Online Resources for Students

Your new textbook provides 18-month access to online study resources that may include videos, code, web chapters, and more. Refer to the preface in the textbook for a detailed list of resources.

Follow the instructions below to register for the Companion Website for James K. Wight's *Reinforced Concrete: Mechanics and Design*, Seventh Edition.

1. Go to www.pearsonhighered.com/eng-resources
2. Either browse by author name or use the search feature to locate the resource page that corresponds with your textbook.
3. Click Companion Website
4. Click Register and follow the on-screen instructions to create a login name and password.

Use a coin to scratch off the coating and reveal your access code.
Do not use a sharp knife or other sharp object as it may damage the code.

Use the login name and password you created during registration to start using the online resources that accompany your textbook.

IMPORTANT:

This access code can only be used once. This subscription is valid for 18 months upon activation and is not transferrable. If the access code has already been revealed it may no longer be valid. If this is the case you can purchase a subscription on the login page for the Companion Website.

For technical support go to http://247pearsoned.custhelp.com

REINFORCED CONCRETE
Mechanics and Design

About the Cover

The signature feature of the fifty-five story Emirates Pearl Hotel is the form of its tower that projects 45 stories above the podium. Floor levels are partial ellipses that spiral around a central elliptical core, causing perimeter columns to be inclined 15 degrees from vertical and shift 19 inches in the circumferential direction over each story. The tower includes luxury apartments and a five-star hotel. The podium includes five levels of underground parking plus five levels of restaurants, retail areas, spas, swimming pools, and a gymnasium.

REINFORCED CONCRETE
Mechanics and Design

SEVENTH EDITION

James K. Wight

F. E. Richart, Jr. Collegiate Professor
Department of Civil & Environmental Engineering
University of Michigan

PEARSON

Boston Columbus Hoboken Indianapolis New York San Francisco
Amsterdam Cape Town Dubai London Madrid Milan Munich Paris Montréal Toronto
Delhi Mexico City São Paulo Sydney Hong Kong Seoul Singapore Taipei Tokyo

Vice President and Editorial Director: *Marcia J. Horton*
Executive Editor: *Holly Stark*
VP of Marketing: *Christy Lesko*
Director of Field Marketing: *Tim Galligan*
Product Marketing Manager: *Bram van Kempen*
Field Marketing Manager: *Demetrius Hall*
Marketing Assistant: *Jon Bryant*
Senior Managing Editor: *Scott Disanno*
Production Project Manager: *Greg Dulles*
Program Manager: *Erin Ault*
Global HE Director of Vendor Sourcing and Procurement: *Diane Hynes*
Operations Specialist: *Maura Zaldivar-Garcia*
Cover Designer: *Black Horse Designs*
Manager, Rights and Permissions: *Rachel Youdelman*
Associate Project Manager, Rights and Permissions: *William Opaluch*
Full-Service Project Management: *Revathi Viswanathan, Lumina Datamatics Inc.*
Composition: *Lumina Datamatics Inc.*
Printer/Binder: *RR Donnelley*
Cover Printer: *Phoenix Color/Hagerstown*
Cover Image: *Project name "Emirates Pearl Hotel in Abu Dhabi".* © *DeSimone Consulting Engineers*

Library of Congress Cataloging-in-Publication Data: Available upon request.

10 9 8 7 6 5 4 3 2 1

PEARSON

ISBN-13: 978-0-13-348596-7
ISBN-10: 0-13-348596-X

Contents

PREFACE xiii

ABOUT THE AUTHOR xix

CHAPTER 1 **INTRODUCTION** 1

1-1 Reinforced Concrete Structures 1
1-2 Mechanics of Reinforced Concrete 1
1-3 Reinforced Concrete Members 2
1-4 Factors Affecting Choice of Reinforced Concrete for a Structure 6
1-5 Historical Development of Concrete and Reinforced Concrete as Structural Materials 7
1-6 Building Codes and the ACI Code 10
References 10

CHAPTER 2 **THE DESIGN PROCESS** 12

2-1 Objectives of Design 12
2-2 The Design Process 12
2-3 Limit States and the Design of Reinforced Concrete 13
2-4 Structural Safety 17
2-5 Probabilistic Calculation of Safety Factors 19
2-6 Design Procedures Specified in the ACI Building Code 20
2-7 Load Factors and Load Combinations in the 2014 ACI Code 22
2-8 Loadings and Actions 27

2-9 Design for Economy 37
2-10 Sustainability 38
2-11 Customary Dimensions and Construction Tolerances 40
2-12 Inspection 40
2-13 Accuracy of Calculations 40
2-14 Handbooks and Design Aids 41
References 41

CHAPTER 3 **MATERIALS** **43**

3-1 Concrete 43
3-2 Behavior of Concrete Failing in Compression 43
3-3 Compressive Strength of Concrete 46
3-4 Strength Under Tensile and Multiaxial Loads 59
3-5 Stress–Strain Curves for Concrete 67
3-6 Time-Dependent Volume Changes 73
3-7 High-Strength Concrete 85
3-8 Lightweight Concrete 87
3-9 Fiber Reinforced Concrete 88
3-10 Durability of Concrete 90
3-11 Behavior of Concrete Exposed to High and Low Temperatures 91
3-12 Shotcrete 93
3-13 High-Alumina Cement 93
3-14 Reinforcement 93
3-15 Fiber-Reinforced Polymer (FRP) Reinforcement 99
3-16 Prestressing Steel 100
References 102

CHAPTER 4 **FLEXURE: BEHAVIOR AND NOMINAL STRENGTH OF BEAM SECTIONS** **105**

4-1 Introduction 105
4-2 Flexure Theory 108
4-3 Simplifications in Flexure Theory for Design 119
4-4 Analysis of Nominal Moment Strength for Singly Reinforced Beam Sections 124
4-5 Definition of Balanced Conditions 131
4-6 Code Definitions of Tension-Controlled and Compression-Controlled Sections 132
4-7 Beams with Compression Reinforcement 142
4-8 Analysis of Flanged Sections 152
4-9 Unsymmetrical Beam Sections 165
References 172

CHAPTER 5 **FLEXURAL DESIGN OF BEAM SECTIONS** **173**

5-1 Introduction 173
5-2 Analysis of Continuous One-Way Floor Systems 173
5-3 Design of Singly Reinforced Beam Sections with Rectangular Compression Zones 195
5-4 Design of Doubly Reinforced Beam Sections 220
5-5 Design of Continuous One-Way Slabs 228
References 242

CHAPTER 6 **SHEAR IN BEAMS** **243**

6-1 Introduction 243
6-2 Basic Theory 245
6-3 Behavior of Beams Failing in Shear 250
6-4 Analysis and Design of Reinforced Concrete Beams for Shear—ACI Code 260
6-5 Other Shear Design Methods 287
6-6 Hanger Reinforcement 292
6-7 Tapered Beams 294
6-8 Shear in Axially Loaded Members 295
References 302

CHAPTER 7 **TORSION** **304**

7-1 Introduction and Basic Theory 304
7-2 Behavior of Reinforced Concrete Members Subjected to Torsion 315
7-3 Thin-Walled Tube Analogies 316
7-4 Design for Torsion and Shear—ACI Code Approach 323
7-5 ACI Code Design Method for Torsion 337
References 358

CHAPTER 8 **DEVELOPMENT, ANCHORAGE, AND SPLICING OF REINFORCEMENT** **359**

8-1 Introduction 359
8-2 Mechanism of Bond Transfer 364
8-3 Development Length 365
8-4 Hooked Anchorages 373
8-5 Headed and Mechanically Anchored Bars in Tension 378
8-6 Design for Anchorage 379
8-7 Bar Cutoffs and Development of Bars in Flexural Members 385
8-8 Reinforcement Continuity and Structural Integrity Requirements 395
8-9 Splices 413
References 417

CHAPTER 9 **SERVICEABILITY** **419**

9-1 Introduction 419
9-2 Elastic Analysis of Stresses in Beam Sections 420
9-3 Cracking 430
9-4 Deflections of Concrete Beams 439
9-5 Consideration of Deflections in Design 447
9-6 Frame Deflections 458
9-7 Vibrations 458
9-8 Fatigue 460
References 462

CHAPTER 10 **CONTINUOUS BEAMS AND ONE-WAY SLABS** **464**

10-1 Introduction 464
10-2 Continuity in Reinforced Concrete Structures 464
10-3 Continuous Beams 468
10-4 Design of Girders 489
10-5 Joist Floors 490
10-6 Moment Redistribution 492
References 494

CHAPTER 11 **COLUMNS: COMBINED AXIAL LOAD AND BENDING** **495**

11-1 Introduction 495
11-2 Tied and Spiral Columns 496
11-3 Interaction Diagrams 502
11-4 Interaction Diagrams for Reinforced Concrete Columns 504
11-5 Design of Short Columns 523
11-6 Contributions of Steel and Concrete to Column Strength 544
11-7 Biaxially Loaded Columns 545
References 558

CHAPTER 12 **SLENDER COLUMNS** **559**

12-1 Introduction 559
12-2 Behavior and Analysis of Pin-Ended Columns 564
12-3 Design of Columns in Nonsway Frames 582
12-4 Behavior of Restrained Columns in Sway Frames 593
12-5 Calculation of Moments in Sway Frames Using Second-Order Analyses 596
12-6 Design of Columns in Sway Frames 601
12-7 General Analysis of Slenderness Effects 619
12-8 Torsional Critical Load 619
References 622

CHAPTER 13 **TWO-WAY SLABS: BEHAVIOR, ANALYSIS, AND DESIGN** **624**

13-1 Introduction 624
13-2 History of Two-Way Slabs 626
13-3 Behavior of Slabs Loaded to Failure in Flexure 626
13-4 Analysis of Moments in Two-Way Slabs 629
13-5 Distribution of Moments in Slabs 633
13-6 Design of Slabs 639
13-7 The Direct-Design Method 644
13-8 Equivalent-Frame Methods 659
13-9 Use of Computers for an Equivalent-Frame Analysis 681
13-10 Shear Strength of Two-Way Slabs 687
13-11 Combined Shear and Moment Transfer in Two-Way Slabs 705
13-12 Details and Reinforcement Requirements 724
13-13 Design of Slabs Without Beams 729
13-14 Design of Slabs with Beams in Two Directions 755
13-15 Construction Loads on Slabs 765
13-16 Deflections in Two-Way Slab Systems 767
13-17 Use of Post-Tensioning 771
References 776

CHAPTER 14 **TWO-WAY SLABS: ELASTIC AND YIELD-LINE ANALYSES** **779**

14-1 Review of Elastic Analysis of Slabs 779
14-2 Design Moments from a Finite-Element Analysis 781
14-3 Yield-Line Analysis of Slabs: Introduction 783
14-4 Yield-Line Analysis: Applications for Two-Way Slab Panels 790
14-5 Yield-Line Patterns at Discontinuous Corners 800
14-6 Yield-Line Patterns at Columns or at Concentrated Loads 801
References 805

CHAPTER 15 **FOOTINGS** **806**

15-1 Introduction 806
15-2 Soil Pressure Under Footings 806
15-3 Structural Action of Strip and Spread Footings 814
15-4 Strip or Wall Footings 821
15-5 Spread Footings 824
15-6 Combined Footings 838
15-7 Mat Foundations 848
15-8 Pile Caps 848
References 851

CHAPTER 16 **SHEAR FRICTION, HORIZONTAL SHEAR TRANSFER, AND COMPOSITE CONCRETE BEAMS** **852**

16-1 Introduction 852
16-2 Shear Friction 852
16-3 Composite Concrete Beams 863
References 872

CHAPTER 17 **DISCONTINUITY REGIONS AND STRUT-AND-TIE MODELS** **873**

17-1 Introduction 873
17-2 Struts 876
17-3 Ties 882
17-4 Nodes and Nodal Zones 883
17-5 Other Strut-and-Tie Elements 890
17-6 Layout of Strut-and-Tie Models 892
17-7 Deep Beams 896
17-8 Continuous Deep Beams 908
17-9 Brackets and Corbels 921
17-10 Dapped Ends 932
17-11 Beam–Column Joints 939
17-12 Bearing Strength 951
17-13 T-Beam Flanges 953
References 956

CHAPTER 18 **WALLS AND SHEAR WALLS** **959**

18-1 Introduction 959
18-2 Bearing Walls 962
18-3 Retaining Walls 966
18-4 Tilt-Up Walls 966
18-5 Shear Walls 966
18-6 Lateral Load-Resisting Systems for Buildings 967
18-7 Shear-Wall–Frame Interaction 969
18-8 Coupled Shear Walls 970
18-9 Design of Structural Walls—General 975
18-10 Flexural Strength of Shear Walls 985
18-11 Shear Strength of Shear Walls 991
18-12 Critical Loads for Axially Loaded Walls 1003
References 1011

CHAPTER 19 **DESIGN FOR EARTHQUAKE RESISTANCE** **1013**

19-1 Introduction 1013
19-2 Seismic Response Spectra 1014
19-3 Seismic Design Requirements 1019
19-4 Seismic Forces on Structures 1023

19-5 Ductility of Reinforced Concrete Members 1027
19-6 General ACI Code Provisions for Seismic Design 1028
19-7 Beams in Special Moment Frames 1032
19-8 Columns in Special Moment Frames 1046
19-9 Joints of Special Moment Frames 1056
19-10 Structural Diaphragms 1058
19-11 Structural Walls 1060
19-12 Frame Members Not Proportioned to Resist Forces Induced by Earthquake Motions 1068
19-13 Special Precast Structures 1068
19-14 Foundations 1068
References 1069

APPENDIX A ***DESIGN AIDS*** ***1071***

APPENDIX B ***NOTATION*** ***1121***

INDEX ***1129***

Preface

Reinforced concrete design encompasses both the art and science of engineering. This book presents the theory of reinforced concrete design as a direct application of the laws of statics and mechanics of materials. It emphasizes that a successful design not only satisfies design rules, but is capable of being built in a timely fashion for a reasonable cost and should provide a long service life.

Philosophy of Reinforced Concrete: Mechanics and Design

A multitiered approach makes *Reinforced Concrete: Mechanics and Design* an outstanding textbook for a variety of university courses on reinforced concrete design. Topics are normally introduced at a fundamental level, and then move to higher levels where prior educational experience and the development of *engineering judgment* will be required. The analysis of the flexural strength of beam sections is presented in Chapter 4. Because this is the first significant design-related topic, it is presented at a level appropriate for new students. Closely related material on the analysis of column sections for combined axial load and bending is presented in Chapter 11 at a somewhat higher level, but still at a level suitable for a first course on reinforced concrete design. Advanced subjects are also presented in the same chapters at levels suitable for advanced undergraduate or graduate students. These topics include, for example, the complete moment versus curvature behavior of a beam section with various tension reinforcement percentages and the use of strain-compatibility to analyze either over-reinforced beam sections, or column sections with multiple layers of reinforcement. More advanced topics are covered in the later chapters, making this textbook valuable for both undergraduate and graduate courses, as well as serving as a key reference in design offices. Other features include the following:

1. Extensive figures are used to illustrate aspects of reinforced concrete member behavior and the design process.

2. Emphasis is placed on logical order and completeness for the many design examples presented in the book.

3. Guidance is given in the text and in examples to help students develop the engineering judgment required to become a successful designer of reinforced concrete structures.

4. Chapters 2 and 3 present general information on various topics related to structural design and construction, and concrete material properties. Frequent references are made back to these topics throughout the text.

Overview—Reorganization of the ACI Building Code

The ACI 318-14 Building Code for Structural Concrete addresses design and detailing requirements for concrete members. The 2014 Code has been reorganized to make technical material more accessible by presenting it in a member-based design format. Each member chapter presents the required technical information for design of a specific member in a logical and repeatable format that should be more intuitive for younger and less-experienced engineers. Each member chapter is organized such that when a structural designer has completed the chapter, all of the design requirements for that member will have been satisfied. Also, the sequencing of the member-based design chapters were arranged to parallel a typical design process that follows the flow of forces from slabs, to beams, to columns, and finally to the foundation.

To avoid duplication from chapter to chapter, the code committee came up with the concept of "toolbox" chapters where key sets of equations and tabulated limits are located. For example, the design for shear strength in a beam, one-way slab or column may all use the same set of equations that are contained in a toolbox chapter. So, when designing a beam for shear, the structural engineering will be directed to the toolbox chapter to select the applicable shear strength equations. After completing the shear design, the engineer will return to the Beam Chapter and continue to the next design step.

A key change was to make more extensive use of tables to replace design requirements that were previously presented in a paragraph format. The code committee believed tables permitted the presentation of design requirements in a form that engineers understand, thus leading to less confusion or potential misinterpretations. Another change was to subdivide lengthy and difficult-to-interpret provisions into shorter statements that address a single design or detailing requirement.

The reorganization process identified overlapping provisions, which were consolidated into a single location, and technical gaps, which were filled with either new sections or chapters, for example chapters on Structural System Requirements and design and detailing for structural Diaphragms. During the reorganization effort, only a limited number of technical changes were made to existing design provisions. Thus, the knowledge base contained in prior editions of the ACI Code has been retained and enhanced.

Changes and Features for the Seventh Edition

All chapters of the text have been reviewed and updated to be in compliance with the 2014 edition of the ACI Building Code. New problems were developed for several chapters and all of the examples given in the book were either reworked or checked for accuracy. Other changes and some continuing features include the following.

1. The presentation of technical information in Chapters 6, 7, and 17 was rearranged to provide a better flow from discussions of member behavior to development of design code requirements. In Chapter 7 additional information was given for the equivalent tube analogies used to define member strength and behavior before and after torsional cracking.

2. Changes were made in the earthquake-resistant design requirements in Chapter 19 to be in compliance with updates to seismic provisions in the ACI Building Code. A key change is the requirement for more transverse reinforcement in columns subjected to a high level of axial load.

3. Flexural design procedures for the full spectrum of beam sections are developed in Chapter 5. Although these design procedures are developed for beam sections, they are easily applied to the flexural design of one-way and two-way slab sections.

4. The design of coupled shear walls and coupling beams in seismic regions is given in Chapter 19, including a discussion of coupling beams with moderate span-to-depth ratios, a topic not well-covered in the ACI Building Code.

5. Chapter 2 contains a discussion of sustainability for design and construction of concrete structures. The use of concrete in building construction for reduced CO_2 emissions and life-cycle costs, as well as improved thermal properties and building aesthetics are discussed.

6. Information is provided for structural analysis of both one-way (Chapter 5) and two-way (Chapter 13) continuous floor systems. Typical modeling assumptions for both types of systems and the interplay between analysis and design are discussed.

7. Axial load vs. moment interaction diagrams are given for a variety of column sections in Appendix A. These diagrams include the required strength-reduction factor, and thus, are very useful for section design either in a classroom or a design office.

Video Solution

8. Video solutions are given for a variety of problems assigned in various chapters. These videos, which were developed by the author, provide step-by-step procedures for solving analysis or design problems. Icons in the margin identify problems for which video solutions are provided. Video Solutions along with Pearson's eText version of this book are provided on the Companion Website at http://www.pearsonhighered.com/eng-resources.

Use of Textbook in Undergraduate and Graduate Courses

The following paragraphs give a suggested set of topics and chapters to be covered in the first and second reinforced concrete design courses, normally given at the undergraduate and graduate levels, respectively. It is assumed that these are semester courses.

First Design Course

Chapters 1 through 3 should be assigned, but the detailed information on loading in Chapter 2 can be covered in a second course. The information on concrete material properties in Chapter 3 could be covered with more depth in a separate undergraduate materials course. **Chapters 4 and 5** are extremely important for all students and should form the foundation of the first undergraduate course. The information in Chapter 4 on moment vs. curvature behavior of beam sections is important for all designers, but this topic could be significantly expanded in a graduate course. Chapter 5 presents a variety of design procedures for developing efficient flexural designs of either singly-reinforced or doubly-reinforced sections. The discussion of structural analysis for continuous floor systems in Section 5-2 could be skipped if either time is limited or students are not yet prepared to handle this topic. The first undergraduate course should cover **Chapter 6** information on member behavior in shear and the shear design requirements given in the ACI Code. Discussions of other methods for determining the shear strength of concrete members can be saved for a second design course. Design for torsion, as covered in **Chapter 7**, could be covered in a first design course, but more often is left for a second design course. The reinforcement anchorage provisions of **Chapter 8** are important material for the first undergraduate design course. Students should develop a basic understanding of development length requirements for straight and hooked bars, as well as the procedure to determine bar cutoff points and reinforcement details required at those cutoff points. The serviceability requirements in **Chapter 9** for control of deflections and cracking are also important topics for the first undergraduate course. In particular, the ability to do an elastic section analysis and find moments of inertia for cracked and uncracked sections is an important skill for designers of concrete structures. **Chapter 10** serves to tie together all of the requirements for continuous floor systems introduced in Chapters 5 through 9. The examples include details for flexural and shear design, as well as full-span detailing of longitudinal and transverse reinforcement. This chapter could either be skipped for the first undergraduate course or be used as a source for a more extensive class design project. **Chapter 11** concentrates on the analysis and design of columns sections and should be included in the first undergraduate course. The portion of Chapter 11 that covers column sections subjected to biaxial bending may either be included in a first undergraduate course or saved for a graduate course. **Chapter 12** considers slenderness effects in columns, and the more detailed analysis required for this topic is commonly presented in a graduate course. If time permits, the basic information in **Chapter 15** on the design of typical concrete footings may be included in a first undergraduate course. This material may also be covered in a foundation design course taught at either the undergraduate or graduate level.

Second Design Course

Clearly, the instructor in a graduate design course has many options for topics, depending on his/her interests and the preparation of the students. **Chapter 13** is a lengthy chapter and is intended to be a significant part of a graduate course. The chapter gives extensive coverage of flexural analysis and design of two-way floor systems that builds on the analysis and design of one-way floor systems covered in Chapter 5. The direct design method and the classic equivalent frame method are discussed, along with more modern analysis and modeling techniques. Problems related to punching shear and the combined transfer of shear and moment at slab-to-column connections are covered in detail. The design of slab shear reinforcement, including the use of shear studs, is also presented. Finally, procedures for calculating deflections in two-way floor systems are given. Design for torsion, as given in **Chapter 7**, should be covered in conjunction with the design and analysis of two-way floor systems in Chapter 13. The design procedure for compatibility torsion at the edges of a floor system has a direct impact on the design of adjacent floor members. The presentation of the yield-line method in **Chapter 14** gives students an alternative analysis and design method for two-way slab systems. This topic could also tie in with plastic analysis methods taught in graduate level analysis courses. The analysis and design of slender columns, as presented in **Chapter 12**, should also be part of a graduate design course. The students should be prepared to apply the frame analysis and member modeling techniques required to either directly determine secondary moments or calculate the required moment-magnification factors. Also, if the topic of biaxial bending in Chapter 11 was not covered in the first design course, it could be included at this point. **Chapter 18** covers bending and shear design of structural walls that resist lateral loads due to either wind or seismic effects. A capacity-design approach is introduced for the shear design of walls that resist earthquake-induced lateral forces. **Chapter 17** covers the concept of *disturbed* regions (D-regions) and the use of the strut-and-tie models to analyze the flow of forces through D-regions and to select appropriate reinforcement details. The chapter contains detailed examples to help students learn the concepts and code requirements for strut-and-tie models. If time permits, instructors could cover the design of combined footings in **Chapter 15**, shear-friction design concepts in **Chapter 16**, and design to resist earthquake-induced forces in **Chapter 19**.

Instructor Materials

An Instructor's Solutions Manual and PowerPoints to accompany this text are available for download to instructors only at http://www.pearsonhighered.com/eng-resources.

Acknowledgment

I would like to take this opportunity to acknowledge the pioneering work done for this textbook by my former co-author, Professor James G. MacGregor. Professor MacGregor was the sole author of this textbook through its first three editions. He initiated the layout of the chapters and the presentation style of first developing an understanding of member and structural behavior, followed by derivation of specific design requirements. I used Professor MacGregor's textbook in my reinforced concrete design classes and I was very happy to join him as a co-author for the Fourth Edition. I became the primary author for the Fifth and Sixth Editions, and I owe a great deal of gratitude to Professor MacGregor for creating an outstanding textbook that I have had the privilege of modifying and enhancing. I wish to express sincere thanks to my recently departed colleague, James G. MacGregor, and best wishes to his devoted wife, Barb.

Dedication

This book is dedicated to all of my colleagues and students who have either interacted with me or taken my classes over the years. My knowledge of the behavior of reinforced concrete members and my development of various design procedures for concrete members and structures was significantly enhanced by my numerous interactions with all of you. I also wish to dedicate this book to my wife, Linda, for her support and encouragement through many long evenings and lost weekends.

The manuscript for the fifth edition book was reviewed by Guillermo Ramirez of the University of Texas at Arlington; Devin Harris of Michigan Technological University; Sami Rizkalla of North Carolina State University; Aly Marei Said of the University of Nevada, Las Vegas; and Roberto Leon of Georgia Institute of Technology. Suggested changes for the sixth edition were submitted by Christopher Higgins and Thomas Schumacher of Oregon State University, Dionisio Bernal of Northeastern University, R. Paneer Selvam of the University of Arkansas, Aly Said of the University of Nevada and Chien-Chung Chen of Pennsylvania State University. The book was reviewed for accuracy by Robert W. Barnes and Anton K. Schindler of Auburn University. This book was greatly improved by all of their suggestions.

James K. Wight
F. E. Richart, Jr. Collegiate Professor
University of Michigan

About the Author

James K. Wight received his B.S. and M.S. degrees in civil engineering from Michigan State University in 1969 and 1970, respectively, and his Ph.D. from the University of Illinois in 1973. He has been a professor of structural engineering in the Civil and Environmental Engineering Department at the University of Michigan since 1973. He teaches undergraduate and graduate classes on analysis and design of reinforced concrete structures. He is well known for his work in earthquake-resistant design of concrete structures and spent a one-year sabbatical leave in Japan where he was involved in the construction and simulated earthquake testing of a full-scale reinforced concrete building. Professor Wight has been an active member of the American Concrete Institute (ACI) since 1973 and was named a Fellow of the Institute in 1984. He is a Past-President of ACI and a past Chair of the ACI Building Code Committee 318. He is also past Chair of the ACI Technical Activities Committee and Committee 352 on Joints and Connections in Concrete Structures. He has received several awards from the American Concrete Institute including the Delmar Bloem Distinguished Service Award (1991), the Joe Kelly Award (1999), the Boise Award (2002), the C.P. Siess Structural Research Award (2003 and 2009), the Alfred Lindau Award (2008), the Wason Medal (2012), and the Charles S. Whitney Medal (2015). Professor Wight has received numerous awards for his teaching and service at the University of Michigan, including the ASCE Student Chapter Teacher of the Year Award, the College of Engineering Distinguished Service Award, the College of Engineering Teaching Excellence Award, the Chi Epsilon-Great Lakes District Excellence in Teaching Award, and the Rackham Distinguished Graduate Mentoring Award. He has also received Distinguished Alumnus Awards from the Civil and Environmental Engineering Departments of the University of Illinois (2008) and Michigan State University (2009).

1

Introduction

1-1 REINFORCED CONCRETE STRUCTURES

Concrete and reinforced concrete are used as building construction materials in every country. In many, including the United States and Canada, reinforced concrete is a dominant structural material in engineered construction. The universal nature of reinforced concrete construction stems from the wide availability of reinforcing bars and of the constituents of concrete (gravel or crushed rock, sand, water, and cement), from the relatively simple skills required in concrete construction, and from the economy of reinforced concrete compared with other forms of construction. Plain concrete and reinforced concrete are used in buildings of all sorts (Fig. 1-1), underground structures, water tanks, wind turbine foundations (Fig. 1-2) and towers, offshore oil exploration and production structures, dams, bridges (Fig. 1-3), and even ships.

1-2 MECHANICS OF REINFORCED CONCRETE

Concrete is strong in compression, but weak in tension. As a result, cracks develop whenever loads, restrained shrinkage, or temperature changes give rise to tensile stresses in excess of the tensile strength of the concrete. In the plain concrete beam shown in Fig. 1-4b, the moments about point O due to applied loads are resisted by an internal tension–compression couple involving tension in the concrete. An unreinforced beam fails very suddenly and completely when the first crack forms. In a *reinforced concrete* beam (Fig. 1-4c), reinforcing bars are embedded in the concrete in such a way that the tension forces needed for moment equilibrium after the concrete cracks can be developed in the bars.

Alternatively, the reinforcement could be placed in a longitudinal duct near the bottom of the beam, as shown in Fig. 1-5, and stretched or *prestressed*, reacting on the concrete in the beam. This would put the reinforcement into tension and the concrete into compression. This compression would delay cracking of the beam. Such a member is said to be a *prestressed concrete* beam. The reinforcement in such a beam is referred to as *prestressing tendons* and must be fabricated from high-strength steel.

The construction of a reinforced concrete member involves building a form or mould in the shape of the member being built. The form must be strong enough to support the weight and hydrostatic pressure of the wet concrete, plus any forces applied to it by workers,

Fig. 1-1
Trump Tower of Chicago.
(Photograph courtesy of Larry Novak, Portland Cement Association.)

Completed in 2009, the 92-story Trump International Hotel and Tower is an icon of the Chicago skyline. With a height of 1170 ft (1389 ft to the top of the spire), the Trump Tower is the tallest building built in North America since the completion of Sears Tower in 1974. The all reinforced concrete residential/hotel tower was designed by Skidmore, Owings & Merrill LLP (SOM). The tower's 2.6 million ft² of floor space is clad in stainless steel and glass, providing panoramic views of the City and Lake Michigan. The project utilized high-performance concrete mixes specified by SOM and designed by Prairie Materials Sales. The project includes self-consolidating concrete with strengths as high as 16,000 psi. The Trump Tower is not only an extremely tall structure; it is also very slender with an aspect ratio exceeding 8 to 1 (height divided by structural base dimension). Slender buildings can be susceptible to dynamic motions under wind loads. To provide the required stiffness, damping, and mass to assist in minimizing the dynamic movements, high-performance reinforced concrete was selected as the primary structural material for the tower. Lateral wind loads are resisted by a core and outrigger system. Additional torsional stiffness and structural robustness is provided by perimeter belt walls at the roof and three mechanical levels. The typical residential floor system consists of 9-in. thick flat plates with spans up to 30 ft.

concrete casting equipment, wind, and so on. The reinforcement is placed in the form and held in place during the concreting operation. After the concrete has reached sufficient strength, the forms can be removed.

1-3 REINFORCED CONCRETE MEMBERS

Reinforced concrete structures consist of a monolithic series of "members" that interact to support the loads placed on the structure. The second floor of the building in Fig. 1-6 is built of concrete joist–slab construction. Here, a series of parallel ribs or *joists* support the load from the top slab. The reactions supporting the joists apply loads to the beams, which in turn are supported by columns. In such a floor, the top slab has two functions: (1) it transfers load laterally to the joists, and (2) it serves as the top flange of the joists, which act as T-shaped beams that transmit the load to the beams running at right angles to the joists.

Fig. 1-2
Wind turbine foundation.
(Photograph courtesy of
Invenergy.)

This wind turbine foundation was installed at Invenergy's Raleigh Wind Energy Center in Ontario, Canada, to support a 1.5 MW turbine with an 80 meter hub height. It consists of 313 cubic yards of 4350 psi concrete, 38,000 lbs of reinforcing steel and is designed to withstand an overturning moment of 29,000 kip-ft. Each of the 140 anchor bolts shown in the photo is post-tensioned to 72 kips.

Fig. 1-3
St. Anthony Falls Bridge.
(Photograph courtesy of
FIGG Bridge Engineers, Inc.)

The new I-35W Bridge (St. Anthony Falls Bridge) in Minneapolis, Minnesota, features a 504 ft main span over the Mississippi River. The concrete piers and superstructure were shaped to echo the arched bridges and natural features in the vicinity. The bridge was designed by FIGG Bridge Engineers, Inc. and constructed by Flatiron-Manson Joint Venture in less than 14 months after the tragic collapse of the former bridge at this site. Segmentally constructed post-tensioned box girders with a specified concrete strength of 6500 psi were used for the bridge superstructure. The tapered piers were cast-in-place and used a specified concrete strength of 4000 psi. Also, a new self-cleaning pollution-eating concrete was used to construct two 30-ft gateway sculptures located at each end of the bridge. A total of approximately 50,000 cubic yards of concrete and 7000 tons of reinforcing bars and post-tensioning steel were used in the project.

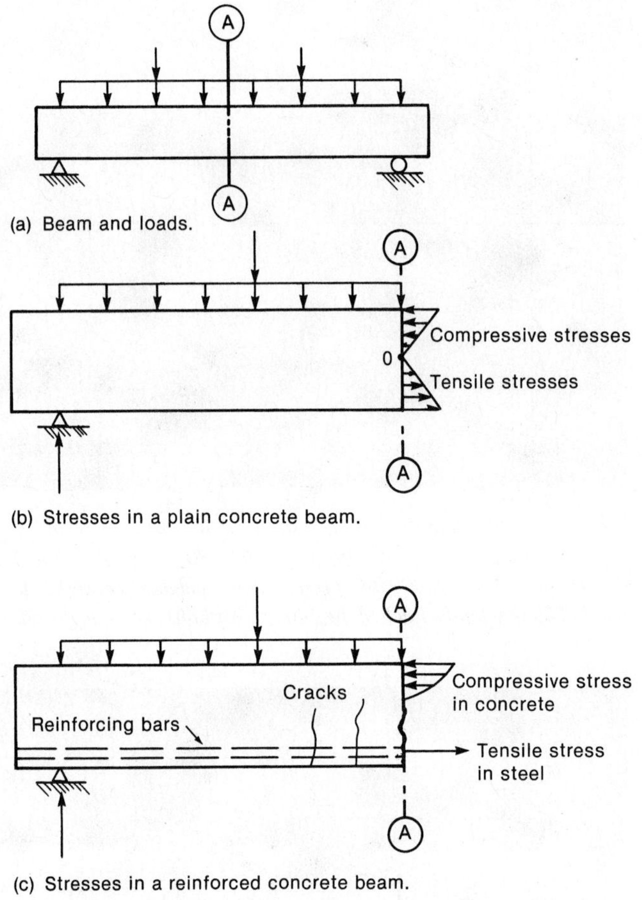

Fig. 1-4
Plain and reinforced concrete beams.

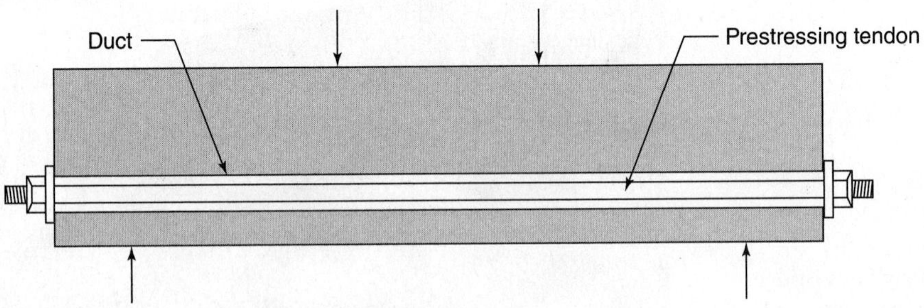

Fig. 1-5
Prestressed concrete beam.

The first floor of the building in Fig. 1-6 has a slab-and-beam design in which the slab spans between beams, which in turn apply loads to the columns. The column loads are applied to *spread footings*, which distribute the load over an area of soil sufficient to prevent overloading of the soil. Some soil conditions require the use of pile foundations or other deep foundations. At the perimeter of the building, the floor loads are supported either directly on the walls, as shown in Fig. 1-6, or on exterior columns, as shown in Fig. 1-7. The walls or columns, in turn, are supported by a basement wall and wall footings.

The first and second floor slabs in Fig. 1-6 are assumed to carry the loads in a north–south direction (see direction arrow) to the joists or beams, which carry the loads in an east–west

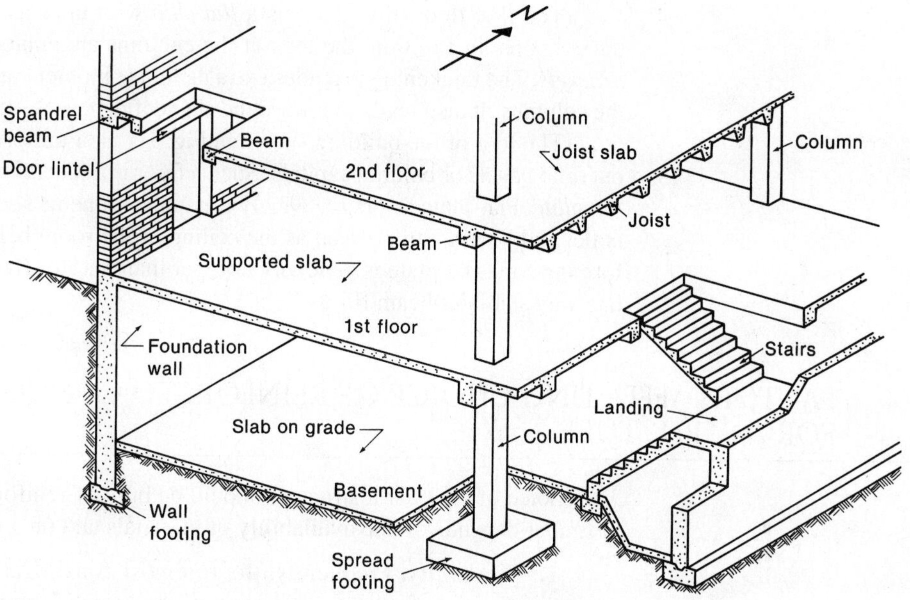

Fig. 1-6
Reinforced concrete building elements. (Adapted from [1-1].)

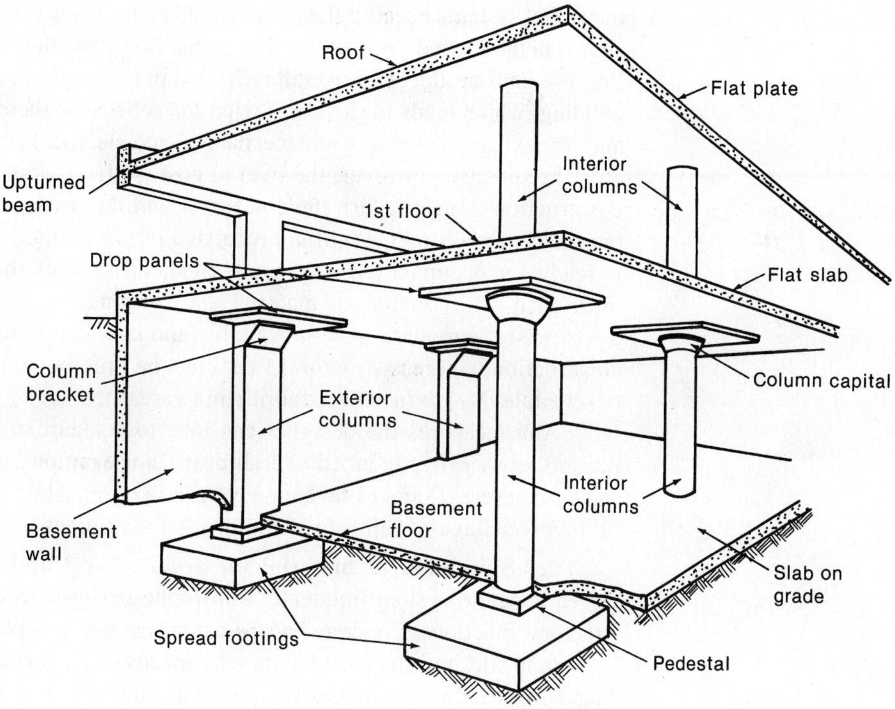

Fig. 1-7
Reinforced concrete building elements. (Adapted from [1-1].)

direction to other beams, girders, columns, or walls. This is referred to as *one-way slab* action and is analogous to a wooden floor in a house, in which the floor decking transmits loads to perpendicular floor joists, which carry the loads to supporting beams, and so on.

The ability to form and construct concrete slabs makes possible the slab or plate type of structure shown in Fig. 1-7. Here, the loads applied to the roof and the floor are transmitted in two directions to the columns by plate action. Such slabs are referred to as *two-way slabs*.

The first floor in Fig. 1-7 is a *flat slab* with thickened areas called *drop panels* at the columns. In addition, the tops of the columns are enlarged in the form of *capitals* or *brackets*. The thickening provides extra depth for moment and shear resistance adjacent to the columns. It also tends to reduce the slab deflections.

The roof of the building shown in Fig. 1-7 is of uniform thickness throughout, without drop panels or column capitals. Such a floor is a special type of *flat slab* referred to as a *flat plate*. Flat-plate floors are widely used in apartments because the underside of the slab is flat and hence, can be used as the ceiling of the room below. Of equal importance, the forming for a flat plate is generally cheaper than that for flat slabs with drop panels or for one-way slab-and-beam floors.

1-4 FACTORS AFFECTING CHOICE OF REINFORCED CONCRETE FOR A STRUCTURE

The choice of whether a structure should be built of reinforced concrete, steel, masonry, or timber depends on the availability of materials and on a number of value decisions.

1. Economy. Frequently, the foremost consideration is the overall cost of the structure. This is, of course, a function of the costs of the materials and of the labor and time necessary to erect the structure. Concrete floor systems tend to be thinner than structural steel systems because the girders and beams or joists all fit within the same depth, as shown in the second floor in Fig. 1-6, or the floors are flat plates or flat slabs, as shown in Fig. 1-7. This produces an overall reduction in the height of a building compared to a steel building, which leads to (a) lower wind loads because there is less area exposed to wind and (b) savings in cladding and mechanical and electrical risers.

Frequently, however, the overall cost is affected as much or more by the overall construction time, because the contractor and the owner must allocate money to carry out the construction and will not receive a return on their investment until the building is ready for occupancy. As a result, financial savings due to rapid construction may more than offset increased material and forming costs. The materials for reinforced concrete structures are widely available and can be produced as they are needed in the construction, whereas structural steel must be ordered and partially paid for in advance to schedule the job in a steel-fabricating yard.

Any measures the designer can take to standardize the design and forming will generally pay off in reduced overall costs. For example, column sizes may be kept the same for several floors to save money in form costs, while changing the concrete strength or the percentage of reinforcement allows for changes in column loads.

2. Suitability of material for architectural and structural function. A reinforced concrete system frequently allows the designer to combine the architectural and structural functions. Concrete has the advantage that it is placed in a plastic condition and is given the desired shape and texture by means of the forms and the finishing techniques. This allows such elements as flat plates or other types of slabs to serve as load-bearing elements while providing the finished floor and ceiling surfaces. Similarly, reinforced concrete walls can provide architecturally attractive surfaces in addition to having the ability to resist gravity, wind, or seismic loads. Finally, the choice of size or shape is governed by the designer and not by the availability of standard manufactured members.

3. Fire resistance. The structure in a building must withstand the effects of a fire and remain standing while the building is being evacuated and the fire extinguished. A concrete building inherently has a 1- to 3-hour fire rating without special fireproofing or other details. Structural steel or timber buildings must be fireproofed to attain similar fire ratings.

4. Rigidity. The occupants of a building may be disturbed if their building oscillates in the wind or if the floors vibrate as people walk by. Due to the greater stiffness and mass of a concrete structure, vibrations are seldom a problem.

5. Low maintenance. Concrete members inherently require less maintenance than do structural steel or timber members. This is particularly true if dense, air-entrained concrete has been used for surfaces exposed to the atmosphere and if care has been taken in the design to provide adequate drainage from the structure.

6. Availability of materials. Sand, gravel or crushed rock, water, cement, and concrete mixing facilities are very widely available, and reinforcing steel can be transported to most construction sites more easily than can structural steel. As a result, reinforced concrete is frequently the preferred construction material in remote areas.

On the other hand, there are a number of factors that may cause one to select a material other than reinforced concrete. These include:

1. Low tensile strength. As stated earlier, the tensile strength of concrete is much lower than its compressive strength (about $\frac{1}{10}$); hence, concrete is subject to cracking when subjected to tensile stresses. In structural uses, the cracking is restrained by using reinforcement, as shown in Fig. 1-4c, to carry tensile forces and limit crack widths to within acceptable values. Unless care is taken in design and construction, however, these cracks may be unsightly or may allow penetration of water and other potentially harmful contaminants.

2. Forms and shoring. The construction of a cast-in-place structure involves three steps not encountered in the construction of steel or timber structures. These are: (a) the construction of the forms, (b) the removal of these forms, and (c) the propping or shoring of the new concrete to support its weight until its strength is adequate. Each of these steps involves labor and/or materials that are not necessary with other forms of construction.

3. Relatively low strength per unit of weight or volume. The compressive strength of concrete is roughly 10 percent that of steel, while its unit density is roughly 30 percent that of steel. As a result, a concrete structure requires a larger volume and a greater weight of material than does a comparable steel structure. As a result, steel is often selected for long-span structures.

4. Time-dependent volume changes. Both concrete and steel undergo approximately the same amount of thermal expansion and contraction. Because there is less mass of steel to be heated or cooled, and because steel is a better conductor than concrete, a steel structure is generally affected by temperature changes to a greater extent than is a concrete structure. On the other hand, concrete undergoes drying shrinkage, which, if restrained, may cause deflections or cracking. Furthermore, deflections in a concrete floor will tend to increase with time, possibly doubling, due to creep of the concrete under sustained compression stress.

1-5 HISTORICAL DEVELOPMENT OF CONCRETE AND REINFORCED CONCRETE AS STRUCTURAL MATERIALS

Cement and Concrete

Lime mortar was first used in structures in the Minoan civilization in Crete about 2000 B.C. and is still used in some areas. This type of mortar had the disadvantage of gradually dissolving when immersed in water and hence could not be used for exposed or underwater joints. About the third century B.C., the Romans discovered a fine sandy volcanic ash that, when mixed with lime mortar, gave a much stronger mortar, which could be used under water.

One of the most remarkable concrete structures built by the Romans was the dome of the Pantheon in Rome, completed in A.D. 126. This dome has a span of 144 ft, a span not exceeded until the nineteenth century. The lowest part of the dome was concrete with aggregate consisting of broken bricks. As the builders approached the top of the dome they used lighter and lighter aggregates, using pumice at the top to reduce the dead-load moments. Although the outside of the dome was, and still is, covered with decorations, the marks of the forms are still visible on the inside [1-2], [1-3].

While designing the Eddystone Lighthouse off the south coast of England just before A.D. 1800, the English engineer John Smeaton discovered that a mixture of burned limestone and clay could be used to make a cement that would set under water and be water resistant. Owing to the exposed nature of this lighthouse, however, Smeaton reverted to the tried-and-true Roman cement and mortised stonework.

In the ensuing years a number of people used Smeaton's material, but the difficulty of finding limestone and clay in the same quarry greatly restricted its use. In 1824, Joseph Aspdin mixed ground limestone and clay from different quarries and heated them in a kiln to make cement. Aspdin named his product Portland cement because concrete made from it resembled Portland stone, a high-grade limestone from the Isle of Portland in the south of England. This cement was used by Brunel in 1828 for the mortar in the masonry liner of a tunnel under the Thames River and in 1835 for mass concrete piers for a bridge. Occasionally in the production of cement, the mixture would be overheated, forming a hard clinker, which was considered to be spoiled and was discarded. In 1845, I. C. Johnson found that the best cement resulted from grinding this clinker. This is the material now known as Portland cement. Portland cement was produced in Pennsylvania in 1871 by D. O. Saylor and about the same time in Indiana by T. Millen of South Bend, but it was not until the early 1880s that significant amounts were produced in the United States.

Reinforced Concrete

W. B. Wilkinson of Newcastle-upon-Tyne obtained a patent in 1854 for a reinforced concrete floor system that used hollow plaster domes as forms. The ribs between the forms were filled with concrete and were reinforced with discarded steel mine-hoist ropes in the center of the ribs. In France, Lambot built a rowboat of concrete reinforced with wire in 1848 and patented it in 1855. His patent included drawings of a reinforced concrete beam and a column reinforced with four round iron bars. In 1861, another Frenchman, Coignet, published a book illustrating uses of reinforced concrete.

The American lawyer and engineer Thaddeus Hyatt experimented with reinforced concrete beams in the 1850s. His beams had longitudinal bars in the tension zone and vertical stirrups for shear. Unfortunately, Hyatt's work was not known until he privately published a book describing his tests and building system in 1877.

Perhaps the greatest incentive to the early development of the scientific knowledge of reinforced concrete came from the work of Joseph Monier, owner of a French nursery garden. Monier began experimenting in about 1850 with concrete tubs reinforced with iron for planting trees. He patented his idea in 1867. This patent was rapidly followed by patents for reinforced pipes and tanks (1868), flat plates (1869), bridges (1873), and stairs (1875). In 1880 and 1881, Monier received German patents for many of the same applications. These were licensed to the construction firm Wayss and Freitag, which commissioned Professors Mörsch and Bach of the University of Stuttgart to test the strength of reinforced concrete and commissioned Mr. Koenen, chief building inspector for Prussia, to develop a method for computing the strength of reinforced concrete. Koenen's book, published in 1886, presented an analysis that assumed the neutral axis was at the midheight of the member.

The first reinforced concrete building in the United States was a house built on Long Island in 1875 by W. E. Ward, a mechanical engineer. E. L. Ransome of California experimented with reinforced concrete in the 1870s and patented a twisted steel reinforcing bar in 1884. In the same year, Ransome independently developed his own set of design procedures. In 1888, he constructed a building having cast-iron columns and a reinforced concrete floor system consisting of beams and a slab made from flat metal arches covered with concrete. In 1890, Ransome built the Leland Stanford, Jr. Museum in San Francisco. This two-story building used discarded cable-car rope as beam reinforcement. In 1903 in Pennsylvania, he built the first building in the United States completely framed with reinforced concrete.

In the period from 1875 to 1900, the science of reinforced concrete developed through a series of patents. An English textbook published in 1904 listed 43 patented systems, 15 in France, 14 in Germany or Austria–Hungary, 8 in the United States, 3 in the United Kingdom, and 3 elsewhere. Most of these differed in the shape of the bars and the manner in which the bars were bent.

From 1890 to 1920, practicing engineers gradually gained a knowledge of the mechanics of reinforced concrete, as books, technical articles, and codes presented the theories. In an 1894 paper to the French Society of Civil Engineers, Coignet (son of the earlier Coignet) and de Tedeskko extended Koenen's theories to develop the working-stress design method for flexure, which was used universally from 1900 to 1950. During the past seven decades, extensive research has been carried out on various aspects of reinforced concrete behavior, resulting in the current design procedures.

Prestressed concrete was pioneered by E. Freyssinet, who in 1928 concluded that it was necessary to use high-strength steel wire for prestressing because the creep of concrete dissipated most of the prestress force if normal reinforcing bars were used to develop the prestressing force. Freyssinet developed anchorages for the tendons and designed and built a number of pioneering bridges and structures.

Design Specifications for Reinforced Concrete

The first set of building regulations for reinforced concrete were drafted under the leadership of Professor Mörsch of the University of Stuttgart and were issued in Prussia in 1904. Design regulations were issued in Britain, France, Austria, and Switzerland between 1907 and 1909.

The American Railway Engineering Association appointed a Committee on Masonry in 1890. In 1903 this committee presented specifications for portland cement concrete. Between 1908 and 1910, a series of committee reports led to the *Standard Building Regulations for the Use of Reinforced Concrete*, published in 1910 [1-4] by the National Association of Cement Users, which subsequently became the American Concrete Institute.

A Joint Committee on Concrete and Reinforced Concrete was established in 1904 by the American Society of Civil Engineers, the American Society for Testing and Materials, the American Railway Engineering Association, and the Association of American Portland Cement Manufacturers. This group was later joined by the American Concrete Institute. Between 1904 and 1910, the Joint Committee carried out research. A preliminary report issued in 1913 [1-5] lists the more important papers and books on reinforced concrete published between 1898 and 1911. The final report of this committee was published in 1916 [1-6]. The history of reinforced concrete building codes in the United States was reviewed in 1954 by Kerekes and Reid [1-7].

1-6 BUILDING CODES AND THE ACI CODE

The design and construction of buildings is regulated by municipal bylaws called *building codes*. These exist to protect the public's health and safety. Each city and town is free to write or adopt its own building code, and in that city or town, only that particular code has legal status. Because of the complexity of writing building codes, cities in the United States generally base their building codes on model codes. Prior to the year 2000, there were three model codes: the *Uniform Building Code* [1-8], the *Standard Building Code* [1-9], and the *Basic Building Code* [1-10]. These codes covered such topics as use and occupancy requirements, fire requirements, heating and ventilating requirements, and structural design. In 2000, these three codes were replaced by the *International Building Code* (*IBC*) [1-11], which is normally updated every three years.

The definitive design specification for reinforced concrete buildings in North America is the *Building Code Requirements for Structural Concrete* (ACI 318-14) and *Commentary* (ACI 318R-14) [1-12]. The code and the commentary are bound together in one volume.

This code, generally referred to as the *ACI Code*, has been incorporated by reference in the *International Building Code* and serves as the basis for comparable codes in Canada, New Zealand, Australia, most of Latin America, and some countries in Asia and the Middle East. The ACI Code has legal status only if adopted in a local building code.

In recent years, the ACI Code has undergone a major revision every three years. Current plans are to publish major revisions on a six-year cycle with interim revisions half-way through the cycle. This book refers extensively to the 2014 ACI Code. It is recommended that the reader have a copy available.

The term *structural concrete* is used to refer to the entire range of concrete structures: from *plain concrete* without any reinforcement; through *ordinary reinforced concrete*, reinforced with normal reinforcing bars; through *partially prestressed concrete*, generally containing both reinforcing bars and prestressing tendons; to *fully prestressed concrete*, with enough prestress to prevent cracking in everyday service. In 1995, the title of the ACI Code was changed from *Building Code Requirements for Reinforced Concrete* to *Building Code Requirements for Structural Concrete* to emphasize that the code deals with the entire spectrum of structural concrete.

The rules for the design of concrete highway bridges are specified in the *AASHTO LRFD Bridge Design Specifications*, American Association of State Highway and Transportation Officials, Washington, D.C. [1-13].

Each nation or group of nations in Europe has its own building code for reinforced concrete. The *fib Model Code for Concrete Structures* 2010 [1-14], published in 2013 by the International Federation for Structural Concrete (fib), Lausanne, Switzerland, is intended to serve as the basis for future attempts to unify European codes.

Another document that will be used extensively in Chapters 2 and 19 is the ASCE standard *ASCE/SEI 7-10*, entitled *Minimum Design Loads for Buildings and Other Structures* [1-16], published in 2010.

REFERENCES

1-1 *Reinforcing Bar Detailing Manual*, Fourth Edition, Concrete Reinforcing Steel Institute, Chicago, IL, 290 pp.

1-2 Robert Mark, "Light, Wind and Structure: The Mystery of the Master Builders," *MIT Press*, Boston, 1990, pp. 52–67.

1-3 Michael P. Collins, "In Search of Elegance: The Evolution of the Art of Structural Engineering in the Western World," *Concrete International*, Vol. 23, No. 7, July 2001, pp. 57–72.

1-4 Committee on Concrete and Reinforced Concrete, "Standard Building Regulations for the Use of Reinforced Concrete," *Proceedings, National Association of Cement Users*, Vol. 6, 1910, pp. 349–361.

1-5 Special Committee on Concrete and Reinforced Concrete, "Progress Report of Special Committee on Concrete and Reinforced Concrete," *Proceedings of the American Society of Civil Engineers*, 1913, pp. 117–135.

1-6 Special Committee on Concrete and Reinforced Concrete, "Final Report of Special Committee on Concrete and Reinforced Concrete," *Proceedings of the American Society of Civil Engineers*, 1916, pp. 1657–1708.

1-7 Frank Kerekes and Harold B. Reid, Jr., "Fifty Years of Development in Building Code Requirements for Reinforced Concrete," *ACI Journal*, Vol. 25, No. 6, February 1954, pp. 441–470.

1-8 *Uniform Building Code*, International Conference of Building Officials, Whittier, CA, various editions.

1-9 *Standard Building Code*, Southern Building Code Congress, Birmingham, AL, various editions.

1-10 *Basic Building Code*, Building Officials and Code Administrators International, Chicago, IL, various editions.

1-11 International Code Council, 2012 *International Building Code*, Washington, D.C., 2012.

1-12 ACI Committee 318, *Building Code Requirements for Structural Concrete (ACI 318-14) and Commentary*, American Concrete Institute, Farmington Hills, MI, 2014.

1-13 *AASHTO LRFD Bridge Design Specifications*, 6th Edition, with 2013 Interim Revisions, American Association of State Highway and Transportation Officials, Washington, D.C., 2013.

1-14 *fib Model Code for Concrete Structures 2010*, International Federation for Structural Concrete (fib), Lausanne, Switzerland, 2013.

1-15 *Minimum Design Loads for Buildings and Other Structures* (ASCE/SEI 7-10), American Society of Civil Engineers, Reston, VA, 2010.

2

The Design Process

2-1 OBJECTIVES OF DESIGN

A structural engineer is a member of a team that works together to design a building, bridge, or other structure. In the case of a building, an architect generally provides the overall layout, and mechanical, electrical, and structural engineers design individual systems within the building.

The structure should satisfy four major criteria:

1. **Appropriateness.** The arrangement of spaces, spans, ceiling heights, access, and traffic flow must complement the intended use. The structure should fit its environment and be aesthetically pleasing.

2. **Economy.** The overall cost of the structure should not exceed the client's budget. Frequently, teamwork in design will lead to overall economies.

3. **Structural adequacy.** Structural adequacy involves two major aspects.
   - **(a)** A structure must be strong enough to support all anticipated loadings safely.
   - **(b)** A structure must not deflect, tilt, vibrate, or crack in a manner that impairs its usefulness.

4. **Maintainability.** A structure should be designed so as to require a minimum amount of simple maintenance procedures.

2-2 THE DESIGN PROCESS

The design process is a sequential and iterative decision-making process. The three major phases are the following:

1. **Definition of the client's needs and priorities.** All buildings or other structures are built to fulfill a need. It is important that the owner or user be involved in determining the attributes of the proposed building. These include functional requirements, aesthetic requirements, and budgetary requirements. The latter include initial cost, premium for rapid construction to allow early occupancy, maintenance, and other life-cycle costs.

2. **Development of project concept.** Based on the client's needs and priorities, a number of possible layouts are developed. Preliminary cost estimates are made, and the

final choice of the system to be used is based on how well the overall design satisfies the client's needs within the budget available. Generally, systems that are conceptually simple and have standardized geometries and details that allow construction to proceed as a series of identical cycles are the most cost effective.

During this stage, the overall structural concept is selected. From approximate analyses of the moments, shears, and axial forces, preliminary member sizes are selected for each potential scheme. Once this is done, it is possible to estimate costs and select the most desirable structural system.

The overall thrust in this stage of the structural design is to satisfy the design criteria dealing with appropriateness, economy, and, to some extent, maintainability.

3. Design of individual systems. Once the overall layout and general structural concept have been selected, the structural system can be designed. Structural design involves three main steps. Based on the preliminary design selected in phase 2, a *structural analysis* is carried out to determine the moments, shears, torques, and axial forces in the structure. The individual members are then *proportioned* to resist these load effects. The proportioning, sometimes referred to as *member design*, must also consider overall aesthetics, the constructability of the design, coordination with mechanical and electrical systems, and the sustainability of the final structure. The final stage in the design process is to prepare construction drawings and specifications.

2-3 LIMIT STATES AND THE DESIGN OF REINFORCED CONCRETE

Limit States

When a structure or structural element becomes unfit for its intended use, it is said to have reached a *limit state*. The limit states for reinforced concrete structures can be divided into three basic groups:

1. Strength limit states. These involve a structural collapse of part or all of the structure. Such a limit state should have a very low probability of occurrence, because it may lead to loss of life and major financial losses. The major strength limit states are as follows:

(a) Loss of equilibrium of a part or all of the structure as a rigid body. Such a failure would generally involve tipping or sliding of the entire structure and would occur if the reactions necessary for equilibrium could not be developed.

(b) Failure of critical parts of the structure, leading to partial or complete collapse. The majority of this book deals with this limit state. Chapters 4 and 5 consider flexural failures; Chapter 6, shear failures; and so on.

(c) Progressive collapse. In some structures, an overload on one member may cause that member to fail. The load acting on it is transferred to adjacent members which, in turn, may be overloaded and fail, causing them to shed their load to adjacent members, causing them to fail one after another, until a major part of the structure has collapsed. This is called a *progressive collapse* [2-1], [2-2]. Progressive collapse is prevented, or at least is limited, by one or more of the following:

(i) Controlling accidental events by taking measures such as protection against vehicle collisions or explosions.

(ii) Providing local resistance by designing key members to resist accidental events.

(iii) Providing minimum horizontal and vertical ties to transfer forces.

(iv) Providing alternative lines of support to anchor the tie forces.

(v) Limiting the spread of damage by subdividing the building with planes of weakness, sometimes referred to as *structural fuses*.

A structure is said to have *general structural integrity* if it is resistant to progressive collapse. For example, a terrorist bomb or a vehicle collision may accidentally remove a column that supports an interior support of a two-span continuous beam. If properly detailed, the structural system may change from two spans to one long span. This would entail large deflections and a change in the load path from beam action to *catenary* or tension membrane action. ACI Code Section 9.7.7 requires continuous tensile reinforcement in beams around the perimeter of a building at each floor to reduce the risk of progressive collapse. The ties provide reactions to anchor the catenary forces and limit the spread of damage. Because such failures are most apt to occur during construction, the designer should be aware of the applicable construction loads and procedures.

(d) Formation of a plastic mechanism. A mechanism is formed when the reinforcement yields to form plastic hinges at enough sections to make the structure unstable.

(e) Instability due to deformations of the structure. This type of failure involves buckling and is discussed more fully in Chapter 12.

(f) Fatigue. Fracture of members due to repeated stress cycles of service loads may cause collapse. Fatigue is discussed in Sections 3-14 and 9-8.

2. Serviceability limit states. These involve disruption of the functional use of the structure, but not collapse per se. Because there is less danger of loss of life, a higher probability of occurrence can generally be tolerated than in the case of a strength limit state. Design for serviceability is discussed in Chapter 9. The major serviceability limit states include the following:

(a) Excessive deflections for normal service. Excessive deflections may cause machinery to malfunction, may be visually unacceptable, and may lead to damage to nonstructural elements or to changes in the distribution of forces. In the case of very flexible roofs, deflections due to the weight of water on the roof may lead to increased depth of water, increased deflections, and so on, until the strength of the roof is exceeded. This is a *ponding failure* and in essence is a collapse brought about by failure to satisfy a serviceability limit state.

(b) Excessive crack widths. Although reinforced concrete must crack before the reinforcement can function effectively, it is possible to detail the reinforcement to minimize the crack widths. Excessive crack widths may be unsightly and may allow leakage through the cracks, corrosion of the reinforcement, and gradual deterioration of the concrete.

(c) Undesirable vibrations. Vertical vibrations of floors or bridges and lateral and torsional vibrations of tall buildings may disturb the users. Vibration effects have rarely been a problem in reinforced concrete buildings.

3. Special limit states. This class of limit states involves damage or failure due to abnormal conditions or abnormal loadings and includes:

(a) damage or collapse in extreme earthquakes,

(b) structural effects of fire, explosions, or vehicular collisions,

(c) structural effects of corrosion or deterioration, and

(d) long-term physical or chemical instability (normally not a problem with concrete structures).

Limit-States Design

Limit-states design is a process that involves

1. the identification of all potential modes of failure (i.e., identification of the significant limit states),

2. the determination of acceptable levels of safety against occurrence of each limit state, and

3. structural design for the significant limit states.

For normal structures, step 2 is carried out by the building-code authorities, who specify the load combinations and the load factors to be used. For unusual structures, the engineer may need to check whether the normal levels of safety are adequate.

For buildings, a limit-states design starts by selecting the concrete strength, cement content, cement type, supplementary cementitious materials, water–cementitious materials ratio, air content, and cover to the reinforcement to satisfy the durability requirements of ACI Code Chapter 19. Next, the minimum member sizes and minimum covers are chosen to satisfy the fire-protection requirements of the local building code. Design is then carried out, starting by proportioning for the ultimate limit states followed by a check of whether the structure will exceed any of the serviceability limit states. This sequence is followed because the major function of structural members in buildings is to resist loads without endangering the occupants. For a water tank, however, the limit state of excessive crack width is of equal importance to any of the strength limit states if the structure is to remain watertight [2-3]. In such a structure, the design for the limit state of crack width might be considered before the strength limit states are checked. In the design of support beams for an elevated monorail, the smoothness of the ride is extremely important, and the limit state of deflection may govern the design.

Basic Design Relationship

Figure 2-1a shows a beam that supports its own dead weight, w, plus some applied loads, P_1, P_2, and P_3. These cause bending moments, distributed as shown in Fig. 2-1b. The bending moments are obtained directly from the loads by using the laws of statics, and for a known span and combination of loads w, P_1, P_2, and P_3, the moment diagram is independent of the composition or shape of the beam. The bending moment is referred to as a *load effect*. Other load effects include shear force, axial force, torque, deflection, and vibration.

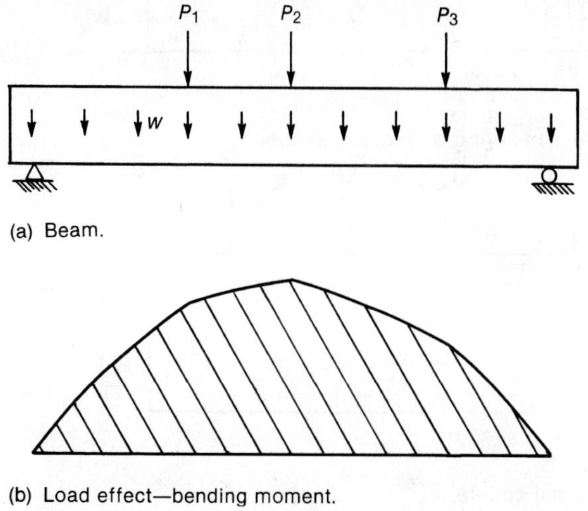

Fig. 2-1
Beam with loads and a load effect.

Figure 2-2a shows flexural stresses acting on a beam cross section. The compressive and tensile stress blocks in Fig. 2-2a can be replaced by forces C and T that are separated by a distance jd, as shown in Fig. 2-2b. The resulting couple is called an *internal resisting moment*. The internal resisting moment when the cross section fails is referred to as the *moment strength* or *moment resistance*. The word *strength* also can be used to describe shear strength or axial load strength.

The beam shown in Fig. 2-2 will support the loads safely if, at every section, the resistance (strength) of the member exceeds the effects of the loads:

$$\text{resistances} \geq \text{load effects} \tag{2-1}$$

To allow for the possibility that the resistances will be less than computed or the load effects larger than computed, *strength-reduction factors*, ϕ, less than 1, and *load factors*, α, greater than 1, are introduced:

$$\phi R_n \geq \alpha_1 S_1 + \alpha_2 S_2 + \cdots \tag{2-2a}$$

Here, R_n stands for nominal resistance (strength) and S stands for load effects based on the specified loads. Written in terms of moments, (2-2a) becomes

$$\phi_M M_n \geq \alpha_D M_D + \alpha_L M_L + \cdots \tag{2-2b}$$

where M_n is the *nominal moment strength*. The word *nominal* implies that this strength is a computed value based on the specified concrete and steel strengths and the dimensions shown on the drawings. M_D and M_L are the bending moments (load effects) due to the specified dead load and specified live load, respectively; ϕ_M is a strength-reduction factor for moment; and α_D and α_L are load factors for dead and live load, respectively.

Similar equations can be written for shear, V, and axial force, P:

$$\phi_V V_n \geq \alpha_D V_D + \alpha_L V_L + \cdots \tag{2-2c}$$

$$\phi_P P_n \geq \alpha_D P_D + \alpha_L P_L + \cdots \tag{2-2d}$$

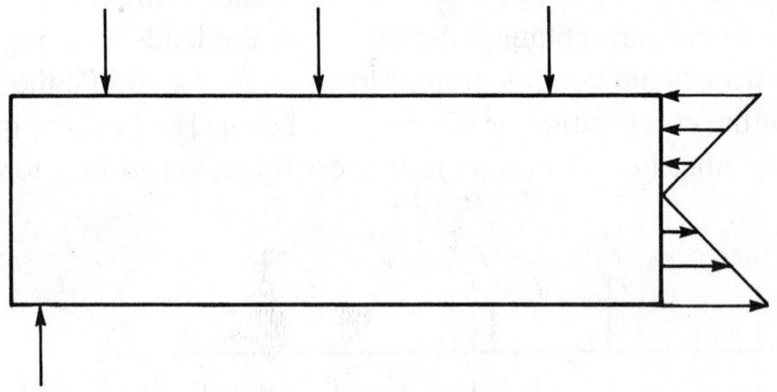

(a) Stresses acting on a cross section.

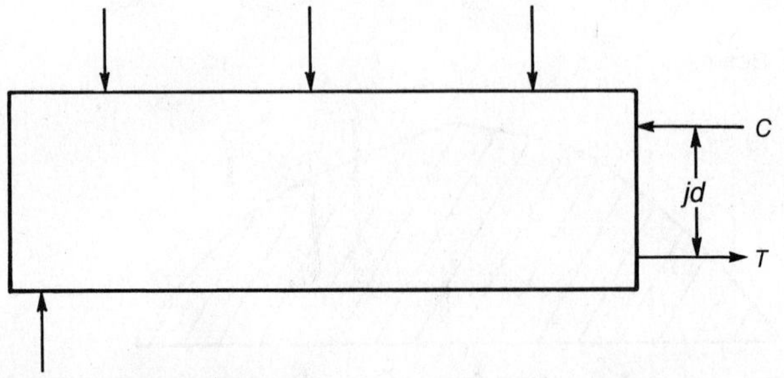

(b) Internal couple.

Fig. 2-2
Internal resisting moment.

Equation (2-1) is the basic limit-states design equation. Equations (2-2a) to (2-2d) are special forms of this basic equation. Throughout the ACI Code, the symbol U is used to refer to the combination $(\alpha_D D + \alpha_L L + \cdots)$. This combination is referred to as the *factored loads*. The symbols M_u, V_u, T_u, and so on, refer to *factored-load effects* calculated from the factored loads.

2-4 STRUCTURAL SAFETY

There are three main reasons why safety factors, such as load and resistance factors, are necessary in structural design:

1. Variability in strength. The actual strengths (resistances) of beams, columns, or other structural members will almost always differ from the values calculated by the designer. The main reasons for this are as follows [2-4]:

(a) variability of the strengths of concrete and reinforcement,

(b) differences between the as-built dimensions and those shown on the structural drawings, and

(c) effects of simplifying assumptions made in deriving the equations for member strength.

A histogram of the ratio of beam moment capacities observed in tests, M_{test}, to the nominal strengths computed by the designer, M_n, is plotted in Fig. 2-3. Although the mean strength is roughly 1.05 times the nominal strength in this sample, there is a definite chance that some beam cross sections will have a lower capacity than computed. The variability shown here is due largely to the simplifying assumptions made in computing the nominal moment strength, M_n.

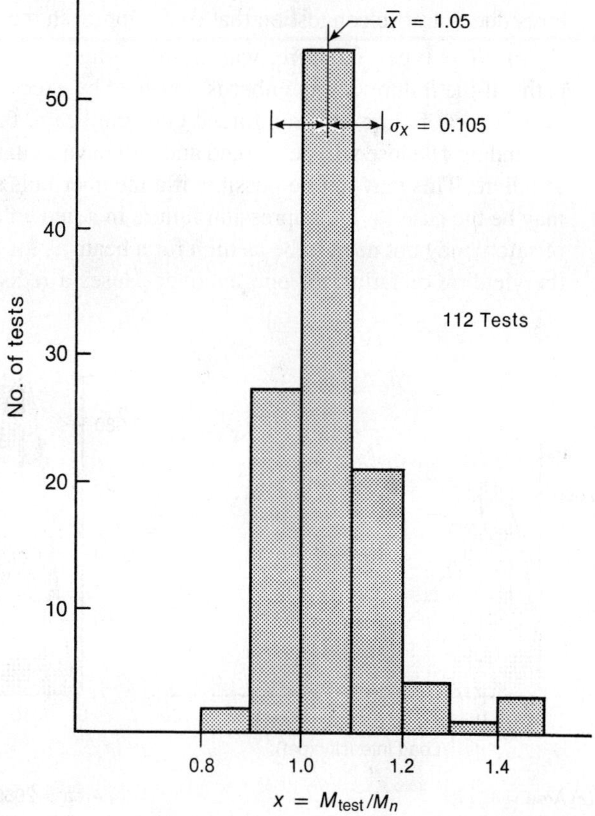

Fig. 2-3
Comparison of measured and computed failure moments, based on data for reinforced concrete beams with $f'_c > 2000$ psi [2-5].

2. Variability in loadings. All loadings are variable, especially live loads and environmental loads due to snow, wind, or earthquakes. Figure 2-4a compares the sustained component of live loads measured in a series of 151-ft^2 areas in offices. Although the average sustained live load was 13 psf in this sample, 1 percent of the measured loads exceeded 44 psf. For this type of occupancy and area, building codes specify live loads of 50 psf. For larger areas, the mean sustained live load remains close to 13 psf, but the variability decreases, as shown in Fig. 2-4b. A transient live load representing unusual loadings due to parties, temporary storage, and so on, must be added to get the total live load. As a result, the maximum live load on a given office will generally exceed the 13 to 44 psf discussed here.

In addition to actual variations in the loads themselves, the assumptions and approximations made in carrying out structural analyses lead to differences between the actual forces and moments and those computed by the designer [2-4]. Due to the variabilities of strengths and load effects, there is a definite chance that a weaker-than-average structure will be subjected to a higher-than-average load, and in this extreme case, failure may occur. The load factors and resistance (strength) factors in Eqs. (2-2a) through (2-2d) are selected to reduce the probability of failure to a very small level.

The consequences of failure are a third factor that must be considered in establishing the level of safety required in a particular structure.

3. Consequences of failure. A number of subjective factors must be considered in determining an acceptable level of safety for a particular class of structure. These include:

(a) The potential loss of life—it may be desirable to have a higher factor of safety for an auditorium than for a storage building.

(b) The cost to society in lost time, lost revenue, or indirect loss of life or property due to a failure—for example, the failure of a bridge may result in intangible costs due to traffic congestion that could approach the replacement cost.

(c) The type of failure, warning of failure, and existence of alternative load paths. If the failure of a member is preceded by excessive deflections, as in the case of a flexural failure of a reinforced concrete beam, the persons endangered by the impending collapse will be warned and will have a chance to leave the building prior to failure. This may not be possible if a member fails suddenly without warning, as may be the case for a compression failure in a tied column. Thus, the required level of safety may not need to be as high for a beam as for a column. In some structures, the yielding or failure of one member causes a redistribution of load to adjacent

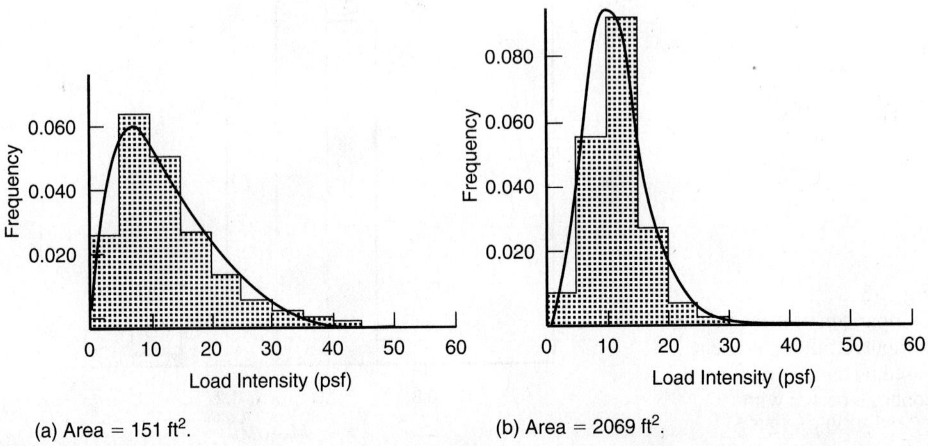

Fig. 2-4
Frequency distribution of sustained component of live loads in offices. (From [2-6].)

members. In other structures, the failure of one member causes complete collapse. If no redistribution is possible, a higher level of safety is required.

(d) The direct cost of clearing the debris and replacing the structure and its contents.

2-5 PROBABILISTIC CALCULATION OF SAFETY FACTORS

The distribution of a population of resistances, R, of a group of similar structures is plotted on the horizontal axis in Fig. 2-5. This is compared to the distribution of the maximum load effects, S, expected to occur on those structures during their lifetimes, plotted on the vertical axis in the same figure. For consistency, both the resistances and the load effects can be expressed in terms of a quantity such as bending moment. The 45° line in this figure corresponds to a load effect equal to the resistance. Combinations of S and R falling above this line correspond to $S > R$ and, hence, failure. Thus, load effect S_1 acting on a structure having strength R_1 would cause failure, whereas load effect S_2 acting on a structure having resistance R_2 represents a safe combination.

For a given distribution of load effects, the probability of failure can be reduced by increasing the resistances. This would correspond to shifting the distribution of resistances to the right in Fig. 2-5. The probability of failure also could be reduced by reducing the dispersion of the resistances.

The term $Y = R - S$ is called the *safety margin*. By definition, failure will occur if Y is negative, represented by the shaded area in Fig. 2-6. The *probability of failure*, P_f, is the chance that a particular combination of R and S will give a negative value of Y. This probability is equal to the ratio of the shaded area to the total area under the curve in Fig. 2-6. This can be expressed as

$$P_f = \text{probability that } [Y < 0] \tag{2-3}$$

The function Y has mean value $\overline{Y}$ and standard deviation σ_Y. From Fig. 2-6, it can be seen that $\overline{Y} = 0 + \beta\sigma_Y$, where $\beta = \overline{Y}/\sigma_Y$. If the distribution is shifted to the right by increasing the resistance, thereby making $\overline{Y}$ larger, β will increase, and the shaded area, P_f, will decrease. Thus, P_f is a function of β. The factor β is called the *safety index*.

If Y follows a standard statistical distribution, and if $\overline{Y}$ and σ_Y are known, the probability of failure can be calculated or obtained from statistical tables as a function of the type of distribution and the value of β. Consequently, if Y follows a normal distribution and β is 3.5, then $\overline{Y} = 3.5\sigma_Y$, and, from tables for a *normal distribution*, P_f is 1/9090, or 1.1×10^{-4}.

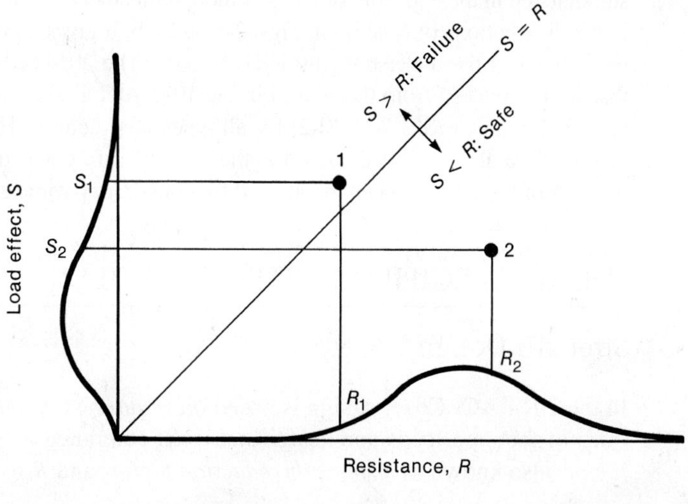

Fig. 2-5
Safe and unsafe combinations of loads and resistances. (From [2-7].)

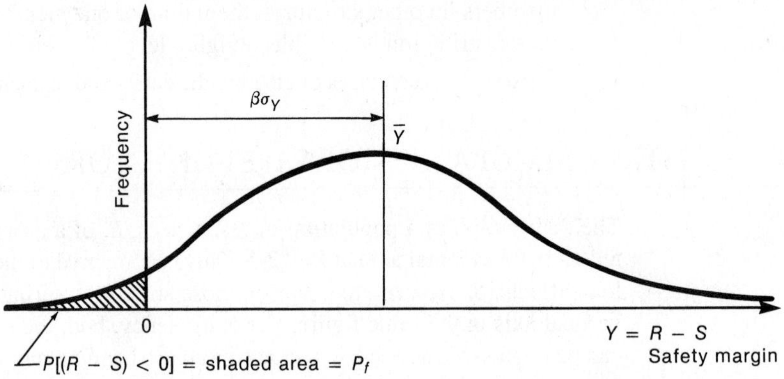

Fig. 2-6
Safety margin, probability of failure, and safety index. (From [2-7].)

This suggests that roughly 1 in every 10,000 structural members designed on the basis that $\beta = 3.5$ will fail due to excessive load or understrength sometime during its lifetime.

The appropriate values of P_f (and hence of β) are chosen by bearing in mind the consequences of failure. Based on current design practice, β is taken between 3 and 3.5 for ductile failures with average consequences of failure and between 3.5 and 4 for sudden failures or failures having serious consequences [2-7], [2-8].

Because the strengths and loads vary independently, it is desirable to have one factor, or a series of factors, to account for the variability in resistances and a second series of factors to account for the variability in load effects. These are referred to, respectively, as *strength-reduction factors* (also called *resistance factors*), ϕ, and *load factors*, α. The resulting design equations are Eqs. (2-2a) through (2-2d).

The derivation of probabilistic equations for calculating values of ϕ and α is summarized and applied in [2-7], [2-8], and [2-9].

The resistance and load factors in the 1971 through 1995 ACI Codes were based on a statistical model, which assumed that if there were a 1/1000 chance of an "overload" and a 1/100 chance of "understrength," the chance that an "overload" and an "understrength" would occur simultaneously is 1/1000 × 1/100 or 1×10^{-5}. Thus, the ϕ factors for ductile beams originally were derived so that a strength of ϕR_n would exceed the load effects 99 out of 100 times. The ϕ factors for columns were then divided by 1.1, because the failure of a column has more serious consequences. The ϕ factors for tied columns that fail in a brittle manner were divided by 1.1 a second time to reflect the consequences of the mode of failure. The original derivation is summarized in the appendix of [2-7]. Although this model is simplified by ignoring the overlap in the distributions of R and S in Figs. 2-5 and 2-6, it gives an intuitive estimate of the relative magnitudes of the understrengths and overloads. The 2014 ACI Code [2-10] uses load factors that were modified from those used in the 1995 ACI Code to be consistent with load factors specified in ASCE/SEI 7-10 [2-2] for all types of structures. However, the strength reduction factors were also modified such that the level of safety and the consideration of the consequences of failure have been maintained for consistency with earlier editions of the ACI Code.

2-6 DESIGN PROCEDURES SPECIFIED IN THE ACI BUILDING CODE

Strength Design

In the 2014 ACI Code, design is based on *required strengths* computed from combinations of factored loads and *design strengths* computed as ϕR_n, where ϕ is a *resistance factor*, also known as a *strength-reduction factor*, and R_n is the nominal resistance. This process is called *strength design*. In the AISC Specifications for steel design, the same design process is known as LRFD (Load and Resistance Factor Design). Strength design and LRFD are methods of limit-states design, except that primary attention is placed on the strength limit states, with the serviceability limit states being checked after the original design is completed. Each member design chapter of the 2014 ACI Code contains a general statement that the design strength (defined here as ϕR_n) at all member sections shall equal or exceed the required strength for the load combination listed in Chapter 5 of the Code. Those load combinations will be defined and discussed in the following section.

Plastic Design

Plastic design, also referred to as *limit design* (not to be confused with limit-states design) or *capacity design*, is a design process that considers the redistribution of moments as successive cross sections yield, thereby forming *plastic hinges* that lead to a plastic mechanism. These concepts are of considerable importance in seismic design, where the amount of ductility expected from a specific structural system leads to a decrease in the forces that must be resisted by the structure.

Plasticity Theorems

Several aspects of the design of statically indeterminate concrete structures are justified, in part, by using the theory of plasticity. These include the (plastic) strength design of continuous frames and two-way slabs for elastically computed loads and moments, and the use of strut-and-tie models for concrete design. Before the theorems of plasticity are presented, several definitions are required:

- A distribution of internal forces (moments, axial forces, and shears) or corresponding stresses is said to be *statically admissible* if it is in equilibrium with the applied loads and reactions.
- A distribution of cross-sectional strengths that equals or exceeds the statically admissible forces, moments, or stresses at every cross section in the structure is said to be a *safe* distribution of strengths.
- A structure is said to be a *collapse mechanism* if there is one more hinge, or plastic hinge, than required for stable equilibrium.
- A distribution of applied loads, forces, and moments that results in a sufficient number and distribution of plastic hinges to produce a collapse mechanism is said to be *kinematically admissible*.

The theory of plasticity is expressed in terms of the following three theorems:

1. Lower-bound theorem. If a structure is subjected to a statically admissible distribution of internal forces and if the member cross sections are chosen to provide a safe distribution of strength for the given structure and loading, the structure either will not collapse or will be just at the point of collapsing. The resulting distribution of internal forces and moments corresponds to a failure load that is a lower bound to the load at failure. This is called a *lower bound* because the computed failure load is less than or equal to the actual collapse load.

2. Upper-bound theorem. A structure will collapse if there is a kinematically admissible set of plastic hinges that results in a plastic collapse mechanism. For any kinematically admissible plastic collapse mechanism, a collapse load can be calculated by equating external and internal work. The load calculated by this method will be greater than or equal to the actual collapse load. Thus, the calculated load is an *upper bound* to the failure load.

3. Uniqueness theorem. If the lower-bound theorem involves the same forces, hinges, and displacements as the upper-bound solution, the resulting failure load is the true or *unique* collapse load.

For the upper- and lower-bound solutions to occur, the structure must have enough ductility to allow the moments and other internal forces from the original loads to redistribute to those corresponding to the bounds of plasticity solutions.

Reinforced concrete design is usually based on elastic analyses. Cross sections are proportioned to have factored nominal strengths, ϕM_n, ϕP_n, and ϕV_n, greater than or equal to the M_u, P_u, and V_u from an elastic analysis. Because the elastic moments and forces are a statically admissible distribution of forces, and because the resisting-moment diagram is chosen by the designer to be a safe distribution, the strength of the resulting structure is a lower bound.

Similarly, the strut-and-tie models presented in Chapter 17 give lower-bound estimates of the capacity of concrete structures if

(a) the strut-and-tie model of the structure represents a statically admissible distribution of forces,

(b) the strengths of the struts, ties, and nodal zones are chosen to be safe, relative to the computed forces in the strut-and-tie model, and

(c) the members and joint regions have enough ductility to allow the internal forces, moments, and stresses to make the transition from the strut-and-tie forces and moments to the final force and moment distribution.

Thus, if adequate ductility is provided the strut-and-tie model will give a so-called safe estimate, which is a lower-bound estimate of the strength of the strut-and-tie model. Plasticity solutions are used to develop the yield-line method of analysis for two-way slabs, presented in Chapter 14.

2-7 LOAD FACTORS AND LOAD COMBINATIONS IN THE 2014 ACI CODE

The 2014 ACI Code presents load factors and load combinations in Chapter 5, which are from ASCE/SEI 7-10, *Minimum Design Loads for Buildings and Other Structures* [2-2], with slight modifications. The load factors from Code Section 5.3.1 are to be used with the strength-reduction factors in Code Sections 21.2.1 and 21.2.2. These load factors and strength reduction factors were derived in [2-8] for use in the design of steel, timber, masonry, and concrete structures and are used in the AISC LRFD Specification for steel structures [2-11]. For concrete structures, resistance factors that are compatible with the ASCE/SEI 7-10 load factors were derived by ACI Committee 318 and Nowak and Szerszen [2-12].

Terminology and Notation

The ACI Code uses the subscript u to designate the *required strength*, which is a load effect computed from combinations of factored loads. The sum of the combination of factored loads is U as, for example, in

$$U = 1.2D + 1.6L \tag{2-4}$$

where the symbol U and subscript u are used to refer to the sum of the factored loads in terms of loads, or in terms of the effects of the factored loads, M_u, V_u, and P_u.

The member strengths computed using the specified material strengths, f'_c and f_y, and the nominal dimensions, as shown on the drawings, are referred to as the *nominal moment strength*, M_n, or *nominal shear strength*, V_n, and so on. The *reduced nominal strength* or design strength is the nominal strength multiplied by a strength-reduction factor, ϕ. The design equation is thus:

$$\phi M_n \geq M_u \tag{2-2b}$$

$$\phi V_n \geq V_u \tag{2-2c}$$

and so on.

Load Factors and Load Combinations

Load Combinations

Structural failures usually occur under combinations of several loads. In recent years these combinations have been presented in what is referred to as the *companion action format*. This is an attempt to model the expected load combinations.

The load combinations in ACI Code Section 5.3 are examples of *companion action load combinations* chosen to represent realistic load combinations that might occur. In principle, each of these combinations includes one or more *permanent loads* (D or F) with load factors of 1.2, plus the dominant or *principal variable load* (L, S, or others) with a load factor of 1.6, plus one or more *companion-action variable loads*. The companion-action loads are computed by multiplying the specified loads (L, S, W, or others) by *companion-action load factors* between 0.2 and 1.0. The companion-action load factors were chosen to provide results for the companion-action load effects that would be likely during an instance in which the principal variable load is maximized.

In the design of structural members in buildings that are not subjected to significant wind or earthquake forces the factored loads are computed from either Eq. (2-5) or Eq. (2-6):

$$U = 1.4D \tag{2-5}$$

(ACI Eq. 5.3.1a)

where D is the specified dead load. Where a fluid load, F, is present, it shall be included with the same load factor as used for D in this and the following equations.

For combinations including dead load; live load, L; and roof loads:

$$U = 1.2D + 1.6L + 0.5(L_r \text{ or } S \text{ or } R) \tag{2-6}$$

(ACI Eq. 5.3.1b)

where

L = live load that is a function of use and occupancy
L_r = roof live load
S = roof snow load
R = roof rain load

The terms in Eqs. (2-5) through (2-11) may be expressed as *direct loads* (such as distributed loads from dead and live weight) or *load effects* (such as moments and shears caused by the given loads). The design of a roof structure, or the columns and footings supporting a roof and one or more floors, would take the roof live load equal to the largest of the three loads (L_r or S or R), with the other two roof loads in the brackets taken as zero. For the common case of a member supporting dead and live load only, Eq. (2-6) is written as:

$$U = 1.2D + 1.6L \tag{2-4}$$

If the roof load exceeds the floor live loads, or if a column supports a total roof load that exceeds the total floor live load supported by the column:

$$U = 1.2D + 1.6(L_r \text{ or } S \text{ or } R) + (1.0L \text{ or } 0.5W) \tag{2-7}$$

(ACI Eq. 5.3.1c)

The roof loads are *principal variable loads* in Eq. (2-7), and they are *companion variable loads* in Eqs. (2-6) and (2-8).

$$U = 1.2D + 1.0W + 1.0L + 0.5(L_r \text{ or } S \text{ or } R) \tag{2-8}$$

(ACI Eq. 5.3.1d)

Wind load, W, is the principal variable load in Eq. (2-8) and is a companion variable load in Eq. (2-7). Wind loads specified in ASCE/SEI 7-10 represent *strength-level* winds, as opposed to the *service-level* wind forces specified in earlier editions of the minimum load standards from ASCE/SEI Committee 7. If the governing building code for the local jurisdiction specifies service-level wind forces, $1.6W$ is to be used in place of $1.0W$ in Eqs. (2-8) and (2-10), and $0.8W$ is to be used in place of $0.5W$ in Eq. (2-7).

Earthquake Loads

If earthquake loads are significant:

$$U = 1.2D + 1.0E + 1.0L + 0.2S \tag{2-9}$$

(ACI Eq. 5.3.1e)

where the load factor of 1.0 for the earthquake loads corresponds to a *strength-level earthquake* that is larger than a *service-load earthquake*. If the loading code used in a jurisdiction is based on the service-load earthquake, the load factor on E is 1.4 instead of 1.0.

Dead Loads That Stabilize Overturning and Sliding

If the effects of dead loads stabilize the structure against wind or earthquake loads,

$$U = 0.9D + 1.0W \tag{2-10}$$

(ACI Eq. 5.3.1f)

or

$$U = 0.9D + 1.0E \tag{2-11}$$

(ACI Eq. 5.3.1g)

Load Factor for Small Live Loads

ACI Code Section 5.3.3 allows that the load factor of 1.0 for L in Eqs. (2-7), (2-8), and (2-9) may be reduced to 0.5 except for

(a) garages,

(b) areas occupied as places of public assembly, and

(c) all areas where the live load is greater than 100 psf.

Lateral Earth Pressure

Lateral earth pressure is represented by the letter H. Where lateral earth pressure adds to the effect of the principal variable load, H should be included in Eqs. (2-6),

(2-10), and (2-11) with a load factor of 1.6. When lateral earth pressure is permanent and reduces the affect of the principal variable load, H should be included with a load factor of 0.9. For all other conditions, H is not to be used in the ACI load combination equations.

Self-Straining Effects

ACI Code Section 5.3.6 uses the letter T to represent actions caused by differential settlement and restrained volume change movements due to either shrinkage or thermal expansion and contraction. Where applicable, these loads are to be considered in combination with other loads. In prior editions of the ACI Code, T was combined with dead load, D, in Eq. (2-6), and thus, the load factor was 1.2. The 2014 edition of the ACI Code states that to establish the appropriate load factor for T the designer is to consider the uncertainty associated with the magnitude of the load, the likelihood that T will occur simultaneously with the maximum value of other applied loads, and the potential adverse effects if the value of T has been underestimated. In any case, the load factor for T is not to be taken less than 1.0. In typical practice, expansion joints and construction pour strips have been used to limit the effects of volume change movements. A recent study of precast structural systems [2-13] gives recommended procedures to account for member and connection stiffnesses and other factors that may influence the magnitude of forces induced by volume change movements.

In the analysis of a building frame, it is frequently best to analyze the structure elastically for each load to be considered and to combine the resulting moments, shears, and so on for each member according to Eqs. (2-4) to (2-11). (Exceptions to this are analyses of cases in which linear superposition does not apply, such as second-order analyses of frames. These must be carried out at the factored-load level.) The procedure used is illustrated in Example 2-1.

EXAMPLE 2-1 Computation of Factored-Load Effects

Figure 2-7 shows a beam and column from a concrete building frame. The loads per foot on the beam are dead load, $D = 1.58$ kips/ft, and live load, $L = 0.75$ kip/ft. Additionally, wind load is represented by the concentrated loads at the joints. The moments in a beam and in the columns over and under the beam due to $1.0D$, $1.0L$, and $1.0W$ are shown in Figs. 2-7b to 2-7d.

Compute the required strengths, using Eqs. (2-4) through (2-11). For the moment at section A, four load cases must be considered

(a) $U = 1.4D$ (2-5)

- Because there are no fluid or thermal forces to consider, $U = 1.4 \times -39 = -54.6$ k-ft.

(b) $U = 1.2D + 1.6L + 0.5(L_r \text{ or } S \text{ or } R)$ (2-6)

- Assuming that there is no differential settlement of the interior columns relative to the exterior columns and assuming there is no restrained shrinkage, the self-equilibrating actions, T, will be taken to be zero.

- Because the beam being considered is not a roof beam, L_r, S, and R are all equal to zero. (Note that the axial loads in the columns support axial forces from the roof load and the slab load.)

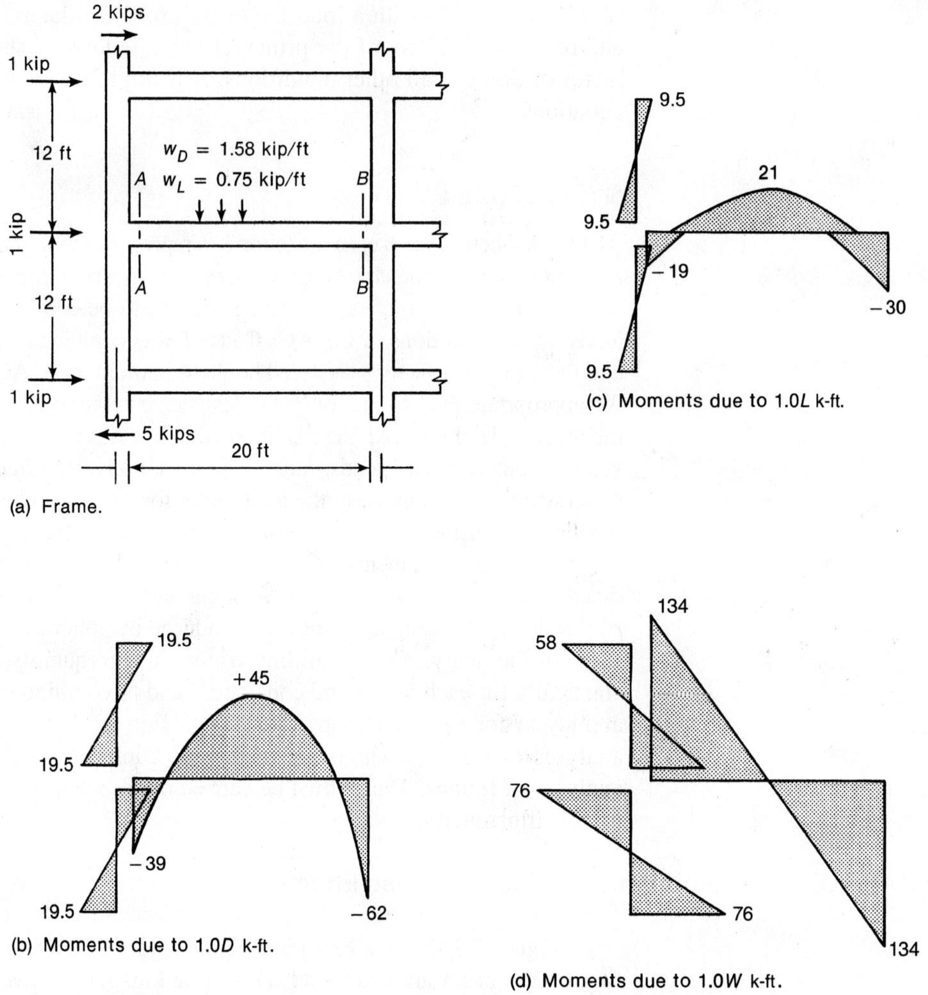

Fig. 2-7
Moment diagrams—
Example 2-1.

Equation (2-6) becomes

$$
\begin{aligned}
U &= 1.2D + 1.6L \qquad (2\text{-}4)\\
&= 1.2 \times -39 + 1.6 \times -19 = -77.2 \text{ k-ft}
\end{aligned}
$$

(c) Equation (2-7) does not govern because this is not a roof beam.

(d) For Eq. (2-8), assume strength-level wind forces have been specified, so the load factor of 1.0 is used for W.

$$U = 1.2D + 1.0W + 1.0L + 0.5(L_r \text{ or } S \text{ or } R) \qquad (2\text{-}8)$$

where ACI Code Section 5.3.3 allows 1.0L to be reduced to 0.5, so,

$$
\begin{aligned}
U &= 1.2D + 1.0W + 0.5L\\
&= 1.2 \times -39 \pm 1.0 \times 134 + 0.5 \times -19\\
&= -56.3 \pm 134\\
&= -190 \text{ or } +77.7 \text{ k-ft}
\end{aligned}
$$

The positive and negative values of the wind-load moment are due to the possibility of winds alternately blowing on the two sides of the building.

(e) The dead-load moments can counteract a portion of the wind- and live-load moments. This makes it necessary to consider Eq. (2-10):

$$\begin{aligned} U &= 0.9D + 1.6W \qquad (2\text{-}10) \\ &= 0.9 \times -39 \pm 1.0 \times 134 = -35.1 \pm 134 \\ &= +98.9 \text{ or } -169 \text{ k-ft} \end{aligned}$$

Thus the required strengths, M_u, at section A–A are +98.9 k-ft and −190 k-ft. ■

This type of computation is repeated for a sufficient number of sections to make it possible to draw shearing-force and bending-moment envelopes for the beam.

Strength-Reduction Factors, ϕ

ACI Code Chapter 21 defines the following set of strength-reduction factors to be used in conjunction with the load combinations given in ACI Code Eqs. (5.3.1a) through (5.3.1g).

Flexure or Combined Flexure and Axial Load

Tension-controlled sections	$\phi = 0.90$
Compression-controlled sections:	
(a) Members with spiral reinforcement	$\phi = 0.75$
(b) Other compression-controlled sections	$\phi = 0.65$

There is a transition region between tension-controlled and compression-controlled sections. The concept of tension-controlled and compression-controlled sections, and the resulting strength-reduction factors, will be presented for beams in flexure, axially loaded columns, and columns loaded in combined axial load and bending in Chapters 4, 5, and 11. The derivation of the ϕ factors will be introduced at that time.

Other actions

Shear and torsion	$\phi = 0.75$
Bearing on concrete	$\phi = 0.65$
Strut-and-tie model	$\phi = 0.75$

2-8 LOADINGS AND ACTIONS

Direct and Indirect Actions

An *action* is anything that gives rise to stresses in a structure. The term *load* or *direct action* refers to concentrated or distributed forces resulting from the weight of the structure and its contents, or pressures due to wind, water, or earth. An *indirect action* or *imposed deformation* is a movement or deformation that does not result from applied loads, but that causes stresses in a structure. Examples are uneven support settlements of continuous beams and shrinkage of concrete if it is not free to shorten.

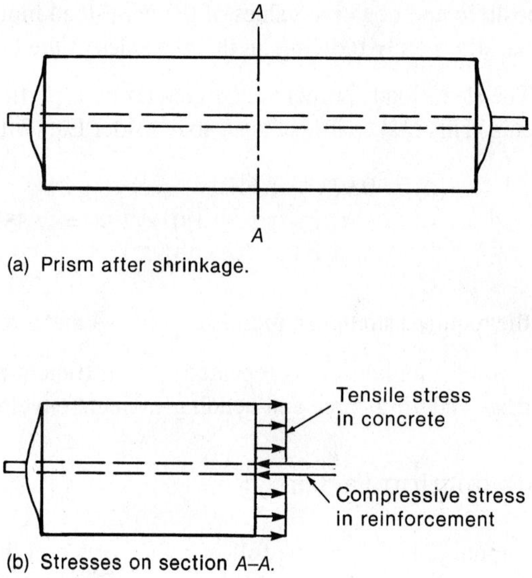

Fig. 2-8
Self-equilibrating stresses due to shrinkage.

Because the stresses due to imposed deformations do not resist an applied load, they are generally *self-equilibrating*. Consider, for example, a prism of concrete with a reinforcing bar along its axis. As the concrete shrinks, its shortening is resisted by the reinforcement. As a result, a compressive force develops in the steel and an equal and opposite tensile force develops in the concrete, as shown in Fig. 2-8. If the concrete cracks from this tension, the tensile force in the concrete at the crack is zero, and for equilibrium, the steel force must also disappear at the cracked section. Section 1.3.3 of ASCE/SEI 7-10 refers to imposed deformations as *self-straining forces*.

Classifications of Loads

Loads may be described by their variability with respect to time and location. A *permanent* load remains roughly constant once the structure is completed. Examples are the self-weight of the structure and soil pressure against foundations. *Variable* loads, such as occupancy loads and wind loads, change from time to time. Variable loads may be *sustained loads* of long duration, such as the weight of filing cabinets in an office, or loads of *short duration*, such as the weight of people in the same office. Creep deformations of concrete structures result from permanent loads and the sustained portion of the variable loads. A third category is *accidental loads*, which include vehicular collisions and explosions.

Variable loads may be *fixed* or *free* in location. Thus, the live loading in an office building is free, because it can occur at any point in the loaded area. A train load on a bridge is not fixed longitudinally, but is fixed laterally by the rails.

Loads frequently are classified as *static loads* if they do not cause any appreciable acceleration or vibration of the structure or structural elements and as *dynamic loads* if they do. Small accelerations are often taken into account by increasing the specified static loads to account for the increases in stress due to such accelerations and vibrations. Larger accelerations, such as those which might occur in highway bridges, crane rails, or elevator supports are accounted for by multiplying the effect of the live load by an *impact factor*. Alternatively, dynamic analyses may be used.

Three levels of live load or wind load may be of importance. The load used in calculations involving the strength limit states should represent the maximum load on the structure in its lifetime. Wherever possible, therefore, the specified live, snow, and wind loadings should represent the mean value of the corresponding maximum lifetime load. A *companion-action load* is the portion of a variable load that is present on a structure when some other variable load is at its maximum. In checking the serviceability limit states, it may be desirable to use a *frequent* live load, which is some fraction of the mean maximum lifetime load (generally, 50 to 60 percent); for estimating sustained load deflections, it may be desirable to consider a *sustained* or *quasi-permanent* live load, which is generally between 20 and 30 percent of the specified live load. This differentiation is not made in the ACI Code, which assumes that the entire specified load will be the load present in service. As a result, service-load deflections and creep deflections of slender columns tend to be overestimated.

Loading Specifications

Most cities in the United States base their building codes on the *International Building Code* [2-14]. The loadings specified in this code are based on the loads recommended in *Minimum Design Loads for Buildings and Other Structures*, ASCE/SEI 7-10 [2-2].

In the following sections, the types of loadings presented in ASCE/SEI 7-10 will be briefly reviewed. This review is intended to describe the characteristics of the various loads. For specific values, the reader should consult the building code in effect in his or her own locality.

Dead Loads

The *dead load* on a structural element is the weight of the member itself, plus the weights of all materials permanently incorporated into the structure and supported by the member in question. This includes the weights of permanent partitions or walls, the weights of plumbing stacks, electrical feeders, permanent mechanical equipment, and so on. Tables of dead loads are given in ASCE/SEI 7-10.

In the design of a reinforced concrete member, it is necessary initially to estimate the weight of the member. Methods of making this estimate for beams and one-way slabs are given in Chapter 5. Once the member size has been computed, its weight is calculated by multiplying the volume by the density of concrete, taken as 145 lb/ft^3 for plain concrete and 150 lb/ft^3 for reinforced concrete (5 lb/ft^3 is added to account for reinforcement). For lightweight concrete members, the density of the concrete must be determined from trial batches or as specified by the producer. In heavily reinforced members, the density of the reinforced concrete may exceed 150 lb/ft^3.

In working with SI units (metric units), the weight of a member is calculated by multiplying the volume by the mass density of concrete and the gravitational constant, 9.81 N/kg. In this calculation, it is customary to take the mass density of normal-density concrete containing an average amount of reinforcement (roughly 2 percent by volume) as 2450 kg/m^3, made up of 2300 kg/m^3 for the concrete and 150 kg/m^3 for the reinforcement. The weight of a cubic meter of reinforced concrete is thus (1 m^3 $\times$ 2450 kg/m^3 $\times$ 9.81 N/kg)/1000 = 24.0 kN, and its weight density is 24 kN/m^3.

The dead load referred to in Eqs. (2-5) to (2-11) is the load computed from the dimensions shown on drawings and the assumed densities. It is therefore close to the mean value of this load. Actual dead loads will vary from the calculated values, because the actual dimensions and densities may differ from those used in the calculations. Sometimes the materials for the roof, partitions, or walls are chosen on the basis of a separate bid document, and their actual weights may be unknown at the time of the design. Tabulated densities of materials frequently tend to underestimate the actual dead loads of the material in place in a structure.

Some types of dead load tend to be highly uncertain. These include pavement on bridges, which may be paved and repaved several times over a period of time, or where a greater thickness of pavement may be applied to correct sag or alignment problems. Similarly, earth fill over an underground structure may be up to several inches thicker than assumed and may or may not be saturated with water. In the construction of thin curved-shell roofs or other lightweight roofs, the concrete thickness may exceed the design values and the roofing may be heavier than assumed.

If dead-load moments, forces, or stresses tend to counteract those due to live loads or wind loads, the designer should carefully examine whether the counteracting dead load will always exist. Thus, dead loads due to soil or machinery may be applied late in the construction process and may not be applied evenly to all parts of the structure at the same time, leading to a potentially critical set of moments, forces, or stresses under partial loads.

It is generally not necessary to checkerboard the self-weight of the structure by using dead-load factors of $\alpha_D = 0.9$ and 1.2 in successive spans, because the structural dead loads in successive spans of a beam tend to be highly correlated. On the other hand, it may be necessary to checkerboard the superimposed dead load by using load factors of $\alpha_D = 0$ or 1.2 in cases where counteracting dead load is absent at some stages of construction or use.

Live Loads Due to Use and Occupancy

Most building codes contain a table of design or specified live loads. To simplify the calculations, these are expressed as uniform loads on the floor area. In general, a building live load consists of a sustained portion due to day-to-day use (see Fig. 2-4) and a variable portion generated by unusual events. The sustained portion changes a number of times during the life of the building—when tenants change, when the offices are rearranged, and so on. Occasionally, high concentrations of live loading occur during periods when adjacent spaces are remodeled, when office parties are held, or when material is stored temporarily. The loading given in building codes is intended to represent the maximum sum of these loads that will occur on a small area during the life of the building. Typical specified live loads are given in Table 2-1.

In buildings where nonpermanent partitions might be erected or rearranged during the life of the building, allowance should be made for the weight of these partitions. ASCE/SEI 7-10 specifies that provision for partition weight should be made, regardless of whether partitions are shown on the plans, unless the specified live load exceeds 80 psf. It is customary to represent the partition weight with a uniform load of 20 psf or a uniform load computed from the actual or anticipated weights of the partitions placed in any probable position. ASCE/SEI 7-10 considers this to be live load, because it may or may not be present in a given case.

As the loaded area increases, the average maximum lifetime load decreases because, although it is quite possible to have a heavy load on a small area, it is unlikely that this would occur in a large area. This is taken into account by multiplying the specified live loads by a *live-load reduction factor*.

TABLE 2-1 Typical Live Loads Specified in ASCE/SEI 7-10

	Uniform, psf	Concentration, lb
Apartment buildings		
Private rooms and corridors serving them	40	
Public rooms and corridors serving them	100	
Office buildings		
Lobbies and first-floor corridors	100	2000
Offices	50	2000
Corridors above first floor	80	2000
File and computer rooms shall be designed for heavier loads based on anticipated occupancy		
Schools		
Classrooms	40	1000
Corridors above first floor	80	1000
First-floor corridors	100	1000
Stairs and exitways	100	
Storage warehouses		
Light	125	
Heavy	250	
Stores		
Retail		
Ground floor	100	1000
Upper floors	75	1000
Wholesale, all floors	125	1000

In ASCE/SEI 7-10, this factor is based on the *influence area*, A_I, for the member being designed. The concept of influence lines and influence areas is discussed in Chapter 5. To figure out the influence area of a given member, one can imagine that the member in question is raised by a unit amount, say, 1 in. as shown in Fig. 2-9. The portion of the loaded area that is raised when this is done is called the *influence area*, A_I, because loads acting anywhere in this area will have a significant impact on the load effects in the member in question. This concept is illustrated in Fig. 2-9 for an interior floor beam and an edge column.

In contrast, the *tributary area*, A_T, extends out from the beam or column to the lines of zero shear in the floor around the member under consideration. For the beam in Fig. 2-9a, the limits on A_T are given by the dashed lines halfway to the next beam on each side. The tributary areas are shown in a darker shading in Figs. 2-9a and 2-9b. An examination of Fig. 2-9a shows that A_T is half of A_I for an interior beam. For the column in Fig. 2-9b, A_T is one-fourth of A_I. Because two-way slab design is based on the total moments in one slab panel, the influence area for such a slab is defined by ASCE/SEI 7-10 as the total panel area.

Previous versions of the ASCE/SEI 7 document allowed the use of reduced live loads, L, in the design of members, based on the influence area A_I. However, the influence-area concept is not widely known compared with that of the tributary area, A_T. In ASCE/SEI 7-10, the influence area is given as $A_I = K_{LL}A_T$, where A_T is the tributary area of the member being designed and K_{LL} is the ratio A_I/A_T. The reduced live load, L, is given by

$$L = L_o\left[0.25 + \frac{15}{\sqrt{K_{LL}A_T}}\right] \tag{2-12}$$

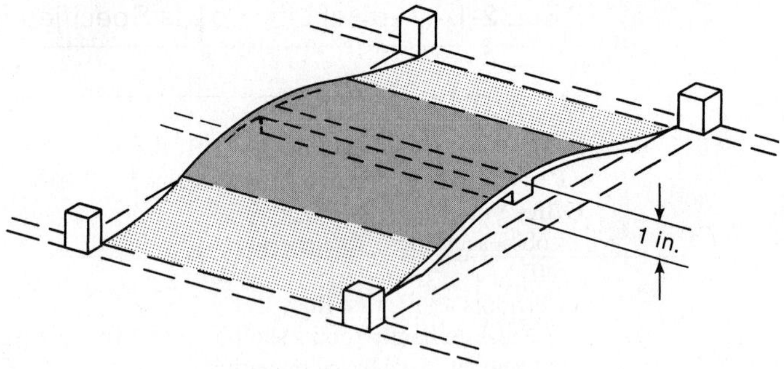

(a) Interior floor beam.

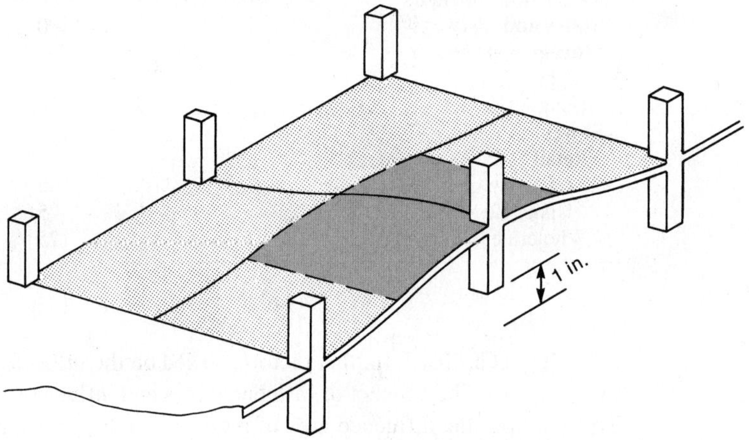

Fig. 2-9
Influence areas.

(b) Edge column.

where L_o is the unreduced live load. Values of K_{LL} are given as follows:

Interior columns and exterior columns without cantilever slabs	$K_{LL} = 4$
Exterior columns with cantilever slabs	$K_{LL} = 3$
Corner columns with cantilever slabs	$K_{LL} = 2$
Interior beams and edge beams without cantilever slabs	$K_{LL} = 2$
All other members, including one-way and two-way slabs	$K_{LL} = 1$

The live-load reduction applies only to live loads due to use and occupancy (not for snow, etc.). No reduction is made for areas used as places of public assembly, for garages, or for roofs. In ASCE/SEI 7-10, the reduced live load cannot be less than 50 percent of the unreduced live load for columns supporting one floor or for flexural members, and no less than 40 percent for other members.

For live loads exceeding 100 psf, no reduction is allowed by ASCE/SEI 7-10, except that the design live load on columns supporting more than one floor can be reduced by 20 percent.

The reduced uniform live loads are then applied to those spans or parts of spans that will give the maximum shears, moments, and so on, at each critical section. This approach is illustrated in Chapter 5.

The ASCE/SEI 7-10 standard requires that office and garage floors and sidewalks be designed to safely support either the reduced uniform design loads or a concentrated load of from 1000 to 8000 lb (depending on occupancy), spread over an area of from 4.5 in. by 4.5 in. to 30 in. by 30 in. The concentrated loads are intended to represent heavy items such as office safes, pianos, car wheels, and so on.

In checking the concentrated load capacity, it generally is necessary to assume an effective width of floor to carry the load to the supports. For one-way floors, this is usually the width of the concentrated load reaction plus one slab effective depth on each side of the load. For two-way slabs, Chapter 13 shows that a concentrated load applied at various points in the slab gives maximum moments (at midspan and near the support columns) that are similar in magnitude to those computed for a complete panel loaded with a uniform load. In many cases, this makes it unnecessary to check the concentrated load effects on maximum moment for two-way slabs.

The live loads are assumed to be large enough to account for the impact effects of normal use and traffic. Special impact factors are given in the loading specifications for supports of elevator machinery, large reciprocating or rotating machines, and cranes.

Classification of Buildings for Wind, Snow, and Earthquake Loads

The ASCE/SEI 7-10 requirements for design for wind, snow, and earthquake become progressively more restrictive as the level of risk to human life in the event of a collapse increases. These are referred to as *risk categories*:

I. Buildings and other structures that represent a low hazard to human life in the event of failure, such as agricultural facilities.

II. Buildings and other structures that do not fall into categories I, III, or IV.

III. Buildings or other structures that represent a substantial hazard to human life in the event of failure, such as assembly occupancies, schools, and detention facilities. Also, buildings and other structures not included in risk category IV that contain a sufficient quantity of highly toxic or explosive substances that pose a significant threat to the general public if released.

IV. Buildings and other structures designated as essential facilities, such as hospitals, fire and police stations, communication centers, and power-generating stations and facilities. Also, buildings and other structures that contain a sufficient quantity of highly toxic or explosive substances that pose a significant threat to the general public if released.

Snow Loads, S

Snow accumulation on roofs is influenced by climatic factors, roof geometry, and the exposure of the roof to the wind. Unbalanced snow loads due to drifting or sliding of snow or uneven removal of snow by workers are very common. Large accumulations of snow often will occur adjacent to parapets or other points where roof heights change. ASCE/SEI 7-10 gives detailed rules for calculating snow loads to account for the effects of snow drifts. It is necessary to design for either a uniform or an unbalanced snow load, whichever gives the worst effect.

Roof Live Loads, L_r, and Rain Loads, R

In addition to snow loads, roofs should be designed for certain minimum live loads (L_r) to account for workers or construction materials on the roof during erection or when repairs are made. Consideration must also be given to loads due to rainwater, R. Because roof drains are rarely inspected to remove leaves or other debris, ASCE/SEI 7-10 requires that roofs be able to support the load of all rainwater that could accumulate on a particular portion of a roof if the primary roof drains were blocked. Frequently, controlled-flow roof drains are used to slow the flow of rainwater off a roof. This reduces plumbing and storm sewage costs but adds to the costs of the roof structure.

If the design snow load is small and the roof span is longer than about 25 ft, rainwater will tend to form ponds in the areas of maximum deflection. The weight of the water in these regions may cause an increase in the deflections, allowing more water to collect, and so on. If the roof is not sufficiently stiff, a *ponding failure* will occur when the weight of ponded water reaches the capacity of the roof members [2-15].

Construction Loads

During the construction of concrete buildings, the weight of the fresh concrete is supported by formwork, which frequently are supported by floors lower down in the structure. In addition, construction materials are often piled on floors or roofs during construction. ACI Code Section 26.11.2.1(c) states the following:

> No construction loads shall be placed on, nor any formwork removed from, any part of the structure under construction except when that portion of the structure in combination with remaining formwork has sufficient strength to support safely its weight and loads placed thereon and without impairing serviceability.

Wind Loads

The pressure exerted by the wind is related to the square of its velocity. Due to the roughness of the earth's surface, the wind velocity at any particular instant consists of an average velocity plus superimposed turbulence, referred to as *gusts*. As a result, a structure subjected to wind loads assumes an average deflected position due to the average velocity pressure and vibrates from this position in response to the gust pressure. In addition, there will generally be deflections transverse to the wind (due to vortex shedding) as the wind passes the building. The vibrations due to the wind gusts are a function of (1) the relationship between the natural energy of the wind gusts and the energy necessary to displace the building, (2) the relationship between the gust frequencies and the natural frequency of the building, and (3) the damping of the building [2-16].

Three procedures are specified in ASCE/SEI 7-10 for the calculation of wind pressures on buildings: the *envelope procedure*, limited in application to buildings with a mean roof height of 60 ft or less; the *directional procedure*, limited to regular buildings that do not have response characteristics making it subject to a cross-wind loading, vortex shedding, or channeling of the wind due to upwind obstructions; and the *wind tunnel procedure*, used for complex buildings. We shall consider the directional procedure. Variations of this method apply to design of the main wind-force-resisting systems of buildings and to the design of components and cladding.

In the directional procedure, the wind pressure on the main wind-force-resisting system is

$$p = qGC_p - q_i(GC_{pi}) \tag{2-13}$$

where either $q = q_z$, the velocity pressure evaluated at height z above the ground on the windward wall, or $q = q_h$, the pressure on the roof, leeward walls, and sidewalls, evaluated at the mean roof height, h, and q_i is the internal pressure or suction on the interior of the walls and roof of the building, also evaluated at the mean roof height.

The total wind pressure p, is the sum of the external pressure on the windward wall and the suction on the leeward wall, which is given by the first term on the right-hand side of Eq. (2-13) plus the second term, p_i, which accounts for the internal pressure. The internal pressure, p_i, is the same on all internal surfaces at any given time. Thus, the internal pressure or suction on the inside of the windward wall is equal but opposite in direction to the internal pressure or suction on the inside of the leeward wall. As a result, the interior wind forces on opposite walls cancel out in most cases, leaving only the external pressure to be resisted by the main wind-force-resisting system. The terms in Eq. (2-13) are defined as:

1. Design pressure, *p*. The *design pressure* is an equivalent static pressure or suction in psf assumed to act perpendicular to the surface in question. On some surfaces, it varies over the height; on others, it is assumed to be constant.

2. Wind Velocity pressure, *q*. The *wind velocity pressure* at height z on the windward wall, q_z, is the pressure (psf) exerted by the wind on a flat plate suspended in the wind stream. It is calculated as

$$q_z = 0.00256 K_z K_{zt} K_d V^2 \tag{2-14}$$

where

V = nominal design 3-sec gust wind speed in miles per hour at a height of 33 ft (10 m) above the ground in Exposure C, open terrain (3 percent probability of exceedance in 50 years; Category III and IV buildings)

K_z = velocity pressure exposure coefficient, which increases with height above the surface and reflects the roughness of the surface terrain

K_{zt} = the topographic factor that accounts for increases in wind speed as it passes over hills

K_d = directionality factor equal to 0.85 for rectangular buildings and 0.90 to 0.95 for circular tanks and the like

The constant 0.00256 reflects the mass density of the air and accounts for the mixture of units in Eq. (2-14).

Several maps and tables for the variables in Eq. (2-14) are given in ASCE/SEI 7-10. Special attention must be given to mountainous terrain, gorges, and promontories subject to unusual wind conditions and regions subject to tornadoes.

At any location, the mean wind velocity is affected by the roughness of the terrain upwind from the structure in question. At a height of 700 to 1500 ft, the wind reaches a steady velocity, as shown by the vertical lines in the plots of K_z in Fig. 2-10. Below this height, the velocity decreases and the turbulence, or gustiness, increases as one approaches the surface. These effects are greater in urban areas than in rural areas, due to the greater surface roughness in built-up areas. The factor K_z in Eq. (2-14) relates the wind pressure at any elevation z feet to that at 33 ft (10 m) above the surface for Exposure C. ASCE/SEI 7-10 gives tables and equations for K_z as a function of the type of exposure (urban, country, etc.) and the height above the surface.

For **side walls**, **leeward wall**, and **roof surfaces**, q_h is a constant suction (negative pressure) on side walls, the leeward wall, and flat roofs, evaluated by using h equal to the average height of the roof. For a sloping roof, q_h must be evaluated as both a positive (downward) pressure and a negative (upward) pressure for the portion of the roof sloping toward the direction of the wind.

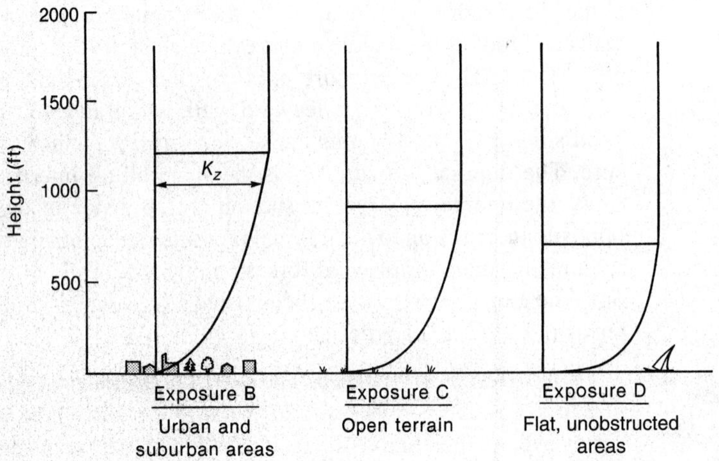

Fig. 2-10
Profiles of velocity pressure exposure coefficient, K_z, for differing terrain.

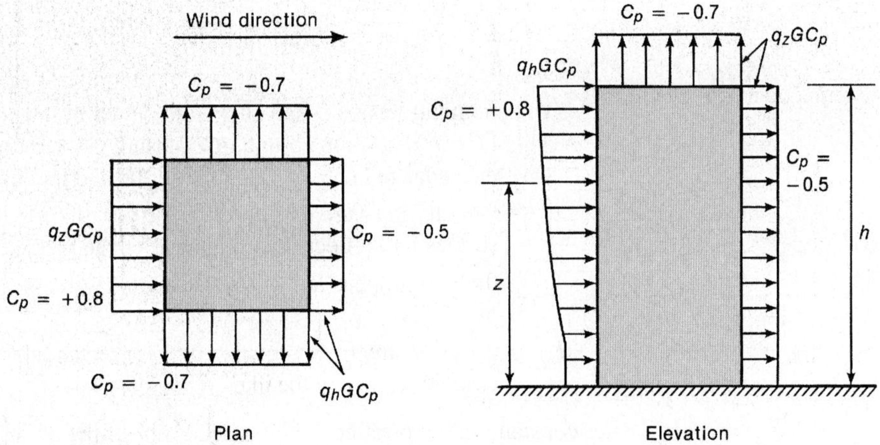

Fig. 2-11
Wind pressures and suctions on a building.

3. Gust-effect factor, *G*. The gust-effect factor, G, in Eq. (2-13) relates the dynamic properties of the wind and the structure. For flexible buildings, it is calculated. For most buildings that tend to be stiff, it is taken to be equal to 0.85.

4. External pressure coefficient, C_p. When wind blows past a structure, it exerts a positive pressure on the windward wall and a negative pressure (suction) on the leeward wall, side walls, and roof as shown in Fig. 2-11. The overall pressures to be used in the design of a structural frame are computed via Eq. (2-13), where C_p is the sum or difference in the pressure coefficients for the windward and leeward walls. Thus, $C_p = +0.80$ (pressure) on the left-hand (windward) wall in Fig. 2-11 and $C_p = -0.50$ (suction) on the right-hand (leeward) wall add together to produce the load on the frame because they have the same direction. Values of the pressure coefficients are given in the ASCE/SEI 7-10. Typical values are shown in Fig. 2-11 for a building having the shape and proportions shown. For a rectangular building with the wind on the narrow side, C_p for the leeward wall varies between −0.5 and −0.2.

Earthquake Loads

Earthquake loads and design for earthquakes are discussed in Chapter 19.

Self-Equilibrating Loadings

Most loads result from things like the weight of the structure or externally applied loads, such as live load or wind load. These loads cause internal forces and moments that are in equilibrium with the external loads. Many structures are subjected to *imposed or restrained deformations*, which are independent of the applied loads. Examples include differential settlements, nonlinear thermal stresses in bridge decks, restrained shrinkage, and prestressing of indeterminate structures. These deformations cause a set of internal forces or moments that are in equilibrium with themselves, as shown in Fig. 2-8. ASCE/SEI 7-10 refers to these as *self-straining forces*. Because these loading cases do not involve applied loads, the magnitude of the internal forces and moments results from

(a) the magnitude of the imposed deformation, and

(b) the resistance of the structure to the deformation (a function of the stiffness of the structure at the time that the deformation occurs).

Consider a two-span beam in which the central support settles relative to the line joining the end supports. The structure resists the differential settlement, setting up internal forces and moments. If the beam is uncracked when it is forced through the differential settlement, the internal forces are larger than they would be if the beam were cracked. If the beam undergoes creep, the magnitude of the internal forces and moments decreases, as shown experimentally by Ghali, Dilger, and Neville [2-17].

Similarly, prestress forces in a two-span continuous beam may tend to lift the center reaction off its support, changing the reactions. This, in turn, causes internal forces and so-called *secondary moments* that are in equilibrium with the change in the reactions. The magnitude of these forces and moments is larger in an uncracked beam than in a cracked beam. They may be partially dissipated by creep.

Other Loads

ASCE/SEI 7-10 also gives soil loads on basement walls, loads due to floods, and loads due to ice accretion.

2-9 DESIGN FOR ECONOMY

A major aim of structural design is economy. The overall cost of a building project is strongly affected by both the cost of the structure and the financing charges, which are a function of the rate of construction.

In a cast-in-place building, the costs of the floor and roof systems make up roughly 90 percent of the total structural costs. The cost of a floor system is divided between the costs of building and stripping the *forms*; providing, bending, and placing the *reinforcement*; and providing, placing, and finishing the *concrete*. In general, material costs increase as the column spacing increases and the cost of the forms is the largest single item, accounting for 40 to 60 percent of the total costs.

Formwork costs can be reduced by reusing the forms from area to area and floor to floor. Beam, slab, and column sizes should be chosen to allow the maximum reuse of the forms. It is generally uneconomical to try to save concrete and steel by meticulously calculating the size of every beam and column to fit the loads exactly, because, although this could save cents in materials, it will cost dollars in forming costs.

Furthermore, changing section sizes often leads to increased design complexity, which in turn leads to a greater chance of design error and a greater chance of construction error. A simple design that achieves all the critical requirements saves design and construction time and generally gives an economical structure.

Wherever possible, haunched beams should be avoided. If practical, beams should be the same width as or a little wider than the columns into which they frame, to simplify the formwork for column-beam joints. Deep spandrel beams along the edge of a building make it difficult to move forms from floor to floor and should be avoided if possible. In one-way floor construction, it is advisable to use the same beam depth throughout rather than switching from deep beams for long spans to shallow beams for short spans. The saving in concrete due to such a change is negligible and generally is more than offset by the extra labor and materials required, plus the need to rent or construct different sizes of beam forms.

If possible, a few standard column sizes should be chosen, with each column size maintained for three or four stories or the entire building. The amount of reinforcement and the concrete strength used can vary as the load varies. Columns should be aligned on a regular grid, if possible, and constant story heights should be maintained.

Economies are also possible in reinforcement placing. Complex or congested reinforcement will lead to higher per-pound charges for placement of the bars. It frequently is best, therefore, to design columns for 1.5 to 2 percent reinforcement and beams for no more than one-half to two-thirds of the maximum allowable reinforcement ratios. Grade-60 reinforcement almost universally is used for column reinforcement and flexural reinforcement in beams. In slabs where reinforcement quantities are controlled by minimum reinforcement ratios, there may be a slight advantage in using Grade-40 reinforcement (only available in smaller bar sizes). The same may be true for stirrups in beams if the stirrup spacings tend to be governed by the maximum allowable spacings. However, before specifying Grade-40 steel, the designer should check whether it is available locally in the sizes needed.

Because the flexural strength of a floor is relatively insensitive to concrete strength, there is no major advantage in using high-strength concrete in floor systems. An exception to this would be a flat-plate system, where the shear capacity may govern the thickness. On the other hand, column strengths are related directly to concrete strength, and the most economical columns tend to result from the use of high-strength concrete.

2-10 SUSTAINABILITY

Sustainability and *green buildings* are currently hot topics in the construction industry, but durability and longevity (major components of sustainability) have always been major reasons for selecting reinforced concrete as the construction material for buildings and other civil-infrastructure systems. The aesthetic qualities and the versatility of reinforced concrete have made it a popular choice for many architects and structural engineers. Both initial and life-cycle economic considerations, as well as the thermal properties of concrete, also play major roles in the selection of reinforced concrete for buildings and other construction projects.

Sustainable/green construction is not easily defined, but an excellent discussion of sustainability issues in concrete construction is given in reference [2-18]. In general, green buildings will be viewed somewhat differently by the owner, designer, and general public, but as noted in reference [2-18], sustainable design is generally accepted as a compromise between economic considerations, social values, and environmental impacts. Reinforced concrete construction fits into this general framework as follows.

Economic impact is one of the three primary components of sustainable construction. Economic considerations include both the initial and life-cycle costs of either a building or component of the civil-infrastructure system. Whether cast-in-place or precast, reinforced concrete is normally produced using local materials and labor, and thus, helps to stimulate the local economy while reducing transportation costs and energy consumption. Efficient structural designs can reduce the total quantity of concrete and reinforcing steel required for different building components and innovative mix designs can include recycled industrial by-products to reduce the consumption of new materials and the amount of cement required per cubic yard of concrete. Concrete's thermal mass and reflective properties can also reduce life-cycle energy consumption, and thus the operating costs for a building.

Aesthetics and occupant comfort are major factors in evaluating the sustainability of a building. A well-designed and aesthetically pleasing building will have a low environmental impact and can be a source of pride for the local community. Concrete's ability to be molded into nearly any form can make it particularly suitable for innovative and aesthetically pleasing architecture. A sustainable building should also provide a comfortable living and working environment for its occupants. Through its thermal mass properties, concrete can play a role in modulating interior temperatures; it can reduce natural lighting requirements because of its reflectance and ability to adapt to various methods of utilizing natural lighting, and it can reduce the use of potentially hazardous interior finishes because it can be used as a finished interior or exterior surface. *Durability* of a structure is an integral part of reducing the long-term costs and use of natural resources in a sustainable building. Many buildings change usage and owners over their service life and the longer a building can perform its required functions without undergoing major renovations, the more it benefits the overall society. Concrete has a long history of providing durable and robust structures, and while a fifty-year service life is typically discussed for most new construction, modern concrete structures are likely to have a service life that exceeds one hundred years.

Reducing the carbon footprint is a major concern for all new construction and is often discussed in terms of CO_2 emissions both during construction and over the life span of a building. Many items that reduce the energy consumption, and thus CO_2 emissions, over the service life of a concrete structure have been noted in the previous paragraphs. One of the commonly noted concerns regarding concrete construction is the emission of green-house gases during the manufacture of cement. The three primary sources of CO_2 emissions in cement production and distribution are: (1) the energy consumed to heat the kilns during cement production, (2) the release of CO_2 from limestone during the physical/chemical process that converts limestone, shale, clay, and other raw materials into calcium silicates, and (3) the transportation of cement from the point of manufacture to concrete production facilities. The cement industry is actively working to reduce CO_2 emissions in all three areas through the use of alternate fuels to fire the kilns, plant modifications to improve energy efficiency, carbon capture and storage systems, and more fuel-efficient cement handling and distribution systems. As noted previously, the carbon footprint per cubic yard of concrete can also be reduced through the use of supplemental cementitious materials, such as fly ash, slag cement, and silica fume, to replace a portion of the cement in a typical mix design.

Sustainability considerations are not typically incorporated into national building codes like the widely used *International Building Code* [2-14]. The American Concrete Institute's *Building Code Requirements for Structural Concrete* [2-10] is the recognized standard for the design of concrete structures and is adopted by reference into the *International Building Code*. The ACI has recently established a sustainability committee (ACI Committee 130) that is tasked to work with other ACI technical committees, including the building code committee, to include sustainability issues in the design requirements for concrete structures. Many ACI documents and standards refer to materials standards developed by the American Society for Testing and Materials (ASTM) and ASTM has also developed a sustainability committee to work with its technical committees to include sustainability considerations in the development and revision of ASTM standards.

2-11 CUSTOMARY DIMENSIONS AND CONSTRUCTION TOLERANCES

The selection of dimensions for reinforced concrete members is based on the size required for strength and for other aspects arising from construction considerations. Beam widths and depths and column sizes are generally varied in whole inch increments, and slab thicknesses in $\frac{1}{2}$-in. increments.

The actual as-built dimensions will differ slightly from those shown on the drawings, due to construction inaccuracies. ACI Standard 347 [2-19] on formwork gives the accepted tolerances on cross-sectional dimensions of concrete columns and beams as $\pm\frac{1}{2}$ in. and on the thickness of slabs and walls as in $\pm\frac{1}{4}$ in. For footings, they recommend tolerances of +2 in. and $-\frac{1}{2}$ in. on plan dimensions and −5 percent of the specified thickness.

The lengths of reinforcing bars are generally given in 2-in. increments. The tolerances for reinforcement placing concern the variation in the effective depth, d, of beams, the minimum reinforcement cover, and the longitudinal location of bends and ends of bars. These are specified in ACI Code Sections 26.6.2.1(a) and (b). ACI Committee 117 has published a comprehensive list of tolerances for concrete construction and materials [2-20].

2-12 INSPECTION

The quality of construction depends in part on the workmanship during construction. Inspection is necessary to confirm that the construction is in accordance with the project drawings and specifications. ACI Code Section 26.13.1 requires that concrete construction be inspected throughout the various work stages by, or under the supervision of, a licensed design professional, or by a qualified inspector. More stringent requirements are given in ACI Code Section 26.13.1.4 for inspection of moment-resisting frames in seismic regions.

The ACI and other organizations certify the qualifications of construction inspectors. Inspection reports should be distributed to the owner, the designer, the contractor, and the building official. The inspecting engineer or architect preserves these reports for at least two years after the completion of the project.

2-13 ACCURACY OF CALCULATIONS

Structural loads, with the exception of dead loads or fluid loads in a tank, are rarely known to more than two significant figures. Thus, although calculator and software output may include several significant figures, it is seldom necessary to use more than three significant figures in design calculations for reinforced concrete structural members. In this book, three significant figures are used.

Most mistakes in structural design arise from three sources: errors in looking up or writing down numbers, errors due to unit conversions, and failure to understand fully the statics or behavior of the structure being analyzed and designed. The last type of mistake is especially serious, because failure to consider a particular type of loading or the use of the wrong statical model may lead to serious maintenance problems or collapse. For this reason, designers are urged to use the limit-states design process to consider all possible modes of failure and to use free-body diagrams to study the equilibrium of parts or all of the structure.

2-14 HANDBOOKS AND DESIGN AIDS

Because a great many repetitive computations are necessary to proportion reinforced concrete members, handbooks containing tables or graphs of the more common quantities are available from several sources. Shortly after the 2014 Code is completed, the American Concrete Institute plans to publish *The Reinforced Concrete Design Manual, SP 17(14)* [2-21]. Also, the Concrete Reinforcing Steel Institute will publish an update of the *CRSI Design Handbook* [2-22].

Once a design has been completed, it is necessary for the details to be communicated to the reinforcing-bar suppliers and placers and to the construction crew. The *ACI Detailing Manual* [2-23] presents drafting standards and is an excellent guide to field practice. ACI Standard 301, *Specifications for Structural Concrete* [2-24] indicates the items to be included in construction specifications. Finally, the ACI publication *Guide to Formwork for Concrete* [2-19] gives guidance for form design.

The *ACI Manual of Concrete Practice* [2-25] collects together most of the ACI committee reports on concrete and structural concrete and is an invaluable reference on all aspects of concrete technology. It is published annually in hard copy and electronic format.

REFERENCES

2-1 Donald Taylor, "Progressive Collapse," *Canadian Journal of Civil Engineering*, Vol. 2, No. 4, December 1975, pp. 517–529.

2-2 *Minimum Design Loads for Buildings and Other Structures* (ASCE/SEI 7-10), American Society of Civil Engineers, Reston, VA, 2010, 608 pp.

2-3 ACI Committee 350, "Code Requirements for Environmental Engineering Concrete Structures and Commentary (ACI 350-06)," *ACI Manual of Concrete Practice*, American Concrete Institute, Farmington Hills, MI, 485 pp.

2-4 C. Allan Cornell, "A Probability Based Structural Code," *ACI Journal*, Vol. 66, No. 12, December 1969, pp. 974–985.

2-5 Alan Mattock, Ladislav Kriz, and Elvind Hognestad, "Rectangular Concrete Stress Distribution in Ultimate Strength Design," *ACI Journal*, Vol. 32, No. 8, February 1961, pp. 875–928.

2-6 Jong-Cherng Pier and C. Allan Cornell, "Spatial and Temporal Variability of Live Loads," *ASCE, Journal of the Structural Division*, Vol. 99, No. ST5, May 1973, pp. 903–922.

2-7 James G. MacGregor, "Safety and Limit States Design for Reinforced Concrete," *Canadian Journal of Civil Engineering*, Vol. 3, No. 4, December 1976, pp. 484–513.

2-8 Bruce Ellingwood, Theodore Galambos, James MacGregor, and C. Allan Cornell, *Development of a Probability Based Load Criterion for American National Standard A58*, NBS Special Publication 577, National Bureau of Standards, US Department of Commerce, Washington, DC, June 1980, 222 pp.

2-9 James G. MacGregor, "Load and Resistance Factors for Concrete Design," *ACI Structural Journal*, Vol. 80, No. 4, July–August 1983, pp. 279–287.

2-10 ACI Committee 318, "Building Code Requirements for Structural Concrete (ACI 318-14) and Commentary," *ACI Manual of Concrete Practice*, American Concrete Institute, Farmington Hills, MI.

2-11 *Specifications for Structural Steel Buildings*, American Institute of Steel Construction, Chicago, IL, 2010.

2-12 Andrzej Nowak and Maria Szerszen, "Calibration of Design Code for Buildings (ACI 318): Part 1—Statistical Models for Resistance," pp. 377–382; and "Part 2—Reliability Analysis and Resistance Factors," pp. 383–389, *ACI Structural Journal*, Vol. 100, No. 3, May–June 2003.

2-13 G.J. Klein and R.J. Linderberg, "Volume Change Response of Precast Concrete Buildings," *PCI Journal*, Precast/Prestressed Concrete Institute, Chicago, IL, Vol. 54, No. 4, Fall 2009, pp. 112–131.

2-14 International Code Council, *2012 International Building Code*, Washington, DC, 2012.

2-15 Donald Sawyer, "Ponding of Rainwater on Flexible Roof Systems," *ASCE, Journal of the Structural Division*, Vol. 93, No. ST1, February 1967, pp. 127–147.

2-16 Allan Davenport, "Gust Loading Factors," *ASCE, Journal of the Structural Division*, Vo. 93, No. ST3, June 1967, pp. 11–34.

2-17 Amin Ghali, Walter Dilger, and Adam Neville, "Time Dependent Forces Induced by Settlement of Supports in Continuous Reinforced Concrete Beams," *ACI Journal*, Vol. 66, No. 12, December 1969, pp. 907–915.

2-18 Andrea Schokker, "The Sustainable Concrete Guide—Strategies and Examples," *US Green Concrete Council*, Farmington Hills, MI, 2010, 86 pp.

2-19 ACI Committee 347, "Guide to Formwork for Concrete (ACI 347-04)," *ACI Manual of Concrete Practice*, American Concrete Institute, Farmington Hills, MI.

2-20 ACI Committee 117, "Specification for Tolerances for Concrete Construction and Materials and Commentary (ACI 117-10)," *ACI Manual of Concrete Practice*, American Concrete Institute, Farmington Hills, MI.

2-21 *The Reinforced Concrete Design Manual, SP 17(14)*, American Concrete Institute, Farmington Hills, MI, 2014.

2-22 *CRSI Design Handbook*, 11th Edition, Concrete Reinforcing Steel Institute, Schaumberg, IL, 2012, 420 pp.

2-23 *ACI Detailing Manual SP-66(04)*, American Concrete Institute, Farmington Hills, MI, 2004.

2-24 ACI Committee 301, "Specifications for Structural Concrete (ACI 301-10)," *ACI Manual of Concrete Practice*, American Concrete Institute, Farmington Hills, MI.

2-25 *Manual of Concrete Practice*, American Concrete Institute, Farmington Hills, MI, published annually.

3

Materials

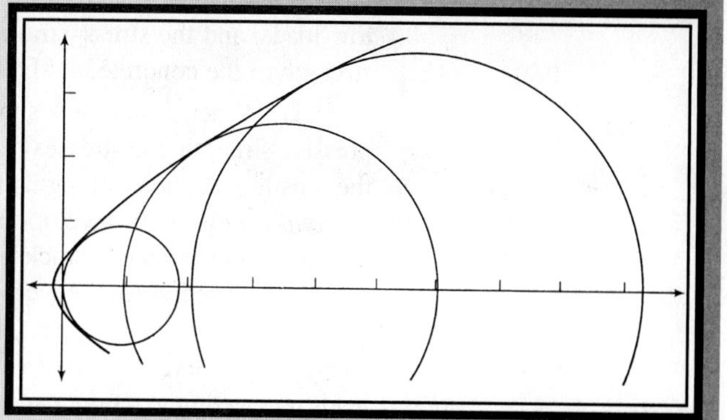

3-1 CONCRETE

Concrete is a composite material composed of aggregate, generally sand and gravel, chemically bound together by hydrated portland *cement*. The aggregate generally is graded in size from sand to gravel, with the maximum gravel size in structural concrete commonly being $\frac{3}{4}$ in., although $\frac{3}{8}$-in. or $1\frac{1}{2}$-in. aggregate may be used.

3-2 BEHAVIOR OF CONCRETE FAILING IN COMPRESSION

Mechanism of Failure in Concrete Loaded in Compression

Concrete is a mixture of cement paste and aggregate, each of which has an essentially linear and brittle stress–strain relationship in compression. Brittle materials tend to develop tensile fractures perpendicular to the direction of the largest tensile strain. Thus, when concrete is subjected to uniaxial compressive loading, cracks tend to develop parallel to the maximum compressive stress. In a cylinder test, the friction between the heads of the testing machine and the ends of the cylinder prevents lateral expansion of the ends of the cylinder and in doing so restrains the vertical cracking in those regions. This strengthens conical regions at each end of the cylinder. The vertical cracks that occur at midheight of the cylinder do not enter these conical regions and the failure surface appears to consist of two cones.

Although concrete is made up of essentially elastic, brittle materials, its stress–strain curve is nonlinear and appears to be somewhat ductile. This can be explained by the gradual development of *microcracking* within the concrete and the resulting redistribution of stress from element to element in the concrete [3-1]. Microcracks are internal cracks $\frac{1}{8}$ to $\frac{1}{2}$ in. in length. Microcracks that occur along the interface between paste and aggregate are called *bond cracks*; those that cross the mortar between pieces of aggregate are known as *mortar cracks*.

There are four major stages in the development of microcracking and failure in concrete subjected to uniaxial compressive loading:

1. Shrinkage of the paste occurs during hydration, and this volume change of the concrete is restrained by the aggregate. The resulting tensile stresses lead to *no-load bond* *cracks*, before the concrete is loaded. These cracks have little effect on the concrete at low loads, and the stress–strain curve remains linear up to 30 percent of the compressive strength of the concrete, as shown by the solid line in Fig. 3-1.

2. When concrete is subjected to stresses greater than 30 to 40 percent of its compressive strength, the stresses on the inclined surfaces of the aggregate particles will exceed the tensile and shear strengths of the paste–aggregate interfaces, and new cracks, known as *bond cracks*, will develop. These cracks are stable; they propagate only if the load is increased. Once such a crack has formed, however, any additional load that would have been transferred across the cracked interface is redistributed to the remaining unbroken interfaces and to the mortar. This redistribution of load causes a gradual bending of the stress–strain curve for stresses above 40 percent of the short-time strength. The loss of bond leads to a wedging action, causing transverse tensions above and below the aggregates.

3. As the load is increased beyond 50 or 60 percent of ultimate, localized *mortar cracks* develop between bond cracks. These cracks develop parallel to the compressive loading and are due to the transverse tensile strains. During this stage, there is stable crack propagation; cracking increases with increasing load but does not increase under constant load. The onset of this stage of loading is called the *discontinuity limit* [3-2].

4. At 75 to 80 percent of the ultimate load, the number of mortar cracks begins to increase, and a continuous pattern of microcracks begins to form. As a result, there are fewer undamaged portions to carry the load, and the stress versus longitudinal-strain curve becomes even more markedly nonlinear. The onset of this stage of cracking is called the *critical stress* [3-3].

If the lateral strains, ϵ_3, are plotted against the longitudinal compressive stress, the dashed curve in Fig. 3-1 results. The lateral strains are tensile and initially increase, as is expected from the Poisson's effect. As microcracking becomes more extensive, these cracks contribute to the apparent lateral strains. As the load exceeds 75 to 80 percent of the ultimate compressive strength, the cracks and lateral strains increase rapidly, and the volumetric strain (relative increase in volume), ϵ_v, begins to increase, as shown by the broken line in Fig. 3-1.

The critical stress is significant for several reasons. The ensuing increase in volume causes an outward pressure on ties, spirals, or other confining reinforcement, and these in turn act to restrain the lateral expansion of the concrete, thus delaying its disintegration.

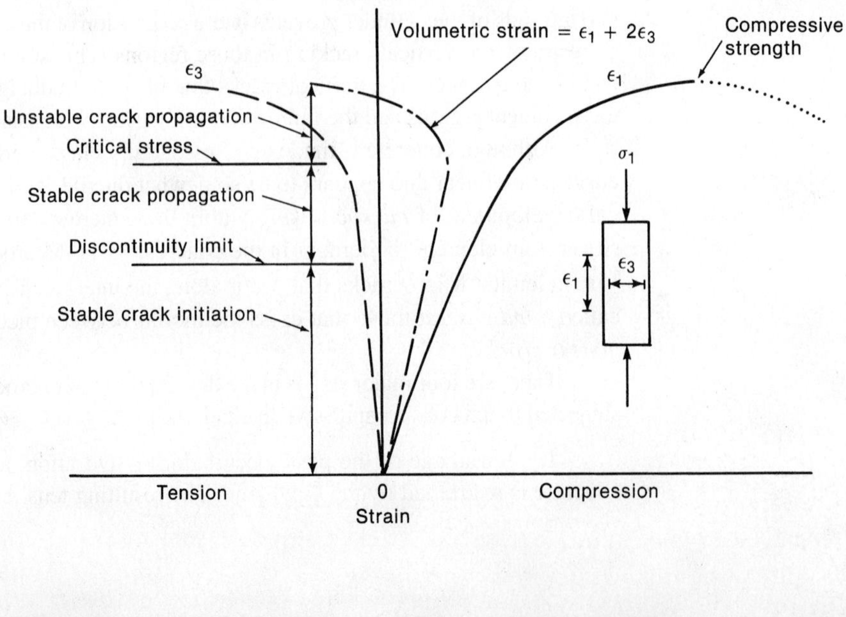

Fig. 3-1
Stress–strain curves for concrete loaded in uniaxial compression. (From [3-2].)

Equally important is the fact that the structure of the concrete tends to become unstable at loads greater than the critical load. Under stresses greater than about 75 percent of the short-time strength, the strains increase more and more rapidly until failure occurs. Figure 3-2a shows the strain–time response of concrete loaded rapidly to various fractions

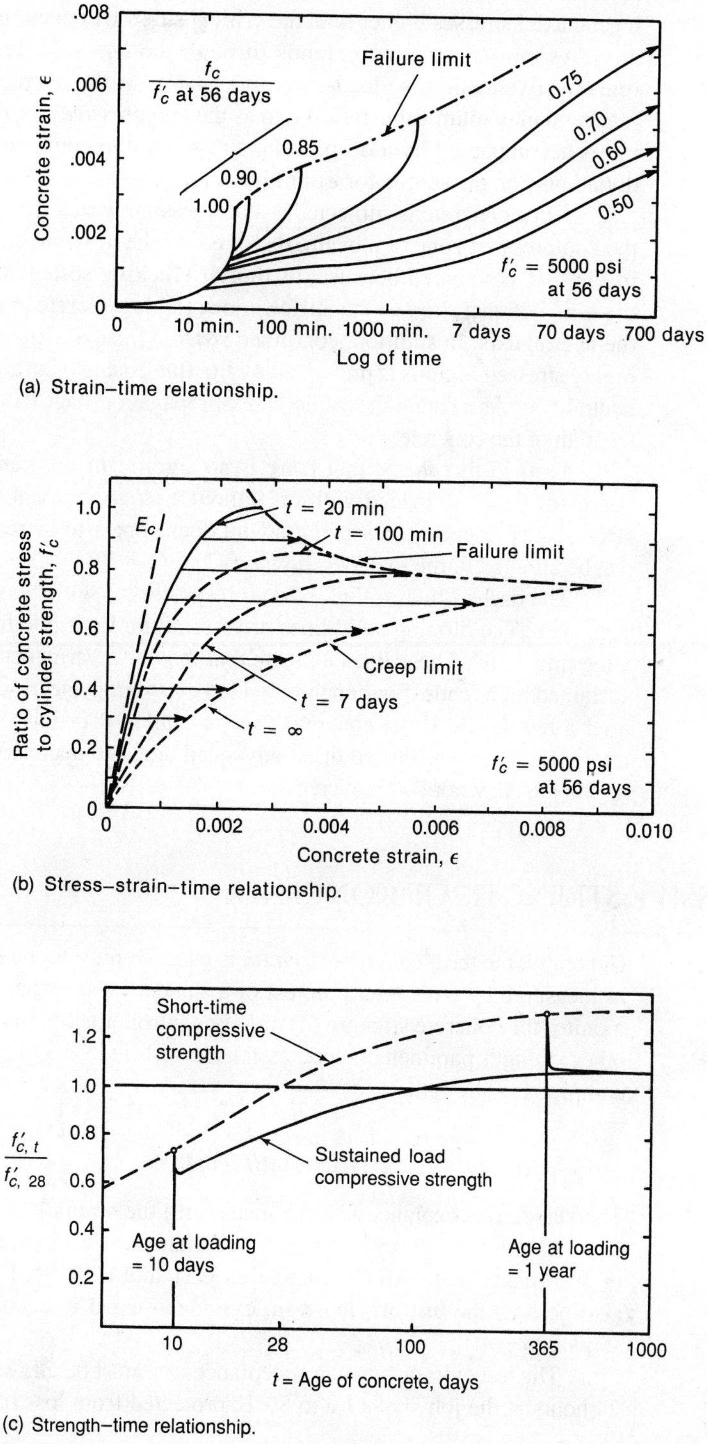

Fig. 3-2
Effect of sustained loads on the behavior of concrete in uniaxial compression.
(From [3-4].)

of its short-time strength, with this load being sustained for a long period of time or until failure occurred. As shown in Fig. 3-2b, concrete subjected to a sustained axial load greater than the critical load will eventually fail under that load. The critical stress is between 0.75 and $0.80f'_c$.

Under cyclic compressive loads, axially loaded concrete has a *shake-down* limit approximately equal to the point of onset of significant mortar cracking at the critical stress. Cyclic axial stresses higher than the critical stress will eventually cause failure.

As mortar cracking extends through the concrete, less and less of the structure remains. Eventually, the load-carrying capacity of the uncracked portions of the concrete reaches a maximum value referred to as the *compressive strength* (Fig. 3-1). Further straining is accompanied by a drop in the stress that the concrete can resist, as shown by the dotted portion of the line for ϵ_1 in Fig. 3-1.

When concrete is subjected to compression with a strain gradient, as would occur in the compression zone of a beam, the effect of the unstable crack propagation stage shown in Fig. 3-1 is reduced because, as mortar cracking softens the highly strained concrete, the load is transferred to the stiffer, more stable concrete at points of lower strain nearer the neutral axis. In addition, continued straining and the associated mortar cracking of the highly stressed regions is prevented by the stable state of strain in the concrete closer to the neutral axis. As a result, the stable-crack-propagation stage extends almost up to the ultimate strength of the concrete.

Tests [3-5] suggest that there is no significant difference between the stress–strain curves of concrete loaded with or without a strain gradient up to the point of maximum stress. The presence of a strain gradient does appear to increase the maximum strains that can be attained in the member, however.

The dashed line in Fig. 3-2c represents the gain in short-time compressive strength with time. The dipping solid lines are the failure limit line from Fig. 3-2b plotted against a log time scale. These lines indicate that there is a permanent reduction in strength due to sustained high loads. For concrete loaded at a young age, the minimum strength is reached after a few hours. If the concrete does not fail at this time, it can sustain the load indefinitely. For concrete loaded at an advanced age, the decrease in strength due to sustained high loads may not be recovered.

3-3 COMPRESSIVE STRENGTH OF CONCRETE

Generally, the term *concrete strength* is taken to refer to the uniaxial compressive strength as measured by a compression test of a standard test cylinder, because this test is used to monitor the concrete strength for quality control or acceptance purposes. For convenience, other strength parameters, such as tensile or bond strength, are expressed relative to the compressive strength.

Standard Compressive-Strength Tests

The standard acceptance test for measuring the strength of concrete involves short-time compression tests on cylinders 6 in. in diameter by 12 in. high, made, cured, and tested in accordance with ASTM Standards C31 and C39. ACI Code Section 26.12.1.1 now also permits the use of 4-by-8-in. cylinders tested in accordance with the same ASTM standards.

The test cylinders for an acceptance test must be allowed to harden in their molds for 24 hours at the job site at 60 to 80°F, protected from loss of moisture and excessive heat, and then must be cured at 73°F in a moist room or immersed in water saturated with lime. The standard acceptance test is carried out when the concrete is 28 days old.

Field-cured test cylinders are frequently used to determine when the forms may be removed or when the structure may be used. These should be stored as near the location of that concrete in the structure as is practical and should be cured in a manner as close as possible to that used for the concrete in the structure.

The standard strength "test" is the average of the strengths of two 6-by-12-in. cylinders or three 4-by-8-in. cylinders from the same concrete batch tested at 28 days (or an earlier age, if specified). These are tested at a loading rate of about 35 psi per second, producing failure of the cylinder at $1\frac{1}{2}$ to 3 minutes. For high-strength concrete, acceptance tests are sometimes carried out at 56 or 90 days, because some high-strength concretes take longer than normal concretes to reach their design strength.

Traditionally, the compressive strength has been tested by using 6-by-12-in. cylinders. For high-strength concretes, the axial stiffness of some testing machines is close to the axial stiffness of the cylinders being tested. In such cases, the strain energy released by the machine at the onset of crushing of the test cylinder leads to a brittle failure of the cylinder. This can cause a decrease in the measured f'_c. This is alleviated by testing 4-by-8-in. cylinders, which have an axial stiffness less than a fifth of that of 6-by-12-in. cylinders. Aïtcin et al. [3-6] report tests on 8-in.-, 6-in.-, and 4-in.-diameter cylinders of concretes with nominal strengths of 5000, 13,000, and 17,500 psi; some of each strength were cured in air, or sealed, or cured in lime-water baths.

The water-cured specimens and the sealed specimens had approximately the same strengths at ages of 7, 28, and 91 days of curing. Aïtcin et al. [3-6] concluded the strengths of the 4-in.- and 6-in.-diameter cylinders were similar. This suggests that the strengths of 4-by-8-in. cylinders will be similar to the strengths of 6-by-12-in. cylinders, and that 4-in. cylinders can be used as control tests.

Other studies quoted in the 1993 report on high-strength concrete by ACI Committee 363 [3-7] gave different conversion factors. The report concluded that 4-by-8-in. control cylinders give a higher strength and a larger coefficient of variation than 6-by-12-in. cylinders.

Statistical Variations in Concrete Strength

Concrete is a mixture of water, cement, aggregate, and air. Variations in the properties or proportions of these constituents, as well as variations in the transporting, placing, and compaction of the concrete, lead to variations in the strength of the finished concrete. In addition, discrepancies in the tests will lead to apparent differences in strength. The shaded area in Fig. 3-3 shows the distribution of the strengths in a sample of 176 concrete-strength tests.

The mean or average strength is 3940 psi, but one test has a strength as low as 2020 psi and one is as high as 6090 psi.

If more than about 30 tests are available, the strengths will generally approximate a normal distribution. The normal distribution curve, shown by the curved line in Fig. 3-3, is symmetrical about the mean value, $\bar{x}$, of the data. The dispersion of the data can be measured by the *sample standard deviation*, s, which is the root-mean-square deviation of the strengths from their mean value:

$$s = \sqrt{\frac{(x_1 - \bar{x})^2 + (x_2 - \bar{x})^2 + (x_3 - \bar{x})^2 + \cdots + (x_n - \bar{x})^2}{n - 1}} \quad (3\text{-}1)$$

The standard deviation divided by the mean value is called the *coefficient of variation*, V:

$$V = \frac{s}{\bar{x}} \quad (3\text{-}2)$$

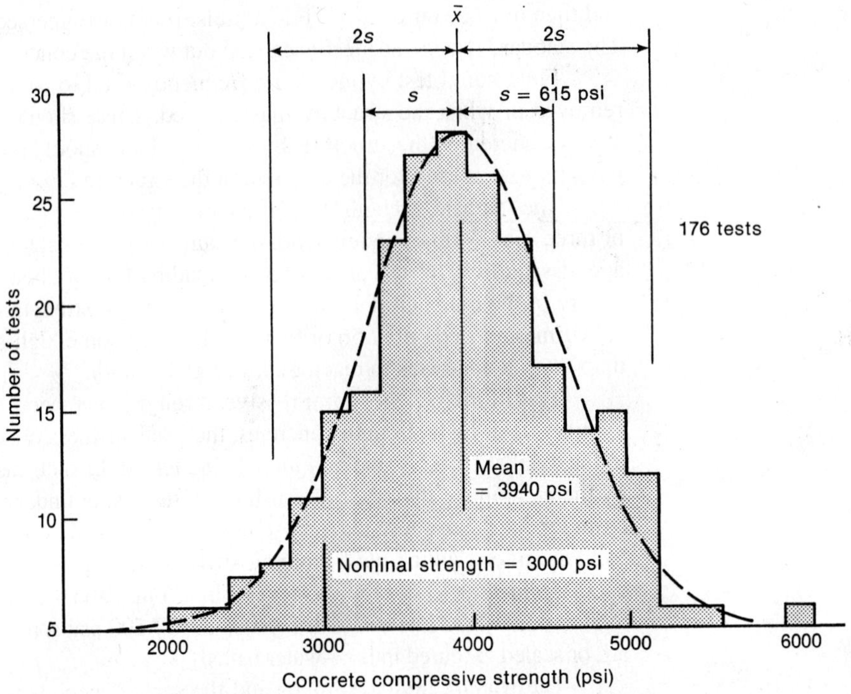

Fig. 3-3
Distribution of concrete strengths.

This makes it possible to express the degree of dispersion on a fractional or percentage basis rather than an absolute basis. The concrete test data in Fig. 3-3 have a standard deviation of 615 psi and a coefficient of variation of $615/3940 = 0.156$, or 15.6 percent.

If the data correspond to a normal distribution, their distribution can be predicted from the properties of such a curve. Thus, 68.3 percent of the data will lie within 1 standard deviation above or below the mean. Alternatively, 15.6 percent of the data will have values less than $(\bar{x} - s)$. Similarly, for a normal distribution, 10 percent of the data, or 1 test in 10, will have values less than $\bar{x}(1 - aV)$, where $a = 1.282$. Values of a corresponding to other probabilities can be found in statistics texts.

Figure 3-4 shows the mean concrete strength, f_{cr}, required for various values of the coefficient of variation if no more than 1 test in 10 is to have a strength less than 3000 psi.

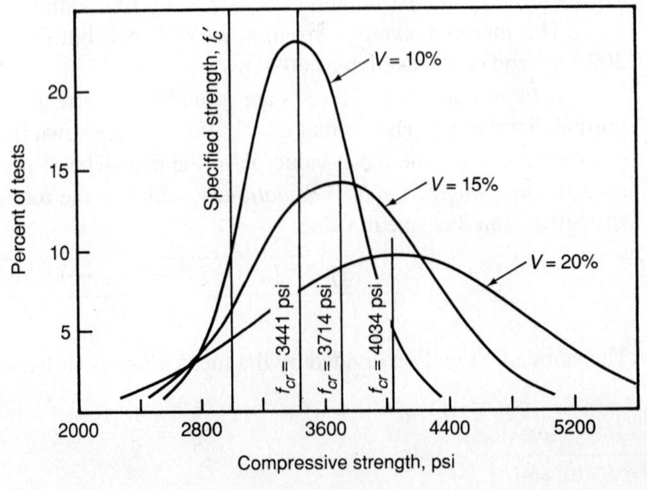

Fig. 3-4
Normal frequency curves for coefficients of variation of 10, 15, and 20 percent. (From [3-9].)

As shown in this figure, as the coefficient of variation is reduced, the value of the mean strength, f_{cr}, required to satisfy this requirement can also be reduced.

Based on the experience of the U.S. Bureau of Reclamation on large projects, ACI Committee 214 [3-8] has defined various standards of control for moderate-strength concretes. A coefficient of variation of 15 percent represents *average control.* About one-tenth of the projects studied had coefficients of variation less than 10 percent, which was termed *excellent control*, and another tenth had values greater than about 20 percent, which was termed *poor control.* Above a mean strength of about 4000 psi, the standard deviation tends to be independent of the mean strength, and for average control s is about 600 psi [3-8]. The test data plotted in Fig. 3-3 correspond to average control, following the Committee 214 definition of *average control.*

In 2001, Nowak and Szerszen [3-9] and [3-10] collected concrete control data from sources around the United States. The data are summarized in Table 3-1. The degree of concrete control was considerably better than that assumed by ACI Committee 214. In particular, the mean of the coefficients of variation reported by Nowak and Szerszen is much lower than the $V = 15$ percent that ACI 214 assumed to be representative of average control. In Table 3-1, the coefficients of variation range from 0.07 to 0.115, with one exception (lightweight concrete). This range of concrete variability appears to be representative of concrete produced in modern ready-mix plants, which represents the vast majority of concrete in North America. Nowak and Szerszen recommend a single value of $V = 0.10$. It would appear that this is a "property" of modern ready-mix concretes.

Nowak and Szerszen suggest that the ratio of mean test strength to specified strength can be taken as 1.35 for 3000 psi concrete, decreasing linearly to 1.14 at $f'_c = 5000$ psi and remaining constant at 1.14 for higher strengths. However, the mean strength ratio cannot be considered a property of modern concrete, because it is easy for a mix designer to increase or decrease this while proportioning the concrete mix.

The data in Table 3-1 suggest the following coefficients of variation for various degrees of concrete control:

Poor control	$V > 0.140$
Average control	$V = 0.105$
Excellent control	$V < 0.070$

TABLE 3-1 Statistical Parameters for f'_c for Concrete

Type of Concrete	Number of Tests	Specified Strengths	Mean Strengths	Mean/ Specified	Coefficient of Variation
Ordinary ready mix concrete	317	3000 to 6000 psi	4060 to 6700 psi	3000 psi—1.38 6000 psi—1.14	3000 psi—0.111 6000 psi—0.080
Ordinary plant-precast concrete	1174	5000 to 6500 psi	6910 to 7420 psi	5000 psi—1.38 6500 psi—1.14	0.10
Lightweight concrete	769	3000 to 5000 psi	4310 to 5500 psi	3000 psi—1.44 5000 psi—1.10	3000 psi—0.185 5000 psi—0.070
High-strength concrete—28 days	2052	7000 to 12,000 psi	8340 to 12,400 psi	7000 psi—1.19 12,000 psi—1.04	7000 psi—0.115 12,000 psi—0.105
High-strength concrete—56 days	914	7000 to 12,000 psi	10,430 to 14,000 psi	7000 psi—1.49 12,000 psi—1.17	7000 psi—0.080 12,000 psi—0.105

Source: From data presented in [3-10] and [3-11].

Building-Code Definition of Compressive Strength

The *specified compressive strength*, f'_c, is measured by compression tests on 6-by-12-in. or 4-by-8-in. cylinders tested after 28 days of moist curing. This is the strength specified on the construction drawings and used in the calculations. As shown in Fig. 3-4, the specified strength is less than the average strength. The required mean strength of the concrete, f_{cr}, must be at least:

For specified compressive strength, f'_c, less than or equal to 5000 psi:

Use the larger value of

$$f_{cr} = f'_c + 1.34s \tag{3-3a}$$

and

$$f_{cr} = f'_c + 2.33s - 500 \tag{3-3b}$$

For specified compressive strength, f'_c, greater than 5000 psi:

Use the larger value of

$$f_{cr} = f'_c + 1.34s \tag{3-4a}$$

and

$$f_{cr} = 0.90f'_c + 2.33s \tag{3-4b}$$

where s is the standard deviation.

Equations (3-3a) and (3-4a) give the lowest average strengths required to ensure a probability of not more than 1 in 100 that the average of any three consecutive strength tests will be below the specified strength. Alternatively, it ensures a probability of not more than 1 in 11 that any one test will fall below f'_c. Equation (3-3b) gives the lowest mean strength to ensure a probability of not more than 1 in 100 that any individual strength test will be more than 500 psi below the specified strength. Lines indicating the corresponding required average strengths, f_{cr}, are plotted in Fig. 3-4. In these definitions, a test is the average of two 6-by-12-in. cylinder tests or three 4-by-8-in. cylinder tests.

For any one test, Eqs. (3-3a and b) and (3-4a and b) give a probability of 0.99 that a single test will fall more than 500 psi below the specified strength, equivalent to a 0.01 chance of understrength. This does not ensure that the number of low tests will be acceptable, however. Given a structure requiring 4000 cubic yards of concrete with 80 concrete tests during the construction period, the probability of a single test falling more than 500 psi below the specified strength is $1 - 0.99^{80}$, or about 55 percent [3-11].

This may be an excessive number of understrength test results in projects where owners refuse to pay for concretes that have lower strengths than specified. Thus, a higher target for the mean concrete strength than that currently required by Eqs. (3-3) and (3-4) will frequently be specified to reduce the probability of low strength tests.

Factors Affecting Concrete Compressive Strength

Among the large number of factors affecting the compressive strength of concrete, the following are probably the most important for concretes used in structures.

1. Water/cement ratio. The strength of concrete is governed in large part by the ratio of the weight of the water to the weight of the cement for a given volume of concrete. A lower water/cement ratio reduces the porosity of the hardened concrete and thus increases the number of interlocking solids. The introduction of tiny, well-distributed air bubbles in the cement paste, referred to as *air entrainment*, tends to increase the freeze–thaw durability of the concrete. When the water in the concrete freezes, pressure is generated in the capillaries and pores in the hardened cement paste. The presence of tiny, well-distributed air bubbles provides a way to dissipate the pressures due to freezing. However, the air voids introduced by air entrainment reduce the strength of the concrete. A water/cement ratio of 0.40 corresponds to 28-day strengths in the neighborhood of 4700 psi for air-entrained concrete and 5700 psi for non-air-entrained concrete. For a water/cement ratio of 0.55, the corresponding strengths are 3500 and 4000 psi, respectively. Voids due to improper consolidation tend to reduce the strength below that corresponding to the water/cement ratio used in design.

2. Type of cement. Traditionally, five basic types of portland cement have been produced:

Normal, Type I: used in ordinary construction, where special properties are not required.

Modified, Type II: lower heat of hydration than Type I; used where moderate exposure to sulfate attack exists or where moderate heat of hydration is desirable.

High early strength, Type III: used when high early strength is desired; has considerably higher heat of hydration than Type I.

Low heat, Type IV: developed for use in mass concrete dams and other structures where heat of hydration is dissipated slowly. In recent years, very little Type IV cement has been produced. It has been replaced with a combination of Types I and II cement with fly ash.

Sulfate resisting, Type V: used in footings, basement walls, sewers, and so on that are exposed to soils containing sulfates.

In recent years, blended portland cements produced to satisfy ASTM C1157 *Standard Performance Specification for Hydraulic Cement* have partially replaced the traditional five basic cements. This in effect allows the designer to select different blends of cement.

Figure 3-5 illustrates the rate of strength gain with different cements. Concrete made with Type III (high early strength) cement gains strength more rapidly than does concrete made with Type I (normal) cement, reaching about the same strength at 7 days as a corresponding mix containing Type I cement would reach at 28 days. All five types tend to approach the same strength after a long period of time, however.

3. Supplementary cementitious materials. Sometimes, a portion of the cement is replaced by materials such as fly ash, ground granulated blast-furnace slag, or silica

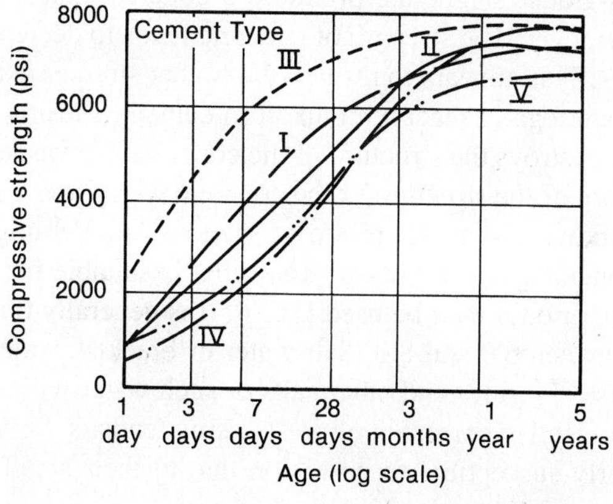

Fig. 3-5
Effect of type of cement on strength gain of concrete (moist cured, water/cement ratio = 0.49). (From [3-12] copyright ASTM; reprinted with permission.)

fume to achieve economy, reduction of heat of hydration, and, depending on the materials, improved workability. Fly ash and silica fume are referred to as *pozzolans*, which are defined as siliceous, or siliceous and aluminous materials that in themselves possess little or no cementitious properties but that will, in the presence of moisture, react with calcium hydroxide to form compounds with such properties. When supplementary cementitious materials are used in mix design, the water/cement ratio, w/c, is restated in terms of the *water/cementitious materials ratio*, w/cm, where cm represents the total weight of the cement and the supplementary cementitious materials, as defined in ACI Code Section 26.4.1.1.1. The design of concrete mixes containing supplementary cementitious materials is discussed in [3-13].

Fly ash, precipitated from the chimney gases of coal-fired power plants, frequently leads to improved workability of the fresh concrete. It often slows the rate of strength gain of concrete, but generally not the final strength, and depending on composition of the fly ash, might reduce or improve the durability of the hardened concrete [3-14]. Fly ashes from different sources vary widely in composition and have different effects on concrete properties. They also affect the color of the concrete.

Ground granulated blast-furnace slag tends to reduce the early-age strength and heat of hydration of concrete. Strengths at older ages will generally exceed those for normal concretes with similar w/cm ratios. Slag tends to reduce the permeability of concrete and its resistance to attack by certain chemicals [3-15].

Silica fume consists of very fine spherical particles of silica produced as a by-product in the manufacture of ferrosilicon alloys. The extreme fineness and high silica content of the silica fume make it a highly effective pozzolanic material. It is used to produce low-permeability concrete with enhanced durability and/or high strength [3-13].

4. Aggregate. The strength of concrete is affected by the strength of the aggregate, its surface texture, its grading, and, to a lesser extent, by the maximum size of the aggregate. Strong aggregates, such as felsite, traprock, or quartzite, are needed to make very-high-strength concretes. Weak aggregates include sandstone, marble, and some metamorphic rocks, while limestone and granite aggregates have intermediate strength. Normal-strength concrete made with high-strength aggregates fails due to mortar cracking, with very little aggregate failure. The stress–strain curves of such concretes tend to have an appreciable declining branch after reaching the maximum stress. On the other hand, if aggregate failure precedes mortar cracking, failure tends to occur abruptly with a very steep declining branch. This occurs in very-high-strength concretes (see Fig. 3-18) and in some lightweight concretes.

Concrete strength is affected by the bond between the aggregate and the cement paste. The bond tends to be better with crushed, angular pieces of aggregate.

A well-graded aggregate produces a concrete that is less porous. Such a concrete tends to be stronger. The strength of concrete tends to decrease as the maximum aggregate size increases. This appears to result from higher stresses at the paste–aggregate interface.

Some aggregates react with alkali in cement, causing a long-term expansion of the concrete that destroys the structure of the concrete. Unwashed marine aggregates also lead to a breakdown of the structure of the concrete with time.

5. Mixing water. There are no standards governing the quality of water for use in mixing concrete. In most cases, water that is suitable for drinking and that has no pronounced taste or odor may be used [3-16]. It is generally thought that the pH of the water should be between 6.0 and 8.0. Salt water or brackish water must not be used as mixing water, because chlorides and other salts in such water will attack the structure of the concrete and may lead to corrosion of prestressing tendons. Strands and wires used as tendons are particularly susceptible to corrosion due to their small diameter and higher stresses compared to reinforcing bars [3-17].

6. Moisture conditions during curing. The development of the compressive strength of concrete is strongly affected by the moisture conditions during curing. Prolonged moist curing leads to the highest concrete strength, as shown in Fig. 3-6.

7. Temperature conditions during curing. The effect of curing temperature on strength gain is shown in Fig. 3-7 for specimens placed and moist-cured for 28 days under the constant temperatures shown in the figure and then moist-cured at 73°F. The 7- and 28-day strengths are reduced by cold curing temperatures, although the long-term strength tends to be enhanced. On the other hand, high temperatures during the first month increase the 1- and 3-day strengths but tend to reduce the 1-year strength.

The temperature during the setting period is especially important. Concrete placed and allowed to set at temperatures greater than 80°F will never reach the 28-day strength of concrete placed at lower temperatures. Concrete that freezes soon after it has been placed will have a severe strength loss.

Occasionally, control cylinders are left in closed boxes at the job site for the first 24 hours. If the temperature inside these boxes is higher than the ambient temperature, the strength of the control cylinders may be affected.

8. Age of concrete. Concrete gains strength with age, as shown in Figs. 3-5 to 3-7. Prior to 1975, the 7-day strength of concrete made with Type I cement was generally 65 to 70 percent of the 28-day strength. Changes in cement production since then have resulted in a more rapid early strength gain and less long-term strength gain. ACI Committee 209 [3-20] has proposed the following equation to represent the rate of strength gain for concrete made from Type I cement and moist-cured at 70°F:

$$f'_{c(t)} = f'_{c(28)}\left(\frac{t}{4 + 0.85t}\right) \tag{3-5}$$

Here, $f'_{c(t)}$ is the compressive strength at age t. For Type III cement, the coefficients 4 and 0.85 become 2.3 and 0.92, respectively.

Concrete cured under temperatures other than 70°F may set faster or slower than indicated by these equations, as shown in Fig. 3-7.

9. Maturity of concrete. Young concrete gains strength as long as the concrete remains about a *threshold temperature* of −10 to −12°C or +11 to +14°F. Maturity is the

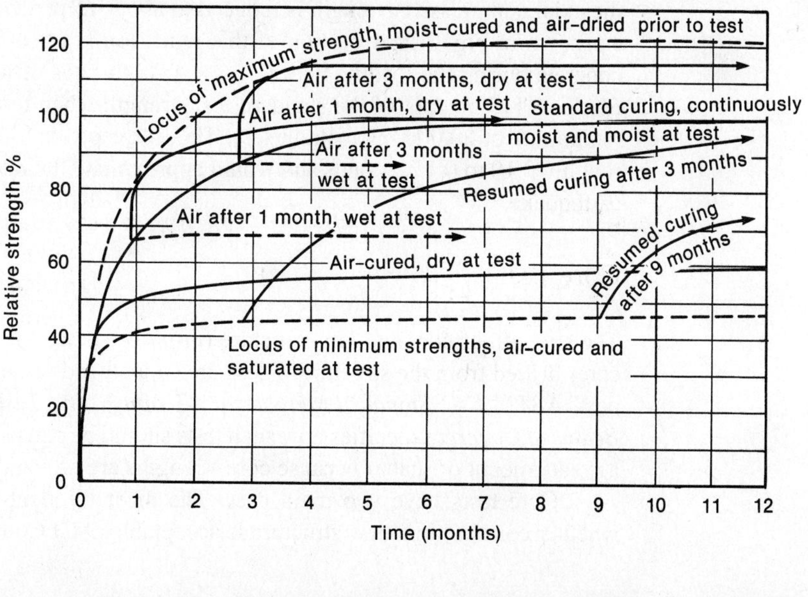

Fig. 3-6
Effect of moist-curing conditions at 70°F and moisture content of concrete at time of test on compressive strength of concrete. (From [3-18].)

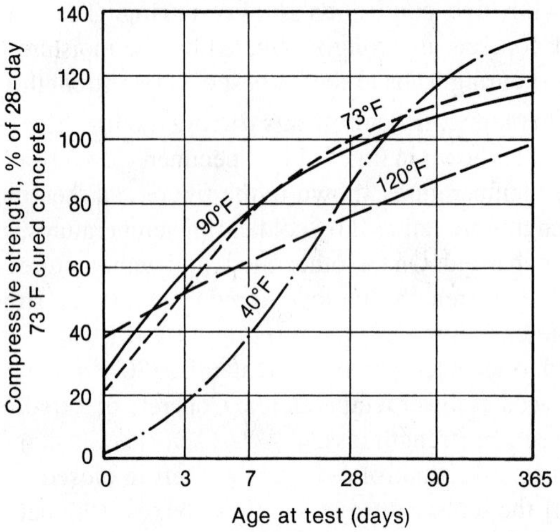

Fig. 3-7
Effect of temperature during the first 28 days on the strength of concrete (water/cement ratio = 0.41, air content = 4.5 percent, Type I cement, specimens cast and moist-cured at temperature indicated for first 28 days, all moist-cured at 73°F thereafter). (From [3-19].)

summation of the product of the difference between the curing temperature and the threshold temperature, and the time the concrete has cured at that temperature, [3-21] and [3-22].

$$\text{Maturity} = M = \sum_{i=1}^{n} (T_i + 10)(t_i) \tag{3-6}$$

In this equation, T_i is the temperature in Celsius during the ith interval and t_i is the number of days spent curing at that temperature. Figure 3-8 shows the form of the relationship between maturity and compressive strength of concrete. Although no unique relationship exists, Fig. 3-8 may be used for guidance in determining when forms can be removed. Maturity should not be used as the sole determinant of adequate strength. It will not detect errors in the concrete batching, such as inadequate cement or excess water, or excessive delays in placing the concrete after batching.

10. Rate of loading. The standard cylinder test is carried out at a loading rate of roughly 35 psi per second, and the maximum load is reached in $1\frac{1}{2}$ to 2 minutes, corresponding to a strain rate of about 10 microstrain/sec. Under very slow rates of loading, the axial compressive strength is reduced to about 75 percent of the standard test strength, as shown in Fig. 3-2. A portion of this reduction is offset by continued maturing of the concrete during the loading period [3-4]. At high rates of loading, the strength increases, reaching 115 percent of the standard test strength when tested at a rate of 30,000 psi/sec (strain rate of 20,000 microstrain/sec). This corresponds to loading a cylinder to failure in roughly 0.10 to 0.15 seconds and would approximate the rate of loading experienced in an earthquake.

Core Tests

The strength of concrete in a structure (*in-place strength*) is occasionally measured from cores drilled from the structure. These are capped and tested in the same manner as cylinders. ASTM C42 *Standard Method for Obtaining and Testing Drilled Cores and Sawed Beams of Concrete* specifies how such tests should be carried out. Core-test strengths show a great amount of scatter because core strengths are affected by a wide range of variables.

Core tests have two main uses. The most frequent use of core tests is to assess whether concrete in a new structure is acceptable. ACI Code Section 26.12.4.1(b) permits

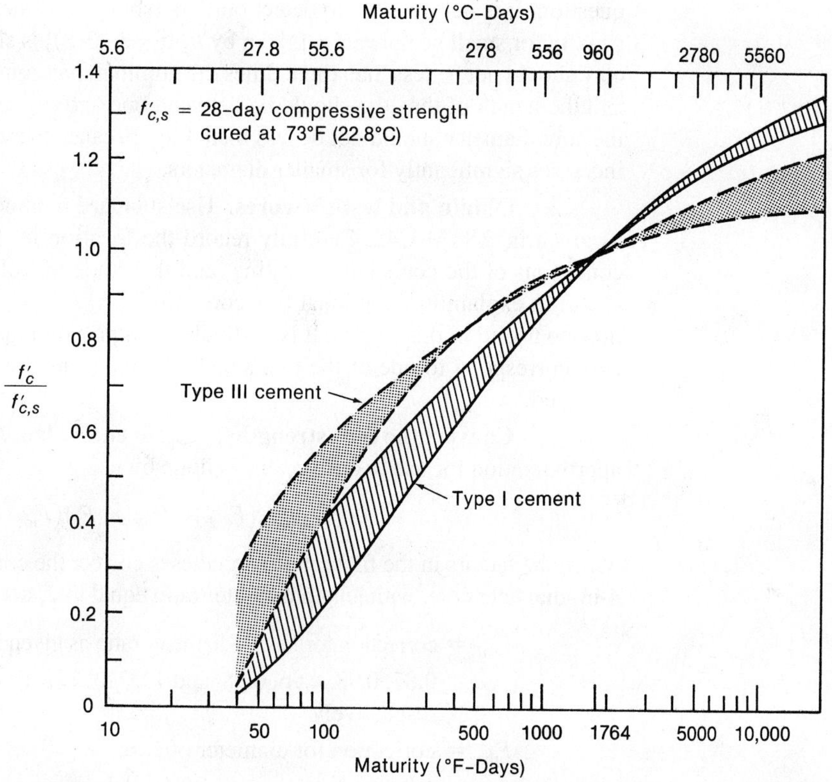

Fig. 3-8 Normalized compressive strength versus maturity. (From [3-22].)

the use of core tests in such cases and requires three cores for each strength test more than 500 psi below the specified value of f'_c. Cores obtained by using a water-cooled bit have a moisture gradient from the wet outside surface to the dry interior concrete. This causes stress gradients that reduce the test strength of the core. ACI Code Section 26.12.4.1(c) requires that cores be prepared for shipping to the testing lab by wrapping them in watertight bags or containers immediately after drilling [3-23]. Cores should not be tested earlier than 48 hours after drilling, nor later than 7 days after drilling. Waiting 48 hours enables the moisture gradients in the cores to dissipate. This reduces the stress gradient in the core. ACI Code Section 26.5.3.2(e) states that concrete evaluated with the use of cores has adequate strength if the average strength of the cores is at least 85 percent of f'_c. Because the 85 percent value tends to be smaller than the actual ratio of core strength to cylinder strength, the widespread practice of taking the in-place strength equal to (core strength)/0.85 overestimates the in-place strength.

Neville [3-23] discusses core testing of in-place concrete and points out the advantages and drawbacks to using cores to estimate the concrete strength in a structure. Bartlett and MacGregor [3-24] and [3-25] suggest the following procedure for estimating the equivalent specified strength of concrete in a structure by using core tests:

1. **Plan the scope of the investigation.** The regions that are cored must be consistent with the information sought. That is, either the member in question should be cored, or, if this is impractical, the regions that are cored should contain the same type of concrete, of about the same age, and cured in the same way as the suspect region. The number of cores taken depends, on one hand, on the cost and the hazard from taking cores out of critical parts of the structure, and on the other hand, on the desired accuracy of the strength estimate. If possible, at least six cores should be taken from a given grade of concrete in

question. It is not possible to detect outliers (spurious values) in smaller samples, and the penalty for small sample sizes [given by k_1 in Eq. (3-8)] is significant. The diameter of the core should not be less than three times the nominal maximum size of the coarse aggregate, and the length of the core should be between one and two times the diameter. If possible, the core diameter should not be less than 4 in., because the variability of the core strengths increases significantly for smaller diameters.

2. Obtain and test the cores. Use standard methods to obtain and test the cores as given in ASTM C42. Carefully record the location in the structure of each core, the conditions of the cores before testing, and the mode of failure. This information may be useful in explaining individual low core strengths. A load–stroke plot from the core test may be useful in this regard. It is particularly important that the moisture condition of the core correspond to one of the two standard conditions prescribed in ASTM C42 and be recorded.

3. Convert the core strengths, f_{core}, to equivalent in-place strengths, f_{cis}. As an approximation for use in design, this is done by using

$$f_{cis} = f_{core}(F_{\ell/d} \times F_{dia} \times F_r)(F_{mc} \times F_d) \tag{3-7}$$

where the factors in the first set of parentheses correct the core strength to that of a standard 4-in.-diameter core, with length/diameter ratio equal to 2, not containing reinforcement:

$F_{\ell/d}$ = correction for length/diameter ratio as given in ASTM C 42
= 0.87, 0.93, 0.96, 0.98, and 1.00 for ℓ/d = 1.0, 1.25, 1.50, 1.75, and 2.0, respectively

F_{dia} = correction for diameter of core
= 1.06 for 2-in. cores, 1.00 for 4-in. cores, and 0.98 for 6-in. cores

F_r = correction for the presence of reinforcing bars
= 1.00 for no bars, 1.08 for one bar, and 1.13 for two bars

It is generally prudent to cut off parts of a core that contain reinforcing bars, provided the specimen that remains for testing has a length/diameter ratio equal to at least 1.0.

The factors in the second set of parentheses account for differences between the condition of the core and that of the concrete in the structure:

F_{mc} = accounts for the effect of the moisture condition of the core at the time of the core test
= 1.09 if the core was soaked before testing, and 0.96 if the core was air-dried at the time of the test

F_d = accounts for damage to the surface of the core due to drilling
= 1.06 if the core is damaged

4. Check for outliers in the set of equivalent in-place strengths. Reference [3-25] gives a technique for doing this. If an outlier is detected via a statistical test, one should try to determine a physical reason for the anomalous strength.

5. Compute the equivalent specified strength from the in-place strengths. The *equivalent specified strength*, f'_{ceq}, is the strength that should be used in design equations when checking the capacity of the member in question. To calculate it, one first computes the mean, $\bar{f}_{cis}$, and sample standard deviation, s_{cis}, of the set of equivalent in-place strengths, f_{cis}, which remains after any outliers have been removed. Bartlett and MacGregor [3-25] present the following equation for f'_{ceq}, which uses the core test data to obtain a lower-bound estimate of the 10 percent fractile of the in-place strength:

$$f'_{ceq} = k_2\left[\bar{f}_{cis} - 1.282\sqrt{\frac{(k_1 s_{cis})^2}{n} + \bar{f}_{cis}^{\,2}(V_{\ell/d}^{\,2} + V_{dia}^{\,2} + V_r^2 + V_{mc}^{\,2} + V_d^2)}\right] \quad (3\text{-}8)$$

Here,

k_1 = a factor dependent on the number of core tests, after removal of outliers, equal to 2.40 for 2 tests, 1.47 for 3 tests, 1.20 for 5 tests, 1.10 for 8 tests, 1.05 for 16 tests, and 1.03 for 25 tests

k_2 = a factor dependent on the number of batches of concrete in the member or structure being evaluated, equal to 0.90 and 0.85, respectively, for a cast-in-place member or structure that contains one batch or many batches, and equal to 0.90 for a precast member or structure

n = number of cores after removal of outliers

$V_{\ell/d}$ = coefficient of variation due to length/diameter correction, equal to 0.025 for $\ell/d = 1$, 0.006 for $\ell/d = 1.5$, and zero for $\ell/d = 2$

V_{dia} = coefficient of variation due to diameter correction, equal to 0.12 for 2-in.-diameter cores, zero for 4-in. cores, and 0.02 for 6-in. cores

V_r = coefficient of variation due to presence of reinforcing bars in the core, equal to zero if none of the cores contained bars, and to 0.03 if more than a third of them did

V_{mc} = coefficient of variation due to correction for moisture condition of core at time of testing, equal to 0.025

V_d = coefficient of variation due to damage to core during drilling, equal to 0.025

The individual coefficients of variation in the second term of Eq. (3-8) are taken equal to zero if the corresponding correction factor, F, is taken equal to 1.0 in Eq. (3-7).

EXAMPLE 3-1 Computation of an Equivalent Specified Strength from Core Tests

As a part of an evaluation of an existing structure, it is necessary to compute the strength of a 6-in.-thick slab. To do so, it is necessary to have an equivalent specified compressive strength, f'_{ceq}, to use in place of f'_c in the design equations. Several batches of concrete were placed in the slab.

1. **Plan the scope of the investigation.** From a site visit, it is learned that five cores can be taken. These are 4-in.-diameter cores drilled vertically through the slab, giving cores that are 6 in. long. They are taken from randomly selected locations around the entire floor in question.

2. **Obtain and test the cores.** The cores were tested in an air-dried condition. None of them contained reinforcing bars. The individual core strengths were 5950, 5850, 5740, 5420, and 4830 psi.

3. **Convert the core strengths to equivalent in-place strengths.** From Eq. (3-7),

$$f_{cis} = f_{core}(F_{\ell/d} \times F_{dia} \times F_r)(F_{mc} \times F_d)$$

The ℓ/d of the cores was 6 in./4 in. = 1.50. For this ratio, $F_{\ell/d} = 0.96$, and we have

$$\begin{aligned} f_{cis} &= f_{core}(0.96 \times 1.0 \times 1.0)(0.96 \times 1.06) \\ &= f_{core} \times 0.977 \end{aligned}$$

The individual strengths, f_{cis}, are 5812, 5715, 5607, 5295, and 4720 psi.

4. Check for low outliers. Although there is quite a difference between the lowest and second-lowest values, we shall assume that all five tests are valid.

5. Compute the equivalent specified strength. From Eq. (3-8),

$$f'_{ceq} = k_2\left[\bar{f}_{cis} - 1.282\sqrt{\frac{(k_1 s_{cis})^2}{n} + \bar{f}_{cis}^{\,2}(V_{\ell/d}^{\,2} + V_{dia}^{\,2} + V_r^2 + V_{mc}^{\,2} + V_d^2)}\right]$$

The mean and sample standard deviation of the f_{cis} values are $\bar{f}_{cis} = 5430$ psi and $s_{cis} = 442$ psi, respectively. Other terms in Eq. (3-8) are $k_1 = 1.20$ for five tests, $k_2 = 0.85$ for several batches, and n = five tests. Because no correction was made in step 3 for the effects of core diameter or reinforcement in the core (F_{dia} and $F_r = 1.0$), V_{dia} and V_r are equal to zero. The terms under the square-root sign in Eq. (3-8) are

$$\frac{(k_1 s_{cis})^2}{n} = \frac{(1.20 \times 442)^2}{5} = 56{,}265$$

$$\begin{aligned} \bar{f}_{cis}^{\,2}(V_{\ell/d}^{\,2} + V_{dia}^{\,2} + V_r^2 + V_{mc}^{\,2} + V_d^2) &= 5430^2(0.006^2 + 0.0^2 + 0.0^2 \\ &\quad + 0.025^2 + 0.025^2) \\ &= 37{,}918 \\ f'_{ceq} &= 0.85(5430 - 1.282\sqrt{56{,}265 + 37{,}918}) \\ &= 4280 \text{ psi} \end{aligned}$$

The concrete strength in the slab should be taken as 4280 psi when calculating the capacity of the slab. ■

Strength of Concrete in a Structure

The strength of concrete in a structure tends to be somewhat lower than the strength of control cylinders made from the same concrete. This difference is due to the effects of different placing, compaction, and curing procedures; the effects of vertical migration of water during the placing of the concrete in deep members; the effects of difference in size and shape; and the effects of different stress regimes in the structure and the cylinders.

The concrete near the top of deep members tends to be weaker than the concrete lower down, probably due to the increased water/cement ratio at the top due to upward water migration after the concrete is placed and by the greater compaction of the concrete near the bottom due to the weight of the concrete higher in the form [3-26].

3-4 STRENGTH UNDER TENSILE AND MULTIAXIAL LOADS

Tensile Strength of Concrete

Under uniaxial tensile loadings, small localized cracks are initiated at tensile–strain concentrations and these relieve these strain concentrations. This initial stage of loading results in an essentially linear stress–strain curve during the stage of stable crack initiation. Following a very brief interval of stable crack propagation, unstable crack propagation and fracture occur. The direction of cracking is perpendicular to the principal tensile stress and strain.

The tensile strength of concrete falls between 8 and 15 percent of the compressive strength. The actual value is strongly affected by the type of test carried out to determine the tensile strength, the type of aggregate, the compressive strength of the concrete, and the presence of a compressive stress transverse to the tensile stress [3-27] and [3-28].

Standard Tension Tests

Two types of tests are widely used. The first of these is the *modulus of rupture* or flexural test (ASTM C78), in which a plain concrete beam, generally 6 in. × 6 in. × 30 in. long, is loaded in flexure at the third points of a 24-in. span until it fails due to cracking on the tension face. The flexural tensile strength or modulus of rupture, f_r, from a modulus-of-rupture test is calculated from the following equation, assuming a linear distribution of stress and strain:

$$f_r = \frac{6M}{bh^2} \tag{3-9}$$

In this equation,

M = moment
b = width of specimen
h = overall depth of specimen

The second common tensile test is the *split cylinder* test (ASTM C496), in which a standard 6-by-12-in. compression test cylinder is placed on its side and loaded in compression along a diameter, as shown in Fig. 3-9a. A simplified force-transfer system is shown in Fig. 3-9b.

In a split-cylinder test, an element on the vertical diameter of the specimen is stressed in biaxial tension and compression, as shown in Fig. 3-9c. The stresses acting across the vertical diameter range from high transverse compressions at the top and bottom to a nearly uniform tension across the rest of the diameter, as shown in Fig. 3-9d. The splitting tensile strength, f_{ct}, from a split-cylinder test is computed as:

$$f_{ct} = \frac{2P}{\pi \ell d} \tag{3-10}$$

where

P = maximum applied load in the test
ℓ = length of specimen
d = diameter of specimen

Various types of tension tests give different strengths. In general, the strength decreases as the volume of concrete that is highly stressed in tension is increased. A third-point-loaded modulus-of-rupture test on a 6-in.-square beam gives a modulus-of-rupture strength f_r that

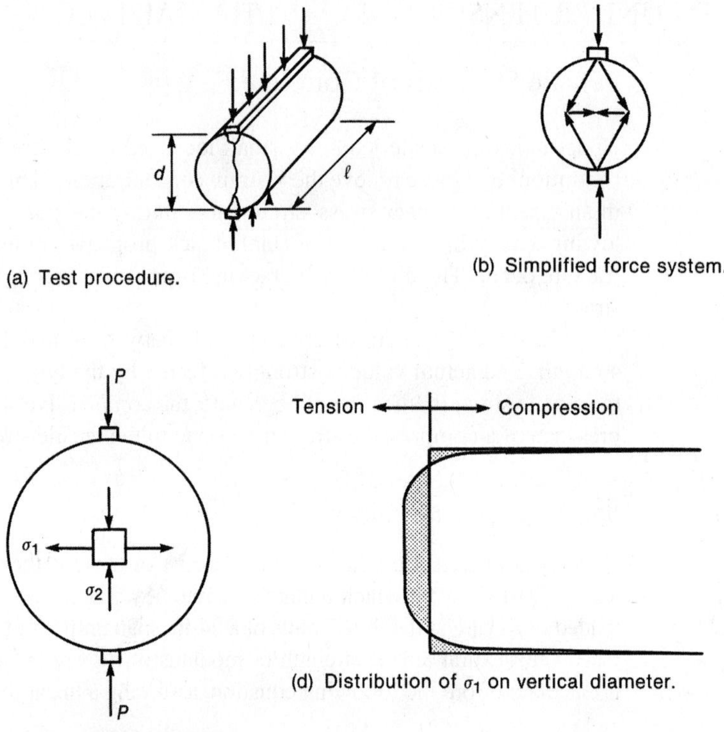

(a) Test procedure.

(b) Simplified force system.

(c) Stresses on element.

(d) Distribution of σ_1 on vertical diameter.

Fig. 3-9
Split-cylinder test.

averages 1.5 times f_{ct}, while a 6-in.-square prism tested in pure tension gives a direct tensile strength that averages about 86 percent of f_{ct}.

Relationship between Compressive and Tensile Strengths of Concrete

Although the tensile strength of concrete increases with an increase in the compressive strength, the ratio of the tensile strength to the compressive strength decreases as the compressive strength increases. Thus, the tensile strength is approximately proportional to the square root of the compressive strength. The mean split cylinder strength, $\bar{f}_{ct}$, from a large number of tests of concrete from various localities has been found to be:

$$\bar{f}_{ct} = 6.4\sqrt{f'_c} \tag{3-11}$$

where $\bar{f}_{ct}$, f'_c, and $\sqrt{f'_c}$ are all in psi. Values from Eq. (3-11) are compared with split-cylinder test data in Fig. 3-10. It is important to note the wide scatter in the test data. The ratio of measured to computed splitting strength is essentially normally distributed.

Similarly, the mean modulus of rupture, $\bar{f}_r$, can be expressed as:

$$\bar{f}_r = 8.3\sqrt{f'_c} \tag{3-12a}$$

Again, there is scatter in the modulus of rupture. Raphael [3-27] discusses the reasons for this, as do McNeely and Lash [3-28]. The distribution of the ratio of measured to computed modulus-of-rupture strength approaches a log-normal distribution.

ACI Code Section 19.2.3.1 defines the modulus of rupture for use in calculating deflections as

$$f_r = 7.5\lambda\sqrt{f'_c} \tag{3-12b}$$

where $\lambda = 1.0$ for normal weight concrete. Lightweight concrete is discussed in Section 3-8.

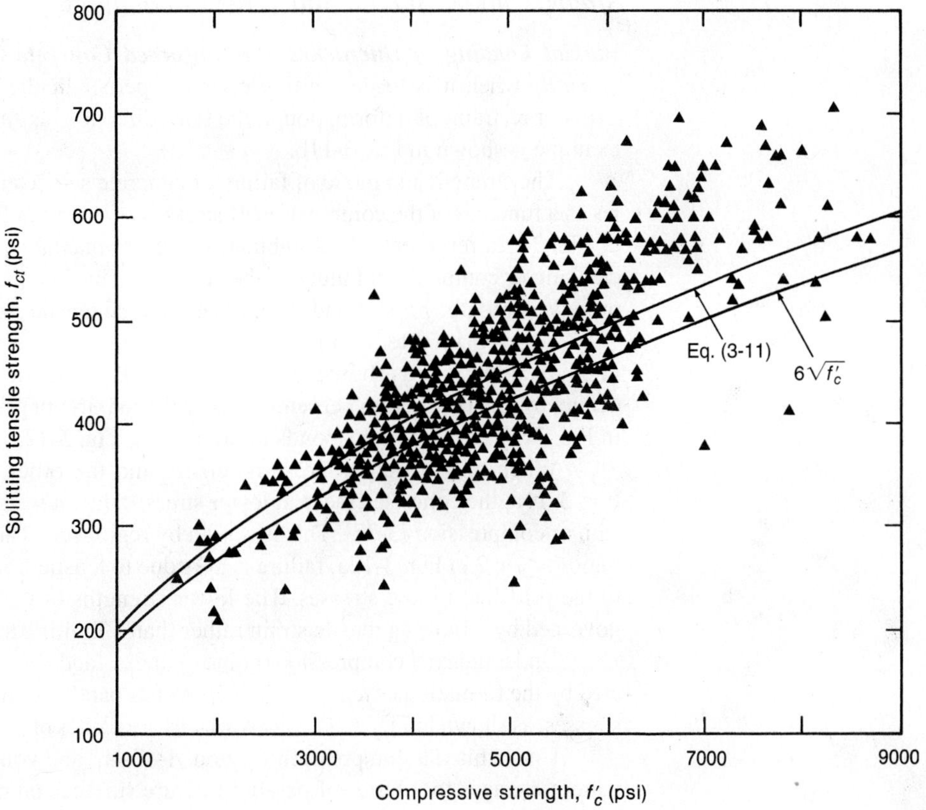

Fig. 3-10
Relationship between splitting tensile strengths and compression strengths. (From [3-27].)

A lower value is used for the average splitting tensile strength (ACI Commentary Section R19.2.4):

$$f_{ct} = 6.7\lambda\sqrt{f_c'} \tag{3-12c}$$

Factors Affecting the Tensile Strength of Concrete

The tensile strength of concrete is affected by the same factors that affect the compressive strength. In addition, the tensile strength of concrete made from crushed rock may be up to 20 percent greater than that from rounded gravels. The tensile strength of concrete made from lightweight aggregate tends to be less than that for normal sand-and-gravel concrete, although this varies widely, depending on the properties of the particular aggregate under consideration.

The tensile strength of concrete develops more quickly than the compressive strength. As a result, such things as shear strength and bond strength, which are strongly affected by the tensile strength of concrete, tend to develop more quickly than the compressive strength. At the same time, however, the tensile strength increases more slowly than would be suggested by the square root of the compressive strength at the age in question. Thus, concrete having a 28-day compressive strength of 3000 psi would have a splitting tensile strength of about $6.7\sqrt{f_c'} = 367$ psi. At 7 days this concrete would have compressive strength of about 2100 psi (0.70 times 3000 psi) and a tensile strength of about 260 psi (0.70 times 367 psi). This is less than the tensile strength of $6.7\sqrt{2100} = 307$ psi that one would compute from the 7-day compressive strength. This is of importance in choosing form-removal times for flat slab floors, which tend to be governed by the shear strength of the column–slab connections [3-29].

Strength under Biaxial and Triaxial Loadings

Biaxial Loading of Uncracked, Unreinforced Concrete Concrete is said to be *loaded biaxially* when it is loaded in two mutually perpendicular directions with essentially no stress or restraint of deformation in the third direction, as shown in Fig. 3-11a. A common example is shown in Fig. 3-11b.

The strength and mode of failure of concrete subjected to biaxial states of stress varies as a function of the combination of stresses as shown in Fig. 3-12. The pear-shaped line in Fig. 3-12a represents the combinations of the biaxial stresses, σ_1 and σ_2, which cause cracking or compression failure of the concrete. This line passes through the uniaxial compressive strength, f'_c, at A and A' and the uniaxial tensile strength, f'_t, at B and B'.

Under biaxial tension (σ_1 and σ_2 both tensile stresses) the strength is close to that in uniaxial tension, as shown by the region B–D–B' (zone 1) in Fig. 3-12a. Here, failure occurs by tensile fracture perpendicular to the maximum principal tensile stress, as shown in Fig. 3-12b, which corresponds to point B' in Fig. 3-12a.

When one principal stress is tensile and the other is compressive, as shown in Fig. 3-11a, the concrete cracks at lower stresses than it would if stressed uniaxially in tension or compression [3-30]. This is shown by regions A–B and A'–B' in Fig. 3-12a. In this region, zone 2 in Fig. 3-12a, failure occurs due to tensile fractures on planes perpendicular to the principal tensile stresses. The lower strengths in this region suggest that failure is governed by a limiting tensile strain rather than a limiting tensile stress.

Under uniaxial compression (points A and A' and zone 3 in Fig. 3-12a), failure is initiated by the formation of tensile cracks on planes parallel to the direction of the compressive stresses, as shown in Fig. 3-12c. These planes are planes of maximum principal tensile strain.

Under biaxial compression (region A–C–A' and zone 4 in Fig. 3-12a), the failure pattern changes to a series of parallel fracture surfaces on planes parallel to the unloaded

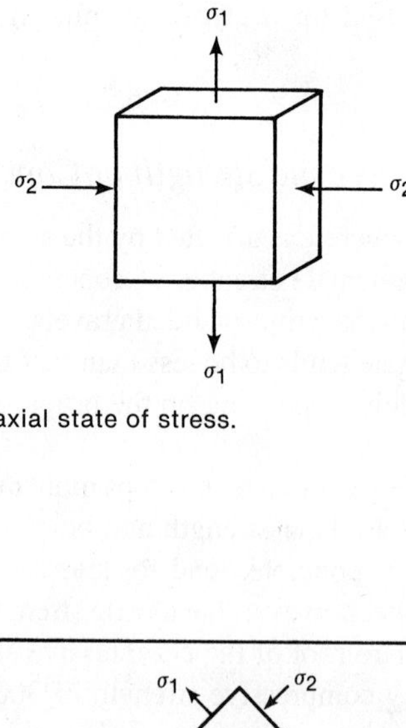

(a) Biaxial state of stress.

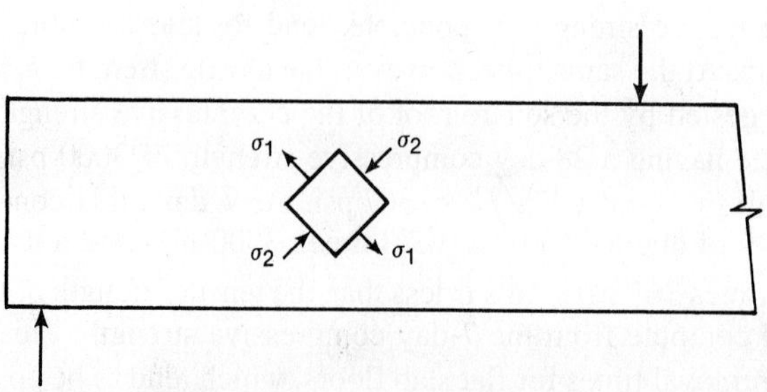

(b) Biaxial state of stress in the web of a beam.

Fig. 3-11
Biaxial stresses.

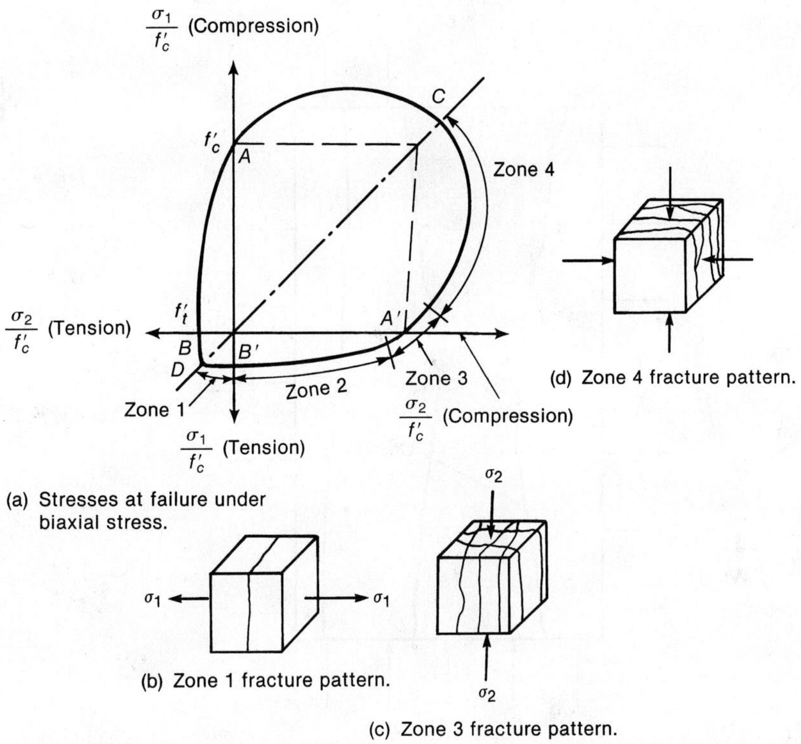

Fig. 3-12
Strength and modes of failure of unreinforced concrete subjected to biaxial stresses. (From [3-30].)

sides of the member, as shown in Fig. 3-12d. Such planes are acted on by the maximum tensile strains. Biaxial and triaxial compression loads delay the formation of bond cracks and mortar cracks. As a result, the period of stable crack propagation is longer and the concrete is more ductile. As shown in Fig. 3-12, the strength of concrete under biaxial compression is greater than the uniaxial compressive strength. Under equal biaxial compressive stresses, the strength is about 107 percent of f'_c, as shown by point C.

In the webs of beams, the principal tensile and principal compressive stresses lead to a biaxial tension–compression state of stress, as shown in Fig. 3-11b. Under such a loading, the tensile and compressive strengths are less than they would be under uniaxial stress, as shown by the quadrant AB or $A'B'$ in Fig. 3-12a. A similar biaxial stress state exists in a split-cylinder test, as shown in Fig. 3-9c. This explains in part why the splitting tensile strength is less than the flexural tensile strength.

In zones 1 and 2 in Fig. 3-12, failure occurred when the concrete cracked, and in zones 3 and 4, failure occurred when the concrete crushed. In a reinforced concrete member with sufficient reinforcement parallel to the tensile stresses, cracking does not represent failure of the member because the reinforcement resists the tensile forces after cracking. The biaxial load strength of cracked reinforced concrete is discussed in the next subsection.

Compressive Strength of Cracked Reinforced Concrete

If cracking occurs in reinforced concrete under a biaxial tension–compression loading and there is reinforcement across the cracks, the strength and stiffness of the concrete under compression parallel to the cracks is reduced. Figure 3-13a shows a concrete element that has been cracked by horizontal tensile stresses. The natural irregularity of the shape of the cracks leads to variations in the width of a piece between two cracks, as shown. The compressive stress

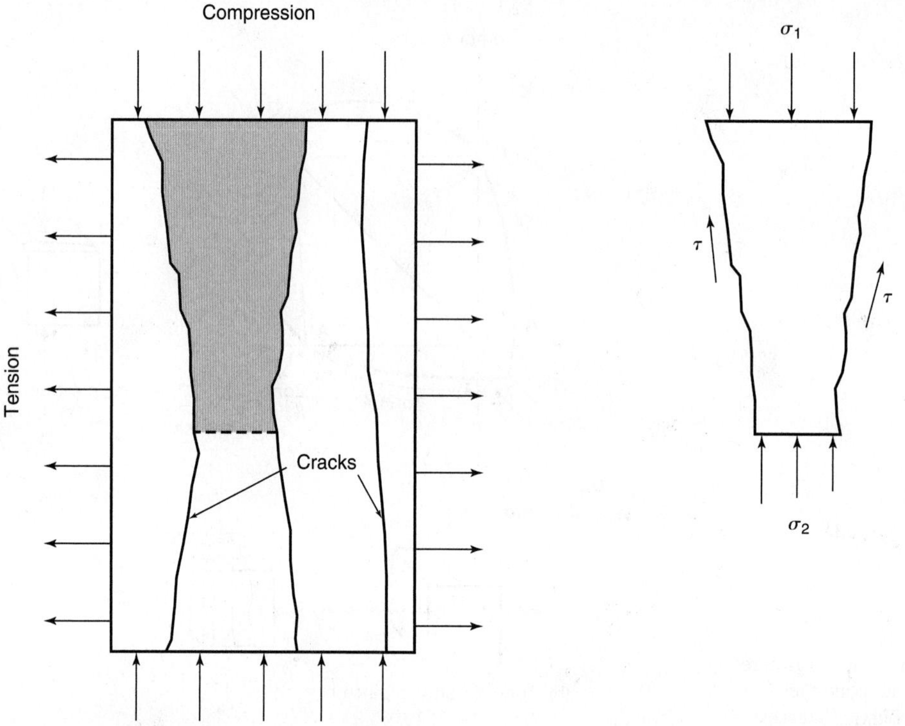

(a) Compression member with cracks.

(b) Free-body diagram of the shaded area in (a).

Fig. 3-13
Stresses in a biaxially loaded, cracked-concrete panel with cracks parallel to the direction of the principal compression stress.

acting on the top of the shaded portion is equilibrated by compressive stresses and probably some bearing stresses on the bottom and shearing stresses along the edges, as shown in Fig. 3-13b. When the crack widths are small, the shearing stresses transfer sufficient load across the cracks that the compressive stress on the bottom of the shaded portion is not significantly larger than that on the top, and the strength is unaffected by the cracks. As the crack widths increase, the ability to transfer shear across them decreases. For equilibrium, the compressive stress on the bottom of the shaded portion must then increase. Failure occurs when the highest stress in the element approaches the uniaxial compressive strength of the concrete.

Tests of concrete panels loaded in in-plane shear, carried out by Vecchio and Collins [3-31], have shown a relationship between the transverse tensile strain, ϵ_1, and the compressive strength parallel to the cracks, f_{2max}:

$$\frac{f_{2max}}{f'_c} = \frac{1}{0.8 + 170\epsilon_1} \tag{3-13}$$

where the subscripts 1 and 2 refer to the major (tensile) and minor (compressive) principal stresses and strains. The average transverse strain, ϵ_1, is the average transverse strain measured on a gauge length that includes one or more cracks. Equation (3-13) is plotted in Fig. 3-14a. An increase in the strain ϵ_1 leads to a decrease in compressive strength. The same authors [3-32] recommended a stress–strain relationship, f_2–ϵ_2, for transversely cracked concrete:

$$f_2 = f_{2max}\left[2\left(\frac{\epsilon_2}{\varepsilon_o}\right) - \left(\frac{\epsilon_2}{\varepsilon_o}\right)^2\right] \tag{3-14}$$

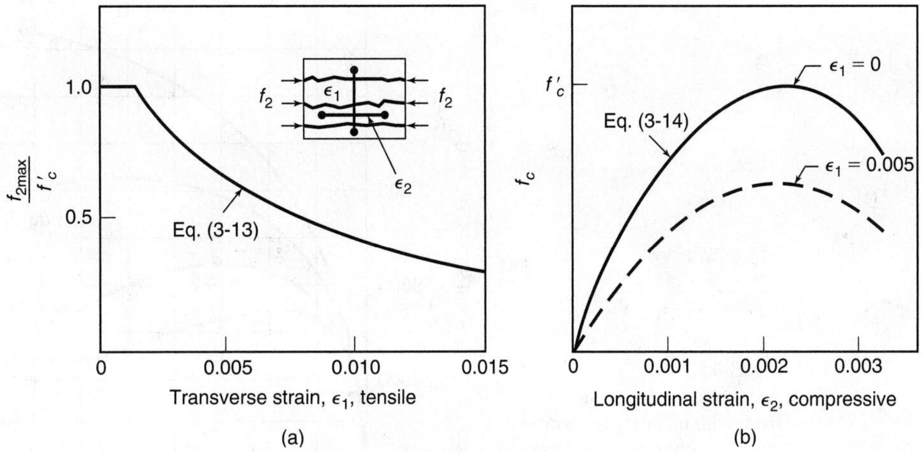

Fig. 3-14
Effect of transverse tensile strains on the compressive strength of cracked concrete.

where $f_{2\max}$ is given by Eq. (3-13), and ε_o is the strain at the highest point in the compressive stress–strain curve, which the authors took as 0.002. The term in brackets describes a parabolic stress–strain curve with apex at ε_o and a peak stress that decreases as ϵ_1 increases.

If the parabolic stress–strain curve given by Eq. (3-14) is used, the strain for any given stress can be computed from

$$\epsilon_c = \epsilon_c'\left(1 - \sqrt{\frac{f_2}{f_c'}}\right) \tag{3-15}$$

If the descending branch of the curve is also assumed to be a parabola, Eq. (3-15) can be used to compute strains on the postpeak portion of the stress–strain curve if the minus sign before the radical is changed to a plus.

The stress–strain relationships given by Eqs. (3-13) and (3-14) represent stresses and strains averaged over a large area of a shear panel or beam web. The strains computed in this way include the widths of cracks in the computation of tensile strains, ϵ_1, as shown in the inset to Fig. 3-14a. These equations are said to represent *smeared* properties. Through smearing, the peaks and hollows in the strains have been attenuated by using the averaged stresses and strains. In this way, Eqs. (3-13) and (3-14) are an attempt to replace the stress analysis of a cracked beam web having finite cracks with the analysis of a continuum. This substitution was a breakthrough in the analysis of concrete structures.

Triaxial Loadings

Under triaxial compressive stresses, the mode of failure involves either tensile fracture parallel to the maximum compressive stress (and thus orthogonal to the maximum tensile strain, if such exists) or a shear mode of failure. The strength and ductility of concrete under triaxial compression exceed those under uniaxial compression, as shown in Fig. 3-15. This figure presents the stress–longitudinal strain curves for cylinders each subjected to a constant lateral fluid pressure $\sigma_2 = \sigma_3$, while the longitudinal stress, σ_1, was increased to failure. These tests suggested that the longitudinal stress at failure was

$$\sigma_1 = f_c' + 4.1\sigma_3 \tag{3-16}$$

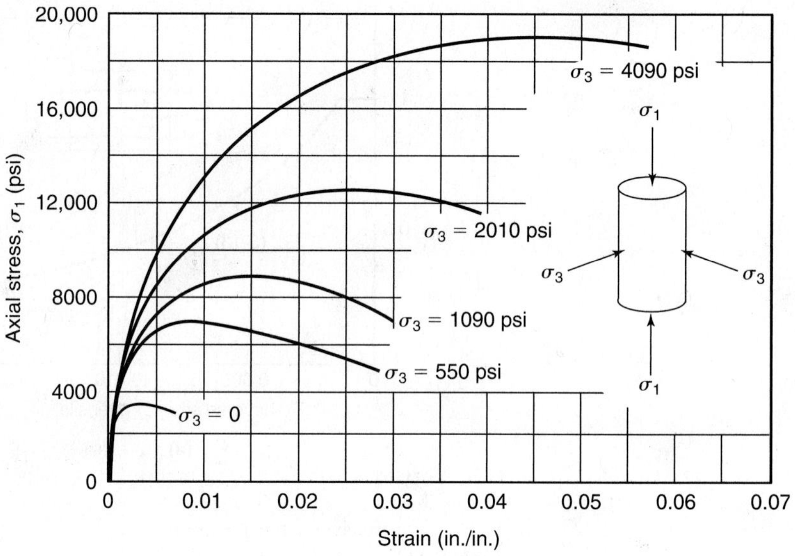

Fig. 3-15
Axial stress–strain curves from triaxial compression tests on concrete cylinders; unconfined compressive strength $f'_c = 3600$ psi. (From [3-3].)

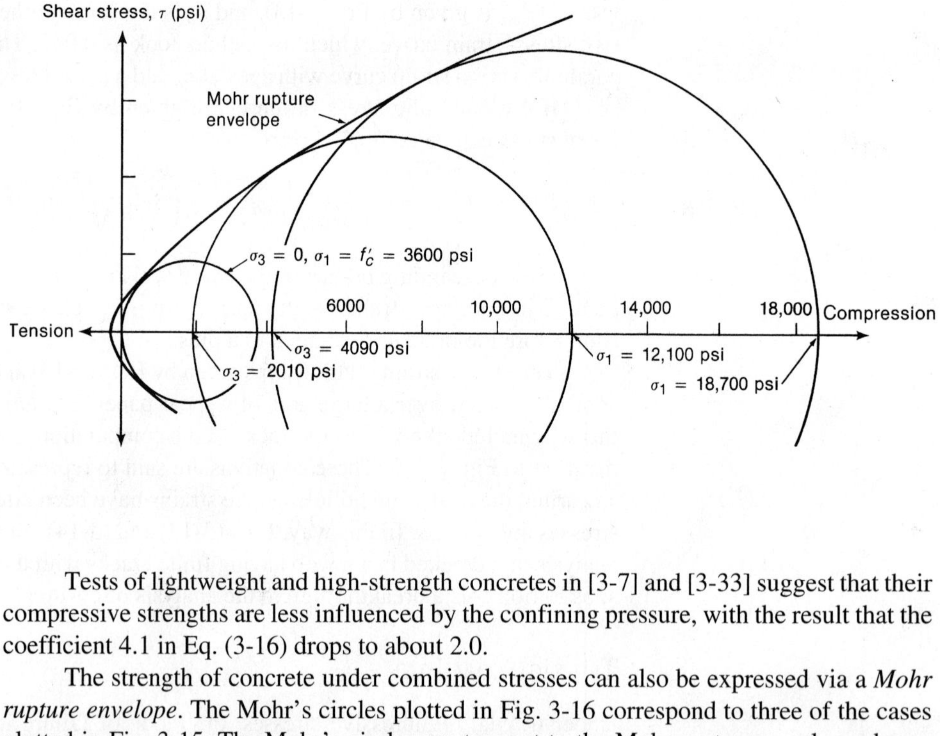

Fig. 3-16
Mohr rupture envelope for concrete tests from Fig. 3-15.

Tests of lightweight and high-strength concretes in [3-7] and [3-33] suggest that their compressive strengths are less influenced by the confining pressure, with the result that the coefficient 4.1 in Eq. (3-16) drops to about 2.0.

The strength of concrete under combined stresses can also be expressed via a *Mohr rupture envelope*. The Mohr's circles plotted in Fig. 3-16 correspond to three of the cases plotted in Fig. 3-15. The Mohr's circles are tangent to the Mohr rupture envelope shown with the outer line.

In concrete columns or in beam–column joints, concrete in compression is sometimes enclosed by closely spaced hoops or spirals. When the width of the concrete element increases due to Poisson's ratio and microcracking, these hoops or spirals are stressed in tension, causing an offsetting compressive stress in the enclosed concrete. The resulting triaxial state of stress in the concrete enclosed or *confined* by the hoops or spirals increases the ductility and strength of the confined concrete. This effect is discussed in Chapters 11 and 19.

3-5 STRESS–STRAIN CURVES FOR CONCRETE

The behavior and strength of reinforced concrete members is controlled by the size and shape of the members and by the stress–strain properties of the concrete and the reinforcement. The stress–strain behavior discussed in this section will be used in subsequent chapters to develop relationships for the strength and behavior of reinforced concrete beams and columns.

Tangent and Secant Moduli of Elasticity

Three ways of defining the modulus of elasticity are illustrated in Fig. 3-17. The slope of a line that is tangent to a point on the stress–strain curve, such as A, is called the *tangent modulus of elasticity*, E_T, at the stress corresponding to point A. The slope of the stress–strain curve at the origin is the *initial tangent modulus of elasticity*. The *secant modulus of elasticity at a given stress* is the slope of a line from the origin and through the point on the curve representing that stress (for example, point B in Fig. 3-17). Frequently, the secant modulus is defined by using the point corresponding to $0.4 f'_c$, representing service-load stresses. The slopes of these lines have units of psi/strain, where strain is unitless, with the result that the units of the modulus of elasticity are psi.

Stress–Strain Curve for Normal-Weight Concrete in Compression

Typical stress–strain curves for concretes of various strengths are shown in Fig. 3-18. These curves correspond to tests lasting about 15 minutes on specimens resembling the compression zone of a beam.

The stress–strain curves in Fig. 3-18 all rise to a maximum stress, reached at a strain between 0.0015 and 0.003, followed by a descending branch. The shape of this curve results from the gradual formation of microcracks within the structure of the concrete, as discussed in Section 3-2.

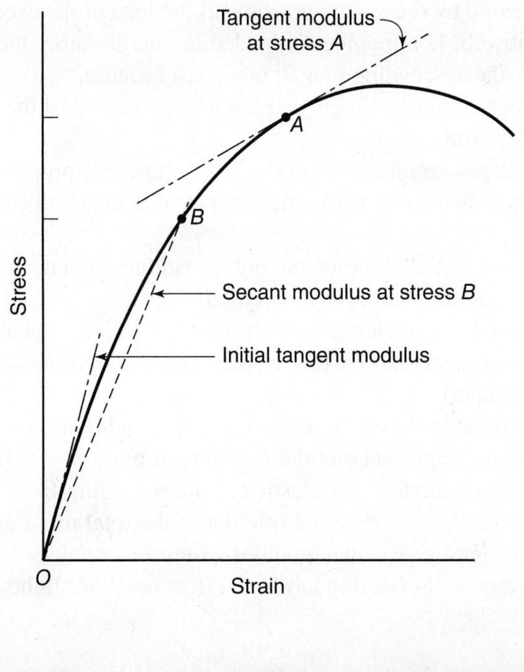

Fig. 3-17
Tangent and secant moduli of elasticity.

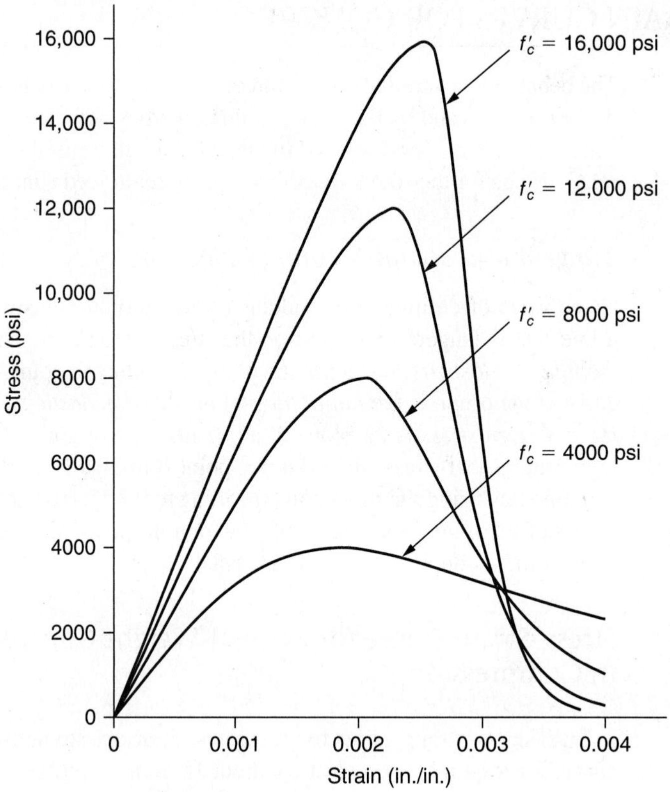

Fig. 3-18
Typical concrete stress–strain curves in compression. [Plotted using Eqs. (3-20) to (3-26).]

The length of the descending branch of the curve is strongly affected by the test conditions. Frequently, an axially loaded concrete test cylinder will fail explosively at the point of maximum stress. This will occur in axially flexible testing machines if the strain energy released by the testing machine as the load drops exceeds the energy that the specimen can absorb. If a member is loaded in compression due to bending (or bending plus axial load), the descending branch may exist because, as the stress drops in the most highly strained fibers, other less highly strained fibers can resist the load, thus delaying the failure of the highly strained fibers.

The stress–strain curves in Fig. 3-18 show five properties used in establishing mathematical models for the stress–strain curve of concrete in compression [3-34]:

1. The initial slope of the curves (initial tangent modulus of elasticity) increases with an increase in compressive strength.

The modulus of elasticity of the concrete, E_c, is affected by the modulus of elasticity of the cement paste and by that of the aggregate. An increase in the water/cement ratio increases the porosity of the paste, reducing its modulus of elasticity and strength. This is accounted for in design by expressing E_c as a function of f'_c.

Of equal importance is the modulus of elasticity of the aggregate. Normal-weight aggregates have modulus-of-elasticity values ranging from 1.5 to 5 times that of the cement paste. Because of this, the fraction of the total mix that is aggregate also affects E_c. Lightweight aggregates have modulus-of-elasticity values comparable to that of the paste; hence, the aggregate fraction has little effect on E_c for lightweight concrete.

The modulus of elasticity of concrete is frequently taken as given in ACI Code Section 19.2.2.1, namely,

$$E_c = 33(w^{1.5})\sqrt{f'_c} \text{ psi} \tag{3-17}$$

where w is the weight of the concrete in lb/ft^3. This equation was derived from short-time tests on concretes with densities ranging from 90 to 155 lb/ft^3 and corresponds to the secant modulus of elasticity at approximately $0.50f'_c$ [3-35]. The initial tangent modulus is about 10 percent greater. Because this equation ignores the type of aggregate, the scatter of data is very wide. Equation (3-17) systematically overestimates E_c in regions where low-modulus aggregates are prevalent. If deflections or vibration characteristics are critical in a design, E_c should be measured for the concrete to be used.

For normal-weight concrete with a density of 145 lb/ft^3, ACI Code Section 19.2.2.1 gives the modulus of elasticity as

$$E_c = 57{,}000\sqrt{f'_c} \text{ psi} \tag{3-18}$$

ACI Committee 363 [3-7] proposed the following equation for high-strength concretes:

$$E_c = 40{,}000\sqrt{f'_c} + 1.0 \times 10^6 \text{ psi} \tag{3-19}$$

2. The rising portion of the stress–strain curve resembles a parabola with its vertex at the maximum stress.

For computational purposes the rising portion of the curves is frequently approximated by a parabola [3-34], [3-36], and [3-37]. This curve tends to become straighter as the concrete strength increases [3-38].

3. The strain, ϵ_0, at maximum stress increases as the concrete strength increases.

4. As explained in Section 3-2, the slope of the descending branch of the stress–strain curve results from the destruction of the structure of the concrete, caused by the spread of microcracking and overall cracking. For concrete strengths up to about 6000 psi, the slope of the descending branch of the stress–strain curve tends to be flatter than that of the ascending branch. The slope of the descending branch increases with an increase in the concrete strength, as shown in Fig. 3-18. For concretes with f'_c greater than about 10,000 psi, the descending branch is a nearly vertical, discontinuous "curve." This is because the structure of the concrete is destroyed by major longitudinal cracking.

5. The maximum strain reached, ϵ_{cu}, decreases with an increase in concrete strength.

The descending portion of the stress–strain curve after the maximum stress has been reached is highly variable and is strongly dependent on the testing procedure. Similarly, the maximum or limiting strain, ϵ_{cu}, is very strongly dependent on the type of specimen, type of loading, and rate of testing. The limiting strain tends to be higher if there is a possibility of load redistribution at high loads. In flexural tests, values from 0.0025 to 0.006 have been measured.

Equations for Compressive Stress–Strain Diagrams

A common representation of the stress–strain curve for concretes with strengths up to about 6000 psi is the *modified Hognestad* stress–strain curve shown in Fig. 3-19a. This consists of a second-degree parabola with apex at a strain of $1.8f''_c/E_c$, where $f''_c = 0.9f'_c$, followed by a downward-sloping line terminating at a stress of $0.85f''_c$ and a limiting strain of 0.0038 [3-36]. The equation given in Fig. 3-19a describes a second-order parabola with

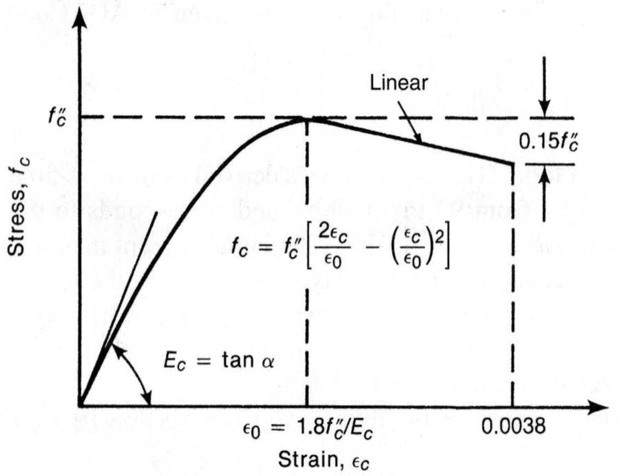

(a) Modified Hognestad. (From [3-37].)

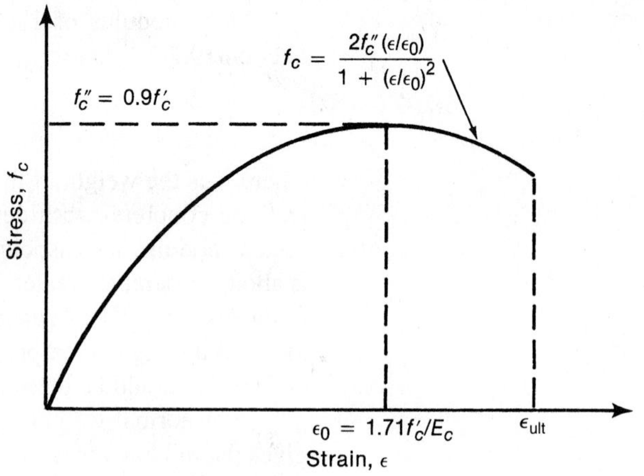

(b) Todeschini. (From [3-39].)

Fig. 3-19
Analytical approximations to the compressive stress–strain curve for concrete.

its apex at the strain ε_o. The reduced strength, $f''_c = 0.9f'_c$, accounts for the differences between cylinder strength and member strength. These differences result from different curing and placing, which give rise to different water-gain effects due to vertical migration of bleed water, and differences between the strengths of rapidly loaded cylinders and the strength of the same concrete loaded more slowly, as shown in Fig. 3-2.

Two other expressions for the stress–strain curve will be presented. The stress–strain curve shown in Fig. 3-19b is convenient for use in analytical studies involving concrete strengths up to about 5000 psi because the entire stress–strain curve is given by one continuous function. The highest point in the curve, f''_c, is taken to equal $0.9f'_c$ to give stress-block properties similar to that of the rectangular stress block of Section 4-3 when $\epsilon_{ult} = 0.003$ for f'_c up to 5000 psi. The strain ε_o, corresponding to maximum stress, is taken as $1.71f'_c/E_c$. For any given strain ϵ, $x = \epsilon/\varepsilon_o$. The stress corresponding to that strain is

$$f_c = \frac{2f''_c x}{1 + x^2} \tag{3-20}$$

For a compression zone of constant width, the average stress under the stress block from $\epsilon = 0$ to ϵ is $\beta_1 f''_c$, where

$$\beta_1 = \frac{\ln(1 + x^2)}{x} \tag{3-21}$$

The center of gravity of the area of the stress–strain curve between $\epsilon = 0$ and ϵ is at $k_2\epsilon$ from the point where ϵ exists, where

$$k_2 = 1 - \frac{2(x - \tan^{-1}x)}{x^2\beta_1} \tag{3-22}$$

where x is in radians when computing $\tan^{-1}x$. The stress–strain curve is satisfactory for concretes with stress–strain curves that display a gradually descending stress–strain curve at strains greater than ϵ_0. Hence, it is applicable for f'_c up to about 5000 psi for normal-weight concrete and about 4000 psi for lightweight concrete.

Expressions for the compressive stress–strain curve for concrete are reviewed by Popovics [3-38]. Thorenfeldt, Tomaszewicz, and Jensen [3-40] generalized two of these expressions to derive a stress–strain curve that applies to concrete strengths from 15 to 125 MPa (2000 to 18,000 psi). The relationship between a stress, f_c, and the corresponding strain, ϵ_c, is

$$\frac{f_c}{f'_c} = \frac{n(\epsilon_c/\varepsilon_o)}{n - 1 + (\epsilon_c/\varepsilon_o)^{nk}} \tag{3-23}$$

where

f'_c = peak stress obtained from a cylinder test
ε_o = strain when f_c reaches f'_c [see Eq. (3-27)]
n = a curve-fitting factor equal to $E_c/(E_c - E'_c)$ [see Eq. (3-24)]
E_c = initial tangent modulus (when $\epsilon_c = 0$)
$E'_c = f'_c/\varepsilon_o$
k = a factor to control the slopes of the ascending and descending branches of the stress–strain curve, taken equal to 1.0 for ϵ_c/ε_o less than 1.0 and taken greater than 1.0 for ϵ_c/ε_o greater than 1.0. [See Eqs. (3-25) and (3-26).]

The four constants ε_o, E_c, n, and k can be derived directly from a stress–strain curve for the concrete if one is available. If not, they can be computed from Eqs. (3-25) to (3-27), given by Collins and Mitchell [3-41]. Equations (3-17) and (3-18) can be used to compute E_c, although they were derived for the secant modulus from the origin and through points representing 0.4 to $0.5f'_c$. For normal-density concrete,

$$n = 0.8 + \left(\frac{f'_c}{2500}\right) \tag{3-24}$$

where f'_c is in psi. For ϵ_c/ε_o less than or equal to 1.0,

$$k = 1.0 \tag{3-25}$$

and for $\epsilon_c/\varepsilon_o > 1.0$,

$$k = 0.67 + \left(\frac{f'_c}{9000}\right) \geq 1.0 \text{ (psi)} \tag{3-26}$$

If n, f'_c, and E_c are known, the strain at peak stress can be computed from

$$\varepsilon_o = \frac{f'_c}{E_c}\left(\frac{n}{n-1}\right) \tag{3-27}$$

A family of stress–strain curves calculated from Eq. (3-23) is shown in Fig. 3-18. Equation (3-23) produces a smooth continuous descending branch. Actually, the descending branch for high-strength concretes tends to drop in a series of jagged steps as the structure of the concrete is destroyed. Equation (3-23) approximates this with a smooth curve, as shown in Fig. 3-18.

Traditionally, equivalent stress blocks used in design are based directly on stress–strain curves that have the peak stress equal to f''_c, which is $0.85f'_c$ to $0.9f'_c$, to allow for differences between the in-place strength and the cylinder strength. For prediction of experimentally obtained behavior, the ordinates of the stress–strain curve should be computed for a strength f'_c and then multiplied by 0.90. For design based on stress–strain relationships, the stress–strain curve should be derived for a strength of f'_c and the ordinates multiplied by 0.90.

As shown in Fig. 3-15, a lateral confining pressure causes an increase in the compressive strength of concrete and a large increase in the strains at failure. The additional strength and ductility of confined concrete are utilized in hinging regions of structures in seismic regions. Stress–strain curves for confined concrete are described in [3-42].

When a compression specimen is loaded, unloaded, and reloaded, it has the stress–strain response shown in Fig. 3-20. The envelope to this curve is very close to the stress–strain curve for a monotonic test. This and the large residual strains that remain after unloading suggest that the inelastic response is due to damage to the internal structure of the concrete, as is suggested by the microcracking theory presented earlier.

Stress–Strain Curve for Normal-Weight Concrete in Tension

The stress–strain response of concrete loaded in axial tension can be divided into two phases. Prior to the maximum stress, the stress–strain relationship is slightly curved. The diagram is linear to roughly 50 percent of the tensile strength. The strain at peak stress is about 0.0001 in pure tension and 0.00014 to 0.0002 in flexure. The rising part of the stress–strain curve may be approximated either as a straight line with slope E_c and a maximum stress equal to the tensile strength f'_t or as a parabola with a maximum strain $\epsilon'_t = 1.8f'_t/E_c$ and a maximum stress f'_t. The latter curve is illustrated in Fig. 3-21a with f'_t and E_c based on Eqs. (3-11) and (3-18).

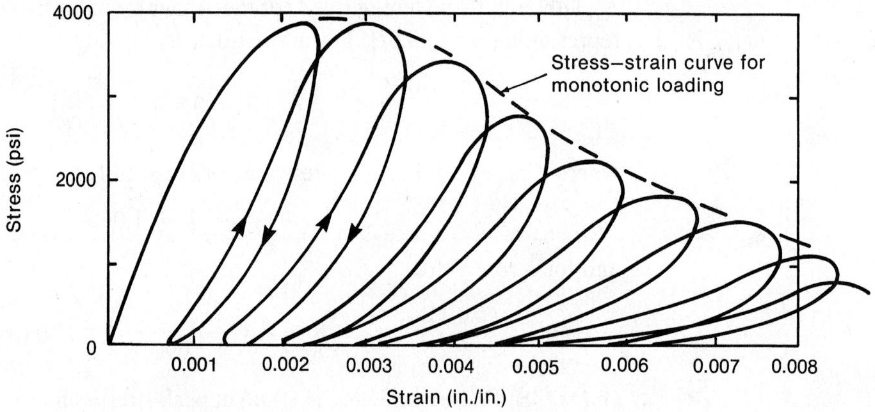

Fig. 3-20
Compressive stress–strain curves for cyclic loads. (From [3-43].)

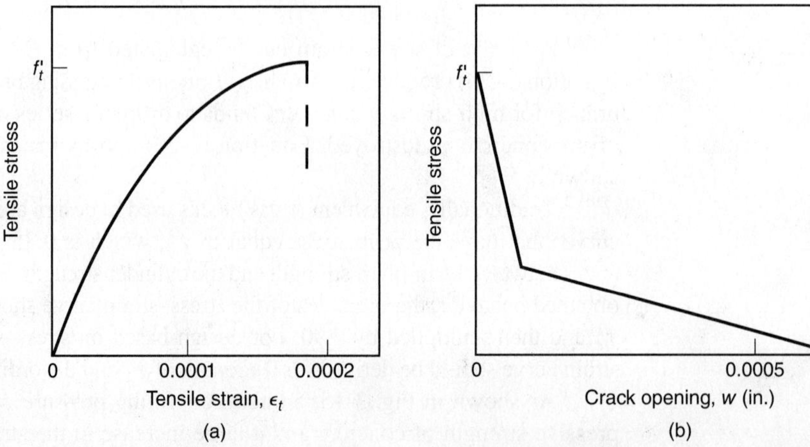

Fig. 3-21
Stress–strain curve and stress–crack opening curves for concrete loaded in tension.

After the tensile strength is reached, microcracking occurs in a *fracture process zone* adjacent to the point of highest tensile stress, and the tensile capacity of this concrete drops very rapidly with increasing elongation. In this stage of behavior, elongations are concentrated in the fracture process zone while the rest of the concrete is unloading elastically. The unloading response is best described by a *stress-versus-crack-opening diagram*, idealized in Fig. 3-21b as two straight lines. The crack widths shown in this figure are of the right magnitude, but the actual values depend on the situation. The tensile capacity drops to zero when the crack is completely formed. This occurs at a very small crack width. A more detailed discussion is given in [3-44].

Poisson's Ratio

At stresses below the critical stress (see Fig. 3-1), Poisson's ratio for concrete varies from about 0.11 to 0.21 and usually falls in the range from 0.15 to 0.20. On the basis of tests of biaxially loaded concrete, Kupfer et al. [3-30] report values of 0.20 for Poisson's ratio for concrete loaded in compression in one or two directions: 0.18 for concrete loaded in tension in one or two directions and 0.18 to 0.20 for concrete loaded in tension and compression. Poisson's ratio remains approximately constant under sustained loads.

3-6 TIME-DEPENDENT VOLUME CHANGES

Concrete undergoes three main types of volume change, which may cause stresses, cracking, or deflections that affect the in-service behavior of reinforced concrete structures. These are shrinkage, creep, and thermal expansion or contraction.

Shrinkage

Shrinkage is the decrease in the volume of concrete during hardening and drying under constant temperature. The amount of shrinkage increases with time, as shown in Fig. 3-22a.

The primary type of shrinkage is called *drying shrinkage* or simply *shrinkage* and is due to the loss of a layer of *adsorbed water* (electrically bound water molecules) from the surface of the gel particles. This layer is roughly one water molecule thick, or about 1 percent of the size of the gel particles. The loss of free unadsorbed water has little effect on the magnitude of the shrinkage.

Shrinkage strains are dependent on the relative humidity and are largest for relative humidities of 40 percent or less. They are partially recoverable upon rewetting the concrete, and structures exposed to seasonal changes in humidity may expand and contract slightly due to changes in shrinkage strains.

The magnitude of shrinkage strains also depends on the composition of the concrete mix and the type of cement used. The hardened cement paste shrinks, whereas the aggregate acts to restrain shrinkage. Thus, the larger the fraction of the total volume of the concrete that is made up of hydrated cement paste, the greater the shrinkage. This may be particularly important with the more common use of self-consolidating concrete, which has significantly higher paste content than normally consolidated concrete of the same strength. An increase in the water/cementitious materials ratio or the total cement content reduces the volume of aggregates, thus reducing the restraint of shrinkage by the aggregate. Also, more finely ground cements have a larger surface area per unit volume, and thus, there is more adsorbed water to be lost during shrinkage. There is less shrinkage in concrete made with quartz or granite aggregates than with sandstone aggregates because quartz and granite have a higher modulus of elasticity.

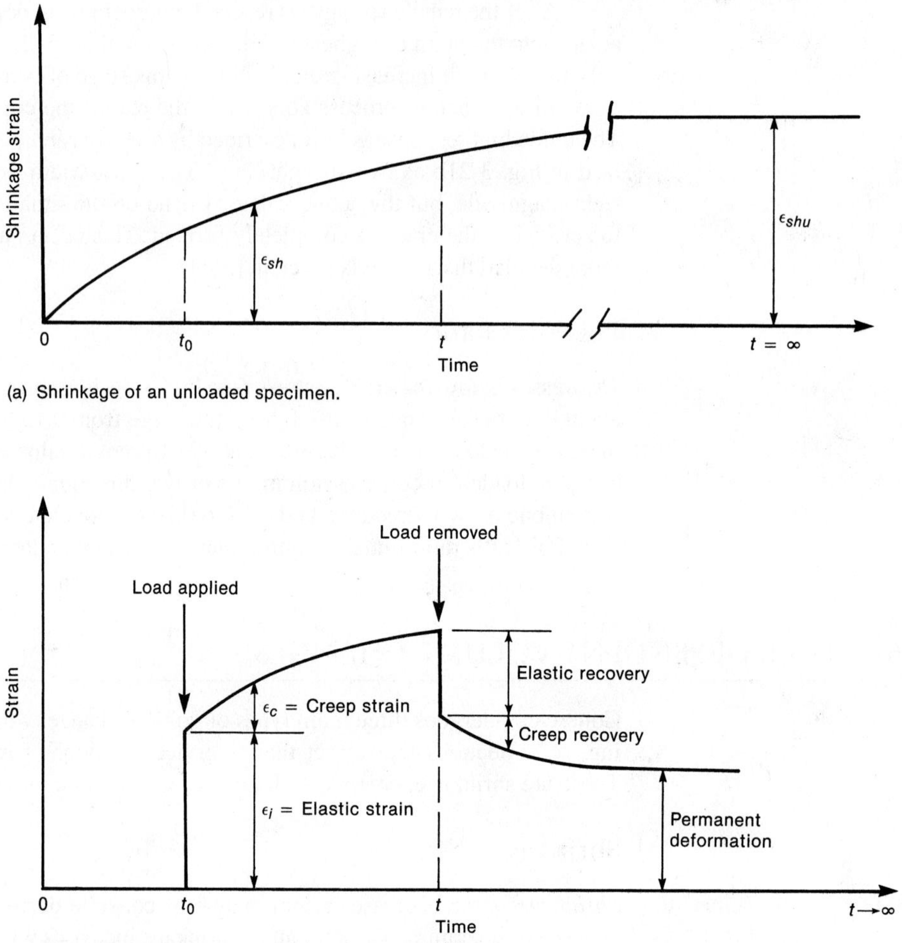

(a) Shrinkage of an unloaded specimen.

(b) Elastic and creep strains due to loading at time, t_0, and unloading at time, t.

Fig. 3-22
Time-dependent strains.

Drying shrinkage occurs as the moisture diffuses out of the concrete. As a result, the exterior shrinks more rapidly than the interior. This leads to tensile stresses in the outer skin of the concrete and compressive stresses in the interior. For large members, the ratio of volume to surface area increases, resulting in less shrinkage because there is more moist concrete to restrain the shrinkage. Shrinkage also develops more slowly in large members.

Autogenous shrinkage occurs without the loss of moisture due to hydration reactions inside the cement matrix. In earlier studies this was considered to be a very small portion of the total shrinkage, but with a greater use of high-performance concretes (water/cement ratio below 0.40), autogenous shrinkage may constitute a more significant percentage of the total shrinkage [3-45].

A final form of shrinkage called *carbonation shrinkage* occurs in carbon-dioxide rich atmospheres, such as those found in parking garages. At 50 percent relative humidity, the amount of carbonation shrinkage can equal the drying shrinkage, effectively doubling the total amount of shrinkage. At higher and lower humidities, the carbonation shrinkage decreases.

The ultimate drying shrinkage strain, ϵ_{shu}, for a 6-by-12-in. cylinder maintained for a very long time at a relative humidity of 40 percent ranges from 0.000400 to 0.001100

(400 to 1100 $\times$ 10^{-6} strain), with an average of about 0.000800 [3-16]. Thus, in a 25-ft bay in a building, the average shrinkage strain would cause a shortening of about $\frac{1}{4}$ in. in unreinforced concrete. In a structure, however, the shrinkage strains will tend to be less for the same concrete, for the following reasons:

1. The ratio of volume to surface area will generally be larger than for the cylinder; as a result, drying shrinkage should be reduced.

2. A structure is built in stages, and some of the shrinkage is dissipated before adjacent stages are completed.

3. The reinforcement restrains the development of the shrinkage.

The *fib Model Code 2010* [3-46] and ACI Committee 209 [3-20] have published procedures for estimating shrinkage strains. Because the fib Model Code procedure is more complicated than is required for typical structural design, the procedure developed by ACI Committee 209 [3-20] with some modifications from Mindess et al. [3-47] will be presented here.

The general expression for the development of shrinkage strain in concrete that is moist-cured for 7 days and then dried in 40 percent relative humidity is:

$$(\varepsilon_{sh})_t = \frac{t}{35 + t}(\varepsilon_{sh})_u \qquad (3\text{-}28)$$

where $(\varepsilon_{sh})_t$ is the shrinkage strain after t days of drying and $(\varepsilon_{sh})_u$ is the ultimate value for drying shrinkage. For concrete that is steam-cured for 1 to 3 days, the constant 35 in Eq. (3-28) is increased to 55. The value for $(\varepsilon_{sh})_u$ may vary between 415 $\times$ 10^{-6} and 1070 $\times$ 10^{-6}. In the absence of detailed shrinkage data for the local aggregates and conditions, $(\varepsilon_{sh})_u$ can be taken as:

$$(\varepsilon_{sh})_u = 780 \times 10^{-6} \qquad (3\text{-}29)$$

Modification for Relative Humidity. Concrete shrinkage strains are reduced in locations with a high ambient relative humidity (RH), as indicated in Fig. 3-23, in which

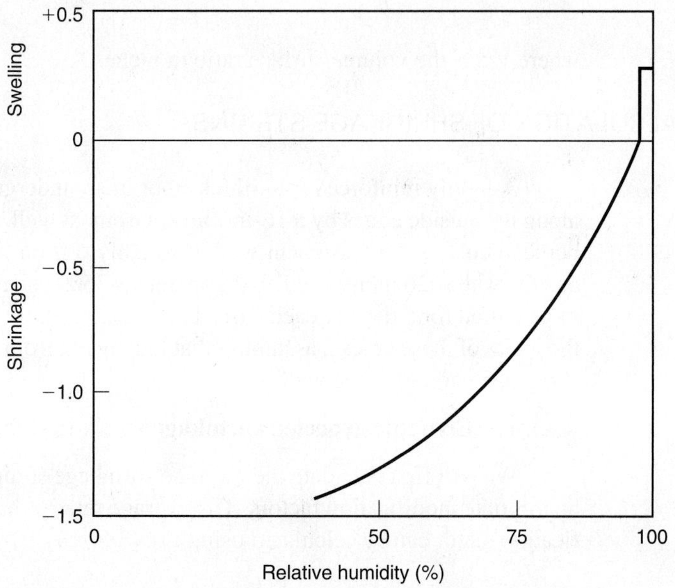

Fig. 3-23
Effect of relative humidity on shrinkage.

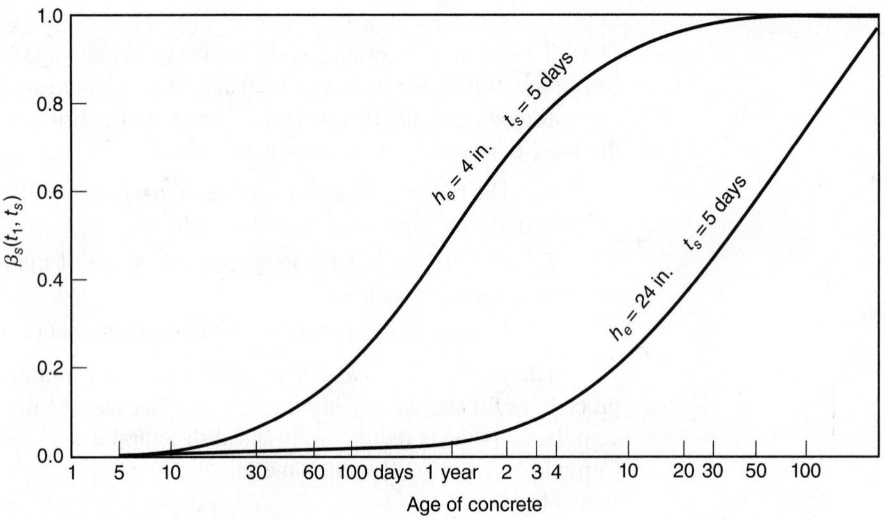

Fig. 3-24
Effect of effective thickness, h_e, on the rate of development of shrinkage.

β_{RH} is a coefficient that accounts for relative humidity. To account for RH values greater that 40 percent, $(\varepsilon_{sh})_u$ can be multiplied by the correction factor, γ_{rh}, given as:

$$\text{for } 40\% \leq \text{RH} \leq 80\%: \gamma_{rh} = 1.40 - 0.01 \times \text{RH} \quad (3\text{-}30a)$$

$$\text{for RH} > 80\%: \gamma_{rh} = 3.00 - 0.03 \times \text{RH} \quad (3\text{-}30b)$$

Modification for Volume/Surface Ratio. Concrete members with a large surface area per unit volume will tend to lose more moisture to the atmosphere, and thus, will exhibit higher shrinkage strains. ACI Committee 209 describes two methods to account for the shape and size of a concrete member using either the average member thickness or the member volume/surface ratio. The effect that the average member thickness, h_e, has on the development of shrinkage over time, β_s, is shown in Fig. 3-24, where t_s indicates the length of time for moist-curing. Members with a large average thickness have a larger volume/surface ratio. The value of $(\varepsilon_{sh})_u$ given in Eq. (3-29) assumes an average member thickness of 6 in., and a volume/surface ratio of 1.5 in. Using the volume/surface approach from ACI Committee 209, the correction factor, γ_{vs}, is given as:

$$\gamma_{vs} = 1.2^{-0.12V/S} \quad (3\text{-}31)$$

where *V/S* is the volume/surface ratio in inches.

EXAMPLE 3-2 CALCULATION OF SHRINKAGE STRAINS

A lightly reinforced 6-in.-thick floor in an underground parking garage is supported along its outside edges by a 16-in.-thick basement wall. Cracks have developed in the slab perpendicular to the basement wall at roughly 6 ft on centers. The slab is 24 months old and the wall is 26 months old. The concrete is 3500 psi, made from Type I cement, and was moist-cured for 5 days in each case. The average relative humidity is 50 percent. Compute the width of these cracks, assuming that they result from the basement wall restraining slab shrinkage parallel to the wall.

1. Compute expected shrinkage strain in slab.

We will first calculate the ultimate shrinkage strain for the slab using Eq. (3-29) and appropriate modification factors. The average relative humidity is 50 percent, so the modification factor can be calculated using Eq. (3-30a).

$$\gamma_{rh} = 1.40 - 0.01 \times RH = 1.40 - 0.50 = 0.90$$

Assume the slab spans 24 ft in one direction and 12 ft in the other direction. Thus, the volume of concrete is 6 in. $\times$ 24 ft $\times$ 12 ft. Assume that due to continuity with adjacent slabs, the edges of the slab are not exposed to the atmosphere. Thus, the exposed surface area on the top and bottom of the slab is 2 $\times$ 24 ft $\times$ 12 ft. With these two values, the volume/surface ratio is, $V/S = 6$ in./2 $= 3$ in. Using this value in Eq. (3-31), the modification factor for volume/surface ratio is,

$$\gamma_{vs} = 1.2^{-0.12V/S} = 1.2^{-0.36} = 0.936 \cong 0.94$$

Rounding this to two significant figures is appropriate because the constants in Eq. (3-31) are only given to two significant figures.

We can now use Eq. (3-29) and the two modification factors to determine the ultimate shrinkage strain and then use Eq. (3-28) to determine the shrinkage strain after 24 months. Putting the modification factors into Eq. (3-29) results in,

$$(\varepsilon_{sh})_u = \gamma_{rh} \times \gamma_{vs} \times 780 \times 10^{-6}$$

$$= 0.90 \times 0.94 \times 780 \times 10^{-6} = 660 \times 10^{-6} \text{ strain}$$

Assume the slab had 7 days of moist-curing before being exposed to the atmosphere. Thus, the number of drying days after 24 months (2 years) is,

$$t = 2 \times 365 - 7 = 730 - 7 = 723 \text{ days}$$

Using this number in Eq. (3-28) results in,

$$(\varepsilon_{sh})_t = \frac{t}{35 + t}(\varepsilon_{sh})_u = \frac{723}{35 + 723} \times 660 \times 10^{-6}$$

$$= 629 \times 10^{-6} \cong 630 \times 10^{-6} \text{ strain}$$

2. Compute expected shrinkage strains in the wall.

We will calculate the total expected shrinkage strain in the wall for 26 months exposure and then subtract from that the expected shrinkage strain during the first 2 months before the slab was cast. The difference will give us the shrinkage strains experienced in the wall from the time the slab was cast up to the 24 months after the slab was cast.

The coefficient, γ_{rh}, is the same as that calculated for the slab (0.90). Assume the portion of the wall under consideration is 10 ft high and has a length of 24 ft. Thus, the volume of concrete in the wall is 16 in. $\times$ 10 ft $\times$ 24 ft. Assume the bottom and edges of the wall are continuous, and thus, not exposed to the atmosphere. The exposed surface area for the first 2 months consists of the front and back of the wall (2 $\times$ 10 ft $\times$ 24 ft) plus the top of the wall (24 ft $\times$ 1.5 ft). After the slab is cast the top of the wall is not exposed, and thus, is not part of the exposed surface area. For the first 2 months after the wall is cast the V/S ratio is,

$$\frac{V}{S} = \frac{16 \text{ in.} \times 240 \text{ ft}^2}{2 \times 240 \text{ ft}^2 + 36 \text{ ft}^2} = 16 \text{ in.}\frac{240 \text{ ft}^2}{516 \text{ ft}^2} = 7.44 \text{ in.}$$

Using Eq. (3-31), the modification factor for the volume/surface ratio during the first 2 months is:

$$\gamma_{vs} = 1.2^{-0.12V/S} = 1.2^{-0.893} \cong 0.85$$

During the following 24 months, $V/S = 8$ in. and the modification factor is,

$$\gamma_{vs} = 1.2^{-0.12V/S} = 1.2^{-0.96} \cong 0.84$$

The difference between these two coefficients is trivial and can be ignored in the calculation of the ultimate shrinkage strain in the wall. Using Eq. (3-29) and the calculated modification factors, the ultimate shrinkage strain expected in the wall is,

$$\begin{aligned}(\varepsilon_{sh})_u &= \gamma_{rh} \times \gamma_{vs} \times 780 \times 10^{-6}\\ &= 0.90 \times 0.84 \times 780 \times 10^{-6} = 590 \times 10^{-6} \text{ strain}\end{aligned}$$

Again assuming 7 days of moist-curing, the shrinkage strain expected in the wall after 26 months can be calculated using the number of drying days equal to,

$$t = 2 \times 365 + 2 \times 30 - 7 = 783 \text{ days}$$

Substituting this and $(\varepsilon_{sh})_u$ into Eq. (3-28) gives,

$$\begin{aligned}(\varepsilon_{sh})_t &= \frac{t}{35 + t}(\varepsilon_{sh})_u = \frac{783}{35 + 783} \times 590 \times 10^{-6}\\ &= 564 \times 10^{-6} \cong 560 \times 10^{-6} \text{ strain}\end{aligned}$$

To calculate the shrinkage strain in the wall during the first 2 months, use $t = 60 - 7 = 53$ days in Eq. (3-28),

$$\begin{aligned}(\varepsilon_{sh})_t &= \frac{53}{35 + 53} \times 590 \times 10^{-6}\\ &= 355 \times 10^{-6} \cong 360 \times 10^{-6}\end{aligned}$$

Thus, the net shrinkage strain expected in the wall after the slab is cast is,

$$\text{Net wall strain} = (560 - 360) \times 10^{-6} = 200 \times 10^{-6} \text{ strain}$$

3. Relative shrinkage strain and expected crack width.

Using the shrinkage strain values calculated in the prior steps for the slab and the wall after the slab was cast, the net differential shrinkage strain between the slab and the wall is,

$$\text{Net differential strain} = (630 - 200) \times 10^{-6} = 430 \times 10^{-6} \text{ strain}$$

With this value, if the observed cracks in the slab are occurring at a spacing of 6 ft, the expected crack widths would be,

$$\begin{aligned}\text{Crack width} &\cong 6 \text{ ft} \times 12 \text{ in./ft} \times 430 \times 10^{-6}\\ &\cong 0.031 \text{ in.}\end{aligned}$$

This is an approximate value for the crack width because it assumes a uniform spacing between the cracks in the slab and does not account for the effect of reinforcement restraining shrinkage strains in the concrete. If reinforcement is present the shrinkage strains would be from 75 to 90 percent of the calculated values. ■

Creep of Unrestrained Concrete

When concrete is loaded in compression, an instantaneous elastic strain develops, as shown in Fig. 3-22b. If this load remains on the member, creep strains develop with time. These occur because the adsorbed water layers tend to become thinner between gel particles transmitting compressive stress. This change in thickness occurs rapidly at first, slowing down with time. With time, bonds form between the gel particles in their new position. If the load is eventually removed, a portion of the strain is recovered elastically and another portion by creep, but a residual strain remains (see Fig. 3-22b), due to the bonding of the gel particles in the deformed position.

Creep strains, ϵ_c, which continue to increase over a period of two to five years, are on the order of one to three times the instantaneous elastic strains. Increased concrete compression strains due to creep will lead to an increase in deflections with time, may lead to a redistribution of stresses within cross sections, and cause a decrease in prestressing forces.

The ratio of creep strain after a very long time to elastic strain, ϵ_c/ϵ_i, is called the *creep coefficient*, ϕ. The magnitude of the creep coefficient is affected by the ratio of the sustained stress to the strength of the concrete, the age of the concrete when loaded, the humidity of the environment, the dimensions of the element, and the composition of the concrete. Creep is greatest in concretes with a high cement–paste content. Concretes containing a large aggregate fraction creep less, because only the paste creeps and because creep is restrained by the aggregate. The rate of development of the creep strains is also affected by the temperature, reaching a plateau at about 160°F. At the high temperatures encountered in fires, very large creep strains occur. The type of cement (i.e., normal or high-early-strength cement) and the water/cement ratio are important only in that they affect the strength at the time when the concrete is loaded.

For creep, as for shrinkage, several calculation procedures exist [3-20], [3-46], and [3-47]. For stresses less than $0.40f'_c$, creep is assumed to be linearly related to stress. Beyond this stress, creep strains increase more rapidly and may lead to failure of the member at stresses greater than $0.75f'_c$, as shown in Fig. 3-2a. Similarly, creep increases significantly at mean temperatures in excess of 90°F.

The total strain, $\epsilon_c(t)$, at time t in a concrete member uniaxially loaded with a constant stress $\sigma_c(t_0)$ at time t_0 is

$$\epsilon_c(t) = \epsilon_{ci}(t_0) + \epsilon_{cc}(t) + \epsilon_{cs}(t) + \epsilon_{cT}(t) \tag{3-32}$$

where

$\epsilon_{ci}(t_0)$ = initial strain at loading = $\sigma_c(t_0)/E_c(t_0)$
$\epsilon_{cc}(t)$ = creep strain at time t where t is greater than t_0
$\epsilon_{cs}(t)$ = shrinkage strain at time t
$\epsilon_{cT}(t)$ = thermal strain at time t
$E_c(t_0)$ = modulus of elasticity at the age of loading

The stress-dependent strain at time t is

$$\epsilon_{cs}(t) = \epsilon_{ci}(t_0) + \epsilon_{cc}(t) \tag{3-33}$$

For a stress σ_c applied at time t_0 and remaining constant until time t, the creep strain ϵ_{cc} between time t_0 and t is

$$\epsilon_{cc}(t, t_0) = \frac{\sigma_c(t_0)}{E_c(28)}C_t \tag{3-34}$$

where $E_c(28)$ is the modulus of elasticity at the age of 28 days, given by Eq. (3-17) or (3-18). Because creep strains involve the entire member, the value for the elastic modulus should be based on the average concrete strength for the full member. It is recommended that the value of mean concrete strength for a member, f_{cm}, be taken as $1.2f'_c$.

From reference [3-20], the creep coefficient as a function of time since load application, C_t, is given as:

$$C_t = \frac{t^{0.6}}{10 + t^{0.6}} \times C_u \tag{3-35}$$

where t is the number of days after application of the load and C_u is the ultimate creep coefficient, which is defined below in Eq. (3-36). The constant, 10, may vary for different concretes and curing conditions, but this value is commonly used for steam-cured concrete and normal concrete that is moist-cured for 7 days.

As with the coefficient for ultimate shrinkage strain, the coefficient C_u consists of a constant multiplied by correction factors.

$$C_u = 2.35 \times \lambda_{rh} \times \lambda_{vs} \times \lambda_{to} \tag{3-36}$$

The constant in this equation can range from 1.30 to 4.15, but the value of 2.35 is commonly recommended. The coefficients λ_{rh} and λ_{vs} account for the ambient relative humidity and the volume/surface ratio, respectively. As with shrinkage strains, a higher value of relative humidity and a larger volume/surface ratio (can also be expressed as a larger effective thickness), will tend to reduce the magnitude of creep strains. For an ambient relative humidity (RH) greater than 40 percent, the modifier for relative humidity is:

$$\lambda_{rh} = 1.27 - 0.0067 \times \text{RH} \tag{3-37}$$

The modifier to account for the volume/surface ratio is:

$$\lambda_{vs} = 0.67\left[1 + 1.13^{-0.54V/S}\right] \tag{3-38}$$

where V/S is the volume/surface area ratio in inches for the member in question.

The coefficient λ_{to} in Eq. (3-36) is used to account for the age of the concrete when load is applied to the member. Early loading of a concrete member will result in higher shrinkage strains, as shown in Fig. 3-25, in which t_o is the time of initial loading in days, h_e is the member effective thickness, and $\phi(t, t_o)$ is the symbol used for the creep coefficient. Values for λ_{to} from ACI Committee 209 [3-20] are:

$$\text{for moist-cured concretes:} \quad \lambda_{to} = 1.25 \times t_o^{-0.118} \tag{3-39a}$$

$$\text{for steam-cured concretes:} \quad \lambda_{to} = 1.13 \times t_o^{-0.094} \tag{3-39b}$$

where t_o is the time in days at initial loading of the member.

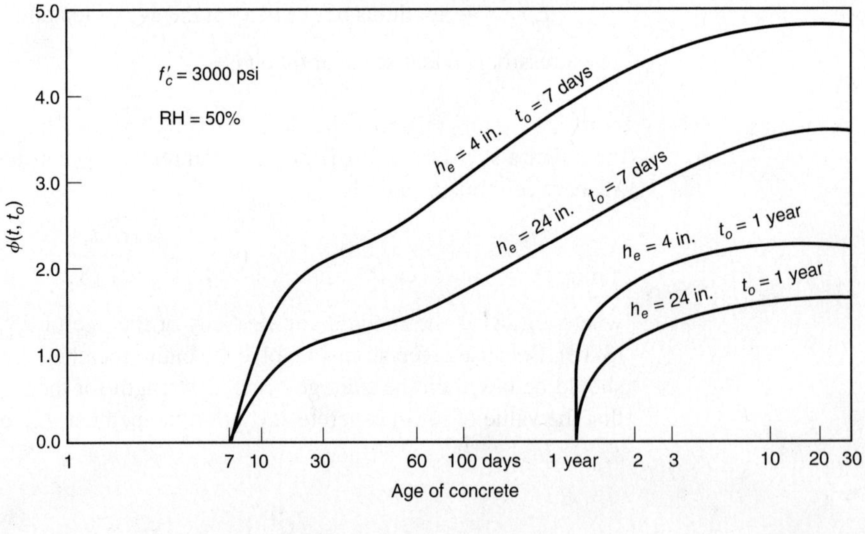

Fig. 3-25
Effect of effective thickness, h_e, and of age at loading, t_o, on creep coefficient.

The expressions given here for creep strains are intended for general use and do not consider significant variations in curing conditions and the types and amounts of aggregates used in the mix design. If creep deflections are anticipated to be a serious problem for a particular structure, consideration should be given to carrying out creep tests on the concrete to be used. Further, a more sophisticated approach is recommended for applications where an accurate calculation of deflection versus time after initial loading is required, such as in segmentally constructed post-tension concrete bridges.

Example 3-3 Calculation of Unrestrained Creep Strains

A plain concrete column 24 in. × 24 in. × 10 ft high is subjected to an average stress of 1000 psi. Compute the total shortening in 5 years if the load is applied 2 weeks after the concrete is cast. The properties of the concrete and the exposure are the same as in Example 3-2.

1. Compute the ultimate shrinkage strain coefficient, C_u.

From Eq. (3-36), the ultimate creep coefficient is,

$$C_u = 2.35 \times \lambda_{rh} \times \lambda_{vs} \times \lambda_{to}$$

For a relative humidity of 50 percent, Eq. (3-37) is used to calculate the modification factor, λ_{rh}.

$$\begin{aligned} \lambda_{rh} &= 1.27 - 0.0067 \times \text{RH} \\ &= 1.27 - 0.0067 \times 50 = 1.27 - 0.33 = 0.94 \end{aligned}$$

The load on the column was applied at $t_o = 14$ days, and it is assumed that the column was moist-cured. Thus, from Eq. (3-39a),

$$\begin{aligned} \lambda_{to} &= 1.25 \times t_o^{-0.118} \\ &= 1.25 \times 14^{-0.118} = 1.25 \times 0.732 \cong 0.92 \end{aligned}$$

The volume of concrete in the column is 2 ft × 2 ft × 10 ft. Assuming that only the sides of the column are exposed to the atmosphere, the exposed surface area is 4 × 2 ft × 10 ft. Thus, the volume/surface ratio is,

$$\frac{V}{S} = \frac{2 \text{ ft} \times 2 \text{ ft} \times 10 \text{ ft}}{4 \times 2 \text{ ft} \times 10 \text{ ft}} = \frac{2 \text{ ft}}{4} = 6 \text{ in.}$$

From Eq. (3-38), the modification factor for volume/surface ratio is,

$$\begin{aligned} \lambda_{vs} &= 0.67\left[1 + 1.13^{-0.54V/S}\right] \\ &= 0.67\left[1 + 1.13^{-3.24}\right] = 0.67\,[1 + 0.673] = 1.12 \end{aligned}$$

Putting these coefficients into Eq. (3-36) results in,

$$C_u = 2.35 \times 0.94 \times 0.92 \times 1.12 = 2.28$$

2. Compute creep coefficient for time since loading.

The time since the load was applied is stated as 5 years minus 2 weeks. Thus, t is,

$$t = 5 \times 365 - 2 \times 7 = 1811 \text{ days}$$

Using this value in Eq. (3-35) to calculate the creep coefficient as a function of time results in,

$$C_t = \frac{t^{0.6}}{10 + t^{0.6}} \times C_u$$

$$= \frac{1811^{0.6}}{10 + 1811^{0.6}} \times 2.28 = \frac{90.1}{10 + 90.1} \times 2.28 = 2.05$$

3. Compute the total stress-dependent strain.

The total stress-dependent strain is a sum of the initial strain plus the creep strain that develops between the time of initial loading, t_o, and the time of interest, t. The creep strain will be calculated using Eq. (3-34). The concrete stress at initial loading, $\sigma_c(t_o)$, is given as 1000 psi. The concrete modulus, $E_c(28)$, will be calculated using Eq. (3-18). The concrete strength to use in Eq. (3-18) is taken as the average concrete strength in the entire member, f_{cm}, which is assumed to be 1.2 f'_c. So,

$$f_{cm} = 1.2 \times f'_c = 1.2 \times 3500 = 4200 \text{ psi}$$

With this value, the elastic modulus for the concrete is,

$$E_c(28) = 57{,}000\sqrt{f_{cm}} = 57{,}000 \times \sqrt{4200} = 3.69 \times 10^6 \text{ psi}$$

Thus, from Eq. (3-34) the creep strain between times t_o and t is,

$$\varepsilon_{cc}(t, t_o) = \frac{\sigma_c(t_o)}{E_c(28)} \times C_t$$

$$= \frac{1000 \text{ psi}}{3{,}690{,}000 \text{ psi}} \times 2.05 = 0.556 \times 10^{-3} \text{ strain}$$

The initial strain at the application of load is to be calculated using the concrete modulus at the time of loading, $E_c(t_o)$. This is to be calculated using the concrete strength at the time of loading, which can be calculated using Eq. (3-5), where t_o (14 days) will be used in place of the symbol t used in Eq. (3-5).

$$f'_c(t_o) = f'_c(28)\left(\frac{t_o}{4 + 0.85t_o}\right)$$

$$= 3500\left(\frac{14}{4 + 0.85 \times 14}\right) = 3080 \text{ psi}$$

Again assuming that f_{cm} is equal to 1.2 f'_c,

$$f_{cm}(t_o) = 1.2f'_c(t_o) = 1.2 \times 3080 = 3700 \text{ psi}$$

From Eq. (3-18),

$$E_c(t_o) = 57{,}000\sqrt{f_{cm}(t_o)}$$

$$= 57{,}000\sqrt{3700} = 3.47 \times 10^6 \text{ psi}$$

From this, calculate the initial concrete strain as,

$$\varepsilon_c(t_o) = \frac{\sigma_c(t_o)}{E_c(t_o)} = \frac{1000 \text{ psi}}{3{,}470{,}000 \text{ psi}} = 0.288 \times 10^{-3} \text{ strain}$$

Thus, the total stress-dependent strain in the concrete is,

$$\varepsilon_c(\text{total}) = \varepsilon_c(t_o) + \varepsilon_{cc}(t, t_o)$$
$$= (0.288 + 0.556) \times 10^{-3} = 0.844 \times 10^{-3} \text{ strain}$$

4. Compute the expected shortening of the column related to stress-dependent strains.

The column is 10 ft long, so the total expected shortening due to stress-dependent strain is,

$$\Delta\ell = \ell \times \varepsilon_c(\text{total}) = 120 \text{ in.} \times 0.844 \times 10^{-3} = 0.101 \text{ in} \cong 0.10 \text{ in.}$$

Thus, the column would be expected to shorten approximately 0.10 in. over 5 years due to the applied load. ■

Restrained Creep

In an axially loaded reinforced concrete column, the creep shortening of the concrete causes compressive strains in the longitudinal reinforcement, increasing the load in the steel and reducing the load, and hence the stress, in the concrete. As a result, a portion of the elastic strain in the concrete is recovered and, in addition, the creep strains are smaller than they would be in a plain concrete column with the same initial concrete stress. A similar redistribution occurs in the compression zone of a beam with compression steel.

This effect can be modeled using an *age-adjusted effective modulus*, $E_{caa}(t, t_0)$, and an *age-adjusted transformed section* in the calculations [3-48], [3-49], and [3-50], where

$$E_{caa}(t, t_0) = \frac{E_c(t_0)}{1 + \chi(t, t_0)[E_c(t_0)/E_c(28)]C_t} \tag{3-40}$$

in which $\chi(t, t_0)$ is an aging coefficient that can be approximated by Eq. (3-41) [3-51]

$$\chi(t, t_0) = \frac{t_0^{0.5}}{1 + t_0^{0.5}} \tag{3-41}$$

The axial strain at time t in a column loaded at age t_0 with a constant load P is

$$\epsilon_c(t, t_0) = \frac{P}{A_{traa} \times E_{caa}(t, t_0)} \tag{3-42}$$

where A_{traa} is the age-adjusted transformed area of the column cross section. The concept of the transformed sections is presented in Section 9-2. For more information on the use of the age-adjusted effected modulus, see [3-48] through [3-50].

EXAMPLE 3-4 Computation of the Strains and Stresses in an Axially Loaded Reinforced Concrete Column

A concrete column 24 in. × 24 in. × 10 ft high has eight No. 8 longitudinal bars and is loaded with a load of 630 kips at an age of 2 weeks. Compute the elastic stresses in the concrete and steel at the time of loading and the stresses and strains at an age of 5 years. The properties of the concrete and the exposure are the same as in Examples 3-2 and 3-3.

In Example 3-3 the following quantities were computed:

$$f_{cm}(14) = 3700 \text{ psi} \qquad f_{cm}(28) = 4200 \text{ psi}$$

$$E_c(14) = 3{,}470{,}000 \text{ psi} \quad E_c(28) = 3{,}690{,}000 \text{ psi}$$

$$C_t = 2.05$$

1. Compute the transformed area at the instant of loading, A_{tr}. (Transformed sections are discussed in Section 9-2.)

$$\text{Elastic modular ratio} = n = \frac{E_s}{E_c(14)} = \frac{29{,}000{,}000}{3{,}470{,}000}$$

$$\cong 8.4$$

The steel area will be "transformed" into an equivalent concrete area, giving the transformed area

$$A_{tr} = A_c + (n - 1)A_s = 576 \text{ in.}^2 + (8.4 - 1) \times 6.32 \text{ in}^2$$

$$= 623 \text{ in.}^2$$

The stress in the concrete is 630,000 lb/623 in.^{2} = 1010 psi. The stress in the steel is n times the stress in the concrete = 8.4 × 1010 psi = 8480 psi.

2. Compute the age-adjusted effective modulus, $E_{caa}(t, t_0)$, and the age-adjusted modular ratio, n_{aa}.

$$E_{caa}(t, t_0) = \frac{E_c(t_0)}{1 + \chi(t, t_0)[E_c(t_0)/E_{cm}(28)]\phi(t, t_0)} \tag{3-40}$$

where

$$\chi(t, t_0) = \frac{t_0^{0.5}}{1 + t_0^{0.5}} = \frac{14^{0.5}}{1 + 14^{0.5}} \tag{3-41}$$

$$= 0.789$$

$$E_{caa}(t, t_0) = \frac{3{,}470{,}000}{1 + 0.789 \times \dfrac{3{,}470{,}000}{3{,}690{,}000} \times 2.05}$$

$$= 1{,}380{,}000 \text{ psi}$$

$$\text{Age-adjusted modular ratio, } n_{aa} = \frac{E_s}{E_{caa}(t, t_0)} = \frac{29{,}000{,}000}{1{,}380{,}000}$$

$$= 21.0$$

3. Compute the age-adjusted transformed area, A_{traa}, the stresses in the concrete and in the steel, and the shortening. Again, the steel will be transformed to concrete.

$$A_{traa} = A_c + (n_{aa} - 1)A_s = 576 \text{ in.}^2 + (21.0 - 1) \times 6.32 \text{ in.}^2$$

$$= 702 \text{ in.}^2$$

$$\text{Stress in concrete} = f_c = \frac{P}{A_{traa}} = \frac{630{,}000 \text{ lb}}{702 \text{ in.}^2}$$

$$= 897 \text{ psi}$$

$$\text{Stress in steel} = n_{aa} \times f_c = 21.0 \times 897 \text{ psi}$$
$$= 18{,}800 \text{ psi}$$
$$\text{Strain} = \frac{f_c}{E_{caa}} = \frac{897}{1{,}380{,}000}$$
$$= 0.000650 \text{ strain}$$
$$\text{Shortening } \epsilon \times \ell = 0.000650 \times 120 \text{ in.}$$
$$= 0.078 \text{ in.}$$

The creep has reduced the stress in the concrete from 1010 psi at the time of loading to 897 psi at 5 years. During the same period, the steel stress has increased from 8480 psi to 18,800 psi. A column with less reinforcement would experience a larger increase in the reinforcement stress. To prevent yielding of the steel under sustained loads, ACI Code Section 10.6.1.1 sets a lower limit of 1 percent on the reinforcement ratio in columns. The reinforcement ratio (A_s/A_g) for this column is 1.1 percent.

The plain concrete column in Example 3-3, which had a constant concrete stress of 1000 psi throughout the 5-year period, shortened 0.10 in. The column in this example, which had an initial concrete stress of 1010 psi but was reinforced, was shortened approximately 80 percent as much. ■

Thermal Expansion

The coefficient of thermal expansion or contraction, α, is affected by such factors as composition of the concrete, moisture content of the concrete, and age of the concrete. Ranges from normal-weight concretes are 5 to 7 $\times$ 10^{-6} *strain*/°F for those made with siliceous aggregates and 3.5 to 5 $\times$ 10^{-6}/°F for concretes made from limestone or calcareous aggregates. Approximate values for lightweight concrete are 3.6 to 6.2 $\times$ 10^{-6}/°F. An all-around value of 5.5 $\times$ 10^{-6}/°F may be used. The coefficient of thermal expansion for reinforcing steel is 6 $\times$ 10^{-6}/°F. In calculations of thermal effects, it is necessary to allow for the time lag between air temperatures and concrete temperatures.

As the temperature rises, so does the coefficient of expansion and at the temperatures experienced in building fires, it may be several times the value at normal operating temperatures [3-52]. The thermal expansion of a floor slab in a fire may be large enough to exert large shear-forces on the supporting columns.

3-7 HIGH-STRENGTH CONCRETE

Concretes with 28-day strengths in excess of 6000 psi are referred to as *high-strength concretes*. Strengths of up to 18,000 psi have been used in buildings. Reference [3-7] presents the state of the art of the production and use of high-strength concrete.

Admixtures such as superplasticizers improve the dispersion of cement in the mix and produce workable concretes with much lower water/cement ratios than were previously possible. The resulting concrete has a lower void ratio and is stronger than normal concretes. Most high-strength concretes have water-to-cementitious-materials ratios (w/cm ratios) of 0.40 or less. Many have w/cm ratios in the range from 0.25 to 0.35. Only the amount of water needed to hydrate the cement in the mix is provided. This results in concrete with a dense amorphous structure without voids. Coarse aggregates should consist of strong fine-grained gravel with a rough surface. Smooth river gravels give a lower paste–aggregate bond strength and a weaker concrete.

Enhanced concrete production control must be enforced at the job site, because all shortcomings in selection of aggregates, amounts of water used in mixes, placing, curing, and the like, lead to weaker concrete. Attention should be given to limiting and controlling the temperature rise due to hydration.

High-Performance Concrete

The term *high-performance concrete* is used to refer to concrete with special properties, such as ease of placement and consolidation, high early-age strength to allow early stripping of forms, durability, and high strength. High-strength concrete is only one type of high-performance concrete.

Mechanical Properties

Many of the mechanical properties of high-strength concretes are reviewed in [3-7], [3-53], and [3-54]. It is important to remember that high-strength concrete is not a unique material with a unique set of properties. For example, the modulus of elasticity is strongly affected by the modulus of elasticity of the coarse aggregate.

As shown in Fig. 3-18, the stress–strain curves for higher-strength concretes tend to have a more linear loading branch and a steep descending branch. High-strength concrete exhibits less internal microcracking for a given strain than does normal concrete. In normal strength concrete, unstable microcracking starts to develop at a compressive stress of about $0.75f'_c$, referred to as the critical stress (see Section 3-2). In high-strength concrete, the critical stress is about $0.85f'_c$. Failure occurs by fracture of the aggregate on relatively smooth planes parallel to the direction of the applied stress. The lateral strains tend to be considerably smaller than in lower-strength concrete. One implication of this is that spiral and confining reinforcement may be less effective in increasing the strength and ductility of high-strength concrete column cores.

Equations (3-17) and (3-18) overestimate the modulus of elasticity of concretes with strengths in excess of about 6000 psi. Reference [3-7] proposes that

$$E_c = 40{,}000\sqrt{f'_c} + 1.0 \times 10^6 \text{ (psi)} \tag{3-19}$$

As noted earlier, E_c varies as a function of the modulus of the coarse aggregate.

The modulus of rupture of high-strength concretes ranges from $(7.5 \text{ to } 12)\sqrt{f'_c}$. A lower bound to the split-cylinder tensile test data is given by $6\sqrt{f'_c}$ (Fig. 3-10).

28-Day and 56-Day Compression Strengths

High-strength concrete frequently contains admixtures that delay the final strength gain. As a result, the concrete is still gaining strength at 56 days, rather than reaching a maximum at about 28 days. In 2001, Nowak and Szerszen [3-9] and [3-10] collected cylinder-test data on high-strength concretes, including tests of companion cylinders at 28 and 56 days. The data is summarized in Table 3-2.

The overall average of measured versus specified cylinder strength was 1.11 at 28 days, increasing to 1.20 at 56 days—an increase of 8.7 percent between 28 days and 56 days.

In the development of resistance factors for the design of reinforced concrete members such as columns, the strength gain after 28 days was generally ignored, giving a

TABLE 3-2 Differences between 28-day and 56-day Concrete Strengths

Specified f'_c	Age	No. of Tests	$f'_{c,\,test}/f'_{c,specified}$	Coefficient of Variation
7000 psi	28 days	210	1.19	0.115
	56 days	58	1.49	0.080
8000 psi	28 days	753	1.09	0.090
	56 days	428	1.09	0.095
10,000 psi	28 days	635	1.13	0.115
	56 days	238	1.18	0.105
12,000 psi	28 days	381	1.04	0.105
	56 days	190	1.17	0.105

strength reserve of about 5 to 9 percent. If the member strength was based on reaching the desired the 56-day concrete strength, some or all of this strength reserve would be lost.

Shrinkage and Creep

Shrinkage of concrete is approximately proportional to the percentage of water by volume in the concrete. High-strength concrete has a higher paste content, but the paste has a lower water/cement ratio. As a result, the shrinkage of high-strength concrete is about the same as that of normal concrete.

Test data suggest that the creep coefficient, C_t, for high-strength concrete is considerably less than that for normal concrete [3-7].

3-8 LIGHTWEIGHT CONCRETE

Structural lightweight concrete is concrete having a density between 90 and 120 lb/ft^3 and containing naturally occurring lightweight aggregates such as pumice; artificial aggregates made from shales, slates, or clays that have been expanded by heating; or sintered blast-furnace slag or cinders. Such concrete is used when a saving in dead load is important. Lightweight concrete costs about 20 percent more than normal concrete. The terms *all-lightweight concrete* and *sand-lightweight concrete* refer to mixes having either lightweight fine aggregates or natural sand, respectively.

The modulus of elasticity of lightweight concrete is less than that of normal concrete and can be computed from Eq. (3-17).

The stress–strain curve of lightweight concrete is affected by the lower modulus of elasticity and relative strength of the aggregates and the cement paste. If the aggregate is the weaker of the two, failure tends to occur suddenly in the aggregate, and the descending branch of the stress–strain curve is very short or nonexistent, as shown by the upper solid line in Fig. 3-26. The fracture surface of those lightweight concretes tends to be smoother than for normal concrete. On the other hand, if the aggregate does not fail, the stress–strain curve will have a well-defined descending branch, as shown by the curved lower solid line in this figure. As a result of the lower modulus of elasticity of lightweight concrete, the strain at which the maximum compressive stress is reached is higher than for normal-weight concrete.

The tensile strength of all-lightweight concrete is 70 to 100 percent of that of normal-weight concrete. Sand-lightweight concrete has tensile strengths in the range from 80 to 100 percent of those of normal-weight concrete.

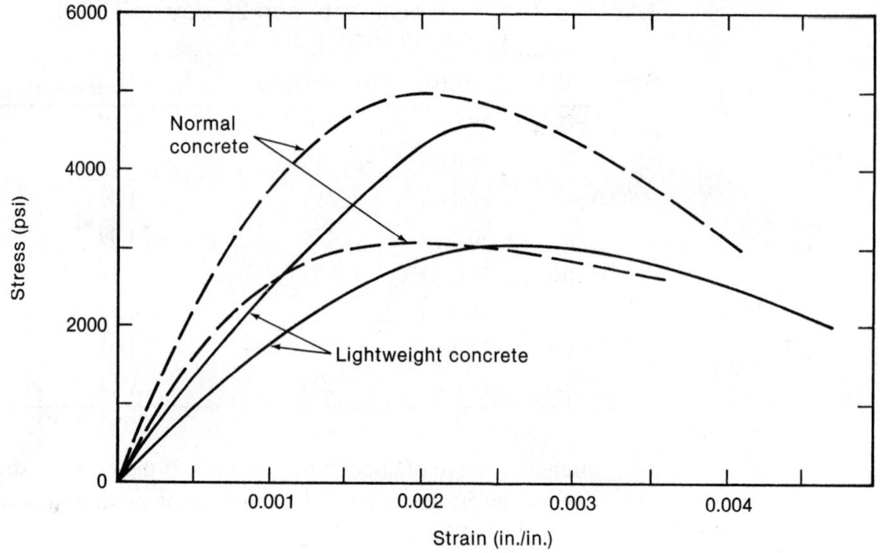

Fig. 3-26
Compressive stress–strain curves for normal-weight and lightweight concretes, $f'_c = 3000$ and 5000 psi. (From [3-55].)

The shrinkage and creep of lightweight concrete are similar to or slightly greater than those for normal concrete. The creep coefficients computed from Eq. (3-35) can be used for lightweight concrete.

3-9 FIBER REINFORCED CONCRETE

Fiber reinforced concrete refers to concrete reinforced with short, randomly oriented fibers. Based on their material, fibers can be divided into four major groups: steel fibers, glass fibers, synthetic fibers, and natural fibers [3-56]. The amount of fibers added to the concrete depends on the type of fiber and target performance, but practical considerations limit the fiber dosage in structural elements to approximately 1.5 percent by volume. Traditional applications of fiber reinforced concrete include slabs on ground, tunnel liners, and architectural elements, where fibers have been primarily used as replacement of minimum reinforcement for cracking control and, to a lesser degree, replacement of minimum shear and/or flexural reinforcement. Applications of fiber reinforced concrete in building structures, on the other hand, have been rather limited. This has been primarily due to limited experimental research on the behavior of structural elements and consequently, the lack of design provisions in building codes. It was not until the 2008 edition when fiber reinforced concrete was recognized as a structural material in the ACI Code.

Fibers are primarily used for their ability to provide post-cracking tension resistance to the concrete and thus, in addition to evaluating the compressive behavior of fiber reinforced concrete, its tensile behavior should also be assessed. The addition of fibers to concrete in low-to-moderate dosages ($\leq$1.5 percent by volume) does not greatly affect compression strength and elastic modulus. Improvements in post-peak behavior, however, have been observed, characterized by an increased compression strain capacity and toughness [3-57].

In tension, the ability of fibers to enhance concrete post-cracking behavior primarily depends on fiber strength, fiber stiffness, and bond with the surrounding concrete matrix. As opposed to reinforcing bars, which are designed to be anchored in the concrete such that their yield strength can be developed, fibers are designed to pullout of the concrete matrix prior to achieving their strength. Thus, the behavior of fiber reinforced concrete is highly dependent on the ability of the fibers to maintain good bond with the concrete as they are pulled out.

Ideally, the tensile behavior of fiber reinforced concrete should be evaluated from direct tension tests. However, difficulties in conducting such a test have led to the use of a four-point flexural test as the most common test method for evaluating the post-cracking behavior of fiber reinforced concrete. In the United States, specifications for this test can be found in ASTM C1609. The size of the flexural test specimens depends on the fiber length and concrete aggregate size but typically, beams with a 6-in. square cross section and an 18-in. span are used. The test is run until a midspan deflection equal to 1/150 of the span length is reached.

Based on its performance under flexure, fiber reinforced concretes can be classified as either deflection softening or deflection hardening [3-58], as shown in Fig. 3-27. Deflection softening implies a drop in the load at first cracking under a flexural test, while deflection hardening fiber reinforced concretes exhibit a flexural strength greater than their first cracking strength. When subjected to direct tension most fiber reinforced concretes will exhibit a drop in stress at first cracking. However, some fiber reinforced concrete with higher fiber contents exhibits a pseudo strain-hardening response with multiple cracking under direct tension. This particular type of fiber reinforced concrete has been referred to as either strain-hardening or high-performance fiber reinforced concrete [3-59]. For structural applications, it is desirable that fiber reinforced concrete exhibits at least a deflection hardening behavior.

Research on the use of fiber reinforcement in structural elements has been primarily limited to steel fibers, as opposed to synthetic or natural fibers. Thus, we shall concentrate on the properties and structural applications of steel fibers. The vast majority of steel fibers used for structural purposes are 1 in. to 2 in. in length and 0.015 in. to 0.04 in. in diameter, with length-to-diameter ratios typically ranging between 50 and 80. The strength of the steel wire used to manufacture fibers has a tensile strength in the order of 170 ksi, although steel wire with strength greater than 350 ksi is sometimes used. In order to improve bond with the concrete matrix as they are pulled out, steel fibers are typically deformed, most commonly through hooks at their ends (Fig. 3-28).

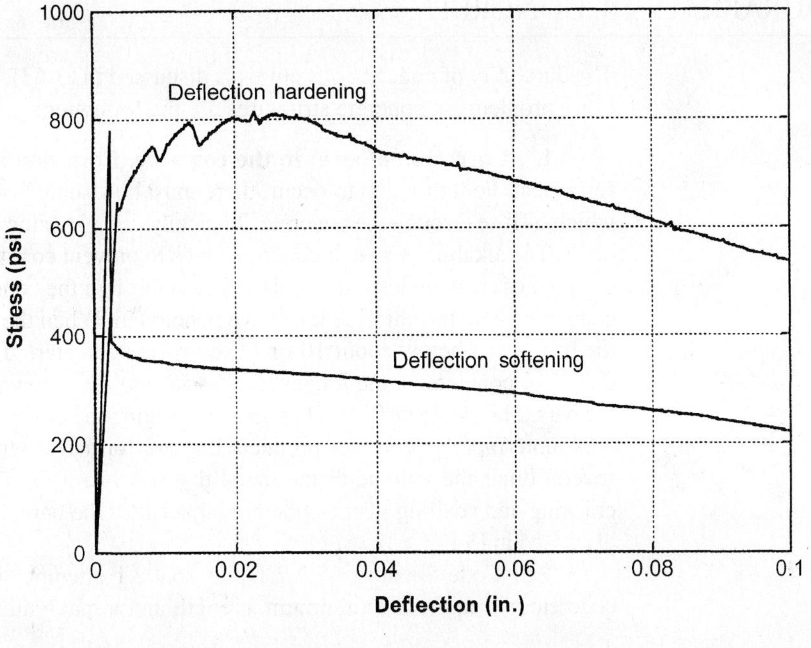

Fig. 3-27
Examples of deflection-hardening and deflection-softening behavior.

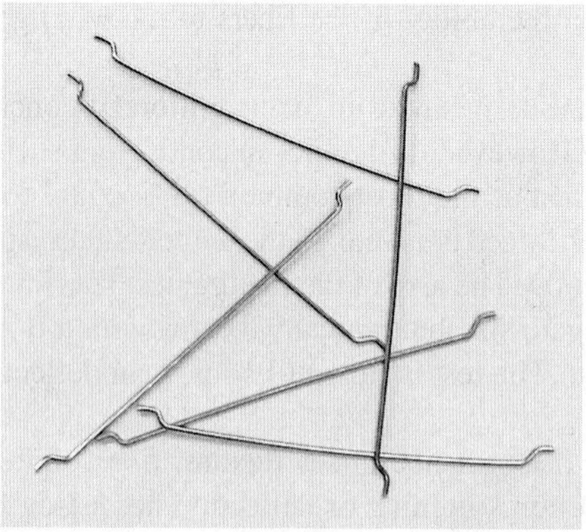

Fig. 3-28
Typical hooked steel fibers used in fiber reinforced concrete.

Research on structural applications of fiber reinforced concrete has primarily focused on their use as shear reinforcement in beams [3-60] and flat plates, and as shear/confinement reinforcement in elements subjected to high shear reversals, such as beam–column connections, structural walls, and coupling beams of earthquake-resistant structures [3-61] and [3-62].

In 2008 the ACI Code allowed for the first time the use of deformed steel fibers as minimum shear reinforcement in beams (see Section 6-3). In order to account for differences in performance between various fibers, as well as fiber contents, performance criteria are used for acceptance of fiber reinforced concrete, based on flexural tests as per ASTM 1609. Based on this material test, fiber reinforced concrete is considered acceptable for shear resistance if the residual strength obtained at deflections of 1/300 and 1/150 of the span length of the beam are greater than or equal to 90 percent and 75 percent of the first peak (cracking) strength, respectively. First peak or cracking strength is determined experimentally, but shall not be taken less than f_r, as defined in ACI Code Eq. (19.2.3.1). In addition, regardless of the performance obtained, fiber dosage shall not be less than 100 lb per cubic yard.

3-10 DURABILITY OF CONCRETE

The durability of concrete structures is discussed in [3-63]. The three most common durability problems in concrete structures are the following:

1. Corrosion of steel in the concrete. Corrosion involves oxidation of the reinforcement. For corrosion to occur, there must be a source of oxygen and moisture, both of which diffuse through the concrete. Typically, the pH value of new concrete is on the order of 13. The alkaline nature of concrete tends to prevent corrosion from occurring. If there is a source of chloride ions, these also diffuse through the concrete, decreasing the pH of the concrete where the chloride ions have penetrated. When the pH of the concrete adjacent to the bars drops below about 10 or 11, corrosion can start. The thicker and less permeable the cover concrete is, the longer it takes for moisture, oxygen, and chloride ions to reach the bars. Shrinkage or flexural cracks penetrating the cover allow these agents to reach the bars more rapidly. The rust products that are formed when reinforcement corrodes have several times the volume of the metal that has corroded. This increase in volume causes cracking and spalling of the concrete adjacent to the bars. Factors affecting corrosion are discussed in [3-64].

ACI Code Sections 19.3.2.1 and 26.4.2.1 attempt to control corrosion of steel in concrete by requiring a minimum strength and a maximum water/cementitious materials ratio to reduce the permeability of the concrete and by requiring at least a minimum cover to the reinforcing bars. The amount of chlorides in the mix also is restricted. Epoxy-coated bars sometimes are used to delay or prevent corrosion.

Corrosion is most serious under conditions of intermittent wetting and drying. Adequate drainage should be provided to allow water to drain off structures. Corrosion is seldom a problem for permanently submerged portions of structures.

2. Breakdown of the structure of the concrete due to freezing and thawing. When concrete freezes, pressures develop in the water in the pores, leading to a breakdown of the structure of the concrete. *Entrained air* provides closely spaced microscopic voids, which relieve these pressures [3-65]. ACI Code Section 19.3.3 requires minimum air contents to reduce the effects of freezing and thawing exposures. The spacing of the air voids is also important, and some specifications specify spacing factors. ACI Code Section 19.3.2.1 sets maximum water/cementitious materials ratios of 0.45 and minimum concrete strengths of 4500 psi for concretes, depending on the severity of the exposure. These can give strengths higher than would otherwise be used in structural design. A water/cement ratio of 0.40 will generally correspond to a strength of 4500 to 5000 psi for air-entrained concrete. This additional strength can be utilized in computing the strength of the structure.

Again, drainage should be provided so that water does not collect on the surface of the concrete. Concrete should not be allowed to freeze at a very young age and should be allowed to dry out before severe freezing.

3. Breakdown of the structure of the concrete due to chemical attack. Sulfates cause disintegration of concrete unless special cements are used. ACI Code Section 19.3.4 specifies cement type, maximum water/cementitious materials ratios, and minimum compressive strengths for various sulfate exposures. Geotechnical reports will generally give sulfate levels. ACI Code Table 19.3.2.1 gives special requirements for concrete in contact with sulfates in soils or in water. In many areas in the western United States, soils contain sulfates.

Some aggregates containing silica react with the alkalies in the cement, causing a disruptive expansion of the concrete, leading to severe random cracking. This *alkali silica reaction* is counteracted by changing the source of the aggregate or by using low-alkali cements [3-66]. It is most serious if the concrete is warm in service and if there is a source of moisture. Reference [3-67] lists a number of other chemicals that attack concretes.

ACI Code Chapter 19 presents requirements for concrete that is exposed to freezing, thawing, deicing chemicals, sulfates, and chlorides. Examples are pavements, bridge decks, parking garages, water tanks, and foundations in sulfate-rich soils.

3-11 BEHAVIOR OF CONCRETE EXPOSED TO HIGH AND LOW TEMPERATURES

High Temperatures and Fire

When a concrete member is exposed to high temperatures such as occur in a building fire, for example, it will behave satisfactorily for a considerable period of time. During a fire, high thermal gradients occur, however, and as a result, the surface layers expand and eventually crack or spall off the cooler, interior part of the concrete. The spalling is aggravated if water from fire hoses suddenly cools the surface.

The modulus of elasticity and the strength of concrete decrease at high temperatures, whereas the coefficient of thermal expansion increases [3-52]. The type of aggregate affects

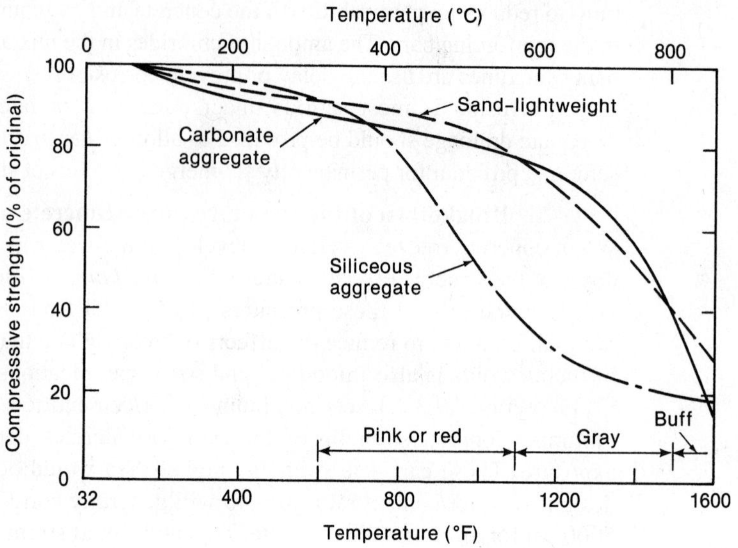

Fig. 3-29
Compressive strength of concretes at high temperatures. (From [3-52].)

the strength reduction, as is shown in Fig. 3-29. Most structural concretes can be classified into one of three aggregate types: carbonate, siliceous, or lightweight. Concretes made with carbonate aggregates, such as limestone and dolomite, are relatively unaffected by temperature until they reach about 1200 to 1300°F, at which time they undergo a chemical change and rapidly lose strength. The quartz in siliceous aggregates, such as quartzite, granite, sandstones, and schists, undergoes a phase change at about 800 to 1000°F, which causes an abrupt change in volume and spalling of the surface. Lightweight aggregates gradually lose their strength at temperatures above 1200°F.

The reduction in strength and the extent of spalling due to heat are most pronounced in wet concrete and, as a result, fire is most critical with young concrete. The tensile strength tends to be affected more by temperature than does the compressive strength.

Concretes made with limestone and siliceous aggregates tend to change color when heated, as indicated in Fig. 3-29, and the color of the concrete after a fire can be used as a rough guide to the temperature reached in the concrete. As a general rule, concrete whose color has changed beyond pink is suspect. Concrete that has passed the pink stage and gone into the gray stage is probably badly damaged. Such concrete should be chipped away and replaced with a layer of new concrete or shotcrete.

Very Cold Temperatures

In low temperatures, the strength of hardened concrete tends to increase, the increase being greatest for moist concrete, as long as the water does not freeze [3-68]. Very cold temperatures are encountered in liquid-natural-gas storage facilities.

Subfreezing temperatures can significantly increase the compressive and tensile strengths and the modulus of elasticity of moist concrete. Dry concrete properties are not affected as much by low temperature. Reference [3-68] reports compression tests of moist concrete with a strength of 5000 psi at 75°F that reached a strength of 17,000 psi at −150°F. The same concrete tested oven-dry or at an interior relative humidity of 50 percent showed a 20 percent increase in compressive strength relative to the strength at 75°F. The split-cylinder tensile strength of the same concrete increased from 600 psi at +75°F to 1350 psi at −75°F.

3-12 SHOTCRETE

Shotcrete is concrete or mortar that is pneumatically projected onto a surface at high velocity. A mixture of sand, water, and cement is sprayed through a nozzle. Shotcrete is used as new structural concrete or as repair material. It has properties similar to those of cast-in-place concrete, except that the properties depend on the skill of the nozzleperson who applies the material. Further information on shotcrete is available in [3-69].

3-13 HIGH-ALUMINA CEMENT

High-alumina cement is occasionally used in structures. Concretes made from this type of cement have an unstable crystalline structure that could lose its strength over time, especially if exposed to moderate to high humidities and temperatures [3-70]. In general, high-alumina cements should be avoided in structural applications.

3-14 REINFORCEMENT

Because concrete is weak in tension, it is reinforced with steel bars or wires that resist the tensile stresses. The most common types of reinforcement for nonprestressed members are hot-rolled deformed bars and wire fabric. In this book, only the former will be used in examples, although the design principles apply with very few exceptions to members reinforced with welded wire mesh or cold-worked deformed bars.

The ACI Code requires that reinforcement be steel bars or steel wires. Significant modifications to the design process are required if materials such as fiber-reinforced-plastic (FRP) rods are used for reinforcement because such materials are brittle and do not have the ductility assumed in the derivation of design procedures for concrete reinforced with steel bars. In addition, special attention must be given to the anchorage of FRP reinforcement.

Hot-Rolled Deformed Bars

Grades, Types, and Sizes

Steel reinforcing bars are basically round in cross section, with lugs or deformations rolled into the surface to aid in anchoring the bars in the concrete (Fig. 3-30). They are produced according to the following ASTM specifications, which specify certain dimensions and certain chemical and mechanical properties.

1. **ASTM A 615:** *Standard Specification for Deformed and Plain Carbon-Steel Bars for Concrete Reinforcement.* This specification covers the most commonly used reinforcing bars. They are available in sizes 3 to 18 in Grade 60 (yield strength of 60 ksi) plus sizes 3 to 6 in Grade 40 and sizes 6 to 18 in Grade 75. The specified mechanical properties are summarized in Table 3-3. The diameters, areas, and weights are listed in Table A-1 in Appendix A. The phosphorus content is limited to ≤ 0.06 percent.

2. **ASTM A 706:** *Standard Specification for Low-Alloy Steel Deformed and Plain Bars for Concrete Reinforcement.* This specification covers bars intended for special applications where weldability, bendability, or ductility is important. As indicated in Table 3-3, the A 706 specification requires a larger elongation at failure and a more stringent bend test than A 615. ACI Code Section 20.2.2.5 requires the use of A 615 bars meeting special requirements or A 706 bars in seismic applications. There is both a lower and an upper limit on the yield strength. A 706 limits the amounts of carbon, manganese, phosphorus, sulfur,

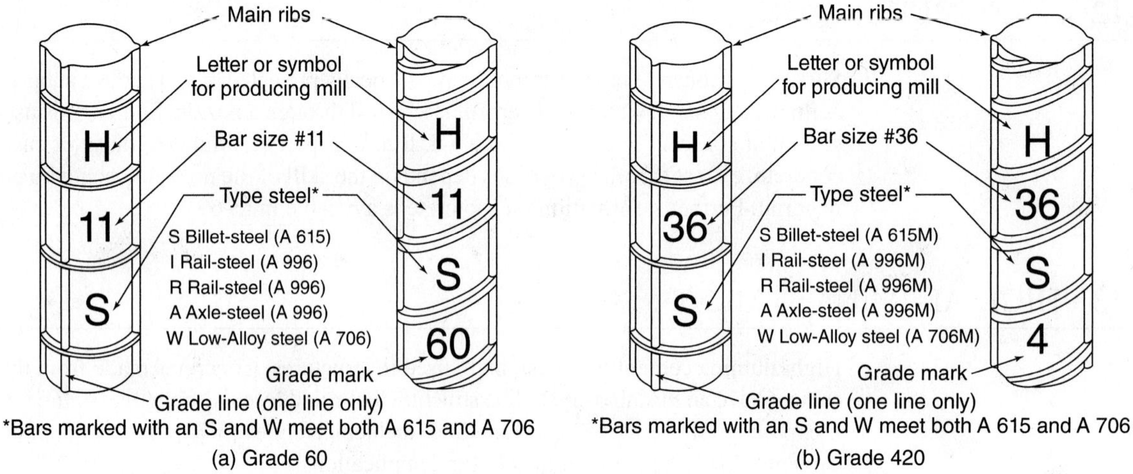

Fig. 3-30
Standard reinforcing-bar markings. (Courtesy of Concrete Reinforcing Steel Institute.)

TABLE 3-3 Summary of Mechanical Properties of Reinforcing Bars from ASTM A 615 and ASTM A 706

	Billet-Steel A 615			Low-Alloy Steel, A 706
	Grade 40	Grade 60	Grade 75	Grade 60
Minimum tensile strength, psi	70,000	90,000	100,000	80,000[a]
Minimum yield strength, psi	40,000	60,000	75,000	60,000
Maximum yield strength, psi	—	—	—	78,000
Minimum elongation in 8-in. gauge length, percent				
No. 3	11	9	—	14
No. 4 and 5	12	9	—	14
No. 6	12	9	7	14
No. 7 and 8	—	8	7	12
No. 9, 10, and 11	—	7	6	12
No. 14 and 18	—	7	6	10
Pin diameter for bend test,[b] where d = nominal bar diameter				
No. 3, 4, and 5	3.5d	3.5d	—	3d
No. 6	5d	5d	5d	4d
No. 7 and 8	—	5d	5d	4d
No. 9, 10, and 11	—	7d	7d	6d
No. 14 and 18	—	9d	9d	8d

[a]But not more than 1.25 times the actual yield.

[b]Bend tests are 180°, except that 90° bends are permitted for No. 14 and 18 A 615 bars.

and silicon and limits the carbon equivalent to ≤0.55 percent. These bars are available in sizes 3 through 18 in Grade 60.

3. ASTM A 996: *Standard Specification for Rail-Steel and Axle-Steel Deformed Bars for Concrete Reinforcement.* This specification covers bars rolled from discarded railroad rails or from discarded train car axles. It is less ductile and less bendable than

A 615 steel. Only Type R rail-steel bars with R rolled into the bar are permitted by the ACI Code. These bars are not widely available.

Reinforcing bars are available in three grades, with yield strengths at 40, 60, and 75 ksi, referred to as Grades 40, 60, and 75, respectively. Grade 60 is the steel most commonly used in buildings and bridges. Other grades may not be available in some areas. Grade 75 is used in large columns. Grade 40 is the most ductile, followed by Grades 60, and 75, in that order.

Grade-60 deformed reinforcing bars are available in the 11 sizes listed in Table A-1. The sizes are referred to by their nominal diameter expressed in eighths of an inch. Thus, a No. 4 bar has a diameter of $\frac{4}{8}$ in. (or $\frac{1}{2}$ in.). The nominal cross-sectional area can be computed directly from the nominal diameter, except for that of the No. 10 and larger bars, which have diameters slightly larger than $\frac{10}{8}$ in., $\frac{11}{8}$ in., and so on. Size and grade marks are rolled into the bars for identification purposes, as shown in Fig. 3-30. Grade-40 bars are available only in sizes 3 through 6. Grade-75 steel is available only in sizes 6 to 18.

ASTM A 615 and A 706 also specify metric (SI) bar sizes. They are available in 11 sizes. Each is the same as an existing inch–pound bar size but is referred to by its nominal diameter in whole millimeters. The sizes are #10, #13, #16, #19, #22, #25, #29, #32, #36, #43, and #57, corresponding to the nominal diameters 10 mm, 13 mm, 16 mm, and so on. The nominal diameters of metric reinforcement are the traditional U.S. Customary unit diameters—$\frac{3}{8}$ in. (9.5 mm), $\frac{4}{8}$ in. (12.7 mm), $\frac{5}{8}$ in. (15.9 mm), and so on—rounded to the nearest whole millimeter. The bar size designation will often include an "M" to denote a metric size bar. The diameters, areas, and weights of SI bar sizes are listed in Table A-1M in Appendix A.

ASTM A 615 defines three grades of metric reinforcing bars: Grades 300, 420, and 520, having specified yield strengths of 300, 420, and 520 MPa, respectively.

For the review of the strength of existing buildings the yield strength of the bars must be known. Prior to the late 1960s, reinforcing bars were available in *structural*, *intermediate*, and *hard* grades with specified yield strengths of 33 ksi, 40 ksi, and 50 ksi (228 MPa, 276 MPa, and 345 MPa), respectively. Reinforcing bars were available in inch–pound sizes 3 to 11, 14, and 18. For sizes 3 to 8, the size number was the nominal diameter of the bar in eighths of an inch, and the cross-sectional areas were computed directly from this diameter. For sizes 9 to 18, the diameters were selected to give the same areas as previously used square bars, and the size numbers were approximately equal to the diameter in eighths of an inch. In the 1970s, the 33-ksi and 50-ksi bars were dropped, and a new 60-ksi yield strength was introduced.

Mechanical Properties

Idealized stress–strain relationships are given in Fig. 3-31 for Grade-40, -60, and -75 reinforcing bars, and for welded-wire fabric. The initial tangent modulus of elasticity, E_s, for all reinforcing bars can be taken as 29×10^6 psi. Grade-40 bars display a pronounced yield plateau, as shown in Fig. 3-31. Although this plateau is generally present for Grade-60 bars, it is typically much shorter. High-strength bars generally do not have a well-defined yield point.

Figure 3-32 is a histogram of mill-test yield strengths of Grade-60 reinforcement having a nominal yield strength of 60 ksi. As shown in this figure, there is a considerable variation in yield strength, with about 10 percent of the tests having a yield strength equal to or greater than 80 ksi—133 percent of the nominal yield strength. The coefficient of variation of the yield strengths plotted in Fig. 3-32 is 9.3 percent.

ASTM specifications base the yield strength on *mill tests* that are carried out at a high rate of loading. For the slow loading rates associated with dead loads or for many live

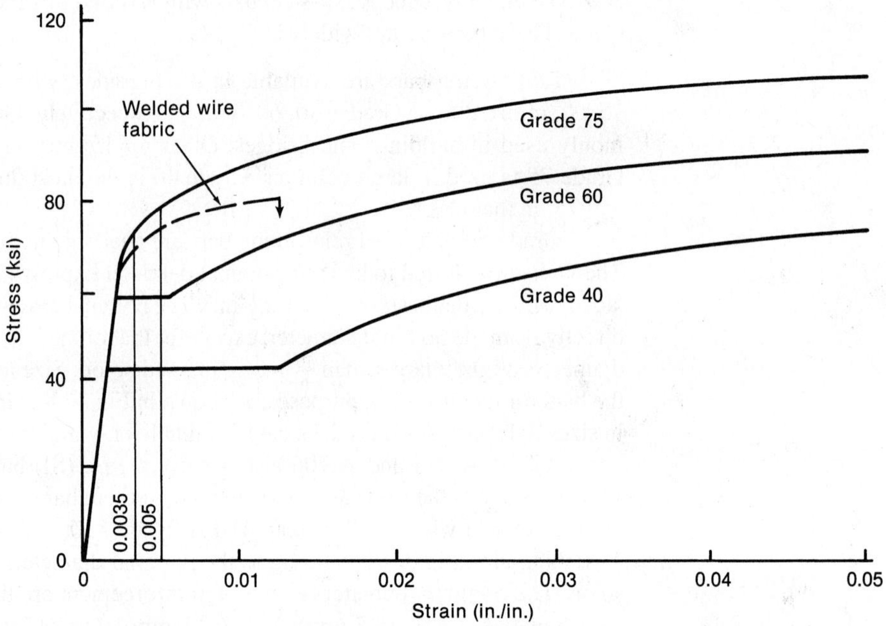

Fig. 3-31
Stress–strain curves for reinforcement.

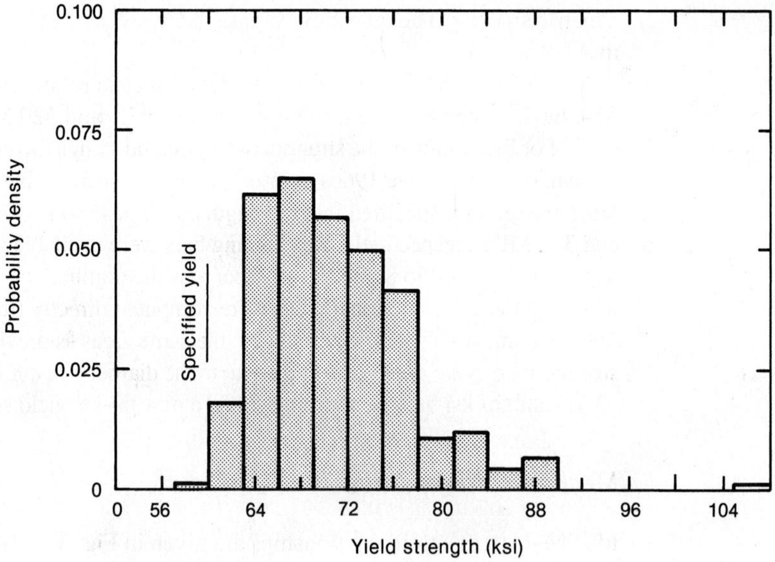

Fig. 3-32
Distribution of mill-test yield strengths for Grade-60 steel. (From [3-71].)

loads, the *static yield strength* is applicable. This is roughly 4 ksi less than the mill-test yield strength [3-71].

Fatigue Strength

Some reinforced concrete elements, such as bridge decks, are subjected to a large number of loading cycles. In such cases, the reinforcement may fail in fatigue. Fatigue failures of the reinforcement will occur only if one or both of the extreme stresses in the stress cycle is tensile. The relationship between the range of stress, S_r, and the number of cycles is shown in Fig. 3-33. For practical purposes, there is a fatigue threshold or *endurance limit* below which fatigue failures will normally not occur. For straight ASTM A 615 bars, this is

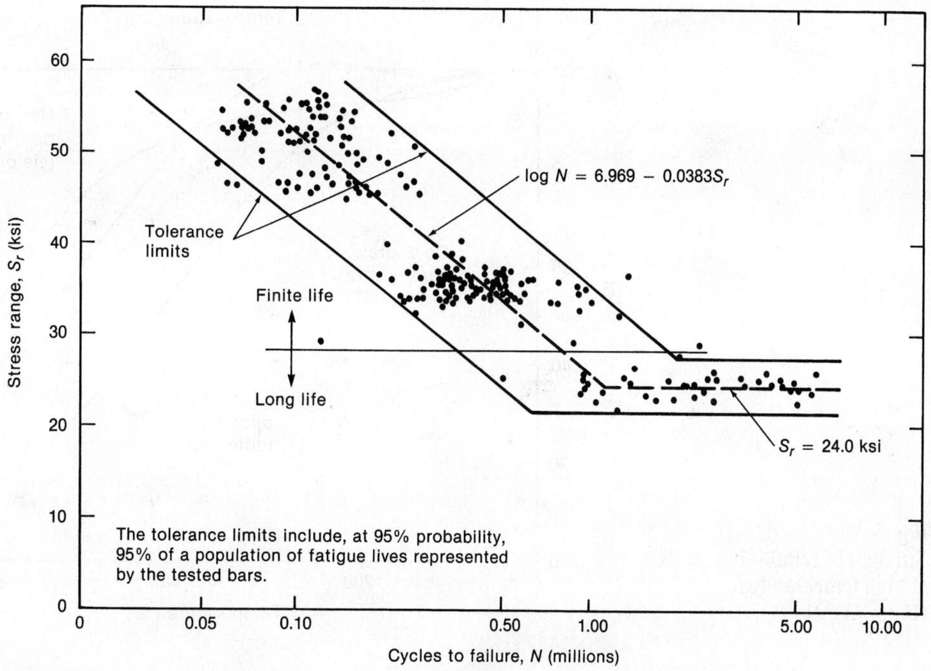

Fig. 3-33
Test data on fatigue of deformed bars from a single U.S. manufacturer. (From [3-72].)

about 24 ksi and is essentially the same for Grade-40 and Grade-60 bars. If there are fewer than 20,000 cycles, fatigue will not be a problem with deformed-bar reinforcement.

The fatigue strength of deformed bars decreases:

(a) as the stress range (the maximum tensile stress in a cycle minus the algebraic minimum stress) increases,

(b) as the level of the lower (less tensile) stress in the cycle is reduced, and

(c) as the ratio of the radius of the fillet at the base of the deformation lugs to the height of the lugs is decreased. The fatigue strength is essentially independent of the yield strength.

(d) In the vicinity of welds or bends, fatigue failures may occur if the stress range exceeds 10 ksi. Further guidance is given in [3-73].

For design, the following rules can be applied: If the deformed reinforcement in a particular member is subjected to 1 million or more cycles involving tensile stresses, or a combination of tension and compression stresses, fatigue failures may occur if the difference between the maximum and minimum stresses under the repeated loading exceeds 20 ksi.

Strength at High Temperatures Deformed-steel reinforcement subjected to high temperatures in fires tends to lose some of its strength, as shown in Fig. 3-34 [3-52]. When the temperature of the reinforcement exceeds about 850°F, both the yield and ultimate strengths drop significantly. One of the functions of concrete cover on reinforcement is to prevent the reinforcement from getting hot enough to lose strength.

Welded-Wire Reinforcement

Welded-wire reinforcement is a prefabricated reinforcement consisting of smooth or deformed wires welded together in square or rectangular grids. Sheets of wires are welded in electric-resistance welding machines in a production line. This type of reinforcement is

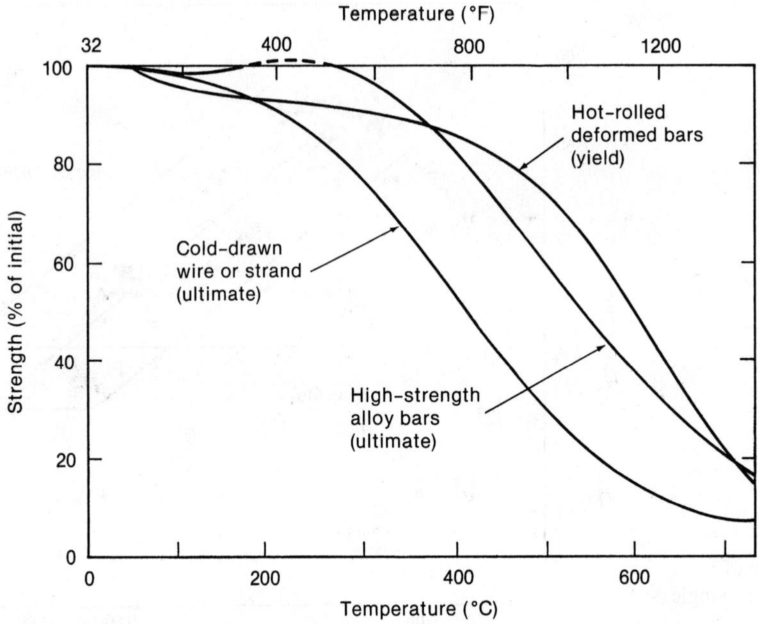

Fig. 3-34
Strength of reinforcing steels at high temperatures. (From [3-52].)

used in pavements, walls or slabs where relatively regular reinforcement patterns are possible. The ability to place a large amount of reinforcement with a minimum of work makes welded-wire fabric economical.

The wire for welded-wire fabric is produced in accordance with the following specifications: ASTM A82 *Standard Specification for Steel Wire, Plain, for Concrete Reinforcement*, and ASTM A496 *Standard Specification for Steel Wire, Deformed, for Concrete Reinforcement*. The deformations are typically two or more lines of *indentations* of about 4 to 5 percent of the bar diameter, rolled into the wire surface. As a result, the deformations on wires are less pronounced than on deformed bars. Wire sizes range from about 0.125 in. diameter to 0.625 in. diameter and are referred to as W or D, for plain or deformed wires, respectively, followed by a number that corresponds to the cross-sectional area of the wire in approximately 0.03-in.^{2} increments. Thus a W2 wire is a smooth wire with a cross-sectional area of 0.06 in.^{2} ACI Code Section 20.2.1.7.1 does not allow wires smaller than size D4. Diameters and areas of typical wire sizes are given in Table A-2a.

Welded-wire fabric satisfies the following specifications: ASTM A185 *Standard Specification for Steel Welded Wire Reinforcement, Plain, for Concrete*, and ASTM A497 *Standard Specification for Steel Welded Wire Reinforcement, Deformed, for Concrete*. Deformed welded-wire fabric may contain some smooth wires in either direction. Welded-wire fabric is available in standard or custom patterns, referred to by a style designation (such as 6 × 6—W4 × W4). The numbers in the style designation refer to: spacing of longitudinal wires × spacing of transverse wires—size of longitudinal wires × size of transverse wires. Thus a 6 × 6–W4 × W4 fabric has W4 wires at 6 in. on centers each way. Areas and weights of common welded-wire fabric patterns are given in Table A-2b.

Welded smooth-wire fabric depends on the crosswires to provide a mechanical anchorage with the concrete, while welded deformed-wire fabric utilizes both the wire deformations and the crosswires for bond and anchorage. In smooth wires, two crosswires are needed to mechanically anchor the bar for its yield strength.

The minimum yield and tensile strength of smooth wire for wire fabric is 65 ksi and 75 ksi. For deformed wires, the minimum yield and tensile strengths are 70 ksi and 80 ksi. According to ASTM A497, these yield strengths are measured at a strain of 0.5 percent.

ACI Code Tables 10.2.2.4a and b define the maximum useable yield strengths of both smooth and deformed wires.

The elongation at failure decreases as the wire size decreases, because the cold-working process used in drawing the small-diameter wires strain-hardens the steel. Reference [3-74] quotes tests indicating that the mean elongation at failure ranges from about 1.25 percent for W1.4 wires (0.133 in. diameter) to about 6 percent for W31 wires (0.628 in. diameter). These are smaller than the elongations at failure of reinforcing bars, given in Table 3-3, which range from 6 to 14 percent. There is no ACI Code limitation on minimum elongation at failure in tension tests. If it is assumed that 3 percent elongation is adequate for moment redistribution in structures reinforced with A 615 bars, wires of sizes W8.5 or D8.5 (0.328 in. diameter) or larger should have adequate ductility. References [3-75] and [3-76] describe tests in which welded-wire fabric showed adequate ductility for use as stirrups or joint ties in members tested under cyclic loads.

3-15 FIBER-REINFORCED POLYMER (FRP) REINFORCEMENT

Since 1990, extensive research has been carried out on structures reinforced with *fiber-reinforced polymer reinforcement* (FRP) in the form of bars or preformed two-dimensional grids. These bars consist of aligned fibers encased in a hardened resin and are made by a number of processes, including pultrusion, braiding, and weaving. FRP reinforcement has been used in structures subject to corrosion and in applications that require non-magnetic bars, such as floors supporting some medical devices (such as MRI machines). Common types are GFRP (made with glass fibers), AFRP (made with Aramid fibers), and CFRP (made with carbon fibers).

Properties of FRP Reinforcement

All types of FRP reinforcement have elastic–brittle stress–strain curves, with ultimate tensile strengths between 60,000 and 300,000 psi [3-77]. The strengths and the moduli of elasticity vary, depending on the type of fibers and on the ratio of the volume of fibers to the total volume of the FRP bars. Typical values of the modulus of elasticity in tension, expressed as a percentage of the modulus for steel reinforcement, range from 20 to 25 percent for GFRP, from 20 to 60 percent for AFRP, and from 60 to 80 percent for CFRP. In a similar manner, compressive strengths on the order of 55 percent, 78 percent, and 20 percent of the tensile strength have been reported [3-77] for GFRP, CFRP, and AFRP, respectively. In some bars, there is a size effect due to shear lag between the surface and the center of the bars, which leads to a lower apparent tensile strength because the interior fibers are not fully stressed at the onset of rupture of the bars.

FRP is susceptible to creep rupture under high, sustained tensile loads. Extrapolated strengths after 500,000 hours of sustained loads vary. They are on the order of 47 to 66 percent of the initial ultimate strength for AFRP and 79 to 91 percent for CFRP, depending on the test method.

The bond strength of FRP bars and concrete is affected by the smooth surface of the resin bars. Some bars are manufactured with windings of FRP cords or are coated with sand to improve bond. There is no standardized deformation pattern, however. FRP bars tend to be susceptible to surface damage during construction. FRP bars cannot be bent once the polymer has set. If bent bars are required, they must be bent during manufacture. The polymer resins used to make FRP bars undergo a phase change between 150 and 250°F, causing a reduction in strength. By the time the temperature of the bar reaches 480°F, the tensile strength has dropped to about 20 percent of the strength at room temperature.

The elastic–brittle stress–strain behavior of FRP bars affects the beam-design philosophy. In the ACI beam-design philosophy, the value of the strength-reduction factor, ϕ, ranges from 0.65 for members in which the strain in the extreme tensile layer of steel is zero or compression to 0.90 for beams in which the bar strain at ultimate exceeds 0.005 strain in tension. For beams designed with FRP reinforcement, proposed values for ϕ vary between 0.5 and 0.7.

3-16 PRESTRESSING STEEL

Prestressing steel is available as individual wires, seven-wire strands and high strength steel bars. A typical seven-wire strand and a ribbed high strength bar are shown in Fig. 3-35. Prestressing wire is produced through a cold-working process, either drawing or rolling. Seven-wire strands are produced by helically winding six peripheral wires around a central wire, which has a slightly larger diameter than the other wires. After they are formed, both individual wires and seven-wire strands are put through a *stress-relieving* process where they are heated to a specified temperature, usually less than 500°C, to improve their ductility. *Wedge anchors* are commonly used to anchor wires and seven-wire strands (Fig. 3-35) at the ends of members or at other intermediate locations. When manufactured to specified lengths wires may have button heads formed at their ends for use in *button-head anchorages*.

High-strength steel bars are composed of various alloys and may be either smooth or ribbed. For a ribbed-bar the ribs are formed to act as threads (Fig. 3-35), and thus, they can be anchored at any point along their length. Smooth bars are typically end-threaded for anchorage with a nut and plate assembly, similar to that shown in Fig. 1-5.

The range of available sizes and grades of prestressing steel, and the governing ASTM standards are given in Table 3-4.

As indicated in Table 3-4, the tensile strength of prestressing steel is significantly larger than that for normal reinforcing bars. This higher strength, and the corresponding high initial prestress are necessary because a significant amount of the initial prestress will be lost (referred to as *prestress losses*) due to elastic shortening of the prestressed member, deformation of the anchorage assembly, relaxation of the prestressing steel, shrinkage and creep of the concrete member, and other load effects.

In addition to the minimum tensile strengths listed in Table 3-4, other important mechanical properties include minimum tensile strain at failure (usually 0.040), the yield point and the elastic modulus. All prestressing steels have a rounded yield point, similar to that shown for Grade-75 steel in Fig. 3-31, as opposed to the sharp yield point that is typical for Grade-40 or Grade-60 reinforcing steel (Fig. 3-31). Thus, the effective yield strength of prestressing steel is defined as the measured stress when a specified tensile strain is reached. For prestressing wire and seven-wire strands the specified strain value is 0.010, and for prestressing bars that strain value is 0.007.

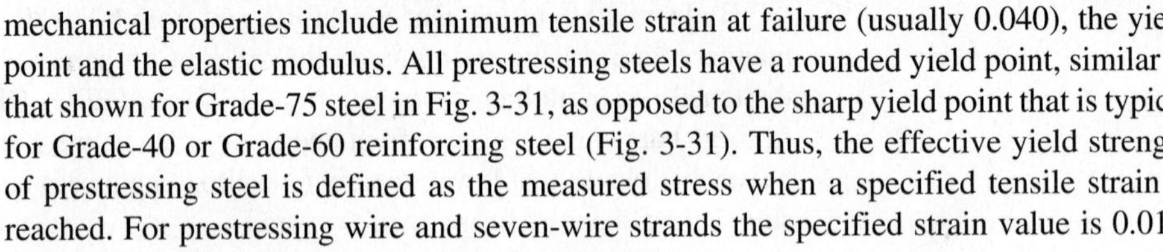

Fig. 3-35
Typical prestressing steel.

(a) Seven-wire strand.

(b) High strength (ribbed) steel bar.

TABLE 3–4 Available Types of Prestressing Steel

Prestressing Steel	Type or Grade	Nominal Diameter (in.)	Minimum Tensile Strength, f_{pu} (ksi)
Stress-relieved wires (ASTM A421)	Wedge-anchor (WA) or Button-anchor (BA)	0.192 to 0.276	235 to 250
Stress-relieved seven-wire strands (ASTM A416)	Grade 250	0.25 to 0.60	250
	Grade 270	0.375 to 0.60	270
High-strength steel bars (ASTM A722)	Grade 145	0.75 to 1.375	145
	Grade 160	0.75 to 1.375	160

The elastic modulus of prestressing wires is the same as that for normal reinforcing steel, 29,000 ksi. Because of the helical winding of seven-wire strands, their effective elastic modulus is normally taken as 27,000 ksi. The various alloys used to produce prestressing bars result in a slightly lower elastic modulus of 28,000 ksi.

PROBLEMS

3-1 What is the significance of the "critical stress"

(a) with respect to the structure of the concrete?

(b) with respect to spiral reinforcement?

(c) with respect to strength under sustained loads?

3-2 A group of 45 tests on a given type of concrete had a mean strength of 4820 psi and a standard deviation of 550 psi. Does this concrete satisfy the strength requirement for 4000-psi concrete?

3-3 The concrete containing Type I cement in a structure is cured for 3 days at 70°F, followed by 6 days at 40°F. Use the maturity concept to estimate its strength as a fraction of the 28-day strength under standard curing.

3-4 Use Fig. 3-12a to estimate the compressive strength σ_2 for biaxially loaded concrete subjected to

(a) $\sigma_1 = 0$.

(b) $\sigma_1 = 0.50$ times the tensile strength, in tension.

(c) $\sigma_1 = 0.75$ times the compressive strength, in compression.

3-5 The concrete in the core of a spiral column is subjected to a uniform confining stress σ_3 of 680 psi. What will the compressive strength σ_1 be? The unconfined uniaxial compressive strength is 5000 psi.

3-6 What factors affect the shrinkage of concrete?

3-7 What factors affect the creep of concrete?

3-8 A structure is made from concrete containing Type I cement. The average ambient relative humidity is 70 percent. The concrete was moist-cured for 7 days. $f'_c = 4000$ psi.

(a) Compute the unrestrained shrinkage strain of a rectangular beam with cross-sectional dimensions 8 in. × 20 in. at 2 years after the concrete was placed.

(b) Compute the axial shortening of a 20 in. × 20 in. × 12 ft plain concrete column at age 3 years. A compression load of 400 kips was applied to the column at age 30 days.

REFERENCES

3-1 Thomas T. C. Hsu, F. O. Slate, G. M. Sturman, and George Winter, "Micro-cracking of Plain Concrete and the Shape of the Stress–Strain Curve," *ACI Journal, Proceedings*, Vol. 60, No. 2, February 1963, pp. 209–224.

3-2 K. Newman and J. B. Newman, "Failure Theories and Design Criteria for Plain Concrete," Part 2 in M. Te'eni (ed.), *Solid Mechanics and Engineering Design*, Wiley-Interscience, New York, 1972, pp. 83/1–83/33.

3-3 F. E. Richart, A. Brandtzaeg, and R. L. Brown, *A Study of the Failure of Concrete under Combined Compressive Stresses*, Bulletin 185, University of Illinois Engineering Experiment Station, Urbana, IL, November 1928, 104 pp.

3-4 Hubert Rüsch, "Research toward a General Flexural Theory for Structural Concrete," *ACI Journal, Proceedings*, Vol. 57, No. 1, July 1960, pp. 1–28.

3-5 Llewellyn E. Clark, Kurt H. Gerstle, and Leonard G. Tulin, "Effect of Strain Gradient on Stress–Strain Curve of Mortar and Concrete," *ACI Journal, Proceedings*, Vol. 64, No. 9, September 1967, pp. 580–586.

3-6 Aïtcin, P-C., Miao, B., Cook, W.D., and Mitchell, D., "Effects of Size and Curing on Cylinder Compressive Strength of Normal and High-Strength Concretes," *ACI Materials Journal*, Vol. 91, No. 4, July–August 1994, pp. 349–354.

3-7 ACI Committee 363, "Report on High-Strength Concrete (ACI 363R-10)," *ACI Manual of Concrete Practice*, American Concrete Institute, Farmington Hills, MI.

3-8 ACI Committee 214, "Evaluation of Strength Test Results of Concrete (ACI 214R-11)," *ACI Manual of Concrete Practice*, American Concrete Institute, Farmington Hills, MI.

3-9 Andrzej Nowak and Maria Szerszen, "Calibration of Design Code for Buildings (ACI 318): Part 1—Statistical Models for Resistance," pp. 377–382; and "Part 2—Reliability Analysis and Resistance Factors," pp. 383–389, *ACI Structural Journal,* Vol. 100, No. 3, May–June 2003.

3-10 Andrzej S. Nowak and Maria Szerszen, "Reliability-Based Calibration for Structural Concrete, Phase 1," Report UMCEE 01–04, University of Michigan, 2001, 73 pp.

3-11 Michael L. Leming, "Probabilities of Low Strength Events in Concrete," *ACI Structural Journal*, Vol. 96, No. 3, May–June 1999, pp. 369–376.

3-12 H. F. Gonnerman and W. Lerch, *Changes in Characteristics of Portland Cement as Exhibited by Laboratory Tests over the Period 1904 to 1950*, ASTM Special Publication 127, American Society for Testing and Materials, Philadelphia, PA, 1951.

3-13 ACI Committee 211, "Standard Practice for Selecting Proportions for Normal, Heavyweight and Mass Concrete (ACI 211.1-91, Reapproved 2009)," *ACI Manual of Concrete Practice*, American Concrete Institute, Farmington Hills, MI.

3-14 ACI Committee 232, "Use of Fly Ash in Concrete (ACI 232.2R-03)," *ACI Manual of Concrete Practice*, American Concrete Institute, Farmington Hills, MI.

3-15 ACI Committee 232, "Report on the Use of Raw or Processed Natural Pozzolans in Concrete (ACI 232.1R-12)," *ACI Manual of Concrete Practice*, American Concrete Institute, Farmington Hills, MI.

3-16 Adam M. Neville, "Water, Cinderella Ingredient of Concrete," *Concrete International*, Vol. 22, No. 9, pp. 66–71.

3-17 Adam M. Neville, "Seawater in the Mixture," *Concrete International*, Vol. 23, No. 1, January 2001, pp. 48–51.

3-18 Walter H. Price, "Factors Influencing Concrete Strength," *ACI Journal, Proceedings*, Vol. 47, No. 6, December 1951, pp. 417–432.

3-19 Paul Klieger, "Effect of Mixing and Curing Temperature on Concrete Strength," *ACI Journal, Proceedings*, Vol. 54, No. 12, June 1958, pp. 1063–1081.

3-20 ACI Committee 209, "Prediction of Creep, Shrinkage and Temperature Effects in Concrete Structures (ACI 209R-92, Reapproved 2008)," *ACI Manual of Concrete Practice*, American Concrete Institute, Farmington Hills, MI.

3-21 V. M. Malhotra, "Maturity Concept and the Estimation of Concrete Strength: A Review," *Indian Concrete Journal*, Vol. 48, No. 4, April 1974, pp. 122–126 and 138; No. 5, May 1974, pp. 155–159 and 170.

3-22 H. S. Lew and T. W. Reichard, "Prediction of Strength of Concrete from Maturity," *Accelerated Strength Testing*, ACI Publication SP-56, American Concrete Institute, Detroit, 1978, pp. 229–248.

3-23 Adam M. Neville, "Core Tests: Easy to Perform, Not Easy to Interpret," *Concrete International*, Vol. 23, No. 11, November, 2001.

3-24 F.M. Bartlett, and J.G. MacGregor, "Effect of Moisture Content on Concrete Core Strengths," *ACI Materials Journal*, Vol. 91, No. 3, May–June 1994, pp. 227–236.

3-25 F. Michael Bartlett and James G. MacGregor, "Equivalent Specified Concrete Strength from Core Test Data," *Concrete International*, Vol. 17, No. 3, March 1995, pp. 52–58.

3-26 F. Michael Bartlett and James G. MacGregor, "Statistical Analysis of the Compressive Strength of Concrete in Structures," *ACI Materials Journal*, Vol. 93, No. 2, March–April 1996, pp. 158–168.

3-27 Jerome M. Raphael, "Tensile Strength of Concrete," *ACI Journal, Proceedings*, Vol. 81, No. 2, March–April 1984, pp. 158–165.

3-28 D. J. McNeely and Stanley D. Lash, "Tensile Strength of Concrete," *Journal of the American Concrete Institute, Proceedings*, Vol. 60, No. 6, June 1963, pp. 751–761.

3-29 H. S. Lew and T. W. Reichard, "Mechanical Properties of Concrete at Early Ages," *ACI Journal, Proceedings*, Vol. 75, No. 10, October 1978, pp. 533–542.

3-30 H. Kupfer, Hubert K. Hilsdorf, and Hubert Rüsch, "Behavior of Concrete under Biaxial Stress," *ACI Journal, Proceedings*, Vol. 66, No. 8, August 1969, pp. 656–666.

3-31 Frank J. Vecchio and Michael P. Collins, *The Response of Reinforced Concrete to In-Plane Shear and Normal Stresses*, Publication 82-03, Department of Civil Engineering, University of Toronto, Toronto, March 1982, 332 pp.

3-32 Frank J. Vecchio and Michael P. Collins, "The Modified Compression Field Theory for Reinforced Concrete Elements Subjected To Shear," *ACI Journal, Proceedings*, Vol. 83, No. 2, March–April 1986, pp. 219–231.

3-33 J. A. Hansen, "Strength of Structural Lightweight Concrete under Combined Stress," *Journal of the Research and Development Laboratories, Portland Cement Association*, Vol. 5, No. 1, January 1963, pp. 39–46.

3-34 Paul H. Kaar, Norman W. Hanson, and H. T. Capell, "Stress–Strain Characteristics of High-Strength Concrete," *Douglas McHenry International Symposium on Concrete and Concrete Structures*, ACI Publication SP-55, American Concrete Institute, Detroit, 1978, pp. 161–186.

3-35 Adrian Pauw, "Static Modulus of Elasticity as Affected by Density," *ACI Journal, Proceedings*, Vol. 57, No. 6, December 1960, pp. 679–683.

3-36 Eivind Hognestad, Norman W. Hanson, and Douglas McHenry, "Concrete Stress Distribution in Ultimate Strength Design," *ACI Journal, Proceedings*, Vol. 52, No. 4, December 1955, pp. 475–479.

3-37 Eivind Hognestad, *A Study of Combined Bending and Axial Load in Reinforced Concrete Members*, Bulletin 399, University of Illinois Engineering Experiment Station, Urbana, Ill., November 1951, 128 pp.

3-38 Popovics, S., "A Review of Stress–Strain Relationships for Concrete, *ACI Journal, Proceedings*, Vol. 67, No. 3, March 1970, pp. 243–248.

3-39 Claudio E. Todeschini, Albert C. Bianchini, and Clyde E. Kesler, "Behavior of Concrete Columns Reinforced with High Strength Steels," *ACI Journal, Proceedings*, Vol. 61, No. 6, June 1964, pp. 701–716.

3-40 Thorenfeldt, E., Tomaszewicz, A. and Jensen, J. J., "Mechanical Properties of High Strength Concrete and Application to Design," *Proceedings of the Symposium: Utilization of High-Strength Concrete*," Stavanger, Norway, June 1987, Tapir, Trondheim, pp. 149–159.

3-41 Collins, M.P. and Mitchell, D., *Prestressed Concrete Structures*, Prentice Hall, Englewood Cliffs, 1991, 766 pp.

3-42 S. H. Ahmad and Surendra P. Shah, "Stress–Strain Curves of Concrete Confined by Spiral Reinforcement," *ACI Journal, Proceedings*, Vol. 79, No. 6. November–December 1982, pp. 484–490.

3-43 B. P. Sinha, Kurt H. Gerstle, and Leonard G. Tulin, "Stress–Strain Relations for Concrete under Cyclic Loading," *ACI Journal, Proceedings*, Vol. 61, No. 2, February 1964, pp. 195–212.

3-44 Surendra P. Shah and V. S. Gopalaratnam, "Softening Responses of Plain Concrete in Direct Tension," *ACI Journal, Proceedings*, Vol. 82, No. 3, May–June 1985, pp. 310–323.

3-45 ACI Committee 209, "Prediction of Creep, Shrinkage, and Temperature Effects in Concrete Structures (ACI 209R-92, Reapproved 2008)," *ACI Manual of Concrete Practice*, American Concrete Institute, Farmington Hills, MI.

3-46 *fib Model Code for Concrete Structures 2010*, International Federation for Structural Concrete (fib), Lausanne, Switzerland, 2013.

3-47 Sidney Mindess, J. Francis Young, and David Darwin, *Concrete*, 2nd Edition, Pearson Educational—Prentice Hall, New Jersey, 2003, 644 pp.

3-48 Zedenek P. Bazant, "Prediction of Concrete Creep Effects Using Age-Adjusted Effective Modulus Method," *ACI Journal, Proceedings*, Vol. 69, No. 4, April 1972, pp. 212–217.

3-49 Walter H. Dilger, "Creep Analysis of Prestressed Concrete Structures Using Creep-Transformed Section Properties," *PCI Journal*, Vol. 27, No. 1, January–February 1982, pp. 99–118.

3-50 Amin Ghali and Rene Favre, *Concrete Structures: Stresses and Deformations*, Chapman & Hall, New York, 1986, 348 pp.

3-51 *Structural Effects of Time-Dependent Behaviour of Concrete*, Bulletin d'Information, 215, Comité Euro-International du Béton, Laussane, March 1993, pp. 265–291.

3-52 Joint ACI/TMS Committee 216, "Code Requirements for Determining Fire Resistance of Concrete and Masonry Construction Assemblies, ACI 216.1-07/TMS-0216-07," *ACI Manual of Concrete Practice*, American Concrete Institute, Farmington Hills, MI.

3-53 Said Iravani, "Mechanical Properties of High-Performance Concrete," *ACI Materials Journal*, Vol. 93, No. 5, September–October 1996, pp. 416–426.

3-54 Said Iravani and James G. MacGregor, "Sustained Load Strength and Short-Term Strain Behavior of High-Strength Concrete," *ACI Materials Journal*, Vol. 95, No. 5, September–October 1998, pp. 636–647.

3-55 Boris Bresler, "Lightweight Aggregate Reinforced Concrete Columns," *Lightweight Concrete*. ACI Publication SP-29, American Concrete Institute, Detroit, 1971, pp. 81–130.

3-56 ACI Committee 544, "Report on Fiber Reinforced Concrete (ACI 544.1R-96, Reapproved 2009)," *ACI Manual of Concrete Practice*, American Concrete Institute, Farmington Hills, MI.

3-57 D.E. Otter and A.E. Naaman, "Fiber Reinforced Concrete Under Cyclic and Dynamic Compressive Loadings," *Report UMCE 88-9*, Department of Civil Engineering, University of Michigan, Ann Arbor, MI, 178 pp.

3-58 A.E. Naaman and H.W. Reinhardt, "Characterization of High Performance Fiber Reinforced Cement Composites–HPFRCC," *High Performance Fiber Reinforced Cement Composites 2 (HPFRCC 2), Proceedings of the Second International RILEM Workshop*, Ann Arbor, USA, June 1995, Ed. A.E. Naaman and H.W. Reinhardt, E & FN Spon, London, UK, pp. 1–24.

3-59 A.E. Naaman, "High-Performance Fiber-Reinforced Cement Composites," *Concrete Structures for the Future*, IABSE Symposium, Zurich, pp. 371–376.

3-60 G.J. Parra-Montesinos, "Shear Strength of Beams with Deformed Steel Fibers," *Concrete International*, Vol. 28, No. 11, pp. 57–66.

3-61 G.J. Parra-Montesinos, "High-Performance Fiber Reinforced Cement Composites: A New Alternative for Seismic Design of Structures," *ACI Structural Journal*, Vol. 102, No. 5, September–October 2005, pp. 668-675.

3-62 R.D. Lequesne, G.J. Parra-Montesinos, and J.K. Wight, J.K., "Seismic Behavior and Detailing of High-Performance Fiber-Reinforced Concrete Coupling Beams and Coupled Wall Systems," *Journal of Structural Engineering, ASCE*, August, 2013, pp. 1362-1370.

3-63 ACI Committee 201, "Guide to Durable Concrete, (ACI 201.2R-08)," *ACI Manual of Concrete Practice*, American Concrete Institute, Farmington Hills, MI.

3-64 ACI Committee 222, "Protection of Metals in Concrete Against Corrosion (ACI 222R–01, Reapproved 2010)," *ACI Manual of Concrete Practice*, American Concrete Institute, Farmington Hills, MI.

3-65 Adam M. Neville, *Properties of Concrete*, 3rd Edition, Pitman, 1981, 779 pp.

3-66 PCI Committee on Durability, "Alkali–Aggregate Reactivity—A Summary," *PCI Journal*, Vol. 39, No. 6, November–December, 1994, pp. 26–35.

3-67 ACI Committee 515, "A Guide to the Use of Waterproofing, Dampproofing, Protective, and Decorative Barrier Systems for Concrete (ACI 515.R-85)," *ACI Manual of Concrete Practice*, American Concrete Institute, Farmington Hills, MI.

3-68 Monfore, G.E. and Lentz, A.E., "Physical Properties of Concrete at Very Low Temperatures," *Journal of the PCA Research and Development Laboratories*, Vol. 4, No. 2, May 1962, pp. 33–39.

3-69 ACI Committee 506, "Guide to Shotcrete (ACI 506R-05)," *ACI Manual of Concrete Practice*, American Concrete Institute, Farmington Hills, MI.

3-70 Neville, Adam M., "A 'New' Look at High-Alumina Cement," *Concrete International*, August 1998, pp. 51–55.

3-71 Sher Al Mirza and James G. MacGregor, "Variability of Mechanical Properties of Reinforcing Bars," *Proceedings ASCE, Journal of the Structural Division*, Vol. 105, No. ST5, May 1979, pp. 921–937.

3-72 T. Helgason and John M. Hanson, "Investigation of Design Factors Affecting Fatigue Strength of Reinforcing Bars—Statistical Analysis," *Abeles Symposium on Fatigue of Concrete*, ACI Publication SP-41, American Concrete Institute, Detroit, 1974, pp. 107–137.

3-73 ACI Committee 215, "Considerations for Design of Concrete Structures Subjected to Fatigue Loading, (ACI 215R-74, revised 1992/Reapproved 1997)," *ACI Manual of Concrete Practice*, American Concrete Institute, Farmington Hills, MI.

3-74 Mirza, S.A. and MacGregor, J.G., "Strength and Ductility of Concrete Slabs Reinforced with Welded Wire Fabric," *ACI Journal, Proceedings*, Vol. 78, No. 5, September–October 1981, pp. 374–380.

3-75 Griezic, A., Cook, W.D., and Mitchell, D., "Tests to Determine Performance of Deformed Welded Wire Fabric Stirrups," *ACI Structural Journal*, Vol. 91, No. 2, March–April 1994, pp. 213–219.

3-76 Guimaraes, G.N. and Kreger, M.E., "Evaluation of Joint-Shear Provisions for Interior Beam-Column Connections Using High-Strength Materials,' *ACI Structural Journal*, Vol. 89, No. 1, January–February 1992, pp. 89–98.

3-77 ACI Committee 440, "Guide for the Design and Construction of Structural Concrete Reinforced with FRP Bars, (440.1R-06)," *ACI Manual of Concrete Practice,* American Concrete Institute, Farmington Hills, MI.

4

Flexure: Behavior and Nominal Strength of Beam Sections

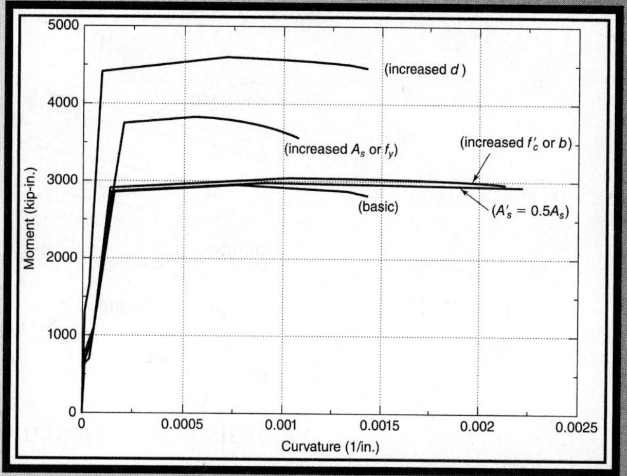

4-1 INTRODUCTION

In this chapter, the stress–strain relationships for concrete and reinforcement from Chapter 3 are used to develop an understanding of the flexural behavior of rectangular beam sections. The effect of changes in material and section properties on the flexural behavior (moment versus curvature relationship) of beam sections will be presented. A good understanding of how changes in these primary design variables affect section behavior will be important for making good design decisions concerning material and section properties, as will be covered in the next chapter.

After gaining a good understanding of the entire range of flexural behavior, a general procedure will be developed to evaluate the *nominal flexural strength*, M_n, for a variety of beam sections. Simplifications for modeling material properties, which correspond to the ACI Code definitions for nominal strength, will be presented. Emphasis will be placed on developing a fundamental approach that can be applied to any beam or slab section.

In Chapter 11, the section analysis procedures developed in this chapter will be extended to sections subjected to combined bending and axial load to permit the analysis and design of column sections.

Most reinforced concrete structures can be subdivided into beams and slabs, which are subjected primarily to flexure (bending), and columns, which are subjected to axial compression and bending. Typical examples of flexural members are the slab and beams shown in Fig. 4-1. The load, P, applied at Point A is carried by the strip of slab shown shaded. The end reactions due to the load P and the weight of the slab strip load the beams at B and C. The beams, in turn, carry the slab reactions and their own weight to the columns at D, E, F, and G. The beam reactions normally cause axial load and bending in the columns. The slab in Fig. 4-1 is assumed to transfer loads in one direction and hence is called a *one-way slab*. The design of such slabs will be discussed in the next chapter. If there were no beams in the floor system shown in Fig. 4-1, the slab would carry the load in two directions. Such a slab is referred to as a *two-way slab*. The design of two-way slabs will be discussed in Chapter 13.

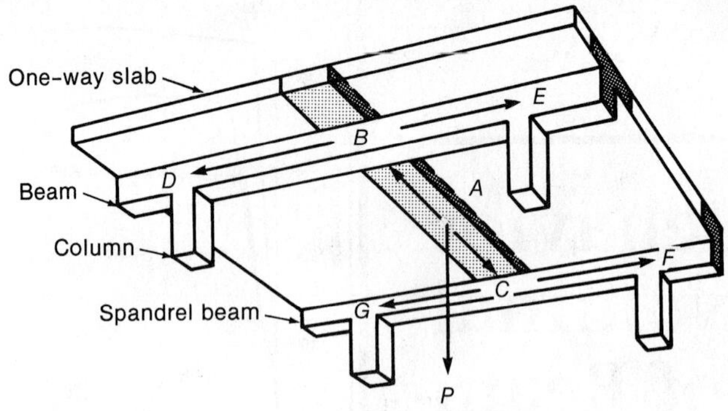

Fig. 4-1
One-way flexure.

Analysis versus Design

Two different types of problems arise in the study of reinforced concrete:

1. ***Analysis.*** Given a cross section, concrete strength, reinforcement size and location, and yield strength, compute the resistance or strength. In analysis there should be one unique answer.

2. ***Design.*** Given a factored design moment, normally designated as M_u, select a suitable cross section, including dimensions, concrete strength, reinforcement, and so on. In design there are many possible solutions.

Although both types of problem are based on the same principles, the procedure is different in each case. Analysis is easier, because all of the decisions concerning reinforcement, beam size, and so on have been made, and it is only necessary to apply the strength-calculation principles to determine the section strength. Design, on the other hand, involves the choice of section dimensions, material strengths, and reinforcement placement to produce a cross section that can resist the moments due to factored loads. Because the analysis problem is easier, this chapter deals with section analysis to develop the fundamental concepts before considering design in the next chapter.

Required Strength and Design Strength

The basic strength requirement from the ACI Code [4-2] is:

$$\text{Reduced nominal strength} \geq \text{factored load effects} \quad (4\text{-}1a)$$

or for flexure,

$$\phi M_n \geq M_u \quad (4\text{-}1b)$$

where M_u is the *moment due to the factored loads*, which commonly is referred to as the *factored design moment*. This is a load effect computed by structural analysis from the governing combination of factored loads given in ACI Code Section 5.3.1. The term M_n refers to the *nominal moment strength* of a cross section, computed from the nominal dimensions and specified material strengths. The factor ϕ in Eq. (4-1b) is a *strength-reduction factor* (ACI Code Tables 21.2.1 and 21.2.2) to account for possible variations in dimensions and material strengths and possible inaccuracies in the strength equations. Since the mid 1990s, the ACI Code has referred to the load factors and load combinations developed by ASCE/SEI Committee 7, which is responsible for the ASCE/SEI Standard for *Minimum Design Loads for Buildings and Other Structures* [4-1]. The load factors and load combinations given in ACI Code Section 5.3.1 are essentially the same as those developed by ASCE/SEI

Committee 7. The strength-reduction factors given in ACI Code Tables 21.2.1 and 21.2.2 are based on statistical studies of material properties [4-3] and were selected to give approximately the same level of safety as obtained with the load and strength-reduction factors used in earlier editions of the code.

For flexure without axial load, ACI Code Table 21.2.2 gives $\phi = 0.90$ for what are called *tension-controlled sections*. Most practical beams will be tension-controlled sections, and ϕ will be equal to 0.90. The concept of tension-controlled sections will be discussed later in this chapter. The product, ϕM_n, commonly is referred to as the *reduced nominal moment strength*.

Positive and Negative Moments

A moment that causes compression on the top surface of a beam and tension on the bottom surface will be called a *positive moment*. The compression zones for positive and negative moments are shown shaded in Fig. 4-2. In this textbook, positive bending-moments will be plotted above the beam centroid and negative moments will be plotted below the centroid.

Symbols and Notation

Although symbols are defined as they are first used and are summarized in Appendix B, several symbols should essentially be memorized because they are commonly used in discussions of reinforced concrete members. These include the terms M_u and M_n (defined earlier) and the cross-sectional dimensions illustrated in Fig. 4-2. The following is a list of common symbols used in this book:

- A_s is the area of reinforcement near the tension face of the beam, tension reinforcement, in.^{2}.
- A_s' is the area of reinforcement on the compression side of the beam, compression reinforcement, in.^{2}.
- b is a general symbol for the width of the compression zone in a beam, in. This is illustrated in Fig. 4-2 for positive and negative moment regions. For flanged sections this symbol will normally be replaced with b_e or b_w.

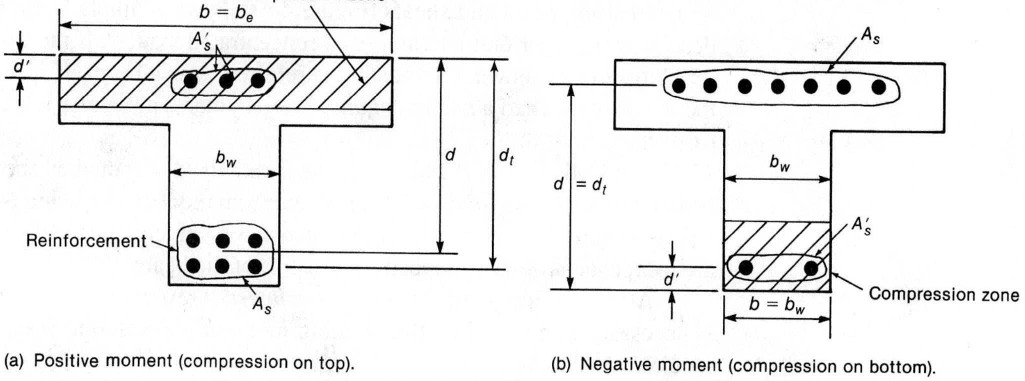

Fig. 4-2
Cross-sectional dimensions.

- b_e is the effective width of a compression zone for a flanged section with compression in the flange, in.
- b_w is the width of the web of the beam (and may or may not be the same as b), in.
- d is the distance from the extreme fiber in compression to the centroid of the longitudinal reinforcement on the tension side of the member, in. In the positive-moment region (Fig. 4-2a), the tension steel is near the bottom of the beam, while in the negative-moment region (Fig. 4-2b) it is near the top.
- d' is the distance from the extreme compression fiber to the centroid of the longitudinal compression steel, in.
- d_t is the distance from the extreme compression fiber to the farthest layer of tension steel, in. For a single layer of tension reinforcement, $d_t = d$, as shown in Fig. 4-2b.
- f'_c is the specified compressive strength of the concrete, psi.
- f_c is the stress in the concrete, psi.
- f_s is the stress in the tension reinforcement, psi.
- f_y is the specified yield strength of the reinforcement, psi.
- h is the overall height of a beam cross section.
- jd is the *lever arm*, the distance between the resultant compressive force and the resultant tensile force, in.
- j is a dimensionless ratio used to define the lever arm, jd. It varies depending on the moment acting on the beam section.
- ε_{cu} is the assumed maximum useable compression strain in the concrete.
- ε_s is the strain in the tension reinforcement.
- ε_t is the strain in the extreme layer of tension reinforcement.
- ρ is the longitudinal tension reinforcement ratio, $\rho = A_s/bd$.

4-2 FLEXURE THEORY

Statics of Beam Action

A *beam* is a structural member that supports applied loads and its own weight primarily by internal moments and shears. Figure 4-3a shows a simple beam that supports its own dead weight, w per unit length, plus a concentrated load, P. If the axial applied load, N, is equal to zero, as shown, the member is referred to as a beam. If N is a compressive force, the member is called a *beam–column*. This chapter will be restricted to the very common case where $N = 0$.

The loads w and P cause *bending moments*, distributed as shown in Fig. 4-3b. The bending moment is a *load effect* calculated from the loads by using the laws of statics. For a simply supported beam of a given span and for a given set of loads w and P, the moments are independent of the composition and size of the beam.

At any section within the beam, the *internal resisting moment*, M, shown in Fig. 4-3c is necessary to equilibrate the bending moment. An internal resisting shear, V, also is required, as shown.

The internal resisting moment, M, results from an internal compressive force, C, and an internal tensile force, T, separated by a lever arm, jd, as shown in Fig. 4-3d. Because there are no external axial loads, summation of the horizontal forces gives

$$C - T = 0 \quad \text{or} \quad C = T \tag{4-2}$$

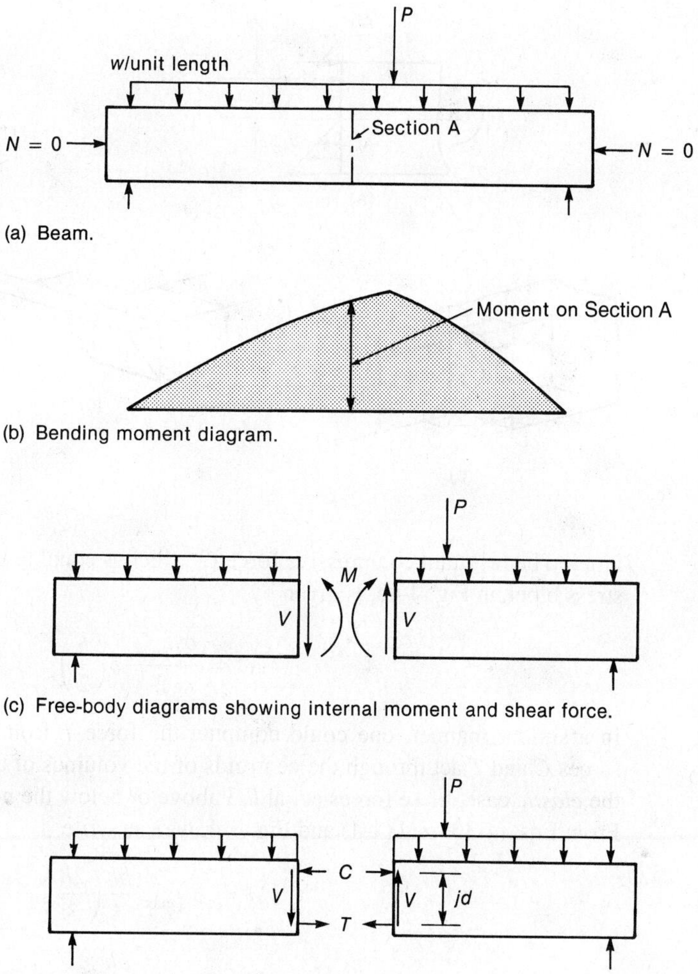

(a) Beam.

(b) Bending moment diagram.

(c) Free-body diagrams showing internal moment and shear force.

(d) Free-body diagrams showing internal moment as a compression–tension force couple.

Fig. 4-3
Internal forces in a beam.

If moments are summed about an axis through the point of application of the compressive force, C, the moment equilibrium of the free body gives

$$M = T \times jd \tag{4-3a}$$

Similarly, if moments are summed about the point of application of the tensile force, T,

$$M = C \times jd \tag{4-3b}$$

Because $C = T$, these two equations are identical. Equations (4-2) and (4-3) come directly from statics and are equally applicable to beams made of steel, wood, or reinforced concrete.

The conventional *elastic* beam theory results in the equation $\sigma = My/I$, which, for an uncracked, homogeneous rectangular beam without reinforcement, gives the distribution of stresses shown in Fig. 4-4a. The stress diagram shown in Fig. 4-4b and 4-4c may be visualized as having a "volume"; hence, one frequently refers to the *compressive stress*

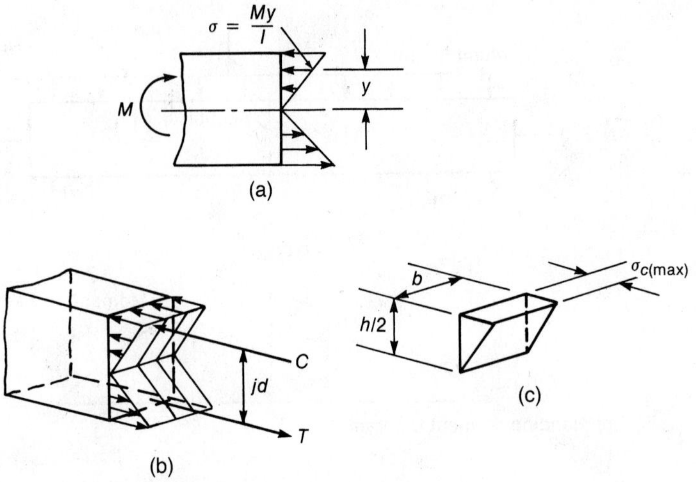

Fig. 4-4
Elastic beam stresses and stress blocks.

block. The resultant compressive force C, which is equal to the volume of the compressive stress block in Fig. 4-4c, is given by

$$C = \frac{\sigma_{c(\max)}}{2} b \left(\frac{h}{2}\right) \tag{4-4}$$

In a similar manner, one could compute the force T from the tensile stress block. The forces C and T act through the centroids of the volumes of the respective stress blocks. In the *elastic* case, these forces act at $h/3$ above or below the neutral axis, so that $jd = 2h/3$. From Eqs. (4-3b) and (4-4) and Fig. 4-4, we can write

$$M = \sigma_{c(\max)} \frac{bh}{4}\left(\frac{2h}{3}\right) \tag{4-5a}$$

$$M = \sigma_{c(\max)} \frac{bh^3/12}{h/2} \tag{4-5b}$$

or, because

$$I = \frac{bh^3}{12}$$

and

$$y_{\max} = h/2$$

it follows that

$$M = \frac{\sigma_{c(\max)} I}{y_{\max}} \tag{4-5c}$$

Thus, for the elastic case, identical answers are obtained from the traditional beam stress equation, Eq. (4-5c), and when the stress block concept is used in Eq. (4-5a).

The elastic beam theory in Eq. (4-5c) is not used in the design of reinforced concrete beams, because the compressive stress–strain relationship for concrete becomes nonlinear at higher strain values, as shown in Fig. 3-18. What is even more important is that concrete cracks at low tensile stresses, making it necessary to provide steel reinforcement to carry the tensile force, T. These two factors are easily handled by the stress-block concept, combined with Eqs. (4-4) and (4-5).

Flexure Theory for Reinforced Concrete

The theory of flexure for reinforced concrete is based on three basic assumptions, which are sufficient to allow one to calculate the moment resistance of a beam. These are presented first and used to illustrate flexural behavior, i.e., the *moment–curvature relationship* for a beam cross section under increasing moment. After gaining an understanding of the general development of a moment–curvature relationship for a typical beam section, a series of moment–curvature relationships will be developed to illustrate how changes in section properties and material strengths affect flexural behavior.

Basic Assumptions in Flexure Theory

Three basic assumptions are made:

1. Sections perpendicular to the axis of bending that are plane before bending remain plane after bending.

2. The strain in the reinforcement is equal to the strain in the concrete at the same level.

3. The stresses in the concrete and reinforcement can be computed from the strains by using stress–strain curves for concrete and steel.

The first of these is the traditional "plane sections remain plane" assumption made in the development of flexural theory for beams constructed with any material. The second assumption is necessary, because the concrete and the reinforcement must act together to carry load. This assumption implies a perfect bond between the concrete and the steel. The third assumption will be demonstrated in the following development of moment–curvature relationships for beam sections.

Flexural Behavior

General moment–curvature relationships will be used to describe and discuss the flexural behavior of a variety of beam sections. The initial discussion will be for *singly reinforced* sections, i.e., sections that have reinforcement only in their tension zone, as shown in Fig. 4-5. After singly reinforced sections have been discussed, a short discussion will be given

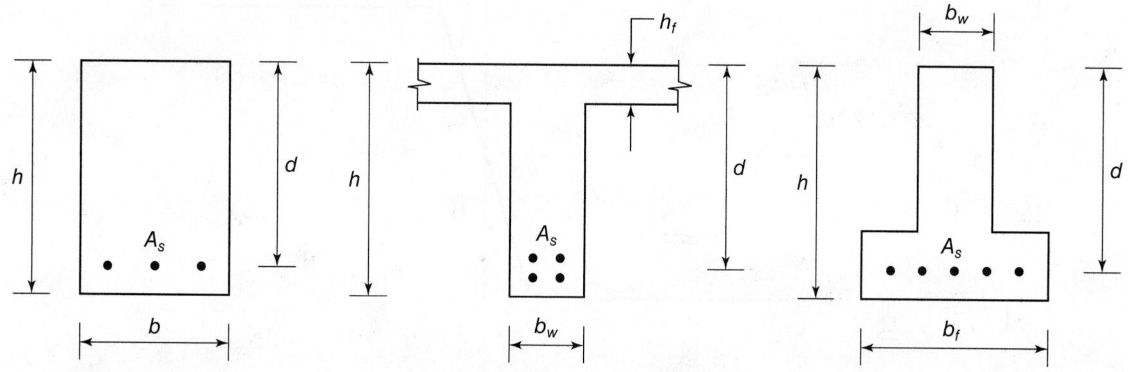

Fig. 4-5
Typical singly reinforced sections in positive bending (tension in bottom).

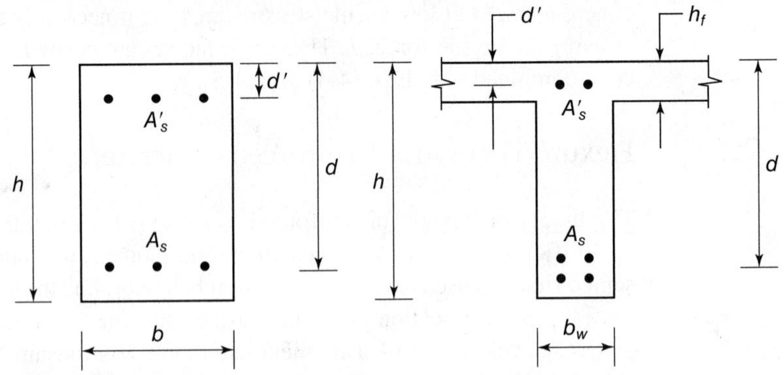

Fig. 4-6
Typical doubly reinforced sections in positive bending.

on how adding steel in the compression zone to create a *doubly reinforced* section, as shown in Fig. 4-6, will affect flexural behavior. All of the sections considered here will be *under-reinforced.* Although this may sound like a bad design, this is exactly the type of cross section we will want to design to obtain the preferred type of flexural behavior. The meaning of an under-reinforced beam section is that, when the section is loaded in bending beyond its elastic range, the tension zone steel will *yield* before the concrete in the compression zone reaches its maximum useable strain, ε_{cu}.

To analytically create a moment–curvature relationship for any beam section, assumptions must be made for material stress–strain relationships. A simple elastic-perfectly plastic model will be assumed for the reinforcing steel in tension or compression, as shown in Fig. 4-7. The steel elastic modulus, E_s, is assumed to be 29,000 ksi.

The stress–strain relationship assumed for concrete in compression is shown in Fig. 4-8. This model consists of a parabola from zero stress to the compressive strength of the concrete, f'_c. The strain that corresponds to the peak compressive stress, ε_o, is often assumed to be 0.002 for normal strength concrete. The equation for this parabola, which was originally introduced by Hognestad [4-4], is

$$f_c = f'_c\left[2\left(\frac{\varepsilon_c}{\varepsilon_o}\right) - \left(\frac{\varepsilon_c}{\varepsilon_o}\right)^2\right] \tag{4-6}$$

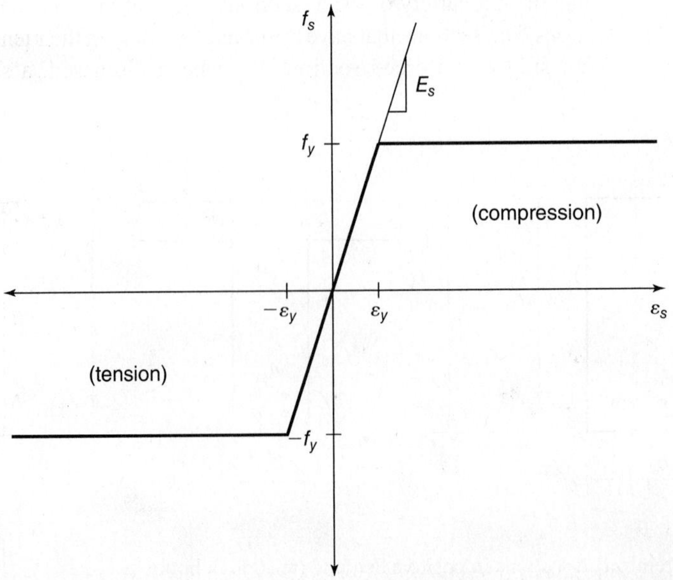

Fig. 4-7
Assumed stress–strain relationship for reinforcing steel.

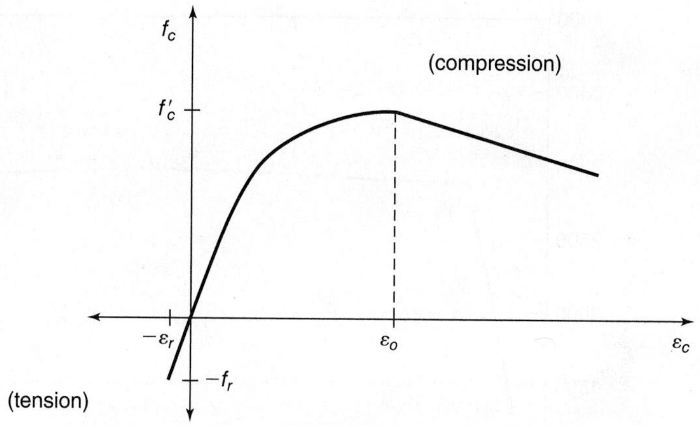

Fig. 4-8
Assumed stress–strain relationship for concrete.

Beyond the strain, ε_o, the stress is assumed to decrease linearly as the strain increases. An equation for this portion of the relationship can be expressed as

$$f_c = f'_c\left[1 - \frac{Z}{1000}\left(\frac{\varepsilon_c - \varepsilon_o}{\varepsilon_o}\right)\right] \tag{4-7}$$

where Z is a constant to control the slope of the line. For this discussion, Z will be set equal to a commonly used value of 150. Lower values for Z (i.e., a shallower unloading slope) can be used if longitudinal and transverse reinforcement are added to confine the concrete in the compression zone.

In tension the concrete is assumed to have a linear stress–strain relationship (Fig. 4-8) up to the concrete modulus of rupture, f_r, defined in Chapter 3.

Consider a singly reinforced rectangular section subjected to positive bending, as shown in Fig. 4-9a. In this figure, A_s represents the total *area of tension reinforcement*, and d represents the *effective flexural depth* of the section, i.e., the distance from the extreme compression fiber to the centroid of the tension reinforcement. A complete moment–curvature relationship, as shown in Fig. 4-10, can be generated for this section by continuously increasing the section curvature (slope of the strain diagram) and using the assumed material stress–strain relationships to determine the resulting section stresses and forces, as will be discussed in the following paragraphs.

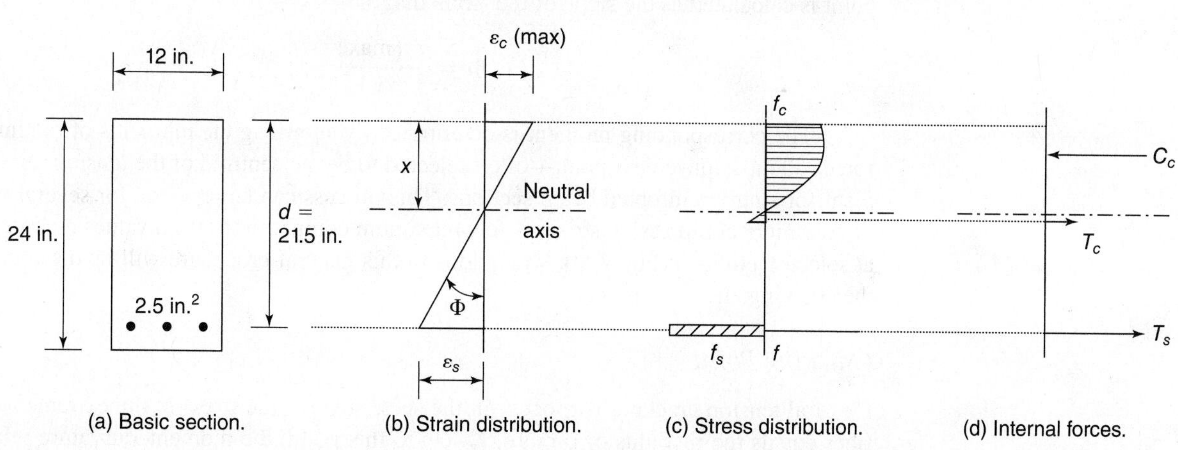

Fig. 4-9
Steps in analysis of moment and curvature for a singly reinforced section.

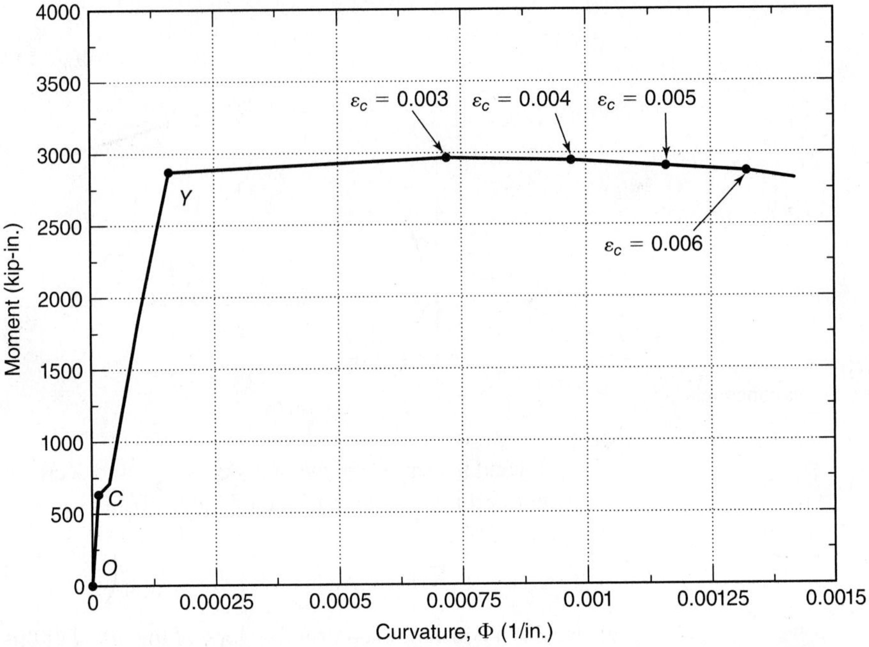

Fig. 4-10
Moment–curvature relationship for the section in Fig. 4-9a using $f'_c = 4000$ psi and $f_y = 60$ ksi.

The calculation of specific points on the moment–curvature curve follows the process represented in Figs. 4-9b through 4-9d. Each point is usually determined by selecting a specific value for the maximum compression strain at the extreme compression fiber of the section, $\varepsilon_c(\max)$. From the assumption that plane sections before bending remain plane, the strain distribution through the depth of the section is linear. From the strain diagram and the assumed material stress–strain relationships, the distribution of stresses is determined. Finally, by integration, the volume under the stress distributions (i.e., the section forces) and their points of action can be determined.

After the section forces are determined, the following steps are required to complete the calculation. First, the distance from the extreme compression fiber to the section neutral axis (shown as x in Fig. 4-9b) must be adjusted up or down until section equilibrium is established, as given by Eq. (4-2). When Eq. (4-2) is satisfied, the curvature, Φ, for this point is calculated as the slope of the strain diagram,

$$\Phi = \frac{\varepsilon_c(\max)}{x} \tag{4-8}$$

The corresponding moment is determined by summing the moments of the internal forces about a convenient point—often selected to be the centroid of the tension reinforcement for singly reinforced beam sections. This process can be repeated for several values of maximum compression strain. A few maximum compression strain values are indicated at selected points in Fig. 4-10. Exceptions to this general procedure will be discussed for the cracking and yield points.

Cracking Point

Flexural tension cracking will occur in the section when the stress in the extreme tension fiber equals the modulus of rupture, f_r. Up to this point, the moment–curvature relationship is linear and is referred to as the *uncracked-elastic* range of behavior (from O to C in Fig. 4-10). The moment and curvature at cracking can be calculated directly from elastic flexural theory, as expressed in Eq. (4-5c). In most cases, the contribution of the reinforcement can be ignored in this range of behavior, and the cracking moment can be calculated using only the concrete section, normally referred to as the *gross section*. If the moment of inertia for the gross section is defined as I_g, and the distance from the section centroid to the extreme tension fiber is defined as y_t, then the stress at the extreme tension fiber in a modified version of Eq. (4-5c) is

$$f = \frac{My_t}{I_g} \tag{4-5d}$$

The *cracking moment* is defined as the moment that causes the stress in the extreme tension fiber to reach the modulus of rupture, i.e.,

$$M_{cr} = \frac{f_r I_g}{y_t} \tag{4-9}$$

This expression is the same as used in ACI Code Section 19.2.3.1. When calculating M_{cr}, it is recommended to take the modulus of rupture, f_r, equal to $7.5\sqrt{f'_c}$. The reinforcement could be included in this calculation by using a transformed section method to define the section properties, but for typical sections, this would result in a relatively small change in the value for M_{cr}.

The section curvature at cracking, Φ_{cr}, can be calculated for this point using the elastic bending theory,

$$\Phi_{cr} = \frac{M_{cr}}{E_c I_g} \tag{4-10}$$

where the elastic concrete modulus, E_c, can be taken as $57{,}000\sqrt{f'_c}$, in psi units.

When a beam section cracks in tension, the crack usually propagates to a point near the centroid of the section and there is a sudden transfer of tension force from the concrete to the reinforcing steel in the tension zone. Unless a minimum amount of reinforcement is present in the tension zone, the beam would fail suddenly. To prevent such a brittle failure, the minimum moment strength for a *reinforced* concrete beam section should be equal to or greater than the cracking moment strength for the *plain* concrete section. Such a specific recommendation is not given in the ACI Code for reinforced concrete beam sections. However, based on a calculation that sets the moment capacity of a reinforced section equal to approximately twice that of a plain section, ACI Code Section 9.6.1.2 specifies the following minimum area of longitudinal reinforcement for beam sections in positive bending as

$$A_{s,\min} = \frac{3\sqrt{f'_c}}{f_y} b_w d \tag{4-11}$$

where the quantity $3\sqrt{f'_c}$ is not to be taken less than 200 psi. The notation for *web width*, b_w, is used here to make the equation applicable for both rectangular and flanged sections. Additional discussion of this minimum area requirement is given in Section 4-8 for flanged sections with the flange in tension.

From the metric version of ACI Code Section 9.6.1.2, using MPa units for f'_c and f_y the expression for $A_{s,\min}$ is:

$$A_{s,\min} = \frac{0.25\sqrt{f'_c}}{f_y} b_w d \geq \frac{1.4 b_w d}{f_y} \tag{4-11M}$$

After cracking but before yielding of the tension reinforcement, the relationship between moment and curvature is again approximately linear, but with a different slope than before cracking. This is referred to as the *cracked-elastic* range of behavior (from C to Y in Fig. 4-10). This linear relationship is important for the calculation of deflections, as will be discussed in Chapter 9.

Yield Point

The yield point represents the end of the elastic range of behavior. As the moment applied to the section continues to increase after the cracking point, the tension stress in the reinforcement and the compression stress in the concrete compression zone will steadily increase. Eventually, either the steel or the concrete will reach its respective capacity and start to yield (steel) or crush (concrete). Because the section under consideration here is assumed to be *under-reinforced*, the steel will yield before the concrete reaches its maximum useable strain.

To calculate moment and curvature values for the yield point, the strain at the level of the tension steel is set equal to the yield strain ($\varepsilon_y = f_y/E_s$). As discussed previously for the general procedure, the neutral axis needs to be adjusted up or down until section equilibrium is established. At this stage of flexural behavior the contribution of the concrete in tension is not significant for section equilibrium and moment calculations, so the vector, T_c, shown in Fig. 4-9d can be ignored. After section equilibrium is established, the section *yield moment*, M_y, is then calculated as the sum of the moments of the internal forces about a convenient point. Referring to Fig. 4-9b, the *yield curvature* is calculated as the slope of the strain diagram, which can be calculated by setting the strain at the level of the tension reinforcement equal to the yield strain,

$$\Phi_y = \frac{\varepsilon_y}{d - x} \tag{4-12}$$

Points beyond the Yield Point

Additional points on the moment–curvature relationship can be determined by steadily increasing the maximum strain in the extreme compression fiber, following the general procedure described previously. Usually, points are generated until some predefined maximum useable compression strain is reached or until the section moment capacity drops significantly below the maximum calculated value. Points representing maximum compression strains of 0.003, 0.004, 0.005, and 0.006 are noted in Fig. 4-10. For each successive point beyond the compression strain of 0.003, the section moment capacity is decreasing at an increasing rate. If a more realistic model was used for the stress–strain properties of the reinforcing steel, i.e., a model that includes strain hardening (Fig. 3-31), the moment capacity would increase beyond the yield point and would hold steady or at least show a less significant decrease in moment capacity for maximum compressive strain values greater than 0.003.

Most concrete design codes specify a maximum useable compression strain at which the *nominal moment strength* of the section is to be calculated. For the ACI Code, this maximum useable strain value is specified as 0.003. For the Canadian Concrete Code [4-5], a maximum useable compressive strain value of 0.0035 is specified. It should be clear from Fig. 4-10 that the calculation of a nominal moment capacity for this section would not be affected significantly by selection of either one of these values. Also, the beam section shown here has considerable deformation capacity beyond the limit corresponding to either of the maximum useable compression strains discussed here.

Any discussion of flexural behavior of a reinforced concrete beam section usually involves a discussion of *ductility*, i.e., the ability of a section to deform beyond its yield point without a significant strength loss. Ductility can be expressed in terms of displacement, rotation, or curvature ratios. For this discussion, section ductility will be expressed in terms of the ratio of the curvature at maximum useable compression strain to the curvature at yield. The maximum useable strain can be expressed as a specific value, as done by most codes, or it could be defined as the strain at which the moment capacity of the section has dropped below some specified percentage of the maximum moment capacity of the section. By either measure, the moment–curvature relationship given in Fig. 4-10 represents good ductile behavior.

Effect of Major Section Variables on Strength and Ductility

In this subsection, a series of systematic changes are made to section parameters for the beam given in Fig. 4-9a to demonstrate the effect of such parametric changes on the moment–curvature response of the beam section. Values of material strengths and section parameters are given for seven different beams in Table 4-1. The first column (Basic Section) represents the original values that correspond to the M–Φ curve given in Fig. 4-10. Each successive beam section (represented by a column in Table 4-1) represents a modification of either the material properties or section properties from those for the *basic section*. Note that for each new beam section (column in table) only one of the parameters has been changed from those used for the basic section.

M–Φ plots that correspond to the first three sections given in Table 4-1 are shown in Fig. 4-11. The only change for these sections is an increase in the area of tension reinforcement, A_s. It is clear that increasing the tension steel area causes a proportional increase in the strength of the section. However, the higher tension steel areas also causes a less ductile behavior for the section. Because of this loss of ductility as the tension steel area is increased, the ACI Code places an upper limit on the permissible area of tension reinforcement, as will be discussed in detail in Section 4-6.

Figure 4-12 shows M–Φ plots for the basic section and for the sections defined in the last four columns of Table 4-1. It is interesting from a design perspective to observe how changes in the different section variables affect the flexural strength, stiffness, and ductility of the beam sections. An increase in the steel yield strength has essentially the same effect as increasing the tension steel area—that is, the section moment strength increases and the section ductility decreases. Increases in either the steel yield strength or the tension steel area have very little effect on the stiffness of the section before yield (as represented by the elastic slope of the M–Φ relationship).

Increasing the effective flexural depth of the section, d, increases the section moment strength without decreasing the section ductility. Increasing the effective flexural depth

TABLE 4-1 Material and Section Properties for Various Beam Sections

Primary Variables	Basic Section	Moderate* A_s	High* A_s	High* f_y	Large* d	High* f'_c	Large* b
A_s (sq.in.)	2.5	4.5	6.5	2.5	2.5	2.5	2.5
f_y (ksi)	60	60	60	80	60	60	60
d (in.)	21.5	21.5	21.5	21.5	32.5	21.5	21.5
f'_c (psi)	4000	4000	4000	4000	4000	6000	4000
b (in.)	12	12	12	12	12	12	18

*Relative to values in the basic section.

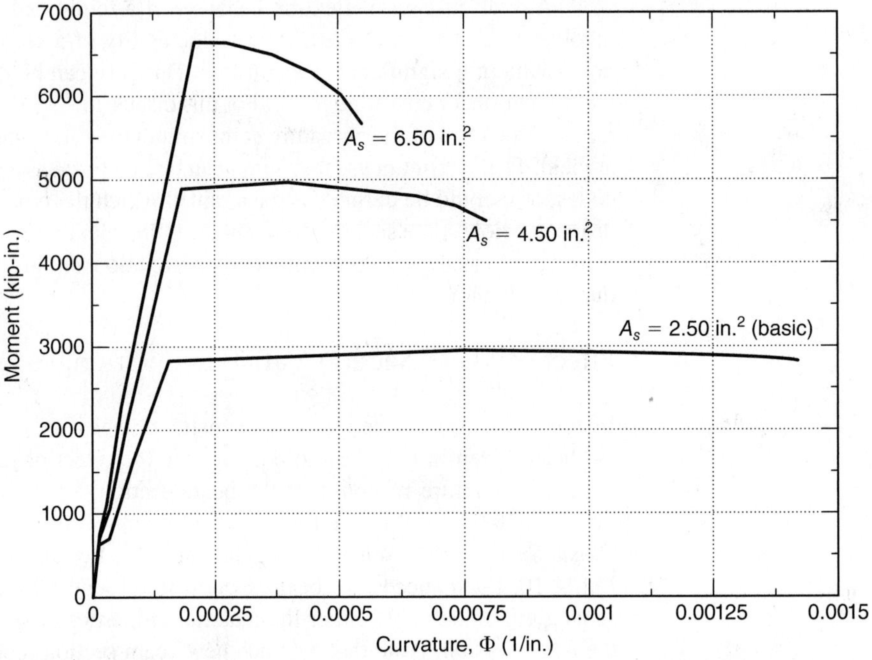

Fig. 4-11
Effect of increasing tension steel area, A_s.

also increases the elastic stiffness of the section, because the section moment of inertia is significantly affected by the depth of the section. These results clearly indicate the importance of the effective flexural depth of a member, so proper placement of reinforcement during construction should be a priority item for field inspectors.

Changes in concrete strength and section width have a smaller effect on moment strength than might be initially expected. These two variables will have only a small affect

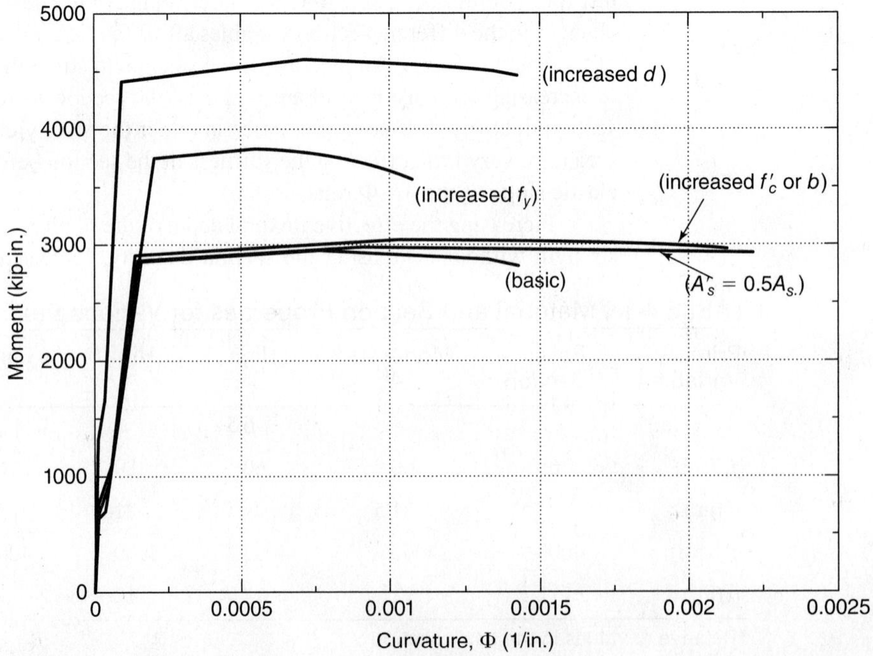

Fig. 4-12
Effect of increasing f_y, d, f'_c, b, and A'_s.

on the moment arm between the tension and compression forces shown in Fig. 4-9d, but they do not affect the value of the tension (and thus the compression) force. Thus, these variables have a significantly smaller effect on moment strength of the section than the tension steel area, steel yield strength, and effective flexural depth. Because final failure in bending for these sections is governed by reaching the maximum useable compression strain in the extreme concrete compression fiber, increases in either the concrete strength or section width do cause a significant increase in curvature at failure, as calculated in Eq. (4-8), by decreasing the neutral axis depth required to balance the tension force from the reinforcing steel.

The last variable discussed here is the addition of compression zone reinforcement, A_s', equal to one-half of the area of tension reinforcement, A_s. This variable is not listed in Table 4-1, because all of the other cross-section values are set equal to those listed for the basic section. As shown in Fig. 4-12, the addition of compression reinforcement has very little effect on the moment strength of the beam section. However, because the compression reinforcement carries part of the compression force that would be carried by the concrete in a singly reinforced beam, the required depth of the neutral axis is decreased and the section reaches a much higher curvature (higher ductility) before the concrete reaches its maximum useable strain. Thus, one of the primary reasons for using compression reinforcement will be to increase the ductility of a given beam section.

4-3 SIMPLIFICATIONS IN FLEXURE THEORY FOR DESIGN

The three assumptions already made are sufficient to allow calculation of the strength and behavior of reinforced concrete elements. For design purposes, however, the following additional assumptions are introduced to simplify the problem with little loss of accuracy.

1. The tensile strength of concrete is neglected in flexural-strength calculations (ACI Code Section 22.2.2.2).

The strength of concrete in tension is roughly one-tenth of the compressive strength, and the tensile force in the concrete below the zero strain axis, shown as T_c in Fig. 4-9d, is small compared with the tensile force in the steel. Hence, the contribution of the tensile stresses in the concrete to the flexural capacity of the beam is small and can be neglected. It should be noted that this assumption is made primarily to simplify flexural calculations. In some instances, particularly shear, bond, deflection, and service-load calculations for prestressed concrete, the tensile resistance of concrete is not neglected.

2. The section is assumed to have reached its nominal flexural strength when the strain in the extreme concrete compression fiber reaches the maximum useable compression strain, ε_{cu}.

Strictly speaking, this is an artificial limit developed by code committees to define at what point on the general moment–curvature relationship the nominal strength of the section is to be calculated. As shown in Fig. 4-10, the moment–curvature relationship for a typical beam section is relatively flat after passing the yield point, so the selection of a specific value for ε_{cu} will not significantly affect the calculated value for the nominal flexural strength of the section. Thus, design calculations are simplified when a limiting strain is assumed.

The maximum compressive strains, ε_{cu}, from tests of beams and eccentrically loaded columns of normal-strength, normal-density concrete are plotted in Fig. 4-13a [4-6], [4-7]. Similar data from tests of normal-density and lightweight concrete are compared in Fig. 4-13b. ACI Code Section 22.2.2.1 specifies a limiting compressive strain, ε_{cu}, equal to 0.003, which approximates the smallest measured values plotted in Fig. 4-13a

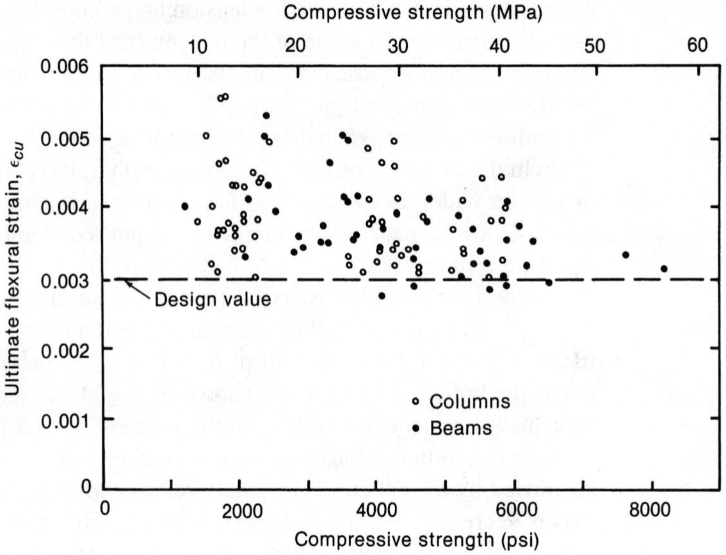

(a) Ultimate strain from tests of reinforced members.

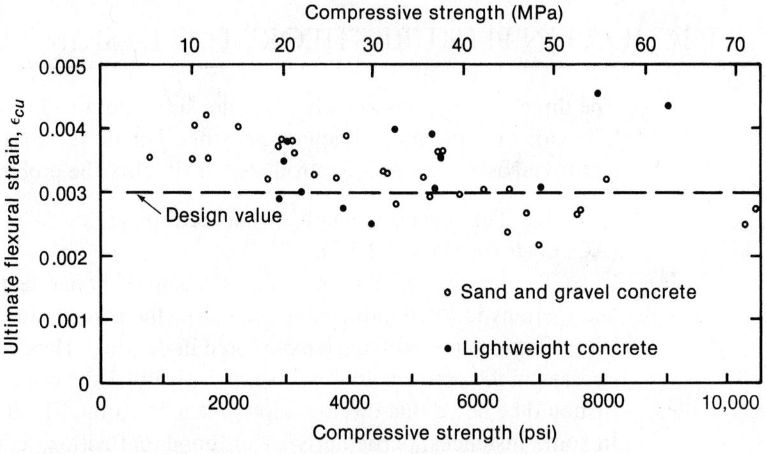

(b) Ultimate strain from tests of plain concrete specimens.

Fig. 4-13
Limiting compressive strain.
(From [4-6] and [4-7].)

and b. In Canada, the CSA Standard [4-5] uses $\varepsilon_{cu} = 0.0035$ for beams and eccentrically loaded columns. Higher limiting strains have been measured in members with a significant moment gradient and in members in which the concrete is confined by spirals or closely spaced hoops [4-8], [4-9]. Throughout this book, however, a constant maximum useable compressive strain equal to 0.003 will be used.

3. The compressive stress–strain relationship for concrete may be based on measured stress–strain curves or may be assumed to be rectangular, trapezoidal, parabolic, or any other shape that results in prediction of flexural strength in substantial agreement with the results of comprehensive tests (ACI Code Section 22.2.2.3).

Thus, rather than using a closely representative stress–strain curve (such as that given in Fig. 4-8), other diagrams that are easier to use in computations are acceptable, provided they adequately predict test results. As is illustrated in Fig. 4-14, the shape of the

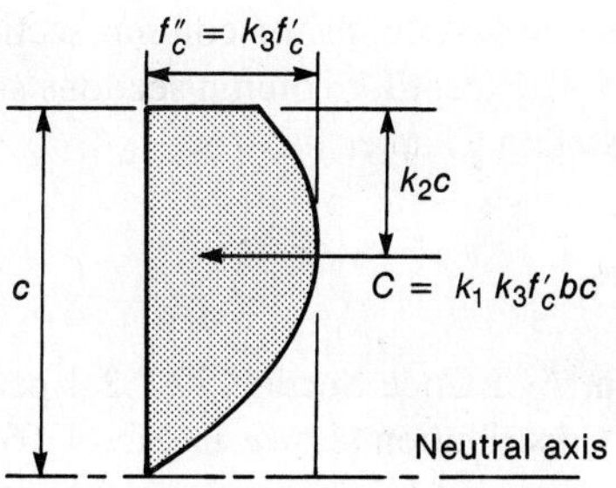

Fig. 4-14
Mathematical description of compression stress block.

stress block in a beam at the ultimate moment can be expressed mathematically in terms of three constants:

k_3 = ratio of the maximum stress, f''_c, in the compression zone of a beam to the cylinder strength, f'_c

k_1 = ratio of the average compressive stress to the maximum stress (this is equal to the ratio of the shaded area in Fig. 4-15 to the area of the rectangle, $c \times k_3 f'_c$)

k_2 = ratio of the distance between the extreme compression fiber and the resultant of the compressive force to the depth of the neutral axis, c, as shown in Figs. 4-14 and 4-15.

For a rectangular compression zone of width b and depth to the neutral axis c, the resultant compressive force is

$$C = k_1 k_3 f'_c bc \tag{4-13a}$$

Values of k_1 and k_2 are given in Fig. 4-15 for various assumed compressive stress–strain diagrams or *stress blocks*. The use of the constant k_3 essentially has disappeared from the flexural theory of the ACI Code. As shown in Fig. 4-12, a large change in the concrete compressive strength did not cause a significant change in the beam section moment capacity. Thus, the use of either f'_c or $f''_c = k_3 f'_c$, with k_3 typically taken equal to 0.85, is not significant for the flexural analysis of beams. The use of f''_c is more significant for column sections subjected to high axial load and bending. Early papers by Hognestad [4-10] and Pfrang, Siess, and Sozen [4-11] recommended the use of $f''_c = k_3 f'_c$ when analyzing

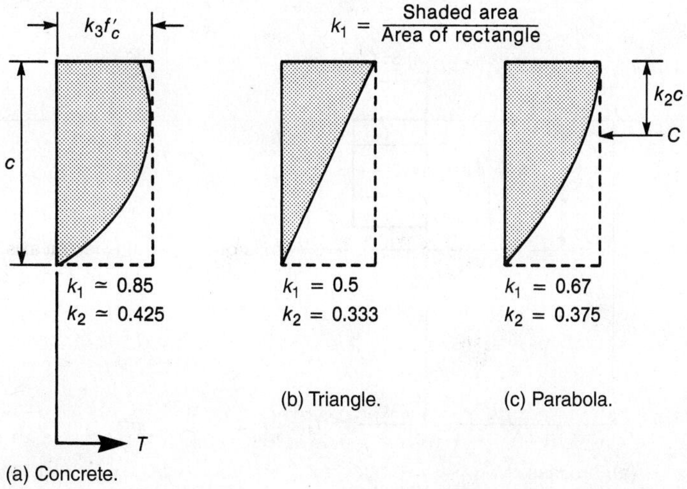

Fig. 4-15
Values of k_1 and k_2 for various stress distributions.

the combined axial and bending strength for column sections. However, the ACI Code does not refer to the use of f''_c except for column sections subjected to pure axial load (no bending), as will be discussed in Chapter 11.

Whitney Stress Block

As a further simplification, ACI Code Section 22.2.2.4 permits the use of an equivalent rectangular concrete stress distribution shown in Fig. 4-16 for nominal flexural strength calculations. This rectangular stress block, originally proposed by Whitney [4-12], is defined by the following:

1. A uniform compressive stress of $0.85\,f'_c$ shall be assumed distributed over an equivalent compression zone bounded by the edges of the cross section and a straight line located parallel to the neutral axis at a distance $a = \beta_1 c$ from the concrete fiber with the maximum compressive strain. Thus, $k_2 = \beta_1/2$, as shown in Fig. 4-16.

2. The distance c from the fiber of maximum compressive strain to the neutral axis is measured perpendicular to that axis.

3. The factor β_1 shall be taken as follows [4-7]:

(a) For concrete strengths, f'_c, up to and including 4000 psi,

$$\beta_1 = 0.85 \tag{4-14a}$$

(b) For 4000 psi $< f'_c \leq$ 8000 psi,

$$\beta_1 = 0.85 - 0.05\frac{f'_c - 4000\text{ psi}}{1000\text{ psi}} \tag{4-14b}$$

(c) For f'_c greater than 8000 psi,

$$\beta_1 = 0.65 \tag{4-14c}$$

For a rectangular compression zone of constant width b and depth to the neutral axis c, the resultant compressive force is

$$C = 0.85 f'_c\, b\beta_1 c = 0.85\,\beta_1 f'_c\, bc \tag{4-13b}$$

Comparing Eqs. (4-13a) and (4-13b), and setting $k_3 = 1.0$, results in $k_1 = 0.85\beta_1$.

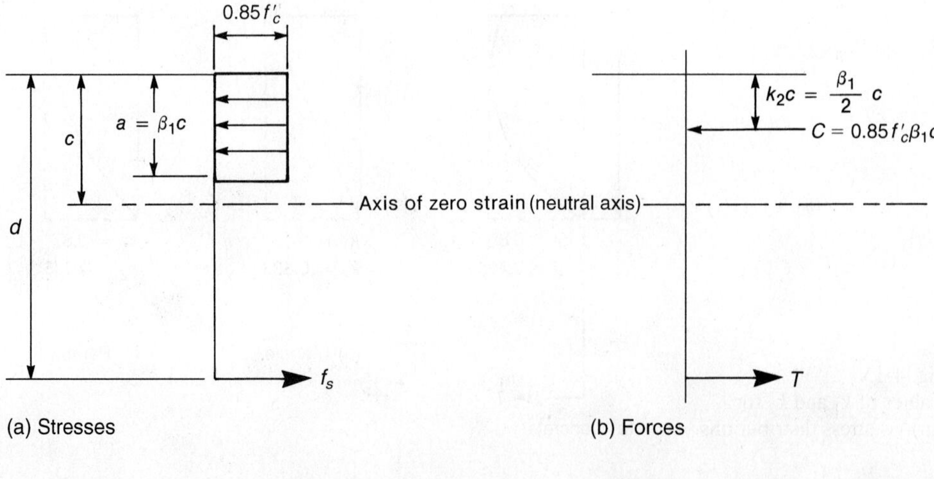

Fig. 4-16
Equivalent rectangular stress block.

In metric units (MPa), the factor β_1 shall be taken as follows:

(a) For concrete strengths, f'_c, up to and including 28 MPa,

$$\beta_1 = 0.85 \tag{4-14Ma}$$

(b) For 28 MPa $< f'_c \leq$ 56 MPa,

$$\beta_1 = 0.85 - 0.05\frac{f'_c - 28\text{ MPa}}{7\text{ MPa}} \tag{4-14Mb}$$

(c) For f'_c greater than 58 MPa,

$$\beta_1 = 0.65 \tag{4-14Mc}$$

The dashed line in Fig. 4-17 is a lower-bound line corresponding to a rectangular stress block with a height of $0.85 f'_c$ and by using β_1 as given by Eq. (4-14). This equivalent rectangular stress block has been shown [4-6], [4-7] to give very good agreement with test data for calculation of the nominal flexural strength of beams. For columns, the agreement is good up to a concrete strength of about 6000 psi. For columns loaded with small eccentricities and having strengths greater than 6000 psi, the moment capacity tends to be overestimated by the ACI Code stress block. This is because Eq. (4-14) for β_1 was chosen as a lower bound on the test data, as indicated by the dashed line in Fig. 4-17. The internal moment arm of the compression force in the concrete about the *centroidal axis* of a rectangular column is $(h/2 - \beta_1 c/2)$, where c is the depth to the neutral axis (axis of zero strain). If β_1 is too small, the moment arm will be too large, and the moment capacity will be overestimated.

To correct this potential error, which can lead to unconservative designs of columns constructed with high-strength concrete, an ACI Task Group [4-13] has recommended the

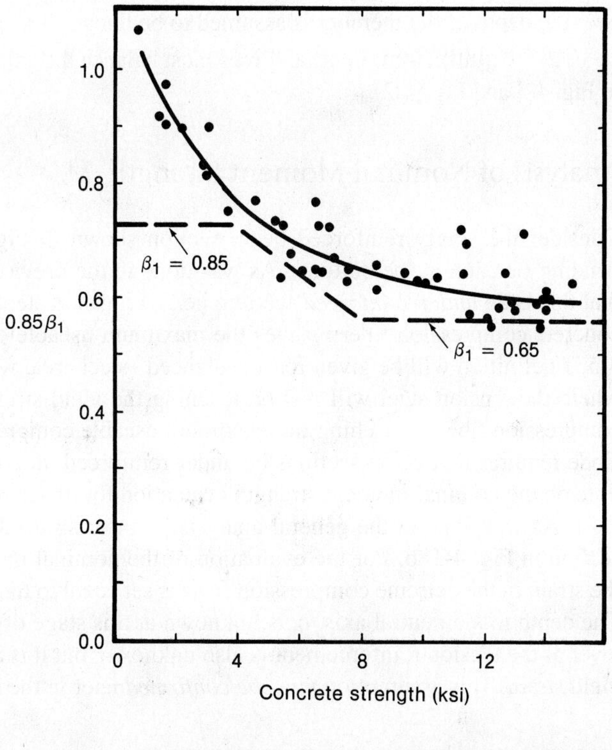

Fig. 4-17
Values of β_1 from tests of concrete prisms. (From [4-7].)

use of a coefficient, α_1, to replace the constant 0.85 as the definition for the height of the stress block shown in Fig. 4-16a. This new coefficient is defined as follows:

(a) For concrete strengths, f'_c, up to and including 8000 psi,

$$\alpha_1 = 0.85 \tag{4-15a}$$

(b) For 8000 psi $< f'_c \leq$ between 8000 and 18,000 psi,

$$\alpha_1 = 0.97 - 0.015\frac{f'_c}{1000} \tag{4-15b}$$

(c) For f'_c greater than 18,000 psi,

$$\alpha_1 = 0.70 \tag{4-15c}$$

Until the ACI Code adopts a modification of the stress block shown in Fig. 4-16a, the authors recommend the use of this coefficient, α_1, when analyzing the flexural strength of columns constructed with concrete strengths exceeding 8000 psi.

4-4 ANALYSIS OF NOMINAL MOMENT STRENGTH FOR SINGLY REINFORCED BEAM SECTIONS

Stress and Strain Compatibility and Section Equilibrium

Two requirements are satisfied throughout the flexural analysis and design of reinforced concrete beams and columns:

1. Stress and strain compatibility. The stress at any point in a member must correspond to the strain at that point. Except for short, deep beams, the distribution of strains over the depth of the member is assumed to be linear.

2. Equilibrium. Internal forces must balance the external load effects, as illustrated in Fig. 4-3 and Eq. (4-2).

Analysis of Nominal Moment Strength, M_n

Consider the singly reinforced beam section shown in Fig. 4-18a subjected to positive bending (tension at the bottom). As was done in the previous section, it will be assumed that this is an *under-reinforced* section, i.e., the tension steel will yield before the extreme concrete compression fiber reaches the maximum useable compression strain. In Section 4-5, a definition will be given for a "balanced" steel area, which results in a beam section where the tension steel will just be reaching the yield strain when the extreme concrete compression fiber is reaching the maximum useable compression strain. Because the ACI Code requires that beam sections be under-reinforced, this initial discussion will concentrate on the nominal moment strength evaluation for under-reinforced sections.

As was done in the general analysis, a linear strain distribution is assumed for the section in Fig. 4-18b. For the evaluation of the nominal moment capacity of the section, the strain in the extreme compression fiber is set equal to the maximum useable strain, ε_{cu}. The depth to the neutral axis, c, is unknown at this stage of the analysis. The strain at the level of the tension reinforcement is also unknown, but it is assumed to be greater than the yield strain. This assumption *must be confirmed* later in the calculation.

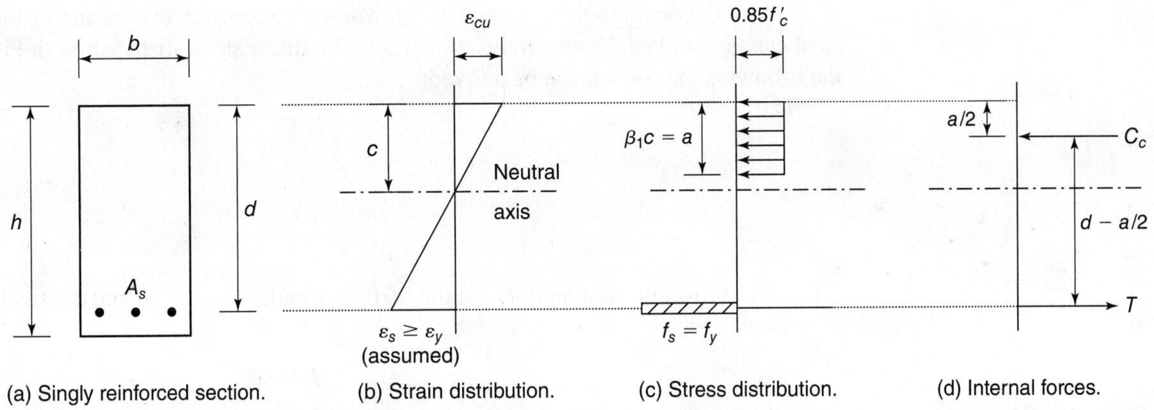

Fig. 4-18
Steps in analysis of M_n for singly reinforced rectangular sections.

The assumed stress distribution is given in Fig. 4-18c. Above the neutral axis, the stress-block model from Fig. 4-16 is used to replace the actual concrete stress distribution. The coefficient β_1 is multiplied by the depth to the neutral axis, c, to get the depth of the stress block, a. The concrete is assumed to carry no tension, so there is no concrete stress distribution below the neutral axis. At the level of the steel, the stress, f_s, is assumed to be equal to the steel yield stress, f_y. This corresponds to the assumptions that the steel strain exceeds the yield strain and that the steel stress remains constant after yielding occurs (Fig. 4-7).

The final step is to go from the stress distributions to the equivalent section forces shown in Fig. 4-18d. The concrete compression force, C_c, is equal to the volume under the stress block. For the rectangular section used here,

$$C_c = 0.85 f'_c b\beta_1 c = 0.85 f'_c ba \tag{4-13b}$$

The compression force in the concrete cannot be evaluated at this stage, because the depth to the neutral axis is still unknown. The tension force shown in Fig. 4-18d is equal to the tension steel area, A_s, multiplied by the yield stress, f_y. Based on the assumption that the steel has yielded, this force is known.

A key step in section analysis is to enforce section equilibrium. For this section, which is assumed to be subject to only bending (no axial force), the sum of the compression forces must be equal to the sum of the tension forces. So,

$$C_c = T \tag{4-2}$$

or

$$0.85 f'_c b\beta_1 c = 0.85 f'_c ba = A_s f_y$$

The only unknown in this equilibrium equation is the depth of the stress block. So, solving for the unknown value of a,

$$a = \beta_1 c = \frac{A_s f_y}{0.85 f'_c b} \tag{4-16}$$

and

$$c = \frac{a}{\beta_1} \tag{4-17}$$

With the depth to the neutral axis known, the assumption of yielding of the tension steel can be checked. From similar triangles in the linear strain distribution in Fig. 4-18b, the following expression can be derived:

$$\frac{\varepsilon_s}{d - c} = \frac{\varepsilon_{cu}}{c}$$

$$\varepsilon_s = \left(\frac{d - c}{c}\right)\varepsilon_{cu} \tag{4-18}$$

To confirm the assumption that the section is under-reinforced and the steel is yielding, show

$$\varepsilon_s \geq \varepsilon_y = \frac{f_y}{E_s} = \frac{f_y \text{ (ksi)}}{29{,}000 \text{ ksi}} \tag{4-19}$$

Once this assumption is confirmed, the nominal moment strength can be calculated by referring back to the section forces in Fig. 4-18d. The compression force is acting at the middepth of the stress block, and the tension force is acting at a distance d from the extreme compression fiber. Thus, the nominal moment strength can be expressed as either the tension force or the compression force multiplied by the moment arm, $d - a/2$:

$$M_n = T\left(d - \frac{a}{2}\right) = C_c\left(d - \frac{a}{2}\right) \tag{4-20}$$

For singly reinforced sections, it is more common to express the nominal moment strength using the definition of the tension force as

$$M_n = A_s f_y\left(d - \frac{a}{2}\right) \tag{4-21}$$

This simple expression can be used for all singly reinforced sections with a rectangular (constant width) compression zone after it has been confirmed that the tension steel is yielding. The same fundamental process as used here to determine M_n for singly reinforced rectangular sections will be applied to other types of beam sections in the following parts of this chapter. However, the reader is urged to concentrate on the process rather than the resulting equations. If the process is understood, it can be applied to any beam section that may be encountered.

EXAMPLE 4-1 Calculation of M_n for a Singly Reinforced Rectangular Section

For the beam shown in Fig. 4-19a, calculate M_n and confirm that the area of tension steel exceeds the required minimum steel area given by Eq. (4-11). The beam section is made of concrete with a compressive strength, $f'_c = 4000$ psi, and has four No. 8 bars with a yield strength of $f_y = 60$ ksi.

For this beam with a single layer of tension reinforcement, it is reasonable to assume that the effective flexural depth, d, is approximately equal to the total beam depth minus 2.5 in. This accounts for a typical concrete clear cover of 1.5 in., the diameter of the stirrup (typically a No. 3 or No. 4 bar) and half the diameter of the beam longitudinal reinforcement. Depending on the sizes of the stirrup and longitudinal bar, the dimension to the center of the steel layer will vary slightly, but the use of 2.5 in. will be accurate enough for most design work unless adjustments in reinforcement location are required to avoid rebar interference at connections with other members. Small bars are often used in the compression

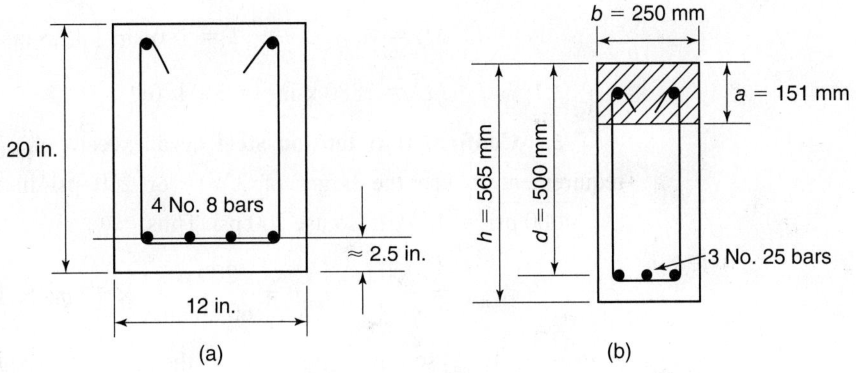

Fig. 4-19
Beam sections for
(a) Example 4-1 and
(b) Example 4-1M.

zone to hold the stirrups in position, but these bars normally are ignored unless they were specifically designed to serve as compression-zone reinforcement.

1. Following the procedure summarized in Fig. 4-18, assume that the steel strain exceeds the yield strain, and thus, the stress f_s in the tension reinforcement equals the yield strength, f_y. Compute the steel tension force:

$$A_s = 4\text{ No. 8 bars} = 4 \times 0.79\text{ in.}^2 = 3.16\text{ in.}^2$$

$$T = A_s f_y = 3.16\text{ in.}^2 \times 60\text{ ksi} = 190\text{ kips}$$

The assumption that $\varepsilon_s > \varepsilon_y$ will be checked in step 3. This assumption generally should be true, because the ACI Code requires that the steel area be small enough in beam sections such that the steel will yield before the concrete reaches the maximum useable compression strain.

2. Compute the area of the compression stress block so that $C_c = T$. This is done for the equivalent rectangular stress block shown in Fig. 4-16a. The stress block consists of a uniform stress of $0.85\,f'_c$ distributed over a depth $a = \beta_1 c$ which is measured from the extreme compression fiber. For $f'_c = 4000$ psi, Eq. (4-14a) gives $\beta_1 = 0.85$. Using Eq. (4-16), which was developed from section equilibrium,

$$a = \beta_1 c = \frac{A_s f_y}{0.85\,f'_c\,b} = \frac{190\text{ kips}}{0.85 \times 4\text{ ksi} \times 12\text{ in.}} = 4.66\text{ in.}$$

3. Check that the tension steel is yielding. The yield strain is

$$\varepsilon_y = \frac{f_y}{E_s} = \frac{60\text{ ksi}}{29{,}000\text{ ksi}} = 0.00207$$

From above, $c = a/\beta_1 = 5.48$ in. Now, use strain compatibility, as expressed in Eq. (4-18), to find

$$\varepsilon_s = \left(\frac{d - c}{c}\right)\varepsilon_{cu} = \left(\frac{17.5 - 5.48}{5.48}\right)0.003 = 0.00658$$

Clearly, ε_s exceeds ε_y, so the assumption used above to establish section equilibrium is confirmed. Remember that you *must make this check* before proceeding to calculate the section nominal moment strength.

4. Compute M_n. Using Eq. (4-21), which was derived for sections with constant width compression zones,

$$M_n = A_s f_y\left(d - \frac{a}{2}\right) = 190 \text{ kips}\left(17.5 \text{ in.} - \frac{4.66 \text{ in.}}{2}\right)$$

$$M_n = 2880 \text{ k-in.} = 240 \text{ k-ft}$$

5. Confirm that tension steel area exceeds $A_{s,\min}$. For Eq. (4-11), there is a requirement to use the larger of $3\sqrt{f'_c}$ or 200 psi in the numerator. In this case, $3\sqrt{4000 \text{ psi}} = 190$ psi, so use 200 psi. Thus,

$$A_{s,\min} = \frac{200 \text{ psi}}{f_y} b_w d = \frac{200 \text{ psi}}{60{,}000 \text{ psi}} \times 12 \text{ in.} \times 17.5 \text{ in.} = 0.70 \text{ in.}^2$$

A_s exceeds $A_{s,\min}$, so this section satisfies the ACI Code requirement for minimum tension reinforcement. ■

EXAMPLE 4-1M Analysis of Singly Reinforced Beams: Tension Steel Yielding—SI Units

Compute the nominal moment strength, M_n, of a beam (Fig. 4-19b) with $f'_c =$ 20 MPa ($\beta_1 = 0.85$), $f_y = 420$ MPa, $b = 250$ mm, $d = 500$ mm, and three No. 25 bars (Table A-1M) giving $A_s = 3 \times 510 = 1530 \text{ mm}^2$. Note that the difference between the total section depth, h, and the effect depth, d, is 65 mm, which is a typical value for beam sections designed with metric dimensions.

1. Compute a (assuming the tension steel is yielding).

$$a = \frac{A_s f_y}{0.85 f'_c b}$$

$$= \frac{1530 \text{ mm}^2 \times 420 \text{ MPa}}{0.85 \times 20 \text{ MPa} \times 250 \text{ mm}} = 151 \text{ mm}$$

Therefore, $c = a/\beta_1 = 151/0.85 = 178$ mm.

2. Check whether the tension steel is yielding. The yield strain for the reinforcing steel is

$$\varepsilon_y = \frac{f_y}{E_s} = \frac{420 \text{ MPa}}{200{,}000 \text{ MPa}} = 0.0021$$

From Eq. (4-18),

$$\varepsilon_s = \left(\frac{500 \text{ mm} - 178 \text{ mm}}{178 \text{ mm}}\right) \times 0.003 = 0.00543$$

Thus, the steel is yielding as assumed in step 1.

3. Compute the nominal moment strength, M_n. From Eq. (4-21), M_n is (where $1 \text{ MPa} = 1 \text{ N/mm}^2$)

$$M_n = A_s f_y\left(d - \frac{a}{2}\right) = 1530 \text{ mm}^2 \times 420 \text{ N/mm}^2\left(500 - \frac{151}{2}\right) \text{mm}$$

$$= 273 \times 10^6 \text{ N-mm} = 273 \text{ kN-m}$$

4. Confirm that the tension steel area exceeds $A_{s,\min}$. For the given concrete strength of 20 MPa, the quantity $0.25\sqrt{f'_c} = 1.12$ MPa, which is less than 1.4 MPa. Therefore, the second part of Eq. (4-11M) governs for $A_{s,\min}$ as

$$A_{s,\min} = \frac{1.4 b_w d}{f_y} = \frac{1.4 \text{ MPa} \times 250 \text{ mm} \times 500 \text{ mm}}{420 \text{ MPa}} = 417 \text{ mm}^2$$

A_s exceeds $A_{s,\min}$, so this section satisfies the ACI Code requirement for minimum tension reinforcement. ■

EXAMPLE 4-2 Calculation of the Nominal Moment Strength for an Irregular Cross Section

The beam shown in Fig. 4-20 is made of concrete with a compressive strength, $f'_c = 3000$ psi and has three No. 8 bars with a yield strength, $f_y = 60$ ksi. This example is presented to demonstrate the general use of strain compatibility and section equilibrium equations for any type of beam section.

1. Initially, assume that the stress f_s in the tension reinforcement equals the yield strength f_y, and compute the tension force $T = A_s f_y$:

$$A_s = 3 \text{ No. 8 bars} = 3 \times 0.79 \text{ in.}^2 = 2.37 \text{ in.}^2$$

$$T = A_s f_y = 2.37 \text{ in.}^2 \times 60 \text{ ksi} = 142 \text{ kips}$$

The assumption that the tension steel is yielding will be checked in step 3.

2. Compute the area of the compression stress block so that $C_c = T$. As in the prior problem, this is done using the equivalent rectangular stress block shown in Fig. 4-16a. The stress block consists of a uniform stress of $0.85\, f'_c$ distributed over a depth $a = \beta_1 c$, which is measured from the extreme compression fiber. For $f'_c = 3000$ psi, Eq. (4-14a) gives $\beta_1 = 0.85$. The magnitude of the compression force is obtained from equilibrium as

$$C_c = T = 142 \text{ kips} = 142{,}000 \text{ lbs}$$

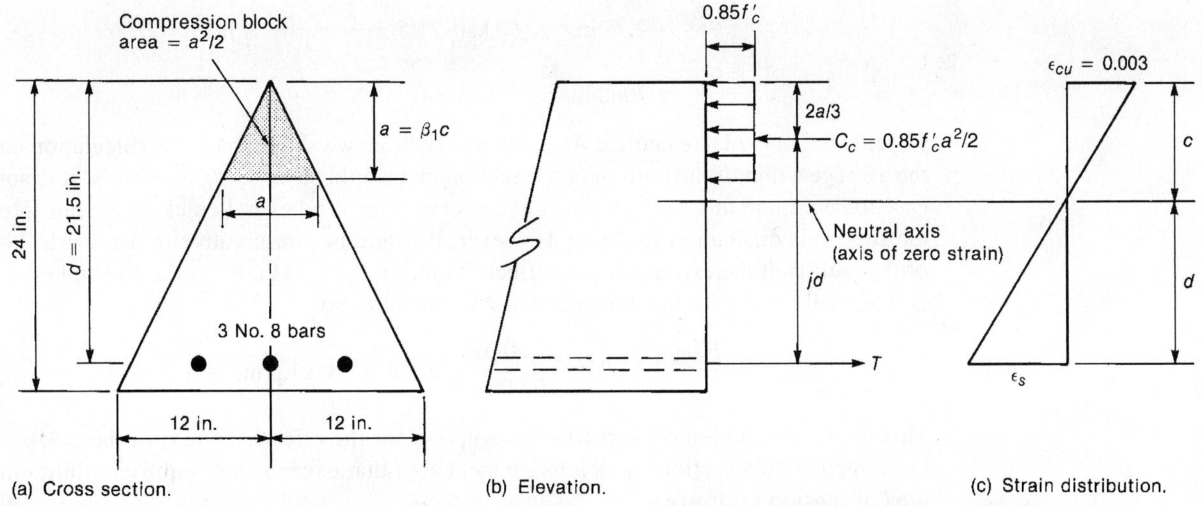

Fig. 4-20
Analysis of arbitrary cross section—Example 4-2.

By the geometry of this particular triangular beam, shown in Fig. 4-20, if the depth of the compression block is a, the width at the bottom of the compression block is also a, and the area is $a^2/2$. This is, of course, true only for a beam of this particular triangular shape. Therefore, $C_c = (0.85 f'_c)(a^2/2)$ and

$$a = \sqrt{\frac{142{,}000 \text{ 1b} \times 2}{0.85 \times 3000 \text{ psi}}} = 10.6 \text{ in.}$$

3. Check whether $f_s = f_y$. This is done by using strain compatibility. The strain distribution at ultimate is shown in Fig. 4-20c. As before,

$$c = \frac{a}{\beta_1} = \frac{10.6 \text{ in.}}{0.85} = 12.4 \text{ in.}$$

Using strain compatibility, as expressed in Eq. (4-18), calculate

$$\varepsilon_s = \left(\frac{21.5 - 12.4}{12.4}\right)0.003 = 0.00220$$

Although this is close to the yield strain, it does exceed the yield strain as calculated in step 3 of Example 4-1 for Grade-60 reinforcement. Thus, the assumption made in step 1 is satisfied.

4. Compute M_n.

$$M_n = C_c jd = Tjd$$

where jd is a general expression for the lever arm, i.e., the distance from the resultant tensile force (at the centroid of the reinforcement) to the resultant compressive force C_c. Because the area on which the compression stress block acts is triangular in this example, C_c acts at $2a/3$ from top of the beam. Therefore,

$$jd = d - \frac{2a}{3}$$

$$M_n = A_s f_y\left(d - \frac{2a}{3}\right)$$

$$= 2.37 \text{ in.}^2 \times 60 \text{ ksi}\left(21.5 - \frac{2 \times 10.6}{3}\right) \text{ in.}$$

$$= 2060 \text{ lb-in.} = 171 \text{ k-ft}$$

Note: If we wanted to calculate $A_{s,\min}$ for this section, we should base the calculation on the average width of the portion of the section that would be cracking in tension. It is not easy to determine this value, because the distance that the flexural crack penetrates into the section is difficult to evaluate. However, it would be conservative to use the width of the section at the extreme tension fiber, 24 in., in Eq. (4-11). As in the Example 4-1, 200 psi will govern for the numerator in this equation. So,

$$A_{s,\min} = \frac{200 \text{ psi}}{f_y} b_w d = \frac{200 \text{ psi}}{60{,}000 \text{ psi}} \times 24 \text{ in.} \times 21.5 \text{ in.} = 1.72 \text{ in.}^2$$

Therefore, even with a conservative assumption for the effective width of the cracked tension zone, this section has a tension steel area that exceeds the required minimum area of tension reinforcement. ■

4-5 DEFINITION OF BALANCED CONDITIONS

The prior sections dealt with under-reinforced beam sections. To confirm that a particular section was under-reinforced, the section was put into equilibrium, and the steel strain evaluated using Eq. (4-18) was shown to be greater than the yield strain. This discussion will concentrate on the condition when the steel strain corresponding to section equilibrium is equal to the yield strain, ε_y, and the strain in the extreme concrete fiber is equal to the maximum useable compression strain, ε_{cu}. The area of tension steel required to cause this strain condition in a beam section will be defined as the *balanced area* of tension reinforcement. The balanced area is an important parameter for design of beam and slab sections and will be referred to in later chapters of this book. The analysis procedure to find the balanced area of tension reinforcement is similar to the analysis for M_n.

The key starting point for the analysis of the balanced area of tension reinforcement is the *balanced strain diagram*, as shown in Fig. 4-21b. This strain diagram corresponds to a *balanced failure*, i.e., the tension reinforcement is just reaching its yield strain, ε_y, at the same time the extreme concrete compression fiber is reaching the maximum useable compression strain, ε_{cu}. Understanding and using the balanced strain diagram is important for the analysis of both beam sections subjected to only bending and column sections subjected to bending plus axial load (Chapter 11).

The balanced strain diagram is being applied to the singly reinforced beam section shown in Fig. 4-21a. There are some important differences between this analysis and the analysis for M_n discussed in the previous sections. First, the major unknown now is the balanced area of steel, A_s(bal). Second, everything is known in the strain diagram, including the depth to the neutral axis, c(bal). This is calculated with the use of similar triangles from the strain diagram:

$$\frac{c(\text{bal})}{\varepsilon_{cu}} = \frac{d}{\varepsilon_{cu} + \varepsilon_y}$$

$$c(\text{bal}) = \left(\frac{\varepsilon_{cu}}{\varepsilon_{cu} + \varepsilon_y}\right) d \tag{4-22}$$

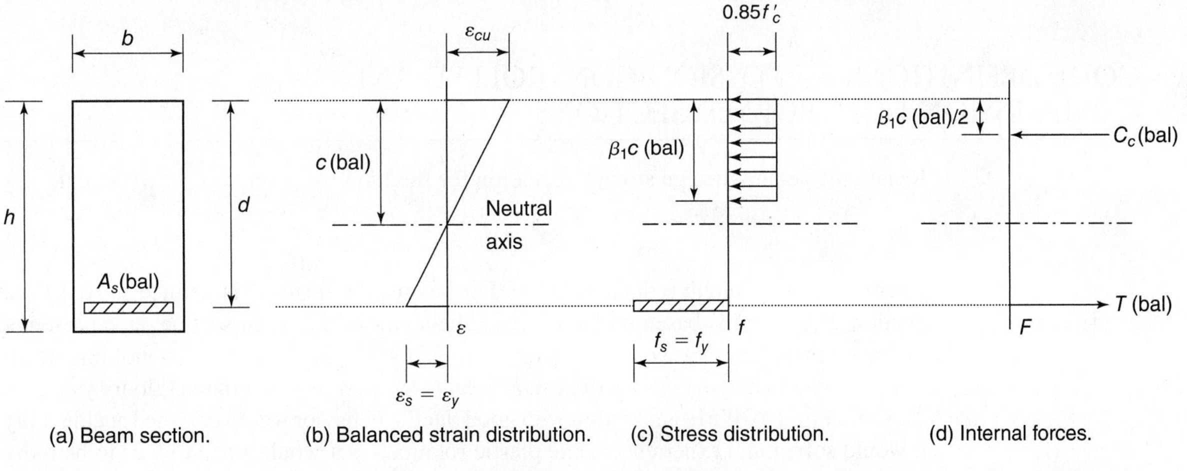

Fig. 4-21
Steps in analysis of A_s(bal), singly reinforced rectangular section.

The next steps through the stress distribution (Fig. 4-21c) and the force diagram (Fig. 4-21d) are similar to what was done for the analysis of M_n in the previous sections. The only difference is that the forces have been labeled as C_c(bal) and T(bal) to distinguish them from the forces in the procedure for the analysis of M_n.

Enforcing section equilibrium,

$$T(\text{bal}) = C_c(\text{bal})$$

$$A_s(\text{bal})f_y = 0.85 f'_c\, b\beta_1 c\,(\text{bal})$$

Solving for the only unknown A_s(bal),

$$A_s(\text{bal}) = \frac{1}{f_y}[0.85 f'_c\, b\beta_1 c(\text{bal})] \tag{4-23}$$

This general expression applies *only* to singly reinforced rectangular sections and will not be used frequently. However, the reinforcement ratio at balanced conditions, ρ_b, is a parameter that often is used in design. Recalling that the reinforcement ratio is the tension steel area divided by the effective area of concrete, bd, and using the definition of c(bal) from Eq. (4-22), we get

$$\rho_b = \frac{A_s(\text{bal})}{bd} = \frac{0.85\,\beta_1 f'_c}{f_y} \times \frac{b}{bd} \times \left(\frac{\varepsilon_{cu}}{\varepsilon_{cu} + \varepsilon_y}\right)d$$

$$\rho_b = \frac{0.85\,\beta_1 f'_c}{f_y}\left(\frac{\varepsilon_{cu}}{\varepsilon_{cu} + \varepsilon_y}\right) \tag{4-24}$$

Although this form is acceptable, the more common form is obtained by substituting in ε_{cu} equal to 0.003 and then multiplying both the numerator and denominator by E_s = 29,000,000 psi to obtain

$$\rho_b = \frac{0.85\,\beta_1 f'_c}{f_y}\left(\frac{87{,}000}{87{,}000 + f_y}\right) \tag{4-25}$$

where f_y and f'_c are used in psi units. Equations (4-24) and (4-25) represent classic definitions for the *balanced reinforcement ratio*. Some references to this reinforcement ratio will be made in later chapters of this book.

4-6 CODE DEFINITIONS OF TENSION-CONTROLLED AND COMPRESSION-CONTROLLED SECTIONS

Recall, the general design strength equation for flexure is

$$\phi M_n \geq M_u \tag{4-1b}$$

where ϕ is the strength reduction factor. For beams, the factor ϕ is defined in ACI Code Section 21.2.2 and is based on the expected behavior of the beam section, as represented by the moment–curvature curves in Figs. 4-11 and 4-12. Because of the monolithic nature of reinforced concrete construction, most beams are part of a continuous floor system, as shown in Fig. 1-6. If a beam section with good ductile behavior was overloaded accidentally, it would soften and experience some plastic rotations that would permit loads to be redistributed to other portions of the continuous floor system. This type of behavior essentially creates an increased level of safety in the structural system, so a higher ϕ-value is permitted for beams designed to exhibit ductile behavior. For beams that exhibit less ductile behavior, as indicated for the sections with larger tension steel areas in Fig. 4-11, the ability to redistribute loads away from an overloaded section is reduced. Thus, to maintain an acceptable level of safety in design, a lower ϕ-value is required for such sections.

Until 2002, the ACI Code defined only a single ϕ-value for the design of reinforced concrete beam sections, but the behavior was controlled by limiting the permitted area of tension reinforcement. The design procedure was to keep the reinforcement ratio, ρ, less than or equal to 0.75 times the *balanced reinforcement ratio* defined in Eq. (4-25). This procedure is easy to apply to singly reinforced rectangular sections, but becomes more complicated for flanged sections and sections that use compression reinforcement. When the same criteria is applied to beam sections that contain both normal reinforcement and prestressing tendons, the definition for the permitted area of tension reinforcement becomes quite complex.

Another method for controlling the ductility of a section is to control the value of tension strain reached at the level of the tension reinforcement when the extreme concrete compression fiber reaches the maximum useable compression strain, i.e., at *nominal strength conditions* (Fig. 4-18b). Requiring higher tension strains at the level of tension steel is a universal method for controlling the ductility of all sections, as initially discussed by Robert Mast [4-14]. Starting with the 2002 edition, this is the procedure used in the ACI Code to control section ductility, and thus, specify the corresponding values for the strength-reduction factor, ϕ.

Definitions of Effective Depth and Distance to Extreme Layer of Tension Reinforcement

The *effective depth*, d, is measured from the extreme compressive fiber to the *centroid* of the longitudinal reinforcement. This is the distance used in calculations of the nominal moment strength, as demonstrated in prior examples. To have consistency in controlling tension strains for a variety of beam and column sections, the ACI Code defines a distance, d_t, which is measured from the extreme compression fiber to the *extreme layer of tension reinforcement*, as shown in Fig. 4-22a. The strain at this level of reinforcement, ε_t, is defined as the net strain at the extreme layer of tension reinforcement at nominal-strength conditions, excluding strains due to effective prestress, creep, shrinkage, and temperature. For beam sections with more than one layer of reinforcement, ε_t will be slightly larger than the strain at the centroid of the tension reinforcement, ε_s, as shown in Fig. 4-22b. The ACI Code uses the strain ε_t to define the behavior of the section at nominal conditions, and thus, to define the value of ϕ.

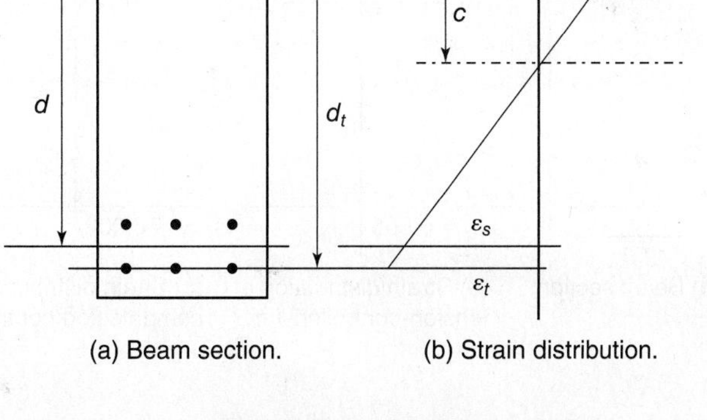

Fig. 4-22
Definitions for d_t and ε_t.

Definitions of Tension-Controlled and Compression-Controlled Sections

A *tension-controlled section* has a tension-reinforcement area such that when the beam reaches its nominal flexural strength, the net tensile strain in the extreme layer of tensile steel, ε_t, is greater than or equal to 0.005. For Grade-60 reinforcement with a yield strength $f_y = 60$ ksi, the tensile yield strain is $\varepsilon_y = 60/29{,}000 = 0.00207$. The tension-controlled limit strain of 0.005 was chosen to be approximately 2.5 times the yield strain of the reinforcement, giving a moment–curvature diagram similar to that shown in Fig. 4-11 for the section with an area of tension reinforcement equal to 4.50 in.[2] The strain diagram corresponding to the tension-controlled limit (TCL) is demonstrated in Fig. 4-23b, with the depth from the extreme compression fiber to the neutral axis defined as c(TCL). From the strain diagram it can be shown that

$$c(\text{TCL}) = \frac{3}{8}d_t = 0.375d_t \tag{4-26}$$

Clearly, if a calculated value of c is *less* that 3/8 d_t, the strain, ε_t, will *exceed* 0.005. Thus, when analyzing the nominal flexural strength of a beam section, demonstrating that the depth to the neutral axis obtained from section equilibrium is less than 3/8 d_t, as given in Eq. (4-26), will be one method to verify that the section is tension-controlled.

A *compression-controlled section* has a tension-reinforcement area such that when the beam reaches its nominal flexural strength, the net tensile strain in the extreme layer of tensile steel, ε_t, is less than or equal to the yield strain. For beams with Grade-60 reinforcement ($\varepsilon_y = 0.00207$) and beams with prestressed reinforcement, ACI Code Section 21.2.2.1 permits the use of 0.002 in place of the yield strain. A beam section with this amount of tension reinforcement would exhibit a moment–curvature relationship similar to that shown in Fig. 4-11 for the section with the largest steel area. The strain diagram corresponding to the compression-controlled limit (CCL) is demonstrated in Fig. 4-23c, with the depth from the extreme compression fiber to the neutral axis defined as c(CCL). From the strain diagram it can be shown that

$$c(\text{CCL}) = \frac{3}{5}d_t = 0.60d_t \tag{4-27}$$

Clearly, if a calculated value of c is *greater* that 3/5 d_t, the strain ε_t will be *less* than 0.002.

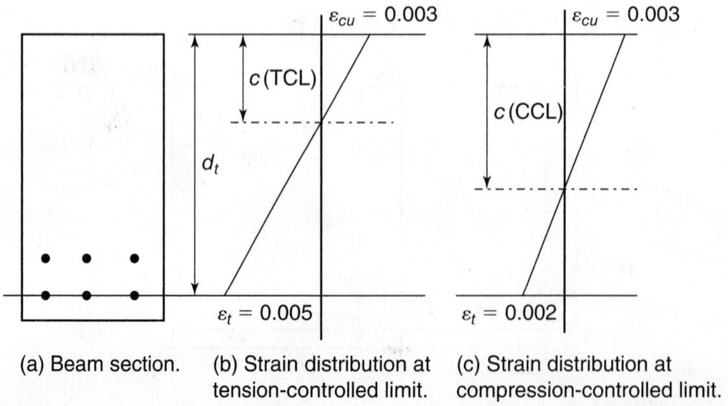

Fig. 4-23
Strain distributions at tension-controlled and compression-controlled limits.

A *transition-zone section* has a tension-reinforcement area such that when the beam reaches its nominal flexural strength, the net tensile strain in the extreme layer of tensile steel, ε_t, is between 0.002 and 0.005. A beam section with this amount of tension reinforcement would exhibit a moment–curvature relationship in between those shown in Fig. 4-11 for sections with tension steel areas of 4.50 and 6.50 in.^{2}

Because tension-controlled sections demonstrate good ductile behavior if overloaded, they are analyzed and designed using a strength-reduction factor, ϕ, of 0.9. Because of their brittle behavior if overloaded, compression-controlled sections are analyzed and designed with ϕ equal to 0.65. (*Note:* This is the value for beams with standard stirrup-tie reinforcement similar to that shown in Fig. 4-19. As will be discussed in Chapter 11, for column sections with *spiral reinforcement*, the value of ϕ is 0.75 if the section is compression-controlled.)

The variation of the strength-reduction factor, ϕ, as a function of either the strain, ε_t, or the ratio, c/d_t, at nominal strength conditions is shown in Fig. 4-24. For beam or column sections that are either compression-controlled ($\varepsilon_t \leq 0.002$) or tension-controlled ($\varepsilon_t \geq 0.005$), the value of ϕ is constant. When analyzing a transition-zone section with stirrup-tie (or hoop) transverse reinforcement, the value of ϕ varies linearly between 0.65 and 0.90 as a function of either ε_t or c/d_t, as given in Eqs. (4-28a) and (4-28b), respectively.

$$\phi = 0.65 + (\varepsilon_t - 0.002) \times \frac{250}{3} \tag{4-28a}$$

$$\phi = 0.65 + 0.25\left(\frac{1}{c/d_t} - \frac{5}{3}\right) \tag{4-28b}$$

For a transition-zone section with spiral transverse reinforcement (column section), the variation of the value of ϕ as a function of either ε_t or c/d_t is given in Eqs. (4-29a) and (4-29b), respectively.

$$\phi = 0.75 + (\varepsilon_t - 0.002) \times 50 \tag{4-29a}$$

$$\phi = 0.75 + 0.15\left(\frac{1}{c/d_t} - \frac{5}{3}\right) \tag{4-29b}$$

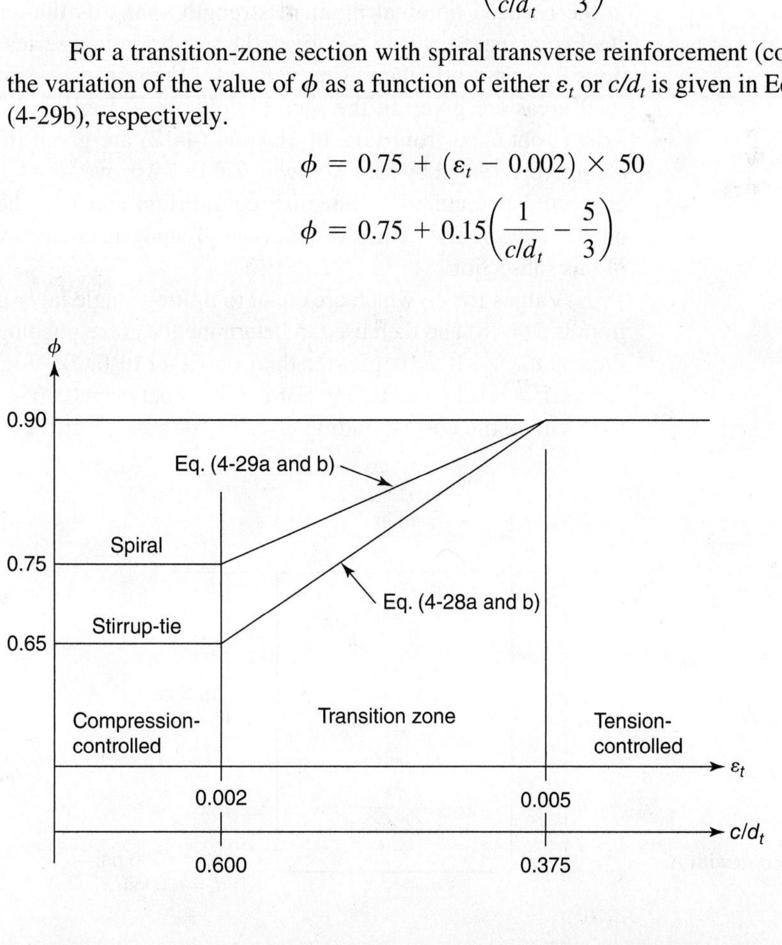

Fig. 4-24
Variation of ϕ-factor with ε_t and c/d_t for spiral and stirrup-tie transverse reinforcement.

In the prior examples, the values of ϕ now can be calculated. In all three examples, there was only one layer of tension steel, so ε_t is equal to ε_s. For the rectangular beams in Examples 4-1 and 4-1M, the value of ε_s exceeded 0.005, so the ϕ-value would be 0.9. For the triangular beam section in Example 4-2, the value of ε_s was 0.00220. Using that as the value for ε_t in Eq. (4-28a) results in a ϕ-value of 0.67 (the author recommends using only two significant figures for ϕ).

Upper Limit on Beam Reinforcement

Prior to the 2002 edition of the ACI Code, the maximum-tension steel area in beams was limited to 0.75 times the steel area corresponding to balanced conditions (Fig. 4-21). In the ACI Code (ACI 318-14), Section 9.3.3.1 requires that for reinforced concrete (non-prestressed) beam sections (stated as members with axial compressive load less than $0.10 f'_c A_g$) the value of ε_t at nominal flexural strength conditions shall be greater than or equal to 0.004. This strain value was selected to approximately correspond to the former ACI Code requirement of limiting the tension steel area to 0.75 times the balanced-tension steel area. A beam section with a tension steel area resulting in $\varepsilon_t = 0.004$ at nominal conditions would have a higher M_n value than a beam section with a lower-tension steel area that resulted in $\varepsilon_t = 0.005$ (the tension-controlled limit) at nominal strength conditions. However, because there are different ϕ-values for these two beam sections, the resulting values of ϕM_n for the two sections will be approximately equal.

The rectangular beam section in Fig. 4-25 will be used to demonstrate the change in the reduced nominal moment strength, ϕM_n, as the amount of tension-reinforcing steel is increased. Table 4-2 gives the results from a series of moment strength calculations for constantly increasing values for the reinforcement ratio, ρ. The corresponding steel areas are given in the second column of Table 4-2, and the depth to the neutral axis, c, obtained from Eqs. (4-16) and (4-17) are given in the third column. The beam section represented by the last row in Table 4-2 is over-reinforced and a strain-compatibility procedure is required to establish equilibrium and find the corresponding depth to the neutral axis, c. The details of that type of analysis procedure will be discussed at the end of this subsection.

Values for ε_t, which are equal to ε_s for a single layer of reinforcement, are obtained from Eq. (4-18) and then used to determine the corresponding values of the strength reduction factor, ϕ. If ε_t is greater than or equal to 0.005 (signifying a tension-controlled section), ϕ is set equal to 0.9. For ε_t values between 0.005 and 0.002, Eq. (4-28a) is used to calculate the corresponding ϕ-value to three significant figures for this comparsion.

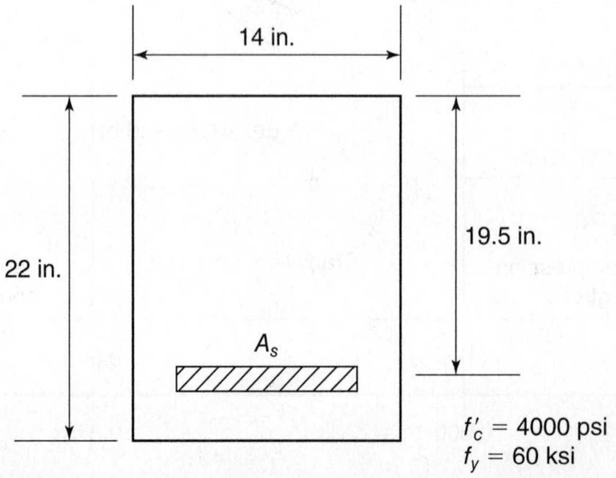

Fig. 4-25
Typical beam section with A_s as a variable.

TABLE 4-2 Relationship between Reinforcement Ratio and Nominal Moment Strength

ρ	A_s (in.^{2})	c (in.)	ε_t	ϕ	M_n (k-ft)	ϕM_n (k-ft)
0.005	1.37	2.03	0.0258	0.900	128	115
0.010	2.73	4.05	0.0115	0.900	243	218
0.015	4.10	6.08	0.00662	0.900	347	312
0.0181	4.93	7.31	0.0050	0.900	404	364
0.0207	5.64	8.36	0.0040	0.817	449	367
0.025	6.83	10.1	0.00278	0.715	519	371
0.0285	7.78	11.5	0.00207	0.656	568	372
0.030	8.19	11.7	0.00200	0.650	575	374

For the largest ρ-value in Table 4-2 (last row), the calculated value of ε_t is equal to the compression-controlled limit of 0.002, so ϕ was set equal to 0.65. Finally, Eq. (4-20) was used to calculate the nominal moment strength, M_n, which was multiplied by ϕ to get the values of ϕM_n given in the last column of the Table 4-2.

Some interesting results can be observed in the plots of ρ versus M_n and versus ϕM_n in Fig. 4-26. There is an almost linear increase in M_n and ϕM_n for increasing values of ρ up to the point where the tension strain, ε_t, reaches the tension-controlled limit of 0.005. Beyond this point, M_n continues to increase almost linearly for increasing values of ρ, but the value of ϕM_n tends to stay almost constant due the decrease in the value of ϕ obtained from Eq. (4-28a). This is a very important result that diminishes the significance of the limit set on ε_t in ACI Code Section 9.3.3.1 ($\varepsilon_t \geq 0.004$). The author believes that the **important**

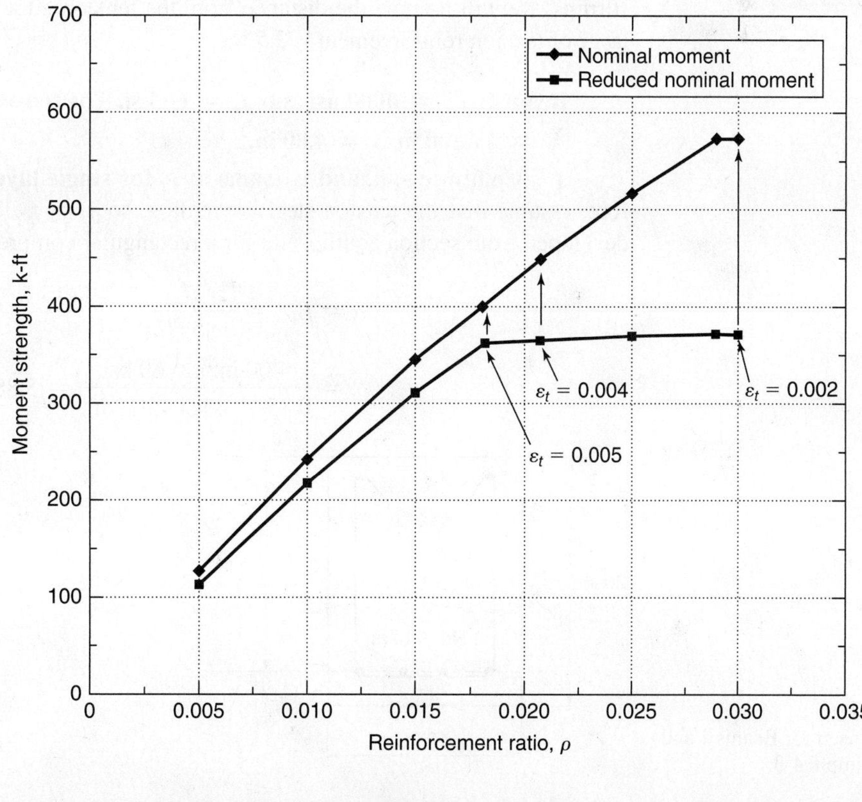

Fig. 4-26
Relationship between ρ and values for M_n and ϕM_n.

limit for the amount of tension steel to use in the design of beam sections will be to keep ε_t at or above the tension-controlled limit of 0.005, because Fig. 4-26 clearly shows that beyond this point it is not economical to add more tension steel to the section. Thus, for the flexural design procedures discussed in Chapter 5, the author will always check that the final section design is classified as a tension-controlled section ($\varepsilon_t \geq 0.005$), and thus, the ϕ-value always will be 0.90.

One final point of interest in Fig. 4-26 occurs in the plot of ρ versus M_n for a steel area larger that the balanced steel area given in Eq. (4-25). This section (last row of values in Table 4-2) is referred to as being *over-reinforced*, but the value for M_n does not increase for this larger area of tension steel because the concrete compression zone will start to fail before the steel reaches its yield stress. Thus, the values for the compression force, C_c, and therefore the tension force, T, tend to stay relatively constant. Exact values for the steel stress and strain can be determined using the fundamental procedure of satisfying section equilibrium [Eq. (4-2)] and strain compatibility [Eq. (4-18)]. Then the section nominal moment strength, M_n, can be calculated by the more general expression in Eq. (4-20). For over-reinforced sections, the nominal moment strength will tend to decrease as more tension steel is added to the section, because the moment arm $(d - a/2)$ will decrease as A_s is increased. An analysis of an over-reinforced beam section is presented as Beam 3 in the following example.

EXAMPLE 4-3 Analysis of Singly Reinforced Rectangular Beams

Compute the nominal moment strengths, M_n, and the strength reduction factor, ϕ, for three singly reinforced rectangular beams, each with a width b = 12 in. and a total height h = 20 in. As shown in Fig. 4-27 for the first beam section to be analyzed, a beam normally will have small longitudinal bars in the compression zone to hold the stirrups (shear reinforcement) in place. These bars typically are ignored in the calculation of the section nominal moment strength. Assuming that the beam has $1\frac{1}{2}$ in. of clear cover and uses No. 3 stirrups, we will assume the distance from the tension edge to the centroid of the lowest layer of tension reinforcement is 2.5 in.

Beam 1: f'_c = 4000 psi and f_y = 60 ksi. The tension steel area, $A_s = 4\,(1.00\text{ in.}^2) = 4.00\text{ in.}^2$

1. Compute a, c, and ε_s (same as ε_t for single layer of reinforcement). As before, assume that the tension steel is yielding, so $f_s = f_y$. Using Eq. (4-16), which was developed from section equilibrium for a rectangular compression zone,

$$a = \beta_1 c = \frac{A_s f_y}{0.85 f'_c b}$$

$$= \frac{4.00\text{ in.}^2 \times 60\text{ ksi}}{0.85 \times 4\text{ ksi} \times 12\text{ in.}} = 5.88\text{ in.}$$

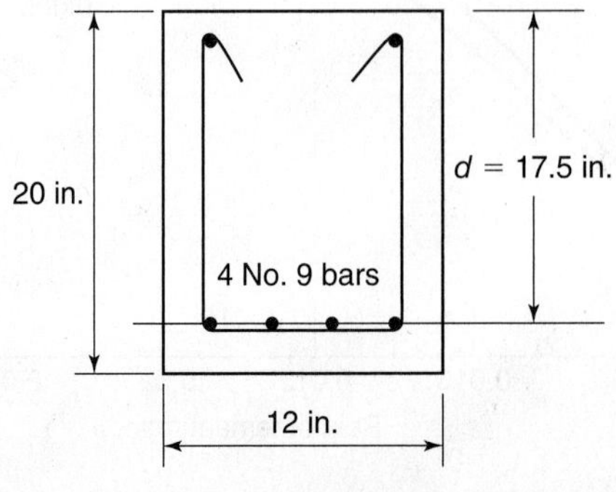

Fig. 4-27
Section used for Beams 1 and 2 of Example 4-3.

For $f'_c = 4000$ psi, β_1 is equal to 0.85. Thus, $c = a/\beta_1 = 6.92$ in., and using strain compatibility as expressed in Eq. (4-18), find

$$\varepsilon_s = \left(\frac{d - c}{c}\right)\varepsilon_{cu} = \left(\frac{17.5 - 6.92}{6.92}\right)0.003 = 0.00459$$

This exceeds the yield strain for Grade-60 steel ($\varepsilon_y = 0.00207$, previously calculated), so the assumption that the tension steel is yielding is confirmed.

2. Compute the nominal moment strength, M_n. As in Example 4-1, use Eq. (4-21), which applies to sections with rectangular compression zones for

$$M_n = A_s f_y\left(d - \frac{a}{2}\right) = 4.0 \text{ in.}^2 \times 60 \text{ ksi}\left(17.5 \text{ in.} - \frac{5.88 \text{ in.}}{2}\right)$$

$$M_n = 3490 \text{ k-in.} = 291 \text{ k-ft}$$

3. Confirm that tension steel area exceeds $A_{s,\text{min}}$. Although this is seldom a problem with most beam sections, it is good practice to make this check. The expression for $A_{s,\text{min}}$ is given in Eq. (4-11) and includes a numerator that is to be taken equal to $3\sqrt{f'_c}$, but not less than 200 psi. As was shown in Example 4-1, the value of 200 psi governs for beams constructed with 4000 psi concrete. Thus,

$$A_{s,\text{min}} = \frac{200 \text{ psi}}{f_y} b_w d = \frac{200}{60{,}000} \times 12 \text{ in.} \times 17.5 \text{ in.} = 0.70 \text{ in.}^2$$

Clearly, A_s for this section satisfies the ACI Code requirement for minimum tension reinforcement.

4. Compute the strength reduction factor, ϕ, and the resulting value of ϕM_n. As stated previously, for a single layer of tension reinforcement, ε_t is equal to ε_s, which was calculated in step 1. Because ε_t is between 0.002 and 0.005, this is a transition-zone section. Thus, Eq. (4-28a) is used to calculate ϕ:

$$\phi = 0.65 + (0.00459 - 0.002)\frac{250}{3} = 0.87$$

Then,

$$\phi M_n = 0.87 \times 291 \text{ k-ft} = 253 \text{ k-ft}$$

Beam 2: Same as Beam 1, except that $f'_c = 6000$ psi. As shown in Fig. 4-12, changing the concrete compressive strength will not produce a large change in the nominal moment strength, but it does increase the ductility of the section. Thus, increasing the concrete compressive strength might change the beam section in Fig. 4-27 from a *transition-zone section* to a *tension-controlled section.*

1. Compute a, c, and ε_s. Again, assume that the tension steel is yielding, so $f_s = f_y$. For this compressive strength, Eq. (4-14b) is used to determine that $\beta_1 = 0.75$. Then, using Eq. (4-16),

$$a = \beta_1 c = \frac{A_s f_y}{0.85 f'_c b}$$

$$= \frac{4.00 \text{ in.}^2 \times 60 \text{ ksi}}{0.85 \times 6 \text{ ksi} \times 12 \text{ in.}} = 3.92 \text{ in.}$$

Thus, $c = a/\beta_1 = 5.23$ in., and using strain compatibility as expressed in Eq. (4-18), find

$$\varepsilon_s = \left(\frac{d - c}{c}\right)\varepsilon_{cu} = \left(\frac{17.5 - 5.23}{5.23}\right)0.003 = 0.00704$$

This exceeds the yield strain for Grade-60 steel ($\varepsilon_y = 0.00207$), confirming the assumption that the tension steel is yielding.

2. Compute the nominal moment strength, M_n. As in Example 4-1, use Eq. (4-21), which applies to sections with rectangular compression zones:

$$M_n = A_s f_y\left(d - \frac{a}{2}\right) = 4.0\text{ in.}^2 \times 60\text{ ksi}\left(17.5\text{ in.} - \frac{3.92\text{ in.}}{2}\right)$$

$$M_n = 3730\text{ k-in.} = 311\text{ k-ft (7 percent increase from Beam 1)}$$

3. Confirm that tension steel area exceeds $A_{s,\min}$. For this beam section with 6000 psi concrete, the value of $3\sqrt{f'_c}$ exceeds 200 psi, and will govern in Eq. (4-11). Thus,

$$A_{s,\min} = \frac{3\sqrt{f'_c}}{f_y}b_w d = \frac{3\sqrt{6000}}{60{,}000} \times 12\text{ in.} \times 17.5\text{ in.} = 0.81\text{ in.}^2$$

Again, A_s for this section easily satisfies the ACI Code requirement for minimum tension reinforcement.

4. Compute the strength reduction factor, ϕ, and the resulting value of ϕM_n. As before, ε_t is equal to ε_s, which was calculated in step 1. By increasing the concrete strength, this beam section is now a tension-controlled section, so $\phi = 0.9$. Then,

$$\phi M_n = 0.9 \times 311\text{ k-ft} = 280\text{ k-ft (11 percent increase from beam 1)}$$

Beam 3: Same as Beam 1, except increase tension steel to six No. 9 bars in two layers (Fig. 4-28). For this section, ε_t will be larger than ε_s and will be calculated using the distance to the extreme layer of tension reinforcement, d_t. Assuming the same cover and size of stirrup, $d_t = 17.5$ in., as used for d in Beams 1 and 2. The value of d for this section involves a centroid calculation for the six No. 9 bars. ACI Code Section 25.2.2 requires a clear spacing between layers of reinforcement greater than or equal to 1 in. Thus, we can assume that the second layer of steel (two bars) is one bar diameter plus 1 in. above the lowest layer—or a total of 2.5 in. + 1.128 in. + 1 in. ≈ 4.63 in. from the extreme tension fiber. A simple calculation is used to find the distance from the bottom of the beam to the centroid of the tension reinforcement, g, and then find the value of $d = h - g$.

$$g = \frac{4.0\text{ in.}^2 \times 2.5\text{ in.} + 2.0\text{ in.}^2 \times 4.63\text{ in.}}{6.0\text{ in.}^2} = 3.21\text{ in.}$$

$$d = h - g = 20\text{ in.} - 3.21\text{ in.} \approx 16.8\text{ in.}$$

1. Compute a, c, and ε_s. Again, assume that the tension steel is yielding, so $f_s = f_y$. Then, using Eq. (4-16):

$$a = \beta_1 c = \frac{A_s f_y}{0.85 f'_c\, b}$$

$$= \frac{6.00\text{ in.}^2 \times 60\text{ ksi}}{0.85 \times 4\text{ ksi} \times 12\text{ in.}} = 8.82\text{ in.}$$

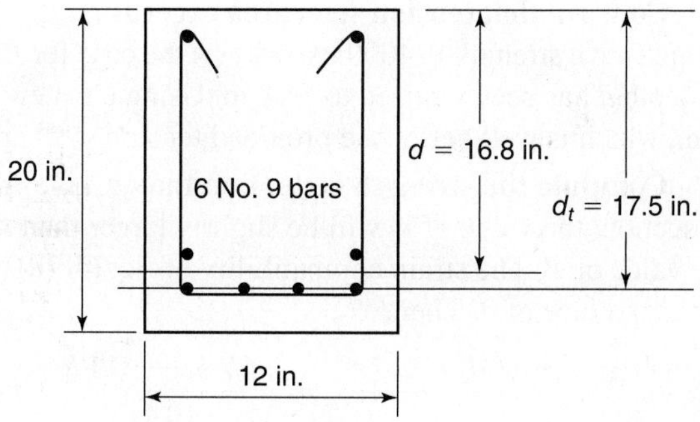

Fig. 4-28
Section used for Beam 3 of Example 4-3.

As for Beam 1, $\beta_1 = 0.85$. Thus, $c = a/\beta_1 = 10.4$ in. and using strain compatibility, as expressed in Eq. (4-18), find

$$\varepsilon_s = \left(\frac{d - c}{c}\right)\varepsilon_{cu} = \left(\frac{16.8 - 10.4}{10.4}\right)0.003 = 0.00186$$

This is less than the yield strain for Grade-60 steel ($\varepsilon_y = 0.00207$), so the assumption that the tension steel is yielding is *not confirmed*. Because the tension steel is not yielding, this is referred to as an *over-reinforced* section, and the previously developed procedure for calculating the nominal moment strength *does not apply*. A procedure that enforces strain compatibility and section equilibrium will be demonstrated in the following.

2. Compute the nominal moment strength, M_n, by enforcing strain compatibility and section equilibrium. Referring to Fig. 4-18, we must now assume that the steel stress, f_s, is an unknown but is equal to the steel steel strain, ε_s, multiplied by the steel modulus, E_s. Strain compatibility as expressed in Eq. (4-18) still applies, so the steel stress and thus the tension force can be expressed as a function of the unknown neutral axis depth, c.

$$T = A_s f_s = A_s E_s \varepsilon_s = A_s E_s\left(\frac{d - c}{c}\right)\varepsilon_{cu}$$

Similarly, the concrete compression force can be expressed as a function of the neutral axis depth, c.

$$C_c = 0.85 f'_c b \beta_1 c$$

Enforcing equilibrium by setting $T = C_c$, we can solve a second degree equation for the unknown value of c. The solution normally results in one positive and one negative value for c; the positive value will be selected. Using all of the given section and material properties and recalling that $E_s = 29{,}000$ ksi and $\varepsilon_{cu} = 0.003$, the resulting value for c is 10.1 in. Using this value, the author obtained

$$T = 346 \text{ kips} \cong C_c = 350 \text{ kips}$$

An average value of $T = C_c = 348$ kips will be used to calculate M_n. Then, using $a = \beta_1 c = 0.85 \times 10.1 \text{ in.} = 8.59$ in., calculate M_n using the more general expression in Eq. (4-20).

$$M_n = T\left(d - \frac{a}{2}\right) = 348 \text{ kips}\left(16.8 \text{ in.} - \frac{8.59 \text{ in.}}{2}\right)$$

$$M_n = 4350 \text{ k-in.} = 363 \text{ k-ft}$$

3. Confirm that tension steel area exceeds $A_{s,\min}$. For this beam section, the concrete compressive strength is 4000 psi, as was the case for Beam 1. However, the effective flexural depth d has been reduced to 16.8 in. Using this new value of d, the value for $A_{s,\min}$ is 0.67 in., which is well below the provided tension steel area A_s.

4. Compute the strength reduction factor, ϕ, and the resulting value of ϕM_n. For this section, the value of ε_t will be slightly larger than ε_s and should be used to determine the value of ϕ. The strain compatibility of Eq. (4-18) can be modified to calculate ε_t by using d_t in place of d. Then,

$$\varepsilon_t = \left(\frac{d_t - c}{c}\right)\varepsilon_{cu} = \left(\frac{17.5 - 10.1}{10.1}\right)0.003 = 0.00220$$

This is an interesting result, because we have previously considered this to be an over-reinforced section based on the strain, ε_s, calculated at the centroid of the tension reinforcement. However, because of the difference between d and d_t, we now have found the value of ε_t to be between 0.002 and 0.005. Thus, this is a transition-zone section, and we must use Eq. (4-28a) to calculate ϕ.

$$\phi = 0.65 + (0.00220 - 0.002)\frac{250}{3} = 0.67$$

Then,

$$\phi M_n = 0.67 \times 363 = 243 \text{ k-ft}$$

It should be noted that even though this section has 50 percent more steel than that of Beam 1, the reduced nominal moment strength is smaller for this beam section than for Beam 1. This demonstrates a very important result for heavily reinforced sections—the only way to increase the reduced nominal moment strength is to add steel to *both* the tension and compression zones of the member. The next part of this chapter deals with the analysis of doubly reinforced beam sections, i.e., beams with longitudinal steel in both the tension and compression zones. ■

4-7 BEAMS WITH COMPRESSION REINFORCEMENT

Occasionally, beam sections are designed to have both tension reinforcement and compression reinforcement. These are referred to as *doubly reinforced* sections. Two cases where compression reinforcement is used frequently are the negative bending region of continuous beams and midspan regions of long-span or heavily loaded beams where deflections need to be controlled. The effect of compression reinforcement on the behavior of beams and the reasons it is used are discussed in this section, followed by a method to analyze such beam sections.

Effect of Compression Reinforcement on Strength and Behavior

The resultant internal forces at nominal-strength conditions in beams with and without compression reinforcement are compared in Fig. 4-29. As was done in the analysis of singly reinforced beam sections, we initially will assume that the tension steel is yielding, so $f_s = f_y$. The beam in Fig. 4-29b has a compression steel of area A_s' located at d' from the extreme compression fiber. The area of the tension reinforcement, A_s, is the same in

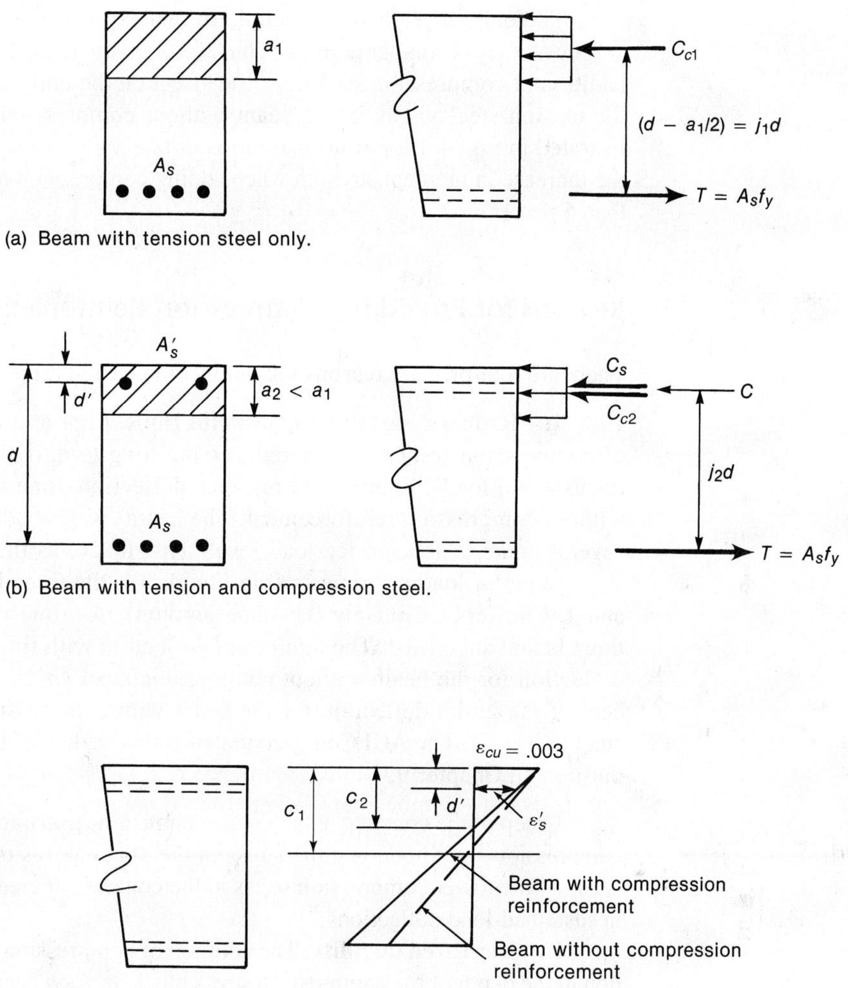

(c) Effect of compression reinforcement on strain distribution in two beams with the same area of tension reinforcement.

Fig. 4-29
Effect of compression reinforcement on moment strength.

both beams. In both beams, the total compressive force is equal to the tension force, where $T = A_sf_y$. In the beam without compression reinforcement (Fig. 4-29a), this compressive force, C_{c1}, is resisted entirely by concrete. In the other case (Fig. 4-29b), C is the sum of C_{c2} provided by the concrete and C_s provided by the compression steel. Because some of the compression is resisted by compression reinforcement, C_{c2} will be less than C_{c1}, with the result that the depth of the compression stress block, a_2, in Fig. 4-29b is less than a_1 in Fig. 4-29a. The change in the required depth of the stress block causes a related change in the depth to the neutral axis, c, as shown in Fig. 4-29c.

Summing moments about the centroid of the resultant compressive force gives the following results:

For the beam without compression steel,

$$M_n = A_sf_y(j_1\,d)$$

For the beam with compression steel,

$$M_n = A_sf_y(j_2\,d)$$

The only difference between these two expressions is that j_2 is a little larger than j_1, because a_2 is smaller than a_1. Thus, for a given amount of tension reinforcement, the addition of compression steel has little effect on the nominal moment strength, provided the tension steel yields in the beam without compression reinforcement. This was illustrated in Fig. 4-12. For normal ratios of tension reinforcement ($\rho = A_s/bd \leq 0.015$), the increase in moment strength when adding compression reinforcement generally is less than 5 percent.

Reasons for Providing Compression Reinforcement

There are four primary reasons for using compression reinforcement in beams:

1. Reduced sustained-load deflections. First and most important, the addition of compression reinforcement reduces the long-term deflections of a beam subjected to sustained loads. Figure 4-30 presents deflection–time diagrams for beams with and without compression reinforcement. The beams were loaded gradually over a period of several hours to the service-load level. This load was then maintained for two years. At the time of loading (time = 0 in Fig. 4-30), the three beams deflected between 1.6 and 1.9 in. (approximately the same amount). As time passed, the deflections of all three beams increased. The *additional* deflection with time is 195 percent of the initial deflection for the beam without compression steel ($\rho' = A_s'/bd = 0$) but only 99 percent of the initial deflection for the beam with compression steel equal to the tension steel ($\rho' = \rho$). The ACI Code accounts for this in the deflection-calculation procedures outlined in Chapter 9.

Creep of the concrete in the compression zone transfers load from the concrete to the compression steel, reducing the stress in the concrete (as occurred in Example 3-4). Because of the lower compression stress in the concrete, it creeps less, leading to a reduction in sustained-load deflections.

2. Increased ductility. The addition of compression reinforcement causes a reduction in the depth of the compression stress block, a. As a decreases, the strain in the tension reinforcement at failure increases, as shown in Fig. 4-29c, resulting in more ductile behavior, as was shown in Fig. 4-12 for $A_s' = 0.5\,A_s$. Figure 4-31 compares moment–curvature

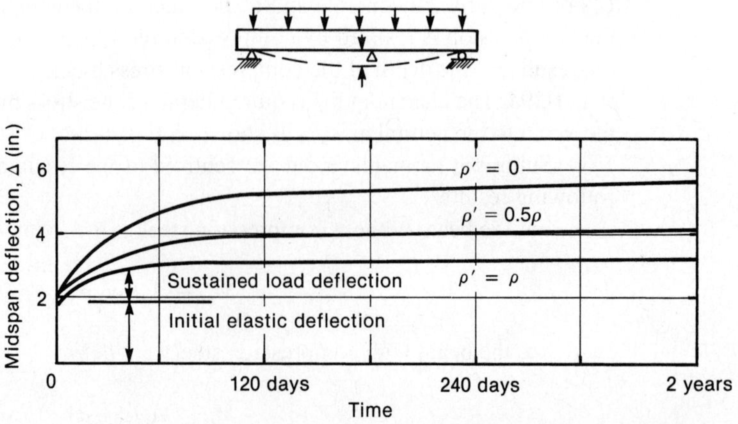

Fig. 4-30
Effect of compression reinforcement on deflections under sustained loading. (From [4-15].)

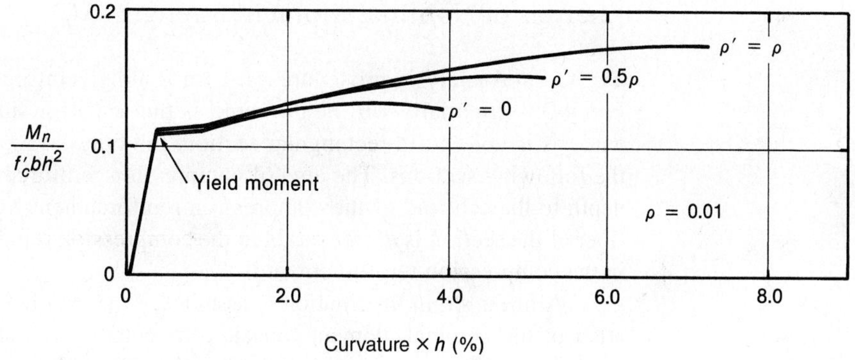

Fig. 4-31
Effect of compression reinforcement on strength and ductility of under-reinforced beams. (From [4-16].)

diagrams for three beams with $\rho < \rho_b$, as defined in Eq. (4-25), and varying amounts of compression reinforcement, ρ'. The moment at first yielding of the tension reinforcement is seen to change very little when compression steel is added to these beams. The increase in moment after yielding in these plots is largely due to strain hardening of the reinforcement. Because this occurs at very high curvatures and deflections, it is ignored in design. On the other hand, the ductility increases significantly when compression reinforcement is used, as shown in Fig. 4-31. This is particularly important in seismic regions or if moment redistribution is desired.

3. Change of mode of failure from compression to tension. When $\rho > \rho_b$, a beam fails in a brittle manner through crushing of the compression zone before the steel yields. A moment–curvature diagram for such a beam is shown in Fig. 4-32 ($\rho' = 0$). When enough compression steel is added to such a beam, the compression zone is strengthened sufficiently to allow the tension steel to yield before the concrete crushes. The beam then displays a ductile mode of failure. For earthquake-resistant design, all beam sections are required to have $\rho' \geq 0.5\rho$.

4. Fabrication ease. When assembling the reinforcing cage for a beam, it is customary to provide small bars in the corners of the stirrups to hold the stirrups in place in the form and also to help anchor the stirrups. If developed properly, these bars in effect are compression reinforcement, although they generally are disregarded in design, because they have only a small affect on the moment strength.

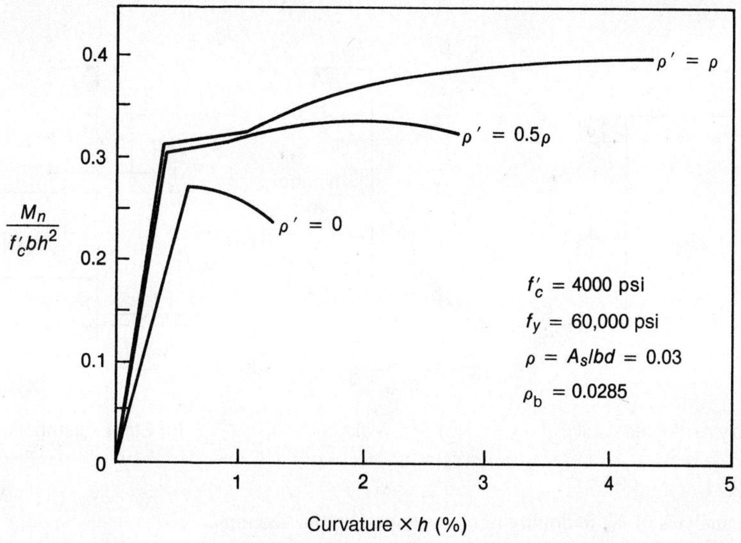

Fig. 4-32
Moment–curvature diagram for beams, with and without compression reinforcement. (From [4-16].)

Analysis of Nominal Moment Strength, M_n

The flexural analysis procedure used for doubly reinforced sections, as illustrated in Fig. 4-33, essentially will be the same as that used for singly reinforced sections. The analysis is done for a rectangular section, but other section shapes will be included in the following sections. The area of compression reinforcement is referred to as A'_s, the depth to the centroid of the compression reinforcement from the extreme compression fiber of the section is d', the strain in the compression reinforcement is ε'_s, and the stress in the compression reinforcement is f'_s.

A linear strain distribution is assumed, as shown in Fig. 4-33b, and for the evaluation of the nominal moment capacity, the compression strain in the extreme concrete compression fiber is set equal to the maximum useable concrete compressive strain, ε_{cu}. As was done for singly reinforced sections, the section is assumed to be under-reinforced, so the strain in the tension reinforcement is assumed to be larger than the yield strain. The exact magnitude of that strain is not known, and thus, the depth to the neutral axis, c, also is unknown. An additional unknown for a doubly reinforced section is the strain in the compression reinforcement, ε'_s. Unlike the tension-reinforcement strain, it is not reasonable to assume that this strain exceeds the yield strain when analyzing the nominal moment strength of a beam section. The following relationship can be established from similar triangles in the strain diagram:

$$\frac{\varepsilon'_s}{c - d'} = \frac{\varepsilon_{cu}}{c}$$

or

$$\varepsilon'_s = \left(\frac{c - d'}{c}\right)\varepsilon_{cu} \tag{4-30}$$

The assumed distribution of stresses is shown in Fig. 4-33c. As before, the real concrete compression stress distribution is replaced by Whitney's stress block. The stress in the compression reinforcement, f'_s, is not known and cannot be determined until the depth to the neutral axis has been determined. As was done in the analysis of a singly reinforced section, the stress in the tension reinforcement is set equal to the yield stress, f_y.

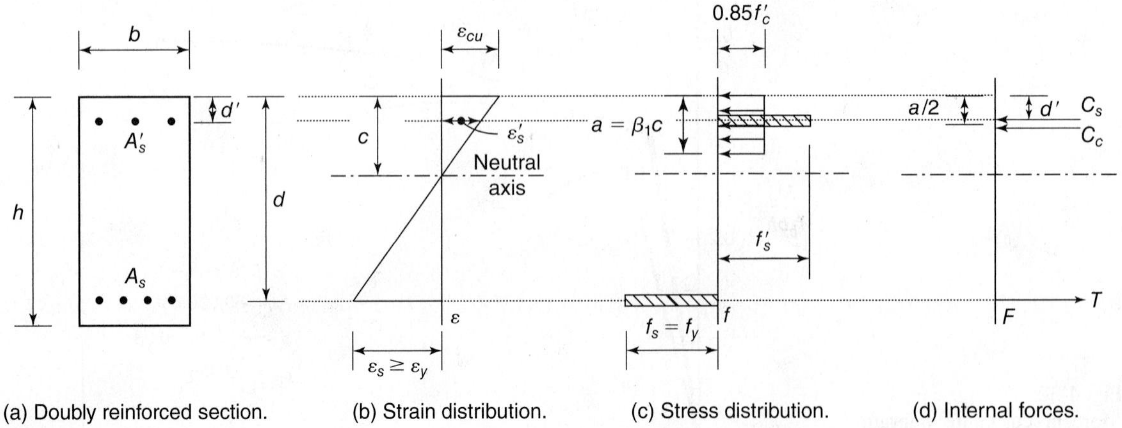

Fig. 4-33
Steps in analysis of M_n in doubly reinforced rectangular sections.

The internal section forces (stress resultants) are shown symbolically in Fig. 4-33d. The concrete compression force, C_c, is assumed to be the same as that calculated for a singly reinforced section.

$$C_c = (0.85)f'_c b\beta_1 c = (0.85)f'_c ba \tag{4-13b}$$

This expression contains a slight error, because part of the compression zone is occupied by the compression reinforcement. Some designers elect to ignore this error, but in this presentation, the error will be corrected in the calculation of the force in the compression steel by subtracting the height of the compression stress block from the stress in the compression reinforcement, f'_s. By correcting this error at the level of the compression reinforcement, the locations of the section forces are established easily. So, the force in the compression reinforcement is expressed as

$$C_s = A'_s(f'_s - 0.85f'_c) \tag{4-31}$$

The stress in the compression reinforcement is not known, but can be expressed as

$$f'_s = E_s\varepsilon'_s \leq f_y \tag{4-32}$$

The tension force is simply the area of tension reinforcement multiplied by the yield stress. Thus, establishing section equilibrium results in the following:

$$T = C_c + C_s$$

or

$$A_s f_y = (0.85)f'_c b\beta_1 c + A'_s(f'_s - 0.85f'_c) \tag{4-33}$$

In this expression, there are two unknowns: the neutral axis depth, c, and the stress in the compression reinforcement, f'_s. The compression steel stress can be assumed to be linearly related to the compression steel strain, ε'_s, as expressed in the first part of Eq. (4-32). Also, the compression steel strain is linearly related to the neutral axis depth, as given in Eq. (4-30). Thus, the section equilibrium expressed in Eq. (4-33) could be converted to a quadratic equation in terms of one unknown, c.

However, the solution of such a quadratic equation has two potential problems. First, after a value has been found for the neutral axis depth, c, a check will be required to confirm the assumption that the compression steel is not yielding in Eq. (4-32). If the compression steel is yielding, Eq. (4-33) would need to be solved a second time (linear solution) starting with the assumption that $f'_s = f_y$. The second, and more important potential problem, is that the engineer does not develop any "feel" for the correct answer. What is a reasonable value for c? What should be done if c is less that the depth to the compression reinforcement, d'?

To develop some "feel" for the correct solution, the author prefers an iterative solution for the neutral axis depth, c. With some experience this process converges quickly and allows for modifications during the solution. The recommended steps are listed below and described in a flowchart in Fig. 4-34.

1. Assume the tension steel is yielding, $\varepsilon_s \geq \varepsilon_y$.
2. Select a value for the neutral axis depth, c (start with a value between *d/4* and *d/3*).
3. Calculate the compression steel strain, ε'_s, Eq. (4-30).

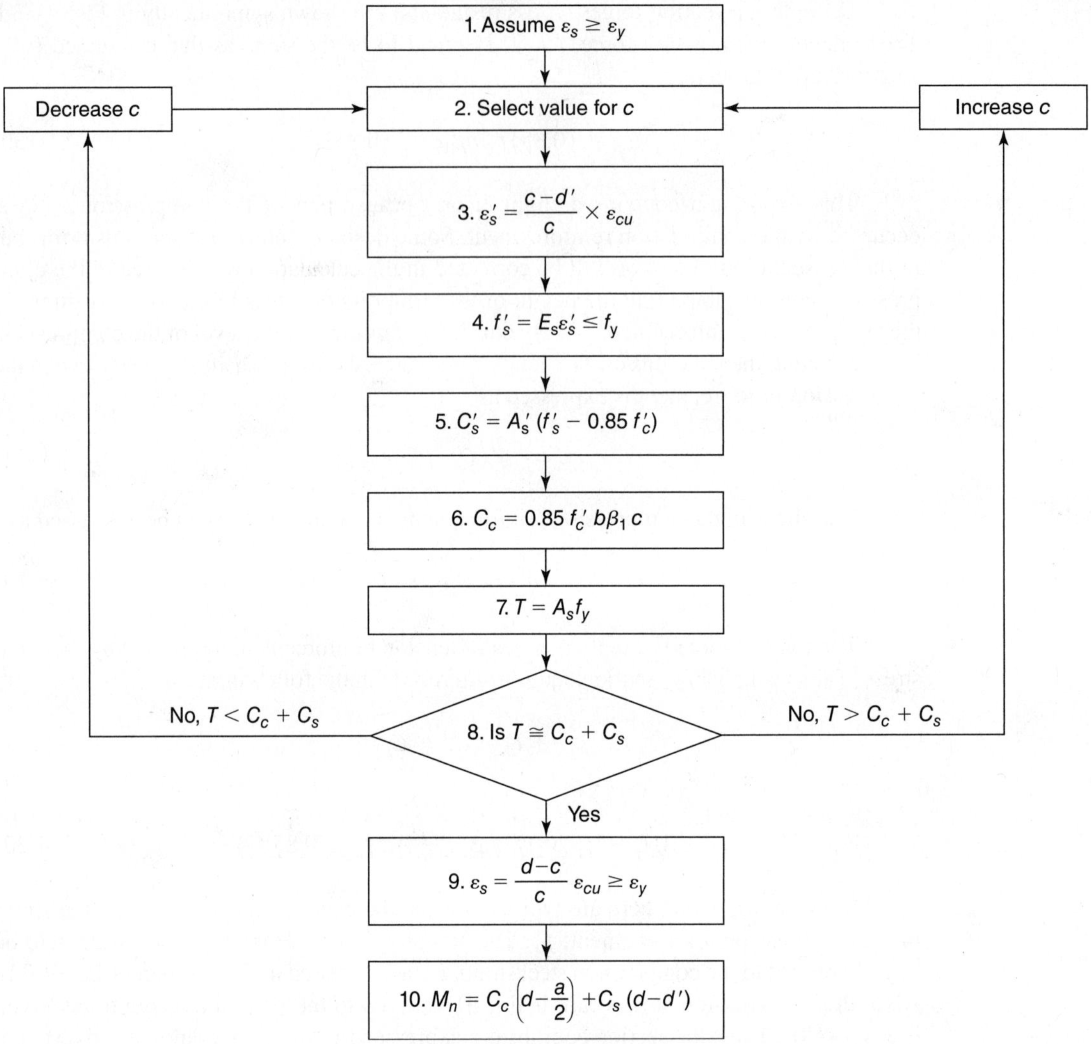

Fig. 4-34
Flowchart for analysis of doubly reinforced beam sections.

4. Calculate the compression steel stress, f_s', Eq. (4-32).
5. Calculate the compression steel force, C_s, Eq. (4-31).
6. Calculate the concrete compression force, C_c, Eq. (4-13b).
7. Calculate the tension steel force, $T = A_s f_y$.
8. Check section equilibrium, Eq. (4-33). If $T \cong C_c + C_s$ (difference less than 5 percent of T), then go to step 9.
   - **(a)** If $T > C_c + C_s$, increase c and return to step 3.
   - **(b)** If $T < C_c + C_s$, decrease c and return to step 3.
9. Confirm that tension steel is yielding in Eqs. (4-18) and (4-19).
10. Calculate nominal moment strength, M_n, as given next.

As stated previously, this process quickly converges and gives the engineer control of the section analysis process. To answer the one question raised previously, if during this process it is found that c is less that d', the author recommends removing C_s from the calculation because the compression steel is not working in compression. Thus, the section should be analyzed as if it is singly reinforced, following the procedure given in Section 4-4. This will often happen when a beam section includes a compression flange, as will be discussed in the next section of the text.

Once the process has converged and section equilibrium is established (step 8), and it has been confirmed that the tension steel is yielding (step 9), the section nominal moment strength can be calculated by multiplying the section forces times their moment arms about a convenient point in the section. For the analysis presented here, that point is taken at the level of the tension reinforcement. Thus, T is eliminated from the calculation and the resulting expression for M_n from Fig. 4-33d is

$$M_n = C_c\left(d - \frac{a}{2}\right) + C_s(d - d') \tag{4-34}$$

where $a = \beta_1 c$, with β_1 defined previously in Eq. (4-14).

Analysis of Strength-Reduction Factor, ϕ

The next step in the flexural analysis of a doubly reinforced beam section is to determine a value for the strength-reduction factor, ϕ, so the value of ϕM_n can be compared with the factored design moment, M_u, that must be resisted by the section. The general procedure is the same for all beam sections. The value of the tension strain in the extreme layer of tension reinforcement, ε_t, can be determined from a strain compatibility expression similar to Eq. (4-18) with the distance to the extreme layer of tension reinforcement, d_t, used in place of d.

$$\varepsilon_t = \left(\frac{d_t - c}{c}\right)\varepsilon_{cu} \tag{4-35}$$

If the value of ε_t is greater than or equal to 0.005, the section is tension-controlled, and $\phi = 0.90$. If ε_t is less than or equal to 0.002, the section is compression controlled, and $\phi = 0.65$. If ε_t is between these two limits, the section is in the transition zone, and Eq. (4-28a) can be used to determine the value for ϕ. For tension-controlled sections, this process can be shortened if the value of ε_s, calculated in step 9 of the section analysis process described previously, is found to be greater than or equal to 0.005. Because the value of d_t is always greater than or equal to d, then ε_t will always equal or exceed ε_s and thus would be greater than 0.005.

Minimum Tension Reinforcement and Ties for Compression Reinforcement

Minimum tension reinforcement, which is seldom an issue for doubly reinforced beam sections, is the same as that for singly reinforced rectangular sections, as given in Eq. (4-11).

As a beam section reaches in maximum moment capacity, the compression steel in the beam may buckle outward, causing the surface layer of concrete to spall off. For this reason, ACI Code Section 9.7.6.4.1 requires compression reinforcement to be enclosed within stirrups or ties over the length that the bars are needed in compression. The spacing and size of the ties is similar to that required for columns ties, as will be discussed in

Chapter 11. Frequently, longitudinal reinforcement, which has been detailed to satisfy bar cutoff rules in Chapter 8, is stressed in compression near points of maximum moment. These bars normally are not enclosed in ties if the compression in them is not included in the calculation of the section nominal moment strength. Ties are required throughout the portion of the beam where the compression steel is used in compression when determining the nominal moment strength of a beam section. If the compression steel will be subjected to stress reversals, or if this steel is used to resist torsion, closed stirrups must be used to confine these bars. Details for closed stirrups will be discussed in Chapters 6 and 7 on design to resist shear and torsion.

EXAMPLE 4-4 Analysis of Doubly Reinforced Rectangular Beam Section

Compute the nominal moment strength, M_n, and the strength-reduction factor, ϕ, for the doubly reinforced rectangular beam shown in Fig. 4-35. This beam section is very similar to the section for Beam 3 of Example 4-3. For the beam section in Fig. 4-35, three No. 9 bars have been used as compression reinforcement, and a closed No. 3 stirrup-tie is used to help hold the top bars in place during casting. This example will demonstrate how beam section behavior can be changed by adding compression reinforcement. Assuming that the beam has a $1\frac{1}{2}$-in. clear cover, we will assume the distance from the compression edge to the centroid of the compression reinforcement, d', is 2.5 in. The values for d and d_t are the same as used for Beam 3 of Example 4-3. Assume the material properties are $f'_c = 4000$ psi and $f_y = 60$ ksi. Recall that for the given concrete compressive strength $\beta_1 = 0.85$, and that the steel modulus $E_s = 29{,}000$ ksi.

1. Use the iterative procedure discussed in the prior paragraphs to establish section equilibrium and find the depth to the neutral axis, *c*.

1. Assume the tension steel is yielding, so $f_s = f_y$. (Before adding the compression reinforcement, this was an over-reinforced beam section. We will assume that it is now an under-reinforced section).

2. Select an initial value for c. $d/4 = 4.20$ in. and $d/3 = 5.60$ in. **Try $c = 5$ in.**

3. Find the strain in the compression reinforcement.

$$\varepsilon'_s = \left(\frac{c - d'}{c}\right)\varepsilon_{cu} = \left(\frac{5 - 2.5}{5}\right)0.003 = 0.0015$$

4. Find stress in compression reinforcement, $f'_s = E_s\varepsilon'_s = 29{,}000 \text{ ksi} \times 0.0015 = 43.5 \text{ ksi}$ (<60 ksi, o.k.).

5. Find force in compression reinforcement, $C_s = A'_s(f'_s - 0.85f'_c) = 3 \times 1.00 \text{ in.}^2 \times (43.5 - 3.4) \text{ ksi} = 120$ kips.

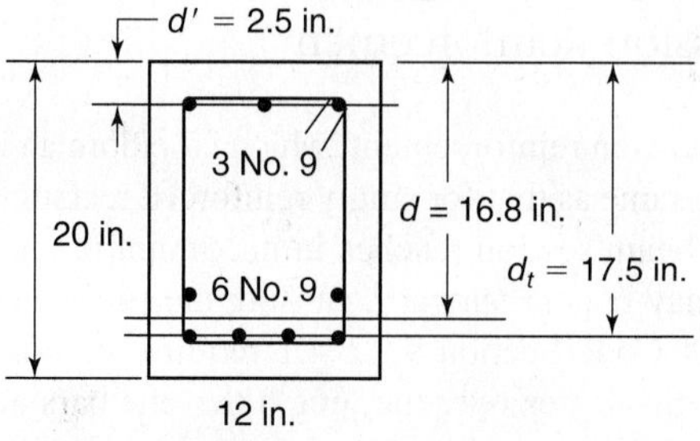

Fig. 4-35
Beam section used for Example 4-4.

6. Find concrete compression force, $C_c = 0.85 f'_c b\beta_1 c = 0.85 \times 4\text{ ksi} \times 12\text{ in.} \times 0.85 \times 5\text{ in.} = 173\text{ kips}$.

7. Find force in tension reinforcement, $T = A_s f_y = 6 \times 1.00\text{ in.}^2 \times 60\text{ ksi} = 360\text{ kips}$.

8. Check section equilibrium, $C_s + c_c = 293\text{ kips} < T = 360\text{ kips}$, thus must increase c.

The step size for the next iteration is not easy to specify. Some judgment must be developed and the only way to develop that judgment is to use this method for a variety of sections.

Try c = 5.5 in. For this trial, the compression steel is still not yielding. $C_s = 132$ kips and $C_c = 191$ kips, so the sum of C_s and C_c is 323 kips, which is 37 kips (approximately 10 percent) less than T.

Try c = 6.0 in. Again, the compression steel is not yielding. $C_s = 142$ kips and $C_c = 208$ kips, so the sum of C_s and C_c is 350 kips, which is only 10 kips (approximately 3 percent) less than T. **This is close enough.** (*Note*: This series of simple calculations can be handled easily with a spreadsheet or Mathcad procedure.)

9. Confirm that the tension steel is yielding using Eq. (4-18):

$$\varepsilon_s = \left(\frac{d - c}{c}\right)\varepsilon_{cu} = \left(\frac{16.8 - 6}{6}\right)0.003 = 0.00540 > \varepsilon_y = 0.00207$$

2. Calculate the nominal moment strength, M_n (step 10).
The depth of the compression stress block, $a = \beta_1 c = 0.85 \times 6\text{ in.} = 5.10\text{ in.}$ Using this in Eq. (4-34),

$$M_n = C_c\left(d - \frac{a}{2}\right) + C_s(d - d')$$

$$= 208\text{ k}\left(16.8\text{ in.} - \frac{5.10\text{ in.}}{2}\right) + 142\text{ k}\,(16.8\text{ in.} - 2.5\text{ in.})$$

$$= 2960\text{ k-in.} + 2030\text{ k-in.} = 4990\text{ k-in.} = 416\text{ k-ft}$$

In case the reader is concerned about more accuracy in satisfying section equilibrium in step 8, the following information is presented. Using a spreadsheet, the author found a closer section equilibrium with $c = 6.2$ in. With this, it can be shown easily that the tension steel is yielding (as assumed) and that the final value for M_n is 428 k-ft. This small increase (approximately 3 percent) will be relatively unimportant in most design situations, as will be discussed in the next chapter on design of beam sections.

3. Confirm that tension steel area exceeds $A_{s,\min}$. This requirement from the ACI Code does not change when compression reinforcement is used. Thus, the required minimum area for this beam section is the same as that for Beam 3 of Example 4-2. Thus, $A_{s,\min} = 0.67$ in., which is well below the provided tension steel area, A_s.

4. Compute the strength reduction factor, ϕ, and the resulting value of ϕM_n. For this section, the value of d_t exceeds d, and thus $\varepsilon_t > \varepsilon_s > 0.005$ (step 9). So, $\phi = 0.9$ and $\phi M_n = 374$ k-ft (using $c = 6.0$ in.). ■

4-8 ANALYSIS OF FLANGED SECTIONS

In the floor system shown in Fig. 4-36, the slab is assumed to carry the loads in one direction to beams that carry them in the perpendicular direction. During a normal construction sequence, concrete in the columns is placed and allowed to harden before the concrete in the floor is placed (ACI Code Section 26.5.7.2a). In the next construction step, concrete is placed in the beams and slab in a monolithic pour (ACI Code Section 26.5.7.2b). As a result, the slab serves as the top flange of the beams, as indicated by the shading in Fig. 4-36. Such a beam is referred to as a *T-beam*. The interior beam, *AB*, has a flange on both sides. The *spandrel beam, CD*, with a flange on one side only, is often referred to as an *inverted L-beam*.

An exaggerated deflected view of the interior beam is shown in Fig. 4-37. This beam develops positive moments at midspan (section *A–A*) and negative moments over

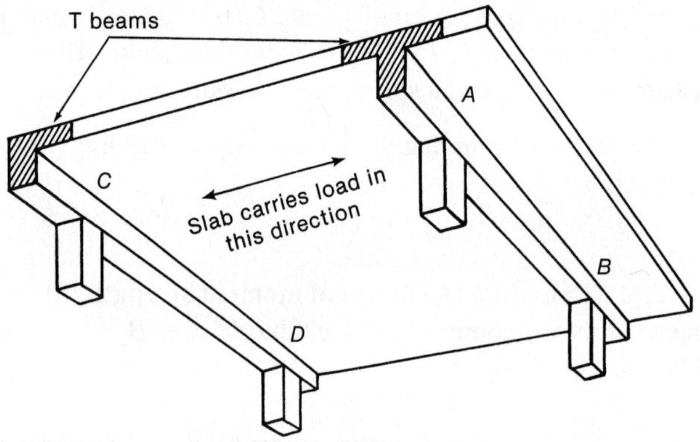

Fig. 4-36
T-beams in a one-way beam and slab floor.

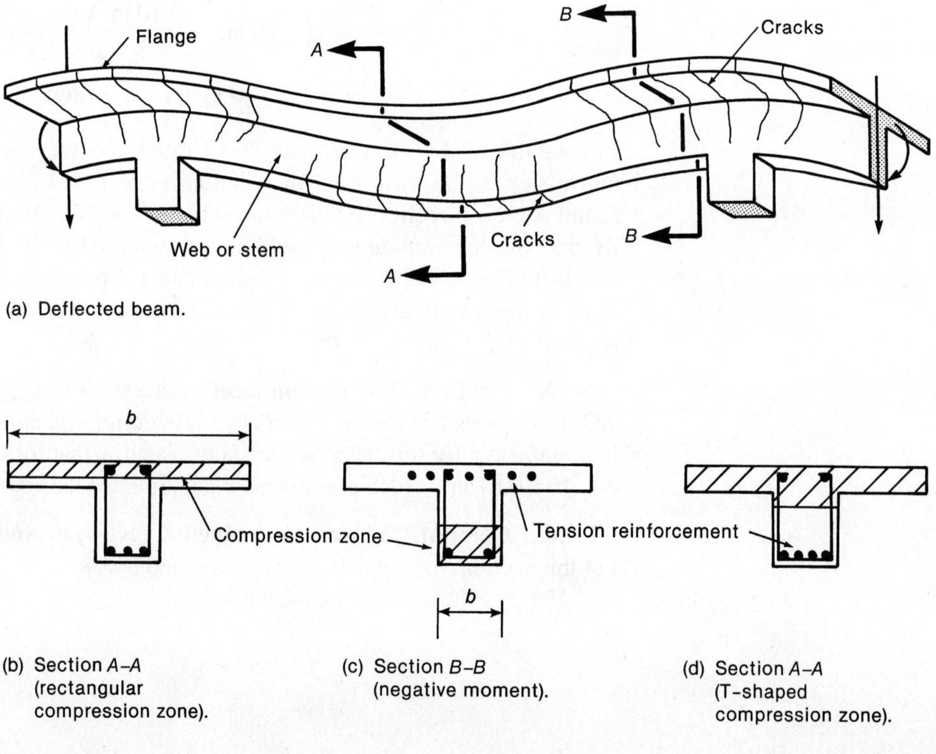

Fig. 4-37
Positive and negative moment regions in a T-beam.

the supports (section *B–B*). At midspan, the compression zone is in the flange, as shown in Figs. 4-37b and 4-37d. Generally, it is rectangular, as shown in 4-37b, although, in very rare cases for typical reinforced concrete construction, the neutral axis may shift down into the web, giving a T-shaped compression zone, as shown in Fig. 4-37d. At the support, the compression zone is at the bottom of the beam and is rectangular, as shown in Fig. 4-37c.

Frequently, a beam-and-slab floor involves slabs supported by beams which, in turn, are supported by other beams referred to as *girders* (Fig. 4-38). Again, all of the concrete above the top of the column is placed at one time, and the slab acts as a flange for both the beams and girders.

Effective Flange Width and Reinforcement in the Transverse Direction

The forces acting on the flange of a simply supported T-beam are illustrated in Fig. 4-39. At the support, there are no longitudinal compressive stresses in the flange, but at midspan, the full width is stressed in compression. The transition requires horizontal shear stresses on the web–flange interface as shown in Fig. 4-39. As a result there is a "shear-lag" effect, and the portions of the flange closest to the web are more highly stressed than those portions farther away, as shown in Figs. 4-39 and 4-40.

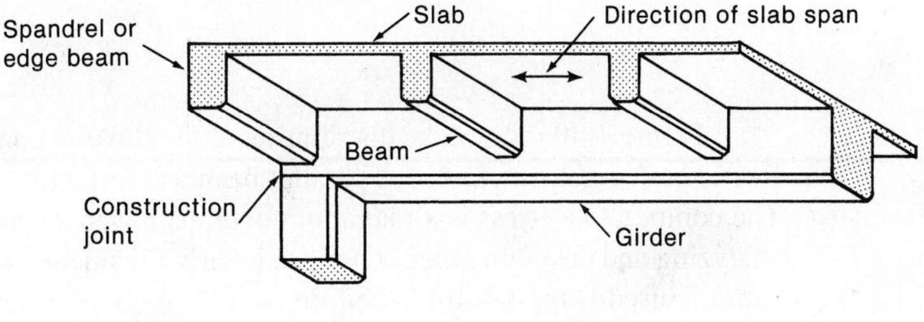

Fig. 4-38
Slab, beam, and girder floor.

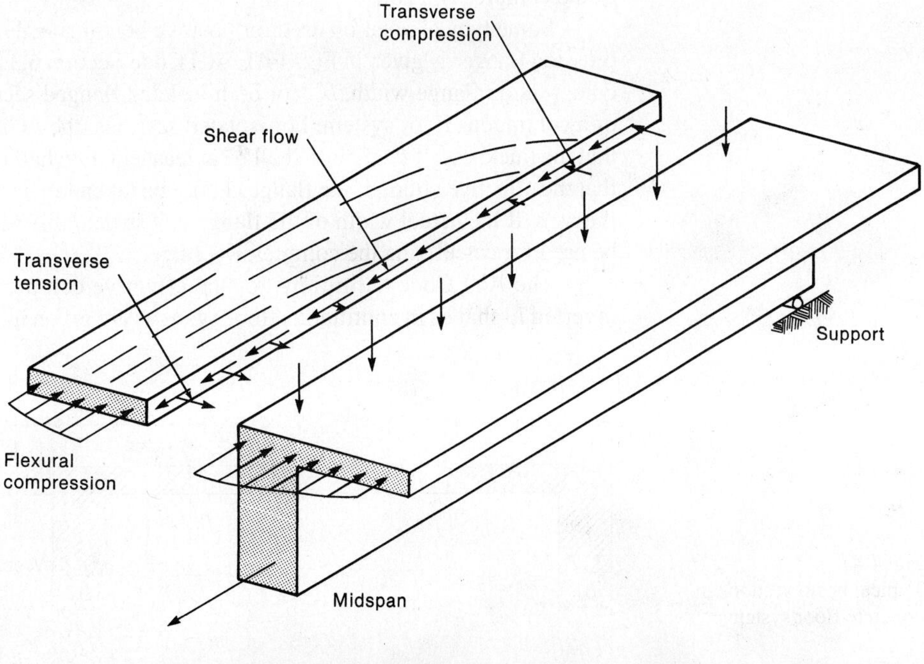

Fig. 4-39
Actual flow of forces on a
T-beam flange.

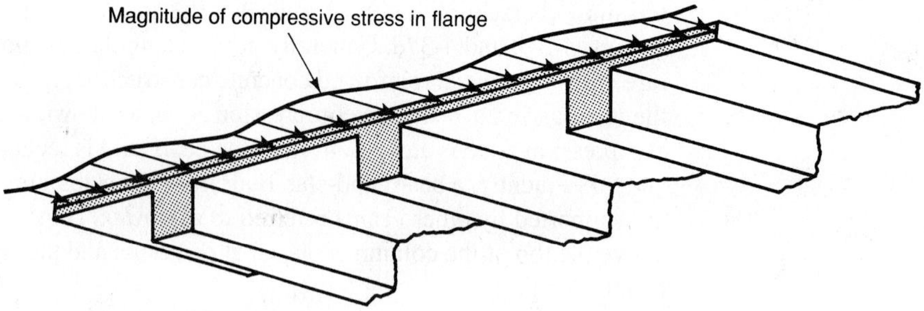

(a) Distribution of maximum flexural compressive stresses.

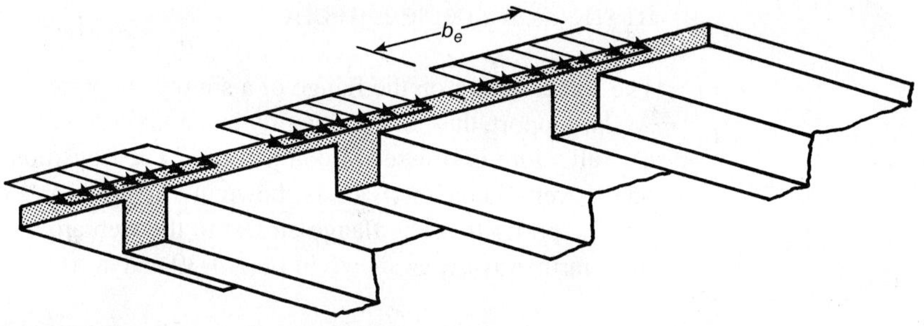

(b) Flexural compressive stress distribution assumed in design.

Fig. 4-40
Effective width of T-beams.

Figure 4-40a shows the distribution of the flexural compressive stresses in a slab that forms the flanges of a series of parallel beams at a section of maximum positive moment. The compressive stress is a maximum over each web, dropping between the webs. When analyzing and designing the section for positive moments, an *effective compression flange width* is used (Fig. 4-40b). When this width, b_e, is stressed uniformly to $0.85 f'_c$, it will give approximately the same compression *force* that actually is developed in the full width of the compression zone.

Some typical notation used for positive bending analysis of beam sections with compression flanges is given in Fig. 4-41. ACI Code Section 6.3.2 gives definitions for effective compression flange width, b_e, for both isolated flanged sections and sections that are part of a continuous floor system. For isolated sections, the ACI Code Section 6.3.2.2 requires that the thickness of the flange shall be at least equal to half of the thickness of the web, and that the effective width of the flange cannot be taken larger than four times the thickness of the web. If the actual width of the flange is less than this value, then the actual width is to be used for calculating the compression force.

The ACI Code definitions for the effective compression flange width for T- and inverted L-shapes in continuous floor systems are given in Section 6.3.2.1 and illustrated

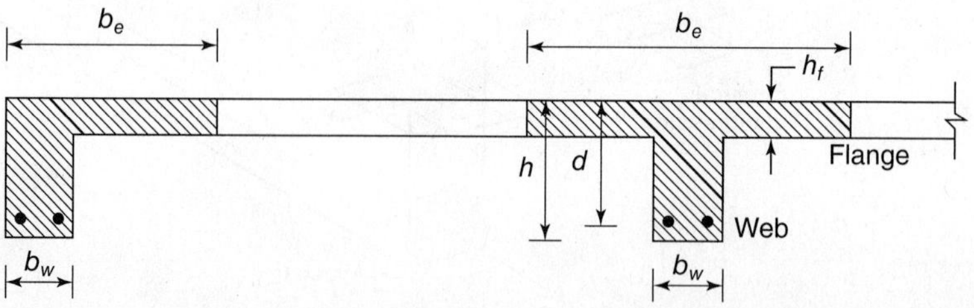

Fig. 4-41
Typical beam sections in concrete floor systems.

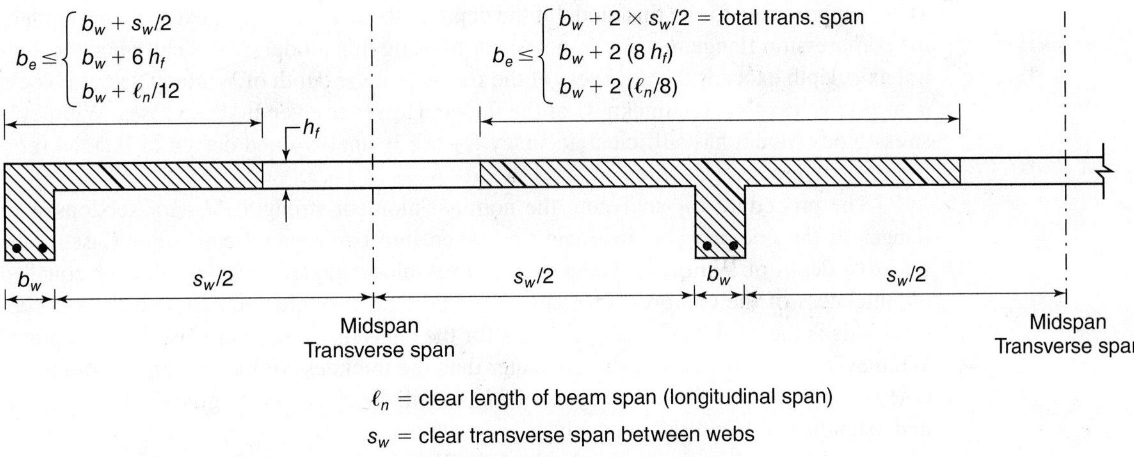

Fig. 4-42
ACI Code definitions for effective width of compression flange, b_e.

in Fig. 4-42. For T-shapes, the following three limits are given for the effective width of the *overhanging portion* of the compression flange:

(a) one-half of the clear span between webs, $s_w/2$
(b) eight times the thickness of the flange (slab), 8 h_f
(c) one-eighth of the clear span length of the beam, $\ell_n/8$

For L-shapes, the following three limits are given for the effective width of the *overhanging portion* of the compression flange:

(a) one-half of the clear span between webs, $s_w/2$
(b) six times the thickness of the flange (slab), 6 h_f
(c) one-twelfth of the clear span length of the beam, $\ell_n/12$

In the following sections for the analysis of section nominal moment strength, M_n, it will be assumed that b_e has already been determined. A sample evaluation of the effective width of the effective compression flange width for a T-section will be given at the start of Example 4-5.

Loads applied to the flange will cause negative moments in the flange where it joins the web. If the one-way floor slab is continuous and spans perpendicular to the beam, as in Fig. 4-36, the slab flexural reinforcement will be designed to resist these moments. If, however, the slab is not continuous (as in an isolated T-beam) or if the slab flexural reinforcement is parallel to the beam web (as is the case of the "girders" in Fig. 4-38) additional reinforcement is required at the top of the slab, perpendicular to the beam web. ACI Code Section 7.5.3.2 states that this reinforcement is to be designed by assuming that the flange acts as a cantilever loaded with the factored dead and live loads. For an isolated T-beam, the full overhanging flange width is considered. For a girder in a monolithic floor system (Fig. 4-38), the overhanging part of the effective flange width (defined above) is used in this calculation.

Analysis of Nominal Moment Strength for Flanged Sections in Positive Bending

As was done for rectangular sections, Whitney's stress block will be used to model the distribution of concrete compression stresses. This model was derived for a unit width, so it theoretically applies only for constant width compression zones. Therefore, it would seem to be inappropriate to use this model if the depth to the neutral axis, c, exceeds the depth of the compression flange, h_f. The largest error in using this model may occur when the neutral axis depth exceeds the thickness of the flange, but the depth of Whitney's stress block, $a = \beta_1 c$, is less than the thickness of the flange. However, even in those cases, Whitney's stress-block model has sufficient accuracy for use in analysis and design of flanged reinforced concrete beam sections.

The procedure for analyzing the nominal moment strength, M_n, for sections with flanges in the compression zone can be broken into two general cases. For Case 1, the effective depth of Whitney's compressive stress block model, a, is less than or equal to the thickness of the compression flange, h_f. For normal reinforced concrete flanged sections, this is the case that usually governs for the analysis of M_n. For Case 2, the depth of Whitney's stress block model, a, is greater than the thickness of the flange. Although this case seldom governs for the analysis of M_n, it will be discussed to give the reader a full understanding of beam section analysis procedures.

For a **Case 1** analysis, the depth of the Whitney stress-block model is less than or equal to the thickness of the compression flange, as shown in Fig. 4-43. Because this case usually will govern for reinforced concrete sections, it is recommended to start the analysis for nominal moment strength with this case and only switch to Case 2 if it is shown that the depth of Whitney's stress block exceeds the depth of the flange. The analysis of M_n is essentially the same as that covered in Section 4-4 for singly reinforced rectangular sections, except now the width of the compression zone is equal to the effective compression flange width b_e. The recommended steps for Case 1 are:

1. Assume $a = \beta_1 c \le h_f$
2. Assume $\varepsilon_s \ge \varepsilon_y$
3. From section equilibrium, use Eq. (4-16) to calculate a with b_e used in place of b:

$$a = \frac{A_s f_y}{0.85 f'_c\, b_e}$$

4. Show $a \le h_f$ (if yes continue; if not, go to Case 2)
5. Confirm $\varepsilon_s \ge \varepsilon_y$ (should be true, by inspection)

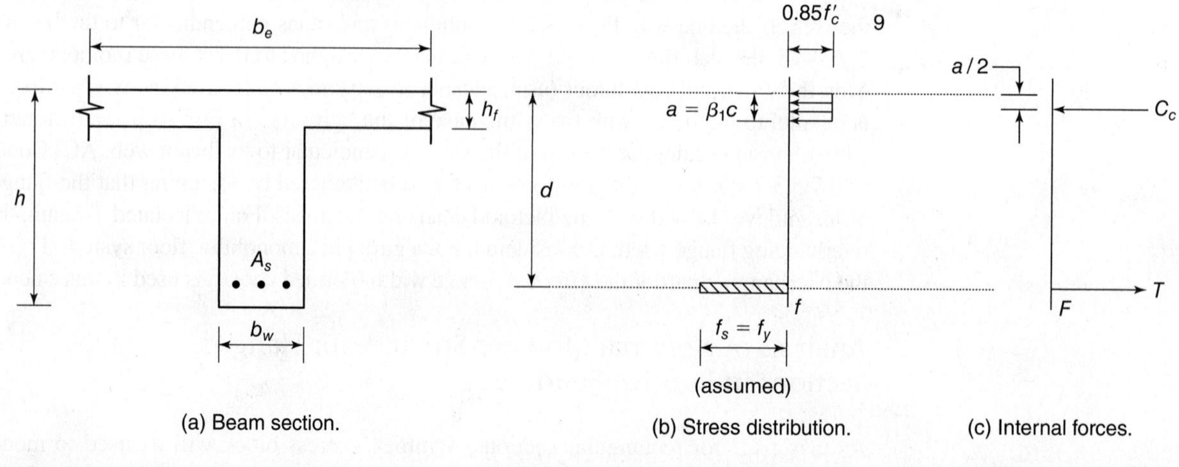

Fig. 4-43
Case 1 analysis ($\beta_1 c \le h_f$) for M_n in T-section.

6. Calculate M_n using Eq. (4-21):

$$M_n = A_s f_y\left(d - \frac{a}{2}\right) \tag{4-21}$$

The **Case 2** analysis procedure must be used if in step 4 of the Case 1 procedure the depth of Whitney's stress block exceeds the thickness of the flange. The assumption that the tension steel is yielding is retained. For Case 2 analysis, the section is artificially divided into two parts (i.e., the overhanging flanges and the full depth of the web, as shown in Fig. 4-44). The total area of tension reinforcement also is divided into two parts, but it is not important in this analysis to find a specific value for A_{sf} and A_{sw}. In Part 1 (Fig. 4-44b), the compression force in the overhanging portion of the flange is given as

$$C_{cf} = 0.85 f'_c (b_e - b_w) h_f \tag{4-36}$$

Every term in Eq. (4-36) is known. In Part 2 (Fig. 4-44c), the compression force in the web is given as

$$C_{cw} = 0.85 f'_c b_w a \tag{4-37}$$

In this equation, the depth of Whitney's stress block, a, is unknown. As before, we can find this by enforcing section equilibrium:

$$T = A_s f_y = C_{cf} + C_{cw}$$

And from this, solve for the depth of Whitney's stress block:

$$a = \frac{T - C_{cf}}{0.85 f'_c b_w} \tag{4-38}$$

As was done in the analysis of other beam sections at this stage, we will solve for the neutral axis depth, $c = a/\beta_1$, and confirm that the tension steel strain, ε_s, calculated using Eq. (4-18), is larger than the yield strain. Then, the nominal moment strength can be found by summing the moments from the two beam parts shown in Fig. 4-44. In this case with two compression forces, it is convenient to sum the moments caused by those two forces acting about the level of the tension reinforcement as

$$M_n = C_{cf}\left(d - \frac{h_f}{2}\right) + C_{cw}\left(d - \frac{a}{2}\right) \tag{4-39}$$

For both the Case 1 and Case 2 analysis procedures described, it was assumed that no compression reinforcement was used in the section. Although such reinforcement usually will have very little effect on the nominal moment capacity of a beam section with a compression flange, its contribution could be included following a procedure similar to those described in Section 4-7. If it is determined that this is a Case 1 analysis (step 4), then definitely ignore the compression reinforcement and calculate M_n using Eq. (4-21). However, if it is determined to be a Case 2 analysis, the compression reinforcement can be included in Part 2 of the Case 2 analysis procedure described, following the steps for compression reinforcement described in Section 4-7. The resulting expression for the nominal moment strength will be

$$M_n = C_{cf}\left(d - \frac{h_f}{2}\right) + C_{cw}\left(d - \frac{a}{2}\right) + C_s(d - d') \tag{4-40}$$

where the term C_s is defined in Eq. (4-31).

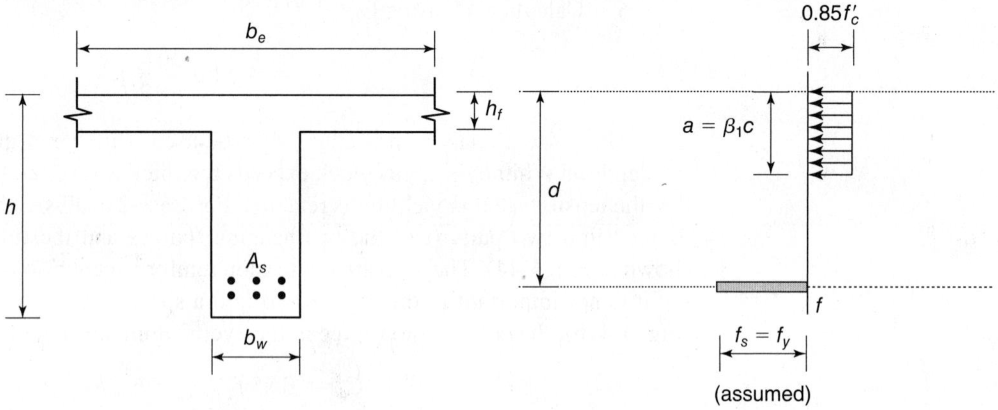

(a) Total T-section and stress distribution.

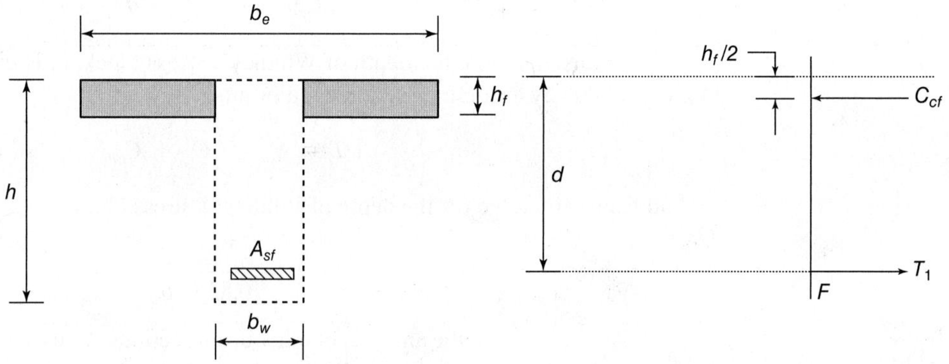

(b) Part 1: Overhanging flange(s) and corresponding internal forces.

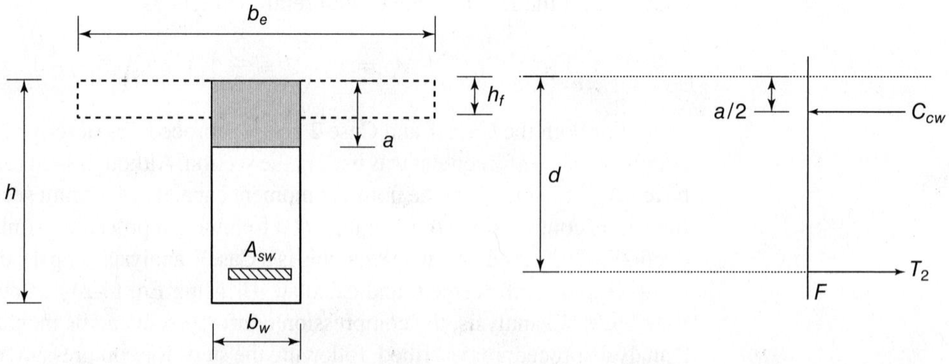

(c) Part 2: Web of section and corresponding internal forces.

Fig. 4-44
Case 2 analysis ($\beta_1 c > h_f$) for M_n in T-section.

Determination of Strength-Reduction Factor, ϕ

Most flanged sections will be *tension-controlled sections* with the strength-reduction factor, ϕ, equal to 0.90. This quickly can be shown by comparing the calculated value of the depth to the neutral axis, c, to the value of the neutral axis depth for the tension-controlled limit, 3/8 of d_t, as given by Eq. (4-26). If there is any doubt, Eq. (4-35) can be used to calculate the strain at the level of the extreme layer of tension reinforcement, ε_t, and show that it is greater than or equal to 0.005. If the calculated value of ε_t is less than 0.005 but more than 0.002, than this is a *transition zone section* and Eq. (4-28a) should be used to calculate the strength-reduction factor, ϕ. For a section with a compression flange, it would be very difficult to put in enough tension reinforcement to make the section over-reinforced ($\varepsilon_t < 0.002$), and thus have $\phi = 0.65$.

Evaluation of $A_{s,\,\min}$ in Flanged Sections

The general expression for $A_{s,\min}$ was given in Eq. (4-11). However, it is not unusual for some confusion to develop when applying this equation to flanged sections. The primary question is, which section width, b_w or b_e, should be used in Eq. (4-11). The reader should recall that the specification of a minimum area of tension reinforcement is used to prevent a sudden flexural failure at the onset of flexural tension cracking. For a typical T-section subjected to positive bending, flexural tension cracking will initiate at the bottom of the section, and thus, the use of b_w is appropriate.

The answer for bending moments that put the flange portion of the section in tension is not quite as clear. It is reasonable that some consideration should be given to the potentially larger tension force that will be released when cracking occurs in the flange portion of the section. Based on several years of satisfactory performance using Eq. (4-11) for the design of *continuous reinforced concrete floor systems*, the ACI Code does not recommend any modification of Eq. (4-11) when used in bending zones that put the flanged portion of the beam sections in tension. However, for *statically determinate* beams where the flange portion of the section is in tension, ACI Code Section 9.6.1.2 recommends that b_w in Eq. (4-11) be replaced by the *smaller* of $2b_w$ or b_e, but need not exceed the actual flange width. Members that fit into this category could be a T-section used in a cantilever span or an inverted T-section (Fig. 4-5), sometimes referred to as a ledger beam, used to span between simple supports.

EXAMPLE 4-5 Analysis of T-Sections in Positive and Negative Bending

1. Determine b_e for a beam T-section that is part of a continuous floor system. Consider the portion of the continuous floor system shown in Fig. 4-45 and the central floor beam spanning in the horizontal direction. The beam sections corresponding to section lines *A–A* and *B–B* in Fig. 4-45 are given in Figs. 4-46 and 4-47, respectively. Assume the clear span, ℓ_n, is 23 ft. The limits given in ACI Code Section 6.3.2.1 for determining the effective width of the compression flange for a beam section in a continuous floor system are

$$b_e \le b_w + 2\left(\frac{\ell_n}{8}\right) = 12\text{ in.} + 2(34.5\text{ in.}) = 81\text{ in.}$$

$$b_e \le b_w + 2(8h_f) = 12\text{ in.} + 16(5\text{ in.}) = 92\text{ in.}$$

$$b_e \le b_w + 2\left(\frac{10\text{ ft} - b_w}{2}\right) = 10\text{ ft} = 120\text{ in.}$$

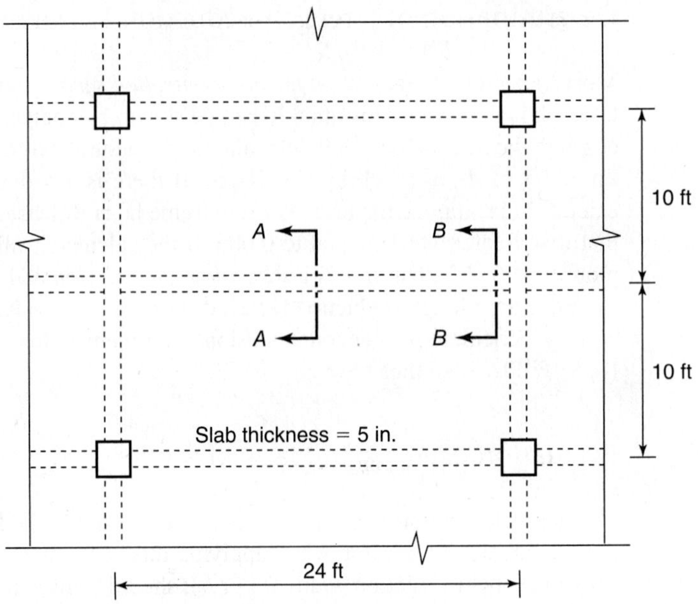

Fig. 4-45
Continuous floor system for Example 4-5.

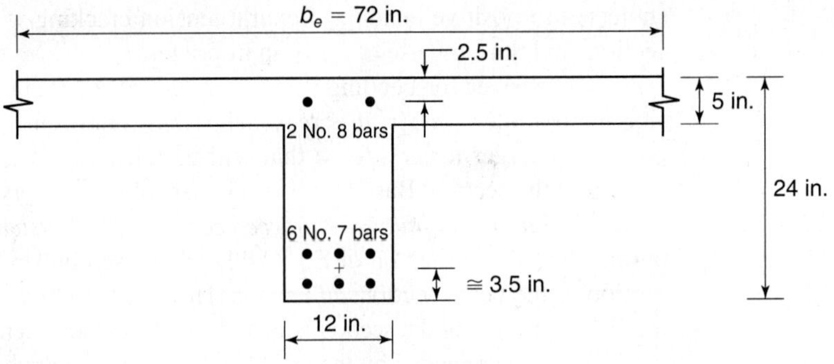

Fig. 4-46
Section *A–A* from continuous floor system in Fig. 4-45.

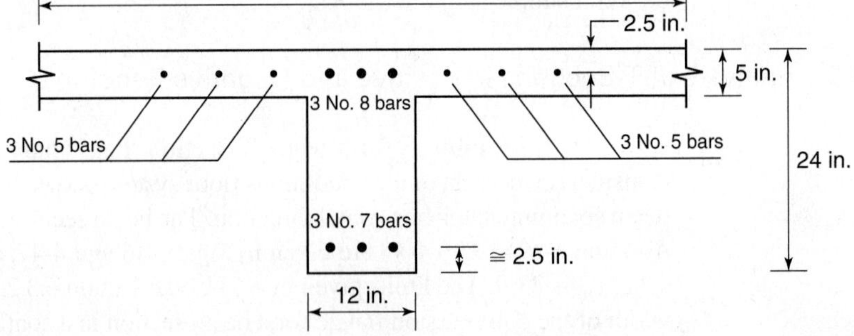

Fig. 4-47
Section *B–B* from continuous floor system in Fig. 4-45.

It should be noted that for a floor system with a uniform spacing between beams, the third limit defined above should always result in a value equal to the center-to-center spacing between the beams. The first limit governs for this section, so in the following parts of this example it will be assumed that $b_e = 81$ in.

For Parts (1) and (2) use the following material properties:

$$f'_c = 4000 \text{ psi } (\beta_1 = 0.85) \text{ and } f_y = 60 \text{ ksi}$$

2. For the T-section in Fig. 4-46, calculate ϕM_n and $A_{s,min}$. For the given section, $A_s = 6(0.60 \text{ in.}^2) = 3.60 \text{ in.}^2$ and $A'_s = 2(0.79 \text{ in.}^2) = 1.58 \text{ in.}^2$

For a typical floor system, midspan sections are subjected to positive bending, and sections near the end of the span are subjected to negative bending. The beam section in Fig. 4-46 represents the midspan section of the floor beam shown in Fig. 4-45 and thus is subjected to positive bending. The tension reinforcement for this section is provided in two layers. The minimum spacing required between layers of reinforcement is 1 in. (ACI Code Section 25.2.2). Thus, the spacing between the centers of the layers is approximately 2 in. Assuming the section will include a No. 3 or No. 4 stirrup, it is reasonable to assume that the distance from the extreme tension edge of the section to the centroid of the lowest layer of steel is approximately 2.5 in. So, the distance from the tension edge to the centroid of the total tension reinforcement is approximately 3.5 in. Thus, the effective flexural depth, d, and the distance from the top of the section (compression edge) to the extreme layer of tension reinforcement, d_t, can be calculated to be

$$d = 24 \text{ in.} - 3.5 \text{ in.} = 20.5 \text{ in.}$$

$$d_t = 24 \text{ in.} - 2.5 \text{ in.} = 21.5 \text{ in.}$$

Calculation of ϕM_n Assume this is a Case 1 analysis ($a \leq h_f$) and assume that the tension steel is yielding ($\varepsilon_s \geq \varepsilon_y$). For section equilibrium, use Eq. (4-16) with b_e substituted for b, giving

$$a = \frac{A_s f_y}{0.85 f'_c b_e} = \frac{(3.60 \text{ in.}^2)(60 \text{ ksi})}{0.85(4 \text{ ksi})(81 \text{ in.})} = 0.78 \text{ in.}$$

This is less than h_f, as expected. This value also is less than d', so we can ignore the compression reinforcement for the analysis of M_n. This is a very common result for a T-section in positive bending. For such beams with large compression zones, compression steel is not required for additional moment strength. The compression steel in this beam section may be present for reinforcement continuity requirements (Chapter 8), to reduce deflections (Chapter 9), or to simply support shear reinforcement.

The depth to the neutral axis, c, which is equal to a/β_1, will be approximately equal to 1 in. Comparing this to the values for d and d_t, it should be clear without doing calculations that the tension steel strain, ε_s, easily exceeds the yield strain (0.00207) and the strain at the level of the extreme layer of tension reinforcement, ε_t, easily exceeds the limit for tension-controlled sections (0.005). Thus, $\phi = 0.9$, and we can use Eq. (4-21) to calculate M_n as

$$M_n = A_s f_y\left(d - \frac{a}{2}\right) = (3.60 \text{ in.}^2)(60 \text{ ksi})\left(20.5 \text{ in.} - \frac{0.78 \text{ in.}}{2}\right)$$

$$M_n = 4340 \text{ k-in.} = 362 \text{ k-ft}$$

$$\phi M_n = 0.9 \times 362 = 326 \text{ k-ft}$$

Check of $A_{s,min}$ Tension is at the bottom of this section, so it is clear that we should use b_w in Eq. (4-11). Also, $3\sqrt{f'_c}$ is equal to 190 psi, so use 200 psi in the numerator:

$$A_{s,\min} = \frac{200 \text{ psi}}{f_y} b_w d = \frac{200 \text{ psi}}{60{,}000 \text{ psi}}(12 \text{ in.})(20.5 \text{ in.}) = 0.82 \text{ in.}^2 < A_s \text{ (o.k.)}$$

3. For the T-section shown in Fig. 4-47, calculate ϕM_n and $A_{s,min}$. Because this section is subjected to negative bending, flexural tension cracking will develop in the top flange and the compressive zone is at the bottom of the section. Note that ACI Code Section 24.3.4 requires that a portion of the tension reinforcement be distributed into the flange, which coincidentally allows all the negative-moment tension reinforcement to be placed in one layer. Thus, assume that the No. 5 bars in the flange are part of the tension reinforcement. So, for the given section,

$$A_s = 3 \times 0.79 \text{ in.}^2 + 6 \times 0.31 \text{ in.}^2 = 4.23 \text{ in.}^2$$

$$A_s' = 3 \times 0.60 \text{ in.}^2 = 1.80 \text{ in.}^2$$

Using assumptions similar to those used in prior examples, d' is approximately equal to 2.5 in. and $d = d_t$ is approximately equal to the total beam depth, h, minus 2.5 in., i.e., 21.5 in.

Calculation of ϕM_n Because this is a doubly reinforced section, we initially will assume the tension steel is yielding and use the trial-and-error procedure described in Section 4-7 to find the neutral axis depth, c.

Try $c = d/4 \approx 5.5$ in.

$$\varepsilon_s' = \left(\frac{c - d'}{c}\right)\varepsilon_{cu} = \left(\frac{5.5 \text{ in.} - 2.5 \text{ in.}}{5.5 \text{ in.}}\right)(0.003) = 0.00164$$

$$f_s' = E_s\varepsilon_s' = 29{,}000 \text{ ksi} \times 0.00164 = 47.5 \text{ ksi } (\leq f_y)$$

$$C_s = A_s'(f_s' - 0.85 f_c') = 1.80 \text{ in.}^2 (47.5 \text{ ksi} - 3.4 \text{ ksi}) = 79.3 \text{ kips}$$

$$C_c = 0.85 f_c' b_w \beta_1 c = 0.85 \times 4 \text{ ksi} \times 12 \text{ in.} \times 0.85 \times 5.5 \text{ in.} = 191 \text{ kips}$$

$$T = A_s f_y = 4.23 \text{ in.}^2 \times 60 \text{ ksi} = 254 \text{ kips}$$

Because $T < C_c + C_s$, we should decrease c for the second trial.

Try $c = 5.1$ in.

$$\varepsilon_s' = 0.00153$$
$$f_s' = 44.4 \text{ ksi } (\leq f_y)$$
$$C_s = 73.7 \text{ kips}$$
$$C_c = 177 \text{ kips}$$
$$T = 254 \text{ kips} \cong C_s + C_c = 251 \text{ kips}$$

With section equilibrium established, we must confirm the assumption that the tension steel is yielding. Because $d = d_t$ for this section, we can confirm that this is a tension-controlled section in the same step. Using Eq. (4-18):

$$\varepsilon_s(= \varepsilon_t) = \frac{d - c}{c}\varepsilon_{cu} = \left(\frac{21.5 \text{ in.} - 5.1 \text{ in.}}{5.1 \text{ in.}}\right)0.003 = 0.00965$$

Clearly, the steel is yielding ($\varepsilon_s > \varepsilon_y = 0.00207$) and this is a tension-controlled section ($\varepsilon_t > 0.005$). So, using $a = \beta_1 c = 0.85 \times 5.1 \text{ in.} = 4.34$ in., use Eq. (4-34) to calculate M_n as

$$M_n = C_c\left(d - \frac{a}{2}\right) + C_s(d - d') = 177 \text{ k} \times 19.3 \text{ in.} + 73.7 \text{ k} \times 19.0 \text{ in.}$$

$$M_n = 3420 \text{ k-in.} + 1400 \text{ k-in.} = 4820 \text{ k-in.} = 401 \text{ k-ft}$$

$$\phi M_n = 0.9 \times 401 = 361 \text{ k-ft}$$

Calculation of $A_{s,min}$ As discussed in Section 4-8, the value of $A_{s,min}$ for beam sections with a flange in the tension zone is a function of the use of that beam. The beam section for this example is used in the negative bending zone of a continuous, statically indeterminate floor system. Thus, the minimum tension reinforcement should be calculated using b_w, as was done in Part (2) of this example. Using an effective depth, d, of 21.5 in., and noting that $3\sqrt{f'_c}$ is less than 200 psi, the following value is calculated using Eq. (4-11):

$$A_{s,\min} = \frac{200 \text{ psi}}{f_y} b_w d = \frac{200 \text{ psi}}{60{,}000 \text{ psi}} (12 \text{ in.})(21.5 \text{ in.}) = 0.86 \text{ in.}^2 < A_s \text{ (o.k.)}$$

If the beam section considered here was used as a statically determinate cantilever beam subjected to gravity loading (all negative bending), then the term b_w should be replaced with the smaller of $2b_w$ or b_e. For this section, $2b_w$ is the smaller value, so for such a case, the value of $A_{s,min}$ would be

$$A_{s,\min} = \frac{200 \text{ psi}}{f_y} (2b_w)d = 1.72 \text{ in.}^2 < A_s \text{ (o.k.)}$$ ■

EXAMPLE 4-6 Analysis of a T-Beam with the Neutral Axis in the Web

Compute the positive moment strength ϕM_n and $A_{s,min}$ for the beam shown in Fig. 4-48. Assume that the concrete and steel strengths are 3000 psi and 60 ksi, respectively. Also assume the beam contains No. 3 stirrups as shear reinforcement, which are not shown in Fig. 4-48.

1. Compute b_e. Assume this beam is an isolated T-beam in which the flange is used to increase the area of the compression zone. For such a beam, ACI Code Section 6.3.2.2 states that the flange thickness shall not be less than one-half the width of the web and that the effective flange width shall not exceed four times the width of the web. By observation, the given flange dimensions satisfy these limits. Thus, $b_e = 18$ in.

2. Compute *d*. As in the prior example with two layers of tension reinforcement, assume $d \approx h - 3.5 \text{ in.} = 24.5$ in., as shown in Fig. 4-48a.

3. Compute *a*. Assume this is a Case 1 analysis ($a \le h_f$), and thus, the compression zone will be rectangular. Accordingly,

$$a = \frac{A_s f_y}{0.85 f'_c b_e} = \frac{4.74 \text{ in.}^2 \times 60 \text{ ksi}}{0.85 \times 3 \text{ ksi} \times 18 \text{ in.}} = 6.20 \text{ in.}$$

Because a is greater than the thickness of the flange ($h_f = 5$ in.), our assumption that the compression zone is rectangular is wrong, and our calculated value of a is *incorrect*. It therefore is necessary to use the Case 2 analysis procedure discussed in Section 4-8 and artificially break the section into two beams shown as beam F and beam W in Figs. 4-48c and 4-48d, respectively.

4. Analysis of M_n for the flanged section with $a > h_f$. The compression force in beam F is given by Eq. (4-36) as

$$C_{cf} = 0.85 f'_c (b_e - b_w) h_f = 0.85 \times 3 \text{ ksi}(18 \text{ in.} - 10 \text{ in.}) \times 5 \text{ in.} = 102 \text{ kips}$$

The compression force in beam W is given by Eq. (4-37) as

$$C_{cw} = 0.85 f'_c b_w a$$

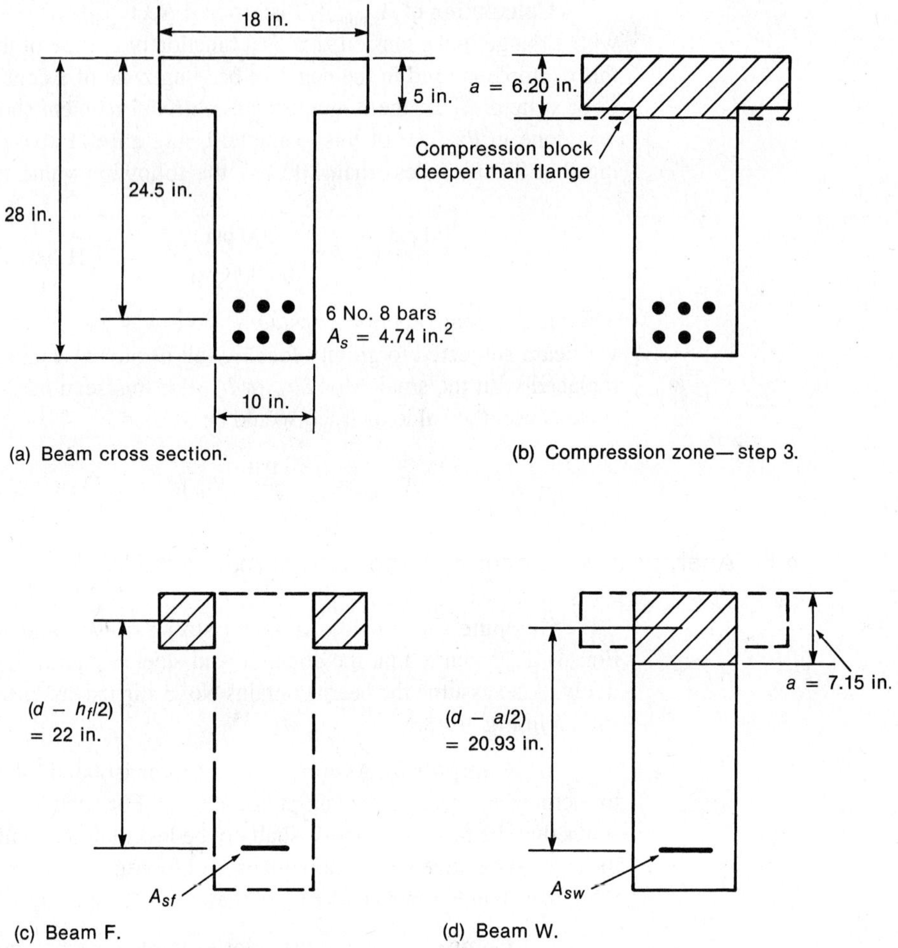

Fig. 4-48
Beam sections for Example 4-6.

Of course, the depth of Whitney's stress block, a, is the major unknown for this section analysis procedure. It is found by setting the tension force, $T = A_s f_y = 4.74 \text{ in.}^2 \times 60 \text{ ksi} = 284$ kips, equal to the sum of the compression forces, as was done to derive Eq. (4-38):

$$a = \frac{T - C_{cf}}{0.85 f'_c b_w} = \frac{284 \text{ k} - 102 \text{ k}}{0.85 \times 3 \text{ ksi} \times 10 \text{ in.}} = 7.15 \text{ in.}$$

With this value of a, $C_{cw} = 182$ kips. Before calculating M_n, we must confirm that the tension steel is yielding. Using $c = a/\beta_1 = 7.15/0.85 = 8.42$ in., Eq. (4-18) can be used to calculate the tension steel strain as

$$\varepsilon_s = \frac{d - c}{c}\varepsilon_{cu} = \left(\frac{24.5 \text{ in.} - 8.42 \text{ in.}}{8.42 \text{ in.}}\right)0.003 = 0.00573$$

This clearly exceeds the yield strain (0.00207), so the assumption that the tension steel is yielding is confirmed. It should be noted that the distance to the extreme layer of tension steel, d_t, will exceed d for this section, so the strain at the extreme layer of tension steel, ε_t, will exceed ε_s. Thus, ε_t exceeds 0.005, making this a tension-controlled section with ϕ equal to 0.9. Using Eq. (4-39) to calculate M_n,

$$M_n = C_{cf}\left(d - \frac{h_f}{2}\right) + C_{cw}\left(d - \frac{a}{2}\right)$$

$$M_n = 102\text{ k}\left(24.5\text{ in.} - \frac{5\text{ in.}}{2}\right) + 182\text{ k}\left(24.5\text{ in.} - \frac{7.15\text{ in.}}{2}\right)$$

$$M_n = 2240\text{ k-in.} + 3820\text{ k-in.} = 6060\text{ k-in.} = 505\text{ k-ft}$$

$$\text{and, } \phi M_n = 0.9 \times 505 = 455\text{ k-ft}$$

5. Check whether $A_s \geq A_{s,\min}$. Assuming this beam is continuous over several spans, we can use $b = b_w$ for this calculation. For a concrete strength of 3000 psi, note that $3\sqrt{f'_c}$ is less than 200 psi, so use 200 psi in Eq. (4-11), giving

$$A_{s,\min} = \frac{200\text{ psi}}{f_y} b_w d = \frac{200\text{ psi}}{60{,}000\text{ psi}}(10\text{ in.})(24.5\text{ in.}) = 0.82\text{ in.}^2 < A_s\text{ (o.k.)}$$ ■

4-9 UNSYMMETRICAL BEAM SECTIONS

Figure 4-49 shows half of the span of a simply supported beam with an unsymmetrical cross section. The loads lie in a plane referred to as the *plane of loading*, and it is assumed that this passes through the shear center of the unsymmetrical section. This beam

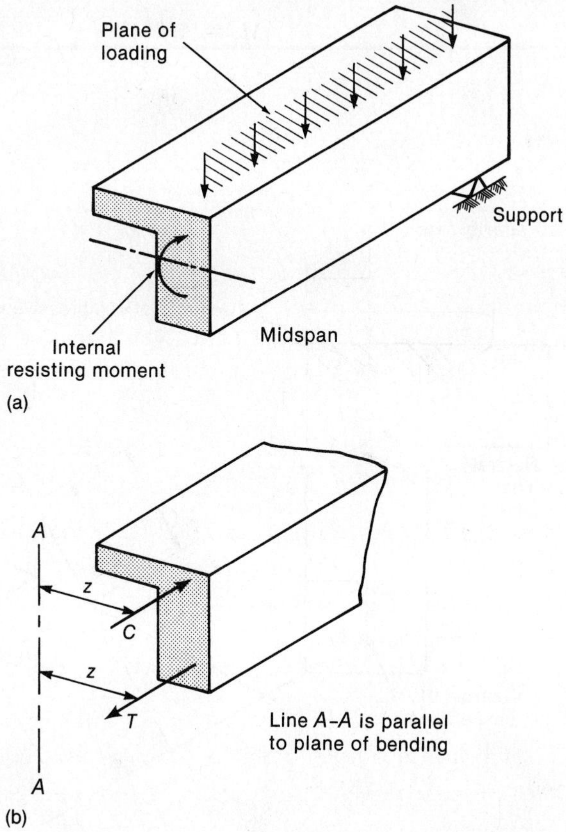

Fig. 4-49
Location of C and T forces in an unsymmetrical beam section.

is free to deflect vertically and laterally between its supports. The applied loads cause moments that must be resisted by an internal resisting moment about a horizontal axis, shown by the moment in Fig. 4-49a. This internal resisting moment results from compressive and tensile forces C and T, as shown in Fig. 4-49b. Because the applied loads do not cause a moment about an axis parallel to the plane of loading (such as section A–A), the internal force resultants C and T cannot do so either. As a result, C and T both must lie in the plane of loading or in a plane parallel to it. The distances z in Fig. 4-49b must be equal.

Figure 4-50 shows a cross section of an inverted L-shaped beam loaded with gravity loads. Because this beam is loaded with vertical loads, leading to moments about a horizontal axis, the line joining the centroids of the compressive and tensile forces must be vertical as shown (both are a distance f from the right-hand side of the beam). As a result, the shape of the compression zone must be triangular and the neutral axis must be inclined, as shown in Fig. 4-50.

Because $C = T$, and assuming that $f_s = f_y$,

$$\frac{1}{2}(3f \times g \times 0.85f'_c) = A_s f_y \tag{4-41}$$

Because the moment is about a horizontal axis, the lever arm must be vertical. Thus, for the case shown in Fig. 4-50,

$$jd = d - \frac{g}{3} \tag{4-42}$$

and

$$M_n = A_s f_y\left(d - \frac{g}{3}\right) \tag{4-43}$$

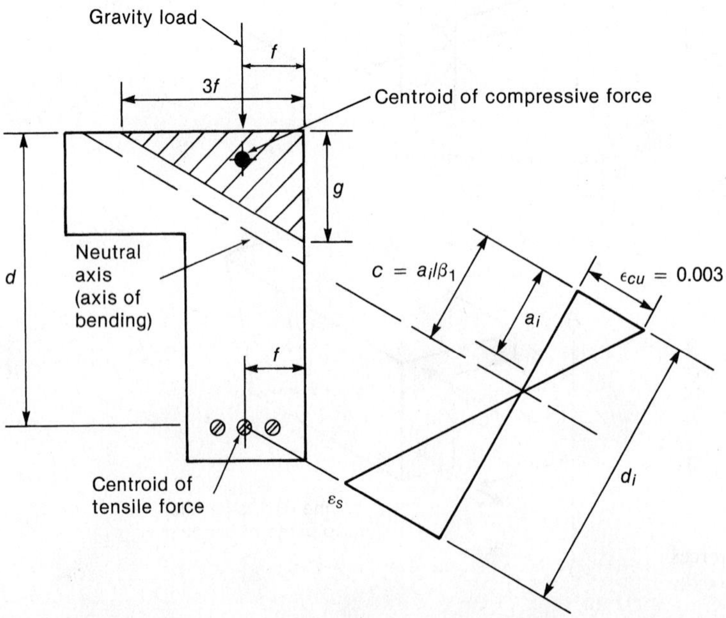

Fig. 4-50
Unsymmetrical beam.

These equations apply only to the triangular compression zone shown in Fig. 4-50. Different equations or a trial-and-error solution generally will be necessary for other shapes.

The rectangular stress block was derived for rectangular beams with the neutral axis parallel to the compression face of the beam. This assumes the resultant compression force C can be reached without crushing of the extreme fibers. Rüsch [4-17] and Mattock, et al. [4-6] have studied this and conclude the rectangular stress block is applicable to a wide variety of shapes of compression zones.

The checks of whether $f_s = f_y$ or whether the section is tension-controlled, respectively, are done by checking steel strains using the inclined strain diagram in Fig. 4-50. For one layer of tension steel, we can assume that $\varepsilon_s = \varepsilon_t$ and use Eq. (4-18) to calculate ε_s, using d_i in place of d.

The discussion to this point has dealt with *isolated* beams which are free to deflect both vertically and laterally. Such a beam would deflect perpendicular to the axis of bending, that is, both vertically and laterally. If the beam in Fig. 4-50 were the edge beam for a continuous slab that extended to the left, the slab would prevent lateral deflections. As a result, the neutral axis would be forced to be very close to horizontal and the beam could be analyzed in the normal fashion.

EXAMPLE 4-7 Analysis of an Unsymmetrical Beam

The beam shown in Fig. 4-51 has an unsymmetrical cross section and an unsymmetrical arrangement of reinforcement. This beam is subjected to vertical loads only. Compute ϕM_n and $A_{s,\min}$ for this cross section if $f'_c = 3000$ psi ($\beta_1 = 0.85$) and $f_y = 60{,}000$ psi.

1. Assume that $f_s = f_y$ and compute the size of the compression zone. The centroid of the three bars is computed to be at 6.27 in. from the right side of the web. The centroid of the compression zone also must be located this distance from the side of the web. Thus, the width of the compression zone is $3 \times 6.27 = 18.8$ in.

Because $C = T$,

$$\frac{1}{2}(18.8 \times g \times 0.85 f'_c) = A_s f_y$$

or

$$g = \frac{2.58 \text{ in.}^2 \times 60 \text{ ksi} \times 2}{18.8 \text{ in.} \times 0.85 \times 3 \text{ ksi}} = 6.46 \text{ in.}$$

The compression zone is shown shaded in Fig. 4-51. If the compression zone were deeper than shown and cut across the reentrant corner, a more complex trial-and-error solution would be required.

2. Check if $f_s = f_y$ and whether the section is tension-controlled. From Fig. 4-51a, it can be found that $\alpha = 19°$, $a_i = 6.11$ in., and $d_i = 22.4$ in. Find $c = a_i/\beta_1 = 7.19$ in., then use Eq. (4-18) with d_i replacing d to find

$$\varepsilon_s = \left(\frac{d_i - c}{c}\right)\varepsilon_{cu} = \frac{22.4 \text{ in.} - 7.19 \text{ in.}}{7.19 \text{ in.}} \times 0.003 = 0.00635$$

Clearly, the tension steel is yielding ($\varepsilon_s > \varepsilon_y$), and this is a tension-controlled section ($\varepsilon_t = \varepsilon_s > 0.005$), so $\phi = 0.9$.

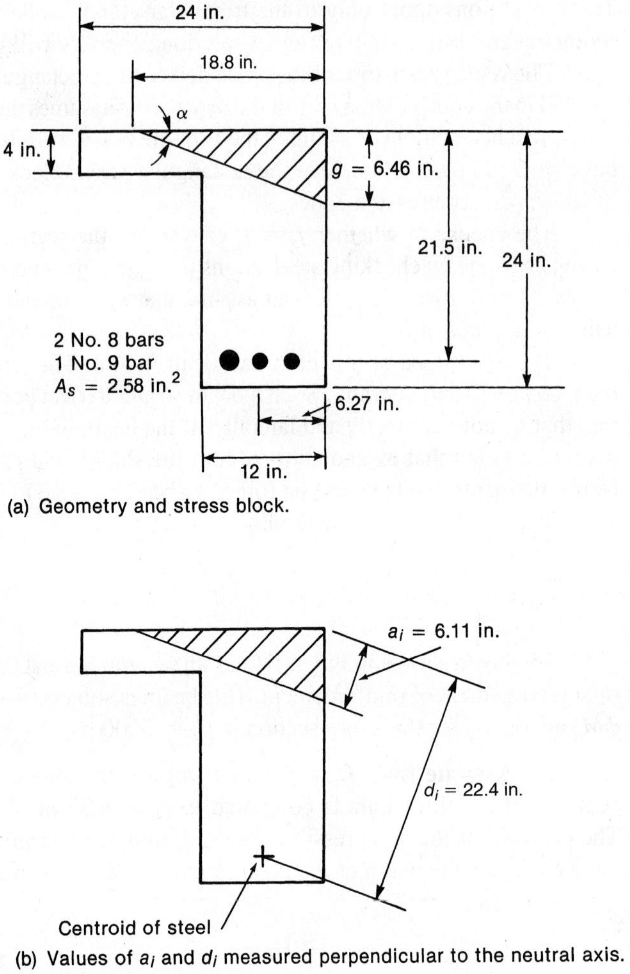

Fig. 4-51
Beam section for
Example 4-7.

3. Compute ϕM_n by applying ϕ to Eq. (4-43):

$$\phi M_n = \phi\left[A_s f_y\left(d - \frac{g}{3}\right)\right]$$

$$\phi M_n = 0.9\left[2.58\text{ in.}^2 \times 60\text{ ksi}\left(21.5\text{ in.} - \frac{6.46\text{ in.}}{3}\right)\right]$$

$$= 2700\text{ k-in.} = 225\text{ k-ft}$$

Note that the moment calculation is based on the lever arm measured *vertically* (parallel to the plane of loading).

4. Check if $A_s \geq A_{s,\min}$. Again, for 3000 psi concrete, $3\sqrt{f'_c}$ is less than 200 psi. So, the value of $A_{s,\min}$ from Eq. (4-11) is

$$A_{s,\min} = \frac{200\text{ psi}}{f_y} b_w d = \frac{200\text{ psi}}{60{,}000\text{ psi}}(12\text{ in.})(21.5\text{ in.}) = 0.86\text{ in.}^2 < A_s\text{ (o.k.)} \quad \blacksquare$$

PROBLEMS

Note: For all problems use load factors from ACI Code Section 5.3.1 and strength reduction factors from ACI Code Tables 21.2.1 and 21.2.2.

4-1 Figure P4-1 shows a simply supported beam and the cross section at midspan. The beam supports a uniform service (unfactored) dead load consisting of its own weight plus 1.5 kips/ft and a uniform service (unfactored) live load of 1.2 kips/ft. The concrete strength is 4500 psi, and the yield strength of the reinforcement is 60,000 psi. The concrete is normal-weight concrete. For the midspan section shown in Fig. P4-1b, compute ϕM_n and show that it exceeds M_u.

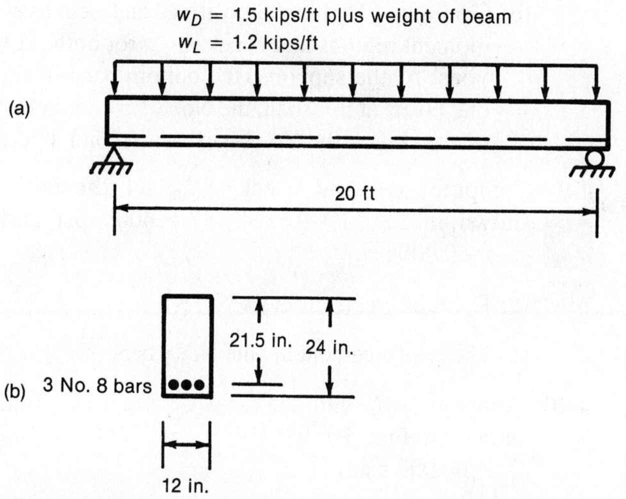

Fig. P4-1

4-2 A cantilever beam shown in Fig. P4-2 supports a uniform service (unfactored) dead load of 1 kip/ft plus its own weight and a concentrated service (unfactored) live load of 12 kips, as shown. The concrete is normal-weight concrete with f'_c = 4000 psi and the steel is Grade 60. For the end section shown in Fig. P4-2b, compute ϕM_n and show that it exceeds M_u.

Video Solution

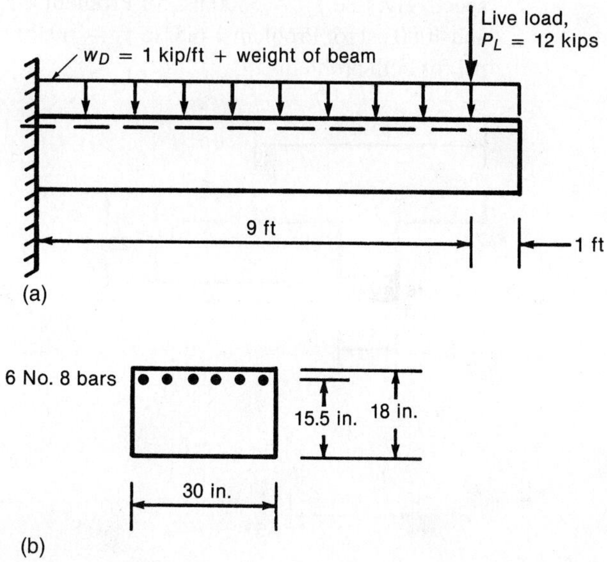

Fig. P4-2

4-3 (a) Compare ϕM_n for singly reinforced rectangular beams having the following properties.

(b) Taking beam 1 as the reference point, discuss the effects of changing A_s, f_y, f'_c, and d on ϕM_n. (Note that each beam has the same properties as beam 1 except for the italicized quantity.)

(c) What is the most effective way of increasing ϕM_n? What is the least effective way?

Beam No.	b (in.)	d (in.)	Bars	f'_c (psi)	f_y (psi)
1	12	22	3 No. 7	4000	60,000
2	12	22	*2 No. 9 plus 1 No. 8*	4000	60,000
3	12	22	3 No. 7	4000	*80,000*
4	12	22	3 No. 7	*6000*	60,000
5	12	*33*	3 No. 7	4000	60,000

4-4 A 12-ft-long cantilever supports its own weight plus an additional uniform service (unfactored) dead load of 0.5 kip/ft. The beam is made from normal-weight 4000-psi concrete and has b = 16 in., d = 15.5 in., and h = 18 in. It is reinforced with four No. 7 Grade-60 bars. Compute the maximum service (unfactored) concentrated live load that can be applied at 1 ft from the free end of the cantilever.

4-5 and 4-6 Compute ϕM_n and check $A_{s,\min}$ for the beams shown in Figs. P4-5 and P4-6, respectively. Use $f'_c = 5000$ psi for Problem 4-5 and 4000 psi for Problem 4-6. Use $f_y = 60{,}000$ psi for both problems.

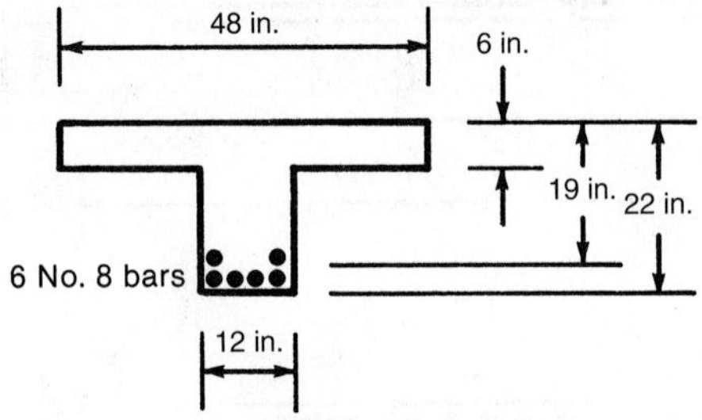

Fig. P4-5

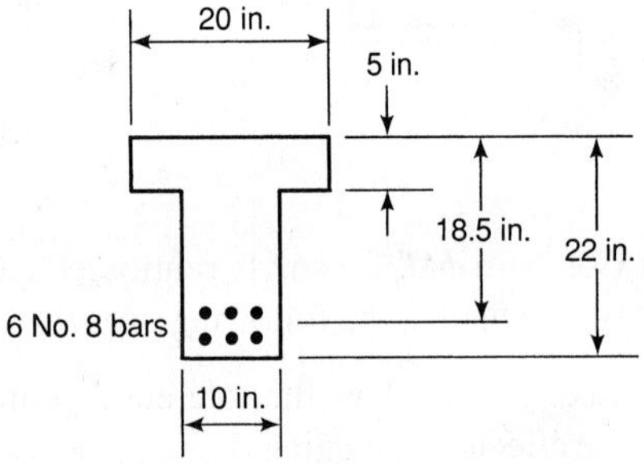

Fig. P4-6

4-7 Compute the negative-moment capacity, ϕM_n, and check $A_{s,\min}$ for the beam shown in Fig. P4-7. Use $f'_c = 4500$ psi and $f_y = 60{,}000$ psi.

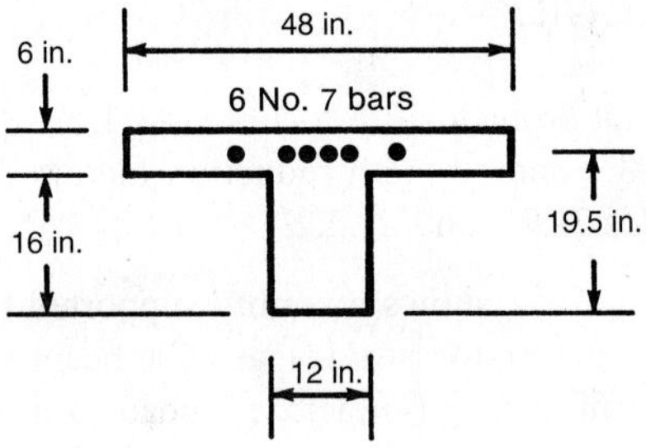

Fig. P4-7

4-8 For the beam shown in Fig. P4-8, $f'_c = 4000$ psi and $f_y = 60{,}000$ psi.

Video Solution

(a) Compute the effective flange width at midspan.

(b) Compute ϕM_n for the positive- and negative-moment regions and check $A_{s,\min}$ for both sections. At the supports, the bottom bars are in one layer; at midspan, the No. 8 bars are in the bottom layer, the No. 7 bars in a second layer.

4-9 Compute ϕM_n and check $A_{s,\min}$ for the beam shown in Fig. P4-9. Use $f'_c = 4000$ psi and $f_y = 60{,}000$ psi.

Video Solution

(a) The reinforcement is six No. 8 bars.

(b) The reinforcement is nine No. 8 bars.

4-10 Compute ϕM_n and check $A_{s,\min}$ for the beam shown in Fig. P4-10. Use $f'_c = 5000$ psi and $f_y = 60{,}000$ psi.

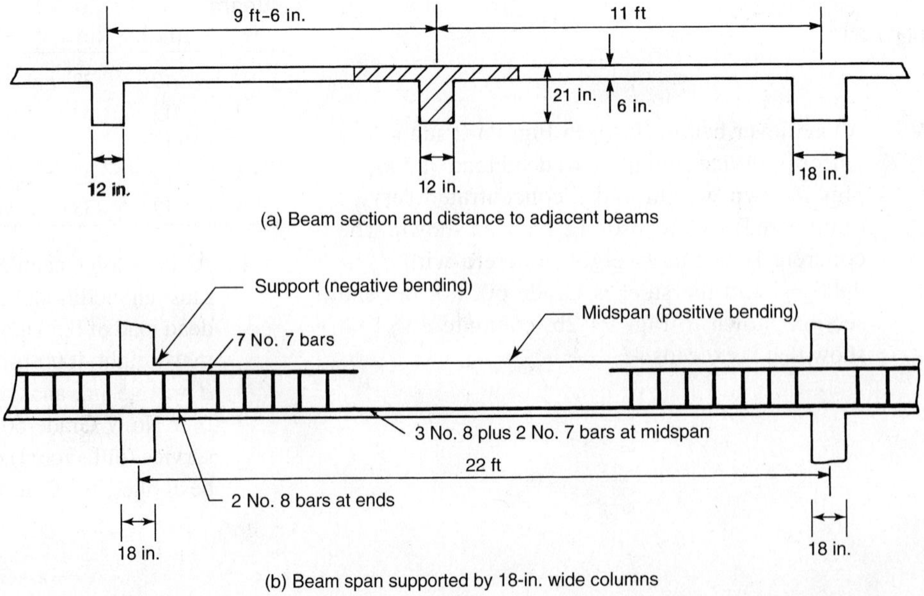

Fig. P4-8

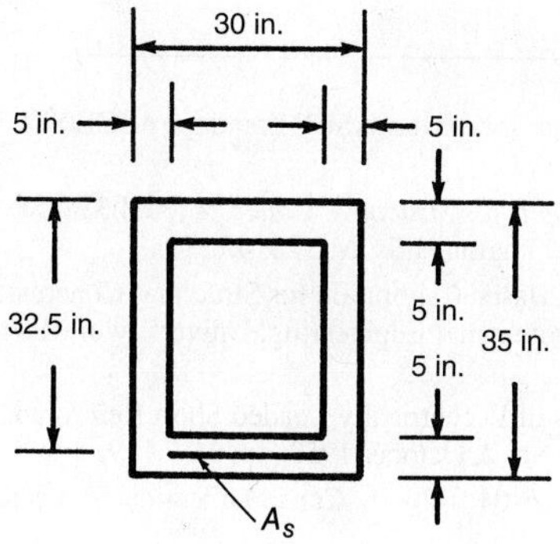

Fig. P4-9

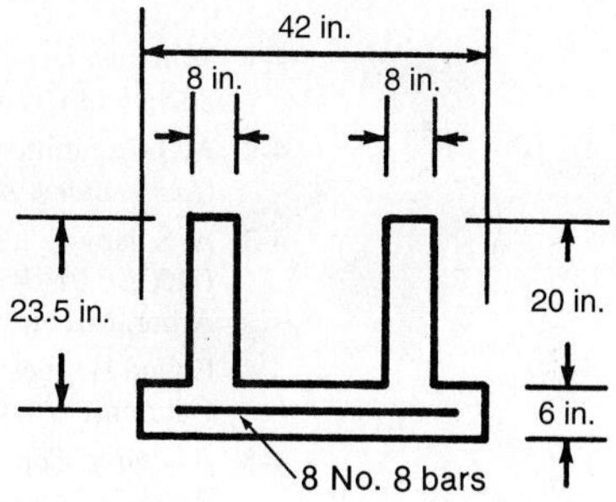

Fig. P4-10

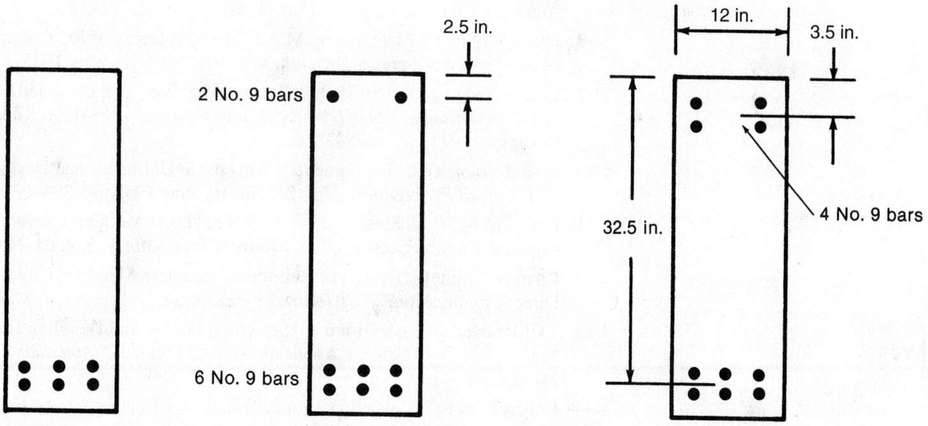

Fig. P4-11

4-11 (a) Compute ϕM_n for the three beams shown in Fig. P4-11. In each case, $f'_c = 4000$ psi, $f_y = 60$ ksi, $b = 12$ in., $d = 32.5$ in., and $h = 36$ in.

(b) From the results of part (a), comment on whether adding compression reinforcement is a cost-effective way of increasing the strength, ϕM_n, of a beam.

4-12 Compute ϕM_n for the beam shown in Fig. P4-12. Use $f'_c = 4500$ psi and $f_y = 60{,}000$ psi. Does the compression steel yield in this beam at nominal strength?

Video Solution

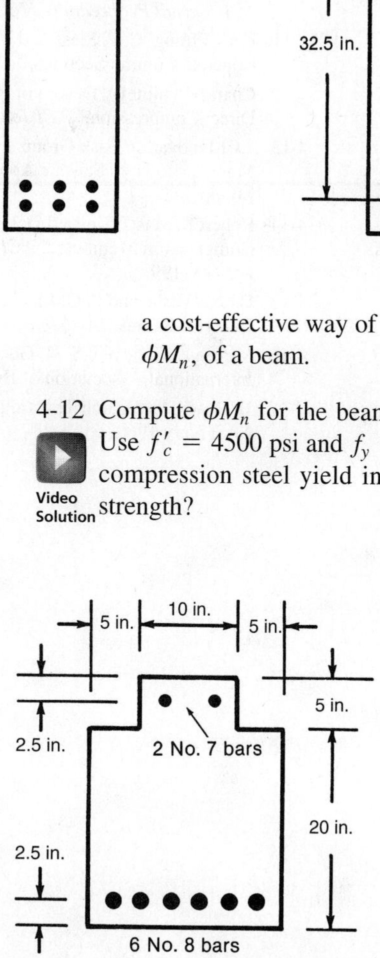

Fig. P4-12

REFERENCES

4-1 *Minimum Design Loads for Buildings and Other Structures*, ASCE Standard ASCE/SEI 7-10, American Society of Civil Engineers, Reston, VA, 2010.

4-2 ACI Committee 318, *Building Code Requirements for Structural Concrete* (ACI 318-14) and *Commentary*, American Concrete Institute, Farmington Hills, MI, 2014.

4-3 A. S. Nowak and M. M. Szerszen, "Reliability-Based Calibration for Structural Concrete," *Report UMCEE 01-04*, Department of Civil and Environmental Engineering, University of Michigan, Ann Arbor, MI, November. 2001, 120 pp.

4-4 Eivind Hognestad, "Inelastic Behavior in Tests of Eccentrically Loaded Short Reinforced Concrete Columns," *ACI Journal Proceedings*, Vol. 24, No. 2, October 1952, pp. 117–139.

4-5 *Design of Concrete Structures*, CAN/CSA-A23.3-04 (R2010), Canadian Standards Association, Rexdale, Ontario, Canada.

4-6 Alan H. Mattock, Ladislav B. Kriz, and Eivind Hognestad, "Rectangular Concrete Stress Distribution in Ultimate Strength Design," *ACI Journal, Proceedings*, Vol. 57, No. 8, February 1961, pp. 875–926.

4-7 Paul H. Kaar, Norman W. Hanson, and H. T. Capell, "Stress–Strain Characteristics of High Strength Concrete," *Douglas McHenry International Symposium on Concrete Structures*, ACI Publication SP-55, American Concrete Institute, Detroit, MI, 1978, pp. 161–185.

4-8 James Wight and Mete Sozen, M.A., "Strength Decay of RC Columns Under Shear Reversals," *ASCE Journal of the Structural Division*, Vol. 101, No. ST5, May 1975, pp. 1053–1065.

4-9 Dudley Kent and Robert Park, "Flexural Members with Confined Concrete," *ASCE, Journal of the Structural Division*, Vol. 97, No. ST7, July 1971, pp. 1969–1990; Closure to Discussion, Vol. 98, No. ST12, December 1972, pp. 2805–2810.

4-10 Eivind Hognestad, "Fundamental Concepts in Ultimate Load Design of Reinforced Concrete Members," *ACI Journal Proceedings*, Vol. 23, No. 10, June 1952, pp. 809–830.

4-11 E. O. Pfrang, C. P. Siess and M. A. Sozen, "Load-Moment-Curvature Characteristics of Reinforced Concrete Column Sections," *ACI Journal Proceedings*, Vol. 61, No. 7, July 1964, pp. 763–778.

4-12 Charles Whitney, "Design of Reinforced Concrete Members Under Flexure or Combined Flexure and Direct Compression," *ACI Journal Proceedings*, Vol. 8, No. 4, March–April 1937, 483–498.

4-13 ACI Innovation Task Group 4, "Structural Design and Detailing for High-Strength Concrete in Moderate to High Seismic Applications (ACI ITG 4.3)," *American Concrete Institute*, Farmington Hills, MI, pp. 212.

4-14 Robert F. Mast, "Unified Design Provisions for Reinforced and Prestressed Concrete Flexural and Compression Members," *ACI Structural Journal, Proceedings*, Vol. 89, No. 2, March–April 1992, pp. 185–199.

4-15 G. W. Washa and P. G. Fluck, "Effect of Compressive Reinforcement on the Plastic Flow of Reinforced Concrete Beams," *ACI Journal, Proceedings*, Vol. 49, No. 4, October 1952, pp. 89–108.

4-16 Mircea Z. Cohn and S. K. Ghosh, "Flexural Ductility of Reinforced Concrete Sections," *Publications*, International Association of Bridge and Structural Engineers, Zurich, Vol. 32-II, 1972, pp. 53–83.

4-17 H. Rusch, "Research Toward a General Flexural Theory for Structural Concrete," *ACI Journal*, Vol. 57, No. 1, July 1960, pp. 1–28.

5

Flexural Design of Beam Sections

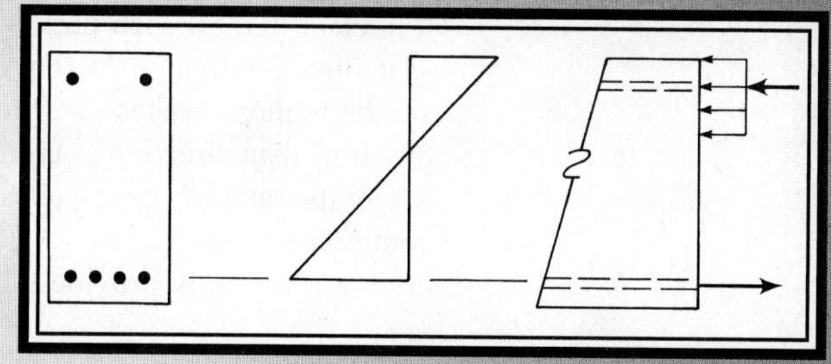

5-1 INTRODUCTION

Using the information provided in Chapter 4, the reader should have the ability to find the nominal moment strength, M_n, for any beam section and the corresponding strength-reduction factor, ϕ, for that section. So, if the factored design moment, M_u, is known for any beam section, he/she should be able to determine if ϕM_n equals or exceeds M_u. The primary topic to be discussed in this chapter is to start with a known value of M_u, design a beam cross section capable of resisting that moment (i.e., $\phi M_n \geq M_u$), and also have that section satisfy all of the ACI Code requirements for flexural reinforcement and section detailing. It probably is clear to the reader that the final value for M_u cannot be determined until the size of the beam section, and thus the self-weight of the beam, is known. This sets up the normal interaction cycle between analysis and design, where there will be an initial analysis based on assumed section sizes, followed by member design based on that analysis, then reanalysis based on updated section sizes, and some final design modifications based on the updated analysis.

Because of this interplay between analysis and design, the next section of this chapter (Section 5-2) will deal with the analysis of continuous one-way floor systems. This will give the reader a good understanding of how such floor systems carry and distribute loads from the slabs to the floor beams, girders, and columns and how the floor system can be analyzed following procedures permitted by the ACI Code. Once we have fully discussed how factored design moments can be determined at various sections in a continuous floor system, including slab sections, section design procedures will be developed (Sections 5-3 and beyond).

If the reader prefers to move directly to section design procedures, Section 5-2 can be skipped at this time.

5-2 ANALYSIS OF CONTINUOUS ONE-WAY FLOOR SYSTEMS

Reinforced concrete floor systems are commonly referred to as one-way or two-way systems based on the ratio of the span lengths for the floor slab in the two principal horizontal directions. Referring to the two floor plans shown in Figs. 5-1a and 5-1c, it is clear that the slab panels between the supporting beams have relatively short span lengths in one horizontal direction compared to their span lengths in the perpendicular direction. Recalling from frame analysis that flexural stiffness is inversely related to span length, it is clear that the slab panels shown in Fig. 5-1 would be much stiffer in their shorter span direction than in the longer span direction. Thus, for any load applied to floor panels similar to those in Figs. 5-1a and 5-1c, a higher percentage of the load would be carried in the short span direction as compared to the long span direction. In a concrete floor system where the ratio of the longer to the shorter span length is greater than or equal to 2, it is common practice to provide flexural reinforcement to resist the entire load in the short direction and only provide minimum steel for temperature and shrinkage effects in the

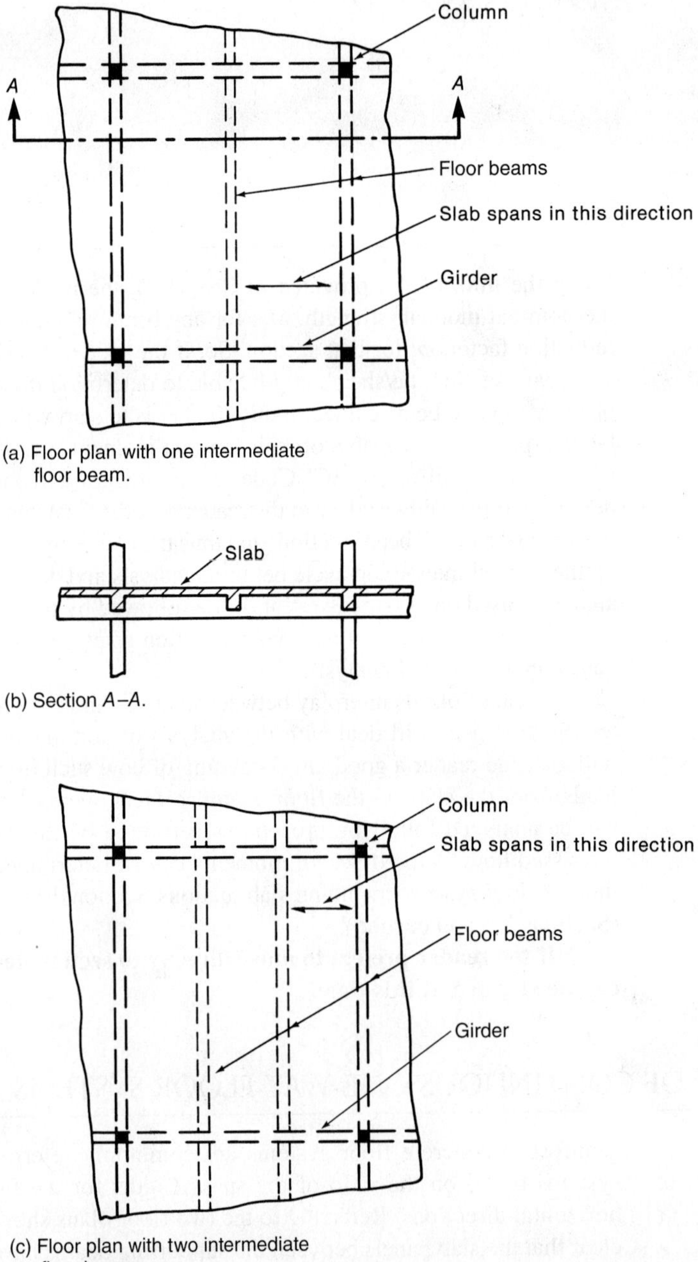

(a) Floor plan with one intermediate floor beam.

(b) Section A–A.

(c) Floor plan with two intermediate floor beams.

Fig. 5-1
Typical one-way floor systems.

long direction. Such slabs are referred to as *one-way slabs* because they are designed to carry applied loads in only one direction. A floor system consisting of one-way slabs and supporting beams, as shown in Fig. 5-1, is referred to as a *one-way floor system.*

If the floor systems in Fig. 5-1 were modified such that the only beams were those that spanned between the columns, the remaining slab panel would have a long span to short span ratio of less than 2. For such a case, flexural reinforcement would be provided in the two principal horizontal directions of the slab panel to enable it to carry applied loads in two directions. Such slabs are referred to as *two-way slabs*. The analysis and design of *two-way floor systems* will be discussed in Chapter 13.

Load Paths in a One-Way Floor System

Consider the idealized one-way floor system shown in Fig. 5-2. The floor system is not realistic, because it does not have openings for stairwells, elevators, or other mechanical systems. However, this floor system will be useful as a teaching tool to discuss load paths and the analysis of bending moments and shear forces in the various structural members.

To study load paths in a one-way floor system, assume a concentrated load is applied at the point p in the central slab panel of the floor system shown in Fig. 5-2. This concentrated load could represent part of a uniformly distributed live load or dead load acting on a specified portion (e.g., 1 ft by 1 ft) of the floor area. The one-way slab panel is assumed to initially carry the concentrated load in the north–south direction to the points m and n on the two adjacent *floor beams* supporting the one-way slab. The *floor beams* then carry the loads in the east–west direction to the points h, i, j, and k on the *girders* that support the floor beams. *Girder* is the name given to a primary support member (beam) that spans from column to column and supports the floor beams. A schematic sketch of a slab, floor beam, and girder system is given in Fig. 5-3. Girders normally have a total member depth that is greater than or equal to the depth of the floor beams that it supports. The final step

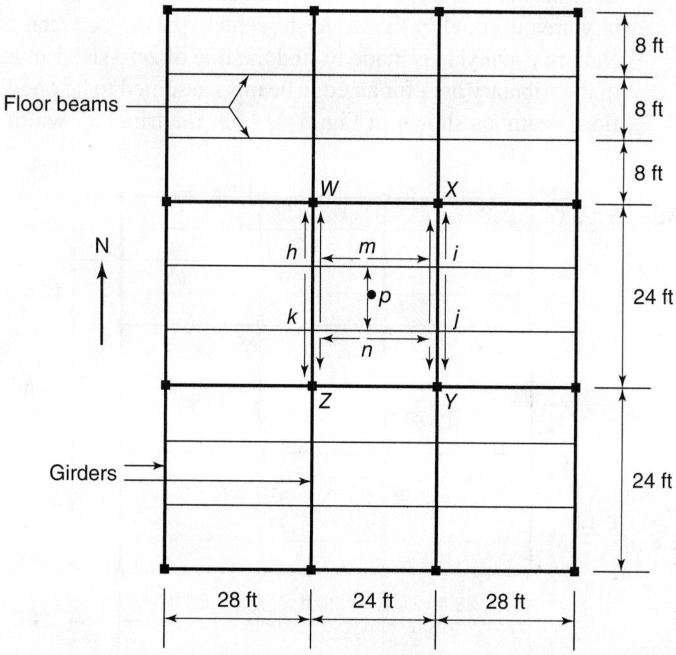

Fig. 5-2
Load paths in a one-way floor system.

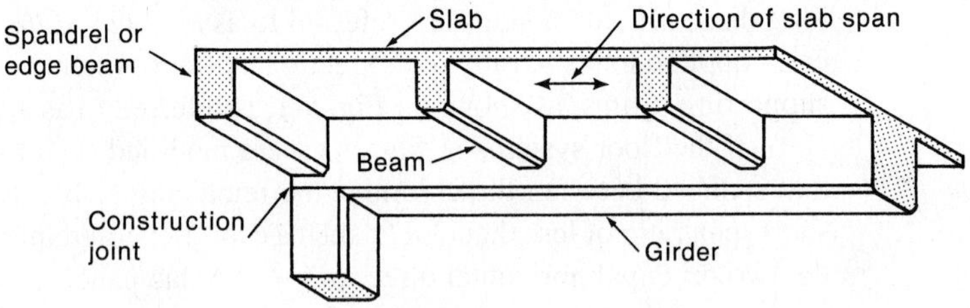

Fig. 5-3
Slab, beam, and girder floor system.

on the load path for the floor system in Fig. 5-2 is the transfer of loads from the girders to the columns at *W*, *X*, *Y*, and *Z*. It should be noted that some of the floor beams in a typical floor system will connect directly to columns, and thus, they transfer their loads directly to those columns, as is the case for the floor beam between the columns at *W* and *X*.

Tributary Areas, Pattern Loadings, and Live Load Reductions

Floor systems in almost all buildings are designed for uniformly distributed dead and live loads, normally given or calculated in unit of pounds per square foot (psf). The symbol q will be used to represent these loads with subscripts L for live load and D for dead load. Total dead load normally is composed of dead loads superimposed on the floor system as well as the self-weight of the floor members. Typical live load values used in design of various types of structures were given in Table 2-1. The analysis procedure for concentrated loads will be presented later in this section.

Floor beams typically are designed to resist area loads acting within the *tributary area* for that beam, as shown by the shaded regions in Fig. 5-4. As discussed in Chapter 2, the tributary area extends out from the member in question to the lines of zero shear on either side of the member. The zero shear lines normally are assumed to occur halfway to the next similar structural member (floor beam in this case). Thus, the width of the tributary area for a typical floor beam is equal to the sum of one-half the distances to the adjacent floor beams. For a floor system with uniformly spaced floor beams, the width of the tributary area is equal to the center-to-center spacing between the floor beams. Unless a more elaborate analysis is made to find the line of zero shear in an exterior slab panel, the width of the tributary area for an edge beam is assumed to be one-half the distance to the adjacent floor beam, as shown in Fig. 5-4. After the tributary width has been established, the area

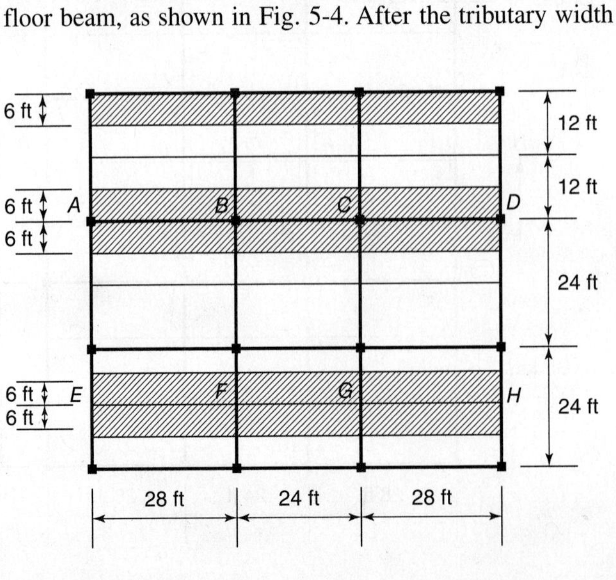

Fig. 5-4
Tributary areas for floor beams.

load, q, is multiplied by the tributary width to obtain a line load, w (lbs/ft or kips/ft), that is applied to the floor beam. This will be demonstrated in Example 5-1.

For one-way slabs, the width of the tributary area is set equal to the width of the *analysis strip*, which is commonly taken as 1 foot. Thus, the cross-hatched area in Fig. 5-5 represents both the tributary area and the width of the analysis strip for the continuous one-way slab portion of this floor system. The effective line load, w, is found by multiplying the area load, q, times the width of the analysis strip (usually 1 ft).

Pattern Loadings for Live Load

The largest moments in a continuous beam or a frame occur when some spans are loaded with live load and others are not. Diagrams, referred to as *influence lines*, often are used to determine which spans should and should not be loaded. An influence line is a graph of the variation in the moment, shear, or other effect at *one particular point* in a beam due to a unit load that moves across the beam.

Figure 5-6a is an influence line for the moment at point C in the two-span beam shown in Fig. 5-6b. The horizontal axis refers to the *position* of a unit (1 kip) load on the beam, and the vertical ordinates are the *moment at C* due to a 1-kip load acting at the point in question. The derivation of the ordinates at B, C, and E is illustrated in Figs. 5-6c to 5-6e. When a unit load acts at B, it causes a moment of 1.93 k-ft at C (Fig. 5-6c). Thus, the ordinate at B in Fig. 5-6a is 1.93 k-ft. Figures 5-6d and 5-6e show that the moments at C due to loads at C and E are 4.06 and -0.90 k-ft, respectively. These are the ordinates at C and E in Fig. 5-6a and are referred to as *influence ordinates*. If a concentrated load of P kips acted at point E, the moment at C would be P times the influence ordinate at E, or $M = -0.90P$ k-ft. If a uniform load of w acted on the span A–D, the moment at C would be w times the area of the influence diagram from A to D.

Figure 5-6a shows that a load placed anywhere between A and D will cause positive moment at point C, whereas a load placed anywhere between D and F will cause a negative moment at C. Thus, to get the maximum positive moment at C, we must load span A–D only.

Two principal methods are used to calculate influence lines. In the first, a 1-kip load is placed successively at evenly spaced points across the span, and the moment (or shear) is calculated at the point for which the influence line is being drawn, as was done in Figs. 5-6c to 5-6e.

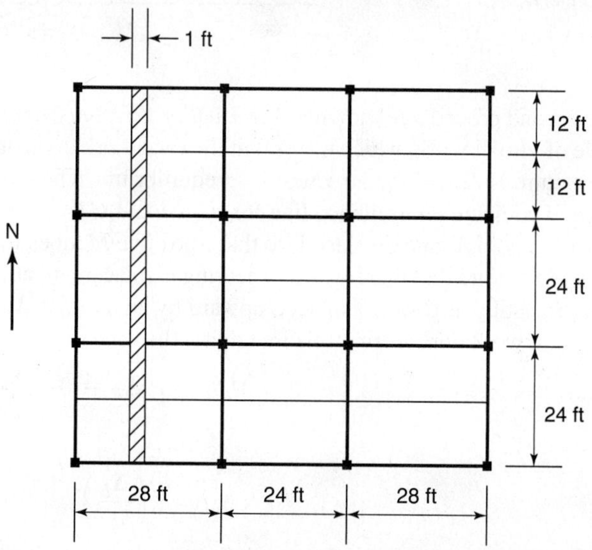

Fig. 5-5
Width of analysis strip and tributary area for one-way slab strip.

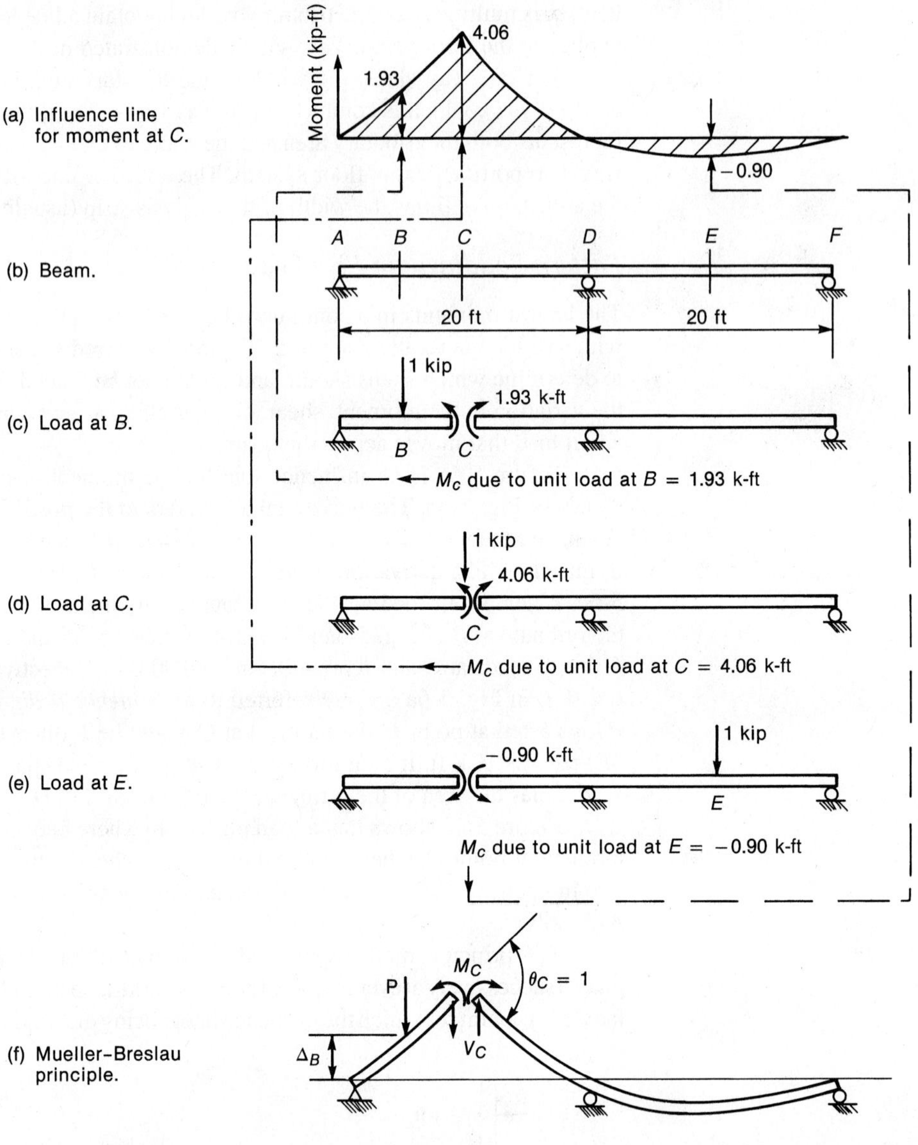

Fig. 5-6
Concept of influence lines.

The second procedure, known as the *Mueller–Breslau principle* [5-1], is based on the principle of virtual work, which states that the total work done during a virtual displacement of a structure is zero if the structure is in equilibrium. The use of the Mueller–Breslau principle to compute an influence line for moment at C is illustrated in Fig. 5-6f. The beam is broken at point C and displaced, so that a positive M_c does work by acting through an angle change θ_c. Note that there was no shearing displacement at C, so V_c does not do work. The load, P, acting at B was displaced upward by an amount Δ_B and hence did negative work. The total work done during this imaginary displacement was

$$M_c\theta_c - P\,\Delta_B = 0$$

so

$$M_c = P\left(\frac{\Delta_B}{\theta_c}\right) \tag{5-1}$$

where Δ_B/θ_c is the influence ordinate at B. Thus, the *deflected shape* of the structure for such a displacement has the same shape and is proportional to the influence line for moment at *C*. (See Figs. 5-6a and 5-6f.)

The Mueller–Breslau principle is presented here as a *qualitative guide* to the shape of influence lines to determine where to load a structure to cause maximum moments or shears at various points. The ability to determine the critical loading patterns rapidly by using sketches of influence lines expedites the structural analysis considerably, even for structures that will be analyzed via computer software packages.

Influence lines can be used to establish loading patterns to maximize the moments or shears due to live load. Figure 5-7 illustrates influence lines drawn in accordance with the Mueller–Breslau principle. Figure 5-7a shows the qualitative influence line for moment at *B*. The loading pattern that will give the largest positive moment at *B* consists of loads on all spans having positive influence ordinates. Such a loading is shown in Fig. 5-7b and is referred to as an *alternate span* loading or a *checkerboard* loading. This is the common loading pattern for determining maximum midspan positive moments due to live load.

The influence line for moment at the support *C* is found by breaking the structure at *C* and allowing a positive moment, M_c, to act through an angle change θ_c. The resulting deflected shape, as shown in Fig. 5-7c, is the qualitative influence line for M_c. The maximum *negative* moment at *C* will result from loading all spans having negative influence ordinates, as shown in Fig. 5-7d. This is referred to as an *adjacent span* loading with alternate span loading occurring on more distant spans. Adjacent span loading is the common loading pattern for determining maximum negative moments at supports due to live load.

Qualitative influence lines for shear can be drawn by breaking the structure at the point in question and allowing the shear at that point to act through a unit shearing displacement, Δ, as shown in Fig. 5-8. During this displacement, the parts of the beam on the two sides of the break must remain parallel so that the moment at the section does not

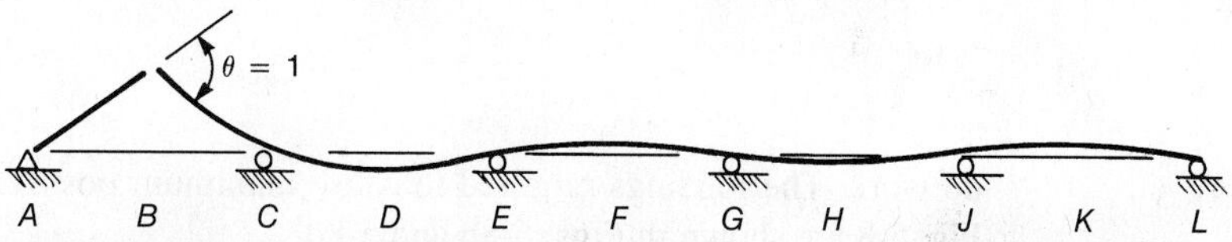

(a) Qualitative influence line for moment at *B*.

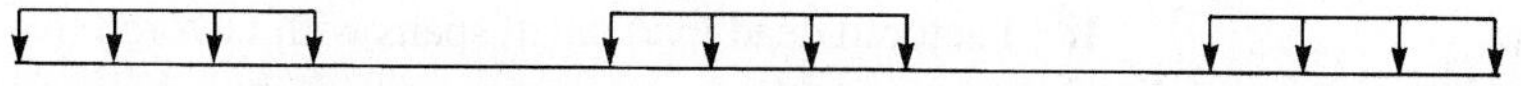
(b) Loading for maximum positive moment at *B*.

(c) Qualitative influence line for moment at *C*.

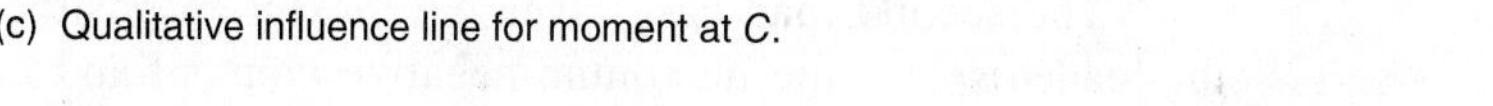
(d) Loading for maximum negative moment at *C*.

Fig. 5-7
Qualitative influence lines for moments and loading patterns.

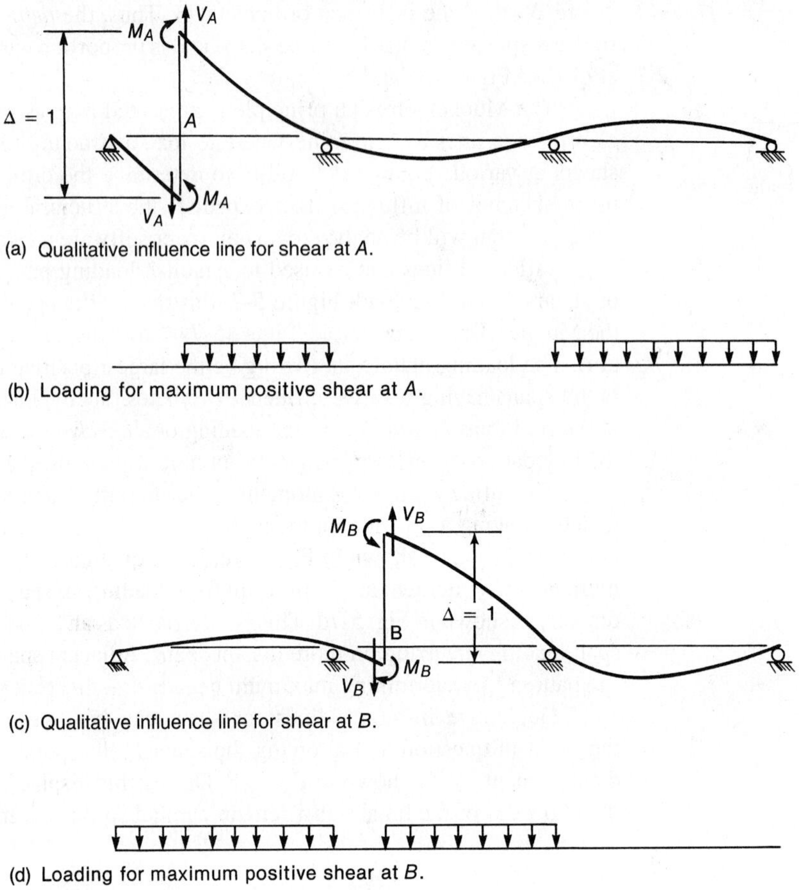

Fig. 5-8
Qualitative influence lines for shear.

do work. The loadings required to cause maximum positive shear at sections *A* and *B* in Fig. 5-8 are shown in Figs. 5-8b and 5-8d.

Using this sort of reasoning, ACI Code Section 6.4.2 defines loading patterns to determine maximum design moments for continuous beams and one-way slabs:

1. Factored dead load on all spans with factored live load on two adjacent spans and no live load on any other spans.

2. Factored dead load on all spans with factored live load on alternate spans.

The first case will give the maximum negative moment and maximum shear at the supports between the two loaded spans. Alternate span loading could be used for spans further from the support section, as shown in Fig. 5-7d. For simplicity, the ACI Code does not require this additional loading, because the influence ordinates are relatively small for those spans, and thus the effect of loading those spans is small compared to the effect of loading the adjacent spans.

The second load case gives the maximum positive moments at the midspan of the loaded spans, the maximum negative moment and maximum shear at the exterior support, and the minimum positive moment, which could be negative, at the midspan of the unloaded spans. Using factored dead load and live load on all spans will represent the maximum vertical loading to be transferred to the columns supporting the floor system. The use of pattern loading will be demonstrated in Example 5-2.

Live Load Reductions

Most standard building codes permit a reduction in the live loads used for member design based on a multiple of the tributary area for that member. In Chapter 2, this was referred to as the *influence area*, A_I, and was defined as

$$A_I = K_{LL} \times A_T \tag{5-2}$$

where A_T is the tributary area for the member in question and K_{LL} is the multiplier based on the type of member under consideration. For edge beams and interior beams, the ASCE/SEI Standard [5-2] states that K_{LL} shall be taken as 2.0. For one-way slabs, there is no need to define either K_{LL} or an influence area because no live load reduction is permitted for those members. The appropriate tributary area for floor beams is based partially on the previously discussed live load patterns and thus can be different for different locations of the same continuous beam. Figure 5-9 shows some different tributary areas to be used for the analysis and design of different beam sections (shown on different floor beams for clarity). For midspan sections (positive bending), as represented by $M1$ and $M2$ in Fig. 5-9, the tributary area is equal to the tributary width discussed previously multiplied by the length of the span in question. Clearly, the tributary areas for $M1$ and $M2$ will be different because of the different span lengths. For sections near a support (negative bending), as represented by $M3$ in Fig. 5-9, the tributary area is equal to the tributary width multiplied by the total length of the two adjacent spans. This is related to the *adjacent span loading pattern* used to maximize the negative moment at this section. Essentially, loads in the two adjacent spans will significantly *influence* the moment at this section.

After the influence area has been determined, the reduced live load, defined here as L_r, can be determined from the following expression, which is a slight modification of Eq. (2-12):

$$L_r = L\left[0.25 + \frac{15}{\sqrt{A_I}}\right] \tag{5-3}$$

where L is the unreduced live load and the influence area, A_I, is to be given in square feet. No reduction is permitted if the influence area does not exceed 400 ft^2. Also, the maximum permissible reduction of live load is 50 percent for any floor beam or girder in a floor system.

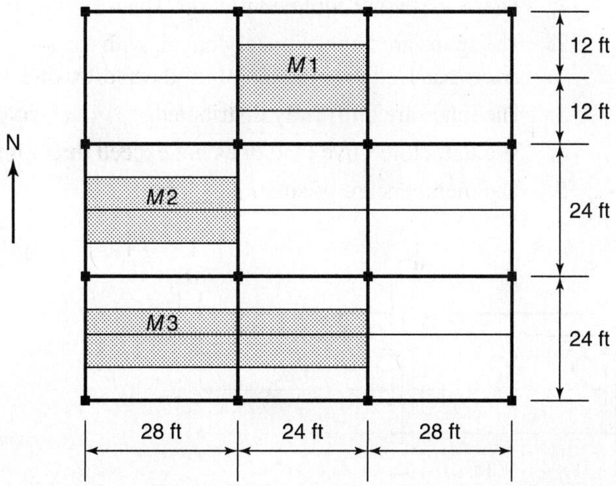

Fig. 5-9
Tributary areas for analysis and design of different beam sections.

As noted earlier, the influence area will change for different locations in a continuous beam, so a different reduced live load can be used when analyzing and designing those different sections. Of course, the designer has the option of using only one value, the most conservative (largest), for the reduced live load when analyzing the moments at various sections along the continuous beam. Examples 5-1 and 5-2 will demonstrate the use of Eq. (5-3).

ACI Moment and Shear Coefficients

Based on the prior discussions of pattern loading and live load reductions, it should be clear that finding the maximum moments and shears at various sections of continuous beams and one-way slabs will require a full structural analysis for at least three and maybe several load cases. Because large parts of the ACI Code were developed and written before the broad accessibility to structural analysis software, a set of approximate moment and shear coefficients were developed for the analysis and design of *non-prestressed* continuous beams and one-way slabs subjected to distributed loading and having relatively uniform span lengths. Because continuous beams and slabs are permitted to be designed for the moments and shears at the faces of their supports, the ACI moment and shear coefficients are based on the clear span, ℓ_n, as opposed to the center-to-center span length, ℓ, which are illustrated in Fig. 5-10. The average span length, ℓ_n(avg), shown in Fig. 5-10 will be used for the negative moments at interior supports, because those moments are influenced by the lengths of the two adjacent spans. Moment coefficients are given at midspan and at the faces of supports, while shear coefficients are given only at the faces of the supports.

The moment and shear coefficients all are based on the total distributed factored load, w_u, which normally is equal to the sum of the factored dead and live loads. For common floor systems, the following load combination will typically govern:

$$w_u = 1.2w_D + 1.6w_L \tag{5-4}$$

If the load case, $w_u = 1.4\, w_D$, happens to govern, then the ACI moment coefficients should not be used, because they assume a pattern live loading that is not appropriate if only dead load is considered. For this condition, a structural analysis would be required for this single loading case.

The requirements for using the ACI moment and shear coefficients are given in ACI Code Section 6.5.1 as:

1. There are two or more continuous spans.
2. The spans are approximately equal, with the longer of the two adjacent spans not more that 1.2 times the length of the shorter one.
3. The loads are uniformly distributed.
4. The unfactored live load does not exceed three times the unfactored dead load.
5. The members are prismatic.

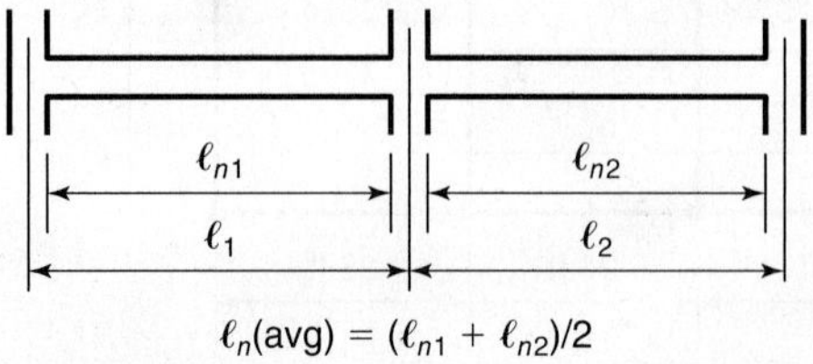

Fig. 5-10
Definitions of clear span and average clear span for use with ACI moment and shear coefficients.

If any of these conditions are violated, then a structural analysis of the continuous member is required, as will be discussed later in this section. Also, it is implicitly assumed that the continuous floor members in question are not resisting any significant moments or shears due to lateral loads.

The maximum positive and negative moments and shears are computed from the following expressions:

$$M_u = C_m\,(w_u \ell_n^2) \tag{5-5}$$

$$V_u = C_v\left(\frac{w_u \ell_n}{2}\right) \tag{5-6}$$

where C_m and C_v are moment and shear coefficients from ACI Code Sections 6.5.2 and 6.5.4 are given in Fig. 5-11. For all positive midspan moments, all shears and the negative moment at exterior supports, ℓ_n, is for the span under consideration. For the negative moment at interior supports, ℓ_n shall be taken as ℓ_n(avg), as defined in Fig. 5-10. The terminology used in the ACI Code to identify various critical design sections is illustrated in Fig. 5-11a. Midspan shear coefficients are not given here, but will be discussed in Chapter 6. At most locations, the shear coefficient is 1.0, except at the exterior face of

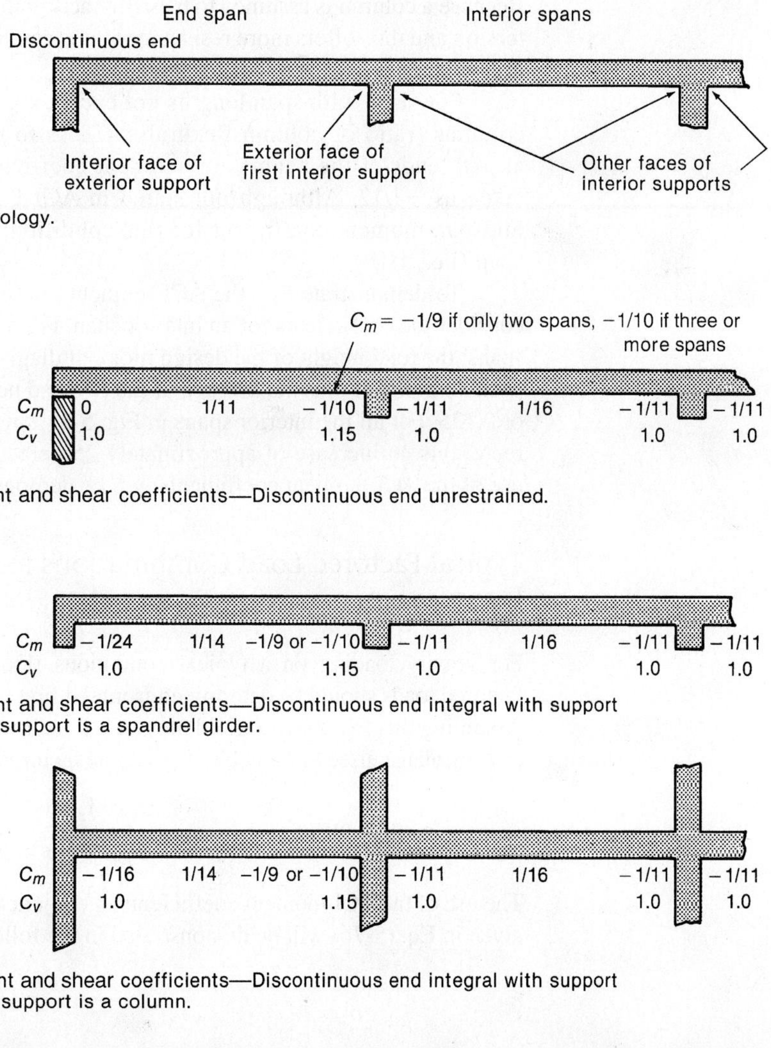

(a) Terminology.

(b) Moment and shear coefficients—Discontinuous end unrestrained.

(c) Moment and shear coefficients—Discontinuous end integral with support where support is a spandrel girder.

(d) Moment and shear coefficients—Discontinuous end integral with support where support is a column.

Fig. 5-11
ACI moment and shear coefficients.

the first interior support where it is increased to 1.15. This increase is to account for the fact that the zero shear point is probably closer to the exterior support, and thus, more load for the exterior span is likely to be carried at the first interior support. It should be noted that no corresponding reduction is made for the load resisted by the exterior support.

The moment coefficients are always the same for an interior span, but they vary in the exterior span depending on the type of rotational resistance provided at the exterior support. The exterior support in Fig. 5-11b can be assumed to be masonry wall that is not built integrally with the beam or slab and thus offers no resistance to rotations at the end of the member ($C_m = 0$). For such a case, higher positive moments would be expected at midspan of this exterior span than if the exterior support offered some resistance to rotation, as indicated in Figs. 5-11c and 5-11d. In Fig. 5-11c, the exterior support is assumed to be a *spandrel beam*, which is a word that is often used for a beam or girder at the exterior of at floor system. Thus, this case would represent the exterior spans of the continuous floor beam, *E–F–G–H*, in Fig. 5-4 or the one-way slab in Fig. 5-5. These end moments will put torsion into the spandrel support beams, so this particular moment coefficient, $C_m = -1/24$, will be discussed again in Chapter 7 on design for torsion. Finally, in Fig. 5-11d, the exterior support is assumed to be a column. This case would represent the exterior spans of the continuous floor beam, *A–B–C–D*, in Fig. 5-4. Because a column is assumed to be stiffer acting in bending than a spandrel beam acting in torsion and thus offers more resistance to end rotation of the continuous beam, the exterior moment coefficient is larger for this case.

For slabs with span lengths not exceeding 10 ft and for beams framing into stiff columns (ratio of column flexural stiffness to beam flexural stiffness exceeds eight at both ends of the beam), the moment coefficient at the face of the supports can be taken as $-1/12$. Although not stated in ACI Code Section 6.5.2, the corresponding midspan moment coefficient for this condition should be the same as for an interior span (i.e., 1/16).

To demonstrate that the ACI moment coefficients do account for pattern loadings, consider the coefficients for an interior span. For a span not affected by loading in adjacent spans, the total height of the design moment diagram (i.e., the absolute sum of the midpan positive moment plus the average of the two end negative moments) should be equal to 1/8 or 0.125. For all the interior spans in Fig. 5-11, this sum is 1/16 plus 1/11, or 0.153, which represents an increase of approximately 25 percent due to potential pattern loading. The use of the ACI moment coefficients will be demonstrated in Example 5-1.

Typical Factored Load Combinations for a Continuous Floor System

For gravity loading on a typical continuous floor system, the required combination of factored loads should be determined from the first two equations in ACI Code Section 5.3.1. Assuming that loads due to fluid pressure, F; soil weight or pressure, H; and thermal, creep, and shrinkage effects, T can be ignored, the factored-load combinations to be considered are

$$U = 1.4D \tag{5-7a}$$

$$U = 1.2D + 1.6L \tag{5-7b}$$

The use of the ACI moment coefficients in conjunction with the factored-load combinations given in Eq. (5-7b) will be demonstrated in the following example.

EXAMPLE 5-1 Use of ACI Moment Coefficients for Continuous Floor Beams

Consider the continuous floor beam *A–B–C–D* in Fig. 5-4. Use the ACI moment coefficients to find the design moments at the critical sections for one exterior span and the interior span. Then, repeat these calculations for the floor beam *E–F–G–H* in Fig. 5-4. Assume the floor slab has a total thickness of 6 in. and assume the floor beams have a total depth of 24 in. and a web width of 12 in. Assume the columns are 18 in. by 18 in. Finally, assume the floor is to be designed for a live load of 60 psf and a superimposed dead load (SDL) of 20 psf.

1. Confirm that the ACI moment coefficients can be used. There are two or more spans, the loads are uniformly distributed, and the members are prismatic. The ratio of the longer span to the shorter span is 28/24 = 1.17, which is less than 1.2. The floor slab is 6 in. thick and thus weighs 75 psf. Therefore, the unfactored live load does not exceed three times the dead load.

2. Determine live load reductions.

(a) Exterior span *A–B:* For the exterior negative moment and the midspan positive moment, the tributary area is equal to the tributary width times the span length. Thus,

$$A_T = 12\text{ ft} \times 28\text{ ft} = 336\text{ ft}^2$$

$$A_I = K_{LL} \times A_T = 2 \times 336 = 672\text{ ft}^2$$

$$L_r = L\left[0.25 + \frac{15}{\sqrt{A_I}}\right] = 60\text{ psf}\left[0.25 + \frac{15}{\sqrt{672}}\right]$$

$$= 60[0.25 + 0.579] = 49.7\text{ psf} > 0.5 \times 60\text{ psf (o.k.)}$$

(b) Interior span *B–C:* For the midspan positive moment, the following applies:

$$A_T = 12\text{ ft} \times 24\text{ ft} = 288\text{ ft}^2$$

$$A_I = K_{LL} \times A_T = 2 \times 288 = 576\text{ ft}^2$$

$$L_r = L\left[0.25 + \frac{15}{\sqrt{A_I}}\right] = 60\text{ psf}\,[0.25 + 0.625]$$

$$= 52.5\text{ psf} > 0.5 \times 60\text{ psf (o.k.)}$$

(c) Negative moments at *B*: For the interior support, the combined lengths of the two adjacent spans are used to find the tributary area. Thus,

$$A_T = 12\text{ ft} \times (28\text{ ft} + 24\text{ ft}) = 624\text{ ft}^2$$

$$A_I = K_{LL} \times A_T = 2 \times 624 = 1250\text{ ft}^2$$

$$L_r = L\left[0.25 + \frac{15}{\sqrt{A_I}}\right] = 60\text{ psf}\,[0.25 + 0.424]$$

$$= 40.5\text{ psf} > 0.5 \times 60\text{ psf (o.k.)}$$

3. Total factored loads.

(a) Exterior negative moment and midspan positive moment of span *A–B*: The distributed live load acting on the beam is

$$w_L = q_L\,(\text{reduced}) \times \text{tributary width}$$
$$= 49.7 \text{ psf} \times 12 \text{ ft} = 596 \text{ lb/ft} = 0.596 \text{ k/ft}$$

The distributed dead loads from the slab and superimposed dead load are

$$q(\text{slab}) = \frac{6 \text{ in.}}{12 \text{ in./ft}} \times 150 \text{ lb/ft}^3 = 75 \text{ psf}$$
$$w(\text{slab} + SDL) = (75 \text{ psf} + 20 \text{ psf}) \times 12 \text{ ft} = 1140 \text{ lb/ft} = 1.14 \text{ k/ft}$$

The dead load of the beam also needs to be included, but we need to avoid double counting the weight of the slab where it passes over the top of the beam web. The weight of the beam web is calculated from the shaded region shown in Fig. 5-12.

$$w(\text{beam web}) = \frac{(24 \text{ in.} - 6 \text{ in.}) \times 12 \text{ in.}}{144 \text{ in.}^2/\text{ft}^2} \times 150 \text{ lb/ft}^3 = 225 \text{ lb/ft} = 0.225 \text{ k/ft}$$

The total dead load is

$$w_D = w(\text{slab} + \text{SDL}) + w(\text{beam}) = 1.14 + 0.225 = 1.37 \text{ k/ft}$$

Now, the total factored load is

$$w_u = 1.2w_D + 1.6w_L = 1.2 \times 1.37 + 1.6 \times 0.596 = 2.60 \text{ k/ft}$$
$$\text{or, } w_u = 1.4w_D = 1.4 \times 1.37 = 1.92 \text{ k/ft (does not govern)}$$

(b) Midspan positive moment for span *B–C*: The distributed live load acting on the beam is

$$w_L = q_L\,(\text{reduced}) \times \text{tributary width}$$
$$= 52.5 \text{ psf} \times 12 \text{ ft} = 630 \text{ lb/ft} = 0.63 \text{ k/ft}$$

So, the total factored load is

$$w_u = 1.2w_D + 1.6w_L = 1.2 \times 1.37 + 1.6 \times 0.63 = 2.65 \text{ k/ft}$$

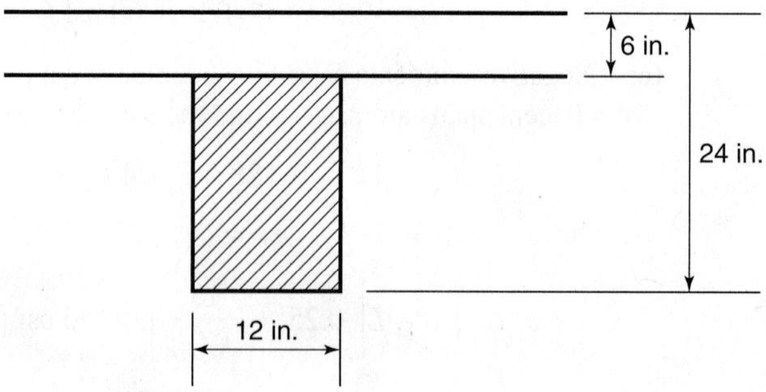

Fig. 5-12
Web area to be included in dead weight calculation.

(c) Negative moment at support *B*: The distributed live load is

$$w_L = q_L\,(\text{reduced}) \times \text{tributary width}$$
$$= 40.5 \text{ psf} \times 12 \text{ ft} = 486 \text{ lb/ft} = 0.486 \text{ k/ft}$$

So, the total factored load is

$$w_u = 1.2w_D + 1.6w_L = 1.2 \times 1.37 + 1.6 \times 0.486 = 2.42 \text{ k/ft}$$

4. **Calculate design moments.**

(a) Negative moment at face of support *A*: From Fig. 5-11d, the coefficient is negative 1/16.

$$\ell_n(A\text{–}B) = 28 \text{ ft} - \frac{18 \text{ in.}}{12 \text{ in./ft}} = 26.5 \text{ ft}$$
$$M_u = -1/16 \times 2.60 \text{ k/ft} \times (26.5 \text{ ft})^2 = -114 \text{ k-ft}$$

(b) Positive moment at midspan of beam *A–B*: From Fig. 5-11d, the coefficient is positive 1/14.

$$M_u = 1/14 \times 2.60 \text{ k/ft} \times (26.5 \text{ ft})^2 = 130 \text{ k-ft}$$

(c) Positive moment at midspan of beam *B–C*: From Fig. 5-11d, the appropriate moment coefficient is positive 1/16, and from step 3, the total distributed load is 2.65 k/ft.

$$\ell_n(B\text{–}C) = 24 \text{ ft} - \frac{18 \text{ in.}}{12 \text{ in./ft}} = 22.5 \text{ ft}$$
$$M_u = 1/16 \times 2.65 \text{ k/ft} \times (22.5 \text{ ft})^2 = 83.8 \text{ k-ft}$$

(d) Negative moment at face of support *B*: Because the beam section design will not change from one side of the column to the other, the final design at both faces of support *B* will need to be for the larger of the two negative moments from the interior and exterior spans. The calculation of both moments will use the average clear span, so the larger of the two moment coefficients will govern. From Fig. 5-11d, it can be seen that the coefficient from the exterior span (negative 1/10 for more than two spans) will govern. Using a total distributed load of 2.42 k/ft,

$$\ell_n(\text{avg}) = 0.5(26.5 + 22.5) = 24.5 \text{ ft}$$
$$M_u = -1/10 \times 2.42 \text{ k/ft} \times (24.5 \text{ ft})^2 = -145 \text{ k-ft}$$

5. **Calculate design moments for beam *E–F–G–H*.** A quick review of Figs. 5-11c and 5-11d indicates that the only change in moment coefficient occurs at the exterior end of the exterior span (coefficient changes to negative 1/24). However, depending on the size of the girders used in this floor system, the clear spans also will change. Assuming the widths of the girders are 12 in., the clear spans and the average clear span are

$\ell_n(E\text{–}F) = 27$ ft, $\ell_n(F\text{–}G) = 23$ ft, and for the exterior and interior spans

$\ell_n(\text{avg}) = 25$ ft

Then, the resulting factored design moments are

$$M_u(E) = -1/24 \times 2.60 \text{ k/ft} \times (27 \text{ ft})^2 = -79.0 \text{ k-ft}$$
$$M_u(\text{midspan } E\text{–}F) = 1/14 \times 2.60 \text{ k/ft} \times (27 \text{ ft})^2 = 135 \text{ k-ft}$$
$$M_u(\text{midspan } F\text{–}G) = 1/16 \times 2.65 \text{ k/ft} \times (23 \text{ ft})^2 = 87.6 \text{ k-ft}$$
$$M_u(F) = -1/10 \times 2.42 \text{ k/ft} \times (25 \text{ ft})^2 = -151 \text{ k-ft}$$ ■

Structural Analysis of Continuous Beams and One-Way Slabs

In many one-way floor systems, the span length and loading limitations given in ACI Code Section 6.5.1 are not satisfied. Common situations include a continuous girder subjected to concentrated loads from floor beams and continuous beams with adjacent span lengths that vary by more than 20 percent. For all such cases, a structural analysis of the continuous beam or one-way slab is required to find the design moments and shears at critical sections (commonly midspan and faces of supports). Of course, a structural analysis can be performed for any continuous member, even if it satisfies the limitations given in ACI Code Section 6.5.1.

In general, a two-dimensional analysis is permitted for determining design moments in a typical continuous beam and column frame system. Further, when finding design moments and shears in a floor system subjected to only gravity loads, ACI Code Section 6.3.1.2 states that it is permitted to isolate the analysis to the particular floor level in question. Thus, for floor beams or girders that frame directly into columns, the analysis model can consist of the beams or girders plus the columns immediately above and below the floor level, with the far ends of those columns fixed against rotations. For gravity loading on the continuous floor beam *A–B–C–D* in Fig. 5-4, an acceptable analysis model is shown in Fig. 5-13, where ℓ_a and ℓ_b represent the column lengths above and below the floor system being analyzed. A vertical roller support should be added at either joint *A* or joint *D* to prevent horizontal displacements at the floor level.

The structural model in Fig. 5-13 can be used for the analysis of any combination of distributed or concentrated loads on the continuous beam. For beams built integrally with supports, ACI Code Section 9.4.2.1 permits the calculation of design moments (and shears) at the face of the support. Most frame analysis software packages allow for the specification of a rigid zone at each end of a frame member, as shown in Fig. 5-14. Because nodes are typically located at the center of the supporting member, the length of the rigid end zones, x_i and x_j, are taken as one-half the total width of the supporting member. For the frame shown in

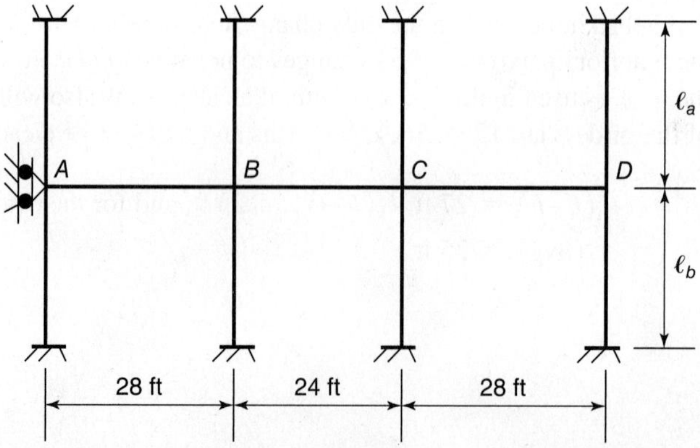

Fig. 5-13
Permissible analysis model for continuous beams subjected to only gravity loading.

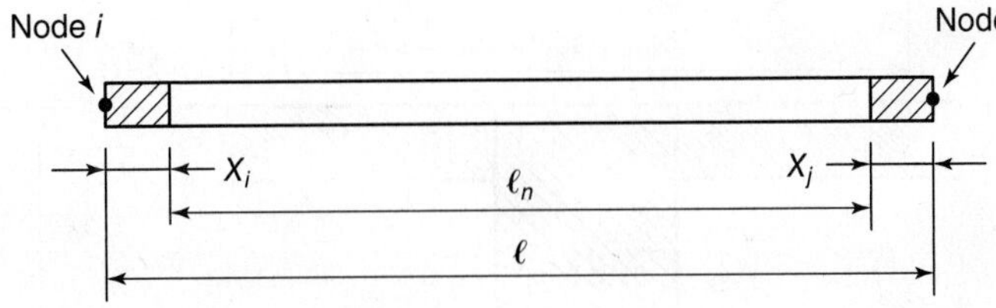

Fig. 5-14
Rigid end zones in frame elements.

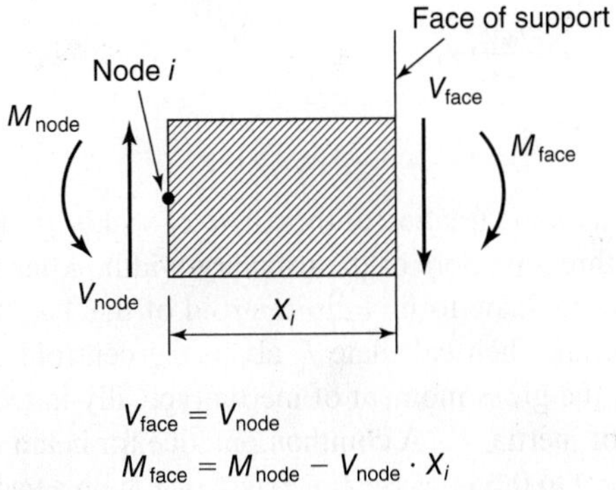

Fig. 5-15
Final design shear and moment at face of support.

Fig. 5-13, the length of the rigid end zones for each beam would be equal to one-half the width of the column at each end of the beam. The major advantage of specifying a rigid end zone is that output from analysis software will give moments and shears at the end of the rigid zones (i.e., at the faces of the support) as opposed to the node points. If either the available software does not permit the designation of a rigid end zone or if a hand-calculation procedure was used, a simple calculation similar to that shown in Fig. 5-15 will be required to find the moment and shear at the face of the support at each end of the beam. The use of rigid end zones is not required for the column elements in Fig. 5-13 unless the output from the analysis also is being used to determine design moments in the columns. In general, a full frame analysis will be used to determine the column design moments, and the rigid end zones at the top and bottom of the column should represent the distances from the selected node points to the sections where the column intersects with either the bottom or the top of the beams on adjacent floor levels, respectively.

For the initial analysis—design cycle, preliminary member sizes can be selected based on prior experience with similar floor systems. Total beam depths, h, are typically in the range of $\ell/18$ to $\ell/12$, where ℓ is the center-to-center span length of the beam. In typical U.S. practice, beam depths are rounded to a whole inch unit and often to an even number of inches. Beam width, b, or web width, b_w, commonly are taken to be approximately one-half of the total beam depth and are rounded to a whole inch unit. Architectural limitations on permissible dimensions and required clearances also may affect the selection of preliminary beam sizes.

After the initial member sizes are selected, most designers will use the *gross moment of inertia*, I_g, for determining the flexural stiffness of the column sections and the *cracked moment of inertia*, I_{cr}, for determining the flexural stiffness of the beam sections. The gross moment of inertia for a column or beam is based on the dimensions of the concrete section, ignoring the contribution from reinforcement. For beams, the concrete section will include some part of the floor slab, as indicated in Fig. 5-16. The slab width (flange width) that should be used to determine the gross moment of inertia, I_g, for a floor beam

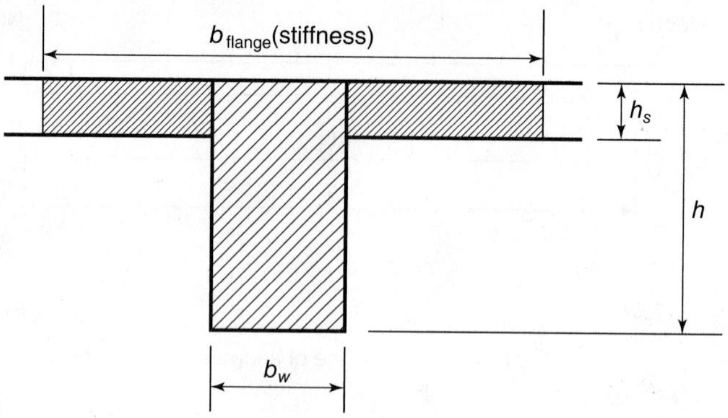

Fig. 5-16
Effective beam section for flexural stiffness analysis of a floor beam carrying gravity loading.

normally is taken as some fraction of the tributary width for that beam. Typical values range from one-half to three-quarters of the tributary width. After the flange width is selected, a calculation should be made to find the centroid of the T-section (or inverted L-section for a spandrel beam), and then calculate I_g about the centroid of the section. To account for flexural cracking, the gross moment of inertia typically is reduced to obtain a value for the cracked moment of inertia, I_{cr}. A common practice for beam sections is to assume that I_{cr} is approximately equal to 0.5 I_g. Based on experience with a variety of concrete floor systems, the author recommends that a good approximation for the cracked moment of inertia of a T-section can be obtained by calculating the gross moment of inertia for the extended web of the section, as shown by the heavily shaded region in Fig. 5-16. Using this procedure,

$$I_{cr}(\text{T-beam}) \cong \frac{1}{12} b_w h^3 \tag{5-8}$$

This procedure eliminates the need to define the effective flange width and the resulting gross moment of inertia for a flanged beam section.

ACI Code Section 6.2.2 states that "continuous construction shall be analyzed for the maximum effects of factored loads," so the pattern live loading cases discussed earlier will need to be used. The minimum number of factored live load patterns to be used in combination with factored dead loads is specified in ACI Code Section 6.4.2. A combination of factored dead load and factored live load on all spans also should be included to determine maximum shear forces at beam ends and the maximum loads transferred to the columns. Example 5-2 will demonstrate the analysis for maximum moments in the continuous floor beam shown in Fig. 5-13 using appropriate combinations of factored dead load and patterns of factored live load.

For the continuous floor beam *E–F–G–H* in Fig. 5-4, a different analysis model must be used. No guidance is given in the ACI Code for the analysis of continuous beams and one-way slabs supported by other beams. In general, these beam supports will not provide much restraint to rotations (i.e., their torsional stiffnesses are relatively small), and thus, the author recommends the use of an analysis model similar to that shown in Fig. 5-17. As stated for the previous model, the model in Fig. 5-17 is only to be used for gravity load analysis and should be subjected to combinations of factored dead load and appropriate patterns of factored live loads. As before, rigid end zones can be used at the ends of the beam elements to directly get output of the moments and shears at the faces of the supports. For this case, the length of the rigid end zone at both ends of the beam element should equal one-half of the width of the supporting beam.

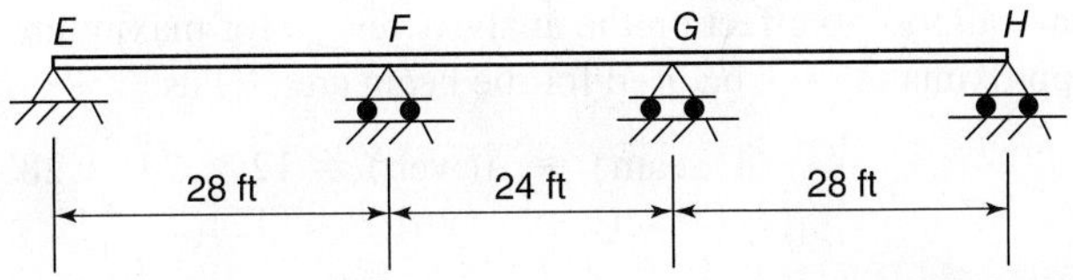

Fig. 5-17
Recommended analysis model for continuous beam or one-way slab supported by beams.

The model in Fig. 5-17 will give reasonable results for design moments and shears at all the critical sections, except for the midspan positive moments in the exterior spans and the zero moments at the exterior supports. Clearly, the spandrel beam supports at the edge of the floor system will have some torsional stiffness, and thus, there should be some negative moment at the exterior supports. Rather than attempt to define a reasonable torsional stiffness for these spandrel beams, which may or may not be cracked due to a combination of bending, shear, and torsion, the author simply recommends that the ACI moment coefficient given in Fig. 5-11c for a floor beam supported by a spandrel beam $(-w_u\ell_n^2/24)$ be used for design moments at supports E and H in Fig. 5-17. The addition of this end moment to the analysis results obtained for the model in Fig. 5-17 will result in an over-design for the total moment capacity of the exterior span unless a corresponding adjustment is made to the midspan positive moment. This over-design, however, could prove to be beneficial when designing the spandrel beam for torsion, as will be discussed in Chapter 7. The torsional design process for the spandrel beam often will require a redistribution of moments away from the spandrel beam and into the floor system—a step that *would not be required* if the analysis model in Fig. 5-17 is used to find the design moments in the exterior spans of the continuous floor beam. A demonstration of the use of the analysis model in Fig. 5-17 will be given in Example 5-2.

EXAMPLE 5-2 Use of Structural Analysis to Find Design Moments in Continuous Floor Beams

As was done in Example 5-1, we will first consider the continuous floor beam *A–B–C–D* in Fig. 5-4. To determine the factored design moments at critical locations along this continuous beam, we will use the analysis model in Fig. 5-13. We will use all of the same member dimensions, dead loads, and reduced live loads calculated in Example 5-1. For our analysis, we will use the appropriate pattern live loads to maximize the moments at the critical locations. After we have finished the analysis of floor beam *A–B–C–D*, we will make similar calculations for floor beam *E–F–G–H* in Fig. 5-4. To determine the factored moments at critical locations along this continuous floor beam, we will use the analysis model in Fig. 5-17.

1. **Analysis model for floor beam *A–B–C–D*.** The beam span lengths are given in Fig. 5-13, and we will assume that the column lengths above and below this floor system are 11 ft. The gross section properties for the columns are

$$A_g = 18 \times 18 = 324 \text{ in.}^2$$
$$I_g = (18)^4/12 = 8750 \text{ in.}^4$$

As discussed previously, we will assume that the approximate cracked moment of inertia for the beam can be taken as the gross moment of inertia for the extended beam web with previously assumed dimensions of 12 in. by 24 in. Because axial stiffness of the beam will have almost no effect on the analysis results for maximum moments and shears, the same approximation can be used for the beam area. Thus,

$$A(\text{beam}) \cong A(\text{web}) = 12 \times 24 = 288 \text{ in.}^2$$
$$I_{cr}(\text{beam}) \cong I_g(\text{web}) = (12)(24)^3/12 = 13{,}800 \text{ in.}^4$$

For all of the beams, we will assume that there is a rigid end zone at each end of the beams (Fig. 5-14) equal to one-half of the column dimension, i.e., 9 inches. Assuming a concrete compressive strength of 4000 psi, the elastic modulus for the beam and column sections will be taken as

$$E_c \cong 57{,}000\sqrt{4000 \text{ psi}} = 3.60 \times 10^6 \text{ psi} = 3600 \text{ ksi}$$

This should be all of the information required for input into an appropriate structural analysis software program.

2. **Analysis for maximum moment at *A* and midspan of member *A–B*.** The appropriate live load pattern to maximize the moments at *A* and at the midspan of the member *A–B* is given in Fig. 5-18a. As determined in Example 5-1, the distributed dead load for all spans is 1.37 k/ft, and the reduced live load for this loading pattern (as determined for span ***A–B***) is 0.596 k/ft. Using the load factors of 1.2 for dead load and 1.6 for live load, the analysis results for the model and loading shown in Fig. 5-18a are $M_A = -102$ k-ft and $M(\text{midspan}) = 105$ k-ft. These results are compared to those obtained using the ACI Moment Coefficients in Table 5-1. All of those results will be discussed in step 5 of this example.

3. **Analysis for maximum moment at midspan of member *B–C*.** The appropriate live load pattern to maximize the moment at the midspan of the member *B–C* is given in Fig. 5-18b. The distributed dead load is unchanged, and the reduced live load for this loading pattern (as determined in Example 5-1 for span *B–C*) is 0.63 k/ft. Using the load factors of 1.2 for dead load and 1.6 for live load, the analysis result for the model and loading shown in Fig. 5-18b is $M(\text{midspan}) = 56.9$ k-ft. Again, this result is compared to that obtained using the ACI Moment Coefficients in Table 5-1.

4. **Analysis for maximum moment at faces of support *B*.** The appropriate live load pattern to maximize the moment at the faces of support *B* is given in Fig. 5-18c. The distributed dead load is unchanged, and the reduced live load for this loading pattern (as determined in Example 5-1 for spans *A–B* and *B–C*) is 0.486 k/ft. Using the load factors of 1.2 for dead load and 1.6 for live load, the analysis result for the model and loading shown in Fig. 5-18c is $M_B(\text{exterior face}) = -154$ ft-kips and $M_B(\text{interior face}) = -123$ k-ft.

TABLE 5-1 Comparison of Factored Design Moments for Continuous Floor Beam with Column Supports

Moment (k-ft)	Face of Support *A*	Midspan of Member *A–B*	Faces of Support *B*	Midspan of Member *B–C*
Results using ACI Moment Coefficients	−114	130	−145	83.8
Results from Structural Analysis	−102	105	−154	56.9

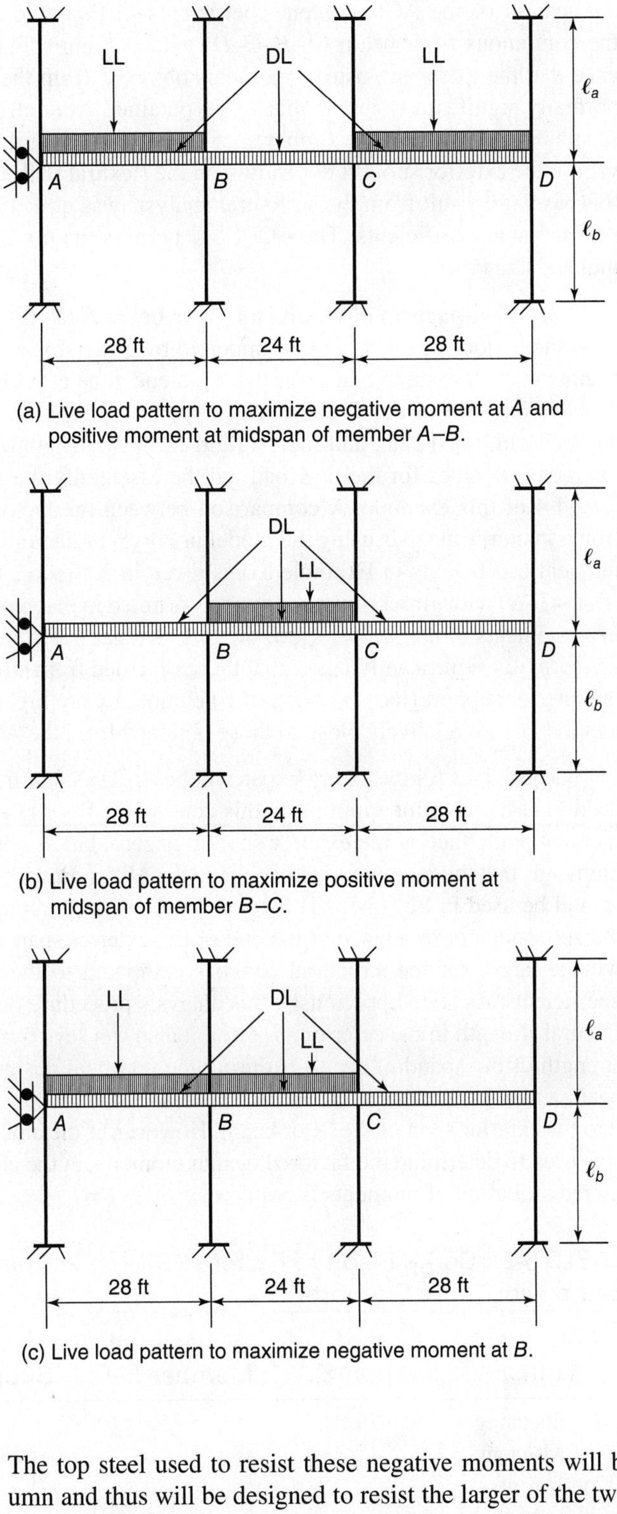

Fig. 5-18
Live load patterns to maximize positive and negative moments.

The top steel used to resist these negative moments will be continuous through the column and thus will be designed to resist the larger of the two moments. So, only the larger moment is compared to that obtained using the ACI Moment Coefficients in Table 5-1.

5. Comparison of results for floor beam *A–B–C–D*. A comparison between the factored design moments obtained from structural analysis in this example and from application of the ACI moment coefficients in Example 5-1 is given in Table 5-1 for the continuous floor beam (*A–B–C–D*) with column supports. It is the author's experience that the midspan positive moments obtained from the ACI moment coefficients are normally significantly larger that those obtained from structural analysis. The negative moments at the face of the supports are usually quite similar for interior supports but can vary at the exterior support depending on the flexural stiffness of the exterior column. For this case, the result from the structural analysis was quite close to that obtained using the ACI moment coefficients. The ACI Code permits the use of either set of factored design moments.

6. Comparison of results for floor beam *E–F–G–H*. The analysis model for this continuous floor beam, which is supported by girders, is given in Fig. 5-17. For all of the beams, we will assume that there is a rigid end zone at each end of the beams (Fig. 5-14) equal to one-half of the width of the supporting girder. For this analysis we will assume a girder width of 12 in., and thus, a rigid end zone length 6 in. The pattern live loads and the reduced values for the live load will be essentially the same as those used in parts 2, 3, and 4 of this example. A comparison between the factored design moments obtained from structural analysis using the model in Fig. 5-17 and those from application of the ACI moment coefficients in Example 5-1 is given in Table 5-2 for the continuous floor beam (*E–F–G–H*) with girder (beam) supports. As noted in the previous step for the midspan positive moments of an *interior span*, the results obtained from the ACI moment coefficients are normally significantly larger that those obtained from structural analysis. The results at the interior support (faces of support F) commonly are higher from the structural analysis method but are relatively close to those obtained from the ACI moment coefficients.

The results for the exterior span will be affected significantly by the assumed pin connection at the exterior support for this continuous floor beam (Fig. 5-17). The calculated moment at the face of the exterior support is zero, but as noted previously, the author recommends that the moment obtained using the ACI moment coefficient, C_m, equal to $-1/24$ should be used in Eq. (5-5). This result is shown in parenthesis in Table 5-2. Because of the zero-moment resistance at the end of the exterior span, the midspan positive moment will be larger for the structural analysis compared to the result from the ACI moment coefficients. As stated previously, this analysis procedure does result in an overdesign for flexural strength in the exterior span, but it also can save time when checking the torsional strength of the spandrel beam. If during the torsional design it is found that the spandrel beam will crack under factored torsion, the ACI code requires a redistribution of moments into the exterior span of the floor beam. However, if the analysis procedure discussed here was used to determine the factored design moments in the exterior span of the floor beam, no redistribution of moments is required. ■

TABLE 5-2 Comparison of Factored Design Moments for Continuous Floor Beam with Beam Supports

Moment (k-ft)	Face of Support *E*	Midspan of Member *E–F*	Faces of Support *F*	Midspan of Member *F–G*
Results using ACI Moment Coefficients	−79	135	−151	87.6
Results from Structural Analysis	(−79)	179	−172	42.4

5-3 DESIGN OF SINGLY REINFORCED BEAM SECTIONS WITH RECTANGULAR COMPRESSION ZONES

General Factors Affecting the Design of Rectangular Beams

Location of Reinforcement

Concrete cracks due to tension and (as a result) reinforcement is required where flexure, axial loads, or shrinkage effects cause tensile stresses. A uniformly loaded, simply supported beam deflects as shown in Fig. 5-19a and has the moment diagram shown in Fig. 5-19b. Because this beam is in positive moment throughout, tensile flexural stresses and cracks are developed along the bottom of the beam. Longitudinal reinforcement is required to resist these tensile stresses and is placed close to the bottom side of the beam, as shown in Fig. 5-19c. Because the moments are greatest at midspan, more reinforcement is required at the midspan than at the ends, and it may not be necessary to extend all the bars into the supports. In Fig. 5-19c, some of the bars are *cut off* within the span.

A cantilever beam develops negative moment throughout and deflects as shown in Fig. 5-20 with the concave surface downward, so that flexural tensions and cracks develop

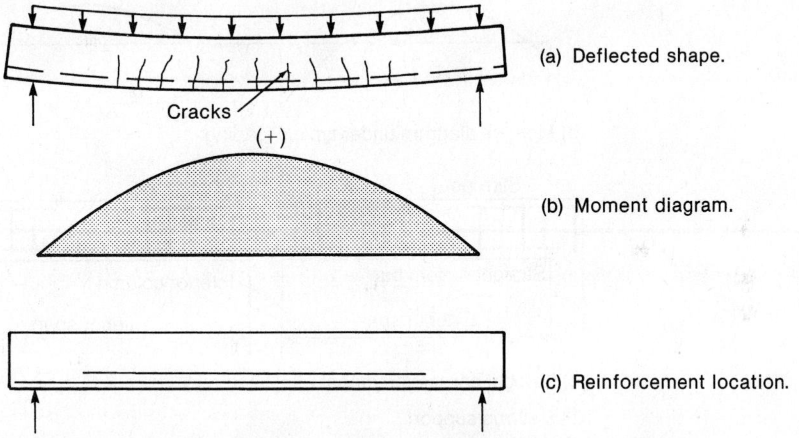

Fig. 5-19
Simply supported beam.

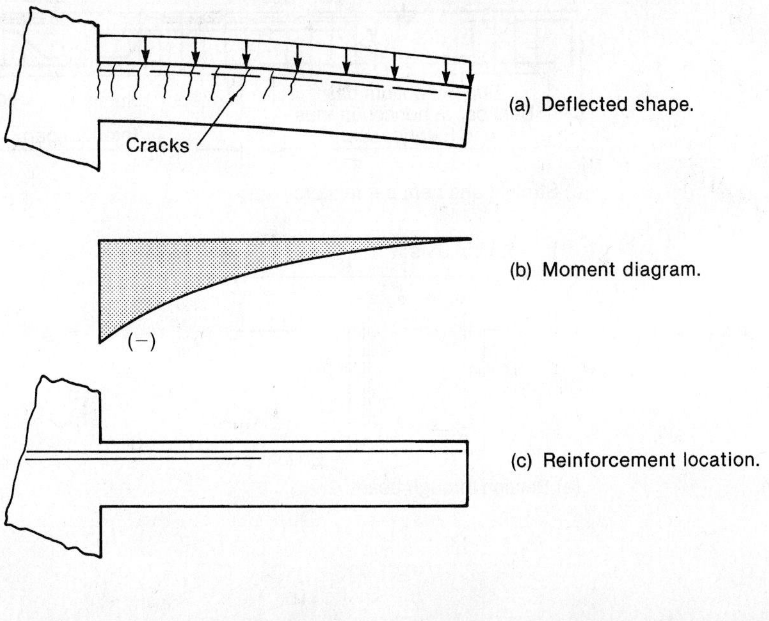

Fig. 5-20
Cantilever beam.

on the top surface. In this case, the reinforcement is placed near the top surface, as shown in Fig. 5-20c. Because the moments are largest at the fixed end, more reinforcement is required there than at any other point. In some cases, some of the bars may be terminated before the free end of the beam. Note that the bars must be anchored into the support.

Commonly, reinforced concrete beams are continuous over several supports, and under gravity loads, they develop the moment diagram and deflected shape shown in Fig. 5-21. Again, reinforcement is needed on the tensile face of the beam, which is at the top of the beam in the negative-moment regions near the supports and at the bottom in the positive-moment regions near the midspans. Two possible arrangements of this reinforcement are shown in Figs. 5-21c and 5-21d. Prior to 1965, it was common practice to bend the bottom reinforcement up to the top of the beam when it was no longer required at

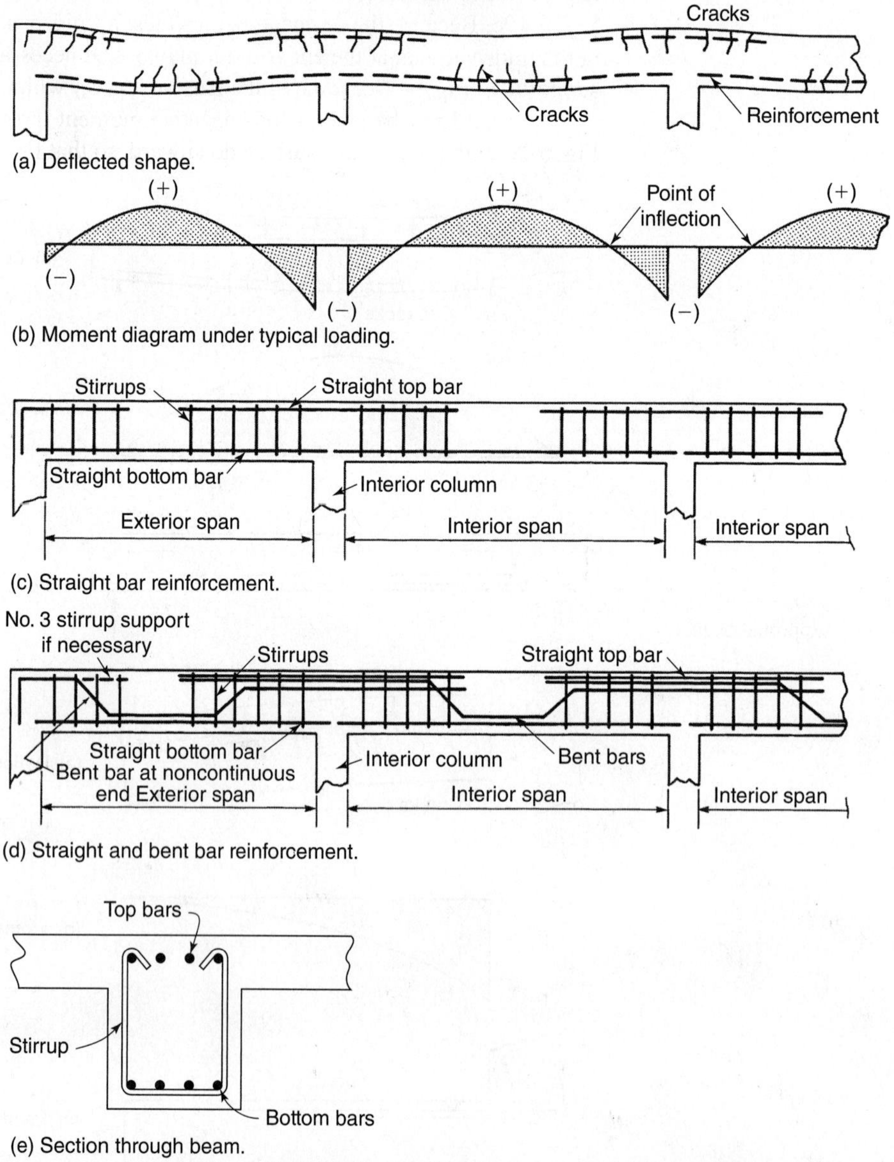

Fig. 5-21
Continuous beam.

the bottom. In this way, a *bent-up* or *truss bar* could serve as negative and positive reinforcement in the same beam. Such a system is illustrated in Fig. 5-21d. Today, the straight bar arrangement shown in Fig. 5-21c is used almost exclusively. In some cases, a portion of the positive-moment or negative-moment reinforcement is terminated or cut off when no longer needed. However, it should be noted that a portion of the steel is extended past the points of inflection, as shown. This is done primarily to account for shifts in the points of inflection due to shear cracking and to allow for changes in loadings and loading patterns. The calculation of bar-cutoff points is discussed in Chapter 8.

In addition to longitudinal reinforcement, transverse bars (referred to as *stirrups*) are provided to resist shear forces and to hold the various layers of bars in place during construction. These are shown in the cross section in Fig. 5-21e. The design of shear reinforcement is discussed in Chapter 6.

In conclusion, it is important that designers be able to visualize the deflected shape of a structure. The reinforcing bars for flexure are placed on the tensile face of the member, which corresponds to the convex side of the deflected shape.

Construction of Reinforced Concrete Beams and Slabs

The simplest concrete flexural member is the one-way slab shown in Fig. 5-1. The form for such a slab consists of a flat surface generally built of plywood supported on wooden or steel joists. Whenever possible, the forms are constructed in such a way that they can be reused on several floors. The forms must be strong enough to support the weight of the wet concrete plus construction loads, such as workers and construction equipment used in the casting and finishing process. In addition, the forms must be aligned correctly and *cambered* (arched upward), if necessary, so that the finished floor is flat after the forms are removed.

The reinforcement is supported in the forms on wire or plastic supports referred to as *bolsters* or *chairs*, which hold the bars at the correct distance above the forms until the concrete has hardened. If the finished slab is expected to be exposed to moisture, wire bolsters may rust, staining the surface. In such a case, small, precast concrete blocks or plastic bar chairs may be used instead. Wire bolsters can be seen in the photograph in Fig. 5-22.

Beam forms most often are built of plywood supported by scaffolding or by wooden supports. The size of beam forms generally is chosen to allow maximum reuse of the forms, because the cost of building the forms is a significant part of the total cost of a concrete floor system, as was discussed in Section 2-9.

Reinforcement for two beams and some slabs is shown in Fig. 5-22. Here, closed stirrups have been used and the top beam bars are supported by the top of the closed stirrups. The negative-moment bars in the slabs still must be placed. Frequently, the positive-moment steel, stirrups, and stirrup-support bars for a beam are preassembled into a cage that is dropped into the form.

Preliminary Beam and Slab Depths for Control of Deflections

The deflections of a beam can be calculated from equations of the form

$$\Delta_{max} = C\frac{w\ell^4}{EI} \tag{5-9a}$$

Fig. 5-22
Intersection of a column and two beams.

Rearranging this and making assumptions concerning strain distribution and neutral-axis depth eventually gives an equation of the form

$$\frac{\Delta}{\ell} = C\frac{\ell}{h} \tag{5-9b}$$

Thus, for any acceptable ratio of deflection to span lengths, Δ/ℓ, it should be possible to specify span-to-depth ratios, ℓ/h, which if exceeded may result in unacceptable deflections. In the previous section on analysis, the author suggested that typical beam depths range between $\ell/12$ and $\ell/18$. The selected beam depth, h, will need to be checked against the minimum member thicknesses (depth, h) given in ACI Table 9.3.1.1 for members *not supporting partitions* or other construction that are *likely to be damaged* by deflection. The reader should note that the minimum member depths given in ACI Table 9.3.1.1 for continuous construction are less than the range of member depths suggested by the author.

In contrast, the minimum thicknesses given for solid slabs in ACI Table 7.3.1.1 are used frequently in selecting the overall depth of slabs. In general, thicknesses calculated in row 1 of the table should be rounded up to the next one-quarter inch for slabs less than 6 in. thick and to the next one-half inch for thicker slabs. The calculation of deflections will be discussed in Chapter 9.

Concrete Cover and Bar Spacing

It is necessary to have cover (concrete between the surface of the slab or beam and the reinforcing bars) for four primary reasons:

1. To bond the reinforcement to the concrete so that the two materials act together. The efficiency of the bond increases as the cover increases. A cover of at least one bar diameter is required for this purpose in beams and columns. (See Chapter 8.)

2. To protect the reinforcement against corrosion. Depending on the environment and the type of member, varying amounts of cover ranging from $\frac{3}{8}$ to 3 in. are required (ACI Code Table 20.6.1.3.1). In highly corrosive environments, such as slabs or bridges exposed to deicing salts or ocean spray, ACI Code Section 20.6.1.4.1 states that cover should be increased and the concrete mix design should satisfy requirements in Code Section 19.3.1 for *Exposure Category C*. Alternative protection methods, such as the use of a waterproof membrane on the exposed surface, may also be used.

3. To protect the reinforcement from strength loss due to overheating in the case of fire. The cover for fire protection is specified in the local building code. Generally speaking, $\frac{3}{4}$-in. cover to the reinforcement in a structural slab will provide a 1-hour fire rating, while a $1\frac{1}{2}$-in. cover to the stirrups or ties of beams corresponds to a 2-hour fire rating.

4. Additional cover sometimes is provided on the top of slabs, particularly in garages and factories, so that abrasion and wear due to traffic will not reduce the cover below that required for structural and other purposes.

In this book, the amounts of clear cover will be based on ACI Code Table 20.6.1.3.1 unless specified otherwise. The arrangement of bars within a beam must allow sufficient concrete on all sides of each bar to transfer forces into or out of the bars; sufficient space so that the fresh concrete can be placed or consolidated around all the bars; and sufficient space to allow an internal vibrator to reach through to the bottom of the beam. Pencil-type concrete immersion vibrators used in consolidation of the fresh concrete are $1\frac{1}{2}$ to $2\frac{1}{2}$ in. in diameter. Enough space should be provided between the beam bars to allow a vibrator to reach the bottom of the form in at least one place in the beam width.

The photo in Fig. 5-22 shows the reinforcement at an intersection of two beams and a column. The longitudinal steel in the beams is at the top of the beams because this is a negative-moment region. Although this region looks congested, there are adequate openings to place and vibrate the concrete. Reference [5-3] discusses the congestion of reinforcement in regions such as this and recommends design measures to reduce the congestion.

ACI Code Sections 25.2.1 and 25.2.2 specify the spacings and arrangements shown in Fig. 5-23. When bars are placed in two or more layers, the bars in the top layer must be directly over those in the other layers to allow the concrete and vibrators to pass through the layers. Potential conflicts between the reinforcement in a beam and the bars in the columns or other beams should be considered. Figure 5-24, based on an actual case history [5-4], shows what can happen if potential conflicts at a joint are ignored. The left-hand side shows how the design was envisioned, and the right-hand side shows the way the joint was built. Placement tolerances and the need to resolve the interference problems have reduced the effective depth of the negative-moment reinforcement from $9\frac{1}{2}$ in. to $7\frac{3}{4}$ in.— an 18 percent reduction in depth and thus a corresponding reduction in moment strength. To identify and rectify bar conflicts, it sometimes is necessary to draw the joint to scale, showing the actual width of the bars. Conflicts between bars in the columns and other beams must be considered.

Calculation of Effective Depth and Minimum Web Width for a Given Bar Arrangement

The effective depth, d, of a beam is defined as the distance from the extreme compression fiber to the centroid of the longitudinal tensile reinforcement.

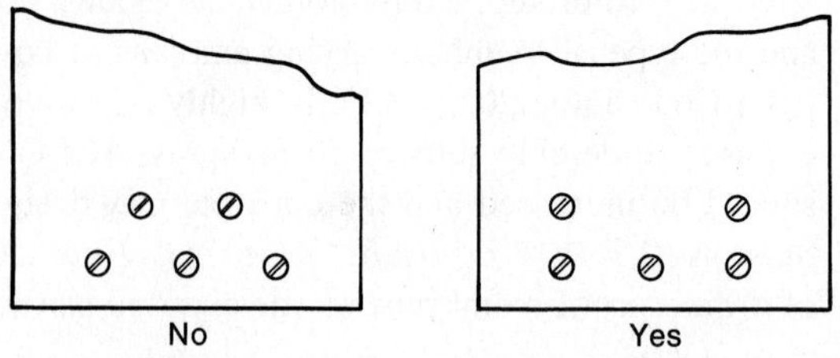

(a) Arrangement of bars in two layers (ACI Section 25.2.2).

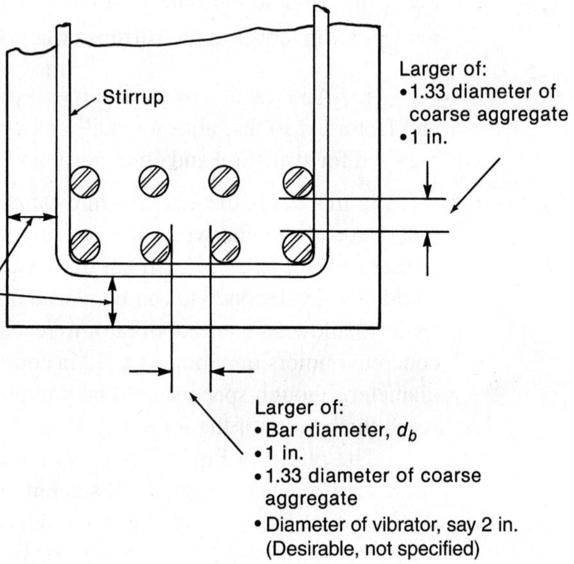

(b) Minimum bar spacing and cover limits in ACI Code.

Fig. 5-23
Bar spacing limits in ACI Code.

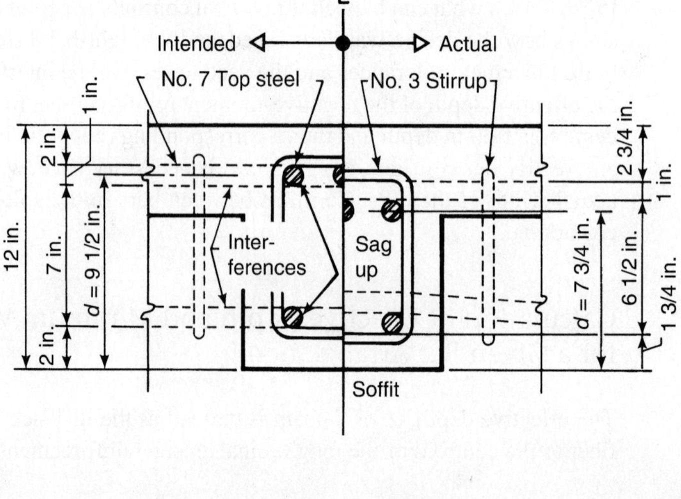

Fig. 5-24
Bar placing problems at the intersections of two beams. (From [5-4].)

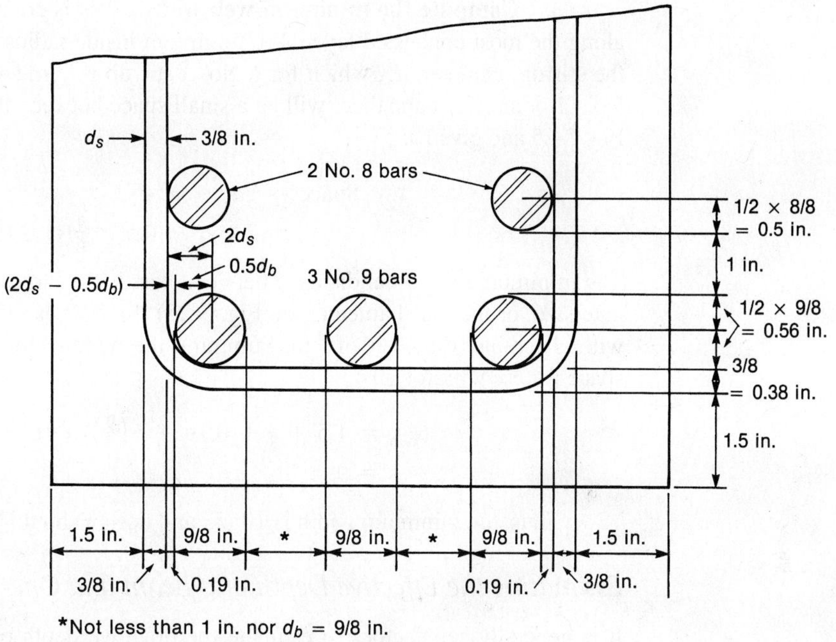

Fig. 5-25
Beam section for Example 5-3.

EXAMPLE 5-3 Calculation of *d* and of Minimum *b*

Compute d and the minimum value of b for a beam having bars arranged as shown in Fig. 5-25. The maximum size of coarse aggregate is specified as $\frac{3}{4}$ in. The overall depth, h, of the beam is 24 in.

This beam has two different bar sizes. The larger bars are in the bottom layer to maximize the effective depth and hence the moment lever arm. Also, notice that the bars are symmetrically arranged about the centerline of the beam. The bars in the upper layer are directly above those in the lower layer. Placing the top bars on the outside of the section allows those bars to be supported by tying them directly to the stirrups.

1. Compute clear cover. From ACI Code Table 20.6.1.3.1, the clear cover to the stirrups is 1.5 in. (Fig. 5-23). From ACI Code Sections 25.2.1 and 25.2.2, the minimum distance between layers of bars is the larger of 1 in. or 4/3 times the aggregate size, which in this case gives $\frac{4}{3} \times \frac{3}{4} = 1$ in.

2. Compute the centroid of the bars.

Layer	Area, A(in.^{2})	Distance from Bottom, y (in.)	Ay in.^{3}
Bottom	$3 \times 1.00 = 3.00$	$1.5 + \frac{3}{8} + \left(\frac{1}{2} \times \frac{9}{8}\right) = 2.44$	7.31
Top	$2 \times 0.79 = \underline{1.58}$	$2.44 + \left(\frac{1}{2} \times \frac{9}{8}\right) + 1 + \left(\frac{1}{2} \times \frac{8}{8}\right) = 4.50$	$\underline{7.11}$
	Total $A = 4.58$		Total $Ay = 14.42$

The centroid is located at $Ay/A = \bar{y} = 14.42/4.58 = 3.15$ in. from the bottom of the beam. The effective depth $d = 24 - 3.15$ in. $= 20.85$ in.—say, $d = 20.8$ in. It is conservative to round the value of d down, not up.

3. **Compute the minimum web width.** This is computed by summing the widths along the most congested layer. The minimum inside radius of a stirrup bend is two times the stirrup diameter, d_s, which for a No. 3 stirrup is $\frac{3}{4}$ in. (ACI Code Section 25.3.2). For No. 11 or smaller bars, there will be a small space between the bar and the tie, as shown in Fig. 5-25 and given as

$$\begin{aligned}\text{Space} &= 2d_s - 0.5d_b \\ &= 2 \times \tfrac{3}{8} - 0.5 \times \tfrac{9}{8} = 0.19 \text{ in.}\end{aligned}$$

The minimum horizontal distance between bars is the largest of 1 in., 4/3 times the aggregate size, or the bar diameter (see Fig. 5-25). In this case, the largest bars are No. 9 bars with a nominal diameter of $\frac{9}{8}$ in. Summing the widths along a section at A and ignoring space for the vibrator gives

$$\begin{aligned}b_{\text{min}} &= 1.5 + \tfrac{3}{8} + 0.19 + 5\left(\tfrac{9}{8}\right) + 0.19 + \tfrac{3}{8} + 1.5 \\ &= 9.76 \text{ in.}\end{aligned}$$

Thus, the minimum width is 10 in., and design should be based on $d = 20.8$ in. ■

Estimating the Effective Depth of a Beam and One-Way Slab

It is generally satisfactory to estimate the effective depth of a beam using the following approximations:

For beams with one layer of reinforcement,

$$d \simeq h - 2.5 \text{ in.} \tag{5-10a}$$

For beams with two layers of reinforcement,

$$d \simeq h - 3.5 \text{ in.} \tag{5-10b}$$

The value 3.5 in. given by Eq. (5-10b) corresponds to the 3.15 in. computed in Example 5-3. The error introduced by using Eq. (5-10b) to compute d is in the order

$$\frac{(24 - 3.5)}{(24 - 3.15)} = 0.983$$

Thus, (5-10b) underestimates d by 1.7 percent. This is acceptable.

For reinforced concrete slabs, the minimum clear cover is $\frac{3}{4}$ in. rather than $1\frac{1}{2}$ in., and the positive moment steel is all in one layer, with the negative moment steel in another layer. This steel generally will be No. 3, 4, or 5 bars. Stirrups are seldom, if ever, used in one-way slabs in buildings. For this case, Eqs. (5-10a) and (5-10b) can be rewritten as follows:

For one-way slab spans up to 12 ft,

$$d \simeq h - 1 \text{ in.} \tag{5-10c}$$

For one-way slab spans over 12 ft,

$$d \simeq h - 1.1 \text{ in.} \tag{5-10d}$$

In SI units and rounding to a 5 mm value, Eq. (5-10) works out to be the following:

For beams with one layer of tension reinforcement,

$$d \simeq h - 65 \text{ mm} \tag{5-10aM}$$

For beams with two layers of tension reinforcement,

$$d \simeq h - 90 \text{ mm} \tag{5-10bM}$$

For one-way slabs with spans up to 3.5 m,

$$d \simeq h - 25 \text{ mm} \tag{5-10cM}$$

For one-way slabs with spans over 3.5 m,

$$d \simeq h - 30 \text{ mm} \tag{5-10dM}$$

It is important not to overestimate d, because normal construction practices may lead to smaller values of d than are shown on the drawings. Studies of construction accuracy show that, on the average, the effective depth of the negative-moment reinforcement in slabs may be 0.75 in. less than specified [5-5]. In thin slabs, this error in the steel placement will cause a significant reduction in the nominal moment strength.

Generally speaking, beam width b should not be less than 10 in., although with two bars, beam widths as low as 7 in. can be used in extreme cases. The use of a layer of closely spaced bars may lead to a splitting failure along the plane of the bars, as will be explained in Chapter 8. Because such a failure may lead to a loss of bar anchorage or to corrosion, care should be taken to have at least the required minimum bar spacings. Where there are several layers of bars, a continuous vertical opening large enough for the concrete vibrator to pass through should be provided. Minimum web widths for multiple bars per layer are given in Table A-5 (see Appendix A).

Minimum Reinforcement

As was discussed in Chapter 4, to prevent a sudden failure with little or no warning when the beam cracks in flexure, ACI Code Section 9.6.1.2 requires a minimum amount of flexural reinforcement equal to that in Eq. (4-11) and repeated here:

$$A_{s,\min} = \frac{3\sqrt{f'_c}}{f_y} b_w d, \text{ and } \geq \frac{200 b_w d}{f_y} \tag{5-11}$$

where f'_c and f_y are in psi. In SI units, this becomes

$$A_{s,\min} = \frac{0.25\sqrt{f'_c}}{f_y} b_w d, \text{ and } \geq \frac{1.4 b_w d}{f_y} \tag{5-11M}$$

where f'_c and f_y are in MPa.

An evaluation of $A_{s,\min}$ in flanged sections was discussed in Section 4-8.

General Strength Design Requirements for Beams

In the design of beam cross sections, the general strength requirement is

$$\phi M_n \geq M_u \tag{5-12}$$

Here, M_u represents the factored moments at the section due to factored loads. Referring to ACI Code Table 5.3.1 and the assumptions made for Eq. (5-7) that the effects of fluid pressure, soil pressure, and thermal effects can be ignored, the factored-load combinations commonly considered in beam design are

$$M_u = 1.4M_D \tag{5-13}$$

$$M_u = 1.2M_D + 1.6M_L \tag{5-14}$$

where M_D and M_L are the moments due to the unfactored dead and live loads, respectively.

We normally will design beam sections to be tension-controlled, and thus, the strength reduction factor, ϕ, initially is assumed to be equal to 0.9. This will need to be confirmed at the end of the design process. The easiest expression for the analysis of the nominal moment strength of a singly reinforced beam section with a rectangular compression zone is

$$M_n = A_s f_y\left(d - \frac{a}{2}\right) \tag{5-15}$$

where $(d - a/2)$ is referred to as the moment arm and sometimes is denoted as jd. Typical values for this moment arm will be discussed in the following section.

Design of Tension Reinforcement When Section Dimensions Are Known

In this case, b and h (and thus, d) are known, and it only is necessary to compute A_s. This is actually a very common case for continuous members where the same section size will be used in both positive and negative bending regions and may be used for several of the typical beam spans in a floor system. These dimensions may be established by architectural limits on member dimensions or estimated using guidelines such as total depth, h, between $\ell/18$ and $\ell/12$ and beam (web) width between 0.4 and $0.8h$. Beam dimensions may also be established by designing the section of the beam that is resisting the largest bending moment. The design of that section will then establish the beam dimensions to be used throughout at least one span—probably for several spans. The initial design of a beam section for which dimensions are not known will be covered in the following subsection.

For the most common steel percentages in beams, the value of the moment arm, jd, generally is between $0.87d$ and $0.91d$. For slab sections and beam sections with wide compression zones (T-beam in positive bending), the value of jd will be close to $0.95d$. Thus, for design problems in this book where section dimensions are known, j will initially be assumed to be equal to 0.9 for beams with narrow compression zones (width of compression zone equal to width of member at mid-depth) and 0.95 for slabs and beam sections that have wide compression zones.

Combining the strength requirement in Eq. (5-12) with the section nominal moment strength expression in Eq. (5-15) leads to an important equation for determining the required steel area in a singly reinforced section.

$$A_s \geq \frac{M_u}{\phi f_y\left(d - \frac{a}{2}\right)} \cong \frac{M_u}{\phi f_y(jd)} \tag{5-16}$$

Using the suggested values for j given above, this equation will give a good approximation of the required area of tension reinforcement. One quick iteration can be used to refine the value for A_s by enforcing section equilibrium to determine the depth of the compression stress block, a (as was done in Chapter 4 for a compression zone with a constant width, b):

$$a = \frac{A_s f_y}{0.85 f'_c\, b} \tag{5-17}$$

and then putting that value of a into Eq. (5-16) to calculate an improved value for A_s. The iterative process represented by Eqs. (5-16) and (5-17) will be used extensively throughout this book and is illustrated now in the following examples.

EXAMPLE 5-4 Design of Reinforcement When Section Dimensions Are Known

We will design tension reinforcement for one positive bending section and one negative bending section of the continuous floor beam *A–B–C–D* shown in Fig. 5-4 and analyzed in Examples 5-1 and 5-2. The section dimensions for this flanged-beam section are shown in Fig. 5-26a. Assume a concrete compressive strength of 4000 psi and a steel yield strength of 60 ksi.

1. Design the midspan section of beam *A–B*. The factored design moment at this section was found to be 130 k-ft using the ACI moment coefficients and 105 k-ft using structural analysis software. We will use the larger value for this example. Because this is a T-section in positive bending, we initially will assume a moment arm, jd, equal to $0.95d$ (wide compression zone). Assume we will use a single layer of reinforcement, so d can be taken as $h - 2.5$ in. or 21.5 in. Assuming that this will be a tension-controlled section ($\phi = 0.9$), we will use Eq. (5-16) to get the first estimate for the required area of tension steel:

$$A_s \geq \frac{M_u}{\phi f_y\left(d - \dfrac{a}{2}\right)} \cong \frac{M_u}{\phi f_y (jd)} = \frac{130 \text{ k-ft} \times 12 \text{ in./ft}}{0.9 \times 60 \text{ ksi} \times 0.95 \times 21.5 \text{ in.}} = 1.41 \text{ in.}^2$$

Because this is a small value, we should check $A_{s,\min}$ from Eq. (5-11). For the given concrete strength, $3\sqrt{f'_c} = 190$ psi, so use 200 psi in Eq. (5-11):

$$A_{s,\min} = \frac{200 b_w d}{f_y} = \frac{200 \text{ psi} \times 12 \text{ in.} \times 21.5 \text{ in.}}{60{,}000 \text{ psi}} = 0.86 \text{ in.}^2$$

Thus, the minimum area will not govern, and we will do one iteration to improve the value of A_s using Eqs. (5-17) and (5-16). To determine the depth of the compression stress block, a, we must determine the effective width of the compression zone to use in Eq. (5-17). Referring to Section 4-8 of this book and ACI Code Section 6.3.2.1, the limits for the effective width of the compression flange are (using $\ell_n = 26.5$ ft):

$$b_e \leq b_w + 2(\ell_n/8) = 12 \text{ in.} + \frac{26.5 \text{ ft}}{4} \times 12 \text{ in./ft} = 91.5 \text{ in.}$$

$$b_e \leq b_w + 2(8h_f) = 12 \text{ in.} + 2 \times 8 \times 6 \text{ in.} = 108 \text{ in.}$$

$$b_e \leq \text{spacing between beams} = 12 \text{ ft} = 144 \text{ in.}$$

The last limit is the result of adding the web width to one-half of the clear spans to adjacent beam webs on each side of the beam under consideration. The first limit governs, so we will use a compression zone width of 91.5 in. in Eq. (5-17):

$$a = \frac{A_s f_y}{0.85 f'_c\, b} = \frac{1.41 \text{ in.}^2 \times 60 \text{ ksi}}{0.85 \times 4 \text{ ksi} \times 91.5 \text{ in.}} = 0.272 \text{ in.}$$

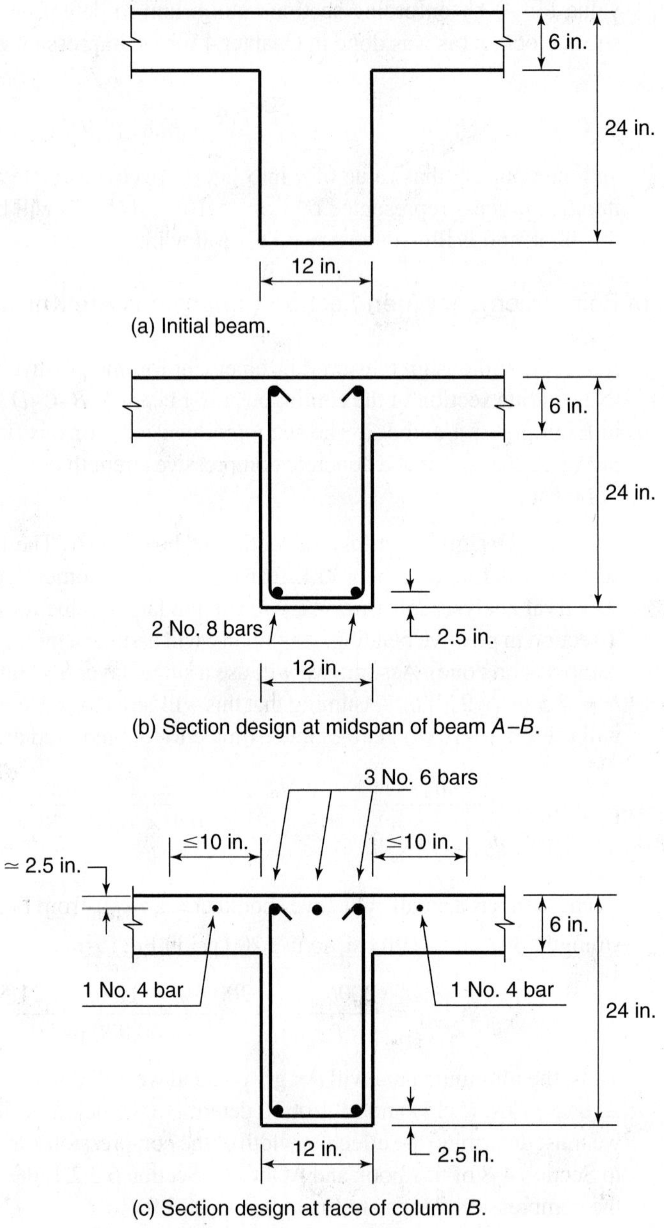

Fig. 5-26
Beam sections—Example 5-4.

At first one might think there is an error in this calculation, but it is not unusual to calculate very small values for the stress-block depth for a T-section in positive bending. Using this value of a in Eq. (5-16) gives

$$A_s \geq \frac{M_u}{\phi f_y\left(d - \dfrac{a}{2}\right)} = \frac{130\text{ k-ft} \times 12\text{ in./ft}}{0.9 \times 60\text{ ksi} \times (21.5\text{ in.} - 0.136\text{ in.})} = 1.35\text{ in.}^2$$

For this required area, **select two No. 8 bars**, which results in an area, A_s, equal to 1.58 in.^{2} (Table A-4) and requires a web width of 7.5 in. (Table A-5). It is possible to select some combination of bar sizes to get closer to the required tension steel area, but the use of multiple bar sizes in a single layer of reinforcement is not preferred, because it could lead to errors during construction. So, we will stay with two No. 8 bars.

2. Detailing Check. Now that we have selected the size of the longitudinal reinforcing bar, we could refine the assumed value of d. However, as stated previously, using $d \approx h - 2.5$ in. will be accurate enough for most beam designs. Only when using large longitudinal bars (>No. 10) or large stirrups (>No. 4) would an adjustment be required to prevent a significant overestimate the value for d.

To limit the widths of flexural cracks in beams and slabs, ACI Code Section 24.3.2 defines an upper limit on the center-to-center spacing between bars in the layer of reinforcement closest to the tension face of a member. In some cases, this requirement could force a designer to select a larger number of smaller bars in the extreme layer of tension reinforcement. The spacing limit is:

$$s \leq 15\left(\frac{40{,}000}{f_s}\right) - 2.5c_c \tag{5-18}$$

but,

$$s \leq 12\left(\frac{40{,}000}{f_s}\right)$$

In Eq. (5-18), c_c is the least distance from the *surface* of the reinforcement bar to the tension face. For the tension zone in a typical beam, as shown in Fig. 5-25, this would include the clear cover to the stirrups (1.5 in.) plus the diameter of the stirrup bar (usually 3/8 or 4/8 in.). Thus, for a typical beam the value of c_c can be taken as 2.0 in.

The term f_s in Eq. (5-18) represents the stress in the flexural reinforcement closest to the tension face due to acting loads (not factored loads). Procedures for calculating f_s will be discussed in Chapter 9, but ACI Code Section 24.3.2.1 permits the value of f_s to be taken as two-thirds of the yield stress, f_y (in psi units). Thus, for Grade-60 steel, f_s can be set equal to 40,000 psi.

Using $c_c = 2$ in. and $f_s = 40{,}000$ psi, the limit on the center-to-center spacing between the two No. 8 bars in the extreme layer of tension reinforcement (Fig. 5-26b) is:

$$s \leq 15(\text{in.})\left(\frac{40{,}000}{40{,}000}\right) - 2.5(2\text{ in.}) = 10\text{ in.}$$

and,

$$s \leq 12(\text{in.})\left(\frac{40{,}000}{40{,}000}\right) = 12\text{ in.}$$

Assuming the distance from the sides of the beam to the center of each No. 8 bar is 2.5 in. (Fig. 5-25), the center-to-center spacing between the No. 8 bars is:

$$s = 12\text{ in.} - 2(2.5\text{ in.}) = 7\text{ in.} < 10\text{ in.}$$

Thus, the spacing between the bars satisfies the Code requirement. If the spacing was too large we would need to use three (smaller) bars to reduce the center-to-center spacing between the bars to a satisfactory value.

3. Required strength check. We already have calculated the required minimum steel area, and it is less than the selected area of steel. Because we have used $\phi = 0.9$, we must confirm that this is a tension-controlled section. For a T-section where we have already calculated a very small value for the depth of the compression stress block, one simply might say that this is clearly a tension-controlled section, because the depth to the neutral axis, c, will be significantly less that the tension-controlled limit of 3/8 of d, as discussed in Section 4-6. For completeness in this first design example, we will calculate c and compare it to 3/8 of d, or 0.375×21.5 in. $= 8.06$ in. For the selected area of steel, use Eq. (5-17) to find

$$a = \frac{1.58 \text{ in.}^2 \times 60 \text{ ksi}}{0.85 \times 4 \text{ ksi} \times 91.5 \text{ in.}} = 0.305 \text{ in.}$$

For a concrete compressive strength of 4000 psi, the factor β_1 is equal to 0.85. Thus, the depth to the neutral axis is

$$c = a/\beta_1 = 0.305 \text{ in.}/0.85 = 0.36 \text{ in.}$$

This value for c is clearly less than 3/8 of d, so this is a tension-controlled section. Also note that the final of jd $(d - a/2)$ is 21.3 in., which is approximately 4 percent larger than the assumed value of 0.95 d.

The final check is to confirm the strength of the section using Eq. (5-15), including the strength reduction factor, ϕ:

$$\phi M_n = \phi A_s f_y\left(d - \frac{a}{2}\right) = 0.9 \times 1.58 \text{ in.}^2 \times 60 \text{ ksi}\left(21.5 \text{ in.} - \frac{0.305 \text{ in.}}{2}\right)$$

$$= 1820 \text{ k-in.} = 152 \text{ k-ft} \geq M_u = 130 \text{ k-ft}$$

The strength is adequate without being too excessive. So, **use two No. 8 bars**, as shown in Fig. 5-26b.

4. Design for factored moment at face of support *B*. The design here represents the design at both faces of support B. To be consistent with the design of the midspan section, we will use the factored design moment obtained from the ACI Moment Coefficients, i.e., a negative moment of 145 k-ft. Because this is a negative moment, compression will occur in the bottom of the section, and thus, we have a relatively narrow compression zone. Recall that for this case the author recommends the use of a moment arm, jd, equal to $0.9d$. Assuming that this will be a tension-controlled section ($\phi = 0.9$), Eq. (5-16) is used to get the first estimate for the tension steel area.

$$A_s \geq \frac{M_u}{\phi f_y\left(d - \frac{a}{2}\right)} \cong \frac{M_u}{\phi f_y(jd)} = \frac{145 \text{ k-ft} \times 12 \text{ in./ft}}{0.9 \times 60 \text{ ksi} \times 0.9 \times 21.5 \text{ in.}} = 1.67 \text{ in.}^2$$

As before, we will do one iteration using Eqs. (5-17) and (5-16) to improve the value of A_s. For this case, the width of the compression zone, b, is equal to the web width, $b_w = 12$ in.

$$a = \frac{A_s f_y}{0.85 f'_c b} = \frac{1.67 \text{ in.}^2 \times 60 \text{ ksi}}{0.85 \times 4 \text{ ksi} \times 12 \text{ in.}} = 2.46 \text{ in.}$$

Using this value of a in Eq. (5-16) gives

$$A_s \geq \frac{M_u}{\phi f_y\left(d - \frac{a}{2}\right)} = \frac{145 \text{ k-ft} \times 12 \text{ in./ft}}{0.9 \times 60 \text{ ksi}(21.5 \text{ in.} - 1.23 \text{ in.})} = 1.59 \text{ in.}^2$$

The selection of reinforcing bars for this negative bending section is complicated by ACI Code Section 24.3.4, which reads in part, "Where flanges of T-beam construction are in tension, part of the flexural tension reinforcement shall be distributed over an effective flange width . . . ." The definition of the word *part* and the intention of the Code are not clarified by reading the Commentary to the Code. The author's interpretation of this Code requirement is that the majority of the tension reinforcement should be placed above the web of the beam section, and the remainder of the required tension steel should be placed in a region of the flange (slab) close to the web of the beam. ACI Code Section 24.3.4

states that tension reinforcement used in the flange must be placed within the effective flange width defined in Code Section 6.3.2.1, but not wider than $\ell_n/10$. The limit of $\ell_n/10$ (approximately 32 in.) governs for this case. For the tension steel area required in this case, **use three No. 6 bars above the web and place two No. 4 bars in the flanges,** one on each side of the web (Fig. 5-26c). Thus,

$$A_s = 3 \times 0.44 \text{ in.}^2 + 2 \times 0.20 \text{ in.}^2 = 1.72 \text{ in.}^2$$

The minimum web width for three No. 6 bars is 9.0 in. (Table A-5). Although the three No. 6 bars are at the level of the flange, we want to make this minimum width check to be sure the shear reinforcement (to be designed in Chapter 6) fits the bars to be used at both the top and bottom of the web. The bar spacing check from ACI Code Section 24.3.2 is the same for this layer of bars as it was for the bottom layer of bars. Because three bars are used in this layer, the spacing limit is easily satisfied.

Because of the thinner cover permitted in a slab, the No. 4 bars in the flange generally will be higher in the T-beam section than the bars placed above the web. However, for strength calculations, we can achieve sufficient accuracy by assuming that all of the tension reinforcement is approximately 2.5 in. from the top of the section. Because the actual d is a little larger, this approach is conservative.

5. Required strength check. Because this T-section is part of a continuous floor beam, the value for $A_{s,\min}$ is the same as that calculated for the midspan section (0.86 in.^{2}). Thus, the provided A_s exceeds the required minimum tension steel area.

To check if this is a tension-controlled section, we can compare the depth to the neutral axis, c, to 3/8 of d, which is the limit on the neutral axis depth for tension-controlled sections. Equation (5-17) will be used to determine the depth of the compression stress block for the selected tension steel area.

$$a = \frac{A_s f_y}{0.85 f'_c\, b} = \frac{1.72 \text{ in.}^2 \times 60 \text{ ksi}}{0.85 \times 4 \text{ ksi} \times 12 \text{ in.}} = 2.53 \text{ in.}$$

Then, using $\beta_1 = 0.85$, the depth to the neutral axis is

$$c = a/\beta_1 = 2.53 \text{ in.}/0.85 = 2.98 \text{ in.}$$

This is less than $3/8\, d = 3/8 \times 21.5$ in. $= 8.06$ in. Thus, this is a tension-controlled section, and $\phi = 0.9$.

Finally, we should use Eq. (5-15), including the use of the strength-reduction factor, ϕ, to check the strength of the final section design.

$$\phi M_n = \phi A_s f_y \left(d - \frac{a}{2} \right) = 0.9 \times 1.72 \text{ in.}^2 \times 60 \text{ ksi}(21.5 \text{ in.} - 1.27 \text{ in.})$$

$$= 1880 \text{ k-in.} = 157 \text{ k-ft} > M_u = 145 \text{ k-ft}$$

The strength is adequate without being too excessive. So, **use three No. 6 bars and two No. 4 bars**, placed as shown in Fig. 5-26c. ■

Design of Beams When Section Dimensions Are Not Known

The second type of section design problem involves finding b, d, and A_s. Three decisions not encountered in Example 5-4 must be made here, that is, a preliminary estimate of the self weight of the beam, selection of a target steel percentage, and final selection of the section dimensions b and h (and d).

Although no dependable rule exists for guessing the weight of beams prior to selection of the dimensions, the weight of a rectangular beam will be roughly 10 to 15 percent of the unfactored loads it must carry. Alternatively, one can estimate h as being between

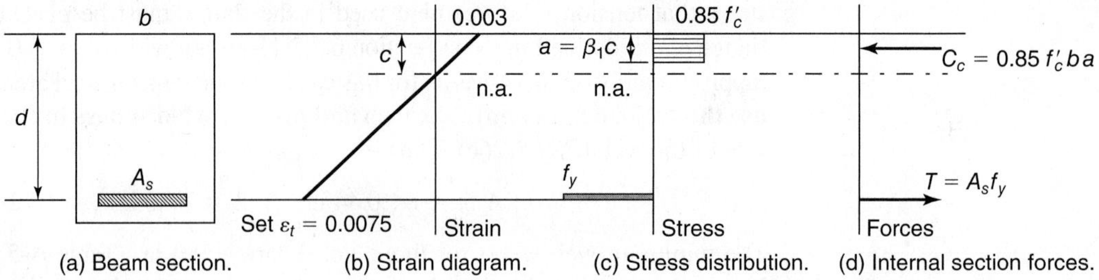

Fig. 5-27 Assumptions for design of singly reinforced beam section.

1/18 and 1/12 of the span, as discussed previously. Past practice at this stage was to estimate b as approximately $0.5h$. However, to save form-work costs, it is becoming more common to select the beam width equal to the column width if that dimension is known at this stage of the design. Even if the column width has not yet been determined, it is probably better to assume a wider beam width—say b at approximately $0.8\,h$—when estimating the weight of the beam. The dead load estimated at this stage will be corrected when the section dimensions are finally chosen, if necessary.

The next step in the process is to select a reasonable starting value for the reinforcement ratio, $\rho = A_s/bd$. Since 2002, for sections subjected to only bending or bending plus axial load, the ACI Code has used a direct relationship between the strength-reduction factor, ϕ, and the strain at the extreme layer of tension reinforcement, ε_t. To be consistent with the ACI Code, the author will use a procedure to select an initial value for ρ that will result in a tension-controlled section (i.e., a section with adequate ductility to justify the use of $\phi = 0.9$).

Assume the singly reinforced beam section shown in Fig. 5-27a is subjected to positive bending. At this stage, the section dimensions and area of tension reinforcement are not known. To start the design process, we will select a target strain diagram, as shown in Fig. 5-27b. Because there is a single layer of tension steel, the strain at the centroid of the tension reinforcement, ε_s, is equal to ε_t. To justify the use of $\phi = 0.9$, ε_t must equal or exceed 0.005 for the final section design. To get a final design that is similar to past practice, as will be demonstrated next, the author recommends setting $\varepsilon_t = 0.0075$ at this initial stage of section design.

The stress and force diagrams shown in Figs. 5-27c and 5-27d are similar to those discussed in Chapter 4. From the strain distribution in Fig. 5-27b, the following value is obtained for the distance to the neutral axis.

$$c = \left(\frac{0.003}{0.003 + 0.0075}\right)d = 0.286d$$

Using this value of c, the expression for the concrete compression force, C_c, is

$$C_c = 0.85 f'_c\, b\beta_1 c = 0.85(0.286) f'_c\, \beta_1 (bd)$$
$$= 0.24\, \beta_1 f'_c\, (bd)$$

Enforcing section equilibrium, $T = C_c$, we can solve for an initial value of the reinforcement ratio, ρ:

$$T = C_c$$
$$A_s f_y = 0.24 \beta_1 f'_c\, (bd)$$
$$\rho(\text{initial}) = \frac{A_s}{bd} = \frac{0.24\beta_1 f'_c}{f_y} \cong \frac{\beta_1 f'_c}{4 f_y} \tag{5-19}$$

Equation (5-19) gives an initial target reinforcement ratio that will be used for the design of singly reinforced rectangular sections.

As an extra discussion, the author would like to present a comparison between the initial ρ value given in Eq. (5-19) and prior procedures for making an initial selection of ρ, which were based on the balanced reinforcement ratio. An expression for the balanced reinforcement ratio was given in Eq. (4-24).

$$\rho_b = \frac{0.85\,\beta_1 f'_c}{f_y}\left(\frac{\varepsilon_{cu}}{\varepsilon_{cu} + \varepsilon_y}\right) \tag{4-24}$$

Recall that the maximum useable compression strain, ε_{cu}, is equal to 0.003. In the following, we will assume that the steel yield strain, ε_y, for Grade-60 steel can be taken as approximately 0.002. Thus, the strain ratio Eq. (4-24) can be taken approximately equal to 3/5. Prior design practice was to select an initial ρ value equal to 45 or 50 percent of the balanced reinforcement ratio given in Eq. (4-24). Using 50 percent of the balanced reinforcement ratio as a target value for ρ results in the following:

$$\rho(\text{target}) = 0.5 \times \rho_b = 0.5 \times \frac{0.85\,\beta_1 f'_c}{f_y} \times \frac{3}{5} = 0.255\frac{\beta_1 f'_c}{f_y} \tag{5-20}$$

The target ρ value from Eq. (5-20) is very similar to that given in Eq. (5-19). Thus, we would expect a section designed with an initial ρ value from Eq. (5-19) will be similar to those obtained from prior practice.

Having selected an initial ρ value, we must now develop a procedure that results in section dimensions and a reinforcement area that satisfy the basic strength requirement, $\phi M_n \geq M_u$. As part of this process, use the following definition for the reinforcement index, ω:

$$\omega = \rho\frac{f_y}{f'_c} \tag{5-21}$$

The nominal flexural strength of a singly reinforced rectangular section was given in Eq. (5-15) as

$$M_n = A_s f_y\left(d - \frac{a}{2}\right) \tag{5-15}$$

Also, the expression for the depth of the compression stress block, a, was given in Eq. (5-17), which can be modified to be

$$a = \frac{A_s f_y}{0.85\,f'_c\,b} \times \frac{d}{d} = \frac{A_s}{bd} \times \frac{f_y}{f'_c} \times \frac{d}{0.85} = \rho\frac{f_y}{f'_c} \times \frac{d}{0.85} = \frac{\omega d}{0.85}$$

Putting that expression for a into Eq. (5-15) and making some notation substitutions results in

$$M_n = A_s f_y\left(d - \frac{\omega d}{2 \times 0.85}\right) = A_s f_y\, d(1 - 0.59\omega) \times \frac{(bd)f'_c}{(bd)f'_c}$$

$$= \frac{A_s}{bd} \times \frac{f_y}{f'_c} \times f'_c(1 - 0.59\omega)(bd^2) = \omega f'_c(1 - 0.59\omega)(bd^2)$$

The symbol, R, commonly referred to as the *flexural-resistance factor*, is used to represent the first part of this expression, i.e.,

$$R = \omega f'_c(1 - 0.59\omega) \tag{5-22}$$

Because this factor can be used either as a convenient starting point for the flexural design of beam sections or to select required reinforcement for an existing beam section,

R-factor design aides for various concrete strengths are given in Table A-3 (U.S. Customary units) and Table A-3M (SI Metric units). The use of these tables will be demonstrated in Examples 5-5 and 5-5M.

After calculating the *R*-factor, the strength requirement in Eq. (5-12) becomes

$$\phi M_n \geq M_u$$
$$\phi R(bd^2) \geq M_u$$
$$(bd^2) \geq \frac{M_u}{\phi R} \tag{5-23a}$$

Equation (5-23a) represents an expression for obtaining a quasi-section modulus for the beam section. As was discussed earlier, the section width eventually may be set equal to the column width to save on formwork costs. So, if the desired beam width is known, Eq. (5-23a) can be used to solve directly for the required effective flexural depth, *d*. Also, if for architectural reasons the total beam depth, *h*, has been limited to a specific value and we assume $d \approx h - 2.5$ in., Eq. (5-23a) can be used to solve directly for the required beam width, *b*. If there are no restrictions on the beam dimensions, the beam width, *b*, can be set equal to some percentage, α, of the effective flexural depth, *d*. The author suggests that α can be set equal to a value between 0.5 and 0.8. With *b* now expressed as a ratio of *d*, Eq. (5-23a) can be solved for a value of *d*:

$$(\alpha d^3) \geq \frac{M_u}{\phi R}$$
$$d \geq \left(\frac{M_u}{\alpha \phi R}\right)^{1/3} \tag{5-24}$$

This value of *d* is normally rounded to a *half-inch* value, because the difference between *h* and *d* commonly is taken as 2.5 in. for one layer of steel and 3.5 in. for two layers of tension steel. Thus, rounding *d* to a half-inch value effectively rounds *h* to a whole-inch value. To avoid possible deflection calculations, the height of the beam *h*, should be taken as greater than or equal to the minimum values given for beams in ACI Code Table 9.3.1.1. Deflection calculations will be discussed in Chapter 9.

After a value for *d* has been selected, the required value of the section width *b* can be found using Eq. (5-23a) and then rounding up to a whole inch value. If the section dimensions were selected without significant changes from the calculated values, the required value for A_s can be estimated by multiplying the value of ρ selected in Eq. (5-19) times the product, $b \times d$ (using the calculated value for *b*, not the rounded value). If significant changes were made to the section dimensions, the value for the moment arm, *jd*, can be assumed (as was done previously) and Eq. (5-16) can be used to determine an initial value for A_s. After an initial value for A_s has been determined by either procedure, one iteration using Eqs. (5-17) and (5-16) can then be used to reach a final value for A_s that provides adequate strength for the selected section dimensions *b* and *d*. A demonstration of this design process will be given in the following example.

EXAMPLE 5-5 Design of a Beam Section for Which *b* and *d* Are Not Known

In this example, we will go through the steps for a complete beam section design for the continuous floor beam *A–B–C–D* in Fig. 5-4. When determining the section size, the design process should start at the location of the largest factored design moment. From the analyses in Examples 5-1 and 5-2, the largest design moment occurs at the face of column *B*. As was done in those examples, we will assume a slab thickness of 6 in., a superimposed dead load of 20 psf, and a live load of 60 psf.

We will make a new estimate for the weight of the beam web and then use the ACI moment coefficients to determine the factored design moment at the face of column *B*. We will assume a normal weight concrete with $f'_c = 4000$ psi and reinforcing steel with $f_y = 60$ ksi.

1. Estimate weight of the beam web. Assume the total depth of the beam will be between 1/18 and 1/12 of the span length from *A* to *B*, which is the longest span length for beam *A–B–C–D*. Thus,

$$h \approx \ell/12 \text{ to } \ell/18$$

$$h \approx \frac{28\text{ ft} \times 12\text{ in./ft}}{12} = 28\text{ in. to } \frac{28\text{ ft} \times 12\text{ in./ft}}{18} = 18.7\text{ in.}$$

Select $h = 24$ in., as was done in Example 5-1. We now could select the beam width, b, as some percentage of h, or we could set the beam width equal to the column width. In Example 5-1, the column dimensions were given as 18 in. × 18 in. Thus, set b equal to 18 in., which is 75 percent of h (a reasonable percentage for estimating the beam weight). For a slab thickness of 6 in., the weight of the beam web is

$$w(\text{beam web}) = \frac{(24\text{ in.} - 6\text{ in.}) \times 18\text{ in.}}{144\text{ in.}^2/\text{ft}^2} \times 150\text{ lb/ft}^3$$

$$w(\text{beam web}) = 338\text{ lb/ft} = 0.338\text{ k/ft}$$

2. Compute total factored load and factored design moment, M_u. Additional dead load includes the weight of the slab (75 psf) and the superimposed dead load (SDL) of 20 psf. For the 12-ft. tributary width, this results in

$$w(\text{slab and SDL}) = (75\text{ psf} + 20\text{ psf}) \times 12\text{ ft} = 1140\text{ lb/ft} = 1.14\text{ k/ft}$$

Combining this with the weight of the beam web, the total dead load is 1.48 k/ft. From Example 5-1, the reduced live load on this beam for calculation of the maximum negative moment at *B* is 0.486 k/ft. Using those values, the total factored load is

$$w_u = 1.2w_D + 1.6w_L, \text{ or } 1.4w_D$$

$$w_u = 1.2(1.48\text{ k/ft}) + 1.6(0.486\text{ k/ft}), \text{ or } 1.4(1.48\text{ k/ft})$$

$$w_u = 2.55\text{ k/ft, or } 2.07\text{ k/ft}$$

The larger value will be used with the appropriate ACI moment coefficient $(-1/10)$ and the average clear span length for spans *A–B* and *B–C* (24.5 ft from Example 5-1). Thus,

$$M_u = (-1/10) \times 2.55\text{ k/ft} \times (24.5\text{ ft})^2 = -153\text{ k-ft}$$

3. Selection of ρ value and corresponding *R-factor*. Equation (5-19) will be used to select an initial value for ρ. For $f'_c \leq 4000$ psi, $\beta_1 = 0.85$. Thus,

$$\rho \cong \frac{\beta_1 f'_c}{4f_y} = \frac{0.85 \times 4\text{ ksi}}{4 \times 60\text{ ksi}} = 0.0142$$

Rounding this off to 0.014 and using Eq. (5-21),

$$\omega = \rho\frac{f_y}{f'_c} = 0.014\frac{60\text{ ksi}}{4\text{ ksi}} = 0.210$$

Then, from Eq. (5-22),

$$R = \omega f'_c(1 - 0.59\omega) = 0.21 \times 4\text{ ksi}(1 - 0.59 \times 0.21) = 0.736\text{ ksi}$$

It is important for a designer to have some judgment about reasonable values for the R-factor. Many experienced designers will select the R-factor directly (between 0.70 and 1.0 ksi) without going through the intermediate step of selecting a reasonable starting value for ρ. As noted previously, R-factors for the flexural design of a singly reinforced beam section are given in Table A-3.

4. Selection of section dimensions, *b* and *h*. Putting the calculated value for R into Eq. (5-23a) results in

$$bd^2 \geq \frac{M_u}{\phi R} = \frac{153 \text{ k-ft} \times 12 \text{ in./ft}}{0.9 \times 0.736 \text{ ksi}} = 2770 \text{ in.}^3$$

We now need to select a value for the ratio between b and d, which we have defined as α. Using $\alpha = 0.7$, which is within the range suggested by the author, leads to

$$d \geq \left(\frac{M_u}{\alpha \phi R}\right)^{0.333} = \left(\frac{2770 \text{ in.}^3}{0.7}\right)^{0.333} = 15.8 \text{ in.}$$

Rounding this up to the next half-inch value results in $d = 16.5$ in. and $h \approx d + 2.5$ in. $= 19$ in. (Note: some designers prefer the use of an even number of inches for h, which could be taken as 20 in. in this case). Before proceeding, the author recommends that the selected value for total beam depth, h, should be checked against the value given in ACI Code Table 9.3.1.1 for minimum thicknesses to avoid deflection calculations in most cases. From that table, the minimum thickness (h) for an exterior span of a continuous floor beam is $\ell/18.5$, or

$$h(\text{min}) = \frac{\ell}{18.5} = \frac{28 \text{ ft} \times 12 \text{ in./ft}}{18.5} = 18.2 \text{ in.}$$

Because the moment of inertia for a beam section is approximately proportional to h^3, it would be a good idea to select a beam depth that is at least a full inch greater than this minimum value. Thus, **select *h* = 20 in.**, and then $d \approx h - 2.5$ in. $= 17.5$ in. This value for d then is used with the previously calculated value for bd^2 to determine b as

$$b \geq \frac{2770 \text{ in.}^3}{(17.5 \text{ in.})^2} = 9.1 \text{ in.}$$

Clearly, this value is much smaller than the value of 18 in. that was assumed for estimating the beam self-weight. If we did not want to match the column dimension or if this was a floor beam that did not frame into a column (i.e., a beam supported by girders), then we would select b equal to either 10 in. or a little larger value depending on the beam width required at midspan to place the positive bending reinforcement in a single layer.

Because we have assumed that we want to have the width of this floor beam equal to the column width that it connects to, **select *b* = 18 in.** Recall from the discussion of flexural behavior in Chapter 4 that a wider beam will not be much stronger than a narrow beam, but it will be more ductile. Thus, this should be a very ductile section that is easily within the tension-controlled region of behavior.

5. Determination of A_s and selection of reinforcing bars. With the selected dimensions of b and h, we can recalculate the beam weight and the resulting factored design moment. Because the dimensions are smaller and the factored moment will be lower, we could safely ignore this step. However, for completeness,

$$w(\text{beam web}) = \frac{(20\text{ in.} - 6\text{ in.}) \times 18\text{ in.}}{144\text{ in.}^2/\text{ft}^2} \times 150\text{ lb/ft}^3 = 263\text{ lb/ft} = 0.263\text{ k/ft}$$

With this value, the revised values for w_D, w_u, and M_u are

$$w_D = 1.40\text{ k/ft}$$
$$w_u = 2.46\text{ k/ft}$$
$$M_u = -148\text{ k-ft}$$

Because the selected beam width is significantly larger than the calculated value (9.1 in.), use Eq. (5-16) to estimate the required area of tension reinforcement. Assming a moment arm, *jd*, approximately equal to 0.9 *d* (narrow compression zone), Eq. (5-16) gives

$$A_s \geq \frac{M_u}{\phi f_y\left(d - \frac{a}{2}\right)} \cong \frac{M_u}{\phi f_y(jd)} = \frac{148\text{ k-ft} \times 12\text{ in./ft}}{0.9 \times 60\text{ ksi} \times 0.9 \times 17.5\text{ in.}} = 2.09\text{ in.}^2$$

Then, proceeding through one iteration using Eq. (5-17) to get a value for the depth of the compression stress block, we have

$$a = \frac{A_s f_y}{0.85 f'_c\, b} = \frac{2.09\text{ in.}^2 \times 60\text{ ksi}}{0.85 \times 4\text{ ksi} \times 18\text{ in.}} = 2.05\text{ in.}$$

Then using Eq. (5-16) to get a revised steel area, we have

$$A_s \geq \frac{M_u}{\phi f_y\left(d - \frac{a}{2}\right)} = \frac{148\text{ k-ft} \times 12\text{ in./ft}}{0.9 \times 60\text{ ksi}(17.5\text{ in.} - 1.02\text{ in.})} = 2.00\text{ in.}^2$$

As was done in Example 5-4 to be in compliance with ACI Code Section 24.3.4, **select three No. 7 bars for over the web of the section and select two No. 4 bars to be placed in the flanges**, one on each side of the web, as shown in the final section design given in Fig. 5-28. The resulting steel area is

$$A_s = 3 \times 0.60\text{ in.}^2 + 2 \times 0.20\text{ in.}^2 = 2.20\text{ in.}^2$$

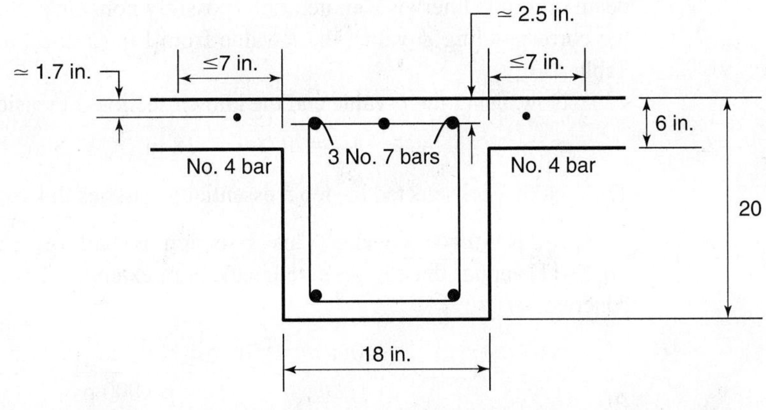

Fig. 5-28
Final beam section design at face of column—Example 5-5.

Assuming the beam will have normal cover (1.5 in.) and that a No. 3 or No. 4 bar will be used as shear reinforcement, the center-to-center spacing between the No. 7 bars in the top tension layer is:

$$s \cong \frac{18 \text{ in.} - 2(2.5 \text{ in.})}{2 \text{ spaces}} = 6.5 \text{ in.} < 10 \text{ in.}$$

This value satisfies the spacing limits in ACI Code Section 24.3.2.1, which were defined and discussed in Example 5-4, and easily represents a clear spacing between the bars that exceeds the minimum required values from ACI Code Section 25.2.1.

6. **Determination of required A_s using Table A-3.** After the section dimensions have been selected, Table A-3 can be used to directly solve for the required area of tension reinforcement in a singly reinforced beam section without going through the iteration process in step 5. Also, when using Table A-3, we can determine if the reinforcement ratio exceeds the minimum value required by the ACI Code and if this is a tension-controlled section. This normally would allow us to eliminate the checks completed in the first part of step 7.

If b and d are known, Eq. (5-23a) can be solved for the required R-value as

$$R \geq \frac{M_u}{\phi b d^2} \tag{5-23b}$$

Using $b = 18$ in. and $d = 17.5$ in. from step 4, the required R-value is

$$R \geq \frac{148 \text{ k-ft} \times 12 \text{ in./ft}}{0.9 \times 18 \text{ in.} \times (17.5 \text{ in.})^2} = 0.358 \text{ ksi} = 358 \text{ psi}$$

Using Table A-3 for 4000 psi concrete, we can read that the required ρ-value is 0.007. Note, instead of interpolating in Table A-3, a designer normally will select the smallest ρ-value that corresponds to an R-value *greater than or equal to* the required R-value. The smallest R-values given at the top of the columns in Table A-3 correspond to the minimum-reinforcement ratio required by the ACI Code, so that Code requirement is satisfied by using this table. Also, if you are reading values in the table that are not printed with boldface type, then your beam section will be a *tension-controlled* section, and ϕ will be equal to 0.9. The boldfaced numbers represent sections in the *transition zone* where the ϕ-value will be between 0.65 and 0.9. As noted previously, the author recommends that beam sections be designed as tension-controlled sections. So, if your required R-value is located in the boldface part of the table, the author recommends that you increase the size of your beam section. Otherwise, an iteration (possibly nonconverging) will be required to find ε_t, the corresponding ϕ-value, the R-value from Eq. (5-23b), and the required ρ-value from Table A-3.

Now, using the ρ-value and the known section dimensions, the required steel area is

$$A_s \geq \rho b d = 0.007 \times 18 \text{ in.} \times 17.5 \text{ in.} = 2.21 \text{ in.}^2$$

The area of steel selected in step 5 essentially satisfies this requirement.

7. **Required checks.** This T-section is part of a continuous floor system, so Eq. (5-11) applies directly. As noted in the prior example, 200 psi exceeds $3\sqrt{f'_c}$ for 4000 psi concrete, so

$$A_{s,\min} = \frac{200 b_w d}{f_y} = \frac{200 \text{ psi} \times 18 \text{ in.} \times 17.5 \text{ in.}}{60{,}000 \text{ psi}} = 1.05 \text{ in.}^2$$

The selected A_s exceeds this value, so it is o.k.

To confirm that the tension steel is yielding and that this is a tension-controlled section, we can start by using Eq. (5-17) to calculate the stress block depth, a, for the selected steel area as

$$a = \frac{A_s f_y}{0.85 f'_c b} = \frac{2.20 \text{ in.}^2 \times 60 \text{ ksi}}{0.85 \times 4 \text{ ksi} \times 18 \text{ in.}} = 2.16 \text{ in.}$$

Using $\beta_1 = 0.85$ for 4000 psi concrete,

$$c = \frac{a}{\beta_1} = \frac{2.16 \text{ in.}}{0.85} = 2.54 \text{ in.}$$

Then, using Eq. (4-18) to calculate the steel strain (ε_s equal to ε_t for single steel layer) for the assumed linear strain distribution, we have

$$\varepsilon_s(= \varepsilon_t) = \frac{d - c}{c} \times \varepsilon_{cu} = \frac{17.5 \text{ in.} - 2.54 \text{ in.}}{2.54 \text{ in.}} \times 0.003 = 0.0177$$

This value is both greater than the yield strain for Grade-60 steel ($\varepsilon_y = 0.00207$) and the strain limit for tension-controlled sections (0.005), so the assumptions that the steel is yielding and that $\phi = 0.9$ are valid.

Finally, using Eq. (5-15) to calculate the nominal flexural strength and including the strength reduction factor ϕ, we have

$$\phi M_n = \phi A_s f_y \left(d - \frac{a}{2} \right) = 0.9 \times 2.20 \text{ in.}^2 \times 60 \text{ ksi}(17.5 \text{ in.} - 1.08 \text{ in.})$$
$$= 1950 \text{ k-in.} = 163 \text{ k-ft} > M_u = 146 \text{ k-ft}$$

The strength is adequate without being too excessive. So, **select three No. 7 and two No. 4 bars**, placed as shown in Fig. 5-28. ■

EXAMPLE 5-5M Design of Beam Section When *b* and *d* Are Not Known—SI Units

Assume that beam dimensions were estimated ($b = 300$ mm and $h = 650$ mm) and an analysis of a continuous floor beam, similar to that completed for Example 5-5, has been completed. Also assume that the factored design moment, M_u, was found to be a negative 220 kN-m. Design the final section dimensions b, h, and d and find the required area of tension reinforcement, A_s, assuming $f'_c = 25$ MPa ($\beta_1 = 0.85$) and $f_y = 420$ MPa.

1. Compute *b*, *d*, and *h*. For the given material strengths, use Eqs. (5-19), (5-21), and (5-22) to calculate ρ, ω, and R.

$$\rho(\text{initial}) = \frac{\beta_1 f'_c}{4 f_y} = 0.0126$$
$$\omega = \rho f_y / f'_c = 0.212$$
$$R = \omega f'_c (1 - 0.59\omega) = 4.63 \text{ MPa}$$

Because this R-factor is for tension-controlled beam section, use $\phi = 0.9$ in Eq. (5-23a) to calculate a value for (bd^2):

$$bd^2 \geq \frac{M_u}{\phi R} = \frac{220 \text{ kN-m}}{0.9 \times 4.63 \text{ N/mm}^2} = \frac{220 \times 10^6 \text{ N-mm}}{4.17 \text{ N/mm}^2} = 52.8 \times 10^6 \text{ mm}^3$$

Select $b \approx 0.6\,d$, and then solve for d:

$$d \cong \left(\frac{52.8 \times 10^6\ \text{mm}^3}{0.6}\right)^{0.333} = 444\ \text{mm}$$

With this, $h \approx d + 65$ mm $= 509$ mm. We will **set h = 500 mm**, and then **$d \approx$ 435 mm**. Using that value, we can go back to the required value of (bd^2) to solve for b:

$$b \cong \frac{52.8 \times 10^6\ \text{mm}^3}{(435\ \text{mm})^2} = 279\ \text{mm}$$

Round this value and **set b = 300 mm**. This section is a little smaller than the assumed section size for calculating dead load, so there is no need to recalculate M_u.

2. Determination of required A_s and selection of reinforcing bars. We are designing a T-section in negative bending (narrow compression zone), so select a moment arm, jd, approximately equal to 0.9 d. Use Eq. (5-16) to estimate the required value of A_s.

$$A_s \geq \frac{M_u}{\phi f_y(d - a/2)} \cong \frac{220\ \text{kN-m}}{\phi f_y\,(jd)} = \frac{220 \times 10^6\ \text{N-mm}}{0.9 \times 420\ \text{N/mm}^2 \times 0.9 \times 435\ \text{mm}} = 1490\ \text{mm}^2$$

Use one iteration to refine this value by first inserting it in Eq. (5-17) to find the depth of the equivalent stress block, so that

$$a = \frac{A_s f_y}{0.85\,f'_c\,b} = \frac{1490\ \text{mm}^2 \times 420\ \text{MPa}}{0.85 \times 25\ \text{MPa} \times 300\ \text{mm}} = 98.2\ \text{mm}$$

Then, use that value of a in Eq. (5-16) to get an improved value for A_s:

$$A_s \geq \frac{M_u}{\phi f_y(d - a/2)} = \frac{220 \times 10^6\ \text{N-mm}}{0.9 \times 420\ \text{N/mm}^2\,(435\ \text{mm} - 49.1\ \text{mm})} = 1510\ \text{mm}^2$$

As before, we need to be aware that ACI Code Section 24.3.4 requires that some of the tension-zone reinforcement for a flanged section be distributed into the flange. For this section, **select three No. 22 bars for over the web and place two No. 16 bars in the flanges**, as shown in Fig. 5-29. These bars give a total tension steel area, $A_s = 3 \times 387 + 2 \times 199 = 1560\ \text{mm}^2$.

3. Use of Table A-3M to select required A_s. As was done in Example 5-5, we can use Table A-3M to find the required area of tension reinforcement after the section dimensions have been selected. Also, when using Table A-3M, we will know that the selected reinforcement ratio exceeds the minimum ratio required by the ACI Code, and

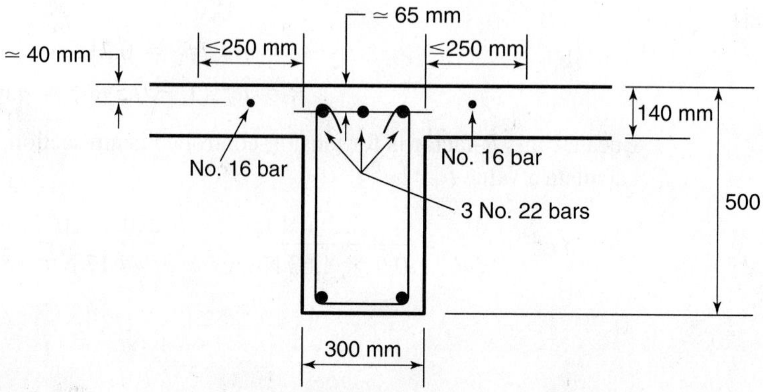

Fig. 5-29
Final beam section design at face of column—Example 5-5M.

that this is a tension-controlled section. This would eliminate the need for the checks at the beginning of step 4, which are given for completeness in this example.

Assuming $\phi = 0.9$ (tension-controlled section) and using $b = 300$ mm and $d = 435$ mm, we can use Eq. (5-23b) to determine the required R-value.

$$R \geq \frac{M_u}{\phi\, bd^2} \tag{5-23b}$$

$$\geq \frac{220 \times 10^6 \text{ N-mm}}{0.9 \times 300 \text{ mm} \times (435 \text{ mm})^2} = 4.31 \text{ N/mm}^2 = 4.31 \text{ MPa}$$

Using Table A-3M for $f'_c = 25$ MPa, we can read that the required ρ-value is 0.012. This clearly is greater than the minimum required reinforcement ratio, and it does correspond to a tension-controlled section (non-boldfaced number), as assumed. Thus, using this ρ-value and the known section dimensions, the required tension steel area is

$$A_s \geq \rho bd = 0.012 \times 300 \text{ mm} \times 435 \text{ mm} = 1570 \text{ mm}^2$$

The selected reinforcement in step 2 essentially satisfies this requirement, and we could proceed directly to the strength check at the end of step 4.

4. **Required checks.** This T-section is part of a continuous floor system, so Eq. (5-11M) applies directly. For the given concrete strength, $0.25\sqrt{f'_c}$ is less than 1.4 MPa. So, use 1.4 MPa in Eq. (5-11M) to give

$$A_{s,\min} = \frac{1.4 b_w d}{f_y} = \frac{1.4 \text{ MPa} \times 300 \text{ mm} \times 435 \text{ mm}}{420 \text{ MPa}} = 435 \text{ mm}^2$$

The selected A_s exceeds this value, so it is o.k.

To confirm that the tension steel is yielding and that this is a tension-controlled section, we can start by using Eq. (5-17) to calculate the stress block depth, a, for the selected steel area as

$$a = \frac{A_s f_y}{0.85 f'_c\, b} = \frac{1560 \text{ mm}^2 \times 420 \text{ MPa}}{0.85 \times 25 \text{ MPa} \times 300 \text{ mm}} = 103 \text{ mm}$$

Using $\beta_1 = 0.85$ for 25 MPa concrete,

$$c = \frac{a}{\beta_1} = \frac{103 \text{ mm}}{0.85} = 121 \text{ mm}$$

Then, using Eq. (4-18) to calculate the steel strain at the one level of tension steel,

$$\varepsilon_s (= \varepsilon_t) = \frac{d - c}{c} \times \varepsilon_{cu} = \frac{435 \text{ mm} - 121 \text{ mm}}{121 \text{ mm}} \times 0.003 = 0.0078$$

This value is both greater that the yield strain for Grade-420 steel ($\varepsilon_y = 0.0021$) and the strain limit for tension-controlled sections (0.005), so the assumptions that the steel is yielding and that $\phi = 0.9$ are valid.

Finally, using Eq. (5-15) to calculate the nominal moment strength and including the strength reduction factor ϕ,

$$\phi M_n = \phi A_s f_y \left(d - \frac{a}{2} \right) = 0.9 \times 1560 \text{ mm}^2 \times 420 \text{ N/mm}^2 (435 \text{ mm} - 52 \text{ mm})$$

$$\phi M_n = 226 \times 10^6 \text{ N-mm} = 226 \text{ kN-m} > M_u = 220 \text{ kN-m}$$

The strength of the section is adequate without being too excessive. So, **select three No. 22 bars and two No. 16 bars**, placed as shown in Fig. 5-29. ■

5-4 DESIGN OF DOUBLY REINFORCED BEAM SECTIONS

As discussed in Chapter 4, the addition of compression reinforcement to an existing beam section (Fig. 4-12) does not significantly increase the nominal moment strength of the section. However, the addition of compression reinforcement does increase the ductility of a beam section (Fig. 4-31). Thus, the use of compression reinforcement will permit the use of more tension reinforcement to increase the strength of a given beam section while keeping the section in the tension-controlled region of behavior. That is, when evaluating the nominal moment strength, M_n, for the section, it can be shown that the strain in the extreme layer of tension reinforcement, ε_t, will exceed the tension-control limit of 0.005, and thus, the strength-reduction factor, ϕ, can be set equal to 0.9.

Two common cases may result in the need to use compression reinforcement to achieve the required nominal moment strength while keeping the section in the tension-controlled region of behavior. If a designer wants to reduce the size, and thus the weight of a beam, he/she could design for a larger percentage of tension reinforcement and use compression reinforcement to keep the section in the tension-controlled region of behavior. Also, when faced with severe architectural restrictions on the dimensions of a beam, a designer may be forced to use a doubly reinforced section. For both of these cases, the final beam section design likely would be classified as being in the transition region $(0.005 > \varepsilon_t > 0.002)$ or the compression-controlled region $(\varepsilon_t \leq 0.002)$ of behavior without the addition of compression reinforcement. This fact leads to the following design procedure for doubly reinforced beam sections, which is a modification of the procedure for singly reinforced sections.

The important first step in the design procedure for doubly reinforced beam sections is the selection of a target value for the tension reinforcement ratio, ρ. A procedure similar to that shown in Fig. 5-27 will be followed, but to obtain a larger initial ρ value (and thus a smaller beam section), the strain in the tension steel layer, ε_t, shown in Fig. 5-27b, will be set to a lower value. The author believes that a reasonable doubly reinforced beam section with respect to ductility and deflection control will result if ε_t is initially set equal to 0.004 (Note: at the end of this design procedure compression steel will be added to make this into a tension-controlled section). Different initial values could be selected to result in somewhat larger (set ε_t to a larger initial value) or somewhat smaller (set ε_t to a smaller initial value) beam sections. However, large variations from the target ε_t value suggested here can result in a beam section that is either too large (i.e., could be a singly reinforced section) or too small (cannot fit all of the tension and compression reinforcements into the section practically). Whichever initial value is selected, the final value of ε_t will need to be checked after the design of the doubly reinforced beam section is completed.

Changing the value of ε_t to 0.004 in Fig. 5-27b results in the following value for the distance to the neutral axis:

$$c = \left(\frac{0.003}{0.003 + 0.004}\right)d = \frac{3}{7}d$$

For this value of c, the expression for the concrete compression force in Fig. 5-27d becomes

$$C_c = 0.85 f'_c\, b\beta_1 c = 0.85(3/7)\beta_1 f'_c(bd) \cong 0.36\beta_1 f'_c(bd)$$

As before, enforcing section equilibrium ($T = C_c$), a solution for a target tension reinforcement ratio can be obtained as

$$T = C_c$$

$$A_s f_y = 0.36\,\beta_1 f'_c(bd)$$

$$\rho = \frac{A_s}{bd} = \frac{0.36\,\beta_1 f'_c}{f_y} \cong \frac{\beta_1 f'_c}{3f_y} \tag{5-25}$$

After this target tension-reinforcement ratio has been established, the procedure used for the design of singly reinforced beam sections can be followed. The reinforcement index, ω, is defined in Eq. (5-21), and the resulting R value is given by Eq. (5-22). The R value then can be used in Eq. (5-23a) to establish the required value of the quantity, bd^2. Then either by selecting a specific value for the section width, b (e.g., equal to the column width), or setting b equal to some percentage, α, times the effective depth, d, values can be determined for b, d, and the total section depth, h. After these section dimensions have been established, the procedure used earlier in this chapter to determine the required area of tension reinforcement, A_s, when section dimension are known will be followed.

Using this described procedure will result in a beam section with a relatively large amount of tension reinforcement. *Compression reinforcement now must be added* to give the section more ductility and thus have it classified as tension-controlled ($\varepsilon_t \geq 0.005$). The only guidance on how much compression reinforcement should be added to obtain reasonable section ductility can be found in the earthquake-resistant design requirements for intermediate and special moment frames in ACI Code Chapter 18. ACI Code Section 18.6.3.2 requires that the area of compression reinforcement at a column face should be greater than or equal to one-half of the area of tension reinforcement. The author recommends using this requirement ($A'_s \geq 0.5\,A_s$) to select the area of compression reinforcement in any doubly reinforced beam section. After this step has been completed and all of the bars for the tension and compression reinforcement have been selected and placed in the beam section, that section must be checked to show that $\phi M_n \geq M_u$ and $\varepsilon_t \geq 0.005$, so ϕ can be set equal to 0.9. This process for the design of doubly reinforced beam sections, including the required checks, will be demonstrated in the following example.

EXAMPLE 5-6 Design of a Doubly Reinforced Beam Section

For this example, we will design a doubly reinforced section for the maximum negative moment in the continuous girder $C1$–$C2$–$C3$–$C4$ in Fig. 5-30a, which is extracted from the floor system in Fig. 5-4. For this girder, the maximum negative design moment will occur at the exterior face of the first interior column ($C2$ or $C3$). The slab thickness and dimensions of the floor beams are given. Use $f'_c = 4000$ psi and $f_y = 60$ ksi. As in prior examples dealing with this floor system, assume a superimposed dead load of 20 psf and a design live load of 60 psf. If the reader prefers to **skip the initial steps** that demonstrate the analysis of the maximum design moment, he/she can jump to step 7 to proceed with the design of a doubly reinforced beam section in negative bending.

1. **Analysis model.** As discussed earlier in this chapter, loads on a typical floor system are assumed to flow from the slab to the floor beams and then to the girders and columns. Thus, the girders will be carrying concentrated loads from the floor beams and cannot be analyzed using the ACI moment coefficients. ACI Code Section 6.3.1.2 permits the analysis of an isolated floor plus the columns above and below the floor in question. The columns are assumed to be fixed at their far ends, so the structural analysis model to be used in this example is shown in Fig. 5-31. It is assumed that the story heights above and below

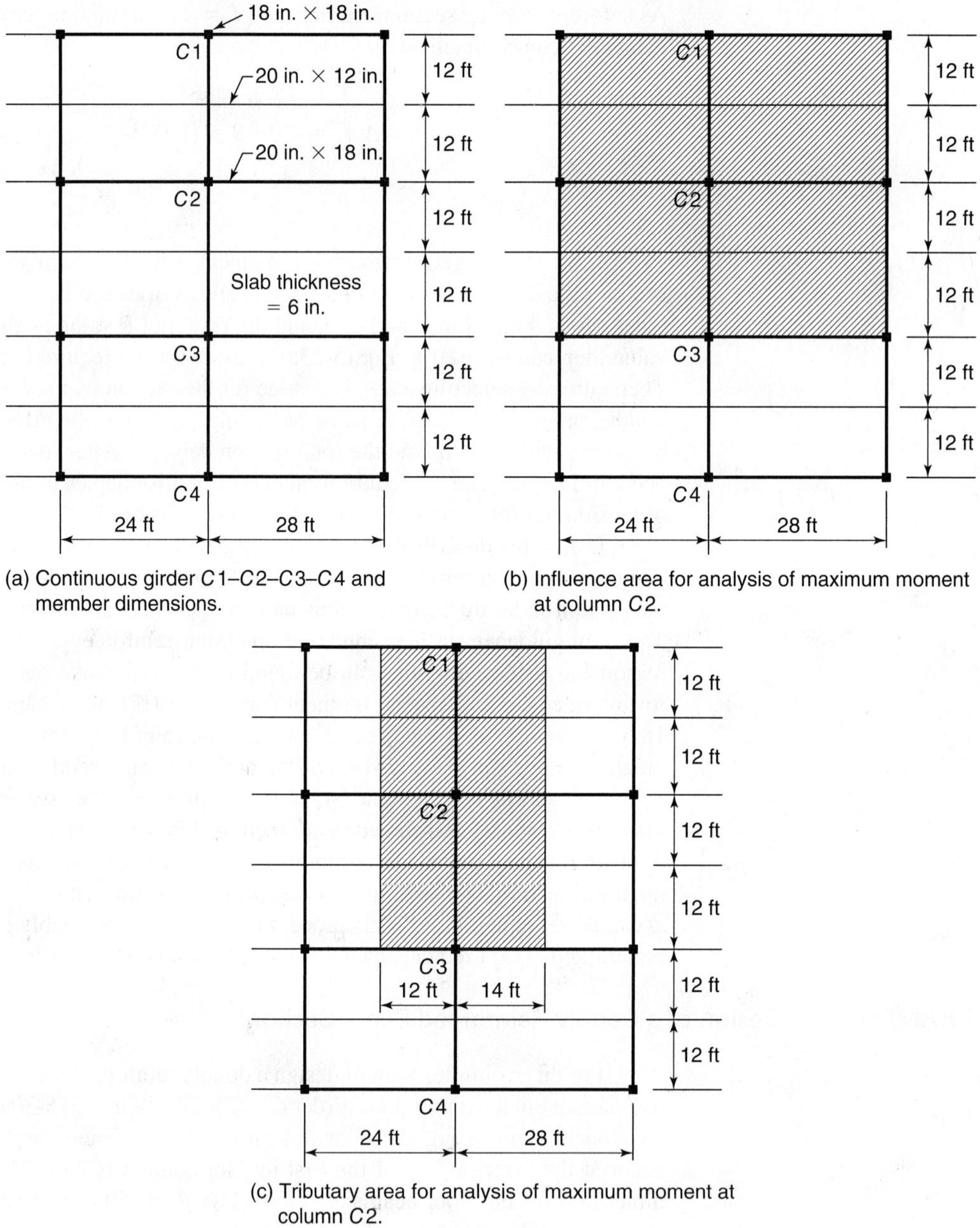

Fig. 5-30
Part of floor system used for design for Example 5-6.

the floor level to be analyzed are 11 ft. The continuous girder *C1–C2–C3–C4* is loaded at midspan by concentrated loads from the floor beams and by a distributed load due to its own weight. The values for those loads are given in the following sections.

2. Reduced live load. The appropriate *influence area*, A_I, for maximizing the negative moment in the girder at the face of column C2 is shown in Fig. 5-30b. Recall that the influence area is a multiple $(K_{LL} = 2)$ of the *tributary area*, A_T, which is shown in Fig. 5-30c. Thus, from Eq. (5-2),

$$A_I = K_{LL}A_T = 2\left[(24\text{ ft} + 24\text{ ft}) \times \left(\frac{24\text{ ft}}{2} + \frac{28\text{ ft}}{2}\right)\right] = 2500\text{ ft}^2$$

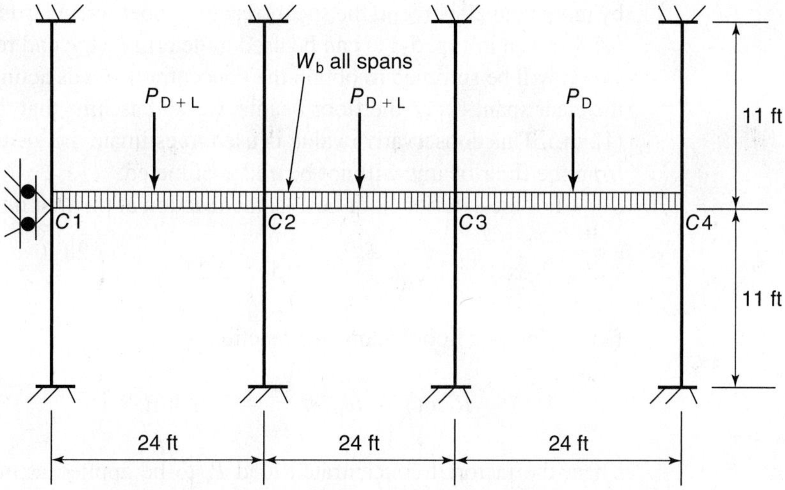

Fig. 5-31
Analysis model used for Example 5-6.

With this area, the reduced live load is calculated using Eq. (5-3):

$$L_r = L\left[0.25 + \frac{15}{\sqrt{A_I}}\right] = 60\text{ psf}\left[0.25 + \frac{15}{\sqrt{2500}}\right] \tag{5-3}$$

$$= 60\text{ psf }[0.25 + 0.30] = 33\text{ psf }(>0.5 \times 60\text{ psf, o.k.})$$

3. Concentrated loads from floor beams. Distributed loads acting on the floor beams will be transferred as concentrated load to the girders. The distributed dead load from the 12-ft wide tributary width for the floor beams consists of superimposed dead load (SDL), the weight of the floor slab, and the weight of the web of the floor beam.

$$w(\text{SDL}) = 20\text{ psf} \times 12\text{ ft} = 240\text{ lb/ft} = 0.24\text{ k/ft}$$

$$w(\text{slab}) = \frac{6\text{ in.}}{12\text{ in./ft}} \times \frac{0.15\text{ k}}{\text{ft}^3} \times 12\text{ ft} = 0.90\text{ k/ft}$$

$$w(\text{web}) = \frac{(20\text{ in.} - 6\text{ in.}) \times 12\text{ in.}}{144\text{ in.}^2/\text{ft}^2} \times \frac{0.15\text{ k}}{\text{ft}^3} = 0.175\text{ k/ft}$$

Thus, the total distributed dead load is

$$w_D = w(\text{SDL}) + w(\text{slab}) + w(\text{web}) = 1.32\text{ k/ft}$$

Similarly, the distributed live load is

$$w_L = L_r \times 12\text{ ft} = 33\text{ psf} \times 12\text{ ft} = 396\text{ lb/ft} \cong 0.40\text{ k/ft}$$

To maximize the negative moment at the face of column $C2$, we should assume that the live load is acting only in the shaded portion of the floor system shown in Fig. 5-30b, and only dead load is assumed to act in the unshaded portion of the floor system. In the portions of the floor where the live load is acting, the total distributed load in the floor beams is

$$w_u = 1.2w_D + 1.6w_L;\text{ or }1.4w_D$$
$$= 1.2(1.32\text{ k/ft}) + 1.6(0.40\text{ k/ft});\text{ or }1.4(1.32\text{ k/ft})$$
$$= 2.22\text{ k/ft; or }1.85\text{ k/ft}$$

The larger value governs and will be used to determine the concentrated loads acting at midspan for spans $C1$–$C2$ and $C2$–$C3$. Because the lengths for the floor beams do not vary by more than 20 percent, the shear (reaction) coefficients from ACI Code Sections 6.5.2 and 6.5.4 (given in Fig. 5-11) can be used to determine the end reactions from the floor beams, which will be summed to obtain the concentrated loads acting on the girders. To determine the clear span, ℓ_n, of the floor beams, we will assume that the girders have a width of 1 ft (12 in.). This conservative value is used to estimate the clear span, so the calculated loads from the floor beams will not be underestimated.

For the exterior floor beams, the reaction at the first interior support is

$$R(\text{ext}) = w_u \times \frac{\ell_n}{2} \times 1.15 = 2.22 \text{ k/ft} \times \left(\frac{28 \text{ ft} - 1 \text{ ft}}{2}\right) \times 1.15 = 34.5 \text{ k}$$

For the interior floor beams, the reaction is

$$R(\text{int}) = w_u \times \frac{\ell_n}{2} = 2.22 \text{ k/ft} \times \left(\frac{24 \text{ ft} - 1 \text{ ft}}{2}\right) = 25.5 \text{ k}$$

Then, the factored concentrated load P_u to be applied at midspan of girders $C1$–$C2$ and $C2$–$C3$ is

$$P_{D+L} = R(\text{ext}) + R(\text{int}) = 60.0 \text{ k}$$

The subscript $D + L$ was used here and in Fig. 5-31 because both the factored dead and live load are acting on these two spans. In the last girder span $C3$–$C4$, it is assumed that only the factored dead load is acting. To be consistent over the total length of this continuous girder, the same load factors that governed in spans $C1$–$C2$ and $C2$–$C3$ should be used in span $C3$–$C4$. Thus, the factored distributed dead load from the floor beam is

$$w_u = 1.2 w_D = 1.2 \times 1.32 \text{ k/ft} = 1.58 \text{ k/ft}$$

Using this distributed dead load, the factored concentrated load P_u to be applied at the midspan of girder $C3$–$C4$ is

$$P_D = w_u\left[\frac{28 \text{ ft} - 1 \text{ ft}}{2} \times 1.15 + \frac{24 \text{ ft} - 1 \text{ ft}}{2}\right] = 1.58 \text{ k/ft} \times 27.0 \text{ ft} = 42.7 \text{ k}$$

4. Distributed load on the girder. Assuming that all of the other loads have been accounted for, only the weight of the web of the girder must be included as a distributed load acting on the girder. The girder should have a total depth greater than or equal to the depth of the floor beams, so assume $h = 22$ in. For estimating the weight of the web, we will assume that the width of the girder is equal to the width of the column, so $b_w = 18$ in. Thus, the estimated weight of the girder web is

$$w(\text{web}) = \frac{(22 \text{ in.} - 6 \text{ in.}) \times 18 \text{ in.}}{144 \text{ in.}^2/\text{ft}^2} \times \frac{0.15 \text{ k}}{\text{ft}^3} = 0.30 \text{ k/ft}$$

Using a dead load factor of 1.2 to be consistent with load factors used for other loads on the floor system;

$$w_u = 1.2 \times w(\text{web}) = 1.2 \times 0.30 \text{ k/ft} = 0.36 \text{ k/ft}$$

5. Member stiffness for analysis model. As discussed in Section 5-3, in a typical structural analysis of a reinforced concrete frame, we will assume the beams are cracked in flexure and the columns are not cracked. Thus, for the 18 in. by 18 in. column we will use

$$I(\text{col}) = I_g = \frac{1}{12}(18 \text{ in.})^4 = 8750 \text{ in.}^4$$

For reinforced concrete beam sections, it is common to assume that the cracked moment of inertia is approximately one-half of the gross moment of inertia. However, to determine the gross moment of inertia for a typical beam (or girder) section that includes a portion of the floor slab, we need to assume how much of the slab is effective in contributing to the flexural stiffness of the slab-beam section. To avoid this issue, the author previously recommended that the cracked stiffness of the slab–beam section can be approximated as being equal to the gross moment of inertial for the full-depth web of the beam, that is,

$$I_{cr}(\text{slab–beam}) \cong I_g(\text{web}) = \frac{1}{12} b_w h^3 \tag{5-26}$$

For the girder $C1$–$C2$–$C3$–$C4$ in this example,

$$I_{cr}(\text{slab–beam}) = \frac{1}{12} \times 18 \text{ in.} \times (22 \text{ in.})^3 = 16{,}000 \text{ in.}^4$$

6. Results of structural analysis. Using the member stiffnesses and applying the loads discussed here to the analysis model in Fig. 5-31 resulted in a maximum negative moment of −229 k-ft at the exterior face of column C2. This moment will be used to design the girder section at column C2.

7. Section design. We will use Eq. (5-25) to calculate a target value for the reinforcement ratio, ρ:

$$\rho = \frac{\beta_1 f'_c}{3 f_y} \tag{5-25}$$

$$= \frac{0.85 \times 4 \text{ ksi}}{3 \times 60 \text{ ksi}} = 0.0189$$

This will be rounded to 0.02, and then use Eqs. (5-21) and (5-22) to calculate ω and R.

$$\omega = \rho \frac{f_y}{f'_c} \tag{5-21}$$

$$= 0.02 \frac{60 \text{ ksi}}{4 \text{ ksi}} = 0.30$$

and

$$R = \omega f'_c (1 - 0.59\omega) \tag{5-22}$$

$$= 0.30 \times 4 \text{ ksi}(1 - 0.59 \times 0.30) = 0.988 \text{ ksi}$$

Assuming $\phi = 0.9$, use Eq. (5-23a) and $M_u = 229$ k-ft to calculate the required value for bd^2.

$$bd^2 \geq \frac{M_u}{\phi R}$$

$$\geq \frac{229 \text{ k-ft} \times 12 \text{ in./ft}}{0.9 \times 0.988 \text{ ksi}} = 3090 \text{ in.}^3$$

As stated earlier, we want the total depth of the girder to equal or exceed that of the floor beams supported by the girder. Thus, assume $h = 20$ in. and $d \approx h - 2.5$ in. $= 17.5$ in. Use this value of d to calculate the required width of the section, b (same as b_w).

$$b \geq \frac{3090 \text{ in.}^3}{d^2} = \frac{3090 \text{ in.}^3}{(17.5 \text{ in.})^2} = 10.1 \text{ in.}$$

As before, we could select a value of b equal to the column width (18 in.), but to better demonstrate the need for compression reinforcement, select $b = 12$ in. These dimensions for the girder ($h = 20$ in. and $b = 12$ in.) are smaller than assumed in step 4, but the resulting reduction in the weight of the girder would not cause much of a reduction in the factored design moment. Thus, a recalculation of M_u is not required.

With section dimensions selected, we can use the iterative design procedure for known section dimensions to determine the required area of tension reinforcement, A_s. Because this is a negative bending region and the compression zone will be at the bottom of the girder section ($b = 12$ in.), we will assume the moment arm, jd, in Eq. (5-16) will be set equal to 0.9 d, as assumed for narrow compression zones.

$$A_s \geq \frac{M_u}{\phi f_y(d - a/2)} \cong \frac{M_u}{\phi f_y(jd)}$$

$$\geq \frac{229 \text{ k-ft} \times 12 \text{ in./ft}}{0.9 \times 60 \text{ ksi} \times 0.9 \times 17.5 \text{ in.}} = 3.23 \text{ in.}^2$$

We will make one iteration using Eqs. (5-17) and (5-16) to improve the required value for A_s.

$$a = \frac{A_s f_y}{0.85 f'_c b} \tag{5-17}$$

$$= \frac{3.23 \text{ in.}^2 \times 60 \text{ ksi}}{0.85 \times 4 \text{ ksi} \times 12 \text{ in.}} = 4.75 \text{ in.}$$

and

$$A_s \geq \frac{M_u}{\phi f_y(d - a/2)} \tag{5-16}$$

$$\geq \frac{229 \text{ k-ft} \times 12 \text{ in./ft}}{0.9 \times 60 \text{ ksi}(17.5 \text{ in.} - 2.38 \text{ in.})} = 3.36 \text{ in.}^2$$

Try three No. 8 bars over the web of the girder section and **four No. 5 bars** in the flanges adjacent to the web as shown in Fig. 5-32. Recall that the permissible flange dimension for placement of the tension reinforcement is $\ell_n/10$ (2.4 ft or approximately 28 in. for this girder section). The resulting area of tension reinforcement is

$$A_s = 3 \times 0.79 \text{ in.}^2 + 4 \times 0.31 \text{ in.}^2 = 3.61 \text{ in.}^2$$

To complete the design of the section, select a compression steel area, A'_s, greater than or equal to one-half of the tension steel area. **Try three No. 7 bars** ($A'_s = 3 \times 0.60 \text{ in.}^2 = 1.80 \text{ in.}^2$) to be placed in the bottom of the girder web, as shown in Fig. 5-32.

8. Check section strength, ϕ-value and $A_{s,min}$. As discussed in Chapter 4, the author recommends the use of a trial-and-error procedure to establish section equilibrium

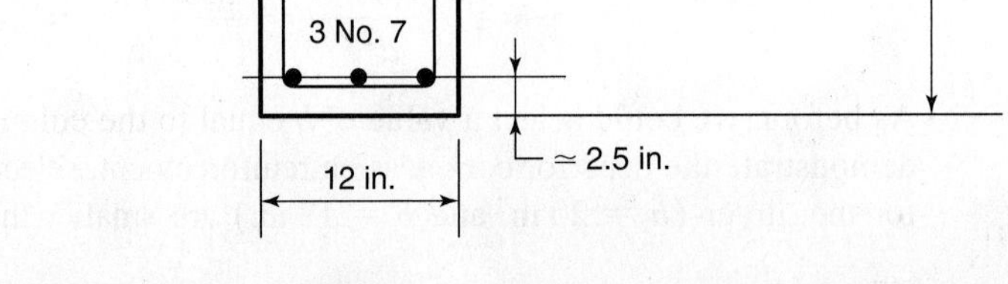

Fig. 5-32
Final girder section design at face of column $C2$ for Example 5-6.

and calculate the section nominal moment capacity, M_n, for a doubly reinforced section. Following that procedure, section equilibrium is satisfied for c = **4.45 in.** For the single layer of tension reinforcement, use Eq. (4-18) to determine the steel strain, ε_s $(= \varepsilon_t)$.

$$\varepsilon_s(= \varepsilon_t) = \left(\frac{d - c}{c}\right) \varepsilon_{cu}$$

$$= \left(\frac{17.5\text{ in.} - 4.45\text{ in.}}{4.45\text{ in}}\right) 0.003 = 0.00880$$

This confirms that the tension steel is yielding ($\varepsilon_s > \varepsilon_y = 0.00207$) and that this section is tension-controlled ($\varepsilon_t \geq 0.005$). Thus, the use of $\phi = 0.9$ is confirmed.

As part of the iteration process, all of the section forces are calculated and put into equilibrium. Those values were not shown here, but we will use the depth to the neutral axis, c, to calculate the other section forces and the nominal moment strength. Equation (4-29) is used to calculate the strain in the compression steel,

$$\varepsilon_s' = \left(\frac{c - d'}{c}\right) \varepsilon_{cu} \tag{4-30}$$

$$= \left(\frac{4.45\text{ in.} - 2.5\text{ in.}}{4.45\text{ in.}}\right) 0.003 = 0.00131$$

Then,

$$f_s' = E_s\varepsilon_s' = 29{,}000\text{ ksi} \times 0.00131 = 38.1\text{ ksi} \leq f_y \tag{4-32}$$

With this,

$$C_s = A_s'(f_s' - 0.85 f_c') \tag{4-31}$$

$$= 1.80\text{ in.}^2\,(38.1\text{ ksi} - 0.85 \times 4\text{ ksi}) = 62.5\text{ kips}$$

The concrete compression force is

$$C_c = 0.85 f_c' b\beta_1 c \tag{4-13b}$$

Recall that $a = \beta_1 c = 0.85 \times 4.45\text{ in.} = 3.78\text{ in.}$, which will be used to calculate M_n. Calculating C_c:

$$C_c = 0.85 \times 4\text{ ksi} \times 12\text{ in.} \times 3.78\text{ in.} = 154\text{ kips}$$

Finally, the tension force is

$$T = A_s f_y = 3.61\text{ in.}^2 \times 60\text{ ksi} = 217\text{ kips}$$

Because this is essentially equal to the sum of the compression forces ($C_c + C_s' = 217$ k), section equilibrium is verified. Finally, use Eq. (4-34) to calculate the section nominal moment strength.

$$M_n = C_c(d - a/2) + C_s(d - d')$$

$$= 154\text{ k}(17.5\text{ in.} - 3.78\text{ in.}/2) + 62.5\text{ k}(17.5\text{ in.} - 2.5\text{ in.})$$

$$= 2400\text{ k-in.} + 938\text{ k-in.} = 3340\text{ k-in.} = 278\text{ k-ft}$$

Using $\phi = 0.9$ to check the moment strength of the section,

$$\phi M_n = 0.9 \times 278\text{ k-ft} = 251\text{ k-ft} > M_u = 229\text{ k-ft}$$

Thus, the section has adequate strength without being significantly stronger than required.

For completeness, we will confirm that the provided area of tension steel exceeds $A_{s,\min}$ in Eq. (5-11). For concrete with $f'_c = 4000$ psi, the minimum value of 200 psi exceeds $3\sqrt{f'_c}$, so

$$A_{s,\min} = \frac{200 \text{ psi}}{f_y} b_w d = \frac{200 \text{ psi}}{60{,}000 \text{ psi}} \times 12 \text{ in.} \times 17.5 \text{ in.} = 0.70 \text{ in.}^2 < A_s \text{ (o.k.)}$$

Thus, **use three No. 8 and four No. 5 bars as tension reinforcement and three No. 7 bars as compression reinforcement**, as shown in Fig. 5-32. ■

5-5 DESIGN OF CONTINUOUS ONE-WAY SLABS

One-way slab-and-beam systems having plans similar to those shown in Fig. 5-1 are used commonly, especially for spans of greater than 20 ft and for heavy live loads. Generally, the ratio of the long side to the short side of the slab panels exceeds 2.0.

For design purposes, a one-way slab is assumed to act as a series of parallel, independent 1-ft wide strips of slab, continuous over the supporting beams. The slab strips span the short direction of the panel, as shown in Fig. 5-5. Near the girders, which are parallel to the one-way slab strips, the floor load is supported by two-way slab action, which is discussed more fully in Chapter 13. This is ignored during the design of the one-way slab strips but is accounted for in ACI Code Section 7.5.2.3, which requires reinforcement extending into the top of the slabs on each side of the girders across the ends of the panel. If this reinforcement is omitted, wide cracks parallel to the webs of the girders may develop in the top of the slab.

Thickness of One-Way Slabs

Except for very heavily loaded slabs, such as slabs supporting several feet of earth, the slab thickness is chosen so that deflections will not be a problem. Occasionally, the thickness will be governed by shear or flexure, so these are checked in each design. Table 7.3.1.1 of the ACI Code gives minimum thicknesses of slabs *not supporting or attached to* partitions or other construction liable to be damaged by large deflections. No guidance is given for other cases. Table A-9 gives recommended minimum thicknesses for one-way slabs that do and do not support such partitions.

Sometimes, slab thicknesses are governed by the danger of heat transmission during a fire. For this criterion the fire rating of a floor is the number of hours necessary for the temperature of the unexposed surface to rise by a given amount, generally 250°F. For a 250°F temperature rise, a 3-1/2 in. thick slab will give a 1-hour fire rating, a 5-in. slab will give a 2-hour fire rating, and a 6-1/4 in. slab will give a 3-hour fire rating [5-6]. Generally, slab thicknesses are selected in $\frac{1}{4}$-in. increments up to 6 in. and in $\frac{1}{2}$-in. increments for thicker slabs. Slab reinforcement is supported at the correct height above the forms on bent wire or plastic supports called *chairs*. The height of available chairs may control the slab thickness.

Cover

Concrete cover to the reinforcement provides corrosion resistance, fire resistance, and a wearing surface and is necessary to develop a bond between steel and concrete. ACI Code Table 20.6.1.3.1 gives the following minimum covers for corrosion protection in slabs:

1. For concrete *not exposed* to weather or in contact with the ground; No. 11 bars and smaller, **$^3/_4$ in.**

2. For concrete *exposed* to weather or in contact with the ground; No. 5 bars and smaller, **$1\frac{1}{2}$ in.**; No. 6 bars and larger, **2 in.**

3. For concrete *cast against* or *permanently exposed* to ground, **3 in.**

The words *exposed to weather* imply direct exposure to moisture changes. The undersides of exterior slabs are not considered exposed to weather unless subject to alternate wetting and drying, including that due to condensation, leakage from the exposed top surface, or runoff. ACI Commentary Section R20.6.1.4.1 recommends a minimum cover of 2 in. for slabs exposed to chlorides, such as deicing salts (as in parking garages).

The structural endurance of a slab exposed to fire depends (among other things) on the cover to the reinforcement. Building codes give differing fire ratings for various covers. Reference [5-6] states that, for normal ratios of service-load moment to ultimate moment, $\frac{3}{4}$-in. cover will give a $1\frac{1}{4}$-hour fire rating, 1-in. cover about $1\frac{1}{2}$ hours, and $1\frac{1}{2}$-in. cover about 3 hours.

Reinforcement

Reinforcement details for one-way slabs are shown in Fig. 5-33. The straight-bar arrangement in Fig. 5-31a is almost always used in buildings. Prior to the 1960s, slab reinforcement was arranged by using the bent-bar and straight-bar arrangement shown in Fig. 5-33b. The cut-off points shown in Fig. A-5 can be used if the slab satisfies the requirements for use of the moment coefficients in ACI Code Section 6.5.1. Cut-off points in slabs not satisfying this clause are obtained via the procedure given in Chapter 8. One-way slabs normally are designed by assuming a 1-ft-wide strip. The area of reinforcement then is computed as A_s/ft of width. The area of steel is the product of the area of a bar times the number of bars per foot, or

$$A_s/ft = A_b\left(\frac{12\text{ in.}}{\text{bar spacing in inches, } s}\right) \tag{5-27}$$

where A_b is the area of one bar. In SI units, Eq. (5-27) becomes

$$A_s/m = A_b\left(\frac{1000\text{ mm}}{\text{bar spacing in mm, } s}\right) \tag{5-27M}$$

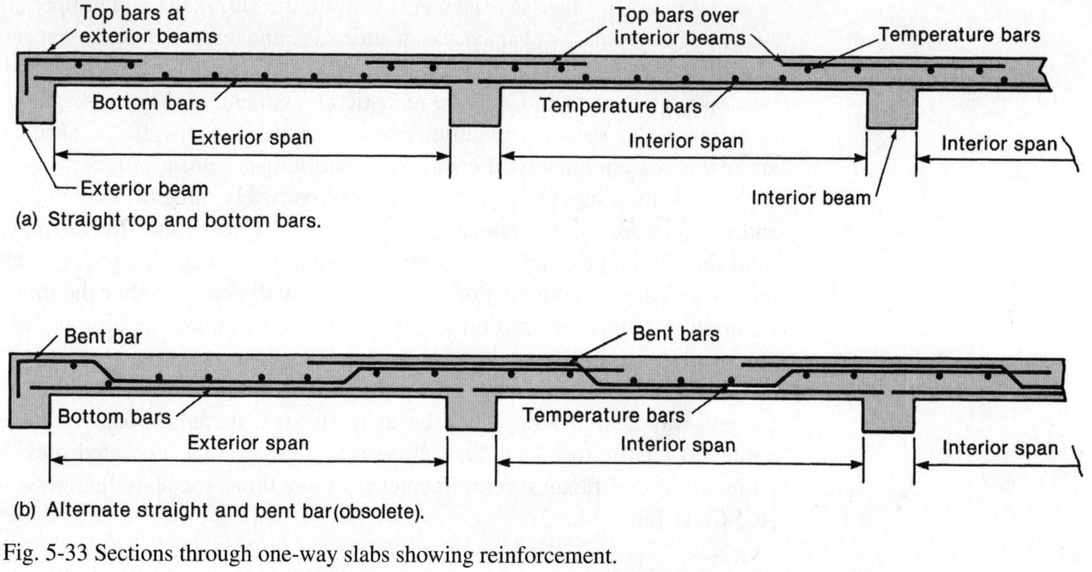

Fig. 5-33 Sections through one-way slabs showing reinforcement.

In most cases, the required steel area has been determined, so these expressions can be used to solve for the maximum spacing to achieve the required steel area:

$$s(\text{inches}) \leq \frac{A_b \times 12 \text{ in.}}{\text{required } A_s/ft} \tag{5-28}$$

and in metric units:

$$s(\text{mm}) = \frac{A_b \times 1000 \text{ mm}}{\text{required } A_s/m} \tag{5-28M}$$

The maximum spacing of bars used as primary flexural reinforcement in one-way slabs is three times the slab thickness or 18 in., whichever is smaller (ACI Code Section 7.7.2.3). The maximum bar spacing also is governed by crack-control provisions (ACI Code Table 24.3.2), as will be shown in Example 5-7 and then discussed more completely in Chapter 9.

Because a slab is thinner than the beams supporting it, the concrete in the slab shrinks more rapidly than the concrete in the beams. This may lead to shrinkage cracks in the slab. Shrinkage cracks perpendicular to the span of the one-way design strips will be crossed by flexural reinforcement, which will limit the width of these cracks. To limit the width of potential shrinkage cracks parallel to the one-way design strips, *shrinkage and temperature reinforcement* is placed perpendicular to the primary flexural reinforcement. The amount of reinforcement required is specified in ACI Code Table 7.6.1.1, which requires the following ratios of reinforcement area to gross concrete area:

1. Slabs with Grade-40 deformed bars: 0.0020
2. Slabs with Grade-60 deformed bars or welded-wire fabric (smooth or deformed): 0.0018
3. Slabs with reinforcement with a yield strength, f_y, in excess of 60,000 psi: $(0.0018 \times 60{,}000 \text{ psi})/f_y$, but not less than 0.0014

Shrinkage and temperature reinforcement is spaced not farther apart than the smaller of five times the slab thickness and 18 in. Splices of such reinforcement (covered in Chapter 8) must be designed to develop the full yield strength of the bars in tension.

It should be noted that shrinkage cracks could be wide even when this amount of shrinkage reinforcement is provided [5-7]. In buildings, this may occur when shear walls, large columns, or other stiff elements restrain the shrinkage and temperature movements. ACI Code Section 24.4.2 states that if shrinkage and temperature movements are restrained significantly, the requirements of ACI Code Section 5.3.6 shall be considered. This section asks the designer to make a realistic assessment of the shrinkage deformations and to estimate the stresses resulting from these movements. If the shrinkage movements are restrained completely, the shrinkage and temperature reinforcement may yield at the cracks, resulting in a few wide cracks. Approximately three times the minimum shrinkage and temperature reinforcement specified in ACI Code Table 7.6.1.1 may be required to limit the shrinkage cracks to reasonable widths. Alternatively, unconcreted control strips may be left during construction to be filled in with concrete after the initial shrinkage has occurred. Methods of limiting shrinkage and temperature cracking in concrete structures are reviewed in [5-8].

ACI Code Section 7.6.1.1 specifies that the minimum flexural reinforcement in the one-way design strips shall be at least equal to the amount required in ACI Code Table 7.6.1.1 for shrinkage and temperature, except that, as stated previously, the maximum spacing of flexural reinforcement is three times the slab thickness, or as limited by ACI Code Table 24.3.2.

Generally, No. 4 and larger bars are used for flexural reinforcement in slabs, because smaller bars or wires tend to be bent out of position by workers walking on the reinforcement during construction. This is more critical for top reinforcement than for bottom reinforcement, because the effective depth, *d*, of the top steel is reduced if it is pushed down, whereas that of the bottom steel is increased.

EXAMPLE 5-7 Design of a One-Way Slab

Design the eight-span floor slab spanning east–west in Fig. 5-34. A typical 1-ft-wide design strip is shown shaded. A partial section through this strip is shown in Fig. 5-35. The underside of a typical floor is shown in Fig. 5-36. The interior beams are assumed to be 14-in. wide and the exterior (spandrel) beams are 16-in. wide. The concrete strength is 4000 psi, and the reinforcement strength is 60 ksi. Assume a superimposed dead load of 20 psf to account for floor covering, the ceiling, and mechanical equipment. In consultation

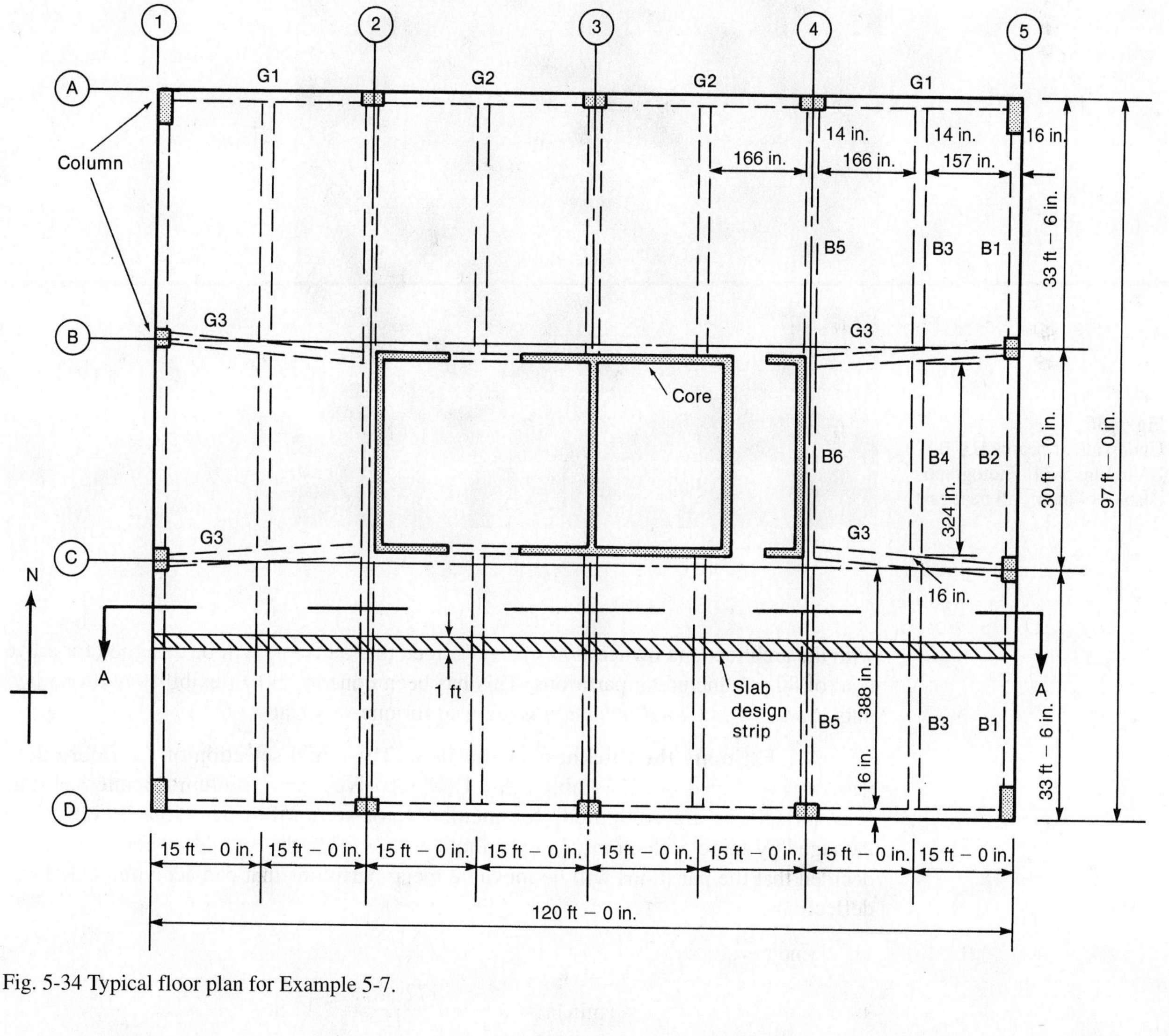

Fig. 5-34 Typical floor plan for Example 5-7.

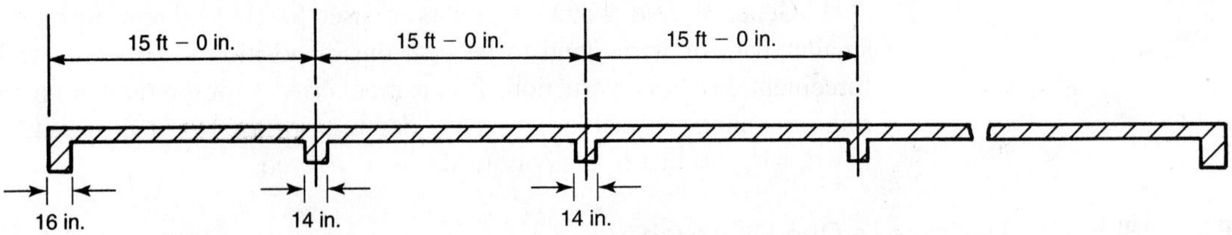

Fig. 5-35 Section A–A for Example 5-7.

Fig. 5-36
Underside of beams B3–B4–B3 in Fig. 5-34. Photograph taken looking north near east wall.

with the architect and owner, it has been decided that all floors will be designed for a live load of 80 psf, including partitions. This has been done to allow flexibility in office layouts. *Note*: *No live-load reduction* is allowed for one-way slabs.

1. Estimate the thickness of the floor. The initial selection of the floor thickness will be based on ACI Table 7.3.1.1, which gives the minimum thicknesses (unless deflections are computed) for members not supporting partitions likely to be damaged by large deflections. In consultation with the architect and owner, it has been decided that the partitions will be movable metal partitions that can accommodate floor deflections.

End bay:

$$(\min)h = \frac{\ell}{24} = \frac{172 \text{ in.}}{24} = 7.17 \text{ in.}$$

Interior bays:

$$(\min)h = \frac{\ell}{28} = \frac{180}{28} = 6.43 \text{ in.}$$

Therefore, **try a 7-in. slab** and assume that we will check deflections in the exterior span (Chapter 9). **Using $\frac{3}{4}$-in. clear cover and No. 4 bars,**

$$d = h - \text{clear cover} - d_b/2 = 7 \text{ in.} - 0.75 \text{ in.} - 0.5 \text{ in.}/2 = 6 \text{ in.}$$

Before the thickness is finalized, it will be necessary to check whether it is adequate for moment and shear. Shear strength is discussed in Chapter 6, but a short check on the slab thickness will be made in step 5 of this example.

2. Compute the unfactored loads. Given the thickness selected in step 1, it is now possible to compute the unfactored uniform loads. The dead load is as follows:

Slab:

$$w(\text{slab}) = \frac{7 \text{ in.}}{12 \text{ in./ft.}} \times 150 \text{ lb/ft}^3$$

$$= 87.5 \text{ lb/ft}^2 \text{ of floor surface}$$

Superimposed dead loads: $w(\text{SDL}) = 20$ psf

Total dead load: $w_D = 108$ psf

Live load: $w_L = 80$ psf

3. Select load and strength-reduction factors. Check the two-factored load combinations as used in the prior examples for a continuous floor system.

$$w_u = 1.4 \times w_D = 1.4 \times 108 \text{ psf} = 151 \text{ psf}$$

$$w_u = 1.2 \times w_D + 1.6 \times w_L$$

$$= 1.2 \times 108 \text{ psf} + 1.6 \times 80 \text{ psf} = 258 \text{ psf}$$

The second load combination governs and will be used in the following. Assume the slab is tension-controlled, and thus, $\phi = 0.9$.

4. Check whether the slab thickness is adequate for the maximum moment. Because $w_L < 3w_D$ and the other requirements of ACI Code Section 6.5.1 are met, use the ACI moment coefficients (Fig. 5-11) to calculate the design moments. If the slab thickness is adequate for the largest design moment, it will be acceptable at all other locations. The maximum moment, M_u, will occur at the first or second interior support.

From ACI Code Table 6.5.2, the moment at the exterior face of the first interior support is

$$M_u = -\frac{w_u \ell_n^2}{10}$$

where ℓ_n for computation of the negative moment at *interior supports* is the average of the clear spans of the adjacent spans. From Fig. 5-35,

$$\ell_n(\text{avg}) = \frac{(157 \text{ in.} + 166 \text{ in.})}{2} \times \frac{1 \text{ ft}}{12 \text{ in.}} = 13.5 \text{ ft}$$

$$M_u = \frac{258 \text{ psf} \times 1 \text{ ft} \times (13.5 \text{ ft})^2}{10} = 4700 \text{ lb-ft/ft of width} = 4.70 \text{ k-ft/ft}$$

For the second interior support,

$$M_u = -\frac{w_u \ell_n^2}{11}$$

where

$$\ell_n(\text{avg}) = 166 \text{ in.} = 13.8 \text{ ft}$$

$$M_u = \frac{258 \text{ psf} \times 1 \text{ ft} \times (13.8 \text{ ft})^2}{11} = 4470 \text{ lb-ft/ft} = 4.47 \text{ k-ft/ft}$$

Therefore, maximum negative design moment is $M_u = 4.70$ k-ft/ft.

One-way slabs normally have a very low reinforcement ratio, ρ. Thus, assume the flexural reinforcement moment arm, jd, is equal to 0.95 d, and use Eq. (5-16) to obtain an initial value for the required steel area, A_s, per one-foot width of slab.

$$A_s \geq \frac{M_u}{\phi f_y\left(d - \frac{a}{2}\right)} \cong \frac{M_u}{\phi f_y(jd)} = \frac{4.70 \text{ k-ft/ft} \times 12 \text{ in./ft}}{0.9 \times 60 \text{ ksi} \times 0.95 \times 6 \text{ in.}} = 0.183 \text{ in.}^2\text{/ft}$$

As was done for the design of beam sections, we can go through one iteration with Eqs. (5-16) and (5-17) to improve this value. Using $b = 1$ ft $= 12$ in. in Eq. (5-17),

$$a = \frac{A_s f_y}{0.85 f'_c b} = \frac{0.183 \text{ in.}^2 \times 60 \text{ ksi}}{0.85 \times 4 \text{ ksi} \times 12 \text{ in.}} = 0.269 \text{ in.}$$

Because of this very small value for a, it is clear that $c = a/\beta_1$ is significantly less than 3/8 of d, and thus, this is a tension-controlled section for which $\phi = 0.9$. Then, from Eq. (5-16),

$$A_s \geq \frac{M_u}{\phi f_y\left(d - \frac{a}{2}\right)} = \frac{4.70 \text{ k-ft/ft} \times 12 \text{ in./ft}}{0.9 \times 60 \text{ ksi}(6 \text{ in.} - 0.135 \text{ in.})} = 0.178 \text{ in.}^2\text{/ft}$$

For this required steel area, the steel reinforcement ratio is

$$\rho = \frac{A_s/ft}{bd} = \frac{0.178 \text{ in.}^2}{12 \text{ in.} \times 6 \text{ in.}} = 0.00247$$

This is a very low reinforcement ratio, as is common in most slabs—so clearly, the selected slab thickness is adequate for the design bending moments. Before selecting reinforcement, we will quickly check to be sure that this slab thickness is also adequate for shear strength.

5. Check whether thickness is adequate for shear. The topic of shear strength in beams and slabs will be covered in Chapter 6. For simplicity here, we will show a check of the shear strength at the exterior face of the first interior support. Using the ACI coefficients from Code Table 6.5.2 (Fig. 5-11) the shear at this section is increased by 15 percent to account for an unsymmetrical moment diagram in the exterior span:

$$V_u = \frac{1.15 w_u \ell_n}{2} = \frac{1.15 \times 258 \text{ lb/ft} \times {}^{157 \text{ in.}}/_{12 \text{ in./ft}}}{2}$$

$$= 1940 \text{ lb per 1-ft width of slab}$$

In Chapter 6, we will introduce an equation (Eq. 6-8b) that is commonly used to determine the shear that can be resisted by the concrete in a beam or one-way slab:

$$V_c = 2\lambda\sqrt{f'_c}\, b_w d \qquad \text{(6-8b)}$$

where λ is taken as 1.0 for normal weight concrete. For this equation, the concrete strength must be given in psi units and the units obtained from the first term of the equation, $2\lambda\sqrt{f'_c}$, also is to be in psi units. Using a 1-ft-wide strip of slab ($b = b_w = 12$ in.),

$$V_c = 2 \times 1\sqrt{4000} \times 12 \text{ in.} \times 6 \text{ in.} = 9110 \text{ lb per 1-ft width of slab}$$

ACI Code Table 21.2.1 gives $\phi = 0.75$ for shear and torsion. Thus, $\phi V_c = 0.75 \times 9110$ lb/ft = 6830 lb/ft $> V_u$.

Therefore, use $\boldsymbol{h} = \mathbf{7}$ **in.** When slab thickness is selected on the basis of deflection control, moment and shear strength seldom require an increase in slab thickness.

6. Design of reinforcement. The calculations shown in Table 5-3 are based on a 1-ft strip of slab. First, several constants used in that table must be calculated.

Line 1—The clear spans, ℓ_n, can be computed from Fig. 5-35.

- End bay, $\ell_n = 15 \text{ ft} \times 12 \text{ in./ft} - 16 \text{ in.} - 14 \text{ in./2} = 157 \text{ in.} = 13.1 \text{ ft.}$
- Interior bay, $\ell_n = 15 \text{ ft.} \times 12 \text{ in./ft} - 2(14 \text{ in./2}) = 166 \text{ in.} = 13.8 \text{ ft.}$
- For the calculation of design moments at interior supports, ℓ_n is taken as the average of the two adjacent spans, so $\ell_n(\text{avg})$ at the first interior support is 161.5 in. = 13.5 ft.

Lines 2, 3, and 4—For line 2, the value of w_u comes from step 3. The moment coefficients in line 3 come from Fig. 5-11. At the first interior support the coefficient of –1/10 is selected because it will govern over the alternate coefficient of –1/11 in Fig. 5-11. The moments in line 4 are computed as $M_u = w_u \ell_n^2 \times$ moment coefficient from line 3.

Line 5—The maximum factored moment calculated in line 4 occurs at the first interior support, as previously determined in step 4. The required reinforcement at this section also was calculated in step 4 as 0.178 in.^{2}/ft. Because the moment arm, $(d - a/2)$, will be similar at all design sections, we can use the solution at this section to develop a scaling factor that can be applied at all other sections.

$$\frac{A_s \text{ (in.}^2\text{/ft)}}{M_u \text{ (k-ft/ft)}} = \frac{0.178 \text{ in.}^2\text{/ft}}{4.70 \text{ k-ft/ft}} = 0.0379 \frac{\text{in.}^2\text{/ft}}{\text{k-ft/ft}}$$

This value was used to calculate all of the other required steel areas in line 5 of Table 5-3.

TABLE 5-3 Calculations for One-Way Slab for Example 5-7

Line No./ Item	External Support	Exterior Midspan	First Interior Support	Interior Midspan	Second Interior Support
1. ℓ_n, (ft)	13.1	13.1	13.5	13.8	13.8
2. $w_u \ell_n^2$, (k-ft/ft)	44.3	44.3	47.0	49.1	49.1
3. M coefficient	–1/24	1/14	–1/10	1/14	–1/11
4. M_u, (k-ft/ft)	1.84	3.16	4.70	3.51	4.47
5. A_s(req'd), (in.^{2}/ft)	0.070	0.120	0.178	0.133	0.169
6. A_s(min), (in.^{2}/ft)	0.151	0.151	0.151	0.151	0.151
7. Select bars	No. 4 at 12 in.	No. 4 at 12 in.	No. 4 at 12 in.	No. 4 at 12 in.	No. 4 at 12 in.
8. Final A_s, (in.^{2}/ft)	0.20	0.20	0.20	0.20	0.20

Line 6—Minimum reinforcement often governs for one-way slabs. Compute the minimum flexural reinforcement using ACI Code Section 7.6.1.1, which refers to ACI Code Table 7.6.1.1 for the amount of minimum reinforcement for temperature and shrinkage effects, which for Grade-60 reinforcement is:

$$A_{s,\min} = 0.0018 \times b \times h = 0.0018 \times 12\text{ in.} \times 7\text{ in.} = 0.151\text{ in.}^2/\text{ft}$$

For one-way slab sections where there is a required area of flexural reinforcement, as indicated in line 5 for all slab sections, ACI Code Section 7.7.2.3 limits the spacing between that reinforcement to the smaller of $3h$ and 18 in. For this case, the 18 in. value governs.

7. Check reinforcement spacing for crack control. To control the width of cracks on the tension face of the slab, ACI Code Table 24.3.2 limits the maximum spacing of the flexural reinforcement closest to the tension face of the slab to

$$s = 15\left(\frac{40{,}000}{f_s}\right) - 2.5c_c, \text{ but not greater than, } 12\left(\frac{40{,}000}{f_s}\right)$$

where f_s is the stress in the tension steel in psi, which can be taken as $2/3\, f_y = 40{,}000$ psi for Grade-60 steel, and c_c is the clear cover from the tension face of the slab to the *surface* of the reinforcement nearest to it, taken as 0.75 in. for this one-way slab example. Thus,

$$s = 15(\text{in.})\left(\frac{40{,}000}{40{,}000}\right) - 2.5 \times 0.75\text{ in.} = 13.1\text{ in., but} \leq 12(\text{in.})\left(\frac{40{,}000}{40{,}000}\right) = 12\text{ in.}$$

This result overrides the 18-in. maximum spacing from the previous step. Thus, maximum bar spacing is 12 in., which was used in line 7 of Table 5-3.

8. Select the top and bottom flexural steel. The choice to use **No. 4 bars at a spacing of 12 in.** in line 7 of Table 5-3 satisfies the strength, minimum area, and spacing requirements at all design locations. If a wider variety of required steel areas had been calculated, the choice of the reinforcement in line 7 could have been made by using Eq. (5-27). The resulting steel arrangement is shown in Fig. 5-37. The cut-off points, which will be discussed in Chapters 8 and 10, have been determined by using Fig. A-5c because the slab geometry permitted the use of the ACI moment coefficients.

9. Determine the shrinkage and temperature reinforcement for transverse direction. ACI Code Table 7.6.1.1 specifies the amount of shrinkage and temperature reinforcement to be provided perpendicular to the span of the one-way slab:

$$A_s(\text{S \& T}) = 0.0018 \times b \times h = 0.0018 \times 12\text{ in.} \times 7\text{ in.} = 0.151\text{ in.}^2/\text{ft}$$

$$\text{Maximum spacing} \leq 5 \times h, \text{ and } \leq 18\text{ in. (ACI Code Section 7.7.2.4)}$$

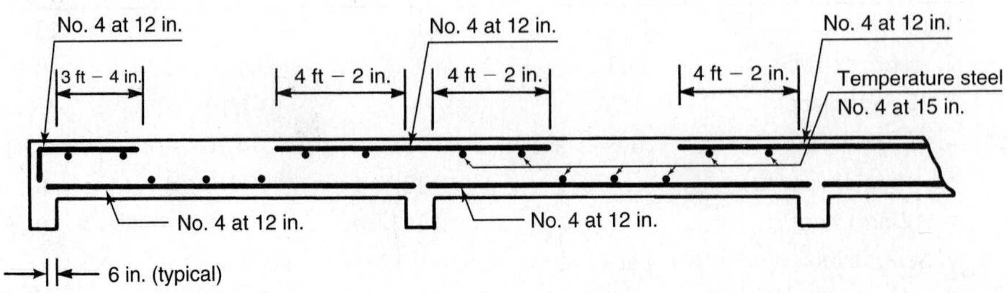

Fig. 5-37
Slab reinforcement for Example 5-7.

Therefore, provide **No. 4 bars at 15 in. o.c., as shrinkage and temperature reinforcement**. Using Eq. (5-27), this results in a steel area equal to 0.160 in.^{2}/ft. These bars can be placed either in the top or bottom of the slab. If they are placed at the top, they should be placed below the top flexural reinforcement to permit the larger effective depth for that flexural reinforcement, and similarly, they should be placed on top of the bottom layer of flexural reinforcement, as shown in Fig. 5-37. *Chairs* will be used to support the flexural steel during placement of concrete.

10. Design the transverse top steel at girders. Due to localized two-way action adjacent to the girders (G1, G2, G3, etc., in Fig. 5-34), ACI Code Section 7.5.2.3 requires that *top* transverse reinforcement be designed for the slab to carry the factored floor load acting on the effective width of the overhanging flange (slab), which is assumed to act as a cantilevered beam. The definitions for the width of the overhanging slab are given in ACI Code Table 6.3.2.1 for interior and exterior girders.

For this floor system, the overhang length for the interior girders (G3) is more critical and can be determined to have an effective cantilevered length of approximately 1.75 ft. Calling this length ℓ_o and using the factored load for the floor calculated in step 3, the factored design moment for this cantilever is

$$M_u = w_u\left(\frac{\ell_o{}^2}{2}\right) = 0.258\text{ ksf}\frac{(1.75\text{ ft})^2}{2} \times 1\text{ ft} = 0.40\text{ k-ft/ft}$$

Because the steel to be provided will be flexural reinforcement (not temperature and shrinkage reinforcement), the maximum spacing for these bars will be 12 in., as determined in step 7. Thus, it is reasonable to use No. 4 at 12 in., as was used for the top flexural reinforcement in the direction of the one-way slab strips. To determine the nominal moment capacity for this reinforcement, we will need to use a smaller effective flexural depth, d, because these bars will be placed below the primary flexural reinforcement, as shown in Fig. 5-37. The effective depth for this transverse steel essentially will be one bar diameter smaller than the 6 in. value determined for the primary flexural reinforcement in step 1 (i.e., $d = 6 - 0.5$ in. $= 5.5$ in.). Equation (5-17) can be used to determine the depth of the compression stress block:

$$a = \frac{A_s f_y}{0.85 f'_c\, b} = \frac{0.20\text{ in.}^2 \times 60\text{ ksi}}{0.85 \times 4\text{ ksi} \times 12\text{ in.}} = 0.294\text{ in.}$$

Because this is very low, it is clear that c/d is less that 3/8, so this is a tension-controlled section with $\phi = 0.9$. Then, the reduced nominal moment capacity is

$$\phi M_n = \phi A_s f_y\,(d - a/2) = 0.9 \times 0.20\text{ in.}^2 \times 60\text{ ksi}(5.5\text{ in.} - 0.147\text{ in.})$$
$$= 57.8\text{ k-in./ft} = 4.82\text{ k-ft/ft}$$

Because this exceeds M_u, the use of **No. 4 bars at 12 in.** as top reinforcement in the transverse direction over all of the girders will satisfy ACI Code Section 7.5.2.3. ■

This completes the design of the one-way slab in Fig. 5-34. The slab thickness was selected in step 1 to limit deflections. A 6.5-in. thickness would be acceptable for the six interior spans, but a larger thickness was required in the end spans. If the entire floor slab was decreased to a 6-in. thickness, instead of the 7-in. thickness used in the example, about 36 cubic yards of concrete could be saved per floor, with a resultant saving of 145 kips of dead load per floor. In a 20-story building, that amount represents a considerable saving. With this in mind, the floor could be redesigned with a 6-in. thickness, and the computed deflections can be shown to be acceptable. The calculations are not given here because deflections will be discussed in Chapter 9.

EXAMPLE 5-7M Design of a One-Way Slab Section in SI Units

Because the prior example contains many details regarding minimum slab thickness for deflection control and shear strength, this example will only concentrate on the flexural design of a one slab section and the requirement for shrinkage and temperature reinforcement in the transverse direction. Assume we have a slab thickness of 160 mm and a factored design moment, $M_u = 35$ kN-m. *Note*: In SI units one-way slabs are typically designed using a 1-meter strip. Assume the material strengths are, $f'_c = 25$ MPa ($\beta_1 = 0.85$) and $f_y = 420$ MPa.

1. Effective flexural depth, *d*. ACI Metric Code Table 20.6.1.3.1 requires a minimum clear cover of 20 mm for slabs using Grade-420 reinforcement of sizes No. 36 and smaller. Assuming that we will use a bar size close to a No. 16 bar, the effective depth is

$$d \cong h - \text{clear cover} - d_b/2 = 160\text{ mm} - 20\text{ mm} - 16\text{ mm}/2 = 132\text{ mm}$$

2. Select flexural reinforcement. With the slab depth selected, we can treat this as a section design where the member dimensions are known and solve directly for the required area of tension reinforcement, A_s. Because slabs are usually lightly reinforced, we can assume that this will be a tension-controlled section ($\phi = 0.9$) and that the flexural moment arm, jd, in Eq. (5-16) is 0.95 d. Thus,

$$A_s \geq \frac{M_u}{\phi f_y(d - a/2)} \cong \frac{35\text{ kN-m}}{\phi f_y(jd)} = \frac{35 \times 10^6\text{ N-mm}}{0.9 \times 420\text{ N/mm} \times 0.95 \times 132\text{ mm}} = 738\text{ mm}^2$$

As has been done in previous examples, we will go through one iteration using Eqs. (5-16) and (5-17) to improve this value. Using $b = 1\text{ m} = 1000$ mm in Eq. (5-17),

$$a = \frac{A_s f_y}{0.85 f'_c b} = \frac{738\text{ mm}^2 \times 420\text{ MPa}}{0.85 \times 25\text{ MPa} \times 1000\text{ mm}} = 14.6\text{ mm}$$

The depth to the neutral axis, $c = a/\beta_1 = 14.6/0.85 = 17.2$ mm, is well below 3/8 of d_t ($d = d_t$), so this is a tension-controlled section and we can use $\phi = 0.9$. Using the calculated value of a, Eq. (5-16) gives

$$A_s = \frac{M_u}{\phi f_y(d - a/2)} = \frac{35 \times 10^6\text{ N-mm/m}}{0.9 \times 420\text{ N/mm}^2(132\text{ mm} - 7.3\text{ mm})} = 743\text{ mm}^2$$

Before selecting bars, we must check if the requirement for minimum reinforcement to control cracking due to temperature and shrinkage effects governs for this section. For slabs using reinforcement with $f_y = 420$ MPa, ACI Metric Code Table 7.6.1.1 requires

$$A_{s,\min} = 0.0018 \times b \times h = 0.0018 \times 1000\text{ mm} \times 160\text{ mm} = 288\text{ mm}^2/\text{m}$$

Clearly, this does not govern. So, using the required area of reinforcement calculated above and assuming that we will use a No. 16 bar, Eq. (5-28M) can be used to solve for

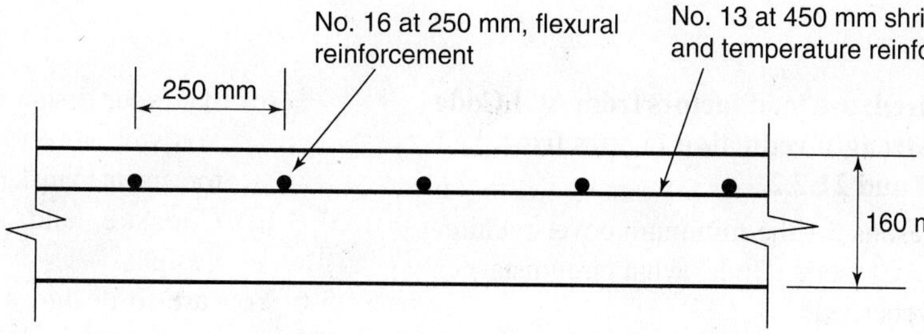

Fig. 5-38
Final slab section design for Example 5-7M.

the maximum permissible spacing between bars to satisfy the nominal moment strength requirement.

$$s \leq \frac{A_b \times 1000 \text{ mm}}{(\text{required})A_s/\text{m}} = \frac{199 \text{ mm}^2 \times 1000 \text{ mm}}{743 \text{ mm}^2} = 268 \text{ mm}$$

Before selecting the final bar spacing, we must also check the maximum spacing limit to control flexural crack widths given in ACI Metric Code Table 24.3.2. The value for maximum bar spacing in that section is

$$s = 380\left(\frac{280}{f_s}\right) - 2.5c_c \leq 300\left(\frac{280}{f_s}\right)$$

where f_s can be taken as 2/3 of f_y (2/3 × 420 MPa = 280 MPa) and c_c is the cover to the bar in question. For a slab design, this is the same as the clear cover (20 mm). So,

$$s = 380 \text{ mm}\left(\frac{280 \text{ MPa}}{280 \text{ MPa}}\right) - 2.5 \times 20 \text{ mm} = 330 \text{ mm} \leq 300 \text{ mm}\left(\frac{280 \text{ MPa}}{280 \text{ MPa}}\right)$$

The upper limit of 300 mm governs here, but the flexural strength requirement governs overall ($s \leq 268$ mm). Thus, use s **= 250 mm** as shown for a cross section of the slab in Fig. 5-38. This also satisfies the upper limit on spacing for flexural reinforcement given in ACI Metric Code Section 7.7.2.3, which states that the maximum spacing shall be less than the smaller of $3h$ and 450 mm.

3. Temperature and shrinkage reinforcement. For one-way slabs, reinforcement must be placed perpendicular to the primary flexural reinforcement to control cracking due to temperature and shrinkage effects. The required area calculated in step 2 can be used in Eq. (5-27M) to determine the maximum permissible spacing. Assuming that we will use a No. 13 bar for this reinforcement,

$$s \leq \frac{A_b \times 1000 \text{ mm}}{(\text{required})A_s/\text{m}} = \frac{129 \text{ mm}^2 \times 1000 \text{ mm}}{288 \text{ mm}^2} \cong 450 \text{ mm}$$

ACI Metric Code Section 7.7.2.4 states that the spacing of temperature and shrinkage reinforcement shall not exceed the smaller of $5h$ and 450 mm. Thus, **use No. 13 bars at 450 mm**, as shown in Fig. 5-38, to satisfy these requirements. ■

PROBLEMS

Note: Where required, use load factors from ACI Code Section 5.3.1 and strength reduction factors from ACI Code Tables 21.2.1 and 21.2.2.

5-1 Give three reasons for the minimum cover requirements in the ACI Code. Under what circumstances are larger covers used?

5-2 Give three reasons for using compression reinforcement in beams.

5-3 Design a rectangular beam section (i.e., select b, d, h, and the required tension reinforcement) at midspan for a 22-ft-span simply supported rectangular beam that supports its own dead load, a superimposed service dead load of 1.25 kip/ft, and a uniform service live load of 2 kip/ft. Use the procedure in Section 5-3 for the design of beam sections when the dimensions are unknown. Use $f'_c = 4500$ psi and $f_y = 60$ ksi.

Video Solution

5-4 The rectangular beam shown in Fig. P5-4 carries its own dead load (you must guess values for b and h) plus an additional uniform service dead load of 0.5 kip/ft and a uniform service live load of 1.5 kip/ft. The dead load acts on the entire beam, of course, but the live load can act on parts of the span. Three possible loading cases are shown in Fig. P5-4.

(a) Draw factored bending-moment diagrams for the three loading cases shown and superimpose them to draw a bending-moment envelope.

(b) Design a rectangular beam section for the maximum positive bending moment between the supports, selecting b, d, h, and the reinforcing bars. Use the procedure in Section 5-3 for the design of beam sections when the dimensions are unknown. Use $f'_c = 5000$ psi and $f_y = 60$ ksi.

(c) Starting with the beam section from part (b), use the procedure for design of tension reinforcement when section dimensions are known to design flexural reinforcement for the maximum negative moment over the roller support.

5-5 Design three rectangular beam sections (i.e., select b and d and the tension steel area A_s) to resist a factored design moment, $M_u = 260$ k-ft. For all three cases, select a section with $b = 0.5d$ and use $f'_c = 4000$ psi and $f_y = 60$ ksi.

(a) Start your design by assuming that $\varepsilon_t = 0.0075$ (as was done in Section 5-3).

(b) Start your design by assuming that $\varepsilon_t = 0.005$.

(c) Start your design by assuming that $\varepsilon_t = 0.0035$. You will probably need to add compression reinforcement to make this a tension-controlled section.

(d) Compare and discuss your three section designs.

5-6 You are to design a rectangular beam section to resist a negative bending moment of 320 k-ft. Architectural requirements will limit your beam dimensions to a width of 12 in. and a total depth of 20 in. Using those maximum permissible dimensions, select reinforcement to provide the required moment strength. You may need to add compression steel to make this a tension-controlled section. Use $f'_c = 5000$ psi and $f_y = 60$ ksi.

Video Solution

All of the following problems refer to the floor plan in Fig. P5-7 and use $f'_c = 4000$ psi.

5-7 For column line 2, use the ACI moment coefficients given in ACI Code Table 6.5.2 to determine the maximum positive and negative factored moments at the support faces for columns A2 and B2, and at the midspans of an exterior span and the interior span.

5-8 Repeat Problem 5-7, but use structural analysis software to determine the maximum positive and negative moments described. The assumed beam, slab, and column dimensions are given in the figure. Assume 12-ft story heights above and below this floor level. You must use appropriate live load patterns to maximize the various factored moments. Use a table to compare the answers from Problems 5-7 and 5-8.

5-9 Repeat Problems 5-7 and 5-8 for column line 1.

5-10 Repeat Problems 5-7 and 5-8 for the beam *m–n–o–p* in Fig. P5-7. Be sure to comment on the factored design moment at the face of the spandrel beam support at point *m*.

5-11 Repeat Problems 5-7 and 5-8 for the one-way slab strip shown in Fig. P5-7. For this problem, find the factored design moments at all of the points, *a* through *i*, indicated in Fig. P5-7.

5-12 Use structural analysis software to find the maximum factored moments for the girder on column line C. Find the maximum factored positive moments at *o* and *y*, and the maximum factored negative moments at columns C1, C2, and C3.

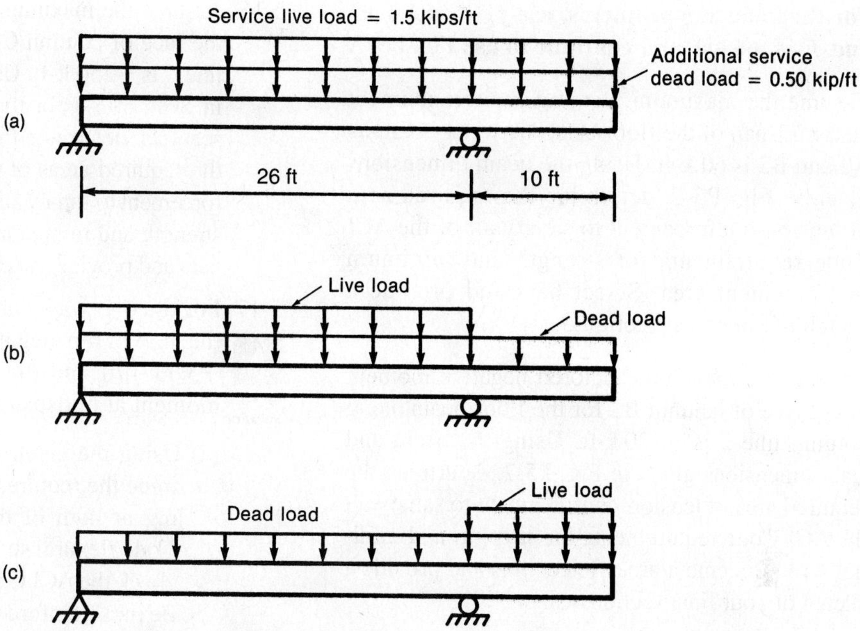

Fig. P5-4

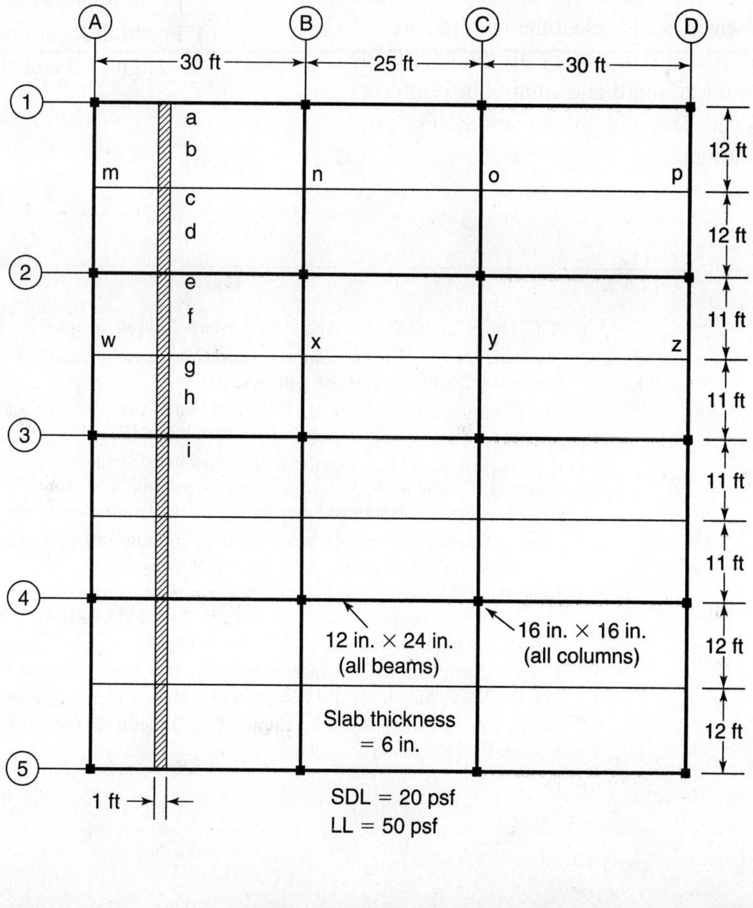

Fig. P5-7
Floor plan for various problems in Chapter 5.

For the following problems, use $f'_c = 4000$ psi and $f_y = 60$ ksi, and continue to use Fig. P5-7.

5-13 Assume the maximum factored positive moment near midspan of the floor beam between columns A2 and B2 is 60 k-ft. Using the beam dimensions given in Fig. P5-7, determine the required area of tension reinforcement to satisfy all of the ACI Code requirements for strength and minimum reinforcement area. Select bars and provide a sketch of your final section design.

Video Solution

5-14 Assume the maximum factored negative moment at the face of column B2 for the floor beam along column line 2 is −120 k-ft. Using the beam and slab dimensions given in Fig. P5-7, determine the required area of tension reinforcement to satisfy all the ACI Code requirements for strength and minimum reinforcement area. Select bars and provide a sketch of your final section design.

5-15 Assume the maximum factored negative moment at support *n* of the floor beam *m–n–o–p* is −150 k-ft. Using the design procedure for singly reinforced beam sections given in Section 5-3 (design of beams when section dimensions are not known), determine the beam dimensions and select the required area of tension reinforcement to satisfy all the ACI Code requirements for strength and minimum reinforcement area. Select bars and provide a sketch of your final section design.

Video Solution

5-16 Assume the maximum factored negative moment at the face of column C2 for the girder along column line C is −250 k-ft. Using the design procedure given in Section 5-4 for the design of doubly reinforced sections, determine the beam dimensions and select the required areas of tension and compression reinforcement to satisfy all the ACI Code requirements for strength and minimum reinforcement area. Select all bars and provide a sketch of your final section design.

5-17 For the one-way slab shown in Fig. P5-7, assume the maximum negative moment at support *c* is −3.3 k-ft/ft and the maximum factored positive moment at midspan point *b* is 2.4 k-ft/ft.

Video Solution

(a) Using the given slab thickness of 6 in., determine the required reinforcement size and spacing at both of these locations to satisfy ACI Code flexural strength requirements. Be sure to check the ACI Code requirements for minimum flexural reinforcement in slabs.

(b) At both locations, determine the required bar size and spacing to be provided in the transverse direction to satisfy ACI Code Table 7.6.1.1 requirements for minimum shrinkage and temperature reinforcement.

(c) For both locations, provide a sketch of the final design of the slab section.

REFERENCES

5-1 R. C. Hibbeler, "Structural Analysis," Seventh Edition, Pearson Prentice Hall, 2009, p. 224.

5-2 *Minimum Design Loads for Buildings and Other Structures*, ASCE Standard, ASCE/SEI 7-10, American Society of Civil Engineers, Reston, VA, 2010.

5-3 ACI Committee 309, "Guide for Consolidation of Concrete," ACI 309R-05, *ACI Manual of Concrete Practice*, American Concrete Institute, Farmington Hills, MI.

5-4 P. W. Birkeland and L. J. Westhoff, "Dimensional Tolerance—Concrete," State-of-Art Report 5, Technical Committee 9, *Proceedings of International Conference on Planning and Design of Tall Buildings*, Vol. Ib, American Society of Civil Engineers, New York, 1972, pp. 845–849.

5-5 Sher-Ali Mirza and James G. MacGregor, "Variations in Dimensions of Reinforced Concrete Members," *Proceedings ACSE, Journal of the Structural Division*, Vol. 105, No. ST4, April 1979, pp. 751–766.

5-6 Joint ACI/TMS Committee 216, "Code Requirements for Determining Fire Resistance of Concrete and Masonry Construction Assemblies, ACI 216.1-07/TMS-0216-07," *ACI Manual of Concrete Practice*, American Concrete Institute, Farmington Hills, MI.

5-7 R. Ian Gilbert, "Shrinkage Cracking in Fully-Restrained Concrete Members," *ACI Structural Journal*, Vol. 89, No. 2, March–April 1992, pp. 141–150.

5-8 *Building Movements and Joints*, Engineering Bulletin EB 086.10B, Portland Cement Association, Skokie, IL, 1982, 64 pp.

6

Shear in Beams

6-1 INTRODUCTION

A beam resists loads primarily by means of internal moments, M, and shears, V, as shown in Fig. 6-1. In the design of a reinforced concrete member, flexure is usually considered first, leading to the size of the section and the arrangement of reinforcement to provide the necessary moment resistance. Limits are placed on the amounts of flexural reinforcement to ensure that if failure was to occur, it would develop gradually, giving warning to the occupants. The beam is then proportioned for shear. Because a shear failure is frequently sudden and brittle, as suggested by the damage sustained by the building in Fig. 6-2 [6-1], the design for shear must ensure that the shear strength equals or exceeds the flexural strength at all points in the beam.

The manner in which shear failures can occur varies widely with the dimensions, geometry, loading, and properties of the members. For this reason, there is no unique way to design for shear. In this chapter, we deal with the internal shear force, V, in relatively slender beams and the effect of shear on the behavior and strength of beams. Examples of the design of such beams for shear are given in this chapter. Footings and two-way slabs supported on isolated columns develop shearing stresses on sections around the circumference of the columns, leading to failures in which the column and a conical piece of the slab punch through the slab (Chapter 13). Short, deep members such as brackets, corbels, and deep beams transfer shear to the support by in-plane compressive stresses rather than shear stresses. Such members are considered in Chapter 17.

Chapter 18 of the ACI Code gives special rules for shear reinforcement in members resisting seismic loads. These are reviewed in Chapter 19.

This chapter uses three different models of the shear strength of beams.

1. The stresses in uncracked beams are presented to explain the onset of shear cracking.

2. The ACI Code design procedure for shear in beams is presented and illustrated by examples.

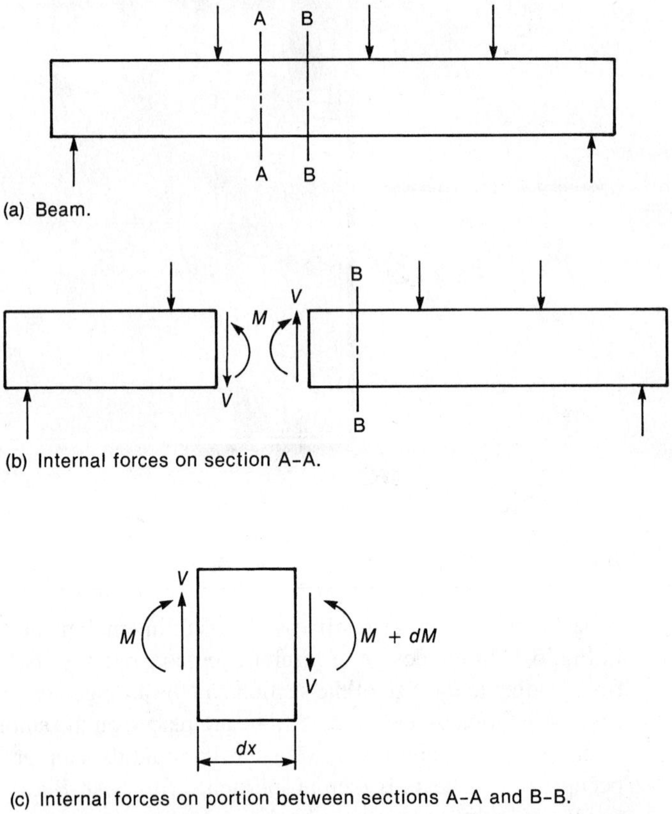

Fig. 6-1
Internal forces in a beam.

Fig. 6-2
Shear failure: U.S. Air Force warehouse. Note the small size and large spacing of the vertical web reinforcement that has fractured. (Photograph courtesy of C. P. Siess.)

3. Several comprehensive models of cracked beams loaded in shear are reviewed, on the basis of recent revisions to shear design theory. These models are mentioned because they give a more comprehensive explanation of the shear strength of reinforced concrete members [6-2].

Items 1 and 2 in this list are included to provide background for the ACI Code design methods. Item 3 shows the effects of other variables that affect the shear strength of slender beams.

6-2 BASIC THEORY

Stresses in an Uncracked Elastic Beam

From the free-body diagram in Fig. 6-1c, it can be seen that $dM/dx = V$. Thus shear forces and shear stresses will exist in those parts of a beam where the moment changes from section to section. By the traditional theory for *homogeneous*, *elastic*, *uncracked* beams, we can calculate the shear stresses, v, on elements 1 and 2 cut out of a beam (Fig. 6-3a), using the equation

$$v = \frac{VQ}{Ib} \tag{6-1}$$

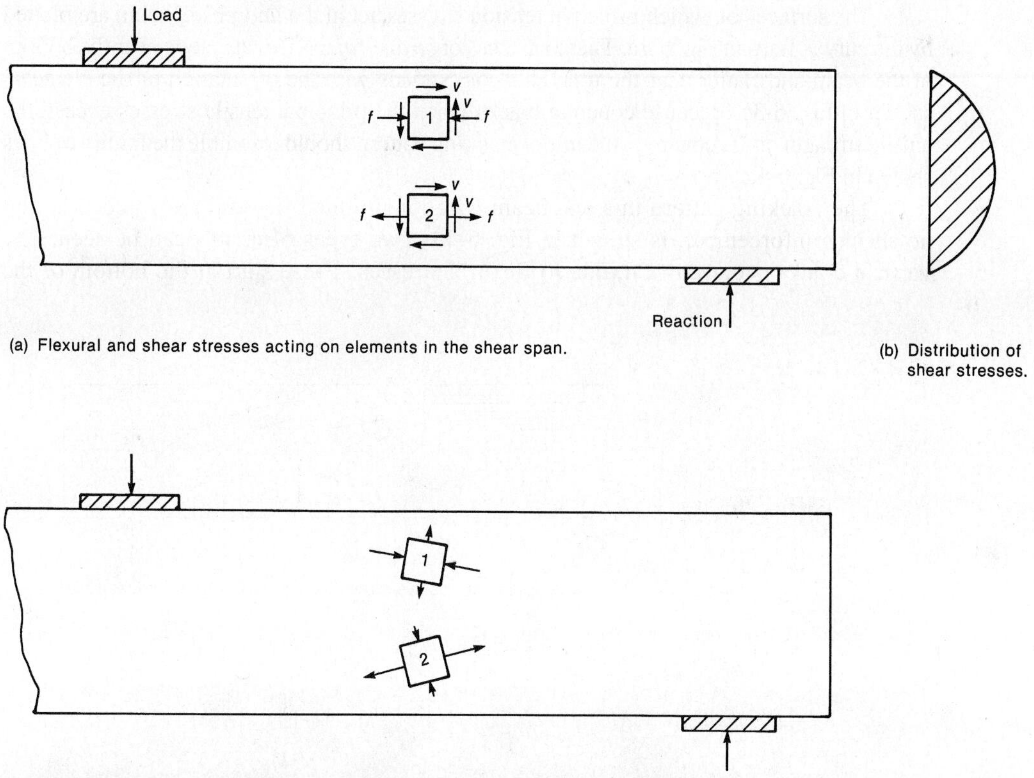

(a) Flexural and shear stresses acting on elements in the shear span.

(b) Distribution of shear stresses.

(c) Principal stresses on elements in shear span.

Fig. 6-3
Normal, shear, and principal stresses in a homogeneous uncracked beam.

where

V = shear force on the cross section

I = moment of inertia of the cross section

Q = first moment about the centroidal axis of the part of the cross-sectional area lying farther from the centroidal axis than the point where the shear stresses are being calculated

b = width of the member at the depth within the section where the stresses are being calculated

Equal shearing stresses exist on both the horizontal and vertical planes through an element, as shown in Fig. 6-3a. The shear stresses on the top and bottom of the elements cause a clockwise couple, and those on the vertical sides of the element cause a counterclockwise couple. These two couples are equal and opposite in magnitude and hence cancel each other out. The horizontal shear stresses are important in the design of construction joints, web-to-flange joints, and regions adjacent to holes in beams. For an *uncracked rectangular* beam, Eq. (6-1) gives the distribution of shear stresses shown in Fig. 6-3b.

The elements in Fig. 6-3a are subjected to combined normal stresses due to flexure, f, and shearing stresses, v. The largest and smallest normal stresses acting on such an element are referred to as *principal stresses*. The principal stresses and the planes they act on are found by using a Mohr's circle for stress, as explained in any mechanics-of-materials textbook. The orientations of the principal stresses on the elements in Fig. 6-3a are shown in Fig. 6-3c.

The surfaces on which principal tension stresses act in the *uncracked* beam are plotted by the curved lines in Fig. 6-4a. These surfaces or *stress trajectories* are steep near the bottom of the beam and flatter near the top. This corresponds with the orientation of the elements shown in Fig. 6-3c. Because concrete cracks when the principal tensile stresses exceed the tensile strength of the concrete, the initial cracking pattern should resemble the family of lines shown in Fig. 6-4a.

The cracking pattern in a test beam with longitudinal flexural reinforcement, but no shear reinforcement, is shown in Fig. 6-4b. Two types of cracks can be seen. The vertical cracks occurred first, due to flexural stresses. These start at the bottom of the

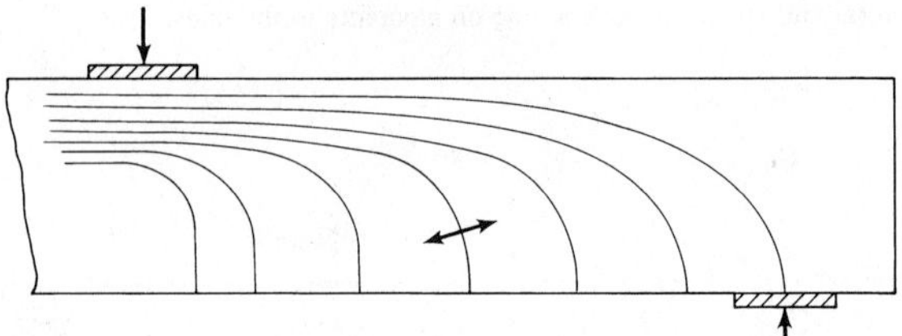

(a) Principal compressive stress trajectories in an uncracked beam.

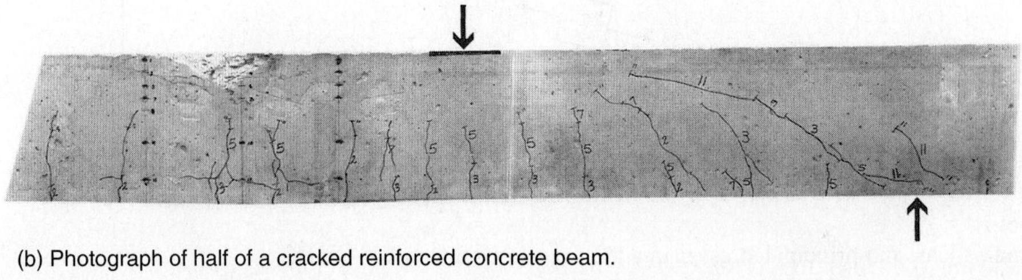

(b) Photograph of half of a cracked reinforced concrete beam.

Fig. 6-4
Principal compressive stress trajectories and inclined cracks. (Photograph courtesy of J. G. MacGregor.)

beam where the flexural stresses are the largest. The inclined cracks near the ends of the beam are due to combined shear and flexure. These are commonly referred to as *inclined cracks*, *shear cracks*, or *diagonal tension cracks*. Such a crack must exist before a beam can fail in shear.

Although there is a similarity between the planes of maximum principal tensile stress and the cracking pattern, this relationship is by no means perfect. In reinforced concrete beams, flexural cracks generally occur before the principal tensile stresses at midheight become critical. Once a flexural crack has occurred, the tensile stress perpendicular to the crack drops to zero. To maintain equilibrium, a major redistribution of stresses is necessary. As a result, the onset of inclined cracking in a beam cannot be predicted from the principal stresses unless shear cracking precedes flexural cracking. This rarely happens in reinforced concrete, but it does occur in some prestressed concrete beams.

Average Shear Stress between Cracks

The initial stage of cracking starts with vertical cracks which, with increasing load, extend in a diagonal manner, as shown in Fig. 6-4b. The equilibrium of the section of beam between two such cracks (Fig. 6-5b) can be written as

$$T = \frac{M}{jd} \qquad \text{and} \qquad T + \Delta T = \frac{M + \Delta M}{jd}$$

or

$$\Delta T = \frac{\Delta M}{jd}$$

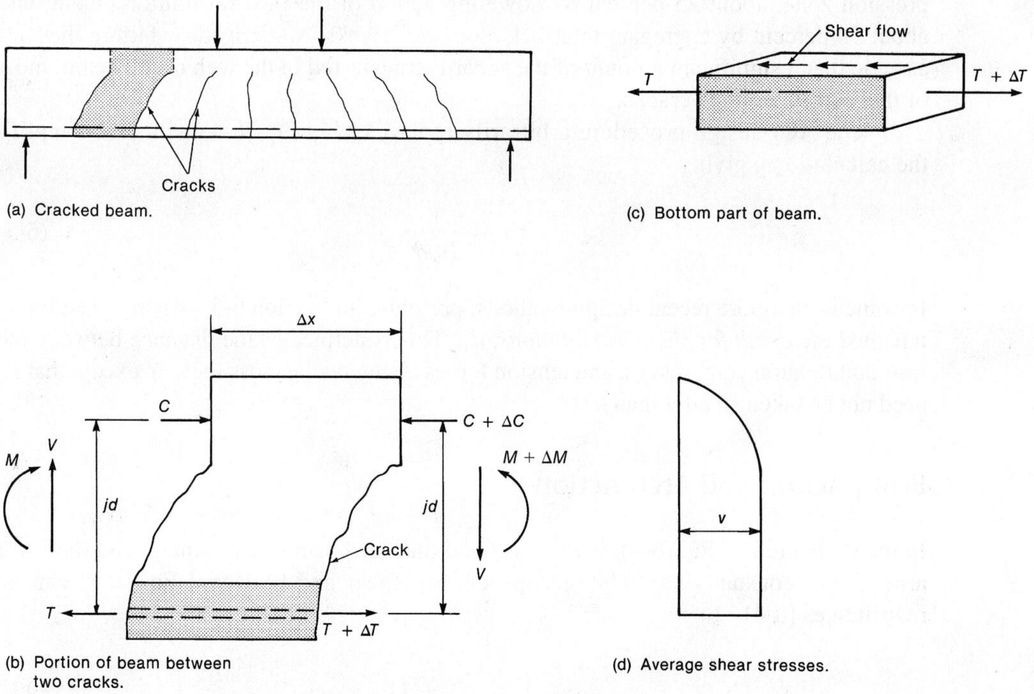

Fig. 6-5
Calculation of average shear stress between cracks.

where jd is the flexural lever arm, which is assumed to be constant. For moment equilibrium of the element,

$$\Delta M = V\Delta x \tag{6-2}$$

and

$$\Delta T = \frac{V\Delta x}{jd} \tag{6-3}$$

If the shaded portion of Fig. 6-5b is isolated as shown in Fig. 6-5c, the force ΔT must be transferred by horizontal shear stresses on the top of the element. The *average* value of these stresses below the top of the crack is

$$v = \frac{\Delta T}{b_w \Delta x}$$

or

$$v = \frac{V}{b_w jd} \tag{6-4}$$

where $jd \simeq 0.9d$ and b_w is the thickness of the web. The distribution of *average* horizontal shear stresses is shown in Fig. 6-5d. Because the vertical shear stresses on an element are equal to the horizontal shear stresses on the same element, the distribution of vertical shear stresses will be as shown in Fig. 6-5d. This assumes that about 30 percent of the shear is transferred in the compression zone. The rest of the shear is transferred across the cracks. In 1970, Taylor [6-3] reported tests of beams without web reinforcement in which he found that about 25 percent of the shear was transferred by the compression zone, about 25 percent by doweling action of the flexural reinforcement, and about 50 percent by aggregate interlock along the cracks. Modern shear failure theories assume that a significant amount of the shear is transferred in the web of the beam, most of this across inclined cracks.

The ACI design procedure arbitrarily replaces jd in Eq. (6-4) with d to simplify the calculations, giving

$$v = \frac{V}{b_w d} \tag{6-5}$$

In some of the more recent design methods, presented in Section 6-5, jd is retained but is renamed the *depth for shear calculations*, d_v. This is defined as the distance between the resultant flexural compression and tension forces acting on the cross section, except that d_v need not be taken smaller than $0.9d$.

Beam Action and Arch Action

In the derivation of Eq. (6-4), it was assumed that the beam was prismatic and the lever arm jd was constant. The relationship between shear and bar force Eq. (6-3) can be rewritten as [6-4]

$$V = \frac{d}{dx}(Tjd) \tag{6-6}$$

which can be expanded as

$$V = \frac{d(T)}{dx}jd + \frac{d(jd)}{dx}T \tag{6-7}$$

Two extreme cases can be identified. If the lever arm, *jd*, remains constant, as assumed in normal elastic beam theory, then

$$\frac{d(jd)}{dx} = 0 \quad \text{and} \quad V = \frac{d(T)}{dx}jd$$

where $d(T)/dx$ is the *shear flow* across any horizontal plane between the reinforcement and the compression zone, as shown in Fig. 6-5c. For beam action to exist, this shear flow must exist.

The other extreme occurs if the shear flow, $d(T)/dx$, equals zero, giving

$$V = T\frac{d(jd)}{dx}$$

or

$$V = C\frac{d(jd)}{dx}$$

This occurs if the shear flow cannot be transmitted, because the steel is unbonded, or if the transfer of shear flow is disrupted by an inclined crack extending from the load to the reactions. In such a case, the shear is transferred by *arch action* rather than beam action, as illustrated in Fig. 6-6. In this member, the compression force *C* in the inclined strut and the tension force *T* in the reinforcement are constant over the length of the shear span.

Shear Reinforcement

In Chapter 4, we saw that horizontal reinforcement was required to restrain the opening of a vertical flexural crack, as shown in Fig. 6-7a. An inclined crack opens approximately perpendicular to itself, as shown in Fig. 6-7b, and either a combination of horizontal flexural reinforcement and inclined reinforcement (Fig. 6-7c) or a combination of horizontal and vertical reinforcement (Fig. 6-7d) is required to restrain it from opening too wide. The inclined or vertical reinforcement is referred to as *shear reinforcement* or *web reinforcement* and may be provided by inclined or vertical *stirrups*. Most often, vertical stirrups are used in North America. Inclined stirrups are not effective in beams resisting shear reversals, such as seismic loads, because the reversals of shear will cause cracking parallel to the inclined reinforcement, rendering it ineffective.

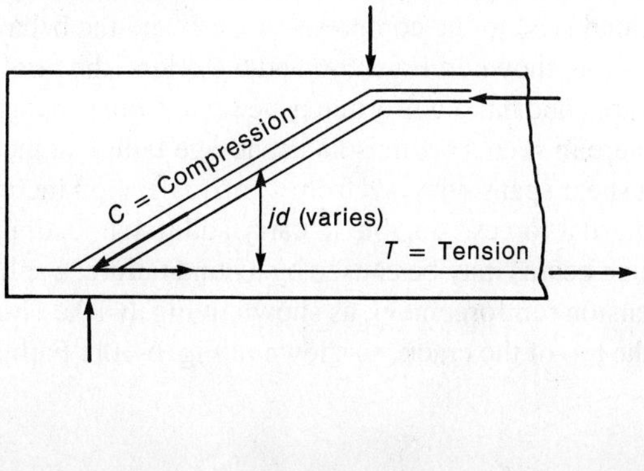

Fig. 6-6
Arch action in a beam.

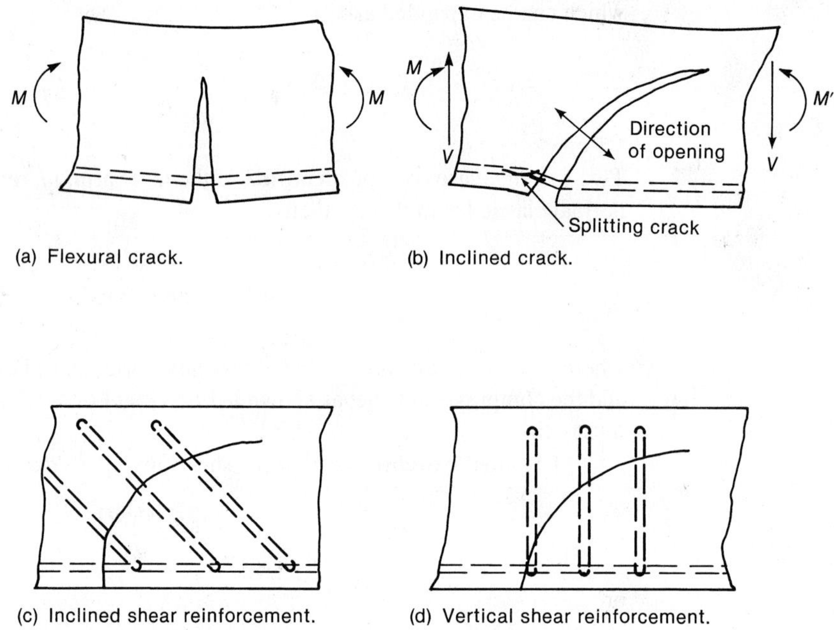

Fig. 6-7
Inclined cracks and shear reinforcement.

6-3 BEHAVIOR OF BEAMS FAILING IN SHEAR

The behavior of beams failing in shear varies widely, depending on the relative contributions of beam action and arch action and the amount of web reinforcement.

Behavior of Beams without Web Reinforcement

The moments and shears at inclined cracking and failure of rectangular beams without web reinforcement are plotted in Figs. 6-8b and 6-8c as a function of the ratio of the shear span a to the effective depth d. (See Fig. 6-8a.) The beam cross section remains constant as the span is varied. The maximum moment (and shear) that can be developed correspond to the nominal moment capacity, M_n, of the cross section plotted as a horizontal line in Fig. 6-8b. The shaded areas in this figure show the reduction in strength due to shear. Web reinforcement is normally provided to ensure that the beam reaches the full flexural capacity, M_n [6-5].

Figure 6-8b suggests that the shear spans can be divided into three types: short, slender, and very slender shear spans. The term *deep beam* is also used to describe beams with short shear spans. Very short shear spans, with a/d from 0 to 1, develop inclined cracks joining the load and the support. These cracks, in effect, destroy the horizontal shear flow from the longitudinal steel to the compression zone, and the behavior changes from beam action to arch action, as shown in Figs. 6-6 and 6-9. Here, the reinforcement serves as the tension tie of a tied arch and has a uniform tensile force from support to support. The most common mode of failure in such a beam is an anchorage failure at the ends of the tension tie.

Short shear spans with a/d from 1 to 2.5 develop inclined cracks and, after a redistribution of internal forces, are able to carry additional load, in part by arch action. The final failure of such beams may be caused by a bond failure, a splitting failure, or a dowel failure along the tension reinforcement, as shown in Fig. 6-10a, or by crushing of the compression zone over the top of the crack, as shown in Fig. 6-10b. Failures in the flexural tension zone

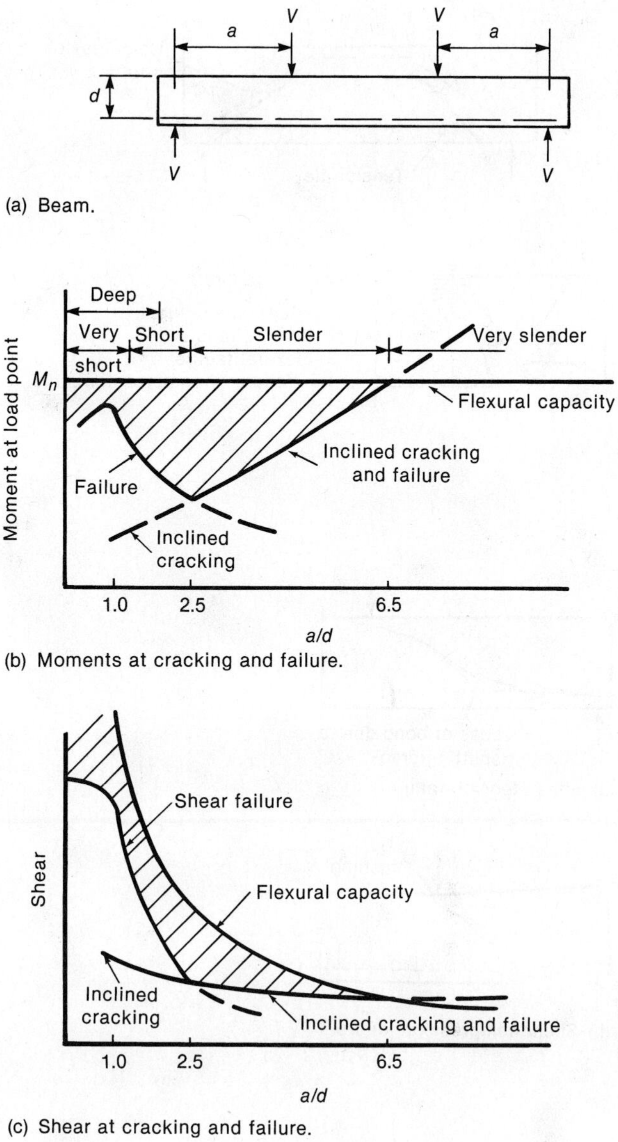

Fig. 6-8
Effect of a/d ratio on shear strength of beams without stirrups.

are referred to as *shear-tension* failures (Fig. 6-10a) and those in the flexural compression zone are referred to as a *shear-compression* failures (Fig. 6-10b). Because the inclined crack generally extends higher into the beam than does a flexural crack, failure occurs at less than the flexural moment capacity.

In slender shear spans, those having a/d from 2.5 to about 6, the inclined cracks disrupt equilibrium to such an extent that the beam fails at the inclined cracking load, as shown in Fig. 6-8b. Very slender beams, with a/d greater than about 6, will fail in flexure prior to the formation of inclined cracks.

Figures 6-9 and 6-10 come from [6-5], which presents an excellent discussion of the behavior of beams failing in shear and the factors affecting their strengths. It is important to note that, for short and very short beams, a major portion of the load capacity after inclined cracking is due to load transfer by the compression struts shown

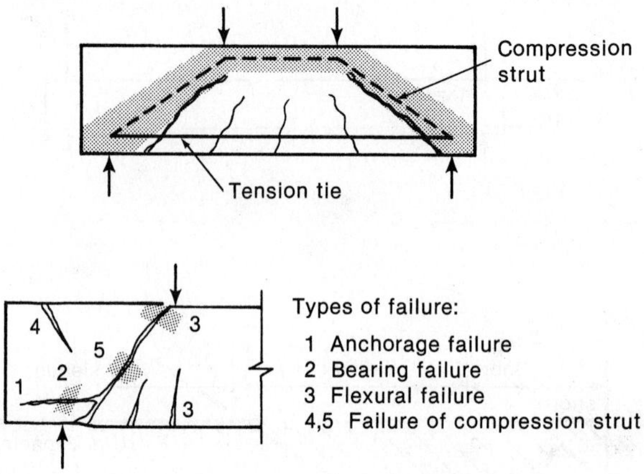

Fig. 6-9
Modes of failure of deep beams, a/d = 0.5 to 2.0. (Adapted from [6-5].)

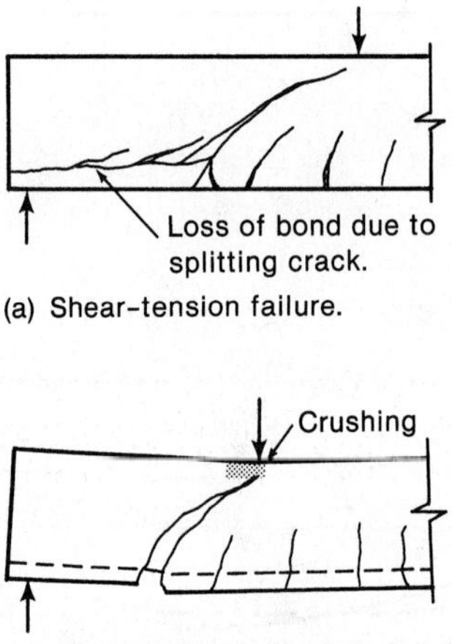

(a) Shear-tension failure.

(b) Shear compression failure.

Fig. 6-10
Modes of failure of short shear spans, a/d = 1.5 to 2.5. (Adapted from [6-5].)

in Fig. 6-9. If the beam is not loaded on the top and supported on the bottom in the manner shown in Fig. 6-9, these compression struts will not form and failure occurs at, or close to, the inclined cracking load.

Because the moment at the point where the load is applied is $M = Va$ for a beam loaded with concentrated loads, as shown in Fig. 6-8a, Fig. 6-8b can be replotted in terms of shear capacity, as shown in Fig. 6-8c. The shear corresponding to a flexural failure is the upper curved line. If stirrups are not provided, the beam will fail at the shear given by the "shear failure" line. This is roughly constant for a/d greater than about 2. Again, the shaded area indicates the loss in capacity due to shear. Note that the inclined cracking loads of the short shear spans and slender shear spans are roughly constant. This is recognized in typical designs by ignoring a/d in the equations for the shear at inclined cracking. In the case of slender beams, inclined cracking causes an immediate shear failure if no web reinforcement is provided.

Inclined Cracking

Inclined cracks must exist before a shear failure can occur. Inclined cracks form in the two different ways shown in Fig. 6-11. In thin-walled I beams in which the a/d ratio is small, the shear stresses in the web are high, while the flexural stresses are low. In a few extreme cases and in some prestressed beams, the principal tension stresses at the neutral axis may exceed those at the bottom flange. In such a case, a *web-shear crack* occurs (Fig. 6-11a). The corresponding inclined cracking shear can be calculated as the shear necessary to

(a) Web-shear cracks.

(b) Flexure-shear cracks.

Fig. 6-11
Types of inclined cracks.
(Photographs courtesy of J. K. Wight.)

cause a principal tensile stress equal to the tensile strength of the concrete at the centroid of the beam.

In most reinforced concrete beams, however, flexural cracks occur first and extend more or less vertically into the beam, as shown in Fig. 6-4b or 6-11b. These alter the state of stress in the beam, causing a stress concentration near the head of the crack. In time, either

1. the flexural cracks extend to become *flexure-shear cracks* (Fig. 6-11b), or

2. flexure-shear cracks extend into the uncracked region above the flexural cracks (Fig. 6-4b).

Flexure-shear cracking *cannot* be predicted by calculating the principal stresses in an uncracked beam. For this reason, empirical equations have been derived to calculate the flexure-shear cracking load.

Flexural-shear cracks in a T beam loaded to produce positive and negative moments are shown in Fig. 5-21a. The slope of the inclined cracks in the negative-moment regions changes direction over the intermediate support because the shear force changes sign from one side of the support to the other side.

Internal Forces in a Beam without Stirrups

The forces transferring shear across an inclined crack in a beam without stirrups are illustrated in Fig. 6-12. Shear is transferred across line *A–B–C* by V_{cy}, the shear in the compression zone, by V_{ay}, the vertical component of the shear transferred across the crack by interlock of the aggregate particles on the two faces of the crack, and by V_d, the dowel action of the longitudinal reinforcement. Immediately after inclined cracking, as much as 40 to 60 percent of the total shear is carried by V_d and V_{ay} together [6-3].

Considering the *D–E–F* portion of the beam below the crack and summing moments about the reinforcement at point *E* shows that V_d and V_a cause a moment about *E* that must be equilibrated by a compression force C'_1. Horizontal force equilibrium on section *A–B–D–E* shows that $T_1 = C_1 + C'_1$, and finally, T_1 and $C_1 + C'_1$ must equilibrate the external moment at this section.

As the crack widens, V_a decreases, increasing the fraction of the shear resisted by V_{cy} and V_d. The dowel shear, V_d, leads to a splitting crack in the concrete along the reinforcement (Fig. 6-10a). When this crack occurs, V_d drops, approaching zero. When V_a and V_d disappear, so do V'_{cy} and C'_1, with the result that all the shear and compression are transmitted in the depth *AB* above the crack. At this point in the life of the beam, the section *A–B* is often too shallow to resist the compression forces needed for equilibrium. As a result, this region crushes or buckles upward (shear-compression failure).

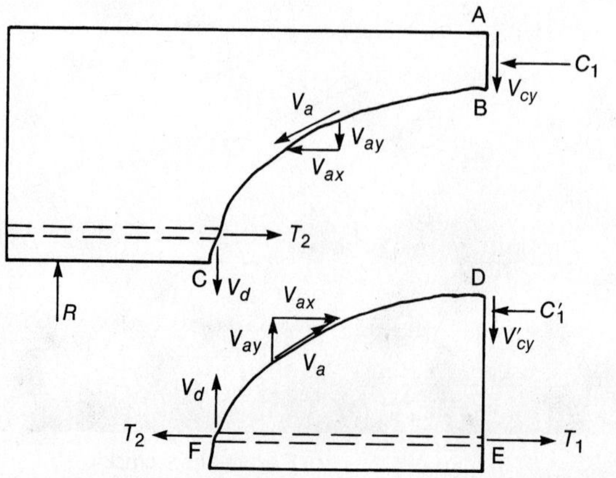

Fig. 6-12
Internal forces in a cracked beam without stirrups.

Note also that if $C_1' = 0$, then $T_2 = T_1$, and as a result, $T_2 = C_1$. In other words, the inclined crack has made the tensile force at point C a function of the moment at section A–B–D–E. This shift in the tensile force must be considered in detailing the bar cutoff points and in anchoring the bars.

The shear failure of a slender beam without stirrups is sudden and dramatic. This is evident from Fig. 6-2. Although this beam had stirrups (which have broken and are hanging down from the upper part of the beam), they were too small to alter the failure mode.

Factors Affecting the Shear Strength of Beams without Web Reinforcement

Beams without web reinforcement will fail when inclined cracking occurs or shortly afterward. For this reason, the shear capacity of such members is taken equal to the inclined cracking shear. The inclined cracking load of a beam is affected by five principal variables, some included in design equations and others not.

Tensile Strength of Concrete. The inclined cracking load is a function of the tensile strength of the concrete, f_{ct}. The stress state in the web of the beam involves biaxial principal tension and compression stresses, as shown in Fig. 6-3. A similar biaxial state of stress exists in a split-cylinder tension test (Fig. 3-9), and the inclined cracking load is frequently related to the strength from such a test. As discussed earlier, the flexural cracking that precedes the inclined cracking disrupts the elastic-stress field to such an extent that inclined cracking occurs at a principal tensile stress roughly half of f_{ct} for the uncracked section.

Longitudinal Reinforcement Ratio, ρ_w. Figure 6-13 presents the shear capacities (psi units) of simply supported beams without stirrups as a function of the steel ratio, $\rho_w = A_s/b_w d$. The practical range of ρ_w for beams developing shear failures is about 0.0075 to 0.025. In this range, the shear strength is approximately

$$V_c = 2\sqrt{f_c'}\,b_w d \quad \text{(lb)} \tag{6-8a}$$

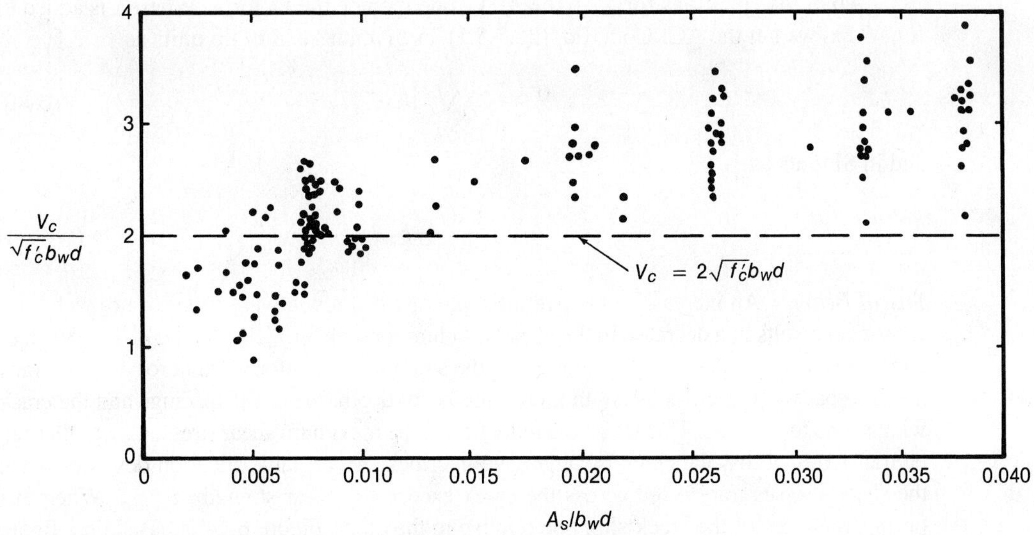

Fig. 6-13
Effect of reinforcement ratio, ρ_w, on shear capacity, V_c, of beams constructed with normal-weight concrete and without stirrups [6-6].

or, in SI units,

$$V_c = \frac{\sqrt{f'_c}\, b_w\, d}{6} \quad \text{(N)} \tag{6-8Ma}$$

as indicated by the horizontal dashed line in Fig. 6-13. This equation tends to overestimate V_c for beams with small longitudinal steel percentages [6-6].

When the steel ratio, ρ_w, is small, flexural cracks extend higher into the beam and open wider than would be the case for large values of ρ_w. An increase in crack width causes a decrease in the maximum values of the components of shear, V_d and V_{ay}, that are transferred across the inclined cracks by dowel action or by shear stresses on the crack surfaces. Eventually, the resistance along the crack drops below that required to resist the loads, and the beam fails suddenly in shear.

Shear Span-to-Depth Ratio, a/d. The shear span-to-depth ratio, *a/d* or *M/Vd*, affects the inclined cracking shears and ultimate shears of portions of members with *a/d* less than 2, as shown in Fig. 6-8c. Such shear spans are "deep beam" regions and are discussed in Chapter 17. For longer shear spans, *a/d* has little effect on the inclined cracking shear (Fig. 6-8c) and can be neglected.

Lightweight Aggregate Concrete. As discussed in Chapter 3, lightweight aggregate concrete has a lower tensile strength than normal weight concrete for a given concrete compressive strength. Because the shear strength of a concrete member without shear reinforcement is directly related to the tensile strength of the concrete, equations for V_c similar to Eq. (6-8a) must be modified for members constructed with lightweight aggregate concrete. This is handled in the ACI Code through the introduction of the factor, λ, which accounts for the difference for the tensile strength of lightweight concrete. ACI Code Table 19.2.4.2 states that for sand-lightweight concrete (i.e., concrete with normal weight small aggregates, and lightweight large aggregates), λ is to be taken as 0.85. For all lightweight concrete (i.e., both the large and small aggregate are lightweight materials), λ is taken as 0.75. Alternatively, if the splitting strength, f_{ct}, for the lightweight concrete is specified, λ can be taken as $f_{ct}/6.7\sqrt{f'_c} \leq 1.0$.

All of the equations for V_c in the ACI Code include the factor λ. When λ is added to Eq. (6-8a) we get the ACI Code Eq. (22.5.5.1), as given here in in.-lb units as

$$V_c = 2\lambda\sqrt{f'_c}\, b_w d \tag{6-8b}$$

and in SI units as

$$V_c = \frac{\lambda\sqrt{f'_c}}{6} b_w d \tag{6-8bM}$$

Size of Beam. An increase in the overall depth of a beam with very little (or no) web reinforcement results in a decrease in the shear at failure for a given f'_c, ρ_w, and *a/d*. The width of an inclined crack depends on the product of the strain in the reinforcement crossing the crack and the spacing of the cracks. With increasing beam depth, the crack spacings and the crack widths tend to increase. This leads to a reduction in the maximum shear stress, $v_{ci\,\max}$, that can be transferred across the crack by aggregate interlock. An unstable situation develops when the shear stresses transferred across the crack exceed the shear strength, $v_{ci\,\max}$. When this occurs, the faces of the crack slip, one relative to the other. Figure 6-14 is based on a figure presented by Collins and Kuchma [6-7]. It shows a significant decrease in the shear strengths of geometrically similar, uniformly loaded beams with effective depths *d* ranging from 4 in. to 118 in. and made with 0.1-in., 0.4-in., and 1-in. maximum size coarse aggregate.

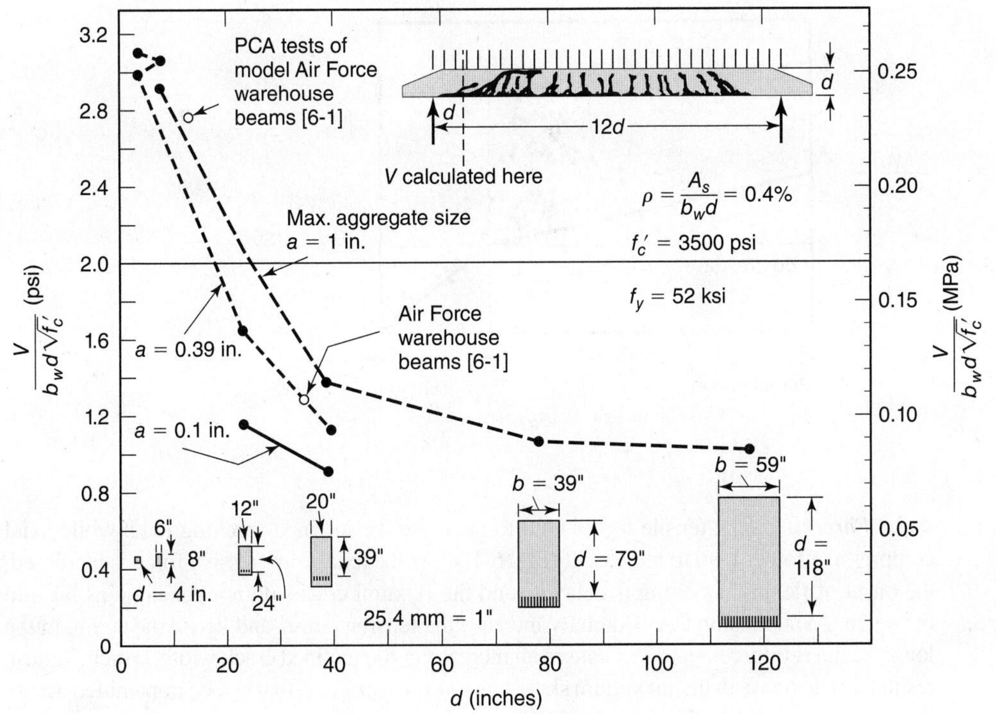

Fig. 6-14
Effect of beam depth, d, on failure shear for beams of various sizes. (From [6-7].)

The dashed lines in Fig. 6-14 show the variation in shear strength of beams without stirrups in tests. The beams were uniformly loaded and simply supported as shown in the inset. Each black circular dot in the figure corresponds to the strength of a beam having the section plotted directly below it.

The open circle labelled Air Force Warehouse Beams refers to the beam in Fig. 6-2. There is very good agreement between the shears at failure of this beam and the dashed lines in Fig. 6-14. The horizontal line at $V_u/\sqrt{f'_c}b_w d = 2.0$ shows the shear V_c that the ACI Code assumes to be carried by the concrete. Figure 6-14 shows a very strong size dependency in uniformly loaded beams *without* web reinforcement [6-7].

In beams *with* at least the minimum required web reinforcement, the web reinforcement holds the crack faces together so that the shear transfer across the cracks by aggregate interlock is not lost. As a result, the reduction in shear strength due to size shown in Fig. 6-14 is not observed in beams with web reinforcement [6-7].

Using fracture mechanics, Bazant [6-8] has explained the size effect on the basis of energy release at cracking. The amount of energy released increases with an increase in size, particularly in the depth. Although a portion of the size effect results from energy release, the author believes crack-width explanation of size effects fits the test data trends more closely.

The two open circles plotted in Fig. 6-14 represent the shear stress at failure in the roof beams in the U.S. Air Force warehouse collapse [6-1] shown in Fig. 6-2, and the shear stress at failure of a scale model of these beams, respectively. The fact that these points fall close to the experimental data suggests that the U.S. Air Force warehouse failures were strongly affected by their size [6-7].

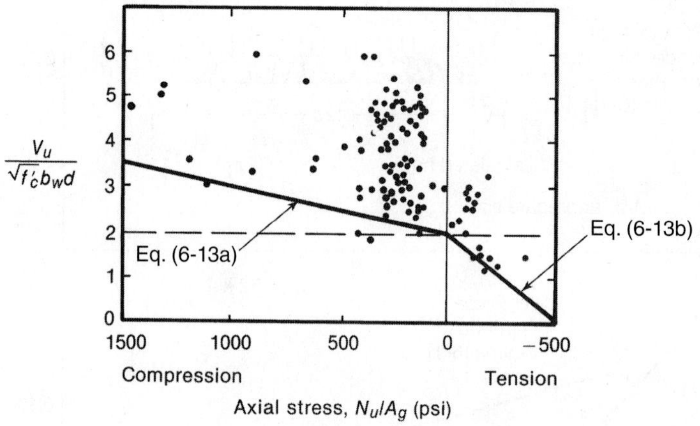

Fig. 6-15
Effect of axial loads on inclined cracking shear. From [6-6].

Axial Forces. Axial tensile forces tend to decrease the inclined cracking load, while axial compressive forces tend to increase it (Fig. 6-15). As the axial compressive force is increased, the onset of flexural cracking is delayed, and the flexural cracks do not penetrate as far into the beam. Axial tension forces directly increase the tension stress, and hence the strain, in the longitudinal reinforcement. This causes an increase in the inclined crack width, which, in turn, results in a decrease in the maximum shear tension stress, $v_{ci\,\max}$, that can be transmitted across the crack. This reduces the shear failure load.

A similar increase is observed in prestressed concrete beams. The compression due to prestressing reduces the longitudinal strain, leading to a higher failure load.

Coarse Aggregate Size. As the size (diameter) of the coarse aggregate increases, the roughness of the crack surfaces increases, allowing higher shear stresses to be transferred across the cracks. As shown in Fig. 6-14, a beam with 1-in. coarse aggregate and 40-in. effective depth failed at about 150 percent of the failure load of a beam with $d = 40$ in. and 0.1-in. maximum aggregate size. In high-strength concrete beams and some lightweight concrete beams, the cracks penetrate pieces of the aggregate rather than going around them, resulting in a smoother crack surface. This decrease in the shear transferred by aggregate interlock along the cracks reduces V_c [6-7].

Behavior of Beams with Web Reinforcement

Inclined cracking causes the shear strength of beams to drop below the flexural capacity, as shown in Figs. 6-8b and 6-8c. The purpose of web reinforcement is to ensure that the full flexural capacity can be developed.

Prior to inclined cracking, the strain in the stirrups is equal to the corresponding strain of the concrete. Because concrete cracks at a very small strain, the stress in the stirrups prior to inclined cracking will not exceed 3 to 6 ksi. Thus, stirrups do not prevent inclined cracks from forming; they come into play only after the cracks have formed.

The forces in a beam with stirrups and an inclined crack are shown in Fig. 6-16. The terminology is the same as in Fig. 6-12. The shear transferred by tension in the stirrups, V_s, does not disappear when the crack opens wider, so there will always be a compression force C'_1 and a shear force V'_{cy} acting on the part of the beam below the crack. As a result, T_2 will be less than T_1, the difference depending on the amount of web reinforcement. The force T_2 will, however, be larger than the flexural tension $T = M/jd$ based on the moment at C.

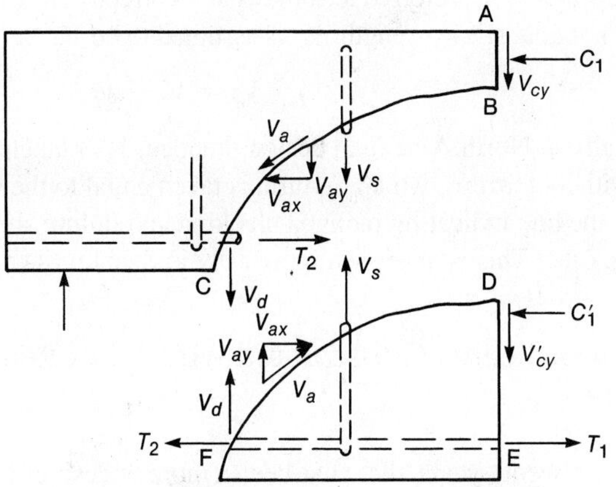

Fig. 6-16
Internal forces in a cracked beam with stirrups.

The loading history of such a beam is shown qualitatively in Fig. 6-17. The components of the internal shear resistance must equal the applied shear, indicated by the upper 45° line. Prior to flexural cracking, all the shear is carried by the uncracked concrete. Between flexural and inclined cracking, the external shear is resisted by V_{cy}, V_{ay}, and V_d. After inclined cracking, the stirrups (V_s) start to participate in resisting applied shear. Eventually, the stirrups crossing the crack yield, and V_s stays constant for higher applied shears. Once the stirrups yield, the inclined crack opens more rapidly. As the inclined crack widens, V_{ay} decreases further, forcing V_d and V_{cy} to increase at an accelerated rate, until either a splitting (dowel) failure occurs, the compression zone crushes due to combined shear and compression, or the web crushes.

Each of the components of this process except V_s has a brittle load–deflection response. As a result, it is difficult to quantify the contributions of V_{cy}, V_d, and V_{ay}. In design, these are

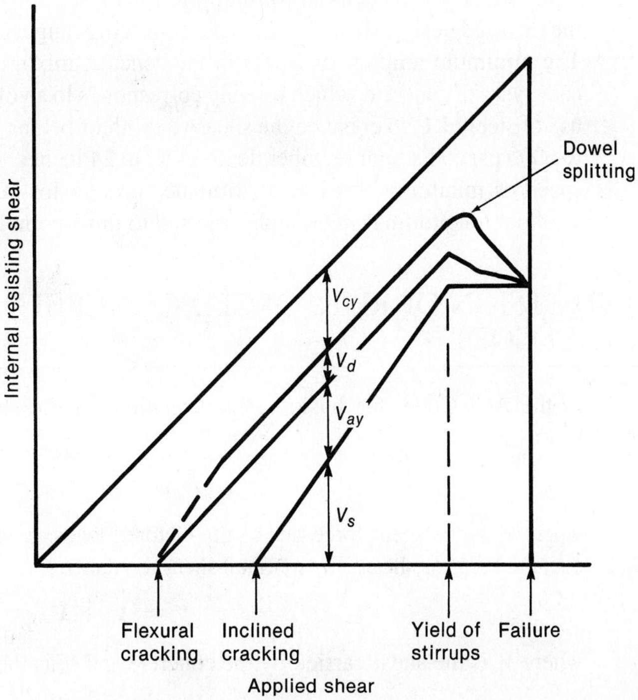

Fig. 6-17
Distribution of internal shears in a beam with web reinforcement. (From [6-5].)

lumped together as V_c, referred to somewhat incorrectly as "the shear carried by the concrete." Thus, the nominal shear strength, V_n, is assumed to be

$$V_n = V_c + V_s \tag{6-9}$$

Traditionally in North American design practice, V_c is taken equal to the failure capacity of a beam without stirrups, which, in turn, is taken equal to the inclined cracking shear, as suggested by the line indicating inclined cracking and failure shear are equal for *a/d* from 2.5 to 6.5 in Fig. 6-8c. This is discussed more fully in Section 6-4.

Behavior of Beams Constructed with Fiber Reinforced Concrete

Over the last twenty years, there has been a more widespread use of fiber reinforced concrete (FRC) in the building construction industry. Engineers are finding a variety of applications for the enhanced tensile properties and toughness of FRC. For most industrial floors and in some other structural applications, FRC with a minimum volume fraction of 0.5 percent steel fibers has been used to replace the need for minimum shrinkage and temperature reinforcement, similar to that specified in ACI Code Section 24.4. ACI Committee 544 [6-9] also has investigated other structural applications for FRC.

A large database recently was collected [6-10] to evaluate the ability of steel FRC to enhance the shear strength and deformation capacity of concrete beams that did not have normal shear reinforcement. The steel fibers, which typically are crimped or hooked, tend to increase the shear capacity by providing post-cracking tensile resistance across an inclined crack, resulting in higher aggregate interlock forces (Fig. 6-16) in a manner similar to that observed for beams with normal stirrup-type shear reinforcement. The use of steel FRC also leads to the formation of multiple inclined cracks and a less sudden shear failure than commonly is observed in tests of concrete beams without shear reinforcement.

Based on a review of the database [6-10], ACI Code Table 9.6.3.1 permits the use of steel FRC as a replacement for minimum shear reinforcement in concrete members when the factored design shear, V_u, is in the following range, $0.5\,\phi V_c < V_u \leq \phi V_c$ ($\phi = 0.75$). The minimum amount of fibers in the concrete mixture is 100 pounds of steel fibers per cubic yard of concrete, which roughly corresponds to a volume fraction of 0.75 percent. This use of steel FRC to enhance the shear strength of beams is limited to concrete strengths up to 6000 psi and a total member depth of up to 24 inches. Also, ACI Code Section 26.12.5.1 specifies minimum flexural performance criteria for beams constructed with steel FRC (without longitudinal steel) and subjected to third-point bending tests (ASTM C1609).

6-4 ANALYSIS AND DESIGN OF REINFORCED CONCRETE BEAMS FOR SHEAR—ACI CODE

In the ACI Code, the basic design equation for the shear capacity of slender concrete beams is

$$\phi V_n \geq V_u \tag{6-10}$$

where V_u is the shear force due to the factored loads, ϕ is a strength-reduction factor, taken equal to 0.75 for shear. The nominal shear resistance is

$$V_n = V_c + V_s \tag{6-9}$$

where V_c is the shear carried by the concrete and V_s is the shear carried by the stirrups.

A shear failure is said to occur when one of several shear limit states is reached. The following paragraphs list the principal shear limit states and describe how these are accounted for in the ACI Code.

Shear-Failure Limit States: Beams without Web Reinforcement

Slender beams without web reinforcement will fail when inclined cracking occurs or shortly afterward. For this reason, the shear strength of such members is taken equal to the inclined cracking shear. The factors affecting the inclined cracking load were discussed in Section 6-3.

Design Equations for the Shear Strength of Members without Web Reinforcement

In 1962, the ACI–ASCE Committee on Shear and Diagonal Tension [6-11] presented an equation for calculating the shear at inclined cracking in beams without web reinforcement:

$$V_c = \left(1.9\lambda\sqrt{f'_c} + \frac{2500\rho_w V_u d}{M_u}\right) b_w d \text{ (psi)} \qquad (6\text{-}11)$$

The symbol λ has been added to the original equation from [6-11] to make it compatible with the ACI Code equation given in Table 22.5.5.1. The derivation of this equation followed two steps. First, a rudimentary analysis of the stresses at the head of a flexural crack in a shear span was carried out to identify the significant parameters. Then, the existing test data were statistically analyzed to establish the constants, 1.9 and 2500, and to drop other terms. The data used in the statistical analysis included "short" and "slender" beams, thereby mixing data from two different behavior types. In addition, most of the beams had high longitudinal reinforcement ratios, ρ_w. Studies have suggested that Eq. (6-11) underestimates the effect of ρ_w for beams without web reinforcement and is not entirely correct in its treatment of the variable *a/d*, expressed as $V_u d/M_u$ in the equation [6-6].

On the basis of statistical studies of beam data for slender beams without web reinforcement, Zsutty [6-12] derived the following equation, which more closely models the actual effects of f'_c, ρ_w, and a/d than does Eq. (6-11):

$$v_c = 59\left(f'_c \rho_w \frac{d}{a}\right)^{1/3} \text{ (psi)} \qquad (6\text{-}12)$$

For the normal range of variables, the second term in the parentheses in Eq. (6-11) will be equal to about $0.1\sqrt{f'_c}$. If this value is substituted into Eq. (6-11), then the result is ACI Code Eq. (22.5.5.1):

$$V_c = 2\lambda\sqrt{f'_c}\, b_w d \qquad (6\text{-}8b)$$

For design, the ACI Code presents both Eqs. (6-8b) and (6-11) for computing V_c. For axially loaded members, the ACI Code modifies (6-8b) as follows:

Axial compression (ACI Code Section 22.5.6.1):

$$V_c = 2\left(1 + \frac{N_u}{2000A_g}\right)\lambda\sqrt{f'_c}\, b_w d \qquad (6\text{-}13a)$$

Axial tension (ACI Code Section 22.5.7.1):

$$V_c = 2\left(1 + \frac{N_u}{500A_g}\right)\lambda\sqrt{f'_c}b_w d \quad \text{(6-13b)}$$

In both of these equations, N_u is positive in compression (negative in tension) and $\sqrt{f'_c}$, N_u/A_g, 500, and 2000 all have units of psi. Axially loaded members are discussed more fully in Section 6-7.

Circular Cross Sections. Members with circular cross sections, such as some columns, may have to be designed for shear. When circular ties or spirals are used as web reinforcement, the calculation of V_c can be based on Eqs. (6-8b), (6-13a), and (6-13b), with b_w taken equal to the diameter of the circular section and d taken equal to the distance from the extreme compression fiber to the centroid of the longitudinal tension reinforcement, but may be taken as $0.8h$. For the contribution from ties or spirals, A_v is taken equal to twice the area, A_b, of the bar used as a circular tie or as a spiral. These definitions are based on tests reported in [6-13] and [6-14].

There is a tendency for ties located close to the inside surface of a hollow member to straighten and pull out through the inside surface of the tube. Means of preventing this must be considered in the design of hollow members.

In SI units, Eqs. (6-8b), (6-13a), and (6-13b) become, respectively,

$$V_c = \frac{\lambda\sqrt{f'_c}}{6}b_w d \quad \text{(6-8bM)}$$

For axial compression,

$$V_c = \left(1 + \frac{N_u}{14A_g}\right)\left(\frac{\lambda\sqrt{f'_c}}{6}\right)b_w d \quad \text{(6-13aM)}$$

and for axial tension,

$$V_c = \left(1 + \frac{N_u}{3.33A_g}\right)\left(\frac{\lambda\sqrt{f'_c}}{6}\right)b_w d \quad \text{(6-13bM)}$$

where N_u is positive in compression and the terms $\sqrt{f'_c}$, N_u/A_g, 14, and 3.33 all have units of MPa.

Shear Failure Limit States: Beams with Web Reinforcement

1. Failure due to yielding of the stirrups. In Fig. 6-16, shear was transferred across the surface A–B–C by shear in the compression zone, V_{cy}, by the vertical component of the aggregate interlock, V_{ay}, by dowel action, V_d, and by stirrups, V_s. In the ACI Code V_{cy}, V_{ay}, and V_d are lumped together as V_c, which is referred to as the "shear carried by the concrete." Thus, the nominal shear strength, V_n, is assumed to be

$$V_n = V_c + V_s \quad \text{(6-9)}$$

The ACI Code further assumes that V_c is equal to the shear strength of a beam without stirrups, which, in turn, is taken equal to the inclined cracking load, as given by Eqs. (6-8b), (6-11), or (6-13). It should be emphasized that taking V_c equal to the shear at inclined cracking is an *empirical* observation from tests, which is *approximately true* if it is assumed that the horizontal projection of the inclined crack is d, as shown in Fig. 6-18. If a flatter crack is used, so that $jd \cot \theta$ is greater than d, a smaller value of V_c must be used.

Figure 6-18a shows a free body between the end of a beam and an inclined crack. The horizontal projection of the crack is taken as d, suggesting that the crack is slightly flatter than 45°. If s is the stirrup spacing, the number of stirrups cut by the crack is d/s. Assuming that all the stirrups yield at failure, the shear resisted by the stirrups is given in ACI Code Eq. (22.5.10.5.3) as:

$$V_s = \frac{A_v f_{yt} d}{s} \tag{6-14}$$

If the stirrups are inclined at an angle α to the horizontal, as shown in Fig. 6-18b, the number of stirrups crossing the crack is approximately $d(1 + \cot \alpha)/s$, where s is the horizontal spacing of the stirrups. The inclined force is

$$F = A_v f_{yt}\left[\frac{d(1 + \cot \alpha)}{s}\right] \tag{6-15}$$

The shear resisted by the stirrups, V_s, is the vertical component of F, which is $F \sin \alpha$, resulting in ACI Code Eq. (22.5.10.5.4):

$$V_s = A_v f_{yt}\left(\sin \alpha + \cos \alpha\right)\frac{d}{s} \tag{6-16}$$

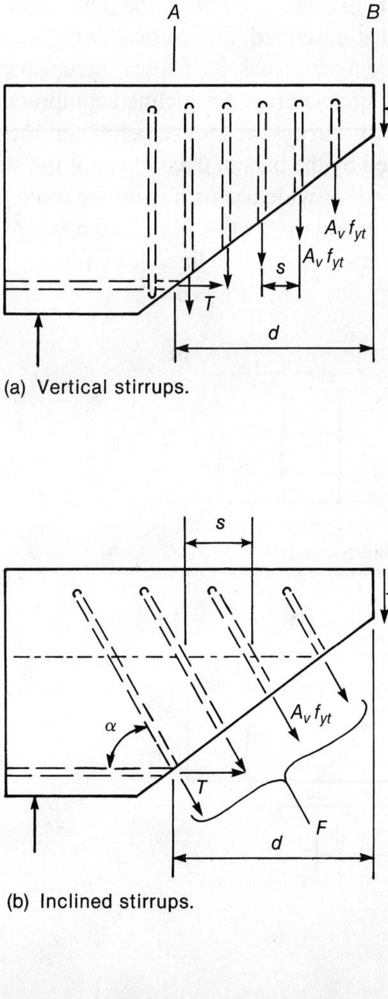

Fig. 6-18
Shear resisted by stirrups.

Figures 6-18 and 6-16 also show that the inclined crack affects the longitudinal tension force, T, making it larger than the moment diagram would suggest.

If V_u exceeds ϕV_c, stirrups must be provided so that

$$V_u \leq \phi V_n \tag{6-10}$$

where V_n is given by Eq. (6-9). In design, this is generally rearranged to the form

$$\phi V_s \geq V_u - \phi V_c \quad \text{or} \quad V_s \geq \frac{V_u}{\phi} - V_c$$

Introducing Eq. (6-14) gives the required stirrup spacing:

$$s \leq \frac{A_v f_{yt} d}{V_u/\phi - V_c} \tag{6-17}$$

This equation applies only to vertical stirrups.

Stirrups are unable to resist shear unless they are crossed by an inclined crack. For this reason, ACI Code Table 9.7.6.2.2 sets the maximum spacing of vertical stirrups as the smaller of $d/2$ or 24 in., so that each 45° crack will be intercepted by at least one stirrup (Fig. 6-19a). The maximum spacing of inclined stirrups is such that a 45° crack extending from midheight of the member to the tension reinforcement will intercept at least one stirrup, as shown in Fig. 6-19b.

If $V_u/\phi - V_c = V_s$ exceeds $4\sqrt{f'_c}b_w d$, the maximum allowable stirrup spacings are reduced to half those just described. For vertical stirrups, the maximum is the smaller of $d/4$ or 12 in. This is done for two reasons. Closer stirrup spacing leads to narrower inclined cracks and provides better anchorage for inclined compression forces.

In a wide beam with stirrups around the perimeter, inclined compression forces in the web tend to be supported by the bars in the corners of the stirrups, as shown in Fig. 6-20a. The situation is improved for wide beams if there are more than two stirrup legs, as shown in Fig. 6-20b. ACI Commentary Section R22.5.10.5 suggests that the transverse spacing of stirrup legs in wide beams should be limited by placing extra stirrups. Based primarily

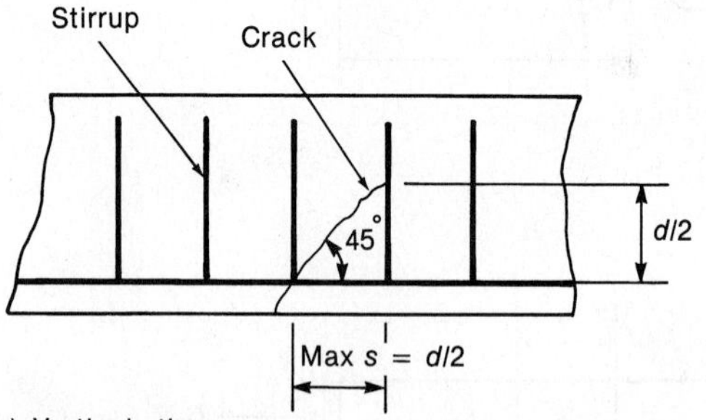

(a) Vertical stirrups.

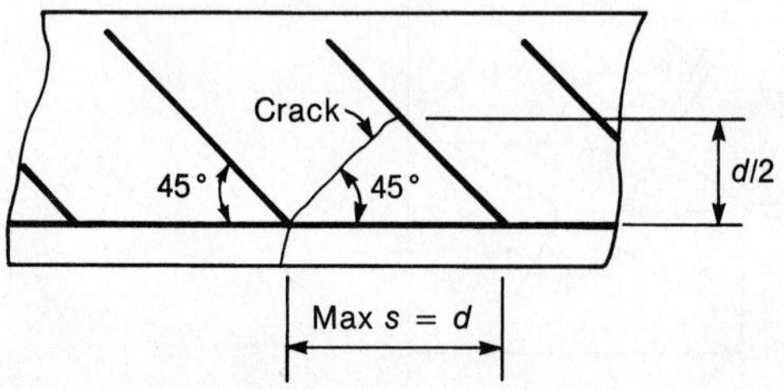

(b) Inclined stirrups.

Fig. 6-19
Maximum spacing of stirrups.

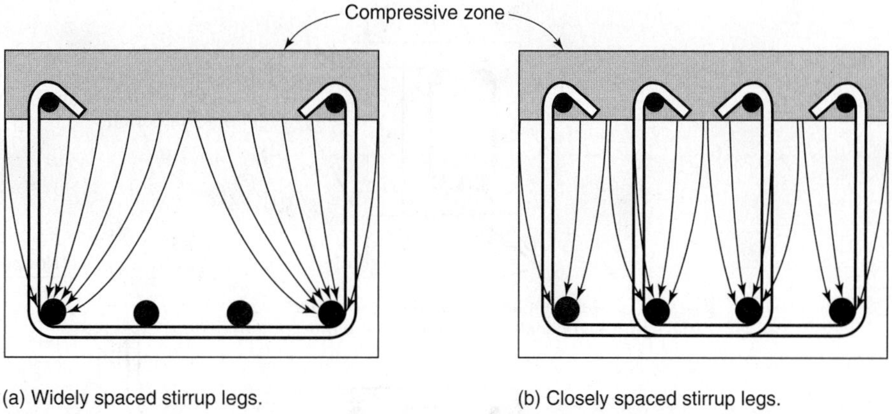

Fig. 6-20
Flow of diagonal compression force in the cross sections of beams with stirrups.

on a report by Leonhardt and Walther [6-15], the fib Model Code [6-16] suggests that the maximum transverse spacing of stirrup legs should be limited to $2d/3$ or 400 mm (16 in.), whichever is smaller.

2. Shear failure initiated by failure of the stirrup anchorages. Equations (6-17) and (6-14) are based on the assumption that the stirrups will yield at ultimate. This will be true only if the stirrups are well anchored. Generally, the upper end of the inclined crack approaches very close to the compression face of the beam, as shown in Figs. 6-4 and 6-21. At ultimate, the stress in the stirrups approaches or equals the yield strength, f_{yt}, at every point where an inclined crack intercepts a stirrup. Thus the portions of the stirrups shown shaded in Fig. 6-21 must be able to anchor f_{yt}. For this reason, ACI Code Section 25.7.1.1 requires that the stirrups extend as close to the compression and tension faces as cover and bar spacing requirements permit and, in addition, specifies certain types of hooks to anchor the stirrups. The ACI Code requirements for stirrup anchorage are illustrated in Fig. 6-22:

(a) ACI Code Section 25.7.1.2 requires that each bend in a simple U– or multiple U–stirrup shall enclose a longitudinal bar, as shown in Fig. 6-22a.

(b) For No. 5 bar or D31 wire stirrups and smaller with any yield strength, ACI Code Section 25.7.1.3 allows the use of a standard hook around longitudinal reinforcement without any specified embedment length, as shown in Fig. 6-22b. Either 135° or 180° hooks are preferred.

(c) For No. 6, 7, and 8 stirrups with f_{yt} of 40,000 psi, ACI Section 25.7.1.3 allows the details shown in Fig. 6-22b.

(d) For No. 6, 7, and 8 stirrups with f_{yt} greater than 40,000 psi, ACI Section 25.7.1.3 requires a standard hook around a longitudinal bar plus embedment between midheight of the member and the outside of the hook of at least $0.014 d_b f_{yt}/\lambda\sqrt{f'_c}$.

(e) Requirements for welded-wire fabric stirrups formed of sheets bent in U shape or vertical flat sheets are illustrated in ACI Commentary Fig. R25.7.1.4.

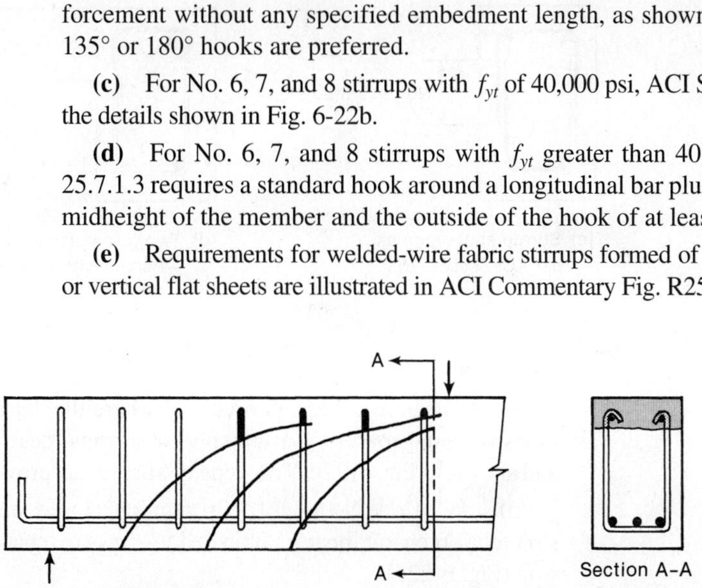

Fig. 6-21
Anchorage of stirrups.

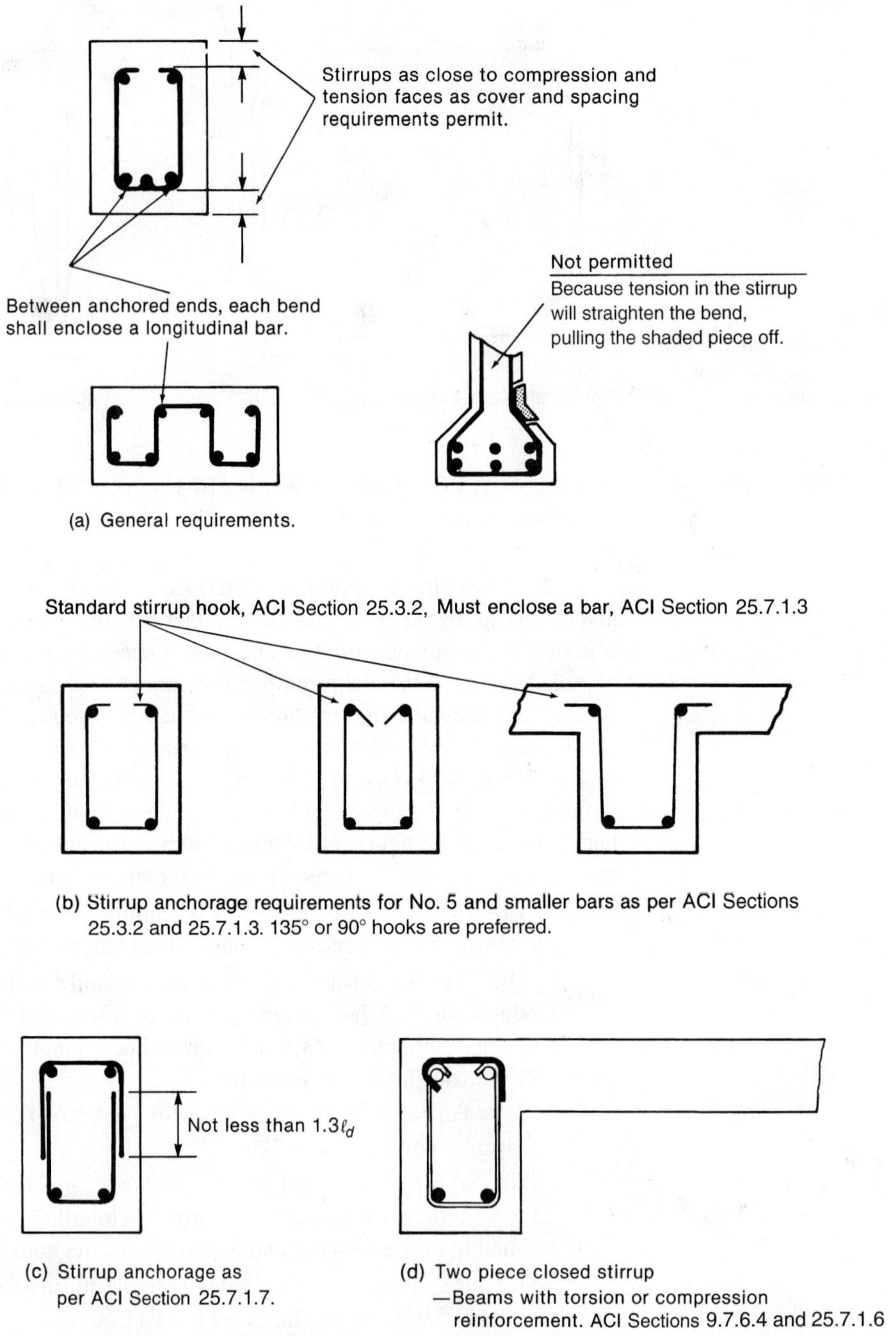

Fig. 6-22
Stirrup detailing requirements.

(f) In deep members, particularly where the depth varies gradually, it is sometimes advantageous to use lap-spliced stirrups, described in ACI Section 25.7.1.7 and shown in Fig. 6-22c. This type of stirrup has proven unsuitable in seismic areas.

(g) ACI Code Section 9.7.6.4 requires closed stirrups in beams with compression reinforcement, beams subjected to stress reversals, and beams subjected to torsion (Fig. 6-22d).

(h) Structural integrity provisions in ACI Code Section 9.7.7.1 require closed stirrups around longitudinal reinforcement in all perimeter beams. These either may be formed from one continuous bar or constructed in two pieces, as shown in Fig. 6-22. Note that the 135° hook is to be placed at the exterior corner of the beam section where the corner concrete is more likely to spall in case of an overload condition.

(i) ACI Code Section 25.4.4 contains requirements for the use of headed and mechanically anchored deformed bars. This type of anchorage detail may be useful for reducing reinforcement interference in regions where large amounts of longitudinal and transverse (shear) reinforcement are required and for through-depth reinforcement in walls of large structures.

Standard hooks, headed bars, and the development length ℓ_d are discussed in Chapter 8 of this book and in ACI Code Chapter 25.

Standard stirrup hooks are bent around a smaller-diameter pin than normal bar bends. Very-high-strength steels may develop small cracks during this bending operation. These cracks may in turn lead to fracture of the bar before the yield strength can be developed. For this reason, ACI Code Table 20.2.2.4(a) limits the yield strength used in design calculations to 60,000 psi, except for welded-wire fabric stirrups, for which the limit is 80,000 psi. This difference is justified by the fact that the bend test for the wire used to make welded-wire fabric is more stringent than that for bars. In addition, welded-wire stirrups tend to be more closely spaced than stirrups made from reinforcing bars and thus give better control of inclined-crack widths.

Because the anchorage length available between the inclined crack and the top of the stirrup is generally very short, as shown in Fig. 6-21, the author recommends both the use of the smallest-diameter stirrups possible and the use of 40,000-psi steel in stirrups where this grade is available. This has the added advantage that the more closely spaced stirrups will help prevent excessively wide inclined cracks, and the lower-yield strength is easier to develop. In addition, it is recommended that 135° stirrup hooks be used throughout, except in very narrow beams, where 180° stirrups may be needed to avoid having the tails of the hooks cross, which would make it difficult to place the longitudinal bars inside the stirrups.

For a beam with stirrup hooks, as shown in the middle cross section in Fig. 6-22b, ACI Code Table 25.3.2 defines stirrup hooks for No. 5 and smaller bars, the usual sizes of bars used for stirrups, as either a 90° bend plus $6d_b$ extension at the free end of the bar or a 135° bend plus $6d_b$ extension. ACI Code Table 25.3.2 also sets the minimum inside diameter of a stirrup bend as $4d_b$ for No. 5 and smaller bars. The minimum beam width to avoid crossing of the tails of No. 3 stirrups with 135° stirrup hooks can be computed as follows:

- cover to outside of the stirrups $= 1.5$ in.
- outside radius of a No. 3 stirrup bend $= (4 + 2) \times 0.375$ in./2 $= 1.125$ in.
- width of bend from 90° to 135° $= 1.125/\sqrt{2} = 0.80$ in.
- bar extension $= 6 \times 0.375$ in./$\sqrt{2} = 1.60$ in.

This totals 5.03 in. on each side of the beam. Leaving a clear gap of about 2 in. to lower the longitudinal steel through gives a minimum beam width of 12 in. when using 135° stirrup hooks. For 90° stirrup hooks, the corresponding total is 4.88 in. each side, also giving a minimum beam width of 12 in.

3. Serviceability failure due to excessive crack widths at service loads. Wide inclined cracks in beams are unsightly and may allow water to penetrate the beam, possibly causing corrosion of the stirrups. ACI Code Section 22.5.1.2 attempts to guard against excessive crack widths by limiting the maximum shear that can be transmitted by stirrups to $V_{s(\max)} = 8\sqrt{f'_c}b_w d$. In a beam with $V_{s(\max)}$, the stirrup stress will be approximately 34 ksi at service loads, corresponding to a maximum crack width of about 0.014 in. [6-5]. Although this limit generally gives satisfactory crack widths, the use of closely spaced stirrups and horizontal steel near the faces of beam webs is also effective in reducing crack widths.

4. Shear failure due to crushing of the web. As indicated by the truss analogies in Figs. 6-6 and 6-9, compression stresses exist in compression diagonals in the web of a beam. In very thin-walled beams, these stresses may lead to crushing of the web. Because the diagonal compression stress is related to the shear stress, a number of codes limit the ultimate shear stress to 0.2 to 0.25 times the compression strength of the concrete. The ACI Code limit on V_s for crack control ($V_{s(\max)} = 8\sqrt{f'_c}b_w d$) provides adequate safety against web crushing in reinforced concrete beams.

5. Shear failure initiated by failure of the tension chord. Figures 6-12 and 6-16 show that the force in the longitudinal tensile reinforcement at a given point in the shear span is a function of the moment at a section located approximately d closer to the section of maximum moment. Partly for this reason, ACI Code Section 9.7.3.3 requires that beam flexural reinforcement extend the larger of d or 12 bar diameters past the point where it is no longer needed to resist tension (except at the supports of simple spans or at the ends of cantilevers).

The exemption of the free ends of cantilevers recognizes that it is difficult to extend the bars at these locations. At a simple support, however, there is a potential force in the bars at the supports.

Figure 6-23 shows a free-body diagram of a simple support. Summing moments about point 0 and ignoring any moment resulting from the shearing stresses acting along the inclined crack gives

$$\frac{V_u}{\phi} \times d_v = T_n \times d_v + V_s \times 0.5d_v$$

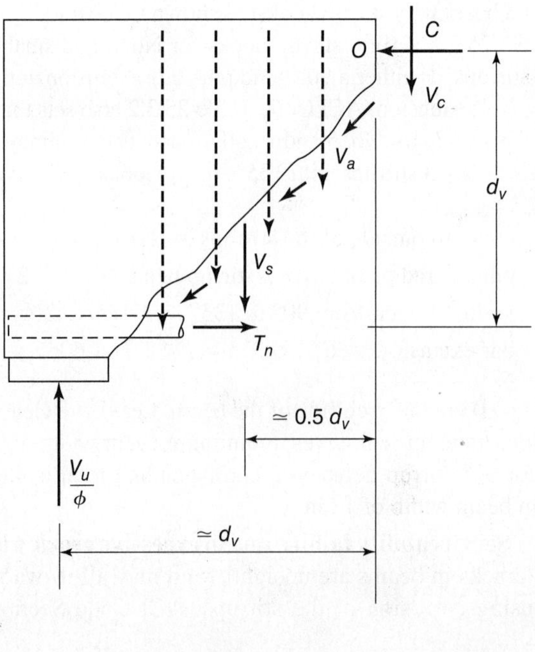

Fig. 6-23
Free-body diagram of beam reaction zone.

or

$$T_n = \frac{V_u}{\phi} - 0.5V_s \tag{6-18}$$

where V_s is the shear resisted by the stirrups near the face of the support and T_n is the longitudinal bar force to be anchored. Frequently, there are more stirrups than needed as a result of satisfying maximum spacing requirements; in such a case, V_s would be less than the value given by Eq. (6-14), because the stress in the stirrups would be less than yield. From Eq. (6-9),

$$V_s = \frac{V_u}{\phi} - V_c$$

Substituting this result into Eq. (6-18) gives

$$T_n = 0.5\left(\frac{V_u}{\phi} + V_c\right) \tag{6-19}$$

This issue is addressed by ACI Code Section 9.7.3.8.3, which gives specific requirements for positive moment reinforcement at simple supports. Anchorage of positive reinforcement at supports is covered in Section 8-7.

Types of Web Reinforcement

ACI Code Sections 22.5.10.5 and 22.5.10.6 allow shear reinforcement for nonprestressed beams to consist of:

(a) Stirrups or ties perpendicular to the axis of the member. This is by far the most common type of web reinforcement.

(b) Welded wire fabric with the wires perpendicular to the axis of the member serving as shear reinforcement. The welded wire fabric may be in the form of a sheet bent into a U shape or may be a ladderlike sheet with the rungs of the ladder transverse to the member. The latter are widely used in precast members. ACI Code Sections 25.7.1.4 and 25.7.1.5 describe how welded wire fabric stirrups are to be anchored.

(c) Stirrups inclined at an angle of 45° or more to the longitudinal tension reinforcement. The inclination of these stirrups must be such that they intercept potential inclined cracks. Inclined stirrups are difficult to detail near the ends of a beam and are not widely used. It is believed that stirrups flatter than 45° may slip along the longitudinal bars at high loads.

(d) A portion of the longitudinal flexural reinforcement may be bent up where no longer needed for flexure, as shown in Fig. 5-21d. Bent-up bars are no longer widely used because of the added cost of the labor needed to bend the bars. Because major inclined cracks tend to occur at the bend points, only the center 75 percent of the inclined portion of the bent bar is considered to be effective as shear reinforcement.

(e) Combinations of spirals, circular ties, and hoops.

Minimum Web Reinforcement

Because a shear failure of a beam without web reinforcement is sudden and brittle, and because shear-failure loads vary widely around the values given by the design equations, ACI Code Section 9.6.3.1 requires a minimum amount of web reinforcement to be provided in beams if the applied shear force, V_u, exceeds half of the factored inclined cracking shear, $\phi(0.5V_c)$, except in

1. concrete joist construction defined by ACI Code Section 9.8;

2. isolated beams with h not greater than 10 in.;

3. beams integral with slabs with both h not greater than 24 in. and h not greater than the larger of 2.5 times the flange thickness or 0.5 time the width of the web; and

4. beams constructed of steel fiber-reinforced, normal-weight concrete with f'_c not exceeding 6000 psi, h not greater that 24 in., and V_u not greater than $2\phi\sqrt{f'_c}\,b_w d$.

The first exception represent a type of member or structural system where load redistributions can occur in the transverse direction. The next two exceptions relate to the *size-effect* issue discussed earlier in this chapter [6-7] and [6-8]. The final exception for beams constructed with steel fiber-reinforced concrete was discussed at the end of Section 6-3.

Where required, the minimum web reinforcement shall be (ACI Code Table 9.6.3.3)

$$A_{v,\min} = 0.75\sqrt{f'_c}\frac{b_w s}{f_{yt}} \tag{6-20}$$

but not less than

$$A_{v,\min} = \frac{(50\text{ psi})b_w s}{f_{yt}} \tag{6-21}$$

From Eqs. (6-20) and (6-21), the minimum web reinforcement provides web reinforcement to transmit shear stresses of $0.75\sqrt{f'_c}$ psi and 50 psi, respectively. From Eq. (6-8a), for $f'_c = 2500$ psi, 50 psi is half of the shear stress at inclined cracking. Equation (6-20) governs when f'_c exceeds 4440 psi and reflects the fact that the tensile stress at cracking increases as the compressive strength increases [6-17].

In SI units, Eqs. (6-20) and (6-21) respectively become

$$A_{v,\min} = \frac{1}{16}\sqrt{f'_c}\frac{b_w s}{f_{yt}} \tag{6-20M}$$

but not less than

$$A_{v,\min} = \frac{b_w s}{3f_{yt}} \tag{6-21M}$$

In seismic regions, web reinforcement is required in most beams, because V_c is taken equal to zero if earthquake-induced shear exceeds half the total shear.

Strength-Reduction Factor for Shear

The strength-reduction factor, ϕ, for shear and torsion is 0.75 (ACI Code Table 21.2.1). This value is lower than for flexure because shear-failure loads are more variable than flexure-failure loads. Special strength-reduction factors are required for shear in some members subjected to seismic loads.

Location of Maximum Shear for the Design of Beams

In a beam loaded on the top flange and supported on the bottom as shown in Fig. 6-24a, the closest inclined cracks that can occur adjacent to the supports will extend outward from the supports at roughly 45°. Loads applied to the beam within a distance d from the support in such a beam will be transmitted directly to the support by a compression fan above the 45° cracks and will not affect the stresses in the stirrups crossing the cracks shown in Fig. 6-24. As a result, ACI Code Section 9.4.3.2 states,

> Sections between the face of the support and a critical section located $\boldsymbol{d}$ from the face of support for nonprestressed beams . . . shall be permitted to be designed for V_u at that critical section.

This is permitted only when

1. the support reaction, in the direction of the applied shear, introduces compression into the end regions of a member,

2. the loads are applied at or near the top of the beam, and

3. no concentrated load occurs within d from the face of the support.

Thus, for the beam shown in Fig. 6-24a, the values of V_u used in design are shown shaded in the shear force diagram of Fig. 6-24b.

This allowance must be applied carefully because it is not applicable in all cases. Figure 6-25 shows five other typical cases that arise in design. If the beam shown in Fig. 6-24 was loaded on the *lower* flange, as indicated in Fig. 6-25a, the critical section for design would be at the face of the support, because loads applied within d of the support must be transferred across the inclined crack before they reach the support.

A typical beam-to-column joint is shown in Fig. 6-25b. Here the critical section for design is d away from the section as shown.

If the beam is supported by a girder of essentially the same depth, as shown in Fig. 6-25c, the compression fans that form in the supported beams will tend to push the bottom off the

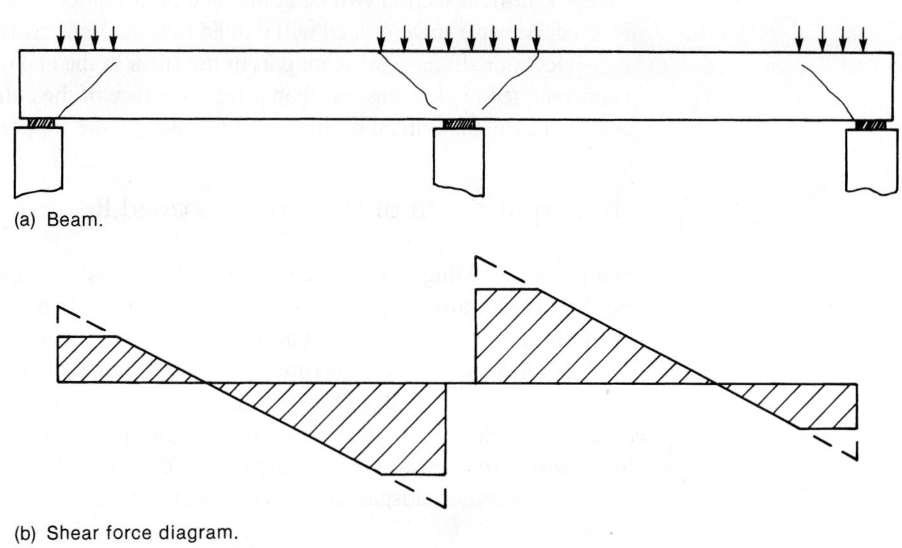

Fig. 6-24
Shear force diagram for design.

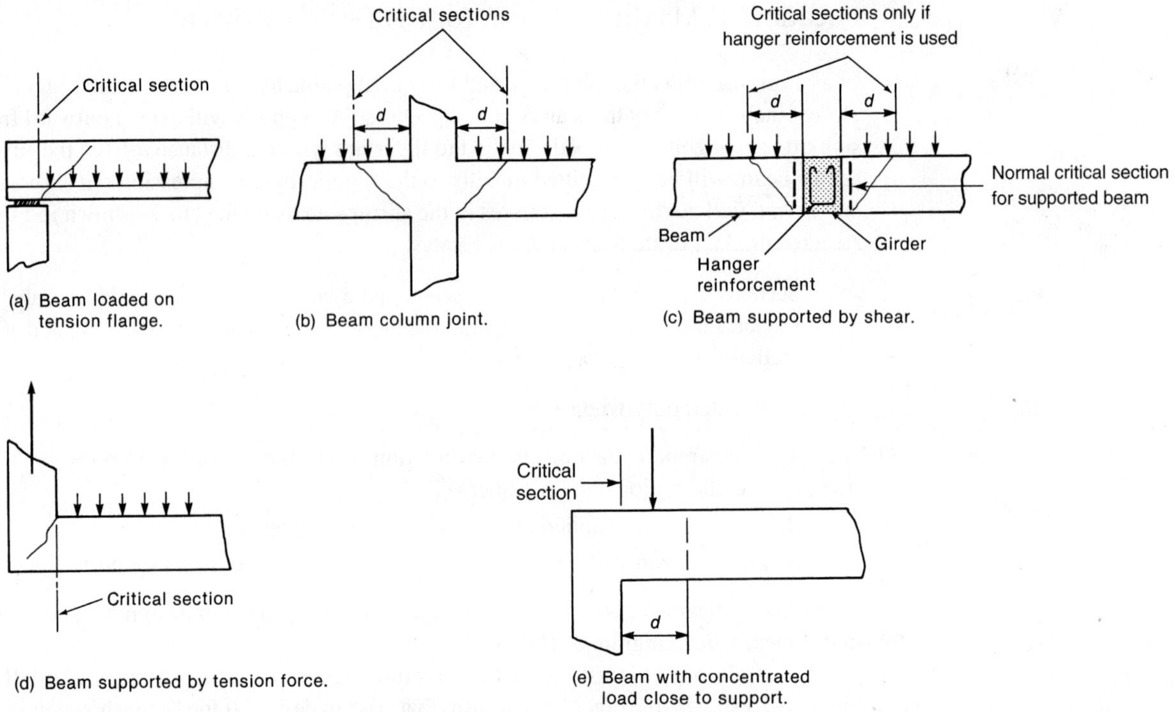

Fig. 6-25
Application of ACI Section 9.4.3.2.

supporting beam. Thus, the critical shear design sections in the supported beams normally are taken at the face of the supporting beam. The critical section may be taken at d from the end of the beam if *hanger reinforcement* is provided to support the reactions from the compression fans. The design of hanger reinforcement is discussed in Section 6-6.

Generally, if the beam is supported by a tensile force rather than a compressive force, the critical section will be at the face of the support, and the joint must be carefully detailed, because shear cracks will extend into the joint, as shown in Fig. 6-25d.

Occasionally, a significant part of the shear at the end of the beam will be caused by a concentrated load acting less than d from the face of the column, as shown in Fig. 6-25e. In such a case, the critical section must be taken at the support face.

Shear at Midspan of Uniformly Loaded Beams

In a normal building, the dead and live loads are assumed to be uniform loads. Although the dead load is always present over the full span, the live load may act over the full span, as shown in Fig. 6-26c, or over part of the span, as shown in Fig. 6-26d. Full uniform load over the full span gives the maximum shear at the ends of the beam. Full uniform load over half the span plus only dead load on the remaining half is used to find the maximum shear at midspan. The maximum shears at other points in the span are closely approximated by a linear *shear-force envelope* resulting from these cases (Fig. 6-26e).

The shear at midspan due to a uniform live load on half the span is

$$V_{u,\text{midspan}} = \frac{w_{Lu}\ell}{8} \tag{6-22}$$

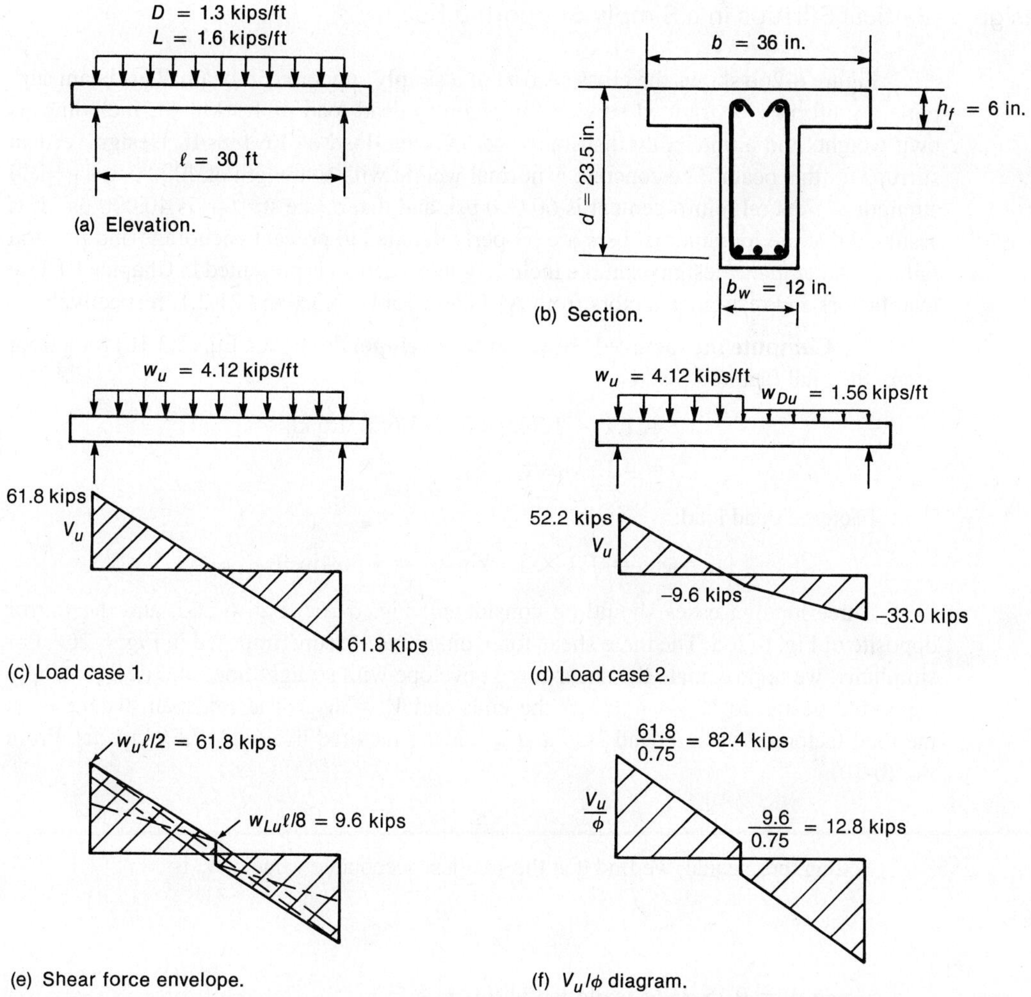

Fig. 6-26
Beam and shear force envelope—Example 6-1.

This can be positive or negative. Although this has been derived for a simple beam, it is also acceptable to apply Eq. (6-22) to continuous beams.

High-Strength Concrete

Tests suggest that the inclined cracking load of beams increases less rapidly than $\sqrt{f'_c}$ for f'_c greater than about 8000 psi. This observation was offset by an increased effectiveness of stirrups in high-strength concrete beams [6-18], [6-19]. Other tests suggest that the required amount of minimum web reinforcement increases as f'_c increases. ACI Code Section 22.5.3.2 allows the use of $\sqrt{f'_c}$ in excess of 100 psi when computing the shear carried by the concrete (V_c, V_{ci}, and V_{cw}) for reinforced concrete and prestressed concrete beams and reinforced concrete joists, *provided* that they have minimum web reinforcement in accordance with ACI Sections 9.6.3.3 or 9.6.4.2. For high-strength concrete in two-way slabs, $\sqrt{f'_c}$ is limited to 100 psi by the lack of test data on two-way slabs with higher strength concrete.

EXAMPLE 6-1 Design of Vertical Stirrups in a Simply Supported Beam

Figure 6-26b shows the cross section of a simply supported T-beam. This beam supports a uniformly distributed service (unfactored) dead load of 1.3 kips/ft, including its own weight, and a uniformly distributed service live load of 1.6 kips/ft. Design vertical stirrups for this beam. The concrete is normal weight with a strength of 4000 psi, the yield strength of flexural reinforcement is 60,000 psi, and that of the stirrups is 40,000 psi. It is assumed that the longitudinal bars are properly detailed to prevent anchorage and flexural failures. A complete design example including those details is presented in Chapter 10. Use load factors and resistance factors from ACI code Tables 5.3.1 and 21.2.1, respectively.

1. Compute the factored shear-force envelope. From ACI Eq. (5.3.1b) for a floor beam, the total factored load is:

$$\begin{aligned} w_u &= 1.2 \times 1.3 \text{ kips/ft} + 1.6 \times 1.6 \text{ kips/ft} \\ &= 4.12 \text{ kips/ft} \end{aligned}$$

Factored dead load:

$$w_{Du} = 1.2 \times 1.3 \text{ kips/ft} = 1.56 \text{ kips/ft}$$

Three loading cases should be considered: Fig. 6-26c, Fig. 6-26d, and the mirror opposite of Fig. 6-26d. The three shear-force diagrams are superimposed in Fig. 6-26e. For simplicity, we approximate the shear-force envelope with straight lines and design simply supported beams for $V_u = w_u\ell/2$ at the ends and $V_u = w_{Lu}\ell/8$ at midspan, where w_u is the total factored live and dead load and w_{Lu} is the factored live load (2.56 kips/ft). From Eq. (6-10),

$$V_u \leq \phi V_n$$

Setting these equal, we find that the smallest acceptable value of V_n is

$$V_n = \frac{V_u}{\phi}$$

Using $\phi = 0.75$, V_u/ϕ is plotted in Fig. 6-26f.

Because this beam is loaded on the top and supported on the bottom, the critical section for shear is located at $d = 23.5$ in. $= 1.96$ ft from the support. From Fig. 6-26f and similar triangles, the shear at d from the support is

$$\begin{aligned} \frac{V_u}{\phi} \text{ at } d &= 82.4 \text{ kips} - \frac{1.96 \text{ ft}}{15 \text{ ft}}(82.4 - 12.8) \text{ kips} \\ &= 73.3 \text{ kips} \end{aligned}$$

Therefore,

$$\frac{V_u}{\phi} \text{ at } d = 73.3 \text{ kips and min. } V_n = 73.3 \text{ kips}$$

2. Are stirrups required by ACI Code Section 9.6.3.1? No stirrups are required if $V_u/\phi \leq V_c/2$, where

$$\begin{aligned} V_c &= 2\lambda\sqrt{f'_c}b_w d \qquad (6\text{-}8b) \\ &= 2 \times 1\sqrt{4000} \text{ psi} \times 12 \text{ in.} \times 23.5 \text{ in.} \\ &= 35{,}700 \text{ lbs} = 35.7 \text{ kips} \end{aligned}$$

Because $V_u/\phi = 73.1$ kips exceeds $V_c/2 = 17.8$ kips, stirrups are required.

3. Is the cross section large enough? ACI Code Section 22.5.1.2 gives the maximum shear in the stirrups as

$$V_{s,\max} = 8\sqrt{f_c'}\, b_w d$$

Thus, in this case the maximum V_u/ϕ is

$$(V_u/\phi)_{\max} = V_c + V_{s,\max} = 10\sqrt{f'_c} b_w d = 5V_c$$
$$= 178 \text{ kips}$$

Because V_u/ϕ at $d = 73.3$ kips is less than 178 kips, the section is large enough.

4. Check the anchorage of stirrups and maximum spacing. Try No. 3 double-leg stirrups, $f_{yt} = 40{,}000$ psi:

$$A_v = 2 \times 0.11 \text{ in.}^2 = 0.22 \text{ in.}^2$$

(a) Check the anchorage of the stirrups. Because the bar size of the stirrups is less than No. 6, ACI Code Section 25.7.1.3 states that such stirrups can be anchored by a 90° or 135° stirrup hook around a longitudinal bar. Provide a No. 4 bar in the upper corners of the stirrups to anchor them.

(b) Find the maximum stirrup spacing.

Based on the beam depth: ACI Code Table 9.7.6.2.2 sets the maximum spacing as the smaller of $0.5d \cong 12$ in. or 24 in. The same code table also requires half this spacing in regions where V_s exceeds $4\sqrt{f'_c}\, b_w d$. Thus, the stirrup spacing must be cut in half if V_u/ϕ exceeds

$$V_c + 4\sqrt{f'_c} b_w d = 6\sqrt{f'_c} b_w d = 107 \text{ kips}$$

Because the maximum V_u/ϕ is less than 107 kips, the maximum spacing based on the beam depth is $s = 12$ in.

Based on minimum A_v: In Eq. (6-20),

$$A_{v,\min} = 0.75\sqrt{f'_c}\frac{b_w s}{f_{yt}} \tag{6-20}$$

but not less than

$$A_{v,\min} = \frac{50 b_w s}{f_{yt}} \tag{6-21}$$

Because $0.75\sqrt{f'_c} = 47.4 \text{ psi} < 50 \text{ psi}$, Eq. (6-21) governs. Rearranging Eq. (6-21) gives

$$s_{\max} = \frac{A_v f_{yt}}{50 b_w}$$
$$= \frac{0.22 \text{ in.}^2 \times 40{,}000 \text{ psi}}{50 \text{ psi} \times 12 \text{ in.}} = 14.7 \text{ in.}$$

Therefore, the maximum spacing based on the beam depth governs. **Maximum $s = 12$ in.**

5. Compute the stirrup spacing required to resist the shear forces. For vertical stirrups, from Eq. (6-17),

$$s \leq \frac{A_v f_{yt} d}{V_u/\phi - V_c}$$

where $V_c = 35.7$ kips.

At d from the support, $V_u/\phi = 73.3$ kips, thus

$$s \leq \frac{0.22 \text{ in.}^2 \times 40 \text{ ksi} \times 23.5 \text{ in.}}{73.3 \text{ k} - 35.7 \text{ k}} = 5.53 \text{ in.}$$

Because this is a tight spacing for a beam with a total depth of 26 in., switch to No. 4 double-leg stirrups ($A_v = 2 \times 0.20 \text{ in.}^2 = 0.40 \text{ in.}^2$). The required spacing for strength is:

$$s \leq \frac{0.40 \text{ in.}^2 \times 40 \text{ ksi} \times 23.5 \text{ in.}}{73.3 \text{ k} - 35.7 \text{ k}} = 10.0 \text{ in.}$$

This is a reasonable result, so use No. 4 double-leg stirrups at a spacing of 10 in. near the support. However, it normally is assumed that each stirrup reinforces a length of beam extending $s/2$ on each side of the stirrup. For this reason, the first stirrup should be placed at $s/2 = 5$ in. from the face of the support.

We now will find the point where we can change to No. 3 double-leg stirrups ($A_v = 0.22 \text{ in.}^2$) at the maximum permissible spacing of 12 in. The selection of more intermediate stirrup size and spacing combinations is up to the designer. Generally, no more than three different size/spacing combinations are used in a single span.

Using No. 3 double-leg stirrups at 12 in., from Eq. (6-14) we have

$$V_s = \frac{A_v f_{yt} d}{s} = \frac{0.22 \text{ in.}^2 \times 40 \text{ ksi} \times 23.5 \text{ in.}}{12 \text{ in.}} = 17.2 \text{ kips}$$

Then, using Eq. (6-9),

$$V_n = V_c + V_s = 35.7 \text{ k} + 17.2 \text{ k} = 52.9 \text{ kips}$$

Setting $V_n = V_u/\phi = 52.9$ kips and using similar triangles from the shear envelope in Fig. 6-26f, it can be shown that this shear occurs at

$$x_1 = \frac{82.4 \text{ k} - 52.9 \text{ k}}{82.4 \text{ k} - 12.8 \text{ k}} \times 15 \text{ ft}$$

$$= 6.36 \text{ ft} = 76.3 \text{ in. (from center of support)}$$

Stirrups must be continued to the point where $V_u/\phi = V_c/2 = 17.8$ kips. Using similar triangles from Fig. 6-26f, this shear occurs at

$$x_2 = \frac{82.4 \text{ k} - 17.8 \text{ k}}{82.4 \text{ k} - 12.8 \text{ k}} \times 180 \text{ in.}$$

$$= 167 \text{ in. (from center of the support)}$$

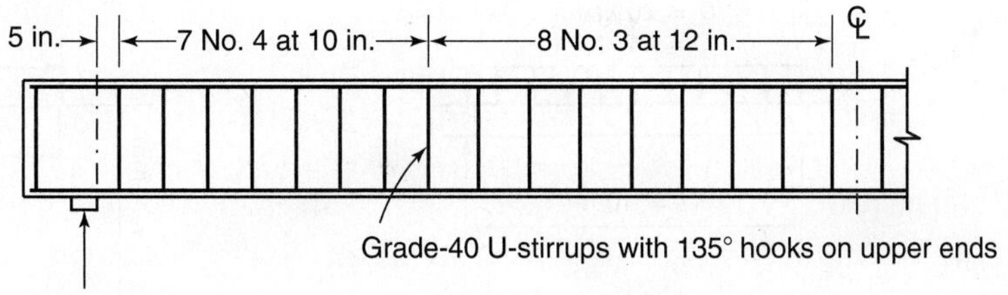

Fig. 6-27 Stirrups in beam for Example 6-1.

To finalize the stirrup design for this beam, we will assume the beam rests on a 6-in. bearing plate. The first No. 4 stirrup must be placed 5 in. from the edge of the bearing plate (a total of 8 in. from the center of the support). Then use seven more No. 4 stirrups at a 10-in. spacing (extends to 78 in. from the center of the support). After that, use eight No. 3 stirrups at a 12-in. spacing (extends to 174 in. from center of support). By coincidence, the last stirrup is only 6 in. from the midspan of the beam, and thus only 12 in. from the next No. 3 stirrup that will be placed symmetrically in the other half of the beam span. For other beam designs there may be a larger region near the midspan of a beam where no stirrups are required to satisfy the ACI Code requirements. It is the designer's option to put stirrups in that region either for enhanced shear strength or to hold longitudinal reinforcement in the proper position. The final stirrup design, as measured from the face of the support plate is:

One No. 4 at 5 in. (8 in. from center of support)
Seven No. 4 at 10 in. (extends to 78 in. from center of support)
Eight No. 3 at 12 in. (extends to 174 in. from center of support)
Symmetrically placed in both halves of the beam span.

The final stirrup design is shown for half of the beam in Fig. 6-27, and a typical beam section is shown in Fig. 6-26b. ■

EXAMPLE 6-1M Design of Vertical Stirrups in a Simply Supported Beam—SI Units

Figure 6-28 shows the elevation and cross section of a simply supported T-beam. This beam supports a uniformly distributed service (unfactored) dead load of 20 kN/m, including its own weight, and a uniformly distributed service live load of 24 kN/m. Design vertical stirrups for this beam. The concrete is normal weight with a strength of 25 MPa, the yield strength of the flexural reinforcement is 420 MPa, and the yield strength of the stirrups is 300 MPa. Use load and resistance factors from ACI Code Tables 5.3.1 and 21.2.1.

1. Compute the design factored shear-force envelope. For ACI (Eq. 5.3.1b) for a floor beam, total factored load:

$$w_u = 1.2 \times 20 \text{ kN/m} + 1.6 \times 24 \text{ kN/m}$$
$$= 62.4 \text{ kN/m}$$

Factored dead load:

$$w_{Du} = 1.2 \times 20 \text{ kN/m} = 24.0 \text{ kN/m}$$

Three loading cases should be considered: Fig. 6-28c, Fig. 6-28d, and the mirror opposite of Fig. 6-28d. The three shear-force diagrams are superimposed in Fig. 6-28e. For simplicity, we approximate the shear-force envelope with straight lines and design simply

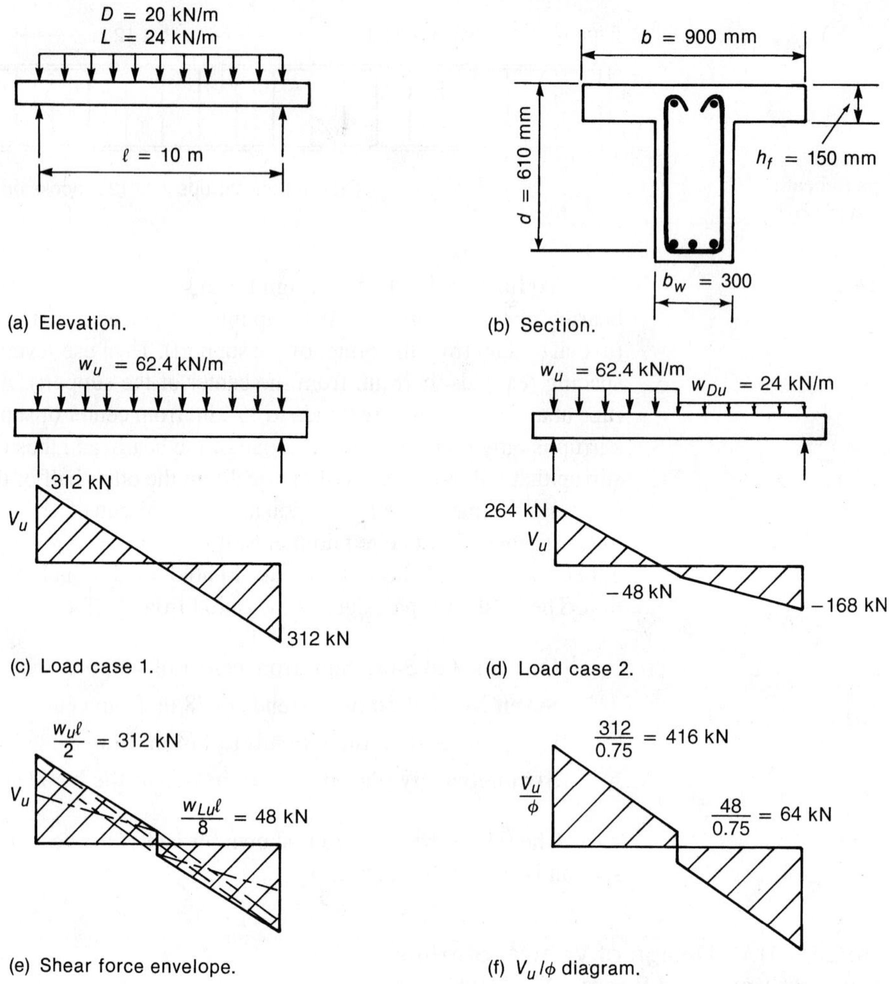

Fig. 6-28
Beam and shear force envelope for Example 6-1M.

supported beams for $V_u = w_u\ell/2$ at the ends and $V_u = w_{Lu}\ell/8$ at midspan, where w_u is the total factored live and dead load and w_{Lu} is the factored live load (38.4 kN/m). From Eq. (6-10),

$$V_u \le \phi V_n$$

Setting these equal, we obtain the smallest value of V_n that satisfies Eq. (6-10):

$$V_n = \frac{V_u}{\phi}$$

V_u/ϕ is plotted in Fig. 6-28f.

Because this beam is loaded on the top flange and supported on the bottom flange, the critical section is located at $d = 0.61$ m from the support. From Fig. 6-28f and similar triangles, the shear at d from the support is

$$\frac{V_u}{\phi} \text{ at } d = 416 \text{ kN} - \frac{0.61 \text{ m}}{5 \text{ m}}(416 - 64)$$

$$= 373 \text{ kN}$$

2. Are stirrups required by ACI Code Section 9.6.3.1? No stirrups are required if $V_n \leq V_c/2$, where

$$V_c = \lambda \frac{\sqrt{f'_c}\, b_w d}{6} \tag{6-8Mb}$$

$$= 1 \times \frac{\sqrt{25 \text{ MPa}} \times 300 \text{ mm} \times 610 \text{ mm}}{6}$$

$$= 153{,}000 \text{ N} = 153 \text{ kN}$$

Because $V_u/\phi = 373$ kN exceeds $V_c/2 = 76.3$ kN, stirrups are required.

3. Is the cross section large enough? ACI Metric Code Section 22.5.1.2 gives the maximum shear in the stirrups as

$$V_{s,\max} = \frac{2}{3}(\sqrt{f'_c}\, b_w d)$$

Thus, the maximum allowable V_u/ϕ is:

$$\left(\frac{V_u}{\phi}\right)_{\max} = V_c + V_{s,\max} = \left(\frac{2}{3} + \frac{1}{6}\right)\sqrt{f'_c}\, b_w d = 5V_c$$

$$= 765 \text{ kN}$$

Because V_u/ϕ at $d = 373$ kN is less than 765 kN, the section is large enough.

4. Check anchorage of stirrups and maximum spacing. Try No. 10M double-leg stirrups, $f_{yt} = 300$ MPa:

$$A_v = 2 \times 71 = 142 \text{ mm}^2$$

(a) Check the anchorage of the stirrups. ACI Code Section 25.7.1.3 allows No. 25M and smaller stirrups to be anchored by a standard 90° or 135° hook stirrup hook around a longitudinal bar. Provide a No. 10M or larger bar in each of the upper corners of the stirrup to anchor them.

(b) Find the maximum stirrup spacing.

Based on the beam depth: ACI Code Table 9.7.6.2.2 requires the smaller of $0.5d = 305$ mm or 600 mm. The same code table also requires half these spacings if V_s exceeds $(\frac{1}{3})\sqrt{f'_c}\, b_w d$:

$$V_c + \frac{\sqrt{f'_c}\, b_w d}{3} = 153 \text{ kN} + 306 \text{ kN} = 459 \text{ kN}$$

Because the maximum V_u/ϕ is less than 459 kN, the maximum spacing is 305 mm.

Based on minimum allowed A_v: In Eq. (6-20M),

$$A_{v,\min} = \frac{1}{16}\sqrt{f'_c}\, \frac{b_w s}{f_{yt}} \tag{6-20M}$$

but not less than

$$A_{v,\min} = \frac{1}{3}\frac{b_w s}{f_{yt}} \tag{6-21M}$$

Because $= \frac{1}{16}\sqrt{f'_c} = 0.313 \text{ MPa} < \frac{1}{3}$ MPa, Eq. (6-21M) governs.

Rearranging Eq. (6-21M) to solve for (s),

$$s_{\max} = \frac{3A_v f_{yt}}{b_w} = \frac{3 \times (2 \times 71) \times 300}{300} = 426 \text{ mm}$$

Therefore, the maximum spacing based on the beam depth governs. **Maximum s = 305 mm.**

5. Compute the stirrup spacing required to resist the shear forces. For vertical stirrups, Eq. (6-17) applies; that is,

$$s \leq \frac{A_v f_{yt} d}{V_u/\phi - V_c}$$

where $V_c = 153$ kN. At d from the support, $V_u/\phi = 373$ kN and

$$s = \frac{2 \times 71 \text{ mm}^2 \times 300 \text{ MPa} \times 610 \text{ mm}}{(373 \text{ kN} - 153 \text{ kN}) \times 1000 \text{ N/kN}} = 118 \text{ mm}$$

As in Example 6-1, this spacing is unreasonably small compared to the effective depth, $d = 610$ mm. Therefore, switch to No. 13M double-leg stirrups ($A_v = 2 \times 129 \text{ mm}^2 = 258 \text{ mm}^2$). The required spacing for strength is

$$s \leq \frac{258 \text{ mm}^2 \times 300 \text{ MPa} \times 610 \text{ mm}}{(373 \text{ kN} - 153 \text{ kN}) \times 1000 \text{ N/kN}} = 215 \text{ mm}$$

This is a reasonable result, so use No. 13M double-leg stirrups at a spacing of 200 mm near the support. As in Example 6-1, the first stirrup should be placed at $s/2 = 100$ mm from the face of the support.

Next, find the point where we can change to No. 10M double-leg stirrups ($A_v = 142 \text{ mm}^2$) at the maximum permissible spacing of 300 mm. The selection of more intermediate stirrup size and spacing combinations is up to the designer. Generally, no more than three different size/spacing combinations are used in a single span.

Using No. 10M double-leg stirrups at 300 mm, from Eq. (6-14), we have

$$V_s = \frac{A_v f_{yt} d}{s} = \frac{142 \text{ mm}^2 \times 300 \text{ MPa} \times 610 \text{ mm}}{300 \text{ mm}} = 86{,}600 \text{ N} = 86.6 \text{ kN}$$

Then, using Eq. (6-9),

$$V_n = V_c + V_s = 153 \text{ kN} + 86.6 \text{ kN} = 240 \text{ kN}$$

Setting $V_n = V_u/\phi = 240$ kN and using similar triangles from the shear envelope in Fig. 6-28f, it can be shown that this shear occurs at

$$x_1 = \frac{416 \text{ kN} - 240 \text{ kN}}{416 \text{ kN} - 64 \text{ kN}} \times 5000 \text{ mm}$$

$$= 2500 \text{ mm (from center of support)}$$

Stirrups must be continued to the point where $V_u/\phi = V_c/2 = 76.5$ kN. Using similar triangles from Fig. 6-28f, this shear occurs at

$$x_2 = \frac{416 \text{ kN} - 76.5 \text{ kN}}{416 \text{ kN} - 64 \text{ kN}} \times 5000 \text{ mm}$$

$$= 4820 \text{ mm (from center of the support)}$$

When choosing the number of stirrups at each spacing, we will assume that the beam rests on a 150-mm bearing plate. The final stirrup design is:

Place the first No. 13M stirrup at 100 mm from the edge of the bearing plate (total of 175 from the center of the support).
Use twelve more No. 13M stirrups at a 200-mm spacing (extends to 2575 mm from the center of the support).
Use eight No. 10M stirrups at a 300-mm spacing (extends to 4975 mm from center of support).

This puts the last stirrup only 25 mm from the beam midspan. In this case, it would be prudent to move this last stirrup to the midspan (spacing becomes 325 mm) and then place stirrups symmetrically in the other half of the beam span. ■

EXAMPLE 6-2 Design of Stirrups in a Continuous Beam

Figure 6-29 shows the elevation and cross section of an interior span of a continuous T-beam that frames into column supports. This beam supports a floor supporting an unfactored dead load of 2.45 kips/ft (including its own weight) and an unfactored live load of 2.4 kips/ft. The concrete is of normal density, with $f'_c = 4500$ psi. The yield strength of the web reinforcement is 40 ksi. Design vertical stirrups to satisfy the strength requirements of the ACI Code. Use the load and resistance factors from Tables 5.3.1 and 21.2.1 of the ACI Code.

1. Compute the factored shear-force envelope. ACI Code Table 5.3.1 gives the following load combinations for beams loaded with dead and live loads:

$$U = 1.4D \qquad \text{(ACI Eq. 5.3.1a)}$$

$$U = 1.2D + 1.6L + 0.5(L_r \text{ or } S \text{ or } R) \qquad \text{(ACI Eq. 5.3.1b)}$$

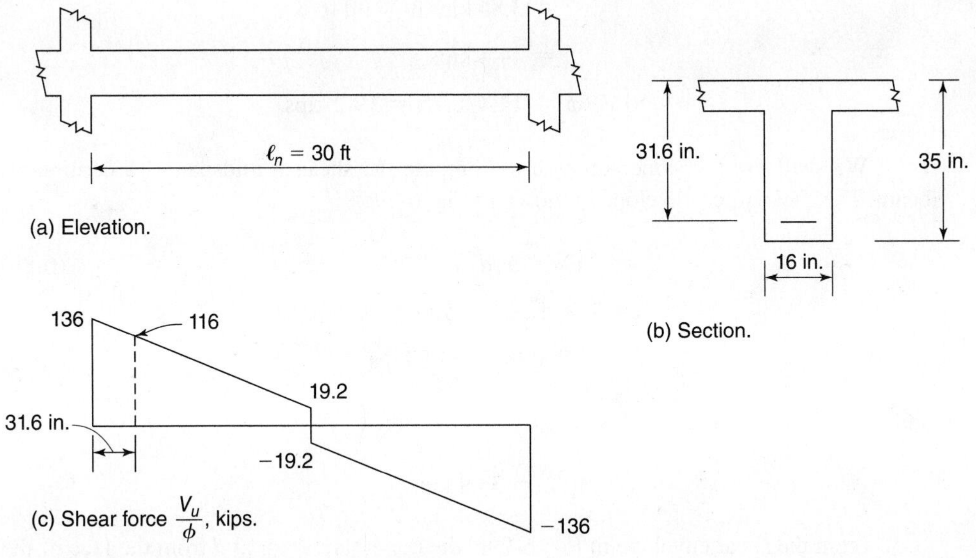

Fig. 6-29
Beam and shear force envelope—Example 6-2.

L_r, S, and R are live load, snow load, and rain load on a roof. Because this beam supports a floor, these loads are not applicable.

Thus,

$$w_u = 1.4D = 1.4 \times 2.45 = 3.43 \text{ kip/ft}$$

and

$$w_u = 1.2D + 1.6L = 1.2 \times 2.45 + 1.6 \times 2.4 = 6.78 \text{ kip/ft}$$

The larger of these governs, so we have

$$\text{Factored total load } w_u = 6.78 \text{ kip/ft}$$

and

$$\text{Factored live load } w_{Lu} = 3.84 \text{ kip/ft.}$$

This is an interior span, and ACI Code Table 6.5.4 gives the shear at the face of the support as

$$V_u = w_u \ell_n / 2$$

where $\ell_n = 30$ ft is the clear span.

$$\begin{aligned} V_u &= 6.78 \text{ kip/ft} \times 30 \text{ ft}/2 \\ &= 102 \text{ kips} \end{aligned}$$

$V_u/\phi = 102/0.75 = 136$ kips at the faces of the supports.

For simply supported beams, we have taken the shear at midspan as

$$\begin{aligned} V_u &= w_{Lu} \ell_n / 8 \qquad (6\text{-}22) \\ &= 3.84 \text{ kip/ft} \times 30 \text{ ft}/8 \\ &= 14.4 \text{ kips} \\ \text{so } V_u/\phi &= 14.4/0.75 = 19.2 \text{ kips} \end{aligned}$$

We shall use the same equation to compute the shear at midspan of a continuous beam. The shear-force envelope is shown in Fig. 6-29c.

$$\begin{aligned} V_c &= 2\lambda\sqrt{f'_c} b_w d \qquad (6\text{-}8b) \\ &= 2 \times 1 \times \sqrt{4500} \times 16 \times 31.6 \\ &= 67{,}800 \text{ lb} = 67.8 \text{ kips} \end{aligned}$$

and

$$V_c/2 = 33.9 \text{ kips}$$

From the shear envelope in Fig. 6-29c, the maximum shear at d from the face of the support is $V_u/\phi = 116$ kips. Because 116 kips exceeds 33.9 kips, web reinforcement is required.

2. Anchorage of stirrups. Try No. 3 Grade-40 U stirrups with 135° hooks around a longitudinal bar in each upper corner. ACI Code Section 25.7.1.3 considers such stirrups to be anchored.

3. Find the maximum stirrup spacing.

Based on beam depth: ACI Code Table 9.7.6.2.2 gives a maximum spacing of $d/2 = 31.6\text{ in.}/2 = 15.8\text{ in.}$ or 24 in. If V_s exceeds $4\sqrt{f'_c}\,b_w d = 136$ kips, which corresponds to $V_u/\phi = (4 + 2)\sqrt{f'_c}\,b_w d = 204$ kips, ACI Code Table 9.7.6.2.2 requires the maximum spacing to be reduced to $d/4$ or 12 in.

The maximum shear at d from the face of a support is $V_u/\phi = 116$ kips. Because this value is less than 204 kips, the maximum spacing based on beam depth is $d/2 = 15.8$ in.

Based on minimum A_v:

$$A_{v,\min} = 0.75\sqrt{f'_c}\,\frac{b_w s}{f_{yt}} \tag{6-20}$$

but not less than

$$A_{v,\min} = \frac{50 b_w s}{f_{yt}} \tag{6-21}$$

For the given concrete strengths, Eq. (6-20) will govern, so

$$s_{\max} = \frac{A_v f_{yt}}{50.3 b_w} = \frac{2 \times 0.11\text{ in.}^2 \times 40{,}000\text{ psi}}{50.3\text{ psi} \times 16\text{ in.}} = 10.9\text{ in. (governs)}$$

4. Compute the stirrup spacing required to resist the factored shear force.

$$s \le \frac{A_v f_{yt} d}{\dfrac{V_u}{\phi} - V_c} \tag{6-17}$$

where, from step 1, $V_c = 67.8$ kips.

At d from support,

$$\frac{V_u}{\phi} = 116\text{ kips and}$$

$$s \le \frac{2 \times 0.11\text{ in.}^2 \times 40\text{ ksi} \times 31.6\text{ in.}}{(116\text{ kips} - 67.8\text{ kips})}$$

$$\le 5.77\text{ in.}$$

This is a very tight spacing, so change to No. 4 double-leg stirrups ($A_v = 2 \times 0.20\text{ in.}^2 = 0.40\text{ in.}^2$). Use Eq. (6-17) to calculate a new spacing required for strength.

$$s \le \frac{0.40\text{ in.}^2 \times 40\text{ ksi} \times 31.6\text{ in.}}{116\text{ k} - 67.8\text{ k}} = 10.5\text{ in.}$$

Thus, use No. 4 double-leg stirrups at a spacing of 10 in. near the support with the first stirrup placed at $s/2 = 5$ in. from the face of the support.

We now will find the point where we can change to No. 3 double-leg stirrups ($A_v = 0.22$ in.^{2}) at a spacing of 10 in., which satisfies the maximum permissible spacing of 10.9 in. Using Eq. (6-14),

$$V_s = \frac{A_v f_{yt} d}{s} = \frac{0.22 \text{ in.}^2 \times 40 \text{ ksi} \times 31.6 \text{ in.}}{10 \text{ in.}} = 27.8 \text{ kips}$$

Then, using Eq. (6-9),

$$V_n = V_c + V_s = 67.8 \text{ k} + 27.8 \text{ k} = 95.6 \text{ kips}$$

Using this value for V_n and the shear envelope for V_u/ϕ in Fig. 6-29c, we can determine where we can change to No. 3 stirrups at 10 in.:

$$x_1 = \frac{136 \text{ k} - 95.6 \text{ k}}{136 \text{ k} - 19.2 \text{ k}} \times 15 \text{ ft} \times 12 \text{ in./ft} = 62.3 \text{ in.}$$

We also can find the point where $V_u/\phi = V_c/2 = 33.9$ kips, and thus, stirrups are no longer required.

$$x_2 = \frac{136 \text{ k} - 33.9 \text{ k}}{136 \text{ k} - 19.2 \text{ k}} \times 180 \text{ in.} = 157 \text{ in.}$$

We can use these values to finalize the selection of stirrups for this interior span of a continuous beam:

One No. 4 at 5 in.

Six No. 4 at 10 in. (extends to 65 in. from face of support)

Nine No. 3 at 10 in. (extends to 155 in. from face of support)

This leaves a beam segment of approximately 15 in. on each side of the beam midspan without stirrups. The ACI Code does not require any stirrups in this region, but some designers will provide two or three stirrups in this region to support the longitudinal reinforcement. ■

EXAMPLE 6-3 Design of Shear Reinforcement for a Girder Subjected to Concentrated Loads

Figure 6-30a shows an elevation view for a continuous girder (similar to those in Fig. 5-1c) subjected to concentrated loads from floor beams that frame into the girder at each of the third-points of its span. These concentrated loads represent all of the dead and live loads acting in the portion of the floor system supported by this girder. From a design of the floor beams (not given here), the concentrated dead load is $P_D = 32.2$ kips, and the concentrated live load is $P_L = 14.9$ kips. The girder also is subjected to a distributed dead load due to its self-weight, which is assumed to include only the portion of the web below the floor slab (Fig. 6-30c).

$$w_D = \frac{16 \text{ in.} \times 20 \text{ in.}}{144 \text{ in.}^2/\text{ft}^2} \times 0.15 \text{ k/ft}^3 = 0.33 \text{ k/ft}$$

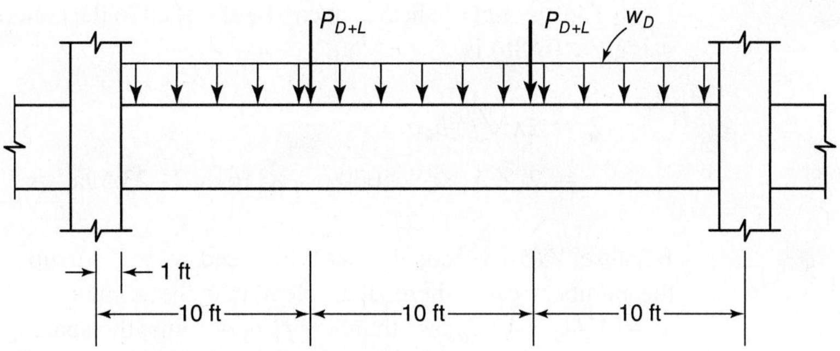

(a) Loads on girder for maximum shear at face of support.

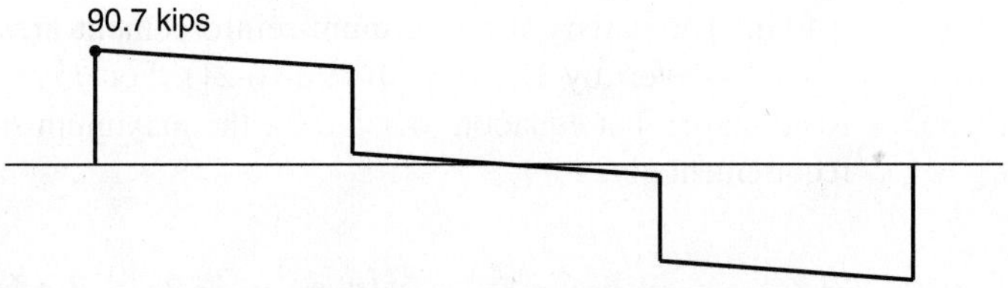

(b) Shear design envelope, V_u/ϕ, for loading in part (a).

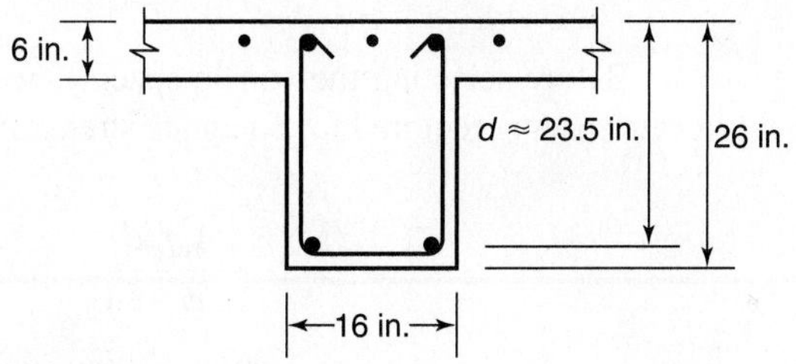

(c) Girder section at support.

Fig. 6-30
Loading, shear envelope, and girder section for Example 6-3.

The girder is constructed with normal-weight concrete with $f'_c = 5000$ psi. The yield strength of the steel reinforcement for the stirrups is $f_{yt} = 40$ ksi.

1. Design shear reinforcement for the exterior thirds of the girder. Assuming similar span lengths for the adjacent girders and similar column stiffnesses at each end of the girder, the general shape of the shear design envelope, V_u/ϕ, is given in Fig. 6-30b for the loading condition in Fig. 6-30a. It is clear that there is a relatively large and almost constant design shear in the exterior thirds of the girder span. Thus, we will not consider finding the design shear at a distance, d, from the face of the support, but rather we will design for the shear at the face of the support. Using a clear span, ℓ_n, of 28 ft (Fig. 6-30a), the factored shear force at the support face is

$$V_u = 1.2P_D + 1.6P_L + 1.2\frac{w_D\ell_n}{2}$$

$$= 1.2 \times 32.2\text{ k} + 1.6 \times 14.9\text{ k} + 1.2\frac{0.33\text{ k/ft} \times 28\text{ ft}}{2} = 68.0\text{ kips}$$

Then, the value for the shear design envelope at this point is

$$(V_u/\phi)_{\max} = 68.0\text{ k}/0.75 = 90.7\text{ kips}$$

The amount of shear that can be assigned to the concrete for this girder can be found using Eq. (6-8b) is

$$V_c = 2\lambda\sqrt{f'_c}\,b_w d$$
$$= 2 \times 1 \times \sqrt{5000}\text{ psi} \times 16\text{ in.} \times 23.5\text{ in.} = 53{,}200\text{ lbs} = 53.2\text{ kips}$$

Because V_u/ϕ exceeds V_c, we will need to use stirrups for additional strength. From the numbers given here, it is clear that the amount of shear assigned to the stirrups, $V_s = V_u/\phi - V_c$, is less than $4\sqrt{f'_c}\,b_w d$. Thus, the spacing limit in the first row of Table 9.7.6.2.2 applies, and the maximum stirrup spacing is $d/2 = 11.8$ in.

Assuming that we will use a two-leg No. 4 stirrup ($A_v = 2 \times 0.20\text{ in.}^2 = 0.40\text{ in.}^2$) to satisfy the minimum-reinforcement area requirement in ACI Code Table 9.6.3.3—given by Eqs. (6-20) and (6-21). For $f'_c = 5000$ psi, Eq. (6-20) will govern. Rearranging that equation to solve for the maximum spacing to satisfy the minimum area requirement gives

$$s_{\max} = \frac{A_v f_{yt}}{0.75\sqrt{f'_c}\,b_w} = \frac{0.40\text{ in.}^2 \times 40{,}000\text{ psi}}{0.75 \times \sqrt{5000}\text{ psi} \times 16\text{ in.}} = 18.9\text{ in.}$$

Before selecting the stirrup spacing, we also must use Eq. (6-17) to determine the stirrup spacing required for adequate strength:

$$s \le \frac{A_v f_{yt} d}{V_u/\phi - V_c}$$

$$s \le \frac{0.40\text{ in.}^2 \times 40\text{ ksi} \times 23.5\text{ in.}}{90.7\text{ k} - 53.2\text{ k}} = 10.0\text{ in.}$$

The strength requirement governs, so use a stirrup spacing of 10 in. with the first stirrup placed at 5 in. from the face of the support. A typical beam section near the support is shown in Fig. 6-30c, and the final stirrup spacing in the exterior thirds of the girder is:

One No. 4 stirrup at 5 in. (from the face of the support)
Eleven No. 4 stirrups at 10 in. (extends to 115 in. from the face of the support)

2. Select stirrup size and spacing for the interior third of the girder span. If the loads are applied symmetrically to this girder, the values for the design shear envelope in the middle third are very low (Fig. 6-30b). To properly design this region of the girder, we must assume an unsymmetrical placement of the live load, as shown in Fig. 6-31a. Assuming that the ends of the girder are fixed, fixed-end moments for this loading case [6-20] can be used to calculate the resulting shear force diagram in Fig. 6-31b. Combining this shear force with the contribution from the distributed dead load (the weight of the girder web) results in a shear design envelope similar to that in Fig. 6-31c. The maximum value at the beginning of the middle third of the girder is

$$(V_u)_{\max} = 1.6 \times 0.24 \times P_L + 1.2 \times w_D \times 5\text{ ft}$$
$$= 1.6 \times 0.24 \times 14.9\text{ k} + 1.2 \times 0.33\text{ k/ft} \times 5\text{ ft} = 7.7\text{ kips}$$

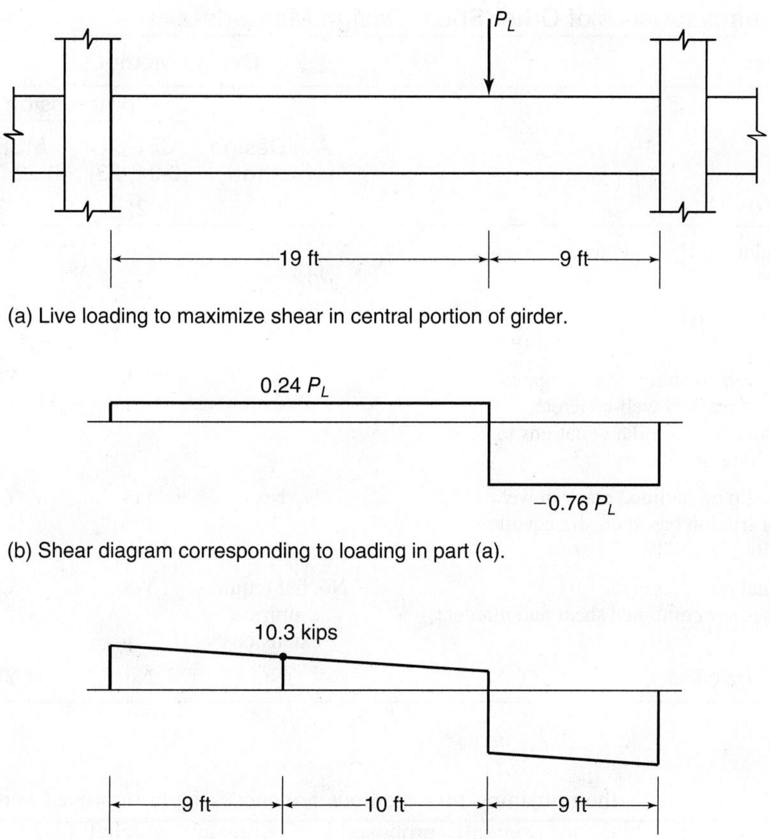

Fig. 6-31
Alternate loading case for Example 6-3.

(c) Shear design envelope, V_u/ϕ, for alternate loading case.

Dividing this by ϕ, we get

$$(V_u/\phi)_{\max} = 7.7 \text{ k}/0.75 = 10.3 \text{ kips}$$

This value is less than $V_c/2 = 26.6$ kips, so no stirrups are required in the middle portion (between the concentrated loads) of this girder. However, it may be useful to place No. 3 stirrups at a 12-in. spacing (approximately $d/2$) to hold the longitudinal bars in place. ■

6-5 OTHER SHEAR DESIGN METHODS

In most modern concrete codes, shear design is based on Eq. (6-9),

$$V_n = V_c + V_s$$

where V_c is the "shear carried by the concrete." The ACI Code value for V_c was empirically derived from test results. Ever since reinforced concrete began to be studied, a mechanical explanation of V_c has been the *Holy Grail* sought by researchers. Today, the underlying mechanisms of shear transfer are understood enough to allow shear design based at least partially on theory, instead of on equations extrapolated from test results. An excellent review of the current status of modern shear design methods is given in [6-2].

The details of five methods of designing for shear in slender beams are compared in Table 6-1. The first method is the *traditional ACI method*, discussed in Section 6-4. Three of

TABLE 6-1 Characteristics of Other Shear Design Methods

Features	Design Methods				
		Compression Field Theories			Shear Friction [6-27, 28]
	ACI Design Method	CFT-84 [6-21, 23]	MCFT-94 [6-22, 24, 25]	MCFT-04 [6-26]	
Notes	1.	2.	3.	4.	5.
1. Includes additive V_c and V_s terms giving $V_n = V_c + V_s$	Yes	No	Yes	Yes	Yes
2. Uses a Mohr's circle for strain to derive equations used to check compatibility	No	Yes	Yes	Yes	No
3. Considers web crushing. Uses smeared properties of cracked web-concrete, based on [6-22], or similar equations to check crushing of the web	No	Yes	Yes	Yes	No
4. Considers slip on inclined crack in web. Uses shear friction based on slip equations proposed [6-27], [6-28]	No	Yes	Yes	Yes	Yes
5. Longitudinal reinforcement proportioned for combined shear and moment	No, but requires empirical extensions	Yes	Yes	Yes	Yes
6. Considers size effect	No	No	Yes	Yes	No

the remaining four methods are successively improved versions of the compression field theory originally proposed by Collins and Mitchell [6-21] and later modified by Vecchio and Collins [6-22]. The original compression field theory served as the basis for shear design in the 1984 edition of the Canadian Concrete Code [6-23] and is labeled as CFT-84 in column 2 of Table 6-1. The modified compression field theory, labeled as MCFT-94 in column 3 of Table 6-1, served as the basis for shear design in both the 1994 edition of the Canadian Concrete Code [6-24] and the 1998 LRFD Bridge Specifications of the American Association of State Highway and Transportation Officials [6-25]. After some further adjustment, the modified compression field theory continues to serve as the basis for shear design in the most recent edition of the Canadian Concrete Code [6-26] and is labeled as MCFT-04 in column 4 of Table 6-1.

The final procedure listed in column 5 of Table 6-1 is based on an application of shear friction to slender beams failing in shear. Except for the CFT-84 method, the design methods are based on Eq. (6-9), although each of the methods has a unique way of calculating V_c, as shown in Table 6-1.

Definitions and Design Methods

Design procedures for shear fall into two classifications:

1. *Whole-member design methods*, in which the member is idealized as a *truss* or a *strut-and-tie model* that represents the complete load-resisting mechanism for one load case.

2. *Sectional design methods*, in which shear-force and bending-moment envelopes are derived for all significant load cases and design is carried out cross section by cross section for the envelope values at sections along the length of the beam.

Both methods are often required for a successful design.

Two other concepts that need to be defined are as follows:

1. *Smeared-crack models* replace cracked reinforced concrete with a hypothetical new material that does not crack, as such. The material has properties that are *averaged* or *smeared* over a long gauge length that includes the width of one or more cracks. This approach reduces the discontinuities in post-cracking behavior.

2. A shear-sensitive region must satisfy both equilibrium and *compatibility*. Compatibility relationships for shear cracking of concrete and post-cracking deformations and strengths have been derived from Mohr's circles. The Mohr's circle of strain is used to determine relationships between deformations [6-21] and [6-22]. The Mohr's circle of stress is used to determine relationships between stresses.

Traditional ACI Design Procedure

The traditional ACI design method uses V_c, which we shall write as

$$V_c = \beta\lambda\sqrt{f'_c}b_w d \tag{6-8c}$$

where β is an empirically derived function whose value is taken to be 2 in Eq. (6-8b), λ is a modification factor used to allow for lightweight concrete, and d is the distance from the extreme compression face to the centroid of the longitudinal tension reinforcement. In the ACI Code, $V_s = A_v f_{yt}\, d/s$, which implies that the cracks are at an angle close to 45°. Other design expressions have similar equations for V_s, except for the shear friction method, which computes V_s on the basis of the number of stirrups and other bars crossed by an inclined shear plane or crack crossing the web at a given slope.

Compression Field Theories

Three of the design procedures in Table 6-1 use the compression field theory to model the given structure. This theory is the inverse of the *tension field theory* developed by Wagner [6-29] in 1929 for the design of light-gauge plates in metal airplane fuselages. If a light-gauge metal web is loaded in shear, it buckles due to the diagonal compression in the web. Once buckling has occurred, further increases in shear require the web shear mechanism to be replaced by a truss or a field of inclined tension forces between the buckles in the web. This diagonal *tension field* in turn requires a truss that includes vertical compression struts and longitudinal compression chords to resist the reactions from the tension diagonals in the web.

The compression field theory (CFT) is just the opposite of the tension field theory. In the CFT, the web of the beam cracks due to the principal tension stresses in the web. Cracking reduces the ability of the web to transmit diagonal tension forces across the web. After cracking, loads are carried by a truss-like mechanism that consists of a field of hypothetical diagonal compression members between the cracks combined with the stirrups and the longitudinal bars serving as tension members.

The CFT from the 1984 Canadian Concrete Code [6-23] did not have a V_c term. Instead, stirrups were provided for the full shear. In designing a structure, it was necessary to check web crushing by using stresses and strains derived from Mohr's circles. In design, the principal compression angle θ was assumed and was used to compute web stresses and the capacity of the concrete. In the CFT-84, the angle θ could have any value between 15 and 75°, as long as the same angle was used for all the calculations at a given section.

In the 1994 version of the modified compression field theory, [6-24], [6-25]

$$V_c = \beta\lambda\sqrt{f'_c}b_w\,d_v \tag{6-8d}$$

where, for beams with at least the minimum shear reinforcement required by the code, λ is a modifier for lightweight concrete, similar to that used in the ACI Code, and the factor β is a function of the strut angle θ. Tables of values of β and θ, which had been computed iteratively, are used to select β and θ to minimize the total reinforcement costs. The resulting set of θ and β values oscillated widely for practical beams.

For members with less than minimum web reinforcement, β and θ were functions of the widths of shear cracks, taken as the product of the longitudinal strain and the crack spacing. The value of β was a function of:

(a) an *index value* of the *shear stress ratio* v/f'_c.

(b) an *index value* of the *longitudinal strain* ε_x, computed as the strain in the longitudinal reinforcement due to flexure, shear, and normal force.

(c) a *crack spacing factor* s_{ze}, related to the distance s_z between reinforcement crossing the cracks, and

(d) the crack angle θ.

Special values of s_{ze} were given for members without stirrups or for members with less than the minimum stirrups specified in ACI Code Table 9.6.3.3. Items (a) and (b) are referred to as *index* values because the exact values of v/f'_c and ε_x are less important in selecting β and θ than the trends in these variables. The longitudinal strain ε_x was taken to act at the level of the longitudinal tension reinforcement.

The 2004 version of the MCFT (called MCFT-04 in Table 6-1) is used in the 2004 Canadian Concrete Code [6-26]. In it, equations are given for θ, from which β can be calculated. The objective function used to determine the most economical value was relaxed, allowing much simpler equations for β.

Shear Friction Method

The shear friction method for beam shear takes V_c to be the shear force transferred across each of a series of inclined sections cut through the entire beam, representing inclined cracks or shear slip planes. For simplicity, the inclined sections are taken to be straight. The shear force V_{ci} on the inclined plane is based on one of several equations derived by Walraven [6-27] or Loov and Patnaik [6-28] for the shear transferred across the shear plane in a composite beam at the onset of slip.

The clamping force needed to mobilize shear friction stress v_{ci} along the crack is taken as the sum of the components, perpendicular to the slip plane, of the yield forces A_sf_y from the longitudinal reinforcement and the stirrups A_vf_{yt} crossed by the crack.

Equivalence of Shears on Inclined and Vertical Planes

For an element from the web, the shear force V_{ci} on the inclined slip plane can be related to the shear on a vertical plane as shown in Fig. 6-32. On the vertical plane,

$$V_c = v_c\,b_w\,d_v. \tag{6-23a}$$

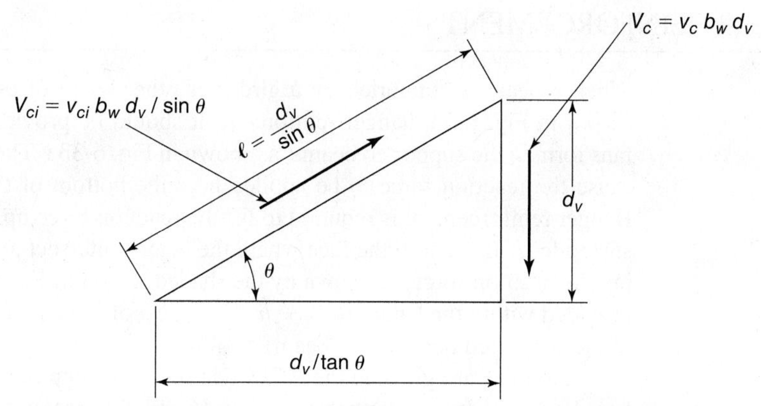

Fig. 6-32
Equivalence of shear stresses acting on an element.

The length of the inclined section is

$$\ell = (d_v/\sin\theta)$$

The inclined shear force on the sloping side of the element is

$$V_{ci} = v_{ci}b_w(d_v/\sin\theta) \qquad (6\text{-}24)$$

where v_{ci} is the average shear stress parallel to the inclined plane, b_w is the width of the web of the beam, d_v is the effective depth for shear, and θ is the angle between the slip plane and the longitudinal flexural reinforcement. The vertical component of V_{ci} is $V_c = V_{ci}\sin\theta$. Substituting this expression into Eq. (6-23a) gives

$$V_c = (v_{ci}b_w d_v \sin\theta/\sin\theta) \qquad (6\text{-}23b)$$

The $\sin\theta$ terms cancel out, leaving

$$V_c = v_{ci}b_w d_v$$

From this equation and Eq. (6-23a), it follows that $v_{ci} = v_c$. Thus, for a given shear force, the average shear stresses v_{ci} on an inclined plane are the same as the vertical shear stresses v_c on the vertical plane due to the same loading.

Two relationships for the shear stress that can be transferred across an inclined crack are, from Walraven [6-27]

$$V_{ci} = \frac{2.16\sqrt{f'_c}}{0.3 + \dfrac{24w}{a + 0.63}} \qquad (6\text{-}25)$$

and, from Loov and Patnaik [6-28],

$$v_{ci} = 0.6\lambda\sqrt{\alpha f'_c} \qquad (6\text{-}26)$$

For a given set of loads, concrete stresses, and crack widths, taking into account the different assumptions, it can be shown that the magnitude of V_{ci} from the MCFT-04 and that from the shear friction method are similar. This suggests that in many cases the "shear carried by the concrete" is closely related to shear friction on the crack surfaces and V_c is due to shear friction along the crack.

6-6 HANGER REINFORCEMENT

When a beam is supported by a girder or other beam of essentially the same depth, as shown in Fig. 6-33, hanger reinforcement should be provided in the joint. Compression fans form in the supported beams, as shown in Fig. 6-33a. The inclined compression forces cause the reaction force to be applied near the bottom of the supporting beam (girder). Hanger reinforcement is required to lift the reaction force up to the top of the girder. Only stirrup legs adjacent to the face where the beams intersect are considered effective in lifting the reaction force, as shown by the shaded region in Fig. 6-33b. Stirrup legs should be provided within the length $(b_{w2} + h_2/2 + 2h_b)$ of the girder and within the length $(d_2/4)$ of the supported beam (floor beam).

No rules are given in the ACI Code for the design of such reinforcement. However, based on a study by Mattock and Shen [6-30] and provisions in the 1984 edition of the Canadian Concrete Code [6-23], the author recommends that the total vertical reinforcement area, A_h, (sum of areas of stirrup legs) have a tensile capacity of:

$$\phi A_h f_{yt} \geq \left(1 - \frac{h_b}{h_1}\right) V_{u2} \tag{6-27}$$

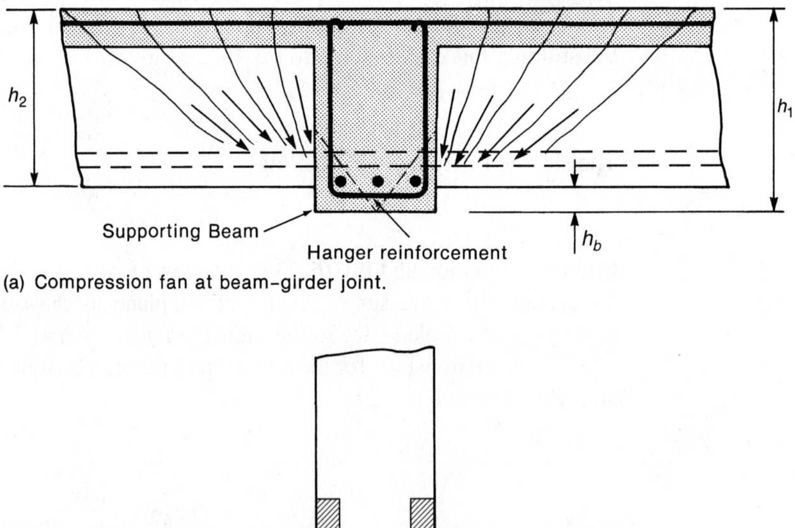

(a) Compression fan at beam-girder joint.

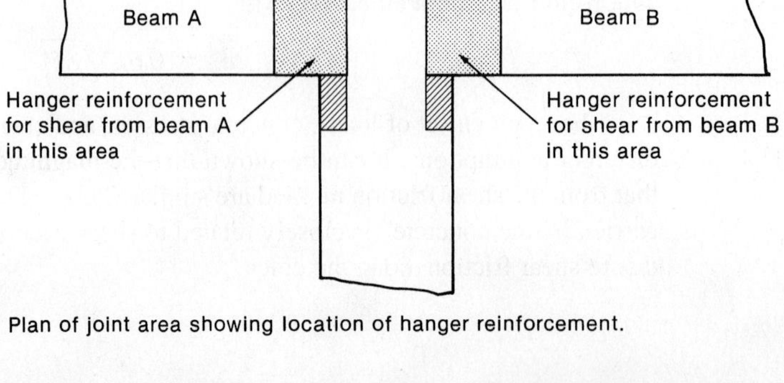

(b) Plan of joint area showing location of hanger reinforcement.

Fig. 6-33
Hanger reinforcement.

where,

A_h = the area of hanger reinforcement adjacent to one face of the supporting beam,
b_{w2} = the width of the supported beam,
d_2 = the effective depth of the supported beam,
h_b = the vertical distance from the bottom of the supporting beam to the bottom of the supported beam,
h_1 = the overall depth of the supporting beam,
h_2 = the overall depth of the supported beam, and
V_{u2} = the factored shear at the end of the supported beam.

If shears are transferred to both side faces of the supporting beam, Eq. (6-31) is to be satisfied separately for each face. These provisions can be waived if the shear, V_{u2}, at the end of the supported beam is less than $3\sqrt{f'_c}b_{w2}d_2$, because inclined cracking is not fully developed for this value of shear force.

The hanger reinforcement should be well anchored, top and bottom. Also, the lower layer of reinforcement in the supported beam should always be placed above the reinforcement in the supporting beam.

EXAMPLE 6-4 Design of Hanger Reinforcement in a Beam-Girder Junction

Figure 6-34a shows a floor beam-girder joint. Each floor beam transfers a factored end shear of 45 kips to the girder. If needed, design hanger reinforcement assuming a concrete compressive strength of 4000 psi and a steel yield strength of 60 ksi. Assume the shear reinforcement in both the floor beam and the girder consists of No. 3 double-leg stirrups.

Do we need to check design for hanger reinforcement?

$$3\sqrt{f'_c}\, b_{w2}d_2 = 42.1 \text{ k} < 45 \text{ k (so check hanger reinforcement)}$$

Design of hanger reinforcement:

From Eq. (6-27), and referring to Fig. 6-34a, we can find the required area of hanger reinforcement for each side of the girder.

$$\phi A_h f_{yt} \geq \left(1 - \frac{h_b}{h_1}\right) V_{u2} = \left(1 - \frac{9 \text{ in.}}{30 \text{ in.}}\right) \times 45 \text{ k} = 31.5 \text{ k}$$

So,

$$A_h \geq \frac{31.5 \text{ k}}{0.75 \times 60 \text{ ksi}} = 0.70 \text{ in.}^2$$

Use seven No. 3 legs, giving $A_h = 7 \times 0.11 \text{ in.}^2 = 0.77 \text{ in.}^2$, on each face of the girder. The stirrup legs could be placed as shown in Fig. 6-34b. The author believes hanger reinforcement should be added as a *supplement* to existing shear reinforcement. ■

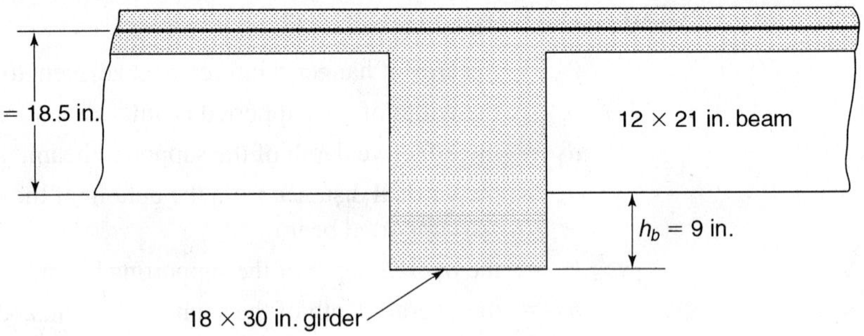

(a) Beams.

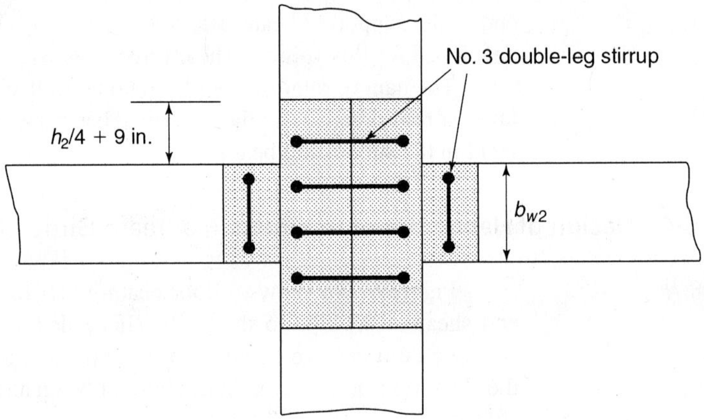

Fig. 6-34
Example 6-4.

(b) Plan view of joint zone.

6-7 TAPERED BEAMS

In a prismatic beam, the average shear stress between two cracks is calculated as

$$v = \frac{V}{b_w jd} \tag{6-4}$$

which is simplified to

$$v = \frac{V}{b_w d} \tag{6-5}$$

In the derivation of Eq. (6-4), it was assumed that jd was constant. If the depth of the beam varies, the compressive and tensile forces due to flexure will have vertical components. A segment of a tapered beam is shown in Fig. 6-35. The moment, M_1, at the left end of the section can be represented by two horizontal force components, C and T, separated by the lever arm jd. The tension force actually acts parallel to the centroid of the reinforcement and hence has a vertical component $T \tan \alpha_T$, where α_T is the angle between the tensile force and horizontal. Similarly, the compressive force acts along a line joining the centroids

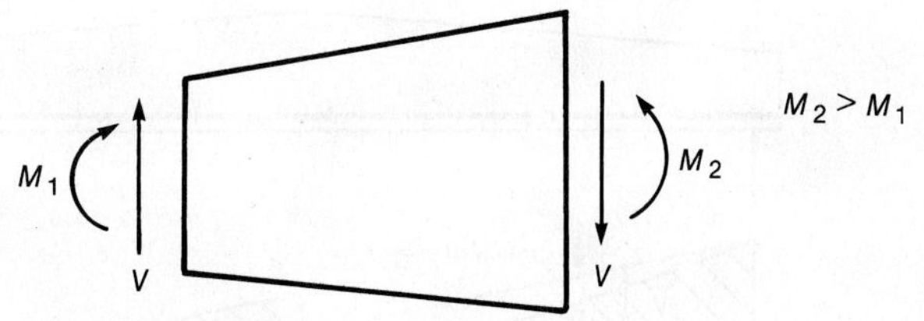

(a) Forces on segment of beam.

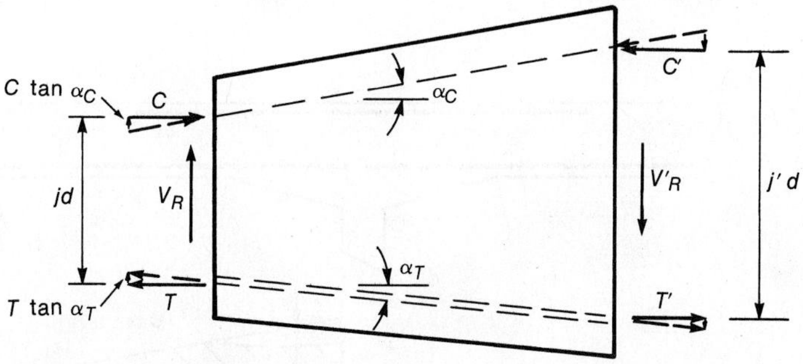

(b) Internal forces and shears.

Fig. 6-35
Reduced shear force in nonprismatic beam.

of the stress blocks at the two sections and hence has a vertical component $C \tan \alpha_c$. The shear force on the left end of the element can be represented as

$$V = V_R + C \tan \alpha_c + T \tan \alpha_T$$

where V_R is the reduced shear force resisted by the stirrups and the concrete. Substituting $C = T = M/jd$ and letting $\alpha = \alpha_c + \alpha_T$ gives

$$V_R = V - \frac{|M|}{jd} \tan \alpha \tag{6-28}$$

where $|M|$ represents the absolute value of the moment and α is positive if the lever arm jd increases in the same direction as $|M|$ increases.

The shear stresses in a tapered beam then become

$$v = \frac{V_R}{b_w d} \tag{6-29}$$

Examples for the use of Eq. (6-28) are shown in Fig. 6-36.

6-8 SHEAR IN AXIALLY LOADED MEMBERS

Reinforced concrete beams can be subjected to shear plus axial tensile or compressive forces due to such causes as gravity load effects in inclined members and stresses resulting from restrained shrinkage or thermal deformations. Similarly, wind or seismic forces cause

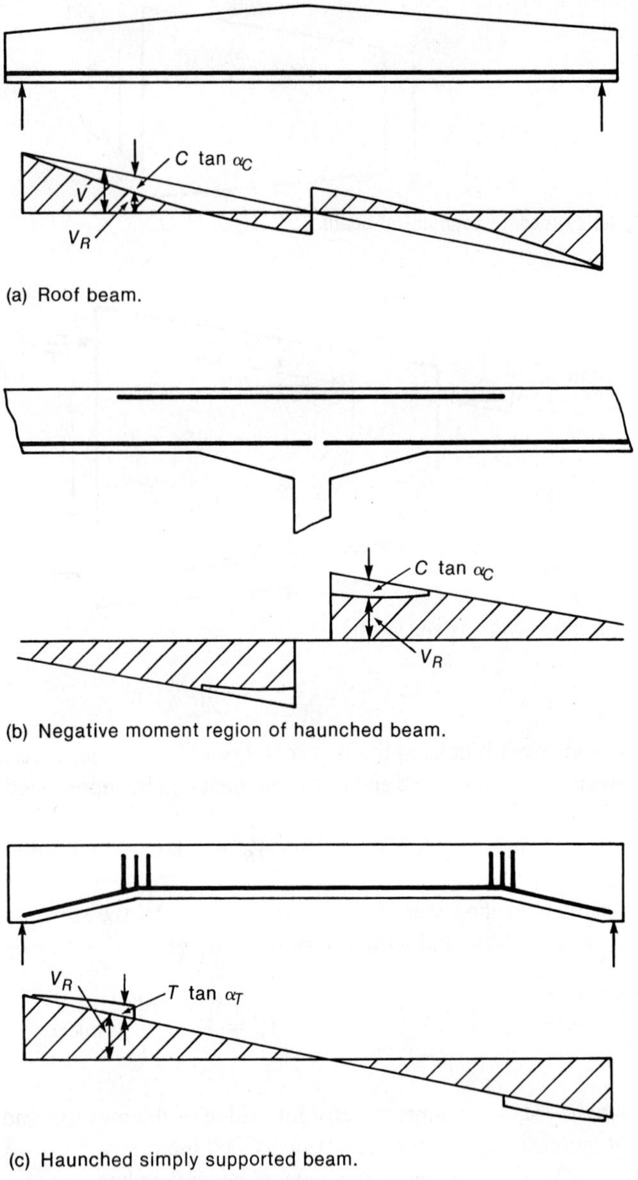

Fig. 6-36
Examples of V_R.

shear forces in axially loaded columns. Figure 6-37 shows a tied column that failed in shear during an earthquake. In seismic regions, columns are particularly vulnerable to shear failures. Reversed loading cycles cause crisscrossing inclined cracks (Fig. 6-45), which cause V_c to decrease to zero. As a result, special calculation procedures and special details are required in seismic regions. (See Chapter 19.)

Axial forces have three major effects on the shear strength. An axial compressive or tensile force will increase or reduce, respectively, the load at which flexural and inclined cracks occur. If V_c is assumed to be related to the inclined-cracking shear, as is done in the ACI Code, this will directly affect the design. If they have not been considered in the design, axial tensile forces may lead to premature yielding of the longitudinal reinforcement, which in turn will effectively do away with any transfer of shear by aggregate interlock.

Fig. 6-37
Shear failure in a tied column, 1971 San Fernando earthquake. (Photograph courtesy of U.S. National Bureau of Standards.)

Axial Tension

For axial tensile loadings, the nominal shear carried by the concrete is given by

$$V_c = 2\left(1 + \frac{N_u}{500A_g}\right)\lambda\sqrt{f'_c}b_w d \qquad \text{(6-13b)}$$

where N_u/A_g is expressed in psi and is negative in tension. The term inside the parentheses becomes zero when the factored axial stress on the section reaches or exceeds 500 psi in tension, which is roughly the tensile strength of concrete.

This expression tends to be conservative, especially for high tensions, as shown by the data points in the right-hand portion of Fig. 6-15. Although the evidence is ambiguous, tests have shown that beams subjected to tensions large enough to crack them completely through can resist shears approaching those for beams not subjected to axial tensions [6-6]. This shear capacity results largely from aggregate interlock along the tension cracks. It should be noted, however, that if the longitudinal tension reinforcement yields under the action of shear, moment, and axial tension, the shear capacity drops very significantly. This is believed to have affected the failure of the beam shown in Fig. 6-2.

Axial Compression

Axial compression tends to increase the shear strength. The ACI Code presents the following equation for calculating V_c for members subjected to combined shear, moment, and axial compression:

$$V_c = 2\left(1 + \frac{N_u}{2000A_g}\right)\lambda\sqrt{f'_c}\,b_w d \text{ lb} \tag{6-13a}$$

Here, N_u/A_g is positive in compression and has units of psi.

The design of a beam subjected to axial compression or tension is identical to that for a beam without such forces, except that the value of V_c is modified.

EXAMPLE 6-5 Checking the Shear Capacity of a Column Subjected to Axial Compression Plus Shear and Moments

A 12-in. × 12-in. column with normal-weight concrete $f'_c = 5000$ psi and ties having $f_{yt} = 40$ ksi is subjected to factored axial forces, moments, and shears, as shown in Fig. 6-38.

1. Compute the nominal shear force in the column. Summing moments about the centroid at one end of the column, we find that the factored shear is

$$V_u = \frac{(42 + 21)\text{ kip-ft}}{10\text{ ft}} = 6.3\text{ kips}$$

$$\frac{V_u}{\phi} = \frac{6.3}{0.75} = 8.40\text{ kips}$$

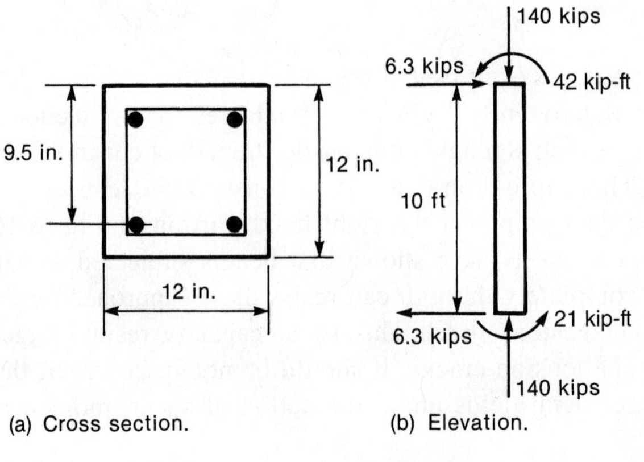

Fig. 6-38
Shear in column—
Example 6-5.

2. Are stirrups required by ACI Code Section 10.6.2.1? No stirrups are required if $V_n < V_c/2$, where

$$V_c = 2\left(1 + \frac{N_u}{2000A_g}\right)\lambda\sqrt{f'_c}b_w d \tag{6-13}$$

$$= 2\left(1 + \frac{140{,}000}{2000 \times 144}\right)1 \times \sqrt{5000} \times 12 \times 9.5 = 24{,}000 \text{ lbs} = 24.0 \text{ kips}$$

and $V_c/2 = 12.0$ kips. Because V_u/ϕ is less than $V_c/2$, shear reinforcement is not necessary. Minimum tie requirements for columns will be discussed in Chapter 11. ■

PROBLEMS

6-1 For the rectangular beam shown in Fig. P6-1,

(a) Draw a shear-force diagram.

(b) Assuming the beam is uncracked, show the direction of the principal tensile stresses at middepth at points A, B, and C.

(c) Sketch, on a drawing of the beam, the inclined cracks that would develop at A, B, and C.

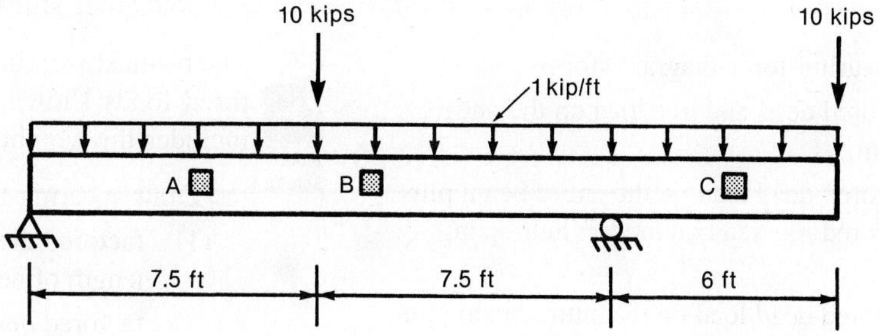

Fig. P6-1

6-2, 6-3, 6-4, and 6-5 Compute ϕV_n for the cross sections shown in Figs. P6-2, P6-3, P6-4, and P6-5. In each case, use $f'_c = 4000$ psi and $f_{yt} = 40{,}000$ psi.

Video Solution P6-5

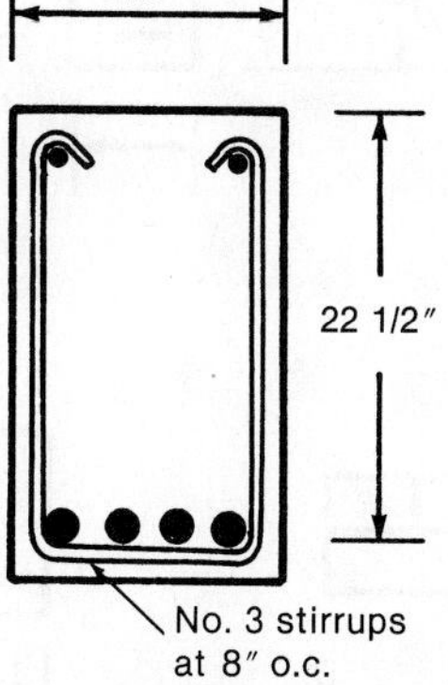

Fig. P6-2

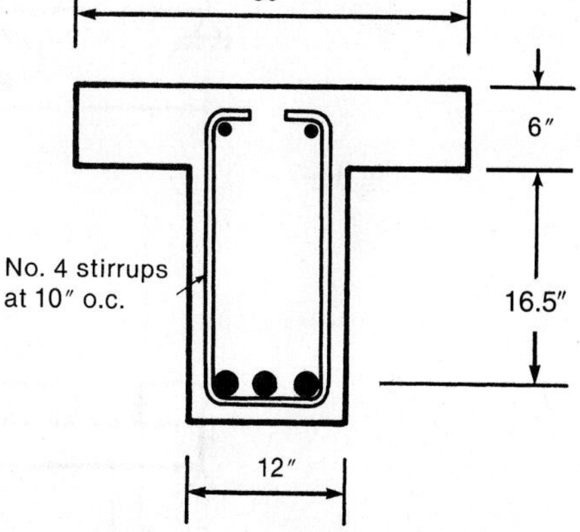

Fig. P6-3

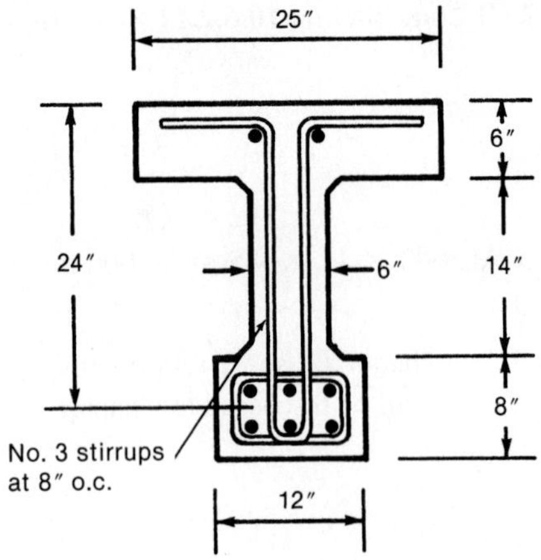

Fig. P6-4

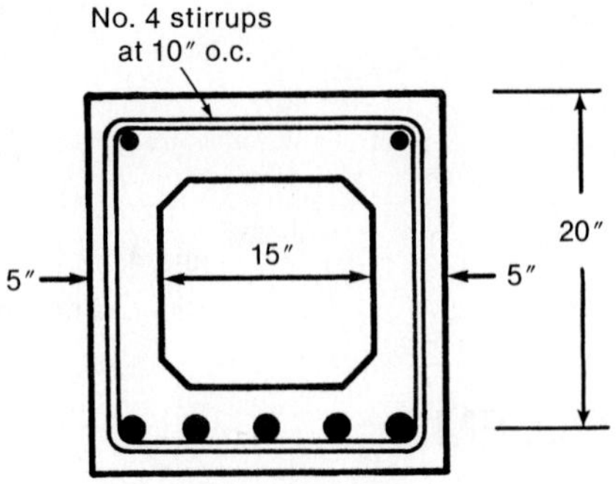

Fig. P6-5

6-6 ACI Code Table 9.7.6.2.2 sets the maximum spacing of vertical stirrups at $d/2$. Explain why.

6-7 The beam shown in Fig. P6-7 supports the unfactored loads shown. The dead load includes the weight of the beam.

(a) Draw shearing-force diagrams for

(1) factored dead and live load on the entire beam;

(2) factored dead load on the entire beam plus factored live load on the left half-span; and

(3) factored dead load on the entire beam plus factored live load on the right half-span.

(b) Superimpose the diagrams to get a shear-force envelope. Compare the shear at midspan to that from Eq. (6-22).

(c) Design stirrups. Use $f'_c = 4500$ psi and No. 3 double-leg stirrups with $f_{yt} = 60{,}000$ psi.

6-8 The beam shown in Fig. P6-8 supports the unfactored loads shown in the figure. The dead load includes the weight of the beam.

Video Solution

(a) Draw shearing-force diagrams for

(1) factored dead and live load on the entire length of beam;

(2) factored dead load on the entire beam plus factored live load between B and C; and

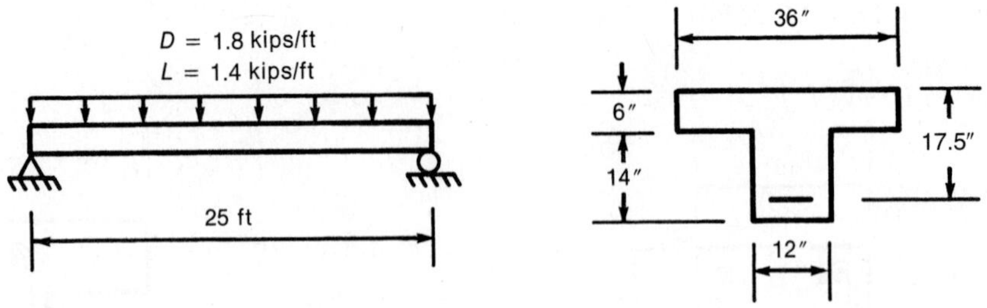

Fig. P6-7

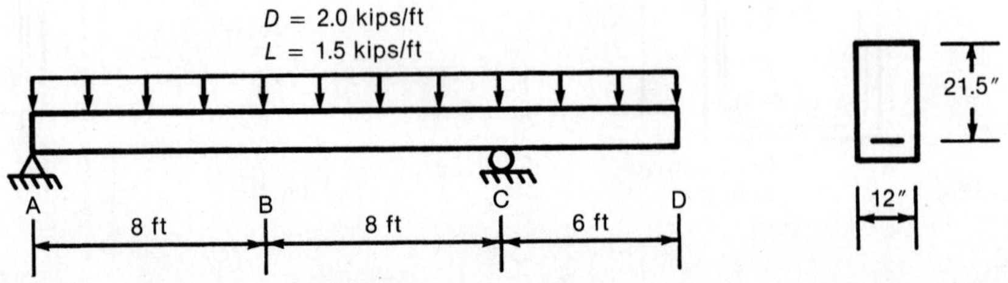

Fig. P6-8

(3) factored dead load on the entire beam plus factored live load between A and B and between C and D. Loadings (2) and (3) will give the maximum positive and negative shears at B.

(b) Draw the factored shear-force envelope.

(c) Design stirrups. Use $f'_c = 5000$ psi and $f_{yt} = 40{,}000$ psi.

6-9 Figure P6-9 shows an interior span of a continuous beam. The shears at the ends are $\pm w_u \ell_n / 2$. The shear at midspan is from Eq. (6-22).

(a) Draw a shear-force envelope.

(b) Design stirrups, using $f'_c = 4000$ psi and $f_{yt} = 40{,}000$ psi.

6-10 Design shear reinforcement for the C1–C2 span of the girder designed in Example 5-6 (final section given in Fig. 5-32). From the structural analysis discussed in Example 5-6, the factored design end shears for this girder are 28.9 kips at the face of column C1 and 39.2 kips at the face of column C2. Use $f'_c = 4000$ psi and $f_{yt} = 40{,}000$ psi.

6-11 Figure P6-11 shows a rigid frame and the factored loads acting on the frame. The 7-kip horizontal load can act from the left or the right. $f'_c = 5000$ psi and $f_{yt} = 40{,}000$ psi.

Video Solution

(a) Design stirrups in the beam.

(b) Are stirrups required in the columns? If so, design the stirrups for the columns.

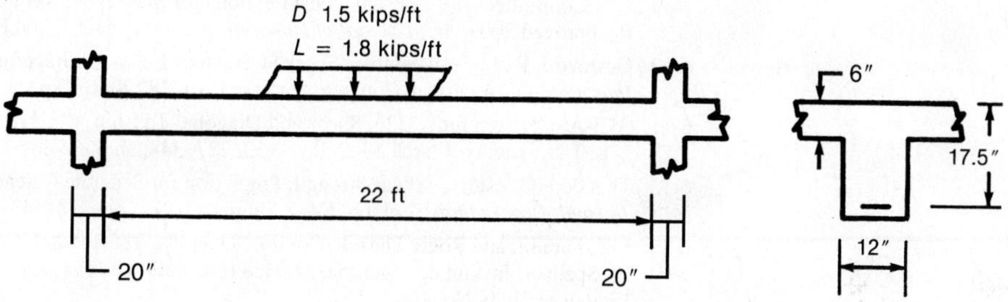

Fig. P6-9

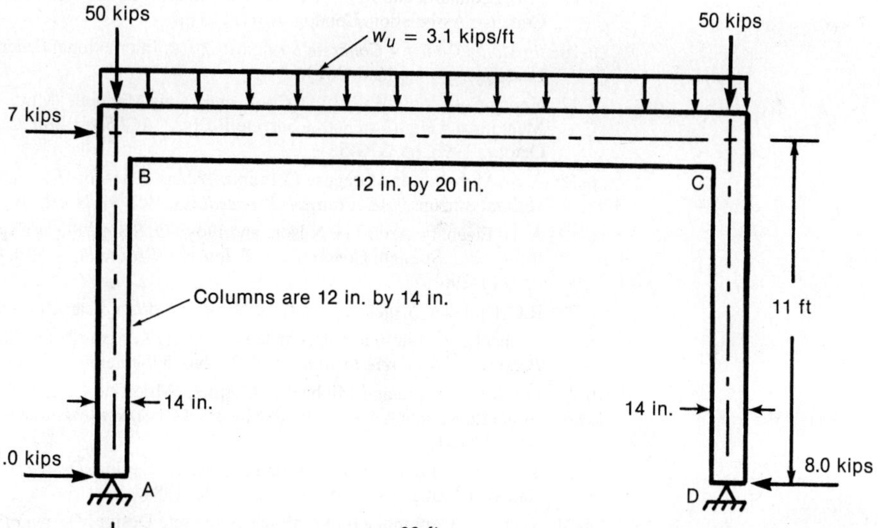

Fig. P6-11

REFERENCES

6-1 Boyd G. Anderson, "Rigid Frame Failures," *ACI Journal, Proceedings*, Vol. 53, No. 7, January 1957, pp. 625–636.

6-2 ACI Committee 445, "Recent Approaches to Shear Design of Structural Concrete," *Journal of Structural Engineering*, ASCE, Vol. 124, No. 12, December 1998.

6-3 Howard P. J. Taylor, "Investigation of Forces Carried across Cracks in Reinforced Concrete Beams in Shear by Interlock of Aggregate," TRA 42.447, Cement and Concrete Association, London, 1970, 22 pp.

6-4 Robert Park and Thomas Paulay, *Reinforced Concrete Structures*, A Wiley-Interscience Publication, Wiley, New York, 1975, 769 pp.

6-5 ACI-ASCE Committee 426, "The Shear Strength of Reinforced Concrete Members—Chapters 1 to 4," *Proceedings ASCE, Journal of the Structural Division*, Vol. 99, No. ST6, June 1973, pp. 1091–1187.

6-6 ACI-ASCE Committee 426, *Suggested Revisions to Shear Provisions for Building Codes*, American Concrete Institute, Detroit, 1978, 88 pp; abstract published in *ACI Journal, Proceedings*, Vol. 75, No. 9, September 1977, pp. 458–469; Discussion, Vol. 75, No. 10, October 1978, pp. 563–569.

6-7 Michael P. Collins and Dan Kuchma, "How Safe Are Our Large, Lightly Reinforced Concrete Beams, Slabs, and Footings?" *ACI Structural Journal, Proceedings*, Vol. 96, No. 4, July–August 1999, pp. 482–490.

6-8 Z.P. Bazant and J.K. Kim, "Size Effect in Shear Failure of Longitudinally Reinforced Beams," *ACI Journal, Proceedings*, Vol. 81, No. 5, April 1984, pp. 456–468.

6-9 ACI Committee 544, "Design Considerations for Steel Fiber Reinforced Concrete," 544.4 R-88, Reapproved 2009, *ACI Manual of Concrete Practice*, Farmington Hills, MI.

6-10 Gustavo J. Parra-Montesinos, "Shear Strength of Beams with Deformed Steel Fibers," *Concrete International*, American Concrete Institute, Vol. 28 (2006), No. 11, pp. 57–66.

6-11 ACI-ASCE Committee 326, Shear and Diagonal Tension," *ACI Journal, Proceedings*, Vol. 59, Nos. 1–3, January–March 1962, pp. 1–30, 277–344, and 352–396.

6-12 Theodore C. Zsutty, "Shear Strength Prediction for Separate Categories of Simple Beam Tests," *ACI Journal, Proceedings*, Vol. 68, No. 2 February 1971, pp. 138–143.

6-13 M.J. Faradji, and Roger Diaz de Cossio, "Diagonal Tension in Concrete Members of Circular Section," (in Spanish) Institut de Ingeniria, Mexico (translation by Portland Cement Association, Foreign Literature Study No. 466).

6-14 J.U. Khalifa, and Michael P. Collins, "Circular Members Subjected to Shear," Publications No. 81–08, Department of Civil Engineering, University of Toronto, December 1981.

6-15 Fritz Leonhardt and Rene Walther, *The Stuttgart Shear Tests, 1961*, Translation 111, Cement and Concrete Association, London, 1964, 110 pp.

6-16 *fib Model Code for Concrete Structures 2010*, International Federation for Structural Concrete (fib), Lausanne, Switzerland, 2013.

6-17 Young-Soo Yoon, William D. Cook, and Denis Mitchell, "Minimum Shear Reinforcement in Normal, Medium and High-Strength Concrete Beams," *ACI Structural Journal*, Vol. 93, No. 5, September–October 1996, pp. 576–584.

6-18 A. G. Mphonde and Gregory C. Frantz, "Shear Tests for High- and Low-Strength Concrete Beams without Stirrups," *ACI Journal, Proceedings*, Vol. 81, No. 4, July–August 1984, pp. 350–357.

6-19 A. H. Elzanaty, Arthur H. Nilson, and Floyd O. Slate, "Shear Capacity of Reinforced Concrete Beams Using High Strength Concrete," *ACI Journal, Proceedings*, Vol. 83, No. 2, March–April 1986, pp. 290–296.

6-20 R.C. Hibbeler, *Structural Analysis, Seventh Edition*, Pearson Prentice Hall, New Jersey, 2009.

6-21 Michael P. Collins and Denis Mitchell, "Design Proposals for Shear and Torsion," *Journal of the Prestressed Concrete Institute*, Vol. 25, No. 5, September–October 1980, 70 pp.

6-22 Frank J. Vecchio and Michael P. Collins, "Modified Compression Field Theory for Reinforced Concrete Elements Subjected to Shear," *ACI Structural Journal*, Vol. 83, No. 2, March–April 1986, pp. 219–231.

6-23 Technical Committee on Reinforced Concrete Design, *Design of Concrete Structures, A23.3-M84*, Canadian Standards Association, Rexdale, 1984.

6-24 Technical Committee on Reinforced Concrete Design, *Design of Concrete Structures, A23.3–94*, Canadian Standards Association, Rexdale, Ontario, 1994.

6–25 *LRFD Bridge Specifications and Commentary*, 2nd Edition, American Association of State Highway and Transportation Officials, Washington, 1998, 1216 pp.

6-26 *Design of Concrete Structures*, CAN/CSA-A23.3-04 (R2010), Canadian Standards Association, Rexdale, Ontario, Canada.

6-27 Joost C. Walraven, "Fundamental Analysis of Aggregate Interlock," *Journal of the Structural Division*, American Society of Civil Engineers, Vol. 107, No. ST11, November 1981, pp. 2245–2270.

6-28 Robert E. Loov and Anil K. Patnaik, "Horizontal Shear Strength of Composite Concrete Beams with a Rough Interface," *PCI Journal*, January–February 1994, Vol. 39, No. 1, pp. 369–390.

6-29 H. Wagner, "Metal Beams with Very Thin Webs," *Zeitscrift für Flugteknik und Motorluftschiffahr*, Vol. 20, No. 8 to 12, 1929.

6-30 Alan H. Mattock and J.F. Shen, "Joints between Reinforced Concrete Members of Similar Depth," *ACI Structural Journal*, Proceedings Vol. 89, No. 3, May–June 1992, pp. 290–295.

7

Torsion

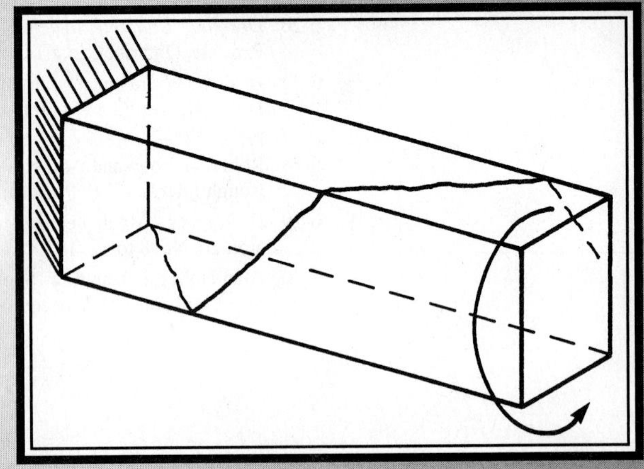

7-1 INTRODUCTION AND BASIC THEORY

A moment acting about the longitudinal axis of a member is called a *twisting moment*, a *torque*, or a *torsional moment, T*. In structures, torsion results from eccentric loading of beams, as shown in Fig. 7-21 (which will be discussed later), or from deformations resulting from the continuity of beams or similar members that join at an angle to each other, as shown in Fig. 7-23 (also discussed later).

Shearing Stresses due to Torsion in Uncracked Members

Solid Members

In a member subjected to torsion, a torsional moment causes shearing stresses on cross-sectional planes and on radial planes extending from the axis of the member to the surface. The element shown in Fig. 7-1 is stressed in shear, τ, by the applied torque, T. In a circular member, the shearing stresses are zero at the axis of the bar and increase linearly to a maximum stress at the outside of the bar, as shown in Fig. 7-2a. In a rectangular bar, the shearing stresses vary from zero at the center to a maximum at the centers of the long sides. Around the perimeter of a square bar, the shearing stresses vary from zero at the corners to a maximum at the center of each side, as shown in Fig. 7-2b.

The distribution of shearing stresses on a cross section can be visualized by using the *soap-film analogy*. The equations for the slope of an inflated membrane are analogous to the equations for shearing stress due to torsion. Thus, the distribution of shearing stresses can be visualized by cutting an opening in a plate that is proportional to the shape of the cross section loaded in torsion, stretching a membrane or soap film over this opening, and inflating the membrane. Figure 7-3 shows an inflated membrane over a circular opening, representing a circular shaft. The slope at each point in the membrane is proportional to the shearing stress at that point. The shearing stress acts perpendicular to the direction of a line tangent to the slope. Such a line is tangent to the slope at point A in Fig. 7-3. A section through the membrane along a diameter is parabolic. Its slope varies linearly from zero at the center to a maximum at the edge in the same way as the stress distribution plotted

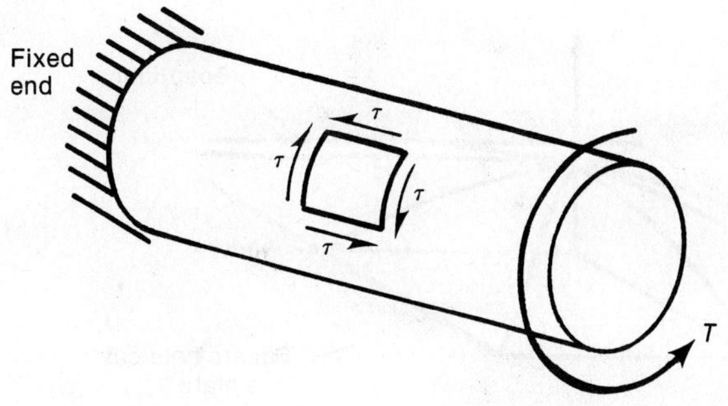

Fig. 7-1
Shear stresses due to torsion.

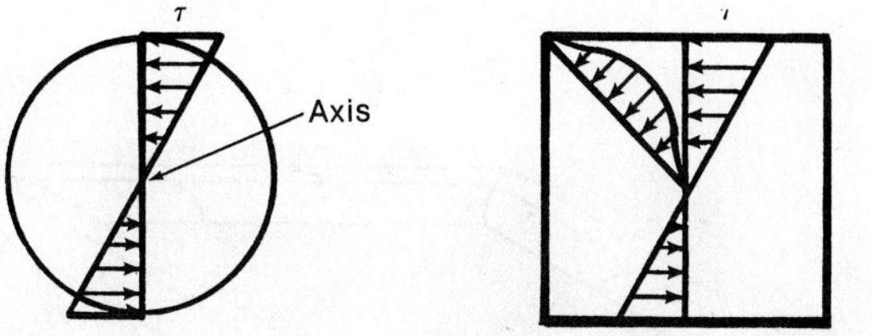

(a) Stress distribution in a circular bar.

(b) Stress distribution in a square bar.

Fig. 7-2
Distribution of torsional shear stresses in a circular bar and in a square bar.

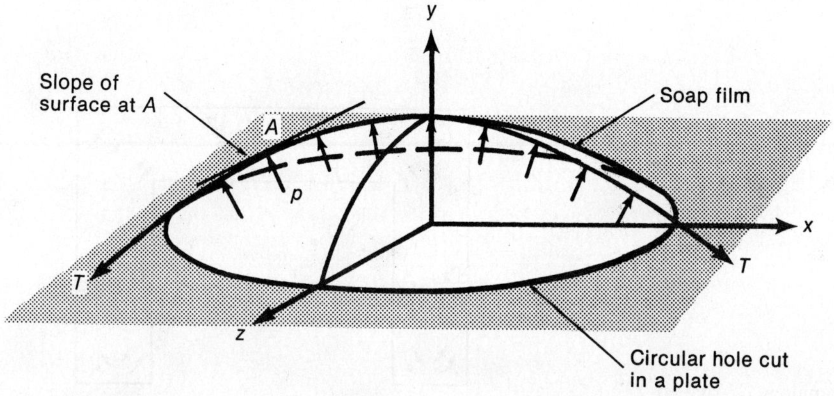

Fig. 7-3
Soap-film analogy: circular bar.

in Fig. 7-2a. Figure 7-4 shows a membrane over a square opening. Here, the slopes of the radial lines correspond to the stress distribution shown in Fig. 7-2b. A similar membrane for a U-shaped cross section made up from a series of rectangles is shown in Fig. 7-5a. The corresponding stress distribution is shown in Fig. 7-5b. For a hollow member with continuous walls, the membrane covers the entire section shape and is similar to Fig. 7-3 or 7-4, except that the region inside the hollow part is represented by an elevated plane having the shape of the hole.

The torsional moment is proportional to the volume under the membrane. A comparison of Figs. 7-4 and 7-5a shows that, for a given slope corresponding to the maximum shearing stress, the volume under a solid section or full hollow section is much greater than that under an open section. Thus, for a given maximum shearing stress, a solid rectangular cross section or full hollow section can transmit a much higher torsional moment than an open section can.

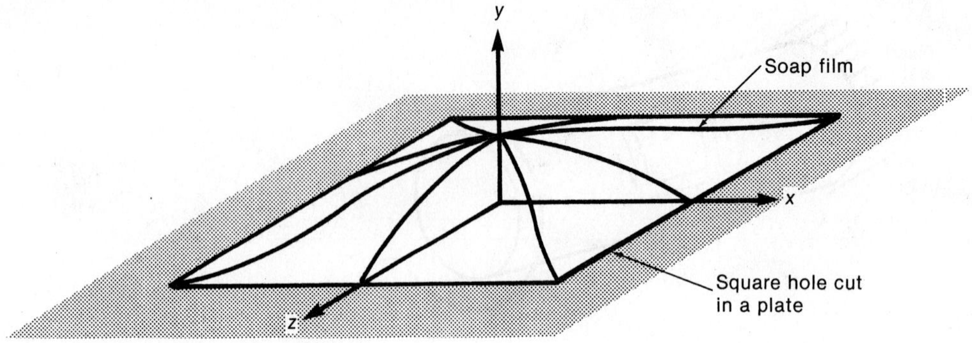

Fig. 7-4
Soap-film analogy: square bar.

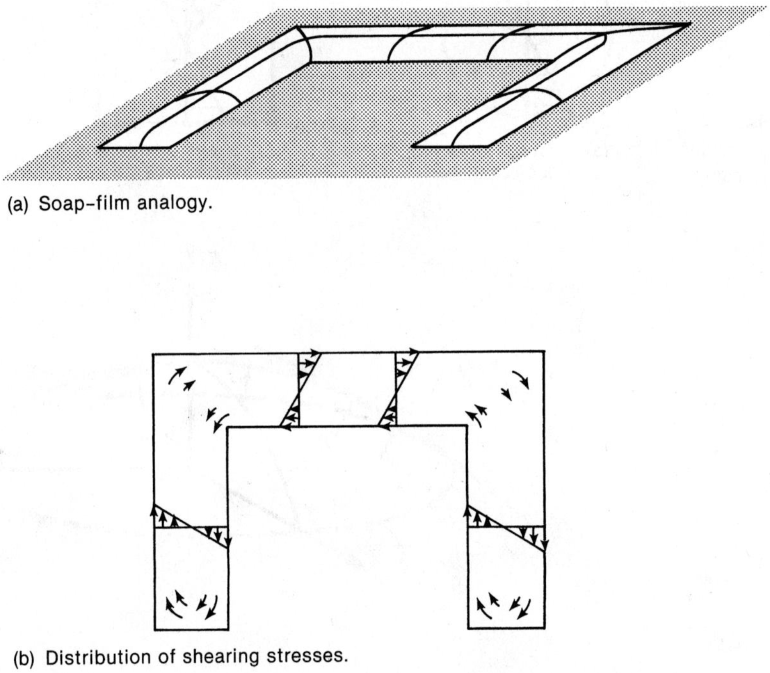

Fig. 7-5
Soap-film analogy: U-shaped member.

The maximum shearing stress in an elastic circular shaft is

$$\tau_{max} = \frac{Tr}{J} \tag{7-1}$$

where

τ_{max} = maximum shearing stress
T = torsional moment
r = radius of the bar
J = polar moment of inertia, $\pi r^4/2$

In a similar manner, the maximum shearing stress in a rectangular elastic shaft occurs at the center of the long side and can be written as

$$\tau_{max} = \frac{T}{\alpha x^2 y} \tag{7-2}$$

where x is the shorter overall dimension of the rectangle, y is the longer overall dimension, and α varies from 0.208 for $y/x = 1.0$ (square bar) to 0.333 for $y/x = \infty$ (an infinitely wide plate) [7-1]. An approximation to α is

$$\alpha = \frac{1}{3 + 1.8x/y} \tag{7-3}$$

For a cross section made up of a series of thin rectangles, such as that in Fig. 7-5,

$$\tau_{\max} \simeq \frac{T}{\Sigma(x^2y/3)} \tag{7-4}$$

where the term $x^2y/3$ is evaluated for each of the component rectangles.

The soap-film analogy and Eqs. (7-1) to (7-4) apply to elastic materials. For fully plastic materials, the shearing stress will be the same at all points. Thus, the soap-film analogy must be replaced by a figure having a slope that is constant at the value corresponding to the fully plastic case, producing a cone for a circular shaft or a pyramid for a square member. Such a figure could be formed by pouring sand onto a plate having the same shape as the cross section. This is referred to as the *sand-heap analogy*. For a solid rectangular cross section, the fully plastic shearing stress is

$$\tau_p = \frac{T}{\alpha_p x^2 y} \tag{7-5}$$

where α_p varies from 0.33 for $y/x = 1.0$ to 0.5 for $y/x = \infty$.

The behavior of uncracked concrete members in torsion is neither perfectly elastic nor perfectly plastic, as assumed by the soap-film and sand-heap analogies. However, solutions based on each of these models have been used successfully to predict torsional strength.

Hollow Members

Figure 7-6a shows a thin-walled tube with continuous walls subjected to a torque about its longitudinal axis. An element *ABCD* cut from the wall is shown in Fig. 7-6b. The thicknesses of the walls along sides *AB* and *CD* are t_1 and t_2, respectively. The applied torque causes shearing forces V_{AB}, V_{BC}, V_{CD}, and V_{DA} on the sides of the element as shown, each equal to the shearing stress on that side of the element times the area of the side. From $\Sigma F_x = 0$, we find that $V_{AB} = V_{CD}$; but $V_{AB} = \tau_1 t_1 dx$ and $V_{CD} = \tau_2 t_2 dx$, which together give $\tau_1 t_1 = \tau_2 t_2$, where τ_1 and τ_2 are the shearing stresses acting on sides *AB* and *CD*, respectively. The product τt is referred to as the *shear flow, q*.

For equilibrium of a smaller element at corner *B* of the element in Fig. 7-6b, $\tau_1 = \tau_3$; similarly, at *C*, $\tau_2 = \tau_4$, as shown in Figs. 7-6c and 7-6d. Thus, at points *B* and *C* on the perimeter of the tube, $\tau_3 t_1 = \tau_4 t_2$. This shows that, for a given applied torque, *T*, the shear flow, *q*, is constant around the perimeter of the tube. The shear flow has units of (stress $\times$ length) pounds force per inch (or N/mm). The name *shear flow* comes from an analogy to water flowing around a circular flume. The volume of water flowing past any given point in the flume is constant at any given period of time.

Figure 7-6e shows an end view of the tube. The torsional shear force acting on the length *ds* of wall is *q ds*. The perpendicular distance from this force to the centroidal axis of the tube is *r*, and the moment of this force about the axis is *rq ds*, where *r* is measured from

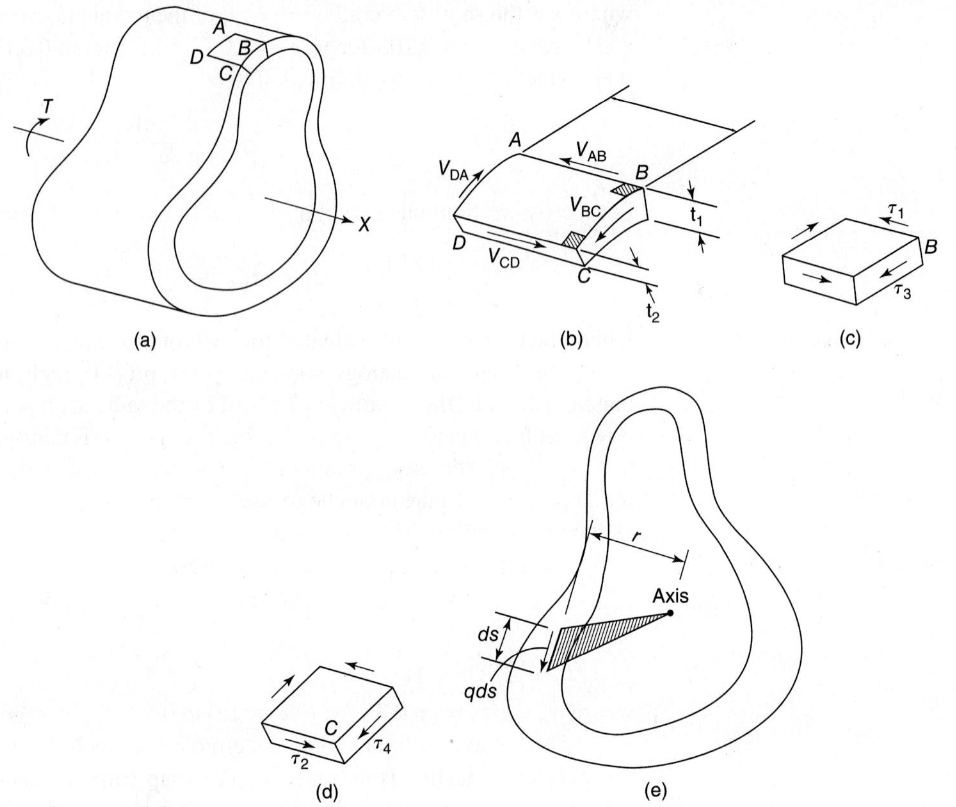

Fig. 7-6
Shear stresses in a thin-walled tube. (From [7-2] Popov, E. P., *Mechanics of Materials*, SI Version, 2/e © 1978, p. 80. Reprinted by permission of Prentice Hall, Upper Saddle River, New Jersey.)

midplane of the wall because that is the line of action of the force $q\ ds$. Integrating around the perimeter gives the torque in the tube:

$$T = \int_p rq\ ds \tag{7-6}$$

where $\int_p$ denotes integration around the perimeter of the tube. However, q is constant around the perimeter of the tube, so q can be moved outside the integral giving

$$T = q \int_p r\ ds \tag{7-7}$$

The shaded triangle in Fig. 7-6e has an area of $rds/2$. Thus, rds in Eq. (7-7) is two times the area of the shaded triangle stretching between the elemental length of perimeter, ds, and the axis of the tube. Furthermore, $\int_p rds$ is equal to two times the area enclosed by the centerline of the wall thickness. This area is referred to as the *area enclosed by the shear flow path*, A_o. For the cross section shown in Fig. 7-7, A_o is the area inside the dashed line, including the area of the hole in the center of the tube. Equation (7-7) becomes

$$T = 2qA_o \tag{7-8}$$

where $q = \tau t$. Rearranging gives

$$\tau = \frac{T}{2A_o t} \tag{7-9}$$

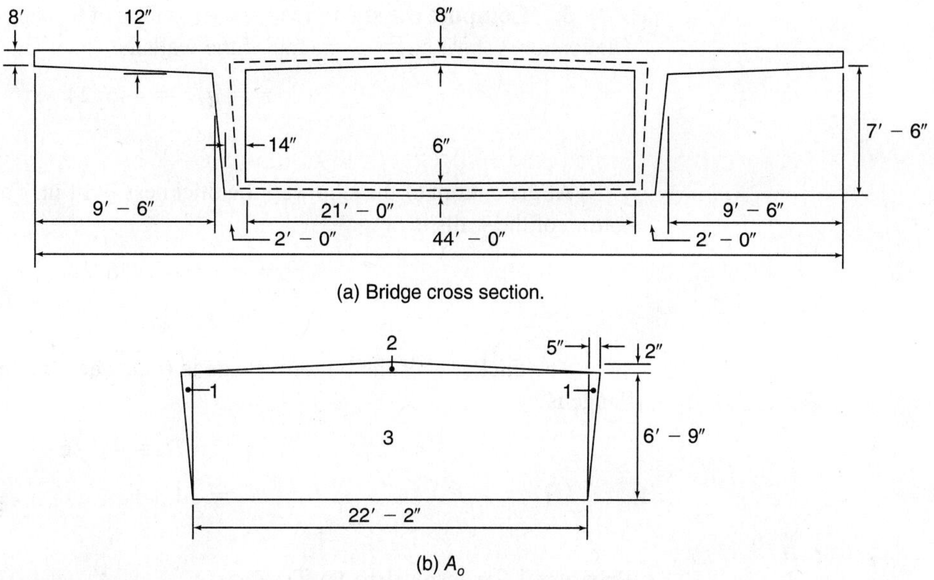

Fig. 7-7
Cross section of a bridge—Example 7-1.

where t is the wall thickness at the point where the shear stress, τ, due to torsion is being computed. The maximum torsional shear stress occurs where the wall thickness is the least. For the hollow trapezoidal bridge cross section shown in Fig. 7-7, for example, this would be in the lower flange.

This analysis applies only if the walls of the tube are continuous (no slits parallel to the axis of the tube) or if, in the case of a solid member, the member can be approximated as a tube with continuous walls. Equation (7-9) can be applied to either elastic or inelastic sections. Consider the shape of the membrane in Fig. 7-4. A tube can be called thin walled if the change in slope of the membrane is small across the thickness of the wall and can be ignored without serious loss of accuracy.

EXAMPLE 7-1 Compute Torsional Shear Stresses in a Bridge Cross Section, Using Thin-Walled Tube Theory

Figure 7-7a shows the cross section of a box girder bridge. When computing the torsional strength and stiffness of such a section, the contribution of the flanges is often ignored. They will be ignored for this example. Compute the shear stresses, τ, at the top and bottom of the side walls and in the lower flange that are due to an applied torque of 1650 kip-ft.

1. Compute A_o. A_o is the area enclosed by the midplane of the walls of the tube. The dashed line in Fig. 7-7a is the perimeter of A_o. Divide A_o into triangles and a rectangle as shown in Fig. 7-7b. Then

$$A_o = (2 \times 6'9'' \times 5''/2) + (23' \times 2''/2) + (22'2'' \times 6'9'')$$

$$= 405 + 276 + 21{,}546$$

$$= 22{,}200 \text{ in.}^2$$

2. Compute the shear flow, q. From Eq. (7-8a),

$$q = \frac{T}{2A_o} = \frac{1650 \times 12{,}000}{2 \times 22{,}200}$$

$$= 446 \text{ lb/in.}$$

3. Compute the shear stresses. At the top of the side wall the thickness, t, is 24 in. The torsional shear stress at the top of the walls is

$$\tau = q/t = 446/24$$
$$= 18.6 \text{ psi}$$

At the bottom of the side wall the thickness is 14 in. The torsional shear stress at the bottom of the walls is

$$\tau = q/t = 446/14$$
$$= 31.9 \text{ psi}$$

The thickness of the bottom flange is 6 in. The torsional shear stress in the bottom flange is

$$\tau = q/t = 446/6$$
$$= 74.3 \text{ psi}$$

■

Principal Stresses due to Torsion

When the beam shown in Fig. 7-8 is subjected to a torsional moment, T, shearing stresses develop on the top and front faces, as shown by the elements in Fig. 7-8a. The principal stresses on these elements are shown in Fig. 7-8b. The principal tensile stress equals the principal compressive stress, and both are equal to the shear stress if T is the only loading. The principal tensile stresses eventually cause cracking that spirals around the body, as shown by the line *A–B–C–D–E* in Fig. 7-8c.

In a reinforced concrete member, such a crack would cause failure unless it was crossed by reinforcement. This generally takes the form of longitudinal bars in the corners and closed stirrups. Because the crack spirals around the body, four-sided (closed) stirrups are required.

Stresses due to Torsion, Shear and Moment

If a beam is subjected to combined shear and torsion as shown in Fig. 7-9, the two shearing-stress components add on one side face (front face in this case) and counteract each other on the other, as shown in Figs. 7-9a and 7-9b. As a result, inclined cracking starts on the face where the stresses add (crack *AB*) and extends across the flexural tensile face of the beam (in this case the top because this is a cantilever beam). If the bending moments are sufficiently large, the cracks will extend almost vertically across the back face, as shown by crack *CD* in Fig. 7-9c. The flexural compression zone near the bottom of the beam prevents the cracks from extending the full height of the front and back faces.

Circulatory Torsion and Warping Torsion

Members subjected to torsion can be divided into two families, distinguished by how the torsion is resisted. If the cross section is solid (either square, rectangular, circular, or polygonal) or is a closed tube, torsion is resisted by torsional stresses, τ, which act in a continuous manner around the section, as shown in Figs. 7-2, 7-5, and 7-6. In the closed tube, these stresses can be represented by a shear flow, $q = \tau t$, which is constant around the circumference. This is referred to as *circulatory torsion* or *St. Venant torsion* after the French mathematician who derived the equations for torsional stresses in noncircular cross sections in 1853 and developed the soap-film analogy.

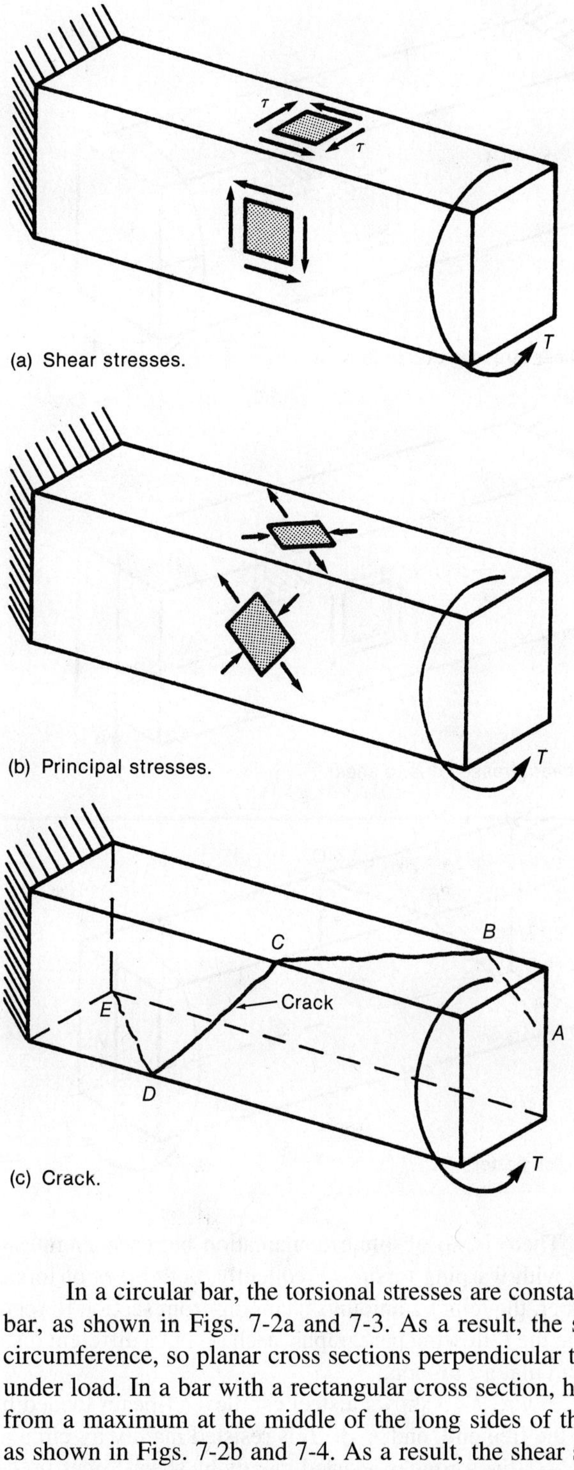

Fig. 7-8
Principal stresses and cracking due to pure torsion.

In a circular bar, the torsional stresses are constant around the circumference of the bar, as shown in Figs. 7-2a and 7-3. As a result, the shear strain is constant around the circumference, so planar cross sections perpendicular to the axis of the bar remain planar under load. In a bar with a rectangular cross section, however, the torsional stresses vary from a maximum at the middle of the long sides of the rectangle to zero at the corners, as shown in Figs. 7-2b and 7-4. As a result, the shear strain varies around the circumference of the section, causing the section to deform in such a manner that plane sections through the bar do not remain plane. This distortion is referred to as *warping*. If the warping deformations are restrained, a part (or all) of the torsion is resisted by *warping torsion*. Warping torsion generally occurs in a rectangular cross section and sections with flanges.

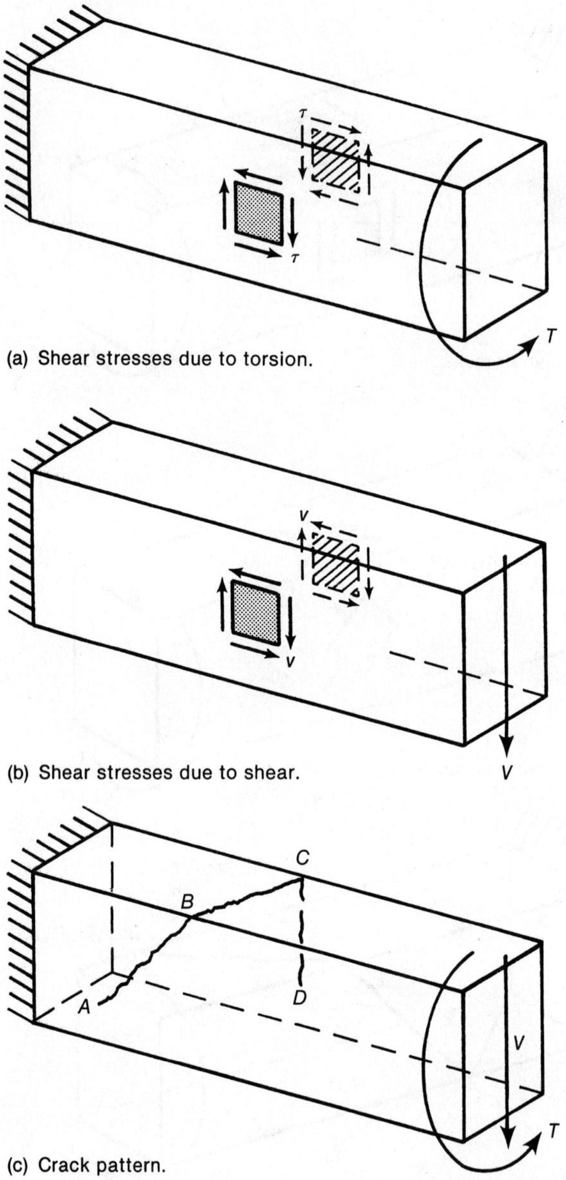

Fig. 7-9
Combined shear, torsion, and moment.

There is no absolute demarcation between members with circulatory torsion and those with warping torsion. Frequently, both types of torsion will be present in the same member, the relative amounts changing from section to section. Two cases will be considered in the following paragraphs, each giving a different distribution of the warping torsion and circulatory torsion.

Figure 7-10 shows a steel cantilever I-beam loaded by a torque, T_u, at the free end. Near the free end, end A–B, T_u is resisted mainly by circulatory torsion. Near the support end C–D, the torque is resisted mainly by shear forces in the flanges:

$$V_{fu} = T_u/h$$

These forces act at the midthickness of the flanges, with h the distance between the resultant forces in the two flanges. The forces V_{fu} can be idealized as acting at a distance a

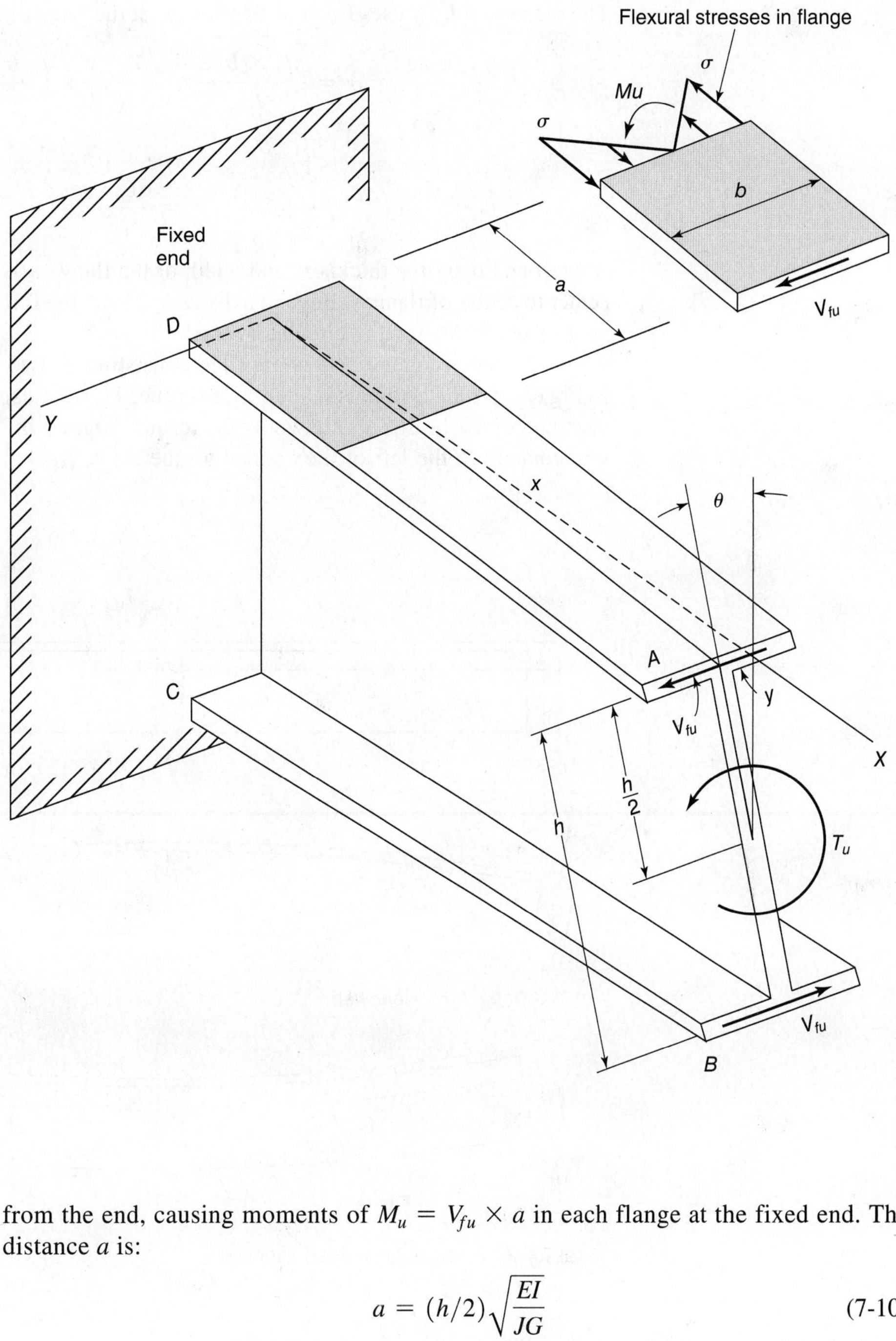

Fig. 7-10
Warping torsion of an I-beam.

from the end, causing moments of $M_u = V_{fu} \times a$ in each flange at the fixed end. The distance a is:

$$a = (h/2)\sqrt{\frac{EI}{JG}} \tag{7-10}$$

where

E and G are the modulus of elasticity and the shearing modulus, respectively

I is the moment of inertia of the entire section about a plane of symmetry in the web so that the moment of inertia of one flange is approximately $I/2$

J is the polar moment of inertia of the cross section

The moment, M_u, causes flexural stresses, σ, at the fixed end of the flanges equal to

$$\sigma = \frac{M_u \times b/2}{I_{f\ell}} = \left(\frac{T_u a}{h}\right) \times \left(\frac{b/2}{tb^3/12}\right) \tag{7-11}$$

or

$$\sigma = \frac{6T_u a}{thb^2} \tag{7-12}$$

where t and b are the thickness and width of the flange and h is the height of the I-beam, center to center of flanges. Beyond a distance a from the fixed end, all of the torsion can be assumed to be resisted by circulatory torsion.

Another frequent case is a bridge consisting of two girders and a slab, as shown in Fig. 7-11. The bridge is loaded by a torque, T_u, at midspan. Half of this is resisted by each end of the bridge, as shown by the torque diagram in Fig. 7-11b. The cross sections immediately to the left of the applied torque are restrained against warping by the cross

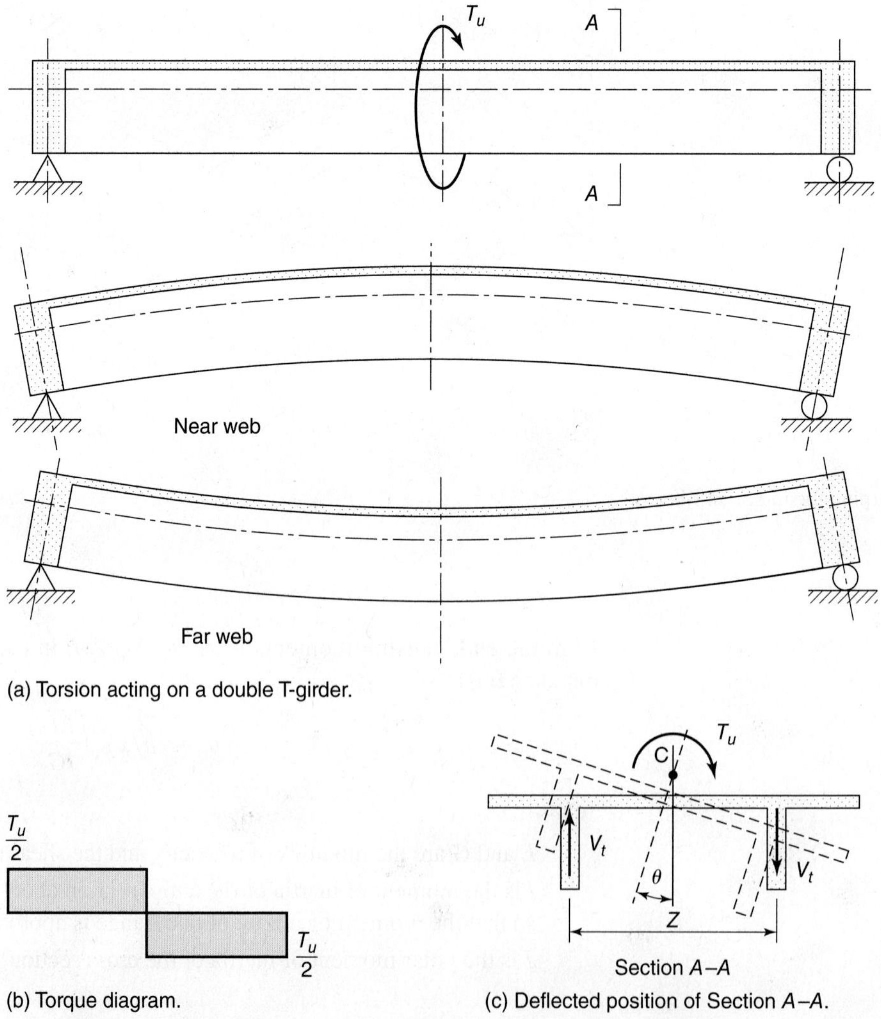

Fig. 7-11
Warping torsion on a bridge.
(From [7-3].)

sections on the right of the loading point, which have torsional stresses of the opposite sign. The cross sections near the ends of the beam are free to warp.

The balance of this chapter will consider only circulatory torsion. Analyses of structures subjected to warping torsion are presented in books on advanced strength of materials or on bridge design [7-3].

7-2 BEHAVIOR OF REINFORCED CONCRETE MEMBERS SUBJECTED TO TORSION

Prior to Cracking

When a concrete member is loaded in pure torsion, shearing stresses develop in a solid or hollow cross section as discussed in the prior section. Reinforcement does not play a significant role at this stage of behavior. Thus, if a value is selected for the maximum torsional shear stress at the onset of cracking, elastic theory, as represented by Eq. (7-2) for solid rectangular sections and Eq. (7-9) for hollow sections, can be used to predict the cracking torsional moment, T_{cr}. A typical value that has been used for the torsional shear stress (principal tension stress) at cracking is $4\sqrt{f'_c}$ in psi units ($0.33\sqrt{f'_c}$ in SI units).

In earlier editions of the ACI Code, the elastic theory [Eqs. (7-2) and (7-4)] was used to determine the torsional cracking moment, and thus, the concrete contribution, T_c, for the overall torsional strength of a reinforced concrete member with rectangular of flanged cross sections. However, since 1995 the ACI Code has used a *thin-walled tube analogy* [Eq. (7-9)] to determine the cracking moment for members with either solid or hollow cross sections. A development of the current ACI Code equations for cracking torque will be given in Section 7-3.

After Cracking

Because torsion cracking will tend to spiral around a member, as shown in Fig. 7-8, a combination of longitudinal steel and closed transverse reinforcement, both placed around the perimeter of the section, will be required to control the growth and opening of torsional cracks. Two very different theories have been used to determine the strength of reinforced concrete members after cracking. The first, based on a skew bending theory developed by Lessig [7-4] and extended by Hsu [7-5], was the basis for the torsion design provisions in the ACI Code from 1971 through 1992. This theory assumes that a portion of the design shear and torsion is resisted by the concrete and the remainder is resisted by shear and torsion reinforcement. The modes of failure are assumed to involve bending on a surface that is skewed with respect to the principal axes of the cross section (Fig. 7-12) due to a crack spiraling around the cross section.

The second theory is based on a post-cracking thin-walled tube analogy or a plastic space truss model that is similar to the truss analogy discussed in Chapter 6 for shear. The dimensions of both the thin-walled tube and plastic space truss are a function of the size and strength of the closed stirrups and longitudinal reinforcement used in the member. The development of the ACI Code equations for torsional strength will be given in Section 7-4. A thorough discussion of the development of the various theories used to analyze reinforced concrete members before and after cracking is given in a report from ACI/ASCE Committee 445 [7-6].

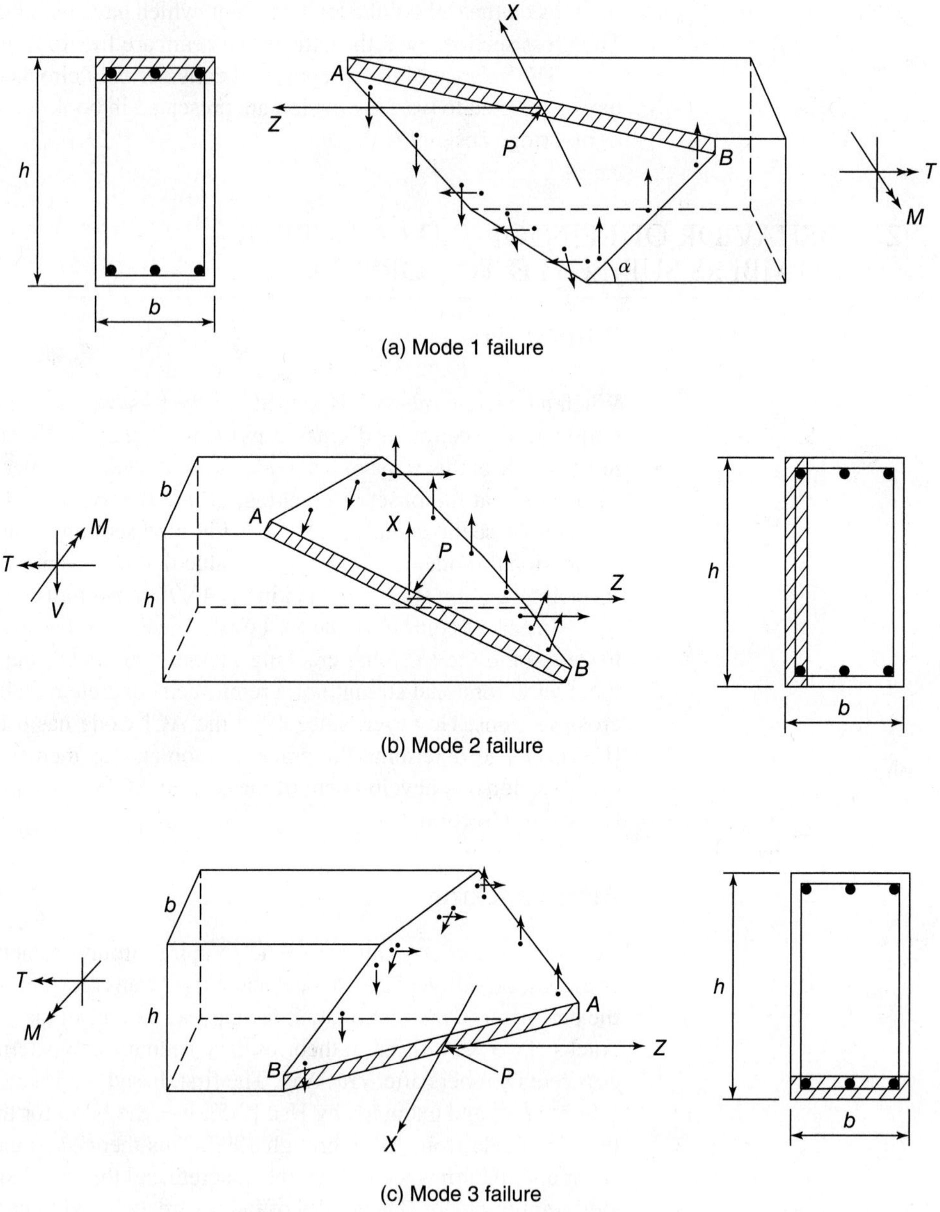

Fig. 7-12
Three failure modes in skew-bending theories. (From [7-6].)

7-3 THIN-WALLED TUBE ANALOGIES

Prior to cracking the largest torsional shear stresses in a member subjected to pure torsion occur at the outer edges of the member, as shown in Fig. 7-2. Due to this fact, the torsion carried by a solid and hollow section with the same outer dimensions at the onset of torsional cracking is approximately the same [7-7]. Since 1995, the ACI Code has used an equivalent thin-walled tube analogy to analyze the torsional cracking strength of a solid or hollow section.

As noted in the prior section, a different thin-walled tube analogy than the one used prior to cracking can be used to determine the strength of a reinforced concrete section after cracking. This second thin-walled tube analogy can be analyzed using a model referred to as a plastic-truss analogy. A discussion of the pre-cracking and post-cracking thin-walled analogies will be given in this section.

Pre-Cracking Thin Tube Analogy

Since 1995, the ACI Code has used the thin-walled tube analogy to model the behavior and strength of solid and hollow sections subjected to torsion. A typical solid cross section for a *ledger beam*, which is often used to support precast beams or slabs, is shown in Fig. 7-13. As shown in Fig. 7-13(a), the total area of the section is defined as A_{cp} and the perimeter of the section is defined as p_{cp}. Figure 7-13(b) shows the thin-walled tube analogy for this section. ACI Code Section 22.7.4 is based on the assumption that prior to cracking the wall thickness, t, can be taken as approximately $3\,A_{cp}/4\,p_{cp}$. As discussed in Section 7-1, when torsion is applied to a thin-walled tube section, a shear flow, q, defined as torsional shear stress, τ, multiplied by the wall thickness, t, develops in the walls of the section. The area enclosed by the shear flow is defined as A_o, and the perimeter of the shear flow is p_o. For a typical concrete section, A_o can be taken as approximately $2A_{cp}/3$.

Torsional cracking for a concrete member subjected to pure torsion is assumed to occur when the principal tensile stress in the concrete reaches the tensile strength of the concrete in biaxial tension-compression (Fig. 3-12a), taken as $4\sqrt{f'_c}$ (psi units). Thus, from Eq. (7-8) and using the approximations discussed above for t and A_o, the cracking torque, T_{cr}, for a solid section subjected to pure torsion can be taken as,

$$T_{cr} = 2qA_o = 2\tau t \times \frac{2A_{cp}}{3}$$

$$= 2 \times 4\sqrt{f'_c} \times \frac{3A_{cp}}{4p_{cp}} \times \frac{2A_{cp}}{3}$$

$$= 4\sqrt{f'_c}\left(\frac{A_{cp}^2}{p_{cp}}\right) \tag{7-13}$$

This is the expression given in ACI Code Table 22.7.5.1 for the cracking torque in a non–prestressed concrete beam.

For a concrete member with a solid cross section that is subjected to combined shear and torsion, test results (Fig. 7-14) indicate that the load at the onset of inclined cracking can be represented by a circular interaction diagram,

$$\left(\frac{T}{T_{cr}}\right)^2 + \left(\frac{V}{V_{cr}}\right)^2 = 1 \tag{7-14}$$

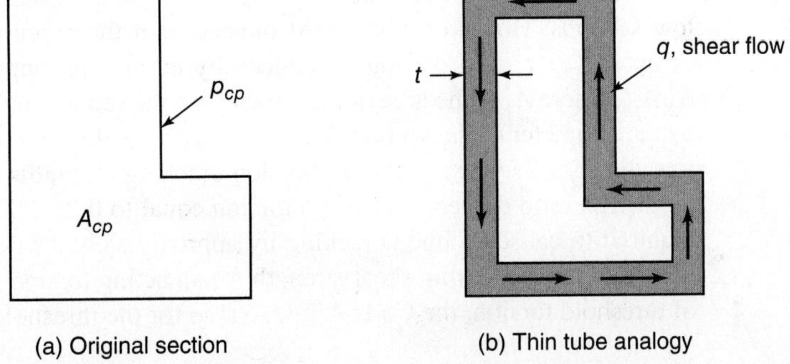

(a) Original section (b) Thin tube analogy

Fig. 7-13
Solid section modeled as thin-walled tube.

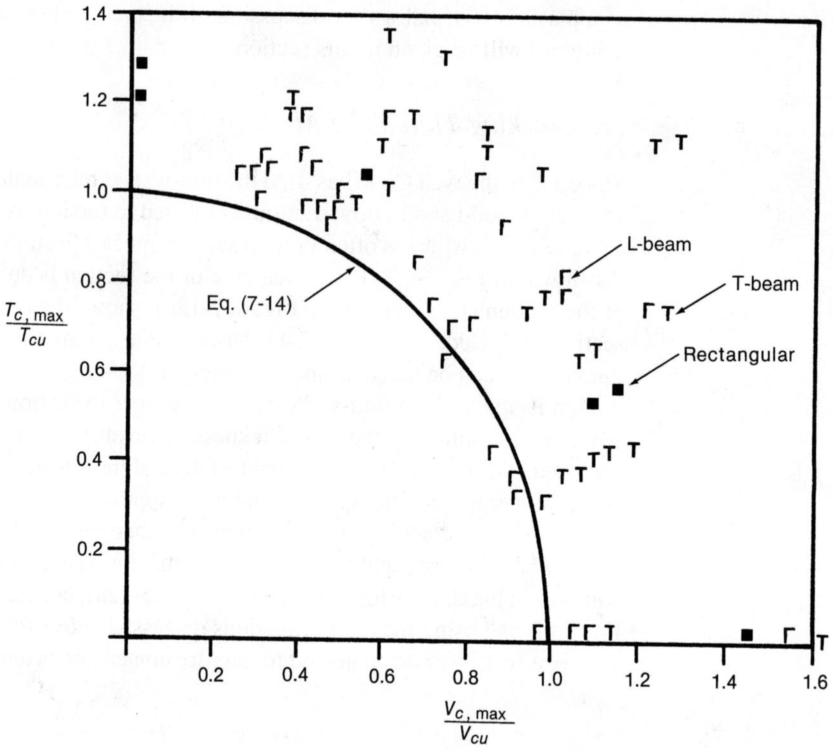

Fig. 7-14
Interaction of torsion and shear. (From [7-8].)

where V_{cr} is the inclined cracking shear in the absence of torsion and T_{cr} is the inclined cracking torque in the absence of shear. For this circular interaction diagram, the existence of a torque equal to a quarter of the inclined cracking torque will reduce the inclined cracking shear by only 3 percent. This can be considered to be a negligible difference. Thus, ACI Code Table 22.7.4.1(a) defines a *threshold torsion*, below which torsion can be ignored in a non–prestressed concrete beam with a solid cross section as,

$$T_{th} = \phi\sqrt{f'_c}\left(\frac{A_{cp}^2}{p_{cp}}\right) \tag{7-15}$$

where ϕ is the capacity reduction factor for torsion ($\phi = 0.75$). In SI units, Eq. (7-15) becomes,

$$T_{th} = \frac{\phi\sqrt{f'_c}}{12}\left(\frac{A_{cp}^2}{p_{cp}}\right) \tag{7-15M}$$

The same thin-walled tube analogy used for solid sections can also be used for hollow sections. However, tests [7-5] indicate that the cracking load for a hollow section compared to a solid section is reduced by an amount approximately equal to the ratio A_g/A_{cp}, where A_g is the area of the concrete in the section and A_{cp} is the total area enclosed by the perimeter of the section (Fig. 7-15). Test results have also indicated that the interaction diagram between shear and torsion cracking strengths approaches a straight line as the A_g/A_{cp} ratio decreases. Thus, a torsion equal to 0.25 T_{cr} would reduce the shear force required to cause inclined cracking by approximately 25 percent. To avoid such a large decrease in the cracking shear strength for an acting torsion near the ACI Code definition of threshold torsion, the Code-defined value for the threshold torsion for a hollow section

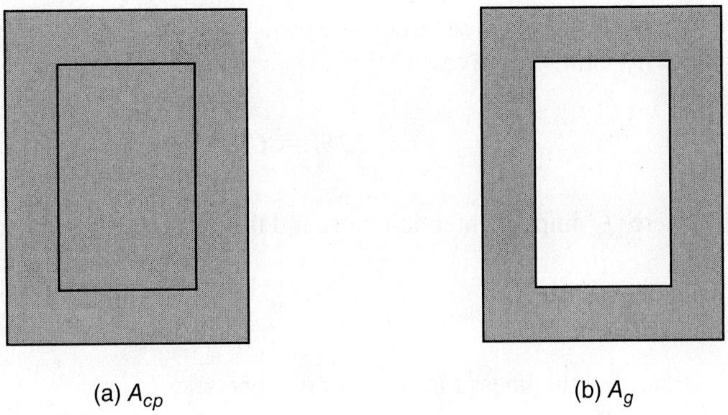

Fig. 7-15
Definition of A_{cp} and A_g for a hollow section.

was reduced by the ratio A_g/A_{cp}. When this reduction is combined with the reduction of the cracking torque (by the same ratio), the resulting definition of the threshold torsion, as given in ACI Code Table 22.7.4.1(b) for a non–prestressed concrete beam with a hollow section is,

$$T_{th}(\text{hollow}) = \frac{A_g}{A_{cp}} \times T_{th}(\text{solid})$$

$$= \frac{A_g}{A_{cp}} \times \left[\phi\sqrt{f'_c}\left(\frac{A_{cp}^2}{p_{cp}}\right)\right] \times \frac{A_g}{A_{cp}}$$

$$T_{th}(\text{hollow}) = \phi\sqrt{f'_c}\left(\frac{A_g^2}{p_{cp}}\right) \tag{7-16}$$

Torsional Stiffness

The torsional stiffness, K_t, of a member of length ℓ is defined as the torsional moment, T, required to cause a unit twist in the length ℓ; that is,

$$K_t = \frac{T}{\phi_t \ell} \tag{7-17}$$

or

$$K_t = \frac{T}{\theta_t} \tag{7-18}$$

where ϕ_t is the angle of twist per unit length and $\theta_t = \phi_t\ell$ is the total twist in the length ℓ. For a thin-walled tube of length ℓ, the total twist can be found by virtual work by equating the external work done when the torque, T, acts through a virtual angle change θ_t to the internal work done when the shearing stresses due to torsion, τ, act through a shear strain $\gamma = \tau/G$. The resulting integral equation is

$$T\theta_t = \int_V \tau\gamma dV$$

where $\int_V$ implies integration over the volume. Replacing γ with τ/G, τ with Eq. (7-9), and d_V with $\ell t\, ds$ gives

$$T\theta_t = \ell \int_p \frac{T}{2A_o t}\left(\frac{T}{2A_o tG}\right) t\, ds$$

where $\int_p$ implies integration around the perimeter of the tube. This reduces to

$$\theta_t = \ell \frac{T}{4A_o^2 G}\int_p \frac{ds}{t} \qquad (7\text{-}19)$$

If the wall thickness t is constant this becomes,

$$\theta_t = \frac{Tp_o}{4A_o^2 tG} \qquad (7\text{-}20)$$

where p_o is the perimeter of A_o. Substituting Eq. (7-20) into Eq. (7-18) gives the torsional stiffness as

$$K_t = \left(\frac{4A_o^2 t}{p_o}\right)\frac{G}{\ell} = \frac{CG}{\ell} \qquad (7\text{-}21)$$

where C, the torsional constant, refers to the term in parentheses. Equation (7-21) is similar to the equation for the flexural stiffness of a beam that is fixed at the far end:

$$K = \frac{4EI}{\ell}$$

In this equation, EI is the *flexural rigidity* of the section. The term CG in Eq. (7-21) is the *torsional rigidity* of the section.

Prior to cracking for a concrete beam with a rectangular section, the value of GC approximately corresponds to the uncracked value for a member constructed with an elastic material, defined as $G \times \beta_t x^2 y$ [7-1], where x and y are the smaller and larger dimensions, respectively, for the section. For ratios of y/x less than 3, an approximate value for β_t is,

$$\beta_t \cong \frac{1}{3}\left(1 - 0.63\frac{x}{y}\right)$$

Thus,

$$C \cong \left(1 - 0.63\frac{x}{y}\right)\frac{x^2 y}{3} \qquad (7\text{-}22a)$$

This expression for C is used in the ACI Code for rectangular sections and sections that can be divided into rectangular components, as shown in Fig. 7-16. The ACI Code expression is,

$$C = \sum_{i=1}^{n}\left(1 - 0.63\frac{x_i}{y_i}\right)\frac{x_i^2 y_i}{3} \qquad (7\text{-}22b)$$

where n is the number of rectangular components for the section.

Figure 7-17 shows a torque versus twist curve for a moderately reinforced beam subjected to pure torsion. From this figure it is clear that the post-cracking torsional stiffness of a torsionally cracked beam is typically equal to 10–15 percent of the torsional stiffness of the uncracked beam. Thus, if such a beam is part of a statically determinate floor system,

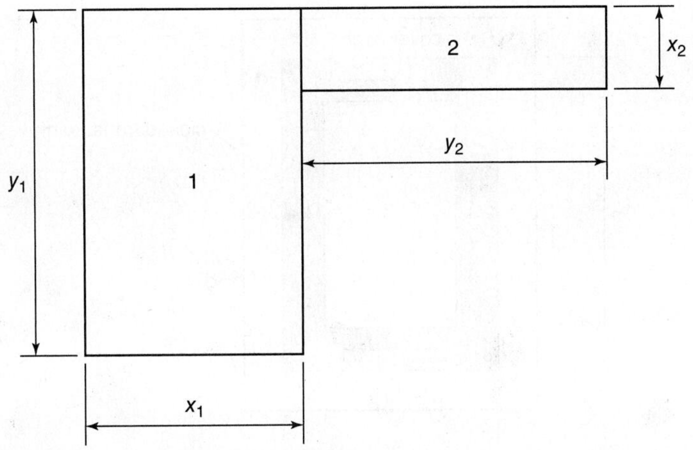

Fig. 7-16
Beam section broken into two rectangular components.

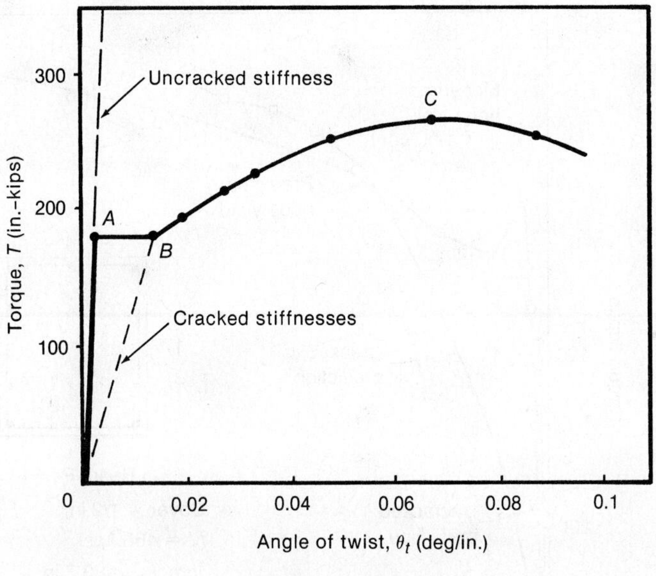

Fig. 7-17
Torque twist curve for a rectangular beam.

a structural analysis of that floor system will indicate that loads will redistribute away from the beam, after it cracks in torsion, to stiffer elements in the floor system.

Post-Cracking Thin Tube Analogy

A different thin-walled tube analogy, as shown in Fig. 7-18, can be used to evaluate the strength and stiffness of a reinforced concrete beam section after cracking [7-7], [7-9]. As noted earlier, the reinforcement to resist torsion must consist of closed transverse reinforcement and longitudinal reinforcement distributed around the perimeter of the transverse reinforcement. It is assumed that most of the cover concrete either spalls off or is loosened after the onset of torsional cracking and thus, is not effective in resisting torsion. Torsion is assumed to be resisted by a thin-walled tube section that forms inside the closed transverse reinforcement, shown as the shaded region in Fig. 7-18. The outer dimensions of the tube are determined by the centerline dimensions of the transverse reinforcement, shown as x_o and y_o for the rectangular hoop in Fig. 7-18. The area enclosed by the closed transverse reinforcement is referred to as A_{oh} and the perimeter is referred to as p_h. As with all thin-walled

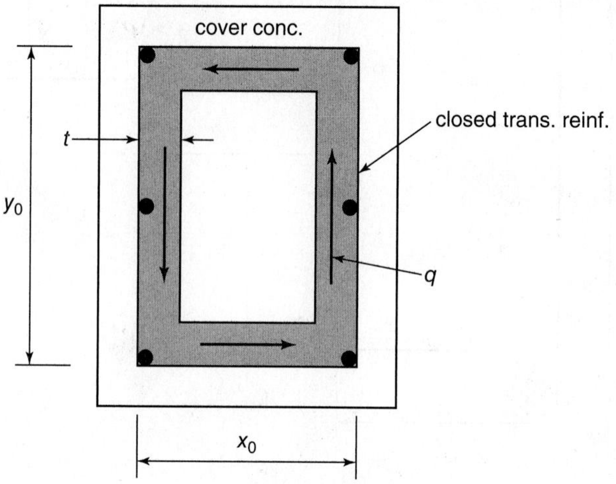

Fig. 7-18
Thin-walled tube analogy after cracking.

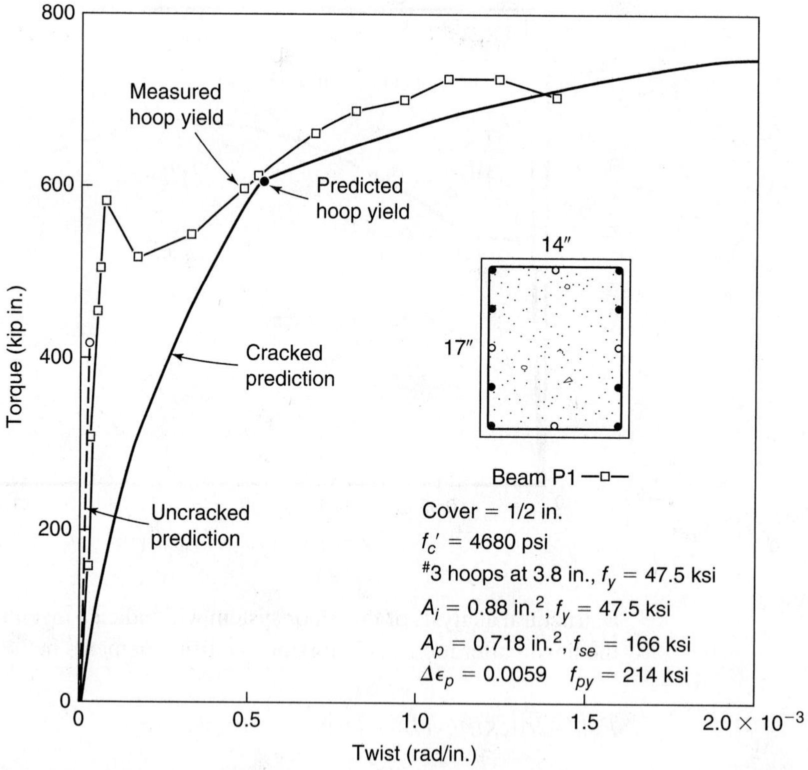

Fig. 7-19
Measured and predicted torque-twist response for a prestressed beam.
(From [7-9].)

tube analogies, a shear flow, q, is assume to act at the centerline of the tube walls and the area enclosed by the shear flow is defined as A_o. If q and A_o can be defined, Eq. (7-8) can be used to calculate the applied torsion acting on the section. In a classic paper by Mitchel and Collins [7-9], the diagonal compression field theory was used to define the tube wall thickness, t, as well as q and A_o for any applied torque, resulting in the development of a complete torque versus twist curve for a torsionally cracked member (Fig. 7-19).

If one assumes that the torsional reinforcement will yield before the concrete will crush in principal compression, the maximum shear flow that can be developed in the tube walls is a function of the amount and strength of the longitudinal and transverse reinforcement used in the member. The ACI Code approach for determining the nominal

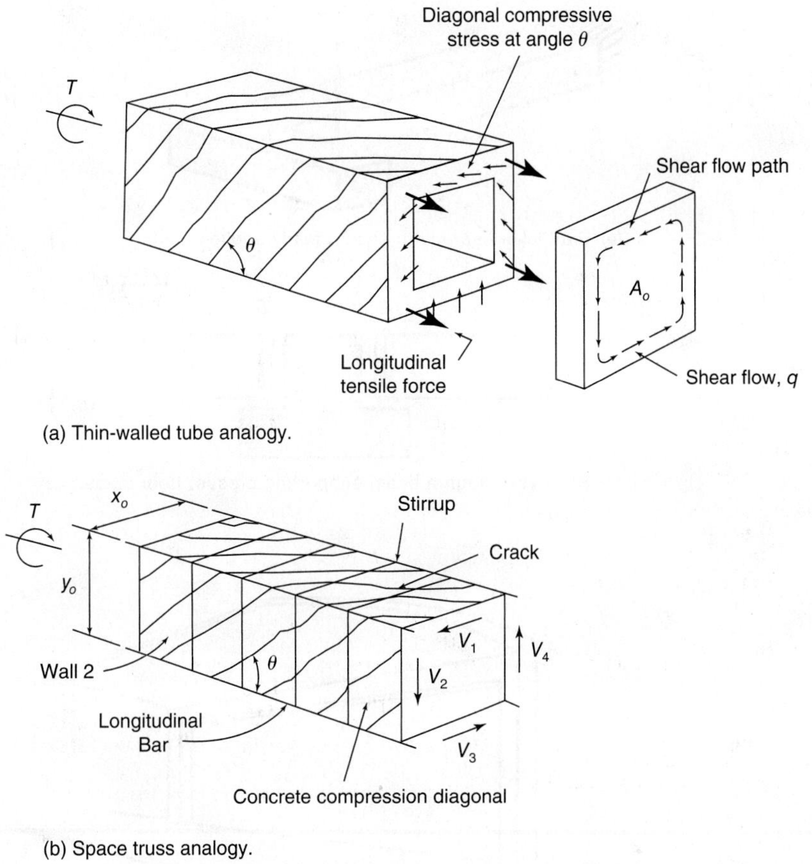

Fig. 7-20
Thin-walled tube analogy and space truss analogy.

torsional strength, T_n, of a reinforced concrete member includes a combination of the thin-walled tube analogy and a space truss analogy, as shown in Fig. 7-20. The ACI Code does not define a procedure for calculating the thin-walled tube thickness, t, and thus, the area enclosed by the shear flow, A_o. Instead, the ACI Code assumes that A_o can be taken as 85 percent of the area enclosed by the closed transverse reinforcement, A_{oh}. The ACI Code then uses the space truss analogy to calculate the shear flow, q, as a function of the yield strength of the longitudinal and transverse reinforcement provided in the member to resist torsion. The ACI Code procedure for calculating torsional strength, T_n, including reinforcement limits to prevent a concrete compression failure prior to reinforcement yielding, is presented in the following section.

7-4 DESIGN FOR TORSION AND SHEAR—ACI CODE APPROACH

Equilibrium and Compatibility Torsion

Torsional loadings can be separated into two basic categories: *equilibrium torsion*, where the torsional moment is required for the equilibrium of the structure, and *compatibility torsion*, where the torsional moment results from the compatibility of deformations between members meeting at a joint.

For equilibrium torsion the ACI Code requires that members resisting torsion must be designed to resist the full factored torsion, T_u. Figure 7-21 shows three examples of

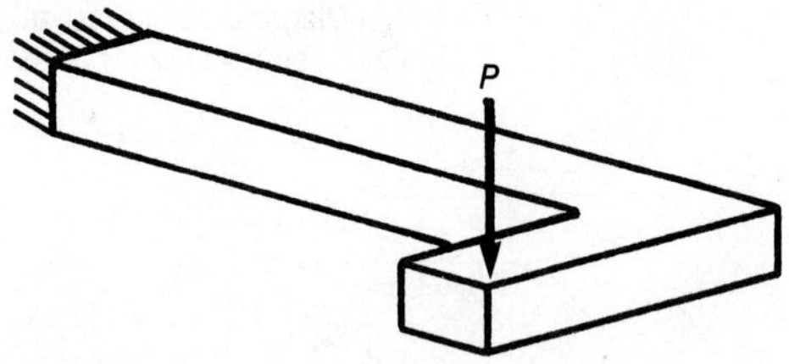

(a) Cantilever beam with eccentrically applied load.

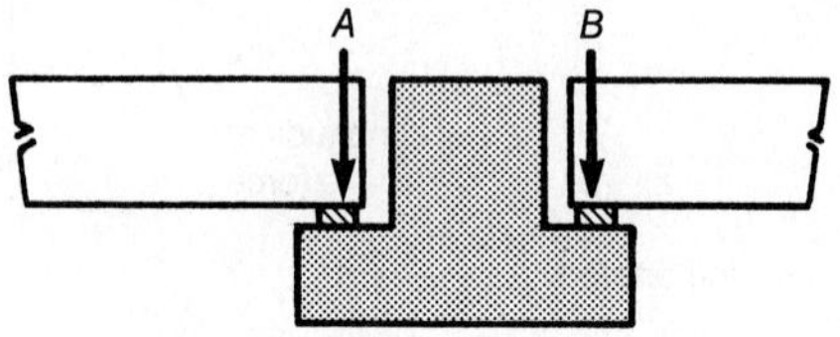

(b) Section through a beam supporting precast floor slabs.

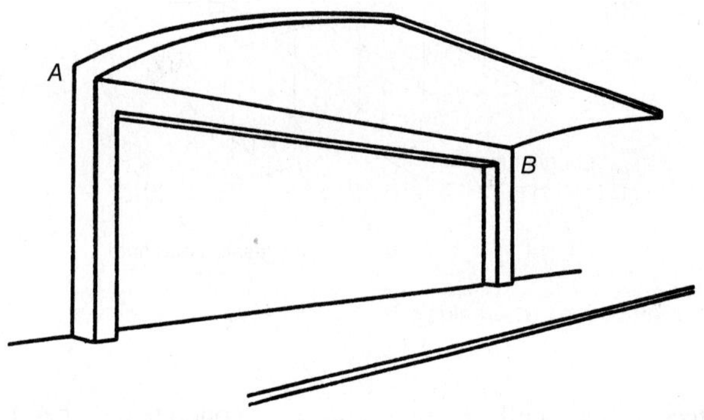

Fig. 7-21
Examples of equilibrium torsion.

(c) Canopy.

equilibrium torsion. Figure 7-21a shows a cantilever beam supporting an eccentrically applied load P that causes torsion. This is a statically determinate member and the torsional moment at any section can be calculated by cutting a free-body diagram at that section. Figure 7-21b shows the cross section of a beam supporting precast floor slabs. Torsion will result if the dead loads of slabs A and B differ, or if one supports a live load and the other does not. The torsion in the beams in Fig. 7-21a and b must be resisted by the structural system if the beam is to remain in equilibrium. If the applied torsion is not resisted, the beam will rotate about its axis until the structure collapses. Similarly, the canopy shown in Fig. 7-21c applies a torsional moment, t, per foot to the beam A–B where t is the torsional moment due to the loads per 1-ft strip of the projecting canopy taken about the centerline of beam A–B. For this structure to stand, the beam must resist the torsional moment, and the columns must resist the resulting bending moments. If the flexural stiffnesses of the columns at ends A and B are the same and if the torsional stiffness, GC, is constant along member A–B, the torque will vary linearly along the length of the member and at ends A and B will be equal to $\pm(t \times \ell/2)$, as shown in Fig. 7-22. If the flexural stiffnesses of the columns at A and B were not equal, the distribution of the torsion along member A–B would still be linear, but would have a higher value at the end with the stiffer column support. However, if the beam is detailed to have adequate

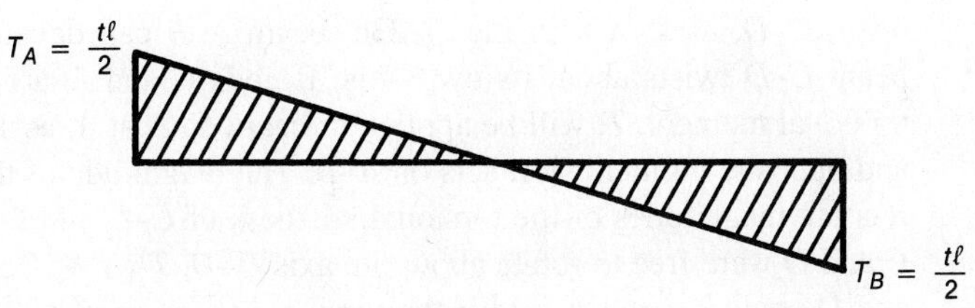

Fig. 7-22
Distribution of torsion in beam *A–B* of the structure shown in Fig. 7-21c.

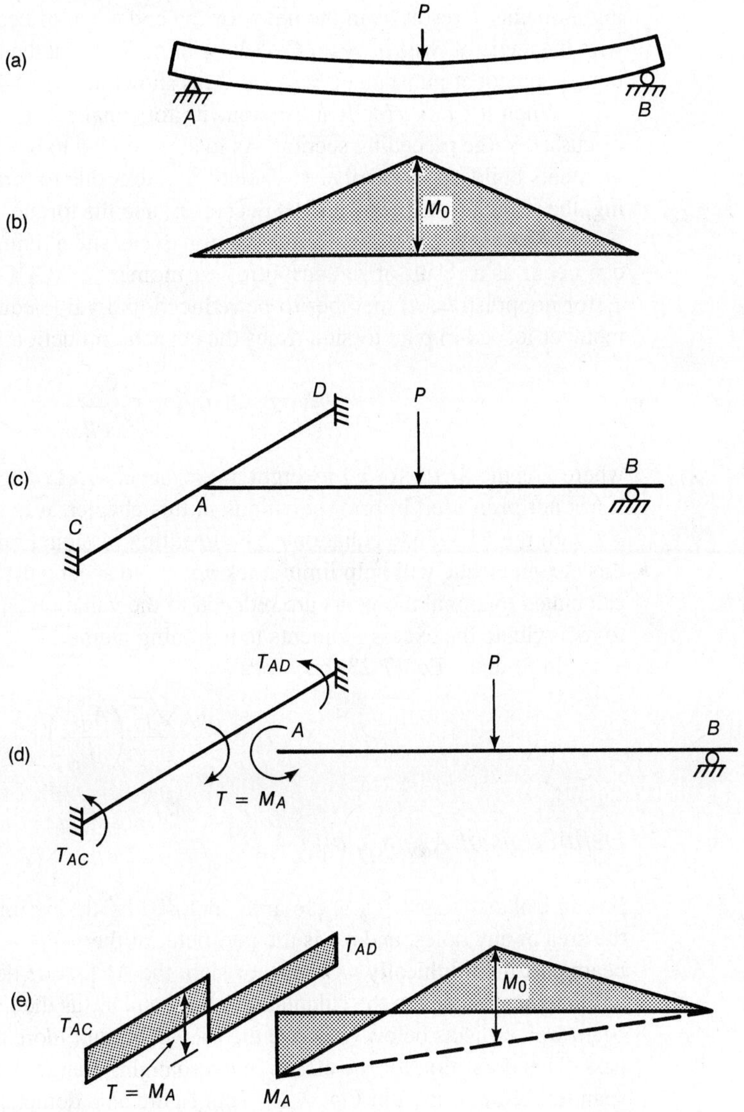

Fig. 7-23
Compatibility torsion.

ductility in torsion, a safe design will result for any reasonable distribution of torsion that accounts for the total torsion applied to the member.

By contrast, Fig. 7-23 shows an example of compatibility torsion. The beam *A–B* in Fig. 7-23a develops a slope at each end when loaded and develops the bending moment diagram shown in Fig. 7-23b. If, however, end *A* is built monolithically with a cross beam $C–D$, as shown in Fig. 7-23c, beam $A–B$ can develop an end slope at A only if beam $C–D$ twists about its own axis. If ends C and D are restrained against rotation, a torsional moment, T, will be applied to beam $C–D$ at A, as shown in Fig. 7-23d. An equal and opposite moment, M_A, acts on $A–B$. The magnitude of these moments depends on the relative magnitudes of the torsional stiffness of $C–D$ and the flexural stiffness of $A–B$. If C and D were free to rotate about the axis $C–D$, T would be zero. On the other hand, if C and D could not rotate, and if the torsional stiffness of $C–D$ was very much greater than the flexural stiffness of $A–B$, the moment M_A would approach a maximum equal to the moment that would be developed if A were a fixed end. Thus, the moment M_A and the twisting moments T result from the need for the end slope of beam $A–B$ at A to be *compatible* with the angle of twist of beam $C–D$ at point A. Note that the moment M_A causes a reduction in the moment at midspan of beam $A–B$, as shown in Fig. 7-23e.

When a beam cracks in torsion, its torsional stiffness drops significantly, as was discussed in the preceding section. As load is applied to beam $A–B$ in Fig. 7-23c, torsional moments build up in member $C–D$ until it cracks due to torsion. With the onset of cracking, the torsional stiffness of $C–D$ decreases, and the torque, T, and the moment, M_A, drop.

In cases where compatibility torsion exists and a reduction of the torsional moment can occur as a result of redistribution of moments, ACI Code Section 22.7.3.2 permits T_u for nonprestressed member to be reduced to a value equal to the cracking torque of a member loaded in pure torsion times the capacity reduction factor, ϕ.

$$T_u = \phi 4\lambda\sqrt{f'_c}\left(\frac{A_{cp}^2}{p_{cp}}\right) \tag{7-23}$$

where λ is the factor for lightweight aggregate concrete discussed in Chapter 6. Although it has not been used in prior equations in this chapter, it is introduced here to be consistent with the ACI Code equations. The resulting torsional reinforcement required to resist this design torque will help limit crack widths to acceptable values at service loads. If the calculated torsional moments are reduced to the value given by Eq. (7-23), it is necessary to redistribute the excess moments to adjoining members.

In SI units, Eq. (7-23) becomes

$$T_u = \frac{\phi\lambda\sqrt{f'_c}}{3}\left(\frac{A_{cp}^2}{p_{cp}}\right) \tag{7-23M}$$

Definitions of A_{cp}, p_{cp}, and A_{oh}

For an isolated beam, A_{cp} is the area enclosed by the perimeter of the section, including the area of any holes, and p_{cp} is the perimeter of the section, as shown in Fig. 7-13a. For a beam cast monolithically with a floor slab, the ACI Code defines the overhanging flange width to be included in the calculation of A_{cp} and p_{cp} as the greater of the distances that the beam web projects below or above the flange, but not more than four times the slab thickness. This definition for the effective overhanging flange in torsion is shown for a typical spandrel beam section in Fig. 7-24. This figure also demonstrates the definition of b_t, the width of that part of the cross section containing the closed stirrup resisting torsion. This dimension will be used in the torsion design examples given later in this chapter.

A_{oh} is defined in the ACI Code as the area enclosed by the outermost perimeter of closed transverse torsional reinforcement, measured at the centerline of that reinforcement. Examples of A_{oh} for a variety of solid and hollow beam sections are shown in Fig. 7-25. As stated in the prior section, the ACI Code states that for a cracked section, the area, A_o, enclosed by the path of the shear flow can be taken as 85 percent of A_{oh}.

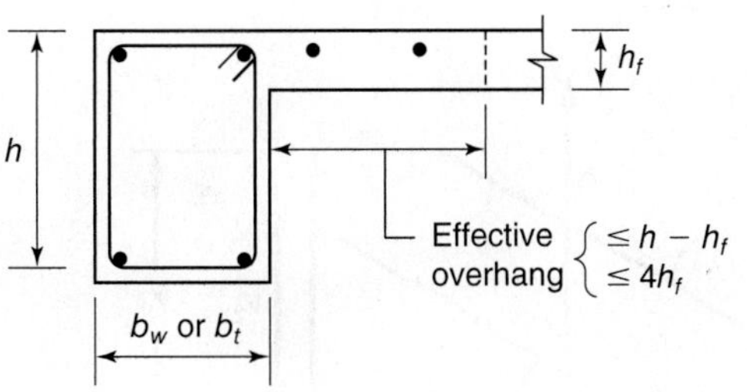

Fig. 7-24
Part of overhanging flange effective for torsion.

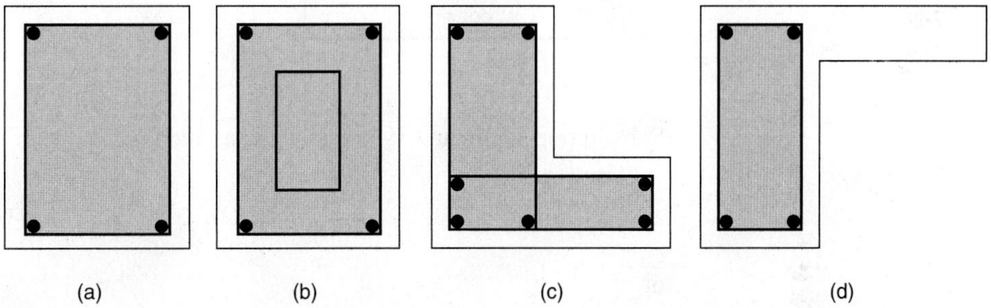

Fig. 7-25
Examples of A_{oh}.

Area of Stirrups for Torsion

A cracked beam subjected to pure torsion can be modeled as shown in Figs. 7-20a and 7-20b. A rectangular beam will be considered for simplicity, but a similar derivation could be applied to any cross-sectional shape. The beam is idealized as a space truss consisting of longitudinal bars in the corners, closed stirrups, and diagonal concrete compression members that spiral around the beam between the cracks. The height and width of the truss are y_o and x_o, which are approximately equal to the distances between the centers of the longitudinal corner bars. The angle of the cracks is θ, which initially is close to 45°, but may become flatter at high torques.

To calculate the required area of stirrups, it is necessary to resolve the shear flow into shear forces acting on the four walls of the tube, as shown in Fig. 7-20b. From Eq. (7-8), the shear force per unit length of the perimeter of the tube or truss, referred to as the shear flow, q, is given by

$$q = \frac{T}{2A_o} \tag{7-24}$$

The total shear force due to torsion along each of the top and bottom sides of the truss is

$$V_1 = V_3 = \frac{T}{2A_o} x_o \tag{7-25a}$$

Similarly, the shear forces due to torsion along each of the two vertical sides are

$$V_2 = V_4 = \frac{T}{2A_o} y_o \tag{7-25b}$$

Summing moments about one corner of the truss, we find that the internal torque is

$$T = V_1 y_o + V_2 x_o$$

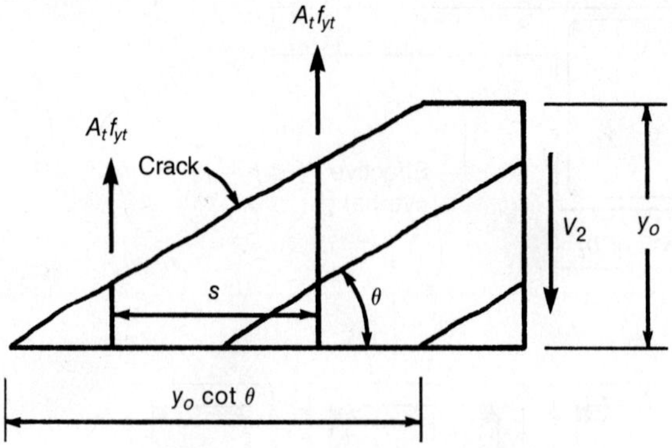

Fig. 7-26
Forces in stirrups.

Substituting for V_1 and V_2 in Eqs. (7-25a) and (7-25b) gives

$$T = \left(\frac{T}{2A_o}x_o\right)y_o + \left(\frac{T}{2A_o}y_o\right)x_o \tag{7-26}$$

or

$$T = \frac{2T(x_o y_o)}{2A_o} \tag{7-27}$$

By definition, however, $x_o y_o = A_o$. Thus, we have shown that the internal forces V_1 through V_4 equilibrate the applied torque, T.

A portion of one of the vertical sides is shown in Fig. 7-26. The inclined crack cuts

$$n_2 = \frac{y_o \cot \theta}{s}$$

stirrups, where s is the spacing of the stirrups. The force in the stirrups must equilibrate V_2. Assuming that all the stirrups yield at nominal strength, we have

$$V_2 = \frac{A_t f_{yt} y_o}{s} \cot \theta \tag{7-28}$$

where f_{yt} is the yield strength of the stirrups. Replacing V_2 with Eq. (7-25b) and taking T equal to the nominal torsion capacity, T_n, gives

$$T_n = \frac{2A_o A_t f_{yt}}{s} \cot \theta \tag{7-29}$$

(ACI Eq. 22.7.6.1a)

where θ is the assumed angle of principal (diagonal) compression with respect to the longitudinal axis of the member. Taking $A_o = 0.85\, A_{oh}$ and substituting T_u/ϕ for T_n gives the following strength requirement for transverse reinforcement.

$$\frac{A_t}{s} \geq \frac{T_u/\phi}{1.7\, A_{oh} f_{yt}} \tan \theta \tag{7-30}$$

Area of Longitudinal Reinforcement

The longitudinal reinforcement must be proportioned to resist the longitudinal tension forces that occur in the space truss. As shown by the force triangle in Fig. 7-27, the shear

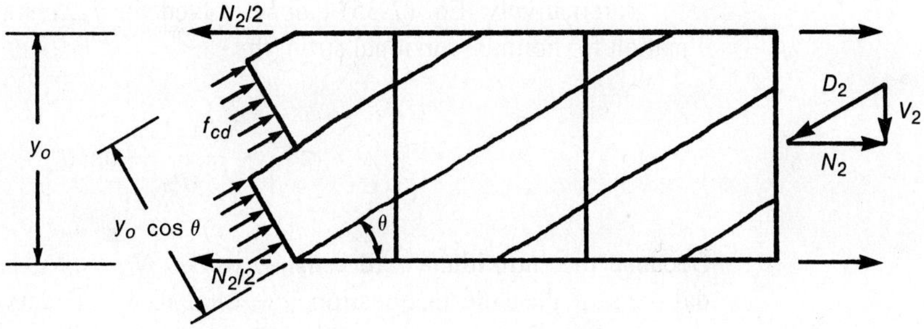

Fig. 7-27
Side of space truss—replacement of shear force V_2.

force V_2 can be replaced with a diagonal compression force, D_2, parallel to the concrete struts and an axial tension force, N_2, where D_2 and N_2 are respectively given by

$$D_2 = \frac{V_2}{\sin \theta} \tag{7-31}$$

and

$$N_2 = V_2 \cot \theta \tag{7-32}$$

Because the shear flow, q, is constant from point to point along side 2, the force N_2 acts along the centroidal axis of side 2. For a beam with longitudinal bars in the top and bottom corners of side 2, half of N_2 will be resisted by each corner bar. A similar resolution of forces occurs on each side of the truss (e.g. $N_1 = V_1 \cot \theta$). For a rectangular member, as shown in Fig. 7-20b, the total longitudinal force is

$$N = 2(N_1 + N_2) \tag{7-33}$$

Substituting Eqs. (7-25a and 7-25b) into (7-33) and taking T equal to T_n gives

$$N = \frac{T_n}{2A_o} 2(x_o + y_o) \cot \theta \tag{7-34}$$

where $2(x_o + y_o)$ is approximately equal to the perimeter of the closed stirrup, p_h. Longitudinal reinforcement with a total area of A_ℓ must be provided for the longitudinal force, N. Assuming that this reinforcement yields at nominal strength, with a yield strength of f_y, produces

$$A_\ell f_y = N$$

or

$$A_\ell = \frac{T_n p_h}{2A_o f_y} \cot \theta \tag{7-35}$$

and again taking $A_o = 0.85\,A_{oh}$ and T_u/ϕ for T_n gives

$$A_\ell = \frac{(T_u/\phi) p_h}{1.7 A_{oh} f_y} \cot \theta \tag{7-36}$$

Alternatively, Eq. (7-35) can be solved for T_n, resulting in the second ACI Code equation for nominal torsional strength,

$$T_n = \frac{2A_oA_\ell f_y}{p_h} \tan\theta \tag{7-37}$$

(ACI Code Eq. 22.7.6.1b)

Because the individual wall tension forces N_1, N_2, N_3, and N_4 act along the centroidal axes of the side in question, the total force, N, acts along the centroidal axis of the member. For this reason, the required longitudinal torsional reinforcement from Eq. (7-36) must be distributed evenly around the perimeter of the cross section so that the centroid of the bar areas coincides approximately with the centroid of the member. One bar must be placed in each corner of the stirrups to anchor the compression struts where the compressive forces change direction around the corner.

Value of θ

The ACI Code allows θ to be taken as any value from 30° to 60°. ACI Code Section 22.7.6.1 requires that the value of θ used to calculate the required area of longitudinal steel, A_ℓ, be the same value as used to calculate the area, A_t, and spacing, s, of transverse reinforcement. This is because a reduction in θ leads to a reduction in the required area of stirrups at a given spacing, as indicated by Eq. (7-29), and an increase in the required area of longitudinal steel, as indicated by Eq. (7-36). ACI Code Section 22.7.6.1.2 suggests using a default value of $\theta = 45°$ for nonprestressed reinforced concrete members. This value will be used in the examples for this chapter.

Limits on Concrete and Steel Strengths for Design

Due to a lack of experimental data for beams constructed with high-strength concrete and tested in torsion, the ACI Code limits the value of $\sqrt{f'_c}$ to 100 psi in all torsion design calculations. Also, to control crack widths at service loads, the ACI Code limits the reinforcement yield strengths (f_y and f_{yt}) *used in torsion design calculations* to 60 ksi.

Combined Shear and Torsion

In ACI Codes prior to 1995, a portion T_c of the torsion was carried by concrete and a portion V_c of the shear was carried by concrete. When both shear and torsion acted, an elliptical interaction diagram was assumed between T_c and V_c, and stirrups were provided for the rest of the torsion and shear. The derivation of Eqs. (7-30) and (7-36) from the space truss analogy assumed that all the torsion was carried by reinforcement, T_s, without any "torsion carried by concrete," T_c. When shear and torsion act together, the 1995 and subsequent ACI Codes assume that V_c remains constant and T_c remains equal to zero, so that

$$V_n = V_c + V_s \tag{6-9}$$

$$T_n = T_s \tag{7-38}$$

where V_c is normally given by Eq. (6-8b). The assumption that there is no interaction between V_c and T_c greatly simplifies the calculations, compared with those required by the

ACI Codes prior to 1995. Design comparisons carried out by ACI Committee 318 showed that, for combinations of low V_u and high T_u, with v_u less than about $0.8(\phi 2\sqrt{f'_c})$ psi, the current ACI Code method requires more stirrups than are required by previous ACI Codes. For v_u greater than this value, the thin-walled-tube method requires the same or marginally fewer stirrups than required by earlier editions of the ACI Code.

Amounts of Torsional and Shear Reinforcement

Torsional reinforcement consists of closed stirrups and longitudinal bars. According to ACI Code Section 9.5.4.3, these are added to the longitudinal bars and stirrups provided for flexure and shear.

In designing for shear, a given size of stirrup, with the area of the two outer legs being A_v, is chosen, and the required spacing, s, is computed. In considering combined shear and torsion, it is necessary to add the stirrups required for shear to those required for torsion. The area of stirrups required for shear and torsion will be computed in terms of A_v/s and A_t/s, both with units of in.^{2}/in. of length of beam. Because A_v refers to all legs of a stirrup (usually two), while A_t refers to only one perimeter leg, the total required stirrup area is

$$\frac{A_{v+t}}{s} = \frac{A_v}{s} + \frac{2A_t}{s} \tag{7-39}$$

where A_{v+t} refers to the cross-sectional area of both legs of a stirrup. Using Eq. (7-39) it is possible to select a stirrup size and compute the required spacing s. If a stirrup in a wide beam has more than two legs for shear, only the outer legs should be included in the summation in Eq. (7-39).

Types of Torsional Reinforcement and Its Anchorage

Because the inclined cracks can spiral around the beam, as shown in Figs. 7-8c and 7-20, stirrups are required in all four faces of the beam. For this reason, ACI Code Section 9.7.6.3 requires the use of closed transverse reinforcement perpendicular to the axis of the member that satisfies detail requirements in ACI Code Section 25.7.1.6. These should extend as close to the perimeter of the member as cover requirements will allow, so as to make A_{oh} as large as possible.

Tests by Mitchell and Collins [7-10] have examined the types of stirrup anchorages required. Figure 7-28a shows one corner of the space truss model shown in Fig. 7-20b. The inclined compressive stresses in the concrete, f_{cd}, have components parallel to the top and side surfaces, as shown in Fig. 7-28b. The components acting toward the corner are balanced by tensions in the stirrups. The concrete outside the reinforcing cage is not well anchored, and the shaded region will spall off if the compression in the outer shell is large. For this reason, ACI Code Section 25.7.1.6 requires that stirrups be anchored with 135° hooks around a longitudinal bar if the corner can spall. If the concrete around the stirrup anchorage is restrained against spalling by a flange or slab or similar member, the ACI Code allows the use of the anchorage details shown in Fig. 7-29a.

ACI Code Section 9.7.5.4 requires that longitudinal reinforcement for torsion be developed at both ends of a beam. Because the maximum torsions generally act at the ends of a beam, it is generally necessary to anchor the longitudinal torsional reinforcement for its yield strength at the face of the support. This may require hooks or horizontal U-shaped bars lap spliced with the longitudinal torsion reinforcement. A common error is to extend the bottom reinforcement in spandrel beams loaded in torsion 6 in. into the support, as allowed in ACI Code Section 9.7.3.8.1 for positive moment reinforcement. Generally, this is not adequate to develop the longitudinal bars needed to resist torsion.

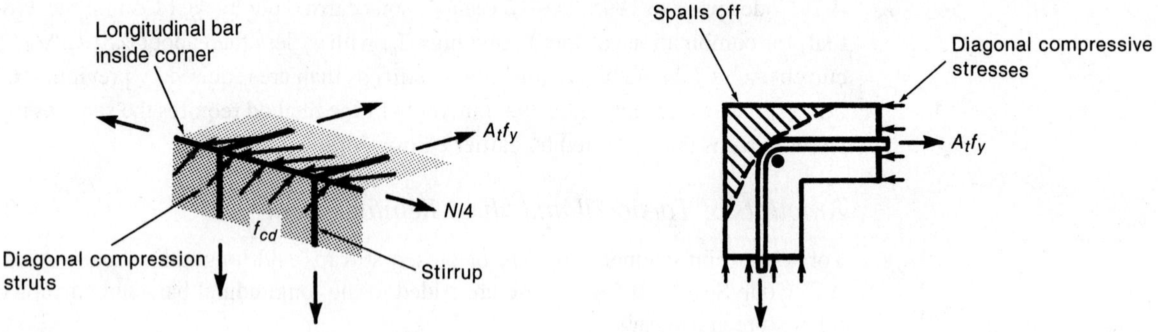

(a) Forces at a corner of the space truss.

(b) Spalling at corner of space truss.

Fig. 7-28
Compressive strut forces at a corner of a torsional member.

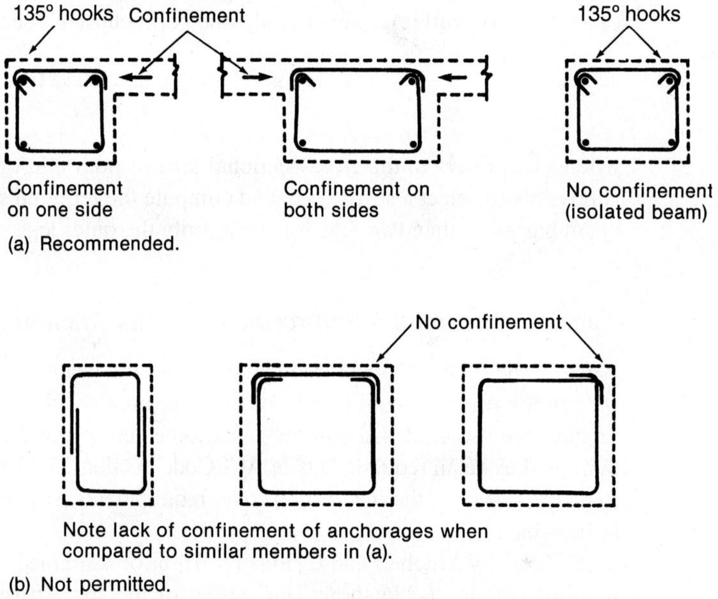

Fig. 7-29
Anchorage of closed stirrups.
(From [7-10].)

Minimum Torsional Reinforcement

When the factored torsional moment exceeds the threshold torsion the larger of: (a) the torsional reinforcement satisfying the strength requirements of ACI Code Section 22.7.6.1, and (b) the minimum reinforcement required by ACI Code Section 9.6.4 must be provided. ACI Code Section 9.6.4.2 specifies that the minimum area of closed stirrups shall be

$$A_v + 2A_t = 0.75\sqrt{f'_c}\,\frac{b_w s}{f_{yt}}, \text{ and } \geq \frac{50 b_w s}{f_{yt}} \tag{7-40}$$

In SI units, this becomes

$$A_v + 2A_t = 0.062\sqrt{f'_c}\,\frac{b_w s}{f_{yt}}, \text{ and } \geq \frac{0.35 b_w s}{f_{yt}} \tag{7-40M}$$

In Hsu's tests [7-5] of rectangular reinforced concrete members subjected to pure torsion, two beams failed at the torsional cracking load. In these beams, the total ratio of the volume of the stirrups and longitudinal reinforcement to the volume of the concrete was 0.802 and 0.88 percent. A third beam, with a volumetric ratio of 1.07 percent, failed at 1.08 times the cracking torque. All the other beams tested by Hsu had volumetric ratios of 1.07 percent or greater and failed at torques in excess of 1.2 times the cracking torque. This suggests that beams with similar concrete and steel strengths loaded in pure torsion should have a minimum volumetric ratio of torsional reinforcement in the order of 0.9 to 1.0 percent. Thus, the minimum volumetric ratio should be set at about 1 percent; that is,

$$\frac{A_{\ell,\,\min}s}{A_{cp}s} + \frac{A_t p_h}{A_{cp}s} \geq 0.01 \tag{7-41a}$$

or

$$A_{\ell,\,\min} = 0.01A_{cp} - \frac{A_t p_h}{s} \tag{7-41b}$$

If the constant 0.01 is assumed to be a function of the material strengths in the test specimens, the constant in the first term on the right-hand side of Eq. (7-41b) can be rewritten as $7.5\sqrt{f'_c}/f_y$. In the 1971 to 1989 ACI Codes, a transition was provided between the total volume of reinforcement required by the equation for $A_{\ell,\,\min}$ for pure torsion and the much smaller amount of minimum reinforcement required in beams subjected to shear without torsion. This was accomplished by multiplying the same term by $\tau/(\tau + v)$, giving

$$A_{\ell,\,\min} = \frac{7.5\sqrt{f'_c}}{f_y}A_{cp}\left(\frac{\tau}{\tau + v}\right) - \left(\frac{A_t}{s}\right)p_h\left(\frac{f_{yt}}{f_y}\right) \tag{7-42}$$

During the development of the 1995 torsion provisions, it was assumed that a practical limit on $\tau/(\tau + v)$ was 2/3 for beams that satisfied Eq. (7-36). When this was introduced, Eq. (7-42) became the following from ACI Code Section 9.6.4.3.

$$A_{\ell,\,\min} = \frac{5\sqrt{f'_c}}{f_y}A_{cp} - \left(\frac{A_t}{s}\right)p_h\left(\frac{f_{yt}}{f_y}\right) \tag{7-43}$$

This equation was derived for the case of pure torsion. When it is applied to combined shear, moment, and torsion, it is not clear how much of the area of the stirrups should be included in A_t/s. In this book, we shall assume that A_t/s in Eq. (7-43) is the actual amount of transverse reinforcement provided for *torsion strength* in Eq. (7-30), where A_t is for one leg of a closed stirrup. The value of A_t/s should not be taken less than $25b_w/f_{yt}$, half of the minimum amount corresponding to Eq. (7-40).

In SI units, (7-43) becomes

$$A_{\ell,\,\min} = \frac{5\sqrt{f'_c}}{12f_y}A_{cp} - \left(\frac{A_t}{s}\right)p_h\left(\frac{f_{yt}}{f_y}\right) \tag{7-43M}$$

Spacing of Torsional Reinforcement

Figure 7-28 shows that the corner longitudinal bars in a beam help to anchor the compressive forces in the struts between cracks. If the stirrups are too far apart, or if the longitudinal bars in the corners are too small in diameter, the compressive forces will tend to bend the longitudinal bars outward, weakening the beam. ACI Code Section 9.7.6.3.3 limits the stirrup spacing to the smaller of $p_h/8$ or 12 in., where p_h is the perimeter of the outermost closed stirrups.

As stated earlier, longitudinal torsional reinforcement should be distributed around the perimeter of the closed stirrups, with the centroid of the steel approximately at the centroid of the cross section. ACI Code Section 9.7.5.1 states that the maximum spacing between longitudinal bars is 12 in. The longitudinal reinforcement should be inside the stirrups, with a bar inside each corner of the stirrups. ACI Code Section 9.7.5.2 requires the diameter of the longitudinal bars to be at least 0.042 times the stirrup spacing, but not less than 0.375 in. In tests, [7-10] corner bars with a diameter of 1/31 of the stirrup spacing bent outward at failure.

ACI Code Section 9.7.5.3 requires that torsional reinforcement continue a distance $(b_t + d)$ past the point where the torque is less than the thereshold torsion, where b_t is the width of that part of the cross section containing the closed stirrups (Fig. 7-24). This length takes into account the fact that torsional cracks spiral around the beam.

Maximum Shear and Torsion

A member loaded by torsion or by combined shear and torsion may fail by yielding of the stirrups and longitudinal reinforcement, as assumed in the derivation of Eqs. (7-30) and (7-36), or by crushing of the concrete due to the diagonal compressive forces. Also, a serviceability failure may occur if the inclined cracks are too wide at service loads. The limit on combined shear and torsion in ACI Code Section 22.7.7.1 was derived to limit service-load crack widths, but as is shown later, it also gives a lower bound on the web's crushing capacity.

Crack Width Limit

As was explained in Section 6-4, ACI Code Section 22.5.1.2 attempts to guard against excessive crack widths by limiting the maximum shear, V_s, that can be transferred by stirrups to an upper limit of $8\sqrt{f'_c}b_w d$. In ACI Code Section 22.7.7.1, the same concept is used, expressed in terms of stresses. The shear stress, v, due to direct shear is $V_u/b_w d$. From Eq. (7-9), with A_o after torsional cracking taken as $0.85A_{oh}$ and $t = A_{oh}/p_h$, the shear stress, τ, due to torsion is $T_u p_h/(1.7A_{oh}^2)$. In a *hollow section*, these two shear stresses are additive on one side, at point A in Fig. 7-30a, and the limit is given by

$$\frac{V_u}{b_w d} + \frac{T_u p_h}{1.7A_{oh}^2} \leq \phi\left(\frac{V_c}{b_w d} + 8\sqrt{f'_c}\right) \qquad (7\text{-}44)$$

(ACI Eq. 22.7.7.1b)

If the wall thickness varies around the cross section, as for example in Fig. 7-7, ACI Code Section 22.7.7.1.2 states that Eq. (7-44) is evaluated at the location where the left-hand side is the greatest.

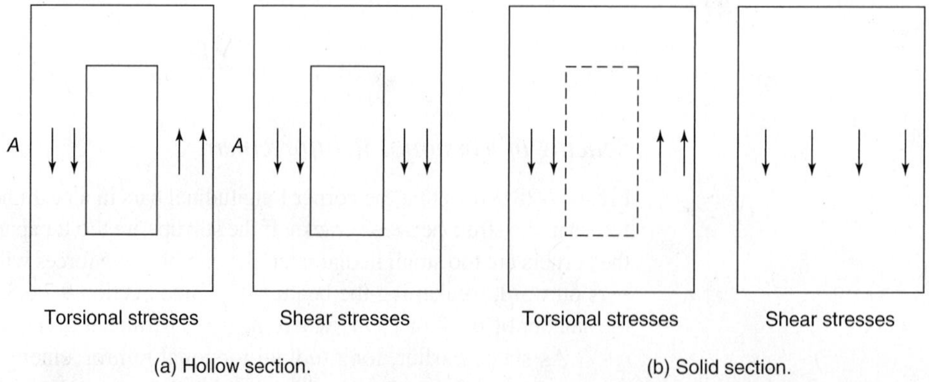

Fig. 7-30
Addition of shear stresses due to torsion and shear. (From [7-11].)

If a hollow section has a wall thickness, t, less than A_{oh}/p_h, ACI Code Section 22.7.7.2 requires that the actual wall thickness be used. Thus, the second term of Eq. (7-44) becomes $T_u/(1.7A_{oh}t)$. Alternatively, the second term on the left-hand side of Eq. (7-44) can be taken as $T_u/(A_o t)$, where A_o and t are computed as in Example 7-1.

In a *solid section*, the shear stresses due to direct shear are assumed to be distributed uniformly across the width of the web, while the torsional shear stresses exist only in the walls of the equivalent thin-walled tube, as shown in Fig. 7-30b. In this case, a direct addition of the two terms tends to be conservative, so a root-square summation is used instead:

$$\sqrt{\left(\frac{V_u}{b_w d}\right)^2 + \left(\frac{T_u p_h}{1.7A_{oh}^2}\right)^2} \le \phi\left(\frac{V_c}{b_w d} + 8\sqrt{f'_c}\right) \qquad (7\text{-}45)$$

(ACI Eq. 22.7.7.1a)

The right-hand sides of Eqs. (7-44) and (7-45) include the term $V_c/b_w d$; hence, the same equation can be used for prestressed concrete members and for members with axial tension or compression that have different values of V_c.

In SI units, the right-hand sides of Eqs. (7-44) and (7-45) become

$$\phi\left(\frac{V_c}{b_w d} + \frac{8\sqrt{f'_c}}{12}\right)$$

Web Crushing Limit

Failure can also occur due to crushing of the concrete in the walls of the tube due to the inclined compressive forces in the struts between cracks. The diagonal compressive force in a vertical side of the member shown in Fig. 7-27 is given by Eq. (7-31). This force acts on a width $y_o \cos\theta$, as shown in Fig. 7-27. The resulting compressive stress due to torsion is

$$f_{cd} = \frac{V_2}{t y_o \cos\theta \sin\theta} \qquad (7\text{-}46)$$

Substituting Eq. (7-9), again taking A_o equal to $0.85A_{oh}$ and approximating t as A_{oh}/p_h, gives

$$f_{cd} = \frac{T_u p_h}{1.7A_{oh}^2 \cos\theta \sin\theta} \qquad (7\text{-}47)$$

The diagonal compressive stresses due to shear may be calculated in a similar manner as

$$f_{cd} = \frac{V_u}{b_w d \cos\theta \sin\theta} \qquad (7\text{-}48)$$

For a solid section, these will be “added” via square root, as explained in the derivation of Eq. (7-45), giving

$$f_{cd} = \sqrt{\left(\frac{V_u}{b_w d \cos\theta \sin\theta}\right)^2 + \left(\frac{T_u p_h}{1.7A_{oh}^2 \cos\theta \sin\theta}\right)^2} \qquad (7\text{-}49)$$

The value of f_{cd} from Eq. (7-47) should not exceed the crushing strength of the cracked concrete in the tube, f_{ce}. Collins and Mitchell [7-10], [7-12] have related f_{ce} to the strains in the longitudinal and transverse reinforcement in the tube. For $\theta = 45°$ and for longitudinal and transverse strains, $\epsilon = 0.002$, equal to the yield strain of Grade-60 steel, Collins and Mitchell predict $f_{ce} = 0.549f'_c$. Setting f_{cd} in Eq. (7-46) equal to $0.549f'_c$ evaluating $\cos\theta \sin\theta$ for $\theta = 45°$ gives the upper limit on the shears and torques, as determined by crushing of the concrete in the walls of the tube:

$$\sqrt{\left(\frac{V_u}{b_w d}\right)^2 + \left(\frac{T_u p_h}{1.7A_{oh}^2}\right)^2} \leq \phi(0.275f'_c) \tag{7-50}$$

The limit in Eqs. (7-44) and (7-45) has been set at $\phi(v_c + 8\sqrt{f'_c})$ to limit crack widths where, for reinforced concrete, v_c can be assumed to be $2\sqrt{f'_c}$, giving a limit of $\phi 10\sqrt{f'_c}$. The limit $\phi(0.275f'_c)$ in Eq. (7-50) will always exceed $\phi 10\sqrt{f'_c}$ for f'_c greater than 1320 psi. Because reinforced concrete members will always have f'_c greater than 1320 psi, only the crack width limits, Eqs. (7-44) and (7-45), are included in the ACI Code. Two simplifications were made in the derivation of Eq. (7-50). First, the calculation of f_{cd} in Eq. (7-48) involved the effective depth, d, while the calculation of f_{cd} in Eq. (7-46) used the height of a wall of the space truss, y_o, which is about $0.9d$. Second, all the shear was assumed to be carried by truss action without a V_c term. These were considered to be reasonable approximations in view of the levels of accuracy of the right-hand sides of Eqs. (7-45) and (7-50).

In [7-11], the code limit, Eq. (7-45), is compared with tests of reinforced concrete beams in pure torsion that failed due to crushing of the concrete in the tube. The limit gave an acceptable lower bound on the test results.

Combined Moment and Torsion

Torsion causes an axial tensile force N, given by Eq. (7-34). Half of this, $N/2$, is assumed to act in the top chord of the space truss, half in the bottom chord, as shown in Fig. 7-31a. Flexure causes a compression–tension couple, $C = T = M_u/jd$, shown in Fig. 7-31b, where $j \approx 0.9$. For combined moment and torsion, these internal forces add together, as shown in Fig. 7-31c. The reinforcement provided for the flexural tension force, T, and that provided for the tension force in the lower chord due to torsion, $N/2$, must be added together, as required by ACI Code Section 9.5.4.3.

In the flexural compression zone, the force C tends to cancel out some, or all, of $N/2$. ACI Code Section 9.5.4.5 allows the area of the longitudinal torsion reinforcement in the compression zone to be reduced by an amount equal to $M_u/(0.9df_y)$, where M_u is the moment that acts in conjunction with the torsion at the section being designed. It is necessary to compute this reduction at a number of sections, because the bending moment varies along the length of the member. If several loading cases must be considered in design, M_u and T_u must be from the same loading case. Normally, the reduction in the area of the compression steel is not significant, as will be shown in Example 7-2.

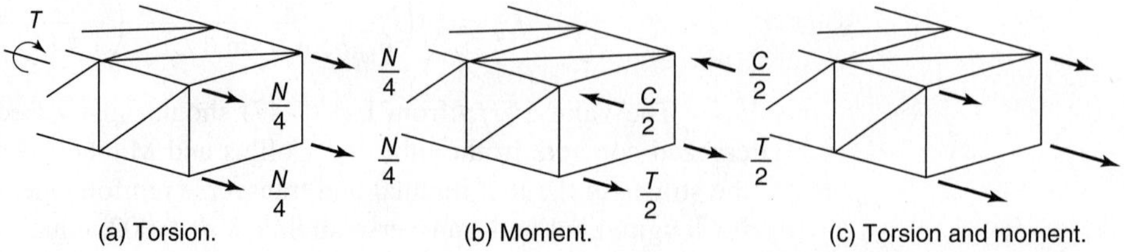

Fig. 7-31
Internal forces due to combined torsion and moment.

7-5 ACI CODE DESIGN METHOD FOR TORSION

The design procedure for combined torsion, shear, and moment involves designing for the moment while ignoring the torsion and shear, and then providing stirrups and longitudinal reinforcement to give adequate shear and torsional strength. The basic design equations are

$$\phi V_n \geq V_u \tag{6-14}$$

$$V_n = V_c + V_s \tag{6-9}$$

and

$$\phi T_n \geq T_u \tag{7-51}$$

where ϕ is the strength reduction factor for shear and torsion, taken equal to 0.75. T_n is given by Eqs. (7-29) and (7-37).

Selection of Cross Section for Torsion

A torsional moment is resisted by shearing stresses in the uncracked member (Figs. 7-2 and 7-5b) and by the shear flow forces (Fig. 7-20b) in the member after cracking. For greatest efficiency, the shearing stresses and shear flow forces should flow around the member in the same circular direction and should be located as far from the axis of the member as possible. Thus the solid square member in Figs. 7-2b and 7-4 is more efficient than the U-shaped member in Fig. 7-5. For equal volumes of material, a closed tube will be much more efficient than a solid section. For building members, solid rectangular sections are generally used for practical reasons. For bridges, box sections like Fig. 7-7a are frequently used. Open U sections like the one shown in Fig. 7-11, made up of beams and a deck, are common, but they are not efficient and lack stiffness in torsion. Much of the torsion is resisted by warping torsion in such cases.

For proportioning hollow sections, some codes require that the distance from the centerline of the transverse reinforcement to the inside face of the wall not be less than $0.5A_{oh}/p_h$. This ensures that the diagonal compression struts will develop their required thickness within the wall, corresponding to the assumption that $A_o = 0.85A_{oh}$. For hollow sections with very thin walls, such as the bridge girder shown in Fig. 7-7, the term $(T_u p_h/1.7A_{oh}^2)$ in Eq. (7-44) is replaced by $(T_u/(1.7A_{oh}t)$, where t is the distance from the centerline of the stirrup to the inside face of the wall. Hollow sections are more difficult to form and to place reinforcement and concrete in than are solid sections. The form for the void must be built; then it must be held in position to prevent it from floating upwards in the fresh concrete; finally, it must be removed after the concrete has hardened. Sometimes, styrofoam void forms are left in place.

Although the ACI Code does not require fillets at the inside corners of a hollow section, it is good practice to provide them. Fillets reduce the stress concentrations where the inclined compressive forces flow around inside corners and also aid in the removal of the formwork. The author suggests that each side of a fillet should be $x/6$ in length if there are fewer than eight longitudinal bars, where x is the smaller dimension of a rectangular cross section, and $x/12$ if the section has eight or more longitudinal bars, but not necessarily more than 4 in.

Location of Critical Section for Torsion

In Section 6-4 and Figs. 6-24 and 6-25, the critical section for shear was often found to be located at a distance d away from the face of the support. For an analogous reason, ACI

Code Section 9.4.4.3 allows sections located at less than d from the support to be designed for the same torque, T_u, that exists at a distance d from the support. This does not apply if a large torque is applied within a distance d from the support.

Steps in the Design Method

1. Calculate the factored bending-moment (M_u) diagram or envelope for the member.

2. Select b, d, and h based on the maximum flexural moment. For problems involving torsion, shallower wide-beam sections are preferable to deep and narrow sections.

3. Given b and h, draw final M_u, V_u, and T_u diagrams or envelopes. Calculate the area of reinforcement required for flexure.

4. Determine whether torsion must be considered. Torsion must be considered if T_u exceeds the threshold torsion given by Eq. (7-15). Otherwise, it can be neglected, and the shear design carried out according to Chapter 6 of this book.

5. Determine whether the case involves equilibrium or compatibility torsion. If it is the latter, the torque may be reduced to the value given by Eq. (7-23) at the sections d from the faces of the supports. If the calculated torsional moment is reduced, the moments and shears in the adjacent members must be adjusted accordingly.

6. Check whether the section is large enough for torsion. If the combination of V_u and T_u exceeds the values given by Eqs. (7-44) or (7-45), enlarge the section.

7. Compute the area of stirrups required for shear. This is done by using Eqs. (6-8b), (6-9), (6-10), and (6-14). To facilitate the addition of stirrups for shear and torsion, calculate

$$\frac{A_v}{s} = \frac{V_s}{f_{yt}d}$$

8. Compute the area of stirrups required for torsion by using Eqs. (7-51) and (7-30). Again, these will be computed in terms of A_t/s.

9. Add the required stirrup amounts together, using Eq. (7-39), and select the stirrups. The area of the stirrups must exceed the minimum given by Eq. (7-40). Their spacing and location must satisfy ACI Code Section 9.7.6.3. The stirrups must be closed.

10. Design the longitudinal reinforcement for torsion using Eq. (7-36) and add it to that provided for flexure. The longitudinal reinforcement for torsion must exceed the minimum given by Eq. (7-43) and must satisfy ACI Code Section 9.7.5.

EXAMPLE 7-2 Design for Torsion, Shear, and Moment: Equilibrium Torsion

The cantilever beam shown in Fig. 7-32a supports its own dead load plus a concentrated load as shown. The beam is 54 in. long, and the concentrated load acts at a point 6 in. from the end of the beam and 6 in. away from the centroidal axis of the member. The *unfactored* concentrated load consists of a 20-kip dead load and a 20-kip live load. Use normal weight concrete with $f'_c = 4000$ psi and both f_y and $f_{yt} = 60{,}000$ psi. Use load combination and strength-reduction factors from the ACI Code.

1. Compute the bending-moment diagram. Estimate the size of the member. The minimum depth of control flexural deflections is (ACI Table 9.3.1.1). This seems

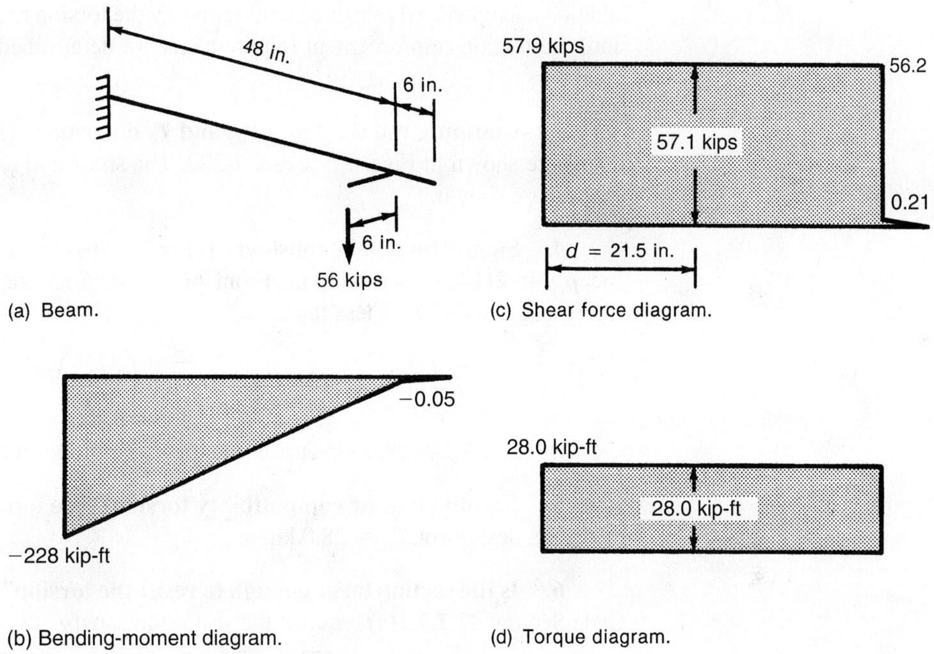

Fig. 7-32
Cantilever beam—
Example 7-2.

too small, in view of the loads involved. As a first trial, use a 14-in.-wide-by-24-in. deep section, with $d = 21.5$ in.:

$$w = \frac{14 \times 24}{144} \times 0.15 = 0.35 \text{ kip/ft}$$

$$\text{Factored uniform dead load} = 1.2 \times 0.35 = 0.42 \text{ kip/ft}$$

$$\text{Factored concentrated load} = 1.2 \times 20 + 1.6 \times 20 = 56 \text{ kips}$$

The bending-moment diagram is as shown in Fig. 7-32b, with the maximum $M_u = 228$ kip-ft.

2. Select A_s for flexure. This is essentially the design of a rectangular section for which the dimensions are known, so we can use Table A-3 to expedite this process. From Eq. (5-23), we can find the required value for the flexural resistance factor, R:

$$R = \frac{M_u}{\phi b d^2} = \frac{228 \text{ k-ft} \times 12 \text{ in./ft}}{0.9 \times 14 \text{ in.} \times (21.5 \text{ in.})^2} = 0.470 \text{ ksi} = 470 \text{ psi}$$

Referring to Table A-3 for $f'_c = 4000$ psi, we see that this section is a tension-controlled section with a reinforcement ratio well above the ACI Code minimum value. Doing a linear interpolation of the given tabular values results in a required reinforcement ratio, $\rho \cong 0.0085$. From this, the required area of tension reinforcement is

$$A_s = \rho b d = 0.0085 \times 14 \text{ in.} \times 21.5 \text{ in.} = 2.56 \text{ in.}^2$$

At this stage in a flexural design, we normally would select a bar size and the number of bars for the required area of steel. In this case, we probably will need to provide some additional longitudinal reinforcement to satisfy the torsion requirements. Thus, we will wait until the torsion reinforcement requirements are determined before selecting the longitudinal bars.

3. Compute the final M_u, V_u, and T_u diagrams. The shear force and torque diagrams are shown in Figs. 7-32c and 7-32d. The shear and torque at d from the face of the support are shown.

4. Should torsion be considered? For the cross section, $A_{cp} = 14 \times 24 = 336$ in.^{2} and $p_{cp} = 2(14 + 24) = 76$ in. From ACI Code Sections 9.5.4.1 and 22.7.4.1, torsion can be neglected if T_u is less than

$$T_{th} = \phi\lambda\sqrt{f'_c}\left(\frac{A_{cp}{}^2}{p_{cp}}\right) = 0.75 \times 1\sqrt{4000}\left(\frac{336^2}{76}\right) = 70{,}500 \text{ in.-lb} = 5.87 \text{ kip-ft}$$

Because $T_u = 28.0$ kip-ft exceeds the threshold torque, torsion must be considered.

5. Equilibrium or compatibility torsion? The torsion is needed for equilibrium; therefore, design for $T_u = 28.0$ kip-ft.

6. Is the section large enough to resist the torsion? For a solid cross section, ACI Code Section 22.7.7.1(a) requires the section to satisfy

$$\sqrt{\left(\frac{V_u}{b_w d}\right)^2 + \left(\frac{T_u p_h}{1.7A_{oh}^2}\right)^2} \le \phi\left(\frac{V_c}{b_w d} + 8\sqrt{f'_c}\right) \tag{7-45}$$

where ϕ is 0.75 for shear and torsion.

From ACI Code Section 22.5.5.1, take $V_c = 2\lambda\sqrt{f'_c}b_w d$. A_{oh} = area within centerline of closed stirrups. Assume 1.5 in. of cover and No. 4 stirrups, as shown in Fig. 7-33. Then

$$A_{oh} = (14 - 2 \times 1.5 - 0.5)(24 - 2 \times 1.5 - 0.5) = 215 \text{ in.}^2$$

$$p_h = 2(10.5 + 20.5) = 62 \text{ in.}$$

$$\sqrt{\left(\frac{57{,}100}{14 \times 21.5}\right)^2 + \left(\frac{28.0 \times 12{,}000 \times 62}{1.7 \times 215^2}\right)^2} \le 0.75(2 \times 1\sqrt{4000} + 8\sqrt{4000})$$

$$\sqrt{36{,}000 + 70{,}300} = 326 \text{ psi} \le 0.75 \times 10\sqrt{4000} = 474 \text{ psi}$$

Because 326 psi is less than 474 psi, the cross section is large enough.

7. Compute the stirrup area required for shear. From Eqs. (6-8b), (6-9), and (6-10),

$$V_u \le \phi(V_c + V_s)$$

$$V_c = 2\lambda\sqrt{f'_c}b_w d = 2 \times 1\sqrt{4000} \times 14 \times 21.5$$

$$= 38{,}100 \text{ lbs} = 38.1 \text{ kips}$$

$$V_s \ge \frac{57.1}{0.75} - 38.1 = 38.0 \text{ kips}$$

From Eq. (6-14),

$$V_s = \frac{A_v f_{yt} d}{s} \quad \text{or} \quad \frac{A_v}{s} = \frac{V_s}{f_{yt} d}$$

$$\frac{A_v}{s} \ge \frac{38.0 \text{ k}}{60 \text{ ksi} \times 21.5 \text{ in.}} = 0.0295 \text{ in.}^2\text{/in.}$$

For shear, we require stirrups with $A_v/s = 0.0295$ in.^{2}/in.

8. Compute the stirrup area required for torsion. From Eq. (7-51), $\phi T_n \geq T_u$. Therefore,

$$T_n \geq \frac{28.0 \times 12 \text{ in./ft}}{0.75} = 448 \text{ k-in.}$$

From Eq. (7-29),

$$T_n = \frac{2A_o A_t f_{yt}}{s} \cot\theta \quad \text{or} \quad \frac{A_t}{s} = \frac{T_n}{2A_o f_{yt}} \tan\theta$$

From ACI Code Section 22.7.6.1.1,

$$A_o \cong 0.85 A_{oh} = 0.85 \times 215$$
$$= 183 \text{ in.}^2$$

Taking $\theta = 45^\circ$

$$\frac{A_t}{s} = \frac{448}{2 \times 183 \times 60} = 0.0204 \text{ in.}^2/\text{in.}$$

For torsion, we require stirrups with $A_t/s = 0.0204$ in.^{2}/in.

9. Add the stirrup areas and select stirrups. From Eq. (7-39),

$$\frac{A_{v+t}}{s} = \frac{A_v}{s} + \frac{2A_t}{s}$$

$$= 0.0295 + 2 \times 0.0204 = 0.0703 \text{ in.}^2/\text{in.}$$

Check minimum stirrups: From Eq. (7-40),

$$\frac{A_v + 2A_t}{s} \geq 0.75\sqrt{f'_c}\frac{b_w}{f_{yt}}, \text{ and } \geq \frac{50b_w}{f_{yt}}$$

$$\text{Minimum}\frac{A_{v+t}}{s} = \frac{0.75\sqrt{4000} \times 14}{60{,}000} = 0.011 \text{ in.}^2/\text{in., and } \geq 0.012 \text{ in.}^2/\text{in.}$$

Because 0.0703 in.^{2}/in. exceeds 0.012 in.^{2}/in., the minimum does not govern.

For No. 3 stirrups, A_{v+t}(two legs) $= 0.22$ in.^{2}, and the required $s = 3.13$ in. For No. 4 stirrups, A_{v+t}(two legs) $= 0.40$ in.^{2}, and the required $s = 5.69$ in. The minimum stirrup spacing for torsion (ACI Code Section 9.7.6.3.3) is the smaller of $p_h/8 = 62/8 = 7.75$ in., or 12 in.

The maximum stirrup spacing for shear (ACI Code Section 9.7.6.2.2) is the smaller of $d/2 = 10.8$ in. or 24 in.

Use No. 4 closed stirrups at 5 in. on centers, placing the first stirrup at 2.5 in. from the face of the support.

10. Design the longitudinal reinforcement for torsion. From Eq. (7-30),

$$A_\ell = \frac{(T_u/\phi)\, p_h}{1.7\, A_{oh} f_y} \cot\theta$$

$$A_\ell = \frac{(336 \text{ k-in.}/0.75)\ 62 \text{ in.}}{1.7 \times 215 \text{ in.}^2 \times 60 \text{ ksi}} \times 1.0 = 1.27 \text{ in.}^2$$

From Eq. (7-43), the minimum A_ℓ is

$$A_{\ell\,\text{min.}} = \frac{5\sqrt{f'_c}\,A_{cp}}{f_y} - \left(\frac{A_t}{s}\right)p_h\left(\frac{f_{yt}}{f_y}\right)$$

where A_t/s will be based on the stirrup area required for torsion in step 8, i.e., 0.0204 in.^{2}/in.

$$A_\ell = \frac{5\sqrt{4000} \times 336}{60{,}000} - (0.0204) \times 62 \times \frac{60}{60}$$
$$= 0.51 \text{ in.}^2$$

This value is less than the strength requirement, so the minimum A_ℓ does not govern. Provide $A_\ell = 1.27$ in.^{2}.

To satisfy the 12-in. maximum spacing between longitudinal bars specified in ACI Code Section 9.7.5.1, we need at least 6 bars each of area 0.21 in.^{2}. There must be a longitudinal bar in each corner of the stirrups. The minimum bar diameter is 0.042 times the stirrup spacing $= 0.042 \times 5 = 0.21$ in.

Provide four No. 4 bars in the lower portion of the beam (two at mid-depth and two at bottom) and add $1.27 - (4 \times 0.20) = 0.47$ in.^{2} to the flexural steel. The total area of steel required in the top face of the beam is $2.56 + 0.47 = 3.03$ in.^{2}. Thus, we have

six No. 7 bars: $A_s = 3.60$ in.^{2}, will not fit in one layer.

four No. 8 bars: $A_s = 3.16$ in.^{2}, will fit in one layer.

Provide four No. 8 bars at the top of the beam.

ACI Code Section 9.5.4.5 allows a designer to subtract $M_u/(0.9df_y)$ from the area of longitudinal steel in the flexural compression zone. We shall not do this, for three reasons: First, M_u varies along the length of the beam; second, we need two bars of minimum diameter 0.21 in. in the corners of the stirrups; third, we need bars to support the corners of the stirrups.

The final beam design is shown in Fig. 7-33. ■

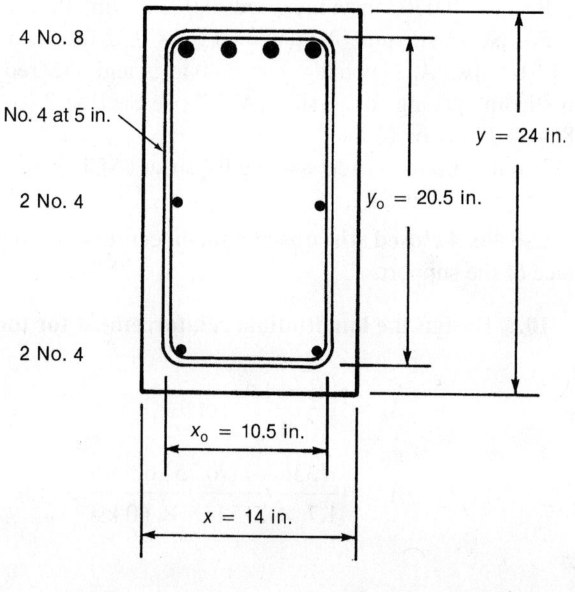

Fig. 7-33
Cross section of a cantilever beam—Example 7-2.

EXAMPLE 7-3 Design for Torsion, Shear, and Moment: Equilibrium Torsion, Hollow Section

Redesign the cantilever beam from Example 7-2, using a hollow cross section. This is a hypothetical example to show the differences between the design for a solid section and that for a hollow section. For a beam of this size, the additional costs of forming the void, holding the void forms in place when the concrete is placed, and removing them when the concrete has hardened would more than offset any material savings due to the void. On the other hand, the cross section of the bridge shown in Fig. 7-7 is clearly more economical if built as a hollow section.

1. Estimate the size and weight of the beam. The most efficient cross section for torsion is a circle; slightly poorer is a rectangular section approaching a square. To estimate the weight of the beam, we shall try a 20-in.-by-24-in. cross section with 5-in.-thick walls on all four sides, leaving a 10-in.-by-14-in. void. The corners of the void have 3-in.-by-3-in. fillets. The area of concrete in the cross section is

$$A_g = (20 \times 24) - (10 \times 14) + 4(3 \times 3/2) = 358 \text{ in.}^2$$

The beam self-weight is

$$w = (358/144) \times 0.150 = 0.373 \text{ kips/ft}$$

The factored uniform self-weight is

$$1.2 \times 0.373 = 0.448 \text{ kips/ft}$$

The factored concentrated load is

$$1.2 \times 20 + 1.6 \times 20 = 56 \text{ kips}$$

The bending-moment diagram is similar to that shown in Fig. 7-32b, with the maximum $M_u = 228$ k-ft.

2. Select A_s for flexure. A hollow section is analyzed like a flanged section in bending. If we can confirm that the compression stress block will stay within the depth of the flange, we will have a constant-width compression zone and can thus use Table A-3 to select an appropriate reinforcement ratio. Using Eq. (5-23), we can find the required value for the flexural resistance factor, R:

$$R = \frac{M_u}{\phi b d^2} = \frac{228 \text{ k-ft} \times 12 \text{ in./ft}}{0.9 \times 20 \text{ in.} \times (21.5 \text{ in.})^2} = 0.330 \text{ ksi} = 330 \text{ psi}$$

Referring to Table A-3, we see that this section is a tension-controlled section with a reinforcement ratio well above the ACI Code minimum value. Table A-3 indicates that a reinforcement ratio, $\rho = 0.006$, will be sufficient. From this, the required area of tension reinforcement is

$$A_s = \rho b d = 0.006 \times 20 \text{ in.} \times 21.5 \text{ in.} = 2.58 \text{ in.}^2$$

For this tension steel area, the depth of the compression stress block, a, can be calculated using Eq. (5-17):

$$a = \frac{A_s f_y}{0.85 f_c' b} = \frac{2.58 \text{ in.}^2 \times 60 \text{ ksi}}{0.85 \times 4 \text{ ksi} \times 20 \text{ in.}} = 2.28 \text{ in.}$$

This value is less than the 5-in. thickness of the lower flange (where the compression is for negative bending), so we can proceed with a required tension steel area of 2.58 in.^{2} and wait on the final selection of longitudinal bars until the torsion reinforcement requirements are determined.

3. Compute the final M_u, V_u, and T_u diagrams. The moment, shear force, and torque diagrams are essentially the same as in Fig. 7-32. The shear and torque at d from the face of the support are shown.

4. Should torsion be considered? Torsion must be considered if T_u exceeds the threshold torque given by

$$\phi\lambda\sqrt{f'_c}\left(\frac{A_{cp}^2}{p_{cp}}\right).$$

Because the cracking torque, and hence the threshold torque, is lower in a hollow section than in a solid section, use Eq. (7-16), including λ.

$$\phi\lambda\sqrt{f'_c}\left(\frac{A_g^2}{p_{cp}}\right)$$

$$A_g = 358 \text{ in.}^2$$

$$p_{cp} = 2(20 + 24) = 88 \text{ in.}$$

$$\phi\lambda\sqrt{f'_c}\left(\frac{A_g^2}{p_{cp}}\right) = 0.75 \times 1\sqrt{4000}\left(\frac{358^2}{88}\right) = 69{,}100 \text{ lb-in.} = 5.76 \text{ k-ft.}$$

Because T_u exceeds 5.76 k-ft, torsion must be considered.

5. Equilibrium or compatibility torsion? The torsion is needed for equilibrium; therefore, design the beam for $T_u = 28.0$ k-ft.

6. Is the section big enough to resist the torsion? For a hollow section, use Eq. (7-44).

$$\left(\frac{V_u}{b_w d}\right) + \left(\frac{T_u p_h}{1.7A_{oh}^2}\right) \le \phi\left(\frac{V_c}{b_w d} + 8\sqrt{f'_c}\right)$$

Assuming that the beam is not exposed to weather and the minimum cover to the No. 4 stirrups is 1.5 in., the center-to-center widths of the stirrups are as follows:

$$\text{Horizontally} = 20 - 2(1.5 + 0.25) = 16.5 \text{ in.}$$

$$\text{Vertically} = 24 - 2(1.5 + 0.25) = 20.5 \text{ in.}$$

$$p_h = 2(16.5 + 20.5) = 74.0 \text{ in.}$$

$$A_{oh} = 16.5 \times 20.5 = 338 \text{ in.}^2$$

$$b_w = 20 - 10 = 10 \text{ in.}$$

ACI Code Section 22.7.7.2 requires that the second term on the left side of Eq. (7-44) be replaced with $T_u/1.7A_{oh}t$ if A_{oh}/p_h exceeds the wall thickness. $A_{oh}/p_h = 338/74 = 4.57$. Therefore, Eq. (7-44) may be used as given. Substituting into Eq. (7-44) gives:

$$\text{left-hand side: } \left(\frac{57.1 \times 1000}{10 \times 21.5}\right) + \left(\frac{28.0 \times 12{,}000 \times 74.0}{1.7 \times 338^2}\right)$$

$$= 266 + 128 = 394 \text{ psi}$$

$$\text{right-hand side: } \phi\left(\frac{V_c}{b_w d} + 8\sqrt{f'_c}\right) = 0.75(10\sqrt{4000}) = 474 \text{ psi}$$

Because 394 psi is less than 474 psi, the section is big enough.

7. Compute the stirrup area required for shear.

$$\phi V_n \geq V_u$$

so,

$$V_n \geq \frac{V_u}{\phi} = \frac{57.1}{0.75} = 76.1 \text{ kips}$$

$$V_n = V_c + V_s$$

$$V_c = 2\lambda\sqrt{f'_c}\, b_w d = 2 \times 1\sqrt{4000} \times 10 \times 21.5$$

$$= 27{,}200 \text{ lbs} = 27.2 \text{ kips}$$

$$V_s \text{ required at } d \text{ from support} = 76.1 - 27.2 = 48.9 \text{ kips}$$

$$V_s = \frac{A_s f_{yt} d}{s}$$

Rearranging gives

$$\frac{A_v}{s} = \frac{V_s}{f_{yt} d} \geq \frac{48.9 \text{ k}}{60 \text{ ksi} \times 21.5 \text{ in.}}$$

$$\geq 0.0379 \text{ in.}^2/\text{in.}$$

For shear, we need stirrups with $A_v/s \geq 0.0379$ in.^{2}/in.

8. Compute the stirrup area required for torsion.

$$\phi T_n \geq T_u \tag{7-51}$$

$$T_n \geq \frac{T_u}{\phi} = \frac{28.0}{0.75} = 37.3 \text{ kip-ft}$$

$$T_n = 2A_o \frac{A_t f_{yt}}{s} \cot\theta \tag{7-29}$$

Rearranging terms and setting θ equal to the default value of 45° from ACI Code Eq. (22.7.6.1a) gives

$$\frac{A_t}{s} = \frac{T_n}{2A_o f_{yt}}$$

where A_o may be taken as 0.85 A_{oh}

$$\frac{A_t}{s} \geq \frac{37.3 \times 12}{2(0.85 \times 338) \times 60}$$

$$\geq 0.0130 \text{ in.}^2/\text{in.}$$

For torsion, we need stirrups with $A_t/s \geq 0.0130$ in.^{2}/in.

9. Add the stirrup areas and select stirrups.

Stirrups required for combined shear and torsion

$$\frac{A_{v+t}}{s} = \frac{A_v}{s} + 2\frac{A_t}{s} \tag{7-39}$$

$$= 0.0379 + 2 \times 0.0130 = 0.0639 \text{ in.}^2/\text{in.}$$

Minimum stirrups

From ACI Code Section 9.6.4.2, and recalling from Example 7-2 that 50 psi $> 3\sqrt{f'_c}$ the minimum stirrups required for combined shear and torsion are:

$$A_{v+t,\,\text{min}} = \frac{50b_w}{f_{yt}} = 0.0083\ \text{in.}^2/\text{in.} \qquad (7\text{-}40)$$

Because the required A_{v+t}/s exceeds this, use the amount required for combined shear and torsion. Try No. 4 closed stirrups, area of two legs $= 2 \times 0.20 = 0.40\ \text{in.}^2$. From the combined strength requirement, the spacing required for the factored shear and torsion is

$$\frac{A_{v+t}}{s} = 0.0639\ \text{in.}^2/\text{in.}$$

so

$$s = \frac{0.40\ \text{in.}^2}{0.0639\ \text{in.}^2/\text{in.}} = 6.26\ \text{in.}$$

From ACI Code Section 9.7.6.2.2, the maximum spacing of stirrups for shear is $d/2$ or 24 in. for V_s less than $4\sqrt{f'_c}\,b_w d = 4\sqrt{4000} \times 10 \times 21.5 = 54{,}400$ lbs $= 54.4$ kips. Because $V_s = 52.5$ kips, the maximum spacing for shear is $d/2 = 10.8$ in. From ACI Code Section 9.7.6.3.3, the maximum spacing of stirrups for torsion is the smaller of $p_h/8 = 74.0/8 = 9.25$ in. and 12 in. Therefore, the maximum stirrup spacing is 6.25 in.

Use No. 4 closed stirrups: One at 3 in. from face of support and eight at 6 in. on centers.

10. Design the longitudinal reinforcement for torsion.

$$A_\ell = \frac{(T_u/\phi)\,p_h}{1.7\,A_{oh} f_y}\cot\theta$$

Using $\theta = 45°$

$$A_\ell = \frac{(336\ \text{k-in.}/0.75) \times 74\ \text{in.}}{1.7 \times 338\ \text{in.}^2 \times 60\ \text{ksi}} \times 1.0 = 0.962\ \text{in.}^2$$

From ACI Code Section 9.6.4.3, the minimum area of longitudinal steel is

$$A_\ell = \frac{5\sqrt{f'_c}}{f_y}A_{cp} - \left(\frac{A_t}{s}\right)p_h\left(\frac{f_{yt}}{f_y}\right)$$

$$= \frac{5\sqrt{4000}}{60{,}000}480 - (0.0130) \times 74 \times \left(\frac{60}{60}\right)$$

$$= 2.53 - 0.962 = 1.57\ \text{in.}^2$$

Therefore, we need to provide 1.57 in.^{2} of longitudinal reinforcement for torsion. From ACI Code Section 9.7.5.1, the maximum spacing of the longitudinal bars is 12.0 in. We will put one longitudinal bar in each corner of the stirrups, one at the middle of the top and bottom sides, and one at midheight of each vertical side, a total of eight bars. The area per bar is $1.57/8 = 0.196$ in.^{2}. From ACI Code Section 9.7.5.2, the minimum diameter of the corner bar is $0.042 \times$ the stirrup spacing $= 0.042 \times 6 = 0.25$ in. We will use No. 4 bars.

For the top of the beam,

for flexure, we need $A_s = 2.28 \text{ in.}^2$

for torsion, we need $3 \times 0.196 \text{ in.}^2 = 0.588 \text{ in.}^2$

total steel at top of beam $= 2.28 + 0.588 = 2.87 \text{ in.}^2$

The possible choices are as follows:

six No. 7 bars, $A_s = 3.60 \text{ in.}^2$. These will fit in the 20 in. width in one layer.

four No. 8 bars, $A_s = 3.16 \text{ in.}^2$. These will fit.

Use four No. 8 bars in the top of the beam plus five additional No. 4 longitudinal bars, two at midheight and three in the bottom flange. ■

EXAMPLE 7-4 Design for Compatibility Torsion

The one-way joist system shown in Fig. 7-34 supports a total factored dead load of 157 psf and a factored live load of 170 psf, totaling 327 psf. Design the end span, *AB*, of the exterior spandrel beam on grid line 1. The factored dead load of the beam and the factored loads applied directly to it total 1.1 kips/ft. The spans and loadings are such that the moments and shears can be calculated by using the coefficients from ACI Code Sections 6.5.2 and 6.5.4 (see Section 5-2 of this book). Use $f_y = f_{yt} = 60{,}000$ psi and normal-weight concrete with $f'_c = 4000$ psi.

1. Compute the bending moments for the beam. In laying out the floor, it was found that joists with an overall depth of 18.5 in. would be required. The slab thickness is 4.5 in. The spandrel beam was made the same depth, to save forming costs. The columns supporting the beam are 24 in. square. For simplicity in forming the joists, the beam overhangs the inside face of the columns by $1\frac{1}{2}$ in. Thus, the initial choice of beam size is $h = 18.5$ in., $b = 25.5$ in., and $d = 16$ in.

Although the joist loads are transferred to the beam by the joist webs, we shall assume a uniform load for simplicity. Very little error is introduced by this assumption. The joist reaction per foot of length of beam is

$$\frac{w\ell_n}{2} = \frac{0.327 \text{ ksf} \times 29.75 \text{ ft}}{2} = 4.86 \text{ kips/ft}$$

The total load on the beam is

$$w_u = 4.86 + 1.1 = 5.96 \text{ kips/ft}$$

Using $\ell_n = 21.5$ ft for the spandrel beam, the moments in the edge beam are as follows:

$$\text{Exterior end negative: } M_u = -\frac{w_u\ell_n^2}{16} = -172 \text{ kip-ft}$$

$$\text{Midspan positive: } M_u = \frac{w_u\ell_n^2}{14} = +197 \text{ kip-ft}$$

$$\text{First interior negative: } M_u = -\frac{w_u\ell_n^2}{10} = -276 \text{ kip-ft}$$

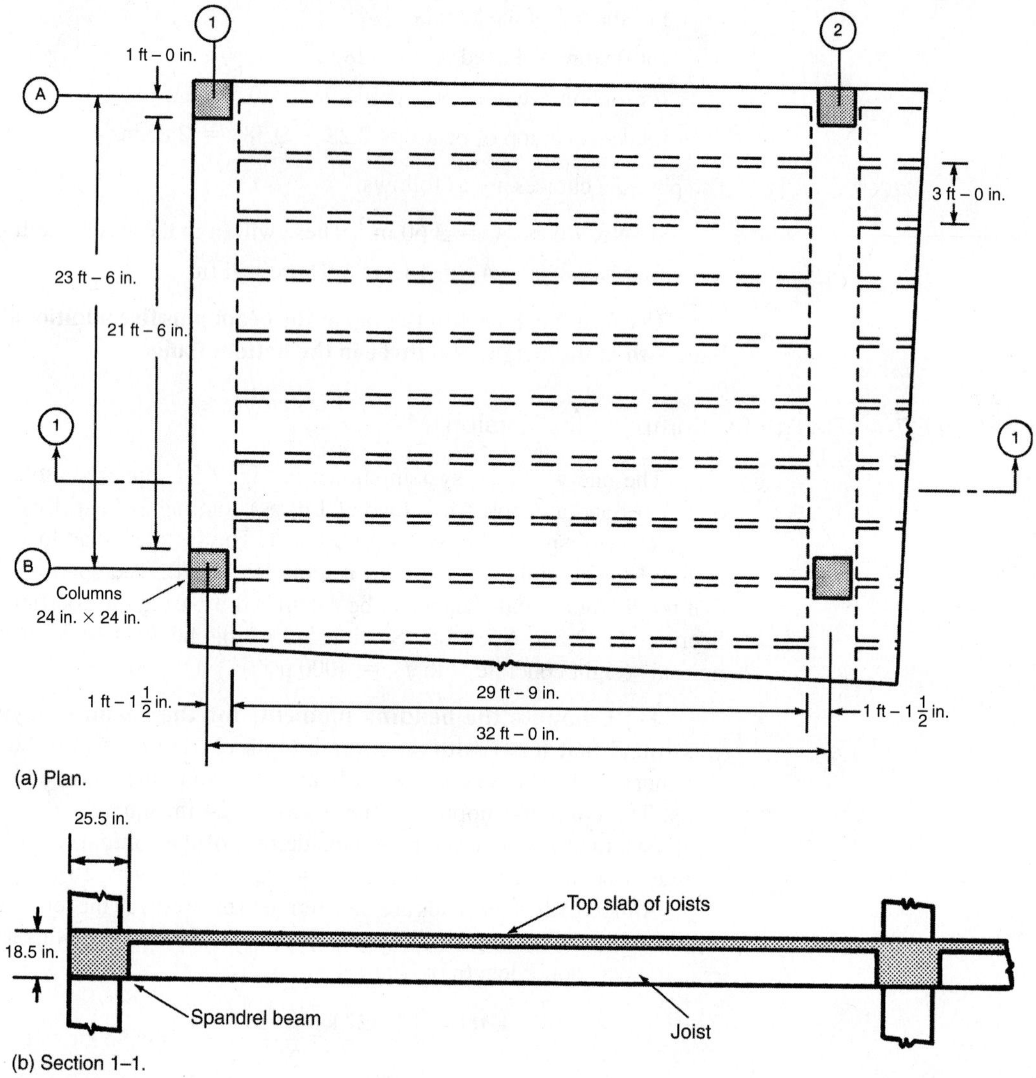

Fig. 7-34
Joist floor—Example 7-4.

2. Compute required A_s for flexure. Because b and d are known, we can check whether the selected section size is sufficiently large enough to ensure a ductile flexural behavior. As was done in the two prior examples, Eq. (5-23) can be used to find the required flexural-resistance factor, R, for the largest design moment, which occurs at the first interior support:

$$R = \frac{M_u}{\phi b d^2} = \frac{276 \text{ k-ft} \times 12 \text{ in./ft}}{0.9 \times 25.5 \text{ in.} \times (16 \text{ in.})^2} = 0.564 \text{ ksi} = 564 \text{ psi}$$

Referring to Table A-3, we can see that this section can be designed as a tension-controlled section ($\phi = 0.9$) with a reinforcement ratio well above the minimum value. Doing a linear interpolation of the given values in Table A-3 results in a required reinforcement ratio, $\rho = 0.0103$. Thus, for this beam section, the required area of tension steel for flexure is

$$A_s = \rho bd = 0.0103 \times 25.5 \text{ in.} \times 16 \text{ in.} = 4.20 \text{ in.}^2$$

Using a similar procedure results in the following required areas of longitudinal reinforcement for flexure at the three critical sections for this beam:

First interior support negative moment: $A_s = 4.20 \text{ in.}^2$

Midspan positive moment: $A_s = 2.92 \text{ in.}^2$

Exterior support negative moment: $A_s = 2.54 \text{ in.}^2$

The actual selection of reinforcing bars will be delayed until the torsion requirements for longitudinal steel have been determined.

3. Compute the final M_u, V_u, and T_u diagrams. The moment and shear diagrams for the edge beam computed from the ACI moment coefficients are plotted in Figs. 7-35a and 7-35b.

The joists are designed as having a clear span of 29.75 ft from the face of one beam to the face of the other beam. Because the exterior ends of the joists are built integrally with a "spandrel beam," ACI Code Section 6.5.2 gives the exterior negative moment in the joists as

$$M_u = -\frac{w_u \ell_n^2}{24}$$

Rather than consider the moments in each individual joist, we shall compute an average moment per foot of width of support:

$$M_u = -\frac{0.327 \text{ ksf} \times (29.75 \text{ ft})^2}{24} = -12.1 \text{ k-ft/ft}$$

Although this is a bending moment in the joist, it acts as a twisting moment on the edge beam. As shown in Fig. 7-36a, this moment and the end shear of 4.86 kips/ft act at the face of the edge beam. Summing moments about the center of the columns (point A in Fig. 7-36a) gives the moment transferred to the edge beam as 17.6 k-ft/ft.

For the design of the edge beam for torsion, we need the torque about the axis of the beam. Summing moments about the centroid of the edge beam (Fig. 7-36b) gives the torque:

$$t = 17.6 \text{ k-ft/ft} - 5.96 \text{ kips/ft} \times \frac{0.75}{12} \text{ ft}$$

$$= 17.2 \text{ k-ft/ft}$$

The forces and torque acting on the edge beam per foot of length are shown in Fig. 7-36b.

If the two ends of the beam A–B are fixed against rotation by the columns, the total torque at each end will be

$$T = \frac{t\ell_n}{2}$$

Using $\ell_n = 21.5$ ft, this gives the torque diagram shown in Fig. 7-35c.

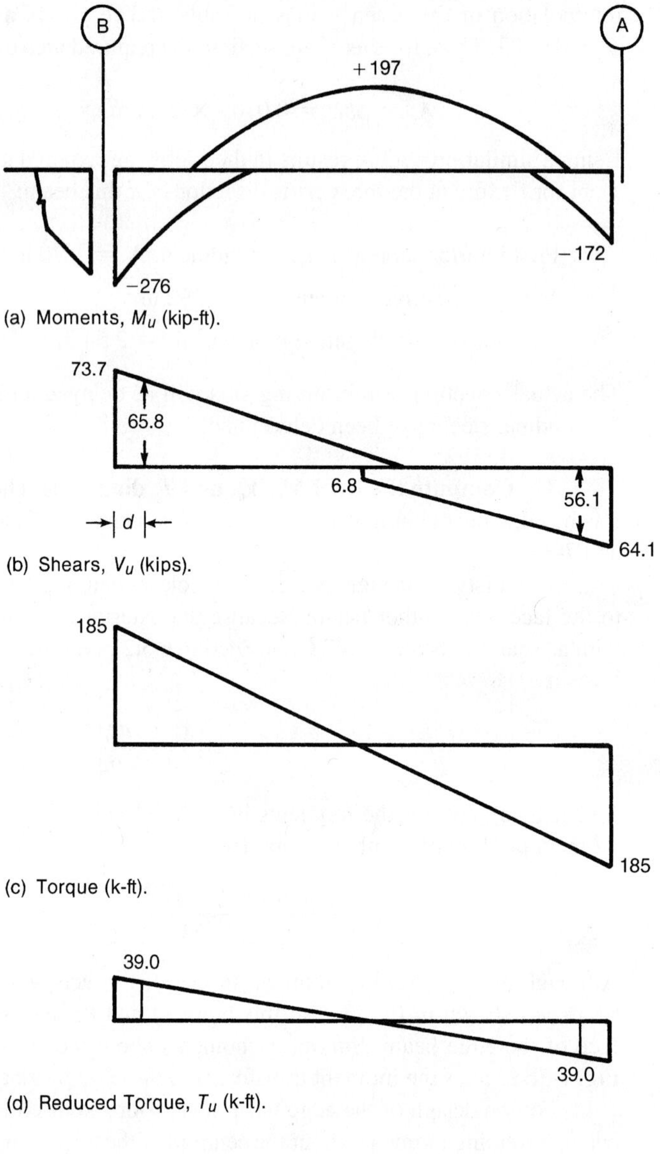

Fig. 7-35
Moments, shears, and torques in end span of edge beam—Example 7-4.

The shear forces in the spandrel beam are:

End A—$V_u = 5.96 \times 21.5/2 = 64.1$ kips

At d from end A, $V_u = 64.1 - 5.96(16/12) = 56.1$ kips

End B—$V_u = 1.15 \times 64.1 = 73.7$ kips

At d from End B—$V_u = 73.7 - 5.96(16/12) = 65.8$ kips

4. Should torsion be considered? If T_u exceeds the following, it must be considered:

$$T_{th} = \phi\lambda\sqrt{f'_c}\left(\frac{A_{cp}{}^2}{p_{cp}}\right)$$

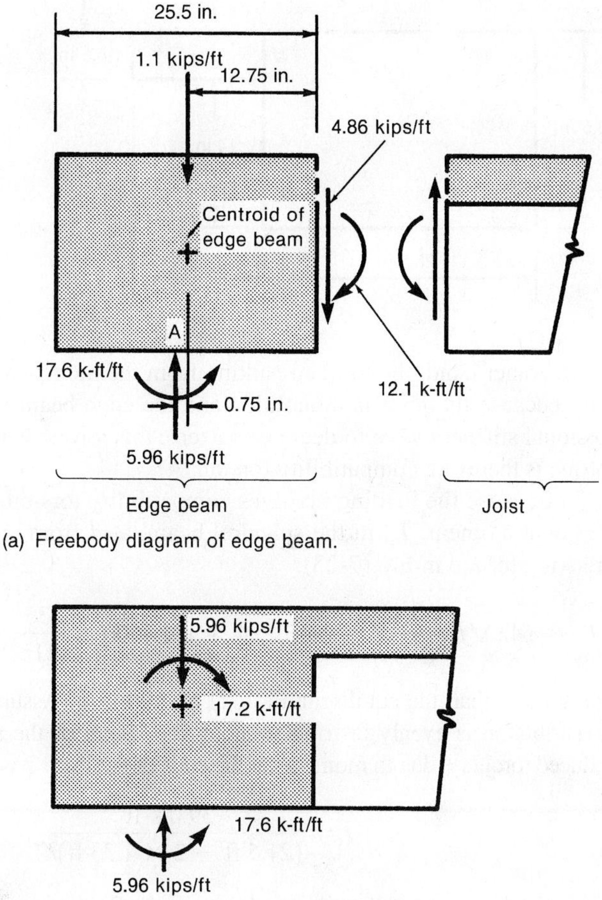

Fig. 7-36
Forces on edge beam—
Example 7-4.

(b) Forces on edge beam resolved through centroid of edge beam.

The effective cross section for torsion is shown in Fig. 7-37. ACI Code Section 9.2.4.4 states that for a monolithic floor system the slab is to be included as part of the section. The projection of the flange is the smaller of the height of the web below the flange (14 in.) and four times the thickness of the flange (18 in.):

$$A_{cp} = 18.5 \times 25.5 + 4.5 \times 14 = 535 \text{ in.}^2$$

$$p_{cp} = 18.5 + 25.5 + 14 + 14 + 4.5 + 39.5 = 116 \text{ in.}$$

$$T_{th} = 0.75 \times 1\sqrt{4000}\left(\frac{535^2}{116}\right) = 117{,}000 \text{ lb-in.}$$

$$= 9.75 \text{ kip-ft}$$

Because the maximum torque of 185 kip-ft exceeds this value, torsion must be considered.

5. (a) Equilibrium or compatibility torsion? The torque resulting from the 0.75-in. offset of the axes of the beam and column (see Fig. 7-36a) is necessary for the equilibrium of the structure and hence is equilibrium torque. The torque at the ends of the beam due to this is

$$5.96 \times \frac{0.75}{12} \times \frac{21.5}{2} = 4.00 \text{ kip-ft}$$

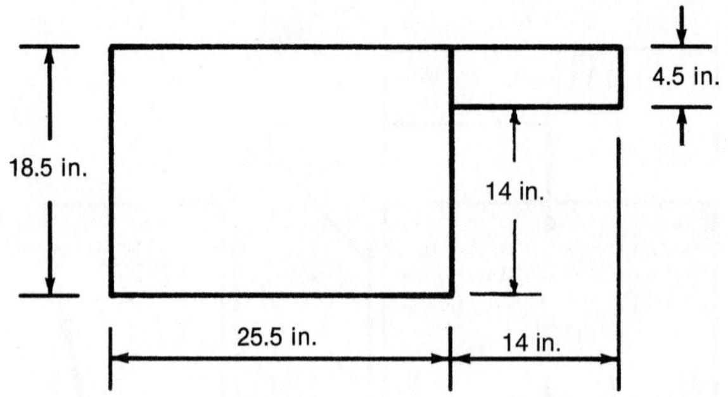

Fig. 7-37
Effective section for torsion—Example 7-4.

On the other hand, the torque resulting from the moments at the ends of the joists exists only because the joint is monolithic and the edge beam has a torsional stiffness. If the torsional stiffness were to decrease to zero, this torque would disappear. This part of the torque is therefore compatibility torsion.

Because the loading involves compatibility torsion, we can reduce the maximum torsional moment, T_u, in the spandrel beam, at d from the faces of the columns to the cracking torque in Eq. (7-23):

$$T_u = \phi 4\lambda\sqrt{f'_c}\left(\frac{A_{cp}^2}{p_{cp}}\right) = 0.75 \times 4 \times 1\sqrt{4000} \times \frac{535^2}{116} = 468{,}000 \text{ lb-in.} = 39.0 \text{ k-ft}$$

but not less than the equilibrium torque of 4.0 k-ft/ft. Assuming the remaining torque after redistribution is evenly distributed along the length of the spandrel beam. The distributed reduced torque, t, due to moments at the ends of the joists has decreased to

$$t = \frac{39.0 \text{ k-ft}}{(21.5 \text{ ft} - 2 \times 1.33 \text{ ft})/2} = 4.14 \text{ k-ft/ft}$$

(b) Adjust moments in the joists. If the floor joists framing into the spandrel beam have been designed using the moment coefficients from ACI Code Section 6.5.2, then the midspan and first interior support sections for those joists will need to be reanalyzed and redesigned, because the spandrel beam is now assumed to be resisting a lower moment than would have been assigned to it by the ACI moment coefficients. This process involves a redistribution of the moment not resisted by the spandrel beam to the positive moment at midspan and the negative moment at the first interior support of the joists. If the joists have been analyzed (and designed) using the structural analysis procedure described in Section 5-2, no load and moment redistribution would be required, because the analysis procedure in Section 5-2 assumed an exterior support (spandrel beam) with zero torsional stiffness.

6. Is the section large enough for the torsion? For a solid section, the limit on shear and torsion is given by

$$\sqrt{\left(\frac{V_u}{b_w d}\right)^2 + \left(\frac{T_u p_h}{1.7A_{oh}^2}\right)^2} \le \phi\left(\frac{V_c}{b_w d} + 8\sqrt{f'_c}\right) \tag{7-45}$$

From Fig. 7-37, and assuming we will use a closed No. 4 stirrup in the web:

$$A_{oh} = (18.5 - 2 \times 1.5 - 0.5)(25.5 - 2 \times 1.5 - 0.5)$$
$$= 330 \text{ in.}^2$$

$$p_h = 2(15.0 + 22.0)$$
$$= 74 \text{ in.}$$

$$\sqrt{\left(\frac{65{,}800}{25.5 \times 16}\right)^2 + \left(\frac{39.0 \times 12{,}000 \times 74}{1.7 \times 330^2}\right)^2} = \sqrt{26{,}000 + 35{,}000}$$
$$= 247 \text{ psi}$$

From Eq. (6-8b),

$$V_c = 2\lambda\sqrt{f'_c}\, b_w d$$
$$\phi(2 \times 1\sqrt{f'_c} + 8\sqrt{f'_c}) = 0.75(10\sqrt{4000})$$
$$= 474 \text{ psi}$$

Because 247 psi is less than 474 psi, the section is large enough.

7. Compute the stirrup area required for shear in the edge beam. From Eqs. (6-9) and (6-14),

$$V_s \geq \frac{V_u}{\phi} - V_c$$

and from Eq. (6-18),

$$\frac{A_v}{s} \geq \frac{V_u/\phi - V_c}{f_{yt} d}$$

where, from Eq. (6-8b),

$$V_c = 2 \times 1\sqrt{4000} \times 25.5 \times \frac{16}{1000} = 51.6 \text{ kips}$$

Thus,

$$\frac{A_v}{s} \geq \frac{V_u/0.75 - 51.6}{60 \times 16}$$

where V_u is in kips.

At d from end B,

$$\frac{A_v}{s} \geq \frac{\dfrac{65.8}{0.75} - 51.6}{60 \times 16} = 0.0376$$

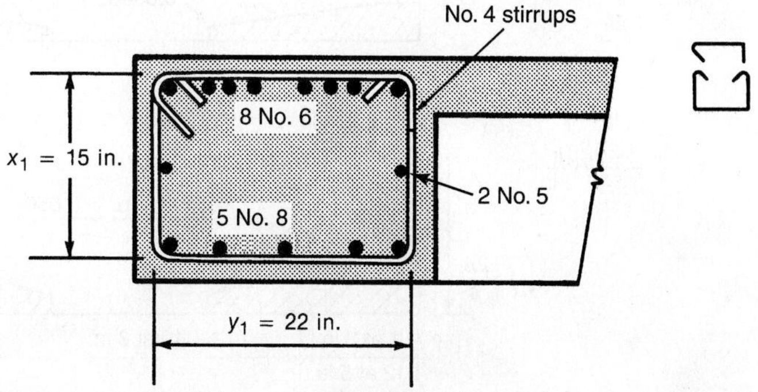

Fig. 7-38
Section *A–A* from Fig. 7-40 through edge beam at exterior negative moment section. Joist reinforcement omitted for clarity.

Figure 7-39a illustrates the calculation of $V_u/\phi - V_c$. Figure 7-39b is a plot of the A_v/s required for shear along the length of the beam. The values of A_v/s for shear and A_t/s for torsion (step 8) will be superimposed in step 9.

8. Compute the stirrups required for torsion. From Eq. (7-30), taking $\theta = 45°$ and $A_o = 0.85A_{oh}$ gives

$$\frac{A_t}{s} \geq \frac{T_u/\phi}{1.7A_{oh}f_{yt}}$$

$$\geq \frac{T_u/\phi \times 12{,}000}{1.7 \times 330 \times 60{,}000}$$

$$\geq 0.000357\frac{T_u}{\phi}$$

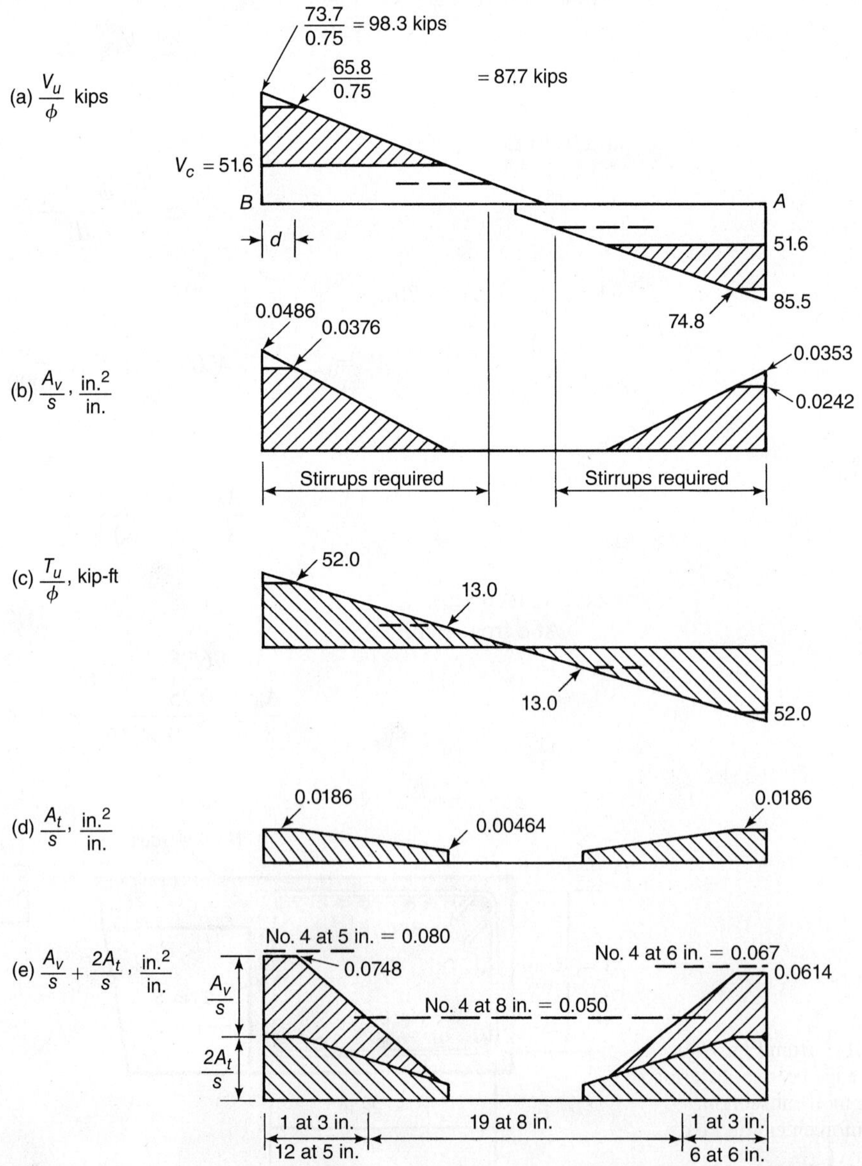

Fig. 7-39
Calculation of stirrups for shear and torsion—Example 7-4.

At d from ends A and B, $T_u = 39$ k-ft, $T_u/\phi = 52.0$ k-ft, and $A_t/s = 0.0186$. T_u/ϕ in kip-ft is plotted in Fig. 7-39c. A_t/s is plotted in Fig. (7-39d). When the factored torque drops below the threshold torque, torsion reinforcement is not required. For $T_u = 9.75$ kip-ft, $T_u/\phi = 13.0$ kip-ft, and $A_t/s = 0.00464$ in.^{2}/in.

9. Add the stirrup areas and select the stirrups.

$$\frac{A_{v+t}}{s} = \frac{A_v}{s} + \frac{2A_t}{s} \tag{7-39}$$

At d from end B,

$$\frac{A_{v+t}}{s} = 0.0376 + 2 \times 0.0186 = 0.0748 \text{ in.}^2\text{/in.}$$

For No. 4 double-leg stirrups, $s = (2 \times 0.20 \text{ in.}^2)/0.0748 \text{ in.}^2\text{/in.} = 5.35$ in.

A_{v+t}/s is plotted in Fig. 7-39e. The maximum allowable spacings are as follows:

for shear (ACI Code Section 9.7.6.2.2), $d/2 = 8$ in.;

for torsion (ACI Code Section 9.7.6.3.3), the smaller of 12 in. and $p_h/8 = 74/8 = 9.25$ in.

The dashed horizontal lines in Fig. 7-39e are the values of A_{v+t}/s for No. 4 closed stirrups at spacings of 5 in.$(= 2 \times 0.20/5.0 = 0.080)$, 6 in. and 8 in. Stirrups must extend to points where $V_u/\phi = V_c/2$, or to $(d + b_t)$ past the points where $T_u/\phi = 9.75/0.75 = 13.0$ kip-ft $(d + b_t = 16 + 25.5 = 41.5$ in.). These points are indicated in Figs. 7-35c to 7-35e. Because the stirrups would be stopped closer than 41.5 in. to midspan, stirrups are required over the entire span.

Provide No. 4 closed stirrups:

End A: One at 3 in., six at 6 in.

End B: One at 3 in., twelve at 5 in., then at 8 in. on centers throughout the rest of the span.

10. Design the longitudinal reinforcement for torsion.

(a) Longitudinal reinforcement required to resist T_n:

$$A_\ell = \frac{(T_u/\phi)\, p_h}{1.7\, A_{oh} f_y} \cot\theta \tag{7-36}$$

This varies along the length of the beam. For simplicity, we shall keep the longitudinal steel constant along the length of the span and shall base it on the maximum $T_u/\phi = 52.0$ k-ft. Again, $\theta = 45°$. We have

$$A_\ell = \frac{(52 \text{ k-ft} \times 12 \text{ in./ft}) \times 74 \text{ in.}}{1.7 \times 330 \text{ in.}^2 \times 60 \text{ ksi}} \times 1.0 = 1.38 \text{ in.}^2$$

The minimum A_ℓ is given by Eq. (7-43) is:

$$A_{\ell,\min} = \frac{5\sqrt{f'_c}A_{cp}}{f_y} - \left(\frac{A_t}{s}\right)p_h\left(\frac{f_{yt}}{f_y}\right)$$

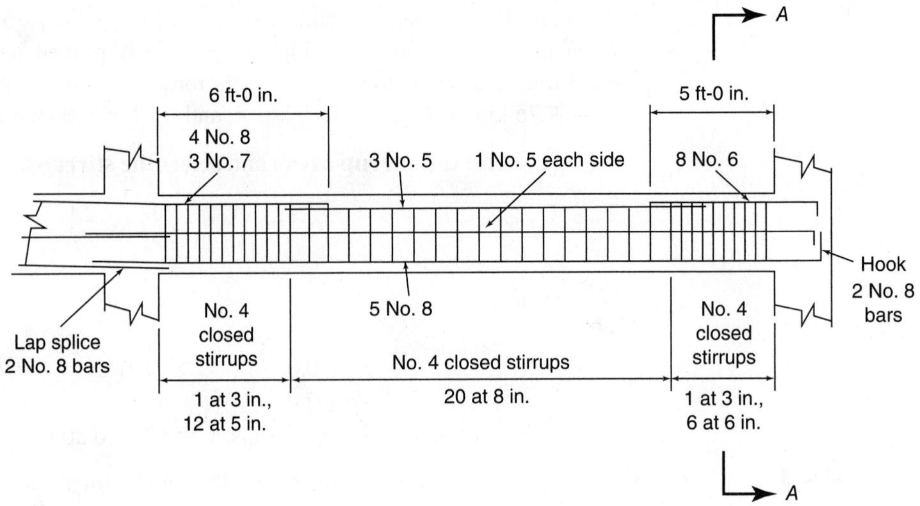

Fig. 7-40
Reinforcement in edge beam—Example 7-4.

where A_t/s shall not be less than $25b_w/f_{yv} = 25 \times 25.5/60{,}000 = 0.0106$. Again, A_t/s varies along the span. The maximum A_ℓ will correspond to the minimum A_t/s. In the center region of the beam, No. 4 stirrups at 8 in. have been chosen. (See Fig. 7-39e.) Assuming half of those stirrups are for torsion, we shall take $A_t/s = 1/2 \times 0.20/8 = 0.0125$ in.^{2}/in.:

$$A_{\ell,\,\min} = \frac{5\sqrt{4000} \times 535}{60{,}000} - 0.0125 \times 74 \times 1.0$$
$$= 2.82 - 0.92 = 1.90 \text{ in.}^2$$

Because $A_\ell = 1.38$ in.^{2} is less than $A_{\ell,\,\min} = 1.90$ in.^{2}, use 1.90 in.^{2}.

From ACI Code Section 9.7.5.1, the longitudinal steel is distributed around the perimeter of the stirrups with a maximum spacing of 12 in. There must be a bar in each corner of the stirrups, and these bars have a minimum diameter of 1/24 of the stirrup spacing, but not less than a No. 3 bar.

The minimum bar diameter corresponds to the maximum stirrup spacing: For 8 in., $8/24 = 0.33$ in.

To satisfy the 12-in.-maximum spacing, we need **three bars at the top and bottom and one bar halfway up each side**. A_s per bar $= 1.90/8 = 0.24$ in.^{2}. **Use No. 5 bars for longitudinal bars.**

The longitudinal torsion steel required at the top of the beam is provided by increasing the area of flexural steel provided at each end and by lap-splicing three No. 5 bars with the negative-moment steel. The lap splices should be at least a Class B tension lap for the larger top bar (see Chapter 8), because all the bars are spliced at the same point.

Exterior end negative moment: $A_s = 2.54 + 3 \times 0.24 = 3.26$ in.^{2}. Use No. 6 bars because smaller bars are more easily anchored in column.

Use eight No. 6 = 3.52 in.^{2}. These fit in one layer.

First interior negative moment: $A_s = 4.20 + 3 \times 0.24 = 4.92$ in.^{2}.

Use four No. 8 and three No. 7 = 4.96 in.^{2}. These fit in one layer, minimum width is 17.5 in.

The longitudinal steel required at the bottom is obtained by increasing the area of steel at midspan. The increased area of steel will be extended from support to support.

Midspan positive moment: $A_s = 2.92 + 3 \times 0.24 = 3.64$ in.^{2}

Use five No. 8 $= 3.95$ in.^{2}. These fit in one layer.

The steel finally chosen is shown in Fig. 7-40. A section through the beam at the exterior support is shown in Fig. 7-38. The cutoff points and lap splices will be discussed in Chapter 8. ■

PROBLEMS

7-1 A cantilever beam 8 ft long and 18 in. wide supports its own dead load plus a concentrated load located 6 in. from the end of the beam and 4.5 in. away from the vertical axis of the beam. The concentrated load is 15 kips dead load and 20 kips live load. Design reinforcement for flexure, shear, and torsion. Use $f_y = 60{,}000$ psi for all steel and $f'_c = 5000$ psi.

7-2 Explain why the torsion in the edge beam *A–B* in Fig. 7-21a is called "equilibrium torsion," while the torsion in the edge beam *A*1–*B*1 in Fig. P7-3 is called "compatibility torsion."

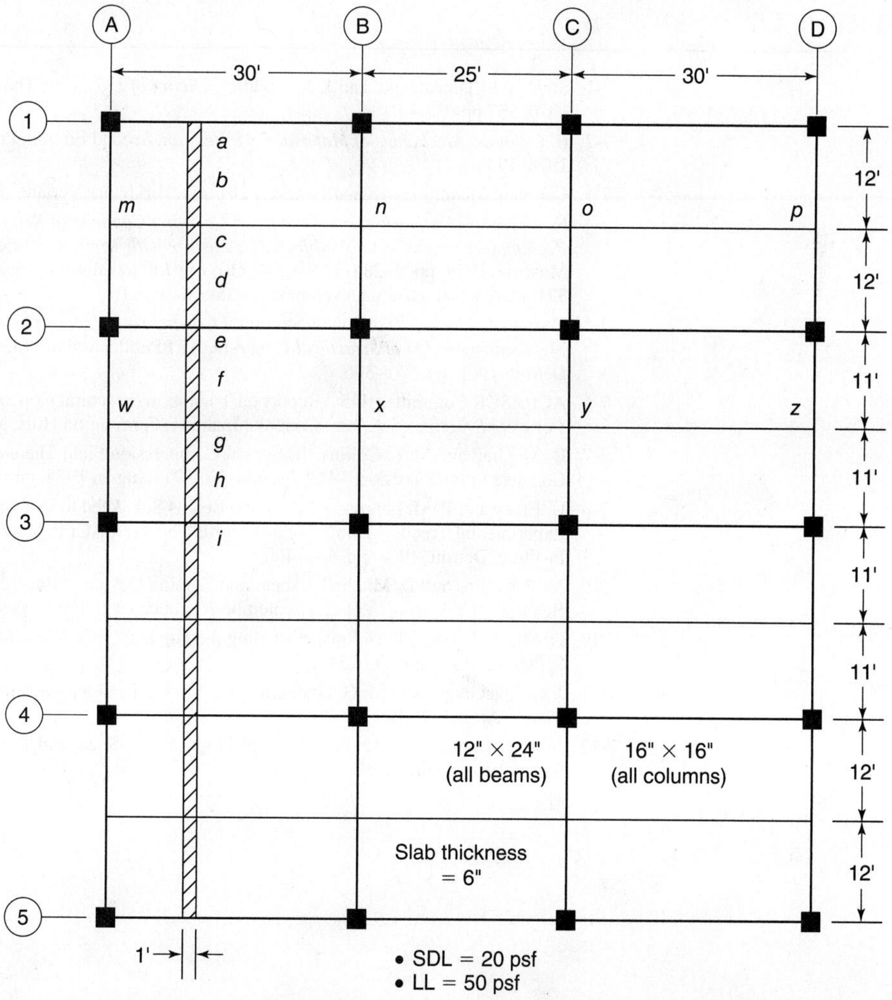

Fig. P7-3
Floor plan for Problems 7-3 and 7-4.

7-3 The two parts of this problem refer to the floor plan shown in Fig. P7-3. Assume that the entire floor system is constructed with normal-weight concrete that has a compressive strength, $f'_c = 4500$ psi. Also assume that the longitudinal steel has a yield strength, $f_y = 60$ ksi, and that the transverse steel has a yield strength of $f_{yt} = 40$ ksi.

(a) Design the spandrel beam between columns $B1$ and $C1$ for bending, shear, and torsion. Check that all of the appropriate ACI Code requirements for strength, minimum-reinforcement area, and reinforcement spacing are satisfied.

Video Solution

(b) Design the spandrel beam between columns $A1$ and $A2$ for bending, shear, and torsion. Check that all of the appropriate ACI Code requirements for strength, minimum-reinforcement area, and reinforcement spacing are satisfied.

7-4 The two parts of this problem refer to the floor plan shown in Fig. P7-3. Assume that the entire floor system is constructed with sand lightweight concrete (115 lb/ft^3) that has a compressive strength, $f'_c = 4000$ psi. Also assume that the longitudinal steel has a yield strength, $f_y = 60$ ksi, and that the transverse steel has a yield strength of $f_{yt} = 60$ ksi.

(a) Design the spandrel beam between columns $B1$ and $C1$ for bending, shear, and torsion. Check that all of the appropriate ACI Code requirements for strength, minimum-reinforcement area, and reinforcement spacing are satisfied.

(b) Design the spandrel beam between columns $A1$ and $A2$ for bending, shear, and torsion. Check that all of the appropriate ACI Code requirements for strength, minimum-reinforcement area, and reinforcement spacing are satisfied.

Video Solution

REFERENCES

7-1 Stephen P. Timoshenko and J. N. Goddier, *Theory of Elasticity*, Third Edition, McGraw-Hill, New York, 1970, 567 pp.

7-2 E. P. Popov, *Mechanics of Materials, S.I. Version*, Second Edition, Prentice Hall, Englewood Cliffs, N. J. 1978, 590 pp.

7-3 Christian Menn, *Prestressed Concrete Bridges*, Birkhäuer Verlang, Basel, 1990, 535 pp.

7-4 N. N. Lessig, *Determination of the Load Carrying Capacity of Reinforced Concrete Elements with Rectangular Cross-Section Subjected to Flexure with Torsion*, Work 5, Institute Betona i Zhelezobetona, Moscow, 1959, pp. 4–28 (in Russian); also English translation available as Foreign Literature Study 371, PCA Research and Development Labs, Skokie, IL.

7-5 Thomas T. C. Hsu, "Torsion of Structural Concrete—Behavior of Reinforced Concrete Rectangular Members," *Torsion of Structural Concrete*, ACI Publication SP-18, American Concrete Institute, Detroit, 1968, pp. 261–306.

7-6 ACI/ASCE Committee 445, "Report on Torsion in Structural Concrete," ACI 445.IR-12, *ACI Manual of Concrete Practice*, American Concrete Institute, Farmington Hills, MI.

7-7 D. Mitchell and M. P. Collins, "Diagonal Compression Field Theory—A Rational Model for Structural Concrete in Pure Torsion," *ACI Journal*, Vol. 71, August 1974, pp. 396–408.

7-8 U. Ersoy and P. M. Ferguson, "Concrete Beams Subjected to Combined Torsion and Shear—Experimental Trends," *Torsion of Structural Concrete*, ACI Publication SP-18, American Concrete Institute, Detroit, 1968, pp. 441–460.

7-9 M. P. Collins and D. Mitchell, "Shear and Torsion Design of Prestressed and Non-Prestressed Concrete Beams," *PCI Journal*, Vol. 25, September/October 1980, Prestressed Concrete Institute, pp. 32–100.

7-10 D. Mitchell and M. P. Collins, "Detailing for Torsion," *ACI Journal Proceedings*, Vol. 73, No. 9, September 1976, pp. 506–511.

7-11 J. G. MacGregor and M. G. Ghoneim, "Design for Torsion," *ACI Structural Journal*, Vol. 92, No. 2, March–April 1995, pp. 211–218.

7-12 M. P. Collins and D. Mitchell, "Design Proposals for Shear and Torsion," *PCI Journal*, Vol. 25, No. 5, September–October 1980, 70 pp.

8

Development, Anchorage, and Splicing of Reinforcement

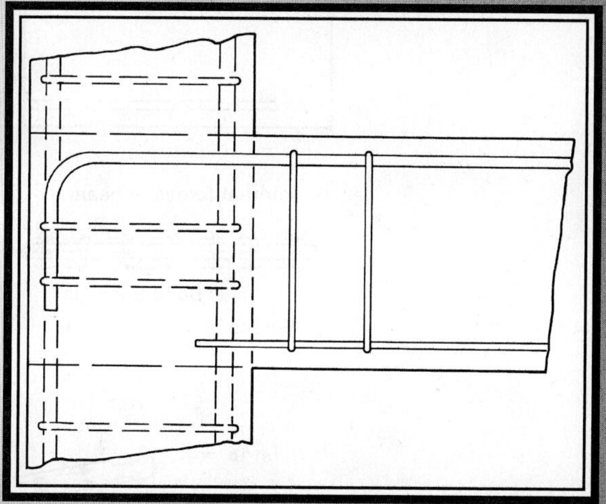

8-1 INTRODUCTION

In a reinforced concrete beam, the flexural compressive forces are resisted by concrete, while the flexural tensile forces are provided by reinforcement, as shown in Fig. 8-1. For this process to exist, there must be a force transfer, or *bond*, between the two materials. The forces acting on the bar are shown in Fig. 8-1b. For the bar to be in equilibrium, bond stresses must exist. If these disappear, the bar will pull out of the concrete and the tensile force, T, will drop to zero, causing the beam to fail.

Bond stresses must be present whenever the stress or force in a reinforcing bar changes from point to point along the length of the bar. This is illustrated by the free-body diagram in Fig. 8-2. If f_{s2} is greater than f_{s1}, bond stresses, μ, must act on the surface of the bar to maintain equilibrium. Summing forces parallel to the bar, one finds that the average bond stress, μ_{avg}, is

$$(f_{s2} - f_{s1})\frac{\pi d_b^2}{4} = \mu_{\text{avg}}(\pi d_b)\ell$$

and taking $(f_{s2} - f_{s1}) = \Delta f_s$ gives

$$\mu_{\text{avg}} = \frac{\Delta f_s d_b}{4\ell} \tag{8-1}$$

If ℓ is taken as a very short length, dx, this equation can be written as

$$\frac{df_s}{dx} = \frac{4\mu}{d_b} \tag{8-2}$$

where μ is the *true bond stress* acting in the length dx.

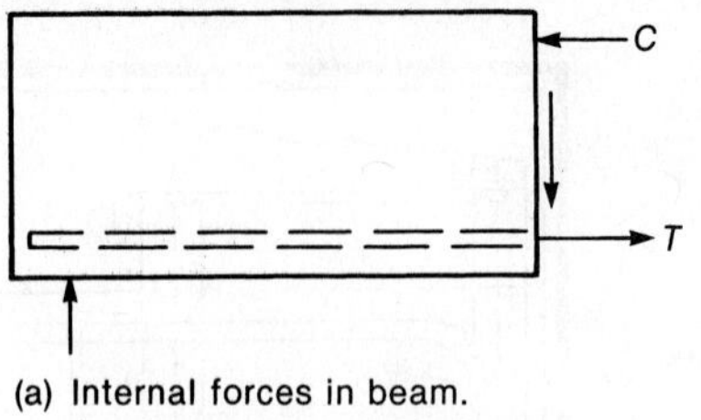

Fig. 8-1
Need for bond stresses.

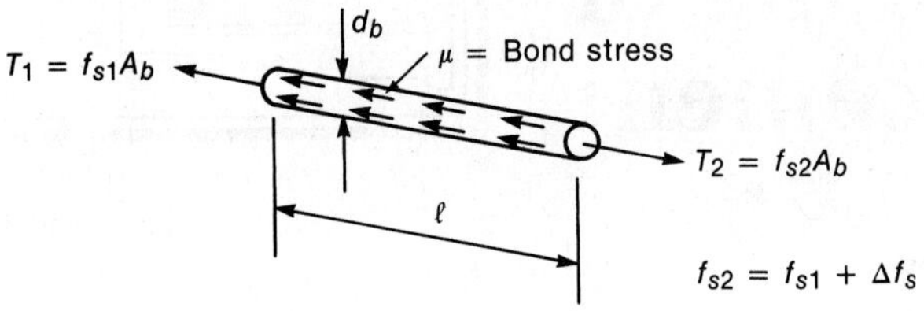

Fig. 8-2
Relationship between change in bar stress and average bond stress.

Average Bond Stress in a Beam

In a beam, the force in the steel at a crack can be expressed as

$$T = \frac{M}{jd} \tag{8-3}$$

where jd is the internal lever arm and M is the moment acting at the section. If we consider a length of beam between two cracks, as shown in Fig. 8-3, the moments acting at the two cracks are M_1 and M_2. If the beam is reinforced with one bar of diameter d_b, the forces on the bar are as shown in Fig. 8-3c. Summing horizontal forces gives

$$\Delta T = (\pi d_b)\mu_{\text{avg}}\,\Delta x \tag{8-4}$$

where d_b is the diameter of the bar, or

$$\frac{\Delta T}{\Delta x} = (\pi d_b)\mu_{\text{avg}}$$

But

$$\Delta T = \frac{\Delta M}{jd}$$

giving

$$\frac{\Delta M}{\Delta x} = (\pi d_b)\mu_{\text{avg}}\,jd$$

From the free-body diagram in Fig. 8-3d, we can see that $\Delta M = V\Delta x$ or $\Delta M/\Delta x = V$. Therefore,

$$\mu_{\text{avg}} = \frac{V}{(\pi d_b)jd} \tag{8-5}$$

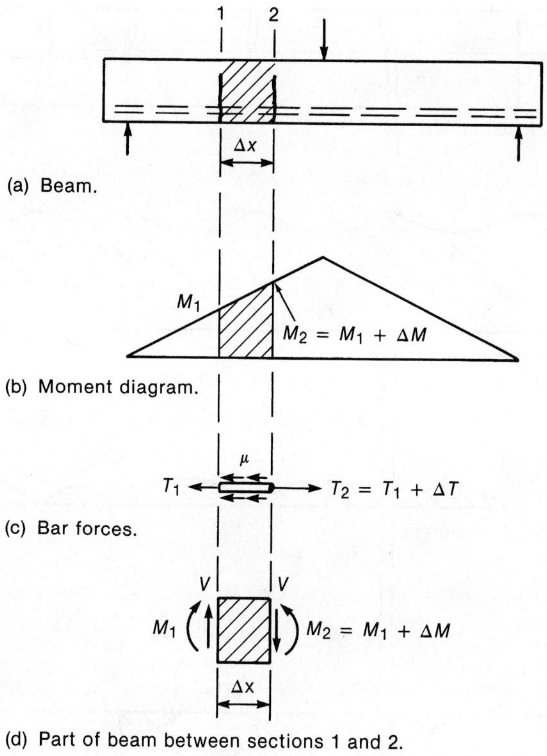

Fig. 8-3
Average flexural bond stress.

If there is more than one bar, the bar perimeter (πd_b) is replaced with the sum of the perimeters, Σo, giving

$$\mu_{avg} = \frac{V}{\Sigma ojd} \tag{8-6}$$

Equations (8-5) and (8-6) give the *average bond stress* between two cracks in a beam. As shown later, the actual bond stresses vary from point to point between the cracks.

Bond Stresses in an Axially Loaded Prism

Figure 8-4a shows a prism of concrete containing one reinforcing bar, which is loaded in tension. At the cracks, the stress in the bar is $f_s = T/A_s$. Between the cracks, a portion of the load is transferred to the concrete by bond, and the resulting distributions of steel and concrete stresses are shown in Figs. 8-4b and 8-4c. From Eq. (8-2), we see that the bond stress at any point is proportional to the slope of the steel stress diagram at that same point. Thus, the bond-stress distribution is shown in Fig. 8-4d. Because the stress in the steel is equal at each of the cracks, the force is also equal, so that $\Delta T = 0$ at the two cracks, and from Eq. (8-4), we see that the average bond stress, μ_{avg}, is also equal to zero. Thus, for the average bond stress to equal zero, the total area under the bond-stress diagram between any two cracks in Fig. 8-4d must equal zero when $\Delta T = 0$.

The bond stresses given by Eq. (8-2) and plotted in Fig. 8-4d are referred to as *true bond stresses* or *in-and-out bond stresses* (they transfer stress into the bar and back out again) to distinguish them from the *average bond stresses* calculated from Eq. (8-1).

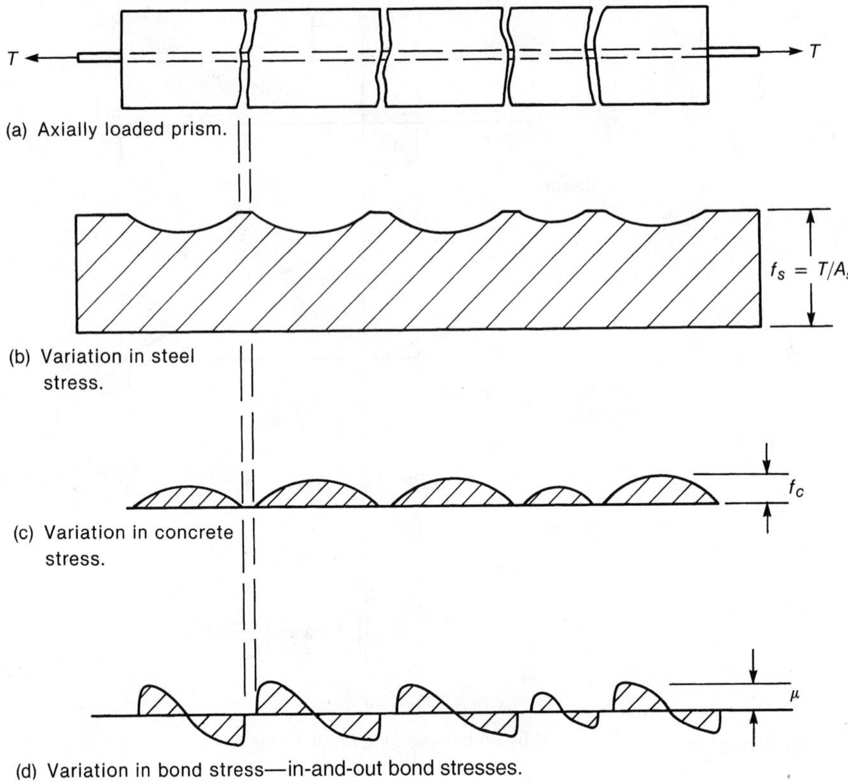

Fig. 8-4
Steel, concrete, and bond stresses in a cracked prism.

True Bond Stresses in a Beam

At the cracks in a beam, the bar force can be computed from Eq. (8-3). If the concrete and the bar are bonded together, a portion of the tensile force will be resisted by the concrete at points between the cracks. As a result, the tensile stresses in the steel and the concrete at the level of the steel will vary, as shown in Figs. 8-5c and 8-5d. This gives rise to the bond-stress distribution plotted in Fig. 8-5e. In the constant-moment region between the two loads in Fig. 8-5a, the shear is zero, and the average bond stress from the in-and-out bond-stress diagram in Fig. 8-5e is zero. Between a support and the nearest load, there is shear. Once again, there are in-and-out bond stresses, but now the total area under the bond-stress diagram is not zero. The average bond stress in Fig. 8-5e must equal the value given by Eq. (8-5).

Bond Stresses in a Pull-Out Test

The easiest way to test the bond strength of bars in a laboratory is by means of the *pull-out test.* Here, a concrete cylinder containing the bar is mounted on a stiff plate and a jack is used to pull the bar out of the cylinder, as shown in Fig. 8-6a. In such a test, the concrete is compressed and hence does not crack. The stress in the bar varies as shown in Fig. 8-6b, and the bond stress varies as shown in Fig. 8-6c. This test does not give values representative of the bond strength of beams because the concrete is not cracked, and hence, there is no in-and-out bond-stress distribution. Also, the bearing stresses of the concrete against the plate cause a frictional component that resists the transverse expansion that would reflect Poisson's ratio. Prior to 1950, pull-out tests were used extensively to evaluate the bond strength of bars. Since then, various types of beam tests have been used to study bond strength [8-1].

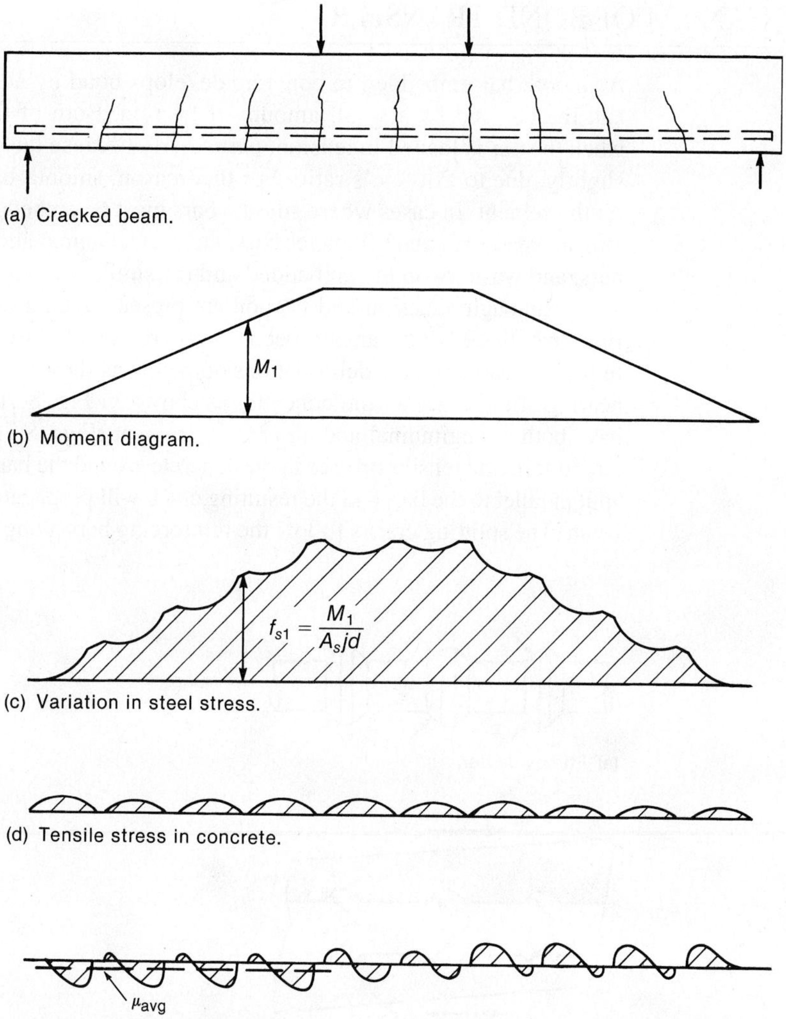

(a) Cracked beam.
(b) Moment diagram.
(c) Variation in steel stress.
(d) Tensile stress in concrete.
(e) In-and-out bond stresses.

Fig. 8-5
Steel, concrete, and bond stresses in a cracked beam.

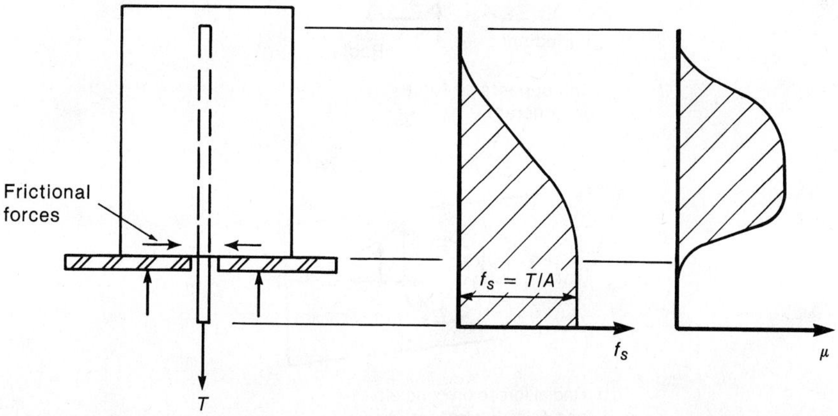

(a) Test method.
(b) Bar stress.
(c) Bond stress.

Fig. 8-6
Stress distribution in a pull-out test.

8-2 MECHANISM OF BOND TRANSFER

A smooth bar embedded in concrete develops bond by adhesion between the concrete and the bar and by a small amount of friction. Both of these effects are quickly lost when the bar is loaded in tension, particularly because the diameter of the bar decreases slightly, due to Poisson's ratio. For this reason, smooth bars are generally not used as reinforcement. In cases where smooth bars must be embedded in concrete (anchor bolts, stirrups made of small diameter bars, etc.), mechanical anchorage in the form of hooks, nuts, and washers on the embedded end (or similar devices) are used.

Although adhesion and friction are present when a deformed bar is loaded for the first time, these bond-transfer mechanisms are quickly lost, leaving the bond to be transferred by bearing on the deformations of the bar as shown in Fig. 8-7a. Equal and opposite bearing stresses act on the concrete, as shown in Fig. 8-7b. The forces on the concrete have both a longitudinal and a radial component (Figs. 8-7c and 8-7d). The latter causes circumferential tensile stresses in the concrete around the bar. Eventually, the concrete will split parallel to the bar, and the resulting crack will propagate out toward the surface of the beam. The splitting cracks follow the reinforcing bars along the bottom or side surfaces of

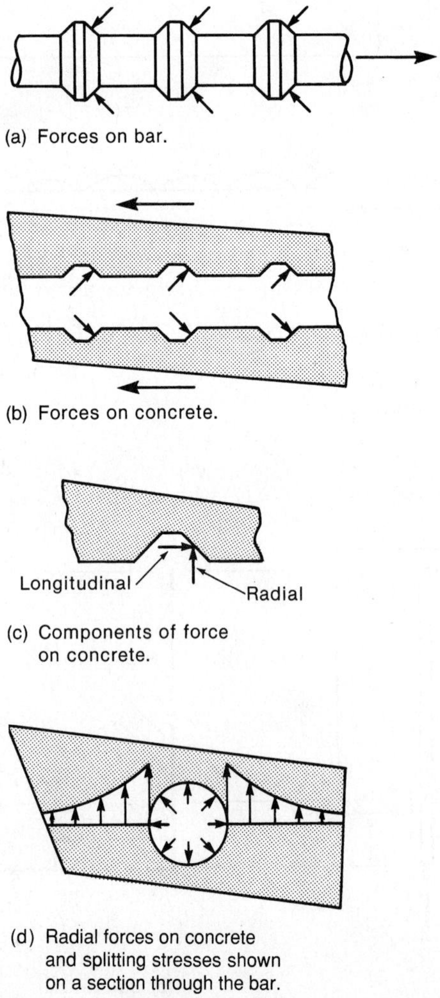

Fig. 8-7
Bond-transfer mechanism.

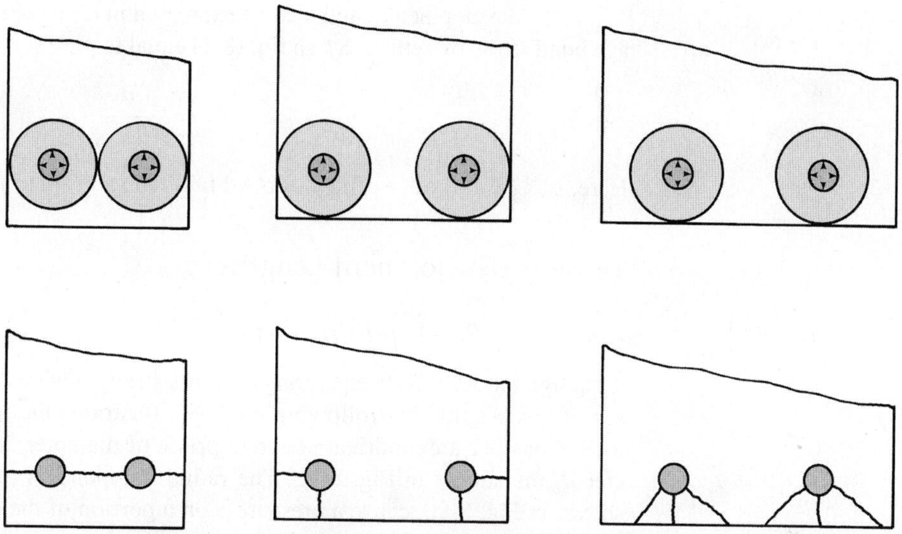

(a) Side cover and half the bar spacing both less than bottom cover.

(b) Side cover = bottom cover, both less than half the bar spacing.

(c) Bottom cover less than side cover and half the bar spacing.

Fig. 8-8
Typical splitting-failure surfaces.

the beam, as shown in Fig. 8-8. Once these cracks develop, the bond transfer drops rapidly unless reinforcement is provided to restrain the opening of the splitting crack.

The load at which splitting failure develops is a function of

1. the minimum distance from the bar to the surface of the concrete or to the next bar—the smaller this distance, the smaller is the splitting load;

2. the tensile strength of the concrete; and

3. the average bond stress—as this increases, the wedging forces increase, leading to a splitting failure.

These factors are discussed more fully in [8-1], [8-2], [8-3], and [8-4]. Typical splitting-failure surfaces are shown in Fig. 8-8. The splitting cracks tend to develop along the shortest distance between a bar and the surface or between two bars. In Fig. 8-8 the circles touch the edges of the beam where the distances are shortest.

If the cover and bar spacings are large compared to the bar diameter, a *pull-out failure* can occur, where the bar and the annulus of concrete between successive deformations pull out along a cylindrical failure surface joining the tips of the deformations.

8-3 DEVELOPMENT LENGTH

Because the actual bond stress varies along the length of a bar anchored in a zone of tension as shown in Fig. 8-5e, the ACI Code uses the concept of *development length* rather than bond stress. The development length, ℓ_d, is the shortest length of bar in which the bar stress can increase from zero to the yield strength, f_y. If the distance from a point where the bar stress equals f_y to the end of the bar is less than the development length, the bar will pull out of the concrete. The development lengths are different in tension and compression, because a bar loaded in tension is subject to in-and-out bond stresses and hence requires a considerably longer development length. Also, for a bar in compression, bearing stresses at the end of the bar will transfer part of the compression force into the concrete.

The development length can be expressed in terms of the ultimate value of the average bond stress by setting Δf_s in Eq. (8-1) equal to f_y:

$$\ell_d = \frac{f_y d_b}{4\mu_{\text{avg},u}} \tag{8-7}$$

Here, $\mu_{\text{avg},u}$ is the value of μ_{avg} at bond failure in a beam test.

Tension-Development Lengths

Analysis of Bond Splitting Load

Although the ACI Code equations for bond strength were derived from statistical analyses of test results, the following analysis illustrates the factors affecting the splitting load. Consider a cylindrical concrete prism of diameter $2c_b$, containing a bar of diameter d_b, as shown in Fig. 8-9a. The radial components of the forces on the concrete, shown in Fig. 8-9b, cause a pressure p on a portion of the cross section of the prism, as shown in Fig. 8-9c. This is equilibrated by tensile stresses in the concrete on either side of the bar. In Fig. 8-9c, the distribution of these stresses has arbitrarily been assumed to be triangular. The circular prism in Fig. 8-9 represents the zones of highest radial tensile stresses, shown by the larger circles in Fig. 8-8. Splitting is assumed to occur when the maximum stress in the concrete is equal to the tensile strength of the concrete, f_{ct}. For equilibrium in the vertical direction in a prism of length equal to ℓ,

$$\frac{p d_b \ell}{2} = K\left(c_b - \frac{d_b}{2}\right) f_{ct}\,\ell$$

where K is the ratio of the average tensile stress to the maximum tensile stress and equals 0.5 for the triangular stress distribution shown in Fig. 8-9c. Rearranging gives

$$p = \left(\frac{c_b}{d_b} - \frac{1}{2}\right) f_{ct}$$

If the forces shown in Fig. 8-7b are assumed to act at 45°, the average bond stress, $\mu_{\text{avg},u}$, at the onset of splitting is equal to p. Taking $f_{ct} = 6\sqrt{f'_c}$ gives

$$\mu_{\text{avg},u} = 6\sqrt{f'_c}\left(\frac{c_b}{d_b} - \frac{1}{2}\right) \tag{8-8}$$

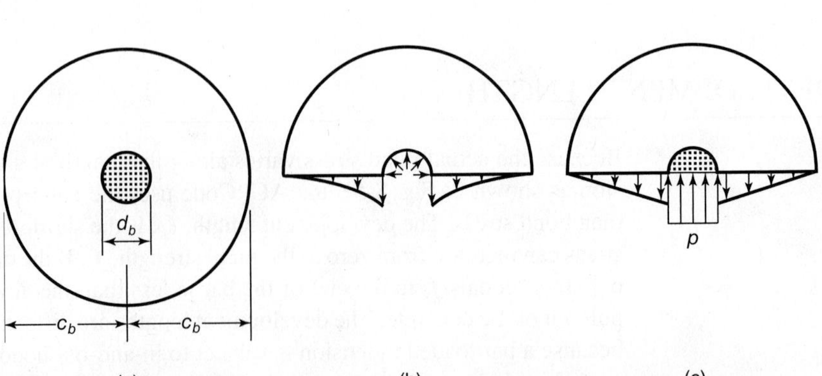

Fig. 8-9
Concrete stresses in a circular concrete prism containing a reinforcing bar that is subjected to bond stresses, shown in section.

The length of bar required to raise the stress in the bar from zero to f_y is called the development length, ℓ_d. From Eq. (8-7),

$$\ell_d = \frac{f_y}{4\mu_{\text{avg},u}} d_b$$

Substituting Eq. (8-8) for $\mu_{\text{avg},u}$ gives

$$\ell_d = \frac{f_y\, d_b}{24\sqrt{f'_c}\left(\dfrac{c_b}{d_b} - \dfrac{1}{2}\right)}$$

Arbitrarily taking $c_b = 1.5d_b$ (for the reasons to be given in the derivation of Eqs. (8-13) and (8-14) later in the chapter) and rearranging yields

$$\ell_d = \frac{f_y}{24\sqrt{f'_c}} d_b \tag{8-9}$$

Four major assumptions were made in this derivation:

(a) The distribution of the tensile stresses in the concrete.

(b) The angle of inclination of the forces on the deformations.

(c) The replacement of the concentrated forces on the deformations with a force that is uniformly distributed along the length and around the circumference of the bar.

(d) The neglect of the effect of in-and-out bond stresses between cracks.

The similarity of the final result here to Eq. (8-11), derived later, reinforces the validity of the splitting model.

Basic Tension-Development Equation

In 1977, Orangun et al. [8-2] fitted a regression equation through the results of a large number of bond and splice tests. The resulting equation for bar development length, ℓ_d, included terms for the bar diameter, d_b; the bar stress to be developed, f_y; the concrete tensile strength, expressed as a coefficient times $\sqrt{f'_c}$; the cover or bar spacing; and the transverse steel ratio. It served as the basis of the development-length provisions in the 1989 ACI Code. These provisions proved difficult to use, however, and between 1989 and 1995, ACI Committee 318 and the ACI bond committee simplified the design expressions, in two stages. First, a basic expression was developed for the development length, ℓ_d, given in ACI Code Section 25.4.2.3 as

$$\ell_d = \frac{3}{40\lambda}\frac{f_y}{\sqrt{f'_c}}\frac{\psi_t \psi_e \psi_s}{\left(\dfrac{c_b + K_{tr}}{d_b}\right)} d_b \tag{8-10}$$

where the confinement term $(c_b + K_{tr})/d_b$ is limited to 2.5 or smaller, to prevent pull-out bond failures, and the length ℓ_d is not to be taken less than 12 in. Also,

ℓ_d is the development length, in.

d_b is the bar diameter, in.

ψ_t is a bar-location factor given in ACI Code Section 25.4.2.4.

ψ_e is an epoxy-coating factor given in ACI Code Section 25.4.2.4.

ψ_s is a bar-size factor given in ACI Code Section 25.4.2.4.

λ is the lightweight concrete factor discussed in Chapters 6 and 7 and defined in ACI Code Section 25.4.2.4.

c_b is the smaller of

(a) the smallest distance measured from the surface of the concrete to the *center* of a bar being developed, and

(b) one-half of the *center-to-center* spacing of the bars or wires being developed.

K_{tr} is a transverse reinforcement factor given in ACI Code Section 25.4.2.3.

Values for these factors will be presented later.

Simplified Tension-Development-Length Equations

Equation (8-10) was simplified by substituting lower limit values of c_b and K_{tr} for common design cases, to get widely applicable equations that did not explicitly include these factors. For deformed bars or deformed wire, ACI Code Section 25.4.2.2 defines the development length, as given in Tables 8-1 and 8-1M.

The Cases 1 and 2 described in the top row of Tables 8-1 and 8-1M are illustrated in Figs. 8-10a and 8-10b. The "code minimum" stirrups and ties mentioned in Case 1 correspond to the minimum amounts and maximum spacings specified in ACI Code Sections 9.6.3.3 and 9.7.6.2.2.

Bar-Spacing Factor, c_b

The factor c_b in Eq. (8-10) is the smaller of two quantities:

1. In the first definition, c_b is the smallest distance from the surface of the concrete to the *center* of the bar being developed. (See Fig. 8-9a.)

ACI Code Table 20.6.1.3.1 gives the minimum cover to principal reinforcement as $1\frac{1}{2}$ in. For a beam stem not exposed to weather, with No. 11 bars enclosed in No. 3 stirrups or ties, c_b will be (1.5-in. cover to the stirrups + 0.375-in. stirrup) + (half of the bar diameter, $1.41/2 = 0.71$ in.) = 2.58 in. (A typical value is approximately 2.5 in.)

2. In the second definition, c_b is equal to one-half of the center-to-center spacing of the bars.

TABLE 8-1 Equations for Development Lengths, ℓ_d[a]

	No. 6 and Smaller Bars and Deformed Wires	No. 7 and Larger Bars
Case 1: Clear spacing of bars being developed or spliced not less than d_b, **and** stirrups or ties throughout ℓ_d not less than the code minimum **or** **Case 2:** Clear spacing of bars being developed or spliced not less than $2d_b$ and clear cover not less than d_b	$\ell_d = \dfrac{f_y \psi_t \psi_e}{25\lambda\sqrt{f'_c}} d_b$ (8-11)	$\ell_d = \dfrac{f_y \psi_t \psi_e}{20\lambda\sqrt{f'_c}} d_b$ (8-12)
Other cases	$\ell_d = \dfrac{3f_y \psi_t \psi_e}{50\lambda\sqrt{f'_c}} d_b$ (8-13)	$\ell_d = \dfrac{3f_y \psi_t \psi_e}{40\lambda\sqrt{f'_c}} d_b$ (8-14)

[a]The length ℓ_d computed using Eqs. (8-11) to (8-14) shall not be taken less than 12 in.

TABLE 8-1M Equations for Development Lengths—SI Units[a]

	No. 20 and Smaller Bars and Deformed Wires	No. 25 and Larger Bars
Case 1: Clear spacing of bars being developed or spliced not less than d_b, **and** stirrups or ties throughout ℓ_d not less than the code minimum **or** **Case 2:** Clear spacing of bars being developed or spliced not less than $2d_b$ and clear cover not less than d_b	$\ell_d = \dfrac{12 f_y \psi_t \psi_e}{25\lambda\sqrt{f'_c}} d_b$ (8-11M)	$\ell_d = \dfrac{12 f_y \psi_t \psi_e}{20\lambda\sqrt{f'_c}} d_b$ (8-12M)
Other cases	$\ell_d = \dfrac{18 f_y \psi_t \psi_e}{25\lambda\sqrt{f'_c}} d_b$ (8-13M)	$\ell_d = \dfrac{18 f_y \psi_t \psi_e}{20\lambda\sqrt{f'_c}} d_b$ (8-14M)

[a]The length ℓ_d computed using Eqs. (8-11M) to (8-14M) shall not be taken less than 300 mm.

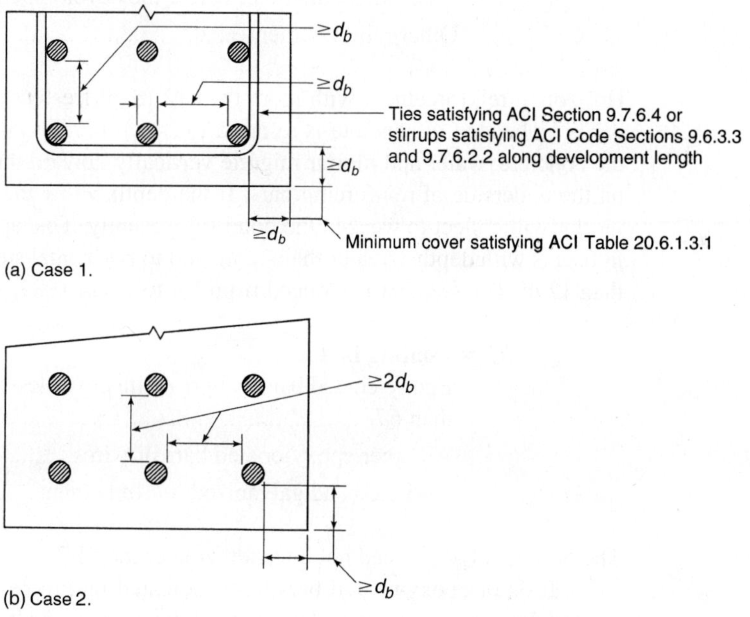

Fig. 8-10
Explanation of Cases 1 and 2 in Table 8-1.

ACI Code Section 25.2.1 gives the minimum clear spacing of parallel bars in a layer as d_b, but not less than 1 in. For No. 11 bars, the diameter is 1.41 in., giving the center-to-center spacing of the bars as $1.41/2 + 1.41 + 1.41/2 = 2.82$ in. and $c_b = 1.41$ in.

The smaller of the two values discussed here is $c_b = 1.41$ in. $= 1.0d_b$. The minimum stirrups or ties given in ACI Code Sections 9.7.6.4 or 9.6.3.3, and 9.7.6.2.2 correspond to K_{tr} between $0.1d_b$ and $0.5d_b$, depending on a wide range of factors. Thus, for this case, $(c_b + K_{tr})/d_b \approx 1.5$. Substituting this and the appropriate bar size factor, ψ_s into Eq. (8-10) gives Eqs. (8-11) and (8-12) in Table 8-1.

For Case 2 in Table 8-1, consider a slab with clear cover to the outer layer of bars of d_b and a clear spacing between the bars of $2d_b$. It is assumed that splitting in the cover will be restrained by bars perpendicular to the bars being developed. As a result, c_b is governed by the bar spacing. If the clear spacing is $2d_b$, the center-to-center spacing is $3d_b$. Thus, $c_b = 3d_b/2 = 1.5d_b$. Substituting this into Eq. (8-10) and taking $K_{tr} = 0$ gives Eqs. (8-11) or (8-12) in Table 8-1.

For the situation where the minimum clear cover to the bar being developed is $1.5d_b$ and the minimum clear spacing is d_b, c_b is the smaller of $1.5d_b$ and d_b. Substituting ψ_s and $c_b = d_b$ into Eq. (8-10), assuming that $K_{tr} = 0$ gives Eqs. (8-13) and (8-14), which apply to all cases other than 1 and 2, as shown in Table 8-1.

Values of ℓ_d computed from Eqs. (8-11) to (8-14) are tabulated in Tables A-6 and A-6M in Appendix A. Typically, ℓ_d is about $30d_b$ for No. 3 to No. 6 bottom bars and about $40d_b$ for No. 7 and larger bottom bars.

Factors in Eqs. (8-10) through (8-14)

The Greek-letter factors in Eqs. (8-10) through (8-14) are defined in ACI Section 25.4.2.4 as follows:

ψ_t = **bar-location factor**

Horizontal reinforcement so placed that more than 12 in. of fresh concrete is cast in the member below the development length or splice...............1.3

Other reinforcement..1.0

Horizontal reinforcement with more than 12 in. of fresh concrete below it at the time the bar is embedded in concrete is referred to as *top reinforcement.* During the placement of the concrete, water and mortar migrate vertically upward through the concrete, collecting on the underside of reinforcing bars. If the depth below the bar exceeds 12 in., sufficient mortar will collect to weaken the bond significantly. This applies to the top reinforcement in beams with depths greater than 12 in. and to horizontal steel in walls cast in lifts greater than 12 in. The factor was reduced from 1.4 to 1.3 in 1989, as a result of tests in [8-5].

ψ_e = **coating factor**

Epoxy-coated bars or wires with cover less than $3d_b$, or clear spacing less than $6d_b$ 1.5

All other epoxy-coated bars or wires..1.2

Uncoated and galvanized reinforcement 1.0

The product of $\psi_t\psi_e$ need not be taken greater than 1.7.

Tests of epoxy-coated bars have indicated that there is negligible friction between concrete and the epoxy-coated bar deformations. As a result, the forces acting on the deformations and the concrete with epoxy-coated bars in Figs. 8-7a and 8-7b act in a direction perpendicular to the surface of the deformations. In a bar without an epoxy coating, friction between the deformation and the concrete allows the forces on the deformation and the concrete to act at an angle flatter than the 45° angle shown in Fig. 8-7. Because of this, the radial-force components are larger in an epoxy-coated bar than in a normal bar for a given longitudinal force component; hence, splitting occurs at a lower longitudinal force [8-6]. The 1.5 value of ψ_e corresponds to cases where splitting failures occur. For larger covers and spacings, pull-out failures tend to occur, and the effect of epoxy coating is smaller.

ψ_s = **bar-size factor**

No. 6 and smaller bars and deformed wires 0.8

No. 7 and larger bars 1.0

Comparison of Eq. (8-10) with a large collection of bond and splice tests showed that a shorter development length was possible for smaller bars.

λ = lightweight-aggregate-concrete factor

When any lightweight-aggregate concrete is used 0.75

However, when the splitting tensile strength f_{ct} is specified, λ shall be permitted to be taken as $f_{ct}/6.7\sqrt{f'_c}$ but not more than 1.0

When normal-weight concrete is used 1.0

The tensile strength of lightweight concrete is generally less than that of normal-weight concrete; hence, the splitting load will be less. In addition, in some lightweight concretes, the wedging forces that the bar deformations exert on the concrete can cause localized crushing, which allows bar slip to occur.

K_{tr} = transverse reinforcement index

$$= \frac{40A_{tr}}{sn}$$

where

A_{tr} = total cross-sectional area of all transverse reinforcement within the spacing s, which crosses the potential plane of splitting along the reinforcement being developed within the development length, in.^{2} (illustrated in Fig. 8-11)

s = maximum center-to-center spacing of transverse reinforcement within ℓ_d, in

n = number of bars or wires being developed or spliced along the plane of splitting

ACI Code Section 25.4.2.3 allows K_{tr} to be taken equal to zero to simplify the calculations, even if there is transverse reinforcement.

Excess Flexural Reinforcement

If the flexural reinforcement provided exceeds the amount required to resist the factored moment, the bar stress that must be developed is less than f_y. In such a case, ACI Code Section 25.4.10 allows ℓ_d to be multiplied by (A_s required/A_s provided). If room is available, the author recommends ignoring this factor, thus ensuring that the steel is fully anchored. In statically indeterminate structures, the increased stiffness resulting from the additional reinforcement can lead to higher moments at the section with excess reinforcement. In such a case, the steel is more highly stressed than would be expected from the ratio of areas. This multiplier is *not applied* in the design of members resisting seismic loads.

The development length, calculated as the product of ℓ_d from ACI Code Table 25.4.2.2 or Section 25.4.2.3 and the factors given here, shall not be taken less than 12 in.

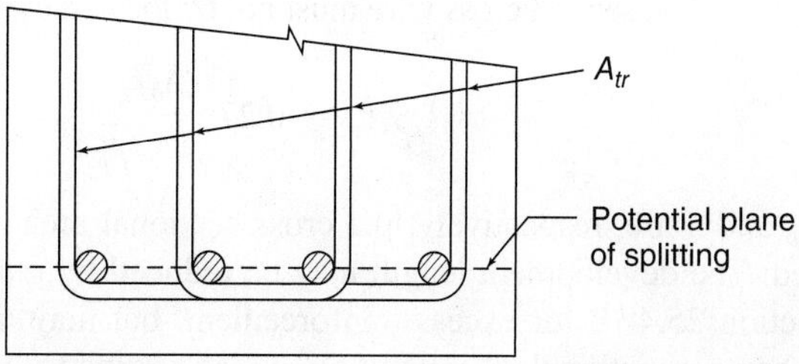

Fig. 8-11
Definition of A_{tr}.

Compression-Development Lengths

Compression-development lengths are considerably shorter than tension-development lengths, because some force is transferred to the concrete by the bearing at the end of the bar and because there are no cracks in such an anchorage region (and hence no in-and-out bond stresses). The basic compression-development length is (ACI Code Section 25.4.9)

$$\ell_{dc} = \frac{\psi_r d_b f_y}{50\lambda\sqrt{f'_c}} \text{ but not less than } 0.0003 d_b f_y \tag{8-15}$$

where ψ_r, as defined in ACI Code Table 25.4.9.3, can be taken as 1.0 and the constant 0.0003 has units of "in.^{2}/lb." Values of ℓ_{dc} are given in Tables A-7 and A-7M. The development length in compression may be reduced by multiplying ℓ_{dc} by the applicable modification factors given in ACI Code Section 25.4.10 for excess reinforcement. The resulting development length shall not be less than 8 in.

Development Lengths for Bundled Bars

Where a large number of bars are required in a beam or column, the bars are sometimes placed in bundles of 2, 3, or 4 bars (ACI Code Section 25.6.1). The effective perimeter for bond failure of bundles is less than the total perimeter of the individual bars in the bundle. ACI Code Section 25.6.1.5 accounts for this by requiring that individual bar development lengths be increased by 1.2 times for bars in a 3-bar bundle and 1.33 times for bars in a 4-bar bundle. To determine the required cover and spacing values for use in ACI Code Table 25.4.2.2, the confinement term in the denominator of Code Eq. 25.4.2.3(a), and the ψ_e factor in Code Table 25.4.2.4, the value of d_b shall be taken as the diameter of a hypothetical single bar having the same area as the bundle.

Development Lengths for Coated Bars

In bridge decks and parking garages, epoxy-coated or galvanized reinforcement are frequently used to reduce corrosion problems. Epoxy-coated bars are covered by the factor ψ_e in ACI Code Table 25.4.2.4. There is no modification factor for zinc-coated bars. The zinc coating on galvanized bars can affect the bond properties via a chemical reaction with the concrete. This effect can be prevented by treating the bars with a solution of chromate after galvanizing. If this is done, the bond is essentially the same as that for normal reinforcement.

Development Lengths for Welded-Wire Reinforcement

ACI Code Section 25.4.7 provides rules for development of welded plain-wire reinforcement in tension. The development of *plain-wire reinforcement* depends on the mechanical anchorage from at least two cross wires. The nearest of these cross wires must be located 2 in. or farther from the critical section. Wires that are closer to the critical section cannot be counted for anchorage. The second cross wire must not be located closer to the critical section than

$$\ell_d = 0.27\frac{A_b f_y}{s\lambda\sqrt{f'_c}} \tag{8-16}$$

where A_b and s are, respectively, the cross-sectional area and spacing of the wire being developed. The development length may be reduced by multiplying by the factor in ACI Code Section 25.4.10 for excess reinforcement, but may not be taken as less than 6 in. except when computing the length of splices according to ACI Code Section 25.5.4.2.

Deformed-wire reinforcement derives anchorage from bond stresses along the deformed wires and from mechanical anchorage from the cross wires. The ASTM specification for welded deformed-wire reinforcement does not require as strong welds for deformed reinforcement as for plain-wire reinforcement. ACI Code Section 25.4.6 gives the basic development length of deformed-wire reinforcement with at least one cross wire in the development length, but not closer than 2 in. to the critical section, as ψ_w times the development length computed from ACI Code Table 25.4.2.2, where ψ_w is the larger of the factors given by Eqs. (8-17a) and (8-17b). From ACI Code Section 25.4.6.3,

$$\psi_w = \frac{f_y - 35{,}000}{f_y} \tag{8-17a}$$

or

$$\psi_w = \frac{5d_b}{s} \tag{8-17b}$$

but ψ_w cannot be taken greater than 1.0. and s is the spacing between wires being developed.

ACI Code Section 25.4.6.4 applies to deformed-wire reinforcement with no cross wires in the development length or with a single cross wire less than 2 in. from the critical section (i.e., the section where f_y must be developed). In this case,

$$\psi_w = 1.0 \tag{8-17c}$$

The development length computed shall not be taken less than 8 in., except when computing the length of splices according to ACI Code Section 25.5.3 or stirrup anchorages according to ACI Code Section 25.7.1.3.

Tests have shown that the development length of deformed-wire reinforcement is not affected by epoxy coating; for this reason, the epoxy-coating factor, ψ_e, is taken equal to 1.0 for epoxy-coated deformed-wire reinforcement.

Deformed welded-wire reinforcement may have some plain wires in one or both directions. For the purpose of determining the development length, such wire reinforcement shall be considered to be plain-wire reinforcement if any of the wires in the direction that development is being considered is a plain wire (ACI Code Section 25.4.6.5).

8-4 HOOKED ANCHORAGES

Behavior of Hooked Anchorages

Hooks are used to provide additional anchorage when there is insufficient straight length available to develop a bar. Unless otherwise specified, the so-called *standard hooks* described in ACI Code Table 25.3.1 are used. Details of 90° and 180° standard hooks and standard stirrup and tie hooks are given in Fig. 8-12. It is important to note that a standard hook on a large bar takes up a lot of room, and the actual size of the hook is frequently quite critical in detailing a structure.

A 90° hook loaded in tension develops forces in the manner shown in Fig. 8-13a. The stress in the bar is resisted by the bond on the surface of the bar and by the bearing on the concrete inside the hook [8-7]. The hook moves inward, leaving a gap between it and the concrete outside the bend. Because the compressive force inside the bend is not collinear with the applied tensile force, the bar tends to straighten out, producing compressive stresses on the outside of the tail. Failure of a hook almost always involves crushing of the concrete inside the hook. If the hook is close to a side face, the crushing will extend

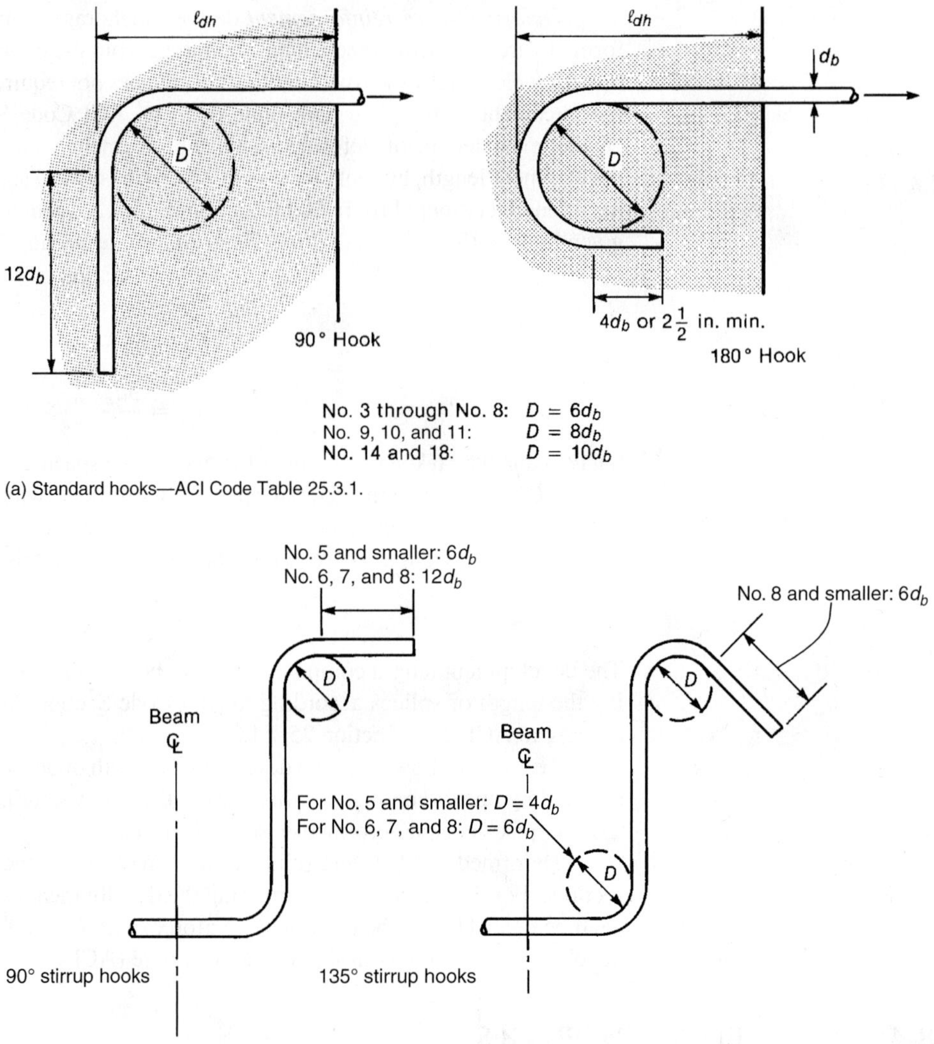

(a) Standard hooks—ACI Code Table 25.3.1.

(b) Stirrup and tie hooks—ACI Code Table 25.3.2.

Fig. 8-12
Standard hooks.

to the surface of the concrete, removing the side cover. Occasionally, the concrete outside the tail will crack, allowing the tail to straighten.

The stresses and slip measured at points along a hook at a bar stress of $1.25f_y$ (75 ksi) in tests of 90° and 180° hooks in No. 7 bars are plotted in Figs. 8-13b and 8-13c [8-7]. The axial stresses in the bar decrease due to the bond on the lead-in length and the bond and friction on the inside of the bar. The magnitude and direction of slip at *A*, *B*, and *C* are shown by the arrows. For the 180° hook, the slip measured at *A* was 1.75 times that measured at *A* in the 90° hook.

The amount of slip depends on, among other things, the angle of the bend and the orientation of the hook relative to the direction of concrete placement. The slip of hooks displays a top-bar effect, which is not recognized in the calculation of ℓ_{dh}. In tests, top-cast hooks, oriented so that weaker mortar was trapped inside the bend during casting, slipped 50 to 100 percent more at a given bar stress than did bottom-cast bars [8-8].

Tests on bars hooked around a corner bar show that tensile stresses can be developed at a given end slip that are 10 to 30 percent larger than can be developed if a bar is not present inside the hook.

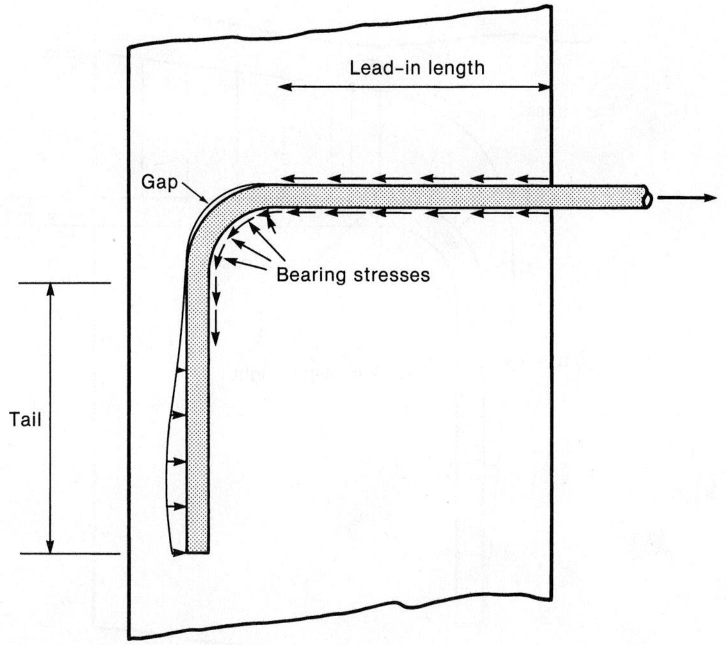

Fig. 8-13
Behavior of hooks.

(a) Forces acting on bar.

Development of Hooked Bars in Tension

The design process described in ACI Code Section 25.4.3 does not distinguish between 90° and 180° hooks or between top and bottom bar hooks. The development length of a hook, ℓ_{dh} (illustrated in Fig. 8-12a), is computed with Eq. (8-18), with the λ and ψ-factors defined in ACI Code Table 25.4.3.2, except as modified in ACI Code Section 25.4.3.3 for hooks at discontinuous ends of members.

$$\ell_{dh} = \left(\frac{f_y \psi_e \psi_c \psi_r}{50\lambda\sqrt{f'_c}}\right) d_b \tag{8-18}$$

The final development length shall not be less than 8 d_b or 6 in., whichever is larger. Values of ℓ_{dh}, before the application of the λ and ψ-factors, are given in Table A-8 and A-8M.

Values for λ and ψ-factors, ACI Code Table 25.4.3.2

λ = lightweight concrete factor; taken as 1.0 for normal weight concrete and 0.75 for a concrete mix that contains any lightweight aggregate.

ψ_e = epoxy-coating factor; taken as 1.0 for uncoated or zinc-coated (galvanized) reinforcement and 1.2 for epoxy-coated reinforcement.

ψ_c = cover factor; taken as 0.7 for 180° hooks of No. 11 and smaller bars with side cover $\geq$2.5 in.; taken as 0.7 for 90° hooks of No. 11 and smaller bars with side cover $\geq$2.5 in. and cover on the back of the tail extension $\geq$2.0 in.; otherwise taken as 1.0.

ψ_r = confining reinforcement factor; taken as 0.8 for 90° and 180° hooks of No. 11 and smaller bars located within ties or stirrups placed along and perpendicular to ℓ_{dh} at $s \leq 3\ d_b$ (of hooked bar); taken as 0.8 for 90° hooks of No. 11 and smaller bars with bar tail extension located within ties or stirrups placed

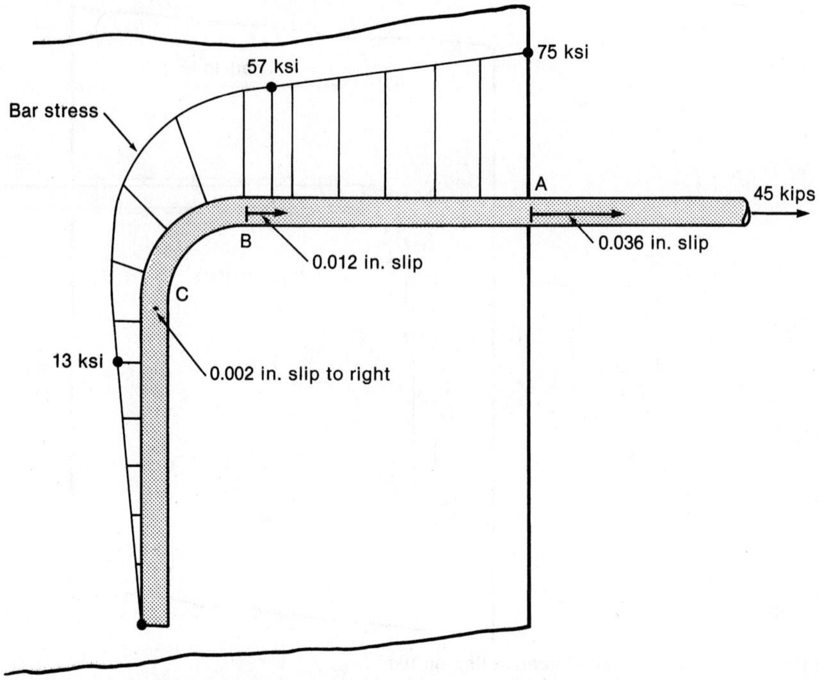

(b) Stresses and slip—90° standard hook.

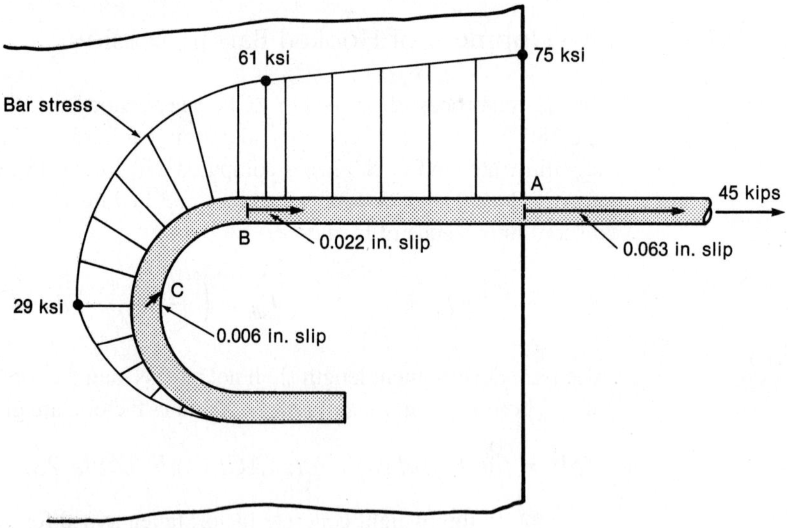

Fig. 8-13 (Continued)

(c) Stresses and slip—180° standard hook.

along and perpendicular to the extension at $s \leq 3\ d_b$; otherwise taken as 1.0. Note: Figure 8-14a shows ties or stirrups placed along and perpendicular to ℓ_{dh}, and Fig. 8-14b shows ties or stirrups placed along and perpendicular to the tail extension. The first tie shall be placed within 2 d_b of the outside bend of the hook.

For cases where excess reinforcement has been provided and anchorage or development of f_y is not specifically required, ACI Code Section 25.4.10.1 permits the calculated hook development length to be reduced by the ratio of A_s(required) to A_s(provided), but it shall

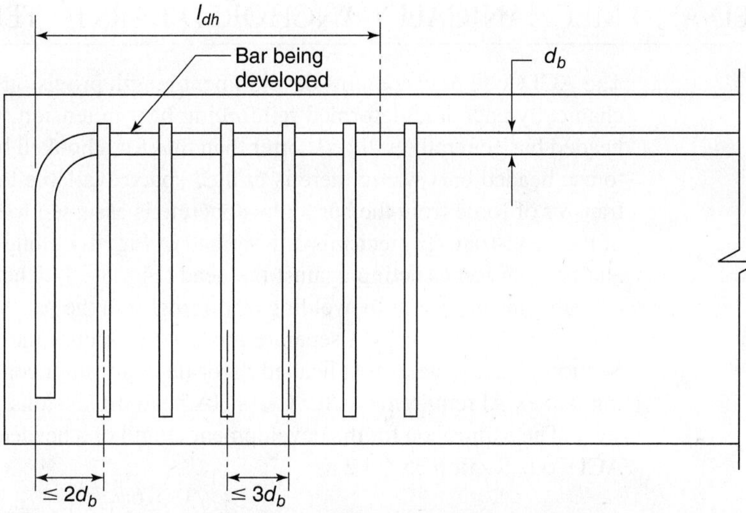

(a) Ties or stirrups placed perpendicular to the bar being developed. (See row 3 in Table 8-2.)

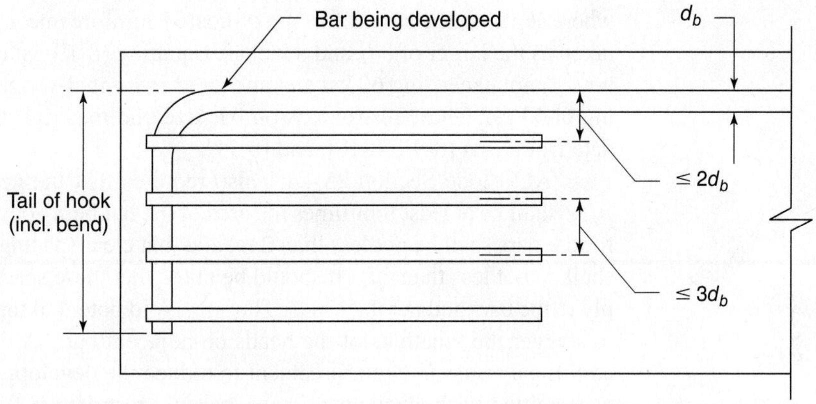

(b) Ties or stirrups placed parallel to the bar being developed. (See row 4 in Table 8-2.)

Fig. 8-14
Confinement of hooks by stirrups and ties.

not be less than 8 d_b or 6 in., whichever is larger. As noted previously, the author does not recommend the use of this reduction if space is available to fully develop the hook anchorage. For No. 14 and 18 bars, all of the ψ-factors are 1.0. However, it is still beneficial to provide stirrups and ample cover along the development length.

ACI Code Section 25.4.3.3 states that for bars being developed by a standard hook at *discontinuous ends* of members with both side cover and top (or bottom) cover over a hook of less than $2\frac{1}{2}$ in., the hooked bar shall be (must be) enclosed within ties or stirrups perpendicular to the bar being developed, spaced not greater than $3d_b$ along the development length of the hook, ℓ_{dh}. In this case, the factors of ACI Code Table 25.4.3.2 shall not apply. ACI Code Section 25.4.3.3 applies at the ends of simply supported beams (particularly if these are deep beams), at the free ends of cantilevers, and at the ends of members that terminate in a joint with less than 2 1/2 in. of both side cover and top (or bottom) cover over the hooked bar. Hooked bars at discontinuous ends of slabs are assumed to have confinement from the slab on each side of the hook; hence, ACI Code Section 25.4.3.3 is not applied.

8-5 HEADED AND MECHANICALLY ANCHORED BARS IN TENSION

The ACI Code now contains development length provisions for the use of headed or mechanically anchored deformed reinforcing bars in tension. The development length for a headed bar generally will be shorter than that for a hooked bar, so it may be advantageous to use headed bars where there is limited space available to develop bars in tension. The transfer of force from the bar to the concrete is assumed to be achieved by a combination of the bond-transfer mechanism described in Fig. 8-7 along the straight portion of the bar and bearing forces acting against the head (Fig. 8-15). The heads can be attached to one or both ends of the bar by welding or forging onto the bar, by internal threads on the head mating to the bar, or by a separate nut used to secure the head onto the bar. ACI Code Section 20.2.1.6 states that headed deformed bars shall conform to ASTM A970, including Annex Al requirements for Class HA head dimensions.

The expression for the development length of a headed bar in tension, ℓ_{dt}, is given in ACI Code Section 25.4.4.2 as

$$\ell_{dt} = \left(\frac{0.016\psi_e f_y}{\sqrt{f'_c}}\right)d_b \qquad (8\text{-}19)$$

where ψ_e is taken as 1.2 for epoxy-coated reinforcement. The minimum value of ℓ_{dt} is taken as the larger of $8d_b$ and 8 inches. Equation (8-19) is limited to No. 11 or smaller bars with f_y not exceeding 60 ksi and anchored in normal-weight concrete with f'_c not exceeding 6000 psi. These restrictions on bar size and material strengths are based on available data from tests [8-9], [8-10], and [8-11].

ACI Code Section 25.4.4.1 also requires that the net bearing area under the head, A_{brg}, shall be at least four times the area of the bar being developed, A_b, that the clear cover for the bars shall be not less than $2d_b$, and that clear spacing between bars being developed shall be not less than $4d_b$. It should be noted that these spacing and cover requirements apply to the bars and not the heads. Thus, to avoid potential interferences, it may be necessary to stagger the locations of the heads on adjacent bars. A designer is not permitted to use additional transverse reinforcement to reduce the development length for a headed bar, as is permitted for hooked anchorages, because test data [8-9], [8-10], and [8-11] has shown that additional transverse reinforcement does not significantly improve the anchorage of headed bars.

Headed reinforcement that does not meet all of the requirements noted here may be used if test data is available to demonstrate the ability of the head or mechanical anchor to fully develop the strength of the bar. No data is available to demonstrate that the use of heads significantly improves the anchorage of bars in compression.

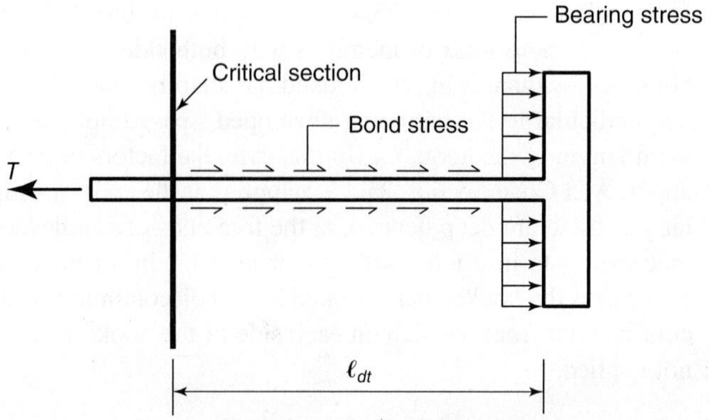

Fig. 8-15
Development length of headed bars in tension.

8-6 DESIGN FOR ANCHORAGE

The basic rule governing the development and anchorage of bars is as follows:

> The calculated tension or compression in reinforcement at each section of reinforced concrete members shall be developed on each side of that section by embedment length, hook, headed deformed bar, or mechanical anchorages, or a combination thereof. (ACI Code Section 25.4.1.1)

This requirement is satisfied in various ways for different types of members. Three examples of bar anchorage, two for straight bars and one for hooks, are presented in this section. The anchorage and cutoff of bars in beams is discussed in Section 8-7.

EXAMPLE 8-1 Anchorage of a Straight Bar

A 16-in.-wide cantilever beam frames into the edge of a 16-in.-thick wall, as shown in Fig. 8-16. To reach M_n, the No. 8 bars at the top of the cantilever are stressed to their yield strength at point A at the face of the wall. Compute the minimum embedment of the bars into the wall and the development length in the beam. The concrete is sand-lightweight concrete with a strength of 4000 psi. The yield strength of the flexural reinforcement is 60,000 psi. Construction joints are located at the bottom and top of the beam, as shown in Fig. 8-16. The beam has closed No. 3 stirrups with $f_{yt} = 40{,}000$ psi at a spacing of 7.5 in. throughout its length. (Stirrups are not shown.) The cover is 1.5 in. to the stirrups. The three No. 8 bars are inside No. 4 bars at 12-in. vertical steel in each face of the wall. The wall steel is Grade 60.

We shall do this problem twice, first using ACI Code Section 25.4.2.2 (Table 8-1) and then using ACI Code Section 25.4.2.3, Eq. (8-10).

1. Find the spacing and confinement case for bars anchored in the wall. The clear side cover to the No. 8 bars in the wall (and beam) is 1.5 + 0.5 = 2 in. = $2d_b$. The clear spacing of the bars is

$$\frac{16 - 2(1.5 + 0.5) - 3 \times 1.0}{2} = 4.5 \text{ in.} = 4.5d_b$$

There are No. 3 stirrups outside of the three No. 8 bars in the beam. Because the clear spacing between the bars is not less than $2d_b$ and the clear cover to the No. 8 bars exceeds d_b, bar development in the wall is governed by Case 2 in Table 8-1, and for No. 8 bars Eq. (8-12) applies.

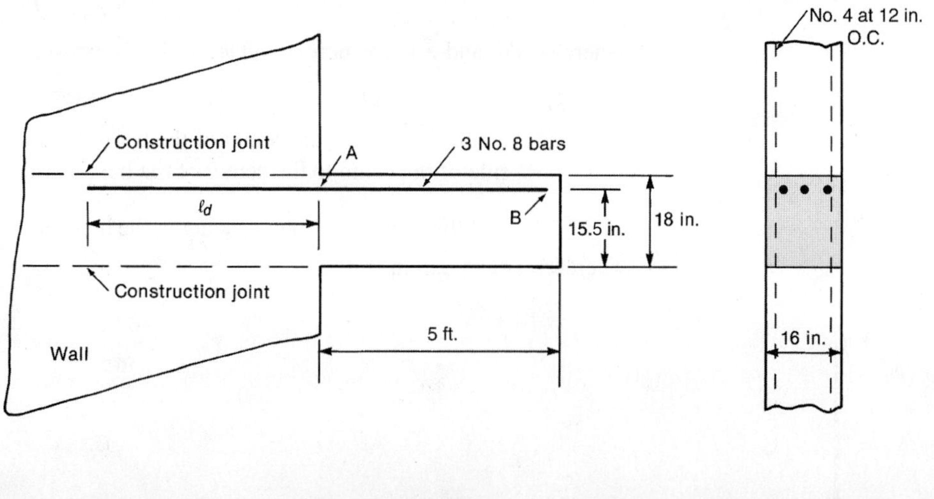

Fig. 8-16
Cantilever beam—
Examples 8-1 and 8-1M.

2. Compute the development length for No. 8 bars in the wall. From Eq. (8-12),

$$\ell_d = \frac{f_y \psi_t \psi_e}{20\lambda\sqrt{f'_c}} d_b$$

where

ψ_t = 1.3 because there will be more than 12 in. of fresh concrete under the bars

ψ_e = 1.0 because the bars are not epoxy coated

λ = 0.75 because the concrete has lightweight aggregates

$$\ell_d = \frac{60{,}000 \times 1.3 \times 1.0}{20 \times 0.75\sqrt{4000}} \times 1.0 = 82.2 \text{ in.}$$

The bars must extend 82.2 in. into the wall to develop the full yield strength. **Extend the bars 7 ft into the wall.**

3. Compute the development length for No. 8 bars in the beam using Table 8-1. We first must determine which case in Table 8-1 governs. From step 1, we know that the clear spacing between the No. 8 bars exceeds d_b. The stirrup spacing of 7.5 in. is less than $d/2$, but we also need to check the minimum area requirement in ACI Code Section 9.6.3.3.

$$A_{v,\min} = \frac{0.75\sqrt{f'_c}\, b_w s}{f_{yt}}, \text{ and } \geq \frac{(50 \text{ psi}) b_w s}{f_{yt}}$$

For f'_c = 4000 psi, the second part of this requirement governs. So,

$$A_{v,\min} = \frac{50 \text{ psi} \times 16 \text{ in.} \times 7.5 \text{ in.}}{40{,}000 \text{ psi}} = 0.150 \text{ in.}^2$$

For a double-leg No. 3 stirrup, $A_v = 2 \times 0.11 = 0.22$ in.^{2}. Therefore, Case 1 in Table 8-1 is satisfied, and Eq. (8-12) governs for No. 8 bars, the same as for the development length in the wall. Thus, the required ℓ_d = 82.2 in., which exceeds the length of the beam. We have two choices: either change to a larger number of smaller bars or use Eq. (8-10) to determine if we can calculate a shorter required development length. We will try the second choice.

4. Compute the development length for No. 8 bars in the beam using Eq. (8-10).

$$\ell_d = \frac{3}{40\lambda} \frac{f_y}{\sqrt{f'_c}} \frac{\psi_t \psi_e \psi_s}{\left(\dfrac{c_b + K_{tr}}{d_b}\right)} d_b$$

where ψ_t, ψ_e, and λ are as before and ψ_s = 1.0 because the bars are No. 8.

c_b = the smaller of

(a) the distance from the center of the bar to the nearest concrete surface: from the side of the beam to the center of the bar is $1.50 + 0.50 + 1.0/2 = 2.5$ in.;

(b) half the center-to-center spacing of the bars, or $0.5\left(\dfrac{16 - 2 \times 2.5}{2}\right) = 2.75$ in.

Therefore, c_b = 2.5 in.

$$K_{tr} = \frac{40A_{tr}}{sn}$$

where

s = spacing of the transverse reinforcement within the development length, ℓ_d = 7.5 in.

A_{tr} = total cross-sectional area of reinforcement crossing the plane of splitting within the spacing s (two No. 3 legs)
= $2 \times 0.11 = 0.22$ in.^{2}

n = number of bars being anchored = 3

Thus,

$$K_{tr} = \frac{40 \times 0.22 \text{ in.}^2}{7.5 \text{ in.} \times 3} = 0.391 \text{ in.}$$

The term

$$\frac{c_b + K_{tr}}{d_b} = \frac{2.5 \text{ in.} + 0.391 \text{ in.}}{1.0 \text{ in.}} = 2.89 \not< 2.5$$

is set equal to 2.5. Substituting into Eq. (8-10) gives

$$\ell_d = \frac{3}{40} \times \frac{60{,}000}{0.75\sqrt{4000}} \times \frac{1.3 \times 1.0 \times 1.0}{2.5} \times 1.0 = 49.3 \text{ in.}$$

Thus, ℓ_d = 49.3 in., so the bars should be extended the full length of the beam. **Extend the bars to 1.5 in. from the end of the beam.**

In this case, there is a large difference between the ℓ_d computed from Eq. (8-12) and the ℓ_d computed from Eq. (8-10). This is because Eq. (8-12) was derived by using $(c_b + K_{tr})/d_b$ equal to 1.5. In this case, it is actually equal to 2.5. ■

EXAMPLE 8-1M Anchorage of a Straight Bar—SI Units

A 400-mm-wide cantilever beam frames into the edge of a 400-mm-thick wall similar to Fig. 8-16. To reach M_n, the three No. 25 bars at the top of the beam are stressed to their yield strength at point A at the face of the wall. Compute the minimum embedment of the bars into the wall and the development length in the beam. The concrete is sand/low-density concrete with a strength of 25 MPa. The yield strength of the flexural reinforcement is 420 MPa. Construction joints are located at the bottom and top of the beam, as shown in Fig. 8-16. The beam has closed No. 10 stirrups with f_{yt} = 420 MPa at a spacing of 180 mm throughout its length. (The stirrups are not shown.) The cover is 40 mm to the stirrups. The three No. 25 bars are inside No. 13 bars used as vertical steel in each face of the wall.

1. Find the spacing and confinement case for bars anchored in wall. The clear side cover to the No. 25 bars in the wall (and beam) is 40 + 13 = 53 mm = $2.1d_b$. The clear spacing of the bars is

$$\frac{400 - 2(40 + 13) - 3 \times 25}{2} = 110 \text{ mm} = 4.4d_b$$

Because the clear spacing between the bars is not less than $2d_b$ and the clear cover to the No. 8 bars exceeds d_b, this is Case 2 in Table 8-1M, and for No. 25 bars, Eq. (8-12M) applies.

2. Compute the development length. From Eq. (8-12M),

$$\ell_d = \frac{12 f_y \psi_t \psi_e}{20\lambda\sqrt{f'_c}} d_b$$

where

ψ_t = 1.3 because there will be more than 300 mm of fresh concrete under the bar when the concrete in the beam covers the bars

ψ_e = 1.0 because the bars are not epoxy coated

λ = 0.75 because the concrete has low-density aggregates

$$\ell_d = \frac{12 \times 420 \times 1.3 \times 1.0}{20 \times 0.75\sqrt{25}} \times 25 = 2180 \text{ mm}$$

The bars must extend 2180 mm into the wall to develop the full yield strength. **Extend the bars 2.2 m into the wall.**

3. Compute the development length for No. 25 bars in the beam using Table 8-1M. Assume the cantilever beam, similar to the one in Fig. 8-16, extends 2 m from the face of the wall and has an effective depth of 380 mm. We must determine which case in Table 8-1M governs. From step 1, we know that the clear spacing between the No. 25 bars exceeds d_b. The stirrup spacing of 180 mm is less than $d/2$, but we also need to check the minimum area requirement in ACI Metric Code Section 9.6.3.3.

$$A_{v,\min} = \frac{0.062\sqrt{f'_c}\, b_w s}{f_{yt}}, \text{ and } \geq \frac{(0.35 \text{ MPa}) b_w s}{f_{yt}}$$

For f'_c = 25 MPa, the second part of this requirement governs. So,

$$A_{v,\min} = \frac{0.35 \text{ MPa} \times 400 \text{ mm} \times 180 \text{ mm}}{420 \text{ MPa}} = 60.0 \text{ mm}^2$$

For a double-leg No. 10 stirrup, $A_v = 2 \times 71 = 142 \text{ mm}^2$. Therefore, Case 1 in Table 8-1M is satisfied, and Eq. (8-12M) governs for No. 25 bars, the same as for the development length in the wall. Thus, the required ℓ_d = 2180 mm, which exceeds the length of the beam. We have two choices: either change to a larger number of smaller bars or use Eq. (8-10M), to determine if we can calculate a shorter required development length. We will try the second choice.

4. Compute the development length for No. 25 bars in the beam using Eq. (8-10M). From ACI Metric Code Section 25.4.2.3, we get Eq. (8-10M):

$$\ell_d = \frac{f_y}{1.1\lambda\sqrt{f'_c}} \frac{\psi_t \psi_e \psi_s}{\left(\dfrac{c_b + K_{tr}}{d_b}\right)} d_b \qquad \text{(8-10M)}$$

where ψ_t, ψ_e, and λ are the same as in step 2. For a No. 25 bar, ψ_s = 1.0. c_b is the smaller of:

(a) the distance from the center of the bar to the nearest concrete surface; measuring from the side face to the center of the bar is 40 + 10 + 25/2 = 62.5 mm, and

(b) half the center to center spacing of the bars; $0.5\left(\frac{400 - 2 \times 62.5}{2}\right) = 69$ mm.

Thus, $c_b = 62.5$ mm.

The metric version of the transverse reinforcement index is the same as used for inch-pound units:

$$K_{tr} = \frac{40A_{tr}}{sn}$$

where s is the spacing of transverse reinforcement along the development length, which is equal to 180 mm in this example. The coefficient n represents the number of bars being anchored, which is three. A_{tr} is the area of transverse reinforcement crossing the potential splitting plane, which is $2 \times 71 = 142$ mm^2. Thus,

$$K_{tr} = \frac{40 \times 142 \text{ mm}^2}{180 \text{ mm} \times 3} = 10.5 \text{ mm}$$

Then,

$$\frac{c_b + K_{tr}}{d_b} = \frac{62.5 \text{ mm} + 10.5 \text{ mm}}{25 \text{ mm}} = 2.92 > 2.5$$

Thus, this term is set equal to 2.5 in Eq. (8-10M) to get

$$\ell_d = \frac{420 \text{ MPa}}{1.1 \times 0.75\sqrt{25} \text{ MPa}} \times \frac{1.3 \times 1.0 \times 1.0}{2.5} \times 25 \text{ mm} = 1320 \text{ mm}$$

Thus, $\ell_d = 1.32$ m, and the bars can be anchored within the cantilever beam. **Extend the No. 25 bars to 40 mm from the end of the beam.** ■

EXAMPLE 8-2 Hooked Bar Anchorage into a Column

The exterior end of a 16-in.-wide-by-24-in.-deep continuous beam frames into a 24-in.-square column, as shown in Fig. 8-17. The column has four No. 11 longitudinal bars. The negative-moment reinforcement at the exterior end of the beam consists of four No. 8 bars. The concrete is 4000-psi normal-weight concrete. The longitudinal steel strength is 60,000 psi. Design the anchorage of the four No. 8 bars into the column. From Example 8-1, it should be clear that a straight No. 8 bar cannot be developed in a 24-in.-deep column. Thus, assume a hooked bar anchorage is required.

1. Compute the development length for hooked beam bars. The basic development length for a Grade-60 hooked bar from Eq. (8-18) is

$$\ell_{dh} = \left(\frac{f_y\psi_e\psi_c\psi_r}{50\lambda\sqrt{f'_c}}\right)d_b$$

Assume we are not using epoxy-coated bars, so ψ_e can be taken as 1.0. For the cover factor, ψ_c, we need to check the side and back cover for this 90° hook. Because of the difference in the size of the column and the width of the beam, it is clear that the side cover on the hook exceeds 2.5 in. (Fig. 8-17b). For the back cover on the tail extension of the hook, we will assume the normal 1.5 in. cover to the stirrups and we will assume that No. 3 ties are used in the column. In this case, the back cover is 1.5 in. + 0.375 in. = 1.875 in. < 2 in., so the back cover requirement is not satisfied and $\psi_c = 1.0$. If we later find out that the hook

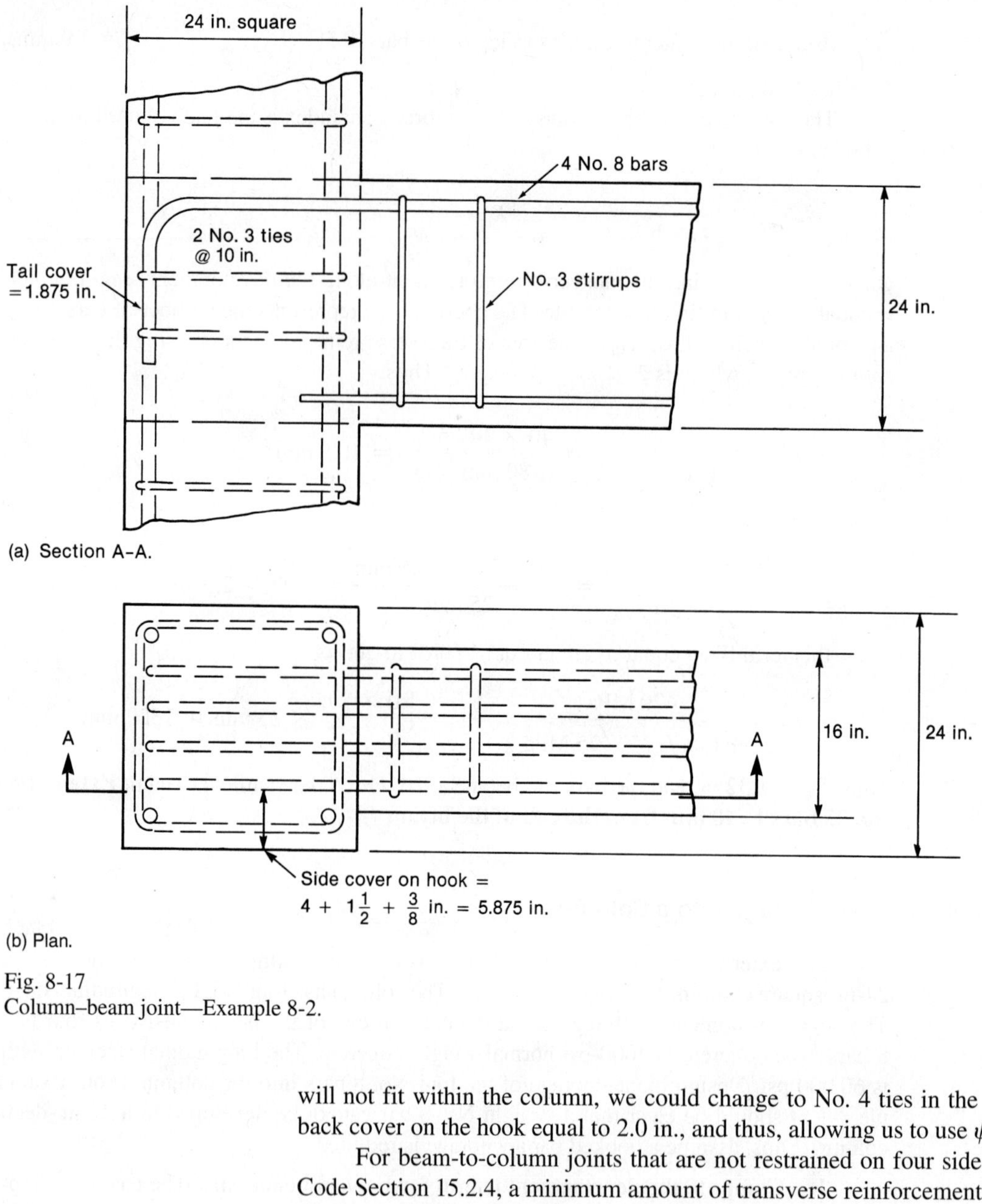

Fig. 8-17
Column–beam joint—Example 8-2.

will not fit within the column, we could change to No. 4 ties in the column, making the back cover on the hook equal to 2.0 in., and thus, allowing us to use $\psi_c = 0.7$.

For beam-to-column joints that are not restrained on four sides, as defined in ACI Code Section 15.2.4, a minimum amount of transverse reinforcement, as specified in ACI Code Section 15.4.2, must be provided. That minimum area is:

$$A_{v,\,\min} = \frac{0.75\sqrt{f'_c}\,b_w s}{f_y}, \text{ but } \geq \frac{50\,b_w s}{f_y}$$

For the given concrete strength, the 50 psi in the second expression will govern. Also, because we have selected No. 3 ties for use in the column, A_v is equal to $2 \times 0.11\text{ in.}^2 = 0.22\text{ in.}^2$, and we can solve for s,

$$s \leq \frac{0.22\text{ in.}^2 \times 60{,}000\text{ psi}}{50\text{ psi} \times 24\text{ in.}} = 11.0\text{ in.}$$

Because of the depth of the beam, two ties need to be provided within the depth of the joint. It is clear that the wide spacing of these ties, as shown in Fig. 8-17a, exceed 3 d_b of the No. 8 hooked bar. Thus, the confining reinforcement factor, ψ_r, is also 1.0. With all of the factors defined for Eq. (8-18), we can calculate ℓ_{dh} as:

$$\ell_{dh} = \left(\frac{60{,}000 \text{ psi} \times 1 \times 1 \times 1}{50 \times 1 \times \sqrt{4000 \text{ psi}}} \right) \times 1.0 \text{ in.} = 19.0 \text{ in.}$$

The hook-development length available is

$$24 \text{ in.} - \text{cover on bar extension-ties} = 24 - 1.875 = 22.1 \text{ in.}$$

Because 22.1 in. exceeds 19.0 in., the hook development length is o.k.

Check the vertical height of a standard hook on a No. 8 bar. From Fig. 8-12a, the vertical height of a 90° standard hook is $4d_b + 12d_b = 16$ in. This will fit into the joint. Therefore, anchor the four No. 8 bars into the joint, as shown in Fig. 8-17. ■

8-7 BAR CUTOFFS AND DEVELOPMENT OF BARS IN FLEXURAL MEMBERS

Why Bars Are Cutoff

In reinforced concrete, reinforcement is provided near the tensile face of beams to provide the tension component of the internal resisting couple. A continuous beam and its moment diagram are shown in Fig. 8-18. At midspan, the moments are positive, and reinforcement is required near the bottom face of the member, as shown in Fig. 8-18a. The opposite is true at the supports. For economy, some of the bars can be *terminated* or *cutoff* where they are no longer needed. The location of the cutoff points is discussed in this section.

Four major factors affect the location of bar cutoffs:

1. Bars can be cutoff where they are no longer needed to resist tensile forces or where the remaining bars are adequate to do so. The location of points where bars are no longer needed is a function of the flexural tensions resulting from the bending moments *and* the effects of shear on these tensile forces.

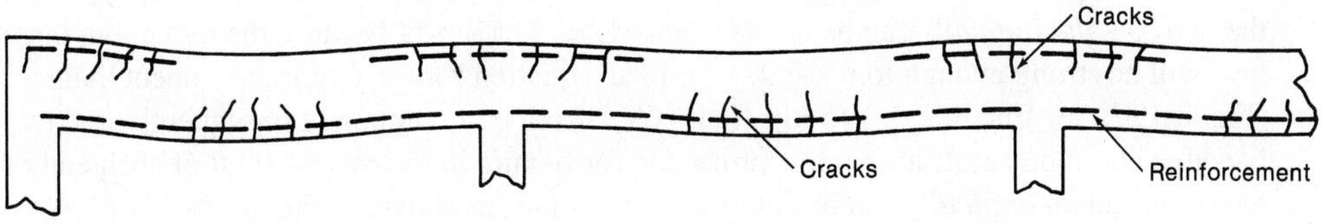

(a) Deflected shape.

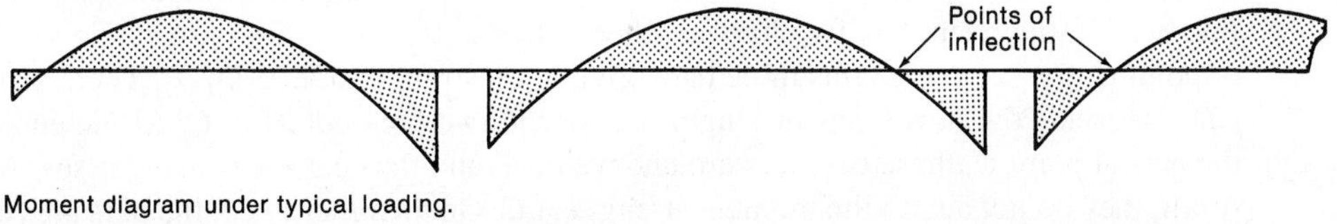

(b) Moment diagram under typical loading.

Fig. 8-18
Moments and reinforcement in a continuous beam.

2. There must be a sufficient extension of each bar, on each side of every section, to develop the force in that bar at that section. This is the basic rule governing the development of reinforcement, presented in Section 8-6 (ACI Code Section 25.4.1.1).

3. Tension bars, cutoff in a region of moderately high shear force cause a major stress concentration, which can lead to major inclined cracks at the bar cutoff.

4. Certain constructional requirements are specified in the code as good practice.

Generally speaking, bar cutoffs should be kept to a minimum to simplify design and construction, particularly in zones where the bars are stressed in tension.

In the following sections, the location of theoretical cutoff points for flexure, referred to as *flexural cutoff points*, is discussed. This is followed by a discussion of how these flexural cutoff locations must be modified to account for shear, development, and constructional requirements to get the *actual cutoff points* used in construction.

Location of Flexural Cutoff Points

The calculation of the flexural cutoff points will be illustrated with the simply supported beam shown in Fig. 8-19a. At midspan, this beam has five No. 8 reinforcing bars, shown in section in Fig. 8-19c. At points C and C', two of these bars are cutoff, leaving three No. 8 bars in the end portions of the beam, as shown in Fig. 8-19b.

The beam is loaded with a uniform factored load of 6.6 kips/ft, including its self-weight, which gives the diagram of ultimate moments, M_u, shown in Fig. 8-19d. This is referred to as the *required-moment diagram*, because, at each section, the beam must have a reduced nominal strength, ϕM_n, at least equal to M_u. The maximum required moment at midspan is ($\ell_n = 20$ ft):

$$M_u = \frac{w_u \ell_n^2}{8} = 330 \text{ kip-ft}$$

Assuming 4000-psi concrete, Grade-60 reinforcement and a tension-controlled section so $\phi = 0.9$, the moment strength, ϕM_n, of the section with five No. 8 bars is 343 kip-ft, which is adequate at midspan. At points away from midspan, the required M_u is less than 330 kip-ft, as shown by the moment diagram in Fig. 8-19d. Thus, less reinforcement (less A_s) is required at points away from midspan. This is accomplished by "cutting off" some of the bars where they are no longer needed. In the example illustrated in Fig. 8-19, it has been arbitrarily decided that two No. 8 bars will be cutoff where they are no longer needed. The remaining three No. 8 bars give a reduced nominal strength $\phi M_n = 215$ kip-ft. Thus, the two bars *theoretically* can be cutoff when $M_u \leq 215$ kip-ft, because the remaining three bars will be strong enough to resist M_u. From an equation for the required-moment diagram (Fig. 8-19d), we find $M_u = 215$ kip-ft at 4.09 ft from each support. Consequently, the two bars that are to be cutoff are no longer needed for flexure in the outer 4.09 ft of each end of the beam and *theoretically* can be cutoff at those points, as shown in Fig. 8-19e.

Figure 8-19f is a plot of the reduced nominal moment strength, ϕM_n, at each point in the beam and is referred to as a *moment-strength diagram*. At midspan (point E in Fig. 8-19e), the beam has five bars and hence has a strength of 343 kip-ft. To the left of point C, the beam contains three bars, giving it a strength of 215 kip-ft. The distance CD represents the development length, ℓ_d, for the two bars cutoff at C. At the ends of the bars at point C, these two bars are undeveloped and thus cannot resist stresses. As a result, they do not add to the moment strength at C. On the other hand, the bars are fully developed at D, and in the region from D to D' they could be stressed to f_y if required.

The three bars that extend into the supports are cutoff at points A and A'. At their ends these bars are undeveloped, and as a result, the moment strength is $\phi M_n = 0$ at

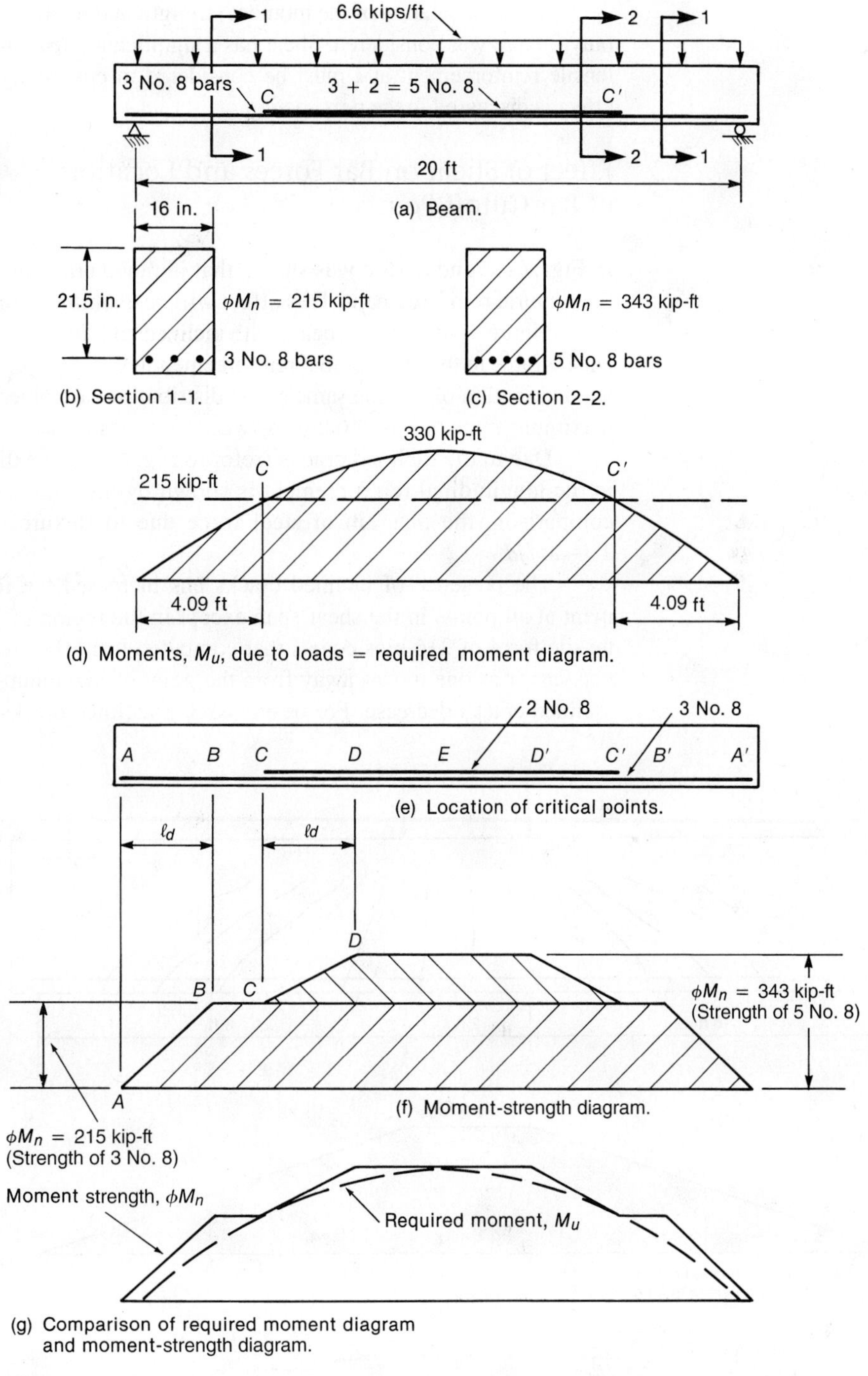

Fig. 8-19
Required moment and moment strength.

A and A'. At points B and B', the bars are fully developed and the moment strength is $\phi M_n = 215$ kip-ft.

In Fig. 8-19g, the moment-strength diagram from Fig. 8-19f and the required moment diagram from Fig. 8-19d are superimposed. Because the moment strength is greater than or equal to the required moment at all points, the beam has adequate strength *for flexure, neglecting the effects of shear.*

In the calculation of the moment strength and required moment diagrams in Fig. 8-19, only flexure was considered. Shear has a significant effect on the stresses in the longitudinal tensile reinforcement and must be considered in computing the actual cutoff points. This effect is discussed in the next section.

Effect of Shear on Bar Forces and Location of Bar Cutoff Points

In Figs. 6-12 and 6-16 it was shown that inclined cracking increased the tension force in the flexural reinforcement. This effect will be examined more deeply later in this chapter.

Figure 8-20a shows a beam with inclined and flexural cracks. From flexure theory, the tensile force in the longitudinal reinforcement is $T = M/jd$. If jd is assumed to be constant, the distribution of T is the same as the distribution of moment, as shown in Fig. 8-20b. The maximum value of T is 216 kips between the loads.

Due to the inclined cracks (refer to Fig. 6-12), the distribution of the tensile force in the longitudinal reinforcement is shown by the solid stepped line in Fig. 8-20c. For comparison, the diagram of steel force due to flexure is shown by the line labeled $T = M/jd$.

The presence of inclined cracks has increased the force in the tension reinforcement at all points in the shear span except in the region of maximum moment, where the tensile force of 216 kips equals that computed from flexure. The increase in tensile force gets larger as one moves away from the point of maximum moment, and the slope of the inclined cracks decrease. For beams having inclined cracks at an angle θ, the increase in

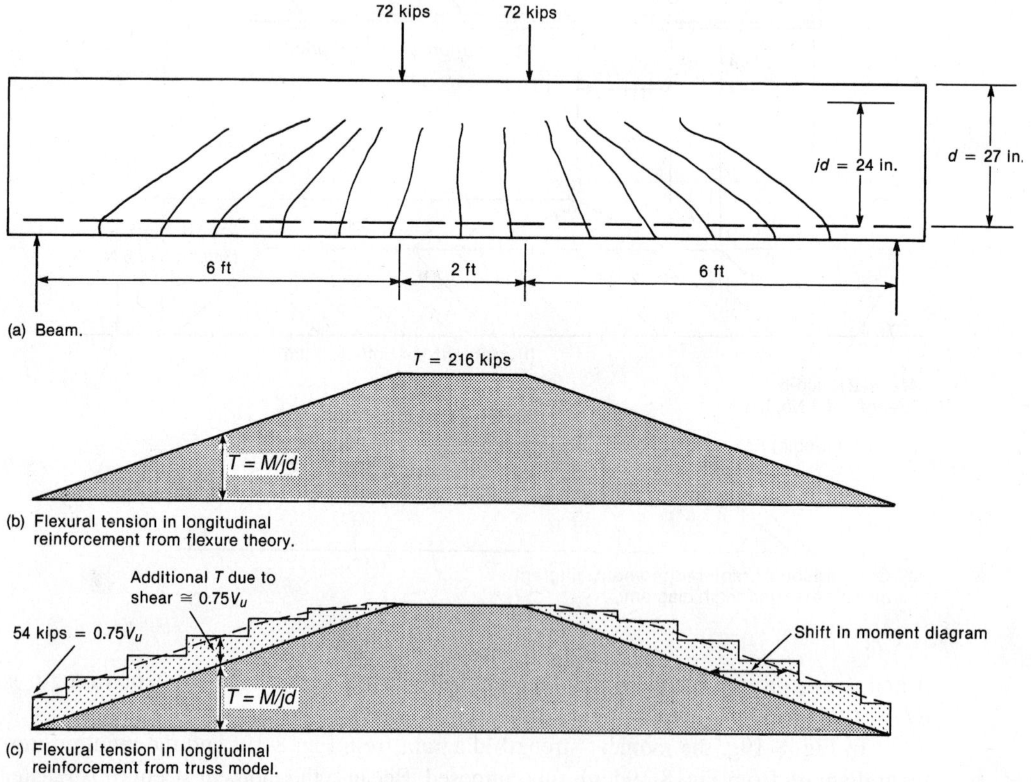

Fig. 8-20
Tension in longitudinal reinforcement.

the tensile force is approximately $V_u/(2 \tan \theta)$, corresponding to a shift of $jd/(2 \tan \theta)$ in the moment diagram.

The ACI Code does not explicitly treat the effect of shear on the tensile force. Instead, ACI Code Section 9.7.3.3 arbitrarily requires that longitudinal tension bars be extended a minimum distance equal to the greater of d or 12 bar diameters past the theoretical cutoff point for flexure. This accounts for the shift in the moment diagram due to shear, plus

> contingencies arising from unexpected loads, yielding of supports, shifting of points of inflection or other lack of agreement with assumed conditions governing the design of elastic structures [8-12] . . . .

In the beam in Fig. 8-20, there is a tensile force of approximately $0.75V_u$ in the tensile reinforcement at the face of the support. If the shear stresses are large enough to cause significant inclined cracking (say, greater than $v_u = 4\sqrt{f'_c}$), it is good practice to anchor these bars for this force. The actual force depends on the angle θ, but $0.75V_u$ is a reasonable value. This is especially important for short, deep beams, as pointed out in ACI Commentary Section R9.9.4.4 and illustrated in Chapter 17 of this book.

Development of Bars at Points of Maximum Bar Force

For reinforcement and concrete to act together, each bar must have adequate embedment on both sides of each section to develop the force in the bar at that section. In beams, this is critical at

1. Points of maximum positive and negative moment, which are points of maximum bar stress.

2. Points where reinforcing bars adjacent to the bar under consideration are cutoff or bent (ACI Code Section 9.7.3.2).

Thus, bars must extend at least a development length, ℓ_d, each way from such points or be anchored with hooks or mechanical anchorages.

It is clear why this applies at points of maximum bar stress, such as point E in Fig. 8-19e, but the situation of bar cutoffs needs more explanation. In Fig. 8-19, the selection of bar cutoffs for flexure alone was discussed. To account for bar forces resulting from shear effects, cutoff bars are then extended away from the point of maximum moment a distance d or 12 bar diameters past the flexural cutoff point. This is equivalent to using the modified bending moment, M'_u, shown in dashed lines in Fig. 8-21a, to select the cutoffs. The beam in Fig. 8-19 required five bars at midspan. If all five bars extended the full length of the beam, the bar-stress diagram would resemble the modified-moment diagram as shown in Fig. 8-21b. Now, two bars will be cutoff at new points C and C', where the moment $M'_u = 215$ kip-ft, which is equal to the reduced nominal moment strength, ϕM_n, of the cross section with three No. 8 bars (Fig. 8-21c). The stresses in the cutoff bars and in the three full-length bars are plotted in Figs. 8-21d and 8-21e, respectively. Points D and D' are located at a development length, ℓ_d, away from the ends of the cutoff bars. By this point in the beam, the cutoff bars are fully effective. As a result, all five bars act to resist the applied moments between D and D', and the stresses in both sets of bars are the same (Figs. 8-21d and 8-21e). Between D and C, the stress in the cutoff bars reduces to zero, while the stress in the remaining three bars increases. At point C, the stress in the remaining three bars reaches the yield strength, f_y, as assumed in selecting the *theoretical* cutoff points. For the bars to reach their yield strength at point C, the distance A–C must not be less than the development length, ℓ_d. If A–C is less than ℓ_d,

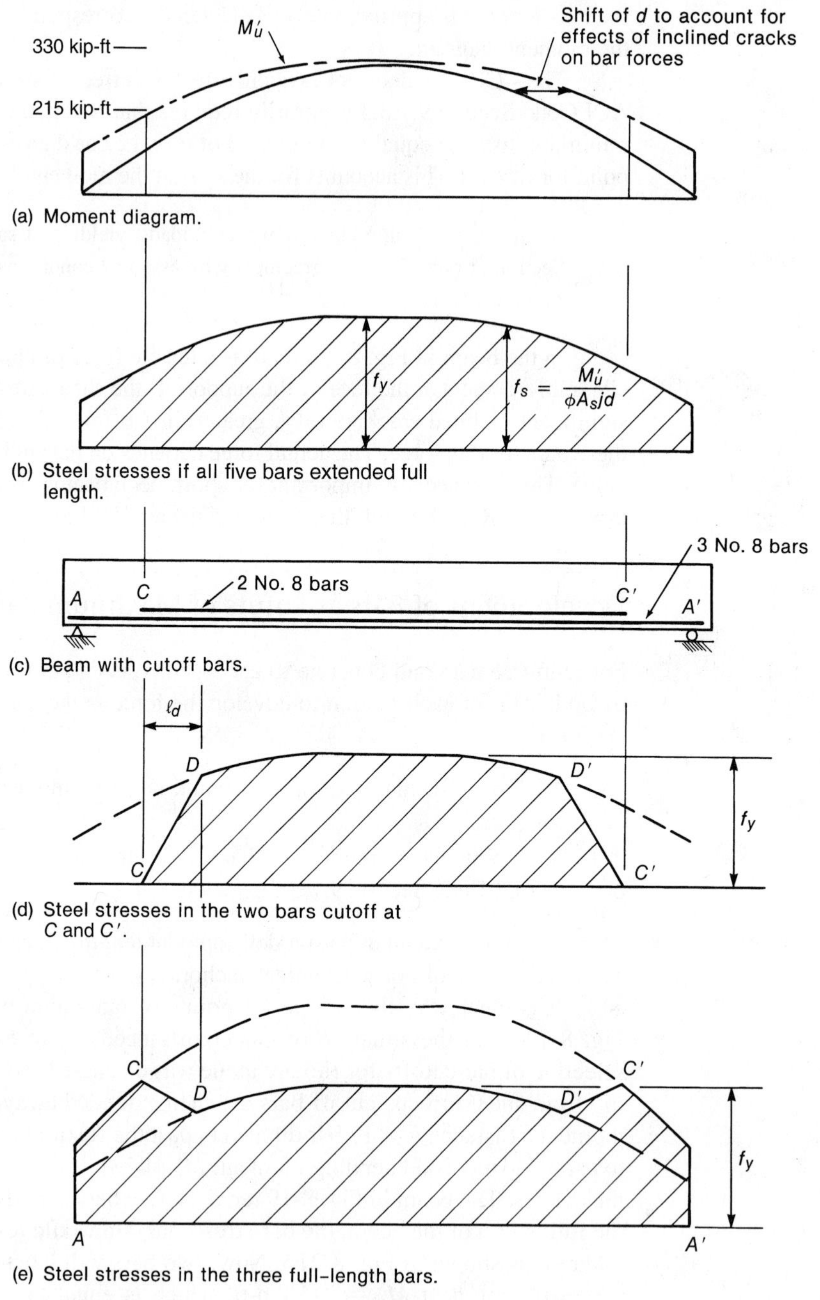

Fig. 8-21
Steel stresses in vicinity of bar cutoff points.

the required anchorage can be obtained by either hooking the three bars at *A*, or by using smaller bars, or by extending all five bars into the support.

Development of Bars in Positive-Moment Regions

Figure 8-22 shows a uniformly loaded, simple beam and its bending-moment diagram. As a first trial, the designer has selected two No. 14 bars as reinforcement. These run the full length of the beam and are enclosed in minimum stirrups. The development length of a

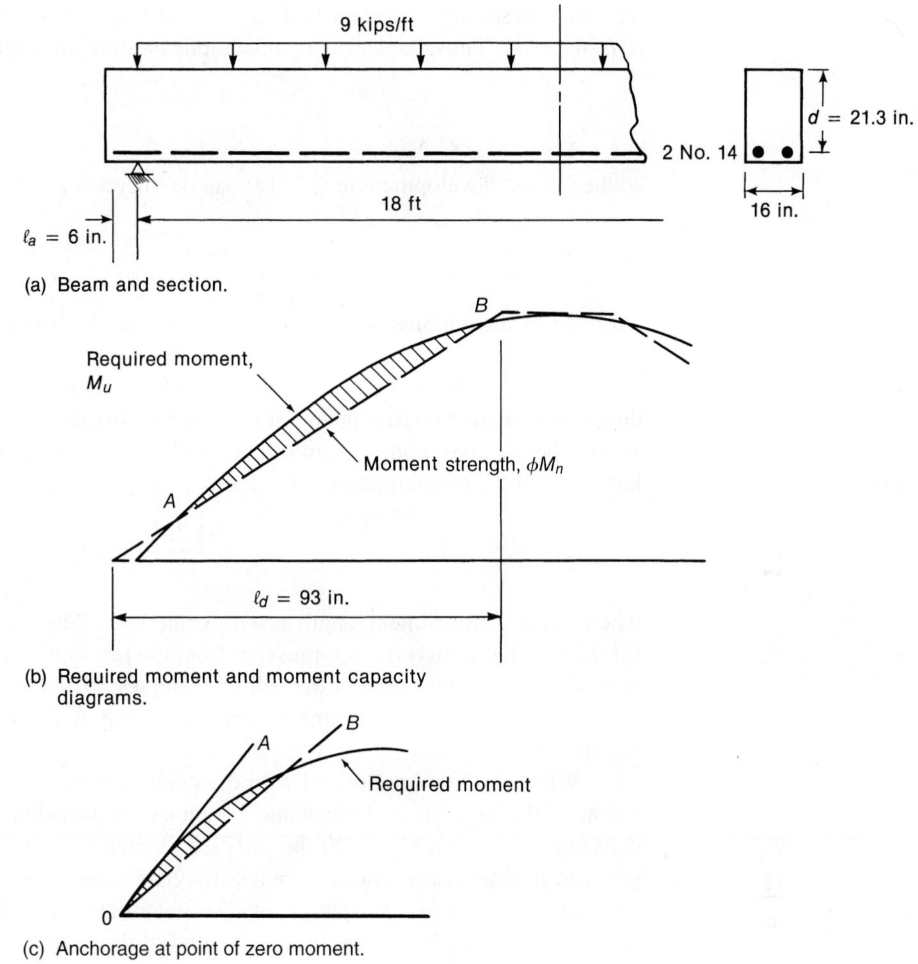

Fig. 8-22
Anchorage of positive-moment reinforcement.

No. 14 Grade-60 bar in 3000-psi concrete is 93 in. The point of maximum bar stress is at midspan, and because the bars extend 9 ft 6 in. = 114 in. each way from midspan, they are developed at midspan.

Because the bending-moment diagram for a uniformly loaded beam is a parabola, it is possible for the bar stress to be developed at midspan but not be developed at, for example, the quarter points of the span, where the moment is three-fourths of the maximum. This is illustrated in Fig. 8-22b, where the moment strength and the required-moment diagrams are compared. The moment strength is assumed to increase linearly, from zero at the ends of the bars to $\phi M_n = 363$ kip-ft at a distance $\ell_d = 93$ in. from the ends of the bars. Between points A and B, the required moment exceeds the moment strength. Stated in a different way, the bar stresses required at points between A and B are larger than those that can be developed in the bar.

Ignoring the extension of the bar into the support for simplicity, it can be seen from Fig. 8-22c that the slope of the rising portion of the moment-strength diagram cannot be less than that indicated by line O–A. If the moment-strength diagram had the slope O–B, the bars would have insufficient development for the required stresses in the shaded region of Fig. 8-22c. Thus, the slope of the moment-strength diagram, $d(\phi M_n)/dx$, cannot be less than that of the tangent to the required-moment diagram, dM_u/dx, at $x = 0$. The slope of the moment-strength diagram is $\phi M_n/\ell_d$. The slope of the required-moment diagram is $dM_u/dx = V_u$. Thus, the least slope the moment-strength diagram can have is

$$\frac{\phi M_n}{\ell_d} = V_u$$

so the longest development length that can be tolerated is

$$\ell_d = \frac{\phi M_n}{V_u}$$

where M_n is the nominal moment strength based on the bars in the beam at 0 and V_u is the shear at 0.

ACI Code Section 9.7.3.8.3 requires that, *at simple supports* where the reaction induces compressive confining stresses in the bars (as would be the case in Fig. 8-22), the size of the positive-moment reinforcement should be small enough that the development length, ℓ_d, satisfies the relation

$$\ell_d \leq \frac{1.3M_n}{V_u} + \ell_a \tag{8-20}$$

where ℓ_a is the embedment length past the centerline of the support. The factor 1.3 accounts for the fact that transverse compression from the reaction force tends to increase the bond strength by offsetting some of the splitting stresses. When the beam is supported in such a way that there are no bearing stresses above the support, the factor 1.3 becomes 1.0 [giving Eq. (8-21)].

When the bars are hooked with the point of tangency of the hook outside the centerline of the support, or if mechanical anchors are provided, ACI Code Section 9.7.3.8.3 does not require that Eq. (8-20) be satisfied. It should be noted that hooked bars at a support can lead to bearing failures unless they are carefully detailed. Figure 8-23a shows, to scale, a support of a simple beam. The potential crack illustrated does not encounter any reinforcement. In precast beams, the end of the beam is often reinforced as shown in Fig. 8-23b.

At positive-moment *points of inflection* (points where the positive-moment envelope passes through zero), a similar situation exists, except that there are no transverse bearing stresses. Here, ACI Code Section 9.7.3.8.3 requires that the diameter of the positive-moment reinforcement should be small enough that the development length, ℓ_d, satisfies

$$\ell_d \leq \frac{M_n}{V_u} + \ell_a \tag{8-21}$$

where ℓ_a is the longer of the effective depth, d, or 12 bar diameters, but not more than the actual embedment of the bar into the negative-moment region past the point of inflection.

Equations (8-20) and (8-21) are written in terms of M_n rather than ϕM_n, because the development length equations were derived on the basis of developing f_y in the bars, not ϕf_y. It should be noted that the derivation of Eqs. (8-20) and (8-21) did not consider the shift in the bar force due to shear. As a result, these equations do not provide a sufficient check of the end anchorage of bars at simple supports of short, deep beams or beams supporting shear forces larger than about $V_u = 4\sqrt{f'_c}b_w d$. In such cases, the bars should be anchored in the support for a force of at least $V_u/2$, and preferably $0.75V_u$, as discussed in the preceding section.

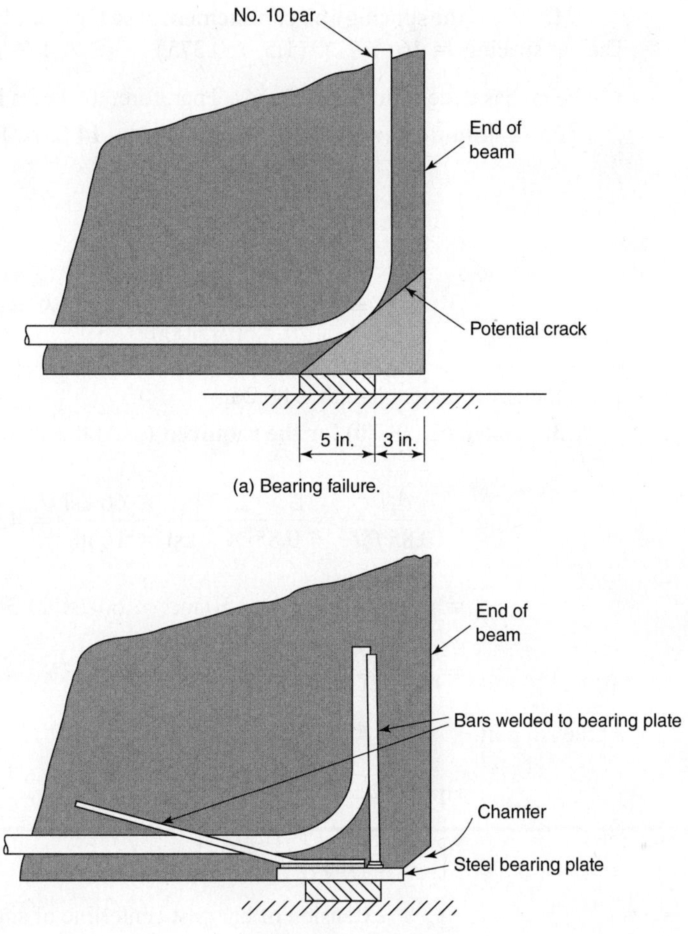

Fig. 8-23
Simple beam supports.

Equations (8-20) and (8-21) are not applied in negative-moment regions, because the shape of the moment diagram is concave downward such that the only critical point for anchorage is the point of maximum bar stress.

EXAMPLE 8-3 Checking the Development of a Bar in a Positive-Moment Region

The beam in Fig. 8-22 has two No. 14 bars and No. 3 U stirrups at 10 in. o.c. The normal-weight concrete has a compressive strength, $f'_c = 4000$ psi and the reinforcing steel has a yield strength, $f_y = 60{,}000$ psi. The beam supports a total factored load of 9.0 kips/ft. Check whether ACI Code Section 9.7.3.8.3 is satisfied. The relevant equation is

$$\ell_d \leq 1.3\frac{M_n}{V_u} + \ell_a \tag{8-20}$$

1. Find the spacing and confinement case for No. 14 bars (diameter = 1.69 in.). The bar spacing $= 16 - 2 \times (1.5 + 0.375) - (2 \times 1.69) = 8.87$ in. $> d_b$.

The beam has code-minimum stirrups. Therefore, the beam is Case 1 in Table 8-1.

2. Compute development length for No. 14 bars. From Eq. (8-12),

$$\ell_d = \frac{f_y \psi_t \psi_e}{20\lambda\sqrt{f'_c}} d_b$$

$$= \frac{60{,}000 \times 1.0 \times 1.0}{20 \times 1\sqrt{4000}} \times 1.69 = 80.2 \text{ in.}$$

Thus, the development length is 80.2 in.

3. Solve Eq. (8-20) for the required ℓ_d. At the support, there are two No. 14 bars:

$$a = \frac{A_s f_y}{0.85 f'_c b} = \frac{2 \times 2.25 \text{ in.}^2 \times 60 \text{ ksi}}{0.85 \times 4 \text{ ksi} \times 16 \text{ in.}} = 4.96 \text{ in.}$$

$$M_n = A_s f_y\left(d - \frac{a}{2}\right) = 4.50 \text{ in.}^2 \times 60 \text{ ksi}(21.3 \text{ in.} - 2.48 \text{ in.})$$

$$= 5080 \text{ k-in.}$$

At the support,

$$V_u = \frac{w_u \ell}{2} = \frac{9.0 \times 18}{2}$$

$$= 81 \text{ kips}$$

$$\ell_a = \text{extension of bar past centerline of support} = 6 \text{ in.}$$

Thus,

$$1.3\frac{M_n}{V_u} + \ell_a = \frac{1.3 \times 5080 \text{ kip-in.}}{81 \text{ kips}} + 6 \text{ in.}$$

$$= 87.5 \text{ in.}$$

In this case, ℓ_d is less than 87.5 in., so we could use two No. 14 bars. However, No. 14 bars are not available from all rebar suppliers. So, investigate the use of six No. 8 bars.

4. Find the spacing and confinement case for six No. 8 bars. Compute the spacing, which works out to 1.20 in. $= 1.2d_b$, and use Table A-5 to find that the minimum web width for six No. 8 bars is 15.5 in. Because 16 in. exceeds 15.5 in., the bar spacing exceeds d_b, and the beam has code-minimum stirrups, this is Case 1.

5. Compute development length for No. 8 bars. From Eq. (8-12),

$$\ell_d = \frac{f_y \psi_t \psi_e}{20\lambda\sqrt{f'_c}} d_b = \frac{60{,}000 \times 1.0 \times 1.0}{20 \times 1\sqrt{4000}} \times 1.0$$

$$= 47.4 \text{ in.}$$

Thus, $\ell_d = 47.4$ in.

6. Solve Eq. (8-20) for the required ℓ_d. $M_n = 5370$ kip-in. (for $d = 21.5$ in.); thus,

$$1.3\frac{M_n}{V_u} + \ell_a = \frac{1.3 \times 5370}{81} + 6 = 92.2 \text{ in.}$$

Because $\ell_d < 92.2$ in., this is acceptable. **Use six No. 8 bars.** ■

Effect of Discontinuities at Bar Cutoff Points in Flexural Tension Zones

The bar-stress diagrams in Figs. 8-21d and 8-21e suggest that a severe discontinuity in bar stresses exists in the vicinity of points where bars are cutoff in a region of flexural tension. One effect of this discontinuity is a reduction in the shear required to cause inclined cracking in this vicinity [8-13]. The resulting inclined crack starts at, or near, the end of the cutoff bars. ACI Code Section 9.7.3.5 prohibits bar cutoffs in a zone of flexural tension unless *one* of the following is satisfied:

1. The factored shear, V_u, at the cutoff point is not greater than:

$$\frac{2}{3}\phi(V_c + V_s) \tag{8-22}$$

2. Extra stirrups are provided over a length of $0.75d$, starting at the end of the cutoff bar and extending along it. The maximum spacing of the extra stirrups is $s = d/8\beta_b$, where β_b is the ratio of the area of the bars that are cutoff to the area of the bars immediately before the cutoff. The area of the stirrups, A_v, is not to be less than $60b_w s/f_{yt}$.

3. For No. 11 bars and smaller, the continuing reinforcement provides twice the area required for flexure at the cutoff point, and V_u is not greater than $0.75\phi(V_c + V_s)$.

Because only one of these need to be satisfied, the author recommends that only the first requirement be considered. To avoid this check, designers frequently will extend all bars into the supports in simple beams or past the points of inflection in continuous beams.

8-8 REINFORCEMENT CONTINUITY AND STRUCTURAL INTEGRITY REQUIREMENTS

For many years, there have been requirements for continuity of beam longitudinal reinforcement in the ACI Code. More recently, requirements for structural integrity were added for cast-in-place reinforced concrete construction, for precast construction, and for prestressed concrete slabs. The primary purpose for both the continuity and structural-integrity reinforcement requirements is to *tie* the structural elements together and prevent localized damage from spreading progressively to other parts of the structure. However, because of the limited amount of calculations required to select and detail this reinforcement, structures satisfying these requirements cannot be said to have been designed to resist *progressive collapse*.

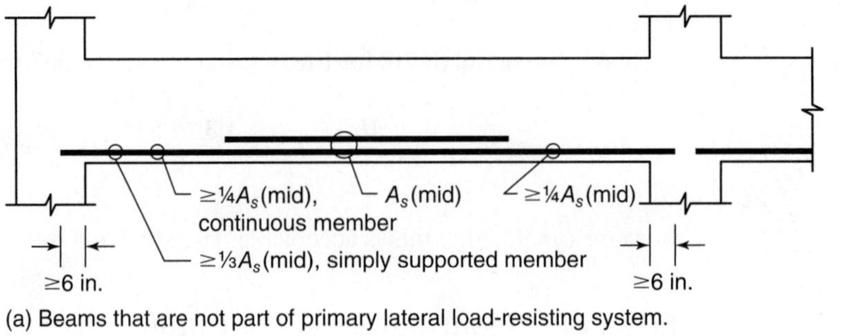

Fig. 8-24
Continuity requirements for positive-moment reinforcement in continuous beams (use at least two bars every where reinforcement is required).

Continuity Reinforcement

Requirements for continuity reinforcement in continuous beams are given in ACI Code Sections 9.7.3.8.1 and 9.7.3.8.2 for positive-moment (bottom) reinforcement and in ACI Code Sections 9.7.3.1 and 9.7.3.8.5 for negative-moment (top) reinforcement. These requirements are summarized in Fig. 8-24 for positive-moment reinforcement and in Fig. 8-25 for negative-moment reinforcement.

ACI Code Sections 9.7.3.8.1 and 9.7.3.8.2 require that at least one-third of the positive-moment reinforcement used at midspan for simply supported members and at least one-fourth of the positive-moment reinforcement used at midspan for continuous members shall be continued at least 6 in. into the supporting member (Fig. 8-24a). If the beam under consideration is part of the primary lateral load-resisting system, ACI Code Section 9.7.3.8.2 also requires that the bottom reinforcement must be continuous through interior supports and fully anchored at exterior supports (Fig. 8-24b).

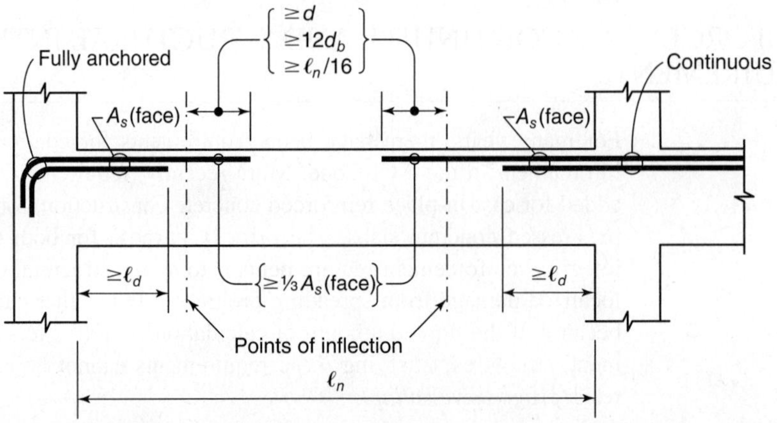

Fig. 8-25
Continuity requirements for negative-moment reinforcement in continuous beams.

ACI Code Section 9.7.3.1 requires that negative-moment reinforcement must be fully anchored at all supports (Fig. 8-25) and must extend the development length, ℓ_d, into the span before being cutoff. Also, ACI Code Section 9.7.3.8.4 requires that at least one-third of the negative-moment reinforcement provided at the face of the support shall be extended beyond the point of inflection a distance greater than or equal to the largest of d, $12d_b$, and $\ell_n/16$. Theoretically, no top steel should be required beyond the point of inflection, where the beam moment changes from negative to positive. The minimum extension given by the ACI Code accounts for possible shifts in the theoretical point of inflection due to changes in loading and for the effect of shear on longitudinal steel requirements, as discussed previously for positive-moment reinforcement.

Structural-Integrity Reinforcement

Requirements for structural-integrity reinforcement in continuous floor members first appeared in the 1989 edition of the ACI Code. These requirements were clarified and strengthened in the 2002 and 2008 editions of the ACI Code. The structural-integrity requirements are supplemental to the continuity requirements discussed previously and were added to better tie the structural members together in a floor system and to provide some resistance to progressive collapse. Similar structural-integrity requirements for reinforced concrete (nonprestressed) continuous slabs, for precast construction, and for prestressed two-way slab systems will not be discussed here.

The structural-integrity requirements for continuous floor members can be divided into requirements for joists, perimeter beams, and interior beams framing into columns. For joist construction, as defined in ACI Code Sections 9.8.1.1 through 9.8.1.4, ACI Code Section 9.8.1.6 requires that at least one bottom bar shall be continuous over all spans and through interior supports and shall be anchored to develop f_y at the face of exterior supports. Continuity of the bar shall be achieved with either a Class B tension lap splice or a mechanical or welded splice satisfying ACI Code Section 25.5.7.1. Class B lap splices are defined in ACI Code Section 25.5.2.1 as having a length of $1.3\ell_d$ (but not less than 12 in.). The value for the development length, ℓ_d, is to be determined in accordance with ACI Code Section 25.4.2, which has been given previously in Table 8-1 and Eq. (8-10). When determining ℓ_d, the 12 in. minimum does not apply, and the reduction for excessive reinforcement cannot be applied.

ACI Code Sections 9.7.7.1, 9.7.7.3, and 9.7.7.4 state that perimeter beams must have continuous top and bottom reinforcement that either passes through or is anchored in the column core, which is defined as the region of the concrete bounded by the column longitudinal reinforcement (Fig. 8-26). The continuous top reinforcement shall consist of at least one-sixth of the negative-moment (top) reinforcement required at the face of the support, but shall not be less than two bars (Fig. 8-27). The continuous bottom reinforcement shall consist of at least one-fourth of the positive-moment (bottom) reinforcement required at midspan, but not less than two bars (Fig. 8-27). At noncontinuous supports (corners), all of these bars must be anchored to develop f_y at the face of the support. Also, all of the continuous longitudinal bars must be enclosed by closed transverse reinforcement (ACI Code Section 25.7.1.6), as specified for torsional or integrity transverse reinforcement, and placed over the full clear span at a spacing not exceeding $d/2$. As before, reinforcement continuity can be achieved through either the use of Class B tension lap splices or a mechanical or welded splice.

For interior beams framing between columns, ACI Code Sections 9.7.7.2 through 9.7.7.6 define two ways to satisfy the structural-integrity requirements for continuous

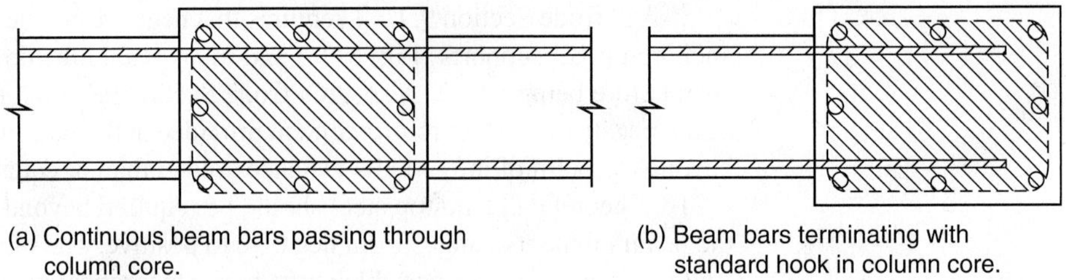

Fig. 8-26
Structural-integrity reinforcement passing through or anchored in column core.

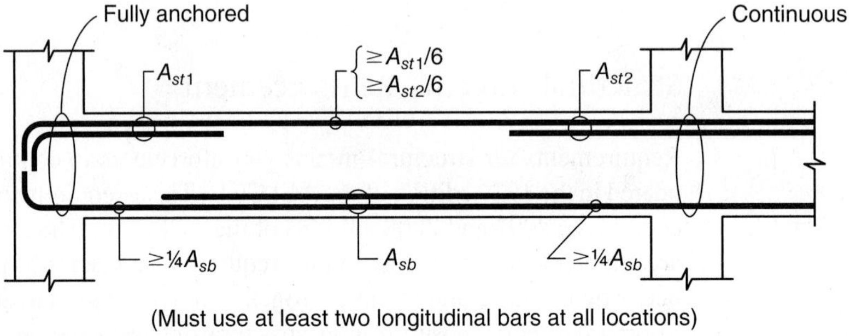

Fig. 8-27
Requirements for longitudinal structural-integrity reinforcement in perimeter beams. (*Note:* required closed transverse reinforcement not shown.)

longitudinal reinforcement. If closed transverse reinforcement is not present, then structural integrity must be achieved by continuous bottom reinforcement similar to that required for perimeter beams (Fig. 8-28a). As before, this reinforcement must pass through or be fully anchored in the column core, and reinforcement continuity can be achieved through either a Class B tension lap splice or a mechanical or welded splice. For interior beams that are not part of the primary system for resisting lateral loads, the bottom reinforcement does not need to be continuous through interior supports or fully anchored at exterior supports, and structural integrity can be achieved by a combination of bottom and top steel that is enclosed by closed transverse reinforcement (Fig. 8-28b). The top steel must satisfy the requirements of ACI Code Sections 9.7.3.1 and 9.7.3.8.4 and must be continuous through the column core of interior supports or fully anchored in the column core of exterior supports. The bottom steel must satisfy the requirements given in ACI Code Sections 9.7.3.8.1 and 9.7.3.8.2. The closed transverse reinforcement (not shown in Fig. 8-28b) must satisfy ACI Code Section 25.7.1.6 and must be provided over the full clear span at a spacing not exceeding $d/2$. How continuity and structural-integrity requirements affect the selection of cutoff points and longitudinal reinforcement detailing are given in the following examples.

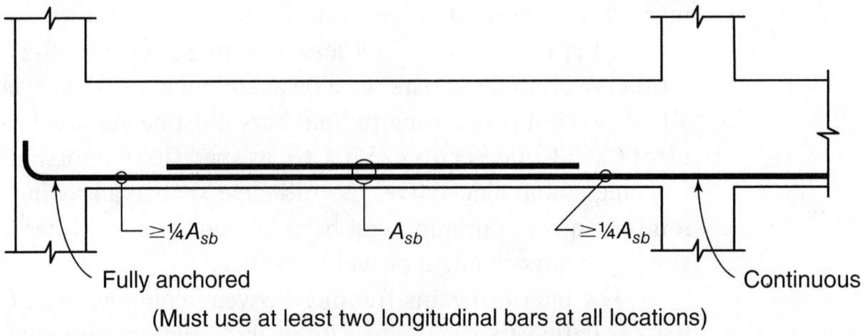

(a) Interior beam without closed transverse reinforcement.

Fig. 8-28
Requirements for longitudinal structural-integrity reinforcement for interior beams framing into columns.

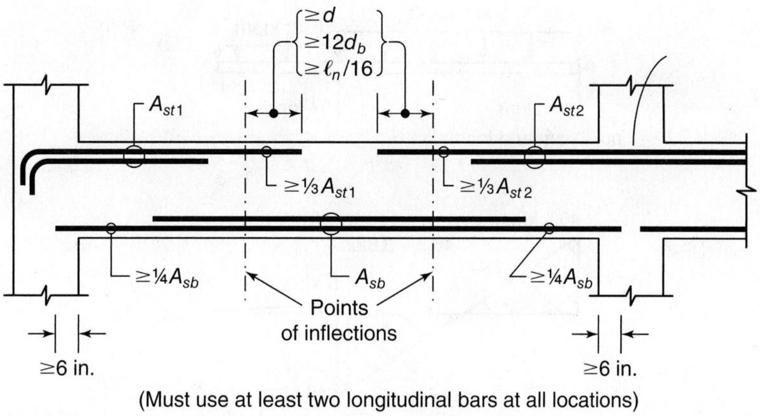

Fig. 8-28 (Continued)

(b) Interior beam with closed transverse reinforcement over total clear span at spacing less than or equal to $d/2$ (transverse reinforcement is not shown).

EXAMPLE 8-4 Calculation of Bar Cutoff Points from Equations of Moment Diagrams

The beam shown in Fig. 8-29a is constructed of normal-weight, 4000-psi concrete and Grade-60 reinforcement. It supports a factored dead load of 0.42 kip/ft and a factored live load of 3.4 kips/ft. The cross sections at the points of maximum positive and negative moment, as given in Figs. 8-30c and 8-32c, are shown in Fig. 8-29b.

1. Locate flexural cutoffs for positive-moment reinforcement. The positive moment in span A–B is governed by the loading case in Fig. 8-30a. From a free-body analysis of a part of span A–B (Fig. 8-30d), the equation for M_u at a distance x from A is

$$M_u = 46.5x - \frac{3.82x^2}{2} \text{ kip-ft}$$

At midspan, the beam has two No. 9 plus two No. 8 bars. The two No. 8 bars will be cutoff. The capacity of the remaining bars is calculated as follows:

$$a = \frac{A_s f_y}{0.85 f'_c b} = \frac{2 \times 1.0 \text{ in.}^2 \times 60 \text{ ksi}}{0.85 \times 4 \text{ ksi} \times 12 \text{ in.}} = 2.94 \text{ in.}$$

$$\phi M_n = \phi A_s f_y (d - a/2) = 0.9 \times 2.0 \text{ in.}^2 \times 60 \text{ ksi } (21.5 \text{ in.} - 1.47 \text{ in.})$$

$$= 2160 \text{ kip-in.} = 180 \text{ k-ft}$$

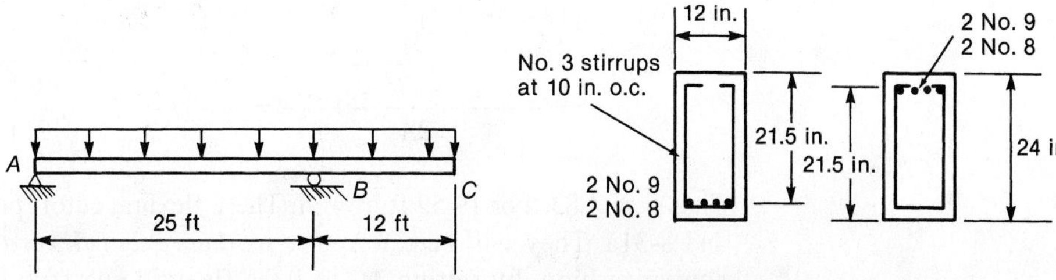

Fig. 8-29
Beam—Example 8-4.

(a) Beam.

(b) Sections of maximum positive and negative moment.

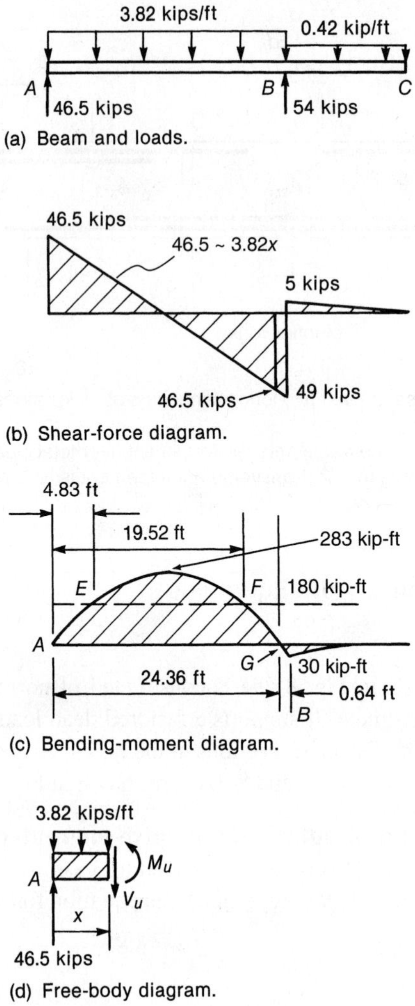

Fig. 8-30
Calculation of flexural cutoff points for a statically determinate beam with an overhang, loaded to produce maximum positive moment—Example 8-4.

Therefore, theoretical flexural cutoff points occur where $M_u = 180$ kip-ft. Setting $M_u = 180$ kip-ft, the equation for M_u can be rearranged to

$$1.91x^2 - 46.5x + 180 = 0$$

This is a quadratic equation of the form

$$Ax^2 + Bx + C = 0$$

and so has the solution

$$x = \frac{-B \pm \sqrt{B^2 - 4AC}}{2A} = \frac{46.5 \pm \sqrt{(-46.5)^2 - 4(1.91)(180)}}{2(1.91)}$$

Thus $x = 4.83$ ft or 19.52 ft from A. These flexural cutoff points are shown in Figs. 8-30c and 8-31a. They will be referred to as *theoretical flexural cutoff points E* and *F*. In a similar fashion, by setting $M_u = 0$ the flexural cutoff point G is found to be 24.36 ft from A or 0.64 ft from B.

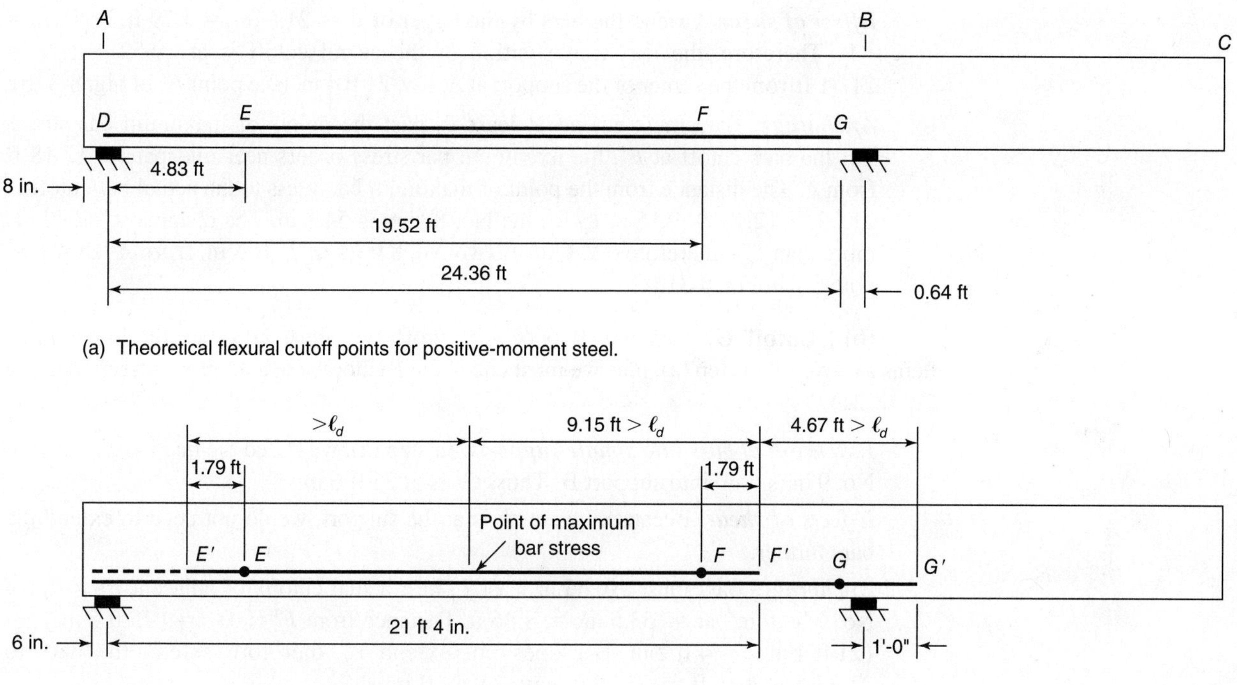

Fig. 8-31
Location of positive-moment cutoff points—Example 8-4.

2. Compute the development lengths for the bottom bars.

$$\text{Bar spacing} = \frac{12 - 2(1.5 + 0.375) - 2 \times 1.128 - 2 \times 1.0}{3} = 1.33 \text{ in.}$$

Because the bar spacing exceeds d_b for both the No. 8 and 9 bars, and because the beam has minimum stirrups, the bars satisfy Case 1 in Table 8-1. From Eq. (8-12),

$$\frac{\ell_d}{d_b} = \frac{f_y \psi_t \psi_e}{20\lambda\sqrt{f'_c}} = \frac{60{,}000 \times 1.0 \times 1.0 \times 1.0}{20 \times 1\sqrt{4000}} = 47.4$$

Thus, for the No. 8 bars, $\ell_d = 47.4 \times 1.0 = 47.4$ in., and for the No. 9 bars, $\ell_d = 47.4 \times 1.128 = 53.5$ in.

3. Locate actual cutoff points for positive-moment reinforcement. The actual cutoff points are determined from the theoretical flexural cutoff points using rules stated earlier. Because the location of cutoffs *G* and *D* are affected by the locations of cutoffs *E* and *F*, the latter are established first, starting with *F*. Because the beam is simply supported it is not included in the ACI Code listing of members that are susceptible to actions requiring structural integrity.

(a) Cutoff *F*. Two No. 8 bars are cutoff. They must satisfy rules for anchorage, extension of bars into the supports and effect of shear on moment diagrams.

Extension of bars into the supports. At least one-third of the positive-moment reinforcement, but not less than two bars, must extend at least 6 in. into the supports. We shall extend two No. 9 bars into each of the supports *A* and *B*.

Effect of shear. Extend the bars by the larger of $d = 21.5$ in. $= 1.79$ ft, or $12d_b =$ 1 ft. Therefore, the first trial position of the actual cutoff is at $19.52 + 1.79 =$ 21.31 ftfrom the center of the support at A, say, 21 ft 4 in. (see point F' in Fig. 8-31b).

Anchorage. Bars must extend at least ℓ_d past the points of maximum bar stress. For the bars cutoff at F', the maximum bar stress occurs near midspan, at 12.18 ft from A. The distance from the point of maximum bar stress to the actual bar cutoff is $21.33 - 12.18 = 9.15$ ft. ℓ_d for the No. 8 bars is 54.8 in. The distance available is more than ℓ_d—therefore o.k. **Cutoff two No. 8 bars at 21 ft 4 in. from *A*** (shown as point F' in Fig. 8-31b).

(b) Cutoff *G*. Two No. 9 bars are cutoff; we must consider the same three items as covered in step (a), plus we must check the anchorage at a point of inflection using Eq. (8-21).

Extension of bars into simple supports. In step (a), we stated the need to extend two No. 9 bars 6 in. into support B. Thus, G' is at 25 ft 6 in.

Effects of shear. Because the cutoff is at the support, we do not need to extend the bars further.

Anchorage. Bars must extend at least ℓ_d past actual cutoffs of adjacent bars. ℓ_d for No. 9 bottom bar $= 53.5$ in. $= 4.46$ ft. Distance from F' to $G' =$ (25 ft 6 in.) $-$ (21 ft 4 in.) $=$ 4 ft 2 in. Bar does not extend ℓ_d, therefore extend the bars to 21.33 ft + 4.46 ft = 25.79 ft, say, 25 ft 10 in.

Anchorage at point of inflection. Must satisfy Eq. (8-21) at point of inflection (point where the moment is zero). Therefore, at G, $\ell_d \leq M_n/V_u + \ell_a$. The point of inflection is 0.64 ft from the support (Fig. 8-30c). At this point V_u is 46.5 kips (Fig. 8-30b) and the moment strength M_n for the bars in the beam at the point of inflection (two No. 9 bars) is

$$M_n = 180 \text{ kip-ft} \times \frac{12}{0.9} = 2400 \text{ kip-in.}$$

$\ell_a =$ larger of d (21.5 in.) or $12d_b$ (13.5 in.) but not more than the actual extension of the bar past the point of inflection $(25.83 - 24.36 = 1.47 \text{ ft} = 17.6 \text{ in.})$. Therefore, $\ell_a = 17.6$ in., and

$$\frac{M_n}{V_u} + \ell_a = \frac{2400}{46.5} + 17.6$$
$$= 69.2 \text{ in.}$$

o.k., because this length exceeds $\ell_d = 53.5$ in. **Cutoff two No. 9 bars 1 ft from *B*** (rounded from 10 in. to 1 ft and shown as point G' in Fig. 8-31b).

(c) Cutoff *E*. Two No. 8 bars are cutoff; we must check for the effect of shear and development length (anchorage).

Effects of shear, positive moment. Extend the bars $d = 1.79$ ft past the flexural cutoff point. Therefore, the actual cutoff E' is at $4.83 - 1.79 = 3.04$ ft (3 ft 1 in.) from A.

Anchorage, positive moment. The distance from the point of maximum moment to the actual cutoff exceeds $\ell_d = 54.8$ in.—therefore o.k. **Cutoff two No. 8 bars at 3 ft 1 in. from *A*** (point E' in Fig. 8-31b; note that this is changed later).

(d) Cutoff *D*. Two No. 9 bars are cutoff; we must consider extension into a support, extension beyond the cutoff point E', and development of bars at a simple support using Eq. (8-20).

Extension into simple support. This was done in step (a).

Bars must extend ℓ_d from the actual cutoff E', where $\ell_d = 53.5$ in. (No. 9 bars). The maximum possible length available is 3 ft 1 in. + 6 in. = 43 in. Because this is less than ℓ_d, we must either extend the end of the beam, hook the ends of the bars, use smaller bars, or eliminate the cutoff E'. We shall do the latter. Therefore, **extend all four bars 6 in. past support *A*.**

Development of bars at simple support. We must satisfy Eq. (8-20) at the support.

$$\ell_d \leq \frac{1.3M_n}{V_u} + \ell_a$$

$$V_u = 46.5 \text{ kips}$$

$$a = \frac{A_s f_y}{0.85 f'_c b} = \frac{3.58 \text{ in.}^2 \times 60 \text{ ksi}}{0.85 \times 4 \text{ ksi} \times 12 \text{ in.}} = 5.26 \text{ in.}$$

$$M_n = A_s f_y (d - a/2) = 3.58 \text{ in.}^2 \times 60 \text{ ksi} (21.5 \text{ in.} - 2.63 \text{ in.})$$

$$= 4050 \text{ kip-in.}$$

$$\ell_a = 6 \text{ in.}$$

$$1.3\frac{M_n}{V_n} + \ell_a = 1.3\frac{4050}{46.5} + 6 = 119 \text{ in.}$$

Because this exceeds ℓ_d, development at the simple support is satisfied. The actual cutoff points are illustrated in Fig. 8-34.

4. Locate flexural cutoffs for negative-moment reinforcement. The negative moment is governed by the loading case in Fig. 8-32a. The equations for the negative bending moments are as follows:

Between *A* and *B*, with *x* measured from *A*,

$$M_u = -5.8x - \frac{0.42x^2}{2} \text{kip-ft}$$

and between *C* and *B*, with x_1 measured from *C*,

$$M_u = \frac{-3.82x_1^2}{2} \text{kip-ft}$$

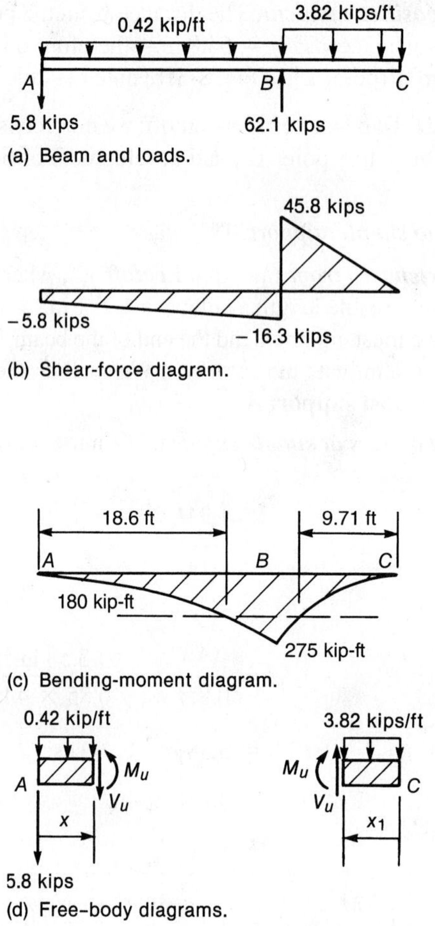

Fig. 8-32
Calculation of flexural cutoff points for negative moment—Example 8-4.

Over the support at B, the reinforcement is two No. 9 bars plus two No. 8 bars. The two No. 8 bars are no longer required when the moment is less than $\phi M_n = 180$ kip-ft (strength of the beam with two No. 9 bars).

So, between A and B,

$$-180 = -5.8x - 0.21x^2$$

$$x = -46.2 \text{ ft or } 18.6 \text{ ft from } A$$

Therefore, the theoretical flexural cutoff point for two No. 8 top bars in span A–B is at 18.6 ft from A. Finally, between B and C,

$$-180 = -1.91x_1^2$$

$$x_1 = 9.71 \text{ ft from } C$$

Therefore, the theoretical flexural cutoff point for two No. 8 top bars in span B–C is at 9.71 ft from C. The flexural cutoff points for the negative-moment steel are shown in Fig. 8-33a and are lettered H, J, K, and L.

5. Compute development lengths for the top bars. Because there is more than 12 in. of concrete below the top bars, $\psi_t = 1.3$. Thus, for the No. 8 bars, $\ell_d = 61.6$ in., and for the No. 9 bars, $\ell_d = 69.6$ in.

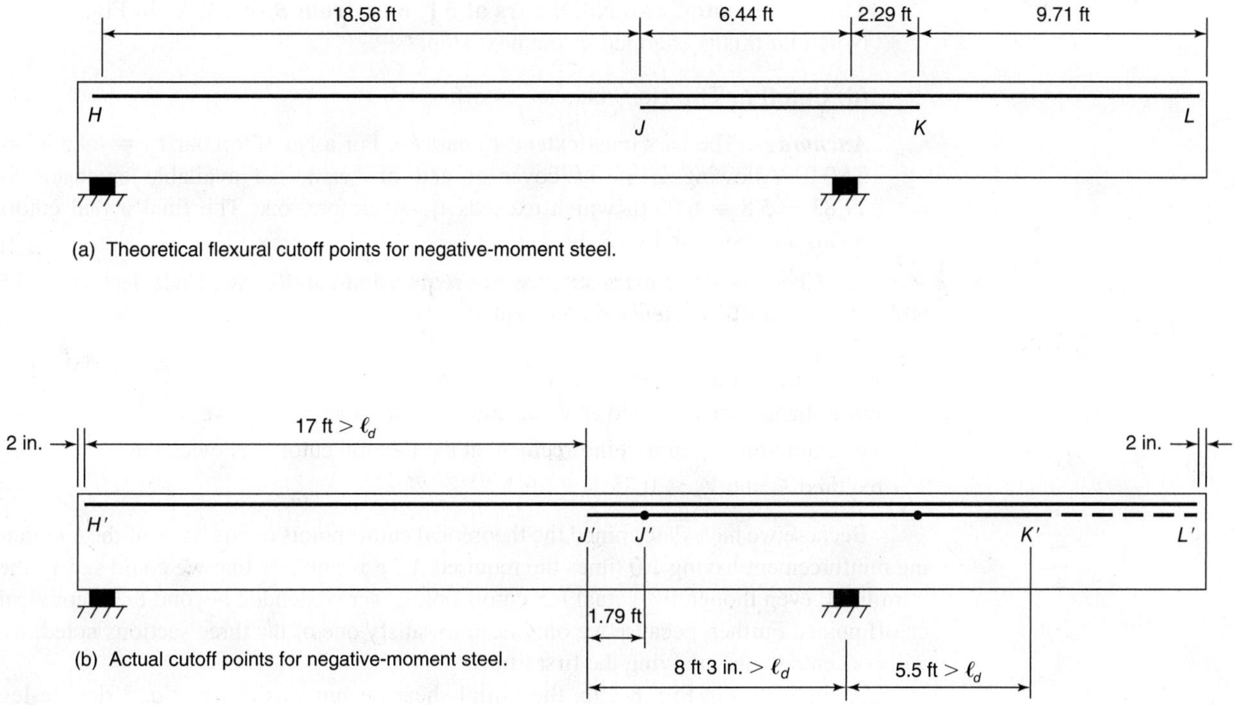

Fig. 8-33
Location of negative-moment cutoff points—Example 8-5.

6. Locate the actual cutoff points for the negative-moment reinforcement. Again, the inner cutoffs will be considered first, because their location affects the design of the outer cutoffs. The choice of actual cutoff points is illustrated in Fig. 8-33b.

(a) Cutoff *J*. Two No. 8 bars are cutoff.

Effects of shear. Extend bars by $d = 1.79$ ft past the theoretical flexural cutoff. Cutoff at $18.56 - 1.79 = 16.8$ ft from *A* and 8.2 ft from *B*, say, 8 ft 3 in.

Anchorage of negative-moment steel. The bars must extend ℓ_d from the point of maximum bar stress. For the two No. 8 top bars, the maximum bar stress is at *B*. The actual bar extension is 8.25 ft $=$ 99 in. This exceeds $\ell_d = 61.9$ in.—therefore o.k. **Cutoff two No. 8 bars at 8 ft 3 in. from *B*** (point *J'* in Fig. 8-33b).

(b) Cutoff *H*. Two No. 9 bars are cutoff.

Anchorage. Bar must extend ℓ_d past *J'*, where $\ell_d = 69.6$ in. Length available $\cong$ 17 ft—therefore o.k. **Extend two No. 9 bars to 2 in. from the end of the beam** (point *H'* in Fig. 8-33b).

(c) Cutoff *K*. Two No. 8 bars are cutoff. The theoretical flexural cutoff is at 9.71 ft from *C* (2.29 ft from *B*).

Effect of shear. Extend bars $d = 1.79$ ft. The end of the bars is at $2.29 + 1.79 =$ 4.08 ft from *B*, say, 4 ft 1 in.

Anchorage. Extend ℓ_d past *B*. ℓ_d for a No. 8 top bar $=$ 61.6 in. The extension of 4 ft 1 in. is thus not enough. Try extending the No. 8 top bars 5 ft 6 in. past *B* to point *K'*.

Therefore, cutoff two No. 8 bars at 5 ft 6 in. from *B* (point K' in Fig. 8-33b). Note that this is changed in the next step.

(d) Cutoff *L*. Two No. 9 bars are cutoff.

Anchorage. The bars must extend ℓ_d past K'. For a No. 9 top bar $\ell_d = 69.6$ in. = 5.80 ft. Allowing 2 in. of cover at end of beam, the available extension is $11.83 - 5.8 = 6.03$ ft, which exceeds ℓ_d—therefore, o.k. The final actual cutoff points are shown in Fig. 8-34.

7. Check whether extra stirrups are required at cutoffs. ACI Code Section 9.7.3.5 prohibits bar cutoffs in a tension zone, unless

V_u at actual cutoff $\leq \frac{2}{3}\phi(V_c + V_s)$ at that point.
Extra stirrups are provided at actual cutoff point.
The continuing flexural reinforcement at the flexural cutoff has twice the required A_s and $V_u \leq 0.75\,\phi(V_c + V_s)$.

Because we have determined the theoretical cutoff points on the basis of the continuing reinforcement having 1.0 times the required A_s, it is unlikely that we could satisfy the third limit, even though the actual bar cutoff points were extended beyond the theoretical cutoff points. Further, because we only need to satisfy one of the three sections noted, we will concentrate on satisfying the first limit.

As indicated in Fig. 8-29b, the initial shear design was to use No. 3 double-leg stirrups throughout the length of the beam. We can use Eqs. (6-9), (6-8b), and (6-18) to determine the value for ϕV_n.

$$\phi V_n = \phi(V_c + V_s) = \phi\left(2\lambda\sqrt{f'_c}\,b_w d + \frac{A_v f_{yt} d}{s}\right)$$

$$= 0.75\left(2 \times 1\sqrt{4000}\text{ psi} \times 12\text{ in.} \times 21.5\text{ in.} + \frac{0.22\text{ in.}^2 \times 60{,}000\text{ psi} \times 21.5\text{ in.}}{10\text{ in.}}\right)$$

$$= 0.75(32{,}600\text{ lb} + 28{,}400\text{ lb}) = 45.8\text{ kips}$$

From this, $\frac{2}{3}\phi V_n = 30.5$ kips. So, if V_u at the *actual* cutoff point exceeds 30.5 kips, we will need to modify the design for the stirrups to increase V_s, and thus, V_n.

(a) Cutoff at F'. This cutoff is at 3 ft 8 in. from B, which is in a flexural tension zone for the bottom reinforcement. From Fig. 8-30b, for load Case 1, the shear at F' is:

$$V_u = -49.0\text{ k} + 3.67\text{ ft} \times 3.82\text{ k/ft} = -35.0\text{ kips}$$

where the sign simply indicates the direction of the shear force. This value for V_u exceeds $\frac{2}{3}\phi V_n$, so we must either decrease the spacing for the No. 3 stirrups or change to a larger (No. 4) stirrup. We will try No. 3 stirrups at a 6-in. spacing:

$$V_s = \frac{A_v f_{yt} d}{s} = \frac{0.22\text{ in.}^2 \times 60\text{ ksi} \times 21.5\text{ in.}}{6\text{ in.}} = 47.3\text{ kips}$$

and then,

$$\tfrac{2}{3}\phi V_n = \tfrac{2}{3} \times 0.75(32.6\text{ k} + 47.3\text{ k}) = 40.0\text{ kips}$$

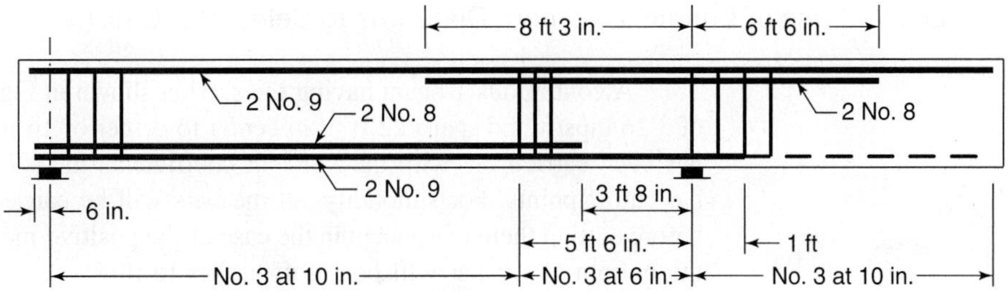

Fig. 8-34 Reinforcement details—Example 8-4.

This value exceeds V_u, so the modified stirrup design is o.k. This tighter spacing should start at the cutoff point and extend at least a distance d toward the maximum positive-moment region. **For simplicity, use a 6-in. stirrup spacing from the center of support B for 5 ft 6 in. toward midspan of span *A–B*** (Fig. 8-34).

(b) Cutoffs at J' and K'. The flexural tension that occurs in these bars is due to load Case 2 (Fig. 8-32). By inspection, V_u at J' is less than $\frac{2}{3}\phi V_n = 30.5$ kips. Therefore, no extra stirrups are required at this cutoff.

The final reinforcement details are shown in Fig. 8-34. For nonstandard beams such as this one, a detail of this sort should be shown in the contract drawings. ■

The calculations just carried out are tedious, and if the underlying concepts are not understood, the detailing provisions are difficult to apply. Several things can be done to simplify these calculations. One is to extend all of the bars past their respective points of inflection so that no bars are cutoff in zones of flexural tension. This reduces the number of cutoffs required and eliminates the need for extra stirrups, on one hand, while requiring more flexural reinforcement, on the other. A second method is to work out the flexural cutoff points graphically. This approach is discussed in the next section.

Graphical Calculation of Flexural Cutoff Points

The flexural strength of a beam is $\phi M_n = \phi A_s f_y jd$, where jd is the internal level arm and is relatively insensitive to the amount of reinforcement. If it is assumed that jd is constant, then ϕM_n is directly proportional to A_s. Because, in design, ϕM_n is set equal to M_u, we can then say that the amount of steel, A_s, required at any section is directly proportional to M_u at that section. If it is desired to cutoff a third of the bars at a particular cutoff point, the remaining two-thirds of the bars would have a strength of two-thirds of the maximum ϕM_n, and hence this cutoff would be located where M_u was two-thirds of the maximum M_u.

Figure A-1 in Appendix A is a schematic graph of the bending-moment envelope for a typical interior span of a multispan continuous beam designed for maximum negative moments of $-w\ell_n^2/11$ and a maximum positive moment of $w\ell_n^2/16$ (as per ACI Code Table 6.5.2). Similar graphs for end spans are given in Figs. A-2 to A-4.

Figure A-1 can be used to locate the flexural cutoff points and points of inflection for typical interior uniformly loaded beams, *provided they satisfy the limitations of ACI Code Section 6.5.1*. Thus, the extreme points of inflection for positive moments (points where the positive-moment diagram equals zero) are at $0.146\ell_n$ from the faces of the two supports, while the corresponding negative-moment points of inflection are at $0.24\ell_n$ from the supports. This means that positive-moment steel must extend from midspan to at least $0.146\ell_n$ from the supports, while negative-moment steel must extend at least $0.24\ell_n$ from the supports. The use of Figs. A-1 through A-4 is illustrated in Example 8-5. A more complete example is given in Chapter 10.

EXAMPLE 8-5 Use of Bending-Moment Diagrams to Select Bar Cutoffs

A continuous T-beam having the section shown in Fig. 8-35a carries a factored load of 3.26 kips/ft and spans 22 ft from center to center of 16-in.-square columns. The design has been carried out with the moment coefficients in ACI Code Table 6.5.2. Locate the bar cutoff points. For simplicity, all the bars will be carried past the points of inflection before cutting them off, except in the case of the positive-moment steel in span *B–C*, where the bottom No. 7 bar will be cutoff earlier to illustrate the use of the bar cutoff graph. Use normal-weight concrete with a compressive strength of 4000 psi, and the steel yield strength is 60,000 psi. Double-leg No. 3 open stirrups are provided at 7.5 in. throughout.

1. Structural integrity provisions. The beam is a continuous interior beam. Design should comply with ACI Code Sections 9.7.7.2 through 9.7.7.6. For open stirrups, these sections require at least one-quarter of the positive-moment flexural reinforcement, but not less than two bars be made continuous by Class B lap splices over, or near, the supports, terminating at discontinuous ends with a standard hook. For span

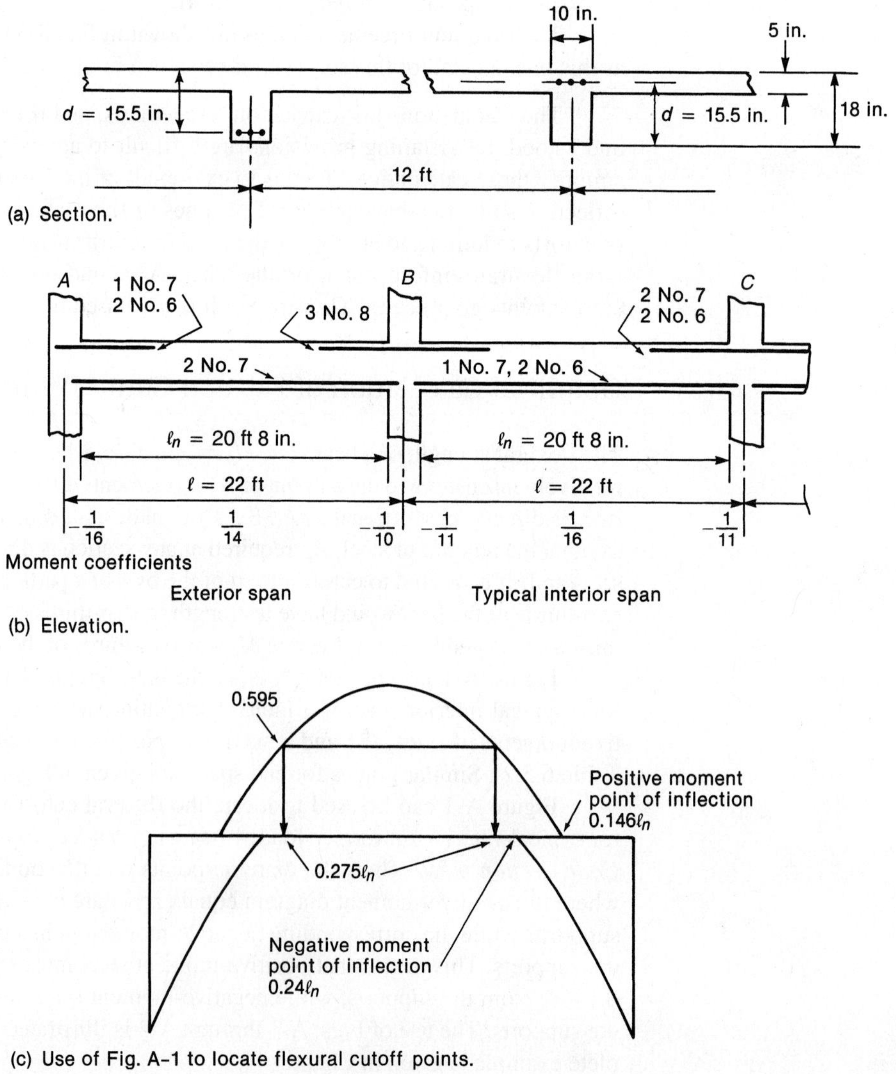

Fig. 8-35
Calculation of flexural cutoff points—Example 8-5.

AB, the two No. 7 bars will serve as the continuous tie. In spans *B–C* (and span *C–D*) we will use the two No. 6 bottom bars as structural-integrity steel. The lap splice length can be taken as 1.3 times ℓ_d for the No. 7 bottom bars.

2. Determine the positive-moment steel cutoff points for a typical interior span—span *B–C*.

(a) Development lengths of bottom bars. From Table A-5, the minimum web width for two No. 7 bars and one No. 6 is 9 in. (checked for three No.7 bars). Because the beam width exceeds this value, the bar spacing is at least d_b, and because the beam has at least code-minimum stirrups, this is Case 1 development. From Table A-6 for bottom bars with $\psi_e = 1.0$ and $\lambda = 1.0$,

No. 6 bar, $\ell_d = 37.9d_b = 28.4$ in. $= 2.37$ ft. Class B splice $= 1.3\ell_d = 3.08$ ft.
No. 7 bar, $\ell_d = 47.4d_b = 41.5$ in. $= 3.46$ ft. Class B splice $= 1.3 \times 3.46 = 4.49$ ft.

(b) Cutoff point for one No. 7 bottom bar—span *B–C*. Cutoff one No. 7 bar when it is no longer needed on each end of beam *BC*, and extend the remaining two No. 6 bars into the supports.

After the No. 7 bar is cutoff, the remaining A_s is 0.88 in.^{2} or $0.88/1.48 = 0.595$ times A_s at midspan. Therefore, the No. 7 bar can be cutoff where M_u is 0.595 times the maximum moment. From Fig. A-1, this occurs at $0.275\ell_n$ from each end for the positive moment in a typical interior span. This is illustrated in Fig. 8-35c.

Therefore, the flexural cutoff point for the No. 7 bar is at

$$\begin{aligned}0.275\ell_n &= 0.275 \times 20.67 \text{ ft}\\ &= 5.68 \text{ ft from the faces of the columns in span } B\text{–}C.\end{aligned}$$

To compute the actual cutoff points, we must satisfy detailing requirements.

Effects of shear. The bar must extend by the longer of $d = 15.5$ in. or $12d_b = 12 \times 0.875$ in. $= 10.5$ in. past the flexural cutoff. Therefore, extend the No. 7 bar 15.5 in. $= 1.29$ ft. The end of the bar will be at $5.68 - 1.29$ ft $= 4.39$ ft, say, 4 ft 4 in. from face of column as shown in Fig. 8-36b.

Anchorage, Positive moment. Bars must extend ℓ_d from the point of maximum bar stress. For No. 7 bottom bar $\ell_d = 41.5$ in. Clearly, the distance from the point of maximum bar stress at midspan to the end of bar exceeds this. Therefore, **cutoff the No. 7 bar at 4 ft 4 in. from the column faces in the interior span.**

Because the No. 7 bar is cutoff in a flexural-tension zone for the bottom steel, the shear should be checked as required in ACI Code Section 9.7.3.5. As in the prior example, we will concentrate on satisfying the first criteria given in that section. The value for $^2/_3\,\phi V_n$ is

$$\begin{aligned}{}^2/_3\,\phi V_n &= {}^2/_3\,\phi(V_c + V_s) = {}^2/_3\,\phi\left(2\lambda\sqrt{f'_c}\,b_w d + \frac{A_v f_{yt} d}{s}\right)\\ &= {}^2/_3 \times 0.75\Bigg(2 \times 1\sqrt{4000}\text{ psi} \times 10\text{ in.} \times 15.5\text{ in.}\\ &\quad + \frac{0.22\text{ in.}^2 \times 60{,}000\text{ psi} \times 15.5\text{ in.}}{7.5\text{ in.}}\Bigg)\\ &= 0.50(19{,}600\text{ lb} + 27{,}300\text{ lb}) = 23.4\text{ kips}\end{aligned}$$

The value for V_u at the *actual* cutoff point is

$$V_u = w_u\left(\frac{\ell_n}{2} - 4.33\text{ ft}\right) = 3.26\text{ k/ft}\left(\frac{20.67\text{ ft}}{2} - 4.33\text{ ft}\right) = 19.6\text{ kips}$$

Because $\frac{2}{3}\phi V_n > V_u$, no extra stirrups are required at the cutoff points for the No. 7 bars.

(c) Detailing of remaining positive-moment steel in span *B–C*. Continuity requirements in ACI Code Section 9.7.3.8.2 state that at least two bars (the remaining No. 6 bars) must be extended at least 6 in. into the column supports at *B* and *C*. Further, the structural-integrity requirements from step 1 state that unless closed transverse reinforcement is used over the full clear span, these bottom bars must be made continuous (spliced) through the supporting columns at *B* and *C*. For a typical interior beam, it is unreasonable to use closed transverse reinforcement, so we will lap splice the two No. 6 bars with bottom bars from the adjacent spans. At column *B*, the lap splice length calculated in step 2(a) for the two No. 7 bars in the exterior span *A–B* will govern (i.e., lap splice length = 4.49 ft). Assume that span *C–D* also has two No. 6 bars to be lapped spliced at column *C*. Then the lap splice length for the No. 6 bars calculated in step 2(a) will govern (i.e., lap splice length = 3.08 ft). These lengths are rounded up to 54 in. and 37 in., as shown in Fig. 8-36b.

3. Determine the negative-moment steel cutoffs for end *B*, span *B–C*. To simplify the calculations, detailing, and construction, all the bars will be extended past the negative-moment point of inflection and cutoff. From Fig. A-1, the negative-moment point of inflection is at $0.24\ell_n = 0.24 \times 20.67\text{ ft} = 4.96\text{ ft}$ from the face of the column. Therefore, the flexural cutoff point is at 4.96 ft from face of column.

Development lengths of top bars. Again, this is Case 1. From Table A-6 for $\psi_e = 1.0$ and $\lambda = 1.0$,

ℓ_d for top No. 6 $= 49.3d_b = 37.0$ in.
ℓ_d for top No. 7 $= 61.7d_b = 54.0$ in.
ℓ_d for top No. 8 $= 61.7d_b = 61.7$ in.

Structural integrity. Has been satisfied by providing continuous bottom reinforcement in step 2(c).

Continuity requirement. At least one-third of the negative-moment reinforcement must extend $d = 15.5$ in. $= 1.29$ ft, $12d_b = 12$ in. (at *B*, less at *C*), or $\ell_n/16 = 20.67/16 = 1.29$ ft past the point of inflection. Therefore, the bars must be extended by $4.96 + 1.29 = 6.25$ ft.

Therefore, try a cutoff point at 6 ft 3 in. from the face of the columns.

Anchorage. Bars must extend ℓ_d past the point of maximum bar stress. For the No. 8 bars at support *B*, $\ell_d = 61.7$ in. The actual bar extension of 6 ft 3 in. is adequate. Also, negative-moment bars must be anchored into the support. This will be accomplished by extending the top bars through to the opposite side of columns *B* and *C*.

Therefore, **cutoff top bars at 6 ft 3 in. from the face of column *B*** (see Fig. 8-36b).

Repeat the calculations at end C of span B–C; the development length of the No. 7 top bars is 54 in. Because this is less than the 6 ft 3 in. chosen earlier, we shall conservatively **cutoff top bars at 6 ft 3 in. from the face of column C.**

4. Detailing positive-moment reinforcement in span A–B. Structural-integrity requirements govern the detailing of this reinforcement. In step 2, we provided continuity of the bottom reinforcement through column B by using a lap splice length that was adequate for the two No. 7 bars in span A–B. The structural-integrity requirements in ACI Code Section 9.7.7.2 also state that at least one-quarter of the reinforcement required at midspan, but not less than two bars be anchored fully into the external support at column A. We must check that the 16-in. square column at A is large enough to anchor a No. 7 bar using a standard hook. Table A-8 gives the development length, ℓ_{dh}, for a hooked bar from the first part of Eq. (8-18) before applying any of the modification factors in ACI Code Table 25.4.3.2. The value of ℓ_{dh} from Table A-8 for a No. 7 bar in 4000 psi concrete is 16.6 in. Thus, to be able to use a No. 7 bar, we must determine if ψ_c equal to 0.7 from Code Table 25.4.3.2 can be applied for this case. Figure 8-36a shows a plan view of the No. 7 reinforcement from beam A–B framing into column A. Clearly, the side cover to the hooked bars within the column exceeds the required minimum value of 2.5 in. Also, if we assume the use of No. 4 ties around the column longitudinal

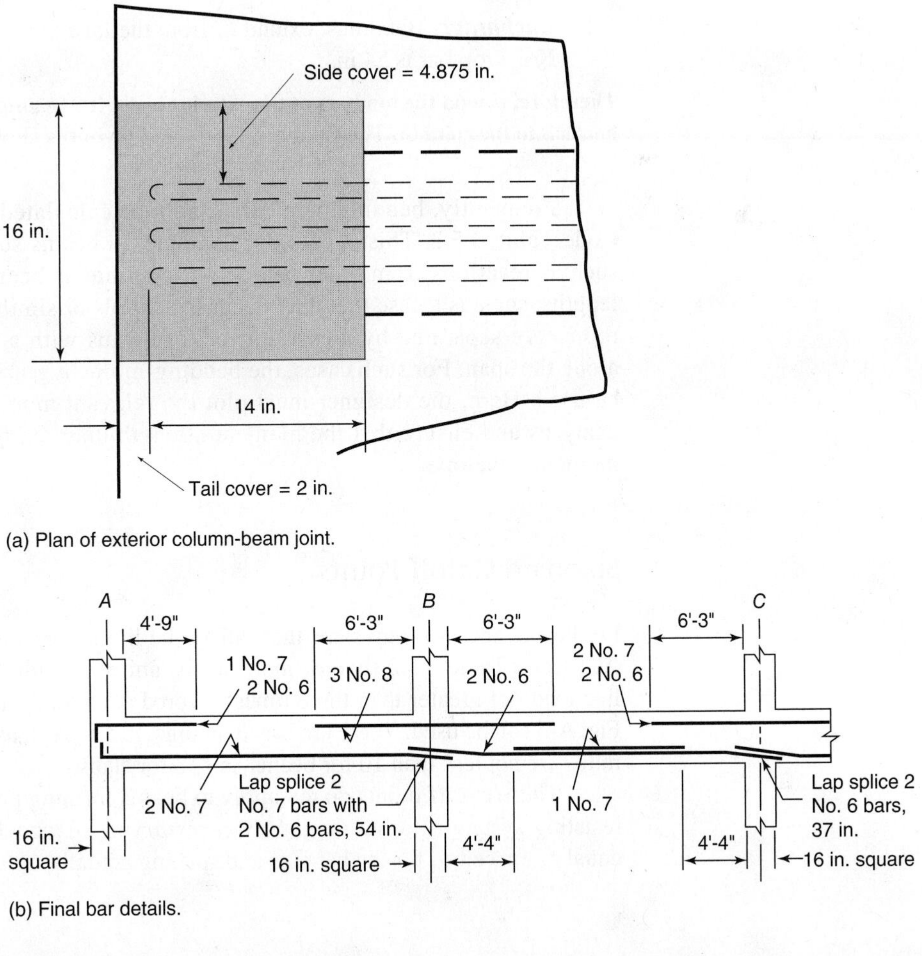

Fig. 8-36
Reinforcement details—
Example 8-5.

reinforcement, the cover on the tail of the hooked bars will be equal to the required minimum value of 2 in. Thus, $\psi_c = 0.7$ and the hooked development length, ℓ_{dh}, is

$$\ell_{dh}\,(\text{final}) = \ell_{dh} \times \text{factors} = 16.6 \text{ in.} \times 0.7 = 11.6 \text{ in.}$$

This value is less than the available length of 14 in. (16 in.–2 in.). **Therefore, the bottom No. 7 bars in beam *A–B* shall be anchored into column *A* using a standard hook.**

5. Determine the negative-moment cutoffs for the interior end of the end span. At the interior end of the end span (at end *B* of span *A–B*), the negative-moment point of inflection is at $0.24\ell_n$ from the face of the interior column (Fig. A-2). Following the calculations for the negative-moment bars in the interior span, we can **cutoff the top bars at end *B* of span *A–B* at 6 ft 3 in. from the face of the column** (see Fig. 8-36b).

6. Determine the negative-moment cutoffs for the exterior end of the end span. In step 4, we demonstrated that there was adequate side cover and tail cover (Fig. 8-36a) to use a standard hook to anchor a No. 7 bar into column *A*. Thus, the top reinforcement in beam *A–B* can be anchored in column *A*, as shown in Fig. 8-36b.

From Fig. A-2, the negative-moment point of inflection is at $0.164\ell_n = 3.39$ ft from the face of the exterior column.

Continuity requirement. Extend the bars by $d = 1.29$ ft, $12d_b = 0.875$ ft, or $\ell_n/16 = 1.29$ ft beyond the point of inflection to $3.39 + 1.29 = 4.68$ ft or 4 ft 9 in. from the support.

Anchorage. Bars must extend ℓ_d from the face of the column. From step 3, ℓ_d for a No. 7 top bar is 54 in.

Therefore, extend the top bars at the exterior end 4 ft 9 in. into the span and provide standard hooks into the column. The final reinforcement layout is shown in Fig. 8-36b. ■

Frequently, bending moments cannot be calculated with the coefficients in ACI Code Table 6.5.2. This occurs, for example, in beams supporting concentrated loads, such as reactions from other beams; for continuous beams with widely varying span lengths, such as a case that may occur in schools or similar buildings, where two wide rooms are separated by a corridor, or for beams with a change in the uniform loads along the span. For such cases, the bending-moment graphs in Figs. A-1 to A-4 cannot be used. Here, the designer must plot the relevant moment envelope from structural analyses and ensure that the moment-strength diagram remains outside the required-moment envelope.

Standard Cutoff Points

For beams and one-way slabs that satisfy the limitations on span lengths and loadings in ACI Code Table 6.5.2 (two or more spans, uniform loads, roughly equal spans, factored live load not greater than three times factored dead load), the bar cutoff points shown in Fig. A-5 can be used. These are based on Figs. A-1 to A-4, assuming that the span-to-depth ratios are not less than 10 for beams and 18 for slabs.

These cutoff points do not apply to beams forming part of the primary lateral load-resisting system of a building. *It is necessary to check* whether the top bar extensions equal or exceed ℓ_d for the bar size and spacing actually used. Figure A-5 can be used as a guide for beams and slabs reinforced with epoxy-coated bars, but *it is necessary to check* the cutoff points selected.

For the beam in Example 8-5, Fig. A-5 would give negative-moment reinforcement cutoffs at 5 ft 2 in. from the face of column *A* and at 6 ft 11 in. from the face of column *B*. Both extensions are a little longer than the computed extension.

8-9 SPLICES

Frequently, reinforcement in beams and columns must be spliced. There are four types of splices: lapped splices, mechanical splices, welded splices, and end-bearing splices. All four types of splices are permitted, as defined in ACI Code Section 25.5.

Tension Lap Splices

In a lapped splice, the force in one bar is transferred to the concrete, which transfers it to the adjacent bar. The force-transfer mechanism shown in Fig. 8-37a is clearly visible from the crack pattern sketched in Fig. 8-37b [8-14]. The transfer of forces out of the

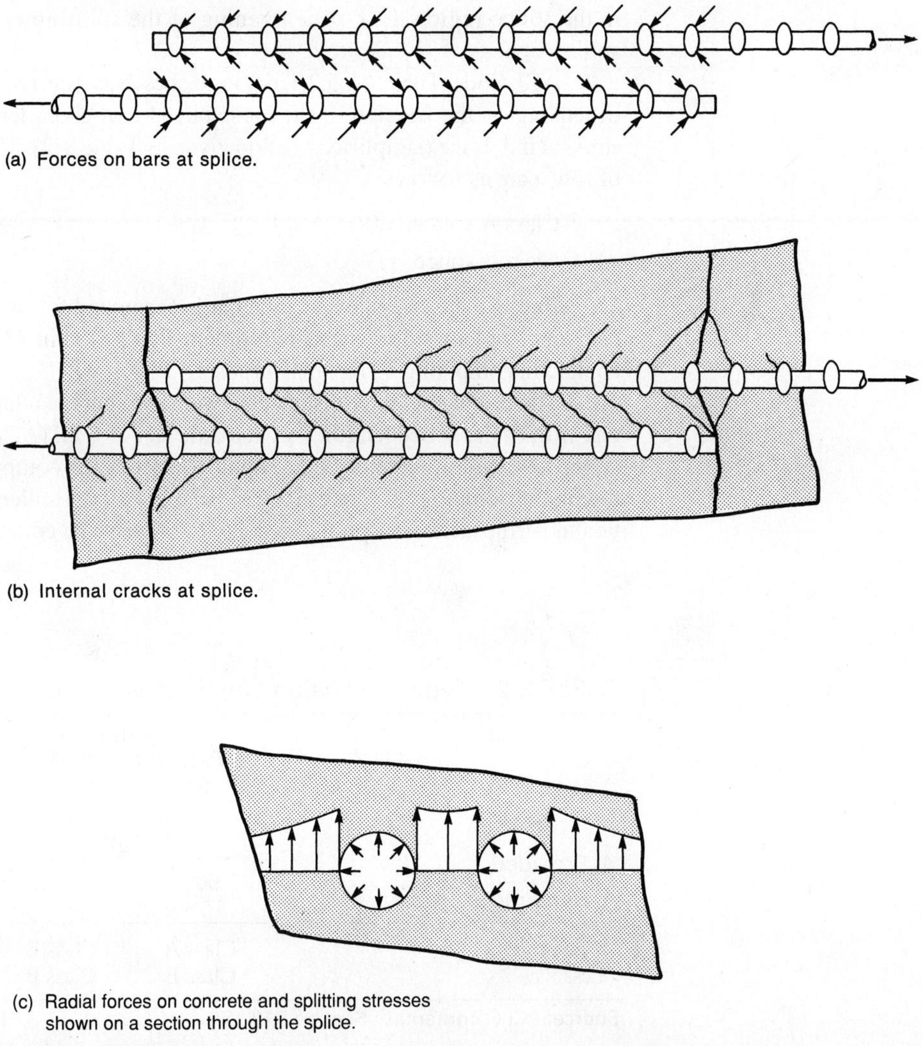

(a) Forces on bars at splice.

(b) Internal cracks at splice.

(c) Radial forces on concrete and splitting stresses shown on a section through the splice.

Fig. 8-37
Tension lap splice.

Fig. 8-38
Failure of a tension lap splice without stirrups enclosing the splice. (Photograph courtesy of J. G. MacGregor.)

bar into the concrete causes radially outward pressures on the concrete, as shown in Fig. 8-37c; these pressures, in turn, cause splitting cracks along the bars similar to those shown in Fig. 8-8a. Once such cracks occur, the splice fails as shown in Fig. 8-38. The splitting cracks generally initiate at the ends of the splice, where the splitting pressures tend to be larger than at the middle, as shown in Fig. 8-37b. Transverse reinforcement in the splice region delays the opening of the splitting cracks and hence improves the splice capacity.

ACI Code Table 25.5.2.1 distinguishes between two types of tension lap splices, depending on the fraction of the bars spliced in a given length and on the reinforcement stress at the splice (simplified version given as Table 8-2). The splice lengths for each class of splice are as follows:

Class A splice: $1.0\ell_d$

Class B splice: $1.3\ell_d$

Because the stress level in the bar is accounted for in Table 8-2, the reduction in the development length for excess reinforcement allowed in ACI Code Section 25.4.10 is *not* applied in computing ℓ_d for this purpose.

The center-to-center distance between two bars in a lap splice cannot be greater than one-fifth of the splice length, with a maximum of 6 in. (ACI Code Section 25.5.1.3). Bars larger than No. 11 cannot be lap spliced, except for compression lap splices at footing-to-column joints (ACI Code Section 16.3.5.4). Lap splices should always be enclosed within stirrups, ties, or spirals, to delay or prevent the complete loss of capacity indicated

TABLE 8-2 Type of Tension Lap Splices Required

$\frac{A_s \text{ Provided}}{A_s \text{ Required}}$	Maximum Percentage of A_s Spliced within Required Lap Length: 50%	100%
Two or more	Class A	Class B
Less than 2	Class B	Class B

Source: ACI Commentary Section R12.15.2.

in Fig. 8-38. As indicated in ACI Code Sections 25.4.2.2 and 25.4.2.3, the presence of transverse steel may lead to shorter ℓ_d and hence shorter splices. ACI Code Section 18.6.3.3 requires tension lap splices of flexural reinforcement in beams resisting seismic loads to be enclosed by hoops or spirals.

Compression Lap Splices

In a compression lap splice, a portion of the force transfer is through the bearing of the end of the bar on the concrete [8-15], [8-16]. This transfer and the fact that no transverse tension cracks exist in the splice length allow compression lap splices to be much shorter than tension lap splices (ACI Code Section 25.5.5). Frequently, a compression lap splice will fail by spalling of the concrete under the ends of the bars. The design of column splices is discussed in Chapter 11.

Welded, Mechanical, and Butt Splices

In addition to lap splices, bars stressed in tension or compression may be spliced by welding or by various mechanical devices, such as sleeves filled with molten cadmium metal, sleeves filled with grout or threaded sleeves. The use of such splices is governed by ACI Code Section 25.5.7. Descriptions of some commercially available splices are given in [8-17].

Most mechanical splices are proprietary steel devices used to provide a positive connection between bars. A common type of mechanical splice consists of a steel sleeve fitted over the joint and filled with either a high strength grout or a molten metallic filler. Other common types are in the form of sleeves that are crimped onto the two bars being spliced. Still others involve tapered threads cut into the end regions of the bars to be spliced. The tapered threads are an attempt to engage the entire area of the bars in the splice, to satisfy the requirement that tests on splices develop 125 percent of the specified yield strength. Common types of mechanical splices are described in [8-17].

ACI Code Section 25.5.7.1 requires a mechanical splice or welded splice to develop a tension or compression force, as applicable, of at least 125 percent of the specified yield strength, f_y, of the bar.

ACI Code Section 25.5.7.4 requires that splices in tension-tie members be made with full mechanical splices or full welded splices and that splices of adjacent bars be staggered by at least 30 in. Special requirements for splices in columns are presented in ACI Code Section 10.7.5.

PROBLEMS

8-1 Figure P8-1 shows a cantilever beam with $b = 12$ in. containing three No. 7 bars that are anchored in the column by standard 90° hooks. $f'_c = 4500$ psi (normal weight) and $f_y = 60{,}000$ psi. If the steel is stressed to f_y at the face of the column, can these bars

Video Solution

(a) be anchored by hooks into the column? The clear cover to the side of the hook is $2\frac{3}{4}$ in. The clear cover to the bar extension beyond the bend is 2 in. The joint is enclosed by ties at 6 in. o.c.

(b) be developed in the beam? The bar ends 2 in. from the end of the beam. The beam has No. 3 double-leg stirrups at 7.5 in.

8-2 Give two reasons why the tension development length is longer than the compression development length.

8-3 Why do bar spacing and cover to the surface of the bar affect bond strength?

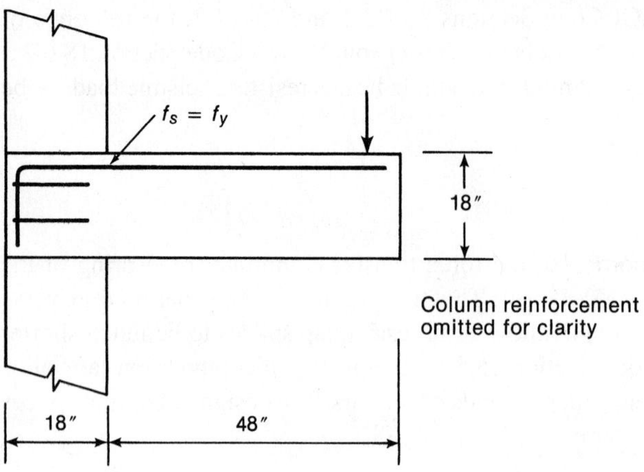

Fig. P8-1

8-4 A simply supported rectangular beam with $b = 14$ in. and $d = 17.5$ in. and with No. 3 Grade-60 stirrups at $s = 8$ in., spans 14 ft and supports a total factored uniform load of 6.5 kips/ft, including its own dead load. It is built of 4000-psi light-weight concrete and contains two No. 9 Grade-60 bars that extend 5 in. past the centers of the supports at each end. Does this beam satisfy ACI Code Section 9.7.3.8.3? If not, what is the largest size bars that can be used?

8-5 Why do ACI Code Sections 9.7.3.3 and 9.7.3.8.4 require that bars extend d past their theoretical flexural cutoff points?

8-6 Why does ACI Code Section 9.7.3.2 define "points within the span where adjacent reinforcement terminates" as critical sections for development of reinforcement in flexural members?

8-7 A rectangular beam with cross section $b = 14$ in., $h = 24$ in., and $d = 21.5$ in. supports a total factored load of 3.9 kips/ft, including its own dead load. The beam is simply supported with a 22-ft span. It is reinforced with five No. 7 Grade-60 bars, two of which are cutoff between midspan and the support and three of which extend 10 in. past the centers of the supports. $f'_c = 4000$ psi. (normal weight). The beam has Grade 60 No. 3 stirrups satisfying ACI Code Sections 9.7.6.2.2 and 9.6.3.3.

(a) Plot to scale the factored moment diagram. $M = w\ell x/2 - wx^2/2$, where x is the distance from the support and ℓ is the span.

(b) Plot the moment-strength diagram and locate the cutoff points for the two cutoff bars such that the moment strength diagram stays above the factored moment.

8-8 Why does ACI Code Section 9.7.3.5 require extra stirrups at bar cutoff points in some cases?

The beam shown in Fig. P8-9 is built of 4000-psi normal-weight concrete and Grade-60 steel. The effective depth $d = 18.5$ in. The beam supports a total factored uniform load of 5.25 kips/ft, including its own dead load. The frame is not part of the lateral load-resisting

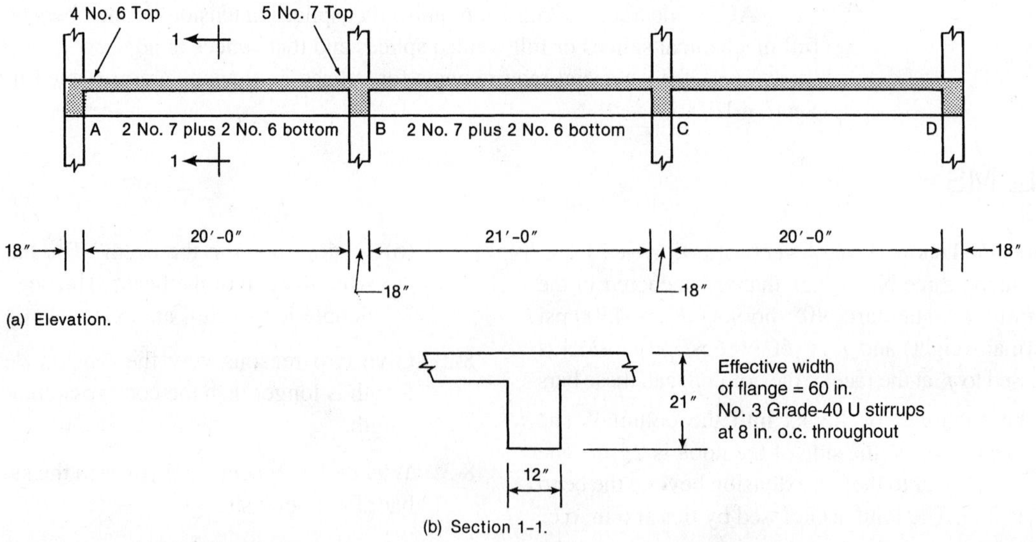

Fig. P8-9

system for the building. Use Figs. A-1 to A-4 to select cutoff points in Problems 8-9 to 8-11.

8-9 Select cutoff points for span *A–B* based on the following requirements:

(a) Cutoff two No. 6 positive-moment bars when no longer needed at each end. Extend the remaining bars into the columns.

(b) Extend all negative-moment bars past the negative-moment point of inflection before cutting them off.

(c) Check the anchorage of the negative-moment bars at *A* and modify the bar size if necessary.

8-10 Repeat Problems 8-9(a) and (b) for span *B–C*.

8-11 Assume the beam is constructed with 4000 psi light weight concrete. Select cutoff points for span *A–B* based on the following requirements:

(a) Extend all negative-moment bars at *A* past the negative-moment point of inflection.

(b) Cutoff the two No. 6 positive-moment bars when no longer needed at each end. Extend the remaining bars into the columns.

(c) Cutoff two of the negative-moment bars near support at *B* when no longer needed. Extend the remaining bars past the point of inflection. Check shear strength requirements at the cutoff point.

REFERENCES

8-1 ACI Committee 408, "Bond and Development of Straight Reinforcing Bars in Tension (ACI 408R-03 reapproved 2012)" *ACI Manual of Concrete Practice*, American Concrete Institute, Farmington Hills, MI.

8-2 C. O. Orangun, J. O. Jirsa, and J. E. Breen, "A Reevaluation of Test Data on Development Length and Splices," *ACI Journal, Proceedings*, Vol. 74, No. 3, March 1977, pp. 114–122; Discussion, pp. 470–475.

8-3 J. O. Jirsa, L. A. Lutz, and P. Gergely, "Rationale for Suggested Development, Splice and Standard Hook Provisions for Deformed Bars in Tension," *Concrete International: Design and Construction*, Vol. 1, No. 7, July 1979, pp. 47–61.

8-4 *Bond Action and Bond Behaviour of Reinforcement, State-of-the-Art Report*, Bulletin d'Information 151, Comité Euro-International du Béton, Paris, April 1982, 153 pp.

8-5 Paul R. Jeanty, Denis Mitchell, and Saeed M. Mirza, "Investigation of Top Bar Effects in Beams," *ACI Structural Journal*, Vol. 85, No. 3, May–June 1988, pp. 251–257.

8-6 Robert A. Treece and James O. Jirsa, "Bond Strength of Epoxy-Coated Reinforcing Bars," *ACI Materials Journal*, Vol. 86, No. 2, March–April 1989, pp. 167–174.

8-7 J. L. G. Marques and J. O. Jirsa, "A Study of Hooked Bar Anchorages in Beam-Column Joints," *ACI Journal, Proceedings*, Vol. 72, No. 5, May 1975, pp. 198–209.

8-8 G. Rehm, "Kriterien zur Beurteilung von Bewehrungsstäben mit hochwertigem Verbund (Criteria for the Evaluation of High Bond Reinforcing Bars)" *Stahlbetonbau-Berichte aus Forschung und Praxis-Hubert Rüsch gewidmet*, Berlin, 1969, pp. 79–85.

8-9 M. K. Thompson, M. J. Ziehl, J. O. Jirsa, and J. E. Breen, "CCT Nodes Anchored by Headed Bars—Part 1: Behavior of Nodes," *ACI Structural Journal*, Vol. 102, No. 6, November–December 2005, pp. 808–815.

8-10 M. K. Thompson, J. O. Jirsa, and J. E. Breen, "CCT Nodes Anchored by Headed Bars—Part 2: Capacity of Nodes," *ACI Structural Journal*, Vol. 103, No. 1, January–February 2006, pp. 65–73.

8-11 M. K. Thompson, A. Ledesma, J. O. Jirsa, and J. E. Breen, "Lap Splices Anchored by Headed Bars," *ACI Structural Journal*, Vol. 103, No. 2, March–April 2006, pp. 271–279.

8-12 Joint Committee on Standard Specifications for Concrete and Reinforced Concrete, *Recommended Practice and Standard Specifications for Concrete and Reinforced Concrete*, American Concrete Institute, Detroit MI, June 1940, 140 pp.

8-13 M. Baron, "Shear Strength of Reinforced Concrete Beams at Point of Bar Cutoff," *ACI Journal, Proceedings*, Vol. 63, No. 1, January 1966, pp. 127–134.

8-14 Y. Goto and K. Otsuka, "Experimental Studies on Cracks Formed in Concrete around Deformed Tension Bars," *Technological Reports of Tohoku University*, Vol. 44, June 1979, pp. 49–83.

8-15 J. F. Pfister and A. H. Mattock, "High Strength Bars as Concrete Reinforcement," Part 5, "Lapped Splices in Concentrically Loaded Columns," *Journal of the Research and Development Laboratories*, Portland Cement Association, Vol. 5, No. 2, May 1963, pp. 27–40.

8-16 F. Leonhardt and K. Teichen, "Druck-Stoesse von Bewehrungsstaeben (Compression Splices of Reinforcing Bars)," *Deutscher Ausschuss fuer Stahlbeton*, Bulletin 222, Wilhelm Ernst, Berlin, 1972, pp. 1–53.

8-17 ACI Committee 439, "Types of Mechanical Splices for Reinforcing Bars (ACI 439.3R-07)," *ACI Manual of Concrete Practice*, American Concrete Institute, Farmington Hills, MI.

9

Serviceability

9-1 INTRODUCTION

In Chapter 2, limit-states design was discussed. The limit states (states at which the structure becomes unfit for its intended function) were divided into two groups: those leading to collapse and those which disrupt the use of structures, but do not cause collapse. These were referred to as *strength limit states* and *serviceability limit states*, respectively. The major serviceability limit states for reinforced concrete structures are caused by excessive crack widths, excessive deflections, and undesirable vibrations. All three are discussed in this chapter. Although fatigue is a failure limit state, it occurs at service loads and so is considered here. The terms *service loads* and *working loads* refer to loads encountered in the everyday use of the structure. Service loads are generally taken to be the specified loads without load factors.

Historically, deflections and crack widths have not been a problem for reinforced concrete building structures. With the advent of strength design and Grade-60 reinforcement, however, the reinforcement stresses at service loads have increased by about 50 percent. Because crack widths, deflections, and fatigue are all related to steel stress, each of these has become more critical.

Serviceability Limit States

In design, a serviceability limit state consists of four parts:

1. A statement of the *limit state* causing the problem. Typical serviceability limit states (SLS) include deflection, cracking, vibrations, and so on.

2. One or more *load combinations* to be used in checking the limit state. Traditionally, the SLS load combinations used a load factor of 1.0 on all service loads. This assumes that all the specified variable loads have the values that they would have at an arbitrary point in time. Generally the variable loads are lower than this, especially for wind loads or snow loads. The selection of the load factors must consider the probability of occurrence of the loads involved, either by themselves or in conjunction with other loads. Generally, service load design codes do not adequately handle offsetting loads. In some structures, the number of loading cycles exceeding the limit state during the service life is important.

3. A *calculation procedure* to be used to check the limit state. Traditionally, serviceability-limit-state calculations for concrete structures are carried out using elastic analysis and the straight-line theory of flexure, based on average material properties.

4. A *criterion* to be used to judge whether the limit state is exceeded. Typically, this is in one of the following forms:

(Calculated deflection or other limit state) $\leq$ (Limiting value)

or, for a prestressed concrete bridge pier,

"Cracks in the concrete should be open less than 100 times in the life of the structure." [9-1]

Major serviceability limit states affecting the design of reinforced concrete buildings are listed in Tables 9-1 through 9-5. These list the four parts of each serviceability limit state and suggest limits. Table 9-1 lists serviceability limit states due to cracking. Table 9-2 lists flexural deflection coefficients for spans with distributed loads. Table 9-3 deals with flexural deflection limit states. Tables 9-4 and 9-5 deal with other serviceability limit states.

9-2 ELASTIC ANALYSIS OF STRESSES IN BEAM SECTIONS

At service loads, the distribution of stresses in the compression zone of a cracked beam is close to being linear, as shown in Fig. 4-4, and the steel is in the elastic range of behavior. As a result, an elastic calculation gives a good estimate of the concrete and steel stresses at service loads. This type of analysis is also referred to as a *straight-line theory* analysis, because a linear stress distribution is assumed. The straight-line theory is used in calculating both the stiffness, EI, at service loads, for deflection calculations, and steel stresses, for use in crack-width or fatigue calculations.

Calculation of *EI*

Modulus of Elasticity and Modular Ratio

Section 19.2.2.1 of the ACI Code assumes that concrete has a modulus of elasticity of

$$E_c = w_c^{1.5}\left(33\sqrt{f'_c}\right) \text{ psi} \tag{9-1}$$

where w_c is the unit weight of concrete, lb/ft^3. For normal-weight concrete, this can be reduced to 57,000 $\sqrt{f'_c}$ psi. As was discussed in Section 3-5, E_c is also affected strongly by the modulus of elasticity of the coarse aggregate. In critical situations, it may be desirable to allow for this. For American concretes E_c will generally lie between 80 percent and 120 percent of the value given by Eq. (9-1). Reinforcing steel has a modulus of elasticity of $E_s = 29 \times 10^6$ psi.

The ratio E_s/E_c is referred to as the *modular ratio*, n, and has values ranging from 9.3 for 3000-psi concrete to 6.6 for 6000-psi concrete. This means that, for a given strain less than the yield strain, the stress in steel will be six to nine times that in concrete subjected to the same strain.

Transformed Section

At service loads, the beam is assumed to act elastically. The basic assumptions in elastic bending are (a) that strains are linearly distributed over the depth of the member and (b) that the stresses can be calculated from the strains by the relationship $\sigma = E \times \epsilon$.

This leads to the elastic bending equation, $\sigma = My/I$. When a beam made of two materials is loaded, the different values of E for the two materials lead to a different stress distribution, because one material is stiffer and accepts more stress for a given strain than the other. However, the elastic-beam theory can be used if the steel–concrete beam is hypothetically *transformed* to either an all-steel beam or an all-concrete beam, customarily the latter. This is done by replacing the area of steel with an area of concrete having the same axial stiffness AE. Because $E_s/E_c = n$, the resulting area of concrete will be nA_s. This transformed area of the steel is assumed to be concentrated at the same point in the cross section as the real steel area.

When the steel is in a compression zone, its transformed area is nA_s', but it displaces an area of concrete equal to A_s'. As a result, compression steel is transformed to an equivalent concrete area of $(n - 1)A_s'$. (In the days of working-stress design, compression steel was transformed to an equivalent concrete area of $(2n - 1)A_s'$, to reflect the effect of concrete creep on the stresses.)

Prior to flexural cracking, the beam shown in Fig. 9-1a has the transformed section in Fig. 9-1b. Because the steel is displacing concrete, which could take stress, the transformed area is $(n - 1)A_s$ for both layers of steel. The cracked transformed section is shown in Fig. 9-1c. Here, the steel in the compression zone displaces stressed concrete and has a transformed area of $(n - 1)A_s'$, while that in the tension zone does not and hence has an area of nA_s.

The neutral axis of the cracked section occurs at a distance $x = kd$ below the top of the section. For an elastic section, the neutral axis occurs at the centroid of the area, which is defined as that point at which

$$\Sigma A_i \bar{y}_i = 0 \tag{9-2}$$

where $\bar{y}_i$ is the distance from the centroidal axis to the centroid of the ith area. The solution of Eq. (9-2) is illustrated in Example 9-1.

EXAMPLE 9-1 Calculation of the Transformed Section Properties

The beam section shown in Fig. 9-1a is built of 4000-psi concrete. Compute the location of the centroid and the moment of inertia for both the uncracked section and the cracked section.

Uncracked Transformed Section

The modulus of elasticity of the concrete is

$$E_c = 57{,}000\sqrt{f_c'} = 3.60 \times 10^6 \text{ psi}$$

The modular ratio is

$$\begin{aligned} n &= \frac{E_s}{E_c} \\ &= \frac{29 \times 10^6}{3.60 \times 10^6} = 8.04 \simeq 8.0 \end{aligned}$$

It is unreasonable to use more than two significant figures for the approximate calculations involving n.

Because all the steel is in uncracked parts of the beam, the transformed areas of the two layers of steel are computed as follows:

$$\text{Top steel:}\quad (8.0 - 1) \times 1.20 = 8.40 \text{ in.}^2$$

$$\text{Bottom steel:}\quad (8.0 - 1) \times 2.40 = 16.8 \text{ in.}^2$$

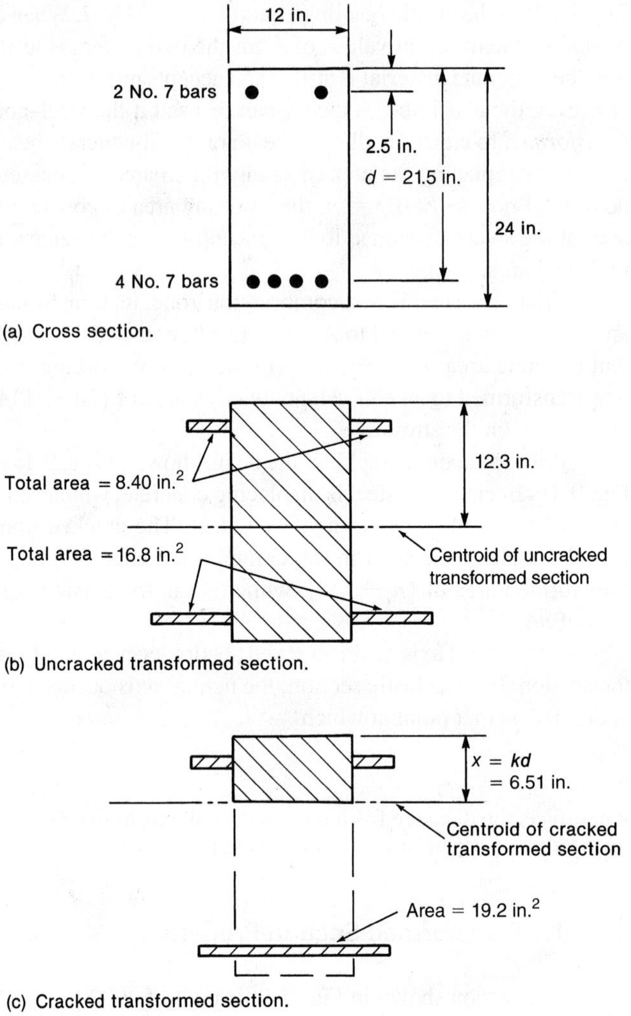

Fig. 9-1
Transformed sections–
Example 9-1.

The centroid of the transformed section is located as follows:

Part	Area (in.^{2})	y_{top} (in.)	Ay_{top} (in.^{3})
Concrete	$12 \times 24 = 288$	12	3456
Top steel	8.40	2.5	21.0
Bottom steel	16.8	21.5	361.2
	313.2		$\Sigma\, Ay_{top} = 3838$

$$\bar{y}_{top} = \frac{3838}{313.2} = 12.3 \text{ in.}$$

Therefore, the centroid is 12.3 in. below the top of the section. The moment of inertia is calculated as follows. (*Note:* The moments of inertia of the steel layers about their centroid are negligible.)

Part	Area (in.^{2})	$\bar{y}$ (in.)	$I_{own\ axis}$ (in.^{4})	$A\bar{y}^2$ (in.^{4})
Concrete	288	−0.3	13,820	25.9
Top steel	8.40	9.8	—	807
Bottom steel	16.8	−9.2	—	1422
				I_{gt} = 16,100 in.^{4}

The uncracked transformed moment of inertia $I_{gt} = 16{,}100$ in.^{4}. This is 16 percent larger than the gross moment of inertia of the concrete alone, referred to as $I_g = bh^3/12$.

Cracked Transformed Section

Assume that the neutral axis is lower than the top steel. The transformed areas are computed as follows:

$$\text{Top steel:}\quad (8.0 - 1) \times 1.20 = 8.40 \text{ in.}^2$$

$$\text{Bottom steel:}\quad 8.0 \times 2.40 = 19.2 \text{ in.}^2$$

Let the depth to the neutral axis be x, as shown in Fig. 9-1c, and sum the moments of the areas about the neutral axis to compute x:

Part	Area (in.^{2})	$\bar{y}$ (in.)	$A\bar{y}$ (in.^{3})
Compression zone	$12x$	$x/2$	$12x^2/2 = 6x^2$
Top steel	8.40	$x - 2.5$	$8.4x - 21.0$
Bottom steel	19.2	$x - 21.5$	$19.2x - 412.8$

But, by definition, x is the distance to the centroid when $\Sigma A\bar{y} = 0$. Therefore, $6x^2 + 27.6x - 433.8 = 0$ and it follows that

$$x = \frac{-27.6 \pm \sqrt{27.6^2 + 4 \times 6 \times 433.8}}{2 \times 6}$$

$$= 6.51 \text{ in. or } -11.1 \text{ in.}$$

The positive value lies within the section. Because the top steel is in the compression zone, the initial assumption is o.k. Therefore, **the centroidal axis (axis of zero strain) is at 6.51 in. below the top of the section.**

Compute the moment of inertia:

Part	Area (in.^{2})	y (in.)	$I_{own\ axis}$ (in.^{4})	Ay^2 (in.^{4})
Compression zone	12 × 6.51 = 78.12	6.51/2	276	828
Top steel	8.40	(6.51 − 2.5)	—	135
Bottom steel	19.2	(6.51 − 21.5)	—	4314
				I = 5550 in.^{4}

Thus, the cracked transformed moment of inertia is I_{cr} = 5550 in.^{4} ■

In this example, I_{cr} is approximately 35 percent of the moment of inertia of the uncracked transformed section and 40 percent of that of the gross moment of inertia. This illustrates the great reduction in stiffness due to cracking.

The procedure followed in Example 9-1 can be used to locate the neutral axis for any shape of cross section bent in uniaxial bending. For the special case of rectangular beams *without* compression reinforcement, Eq. (9-2) gives

$$\frac{bx^2}{2} - nA_s(d - x) = 0$$

Substituting $x = kd$ and $\rho = A_s/bd$ results in

$$\frac{b(kd)^2}{2} - \rho nbd(d - kd) = 0$$

Dividing by bd^2 and solving for k gives

$$k = \sqrt{2\rho n + (\rho n)^2} - \rho n \qquad (9\text{-}3)$$

where $\rho = A_s/bd$ and $n = E_s/E_c$. Equation (9-3) can be used to locate the neutral-axis position directly, thus simplifying the calculation of the moment of inertia. This equation applies only for rectangular beams without compressive reinforcement.

Service-Load Stresses in a Cracked Beam

Because the cross section has been transformed into a single material, we can use the elastic bending equation, $\sigma = M \cdot y/I$, to calculate the acting stresses after cracking in the concrete and the steel reinforcement due the acting moment, M_a. In the following equations we will use f in place of σ to represent stress and for I, we will use the previously calculated moment of inertia for the cracked-transformed section, I_{cr}. Because we transformed the steel into equivalent concrete, Eq. (9-4) is used to find concrete stresses, f_c:

$$f_c = \frac{M_a \cdot y}{I_{cr}} \qquad (9\text{-}4a)$$

To find the maximum concrete stress in the extreme compression fiber, use $y = kd = x$:

$$f_c(\max) = \frac{M_a \cdot x}{I_{cr}} \qquad (9\text{-}4b)$$

Recall that the steel stress in the transformed section will be equal to n-times the concrete stresses at the same location within the depth of the cross-section. So, to find the stress in the tension reinforcement we need to multiply Eq. (9-4) by n, and the value for y at the level of the tension reinforcement will be $d - x$.

$$f_s = \frac{n \cdot M_a(d - x)}{I_{cr}} \qquad (9\text{-}5)$$

Although these equations ignore the effect of creep in the concrete, the change in the calculated stresses would be very small for a singly reinforced beam section. If compression reinforcement was present, it would have a more significant effect on the final concrete stresses and thus should be transformed using $(2n - 1)A_s'$, as noted previously.

This analysis ignores the effects of creep, which will tend to increase the stress in the tension steel by a small amount.

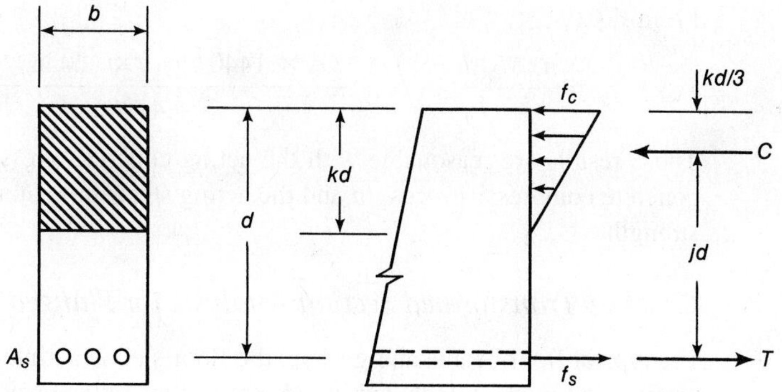

Fig. 9-2
Stress distribution in straight-line theory.

EXAMPLE 9-2 Calculation of the Service-Load Concrete Steel Stresses in a Rectangular Beam

A rectangular beam similar to the one shown in Fig. 9-2 has $b = 10$ in., $d = 20$ in., three No. 8 Grade-60 bars, and $f'_c = 4500$ psi. Compute f_s at service loads if the service live-load moment is 50 kip-ft and the service dead-load moment is 70 kip-ft.

1. Compute k and I_{cr}.

$$E_c = 57{,}000\sqrt{4500} = 3.82 \times 10^6 \text{ psi}$$

$$n = \frac{E_c}{E_s} = \frac{29{,}000 \text{ ksi}}{3820 \text{ ksi}} = 7.6$$

$$\rho = \frac{A_s}{bd} = \frac{3 \times 0.79 \text{ in.}^2}{10 \text{ in.} \times 20 \text{ in.}} = 0.0119$$

$$\rho n = 0.0119 \times 7.6 = 0.0904$$

$$k = \sqrt{2\rho n - (\rho n)^2} - \rho n = 0.344$$

So,

$$kd = x = 0.344 \times 20 \text{ in.} = 6.88 \text{ in.}$$

and,

$$I_{cr} = \frac{bx^3}{12} + (bx)\left(\frac{x}{2}\right)^2 + nA_s(d - x)^2 = \frac{bx^3}{3} + nA_s(d - x)^2$$

$$I_{cr} = \frac{10 \times (6.88)^3}{3} + 7.6 \times 2.37(20 - 6.88)^2$$

$$I_{cr} = 1090 \text{ in.}^4 + 3100 \text{ in.}^4 = 4190 \text{ in.}^4$$

2. Compute f_c(max) and f_s.

$$M_a = 50 \text{ k-ft} + 70 \text{ k-ft} = 120 \text{ k-ft} = 1440 \text{ k-in.}$$

From Eq. (9-4b):

$$f_c(\max) = \frac{M_a \cdot x}{I_{cr}} = \frac{1440 \text{ k-in.} \times 6.88 \text{ in.}}{4190 \text{ in.}^4} = 2.36 \text{ ksi} = 2360 \text{ psi}$$

From Eq. (9-5):

$$f_s = \frac{n \cdot M_a(d - x)}{I_{cr}} = \frac{7.6 \times 1440 \text{ k-in.} \times (20 \text{ in.} - 6.88 \text{ in.})}{4190 \text{ in.}^4} = 34.3 \text{ ksi}$$

These results are reasonable with the acting concrete stress at approximately half of the concrete compressive strength and the acting steel stress at approximately half of its yield strength. ■

Cracked-Transformed Section Analysis for Flanged Sections

For typical floor beams and girders, the floor slab is an integral part of the member cross section and must be included in the elastic flexural analysis of that section. Consider the beam section in Fig. 9-3a. If this section is subjected to negative bending (tension at top), the corresponding cracked-transformed section is shown in Fig. 9-3b. Because this section has a rectangular compression zone, the analysis is similar to that used for the beam section in Fig. 9-2. However, for this case, we will take a more fundamental approach to finding the depth to the neutral axis, x, and the value for the cracked-transformed section moment of inertia, I_{cr}.

For elastic behavior (linear distribution of strain and stress), the centroid of the section should coincide with the location of the neutral axis for a section subjected to pure bending (no axial load). Thus, for the transformed section shown in Fig. 9-3b, the sum of the areas times their centroid distances to the neutral axis should be zero.

$$\sum A_i \cdot y_i = 0$$

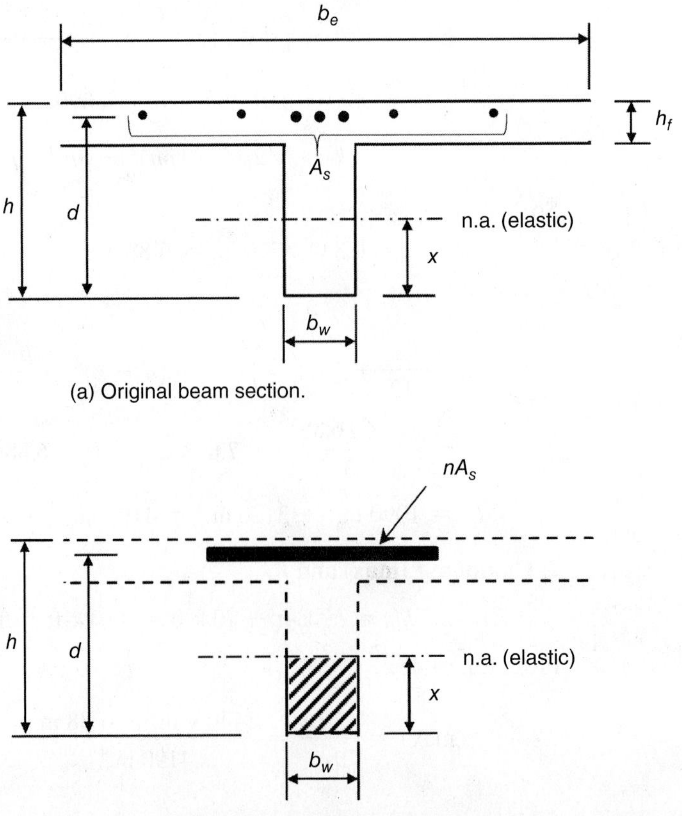

(b) Cracked–transformed beam section.

Fig. 9-3
Singly-reinforced flanged section in negative bending.

For the section in Fig. 9-3b, assuming x is positive toward the compression edge,

$$b_w \cdot x \cdot \left(\frac{x}{2}\right) - n \cdot A_s \cdot (d - x) = 0$$

which can be rearranged as:

$$x^2 + \frac{2nA_s}{b_w} \cdot x - \frac{2nA_s d}{b_w} = 0 \quad \text{(9-6a)}$$

As in Example 9-1, solving this equation will give a positive and negative value for x, and of course, we will select the positive value. With the value for x, we can now use the parallel axis theorem to find the moment of inertia for the cracked-transformed section shown in Fig. 9-3b.

$$I_{cr} = \sum (I_i + A_i \cdot y_i^2)$$

$$I_{cr} = \frac{b_w \cdot x^3}{12} + b_w \cdot x\left(\frac{x}{2}\right)^2 + I(\text{steel layer}) + n \cdot A_s(d - x)^2$$

The moment of inertia of the steel bars about their centroid is very small, and thus, is ignored for this calculation. So, combining the first two terms we have,

$$I_{cr} = \frac{b_w \cdot x^3}{3} + n \cdot A_s(d - x)^2 \quad \text{(9-7a)}$$

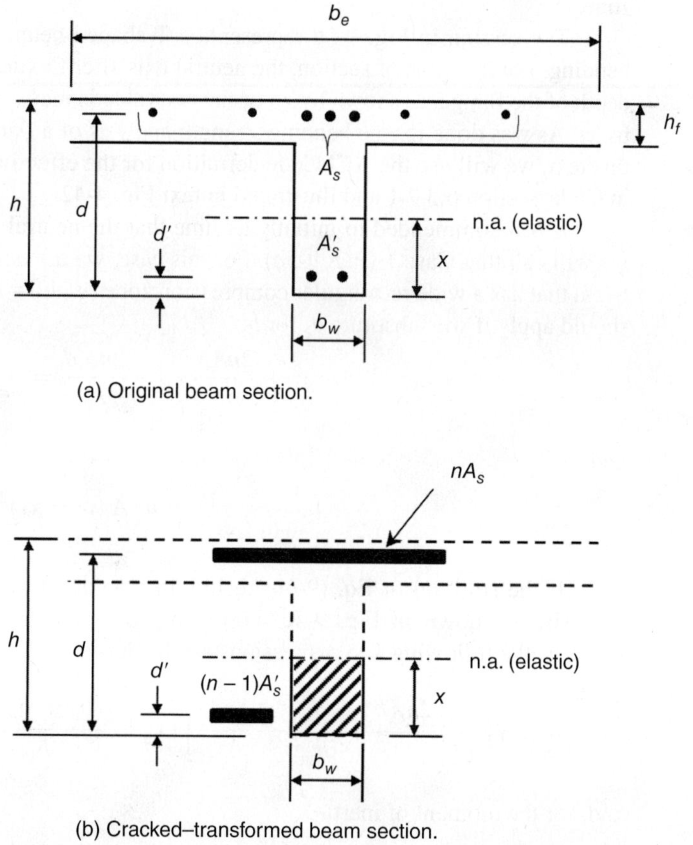

Fig. 9-4
Doubly-reinforced flanged section in negative bending.

This is the same expression that was developed in Example 9-2, and it applies for all singly reinforced sections with a rectangular compression zone.

Let us next consider a similar beam section subjected to negative bending, but for this section we will include compression zone reinforcement, A_s' (see Fig. 9-4a). As before, the cracked-transformed section for this beam section is given in Fig. 9-4b. Note that the transformed compression steel is shown on one side of the section for convenience. As was done previously, setting $\sum A_i \cdot y_i = 0$ about the neutral axis, results in,

$$b_w \cdot x \cdot (x/2) + (n - 1)A_s' \cdot (x - d') - n \cdot A_s \cdot (x - d) = 0$$

which can be rearranged as:

$$x^2 + \left[\frac{2(n - 1)A_s' + 2nA_s}{b_w}\right] \cdot x - \left[\frac{2(n - 1)A_s'd' + 2nA_sd}{b_w}\right] = 0 \tag{9-8}$$

As before, solving this equation will give both a positive and negative solution for x, and we will use the positive value. It is interesting to note the similar handling of both the tension and compression steel in Eq. (9-8). With the value of x, we can use the parallel axis theorem to derive the following value for I_{cr}:

$$I_{cr} = \frac{b_w \cdot x^3}{3} + (n - 1)\, A_s'(x - d')^2 + nA_s(d - x)^2 \tag{9-9}$$

This expression applies for all doubly reinforced sections with a rectangular compression zone.

The section in Fig. 9-5a represents a T-shaped beam section subjected to positive bending. For this type of section, the neutral axis after flexural cracking may be within the depth of the flange, as shown by x_1, or the neutral axis may be located in the web, as shown by x_2. As was done for the bending strength analysis of a flanged section in Section 4-8 of this text, we will use the ACI Code definition for the effective flange width, b_e, as defined in Code Section 6.3.2.1 and illustrated in text Fig. 4-42.

It is recommended to initially assume that the neutral axis is within the flange, and we will call that Case 1 (Fig. 9-5b). For this case, we are dealing with a singly reinforced beam that has a wide rectangular compression zone (width $= b_e$). So, Eqs. (9-6a) and (9-7a) should apply if we substitute b_e for b_w.

$$x_1^2 + \frac{2nA_s}{b_e} \cdot x_1 - \frac{2nA_sd}{b_e} = 0 \tag{9-6b}$$

and,

$$I_{cr} = \frac{b_e \cdot x_1^3}{3} + n \cdot A_s(d - x_1)^2 \tag{9-7b}$$

If the solution of Eq. (9-6b) results in a value of $x_1 > h_f$, then we have a Case 2 analysis, as shown in Fig. 9-5c. Following the same procedure as used for the prior sections, the following two expressions can be derived.

$$x_2^2 + \left[\frac{2h_f(b_e - b_w) + 2nA_s}{b_w}\right] \cdot x_2 - \frac{(b_e - b_w)h_f^2}{b_w} - \frac{2nA_sd}{b_w} = 0 \tag{9-10}$$

and, for the moment of inertia,

$$I_{cr} = \frac{b_w x_2^3}{3} + \frac{(b_e - b_w)h_f^3}{12} + (b_e - b_w)h_f\left(x_2 - \frac{h_f}{2}\right)^2 + nA_s(d - x_2)^2 \tag{9-11}$$

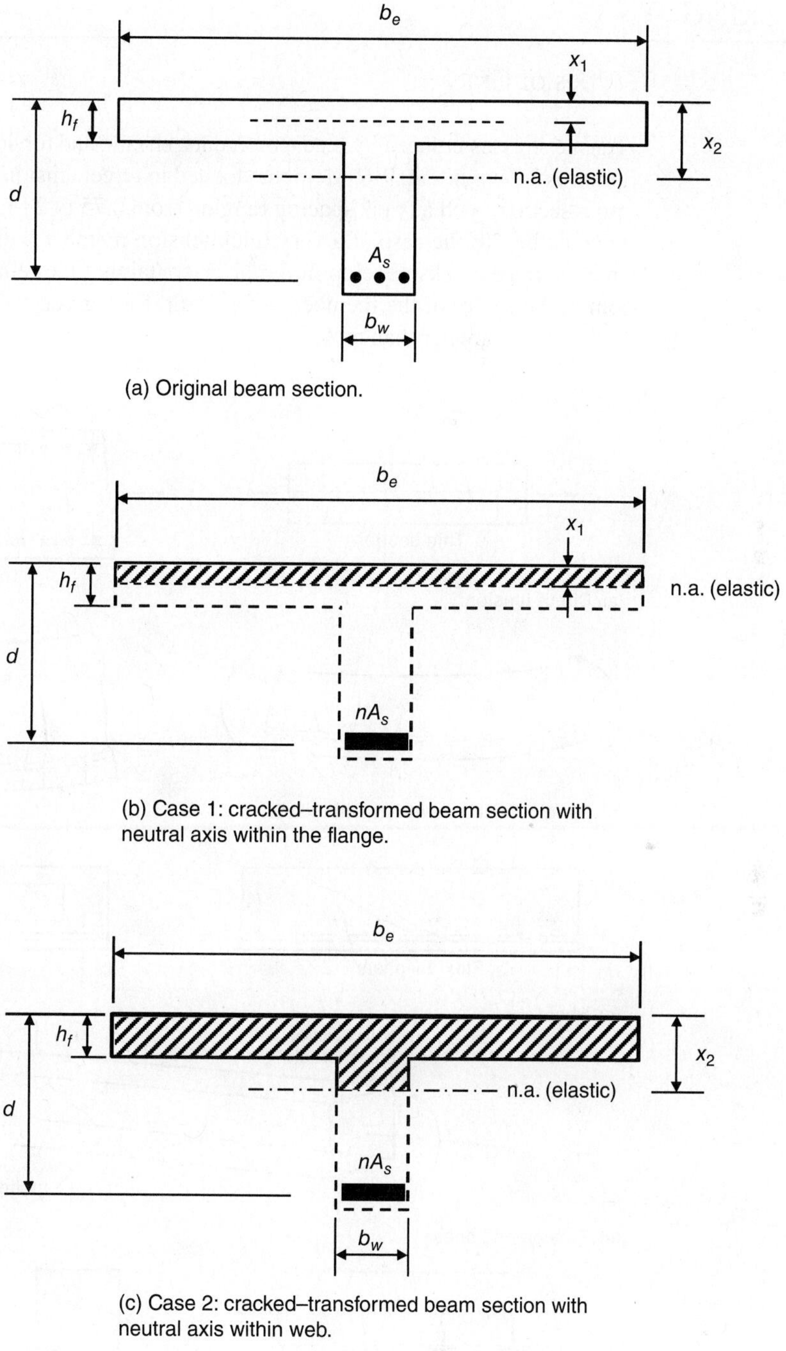

Fig. 9-5
Singly-reinforced flanged section in positive bending.

We will not derive additional formulas for a T-section subjected to positive bending, which contains compression reinforcement. As was done for flexural strength analysis of such sections, we will ignore the presence of the compression reinforcement. Ignoring the compression reinforcement had only a minor effect on the results for strength analysis in Chapter 4, and will have only a minor effect on elastic stress and deflection analyses, as covered in this chapter.

9-3 CRACKING

Types of Cracks

Tensile stresses induced by loads, moments, shears, and torsion cause distinctive crack patterns, as shown in Fig. 9-6. Members loaded in direct tension crack right through the entire cross section, with a crack spacing ranging from 0.75 to 2 times the minimum thickness of the member. In the case of a very thick tension member with reinforcement in each face, small surface cracks develop in the layer containing the reinforcement (Fig. 9-6a). These join in the center of the member. As a result, for a given total change in length, the crack width at B is greater than at A.

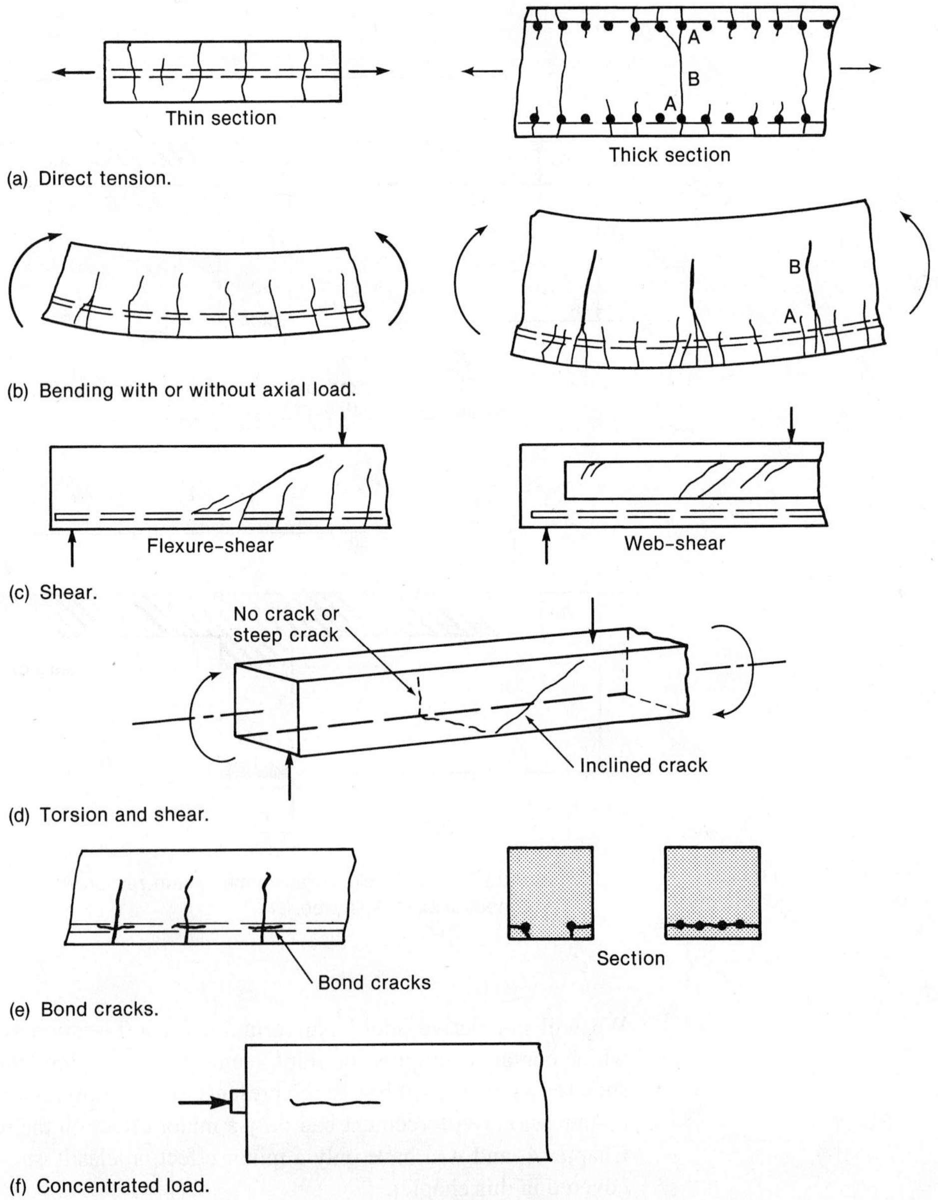

Fig. 9-6
Load-induced cracks.
(From [9-5].)

Various mechanisms of cracking are discussed in detail in [9-6] through [9-9]. The relationship between cracking and corrosion of reinforcement is discussed in [9-10], [9-11], and [9-12]. Crack widths are discussed in [9-13], [9-14], [9-15], and [9-16]. Members subjected to bending moments develop flexural cracks, as shown in Fig. 9-6b. These vertical cracks extend almost to the zero-strain axis (neutral axis) of the member. In a beam with a web that is more than 3 to 4 ft high, the cracking is relatively closely spaced at the level of the reinforcement, with several cracks joining or disappearing above the reinforcement, as shown in Fig. 9-6b. Again, the crack width at B will frequently exceed that at A.

Cracks due to shear have a characteristic inclined shape, as shown in Fig. 9-6c. Such cracks extend upward as high as the neutral axis and sometimes into the compression zone. Torsion cracks are similar. In pure torsion, they spiral around the beam. In a normal beam, where shear and moment also act, they tend to be pronounced on the face where the flexural shear stresses and the shear stresses due to torsion add, and less pronounced (or even absent) on the opposite face, where the stresses counteract (Fig. 9-6d).

Bond stresses lead to splitting along the reinforcement, as shown in Fig. 9-6e. Concentrated loads will sometimes cause splitting cracks or "bursting cracks" of the type shown in Fig. 9-6f. This type of cracking is discussed in Section 17-3. It occurs in bearing areas and in struts of strut-and-tie models.

At service loads, the final cracking pattern has generally not developed completely, with the result that there are normally only a few cracks at points of maximum stress at this load level.

Cracks also develop in response to imposed deformations, such as differential settlements, shrinkage, and temperature differentials. If shrinkage is restrained, as in the case of a thin floor slab attached at each end to stiff structural members, shrinkage cracks may occur. Generally, however, shrinkage simply increases the width of load-induced cracks.

A frequent cause of cracking in structures is restrained contraction resulting from the cooling down to ambient temperatures of very young members that expanded under the *heat of hydration*, which developed as the concrete was setting. This most typically occurs where a length of wall is cast on a foundation that was cast some time before. As the wall cools, its contraction is restrained by the foundation. A typical *heat-of-hydration cracking* pattern is shown in Fig. 9-7. Such cracking can be controlled by limiting the heat rise due to the heat of hydration and the rate of cooling, or both; by placing the wall in short lengths; or by reinforcement considerably in excess of normal shrinkage reinforcement [9-16], [9-17], [9-18].

Plastic shrinkage and slumping of the concrete, which occur as newly placed concrete bleeds and the surface dries, result in settlement cracks along the reinforcement (Fig. 9-8a) or a random cracking pattern, referred to as *map cracking* (Fig. 9-8b). These types of cracks can be avoided by proper mixture design and by preventing rapid drying of the surface during the first hour or so after placing. Map cracking can also be due to alkali–aggregate reaction.

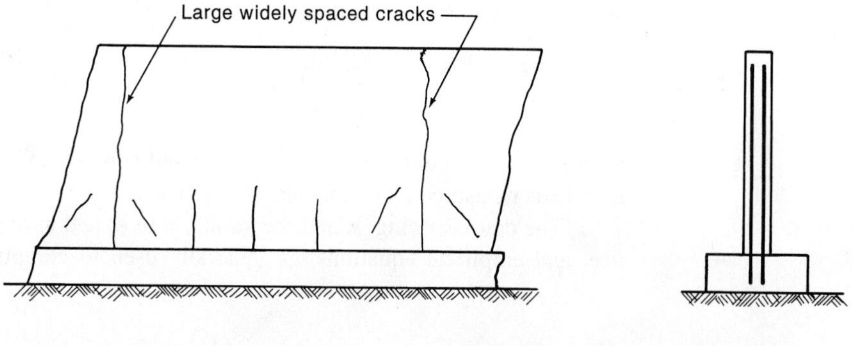

Fig. 9-7
Heat-of-hydration cracking.

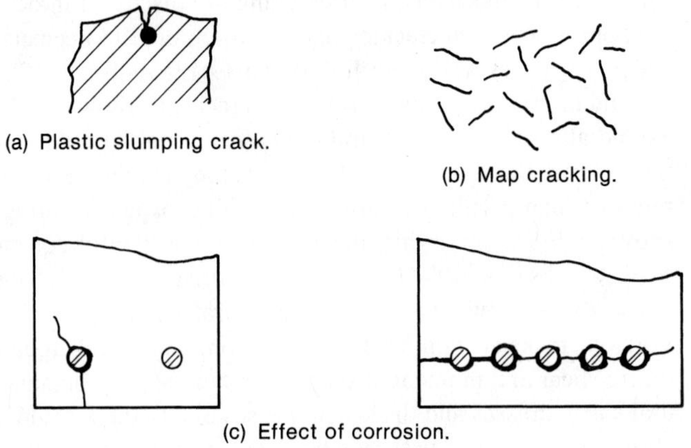

Fig. 9-8
Other types of cracks.

Rust occupies two to three times the volume of the metal from which it is formed. As a result, if rusting occurs, a bursting force is generated at the bar location, which leads to splitting cracks and an eventual loss of cover (Fig. 9-8c). Such cracking looks similar to bond cracking (Fig. 9-6e) and may accompany bond cracking.

Development of Cracks due to Loads

Figure 9-9 shows an axially loaded prism. Cracking starts when the tensile stress in the concrete (shown by the shaded area in Fig. 9-9b) reaches the tensile strength of the concrete (shown by the outer envelope) at some point in the bar. When this occurs, the prism cracks. At the crack, the entire force in the prism is carried by the reinforcement. Bond gradually builds up the stress in the concrete on either side of the crack until, with further loading, the stress reaches the tensile strength at some other section, which then cracks (Fig. 9-9c). With increasing load, this process continues until the distance between the cracks is not large enough for the tensile stress in the concrete to increase enough to cause cracking. Once this stage is reached, the crack pattern has stabilized, and further loading merely widens the existing cracks. The distance between stabilized cracks is a function of the overall member thickness, the cover, the efficiency of the bond, and several other factors. Roughly, however, it is two to three times the bar cover. Cracks that extend completely through the member generally occur at roughly one member thickness apart.

Figures 9-10b and 9-10c show the variation in the steel and concrete stresses along an axially loaded prism with a stabilized crack pattern. At the cracks, the steel stress and strain are at a maximum and can be computed by ignoring the concrete. Between the cracks, there is stress in the concrete. This reaches a maximum midway between two cracks. The total width, w, of a given crack is the difference in the elongation of the steel and the concrete over a length A–B equal to the crack spacing:

$$w = \int_A^B (\epsilon_s - \epsilon_c)\, dx \tag{9-12}$$

where ϵ_s and ϵ_c are the strains in the steel and concrete at a given location between A and B and x is measured along the axis of the prism.

The crack spacing, s, and the variation in ϵ_s and ϵ_c are difficult to calculate in practice, and empirical equations are generally used to compute the crack width. The best

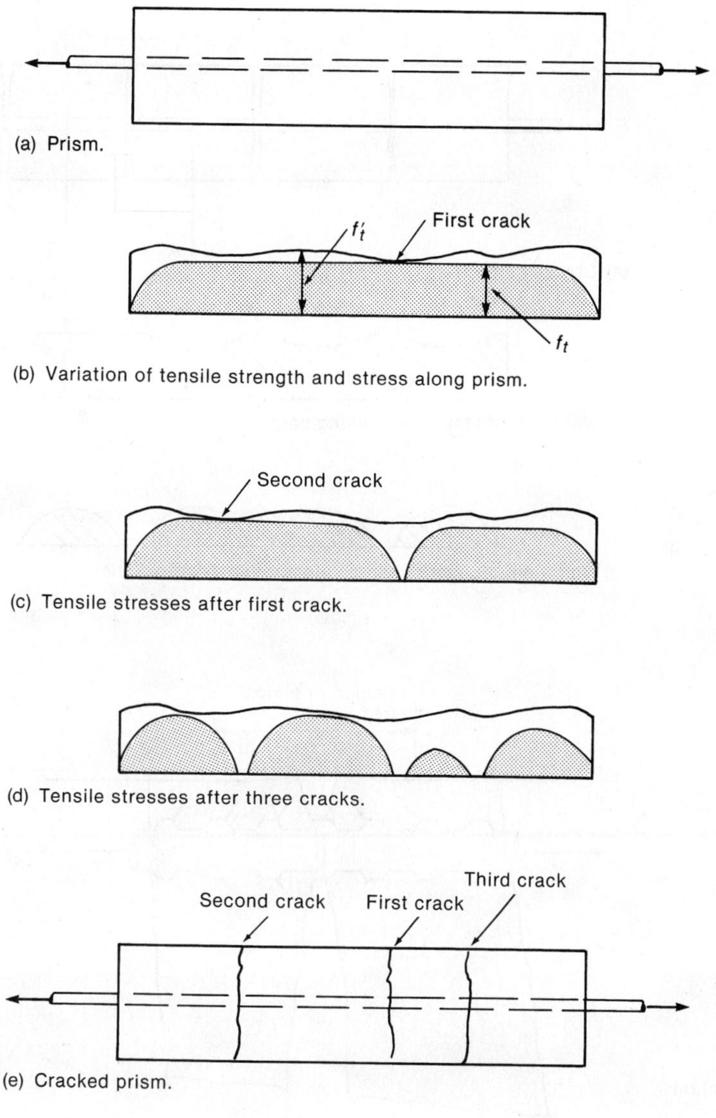

Fig. 9-9
Cracking of an axially loaded prism.

known of these are the Euro-International Concrete Committee (CEB) procedure [9-8], based on Eq. (9-12), and the Gergely–Lutz equation [9-9], derived statistically from a number of test series.

As discussed in Chapter 8, bond stresses are transferred from steel to concrete by means of forces acting on the deformation lugs on the surface of the bar. These lead to cracks in the concrete adjacent to the ribs, as shown in Fig. 9-11. In addition, the tensile stress in the concrete decreases as one moves away from the bar, leading to less elongation of the surface of the concrete than of the concrete at the bar. As a result, the crack width at the surface of the concrete exceeds that at the bar.

In a deep flexural member, the distribution of crack widths over the depth shows a similar effect, particularly if several cracks combine, as shown in Fig. 9-6b. The crack width at B frequently exceeds that at the level of the reinforcement.

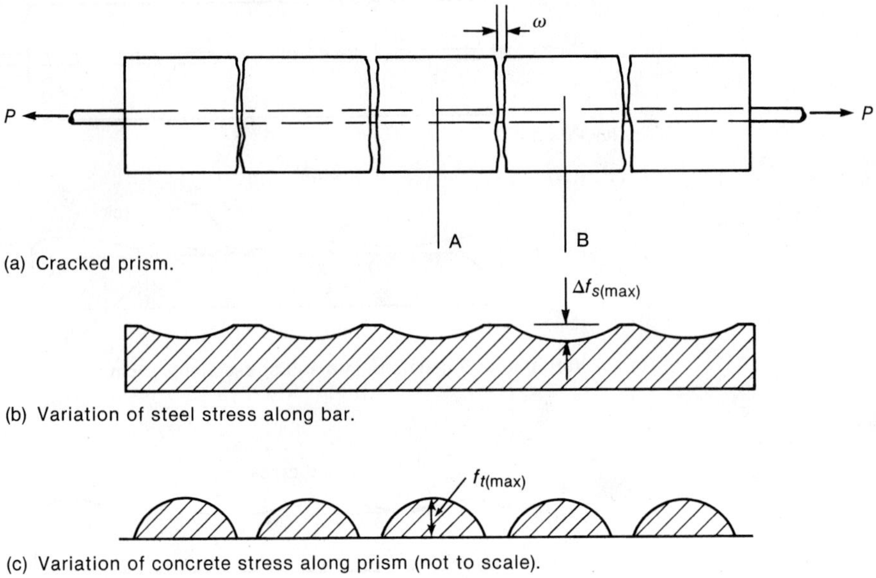

Fig. 9-10
Stresses in concrete and steel in a cracked prism.

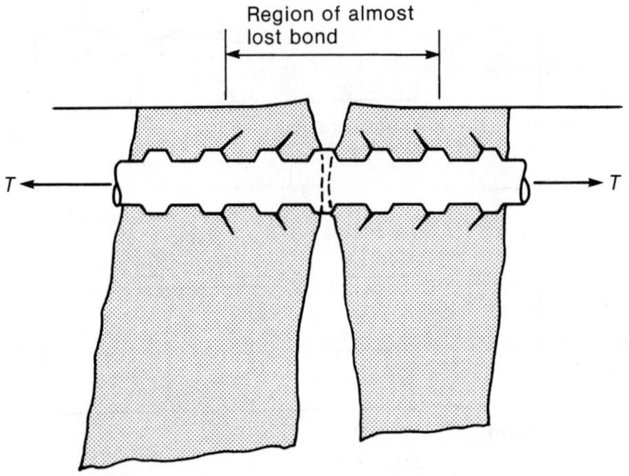

Fig. 9-11
Cracking at bar deformations. (From [9-2].)

Reasons for Controlling Crack Widths

Crack widths are of concern for three main reasons: appearance, leakage, and corrosion. Wide cracks are unsightly and sometimes lead to concern by owners and occupants. Crack width surveys [9-7] suggest that, on easily observed, clean, smooth surfaces, cracks wider than 0.01 to 0.013 in. can lead to public concern. Wider cracks are tolerable if the surfaces are less easy to observe or are not smooth. On the other hand, cracks in exposed surfaces may be accentuated by streaks of dirt or leached materials. Sandblasting the surface will accentuate the apparent width of cracks, making small cracks much more noticeable.

Limit states resulting from cracking of concrete are summarized in Table 9-1. It should be noted that reinforced concrete structures normally crack when carrying service loads. It is only when these cracks affect the usage or the appearance of the structure that cracking limit states are considered.

TABLE 9-1 Loads and Load Combinations for Serviceability—Cracking

(1) Case	(2) Serviceability Limit State and Elements Affected	(3) Serviceability Parameter and Load Combination	(4) Calculation Method and Acceptance Criteria	(5) References
A	**Cracking**			
A-1	**Leakage of Fluids or Gases**	Number and width of cracks. Number of times cracks are open.	The width of cracks must be less than a chosen maximum.	[9-6], [9-8], [9-11], [9-13], [9-18]
A-2	**Corrosion** Water and corrosive solutions reach the reinforcement	Width of cracks that extend to the reinforcement. Number of times crack width exceeds a given limit per year.	ACI Code does not give guidance for how to compute crack widths.	ACI Code Section 20.6.1.4.1. Book Section 9-3. Refs. [9-2], [9-10] to [9-12]
A-3	**Visually obvious cracks**	Cracks that can be seen. This is aggravated by sand-blasting the surface of the concrete.	ACI Code does not give guidance for how to compute crack widths.	ACI Code Sections 24.3.2 and 9.7.2.3. Book Section 9-3. Refs. [9-5], [9-13], [9-14], [9-15], [9-16]
A-4	**Crushing and cracking of prestressed members**	At transfer: 1.0D + 1.0 Prestress In service: 1.0D + 1.0 Prestress + 1.0L	Straight-line flexural stress calculations. Maximum compressive stress vs. long term strength, or tensile stress limit related to concrete tensile strength.	ACI Code Sections 24.3 and 24.5
A-5	**Shrinkage and creep**	Cracks due to restrained shrinkage.	Limits based on resulting damage. 2 to 3 times ACI temperature and shrinkage steel needed to control this cracking.	ACI Code Section 5.3.6. Book Section 3-6. Refs. [9-17], [9-18]

Crack control is important in the design of liquid-retaining structures. Leakage is a function of crack width, provided the cracks extend through the concrete member.

Corrosion of reinforcement has traditionally been related to crack width. More recent studies [9-10], [9-11], [9-12], and [9-13] suggest that the factors governing the eventual development of corrosion are independent of the crack width, although the period of time required for corrosion to start is a function of crack width. Reinforcement surrounded by concrete will not corrode until an electrolytic cell can be established. This will occur when carbonization of the concrete reaches the steel or when chlorides penetrate through the concrete to the bar surface. The time taken for this to occur will depend on a number of factors: whether the concrete is cracked, the environment, the thickness of the cover, and the permeability of the concrete. If the concrete is cracked, the time required for a corrosion cell to be established is a function of the crack width. After a period of 5 to 10 years, however, the amount of corrosion is essentially independent of crack width.

Generally speaking, corrosion is most apt to occur if one or more of the following circumstances occurs:

1. Chlorides or other corrosive substances are present.

2. The relative humidity exceeds 60 percent. Below this value, corrosion rarely occurs except in the presence of chlorides.

3. Ambient temperatures are high, accelerating the chemical reaction.

4. Ponded water and wetting and drying cycles occur that cause the concrete at the level of the steel to be alternately wet and dry. (Corrosion is minimal in permanently saturated concrete, because the water reduces or prevents oxygen flow to the steel.)

5. Stray electrical currents occur in the bars.

Corrosion of steel embedded in concrete is discussed in [9-10], [9-11], and [9-12], and in Section 3-10 of this book. ACI Code Chapter 19 and Section 20.6 are intended to limit or delay corrosion by limiting the permeability of the concrete. This is achieved by setting limits on the composition of the concrete and on the cover to the reinforcement.

Limits on Crack Width

There are no universally accepted rules for maximum crack widths. Prior to 1999, the ACI Code crack-control limits were based on a maximum crack width of 0.016 in. for interior exposure and 0.013 in. for exterior exposure [9-9]. What constitutes interior and exterior exposure was not defined. In addition to crack-control provisions, there are special requirements in ACI Chapter 19 for the composition of concrete subjected to special exposure conditions.

The International Federation for Structural Concrete (fib) [9-8], [9-15] limits the mean crack width (about 60 percent of the maximum crack width) as a function of exposure condition, sensitivity of reinforcement to corrosion, and duration of the loading condition.

ACI Code Sections 24.3.2 and 9.7.2.3 handle crack widths indirectly by limiting the maximum bar spacings and bar covers for beams and one-way slabs. Prior to 1999, these limits were based on the Gergely–Lutz [9-9] equation, which related the maximum crack width w at the tensile surface of a beam or slab and the cover to:

(a) the stress f_s in the steel at service loads,

(b) the distance d_c from the extreme concrete fiber to the centroid of the bar closest to the tension fiber, and

(c) the area A of the prism of concrete concentric with the bar.

The resulting equation tended to give unacceptably small bar spacings for bar covers greater than 2.5 in. For this reason, the Gergely–Lutz equation was replaced in the 1999 ACI Code by Eq. (9-13). This equation was obtained by fitting a straight line to the Gergely–Lutz equation for a flexural crack width of 0.016 in. This resulted in a relationship [9-14] that was modified to be

$$s = 15\left(\frac{40{,}000}{f_s}\right) - 2.5c_c\text{, but not greater than } 12\left(\frac{40{,}000}{f_s}\right) \text{ (psi)} \tag{9-13}$$

$$s = 380\left(\frac{280}{f_s}\right) - 2.5c_c\text{, but not greater than } 300\left(\frac{280}{f_s}\right) \text{ (MPa)} \tag{9-13M}$$

where s is the bar spacing in inches or millimeters, f_s is the service-load bar stress in psi or MPa, and c_c is the clear cover from the nearest surface of the concrete in the tension zone to the surface of the flexural-tension reinforcement, in inches or mm in Eq. (9-13) or (9-13M), respectively.

Equation (9-13) was based on the limiting crack width for interior exposure only. This is because the eventual amount of reinforcement corrosion has been shown to be independent of surface crack width [9-10], [9-11], [9-12].

The steel stress f_s can be computed via Eq. (9-5). Alternatively, ACI Code Section 24.3.2 allows the value of f_s at service loads to be taken as $2/3\,f_y$.

EXAMPLE 9-3 Checking the Distribution of Reinforcement in a Beam

At the point of maximum positive moment, a beam contains the reinforcement shown in Fig. 9-12. The reinforcement has a yield strength of 60,000 psi. Is this distribution satisfactory?

Assume that $f_s = 0.67 f_y = 40{,}000$ psi. Let c_c be the clear cover from the nearest surface of the concrete in tension to the surface of the *flexural-tension* reinforcement in inches. Then

$$c_c = 1.5\text{ in.} + 0.375\text{ in.} = 1.875\text{ in.}$$

$$s = 15\left(\frac{40{,}000}{40{,}000}\right) - 2.5 \times 1.875 = 10.3\text{ in.} < 12 \times \frac{40{,}000}{40{,}000} = 12\text{ in.}$$

Because the bar spacing in Fig. 9-12 is clearly less than 10.3 in., **the steel distribution is acceptable.** ■

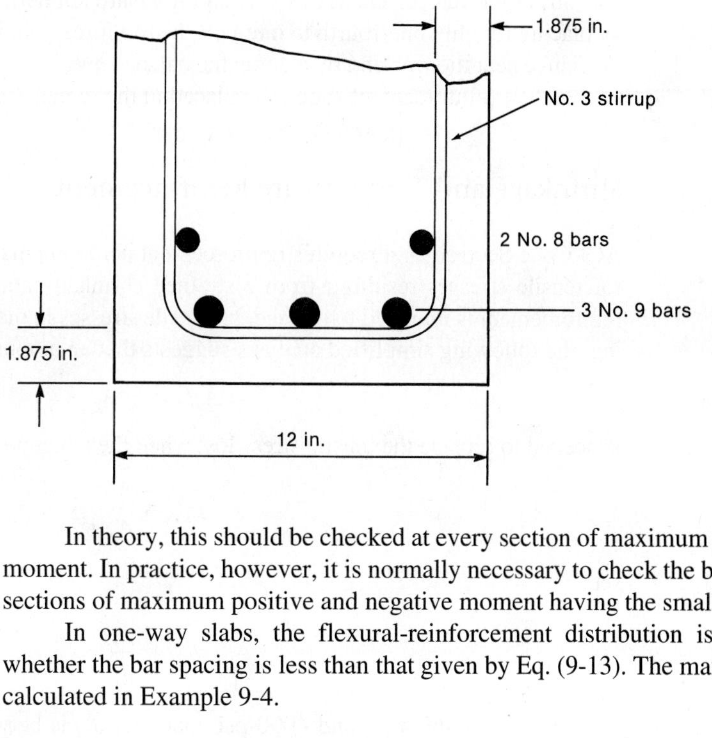

Fig. 9-12
Beam—Example 9-3.

In theory, this should be checked at every section of maximum positive and negative moment. In practice, however, it is normally necessary to check the bar spacing only at the sections of maximum positive and negative moment having the smallest numbers of bars.

In one-way slabs, the flexural-reinforcement distribution is found by checking whether the bar spacing is less than that given by Eq. (9-13). The maximum bar spacing is calculated in Example 9-4.

EXAMPLE 9-4 Calculation of the Maximum Bar Spacing for Bars in a One-Way Slab

An 8-in. thick slab has No. 4 bars at a spacing s. The bars have $f_y = 60{,}000$ psi and a minimum clear cover of $\frac{3}{4}$ in. Compute the maximum value of s.

We have

$$s = 15\left(\frac{40{,}000}{f_s}\right) - 2.5\,c_c \tag{9-13}$$

where we assume:

$$f_s = 0.67 f_y = 40{,}000 \text{ psi}$$

$$c_c = 0.75 \text{ in.}$$

Thus,

$$s = 15\,\frac{40{,}000}{40{,}000} - 2.5 \times 0.75 = 13.1 \text{ in.} > 12 \times \frac{40{,}000}{40{,}000} = 12 \text{ in.}$$

A bar spacing less than or equal to 12 in. satisfies ACI Code Section 24.3.2. The bar spacing must also satisfy ACI Code Section 8.7.2.2, which limits s to the smaller of three times the slab thickness, or 18 in. ■

Concrete covers in excess of 2 in. may be required for durability, for fire resistance, or in members such as footings.

In the negative-moment regions of T-beams, the flanges will be stressed in tension and will crack as shown in Fig. 4-37a. To restrict the width of the cracks in the flanges, ACI Code Section 24.3.4 requires that "part" of the flexural-tension reinforcement be distributed over a width equal to the smaller of the effective flange width and $\ell_n/10$. The same ACI section also requires that "some" longitudinal reinforcement be provided in the outer portions of the flange. The terms *part* and *some* are not defined. This can be accomplished by placing roughly one-fourth to one-half the reinforcement in the overhanging portions of the flange near the web and by placing the balance over the web of the beam. Shrinkage and temperature reinforcement should be placed in the remaining portion of the flange (slab).

Shrinkage and Temperature Reinforcement

ACI Code Section 24.4 requires reinforcement perpendicular to the span in one-way slabs for tensile stresses resulting from restrained shrinkage and temperature changes. If the reinforcement is intended to replace the tensile stresses in the concrete at the time of cracking, the following simplified analysis suggests that an amount equal to

$$A_s f_y = f_t' A_g$$

is needed to replace the tensile stress lost when the concrete cracks. If,

$$\rho_{s,t} = \frac{A_s}{A_g} \tag{9-14}$$

set

$$\rho_{s,t} \geq \frac{f_t'}{f_y}$$

For Grade-60 steel and 4000-psi concrete, $\rho_{s,t}$ is between 0.004 and 0.005. This is about three times the amount of shrinkage and temperature reinforcement specified in ACI Code Table 8.6.1.1. More complete analyses of the required amounts of shrinkage and temperature steel are presented by Gilbert [9-17] and Beeby [9-18].

In a one-way slab the area of shrinkage and temperature steel in a given direction is the sum of the areas of the top and bottom reinforcement in that direction.

Web Face Reinforcement

To control the width of flexural cracks in beams deeper than 3 ft, ACI Code Section 9.7.2.3 requires the use of *skin reinforcement* that is distributed uniformly along both faces of the beam web for a distance of $d/2$ measured from the centroid of the longitudinal tension reinforcement toward the neutral axis. Research [9-14] has shown that the spacing between these reinforcing bars is more important than the size of the bar for controlling the width of potential flexural cracks. Therefore, the spacing between the bars used for skin reinforcement is governed by Eq. (9-13), where c_c is the least dimension from the surface of the skin-reinforcing bar to the side face of the beam. For larger beams, this requirement can result in the addition of a significant amount of reinforcement. If strain compatibility is used to calculate the stresses in the various layers of skin reinforcement, as will be demonstrated in Chapter 11 for column sections with several layers of reinforcement, the effect of this steel on ϕM_n can be included in the section design.

9-4 DEFLECTIONS OF CONCRETE BEAMS

In this section, we deal with the flexural deflections of beams and slabs. Deflections of frames are considered in Section 9-6.

Load–Deflection Behavior of a Concrete Beam

Figure 9-13a traces the load–deflection history of the fixed-ended, reinforced concrete beam shown in Fig. 9-13b. Initially, the beam is uncracked and is stiff (O–A). With further load, flexural cracking occurs when the moment at the ends exceeds the cracking moment.

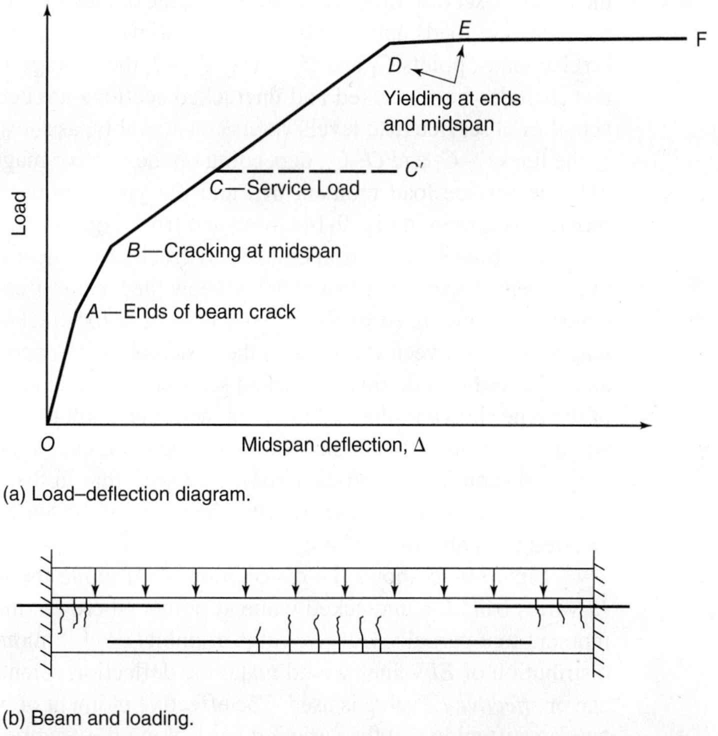

Fig. 9-13
Load–deflection behavior of a concrete beam.

When a section cracks, its moment of inertia decreases, leading to a decrease in the stiffness of the beam. This causes a reduction in stiffness (*A–B*) in the load–deflection diagram in Fig. 9-13a. Flexural cracking in the midspan region causes a further reduction of stiffness (point *B*). Eventually, the reinforcement would yield at the ends or at midspan, an effect leading to large increases in deflection with little change in load (points *D* and *E*). The service-load level is assumed to be at point *C*. The beam is essentially elastic at point *C*, the nonlinear load deflection prior to point *C* being caused by a progressive reduction of flexural stiffness due to increased cracking as the loads are increased.

With time, the service-load deflection would increase from *C* to *C*′, due to creep of the concrete. The short-time or *instantaneous deflection* under service loads (point *C*) and the *long-time deflection* under service loads (point *C*′) are both of interest in design.

Flexural Stiffness and Moment of Inertia

The deflection of a beam is calculated by integrating the curvatures along the length of the beam [9-19]. For an elastic beam, the curvature, $1/r$, is calculated as $1/r = M/EI$, where EI is the flexural stiffness of the cross section. If EI is constant, this is a relatively routine process. For reinforced concrete, however, three different EI values must be considered. These can be illustrated by the moment–curvature diagram for a length of beam, including several cracks, shown in Fig. 9-14d. The slope of any radial line through the origin in such a diagram is $M/\phi = EI$.

Before cracking, the entire cross section shown in Fig. 9-14b is stressed by loads. The moment of inertia of this section is called the *uncracked moment of inertia*, and the corresponding EI can be represented by the radial line *O–A* in Fig. 9-14d. Usually, the gross moment of inertia for the concrete section, I_g, is used for this region of behavior. The uncracked transformed moment of inertia discussed in Section 9-2 is seldom used. The effective cross section at a crack is shown in Fig. 9-14c. As demonstrated in Example 9-1, the *cracked-section EI* is less than the uncracked EI and corresponds relatively well to the curvatures at loads approaching yield, as shown by the radial line *O–B* in Fig. 9-14d. At service loads, points C_1 and C_2 in Fig. 9-14d, the average EI values for this beam segment that includes both cracked and uncracked sections are between these two extremes. The actual EI at service load levels varies considerably, as shown by the difference in the slope of the lines $O–C_1$ and $O–C_2$, depending on the relative magnitudes of the cracking moment M_{cr}, the service load moment M_a, and the yield moment M_y. The variation in EI with moment is shown in Fig. 9-14e, obtained from Fig. 9-14d.

The transition from uncracked to cracked moment of inertia reflects two different phenomena. Figures 9-10b and 9-10c show the tensile stresses in the reinforcement and the concrete in a prism. At loads only slightly above the cracking load, a significant fraction of tensile force between cracks is in the concrete, and hence the member behaves more like an uncracked section than a cracked section. As the loads are increased, internal cracking of the type shown in Fig. 9-11 occurs, with the result that the steel strains increase with no significant change in the tensile force in the concrete. At very high loads, the tensile force in the concrete is insignificant compared with that in the steel, and the member approximates a completely *cracked section*. The effect of the tensile forces in the concrete on EI is referred to as *tension stiffening*.

Figure 9-15 shows the distribution of EI along the beam shown in Fig. 9-13b. The EI varies from the uncracked value at points where the moment is less than the cracking moment to a partially cracked value at points of high moment. Because the use of such a distribution of EI values would make the deflection calculations tedious, an overall average or *effective EI* value is used. The effective moment of inertia must account for both the tension stiffening and the variation of EI along the member.

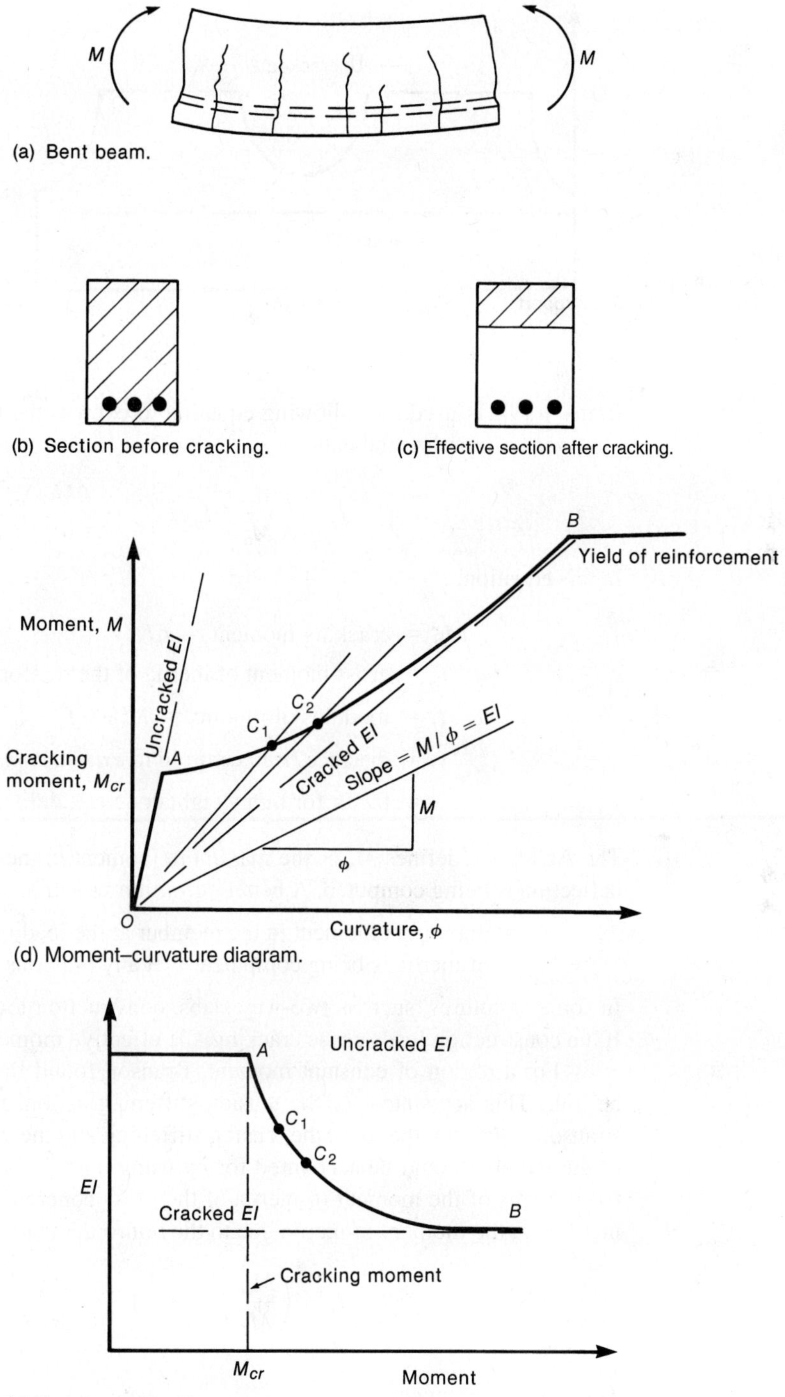

Fig. 9-14
Moment–curvature diagram and variation in *EI*.

Effective Moment of Inertia

The slope of line *OA* in Fig. 9-14d is approximately EI_g, where I_g is referred to as the *gross moment of inertia*, while that of line *OB* is approximately EI_{cr}, where I_{cr} is referred to as the *cracked moment of inertia*. At points between cracking (point *A*) and yielding of the steel (point *B*), intermediate values of *EI* exist. ACI Committee 435 [9-20], and

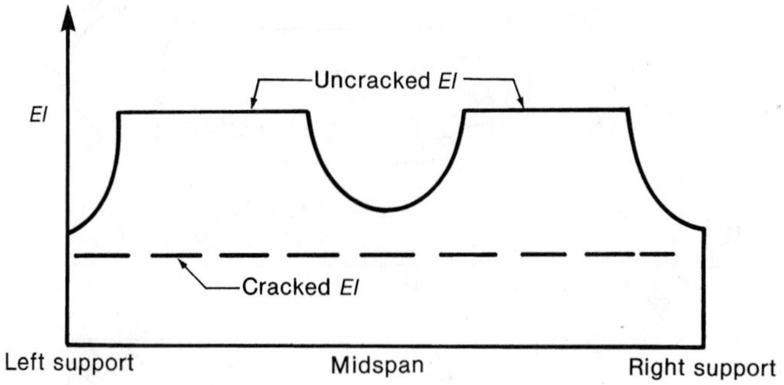

Fig. 9-15
Variation of EI along the length of the beam shown in Fig. 9-13b.

Branson [9-21] used the following equation to express the transition from I_g to I_{cr} that is observed in experimental data:

$$I_e = \left(\frac{M_{cr}}{M_a}\right)^a I_g + \left[1 - \left(\frac{M_{cr}}{M_a}\right)^a\right] I_{cr} \tag{9-15}$$

In this equation,

M_{cr} = cracking moment = $f_r I_g / y_t$
I_g = gross moment of inertia of the reinforced concrete section
f_r = modulus of rupture = $7.5\lambda\sqrt{f'_c}$
y_t = distance from centroid to *extreme tension fiber*
λ = factor for lightweight concrete, defined in ACI Code Table 19.2.4.2

The ACI Code defines M_a as the maximum moment in the member at the stage at which deflection is being computed. A better definition would be

M_a = maximum moment in the member at the loading stage for which the moment of inertia is being computed or at any previous loading stage

In some structures, such as two-way slabs, construction loads may exceed service loads. If the construction loads cause cracking, the effective moment of inertia will be reduced.

For a region of constant moment, Branson found the exponent a in Eq. (9-15) to be four. This accounted for the tension-stiffening action. For a simply supported beam, Branson suggested that both the tension stiffening and the variation in EI along the length of the member could be accounted for by using $a = 3$. The ACI Code equation is written in terms of the moment of inertia of the gross concrete section, I_g, ignoring the small increase in the moment of inertia due to the reinforcement:

$$I_e = \left(\frac{M_{cr}}{M_a}\right)^3 I_g + \left[1 - \left(\frac{M_{cr}}{M_a}\right)^3\right] I_{cr} \tag{9-16a}$$

(ACI Eq. 24.2.3.5a)

This can be rearranged to

$$I_e = I_{cr} + (I_g - I_{cr})\left(\frac{M_{cr}}{M_a}\right)^3 \tag{9-16b}$$

For a continuous beam, the I_e values may be quite different in the negative- and positive-moment regions. In such a case, the positive-moment value may be assumed to apply between the points of contraflexure and the negative-moment values in the end regions. ACI Code Section 24.2.3.6 suggests the use of the average I_e value. A better suggestion is given in [9-20]:

Beams with two ends continuous:

$$\text{Average } I_e = 0.70 I_{em} + 0.15(I_{e1} + I_{e2}) \tag{9-17a}$$

Beams with one end continuous:

$$\text{Average } I_e = 0.85 I_{em} + 0.15(I_{e\text{ continuous end}}) \tag{9-17b}$$

Here, I_{em}, I_{e1}, and I_{e2} are the values of I_e at midspan and the two ends of the beam, respectively. Moment envelopes or moment coefficients should be used in computing M_a and I_e at the positive- and negative-moment sections.

Instantaneous and Additional Sustained Load Deflections

When a concrete beam is loaded, it undergoes a deflection referred to as an *instantaneous deflection*, Δ_i. If the load remains on the beam, additional *sustained-load deflections* occur due to creep. These two types of deflections are discussed separately.

Instantaneous Deflections

Equations for calculating the instantaneous midspan and tip deflections of single-span beams for common support cases are summarized in Fig. 9-16. In this figure, M_{pos} and M_{neg} refer to the maximum positive and negative moments, respectively. The deflections listed

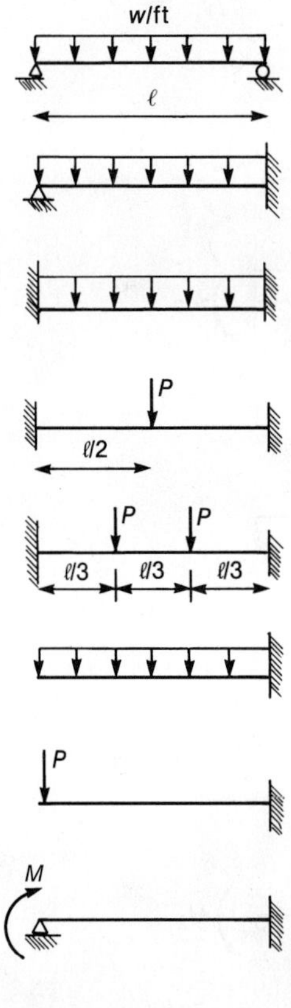

Case 1 $\Delta_{mid} = \frac{5}{384} \cdot \frac{w\ell^4}{EI} = \frac{5}{48} \cdot \frac{M_{pos}\ell^2}{EI}$

Case 2 $\Delta_{mid} = \frac{1}{192} \cdot \frac{w\ell^4}{EI} = \frac{128}{1728} \cdot \frac{M_{pos}\ell^2}{EI}$

Case 3 $\Delta_{mid} = \frac{1}{384} \cdot \frac{w\ell^4}{EI} = \frac{1}{16} \cdot \frac{M_{pos}\ell^2}{EI}$

Case 4 $\Delta_{mid} = \frac{1}{192} \cdot \frac{P\ell^3}{EI} = \frac{1}{24} \cdot \frac{M_{pos}\ell^2}{EI}$

Case 5 $\Delta_{mid} = \frac{5}{684} \cdot \frac{P\ell^3}{EI} = \frac{5}{72} \cdot \frac{M_{pos}\ell^2}{EI}$

Case 6 $\Delta_{tip} = \frac{1}{8} \cdot \frac{w\ell^4}{EI} = \frac{1}{4} \cdot \frac{M_{neg}\ell^2}{EI}$

Case 7 $\Delta_{tip} = \frac{1}{3} \cdot \frac{P\ell^3}{EI} = \frac{1}{3} \cdot \frac{M_{neg}\ell^2}{EI}$

Case 8 $\Delta_{mid} = \frac{1}{32} \cdot \frac{M\ell^2}{EI}$

Fig. 9-16
Deflection equations.

are the maximum deflections along the beam, except for Cases 2 and 8. However, even for these two cases, the midspan deflection is a good estimate of the maximum deflection. Furthermore, it may be necessary to combine different load cases from Fig. 9-16 to get the total beam deflection, so it is appropriate to determine all of these deflections either at midspan of a beam with two supports or at the tip of a cantilever beam.

It should be noted that the author recommends the use of the total span length, ℓ, when calculating deflections, as opposed to the clear span, ℓ_n. This is a conservative decision to avoid underestimating the potential maximum deflections. Also, most concrete beams and slabs are continuous over several spans, so the deflection coefficients for single-span beams may not always be appropriate. The length of adjacent spans and the loads on those spans will affect deflections in the span under consideration. In the following discussions, the length of the adjacent spans will be ignored, but potential occurrence of different live load patterns will be considered.

The midspan deflection for a continuous beam with uniform loads and unequal end moments can be computed with the use of superposition, as shown in Fig. 9-17. Thus,

$$\Delta = \Delta_0 + \Delta_1 + \Delta_2$$

From the right-hand side in Fig. 9-17, assuming that M_0 (the moment at midspan due to uniform loads on a simple beam), M_1, and M_2 are all positive, so that the signs will be compatible, we have

$$\Delta = \frac{5}{48}\frac{M_0\ell^2}{EI} + \frac{3}{48}\frac{M_1\ell^2}{EI} + \frac{3}{48}\frac{M_2\ell^2}{EI}$$

The midspan moment M_m is

$$M_m = M_0 + \frac{M_1}{2} + \frac{M_2}{2}$$

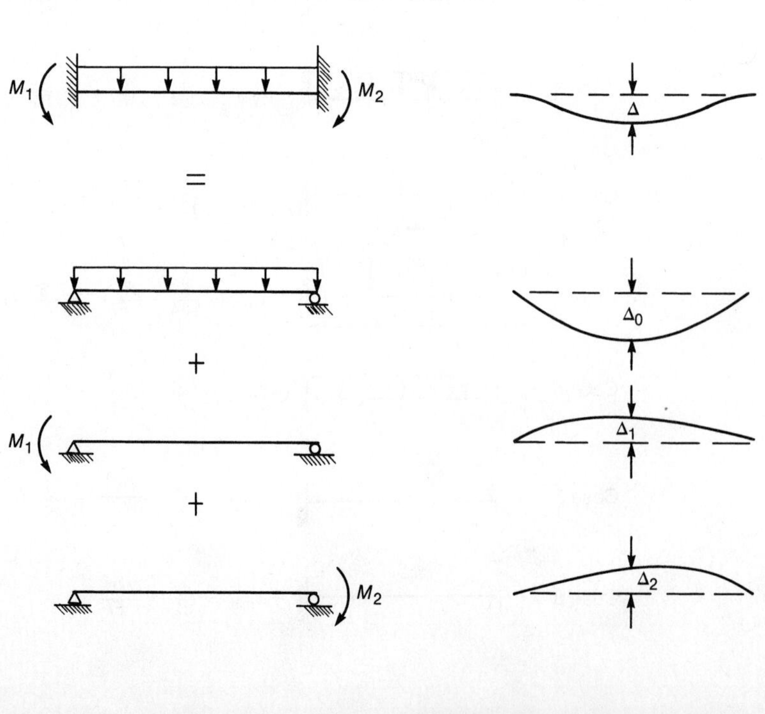

Fig. 9-17
Calculation of deflection for a beam with unequal end moments.

Therefore, expressing the deflection in terms of M_m, M_1, and M_2, we obtain

$$\Delta = \frac{\ell^2}{48\,EI}\left[5M_m - \frac{5}{2}(M_1 + M_2) + 3(M_1 + M_2)\right]$$

$$= \frac{5}{48}\frac{\ell^2}{EI}\left[M_m + 0.1(M_1 + M_2)\right] \tag{9-18}$$

where Δ is the midspan deflection, M_m is the midspan moment, and M_1 and M_2 are the moments at the two ends, which must be given the correct algebraic sign. Generally, M_1 and M_2 will be negative.

In the following, some specific deflection coefficients will be derived for continuous beams or slabs subjected to uniformly distributed loads. For the calculation of deflections due to dead load where adjacent spans will all be loaded with the same uniform dead load, the load cases in Fig. 9-16 can be used. For an interior span, Case 3 in Fig. 9-16 is the appropriate case, and the midspan deflection would be

$$\frac{1}{384}\frac{w\ell^4}{EI} = 2.6 \times 10^{-3}\frac{w\ell^4}{EI}$$

This result is given in the first row of Table 9-2, with w representing the unfactored uniform dead load. Note that only two significant figures are used for the deflection coefficients in the last column of Table 9-2 to represent the expected level of accuracy for deflection calculations in reinforced concrete beams and slabs.

For an exterior span beam that is subjected to uniform dead load, the maximum deflection is a function of the rotational stiffness at the exterior support. If the exterior support is a spandrel beam, it is conservative to ignore the torsional stiffness of the spandrel beam and use Case 2 in Fig. 9-16, as is done in row 2 of Table 9-2. However, if the exterior support is a column, we can assume that the flexural stiffness of the column will tend to reduce the maximum beam deflection. This is handled in row 3 of Table 9-4 by combining Cases 2 and 8 from Fig. 9-16. The correct value of M to use in Case 8 is not known unless a full stiffness analysis is carried out. However, an approximate value of M can be obtained from the ACI moment coefficients in Code Table 6.5.2, which were used for some designs in Chapter 5. The correct coefficient from the external moment in this case is $-w\ell^2/16$. Using this moment coefficient in Case 8 of Fig. 9-16 and combining that result with Case 2 of the same figure results in the deflection coefficient given in row 3 of Table 9-2.

To determine maximum displacements due to live load, we should consider the case of alternate span loading that results in maximum values for the midspan positive moment

TABLE 9-2 Deflection Coefficients for Spans with Distributed Loading

Span and Supports	Loading	Deflection Case Fig. 9-16	M in Eq. (9-18) × $(w\ell^2)$ M_1	M_m	M_2	Deflection Coefficient × $(w\ell^4/EI)$
1. Interior	Uniform *DL*	3	—	—	—	2.6×10^{-3}
2. Exterior: Spandrel Ext. Support	Uniform *DL*	2	—	—	—	5.2×10^{-3}
3. Exterior: Column Ext. Support	Uniform *DL*	2 and 8	—	—	—	3.3×10^{-3}
4. Interior	Pattern *LL*	—	−1/12	1/16	−1/12	4.8×10^{-3}
5. Exterior: Spandrel Ext. Support	Pattern *LL*	—	−1/24	1/14	−1/11	6.1×10^{-3}
6. Exterior: Column Ext. Support	Pattern *LL*	—	−1/16	1/14	−1/11	5.8×10^{-3}

and midspan deflection. To determine deflection coefficients for this loading case, the author will use Eq. (9-18) with the M_m values obtained from the ACI moment coefficients in Code Table 6.5.2. The end moments M_1 and M_2 in Eq. (9-18) also should come from the same loading case. Thus, most of the negative-moment coefficients in ACI Code Table 6.5.2 are not appropriate, because they come from different loading patterns intended to maximize the negative moments at the faces of the supports. Therefore, the author used his engineering judgment to select the M_1 and M_2 moment coefficients listed in rows 4, 5, and 6 of Table 9-2. The values of M_1 for exterior spans (rows 5 and 6) do correspond to the ACI moment coefficients, because they come from the same loading case used to maximize the midspan positive moment in the exterior span. However, the M_1 value in row 4 and all of the M_2 values were selected to be less than the corresponding ACI moment coefficients. When using the deflection coefficients in rows 4 through 6 of Table 9-2, w represents the unfactored distributed live load. Use of the deflection coefficients in Table 9-2 will be demonstrated in Examples 9-5 and 9-6.

Sustained-Load Deflections

Under sustained loads, concrete undergoes creep strains and the curvature of a cross section increases, as shown in Fig. 9-18. Because the lever arm is reduced, there is a small increase in the steel force, but for normally reinforced sections, this will be minimal. At the same time, the compressive stress in the concrete decreases slightly, because the compression zone is larger.

If compression steel is present, the increased compressive strains will cause an increase in stress in the compression reinforcement, thereby shifting some of the compressive force from the concrete to the compression steel. As a result, the compressive stress in the concrete decreases, resulting in reduced creep strains. The larger the ratio of compression reinforcement, $\rho' = A_s'/bd$, the greater the reduction in creep, as shown in Fig. 4-30. From test data, Branson [9-21] derived Eq. (9-19), which gives the ratio, λ_Δ, of the *additional* sustained load deflection to the instantaneous deflection (Δ_i). The total instantaneous and sustained load deflection is $(1 + \lambda_\Delta)\,\Delta_i$, where

$$\lambda_\Delta = \frac{\xi}{1 + 50\rho'} \tag{9-19}$$

(ACI Eq. 24.2.4.1.1)

where $\rho' = A_s'/bd$ at midspan for simple and continuous beams and at the support for cantilever beams and ξ is a factor between 0 and 2, depending on the time period over which sustained load deflections are of interest. Values of ξ are given in ACI Code Table 24.2.4.1.3 and Fig. 9-19.

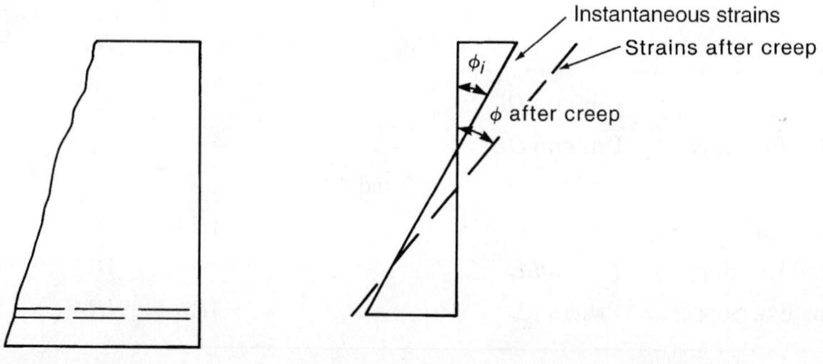

Fig. 9-18
Effect of creep on strains and curvature.

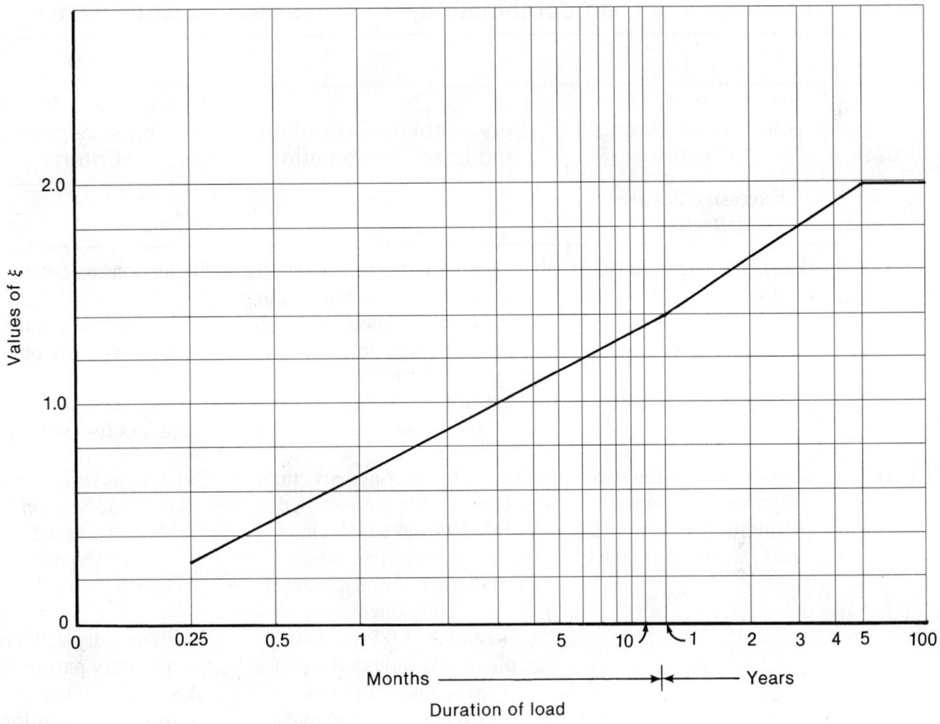

Fig. 9-19
Value of ξ.

In Section 3-6 of this book, Example 3-4 illustrates the internal load transfer from compressed concrete to compression reinforcement in columns carrying sustained axial loads. These calculations were carried out by using transformed sections based on an age-adjusted effective modulus for the concrete [9-2], [9-3], [9-4]. A similar load transfer occurs in the compression zones of beams.

9-5 CONSIDERATION OF DEFLECTIONS IN DESIGN

Serviceability Limit States Resulting from Flexural Deflections

Serviceability limit states arising from excessive flexural deflections are discussed here and are summarized in Table 9-3.

Visual Appearance

Deflections greater than about 1/250 of the span ($\ell/250$) are generally visible for simple or continuous beams, depending on a number of factors, such as floor finish [9-22]. Deflections will be most noticeable if it is possible to compare the deflected member with an undeflected member or a horizontal line. Uneven deflections of the tips of cantilevers can be particularly noticeable. The critical deflection is the total immediate and sustained deflection minus any camber. (See Case B-1 in Table 9-3.)

Damage to Nonstructural Elements

Excessive deflections can cause cracking of partitions or malfunctioning of doors and windows. Such problems can be handled by limiting the deflections that occur after

TABLE 9-3 Loads and Load Combinations for Serviceability Limit States—Flexural Deflections

(1) Case	(2) Serviceability Limit State and Elements Affected	(3) Serviceability Parameter and Load Combination	(4) Calculation Method and Acceptance Criteria	(5) References
B	**Excessive flexural deflections**			
B-1	Unsightly sag of floors, droopy cantilevers	Total sustained load deflections. Deflection visible when building is in service. Camber may hide sag, but large deflections may give nonstructural problems. $1.0D + 1.0$ Prestress plus possibly $0.4L$.	Deflections calculated with ACI Code Section 24.2. Deflection less than limit of visually obvious deflections. $\Delta/\ell \approx 1/250$ and less for cantilevers.	Book Sections 9-4 and 9-5. Refs. [9-8] [9-15], [9-20], [9-21], [9-22], [9-24], [9-28]
B-2	Damage to nonstructural components, drywall partitions, block or brick walls, jamming of doors	Deflection after nonstructural components are installed **taken equal to** sustained load deflection after installation of nonstructural components. $1.0D + 1.0$ Prestress **plus** short-time deflection after installation of nonstructural components. $0.4L$ or $0.6S$.	Deflections calculated with ACI Code Section 24.2. Limits depend on materials. Plaster, drywall, and masonry partitions $\Delta/\ell \leq 1/500$ to 1/1000. Moveable partitions $\Delta/\ell \leq 1/240$.	Book Sections 9-4 and 9-5. Refs. [9-8], [9-15], [9-19], [9-22], [9-24], [9-27]
B-3	Malfunction of equipment	Increase in deflection or increase in curvature of main shaft after machinery is aligned.	Deflections calculated with ACI Code Section 24.2. Limits to be provided by equipment manufacturer.	Book Sections 9-4 and 9-5. Refs. [9-20], [9-25], [9-26]
B-4	Ponding failure	Second-order increase in deflection of a roof or surface holding rain.	Deflections calculated with ACI Code Section 24.2. Short-time deflection under own weight and weight of collected rain.	Book Section 9-5. Refs. [9-23], [9-25]

installation of the nonstructural elements by delaying their installation, or by designing the nonstructural elements to accommodate the required amount of movement. (See Cases B-2 and B-3 in Table 9-3.)

Surveys of partition damage have shown that damage to brittle partitions can occur with deflections as small as $\ell/1000$. A frequent limit on deflections that cause damage is a deflection of $\ell/480$ occurring after attachment of the nonstructural element (ACI Code Table 24.2.2). This is the sum of the instantaneous deflection due to live loads plus the sustained portion of the deflections due to dead load and any sustained live load; that is,

$$\Delta = \lambda_{to,\infty}\Delta_{iD} + \Delta_{iL} + \lambda_{\infty}\Delta_{iLS} \tag{9-20}$$

where Δ_{iD}, Δ_{iL}, and Δ_{iLS} are the instantaneous deflections due to the dead load, the live load, and the sustained portion of the live load, respectively; $\lambda_{to,\infty}$ is the value of λ_{Δ} in Eq. (9-16) based on the value of ξ for 5 years or more, minus the value of ξ at the time t_0 when the partitions and so on are installed. (See Fig. 9-19.) Thus, if the nonstructural elements were installed 3 months after the shores were removed, the value of ξ used in calculating $\lambda_{(t_0,\infty)}$

would be $2.0 - 1.0 = 1.0$. Finally, λ_∞ is the value of λ_Δ based on $\xi = 2.0$, because it is assumed here that all the sustained *live-load* deflection affecting nonstructural damage will occur after the partitions are installed.

Racking or shear deflections of a plastered or brick wall may damage walls or cause doors to jam at very small Δ/ℓ ratios. The maximum slope, Δ/ℓ, will occur near the quarter-points of the spans.

As is evident from Eq. (9-15), the effective moment of inertia, I_e, decreases as the loads increase, once cracking has occurred. As a result, the value of I_e is larger when only dead load acts than when dead and live loads act together. There are two schools of thought about the value of I_e to be used in the calculation of the Δ_{iD} in Eq. (9-20). Some designers use the value of I_e, which is effective when only dead load is supported. Others reason that the portion of the sustained dead-load deflection which is significant is that occurring after the partitions are in place (i.e., that occurring during the live-loading period). This second line of reasoning implies that Δ_{iD} in Eq. (9-20) should be based on the value of I_e corresponding to dead plus live loads. This book supports the second procedure, particularly in view of the fact that construction loads or shoring loads may lead to premature cracking of the structure. The solution of Eq. (9-20) is illustrated in Example 9-6.

Disruption of Function

Excessive deflections may interfere with the use of the structure. This is particularly true if the members support machinery that must be carefully aligned. The corresponding deflection limits will be prescribed by the manufacturer of the machinery or by the owner of the building.

Excessive deflections may also cause problems with drainage of floors or roofs. The deflection due to the weight of water on a roof will cause additional deflections, allowing it to hold more water. In extreme cases, this may lead to progressively deeper and deeper water, resulting in a *ponding failure* of the roof [9-23]. Ponding failures are most apt to occur with long-span roofs in regions where roof design loads are small. (See Case B-4 in Table 9-3.)

Damage to Structural Elements

The deflections corresponding to structural damage are several times larger than those causing serviceability problems. If a member deflects so much that it comes in contact with other members, the load paths may change, causing cracking. Creep may redistribute the reactions of a continuous beam or frame thereby reducing the effect of differential settlements [9-24].

Allowable Deflections

In principle, the designer should select the deflection limit that applies for a given structure, basing it on the characteristics and function of the structure. In this respect, allowable deflections given in codes are only guides.

The ACI Code gives values of maximum permissible computed deflections in ACI Code Table 24.2.2. The ACI Deflection Committee has also proposed a more comprehensive set of allowable deflections [9-25].

Accuracy of Deflection Calculations

The ACI Code deflection-calculation procedure gives results that are within ±20 percent of the true values, provided that M_{cr}/M_a is less than 0.7 or greater than 1.25 [9-26]. Because a

TABLE 9-4 Loads and Load Combinations for Serviceability Limit States—Racking and Relative Vertical Deflections

(1) Case	(2) Serviceability Limit State and Elements Affected	(3) Serviceability Parameter Load Combination	(4) Calculation Method and Acceptance Criteria	(5) References
C	**Racking deflections**			
C-1	Cracking of partitions, jamming of doors, structural damage	Sustained and short-time shear deflections after partitions, door frames, nonstructural elements are installed. $1.0D + 1.0$ Prestress **plus** short-time deflection due to $0.4L$ or $0.6S$	Deflections calculated using ACI Code Section 24.2 Limits on racking deflections.	Book Sections 9-5 and 9-6. Refs. [9-8], [9-22], [9-25], [9-27], [9-30], [9-31]
D	**Relative Vertical Displacements**			
D-1	Crushing and cracking of cladding	Relative vertical displacement of brick or other cladding, and the frame. $1.0T_t + 1.0T_{cs} + 1.0$ Prestress where T_t is thermal expansion, T_{cs} is creep and shrinkage. Crushing strain of cladding.		Book Section 9-6. Refs. [9-2], [9-12] [9-30], [9-31]
D-2	Sloping floors	Relative vertical deflection of columns and shear walls due to creep. $1.0D + 1.0T_{cs} + 1.0$ Prestress, T_{cs} is creep and shrinkage Limit of visually obvious slope.		Book Section 9-6. Refs. [9-22], [9-27], [9-30], [9-31]
D-3	Sloping floors	Relative vertical deflection of interior columns and exterior columns exposed to temperature changes. About $\frac{3}{4}$ in. in 25 ft $1.0D + 1.0T_t$ where T_t is thermal expansion.	Limit of visually obvious slope.	Book Section 9-6. Refs. [9-22], [9-30], [9-31]
D-4	Damage to structure	Differential settlement About $\frac{3}{4}$ in. in 25 ft	Beam and column structures can tolerate differential settlements of about $\frac{3}{4}$ in. in 25 ft	Refs. [9-8], [9-30], [9-31]

small amount of cracking can make a major difference in the value of I_e, as indicated by the big decrease in slope from O–A to O–C in Fig. 9-14d, the accuracy is considerably poorer when the service-load moment is close to the cracking moment. This is because random variations in the tensile strength of the concrete will change M_{cr}, and hence I_e.

For these reasons it is frequently satisfactory to use a simplified deflection calculation as a first trial and then refine the calculations if deflections appear to be a problem.

Deflection Control by Span-to-Depth Limits

The relationship between beam depth and deflections is given by

$$\frac{\Delta}{\ell} = C\frac{\ell}{d} \tag{5-9b}$$

where the term C allows for the support conditions, and the grade, and amount of steel. For a given desired Δ/ℓ, a table of ℓ/d values can be derived. ACI Table 9.3.1.1 is such a table and gives minimum thicknesses of beams and one-way slabs unless deflections are computed. This table is frequently used to select member depths. It should be noted, however, that this table applies only to *members not supporting, or attached to, partitions or other construction likely to be damaged by large deflections*. If the member supports such partitions, and so on, the ACI Code requires calculation of deflections.

Table A-9 in Appendix A summarizes the ACI deflection limits and presents other more stringent limits for the case of beams supporting brittle partitions [9-20], [9-26], [9-27], [9-28].

EXAMPLE 9-5 Calculation of Immediate Deflection

The T-beam shown in Fig. 9-20 supports unfactored dead and live loads of 0.87 kip/ft and 1.2 kips/ft, respectively. It is built of 4000-psi concrete and Grade-60 reinforcement. The moments used in the design of the beam were calculated from the moment coefficients in ACI Code Table 6.5.2. Calculate the immediate midspan deflection. Assume that the construction loads did not exceed the dead load.

1. Is the beam cracked at service loads? The cracking moment is

$$M_{cr} = \frac{f_r I_g}{y_t}$$

Compute I_g for the uncracked T-section (ignore the effect of the reinforcement for simplicity):

Assume flange width = effective flange width from ACI Code Table 6.3.2.1 = 66 in.

(a) Compute the centroid of the cross section.

Part	Area (in.^{2})	y_{top} (in.)	Ay_{top} (in.^{3})
Flange	$66 \times 5 = 330$	2.5	825
Web	$13 \times 10 = 130$	11.5	1495
	Area = 460 in.^{2}		ΣAy_t = 2320 in.^{3}

The centroid is located at $2320/460 = 5.04$ in. below the top of the beam:

$$\bar{y}_{top} = 5.04 \text{ in.} \quad \bar{y}_{bottom} = 12.96 \text{ in.}$$

(b) Compute the moment of inertia, I_g.

Part	Area (in.^{2})	$\bar{y}$ (in.)	$I_{own\ axis}$ (in.^{4})	$A\bar{y}^2$ (in.^{4})
Flange	330	$5.04 - 2.5 = 2.54$	688	2129
Web	130	$5.04 - 11.5 = -6.46$	1831	5425
			I_y = 10,100 in.^{4}	

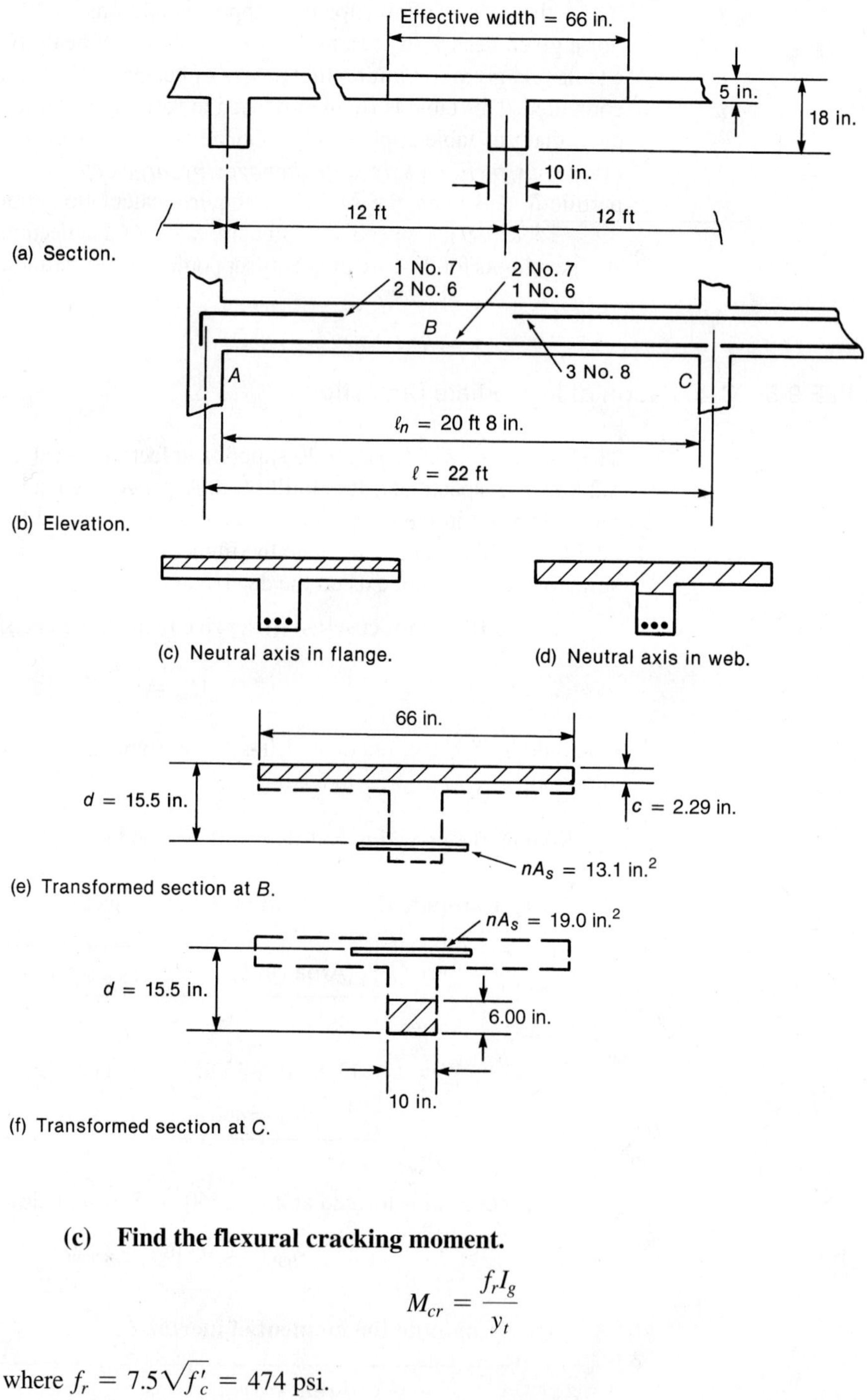

Fig. 9-20
End span of beam—
Examples 9-5 and 9-6.

(c) Find the flexural cracking moment.

$$M_{cr} = \frac{f_r I_g}{y_t}$$

where $f_r = 7.5\sqrt{f'_c} = 474$ psi.

In negative-moment regions at A and C, $y_t = 5.04$ in.:

$$M_{cr} = \frac{474 \text{ psi} \times 10{,}100 \text{ in.}^4}{5.04 \text{ in.}} = 950{,}000 \text{ lb-in.}$$

$$= 79.2 \text{ kip-ft}$$

$$\text{Negative moment at } A = -\frac{w\ell_n^2}{16}$$

$$\text{Dead-load moment} = -\frac{0.87 \times 20.67^2}{16} = -23.2 \text{ kip-ft}$$

$$\text{Dead-plus-live-load moment} = -\frac{(0.87 + 1.20) \times 20.67^2}{16} = -55.3 \text{ kip-ft}$$

Because both of these are less than M_{cr}, section A will not be cracked at service loads.

$$\text{Negative moment at } C = -\frac{w\ell_n^2}{10}$$

$$\text{Dead-load moment} = -\frac{0.87 \times 20.67^2}{10} = -37.2 \text{ kip-ft (not cracked)}$$

$$\text{Dead-plus-live-load moment} = -\frac{2.07 \times 20.67^2}{10} = -88.4 \text{ kip-ft (cracked)}$$

In the positive-moment region, $y_t = 12.96$ in.:

$$M_{cr} = \frac{474 \times 10{,}100}{12.96} = 369{,}000 \text{ lb-in.} = 30.8 \text{ kip-ft}$$

$$\text{Positive moment at } B = \frac{w\ell_n^2}{14}$$

$$\text{Dead-load moment} = \frac{0.87 \times 20.67^2}{14}$$

$$= 26.6 \text{ kip-ft (not cracked)}$$

$$\text{Dead-plus-live-load moment} = \frac{2.07 \times 20.67^2}{14} = 63.2 \text{ kip-ft (cracked)}$$

Therefore, it will be necessary to compute I_{cr} and I_e at midspan and at support C.

2. Compute I_{cr} at midspan. It is not known whether this section will have a rectangular compression zone (Fig. 9-20c) or a T-shaped compression zone (Fig. 9-20d). We shall assume that the compression zone is rectangular for the first trial.

To locate the centroid, either follow the calculation in Example 9-1, or, because the compression zone is assumed to be rectangular and there is no compression steel, use Eq. (9-3) to compute k, where $x = kd$. Either way, we obtain

$$k = \sqrt{2\rho n + (\rho n)^2} - \rho n$$

where

$$n = \frac{E_s}{E_c}$$

$$= \frac{29 \times 10^6}{57{,}000\sqrt{4000}} = 8.0$$

$$\rho \text{ (based on } b_e = 66 \text{ in.)} = \frac{1.64}{66 \times 15.5} = 0.00160$$

$$\rho n = 0.0128$$

$$k = \sqrt{2 \times 0.0128 + 0.0128^2} - 0.0128$$

$$= 0.148$$

Therefore,

$$x = kd = 0.148 \times 15.5$$

$$= 2.29 \text{ in.}$$

Because this is less than the thickness of the flange, this is a Case 1 analysis, as assumed. If it is not, use Eq. (9-10) to find x and Eq. (9-11) to find I_{cr}.

Compute the cracked moment of inertia at *B*. Use Eq. (9-7b) to find I_{cr} with $b_e = 66$ in.

$$I_{cr} = \frac{b_e \cdot x_1^3}{3} + n \cdot A_s(d - x_1)^2$$

$$I_{cr} = \frac{66 \text{ in.}(2.29 \text{ in.})^3}{3} + 8.0 \times 1.64 \text{ in.}^2(15.5 \text{ in.} - 2.29 \text{ in.})^2$$

$$I_{cr} = 264 \text{ in.}^4 + 2290 \text{ in.}^4 = 2550 \text{ in.}^4$$

3. Compute I_{cr} at support *C*. The cross section at support C is shown in Fig. 9-17f. The transformed area of the reinforcement is

$$nA_s = 8.0 \times 2.37 = 19.0 \text{ in.}^2$$

$$\rho = \frac{2.37}{10 \times 15.5} = 0.0153$$

From Eq. (9-3), $k = 0.387$ and

$$x = kd = 6.00 \text{ in.}$$

Assume the bottom reinforcement is not developed for compression at the support and therefore, we will neglect it.

As shown in Fig. 9-20f, this is a singly reinforced section with a rectangular compression zone. Thus, use Eq. (9-7a) to calculate I_{cr}.

$$I_{cr} = \frac{b_w \cdot x^3}{3} + n \cdot A_s(d - x)^2$$

$$I_{cr} = \frac{10 \text{ in.}(6.0 \text{ in.})^3}{3} + 8.0 \times 3 \times 0.79 \text{ in.}^2(15.5 \text{ in.} - 6.0 \text{ in.})^2$$

$$I_{cr} = 720 \text{ in.}^4 + 1710 \text{ in.}^4 = 2430 \text{ in.}^4$$

In summary, I_g is 10,100 in.^{4}, I_{cr} at B is 2550 in.^{4}, and I_{cr} at C is 2430 in.^{4}

4. Compute immediate dead-load deflection.

(a) Compute I_e at A. Because M_a is less than M_{cr}, $I_e = 10{,}100$ in.^{4} at A for dead loads.

(b) Compute I_e at B. M_a is less than M_{cr}, $I_e = 10{,}100$ in.^{4} at B for dead loads.

(c) Compute I_e at C. M_a is less than M_{cr}. Therefore, $I_e = 10{,}100$ in.^{4} at C for dead loads.

(d) Compute the weighted average, I_e. Because the section is not cracked at this loading stage, the average $I_e = I_g = 10{,}100$ in.^{4}

(e) Immediate dead-load deflection. This is an exterior beam span with a column as the exterior support. Therefore, the deflection can be calculated using row 3 of Table 9-2.

$$\Delta = 3.3 \times 10^{-3}\left(\frac{w\ell^4}{EI}\right)$$

where $\ell = 22$ ft $= 264$ in.

$w = 0.87$ kip/ft $= 870$ lb/ft $= 72.5$ lb/in.

$E = 57{,}000\sqrt{4000} = 3.6 \times 10^6$ psi

$I = I_e = I_g = 10{,}100$ in.^{4}

So,

$$\Delta = 3.3 \times 10^{-3}\left(\frac{72.5 \times (264)^4}{3.60 \times 10^6 \times 10{,}100}\right) = 0.032 \text{ in. (down)}$$

So, the total dead-load deflection is 0.032 in. (down)

5. Compute the immediate live-load deflection. When the live load is applied to the beam, the moments will increase, leading to flexural cracking at B and C. As a result, I_e will decrease. The deflection that occurs when the live load is applied can be calculated as follows:

$$\Delta_{iL} = \Delta_{i,L+D} - \Delta_{iD}$$

(a) Compute I_e at A. Because $M_a = -55.3$ kip-ft is less than $M_{cr} = 79.2$ kip-ft, the section is still uncracked and $I_e = 10{,}100$ in.^{4} at A for dead plus live load.

(b) Compute I_e at B. $M_a = 63.2$ kip-ft and $M_{cr} = 30.8$ kip-ft; therefore, $M_{cr}/M_a = 0.487$ and $(M_{cr}/M_a)^3 = 0.116$. Using Eq. (9-10a),

$$\begin{aligned} I_e &= \left(\frac{M_{cr}}{M_a}\right)^3 I_g + \left[1 - \left(\frac{M_{cr}}{M_a}\right)^3\right] I_{cr} \\ &= 0.116 \times 10{,}100 + (1 - 0.116) \times 2550 \\ &= 3{,}430 \text{ in.}^4 \text{ (for dead plus live load at } B) \end{aligned}$$

(c) Compute I_e at C. $M_a = 88.4$ kip-ft and $M_{cr} = 79.2$ kip-ft; therefore, $M_{cr}/M_a = 0.896$ and $(M_{cr}/M_a)^3 = 0.719$. Using Eq. (9-10a):

$$I_e = 0.719 \times 10{,}100 + (1 - 0.719) \times 2430$$
$$= 7{,}940 \text{ in.}^4 \text{ (for dead plus live load at } C\text{)}$$

(d) Compute weighted average value of I_e. Using Eq. (9-11a):

$$I_e(\text{avg}) = 0.70 I_{em} + 0.15(I_{e1} + I_{e2})$$
$$= 0.70 \times 3430 + 0.15(10{,}100 + 7940) = 5{,}110 \text{ in.}^4$$

(e) Compute the immediate dead-plus-live-load deflection. Using the same procedure as in step 4, we will use row 3 of Table 9-2. The changes are: $w = 2.07$ kip/ft $= 173$ lb/in. and $I = I_e = 5110$ in.^{4}

$$\Delta = 3.3 \times 10^{-3}\left(\frac{w\ell^4}{EI}\right) = 3.3 \times 10^{-3}\left(\frac{173 \times (264)^4}{3.60 \times 10^6 \times 5110}\right) = 0.151 \text{ in. (down)}$$

Thus,

$$\Delta_{i,L+D} = 0.151 \text{ in. } \textbf{(down)}$$

and

$$\Delta_{iL} = \Delta_{i,L+D} - \Delta_{iD} = 0.151 - 0.032 = 0.119 \text{ in. } \textbf{(down)}$$

The author believes that this is not the appropriate value for the live-load deflection that is to be compared with allowable values in ACI Code Table 24.2.2. After the live load has been applied and the beam has cracked, the deflection due to dead load will be increased by an amount equal to the ratio of the I_e values used in steps 4 and 5. Thus, an improved value for Δ_{iD} is

$$\Delta_{iD} = 0.032 \times \frac{10{,}100}{5110} = 0.063 \text{ in.}$$

Then, the value for Δ_{iL} would be

$$\Delta_{iL} = 0.151 - 0.063 = 0.088 \text{ in. } \textbf{(down)}$$

ACI Code Table 24.2.2 limits computed immediate live-load deflections to $\ell/360$ for floors that either do not support or are not attached to elements likely to be damaged by large deflections. For this beam, that limit is (22 ft × 12 in./ft)/360 = 0.73 in., which is significantly larger than the value calculated here. ■

If this beam does not support partitions that can be damaged by deflections, ACI Table 9.3.1.1 would have indicated that it was not necessary to compute deflections, because the overall depth of the beam ($h = 18$ in.) exceeds the minimum value of $\ell/18.5 = (22 \times 12)/18.5 = 14.3$ in. given in the table.

If the structure just barely passes the allowable deflection calculations, consideration should be given to means of reducing the deflections, because deflection calculations are of questionable accuracy. The easiest ways to do this are to add compression steel or increase the depth of the section.

EXAMPLE 9-6 Calculation of Deflections Occurring after Attachment of Nonstructural Elements

The beam considered in Example 9-5 will be assumed to support partitions that would be damaged by excessive deflections. Again, it supports a dead load of 0.87 kip/ft and a live load of 1.2 kips/ft, 25 percent of which is sustained. The partitions are installed at least 3 months after the shoring is removed. Will the computed deflections exceed the allowable in the end span?

The controlling deflection is Eq. (9-14)

$$\Delta = \lambda_{to,\infty}\,\Delta_{iD} + \Delta_{iL} + \lambda_{\infty}\Delta_{iLS}$$

In Example 9-5, the following quantities were calculated:

$$\Delta_{iD+L} = 0.151 \text{ in.}$$
$$M_D \text{ (at midspan)} = 26.6 \text{ kip-ft}$$
$$M_{D+L} \text{ (at midspan)} = 63.2 \text{ kip-ft}$$

Therefore, $M_L = (63.2 - 26.6)$ kip-ft $= 36.6$ kip-ft.

From Example 9-5, the immediate live-load deflection is

$$\Delta_{iL} = 0.088$$

Twenty-five percent of this results from the sustained live loads, $\Delta_{iLS} = 0.022$ in.

Also from Example 9-5, the immediate dead-load deflection based on I_e, which is effective when both dead and live loads are acting, is

$$\Delta_{iD} = 0.063 \text{ in.}$$

The long-term deflection multipliers are given by Eq. (9-13):

$$\lambda_{\Delta} = \frac{\xi}{1 + 50\rho'}$$

The beam being checked does not have compression reinforcement; hence, $\rho' = 0$. If it had such reinforcement, the value of ρ' in Eq. (9-13) would be the value at midspan. Note that the bottom steel entering the supports at A and B is not developed for compression in the support and hence does not qualify as compression steel in this calculation, which proceeds as follows:

$$\lambda_{\infty} = \frac{2.0}{1 + 0} = 2.0$$
$$\lambda_{(t_0,\infty)} = \frac{2.0 - 1.0}{1 + 0} = 1.0$$

The deflection occurring after the partitions are in place is

$$\Delta = 1.0 \times 0.063 + 0.088 + 2.0 \times 0.022 = 0.195 \text{ in.}$$

The maximum permissible computed deflection is $\ell/480$ (ACI Table 24.2.2), where ℓ is the span from center-to-center of the supports:

$$\frac{22 \times 12 \text{ in.}}{480} = 0.55 \text{ in.}$$

Therefore, the long-term deflections of this beam are satisfactory. ■

9-6 FRAME DEFLECTIONS

Lateral Deflections

Section 9-4 dealt with vertical flexural deflections of floors and roofs. Several other types of deflection must be considered in the design of concrete-frame structures (Table 9-5). The most obvious of these are lateral deflections of the frame itself. These must be controlled to prevent discomfort to occupants, damage to partitions, and so on in frames, and because lateral deflections lead to second-order, or $P–\Delta$, effects, which are inversely proportional to the lateral stiffness. Guidance in the selection of acceptable lateral-sway indices, Δ/H, is given in [9-29].

Member Stiffnesses

Lateral deflections are important at (a) service loads, and (b) at factoral load levels. Figures 9-14e and 9-15 show a significant reduction in EI as the members of a frame crack. ACI Code Section 6.6 specifies values of E and I for use in computing first-order lateral deflections of a frame at factored load levels and for $P–\Delta$ analyses. These correspond to cases C_1 or C_2 in Figs. 9-14d and 9-14e. At service load levels the cracking will be less pronounced, probably between A and C_1 in Fig. 9-14. ACI Code Section 6.6.3.2.2 suggests using $1/0.70 = 1.4$ times the EI values from ACI Code Section 6.6.3.1 for service load calculations.

Vertical Deflections

If the exterior frames of a building are exposed to atmospheric temperature changes and the interior is maintained at a constant temperature, the thermal contraction and expansion of the exterior gives rise to relative vertical deflections of interior and exterior columns. This problem is discussed in [9-30].

In tall concrete buildings with columns and shear walls, the columns will generally be more highly stressed axially than the walls and will shorten more, due to elastic shortening and creep. The relative shortening may exceed 1 in. after several years. As a result, the floors may slope from the walls down to the columns. Procedures for estimating and dealing with these deflections are discussed in [9-31].

Concrete buildings with brick exteriors occasionally encounter difficulties due to the incompatible long-term deformations of concrete and brick [9-32]. With time, the concrete frame shortens from creep, while the bricks expand with time due to an increase in moisture content above their kiln-dried condition. This, plus thermal deformations of the brick, leads to crushing and cracking of the brick if the joints in the brickwork close so that the bricks begin to carry load. The expansion of bricks due to moisture gain is on the order of 0.0002 in./in., while the shortening of a concrete column due to creep strains may be on the order of 0.0005 to 0.0010 in./in., giving a total relative strain between the brick and the concrete of as much as 0.0012 in./in. In a 12-ft-high story, this amounts to a shortening of the frame by 0.17 in. relative to the brick cladding. This and temperature effects must be considered when designing brick cladding.

9-7 VIBRATIONS

In general, reinforced concrete buildings are not subject to vibration problems. This robustness is due to their mass and stiffness. Occasionally, long-span floors or assembly structures will be susceptible to vibrations induced by persons walking, dancing, or

TABLE 9-5 Loads and Load Combinations for Serviceability—Lateral Deflection, Vibration, and Fatigue-Limit States

(1) Case	(2) Serviceability Limit State and Elements Affected	(3) Serviceability Parameter Load Combination	(4) Calculation Method and Acceptance Criteria	(5) References
D	**Lateral deflection of frames**			
D-1	Wind	Mean downwind deflection plus cyclic deflections relative to mean	Human perception of vibration	Book Sections 9-6 and 9-7. Refs. [9-8] [9-29], [9-32], [9-33]
F	**Floor vibrations**	Discomfort	Human perception of vibration	Book Section 9-7. Refs. [9-22], [9-32], [9-33]
G	**Sway vibrations**	Wind-induced vibrations 1.0D + 1.0 (1-in- 2-year to 1-in-10-year wind)	Human perception of vibration	Book Section 9-7. Refs. [9-19], [9-23], [9-29], [9-32], [9-33]
H	**Fatigue**	Fracture due to large numbers of cycles and stress range. Number of cycles >20,000 Stress range: Compression or zero stress to tension stress		Book Sections 3-9 and 9-8. Refs. [9-34], [9-35].

exercising. These activities induce vibrations of approximately 2 to 4 cycles per second (hertz). If a floor system or supporting structure has a natural frequency of less than five cycles per second, its vibrational properties should be examined more closely to ensure that floor vibrations will not be a problem. Guidance on this problem may be found in [9-32], [9-33].

Fortunately, it is relatively easy to estimate the natural frequency, f, of a floor if its deflection can be calculated. We obtain

$$f = 0.18\sqrt{\frac{g}{\Delta_{is}}}$$

$$= 3.5\sqrt{\frac{1}{\Delta_{is}(\text{in.})}} \quad \text{cycles per second} \tag{9-21}$$

where g is the gravitational constant, and Δ_{is} is the immediate static deflection at the center of the floor due to the loads expected to be on the floor when it is vibrating [9-32]. The deflection, Δ_{is}, should be calculated using the effective moment of inertia, I_e, as defined in Eq. (9-16a) assuming the full dead and live load have acted on the floor. These would include the dead load and that portion of the live load on the floor when vibrating. The factor 3.5 in Eq. (9-21) actually varies from 3.1 to 4.0 for different types of floors (simple beams, fixed slabs, etc.), but the resulting error in f is small compared to inaccuracies in the deflection calculations themselves.

For continuous beams, the first vibration mode involves alternate spans deflecting up and down at the same time. The resulting modal deflections will be larger than those due to downward loads on each span. To use Eq. (9-21) in such a case, the deflections from a special frame analysis with the dead load plus some fraction of the live load applied upward and downward on alternate spans are needed.

9-8 FATIGUE

Structures subjected to cyclic loads may fail in fatigue. This type of failure requires (1) generally in excess of 1 million load cycles, and (2) a change of reinforcement stress in each cycle in excess of about 20 ksi. Because dead-load stresses account for a significant portion of the service-load stresses in most concrete structures, the latter case is infrequent. As a result, fatigue failures of reinforced concrete structures are rare. An overview of the fatigue strength of reinforced concrete structures is given in [9-34].

Fatigue failure of the concrete itself occurs via a progressive growth of microcracking. The fatigue strength of plain concrete in compression or tension for a life of 10 million cycles is roughly 55 percent of the static strength. It is affected by the minimum stress during the cycle and by the stress gradient. Concrete loaded in flexural compression has a 15 to 20 percent higher fatigue strength than axially loaded concrete, because the development of microcracking in highly strained areas is inhibited by the restraint provided by less strained areas. The fatigue strength of concrete is not sensitive to stress concentrations.

The ACI Committee for fatigue of concrete [9-34] recommends that the compressive stress range, f_{cr}, should not exceed

$$f_{cr} = 0.4f'_c + 0.47f_{\min} \tag{9-22}$$

where $f_{\min}$ is the minimum compressive stress in the cycle (positive in compression). The ACI Committee for bridge design [9-35] limits the compressive stress at service load to $0.5f'_c$.

The fatigue strength of deformed reinforcing bars is affected by the stress range and the minimum stress in the stress cycle but is almost independent of the yield strength of the bar.

In tests, bends with the minimum radius of bend permitted by the ACI Code reduced the fatigue strength in the region of the bend by a half, while tack welding of stirrups to longitudinal bars reduced the strength of the latter by more than a third [9-35]. The fatigue strength of bars is also affected by the radius at the bottom of the deformation lugs.

For bars of normal lug geometry, the ACI Committee for bridge design [9-35] recommends that the stress range not exceed

$$f_{sr} = 23.4 - 0.33f_{\min} \quad \text{(ksi)} \tag{9-23}$$

where f_{sr} is the algebraic difference between the maximum and minimum stresses at service loads and $f_{\min}$ is the minimum stress at service loads. In both cases, tension is positive. At bends or near tack welds, the allowable f_{sr} should be taken as half the value given by Eq. (9-23).

The fatigue strength of stirrups is less than that of longitudinal bars [9-35], due in part to a slight kinking of the stirrup where it crosses an inclined crack and, possibly, in part to an increase in the stirrup stress resulting from a reduction in the shear carried by the concrete in cyclically loaded beams. No firm design recommendations are currently available. A reasonable design procedure would be to limit the stirrup stress range to 0.75 times f_{sr} from Eq. (9-23), where the stirrup stress range would be calculated on the basis of the ACI shear design procedures, but using $V_c/2$.

The effect of cyclic loading on bond strength is approximately the same as that on the compressive strength of concrete [9-36]. Hence, a relationship similar to Eq. (9-22) would adequately estimate the allowable bond-stress range. During the initial stages of the cyclic load, the loaded end of the bar will slip relative to the concrete, as a result of internal cracking around the bar. The slip ceases once the cracking has stabilized.

PROBLEMS

9-1 Explain the differences in appearance of flexural cracks, shear cracks, and torsional cracks.

9-2 Why is it necessary to limit the widths of cracks?

9-3 Does the beam shown in Fig. P9-3 satisfy the ACI Code crack control provisions (Section 24.3.2)? $f_y = 60$ ksi.

9-4 Compute the maximum spacing of No. 5 bars in a one-way slab with 1 in. of clear cover that will satisfy the ACI Code crack-control provisions (Section 24.3.2). $f_y = 60$ ksi.

9-5 and 9-6 For the cross sections shown in Figs. P9-5 and P9-6, compute

(a) the gross moment of inertia, I_g;

(b) the location of the neutral axis of the cracked section and I_{cr}; and

(c) I_e for $M_a = 0.55M_n$.

The beams have 1.5 in. of clear cover and No. 3 stirrups. The concrete strength is 4000 psi; $f_y = 60$ ksi.

9-7 Why are deflections limited in design?

9-8 A simply supported beam with the cross section shown in Fig. P9-5 has a span of 28 ft and supports an unfactored dead load of 1.5 kips/ft, including its own self-weight plus an unfactored live load of 1.5 kips/ft. The concrete strength is 4500 psi. Compute

(a) the immediate dead-load deflection.

(b) the immediate dead-plus-live-load deflection.

(c) the deflection occurring after partitions are installed. Assume that the partitions are installed 1 month after the shoring for the beam is removed and assume that 20 percent of the live load is sustained.

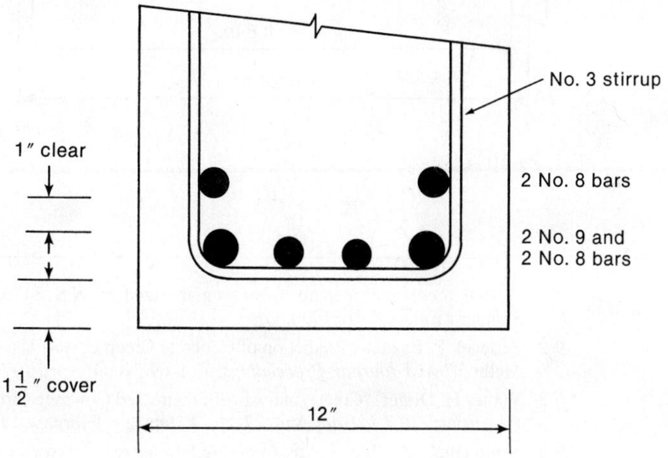

Fig. P9-3

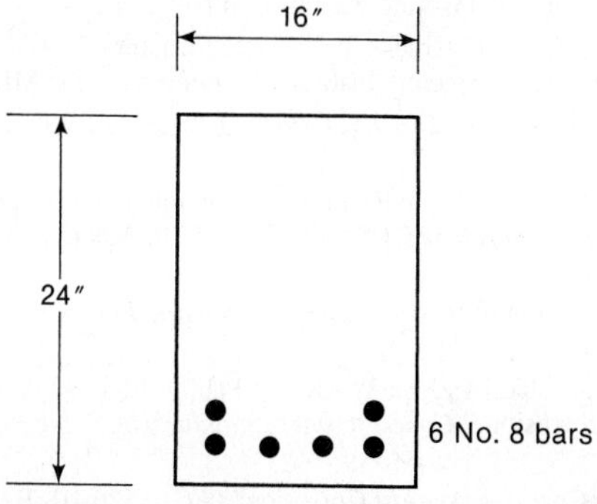

Fig. P9-5

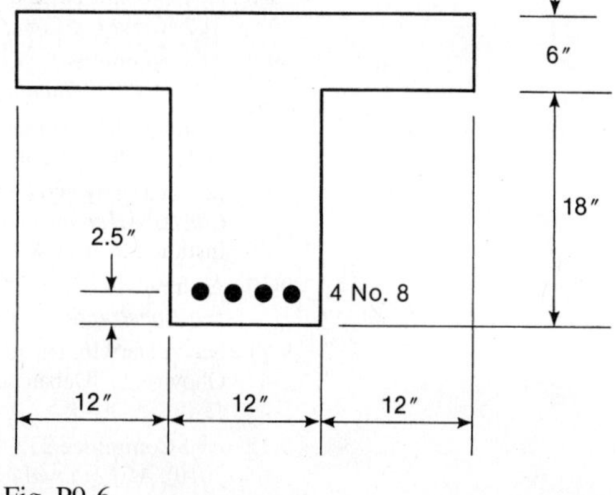

Fig. P9-6

9-9 Repeat Problem 9-8 for the beam section in Fig. P9-6.

Video Solution

9-10 The beam shown in Fig. P9-10 is made of 4000-psi concrete and supports unfactored dead and live loads of 1 kip/ft and 1.1 kip/ft. Compute

(a) the immediate dead-load deflection.

(b) the immediate dead-plus-live-load deflection.

(c) the deflection occurring after partitions are installed. Assume that the partitions are installed 2 months after the shoring is removed and assume that 15 percent of the live load is sustained.

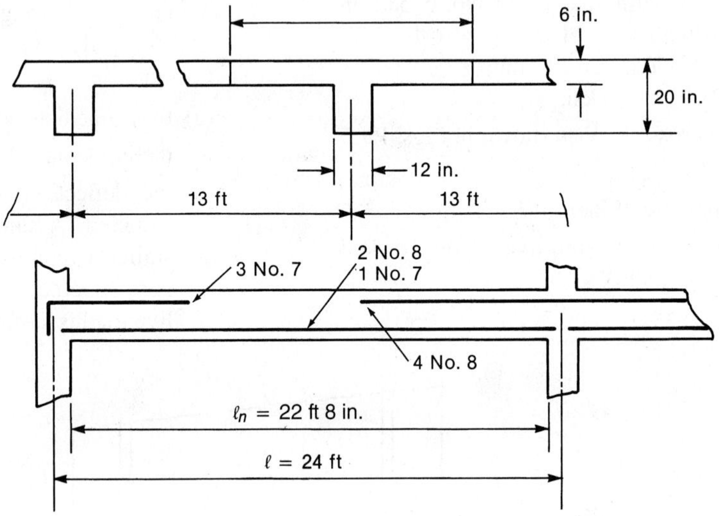

Fig. P9-10

REFERENCES

9-1 *Design of concrete structures*, Norwegian Standard, N.S. 3473, Norwegian Council for Building Standardization, Oslo, 1990.

9-2 Zedenek P. Bazant, "Prediction of Concrete Creep Effects Using Age-Adjusted Effective Modulus Method," *ACI Journal, Proceedings*, Vol. 69, No. 4, April 1972, pp. 212–217.

9-3 Walter H. Dilger, "Creep Analysis of Prestressed Concrete Structures Using Creep-Transformed Section Properties," *PCI Journal*, Vol. 27, No. 1, January–February 1982, pp. 99–118.

9-4 Amin Ghali and Rene Favre, *Concrete Structures: Stresses and Deformations*, Chapman & Hall, New York, 1986, 348 pp.

9-5 Fritz Leonhardt, "Crack Control in Concrete Structures," *IABSE Surveys*, IABSE Periodica, 3/1977, International Association for Bridge and Structural Engineering, Zurich, 1977, 26 pp.

9-6 ACI Committee 224, "Control of Cracking in Concrete Structures" (ACI 224R-01, Reapproved 2008), *ACI Manual of Concrete Practice*, American Concrete Institute, Farmington Hills, MI.

9-7 ACI Committee 224, "Causes, Evaluation, and Repair of Cracks in Concrete Structures" (ACI 224.1R-07), *ACI Manual of Concrete Practice*, American Concrete Institute, Farmington Hills, MI.

9-8 Comité Euro-International du Béton, *CEB Manual—Cracking and Deformations*, Ecole Polytechnique Fédérale de Lausanne, 1985, 232 pp.

9-9 Peter Gergely and Leroy A. Lutz, "Maximum Crack Width in Reinforced Concrete Flexural Members," *Causes, Mechanism and Control of Cracking in Concrete*, ACI Publication SP-20, American Concrete Institute, Detroit, MI, 1973, pp. 87–117.

9-10 Andrew W. Beeby, "Cracking, Cover, and Corrosion of Reinforcement," *Concrete International: Design and Construction*, Vol. 5, No. 2, February 1983, pp. 35–41.

9-11 David Darwin, David G. Manning, Eivind Hognestad, Andrew W. Beeby, Paul F. Rice, Abdoul Q. Ghowrwal, "Debate: crack width, cover, and corrosion," *Concrete International*, Vol. 7, No. 5, May 1985, pp. 20–35.

9-12 ACI Committee 222, "Protection of Metals in Concrete Against Corrosion" (ACI 222R-01, Reapproved 2010), *ACI Manual of Concrete Practice*, American Concrete Institute, Farmington Hills, MI.

9-13 ACI Committee 350, "Code Requirements for Environmental Engineering Concrete Structures and Commentary," (ACI 350–06), *Manual of Concrete Practice*, American Concrete Institute, Farmington Hills, MI.

9-14 R. J. Frosch, "Another Look at Cracking and Crack Control in Reinforced Concrete," *ACI Structural Journal*, Vol. 96, No. 3, May–June 1999, pp. 437–442.

9-15 *fib Model Code for Concrete Structures 2010*, International Federation for Structural Concrete (fib), Lausanne, Switzerland, 2013.

9-16 Gregory C. Frantz and John E. Breen, "Design Proposal for Side Face Crack Control Reinforcement for Large Reinforced Concrete Beams," *Concrete International: Design and Construction*, Vol. 2, No. 10, October 1980, pp. 29–34.

9-17 R. I. Gilbert, "Shrinkage Cracking in Fully Restrained Concrete Members," *ACI Structural Journal*, Vol. 89, No. 2, March–April 1992, pp. 141–149.

9-18 Andrew W. Beeby, Discussion of R. I. Gilbert, "Shrinkage Cracking in Fully Restrained Reinforced Concrete Members," *ACI Structural Journal, Proceedings*, Vol. 90, No. 1. January–February 1993, pp. 123–126.

9-19 Amin Ghali, "Deflection of Reinforced Concrete Members: A Critical Review," *ACI Structural Journal*, Vol. 90, No. 4, July–August 1993, pp. 364–373.

9-20 ACI Committee 435, "Deflections of Reinforced Concrete Flexural Members," *ACI Journal, Proceedings*, Vol. 63 No. 6, June 1966, pp. 637–674.

9-21 Dan E. Branson, "Compression Steel Effect on Long-Time Deflections," *ACI Journal, Proceedings*, Vol. 68, No. 8, August 1971, pp. 555–559.

9-22 H. Mayer and Hubert Rüsch, "Damage to Buildings Resulting from Deflection of Reinforced Concrete Members," *Deutscher Ausschuss für Stahlbeton*, Bulletin 193, Wilhelm Ernst, Berlin, 1967, 90 pp.; English translation: Technical Translation 1412, National Research Council of Canada, Ottawa, 1970, 115 pp.

9-23 D. A. Sawyer, "Ponding of Rainwater on Flexible Roof Systems," *Proceedings ASCE, Journal of the Structural Division*, Vol. 93, No. ST1, February 1967, pp. 127–147.

9-24 Amin Ghali, Walter Dilger, and Adam M. Neville, "Time-Dependent Forces Induced by Settlement of Supports in Continuous Reinforced Concrete Beams," *Journal of the American Concrete Institute*, Vol. 66, No. 11, November 1969, pp. 907–915.

9-25 ACI Committee 435, "Allowable Deflections," *ACI Journal, Proceedings*, Vol. 65, No. 6, June 1968, pp. 433–444.

9-26 Robert Ramsey, Sher-Ali Mirza, and James G. MacGregor, "Variability of Deflections of Reinforced Concrete Beams," *ACI Journal, Proceedings*, Vol. 76, No. 8, August 1979, pp. 897–918.

9-27 Jacob S. Grossman, "Simplified Computations for Effective Moment of Inertia I_e and Minimum Thickness to Avoid Deflection Computations," *ACI Journal, Proceedings*, Vol. 78, No. 6, November–December 1981, pp. 423–440.

9-28 B. S. Choi, B. H. Oh, and Andrew Scanlon, "Probabilistic Assessment of ACI 318 Minimum Thickness Requirements for One-Way Members," *ACI Structural Journal*, May–June 2002, pp. 344–351.

9-29 "Motion Perception and Tolerance," Chapter PC-13, *Planning and Environmental Criteria for Tall Buildings*, Monograph on Planning and Design of Tall Buildings, Vol. PC, American Society of Civil Engineers, New York, 1981, pp. 805–862.

9-30 Fazlur R. Khan and Mark Fintel, "Effects of Column Exposure in Tall Structures," *ACI Journal, Proceedings*, Vol. 62, No. 12, December 1965, pp. 1533–1556; Vol. 63, No. 8, August 1966, pp. 843–864; Vol. 65, No. 2, February 1968, pp. 99–110.

9-31 Mark Fintel and S. K. Ghosh, *Column Shortening in Tall Structures—Prediction and Compensation*, EB108.01D, Portland Cement Association, Skokie, IL., 1986, 35 pp.

9-32 "Commentary A, Serviceability Criteria for Deflections and Vibrations," *Supplement to the National Building Code of Canada 1990*, NRCC 30629, National Research Council of Canada, Ottawa, 1990, pp. 134–140.

9-33 David E. Allen, J. Hans Rainer, and G. Pernica, "Vibration Criteria for Long-Span Concrete Floors," *Vibrations of Concrete Structures*, ACI Publication SP-60, American Concrete Institute, Detroit, 1979, pp. 67–78.

9-34 ACI Committee 215, "Considerations for the Design of Concrete Structures Subjected to Fatigue Loading" (ACI 215R-74, revised 1992, Reapproved 1997), *ACI Manual of Concrete Practice*, American Concrete Institute, Farmington Hills, MI.

9-35 ACI Committee 343, "Analysis and Design of Reinforced Concrete Bridge Structures" (ACI 343R-95, Reapproved 2004), *ACI Manual of Concrete Practice*, American Concrete Institute, Farmington Hills, MI.

9-36 *Bond Action and Bond Behavior of Reinforcement*, Bulletin d'Information 151, Comité Euro-International du Béton, Paris, April 1982, 153 pp.

10

Continuous Beams and One-Way Slabs

10-1 INTRODUCTION

In the design of a continuous beam or slab, it is necessary to consider several strength limit states—failure by flexure, shear, bond, and possibly torsion—and several serviceability limit states—excessive deflections, crack widths, and vibrations. Each of these topics has been covered in a preceding chapter. In this chapter, continuity in reinforced concrete structures is discussed, followed by a design example showing the overall design of a continuous beam.

10-2 CONTINUITY IN REINFORCED CONCRETE STRUCTURES

The construction of two cast-in-place reinforced concrete structures is illustrated in Figs. 10-1 to 10-3. In Fig. 10-1, formwork is under construction for the beams and slabs supporting a floor. The raised platform-like areas are the bottom forms for the slabs. The rectangular openings between them are forms for the beam concrete. The beams meet at right angles at the columns.

Once the formwork is complete and the reinforcement has been placed in the forms (Fig. 10-2), the concrete for the slabs and beams will be placed in one monolithic pour. [See ACI Code Section 26.5.7.2(b).] Following this, the columns for the next story are erected, as shown in Fig. 10-3. ACI Code Section 26.5.7.2(a) requires that the concrete in the columns or walls have set before the concrete in the floor supported by those columns is placed. This sequence is required because the column concrete will tend to settle in the forms while in the plastic state. If the floor concrete had been placed, this would leave a gap between the column concrete and the beam. By placing the floor concrete after the column concrete is no longer plastic, any gap that formed as the concrete settled will be filled. The resulting construction joints can be seen at the bottom and top of each column in Fig. 10-3.

As a result of this placing sequence, each floor acts as a continuous unit. Because the column reinforcement extends through the floor, the columns act with the floors to form a continuous frame.

Fig. 10-1
Formwork for a beam-and-slab floor. (Photograph courtesy of J. G. MacGregor.)

Fig. 10-2
Beam and slab reinforcement in the forms. (Photograph courtesy of J. G. MacGregor.)

Braced and Unbraced Frames

A frame is said to be a *sway* frame (or an *unbraced* frame) if it relies on frame action to resist lateral loads or lateral deformations. Thus, the frame in Fig. 10-4a is unbraced, while that in Fig. 10-4b is braced by the shear wall. If the lateral stiffness of the bracing element in a story exceeds 6 to 10 times the sum of the lateral stiffnesses of all the

Fig. 10-3
Construction of columns in a tall building. (Photograph courtesy of J. G. MacGregor.)

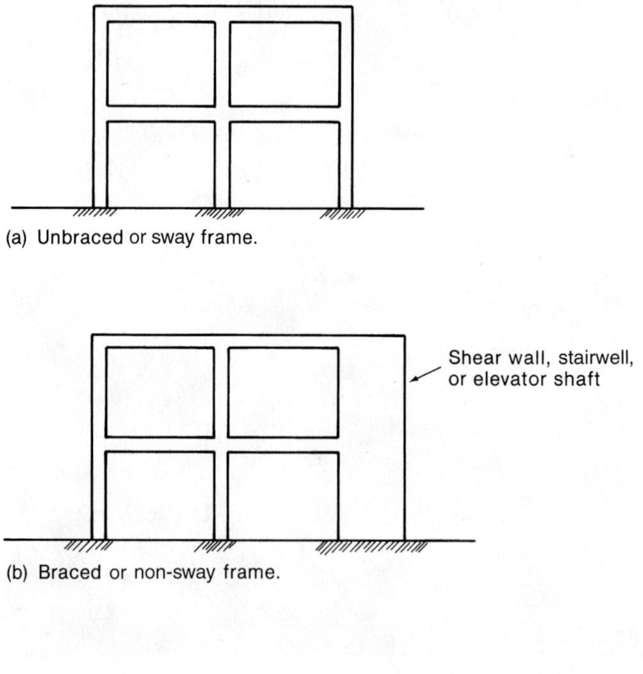

Fig. 10-4
Sway frames and braced frames.

columns in that story, a story can be considered to be braced. ACI Code Sections 6.6.4.3 and 6.6.4.4 give a procedure for classifying stories in a frame as sway or non-sway stories (as will be discussed in Chapter 12). Most concrete buildings are braced by walls, elevator shafts, or stairwells. The floor members in an unbraced frame must resist moments induced by lateral loads as well as gravity loads; in most cases, by contrast, the lateral-load moments can be ignored in the design of beams in a braced frame. The design examples in this chapter are limited to beams in braced frames.

Design Loads for a Continuous Floor System

The design loads are selected in accordance with ASCE/SEI 7-10, *Minimum Design Loads for Buildings and Other Structures* [10-1]. The dead loads include the weight of the floor, 0.5 psf for asphalt tile or carpet, 4 psf for mechanical equipment and lighting fixtures hung below the floor, and 2 psf for a suspended ceiling.

ASCE/SEI 7-10 specifies the minimum live loads for offices as 50 psf and for lobbies as 100 psf. (See Table 2-1.) The live loads for file storage rooms and computer rooms are based on the actual usage of the room. An allowance must be made for the weight of partitions (considered to be live load in ASCE/SEI 7-10) if the floor loading is 80 psf or less. The design live loads can be reduced as a function of the floor area, as discussed in Section 2-8. No live-load reduction is allowed for one-way slabs.

A second loading case required in ASCE/SEI 7-10 for office buildings is a concentrated load of 2000 lb acting on a 30-in.-by-30-in. area placed on the floor to produce the maximum moments or shears in the structural members. The uniform live load is assumed to not be acting on the structure when this load acts. For most floor beams and slabs, the uniform live loads will give greater moments and shears in the beams than the concentrated load.

Transfer of Column Loads through the Floor System

In the building shown in Fig. 10-3, the columns were built of 5000-psi concrete and the floors were built of 3750-psi concrete. These strengths were chosen on the basis of ACI Code Section 15.3, which allows the column concrete to be stronger than the floor concrete.

The floor concrete in the vicinity of the column–beam joint must transfer axial loads from the column above the joint to the column below. It is usual practice for the column concrete placement to be stopped at the elevation of the underside of each floor, as shown in Figs. 10-1 and 10-3, to allow the concrete to settle in the forms before the concrete in the floor over the column is placed. ACI Code Section 15.3.1 allows the concrete in interior, edge, and corner columns to have a strength up to 1.4 times the strength of the concrete in the floor system (slab and beams) without any strengthening of the floor. On the basis of tests reported by Biachini et al. [10-2], the effective strength of the concrete in the floor system is enhanced by the lateral confinement from the floor surrounding the column. For larger ratios of column concrete strength to floor system concrete strength,

(a) ACI Code Section 15.3.1(a) allows the higher strength column concrete to be placed in the floor under and around the column, or

(b) ACI Code Section 15.3.1(b) allows the design of the joint based on the lower strength concrete with dowels provided to make up for the lower strength concrete, or

(c) ACI Code Section 15.3.1(c) allows the joint region to be designed by using an *effective compressive strength* of the concrete in the floor, f'_{ce}, taken as

$$f'_{ce} = 0.75f'_{cc} + 0.35f'_{cs} \qquad (10\text{-}1)$$

In the third case, the maximum ratio is $f'_{cc}/f'_{cs} = 2.5$, where f'_{cc} is the specified strength of the column concrete and f'_{cs} is the specified strength of the slab concrete.

More recent tests, reported by Ospina and Alexander [10-3], showed that the lateral confinement of the floor concrete is reduced when slab loads stress the slab reinforcement in tension. These tests supported the use of the 1.4 strength-ratio threshold for interior columns and edge columns, but suggested that the threshold strength ratio be lowered to 1.2 for corner columns. The joint strengths were also sensitive to the ratio of floor thickness to column width, h/c. The effective concrete strength decreased when h/c increased.

10-3 CONTINUOUS BEAMS

The three major stages in the design of a continuous beam are design for flexure, design for shear, and design of longitudinal reinforcement details. In addition, it is necessary to consider deflections and crack control and, in some cases, torsion. When the area supported by a beam exceeds 400 ft^2, it is usually possible to use a reduced live load in calculating the moments and shears in the beam.

EXAMPLE 10-1 Design of a Continuous T-Beam

Design the floor beam $B3$–$B4$–$B3$ in Fig. 10-5. This beam supports its own dead load plus the load from a 6-in. slab. The beam is supported on girders at lines A, B, C, and D and is symmetrical about the centerline of the building. Use normal-weight concrete with $f'_c = 4000$ psi, $f_y = 60{,}000$ psi for flexural reinforcement, and $f_{yt} = 60{,}000$ psi for stirrups. Base the loadings on the ASCE/SEI 7-10 recommendations and use load factors from ACI Code Table 5.3.1

1. **Compute the trial factored loads on the beam.** Because the size of the beam stem is unknown at this stage, it is not possible to compute the final loads for use in the design. For preliminary purposes, estimate the size of the beam stem. Once the size has been established, the factored load will be corrected and then used in subsequent calculations.

The ASCE/SEI 7-10 recommendations allow live-load reductions based on tributary areas multiplied by a live-load element factor, $K_{LL} = 2$, to convert the tributary area to an influence area. (See Section 2-8.) Three tributary areas will be considered in the design of beam $B3$–$B4$–$B3$, as shown in Fig. 10-6.

(a) To compute the positive moment at midspan of beam $B3$ and negative moment at the exterior end of beam $B3$, it is necessary to load spans AB and CD with live load. Because the majority of the moment in the left span results from loads on that span, we shall take the tributary area A_T to be the area over beam $B3$, extending halfway to the adjacent beams:

$$A_T \cong 33 \text{ ft}\left(\frac{15 \text{ ft}}{2} + \frac{14.3 \text{ ft}}{2}\right) = 483 \text{ ft}^2$$

From Eq. (2-12), the reduced live load is

$$L = L_o\left(0.25 + \frac{15}{\sqrt{K_{LL}A_T}}\right)$$

where L_o, the unreduced live load, is 100 psf and for a typical floor beam $K_{LL} = 2.0$.

$$L = L_o\left(0.25 + \frac{15}{\sqrt{2 \times 483}}\right) = 100 \times 0.732$$
$$= 73.2 \text{ psf}$$

L shall not be less than $0.50L_o$ for members supporting one floor. Thus, $L = 73.2$ psf.

(b) To compute the maximum negative moment at the interior support, B, spans AB and BC should be loaded. The corresponding tributary area, A_T, is the area between lines of zero shear parallel to the beam and on either side of it, as shown in Fig. 10-6b.

$$A_T \cong \left(33 \text{ ft} + 30 \text{ ft}\right)\left(\frac{15 \text{ ft}}{2} + \frac{14.3 \text{ ft}}{2}\right) = 923 \text{ ft}^2$$

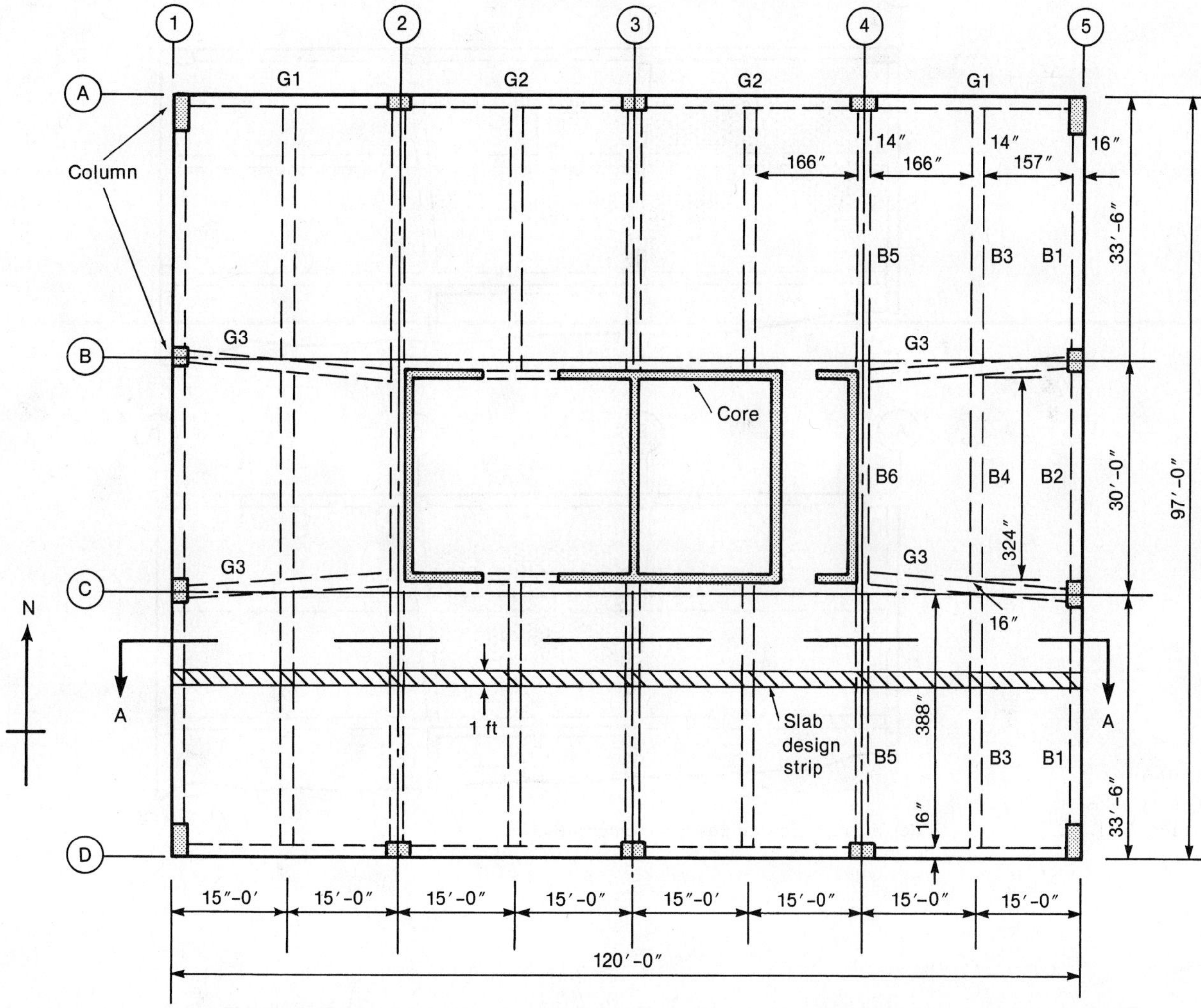

Fig. 10-5
Typical floor plan—Example 10-1.

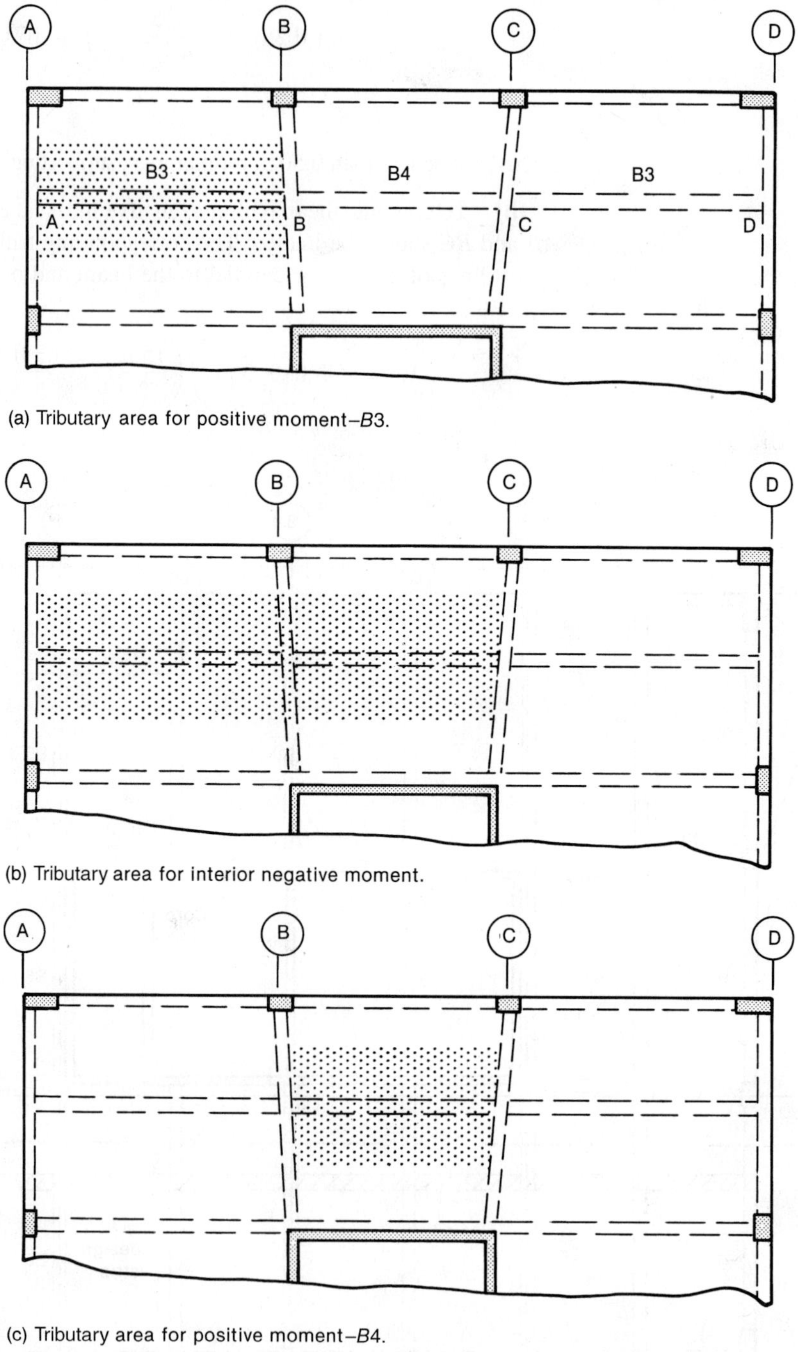

Fig. 10-6
Tributary areas—
Example 10-1.

and

$$K_{LL} = 2$$

$$L = 100\left(0.25 + \frac{15}{\sqrt{2 \times 923}}\right) = 59.9 \text{ psf}$$

where L shall not be less than $0.50L_o$ for members supporting one floor. Thus, $L = 59.9$ psf.

(c) To compute the positive moment in the center span (beam $B4$), span BC should be loaded. The tributary area is shown in Fig. 10-20c. Here,

$$A_T \cong 30 \text{ ft}\left(\frac{15 \text{ ft}}{2} + \frac{14.3 \text{ ft}}{2}\right) = 440 \text{ ft}^2$$

$$L = 100\left(0.25 + \frac{15}{\sqrt{2 \times 440}}\right) = 75.5 \text{ psf}$$

In this example, three reduced live loads were calculated, in part, to illustrate the calculation. To simplify design calculations, one might use a single reduced live load throughout, equal to the largest of the three reduced live loads.

Summary of unfactored reduced live loads.

For negative moment at exterior end, positive moment at midspan, and shear in beam $B3 = 73.2$ psf.
For negative moment at support $B = 59.9$ psf.
For positive moment at midspan and shear in beam $B4 = 75.5$ psf.

2. Select Method of Analysis of Beam and Values of the Strength-Reduction Factor.

If the beam fits the requirements in ACI Code Section 6.5.1, we will use the moment coefficients from that section.

In this example,

(a) there are two or more spans,

(b) the ratio of the longer clear span to the shorter clear span is approximately $33/30 = 1.10$, which is less than 1.20,

(c) the loads are uniformly distributed,

(d) the "unit" (unfactored) live load does not exceed three times the "unit" (unfactored) dead load, and

(e) the members are prismatic.

So far, the beam fits all of these except item (d), which we will check later.

Use ACI Code Table 6.5.2 to compute moments and shears.

3. Select the strength-reduction factors. The design is based on load combinations from ACI Code Table 5.3.1, and the strength reduction factors are taken from ACI Code Table 21.2.1:

For flexure in tension-controlled beam sections, $\phi = 0.90$. (The goal will be to design tension-controlled sections.)

Shear $\phi = 0.75$.

The size of the stem beam $B3$–$B4$–$B3$ will be chosen on the basis of negative moment at the first interior support. For this location, the factored load on the beam is as follows:

(a) Dead load (not including the portion of the beam stem below the slab):

Slab	75 psf
Floor, ceiling, mechanical	6.5 psf
	81.5 psf

Live load (reduced): 59.9 psf
Total factored load from slab:

$$w_u = 1.2 \times 81.5 + 1.6 \times 59.9 = 194 \text{ psf}$$

Assume that the tributary width that loads beam $B3$–$B4$–$B3$ extends halfway from this beam to the adjacent beams on each side of $B3$–$B4$–$B3$, or

$$0.5\,(15\text{ ft} + 14.3\text{ ft}) = 14.7\text{ ft}$$

Therefore, the factored load per foot from the slab is

$$14.7\text{ ft} \times 194\text{ psf} = 2.85\text{ kip/ft}$$

It is now necessary to estimate the weight of the beam stem. Two approximate methods of doing this are:

(1) The factored dead load of the stem is taken as 12 to 20 percent of the other factored loads on the beam. This gives 0.34 to 0.57 kip/ft.

(2) The overall depth of beam h is taken to be 1/18 to 1/12 of the larger span, ℓ, and b_w is taken to be $0.5h$. This gives the overall h as 22 to 33 in., with the stem extending 16 to 27 in. below the slab, and gives b_w as 11 to 17 in. The factored load from stems of such sizes ranges from 0.22 to 0.57 kip/ft.

As a first trial, assume the factored weight of the stem to be 0.50 kip/ft. Then

$$\text{Total trial load per foot} = 2.85 + 0.50\text{ kip/ft}$$
$$= 3.35\text{ kips/ft}$$

say, 3.4 kips/ft.

4. Choose the actual size of the beam stem. The size of the stem is governed by three factors: (a) deflections, (b) moment capacity at the point of maximum negative moment, and (c) shear capacity. In addition, the overall depth of the floor should be minimized to reduce the overall height of the building.

(a) Calculate the minimum depth based on deflections. ACI Table 9.3.1.1 (Table A-9) gives minimum depths, unless deflections are checked. Because the anticipated partitions will be flexible enough to undergo some deflections, this table can be used.

Beam $B3$ is one-end continuous, $f_y = 60{,}000$ psi, and normal-density concrete. Therefore,

$$\text{Minimum } h = \frac{\ell}{18.5}$$

where ℓ shall be taken as the span center-to-center of supports. Consequently,

$$\text{Minimum } h \text{ based on deflections} = \frac{33\text{ ft} \times 12\text{ in./ft}}{18.5} = 21.4\text{ in.}$$

(b) Determine the minimum depth based on the negative moment at the exterior face of the first interior support. Using the appropriate moment coefficient from ACI Code Table 6.5.2,

$$\text{Moment} = -\frac{w_u \ell_n^2}{10}$$

ℓ_n for negative moment is defined as the average of the two adjacent spans. Using dimensions given in Fig. 10-5,

$$\ell_n = \frac{(388 + 324)\text{ in.}}{2} = 356\text{ in.} = 29.7\text{ ft}$$

Therefore,

$$M_u \simeq -\frac{3.4 \times 29.7^2}{10} = -300 \text{ kip-ft}$$

We will use the procedure developed in Chapter 5 for the design of singly reinforced beam sections with rectangular compression zones. Use Eq. (5-19) to select a reinforcement ratio, ρ, that will result in a tension-controlled section.

$$\rho(\text{initial}) \cong \frac{\beta_1 f'_c}{4f_y} = \frac{0.85 \times 4 \text{ ksi}}{4 \times 60 \text{ ksi}} = 0.0142$$

From this, use Eq. (5-21) to find the reinforcement index, ω:

$$\omega = \rho\frac{f_y}{f'_c} = 0.0142\frac{60 \text{ ksi}}{4 \text{ ksi}} = 0.213$$

and then, use Eq. (5-22) to calculate the flexural resistance factor, R.

$$R = \omega f'_c(1 - 0.59\omega)$$
$$= 0.213 \times 4 \text{ ksi}(1 - 0.59 \times 0.213) = 0.745 \text{ ksi}$$

Finally, this value of R can be used to determine the required value of bd^2, using Eq. (5-23a).

$$bd^2 \geq \frac{M_u}{\phi R} = \frac{300 \text{ k-ft} \times 12 \text{ in./ft}}{0.9 \times 0.745 \text{ ksi}} = 5370 \text{ in.}^3$$

Possible choices for b and d are

$$b = 10 \text{ in.}, d = 23.2 \text{ in.}$$
$$b = 12 \text{ in.}, d = 21.2 \text{ in.}$$
$$b = 14 \text{ in.}, d = 19.6 \text{ in.}$$

Select the second of these. With one layer of reinforcement at the supports, $h \simeq 21.2 + 2.5 = 23.7$ in. This exceeds the minimum h based on deflections. **Try a 12-in. wide-by-24-in. deep extending 18 in. below the slab stem, with $d = 21.5$ in.**

(c) Check the shear capacity of the T-beam.

$$V_u = \phi(V_c + V_s)$$

Maximum shear from beam loads at interior end of $B3$ are approximately:

$$V_u = 1.15\, w_u \ell_n/2 = 1.15 \times 3.40 \text{ k/ft} \times 32.3 \text{ ft}/2 = 63.1 \text{ kips}$$

From ACI Code Section 22.5.5.1, the nominal V_c is

$$V_c = 2\lambda\sqrt{f'_c}\,b_w d = 32.6 \text{ kips}$$

ACI Code Section 22.5.1.2 sets the maximum nominal V_s is

$$V_s = 8\sqrt{f'_c}\,b_w d = 131 \text{ kips}$$

Thus, the absolute maximum $\phi V_n = 0.75\,(32.6 + 131 \text{ kips}) = 123$ kips.

The maximum factored V_u due to the applied loads and dead loads is 63.1 kips. **Therefore, the beam stem selected has ample capacity for shear.**

(d) Summary. Use

$$\mathbf{b = 12\ in.}$$

$$\mathbf{h = 24\ in.\ (18\ in.\ below\ slab)}$$

$$\mathbf{d = 21.5\ in.}\text{, assuming one layer of steel at all sections}$$

This section is shown in Fig. 10-7.

5. Compute the dead load of the stem, and recompute the total load per foot.

$$\text{Weight per foot of the stem below the slab} = \frac{12 \times 18}{144} \times 0.15 = 0.225 \text{ kip/ft}$$

Total dead load for *B*3–*B*4–*B*3 is:

$$w_D = 0.0815 \text{ ksf} \times 14.7 \text{ ft} + 0.225 \text{ kip/ft} = 1.42 \text{ kip/ft}$$

Live load for *B*3–*B*4–*B*3 is:

Beam *B*3, $w_L = 0.0732 \text{ ksf} \times 14.7 \text{ ft} = 1.08 \text{ kip/ft}$

Negative moment at *B*, $w_L = 0.0599 \text{ ksf} \times 14.7 \text{ ft} = 0.881 \text{ kip/ft}$

Positive moment for *B*4, $w_L = 0.0755 \text{ ksf} \times 14.7 \text{ ft} = 1.11 \text{ kip/ft}$

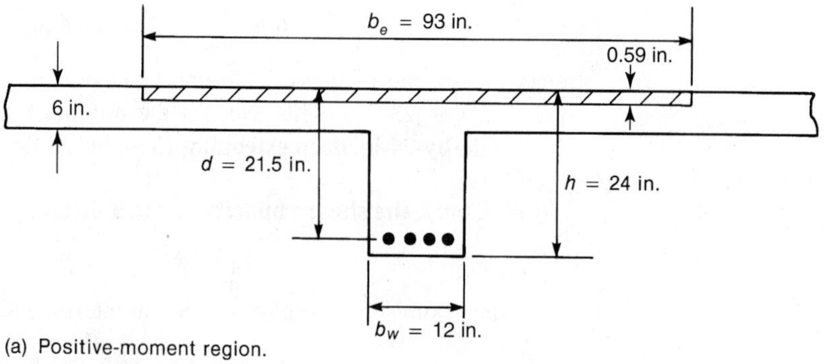

(a) Positive-moment region.

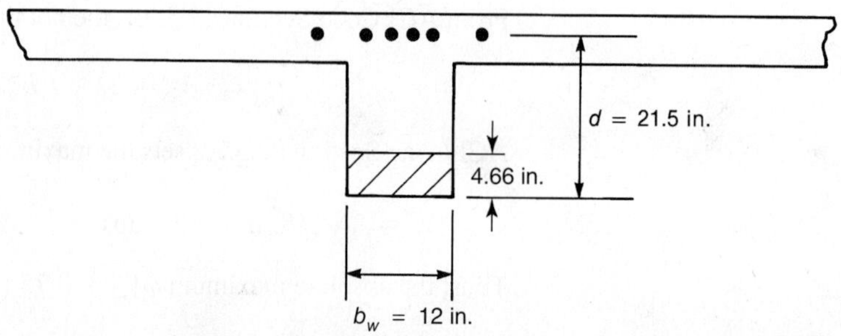

(b) Negative-moment region.

Fig. 10-7
Beam sections.

Summary of factored loads on *B*3–*B*4–*B*3.

Beam *B*3, $w_u = 1.2 \times 1.42 + 1.6 \times 1.08 = 3.43 \text{ kip/ft}$

Negative moment at *B*, $w_u = 1.2 \times 1.42 + 1.6 \times 0.881 = 3.11 \text{ kip/ft}$

Positive moment for *B*4, $w_u = 1.2 \times 1.42 + 1.6 \times 1.11 = 3.48 \text{ kip/ft}$

Note that although the factored weight of the beam stem is only $0.27/0.5 = 0.54$ times the original guess, the error in the original estimate of the total factored load at B is only 9 percent. If the final factored load per foot changed by more than about 10 percent, it may be desirable to recalculate the beam size and load per foot.

6. Calculate the flange width for the positive-moment regions. The beam acts as a T-beam in the positive-moment regions. The effective width of the overhanging flange on each side of the web is the smallest of (ACI Code Section 6.3.2.1) the following:

$$\text{(using shorter } \ell_n\text{); } b_w + 2(\ell_n/8) = 12 \text{ in.} + (324 \text{ in.}/4) = 93 \text{ in.}$$

$$b_w + 2(8 \times 6) = 12 \text{ in.} + 2 \times 48 \text{ in.} = 108 \text{ in.}$$

$$b_w + \frac{157 \text{ in.}}{2} + \frac{166 \text{ in.}}{2} = 174 \text{ in.}$$

Therefore, the effective flange width is 93 in., as shown in Fig. 10-7a.

7. Compute the beam moments. The moments can be computed by using frame analysis, as was discussed in Chapter 5. However, in step 2 it was shown that we could use the ACI moment coefficients from ACI Code Table 6.5.2 if the unfactored unit live load was less than three times the unfactored unit dead load. From the load calculations in step 5, it is clear that this condition is satisfied. Thus, the ACI moment coefficients can be used in this design. The moment calculations are presented in Table 10-1. Because the building is symmetrical about its center, only a part of the beam is shown in this table.

Table 10-1, Line 1 The values of ℓ_n are the clear-span lengths, except that in computing the negative moment at supports *B* and *C*, ℓ_n is the average of the two adjacent spans.

Table 10-1, Line 2 The factored uniform loads differ for each span, due to the different live-load reduction factors for the various sections.

Table 10-1, Line 4 Because the exterior end of span *AB* is supported on a spandrel girder, the moment coefficient at *A* is 1/24. The resulting moment diagram is shown in Table 10-2.

8. Design the flexural reinforcement. The reinforcement will be designed in Table 10-2. This is a continuation of Table 10-1. Prior to entering Table 10-2, however,

TABLE 10-1 Calculation of Moments for Beam *B*3–*B*4–*B*3—Example 10-1

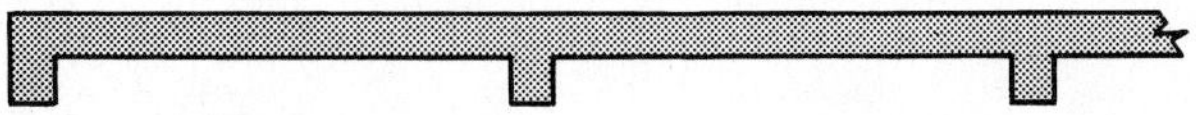

Line		*B*3		*B*4	
1. ℓ_n, ft	32.3	32.3	29.7	27.0	29.7
2. w_u, kip/ft	3.43	3.43	3.11	3.48	3.11
3. $w_u \ell_n^2$	3590	3590	2740	2540	2740
4. C_m	−1/24	1/14	−1/10 −1/11	1/16	−1/11 −1/10
5. $M_u = C_m w_u \ell_n^2$, kip-ft	−149	256	−274	159	−274

TABLE 10-2 Calculation of Reinforcement Required—Example 10-1

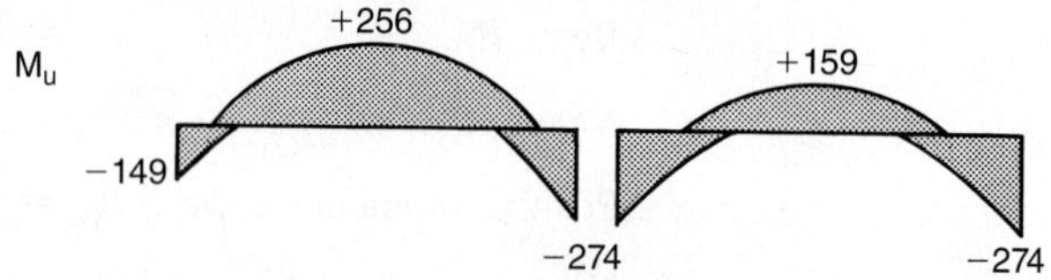

Line				
5. M_u, kip-ft	−149	256	−274	159
6. A_s coefficients	0.0116	0.0105	0.0116	0.0105
7. $A_{s(req'd)}$, in.^{2}	1.72	2.68	3.17	1.67
8. $A_s > A_{s,min}$	Yes	Yes	Yes	Yes
9. Bars selected	3 No. 7	4 No. 8	4 No. 7 2 No. 6	3 No. 7
10. A_s provided, in.^{2}	1.80	3.16	3.28	1.80
11. b_w o.k.	—	Yes	—	Yes

it is necessary to compute some constants for use in subsequent calculations of area of reinforcement.

(a) Compute the area of steel required at the point of maximum negative moment (first interior support). From Eq. (5-16),

$$A_s \geq \frac{M_u}{\phi f_y\left(d - \dfrac{a}{2}\right)} \cong \frac{M_u}{\phi f_y(jd)}$$

Because there is negative moment at the support, the beam acts as a rectangular beam with compression in the web, as shown in Fig. 10-7b. Assume that $j = 0.9$ and $\phi = 0.90$

$$A_s \cong \frac{274 \text{ k-ft} \times 12 \text{ in./ft}}{0.9 \times 60 \text{ ksi} \times (0.9 \times 21.5 \text{ in.})} = 3.15 \text{ in.}^2$$

This value can be improved with one iteration using Eq. (5-17) to find the depth of the compression stress block, a:

$$a = \frac{A_s f_y}{0.85 f'_c b} = \frac{3.15 \text{ in.}^2 \times 60 \text{ ksi}}{0.85 \times 4 \text{ ksi} \times 12 \text{ in.}} = 4.63 \text{ in.}$$

and then reusing Eq. (5-16) with this calculated value of a:

$$A_s \geq \frac{M_u}{\phi f_y\left(d - \dfrac{a}{2}\right)} = \frac{274 \text{ k-ft} \times 12 \text{ in./ft}}{0.9 \times 60 \text{ ksi} \times (21.5 \text{ in.} - 2.32 \text{ in.})} = 3.17 \text{ in.}^2$$

Before proceeding, we need to confirm that this is a tension-controlled section. This could be done by calculating the strain at the level of the tension reinforcement, ε_t, and confirming that it exceeds 0.005 or by simply showing that the depth to the neutral axis, c, is less than 3/8 of d. Using the latter approach with $\beta_1 = 0.85$ for 4000-psi concrete and calculating a revised value of $a = 4.66$ in., which was obtained using the new required area of steel, A_s (3.17 in.^{2}):

$$c = \frac{a}{\beta_1} = \frac{4.66 \text{ in.}}{0.85} = 5.48 \text{ in.}$$

This is less than 3/8 d = 8.06 in., so the section is tension-controlled and can be designed using $\phi = 0.9$. The other negative-moment sections have a lower design moment, so it will be conservative to use the ratio of A_s/M_u obtained here to quickly determine the area of tension steel required at those other locations. That ratio is

$$\frac{A_s}{M_u} = \frac{3.17 \text{ in.}^2}{274 \text{ kip-ft}} = 0.0116 \text{ (in.}^2\text{/kip-ft)}$$

Lines 6 and 7 of Table 10-2 Calculate the area of steel required in negative-moment regions as $A_s = 0.0116M_u$.

(b) Compute the area of steel required at the point of maximum positive moment (point near the middle of the exterior span). In the positive-moment regions, the beam acts as a T-shaped beam with compression in the top flange. Assume that the compression zone is rectangular, as shown in Fig. 10-7a. Take $j = 0.95$ (wide compression zone) for the first calculation of A_s:

$$A_s \cong \frac{256 \text{ k-ft} \times 12 \text{ in./ft}}{0.9 \times 60 \text{ ksi } (0.95 \times 21.5 \text{ in.})} = 2.79 \text{ in.}^2$$

Assume that a is less than h_f; then

$$a = \frac{A_s f_y}{0.85 f'_c b}$$

where b is the effective flange width of 93 in.

$$a = \frac{2.79 \text{ in.}^2 \times 60 \text{ ksi}}{0.85 \times 4 \text{ ksi} \times 93 \text{ in.}} = 0.53 \text{ in.} < h_f = 6.00 \text{ in.}$$

Doing one iteration, as was done for the negative-moment section, results in A_s = 2.68 in.^{2} and $a = 0.51$ in., For this value of a, it is clear that c is less than 3/8 d = 8.06 in., so the section is tension-controlled and can be designed using $\phi = 0.9$. The other positive-moment section has a lower design moment, so it will be conservative to use the ratio of A_s/M_u obtained here to quickly determine the area of tension steel required at that other location. The ratio is

$$\frac{A_s}{M_u} = \frac{2.68 \text{ in.}^2}{256 \text{ kip-ft}} = 0.0105 \text{ (in.}^2\text{/kip-ft)}$$

The area of steel required at positive-moment regions is calculated in lines 6 and 7 of Table 10-2 by using $A_s = 0.0105M_u$.

(c) Calculate the minimum reinforcement. The minimum reinforcement required is, by ACI Code Section 9.6.1.2, and given earlier as Eq. (4-11):

$$A_{s,\min} = \frac{3\sqrt{f'_c}}{f_y} b_w d, \text{ and } \geq \frac{200 b_w d}{f_y}$$

For 4000-psi concrete, the second term governs, so

$$A_{s,\min} \geq \frac{200 \times 12 \times 21.5}{60{,}000}$$

$$\geq 0.86 \text{ in.}^2$$

Table 10-2, Line 8 A_s in line 7 of Table 10-2 exceeds $A_{s,\min}$ in both the positive- and negative-moment regions.

(d) Calculate the area of steel and select the bars. The remaining calculations are done in Table 10-2. The areas computed in line 7 exceed $A_{s,\min}$ in all cases. If they did not, $A_{s,\min}$ should be used.

Lines 9 and 10 of Table 10-2 Give the bars selected at each location, together with their areas. The bars were selected by using Table A-4. Small bars were selected at the exterior support, because they have to be hooked into the support and there may not be enough room for a standard hook on larger bars.

Line 11 of Table 10-2 Check whether the bars will fit into the beam web using Table A-5. In the negative-moment regions, some of the bars can be placed in the slab beside the beams; hence, it is not necessary to check whether they will fit into the web width.

9. Check the distribution of the reinforcement.

(a) Positive-moment region. It is necessary to satisfy ACI Code Section 24.3.2. This will be done at the section with the smallest number of positive-moment bars: the middle of span *B–C*. Here, there are three No. 7 bars, as shown in Fig. 10-8a.

The clear cover to the surface of the flexural steel is

$$c_c = 1.5\text{-in. cover} + 0.375\text{-in. stirrups} = 1.875 \text{ in.}$$

The maximum bar spacing is

$$s = 15\left(\frac{40{,}000}{f_s}\right) - 2.5c_c \leq 12\left(\frac{40{,}000}{f_s}\right) \tag{9-7}$$

where f_s = can be taken as $0.67f_y$ (in which $f_y = 60$ ksi). Thus,

$$s = 15 - 2.5 \times 1.875 = 10.3 \text{ in.} \leq 12 \text{ in.}$$

Because $b_w = 12$ in. and there are three bars, *s* is clearly smaller than 10.3 in.

(b) Negative-moment region. ACI Code Section 24.3.4 says "part" of the negative-moment steel shall be distributed over a width equal to the smaller of the effective flange width (93 in.) and $\ell_n/10 = 324/10 = 32.4$ in. At each of the interior negative-moment regions, there are six top bars. Two of these will be placed in the corners of the stirrups, as shown in Fig. 10-8b, two over the beam web, and the other two in the slab. (Note that, for bars placed in the slab to be

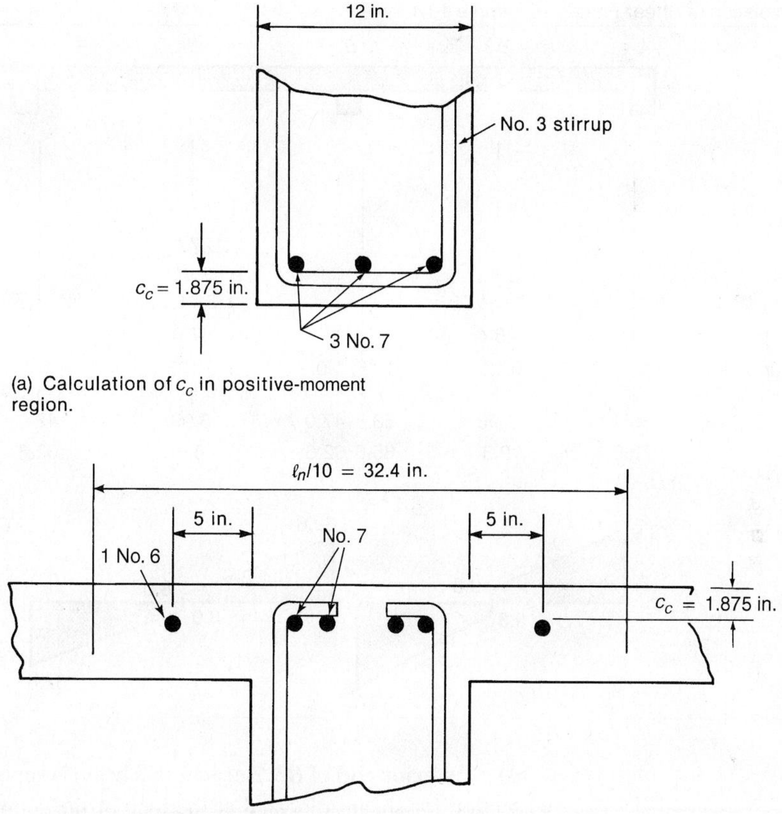

(a) Calculation of c_c in positive-moment region.

(b) Bar spacing at interior support.

Fig. 10-8
Distribution of reinforcement.

completely effective, there should be reinforcement perpendicular to the beam in the slab. In this case the slab reinforcement will serve this purpose.) The two bars placed in the slab will be placed to give spacing, s, approaching the maximum allowed. The maximum spacing allowed is

$$s = 15\left(\frac{40{,}000}{f_s}\right) - 2.5\,c_c$$

$$= 15 - 2.5 \times 1.875$$

$$= 10.3 \text{ in.}$$

Within a width of 32.4 in. we must place six bars. These cannot be farther apart than 10.3 in. We shall arbitrarily place two bars in the slab at 5 in. outside the web of the beam.

ACI Code Section 24.3.4 requires "additional" longitudinal reinforcement in the slab outside this band. We shall assume that the shrinkage and temperature steel already in the slab will satisfy this requirement. The final distribution of negative-moment steel is as shown in Fig. 10-8b.

10. Design the shear reinforcement. The shear-force diagrams are calculated in Table 10-3 and shown at the bottom of that table. The shear coefficients at midspan of the beams (line 5) are based on Eq. (6-26).

TABLE 10-3 Calculation of Shear Forces—Example 10-1

Line	A	B3	B	B	B4	C
1. ℓ_n, ft		32.33			27.0	
2. w_u, kip/ft		3.43			3.48	
3. w_{Lu} kip/ft		1.6 × 1.08			1.6×1.11	
4. $w_u\ell_n/2$		55.4			47.0	
5. C_v at support and midspan	1.0	0.125	1.15	1.0	0.125	1.0
6. V_u	55.4	6.98	63.8	47.0	5.99	47.0
7. V_u/ϕ	73.9	9.3	85.0	62.6	8.0	62.6

$\frac{V_u}{\phi}$ (kips)

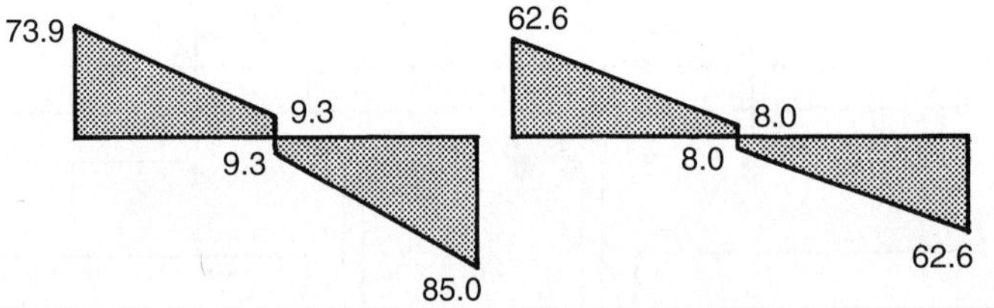

(a) Exterior end of *B*3. Because this beam is supported by other beams, the critical section for shear is located at the face of the support. ACI Code Section 9.6.3.1 requires stirrups if $V_u \geq \phi V_c/2$, where

$$V_c = 2\lambda\sqrt{f'_c}\,b_w d$$

$$= 2 \times 1\sqrt{4000} \times 12 \times 21.5 = 32{,}600 \text{ lbs} = 32.6 \text{ kips}$$

$$\frac{V_c}{2} = 16.3 \text{ kips}$$

Because $V_u/\phi = 73.9$ kips exceeds $V_c/2 = 16.3$ kips, stirrups are required.

Try No. 3 Grade-40 double-leg stirrups with a 90° hook enclosing a No. 7 stirrup-support bar (Fig. 10-8b). Because the required value of $V_s = V_u/\phi - V_c$ does not exceed $4\sqrt{f'_c}\,b_w d$, ACI Code Section 9.7.6.2.2 limits the maximum stirrup spacing to the smaller of $d/2$ and 24 in.

$$\frac{d}{2} = 10.75 \text{ in.}$$

To satisfy the minimum stirrup area requirement in ACI Code Section 9.6.3.3, the stirrup spacing must be less than the smaller of

$$s = \frac{A_v f_{yt}}{50 b_w} = 22.0 \text{ in.}$$

or

$$s = \frac{A_v f_{yt}}{0.75\sqrt{f'_c} \cdot b_w} = 23.2 \text{ in.}$$

Use 10 in. as the maximum spacing.

The spacing required to support the shear forces is

$$s = \frac{A_v f_{yt} d}{V_u/\phi - V_c} \tag{6-21}$$

At end A, $V_u/\phi = 73.9$ kips, and

$$s = \frac{0.22 \text{ in.}^2 \times 60 \text{ ksi} \times 21.5 \text{ in.}}{73.9 \text{ k} - 32.6 \text{ k}}$$

$$= 6.87 \text{ in.—say, 6 in. on centers}$$

At some point in the span, the stirrup spacing will be changed to 10 in. Compute V_n where $s = 10$ in.,

$$V_n = \frac{A_v f_{yt} d}{s} + V_c$$

$$= \frac{0.22 \times 60 \times 21.5}{10} + 32.6$$

$$= 61.0 \text{ kips}$$

Setting $V_n = V_u/\phi$, this value occurs at

$$x = \frac{73.9 - 61.0}{73.9 - 9.3} \times 194 \text{ in.} = 38.7 \text{ in. from end } A$$

Compute where V_u/ϕ is equal to $V_c/2$, and we can stop using stirrups:

$$x = \frac{73.9 - 16.3}{73.9 - 9.3} \times 194 \text{ in.} = 173 \text{ in.}$$

At the exterior end of *B*3, use No. 3 Grade-60 double-leg stirrups: one at 3 in., six at 6 in., and thirteen at 10 in. on centers.

(b) Interior end of *B*3. The shear at support B is 85.0 kips. The spacing required at this point is

$$s = \frac{0.22 \text{ in.}^2 \times 60 \text{ ksi} \times 21.5 \text{ in.}}{85.0 \text{ k} - 32.6 \text{ k}}$$

$$= 5.42 \text{ in.,} \quad \text{so use } s = 5 \text{ in.}$$

Change the stirrup spacing to 10 in. V_n, where $s = 10$ in., is 61.0 kips.

$s = 10$ in. occurs at

$$x = \frac{85.0 - 61.0}{85.0 - 9.3} \times 194 = 61.5 \text{ in. from end } B$$

$V_u/\phi = V_c/2$ at the point where the stirrups can be stopped. This occurs at

$$x = 176 \text{ in. from end } B$$

At the interior end of *B*3, use No. 3 Grade-60 double-leg stirrups: one at 3 in., twelve at 5 in., and eleven at 10 in. on centers.

(c) Ends of beam *B*4. The calculations for beam *B*4 are carried out in the same manner as for beam *B*3. **At each end of beam *B*3, use No. 3 Grade-60 double-leg stirrups: one at 4.5 in. and thirteen at 10 in.**

11. Check the development lengths and design-bar cutoffs.

(a) Perform the preliminary calculations.

Detailing requirements:

$$d = 21.5 \text{ in.}$$

$12d_b$ = 12 in. for No. 8, 10.5 in. for No. 7, and 9 in. for No. 6.

$$\frac{\ell_n}{16} = \frac{388}{16} = 24.3 \text{ in. for } B3$$

$$= \frac{324}{16} = 20.3 \text{ in. for } B4$$

Therefore, d always exceeds $12d_b$, but $\ell_n/16$ exceeds d in beam *B*3.

(b) Select cutoffs for positive-moment steel in *B*3. The reinforcement at midspan is four No. 8 (Table 10-2). Extend two No. 8 bar into the supports and cut off two No. 8 where they are no longer required. The remaining bars have 50 percent of the original area; therefore, the theoretical flexural cutoff is at the point where M_u = 50 percent of the maximum M_u at midspan. From Fig. A-3, this occurs at $0.21\ell_n = 0.21 \times 388 = 81.5$ in. from the exterior end, and $0.26\ell_n = 101$ in. from the interior end. These are points *B* and *C* in Fig. 10-9. To compute the actual cut-off points we must satisfy the detailing rules in various ACI Code Sections. The following discussion will refer to those section.

Before doing so we shall calculate ℓ_d for the bottom bars.

Spacing and confinement case. The bottom bars have clear spacing and cover of at least d_b and are enclosed by at least minimum stirrups. Therefore, this is Case 1 in Table 8-1 (ACI Code Section 25.4.2.2). The bars are No. 8. Thus, we have

$$\ell_d = \frac{f_y \psi_t \psi_e}{20\lambda\sqrt{f'_c}} d_b \qquad (8\text{-}12)$$

$$= \frac{60{,}000 \times 1.0 \times 1.0 \times d_b}{20 \times 1\sqrt{4000}} = 47.4d_b$$

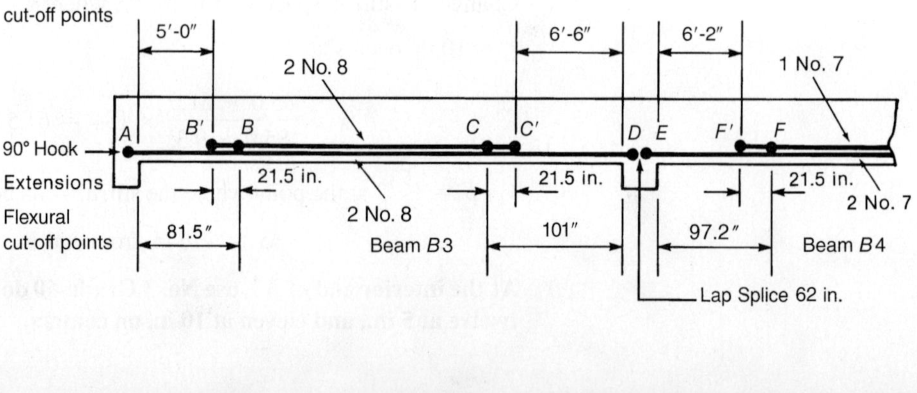

Fig. 10-9
Calculation of bar cut-off points for positive-moment steel.

For the No. 8 bars, $d_b = 1.0$ in. and thus $\ell_d = 47.4$ in. For the No. 7 bars (Beam $B4$), $d_b = 0.875$ in. and thus $\ell_d = 41.5$ in.

Application of detailing rules to beam *B*3

Some or all of each family of detailing rules apply to each region of the beam. The detailing process starts by selecting the rules that apply in each region. It is then easier to select the sub-rules applied in each region. ***Beam B*3** must satisfy rules: 1. Structural integrity; 2. Extension of bars into supports; 3. Effect of shear on bending moment diagrams; and 4. Anchorage, positive moment.

Structural integrity. This beam is a continuous interior beam. ACI Code Sections 9.7.7.2 through 9.7.7.6 apply. In beam line *B*3–*B*4–*B*3, a tension tie is created by lap splicing at least one-quarter of the positive-moment reinforcement, but not less than two bars, from span *B*3 with a similar amount of steel from beam *B*4. The splices will be Class B splices, at or near the supports, with a lap of $1.3\ell_d$. The development length of a No. 8 bottom bar is 47.4 in., so $1.3\ell_d = 61.6$ in.

Lap splice two No. 8 bottom bars from span *B*3, a distance of 62 in. with two No. 7 bottom bars from span *B*4.

The two No. 7 bottom bars which will be part of the longitudinal structural integrity tie must be placed inside the lower corners of U stirrups and must pass through the column core. The longitudinal tie bars must be hooked into the supports at the two outer ends of the beam using standard 90º hooks at the discontinuous ends.

Extension of bars into the supports. ACI Code Section 9.7.3.8.2 requires that at least one-quarter of the positive-moment reinforcement, but not less than two bars from span *B*3 must extend a minimum of 6 in. into the supports. This requirement has already been satisfied.

Effect of shear. The bars must extend d or $12d_b$ past the positive-moment flexural cut-off points. **Extend bars $d = 21.5$ in. $= 1.79$ ft to points B' and C'** (Fig. 10-9).

Anchorage. The bars must extend ℓ_d past the point of maximum bar stress. The maximum bar stress occurs at midspan. By inspection, the distances from midspan to B' and C' exceed $\ell_d = 47.4$ in. Therefore, cut off two No. 8 bars at 81.5 in. − 21.5 in. = 5 ft, from the interior face of the exterior column, and 79.5 in., say 6 ft 6 in., from the exterior face of the interior column.

Anchorage positive moment. Bars must extend ℓ_d past the actual cut-off points of adjacent bars. The actual extensions past B' and C' exceed 60 in. and 78 in., respectively, both of which are greater than $\ell_d = 47.4$ in.

Anchorage at point of inflection. At positive-moment points of inflection, the bars must satisfy

$$\ell_d \leq \frac{M_n}{V_u} + \ell_a \tag{8-21}$$

From Fig. A-3, the positive-moment points of inflection are at $0.098\ell_n = 38$ in. from the exterior end and $0.146\ell_n = 56.6$ in. from the interior end. At each of these points, the remaining steel is two No. 8 bars, $A_s = 1.58$ in.^{2}. Accordingly we have

$$a = \frac{A_s f_y}{0.85 f'_c b_e} = 0.30 \text{ in.}$$

$$M_n = A_s f_y \left(d - \frac{a}{2} \right) = 2020 \text{ kip-in.}$$

Check at the exterior end: from the shear-force diagram in Table 10-3, the V_u at 38 in. from the exterior end is

$$\frac{V_u}{\phi} = 73.9 - \frac{38}{194}(73.9 - 9.3) = 61.2 \text{ kips}$$

$$V_u = 45.9 \text{ kips}$$

and ℓ_a is the larger of d or $12d_b$, but smaller than the actual extension (which exceeds 38 in.), so

$$\ell_a = 21.5 \text{ in.}$$

Thus,

$$\frac{M_n}{V_u} + \ell_a = \frac{2020}{45.9} + 21.5 = 65.5 \text{ in.}$$

This exceeds $\ell_d = 47.4$ in.—therefore, o.k.

Check at interior end:

$$V_u \text{ at 56.6 in. from interior end} = 62.9 \text{ kips}$$

$$\frac{M_n}{V_u} + \ell_a = \frac{2020}{62.9} + 21.5$$

$$= 53.6 \text{ in.—therefore, o.k.}$$

Summary of the bar cut-off points for the positive-moment steel in beam* B3. *This is the end span of a continuous interior beam. Hook two No. 8 bars in the exterior support and lap-splice two No. 8 bars with No. 7 bars in the interior support. Cut off two No. 8 bars at 5 ft from the face of the exterior column and at 6 ft 6 in. from the face of the interior column. These locations are shown as points A, B', C', and D in Fig. 10-9.

(c) Select cutoffs for the positive-moment steel in *B*4. This is a continuous interior beam. At midspan, we have three No. 7 bars. Run two No. 7 bars into the supports and cut off the remaining No. 7 bar. The A_s remaining is $1.20/1.80 = 0.67$ times the original amount. Therefore, the flexural cutoff is located at the point where the moment is 0.67 times the midspan moment. From Fig. A-1, this occurs at $0.30\ell_n = 0.30 \times 324 = 97.2$ in. from the face of the support (Fig. 10-9). Applying the rule for the effect of shear, we find that the bar must extend 21.5 in. past this point, to 75.7 in. from the support. Therefore, cut off one No. 7 bar at 6 ft 2 in. from the support.

The bars have a spacing and cover of at least d_b and are enclosed by at least minimum stirrups. Therefore, for the No. 7 bars, $\ell_d = 47.4 \times 0.875 = 41.5$ in.

Structural integrity. Lap splice the two No. 7 bars with the No. 8 bars from beam *B*3. The length of a Class B lap splice for a No. 8 bar is $1.3\ell_d = 61.6$ in. Therefore, lap splice the bars 62 in.

Anchorage. The positive-moment points of inflection are $0.146\ell_n = 47.3$ in. from supports. At these sections the reinforcement is two No. 7 bars, and we have

$$a = \frac{A_s f_y}{0.85 f'_c b_e} = 0.23 \text{ in.}$$

$$M_n = A_s f_y\left(d - \frac{a}{2}\right) = 1540 \text{ kip-in.}$$

$$V_u = 35.0 \text{ kips}$$

$$\frac{M_n}{V_u} + \ell_a = \frac{1540}{35.0} + 21.5 = 65.5 \text{ in.}$$

This exceeds ℓ_d—therefore, o.k. The cut-off points for the positive-moment steel are as indicated in Fig. 10-9.

(d) Select cutoffs for the negative-moment steel at the exterior end of *B*3. At the exterior end, there are three No. 7 bars. These must be anchored in the spandrel beam, which is 16 in. wide. The basic development lengths of standard hooks are (ACI Code Section 25.4.3.1) and book Section 8-4.

$$\ell_{dh} = \left(\frac{f_y \psi_e \psi_c \psi_r}{50\lambda\sqrt{f'_c}}\right) d_b \tag{8-18}$$

We will assume that λ (lightweight concrete factor), ψ_e (epoxy coating factor), and ψ_r (confining reinforcement factor), are all 1.0. The value of ψ_c (cover factor) can be taken as 0.7 if we can show that the side cover perpendicular to the plane of the hook is at least 2.5 in. and if the tall cover is at least 2.0 in. The hooks in question enter the spandrel beam perpendicular to the axis of the spandrel beam, and thus, the side cover requirement is satisfied. By hooking the No. 7 bars inside the top longitudinal bars for the spandrel beam, we can ensure that the cover on the tail of the hook exceeds 2.0 in. Thus,

$$\ell_{dh} = \left(\frac{60{,}000 \text{ psi} \times 1 \times 0.7 \times 1}{50 \times 1\sqrt{4000 \text{ psi}}}\right) 0.875 \text{ in.} = 11.6 \text{ in.}$$

This can be accommodated in the spandrel beam. **Therefore, anchor the top bars into the spandrel beam with 90° standard hooks.**

Extend all the top bars past the negative-moment point of inflection before cutting them off. From Fig. A-3, the negative-moment point of inflection is at $0.108\ell_n = 41.9$ in. from the face of the exterior support (point *G* in Fig. 10-10). At this cut-off point, we must satisfy the rule for the effect of shear and that the bars extend at least ℓ_d from the support face.

Effects of shear. Requires at least one-third of the bars to extend the longest of d, $12d_b$, and $\ell_n/16$ past the point of inflection. $\ell_n/16 = 24.3$ in. governs here. Therefore, cut off all bars at $41.9 + 24.3 = 66.2$ in., say 5 ft 6 in. from the face of the support.

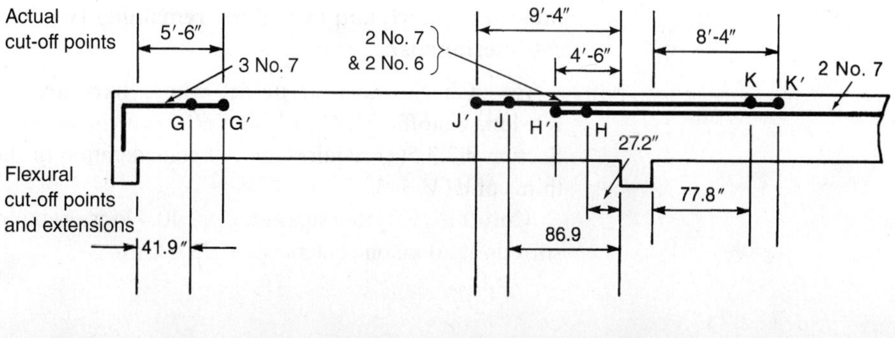

Fig. 10-10
Calculation of bar cut-off points for negative-moment steel.

Anchorage. Bars must extend at least ℓ_d from the face of the support. Using $\psi_t = 1.3$,

$$\ell_d \text{ for No. 7 top bar} = 1.3 \times 41.5 = 54 \text{ in.}$$

Because 5 ft 6 in. exceeds 54 in., this rule is satisfied. **Therefore, cut off all top bars at the exterior support at 5 ft 6 in. from the face of the support.**

(e) Select cutoffs for the negative-moment steel at the interior end of beam *B*3. The negative-moment steel at the first interior support consists of four No. 7 and two No. 6 bars. We will cut off two No. 7 bars when they are no longer required for flexure and extend the remaining bars ($0.63\ A_s$) past the negative point of inflection. From Fig. A-3, $M_u = 0.63\ M_u(\text{max})$ at approximately $0.070\ \ell_n = 27.2$ in. from the face of the support (point *H* in Fig. 10-10).

Effects of shear. Bars must extend $d = 21.5$ in. past *H* to 48.7 in. from the face of the support.

Anchorage. The bars must extend ℓ_d from the point of maximum bar stress at the face of the support. In step 11(d), this distance was found to be 54 in. for a top No. 7 bar.

Therefore, cut off two No. 7 bars at 4 ft 6 in. from the face of the interior support.

The remaining four bars will be extended past the negative-moment point of inflection. From Fig. A-3, this is $0.224\ell_n = 86.9$ in. from the face of the support (point *J* in Fig. 10-10).

Effects of shear. At least one-third of bars must extend $\ell_n/16 = 24.3$ in. past the point of inflection to 111 in. from the face of the support.

Anchorage. Bars must extend $\ell_d = 54$ in. past the actual cutoff H'. Therefore, the bars must extend to 108 in. from the face of the support.

Therefore, cut off two No. 7 and two No. 6 top bars at 9 ft 4 in. from the exterior face of the first interior support.

(f) Select cutoffs for the negative-moment steel in beam *B*4. When the interior span of a three-span continuous beam is less than about 90 percent as long as the end spans, it can have negative moment at midspan under some loads, even though this is not apparent from the ACI moment coefficients. For this reason, we will extend two No. 7 top bars the full length of the beam. The remaining four bars will be extended past the negative-moment point of inflection and will be cut off. From Fig. A-1 the negative-moment point of inflection is $0.24\ell_n = 77.8$ in. from the face of the column (point *K* in Fig. 10-10).

Effects of shear. Says that these bars must extend by the larger of d, $12d_b$, or $\ell_n/16$ past this point to $77.8 + 21.5 = 99.3$ in. from the face of the support.

Anchorage. Says that these bars must extend by $\ell_d = 54$ in. past the face of the support—therefore o.k.

Cut off two No. 7 top and two No. 6 bars at 8 ft 4 in. from the interior face of the interior support, and extend the remaining two No. 7 top bars the full length of the interior span (see Fig. 10-10).

(g) Check shear at points where bars are cut off in a zone of flexural tension. Cutoffs B', C', F', and H' occur in zones of flexural tension. ACI Code Section 9.7.3.5(a) requires special consideration of these regions if V_u exceeds two-thirds of $\phi(V_c + V_s)$.

Cutoff B': 5 ft from support, $V_u = 40.4$ kips. At this point there are No. 3 double-leg stirrups at 10 in. on centers. $V_c = 32.6$ kips.

$$V_s = \frac{0.22 \times 60 \times 21.5}{10} = 28.4 \text{ kips}$$

$$\phi(V_c + V_s) = 0.75(32.6 + 28.4) = 45.8 \text{ kips}$$

Thus, V_u is 0.88 of the shear permitted.

In step 10(a) for the design of shear reinforcement at the exterior end of beam $B3$, we started with a stirrup spacing of 6 in. near the support and increased to a 10-in. spacing at 39 in. from the face of the support. To satisfy ACI Code Section 9.7.3.5(a), we will maintain the 6-in. spacing up to and beyond cut-off point B'. This results in $2/3\ \phi V_n \geq V_u$ at cut-off point B'. The tighter 6-in. stirrup spacing at point B' should be extended at least $d/2$ toward the point of maximum positive moment, which occurs near midspan, because if a flexural-shear crack initiates at B', it will tend to extend toward the maximum positive moment. **Therefore, at the exterior end of beam *B*3, use No. 3 Grade-60 double-leg stirrups: one at 3 in., twelve at 6 in. (extends *d*/2 beyond point *B*′), and ten at 10 in.** (extends beyond the point where $\frac{1}{2}\phi V_c$ exceeds V_u). The final stirrup layout for the external end of beam $B3$ is shown in the left portion of Fig. 10-11a. At least two of the top No. 7 bars, which were shown in Fig. 10-10 to be cut off at 5 ft 6 in. from the support face, should be extended to 14 ft 8 in. from the support face to hold up the stirrups, as shown by the dashed line in Fig. 10-11a.

***Cutoff C′*:** At this point, $V_u = 40.9$ kips, and thus, the same 6-in. stirrup spacing will satisfy ACI Code Section 9.7.3.5(a).

***Cutoff H′*:** At this point, $V_u = 47.9$ kips, so a closer spacing of stirrups will be required. Using a 4-in. spacing:

$$V_s = \frac{0.22 \times 60 \times 21.5}{4} = 71.0 \text{ kips}$$

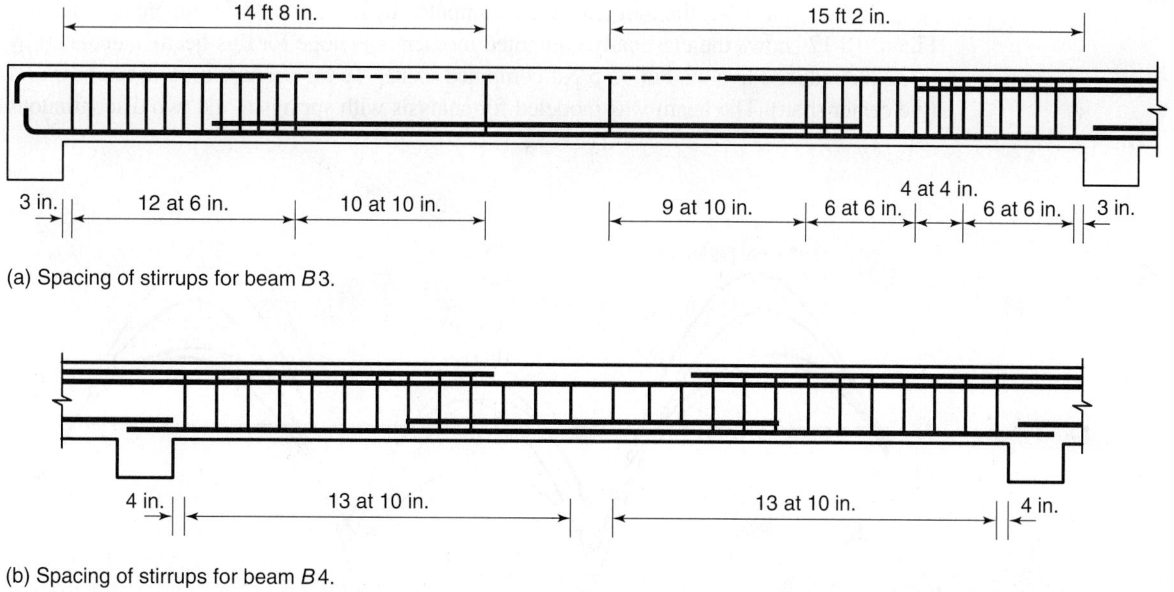

Fig. 10-11
Final layout of stirrups for Example 10-1.

and

$$\frac{2}{3}\phi V_n = 0.667 \times 0.75(32.6 + 71.0) = 51.8 \text{ kips}$$

This exceeds V_u at the cut-off point H'. It should be noted that cut-off point H' is for the negative-moment reinforcement, so the tighter stirrup spacing used at H' should be extended at least $d/2$ toward the point of maximum negative moment, which occurs at the face of the support. The tighter stirrup spacing at C' should extend toward midspan, as was done for cut-off point B'. **Therefore, at the interior end of beam *B*3, use No. 3 Grade-60 stirrups: one at 3 in., six at 6 in., four at 4 in.** (starts at least $d/2$ before point H' and extends just beyond H'), **six at 6 in.** (extends $d/2$ past point C') **and nine at 10 in.** (extends past the point where $\frac{1}{2}\phi V_c$ exceeds V_u). The final stirrup layout for the interior end of beam *B*3 is shown in the right portion of Fig. 10-11a. At least two of the top bars, which were shown in Fig. 10-10 to be cut off at 9 ft 4 in. from the support face, should be extended to 15 ft 2 in. from the support face to hold up the stirrups, as shown by the dashed line in Fig. 10-11a.

Cutoff F': At this point $V_u = 28.2$ kips, so a stirrup spacing of 10 in. will give 2/3 $\phi V_n \geq V_u$. **Therefore, at each end of beam *B*4, use No. 3 Grade-60 double-leg stirrups: one at 4 in. and thirteen more at 10 in.** (extends beyond the point where $\frac{1}{2}\phi V_c$ exceeds V_u). The final stirrup layout for beam *B*4 is shown in Fig. 10-11b. Because two top No. 7 bars already were placed over the full length of beam *B*4, no changes are required to provide support for the stirrups.

12. Design web-face steel. Because d does not exceed 3 ft, no side-face steel is required (ACI Code Section 9.7.2.3).

This concludes the design. The bar detailing in step 11 of this example is quite lengthy. The process could be simplified considerably if all bars were extended past the points of inflection before being cut off. ■

In Example 10-1, the moments were computed by using the ACI moment coefficients. Figure 10-12 shows the elastically computed moment envelope for this beam, calculated by using the reduced live load of 75.5 psf computed in step 1c for positive moment in beam *B*4 (the center span). The beam was modeled for analysis with springs at the two discontinuous

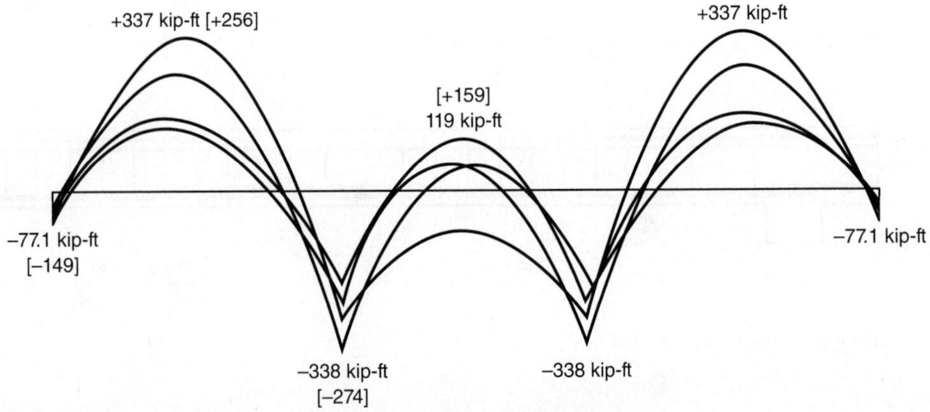

Fig. 10-12
Elastic-moment envelopes—
Example 10-1.

ends to represent torsional stiffness of the spandrel beams. Their spring stiffness was derived from Eq. (7-21).

The moments plotted in Fig. 10-12 can be compared to the moments from the ACI moment coefficients, shown in the moment diagram in the heading of Table 10-2 and given (in brackets) in Fig. 10-12. Moments in the end spans and at the interior supports are relatively close between the two procedures. The negative moments at the exterior ends are directly a function of the assumed torsional stiffness of the spandrel beams. The interior-span positive moment is overestimated by the ACI moment coefficients, which, in addition, fail to predict the possibility of negative moment at midspan of the interior span.

The lengthy example presented in this chapter appears somewhat cumbersome the first time through. With practice and understanding of the principles, the design of continuous beams and slabs becomes straightforward. The beams and slabs designed for one floor can frequently be used many times throughout the building.

10-4 DESIGN OF GIRDERS

Girders support their own weight plus the concentrated loads from the beams they support. It is customary to compute the moments and shears in the girder by assuming that the girder supports concentrated loads equal to the beam reactions, plus a uniform load equal to the self-weight of the girder, plus the live load applied directly over the girder. This approach neglects two-way action in the slab adjacent to the girder.

When the flexural reinforcement in the slab runs parallel to the girder, the two-way action shown in Fig. 10-13 causes negative moments in the slab along the slab–girder connection, as indicated by the curvature of slab strip *B* in Fig. 10-13. To reinforce for these moments, ACI Code Section 7.5.2.3 requires slab reinforcement transverse to the girder, as shown in Fig. 10-14. The reinforcement is designed by assuming that the slab acts as a cantilever, projecting a distance equal to the effective overhanging slab width and carrying the factored dead load and live load supported by this portion of the slab.

Frequently, edge girders are loaded in torsion by beams framing into them from the sides between the ends of the girder. This is *compatibility torsion*, because it exists only because the free-end rotation of the beam is restrained by the torsional stiffness of the girder. In such a case, ACI Code Section 22.7.3.2 allows a reduction in the design torque, as discussed in Chapter 7. If torque is present, the stirrups must be closed stirrups (ACI Code Section 9.7.3.6.1).

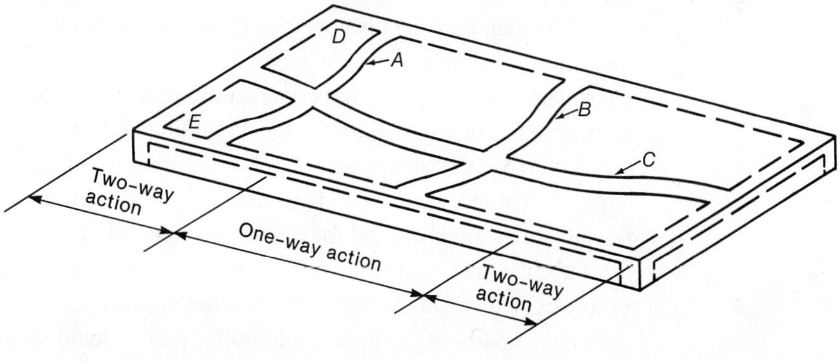

Fig. 10-13
One-way and two-way slab action near beam supports.

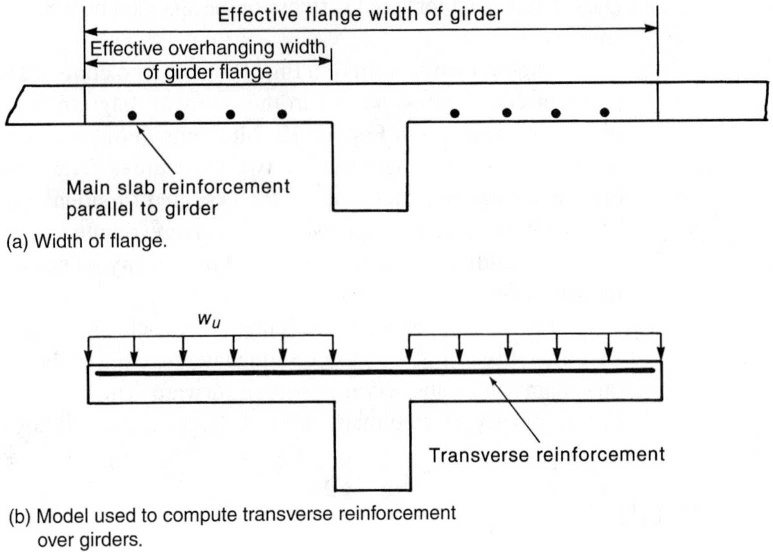

Fig. 10-14
Transverse reinforcement over girders.

10-5 JOIST FLOORS

Long-span floors for relatively light live loads can be constructed as a series of closely spaced, cast-in-place T-beams (or *joists*) with a cross section as shown in Fig. 10-15. The joists span one way between beams. Most often, removable metal forms referred to as *fillers* or *pans* are used to form the joists. Occasionally, joist floors are built by using clay-tile fillers, which serve as forms for the concrete in the ribs that are left in place to serve as the ceiling (ACI Code Section 9.8.2).

When the dimensions of the joists conform to ACI Code Sections 9.8.1.1 to 9.8.1.4, they are eligible for less cover to the reinforcement than for beams (ACI Code Table 20.6.1.3.1) and for a 10 percent increase in the shear, V_c, carried by the concrete (ACI Code Section 9.8.1.5). The principal requirements are that the floor be a monolithic combination of regularly spaced ribs and a top slab with

1. ribs not less than 4 in. in width,
2. depth of ribs not more than $3\frac{1}{2}$ times the minimum web width, and
3. clear spacing between ribs not greater than 30 in.

Ribbed slabs not meeting these requirements are designed as slabs and beams.

The slab thickness is governed by ACI Code Section 9.8.3.1, which requires a thickness of not less than 2 in. for joists formed with 20-in.-wide pans and $2\frac{1}{2}$ in. for 30-in. pans. The slab thickness and cover to the reinforcement are also a function of the fire-resistance rating required by the local building code. For a 1-hour fire rating, a $\frac{3}{4}$ in. cover and a 3- to $3\frac{1}{2}$ in. slab are required. For a 2-hour rating, a 1-in. cover and a $4\frac{1}{2}$ in. slab are required. These vary from code to code and as a function of the type of ceiling used. Joist floor systems are also used for parking garages. Here, the top cover of the slab reinforcement and the quality of concrete used must be chosen to reduce the possibility of corrosion. Epoxy-coated or galvanized bars are often used in such applications. Wheel loads on the slab and abrasion of the concrete in traffic paths should also be considered.

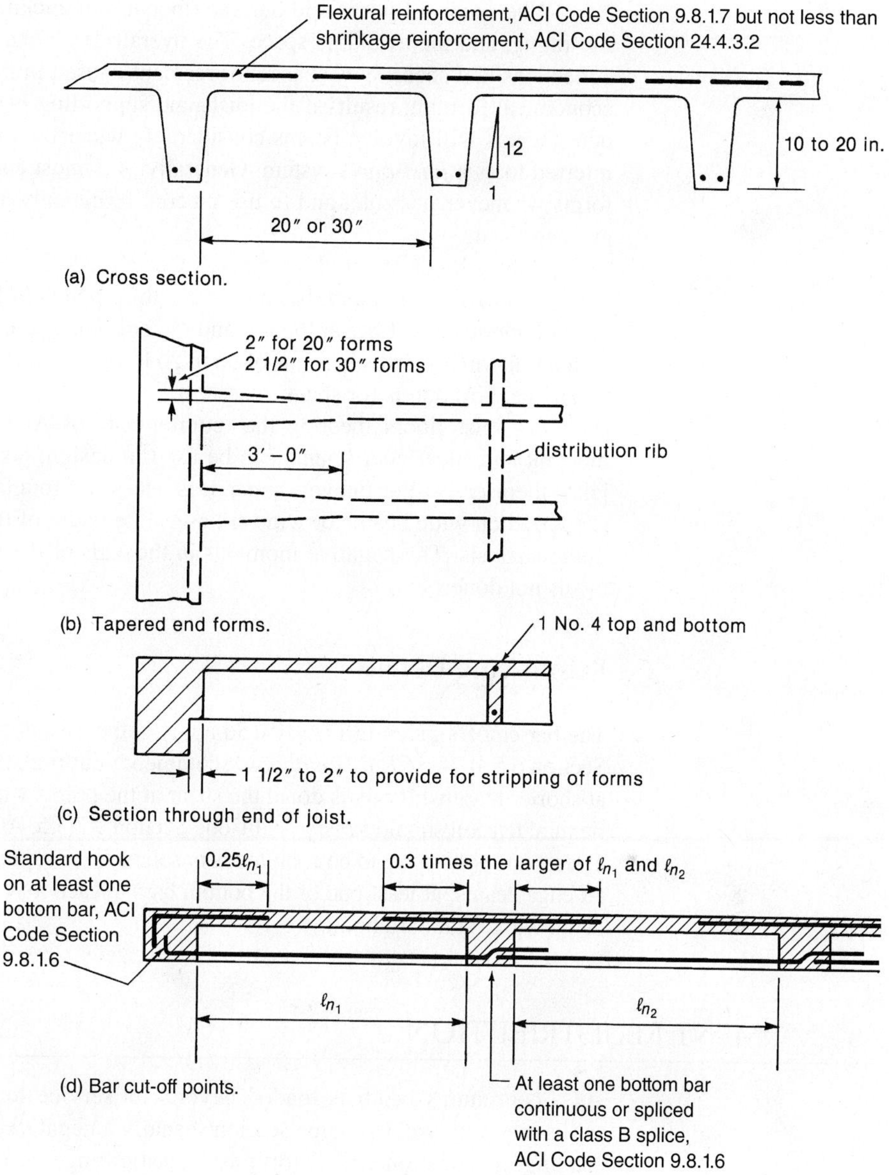

Fig. 10-15
One-way-joist construction.

Standard Dimensions and Layout of One-Way Joists

The dimensions of standard removable pan forms are shown in Fig. 10-15. The forms are available in widths of 20 and 30 in., measured at the bottom of the ribs. Standard depths are 6, 8, 10, 12, 14, 16, and 20 in. End forms have one end filled in to serve as the side form for the supporting beams. Tapered end forms allow the width of the ends of the joists to increase, as shown in Fig. 10-15b, giving additional shear capacity at the end of the joists. The 20-in. tapered end pans reduce in width from 20 in. to 16 in. over a 3-ft length. The 30-in. pans reduce to 25 in., also over a 3-ft length. Pan forms wider than 30 in. also are available, but the resulting floors or roofs must be designed as conventional slabs and beams (ACI Code Section 9.8.1.8).

When such a floor is laid out, the rib and slab thicknesses are governed by strength, fire rating, and the available space. The overall depth and rib thickness are governed by deflections and shear. No shear reinforcement is used in most cases. Generally, the most economical forming results if the joists and supporting beams have the same depth. Frequently, this will involve beams considerably wider than the columns. Such a system is referred to as a *joist-band* system. Generally, it is most economical to use untapered end forms whenever possible, and to use tapered forms only at supports having higher-than-average shears.

Although not required by the ACI Code, load-distributing ribs perpendicular to the joists are provided at the midspan or at the third points of long spans. These have at least one continuous No. 4 bar at the top and the bottom. The *CRSI Handbook* [10-4] suggests no load-distributing ribs in spans of up to 20 ft, one at midspan for spans of 20 to 30 ft, and two at the third points for spans over 30 ft.

For joist floors meeting the requirements of ACI Code Section 6.5.1, the ACI moment and shear coefficients can be used in design, taking ℓ_n as the clear span of the joists themselves. For uneven spans, it is necessary to analyze the floor. If the joists are supported on wide beams or wide columns, the width of the supports should be modeled in the analysis. The negative moments in the ends of the joists will be underestimated if this is not done.

Reinforcing Details

The bar cutoffs given in Fig. 10-15d apply if the joists fit the requirements of ACI Code Section 6.5.1. The *CRSI Handbook* recommends cutting off half of the top and bottom bars at shorter lengths. If this is done, the shear at the points where bars are cut off in a zone of flexural tension should satisfy ACI Code Section 9.7.3.5. ACI Code Section 9.8.1.6 requires at least one bottom bar to be continuous or spliced over supports with a Class B tension splice. At edge beams, at least one of the bottom bars in each joist must be fully anchored into the beam with a standard hook.

10-6 MOMENT REDISTRIBUTION

As a continuous beam is loaded beyond its service loads, the tension reinforcement will eventually yield at some section (usually a negative-moment section). With further loading, this section will deform as a plastic hinge, at a constant moment taken equal to the nominal moment. Increased loads cause an increase in the sum of the positive and negative moments in the beam. Because the moment cannot increase at the plastic hinge, the increase in moments all occurs at the positive-moment location. Thus, the shape of the moment diagram must shift, with more moment going to those sections, which are still elastic. Eventually, a mechanism forms, and the plastic capacity is reached. Procedures for carrying out inelastic analyses of concrete structures are given in [10-5] and [10-6].

For a uniformly loaded fixed-end beam, the maximum elastic positive moments $(w\ell^2/24)$ are half the maximum negative moment $(w\ell^2/12)$. As a result, approximately twice as much reinforcement is required at the supports compared with what is required at midspan. Sometimes, this leads to congestion of the steel at the supports. The plastic-moment distribution results in a much more variable reinforcement layout depending on the actual plastic-moment capacities in the positive- and negative-moment regions, and moments can be redistributed from the elastic distribution, allowing reductions in the higher moments, with corresponding increases in the lower moments. The amount of redistribution that can be tolerated is governed by two aspects. First, the hinging section must be able to undergo the necessary inelastic deformations. Because the inelastic rotational capacity is a function of the reinforcement ratio, as shown in Fig. 4-11, this implies an upper limit on the reinforcement ratio. Second, hinges should not occur at service loads, because wide cracks develop at hinge locations.

ACI Code Section 6.6.5.1 allows the maximum moments at the supports or midspan to be increased or decreased by not more than $1000\epsilon_t$ percent, with a maximum of 20 percent, provided that $\epsilon_t \geq 0.0075$ at the section where moments are being reduced, where ϵ_t is the tensile strain in the layer of steel closest to the tension face of the member.

These moments must have been computed via an elastic analysis, and not the ACI moment coefficients. The modified moments must be used in calculating the moments at all critical sections, shears, and the bar cut-off points. (See ACI Code Sections 6.6.5.4 and 6.6.5.5.)

EXAMPLE 10-2 Moment Redistribution

Compute the design moments for the three-span beam from Example 10-1, using an elastic analysis (Fig. 10-12) and moment redistribution, as permitted in ACI Code Section 6.6.5.1. Step 8 of the example will be repeated. Using $c = a/\beta_1 = 4.66/0.85 = 5.48$ in.,

$$\epsilon_t = \frac{21.5 - 5.48}{5.48} \times 0.003$$

$$= 0.00877$$

Thus, the maximum redistribution that is allowed is approximately $(1000\epsilon_t = 8.8)$ percent. Using the elastic moment shown in Fig. 10-12, the reduced moment is accordingly $(1 - 0.088) \times 338 = 308$ kip-ft. This value exceeds the design moment (274 kip-ft) that was used for this section in Example 10-1. Thus, change the tension reinforcement for the section to be four No. 7 bars and four No. 5 bars ($A_s = 3.64\text{ in}^2$). The four No. 5 bars will be distributed into the flanges, two on each side of the web. Check whether $A_s = 3.64\text{ in.}^2$ is adequate:

$$a = \frac{A_s f_y}{0.85 f'_c b_w} = \frac{3.64\text{ in.}^2 \times 60\text{ ksi}}{0.85 \times 4\text{ ksi} \times 12\text{ in.}} = 5.35\text{ in.}$$

$$\phi M_n = \phi A_s f_y \left(d - \frac{a}{2}\right) = 3700\text{ k-in.} = 308\text{ kip-ft}$$

Thus, the modified section is adequate.

It is now necessary to compute the positive midspan moments corresponding to this reduced value of the negative moment at the interior supports. This step is done to see if the redistribution of negative moments to midspan positive moments will cause these values to exceed those previously calculated for other loading cases (Fig. 10-12). The midspan moments must correspond to the load case that produced the maximum negative moment at the interior support. The appropriate live-load pattern would include live load on the two spans that are adjacent to the interior support. In step 5 of Example 10-1, the total factored distributed load, w_u, for this load case was found to be 3.11 kip/ft. An elastic analysis for this load case resulted in midspan moment values of approximately 260 kip-ft for the exterior span and 65 kip-ft for the interior span. Even if the full distributed moment, which was calculated above to be 338 − 308 = 30 kip-ft, was added to these midspan moments, their values still would be less than the values calculated for other load cases intended to maximize the midspan positive moments (Fig. 10-12). Thus, for this case, very little can be accomplished by redistributing the maximum negative moments. The only change is the reduction in the maximum negative design moment at the interior supports. ■

PROBLEMS

10-1 A five-span one-way slab is supported on 12-in.-wide beams with center-to-center spacing of 16 ft. The slab carries a superimposed dead load of 10 psf and a live load of 60 psf. Using $f'_c = 4000$ psi and $f_y = 60{,}000$ psi, design flexural reinforcement for the slab. Draw a longitudinal section showing the top and bottom reinforcement. Use Fig. A-5 to locate bar cut-off points.

10-2 A four-span one-way slab is supported on 12-in.-wide beams with center-to-center spacing of 14, 16, 16, and 14 ft. The slab carries a superimposed dead load of 20 psf and a live load of 100 psf. Design the slab flexural reinforcement using $f'_c = 3500$ psi and $f_y = 60{,}000$ psi.

Video Solution

10-3 A three-span continuous beam supports 6-in.-thick one-way slabs that span 20 ft center-to-center of beams. The beams have clear spans, face-to-face of 16-in.-square columns, of 27, 30, and 27 ft. The floor supports ceiling, ductwork, and lighting fixtures weighing a total of 10 psf, ceramic floor tile weighing 16 psf, partitions equivalent to a uniform dead load of 20 psf, and a live load of 60 psf. Design the beam, using $f'_c = 4500$ psi. Use $f_y = 60{,}000$ psi for flexural reinforcement and $f_{yt} = 40{,}000$ psi for shear reinforcement. Calculate cut-off points, extending all reinforcement past points of inflection. Draw an elevation view of the beam and enough cross sections to summarize the design.

10-4 Repeat Problem 10-3, but cut off up to 50 percent of the negative-moment bars in each span where they are no longer needed.

REFERENCES

10-1 *Minimum Design Loads for Buildings and Other Structures*, (ASCE/SEI 7-10), American Society of Civil Engineers, Reston, VA, 2010, 608 pp.

10-2 Bianchini, A. C., Woods, Robert E., and Kesler, C. E., "Effect of Floor Concrete Strength on Column Strength," *ACI Journal, Proceedings*, Vol. 56, No. 11, May 1960, pp. 1149–1169.

10-3 Carlos E. Ospina and Scott D. B. Alexander, "Transmission of Interior Concrete Column Loads through Floors," *ASCE Journal of Structural Engineering*, Vol. 124, No. 6, 1998.

10-4 *CRSI Design Handbook*, 11th Edition, Concrete Reinforcing Steel Institute, Schaumberg, IL, 2012.

10-5 Richard W. Furlong, "Design of Concrete Frames by Assigned Limit Moments," *ACI Journal, Proceedings*, Vol. 67, No. 4, April 1970, pp. 341–353.

10-6 Alan H. Mattock, "Redistribution of Design Bending Moments in Reinforced Concrete Continuous Beams," *Proceedings, Institution of Civil Engineers, London*, Vol. 13, 1959, pp. 35–46.

11

Columns: Combined Axial Load and Bending

11-1 INTRODUCTION

A column is a vertical structural member supporting axial compressive loads, with or without moments. The cross-sectional dimensions of a column are generally considerably less than its height. Columns support vertical loads from the floors and roof and transmit these loads to the foundations.

In a typical construction cycle, the reinforcement and concrete for the beams and slabs in a floor system are placed first. Once this concrete has hardened, the reinforcement and concrete for the columns over that floor are placed. This second step in the process is illustrated in Figs. 11-1 and 11-2. Figure 11-1 shows a completed column prior to construction of the formwork for the next floor. This is a *tied column*, so called because the longitudinal bars are tied together with smaller bars at intervals up the column. One set of ties is visible just above the concrete. The longitudinal (vertical) bars protruding from the column will extend through the floor into the next-higher column and will be lap spliced with the bars in that column. The longitudinal bars are bent inward to fit inside the cage of bars for the next-higher column. (Other splice details are sometimes used: see Figs. 11-26 and 11-27). A reinforcement cage that is ready for the column forms is shown in Fig. 11-2. The lap splice at the bottom of the column and the ties can be seen in this photograph. Typically, a column cage is assembled on sawhorses prior to erection and is then lifted into place by a crane.

The more general terms *compression members* and *members subjected to combined axial load and bending* are sometimes used to refer to columns, walls, and members in concrete trusses or frames. These may be vertical, inclined, or horizontal. A column is a special case of a compression member that is vertical.

Stability effects must be considered in the design of compression members. If the moments induced by slenderness effects weaken a column appreciably, it is referred to as a *slender column* or a *long column*. The great majority of concrete columns are sufficiently stocky that slenderness can be ignored. Such columns are referred to as *short columns*. Slenderness effects are discussed in Chapter 12.

Fig. 11-1
Tied column under construction. (Photograph courtesy of J. G. MacGregor.)

Fig. 11-2
Reinforcement cage for a tied column. (Photograph courtesy of J. G. MacGregor.)

Although the theory developed in this chapter applies to columns in seismic regions, such columns require special detailing to resist the shear forces and repeated cyclic loads from earthquakes. In seismic regions, the ties are heavier and much more closely spaced than those shown in Figs. 11-1 and 11-2. This is discussed in Chapter 19.

11-2 TIED AND SPIRAL COLUMNS

Over 95 percent of all columns in buildings in nonseismic regions are tied columns similar to those shown in Figs. 11-1 and 11-2. Tied columns may be square, rectangular, L-shaped, circular, or any other required shape. Occasionally, when high strength and/or high ductility are required, the bars are placed in a circle, and the ties are replaced by a bar bent into a helix or spiral, with a pitch (distance between successive turns of the spiral—distance s in Fig. 11-4a) from $1\frac{3}{8}$ to $3\frac{3}{8}$ in. Such a column, called a *spiral column*, is illustrated in Fig. 11-3. Spiral columns are generally circular, although square or polygonal shapes are sometimes used. The spiral acts to restrain the lateral expansion of the column core under high axial loads and, in doing so, delays the failure of the concrete core, making the column more ductile, as discussed in the next section.

Fig. 11-3
Spiral column. (Photograph courtesy of J. G. MacGregor.)

Behavior of Tied and Spiral Columns

Figure 11-4a shows a portion of the core of a spiral column enclosed by one and a half turns of the spiral. Under a compressive load, the concrete in this column shortens longitudinally under the stress f_1 and so, to satisfy Poisson's ratio, it expands laterally. This lateral expansion is especially pronounced at compression stresses in excess of 70 percent of the cylinder strength, as was discussed in Chapter 3. In a spiral column, the lateral expansion of the concrete inside the spiral (referred to as the *core*) is restrained by the spiral. This stresses the spiral in tension, as is shown in Fig. 11-4b. For equilibrium, the concrete is subjected to lateral compressive stresses, f_2. An element taken from within the core (Fig. 11-4c) is subjected to triaxial compression. In Chapter 3, triaxial compression was shown to increase the strength of concrete:

$$f_1 = f'_c + 4.1f_2 \tag{3-16}$$

Later in this chapter, Eq. (3-16) is used to derive an equation for the amount of spiral reinforcement required in a column.

In a tied column in a nonseismic region, the ties are spaced roughly the width of the column apart and, as a result, provide relatively little lateral restraint to the concrete core. Outward pressure on the sides of the ties due to lateral expansion of the core merely bends them outward, developing an insignificant hoop-stress effect. Hence, normal ties have little effect on the strength of the core in a tied column. They do, however, act to reduce the unsupported length of the longitudinal bars, thus reducing the danger of buckling of those bars as the bar stress approaches yield. The arrangement of ties is discussed in Section 11-5.

Figure 11-5 presents load-deflection diagrams for a tied column and a spiral column subjected to axial loads. The initial parts of these diagrams are similar. As the maximum load is reached, vertical cracks and crushing develop in the concrete shell outside the ties

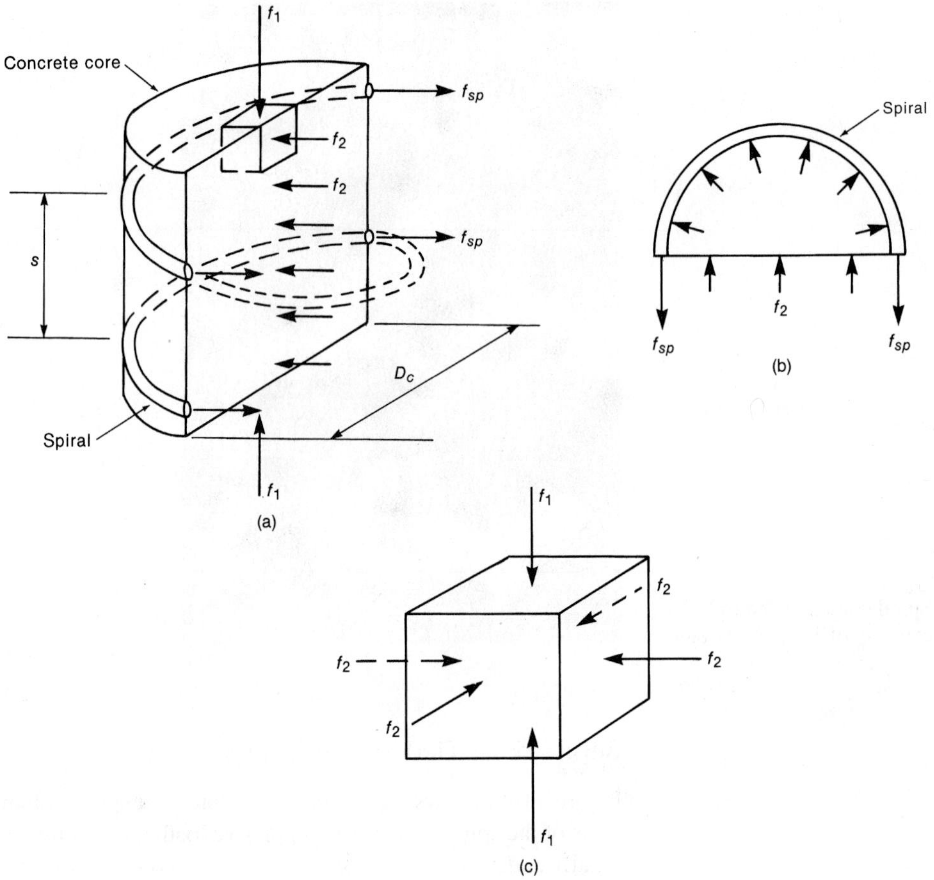

Fig. 11-4
Triaxial stresses in core of spiral column.

or spiral, and this concrete spalls off. When this occurs in a tied column, the capacity of the core that remains is less than the load on the column. The concrete core is crushed, and the reinforcement buckles outward between ties. This occurs suddenly, without warning, in a brittle manner.

When the shell spalls off a spiral column, the column does not fail immediately because the strength of the concrete core has been enhanced by the triaxial stresses resulting from the effect of the spiral reinforcement. As a result, the column can undergo large deformations, eventually reaching a *second maximum load*, when the spirals yield and the column finally collapses. Such a failure is much more ductile than that of a tied column and gives warning of the impending failure, along with possible load redistribution to other members. It should be noted, however, that this is accomplished only at very high strains. For example, the strains necessary to reach the second maximum load correspond to a shortening of about 1 in. in an 8-ft-high column, as shown in Fig. 11-5a.

When spiral columns are eccentrically loaded, the second maximum load may be less than the initial maximum, but the deformations at failure are large, allowing load redistribution (Fig. 11-5b). Because of their greater ductility, compression-controlled failures of spiral columns are assigned a strength-reduction factor, ϕ, of 0.75, rather than the value 0.65 used for tied columns.

Spiral columns are used when ductility is important or where high loads make it economical to utilize the extra strength resulting from the higher ϕ factor. Figures 11-6 and 11-7 show a tied and a spiral column, respectively, after an earthquake. Both columns are in the same building and have undergone approximately the same deformations. The tied

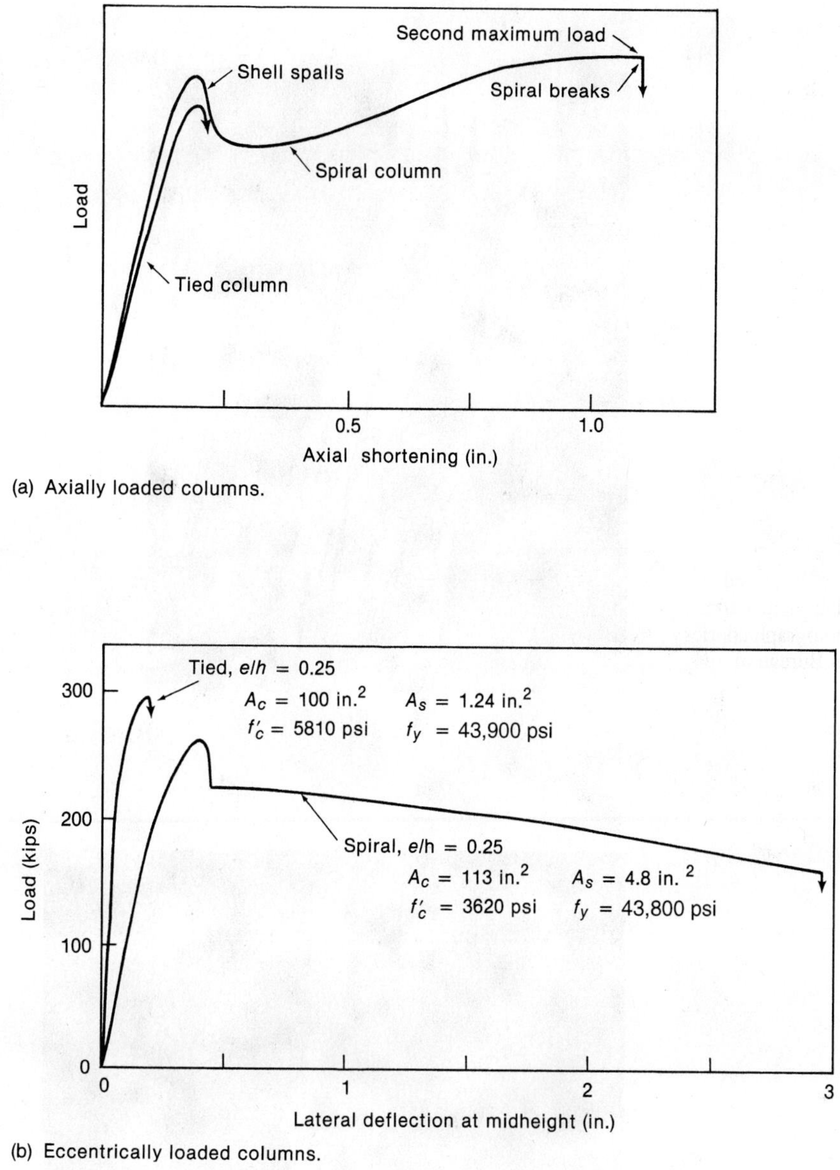

Fig. 11-5
Load-deflection behavior of tied and spiral columns. (Adapted from [11-1].)

column has failed completely, while the spiral column, although badly damaged, is still supporting a vertical load. The very minimal ties in Fig. 11-6 were inadequate to confine the core concrete. Had the column ties been detailed according to ACI Code Section 18.7.5, the column would have performed much better.

Strength of Axially Loaded Columns

When a symmetrical column is subjected to a concentric axial load, P, longitudinal strains, ϵ, develop uniformly across the section, as shown in Fig. 11-8a. Because the steel and concrete are bonded together, the strains in the concrete and steel are equal. For any given strain, it is possible to compute the stresses in the concrete and steel using the stress–strain

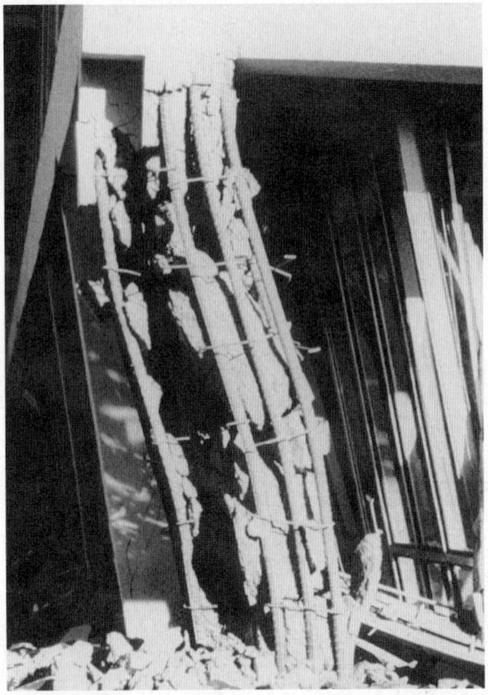

Fig. 11-6
Tied column destroyed in 1971 San Fernando earthquake. (Photograph courtesy of National Bureau of Standards.)

Fig. 11-7
Spiral column damaged by 1971 San Fernando earthquake. Although this column has been deflected sideways 20 in., it is still carrying load. (Photograph courtesy of National Bureau of Standards.)

curves for the two materials. The forces P_c and P_s in the concrete and steel are equal to the stresses multiplied by the corresponding areas. The total load on the column, P, is the sum of these two quantities. Failure occurs when P reaches a maximum. For a steel with a well-defined yield strength (Fig. 11-8c), this occurs when $P_c = f''_c A_c$ and $P_s = f_y A_{st}$, where $f''_c = k_3 f'_c$ is the strength of the concrete loaded as a column. On the basis of tests of 564 columns carried out at the University of Illinois and Lehigh University from 1927 to 1933 [11-2], the ACI Code takes k_3 equal to 0.85. Thus, for a column with a well-defined yield strength, the axial-load capacity is

$$P_o = 0.85 f'_c (A_g - A_{st}) + f_y A_{st} \tag{11-1}$$

where A_g is the gross area and A_{st} is the total area of the longitudinal reinforcement. This equation represents the summation of the fully plastic strength of the steel and the

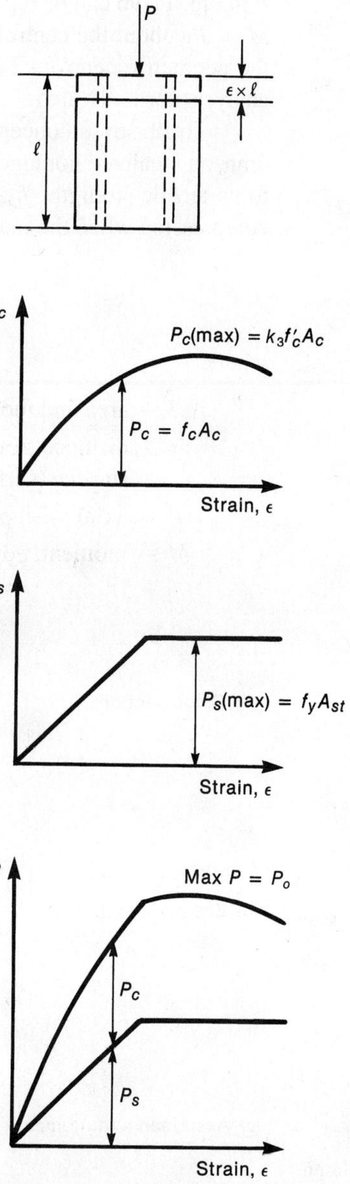

(a) Strains in column.

(b) Load resisted by concrete.

(c) Load resisted by steel.

(d) Total load resisted by column.

Fig. 11-8
Resistance of an axially loaded column.

concrete. If the reinforcement is not elastic–perfectly plastic, failure occurs when P_o reaches a maximum, but this may, or may not, coincide with the strain at which the maximum P_c occurs. As discussed in Section 3-5, the factor k_3 accounts, in part, for the reduction in concrete compressive strength due to: (a) the slow loading or (b) the weakening of the concrete near the top of the column due to upward migration of water in the fresh concrete.

11-3 INTERACTION DIAGRAMS

Almost all compression members in concrete structures are subjected to moments in addition to axial loads. These may be due to misalignment of the load on the column, as shown in Fig. 11-9b, or may result from the column resisting a portion of the unbalanced moments at the ends of the beams supported by the columns (Fig. 11-9c). The distance e is referred to as the *eccentricity* of the load. These two cases are the same, because the eccentric load P in Fig. 11-9b can be replaced by a load P acting along the centroidal axis, plus a moment $M = Pe$ about the centroid. The load P and the moment M are calculated with respect to the geometric centroidal axis because the moments and forces obtained from structural analysis normally are referred to this axis.

To illustrate conceptually the interaction between moment and axial load in a column, an idealized homogeneous and elastic column with a compressive strength, f_{cu}, equal to its tensile strength, f_{tu}, will be considered. For such a column, failure would occur in compression when the maximum stresses reached f_{cu}, as given by

$$\frac{P}{A} + \frac{My}{I} = f_{cu} \tag{11-2}$$

where

A, I = area and moment of inertia of the cross section, respectively

y = distance from the centroidal axis to the most highly compressed surface (surface A–A in Fig. 11-9a), positive to the right

P = axial load, positive in compression

M = moment, positive as shown in Fig. 11-9c

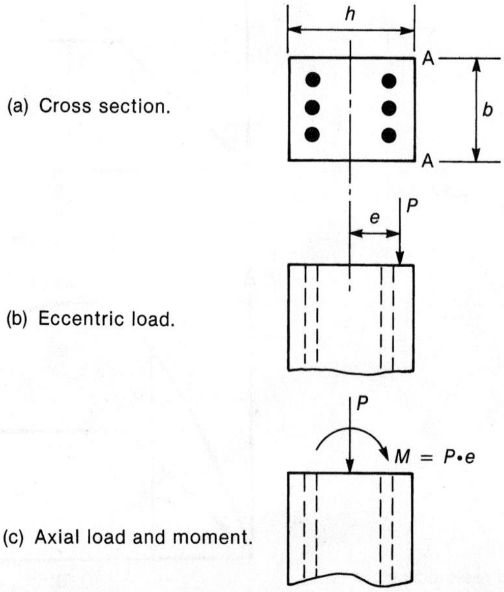

Fig. 11-9
Load and moment on column.

Dividing both sides of Eq. (11-2) by f_{cu} gives

$$\frac{P}{f_{cu}A} + \frac{My}{f_{cu}I} = 1$$

The maximum axial load the column can support occurs when $M = 0$ and is $P_{\max} = f_{cu}A$. Similarly, the maximum moment that can be supported occurs when $P = 0$ and M is $M_{\max} = (f_{cu}I/y)$. Substituting $P_{\max}$ and $M_{\max}$ gives

$$\frac{P}{P_{\max}} + \frac{M}{M_{\max}} = 1 \tag{11-3}$$

This equation is known as an *interaction equation*, because it shows the interaction of, or relationship between, P and M at failure. It is plotted as the line AB in Fig. 11-10. A similar equation for a tensile load, P, governed by f_{tu}, gives the line BC in this figure, and the lines AD and DC result if the moments have the opposite sign.

Figure 11-10 is referred to as an *interaction diagram*. Points on the lines plotted in this figure represent combinations of P and M corresponding to the resistance of the section. A point inside the diagram, such as E, represents a combination of P and M that will not cause failure. Combinations of P and M falling on the line or outside the line, such as point F, will equal or exceed the resistance of the section and hence will cause failure.

Figure 11-10 is plotted for an elastic material with $f_{tu} = -f_{cu}$. Figure 11-11a shows an interaction diagram for an elastic material with a compressive strength f_{cu}, but with the tensile strength, f_{tu}, equal to zero, and Fig. 11-11b shows a diagram for a material with $|-f_{tu}| = 0.5|f_{cu}|$. Lines AB and AD indicate load combinations corresponding to failure initiated by compression (governed by f_{cu}), while lines BC and DC indicate failures initiated by tension. In each case, the points B and D in Figs. 11-10 and 11-11 represent *balanced failures*, in which the tensile and compressive resistances of the material are reached simultaneously on opposite edges of the column.

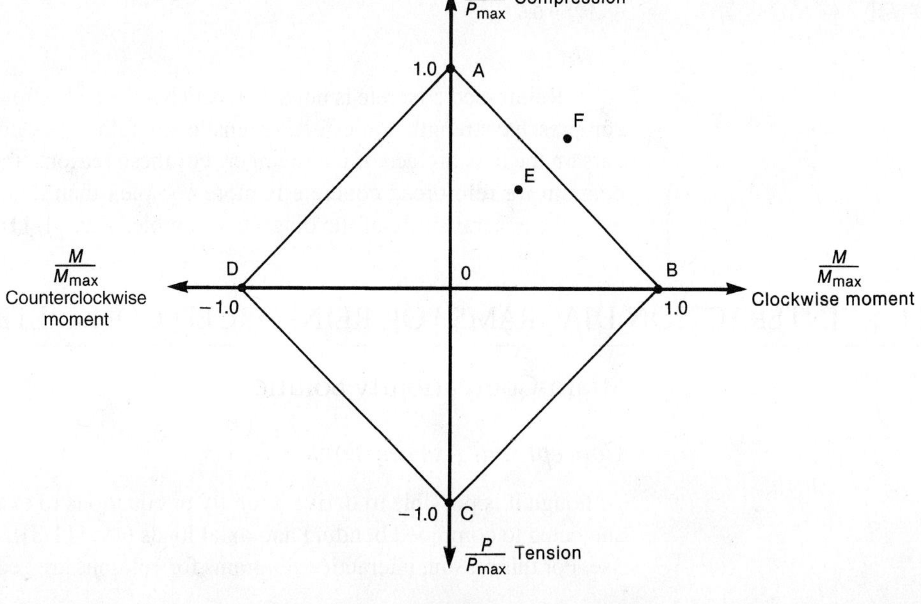

Fig. 11-10
Interaction diagram for an elastic column, $|f_{cu}| = |f_{tu}|$.

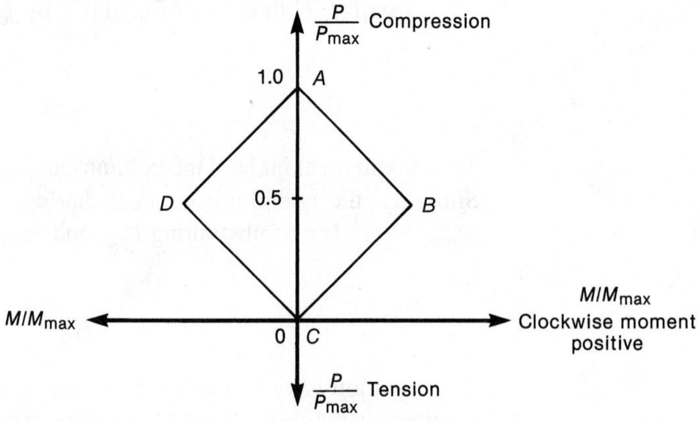

(a) Material with $f_{tu} = 0$.

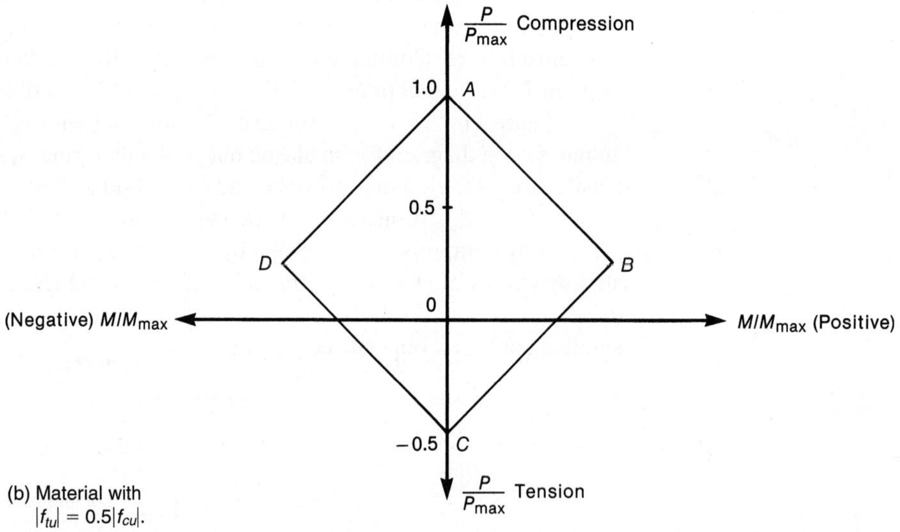

(b) Material with $|f_{tu}| = 0.5|f_{cu}|$.

Fig. 11-11
Interaction diagrams for elastic columns, $|f_{cu}|$ not equal to $|f_{tu}|$.

Reinforced concrete is not elastic and has a tensile strength that is much lower than its compressive strength. An effective tensile strength is developed, however, by reinforcing bars on the tension face of the member. For these reasons, the calculation of an interaction diagram for reinforced concrete is more complex than that for an elastic material. However, the general shape of the diagram resembles Fig. 11-11b.

11-4 INTERACTION DIAGRAMS FOR REINFORCED CONCRETE COLUMNS

Strain-Compatibility Solution

Concept and Assumptions

Although it is possible to derive a family of equations to evaluate the strength of columns subjected to combined bending and axial loads (see [11-3]), these equations are tedious to use. For this reason, interaction diagrams for columns are generally computed by assuming

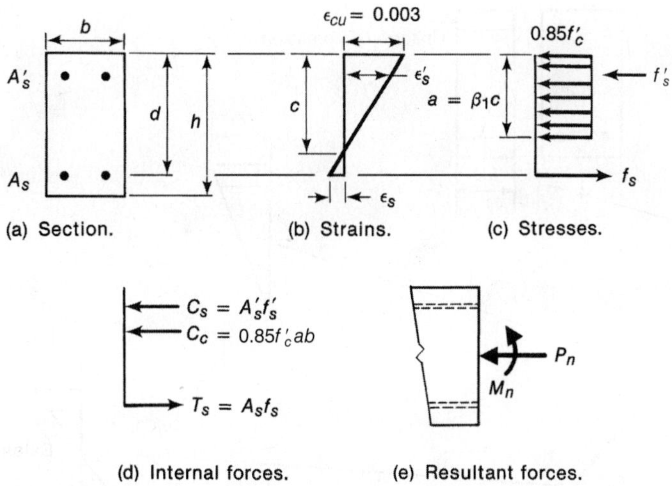

Fig. 11-12
Calculation of P_n and M_n for a given strain distribution.

a series of strain distributions, each corresponding to a particular point on the interaction diagram, and computing the corresponding values of P and M. Once enough such points have been computed, the results are plotted as an interaction diagram.

The calculation process is illustrated in Fig. 11-12 for one particular strain distribution. The cross section is illustrated in Fig. 11-12a, and one assumed strain distribution is shown in Fig. 11-12b. The maximum compressive strain is set at 0.003, corresponding to the ACI Code definition of the maximum useable compression strain. The location of the neutral axis is selected, and the strain in each level of reinforcement are computed from the strain distribution. This information is then used to compute the size of the compression stress block and the stress in each layer of reinforcement, as shown in Fig. 11-12c. The forces in the concrete and the steel layers, shown in Fig. 11-12d, are computed by multiplying the stresses by the areas on which they act. Finally, the axial force P_n is computed by summing the individual forces in the concrete and steel, and the moment M_n is computed by summing the moments of these forces about the geometric centroid of the cross section. These values of P_n and M_n represent one point on the interaction diagram. Other points on the interaction diagram can be generated by selecting other values for the depth, c, to the neutral axis from the extreme compession fiber.

Significant Points on the Column Interaction Diagram

Figure 11-13 and Table 11-1 illustrate a series of strain distributions and the corresponding points on an interaction diagram for a typical tied column. As usual for interaction diagrams, axial load is plotted vertically and moment horizontally. Points on the interaction diagram will have different strength-reduction factors, ϕ factors, for column design in ACI Code Table 21.2.2. The method of computing the strength-reduction factor for columns was changed in the 2002 ACI Code.

1. Point *A*—Pure Axial Load. Point *A* in Fig. 11-13 and the corresponding strain distribution represent uniform axial compression without moment, sometimes referred to as *pure axial load*. This is the largest axial load the column can support. Later in this section the maximum useable axial load will be limited to 0.80 to 0.85 times the pure axial-load capacity.

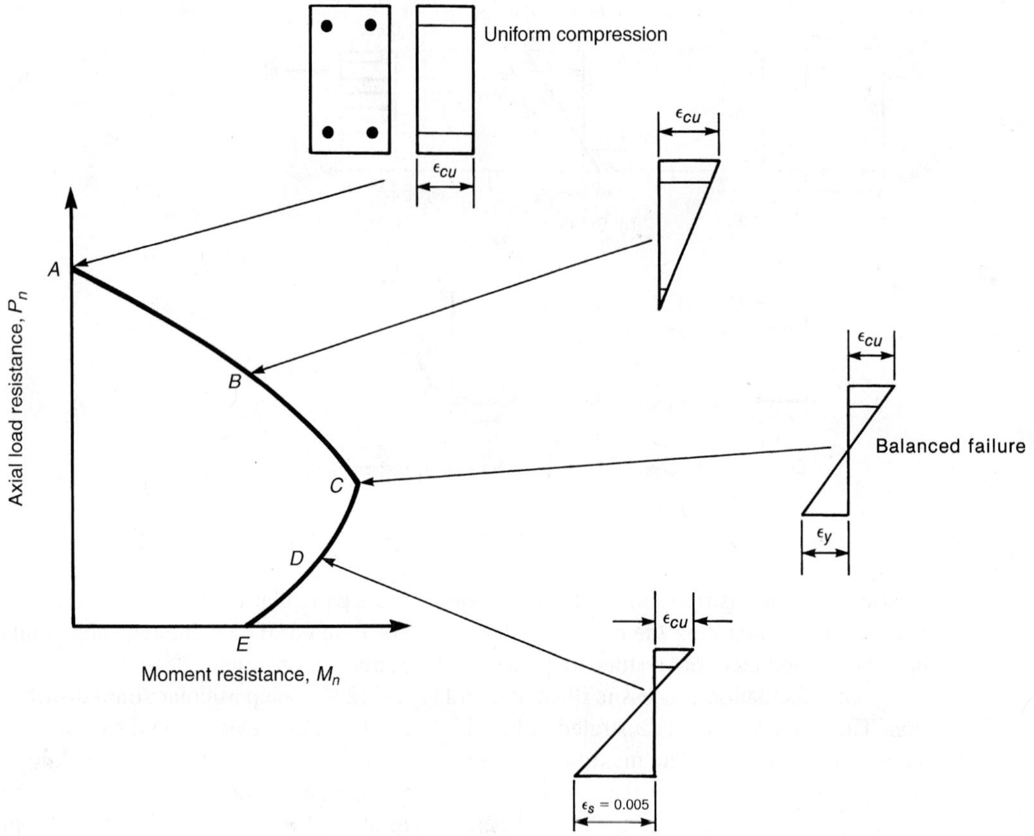

Fig. 11-13
Strain distributions corresponding to points on the interaction diagram.

TABLE 11-1 Strain Regimes and Strength-Reduction Factors, ϕ, for Columns and Beams

Maximum Strain	Compression-Controlled	Transition Region	Tension-Controlled
Max Compressive Strain	$\varepsilon_{cu} = 0.003$ compression	$\varepsilon_{cu} = 0.003$ compression	$\varepsilon_{cu} = 0.003$ compression
Maximum Tensile strain at ultimate	ε_t between 0.003 compression strain and 0.002 tension strain	ε_t between 0.002 tension strain and 0.005 tension strain	ε_t equal to or greater than 0.005 tension
Load Factors $U = 1.2D + 1.6L$	**Values of Strength-Reduction Factor, ϕ**	**Values of Strength-Reduction Factor, ϕ**	**Values of Strength-Reduction Factor, ϕ**
	Tied columns $\phi = 0.65$	*Tied columns* $\phi = 0.65 + (\varepsilon_t - 0.002)\dfrac{250}{3}$ (11-4)	*Tied columns* $\phi = 0.90$
	Spiral columns $\phi = 0.75$	*Spiral columns* $\phi = 0.75 + (\varepsilon_t - 0.002)\,50$ (11-5)	*Spiral columns* $\phi = 0.90$
	All from ACI Code Table 21.2.2		

2. Point *B*—Zero Tension, Onset of Cracking. The strain distribution at *B* in Fig. 11-13 corresponds to the axial load and moment at the onset of crushing of the concrete just as the strains in the concrete on the opposite face of the column reach zero. Case *B* represents the onset of cracking of the least compressed side of the column. Because tensile stresses in the concrete are ignored in the strength calculations, failure loads below point *B* in the interaction diagram represent cases where the section is partially cracked.

3. Region *A–C*—Compression-Controlled Failures. Columns with axial loads P_n and moments M_n that fall on the upper branch of the interaction diagram between points *A* and *C* initially fail due to crushing of the compression face before the extreme tensile layer of reinforcement yields. Hence, they are called *compression-controlled* columns.

4. Point *C*—Balanced Failure, Compression-Controlled Limit Strain. Point *C* in Fig. 11-13 corresponds to a strain distribution with a maximum compressive strain of 0.003 on one face of the section, and a tensile strain equal to the yield strain, ϵ_y, in the layer of reinforcement *farthest* from the compression face of the column. The *extreme tensile strain* ϵ_t occurs in the *extreme tensile layer* of steel located at d_t below the extreme compression fiber. A *balanced failure* occurs when concrete at the extreme compression face reaches the maximum useable compression strain (0.003) simultaneously with yielding of the reinforcement nearest to the opposite face of the column (tensile face). Traditionally, the ACI Code defined a balanced failure as one in which the steel strain at the *centroid* of the tensile reinforcement reached yield in tension when the concrete reached its crushing strain. In the 2002 ACI Code the definition of balanced failure was changed to correspond to the yield of the *extreme tensile layer* of reinforcement rather than the yield at the *centroid* of the tension reinforcement. The two definitions are the same if the tensile reinforcement is all in one layer.

5. Point *D*—Tensile-Controlled Limit. Point *D* in Fig. 11-13 corresponds to a strain distribution with 0.003 compressive strain on the top face and a tensile strain of 0.005 in the *extreme layer* of tension steel (the layer closest to the tensile face of the section). The failure of such a column will be ductile, with steel strains at failure that are about two and a half times the yield strain (for Grade-60 steel). This is defined as the *tension-controlled strain limit*. The strain of 0.005 was chosen to be significantly higher than the yield strain to ensure ductile behavior [11-4].

6. Region *C–D*—Transition Region. Flexural members and columns with loads and moments which would plot between points *C* and *D* in Fig. 11-13 are called *transition failures* because the mode of failure is transitioning from a *brittle* failure at point *C* to a *ductile* failure at point *D*, corresponding respectively to steel strains of 0.002 and 0.005 in the extreme layer of tension steel. This is reflected in the transition of the ϕ-factor, which equals 0.65 (tied column) or 0.75 (spiral column) at point *C* and equals 0.9 at point *D* (see Table 11-1).

7. Region *D–E*—Tension-Controlled Failures. Columns with axial loads, P_n, and moments, M_n, that fall on the lower portion of the interaction diagram between *D* and *E* experience significant tension yielding in the extreme layer of tension reinforcement before the concrete at the extreme compression face reaches the maximum useable compression strain, 0.003. Beam sections designed following the procedures given in Chapter 4 would be represented by the Point *E* ($P_n = 0$).

Strength-Reduction Factor for Columns

In the design of columns, the axial-load and moment capacities must satisfy

$$\phi P_n \geq P_u \quad \phi M_n \geq M_u \tag{11-6a and b}$$

where

P_u and M_u = factored load and moment applied to the column, usually computed from a frame analysis

P_n and M_n = nominal strengths of the column cross section

ϕ = strength-reduction factor; the value of ϕ is the same in both relationships in Eq. (11-6)

Maximum Axial Load

As seen earlier, the strength of a column under truly concentric axial loading can be written as

$$P_{no} = (k_3 f'_c)(A_g - A_{st}) + f_y(A_{st}) \quad (11\text{-}7)$$

where

$k_3 f'_c$ = maximum concrete stress permitted in column design

A_g = gross area of the section (concrete and steel)

f_y = yield strength of the reinforcement

A_{st} = total area of reinforcement in the cross section

The value of $k_3 f'_c$ was derived from tests [11-1], [11-2], [11-3] and normally is taken as $0.85\,f'_c$.

The strength given by Eq. (11-7) cannot normally be attained in a structure because almost always there will be moments present, and, as shown by Fig. 11-13, any moment leads to a reduction in the axial-load capacity. Such moments or eccentricities arise from unbalanced moments in the beams, misalignments of columns from floor to floor, uneven compaction of the concrete across the width of the section, or misalignment of the reinforcement. An examination of Fig. 11-1 will show that the reinforcement has been displaced to the left in this column. Hence in this case, the centroid of the theoretical column resistance does not coincide with the axis of the column, as built. The misalignment of the reinforcement in Fig. 11-1 is considerably greater than the allowable tolerances for reinforcement location [ACI Code Table 26.6.2.1(a)], and such a column should not be acceptable.

To account for the effect of accidental moments, ACI Code Section 22.4.2.1 specifies that the maximum load on a column must not exceed 0.85 times the load from Eq. (11-1) for spiral columns and 0.8 times Eq. (11-1) for tied columns:

Spiral columns:

$$\phi P_{n,\max} = 0.85\phi\,[0.85 f'_c(A_g - A_{st}) + f_y(A_{st})] \quad (11\text{-}8a)$$

Tied columns:

$$\phi P_{n,\max} = 0.80\phi\,[0.85 f'_c(A_g - A_{st}) + f_y(A_{st})] \quad (11\text{-}8b)$$

These limits will be included in the interaction diagram. The difference between the allowable values for spiral and tied columns reflects the more ductile behavior of spiral columns. (See Fig. 11-5.)

Derivation of Computation Method for Interaction Diagrams

In this section, the relationships needed to compute the various points on an interaction diagram are derived by using strain compatibility and mechanics. The calculation of an interaction diagram involves the basic assumptions and simplifying assumptions stated in Section 4-3 of this book and ACI Code Section 22.2. For simplicity, the derivation and the computational example in the next section are limited to rectangular tied columns, as shown in Fig. 11-14a. The extension of this procedure to other cross sections is discussed later in this section.

Throughout the computations, it is necessary to rigorously observe a sign convention for stresses, strains, forces, and directions. Compression has been taken as *positive* in all cases.

Concentric Compressive Axial-Load Capacity and Maximum Axial-Load Capacity

The theoretical top point on the interaction diagram is calculated from Eq. (11-7). For a symmetrical section, the corresponding moment will be zero. Unsymmetrical sections are discussed briefly later in this chapter. The maximum factored axial-load resistances are computed from Eq. (11-8).

General Case

The general case involves the calculation of P_n acting at the centroid and M_n acting about the centroid of the gross cross section, for an assumed strain distribution with $\epsilon_{cu} = 0.003$. The column cross section and the assumed strain distribution are shown in Figs. 11-14a and 11-14b. Four layers of reinforcement are shown, layer 1 having strain ϵ_{s1} and area A_{s1}, and so on. Layer 1 is closest to the "least compressed" surface and is at a distance d_1 from the "most compressed" surface. Layer 1 is called the *extreme tension layer*. It has a depth d_t and a strain ϵ_t.

The strain distribution will be defined by setting $\epsilon_{cu} = 0.003$ and assuming a value for ϵ_{s1}. An iterative calculation will be necessary to consider a series of cases, as shown in Fig. 11-13. The iteration can be controlled by selecting a series of values for the neutral axis depth, c. Large values of c will give points high in the interaction diagram and low

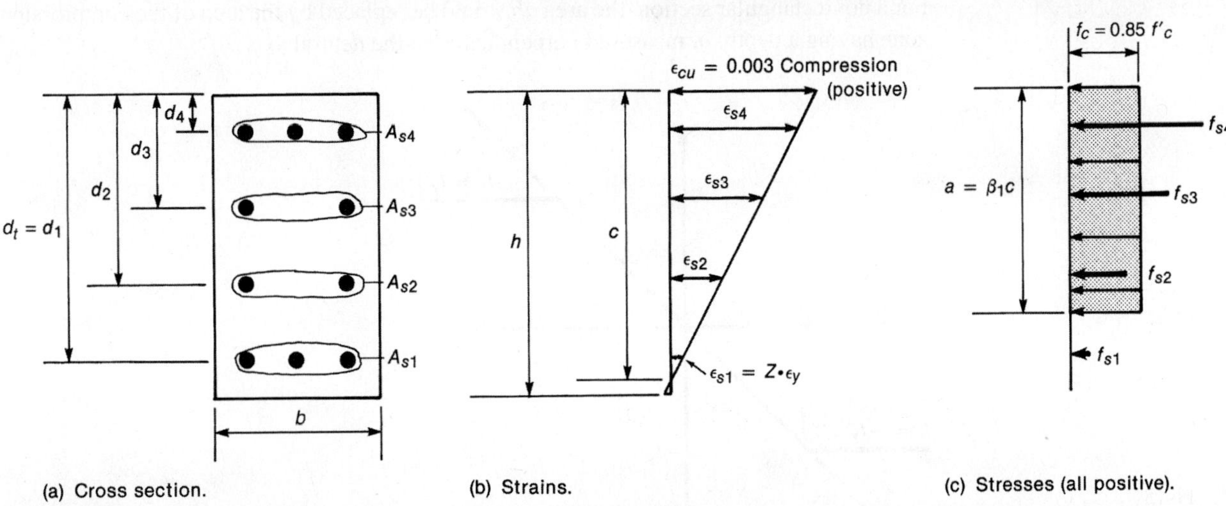

Fig. 11-14
Notation and sign convention for interaction diagram.

values of c will give points low in the interaction diagram. To find points corresponding to specific values of strain in the extreme layer of tension reinforcement, the iteration can be controlled by setting $\epsilon_{s1} = Z\epsilon_y$, where Z is an arbitrarily chosen value. Positive values of Z correspond to positive (compressive) strains (as shown in Fig. 11-14b). For example, $Z = -1$ corresponds to $\epsilon_{s1} = -1\epsilon_y$, the yield strain in tension. Such a strain distribution corresponds to the balanced-failure condition, which will be discussed in Example 11-1.

From Fig. 11-14b, by similar triangles,

$$c = \left(\frac{0.003}{0.003 - Z\epsilon_y}\right)d_1 \tag{11-9}$$

and

$$\epsilon_{si} = \left(\frac{c - d_i}{c}\right)0.003 \tag{11-10}$$

where ϵ_{si} and d_i are the strain in the ith layer of steel and the depth to that layer.

Once the values of c and ϵ_{s1}, ϵ_{s2}, and so on, are known, the stresses in the concrete and in each layer of steel can be computed. For elastic–plastic reinforcement with the stress–strain curve illustrated in Fig. 11-15,

$$f_{si} = \epsilon_{si}E_s \quad \text{but} \quad -f_y \le f_{si} \le f_y \tag{11-11}$$

The stresses in the concrete are represented by the equivalent rectangular stress block introduced in Section 4-3 of this book (ACI Code Section 22.2.2.4). The depth of this stress block is $a = \beta_1 c$, where a, shown in Fig. 11-14c, cannot exceed the overall height of the section, h. The factor β_1 is given by

$$\beta_1 = 0.85 - 0.05\frac{f'_c - 4000 \text{ psi}}{1000 \text{ psi}} \tag{4-14b}$$

but not more than 0.85 nor less than 0.65.

The next step is to compute the compressive force in the concrete, C_c, and the forces in each layer of reinforcement, F_{s1}, F_{s2}, and so on. This is done by multiplying the stresses by corresponding areas. Thus,

$$C_c = (0.85f'_c)(ab) \tag{11-12}$$

For a nonrectangular section, the area ab would be replaced by the area of the compression zone having a depth, a, measured perpendicular to the neutral axis.

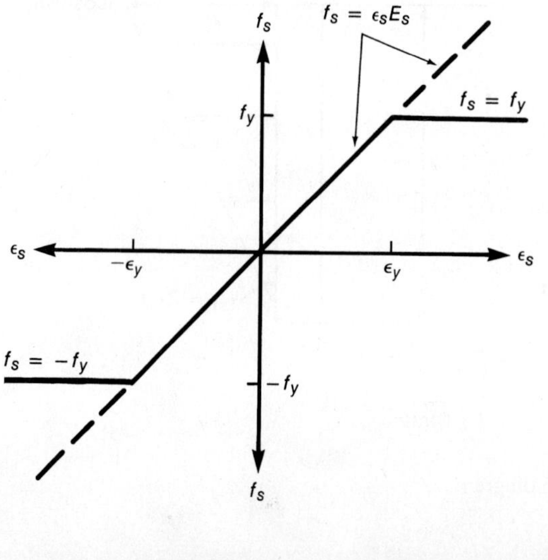

Fig. 11-15
Calculation of stress in steel from Eq. (11-11).

If a is less than d_i,

$$F_{si} = f_{si}A_{si} \text{ (positive in compression)} \tag{11-13a}$$

If a is greater than d_i for a particular layer of steel, the area of the reinforcement in that layer has been included in the area (ab) used to compute C_c. As a result, it is necessary to subtract $0.85f'_c$ from f_{si} before computing F_{si}:

$$F_{si} = (f_{si} - 0.85f'_c)A_{si} \tag{11-13b}$$

The resulting forces C_c and F_{s1} to F_{s4} are shown in Fig. 11-16b.

The nominal axial-load capacity, P_n, for the assumed strain distribution is the summation of the axial forces:

$$P_n = C_c + \sum_{i=1}^{n} F_{si} \tag{11-14}$$

The nominal moment capacity, M_n, for the assumed strain distribution is found by summing the moments of all the internal forces about the *centroid* of the column. The moments are summed about the centroid of the section, because this is the axis about which moments are computed in a conventional structural analysis. In the 1950s and 1960s, moments were sometimes calculated about the *plastic centroid*, the location of the resultant force in a column strained uniformly in compression (case *A* in Fig. 11-13). The centroid and plastic centroid are the same point in a symmetrical column with symmetrical reinforcement.

All the forces are shown positive (compressive) in Fig. 11-16. A positive internal moment corresponds to a compression at the top face, and

$$M_n = C_c\left(\frac{h}{2} - \frac{a}{2}\right) + \sum_{i=1}^{n} F_{si}\left(\frac{h}{2} - d_i\right) \tag{11-15a}$$

Equation (11-15a) is derived by assuming that the centroid of the gross (concrete) section is located at $h/2$ from the extreme compression fiber. If the gross cross section is not symmetrical, the moments would be computed about the centroid of the gross section, and the factored moment resistance would be

$$M_n = C_c\left(\bar{y}_t - \frac{a}{2}\right) + \sum_{i=1}^{n} F_{si}\left(\bar{y}_t - d_i\right) \tag{11-15b}$$

where $\bar{y}_t$ is the distance from the extreme compression fiber to the centroid of the gross section.

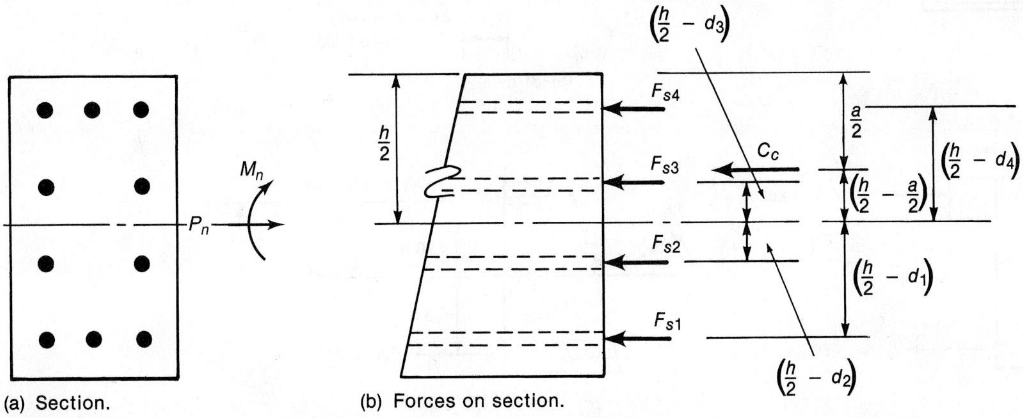

Fig. 11-16
Internal forces and moment arms.

Pure-Axial-Tension Case

The strength under pure axial tension is computed by assuming that the section is completely cracked and subjected to a uniform strain greater than or equal to the yield strain in tension. The stress in all of the layers of reinforcement is therefore $-f_y$ (yielding in tension), and

$$P_{nt} = \sum_{i=1}^{n} -f_y A_{si} \tag{11-16}$$

The axial tensile capacity of the concrete is, of course, ignored. For a symmetrical section, the corresponding moment will be zero. For an unsymmetrical section, Eq. (11-15b) is used to compute the moment.

EXAMPLE 11-1 Calculation of an Interaction Diagram

Compute four points on the interaction diagram for the column shown in Fig. 11-17a. Use $f'_c = 5000$ psi and $f_y = 60{,}000$ psi. $A_g = bh$ is 256 in.^{2}, $A_{s1} = 4$ in.^{2}, $A_{s2} = 4$ in.^{2}, $A_{st} = \Sigma A_{si} = 8$ in.^{2}, and $\rho_g = A_{st}/A_g = 0.031$. The yield strain, $\epsilon_y = f_y/E_s$, is 60,000 psi/29,000,000 psi = 0.00207. Use load factors and strength-reduction factors from ACI Code Table 21.2.2.

1. Compute the concentric axial-load capacity and maximum axial-load capacity. From Eq. (11-7),

$$\begin{aligned} P_o &= (0.85f'_c)(A_g - A_{st}) + f_y(A_{st}) \\ &= (0.85 \times 5 \text{ ksi})(256 - 8) \text{ in.}^2 + 60 \text{ ksi} \times 8 \text{ in.}^2 \\ &= 1050 \text{ kips} + 480 \text{ kips} = 1530 \text{ kips} \end{aligned}$$

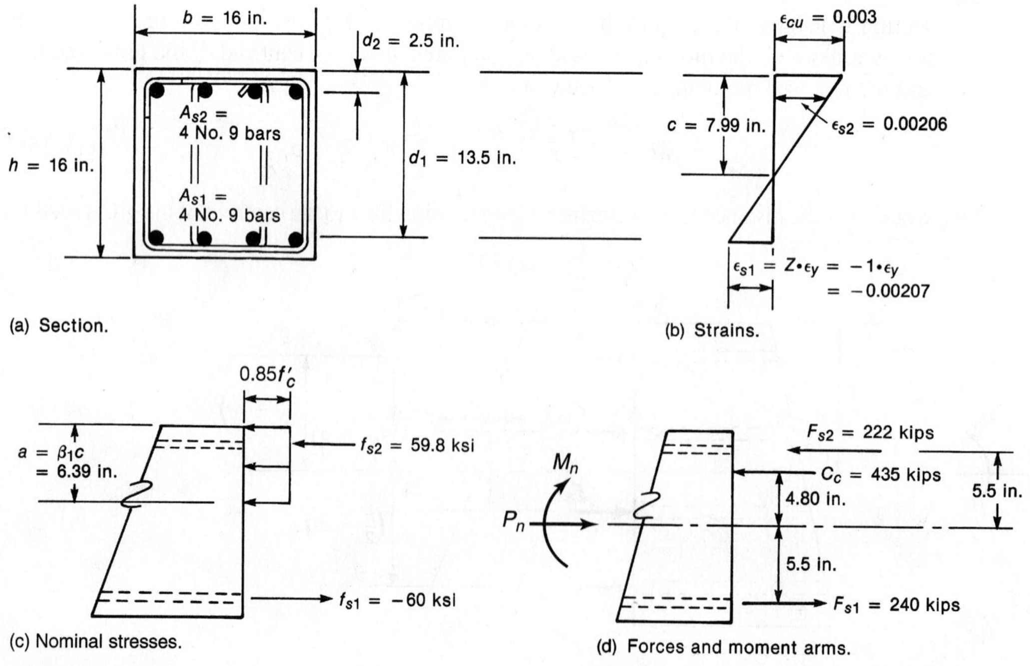

Fig. 11-17
Calculations—Example 11-1, $\epsilon_{s1} = -1 \cdot \epsilon_y$, so $Z = -1$.

This is the nominal concentric axial-load capacity. The value used in drawing a design interaction diagram would be ϕP_o, where $\phi = 0.65$, because (i) the strain in the extreme tension steel, layer 1 in Fig. 11-17a, has a strain that is compressive, and (ii) this is a tied column. Thus,

$$\phi P_o = 997 \text{ kips}$$

P_o and ϕP_o are plotted as points A and A' in Fig. 11-18 and point A in Fig. 11-13.

For this column with $\rho_g = 0.031$ (or 3.1 percent), the 480 kips carried by the reinforcement is roughly 30 percent of the 1530-kip nominal capacity of the column. For axially loaded columns, reinforcement will generally carry between 10 and 35 percent of the total capacity of the column.

The maximum load allowed on this column (ACI Code Sections 22.4.2.1 and 22.4.2.2) is given by Eq. (11-8b):

$$\phi P_{n(\max)} = 0.80\phi[(0.85f'_c)(A_g - A_{st}) + f_y(A_{st})]$$
$$= 0.80\phi P_o = 798 \text{ kips}$$

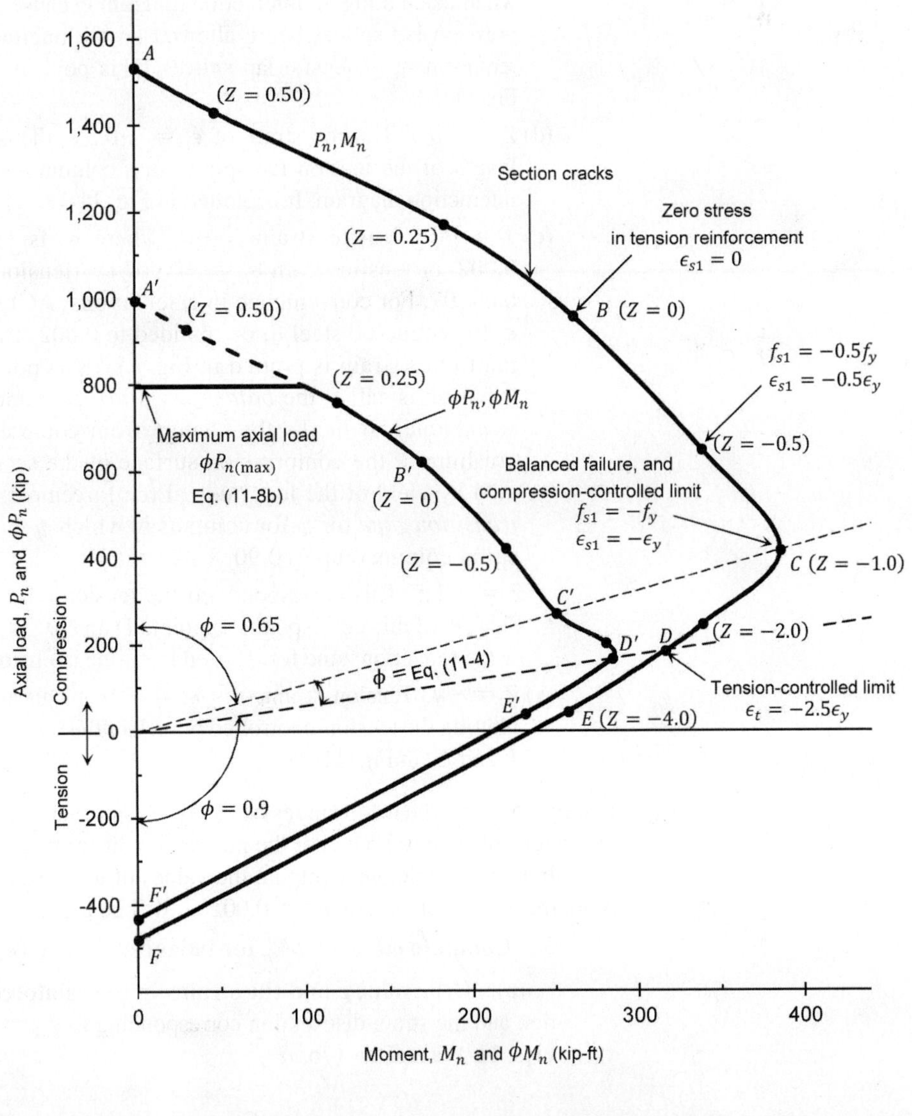

Fig. 11-18
Interaction diagram—
Example 11-1.

This load is plotted as a horizontal solid line in Fig. 11-18. The portion of the ϕP_n, ϕM_n interaction diagram above this line is shown with a dashed line because this capacity cannot be used in design. It should be noted that limiting the factored axial load to $\phi P_{n(\max)}$ gives the column a capacity to resist accidental moments, as represented by the horizontal line in Fig. 11-18.

2. Compute ϕP_n and ϕM_n for the general case. To get a more or less complete interaction diagram, a number of strain distributions must be considered and the corresponding values of P_n, M_n, ϕP_n, and ϕM_n calculated. The incremental step size for Z should become successively larger because the points get closer and closer together as Z gets larger. Ideally the values of Z should be chosen to agree with the *compression-controlled strain limits*, the *tensile-controlled strain limit*, and the limit on beam reinforcement as follows:

(a) $Z = +0.5$. Compression strain of $\epsilon_t = +0.5\ \epsilon_y$ in the extreme tension layer of steel, not plotted in Fig. 11-18.

(b) $Z = +0.25$. Compression strain of $\epsilon_t = +0.25\ \epsilon_y$ in the extreme tension layer of steel. This case is not plotted in Fig. 11-18.

(c) $Z = 0.0$. Strain ϵ_t is zero in extreme layer of tension steel. This case is considered when calculating an interaction diagram because it marks the change from compression lap splices being allowed on all longitudinal bars, to the more severe requirement of tensile lap splices. This point is plotted as points B and B' in Fig 11-18.

(d) $Z = -0.5$. Tension strain of $\epsilon_t = -0.5\ \epsilon_y$. This strain distribution affects the length of the tension lap splices in a column and is customarily plotted on an interaction diagram. It is plotted in Fig. 11-18.

(e) $Z = -1$. Tensile strain $= -\epsilon_y$ where ϵ_y is taken approximately equal to 0.002, or tensile strain $\epsilon_t = -0.967\ \epsilon_y$ (tension) where ϵ_y is taken equal to 0.00207. For convenience in calculations, ACI Code Section 21.2.2.1 allows ϵ_y for Grade-60 steel to be rounded to 0.002 strain. The strength corresponding to this strain is plotted in Fig. 11-18 as points C and C'. This strain distribution is called the *balanced failure* case and the *compression-controlled strain limit*. It marks the change from compression failures originating by crushing of the compression surface of the section, to tension failures initiated by yield of the longitudinal reinforcement. It also marks the start of the *transition zone* for ϕ for columns in which ϕ increases from 0.65 (or 0.75 for spiral columns) up to 0.90.

(f) $Z = -2.5$. This corresponds to the tension-controlled strain limit of 0.005. In Fig. 11-18 this corresponds to points D and D'. It is the strain at the tensile limit of the transition zone for ϕ, used to define a tension-controlled section.

(g) $Z = -4$. Tension strain of $-\epsilon_t = -4\epsilon_y$. This load combination is helpful in plotting the tension-controlled branch of the interaction diagram. It plots as points E and E' in Fig. 11-18.

Finally, two or three more values of Z, say -2 and -6, are desirable to complete plotting of the tension-failure branch of the interaction diagram.

In this example, we compute the values of ϕP_n and ϕM_n for $Z = -1, -2$, and -4. Also, the exact yield strain, $\epsilon_y = 0.00207$, will be used.

3. Compute ϕP_n and ϕM_n for balanced failure ($\epsilon_{s1} = -\epsilon_y$).

(a) Determine c and the strains in the reinforcement. The column cross section and the strain distribution corresponding to $\epsilon_{s1} = -1\epsilon_y (Z = -1)$ are shown in Figs. 11-17a and 11-17b.

The strain in the bottom layer of steel is $-1\epsilon_y = -0.00207$. From similar triangles, the depth to the neutral axis is

$$c = \frac{0.003}{0.003 - (-1 \times 0.00207)} d_1 \tag{11-9}$$

$$= \frac{0.003}{0.003 + 0.00207} \times 13.5 \text{ in.} = 7.99 \text{ in.}$$

Again by similar triangles, the strain in the compression steel is

$$\epsilon_{s2} = \left(\frac{c - d_2}{c}\right) 0.003 \tag{11-10}$$

$$= \left(\frac{7.99 - 2.5}{7.99}\right) 0.003 = 0.00206$$

(b) Compute the stresses in the reinforcement layers. The stress in reinforcement layer 2 is Eq. (11-11)

$$f_{s2} = \epsilon_{s2} E_s \quad \text{but} \quad -f_y \le f_{s2} \le f_y$$

$$\epsilon_{s2} E_s = 0.00206 \times 29{,}000 \text{ ksi} = 59.8 \text{ ksi} < 60 \text{ ksi}$$

Therefore, $f_{s2} = 59.8$ ksi. Because this is positive, it is compressive. Also, $\epsilon_{s1} = -\epsilon_y$ and $f_{s1} = -60$ ksi

(c) Compute *a*. The depth of the equivalent rectangular stress block is $a = \beta_1 c$, where a cannot exceed h. For $f'_c = 5000$ psi,

$$\beta_1 = 0.85 - 0.05 \frac{f'_c - 4000 \text{ psi}}{1000 \text{ psi}} = 0.80$$

and

$$a = \beta_1 c$$

$$= 0.80 \times 7.99 \text{ in.} = 6.39 \text{ in.}$$

This is less than h; therefore, this value can be used. If a exceeded h, $a = h$ would be used. The stresses computed in steps 2 and 3 are shown in Fig. 11-17c.

(d) Compute the forces in the concrete and steel. The force in the concrete, C_c, is equal to the average stress, $0.85f'_c$, times the area of the rectangular stress block, ab:

$$C_c = (0.85 f'_c)(ab) \tag{11-12}$$

$$= 0.85 \times 5 \text{ ksi} \times 6.39 \text{ in.} \times 16 \text{ in.} = 435 \text{ kips}$$

The distance $d_1 = 13.5$ in. to reinforcement layer 1 exceeds $a = 6.39$ in. Hence, this layer of steel lies outside the compression stress block and does not displace concrete included in the area (ab) when computing C_c. Thus,

$$F_{s1} = f_{s1} A_{s1}$$

$$= -60 \text{ ksi} \times 4 \text{ in.}^2 = -240 \text{ kips}$$

(Negative denotes tension.)

Reinforcement layer 2 lies in the compression zone, because $a = 6.39$ in. which exceeds $d_2 = 2.5$ in. Hence, we must allow for the stress in the concrete displaced by the steel when we compute F_{s2}. From Eq. (11-13b),

$$F_{s2} = (f_{cs2} - 0.85f'_c)A_{s2}$$
$$= (59.8 - 0.85 \times 5) \text{ ksi} \times 4 \text{ in.}^2 = 222 \text{ kips}$$

The forces in the concrete and steel are shown in Fig. 11-17d.

(e) Compute P_n. The nominal axial-load capacity, P_n, is found by summing the axial-force components Eq. (11-14):

$$P_n = C_c + \Sigma F_{si}$$
$$= 435 - 240 + 222 = 417 \text{ kips}$$

Because $\epsilon_{s1} = -\epsilon_y$ (yield in tension), this is the balanced-failure condition, and $P_n = P_b$.

(f) Compute M_n. From Fig. 11-17d, the moment of C_c, F_{s1}, and F_{s2} about the centroid of the section is Eq. (11-15a)

$$M_n = C_c\left(\frac{h}{2} - \frac{a}{2}\right) + F_{s1}\left(\frac{h}{2} - d_1\right) + F_{s2}\left(\frac{h}{2} - d_2\right)$$
$$= 435 \text{ kips}\left(\frac{16}{2} - \frac{6.39}{2}\right) \text{in.} + [-240(8 - 13.5)] + 222(8 - 2.5)$$
$$= 2090 + 1320 + 1220 = 4630 \text{ kip-in.}$$

Therefore, $M_n = M_b = 386$ kip-ft.

(g) Compute ϕ, ϕP_n, and ϕM_n. ϕ will be computed according to ACI Code Table 21.2.2. The strain ϵ_t, in the layer of reinforcement farthest from the compression face is $\epsilon_{s1} = -0.00207 = -\epsilon_y$. Thus, this is essentially a compression-controlled section and $\phi = 0.65$.

$$\phi P_n = 0.65 \times 417 = 271 \text{ kips}$$
$$\phi M_n = 0.65 \times 386 = 251 \text{ kip-ft}$$

This completes the calculations for one value of $\epsilon_{s1} = Z(\epsilon_y)$ and gives the points C and C' in Fig. 11-18. Other values of Z are now assumed, and the calculations are repeated until one has enough points to complete the diagram. As noted earlier, other points on the interaction diagram can be found by selecting different values for c.

4. Compute ϕP_n and ϕM_n for $Z = -2$. This point illustrates the calculation of ϕ for cases falling between the compression-controlled limit and the tension-controlled limit.

(a) Determine c and the strains in the reinforcement. From similar triangles, $\epsilon_1 = -2\epsilon_y$ [substituting $Z = -2$ into Eq. (11-9)] gives

$$c = 5.67 \text{ in.}$$

From Eq. (11-10), the strain in the compression steel is

$$\epsilon_{s2} = 0.00168$$

The strain in the tension reinforcement is

$$\epsilon_{s1} = -2 \times 0.00207 = -0.00414$$

(b) Compute the stress in the reinforcement layers.

$$f_{s2} = 0.00168 \times 29{,}000 \text{ ksi} = 48.7 \text{ ksi}$$

This is within the range $\pm f_y$—therefore, o.k.

$$f_{s1} = -60 \text{ ksi, because } |\epsilon_{s1}| > |\epsilon_y|$$

(c) Compute *a*.

$$a = 0.80 \times 5.67 = 4.54 \text{ in.}$$

(d) Compute the forces in the concrete and steel. The compression force in the concrete is

$$C_c = 0.85 \times 5 \text{ ksi} \times 4.54 \text{ in.} \times 16 \text{ in.} = 309 \text{ kips}$$

The force in the tension reinforcement is

$$F_{s1} = -60 \text{ ksi} \times 4 \text{ in.}^2 = -240 \text{ kips}$$

Since $a = 4.54$ in. is greater than $d_2 = 2.5$ in., Eq. (11-13b) is used to compute F_{s2}.

$$F_{s2} = (48.7 - 0.85 \times 5) \times 4 = 178 \text{ kips}$$

(e) Compute P_n. Summing forces perpendicular to the section, Eq. (11-14) gives

$$P_n = 309 - 240 + 178 = 247 \text{ kips}$$

(f) Compute M_n. Summing moments about the centroid of the section, Eq. (11-15a) gives

$$M_n = 309\left(8 - \frac{4.54}{2}\right) + (-240 \times -5.50) + (178 \times 5.50)$$

$$= 4080 \text{ kip-in.} = 340 \text{ kip-ft}$$

(g) Compute ϕ, ϕP_n, and ϕM_n. The strain $\epsilon_t = \epsilon_{s1} = -0.00414$ is between $-\epsilon_y$ and -0.005. Therefore, Eq. (11-4) from Table 11-1 applies, and using $\epsilon_t = 0.00414$:

$$\phi = 0.65 + (\varepsilon_t - 0.002)\frac{250}{3}$$

where ϵ_t is the strain in the extreme tensile layer of steel, ϵ_{s1}. Thus,

$$\phi = 0.65 + (0.00414 - 0.002)\frac{250}{3} = 0.828$$

$$\phi P_n = 0.828 \times 247 = 205 \text{ kips}$$

$$\phi M_n = 0.828 \times 340 \text{ kip-ft} = 282 \text{ kip-ft}$$

P_n, M_n, ϕP_n, and ϕM_n calculated for this strain distribution are not both plotted in Fig. 11-18 but would be located just above the point D and D' in that figure. The peculiar shape of this portion of the ϕP_n, ϕM_n interaction diagram is due to the transition from $\phi = 0.65$ to $\phi = 0.90$.

5. Compute ϕP_n and ϕM_n for the tension-controlled point. Rather than selecting a specific Z-value, we will simply set $\varepsilon_{s1} = \varepsilon_t = -0.005$. From this,

$$c = \frac{0.003}{0.003 + 0.005}d = \frac{3}{8} \times 13.5 \text{ in.} = 5.06 \text{ in.}$$

and

$$a = \beta_1 c = 0.80 \times 5.06 \text{ in.} = 4.05 \text{ in.}$$

From this, the strain in the upper layer of reinforcement is

$$\varepsilon_{s2} = \frac{c - d_2}{c} \times 0.003 = \frac{5.06 - 2.5}{5.06} \times 0.003 = 0.00152$$

and the stress in this layer is

$$f_{s2} = E_s \varepsilon_{s2} = 29{,}000 \times 0.00152 = 44.0 \text{ ksi}$$

Because ϵ_{s1} exceeds the yield strain in tension, $f_{s1} = -60$ ksi.

The section forces are

$$\begin{aligned} C_c &= 0.85 f'_c ba = 0.85 \times 5 \text{ ksi} \times 16 \text{ in.} \times 4.05 \text{ in.} = 275 \text{ kips} \\ F_{s2} &= A_{s2}(f_{s2} - 0.85 f'_c) = 4.0 \text{ in.}^2 (44.0 - 0.85 \times 5)\text{ksi} = 159 \text{ kips} \\ F_{s1} &= A_{s1} f_{s1} = 4.0 \text{ in.}^2 \times (-60 \text{ ksi}) = -240 \text{ kips} \end{aligned}$$

The resultant nominal axial load for this point is

$$P_n = 275 \text{ k} + 159 \text{ k} - 240 \text{ k} = 194 \text{ kips}$$

and, the resultant nominal moment is

$$\begin{aligned} M_n &= 275 \text{ k}\left(8 \text{ in.} - \frac{4.05 \text{ in.}}{2}\right) + 159 \text{ k}(8 \text{ in.} - 2.5 \text{ in.}) - 240 \text{ k}(8 \text{ in.} - 13.5 \text{ in.}) \\ &= 3840 \text{ kip-in.} = 320 \text{ kip-ft} \end{aligned}$$

For this tension-controlled point, $\phi = 0.9$, so

$$\begin{aligned} \phi P_n &= 175 \text{ kips} \\ \phi M_n &= 288 \text{ kip-ft} \end{aligned}$$

These results are plotted as points D and D' in Fig. 11-18.

6. Compute ϕP_n and ϕM_n for $Z = -4$. Repeating the calculations for $Z = -4$ (four times the yield strain) gives

$$P_n = 43.9 \text{ kips} \quad M_n = 257 \text{ kip-ft}$$

Because $\epsilon_t = \epsilon_{s1} = -4 \times 0.00207 = -0.00828$, is more than -0.005, so $\phi = 0.90$ and

$$\phi P_n = 39.5 \text{ kips} \quad \phi M_n = 232 \text{ kip-ft}$$

These are plotted as points E and E' in Fig. 11-18.

7. Compute the capacity in axial tension. The final loading case to be considered in this example is concentric axial tension. The strength under such a loading is equal to the yield strength of the reinforcement in tension, as given by Eq. (11-16):

$$\begin{aligned} P_{nt} &= \sum_{i=1}^{n} (-f_y A_{si}) \\ &= -60 \text{ ksi } (4 + 4) \text{ in.}^2 = -480 \text{ kips} \end{aligned}$$

Because the section is symmetrical, $M = 0$.

The design capacity in pure tension is ϕP_{nt}, where $\phi = 0.9$. Thus,

$$\phi P_{nt} = 0.9 \times -480 \text{ kips} = -432 \text{ kips}$$

Points F and F' in Fig. 11-18 represent the pure-tension case. ■

Interaction Diagrams for Circular Columns

The strain-compatibility solution described in the preceding section can also be used to calculate the points on an interaction diagram for a circular column. As shown in Fig. 11-19b, the depth to the neutral axis, c, is calculated from the assumed strain diagram by using similar triangles [or from Eq. (11-9)]. The depth of the equivalent rectangular stress block, a, is again $\beta_1 c$.

The resulting compression zone is a segment of a circle having depth a, as shown in Fig. 11-19d. To compute the compressive force and its moment about the centroid of the column, it is necessary to be able to compute the area and centroid of the segment. These terms can be expressed as a function of the angle θ shown in Fig. 11-20. The area of the segment is

$$A = h^2\left(\frac{\theta - \sin\theta\cos\theta}{4}\right) \tag{11-17}$$

where θ is expressed in radians (1 radian $= 180°/\pi$). The moment of this area about the center of the column is

$$A\bar{y} = h^3\left(\frac{\sin^3\theta}{12}\right) \tag{11-18}$$

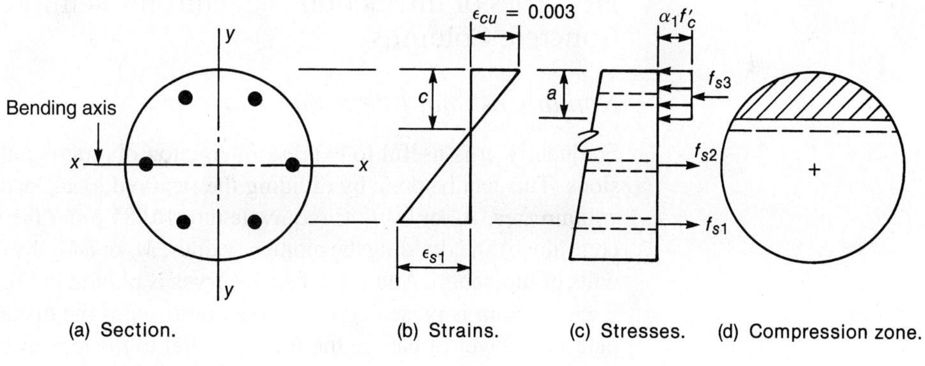

Fig. 11-19
Circular column.

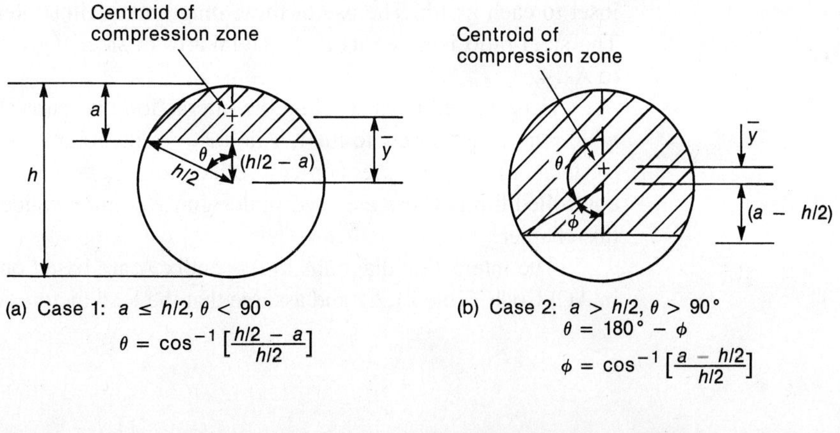

Fig. 11-20
Circular segments.

The shape of the interaction diagram of a circular column is affected by the number of bars and their orientation relative to the direction of the neutral axis. Thus, the moment capacity about axis x–x in Fig. 11-19a is slightly less than that about axis y–y. Because the designer has little control over the arrangement of the bars in a circular column, the interaction diagram should be computed for the least favorable bar orientation. For circular columns with more than eight bars, this problem vanishes, because the bar placement approaches a continuous ring.

The interaction diagrams for the factored cross-sectional strength of a compression-controlled circular tied column and a compression-controlled circular spiral column are comparable, except that

(a) The horizontal cut off at the top of the interaction diagram has a value of $0.80P_{no}$ for tied columns and $0.85P_{no}$ for spiral columns.

(b) The value of the strength-reduction factor for compression-controlled spiral columns is $\phi = 0.75$ for design, compared with 0.65 for compression-controlled tied columns. This requires that the ϕ factors in the transition zone be evaluated using a different equation for ϕ. If the column is a circular spiral column, this change is taken into account by substituting Eq. (11-5) for (11-4).

For extreme tension strains more tensile than -0.005, $\phi = 0.9$ for spiral and tied columns.

It should be noted that the nondimensional interaction diagrams given in Figs. A-12 to A-14 include the ϕ factors for spiral columns. They cannot be used to design circular tied columns unless an adjustment is made to ϕ.

Properties of Interaction Diagrams for Reinforced Concrete Columns

Nondimensional Interaction Diagrams

Frequently, it is useful to express interaction diagrams independently of column dimensions. This can be done by dividing the factored axial-load resistances, P_n or ϕP_n, by the column area, A_g, or by $f'_c A_g$ (equivalent to 1/0.85 times the axial-load capacity of the concrete alone) and dividing the moment values, M_n or ϕM_n, by $A_g h$ or by $f'_c A_g h$ (which has the units of moments). A family of such curves is plotted in Figs. A-6 to A-14 in Appendix A. Each diagram is presented for a given ratio, γ, of the distance between the centers of the outermost layer of bars in the faces parallel to the axis of bending to the overall depth of the section, as shown in the inset to the diagrams. The axis of bending is shown in the inset to each graph. The use of these diagrams is illustrated in Examples 11-2 and 11-3. The steel ratio is given as ρ_g = (total area of steel)/(gross area of section) in Figs. A-6 to A-14.

Four dashed lines crossing the interaction diagrams show the loads and moments at which the layer closest to the tensile face of the column is stressed to $f_s = 0, f_s = 0.5 f_y$ (in tension), the compression-controlled limit $f_s = 1.0 f_y$ (in tension), and the tension-controlled limit. These are used in designing column splices, as will be discussed later in this chapter.

The interaction diagrams in Appendix A are based on the strength-reduction factors in ACI Code Table 21.2.2 and assume that the load factors are from ACI Code Table 5.3.1.

Eccentricity of Load

In Fig. 11-9, it was shown that a load P applied to a column at an eccentricity e was equivalent to a load P acting through the centroid, plus a moment $M = Pe$ about the centroid. A radial line through the origin in an interaction diagram has the slope P/M, or $P/P \times e = 1/e$. For example, the balanced load and moment computed in Example 11-1 correspond to an eccentricity of 386 kip-ft/417 kips = 0.93 ft, and a radial line through this point (point C in Fig. 11-18) would have a slope of 1/0.93. The pure-moment case may be considered to have an eccentricity $M/P = \infty$, since $P = 0$.

In a nondimensional interaction diagram such as Fig. A-6, a radial line through the origin has a slope equal to $(P/A_g)/(M/A_g h)$. Substituting $M = Pe$ shows that the line has the slope h/e or $1/(e/h)$, where e/h represents the ratio of the eccentricity to the column thickness. Radial lines corresponding to several eccentricity ratios are plotted in Figs. A-6 to A-14.

Unsymmetrical Columns

Up to this point, interaction diagrams have been shown only for symmetrical columns. If the column cross section is symmetrical about the axis of bending, the interaction diagram is symmetrical about the vertical $M = 0$ axis, as shown in Fig. 11-21a. For unsymmetrical columns, the diagram is tilted as shown in Fig. 11-21b, provided that the moments are taken about the geometric centroid. The calculation of an interaction diagram for such a member follows the same procedure as Example 11-1, except that for the cases of uniform compressive or tensile strains (axial compression, P_o, and axial tension, P_{nt}), the unsymmetrical bar placement gives rise to a moment of the steel forces about the centroid. Figure 11-21a is drawn for the column shown in Fig. 11-17a and is the same as the interaction diagram in Fig. 11-18. Figure 11-21b is drawn for a similar cross section with four No. 9 bars in one face and two No. 9 bars in the other. For positive moment, the face with four bars is in tension. As a result, the balanced axial load for positive moment is less than that for negative moment.

Simplified Interaction Diagrams for Columns

Generally, designers have access to published interaction diagrams or computer programs to compute interaction diagrams for use in design. Occasionally, this is not true, as, for example, in the design of hollow bridge piers, elevator shafts, or unusually shaped members. Interaction diagrams for such members can be calculated by using the strain-compatibility solution presented earlier. In most cases, it is adequate to represent the interaction diagram by a series of straight lines joining the load and moment values corresponding to the following five strain distributions:

1. a uniform compressive strain of 0.003, giving point 1 in Fig. 11-22;
2. a strain diagram corresponding to incipient cracking, passing through a compressive strain of 0.003 on one face and zero strain on the other (point 2 in Fig. 11-22);
3. the limiting compression-controlled strain distribution, which has a compressive strain of 0.003 on one face and a tensile strain of $-\epsilon_y$ at the centroid of the reinforcement layer nearest to the tensile face (point 3);
4. the limiting tension-controlled strain distribution, which has a compressive strain of 0.003 on one face and a tensile strain of -0.005 in the reinforcement layer nearest to the tensile face (point 4);
5. a uniform tensile strain of $-\epsilon_y$ in the steel with the concrete cracked (point 5).

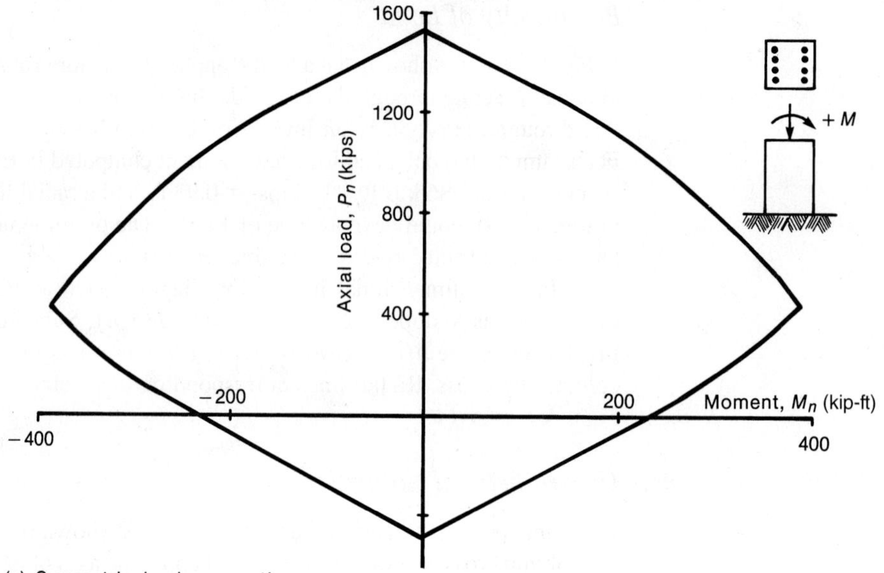

(a) Symmetrical column section.

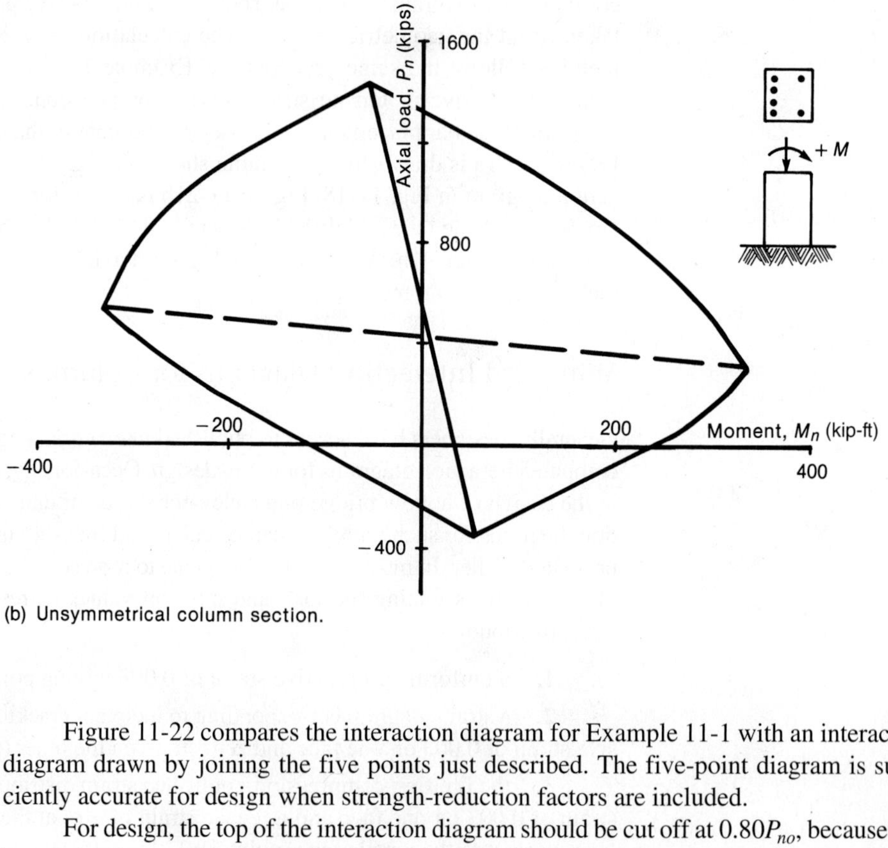

(b) Unsymmetrical column section.

Fig. 11-21
Interaction diagrams for symmetrical and unsymmetrical columns.

Figure 11-22 compares the interaction diagram for Example 11-1 with an interaction diagram drawn by joining the five points just described. The five-point diagram is sufficiently accurate for design when strength-reduction factors are included.

For design, the top of the interaction diagram should be cut off at $0.80P_{no}$, because this is a tied column. The strength-reduction factors, ϕ, have not been included in Fig. 11-22. When ϕ is computed by ACI Code Table 21.2.2 for a tied column, ϕ is 0.65 for points 1, 2, and 3 and ϕ is 0.9 for points 4 and 5.

It is difficult to calculate the pure-moment case directly. If this value is required for a symmetrical section, it can be estimated as the larger of (1) the flexural capacity

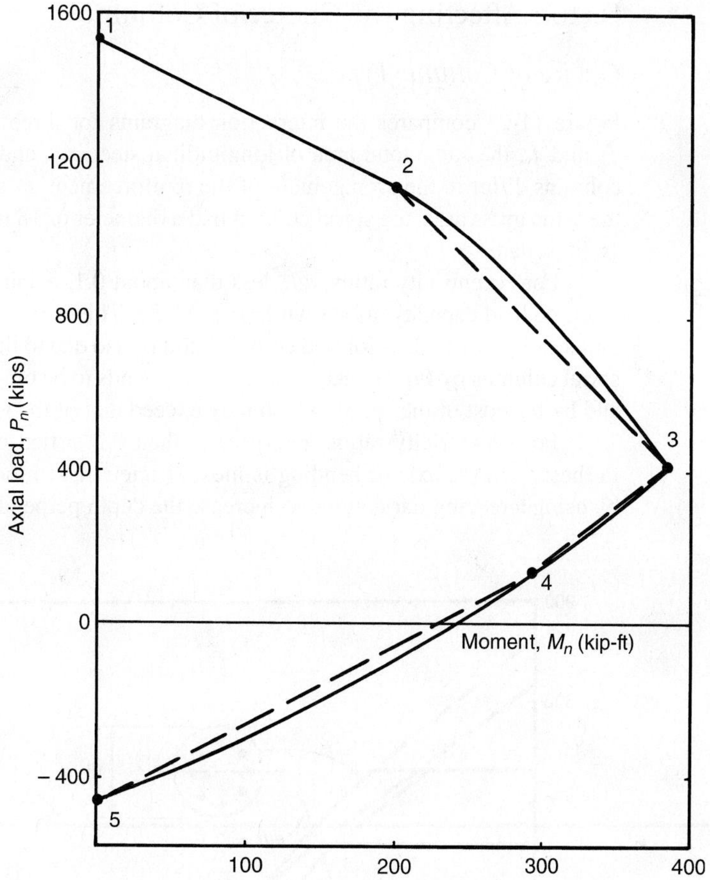

Fig. 11-22
Simplified interaction diagram.

ignoring the reinforcement in the compressive zone, or (2) the moment computed by ignoring the concrete and assuming a strain of $5\epsilon_y$ in the reinforcement adjacent to each face. For the column in Example 11-1, the pure-moment capacity, M_{no}, from a strain-compatibility solution was 236 kip-ft, compared with 227 from the dashed line in Fig. 11-22, 235 computed for it as a beam (ignoring the compression reinforcement), and 220 for it as a steel couple (ignoring the concrete). The column in question has a high steel percentage. The accuracy of these approximations decreases as ρ approaches the minimum allowed.

11-5 DESIGN OF SHORT COLUMNS

Types of Calculations—Analysis and Design

In Chapters 4 and 5, two types of computations were discussed. If the cross-sectional dimensions and reinforcement are known and it is necessary to compute the capacity, a section *analysis* is carried out. On the other hand, if the factored loads and moments are known and it is necessary to select a cross section to resist them, the procedure is referred to as *design* or *proportioning*. A design problem is solved by guessing a section, analyzing whether it will be satisfactory, revising the section, and reanalyzing it. In each case, the analysis portion of the problem for column section design is most easily carried out via interaction diagrams.

Factors Affecting the Choice of Column

Choice of Column Type

Figure 11-23 compares the interaction diagrams for three columns, each with the same f'_c and f_y, the same total area of longitudinal steel, A_{st}, and the same gross area, A_g. The columns differ in the arrangement of the reinforcement, as shown in the figure. To obtain the same gross area, the spiral column had a diameter of 18 in. while the tied columns were 16 in. square.

For eccentricity ratios, e/h, less than about 0.1, a spiral column is more efficient in terms of load capacity, as shown in Fig. 11-23. This is due to ϕ being 0.75 for spiral columns compared to 0.65 for tied columns and is also due to the higher axial load allowed on spiral columns by Eq. (11-8a). This economy tends to be offset by more expensive forming and by the cost of the spiral, which may exceed that of the ties.

For eccentricity ratios, e/h, greater than 0.2, a tied column with bars in the faces farthest from the axis of bending is most efficient. Even more efficiency can be obtained by using a rectangular column to increase the depth perpendicular to the axis of bending.

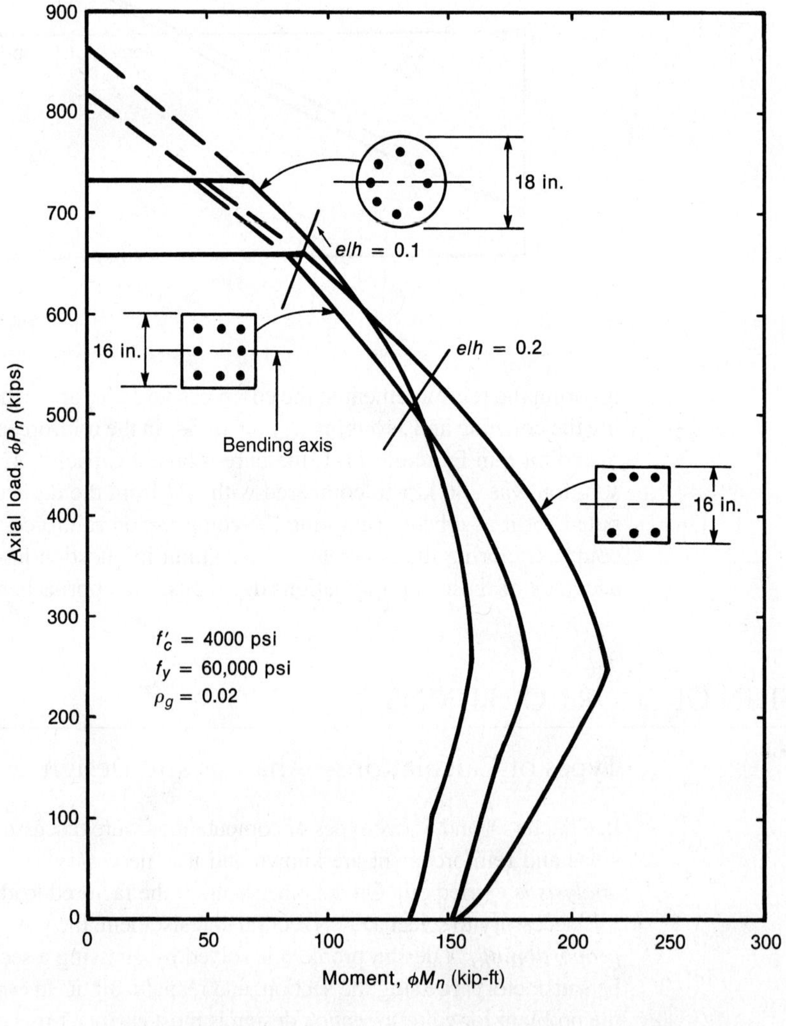

Fig. 11-23
Effect of column type on shape of interaction diagram.

Tied columns with bars in four faces are used for *e/h* ratios of less than about 0.2 and also when moments exist about both axes. Many designers prefer this arrangement because there is less possibility of construction error in the field if there are equal numbers of bars in each face of the column. Spiral columns are used infrequently for buildings in nonseismic areas. In seismic areas or in other situations where ductility is important, spiral columns are used more extensively.

Choice of Material Properties and Reinforcement Ratios

In small buildings, the concrete strength in the columns is selected to be equal to that in the floors, so that one grade of concrete can be used throughout. Frequently, this will be 4000 or 4500 psi.

In tall buildings, the concrete strength in the columns is often higher than that in the floors, to reduce the column size. When this occurs, the designer must consider the transfer of the column loads through the weaker floor concrete, as was discussed in Section 10-2. Tests of interior column–floor junctions subjected to axial column loads have shown that the lateral restraint provided by the surrounding floor members enables a floor slab to transmit the column loads, provided that the strength of the column concrete is not much higher than that of the slab [11-5]. The requirements in ACI Code Section 15.3 came from tests in which the floor slabs were not loaded. Moments resulting from dead and live loads cause the slab to crack around the column. This reduces the strengthening effect of the restraint from the slab [11-6]. Similarly, the strengthening effect is less at edge columns and corner columns that are unrestrained on one or two sides.

In the vast majority of columns, Grade-60 longitudinal reinforcement is used. Spirals are normally made from Grade-60 steel. Ties may be made of either Grade-60 or Grade-40 steel.

Tests of axially loaded tied columns reported by Pfister [11-7] in 1964 included columns with high-strength-alloy-steel longitudinal reinforcement having a specified yield strength of 75 ksi and a strength of 92 ksi at a strain of 0.006. Two columns with ties spaced at the least dimension of the cross section failed more gradually than companion columns with one tie at midheight or no ties. The columns with ties were able to develop a steel stress of 65 ksi. Pfister also found that one single tie around the perimeter of the 10-by-12-in. columns was enough support for the bars.

Hudson found that the influence of ties on the ultimate capacity of axially and eccentrically loaded columns was negligible [11-8]. His columns failed by spalling of the cover on the compression side, immediately followed by buckling of the longitudinal bars.

ACI Code Section 10.6.1.1 limits the area, A_{st}, of longitudinal reinforcement in tied and spiral columns to not less than 0.01 times the gross area, A_g (i.e., $\rho_g = A_{st}/A_g$ not less than 0.01, except as allowed by ACI Code Section 10.3.1.2) and not more than $0.08A_g$ ($0.06A_g$ in columns of *special moment frames* designed to resist earthquake forces). Under sustained loads, creep of the concrete gradually transfers load from the concrete to the reinforcement. In tests of axially loaded columns, column reinforcement yielded under sustained service loads if the ratio of longitudinal steel was less than roughly 0.01 [11-2].

A recently reported reexamination [11-9] of the effect of the creep and shrinkage of modern concretes on the transfer of vertical compression stresses from the concrete to the longitudinal (vertical) bars upheld the lower limit of $\rho_g \geq 0.01$ determined in the 1928 to 1931 tests [11-2].

Although the code allows a maximum steel ratio of 0.08, it is generally very difficult to place this amount of steel in a column, particularly if lapped splices are used. Tables A-10 and A-11 give maximum steel percentages for various column sizes [11-10]. These range from roughly 3 to 5 or 6 percent. In addition, the most economical tied-column section generally involves ρ_g of 1 to 2 percent. As a result, tied columns seldom have ρ_g greater than 3 percent. Exceptions to this are the lower columns in a tall building, where the column size must be limited for architectural reasons. In such a case, the bars may be tied in bundles of two to four bars. Design requirements for bundled bars are given in ACI Code Section 25.6. Because they are often used to resist high axial loads, spiral columns generally have steel ratios between 2.5 and 5 percent.

The minimum number of bars in a rectangular column is four, and that in a circular column or spiral column is six (ACI Code Section 10.7.3.1). Almost universally, an even number of bars is used in a rectangular column, so that the column is symmetrical about the axis of bending, and also almost universally, all the bars are the same size. Table A-12 gives the total area for various even numbers of different bar sizes.

Estimating the Column Size

The initial stage in column design involves estimating the required size of the column. There is no simple rule for doing this, because the axial-load capacity of a given cross section varies with the moment acting on the section. For very small values of M, the column size is governed by the maximum axial-load capacity given by Eq. (11-8). Rearranging, simplifying, and rounding down the coefficients in Eq. (11-8) gives the approximate relationships.

Tied Columns

$$A_{g(\text{trial})} \geq \frac{P_u}{0.40(f'_c + f_y\rho_g)} \tag{11-19a}$$

where $\rho_g = A_{st}/A_g$.

Spiral Columns

$$A_{g(\text{trial})} \geq \frac{P_u}{0.50(f'_c + f_y\rho_g)} \tag{11-19b}$$

Both of these equations will tend to underestimate the column size if there are moments present, because they correspond roughly to the horizontal line portion of the ϕP_n, ϕM_n interaction diagram in Fig. 11-18.

The local fire code usually specifies minimum sizes and minimum cover to the reinforcement. A conservative approximation to these values is 9-in. minimum column thickness for a 1-hour fire rating and 12-in. minimum for 2- or 3-hour ratings. The minimum clear cover to the vertical reinforcement is 1 in. for a 1-hour fire rating and 2 in. for 2- and 3-hour ratings. The tables in Appendix A are based on a 1.5-in. clear cover to the ties and a 2-in. clear cover to the vertical reinforcement. Column widths and depths are generally varied in increments of 2 in. Although the ACI Code does not specify a minimum column size, the minimum dimension of a cast-in-place tied column should not be less than 8 in. and preferably not less than 10 in. The diameter of a spiral column should not be less than 12 in.

Slender Columns

A slender column deflects laterally under load. This increases the moments in the column and hence weakens the column. (See Section 12-1.) Slenderness effects are discussed in Chapter 12. Chapter 11 deals only with short columns in braced frames, the most commonly occurring case. ACI Code Section 6.2.5 states that it is permissible to neglect slenderness effects for columns braced against sidesway if

$$\frac{k\ell_u}{r} \leq 34 + 12\left(\frac{M_1}{M_2}\right) \leq 40 \tag{11-20}$$

where

k = effective length factor, which, for a braced (non-sway) frame, will be less than or equal to 1.0

ℓ_u = unsupported height of column from top of floor to the bottom of the beams or slab in the floor above

r = radius of gyration, equal to 0.3 and 0.25 times the overall depth of rectangular and circular columns, respectively

M_1/M_2 = ratio of the moments at the two ends of the column, which (for compression members in a frame braced against sidesway) will generally be between +0.5 and −0.5

This limit is discussed more fully in Section 12-2. In this chapter, we shall assume that $k = 1.0$ and $M_1/M_2 = -0.5$. This will almost always be conservative. For this combination, columns are short if $k\ell_u/r \leq 28$. For a square column, this corresponds to $\ell_u/h \leq 8.4$.

Bar-Spacing Requirements

ACI Code Table 20.6.1.3.1 requires a clear concrete cover of not less than $1\frac{1}{2}$ in. to the ties or spirals in columns. More cover may be required for fire protection in some cases. The concrete for a column is placed in the core inside the bars and must be able to flow out between the bars and the form. To facilitate this, the ACI Code requires that the minimum clear distance between longitudinal bars shall not be less than the larger of 1.5 times the longitudinal bar diameter, or 1.5 in. (ACI Code Section 25.2.3), or $1\frac{1}{3}$ times the maximum size of the coarse aggregate. These clear-distance limitations also apply to the clear distance between lap-spliced bars and adjacent bars of lap splices (ACI Code Section 25.5.1.2). Because the maximum number of bars occurs at the splices, the spacing of bars at this location generally governs. The spacing limitations at splices in tied and spiral columns are illustrated in Fig. 11-24. Tables A-10 and A-11 give the maximum number of bars that can be used in rectangular tied columns and circular spiral columns, respectively, assuming that the bars are lapped as shown in Figs. 11-24a or 11-24b. These tables also give the area, A_{st}, of these combinations of bars and the ratio $\rho_g = A_{st}/A_g$.

Reinforcement Splices

In most buildings in nonseismic zones, the longitudinal bars in the columns are spliced just above each floor. Lap splices as shown in Fig. 11-25 are the most widely used, although, in large columns with large bars, mechanical splices or butt splices are sometimes used. (See Figs. 11-26 and 11-27.) Lap splices, welded or mechanical splices, and end-bearing splices are covered in ACI Code Sections 10.7.5.2, 25.5.7, and 10.7.5.3, respectively.

The requirements for lap splices vary to suit the state of stress in the bar at the nominal strength. In columns subjected to combined axial load and bending, tensile stresses may occur on one face of the column, as seen in Example 11-1. (See Fig. 11-17c, for example.) Design-interaction charts, such as Fig. A-6, frequently include lines indicating the eccentricities for which various tensile stresses occur in the reinforcement closest to the tensile

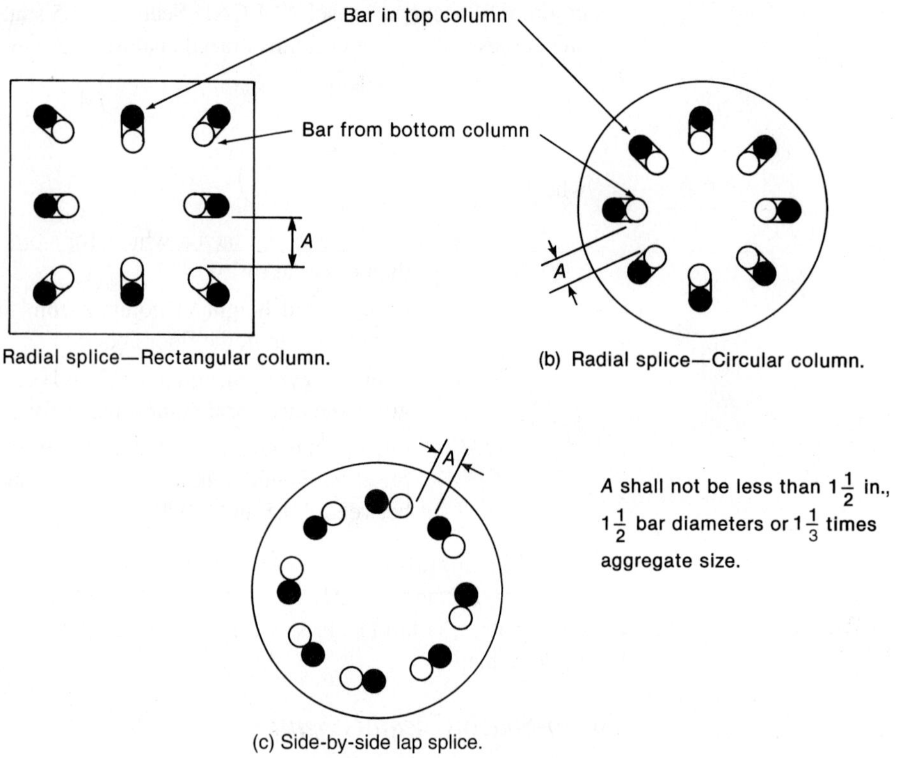

Fig. 11-24
Arrangement of bars at lap splices in columns.

face of the column. In Fig. A-6, these are labeled $f_s = 0$ and $f_s = 0.5f_y$ (in tension). The range of eccentricities for which various types of splices are required is shown schematically in Fig. 11-28.

Column-splice details are important to the designer for two major reasons. First, a compression lap splice will automatically be provided by the reinforcement detailer unless a different lap length is specified by the designer. Hence, if the bar stress at nominal strength is tensile, compression lap splices may be inadequate, and the designer should compute and show the laps required on the drawings. Second, if Class B lap splices are required (see Fig. 11-28), the required splice lengths may be excessive. For closely spaced bars larger than No. 8 or 9, the length of such a splice may exceed 5 ft and thus, may be half or more of the height of the average story. The splice lengths will be minimized by choosing the smallest practical bar sizes and the highest practical concrete strength. Alternatively, welded or mechanical splices should be used. To reduce the chance of field errors, all the bars in a column will be spliced with splices having the same length, regardless of whether they are on the tension or compression face. The required splice lengths are computed via the following steps:

1. Establish whether compression or tension lap splices are required, using the sloping dashed lines in the interaction diagrams. (See Figs. A-6 to A-14.) This is illustrated in Fig. 11-28.

2. If *compression* lap splices are required, compute the basic compression lap-splice length from ACI Code Section 25.5.5. Where applicable, the compression lap-splice length can be multiplied by factors from ACI Code Section 10.7.5.2.1 for bars enclosed within ties or spirals. These factors apply only to compression lap splices.

3. If *tension* lap splices are required, ℓ_d is computed from ACI Code Section 25.4.2.2 or 25.4.2.3. Because ACI Section 25.2.3 requires that longitudinal column bars

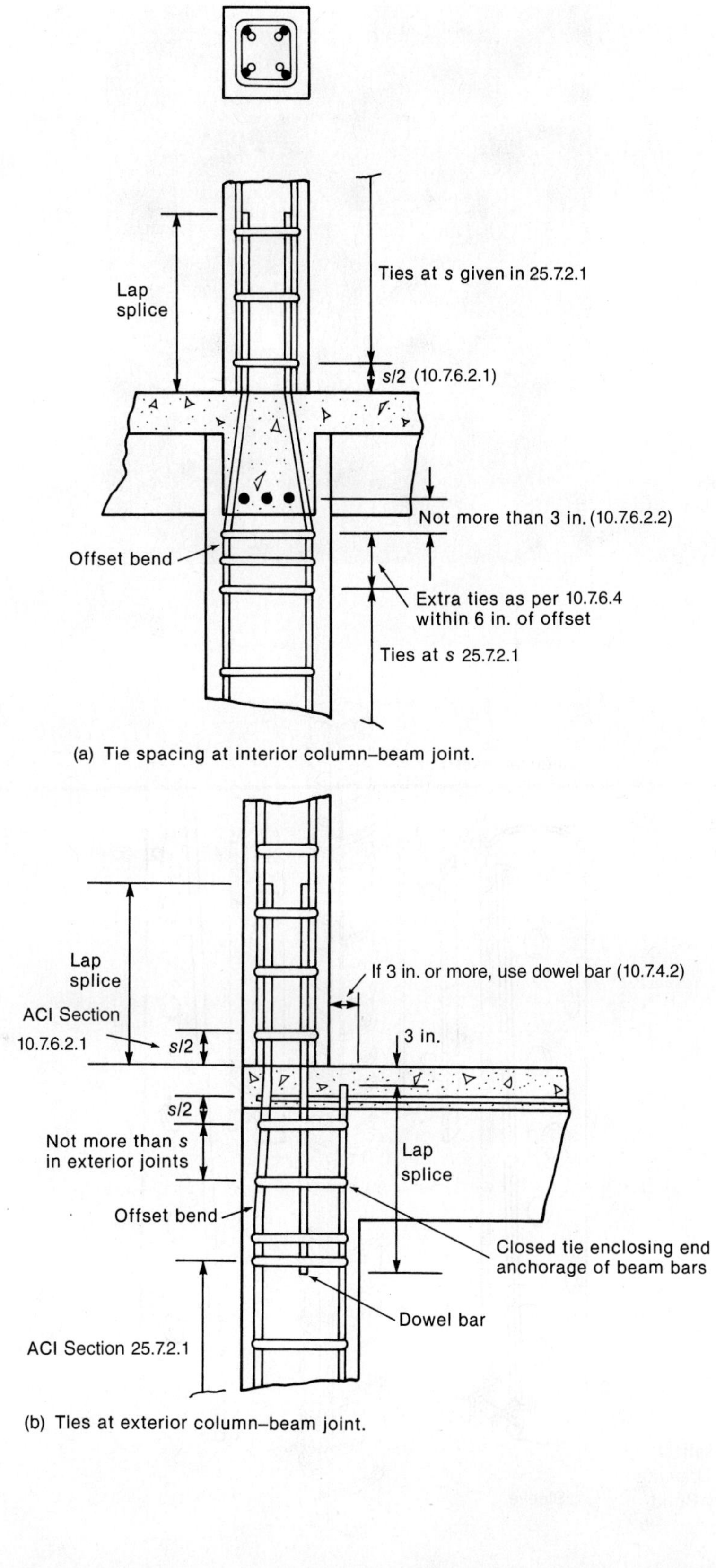

Fig. 11-25
Lap details at column–beam joints following given ACI Code Sections.

Fig. 11-26
Metal-filled bar splice: tension or compression. (Photograph courtesy of Erico Products Inc.)

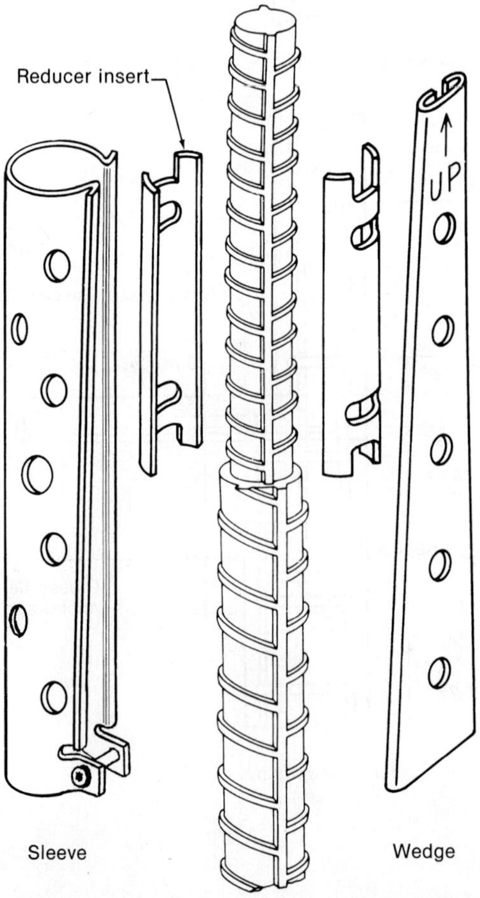

Fig. 11-27
Wedged sleeve bar splice: compression only. (Drawing courtesy of Gateway Building Products.)

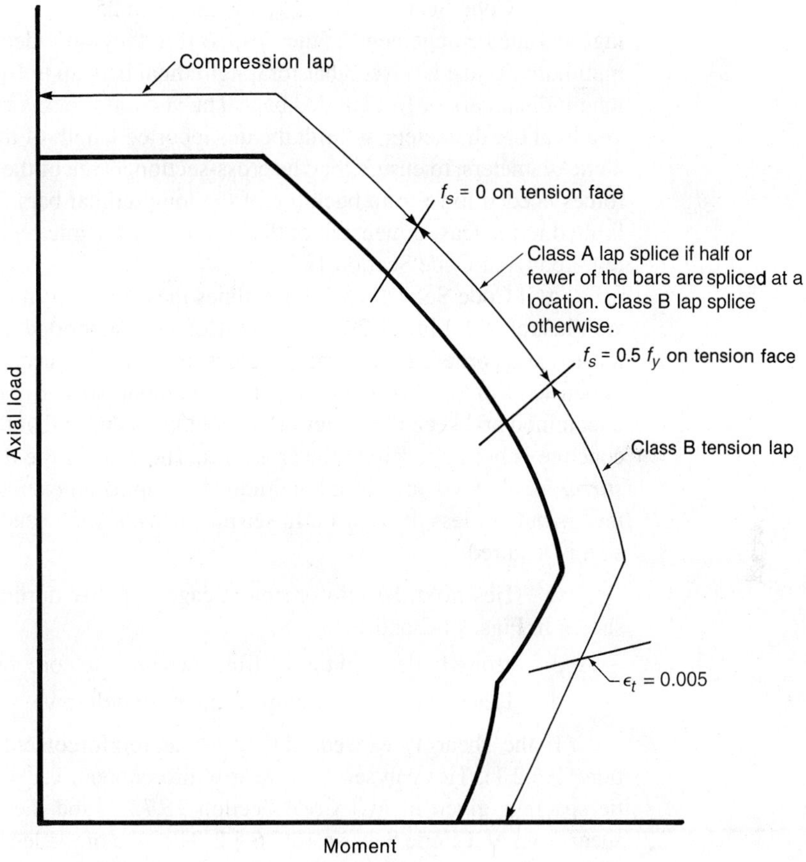

Fig. 11-28
Types of lap splices required if all bars are lap spliced at every floor.

have a minimum clear spacing of $1.5d_b$, and because they will always be enclosed by ties or spirals, ℓ_d is computed for Case 1 in Table 8-1. ℓ_d is then multiplied by 1.0 or 1.3 for a Class A or a Class B splice, respectively.

Bars loaded only in compression can be spliced by end-bearing splices (ACI Code Sections 25.5.6 and 10.7.5.3), provided that the splices are staggered or extra bars are provided at the splice locations so that the continuing bars in each face of the column at the splice location have a tensile strength at least 25 percent of that of all the bars in that face. In an end-bearing splice, the ends of the two bars to be spliced are cut as squarely as possible and held in contact with a wedged sleeve device (Fig. 11-27). End-bearing splices are used only on vertical, or almost vertical, bars enclosed by stirrups or ties. Tests show that the force transfer in end-bearing splices is superior to that in lapped splices, even if the ends of the bars are slightly off square (up to 3° total angle between the ends of the bars).

Spacing and Construction Requirements for Ties

Ties are provided in reinforced concrete columns for four reasons [11-7], [11-8], [11-11]:

1. Ties restrain the longitudinal bars from buckling out through the surface of the column.

ACI Code Sections 25.7.2.1, 25.7.2.2, and 25.7.2.3, give limits on the size, the spacing, and the arrangement of the ties, so that they are adequate to restrain the bars. The minimum tie size is a No. 3 bar for longitudinal bars up to No. 10 and a No. 4 bar for larger longitudinal bars or for bundled bars. The vertical spacing of ties shall not exceed 16 longitudinal bar diameters, to limit the unsupported length of these bars, and shall not exceed 48 tie diameters, to ensure that the cross-sectional area of the ties is adequate to develop the forces needed to restrain buckling of the longitudinal bars. The maximum spacing is also limited to the least dimension of the column. In seismic regions, much closer spacings are required (ACI Code Section 18.7.5).

ACI Code Section 25.7.2.3 outlines the arrangement of ties in a cross section. These are illustrated in Fig. 11-29. A bar is adequately supported against lateral movement if it is located at a corner of a tie where the ties are spaced in accordance with 25.7.2.1 and if the dimension x in Fig. 11-29 is 6 in. or less. Diamond-shaped and octagonal-shaped ties are not uncommon and keep the center of the column open, so that the placing and vibrating of the concrete is not impeded by the cross-ties. The ends of the ties are anchored by a *standard stirrup or tie hook* around a longitudinal bar, plus an extension of at least six tie-bar diameters but not less than $2\frac{1}{2}$ in. In seismic areas, a 135° bend plus a six-tie-diameter extension is required.

2. Ties hold the reinforcement cage together during the construction process, as shown in Figs. 11-2 and 11-3.

3. Properly detailed ties confine the concrete core, providing increased ductility.

4. Ties serve as shear reinforcement for columns.

If the shear V_u exceeds $\phi V_c/2$, shear reinforcement is required (ACI Code Section 10.6.2.1). Ties can serve as shear reinforcement, but must satisfy both the maximum tie spacings given in ACI Code Section 25.7.2.1 and the maximum stirrup spacing for shear from ACI Code Section 10.7.6.5.2. The area of the legs parallel to the direction of the shear force must satisfy ACI Code Section 10.6.2.2.

Specially fabricated welded-wire reinforcement cages incorporating the longitudinal reinforcement and ties are sometimes an economical solution to constructing column cages. Each cage consists of several interlocking sheets bent to form one to three sides of the cage. The bars or wires forming the ties are hooked around longitudinal bars to make the ties continuous. The layout of the cages must be planned carefully, so that bars from one part of the cage do not interfere with those from another when they are assembled in the field.

Ties may be formed from continuously wound wires with a pitch and area conforming to the tie requirements.

ACI Code Sections 10.7.6.2.1 and 10.7.6.2.2 require that the bottom and top ties be placed as shown in Fig. 11-25. ACI Code Section 15.4.3 requires that bar anchorages in connections of beams and columns be enclosed by ties, spirals, or stirrups. Generally, ties are most suitable for this purpose and should be arranged as shown in Fig. 11-25b.

Finally, extra ties are required at the outside (lower) end of offset bends at column splices (see Fig. 11-25) to resist the horizontal force component in the sloping portion of the bar. The design of these ties is described in ACI Code Section 10.7.6.4. If we assume that the vertical bars in Fig. 11-25a are No. 8 Grade-60 bars and that the offset bend is at the maximum allowed slope of 1:6 (ACI Code Code Section 10.7.4.1), then the horizontal force from two bars is $(1/6) \times f_y \times 2 \times 0.79 \text{ in.}^2 = 15.8$ kips, where 0.79 in.^{2} is the area of one No. 8 bar. ACI Code Section 10.7.6.4.1 requires that ties be provided for 1.5 times this force, or 23.7 kips. The number of No. 3 Grade-60 tie legs required to resist this force is $23.7 \text{ kips}/(60 \times 0.11 \text{ in.}^2) = 3.6$ legs. Thus, two, two-leg ties would be required within 6 in. of the bend.

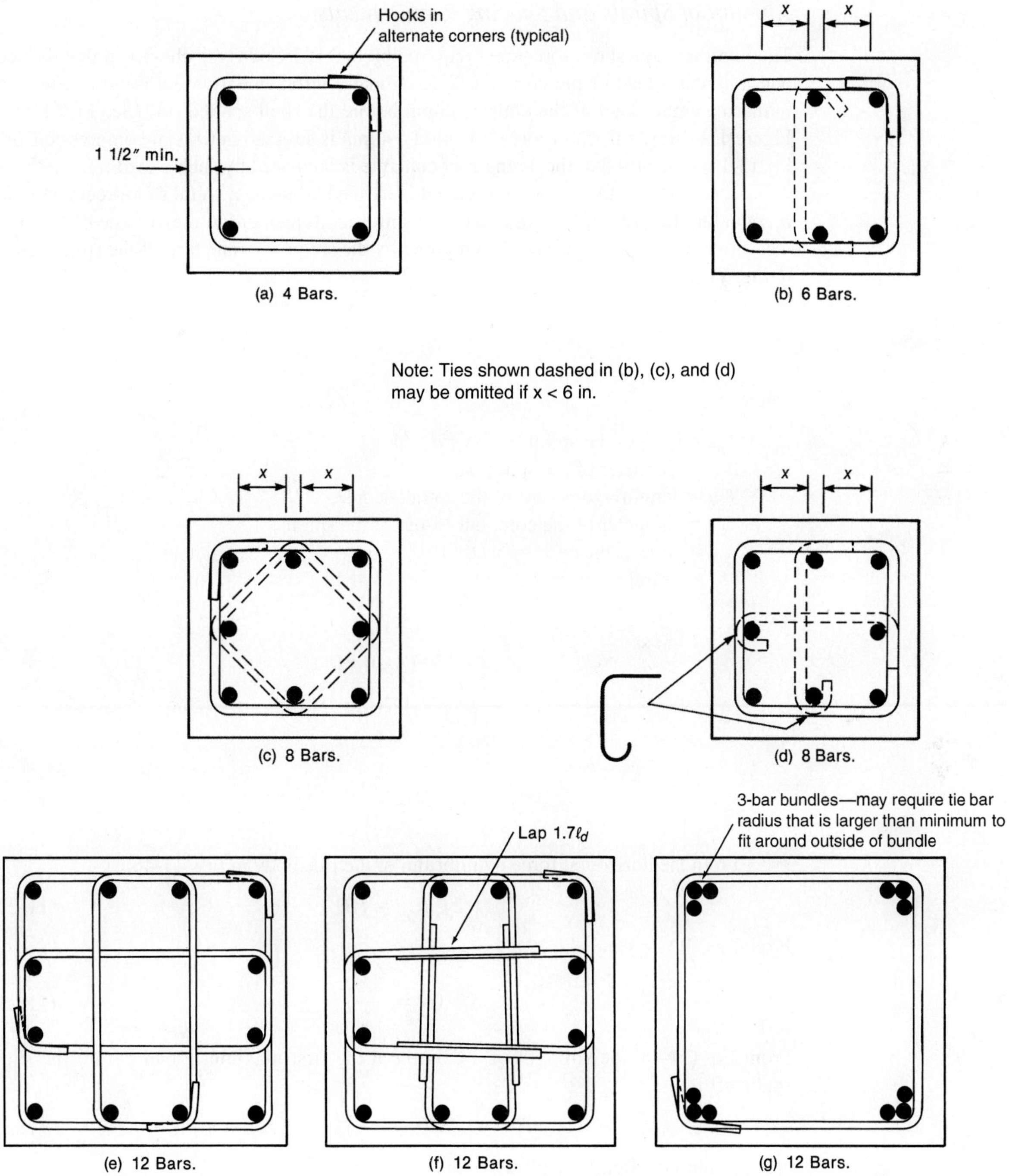

Fig. 11-29
Typical tie arrangements.

Amount of Spirals and Spacing Requirements

The minimum spiral reinforcement required by the ACI Code was chosen so that the second maximum load of the core and longitudinal reinforcement would roughly equal the initial maximum load of the entire column before the shell spalled off. (See Fig. 11-5a.) Figure 11-4 shows that the core of a spiral column is stressed in triaxial compression, and Eq. (3-16) indicates that the strength of concrete is increased by such a loading.

The amount of spiral reinforcement is defined by using a spiral reinforcement ratio, ρ_s, equal to the ratio of the volume of the spiral reinforcement to the volume of the core, measured out-to-out of the spirals, enclosed by the spiral. For one turn of the spiral shown in Fig. 11-4a,

$$\rho_s = \frac{A_{sp}\ell_{sp}}{A_{ch}\ell_c}$$

where

$$\begin{aligned}
A_{sp} &= \text{area of the spiral bar} = \pi d_{sp}^2/4 \\
d_{sp} &= \text{diameter of the spiral bar} \\
\ell_{sp} &= \text{length of one turn of the spiral} = \pi D_c \\
D_c &= \text{diameter of the core, out-to-out of the spirals} \\
A_{ch} &= \text{area of the core} = \pi D_c^2/4 \\
\ell_c &= \text{spiral pitch} = s
\end{aligned}$$

Thus,

$$\rho_s = \frac{(A_{sp})(\pi D_c)}{(\pi D_c^2/4)s}$$

or

$$\rho_s = \frac{4A_{sp}}{sD_c} \tag{11-21}$$

From the horizontal force equilibrium of the free body in Fig. 11-4b,

$$2f_{sp}A_{sp} = f_2 D_c s \tag{11-22}$$

From Eqs. (11-21) and (11-22),

$$f_2 = \frac{f_{sp}\rho_s}{2} \tag{11-23}$$

From Eq. (11-1), the strength of a column at the first maximum load before the shell spalls off is

$$P_o = 0.85f'_c(A_g - A_{st}) + f_y A_{st} \tag{11-1}$$

and the strength after the shell spalls is

$$P_2 = 0.85f_1(A_{ch} - A_{st}) + f_y A_{st}$$

Thus, if P_2 is to equal P_o, $0.85f_1(A_{ch} - A_{st})$ must equal $0.85f'_c(A_g - A_{st})$. Because A_{st} is small compared with A_g or A_{ch}, we can disregard it, giving

$$f_1 = \frac{A_g f'_c}{A_{ch}} \tag{11-24}$$

Substituting Eqs. (3-16) and (11-23) into Eq. (11-24), taking f_{sp} equal to the yield strength of the spiral bar, f_{yt}, rearranging, and rounding down the coefficient gives

$$\rho_s = 0.45\left(\frac{A_g}{A_{ch}} - 1\right)\frac{f'_c}{f_{yt}} \tag{11-25}$$

(ACI Eq. 25.7.3.3)

There is experimental evidence that more spiral reinforcement may be needed in high-strength concrete spiral columns than is given by Eq. (11-25) [11-12], to ensure that ductile behavior precedes any failure.

Design requirements for column spirals are presented in ACI Code Section 25.7.3. ACI Code Section 25.7.3.2 requires that in cast-in-place construction, spirals be at least No. 3 in diameter. The spacing is determined by three rules:

1. Equation (11-25) gives the maximum spacing that will result in a second maximum load that equals or exceeds the initial maximum load. Combining Eqs. (11-21) and (11-25) and rearranging gives the maximum *center-to-center* spacing of the spirals:

$$s \leq \frac{4A_{sp}f_{yt}}{0.45D_c f'_c[(A_g/A_{ch}) - 1]} \tag{11-26}$$

2. For the spirals to confine the core effectively, successive turns of the spiral must be spaced relatively close together. ACI Code Section 25.7.3.1 limits the clear spacing of the spirals to a maximum of 3 in.

3. To avoid problems in placing the concrete in the outer shell of concrete, the spiral spacing must be as large as possible. ACI Code Section 25.7.3.1 limits the minimum clear spacing between successive turns to 1 in., but not less than $1\frac{1}{3}$ times the nominal size of the coarse aggregate, whichever is greater.

The termination of spirals at the tops and bottoms of columns is governed by ACI Code Sections 10.7.6.3.1 and 10.7.6.3.2. Again, joint reinforcement satisfying ACI Code Section 15.4.3 is also required, and generally, ties are provided for this purpose.

EXAMPLE 11-2 Design of a Tied Column for a Given P_u and M_u

Design a tied-column cross section to support $P_u = 450$ kips, $M_u = 120$ kip-ft, and $V_u = 14$ kips. The column is in a braced frame and has an unsupported length of 10 ft 6 in.

1. Select the material properties, trial size, and trial reinforcement ratio. Select $f_y = 60$ ksi and $f'_c = 4$ ksi. The most economical range for ρ_g is from 1 to 2 percent. Assume that $\rho_g = 0.015$ for the first trial value. From (11-19a),

$$A_{g(\text{trial})} \geq \frac{P_u}{0.40(f'_c + f_y\rho_g)}$$

$$\geq \frac{450}{0.40(4 + 60 \times 0.015)}$$

$$\geq 230 \text{ in.}^2 \text{ (or 15.2 in. square)}$$

Because moments act on this column, (11-19a) will underestimate the column size. Choose a 16-in.-square column. *Note*: Column dimensions are normally increased in 2-in. increments.

To determine the preferable bar arrangement, compute the ratio e/h:

$$e = \frac{M_u}{P_u}$$

$$= \frac{120 \text{ kip-ft}}{450 \text{ kips}} = 0.267 \text{ ft} = 3.20 \text{ in.}$$

$$\frac{e}{h} = 0.200$$

For e/h in this range, Fig. 11-23 indicates that a column with bars in two faces will be most efficient. Use a tied column with bars in two faces.

Slenderness can be neglected if

$$\frac{k\ell_u}{r} \leq 34 + 12\left(\frac{M_1}{M_2}\right) \leq 40 \qquad (11\text{-}20)$$

Because this is a braced frame, $k \leq 1.0$, and M_1/M_2 will normally be between +0.5 and −0.5. We shall assume that $k = 1.0$ and $M_1/M_2 = -0.5$. The left-hand side of (11-20) is

$$\frac{k\ell_u}{r} = \frac{1.0 \times 126 \text{ in.}}{0.3 \times 16 \text{ in.}} = 26.3$$

The right-hand side is

$$34 + 12\left(\frac{M_1}{M_2}\right) = 34 + 12(-0.5) = 28$$

Because 26.3 is less than 28, slenderness can be neglected.

Summary for the trial column. A 16-in. × 16-in. tied column with bars in two faces.

2. Compute γ. The interaction diagrams in Appendix A are each drawn for a particular value of the ratio, γ, of the distance between the centers of the outside layers of bars to the overall depth of the column. To estimate γ, assume that the centroid of the longitudinal bars is located 2.5 in. from the edge of the column, as was done in Chapters 4 and 5 for beam sections with nominal 1.5-in. cover.

$$\gamma = \frac{16 - 2(2.5)}{16} \cong 0.69$$

Because the interaction diagrams in Appendix A are given for $\gamma = 0.60$ and $\gamma = 0.75$, it will be necessary to interpolate. Also, because the diagrams only can be read with limited accuracy, it is recommended to express γ with only two significant figures.

3. Use interaction diagrams to determine ρ_g. The interaction diagrams are entered with

$$\frac{\phi P_n}{A_g} = \frac{P_u}{A_g}$$

$$= \frac{450}{16 \times 16} = 1.76 \text{ ksi}$$

$$\frac{\phi M_n}{A_g h} = \frac{M_u}{A_g h}$$

$$= \frac{120 \times 12}{16 \times 16 \times 16} = 0.352 \text{ ksi}$$

From Fig. A-6a (interaction diagram for $\gamma = 0.6$),

$$\rho_g = 0.017$$

From Fig. A-6b (for $\gamma = 0.75$),

$$\rho_g = 0.014$$

Use linear interpolation to compute the value for $\gamma = 0.69$:

$$\rho_g = 0.017 - 0.003 \times \frac{0.09}{0.15} = 0.015$$

If the value of ρ_g computed here exceeds 0.03 to 0.04, a larger section should be chosen. If ρ_g is less than 0.01, either use 0.01 (the minimum allowed by ACI Code Section 10.6.1.1) or recompute, using a smaller cross section. *Note*: Using a smaller section could change this to a slender column, which is something we want to avoid for now.

4. Select the reinforcement.

$$A_{st} = \rho_g A_g$$

$$= 0.015 \times 16 \times 16 \text{ in.}^2 = 3.84 \text{ in.}^2$$

Possible combinations are (Table A-12)

four No. 9 bars, $A_{st} = 4.00$ in.^{2}, two in each face
six No. 8 bars, $A_{st} = 4.74$ in.^{2}, three in each face

An even number of bars will be chosen, so that the reinforcement is symmetrical about the bending axis. From Table A-10, it is seen that none of these violates the minimum bar spacing rules. **Try a 16-in.-square column with six No. 8 bars.**

5. Check the maximum load capacity. P_u should not exceed $\phi P_{n(\max)}$, given by Eq. (11-8b). The upper horizontal lines in the interaction diagrams represent $\phi P_{n(\max)}$, and the section chosen falls below the upper limit. Actually, this check is necessary only if one is using interaction diagrams that do not show this cutoff.

6. Design the lap splices. From Figs. A-6a and A-6b, the stress in the bars adjacent to the tensile face for $P_u/bh = 1.76$ and $M_u/bh^2 = 0.352$ is about $0.2f_y$ in tension. From ACI Code Table 10.7.5.2.2 the splice must be a Class B splice if more than half of the bars are spliced at any section or a Class A splice if half or fewer are spliced at one location. Normally, all the bars would be spliced at the same location. We shall assume that this is done. The splice length is $1.3\ell_d$. From ACI Code Table 25.4.2.2, for No. 8 bars,

$$\ell_d = \left(\frac{f_y \psi_t \psi_e}{20\lambda\sqrt{f'_c}}\right) d_b$$

$$= \left(\frac{60{,}000 \times 1.0 \times 1.0}{20 \times 1 \times \sqrt{4000}}\right) \times 1.00$$

$$= 47.4 \text{ in.}$$

The splice length is

$$1.3\ell_d = 1.3 \times 47.4$$
$$= 61.7 \text{ in.}$$

This is a long splice. It would equal approximately half of the story height.

7. Select the ties. From ACI Code Section 25.7.2.2, No. 3 ties are the smallest allowed. The required spacing (ACI Code Section 25.7.2.1) is the smallest of the following quantities:

$$16 \text{ longitudinal bar diameters} = 16 \times \frac{8}{8} = 16 \text{ in.}$$

$$48 \text{ tie diameters} = 48 \times \frac{3}{8} = 18 \text{ in.}$$

$$\text{least dimension of column} = 16 \text{ in.}$$

If $V_u > 0.5\phi V_c$, the ties must satisfy both ACI Code Sections 10.6.2 and 25.7.2. From ACI Code Section 22.5.6.1, V_c is,

$$V_c = 2\left(1 + \frac{N_u}{2000A_g}\right)\lambda\sqrt{f'_c}b_w d \tag{6-17a}$$

$$= 2\left(1 + \frac{450{,}000}{2000 \times 16 \times 16}\right)1 \times \sqrt{4000} \times 16 \times 13.5$$

$$= 51.3 \text{ kips}$$

$V_u = 14$ kips is less than $0.5\phi V_c = 0.5 \times 0.75$ (for shear) $\times\ 51.3 = 19.3$ kips. Therefore, ACI Code Section 25.7.2 governs. (If $0.5\phi V_c < V_u \leq \phi V_c$, it would be necessary to satisfy ACI Sections 10.6.2.2 and 10.7.6.5.2.) **Use No. 3 ties at 16 in. o.c.** The tie arrangement is shown in Fig. 11-30. ACI Code Section 25.7.2.3 requires that all corner bars and other bars at a clear spacing of more than 6 in. from a corner bar be enclosed by the corner of tie. In this case the clear spacing is approximately 4.5 in., so no extra cross-tie is required.

Summary of the design. Use a 16-in. × 16-in. tied column with eight No. 8 bars, $f_y = 60{,}000$ psi, and $f'_c = 4000$ psi. Use No. 3 closed ties at 16 in. on centers, as shown in Fig. 11-30. ■

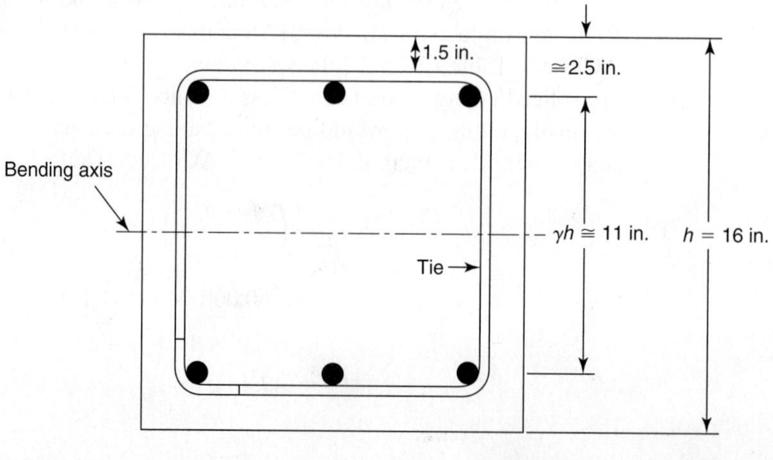

Fig. 11-30
Computation of γ and arrangement of ties—Example 11-2.

EXAMPLE 11-3 Design of a Circular Spiral Column for a Large Axial Load and Small Moment

Design a spiral-column cross section to support factored forces and end moments of $P_u = 1600$ kips and $M_u = 150$ kip-ft.

1. Select the material properties, trial size, and trial reinforcement ratio. Use $f'_c = 4$ ksi and $f_y = 60$ ksi. Try $\rho_g = 0.04$. From (11-19b),

$$A_{g(\text{trial})} = \frac{P_u}{0.50(f'_c + f_y\rho_g)} = 500 \text{ in.}^2$$

This corresponds to a diameter of 25.2 in. We shall try 26 in.

2. Compute γ. Assume that the centroid of each longitudinal bar is 2.5 in. from the edge of the column.

$$\gamma = \frac{26 \text{ in.} - 2(2.5)}{26 \text{ in.}} \cong 0.81$$

3. Use interaction diagrams to determine ρ_g.

$$\frac{P_u}{A_g} = \frac{1600}{\pi \times 13 \text{ in.}^2} = 3.01 \text{ ksi}$$

$$\frac{M_u}{A_g h} = \frac{150 \times 12}{\pi \times 13 \text{ in.}^2 \times 26 \text{ in.}} = 0.130 \text{ ksi}$$

Looking at Figs. A-12b and A-12c (interaction diagrams for spiral columns with $\gamma = 0.75$ and 0.90) indicates that due to the relatively small moment, we are on the flat part of the diagrams. For both γ values read $\rho_g = 0.024$. Because the values of P_u/A_g and $M_u/A_g h$ fall in the upper horizontal part of the diagram, Eq. (11-8a) could be used to solve for A_{st} directly.

4. Select the reinforcement.

$$A_{st} = \rho_g A_g$$
$$= 0.024(13 \text{ in.}^2 \times \pi) = 12.7 \text{ in.}^2$$

From Table A-12, ten No. 10 bars give 12.7 in.^{2}, and Table A-11 shows that these will fit into a 26-in.-diameter column. **Try ten No. 10 bars.**

5. Check the maximum load capacity. From Eq. (11-8a),

$$\phi P_{n,\max} = 0.85 \times 0.75 \text{ (for spiral) } [0.85 \times 4(\pi \times 13 \text{ in.}^2 - 12.7) + 60 \times 12.7]$$
$$= 1610 \text{ kips—therefore, o.k.}$$

6. Select the spiral. The minimum-size spiral is No. 3 (ACI Code Section 25.7.3.2). The center-to-center pitch, s, required to ensure that the second maximum load is equal to the initial maximum load, is given by Eq. (11-26), that is,

$$s \leq \frac{4A_{sp} f_{yt}}{0.45 D_c f'_c[(A_g/A_{ch}) - 1]}$$

where

A_{sp} = area of spiral bar = 0.11 in.^{2}

f_{yt} = yield strength of spiral bar = 60,000 psi (assumed)

D_c = diameter of core, outside to outside of spirals = 26 in. − 2 × 1.5 in. = 23 in.

A_g = gross area = 13 in.^{2} × π = 531 in.^{2}

A_{ch} = area of core = 11.5 in.^{2} × π = 415 in.^{2}

Thus,

$$s \leq \frac{4 \times 0.11 \text{ in.}^2 \times 60{,}000}{0.45 \times 23 \text{ in.} \times 4000[(531/415) - 1]}$$
$$\leq 2.28 \text{ in.}$$

Hence, the center-to-center pitch cannot exceed 2.28 in., but must also satisfy detailing requirements. The maximum clear spacing (ACI Code Section 25.7.3.1) is 3 in. (the maximum pitch is 3.0 + 0.375 = 3.375 in.) The minimum clear spacing (ACI Code Section 25.7.3.1) is 1 in. or $1\frac{1}{3}$ times the size of the coarse aggregate. For $\frac{3}{4}$-in. aggregate, the minimum pitch is 1.375 in.

Use a No. 3 spiral with $2\frac{1}{4}$-in. pitch.

7. Design the lap splices. From the interaction diagrams, f_s is compression. Therefore, use compression lap splices (ACI Code Section 25.5.5). From Table A-13, these must be 38 in. long. Because the bars are enclosed in a spiral, this can be multiplied by 0.75, giving a splice length of 28.5 in. (ACI Code Section 10.7.5.2.1b).

Summary of the design. Use a 26-in.-diameter column with ten No. 10 bars, f_y = 60,000 psi, and f'_c = 4000 psi. Use a No. 3 spiral with a pitch of $2\frac{1}{4}$ in. and f_{yt} = 60,000 psi. Lap-splice all bars 29 in. just above every floor. ■

EXAMPLE 11-4 Design a Tied Column for Combined Axial Load and Bending

Design a tied column considering three different combinations of gravity and lateral wind loads. Use an R-type arrangement of longitudinal reinforcement, starting with

$$\rho_g \cong 0.025. \quad \text{Use } f'_c = 5000 \text{ psi} \quad \text{and} \quad f_y = 60 \text{ ksi.}$$

Given loads: P_D = 300 k, M_D = 60 k-ft
P_L = 200 k, M_L = 120 k-ft
P_W = ±100 k, M_W = ±160 k-ft

Load Combinations from ACI Code Table 5.3.1:

Load Case 1 (Gravity, Code Eq. (5.3.1b)):

$$P_u = 1.2\,(300 \text{ k}) + 1.6\,(200 \text{ k}) = 680 \text{ k}$$
$$M_u = 1.2\,(60 \text{ k-ft}) + 1.6\,(120 \text{ k-ft}) = 264 \text{ k-ft}$$

Design eccentricity, $e = M_u/P_u = 0.388$ ft = 4.66 in.

Load Case 2 (Gravity + Wind, Code Eq. (5.3.1d)):

$$P_u = 1.2\,(300 \text{ k}) + 0.5\,(200 \text{ k}) + 1.0\,(100 \text{ k}) = 560 \text{ k}$$
$$M_u = 1.2\,(60 \text{ k-ft}) + 0.5\,(120 \text{ k-ft}) + 1.0\,(160 \text{ k-ft}) = 292 \text{ k-ft}$$

Design eccentricity, $e = M_u/P_u = 0.521$ ft = 6.26 in.

Load Case 3 (Reduced Gravity + Wind, Code Eq. (5.3.1f)):

$$P_u = 0.9\,(300\text{ k}) - 1.0\,(100\text{ k}) = 170\text{ k}$$
$$M_u = 0.9\,(60\text{ k-ft}) + 1.0\,(160\text{ k-ft}) = 214\text{ k-ft}$$
$$\text{Design eccentricity, } e = M_u/P_u = 1.26\text{ ft} = 15.1\text{ in.}$$

Note: For Load Case 2, we have used a load factor of 0.5 for live load, as permitted in ACI Code Section 5.3.3. For Load Case 3, a reduced acting axial load will result in a reduced moment capacity. Thus, the most critical combination for this load case is obtained by subtracting the wind axial load from the dead axial load, and adding the wind moment to the dead load moment.

1. Initial column section. We will use P_u from Load Case 1 in Eq. (11-19a) to select an initial column area. We will assume primary bending is about one axis and minor about the other axis, so select $b \cong 0.75h$.

$$A_g(\text{trial}) = \frac{P_u}{0.40\left(f_c' + f_y\rho_g\right)}$$

$$A_g(\text{trial}) = \frac{680\text{ k}}{0.40(5\text{ ksi} + 60\text{ ksi} \times 0.025)} = 262\text{ in.}^2$$

Possible section dimensions (using even numbers):

$$b = 14\text{ in. and } h = 20\text{ in.}; A_g = 280\text{ in.}^2 \text{ and } b/h = 0.70$$
$$b = 16\text{ in. and } h = 20\text{ in.}; A_g = 300\text{ in.}^2 \text{ and } b/h = 0.80$$

Because Eq. (11-19a) tends to underestimate the required A_g when significant moments are acting, we will select the second option to continue the design.

2. Design iteration No. 1. We will use the design charts from Appendix A to refine the required section size. Recall that we are using an R-type distribution of longitudinal reinforcement. We need to calculate γ to determine which chart to use. As before, we will assume the distance from the edge of the column to the nearest layer of reinforcement is 2.5 in.

$$\gamma = \frac{20\text{ in.} - 2 \times 2.5\text{ in.}}{20\text{ in.}} = 0.75, \text{ so use Chart R-5-60.75} \quad \text{(Fig. A-10b)}$$

We will enter the design chart with a value for *e*/*h* and ρ_g, and read the value for the ordinate, $(\phi P_n/A_g)$. We do not know P_n, but we know the minimum value for ϕP_n is P_u, which is known. So, from the ordinate value we can calculate the required A_g.

<u>Load Case 1</u>: $e/h = 4.66\text{ in.}/20\text{ in.} = 0.23$ (use just two significant figures) and read:

$$\frac{\phi P_n}{A_g} \cong 2.0\text{ ksi, and using } \phi P_n \geq P_u$$

$$A_g \geq \frac{P_u}{2.0\text{ ksi}} = \frac{680\text{ k}}{2.0\text{ ksi}} = 340\text{ in.}^2$$

Note: It is better to enter these charts using only two significant figures for each value and read the ordinate to only two significant figures. Also, a conservative reading of the ordinate (as was done here) is recommended so as to not underestimate the required area.

<u>Load Case 2</u>: $e/h = 6.26\text{ in.}/20\text{ in.} = 0.31$, and read:

$$\frac{\phi P_n}{A_g} \cong 1.7\text{ ksi, and using } \phi P_n \geq P_u$$

$$A_g \geq \frac{P_u}{1.7\text{ ksi}} = \frac{560\text{ k}}{1.7\text{ ksi}} = 329\text{ in.}^2$$

Load Case 3: $e/h = 15.1 \text{ in.}/20 \text{ in.} = 0.76$, and read:

$$\frac{\phi P_n}{A_g} \cong 0.7 \text{ ksi, and using } \phi P_n \geq P_u$$

$$A_g \geq \frac{P_u}{0.7 \text{ ksi}} = \frac{170 \text{ k}}{0.7 \text{ ksi}} = 243 \text{ in.}^2$$

For this design, Load Case 1 governs. Using an even number for the column dimensions, try a 22 in. by 16 in. section with $A_g = 352 \text{ in.}^2$. This value is relatively close to the area we had at the end of Step 1 ($Ag = 300 \text{ in.}^2$). If there was a more significant difference, we could do another design iteration to converge to a satisfactory cross section.

3. Find required amount of longitudinal reinforcement and select bars. We can now use the design charts to find the required reinforcement ratio, as follows:

$$\frac{\phi P_n}{A_g} \geq \frac{P_u}{bh} = \frac{680 \text{ k}}{16 \text{ in.} \times 22 \text{ in.}} \cong 1.9 \text{ ksi}$$

$$\frac{\phi M_n}{A_g \cdot h} \geq \frac{M_u}{bh^2} = \frac{264 \text{ k-ft} \times 12 \text{ in./ft}}{16 \text{ in.} \times (22 \text{ in.})^2} = 0.41 \text{ ksi}$$

$$\gamma = \frac{22 \text{ in.} - 5 \text{ in.}}{22 \text{ in.}} = 0.77 \cong 0.75 \text{ (no iteration required)}$$

From Chart R-5-60.75 read required $\rho_g \cong 0.015$. So, the required area of steel is, $A_{st} = 0.015 \times 352 \text{ in.}^2 = 5.28 \text{ in.}^2$. Try eight No. 8 bars, so $A_{st} = 8 \times 0.79 \text{ in.}^2 = 6.32 \text{ in.}^2$. Use the reinforcement layout shown in Fig. 11-31.

4. Exact check of section design. Because we can only realistically read the design charts in Appendix A to two significant figures and because we may sometimes be interpolating between charts, it is good practice to do an exact check of the section design using the procedure given in Section 11-4 for developing a nominal strength interaction diagram. We do not need the full diagram, but we need to show $\phi P_n \geq P_u$ at the points on the interaction diagram corresponding to the design eccentricities calculated in Step 1.

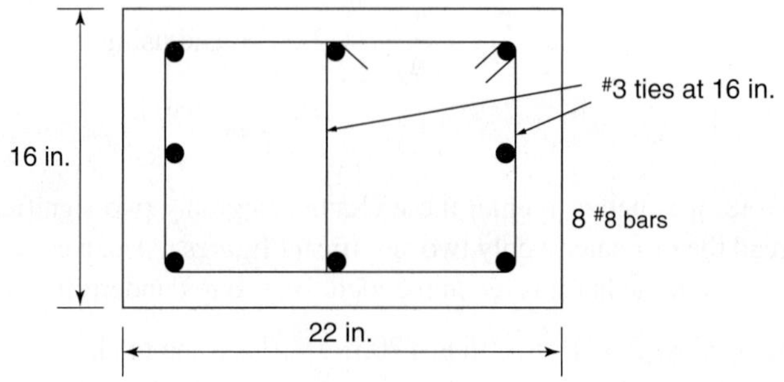

Fig. 11-31
Final column cross section for Example 11-4.

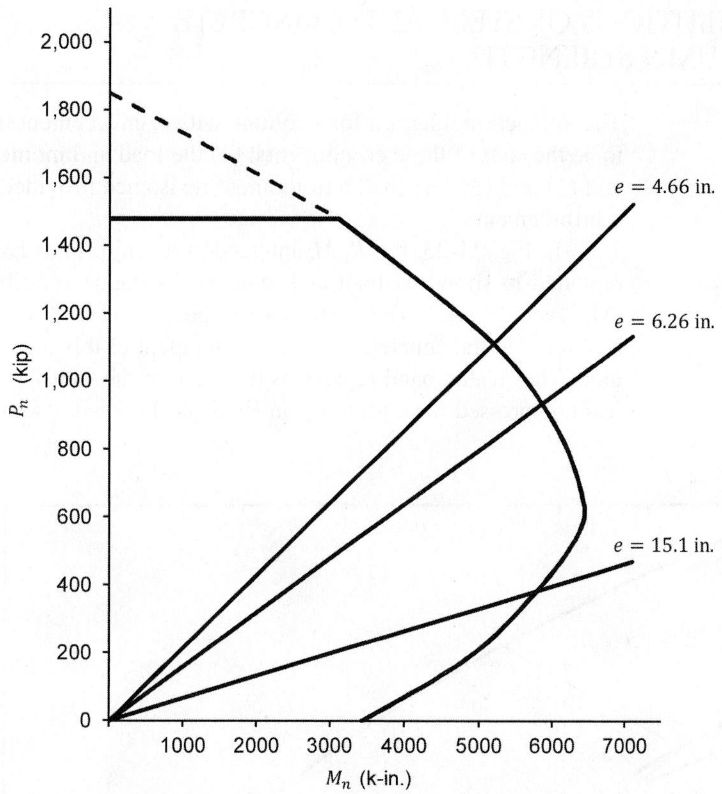

Fig. 11-32
Nominal strength interaction diagram for checking column section in Fig. 11-31.

The interaction diagram in Fig. 11-32 has been generated using the given material properties and the column section in Fig. 11-31. Drawing lines representing the design eccentricities we get the following results:

Load Case 1: $e = 4.66$ in.; read $P_n = 1110$ k. This point is clearly in the compression-controlled region, so $\phi = 0.65$ and $\phi P_n = 722 \text{ k} > 680 \text{ k}$ (o.k.).

Load Case 2: $e = 6.26$ in.; read $P_n = 924$ k. This point is also in the compression-controlled region, so $\phi = 0.65$ and $\phi P_n = 601 \text{ k} > 560 \text{ k}$ (o.k.).

Load Case 3: $e = 15.1$ in.; read $P_n = 383$ k. This point is in the transition region and we need to determine the strain in the extreme layer of tension reinforcement, ε_t, so we can calculate the value for ϕ. Without showing the details, this point corresponds to a neutral axis depth, $c = 7.81$ in., and for that value $\varepsilon_t = 0.00449$. Using this value in Eq. (4-28a) results in $\phi = 0.858$. Thus, $\phi P_n = 328 \text{ k} > 170 \text{ k}$ (o.k.). Load Case 3 will not govern for most column designs, but it is included here for completeness.

5. Detailing of ties. Because we are using No. 8 longitudinal bars, we can use No. 3 ties. The clear spacing between the longitudinal bars along the long face of the column will exceed 6 in., so we will need an extra tie around those bars, as shown in Fig. 11-31. The tie spacing must satisfy the following:

$$s \leq 16\, d_b(\text{longitudinal bar}) = 16 \times 1.0 \text{ in.} = 16 \text{ in.}$$
$$s \leq 48\, d_b(\text{tie bar}) = 48 \times 3/8 \text{ in.} = 18 \text{ in.}$$
$$s \leq \text{least column dimension} = 16 \text{ in.}$$

So, use 16 in. spacing between tie sets, as shown in Fig. 11-31. ■

11-6 CONTRIBUTIONS OF STEEL AND CONCRETE TO COLUMN STRENGTH

The interaction diagram for a column with reinforcement in two faces can be considered to be the sum of three components: (1) the load and moment resistance of the concrete, and (2) and (3), the load and moment resistance provided by each of the two layers of reinforcement.

In Fig. 11-33, the P_n, M_n interaction diagram from Example 11-1 and Fig. 11-18 is replotted to show the load and moment resistance contributed by the concrete and the two layers of steel. The center curved portion, *O–J–A–G–D–L*, represents the compressive force in the concrete, C_c, and the moment of this force about the centroid of the column. The shaded band represents the force in the "tensile" reinforcement adjacent to the less compressed face, force F_{s1} in Example 11-1, and the moment of this force about the

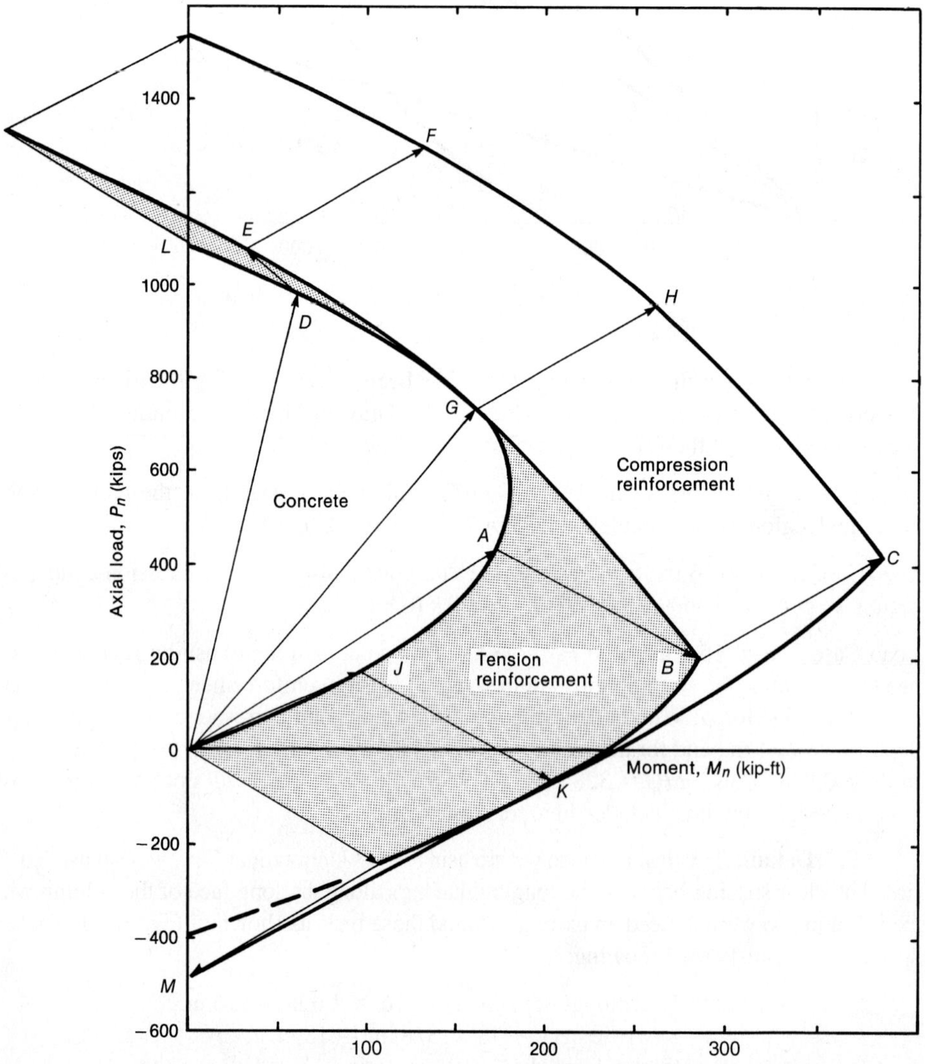

Fig. 11-33
Components of interaction diagram—Example 11-1.

centroid of the column. The outer band is the force F_{s2} in the "compressive" reinforcement adjacent to the most compressed face and its moment about the centroid. A reexamination of Example 11-1 shows that, at the balanced-failure condition, the force C_c was 435 kips and the moment of C_c about the centroid was 174 kip-ft. These are plotted in Fig. 11-33 as the vector *OA*. The force F_{s1} was -240 kips, and its moment about the centroid was 110 kip-ft, plotted as vector *AB*. Finally, the force F_{s2} was 222 kips, and its moment was 102 kip-ft, plotted as vector *BC*. An examination of line *OABC* shows that the majority of the axial-load capacity at the balanced load comes from the concrete, with the forces in the tension and compression steel layers essentially canceling each other out. All three constituents contribute to the moment capacity.

A different case is represented by line *ODEF* in Fig. 11-33. Here, the force F_{s1} is in compression and hence adds to the axial-load capacity and subtracts from the moment capacity, as shown by the vector *DE*. An intermediate case, in which $F_{s1} = 0$, is shown by line *OGH*. A similar case, in which $F_{s2} = 0$, is shown by line *OJK*.

The portion of the interaction diagram due to the concrete is a continuous curve. The discontinuity in the overall interaction diagram at the balanced point is due to the reinforcement. The two branches of the interaction diagram in Figs. 11-18 and 11-33 are relatively straight. This is characteristic of a column with only two layers of steel and a high steel ratio. In the case of low steel ratios, the diagram is much more curved because the concrete portion dominates.

The effects of the individual reinforcement layers can be seen from Fig. 11-33. From the top of the diagram down to about *F*, both layers of steel are effective in increasing the axial capacity. At *H*, only the steel on the compression face is effective. At *C*, both layers are effective in increasing the moment capacity. At *K*, only the reinforcement on the tension face is effective, and finally, at *M*, both layers add to the tensile capacity. To optimize the design of a member subjected to axial load and bending, one may wish to provide different amounts of reinforcement in the two faces. Figure 11-33 can be used as a guide to where the reinforcement should go. Generally speaking, however, columns are built with the same reinforcement in the two faces parallel to the axis of bending: (a) to minimize the chance of putting the bars in the wrong face and (b) because end moments due to wind change sign as a result of winds blowing from alternate sides of the building. In the case of culverts, arch sections, or rigid frame legs, however, it may be desirable to have unsymmetrical reinforcement.

11-7 BIAXIALLY LOADED COLUMNS

Up to this point in the chapter we have dealt with columns subjected to axial loads accompanied by bending about one axis. It is not unusual for columns to support axial forces and bending about two perpendicular axes. One common example is a corner column in a frame.

For a given cross section and reinforcing pattern, one can draw an interaction diagram for axial load and bending about either principal axis. As shown in Fig. 11-34, these interaction diagrams form two edges of a three-dimensional interaction surface for axial load and bending about two axes. The calculation of *each* point on such a surface involves a double iteration: (1) the strain gradient across the section is varied, and (2) the angle of the neutral axis is varied. For the same reasons discussed in Section 4-9, the neutral axis will generally not be parallel to the resultant moment vector [11-13], [11-14]. The calculation of interaction diagrams for biaxially loaded columns is discussed in [11-15].

A horizontal section through such a diagram resembles a quadrant of a circle or an ellipse at high axial loads, and depending on the arrangement of bars, it becomes considerably less circular near the balanced load, as shown in Fig. 11-34.

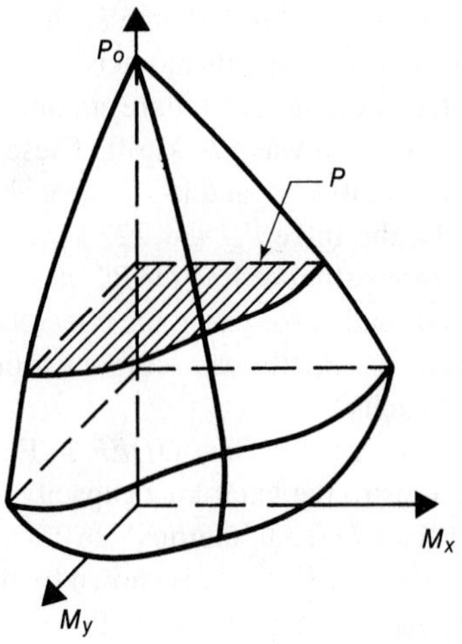

Fig. 11-34
Interaction surface for axial load and biaxial bending.

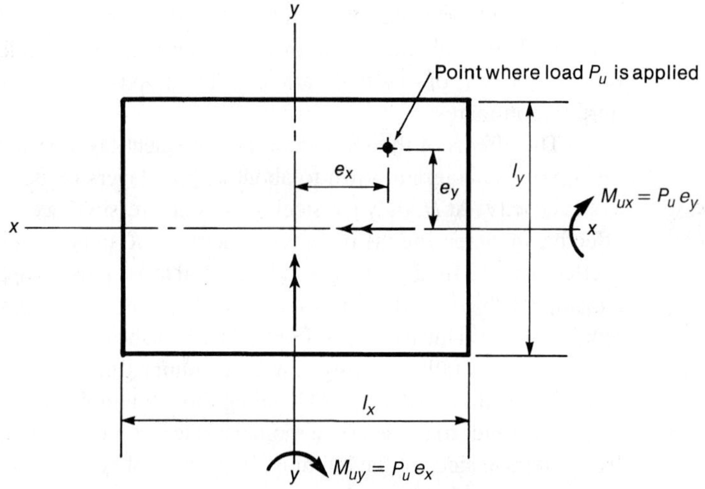

Fig. 11-35
Definition of terms: biaxially loaded columns.

Four procedures commonly used to design rectangular columns subjected to biaxial loads will be illustrated. Notation is defined in Figs. 11-35 and 11-36.

Positive-moment vectors are shown in Fig. 11-35. By this definition of moments, a positive moment M_x causes compression at points with positive y-coordinates, and a positive moment M_y causes compression at points with positive x-coordinates.

In addition to the notation presented earlier, other terms include:

P_u = factored axial load, positive in compression.

e_x = eccentricity of applied load measured parallel to the x-axis, positive to right in the cross section in Fig. 11-35.

e_y = eccentricity of applied load measured parallel to the y-axis, positive upward on cross section in Fig. 11-35.

M_{ux} = factoral moment about x-axis, equal to $P_u e_y$, positive when it causes compression in fibers in the positive y direction.

M_{uy} = factoral moment about y-axis, equal to $P_u e_x$, positive when it causes compression in fibers in the positive x direction.

ϕP_{nx} = reduced nominal axial-load capacity for the moment about the x-axis corresponding to the eccentricity, e_y, and the steel and concrete section provided, with $e_x = 0$.

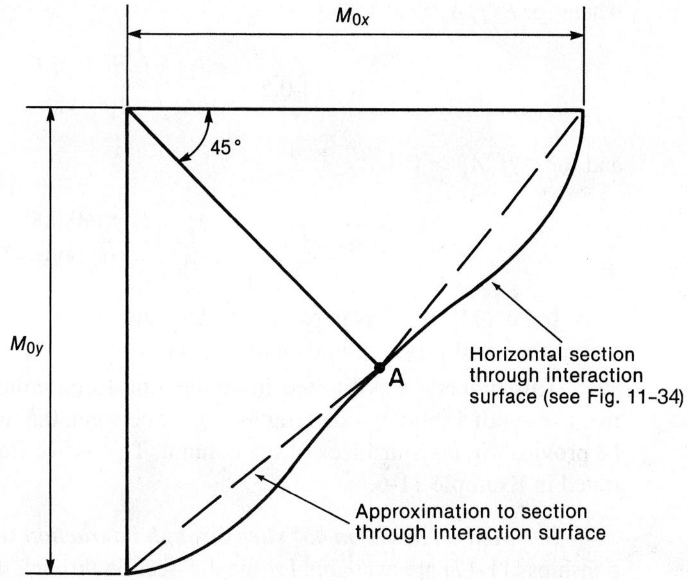

Fig. 11-36
Approximation of section through intersection surface.

ϕP_{ny} = reduced nominal axial-load capacity, for the moment about the y-axis corresponding to the eccentricity, e_x, and the steel and concrete section provided, with $e_y = 0$.

ϕP_{no} = reduced nominal axial-load capacity for the steel and concrete section provided, with both $e_y = 0$ and $e_x = 0$.

ℓ_x = length of side of column parallel to the x-axis.

ℓ_y = length of side of column parallel to the y-axis.

1. ***The strain-compatibility method.*** The use of an iterative strain-compatibility analysis for a biaxially loaded column is illustrated in Example 11-5. This is the most nearly theoretically correct method of solving biaxially loaded column problems presented in this book. Alternatively, design can be based on one of the three widely used approximate design procedures presented after Example 11-5.

2. ***The equivalent eccentricity method.*** The biaxial eccentricities, e_x and e_y, can be replaced by an equivalent uniaxial eccentricity, e_{ox}, and the column designed for uniaxial bending and axial load [11-15], [11-16]. We shall define e_x as the component of the eccentricity parallel to the side l_x and the x-axis, as shown in Fig. 11-35, such that the moments, M_{uy}, and M_{ux}, are

$$M_{uy} = P_u e_x \qquad M_{ux} = P_u e_y \tag{11-27a,b}$$

If

$$\frac{e_x}{\ell_x} \geq \frac{e_y}{\ell_y} \tag{11-28}$$

then the column can be designed for P_u and a factored moment $M_{oy} = P_u e_{ox}$, where

$$e_{ox} = e_x + \frac{\alpha e_y \ell_x}{\ell_y} \tag{11-29}$$

where for $P_u/f'_c A_g \leq 0.4$,

$$\alpha = \left(0.5 + \frac{P_u}{f'_c A_g}\right)\frac{f_y + 40{,}000}{100{,}000} \geq 0.6 \tag{11-30a}$$

and for $P_u/f'_c A_g > 0.4$,

$$\alpha = \left(1.3 - \frac{P_u}{f'_c A_g}\right)\frac{f_y + 40{,}000}{100{,}000} \geq 0.5 \tag{11-30b}$$

In Eq. (11-30), f_y is in psi. If the inequality in Eq. (11-28) is not satisfied, the definition of the x and y axes should be interchanged.

This procedure is limited in application to columns that are symmetrical about two axes with a ratio of side lengths, ℓ_x/ℓ_y, between 0.5 and 2.0. Reinforcement should be provided in all four faces of the column. The use of Eqs. (11-27) to (11-30) is illustrated in Example 11-6.

3. ***Method based on 45° slice through interaction surface.*** Charts [11-16] or relationships [11-17] are available for the 45° section through the interaction surface (M_x and M_y at point A in Fig. 11-36). The design is then based on straight-line approximations to horizontal slices through the interaction surface as shown by the dashed lines in Fig. 11-36.

4. ***Bresler reciprocal load method.*** The following equation, originally presented by Bresler [11-18], is often used for calculating the capacity under biaxial bending:

$$\frac{1}{\phi P_n} = \frac{1}{\phi P_{nx}} + \frac{1}{\phi P_{ny}} - \frac{1}{\phi P_{no}} \tag{11-31}$$

This procedure, widely used to check designs, is illustrated in Example 11-7.

EXAMPLE 11-5 Calculate the Capacity of a Biaxially Loaded Column for a Given Neutral-Axis Location

Compute the nominal axial-load capacity, P_n, and the nominal M_{nx} and M_{ny} moments corresponding to a prescribed strain distribution in the column. Use a 16-in.-by-16-in. tied column with 4000-psi concrete and eight No. 8 Grade-60 bars, with three bars in each face of the column. The neutral-axis position assumed in this example crosses the vertical axis of symmetry of the section (the y-axis) at 10 in. below the top of the section, at an angle of 30° counterclockwise from the x-axis of the cross section. Use the ACI rectangular stress block.

1. Select a sign convention. The cross section of the column is a z-plane. (A z-plane is a plane that is perpendicular to the z-axis.) Positive x and y are to the right and upward, respectively. Positive moments cause compression at positive x locations and positive y locations. Compression is positive.

2. Locate the neutral axis. Figure 11-37a shows the cross section and the specified neutral axis (axis of zero strain). The origin is 8 in. below the top and 8 in. from the left side. The neutral axis crosses the y-axis at 2 in. below the origin and intersects the left and right sides of the section at 14.62 in. and 5.38 in., respectively, below the top of the section. The extreme compression fiber, A, is located in the upper-left corner of the section. The perpendicular distance from this fiber to the neutral axis is

$$c_{\text{incl}} = 14.62 \cos 30° = 12.66 \text{ in.}$$

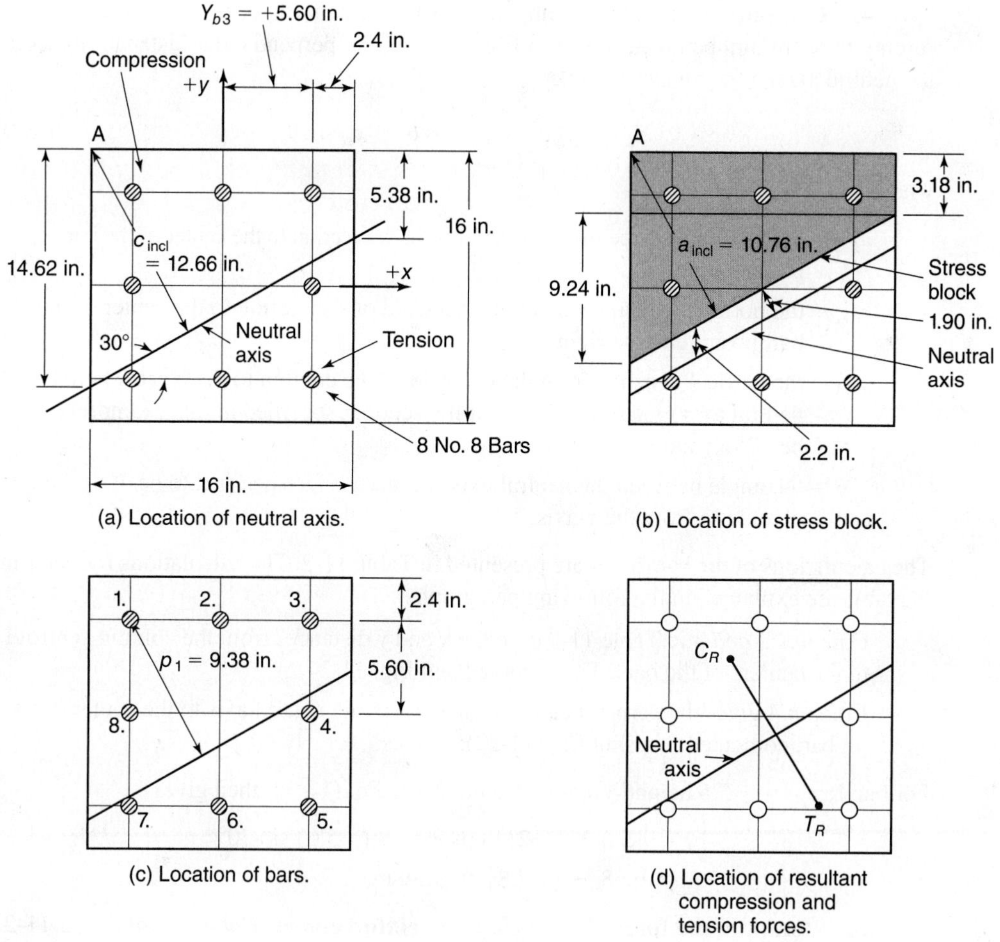

Fig. 11-37
Location of neutral axis and stress block—Example 11-5.

3. Locate the stress block. The stress block extends from corner *A* to a line at $(\beta_1 \times c_{\text{incl}}) = a_{\text{incl}} = 10.76$ in. $(\beta_1 = 0.85)$ from corner *A* measured perpendicular to the neutral axis. The column capacity is assumed to be reached when the strain at *A* reaches a compressive strain of 0.003. A uniform compression stress of $0.85f'_c$ acts on the shaded area in Fig. 11-37b. To simplify the calculation of the forces in the concrete, the area loaded by the stress-block is divided into a rectangle and a triangle, with dimensions as shown in Fig. 11-37b as follows:

(a) Rectangular portion, height = 3.18 in., width = 16 in., area = 50.9 in.^{2} with centroid located at

$$x = 0, \text{ and}$$

$$y = 8.0 - 3.18/2 \text{ in.} = 6.41 \text{ in.}$$

(b) Triangular portion, height = 9.24 in., width = 16 in., area = 73.9 in.^{2}, with centroid located at

$$x = 8 - 16/3 = -2.67 \text{ in. to the left of the origin, and}$$

$$y = 8 - 3.18 - 9.24/3 = 1.74 \text{ in. above origin.}$$

4. Compute the distance from the bars to the neutral axis. The individual reinforcing bars are numbered as shown in Fig. 11-37c. The perpendicular distance, p_i, from the neutral axis to the ith bar is

$$p_i = (y_{si} - y_{NA}) \cos\theta - x_{si} \sin\theta \quad (11\text{-}32)$$

where

y_{si} = the vertical distance from the centroid of the section to the center of the bar, positive upward

x_{si} = the horizontal distance from the centroid of the section to the center of the bar, positive to the right

y_{NA} = the vertical distance from the centroid of the column to the place where the neutral axis crosses the y-axis of the section, specified in this example to be –2 in.; and

θ = the angle between the neutral axis and the x-axis, specified to be 30° counterclockwise from the x-axis

The calculations of the bar forces are presented in Table 11-2. The calculations for bar 1 in the table are explained in the following paragraphs:

Columns 2 *and* 3 in Table 11-2 give the x and y distances from the column centroid to the centers of the bars. These come from Fig. 11-37c.

Column 4 gives the perpendicular distance from the neutral axis to the center of the ith bar, computed by using Eq. (11-32);

For bar 1, $x_{s1} = -5.6$ in. and $y_{s1} = +5.6$ in. Using Eq. (11-32) then gives

$$\begin{aligned} p_1 &= [5.6 - (-2)] \cos 30° - (-5.6) \sin 30° \\ &= 6.58 - (-2.8) = 9.38 \text{ in.} \end{aligned}$$

5. Compute the forces in the column reinforcement. *Column* 5 of Table 11-2, from similar triangles, gives the strain in the reinforcing bar 1 as

$$\varepsilon_{si} = \frac{p_i}{c_{\text{incl}}} \times 0.0030 = \frac{9.38}{12.66} \times 0.0030 = 0.00222$$

TABLE 11-2 Computation of Bar Forces in Biaxially Loaded Column

1	2	3	4	5	6	7
Bar(i)	x_{si} in.	y_{si} in.	Perpendicular Distance = p_i in.	Steel Strain, ϵ_{si}	Steel Stress, ksi	Corrected Steel Force $P_{n,si}$, kips
1	−5.6	+5.6	9.38	0.00222	60	44.7
2	0	+5.6	6.58	0.00156	45.2	33.0
3	+5.6	+5.6	3.78	0.000896	26.0	17.8
4	+5.6	0	−1.07	−0.000254	−7.3	−5.8
5	+5.6	−5.6	−5.92	−0.00140	−40.7	−32.1
6	0	−5.6	−3.12	−0.000739	−21.4	−16.9
7	−5.6	−5.6	−0.32	−0.000076	−2.2	−1.7
8	−5.6	0	4.53	0.00107	31.1	21.9

Column 6 gives the stress in reinforcing bar 1 as

$$f_{si} = \varepsilon_{si} \times E_s = 0.00222 \times 29{,}000$$
$$= 64.5 \text{ ksi}$$

but not more than $f_y = 60$ ksi not less than -60 ksi. Therefore, $f_{s1} = 60$ ksi.

If p_i exceeds the distance from the neutral axis to the lower edge of the rectangular stress block (2.2 in. vertically; see Fig. 11-37b), the bar displaces concrete in the compression stress block, an effect that will be corrected for by subtracting the force in the displaced concrete. Otherwise, $F_{si} = f_{si} \times A_{si}$.

For bar number 1,

$$F_{si} = (f_{si} - 0.85 f'_c)\, A_{si}$$
$$F_{s1} = (60 - 0.85 \times 4 \text{ ksi}) \times 0.79 \text{ in.}$$
$$= 44.7 \text{ kips}$$

6. Compute concrete forces and moments.

Concrete forces:

Area 1—Rectangle: $C_{c1} = 0.85 \times f'_c \times \text{area} = 0.85 \times 4 \times 50.9 = 173$ kips.

Area 2—Triangle: $C_{c2} = 251$ kips.

The calculation of moments is done in Table 11-3. The center of compression is in the upper-left quadrant of the section. From the sign convention,

M_x = positive, because it causes compression stress at positive y-coordinates (in the upper half of the cross section), and

M_y = negative, because it causes compression in the left half of the cross section.

7. Compute ϕ factor, ϕP_n, ϕM_{nx}, and ϕM_{ny}. From step 5, calculate the extreme tension steel strain in bar 5, $\varepsilon_t = -0.00140$, where $-$ is tension. Because this is less than the yield strain in tension, this is a compression-controlled point and $\phi = 0.65$.

The sums of columns 2, 4, and 6 of Table 11-3 are multiplied by $\phi = 0.65$, giving $\phi P_n = 315$ kips, $\phi M_{nx} = 1540$ kip-in, and $\phi M_{ny} = -748$ kip-in.

8. Summary. The nominal load capacity and factored nominal load capacity are

$$\mathbf{P_n = 485 \text{ kips}}$$
$$\mathbf{\phi P_n = 315 \text{ kips}}$$
$$\mathbf{M_{nx} = 2370 \text{ kip-in}}$$
$$\mathbf{\phi M_{nx} = 1540 \text{ kip-in}}$$

M_{nx} causes compression at $+y$ elements.

$$\mathbf{M_{ny} = -1150 \text{ kip-in}}$$
$$\mathbf{\phi M_{ny} = -748 \text{ kip-in}}$$

■

The nominal moments for Example 11-5 act around the x- and y-axes. These moments can be combined into a total moment vector. This vector has a magnitude of

TABLE 11-3 Computation of Moments of Bar Forces and Concrete Forces about *x*- and *y*-Axes of the Column Cross Section

1	2	3	4	5	6
Bar (i) and Concrete	Forces, $P_{n,si}$ and $P_{n,ci}$, kips	x_{si} in.	$M_{ny,si} = P_{n,si} \times x_{si}$ and $M_{ny,ci} = P_{n,ci} \times x_{ci}$ kip-in.	y_{si} in.	$M_{nx,si} = P_{sn,si} \times y_{si}$ and $M_{nx,ci} = P_{cn,si} \times y_{n,ci}$ kip-in.
1	44.7	−5.6	−250	+5.6	250
2	33.0	0	0	+5.6	185
3	17.8	+5.6	100	+5.6	100
4	−5.8	+5.6	−32	0	0
5	−32.1	+5.6	−180	−5.6	180
6	−16.9	0	0	−5.6	95
7	−1.7	−5.6	10	−5.6	10
8	21.9	−5.6	−123	0	0
Concrete Rectangle	173	0	0	6.41	1109
Concrete Triangle	251	−2.67	−670	1.74	437
	Sum = P_n 485 kips		**Sum = M_{ny} −1150 kip-in.**		**Sum = M_{nx} 2370 kip-in.**

ϕM_n = 1710 kip-in and is at an angle of 25.9° with the *x*-axis. In a rectangular column with uniaxial bending about the *x*-axis, the moment vector, M_{nx}, is parallel to the neutral axis. In the general case of biaxial bending of a noncircular, biaxially loaded section, the moment vector is at an angle with the neutral axis. In this example, the difference between the direction of the moment vector and the direction of the assumed neutral axis was small (25.9° compared with 30°). This closeness occurred because the column was square with bars in all four faces.

The internal load components can be combined into a resultant compression force, C_R, by adding all the axial forces, P_n, and the moments M_{nx} and M_{ny} of the individual bars and areas that are in compression. The same is done for the elements stressed in tension forces to get the resultant tension force, T_R. Together, these allow the eccentricities, e_x and e_y, to be calculated.

From column 2 *of Table 11-3*, Σ(compression-resisting forces) $= C_R = 541$ kips
From column 4,

$$\Sigma M_{ny} \text{ for compression-resisting forces} = -943 \text{ kip-in}$$

$$e_x = -943/541 = -1.74 \text{ in.}$$

By inspection of the location of those loads and compressed areas of the column, which are concentrated in the upper left-hand quadrant, e_x must be in the negative *x* direction.

From column 6,

$$\Sigma M_{nx} \text{ for compression-resisting forces} = 2080 \text{ kip-in}$$

$$e_y = 2080/541 = +3.85 \text{ in.}$$

By inspection of the location of those loads and tensile-bar areas of the column, which are located in the lower right-hand quadrant, e_y must be in the positive y direction. Thus, the resultant compression-resisting force, $C_R = 541$ kips, acts at $e_y = 3.85$ in. above and $e_x = -1.74$ in. to the left of the centroid.

From column 2,

$$\Sigma(\text{tension-resisting forces}) = T_R = -56.5 \text{ kips}$$

From column 4,

$$\Sigma M_{ny} \text{ for tension-resisting forces} = -202 \text{ kip-in}$$

$$e_x = -202/-56.5 = 3.58 \text{ in. to the right of the } y\text{-axis}$$

From column 6,

$$\Sigma M_{nx} \text{ for tension-resisting forces} = 285 \text{ kip-in}$$

$$e_y = 285/-56.5 = -5.04 \text{ in. below the } x\text{-axis.}$$

Thus, the resultant tension-resisting force, $T_R = -56.5$ kips, acts at $e_x = 3.58$ in. to the right of the y-axis, and $e_y = -5.04$ in. (below the x-axis).

The line joining the points where C_R and T_R act has a horizontal projection of 1.74 + 3.58 in. = 5.32 in. and a vertical projection of 3.85 in. + 5.04 in. = 8.89 in.

The angle of the line joining these points is arctan (8.89/5.32) = 59.1°. For a square column with bars in all four sides, the line joining the points where the resultants C_R and T_R act should be 60° (perpendicular to the specified 30° neutral axis). The discrepancy is due to round-off errors. The resultant compressive and tensile forces, C_R, and T_R, must lie in a plane with the applied load, P_n.

In our solution for the moments, the moment vector is at an angle of 30° with the neutral axis. If one were using an iterative solution to get the position of the neutral axis for a given set of load and moments, it would be necessary to iterate two variables: (i) the distance, d_n, along the y-axis from the intersection of the neutral axis and the top of the beam and (ii) the angle, θ, between the neutral axis and the x-axis. Warner et al. [11-13] observed that P_n is affected more strongly by the height of the intercept, d_n, than by the angle, θ, and vice-versa for the angle θ and the resultant of the applied moments. They suggest following these steps in solving for the forces in a biaxial-bending problem:

1. Select the cross section.

2. Choose the depth, d_n, to the intersection of the neutral axis with the y-axis of the section and the angle, θ, between the neutral axis and the x-axis of the section.

3. Follow the computations in this example to get an initial set of P_n, M_{nx}, and M_{ny} values.

4. Keep θ constant, and iterate d_n until the computed P_n is close to the target value.

5. Iterate the angle θ until M_{nx} and M_{ny} approach the target values.

EXAMPLE 11-6 Design of a Biaxially Loaded Column: Equivalent Eccentricity Method

Select a tied column cross section to resist factored loads and moments of $P_u = 250$ kips, $M_{ux} = 55$ kip-ft, and $M_{uy} = 110$ kip-ft. Use $f_y = 60$ ksi and $f'_c = 4$ ksi. The first approximate procedure, based on Eqs. (11-27) to (11-30a), will be used.

1. Select a trial section. Assume that $\rho_g = 0.015$. Use a section with bars in the four faces, because the column is loaded biaxially. Then

$$A_{g(\text{trial})} \geq \frac{P_u}{0.40(f'_c + f_y\rho_g)}$$

$$\geq \frac{250}{0.40(4 + 60 \times 0.015)}$$

$$\geq 128 \text{ in.}^2 \text{ or } 11.3 \text{ in. square}$$

Because this column is subjected to biaxial bending, try a 16-in.-square column with eight No. 8 bars, three in each face.

2. Compute γ.

$$\gamma = \frac{(16 - 2 \times 2.5)}{16} = 0.69$$

3. Compute e_x, e_y, and e_{ox}. From the definition of the moments and eccentricities in Fig. 11-35,

$$e_x = \frac{M_{uy}}{P_u}$$

$$= \frac{110 \times 12}{250} = 5.28 \text{ in.}$$

$$e_y = \frac{M_{ux}}{P_u}$$

$$= \frac{55 \times 12}{250} = 2.64 \text{ in.}$$

By inspection, $e_x/\ell_x \geq e_y/\ell_y$; therefore, use Eq. (11-29) as given. If this were not true, you would transpose the x- and y-axes before using it. In any event, we have

$$\frac{P_u}{f'_c A_g} = \frac{250}{4 \times 256} = 0.244 < 0.4$$

Therefore, use (11-30a) to compute α:

$$\alpha = \left(0.5 + \frac{P_u}{f'_c A_g}\right)\left(\frac{f_y + 40{,}000}{100{,}000}\right) \tag{11-30a}$$

$$= (0.5 + 0.244)\left(\frac{60{,}000 + 40{,}000}{100{,}000}\right)$$

$$= 0.744$$

From (11-29),

$$e_{ox} = e_x + \frac{\alpha e_y \ell x}{\ell_y}$$

$$= \left(5.28 + 0.744 \times 2.64 \times \frac{16}{16}\right) = 7.24 \text{ in.}$$

Thus, the equivalent uniaxial moment is

$$M_{oy} = P_u e_{ox}$$

$$= 250 \times 7.24 = 1810 \text{ kip-in.}$$

The column is designed for uniaxial bending for $P_u = 250$ kips and the equivalent moment, $M_{oy} = 1810$ kip-in.

4. Use interaction diagrams to determine ρ_g. Because the column has biaxial bending, we will select a section with bars in four faces. The interaction diagrams are entered with

$$\frac{P_u}{A_g} = \frac{250}{256} = 0.977 \text{ ksi}$$

and

$$\frac{M_{oy}}{A_g h} = \frac{1810}{16^3} = 0.442 \text{ ksi}$$

From Figs. A-9a and A-9b,

$$\text{for } \gamma = 0.60, \rho_g = 0.025$$

$$\text{for } \gamma = 0.75, \rho_g = 0.018$$

By linear interpolation, $\rho_g = 0.021$ for $\gamma = 0.69$.

5. Compute A_{st} and select the reinforcement.

$$A_{st} = \rho_g A_g = 0.021 \times 256 = 5.38 \text{ in.}^2$$

Select eight No. 8 bars, three in each face, A_{st} **= 6.32 in.²**. Design ties and lap splices as in earlier examples. ■

EXAMPLE 11-7 Check Design of a Biaxially Loaded Column: Bresler Reciprocal Load Method

Continue Example 11-6 and use Eq. (11-31) to check the section design.

1. Review trial section. Trial section is a 16-in.-square column, $f'_c = 4$ ksi, $f_y = 60$ ksi, eight No. 8 bars (three in each face), and No. 3 ties.

2. Compute γ. From step 2 in Example 11-6:

$$\gamma = 0.69$$

3. Compute ϕP_{nx}. ϕP_{nx} is the factored axial-load capacity corresponding to e_x and ρ_g. We have

$$\rho_g = \frac{8 \times 0.79}{16 \times 16} = 0.0247$$

$$\frac{e_x}{\ell_x} = \frac{M_{uy}}{P_u \ell_x} = \frac{110 \text{ kip-ft} \times 12}{250 \text{ kips} \times 16 \text{ in.}}$$
$$= 0.330$$

From the interaction diagram, Fig. A-9a, e/h $(e_x/\ell_x) = 0.330$ and $\rho_g = 0.0247$ gives $(\phi P_{nx}/bh) = 1.30$ for $\gamma = 0.60$. From Fig. A-9b, $(\phi P_{nx}/bh) = 1.40$ for $\gamma = 0.75$. Interpolating for $\gamma = 0.69$ gives $(\phi P_{nx}/bh) = 1.36$ and $\phi P_{nx} = 348$ kips.

4. Compute ϕP_{ny}.

$$\frac{e_y}{\ell_y} = \frac{M_{ux}}{P_u \ell_y} = \frac{55 \times 12}{250 \times 16} = 0.165$$

From Fig. A-9a, $(\phi P_{ny}/bh) = 2.10$ for $\gamma = 0.60$. From Fig. A-9b, $(\phi P_{ny}/bh) = 2.20$ for $\gamma = 0.75$. Interpolating for $\gamma = 0.69$ gives $(\phi P_{ny}/bh) = 2.16$ and $\phi P_{ny} = 553$ kips.

5. Compute ϕP_{no}. Use Eq. (11-8b) to calculate $P_{no} = 1230$ kips. Thus, $\phi P_{no} = 0.65 \times 1230 = 798$ kips.

6. Solve for ϕP_n.

$$\frac{1}{\phi P_n} = \frac{1}{\phi P_{nx}} + \frac{1}{\phi P_{ny}} - \frac{1}{\phi P_{no}}$$
$$= \frac{1}{348} + \frac{1}{553} - \frac{1}{798}$$
$$\phi P_n = 292 \text{ kips}$$

The required capacity is 250 kips; therefore, the column design is adequate. **Use eight No. 8 bars, three in each face, $A_{st} = 6.32$ in.^{2}.** ■

Examples 11-5, 11-6, and 11-7 all dealt with the same column and essentially the same loads. The strain-compatibility solution in Example 11-5 computed the axial load and bending strengths about two axes for a specified 16-in.-square column with eight No. 8 bars, three per side, for one assumed neutral-axis position. This gave

$$P_n = 485 \text{ kips}$$
$$\phi P_n = 315 \text{ kips}$$
$$M_{nx} = 2370 \text{ kip-in.}$$
$$\phi M_{nx} = 1540 \text{ kip-in.}$$
$$M_{ny} = -1150 \text{ kip-in.}$$
$$\phi M_{ny} = -713 \text{ kip-in.}$$

The equivalent eccentricity method was used in Example 11-6, to check whether the same cross section could support

$$P_u = 250 \text{ kips}$$
$$M_{ux} = 55 \text{ kip-ft} = 660 \text{ kip-in.}$$
$$M_{ny} = 110 \text{ kip-ft} = 1320 \text{ kip-in.}$$

After the two moments were converted to an equivalent uniaxial moment about the y-axis, interaction diagrams for ϕP_n and ϕM_n were used to compute the strength. This solution indicated that a 16-in.-square column with eight No.8 bars was adequate to resist these loads and moments.

The Bresler reciprocal load method was used in Example 11-7 to check if the selected cross section was safe for the same loading conditions used in Example 11-6. Interaction diagrams in Figs. A-9a and A-9b were used to determine ϕP_{nx} and ϕP_{ny}, and Eq. (11-8b) was used to find ϕP_{no}. The check was then completed using Eq. (11-31), which indicated that the selected column section was adequate.

PROBLEMS

11-1 The column shown in Fig. P11-1 is made of 5000-psi concrete and Grade-60 steel.

(a) Compute the theoretical capacity of the column for pure axial load.
(b) Compute the maximum permissible ϕP_n for the column.

11-2 Why does a spiral improve the behavior of a column?

11-3 Why are tension splices required in some columns?

11-4 Compute the balanced axial-load and moment capacity about the x-axis of the column shown in Fig. P11-1. Use $f'_c = 4000$ psi and $f_y = 60{,}000$ psi.

11-5 For the column shown in Fig. P11-5, use a strain-compatibility solution to compute five points on the interaction diagram corresponding to points 1 to 5 in Fig. 11-22. Plot the interaction diagram. Use $f'_c = 5000$ psi and $f_y = 60{,}000$ psi.

11-6 Use the interaction diagrams in Appendix A to compute the maximum moment, M_{ux}, that can be supported by the column shown in Fig. P11-1 if

(a) $P_u = 583$ kips.
(b) $P_u = 130$ kips.
(c) $e = 4$ in.

Use psi $f'_c = 5000$ and $f_y = 60{,}000$ psi.

11-7 Video Solution Use the interaction diagrams in Appendix A to select tied-column cross sections to support the loads given in the accompanying list. In each case, use $f'_c = 4000$ psi and $f_y = 60{,}000$ psi. Design the ties and calculate the required splice lengths, assuming that the bars extending up from the column below are the same diameter as in the column you have designed. Draw a typical cross section of the column showing the bars and ties.

(a) $P_u = 390$ kips, $M_u = 220$ kip-ft, square column with bars in two faces.

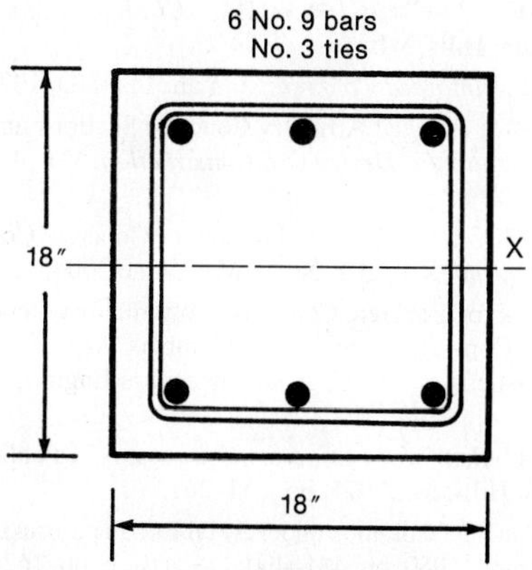

Fig. P11-1

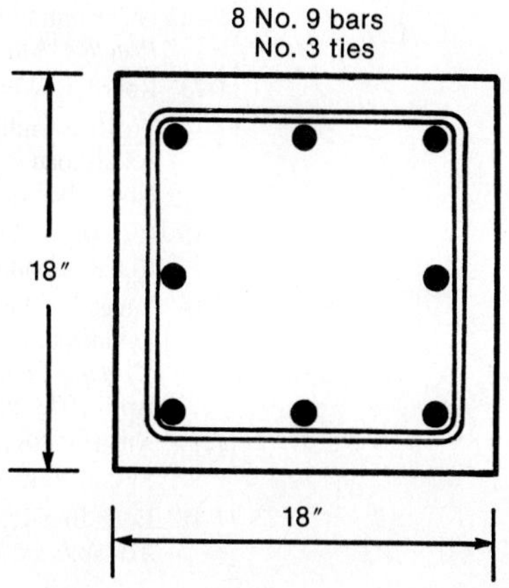

Fig. P11-5

(b) $P_u = 710$ kips, $M_u = 50$ kip-ft, square column with bars in four faces.

(c) $P_u = 200$ kips, $M_u = 240$ kip-ft, square column with bars in four faces.

11-8 Use the interaction diagrams in Appendix A to select spiral-column cross sections to support the loads given in the accompanying list. In each case, use $f'_c = 5000$ psi and $f_y = 60{,}000$ psi, design the spirals and calculate the required splice lengths for the longitudinal bars. Draw a typical cross section of the column showing the bars and spiral.

Video Solution

(a) $P_u = 600$ kips, $M_u = 65$ kip-ft.

(b) $P_u = 200$ kips, $M_u = 150$ kip-ft.

11-9 Design a square cross section and reinforcement to support $P_u = 450$ kips, $M_{ux} = 100$ kip-ft, and $M_{uy} = 130$ kip-ft. Use $f'_c = 5000$ psi and $f_y = 60{,}000$ psi.

REFERENCES

11-1 Eivind Hognestad, *A Study of Combined Bending and Axial Load in Reinforced Concrete Members*, Bulletin 399, University of Illinois Engineering Experiment Station, Urbana, IL., June 1951, 128 pp.

11-2 ACI Committee 105, "Reinforced Concrete Column Investigation," *ACI Journal, Proceedings*, Vol. 26, April 1930, pp. 601–612; Vol. 27, February 1931, pp. 675–676; Vol. 28, November 1931, pp. 157–578; Vol. 29, September 1932, pp. 53–56; Vol. 30, September–October 1933, pp. 78–90; November–December 1933, pp. 153–156.

11-3 ACI-ASCE Committee 327, "Report on Ultimate Strength Design," *Proceedings ASCE*, Vol. 81, October 1955, Paper 809. See also *ACI Journal, Proceedings*, Vol. 2, No. 7, January 1956, pp. 505–524.

11-4 Robert F. Mast, "Unified Design Provision for Reinforced and Prestressed Concrete Members," *ACI Structural Journal*, Vol. 89, No. 2, March–April, pp. 185–199.

11-5 Albert C. Bianchini, R. E. Woods, and Clyde E. Kesler, "Effect of Floor Concrete Strength on Column Strength," *ACI Journal, Proceedings*, Vol. 56, No. 11, May 1960, pp. 1149–1169.

11-6 Carlos E. Ospina and Scott D. B. Alexander, "Transmission of Interior Concrete Column Loads through Floors," *ASCE Journal of Structural Engineering*, Vol. 124, No. 6, 1998.

11-7 J. F. Pfister, "Influence of Ties on the Behavior of Reinforced Concrete Columns," *ACI Journal, Proceedings*, Vol. 61, No. 5, May 1964, pp. 521–537.

11-8 Fred M. Hudson, "Reinforced Concrete Columns: Effect of Lateral Tie Spacing on Ultimate Strength," *ACI Special Publication SP 13*, American Concrete Institute, Farmington Hills, MI, pp. 235–244, 1966.

11-9 P. H. Ziehl, J. E. Cloyd, and Michael E. Kreger, "Investigation of Minimum Longitudinal Reinforcements for Concrete Columns Using Present-Day Construction Materials," *ACI Structural Journal*, Vol. 101, No. 2, March–April 2004, pp. 165–175.

11-10 ACI Committee 340, *Design Handbook Volume 2—Columns* (ACI 340.2R-91), ACI Publication SP-17A(90), American Concrete Institute, Farmington Hills, MI, 250 pp.

11-11 Boris Bresler and P.H. Gilbert, "Tie Requirements for Reinforced Concrete Columns," *ACI Journal*, Vol. 58, No. 5, November 1961, pp. 555–570.

11-12 ACI Committee 363, "Report on High-Strength Concrete" (ACI 363-R10), *ACI Manual of Concrete Practice*, American Concrete Institute, Farmington Hills, MI.

11-13 Robert F. Warner, B. V. Rangan, and A. S. Hall, *Reinforced Concrete*, Pitman Australia, 1976, 475 pp.

11-14 Troels Brondiem-Nielsen, "Ultimate Limit States of Cracked Arbitrary Concrete Sections under Axial Load and Biaxial Loading," *Concrete International: Design and Construction*, Vol. 4, No. 11, November 1982, pp. 51–55.

11-15 Richard W. Furlong, C.-T. T. Hsu, and Sher A. Mirza, "Analysis and Design of Concrete Columns for Biaxial Bending—Overview," *ACI Structural Journal*, Vol. 101, No. 3, May–June 2004, pp. 413–423.

11-16 James G. MacGregor, "Simple Design Procedures for Concrete Columns," Introductory Report, Symposium on Design and Safety of Reinforced Concrete Compression Members, *Reports of the Working Commissions*, Vol. 15, International Association of Bridge and Structural Engineering, Zurich, April 1973, pp. 23–49.

11-17 Albert J. Gouwens, "Biaxial Bending Simplified," *Reinforced Concrete Columns*, ACI Publication SP-50, American Concrete Institute, Farmington Hills, MI, 1975, pp. 233–261.

11-18 Boris Bresler, "Design Criteria for Reinforced Concrete Columns under Axial Load and Biaxial Bending," *ACI Journal, Proceedings*, Vol. 57, No. 5, November 1960, pp. 481–490; Discussion, pp. 1621–1638.

12

Slender Columns

12-1 INTRODUCTION

Definition of a Slender Column

An eccentrically loaded, pin-ended column is shown in Fig. 12-1a. The moments at the ends of the column are

$$M_e = Pe \tag{12-1}$$

When the loads P are applied, the column deflects laterally by an amount δ, as shown. For equilibrium, the internal moment at midheight is (Fig. 12-1b)

$$M_c = P(e + \delta) \tag{12-2}$$

The deflection increases the moments for which the column must be designed. In the symmetrical column shown here, the maximum moment occurs at midheight, where the maximum deflection occurs.

Figure 12-2 shows an interaction diagram for a reinforced concrete column. This diagram gives the combinations of axial load and moment required to cause failure of a column cross section or a very short length of column. The dashed radial line *O–A* is a plot of the end moment on the column in Fig. 12-1. Because this load is applied at a constant eccentricity, e, the end moment, M_e, is a linear function of P, given by Eq. (12-1). The curved, solid line *O–B* is the moment M_c at midheight of the column, given by Eq. (12-2). At any given load P, the moment at midheight is the sum of the end moment, Pe, and the moment due to the deflections, $P\delta$. The line *O–A* is referred to as a *load–moment curve* for the end moment, while the line *O–B* is the load–moment curve for the maximum column moment.

Failure occurs when the load–moment curve *O–B* for the point of maximum moment intersects the interaction diagram for the cross section. Thus the load and moment at failure are denoted by point *B* in Fig. 12-2. Because of the increase in maximum moment due to

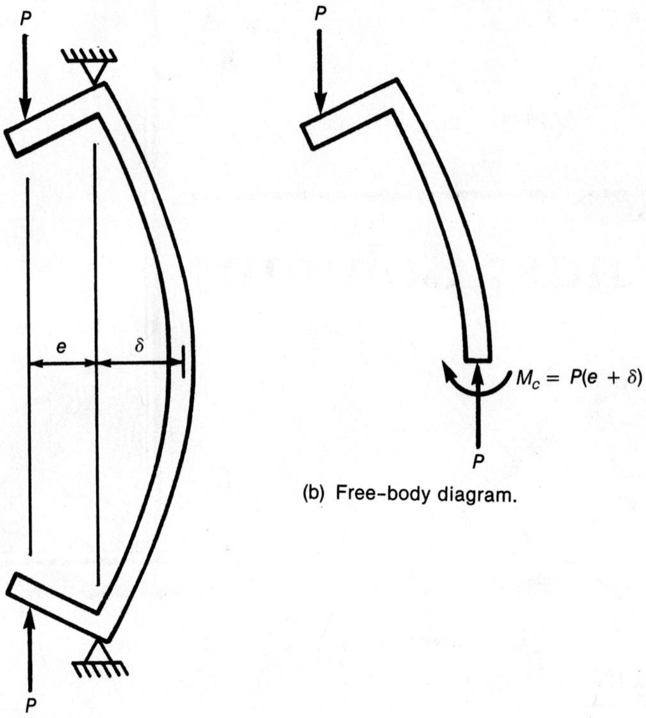

(b) Free-body diagram.

(a) Column.

Fig. 12-1
Forces in a deflected column.

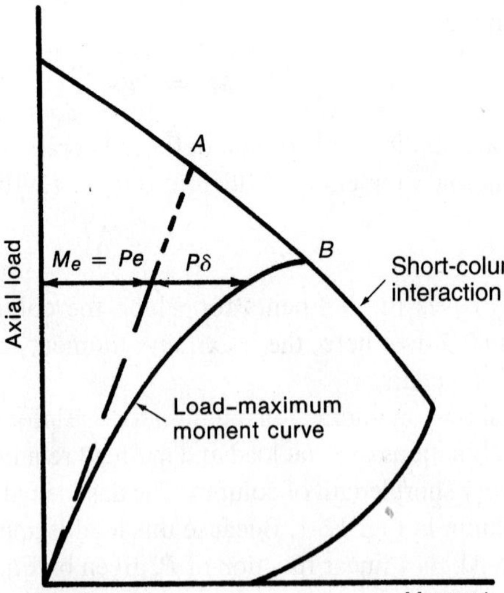

Fig. 12-2
Load and moment in a column.

deflections, the axial-load capacity is reduced from A to B. This reduction in axial-load capacity results from what are referred to as *slenderness effects*.

A *slender column* is defined as a column that has a significant reduction in its axial-load capacity due to moments resulting from lateral deflections of the column. In the derivation of the ACI Code, "a significant reduction" was arbitrarily taken as anything greater than about 5 percent [12-1].

Buckling of Axially Loaded Elastic Columns

Figure 12-3 illustrates three states of equilibrium. If the ball in Fig. 12-3a is displaced laterally and released, it will return to its original position. This is *stable equilibrium.* If the ball in Fig. 12-3c is displaced laterally and released, it will roll off the hill. This is *unstable equilibrium.* The transition between stable and unstable equilibrium is *neutral equilibrium,* illustrated in Fig. 12-3b. Here, the ball will remain in the displaced position. Similar states of equilibrium exist for the axially loaded column in Fig. 12-4a. If the column returns to its original position when it is pushed laterally at midheight and released, it is in stable equilibrium.

Figure 12-4b shows a portion of a column that is in a state of neutral equilibrium. The differential equation for this column is

$$EI\frac{d^2y}{dx^2} = -Py \tag{12-3}$$

In 1744, Leonhard Euler derived Eq. (12-3) and its solution,

$$P_c = \frac{n^2\pi^2EI}{\ell^2} \tag{12-4}$$

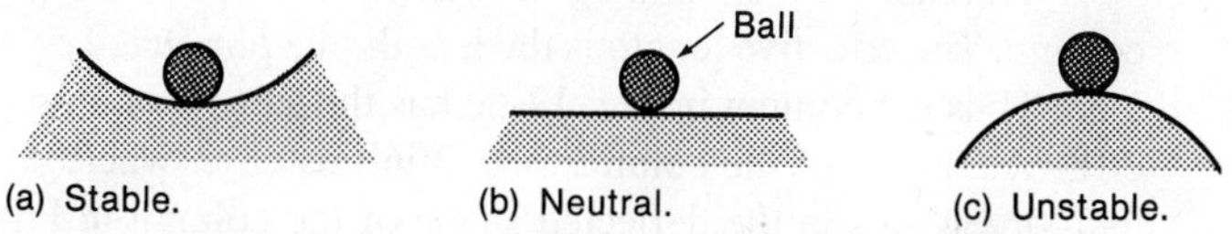

Fig. 12-3
States of equilibrium.

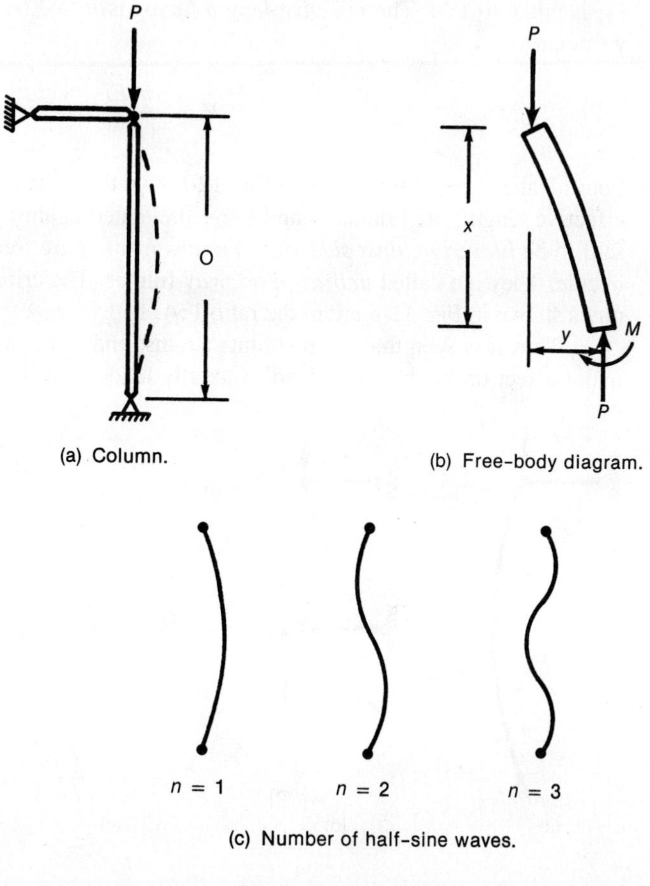

Fig. 12-4
Buckling of a pin-ended column.

where

EI = flexural rigidity of column cross section
ℓ = length of the column
n = number of half-sine waves in the deformed shape of the column

Cases with $n = 1, 2$, and 3 are illustrated in Fig. 12-4c. The lowest value of P_c will occur with $n = 1.0$. This gives what is referred to as the *Euler buckling load*:

$$P_E = \frac{\pi^2 EI}{\ell^2} \tag{12-5}$$

Such a column is shown in Fig. 12-5a. If this column were unable to move sideways at midheight, as shown in Fig. 12-5b, it would buckle with $n = 2$, and the buckling load would be

$$P_c = \frac{2^2 \pi^2 EI}{\ell^2}$$

which is four times the critical load of the same column without the midheight brace.

Another way of looking at this involves the concept of the *effective length* of the column. The effective length is the length of a pin-ended column having the same buckling load. Thus the column in Fig. 12-5c has the same buckling load as that in Fig. 12-5b. The effective length of the column is $\ell/2$ in this case, where $\ell/2$ is the length of each of the half-sine waves in the deflected shape of the column in Fig. 12-5b. The effective length, $k\ell$, is equal to ℓ/n. The *effective length factor* is $k = 1/n$. Equation (12-4) is generally written as

$$P_c = \frac{\pi^2 EI}{(k\ell)^2} \tag{12-6}$$

Four idealized cases are shown in Fig. 12-6, together with the corresponding values of the effective length, $k\ell$. Frames *a* and *b* are prevented against deflecting laterally. They are said to be *braced against sidesway*. Frames *c* and *d* are free to sway laterally when they buckle. They are called *unbraced* or *sway* frames. The critical buckling loads of the columns shown in Fig. 12-6 are in the ratio $1:4:1:\frac{1}{4}$.

Thus it is seen that the restraints against end rotation and lateral translation have a major effect on the buckling load of axially loaded elastic columns. In actual structures,

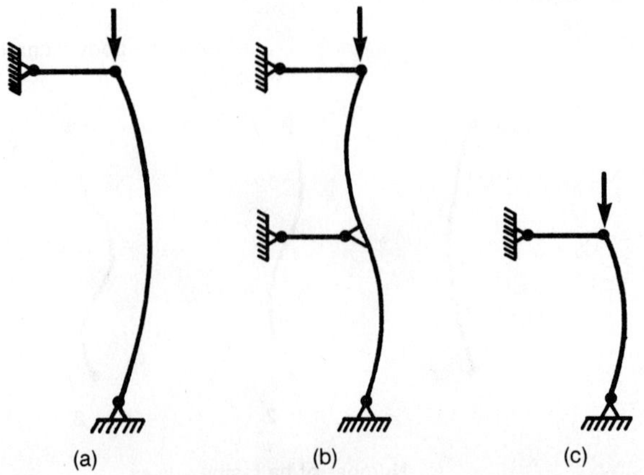

Fig. 12-5
Effective lengths of columns.

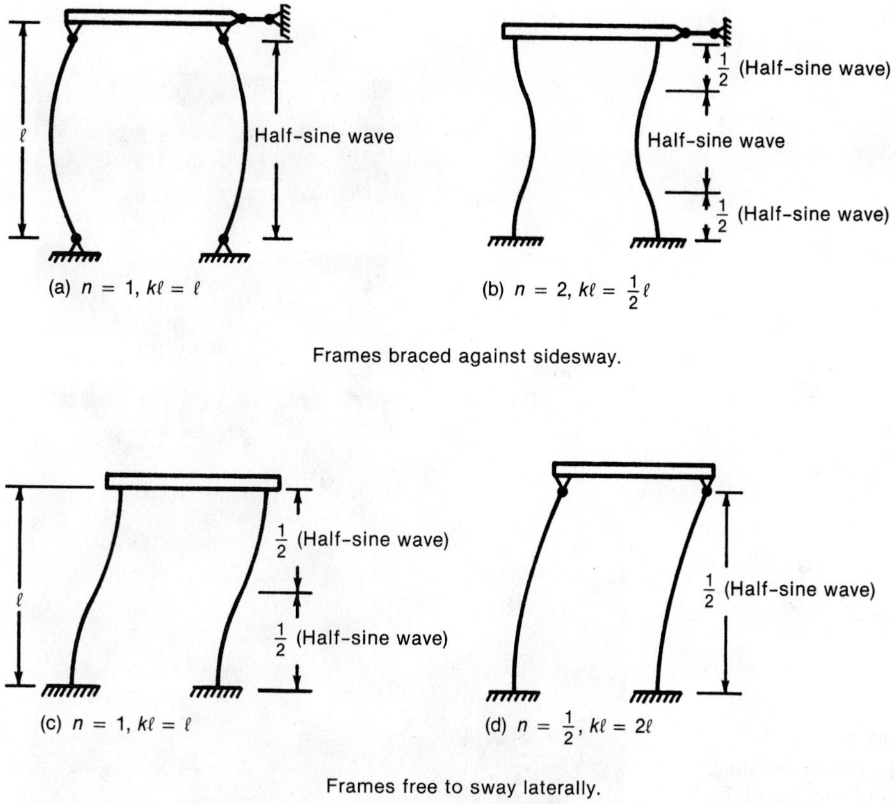

Fig. 12-6
Effective lengths of idealized columns.

fully fixed ends, such as those shown in Figs. 12-6b to 12-6d, rarely, if ever, occur. This is discussed later in the chapter.

Slender Columns in Structures

Pin-ended columns are rare in cast-in-place concrete construction, but do occur in precast construction. Occasionally, these will be slender, as, for example, the columns supporting the back of a precast grandstand.

Most concrete building structures are braced (nonsway) frames, with the bracing provided by shear walls, stairwells, or elevator shafts that are considerably stiffer than the columns themselves (Fig. 10-4). Occasionally, unbraced frames are encountered near the tops of tall buildings, where the stiff elevator core may be discontinued before the top of the building, or in industrial buildings where an open bay exists to accommodate a travelling crane.

Most building columns fall in the short-column category [12-1]. Exceptions occur in industrial buildings and in buildings that have a high first-floor story for architectural or functional reasons. An extreme example is shown in Fig. 12-7. The left corner column has a height of 50 times its least thickness. Some bridge piers and the decks of cable-stayed bridges fall into the slender-column category.

Organization of Chapter 12

The presentation of slender columns is divided into three progressively more complex parts. Slender pin-ended columns are discussed in Section 12-2. Restrained columns in

Fig. 12-7
Bank of Brazil building, Porto Alegre, Brazil. Each floor extends out over the floor below it. (Photograph courtesy of J. G. MacGregor.)

nonsway frames are discussed in Section 12-3. These sections deal with $P\delta$ effects and build on the material in Section 12-2. Finally, columns in sway frames are discussed in Sections 12-4 to 12-6.

12-2 BEHAVIOR AND ANALYSIS OF PIN-ENDED COLUMNS

Lateral deflections of a slender column cause an increase in the column moments, as illustrated in Figs. 12-1 and 12-2. These increased moments cause an increase in the deflections, which in turn lead to an increase in the moments. As a result, the load–moment line *O–B* in Fig. 12-2 is nonlinear. If the axial load is below the critical load, the process will converge to a stable position. If the axial load is greater than the critical load, it will not. This is referred to as a *second-order* process, because it is described by a second-order differential equation (Eq. 12-3).

In a *first-order analysis*, the equations of equilibrium are derived by assuming that the deflections have a negligible effect on the internal forces in the members. In a *second-order analysis*, the equations of equilibrium consider the deformed shape of the structure. Instability can be investigated only via a second-order analysis, because it is the loss of equilibrium of the deformed structure that causes instability. However, because many engineering calculations and computer programs are based on first-order analyses, methods have been derived to modify the results of a first-order analysis to approximate the second-order effects.

P–δ Moments and P–Δ Moments

Two different types of second-order moments act on the columns in a frame:

1. P−δ moments. These result from deflections, δ, of the axis of the bent column away from the chord joining the ends of the column, as shown in Figs. 12-1, 12-10, and 12-11. The slenderness effects in pin-ended columns and in nonsway frames result from P–δ effects, as discussed in Sections 12-2 through 12-4.

2. P−Δ moments. These result from lateral deflections, Δ, of the beam–column joints from their original undeflected locations, as shown in Figs. 12-30 and 12-31. The slenderness effects in sway frames result from P–Δ moments, as discussed in Sections 12-5 through 12-7.

Material Failures and Stability Failures

Load–moment curves are plotted in Fig. 12-8 for columns of three different lengths, all loaded (as shown in Fig. 12-1) with the same end eccentricity, e. The load–moment curve O–A for a relatively short column is practically the same as the line $M = Pe$. For a column of moderate length, line O–B, the deflections become significant, reducing the failure load. This column fails when the load–moment curve intersects the interaction diagram at point B. This is called a *material failure* and is the type of failure expected in most practical columns in braced frames. If a very slender column is loaded with increasing axial load, P, applied at a constant end eccentricity, e, it may reach a deflection δ at which the value of the $\partial M/\partial P$ approaches infinity or becomes negative. When this occurs, the column becomes unstable, because, with further deflections, the axial load capacity will drop. This type of failure is known as a *stability failure* and occurs only with very slender braced columns or with slender columns in sway frames [12-2].

Slender-Column Interaction Curves

In discussing the effects of variables on column strength, it is sometimes convenient to use slender-column interaction curves. Line O–B_1 in Fig. 12-9a shows the load–maximum

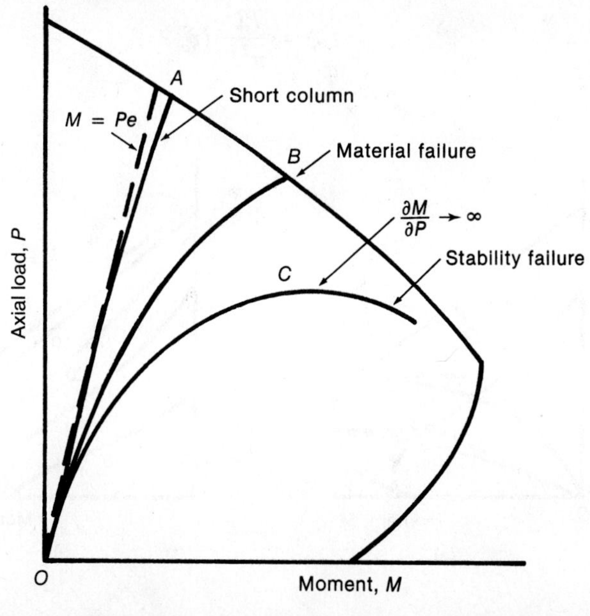

Fig. 12-8
Material and stability failures.

moment curve for a column with slenderness $\ell/h = 30$ and a given end eccentricity, e_1. This column fails when the load–moment curve intersects the interaction diagram at point B_1. At the time of failure, the load and moment at the *end* of the column are given by point A_1. If this process is repeated a number of times, we get the *slender column interaction curve* shown by the broken line passing through A_1 and A_2, and so on. Such curves show the loads and maximum *end* moments causing failure of a given slender column. A family of slender-column interaction diagrams is given in Fig. 12-9b for columns with the same cross section but different slenderness ratios.

Moment Magnifier for Symmetrically Loaded Pin-Ended Beam Column

The column from Fig. 12-1 is shown in Fig. 12-10a. Under the action of the end moment, M_o, it deflects an amount δ_o. This will be referred to as the *first-order* deflection. When the axial loads P are applied, the deflection increases by the amount δ_a. The final deflection at midspan is $\delta = \delta_o + \delta_a$. This *total* deflection will be referred to as the *second-order deflection*. It will be assumed that the final deflected shape approaches a half-sine wave. The primary moment diagram for the member (column) is shown in Fig. 12-10b, and the secondary moments, $P\delta$, are shown in Fig. 12-10c. Because the deflected shape is assumed to be a sine wave, the $P-\delta$ moment diagram is also a sine wave. Using the moment area method and observing that the deflected shape is symmetrical, the deflection δ_a is computed as the moment about the support of the portion of the *M/EI* diagram between the support and midspan, shown shaded in Fig. 12-10c. The area of this portion is

$$\text{Area} = \left[\frac{P}{EI}(\delta_o + \delta_a)\right]\frac{\ell}{2} \times \frac{2}{\pi}$$

and the centroid of the second-order moment diagram is located at ℓ/π from the support. Thus,

$$\delta_o = \left[\frac{P}{EI}(\delta_o + \delta_a)\frac{\ell}{2} \times \frac{2}{\pi}\right]\left(\frac{\ell}{\pi}\right)$$

$$= \frac{P\ell^2}{\pi^2 EI}(\delta_o + \delta_a)$$

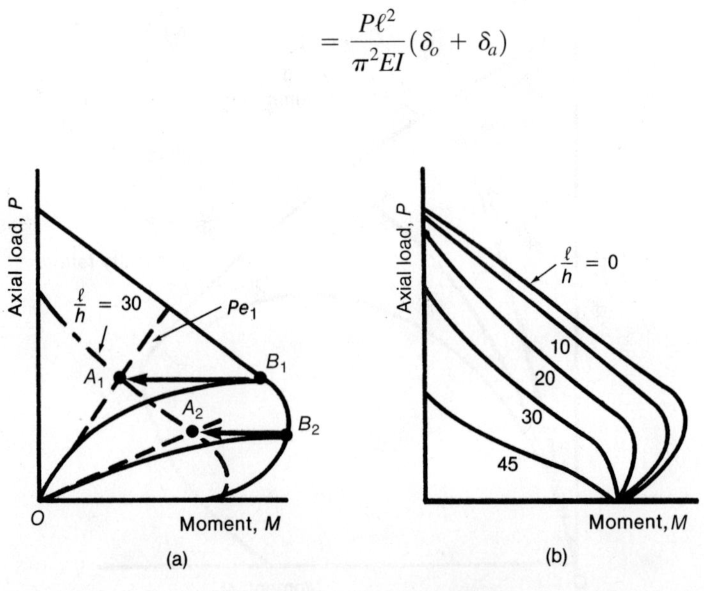

Fig. 12-9
Slender column interaction curves. (From [12-1].)

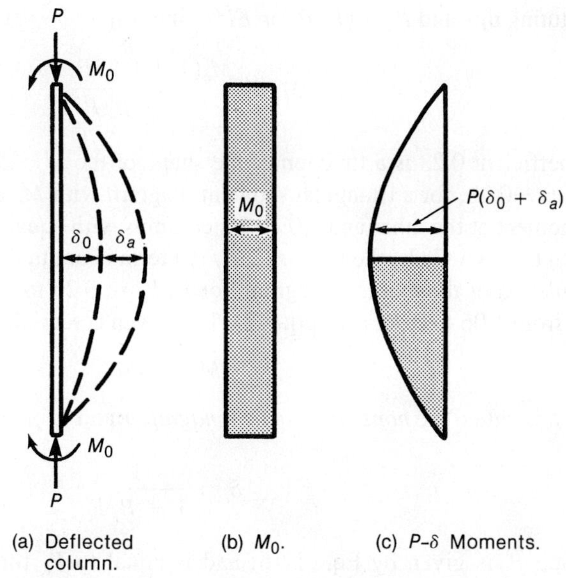

Fig. 12-10
Moments in a deflected column.

This equation can be simplified by replacing $\pi^2 EI/\ell^2$ by P_E, the Euler buckling load of a pin-ended column. Hence,

$$\delta_a = (\delta_o + \delta_a)P/P_E$$

Rearranging gives

$$\delta_a = \delta_o\left(\frac{P/P_E}{1 - P/P_E}\right) \tag{12-7}$$

Because the final deflection δ is the sum of δ_o and δ_a, it follows that

$$\delta = \delta_o + \delta_o\left(\frac{P/P_E}{1 - P/P_E}\right)$$

or

$$\delta = \frac{\delta_o}{1 - P/P_E} \tag{12-8}$$

This equation shows that the second-order deflection, δ, increases as P/P_E increases, reaching infinity when $P = P_E$.

The maximum bending moment is

$$M_c = M_o + P\delta$$

Here M_c is referred to as the *second-order moment*, and M_o is referred to as the *first-order moment*. Substituting Eq. (12-8) gives

$$M_c = M_o + \frac{P\delta_o}{1 - P/P_E} \tag{12-9}$$

For the moment diagram shown in Fig. 12-10b,

$$\delta_o = \frac{M_o \ell^2}{8EI} \tag{12-10}$$

Substituting this and $P = (P/P_E)\pi^2 EI/\ell^2$ into Eq. (12-9) gives

$$M_c = \frac{M_o(1 + 0.23P/P_E)}{1 - P/P_E} \tag{12-11}$$

The coefficient 0.23 is a function of the shape of the M_o diagram [12-3]. For example, it becomes -0.38, for a triangular moment diagram with M_o at one end of the column and zero moment at the other and -0.18 for columns with equal and opposite end moments.

In the ACI Code, the $(1 + 0.23P/P_E)$ term is omitted because the factor 0.23 varies as a function of the moment diagram, for $P/P_E = 0.25$ to -0.18, the term $(1 + C\,P/P_E)$ varies from 1.06 to 0.96, and Eq. (12-11) is given essentially as

$$M_c = \delta M_o \tag{12-12}$$

where δ is called the *nonsway-moment magnifier* and is given by

$$\delta = \frac{1}{1 - P/P_c} \tag{12-13}$$

in which P_c is given by Eq. (12-6) and is equal to P_E for a pin-ended column. Equation (12-13) underestimates the moment magnifier for the column loaded with equal end moments, but approaches the correct solution when the end moments are not equal.

Effect of Unequal End Moments on the Strength of a Slender Column

Up to now, we have considered only pin-ended columns subjected to moments of equal magnitude at the two ends. This is a very special case, for which the maximum deflection moment, $P\delta$, occurs at a section where the applied load moment, Pe, is also a maximum. As a result, these quantities can be added directly, as done in Figs. 12-1 and 12-2.

Before we proceed further in this section, we need to discuss sign conventions for member end moments and member moment diagrams. When conducting a frame analysis (usually using a stiffness-based method), we need to establish a consistent sign convention for member end moments. A common choice is to say counter-clockwise (CCW) end moments are positive. Using that sign convention, the end moment at the top of the column in Fig. 12-10a is positive and the moment at the bottom of the column is negative. However, when we want to plot a member moment diagram, as shown for the same column in Fig. 12-10b, we need to define a sign convention for positive member moments as positive when causing compression on either the left or right side of the column. For the member diagram in Fig. 12-10b, we have defined a positive member moment as that causing compression to occur on the left side of the member. The necessity of using a consistent sign convention for member end moments in a frame analysis and the similar need to have a consistent sign convention for plotting moment diagrams over the length of a member sets up a well-known conflict between these two sign conventions at one end of the member or the other. This sign convention conflict is an issue that structural engineering students must learn to deal with. For the column in Fig. 12-10, the conflict occurs at the bottom of the column where the member end moment is negative (clockwise) and the member moment diagram is plotted as a positive value. In the following discussions, we will assume counter-clockwise (CCW) end moments are positive. Further, end eccentricities are assumed to be positive if the resulting end moment, $P \cdot e$, is CCW.

In the usual case, the end eccentricities, $e_1 = M_1/P$ and $e_2 = M_2/P$, are not equal and so give member moment diagrams as shown shaded in Figs. 12-11b and 12-11c for

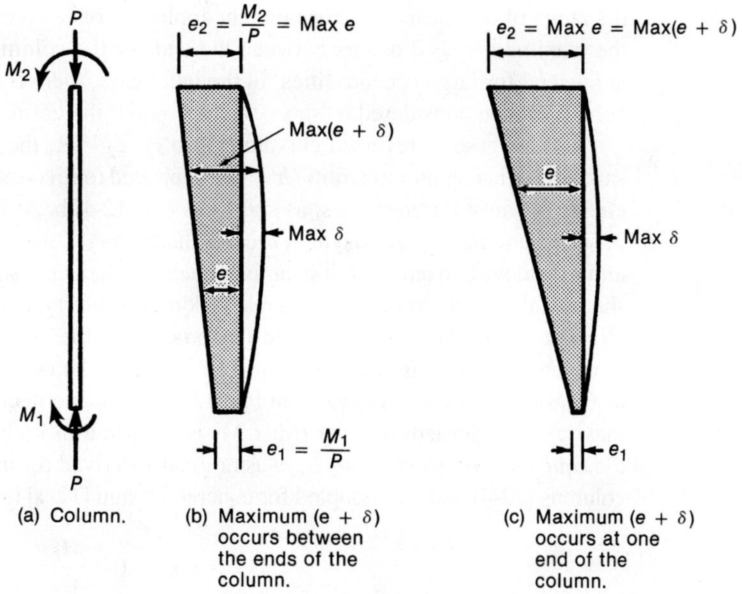

Fig. 12-11
Member moments in columns with unequal end moments.

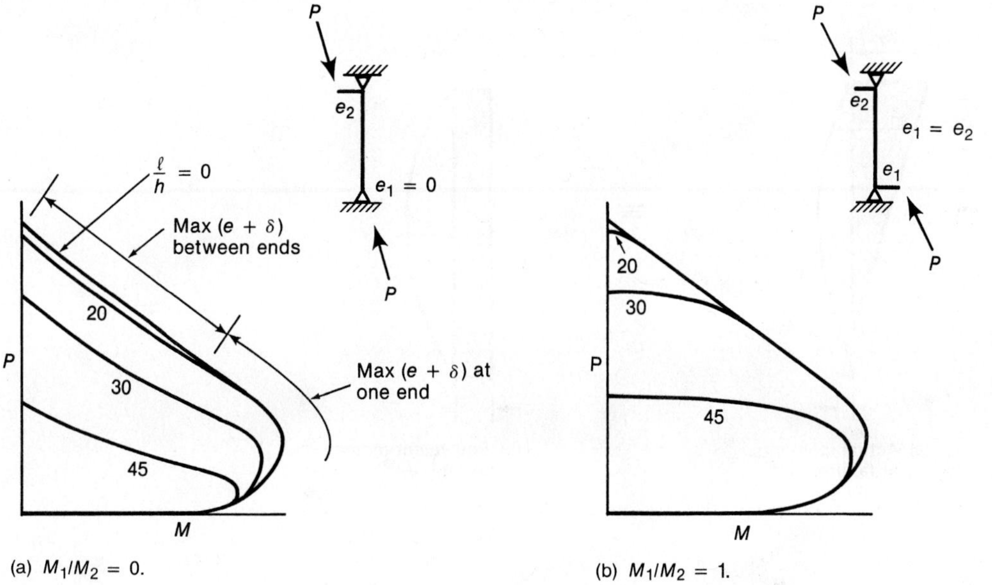

Fig. 12-12
Effect of M_1/M_2 ratio on slender column interaction curves for hinged columns. (From [12-1].)

the column shown in Fig. 12-11a. The maximum value of δ occurs between the ends of the column while the maximum e occurs at one end of the column. As a result, e_{max} and δ_{max} cannot be added directly. Two different cases exist. For a slender column with small end eccentricities, the maximum sum of $e + \delta$ may occur between the ends of the column, as shown in Fig. 12-11b. For a shorter column, or a column with large end eccentricities, the maximum sum of $e + \delta$ will occur at one end of the column, as shown in Fig. 12-11c.

These two types of behavior can be identified in the slender-column interaction diagrams shown in Figs. 12-9b and 12-12. For $e_1 = -e_2$ (Fig. 12-10a), the interaction diagram in Fig. 12-9b for $\ell/h = 20$, for example, shows a reduction in strength throughout the range of eccentricities. For moment applied at one end only ($e_1/e_2 = 0$, Fig. 12-12a), the maximum $e + \delta$ occurs between the ends of the column for small eccentricities and at one end for large eccentricities. In the latter case, there is no slenderness effect, and the column can be considered a "short column" under the definition given in Section 12-1.

In the case of reversed curvature with $e_1/e_2 = 1$, the slender-column range is even smaller, so that a column with $\ell/h = 20$ subjected to reversed curvature has no slenderness effects for most eccentricities, as shown in Fig. 12-12b. At low loads, the deflected shape of such a column is an antisymmetrical S shape. As failure approaches, the column tends to *unwrap*, moving from the initial antisymmetrical deflected shape toward a single-curvature shape. This results from the inevitable lack of uniformity along the length of the column.

In the moment-magnifier design procedure, the column subjected to unequal end moments shown in Fig. 12-13a is replaced with a similar column subjected to equal moments of $C_m M_2$ at both ends, as shown in Fig. 12-13b. The moments $C_m M_2$ are chosen so that the maximum magnified moment (Max M) is the same in both columns. The expression for the *equivalent moment factor* C_m was originally derived for use in the design of steel beam-columns [12-4] and was adopted for concrete design [12-1] (ACI Code Section 6.6.4.5.3):

$$C_m = 0.6 - 0.4\frac{M_1}{M_2} \tag{12-14}$$

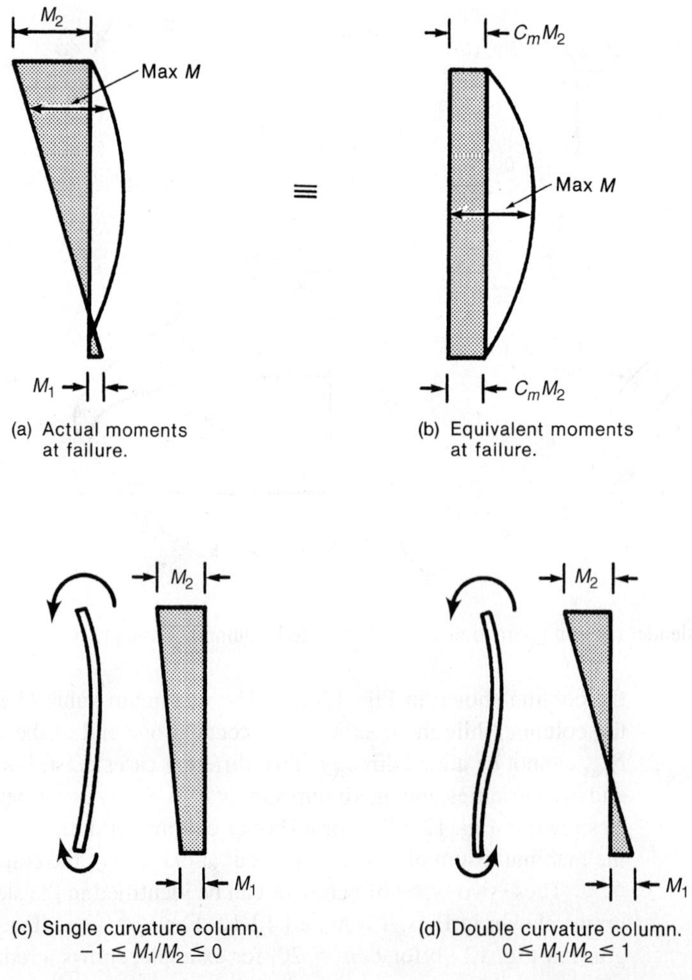

Fig. 12-13
Equivalent moment factor, C_m.

Equation (12-14) was originally derived using the ratio of end eccentricities (e_1/e_2) rather than the ratio of end moments (M_1/M_2). In Eq. (12-14), M_1 and M_2 are the smaller and larger end moments, in terms of magnitude, calculated from a conventional first-order elastic analysis. The sign convention for the ratio M_1/M_2 is illustrated in Figs. 12-13c and 12-13d. If the moments M_1 and M_2 cause single curvature bending without a point of contraflexure between the ends, as shown in Fig. 12-13c, M_1/M_2 is negative. If the moments M_1 and M_2 bend the column in double curvature with a point of zero moment between the two ends, as shown in Fig. 12-13d, that M_1/M_2 is positive.

Equation (12-14) applies only to hinged columns or columns in braced frames, loaded with axial loads and end moments. In all other cases, including columns subjected to transverse loads between their ends and concentrically loaded columns (no end moment), C_m is taken equal to 1.0 (ACI Code Section 6.6.4.5.3). The term C_m is not included in the equation for the moment magnifier for unbraced (sway) frames.

Column Rigidity, *EI*

The calculation of the critical load, P_c, via Eq. (12-6) involves the use of the flexural rigidity, *EI*, of the column section. The value of *EI* chosen for a given column section, axial-load level, and slenderness must approximate the *EI* of the column *at the time of failure*, taking into account the type of failure (material failure or stability failure) and the effects of cracking, creep, and nonlinearity of the stress–strain curves at the time of failure.

Figure 12-14 shows moment–curvature diagrams for three different load levels for a typical column cross section. (P_b is the balanced-failure load.) A radial line in such a diagram has slope $M/\Phi = EI$. The value of *EI* depends on the particular radial line selected. In a material failure, failure occurs when the most highly stressed section fails (point *B* in Fig. 12-8). For such a case, the appropriate radial line should intercept the end of the moment–curvature diagram, as shown for the $P = P_b$ (balanced-load) case in Fig. 12-14. On the other hand, a stability failure occurs before the cross section fails (point *C* in Fig. 12-8). This corresponds to a steeper line in Fig. 12-14 and thus a higher value of *EI*. The multitude of radial lines that can be drawn in Fig. 12-14 suggests that there is no all-encompassing value of *EI* for slender concrete columns.

References [12-1] and [12-5] describe empirical attempts to derive values for *EI*. The following two different sets of stiffness values, *EI*, are given in the slenderness provisions in ACI Code Sections 6.6.3.1.1 and 6.6.4.4.4 and will be discussed separately here:

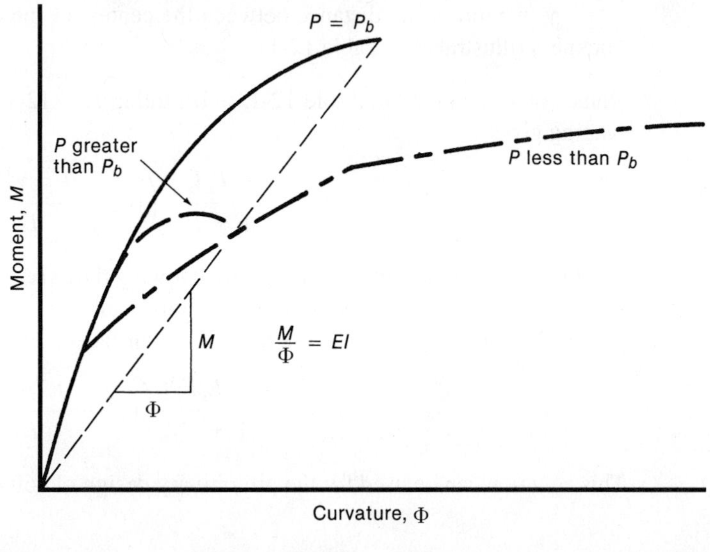

Fig. 12-14
Moment–curvature diagrams for a column cross section.

Design Values of $(EI)_{eff}$ for the Computation of the Critical Loads of Individual Columns

ACI Code Section 6.6.4.4.4 gives the following equations for the computation of *EI* in calculating the critical load of an individual column:

$$(EI)_{eff} = \frac{0.2E_c I_g + E_s I_{se}}{1 + \beta_{dns}} \tag{12-15a}$$

$$(EI)_{eff} = \frac{0.40E_c I_g}{1 + \beta_{dns}} \tag{12-15b}$$

$$(EI)_{eff} = \frac{E_c I}{1 - \beta_{dns}} \tag{12-16}$$

In these equations,

E_c, E_s = moduli of elasticity of the concrete (ACI Code Section 19.2.2.1) and the steel, respectively

I_g = gross moment of inertia of the concrete section about its centroidal axis, ignoring the reinforcement

I_{se} = moment of inertia of the reinforcement about the centroidal axis of the concrete section

I in Eq. (12-16) is calculated according to Table 6.6.3.1.1(b).

The term $(1 + \beta_{dns})$ reflects the effect of creep on the column deflections and is discussed later.

Equation (12-15a) is more "accurate" than Eq. (12-15b) but is more difficult to use because I_{se} is not known until the steel is chosen. It can be rewritten in a more useable form, however. The term I_{se} in Eq. (12-15) can be rewritten as

$$I_{se} = C\rho_g \gamma^2 I_g \tag{12-17}$$

where

C = constant depending on the steel arrangement

ρ_g = total longitudinal-reinforcement ratio

γ = ratio of the distance between the centers of the outermost bars to the column thickness (illustrated in Table 12-1)

Values of C are given in Table 12-1. Substituting Eq. (12-17) into Eq. (12-15a) and rearranging gives

$$(EI)_{eff} = \frac{E_c I_g}{1 + \beta_{dns}}\left(0.2 + \frac{C\rho_g \gamma^2 E_s}{E_c}\right) \tag{12-18}$$

It is then possible to estimate *EI* without knowing the exact steel arrangement by choosing ρ_g, estimating γ from the column dimensions, and using the appropriate value of C from Table 12-1. For the common case of bars in four faces and $\gamma \simeq 0.75$, this reduces to

$$(EI)_{eff} = \frac{E_c I_g}{1 + \beta_{dns}}\left(0.2 + \frac{1.2\rho_g E_s}{E_c}\right) \tag{12-19}$$

This equation can be used for the preliminary design of columns.

TABLE 12-1 Calculation of I_{se}[a]

Type of Column	Number of Bars	I_{se}	$\frac{I_{se}}{I_g} = C\rho_g\gamma^2$
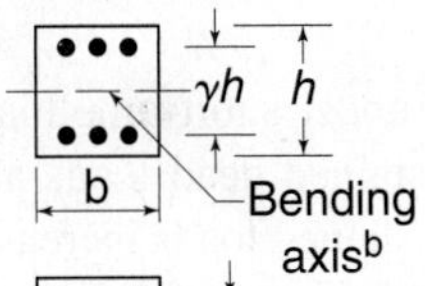	—	$0.25A_{st}(\gamma h)^2$	$3\rho_g\gamma^2$
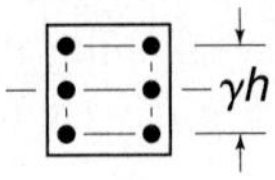	3 bars per face	$0.167A_{st}(\gamma h)^2$	$2\rho_g\gamma^2$
	6 bars per face	$0.117A_{st}(\gamma h)^2$	$1.4\rho_g\gamma^2$
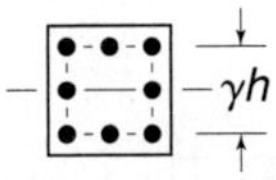	8 bars (3 per face)	$0.187A_{st}(\gamma h)^2$	$2.2\rho_g\gamma^2$
	12 bars (4 per face)	$0.176A_{st}(\gamma h)^2$	$2.10\rho_g\gamma^2$
	16 bars (5 per face)	$0.172A_{st}(\gamma h)^2$	$2.06\rho_g\gamma^2$
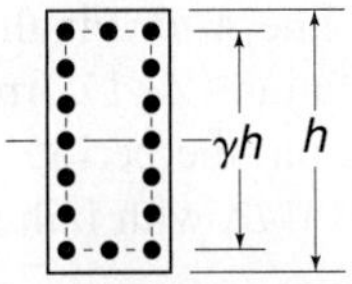	$h = 2b$ 16 bars as shown About strong axis	$0.128A_{st}(\gamma h)^2$	$1.54\rho_g\gamma^2$
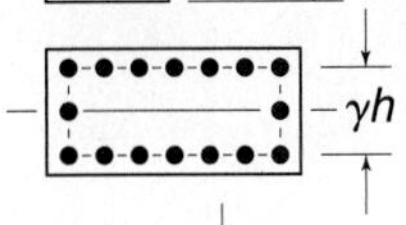	$b = 2h$ About weak axis	$0.219A_{st}(\gamma h)^2$	$2.63\rho_g\gamma^2$
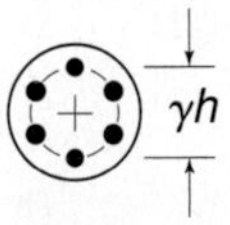	—	$0.125A_{st}(\gamma h)^2$	$2\rho_g\gamma^2$
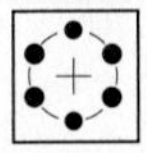	—	$0.125A_{st}(\gamma h)^2$	$1.5\rho_g\gamma^2$

[a]Total area of steel $= A_{st} = \rho_g A_c$
[b]All sections are bent about the horizontal axis of the section that is shown by the dashed line.
Source: [12-5]

EI for the Computation of Frame Deflections and for Second-Order Analyses

Three different sets of EI values are given in the slender-column sections of the ACI Code. ACI Code Section 6.6.4.4.4 gives Eqs. (12-15) and (12-16) for use in Eq. (12-6) to compute P_c when one is using the moment-magnifier method. These represent the behavior of a single, highly loaded column.

ACI Code Table 6.6.3.1.1(b) gives a set of values of the moment of inertia, *I*, for use

(a) in elastic frame analyses, to compute the moments in columns and beams and the lateral deflections of frames, and

(b) to compute the ψ used in computing the effective length factor, *k*. (Both of these topics will be covered in Section 12-4.)

The lateral deflection of a frame is affected by the stiffnesses of all the beams and columns in the frame. For this reason, the moments of inertia in ACI Code Table 6.6.3.1.1(a) are intended to represent an overall average of the moment of inertia values of *EI* for each type of member in a frame. In a similar manner, the effective length of a column in a frame is affected by the flexural stiffnesses of a number of beams and columns. It is incorrect to use the I values from ACI Code Table 6.6.3.1.1(a) when computing the critical buckling load using Eq. (12-16).

Effect of Sustained Loads on Pin-Ended Columns

Up to this point, the discussion has been limited to columns failing under short-time loadings. Columns in structures, on the other hand, are subjected to sustained dead loads and sometimes to sustained live loads. The creep of the concrete under sustained loads increases the column deflections, increasing the moment $M = P(e + \delta)$ and thus weakening the column. The load–moment curve of Fig. 12-2 can be replotted, as shown in Fig. 12-15, for columns subjected to sustained loads.

Effect of Loading History

In Fig. 12-15a, the column is loaded rapidly to the service load (line *O–A*). The service load is assumed to act for a number of years, during which time the creep deflections and resulting second-order effects increase the moment, as shown by line *A–B*. Finally, the column is rapidly loaded to failure, as shown by line *B–C*. The failure load corresponds to point *C*. Had the column been rapidly loaded to failure without the period of sustained service load, the load–moment curve would resemble line *O–A–D*, with failure

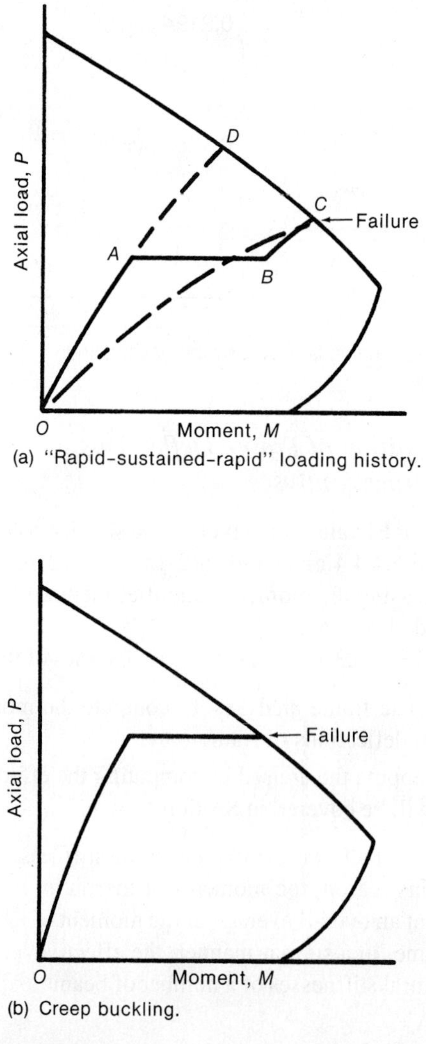

(a) "Rapid-sustained-rapid" loading history.

(b) Creep buckling.

Fig. 12-15
Load–moment behavior for hinged columns subjected to sustained loads.

corresponding to point D. The effect of the sustained loads has been to increase the mid-height deflections and moments, causing a reduction in the failure load from D to C. On the reloading (line B–C), the column deflections are governed by the EI corresponding to rapidly applied loads.

Creep Buckling

The second type of column behavior under sustained loads is referred to as *creep buckling*. Here, as shown in Fig. 12-15b, the column deflections continue to increase under the sustained load, causing failure under the sustained load itself [12-6], [12-7]. This occurs only under high-sustained loads greater than about 70 percent of the short-time capacity represented by point D. Because the sustained load will rarely exceed the strength-reduction factor, ϕ, divided by the dead-load factor, $0.65/1.2 = 0.54$ (for tied columns), times the column capacity, this type of failure is not specifically addressed in the ACI Code design procedures.

Two different design procedures are used to account for creep effects. In the *reduced-modulus procedure* [12-1], [12-7], [12-8], the value of E used to compute P_c is reduced to give the correct failure load. This procedure is illustrated by the broken line O–C in Fig. 12-15a.

The second procedure replaces the column loaded at eccentricity e with one loaded at an eccentricity $e + \delta_{0,cr}$ where $\delta_{0,cr}$ is the creep deflection that would remain in the column in Fig. 12-15a if it were unloaded after reaching point B [12-9].

The ACI Code moment-magnifier procedure uses the reduced-modulus procedure. The value of EI is reduced by dividing by $(1 + \beta_{dns})$, as done in Eqs. (12-15) and (12-16), where, for hinged columns and columns in restrained frames, β_{dns} is defined as the ratio of the factored axial load due to sustained load to the total factored axial load. In a lightly reinforced column, creep of the concrete frequently causes a significant increase in the steel stress, so that the compression reinforcement yields at a load lower than that at which it would yield under a rapidly applied load. This effect is empirically accounted for in Eq. (12-15a) by dividing both $E_c I_g$ and $E_s I_{se}$ by $(1 + \beta_{dns})$.

ACI Code Section 2.2 gives definitions of β_{dns} and β_{ds}, depending on whether the frame is nonsway or sway. To be stable, a pin-ended column must be in a structure that restricts sway of the ends of the column. In addition, it does not develop end moments if the structure sways sideways. In effect, a pin-ended column is always a nonsway column. For columns in a nonsway frame, ACI Code Section 6.6.4.4.4 defines β_{dns} as the ratio of the maximum factored axial sustained load to the total factored axial load for the same load case.

Limiting Slenderness Ratios for Slender Columns

Most columns in structures are sufficiently short and stocky to be unaffected by slenderness effects. To avoid checking slenderness effects for all columns, ACI Code Section 6.2.5 allows slenderness effects to be ignored in the case of columns in sway frames if,

$$\frac{k\ell_u}{r} \le 22 \tag{12-20a}$$

and in nonsway frames if,

$$\frac{k\ell_u}{r} \le 34 + 12\frac{M_1}{M_2} \tag{12-20b}$$

and,

$$\frac{k\ell_u}{r} \le 40 \tag{12-20c}$$

In Eq. (12-20), k refers to the effective length factor, which is 1.0 for a pin-ended column, ℓ_u is the unsupported length of the column, and r is the radius of gyration, taken as $0.3h$ for rectangular sections and $0.25h$ for circular sections (ACI Code Section 6.2.5.1). For other shapes, the value of r can be calculated from the area and moment of inertia of the cross-section. By definition, $r = \sqrt{I/A}$. The sign convention for M_1/M_2 is given in Figs. 12-13c and 12-13d.

Columns having slenderness ratios less than the right-hand side of Eq. (12-20) should have a slender-column strength that is 95 percent or more of the short-column strength.

Definition of Nonsway and Sway Frames

The preceding discussions were based on the assumption that frames could be separated into nonsway (braced) frames and sway (unbraced) frames. A column may be considered to be nonsway in a given direction if the lateral stability of the structure as a whole is provided by walls, bracing, or buttresses designed to resist all lateral forces in that direction.

In actuality, there is no such thing as a "completely braced" frame, and no clear-cut boundary exists between nonsway and sway frames. Some frames are clearly unbraced, as, for example, the frames shown in Figs. 12-25b and 12-25c. Other frames are connected to shear walls, elevator shafts, and so on, which clearly restrict the lateral movements of the frame, as shown in Fig. 12-25a. Because no wall is completely rigid, however, there will always be some lateral movement of a braced frame, and hence some $P\Delta$ moments will result from the lateral deflections.

For the purposes of design, a story or a frame can be considered "nonsway" if horizontal displacements do not significantly reduce the vertical load capacity of the structure. This criterion could be restated as follows: a frame can be considered "nonsway" if the $P\Delta$ moments due to lateral deflections are small compared with the first-order moments due to lateral loads. ACI Code Section 6.6.4.3(a) allows designers to assume that a frame is nonsway if the increase in column-end moments due to second-order effects does not exceed 5 percent of the first-order moments.

Alternatively, ACI Code Section 6.6.4.3(b) allows designers to assume that a story in a frame is nonsway if

$$Q = \frac{\Sigma P_u \Delta_o}{V_{us}\ell_c} \leq 0.05 \tag{12-21}$$

where Q is the *stability index*, ΣP_u is the total vertical load in all the columns and walls in the story in question, V_{us} is the shear in the story due to factored lateral loads, Δ_o is the first-order relative deflection between the top and bottom of that story *due to* V_{us}, and ℓ_c is the height of the story measured from center to center of the joints above and below the story. This concept, which is explained and developed more fully in Section 12-6, results in a limit similar to that from ACI Code Section 6.6.4.3(a). It was originally presented in [12-10] and [12-11] and is discussed in [12-12].

ACI Code Section 6.2.5 states that it shall be permitted to consider compression members braced against sidesway when bracing elements in a story have a total lateral stiffness of at least 12 times the gross lateral stiffness of the columns within the story.

Including Slenderness Effects

Code Section 6.2.5 gives the limiting slenderness ratios, $k\ell_u/r$ values, for classifying a column as either short or slender in nonsway (braced) frames and sway (unbraced) frames. When slenderness effects cannot be ignored (i.e., the column is classified as a slender column), the designer has two options for analyzing the total secondary moments that must be designed for. First, the secondary moments can be determined directly using structural analysis techniques that include secondary effects. This has become a relatively common feature of software packages used in frame analysis. Code Section 6.7 gives extensive guidance for conducting an elastic second-order analysis of frames including slender columns. Code Section 6.8 gives guidance for an inelastic second-order analysis that includes material nonlinearities, the duration of loads, and interactions with the foundation. Suggested values for member moments of inertia to be used in such analysis are defined and will be discussed in Sections 12-5 and 12-6 of this book.

An alternative available to the designer for the analysis of secondary moments in slender columns is the use of the *moment magnifier method*, which has been in the ACI Code for several years. Code Section 6.6.4 is the initial section for the moment magnifier method and Code Section 6.6.4.3 refers to the stability index, Q (Eq. 12-21), which is used to determine if the column is located in a nonsway ($Q \leq 0.05$) or sway ($Q > 0.05$) story of a frame. Code Section 6.6.4.5 gives the procedure to determine secondary moments for columns in nonsway frames, and Code Section 6.6.4.6 gives the procedure to determine secondary moments in sway frames. The analysis of secondary moments using the moment magnifier method will be given in the following paragraphs and in Section 12-4 for nonsway frames, and Sections 12-5 and 12-6 for sway frames.

Summary of ACI Moment Magnifier Design Procedure for Columns in Nonsway Frames

If a column is in a nonsway frame, then design following the moment magnifier method involves several sections in ACI Code Chapter 6. The specific equations and requirements for determining the magnified moments are:

1. Length of column. The unsupported length, ℓ_u, is defined as the clear distance between members capable of giving lateral support to the column. For a pin-ended column it is the distance between the hinges.

2. Effective length. ACI Commentary Section R6.2.5 states that the effective length factor, k, can be taken conservatively as 1.0 for columns in nonsway frames. A procedure for estimating k-values for columns in nonsway and sway frames will be given in Section 12-4.

3. Radius of gyration. For a rectangular section $r = 0.3h$, and for a circular section, $r = 0.25h$. (See ACI Code Section 6.2.5.1.) For other sections, r can be calculated from the area and moment of inertia of the gross concrete section as $r = \sqrt{I_g/A_g}$.

4. Consideration of slenderness effects. For columns in nonsway frames, ACI Code Section 6.2.5 allows slenderness to be neglected if $k\ell_u/r$ satisfies Eqs. (12-20b) and (12-20c). The sign convention for M_1/M_2 is given in Fig. 12-13.

5. Minimum moment. ACI Code Section 6.6.4.5.4 requires that the maximum end moment on the column, M_2, not be taken less than

$$M_{2,\min} = P_u(0.6 + 0.03h) \tag{12-22}$$

where the 0.6 and h are in inches. When M_2 is less than $M_{2,\min}$, C_m shall either be taken equal to 1.0 or evaluated from the actual end moments.

6. Moment-magnifier equation. ACI Code Section 6.6.4.5.1 states that the columns shall be designed for the factored axial load, P_u, and the magnified moment, M_c, defined by

$$M_c = \delta M_2 \tag{12-23}$$

The moment M_2 is defined as the larger (magnitude) end moment acting on the column. ACI Code Section 6.6.4.5.2 goes on to define δ as

$$\delta = \frac{C_m}{1 - P_u/(0.75P_c)} \geq 1.0 \tag{12-24}$$

where

$$C_m = 0.6 - 0.4\frac{M_1}{M_2} \tag{12-14}$$

$$P_c = \frac{\pi^2(EI)_{eff}}{(k\ell_u)^2} \tag{12-25}$$

and

$$(EI)_{eff} = \frac{0.2E_cI_g + E_sI_{se}}{1 + \beta_{dns}} \tag{12-15a}$$

or

$$(EI)_{eff} = \frac{0.40E_cI_g}{1 + \beta_{dns}} \tag{12-15b}$$

or

$$(EI)_{eff} = \frac{E_cI}{1 - \beta_{dns}} \tag{12-16}$$

Equation (12-24) is Eq. (12-13) rewritten to include the equivalent moment factor, C_m, and to include a *stiffness-reduction factor*, ϕ_k, taken equal to 0.75 for all slender columns [12-13]. The number 0.75 is used in Eq. (12-24) rather than the symbol ϕ_k, to avoid confusion with the *capacity-reduction factor*, ϕ, used in design of the column cross-section.

There are two β_d terms. The one that applies to columns in nonsway frames is

$$\beta_{dns} = \frac{\text{maximum factored sustained (dead) axial load in column}}{\text{total factored axial load in column}} \tag{12-26a}$$

Equations (12-18) and (12-19) also could be used to compute $(EI)_{eff}$ for use in Eq. (12-25). However, the *EI* values given in ACI Code Table 6.6.3.1.1(a) should not be used to compute *EI* for use in Eq. (12-25) because those values are assumed to be average values for an entire story in a frame and are intended for use in first- or second-order frame analyses.

If P_u exceeds $0.75P_c$ in Eq. (12-24), δ will be negative, and the column would be unstable. Hence, if P_u exceeds $0.75P_c$, the column cross-section must be enlarged. Further, if δ exceeds 1.4, ACI Code Section 6.2.6 requires that the column cross-section be enlarged.

EXAMPLE 12-1 Design of a Slender Pin-Ended Column (Nonsway)

Design a 20-ft-tall column to support an unfactored dead load of 90 kips and an unfactored live load of 75 kips. The loads act at an eccentricity of 3 in. at the top and 2 in. at the bottom, as shown in Fig. 12-16. Use $f'_c = 4000$ psi and $f_y = 60{,}000$ psi. Use the load combinations and strength-reduction factors from ACI Code Tables 5.3.1 and 21.2.1.

1. **Compute the factored loads and moments and M_1/M_2.**

$$P_u = 1.2D + 1.6L$$
$$= 1.2 \times 90 \text{ kips} + 1.6 \times 75 \text{ kips} = 228 \text{ kips}$$

The moment at the top is

$$M = P_u \times e$$
$$= 228 \text{ kips} \times \frac{3 \text{ in.}}{12 \text{ in.}} = 57.0 \text{ kip-ft}$$

The moment at the bottom is

$$M = 228 \text{ kips} \times \frac{-2 \text{ in.}}{12 \text{ in.}}$$
$$= -38.0 \text{ kip-ft}$$

By definition, M_2 is the larger magnitude end moment in the column. Therefore, $M_2 = 57.0$ kip-ft and $M_1 = -38.0$ kip-ft. The ratio M_1/M_2 is taken to be negative, because the column is bent in single curvature (see Fig. 12-16c). Thus $M_1/M_2 = -0.667$.

2. **Estimate the column size.** From (11-19a), assuming that $\rho_g = 0.015$,

$$A_{g(\text{trial})} \geq \frac{P_u}{0.40(f'_c + f_y\rho_g)}$$

$$A_{g(\text{trial})} \geq \frac{228 \times 1000}{0.40(4000 + 60{,}000 \times 0.015)}$$
$$= 116 \text{ in.}^2$$

This suggests that a 12 in. × 12 in. column would be satisfactory. It should be noted that Eq. (11-19a) was derived for short columns and will underestimate the required sizes of slender columns.

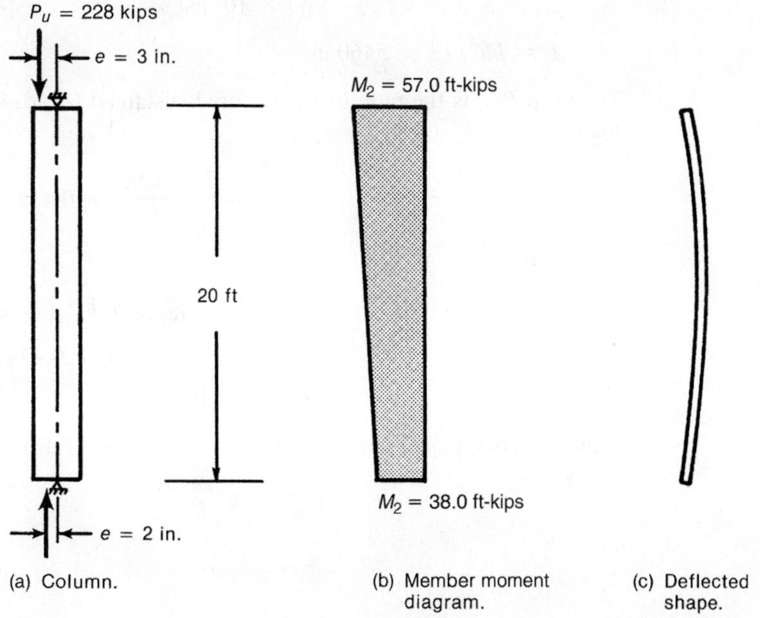

Fig. 12-16
Column—Example 12-1.

3. Is the column slender? From Eq. (12-20b), a column in a nonsway frame is short if

$$\frac{k\ell_u}{r} \leq 34 + 12\frac{M_1}{M_2} \qquad \text{(12-20b)}$$

For the 12 in. × 12 in. section selected in step 2, $k = 1.0$ because the column is pin-ended, and where $r = 0.3h = 0.3 \times 12$ in. $= 3.6$ in.,

$$\frac{k\ell_u}{r} = \frac{1.0 \times 240 \text{ in.}}{3.6 \text{ in.}} = 66.7$$

For $M_1/M_2 = -0.667$,

$$34 + 12\frac{M_1}{M_2} = 34 - 12 \times 0.667 = 26$$

Because $k\ell_u/r = 66.7$ exceeds 26, the column is quite slender. This suggests that the 12 in. × 12 in. section probably is inadequate.

Using increments of 2 in., we shall select a 16-in.-by-16 in.-section for the first trial.

4. Check whether the moments are less than the minimum. ACI Code Section 6.6.4.5.4 requires that a braced column be designed for a minimum eccentricity of $0.6 + 0.03h = 1.08$ in. Because the maximum end eccentricity exceeds this, design for the moments from step 1.

5. Compute $(EI)_{eff}$. At this stage, the area of reinforcement is not known. Additional calculations are needed before it is possible to use Eq. (12-15a) to compute EI, but Eq. (12-15b) can be used.

$$(EI)_{eff} = \frac{0.40E_cI_g}{1 + \beta_{dns}} \qquad \text{(12-15b)}$$

where

$$E_c = 57{,}000\sqrt{f'_c} = 3.60 \times 10^6 \text{ psi}$$

$$I_g = bh^3/12 = 5460 \text{ in.}^4$$

The term β_{dns} is the ratio of the factored sustained (dead) load to the total factored axial load:

$$\beta_{dns} = \frac{1.2 \times 90}{228} = 0.474 \qquad \text{(12-26a)}$$

Thus,

$$(EI)_{eff} = 0.4 \times \frac{3.60 \times 10^6 \text{ psi} \times 5460 \text{ in.}^4}{(1 + 0.474)}$$

$$= 5.33 \times 10^9 \text{ lb-in.}^2$$

Generally, one would use Eq. (12-15a) or Eq. (12-19) if ρ_g exceeded about 0.02.

6. Compute the magnified moment. From Eq. (12-23),

$$M_c = \delta M_2$$

where

$$\delta = \frac{C_m}{1 - P_u/0.75P_c} \geq 1.0 \tag{12-24}$$

$$C_m = 0.6 - 0.4\frac{M_1}{M_2} \tag{12-14}$$

$$= 0.6 + 0.4 \times 0.667 = 0.867$$

$$P_c = \frac{\pi^2 (EI)_{eff}}{(k\ell_u)^2} \tag{12-25}$$

where $k = 1.0$ because the column is pin-ended.

$$P_c = \frac{\pi^2 \times 5.33 \times 10^9 \text{ lb-in.}^2}{(1.0 \times 240 \text{ in.})^2} = 913{,}000 \text{ lb}$$
$$= 913 \text{ kips}$$

and

$$\delta = \frac{0.867}{1 - 228/(0.75 \times 913)} \geq 1.0$$
$$= 1.30$$

If δ exceeds 1.4, a larger cross-section should be selected. Continuing without selecting a larger column, the magnified moment is

$$M_c = 1.30 \times 57.0 \text{ kip-ft} = 74.1 \text{ kip-ft}$$

We will continue with a **16-in.-by-16-in. square section.**

7. Select the column reinforcement. We will use the tied-column interaction diagrams in Appendix A assuming an equal distribution of longitudinal bars in two opposite faces of the column. The parameters required for entering the interaction diagrams are

$$\gamma \cong \frac{16 \text{ in.} - 2 \times 2.5 \text{ in.}}{16 \text{ in.}} = 0.69$$

Assuming $\phi P_n = P_u = 228$ kips,

$$\frac{\phi P_n}{bh} = \frac{228 \text{ kips}}{16 \text{ in.} \times 16 \text{ in.}} = 0.89 \text{ ksi}$$

Assuming $\phi M_n = M_c = 74.1$ kip-ft $= 889$ kip-in.,

$$\frac{\phi M_n}{bh^2} = \frac{889 \text{ kip-in.}}{(16 \text{ in.})^3} = 0.22 \text{ ksi}$$

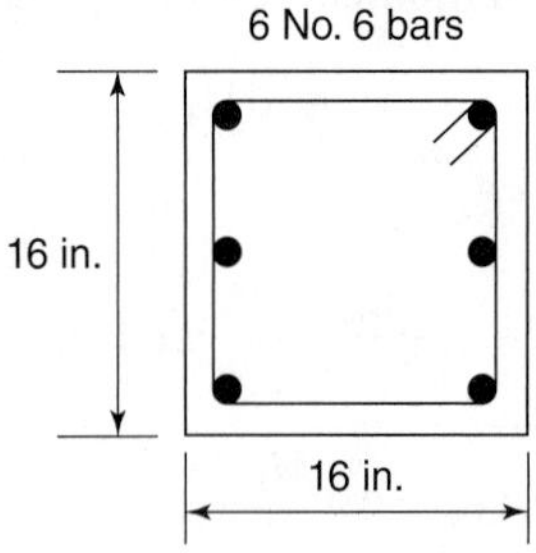

Fig. 12-17
Final column section for Example 12-1.

From both Fig. A-6a ($\gamma = 0.60$) and Fig. A-6b ($\gamma = 0.75$), the required value for ρ_g is less than 0.01. Therefore, to satisfy the minimum column longitudinal-reinforcement ratio from ACI Code Section 10.6.1.1, use $\rho_g = 0.01$. Thus,

$$A_{st}(\text{req'd}) = 0.01 \times A_g = 0.01 \times (16 \text{ in.})^2 = 2.56 \text{ in.}^2$$

Use six No. 6 bars, $A_{st} = 6 \times 0.44 \text{ in.}^2 = 2.64 \text{ in.}^2$ In summary, **use a 16 in. × 16 in. column section with six No. 6 bars**, as shown in Fig. 12-17. This section design would be very conservative if we were designing a short column, but the slenderness of the column has required the use of this larger section. ■

12-3 DESIGN OF COLUMNS IN NONSWAY FRAMES

This section deals with columns in continuous frames with deformations restrained in two ways. First, the frames are "nonsway" or "braced," so *the horizontal deflection of one end of a column relative to the other end* is prevented, or at least restrained, by walls or other bracing elements. Second, the columns are attached to beams that *restrain the rotations of the ends of the column.* This section deals primarily with how the rotational restraints provided by the beams are accounted for in design.

Design Approximation for the Effect of End Restraints in Nonsway Frames

Figure 12-18a shows a restrained column in a frame. The curved solid line in Fig. 12-18b is the member moment diagram (including slenderness effects) for this column at failure. Superimposed on this is the corresponding first-order moment diagram for the same load level. In design, it is convenient to replace the restrained column with an equivalent hinged end column of length ℓ_i, the distance between the points on the second-order moment diagram where the moments are equal to the end moments in the first-order diagram (Fig. 12-18c). This equivalent hinged column is then designed for the axial load, P, and the end moments, M_2, from the first-order analysis.

Unfortunately, the length ℓ_i is difficult to compute. In all modern concrete and steel design codes, the *empirical assumption* is made that ℓ_i can be taken equal to the effective length for elastic buckling, $k\ell$. The accuracy of this assumption is discussed in [12-14], which concludes that $k\ell$ slightly underestimates ℓ_i for an elastically restrained, elastic column.

The concept of effective lengths was discussed earlier in this chapter for the four idealized cases shown in Fig. 12-6. The effective length of a column, $k\ell_u$, is defined as the length of an equivalent pin-ended column having the same buckling load. When a

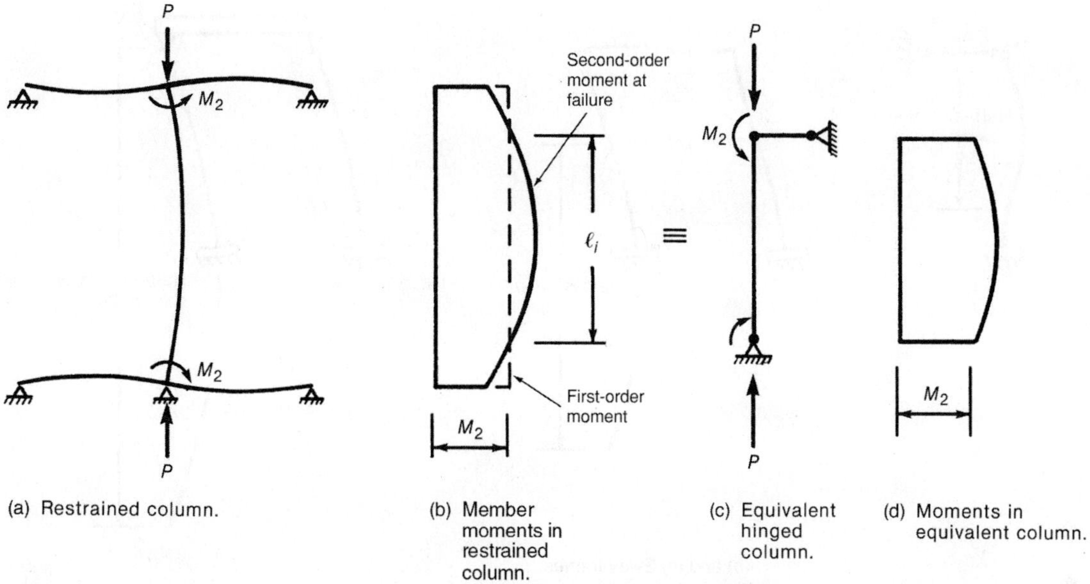

Fig. 12-18
Replacement of restrained column with an equivalent hinged column for design.

pin-ended column buckles, its deflected shape is a half-sine wave, as shown in Fig. 12-6a. The effective length of a restrained column is taken equal to the length of a complete half-sine wave in the deflected shape.

Figures 12-6b to 12-6d are drawn by assuming truly fixed ends. This condition seldom, if ever, actually exists. In buildings, columns are restrained by beams or footings, which always allow some rotation of the ends of the column. Thus the three cases considered in Fig. 12-6 will actually deflect as shown in Fig. 12-19, and the effective lengths will be greater than the values for completely fixed ends. The actual value of k for an elastic column is a function of the relative stiffnesses, ψ, of the beams and columns at each end of the column, where ψ is

$$\psi = \frac{\Sigma\,(E_c I_c/\ell_c)}{\Sigma\,(E_b I_b/\ell_b)} \tag{12-27a}$$

where the subscripts b and c refer to beams and columns, respectively, and the lengths ℓ_b and ℓ_c are measured center-to-center of the joints. The summation sign in the numerator refers to all the compression members meeting at a joint; the summation sign in the denominator refers to all the beams or other restraining members at the joint.

If $\psi = 0$ at one end of the column, the column is fully fixed at that end. Similarly, $\psi = \infty$ denotes a perfect hinge. Thus, as ψ approaches zero at the two ends of a column in a braced frame, k approaches 0.5, the value for a fixed-ended column. Similarly, when ψ approaches infinity at the two ends of a braced column, k approaches 1.0, the value for a pin-ended column. This is illustrated in Table 12-2. The value for columns that are fully fixed at both ends is 0.5, found in the lower-left corner of the table. The value for columns that are pinned at both ends is 1.0, found in the upper-right corner.

In practical structures, there is no such thing as a truly fixed end or a truly hinged end. Reasonable upper and lower limits on ψ are 20 and 0.2. It is recommended for columns in nonsway frames that k should not be taken less than 0.6.

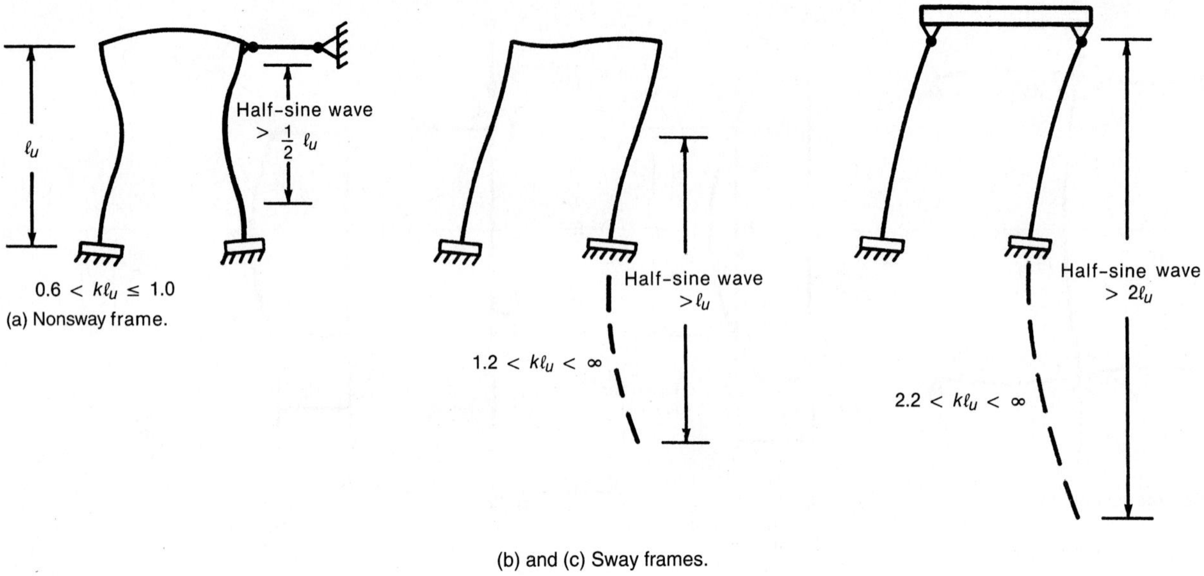

Fig. 12-19
Effective lengths of columns in frames with foundation rotations.

TABLE 12-2 Effective-Length Factors for Nonsway (Braced) Frames

Top	k				
Hinged	0.70	0.81	0.91	0.95	1.00
Elastic $\psi = 3.1$	0.67	0.77	0.86	0.90	0.95
Elastic, flexible $\psi = 1.6$	0.65	0.74	0.83	0.86	0.91
Stiff $\psi = 0.4$	0.58	0.67	0.74	0.77	0.81
Fixed	0.50	0.58	0.65	0.67	0.70
Bottom	Fixed	Stiff	Elastic, flexible	Elastic	Hinged

Calculation of k from Tables

Table 12-2 can be used to select values of k for the design of nonsway frames. The shaded areas correspond to one or both ends truly fixed. Because such a case rarely, if ever, occurs in practice, this part of the table should not be used. The column and row labeled "Hinged," "elastic" through to "fixed" represent conservative practical degrees of end fixity. Because k values for sway frames vary widely, no similar table is given for such frames.

Calculation of k via Nomographs

The nomographs given in Fig. 12-20 are also used to compute k. To use these nomographs, ψ is calculated at both ends of the column, from Eq. (12-27a), and the appropriate value of k is found as the intersection of the line labeled k and a line joining the values of ψ at the two ends of the column. The calculation of ψ is discussed in a later section.

The nomographs in Fig. 12-20 were derived [12-15], [12-16] by considering a typical interior column in an infinitely high and infinitely wide frame, in which all of the columns have the same cross section and length, as do all beams. Equal loads are applied at the tops of each of the columns, while the beams remain unloaded. All columns are assumed to buckle at the same moment. As a result of these very idealized and quite unrealistic assumptions, the nomographs tend to underestimate the value of the effective length factor k for elastic frames of practical dimensions by up to 15 percent [12-14]. This then leads to an underestimate of the magnified moments, M_c.

The lowest practical value for k in a sway frame is about 1.2 due to the lack of truly fixed connections at both ends of a column. When smaller values are obtained from the nomographs, it is good practice to use 1.2.

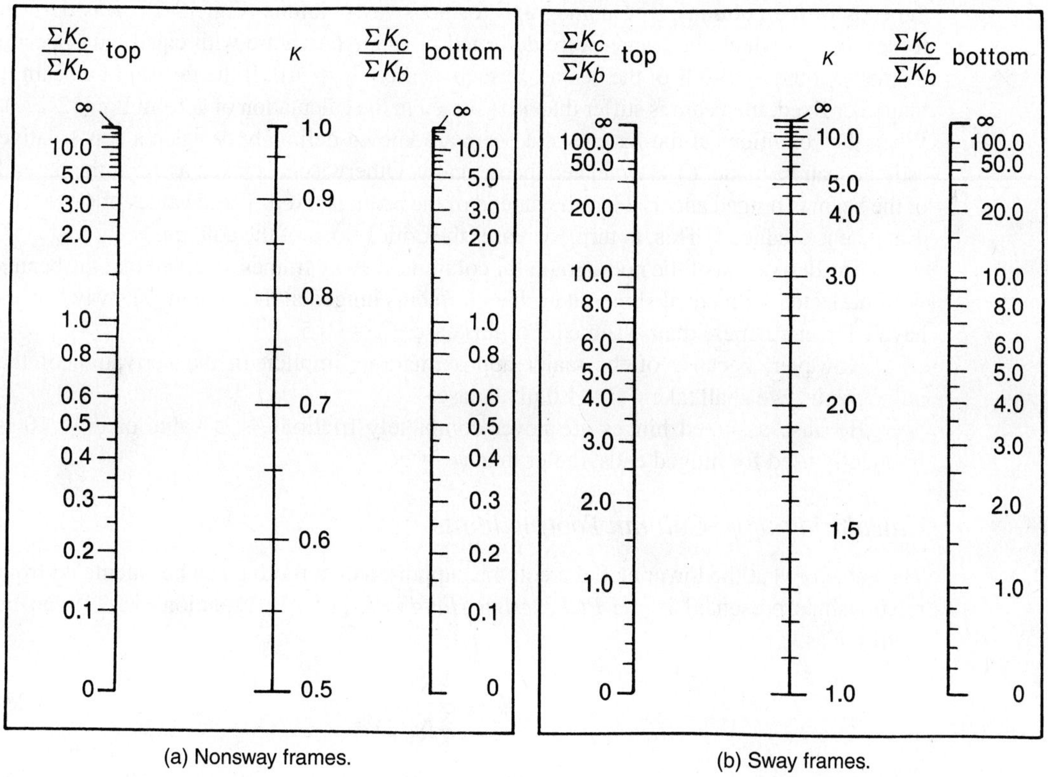

Fig. 12-20
Nomographs for effective length factors.

Calculation of ψ, Column-Beam Frames

The stiffness ratio, ψ, is calculated from Eq. (12-27a). The values of E_cI_c and E_bI_b should be realistic for the state of loading immediately prior to failure of the columns. Generally, at this stage of loading, the beams are extensively cracked and the columns are uncracked or slightly cracked. Ideally, the values of EI should reflect the degree of cracking and the actual reinforcement present. This is not practical, however, because this information is not known at this stage of design. The calculation of k should be based on a ψ based on I values given in ACI Code Section 6.6.3.1.1. In most cases the I values given in ACI Code Table 6.6.3.1.1(a) are used for the evaluation of ψ.

In calculating I_b for a T-beam, the flange width can be taken as defined in ACI Code Table 6.3.2.1. For common ratios of flange thickness to overall depth, h, and flange width to web width, b_w, the gross moment of inertia, I_g, is approximately twice the moment of inertia of a rectangular section with dimensions b_w and h.

In structural-steel design, the term ψ varies to account for the effects of fixity of the far ends of the beams, as expressed in the equation

$$\psi = \frac{\Sigma E_cI_c/\ell_c}{\Sigma C_bE_bI_b/\ell_b} \tag{12-27b}$$

where C_b accounts for the degree of fixity of the far ends of the beams that are attached to the ends of the columns. The nomograph for nonsway columns (Fig. 12-20) was derived for a frame in which the beams were deflected in single curvature with equal but opposite moments at the two ends of the beams. In such a case, $C_b = 1.0$. If the far end of a beam is hinged or fixed, the beam is stiffer than is assumed in the calculation of ψ from Eq. (12-27a). When the conditions at the far end of a beam are known definitely or when a conservative estimate can be made, C_b is evaluated theoretically. Otherwise it is taken as 1.5 if the far end of the beam is hinged and as 2.0 if the far end of the beam is fixed. These values of C_b reduce ψ and hence reduce k. This, in turn, increases the critical load of the column.

The derivation of the nomograph for columns in sway frames assumed that the beams were deflected with equal slopes at each end. Beams hinged at the far ends in sway frames have a lower stiffness than assumed. For this case, $C_b = 0.5$.

However, because of the major approximations implicit in the derivation of the nomographs, we shall take $C_b = 1.0$ in all cases.

Because assumed hinges are never completely frictionless, a value of $\psi = 10$ is frequently used for hinged ends, rather than $\psi = \infty$.

Calculation of ψ, Column Footing Joints

The value of ψ at the lower end of a column supported on a footing can be calculated from relationships presented in the *PCI Design Handbook* [12-17]. Equation (12-27) can be rewritten as

$$\psi = \frac{\Sigma K_c}{\Sigma K_b} \tag{12-28}$$

where ΣK_c and ΣK_b are the sums of the flexural stiffnesses of the columns and the restraining members (beams) at a joint, respectively. At a column-to-footing joint, $\Sigma K_c = 4E_cI_c/\ell_c$ for a braced column restrained at its upper end, and ΣK_b is replaced by the rotational stiffness of the footing and soil, taken equal to

$$K_f = \frac{M}{\theta_f} \tag{12-29}$$

where M is the moment applied to the footing and θ_f is the rotation of the footing. The stress under the footing is the sum of $\sigma = P/A$, which causes a uniform downward settlement, and $\sigma = My/I$, which causes a rotation. The rotation θ_f is

$$\theta_f = \frac{\Delta}{y} \tag{12-30}$$

where y is the distance from the centroid of the footing area and Δ is the displacement of that point y-distance from the centroid relative to the displacement of the centroid of the footing area. If k_s is the coefficient of subgrade reaction, defined as the stress required to compress the soil by a unit amount ($k_s = \sigma/\Delta$), then θ_f is

$$\theta_f = \frac{\sigma}{k_s y} = \frac{My}{I_f} \times \frac{1}{k_s y}$$

Substituting this into Eq. (12-29) gives

$$K_f = I_f k_s \tag{12-31}$$

where I_f is the moment of inertia of the contact area between the bottom of the footing and the soil and k_s is the coefficient of subgrade reaction, which can be taken from Fig. 12-21. Thus, the value of ψ at a footing-to-column joint for a column restrained at its upper end is

$$\psi = \frac{4E_c I_c/\ell_c}{I_f k_s} \tag{12-32}$$

The axis about which the footing rotates is in the plane of the footing–soil interface. As a result, the length ℓ_c used to compute ψ should include the depth of the footing.

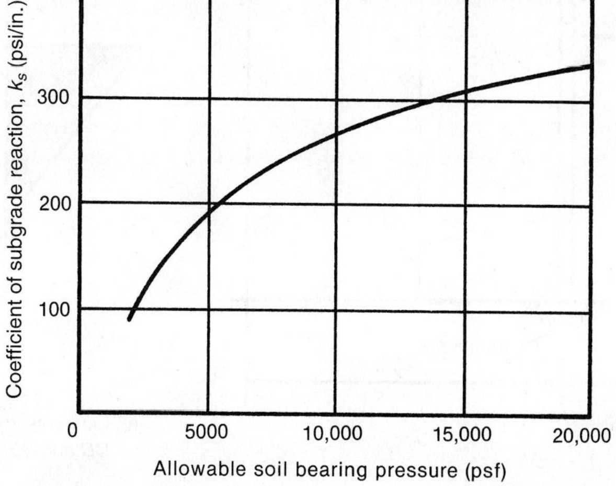

Fig. 12-21
Approximate relationship between allowable soil bearing pressure and the coefficient of subgrade reaction k_s. (From [12-17].)

EXAMPLE 12-2 Design of the Columns in a Braced Frame

Figure 12-22 shows part of a typical frame in an industrial building. The frames are spaced 20 ft apart. The columns rest on 4-ft-square footings. The soil bearing capacity is 4000 psf. Design columns *C–D* and *D–E*. Use $f'_c = 4000$ psi and $f_y = 60{,}000$ psi for beams and columns. Use load combinations and strength-reduction factors from ACI Code Tables 5.3.1 and 21.2.1.

1. Calculate the column loads from a frame analysis.
A first-order elastic analysis of the frame shown in Fig. 12-22 (loads acting on frame members are not shown) gave the forces and moments in the following table. The member end moments given in the table assume CCW is positive.

	Column *CD*	Column *DE*
Service loads, P	Dead = 80 kips	Dead = 50 kips
	Live = 24 kips	Live = 14 kips
Service moments at	Dead = 60 kip-ft	Dead = −42.4 kip-ft
tops of columns	Live = 14 kip-ft	Live = −11.0 kip-ft
Service moments at	Dead = 21 kip-ft	Dead = 32.0 kip-ft
bottoms of columns	Live = 8 kip-ft	Live = 8 kip-ft

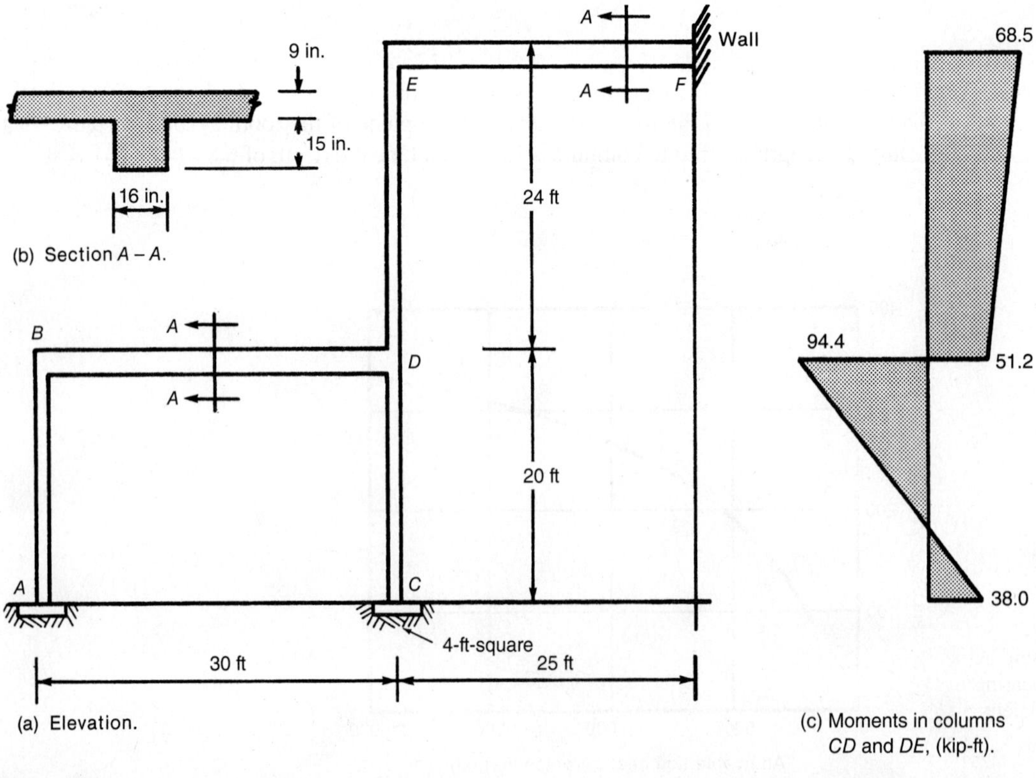

Fig. 12-22
Braced frame—Example 12-2.

The column moment diagrams in Fig. 12-22c assume moments that cause compression on the left side of the column are positive. All wind forces are assumed to be resisted by the end walls of the building.

2. Determine the factored loads.

(a) Column *CD*:

$$P_u = 1.2 \times 80 + 1.6 \times 24 = 134 \text{ kips}$$

$$\text{Moment at top} = 1.2 \times 60 + 1.6 \times 14 = 94.4 \text{ kip-ft}$$

$$\text{Moment at bottom} = 1.2 \times 21 + 1.6 \times 8 = 38.0 \text{ kip-ft}$$

The factored-moment diagram is shown in Fig. 12-22c. By definition, M_2 is always positive, and M_1 is negative if the column is bent in single curvature (Figs. 12-13c and 12-13d). Because column *CD* is bent in double curvature (Fig. 12-22c), M_1 is positive. Thus, for slender column design, $M_2 = +94.4$ kip-ft and $M_1 = +38.0$ kip-ft.

(b) Column *DE*:

$$P_u = 82.4 \text{ kips}$$

$$\text{Moment at top} = -68.5 \text{ kip-ft}$$

$$\text{Moment at bottom} = -51.2 \text{ kip-ft}$$

Thus, set $M_2 = +68.5$ kip-ft and $M_1 = -51.2$ kip-ft. M_1 is negative because the column is in single curvature.

3. Make a preliminary selection of the column size. From Eq. (11-19a) for $\rho_g = 0.015$,

$$A_{g(\text{trial})} \geq \frac{P_u}{0.40(f'_c + f_y\rho_g)}$$

$$= \frac{134}{0.40(4 + 0.015 \times 60)}$$

$$= 68.4 \text{ in.}^2 \text{ or } 8.3 \text{ in. square}$$

Because of the anticipated slenderness effects and because of the large moments, we will take a larger column. Try 14 in. × 14 in. columns throughout.

4. Are the columns slender? From ACI Code Section 6.2.5, a column in a braced frame is short if $k\ell_u/r$ is less than $34 + 12M_1/M_2$, which has an upper limit of 40.

(a) Column *CD*:

$$\ell_u = 20 \text{ ft} - 2 \text{ ft} = 18 \text{ ft}$$

$$= 216 \text{ in.}$$

$$r = 0.3 \times 14 \text{ in.} = 4.2 \text{ in.}$$

From Table 12-2, $k \cong 0.77$. Thus,

$$\frac{kl_u}{r} \cong \frac{0.77 \times 216}{4.2} = 39.6$$

From Eq. (12-20b),

$$34 + 12\left(\frac{M_1}{M_2}\right) = 34 + 12\left(\frac{38.0}{94.4}\right) = 38.8$$

Because 39.6 > 38.8, column *CD* is just slender.

(b) Column *DE*:

$$\ell_u = 24 \text{ ft} - 2 \text{ ft} = 264 \text{ in.}$$
$$k \cong 0.86$$
$$\frac{kl_u}{r} \cong \frac{0.86 \times 264}{4.2} = 54.1 > 40$$

Thus, column DE is also slender.

5. Check whether the moments are less than the minimum. ACI Code Section 6.6.4.5.4 requires that braced slender columns be designed for a minimum eccentricity of $(0.6 + 0.03h)$ in. For 14-in. columns, this is 1.02 in. Thus, column *CD* must be designed for a moment M_2 of at least

$$P_u e_{\min} = 134 \text{ kips} \times 1.02 \text{ in.} = 11.4 \text{ kip-ft}$$

and column *DE* for a moment of at least 7.0 kip-ft. Because the actual moments exceed these values, the columns shall be designed for the actual moments.

6. Compute *EI*. Because the reinforcement is not known at this stage of the design, we can use Eq. (12-15b) to compute *EI*. From Eq. (12-15b),

$$(EI)_{eff} = \frac{0.40\, E_c I_g}{1 + \beta_{dns}}$$

where

$$E_c = 57{,}000\sqrt{f'_c} = 3.60 \times 10^6 \text{ psi}$$
$$I_g = 14^4/12 = 3200 \text{ in.}^4$$
$$0.40\, E_c I_g = 4.61 \times 10^9 \text{ lb-in}^2.$$

(a) Column *CD*:

$$\beta_{dns} = \frac{1.2 \times 80}{134} = 0.716 \qquad (12\text{-}26a)$$

$$EI = \frac{4.61 \times 10^9}{1 + 0.716} = 2.69 \times 10^9 \text{ lb-in.}^2$$

(b) Column *DE*:

$$\beta_{dns} = \frac{1.2 \times 50}{82.4} = 0.728$$

$$(EI)_{eff} = \frac{4.61 \times 10^9}{1.728} = 2.67 \times 10^9 \text{ lb-in}^2.$$

7. Compute the effective-length factors. Two methods of estimating the effective-length factors, *k*, have been presented. In this example, we calculate *k* by both methods, to illustrate their use. In practice, only one of these procedures would be used in a given set of calculations. We begin with

$$\psi = \frac{\Sigma E_c I_c/\ell_c}{\Sigma E_b I_b/\ell_b} \qquad (12\text{-}27a)$$

where values for I_c and I_b can be taken as those given in ACI Code Table 6.6.3.1.1(a). Thus, $I_c = 0.70\, I_g(\text{col})$, and $I_b = 0.35\, I_g(\text{beam})$. For the assumed column section, use $I_c = 0.70 \times 14^4/12 = 2240\ \text{in.}^4$. For the beam section shown in Fig. 12-22b, the governing condition in ACI Code Table 6.3.2.1 gives the effective flange width as being equal to 103 in. Using that width gives $I_g = 38{,}200\ \text{in.}^4$, so $I_b = 0.35 \times 38{,}200 = 13{,}400\ \text{in.}^4$. As noted earlier in this text, I_g for the web portion of the beam section is equal to approximately one-half of I_g for the full-beam section (web and flange), and it often is used as an estimate of the cracking moment of inertia, I_{cr}, for the beam section. Thus, for the analysis used here, $0.35I_g(\text{beam})$ is equal to approximately $0.70\, I_g(\text{web}) = 0.7 \times 16 \times 24^3/12 = 12{,}900\ \text{in.}^4$. This approximation could be used at this stage to save the time required to determine the effective flange width and then calculate the moment of inertia of a flanged section.

In Eq. (12-27a), ℓ_c and ℓ_b are the spans of the column and beam, respectively, measured center-to-center from the joints in the frame.

(a) Column *DE*: The value of ψ at *E* is

$$\psi_E = \frac{E_c \times 2240/(24 \times 12)}{E_b \times 13{,}400/300} = 0.174\,\frac{E_c}{E_b}$$

where $E_c = E_b$. Thus, $\psi_E = 0.174$. The value of ψ at *D* is (for $E_c = E_b$):

$$\psi_D = \frac{E_c \times 2240/288 + E_c \times 2240/240}{E_b \times 13{,}400/360} = 0.460$$

The value of k from Fig. 12-20a is 0.63. The value of k from Table 12-2 was approximately 0.86.

As was pointed out in the discussion of Fig. 12-20, the effective-length nomographs tend to underestimate the values of k for beam columns in practical frames [12-14]. Because Table 12-2 gives reasonable values without the need to calculate ψ, it has been used to compute k in this example. Thus, we shall use $k = 0.86$ for column *DE*.

(b) Column *CD*: The value of ψ at *D* is

$$\psi_D = 0.460$$

The column is restrained at *C* by the rotational resistance of the soil under the footing and is continuous at *D*. From Eq. (12-32),

$$\psi = \frac{4E_c I_c/\ell_c}{I_f k_s}$$

where I_f is the moment of inertia of the contact area between the footing and the soil and k_s is the coefficient of subgrade reaction, which is 160 psi/in. from Fig. 12-21. Using , $I_c = 1.0\, I_g$ (not $0.7I_g$) and assuming a footing depth of 2 ft, so ℓ_c in Eq. (12-32) is 22 ft = 264 in., results in

$$I_f = \frac{48^4}{12} = 442{,}000\ \text{in.}^4$$

$$\psi_c = \frac{4 \times 3.60 \times 10^6\ \text{lb/in.}^2 \times 3200\ \text{in.}^4/264\ \text{in.}}{442{,}000\ \text{in.}^4 \times 160\ \text{lb/in.}^3}$$

$$= 2.47$$

From Fig. 12-20: $k = 0.77$
From Table 12-2: $k \cong 0.77$

Use $k = 0.77$ for column *CD*.

8. Compute the magnified moments. From Eq. (12-23),

$$M_c = \delta M_2 \tag{12-23}$$

where

$$\delta = \frac{C_m}{1 - (P_u/0.75P_c)} \geq 1.0 \tag{12-24}$$

(a) Column *CD*:

$$C_m = 0.6 - 0.4\left(\frac{38.0}{94.4}\right) \tag{12-14}$$

$$= 0.438$$

$$P_c = \frac{\pi^2(EI)_{eff}}{(k\ell_u)^2} \tag{12-25}$$

$$\ell_u = 18 \text{ ft} = 216 \text{ in}$$

Using EI from step 6 and $k = 0.77$,

$$P_c = \frac{\pi^2 \times 2.69 \times 10^9 \text{ lb-in.}^2}{(0.77 \times 216 \text{ in.})^2}$$

$$= 960{,}000 = 960 \text{ kips}$$

$$\delta = \frac{0.438}{1 - 134/(0.75 \times 960)}$$

$$= 0.538$$

Therefore, $\delta = 1.0$. This means that the section of maximum moment remains at the end of the column, so that

$$M_c = 1.0 \times 94.4 = 94.4 \text{ kip-ft}$$

Column *CD* is designed for $P_u = 134$ kips and $M_u = M_c = 94.4$ kip-ft.

(b) Column *DE*:

$$C_m = 0.6 - 0.4\left(-\frac{51.2}{68.5}\right) = 0.900$$

$$P_c = \frac{\pi^2 \times 2.67 \times 10^9}{(0.86 \times 264)^2} = 511 \text{ kips}$$

$$\delta = \frac{0.900}{1 - \dfrac{82.4}{0.75 \times 511}} \geq 1.0$$

$$= 1.15$$

This column is affected by slenderness, so

$$M_c = 1.15 \times 68.5 = 78.8 \text{ kip-ft}$$

Column *DE* is designed for $P_u = 82.4$ kips and $M_u = M_c = 78.8$ kip-ft.

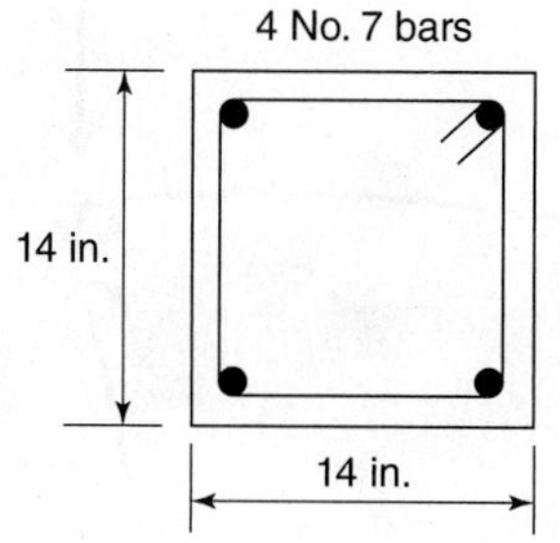

Fig. 12-23
Final column section for Example 12-2.

For both columns *CD* and *DE* the magnified moments are less than 1.4 times the first-order moments, as required by ACI Code Section 6.2.6.

9. Select the column reinforcement. Column *CD* is carrying both a higher axial load and a higher moment, and thus, it will govern the design of the column section. We will use the tied-column interaction diagrams in Appendix A, assuming an equal distribution of longitudinal bars in two opposite faces of the column. The parameters required for entering the interaction diagrams are

$$\gamma \cong \frac{14 \text{ in.} - 2 \times 2.5 \text{ in.}}{14 \text{ in.}} = 0.64 \cong 0.60$$

Assuming $\phi P_n = P_u = 134$ kips,

$$\frac{\phi P_n}{bh} = \frac{134 \text{ kips}}{14 \text{ in.} \times 14 \text{ in.}} = 0.68 \text{ ksi}$$

Assuming $\phi M_n = M_c = 94.4$ kip-ft $= 1130$ kip-in.,

$$\frac{\phi M_n}{bh^2} = \frac{1130 \text{ kip-in.}}{(14 \text{ in.})^3} = 0.41 \text{ ksi}$$

From Fig. A-6a ($\gamma = 0.60$), the required value for ρ_g is approximately 0.012. Thus,

$$A_{st}(\text{req'd}) = 0.012 \times A_g = 0.012 \times (14 \text{ in.})^2 = 2.35 \text{ in.}^2$$

Use four No. 7 bars, $A_{st} = 4 \times 0.60 \text{ in.}^2 = 2.40 \text{ in.}^2$ In summary, **use a 14 in. × 14 in. column section with four No. 7 bars**, as shown in Fig. 12-23. ■

12-4 BEHAVIOR OF RESTRAINED COLUMNS IN SWAY FRAMES

Statics of Sway Frames

A sway (unbraced) frame is one that depends on moments in the columns to resist lateral loads and lateral deflections. Such a frame is shown in Fig. 12-24a. The sum of the moments at the tops and bottoms of all the columns must equilibrate the applied lateral-load moment, $V\ell$, plus the moment due to the vertical loads, $\Sigma P\Delta$. Thus,

$$\Sigma\,(M_{top} + M_{btm}) = V\ell + \Sigma P\Delta \tag{12-33}$$

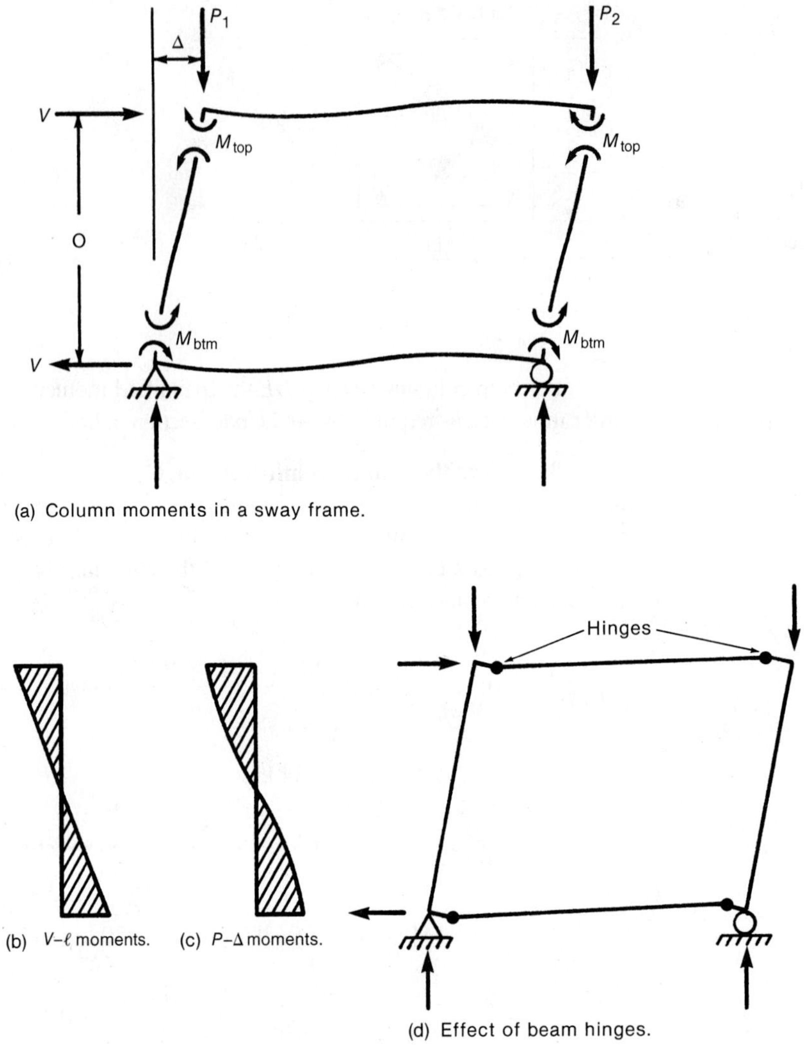

Fig. 12-24
Column moments in a sway frame.

It should be noted that both columns have deflected laterally by the same amount Δ. For this reason, it is not possible to consider columns independently in a sway frame.

If a sway frame includes some pin-ended columns, as might be the case in a precast concrete building, the vertical loads in the pin-ended columns are included in ΣP in Eqs. (12-21) and (12-33). Such columns are referred to as *leaning columns*, because they depend on the frame for stability.

The V–ℓ moment diagram due to the lateral loads is shown in Fig. 12-24b and that due to the P–Δ moments in Fig. 12-24c. Because the maximum lateral-load moments and the maximum P–Δ moments both occur at the ends of the columns, and hence can be added directly, the equivalent moment factor, C_m, given by Eq. (12-14) does not apply to sway columns. On the other hand, Eq. (12-11) becomes

$$M_c = \frac{M_o(1 - 0.18P/P_E)}{1 - P/P_E} \tag{12-34}$$

The term in brackets in Eqs. (12-11) and (12-34) has been left out of the Moment-Magnifier equation in the ACI Code because the resulting change in the magnification does not vary significantly.

It is also important to note that if hinges were to form at the ends of the beams in the frame as shown in Fig. 12-24d, the frame would be unstable. Thus, the beams must resist the *full magnified end moment* from the columns for the frame to remain stable (ACI Code Section 6.6.4.6.3).

Loads causing sway are seldom sustained, although such cases can be visualized, such as a frame that resists the horizontal reaction from an arch roof or a frame resisting lateral earth loads. If a sustained load acts on an unbraced frame, the deflections increase with time, leading directly to an increase in the P–Δ moment. This process is very sensitive to small variations in material properties and loadings. As a result, structures subjected to sustained lateral loads should always be braced. Indeed, braced frames should be used wherever possible, regardless of whether the lateral loads are short time or sustained.

M_{ns} and M_s Moments

Two different types of moments occur in sway frames:

1. Moments due to loads not causing appreciable sway, M_{ns}
2. Moments due to loads causing appreciable sway, M_s

The slenderness effects of these two kinds of moments are considered separately in the ACI Code design process because each is magnified differently as the individual columns deflect and as the entire frame deflects [12-18]. Column moments that cause no appreciable sway are magnified when the column deflects by an amount δ relative to its original straight axis such that the moments at points along the length of the column exceed those at the ends. In Section 12-2, this is referred to as the *member stability* effect or P–δ effect, where the lower case δ refers to deflections relative to the chord joining the ends of the column.

On the other hand, the column moments due to lateral loads can cause appreciable sway. They are magnified by the P–Δ moments resulting from the sway deflections, Δ, at joints in the frame, as indicated by Eq. (12-33). This is referred to as the P–Δ effect or *lateral drift* effect.

ACI Code Section 2.2 defines the *nonsway moment*, M_{ns}, as the factored end moment on a column due to loads that cause *no appreciable sidesway*, as computed by a first-order elastic frame analysis. These moments normally result from gravity loads. In general, "no appreciable sway" is defined as a lateral deflection of less than $1/1500$ of the story height at factored load levels. Gravity loads will cause small lateral deflections, except in the case of symmetrically loaded, symmetrical frames. The *sway moment*, M_s, is defined in ACI Code Section 2.1 as the factored end moment on a column due to loads that cause appreciable sidesway, calculated by a first-order elastic frame analysis. These moments result from either lateral loads or large unsymmetrical gravity loads, or gravity loads on highly unsymmetrical frames.

Treating the P–δ and P–Δ moments separately simplifies design. The nonsway moments frequently result from a series of pattern loads, as was discussed in Chapter 10. The pattern loads can lead to a moment envelope for the nonsway moments. The maximum end moments from the moment envelope are then combined with the magnified sway moments from a second-order analysis or from a sway moment-magnifier analysis.

12-5 CALCULATION OF MOMENTS IN SWAY FRAMES USING SECOND-ORDER ANALYSIS

First-Order and Second-Order Analysis

A *first-order frame analysis* is one in which the effect of lateral deflections on bending moments, axial forces, and lateral deflections is ignored. The resulting moments and deflections are linearly related to the loads. In a *second-order frame analysis*, the effect of deflections on moments, and so on, is considered. The resulting moments and deflections include the effects of slenderness and hence are nonlinear with respect to the load. Because the moments are directly affected by the lateral deflections, as shown in Eq. (12-33), it is important that the stiffnesses, EI, used in the analysis be representative of the stage immediately prior to yielding of the flexural reinforcement.

Second-Order Analysis

Load Level for the Analysis

In a second-order analysis, column moments and lateral frame deflections increase more rapidly than do loads. Thus, it is necessary to calculate the second-order effects at the factored load level. The load combinations, load factors, and strength-reduction factors from ACI Code Tables 5.3.1 and 21.2.1 can be used.

Lateral Stiffness-Reduction Factor

The term 0.75 in the denominator of Eq. (12-24) is the lateral stiffness-reduction factor, ϕ_K, which accounts for variability in the critical load, P_c, and for variability introduced by the assumptions in the moment-magnifier calculation. This factor leads to an increase in the magnified moments. A similar but higher stiffness-reduction factor appears in a second-order analysis. Two things combine to allow the use of a value of ϕ_K larger than 0.75 when a second-order analysis is carried out. First, the modulus of elasticity, E_c, used in the frame analysis is based on the specified strength, f'_c, even though the deflections are a function of the E_c that applies to the mean concrete strength, that is, a concrete strength which is 600 to 1400 psi higher than f'_c. Second, the second-order analysis is a better representation of the behavior of the frame than the sway magnifier given later by Eq. (12-39). The moments of inertia given in ACI Code Table 6.6.3.1.1(a) have been multiplied by 0.875, which, when combined with the underestimate of E_c, lead to an overestimation of the second-order deflections on the order of from 20 to 25 percent, corresponding to an implicit value for ϕ_K of 0.80 to 0.85 in design based on second-order analyses.

Stiffnesses of the Members

Strength Limit State. The stiffnesses appropriate for strength calculations must estimate the lateral deflections accurately at the factored load level. They must be simple to apply, because a frame consists of many cross-sections, with differing reinforcement ratios and differing degrees of cracking. Furthermore, the reinforcement amounts and distributions are not known at the time the analysis is carried out. Using studies of the flexural stiffness of beams with cracked and uncracked regions, MacGregor and Hage [12-12] recommended that the beam stiffnesses be taken as $0.4E_cI_g$ when carrying out a second-order analysis. In

ACI Code Table 6.6.3.1.1(a), this value has been multiplied by a stiffness-reduction factor of 0.875, giving $I = 0.35I_g$.

Two levels of behavior must be distinguished in selecting the *EI* of columns. The lateral deflections of the frame are influenced by the stiffness of all the members in the frame and by the variable degree of cracking of these members. Thus, the *EI* used in the frame analysis should be an average value. On the other hand, in designing an individual column in a frame in accordance with Eq. (12-24), the *EI* used in calculating δ must be for that column. This *EI* must reflect the greater chance that a particular column will be more cracked, or weaker, than the overall average; hence, this *EI* will tend to be smaller than the average *EI* for all the columns acting together. Reference [12-12] recommends the use of $EI = 0.8E_cI_g$ in carrying out second-order analyses of frames. ACI Code Table 6.6.3.1.1(a) gives this value multiplied by 0.875, or $EI = 0.70E_cI_g$ for this purpose. On the other hand, in calculating the moment magnifiers δ and δ_s from Eqs. (12-24) and (12-39), *EI* must be taken as given by Eqs. (12-15) or (12-16).

The value of *EI* for shear walls may be taken equal to the value for beams in those parts of the structure where the wall is cracked by flexure or shear and equal to the value for columns where the wall is uncracked. If the factored moments and shears from an analysis based on $EI = 0.70E_cI_g$ for the walls indicate that a portion of the wall will crack due to stresses reaching the modulus of rupture of the wall concrete, the analysis should be repeated using $EI = 0.35E_cI_g$ for the cracked parts of the wall.

Serviceability Limit State. The moments of inertia given in ACI Code Table 6.6.3.1.1(a) are for the strength limit state. At service loads, the members are cracked less than they are at nominal strength. In computing deflections or vibrations at service loads, the values of *I* should be representative of the degree of cracking at service loads. ACI Code Table 6.6.3.1.1(b) includes two other equations to calculate a refined value for the section moment of inertia, *I*, when all of the member loads and longitudinal reinforcement are known. These equations can be used to calculate deflections under either service loads or strength-level loads [12-19], [12-20]. The equation for compression members (columns) is

$$I = \left(0.8 + \frac{25A_{st}}{A_g}\right)\left(1 - \frac{M_u}{P_uh} - \frac{0.5P_u}{P_o}\right)I_g \leq 0.875I_g \qquad \text{(12-35)}$$

where P_u and M_u are either for the particular load combination under consideration or are from the combination of loads that results in the smallest value of *I* in Eq. (12-35). For service loads, the P_u and M_u values can be replaced by the acting loads. The value of *I* from Eq. (12-35) does not need to be taken less than $0.35\,I_g$.

The equation for flexural members (beams and slabs) is

$$I = (0.10 + 25\rho)\left(1.2 - \frac{0.2b_w}{d}\right)I_g \leq 0.5I_g \qquad \text{(12-36)}$$

where ρ is the tension-reinforcement ratio at the section under consideration. For continuous flexural members, *I* can be taken as the average of the values obtained from Eq. (12-36) for the sections resisting the maximum positive and the maximum negative moments. The value of *I* from Eq. (12-36) does not need to be taken less than $0.25\,I_g$.

If the frame members are subjected to sustained lateral loads, the *I*-value in Eq. (12-35) for compression members must be divided by $1 + \beta_{ds}$, where β_{ds} is defined in Eq. (12-26b), which is given in the following paragraphs.

Foundation Rotations

The rotations of foundations subjected to column end moments reduce the fixity at the foundations and lead to larger sway deflections. These are particularly significant in the case of shear walls or large columns, which resist a major portion of the lateral loads. The effects of foundation rotations can be included in the analysis by modeling each foundation as an equivalent beam having the stiffness given by Eq. (12-31).

Effect of Sustained Loads

Loads causing appreciable sidesway are generally short-duration loads, such as wind or earthquake, and, as a result, do not cause creep deflections. In the unlikely event that sustained lateral loads act on an unbraced structure, the *EI* values used in the frame analysis should be reduced. ACI Code Section 6.6.3.1.1 states that in such a case, *I* shall be divided by $(1 + \beta_{ds})$, where, for this case, β_{ds} is defined in ACI Code Section 2.2 as

$$\beta_{ds} = \frac{\text{maximum factored sustained shear within a story}}{\text{total factored shear in that story}} \tag{12-26b}$$

Methods of Second-Order Analysis

Computer programs that carry out second-order analyses are widely available. The principles of such an analysis are presented in the following paragraphs. Methods of second-order analysis are reviewed in [12-12] and [12-21].

Iterative P–Δ Analysis

When a frame is displaced sideways under the action of lateral and vertical loads, as shown in Figs. 12-24 and 12-25, the column end moments must equilibrate the lateral loads and a moment equal to $(\Sigma P)\Delta$; that is,

$$\Sigma\,(M_{\text{top}} + M_{\text{btm}}) = V\ell_c + \Sigma P\Delta \tag{12-33}$$

where Δ is the lateral deflection of the top of the story relative to the bottom of the story. The moment $\Sigma P\Delta$ in a given story can be represented by shear forces, $(\Sigma P)\,\Delta/\ell_c$, where ℓ_c is the story height, as shown in Fig. 12-25b. These shears give an overturning moment of $(\Sigma P\Delta/\ell_c) \times (\ell_c) = (\Sigma P)\Delta$. Figure 12-25c shows the story shears in two different stories. The algebraic sum of the story shears from the columns above and below a given floor gives rise to a *sway force* acting on that floor. At the *j*th floor, the sway force is

$$\text{Sway force } j = \frac{(\Sigma P_i)\Delta_i}{\ell_i} - \frac{(\Sigma P_j)\Delta_j}{\ell_j} \tag{12-37}$$

where a positive $P\Delta/\ell$ moment and a positive sway force both correspond to forces that would overturn the structure in the same direction. The sway forces are added to the applied lateral loads at each floor level, and the structure is reanalyzed, giving new lateral deflections and larger column moments. If the deflections increase by more than about 2.5 percent, new $\Sigma P\Delta/\ell$ forces and sway forces are computed and the structure is reanalyzed for the sum of the applied lateral loads and the new sway forces. This process is continued until convergence is obtained.

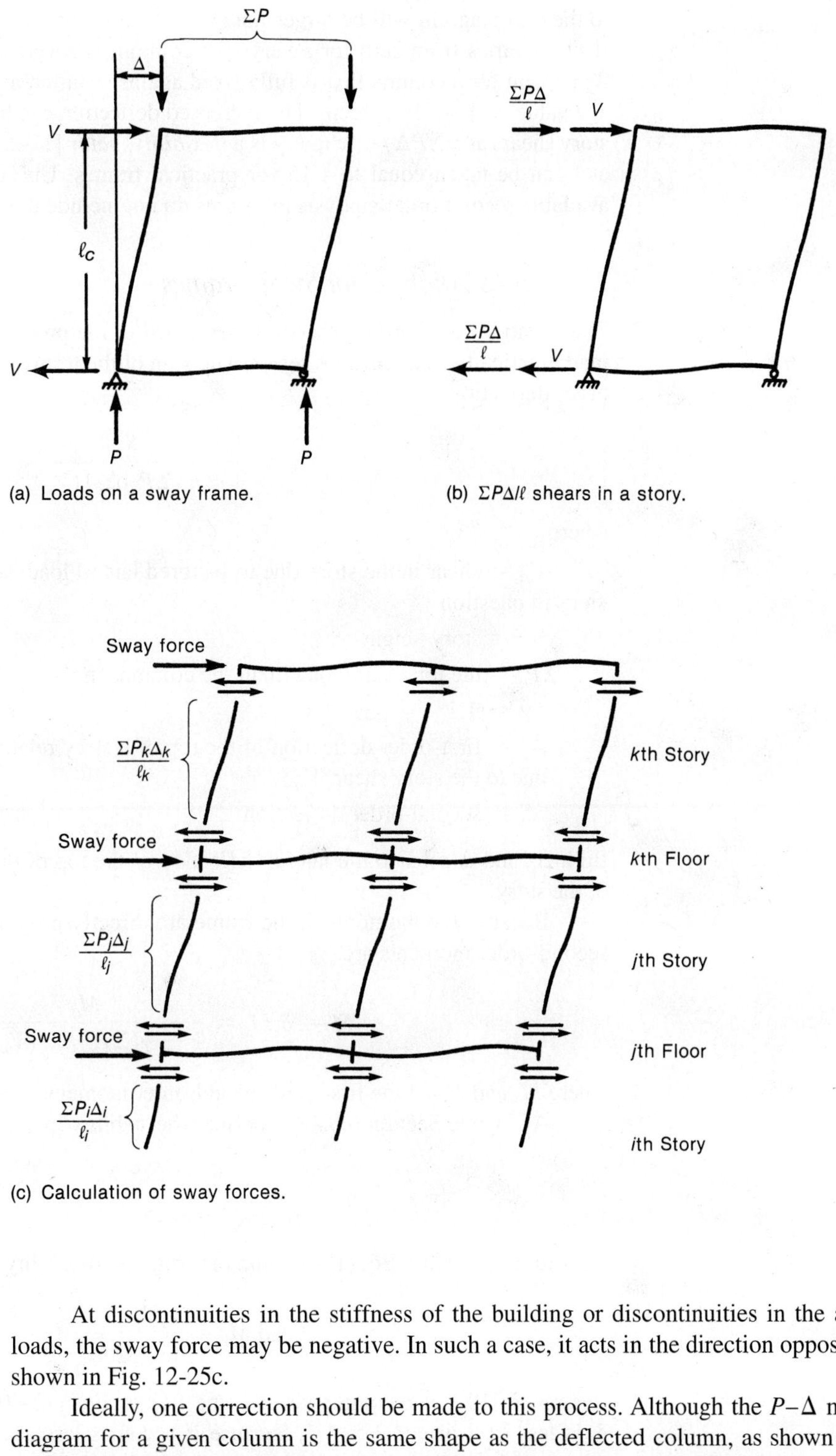

Fig. 12-25
Iterative P–Δ analyses.

At discontinuities in the stiffness of the building or discontinuities in the applied loads, the sway force may be negative. In such a case, it acts in the direction opposite that shown in Fig. 12-25c.

Ideally, one correction should be made to this process. Although the P–Δ moment diagram for a given column is the same shape as the deflected column, as shown in Fig. 12-24c, the moment diagram due to the $P\Delta/\ell$ shears is a straight-line diagram, similar to the $V\ell$ moments shown in Fig. 12-24b, compared to the curved-moment diagram. As a result, the area of the real P–Δ moment diagram is larger than that of the straight-line moment diagram. It can be shown by the moment–area theorems that the deflections due to the real diagram will be larger than those due to the $P\Delta/\ell$ shears. The increase in deflection varies from zero for a very stiff column with very flexible restraining beams, to 22 percent for a column that is fully fixed against rotation at each end. A reasonable average value is about 15 percent. The increased deflection can be accounted for by taking the story shears as $\gamma\Sigma P\Delta/\ell$, where γ is a *flexibility factor* [12-21] that ranges from 1.0 to 1.22 and can be taken equal to 1.15 for practical frames. Unfortunately, most commercially available second-order analysis programs do not include this correction.

Direct P–Δ Analysis for Sway Frames

The iterative calculation procedure described in the preceding section can be described mathematically as an infinite series. The sum of the terms in this series gives the second-order deflection

$$\Delta = \frac{\Delta_o}{1 - \gamma(\Sigma P_u)\Delta_o/(V_{us}\ell_c)} \tag{12-38}$$

where

V_{us} = shear in the story due to factored lateral loads acting on the frame above the story in question

ℓ_c = story height

ΣP_u = the total axial load in all the columns in the story

$\gamma \simeq 1.15$

Δ_o = first-order deflection of the top of a story relative to the bottom of the story, due to the story shear, V_{us}

Δ = second-order deflection

Both Δ_o and Δ refer to the lateral deflection of the top of the story relative to the bottom of the story.

Because the moments in the frame are directly proportional to the deflections, the second-order moments are

$$M = \delta_s M_s = \frac{M_o}{1 - \gamma(\Sigma P_u)\Delta_o/(V_{us}\ell_c)} \tag{12-39}$$

where M_o and M are the first- and second-order moments, respectively.

ACI Code Section 6.6.4.4.1 defines the stability index for a story as

$$Q = \frac{\Sigma P_u \Delta_o}{V_{us}\ell_c} \tag{12-21}$$

Substituting this into Eq. (12-39) and omitting the flexibility factor γ gives

$$\delta_s M_s = \frac{M_s}{1 - Q} \geq M_s \tag{12-40}$$

Reference [12-21] recommends that the use of Eq. (12-40) be limited to cases where δ_s is less than or equal to 1.5, because it becomes less accurate for higher values. This corresponds to $Q \leq 1/3$. For this reason, ACI Code Section 6.6.4.6.2 limits the use of Eq. (12-40) to $\delta_s \leq 1.5$. Example 12-3 illustrates the impact of Eq. (12-40) on magnified moments in a frame.

12-6 DESIGN OF COLUMNS IN SWAY FRAMES

Overview

Starting with the 2008 edition of the ACI Building Code, sections dealing with the design of slender columns were rewritten and rearranged. Because of the general availability of structural analysis software that includes secondary bending effects in sway frames, recommendations for the use of either elastic or inelastic second-order analyses are presented in ACI Code Sections 6.7 and 6.8, respectively. Moment magnification procedures, which are used in conjunction with first-order analysis methods, are given in ACI Code Section 6.6.4.6 for sway frames.

A separate stability check for sway frames, which was required in prior editions of the ACI Code, is now covered by ACI Code Section 6.2.6, which requires that the total secondary moments in compression members shall not exceed 1.4 times the first-order moments calculated for that member. Prior analytical studies [12-12] have shown that the probability of a stability failure increases when the stability index, Q, calculated in Eq. (12-21), exceeds 0.2. This is similar to the limit of 0.25 set by ASCE/SEI 7-10 [12-22] for their definition of a stability coefficient denoted as θ. Using a value of 0.25 results in a secondary-to-primary moment ratio of 1.33, which is the basis for the ACI Code limit of 1.4 for that ratio. If that limit is satisfied, a separate stability check is not required.

ACI Code Section 6.6.4.6.4 states that secondary bending effects shall be considered along the length of a compression member, including columns that are part of a sway frame (sway story). Normally, structural analysis software that includes second-order effects will give the maximum moments at the ends of a column. For slender columns within a sway frame, it is possible that the maximum secondary moment will occur between the ends of the column. This could be accounted for by either subdividing the column (adding extra nodes) along its length or by applying the moment magnification procedure for nonsway frames, which was covered earlier in Sections 12-2 and 12-3.

Computation of $\delta_s M_s$

The following subsections discuss the ACI Code procedures for computing the magnified sway moments, $\delta_s M_s$, in a sway frame.

1. Computation of $\delta_s M_s$ Using Second-Order Analyses

ACI Code Sections 6.7 and 6.8 allow the use of second-order analyses to compute $\delta_s M_s$. Elastic second-order analyses were discussed in Section 12-5. If torsional displacements of the frame are significant, a three-dimensional second-order analysis should be used.

2. Computation of $\delta_s M_s$ Using Direct P–Δ Analysis

As part of the moment magnification method for sway frames, ACI Code Section 6.6.4.6 permits the use of a direct calculation of P–Δ moments via an equation similar to Eq. (12-40). In ACI Code Section 6.6.4.6.2, this is written as

$$\delta_s = \frac{1}{1 - Q} \geq 1 \tag{12-41}$$

where

$$Q = \frac{\Sigma P_u \Delta_o}{V_{us} \ell_c} \tag{12-21}$$

Although there is no easy way to incorporate torsional effects into the calculation of $\delta_s M_s$, Section 12-8 discusses the extension of the Q method to second-order torsion.

3. Computation of $\delta_s M_s$ Using Sway-Frame Moment Magnifier

ACI Code Section 6.6.4.6.2 also allows the use of the traditional sway-frame moment magnifier for computing the magnified sway moments

$$\delta_s = \frac{1}{1 - \Sigma P_u / (0.75\ \Sigma P_c)} \geq 1 \tag{12-42}$$

Here, ΣP_u and ΣP_c refer to the sums of the axial loads and critical buckling loads, respectively, for all the columns in the story being analyzed. In this case, the values of P_c are calculated by using the effective lengths, $k\ell_u$, evaluated for columns in a sway frame, with β_{ds} defined as

$$\beta_{ds} = \frac{\text{maximum factored sustained shear in the story}}{\text{total factored shear in the story}} \tag{12-26b}$$

In most sway frames, the story shear is due to wind or seismic loads and hence is not sustained, resulting in $\beta_{ds} = 0$. The use of the summation terms in Eq. (12-42) accounts for the fact that sway instability involves all the columns and bracing members in the story. (See Fig. 12-25.) The format including the summations was first presented in [12-1].

If $(1 - \Sigma P_u / 0.75\ \Sigma P_c)$ is negative, the load on the frame, ΣP_u, exceeds the buckling load for the story, ΣP_c, indicating that the frame is unstable. A stiffer frame is required.

Moments at the Ends of the Columns

The unmagnified nonsway moments, M_{ns}, are added to the magnified sway moments, $\delta_s M_s$, at each end of each column to get the total design moments:

$$M_1 = M_{1ns} + \delta_s M_{1s} \tag{12-43a}$$

$$M_2 = M_{2ns} + \delta_s M_{2s} \tag{12-43b}$$

The addition is carried out for the moments at the top and bottom of each column. The larger absolute sum of the resulting end moments for a given column is called M_2, and the smaller is called M_1. By definition, M_2 is always taken as positive, and M_1 is taken as negative if the column is bent in single curvature.

Maximum Moment between the Ends of the Column

In most columns in sway frames, the maximum moment will occur at one end of the column and will have the value given by Eqs. (12-43a) and (12-43b). *Occasionally, for very slender, highly loaded columns*, the deflections of the column can cause the maximum column P–δ moment to exceed the P–Δ moment at one or both ends of the column. As stated in the overview for this section, ACI Code Section 6.6.4.6.4 calls attention to this potential problem and ACI Commentary Section R6.6.4.6.4 suggests that this can be accounted for in structural analysis by adding nodes along the length of the column.

This is a rare occurrence and prior editions of the ACI Code used Eq. (12-44) to identify columns that may have $P-\delta$ moments between the ends of the column that exceed the $P-\Delta$ moments at the ends.

$$\frac{\ell_u}{r} > \frac{35}{\sqrt{\frac{P_u}{f'_c A_g}}} \tag{12-44}$$

If ℓ_u/r exceeds the value given by Eq. (12-44), there is a chance that the maximum moment on the column will exceed the larger end moment, M_2 [12-23]. This would occur if M_c, computed from Eq. (12-23), was larger than the end moments M_1 and M_2 from Eqs. (12-43a) and (12-43b). If $M_c < M_2$, the maximum design moment is at the end of the column and is equal to M_2. If $M_c \geq M_2$, the maximum design moment occurs between the ends of the column and is equal to M_c.

Sidesway Buckling under Gravity Loads

Another very rare event is the classical case of sidesway buckling under gravity loads alone. As stated in the overview, ACI Code Section 6.2.6 guards against this by requiring that the secondary-to-primary moment ratio shall not exceed 1.4.

Minimum Moment

The ACI Code specifies a minimum moment $M_{2,\min}$ (Eq. 12-22) to be considered in the design of columns in nonsway frames, but not for columns in sway frames. This will be a problem only for the gravity load combination $U = 1.2D + 1.6L$ acting on a sway frame, because this load combination does not involve $\delta_s M_s$. For this load combination, we should design for the larger of M_2 and $M_{2,\min}$.

EXAMPLE 12-3 Design of the Columns in a Sway Frame

Figure 12-26 shows the elevation and main floor plan for a five-story building. The building is clad with nonstructural precast panels. There are no structural walls or other bracing. The floor beams in the north–south direction are all 18 in. wide with an overall depth of 30 in. The floor slabs are 6 in. thick. Assume all the floors are to be designed for a superimposed dead load of 20 psf plus a live load of 80 psf, which includes a 20 psf partition loading. The roof is assumed to have similar-sized structural members, carries a superimposed dead load of 25 psf, and has a reduced live load, L_r, of 30 psf. Design an interior and exterior column in the first-story level for dead load, live load, and wind forces in the north–south direction. Assume we are using "service-level" wind forces, so the appropriate load factor is 1.6. Use $f'_c = 4000$ psi and $f_y = 60$ ksi.

1. Make a preliminary selection of the column sizes. The preliminary sizing of the column sections will be based on the axial loads in the first-story columns for the gravity-load case $(1.2D + 1.6L + 0.5L_r)$. Based on the tributary areas on the roof and

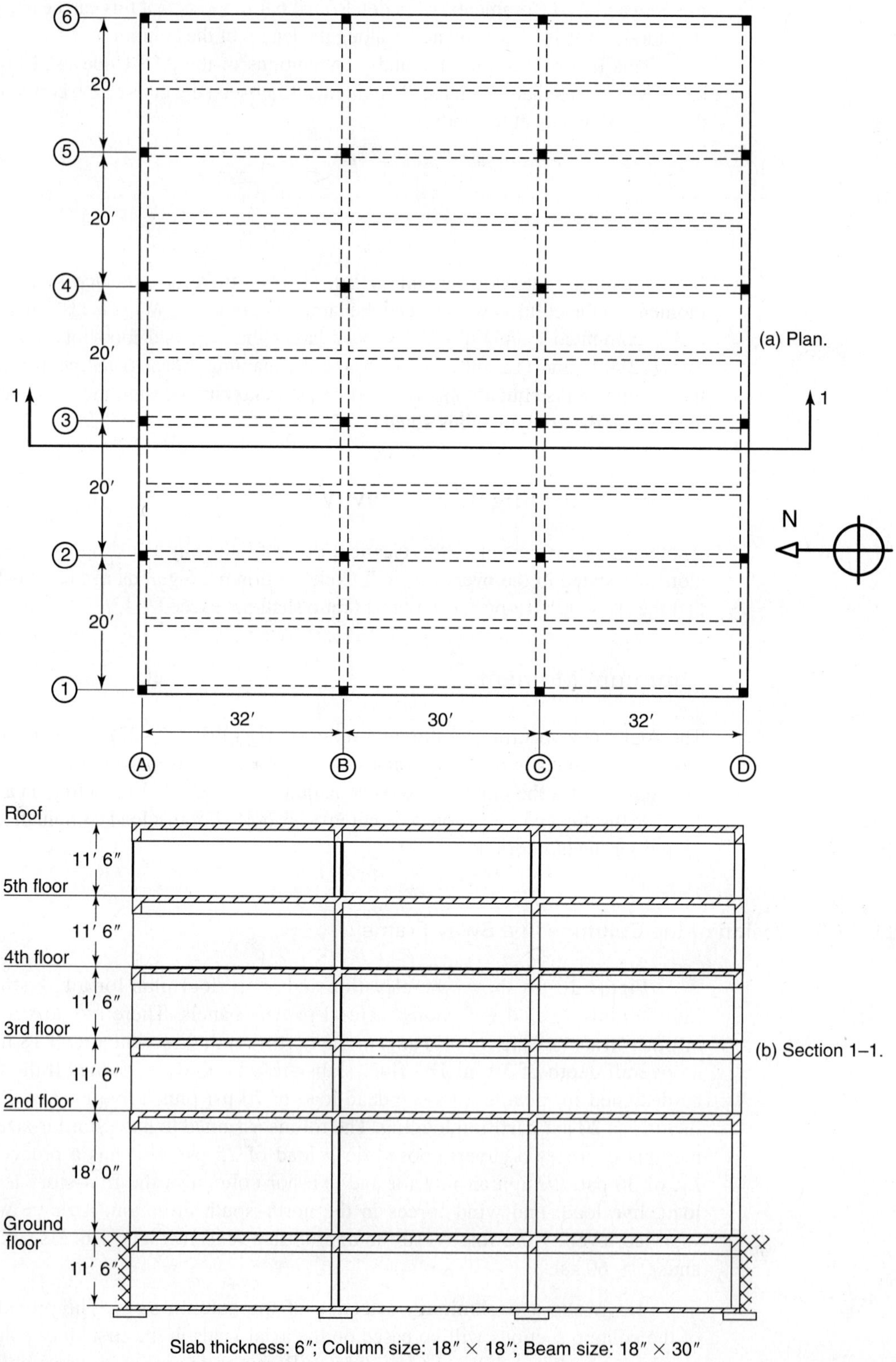

Fig. 12-26
Plan and elevation of sway frame—Example 12-3.

each floor above the main floor, the following unfactored dead and (unreduced) live loads were calculated for the first story columns:

Exterior Column:

Dead load = 284 kips
Live load = 10.1 kips from the roof and 93.3 kips from floor loads

Interior Column:

Dead load = 486 kips
Live load = 18.4 kips from the roof and 173 kips from floor loads

ASCE/SEI 7 [12-22] allows the floor live loads to be reduced as a function of the influence area, A_I, which is taken as a multiple of the tributary area, A_T. For columns, the influence area is taken as four times the tributary area. (See Section 2-8, Fig. 2-9.) For a first-story exterior column in this example, the total tributary area is

$$A_T = (20 \text{ ft} \times 32 \text{ ft}/2) \times 4 \text{ floors} = 1280 \text{ ft}^2$$

Then, from Eq. (2-12), the live load-reduction factor is

$$\frac{L}{L_o} = 0.25 + \frac{15}{\sqrt{4 \times A_T}} \cong 0.46$$

but cannot be less than 0.5 for columns supporting one floor, nor less than 0.4 for columns supporting two or more floors. For this problem, we can use the calculated reduction factor of 0.46 for the first-story exterior columns. Using a similar procedure, the calculated live-load reduction factor for a first-story interior column is equal to approximately 0.40, which is the minimum permissible value. Thus, the modified factored design loads for the exterior and interior columns are

Exterior Column:

Reduced floor live load = 0.46 × 93.3 k = 42.9 kips
Total factored load = 1.2 × 284 k + 1.6 × 42.9 k + 0.5 × 10.1 k = 414 kips

Interior column:

Reduced floor live load = 0.40 × 173 k = 69.2 kips
Total factored load = 1.2 × 486 k + 1.6 × 69.2 k + 0.5 × 18.4 k = 703 kips

For an exterior column, use Eq. (11-19a) with $\rho_g = 0.015$ to get a preliminary size for the column section.

$$A_{g(\text{trial})} \geq \frac{P_u}{0.40(f'_c + f_y\rho_g)}$$

$$\geq \frac{414 \text{ kips}}{0.40(4 \text{ ksi} + 0.015 \times 60 \text{ ksi})} = 211 \text{ in.}^2 \; (\cong 14.5 \text{ in.})^2$$

We will try an 18 in. × 18 in. section to allow for both the probable slenderness considerations and the moments carried by the exterior columns for the gravity-load case.

For an interior column, again assuming $\rho_g = 0.015$, a preliminary section size is

$$A_{g(\text{trial})} \geq \frac{703 \text{ kips}}{0.40(4 \text{ ksi} + 0.015 \times 60 \text{ ksi})} = 359 \text{ in.}^2 \cong (18.9 \text{ in.})^2$$

For this column, we will try a 18 in. × 18 in. section size and increase the reinforcement percentage, if necessary.

For upcoming analyses, we will assume the same column section sizes are used throughout the height of the structural frame.

2. Factored load combinations to be considered. The load combinations from ACI Code Table 5.3.1 will be used, and thus, the different load cases to be considered are

Case 1: Gravity loads, $U = 1.2D + 1.6L + 0.5L_r$

Case 2: Gravity plus wind loads, $U = 1.2D + 1.6W + 1.0L + 0.5L_r$. Assuming that we are designing a typical office building, and not a place of public assembly, ACI Code Section 5.3.3 permits the load factor for live load to be reduced to 0.5. Also, ACI Code Section 5.3.5 requires that the load factor for the wind load shall be 1.6 if the calculated wind loads are for service-level loads.

Case 3: Low gravity plus wind, $U = 0.9D + 1.6W$.
Again, for this load case we will use the load factor of 1.6.

Because this is a symmetrical building, the gravity loads will not cause appreciable sidesway, and thus, the dead and live loads will only give rise to nonsway moments, M_{ns}. The wind loads will cause sway moments, M_s. For clarity, all of the calculations for one load case will be completed before checking the next load case. We will start with Load Case 1.

Load Case 1: Gravity Loads Only

3. Is the story being designed sway or nonsway? It should be noted that even if the first story is found to be a sway story, the column end moment for this gravity loading will be multiplied by the nonsway magnifier, δ, because the gravity loads will not tend to cause sway deflections in this symmetrical frame. ACI Code Section 6.6.4.3(b) defines a story in a frame as being nonsway if

$$Q = \frac{\Sigma P_u \times \Delta_o}{V_{us} \times \ell_c} \leq 0.05 \tag{12-21}$$

where for the first story,

ΣP_u = total factored load in all 24 first-story columns = 11,400 kips (assuming all of the interior and exterior columns carry the same loads as previously calculated and that the corner columns carry one-half of the calculated load for an exterior column)
ℓ_c = total story height = 18 ft = 216 in.

There is no lateral load for this loading case, so to use Eq. (12-21), we need to apply an arbitrary lateral load representing V_{us} at the top of the first story and then calculate the resulting lateral deflection in that story, Δ_o. This lateral load must be applied in conjunction with the factored dead and live loads used for the gravity-load case. The loading and analytical model for this analysis is shown in Fig. 12-27a. A value of 20 kips was selected for the lateral loading, V_{us}.

For this analysis and all subsequent analyses for this frame, we need to use the member moments of inertia recommended in ACI Code Table 6.6.3.1.1(a). Thus, the moment of inertia for the column members shall be taken as

$$I_c = 0.7 \times I_g = 0.7 \times (18\text{ in.})^4/12 = 6120\text{ in.}^4$$

For the beams, ACI Code Table 6.6.3.1.1(a) requires that we use 0.35 I_g. Rather than going through the process of determining the effective flange width for the floor beams and then calculating the gross moment of inertia of the section, we will use the assumption that the gross moment of inertia for the beam's web (full depth) is approximately equal to one-half

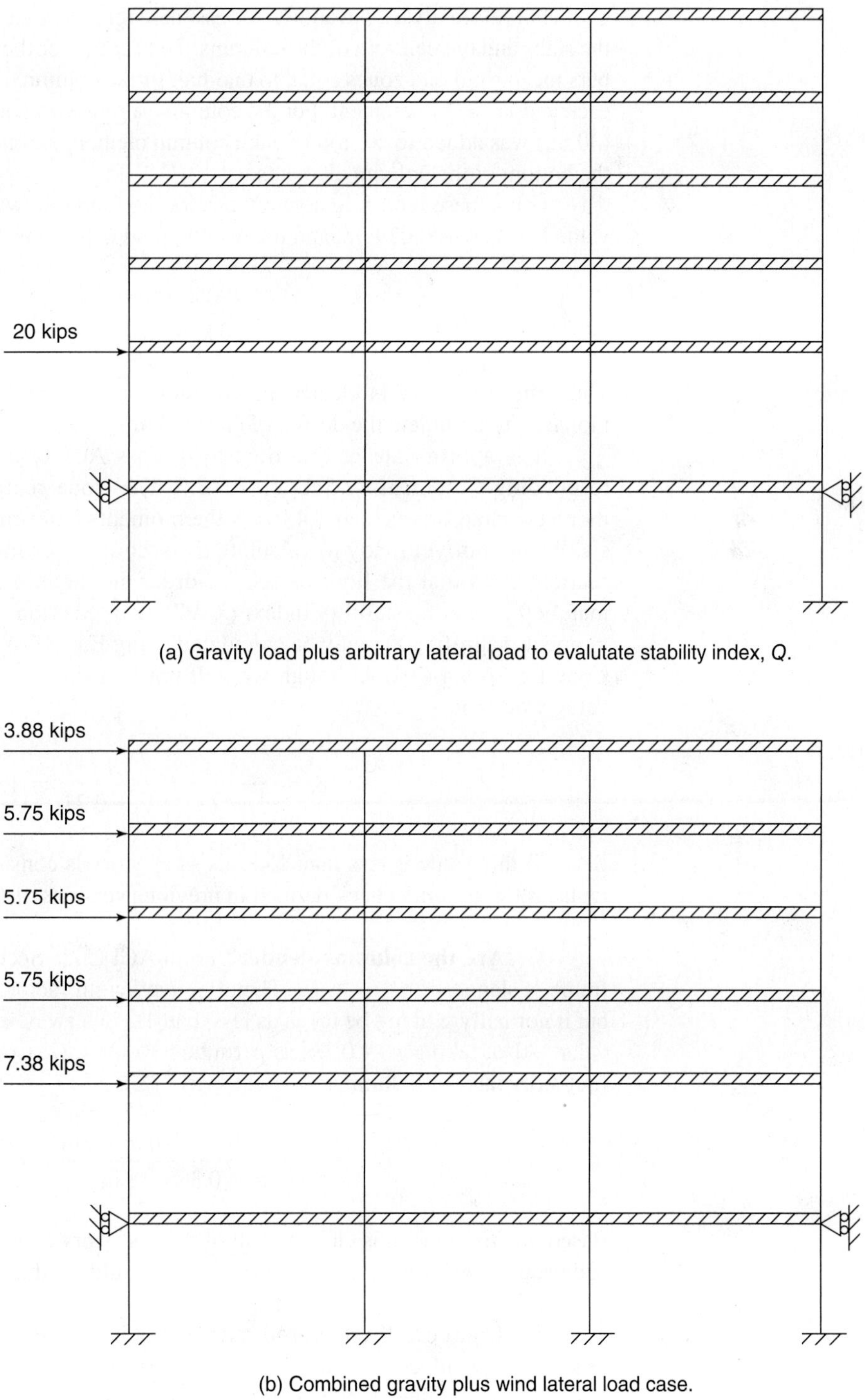

Fig. 12-27
Frame analysis model for combined gravity and lateral loading.

of the moment of inertia for the full T-beam section. With this assumption, the moment of inertia for the beam members was taken as

$$0.35 \times I_g\,(\text{T-beam}) \cong 0.7 \times I_g\,(\text{beam-web})$$
$$\cong 0.7 \times 18\text{ in.} \times (30\text{ in.})^3/12 = 28{,}400\text{ in.}^4$$

For this analysis, the author has assumed that the node points are located at the top of the slabs and the centroid of the columns. To account for the rigid joints where these members meet, rigid end zones equal to one-half of the column dimension (9 in.) were added to each end of a beam element. For the columns, a rigid end zone equal to the full beam depth (30 in.) was added to the top of each column element and no rigid end zone was added to the bottom of the column elements.

Using these modeling assumptions for the frame shown in Fig. 12-27a, the calculated value for Δ_o was 0.079 in., and the resulting value for Q is

$$Q = \frac{11{,}400 \text{ k} \times 0.079 \text{ in.}}{20 \text{ k} \times 216 \text{ in.}} \cong 0.21$$

Thus, the first story is clearly a sway story, and we will need to calculate magnified moments to complete the design of each column.

It is appropriate at this time to discuss ACI Code Section 6.2.6 and the corresponding Commentary Section R6.2.6. The Code section states that the secondary moments shall not exceed 1.4 times the moments determined from a first-order analysis. We are not yet ready to calculate the second-order moments, but the Commentary section states that the limit on second-order moments is related to a limit of approximately 0.25 for the stability index, Q. ACI Code Section 6.6.4.6.2 states that the sway-moment magnifier, δ_s, may be calculated using Eq. (12-41), which corresponds to ACI Code Eq. (6.6.4.4.1). Although we will not be using δ_s for this gravity-load case, the value would be

$$\delta_s = \frac{1}{1 - Q} = \frac{1}{1 - 0.21} = 1.27$$

Because this value is less than 1.4, this sway story is considered to be *stable*. This check replaces the stability check defined in previous versions of the ACI Code.

4. Are the columns slender? From ACI Code Section 6.2.5, a column in a sway frame is slender if $k\ell_u/r \geq 22$. The effective length factor, k, is not known at this stage, but it normally will not be taken as less than 1.2 in a sway frame. We initially will use that value. Also, taking $r = 0.3h$, as permitted by the ACI Code, the slenderness ratio for a first-story interior column is

$$\frac{k\ell_u}{r} = \frac{1.2 \times (216 - 30)\text{in.}}{0.3 \times 18 \text{ in.}} = 41.3$$

Based on this value, it is clear that all of the first-story columns are slender. If this number had been closer to 22, a more accurate value would need to be calculated for k.

5. Compute the factored axial loads and moments from a first-order frame analysis. The unfactored axial loads were computed in step 1 and are given in Table 12-3. The calculated values for the unfactored dead-load moments at the top and bottom of an exterior and an interior column also are given in Table 12-3 (assuming CCW+). The signs indicate that both columns are put into double curvature due the dead load.

For the column end moments due to live load, two loading patterns were considered, as shown in Fig. 12-28. The pattern shown in part (a) of Fig. 12-28 will maximize the end

TABLE 12-3 Axial Load and Moments for Load Case 1, Gravity Load

	Exterior Column	Interior Column
P_D	283 kips	486 kips
P_L (reduced)	42.9 kips	69.2 kips
P_{Lr}	10.1 kips	18.4 kips
M_D (top)	−34.9 kip-ft	4.8 kip-ft
M_D (bottom)	−36.8 kip-ft	6.3 kip-ft
M_L (top, reduced)	−11.2 kip-ft	4.4 kip-ft
M_L (bottom, reduced)	−11.8 kip-ft	−2.0 kip-ft

moments in the exterior columns and develops a double curvature pattern in those columns, which is similar to that for the dead-load case. The end moments calculated for this loading pattern were multiplied by the live-load reduction factor corresponding to the influence area for an exterior column supporting a single floor (0.67). This generally corresponds to the live-load reduction factor that would be used for the beam moments in a typical floor.

The live-load pattern shown in part (b) of Fig. 12-28 was used to find the maximum end moments for the interior columns. This load pattern will put those columns into a single-curvature deformation pattern, which is more critical when designing slender columns. Because this load pattern has a lower probability of occurrence and does not correspond to the loading case used to find the column axial load, the authors believe it is appropriate to use a live-load reduction factor that corresponds to an interior column that supports *two* floor levels (0.46). Because these end moments are very small, they will be (conservatively) combined with the axial loads from the case of live load acting on all spans.

The axial loads and moments from Table 12-3 now will be combined with the appropriate load factors to determine the factored design loads for the column. It should be noted that the roof live load has essentially no effect on the end moments for the columns in the first story of this frame structure.

Exterior Column:

$$\begin{aligned} P_u &= 1.2P_D + 1.6P_L + 0.5P_{Lr} \\ &= 1.2 \times 283 + 1.6 \times 42.9 + 0.5 \times 10.1 = 414 \text{ kips} \\ M_2 &= 1.2 \times (-36.8 \text{ k-ft}) + 1.6 \times (-11.8 \text{ k-ft}) = -63.0 \text{ k-ft} \\ M_1 &= 1.2 \times (-34.9 \text{ k-ft}) + 1.6 \times (-11.2 \text{ k-ft}) = -59.8 \text{ k-ft} \end{aligned}$$

Interior Column:

$$\begin{aligned} P_u &= 1.2P_D + 1.6P_L + 0.5P_{Lr} \\ &= 1.2 \times 486 + 1.6 \times 69.2 + 0.5 \times 18.4 = 703 \text{ kips} \\ M_2 &= 1.2 \times (4.8 \text{ k-ft}) + 1.6 \times (4.4 \text{ k-ft}) = 12.8 \text{ k-ft} \\ M_1 &= 1.2 \times (6.3 \text{ k-ft}) + 1.6 \times (-2.0 \text{ k-ft}) = 4.4 \text{ k-ft} \end{aligned}$$

It should be noted that although the live load tended to put the interior column into a single-curvature mode, the final factored moments put the column into a double-curvature mode of deformation.

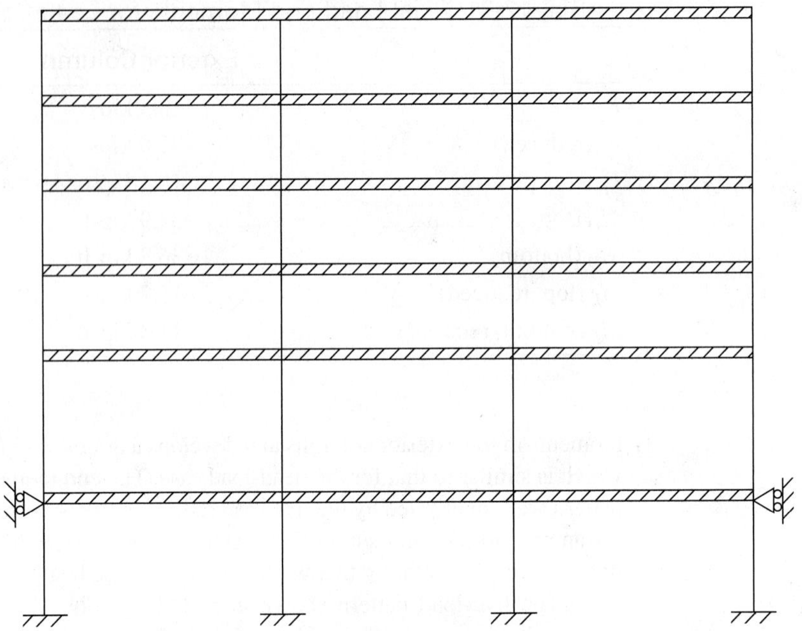

(a) All spans loaded with live load.

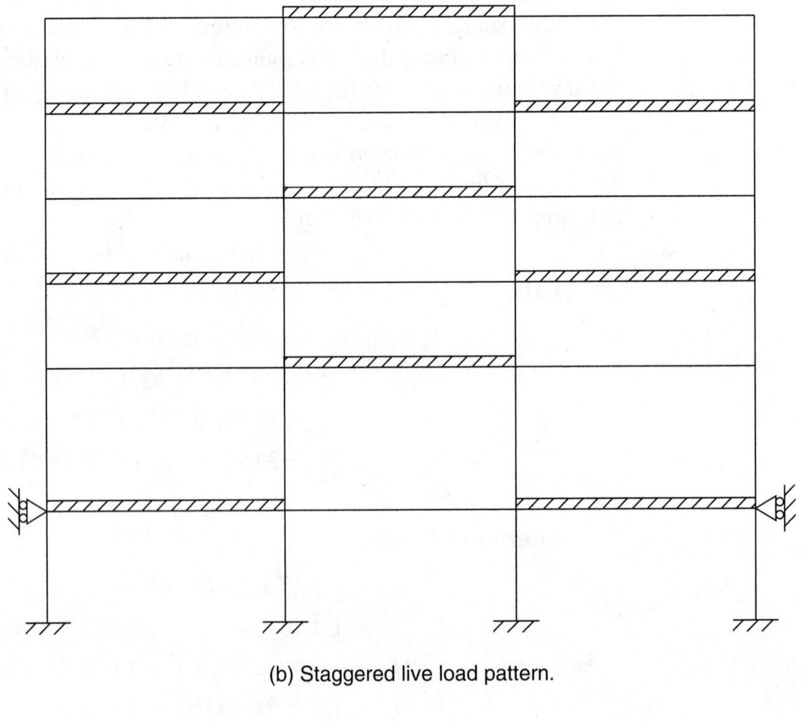

(b) Staggered live load pattern.

Fig. 12-28
Live-load cases considered for column moment in Example 12-3.

6. **Find δ for the interior and exterior columns.** From Eq. (12-24),

$$\delta = \frac{C_m}{1 - P_u/(0.75P_c)} \geq 1.0$$

We now will go through the process of finding C_m and P_c for the exterior and interior columns. From Eq. (12-14),

$$C_m = 0.6 - 0.4\left(\frac{M_1}{M_2}\right)$$

so

$$C_m(\text{ext.}) = 0.6 - 0.4\left(\frac{-59.8}{-63.0}\right) = 0.22$$

and

$$C_m(\text{int.}) = 0.6 - 0.4\left(\frac{4.4}{12.8}\right) = 0.46$$

We will use Eq. (12-25) to determine the critical buckling load, P_c:

$$P_c = \frac{\pi^2 EI}{(k\ell_u)^2}$$

Because the C_m values are quite low, it is likely that δ will not exceed 1.0. Thus, an exact calculation of the effective-length factor, k, is not warranted. Because the gravity loads do not cause sway, we will take $k = 1.0$, and avoid the need to calculate ψ-values at the top and bottom connection for each column.

(a) Calculation of $(EI)_{eff}$ values. The EI values for use in Eq. (12-25) can be calculated using Eq. (12-15a) as

$$(EI)_{eff} = \frac{0.2E_cI_g + E_sI_{se}}{1 + \beta_{dns}}$$

For both of the interior and exterior columns, I_g is

$$I_g = (18\text{ in.})^4/12 = 8750\text{ in.}^4$$

and

$$E_c = 57{,}000\sqrt{f'_c} = 57{,}000\sqrt{4000}\text{ psi}$$

$$= 3.60 \times 10^6\text{ psi} = 3600\text{ ksi}$$

For the steel bars, which have not yet been selected, $E_s = 29{,}000$ ksi. The value of I_{se} can be estimated using Table 12-1. Assuming that we will use three bars per face (eight longitudinal bars) in these square columns, the last column in Table 12-1 gives

$$I_{se} \cong 2.2\rho_g\gamma^2 \times I_g$$

We have assumed a total steel ratio, ρ_g, equal to 0.015, and for an 18-in. column,

$$\gamma \cong \frac{18\text{ in.} - 2 \times 2.5\text{ in.}}{18\text{ in.}} = 0.72$$

Thus,

$$I_{se} \cong 2.2 \times 0.015 \times (0.72)^2 \times 8750\text{ in.}^4 = 150\text{ in.}^4$$

Assuming that only the dead load is considered to cause a sustained axial load on the columns, Eq. (12-26a) will be used with the axial dead loads from Table 12-3 and the total factored axial loads calculated in step 5 to determine values for β_{dns} in the exterior and interior columns:

$$\beta_{dns} = \frac{\text{maximum factored sustained axial load}}{\text{total factored axial load}}$$

$$\beta_{dns}(\text{ext}) = \frac{1.2 \times 283\text{ kips}}{414\text{ kips}} = 0.82$$

and

$$\beta_{dns}(\text{int}) = \frac{1.2 \times 486\text{ kips}}{703\text{ kips}} = 0.83$$

Because these values are so close, only one value (0.83) will be used in Eq. (12-15) to determine a single EI value for both the exterior and interior columns:

$$(EI)_{eff} = \frac{0.2 \times 3600\text{ ksi} \times 8750\text{ in.}^4 + 29{,}000\text{ ksi} \times 150\text{ in.}^4}{1 + 0.83}$$
$$= 5.82 \times 10^6\text{ kip-in.}^2$$

(b) Calculation of P_c. Returning to Eq. (12-25) and using $\ell_u = 216 - 30 = 186$ in., we can calculate a single critical-buckling load for the exterior and interior columns as

$$P_c(\text{ext and int}) = \frac{\pi^2 (EI)_{eff}}{(k\ell_u)^2} = \frac{\pi^2(5.82 \times 10^6\text{ kip-in.}^2)}{(1 \times 186\text{ in.})^2} = 1660\text{ kips}$$

(c) Calculation of δ. We are now ready to use Eq. (12-24) to calculate δ for the exterior and interior columns.

$$\delta(\text{ext}) = \frac{C_m}{1 - \dfrac{P_u}{0.72P_c}} = \frac{0.22}{1 - \dfrac{414\text{ k}}{0.75 \times 1660\text{ k}}} = 0.33\text{, use } 1.0$$

and

$$\delta(\text{int}) = \frac{0.46}{1 - \dfrac{703\text{ k}}{0.75 \times 1660\text{ k}}} = 1.06$$

Thus, $M_c = 1.06 \times 12.8 = 13.5$ k-ft.

For these δ values the magnified moments are clearly less than 1.4 times the first-order moments, as required in ACI Code Section 6.2.6.

7. Check initial column sections for gravity-load case. For this check, we will use the axial load versus moment strength interaction diagrams in Appendix A to check the adequacy of the selected column sections. Starting with the factored axial load and moment acting on the exterior column, the design eccentricity is

$$e = \frac{M_c}{P_u} = \frac{63.0 \text{ kip-ft} \times 12 \text{ in./ft}}{414 \text{ kips}} = 1.83 \text{ in.}$$

and

$$e/h = 1.83 \text{ in.}/18 \text{ in.} \cong 0.10$$

We previously calculated $\gamma = 0.72$, so using the interaction diagram in Fig. A-9b ($\gamma = 0.75$) and the point where $e/h = 0.10$ and $\rho_g = 0.015$, we can extend a line over to the vertical axis and read an approximate value of $\phi P_n/A_g = 2.15$ ksi. Because the minimum value of ϕP_n is P_u, we will use the value read from Fig. A-9b to check the required column area as

$$A_g \geq \frac{P_u}{2.15 \text{ ksi}} = \frac{414 \text{ kips}}{2.15 \text{ ksi}} = 193 \text{ in.}^2$$

Because this is well below the provided area of 324 in.^{2}, we can try a lower longitudinal steel-reinforcement ratio. Using $\rho_g = 0.010$ and returning to Fig. A-9b, we now can read $\phi P_n/A_g = 2.00$ ksi. For this value, the required column area is $A_g = 207$ in.^{2}, which still is well below the provided area. Also, because the provided area is significantly larger than the required area, we do not need to consider interpolation between the interaction diagrams in Appendix A for different γ-values.

For the interior column, the design moments are very small, so the minimum moment value given in ACI Code Section 6.6.4.5.4 will probably govern, i.e.,

$$\begin{aligned} M_{2,\min} &= P_u(0.6 + 0.03h) \\ &= 703 \text{ kips} \times (0.6 \text{ in.} + 0.03 \times 18 \text{ in.}) = 801 \text{ kip-in.} \cong 67 \text{ kip-ft} \end{aligned}$$

This clearly governs, but even for this moment, the value of e/h is very low and we essentially will be reading values from the *top plateau* of the interaction diagrams in Fig. A-9b. This essentially means that the interior axial-load capacity is governed by the value for $\phi P_n(\max)$ given in Eq. (11-8b). Before using that equation, we will select the longitudinal bars for this column. The required area of longitudinal reinforcement is

$$A_{st} = \rho_g A_g = 0.015 \times 324 \text{ in.}^2 = 4.86 \text{ in.}^2$$

To be a little conservative in our design of the interior columns, we will select eight No. 8 longitudinal bars. This results in a total steel area, $A_{st} = 6.32$ in.^{2}, and a longitudinal reinforcement ratio, $\rho_g \approx 0.020$. Using that steel area in Eq. (11-8b) results in

$$\begin{aligned} \phi P_n(\max) &= \phi \times 0.80[0.85 f'_c(A_g - A_{st}) + f_y A_{st}] \\ &= 0.65 \times 0.80[0.85 \times 4 \text{ ksi}(324 - 6.32)\text{in.}^2 + 60 \text{ ksi} \times 6.32 \text{ in.}^2] \\ &= 759 \text{ kips} > P_u = 703 \text{ kips (o.k.)} \end{aligned}$$

This completes the check for Load Case 1. Longitudinal bars for the exterior column will be selected after investigating Load Case 2.

Load Case 2: Gravity Plus Lateral (Wind) Loads

8. Factored axial loads. The axial loads due to unfactored gravity loads were given in Table 12-3. The unfactored axial loads and end moments due to lateral wind loads were determined using the forces shown in Fig. 12-27b. To obtain these forces, the total lateral wind forces acting in the north–south direction (refer to Fig. 12-26a), which were calculated using the minimum design loads from reference [12-22], were divided by seven and then applied to the frame, as shown in Fig. 12-27b. It is typical to assume that each of the seven frame lines shown in Fig. 12-26a will participate equally in resisting the lateral loads. This would not be the case if some of the frame lines included stiffer elements, such as a structural wall. In those cases, the lateral loads would either be allocated to each frame line depending on their relative stiffnesses or would be allocated to the structural wall and frame members in accordance with governing building code requirements—for example, those given in reference [12-24].

The lateral loads shown in Fig. 12-27b were used to obtain the column axial loads and end moments given in Table 12-4.

Using the axial loads from Tables 12-3 and 12-4 along with the load factors given in step 2 for Loading Case 2, the following factored axial loads can be calculated:

$$P_u(\text{ext}) = 1.2P_D + 1.6P_W + 0.5(P_L + P_{Lr})$$
$$= 1.2 \times 283 + 1.6 \times 9 + 0.5(42.9 + 10.1) = 381 \text{ kips}$$

and

$$P_u(\text{int}) = 1.2 \times 486 + 1.6 \times 1.2 + 0.5(69.2 + 18.4) = 629 \text{ kips}$$

9. Calculate magnified moments due to sway. Equation (12-43b), which comes from ACI Code Section 6.6.4.6, will be used to define the larger factored end moment for design of the columns in this sway story.

$$M_2 = M_{2ns} + \delta_s \times M_{2s} \tag{12-43a}$$

The gravity loads, which do not cause sway, are combined with the appropriate load factors from step 2 to calculate the values for M_{2ns}. Using the moments from Table 12-3, and dropping the minus sign:

$$M_{2ns}(\text{ext}) = 1.2M_D + 0.5M_L$$
$$= 1.2 \times 34.9 + 0.5 \times 11.2 = 47.5 \text{ kip-ft}$$

and

$$M_{2ns}(\text{int}) = 1.2 \times 4.8 + 0.5 \times 4.4 \cong 8 \text{ kip-ft}$$

The lateral wind loads, which cause sway deflections, are used to calculate M_{2s}. Using the moments from Table 12-4 and the 1.6 load factor discussed in step 2, the values for M_{2s} are (again dropping the minus sign):

$$M_{2s}(\text{ext}) = 1.6M_W = 1.6 \times 47.8 = 76.5 \text{ kip-ft}$$

and

$$M_{2s}(\text{int}) = 1.6 \times 63.9 = 102 \text{ kip-ft}$$

The value of the sway-moment magnifier, δ_s, will be obtained using Eq. (12-41), which corresponds to ACI Code Section 6.6.4.6.2:

$$\delta_s = \frac{1}{1 - Q} \geq 1$$

TABLE 12-4 Axial Load and End Moments due to Lateral Wind Loads

	Exterior Column	Interior Column
P_W	9.0 kips	1.2 kips
M_W (top)	−47.8 kip-ft	−63.9 kip-ft
M_W (bottom)	−46.1 kip-ft	−63.5 kip-ft

To make this calculation, a new value for the stability index, Q, must be calculated for the loads corresponding to this load case. Using the lateral loads shown in Fig. 12-27b and the corresponding gravity loads, the value for the lateral shear in the first story is $V_{us} = 28.5$ kips, and the resulting lateral deflection in that story is $\Delta_o = 0.075$ in. To complete the calculation for the stability index, we also need the sum of the column axial loads in the first story corresponding to this load case. Using the axial loads from step 8 (without the wind loads, which would cause compression in some columns and tension in others and thus would cancel out), assuming (as before) that all of the interior columns and exterior columns carry the same axial loads as computed in step 8, and assuming that the corner columns carry one-half of the axial load carried by the exterior columns,

$$\Sigma P_u = 8 \times 627 + 12 \times 366 + 4 \times 0.5 \times 366 = 10{,}100 \text{ kips}$$

Now, using Eq. (12-21), we get

$$Q = \frac{\Sigma P_u \times \Delta_o}{V_{us} \times \ell_c} = \frac{10{,}100 \text{ kips} \times 0.075 \text{ in.}}{28.5 \text{ kips} \times 216 \text{ in.}} = 0.123$$

Then, from Eq. (12-41),

$$\delta_s = \frac{1}{1 - Q} = \frac{1}{1 - 0.123} = 1.14$$

With this value for δ_s, we now can calculate the total factored moments for this load case as

$$M_2(\text{ext}) = M_{2ns} + \delta_s M_{2s} = 47.5 + 1.14 \times 76.5 = 135 \text{ kip-ft}$$

and

$$M_2(\text{int}) = 8 + 1.14 \times 102 = 124 \text{ kip-ft}$$

These magnified moments are less than 1.4 times the first-order moments, as required in ACI Code Section 6.2.6.

10. Check column sections for axial loads and moments of Load Case 2. As in step 7, we will use the interaction diagrams in Fig. A-9b, which correspond to a γ-value (0.75) that is slightly larger than the actual γ-value (0.72). For the exterior column, the design eccentricity for this load case is

$$e = \frac{M_2}{P_u} = \frac{135 \text{ k-ft} \times 12 \text{ in./ft}}{381 \text{ k}} = 4.25 \text{ in.}$$

so

$$e/h = 4.25 \text{ in./18 in.} \cong 0.24$$

Entering Fig. A-9b with this value and $\rho_g = 0.010$ (which was selected in step 7), we read $\phi P_n/A_g = 1.4$ ksi. As before, assuming the minimum value for ϕP_n is equal to P_u, we can solve for the required column area:

$$A_g \geq \frac{P_u}{1.4\text{ ksi}} = \frac{381\text{ kips}}{1.4\text{ ksi}} = 272\text{ in.}^2$$

This is less than the provided area of 324 in.^{2}, so the design of the exterior column is acceptable for this load case. Because the selected section for the exterior columns has now been shown to satisfy Load Cases 1 and 2, we will complete the section design by selecting the longitudinal bars for this column. The required area of longitudinal reinforcement is

$$A_{st} = \rho_g A_g = 0.010 \times 324\text{ in.}^2 = 3.24\text{ in.}^2$$

The selection of eight No. 6 longitudinal bars results in a total steel area, $A_{st} = 3.52$ in.^{2}, and a longitudinal reinforcement ratio, $\rho_g \approx 0.011$.

For the interior column, the design eccentricity for this load case is

$$e = \frac{M_2}{P_u} = \frac{124\text{ k-ft} \times 12\text{ in./ft}}{629\text{ k}} = 2.37\text{ in.}$$

so

$$e/h = 2.37\text{ in./18 in.} \cong 0.13$$

Entering Fig. A-9b with this value and $\rho_g = 0.020$ (which was selected in step 7), we read $\phi P_n/A_g = 2.1$ ksi. As before, assuming the minimum value for ϕP_n is equal to P_u, we can solve for the required column area:

$$A_g \geq \frac{P_u}{2.1\text{ ksi}} = \frac{629\text{ kips}}{2.1\text{ ksi}} = 300\text{ in.}^2$$

This is below (but relatively close to) the provided area of 324 in.^{2}. In such cases, the designer should make a more detailed check by comparing the factored axial loads and moments to the nominal axial load versus moment-strength envelope for the column section, as was done in Example 11-4.

Figure 12-29 shows the nominal axial load versus moment-strength envelopes, reduced with the appropriate ϕ-factors, for the exterior and interior columns sections selected to satisfy Load Cases 1 and 2. Also shown in Fig. 12-29 are the combinations of factored axial loads and moments from those two loading cases. From this figure, it is clear that the selected column sections and reinforcement are satisfactory.

Load Case 3: Low Gravity Load plus Lateral Load

11. Summary of results. This load case will seldom govern the design of either interior or exterior columns, except for structures with a large height-to-width ratio. For that type of structure, it is possible that the overturning moments due to the lateral forces could result in a very low compression load, or even a tension load, in the exterior columns. Because the moment capacity could be significantly lower for such axial loads, this load case might govern for the exterior columns.

For the structure under consideration in this example, the factored axial loads and moments, including magnification of sway moments, are plotted in Fig. 12-29. It is clear from this plot that this load case does not govern the design of the column sections.

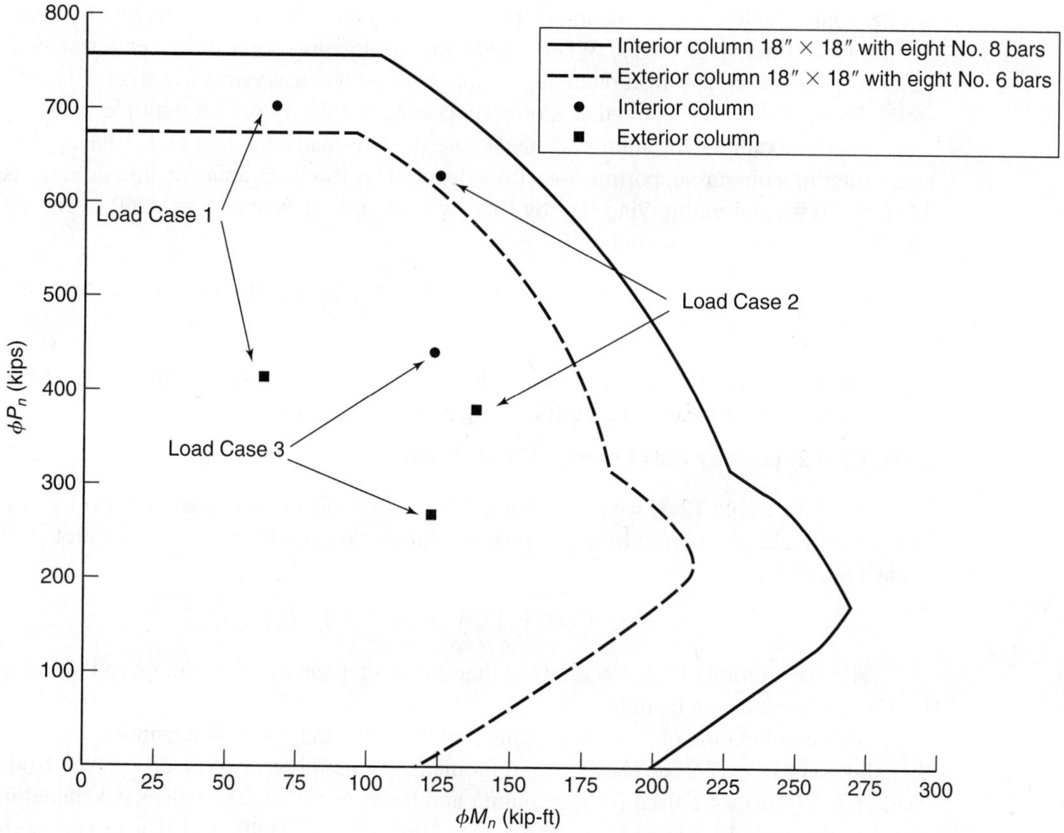

Fig. 12-29
Comparison of column strength and factored loads.

Thus, for the exterior columns, use an 18-in.-by-18-in.-section with eight No. 6 longitudinal bars (arranged with three bars per face). For the interior column, use an 18-in.-by-18-in.-section with eight No. 8 longitudinal bars (three bars per face). ■

EXAMPLE 12-4 Sway Frame: Using Direct P–Δ Analysis

Reconsider the design of the first-story columns for the frame in Example 12-3. For brevity, we only will consider Load Case 2, which was for the combination of gravity and lateral (wind) loading. If secondary bending effects were to be considered for Load Case 1, where the acting gravity loads do not cause sway, extra nodes would need to be added along the length of the columns to determine if moments at those node points exceeded the moments at the ends of the columns. Because the values for δ calculated in Example 12-3 were found to be well below and just above 1.0 for the exterior and interior columns, respectively, there was no apparent need to make such an analysis. Also, as indicated in Fig. 12-29, Load Case 3 was unlikely to govern the design of these first-story columns.

For Example 12-3, the dead-load, live-load, and wind-loading cases could all be determined separately. Then the results would be combined using the appropriate load factors. This procedure permits the application of different live-load reduction factors for the exterior and interior columns and can be used to apply different reduction factors for columns in other stories throughout the frame. When using structural analysis software to directly calculate the total column moments, including P–Δ effects, the structural designer will need to select only one live-load reduction factor to apply for all of the columns within the frame (or at least within a single story). When faced with this decision, some design engineers will elect to either use no live-load reduction factor or to use a conservative live-load reduction factor that applies for a column that is only supporting a single floor (not multiple floors).

For this example, the author decided to use the live-load reduction factor that applies to an interior column supporting one floor level. The tributary area for this column is 32 ft. by 20 ft., and multiplying that by four, we get an influence area of 2560 ft.^{2} Using Eq. (2-15), the live-load reduction factor is

$$\frac{L}{L_o} = 0.25 + \frac{15}{\sqrt{A_I}} \cong 0.55$$

This factor was multiplied by the given floor live load of 80 psf, resulting in a reduced live load of 44 psf to be used in the following structural analysis.

Load Case 2: Gravity plus Lateral (Wind) Loads

As in Example 12-3, we will assume the given wind loads are service-level loads. Thus, we should use a wind load factor of 1.6 for the load combination in row d of ACI Code Table 5.3.1.

$$U = 1.2D + 1.6W + 1.0L + 0.5L_r$$

Also in Example 12-3, we assumed that this is a typical office building, so the factor for the live load, L, can be reduced to 0.5.

This combination of loads was applied to the structural frame in a manner similar to that shown in Fig. 12-27b. The effective moments of inertia recommended by ACI Code Table 6.6.3.1.1(a) were used for the column and beam elements. The values calculated in step 3 of Example 12-3 were $I_{\text{column}} = 6120$ in.^{4} and $I_{\text{beam}} = 28{,}400$ in.^{4}. The elastic modulus for concrete calculated in Example 12-3, 3600 ksi, was used for all of the members. The structural analysis software package entitled RISA-3D [12-25] was used to conduct the second-order analysis of the frame for this example.

The results of that analysis are given in Table 12-5 along with the results obtained in steps 8 and 9 of Example 12-3. For completeness, the author included the calculated moments at both ends of the column, while only the larger moment was used in Example 12-3. The results from the direct second-order analysis are very similar to those obtained from the sway-multiplier procedure applied to the results from a first-order analysis.

Although the results are very close in this case, there may be more substantial differences in other design/analysis cases. In general, the moment-magnifier procedure for sway frames in the ACI Code, which combines first-order analyses and sway multipliers for gravity and wind loads, is more conservative than using structural analysis software that includes a direct second-order analysis for column end moments. ■

TABLE 12-5 Axial Load and End Moments due to Gravity plus Lateral Wind Loads

	Structural Analysis Including Second-Order Effects		First-Order Analysis Results and Application of Sway Multiplier	
	Exterior Column	Interior Column	Exterior Column	Interior Column
P_u	387 kips	640 kips	381 kips	629 kips
M_u (top)	130 k-ft	118 k-ft	135 k-ft	124 k-ft
M_u (bottom)	130 k-ft	120 k-ft	134 k-ft	122 k-ft

12-7 GENERAL ANALYSIS OF SLENDERNESS EFFECTS

ACI Code Section 6.2.6 presents requirements for a general analysis of column and frame stability. It lists a number of things the analysis must include. ACI Code Section 6.8.1 states that any nonlinear second-order analysis procedure shall have been shown to predict failure loads to within 15 percent of the measured failure loads. After the analysis, the structure must be reanalyzed if the final member sizes are more than 10 percent different from those assumed in the initial analysis.

The ACI moment-magnifier method for designing columns is limited by its derivation to prismatic columns with the same reinforcement from end to end. Tall bridge piers or other tall compression members often vary in cross-section along their lengths. In such a case, the design can proceed as follows:

1. Derive moment–curvature diagrams for the various sections along the column for the factored axial loads at each section.

2. Use the Vianello method combined with Newmark's numerical method [12-26] to solve for the deflected shape and the magnified moments at each section, taking into account the loadings and end restraints. A computer program to carry out such a solution is described in [12-27].

These procedures tend to underestimate the magnified moments slightly, because the solution is discretized and shear deformations are ignored.

12-8 TORSIONAL CRITICAL LOAD

Figure 12-30 shows a building subjected to a torque, T, that causes the floors and roof to undergo a rotation, ϕ, per unit of height, about the center of rotation, C_R. The torsional stiffness of the building using θ_1 to represent first-order building rotation is

$$K_\theta = T/\theta_1 \tag{12-45}$$

At height h_i, the angle of rotation is

$$\theta_i = \phi h_i$$

The horizontal deflection of a column-to-floor joint at height h is

$$\Delta = \theta \times r_j$$

where r_j is the radial distance from the center of rotation to the point in question. A gravity load P_j at this point exerts an overturning moment $P_j\Delta = P_j(\theta \times r_j)$ on the building. The moment $P_j\Delta$ can be replaced with an equivalent horizontal force of

$$P_j\Delta/h = P_j(\theta \times r_j)/h$$

This force has a torque about C_R of

$$\frac{P_j(\theta \times r_j)}{h} \times r_j$$

Summing over all the columns in the story, the second-order torque is

$$T_2 = \frac{\Sigma(P_j r_j^2)\theta}{h} \tag{12-46}$$

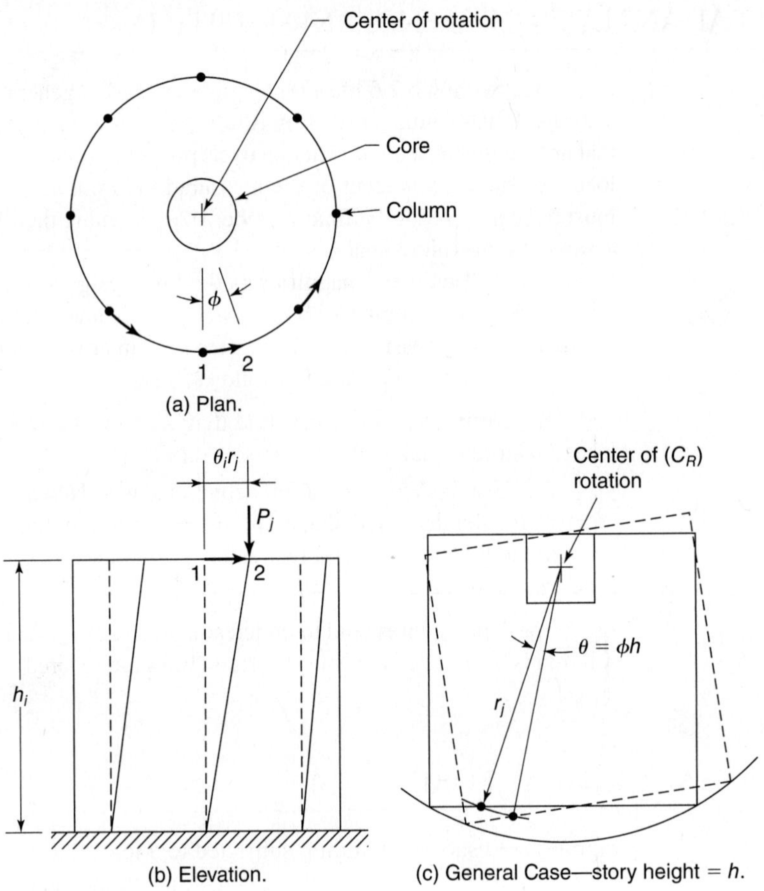

Fig. 12-30
Torsional critical load.

The applied torque, T, and the vertical loads, P_j, cause a total second-order rotation of

$$\theta_2 = \frac{(T + T_2)}{K_\theta} \tag{12-47}$$

but

$$T_2 = \Sigma(P_j r_j^2)\theta_2/h$$

Therefore, using column length, ℓ_c in place of h,

$$\theta_2 = \left(T + \frac{\Sigma(P_j r_j^2)}{\ell_c}\theta_2\right)\frac{1}{K_\theta}$$

and

$$\frac{T}{K_\theta} = \theta_2\left(1 - \frac{\Sigma(P_j r_j^2)}{K_\theta \ell_c}\right)$$

Substituting $K_\theta = T/\theta_1$ gives

$$\theta_2 = \frac{\theta_1}{1 - \dfrac{\Sigma(P_j r_j^2)\theta_1}{T_1 \ell_c}} \tag{12-48}$$

where T_1 is the torque from a first-order analysis. At the critical load, the building is in a state of neutral equilibrium, and the rotation can have any value, including zero. Instability occurs when the denominator of Eq. (12-48) equals zero. Thus, the critical load is

$$\Sigma(P_j r_j^2) = \frac{T_1 \ell_c}{\theta_1} \tag{12-49}$$

where the *stability index for torsional buckling* is

$$Q_T = \frac{\Sigma(P_j r_j^2)}{T_1} \times \frac{\theta_1}{\ell_c} \tag{12-50}$$

Implications of Eq. (12-50)

The stability index for torsional buckling, Q_T, is similar to Q in Eq. (12-21) for sway buckling, except that the torsional stability of a story is affected both by the total vertical load, ΣP_j, and by the square of the distance, r_j^2, from the center of rotation, C_R, to the tops of the columns on which the vertical loads act. The larger ΣPr^2 is, the lower the torsional critical load is. In design, gravity loads should be located such that ΣPr^2 is as small as possible. On the other hand, the elements resisting torsion should be placed as far away from C_R as possible, to increase the torsional stiffness. To estimate the torsional critical load of a building, the stiffnesses of the frames, walls, and cores in the x- and y-directions are computed, as will be discussed in Chapter 18.

PROBLEMS

12-1 A hinged end column 18-ft tall supports unfactored loads of 100 kips dead load and 60 kips live load. These loads are applied at an eccentricity of 3 in. at the bottom and 5 in. at the top. Both eccentricities are on the same side of the centerline of the column. Design a tied column with at least three bars per face using $f'_c = 4000$ psi and $f_y = 60{,}000$ psi.

Video Solution

12-2 Repeat Problem 12-1, but with the top eccentricity to the right of the centerline and the bottom eccentricity to the left.

12-3 Figure P12-3 shows an exterior column in a multistory frame. The dimensions are center-to-center of the joints. The beams are 12 in. wide by 18 in. in over-all depth. The floor slab is 6 in. thick. The building includes a service core that is stiff enough to brace all of the columns in each story. Use $f'_c = 5000$ psi and $f_y = 60{,}000$ psi. The loads and moments (CCW+) on column AB are:

Video Solution

Factored dead load:

Axial force = 300 kips
Moment at top = 80 kip-ft
Moment at bottom = 80 kip-ft

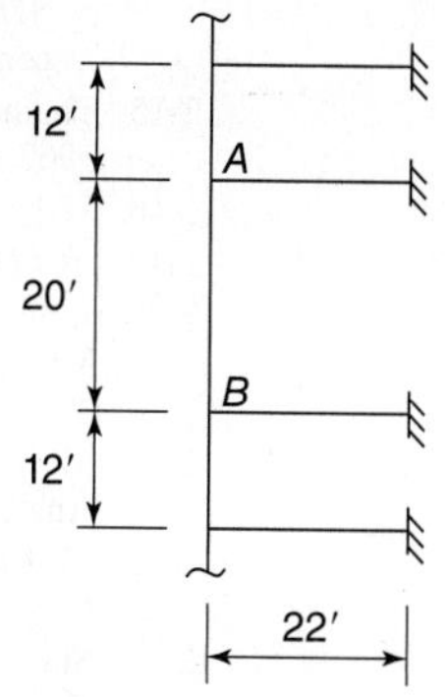

Fig. P12-3

Factored live load:

Axial force = 200 kips

Moment at top = 50 kip-ft

Moment at bottom = 80 kip-ft

Design column *A–B* using a square cross-section with at least three bars per face.

12-4 Use the ACI moment-magnifier method to redesign the columns in the main floor of Example 12-3 assuming that the floor-to-floor height of the first story is 16 ft 0 in. rather than 18 ft 0 in. Also, assume the lateral wind forces are 15 percent larger than those used in Example 12-3.

REFERENCES

12-1 James G. MacGregor, John E. Breen, and Edward O. Pfrang, "Design of Slender Columns," *ACI Journal, Proceedings*, Vol. 67, No. 1, January 1970, pp. 6–28.

12-2 Bengt Broms and Ivan M. Viest, "Long Reinforced Concrete Columns—A Symposium," *Transactions ASCE*, Vol. 126, Part 2, 1961, pp. 308–400.

12-3 *Specifications for Structural Steel Buildings (Commentary)*, American Institute of Steel Construction, Chicago, IL, 2010.

12-4 Walter J. Austin, "Strength and Design of Metal Beam-Columns," *Proceedings ASCE, Journal of the Structural Division*, Vol. 87, No. ST4, April 1961, pp. 1–34.

12-5 James G. MacGregor, Urs H. Oelhafen, and Sven E. Hage, "A Reexamination of the *EI* Value for Slender Columns," *Reinforced Concrete Columns*, ACI Publication SP-50, American Concrete Institute, Farmington Hills, MI, 1975, pp. 1–40.

12-6 A. A. Gvozdev and E. A. Chistiakov, "Effect of Creep on Load Capacity of Slender Compressed Elements," State-of-Art Report 2, Technical Committee 23 ASCE-IABSE International Conference on Tall Buildings, *Proceedings*, Vol. III-23, August 1972, pp. 537–554.

12-7 Roger Green and John E. Breen, "Eccentrically Loaded Columns under Sustained Load," *ACI Journal, Proceedings*, Vol. 66, No. 11, November 1969, pp. 866–874.

12-8 Adam M. Neville, Walter H. Dilger, and J. J. Brooks, *Creep of Plain and Structural Concrete*, Construction Press, New York, 1983, pp. 347–349.

12-9 Robert F. Warner, "Physical-Mathematical Models and Theoretical Considerations," *Introductory Report, Symposium on Design and Safety of Reinforced Concrete Compression Members*, International Association on Bridge and Structural Engineering, Zurich, 1974, pp. 1–21.

12-10 Emilio Rosenblueth, "Slenderness Effects in Buildings," *Journal of the Structural Division*, American Society of Civil Engineers, Vol. 91, No. ST1, Proceedings Paper 4235, pp. 229–252.

12-11 L.K. Stevans, "Elastic Stability of Practical Multi-Storey Frames," *Proceedings*, Institution of Civil Engineers, London, Vol. 36, pp. 99–117.

12-12 James G. MacGregor and Sven E. Hage, "Stability Analysis and Design of Concrete Frames," *Proceedings ASCE, Journal of the Structural Division*, Vol. 103, No. ST10, October 1977, pp. 1953–1970.

12-13 Sher Ali Mirza, P. M. Lee, and D. L. Morgan, "ACI Stability Resistance Factor for RC Columns," *ASCE Structural Engineering*, Vol. 113, No. 9, September 1987, pp. 1963–1976.

12-14 Shu-Ming Albert Lai, James G. MacGregor, and Jostein Hellesland, "Geometric Non-linearities in Nonsway Frames", *Proceedings ASCE, Journal of the Structural Division*, Vol. 109, No. ST12, December 1983, pp. 2770–2785.

12-15 Thomas C. Kavanagh, "Effective Length of Framed Columns," *Transactions ASCE*, Vol. 127, Part 2, 1962, pp. 81–101.

12-16 William McGuire, *Steel Structures*, Prentice Hall, Englewood Cliffs, N.J., 1968, 1110 pp.

12-17 *PCI Design Handbook*, 6th ed., Prestressed Concrete Institute, Chicago, 2004, 740 pp.

12-18 J. Steven Ford, D. C. Chang, and John E. Breen, "Design Implications from Tests of Unbraced Multipanel Concrete Frames," *Concrete International: Design and Construction*, Vol. 3, No. 3, March 1981, pp. 37–47.

12-19 M. Khuntia and S. K. Ghosh, "Flexural Stiffness of Reinforced Concrete Columns and Beams: Analytical Approach," *ACI Structural Journal*, Vol. 101, No. 3, May 2004, pp. 351–363.

12-20 M. Khuntia and S. K. Ghosh, "Flexural Stiffness of Reinforced Concrete Columns and Beams: Experimental Verification," *ACI Structural Journal*, Vol. 101, No. 3, May 2004, pp. 364–374.

12-21 Shu-Ming Albert Lai and James G. MacGregor, "Geometric Non-Linearities in Multi-story Frames," *Proceedings ASCE, Journal of the Structural Division*, Vol. 109, No. ST11, November 1983, pp. 2528–2545.

12-22 *Minimum Design Loads for Buildings and Other Structures* (ASCE/SEI 7–10), American Society of Civil Engineers, Reston, VA, 608 pp.

12-23 James G. MacGregor, "Design of Slender Columns—Revisited," *ACI Structural Journal*, Vol. 90, No. 3, May–June 1993, pp. 302–309.

12-24 *International Building Code*, International Code Council, Washington, D.C. 2009.

12-25 *RISA-3D*, RISA Technologies, *http://www.risatech.com.*

12-26 W. G. Godden, *Numerical Analysis of Beam and Column Structures*, Prentice Hall, Englewood Cliffs, N. J., 1965, 320 pp.

12-27 Randal W. Poston, John E. Breen, and Jose M. Roesset, "Analysis of Nonprismatic or Hollow Slender Concrete Bridge Piers," *Journal of the American Concrete Institute*, Vol. 82, No. 5, September–October 1985, pp. 731–739.

13

Two-Way Slabs: Behavior, Analysis, and Design

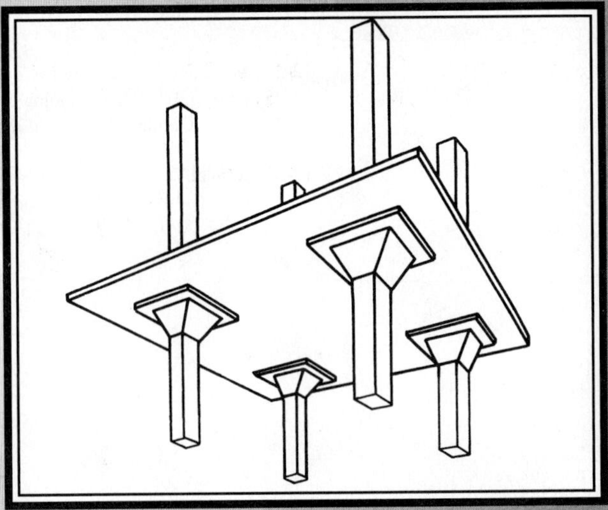

13-1 INTRODUCTION

The structural systems designed in Chapter 10 involved one-way slabs that carried loads to beams, which, in turn, transmitted the loads to columns, as shown in Fig. 4-1. If the beams are incorporated within the depth of the slab itself, the system shown in Fig. 13-1 results. Here, the slab carries load in two directions. The load at A may be thought of as being carried from A to B and C by one strip of slab, and from B to D and E, and so on, by other slab strips. Because the slab must transmit loads in two directions, it is referred to as a *two-way slab*.

Two-way slabs are a form of construction unique to reinforced concrete among the major structural materials. It is an efficient, economical, and widely used structural system. In practice, two-way slabs take various forms. For relatively light loads, as experienced in apartments or similar buildings, *flat plates* are used. As shown in Fig. 13-2a, such a plate is simply a slab of uniform thickness supported on columns. In an apartment building, the top of the slab would be carpeted, and the bottom of the slab may be finished as the ceiling for the story below. Flat plates are most economical for spans from 15 to 20 ft.

For longer spans, the thickness required for the shear transfer of vertical loads to the columns exceeds that required for flexural strength. As a result, the concrete at the middle of the panel is not used efficiently. To lighten the slab, reduce the slab moments, and save material, the slab at midspan can be replaced by intersecting ribs, as shown in Fig. 13-2b. Note that, near the columns, the full depth is retained for shear transfer of loads from the slab to the columns. This type of slab is known as a *waffle slab* (or a *two-way joist* system) and is formed with fiberglass or metal "dome" forms. Waffle slabs are used for spans from 25 to 40 ft.

For heavy industrial loads, the *flat slab* system shown in Fig. 13-2c may be used. Here, the shear transfer to the column is accomplished by thickening the slab near the column with *drop panels* or by flaring the top of the column to form a *column capital.* Drop panels commonly extend about one-sixth of the span in each direction away from the column, giving extra strength and stiffness in the column region while minimizing the amount of concrete at midspan. *Flat slabs* are used for loads in excess of 100 psf and for spans of 20 to 30 ft. Capitals of the type shown in Fig. 13-2c are less common today than they were in the first half of the Twentieth Century, due to the cost of forming them. Slab systems may incorporate beams between some or all of the columns. If the resulting panels are roughly square, the structure is referred to as a *two-way slab with beams* (Fig. 13-2d).

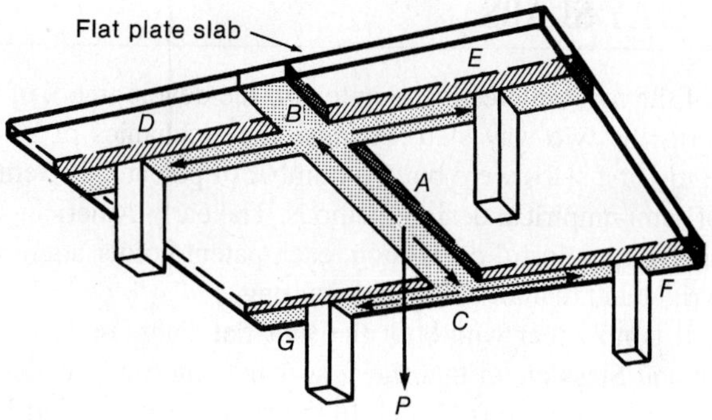

Fig. 13-1
Two-way flexure.

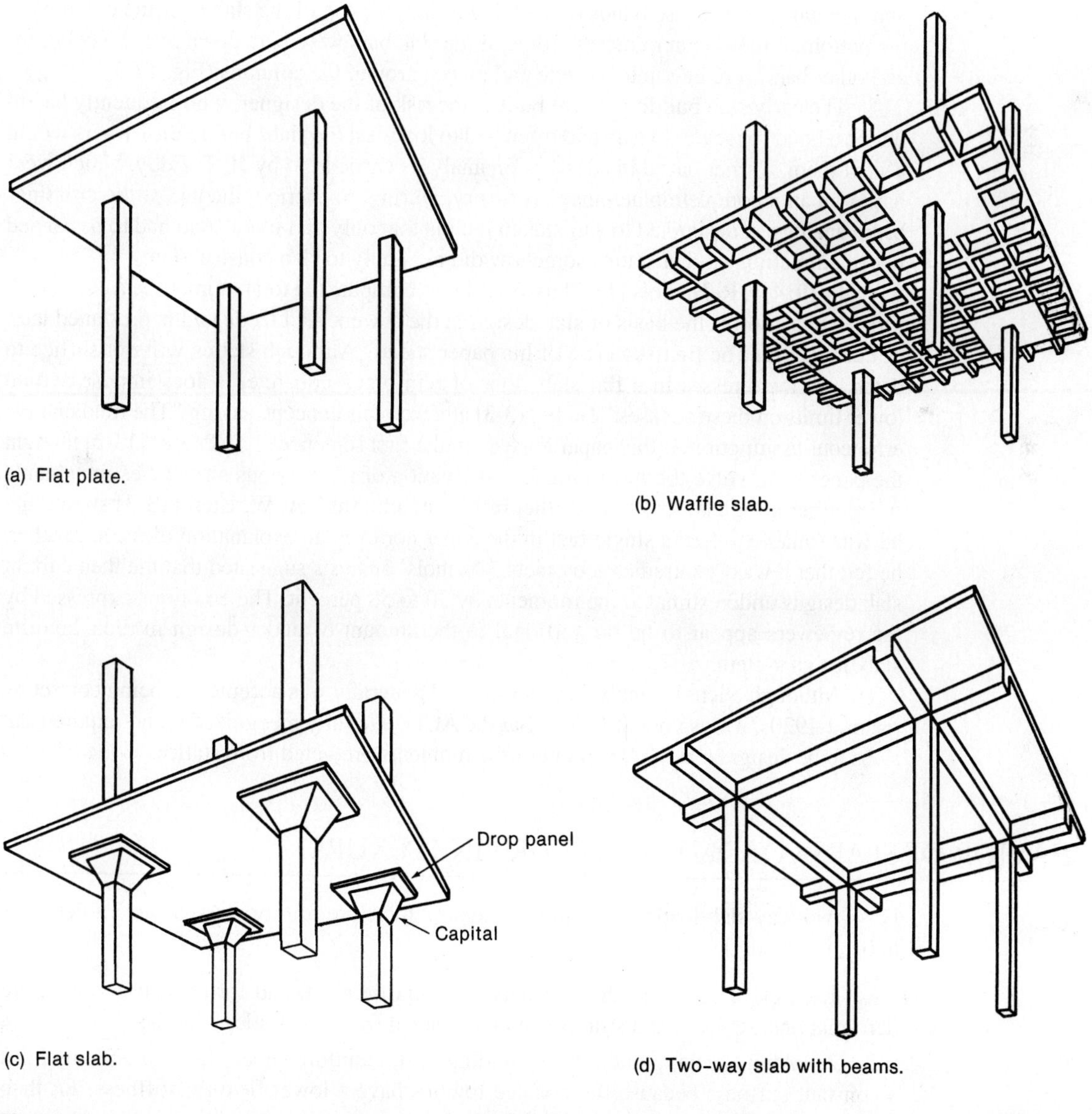

(a) Flat plate.

(b) Waffle slab.

(c) Flat slab.

(d) Two-way slab with beams.

Fig. 13-2
Types of two-way slabs.

13-2 HISTORY OF TWO-WAY SLABS

One of the most interesting chapters in the development of reinforced concrete structures concerns the two-way slab. Because the mechanics of slab action were not understood when the first slabs were built, a number of patented systems developed alongside a number of semi-empirical design methods. The early American papers on slabs attracted copious and very colorful discussion, each patent holder attempting to prove that his theories were right and that all others were wrong.

It is not clear who built the first flat slabs. In their review of the history of slabs, Sozen and Siess claim that the first American true flat slab was built by C. A. P. Turner in 1906 in Minneapolis [13-1]. In the same year, Maillart built a flat slab in Switzerland. Turner's slabs were known as mushroom slabs because the columns flared out to join the slab, which had steel running in bands in four directions (i.e., the two orthogonal directions and the diagonals). These bands draped down from the top of the slab over the columns to the bottom of the slab at midspan. Some of the slab bars were bent down into the columns, and other bars were bent into a circle and placed around the columns (Fig. 13-3).

The early slab buildings were built at the risk of the designer, who frequently had to put up a bond for several years and often had to load-test the slabs before the owners would accept them. Turner based his designs on analyses carried out by H. T. Eddy, which were based on an incomplete plate-analysis theory. During this period, the use of the crossing-beam analogy in design led to a mistaken feeling that only part of the load had to be carried in each direction, so that statics somehow did not apply to slab construction.

In 1914, J. R. Nichols [13-2] used statics to compute the total moment in a slab panel. This analysis forms the basis of slab design in the current ACI Code and is presented later in this chapter. The first sentence of his paper stated "Although statics will not suffice to determine the stresses in a flat slab floor of reinforced concrete, it does impose certain lower limits on these stresses." Eddy [13-3] attacked this concept, saying "The fundamental erroneous assumption of this paper appears in the first sentence . . . " Turner [13-3] thought the paper "to involve the most unique combination of multifarious absurdities imaginable from either a logical, practical or theoretical standpoint." A. W. Buel [13-3] stated that he was "unable to find a single fact in the paper nor even an explanation of facts." Rather, he felt that it was "contradicted by facts." Nichols' analysis suggested that the then current slab designs underestimated the moments by 30 to 50 percent. The emotions expressed by the reviewers appear to be proportional to the amount of under-design in their favorite slab design system.

Although Nichols' analysis is correct and generally was accepted as being correct by the mid-1920s, it was not until 1971 that the ACI Code fully recognized it and required flat slabs to be designed for 100 percent of the moments predicted from statics.

13-3 BEHAVIOR OF SLABS LOADED TO FAILURE IN FLEXURE

For a two-way slab loaded to failure in flexure, there are four or more stages of behavior to be discussed:

1. Before cracking, the slab acts as an elastic plate, and for short-time loads, the deformations, stresses, and strains can be predicted from an elastic analysis.

2. After cracking and before yielding of the reinforcement, the slab no longer has a constant stiffness because the cracked regions have a lower flexural stiffness, *EI*, than the uncracked regions; and the slab is no longer isotropic because the crack pattern may

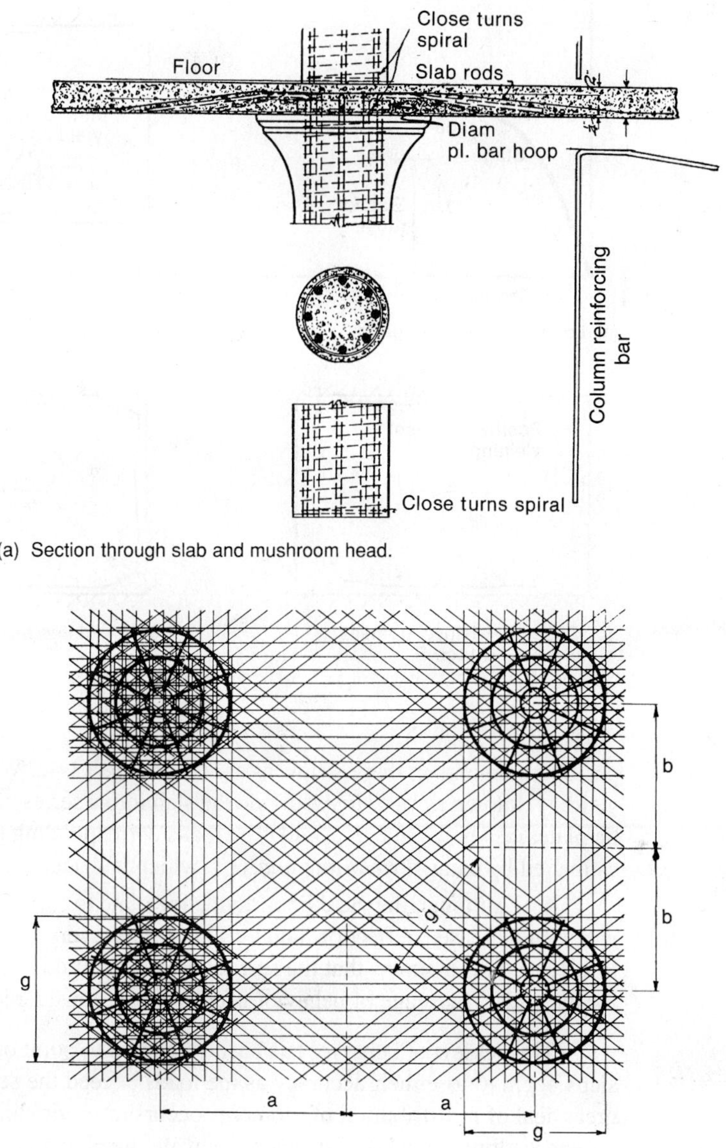

(a) Section through slab and mushroom head.

(b) Plan of reinforcement.

Fig. 13-3
C. A. P. Turner mushroom slab.

differ in the two directions. Although these conditions violate the assumptions in the elastic theory, tests indicate that the elastic theory still predicts the moments adequately. Generally, normal building slabs are partially cracked under service loads.

3. Yielding of the reinforcement eventually starts in one or more region of high moment and spreads through the slab as moments are redistributed from yielded regions to areas that are still elastic. The progression of yielding through a slab fixed on four edges is illustrated in Fig. 13-4. In this case, the initial yielding occurs in response to negative moments that form localized plastic hinges at the centers of the long sides (Fig. 13-4b). These hinges spread along the long sides, and eventually, new hinges form along the short sides of the slab (Fig. 13-4c). Meanwhile, the positive moments increase in slab strips across the center of the slab in the short direction, due to the moment redistribution caused by the plastic hinges at the ends of these strips. Eventually, the reinforcement yields due to

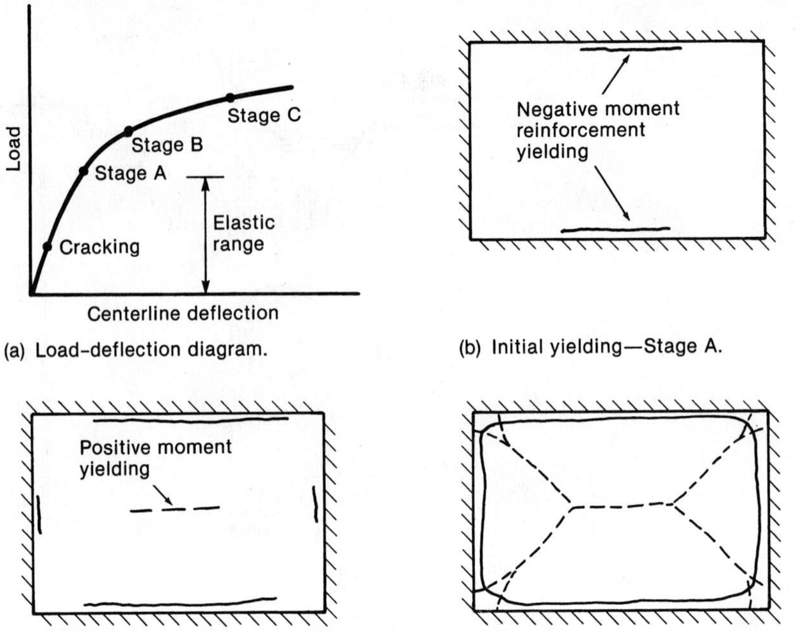

(a) Load-deflection diagram.

(b) Initial yielding—Stage A.

(c) Onset of positive moment yielding—Stage B.

(d) Yield line mechanism—Stage C.

Fig. 13-4
Inelastic action in a slab fixed on four sides.

positive moments in these strips, as shown in Fig. 13-4c. With further load, the regions of yielding, known as *yield lines*, divide the slab into a series of trapezoidal or triangular elastic plates, as shown in Fig. 13-4d. The loads corresponding to this stage of behavior can be estimated by using a *yield-line analysis*, which is discussed in Chapter 14.

4. Although the yield lines divide the plate to form a plastic mechanism, the hinges jam with increased deflection, and the slab forms a very flat compression arch, as shown in Fig. 13-5. This assumes that the surrounding structure is stiff enough to provide reactions for the arch. This stage of behavior usually is not considered in design.

This review of behavior has been presented to point out, first, that elastic analyses of slabs begin to lose their accuracy as the loads exceed the service loads, and second, that a great deal of redistribution of moments occurs after yielding first starts. A slab supported on and continuous with stiff beams or walls has been considered here. In the case of a slab supported on isolated columns, as shown in Fig. 13-2a, similar behavior would be observed, except that the first cracking would be on the top of the slab around the column, followed by a cracking of the bottom of the slab midway between the columns.

Slabs that fail in flexure are extremely ductile. Slabs, particularly flat-plate slabs, also may exhibit a brittle shear mode of failure. The shear behavior and strength of two-way slabs is discussed in Sections 13-10 and 13-11.

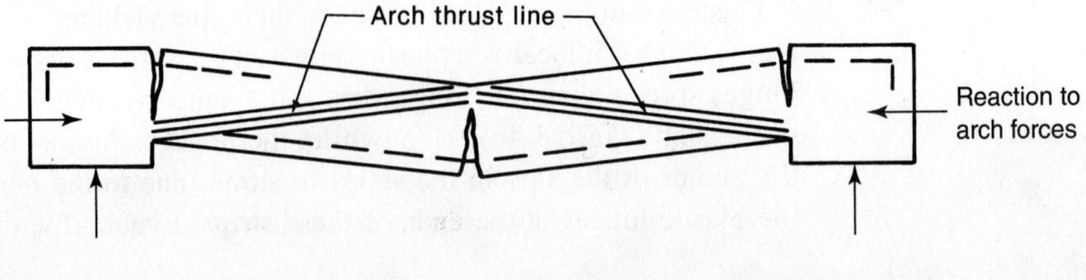

Fig. 13-5
Arch action in slab.

13-4 ANALYSIS OF MOMENTS IN TWO-WAY SLABS

Figure 13-6 shows a floor made up of simply supported planks supported by simply supported beams. The floor carries a load of q lb/ft^2. The moment per foot of width in the planks at section A–A is

$$m = \frac{q\ell_1^2}{8} \text{ kip-ft/ft}$$

The total moment in the entire width of the floor is

$$M_{A\text{-}A} = (q\ell_2)\frac{\ell_1^2}{8} \text{ kip-ft} \tag{13-1}$$

This is the familiar equation for the maximum moment in a simply supported floor of width ℓ_2 and span ℓ_1.

The planks apply a uniform load of $q\ell_1/2$ ft on each beam. The moment at section B–B in one beam is thus

$$M_{1b} = \left(\frac{q\ell_1}{2}\right)\frac{\ell_2^2}{8} \text{kip-ft}$$

The total moment in both beams is

$$M_{B\text{-}B} = (q\ell_1)\frac{\ell_2^2}{8} \text{kip-ft} \tag{13-2}$$

It is important to note that the full load was transferred east and west by the planks, causing a moment equivalent to $w\ell_1^2/8$ in the planks, where $w = q\ell_2$. Then the full load was transferred north and south by the beams, causing a similar moment in the beams. Exactly the

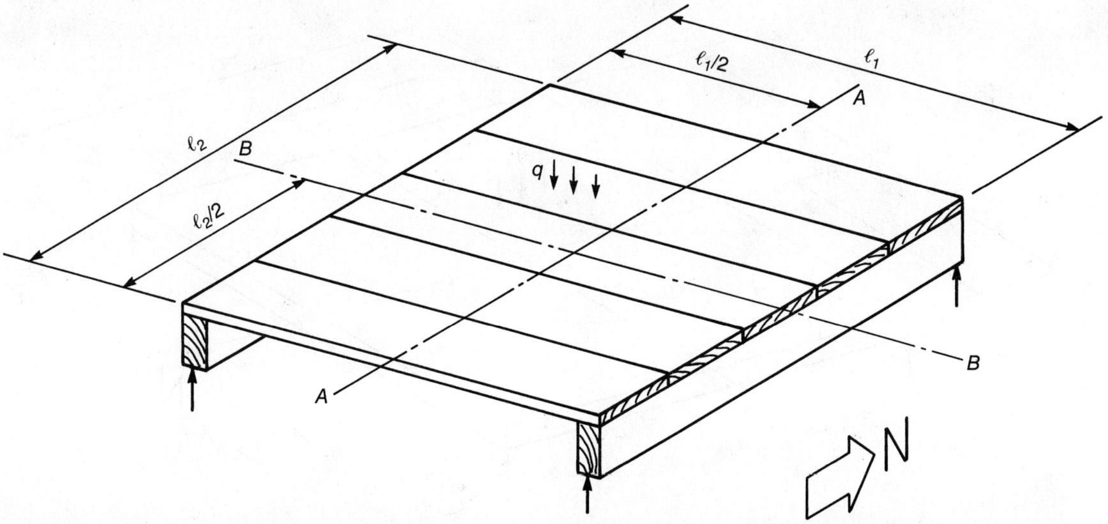

Fig. 13-6
Moments in a plank-and-beam floor.

same thing happens in the two-way slab shown in Fig. 13-7. The total moments required along sections *A–A* and *B–B* are

$$M_{A\text{–}A} = (q\ell_2)\frac{\ell_1^2}{8} \tag{13-1}$$

and

$$M_{B\text{–}B} = (q\ell_1)\frac{\ell_2^2}{8} \tag{13-2}$$

Again, the full load was transferred east and west, and then the full load was transferred north and south—this time by the slab in both cases. This, of course, always must be true regardless of whether the structure has one-way slabs and beams, two-way slabs, or some other system.

Nichols' Analysis of Moments in Slabs

The analysis used to derive Eqs. (13-1) and (13-2) was first published in 1914 by Nichols [13-2]. Nichols' original analysis was presented for a slab on round columns, rather than on the point supports assumed in deriving Eqs. (13-1) and (13-2). Because rectangular columns are more common today, the derivation that follows considers that case. Assume:

1. A typical rectangular, interior panel in a large structure.
2. All panels in the structure are uniformly loaded with the same load.

These two assumptions are made to ensure that the lines of maximum moment, and hence the lines on which the shears and twisting moments are equal to zero, will be lines of symmetry in the structure. This allows one to isolate the portion of the slab shown shaded in Fig. 13-8a. This portion is bounded by lines of symmetry.

The reactions to the vertical loads are transmitted to the slab by shear around the face of the columns. It is necessary to know, or assume, the distribution of this shear to compute

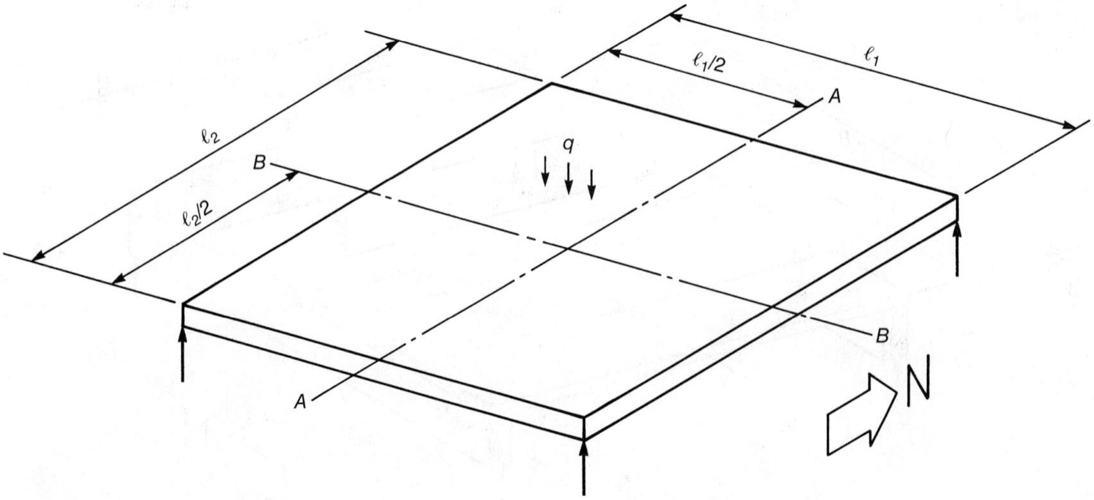

Fig. 13-7
Moments in a two-way slab.

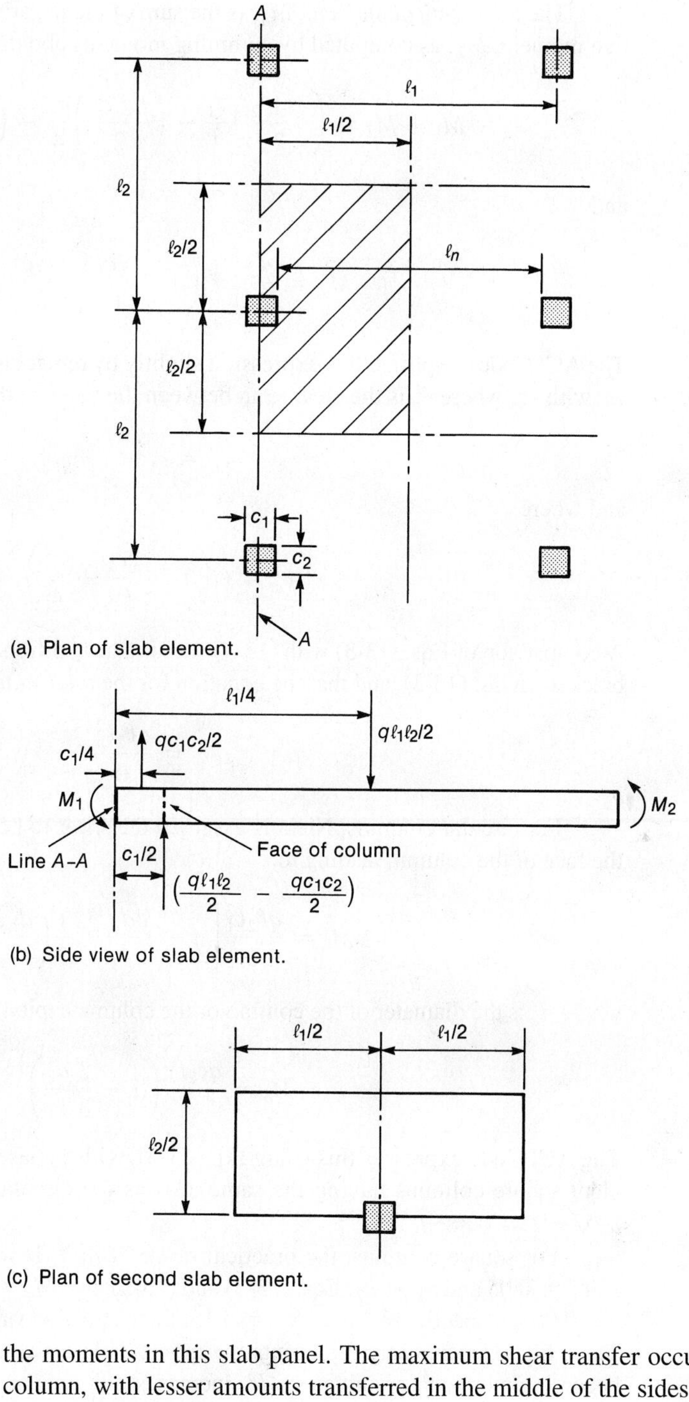

Fig. 13-8
Slab considered in Nichols' analysis.

the moments in this slab panel. The maximum shear transfer occurs at the corners of the column, with lesser amounts transferred in the middle of the sides of the column. For this reason, we shall assume that

3. The column reactions are concentrated at the four corners of each column.

Figure 13-8b shows a side view of the slab element with the forces and moments acting on it. The applied load is $(q\ell_1\ell_2/2)$ at the center of the shaded panel, minus the load on the area occupied by the column $(qc_1c_2/2)$. This is equilibrated by the upward reaction at the corners of the column.

The total *statical moment*, M_o, is the sum of the negative moment, M_1, and the positive moment, M_2, as computed by summing moments about line *A–A*:

$$M_o = M_1 + M_2 = \left(\frac{q\ell_1\ell_2}{2}\right)\frac{\ell_1}{4} - \left(\frac{qc_1c_2}{2}\right)\frac{c_1}{4} - \left(\frac{q\ell_1\ell_2}{2} - \frac{qc_1c_2}{2}\right)\frac{c_1}{2}$$

and

$$M_o = \frac{q\ell_2}{8}\left[\ell_1^2\left(1 - \frac{2c_1}{\ell_1} + \frac{c_2c_1^2}{\ell_2\ell_1^2}\right)\right] \qquad (13\text{-}3)$$

The ACI Code simplified this expression slightly by replacing the term in the square brackets with ℓ_n^2, where ℓ_n is the clear span between the faces of the columns, given by

$$\ell_n = \ell_1 - c_1$$

and where

$$\ell_n^2 = \ell_1^2\left(1 - \frac{2c_1}{\ell_1} + \frac{c_1^2}{\ell_1^2}\right) \qquad (13\text{-}4)$$

A comparison of Eqs. (13-3) with (13-4) shows that ℓ_n^2 differs only slightly from the term in brackets in Eq. (13-3), and that the equation for the *total statical moment* can be written as

$$M_o = \frac{q\ell_2\ell_n^2}{8} \qquad (13\text{-}5)$$

For circular columns, Nichols assumed the shear to be uniformly distributed around the face of the column, leading to

$$M_o = \frac{q\ell_2\ell_1^2}{8}\left[1 - \frac{4d_c}{\pi\ell_1} + \frac{1}{3}\left(\frac{d_c}{\ell_1}\right)^3\right] \qquad (13\text{-}6)$$

where d_c is the diameter of the column or the column capital. Nichols approximated this as

$$M_o = \frac{q\ell_2\ell_1^2}{8}\left(1 - \frac{2}{3}\frac{d_c}{\ell_1}\right)^2 \qquad (13\text{-}7)$$

The ACI Code expresses this using Eq. (13-5), with ℓ_n based on the span between equivalent square columns having the same area as the circular columns. In this case, $c_1 = d_c\sqrt{\pi}/2 = 0.886d_c$.

For square columns, the practical range of c_1/ℓ_1 is roughly from 0.05 to 0.15. For $c_1/\ell_1 = 0.05$ and $c_1 = c_2$, Eqs. (13-3) and (13-5) give $M_o = Kq\ell_2\ell_1^2/8$, where $K = 0.900$ and 0.903, respectively. For $c_1/\ell_1 = 0.15$, the respective values of K are 0.703 and 0.723. Thus, Eq. (13-5) closely represents the moments in a slab supported on square columns, becoming more conservative as c_1/ℓ_1 increases.

For circular columns, the practical range of d_c/ℓ_1 is roughly from 0.05 to 0.20. For $d_c/\ell_1 = 0.05$, Eq. (13-6) gives $K = 0.936$, while Eq. (13-5), with ℓ_n defined by using $c_1 = d_c\sqrt{\pi}/2$, gives $K = 0.913$. For $d_c/\ell_1 = 0.2$, the corresponding values of K from Eqs. (13-6) and (13-5) are 0.748 and 0.677, respectively. Thus, for circular columns, Eq. (13-5) tends to underestimate M_o by up to 10 percent, compared with Eq. (13-6).

If the equilibrium of the element shown in Fig. 13-8c were studied, a similar equation for M_o would result as one having ℓ_1 and ℓ_2 interchanged and c_1 and c_2 interchanged. This indicates once again that the total load must satisfy moment equilibrium in both the ℓ_1 and ℓ_2 directions.

13-5 DISTRIBUTION OF MOMENTS IN SLABS

Relationship between Slab Curvatures and Moments

The principles of elastic analysis of two-way slabs are presented briefly in Section 14-1. The basic equation for moments is Eq. (14-6). Frequently, in studies of concrete plates, Poisson's ratio, ν, is taken equal to zero. When this is done, Eq. (14-6) reduces to

$$\begin{aligned} m_x &= -\frac{Et^3}{12}\left(\frac{\partial^2 z}{\partial x^2}\right) \\ m_y &= -\frac{Et^3}{12}\left(\frac{\partial^2 z}{\partial y^2}\right) \\ m_{xy} &= -\frac{Et^3}{12}\left(\frac{\partial^2 z}{\partial x\,\partial y}\right) \end{aligned} \tag{13-8}$$

In these equations, $\partial^2 z/\partial x^2$ represents the curvature in a slab strip in the x-direction, and $\partial^2 z/\partial y^2$ represents the curvature in a strip in the y-direction where the *curvature* of a slab strip is

$$\frac{1}{r} = \frac{-\varepsilon}{y}$$

where r is the *radius of curvature* (the distance from the midplane axis of the slab to the center of the curve formed by the bent slab), $1/r$ is the curvature, and y is the distance from the centroidal axis to the fiber where the strain ε occurs. In an elastic flexural member, the curvature is

$$\frac{1}{r} = \frac{-M}{EI}$$

This shows the direct relationship between curvature and moments. Thus, by visualizing the deflected shape of a slab, one can qualitatively estimate the distribution of moments.

Figure 13-9a shows a rectangular slab that is fixed on all sides by stiff beams. One longitudinal and two transverse strips are shown. The deflected shape of these strips and the corresponding moment diagrams are shown in Figs. 13-9b through 13-9d. Where the deflected shape is concaved downward, the moment causes compression on the bottom—that is, the moment is negative. This may be seen also from Eq. (13-8). Since z was taken as positive downward, a positive curvature $\partial^2 z/\partial x^2$ corresponds to a curve that is concaved downward. From Eq. (13-8), a positive curvature corresponds to a negative moment. The magnitude of the moment is proportional to the curvature.

The largest deflection, Δ_2, occurs at the center of the panel. As a result, the curvatures (and hence the moments in strip B) are larger than those in strip A. The center portion of strip C essentially is straight, indicating that most of the loads in this region are being transmitted by one-way action across the short direction of the slab.

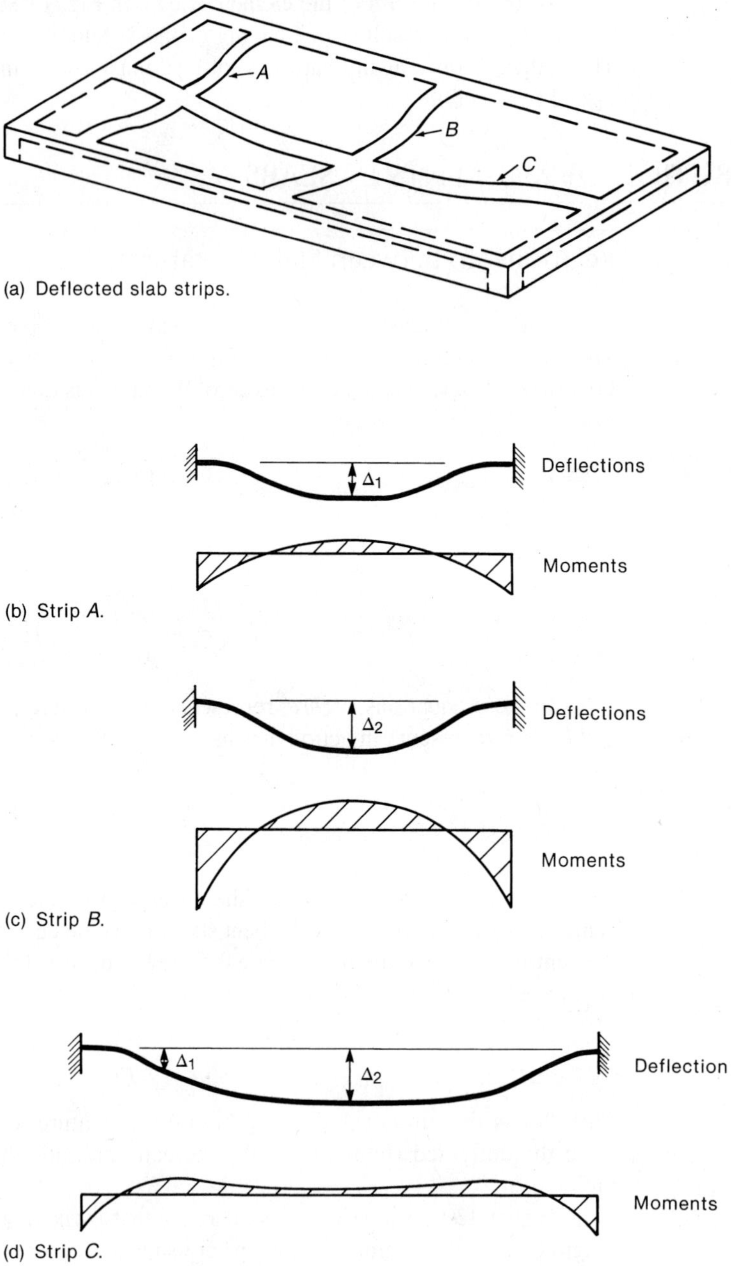

Fig. 13-9
Relationship between slab curvatures and moments.

Fig. 13-10
Deflection of strip B of Fig. 13-9.

The existence of the twisting moments, m_{xy}, in a slab can be illustrated by the crossing-strip analogy. Figure 13-10 shows a section cut through the slab strip B shown in Fig. 13-9. Here, the slab is represented by a series of beam cross sections that run parallel to slab strip C, which also is shown in Fig. 13-9. These slab strips must twist, as shown in Fig. 13-10, due to the m_{xy} twisting moments.

Moments in Slabs Supported by Stiff Beams or Walls

The distributions of moments in a series of square and rectangular slabs will be presented in one of two graphical treatments. The distribution of the negative moments, M_A, and of the positive moments, M_B, along lines across the slab will be as shown in Figs. 13-11a and 13-11b. These distributions may be shown as continuous curves, as shown by the solid lines and shaded areas, or as a series of steps, as shown by the dashed line for M_A. The height of the curve at any point indicates the magnitude of the moment at that point. Occasionally, the distribution of bending moments in a strip *A–B–C* across the slab will be plotted, as shown in Fig. 13-11c or Fig. 13-9.

The moments will be expressed in terms of Cqb^2, where b is the short dimension of the panel. The value of C would be 0.125 in a square, simply supported one-way slab. The units will be lb-ft/ft or kN • m/m of width. In all cases, the moment diagrams are for slabs subjected to a uniform load, q.

It is important to remember that the total load must be transferred from support to support by moments in the slab or beams. Figure 13-12a shows a slab that is simply supported at two ends and is supported by stiff beams along the other two sides. The moments in the slab account for only 19 percent of the total moments. The remaining portion of the moments is divided between the two beams as shown.

Figure 13-12b shows the effect of reducing the stiffness of the two edge beams. Here, the stiffness of the edge beams per unit width has been reduced to be equal to that

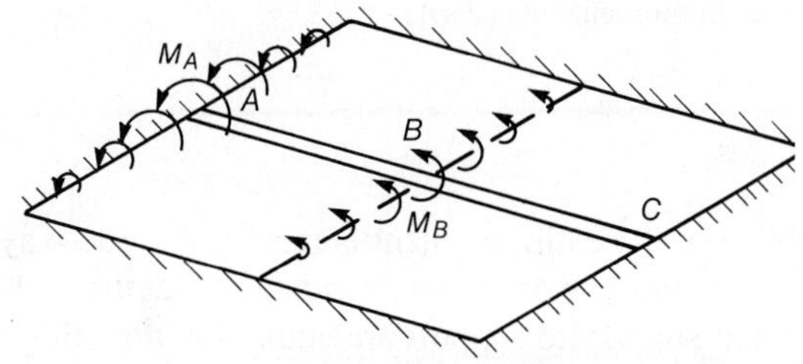

(a) Moments at edge and middle of slab.

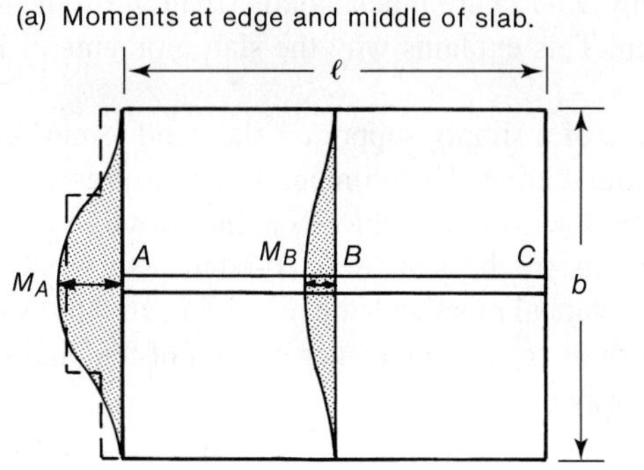

(b) Distribution of moments at edge and middle.

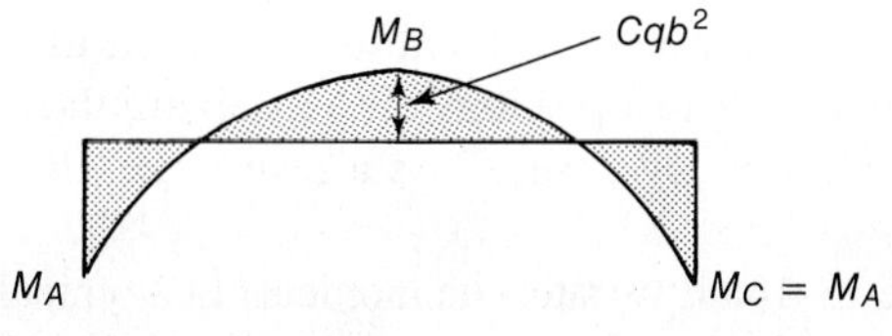

(c) Moments in strip *ABC*.

Fig. 13-11
Types of moment diagrams: four-edged fixed slab.

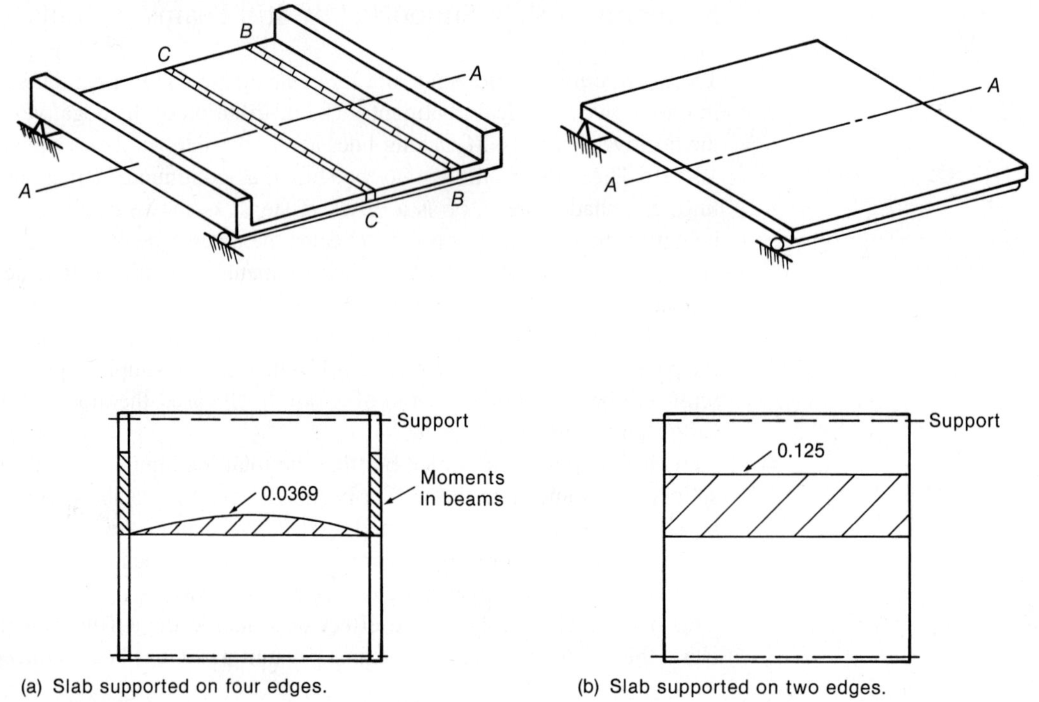

Fig. 13-12
Effect of beam stiffness on moments in slabs.

of the slab. Now the entire moment is resisted by one-way slab action. The fact that the moments about line *A–A* are constant indicates that the midspan curvatures of all the strips spanning from support to support are equal. On the other hand, the curvatures in a strip along *B–B* in Fig. 13-12a are much smaller than those in strip *C–C*, because *B–B* is next to the stiff beam. This explains why the slab moments in Fig. 13-12a decrease near the edge beams.

The corners of a simply supported slab tend to curl up off their supports, as is discussed later in this section. The moment diagrams presented here assume that the corners are held down by downward point loads at the corners.

The moments in the square fixed-edge slab shown in Fig. 13-13a account for 36 percent of the statical moments; the remaining amount is in the edge beams. As the ratio of length to width increases, the center portion of the slab spanning in the short direction approaches one-way action.

Moments in Slabs Supported by Isolated Columns

In a flat plate or flat slab, the slab is supported directly on the columns without any beams. Here, the stiffest portions of the slab are those running from column to column along the four sides of a panel. As a result, the moments are largest in these parts of the slab.

Figure 13-14a illustrates the moments in a typical interior panel of a very large slab in which all panels are uniformly loaded with equal loads. The slab is supported on circular

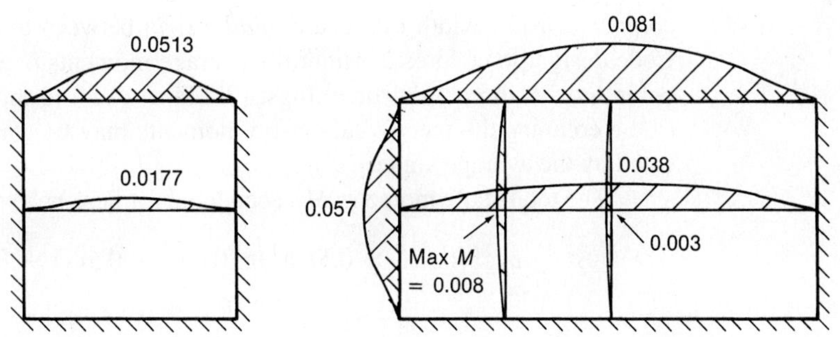

(a) Square fixed slab.

(b) Rectangular fixed slab, $b/a = 0.5$.

Fig. 13-13
Bending moments per unit inch in rectangular slabs with fixed edges. (From [13-4].)

columns with a diameter $c = 0.1\ell$. The largest negative and positive moments occur in the strips spanning from column to column. In Figs. 13-14b and 13-14c, the curvatures and moment diagrams are shown for strips along lines A–A and B–B. Both strips have negative moments adjacent to the columns and positive moments at midspan. In Fig. 13-14d, the moment diagram from Fig. 13-14a is replotted to show the average moments over

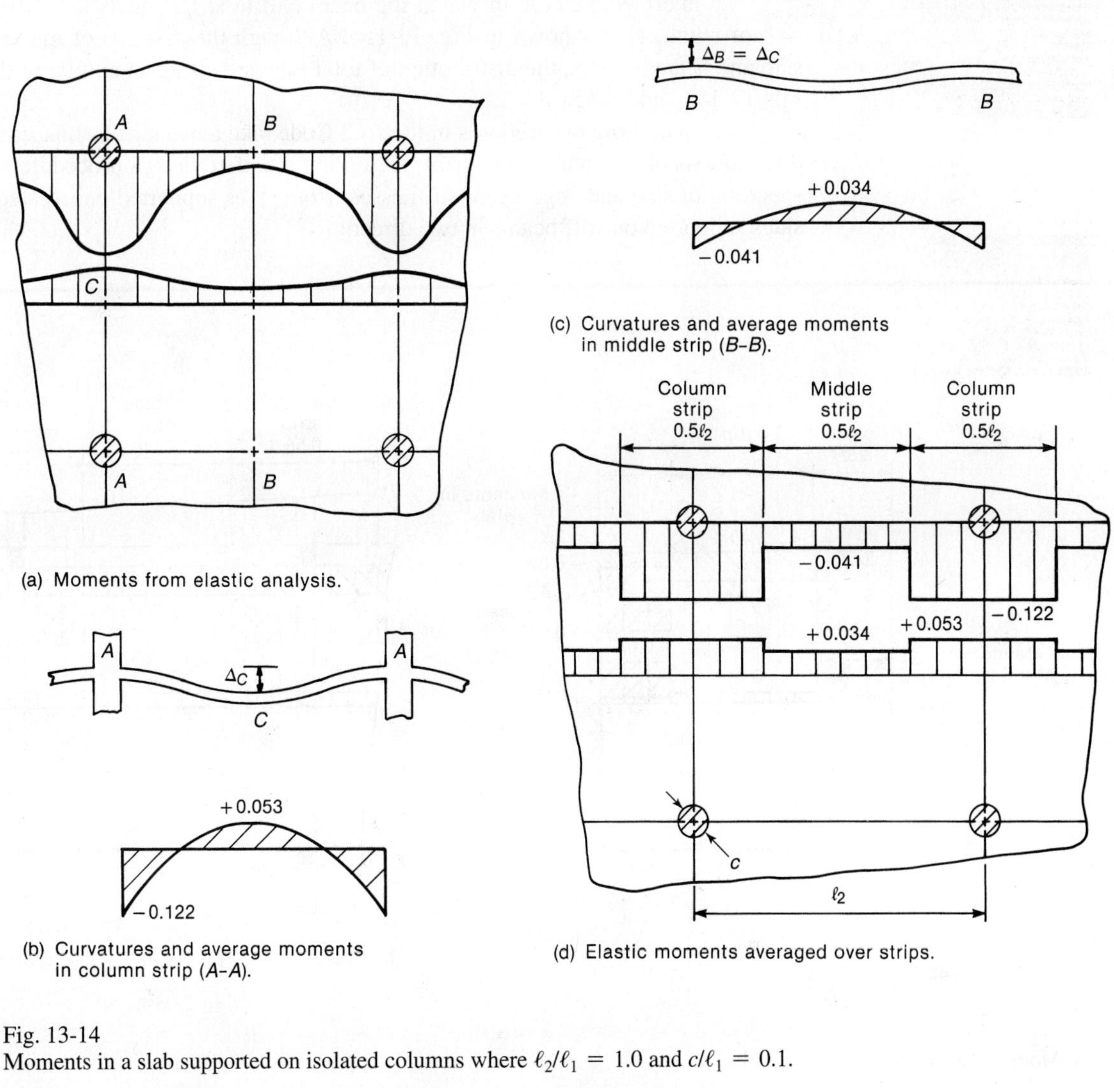

Fig. 13-14
Moments in a slab supported on isolated columns where $\ell_2/\ell_1 = 1.0$ and $c/\ell_1 = 0.1$.

column strips of width $\ell_2/2$ and a *middle strip* between the two column strips. The ACI Code design procedures consider the average moments over the width of the middle and column strips. A comparison of Figs. 13-14a and 13-14d shows that immediately adjacent to the columns the theoretical elastic moments may be considerably larger than is indicated by the average values.

The total static moment, M_o, accounted for here is

$$q\ell_n^2[(0.122 \times 0.5\ell_2) + (0.041 \times 0.5\ell_2) + (0.053 \times 0.5\ell_2) + (0.034 \times 0.5\ell_2)] = 0.125q\ell_2\ell_n^2$$

The distribution of moments given in Fig. 13-13 for a square slab fixed on four edges and supported on rigid beams is replotted in Fig. 13-15a with the moments averaged over column-strip and middle-strip bands in the same way as the flat-plate moments were in Fig. 13-14d. In addition, the sum of the beam moments and the column-strip slab moments has been divided by the width of the column strip and plotted as the *total column-strip moment*. The distribution of moments in Fig. 13-14d closely resembles the distribution of middle-strip and total column-strip moments in Fig. 13-15a.

An intermediate case in which the beam stiffness, I_b, equals the stiffness, I_s, of a slab of width, ℓ_2, is shown in Fig. 13-15b. Although the division of moments between slab and beams differs, the distribution of total moments is again similar to that shown in Fig. 13-14d and 13-15a.

The slab design procedures in the ACI Code take advantage of this similarity in the distributions of the total moments by presenting a unified design procedure for the whole spectrum of slab and edge-beam stiffnesses: from slabs supported on isolated columns to slabs supported on stiff beams in two directions.

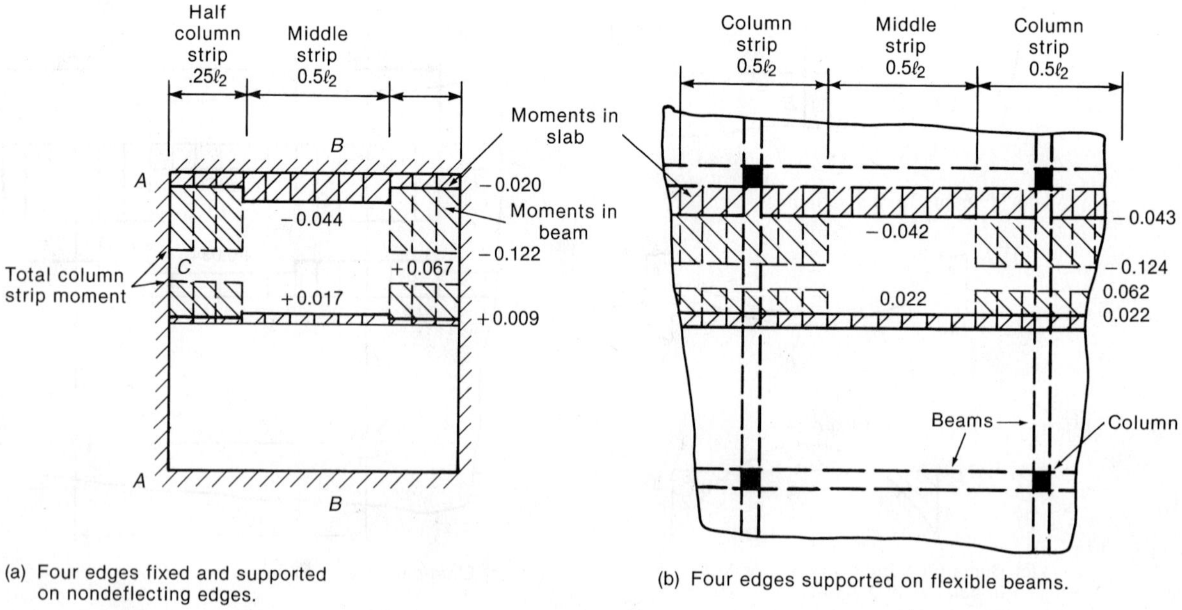

Fig. 13-15
Moments in slabs where $\ell_2/\ell_1 = 1.0$.

13-6 DESIGN OF SLABS

ACI Code Section 8.2.1 allows slabs to be designed by any procedure that satisfies both equilibrium and geometric compatibility, provided that every section has a strength at least equal to the required strength and that serviceability conditions are satisfied. Two procedures for the flexural analysis and design of two-way floor systems are presented in detail in the ACI Code. These are the direct-design method—considered in the following section—and the equivalent-frame design method—presented in Section 13-8. These two methods differ primarily in the way in which the slab moments are computed. The calculation of moments in the direct-design method is based on the *total statical moment*, M_o, presented earlier in this chapter. In this method, the slab is considered panel by panel, and Eq. (13-5) is used to compute the total moment in each panel and in each direction. The statical moment then is divided between positive and negative moments, and these are further divided between middle strips and column strips.

For the equivalent-frame method the slab is divided into a series of two-dimensional frames (in each direction), and the positive and negative moments are computed via an elastic-frame analysis. Once the positive and negative moments are known, they are divided between middle strips and column strips in exactly the same way as in the direct-design method. Similar design examples are given in this chapter to allow a comparison of the design moments obtained from these two methods. Other methods, such as the yield-line method (Chapter 14) and the strip method [13-5], [13-6], are permitted by ACI Code Section 13.5.

The direct-design method is emphasized in this book because an understanding of the method is essential for understanding the concepts of two-way slab design. In addition, it is an excellent method of checking slab design calculations. In design practice, computer programs based on the principles of the equivalent-frame method often are used.

Steps in Slab Design

The steps in the design of a two-way slab are the following:

1. Choose the layout and type of slab to be used. The various types of two-way slabs and their uses have been discussed briefly in Section 13-1. The choice of type of slab is strongly affected by architectural and construction considerations.

2. Choose the slab thickness. Generally, the slab thickness is chosen to prevent excessive deflection in service. Equally important, the slab thickness chosen must be adequate for shear at both interior and exterior columns (see Section 13-10).

3. Choose the method for computing the design moments. Equivalent-frame methods use an elastic-frame analysis to compute the positive and negative moments in the various panels in the slab. The direct-design method uses coefficients to compute these moments.

4. Calculate the distribution of the moments across the width of the slab. The lateral distribution of moments within a panel depends on the geometry of the slab and the stiffness of the beams (if any). This procedure is the same whether the negative and positive moments are calculated from the direct-design method or from an equivalent-frame method.

5. If there are beams, assign a portion of the column-strip moment to the beams.

6. Design reinforcement for the moments from steps 4 and 5. (Note: steps 3 through 6 need to be done for both principal directions.)

7. Check shear strength at a critical section around the columns.

Several of the parameters used in this process will be defined prior to carrying out slab designs.

Beam-to-Slab Stiffness Ratio, α_f

Slabs frequently are built with beams spanning from column to column around the perimeter of the building. These beams act to stiffen the edge of the slab and help to reduce the deflections of the exterior panels of the slab. Very heavily loaded slabs and long-span waffle slabs sometimes have beams joining all of the columns in the structure.

In the ACI Code, the effects of beam stiffness on deflections and the distribution of moments are expressed as a function of α_f, defined as the flexural stiffness, $4EI/\ell$, of the beam divided by the flexural stiffness of a width of slab bounded laterally by the centerlines of the adjacent panels on each side of the beam:

$$\alpha_f = \frac{4E_{cb}I_b/\ell}{4E_{cs}I_s/\ell}$$

Because the lengths, ℓ, of the beam and slab are equal, this quantity is simplified and expressed in the code as

$$\alpha_f = \frac{E_{cb}I_b}{E_{cs}I_s} \tag{13-9}$$

where E_{cb} and E_{cs} are the moduli of elasticity of the beam concrete and slab concrete, respectively, and I_b and I_s are the moments of inertia of the uncracked beams and slabs. The sections considered in computing I_b and I_s are shown shaded in Fig. 13-16. The span perpendicular to the direction being designed is ℓ_2. In Fig. 13-16c, the panels adjacent to

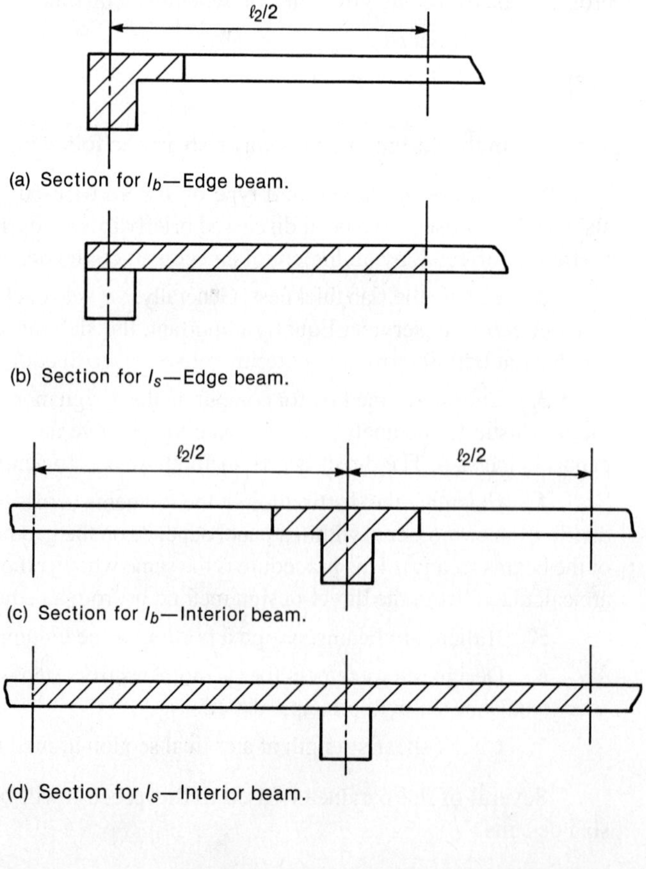

Fig. 13-16
Beam and slab sections for calculations of α_f.

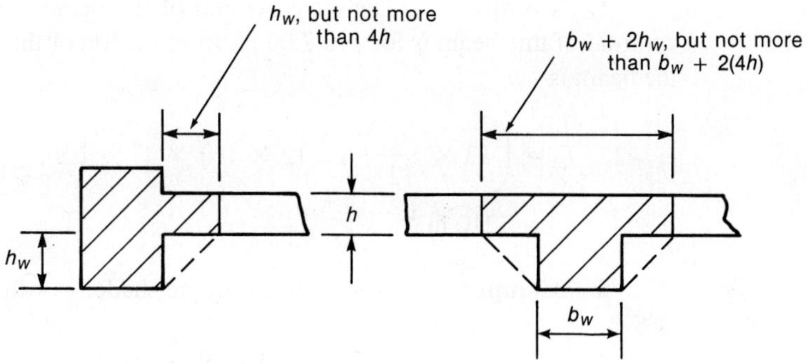

Fig. 13-17
Cross section of beams as defined in ACI Code Section 8.4.1.8.

the beam under consideration have different transverse spans. The calculation of ℓ_2 in such a case is illustrated in Example 13-4. If there is no beam, $\alpha_f = 0$.

ACI Section 8.4.1.8 defines a beam in monolithic or fully composite construction as the beam stem plus a portion of the slab on each side of the beam extending a distance equal to the projection of the beam above or below the slab, whichever is greater, but not greater than four times the slab thickness. This is illustrated in Fig. 13-17. Once the size of the slab and beam have been chosen, values of α_f can be computed from first principles.

EXAMPLE 13-1 Calculation of α_f for an Edge Beam

An 8-in.-thick slab is provided with an edge beam that has a total depth of 16 in. and a width of 12 in., as shown in Fig. 13-18a. The slab and beam were cast monolithically, have the same concrete strength, and thus the same E_c. Compute α_f. Because the slab and beam have the same elastic modulus, Eq. (13-9) reduces to $\alpha_f = I_b/I_s$.

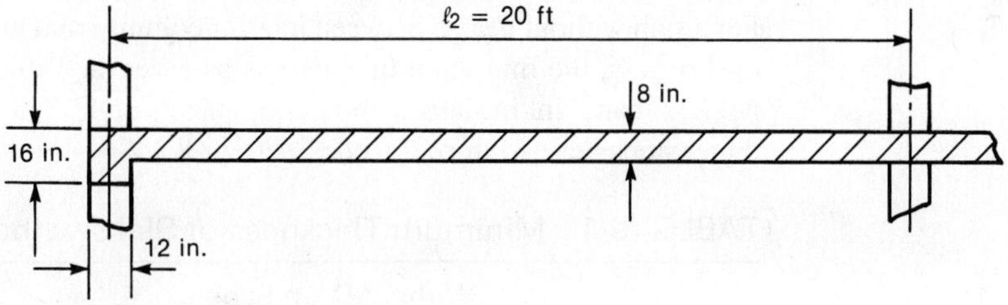

(a) Section through edge of slab.

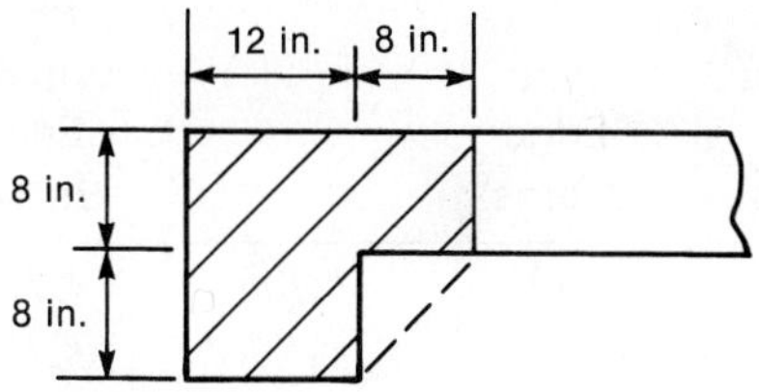

(b) Edge beam.

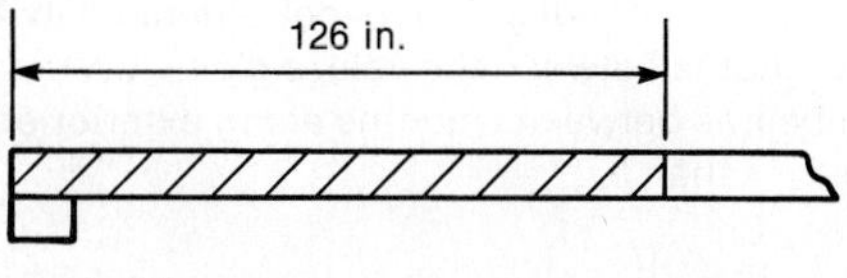

Fig. 13-18
Slab for Example 13-1.

(c) Section of slab.

1. Compute I_b. The cross section of the beam is as shown in Fig. 13-18b. The centroid of this beam is located 7.00 in. from the top of the slab. The moment of inertia of the beam is

$$I_b = \left(12 \times \frac{16^3}{12}\right) + (12 \times 16) \times 1^2 + \left(8 \times \frac{8^3}{12}\right) + (8 \times 8) \times 3^2$$
$$= 5210 \text{ in.}^4$$

2. Compute I_s. I_s is computed for the shaded portion of the slab in Fig. 13-18c:

$$I_s = 126 \times \frac{8^3}{12} = 5380 \text{ in.}^4$$

3. Compute α_f.

$$\alpha_f = \frac{I_b}{I_s} = \frac{5210}{5380} = 0.968$$ ■

Minimum Thickness of Two-Way Slabs

ACI Code Section 8.3.1 defines minimum thicknesses that generally are sufficient to limit slab deflections to acceptable values. Thinner slabs can be used if it can be shown that the computed slab deflections will not be excessive. The computation of slab deflections is discussed in Section 13-16.

Slabs without Beams between Interior Columns

For a slab without beams between interior columns and having a ratio of long to short spans of 2 or less, the minimum thickness is as given in Table 13-1 (ACI Table 8.3.1.1) but is not less than 5 in. in slabs without drop panels or 4 in. in slabs with drop panels having the dimensions defined in ACI Code Sections 8.2.4.

TABLE 13-1 Minimum Thickness of Slabs without Interior Beams

	Without Drop Panels			With Drop Panels[a]		
	Exterior Panels		Interior Panels	Exterior Panels		Interior Panels
Yield Strength, f_y[b] (psi)	Without Edge Beams	With Edge Beams[c]		Without Edge Beams	With Edge Beams[c]	
40,000	$\ell_n/33$	$\ell_n/36$	$\ell_n/36$	$\ell_n/36$	$\ell_n/40$	$\ell_n/40$
60,000	$\ell_n/30$	$\ell_n/33$	$\ell_n/33$	$\ell_n/33$	$\ell_n/36$	$\ell_n/36$
75,000	$\ell_n/28$	$\ell_n/31$	$\ell_n/31$	$\ell_n/31$	$\ell_n/34$	$\ell_n/34$

ℓ_n is the length of the clear span in the longer direction, measured face-to-face of the supports.
[a]The required geometry of a drop panel is defined in ACI Code Section.
[b]For yield strengths between the values given, use linear interpolation.
[c]Slabs with beams between columns along exterior edges. The value of α_f for the edge beam shall not be less than 0.8.

The ACI Code permits thinner slabs to be used if calculated deflections satisfy limits given in ACI Table 24.2.2. (See Section 13-16.)

As noted in the footnote to Table 13-1, a beam must have a stiffness ratio, α_f, of 0.8 or more to be called an edge beam. It can be shown that an edge beam, which has an overall height at least twice the slab thickness, h, and a gross area of at least $4h^2$ always will have α_f greater than 0.8. This rule of thumb can be used to simplify the selection of slab thicknesses.

Excessive slab deflections are a serious problem in many parts of North America. The thickness calculated from Table 13-1 should be rounded up to the next one-quarter inch for slabs less than 6 in. thick and to the next one-half inch for thicker slabs. Rounding the thickness up will give a stiffer slab and hence smaller deflections. Studies of slab deflections presented in [13-7] suggest that slabs without interior beams should be about 10 percent thicker than the ACI minimum values to avoid excessive deflections.

Slabs with Beams between the Interior Supports

For slabs with beams between interior supports, ACI Code Table 8.3.1.2 gives the following minimum thicknesses:

(a) For $\alpha_{fm} \le 0.2$, the minimum thicknesses in Table 13-1 shall apply.

(b) For $0.2 < \alpha_{fm} < 2.0$, the thickness shall not be less than

$$h = \frac{\ell_n[0.8 + (f_y/200{,}000)]}{36 + 5\beta(\alpha_{fm} - 0.2)} \quad \text{but not less than 5 in.} \tag{13-10}$$

(c) For $\alpha_{fm} > 2.0$, the thickness shall not be less than

$$h = \frac{\ell_n[0.8 + (f_y/200{,}000)]}{36 + 9\beta} \quad \text{but not less than 3.5 in.} \tag{13-11}$$

(d) At discontinuous edges, either an edge beam with a stiffness ratio α_f not less than 0.8 shall be provided or the slab thickness shall be increased by at least 10 percent in the edge panel.

In (a), (b), and (c),

h = overall thickness
ℓ_n = clear span of the slab panel under consideration, measured in the longer direction
α_{fm} = the average of the values of α_f for the four sides of the panel
β = longer clear span/shorter clear span of the panel

Note: Minimum thicknesses are computed as the second step of Examples 13-14 and 13-15.

The thickness of a slab also may be governed by shear. This is particularly likely if large moments are transferred to edge columns and for interior columns between two spans that are greatly different in length. The selection of slab thicknesses to satisfy shear requirements is discussed in Section 13-10. Briefly, that section suggests that the trial slab thickness be chosen such that $V_u \simeq 0.5$ to $0.55(\phi V_c)$ at edge columns and $V_u \simeq 0.85$ to $1.0(\phi V_c)$ at interior columns.

13-7 THE DIRECT-DESIGN METHOD

The direct-design method also could have been called *the direct-analysis method*, because this method essentially prescribes values for moments in various parts of the slab panel without the need for a structural analysis. The reader should be aware that this design method was introduced in an era when most engineering calculations were made with a slide rule and computer software was not available to do the repetitive calculations required to analyze a continuous-floor slab system. Thus, for continuous slab panels with relatively uniform lengths and subjected to distributed loading, a series of moment coefficients were developed that would lead to safe flexural designs of two-way floor systems.

Limitations on the Use of the Direct-Design Method

The direct-design method is easier to use than the equivalent-frame method, but can be applied only to fairly regular multipanel slabs. The limitations, given in ACI Code Section 8.10.2, include the following:

1. There must be a minimum of three continuous spans in each direction. Thus, a nine-panel structure (3 by 3) is the smallest that can be considered. If there are fewer than three panels, the interior negative moments from the direct-design method tend to be too small.

2. Rectangular panels must have a long-span/short-span ratio that is not greater than 2. One-way action predominates as the span ratio reaches and exceeds 2.

3. Successive span lengths in each direction shall not differ by more than one-third of the longer span. This limit is imposed so that certain standard reinforcement cut-off details can be used.

4. Columns may be offset from the basic rectangular grid of the building by up to 0.1 times the span parallel to the offset. In a building laid out in this way, the actual column locations are used in determining the spans of the slab to be used in calculating the design moments.

5. All loads must be due to gravity only and uniformly distributed over an entire panel. The direct-design method cannot be used for unbraced, laterally loaded frames, foundation mats, or prestressed slabs.

6. The service (unfactored) live load shall not exceed two times the service dead load. Strip or checkerboard loadings with large ratios of live load to dead load may lead to moments larger than those assumed in this method of analysis.

7. For a panel with beams between supports on all sides, the relative stiffness of the beams in the two perpendicular directions given by $(\alpha_{f1}\ell_2^2)/(\alpha_{f2}\ell_1^2)$ shall not be less than 0.2 or greater than 5. The term α_f was defined in the prior section, and ℓ_1 and ℓ_2 are the spans in the two directions.

Limitations 2 and 7 do not allow use of the direct-design method for slab panels that essentially transmit load as one-way slabs.

Distribution of Moments within Panels—Slabs without Beams between All Supports

Statical Moment, M_o

For design, the slab is considered to be a series of frames in the two directions, as shown in Fig. 13-19. These frames extend to the middle of the panels on each side of the column

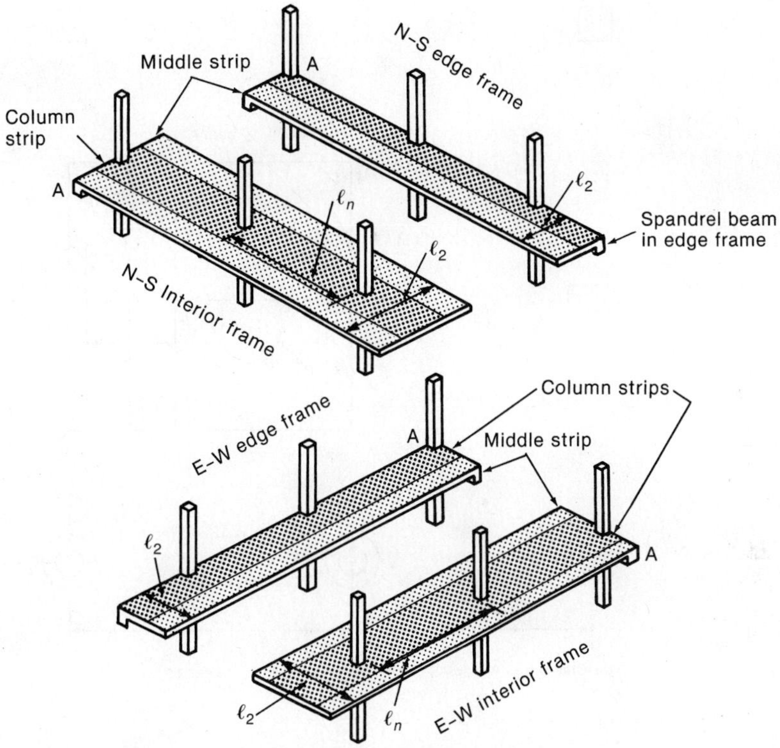

Fig. 13-19
Division of slab into frames for design.

lines. In each span of each of the frames, it is necessary to compute the total statical moment, M_o. We thus have

$$M_o = \frac{q_u \ell_2 \ell_n^2}{8} \tag{13-5}$$

where

q_u = factored load per unit area
ℓ_2 = transverse width of the strip
ℓ_n = clear span between columns

In computing ℓ_n, circular columns or column capitals of diameter d_c are replaced by equivalent square columns with side lengths $0.886d_c$. Values of ℓ_2 and ℓ_n are shown in Fig. 13-19 for panels in each direction. Example 13-2 illustrates the calculation of M_o in a typical slab panel.

EXAMPLE 13-2 Computation of Statical Moment, M_o

Compute the statical moment, M_o, in the slab panels shown in Fig. 13-20. The normal-weight concrete slab is 8 in. thick and supports a live load of 100 psf.

1. Compute the factored uniform loads.

$$q_u = 1.2\left(\frac{8}{12} \times 0.15\right) \text{ksf} + 1.6(0.100) \text{ ksf}$$
$$= 0.280 \text{ ksf}$$

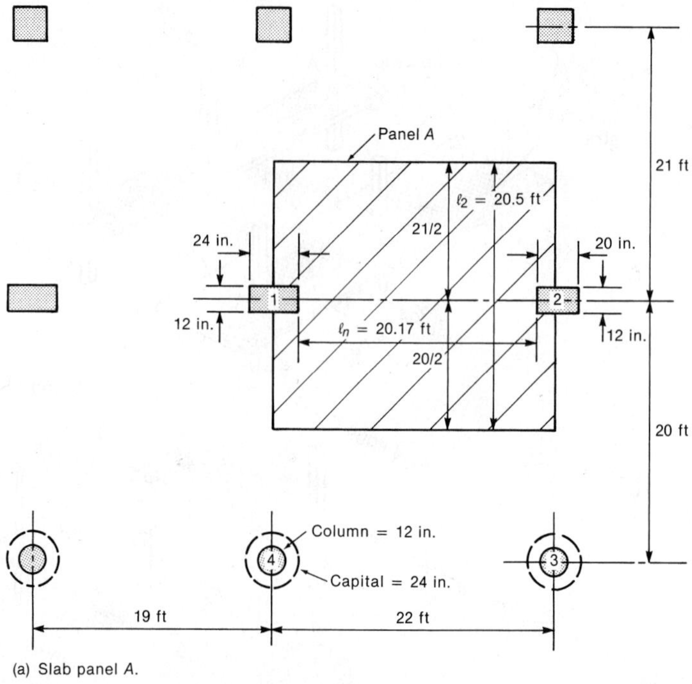

Fig. 13-20
Slab for Example 13-2.

Note that if the local building code allows a live-load reduction, the live load could be multiplied by the appropriate live-load-reduction factor before computing q_u.

2. Consider panel *A* spanning from column 1 to column 2. Slab panel *A* is shown shaded in Fig. 13-20a. The moments computed in this part of the example would be used to design the reinforcement running parallel to lines 1–2 in this panel. From Eq. (13-5),

$$M_o = \frac{q_u \ell_2 \ell_n^2}{8}$$

where

$$\ell_n = \text{clear span of slab panel}$$

$$= 22 \text{ ft} - \frac{1}{2}\left(\frac{20}{12}\right) \text{ ft} - \frac{1}{2}\left(\frac{24}{12}\right) \text{ ft} = 20.17 \text{ ft}$$

$$\ell_2 = \text{width of slab panel}$$

$$= (21/2) \text{ ft} + (20/2) \text{ ft} = 20.5 \text{ ft}$$

Therefore,

$$M_o = \frac{0.280 \text{ ksf} \times 20.5 \text{ ft} \times 20.17^2 \text{ ft}^2}{8}$$

$$= 292 \text{ kip-ft}$$

3. Consider panel *B*, spanning from column 1 to column 4. Slab panel *B* is shown shaded in Fig. 13-20b. The moments computed here would be used to design the reinforcement running parallel to lines 1–4 in this panel.

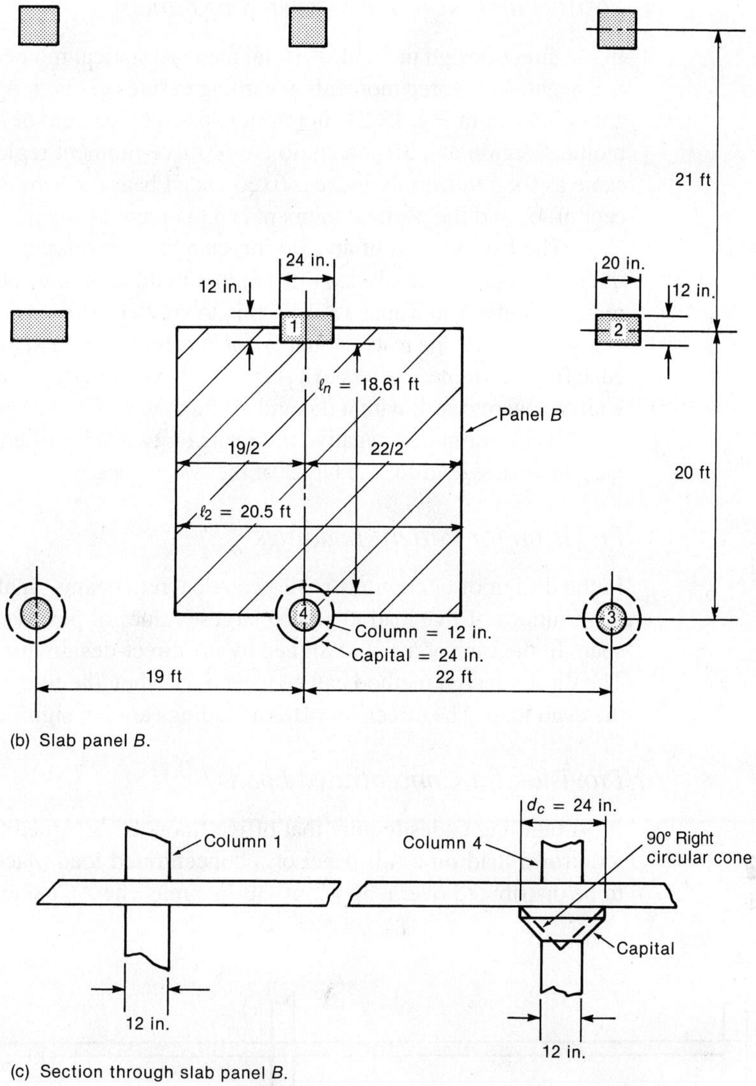

Fig. 13-20
Slab for Example 13-2.

A section through the slab showing columns 1 and 4 is shown in Fig. 13-20c. Column 4 has a column capital. ACI Code Section 8.4.1.4 defines the effective diameter of this capital as the diameter (measured at the bottom of the slab or drop panel) of the largest right circular cone with a 90° vertex that can be included within the column and capital. The outline of such a cone is shown with dashed lines in Fig. 13-20c, and the diameter, d_c, is 24 in. For the purpose of computing ℓ_n, the circular supports are replaced by equivalent square columns having a side length $c_1 = d_c\sqrt{\pi}/2$, or $0.886d_c$. Thus,

$$\ell_n = 20\text{ ft} - \frac{1}{2}\left(\frac{12}{12}\right)\text{ft} - \frac{1}{2}\left(0.886 \times \frac{24}{12}\right)\text{ft} = 18.61\text{ ft}$$

$$\ell_2 = \frac{19}{2}\text{ft} + \frac{22}{2}\text{ft} = 20.5\text{ ft}$$

$$M_o = \frac{0.280\text{ ksf} \times 20.5\text{ ft} \times 18.61^2\text{ ft}^2}{8} = 249\text{ kip-ft}$$

■

Positive and Negative Moments in Panels

In the direct-design method, the total factored statical moment, M_o, is divided into positive and negative factored moments according to rules given in ACI Code Section 8.10.4. These are illustrated in Fig. 13-21. In interior spans, 65 percent of M_o is assigned to the negative-moment region and 35 percent to the positive-moment regions. This is approximately the same as for a uniformly loaded, fixed-ended beam, where the negative moment is 67 percent of M_o and the positive moment is 33 percent.

The exterior end of an exterior span has considerably less fixity than the end at the interior support. The division of M_o in an end span into positive- and negative-moment regions is given in Table 13-2. In this table, "exterior edge unrestrained" refers to a slab whose exterior edge rests on, but is not attached to, for example, a masonry wall. "Exterior edge fully restrained" refers to a slab whose exterior edge is supported by, and is continuous with, a concrete wall with a flexural stiffness as large or larger than that of the slab.

If the computed negative moments on two sides of an interior support are different, the slab is designed for the larger of the two values.

Provision for Pattern Loadings

In the design of continuous-reinforced concrete beams, analyses are carried out for a few distributions of live load to get the largest values of positive and negative moment in each span. In the case of a slab designed by the direct-design method, no such analysis is done. The direct-design method can be used only when the live load does not exceed two times the dead load. The effects of pattern loadings are not significant in such a case [13-8].

Provision for Concentrated Loads

Most building codes require that office floors be designed for the largest effects of either a uniform load on a full panel or a concentrated load placed anywhere on the floor and to be distributed over a 30-in.-by-30-in. area. The ACI slab-design provisions apply only

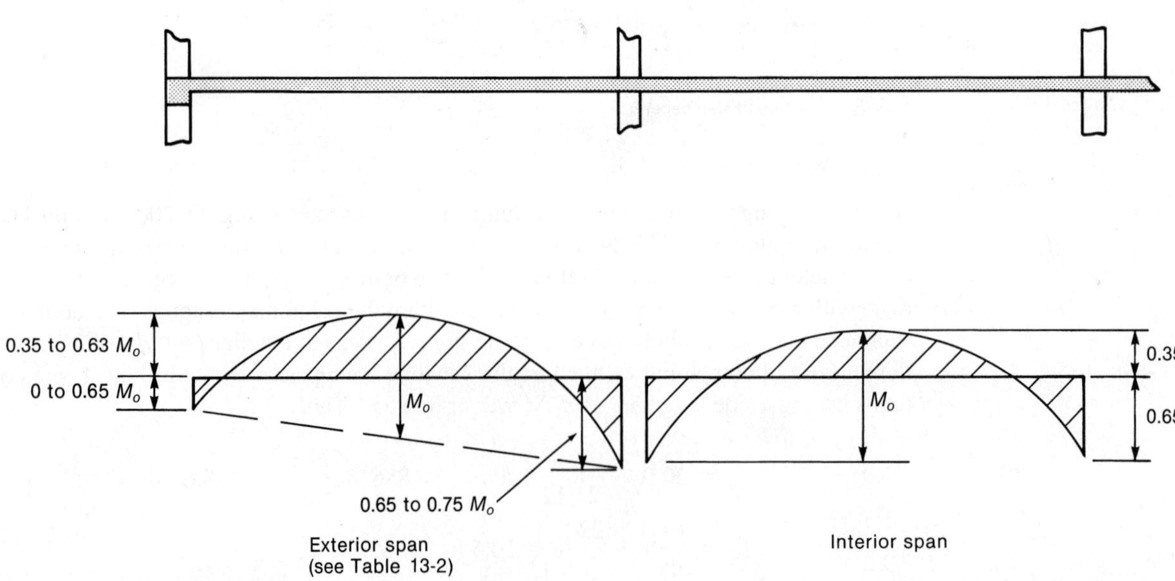

Fig. 13-21
Assignment of positive- and negative-moment regions.

TABLE 13-2 Distribution of Total Factored Static Moment, M_o, in an Exterior Span

	(1) Exterior Edge Unrestrained	(2) Slab *with* Beams between *All* Supports	(3) Slab *without* Beams between *Interior* Supports: Without Edge Beam	(4) Slab *without* Beams between *Interior* Supports: With Edge Beam	(5) Exterior Edge Fully Restrained
Interior Negative Factored Moment	0.75	0.70	0.70	0.70	0.65
Midspan Positive Factored Moment	0.63	0.57	0.52	0.50	0.35
Exterior Negative Factored Moment	0	0.16	0.26	0.30	0.65

Source: ACI Code Section 13.6.3.3.

to the uniformly loaded case. Woodring and Siess [13-9] studied the moments due to concentrated loads acting on square interior slab panels. For 20-, 25-, and 30-ft-square flat plates, the largest moments due to a 2000-lb concentrated load on a 30-in.-square area were equivalent to uniform loads of 39 or −23 psf, 27 or −16 psf, and 20 or −13 psf, respectively, for the three sizes of slab. The positive (downward) equivalent uniform loads are smaller than the 40- to 100-psf loads used in apartment or office floor design and hence do not govern. The negative, equivalent uniform loads indicate that, at some point in the slab, the worst effect of the concentrated load is equivalent to that of an upward uniform load. Because these all are less than the dead load of a concrete floor slab, the concentrated load case does not govern here either. In summary, therefore, design for the uniform-load case will satisfy the 2000-lb concentrated-load case for 20-ft-square and larger slabs. For larger concentrated loads, an equivalent-frame analysis can be used. (See Section 13-8.)

Definition of Column Strips and Middle Strips

As shown in Fig. 13-14a, the moments vary continuously across the width of the slab panel. To aid in steel placement, the design moments are averaged over the width of column strips over the columns and middle strips between the column strips as shown in Fig. 13-14d. The widths of these strips are defined in ACI Code Sections 8.4.1.5 and 8.4.1.6 and are illustrated in Fig. 13-22. The column strips in both directions extend one-fourth of the smaller span, ℓ_{min}, each way from the column line.

Distribution of Moments between Column Strips and Middle Strips

ACI Code Section 8.10.5 defines the fraction of the negative and positive moments assigned to the column strips. The remaining amount of negative and positive moment is assigned to the adjacent half-middle strips (Fig. 13-24). The division is a function of $\alpha_{f1}\ell_2/\ell_1$, which depends on the aspect ratio of the panel, ℓ_2/ℓ_1, and the relative stiffness, α_{f1}, of the beams (if any) spanning parallel to and within the column strip.

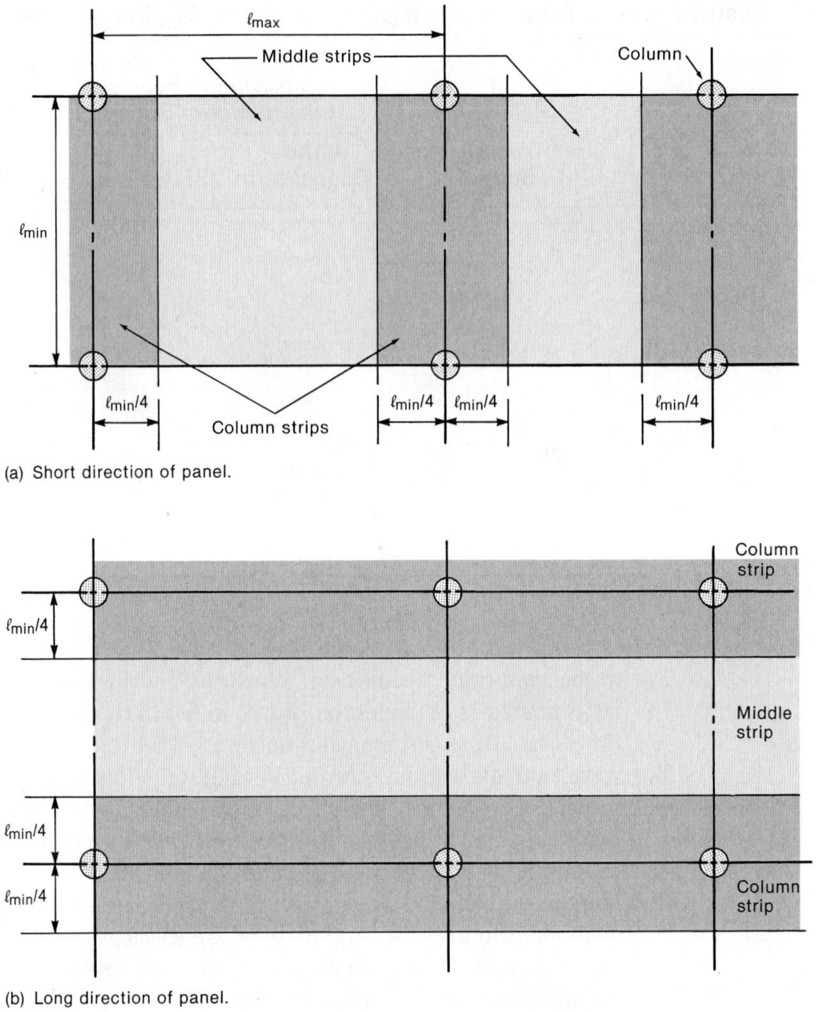

Fig. 13-22
Definitions of columns and middle strips.

Table 13-3 gives the percentage distribution of negative factored moment to the column strip at all interior supports. For floor systems without interior beams, $\alpha_{f1}\ell_2/\ell_1$ is taken to be equal to zero, because $\alpha_{f1} = 0$. In this case, 75 percent of the negative moment is distributed to the column strip, and the remaining 25 percent is divided equally between the two adjacent half-middle strips. For cases where a beam is present in the column strip (spanning in the direction of ℓ_1) and $\alpha_{f1}\ell_2/\ell_1 \geq 1.0$, the second row in Table 13-3 applies. Some linear interpolation may be required based on the ratio ℓ_2/ℓ_1. For cases where $0 \leq \alpha_{f1}\ell_2/\ell_1 \leq 1.0$, linear interpolations will be required between the percentages given in the first and second rows of Table 13-3. Similar procedures are used for the distribution of the factored moments at other locations along the span.

Table 13-4 gives the percentage distribution of positive factored moment to the column strip at midspan for both interior and exterior spans. For floor systems without interior beams, 60 percent of the positive moment is assigned to the column strip and the remaining 40 percent is divided equally between the adjacent half-middle strips. If a beam is present in the column strip (spanning in the direction of ℓ_1), either the percentages in the second row or a linear interpolation between the percentages given in the first and second rows of Table 13-4 will apply.

TABLE 13-3 Percentage Distribution of Interior Negative Factored Moment to Column Strip

ℓ_2/ℓ_1	0.5	1.0	2.0
$(\alpha_{f1}\ell_2/\ell_1) = 0$	75	75	75
$(\alpha_{f1}\ell_2/\ell_1) \geq 1.0$	90	75	45

TABLE 13-4 Percentage Distribution of Midspan Positive Factored Moment to Column Strip

ℓ_2/ℓ_1	0.5	1.0	2.0
$(\alpha_{f1}\ell_2/\ell_1) = 0$	60	60	60
$(\alpha_{f1}\ell_2/\ell_1) \geq 1.0$	90	75	45

At an exterior edge, the division of the exterior-end factored negative moment distributed to the column and middle strips spanning perpendicular to the edge also depends on the torsional stiffness of the edge beam, calculated as the shear modulus, G, times the torsional constant of the edge beam, C, divided by the flexural stiffness of the slab spanning *perpendicular* to the edge beam (i.e., EI for a slab having a width equal to the length of the edge beam from the center of one span to the center of the other span, as shown in Fig. 13-16d). Assuming that Poisson's ratio is zero gives $G = E/2$, then this torsional stiffness ratio is defined as

$$\beta_t = \frac{E_{cb}C}{2E_{cs}I_s} \tag{13-12}$$

where the cross section of the edge beam is as defined in ACI Code Section and Fig. 13-17. If there are no edge beams, β_t can be taken to be equal to zero.

The term C in Eq. (13-12) refers to the torsional constant of the edge beam. This is roughly equivalent to a polar moment of inertia. It is calculated by subdividing the cross section into rectangles and carrying out the summation

$$C = \sum\left[\left(1 - 0.63\frac{x}{y}\right)\frac{x^3y}{3}\right] \tag{13-13}$$

where x is the shorter side of a rectangle and y is the longer side. The subdivision of the cross section of the torsional members is illustrated in Fig. 13-23. Different combinations of rectangles should be tried to get the *maximum* value of C. The maximum value normally is obtained when the wider rectangle is made as long as possible. Thus, the rectangles chosen in Fig. 13-23b will give a larger value of C than will those in Fig. 13-23a.

Table 13-5 gives the percentage distribution of negative factored moment to the column strip at exterior supports. The setup of this table is similar to that used in Tables 13-3 and 13-4, with an addition of two rows to account for presence or absence of an edge beam working in torsion to transfer some of the slab negative moment into the column. When there is no edge beam ($\beta_t = 0$), all of the negative moment is assigned to the column strip. This is reasonable because there is no torsional edge member to transfer moment from the middle strip all the way back to the column. If a stiff edge beam is present ($\beta_t \geq 2.5$), Table 13-5 gives specific percentages to be assigned to the column strip, depending on the value of α_{f1} and the ℓ_2/ℓ_1 ratio, as was done in Tables 13-3 and 13-4. For values of β_t

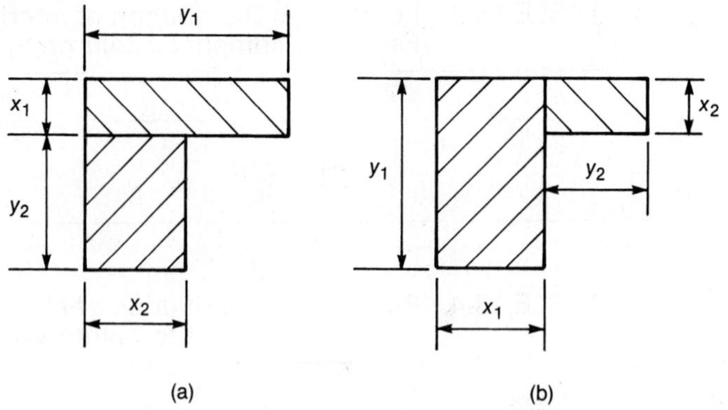

Fig. 13-23
Division of edge members for calculation of torsional constant, C.

between 2.5 and 0.0, and values of $(\alpha_{f1}\,\ell_2/\ell_1)$ between 1.0 and 0.0, two or three levels of linear interpolation may be required to determine the percentage of negative moment assigned to the column strip.

If a beam is present in the column strip (spanning in the direction of ℓ_1), a portion of the column-strip moment is assigned to the beam, as specified in ACI Code Section 8.10.5.7. If the beam has $\alpha_1\ell_2/\ell_1$ greater than 1.0, 85 percent of the column-strip moment is assigned to the beam and 15 percent to the slab. This is discussed more fully in Section 13-14 and Example 13-15.

TABLE 13-5 Percentage Distribution of Exterior Negative Factored Moment to Column Strip

ℓ_2/ℓ_1		0.5	1.0	2.0
$(\alpha_{f1}\ell_2/\ell_1) = 0$	$\beta_t = 0$	100	100	100
	$\beta_t \geq 2.5$	75	75	75
$(\alpha_{f1}\ell_2/\ell_1) \geq 1.0$	$\beta_t = 0$	100	100	100
	$\beta_t \geq 2.5$	90	75	45

EXAMPLE 13-3 Calculation of Moments in an Interior Panel of a Flat Plate

Figure 13-24 shows an interior panel of a flat-plate floor in an apartment building. The slab thickness is 5.5 in. The slab supports a design live load of 50 psf and a superimposed dead load of 25 psf for partitions. The columns and slab have the same strength of concrete. The story height is 9 ft. Compute the column-strip and middle-strip moments in the short direction of the panel.

1. Compute the factored loads.

$$q_u = 1.2\left(\frac{5.5}{12} \times 150 + 25\right) + 1.6(50) = 193 \text{ psf}$$

Note that, if the local building code allows a live-load reduction factor, the 50-psf live load could be multiplied by the appropriate factor. The reduction should be based on a tributary area equal to the total panel area $\ell_1 \times \ell_2$. (Note: $K_{LL} = 1.0$ for two-way slab panels.)

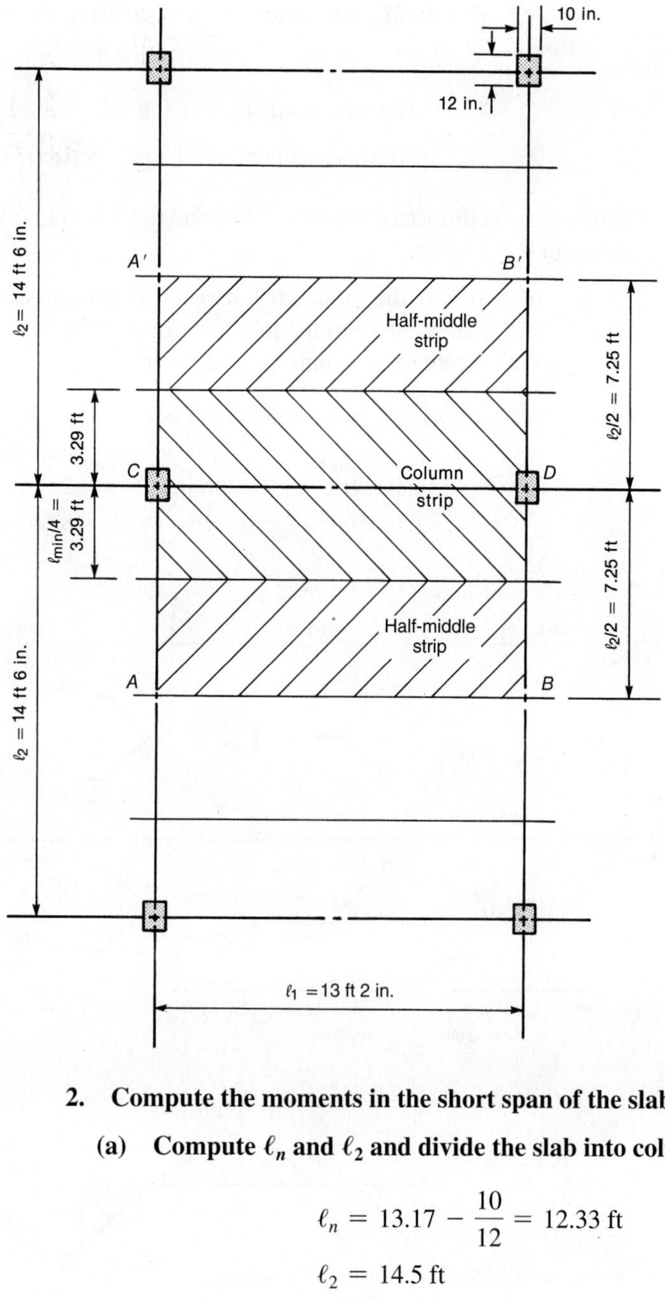

Fig. 13-24
Interior panel of a flat plate for Example 13-3.

2. Compute the moments in the short span of the slab.

(a) Compute ℓ_n and ℓ_2 and divide the slab into column and middle strips.

$$\ell_n = 13.17 - \frac{10}{12} = 12.33 \text{ ft}$$

$$\ell_2 = 14.5 \text{ ft}$$

The column strip extends the smaller of $\ell_2/4$ or $\ell_1/4$ on each side of the column centerline, as shown in Fig. 13-22. Thus, the column strip extends $13.17/4 = 3.29$ ft on each side of column centerline. The total width of the column strip is 6.58 ft. Each half-middle strip extends from the edge of the column strip to the centerline of the panel. The total width of two half-middle strips is $14.5 - 6.58 = 7.92$ ft.

(b) Compute M_o.

$$M_o = \frac{q_u \ell_2 \ell_n^2}{8} = \frac{0.193 \times 14.5 \times 12.33^2}{8}$$

$$= 53.2 \text{ kip-ft}$$

(c) Divide M_o into negative and positive moments. From ACI Code Section 8.10.4.1,

$$\text{Negative moment} = -0.65M_o = -34.6 \text{ kip-ft}$$

$$\text{Positive moment} = 0.35M_o = 18.6 \text{ kip-ft}$$

This process is illustrated in Fig. 13-25a, and the resulting distribution of total moments is shown in Fig. 13-25b.

(d) Divide the moments between the column and middle strips. ***Negative moments:*** From Table 13-3 for $\alpha_{f1}\ell_2/\ell_1 = 0$ ($\alpha_{f1} = 0$ because there are no beams between columns A and B in this panel),

$$\begin{aligned}\text{Column-strip negative moment} &= 0.75 \times -34.6 \text{ kip-ft}\\ &= -26.0 \text{ kip-ft}\\ \text{Middle-strip negative moment} &= 0.25 \times -34.6 \text{ kip-ft}\\ &= -8.66 \text{ kip-ft}\end{aligned}$$

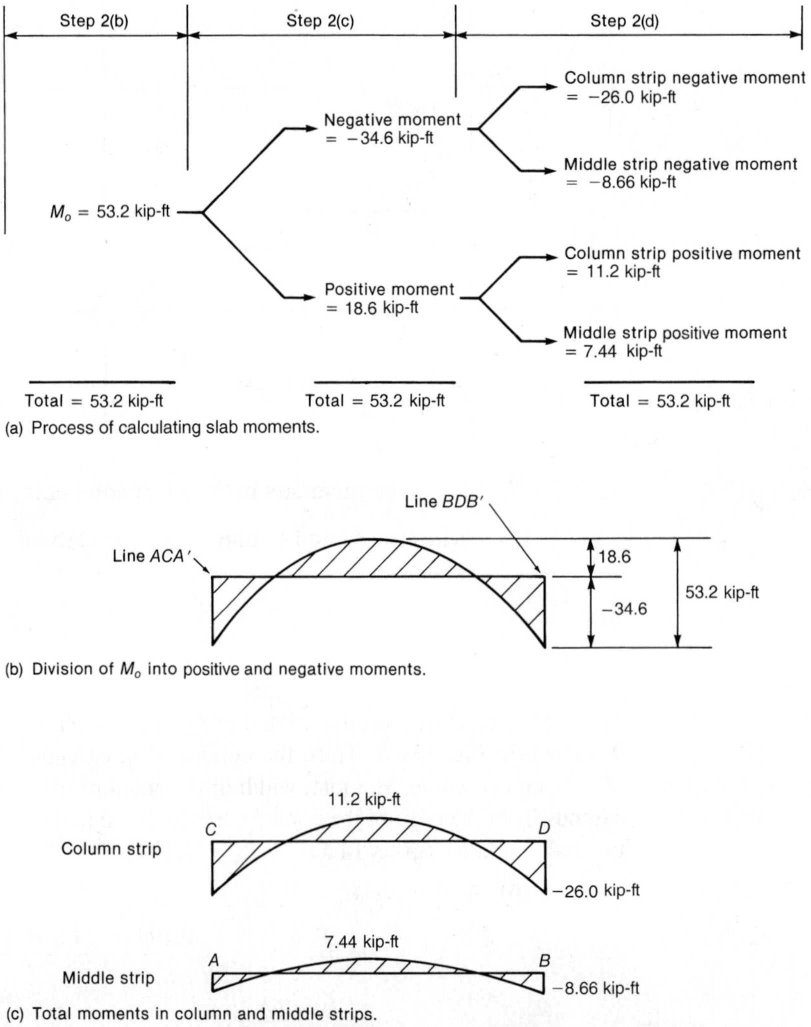

Fig. 13-25
Distribution of moments in an interior panel or flat slab—Example 13-3.

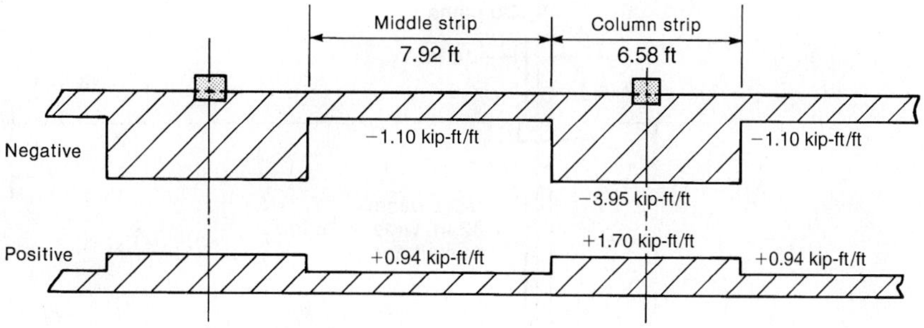

Fig. 13-26
Calculation of moments—
Example 13-3.

Half of this, −4.33 kip-ft, goes to each adjacent half-middle strip. Because the adjacent bays have the same width, ℓ_2, a similar moment will be assigned to the other half of each middle strip so that the total middle-strip negative moment is 8.66 kip-ft. ***Positive moments:*** From Table 13-4, where $a_{f1}\ell_2/\ell_1 = 0$,

$$\text{Column-strip positive moment} = 0.6 \times 18.6 = 11.2 \text{ kip-ft}$$

$$\text{Middle-strip positive moment} = 0.4 \times 18.6 = 7.44 \text{ kip-ft}$$

These calculations are illustrated in Fig. 13-25a. The resulting distributions of moments in the column strip and middle strip are summarized in Fig. 13-25c. In Fig. 13-26, the moments in each strip have been divided by the width of that strip. This diagram is very similar to the theoretical elastic distribution of moments given in Fig. 13-14d.

3. Compute the moments in the long span of the slab. Although it was not asked for in this example, in a slab design, it now would be necessary to repeat steps 2(a) to 2(d) for the long span. ■

EXAMPLE 13-4 Calculation of Moments in an Exterior Panel of a Flat Plate

Compute the positive and negative moments in the column and middle strips of the exterior panel of the slab between columns *B* and *E* in Fig. 13-27. The slab is 8 in. thick and supports a superimposed service dead load of 25 psf and a service live load of 60 psf. The edge beam is 12 in. wide by 16 in. in overall depth and is cast monolithically with the slab.

1. Compute the factored loads.

$$q_u = 1.2\left(\frac{8}{12} \times 150 + 25\right) + 1.6(60) = 246 \text{ psf}$$

As in the prior example, if the local building code allows a live-load reduction factor, the 60-psf live load could be multiplied by the appropriate factor. The reduction should be based on the tributary area $\ell_1 \times \ell_2$.

2. Compute the moments in span *BE*.

(a) Compute ℓ_n and ℓ_2 (avg) and divide the slab into middle and column strips.

$$\ell_n = 21.0 - \frac{1}{2}\left(\frac{14}{12}\right) - \frac{1}{2}\left(\frac{16}{12}\right) = 19.75 \text{ ft}$$

$$\ell_2(\text{avg}) = 19 \text{ ft}$$

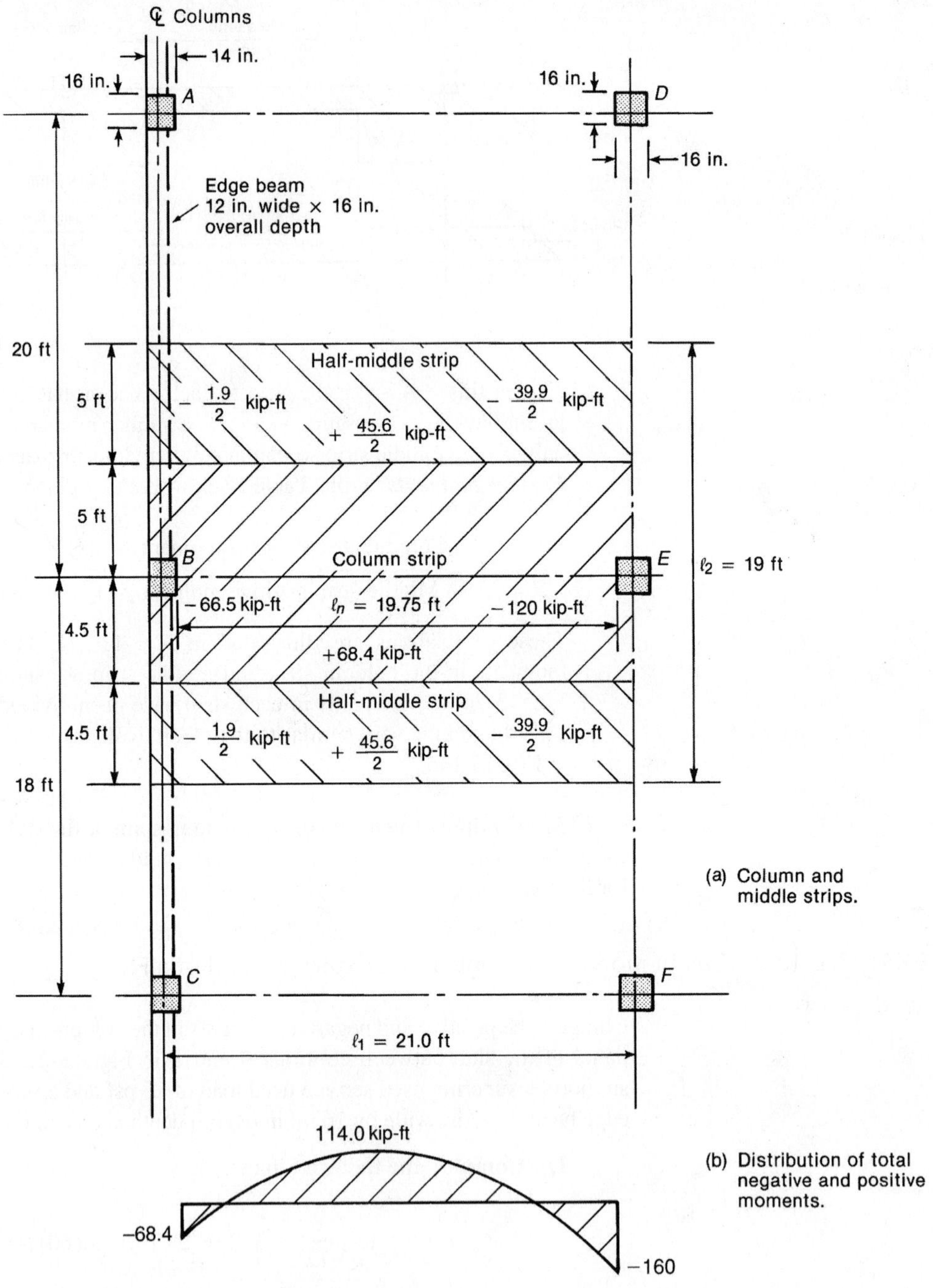

Fig. 13-27
Calculation of moments in an end span—Example 13-4.

The column strip extends the smaller of $\ell_2/4$ or $\ell_1/4$ on each side of the column centerline. Because ℓ_1 is greater than either value of ℓ_2, base this on ℓ_2. The column strip extends $20/4 = 5$ ft toward *AD* and $18/4 = 4.5$ ft toward *CF* from line *BE*, as shown in Fig. 13-27a. The total width of the column strip is 9.5 ft. The half-middle strip between *BE* and *CF* has a width of 4.5 ft, and the other one is 5 ft, as shown.

(b) Compute M_o.

$$M_o = \frac{q_u \ell_2 \ell_n^2}{8} = \frac{0.246 \times 19 \times 19.75^2}{8}$$
$$= 228 \text{ kip-ft}$$

(c) Divide M_o into positive and negative moments. The distribution of the moment to the negative and positive regions is as given in Table 13-2. In the terminology of Table 13-2, this is a "slab without beams between interior supports and with edge beam." From Table 13-2, the total moment is divided as follows:

$$\text{Interior negative } M_u = -0.70M_o = -160 \text{ kip-ft}$$
$$\text{Positive } M_u = 0.50M_o = +114.0 \text{ kip-ft}$$
$$\text{Exterior negative } M_o = -0.30M_o = -68.4 \text{ kip-ft}$$

This calculation is illustrated in Fig. 13-28. The resulting negative and positive moments are shown in Fig. 13-27b.

(d) Divide the moments between the column and middle strips.

Interior negative moments: This division is a function of $\alpha_{f1}\ell_2/\ell_1$, which again equals zero, because there are no beams parallel to *BE*. From Table 13-3,

$$\text{Interior column-strip negative moment} = 0.75 \times -160 = -120 \text{ kip-ft}$$
$$= -12.6 \text{ kip-ft/ft of width of column strip}$$
$$\text{Interior middle-strip negative moment} = -40 \text{ kip-ft}$$

Half of this goes to each of the half-middle strips beside column strip *BE*.

Positive moments: From Table 13-4,

$$\text{Column-strip positive moment} = 0.60 \times 114.0 = 68.4 \text{ kip-ft}$$
$$= 7.20 \text{ kip-ft/ft}$$
$$\text{Middle-strip positive moment} = 45.6 \text{ kip-ft}$$

Half of this goes to each half-middle strip.

Exterior negative moment: From Table 13-5, the exterior negative moment is divided as a function of $\alpha_{f1}\ell_2/\ell_1$ (again equal to zero because there is no beam parallel to ℓ_1) and

$$\beta_t = \frac{E_{cb}C}{2E_{cs}I_s} \tag{13-12}$$

where E_{cb} and C refer to the attached torsional member shown in Fig. 13-29, and E_{cs} and I_s refer to the strip of slab being designed (the column strip and the two half-middle strips shown shaded in Fig. 13-27a). To compute C, divide the edge beam into rectangles. The two possibilities shown in Fig. 13-29 will be considered. For Fig. 13-29a, Eq. (13-13) gives

$$C = \frac{(1 - 0.63 \times 12/16)12^3 \times 16}{3} + \frac{(1 - 0.63 \times 8/8)8^3 \times 8}{3}$$
$$= 5367 \text{ in.}^4$$

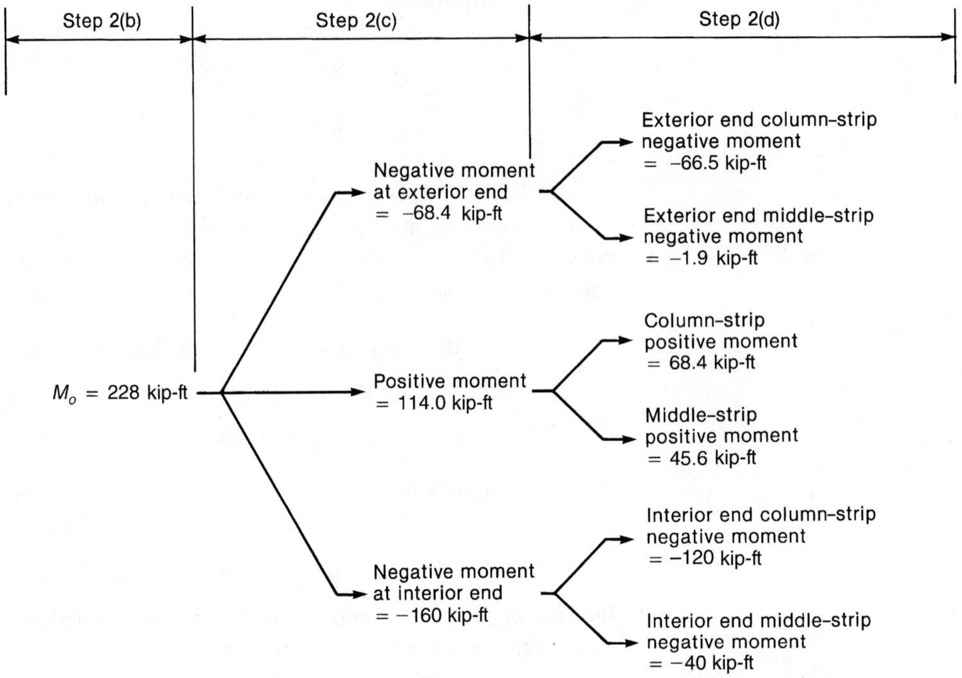

Fig. 13-28
Calculation of moments in an end span for Example 13-4.

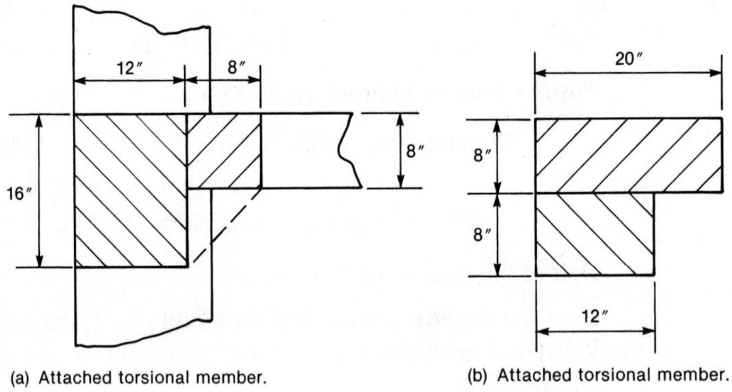

Fig. 13-29
Slab, column, and edge beam for Example 13-4.

For Fig. 13-29b, $C = 3741$ in.^{4} The larger of these two values is used; therefore, $C = 5367$ in.^{4}

I_s is the moment of inertia of the strip of slab section perpendicular to the edge, which has $b = 19$ ft and $h = 8$ in. Then

$$I_s = \frac{(19 \times 12) \times 8^3}{12} = 9728 \text{ in.}^4$$

Because f'_c is the same in the slab and beam, $E_{cb} = E_{cs}$ and

$$\beta_t = \frac{5367}{2(9728)} = 0.276$$

Interpolating in Table 13-5, we have

For $\beta_t = 0$: 100 percent to column strip

For $\beta_t = 2.5$: 75 percent to column strip

Therefore, for $\beta_t = 0.276$, 97.2 percent to column strip, and we have

$$\text{Exterior column-strip negative moment} = 0.972(-68.4) = -66.5 \text{ kip-ft}$$
$$= -7.00 \text{ kip-ft/ft}$$
$$\text{Exterior middle-strip negative moment} = -1.9 \text{ kip-ft}$$

■

Transfer of Moments to Columns

Exterior Columns

When design is carried out by the direct-design method, ACI Code Section 8.10.7.3 specifies that the moment to be transferred from a slab to an edge column is $0.30\,M_o$. This moment is used to compute the shear stresses due to moment transfer to the edge column, as discussed in Section 13-11. Although not specifically stated in the ACI Code, this moment can be assumed to be acting at the centroid of the shear perimeter. The exterior negative moment from the direct-design method calculation is divided between the columns above and below the slab in proportion to the column flexural stiffnesses, $4EI/\ell$. The resulting column moments are used in the design of the columns.

Interior Columns

At interior columns, the moment-transfer calculations and the total moment used in the design of the columns above and below the floor are based on an unbalanced moment resulting from an uneven distribution of live load. The unbalanced moment is computed by assuming that the longer span adjacent to the column is loaded with the factored dead load and half the factored live load, while the shorter span carries only the factored dead load. The total unbalanced negative moment at the joint is thus:

$$M = 0.65\left[\frac{(q_{Du} + 0.5q_{Lu})\ell_2\ell_n^2}{8} - \frac{q'_{Du}\ell'_2(\ell'_n)^2}{8}\right]$$

where q_{Du} and q_{Lu} refer to the factored dead and live loads on the longer span and q'_{Du} refers to the factored dead load on the shorter span adjacent to the column. The values of ℓ_2 and ℓ_n refer to the longer of the adjacent spans and the values ℓ'_2 and ℓ'_n refer to the shorter span. The factor 0.65 is the fraction of the static moment assigned to the negative moment at an interior support from ACI Code Section 8.10.4.1. Using the expression given above, ACI Code Section 8.10.7.2 defines the following moment as being transferred at interior connections from the slab to the column sections above and below the slab in proportion to their flexural stiffnesses:

$$M_{\text{slab}} = 0.07[(q_{Du} + 0.5q_{Lu})\ell_2\ell_n^2 - q'_{Du}\ell'_2\,(\ell'_n)^2] \tag{13-14}$$

This moment is also used to compute the shear stresses due to moment transfer at an interior column.

13-8 EQUIVALENT-FRAME METHODS

The ACI Code presents two general methods for calculating the longitudinal distribution of moments in two-way slab systems. These are the direct-design method (presented in the previous section) and equivalent-frame methods, which are presented in this and the following section. Equivalent-frame methods are intended for use in analyzing moments in any practical slab–column frame. Their scope is thus wider than the direct-design method, which is subject to the limitations presented in Section 13-7 (ACI Code Section 8.10.2). In the direct-design method, the statical moment, M_o, is calculated for each slab span. This moment is then divided between positive- and negative-moment regions using arbitrary moment coefficients, which are adjusted to reflect pattern loadings. For equivalent-frame methods, a stiffness analyses of a slab–column frame is used to determine the longitudinal distribution of bending moments, including possible pattern loadings. The transverse distribution of moments to column and middle strips, as defined in the prior section, is the same for both methods.

The use of frame analyses for two-way slabs was first proposed by Peabody [13-10] in 1948, and a method of slab analysis referred to as "design by elastic analysis" was incorporated in the 1956 and 1963 editions of the ACI Code. In the late 1940s, Siess and Newmark [13-11] studied the application of moment-distribution analyses to two-way slabs on stiff beams. Following extensive research on two-way slabs carried out at the University of Illinois, Corley and Jirsa [13-12] presented a more refined method of frame analysis for slabs. This has been incorporated in the 1971 and subsequent ACI Codes. Corley and Jirsa considered only gravity loads. Studies of the use of frame analyses for laterally loaded slab–column frames [13-13] led to treatment of this problem in the 1983 and subsequent ACI Codes. This will be referred to as the classic equivalent-frame method in the following subsection.

Classic Equivalent-Frame Analysis of Slab Systems for Vertical Loads

The slab is divided into a series of equivalent frames running in the two directions of the building, as shown in Fig. 13-19 (ACI Code Section 8.11.2). These frames consist of the slab, any beams that are present, and the columns above and below the slab. For gravity-load analysis, the code allows analysis of an entire equivalent frame extending over the height of the building, or each floor can be considered separately with the far ends of the columns being fixed. The original derivation of the classic equivalent-frame method assumed that moment distribution would be the calculation procedure used to analyze the continuous-slab system, so some of the concepts in the method are awkward to adapt to other methods of analysis.

Calculation of Stiffness, Carryover, and Fixed-End Moments

In the moment-distribution method, it is necessary to compute *flexural stiffnesses, K*; *carry-over factors, COF*; *distribution factors, DF*; and *fixed-end moments, FEM*, for each of the members in the structure. For a prismatic member fixed at the far end and with negligible axial loads, the flexural stiffness is

$$K = \frac{kEI}{L} \tag{13-15}$$

where $k = 4$ and the carryover factor is ± 0.5, the sign depending on the sign convention used for moments. For a prismatic, uniformly loaded beam, the fixed-end moments are $w\ell^2/12$.

In the equivalent-frame method, the increased stiffness of members within the column–slab joint region is accounted for, as is the variation in cross section at drop panels. As a result, all members have a stiffer section at each end, as shown in Fig. 13-30b. If

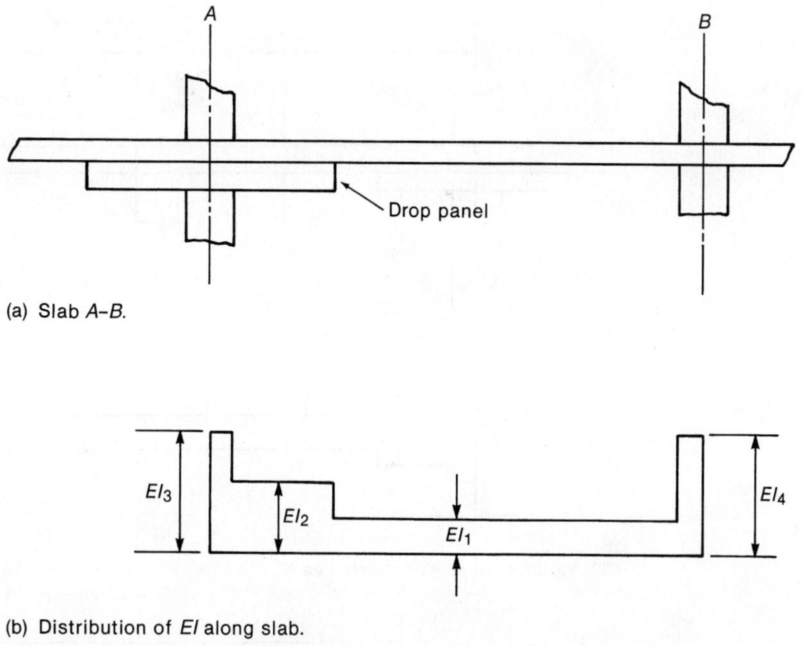

Fig. 13-30
Variation in stiffness along a span.

the *EI* used in Eq. (13-15) is that at the midspan of the slab strip, *k* will be greater than 4; similarly, the carryover factor will be greater than 0.5, and the fixed-end moments for a uniform load w will be greater than $w\ell^2/12$.

Several methods are available for computing values of *k*, *COF*, and *FEM*. Originally, these were computed by using the *column analogy* developed by Hardy Cross. Cross observed an analogy between the equations used to compute stresses in an unsymmetrical column loaded with axial loads and moments, and the equations used to compute moments in a fixed-end beam [13-14].

Tables and charts for computing *k*, *COF*, and *FEM* are given in Appendix A as Tables A-14 through A-17.

Properties of Slab-Beams

The horizontal members in the equivalent frame are referred to as *slab-beams.* These consist of either only a slab, or a slab and a drop panel, or a slab with a beam running parallel to the equivalent frame. ACI Code Section 8.11.3 explains how these nonprismatic beams are to be modeled for analysis:

1. At points outside of joints or column capitals, the moment of inertia may be based on the gross area of the concrete. Variations in the moment of inertia along the length shall be taken into account. Thus, for the slab with a drop panel shown in Fig. 13-31a, the moment of inertia at section *A–A* is that for a slab of width ℓ_2 (Fig. 13-31c). At section *B–B* through the drop panel, the moment of inertia is for a slab having the cross section shown in Fig. 13-31d. Similarly, for a slab system with a beam parallel to ℓ_1, as shown in Fig. 13-32a, the moment of inertia for section *C–C* is that for a slab-and-beam section, as shown in Fig. 13-32c. Section *D–D* is cut through a beam running perpendicular to the page.

2. The moment of inertia of the slab-beams from the center of the column to the face of the column, bracket, or capital (as defined in ACI Code Section 8.4.1.4) shall be taken as the moment of inertia of the slab-beam at the face of the column, bracket, or capital divided by the quantity $(1 - c_2/\ell_2)^2$, where ℓ_2 is the transverse width of the equivalent frame (Fig. 13-19) and c_2 is the width of the support parallel to ℓ_2.

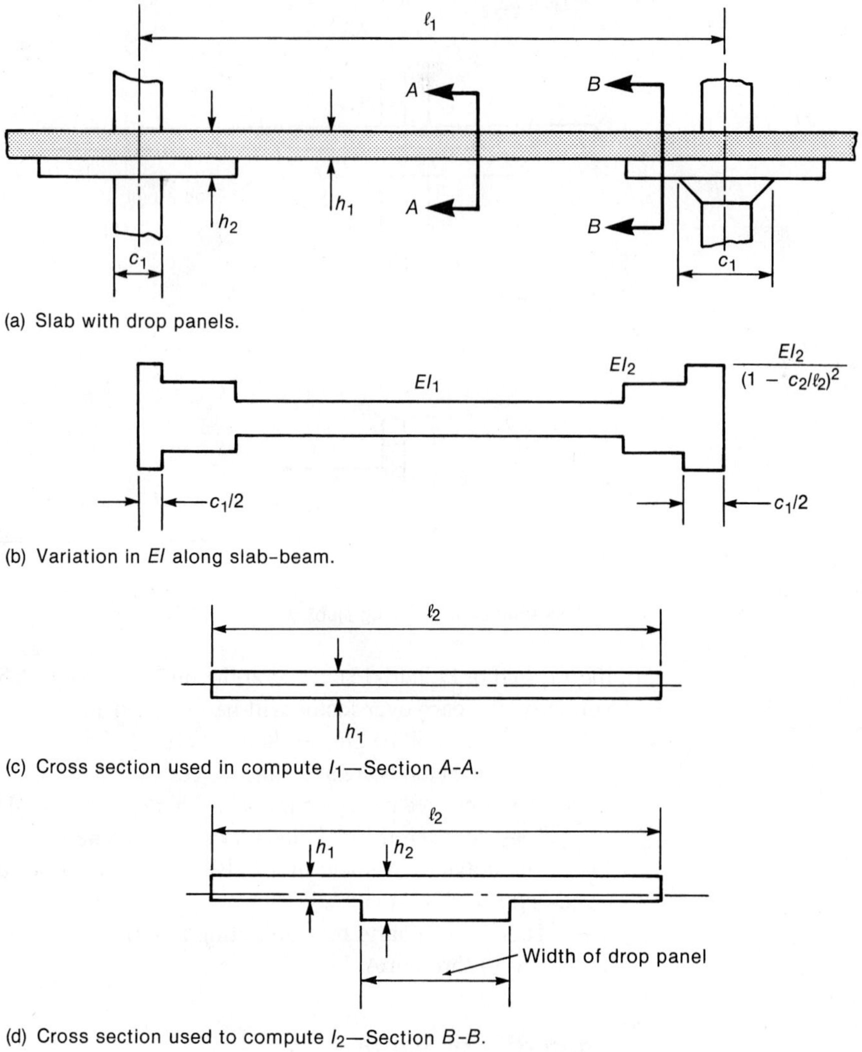

Fig. 13-31
EI values for a slab with a drop panel.

The application of this approach is illustrated in Figs. 13-31 and 13-32. Tables A-14 through A-16 [13-15], [13-16] present moment-distribution constants for flat plates and for slabs with drop panels. For most practical cases, these eliminate the need to use the column-analogy solution to compute the moment-distribution constants.

EXAMPLE 13-5 Calculation of the Moment-Distribution Constants for a Flat-Plate Floor

Figure 13-33 shows a plan of a flat-plate floor without spandrel beams. The floor is 7 in. thick. Compute the moment-distribution constants for the slab-beams in the equivalent frame along line 2, shown shaded in Fig. 13-33.

Span A2–B2. At end A, $c_1 = 12$ in., $\ell_1 = 213$ in., $c_2 = 20$ in., and $\ell_2 = 189$ in. Thus, $c_1/\ell_1 = 0.056$ and $c_2/\ell_2 = 0.106$. Interpolating in Table A-14, we find that the fixed-end moment is

$$M = 0.084 q_u \ell_2 \ell_1^2$$

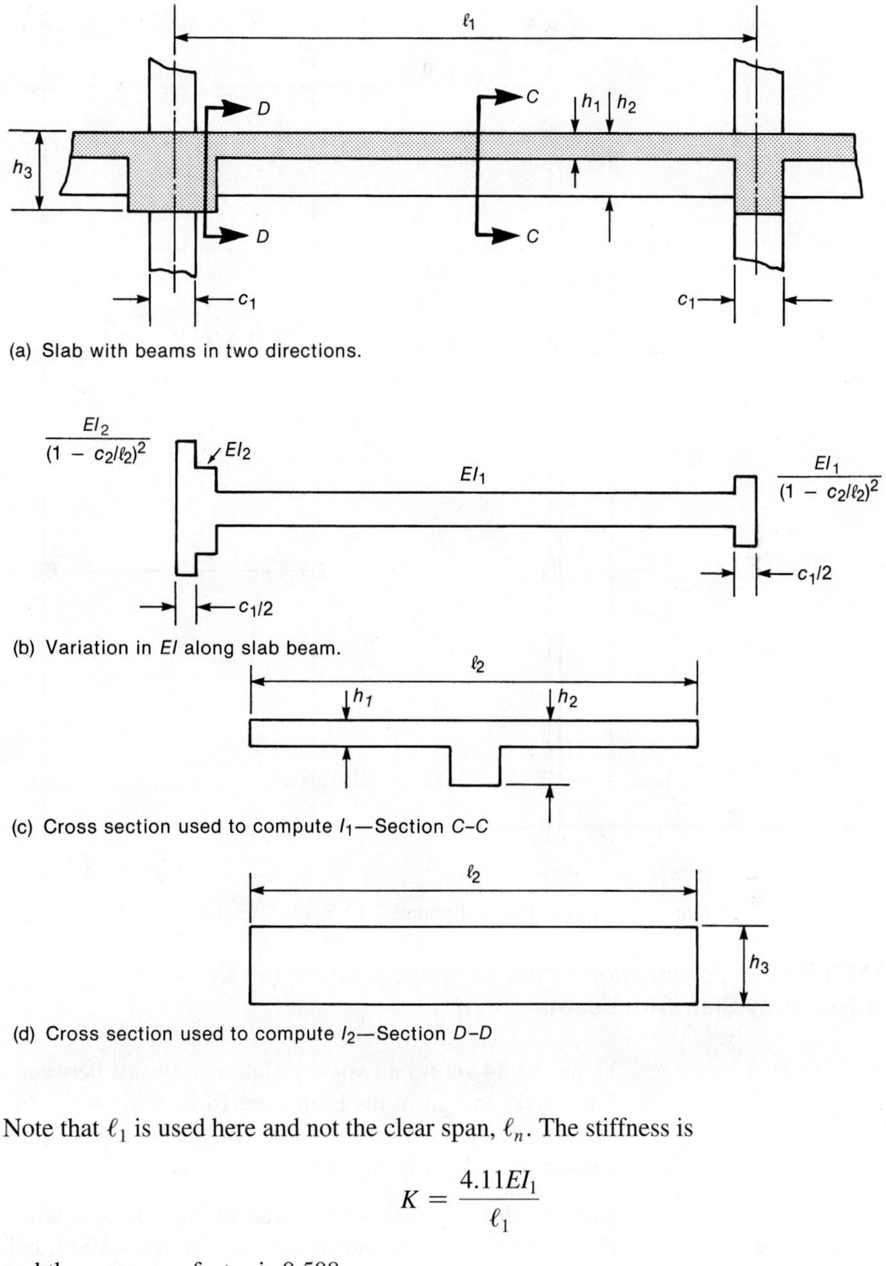

(a) Slab with beams in two directions.

(b) Variation in EI along slab beam.

(c) Cross section used to compute I_1—Section C–C

(d) Cross section used to compute I_2—Section D–D

Fig. 13-32
EI values for a slab and beam.

Note that ℓ_1 is used here and not the clear span, ℓ_n. The stiffness is

$$K = \frac{4.11EI_1}{\ell_1}$$

and the carryover factor is 0.508.

Span B2–C2. For span $B2$–$C2$, $c_1 = 20$ in., $\ell_1 = 240$ in., $c_2 = 12$ in., and $\ell_2 = 189$ in. Thus, $c_1/\ell_1 = 0.083$, $c_2/\ell_2 = 0.064$. From Table A-14, the fixed-end moment is

$$M = 0.084q_u\ell_2\ell_1^2$$

The stiffness is

$$K = \frac{4.10EI_1}{\ell_1}$$

and the carryover factor is 0.507. ■

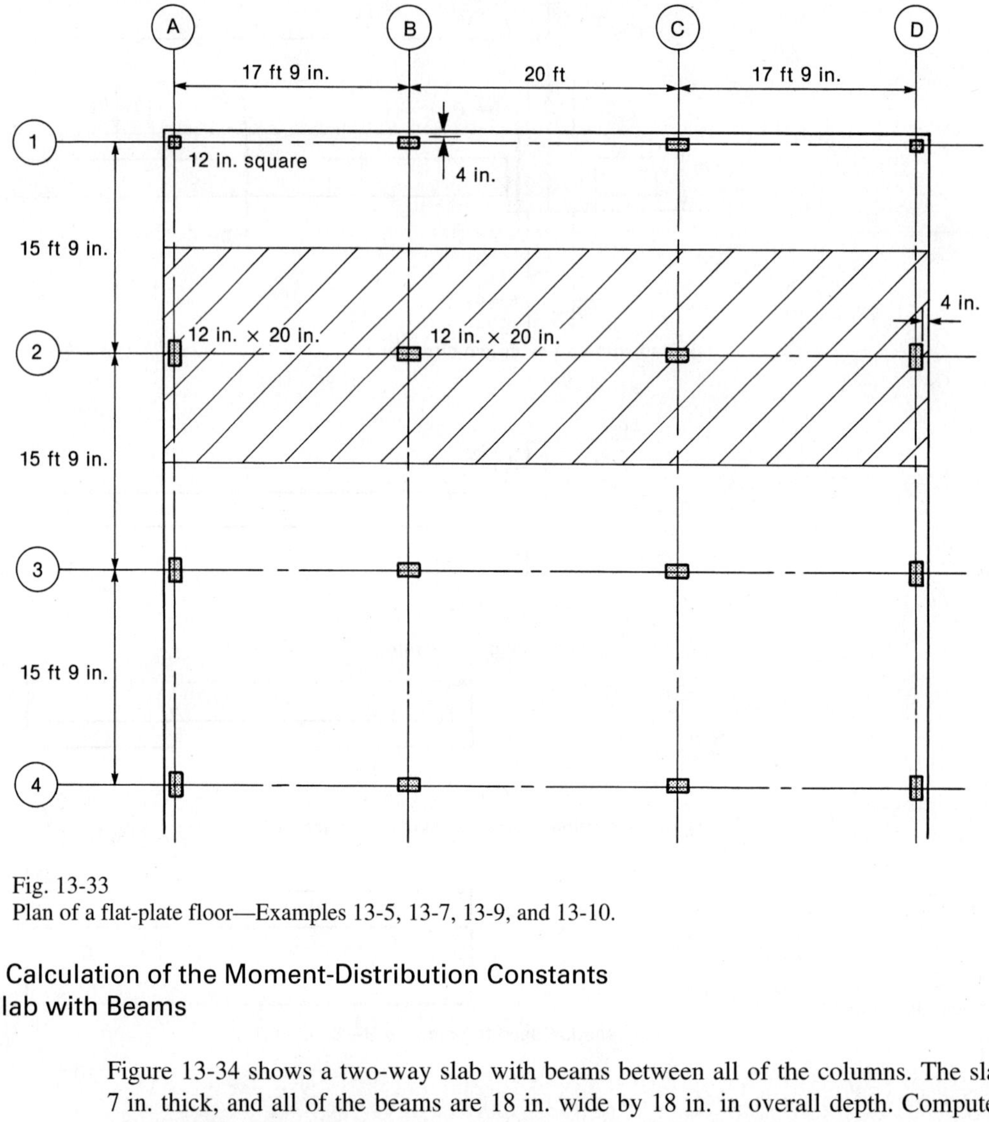

Fig. 13-33
Plan of a flat-plate floor—Examples 13-5, 13-7, 13-9, and 13-10.

EXAMPLE 13-6 Calculation of the Moment-Distribution Constants for a Two-Way Slab with Beams

Figure 13-34 shows a two-way slab with beams between all of the columns. The slab is 7 in. thick, and all of the beams are 18 in. wide by 18 in. in overall depth. Compute the moment-distribution constants for the slab-beams in the equivalent frame along line *B* (shown shaded in Fig. 13-34).

Span B1–B2. Span *B*1–*B*2 is shown in Fig. 13-35. A cross section at midspan is shown in Fig. 13-35b. The centroid of this section lies 4.39 in. below the top of the slab, and its moment of inertia is

$$I_1 = 23{,}800 \text{ in.}^4$$

The columns at both ends are 18 in. square, giving $c_1/\ell_1 = 18/207 = 0.087$ and $c_2/\ell_2 = 0.070$. Because this member has uniform stiffness between the joint regions, we can use Table A-14 to get the moment-distribution constants:

$$\text{Fixed-end moment: } M = 0.084 q_u \ell_2 \ell_1^2$$

$$\text{Stiffness: } K = \frac{4.10 E I_1}{\ell_1}$$

$$\text{Carryover factor} = 0.507$$

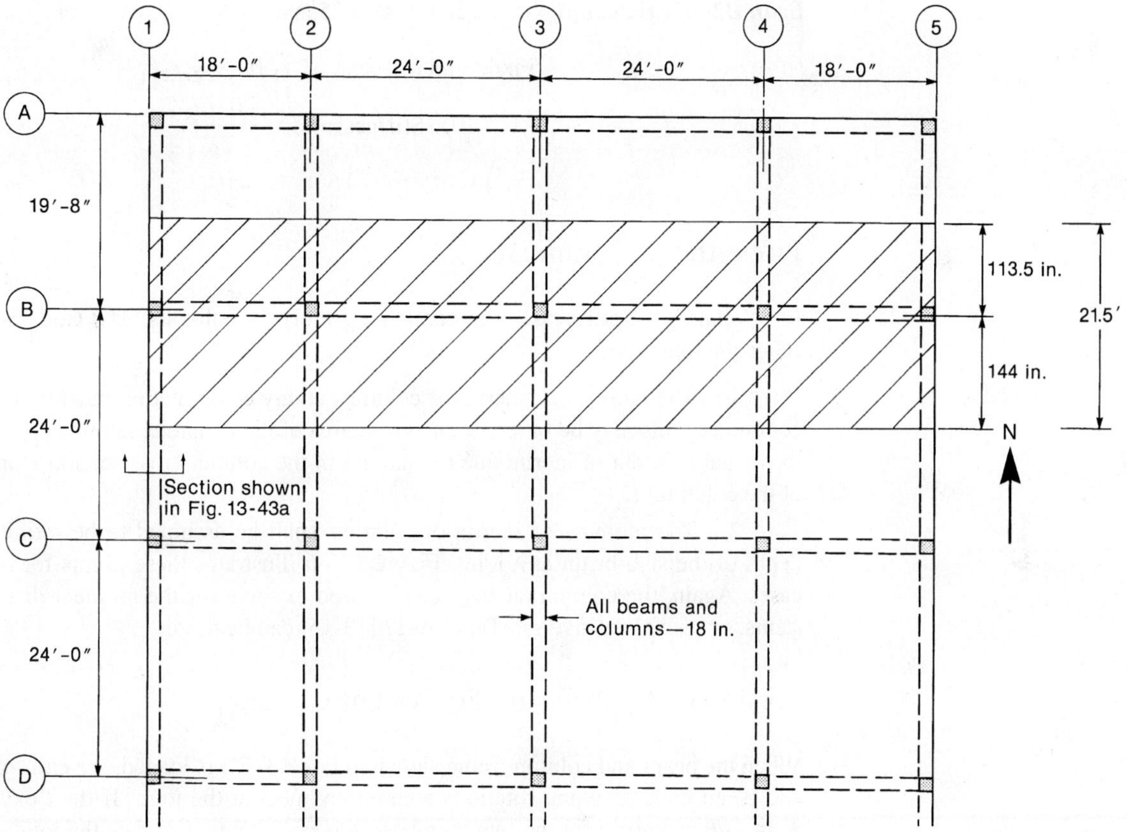

Fig. 13-34
Two-way slabs with beams—Examples 13-6 and 13-8.

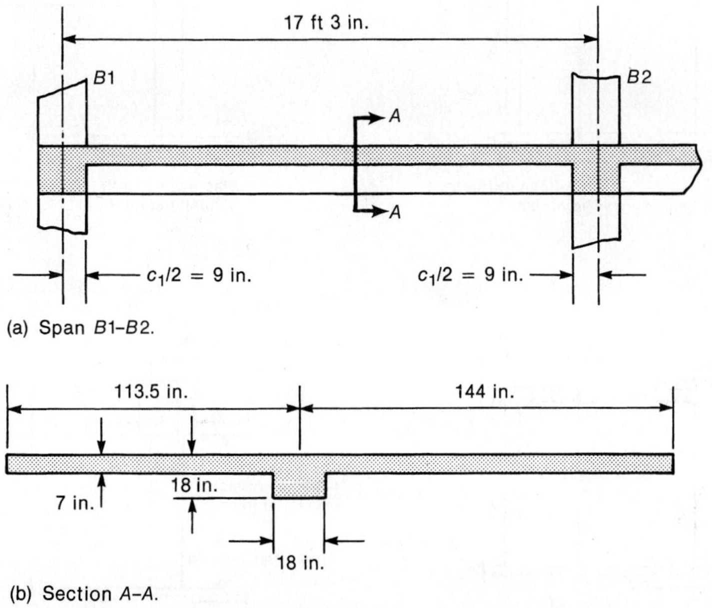

Fig. 13-35
Span $B1$–$B2$—Example 13-6.

Span B2–B3. Here, $c_1/\ell_1 = 18/288 = 0.0625$, and $c_2/\ell_2 = 0.070$. From Table A-14,

$$\text{Fixed-end moment: } M = 0.084 q_u \ell_2 \ell_1^2$$

$$\text{Stiffness: } K = \frac{4.07EI_1}{\ell_1}$$

$$\text{Carryover factor} = 0.505$$

■

Properties of Columns

In computing the stiffnesses and carryover factors for columns, ACI Code Section 8.11.4 states the following:

1. The moment of inertia of columns at any cross section outside of the joints or column capitals may be based on the gross area of the concrete, allowing for variations in the actual moment of inertia due to changes in the column cross section along the length of the column.

2. The moment of inertia of columns shall be assumed to be infinite within the depth of the slab-beam at a joint. Figure 13-36 illustrates these points for four common cases. Again, the column analogy can be used to solve for the moment-distribution constants, or the values given in Table A-17 [13-15] can be used.

Torsional Members and Equivalent Columns

When the beam and column frame shown in Fig. 13-37a is loaded, the ends of the column and beam undergo equal rotations where they meet at the joint. If the flexural stiffness, $K = M/\theta$, is known for the two members, it is possible to calculate the joint rotations and

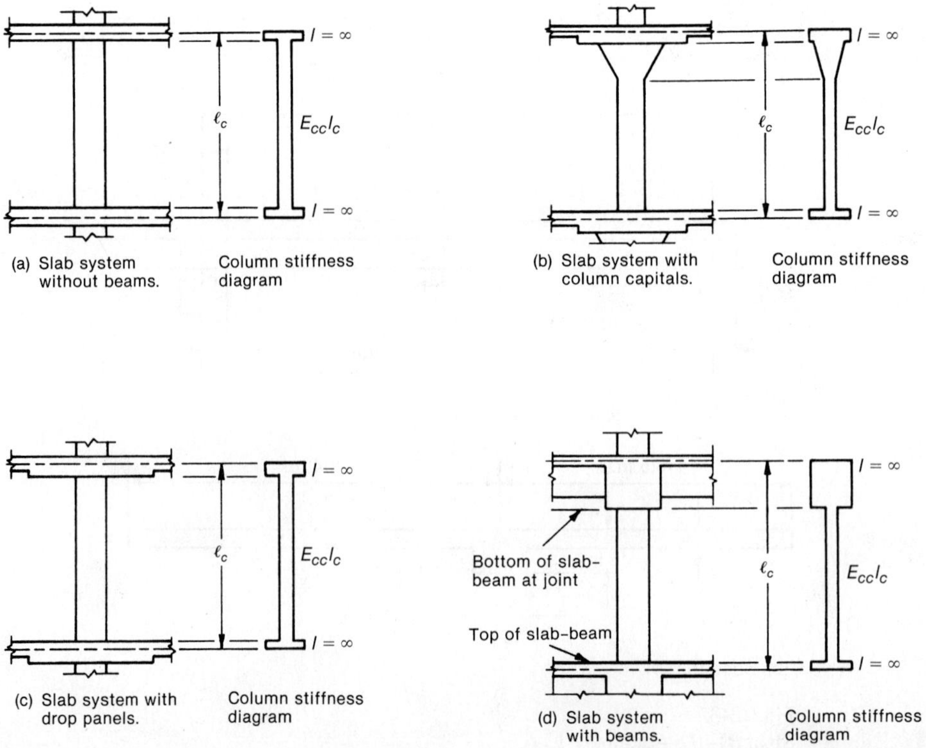

Fig. 13-36
Sections for the calculations of column stiffness, K_c.

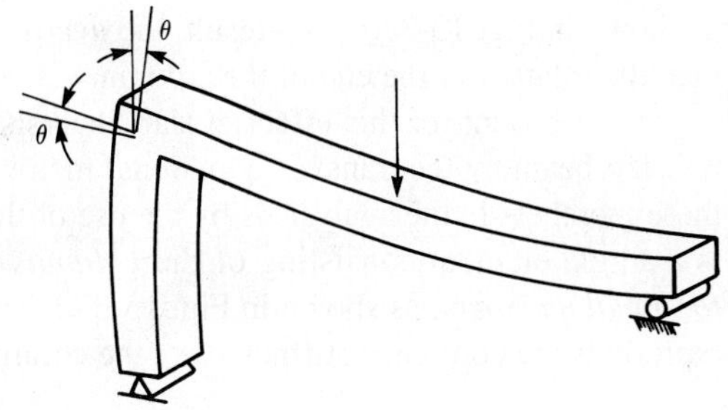

(a) Beam and column frame.

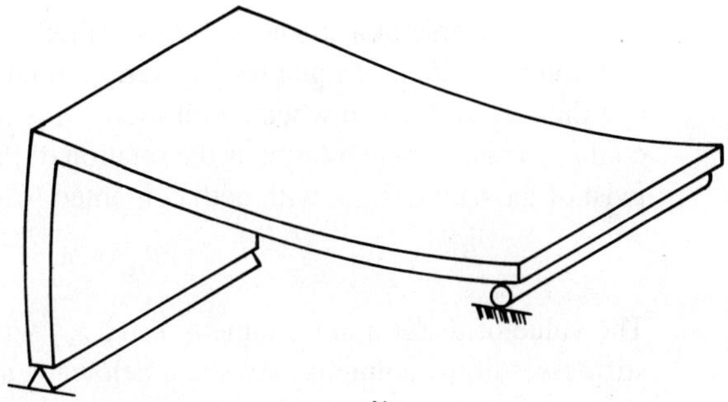

(b) Slab and wall frame.

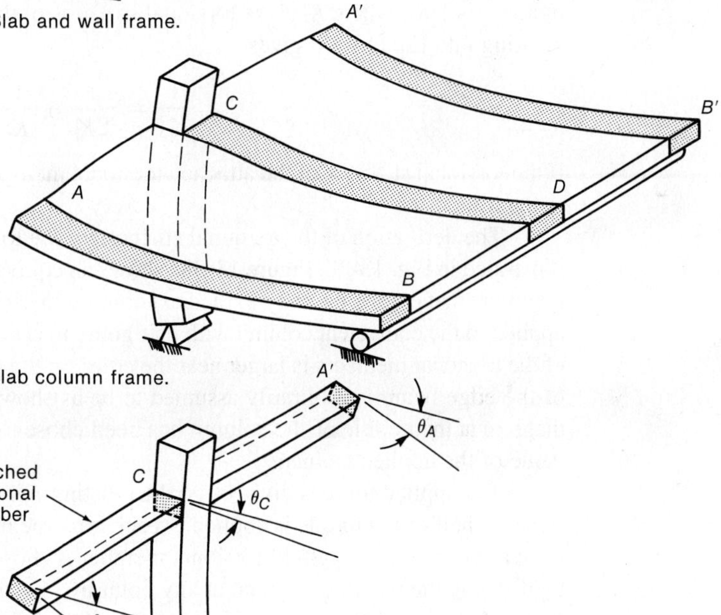

(c) Slab column frame.

(d) Column and edge of slab.

Fig. 13-37
Frame action and twisting of edge member.

the end moments in the members. Similarly, in the case shown in Fig. 13-37b, the ends of the slab and the wall both undergo equal end rotations when the slab is loaded. When a flat plate is connected to a column, as shown in Fig. 13-37c, the end rotation of the column is equal to the end rotation of the strip of slab *C–D*, which is attached to the column. The rotation at *A* of strip *A–B* is greater than the rotation at point *C*, however, because there is less restraint to the rotation of the slab at this point. In effect, the edge of the slab has twisted, as shown in Fig. 13-37d. As a result, the *average* rotation of the edge of the slab is greater than the rotation of the end of the column.

To account for this effect in slab analysis, the column is assumed to be attached to the slab-beam by the transverse torsional members *A–C* and *C–A'*. One way of including these members in the analysis is by the use of the concept of an *equivalent column*, which is a single element consisting of the *columns* above and below the floor and *attached torsional members*, as shown in Fig. 13-37d. The stiffness of the equivalent column, K_{ec}, represents the combined stiffnesses of the columns and attached torsional members:

$$K_{ec} = \frac{M}{\text{average rotation of the edge beam}} \tag{13-16}$$

The inverse of a stiffness, $1/K$, is called the *flexibility*. The flexibility of the equivalent column, $1/K_{ec}$, is equal to the average rotation of the joint between the "edge beam" and the rest of the slab when a unit moment is transferred from the slab to the equivalent column. This average rotation is the rotation of the end of the columns, θ_c, plus the average twist of the beam, $\theta_{t,\text{avg}}$, with both computed for a unit moment:

$$\theta_{ec} = \theta_c + \theta_{t,\text{avg}} \tag{13-17}$$

The value of θ_c for a unit moment is $1/\Sigma K_c$, where ΣK_c refers to the sum of the flexural stiffnesses of the columns above and below the slab. Similarly, the value of $\theta_{t,\text{avg}}$ for a unit moment is $1/K_t$, where K_t is the torsional stiffness of the attached torsional members. Substituting into Eq. (13-17) gives

$$\frac{1}{K_{ec}} = \frac{1}{\Sigma K_c} + \frac{1}{K_t} \tag{13-18}$$

If the torsional stiffness of the attached torsional members is small, K_{ec} will be much smaller than ΣK_c.

The derivation of the torsional stiffness of the torsional members (or edge beams) is illustrated in Fig. 13-38. Figure 13-38a shows an equivalent column with attached torsional members that extend halfway to the next column in each direction. A unit torque, $T = 1$, is applied to the equivalent column with half going to each arm. Because the effective stiffness of the torsional members is larger near the column, the torsional moment, t, per unit length of the edge beam is arbitrarily assumed to be as shown in Fig. 13-38b. The height of this diagram at the middle of the column has been chosen to give a total area equal to 1.0, the value of the applied moment.

The applied torques give rise to the twisting-moment diagram shown in Fig. 13-38c. Because half of the torque is applied to each arm, the maximum twisting moment is $\frac{1}{2}$. The twist angle per unit length of torsional member is shown in Fig. 13-38d. This is calculated by dividing the twisting moment at any point by CG, the product of the torsional constant, C (similar to a polar moment of inertia), and the modulus of rigidity, G. The total twist of the end of an arm relative to the column is the summation of the twists per unit length and is equal to the area of the diagram of twist angle per unit length in Fig. 13-38d. Because this is a parabolic diagram, the angle of twist at the outer end of the arm is one-third of the height times the length of the diagram:

$$\theta_{t,\text{end}} = \frac{1}{3}\frac{(1 - c_2/\ell_2)^2}{2CG}\left[\frac{\ell_2}{2}\left(1 - \frac{c_2}{\ell_2}\right)\right]$$

Replacing G with $E/2$ gives

$$\theta_{t,\text{end}} = \frac{\ell_2(1 - c_2/\ell_2)^3}{6CE}$$

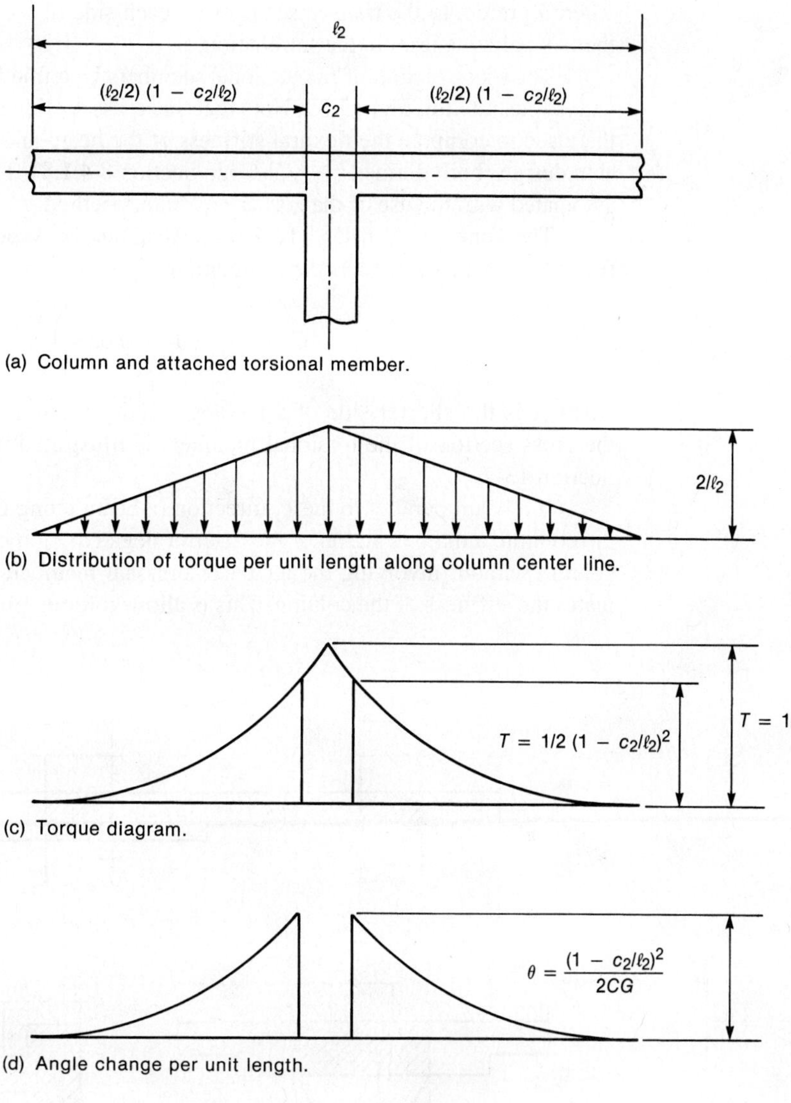

Fig. 13-38
Calculation of K_t.
(From [13-17].)

This is the rotation of the end of the arm. The rotation required for use in Eq. (13-17) is the average rotation of the arm, which is assumed to be a third of the end rotation:

$$\theta_{t,\text{avg}} = \frac{\ell_2(1 - c_2/\ell_2)^3}{18CE} \tag{13-19}$$

Finally, the torsional stiffness of one arm is calculated as $K_t = M/\theta_{t,\text{avg}}$, where the moment resisted by one arm is taken as $\frac{1}{2}$, giving

$$K_t(\text{one-arm}) = \frac{9EC}{\ell_2(1 - c_2/\ell_2)^3}$$

ACI Commentary Section R8.11.5 expresses the torsional stiffness of the two arms as

$$K_t = \sum \frac{9E_{cs}C}{\ell_2(1 - c_2/\ell_2)^3} \tag{13-20}$$

where ℓ_2 refers to the transverse spans on each side of the column. For a corner column, there is only one term in the summation.

The cross section of the torsional members is defined in ACI Code Section 8.11.5.1 and is illustrated in Fig. 13-39. Note that this cross section normally will be different from that used to compute the flexural stiffness of the beam and the beam section used for torsion design (both defined by ACI Code Section 8.4.1.8). This difference always has been associated with the use of the equivalent-frame method.

The constant C in Eq. (13-20) is calculated by subdividing the cross section into rectangles and carrying out the summation

$$C = \sum\left[\left(1 - 0.63\frac{x}{y}\right)\frac{x^3y}{3}\right] \tag{13-21}$$

where x is the shorter side of a rectangle and y is the longer side. The subdivision of the cross section of the torsional members is illustrated in Fig. 13-23 and explained in Section 13-7.

If a beam parallel to the ℓ_1 direction (a beam along C–D in Fig. 13-37) frames into the column, a major fraction of the exterior negative moment is transferred directly to the column without involving the attached torsional member. In such a case, K_{ec} underestimates the stiffness of the column. This is allowed for empirically by multiplying K_t by the

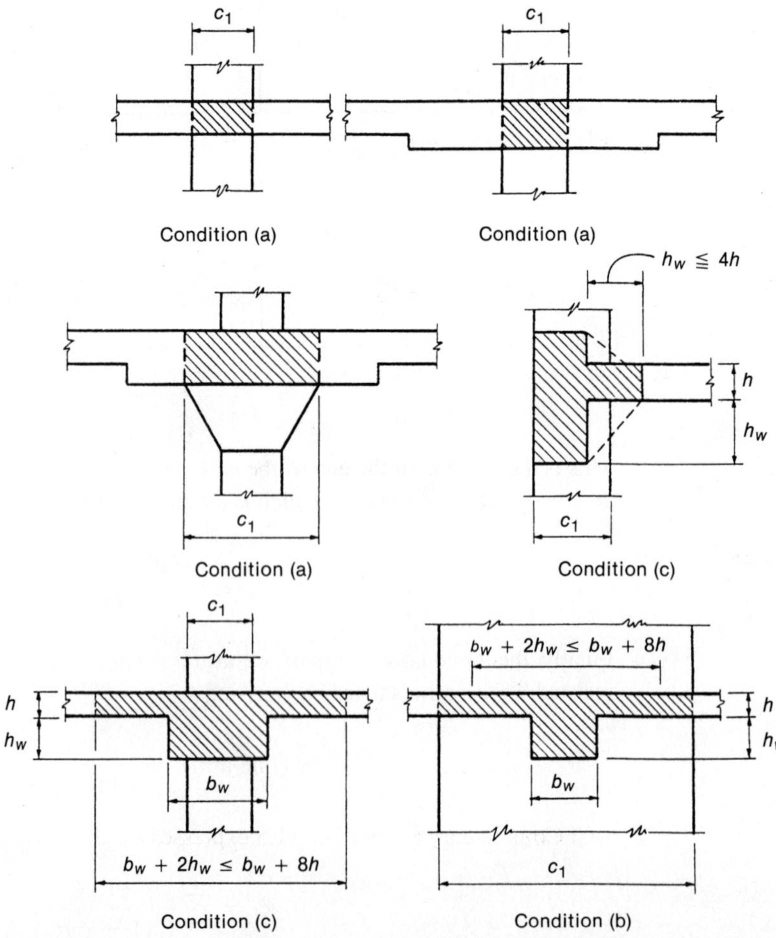

Fig. 13-39
Torsional members.
(From [13-15], courtesy of the Portland Cement Association.)

ratio I_{sb}/I_s, where I_{sb} is the moment of inertia of the slab and beam together and I_s is the moment of inertia of the slab neglecting the beam stem (ACI Code Section 8.11.5.2).

In a moment-distribution analysis, the frame analysis is carried out for a frame with slab-beams having stiffnesses K_s and with equivalent columns having stiffnesses K_{ec}.

EXAMPLE 13-7 Calculation of K_t, ΣK_c, and K_{ec} for an Interior Column

The 7-in.-thick flat plate shown in Fig. 13-33 is attached to 12-in. × 20-in. interior columns oriented with the 20-in. dimension parallel to column line 2. The story-to-story height is 8 ft 10 in. The slab and columns are of 4000-psi concrete. Compute K_t, ΣK_c, and K_{ec} for the interior connection between the slab strip along line 2 and column $B2$.

1. Define the cross section of the torsional members. According to ACI Code Section 8.11.5.1, the attached torsional member at the interior column corresponds to condition (a) in Fig. 13-39, as shown in Fig. 13-40b. Here, $x = 7$ in., and $y = 20$ in.

2. Compute *C*.

$$C = \sum\left(1 - 0.63\frac{x}{y}\right)\frac{x^3y}{3} = \left(1 - 0.63\frac{7}{20}\right)\frac{7^3 \times 20}{3} \qquad (13\text{-}21)$$

$$= 1780 \text{ in.}^4$$

3. Compute K_t.

$$K_t = \sum\frac{9E_{cs}C}{\ell_2(1 - c_2/\ell_2)^3} \qquad (13\text{-}20)$$

where the summation refers to the beams on either side of line 2 and ℓ_2 refers to the length ℓ_2 of the beams on each side of line 2. Because the two beams are similar,

$$K_t = 2\left[\frac{9E_{cs} \times 1780}{15.75 \times 12\left(1 - \dfrac{12}{15.75 \times 12}\right)^3}\right] = 207E_{cs}$$

4. Compute ΣK_c for the columns. The height center-to-center of the floor slabs is 8 ft 10 in. = 106 in. The distribution of stiffnesses along the column is similar to that in Fig. 13-36a. The columns are bent about their strong axis. Thus,

$$I_c = 12 \times \frac{20^3}{12}$$

$$= 8000 \text{ in.}^4$$

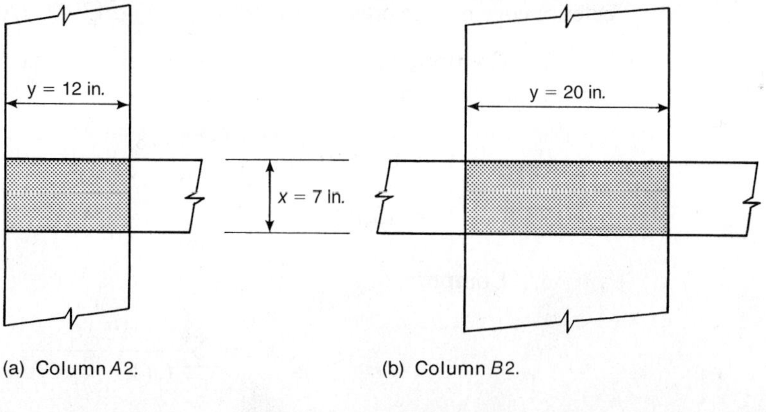

Fig. 13-40
Attached torsional members for Example 13-7.

For this column, the overall height $\ell_c = 106$ in., the unsupported or clear height $\ell_u = 99$ in., and $\ell_c/\ell_u = 1.07$. The distance from the centerline of the slab to the top of the column proper, t_a, is 3.5 in., as is the corresponding distance, t_b, at the bottom of the column. Interpolating in Table A-17 for $\ell_c/\ell_u = 1.07$ and $t_a/t_b = 1.0$ gives

$$K_c = \frac{4.76EI_c}{\ell_c}$$

and the carryover factor is 0.55. Because there are two columns (one above the floor and one under), each with the same stiffness, it follows that

$$\sum K_c = 2\left(\frac{4.76E_{cc} \times 8000}{106}\right) = 718E_{cc}$$

5. Compute the equivalent column stiffness, K_{ec}, for the interior column connection.

$$\frac{1}{K_{ec}} = \frac{1}{\Sigma K_c} + \frac{1}{K_t} = \frac{1}{718E_{cc}} + \frac{1}{207E_{cs}}$$

Because the slab and the columns have the same strength concrete, $E_{cc} = E_{cs} = E_c$. Therefore, $K_{ec} = 161E_c$.

Note that K_{ec} is only 22 percent of ΣK_c. This illustrates the large reduction in effective column stiffness in a flat-slab floor system. ■

EXAMPLE 13-8 Calculation of K_{ec} for an Edge Column in a Two-Way Slab with Beams

The two-way slab with beams between all of the columns in Fig. 13-34 is supported on 18-in.-square columns. The floor-to-floor height is 12 ft above the floor in question and 14 ft below the floor in question, as shown in the elevation of the exterior column in Fig. 13-41. All beams are 18 in. wide by 18 in. in overall depth. The concrete strength is 3000 psi in the slab and beams and 4000 psi in the columns. Compute K_{ec} for the connections between the slab strip along line *B* and column *B*1. A vertical section through the exterior span of the structure is shown in Fig. 13-35a.

1. Define the cross section of the torsional member. The torsional member at the exterior edge has the cross section shown in Fig. 13-42a (ACI Sections 8.11.5.1).

2. Compute *C*. To compute *C*, divide the torsional member into rectangles to maximize *C*, as shown in Fig. 13-42a:

$$C = \left(1 - 0.63\frac{18}{18}\right)\frac{18^3 \times 18}{3} + \left(1 - 0.63\frac{7}{11}\right)\frac{7^3 \times 11}{3}$$
$$= 13{,}700 \text{ in.}^4$$

3. Compute K_t.

$$K_t = \sum \frac{9E_{cs}C}{\ell_2(1 - c_2/\ell_2)^3} \tag{13-20}$$

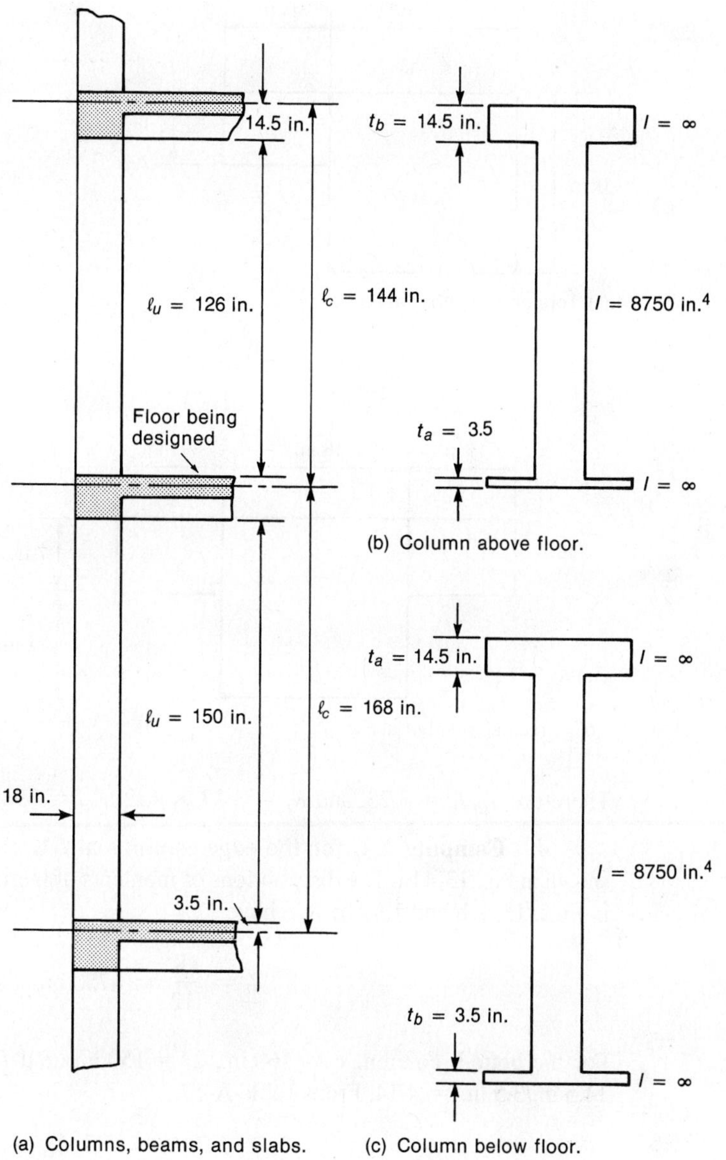

Fig. 13-41
Exterior columns to slab joint at $B1$ for Example 13-8.

For span $A1$–$B1$, $\ell_2 = 18$ ft 11 in. $= 227$ in., while for span $B1$–$C1$, $\ell_2 = 24$ ft 0 in. $=$ 288 in. Thus,

$$K_t = \frac{9E_{cs} \times 13{,}700}{227(1 - 18/227)^3} + \frac{9E_{cs} \times 13{,}700}{288(1 - 18/288)^3}$$
$$= 1220E_{cs}$$

Because the span for which moments are being determined contains a beam parallel to the span, K_t is multiplied by the ratio of the moment of inertia, I_{sb}, of a cross section including the beam stem (Fig. 13-35b) to the moment of inertia, I_s, of the slab alone. From Example 13-6, $I_{sb} = 23{,}800$ in.^{4}, and

$$I_s = \frac{257.5 \times 7^3}{12} = 7360 \text{ in.}^4$$

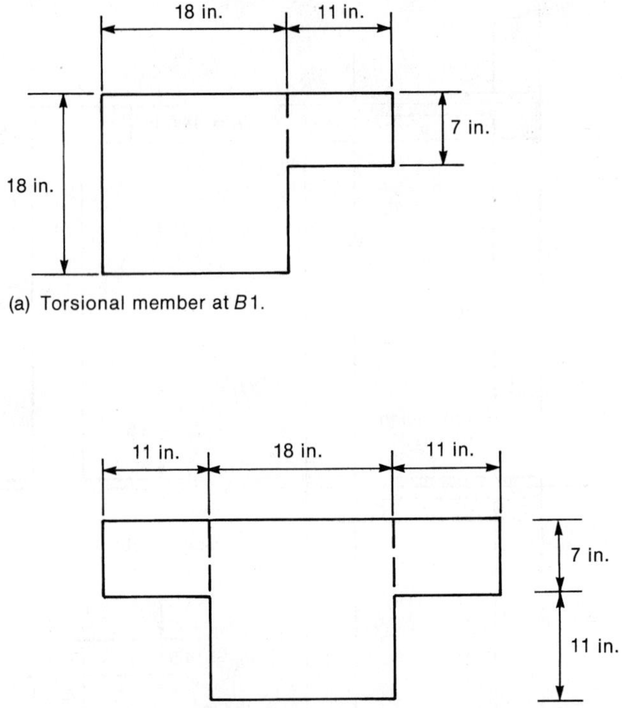

(a) Torsional member at *B*1.

(b) Torsional member at *B*2.

Fig. 13-42
Torsional members at *B*1 and *B*2 for Example 13-8.

Therefore, $I_{sb}/I_s = 3.23$, and $K_t = 3.23 \times 1220E_{cs} = 3940E_{cs}$.

4. Compute ΣK_c for the edge columns at $B1$. The columns and slabs at $B1$ are shown in Fig. 13-41a. The distributions of moments of inertia along the columns are shown in Figs. 13-41b and 13-41c. We have

$$I_c = \frac{18^4}{12} = 8750 \text{ in.}^4$$

For the bottom column, $\ell = 168$ in., $\ell_u = 150$ in., and $\ell/\ell_u = 1.11$. The ratio $t_a/t_b = 14.5 \text{ in.}/3.5 \text{ in.} = 4.14$. From Table A-17,

$$K_{c,\text{lower}} = \frac{5.73EI}{\ell} = \frac{5.73E_{cc} \times 8750}{168}$$
$$= 298E_{cc}$$

For the column above the floor, $\ell_c = 144$ in., $\ell_u = 126$ in., and $\ell_c/\ell_u = 1.14$. Because the stiffness is being evaluated at the lower end, $t_a = 3.5$ in., $t_b = 14.5$ in., and $t_a/t_b = 0.24$. From Table A-17,

$$K_{c,\text{upper}} = \frac{4.93E_{cc} \times 8750}{144} = 300E_c$$
$$\Sigma K_c = 598E_{cc}$$

5. Compute K_{ec}.

$$\frac{1}{K_{ec}} = \frac{1}{\Sigma K_c} + \frac{1}{K_t}$$

$$\frac{1}{K_{ec}} = \frac{1}{598E_{cc}} + \frac{1}{3940E_{cs}}$$

where

$$E_{cc} = 57{,}000\sqrt{4000} = 3.60 \times 10^6 \text{ psi}$$
$$E_{cs} = 57{,}000\sqrt{3000} = 3.12 \times 10^6 \text{ psi}$$

Thus, $E_{cc} = 1.15E_{cs}$ and $K_{ec} = 586E_{cs}$ for the edge column-to-slab connection. ■

Arrangement of Live Loads for Structural Analysis

The placement of live loads to produce maximum moments in a continuous beam or one-way slab was discussed in Section 5-2 and illustrated in Fig. 5-7. Similar loading patterns are specified in ACI Code Section 6.4.3 for the analysis of two-way slabs. If the unfactored live load does not exceed 0.75 times the unfactored dead load, it is not necessary to consider pattern loadings, and only the case of full factored live and dead load on all spans need be analyzed (ACI Code Section 6.4.3.2). This is based on the assumption that the increase in live-load moments due to pattern loadings compared to uniform live loads will be small compared with the magnitude of the dead-load moments. It also recognizes the fact that a slab is sufficiently ductile in flexure to allow moment redistribution.

If the unfactored live load exceeds 0.75 times the unfactored dead load, the pattern loadings described in ACI Code Section 6.4.3.3 need to be considered:

1. For maximum positive moment—factored dead load on all spans, plus 0.75 times the full factored live load on the panel in question and on alternate panels.

2. For maximum negative moment at an interior support—factored dead load on all panels, plus 0.75 times the full-factored live load on the two adjacent panels.

The final design moments shall not be less than for the case of full-factored dead and live load on all panels (ACI Code Section 6.4.3).

ACI Code Section 6.4.3.3 is an empirical attempt to acknowledge the small probability of full live load in a pattern-loading situation. Again, the possibility of moment redistribution is recognized.

Moments at the Face of Supports

The equivalent-frame analysis gives moments at the node point where ends of the members meet in the center of the joint. ACI Code Section 8.11.6 permits the moments at the face of rectilinear supports to be used in the design of the slab reinforcement. The critical sections for negative moments are illustrated in Fig. 13-43. For columns extending more than $0.175\ell_1$ from the face of the support, the moments can be reduced to the values existing at $0.175\ell_1$ from the center of the joint. This limit is necessary because the representation of the slab-beam stiffness in the joint is not strictly applicable for very long, narrow columns.

If the slab meets the requirements for slabs designed by the direct-design method (see Section 13-7 and ACI Code Section 8.10.2) ACI Code Section 8.11.6.5 allows the total design moments in a panel to be reduced such that the absolute sum of the positive

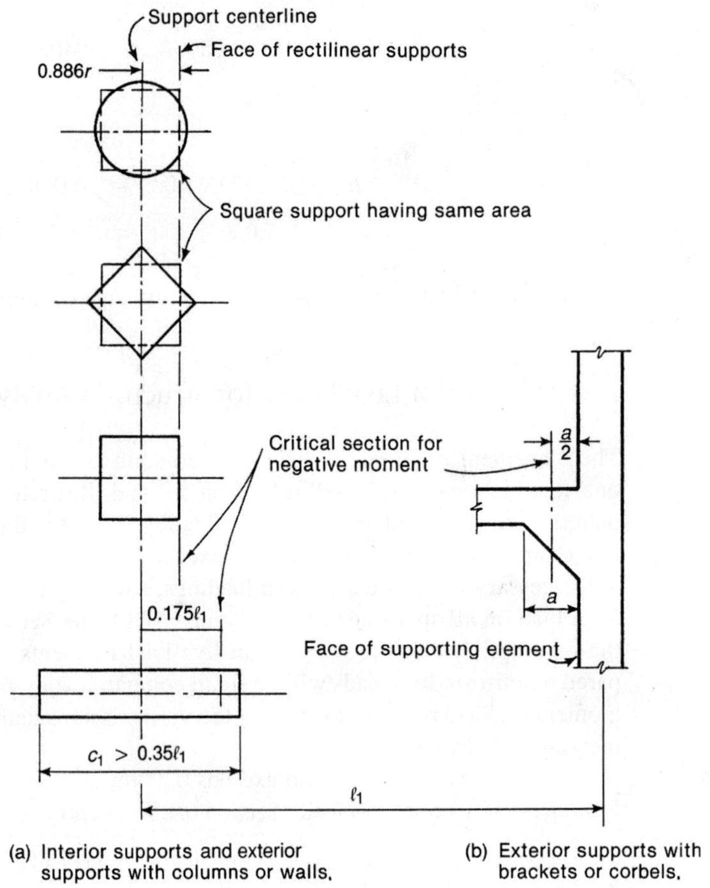

Fig. 13-43
Critical sections for negative moment. (From [13-15], courtesy of the Portland Cement Association.)

moment and the average negative moment does not exceed the statical moment, M_o, for that panel, where M_o is as given by Eq. (13-5). Thus, for the case illustrated in Fig. 13-44, the values of the computed moments M_1, M_2, and M_3 would be limited to

$$\left(\frac{M_1 + M_2}{2} + M_3\right) \leq M_o$$

Distribution of Moments to Column Strips, Middle Strips, and Beams

Once the negative and positive moments have been determined for each equivalent frame, these are distributed to column and middle strips in accordance with ACI Code Sections 8.10.5 and 8.10.6 in exactly the same way as in the direct-design method. This process was discussed more fully in Section 13-7. For panels with beams between the columns on all sides, the distribution of moments to the column and middle strips according to ACI Code Sections 8.10.5 and 8.10.6 is valid only if $\alpha_{f1}\ell_2^2/\alpha_{f2}\ell_1^2$ falls between 0.2 and 5.0. Cases falling outside of this range tend to approach one-way action, and other methods of slab analysis are required.

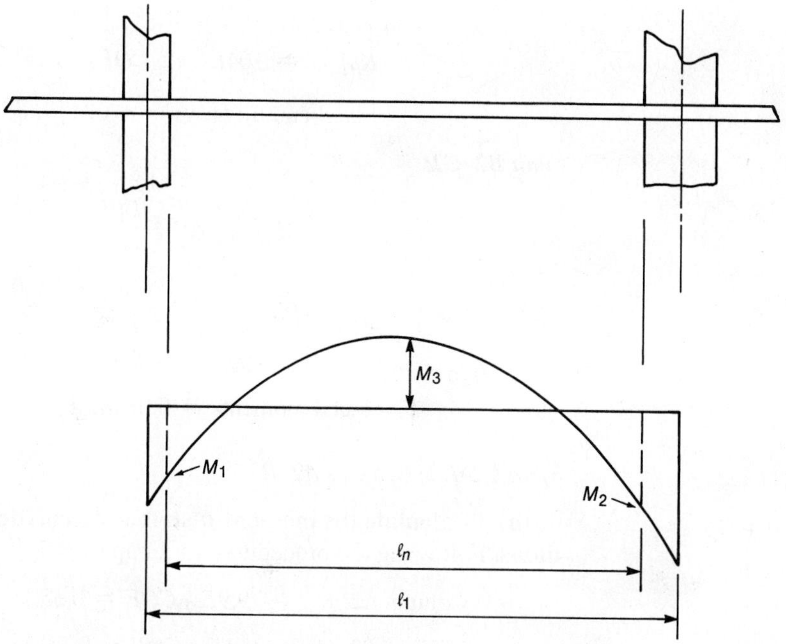

Fig. 13-44
Negative and positive moments in a slab beam.

EXAMPLE 13-9 Analysis of a Flat-Plate Floor Using the Classic Equivalent-Frame Method

Figure 13-33 shows a plan of a flat-plate floor without spandrel beams. Analyze this floor using the classic equivalent-frame method. Use 4000-psi concrete for the columns and slab. The story-to-story height is 8 ft 10 in. The floor supports its own dead load, plus 25 psf for partitions and finishes, and a live load of 40 psf.

This is the same slab that was analyzed in Examples 13-5 and 13-7. Only the parts of the analysis that differ from those examples are discussed here.

1. Select the analysis method. Although the slab satisfies the requirements for the use of the direct-design method, it has been decided to use the classic equivalent-frame method for the analysis of moments.

2. Select the thickness. The selection of thickness is based on Table 13-1 and also on providing adequate shear strength. A 7-in. slab will be used here, but this will be increased to 7.5 in. for Example 13-14.

3. Compute the moments in the equivalent frame along column line 2. The strip of slab along column line 2 acts as a rigid frame spanning between columns *A*2, *B*2, *C*2, and *D*2. For the purposes of analysis, the columns above and below the slabs will be assumed fixed at their far ends.

(a) Calculate the moment-distribution coefficients for the slab-beams. From Example 13-5, we have

Span A2–B2:

$$K_{A2-B2} = \frac{4.11EI_1}{\ell_1}$$

$$= \frac{4.11E_c \times 5400}{213} = 104E_c$$

$$COF_{A2-B2} = 0.508$$

$$K_{B2-A2} = 104E_c \qquad COF_{B2-A2} = 0.508$$

$$\text{Fixed-end moments} = 0.084q\ell_2\ell_1^2$$

Span B2–C2:

$$K_{B2-C2} = \frac{4.10EI_1}{\ell_1}$$

$$= \frac{4.10E_c \times 5400}{288} = 76.9E_c$$

$$COF = 0.507$$

$$\text{Fixed-end moments} = 0.084q\ell_2\ell_1^2$$

Span C2–D2: Same as *A2–B2*.

(b) Calculate the moment-distribution coefficients for the equivalent columns. Following the procedure in Example 13-7,

Column *A2*: $K_{ec} = 79.9E_c$, $COF = 0.55$

Column *B2*: $K_{ec} = 161E_c$, $COF = 0.55$

(c) Compute the distribution factors. The distribution factors are computed in the usual manner:

$$DF_{A2-B2} = \frac{K_{A2-B2}}{K_{A2-B2} + K_{ecA2}}$$

$$= \frac{104E_c}{104E_c + 79.9E_c} = 0.566$$

$$DF_{\text{column }A2} = 0.434$$

Through a similar process the distribution factors can be calculated for the three members meetings at joint B2. The final distribution factors and carryover factors are shown in Fig. 13-45. The cantilever members projecting outward at joints *A2* and *D2* refer to the slab that extends outside the column to support the exterior wall.

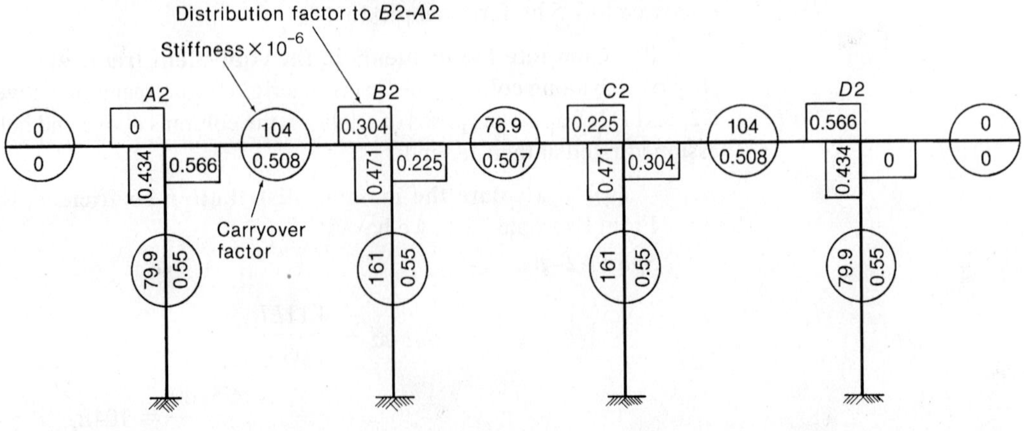

Fig. 13-45
Stiffness, carryover, and distribution factors—Example 13-9.

(d) Select the loading cases and compute the fixed-end moments. Because $q_L = 40$ psf is less than three-fourths of $q_D = 113$ psf, only the case of uniform live load on each panel need be considered (ACI Code Section 6.4.3.2). Because the influence area of each of the panels is less than 400 ft^2, live-load reductions do not apply. Accordingly, we have

$$q_u = 1.2\left(\frac{7}{12} \times 0.15 + 0.025\right) + 1.6(0.040) = 0.199 \text{ ksf}$$

For the slab-beams, $\ell_2 = 15.75$ ft.

Span A2–B2:

$$M = 0.084 q_u \ell_2 \ell_1^2$$
$$= 0.084 \times 0.199 \times 15.75 \times 17.75^2 = 82.9 \text{ kip-ft}$$

(Note that the moment, M, is based on the center-to-center span, ℓ_1, rather than on the clear span, ℓ_n, used in the direct-design method.)

Span B2–C2:

$$M = 0.084 \times 0.199 \times 15.75 \times 20^2 = 105 \text{ kip-ft}$$

The weight of the slab outside line A and the wall load (assumed to be 300 lb/ft) cause a small cantilever moment at joint $A2$. It is assumed that the wall load acts 2 in. outside the exterior face of the column or 8 in. from the center of the column.

$$M = (1.2 \times 15.75 \times 0.300) \times \frac{8}{12} + \left(1.2 \times 15.75 \times \frac{7}{12} \times 0.150\right)\frac{(10/20)^2}{2}$$
$$= 4.0 \text{ kip-ft}$$

Span C2–D2: Same moment as in Span $A2$–$B2$.

(e) Carry out a moment-distribution analysis. The moment-distribution analysis is carried out in Table 13-6. The sign convention used takes clockwise moments *on the joints* as positive. Thus, counterclockwise moments *on the ends of members* are positive. When this sign convention is used, the carryover factors are positive. The resulting moment and shear diagrams are plotted in Figs. 13-46a and 13-46b. (Note that end shears and maximum positive moments are found by putting each slab-beam element into static equilibrium.) The moment diagram is plotted on the compression face of the member, and positive and negative moments in the slab-beam correspond to the normal sign convention for continuous beams.

(f) Calculate the moments at the faces of the supports. The moments at the face of the supports are calculated by subtracting the areas of the shaded parts of Fig. 13-47b from the moments at the ends of the members.

(g) Reduce the moments to M_o. For slabs that fall within the limitations for the use of the direct-design method, the total moments between faces of columns can be reduced to M_o. This slab satisfies ACI Code Section 8.10.2, so this reduction can be used.

Span A2–B2:

$$\text{Total moment} \simeq \frac{24.4 + 80.6}{2} + 55.1 = 108 \text{ kip-ft}$$

$$M_o = \frac{q_u \ell_2 \ell_n^2}{8} = 106 \text{ kip-ft}$$

TABLE 13-6 Moment Distribution—Example 13-9

		A2			*B2*			*C2*			*D2*	
			COF = *0.508*			*COF* = *0.507*			*COF* = *0.508*			
	0	0.434	0.566	0.304	0.471	0.225	0.225	0.471	0.304	0.566	0.434	0
	Cant.	Col.	Slab	Slab	Col.	Slab	Slab	Col.	Slab	Slab	Col.	Cant.
FEM	−4.0	0	+82.9	−82.9	0	+105	−105	0	+82.9	−82.9	0	+4.0
*B*1		−36.0	−46.9	−6.7	−10.4	−5.0	+5.0	+10.4	+6.7	+46.9	+36.0	
*C*1			−3.4	−23.8		+2.5	−2.5		+23.8	+3.4		
*B*2		+1.5	+1.9	+6.5	+10.0	+4.8	−4.8	−10.0	−6.5	−1.9	−1.5	
*C*2			+3.3	+1.0		−2.4	+2.4		−1.0	−3.3		
*B*3		−1.4	−1.9	+0.4	+0.7	+0.3	−0.3	−0.7	−0.4	+1.9	+1.4	
*C*3			+0.2	−0.9		−0.2	+0.2		+0.9	−0.2		
*B*4		−0.1	−0.1	+0.3	+0.5	+0.2	−0.2	−0.5	−0.3	+0.1	+0.1	
Sum	−4.0	−36.0	36.0	−106	+0.80	105	−105	−0.80	+106	−36.0	36.0	+4.0
Sum at joint		0			0			0			0	

Because the results are very close, no adjustment will be made.

Span B2–C2:

$$\text{Total moment} = 80.0 + 50.7 = 131 \text{ kip-ft}$$
$$M_o = 132 \text{ kip-ft}$$

Again, no adjustment will be made. There were no reductions in this example because we have only one load case to consider, with factored dead and live load on all spans. If the live load had exceeded three-fourths of the dead load, it would have been necessary to consider additional load cases (ACI Code Section 6.4.3.3), and reductions of the envelope moment values would be likely.

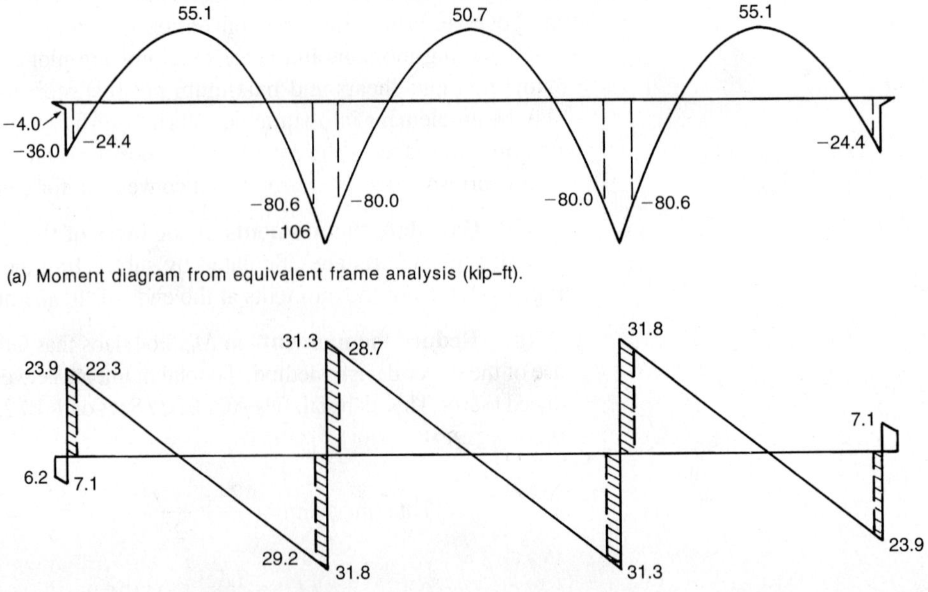

(a) Moment diagram from equivalent frame analysis (kip–ft).

(b) Shear force diagram (kips).

Fig. 13-46
Moments and shears in frame along line 2 for Example 13-9.

(h) Calculate the column moments. The total moment transferred to the columns at $A2$ is 36.0 kip-ft. This moment is distributed between the two columns in the ratio of the stiffnesses, K_c. Because the columns above and below have equal stiffnesses, half goes to each column. At $B2$, only a small moment is transferred to the columns for this loading case.

The analysis results obtained here will be compared with design moments obtained from the direct-design method in Example 13-14. In that example, moments will be distributed to the column and middle strips and reinforcement will be selected to complete the flexural design of the two-way slabs. ■

13-9 USE OF COMPUTERS FOR AN EQUIVALENT-FRAME ANALYSIS

The classic equivalent-frame method was derived by assuming that the structural analysis would be carried out by hand using the moment-distribution method. Thus, tables were developed to evaluate fixed-end moments, stiffnesses, and equivalent-column stiffnesses for use in such an analysis. If standard frame analysis software based on the stiffness method is to be used, the torsional member (and the resulting equivalent-column stiffness) defined in the classic equivalent-frame method will need to be incorporated into the stiffness of either the slab-beam or column elements. The general research direction has been to modify the stiffness of the slab-beam element by defining an *effective slab width* to reduce the element stiffness, particularly at connections. The frame analysis results for gravity loading, obtained using the modified slab-beam elements, should be in reasonable agreement with those obtained from the classic equivalent-frame method.

Several researchers have worked on the development of *effective slab width* models that could be used to define the stiffness of an *equivalent beam* in a standard frame analysis program for the evaluation of moments and shears in a slab–column frame subjected to combined vertical and lateral loading. Pecknold [13-18] and Darvall and Allen [13-19] used classical plate theory and finite-element analyses to define an effective slab width. More recent work by Luo and Durrani [13-20] and [13-21], Hwang and Moehle [13-22], and Dovich and Wight [13-23] based their effective slab width and stiffness models on experimental results for reinforced concrete slabs. Hueste and Wight [13-24] proposed modifications of these effective slab width and stiffness models for post-tensioned slabs, based on observed damaged in a post-tensioned slab–column frame system following a major earthquake. A summary of simple modeling rules for equivalent slab widths and flexural stiffnesses are presented in the following subsection.

The first step in building a slab–column frame analysis model is to select an effective slab width that is a fraction, α, of the total slab width, ℓ_2 (avg), as was shown in Figs. 13-33 and 13-34. A wide range of α values have been suggested by various researchers, but the author prefers to simply use $\alpha = 0.5$ for all positive-bending regions and for negative-bending regions at interior supports. For negative-bending regions at exterior supports, the effective slab width depends on the torsional stiffness at the edge of the slab. If no edge beam is present, then an α value of 0.2 is recommended. If an edge beam is present and has a torsional stiffness such that β_t, as defined in Eq. (13-12), is greater than or equal to 2.5, then the recommended α value is 0.5. If the value of β_t is between 0.0 and 2.5, a linear interpolation can be used to find an α value between 0.2 and 0.5. For low values of α, the effective slab width should not be taken to be less than the column width, c_2, plus one-half the column total depth, c_1, on each side of the column (Fig. 13-47). For slab–column frames along a column line at the edge of a floor plan, the effective slab widths are reduced accordingly. The resulting effective slab-width models for one exterior and one interior column line of the flat-plate floor system analyzed in Example 13-9 is shown in Fig. 13-48.

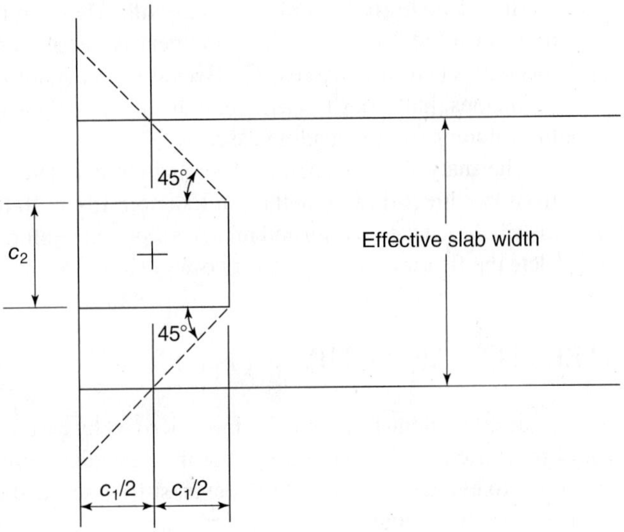

Fig. 13-47
Minimum value for effective slab width at exterior slab-to-column connections.

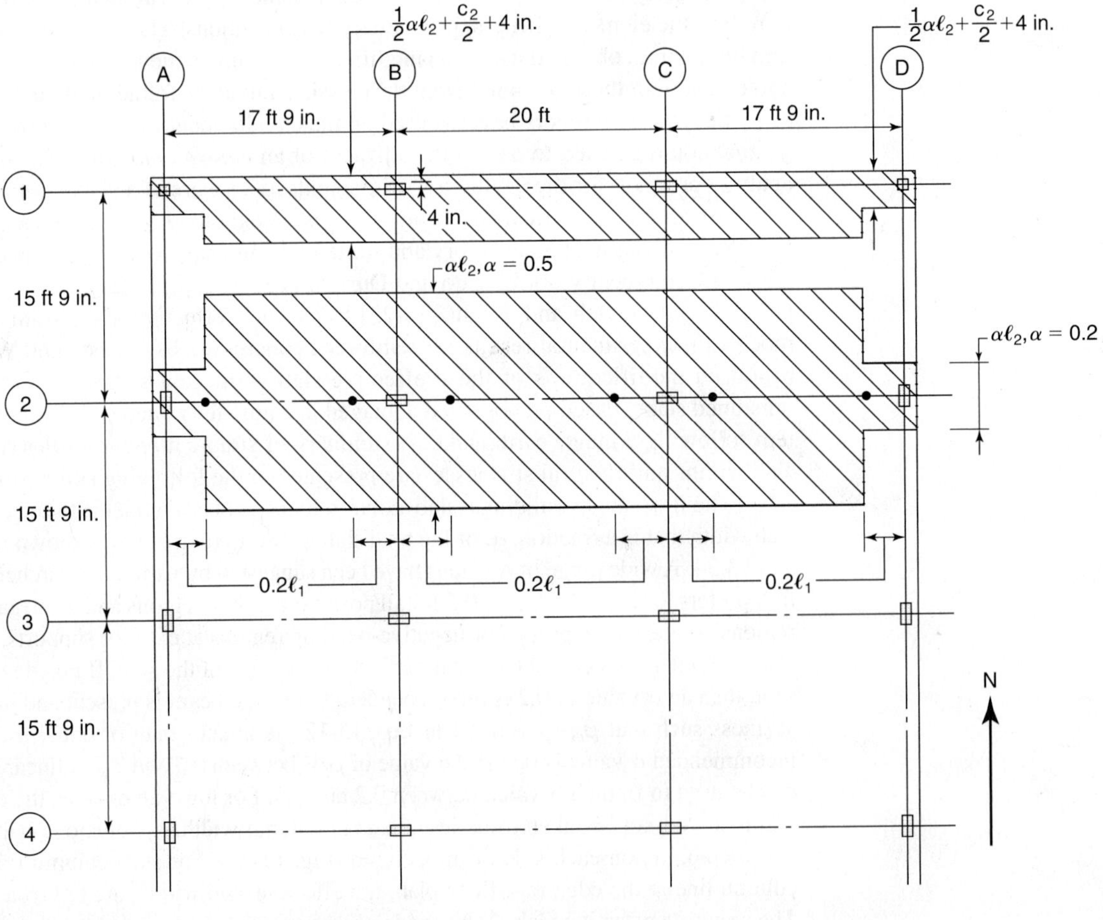

Fig. 13-48
Effective slab width, $\alpha\ell_2$, and locations for intermediate nodes along the span.

As indicated in Fig. 13-48, the negative-bending region at the exterior connection is assumed to extend over 20 percent $(0.2\,\ell_1)$ of the span. The authors recommend that the same assumption be used for negative-bending regions at all interior and exterior connections. This assumption essentially creates extra node points within the span and becomes important when assigning cracked-stiffness values to the positive-and negative-bending regions.

After the effective slab width, $\alpha\ell_2$, has been established, the gross moment of inertia for the slab-beam can be calculated using either a section similar to that in Fig. 13-31c (if no beam is present) or a section similar to that in Fig. 13-32c (if a beam is present). For both cases, the effective slab width, $\alpha\ell_2$, is to be used in place of the ℓ_2 value shown in those figures.

If a *drop panel* is present in the negative-bending region, then a section similar to that used in Fig. 13-31d (with $\alpha\ell_2$ in place of ℓ_2) is to be used. Because a standard drop panel is required to extend for at least 1/6 of the span length (ACI Code Section 8.2.4), in cases where a drop panel is present in the negative-bending region, the authors recommend placing the extra node points at the edge of the drop panel instead of the $0.2\,\ell_1$ value recommended previously. Thus, the negative-bending region is assumed to end at the edge of the drop panel. If a *short drop panel* is used to only provide additional shear strength, as will be discussed in Section 13-10, then two node points should be used in the negative-bending region. One node should be located at the end of the drop panel, and another node should be used at the assumed end of the negative-bending region $(0.2\,\ell_1)$. This will permit the use of different flexural stiffnesses inside and outside the short drop panel.

A final modification is to be made to the slab-beam stiffness to account for flexural cracking. In general, the cracked moment of inertia for a slab-beam section, I_{cr}, is some fraction of the gross moment of inertia for that section. Because slabs normally have lower reinforcement ratios than beams, their cracked-section moment of inertia is usually a smaller fraction of the gross moment of inertia than for a typical beam section. However, because large portions along the slab-beam span will remain uncracked and the flexural cracks that do occur usually will not propagate over the entire width of the slab, an *effective moment of inertia*, I_e, needs to be defined for different portions of the slab-beam span. Commonly, a factor β is used to define the effective moment of inertia as some fraction of the gross moment of inertia $(I_e = \beta\, I_g)$. For all positive-bending regions of the slab, the author recommends using a β factor equal to 0.5. Because larger moments typically occur near interior connections, and in order to not overestimate the slab-to-edge beam- to-column stiffness at an exterior connection, whether or not an edge beam is present, the author recommends a β factor of 0.33 for all negative bending regions.

A summary of the recommended α and β values to be used to represent the flexural stiffness of slab-beam elements is given in Table 13-7. For analysis of post-tensioned slabs, Hueste and Wight [13-24] and Kang and Wallace [13-25] have recommended the use of a β value equal to 0.67 because of the reduced flexural cracking expected in a post-tensioned slab.

For a gravity-load analysis of an equivalent frame representing a two-way floor system, the slab-beam elements can be assembled with column elements that extend one story above and one story below the floor system (Fig. 13-49), as permitted by ACI Code Section 8.11.2.5. The dimensions in Fig. 13-49 are selected to correspond to the equivalent frame analyzed in Example 13-9 (Fig. 13-33). The column lengths should be set equal to the center-to-center dimensions from one floor level to the next, and the gross moment of inertia of the column sections can be used as input to the structural analysis software.

The frame in Fig. 13-49 can be analyzed for various combinations of dead and live load, including pattern live loading as discussed in the previous section (ACI Code Section 6.4.3). When using frame analysis software to analyze a model similar to that shown in

TABLE 13-7 Recommended α and β Values for the Flexural Stiffness of Slab-Beam Elements

Region of the Slab	α-Value (For Effective Width $\alpha\,\ell_2$)	β-Value (For $I_e = \beta\, I_g$)
Positive-bending regions	0.5	0.5
Negative-bending regions (interior columns)	0.5	0.33
Negative-bending regions (exterior columns)	0.2 to 0.5 (function of edge beam stiffness)	0.33

Fig. 13-49, negative moments and shear forces for the ends of the slab-beams will be given at the center of the joint. As discussed in the previous example, these moment and shear forces can be used to determine the moment at the face of the support, which will be used to design slab flexural reinforcement. For interior supports, the larger of the two moments at the opposite faces of the support should be used as the design moment for the slab reinforcement. The results from the frame analysis also can be used to determine the amount of moment transferred from the slab system to the column. Part of this transfer moment will need to be used to check the shear strength of the slab along a *critical perimeter* around the column, as will be discussed in the Section 13-11.

Analysis of Slab–Column Frames for Combined Gravity and Lateral Loads

A frame consisting of columns and either flat plates or flat slabs but lacking shear walls or other bracing elements is inefficient in resisting lateral loads and may be subject to significant lateral drift deflections. As a result, slab–column frame structures of more than two or three stories are generally braced by shear walls.

When unbraced slab–column frames are used, it is necessary to analyze equivalent frame structures for both gravity and lateral loads. The general equivalent-frame analysis method discussed previously can be used by simply extending the slab–column frame over the full height of the structure, as shown in Fig. 13-50. In general, the slab-beams in each floor level are modeled as discussed previously, and the columns are assumed to

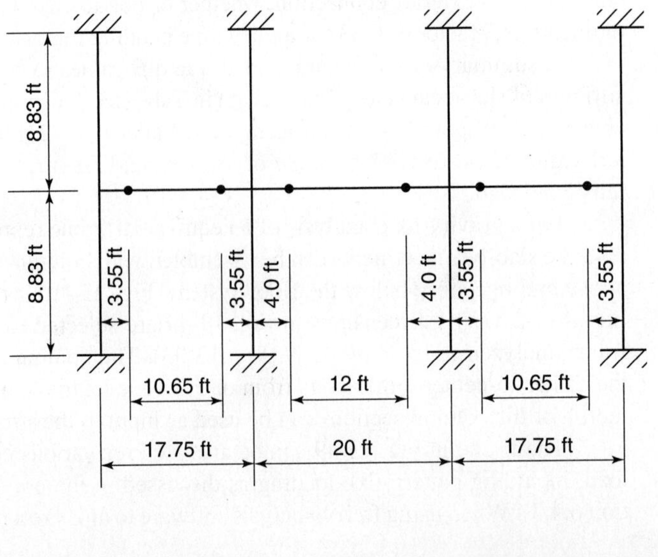

Fig. 13-49
Dimensions and node point locations for slab–column frame analyzed in Example 13-10.

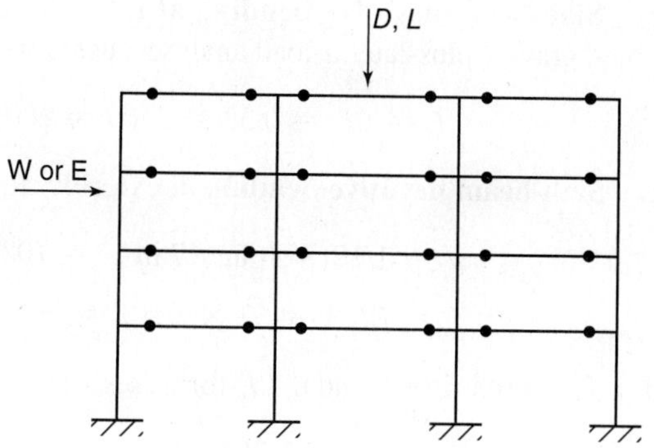

Fig. 13-50
Equivalent-frame model for analysis of slab–column frame subjected to gravity and lateral loads.

have a length equal to the center-to-center distance between floor levels. In order to not overestimate the lateral stiffness of the slab–column frame (and thus underestimate the lateral deflections), the author recommends that the effective moment of inertia of the column sections should be taken as 70 percent of the gross moment of inertia, as required in ACI Code Table 6.6.3.1.1(a) for lateral-stability analysis of frames with slender columns. If a designer uses the slab–column frame similar to that in Fig. 13-50 for the analysis of only gravity loads with the stiffness modifications noted here, then the analysis of the frame for combined gravity and lateral loads simply becomes an additional loading case that can be handled without changing the geometry and stiffness properties of the existing analytical model.

EXAMPLE 13-10 Reanalysis of a Flat-Plate Floor Using Standard Frame Analysis Software

For this problem, the slab–column frame shown in Fig. 13-49 will be analyzed using standard frame analysis software. This is the same frame analyzed in Example 13-9 using the classic equivalent-frame method (CEFM). A plan view of the floor system is given in Fig. 13-33. Two different analysis cases will be considered and the results will be compared to those obtained using the CEFM.

1. **Slab-beam effective widths, $\alpha\ell_2$.** For positive-bending regions and the interior negative-bending regions, use $\alpha = 0.5$. Then, the effective slab width is

$$0.5 \times 15.75 \text{ ft} \times 12 \text{ in./ft} = 94.5 \text{ in.}$$

Because there is not an edge beam, use $\alpha = 0.2$ for the negative-bending region near an exterior support. Thus,

$$0.2 \times 15.75 \text{ ft} \times 12 \text{ in./ft} = 37.8 \text{ in.}$$

This must be checked against the minimum width value at an exterior support:

$$c_2 + c_1 = 20 \text{ in.} + 12 \text{ in.} = 32 \text{ in. (does not govern)}$$

2. **Effective moments of inertia.**

(a) Slab-beam positive-bending region. Use $\beta = 0.5$,

$$I_g = 1/12(94.5 \text{ in.})(7 \text{ in.})^3 = 2700 \text{ in.}^4$$

$$I_e = \beta I_e = 0.5 \times 2700 = 1350 \text{ in.}^4$$

(b) Slab-beam negative bending at interior supports. For gravity-load and combined gravity-plus-lateral-load analyses, use $\beta = 0.33$.

$$I_e = \beta I_g = 0.33 \times 2700 = 900 \text{ in.}^4$$

(c) Slab-beam negative bending at exterior supports. Use $\beta = 0.33$,

$$I_g = 1/12(37.8 \text{ in.})(7 \text{ in.})^3 = 1080 \text{ in.}^4$$

$$I_e = \beta I_g = 0.33 \times 1080 = 360 \text{ in.}^4$$

(d) Columns. Use I_g and 0.7 I_g for Cases 1 and 2. For exterior columns,

$$I_g = 1/12(20)(12)^3 = 2880 \text{ in.}^4$$

$$I_e = 0.7 \times 2880 = 2020 \text{ in.}^4$$

For interior columns,

$$I_g = 1/12(12)(20)^3 = 8000 \text{ in.}^4$$

$$I_e = 0.7 \times 8000 = 5600 \text{ in.}^4$$

3. Analysis Cases; results given in Table 13-8.

(a) Case 1. Use I_e values given above for all slab-beams, and I_g for all columns.

(b) Case 2. Same as Case 1, except use I_e for all columns.

4. Discussion of results. The results for Cases 1 and 2 are essentially the same, so altering the column stiffness by a factor of 0.7 had little effect on the results. When comparing these two cases to the results from the CEFM, the most significant difference occurs for the design moment at the face of the exterior column. The design moment values from the standard frame analysis (Cases 1 and 2) are approximately 75 percent higher that those obtained from the CEFM at this location. Because it would be unreasonable to select even lower values for α and β at this location, it seems that the standard frame analysis did not properly reflect the moment redistribution that occurs due to large rotations of the edge of the slab away from the exterior column. Thus, for Cases 1 and 2, it would be appropriate to increase the calculated design moments at midspan and the face of the first interior support for the external span by approximately 10 percent to account for the redistribution of moments within that span. However, using the moment results from the standard frame analysis software to design the exterior slab–column connection is not unreasonable. As will be discussed in the Section 13-11, the direct-design method requires (ACI Code Section 8.10.7.3) that 30 percent of the total static moment, M_o, in an exterior

TABLE 13-8 Analysis Results: Design Moments from Standard Frame Analysis Software Compared to CEFM

	Exterior Span			Interior Span	
Analysis Case	Moment at Exterior Face (kip-ft)	Moment at Midspan (kip-ft)	Moment at Interior Face (kip-ft)	Moment at Interior Face (kip-ft)	Moment at Midspan (kip-ft)
CEFM	−24.4	55.1	−80.6	−80.0	50.7
Case 1	−43.7	47.0	−74.1	−80.5	51.0
Case 2	−42.6	47.3	−74.9	−80.5	51.0

span be used as the design-transfer moment at the exterior slab-to-column connection (0.30 × 108 kip-ft = 32.4 kip-ft in this case). Note: when an edge beam is present, the results from the standard frame analysis procedure more closely matches the results from the CEFM, as can be seen by solving Problem 13-10 at the end of this chapter. ■

13-10 SHEAR STRENGTH OF TWO-WAY SLABS

A shear failure in a beam results from an inclined crack caused by flexural and shearing stresses. This crack starts at the tensile face of the beam and extends diagonally to the compression zone, as explained in Chapter 6. In the case of a two-way slab or footing, the two shear-failure mechanisms shown in Fig. 13-51 are possible. *One-way shear* or *beam-action shear* (Fig. 13-51a) involves an inclined crack extending across the entire width of the structure. *Two-way shear* or *punching shear* involves a truncated cone or pyramid-shaped surface around the column, as shown schematically in Fig. 13-51b. Generally, the punching-shear capacity of a slab or footing will be considerably less than the one-way shear capacity. In design, however, it is necessary to consider both failure mechanisms. This section is limited to footings and slabs without beams. The shear strength of slabs with beams is discussed in Section 13-14.

Behavior of Slabs Failing in Two-Way Shear—Interior Columns

As discussed in Section 13-5, the maximum moments in a uniformly loaded flat plate occur around the columns and lead to a continuous, flexural crack around each column. After additional loading, the cracks tend to form a fan-shaped yield-line mechanism (see Section 14-6). At about the same time, inclined or shear cracks form on a truncated pyramid-shaped surface (shown in Fig. 13-51b). These cracks can be seen in Fig. 13-52, which shows a slab that has been sawn through along two sides of the column after the slab had failed in two-way shear.

In Chapter 6, truss models are used to explain the behavior of beams failing in shear. Alexander and Simmonds [13-26] have analyzed punching-shear failures using a truss

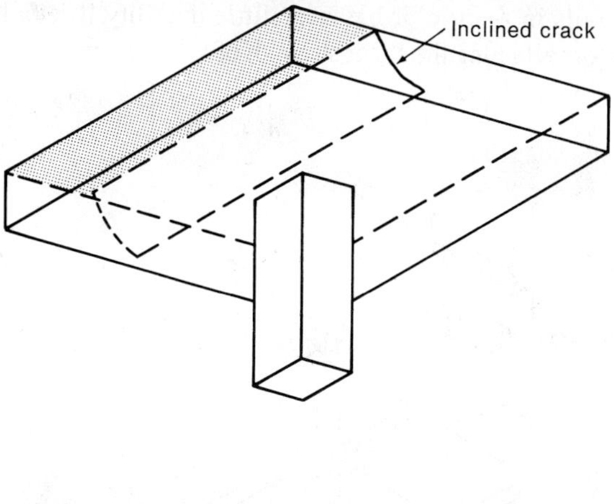

(a) One-way shear.

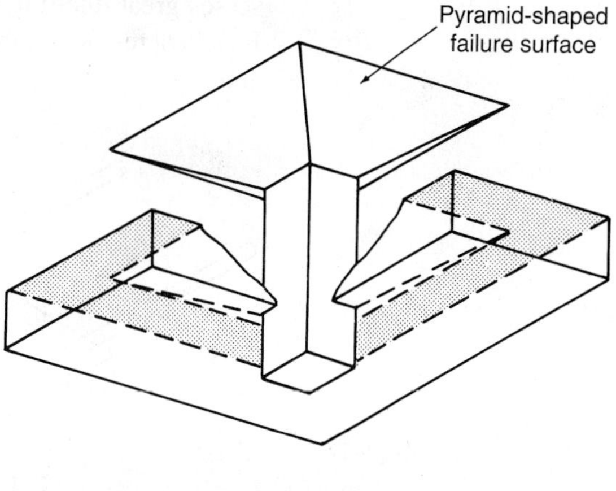

(b) Two-way shear.

Fig. 13-51
Shear failure in a slab.

Fig. 13-52
Inclined cracks in a slab after a shear failure. (Photograph courtesy of J. G. MacGregor.)

model similar to that in Fig. 13-53. Prior to the formation of the inclined cracks shown in Fig. 13-52, the shear is transferred by shear stresses in the concrete. Once the cracks have formed, only relatively small shear stresses can be transferred across them. The majority of the vertical shear is transferred by inclined struts *A–B* and *C–D* extending from the compression zone at the bottom of the slab to the reinforcement at the top of the slab. Similar struts exist around the perimeter of the column. The horizontal component of the force in the struts causes a change in the force in the reinforcement at *A* and *D*, and the vertical component pushes up on the bar and is resisted by tensile stresses in the concrete between the bars. Eventually, this concrete cracks in the plane of the bars, and a punching failure results. Such a failure occurs suddenly, with little or no warning.

Once a punching-shear failure has occurred at a slab–column joint, the shear capacity of that particular joint is almost completely lost. In the case of a two-way slab, as the slab slides down, the column load is transferred to adjacent column-slab connections, thereby possibly overloading them and causing them to fail. Thus, although a two-way slab possesses great ductility if it fails in flexure, it has very little ductility if it fails in shear. Excellent reviews of the factors affecting the shear strength of slabs are presented in [13-27] and [13-28].

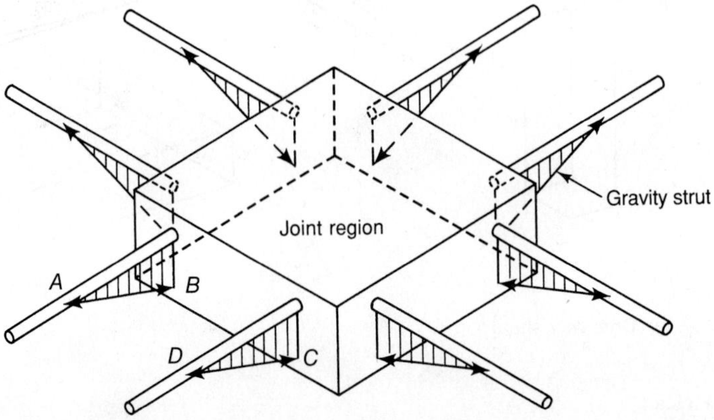

Fig. 13-53
Truss model for shear transfer at an interior column.

Design for Two-Way Shear

Initially, this discussion will consider the case of shear transfer without appreciable moment transfer. The case when both shear and moment are transferred from the slab to the column is discussed in Section 13-11.

From extensive tests, Moe [13-29] concluded that the critical section for shear was located at the face of the column. ACI-ASCE Committee 326 (now 445) [13-30] accepted Moe's conclusions, but showed that a much simpler design equation could be derived by considering a critical section at $d/2$ from the face of the column. This was referred to as the *pseudo-critical section for shear*. This simplification has been incorporated in the ACI Code.

Location of the Critical Perimeter

Two-way shear is assumed to be critical on a vertical section through the slab or footing extending around the column. According to ACI Code Section 22.6.4.1, this section is chosen so that it is never less than $d/2$ from the face of the column and so that its length, b_o, is a minimum. Although this would imply that the corners of the shear perimeter should be rounded, the original derivation of the allowable stresses in punching shear was based on a rectangular perimeter. The intent of the code is that the critical shear section be taken as rectangular if the columns are rectangular, as stated in ACI Code Section 22.6.4.1.1. Several examples of critical shear perimeters are shown in Fig. 13-54.

Critical Sections for Slabs with Drop Panels

When high shear forces are being transferred at a slab–column connection, the slab shear strength can be increased locally by using a *drop panel* to locally increase the thickness of the slab. The term *drop panel* stems from the requirement to "drop" the form-work around the

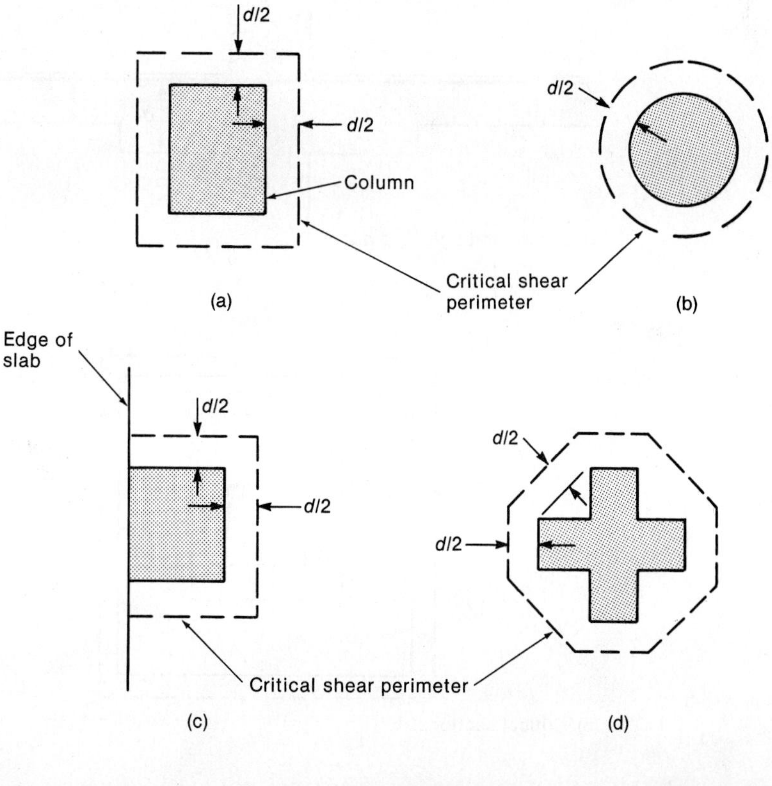

Fig. 13-54
Location of critical shear perimeters.

column to increase the cast-in-place thickness of the slab. ACI Code Section 8.2.4 requires that the total thickness of the slab and drop panel to be at least 1.25 times the thickness of the slab adjacent to the drop panel. In slabs with drop panels, two critical sections should be considered, as shown in Fig. 13-55. When adding a drop panel, the common design objective is to select a depth and length for the drop panel such that the outer critical perimeter shown in Fig. 13-55 does not govern the shear strength of the slab-column connection.

If a drop panel also is being used to control deflections or reduce the amount of flexural reinforcement required in the slab, the drop panel must satisfy the length requirements given in ACI Code Section 8.2.4. Drop panels that do not satisfy those length requirements still can be used for added shear strength and are sometimes referred to as *shear capitals*, or *shear caps*. Both drop panels and shear caps are discussed in Section 13-12.

Critical Sections Near Holes and At Edges

When openings are located at less than 10 times the slab thickness from a column, ACI Code Section 22.6.4.3 requires that the critical perimeter be reduced, as shown in Fig. 13-56a. The critical perimeter for edge or corner columns is not defined specifically in the ACI Code. However, ACI Commentary Section R22.6.4.3 refers to a publication from ACI Committee 326 (now 445) [13-31], which suggested that the side faces of the critical perimeter would extend to the edge of the slab if the distance from the face of the column to the edge of the slab does not exceed the larger of (i) four slab thicknesses, $4h$, or (ii) twice the development length, $2\ell_d$, of the flexural reinforcement perpendicular to the edge, shown by the distances labeled A and B in Fig. 13-56b and c. If the actual distance to the edge of the slab is only slightly smaller that these limits, extending the critical shear section all the way to the edge of the slab may lead to an unreasonably large critical perimeter that would overestimate the shear transfer strength at that connection. The author agrees with the

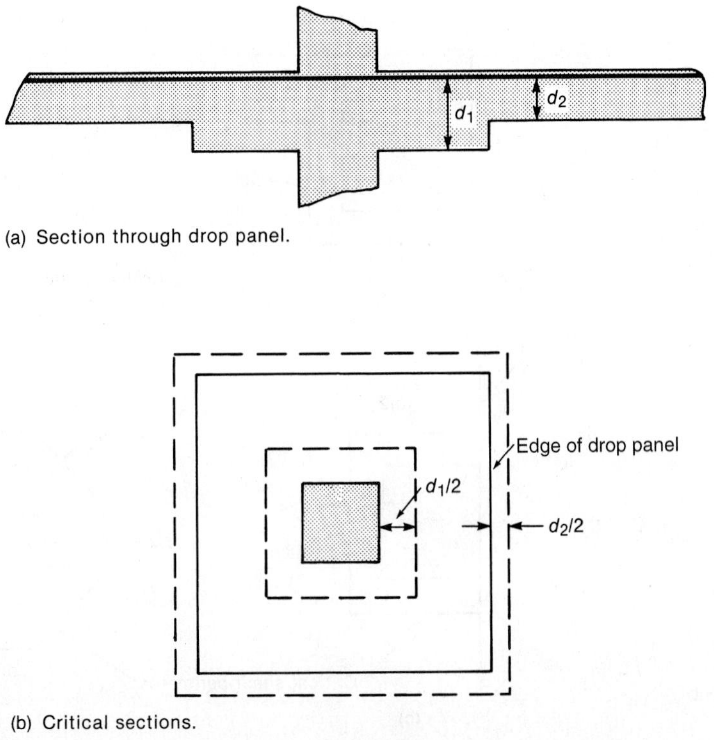

Fig. 13-55
Critical sections in a slab with drop panels.

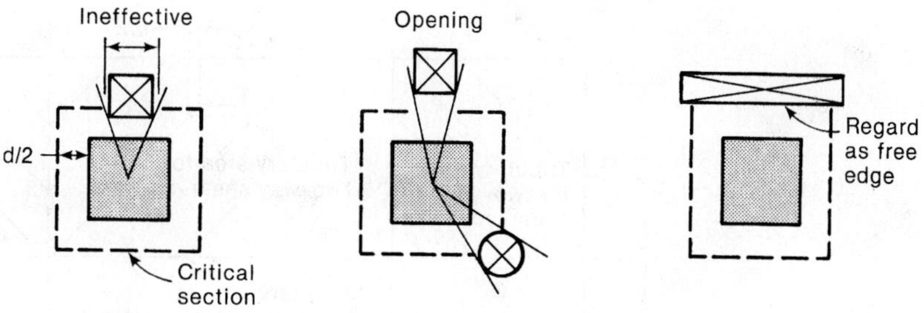

(a) Openings.

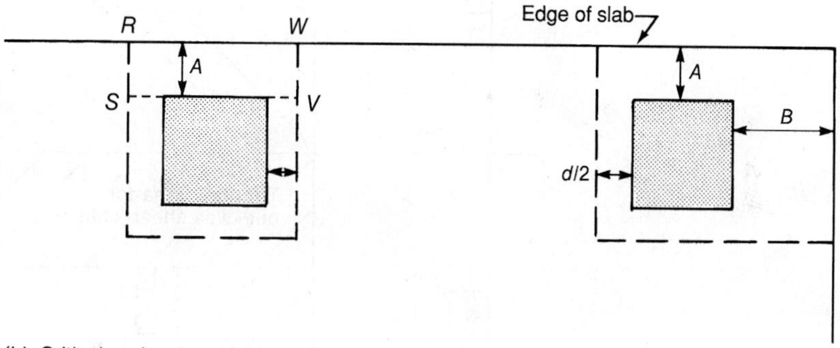

(b) Critical perimeters if A and B do not exceed the greater of $4h$ or $2\ell_d$.

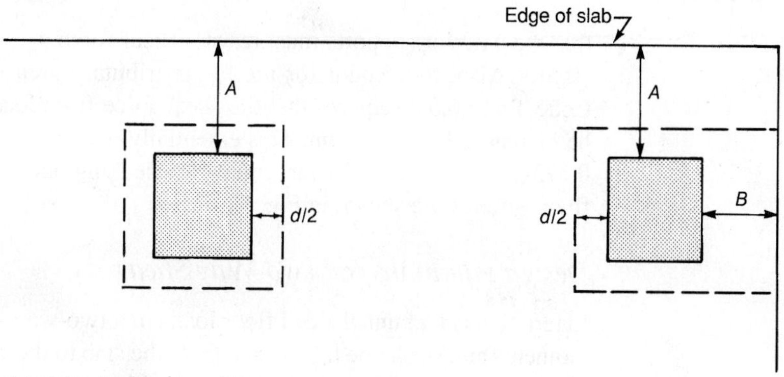

(c) Critical perimeters if A exceeds the greater of $4h$ or $2\ell_d$, but B does not.

Fig. 13-56
Effects of openings and edges on the critical shear perimeter.

guidelines developed by former ACI Committee 326 for deciding when the failure surface should extend *toward* a free edge, as opposed to wrapping around the column. However, the author recommends that the *effective length* of this extension, which is represented by the lines $R - S$ and $V - W$ in Fig. 13-56(b), should be limited to the shorter of $2h$ and the actual dimension to the edge, measured from the face of the column.

Tributary Areas for Shear in Two-Way Slabs

For uniformly loaded two-way slabs, the tributary areas used to calculate V_u are bounded by lines of zero shear. For interior panels, these lines can be assumed to pass through the center of the panel. For edge panels, lines of zero shear are approximately at 0.42ℓ to 0.45ℓ from the center of the exterior column, where ℓ is the span measured from center-to-center of the columns. However, to be conservative in design, ACI Code Section 6.5.4 requires

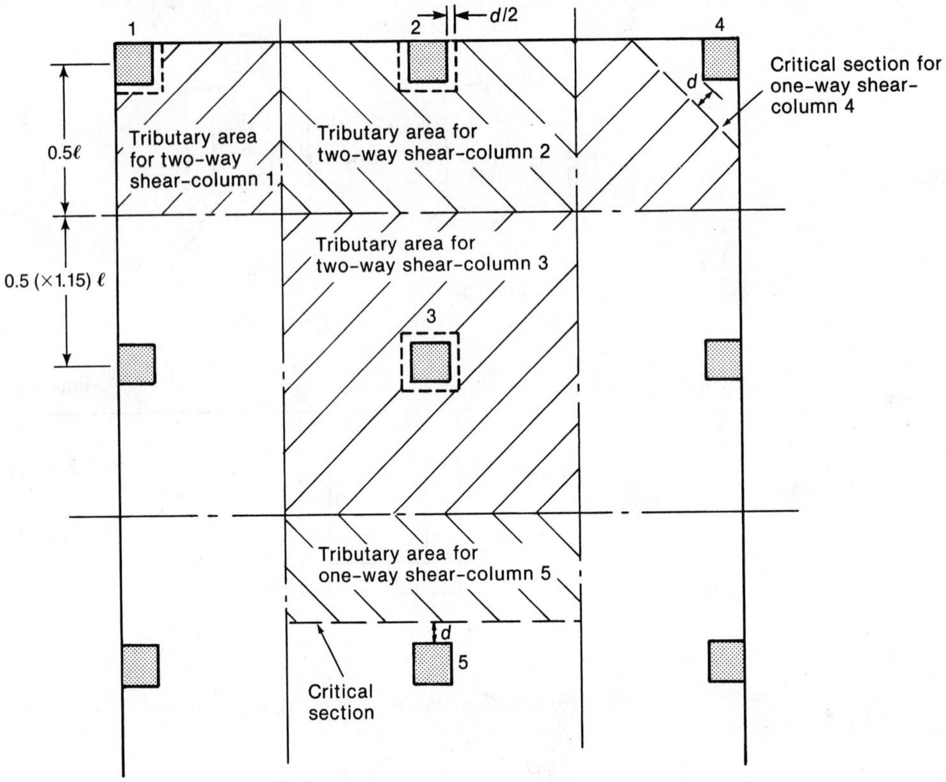

Fig. 13-57
Critical sections and tributary areas for shear in a flat plate.

that the exterior supports must resist a shear force due to loads acting on half of the span (0.5ℓ). Also, to account for the larger tributary area for the first interior support, ACI Code Table 6.5.4 requires that the shear force from loads acting on half of the span must be increased by 15 percent. This essentially results in a tributary length of $1.15 \times 0.5\ell = 0.575\ell$. Examples of tributary areas for checking one-way and two-way shear in flat-plate floor systems are shown in Fig. 13-57.

Design Equations for Two-Way Shear

Lateral loads and unbalanced floor loads in a two-way slab building will require that both moments and shears be transferred from the slab to the columns. In the case of interior columns in a braced frame, the worst loading case for shear generally corresponds to full live load on all spans resulting in a negligible moment being transferred at the slab to column connection. We will initially focus our attention on the case of only transferring shear at an interior slab to column connections and discuss combined shear and moment transfer at interior and exterior connections in the following section. However, when we discuss the combined transfer of shear and moment, we will need to compare the maximum factored shear stress acting on the critical shear section to the nominal shear strength of the slab expressed as a stress acting on the same section area, $b_o \cdot d$. For consistency, even when only shear is being transferred we will compare factored loads to nominal strength in terms of shear stresses on the critical shear section.

Design for two-way shear without moment transfer is carried out using stress-based equations that are similar to the force-based equations used in Chapter 6. The basic shear design expression is:

$$\phi v_n \geq v_u = \frac{V_u}{b_o \cdot d} \tag{13-22}$$

where V_u is the factored shear force due to loads acting across the critical shear section, $b_o \cdot d$, and v_n is the nominal shear strength of the slab, expressed in units of stress. From ACI Code Table 21.2.1, the ϕ-factor for shear is 0.75.

Punching Shear, v_c, Carried by Concrete in Two-Way Slabs

In general, the ACI Code defines v_n as follows:

$$v_n = v_c + v_s \tag{13-23}$$

where v_c and v_s are the shear resistances attributed to the concrete and the shear reinforcement, respectively. For economic reasons, designers prefer to avoid the use of shear reinforcement, and thus, v_s is usually zero. ACI Code Table 22.6.5.2 gives three equations for calculating v_c, with the requirement that the smallest calculated value for v_c must be used in design. For many years, only the first equation was used to determine the punching-shear strength of a two-way slab without shear reinforcement:

$$v_c = 4\lambda\sqrt{f'_c} \tag{13-24}$$

where λ is the factor for lightweight-aggregate concrete introduced in Chapter 6 (equals 0.85 for sand-lightweight concrete and 0.75 for all-lightweight concrete).

The distribution of shear stresses around the column is approximately as shown in Fig. 13-58 with higher shear-stresses transferred in the vicinity of the corners. For rectangular columns with the longer sides significantly larger than the shorter sides, the shear stress along the longer sides at failure conditions can be significantly less than that at the corners [13-32], approaching the nominal strength for one-way shear, $2\sqrt{f'_c}$. Thus, for highly rectangular columns, the following expression was added to the ACI Code:

$$v_c = \left(2 + \frac{4}{\beta}\right)\lambda\sqrt{f'_c} \tag{13-25}$$

where β is the ratio of long-side–short-side dimensions of the column, concentrated load, or reaction area. For nonrectangular columns, this is defined as shown in Fig. 13-59. Equation (13-25) is compared to test data in Fig. 13-60a.

For very large columns, the shear stress at failure conditions along all sides of the critical perimeter will be significantly less than that at the corners. Thus, for columns having a ratio of critical-perimeter–effective-depth, b_o/d, greater than 20, the following expression applies.

$$v_c = \left(\frac{\alpha_s d}{b_o} + 2\right)\lambda\sqrt{f'_c} \tag{13-26}$$

where α_s is 40 for interior columns, 30 for edge columns, and 20 for corner columns.

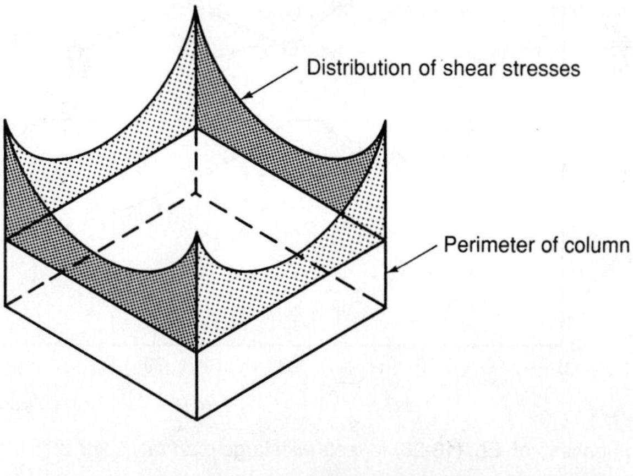

Fig. 13-58
Distribution of shear amount around the perimeter of a square.

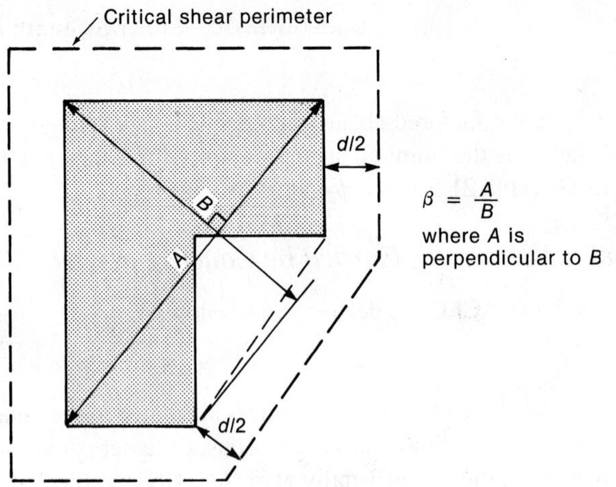

Fig. 13-59
Definition for irregular shaped columns.

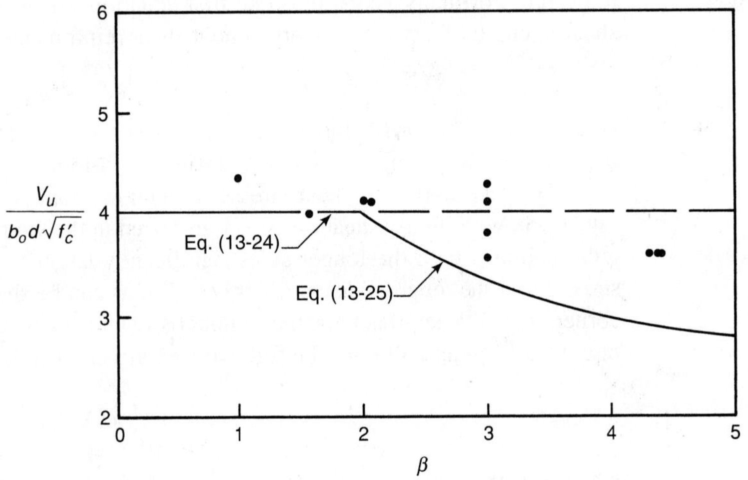

(a) Comparison of Eqs. (13-24) and (13-25) to tests of rectangular columns [13-32].

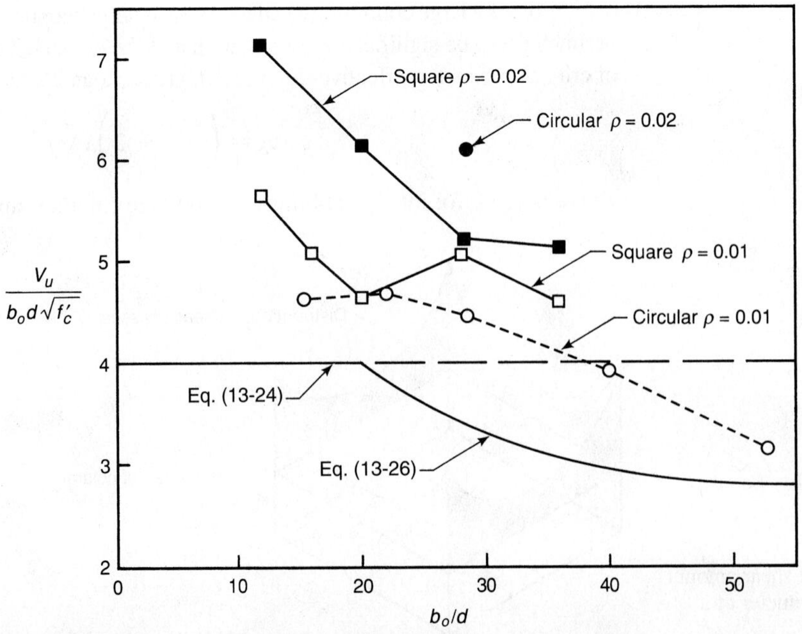

(b) Comparison of Eq. (13-26) to tests with large b_o/d ratios [13-33].

Fig. 13-60
Comparison of design equations to test data.

Equation (13-26) is compared to test data in Fig. 13-60b [13-33]. Equation (13-24) will govern in most cases. For normal size columns where β exceeds 2.0, Eq. (13-25) will govern, and for very large columns or other cases with a large critical shear perimeter, Eq. (13-26) will govern.

One-Way Shear in Slabs

In the case of a uniformly loaded slab, the critical section for one-way shear is located at d from the face of a support (Fig. 13-57), from the face of a drop panel, or from other changes in thickness. The entire width of the slab panel is assumed to be effective in resisting one-way shear (ACI Code Section 8.5.3.1.1). The tributary areas for one-way shear in a slab are illustrated in Fig. 13-57 (for columns 4 and 5). The shear strength on the critical section is computed as was done for beams (Chapter 6) by using the following expression:

$$V_c = 2\lambda\sqrt{f'_c}\,bd \tag{13-27}$$

where b is the width of the critical sections. One-way shear is seldom critical in flat plates or flat slabs, as will be seen from Example 13-11.

EXAMPLE 13-11 Checking One-Way and Two-Way Shear at an Interior Column in a Flat Plate

Figure 13-61 shows an interior column in a large uniform flat-plate slab. The slab is 6 in. thick. An average effective depth, d, as shown in Fig. 13-62, normally is used in shear-strength calculations for two-way slabs. Both one-way and two-way punching shear usually is checked near columns where top reinforcement is used in both principal directions to resist negative-bending moments. ACI Code Table 20.6.1.3.1 states that the minimum clear cover for slab reinforcement is $^3/_4$ in. Thus, assuming No. 4 bars (diameter = 0.5 in.) are used as flexural reinforcement, the average d value for determining shear strength of the slab is

$$d(\text{avg}) = 6\text{ in.} - 0.75\text{ in.} - 0.5\text{ in.} = 4.75\text{ in.}$$

The slab supports a uniform, superimposed dead load of 15 psf and a uniform, superimposed live load of 60 psf. The normal-weight concrete ($\lambda = 1.0$) has a compressive strength of 3000 psi. The moments transferred from the slab to the column (or vice versa) are assumed to be negligible. Check whether the shear capacity is adequate.

1. Determine the factored uniform load.

$$\begin{aligned} q_u &= 1.2\left(\frac{6}{12} \times 150 + 15\right) + 1.6 \times 60 = 204\text{ psf} \\ &= 0.204\text{ ksf} \end{aligned}$$

2. Check the one-way shear. One-way shear is critical at a distance d from the face of the column. Thus, the critical sections for one-way shear are A–A and B–B in Fig. 13-61. The loaded areas causing shear on these sections are cross hatched. Their outer boundaries are lines of symmetry on which $V_u = 0$. Because the tributary area for section A–A is larger, this section will be more critical.

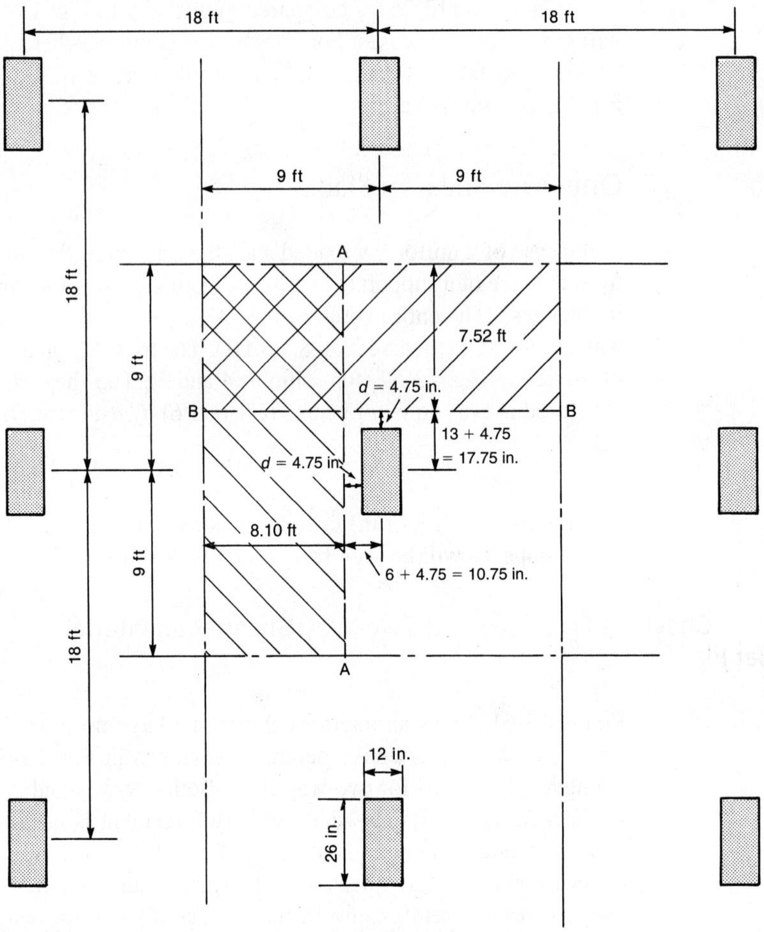

Fig. 13-61
Critical sections for one-way shear at interior column for Example 13-11.

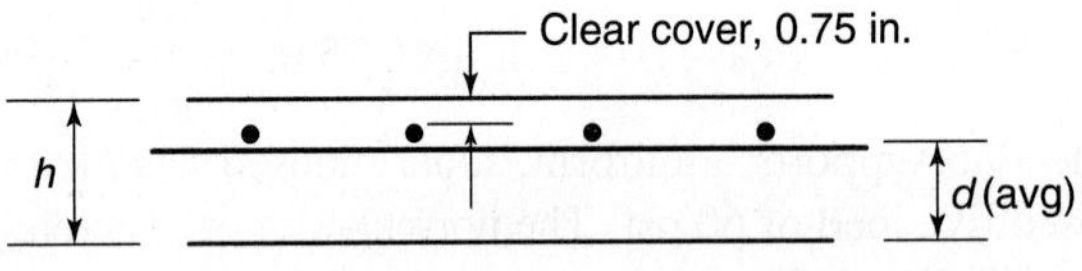

Fig. 13-62
Determination of d(avg) for use in shear strength evaluation of two-way slabs.

(a) Compute V_u at section A–A.

$$V_u = 0.204 \text{ ksf} \times 8.10 \text{ ft} \times 18 \text{ ft} = 29.7 \text{ kips}$$

(b) Compute ϕV_n for one-way shear. Because there is no shear reinforcement, we have

$$\phi V_n = \phi V_c$$

where V_c for one-way shear is given by Eq. (13-27) as

$$\phi V_c = 0.75(2\lambda\sqrt{f'_c}\, bd)$$

$$= 0.75\left(2 \times 1\sqrt{3000} \times \left(18 \text{ ft} \times \frac{12 \text{ in.}}{\text{ft}}\right) \times 4.75 \text{ in.} \times \frac{1 \text{ kip}}{1000 \text{ lb}}\right)$$

$$= 84.3 \text{ kips}$$

Because $\phi V_c > V_u$, the slab is o.k. in one-way shear.

3. Check the two-way punching shear. Punching shear is critical on a rectangular section located at $d/2$ away from the face of the column, as shown in Fig. 13-63. The load on the cross-hatched area causes shear on the critical perimeter. Once again, the outer boundaries of this area are lines of symmetry, where V_u is assumed to be zero.

(a) Compute V_u on the critical perimeter for two-way shear.

$$V_u = 0.204 \text{ ksf}\left[(18 \text{ ft} \times 18 \text{ ft}) - \left(\frac{16.75 \text{ in.}}{12} \times \frac{30.75 \text{ in.}}{12}\right)\text{ft}^2\right]$$

$$= 65.4 \text{ kips}$$

(b) Compute v_u and compare to ϕv_c for the critical section. The length of the critical perimeter is:

$$b_o = 2(16.75'' + 30.75'') = 95 \text{ in.}$$

and,

$$v_u = \frac{V_u}{b_o \cdot d} = \frac{65.4 \text{ k}}{95'' \times 4.75''} = 0.145 \text{ ksi} = 145 \text{ psi}$$

To find the smallest value of v_c, we can calculate the coefficients in Eqs. (13-25) and (13-26) to see how they compare to the coefficient 4.0 in Eq. (13-24). For Eq. (13-25):

$$\beta = \frac{26''}{12''} = 2.17, \text{ and}$$

$$\left(2 + \frac{4}{\beta}\right) = 2 + \frac{4}{2.17} = 3.85 < 4.0$$

Similarly, for Eq. (13-26), using α_s equal to 40 for an interior connection:

$$\left(\frac{\alpha_s \cdot d}{b_o} + 2\right) = \left(\frac{40 \cdot 4.75''}{95''} + 2\right) = 2.11 + 2.0 = 4.11$$

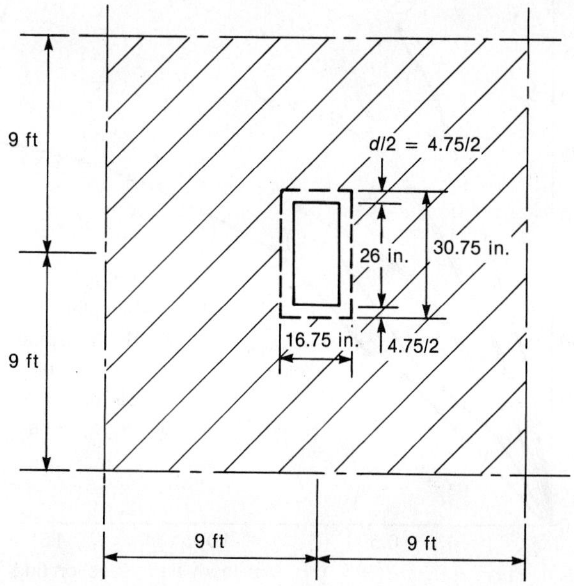

Fig. 13-63
Loaded area and critical sections for a two-way shear—Example 13-11.

Therefore, Eq. (13-25) will govern for determining ϕv_c:

$$\phi v_c = 0.75 \times 3.85 \times \lambda\sqrt{f'_c}$$
$$\phi v_c = 0.75 \times 3.85 \times 1 \times \sqrt{3000}\text{ psi} = 158\text{ psi}$$

Because ϕv_c exceeds v_u, the slab is o.k. in two-way shear. ■

Augmenting the Shear Strength

If ϕv_c is less than v_u, the shear capacity can be increased by any of the following four methods:

1. Thicken the slab over the entire panel. (Note: this may be counterproductive because the weight of the slab may increase v_u significantly.)

2. Use a drop panel to thicken the slab adjacent to the column.

3. Increase b_o by increasing the column size or by adding a fillet or shear capital around the column.

4. Add shear reinforcement.

There are a number of schemes for increasing the two-way shear strength. Figure 13-64 shows load-deflection diagrams for five slab–column connections tested by Ghali and Megally [13-34] to compare the strength and behavior of three of the four methods of providing additional shear strength. Specimen 1 was a flat plate without any shear strengthening, Specimen 2 had a shear capital, Specimen 3 had a drop panel, Specimen 4 had stirrups (note: slab effective depth for this specimen was less than that permitted for stirrups by the ACI Code), and Specimen 5 had headed shear studs. For Specimen 2, the outer corners of the capital spalled off, leaving an essentially round support region with a perimeter that was shorter than would be assumed for the original square capitals.

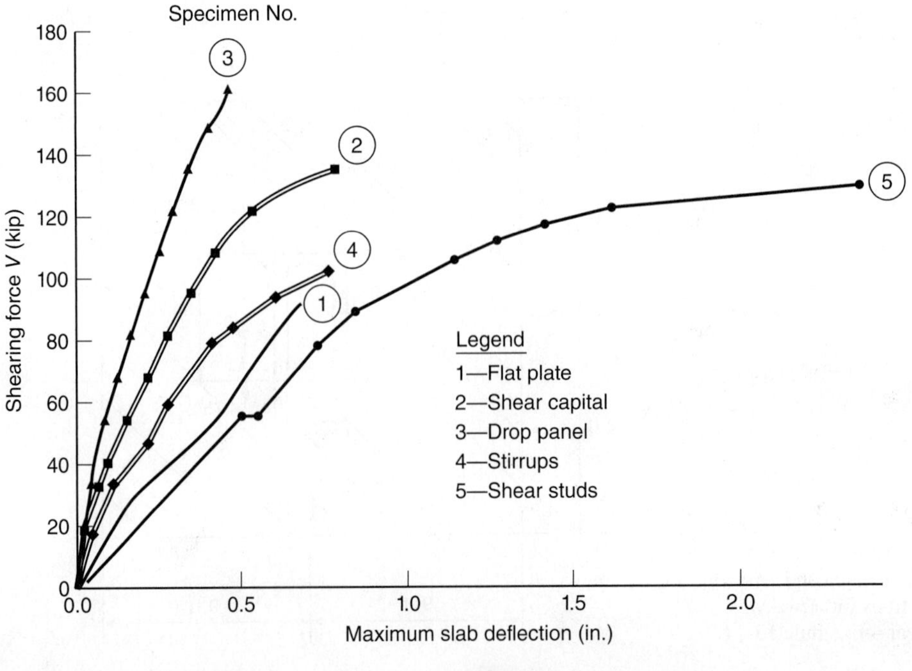

Fig. 13-64
Load deflection curves for slabs strengthened against two-way shear failure.

Shear Reinforcement for Two-Way Slabs

Stirrups

ACI Code Section 22.6.7.1 allows the use of single-leg, multiple-leg, and closed stirrups, provided there are longitudinal bars in all corners of the stirrups, as shown in Fig. 13-65. Stirrups are allowed in slabs with effective depths, d, that exceed the larger of 6 in. or 16 times the stirrup diameter. The precision required to bend and place either the closed stirrups or the multiple-leg stirrups makes these types of shear reinforcement labor-intensive and expensive. As a result, shear reinforcement consisting of stirrups or bent reinforcement is not used widely.

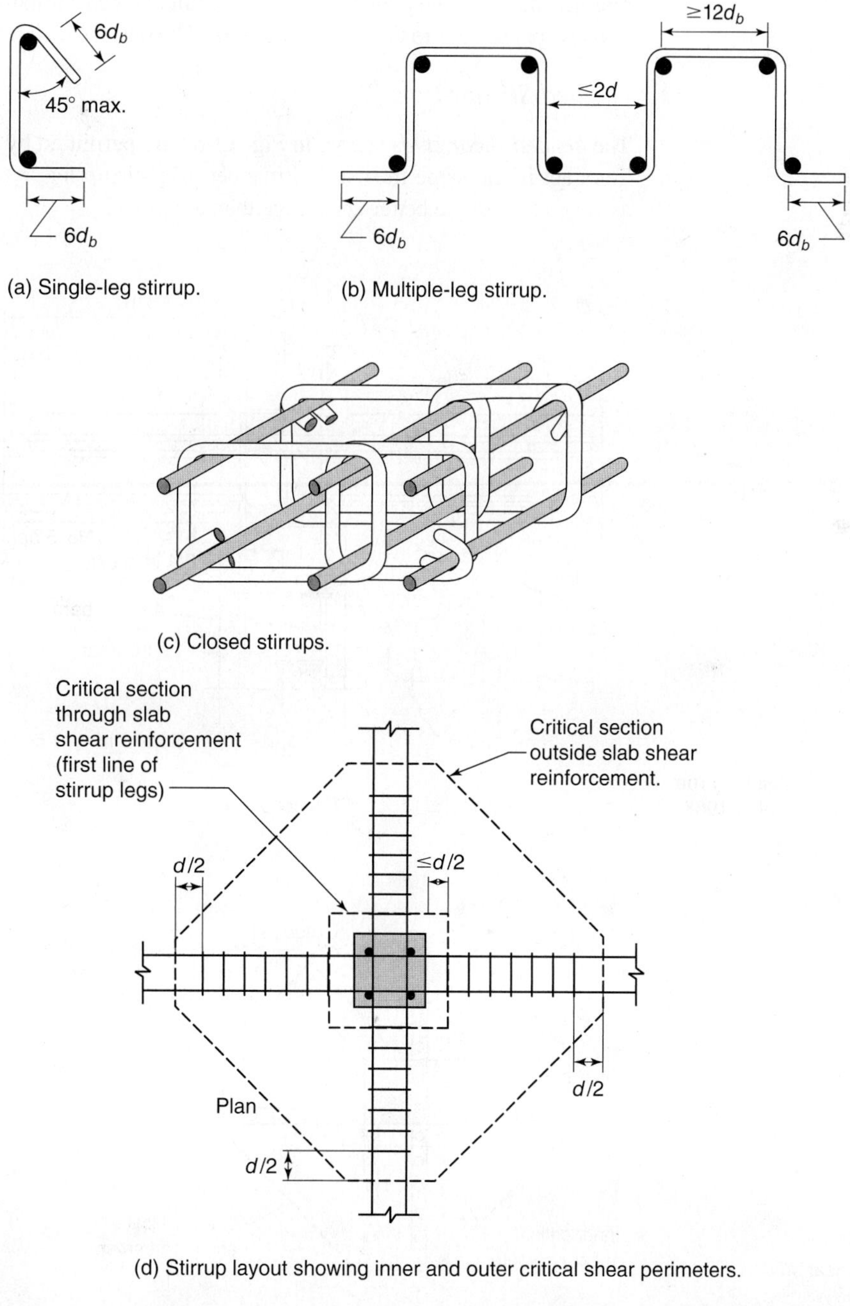

Fig. 13-65
Stirrup-type shear reinforcement in two-way slabs.

Based on two test series with a total of 15 slabs, the 1962 report of ACI-ASCE Committee 326, *Shear and Diagonal Tension* [13-30], stated that the measured shear strength of slabs with stirrups as shear reinforcement exceeded the shear strength of the concrete section, v_c, and also exceeded v_s from the stirrups alone. However, the report goes on to say that in most cases the measured shear strength was less than the sum of v_c and v_s. The authors of the 1962 report were wary about the use of stirrups in slabs with a total thickness less than 10 in.

Structural Steel Shearheads

Structural steel *shearheads*, shown in Fig. 13-66, are designed to resist shear and some of the slab moment according to ACI Code Section 22.6.9. Shearheads at exterior columns require special provisions to transfer the unbalanced moment resulting from loads on the arms perpendicular to the edge of the slab. The design of shearheads is discussed in [13-35].

Headed Shear Studs

The *headed shear studs* shown in Fig. 13-67 are permitted by the ACI Code Section 22.6.8. They act in the same mechanical manner as a stirrup leg, but the head of the shear stud is assumed to provide better anchorage than a bar hook.

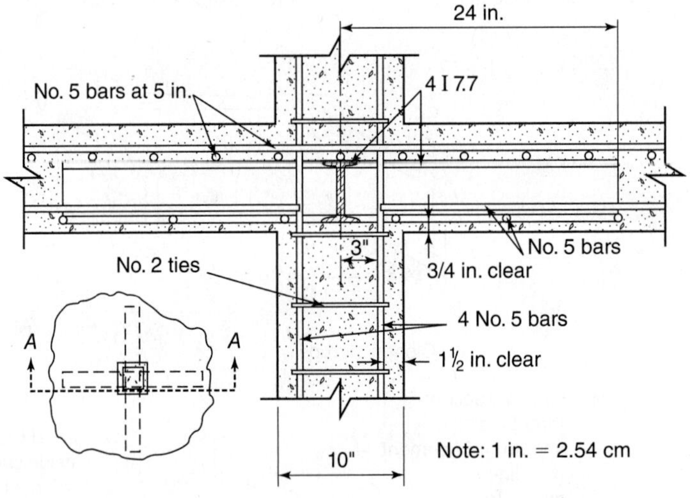

Fig. 13-66
Structural shearheads. (From *ACI Journal*, October 1968.)

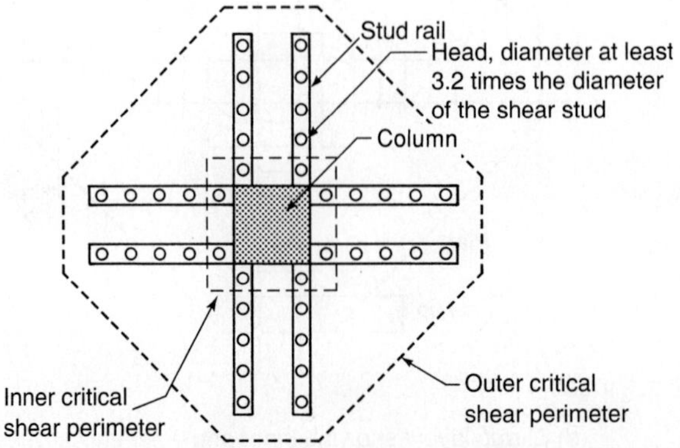

Fig. 13-67
Headed shear studs.

Headed shear-stud reinforcement at a slab–column connection consists of rows of vertical rods, each with a circular head or plate welded or forged on the top end, as shown in Fig 13-67. These rows are placed to extend out from the corners of the column. To aid in the handling and placement of the shear studs and to anchor the lower ends of the studs, they generally are shop-welded to flat steel bars at the desired spacing. The vertical rods are referred to as *headed shear reinforcement* or *headed shear studs*. The assembly of studs plus the bar is called a *stud rail*. Tests suggest the area of the bearing surface of the head should be at least 10 times the area of the vertical rod to enable the yield strength of the rod to be developed prior to crushing of the concrete under the head [13-36].

Recent experimental research [13-37] has indicated that the nominal shear strength at slab–column connections with shear studs provided in the cruciform pattern shown in Fig. 13-67 may be less than the sum of v_c and v_s, as was previously noted for stirrup-type shear reinforcement [13-30]. Adding additional stud rails radiating diagonally from the corners of the column are recommended for obtaining the full nominal shear strength at slab–column connections [13-38].

Design of Shear Reinforcement

Shear reinforcement is required if the shear stress at the *inner critical section* at $d/2$ from the face of the column exceeds the smallest of the values of v_c given by Eqs. (13-24), (13-25), and (13-26). The following design requirements apply to stirrup-type reinforcement and headed shear-studs. Stirrup-type and headed shear-stud reinforcement are designed using Eqs. (13-22) and (13-23). The critical sections for design are

(a) The normal critical section for two-way shear is located at $d/2$ away from the face and corners of the column. We shall refer to this as the *inner critical section* (Figs. 13-65d and 13-67).

(b) A second peripheral line is located at $d/2$ outside the outermost set of stirrup legs or headed shear studs, as shown in Figs. 13-65d and 13-67. This is the *outer critical section*.

ACI Code Table 22.6.6.1 only permits a shear stress of $v_c = 2\lambda\sqrt{f'_c}$ at the inner critical-shear section for a slab with stirrup-type shear reinforcement. This is less than the $v_c = 4\lambda\sqrt{f'_c}$ allowed by ACI Code Table 22.6.5.2 on the same inner critical-shear section in a slab *without shear reinforcement*. The reduction from $4\lambda\sqrt{f'_c}$ to $2\lambda\sqrt{f'_c}$ on the inner critical-shear section is based (in part) on tests of slab–column connections with stirrups, quoted by ACI Committee 326 (now 445) [13-30], in which the measured shear strength, v_n, was less than the sum of v_c and v_s. Hawkins [13-39] noted similar results in tests of slabs with stirrups as shear reinforcement, in which the punching strength of slabs reinforced for shear was about $v_c/2 + v_s$, where v_c was obtained using Eq. (13-24).

Therefore, the ACI Code expression for the nominal shear strength of a two-way slab with stirrup-type shear reinforcement is

$$v_n = v_c + v_s \tag{13-23}$$

where

$$v_c = 2\lambda\sqrt{f'_c}\, b_o d \tag{13-28}$$

The ACI Code assumes that potential punching-shear cracks have a horizontal projection equal to the slab effective depth, d, as shown in Fig. 13-68. In this figure, the critical shear crack nearest the column is shown as a solid line and potential shear cracks that are further from the column face are shown as broken lines. The cracks are assumed to circumscribe the column along a line similar in plan to the perimeter of the

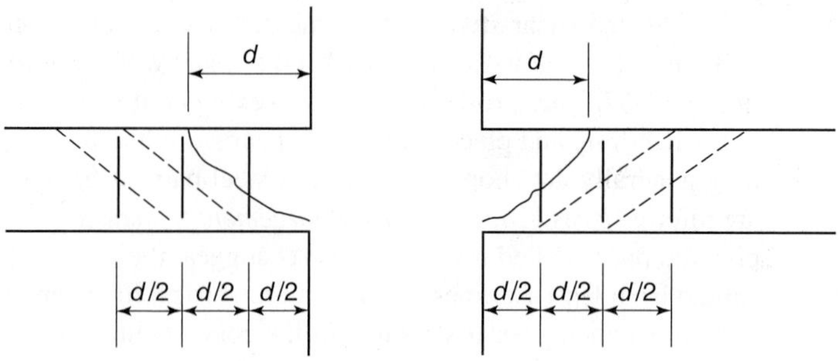

Fig. 13-68
Potential shear cracks and stirrup spacing at slab–column connection.

column. To be sure that stirrup legs cross the potential shear crack nearest the column, ACI Code Table 8.7.6.3 requires that the distance from the face of the column to the first line of stirrup legs that surround the column shall not exceed $d/2$. To intercept potential shear-failure cracks further from the column, ACI Code Table 8.7.6.3 requires that the spacing, s, between successive lines of stirrup legs surrounding the column shall not exceed $d/2$, as shown in Fig. 13-68.

The shear strength provided by the stirrup-type shear reinforcement is

$$v_s = \frac{A_v f_{yt}}{b_o s} \tag{13-29}$$

where A_v is the sum of the areas of all stirrup legs on a peripheral line of stirrups that is geometrically similar to the perimeter of the column section, and f_{yt} is the specified yield strength of the stirrup reinforcement. When using stirrup-type shear reinforcement, the nominal shear strength given by Eq. (13-23) shall not be taken to be larger than $6\sqrt{f'_c}$.

In normal design practice, the shear reinforcement would be extended to a point where the slab shear strength at the outer perimeter shown in Fig. 13-65d does not govern the shear design of the slab–column connection. The nominal shear strength, v_n, at the outer critical section is set equal to the value of v_c given by Eq. (13-28) with b_o taken as the perimeter of the outer critical section.

The design of headed-shear reinforcement in slabs is similar to that for stirrup-type shear reinforcement with a few significant changes. The bearing stresses under the heads anchoring the headed-shear reinforcement confine the slab around the column more effectively than stirrup-type reinforcement. Thus, the maximum shear stress permitted in ACI Code Table 22.6.6.1 is increased from $2\lambda\sqrt{f'_c}$ to $3\lambda\sqrt{f'_c}$ to reflect this better confinement. Also, the maximum value for the nominal shear strength in Eq. (13-23) is increased to $8\sqrt{f'_c}\,b_o d$. The experimental evidence to support these increased values are described in a report from ACI Committee 421 [13-35].

Adjacent to the column face, the maximum permissible spacing between shear studs is measured perpendicular to the face of the column as $d/2$—the same as for stirrup-type shear reinforcement. However, ACI Code Table 8.7.7.1.2 permits the maximum spacing between shear studs to be increased to $3d/4$ at any intermediate critical section where the calculated shear stress due to factored loads, v_u, drops below $\phi\, 6\sqrt{f'_c}$. As with stirrup-type shear reinforcement, the shear studs should be extended to a point where the slab shear strength (as governed by Eq. (13-28) with $v_c = 2\lambda\sqrt{f'_c}$) at the outer critical perimeter shown in Fig. 13-67 does not govern the shear design of the slab–column connection.

The shear strength, v_s, carried by the shear studs is calculated using Eq. (13-29), where A_v is equal to the sum of the cross-sectional areas of all the shear studs on one peripheral line of studs that are crossed by an assumed 45° crack similar to the crack shown in Fig. 13-68.

EXAMPLE 13-12 Design an Interior Slab–Column Connection with Headed Shear Reinforcement—No Appreciable Moment Transfer

A 7.5-in.-thick flat-plate slab with No. 5 flexural reinforcement is supported by 14-in.-square columns spaced at 20 ft on centers *N–S* and 22 ft on centers *E–W*. The service loads on the slab are dead load = 120 psf (including self-weight) and live load = 60 psf. The normal-weight concrete has a compressive strength of 4000 psi. Check the capacity of an interior slab–column connection. If necessary, design shear reinforcement using headed-shear studs. Use the load and resistance factors in ACI Code Tables 5.3.1 and 21.2.1.

1. Select the critical section for two-way shear around the column. At this stage in the calculations, the designer does not know whether shear reinforcement will be required. We will assume it is not and will redo the calculations if we are wrong. Assuming No. 5 flexural reinforcement, the average effective depth of the slab is

$$d(\text{avg}) = (7.5\text{ in.} - 0.75\text{ in. cover} - 0.625\text{ in.}) = 6.13\text{ in.}$$

The *inner critical shear section* for two-way shear in a flat plate extends around the column at $d/2 = 3.06$ in. from the face of the column, as shown in Fig. 13-69.

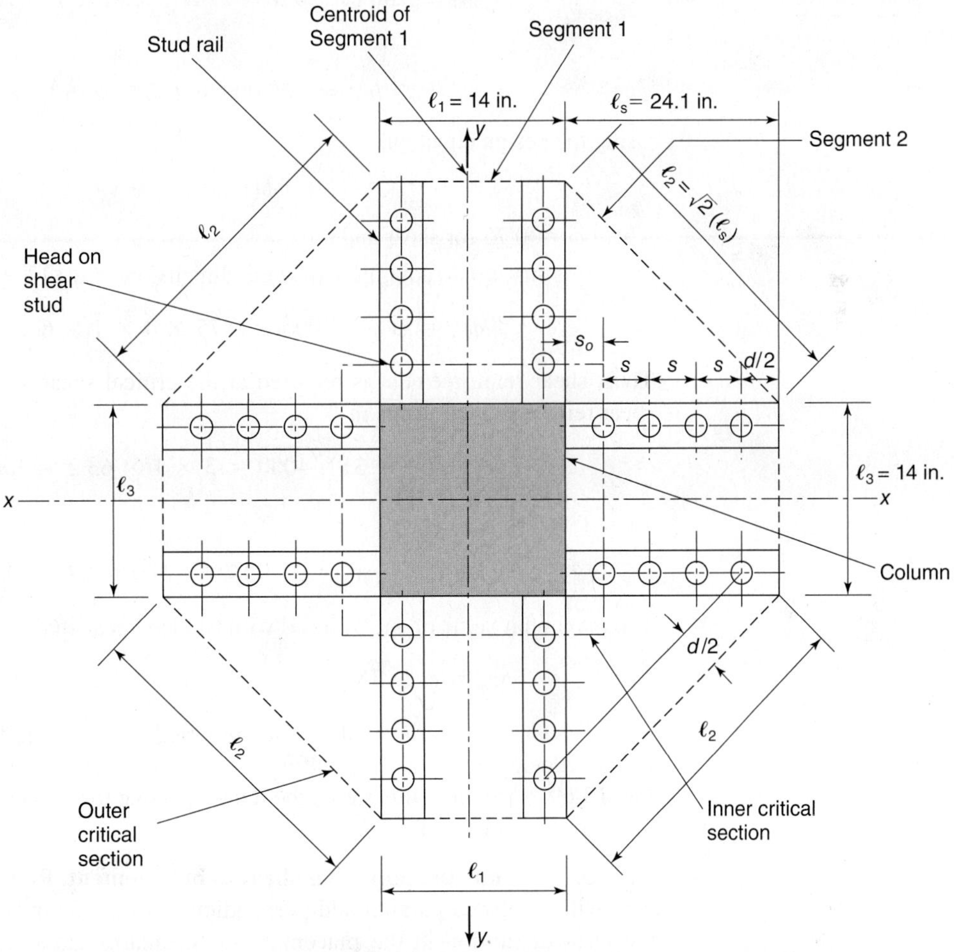

Fig. 13-69
Eight-sided critical shear section for Example 13-12.

The length of one side of the critical shear section around the column is

$$[14 + 6.13] = 20.1 \text{ in.},$$

giving a shear perimeter $b_o = 4 \times 20.1 = 80.5$ in.

The area enclosed within the critical shear section is $(20.1 \times 20.1)/144 = 2.81 \text{ ft}^2$.

2. Compute the shear acting on the critical shear section. The load combinations from ACI Code Table 5.3.1 will be used to compute the total factored dead and live loads. The basic combination for gravity load is

$$U = 1.2(D) + 1.6(L) + 0.5(L_r \text{ or } S \text{ or } R)$$

The slab is not a roof slab, and hence, the roof loads in the last brackets in this load combination do not apply, leaving

$$U = 1.2D + 1.6L$$

$$U = 1.2 \times 120 \text{ psf} + 1.6 \times 60 \text{ psf} = 240 \text{ psf}(>1.4\text{D})$$

The factored shear force and stress on the critical shear section are:

$$V_u = 240 \text{ psf}(20 \text{ ft} \times 22 \text{ ft} - 2.81 \text{ ft}^2) = 105{,}000 \text{ lb}$$

and

$$v_u = V_u/b_o d = 105{,}000 \text{ lb}/(80.5'' \times 6.13'') = 213 \text{ psi}$$

Using the design equation,

$$\phi(v_c + v_s) \geq v_u$$

where $\phi = 0.75$ for shear and torsion.

For this square column with small dimensions, Eq. (13-24) governs:

$$\phi v_c = \phi 4\lambda\sqrt{4000} = 0.75 \times 4 \times 1 \times 63.2 = 190 \text{ psi}$$

Thus, shear reinforcement is required at the critical shear section. Using shear studs as shear reinforcement results in

$$v_c = 3\lambda\sqrt{4000} = 3 \times 1 \times 63.2 = 190 \text{ psi}$$

This gives

$$v_s \geq v_u/\phi - v_c = 213/0.75 - 190 = 284 - 190 = 94 \text{ psi}$$

The maximum value of ϕv_n allowed with headed-shear studs is

$$\phi v_n = \phi \times 8\sqrt{f'_c} = 0.75 \times 8 \times 63.2 = 379 \text{ psi} > 284 \text{ psi} \qquad \text{(o.k.)}$$

Based on experimental results reported in [13-37], the author would recommend reducing v_c to $2\lambda\sqrt{f'_c}$ (126 psi) for the arrangement of shear studs shown in Fig. 13-69. With this value for v_c, the required value for v_s becomes **158 psi**. In the following steps this value will be used.

3. Lay out the punching shear reinforcement. Rows of shear studs welded to bars will be placed parallel and perpendicular to the main slab reinforcement to cause the least disruption in the placement of the main slab steel. Using a trial and error process:

Try eight stud rails, each with seven 1/2-in.-diameter studs ($A_b = 0.20$ in.^{2}) with 1.6-in.-diameter heads, and $f_{yt} = 51$ ksi (typical for shear studs).

The spacing to the first set of shear studs, s_o, is to be taken less than or equal to $d/2$. Thus, select $s_o = 3$ in. (Fig. 13-69). Before proceeding to layout the subsequent rows of shear studs, the shear strength should be checked at the inner critical section. In Eq. (13-29), the area provided by the inner row of shear studs is

$$A_v = 8 \times A_b = 8 \times 0.20 \text{ in.}^2 = 1.60 \text{ in.}^2$$

Then, assuming only one line of shear studs are crossed by the potential critical shear crack nearest the column (Fig.13-68)

$$V_s = A_v f_{yt} = 1.60 \text{ in.}^2 \times 51 \text{ ksi} = 81.6 \text{ k} = 81{,}600 \text{ lb}$$

and

$$v_s = V_s/b_o d = 81{,}600 \text{ lb}/(80.5'' \times 6.13'') = 165 \text{ psi}$$

This exceeds the required value of 158 psi.

Although the spacing between subsequent shear studs along the strips perpendicular to the column faces depends on the shear stress at each successive critical section, the author prefers to use a constant spacing of 3 in. (approximately $d/2$).

Try eight stud rails with the first stud located at $s_o = 3$ in. from the column face. Subsequent studs are at a spacing of $s = 3$ in. with seven 1/2-in.-diameter headed shear studs per rail.

The outermost studs are at 3 in. + 6 × 3 in. = 21 in. from the face of the column, and the outer critical section is $(21 + d/2) = 24.1$ in. from the face of the column.

The outer critical-shear section is a series of straight line segments passing through points located $d/2$ outside the outer shear studs, as shown in Fig. 13-69. The perimeter of this *peripheral line* is

$$b_o = 4[(\sqrt{2} \times 24.1) + 14] = 192 \text{ in.}$$

The area inside this line $= [4(24.1 \times 24.1)/2 + 4 \times (14 \times 24.1) + (14 \times 14)] \text{ in.}^2$
$= 2710 \text{ in.}^2 = 18.8 \text{ ft}^2$

4. Check the shear stresses on the outer critical section. The factored shear force on the concrete at the outer critical section is

$$V_u = 240 \text{ psf} \times (20 \text{ ft} \times 22 \text{ ft} - 18.8 \text{ ft}^2) = 101{,}000 \text{ lb}$$

and

$$v_u = \frac{101{,}000}{192 \times 6.13} = 85.9 \text{ psi}$$

The shear stress on the outer critical section is limited to $\phi v_c = \phi 2\lambda\sqrt{4000} = 94.9$ psi. **Because this is larger than v_u, the design is o.k.** ■

13-11 COMBINED SHEAR AND MOMENT TRANSFER IN TWO-WAY SLABS

Slab–Column Connections Loaded with Shear and Moment

When lateral loads or unbalanced gravity loads cause a transfer of moment between the slab and column, the behavior is complex—involving flexure, shear, and torsion in the portion of the slab attached to the column, as shown in Figs. 13-70a and 13-71a. Depending on the relative strengths in these three modes, failures can take various forms. Research on moment and shear transfer in slabs is reviewed in [13-27] and [13-39].

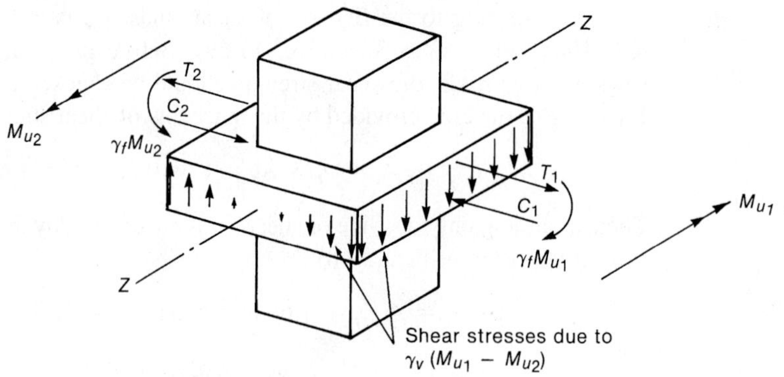

(a) Transfer of unbalanced moments to column.

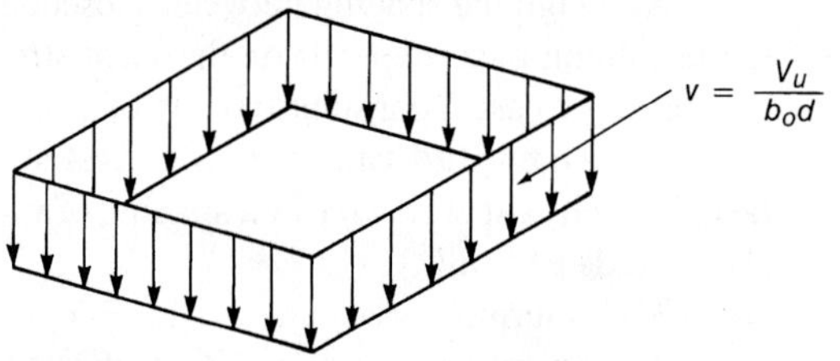

(b) Shear stresses due to V_u.

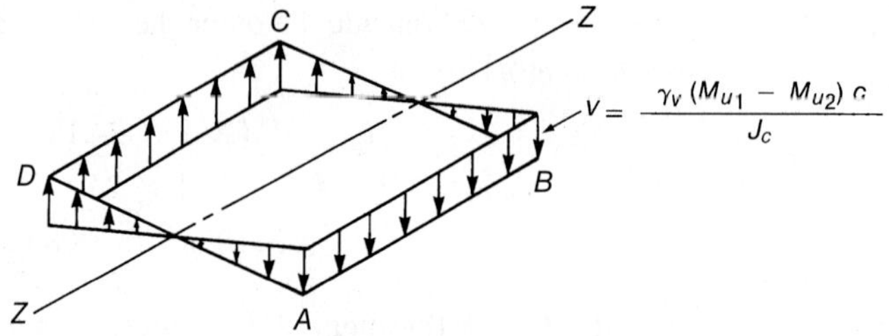

(c) Shear due to unbalanced moment.

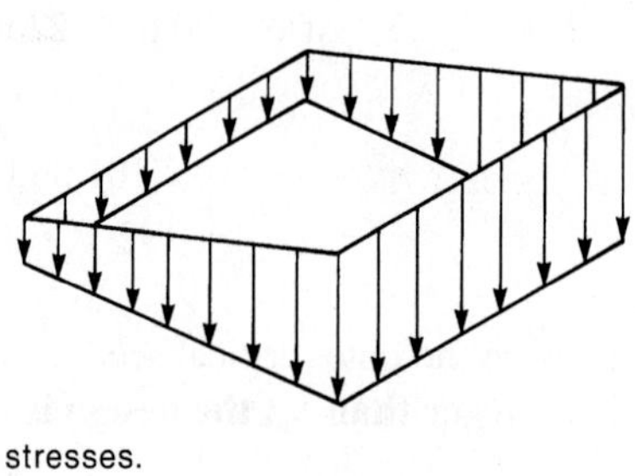

(d) Total shear stresses.

Fig. 13-70
Shear stresses clue to shear and moment transfer.

Methods for designing slab–column connections transferring shear and moment are reviewed briefly, followed by an example using the design method presented in the ACI Commentary.

1. Beam method. The most fundamental method of designing for moment and shear transfer to columns in the literature considers the connection to consist of the joint region plus short portions of the slab strips and beams projecting out from each face of the joint where there is an adjacent slab [13-39]. These strips are loaded by moments, shears, and torsions, and they fail when the combined stresses correspond to one or more of the individual failure modes of the members entering the joint. For an edge beam, the model

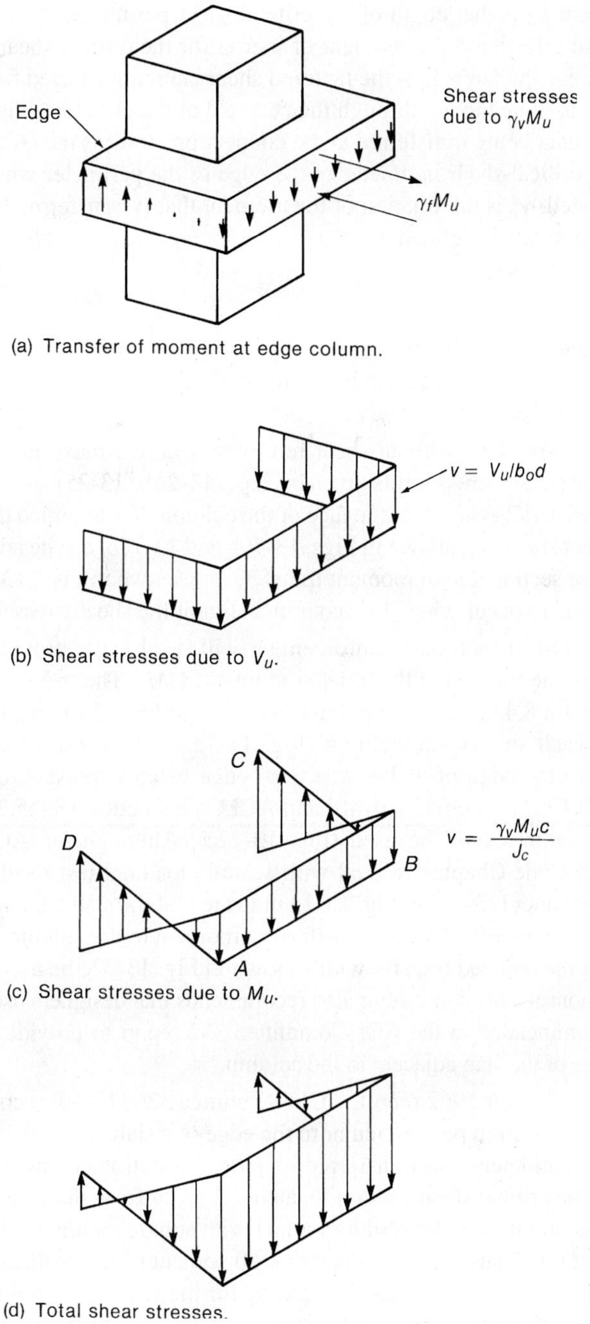

Fig. 13-71
Shear stresses due to shear and moment transfer at an edge column.

gives an estimate of the final distribution of the moments to the column. The beam model is conceptual only and has not been developed sufficiently for use in design.

2. Traditional ACI Commentary Design Method (J_c Method). The traditional ACI method used to calculate the maximum shear stress on the critical section surrounding the column for slab–column connections transferring both shear and moment is given by

$$v_u = \frac{V_u}{b_o d} \pm \frac{\gamma_v M_u c}{J_c} \tag{13-30}$$

where b_o is the length of the critical shear perimeter, d is the effective slab depth, and J_c is an effective polar moment of inertia for the critical shear section, as will be defined in more detail later. V_u is the factored shear being transferred from the slab to the column, and it is assumed to act through the centroid of the critical section for shear. M_u is the factored moment being transferred at the connection, and c is the measurement from the centroid of the critical-shear perimeter to the edge of the perimeter where the stress, v_u, is being calculated. γ_v is the fraction of the moment that is transferred by shear stresses on the critical section and is defined as

$$\gamma_v = 1 - \gamma_f \tag{13-31}$$

where γ_f is the fraction of the moment that is transferred by direct flexure. Reinforcement already designed for flexure in this region can be used to satisfy all or part of this strength requirement.

For slabs without shear reinforcement, the maximum value of v_u from Eq. (13-30) must satisfy stress limits given by Eqs. (13-24), (13-25), and (13-26) on a critical section located at $d/2$ away from the face of the column. It is assumed that the shear stresses due to the direct shear, V_u, shown in Figs. 13-70b and 13-71b, can be added to the shear stresses on the same section due to moment transfer, M_u, shown in Figs. 13-70c and 13-71c. Failure is assumed to occur when the maximum sum of the shear stresses reaches a limiting value ϕv_c.

Sufficient slab reinforcement will need to be provided within a *transfer width* to carry the fraction of the transfer moment, $\gamma_f M_u$. The *transfer width* is defined in ACI Code Section 8.4.2.3.3 as the column width, c_2, plus 1.5 times the slab or drop-panel thickness on each side of the column (Fig. 13-72a and b). For moments transferred perpendicular to the edge of a slab where no edge beam is present, design recommendations from ACI Committee 352 [13-40] and ACI Code Section 18.4.5.2 require that a narrower transfer width should be used (Fig. 13-72c). Although the ACI Committee 352 report and ACI Code Chapter 18 deal with seismic loading, test results reported by Simmonds and Alexander [13-41] in Fig. 13-73 indicate that the slab bars not anchored in the column will not be very effective for transferring moment to the column. Thus, the author recommends that the reduced transfer width shown in Fig. 13-72c be used for all design cases—seismic or nonseismic. The author also recommends that designers use the slab reinforcement details recommended in the ACI Committee 352 report to provide some torsional strength to the edge of the slab adjacent to the column.

In their 1962 report, ACI Committee 326 [13-30] recommended that when moments are transferred perpendicular to the edge of a slab, $\gamma_v = 0.20$. This implied that 80 percent of the moment was transferred by moment reinforcement and 20 percent by shear stresses on the critical shear section located $d/2$ away from the face of the column. Based on their tests on interior slab–column joints with square columns, Hanson and Hanson [13-42] set γ_f at 0.6. Thus, they assumed that 60 percent of the moment was transferred by slab flexural reinforcement. This doubled γ_v for the slab compared to that recommended for edge connections, thereby doubling the shear stresses due to moment transfer.

ACI Code Section 8.4.2.3.2 defines the fraction of the moment transferred by flexure, γ_f, as

$$\gamma_f = \frac{1}{1 + \left(\frac{2}{3}\right)\sqrt{b_1/b_2}} \tag{13-32}$$

where b_1 is the total width of the critical section measured perpendicular to the axis about which the moment acts, and b_2 is the total width parallel to the axis (Fig. 13-74).

For a square critical section where $b_1 = b_2$, γ_f is 0.60 and $\gamma_v = 1 - \gamma_f$ is 0.40, meaning that 60 percent of the unbalanced moment is transferred to the column by slab

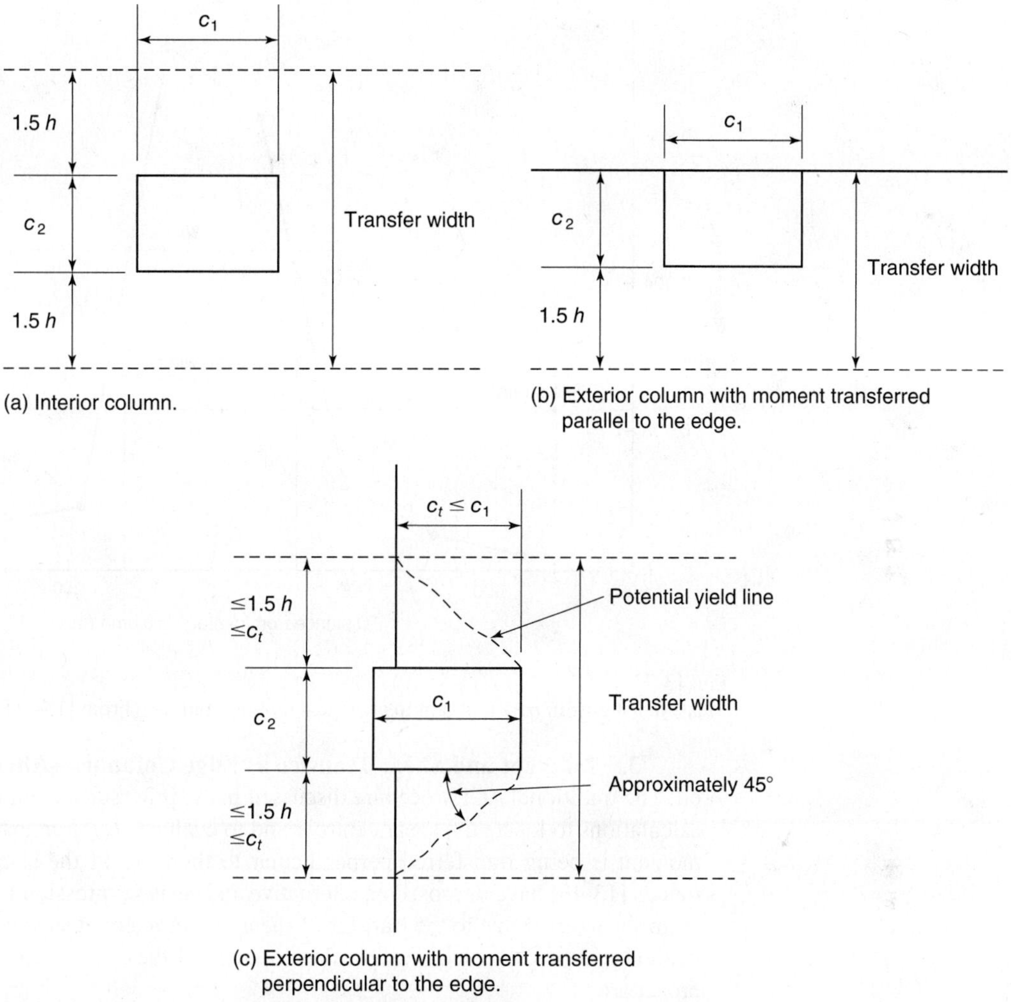

Fig. 13-72
Definition of transfer width at slab–column connections.

flexural reinforcement and 40 percent by eccentric shear stresses. Equation (13-32) was derived to give $\gamma_f = 0.6$ for $b_1 = b_2$, as proposed by Hanson and Hanson [13-42] to provide a transition to $\gamma_f = 1.0$ for a slab attached to the side of a wall and to γ_f approaching zero for a slab attached to the end of a long wall.

Based on test results, ACI Committee 352 [13-40] showed that a lower-bound interaction diagram for combined moment, M, and shear, V, can be represented by the two dashed lines in Fig. 13-75. The area inside those lines does not include any failures. This relationship can be used to modify the value of γ_f so that the fraction of the moment assumed to be transferred by flexure is increased while the fraction transferred by shear is reduced.

ACI Code Section 8.4.2.3.4 allows the designer to increase γ_f from 0.6 to 1.0 if the factored shear stress, v_u, on the connection is less than 0.75 times the design shear stress, ϕv_c, found using the minimum v_c from Eqs. (13-24), (13-25), and (13-26). The same increase of γ_f is permitted at corner connections when v_u is less than or equal to $0.5 \times \phi v_c$. A 25 percent increase of γ_f is permitted at interior connections when v_u is less than or equal to $0.4 \times \phi v_c$.

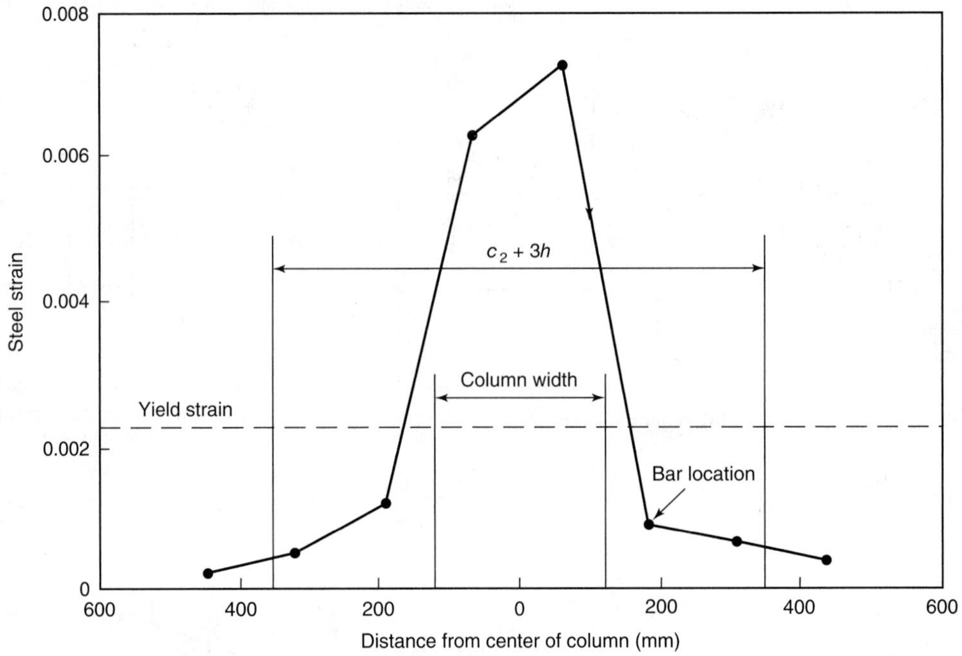

Fig. 13-73
Measured strains in bars adjacent to an edge column at failure. (From [13-41].)

3. Moment and Shear Transfer at Edge Columns—Alternate Analysis Method. The traditional ACI procedure discussed in the prior subsection involves some detailed calculations to locate the shear centroid and to evaluate J_c for an exterior connection when moment is being transferred perpendicular to the edge of the slab. Alexander and Simmonds [13-43] have proposed an alternative and easier expression for evaluating the maximum shear stress due to the transfer of shear and moment at such edge connections. Their method will be presented after the development of the expressions for the traditional ACI procedure.

Properties of Critical Shear Perimeters

The ACI design procedure for analysis of shear and moment transfer at slab–column connections applies to the interior, edge, and corner connections shown in Fig. 13-74 and uses Eq. (13-30) to calculate the stresses on the critical shear section. When calculating the section properties of various shear perimeters, the perimeters generally are subdivided into a number of individual sides similar to the one shown in Fig. 13-76.

The term, J_c, is defined by the ACI Commentary as a property analogous to a *polar moment of inertia* of the shear perimeter at the connection. This term is used to account for torsions and shears on the faces of the shear perimeter. To compute J_c, the critical shear section is divided into two, three, or four individual sides, as shown in Fig. 13-74. The term J_c is the sum of the individual, effective polar moment of inertia values computed for each of the sides.

Polar Moment of Inertia, J_c, for an Isolated Side of a Critical Shear Perimeter

Textbooks on statics or mechanics of material discuss the calculation of the moments of inertia and polar moments of inertia of rectangular areas that are similar to the individual

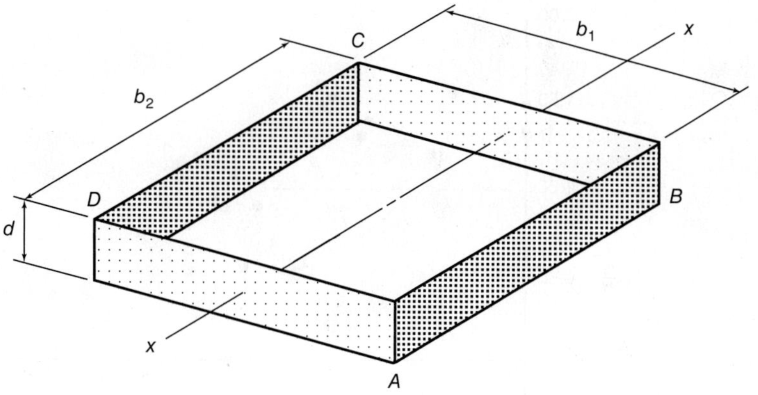

(a) Critical perimeter of an interior column.

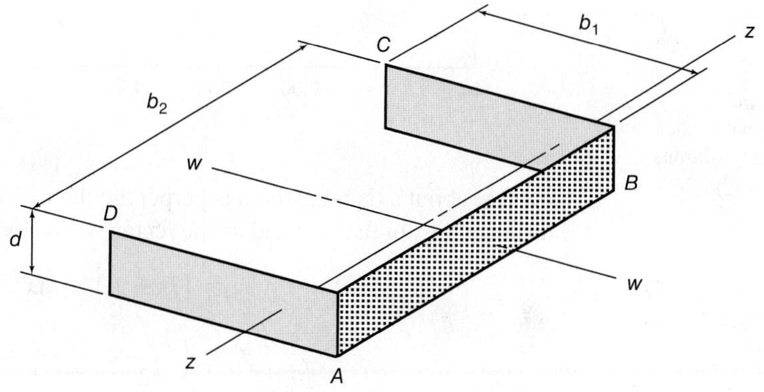

(b) Critical perimeter of edge column.

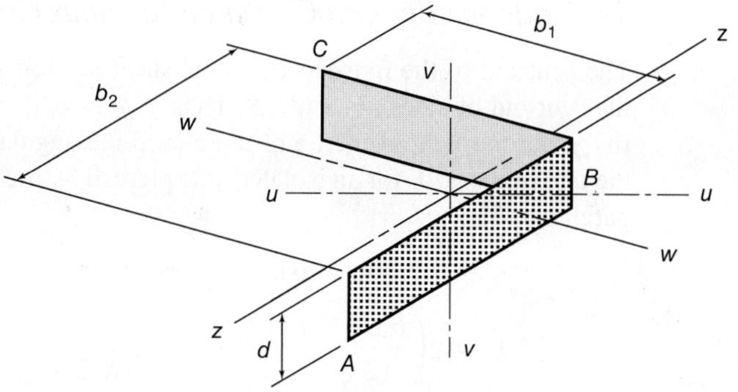

(c) Critical shear perimeter of corner column.

Fig. 13-74
Critical shear perimeters for bending about the indicated axes.

sides of the critical section around a column. Normally, these discussions of moments of inertia are limited to cases where the surface of the area under consideration can be plotted on a flat sheet of paper. The concept becomes ill-defined when it is applied to the open, three-dimensional critical shear sections shown in Fig. 13-74.

To compute the moments of inertia of a critical shear perimeter, the critical section is broken down into two to four individual plates. The rectangle in Fig. 13-76 represents side *A–D* of the three-sided shear perimeter in Fig. 13-74b. The polar moment of inertia of this

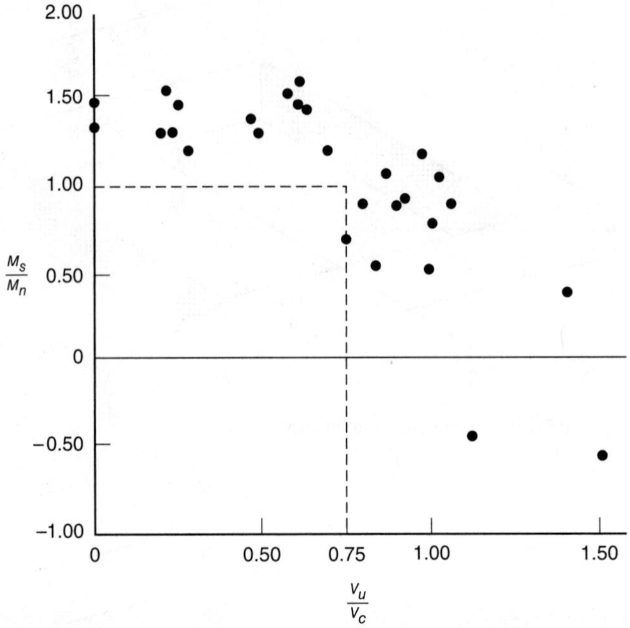

Fig. 13-75
Interaction between shear and moment at edge columns.

rectangle about axis z–z, which is perpendicular to the plane of the rectangle and displaced a distance $\bar{x}$ from the centroid of the rectangle, is given by

$$J_c = (I_x + I_y) + A\bar{x}^2$$

$$J_c = \left(\frac{bd^3}{12} + \frac{db^3}{12}\right) + (bd)\bar{x}^2 \tag{13-33}$$

Section Properties of a Closed Rectangular Critical Shear Section

The centroid of the four-sided critical shear section shown in Fig. 13-74a passes through the centroid of sides DA and CB. Thus, $c_{AB} = c_{CD} = b_1/2$. The calculation for the effective polar moment of inertia of the closed rectangular section in Fig. 13-74a is similar to the derivation of J_c for an isolated side pierced by the axis x–x, plus terms for the two sides parallel to the axis.

$$J_c = J_c(\text{about } x\text{-axis for faces } AD \text{ and } BC) + A\bar{x}^2(\text{faces } AB \text{ and } CD)$$

$$J_c = 2\left(\frac{b_1 d^3}{12} + \frac{db_1^3}{12}\right) + 2(b_2 d)\left(\frac{b_1}{2}\right)^2 \tag{13-34}$$

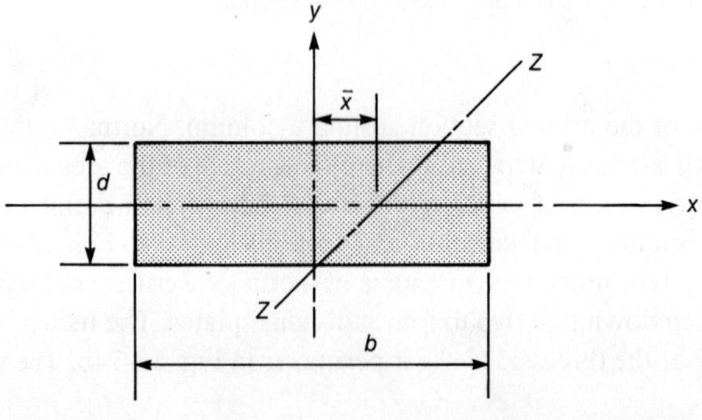

Fig. 13-76
Rectangle considered in computing polar inertia of a side.

where

$$b_1 = c_1 + 2(d/2)$$

$= $ length of the sides of the shear perimeter perpendicular to the axis of bending

$$b_2 = c_2 + 2(d/2)$$

$= $ length of the sides of the shear perimeter parallel to the axis of bending

$c_1 = $ width of column perpendicular to the axis of bending

$c_2 = $ width of column parallel to the axis of bending

If large openings are present adjacent to the column, the shear perimeter will be discontinuous, as shown in Fig. 13-56a. If this occurs, the calculation of the location of the centroid and J_c should include the effect of the openings.

Edge Columns

In the case of an edge column with the moment acting about an axis parallel to the edge, the centroid of the critical perimeter lies closer to the inside face of the column than to the outside face (Fig. 13-74b). As a result, the shear stresses due to the moment shown in Fig. 13-71c are largest at the outside corners of the shear perimeter. If M_u is large and V_u is small, a negative shear stress may occur at these points. If M_u due to the combination of lateral loads and gravity loads is positive on this joint (rather than negative as shown in Fig. 13-71c), the largest shear stress will occur at the outside corners.

Moments about an axis parallel to the edge (axis z–z in Fig. 13-74b). For the three-sided perimeter shown in Fig. 13-74b and taking b_1 as the length of the side perpendicular to the edge, as shown, the location of the centroidal axis z–z is calculated as follows:

$$c_{AB} = \frac{\text{moment of area of the sides about } AB}{\text{area of the sides}}$$

$$= \frac{2(b_1 d)b_1/2}{2(b_1 d) + b_2 d} \tag{13-35}$$

The polar moment J_c of the shear perimeter is given by

$$J_c = J_c(\text{about } z\text{-axis for faces } AD \text{ and } BC) + A\bar{x}^2(\text{face } AB)$$

$$J_c = 2\left[\frac{b_1 d^3}{12} + \frac{d b_1^3}{12} + (b_1 d)\left(\frac{b_1}{2} - c_{AB}\right)^2\right] + b_2 d c_{AB}^2 \tag{13-36}$$

Moments about an axis perpendicular to the edge (axis w–w in Fig. 13-74b). When considering this case, the definitions of b_1 and b_2, as shown in Fig. 13-74b, should be reversed for use in Eq. (13-32). In the following, however, the values of b_1 and b_2 will be as shown in Fig. 13-74b. For this case, the location of the centroid is $c_{BC} = c_{AD} = c = b_2/2$. The effective polar moment of inertia about the w-axis for the three-sided section in Fig. 13-74b is

$$J_c = J_c(\text{about } w\text{-axis for face } AB) + A\bar{x}^2(\text{faces } BC \text{ and } AD)$$

$$J_c = \left(\frac{b_2 d^3}{12} + \frac{d b_2^3}{12}\right) + 2(b_1 d)c^2 \tag{13-37}$$

Corner Columns

For the two-sided perimeter shown in Fig. 13-74c with sides b_1 and b_2, the location of the centroidal axis z–z is

$$c_{AB} = \frac{(b_1 d) b_1/2}{b_1 d + b_2 d} \tag{13-38}$$

The polar moment of inertia of the shear perimeter is

$$J_c = J_c(\text{about } z\text{-axis for face } BC) + A\bar{x}^2(\text{face } AB)$$

$$J_c = \left[\frac{b_1 d^3}{12} + \frac{d b_1^3}{12} + (b_1 d)\left(\frac{b_1}{2} - c_{AB}\right)^2\right] + (b_2 d) c_{AB}^2 \tag{13-39}$$

Circular Columns

For combined shear and moment calculations, ACI–ASCE Committee 426 [13-27] recommended that the shear perimeter of circular columns be based on that of a square column with the same centroid and the same area. In this case, the equivalent square column would have sides of length $c = \sqrt{\pi}\, d_c/2 = 0.886 d_c$, where d_c is the diameter of the column. Two cases are illustrated in Fig. 13-77.

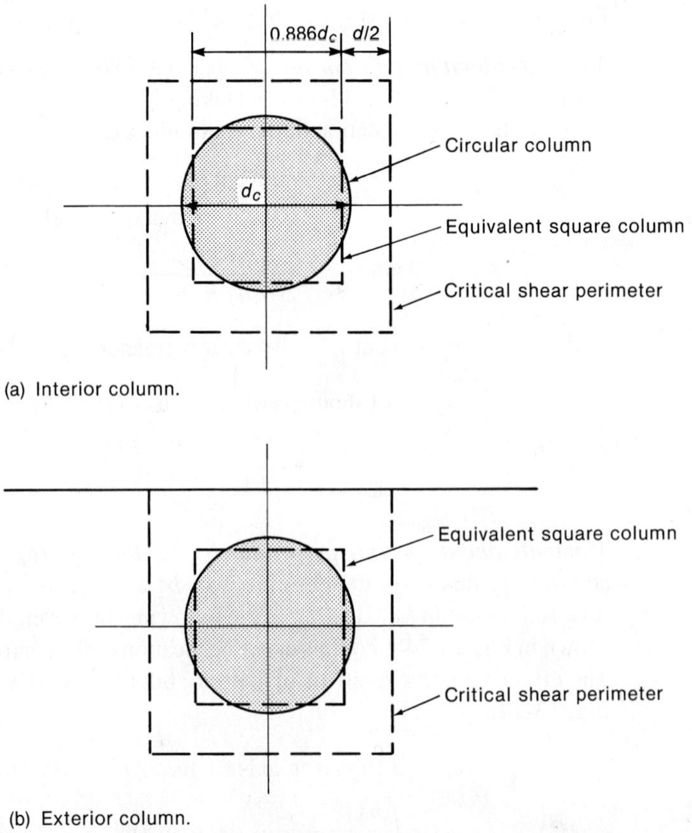

Fig. 13-77
Critical shear perimeters for moments and shear transfer at circular columns.

Load Patterns for Maximum Shear Stress due to Combined Shear and Moment Transfer

The maximum shear stress on the critical sections around columns supporting a slab should be computed from a consistent load case that produces a sum of shear stresses due to gravity load and moment that is likely to be a maximum. The words *consistent load case* imply a combination of loads that are likely to occur together.

For interior columns. The maximum shear stress generally occurs when all of the panels of the floor or roof surrounding a column are loaded with the full-factored dead and live loads. However, if two adjacent spans have significantly different lengths, a second loading case where only the longer, adjacent span is loaded with half of the factored live load should be considered. This loading case will produce a larger transfer moment, but a lower shear force.

For edge columns. (a) For the case of moments acting about an axis parallel to the edge, the critical loading for maximum shear stress occurs with full-factored dead and live loads acting simultaneously on both edge panels adjoining the column. (b) For the case of moments acting about an axis perpendicular to the edge, two cases may need to be considered. For one case, full-factored dead and live loads should be applied to both adjacent panels. This case will produce the maximum shear force, but a smaller transfer moment. For the other case, half of the factored live load should be applied only to the longer of the two adjacent panels. This case will produce a larger transfer moment, but a lower shear force.

For corner columns. The maximum shear stress normally occurs due to the load case where full-factored dead and live load are applied to all panels of the floor system.

EXAMPLE 13-13 Recheck Interior Column Connection from Example 13-11 for Combined Shear and Moment Transfer

Example 13-11 evaluated the shear strength of the interior slab-column connection shown in Fig. 13-61. For that problem, live load was assumed to act on the entire tributary area to give the maximum shear force to be transferred. For this problem, we will include the slab-transfer moment calculated using Eq. (13-14), which assumes a pattern live load. To be consistent with that assumed load case, the shear force to be transferred at this connection will need to be recalculated.

(a) Calculate factored shear and moment to be transferred at this connection. We will check the case when the unbalanced moment is transferred about the weaker column axis. A summary for moment transfer about the strong column axis is given at the end. When using Eq. (13-14), it is assumed that the factored dead load is acting on both sides of the column and fifty percent of the live load is acting only on the longer span adjacent to the column. For this connection, the span lengths are the same (18 ft) on both sides of the column. Using the notation from Eq. (13-14), the factored dead and live loads, and the factored shear force to be transferred are:

$$q_{Du} = q_{Du}' = 1.2\left(\frac{6}{12} \times 150 + 15\right) = 108 \text{ psf}$$

$$q_{Lu} = q_{Lu}' = 1.6 \times 60 = 96 \text{ psf}$$

and ignoring the area inside the critical perimeter,

$$V_u = q_{Du}(18' \times 18') + q_{Lu}(18' \times 9')$$

$$V_u = 35{,}000 \text{ lb} + 15{,}600 \text{ lb} = 50{,}600 \text{ lb}$$

The transfer moment is calculated using Eq. (13-14), with $\ell_2 = \ell_2'$ and $\ell_n = \ell_n' = 18 - 1 = 17$ ft:

$$M_u = M_{\text{slab}} = 0.07[(q_{Du} + 0.5q_{Lu})\ell_2\ell_n^2 - q_{Du}\ell_2'(\ell_n')^2]$$

Because the spans are equal, this reduces to:

$$M_u = 0.07(0.5q_{Lu})\ell_2\ell_n^2 = 0.07 \times 0.5 \times 96 \text{ psf} \times 18 \text{ ft} \times (17 \text{ ft})^2$$

$$M_u = 17{,}500 \text{ lb-ft} = 210{,}000 \text{ lb-in.}$$

(a) Calculate b_1, b_2, b_o, γ_f, γ_v, J_c and c. From Fig. 13-63:

$$b_1 = 16.75 \text{ in.}$$

$$b_2 = 30.75 \text{ in.}$$

$$b_o = 2(b_1 + b_2) = 2 \times 47.5 = 95.0 \text{ in.}$$

From Eq. (13-32):

$$\gamma_f = \frac{1}{1 + (2/3)\sqrt{b_1/b_2}} = \frac{1}{1 + 0.667\sqrt{16.75/30.75}} = 0.67$$

From Eq. (13-31):

$$\gamma_v = 1 - \gamma_f = 1 - 0.67 = 0.33$$

From Fig. 13-74a, $c = b_1/2 = 8.38$ in., and from Eq. (13-34) using $d = 4.75$ in.:

$$J_c = 2\left(\frac{b_1d^3}{12} + \frac{db_1^3}{12}\right) + 2(b_2d)\left(\frac{b_1}{2}\right)^2$$

$$J_c = 2(150 + 1860) + 20{,}500 = 24{,}500 \text{ in.}^4$$

(c) Calculate the maximum v_u and compare it to ϕv_c. The maximum shear stress acting on the critical section is found using Eq. (13-30):

$$v_u = \frac{V_u}{b_od} + \frac{\gamma_v M_u c}{J_c} = \frac{50{,}600 \text{ lb}}{95'' \times 4.75''} + \frac{0.33 \times 210{,}000 \text{ lb-in.} \times 8.38 \text{ in.}}{24{,}500 \text{ in.}^4}$$

$$v_u = 112 \text{ psi} + 24 \text{ psi} = 136 \text{ psi}$$

In step (b) of Example 13-11, it was found that $\phi v_c = 158$ psi

(d) Check for combined and moment transfer about other principal axis. These calculations are very similar to those just presented. Because the larger column dimension in this direction, ℓ_n is shorter and M_u is smaller. However, the b_1/b_2 ratio changes, making γ_f smaller and γ_v larger. The final result is very similar to that found in part (c) with $\phi v_c > v_u$.

So, because ϕv_c exceeds v_u in both principal directions, the slab is o.k. for transfer of shear and moment.

Calculation of Moment about Centroid of Shear Perimeter

The distribution of stresses calculated from Eq. (13-30) and illustrated in Fig. 13-71 assumes that V_u acts through the centroid of the shear perimeter and that M_u acts about the centroidal axis of this perimeter. When using structural analysis software, the values of V_u and M_u may be calculated either at a node point located at the centroid of the column or at the face of the column supporting the slab. For the direct-design method, V_u and M_u normally are calculated at the face of the column. The only exception is for the transfer moment at an edge column where ACI Code Section 8.10.7.3 sets that value equal to 0.3 M_o and it is assumed to act at the centroid of the critical shear perimeter.

After the centroid of the critical shear perimeter has been determined, equilibrium of a simple free body diagram can be used to determine the values for V_u and M_u acting at the centroid. Figure 13-78a represents the case of an exterior column with the shear and moment calculated initially to be acting at the face of the column. An equilibrium calculation can be used to determine the resulting moment and shear at the centroid of the critical perimeter, as indicated in Fig. 13-78a. A similar calculation can be used if the shear and moment are initially calculated to act at the centroid of the column, as shown in Fig. 13-78b. For an interior column, the centroid axis for the critical shear perimeter should pass through the column centroid. Thus, if the shear and moment are initially determined at the faces of an interior column, the equilibrium calculation shown in Fig. 13-78c can be used to determine the shear and moment at the centroid of the shear perimeter. If, due to openings near an interior column the centroid of the shear perimeter does not coincide with the centroid of the column, simple adjustments can be made to the calculations shown in Fig. 13-78c.

Consideration of Moment Transfer in Both Principal Directions

Equation (13-30) for checking combined transfer of shear and moment at a slab–column connection initially was developed in the 1960s. This expression was derived assuming the transfer of moment from one principal direction at a time. For typical building applications, the common approach is for a structural engineer to use this equation to check the maximum shear stress on the perimeter of the critical shear section resulting from the transfer shear and moment in only one principal direction at a time. Of course, this check needs to be made in both orthogonal principal directions using consistent load cases, as was discussed in Example 13-13.

Occasionally, due to unusual span lengths and columns layouts, a structural designer may decide that it is appropriate to check the maximum shear stress due to moments acting simultaneously about both principal axes. For such a case, a third term would be added to Eq. (13-30) as shown in the following expression.

$$v_u = \frac{V_u}{b_o d} \pm \frac{\gamma_v M_{u1} c}{J_{c1}} \pm \frac{\gamma_v M_{u2} c}{J_{c2}} \tag{13-40}$$

Where V_u, M_{u1}, and M_{u2} refer to the shear and moments about the two principal axes for a *consistent load case*, and J_{c1} and J_{c2} (and the corresponding values of c) refer to properties of the critical shear perimeter, as defined previously (Fig. 13-74). It should be noted that this procedure would result in the calculation of a maximum shear stress at a point, while the ACI Code requirements refer only to stress limits over some defined area. Thus, if this calculation procedure is used, a structural designer will need to use engineering judgment

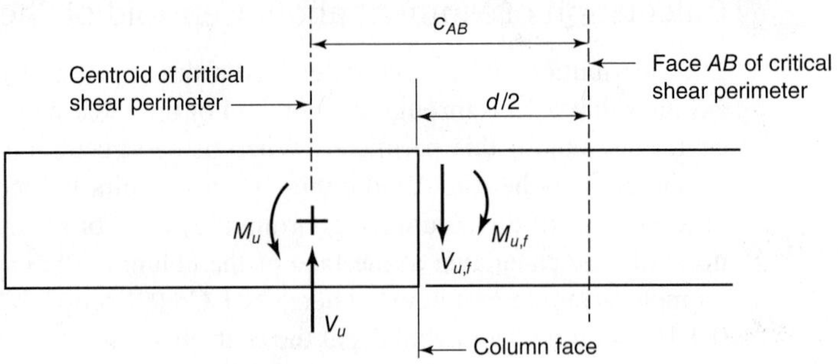

$V_u = V_{u,f}$

$M_u = M_{u,f} + V_{u,f}\,(c_{AB} - d/2)$

(a) Exterior column, starting from moment and shear at face.

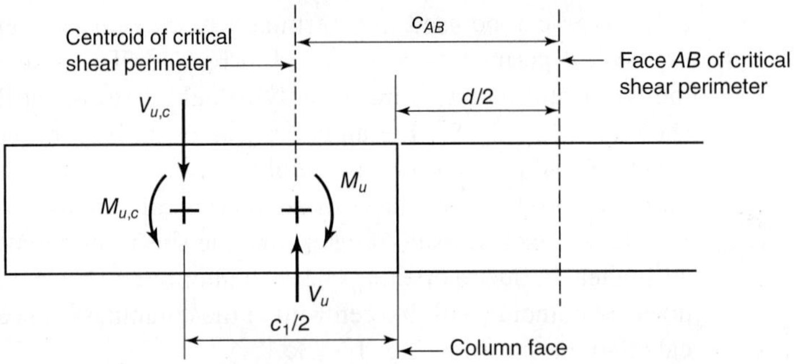

$V_u = V_{u,c}$

$M_u = M_{u,c} + V_{u,c}\,[c_1/2 - (c_{AB} - d/2)]$

(b) Exterior column, starting from moment and shear at column centroid.

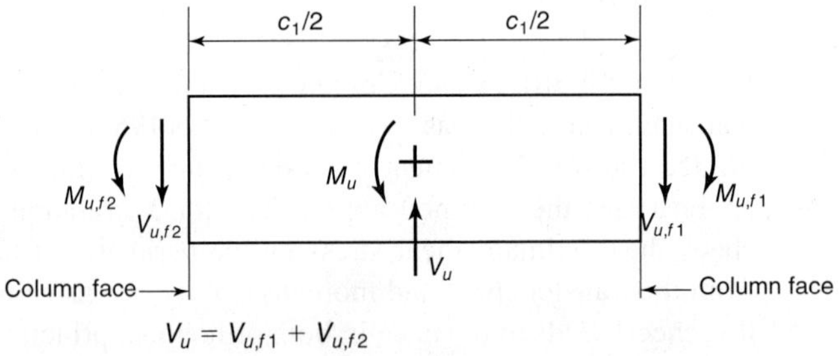

$V_u = V_{u,f1} + V_{u,f2}$

$M_u = M_{u,f1} - M_{u,f2} + (V_{u,f1} - V_{u,f2}) \times c_1/2$

(c) Interior column, starting from moments and shears at column faces.

Fig. 13-78 Calculation of transfer moment and shear at centroid of critical perimeter.

when comparing the calculated maximum shear stress to the ACI Code limit of ϕv_n (or more typically ϕv_c). In Appendix B of the ACI Committee 421 report [13-36], an overstress of 15 percent is assumed to be acceptable due to expected stress redistribution away from the most highly stressed corner of the critical perimeter.

Alternative Analysis of Maximum Shear Stress due to Combined Shear and Moment Transfer at Exterior Connections

In a recent article [13-43], Alexander and Simmonds presented the following expression for determining the maximum shear stress on a critical section at the front face of an exterior column resulting from transfer of shear plus moment acting about an axis parallel to the edge of the slab.

$$v_u = 0.65 \times \frac{V_u + \dfrac{M_{u,\text{face}}}{4d}}{(c_1 + c_2)d} \tag{13-41}$$

V_u is the factored shear to be transferred, $M_{u,\text{face}}$ is the factored moment at the face of the column, d is the effective depth of the slab, and c_1 and c_2 are the column dimensions perpendicular and parallel to the edge of the slab (Fig. 13-72c). Alexander and Simmonds state that Eq. (13-41) is simple enough to be used for preliminary sizing of the slab or drop panel, and it provides relatively close agreement with the maximum stresses obtained from Eq. (13-30). A comparison of the results from Eqs. (13-41) and (13-30) will be given in Example 13-14.

Shear Reinforcement for Slab–Column Connections Transferring Shear and Moment

ACI Code Section 8.4.4.2.3 states that the shear stresses resulting from combined shear and moment transfer shall be assumed to vary linearly about the centroid of the critical section defined in ACI Code Section 22.6.4.1. ACI Code Tables 22.6.5.2 and 22.6.6.1 limit the maximum shear stress on the critical perimeter as follows:

(a) In slabs without shear reinforcement, the factored shear stress shall not exceed

$$\phi v_n = \phi v_c \tag{13-42}$$

where v_c is computed as the smallest of the shear strengths from Eqs. (13-24), (13-25), and (13-26).

(b) In slabs having shear reinforcement, the factored shear stress shall not exceed

$$\phi v_n = \phi(v_c + v_s) \tag{13-18}$$

where v_c is limited to $2\lambda\sqrt{f'_c}$ when using stirrup-type shear reinforcement and to $3\lambda\sqrt{f'_c}$ when using stud-type shear reinforcement. The shear v_s was previously defined in Eq. (13-29). ACI Code Table 22.6.6.1 limits the maximum shear stress to $2\lambda\sqrt{f'_c}$ at a critical section located $d/2$ outside the outermost line of stirrup legs or shear studs extending out from the column. As shown in Fig. 13-67, this outer critical shear perimeter may take on an unusual shape that is not covered by the J_c expressions for the critical perimeters shown in Fig. 13-74. Elgabry and Ghali [13-44] developed expressions for handling such sections that subsequently were used in the report from ACI Committee 421 [13-36].

EXAMPLE 13-14 Checking Combined Shear and Moment Transfer at an Edge Column

A 12-in.-by-16-in. column is located 4 in. from the edge of a flat slab without edge beams, as shown in Fig. 13-79. The slab is 6.5 in. thick, with an average effective depth of 5.5 in. The normal-weight concrete has a compressive strength of 3500 psi and the reinforcement

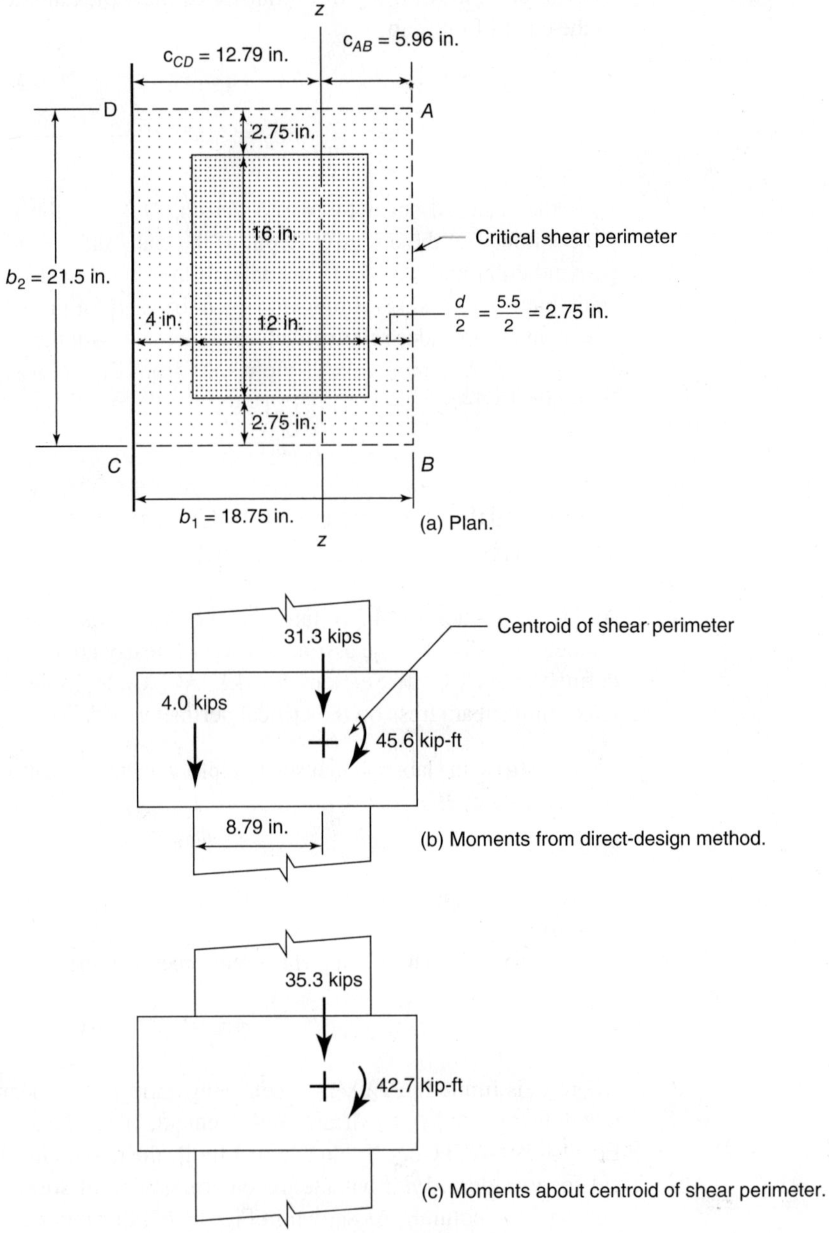

Fig. 13-79
Slab–column joint for Example 13-14.

yield strength is 60,000 psi. The direct-design method gives the statical moment, M_o, in the edge panel as 152 kip-ft. The shear from the edge panel is 31.3 kips. The portion of the slab outside the centerline of the column produces a factored shear force of 4.0 kips acting at 6 in. outside the center of the column. The loading that causes moment about an axis perpendicular to the edge is less critical than the moments about the *z–z* axis (axis parallel to the edge). Use load factors from ACI Code Table 5.3.1.

1. **Locate the critical shear perimeter.** The critical shear perimeter is located at $d/2$ from the sides of the column. Because the edge of the slab is less than $2h$ away from the outside face of the column, a perimeter similar to that shown in Fig. 13-56b is assumed to be critical.

2. Compute the centroid of the shear perimeter.

$$c_{AB} = \frac{\Sigma Ay}{A} \qquad \text{(where } y \text{ is measured from } AB\text{)}$$

$$= \frac{2(18.75 \times 5.5)(18.75/2)}{2(18.75 \times 5.5) + 21.5 \times 5.5} = 5.96 \text{ in.}$$

Therefore, $c_{AB} = 5.96$ in. and $c_{CD} = 12.79$ in.

3. Compute the moment about the centroid of the shear perimeter. For the portion of the slab that spans between the centerline of the edge column and the first interior column, ACI Code Section 8.10.7.3 defines the moment to be transferred at the edge column as $0.3M_o = 0.3 \times 152 = 45.6$ kip-ft. We shall assume that this is about the centroid of the shear perimeter and that V_u from the edge panel acts through this point. The portion of the slab outside the column centerline has a shear of $V_{uc} = 4$ kips acting at 6 in. + 2.79 in. from the centroid of the shear perimeter. The total moment about the centroid of the shear perimeter is

$$M_u = 45.6 \text{ kip-ft} - 4 \text{ kips} \times 8.79/12 \text{ ft} = 42.7 \text{ kip-ft}$$

The total shear to be transferred is

$$V_u = 31.3 \text{ kips} + 4 \text{ kips} = 35.3 \text{ kips}$$

4. Compute ϕv_c and $v_u/\phi v_c$. v_c is the smallest value given by Eqs. (13-24) through (13-26). From ACI Code Table 21.2.1, we will use the strength-reduction factor $\phi =$ 0.75. Check which equation for v_c governs. Consider the column shape and start with Eq. (13-25):

$$v_c = \left(2 + \frac{4}{\beta}\right)\sqrt{f_c} \tag{13-25}$$

where

$$\beta = \frac{\text{long side of the column}}{\text{short side of the column}} = \frac{16}{12} = 1.33$$

$$(2 + 4/1.33) = 5.01 > 4 \text{ (does not govern)}$$

Consider the size of the critical perimeter, so check Eq. (13-26):

$$v_c = \left(\frac{\alpha_s d}{b_o} + 2\right)\sqrt{f'_c} \tag{13-26}$$

where $\alpha_s = 30$ for an edge column and $b_o = 2 \times 18.75$ in. + 21.5 in. = 59 in.

Thus, $(30 \times 5.5/59) + 2 = 4.80 > 4$ (does not govern)

So, use Eq. (13-24) with a coefficient of 4:

$$v_c = 4\lambda\sqrt{f'_c}$$

$$\phi v_c = 0.75 \times 4 \times 1\sqrt{3500} = 177 \text{ psi}$$

$$v_u = V_u/b_od = 35{,}300 \text{ lb}/(59'' \times 5.5'') = 109 \text{ psi}$$

and

$$v_u/\phi v_c = 109/177 = 0.616$$

5. Determine the fraction of the moment transferred by flexure, γ_f.

$$\gamma_f = \frac{1}{1 + \frac{2}{3}\sqrt{\frac{b_1}{b_2}}} \tag{13-32}$$

$$= \frac{1}{1 + \frac{2}{3}\sqrt{\frac{18.75}{21.5}}}$$

$$= 0.616$$

ACI Code Section 8.4.2.3.4 allows γ_f to be increased up to 1.0 if $v_u/\phi v_c$ does not exceed 0.75 at an exterior column. From step 4, $v_u/\phi v_c = 0.616$. Therefore, try γ_f equal to 1.0 and check the reinforcement needed.

6. Design the reinforcement for moment transfer by flexure. From Fig. 13-79, $1.5h < c_t$ (12 in.). So,

$$\text{Width effective for flexure: } c_2 + 3h = 16 \text{ in.} + 3 \times 6.5 \text{ in.}$$

$$= 35.5 \text{ in.}$$

$$\text{Moment: } 1.0 \times 42.7 \text{ kip-ft} = 42.7 \text{ kip-ft} = 42{,}700 \text{ lb-ft}$$

Assume that $jd = 0.95d$. Then

$$A_s = \frac{M_u}{\phi f_y jd}$$

$$= \frac{42{,}700 \text{ lb-ft} \times 12 \text{ in./ft}}{0.9 \times 60{,}000 \text{ psi} \times 0.95 \times 5.5 \text{ in.}}$$

$$= 1.82 \text{ in.}^2$$

Try ten No. 4 bars = 2.00 in.^{2} Because this calculation is based on a guess of jd, we should compute a for $A_s = 2.00$ in.^{2}; then recompute A_s using that value of a:

$$a = \frac{A_S f_y}{0.85 f'_c b}$$

$$= \frac{2.00 \times 60{,}000}{0.85 \times 3500 \times 35.5}$$

$$= 1.14 \text{ in.} < 3d/8 = 2.06 \text{ in. (so, } \phi = 0.9)$$

$$A_s = \frac{M_u}{\phi f_y(d - a/2)} = \frac{42{,}700 \times 12}{0.9 \times 60{,}000(5.5 - 1.14/2)}$$

$$= 1.92 \text{ in.}^2 \quad \text{(o.k.)}$$

Although the use of ten No. 4 bars represents a possible design for the slab reinforcement at this exterior slab–column connection, the author believes the requirement in ACI Code Table 8.4.2.3.4, which states that ε_t must exceed 0.010 for the slab flexural design when γ_f is increased at an interior connection, *should be applied* to all slab–column connections where γ_f is increased. This requirement ensures that the slab transfer zone has adequate flexural ductility to allow load redistributions away from the connection in case of an accidental overload. Thus, before accepting this design, we will calculate ε_t (ε_s) for this design.

$$c = \frac{a}{\beta_1} = \frac{1.14 \text{ in}}{0.85} = 1.34 \text{ in.}$$

Using a linear strain distribution,

$$\varepsilon_t = \varepsilon_s = \frac{d - c}{c} \times \varepsilon_{cu}$$

$$= \frac{5.5 - 1.34}{1.34} \times 0.003 = 0.0093$$

Because ε_t is less than 0.010, we should consider selecting a value of γ_f less than 1.0 but greater than the value calculated in step 5 (0.616). We will select a value of γ_f equal to 0.8, and thus provide eight No. 4 bars to get $A_s = 1.60$ in.^{2} For this steel area, $a = 0.91$ in., $c = 1.07$ in., and $\varepsilon_t = 0.0124$ (o.k.). With these values,

$$\phi M_n = \phi A_s f_y (d - a/2) = 0.9 \times 1.60 \text{ in.}^2 \times 60 \text{ ksi} \times (5.5 \text{ in.} - 0.46 \text{ in.})$$

$$= 436 \text{ k-in.} = 36.3 \text{ k-ft} = \gamma_f M_u$$

Thus, the moment to be transferred by eccentric shear stress is,

$$\gamma_v M_u = 42.7 \text{ k-ft} - 36.3 \text{ k-ft} = 6.4 \text{ k-ft}$$

7. Compute the polar moment of inertia, J_c. Equation (13-36) applies to moments about the *z–z* axis. From that equation,

$$J_c = 2\left(\frac{b_1 d^3}{12}\right) + 2\left(\frac{d b_1^3}{12}\right) + 2(b_1 d)\left(\frac{b_1}{2} - c_{AB}\right)^2 + (b_2 d)c^2_{AB}$$

$$= \frac{2 \times 18.75 \times 5.5^3}{12} + \frac{2 \times 5.5 \times 18.75^3}{12}$$

$$+ 2(18.75 \times 5.5)\left(\frac{18.75}{2} - 5.96\right)^2 + (21.5 \times 5.5)5.96^2$$

$$= 13{,}200 \text{ in.}^4$$

8. Compute the shear stresses.

$$v_u = \frac{V_u}{b_o d} \pm \frac{\gamma_v M_u c}{J_c} \qquad (13\text{-}30)$$

$$v_u = \frac{35{,}300}{59 \times 5.5} \pm \left(\frac{6400 \times 12}{13{,}200}\right)c$$

$$= 109 \pm 5.82c$$

The shear stress at AB is ($c_{AB} = 5.96$ in.)

$$v_{u,AB} = 109 + 5.82 \times 5.96$$
$$= 144 \text{ psi}$$

The shear stress at CD is ($c_{CD} = 12.79$ in.)

$$v_{u,CD} = 109 - 5.82 \times 12.79$$
$$= 34.6 \text{ psi}$$

From step 4, $\phi v_c = 177$ psi. Because ϕv_c exceeds v_u, the shear is o.k. **Use a 12-in.-by-16-in. column, as shown in Fig. 13-79, with eight No. 4 bars at 5 in. on centers, centered on the column.**

9. Check accuracy of approximate method. This step is made just to check the approximate method from Alexander and Simmonds [13-43] for checking the maximum shear stress due to the transfer of shear and moment at an exterior connection. Assume $\gamma_v M_u$ from step 6 can be used for $M_{u,\text{ face}}$, then from Eq. (13-41);

$$v_u = 0.65 \times \frac{V_u + \dfrac{M_{u,\text{ face}}}{4d}}{(c_1 + c_2)d}$$

$$v_u = 0.65 \times \frac{35{,}300 + \dfrac{6400 \times 12}{4 \times 5.5}}{(12 + 16)5.5} = 164 \text{ psi}$$

This is approximately 15 percent higher than the stress calculated on face AB in step 8. For a quick design check of the slab thickness, this certainly would be accurate enough. Also, the accuracy of Eq. (13-41) may have been affected by the relatively low percentage of the moment transferred by eccentric shear stresses. ■

13-12 DETAILS AND REINFORCEMENT REQUIREMENTS

Drop Panels

Drop panels are thicker portions of the slab adjacent to the columns, as shown in Fig. 13-80 or Fig. 13-2c. They are provided for three main reasons:

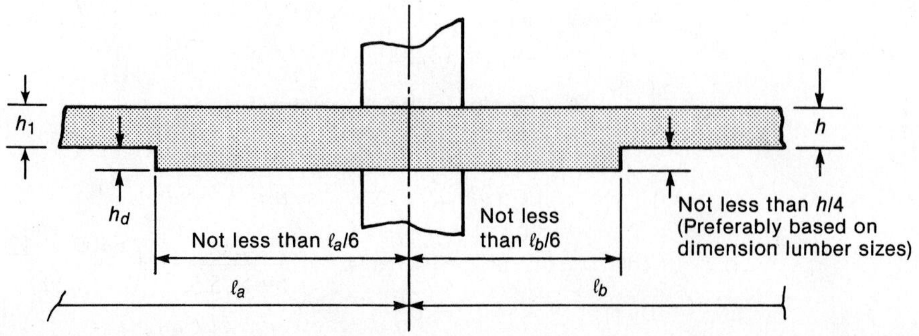

Minimum size of drop panels—ACI Section 13.2.5

Fig. 13-80 Drop panels.

1. The minimum thickness of slab required to limit deflections (see Section 13-6) may be reduced by 10 percent if the slab has drop panels conforming to ACI Code Section 8.2.4. The drop panel stiffens the slab in the region of highest moments and hence reduces the deflection.

2. A drop panel with dimensions conforming to ACI Code Section 8.2.4 can be used to reduce the amount of negative-moment reinforcement required over a column in a flat slab. By increasing the overall depth of the slab, the lever arm, jd, used in computing the area of steel is increased, resulting in less required reinforcement in this region.

3. A drop panel gives additional slab depth at the column, thereby increasing the area of the critical shear section.

The minimum size of a drop panel from ACI Code Section 8.2.4 is illustrated in Fig. 13-80. ACI Code Section 8.5.2.2 states that, in computing the negative-moment flexural reinforcement, the thickness of the drop panel below the slab used in the calculations shall not be taken greater than one-fourth of the distance from the edge of the drop panel to the face of the column or column capital. If the drop panel were deeper than this, it is assumed that the maximum compression stresses would not flow down to the bottom of the drop panel, and thus, the full depth would not be effective.

For economy in form construction, the thickness of the drop, shown as h_d in Fig. 13-80, should be related to actual lumber dimensions, such as $\frac{3}{4}$ in., $1\frac{1}{2}$ in., $3\frac{1}{2}$ in., $5\frac{1}{2}$ in. (nominal 1-in., 2-in., 4-in., or 6-in. lumber sizes) or some combination of these, plus the thickness of plywood used for forms. The drop panel always should be underneath the slab (in a slab loaded with gravity loads), so that the negative-moment steel will be straight over its entire length.

Column Capitals

Occasionally, the top of a column will be flared outward, as shown in Fig. 13-81a or Fig. 13-2c. This is done to provide a larger shear perimeter at the column and to reduce the clear span, ℓ_n, used in computing moments.

ACI Code Section 26.5.7.2 requires that the capital concrete be placed at the same time as the slab concrete. As a result, the floor forming becomes considerably more complicated and expensive. For this reason, other alternatives, such as drop panels or shear reinforcement in the slab, should be considered before capitals are selected. If capitals must be used, it is desirable to use the same size throughout the project.

The diameter or effective dimension of the capital is defined in ACI Code Section 8.4.1.4 as that part of the capital lying within the largest right circular cone or pyramid with a 90° vertex that can be included within the outlines of the supporting column. The diameter is measured at the bottom of the slab or drop panel, as illustrated in Fig. 13-81a. This effective diameter is used to define the effective width for moment transfer, $c_2 + 3h$, and to define the clear span, ℓ_n.

Shear Caps

Shear caps are a projection below the slab, similar to a drop panel, but they do not necessarily satisfy the dimensional limits given in ACI Code Section 8.2.4. Shear caps are used to increase the shear strength at a slab–column connection by locally increasing the effective depth of the slab. They essentially serve the same purpose as a column capital, but because of their rectilinear shape (Fig. 13-81b), they are easier to form. The only dimensional restriction on shear caps is that their horizontal extension from the face of the column must be greater than or equal to their vertical projection below the slab (ACI Code Section 8.2.5).

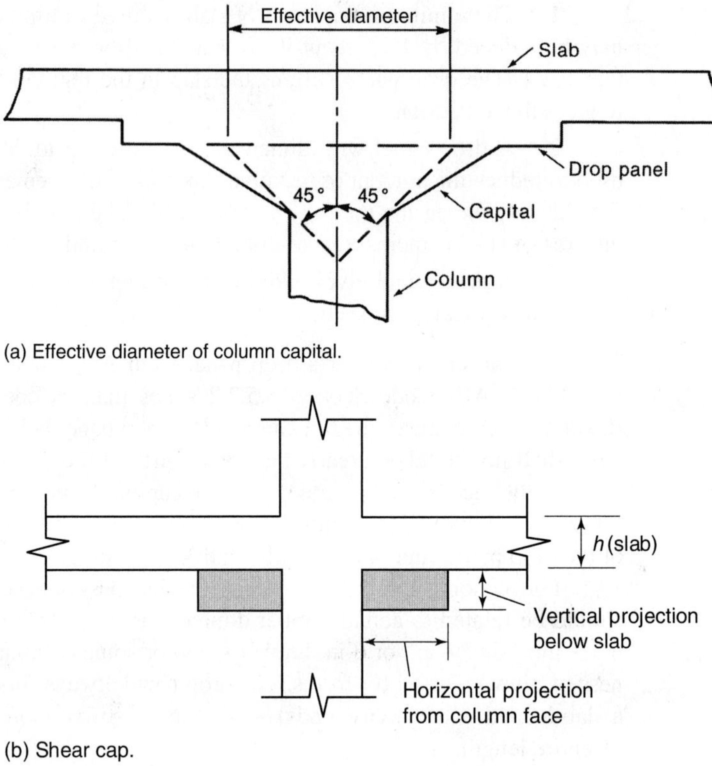

Fig. 13-81
Column capitals and shear caps for increased shear strength at slab–column connections.

In general, the vertical projection of a shear cap below the slab will be from 50 percent to greater than 100 percent of the slab thickness adjacent to the cap. Also, they typically will extend far enough from the column face to ensure that the shear capacity on the critical perimeter outside the shear cap does not govern the nominal shear strength of the connection. Equations (13-24) through (13-26) should be used to evaluate the nominal shear strength for both the critical perimeter within the shear cap and the one outside the shear cap.

Reinforcement

Placement Sequence

In a flat plate or flat slab, the moments are larger in the slab strips spanning the long direction of the panels. As a result, the reinforcement for the long span generally is placed closer to the top and bottom of the slab than is the short-span reinforcement. This gives the larger effective depth for the larger moment. For slabs supported on beams having α_f greater than about 1.0, the opposite is true, and the reinforcing pattern should be reversed. If a particular placing sequence has been assumed in the reinforcement design, it should be shown or noted on the drawings. It also is important to maintain the same arrangements of layers throughout the entire floor, to avoid confusion in the field. Thus, if the east–west reinforcement is nearer the top and bottom surfaces in one area, this arrangement should be maintained over the entire slab, if at all possible.

Cover and Effective Depth

ACI Code Table 20.6.1.3.1 specifies the minimum clear cover to the surface of the reinforcement in slabs as $\frac{3}{4}$ in. for No. 11 and smaller bars, provided that the slab is not exposed to earth or to weather. For concrete exposed to weather, the minimum clear cover is $1\frac{1}{2}$ in. for No. 5 and smaller bars and 2 in. for larger bars. Concrete parking decks exposed to deicing salts should have greater cover, and the use of epoxy-coated bars is often recommended. ACI Commentary Section 20.6.1.4.1 suggests 2 in. of cover in such a case. It may be necessary to increase the cover for fire resistance, as normally would be specified in a local building code.

The reinforcement in a two-way slab with spans of up to 25 ft will generally be No. 5 bars or smaller; and for spans over 25 ft, it will be No. 5 or No. 6 bars. For the longer span of a flat plate or flat slab, $d = h - 3/4 - 0.5d_b$, and for the shorter span, $d = h - 3/4 - 1.5d_b$. For preliminary design, these can be taken as follows:

Flat-plate or flat-slab spans up to 25 ft (7.5 m):

$$\text{Long span } d \simeq h - 1.1 \text{ in. } (30 \text{ mm}) \tag{13-43a}$$

$$\text{Short span } d \simeq h - 1.7 \text{ in. } (45 \text{ mm}) \tag{13-43b}$$

Flat-plate or flat-slab spans over 25 ft (7.5 m):

$$\text{Long span } d \simeq h - 1.15 \text{ in. } (30 \text{ mm}) \tag{13-43c}$$

$$\text{Short span } d \simeq h - 1.9 \text{ in. } (50 \text{ mm}) \tag{13-43d}$$

It is important to conservatively estimate d in slabs, because normal construction inaccuracies tend to result in values of d that are smaller than those shown on the drawings.

Spacing Requirements, Minimum Reinforcement, and Minimum Bar Size

ACI Code Section 8.6.1.1 requires that the minimum area of reinforcement provided for flexure, $A_{s,\min}$, should not be less than:

$0.0020bh$ if Grade-40 or -50 deformed bars are used
$0.0018bh$ if Grade-60 deformed bars or welded-wire fabric is used

The maximum spacing of reinforcement at critical design sections for positive and negative moments in both the middle and column strips shall not exceed two times the slab thickness (ACI Code Section 8.7.2.2), and the bar spacing shall not exceed 18 in. at any location.

Although there is no code limit on bar size, the Concrete Reinforcing Steel Institute recommends that top steel in slab should not be less than No. 4 bars at 12 in. on centers, to give adequate rigidity to prevent displacement of the bars under ordinary foot traffic before the concrete is placed.

Calculation of the Required Area of Steel

The calculation of the steel required is based on Eqs. (5-16) and (13-44), as illustrated in Examples 5-4 and 13-14, namely,

$$A_s = \frac{M_u}{\phi f_y jd} \tag{13-44}$$

where $jd \approx 0.95d$ for slabs of normal proportions. Once a trial value of A_s has been computed for the section of maximum moment, the depth of the compression stress block, a, will be computed and used to compute an improved value of $jd = d - a/2$. This procedure will be used to compute A_s at all maximum moment sections in the slab. It also is necessary to check if the section is tension-controlled and whether A_s exceeds $A_{s,\min}$, defined in ACI Code Section 8.6.1.1.

Bar Cutoffs and Anchorages

For slabs without beams, ACI Code Section 8.7.4.1.3 allows the bars to be cut off as shown in Fig. 13-82 (ACI Code Fig. 8.7.4.1.3). Where adjacent spans have unequal lengths, the extension of the negative-moment bars past the face of the interior support is based on the length of the longer span.

Prior to 1989, ACI Code Fig. 8.7.4.1.3 also showed details for slabs with alternate straight and bent bars. While such a bar arrangement is still allowed, the bent-bar details have been deleted from the figure, because those details rarely are used today. If such a slab were designed or checked, the bar details should be based on Fig. 13.4.8 of the 1983 ACI Code and modified in accordance with ACI Code Sections 8.7.4.2.1 and 8.7.4.2.2.

ACI Code Section 8.7.4.1.1(b) requires that all negative-moment steel perpendicular to an edge be bent, hooked, or otherwise anchored in spandrel beams, columns, and walls along the edge to develop f_y in tension. If there is no edge beam, this steel still should be hooked to act as torsional reinforcement and should extend to the minimum cover thickness from the edge of the slab.

Detailing Slab Reinforcement at an Edge Column

The shear and moment transfer from the slab to an exterior or corner column assumes that the edge of the slab will act as a torsional member. ACI-ASCE Committee 352 [13-40]

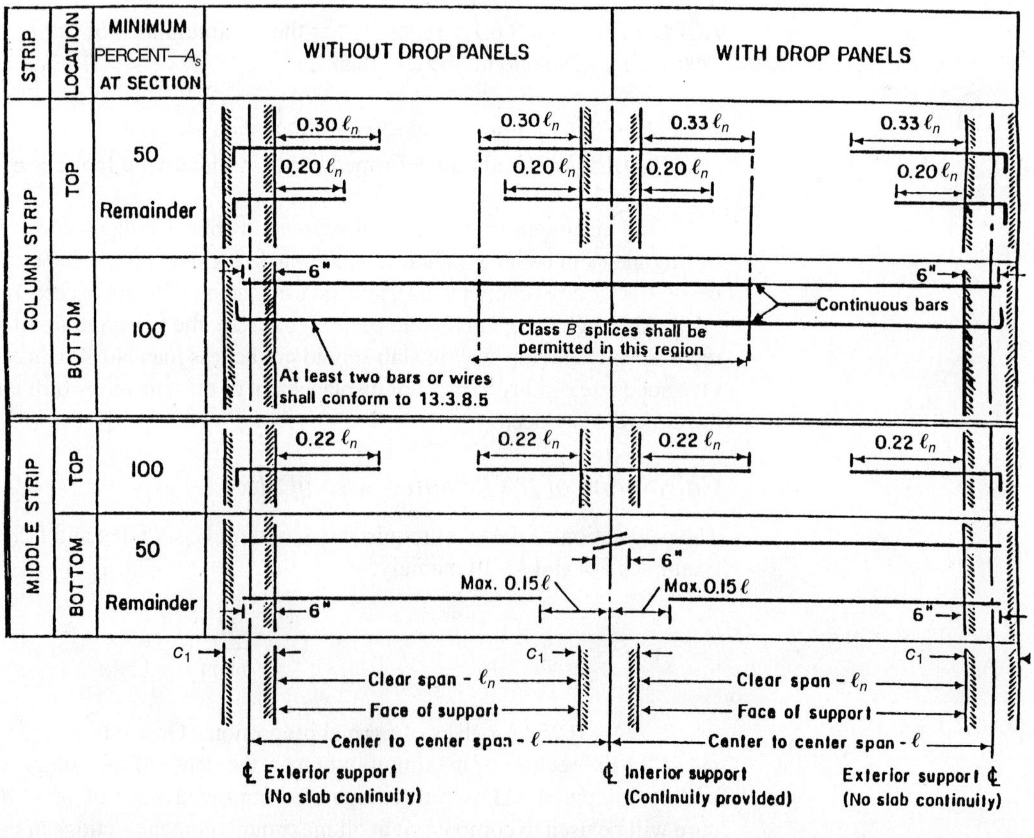

Fig. 13-82
Minimum cut-off points for slabs without beams. (From ACI Code Section 8.7.4.1.3.)

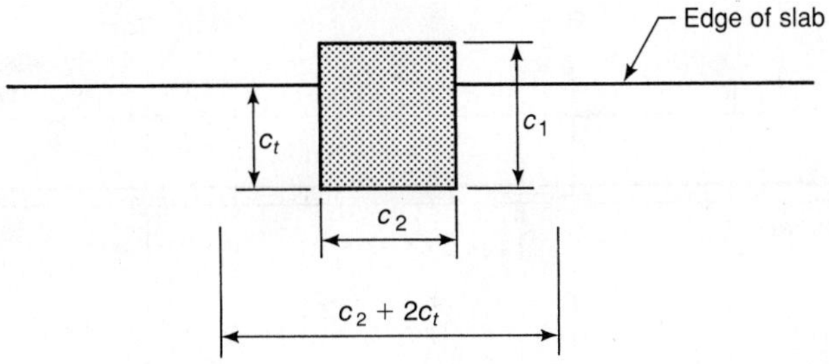

Fig. 13-83
Width effective for moment transfer at exterior columns.

has recommended details for edge–column–slab connections. Working in part from these recommendations, the author recommends the following reinforcing details:

(a) The top steel required to transfer the moment $\gamma_f M_u$ according to ACI Code Section 8.4.2.3.3 should be placed in a width equal to the smaller of $2(1.5h) + c_2$ or $2c_t + c_2$ centered on the column (Fig. 13-83), where c_t is the distance from the inner face of the column to the edge of the slab but not more than the depth of the column, c_1, and c_2 is the width of the column.

(b) Equivalent torsion reinforcement, as described in reference [13-40], should be provided along the edge of the slab within the dimensions defined in (a) and extending at least two slab thicknesses away from the side face of the column.

Structural Integrity Reinforcement

When a punching-shear failure occurs, it completely removes the shear capacity at a column and the slab will drop, pulling the top reinforcement out through the top of the slab. If the slab impacts on the slab below, that slab probably will fail also, causing a progressive type of failure. Research reported by Mitchell and Cook [13-45] suggests that this can be prevented by providing reinforcement through the column at the bottom of the slab. ACI Code Section 8.7.4.2 requires that all bottom steel in the column strip in each direction either be continuous or be lap-spliced or mechanically spliced with a Class B splice (see Fig. 13-82). At least two of the bottom bars in each direction must pass through the column core. In this context, the words *column core* mean that these bars should be between the corner reinforcing bars of the column. At exterior and corner columns, at least two bars perpendicular to the edge should be bent, hooked, or otherwise anchored within the column core. The bars passing through the column are referred to as *integrity steel.* The ACI-ASCE Committee 352 report [13-40] recommends that a specific amount of integrity steel pass through the column core to prevent progressive collapse. In most cases, the Committee 352 requirement will result in a slightly larger area of integrity steel than required by the ACI Code.

13-13 DESIGN OF SLABS WITHOUT BEAMS

EXAMPLE 13-15 Design of a Flat-Plate Floor without Spandrel Beams—Direct-Design Method

Figure 13-84 shows a plan of a part of a flat-plate floor. There are no spandrel beams. The floor supports its own dead load, plus 25 psf for partitions and finishes, and a live load of 50 psf. The slab extends 4 in. past the exterior face of the column to support an exterior wall that weighs 300 lb/ft of length of wall. The story-to-story height is 8 ft

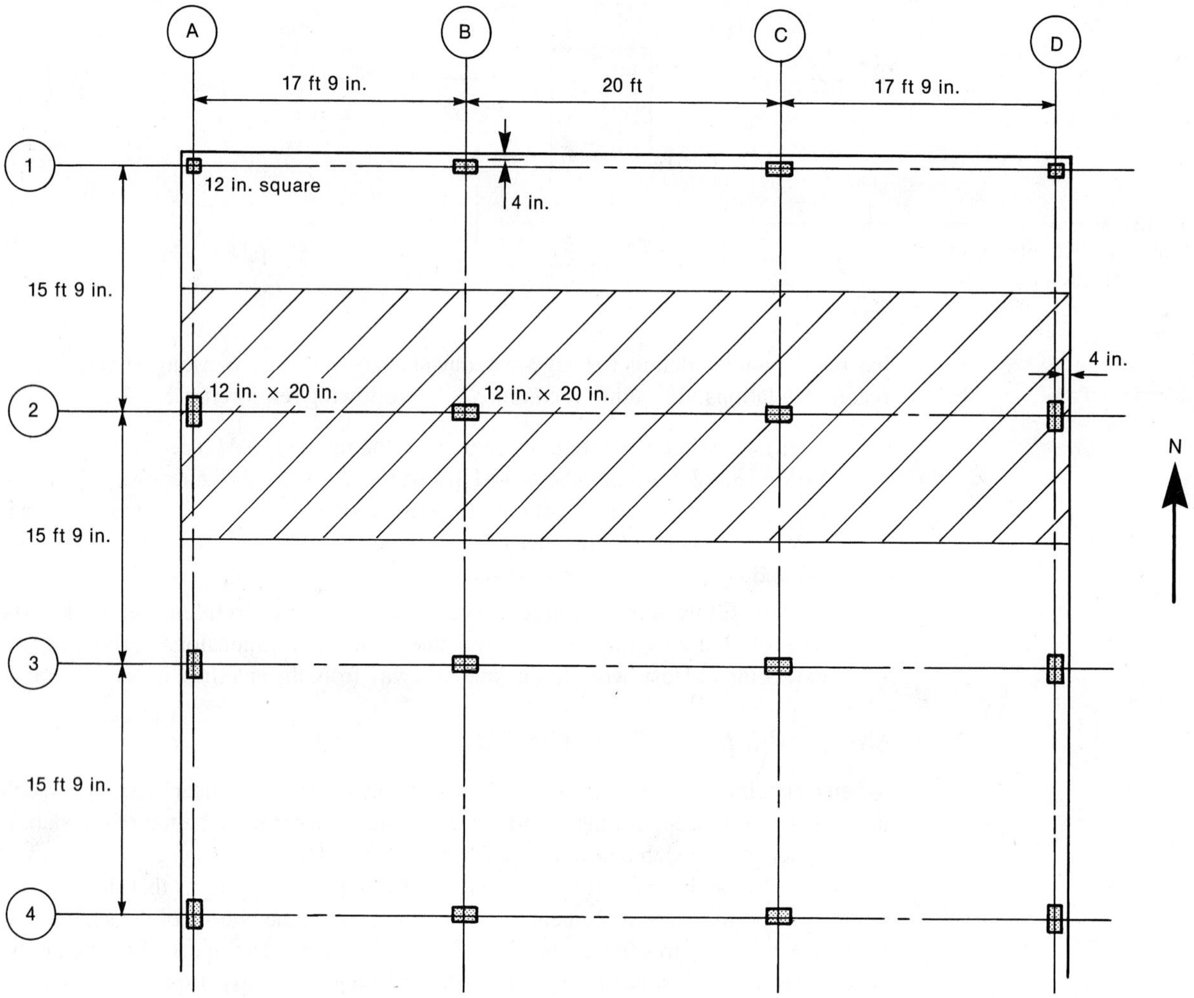

Fig. 13-84
Plan of flat-plate floor for Example 13-15.

10 in. Use 4000-psi normal-weight concrete and Grade-60 reinforcement. The columns are 12 in. × 12 in. or 12 in. × 20 in. and are oriented as shown. Select the thickness, compute the design moments, and select the reinforcement in the slab. Use the load combinations from Table 5.3.1 of the ACI Code.

To fully design the portion of slab shown, it is necessary to consider two east–west strips (Fig. 13-85a) and two north–south strips (Fig. 13-85b). In steps 3 and 4, the computations are carried out for the east–west strip along column line 2 and the edge strip running east–west along column line 1. Steps 5 and 6 deal similarly with the north–south strips.

1. Select the design method, load combinations, and strength-reduction factor. ACI Code Section 8.10.2 allows the use of the direct-design method if the following conditions are satisfied.

(a) Minimum of three consecutive spans each way—therefore, o.k.

(b) Ratio of longer to shorter span within a panel shall be less than 2. In this slab the largest ratio is 1.27—therefore, o.k.

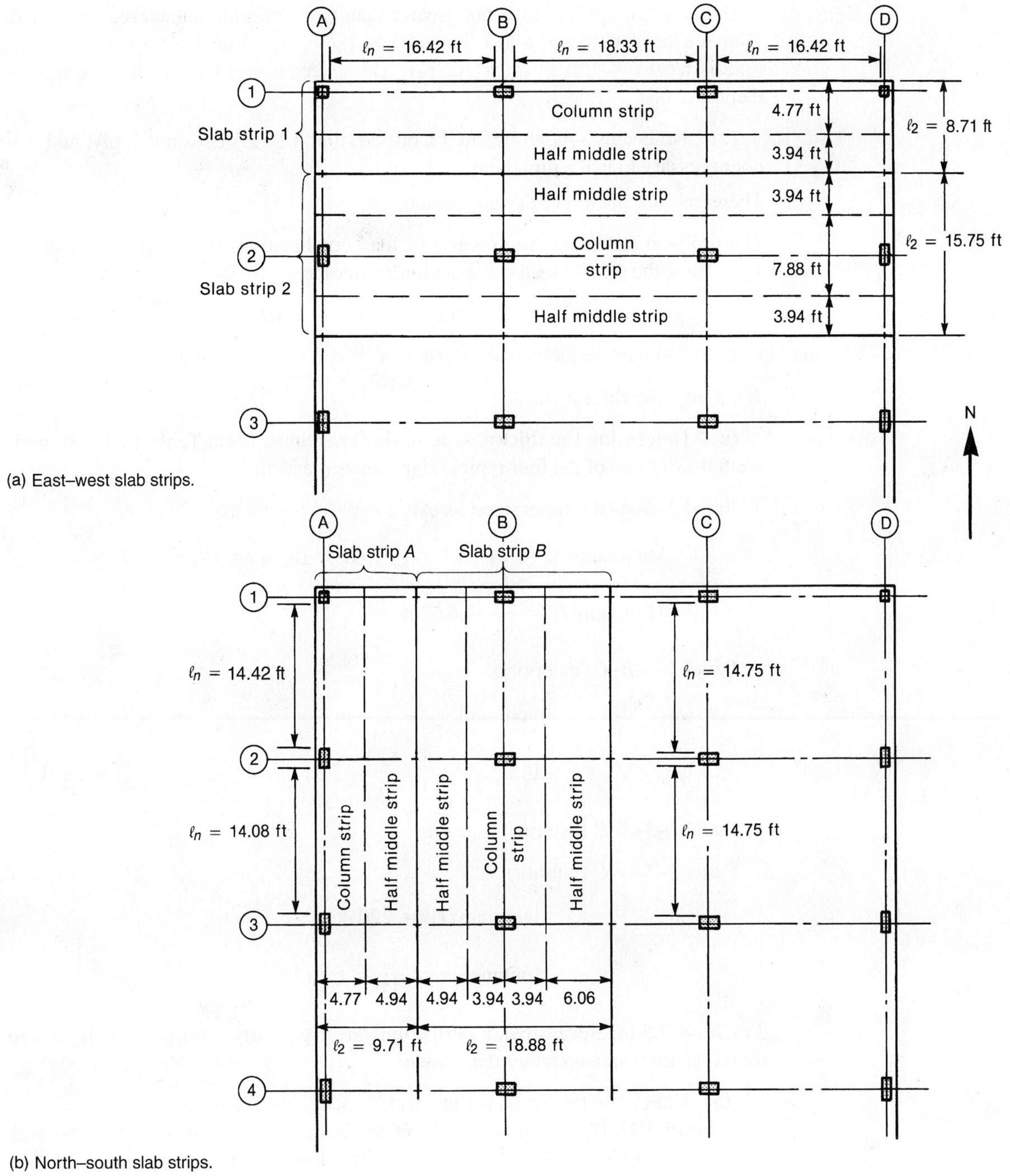

Fig. 13-85
Division of slab into strips for design in Example 13-15.

(c) Successive span lengths differ by not more than one-third of the longer span. Thus, short span/long span $\geq$ 0.667: 17.75/20 = 0.89—therefore, o.k.

(d) Columns offset up to 10 percent—o.k.

(e) All loads are uniformly distributed gravity loads. Strictly speaking, the wall load is not uniformly distributed, but use DDM anyway.

(f) Unfactored live load not greater than two times the unfactored dead load. Estimate the slab thickness as $\ell/36 = 20 \times 12/36 \simeq 6.5$ in. The approximate dead load is $6.5/12 \times 150 + 25 = 106$ psf. This exceeds one-half of the live load—therefore, o.k.

(g) No beams—therefore, ACI Code Section 8.10.2.7 does not apply, and it is not necessary to make this check.

Therefore, use the direct-design method.

The problem statement requires use of load combinations from Table 5.3.1 of the ACI Code. Thus, the gravity load combination reduces to

$$U = 1.2D + 1.6L, \text{ or } 1.4D$$

and the strength-reduction factors are: flexure, $\phi = 0.90$; shear, $\phi = 0.75$

2. Select the thickness.

(a) Determine the thickness to limit deflections. From Table 13-1, the minimum thicknesses of the four typical slab panels are as follows:

Panel 1–2–*A*–*B* (corner; treat as exterior):

$$\text{Maximum: } \ell_n = (17 \text{ ft } 9 \text{ in.}) - (6 + 10)\text{in.} = 197 \text{ in.}$$

$$\text{Minimum: } h = \frac{\ell_n}{30} = 6.57 \text{ in.}$$

Panel 1–2–*B*–*C* (exterior):

$$\text{Maximum: } \ell_n = 220 \text{ in.}$$

$$\text{Minimum: } h = \frac{\ell_n}{30} = 7.33 \text{ in.}$$

Panel 2–3–*A*–*B* (exterior); same as 1-2-A-B.

Panel 2–3–*B*–*C* (interior):

$$\text{Maximum: } \ell_n = 220$$

$$\text{Minimum: } h = \frac{\ell_n}{33} = 6.67 \text{ in.}$$

Try $h = 7.5$ in. Because slab deflections are frequently a problem, it is best to round the thickness up rather than down.

(b) Check the thickness for shear. Check at columns $B2$ and $B1$.

***Column B*1:** The maximum moment about the z–z axis occurs when the area *E–F–G–H–E* in Fig. 13-86 is all loaded with live load. The maximum moment about the axis perpendicular to the edge occurs with live loads on one side of the column line *B*. This loading will not be used, because it corresponds to a live load shear that is half that for loading of *E–F–G–H–E*. Assume the tributary area extends 0.5ℓ from the centers of the edge or corner columns.

Tributary areas for slab strip A (Fig. 13-85b):

***Span A*1–*A*2**

Width is 17.75 ft from centerline of columns in line *A* to centerline of columns line *B*.

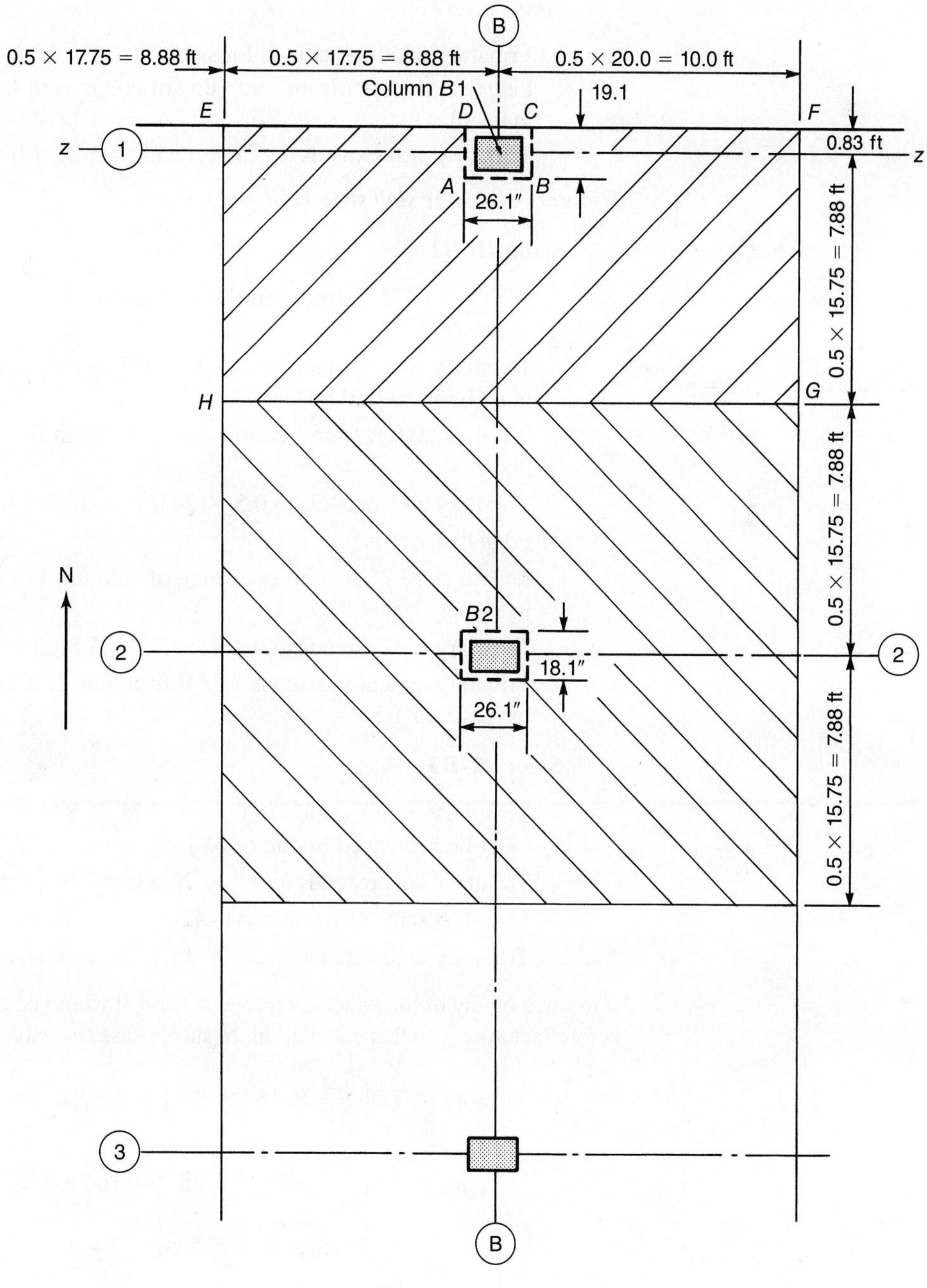

Fig. 13-86
Initial critical shear perimeters and tributary areas for columns *B*1 and *B*2.

Tributary area extends to $0.5 \times 17.75 = 8.88$ ft from line *A*.

Edge distance is $4 + 6$ in. $= 0.833$ ft.

Length is 15.75 ft from centerline of columns in line 1 to centerline of columns in line 2.

Tributary area extends to $0.5 \times 15.75 = 7.88$ ft from line 1.

Tributary area also extends 7.88 ft from line 2, but shear force must be increased by factor of 1.15.

Edge distance is $4 + 6$ in. $= 0.833$ ft.

Span A2–A3

Tributary width is same as for span $A1$–$A2$.

Length is 15.75 ft from centerline of columns in line 2 to centerline of columns in line 3.

Tributary area extends to $0.5 \times 15.75 = 7.88$ ft from lines 2 and 3.

Tributary areas for slab strip B (Fig. 13-86):

Span B1–B2

Width is 17.75 ft from centerline of columns in line A to centerline of columns in line B.

Tributary area extends to $0.5 \times 17.75 = 8.88$ ft from line B, but shear force must be increased by factor of 1.15.

Width is 20.0 ft from centerline of columns in line B to centerline of columns in line C.

Tributary area extends to 0.5×20.0 ft $= 10.0$ ft from lines B and C.

Edge distance is 0.833 ft.

Length is 15.75 ft from centerline of columns in line 1 to centerline of columns in line 2.

Tributary area extends to $0.5 \times 15.75 = 7.88$ ft from line 1.

Tributary area also extends 7.88 ft from line 2, but shear force must be increased by factor of 1.15.

Span B2–B3

Tributary area extends to $0.5 \times 17.75 = 8.88$ ft from line B, but shear force must be increased by factor of 1.15.

Tributary area extends to 0.5×20.0 ft $= 10.0$ ft from line(s) B (and C).

Length is same as for span $A1$–$A2$

Tributary area extends to $0.5 \times 15.75 = 7.88$ ft from lines 2 and 3.

If the area of any of the panels exceeded 400 ft^2, it would be possible to reduce the live load before factoring it in this calculation. In such a case, q_u may differ from panel to panel.

$$q_u = 1.2\left(\frac{7.5}{12} \times 150 + 25\right) + 1.6(50) = 223 \text{ psf}$$

$$q_u = 1.4\left(\frac{7.5}{12} \times 150 + 25\right) = 166 \text{ psf (does not govern)}$$

$$\text{Average } d \cong 7.5 \text{ in.} - 1.4 \text{ in.} \cong 6.1 \text{ in.}$$

Column B2: The long side of the critical shear perimeter is $20 + d = 26.1$ in., and the short side is $12 + d = 18.1$ in. In the following calculation for the factored shear force transferred to column $B2$, the shear force multiplier of 1.15 required for the first interior support will be applied directly to the appropriate tributary lengths. Then,

$$b_o = 2(26.1 + 18.1) = 88.4 \text{ in.}$$

$$V_u = 223\left[(8.88 \times 1.15 + 10.0)(7.88 \times 1.15 + 7.88) - \frac{26.1 \times 18.1}{144}\right]$$

$$= 75{,}600 \text{ lbs} = 75.6 \text{ kips}$$

$$v_u = V_u/b_o d = 75{,}600 \text{ lb}/(88.4'' \times 6.1'') = 140 \text{ psi}$$

From Eq. (13-25),

$$\beta = \frac{c(\text{long})}{c(\text{short})} = \frac{20}{12} = 1.67$$

$$\left(2 + \frac{4}{\beta}\right) = \left(2 + \frac{4}{1.67}\right) = 4.40 > 4.0 \text{ (does not govern)}$$

From Eq. (13-26),

$$\alpha_s = 40, \text{ for an interior slab–column connection}$$

$$\left(\frac{\alpha_s d}{b_o} + 2\right) = \left(\frac{40 \times 6.1}{88.4} + 2\right) = 4.76 > 4.0 \text{ (does not govern)}$$

Thus, using Eq. (13-24),

$$\phi v_c = \phi 4\lambda\sqrt{f'_c} = 0.75 \times 4 \times 1\sqrt{4000}$$
$$= 190 \text{ psi} > v_u$$

Note: $v_u/\phi v_c = 0.737 > 0.4$.

For this ratio, no adjustment will be permitted in the ratio of unbalanced moment resisted by eccentric shear stresses, $\gamma_v = 1 - \gamma_f$, as discussed in ACI Code Section 8.4.2.3.4. However, because this ratio is below 0.8, the slab thickness probably will be sufficient when checking for combined shear and moment transfer at this interior connection.

***Column B*1:**

$$b_o = 26.1 + 2 \times 19.1 = 64.3 \text{ in.}$$

$$V_u = 223\,psf\left[(8.88 \times 1.15 + 10.0)(7.88 + 0.833) - \left(\frac{19.1 \times 26.1}{144}\right)\right]$$
$$+ [1.2 \times 300 \times (8.88 \times 1.15 + 10)]$$
$$= 45{,}900 \text{ lbs. (The last term is the factored wall load.)}$$

$$v_u = V_u/b_o d = 45{,}900 \text{ lb}/(64.3'' \times 6.1'') = 117 \text{ psi}$$

Determine ϕv_c using the following:

Column $B1$ has same dimensions as $B2$, so Eq. (13-25) will not govern.

For Eq. (13-26):

$$\alpha_s = 30, \text{ for an exterior slab–column connection.}$$

$$\left(\frac{\alpha_s d}{b_o} + 2\right) = \left(\frac{30 \times 6.1}{64.3} + 2\right) = 4.85 > 4.0 \text{ (does not govern)}$$

Thus, using Eq. (13-24):

$$\phi v_c = 0.75 \times 4 \times 1\sqrt{4000}$$
$$= 190 \text{ psi}$$

Note: $v_u/\phi v_c = 0.62 < 0.75$.

For this ratio, ACI Code Section 8.4.2.3.4 permits a modification of γ_f for moment transfer about an axis parallel to the edge of the slab. With that information and because this ratio is below 0.8, the slab thickness at this connection should be sufficient for checking shear and moment transfer about an axis perpendicular to the edge of the slab. **Thus, use a 7.5-in. slab throughout.** Final shear checks will be made after completing the flexural design of the slab.

(c) Compute α_f for the exterior beams. Because there are no edge beams, $\alpha_f = 0$.

3. Compute the moments in a slab strip along column line 2 (Fig. 13-85a). This piece of slab acts as a rigid frame spanning between columns *A*2, *B*2, *C*2, and *D*2. In this slab strip, slab panels *A*2–*B*2 and *C*2–*D*2 are "end panels," and *B*2–*C*2 is an "interior panel." Columns *A*2 and *D*2 are "exterior columns," and *B*2 and *C*2 are "interior columns."

These calculations are generally tabulated as shown in Tables 13-9 to 13-12. The individual calculations will be followed in detail for slab strips 2 and 1 shown in Fig. 13-85a.

Line 1. *(Table 13-9).* ℓ_1 is the center-to-center span in the direction of the strip being designed.

Line 2. ℓ_n is the clear span in the direction of the strip.

Line 3. ℓ_2 is the width perpendicular to ℓ_1. The lengths ℓ_2 and ℓ_n are shown in Fig. 13-85a.

Line 4. q_u may differ from panel to panel, especially if panel areas are large enough to be affected by live-load reductions.

Line 5. Total static moment for each span.

Line 6. The moment in the exterior panels is divided by using ACI Code Table 8.10.4.2:

$$\text{Negative moment at exterior end of end span} = -0.26M_o$$

$$\text{Positive moment in end span} = 0.52M_o$$

$$\text{Negative moment at interior end of end span} = -0.70M_o$$

For the interior panel, from ACI Code Section 8.10.4.1,

$$\text{Negative} = -0.65M_o, \text{ and positive} = 0.35M_o$$

Line 7. This is the product of lines 5 and 6. At interior columns, design the slab at the joint for the larger negative moment.

Line 8. ACI Code Section 8.10.7.2 requires that interior columns be designed for the moment given by Eq. (13-14):

$$M_{\text{slab}} = 0.07[(1.2 \times 119 + 0.5 \times 1.6 \times 50) \times 15.75 \times 18.33^2 - (1.2 \times 119) \times 15.75 \times 16.42^2] = 25.2 \text{ kip-ft}$$

Therefore, design the interior columns for a total moment of 25.2 kip-ft divided between the columns above and below the joint in the ratio of their stiffnesses.

The unbalanced moment at the exterior edge (-30.8 kip-ft) is divided between the columns above and below the joint in the ratio of their stiffnesses. The moment due to the wall load acting on the overhanging portion of the slab would partially counterbalance the moment transferred from the slab to the columns, but that moment has been conservatively ignored for this example.

4. Compute the moments in the slab strip along column line 1 (Fig. 13-85a). This strip of slab acts as a rigid frame spanning between columns *A*1, *B*1, *C*1, and *D*1. Spans *A*1–*B*1 and *C*1–*D*1 are "end panels," and span *B*1–*C*1 is an "interior panel" in this strip. Similarly, columns *A*1 and *D*1 are "exterior columns" and *B*1 and *C*1 are "interior columns" in this slab strip.

TABLE 13-9 Calculation of Negative and Positive Moments for East–West Slab Strip 2—Example 13-15

	A2			B2			C2		D2
1. ℓ_1 (ft)		17.75			20			17.75	
2. ℓ_n (ft)		16.42			18.33			16.42	
3. ℓ_2 (ft)		15.75			15.75			15.75	
4. q_u (ksf)		0.223			0.223			0.223	
5. M_o (kip-ft)		118			148			118	
6. Moment coefficients	−0.26	0.52	−0.70	−0.65	0.35	−0.65	−0.70	0.52	−0.26
7. Negative and positive moments (kip-ft)	−30.8	61.6	−82.9	−95.9	51.6	−95.9	−82.9	61.6	−30.8
8. Sum of column moments (kip-ft)	30.8		25.2			25.2			30.8

TABLE 13-10 Calculation of Negative and Positive Moments for East–West Slab Strip 1—Example 13-15

	A1			B1			C1		D1
1. ℓ_1 (ft)		17.75			20			17.75	
2. ℓ_n (ft)		16.42			18.33			16.42	
3. ℓ_2 (ft)		8.71			8.71			8.71	
4. q_u (ksf)		0.223			0.223			0.223	
5. M_o (kip-ft)		65.5			81.6			65.5	
6. Moment coefficients	−0.26	0.52	−0.70	−0.65	0.35	−0.65	−0.70	0.52	−0.26
7. Negative and positive moments (kip-ft)	−17.0	34.1	−45.9	−53.0	28.6	−53.0	−45.9	34.1	−17.0
8. Column moments from slab (kip-ft)	17.0		14.0			14.0			17.0
9. Wall load (kip/ft)		0.36			0.36			0.36	
10. Wall M_o (kip-ft)		12.1			15.1			12.1	
11. Negative and positive moments from wall load (kip-ft)	−3.2	6.3	−8.5	−9.8	5.3	−9.8	−8.5	6.3	−3.2
12. Column moments from wall load (kip-ft)	3.2		1.3			1.3			3.2
13. Total column moment (kip-ft)	20.2		15.3			15.3			20.2

TABLE 13-11 Calculation of Negative and Positive Moments for North–South Slab Strip *B*—Example 13-15

	*B*1			*B*2		*B*3	
1. ℓ_1 (ft)		15.75				15.75	
2. ℓ_n (ft)		14.75				14.75	
3. ℓ_2 (ft)		18.88				18.88	
4. q_u (ksf)		0.223				0.223	
5. M_o (kip-ft)		114				114	
6. Moment coefficients	−0.26	0.52	−0.70		−0.65	0.35	−0.65
7. Negative and positive moments (kip-ft)	−29.8	59.5	−80.1		−74.4	40.1	−74.4
8. Sum column moments (kip-ft)	29.8			11.5			11.5

TABLE 13-12 Calculation of Negative and Positive Moments for North–South Slab Strip *A*—Example 13-15

	*A*1			*A*2		*A*3	
1. ℓ_1 (ft)		15.75				15.75	
2. ℓ_n (ft)		14.42				14.08	
3. ℓ_2 (ft)		9.71				9.71	
4. q_u (ksf)		0.223				0.223	
5. M_o (kip-ft)		56.3				53.7	
6. Moment coefficients	−0.26	0.52	−0.70		−0.65	0.35	−0.65
7. Negative and positive moments (kip-ft)	−14.6	29.3	−39.4		−34.9	18.8	−34.9
8. Column moments from slab (kip-ft)	14.6			6.6			6.6
9. Wall load (kip/ft)		0.36				0.36	
10. Wall M_o (kip-ft)		9.4				8.9	
11. Negative and positive moments from wall load (kip-ft)	−2.4	4.9	−6.6		−5.8	3.1	−5.8
12. Column moments from wall load (kip-ft)	2.4			0.8			0.0
13. Total column moment (kip-ft)	17.0			7.4			6.6

The calculations are carried out in Table 13-10. These are similar to those for slab strip 2, except that the panel is not as wide and there is a uniformly distributed wall load that causes additional moments in the slab. Because the wall load occurs along the edge of the slab, these moments will be assumed to be resisted entirely by the exterior column strip. The wall-load moments will be calculated separately (lines 9 through 12 in Table 13-10) and added in at a later step (line 4 of Table 13-13).

Line 3. ℓ_2 is the transverse span measured from center-to-center of the supports. ACI Code Section 8.10.3.2.3 states that, for edge panels, ℓ_2 in the equation for M_o should be replaced by $\ell_{2,\,\text{edge}}$, which is the width from the edge of the slab to a line halfway between column lines 1 and 2.

Line 8. The interior columns (*B*1 and *C*1) are designed for the moment from the slab given by Eq. (13-14):

$$M_{\text{slab}} = 0.07[(1.2 \times 119 + 0.5 \times 1.6 \times 50)8.71 \times 18.33^2 - (1.2 \times 119)8.71 \times 16.42^2] = 14.0 \text{ kip-ft}$$

Line 9. The factored dead load of the wall is $1.2 \times 0.3 = 0.36$ kip/ft.

TABLE 13-13 Division of Moment to Column and Middle Strips: East–West Strips—Example 13-15

		Column Strip		Middle Strip		Column Strip		Middle Strip		Edge Column Strip
Strip Width, ft		7.88		7.88		7.88		7.88		4.77
<u>Exterior Negative Moments</u>		**A3**				**A2**				**A1**
1. Slab moment (kip-ft)		−30.8				−30.8				−17.0
2. Moment coefficients	0.0	1.0	0.0		0.0	1.0	0.0		0.0	1.0
3. Distributed moments to strips		−30.8		0.0		−30.8		0.0		−17.0
4. Wall moment (kip-ft)										−3.2
5. Total strip moment (kip-ft)		−30.8		0.0		−30.8		0.0		−20.2
6. Required A_s (in.^{2})		1.11		0		1.11		0		0.73
7. Minimum A_s (in.^{2})		1.28		1.28		1.28		1.28		0.77
8. Selected steel		7 No. 4 bars		7 No. 4 bars		7 No. 4 bars		7 No. 4 bars		5 No. 4 bars
9. A_s provided (in.^{2})		1.40		1.40		1.40		1.40		1.00
<u>End Span Positive Moments</u>										
1. Slab moment (kip-ft)		61.6				61.6				34.1
2. Moment coefficients	0.2	0.6	0.2		0.2	0.6	0.2		0.4	0.6
3. Distributed moments to strips		37.0	12.3		12.3	37.0	12.3		13.6	20.5
4. Wall moment (kip-ft)										6.3
5. Total strip moment (kip-ft)		37.0		24.6		37.0		25.9		26.8
6. Required A_s (in.^{2})		1.33		0.89		1.33		0.93		0.97
7. Minimum A_s (in.^{2})		1.28		1.28		1.28		1.28		0.77
8. Selected steel		7 No. 4 bars		7 No. 4 bars		7 No. 4 bars		7 No. 4 bars		5 No. 4 bars
9. A_s provided (in.^{2})		1.40		1.40		1.40		1.40		1.00

(*Continued*)

TABLE 13-13 Division of Moment to Column and Middle Strips: East–West Strips—Example 13-15 (*Continued*)

		Column Strip	Middle Strip		Column Strip	Middle Strip		Edge Column Strip
Strip Width, ft		7.88	7.88		7.88	7.88		4.77
First Interior Negative Moment		***B3***			***B2***			***B1***
1. Slab moment (kip-ft)		−95.9			−95.9			−53.0
2. Moment coefficients	0.125	0.75	0.125	0.125	0.75	0.125	0.25	0.75
3. Distributed moments to strips		−71.9	−12.0	−12.0	−71.9	−12.0	−13.3	−39.8
4. Wall moment (kip-ft)								−9.8
5. Total strip moment (kip-ft)		−71.9	−24.0		−71.9	−25.3		−49.6
6. Required A_s (in.^{2})		2.59	0.87		2.59	0.91		1.79
7. Minimum A_s (in.^{2})		1.28	1.28		1.28	1.28		0.77
8. Selected steel		9 No. 5 bars	7 No. 4 bars		9 No. 5 bars	7 No. 4 bars		6 No. 5 bars
9. A_s provided (in.^{2})		2.79	1.40		2.79	1.40		1.86
Interior Positive Moment								
1. Slab moment (kip-ft)		51.6			51.6			28.6
2. Moment coefficients	0.20	0.60	0.20	0.20	0.60	0.20	0.40	0.60
3. Distributed moments to strips		31.0	10.3	10.3	31.0	10.3	11.4	17.2
4. Wall moment (kip-ft)								5.3
5. Total strip moment (kip-ft)		31.0	20.6		30.1	21.7		22.5
6. Required A_s (in.^{2})		1.12	0.74		1.12	0.78		0.81
7. Minimum A_s (in.^{2})		1.28	1.28		1.28	1.28		0.77
8. Selected steel		7 No. 4 bars	7 No. 4 bars		7 No. 4 bars	7 No. 4 bars		5 No. 4 bars
9. A_s provided (in.^{2})		1.40	1.40		1.40	1.40		1.00

Line 10. The total static moment of the wall load is $w\ell_n^2/8$. Therefore, in span $A1$–$B1$,

$$M_o(\text{wall}) = 0.36 \times \frac{16.42^2}{8} = 12.1 \text{ kip-ft}$$

Line 11. The negative and positive moments due to the wall load are found by multiplying the moments in line 10 by the coefficients in line 6.

Line 12. The column moments at the interior columns are assumed to be equal to the unbalanced moments at the joint. Equation (13-14) is not applied here, because there is no unbalanced live load from the wall.

Line 13. The total column moments are the sum of lines 8 and 12.

5. Compute the moments in the slab strip along column line *B* (Fig. 13-85b). This strip of slab acts as a rigid frame spanning between columns *B*1, *B*2, *B*3, and so on. The calculations are carried out in Table 13-11 and follow a process similar to that in step 3 of this example.

6. Compute the moments in the slab strip along column line A (Fig. 13-85b). This strip of slab acts as a rigid frame spanning between columns $A1$, $A2$, $A3$, and so on. In this strip, span $A1$–$A2$ is an "end span," and $A2$–$A3$ and so on are "interior spans." Column $A1$ is an "exterior column," and columns $A2$ and $A3$ are "interior columns," in the slab strip. The calculations are carried out in Table 13-12 and follow a process similar to that in step 4 of this example.

7. Distribute the negative and positive moments to the column and middle strips and design the reinforcement: strips spanning east and west (strips 1 and 2). In steps 3 and 4, the moments in the strips along column lines 1 and 2 were computed. These moments must be distributed to the various middle and column strips to enable the east–west reinforcement to be designed.

(a) Divide the slab strips into middle and column strips. In each panel, the column strip extends 0.25 times the smaller of ℓ_1 and ℓ_2 from the line joining the columns, as shown in Fig. 13-23 and (for this example) in Fig. 13-85a. Thus, the column strips extend 15 ft 9 in./4 = 47.25 in. on each side of the column lines. The total width of the column strip is 2×47.25 in. = 7.87 ft. The width of the middle strip is 7.88 ft. The edge column strip has a width of 47.25 + 10 in. = 4.77 ft.

(b) Divide the moments between the column and middle strip and design the reinforcement. The calculations are carried out in Table 13-13 for the east–west strips. This table is laid out to resemble a plan of the column and middle strips shown in Fig. 13-85a. The columns are indicated by bold numbering. The computations leading up to the entries in each line of this table are summarized below. The calculations are repeated for each of the negative- and positive-moment regions. Because this slab has only three spans, it is not necessary to consider the typical interior negative moment.

Line 1. These moments are taken from line 7 in Tables 13-9 and 13-10. At the first interior support, the larger of the two adjacent negative moments is used (ACI Code Section 8.10.4.5).

Line 2. These are the moment coefficients.

Exterior negative moments. Table 13-5 gives the fraction of the exterior negative moment resisted by the column strip. This section is entered by using β_t, $\alpha_{f1}\ell_2/\ell_1$, and ℓ_2/ℓ_1. For a slab without edge beams, $\beta_t = 0$; for a slab without beams parallel to ℓ_1 for the span being designed, $\alpha_{f1} = 0$. Thus, from Table 13-5, 100 percent of the exterior negative moment is assigned to the column strips.

Positive-moment regions. Table 13-4 gives the fraction of positive moments assigned to the column strips. Because $\alpha_{f1} = 0$, 60 percent is assigned to the column strip. The remaining 40 percent is assigned to the adjacent half-middle strips. At the edge, there is only one adjacent half-middle strip, so 40 percent of the panel moment goes to it. At an interior column strip, there are two adjacent half-middle strips, and $0.5 \times 40 = 20$ percent goes to each.

Interior negative-moment regions. Table 13-3 specifies the fraction of interior negative moments assigned to the column strips. Because $\alpha_{f1} = 0$, 75 percent goes to the column strip and the remaining 25 percent to the middle strips.

Line 3. This is the product of lines 1 and 2. The arrows in Table 13-13 illustrate the distribution of the moments between the various parts of the slab.

Line 4. The weight of the east–west wall along column line 1 causes moments in the edge column strip. The moments given here come from line 11 of Table 13-10.

Line 5. The moments are the sum of lines 3 and 4.

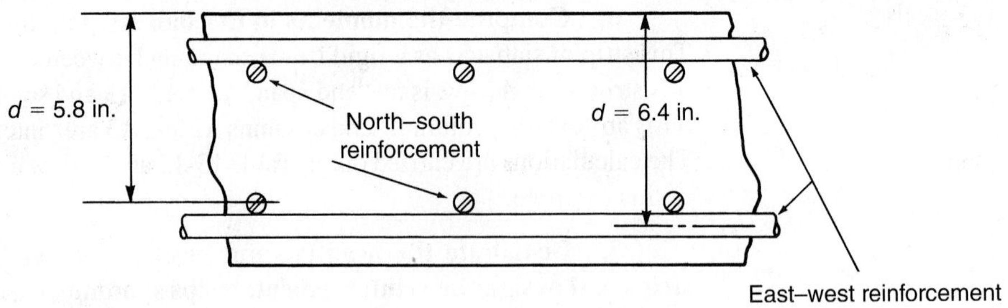

Fig. 13-87
Arrangement of bars in a slab for Example 13-15.

Line 6. The area of steel will be calculated from Eq. (13-44):

$$A_s = \frac{M_u}{\phi f_y jd} \tag{13-44}$$

Compute d. Because the largest moments in the panel occur at support $B2$ in slab strip 2 (see Table 13-9), place reinforcement as shown in Fig. 13-87. Thus,

$$d(\text{east–west}) \cong h - 1.1 \text{ in.} = 7.5 - 1.1 = 6.4 \text{ in.} \tag{13-43a}$$

Compute trial A_s required at the section of maximum moment (column strip at $B2$). The largest M_u is -71.9 kip-ft. Assume that $j = 0.95$. Then,

$$A_s\,(\text{req'd}) = \frac{71.9 \times 12{,}000}{0.9 \times 60{,}000 \times 0.95 \times 6.4} = 2.63 \text{ in.}^2$$

Compute a and check whether the section is tension controlled:

$$a = \frac{A_s f_y}{0.85 f'_c\, b} = \frac{2.63 \times 60{,}000}{0.85 \times 4000 \times (7.88 \times 12)} = 0.49 \text{ in.}$$

$$c = a/\beta_1 = 0.49/0.85 = 0.576 \text{ in.}$$

This is less than $3d/8 = 2.40$ in.; therefore, $\phi = 0.9$.

Compute jd and a constant for computing A_s. (Note: this calculation makes a conservative assumption that the value of a is constant for all sections):

$$jd = d - \frac{a}{2} = 6.4 - \frac{0.49}{2} = 6.16 \text{ in.}$$

$$A_s\,(\text{in.}^2) = \frac{M_u\,(\text{kip-ft}) \times 12{,}000}{0.9 \times 60{,}000 \times 6.16}$$

Thus,

$$A_s\,(\text{in.}^2) = 0.0361 M_u\,(\text{kip-ft}) \tag{A}$$

The values of A_s in lines 6 of Table 13-13 are computed from (A).

Line 7. The minimum A_s is specified in ACI Code Section 8.6.1.1. $A_{s,\min} = 0.0018bh$ for Grade-60 reinforcement. Maximum bar spacing is $2h$ (ACI Code Section 8.7.2.2), but not more than 18 in. Therefore, the maximum spacing is 15 in.

Edge column strip:

$$A_{s,\min} = 0.0018bh = 0.0018(4.77 \times 12) \times 7.5 = 0.77 \text{ in.}^2$$

$$\text{Minimum number of bar spaces} = \frac{4.77 \text{ ft} \times 12 \text{ in./ft}}{15 \text{ in.}} = 3.82$$

Therefore, the minimum number of bars is 4.
Other strips:

$$A_{s,\min} = 0.0018(7.88 \text{ ft} \times 12 \text{ in./ft}) \times 7.5 \text{ in.} = 1.28 \text{ in.}^2$$

The minimum number of bars is 7.

Line 8. The final bar choices are given in line 8.

Line 9. The actual areas of reinforcement provided are given in line 9. The reinforcement chosen for each east–west strip is shown in the section in Fig. 13-88. In the calculations for line 6, it was assumed that No. 5 bars would be used, giving $d = 6.4$ in. In some cases, the bars chosen were No. 4 bars. As a result, $d = 6.5$ in., could be used in some parts of the slab. This will not make enough difference to repeat the calculation.

8. Distribute the negative and positive moments to the column and middle strips and design the reinforcement: strips spanning north and south (strips *A* and *B*). In steps 5 and 6, the moments were computed for the strips along column lines *A* and *B*. These moments must be assigned to the middle and column strips so that the

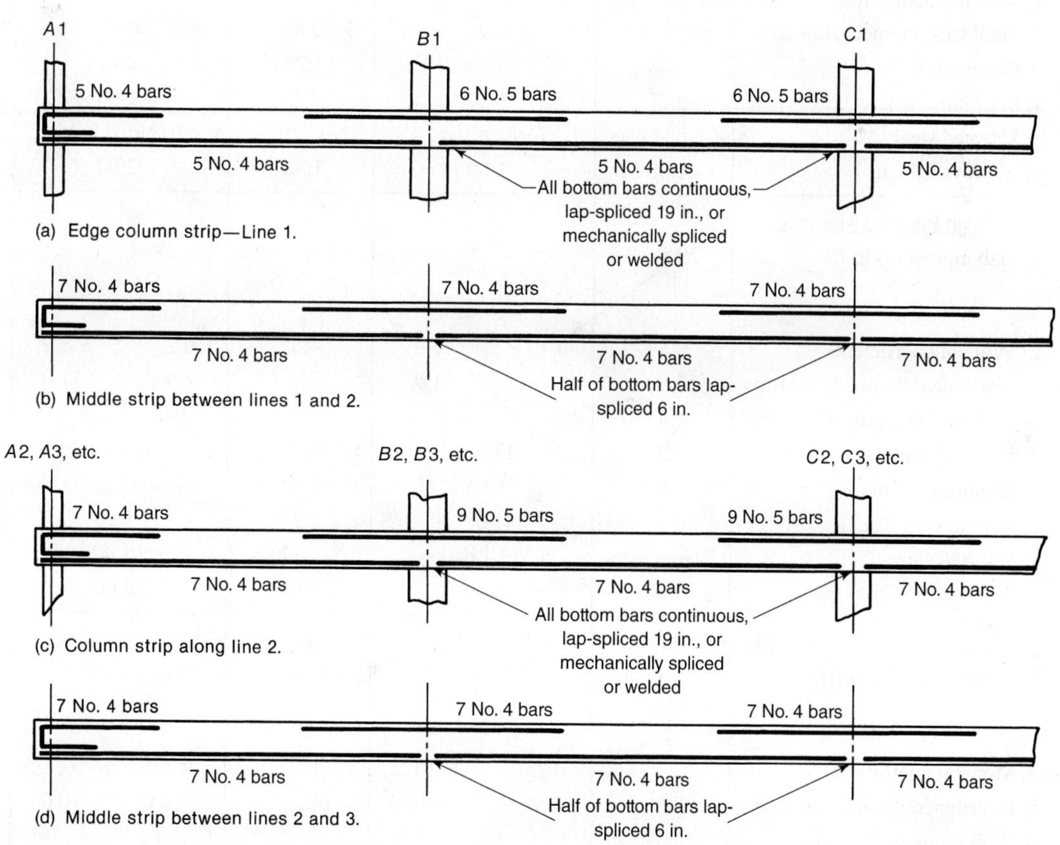

Fig. 13-88
Schematic diagram of reinforcement in east–west strips for Example 13-15.

north–south reinforcement can be designed. This proceeds in the same manner as did step 7, using the moments from lines 7 and 10 of Tables 13-11 and 13-12.

(a) Divide the slab strips into middle and column strips. See Fig. 13-85b.

(b) Divide the moments between the column and middle strips, and design the reinforcement. For the north–south strips, the calculations are carried out in Table 13-14. This table is laid out to represent a plan view of the column and middle strips shown in Fig. 13-85b.

TABLE 13-14 Division of Moment to Column and Middle Strips: North–South Strips—Example 13-15

	Column Strip	Middle Strip		Column Strip	Middle Strip		Edge Column Strip	
Strip Width, ft	4.77	9.88		7.88	12.12		7.88	
Exterior Negative Moments	***A1***			***B1***			***C1***	
1. Slab moment (kip-ft)	−14.6			−29.8			−29.8	
2. Moment coefficients	1.0	0.0	0.0	1.0	0.0	0.0	1.0	0.0
3. Distributed moments to strips	−14.6	0.0		−29.8	0.0		−29.8	
4. Wall moment (kip-ft)	−2.4							
5. Total strip moment (kip-ft)	−17.0	0.0		−29.8	0.0		−29.8	
6. Required A_s (in.^{2})	0.68	0		1.20	0		1.20	
7. Minimum A_s (in.^{2})	0.77	1.60		1.28	1.96		1.28	
8. Selected steel	5 No. 4 bars	8 No. 4 bars		7 No. 4 bars	10 No. 4 bars		7 No. 4 bars	
9. A_s provided (in.^{2})	1.00	1.60		1.40	2.00		1.40	
End Span Positive Moments								
1. Slab moment (kip-ft)	29.3			59.5			59.5	
2. Moment coefficients	0.6	0.4	0.2	0.6	0.2	0.2	0.6	0.2
3. Distributed moments to strips	17.6	11.7	11.9	35.7	11.9	11.9	35.7	
4. Wall moment (kip-ft)	4.9							
5. Total strip moment (kip-ft)	22.5	23.6		35.7	23.8		35.7	
6. Required A_s (in.^{2})	0.90	0.94		1.42	0.95		1.42	
7. Minimum A_s (in.^{2})	0.77	1.60		1.28	1.96		1.28	
8. Selected steel	5 No. 4 bars	8 No. 4 bars		8 No. 4 bars	10 No. 4 bars		8 No. 4 bars	
9. A_s provided (in.^{2})	1.00	1.60		1.60	2.00		1.60	
First Interior Negative Moment	***A2***			***B2***			***C2***	
1. Slab moment (kip-ft)	−39.4			−80.1			−80.1	
2. Moment coefficients	0.75	0.25	0.125	0.75	0.125	0.125	0.75	0.125
3. Distributed moments to strips	−29.6	−9.9	−10.0	−60.1	−10.0	10.0	−60.1	
4. Wall moment (kip-ft)	−6.6							

(Continued)

TABLE 13-14 Division of Moment to Column and Middle Strips: North–South Strips—Example 13-15 (*Continued*)

	Column Strip	Middle Strip	Column Strip	Middle Strip	Edge Column Strip	
Strip Width, ft	4.77	9.88	7.88	12.12	7.88	
5. Total strip moment (kip-ft)	−36.2	−19.9	−60.1	−20.0	−60.1	
6. Required A_s (in.^{2})	1.44	0.79	2.40	0.80	2.40	
7. Minimum A_s (in.^{2})	0.77	1.60	1.28	1.96	1.28	
8. Selected steel	5 No. 5 bars	8 No. 4 bars	8 No. 5 bars	10 No. 4 bars	8 No. 5 bars	
9. A_s provided (in.^{2})	1.55	1.60	2.48	2.00	2.48	
Interior Positive Moment						
1. Slab moment (kip-ft)	18.8		40.1		40.1	
2. Moment coefficients	0.60	0.40 0.20	0.60	0.20 0.20	0.60	0.20
3. Distributed Moments to strips	11.3	7.5 8.0	24.1	8.0 8.0	24.1	
4. Wall moment (kip-ft)	3.1					
5. Total strip moment (kip-ft)	14.4	15.5	24.1	16.0	24.1	
6. Required A_s (in.^{2})	0.58	0.62	0.96	0.64	0.96	
7. Minimum A_s (in.^{2})	0.77	1.60	1.28	1.96	1.28	
8. Selected steel	5 No. 4 bars	8 No. 4 bars	7 No. 4 bars	10 No. 4 bars	7 No. 4 bars	
9. A_s provided (in.^{2})	1.00	1.60	1.40	2.00	1.40	
Typical Interior Negative Moment	***A3***		***B3***		***C3***	
1. Slab moment (kip-ft)	−34.9		−74.4		−74.4	
2. Moment coefficients	0.75	0.25 0.125	0.75	0.125 0.125	0.75	0.125
3. Distributed moments to strips	−26.2	−8.7 −9.3	−55.8	−9.3 −9.3	−55.8	
4. Wall moment (kip-ft)	−5.8					
5. Total strip moment (kip-ft)	−32.0	−18.0	−55.8	−18.6	−55.8	
6. Required A_s (in.^{2})	1.28	0.72	2.23	0.74	2.23	
7. Minimum A_s (in.^{2})	0.77	1.60	1.28	1.96	1.28	
8. Selected steel	5 No. 5 bars	8 No. 4 bars	8 No. 5 bars	10 No. 4 bars	8 No. 5 bars	
9. A_s provided (in.^{2})	1.55	1.60	2.48	2.00	2.48	

Line 1. These moments are from line 7 of Tables 13-11 and 13-12.

Line 6. The area of steel is computed as in Table 13-13, except that d is smaller, as shown in Fig. 13-87.

$$d(\text{north–south}) \cong h - 1.7\text{ in.} = 7.5 - 1.7 = 5.8\text{ in.} \qquad (13\text{-}43b)$$

Compute a trial A_s:

$$\text{Largest } M_u = -60.1 \text{ kip-ft}$$

$$A_s\ (\text{req'd}) = \frac{60.1 \times 12{,}000}{0.9 \times 60{,}000 \times 0.95 \times 5.8} = 2.42 \text{ in.}^2$$

Compute a and check whether the section is tension-controlled:

$$a = \frac{A_s f_y}{0.85 f'_c\, b} = \frac{2.42 \times 60{,}000}{0.85 \times 4000(7.88 \times 12)} = 0.452 \text{ in.}$$

$$c = a/\beta_1 = 0.452/0.85 = 0.531 \text{ in.}$$

This is less than $3d/8 = 2.18$ in.; therefore, $\phi = 0.9$.

As before, compute jd and a constant for computing A_s:

$$jd = d - \frac{a}{2} = 5.8 - \frac{0.452}{2} = 5.57 \text{ in.}$$

$$A_s\ (\text{in.}^2) = \frac{M_u(\text{kip-ft}) \times 12{,}000}{0.9 \times 60{,}000 \times 5.57}$$

Therefore,

$$A_s\ (\text{in.}^2) = 0.0399 M_u\ (\text{kip-ft}) \qquad \text{(B)}$$

The values of A_s in lines 6 of Table 13-14 are computed from (B).

Line 7. For the column strips, the minimum A_s and minimum number of bars are as given in Table 13-13, because the column-strip width is the same in both directions. In the middle strips, different amounts of steel are required. We have

Middle strip between lines A and B:

$$A_{s,\min} = 0.0018(9.88 \times 12) \times 7.5 = 1.60 \text{ in.}^2$$

$$\text{Minimum number of bar spaces} = 7.9$$

$$\text{Minimum number of bars} = 8$$

Middle strip between lines B and C:

$$A_{s,\min} = 0.0018(12.12 \times 12) \times 7.5 = 1.96 \text{ in.}^2$$

$$\text{Minimum number of bars} = 10$$

Line 8. The final bar choices are given in line 8.

Line 9. The reinforcement chosen for each north–south strip is given in line 9 and shown in the sections in Fig. 13-89.

9. Check the shear at the exterior columns for combined shear and moment transfer. Either column $A2$ or column $B1$ will be the most critical. Because the tributary area is largest for $B1$, we shall limit our check to that column, although both joints should be checked.

Because the reinforcement is No. 4 bars at all exterior ends, the effective depths will be a little larger than are shown in Fig. 13-87. Thus, $d(\text{avg}) = h - {}^3\!/_4 \text{ in.} - d_b = 6.25$ in.

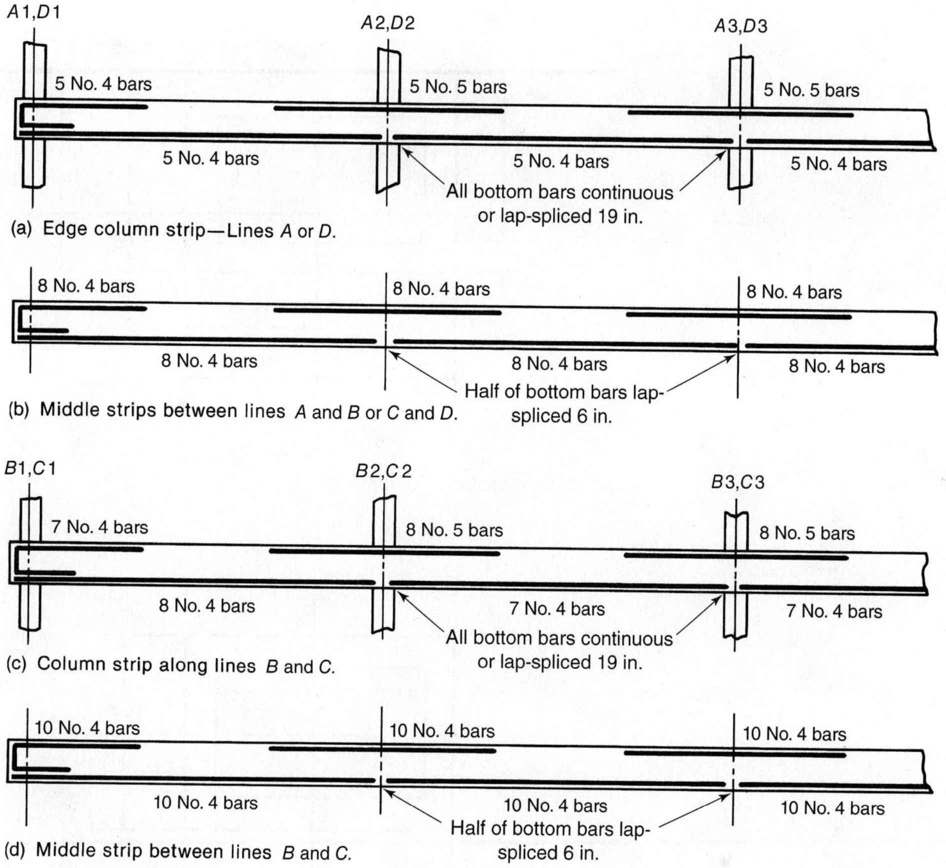

Fig. 13-89
Schematic diagram of reinforcement in north–south strips for Example 13-15.

(a) Locate the critical shear perimeter. The critical shear perimeter is at $d/2$ from the face of the column, where d is the average depth. Because the shortest perimeter results from the section shown in Fig. 13-90a, the perimeter dimensions are

$$b_1 = 16 \text{ in.} + d/2 = 19.1 \text{ in.}$$

$$b_2 = 20 \text{ in.} + d = 26.3 \text{ in.}$$

(b) Locate the centroid of the shear perimeter. For moments about the z–z axis,

$$y_{AB} = \frac{2 \times (19.1 \times 6.25) \times 19.1/2}{2(19.1 \times 6.25) + (26.3 \times 6.25)} = 5.66 \text{ in.}$$

Therefore, $c_{AB} = 5.66$ in. and $c_{CD} = 13.4$ in. For moments about the w–w axis,

$$c_{CB} = c_{AD} = \frac{26.3}{2} = 13.1 \text{ in.}$$

(c) Compute the shear and the moment about the centroid of the shear perimeter. As calculated in step 2(b), $V_u = 45.9$ kips.

For slabs designed by the direct-design method, the moment transferred from the slab to the column about axis z–z in Fig. 13-90a is $0.3M_o$, where, from line 5 in

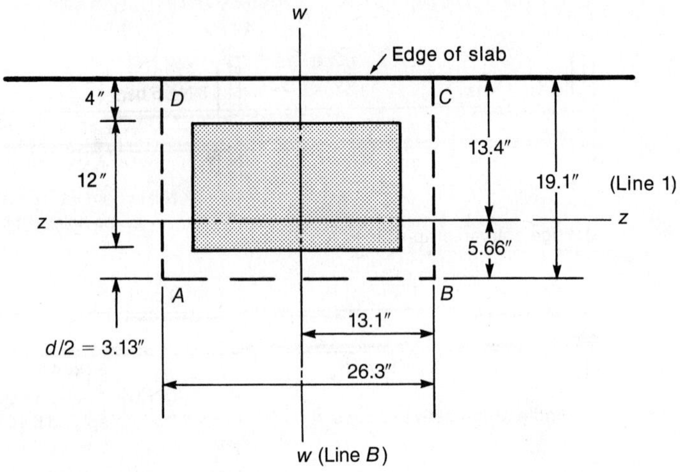

(a) Critical section—Column $B1$.

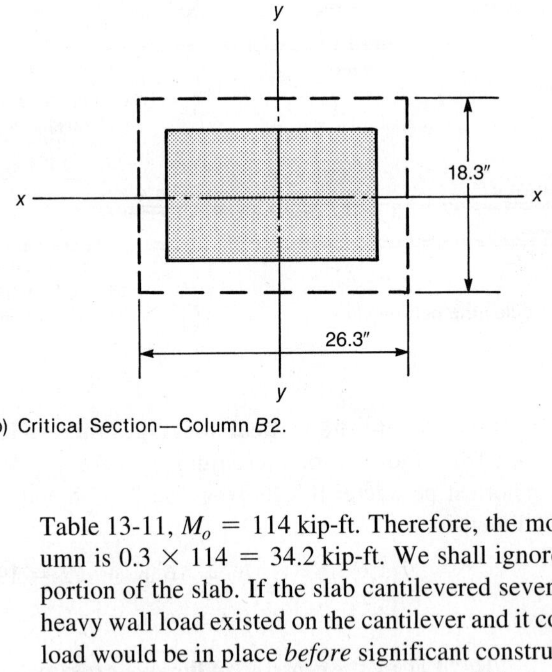

(b) Critical Section—Column $B2$.

Fig. 13-90
Final critical shear perimeters for columns $B1$ and $B2$—Example 13-15.

Table 13-11, $M_o = 114$ kip-ft. Therefore, the moment to be transferred to the column is $0.3 \times 114 = 34.2$ kip-ft. We shall ignore the moment from the cantilever portion of the slab. If the slab cantilevered several feet outside the column, or if a heavy wall load existed on the cantilever and it could be shown that this heavy wall load would be in place *before* significant construction or live loads were placed on the slab itself, the moment about z–z could be reduced.

Moments about axis w–w in Fig. 13-90a come from the slab strips parallel to the edge. Line 7 of Table 13-10 gives the moments for all panels loaded. Line 8 gives the moments transferred to the columns when one adjacent panel is loaded with half of the live load and the other panel is not. Because the critical shear check normally involves the shear due to all of the panels loaded, we shall use the moments from line 7. From line 7 in Table 13-10, we find that the moment of the edge-panel loads about axis w–w of column $B1$ is $53.0 - 45.9 = 7.1$ kip-ft. From line 13, we find that the unbalanced moment due to the wall load is 1.3 kip-ft. The total moment to be transferred is 8.4 ft-kips.

In summary, the connection at $B1$ must be designed for $V_u = 45.9$ kips, and $M_{zz} = 34.2$ kip-ft and $M_{ww} = 8.4$ kip-ft. These moments act about the centroids of the shear perimeter in the two directions.

(d) Determine the fraction of the moment transferred by flexure, γ_f.
Moments about the *z–z* axis:

$$\gamma_f = \frac{1}{1 + \frac{2}{3}\sqrt{b_1/b_2}} \tag{13-32}$$

where, for moment about the *z–z* axis, $b_1 = 19.1$ in. and $b_2 = 26.3$ in. Thus,

$$\gamma_{f1} = \frac{1}{1 + \frac{2}{3}\sqrt{\frac{19.1}{26.3}}} = 0.638$$

ACI Code Section 8.4.2.3.4 allows γ_f to be increased up to 1.0 if $v_u/\phi v_c$ does not exceed 0.75. From step 2 (b), $v_u/\phi v_c = 0.62$. Therefore, take $\gamma_{f1} = 1.0$, and check the strain, ε_t.

Moments about the *w–w* axis: Exchanging b_1 and b_2 in Eq. (13-32) gives $\gamma_{f2} = 0.561$. ACI Code Section 8.4.2.3.4 allows γ_f to be increased by up to 25 percent if $v_u/\phi v_c \leq 0.4$. Because $v_u/\phi v_c = 0.62$, no increase can be made in γ_{f2}. The moment transferred by flexure is $\gamma_{f2}M_u = 0.561 \times 8.4$ kip-ft $= 4.7$ kip-ft.

(e) Design the reinforcement required for moment transfer by flexure.
Moments about the *z–z* axis:

$$\text{Width effective for flexure: } c_2 + 3h = 20 \text{ in.} + 3 \times 7.5 \text{ in.} = 42.5 \text{ in.}$$

$$\text{Effective width} \leq c_2 + 2\,c_t \leq c_2 + 2\,c_1 \text{ (second expression governs)}$$

$$\text{Effective width} \leq 20 \text{ in.} + 2 \times 12 \text{ in.} = 44 \text{ in. (o.k.)}$$

Assume that $jd = 0.95d$ with $d \approx 5.8$ in.

$$A_s = \frac{M_u}{\phi f_y jd} = \frac{34.2 \times 12{,}000}{0.9 \times 60{,}000 \times 0.95 \times 5.8} = 1.38 \text{ in.}^2$$

The steel provided in step 8 is seven No. 4 bars (see Fig. 13-89c) in a column-strip width of 7.88 ft = 94.5 in., or roughly 13.5 in. on centers. The bars within the 42.5 in. effective width can be used for the moment transfer. Place four column-strip bars into this region and add three additional bars in this region, giving $A_s = 1.40$ in.^{2} in the effective width.

Because the computed A_s was based on a guess for jd, we shall compute a for the A_s chosen and recompute A_s as

$$a = \frac{A_s f_y}{0.85 f'_c\, b} = \frac{1.40 \times 60{,}000}{0.85 \times 4000 \times 42.5} = 0.581 \text{ in.}$$

$$A_s = \frac{M_u}{\phi f_y(d - a/2)} = \frac{34.2 \times 12{,}000}{0.9 \times 60{,}000(5.8 - 0.581/2)} = 1.38 \text{ in.}^2$$

Therefore, the steel chosen has adequate capacity.

$$c = \frac{a}{\beta_1} = \frac{0.581 \text{ in.}}{0.85} = 0.68 \text{ in.}$$

From linear strain diagram,

$$\varepsilon_t = \varepsilon_s = \frac{d - c}{c} \times \varepsilon_{cu}$$

$$= \frac{5.8 - 0.68}{0.68} \times 0.003 = 0.0225$$

Because ε_t is larger than 0.010, and we can use $\gamma_{f1} = 1.0$. As a result, it is not necessary to transfer any of the moment about axis z–z by eccentric shear stresses.

Moments about the *w*–*w* axis: Effective width for moment transfer is 12 in. + 4 in. + 1.5 × 7.5 in. = 27.3 in. For the portion of the transfer moment resisted by flexure (4.7 kip-ft), one No. 5 bar is required in this width. This easily will be accomplished if the bars are uniformly distributed in this region of the column strip.

(f) Compute the shear stresses.

$$v_u = \frac{V_u}{b_o d} \pm \frac{\gamma_{v2} M_{u2} c}{J_{c2}} \qquad (13\text{-}30)$$

where M_{u2}, and so on, refer to axis w–w. Here,

$$b_o = 2 \times 19.1 + 26.3 = 64.5 \text{ in.}$$

$$\gamma_{v2} = 1 - \gamma_{f2} = 0.439$$

Refer to Fig. 13-74b and Eq. (13-37) for the calculation of J_{c2}:

$$J_{c2} = \left(\frac{b_2 d^3}{12} + \frac{d b_2^3}{12}\right) + 2(b_1 d)c^2$$

$$= \frac{26.3 \times 6.25^3}{12} + \frac{6.25 \times 26.3^3}{12} + 2(19.1 \times 6.25) \times 13.1^2$$

$$= 51{,}000 \text{ in.}^4$$

The maximum stresses along sides *CB* and *AD* are

$$v_u = \frac{45.9 \times 1000}{64.5 \times 6.25} \pm \frac{0.439 \times 8.4 \times 12{,}000 \times 13.1}{51{,}000}$$

$$= 114 \pm 11 = 125 \text{ psi (at edge } BC \text{ and 103 psi at edge } AD)$$

Because there is no shear reinforcement in the slab, $\phi v_n = \phi v_c$, where, from step 2(b), $\phi v_c = 190$ psi.

Because $v_u < \phi v_n$, the shear is o.k. in this column–slab connection.

If $v_u > \phi v_c$, it would be necessary to modify the connection. Solutions would be to use a drop panel or shear cap, use stronger concrete, enlarge the column, or use shear reinforcement. The choice of solution should be based on a study of the extra costs involved.

10. Check the shear at an interior column for combined shear and moment transfer. Check the shear at column $B2$.

(a) Locate the critical shear perimeter. See Fig. 13-90b.

$$b_o = 2(18.3 + 26.3) = 89.2 \text{ in.}$$

(b) Locate the centroid of the shear perimeter. Because the perimeter is continuous, the centroidal axes pass through the centers of the sides.

(c) Compute the forces to be transferred. From step 2(b), $V_u = 75.6$ kips. The moments to be transferred come from Tables 13-9 and 13-11. Line 7 of these tables gives the moments for all panels loaded. Line 8 of these tables gives moments transferred to the columns when one adjacent panel is loaded with reduced live load and the other panel is not. Because the shear check involves the shear due to all panels loaded, we shall use the moments from line 7. In line 7 of Table 13-9, the difference between the negative moments on the two sides of column $B2$ is $95.9 - 82.9 = 13.0$ kip-ft. From Table 13-11, it is $80.1 - 74.4 = 5.7$ kip-ft.

(d) Compute the fraction of the moment transferred by flexure.

$$\gamma_{f1}(\text{about } x\text{–}x \text{ axis}) = 0.643$$

$$\gamma_{f2}(\text{about } y\text{–}y \text{ axis}) = 0.556$$

ACI Code Section 8.4.2.3.4 allows γ_f to be increased by up to 25 percent if $v_u/\phi v_c \leq 0.4$. As shown in step 2(b), $v_u/\phi v_c = 0.741$, so γ_f cannot be increased about either axis.

(e) Compute the torsional moment of inertia, J_c. Bending about x–x axis z–z using Eq. (13-34):

$$J_{c1} = 2\left(\frac{b_1 d^3}{12} + \frac{d b_1^3}{12}\right) + 2(b_2 d)\left(\frac{b_1}{2}\right)^2$$

where $d = 6.25$ in., $b_1 = 18.3$ in., and $b_2 = 26.3$ in. (see Fig. 13-90b). Thus,

$$J_{c1} = 34{,}700 \text{ in.}^4$$

For bending about axis y–y of column $B2$, $b_1 = 26.3$ in., $b_2 = 18.3$ in., and $J_{c2} = 59{,}600$ in.^{4}

(f) Design the reinforcement for moment transfer. By inspection, the reinforcement already in the slab is adequate.

(g) Compute the shear stresses. The combined shear stress calculation, represented by Eq. (13-30), assumes bending about one principal axis at a time and thus the calculation of a maximum shear stress along one edge of the critical section. In this case, it is not clear which moment will cause the largest combined shear stress, so both will be calculated to determine which one is critical. For bending about axis x–x,

$$M_u(\text{shear transfer}) = \gamma_{v1} \times 5.7 \text{ kip-ft} = 0.357 \times 5.7$$

$$= 2.03 \text{ kip-ft} = 24{,}400 \text{ lb-in.}$$

Then,

$$\frac{\gamma_{v1} M_u c}{J_{c1}} = \frac{24{,}400 \times 9.15}{34{,}700} = 6.4 \text{ psi}$$

For bending about axis *y–y*,

$$M_u(\text{shear transfer}) = \gamma_{v1} \times 13.0 \text{ kip-ft} = 0.444 \times 13.0$$
$$= 5.77 \text{ kip-ft} = 69{,}300 \text{ lb-in.}$$

Then,

$$\frac{\gamma_{v2} M_u c}{J_{c2}} = \frac{69{,}300 \times 13.1}{59{,}600} = 15.2 \text{ psi (use this)}$$

So

$$v_u(\max) = \frac{75{,}600 \text{ lbs.}}{89.2 \text{ in.} \times 6.25 \text{ in.}} + 15.2 \text{ psi}$$
$$= 136 \text{ psi} + 15.2 \text{ psi} = 151 \text{ psi} < \phi v_c = 190 \text{ psi}$$

Because $\phi v_c > v_u$, the shear is o.k. at this column. In this case, the moment transfer increased the shear stress by only 11 percent.

11. Check the shear at the corner column. Tables 13-10 and 13-12 indicate that moments of 20.2 kip-ft and 17.0 kip-ft are transferred from the slab to the corner column, *A*1, from strip 1 and strip *A*, respectively. Two shear perimeters will be considered. Two-way shear may be critical on the perimeter shown in Fig. 13-91a, while one-way shear may be critical on the section shown in Fig. 13-91b.

Two-Way Shear

(a) Locate the critical perimeter. The critical perimeter is as shown in Fig. 13-91a. For two-way shear, we shall base the calculations of the stresses on the orthogonal *x*- and *y*-axes.

(b) Locate the centroid of the perimeter.

$$\bar{x} = \frac{(19.1 \times 6.25)(19.1/2)}{2 \times 19.1 \times 6.25}$$
$$= 4.78 \text{ in. from inside corner}$$

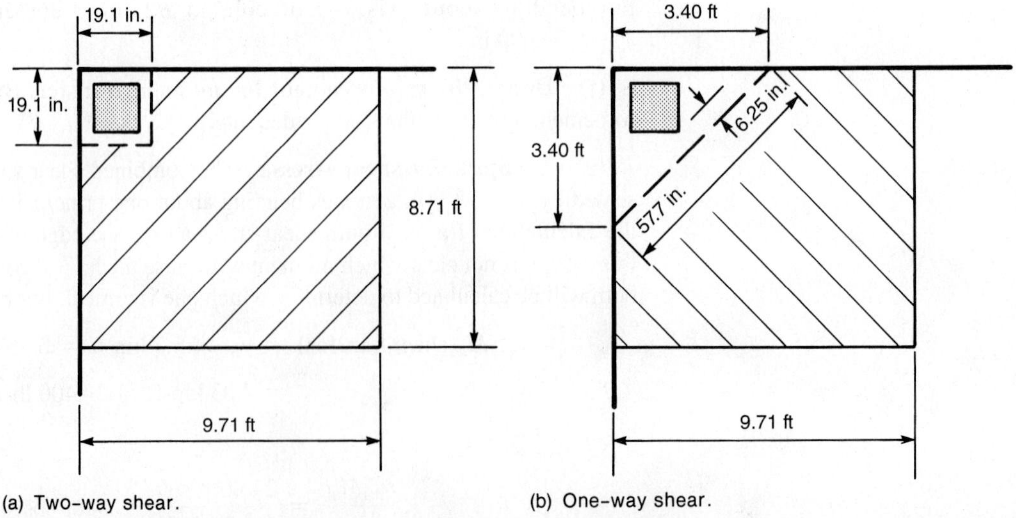

Fig. 13-91
Critical shear perimeters and tributary areas for column *A*1—Example. 13-15.

(c) Compute the forces to be transferred. Factored shear on the area bounded by lines of zero shear at 0.5 times the span length from the exterior end of the panel plus the distance from the center of the column to the edge:

$$V_u = 0.223 \text{ ksf} \times \left[(0.5 \times 17.75 + 0.83) \times (0.5 \times 15.75 + 0.83) - \frac{19.1 \times 19.1}{144}\right]$$
$$= 18.3 \text{ kips}$$

$$\text{Shear due to wall load} = \left[(8.88 + 0.83) + (7.88 + 0.83) - \frac{2 \times 19.1}{12}\right] \times 1.2 \times 0.30 \text{ kips/ft}$$
$$= 5.5 \text{ kips}$$

$$\text{Total shear} = 18.3 + 5.5 = 23.8 \text{ kips}$$

ACI Code Section 8.10.7.3 states that the amount of moment to be transferred from a slab to an edge column is $0.3M_o$. Although no similar statement is made for a corner column, we shall assume that the same moment should be transferred.

$$\text{For strip 1: } M_o = 65.5 \text{ kip-ft}, 0.3M_o = 19.7 \text{ kip-ft}$$

$$\text{For strip } A\text{: } M_o = 56.3 \text{ kip-ft}, 0.3M_o = 16.9 \text{ kip-ft}$$

Again, it will be assumed that these moments and the shears all act through the centroid of the shear perimeter.

(d) Determine the fraction of the moment transferred by flexure. Because the column is square, we shall take $\gamma_f = 0.60$, which is the value for a square interior column. The ACI Code does not give any values of γ_f for corner columns, but we shall assume the same equation can be used that is used for interior columns. ACI Code Section 8.4.2.3.4 allows γ_f to be increased to 1.0, provided that $v_u/\phi v_c \leq 0.5$ in the strip of slab that is effective for moment transfer. In this case, the transfer width = 4 in. + 12 in. + 1.5 × 7.5 in. = 27.3 in.

$$b_o = 2 \times 19.1 = 38.2 \text{ in.}$$

For shear strength, Eq. (13-24) governs and $\phi v_c = 190$ psi.

$$v_u = V_u/b_o d = 23{,}800 \text{ lb}/(38.2'' \times 6.25'') = 99.7 \text{ psi}$$

and

$$v_u/\phi v_c = 99.7/190 = 0.525 > 0.5$$

Thus, we will use $\gamma_f = 0.6$.

(e) Design the reinforcement for moment transfer by flexure. In Table 13-13 for design of the slab strips in the east–west direction (Fig. 13-85a), the column-strip reinforcement in slab strip 1 at the exterior connection was designed to resist $0.26\,M_o$. The required slab flexural reinforcement was equal to 0.73 in.^{2}. The design transfer moment is equal to $0.30\,M_o$, but only 60 percent of that moment is to be resisted by slab flexural reinforcement. Thus, the required area of slab flexural reinforcement in the transfer width is,

$$A_s \text{ (req'd)} = 0.73 \text{ in.}^2 \times \frac{0.30}{0.26} \times 0.60 = 0.51 \text{ in.}^2$$

To meet this requirement, place three No. 4 bars ($A_s = 0.60$ in.^{2}) within the 27.3 in. transfer width for slab strip 1. The other two required No. 4 bars should be distributed in the remaining part of the column strip.

Through a similar calculation for the column strip of slab strip A in the north–south direction (Fig. 13-85b), the required area of slab reinforcement within the transfer width was 0.47 in.^{2}. To meet this requirement place three No. 4 bars within the 27.3 in. transfer width.

(f) Compute the torsional moment of inertia, J_c. Bending about either principal axes, use Eq. (13-39):

$$J_c = \frac{b_1 d^3}{12} + \frac{d b_1^3}{12} + (b_1 d)\left(\frac{b_1}{2} - c\right)^2 + (b_2 d)c^2$$

$$J_c = \frac{19.1 \times 6.25^3}{12} + \frac{6.25 \times 19.1^3}{12}$$

$$+ (19.1 \times 6.25)\left(\frac{19.1}{2} - 4.78\right)^2 + (6.25 \times 19.1)(4.78)^2$$

$$= 9460 \text{ in.}^4$$

(g) Compute the shear stresses. As stated earlier, the ACI eccentric shear-stress equations were developed for bending about only one principal axis. However, for a corner column we may want to consider simultaneous bending about both principal axes. Because it is quite conservative to be checking a critical shear stress at a single point (the corner of the critical section), we should use some engineering judgment when determining the appropriate stress limit. In this case, we shall use 1.15 times ϕv_c, as suggested by ACI Committee 421 [13-36], as the upper limit for v_u in the corner of the critical section.

$$v_u = \frac{v_u}{b_o d} \pm \frac{\gamma_{v1} M_{u1} c}{J_{c1}} \pm \frac{\gamma_{v2} M_{u2} c}{J_{c2}}$$

All of these terms add at the inside corner to give the maximum v_u:

$$v_u = \frac{23.8 \times 1000}{38.2 \times 6.25} + \frac{0.4 \times 19.7 \times 12{,}000 \times 4.78}{9460}$$

$$+ \frac{0.4 \times 16.9 \times 12{,}000 \times 4.78}{9460}$$

$$= 100 + 48 + 41 = 189 \text{ psi}$$

$$1.15 \times \phi v_c = 1.15 \times \phi 4\lambda\sqrt{f'_c} = 1.15 \times 190 \text{ psi} = 219 \text{ psi}$$

Thus, $1.15\phi v_c$ is greater than v_u, and the corner column *is adequate in two-way shear.*

One-Way Shear. This check of one-way shear is not discussed in the ACI Code and thus is not made by all designers. The author, however, believes that it is appropriate to make this check at corner columns.

(a) Locate the critical section. The critical section is as shown in Fig. 13-91b. This section is located $d_{avg} = 6.25$ in. from the corner of the column.

(b) Compute the shear on the critical section.

$$V_u = 0.223[(8.88 + 0.83) \times (7.88 + 0.83) - 3.40^2/2] = 17.6 \text{ kips}$$

Shear due to wall load: $[(9.71 - 3.40) + (8.71 - 3.40)] \times 1.2 \times 0.30 =$ 4.2 kips

Total shear $= 17.6 + 4.2 = 21.8$ kips

(c) ϕV_c for the critical section is

$$\phi V_c = \phi 2\lambda \sqrt{f'_c}\, bd = 0.75 \times 2 \times 1\sqrt{4000} \times 57.7 \text{ in.} \times 6.25 \text{ in.}$$
$$= 34{,}200 \text{ lbs} = 34.2 \text{ kips}$$

Therefore, the slab is o.k. in one-way shear.

This completes the design of the slab. The bar cutoffs are calculated by using Fig. 13-84. ■

13-14 DESIGN OF SLABS WITH BEAMS IN TWO DIRECTIONS

Because of its additional depth, a beam is stiffer than the adjacent slab, and thus, it attracts additional loads and moments. This was discussed in Section 13-5 and illustrated in Figs. 13-12 and 13-15. The average moments in the column strip are almost the same in a flat plate (Fig. 13-14d) and in a slab with beams between all columns (Fig. 13-15b). In the latter case, the column-strip moment is divided between the slab and the beam. This reduces the reinforcement required for the slab in the column strip because the beam must be reinforced to carry most of the load.

The greater stiffness of the beams reduces the overall deflections, allowing a thinner slab to be used than in the case of a flat plate. Thus, an advantage of slabs with beams in two directions lies in their reduced weight. Also, two-way shear does not govern for most two-way slabs with beams, again allowing thinner slabs. This is offset by the increased overall depth of the floor system and increased forming and reinforcement-placing costs.

The direct-design method for computing moments in the slab and beams is the same as the procedure used in slabs without beams, with one additional step. Thus, the designer will, as usual:

1. Compute M_o.

2. Divide M_o between the positive- and negative-moment regions.

3. Divide the positive and negative moments between the column and middle strips.

The additional step needed is

4. Divide the column-strip moments between the beam and the slab.

The amount of moment assigned to the column and middle strips in step 3 and the division of moments between the beam and slab in step 4 are a function of $\alpha_{f1}\ell_2/\ell_1$, where is α_{f1} is the beam–slab stiffness ratio in the direction in which the reinforcement is being designed. (See Example 13-1.)

When slabs are supported on beams having $\alpha_{f1}\ell_2/\ell_1 \geq 1.0$, the beams must be designed for shear forces computed by assuming tributary areas bounded by 45° lines at the corners of the panels and the centerlines of the panels, as is shown in Fig. 13-92. If the beams have $\alpha_{f1}\ell_2/\ell_1$ between 0 and 1.0, the shear forces computed from these tributary areas are multiplied by $\alpha_{f1}\ell_2/\ell_1$. In such a case, the remainder of the shear must be transmitted to the column by shear in the slab. The ACI Code is silent on how this is to be done. The most common interpretation involves using two-way shear in the slab between the beams and one-way shear in the beams, as shown in Fig. 13-93. Frequently, problems are encountered when $\alpha_{f1}\ell_2/\ell_1$ is less than 1.0, because the two-way shear perimeter is inadequate to transfer the portion of the shear not transferred by the beams. Thus it is recommended to select beam sizes such that $\alpha_{f1}\ell_2/\ell_1$ exceeds 1.0 for a two-way slab.

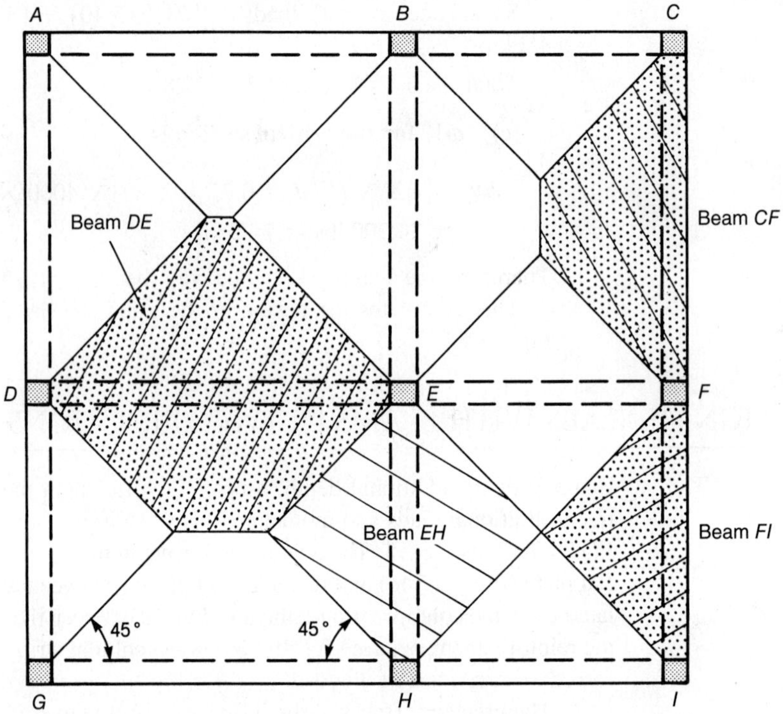

Fig. 13-92
Tributary areas for computing shear in beams supporting two-way slabs.

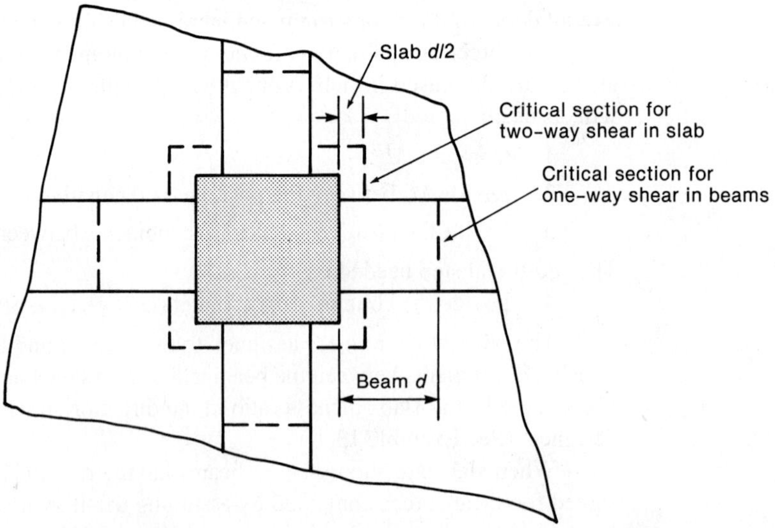

Fig. 13-93
Shear perimeters in slabs with beams.

The size of the beams also is governed by their shear and flexural strengths. The cross section should be large enough so that $V_u \leq \phi(V_c + V_s)$, where an upper practical limit on $V_c + V_s$ would be about $(6\sqrt{f'_c}\, b_w d)$. The critical location for flexure is the point of maximum negative moment, where the reinforcement ratio, ρ, should be approximately equal to the value obtained from Eq. (5-19).

EXAMPLE 13-16 Design of a Two-Way Slab with Beams in Both Directions—Direct-Design Method

Figure 13-94 shows the plan for a part of a floor with beams between all columns. For simplicity in forming, the beams have been made the same width as the columns. The floor supports its own weight, superimposed dead loads of 5 psf for ceiling and mechanical fixtures and 25 psf for future partitions, plus a live load of 80 psf. The exterior wall weighs 300 lb/ft and is supported by the edge beam. The heights of the stories above and below the floor in question are 12 ft and 14 ft, respectively. Lateral loads are resisted by an elevator shaft not shown in the plan. Design the east–west strips of slab along column lines *A* and *B*, using normal-weight 4000-psi concrete and Grade-60 reinforcement. Use load factors and strength-reduction factors from Table 5.3.1 of the ACI Code.

This example illustrates several things not included in Example 13-15, including

(a) the effect of edge beams on the required thickness,

(b) the use of live-load reduction factors,

(c) the effect of beams on the division of moments between the slab and beam in the column strip,

(d) the distribution of negative moments where the direct-design method gives large differences between the moments on each side of a support, and

(f) the calculation of shear in the beam and slab.

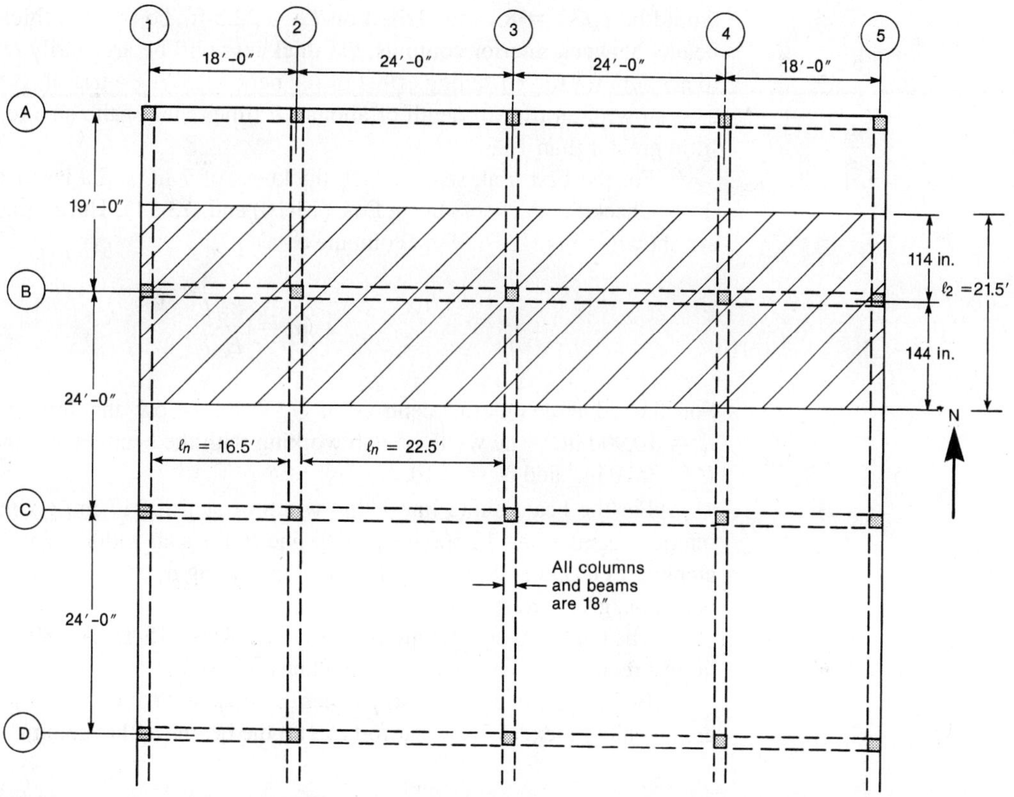

Fig. 13-94
Plan of a two-way slab with beams for Example 13-16.

For live loads of 100 psf or less, on panels having an area greater than 400 ft^2, the live-load may be reduced based on Eq. (2-12):

$$L = L_o\left(0.25 + \frac{15}{\sqrt{K_{LL}A_T}}\right)$$

where L and L_o are the reduced and specified live loads, respectively, A_T is the tributary area and K_{LL} is a factor to convert the tributary area to an influence area. In the case of a two-way slab, the tributary area, A_T, normally is taken as the panel area measured from center-to-center of the columns and K_{LL} is equal to 1.0.

1. Select the design method, load, and resistance factors. ACI Code Section 8.10.2 sets limits on the use of the direct-design method. These are checked as in step 1 of Example 13-15. One additional check is necessary (ACI Code Section 8.10.2.7): The beams will be selected so that all are the same width and depth. As a result, the ratio $\alpha_{f1}\ell_2^2/\alpha_{f2}\ell_1^2$ should fall within the bounds given. Thus, the direct-design method could be used, but the author will use a frame analysis model, so wind loads could be included in a later design step.

2. Select the slab thickness and beam size. The slab thickness is chosen to satisfy deflection requirements once the beam size is known. If $\alpha_{f1}\ell_2/\ell_1$ exceeds 1.0 for all beams, all the shear is transferred to the columns by the beams, making it unnecessary to check shear while selecting the slab thickness. If there were only edge beams, the minimum slab thickness for deflection would be governed by Table 13-1 and would be $\ell_n/33 = 8.18$ in., based on $\ell_n = 22.5$ ft. To select a thickness for a slab with beams between interior columns, the thickness will be arbitrarily reduced by 15 percent to account for the stiffening effect of the beams, giving a trial thickness of 7 in. Assume a beam with an overall depth of about 2.5 times that of the slab to give a value of α_f a little greater than 1.0.

For the first trial, select a slab thickness of 7 in. and a beam 18 in. wide by 18 in. deep. Check the thickness using Eqs. (13-10) and (13-11). The cross sections of the beams are shown in Fig. 13-95. First compute α_f as:

$$\alpha_f = \frac{E_{cb}I_b}{E_{cs}I_s} \tag{13-9}$$

For the edge beam, the centroid is 7.94 in. below the top of the section, giving $I_b = 10{,}900$ in.^{4}. The width of slab working with the beam along line A is 123 in., giving $I_s = 3520$ in.^{4} and $\alpha_f = 3.10$.

For the beam along line 1, the width of slab is 117 in., giving $\alpha_f = 3.26$. For the interior beams, $I_b = 12{,}500$ in.^{4}; along line B, the slab width is 258 in., giving $\alpha_f = 1.70$; along lines C and 3, the slab width is 288 in., giving $\alpha_f = 1.52$; along line 2, the slab width is 252 in., giving $\alpha_f = 1.74$.

The thickness computations are given in Table 13-15. A 6.89-in. slab thickness is the largest required value, so the 7-in. thickness chosen is satisfactory.

Before continuing, check whether the shears at the column with the largest tributary area would put excessive shear forces into the beams at that column:

$$q_u = 1.2\left(\frac{7}{12} \times 0.15 + 0.005 + 0.025\right) + 1.6 \times 0.080 = 0.269 \text{ ksf}$$

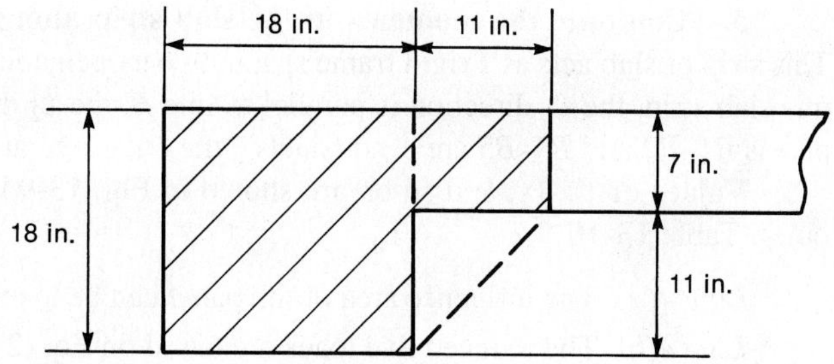

(a) Edge beam.

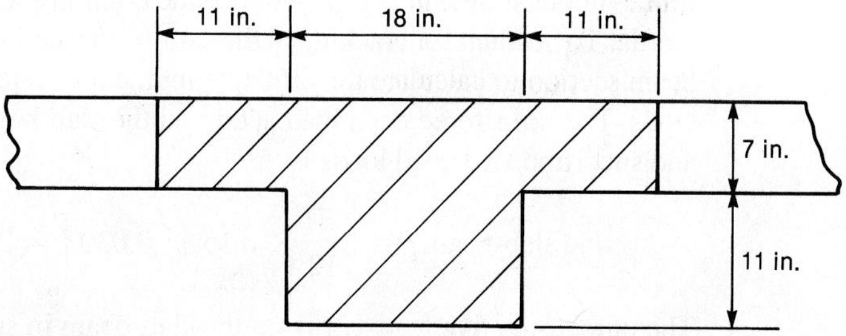

(b) Interior beam.

Fig. 13-95
Cross sections of beams for Example 13-16.

Ignoring the weight of the beam and assuming that the value of d is $(18 - 2.5) = 15.5$ in. Then,

$$V_u = 0.269\left[24 \text{ ft} \times 24 \text{ ft} - \left(\frac{18 + d(\text{beam})}{12}\right)^2\right] = 153 \text{ kips}$$

ϕV_c for the four beams is (using $\lambda = 1.0$):

$$\Sigma\, \phi V_c = 4\left[0.75 \times 2 \times 1\sqrt{4000} \times 18 \times 15.5\right] \times \frac{1 \text{ kip}}{1000 \text{ lbs}} = 106 \text{ kips}$$

Thus, $V_u \le 2\Sigma\phi V_c$, which will not be excessive if stirrups are used. Therefore, use the beams selected earlier.

TABLE 13-15 Computation of Minimum Thickness—Example 13-16

	Panel			
	A–B–1–2 (Corner)	*A–B*–2–3	*B–C*–1–2	*B–C*–2–3 (Interior)
Maximum ℓ_n	19 ft 18 in.	24 ft 18 in.		
	210 in.	270 in.	270 in.	270 in.
Minimum ℓ_n	18 ft 18 in.			
	198 in.	210 in.	198 in.	270 in.
β	$\frac{210}{198} = 1.06$	1.29	1.36	1.00
α_{fm}	$\frac{3.10 + 1.74 + 1.70 + 3.26}{4}$	2.02	2.06	1.62
	$= 2.45$			
Applicable equation	(13-11)	(13-11)	(13-11)	(13-10)
h from equation	5.07	6.24	6.16	6.89
Minimum h	3.5	3.5	3.5	5.0

3. Compute the moments in the slab strip along column line *B* (Fig. 13-94). This strip of slab acts as a rigid frame spanning between columns *B*1, *B*2, *B*3, and so on. In this slab strip, the ℓ_1 direction is parallel to line *B*, the ℓ_2 direction is perpendicular. Slab panels *B*1–*B*2 and *B*4–*B*5 are "end panels"; the other two are "interior panels."

Values of ℓ_n, ℓ_2, and so on are shown in Fig. 13-94. The calculations are carried out in Table 13-16.

Line 4(a). The influence area of the panel can be taken as $\ell_1 \times \ell_2$.

Line 4(b). The reduced live loads are based on Eq. (2-12).

Line 5. Design moments at the ends of the spans and midspan are calculated using a frame model shown in Fig. 13-96. For the column sections, use the gross moment of inertia. To account for cracking in the slab-beam section, use just the 18-in.-by-18-in.-beam section to calculate the effective moment of intertia for this analysis of gravity loads. The unfactored dead load acting on the slab-beam in each span due to the slab and superimposed dead loads is

$$w_D(\text{slab-beam}) = \left[\frac{7}{12} \times 0.15 + 0.030\right] \times 21.5 \text{ ft} = 2.53 \text{ k/ft}$$

The unfactored live load acting on the slab-beam in span *B*1-*B*2 is:

$$w_L = 0.080 \text{ ksf} \times 21.5 \text{ ft} = 1.72 \text{ k/ft}$$

The unfactored reduced live load for slab-beam span *B*2–*B*3 is 1.57 k/ft.

TABLE 13-16 Calculation of Negative and Positive Moments for East–West Slab Strip *B*—Example 13-16

	*B*1		*B*2			*B*3
1. ℓ_1 (ft)		18			24	
2. ℓ_n (ft)		16.5			22.5	
3. ℓ_2 (ft)		21.5			21.5	
4. (a) Area, A_I (ft^2)		387			516	
(b) Reduced live load (ksf)		0.080			0.073	
	End	Mid	End	End	Mid	End
5. Slab-beam moments (kip-ft)	−76.6	74.4	−175	−219	121	−242
6. Beam moments (kip-ft)	−3.3	3.2	−7.7	−9.9	5.5	−11.0

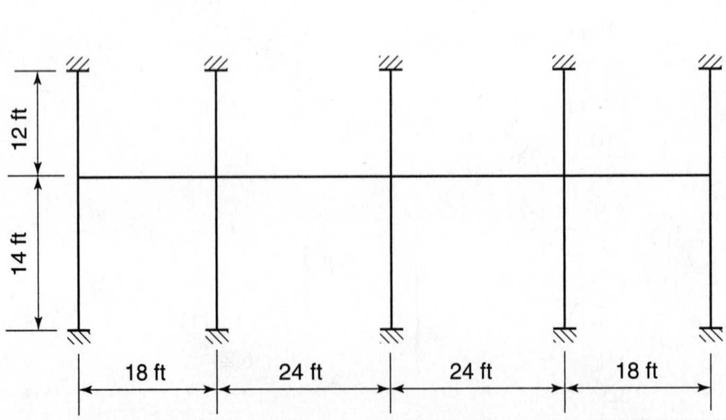

Fig. 13-96
Frame used for the analysis of the two-way slab in Example 13-16.

Because these values are less than three-quarters of the dead load, ACI Code Section 6.4.3.2 states that only the load case of full-factored live and dead load on all spans needs to be considered (i.e., can ignore pattern loading). The design moments that result from applying this load case to the frame shown in Fig. 13-96, using reduced live loads in the appropriate spans, are given on line 5 of Table 13-16. Column end moments also are obtained from this analysis, but they are not included in this example.

Line 6. Because the beam weight will be resisted only by the beam, a separate analysis was used for this dead load that acts directly on the beam.

$$w_D\,(\text{beam}) = \frac{18 \times 11}{144}(\text{ft}^2) \times 0.150\ \text{k/ft}^3 = 0.21\ \text{k/ft}$$

4. Compute the moments in the slab strip along column line *A* (Fig. 13-94). This strip of slab acts as a rigid frame spanning between columns *A*1, *A*2, *A*3 and so on. The slab strip includes an edge beam (*A*1–*A*2, *A*2–*A*3, etc.) parallel to the slab spans. In this slab strip, panels *A*1–*A*2 and *A*4–*A*5 are "end panels"; the other two are "interior panels." The calculations are carried out in Table 13-17 and proceed in the same way as the calculations in Table 13-16, except as noted below:

(a) Because of the small influence areas for this span along the edge of the floor system, there are no live load reductions for any of the spans.

(b) For the structural analysis, the unfactored dead and live loads acting on the slab-beam are:

$$w_D(\text{slab-beam}) = \left[\frac{7}{12} \times 0.15 + 0.030\right] \times 10.25\ \text{ft} = 1.21\ \text{k/ft}$$

$$w_L = 0.080\ \text{ksf} \times 10.25\ \text{ft} = 0.82\ \text{k/ft}$$

As before, a separate analysis will be done for the dead load applied directly to the beam, which must also include the weight of the wall along the edge of the slab:

$$w_D\,(\text{beam}) = \frac{18 \times 11}{144} \times 0.15 + 0.300 = 0.51\ \text{k/ft}$$

5 and 6. Compute the moments in the north–south slab strips. Although a complete solution for this slab would include computation of moments in north–south strips, as well as in east–west strips, these will be omitted here.

TABLE 13-17 Calculation of Negative and Positive Moments for East–West Slab Strip A—Example 13-16

	*A*1		*A*2			*A*3
1. ℓ_1 (ft)		18			24	
2. ℓ_n (ft)		16.5			22.5	
3. ℓ_2 (ft)		10.25			10.25	
4. (a) Area, A_I (ft^2)		185			246	
(b) Reduced live load (ksf)		0.080			0.080	
	End	Mid	End	End	Mid	End
5. Slab-beam moments (kip-ft)	−35.9	35.1	−84.3	−108	60.1	−120
6. Beam moments (kip-ft)	−8.0	7.8	−18.6	−24.0	13.4	−26.8

7. For strips spanning east–west, distribute the negative and positive moments to the column strips, the middle strips, and to the beams. In steps 3 and 4, the moments along column lines *A* and *B* were computed. These moments now must be distributed to column and middle strips, and the column-strip moment must be divided between the slab and the beam.

(a) Divide the slabs into column and middle strips. The widths of the column strips as defined in ACI Code Section 8.4.1.5 will vary from span to span. Along line *B*, the column-strip width is 9.0 ft in span *B*1–*B*2 and 10.75 ft in span *B*2–*B*3. For consistency in design of slab reinforcement from span to span, **assume a constant-width column strip of width 9 ft.**

(b) Divide the moments between the column strips and middle strips. These calculations are carried out in Table 13-18, which is laid out to resemble a plan of the slab. Arrows illustrate the flow of moments to the various parts of the slab. The division of moments is a function of both the beam-stiffness ratio for the beam parallel to the strip being designed and the aspect ratio of the panel. For the east–west strips, these terms are summarized subsequently.

Values of α_f were calculated in step 2, and values of ℓ_1 and ℓ_2 are given in Tables 13-16 and 13-17. For the interior strips, ℓ_2 is taken equal to the value used in calculating the moments in the strips. We have the following calculations:

$$\text{Panel } B1\text{–}B2\text{: } \ell_1 = 18 \text{ ft}, \ell_2 = 21.5 \text{ ft}, \alpha_{f1} = 1.70, \frac{\ell_2}{\ell_1} = 1.19, \frac{\alpha_{f1}\ell_2}{\ell_1} = 2.03$$

$$\text{Panel } B2\text{–}B3\text{: } \ell_1 = 24.0 \text{ ft}, \ell_2 = 21.5 \text{ ft}, \alpha_{f1} = 1.70, \frac{\ell_2}{\ell_1} = 0.90, \frac{\alpha_{f1}\ell_2}{\ell_1} = 1.52$$

$$\text{Panel } C1\text{–}C2\text{: } \ell_1 = 18 \text{ ft}, \ell_2 = 24.0 \text{ ft}, \alpha_{f1} = 1.52, \frac{\ell_2}{\ell_1} = 1.33, \frac{\alpha_{f1}\ell_2}{\ell_1} = 2.03$$

$$\text{Panel } C2\text{–}C3\text{: } \ell_1 = 24.0 \text{ ft}, \ell_2 = 24.0 \text{ ft}, \alpha_{f1} = 1.52, \frac{\ell_2}{\ell_1} = 1.0, \frac{\alpha_{f1}\ell_2}{\ell_1} = 1.52$$

The ACI Code is not clear about which ℓ_2 should be used when considering an edge panel. In current practice, most designers take ℓ_2 equal to the total width of the edge panel. Then

$$\text{Panel } A1\text{–}A2\text{: } \ell_1 = 18 \text{ ft}, \ell_2 = 19 \text{ ft}, \alpha_{f1} = 3.10, \frac{\ell_2}{\ell_1} = 1.06, \frac{\alpha_{f1}\ell_2}{\ell_1} = 3.27$$

$$\text{Panel } A2\text{–}A3\text{: } \ell_1 = 24.0 \text{ ft}, \ell_2 = 19 \text{ ft}, \alpha_{f1} = 3.10, \frac{\ell_2}{\ell_1} = 0.79, \frac{\alpha_{f1}\ell_2}{\ell_1} = 2.45$$

Line 1. These moments come from lines 5 of Tables 13-16 and 13-17. The moments in strip *C* are from a similar set of calculations. **Note** that the larger of the two slab moments is used at the first interior support.

Line 2. *Exterior negative moments:* Table 13-5 gives the fraction of the exterior negative moment resisted by the column strip. This is computed from values for ℓ_2/ℓ_1 and $\alpha_{f1}\ell_2/\ell_1$ just calculated, plus

$$\beta_t = \frac{E_{cb}C}{2E_{cs}I_s} \tag{13-12}$$

where C is the torsional constant for the beam, computed from Eq. (13-13). The edge-beam cross section effective for torsion is defined in ACI Code Section 8.11.5.

TABLE 13-18 Division of Moment to Column and Middle Strips: East–West Strips—Example 13-16

Strip Width, ft	Column Strip 9.0		Middle Strip 15.0		Column Strip 9.0		Middle Strip 10.3		Edge Column Strip 5.0	
<u>Exterior Negative Moments</u>	***C1***				***B1***				***A1***	
1. Slab moment (kip-ft)	−80.3				−76.6				−35.9	
2. Moment coefficients	0.88		0.06	0.06	0.88		0.06	0.21	0.79	
3. Moments to columns and middle strips (kip-ft)	−70.1		−4.8	−4.6	−67.4		−4.6	−7.5	−28.4	
4. (a) Column-strip moments to slabs and beams (kip-ft)	Slab	Beam			Slab	Beam			Slab	Beam
	−10.0	−60.1			−10.1	−57.3			−4.3	−24.1
(b) Moment due to load on beams (kip-ft)		−3.3				−3.3				−8.0
5. Total moments in slabs and beams (kip-ft)	−10.0	−63.4	−9.4		−10.1	−60.6	−12.1		−4.3	−32.1
<u>End Span Positive Moments</u>										
1. Slab moment (kip-ft)	78.0				74.4				35.1	
2. Moment coefficients	0.65		0.175	0.155	0.69		0.155	0.27	0.73	
3. Moments to columns and middle strips (kip-ft)	50.7		13.7	11.5	51.3		11.5	9.5	25.6	
4. (a) Column-strip moments to slabs and beams (kip-ft)	Slab	Beam			Slab	Beam			Slab	Beam
	7.6	43.1			7.7	43.6			3.8	21.8
(b) Moment due to load on beams (kip-ft)		3.2				3.2				7.8
5. Total moments in slabs and beams (kip-ft)	7.6	46.3	25.2		7.7	46.8	21.0		3.8	29.6
<u>First Interior Negative Moment</u>	***C2***				***B2***				***A2***	
1. Slab moment (kip-ft)	−230				−219				−108	
2. Moment coefficients	0.75		0.125	0.11	0.78		0.11	0.19	0.81	
3. Moments to columns and middle strips (kip-ft)	−173		−28.8	−24.1	−171		−24.1	−20.5	−87.5	
4. (a) Column-strip moments to slabs and beams (kip-ft)	Slab	Beam			Slab	Beam			Slab	Beam
	−26	−147			−26	−145			−13.1	−74.4
(b) Moment due to load on beams (kip-ft)		−9.9				−9.9				−24.0
5. Total moments in slabs and beams (kip-ft)	−26	−157	−52.9		−26	−155	−44.6		−13.1	−98.4

The effective cross section is shown in Fig. 13-95a. To compute C, the beam is divided into rectangles. The maximum value of C corresponds to the rectangles shown in Fig. 13-95a, and $C = 13{,}700 \text{ in.}^4$

The I_s in Eq. (13-12) is the moment of inertia of the slab span framing into the edge beam. Thus, $I_s = \ell_2 h^3/12$. Because the slab and beam are cast at the same time, $E_{cb} = E_{cs}$. We then have the following results:

(a) ***Slab strip A:*** $I_s = (10.25 \times 12)\dfrac{7^3}{12} = 3520 \text{ in.}^4$, $\beta_t = \dfrac{13{,}700}{2 \times 3520} = 1.95$

Interpolating in Table 13-5 for $\ell_2/\ell_1 = 1.06$, $\alpha_{f1}\ell_2/\ell_1 = 3.27$ and $\beta_t = 1.95$ gives 0.79 of the exterior negative moment to the column strip (at column $A1$).

(b) ***Slab strip B:*** $I_s = 7370 \text{ in.}^4$, $\beta_t = 0.93$, $\ell_2/\ell_1 = 1.19$, and $\alpha_{f1}\ell_2/\ell_1 = 2.03$. Interpolating in Table 13-5 gives 0.88 times the exterior negative moment to the column strip (at column $B1$).

(c) ***Slab strip C:*** $I_s = 8230 \text{ in.}^4$, $\beta_t = 0.83$, $\ell_2/\ell_1 = 1.19$, and $\alpha_{f1}\ell_2/\ell_1 = 2.03$. Interpolating in Table 13-5 gives 0.88 times the exterior moment to the column strip.

Positive moments:

(a) ***Slab strip A:*** $\ell_2/\ell_1 = 1.06$, $\alpha_{f1}\ell_2/\ell_1 > 1.0$. From Table 13-4, 73 percent of the positive moments go to the column strips.

(b) ***Slab strip B:*** $\ell_2/\ell_1 = 1.19$, $\alpha_{f1}\ell_2/\ell_1 > 1.0$. From Table 13-4, 69 percent of the positive moments go to the column strips.

(c) ***Slab strip C:*** $\ell_2/\ell_1 = 1.33$, $\alpha_{f1}\ell_2/\ell_1 > 1.0$. From Table 13-4, 65 percent of the positive moments go to the column strip.

Interior negative moments: The ℓ_2/ℓ_1 ratio is different for the exterior and interior spans that are adjacent to the first interior columns. Because the larger negative moments at the first interior columns all come from the longer interior spans, the ℓ_2/ℓ_1 ratios for those spans will be used to find the percentage of moment assigned to the column strips. The ℓ_2/ℓ_1 ratios for spans $A2$–$A3$, $B2$–$B3$, and $C2$–$C3$ are 0.792, 0.896, and 1.0, respectively. For all of these spans $\alpha_{f1}\ell_2/\ell_1 > 1.0$. Thus, interpolating in Table 13-3, 81, 78, and 75 percent of the negative moments in slab strips A, B, and C go to the column strip.

Line 4(a). The column-strip moments are divided between the slab and the beam, following the rules given in ACI Code Section 8.10.5.7.1. If $\alpha_{f1}\ell_2/\ell_1 \geq 1.0$, 85 percent of the column-strip moment is assigned to the beam. For this slab, this is true in all cases.

Line 4(b). These moments come from an analysis of the frame shown in Fig. 13-96 subjected to the weight of the beam web for column lines A, B, and C, plus the weight of the exterior wall along column line A.

Line 5. The final moments in the middle strip between lines A and B, in the slab in the column strip along line B, and in the beam along line B are plotted in Fig. 13-97. Note: similar detailed calculations can be completed for all the spans in the east–west direction.

8. Design the slab reinforcement. The design of slab reinforcement is carried out in the same way as in the previous example.

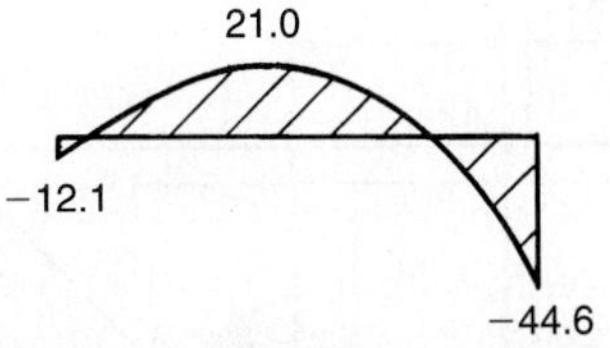

(a) Moments in middle strip between lines *A* and *B*.

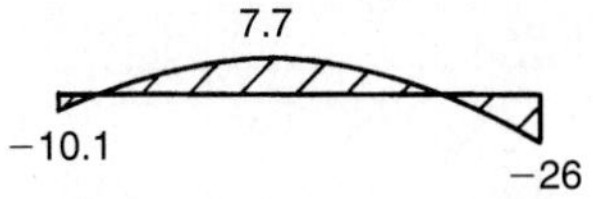

(b) Moments in slab in column strip along line *B*.

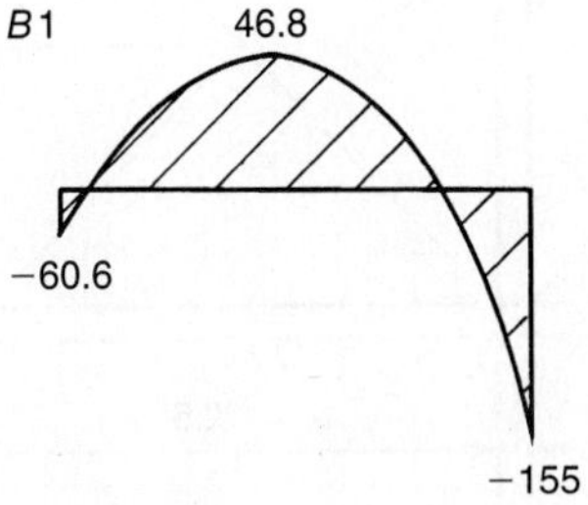

(c) Moments in beam along line *B*.

Fig. 13-97
Moments (kip-ft) in slab strips and beams on line *B* for Example 13-16.

9. Design the beams. The beams must be designed for moment, shear, and bar anchorage, as in the examples in Chapters 5 and 10. The edge beams also are subjected to a distributed torque, which may be considered to be uniformly distributed, as permitted by ACI Code Section 9.4.4.1. Design of edge beams subjected to shear and torsion was covered in Chapter 6 and 7.

Because $\alpha_{f1}\ell_2/\ell_1 \geq 1.0$ for all beams, ACI Code Section 8.10.8.1 requires that the beams be designed for the shear caused by loads on the tributary areas shown in Fig. 13-98a (next page). For the beams along line *A*, the corresponding beam loads and shear-force diagrams are shown in Figs.13-98b and 13-98c. Design of stirrups follows the procedures given in Chapter 6. Note: Some of the end moments in Fig. 13-98b came from structural analysis results that are not recorded in Tables 13-16 and 13-17. ■

13-15 CONSTRUCTION LOADS ON SLABS

Most two-way slab buildings are built by using *flying forms*, which can be removed sideways out of the building and are then lifted or *flown* up to form a higher floor. When the flying form is removed from under a slab, the weight of the slab is taken by posts or *shores*, which are wedged into place to take the load. Sets of flying forms and shores can be seen in Fig. 10-3. To save on the number of shores needed, it is customary to have only three to six floors of shoring below a slab at the time the concrete is placed. As a result, the weight of the fresh concrete is supported by the three to six floors below it. Because these floors are of different ages, they each take a different fraction of the load of the new slab.

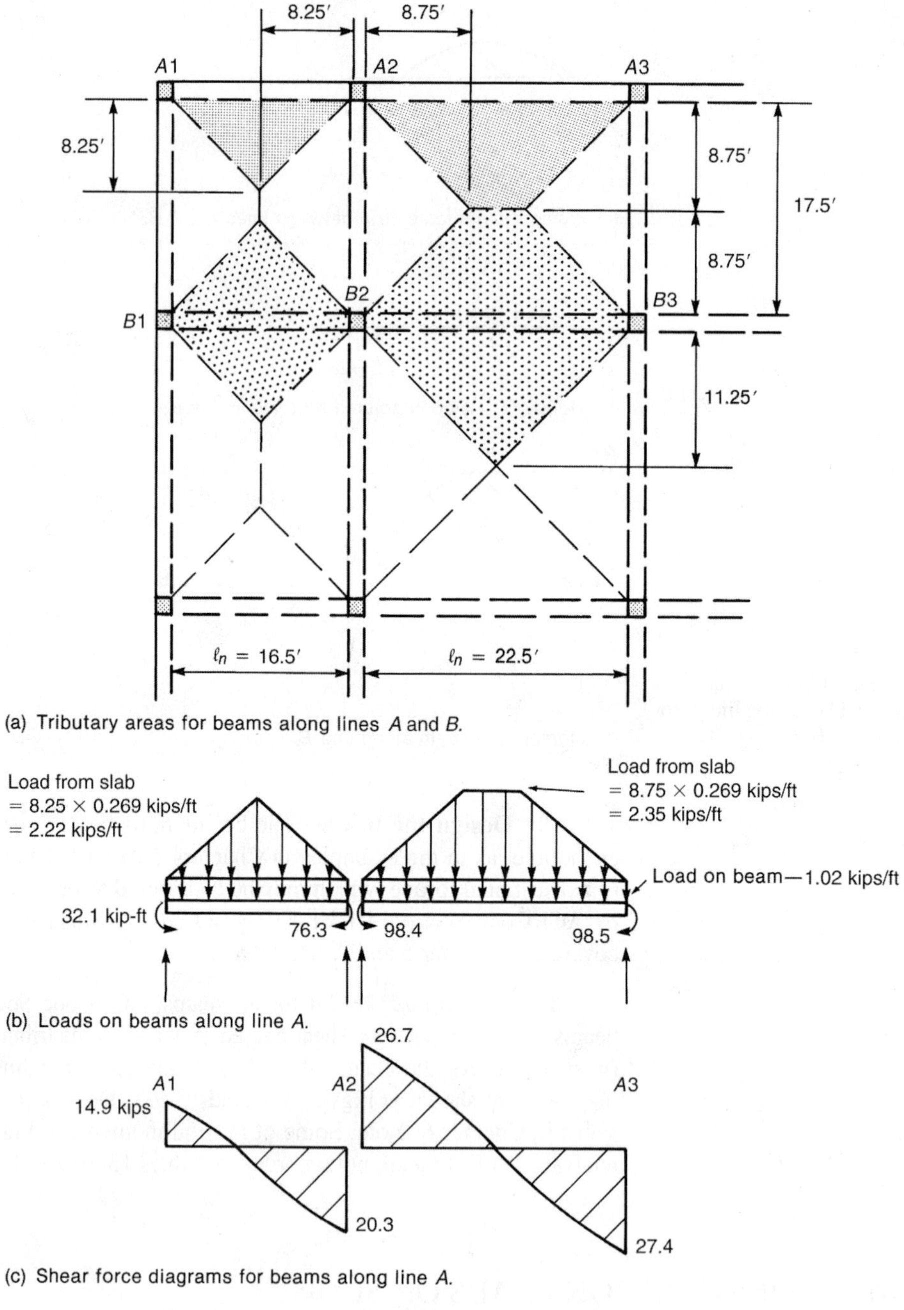

Fig. 13-98
Factored shear on edge beam for Example 13-16.

The calculation of the construction loads on slabs is discussed in [13-46]. Depending on the number of floors that are shored, the sequence of casting and form removal, and the weight of stacked building material, the maximum construction load on a given slab may reach 1.8 to 2.2 times the dead load of the slab. This can approach the capacity of the slab, particularly if, as is the usual case, the slab has not reached its full strength when the construction loads occur. These high loads cause cracking of the less than fully cured concrete slabs and lead to larger short- and long-time deflections than would otherwise be expected [13-47].

13-16 DEFLECTIONS IN TWO-WAY SLAB SYSTEMS

Excessive deflections are potentially a significant problem for two-way slab systems, causing sagging floors and damage to partitions, doors, and windows. ACI Code Section 8.3.1.1 gives minimum thicknesses of two-way slabs to avoid excessive deflections (see Table 13-1). As noted previously, it is good practice to round up the minimum thicknesses in Table 13-1 to the next larger quarter-inch for slabs less than 6 in. thick and the next larger half-inch for thicker slabs. Smaller slab thicknesses are permitted, but calculated deflections must be less than the allowable deflections given in ACI Code Table 24.2.2, found by using the larger value of ℓ for the panel.

The calculations of deflections in a two-way slab panel are approximate at best. This is a complex calculation that should involve a consideration of boundary conditions along the edge of the slab panel, loading history, live load patterns, cracking due to flexure and shrinkage, and the increase in deflections due to creep. Based on reported deflection problems with two-way slab systems, the long-term deflection coefficient, λ_Δ, of 2.0 given in the ACI Code is probably not sufficient. (Note: the value of λ_Δ equals the coefficient ξ because no compression reinforcement is used at midspan of one-way or two-way slabs.) The author recommends that this value be increased to 3.0 for two-way slab systems.

The crossing beam analogy generally is accepted as giving reasonable values for design decisions, although actual deflections could vary considerably from these calculated values. For this analogy, the average of the midspan deflections of the column strips along the edges of the panel are added to the midspan deflection of the perpendicular middle strip, as shown in Fig. 13-99. For interior panels, the columns strips are taken along the longer edge, and for exterior panels, the column strips are taken perpendicular to the exterior edge of the panel. The calculations generally follow the procedures given in Chapter 9.

The effective moment of inertia for a beam or slab span is calculated from Eqs. (9-10) and (9-11). To use these equations, one needs to evaluate M_a, which is defined in the ACI Code as the maximum unfactored moment acting in the member at the stage for which deflection is to be calculated. A better definition would be the maximum unfactored moment acting in the member at the stage in question or at any previous stage. In Section 13-15, it was pointed out that construction loads frequently reach two times the dead load of the slab. Thus, in computing slab deflections, it is recommended that M_a used in Eq. (9-10) should be based on an acting load equal to the larger of either the sum of the unfactored dead and live loads or 2.0 times the dead load of the slab. Furthermore, the cracking moment, M_{cr}, should be based on the age of the concrete when loaded and should consider tensile stresses developed due to restraint of shrinkage. Scanlon and Bischoff [13-48] recommend using

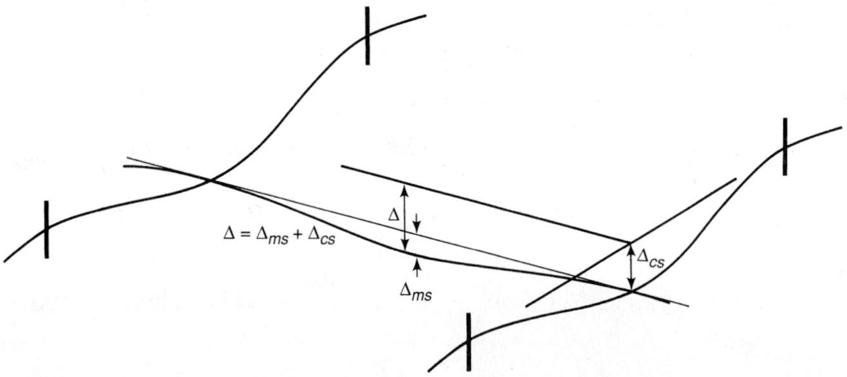

Fig. 13-99
Superposition of column-strip and middle-strip deflections.

two-thirds of the modulus or rupture, f_r, given in ACI Code Section 19.2.3.1, when calculating M_{cr} (ACI Code Section 24.2.3.5).

Drop panels satisfying the dimension limitations in ACI Code Section 8.2.4 should be included when calculating the I_g and I_{cr} values at the ends of the column strip. Smaller drop panels (shear caps) and column capitals should be ignored unless a more sophisticated procedure is used to determine the effective moment of inertia for the column strip.

EXAMPLE 13-17 Calculation of Deflections of an Interior Panel of a Flat-Plate Floor

Compute deflections for interior panel *B–C–2–3* of the slab panel shown in Fig. 13-84 and compare calculated values to limits in ACI Code Table 24.2.2. Assume that this floor panel supports nonstructural elements that are *not* likely to be damaged by large deflections. For this situation, the immediate deflection due to live load and the total deflection after the attachment of partitions need to be checked. In Example 13-15, the thickness of this floor system was initially selected as 7.5 inches to satisfy the minimum thickness criteria in Table 13-1. For this example, assume that the designer has selected a thickness of 7.0 inches, and thus, slab deflections must be checked.

1. Compute the immediate and long-term deflection of the column strip between *B2* and *C2* (assume deflections between *B3* and *C3* are the same).

(a) Compute M_a. Prior factored load on slab = 223 psf

Loads for deflection calculations (slab thickness of 7 in.):

Dead load, 1.0 (dead) = 87.5 + 25 = 113 psf

Service load, 1.0 (dead + live) = 113 + 50 = 163 psf

Construction load, 2.0 (slab dead load) = 2 × 87.5 = 175 psf

Thus, cracking will be governed by construction loads. Take M_a as 175/223 = 0.785 times the column-strip moments from Table 13-13. Then,

Negative moment at B2 and C2: 0.785 × 71.9 = 56.4 kip-ft

Positive moment at midspan: 0.785 × 31.0 = 24.3 kip-ft

(b) Compute M_{cr}. Assume that the maximum construction load on a given slab occurs when it is 14 days old. From Eq. (3-5), a 14-day old slab has a strength of $0.88f'_c$. From ACI Code Section 24.2.3.5,

$$M_{cr} = \frac{f_r I_g}{y_t}$$

using

$$f_r = (2/3)7.5\lambda\sqrt{f'_c} = (2/3)7.5 \times 1.0 \times \sqrt{0.88 \times 4000} = 297 \text{ psi}$$

$$I_g = \frac{(7.875 \text{ ft} \times 12 \text{ in./ft}) \times (7 \text{ in.})^3}{12} = 2700 \text{ in.}^4$$

$$y_t = 3.5 \text{ in.}$$

$$M_{cr} = \frac{297 \times 2700}{3.5} = 229{,}000 \text{ lb-in.} = 19.1 \text{ kip-ft}$$

Because we included the effect of tension stresses in the slab due to shrinkage, the slab in the column strip is cracked in both the negative- and positive-bending regions.

(c) Compute I_{cr} and I_e.

Negative-moment region. Although the slab reinforcement would likely change for this thinner slab, referring to Fig. 13-88c, A_s = 9 No. 5 bars. For this slab, assume the effective depth, $d(\text{east–west}) \approx h - 1.1 \text{ in.} = 5.9 \text{ in.}$, so

$$\rho = \frac{9 \times 0.31}{7.875 \times 12 \times 5.9} = 0.00500$$

$$n \text{ (at time of cracking)} = \frac{29{,}000{,}000}{57{,}000\sqrt{0.88 \times 4000}} = 8.58$$

Thus, $n\rho = 0.0429$, and from Eq. (9-3), $k = 0.253$. It then follows that

$$I_{cr} = \frac{1}{3} \times (7.875 \times 12) \times (0.253 \times 5.9)^3 + 9 \times 0.31 \times 8.58(5.9 - 0.253 \times 5.9)^2$$

$$I_{cr} = 104 + 466 = 570 \text{ in.}^4$$

From Eq. (9-10b),

$$I_e = I_{cr} + (I_g - I_{cr})\left(\frac{M_{cr}}{M_a}\right)^3$$

$$I_e = 570 + (2700 - 570) \times \left(\frac{19.1}{56.4}\right)^3 = 653 \text{ in.}^4$$

Positive-moment region. Following a similar procedure for the seven No. 4 bars used as positive reinforcement at midspan of the column strip (Fig. 13-88c), find:

$$\rho = 0.00251$$

$$\rho n = 0.0215$$

$$x = kd = 1.10 \text{ in. (depth to neutral axis)}$$

$$I_{cr} = 319 \text{ in.}^4$$

$$I_e = 1480 \text{ in.}^4$$

Weighted average value of I_e. From Eq. (9-11a),

$$I_e(\text{average}) = 0.70 I_{em} + 0.15(I_{e1} + I_{e2})$$

$$I_e(\text{average}) = 0.7 \times 1480 + 0.15(653 + 653) = 1230 \text{ in.}^4$$

(d) Compute the immediate and long-term deflections due to dead and live loads. To calculate the column-strip (and middle-strip) deflections, we need to determine what percentage of the dead and live loads are carried by the column and middle strips. This percentage commonly is taken to be the same as the percentage of the moments assigned to the column and middle strips. Because different percentages commonly are assigned for the negative- and positive-bending regions, a simple average is used here to determine the percentage of load carried by each strip. For this interior panel in Example 13-15, 75 percent of the negative moment at the end of the span and 60 percent of the midspan positive moment were assigned to the column strip. Thus, we will assume that 67.5 percent of the distributed dead and live loads are to be carried by the column strip (32.5 percent of the distributed loads will be assigned to the middle strip in the second part of this panel deflection calculation). So,

$$w_D = 113 \text{ psf} \times 15.75 \text{ ft} \times 0.675 = 1200 \text{ lb/ft} = 1.20 \text{ k/ft}$$
$$w_L = 50 \text{ psf} \times 15.75 \text{ ft} \times 0.675 = 532 \text{ lb/ft} = 0.53 \text{ k/ft}$$

From Table 9-2,

$$\Delta_D\,(\text{max}) = 0.0026\frac{w_D\ell^4}{EI}\ (\text{row 1})$$

$$\Delta_L\,(\text{max}) = 0.0048\frac{w_L\ell^4}{EI}\ (\text{row 4})$$

where $\ell = 20$ ft, $I = I_e = 1230$ in.^{4} and $E = E_c$ for $f'_c = 4000$ psi

$$E_c = 57{,}000\sqrt{4000} = 3{,}600{,}000 \text{ psi} = 3600 \text{ ksi}$$

Then, the immediate deflections due to dead and live load are

$$\Delta_D(\text{max}) = \frac{0.0026 \times 1.20 \times 20^4}{3600 \times 1230} \times 12^3 = 0.195 \text{ in.}$$

$$\Delta_L(\text{max}) = \frac{0.0048 \times 0.53 \times 20^4}{3600 \times 1230} \times 12^3 = 0.159 \text{ in.}$$

To determine long-term deflections, assume that 25 percent of the live load is sustained for long periods of time, i.e., it essentially acts like a dead load. Then, using the suggested long-term deflection coefficient of 3.0, the long-term deflection at midspan of the column strip is

$$\Delta(\text{long term}) = 3.0(0.195 + 0.25 \times 0.159) = 0.704 \text{ in.}$$

2. Compute the deflection of the middle strip between column lines 2 and 3.

(a) Compute M_a and M_{cr}. Again, the governing M_a is due to construction loads:

$$\text{Negative at line 2: } 0.785 \times 20.0 = 15.7 \text{ kip-ft}$$
$$\text{Positive at midspan: } 0.785 \times 16.0 = 12.6 \text{ kip-ft}$$
$$\text{Negative at line 3: } 0.785 \times 18.6 = 14.6 \text{ kip-ft}$$

As before,

$$I_g = \frac{12.12 \times 12 \times 7^3}{12} = 4160 \text{ in.}^4$$

and

$$M_{cr} = 29.4 \text{ k-ft}$$

(b) Compute I_e. Because $M_a < M_{cr}$ at all points, $I_e = I_g = 3330$ in.^{4}

(c) Compute the immediate and long-term deflections due to dead and live load.

As before,

$$w_D = 113 \text{ psf} \times 20 \text{ ft} \times 0.325 = 735 \text{ lb/ft} = 0.74 \text{ k/ft}$$
$$w_L = 50 \text{ psf} \times 20 \text{ ft} \times 0.325 = 325 \text{ lb/ft} = 0.33 \text{ k/ft}$$

Using the same deflection formulas with $\ell = 15.75$ ft,

$$\Delta_D(\text{max}) = \frac{0.0026 \times 0.74 \times 15.75^4}{3600 \times 4160} \times 12^3 = 0.014 \text{ in.}$$

$$\Delta_L(\text{max}) = \frac{0.0048 \times 0.33 \times 15.75^4}{3600 \times 4160} \times 12^3 = 0.011 \text{ in.}$$

Also, as before,

$$\Delta(\text{long term}) = 3.0(0.014 + 0.25 \times 0.011) = 0.050 \text{ in.}$$

3. Check deflections against ACI Code limits.

(a) Check immediate deflection due to live load.

$$\Delta_L(\text{max for panel}) = 0.159 + 0.011 = 0.170 \text{ in.}$$

$$\text{ACI Code limit} = \ell/360 = (20 \text{ ft} \times 12 \text{ in./ft})/360 = 0.667 \text{ in. (o.k.)}$$

(b) Check total deflection after attachment of partitions. We will assume that 90 percent of the immediate deflection due to dead load will have occurred at the time the partitions are attached. Thus, the deflections experienced by the partitions should be equal to the sum of 10 percent of the immediate deflection due to dead load, the immediate deflection due to live load, plus 100 percent of the long-term deflections. So,

$$\Delta(\text{max for partitions}) = 0.1(0.195 + 0.014) + 0.170 + (0.704 + 0.050)$$

$$= 0.021 + 0.173 + 0.754 = 0.95 \text{ in.}$$

$$\text{ACI Code limit} = \ell/240 = (20 \text{ ft} \times 12 \text{ in./ft})/240 = 1.00 \text{ in.}$$

The calculated deflection (with conservative assumptions) is less than the ACI limit, so the slab thickness of 7 in. should give satisfactory performance. ■

13-17 USE OF POST-TENSIONING

The use of post-tensioning is quite common in two-way floor systems. In general, post-tensioning allows the designer to control cracking and deflections under service loads and thus, leads to thinner slabs and longer spans. Although the design of post-tensioned slabs will not be covered in detail here, some of the fundamental considerations for using post-tensioning in two-way floor systems will be discussed. More in-depth coverage is provided in textbooks for design of prestressed concrete structures [13-49], [13-50], [13-51] and in manuals from the Post-Tensioning Institute [13-52].

Two-way post-tensioned slabs usually are constructed with unbonded tendons. The post-tensioning tendons (strands) are sheathed in plastic or covered with grease for corrosion protection and to break the bond between the strands and the concrete. In typical two-way slabs, the tendons are grouped together in one direction in a narrow band within the column strip and uniformly distributed in the other direction. The banded tendons normally are used in the longer span direction and the distributed tendons in the shorter direction (Fig. 13-100).

A fundamental design concept for use of post-tensioning in two-way slabs is that the axial tension in the post-tensioning tendons puts an eccentric compression force into a concrete slab, and thus provides a moment that counteracts the moment caused by external loads, as shown in Fig 13-101. A general expression for calculation of stresses in a post-tensioned slab section is given in Eq. (13-45), where P is the effective post-tensioning force in the tendon, e is the eccentricity of the tendon with respect to the centroid of the section, A_g and I_g

Fig. 13-100
Use of banded tendons in one direction and distributed tendons in other direction. (Photo courtesy of Kenneth Bondy)

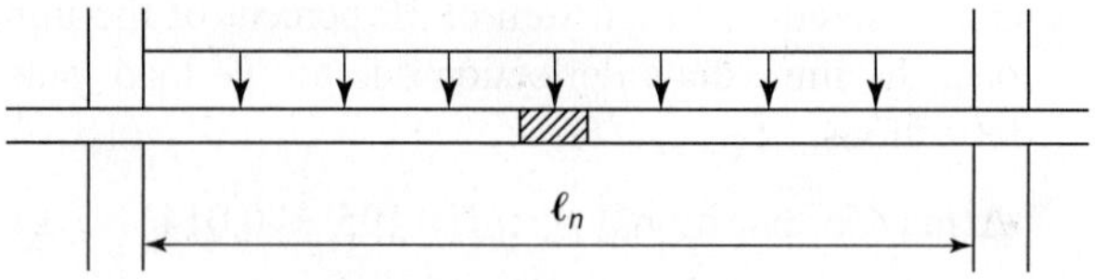

(a) Distributed loading on a continuous floor slab.

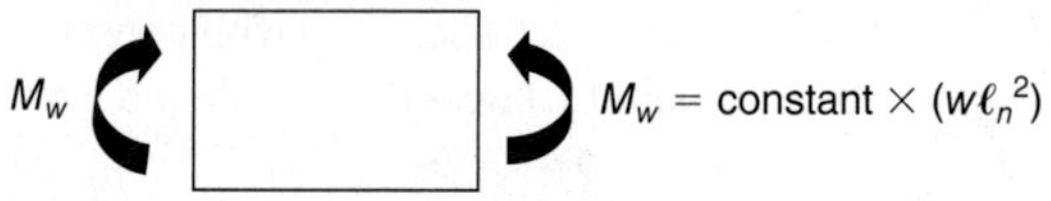

(b) Positive midspan moment due to distributed loading.

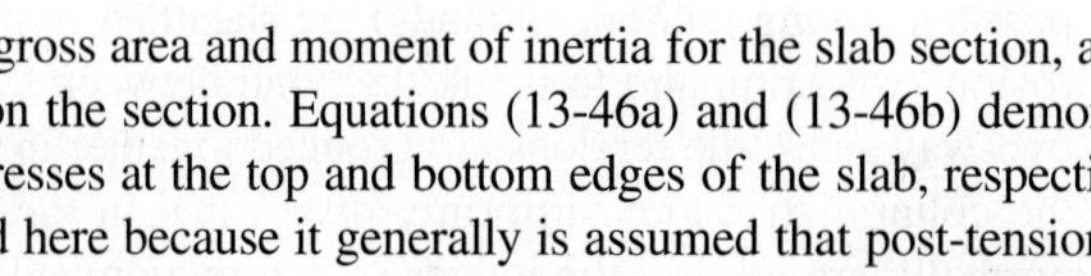

(c) Effective prestressing force, P_e, acting at an eccentricity, e, to develop a negative moment at midspan.

Fig. 13-101
Use of eccentric prestressing force to counterbalance moment due to acting loads.

are the gross area and moment of inertia for the slab section, and M_a is the external moment acting on the section. Equations (13-46a) and (13-46b) demonstrate the calculation of concrete stresses at the top and bottom edges of the slab, respectively. Gross section properties are used here because it generally is assumed that post-tensioned slabs are uncracked under service loads.

$$f = \frac{P}{A_g} \pm \frac{(P \cdot e)\, y}{I_g} \pm \frac{M_a \cdot y}{I_g} \tag{13-45}$$

$$f(\text{top}) = \frac{P}{A_g} - \frac{(P \cdot e)\, y_t}{I_g} + \frac{M_a \cdot y_t}{I_g} \tag{13-46a}$$

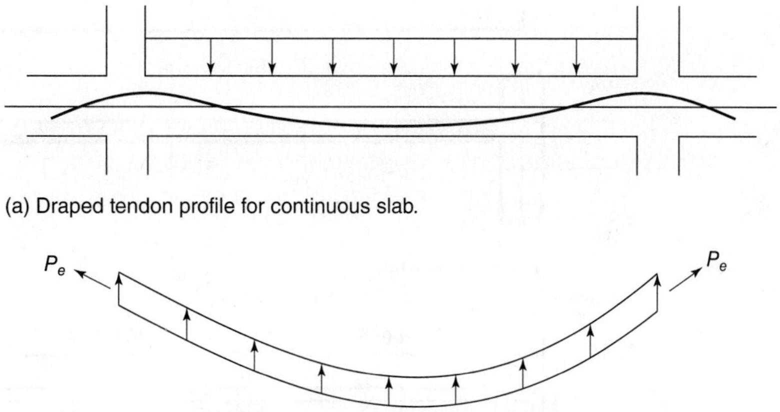

(a) Draped tendon profile for continuous slab.

(b) Effective upward load from parabolically draped tendon.

Fig. 13-102
Use of parabolically draped tendon profile to balance the effect of distributed gravity loading.

$$f(\text{bottom}) = \frac{P}{A_g} + \frac{(P \cdot e)\, y_b}{I_g} - \frac{M_a \cdot y_b}{I_g} \qquad (13\text{-}46b)$$

For a two-way slab system, the acting moment varies over the span, so it is reasonable to also vary the position of the post-tensioning tendons to counteract the effect of the external moment. This results in the use of draped tendon profiles, similar to that shown in Fig. 13-102a. A designer typically would use expressions similar to Eqs. (13-46a) and (13-46b) to determine a tendon profile that would keep the extreme fiber stresses between the maximum permissible tension and compression stresses at the time of tensioning the tendons and under service load conditions.

One method to envision the design process is to assume that if the tendon is draped into a parabolic profile, as shown in Fig. 13-102b, it will provide a distributed upward load in the beam (slab). Based on the magnitude of the post-tensioning force and the profile of the strand, the post-tensioning can be used to "balance" the external loads acting on the slab. In general, this load-balancing design procedure is used to balance most, if not all of the dead load acting on the two-way slab system. Because the live loads usually are not large enough to cause flexural cracking, deflections then can be evaluated using gross section properties. Further, because the dead loads are essentially "balanced," only deflections due to live load commonly need to be checked. Deflections due to sustained load, which usually is approximately equal to the dead load, will not be significant unless a large percentage of the live load is sustained for long periods of time. In such a case, either the tendon profile or the post-tensioning force could be adjusted to "balance" the dead load plus some percentage of the sustained live load.

PROBLEMS

13-1 Compute α_f for the edge beam shown in Fig. P13-1. The concrete for the slab and beam was placed in one pour.

13-2 Compute the column-strip and middle-strip moments in the long-span direction for an interior panel of the flat-slab floor shown in Fig. 13-24. Assume the slab is 6 in. thick, the design live load is 40 psf, and the superimposed dead load is 5 psf for ceiling, flooring, and so on, plus 25 psf for partitions. The columns are 10 in. × 12 in., as shown in Fig. 13-24.

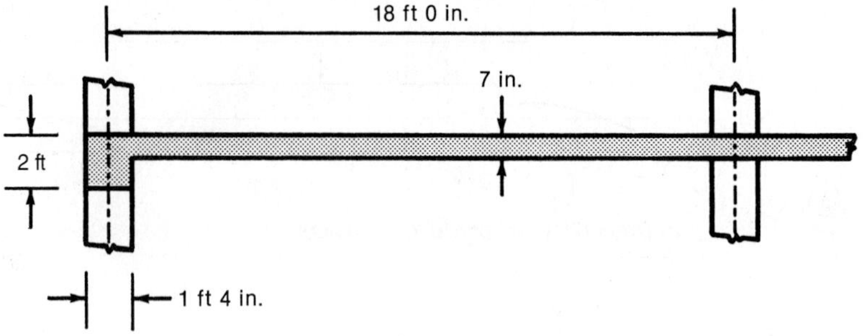

Fig. P13-1

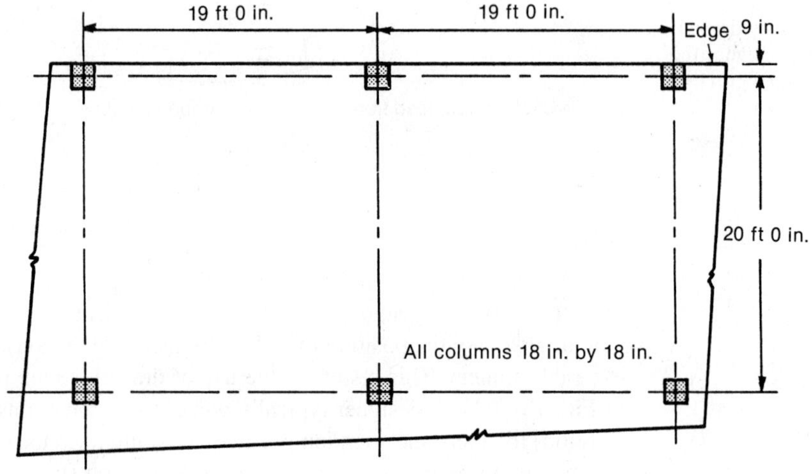

Fig. P13-3

13-3 Use the direct-design method to compute the moments for the column-strip and middle-strip spanning perpendicular to the edge of the exterior bay of the flat-plate slab shown in Fig. P13-3. Assume the slab is $7\frac{1}{2}$ in. thick and supports a superimposed dead load of 25 psf and a live load of 50 psf. There is no edge beam. The columns are all 18 in. square.

Video Solution

13-4 For the slab configuration and loading conditions in P13-3, use the direct-design method to compute moments for the edge-column strip and the middle strip spanning parallel to the edge of the slab.

13-5 A 7-in.-thick flat-plate slab with spans of 20 ft in each direction is supported on 16 in. × 16 in. columns. The average effective depth is 5.6 in. Assume the slab supports its own dead load, plus 25 psf superimposed dead load and 40 psf live load. The concrete strength is 4000 psi. Check two-way shear at a typical interior support. Assume unbalanced moments are negligible.

13-6 Assume the slab described in Problem 13-5 is supported on 10 in. × 24 in. columns. Check two-way shear at a typical interior support. Assume unbalanced moments are negligible.

13-7 The slab shown in Fig. P13-7 supports a superimposed dead load of 25 psf and a live load of 60 psf. The slab extends 4 in. past the exterior face of the column to support an exterior wall that weighs 400 lb/ft of length of wall. The story-to-story height is 9.5 ft. Use 4500-psi concrete and Grade-60 reinforcement.

Video Solution

(a) Select slab thickness.

(b) Use the direct-design method to compute moments, and then design the reinforcement for column and middle strips running north and south along column line 2.

(c) Check two-way shear and moment transfer at columns $A2$ and $B2$. Neglect unbalanced moments about column line 2.

13-8 Refer to the slab shown in Fig. P13-7 and the loadings and material strengths given in Problem 13-7.

(a) Select slab thickness.

(b) Use the direct-design method to compute

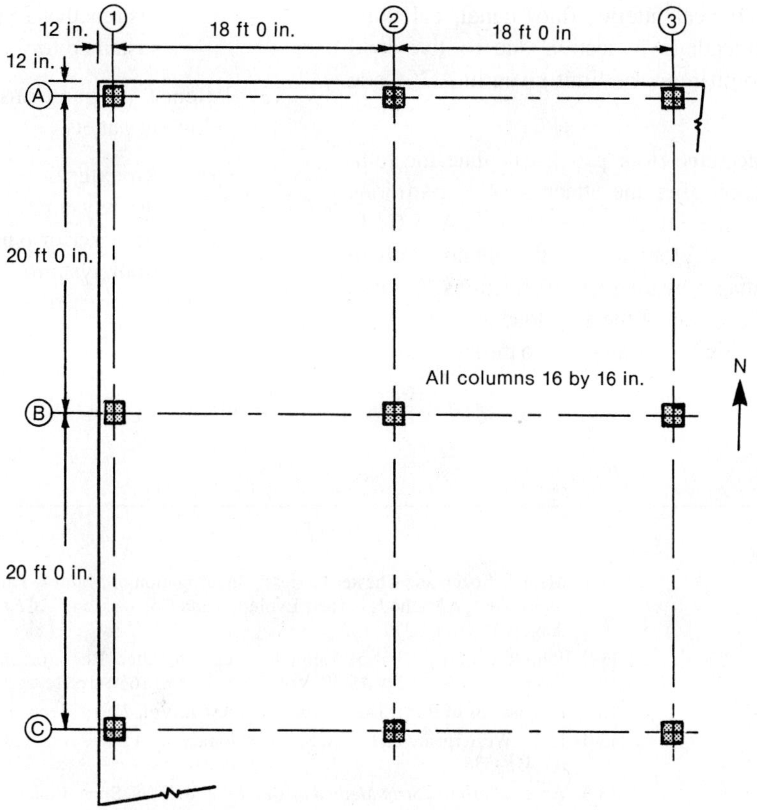

Fig. P13-7

moments, and then design the reinforcement for column and middle strips running east and west along column line *A*.

13-9 For the corner column (*A*1) in Fig. P13-7 and the loadings and material strengths given in Problem 13-7, select a slab thickness to satisfy ACI Code strength requirements for two-way shear and moment transfer, and deflection control.

(a) Make the check for moment transfer in only one principal direction (use the more critical direction).

(b) Make the check for moment transfer in both principal directions, but permit a 15 percent increase in the maximum permissible shear stress calculated at the corner of the critical shear perimeter.

(c) Check one-way shear for a critical diagonal section across the corner near the corner column.

13-10 For the floor system shown in Fig. P13-7, assume the slab has four equal spans in the north–south direction and three equal spans in the east–west direction. Assume story heights above and below this floor are 10 ft – 0 in. Use the loading and material strengths given in Problem 13-7 and assume a slab thickness of 7.5 in.

(a) Use an equivalent-frame method to analyze the factored design moments along column line 2 and compare the results with the moments used in part (b) of Problem 13-7.

(b) Use an equivalent-frame method to analyze the factored design moments along column line *A* and compare the results with the moments used in part (b) of Problem 13-8.

13-11 For the same floor system described in Problem 13-10 and the loading and material strengths given in Problem 13-7, assume the slab thickness has been selected to be 7 in.

(a) For a typical interior floor panel, calculate the immediate deflection due to live load and compare to the limit given in ACI Code Table 24.2.2.

(b) For the same floor panel, calculate the total deflection after the attachment of partitions and compare to the limit given in ACI Code Table 24.2.2 for partitions that are not likely to be damaged by long-term deflections. Assume that 85 percent of the dead load is acting when the partititions are attached to the structure and assume that 25 percent of the live load will be sustained for a period of 1 year.

13-12 Repeat the questions in Problem 13-11 for the following panels.

(a) An exterior panel along the west side of the floor system.

(b) An exterior panel along the north side of the floor system.

(c) A corner panel.

REFERENCES

13-1 Mete A. Sozen and Chester P. Siess, "Investigation of Multiple Panel Reinforced Concrete Floor Slabs: Design Methods—Their Evolution and Comparison," *ACI Journal, Proceedings*, Vol. 60, No. 8, August 1963, pp. 999–1027.

13-2 John R. Nichols, "Statical Limitations upon the Steel Requirements in Reinforced Concrete Flat Slab Floors," *Transactions ASCE*, Vol. 77, 1914, pp. 1670–1681.

13-3 Discussions of Ref. 13–2, *Transactions ASCE*, Vol. 77, 1914, pp. 1682–1736.

13-4 H. M. Westergaard and N. A. Slater, "Moments and Stresses in Slabs," *ACI Proceedings*, Vol. 17, 1921, pp. 415–538.

13-5 Arne Hillerborg, *Strip Method of Design*, E. & F., N. Spon, London, 1975, 258 pp.

13-6 Randal H. Wood and G. S. T. Armer, "The Theory of the Strip Method for Design of Slabs," *Proceedings, Institution of Civil Engineers, London*, Vol. 41, October 1968, pp. 285–313.

13-7 David P. Thompson and Andrew Scanlon, "Minimum Thickness Requirements for Control of Two-Way Slab Deflections," *ACI Structural Journal*, Vol. 85, No. 1, January–February 1988, pp. 12–22.

13-8 James O. Jirsa, Mete A. Sozen, and Chester P. Siess, "Pattern Loadings on Reinforced Concrete Floor Slabs," *Proceedings ASCE, Journal of the Structural Division*, Vol. 95, No. ST6, June 1969, pp. 1117–1137.

13-9 R. E. Woodring and Chester P. Siess, *An Analytical Study of the Moments in Continuous Slabs Subjected to Concentrated Loads*, Structural Research Series 264, Department of Civil Engineering, University of Illinois, Urbana, IL, May 1963, 151 pp.

13-10 Dean Peabody, Jr., "Continuous Frame Analysis of Flat Slabs," *Journal, Boston Society of Civil Engineers*, January 1948.

13-11 Chester P. Siess and Nathan M. Newmark, "Rational Analysis and Design of Two-Way Concrete Slabs," *ACI Journal, Proceedings*, Vol. 45, 1949, pp. 273–315.

13-12 W. Gene Corley and James O. Jirsa, "Equivalent Frame Analysis for Slab Design," *ACI Journal, Proceedings*, Vol. 67, No. 11, November 1970, pp. 875–884.

13-13 M. Daniel Vanderbilt, "Equivalent Frame Analysis of Unbraced Concrete Frames," *Significant Developments in Engineering Practice and Research—A Tribute to Chester P. Siess*, ACI Publication SP-72, American Concrete Institute, Detroit, MI, 1981, pp. 219–246.

13-14 C. K. Wang, *Intermediate Structural Analsyis*, McGraw-Hill, New York, 1983.

13-15 *Notes on ACI-318-05, Building Code Requirements for Reinforced Concrete with Design Applications*, Portland Cement Association, Skokie, IL., 2005.

13-16 Sidney H. Simmonds and Janko Misic, "Design Factors for the Equivalent Frame Method," *ACI Journal, Proceedings*, Vol. 68, No. 11, November 1971, pp. 825–831.

13-17 Corley, W. G. and Jirsa, J.O., "Equivalent Frame Analysis for Slab Design," *ACI Journal, Proceedings*, Vol. 67, No. 11, November 1970, pp. 875–884.

13-18 Pecknold, D., "Slab Effective Width for Equivalent Frame Analysis," *ACI Journal, Proceedings*, Vol. 72, No. 4, April 1975, pp. 135–137.

13-19 Darvall, P. and Allen, F., "Lateral Load Equivalent Frame," *ACI Journal, Proceedings*, Vol. 74, No. 7, July 1977, pp. 294–299.

13-20 Luo, Y. H. and Durrani, A. J., "Equivalent Beam Model for Flat-Slab Buildings—Part I: Interior Connections," *ACI Structural Journal*, Vol. 92, No. 1, January–February 1995, pp. 115–124.

13-21 Luo, Y. H. and Durrani, A. J., "Equivalent Beam Model for Flat-Slab Buildings—Part II: Exterior Connections," *ACI Structural Journal*, Vol. 92, No. 2, March–April 1995, pp. 250–257.

13-22 Hwang, S. J. and Moehle, J. P., "An Experimental Study of Flat-Plate Structures Under Vertical and Lateral Loads," Report No. UCB/SEMM-90/11, Department of Civil Engineering, University of California at Berkeley, January 1996.

13-23 Dovich, L. and Wight, J. K., "Effective Width Model for Seismic Analysis of Flat Slab Systems," *ACI Structural Journal*, Vol. 102, No. 6, November–December 2005, pp. 868–875.

13-24 Hueste, M. B. and Wight, J. K., "Evaluation of a Four-Story Reinforced Concrete Building Damaged During the Northridge Earthquake," *Earthquake Spectra*, EERI, Vol. 13, No. 3, pp. 387–414.

13-25 Kang, T. H. K. and Wallace, J. W., "Dynamic Responses of Flat Plate Systems with Shear Reinforcement," *ACI Structural Journal*, Vol. 102, No. 5, September–October. 2005, pp. 763–773.

13-26 Scott D. B. Alexander and Sidney H. Simmonds, "Ultimate Strength of Column–Slab Connections," *ACI Structural Journal, Proceedings*, Vol. 84, No. 3, May–June 1987, pp. 255–261.

13-27 ACI-ASCE Committee 426, "The Shear Strength of Reinforced Concrete Members," Chapter 5, "Shear Strength of Slabs," *Proceedings ASCE, Journal of the Structural Division*, Vol. 100, No. ST8, August 1974, pp. 1543–1591.

13-28 Widianto O. Bayrak, and J.O. Jirsa, "Two-Way Shear Strength of Slab-Column Connections: Reexamination of ACI 318 Provisions," *ACI Structural Journal*, Vol. 106, No. 2, March-April 2009, pp. 160–170.

13-29 Johannes Moe, *Shearing Strength of Reinforced Concrete Slabs and Footings under Concentrated Loads*, Development Department Bulletin D47, Portland Cement Association, Skokie, IL., April 1961.

13-30 ACI-ASCE Committee 326, "Shear and Diagonal Tension, Slabs," *ACI Journal, Proceedings*, Vol. 59, No. 3, March 1962, pp. 353–396.

13-31 Joint ACI-ASCE Committee 326, "Shear and Diagonal Tension, Part 3—Slabs and Footings," *ACI Journal, Proceedings*, Vol. 59, No. 3, March 1962, pp. 353–396.

13-32 Neil M. Hawkins, H. B. Fallsen, and R. C. Hinojosa, "Influence of Column Rectangularity on the Behavior of Flat Plate Structures," *SP-30 Cracking, Deflection and Ultimate Load of Concrete Slab Systems*, American Concrete Institute, Detroit, 1971, pp. 127–146.

13-33 M. Daniel Vanderbilt, "Shear Strength of Continuous Plates," *Proceedings ASCE, Journal of the Structural Division*, Vol. 98, No. ST5, May 1972, pp. 961–973.

13-34 Sami Megally and Amin Ghali, "Cautionary Note on Shear Capitals," *Concrete International*, Vol. 24, No. 3, March 2002, pp 75–82.

13-35 W. Gene Corley and Neil M. Hawkins, "Shearhead Reinforcement for Slabs," *ACI Journal, Proceedings*, Vol. 65, No. 10, October 1968, pp. 811–824.

13-36 ACI Committee 421, "Shear Reinforcement for Slabs" (ACI 421.1R), *ACI Structural Journal*, Vol. 89, No. 5, September–October 1992, pp. 587–589.

13-37 M.-Y. Cheng, G. J. Parra-Montesinos, and C. K. Shield, "Shear Strength and Drift Capacity of Fiber Reinforced Concrete Slab–Column Connections Subjected to Bi-Axial Displacements," *Journal of Structural Engineering*, ASCE, Vol. 136, No. 9, September 2010, pp. 1078–1088.

13-38 European Committee for Standardization, "Eurocode 2: Design of concrete structures—Part 1-1: General—Common rules for design of buildings and civil engineering structures," English version (EN 1992-1-1: 2004: E), 225 pp.

13-39 Neil M. Hawkins, "Shear Strength of Slabs with Moments Transferred to Columns," *Shear in Reinforced Concrete*, Vol. 2, ACI Publication SP-42, American Concrete Institute, Detroit, MI, 1974, pp. 817–846.

13-40 ACI Committee 352, "Recommendations for Design of Slab-Column Connections in Monolithic Reinforced Concrete Structures," *ACI Structural Journal*, Vol. 85, No. 6, November–December 1988, pp. 675–696.

13-41 Sidney H. Simmonds and Scott D. B. Alexander, "Truss Model for Edge Column–Slab Connections," *ACI Structural Journal*, Vol. 84, No. 4, July–August 1987, pp. 296–303.

13-42 Norman W. Hanson and John M. Hanson, "Shear and Moment Transfer between Concrete Slabs and Columns," *Journal of the Research and Development Laboratories, Portland Cement Association*, Vol. 10, No. 1, January 1968, pp. 2–16.

13-43 S. D. B. Alexander, and S. H. Simmonds, "Shear and Moment Transfer at an Edge Connection," *Concrete International*, Vol. 27, No. 8, August 2005, pp. 69–74.

13-44 Adel E. Elgabry and Amin Ghali, "Design of Stud-Shear Reinforcement for Slabs," *ACI Structural Journal*, Vol. 87, No. 3, May–June 1990, pp. 350–361.

13-45 Denis Mitchell and William D. Cook, "Preventing Progressive Collapse of Slab Structures," *Proceedings ASCE, Journal of the Structural Division*, Vol. 110, No. ST7, July 1984, pp. 1513–1532.

13-46 R. K. Agerwal and Noel J. Gardner, "Form and Shore Requirements for Multistory Flat Plate Type Buildings," *ACI Journal, Proceedings*, Vol. 71, No. 11, November 1974, pp. 559–569.

13-47 John A. Sbarounis, "Multistory Flat Plate Buildings," *Concrete International: Design and Construction*, Vol. 81, No. 2, February 1984, pp. 70–77; No. 4, April 1984, pp. 62–70; No. 8, August 1984, pp. 31–35.

13-48 A. Scanlon and Bischoff, P. H., "Shrinkage Restraint and Loading History Effects on Deflection of Flexural Members," *ACI Structural Journal*, V. 105, No. 4, July–August, 2008, pp. 498–506

13-49 Arthur H. Nilson, *Design of Prestressed Concrete*, John Wiley and Sons, Inc., New York, 1978, 526 pp.

13-50 Michael P. Collins and Denis Mitchell, *Prestressed Concrete Structures*, Prentice-Hall Inc., New Jersey, 1991, 766 pp.

13-51 Antoine E. Naaman, *Prestressed Concrete Analysis and Design*, Second Edition, Techno Press 3000, Ann Arbor, 2004, 1071 pp.

13-52 Post-Tensioning Institute, *Design of Post-Tensioned Slabs*, Third edition, Post Tensioning Institute, Phoenix, 2003, 54 pp.

14

Two-Way Slabs: Elastic and Yield-Line Analyses

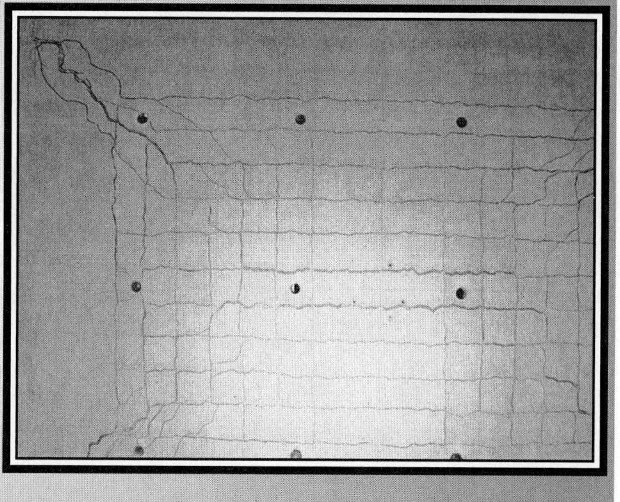

14-1 REVIEW OF ELASTIC ANALYSIS OF SLABS

The concepts involved in the elastic analysis of slabs are reviewed briefly here, to show the relationship between the loads and the internal moments in the slab. In addition, and even more important, this will show the relationship between moments and slab curvatures.

Slabs may be subdivided into *thick* slabs (with a thickness greater than about one-tenth of the span), *thin* slabs (with a thickness less than about one-fortieth of the span), and *medium-thick* slabs. Thick slabs transmit a portion of the loads as a flat arch and have significant in-plane compressive forces, with the result that the internal resisting compressive force C is larger than the internal tensile force T. Thin slabs transmit a portion of the loads by acting as a tension membrane—hence, T is larger than C. A medium-thick slab does not exhibit either arch action or membrane action and thus has $T = C$.

Figure 14-1 shows an element cut from a medium-thick, two-way slab. This element is acted on by the moments shown in Fig. 14-1a and by the shears and loads shown in Fig. 14-1b. (The figures are separated for clarity.)

Two types of moment exist on each edge: bending moments m_x and m_y about axes parallel to the edges, and twisting moments m_{xy} and m_{yx} about axes perpendicular to the edges. The moments are shown by moment vectors represented by double-headed arrows. The moment in question acts about the arrow according to the right-hand-screw rule. The length of vector represents the magnitude of the moment. Vectors can be added graphically or numerically. The moments m_x, and so on, are defined for a unit width of the edge they act on, as are the shears V_x, and so on. The m_x and m_y moments are positive, corresponding to compression on the top surface. The twisting moments on adjacent edges both act to cause compression on the same surface of the slab, at the corner between the two edges, as shown in Fig. 14-1a.

Summing vertical forces gives

$$\frac{\partial V_x}{\partial x} + \frac{\partial V_y}{\partial y} = -q \tag{14-1}$$

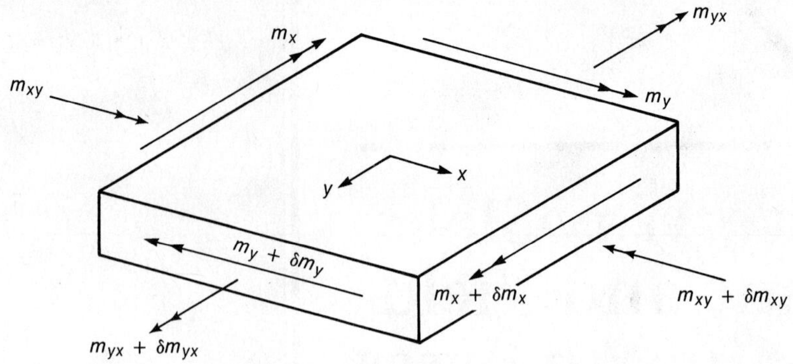

(a) Bending and twisting moments on a slab element.

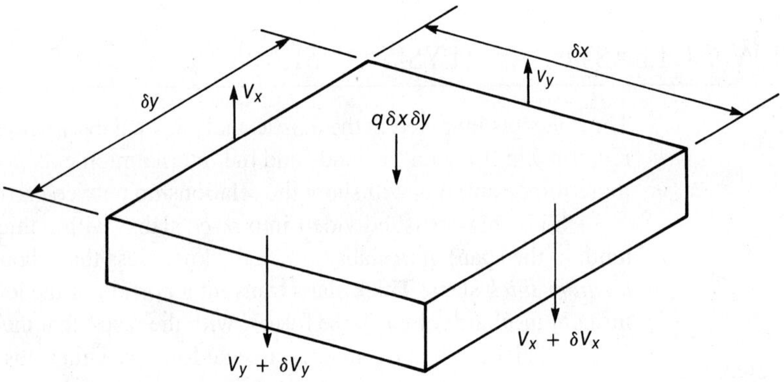

(b) Shears and loads on a slab element.

Fig. 14-1
Moments and forces in a medium-thick plate.

Summing moments about lines parallel to the x and y axes and neglecting higher order terms gives, respectively,

$$\frac{\partial m_y}{\partial y} + \frac{\partial m_{xy}}{\partial x} = V_y \quad \text{and} \quad \frac{\partial m_x}{\partial x} + \frac{\partial m_{yx}}{\partial y} = V_x \tag{14-2}$$

It can be shown that $m_{xy} = m_{yx}$. Differentiating Eq. (14-2) and substituting into Eq. (14-1) gives the basic equilibrium equation for medium-thick slabs:

$$\frac{\partial^2 m_x}{\partial x^2} + \frac{2\partial^2 m_{xy}}{\partial x\,\partial y} + \frac{\partial^2 m_y}{\partial y^2} = -q \tag{14-3}$$

This is purely an equation of statics and applies regardless of the behavior of the plate material. For an elastic plate, the deflection, z, can be related to the applied load by means of

$$\frac{\partial^4 z}{\partial x^4} + 2\frac{\partial^4 z}{\partial x^2\,\partial y^2} + \frac{\partial^4 z}{\partial y^4} = -\frac{q}{D} \tag{14-4}$$

where the *plate rigidity* is

$$D = \frac{Et^3}{12(1 - \nu^2)} \tag{14-5}$$

in which ν is Poisson's ratio. The term D is comparable to the EI value of a unit width of slab.

In an elastic plate analysis, Eq. (14-4) is solved to determine the deflections, z, and the moments are calculated from

$$m_x = -D\left(\frac{\partial^2 z}{\partial x^2} + \frac{v\partial^2 z}{\partial y^2}\right)$$
$$m_y = -D\left(\frac{\partial^2 z}{\partial y^2} + \frac{v\partial^2 z}{\partial x^2}\right) \tag{14-6}$$
$$m_{xy} = -D(1 - v)\frac{\partial^2 z}{\partial x\,\partial y}$$

where z is positive downward.

Solutions of Eq. (14-6) can be found in books on elastic plate theory.

14-2 DESIGN MOMENTS FROM A FINITE-ELEMENT ANALYSIS

The most common way of solving Eq. (14-6) is to use a finite-element analysis. Such an analysis gives values of m_x, m_y, and m_{xy}, in each element, where m_x, m_y, and m_{xy} are moments per unit width. A portion of an element bounded by a diagonal crack is shown in Fig. 14-2. The moments on the x and y faces, from the finite-element analysis, are shown in Fig. 14-2b. The moment about an axis parallel to the crack is m_c, given by

$$m_c\,ds = (m_x\,dy + m_{xy}k\,dy)\cos\theta + (m_y k\,dy + m_{xy}\,dy)\sin\theta$$

or

$$m_c = \left(\frac{dy}{ds}\right)^2 (m_x + k^2 m_y + 2km_{xy}) \tag{14-7}$$

This slab is to be reinforced by bars in the x and y directions having positive-moment capacities m_{rx} and m_{ry} per unit width. The corresponding moment capacity about an axis parallel to the assumed crack is

$$m_{rc} = \left(\frac{dy}{ds}\right)^2 (m_{rx} + k^2 m_{ry}) \tag{14-8}$$

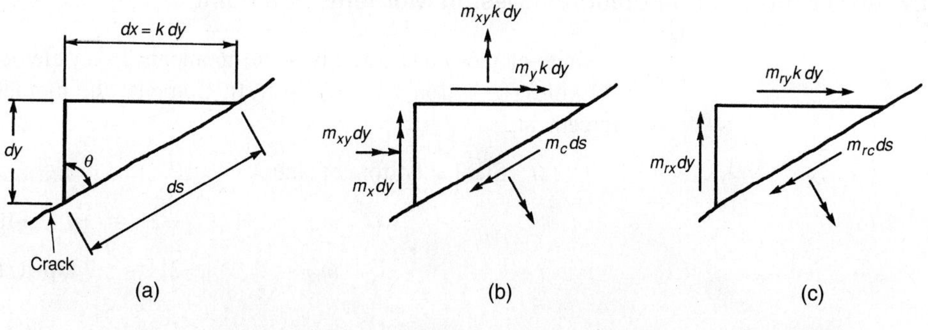

Fig. 14-2
Resolution of moments.
(From [14-1].)

where m_{rc} must equal or exceed m_c to provide adequate strength. Equating these and solving for the minimum, we get

$$m_{ry} = m_y + \frac{1}{k}m_{xy}$$

Because m_{ry} must equal or exceed m_y to account for the effects of m_{xy}, $(1/k)m_{xy} \geq 0$, which gives

$$m_{ry} = m_y + \frac{1}{k}|m_{xy}|$$
$$m_{rx} = m_x + k|m_{xy}| \quad (14\text{-}9)$$

where k is a positive number and $|m_{xy}|$ implies the absolute value of m_{xy}. This must be true for all crack orientations (i.e., for all values of k). As k is increased, m_{ry} goes down and m_{rx} goes up. The smallest of the sums of the two (i.e., the smallest total reinforcement) depends on the slab in question, but $k = 1$ is the best choice for a wide range of moment values [14-1], [14-2].

The reinforcement at the bottom of the slab in each direction is designed to provide positive-moment resistances of

$$m_{ry} = m_y + |m_{xy}| \quad (14\text{-}10a)$$

and

$$m_{rx} = m_x + |m_{xy}| \quad (14\text{-}10b)$$

Reinforcement at the top of the slab in each direction is designed to provide negative-moment resistances of

$$m_{ry} = m_y - |m_{xy}| \quad (14\text{-}10c)$$

and

$$m_{rx} = m_x - |m_{xy}| \quad (14\text{-}10d)$$

Calculation of m_{rx} and m_{ry}—positive moment. When the values of m_{rx} and m_{ry} are used to calculate the areas of positive-moment reinforcement (bottom steel) in a slab, the calculations are based on Eq. (14-10a) or Eq. (14-10b). The + sign and absolute number in these equations combine to make the term "$+ |m_{xy}|$" positive. If either m_{ry} or m_{rx} are negative in these two equations, they are set equal to zero for designing bottom reinforcement.

Calculation of m_{rx} and m_{ry}—negative moment. For negative-moment reinforcement (top steel) the calculations are based on Eqs. (14-10c) and (14-10d). The − sign and absolute number combine to make the term "$- |m_{xy}|$" negative. If either m_{ry} or m_{rx} are positive in these two equations, they are set to zero for designing top reinforcement.

EXAMPLE 14-1 Computation of Design Moments at a Point

A finite-element analysis gives the moments in an element as $m_x = 5$ kip-ft/ft, $m_y = -1$ kip-ft/ft, and $m_{xy} = -2$ kip-ft/ft. Compute the moments to be used to design reinforcement.

(a) Steel at bottom of slab:

$$m_{ry} = -1 + |-2| = +1 \text{ kip-ft/ft}$$
$$m_{rx} = 5 + |-2| = +7 \text{ kip-ft/ft}$$

(b) Steel at top of slab:

$$m_{ry} = -1 - |-2| = -3 \text{ kip-ft/ft}$$

$$m_{rx} = 5 - |-2| = +3 \text{ kip-ft/ft}$$

Because m_{rx} is positive, it is set equal to zero in designing steel at the top of the slab. ■

Generally, the reinforcement is uniformly spaced in orthogonal bands such that

(a) the total reinforcement provided in a band is sufficient to resist the total factored moment computed for that band, and

(b) the moment resistance per unit width in the band is at least two-thirds of the maximum moment per unit width in the band, as computed in the finite-element analysis.

In the direct design method the widths of the bands of reinforcement were taken equal to the widths of the column strip and middle strips defined in ACI Code Sections 8.4.1.5 and 8.4.1.6. The use of such band (strip) widths is generally acceptable.

14-3 YIELD-LINE ANALYSIS OF SLABS: INTRODUCTION

Under overload conditions in a slab failing in flexure, the reinforcement will yield first in a region of high moment. When that occurs, this portion of the slab acts as a plastic hinge, only able to resist its plastic moment strength. When the load is increased further, the hinging region rotates plastically, and the moments due to additional loads are redistributed to adjacent sections, causing them to yield, as shown in Fig. 13-4. The bands in which yielding has occurred are referred to as *yield lines* and divide the slab into a series of elastic plates. Eventually, enough yield lines exist to form a plastic mechanism in which the slab can deform plastically without an increase in the applied load.

The *yield-line analysis method* for slabs actually brings more continuity between the analysis of load effects and analysis of member strengths. In general design, moments and shears from an *elastic analysis* are compared to *plastic member strengths*, using appropriate load factors and strength-reduction factors. In the yield-line method for slabs, the loads required to develop a *plastic mechanism* are compared directly to the *plastic resistance* (nominal strength) of the member. Load factors and strength-reduction factors can be incorporated into the procedure, as will be shown in some later examples.

A yield-line analysis uses rigid plastic theory to compute the failure loads corresponding to given plastic moment resistances in various parts of the slab. It does not give any information about deflections or about the loads at which yielding first starts. Although the concept of yield-line analysis was first presented by A. Ingerslev in 1921–1923, K. W. Johansen developed modern yield-line theory [14-3]. This type of analysis is widely used for slab design in the Scandinavian countries. The yield-line concept is presented here to aid in the understanding of slab behavior between service loads and failure. Further details are given in [14-4] to [14-7].

Yield Criterion

To limit deflections, floor slabs are generally considerably thicker than is required for flexure, and as a result, they seldom have reinforcement ratios exceeding 0.3 to 0.4 times the balanced reinforcement ratio defined in Eq. (4-25). In this range of reinforcement ratios, the

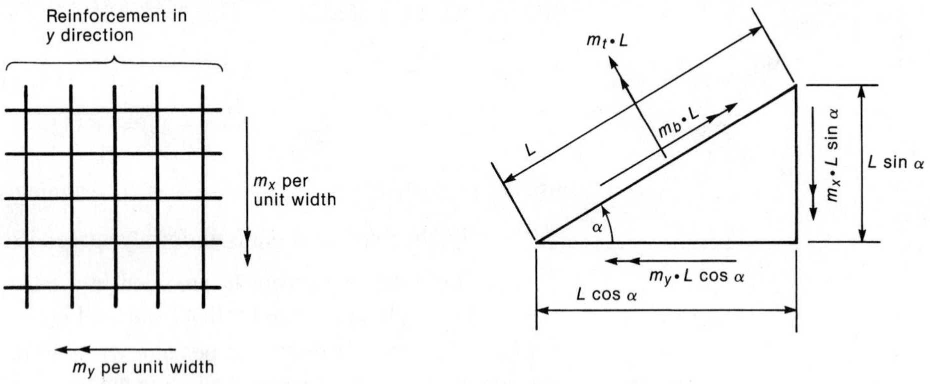

(a) Reinforcement pattern.

(b) Moments on an element.

Fig. 14-3
Yield criterion.

moment–curvature response is essentially elastic–plastic with a plastic moment strength conservatively assumed to be equal to ϕM_n, the design flexural strength of the section.

If yielding occurs along a line at an angle α to the reinforcement, as shown in Fig. 14-3, the bending and twisting moments are assumed to be distributed uniformly along the yield line and are the maximum values provided by the flexural capacities of the reinforcement layers crossed by the yield line. It is further assumed that there is no kinking of the reinforcement as it crosses the yield line. In Fig. 14-3, the reinforcements in the x and y directions provide moment capacities of m_x and m_y per unit width (k-ft/ft). The bending moment, m_b, and the twisting moment, m_t, per unit length of the yield line can be calculated from the moment equilibrium of the element. In this calculation, α will be measured counterclockwise from the x-axis; the bending moments m_x, m_y, and m_b will be positive if they cause tension in the bottom of the slab; and the twisting moment, m_t, will be positive if the moment vector points away from the section, as shown.

Consider the equilibrium of the element in Fig. 14-3b:

$$m_b L = m_x(L \sin \alpha) \sin \alpha + m_y(L \cos \alpha) \cos \alpha$$

This equation gives the bending moment m_b as

$$m_b = m_x \sin^2 \alpha + m_y \cos^2 \alpha \tag{14-11}$$

The twisting moment m_t is

$$m_t = \left(\frac{m_x - m_y}{2}\right) \sin 2\alpha \tag{14-12}$$

These equations apply only for *orthogonal reinforcement*. If $m_x = m_y$, these two equations reduce to $m_b = m_x = m_y$ and $m_t = 0$, regardless of the angle of the yield line. This is referred to as *isotropic reinforcement*.

Locations of Axes and Yield Lines

Yield lines form in regions of maximum moment and divide the plate into a series of elastic plate segments. When the yield lines have formed, all further deformations are concentrated at the yield lines, and the slab deflects as a series of stiff plates joined together by long hinges, as shown in Fig. 14-4. The pattern of deformation is controlled by *axes* that pass along line supports and over columns, as shown in Fig. 14-5, and by the yield lines. Because the individual plates rotate about the axes and/or yield lines, these axes and lines must be straight. To satisfy compatibility of deformations at points such as A and B in Fig. 14-4, the

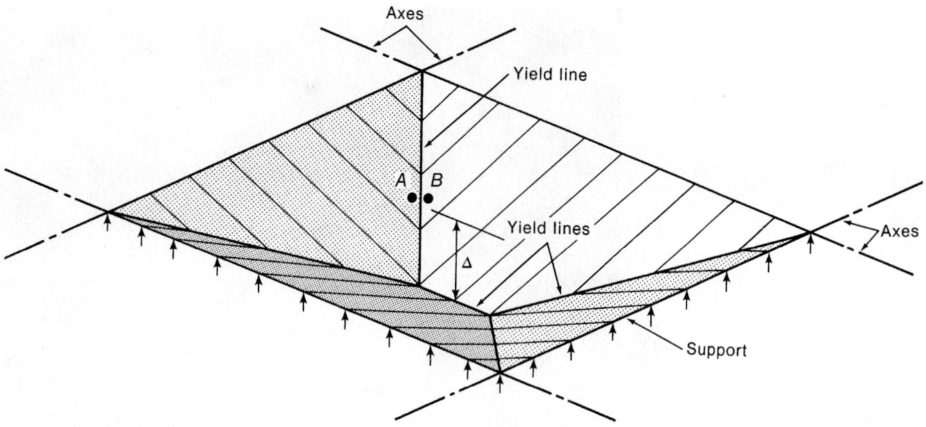

Fig. 14-4
Deformations of a slab with yield lines.

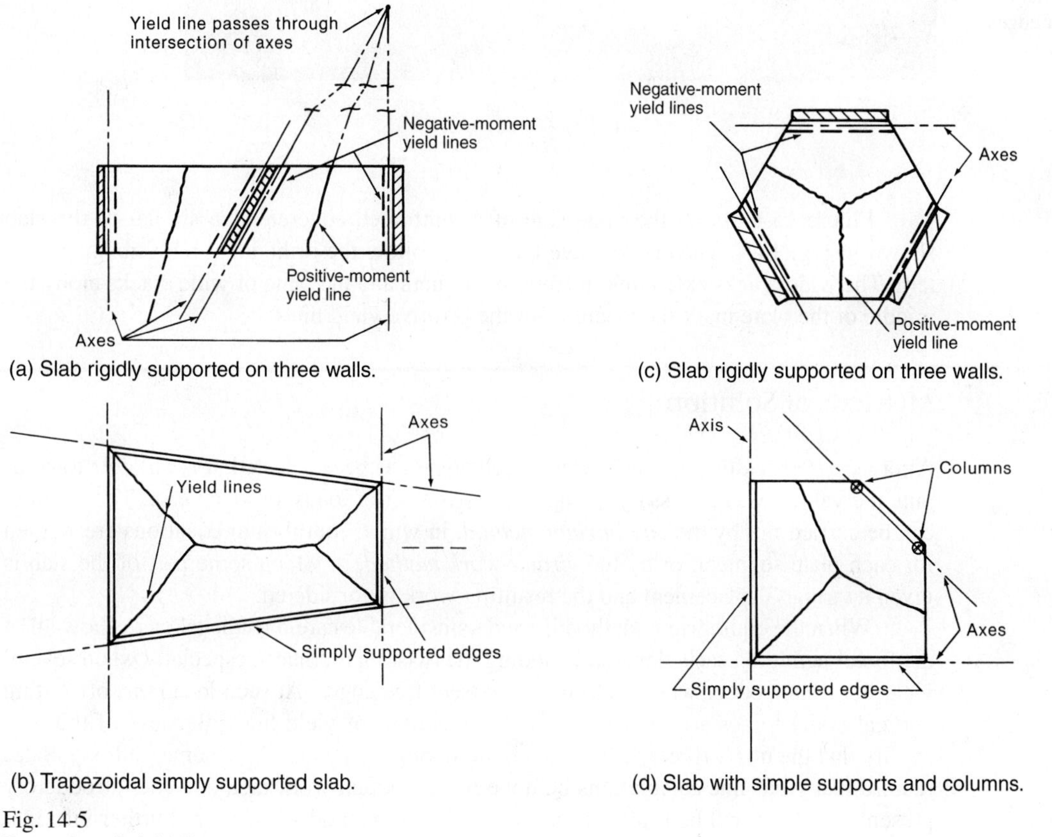

Fig. 14-5
Examples of yield-line patterns.

yield line dividing two plates must intersect the intersection of the axes about which those plates are rotating. Figure 14-5 shows the locations of axes and yield lines in a number of slabs subjected to uniform loads. The yield mechanisms in Figs. 14-4 and 14-5 are referred to as *kinematically admissible mechanisms* because the displacements and rotations of adjacent plate segments are compatible. If more than one kinematically admissible mechanism can be defined for a given slab panel, each mechanism should be investigated to determine which one results in the minimum panel resistance to a given loading.

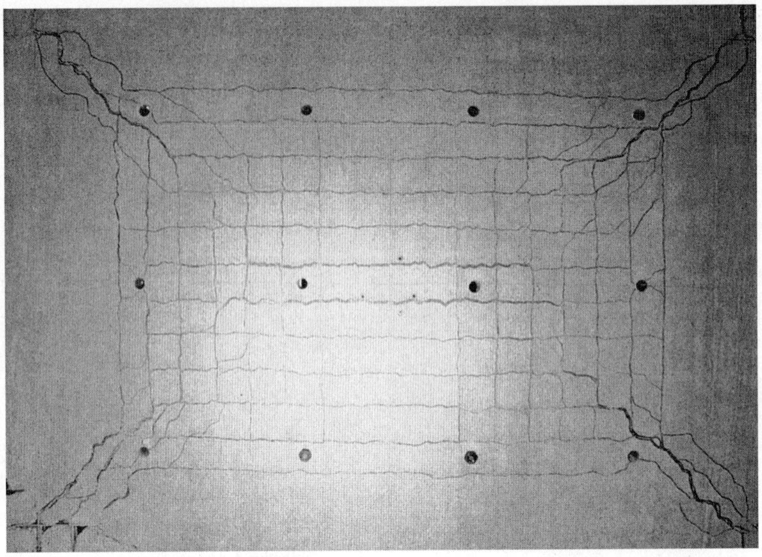

Fig. 14-6
Yield lines in a slab that is simply supported on four edges. (Photograph courtesy of J. G. MacGregor.)

Figure 14-6 shows the underside of a reinforced concrete slab similar to the slab shown in Fig. 14-4. The cracks have tended to follow the orthogonal reinforcement pattern. The wide cracks extending in from the corners and the band of wide cracks along the middle of the plate mark the locations of the positive yield lines.

Methods of Solution

After a kinematically admissible yield mechanism has been selected, it is possible to compute the values of m necessary to support a given set of loads, or vice versa. The solution can be carried out by the *equilibrium method*, in which equilibrium equations are written for each plate segment, or by the *virtual-work method*, in which some part of the slab is given a virtual displacement and the resulting work is considered.

When the equilibrium method is used, considerable care must be taken to show all of the forces acting on each element, including the twisting moments, especially when several yield lines intersect or when yield lines intersect free edges. At such locations, off setting vertical nodal forces are required at the intersections of yield lines. Because of the possibility that the nodal forces will be given the wrong sign or location, some building codes require that yield-line calculations be done by the virtual-work method. The introductory presentation here will be limited to the virtual-work method of solution. Further information on the equilibrium method can be obtained in [14-3] to [14-7].

Virtual-Work Method

Once the yield lines have been chosen, some point on the slab is given a virtual displacement, δ, as shown in Fig. 14-7c. The external work done by the loads when displaced is

$$\begin{aligned}\text{external work} &= \iint q\delta\, dx\, dy \\ &= \Sigma\,(W\Delta_c) \qquad (14\text{-}13)\end{aligned}$$

where q = uniformly distributed load on an element of area
δ = deflection of that element
W = total load on a plate segment
Δ_c = deflection of the centroid of that segment

In Example 14-2, it will be shown that the right-hand side of Eq. (14-13) can be expressed as q times the displaced volume for the slab yield mechanism.
The total external work is the sum of the separate work done by each plate.

The internal work done by rotating the yield lines is

$$\text{internal work} = \sum (m_b \ell \theta) \tag{14-14}$$

where m_b = bending moment per unit length of yield line
ℓ = length of the yield line
θ = angle change at that yield line corresponding to the virtual displacement, δ

The total internal work done during the virtual displacement is the sum of the internal work done on each separate yield line. Because the yield lines are assumed to have formed before the virtual displacement is imposed, no elastic deformations occur during the virtual displacement.

The principle of virtual work states that, for conservation of energy,

$$\text{external work} = \text{internal work}$$

or

$$\sum (W\Delta_c) = \sum (m_b \ell \theta) \tag{14-15}$$

The virtual work solution is an *upper-bound solution*—that is, the load W is equal to or *higher* than the true failure load. If an incorrect set of yield lines is chosen, the resisted W will be too large for a given m, or conversely, the value of m required to resist a given W will be too small.

EXAMPLE 14-2 Yield-Line Analysis of One-Way Slab by Virtual Work

The one-way slab shown in Fig. 14-7 will be analyzed to determine the area load, q_f, that will cause failure. The slab reinforcement is shown in Fig. 14-7a, and the assumed failure mechanism is shown in Fig. 14-7b. When using plan views of all of the slabs in this chapter, cross-hatched notation at the supports indicates that either the slab is fixed at the supports or is continuous across the supports, and the slab is capable of resisting negative bending. A single-hatched notation will be used to indicate supports that are not capable of resisting a negative moment (i.e., a simple support). It is assumed that the negative-bending capacity at the left support, m_{n1}, is equal to 6 k-ft/ft and the negative-bending capacity at the right support, m_{n2}, is equal to 10 k-ft/ft. The positive-bending capacity of the slab reinforcement near midspan, m_p, is equal to 4 k-ft/ft.

1. Select the axes and yield lines. The negative yield lines (shown as dashed lines) and axes of rotation for the two slab segments must be at the supports, as shown in Fig. 14-7b. Because the two axes at the supports are parallel to each other, the positive yield line, shown by the wavy solid line near midspan, must be parallel to the supports for this to be a kinematically admissible mechanism. The distance x that results in the minimum value for q_f is to be determined.

2. Give the slab a virtual displacement. Assume that the positive yield line, C–D, is displaced downward a distance equal to the virtual displacement, δ, as shown in Fig. 14-7c. This causes slab segment I to rotate an angle, θ_{I}, and slab segment II to rotate an angle, θ_{II}. The total rotation across the positive yield line near midspan is the sum, $\theta_{\mathrm{I}} + \theta_{\mathrm{II}}$. For small displacements, θ is approximately equal to *tan* θ, so

$$\theta_{\mathrm{I}} = \frac{\delta}{x}, \text{ and } \theta_{\mathrm{II}} = \frac{\delta}{12\text{ ft} - x}$$

3. Compute the internal work. The internal work is the sum of the work done by the moments in the plastic hinging zones as the slab segments deflect. All of the moment capacities are given per foot of length, so those values will need to be multiplied by the length of the plastic hinges, which equals the slab width, b.

$$\begin{aligned} IW &= m_{n1} \times b \times \theta_{\mathrm{I}} + m_{n2} \times b \times \theta_{\mathrm{II}} + m_p \times b \times (\theta_{\mathrm{I}} + \theta_{\mathrm{II}}) \\ &= (6\text{ k-ft/ft}) \cdot b \cdot \frac{\delta}{x} + (10\text{ k-ft/ft}) \cdot b \cdot \frac{\delta}{12 - x} + (4\text{ k-ft/ft}) \cdot b \cdot \left(\frac{\delta}{x} + \frac{\delta}{12 - x}\right) \\ &= \left[(10\text{ k})\frac{1}{x} + (14\text{ k})\frac{1}{12 - x}\right] b \cdot \delta \end{aligned}$$

Note that the moment values are now expressed in kips, as opposed to k-ft/ft.

4. Compute the external work. As noted after Eq. (14-13), this can be calculated as either a sum of vector forces for each slab segment (plate) times the corresponding displacement of the centroid of the plate or as the area force, q_f, multiplied times the displaced volume for the assumed mechanism. Both procedures will be used here.

For the first procedure, both slab segments are rectangular, so their centroids will be at their midlength. Also, because each segment is supported at one end and deflects a distance, δ, at the other end, each of the centroids will deflect a distance of $\delta/2$. The equivalent vector forces acting at the centroids is equal to the area load, q_f, times the area of the segment. Thus,

$$EW = q_f[x \cdot b + (12 - x)b]\frac{\delta}{2} = \frac{q_f \cdot b \cdot \delta \cdot 12\text{ ft}}{2} = q_f \cdot b \cdot \delta \cdot 6\text{ ft}$$

For the second procedure, the displaced volume of the slab mechanism is equal to the triangular area of the displaced shape in Fig. 14-7c, times the width of the slab. Thus, the expression for external work is

$$EW = q_f(\tfrac{1}{2} \times \delta \times 12\text{ ft})b = q_f \cdot b \cdot \delta \cdot 6\text{ ft}$$

As expected, these two procedures give the same result. For some slab-analysis cases, one procedure may be easier than the other, and thus, it is recommended that a designer be familiar with both procedures.

5. Equate the external and internal work. Set $EW = IW$ and solve for q_f. Note that the virtual deflection, δ, always will cancel across the equal sign. In this case, the slab width also cancels out of the expressions.

$$q_f \cdot b \cdot \delta \cdot 6\text{ ft} = \left(\frac{10\text{ k}}{x} + \frac{14\text{ k}}{12\text{ ft} - x}\right) b \cdot \delta$$

$$q_f = \frac{10\text{ k}}{x \cdot 6\text{ ft}} + \frac{14\text{ k}}{(12\text{ ft} - x) \cdot 6\text{ ft}}$$

In this form, the solution for q_f will be in units of ksf.

6. Solve for the minimum value of q_f. The minimum value of q_f can be found by setting the derivative of q_f, $d(q_f)/dx$, equal to zero and solving for x. Such a solution is easy for this problem, but for more complicated two-way slab problems an iterative solution procedure may be more efficient. A spreadsheet can be used for the iterations. Because of the different end-moment strengths for this one-way slab, the minimum q_f will not occur for $x = 6$ ft. The plastic-moment capacity is larger at the right support, so a reasonable strategy will be to select x-values less than 6 ft to decrease the amount of rotation in slab segment II, θ_{II}, and thus reduce the amount of work done by the plastic hinge at the right support. Of course, this strategy increases the rotation in slab segment I, θ_I, and thus increases the work done by the plastic hinge at the left support. At some point, the increase in work at the left support will be larger than the decrease in work done at the right support. The following table starts with $x = 6$ ft and moves to lower x-values:

x (ft)	q_f (ksf)
6.0	0.667
5.8	0.664
5.6	0.662
5.4	0.662
5.2	0.664
5.0	0.667

From this table, it appears that the minimum value of q_f may occur at $x = 5.5$ ft. Using that value results in $q_f = 0.662$ ksf, which will be accepted as the final solution for this problem. ■

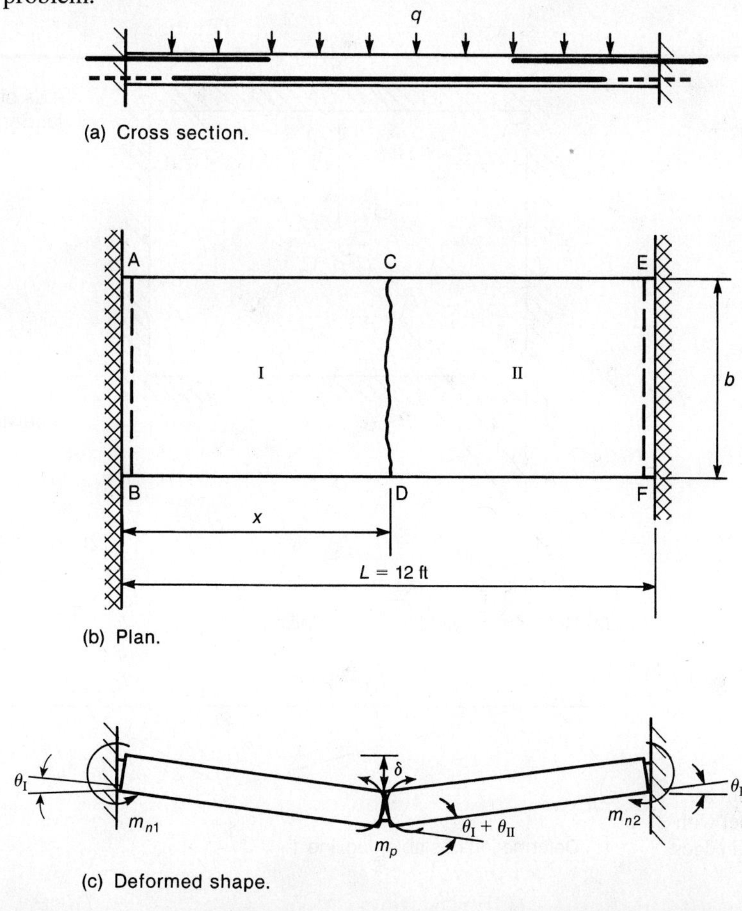

(a) Cross section.

(b) Plan.

(c) Deformed shape.

Fig. 14-7
Slab—Example 14-2.

14-4 YIELD-LINE ANALYSIS: APPLICATIONS FOR TWO-WAY SLAB PANELS

When analyzing the yield mechanisms for two-way slab panels, a designer commonly will need to deal with positive yield lines that run at a skew angle across an orthogonal reinforcement pattern (Figs. 14-5 and 14-6). The most general procedure for determining the internal work generated along this yield line as the slab mechanism deforms is to first use Eq. (14-11) to determine the moment resistance acting along the yield line, assuming the moment strengths are known for the orthogonal reinforcement grid crossed by the yield line. The moment strength along the yield line then is to be multiplied by the length of the yield line and the relative rotation between the two slab segments adjacent to the yield line to obtain the internal work.

The plastic mechanism for a rectangular two-way slab panel with simply supported edges (Fig. 14-8) will be used to demonstrate this procedure. We will concentrate on the internal work done along the positive yield line *O–P*, assuming that there is a virtual displacement of δ at point *P*. From symmetry, the internal work found for yield line *O–P* can be multiplied by four to obtain the total internal work for the slab panel.

Assume the yield line *O–P* has a length of L_d and forms an angle, α, with respect to the longer edge of the panel. The line *A–B* is drawn perpendicular to the yield line *O–P* and extends either to an edge of the panel (point *B*) or to an extended edge of the panel (point *A*), which represents the axis of rotation for slab Segment I (Fig. 14-8a). Section line 1–1 is cut along line *A–B* to show the deformed shape of the panel and the relative rotation between the

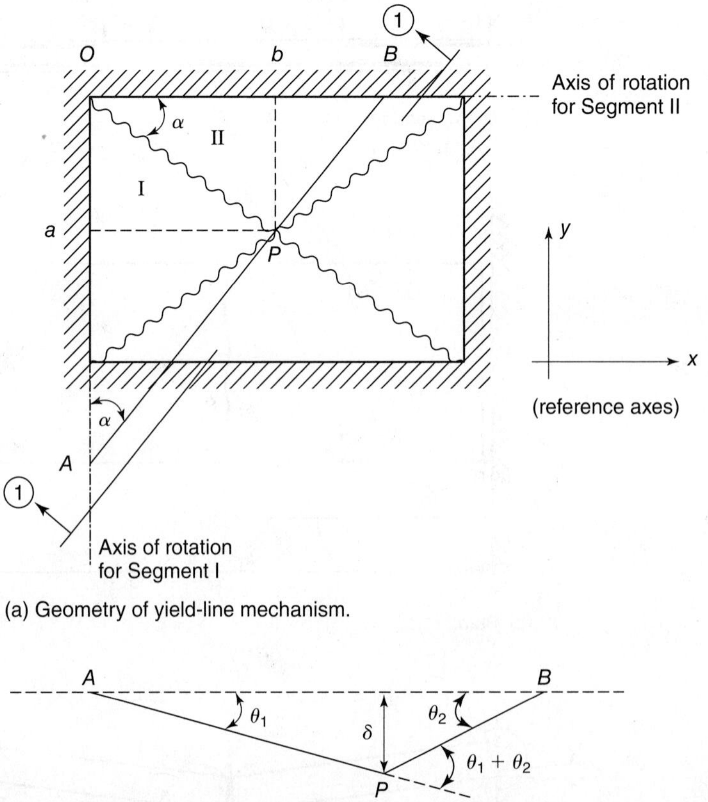

(a) Geometry of yield-line mechanism.

(b) Deformation of slab along line 1–1.

Fig. 14-8
Yield-line mechanism for rectangular slab panel with hinge supports on all edges.

slab segments that meet at yield line *O–P* (Fig. 14-8b). The length of line segment *P–B* is $L_d \cdot \tan \alpha$, and the length of line segment *P–A* is $L_d \cdot \cot \alpha$. Thus, the rotation angles shown in Fig. 14-8b are

$$\theta_1 = \frac{\delta}{L_d \cot \alpha} \text{ and } \theta_2 = \frac{\delta}{L_d \tan \alpha}$$

Assuming the moment resistance along yield line *O–P* has the value, m_b, obtained from Eq. (14-11), the internal work for this yield-line segment is

$$IW(O\text{–}P) = m_b \cdot L_d\,(\theta_1 + \theta_2) = m_b \cdot \delta(\tan \alpha + \cot \alpha) \qquad \text{(14-16a)}$$

An alternative approach can be used if the supported edges of the panel (i.e., the axes of rotation for the slab segments shown in Fig. 14-8a) are orthogonal to the reinforcement pattern used in the slab panel. For such cases, it is possible to project the work done by the slab reinforcement back to the axes of rotation.

The corner of the slab panel in Fig. 14-8 is reproduced in Fig. 14-9 with the yield line *O–P* replaced by a series of short line segments that run parallel and perpendicular to the slab's positive reinforcement. Assuming that a virtual deflection δ occurs at point *P*,

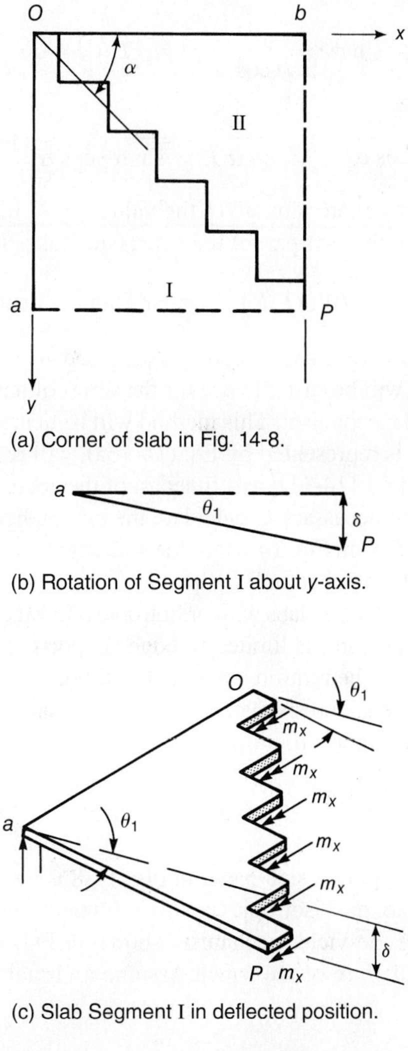

(a) Corner of slab in Fig. 14-8.

(b) Rotation of Segment I about *y*-axis.

(c) Slab Segment I in deflected position.

Fig. 14-9
Rotation of slab Segment I about *y*-axis.

the rotation of slab Segment I about its supported edge is shown in Fig. 14-9b. As indicated in Fig. 14-9c, only the m_x component from the orthogonal slab reinforcement will do work when slab Segment I is rotated about the y-axis. Similarly, for slab Segment II, only the m_y component of the slab resistance will do work when that segment is rotated about its supported edge, which is parallel to the x-axis. Thus, the total internal work along yield line O–P can be calculated as the sum of the resistances from the two orthogonal components of the slab resistance, with each multiplied by the corresponding rotations and projected lengths of the slab segments that are adjacent to the yield line.

This alternative procedure for finding internal work will be demonstrated for yield line O–P of the slab panel in Fig. 14-8, which was assumed to have a length of L_d. Line segments a–P and b–P in Fig. 14-8 have lengths of $L_d \cdot \cos\alpha$ and $L_d \cdot \sin\alpha$, respectively. Thus, the rotations of slab Segments I and II are

$$\theta_{\text{I}} = \frac{\delta}{L_d \cdot \cos\alpha} \text{ and } \theta_{\text{II}} = \frac{\delta}{L_d \cdot \sin\alpha}$$

Line segments O–a and O–b represent the projections of yield line O–P onto the y- and x-axes, and have lengths of $L_d \cdot \sin\alpha$ and $L_d \cdot \cos\alpha$, respectively. Using these lengths, the internal work along yield line O–P can be expressed as

$$IW(O\text{–}P) = m_x \cdot L_d \cdot \sin\alpha \cdot \frac{\delta}{L_d \cdot \cos\alpha} + m_y \cdot L_d \cdot \cos\alpha \cdot \frac{\delta}{L_d \cdot \sin\alpha} \tag{14-16b}$$

$$= \delta\left(m_x \frac{\sin\alpha}{\cos\alpha} + m_y \frac{\cos\alpha}{\sin\alpha}\right) = \frac{\delta}{\sin\alpha \cdot \cos\alpha}(m_x \sin^2\alpha + m_y \cos^2\alpha)$$

The second part of this expression gives the value for m_b in Eq. (14-11), and using a geometric substitution for the first part of the expression takes us back to Eq. (14-16a):

$$IW(O\text{–}P) = m_b \cdot \delta(\tan\alpha + \cot\alpha)$$

Although this alternative method as expressed in Eq. (14-16b) may appear to be more complicated, it will be easier to use for the series of iterations that may be required to find the critical-yield mechanism. This method will be demonstrated in Example 14-4. The prior method, which is represented by Eq. (14-16a), will require a new calculation during each iteration for m_b [Eq. (14-11)] as a function of the angle α. For slab panels without 90° corners, it also will be necessary to calculate the perpendicular measurements to the axes of rotation (P–A and P–B in Fig. 14-8) for the slab segments adjacent to the yield line. This procedure will be demonstrated in Example 14-5.

It must be noted that for slabs with orthotropic reinforcement the alternative procedure, represented by Eq. (14-16b), is limited to edge supports (axes of rotation) that are parallel and perpendicular to the reinforcement pattern. For slabs with isotropic reinforcement, where by definition the slab resistance is the same for any crack orientation, Eq. (14-16b) can be used for slab segments rotating about any axis.

EXAMPLE 14-3 Yield-Line Analysis of a Square Two-Way Slab Panel

Assume the simply supported slab panel in Fig. 14-8 is a square with sides equal to 15 ft. Also assume the slab has isotropic positive-moment reinforcement with $m_x = m_y = m_b = 4$ k-ft/ft. Using the yield mechanism shown in Fig. 14-8, determine the area load, q_f, required to cause failure of the panel. Assume a virtual downward deflection equal to δ at point P.

1. Compute the internal work. The total internal work is four times that expressed by Eq. (4-16a). For a 45° diagonal, both tan α and cot α are equal to 1.0. Thus,

$$IW = 4 \cdot m_p \cdot \delta(1 + 1) = 4(4\text{ k-ft/ft}) \cdot \delta \cdot 2 = (32\text{ k}) \cdot \delta$$

2. Compute the external work. The external work can be expressed at the area load, q_f, multiplied by the displaced volume. In this case, the displaced volume has the shape of an inverted pyramid, and thus, the volume is equal to the total slab area times one-third of δ.

$$EW = q_f\left(L^2 \cdot \frac{\delta}{3}\right) = q_f \cdot \delta\left(\frac{225\text{ ft}^2}{3}\right)$$

3. Equate the external and internal work and solve for q_f. As before, the virtual displacement, δ, will cancel out, so

$$q_f = \frac{32\text{ k}}{(225/3)\text{ ft}^2} = 0.427\text{ ksf} = 427\text{ psf}$$ ■

EXAMPLE 14-4 Yield-Line Analysis of Two-Way Rectangular Slab Panel

The two-way slab shown in Fig. 14-10 will be analyzed to determine the area load, q_f, that will cause failure. The bottom slab reinforcement is orthotropic and has moment strengths of $m_{px} = 3$ k-ft/ft and $m_{py} = 2$ k-ft/ft. The subscript p is used to designate positive moment. The negative-bending resistance is isotropic with $m_{nx} = m_{ny} = m_n = 4$ k-ft/ft.

1. Select the general plastic mechanism. The assumed plastic-collapse mechanism is shown in Fig. 14-10. There are negative yield lines along the supported edges and positive yield lines running diagonally out of each corner where the axes of rotation for adjacent slab segments intersect. The locations of point A and the symmetrical point B are defined by the coefficient β. We will need to iterate to find the value of β that results in the minimum value for q_f.

2. Give the slab a virtual displacement. Assume that the segment of the positive yield line from A to B is displaced downward a virtual displacement, δ. This causes the

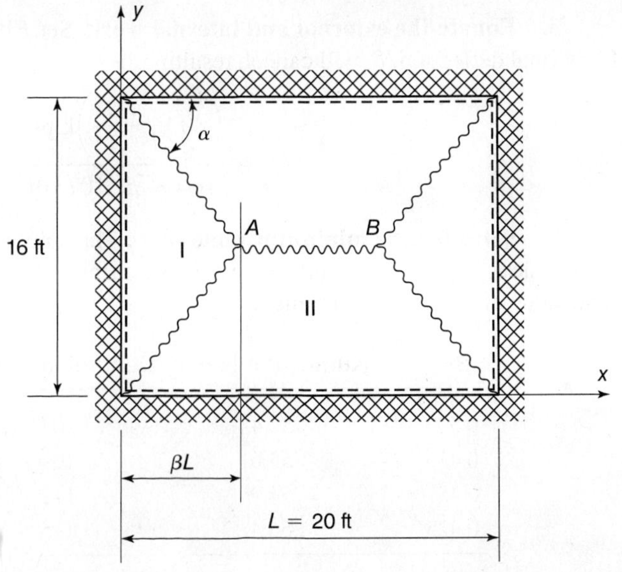

Fig. 14-10
Slab panel and plastic mechanism for Example 14-4.

slab Segment I to rotate an angle, θ_{I}, and the slab Segment II to rotate an angle, θ_{II}. The rotations of the slab segments are

$$\theta_{\text{I}} = \frac{\delta}{\beta \cdot 20\text{ ft}} \text{ and } \theta_{\text{II}} = \frac{\delta}{8\text{ ft}}$$

3. Compute the internal work. The total internal work is equal to two times the sum of the work done by the plastic moments acting on slab Segments I and II. For slab Segment I the internal work is

$$\begin{aligned} IW(\text{I}) &= m_{px} \cdot 16\text{ ft} \cdot \theta_{\text{I}} + m_n \cdot 16\text{ ft} \cdot \theta_{\text{I}} \\ &= (3 + 4)\frac{\text{k-ft}}{\text{ft}} \cdot 16\text{ ft} \cdot \frac{\delta}{\beta \cdot 20\text{ ft}} = \frac{5.6\text{ k} \cdot \delta}{\beta} \end{aligned}$$

Similarly, the internal work for slab Segment II is

$$\begin{aligned} IW(\text{II}) &= m_{py} \cdot 20\text{ ft} \cdot \theta_{\text{II}} + m_n \cdot 20\text{ ft} \cdot \theta_{\text{II}} \\ &= (2 + 4)\frac{\text{k-ft}}{\text{ft}} \cdot 20\text{ ft} \cdot \frac{\delta}{8\text{ ft}} = 15\text{ k} \cdot \delta \end{aligned}$$

From these, the sum of the internal work is

$$\sum IW = 2\left(15\text{ k} + \frac{5.6\text{ k}}{\beta}\right) \cdot \delta$$

4. Compute the external work. The two end regions outside of the points A and B are essentially two half-pyramids that can be combined into a single pyramid with a base area 16 ft × 2β × 20 ft. The central region between points A and B has a triangular cross section and extends over a length or 20 ft × $(1 - 2\beta)$. Thus, the external work is

$$\begin{aligned} EW &= q_f\left[2\beta \cdot 20\text{ ft} \cdot 16\text{ ft} \cdot \frac{\delta}{3} + \frac{1}{2} \cdot 16\text{ ft} \cdot \delta \times (1 - 2\beta) \cdot 20\text{ ft}\right] \\ &= q_f \cdot \delta(\beta \cdot 213\text{ ft}^2 + 160\text{ ft}^2 - \beta \cdot 320\text{ ft}^2) \\ &= q_f \cdot \delta(160 - \beta \cdot 107)\text{ ft}^2 \end{aligned}$$

5. Equate the external and internal work. Set $EW = \Sigma(IW)$ and solve for q_f. The virtual deflection, δ, will cancel, resulting in

$$q_f = \frac{\left(30 + \frac{11.2}{\beta}\right)\text{kips}}{(160 - \beta \cdot 107)\text{ ft}^2}$$

6. Solve for the minimum value of q_f. The solution table below starts with β = 0.5 and slowly decreases β until the value of q_f starts to increase. At that point, the minimum value of q_f has been obtained.

β	Numerator (kips)	Denominator (ft²)	q_f (ksf)
0.50	52.4	107	0.491
0.475	53.6	109	0.490
0.45	54.9	112	0.490
0.425	56.4	115	0.491

From this table, the minimum value of q_f is 0.490 ksf, and it occurs for a β value of approximately 0.46. It is not a practical structural engineering decision to make a finer adjustment of β in an attempt to find that the minimum value of q_f is slightly less than 0.490 ksf.

7. Recompute the internal work using Eq. (14-16a). We will calculate the positive-moment internal work using Eq. (14-16a) for one of the diagonal yield lines in Fig. 14-10, and then multiply that value by four. For β values less than 0.5, we also will need to include the positive-moment internal work along the yield-line segment A–B. Finally, we will need to calculate the negative-moment internal work for the four slab segments in Fig. 14-10. For this single iteration, select $\beta = 0.475$.

The internal work for line segment A–B is

$$IW(A\text{–}B) = m_y \cdot (1 - 2\beta) \cdot 20\text{ ft} \cdot 2 \cdot \theta_{\text{II}}$$

$$= 2\frac{\text{k-ft}}{\text{ft}}(1 - 0.95)\,20\text{ ft} \cdot \frac{2\delta}{8\text{ ft}} = 0.50\text{ k} \cdot \delta$$

To find the value of m_b in Eq. (14-11), which represents the positive-moment resistance along the diagonal yield line extending from the corner, we need to find the value of the angle, α:

$$\tan\alpha = \frac{8\text{ ft}}{\beta \cdot 20\text{ ft}} = \frac{8}{9.5}$$

$$\text{thus, } \alpha = \tan^{-1}\frac{8}{9.5} = 40.1°$$

Then, from Eq. (14-11),

$$m_b = m_{px}\sin^2\alpha + m_{py}\cos^2\alpha$$

$$= 3\frac{\text{k-ft}}{\text{ft}} \cdot 0.415 + 2\frac{\text{k-ft}}{\text{ft}} \cdot 0.585 = 2.41\frac{\text{k-ft}}{\text{ft}}$$

Putting this into Eq. (14-16b), we get

$$IW(\text{diag}) = m_b \cdot \delta(\tan\alpha + \cot\alpha)$$

$$= 2.41\text{ k}\left(\frac{8}{9.5} + \frac{9.5}{8}\right)\delta = 4.90\text{ k} \cdot \delta$$

Thus, the sum of positive-moment internal work is

$$\sum IW(\text{pos}) = (4 \times 4.90\text{ k} + 0.50\text{ k})\delta = 20.1\text{ k} \cdot \delta$$

The sum of the negative-moment internal work is calculated using the θ_{I} and θ_{II} values from step 2:

$$IW(\text{neg}) = m_n(16\text{ ft} \cdot \theta_{\text{I}} + 20\text{ ft} \cdot \theta_{\text{II}}) \times 2$$

$$= 4\frac{\text{k-ft}}{\text{ft}}\left(16\text{ ft}\frac{\delta}{\beta \cdot 20\text{ ft}} + 20\text{ ft}\frac{\delta}{8\text{ ft}}\right) \times 2$$

$$= 8\text{ k} \cdot \delta\left(\frac{16}{9.5} + \frac{20}{8}\right) = 33.5\text{ k} \cdot \delta$$

Now, the total internal work is

$$\sum IW = (20.1\text{ k} + 33.5\text{ k})\delta = 53.6\text{ k} \cdot \delta$$

This value is the same as the value calculated for the total internal work in row 2 of the solution table in step 6. ■

EXAMPLE 14-5 Yield-Line Analysis of a Triangular Slab Panel

The triangular-shaped two-way slab shown in Fig. 14-11 will be analyzed to determine the minimum area load, q_f, that will cause failure. Two edges of the slab are simply supported, and the third edge is free (unsupported). The bottom slab reinforcement is orthotropic with respect to the x- and y-axes, and has moment strengths of $m_{px} = 3.5$ k-ft/ft and $m_{py} = 2.5$ k-ft/ft. For the given shape of the slab panel, the angle β is equal to 53.1°.

1. Select the plastic mechanism. The assumed plastic-collapse mechanism is shown in Fig. 14-11a. There is a single, positive yield line coming out of the corner (point A) where the two supported edges (two axes of rotation) meet. When a slab has a free edge, positive yield lines tend to propagate toward that free edge. The location of the point D on the free edge is not known. Thus, we will select different values for the variable, x, to determine the minimum value for q_f.

2. Give the slab a virtual displacement. Assume the point D is displaced downward a virtual displacement, δ. The line F–D–E is drawn perpendicular to the positive yield line A–D, and Fig. 14-11b shows the deflected shape of the slab along the line F–D–E. Defining the length of yield line A–D as L_d, and referring to the angles α and β in Fig. 14-11a, the values for the length of the line segments represented by the symbols a and b are

$$a = L_d \tan \alpha \text{ and } b = L_d \tan(\beta - \alpha)$$

With these values, the relative rotation between the two slab segments on each side of the yield line is calculated as

$$\theta_d = \frac{\delta}{a} + \frac{\delta}{b}$$

3. Compute the internal work. Because the supported edge A–B is neither perpendicular nor parallel to the orthotropic reinforcement, Eq. (14-16a) cannot be used for

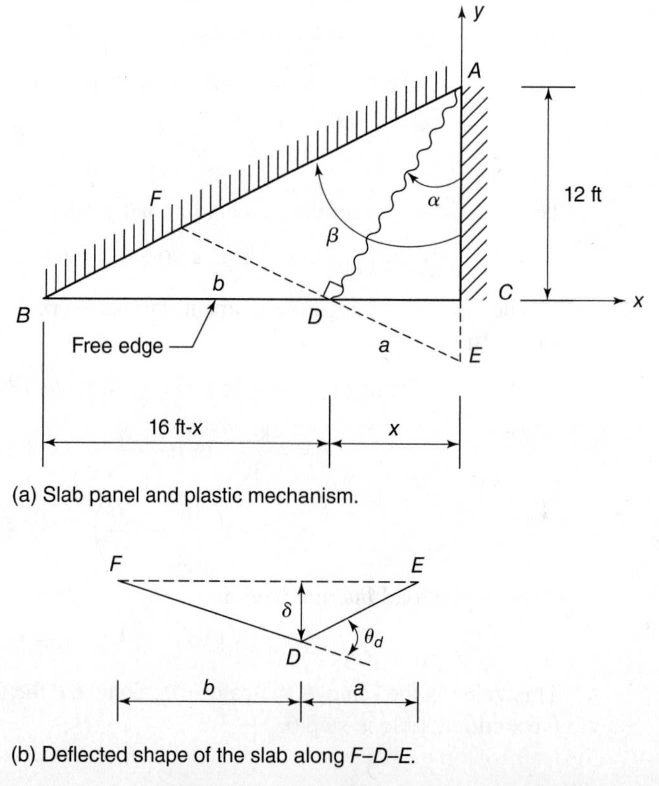

Fig. 14-11
Triangular slab panel with orthotropic reinforcement.

this slab panel. Thus, we will return to the basic procedure of calculating the internal work as being equal to the moment resistance acting along the yield line, found with Eq. (14-11), times the length of the yield line, L_d, and then times the rotation between the slab segments that meet at the yield line, θ_d. From Fig. 14-11a, the length of the yield line is

$$L_d = \sqrt{(12\text{ ft})^2 + x^2}$$

Equation (14-11) is used to determine the moment resistance, m_b, along yield line *A–D*. It should be noted that because the definition of the angle α is different here than was used in the derivation of Eq. (14-11), the $\sin^2$ and $\cos^2$ terms are interchanged.

$$m_b = m_{px}\cos^2\alpha + m_{py}\sin^2\alpha$$

where

$$\tan\alpha = \frac{x}{12\text{ ft}}$$

which implies

$$\alpha = \tan^{-1}\left(\frac{x}{12\text{ ft}}\right)$$

Then

$$IW = m_b \cdot L_d \cdot \theta_d = m_b \cdot L_d \cdot \delta\left(\frac{1}{a} + \frac{1}{b}\right)$$

4. Compute the external work. The deflected shape is similar to a half-pyramid, so the external work is

$$EW = q_f\left[\frac{1}{2} \cdot 16\text{ ft} \cdot 12\text{ ft} \cdot \frac{\delta}{3}\right] = q_f \cdot 32\text{ ft}^2 \cdot \delta$$

5. Equate the external and internal work. Set $EW = IW$ and solve for q_f. The virtual deflection, δ, will cancel, resulting in:

$$q_f = \frac{m_b \cdot L_d\left(\frac{1}{a} + \frac{1}{b}\right)}{32\text{ ft}^2}$$

6. Solve for the minimum value of q_f. The solution table below starts with $x = 5.0$, and then x is increased in 0.5-ft increments to find the minimum value of q_f.

x (ft)	α (deg.)	$\beta - \alpha$ (deg.)	L_d (ft)	a (ft)	b (ft)	m_b (k-ft/ft)	q_f (ksf)
5.0	22.6	30.5	13.0	5.42	7.66	3.35	0.429
5.5	24.6	28.5	13.2	6.04	7.17	3.33	0.419
6.0	26.6	26.5	13.4	6.71	6.71	3.30	0.412
6.5	28.4	24.7	13.6	7.39	6.26	3.27	0.412
7.0	30.3	22.8	13.9	8.10	5.86	3.25	0.415

From this table, the minimum value of q_f is 0.412 ksf, and it occurs for $x = 6.5$ ft. As noted in the prior example, one might be able to find a slightly smaller value of q_f for an x-value between 6.0 and 6.5 ft, but such a search is not practical. ■

EXAMPLE 14-6 Design of a Two-Way Slab Panel Using the Yield-Line Method

The yield-line method will be used to design isotropic top and bottom reinforcement for the rectangular two-way slab panel shown in Fig. 14-12. Assume the slab is 8 in. thick (100 psf), carries a superimposed dead load of 25 psf, and carries a live load of 80 psf. Assume the concrete compressive strength, f'_c, is 4000 psi and the steel yield strength, f_y, is 60 ksi.

1. Determine the factored load for design. From ACI Code Table 5.3.1, the two factored design loads to be considered are

$$q_u = 1.4q_D = 1.4(100 + 25)\text{psf} = 175 \text{ psf}$$

or

$$q_u = 1.2q_d + 1.6q_L = 1.2(125 \text{ psf}) + 1.6(80 \text{ psf}) = 278 \text{ psf}$$

The governing factored design load is $q_u = 278$ psf $= 0.278$ ksf.

2. Select the plastic mechanism and impose a virtual displacement. The assumed plastic-collapse mechanism is shown in Fig. 14-12. There are negative yield lines adjacent to the three fixed supports. Positive yield lines will come out of the corners where two supported edges (two axes of rotation) intersect. Those two positive yield lines will join at point *A* to form a single positive yield line *A–B*, which intersects the free edge of the slab panel. For the assumed symmetrical reinforcement pattern, the yield line *A–B* will be located at midlength of the panel. The location of the point *A* is not known, so the variable β is used to define its location. For this design problem, we will select different values for the variable, β, to *maximize* the required moment strengths of the isotropic top and bottom reinforcement, $m_n + m_p$. Note: It is possible for some slab panel geometries and moment strengths that the point *A* could move outside the slab panel, i.e., $\beta \geq 1.0$.

Assume the yield-line segment *A–B* is displaced downward a virtual displacement, δ. For this displaced shape, the rotations of the slab Segments I and the slab Segment II are

$$\theta_{\text{I}} = \frac{\delta}{10 \text{ ft}} \text{ and } \theta_{\text{II}} = \frac{\delta}{\beta \cdot 18 \text{ ft}}$$

These values still would be correct even if the point *A* was outside the slab panel.

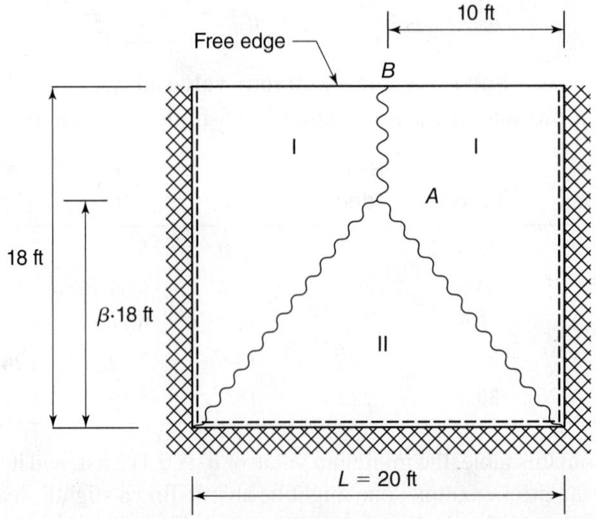

Fig. 14-12
Slab panel and plastic mechanism for Example 14-6.

3. Compute the internal work. Because the bottom (positive bending) reinforcement is isotropic, the internal work done along the positive yield lines can be projected back to the axis of rotation for the corresponding slab segment. Thus, the total internal work for slab Segments I and II is:

$$IW(I) = (m_n + m_p) \cdot 18 \text{ ft} \cdot \theta_{\text{I}} = (m_n + m_p) \cdot 1.8 \cdot \delta$$

$$IW(II) = (m_n + m_p) \cdot 20 \text{ ft} \cdot \theta_{\text{II}} = (m_n + m_p) \frac{1.11}{\beta} \cdot \delta$$

From these, the sum of the internal work is

$$\sum IW = 2 \cdot IW(I) + IW(II) = (m_n + m_p)\left(3.6 + \frac{1.11}{\beta}\right) \cdot \delta$$

4. Compute the external work. From the longer supported edge of the panel out to the point A, the displaced volume will be equal to the base area multiplied by one-third of δ. For the portion of the slab panel from A–B, the displaced volume will be equal to the base area multiplied by one-half of δ. Thus,

$$EW = q_u\left[20 \text{ ft} \cdot \beta \cdot 18 \text{ ft} \cdot \frac{\delta}{3} + 20 \text{ ft} \cdot (1 - \beta) \cdot 18 \text{ ft} \cdot \frac{\delta}{2}\right]$$

$$= q_u \cdot \delta(120 \text{ ft}^2 \cdot \beta + 180 \text{ ft}^2 - 180 \text{ ft}^2 \cdot \beta)$$

$$= q_u \cdot \delta(180 \text{ ft}^2 - 60 \text{ ft}^2 \cdot \beta)$$

5. Set ϕ times the internal work greater than or equal to the external work. The strength-reduction factor, ϕ, must be included at this point to satisfy the ACI Code strength requirements. Because of the expected low percentage of slab reinforcement, we can assume the appropriate flexural ϕ-factor from ACI Code Table 21.2.2 is $\phi = 0.9$ (tension-controlled section). As before, the virtual deflection, δ, will cancel, resulting in

$$\phi(m_n + m_p)\left(3.6 + \frac{1.11}{\beta}\right) \geq q_u(180 \text{ ft}^2 - 60 \text{ ft}^2 \cdot \beta)$$

so

$$(m_n + m_p) \geq \frac{q_u}{\phi}\left(\frac{180 \text{ ft}^2 - 60 \text{ ft}^2 \cdot \beta}{3.6 + \frac{1.11}{\beta}}\right)$$

It should be noted that the term, q_u/ϕ, essentially is equivalent to the q_f term used in all of the prior analysis examples. For this problem, q_u/ϕ is equal to 0.309 ksf.

6. Solve for the maximum required value of $m_n + m_p$. The solution table below starts with $\beta = 0.5$, and then β is increased in 0.05 increments to find the maximum required value of $m_n + m_p$.

β	Numerator (ft^2)	Denominator	$m_n + m_p$ (k-ft/ft)
0.50	150	5.82	7.96
0.55	147	5.62	8.08
0.60	144	5.45	8.16
0.65	141	5.31	8.21
0.70	138	5.19	8.22
0.75	135	5.08	8.21

From this table, the required flexural strength from the isotropic positive and negative reinforcement is 8.22 k-ft/ft.

7. Select top and bottom slab reinforcement. We will assume that the average flexural depth, d, for both the top and bottom slab reinforcement is 6.5 in. For this type of slab design, we normally would want to have the negative-moment strength greater than or equal to the positive-moment strength. Recalling that the maximum bar spacing is two times the slab thickness, the author selected the following reinforcement.

Top steel, No. 4 at 14 in. on centers in each directions; $m_n = 5.5$ k-ft/ft

Bottom steel, No. 3 at 14 in. on centers in each direction: $m_p = 3.0$ k-ft/ft

The sum of m_n and m_p exceeds the required strength, and it can be shown that both the positive- and negative-bending designs are tension-controlled. ■

It should be noted that the *yield-line* method only considers strength requirements and does not address serviceability criteria, such as deflections and control of cracking. Therefore, designers using this method for strength design of slabs will also need to satisfy the ACI Code serviceability criteria previously discussed in Chapters 9 and 13.

14-5 YIELD-LINE PATTERNS AT DISCONTINUOUS CORNERS

At discontinuous corners where the slab is supported on relatively stiff beams, as shown in Fig. 14-13a, there is a tendency for the corner of the slab to lift off its support unless a downward reaction is provided at the corner. In Fig. 14-13a, point B has deflected downward relative to the simply supported edges A–D and C–D. If corner D is held down, strips A–C and B–D develop the curvatures shown. Strip A–C develops positive moments and, ideally, should be reinforced with bars parallel to A–C at the bottom of the slab, as shown in Fig. 14-13b. Strip B–D develops negative moments and should be reinforced with bars parallel to B–D at the top of the slab, as shown in Fig. 14-13c. Diagonal bars are awkward to place and hence may be replaced by reinforcement in two orthogonal directions,

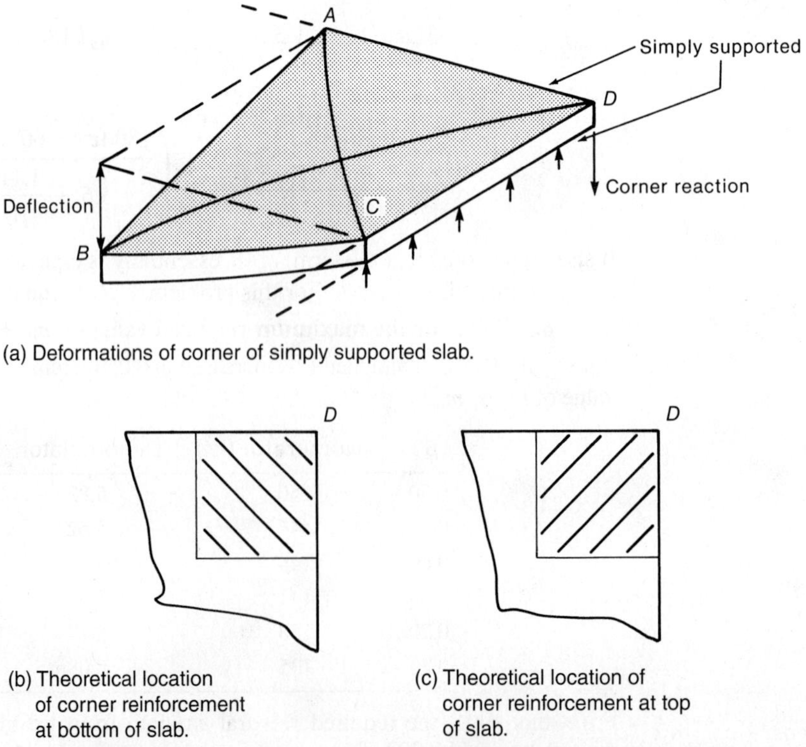

Fig. 14-13
Moments in corner of slab supported on stiff beams.

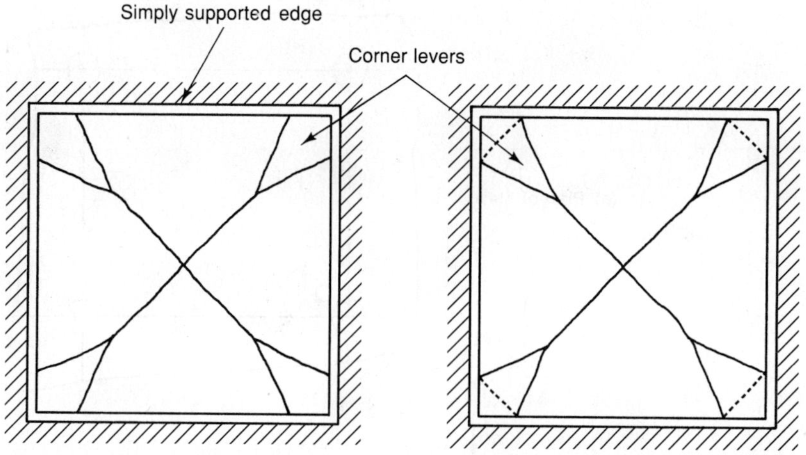

Fig. 14-14
Corner levers in simply supported slabs.

i.e., parallel to the two sides of the slab. Two such orthogonal reinforcing mats would be required, one at the top of the slab and one at the bottom.

In Examples 14-3 to 14-6, the yield lines were assumed to extend along the diagonals into the corners of the slabs. As a result of the localized bending discussed previously, the yield-line patterns in the corners of such a slab tend to fork out to the sides of the slab, as shown in Fig. 14-14a. If the corner of the slab is free to lift, it will do so. If the corner is held down, the slab will form a crack or form a yield line across the corner as shown in Fig. 14-14b. References [14-3] and [14-5] show that inclusion of the corner segments or *corner levers* will reduce the computed uniform load capacity of a simply supported slab by up to 9 percent from the value in an analysis that ignores them.

To counteract the *corner effects* and possible loss of capacity noted above, ACI Code Section 8.7.3.1 requires special corner reinforcement at the exterior corners of all slabs with edge beams having α_f greater than 1.0. The reinforcement may be placed either diagonally (Figs. 14-13b and 14-13c) or in an orthogonal pattern, as noted previously. Both of these options are shown in ACI Commentary Fig. R8.7.3.1. This reinforcement is to be designed for a moment strength-per-unit-width equal to the largest positive moment-per-unit-width acting in the slab panel and is to extend one-fifth of the longer span in each direction from the corner.

14-6 YIELD-LINE PATTERNS AT COLUMNS OR AT CONCENTRATED LOADS

Figure 14-15 shows a slab panel supported on circular columns. The element at A is subjected to bending and twisting moments. If this element is rotated, however, it is possible to orient it so that the element is acted on by bending moments only, as shown in Fig. 14-15c. The moments calculated in this way are known as the principal moments and represent the maximum and minimum moments on any element at this point. On the element shown near the face of the column, the m_1 moment is the largest moment. As a result, a circular crack develops around the column, and eventually the reinforcement crossing the circumference of the column yields. The moments about lines radiating out from the center of the column lead to radial cracks, and as a result, a circular *fan-shaped* yield pattern develops, as shown in Fig. 14-16a. Under a concentrated load, a similar fan pattern also develops, as shown in Fig. 14-16b. Note that these fans involve both positive- and negative-moment yield lines, shown by solid lines and dashed lines, respectively.

(a) Plan of slab.

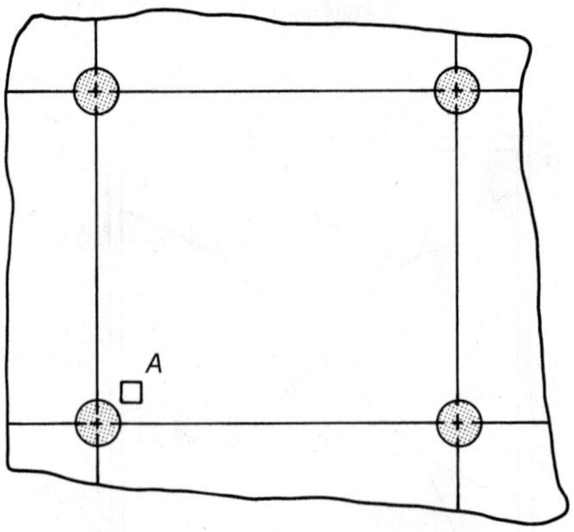

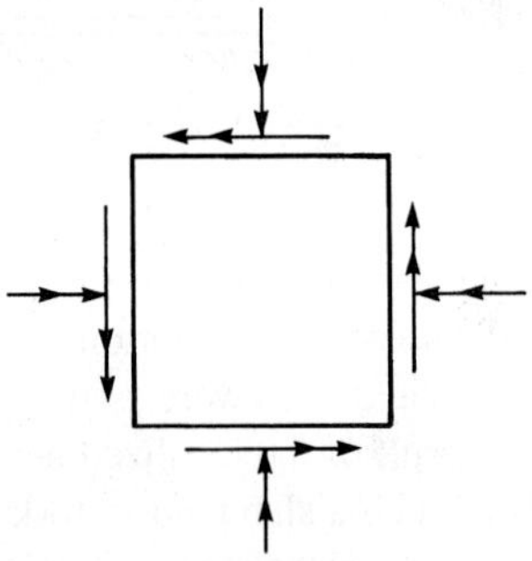

(b) Bending and twisting moments on element at *A*.

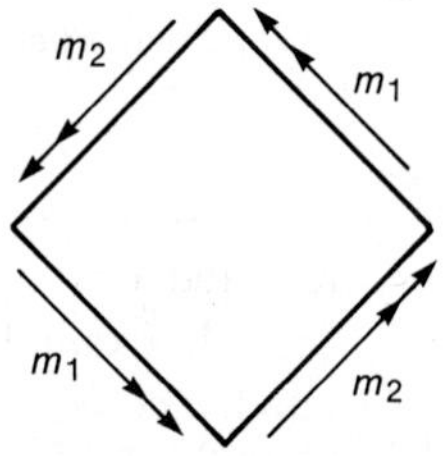

(c) Principal moments on element at *A*.

Fig. 14-15
Principal moments adjacent to a column in a flat plane.

EXAMPLE 14-7 Calculation of the Capacity of a Fan Yield Pattern

Compute the moments required to resist a concentrated load P if a fan mechanism develops. Assume that the negative-moment capacity is m_n and the positive-moment capacity is m_p per unit width in all directions. Figure 14-16c shows a triangular segment from the fan in Fig. 14-16b. This segment subtends an angle α and has an outside radius r.

Consider segment A–B–C.

1. **Give point A a virtual displacement of δ.**
2. **Compute the external work.** For this problem, the external work is simply,

$$EW = P \cdot \delta$$

3. **Compute the internal work.** Consider the typical triangular slab segment A–B–C in Fig. 14-16b. As before, the work done along the positive yield lines can be projected back to the axis of rotation for the slab segment, which is represented by the negative yield line from B–C. Assuming that there are many similar triangular slab segments surrounding the loading point (and thus α is small), then the length of line B–C is equal to $\alpha \cdot r$. As shown in Fig. 14-16c, the rotation of this slab segment, θ_x, is equal to δ/r.

Therefore, the internal work done by slab segment A–B–C is

$$IW(A\text{–}B\text{–}C) = (m_n + m_p)\alpha \cdot r \cdot \frac{\delta}{r} = (m_n + m_p)\alpha \cdot \delta$$

The total number of triangular slab segments surrounding the loading point is $2\pi/\alpha$. So, the total internal work is

$$\sum IW = 2\pi(m_n + m_p) \cdot \delta$$

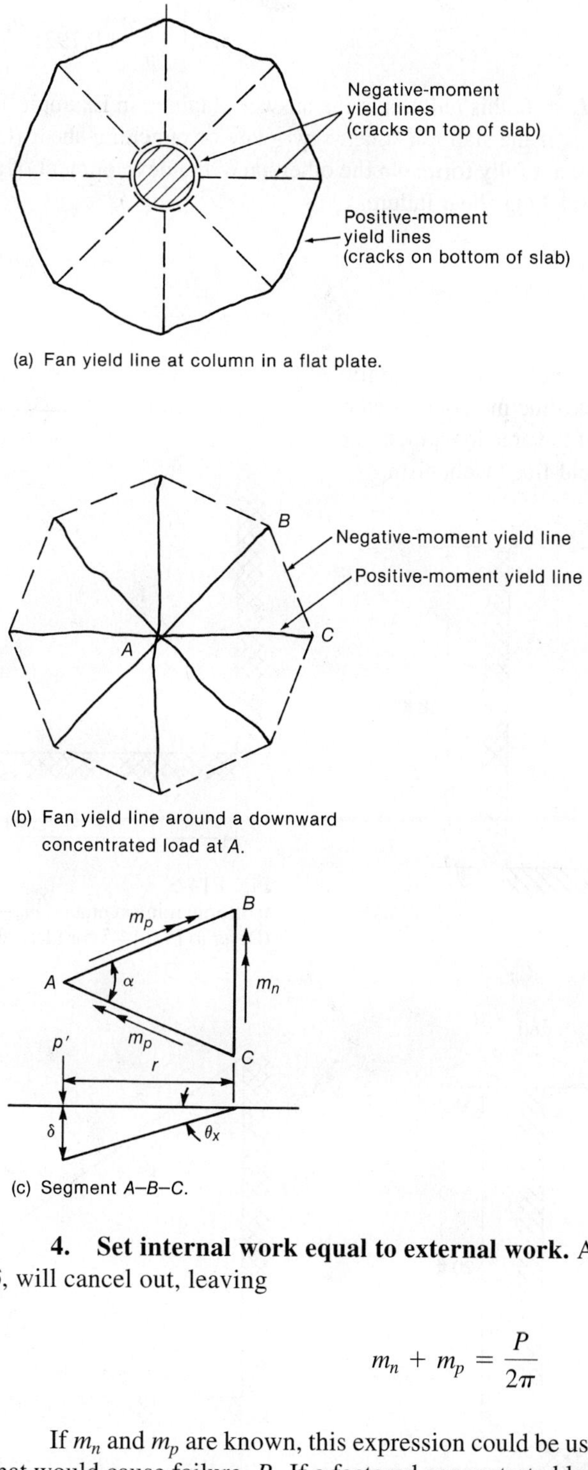

Fig. 14-16
Fan yield lines.

4. Set internal work equal to external work. As before the virtual deflection, δ, will cancel out, leaving

$$m_n + m_p = \frac{P}{2\pi}$$

If m_n and m_p are known, this expression could be used to find the concentrated load that would cause failure, P_f. If a factored concentrated load, P_u, was known, the left side of this expression could be multiplied by ϕ and then used to find the required positive- and negative-moment strengths for the slab reinforcement. ■

Using a similar analysis, Johnson [14-6] computed the yield-line moments for a circular fan around a column of diameter d_c in a uniformly loaded square plate with spans ℓ_1. He found that

$$m_n + m_p = w\ell_1^2\left[\frac{1}{2\pi} - 0.192\left(\frac{d_c}{\ell_1}\right)^{2/3}\right] \tag{14-17}$$

If $d_c = 0$, this reduces to the answer obtained in Example 14-7.

If the slab fails due to two-way or punching shear (Chapter 13) the fan mechanism may not fully form. On the other hand, the development of a fan mechanism may bring on a punching-shear failure.

PROBLEMS

14-1 through 14-4. For the slab panels shown in the following figures, use the yield-line method to determine the minimum value of the area load, q_f, at the formation of the critical yield-line mechanism.

Video Solution

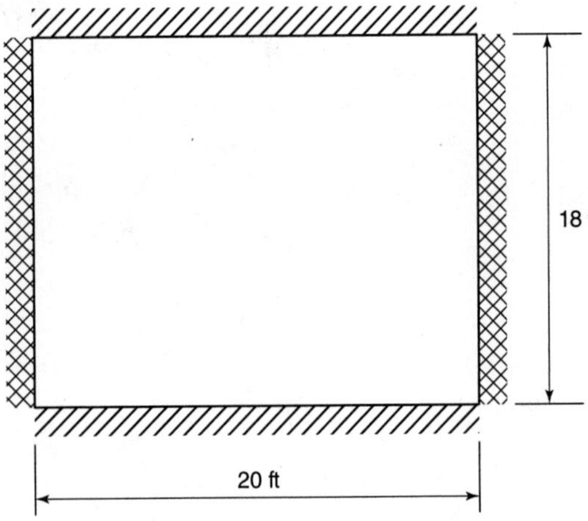

Fig. P14-1
Isotropic reinforcement, $m_n = 6$ k-ft/ft and $m_p = 3.5$ k-ft/ft.

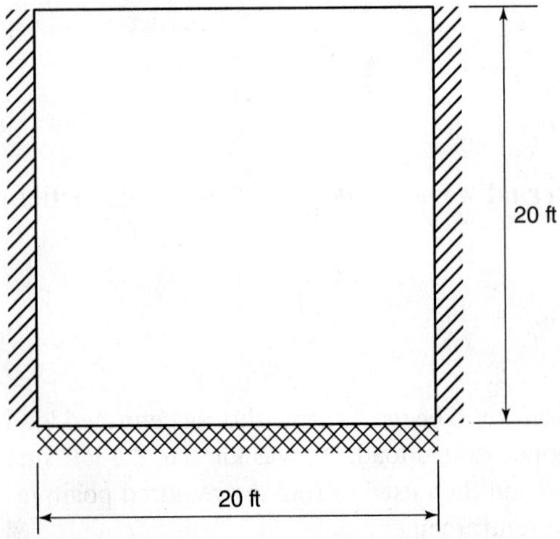

Fig. P14-2
Isotropic reinforcement, $m_n = 8$ k-ft/ft and $m_p = 4.5$ k-ft/ft.

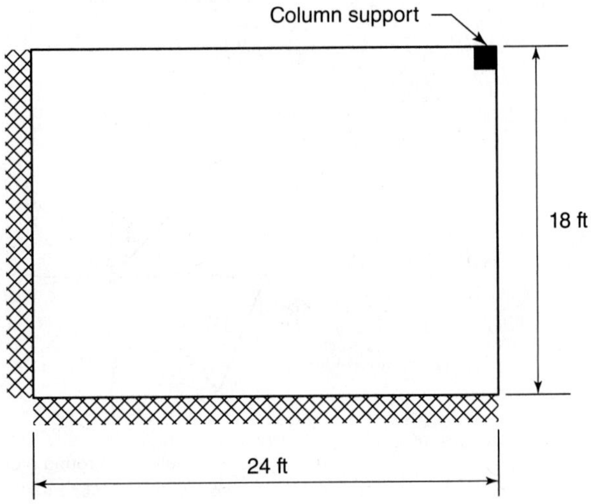

Fig. P14-3
Isotropic reinforcement, $m_n = 8$ k-ft/ft and $m_p = 6$ k-ft/ft.
(Refer to Fig. 14-5 for ideas about proper yield-line mechanism.)

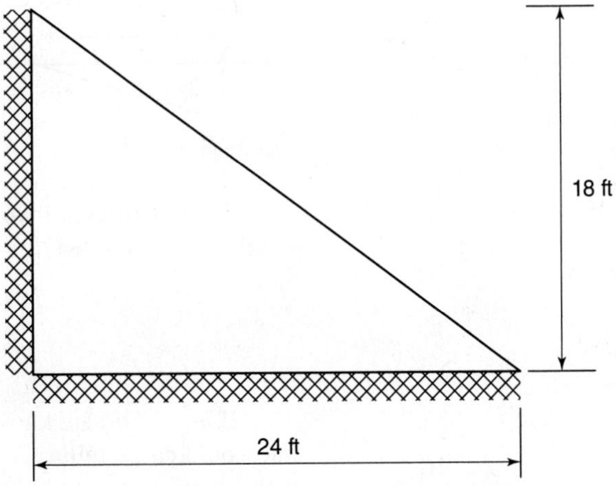

Fig. P14-4
Orthotropic positive reinforcement, $m_{px} = 3$ k-ft/ft and $m_{py} = 4$ k-ft/ft. and isotropic negative reinforcement with $m_n = 5$ k-ft/ft.

14-5 The slab panel shown in Fig. P14-5 is 7.5 in. thick and is made of normal-weight concrete. The slab will need to support a superimposed dead load of 25 psf and a live load of 60 psf. Assume that the slab will be designed with isotropic top and bottom reinforcement.

(a) Find the critical yield-line mechanism and then use the appropriate load factors and strength-reduction factors to determine the required sum of the nominal negative- and positive-moment capacities $(m_n + m_p)$ in units of k-ft/ft.

(b) Start with m_n approximately equal to $1.6 \times m_p$, and then select top and bottom isotropic reinforcement to provide the required combined flexural strength calculated in part (a). Be sure to check reinforcement spacing requirements given in ACI Code Section 8.7.2.2. Use $f'_c = 4000$ psi and $f_y = 60$ ksi.

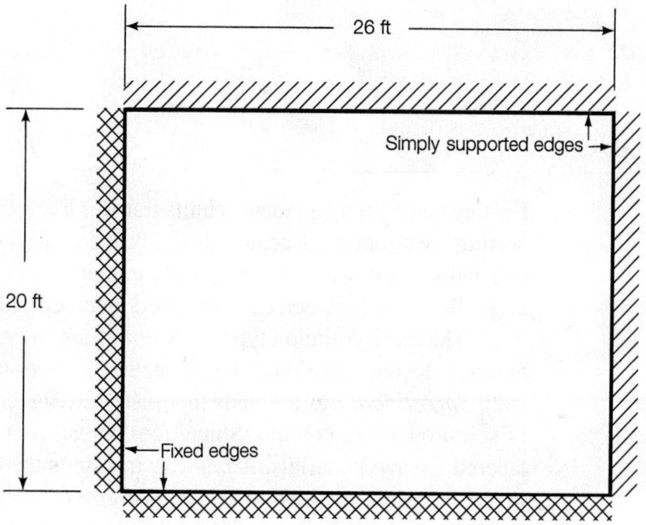

Fig. P14-5

REFERENCES

14-1 Arne Hillerborg, *Strip Method of Design*, E. & F., N. Spon, London, 1975, 258 pp.

14-2 Randal H. Wood and G. S. T. Armer, "The Theory of the Strip Method for Design of Slabs," *Proceedings, Institution of Civil Engineers, London*, Vol. 41, October 1968, pp. 285–313.

14-3 Kurt W. Johansen, *Yield-Line Theory* (English translation from German), Cement and Concrete Association, London, 1962, 181 pp.

14-4 Eivind Hognestad, "Yield-Line Theory for Ultimate Flexural Strength of Reinforced Concrete Slabs," *Journal of the American Concrete Institute*, Vol. 49, No. 7, March 1953, pp. 637–656.

14-5 Leonard L. Jones and Randal H. Wood, *Yield Line Analysis of Slabs*, Elsevier, New York, 1967.

14-6 Roger P. Johnson, *Structural Concrete*, McGraw-Hill, London, 1967, 271 pp.

14-7 Robert Park and William L. Gamble, *Reinforced Concrete Slabs*, Wiley-Interscience, New York, 1980, 618 pp.

15

Footings

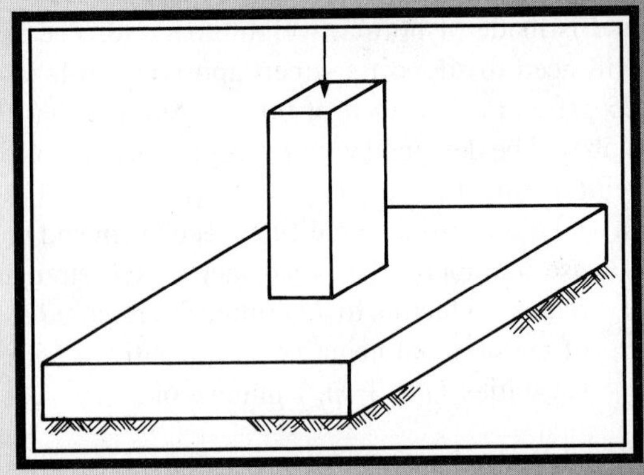

15-1 INTRODUCTION

Footings and other foundation units transfer loads from the structure to the soil or rock supporting the structure. Because the soil is generally much weaker than the concrete columns and walls that must be supported, the contact area between the soil and the footing is much larger than that between the supported member and the footing.

The more common types of footings are illustrated in Fig. 15-1. *Strip footings* or *wall footings* display essentially one-dimensional action, cantilevering out on each side of the wall. *Spread footings* are pads that distribute the column load in two directions to an area of soil around the column. Sometimes spread footings have pedestals, are stepped, or are tapered to save materials. A *pile cap* transmits the column load to a series of *piles*, which, in turn, transmit the load to a strong soil layer at some depth below the surface. *Combined footings* transmit the loads from two or more columns to the soil. Such a footing is often used when one column is close to a property line. A *mat or raft foundation* transfers the loads from all the columns in a building to the underlying soil. Mat foundations are used when very weak soils are encountered. *Caissons* 2 to 5 ft in diameter are sometimes used instead of piles to transmit heavy column loads to deep foundation layers. Frequently, these are enlarged at the bottom (*belled*) to apply load to a larger area.

The choice of foundation type is selected in consultation with the geotechnical engineer. Factors to be considered are the soil strength, the soil type, the variability of the soil type over the area and with increasing depth, and the susceptibility of the soil and the building to deflections.

Strip, spread, and combined footings are considered in this chapter because these are the most basic and most common types.

15-2 SOIL PRESSURE UNDER FOOTINGS

The distribution of soil pressure under a footing is a function of the type of soil and the relative rigidity of the soil and the foundation pad. A concrete footing on sand will have a pressure distribution similar to Fig. 15-2a. The sand near the edges of the footing tends to displace laterally when the footing is loaded, causing a decrease in soil pressure

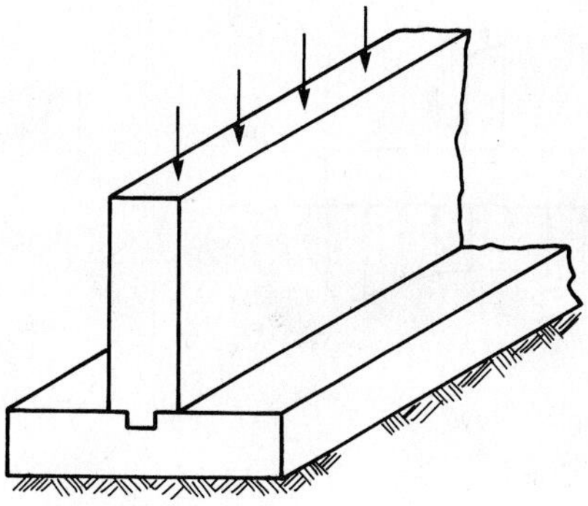

(a) Strip or wall footing.

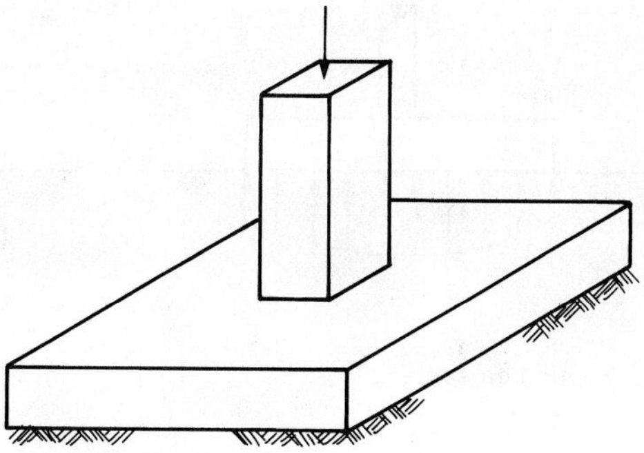

(b) Spread footing.

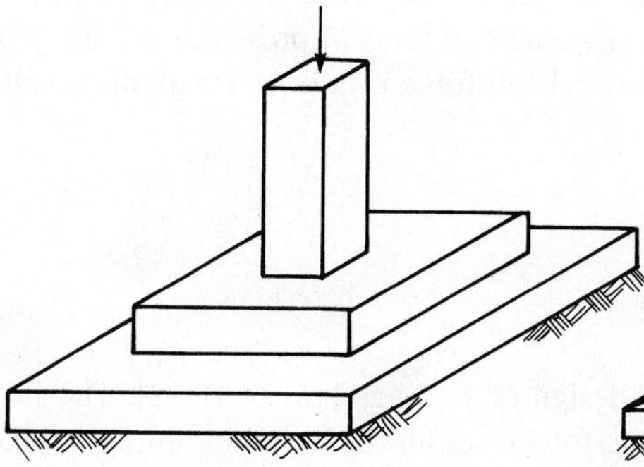

(c) Stepped footing.

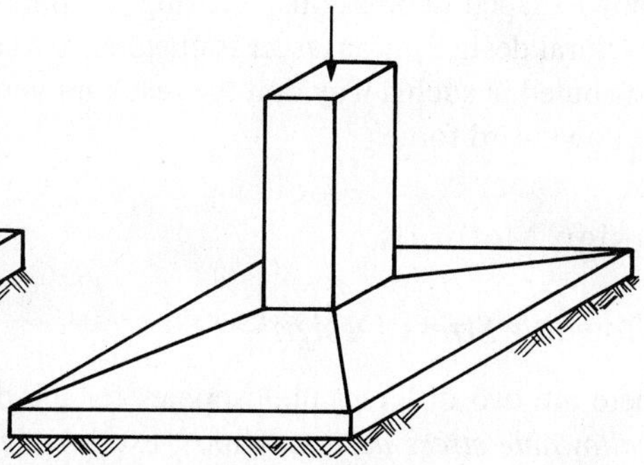

(d) Tapered footing.

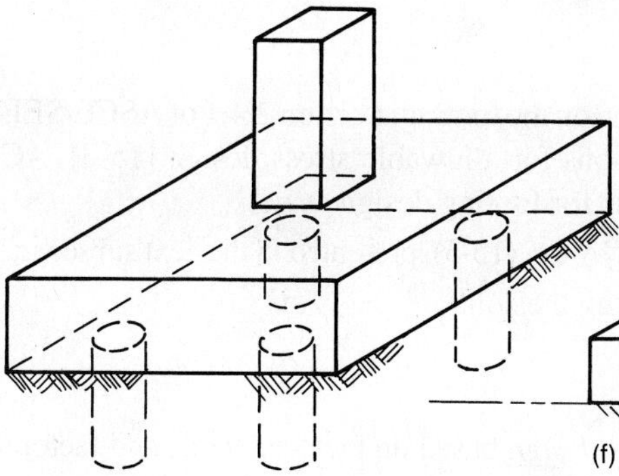

(e) Pile cap.

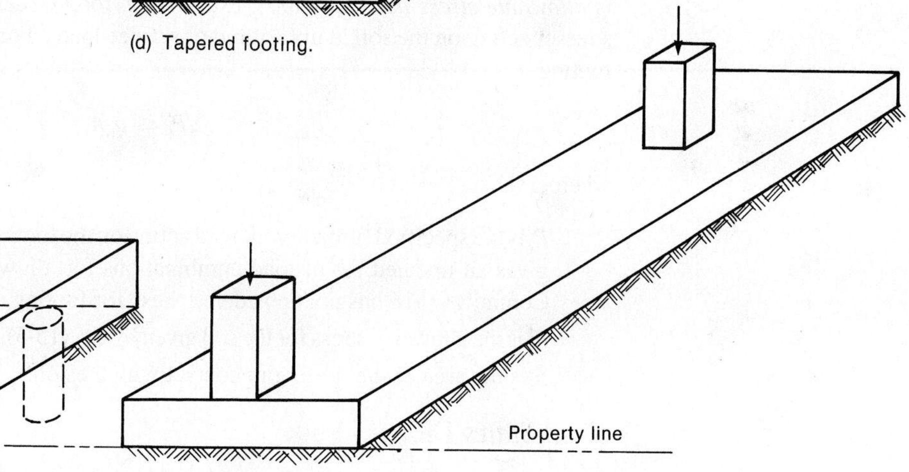

(f) Combined footing.

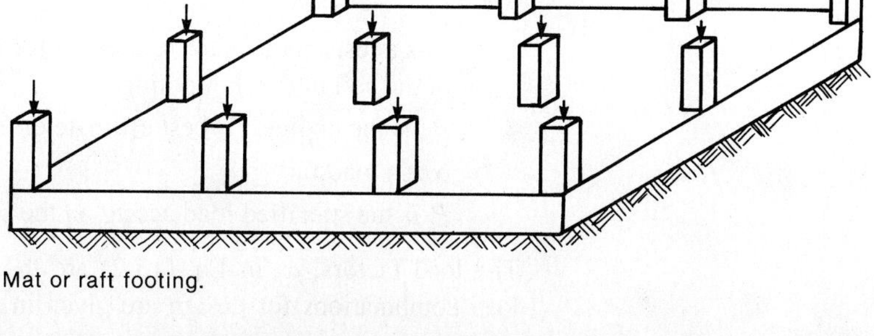

(g) Mat or raft footing.

Fig. 15-1
Types of footings.

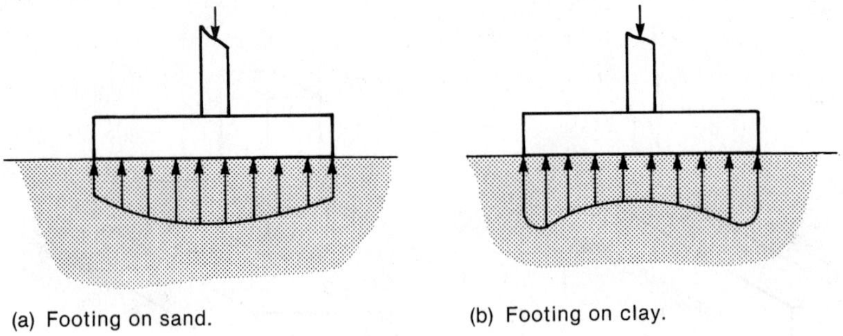

Fig. 15-2
Pressure distribution under footings.

near the edges. On the other hand, the pressure distribution under a footing on clay is similar to Fig. 15-2b. As the footing is loaded, the soil under the footing deflects in a bowl-shaped depression, relieving the pressure under the middle of the footing. For structural design purposes, it is customary to assume that the soil pressures are linearly distributed in such a way that the resultant vertical soil force is collinear with the resultant downward force.

Design Methods

Allowable Stress Design

There are two different philosophies for the design of footings [15-1], [15-2]. The first is *allowable stress design.* Almost exclusively, footing design is based on the allowable stresses acting on the soil at unfactored or service loads. For a concentrically loaded spread footing,

$$\Sigma P_s \leq q_a A \tag{15-1}$$

where

P_s is the specified (unfactored) load acting on the footing. Section 2.4.1 of ASCE/SEI 7 gives an updated set of load combinations for allowable stress design [15-3]. ACI Committee 318 has not considered these for footing design as yet.

q_a is the allowable stress for the soil given by Eq. (15-3), presented in the next subsection.

A is the area of the footing in contact with the soil.

Limit-States Design

The second design philosophy is a *limit-states design* based on factored loads and factored resistances, given by

$$\phi R_n \geq \Sigma \alpha P_s \tag{15-2}$$

where

ϕ is a resistance factor to account for the variability of the load-resisting mechanism of the soil under the footing.

R_n is the engineer's best estimate of the resistance of the soil under the footing.

α is a load factor.

P_s is the specified load acting on the soil at the base of the footing.

The load factors, α, in Eq. (15-2) are those used in building design. Load factors and load combinations for design are given in ACI Code Table 5.3.1. Resistance factors for limit-states design of footings are still being developed. Current estimates of ϕ values for shallow footings are as follows:

Vertical resistance, $\phi = 0.5$.

Sliding resistance dependent on friction, with cohesion equal to zero, $\phi = 0.8$.

Sliding resistance dependent on cohesion, with friction equal to zero, $\phi = 0.6$.

Serviceability limit states should also be checked [15-1], [15-2], [15-3].

At the time of writing, virtually all building footings in North America are designed by using allowable-stress design applied to failures of the soil. The rest of this chapter will apply allowable-stress design to the soil and then use strength design for the reinforced concrete foundation structure.

Limit States for the Design of Foundations

Limit States Governed by the Soil

Three primary limit states of the soil supporting an isolated foundation are [15-1], [15-2], and [15-3]:

1. a bearing failure of the soil under the footing (Fig. 15-3),

2. a serviceability failure in which excessive differential settlement between adjacent footings causes architectural or structural damage to the structure, or

3. excessive total settlement.

Settlement occurs in two stages: *immediate settlement* as the loads are applied, and a long-term settlement known as *consolidation.*

Procedures for minimizing differential settlements involve a degree of geotechnical engineering theory outside the scope of this book.

Bearing failures are controlled by limiting the service-load stress under the footing to less than an allowable stress, q_a, as in Eq. (15-1).

Limit States Governed by the Structure

Similarly, there are four primary structural limit states for the foundations themselves [15-1], [15-2]:

1. flexural failure of the portions of the footing that project from the column or wall,

2. shear failure of the footing,

3. bearing failure at member interfaces, and

4. inadequate anchorage of the flexural reinforcement in the footing.

As stated earlier, bearing failures of the soil supporting the foundation are prevented by limiting the service-load stress under the footing to less than an allowable stress

$$q_a = \frac{q_{\text{ult}}}{\text{FS}} \tag{15-3}$$

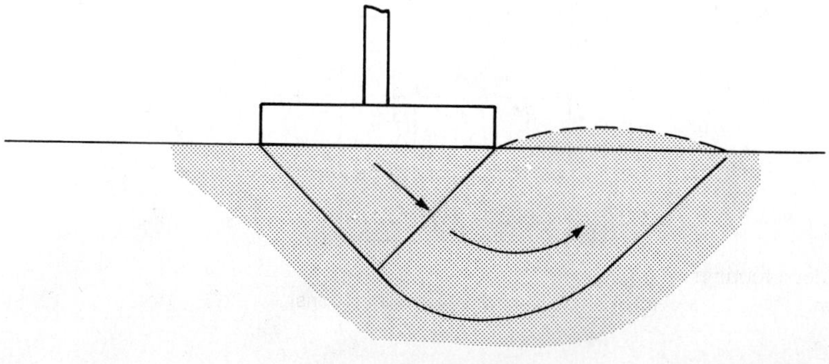

Fig. 15-3
Bearing failure of a footing.

where q_{ult} is the stress corresponding to the failure of the soil under the footing and FS is a factor of safety in the range of 2.5 to 3. Values of q_a are obtained from the principles of geotechnical engineering and depend on the shape of the footing, the depth of the footing, the overburden or surcharge on top of the footing, the position of the water table, and the type of soil. When using a value of q_a provided by a geotechnical engineer, it is necessary to know what strengths were measured and in what kind of tests, and what assumptions have been made in arriving at this allowable soil pressure, particularly with respect to overburden and depth to the base of the footing.

It should be noted that q_a in Eq. (15-3) is a service-load stress, whereas the rest of the structure usually is designed by using factored loads corresponding to the strength limit states. The method of accounting for this difference in philosophy is explained later.

Elastic Distribution of Soil Pressure under a Footing

The soil pressure under a footing is calculated by assuming linearly elastic action in compression, but no tensile strength across the contact between the footing and the soil. If the column load is applied at, or near, the middle of the footing, as shown in Fig. 15-4, the stress, q, under the footing is

$$q = \frac{P}{A} \pm \frac{My}{I} \tag{15-4}$$

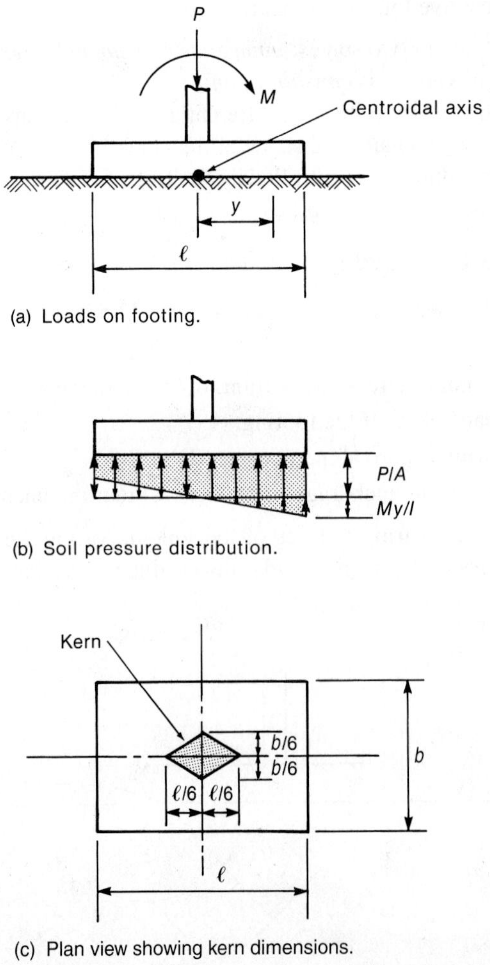

Fig. 15-4
Soil pressure under a footing: loads within kern.

where

P = vertical load, positive in compression

A = area of the contact surface between the soil and the footing

I = moment of inertia of this area

M = moment about the centroidal axis of the footing area

y = distance from the centroidal axis to the point where the stresses are being calculated

The moment, M, can be expressed as Pe, where e is the eccentricity of the load relative to the centroidal axis of the area A. The maximum eccentricity e for which Eq. (15-4) applies is that which first causes $q = 0$ at some point. Larger eccentricities will cause a portion of the footing to lift off the soil, because the soil–footing interface cannot resist tension. For a rectangular footing, this occurs when the eccentricity exceeds

$$e_k = \frac{\ell}{6} \text{ or } e_k = \frac{b}{6} \tag{15-5}$$

This is referred to as the *kern distance*. Loads applied within the *kern*, the shaded area in Fig. 15-4c, will cause compression over the entire area of the footing, and Eq. (15-4) can be used to compute q.

Various pressure distributions for rectangular footings are shown in Fig. 15-5. If the load is applied concentrically, the soil pressure q is $q_{avg} = P/A$. If the load acts through the kern point (Fig. 15-5c), $q = 0$ at one side and $q = 2q_{avg}$ at the other. If the load falls outside the kern point, the resultant upward load is equal and opposite to the resultant downward load, as shown in Fig. 15-5d. Generally, such a pressure distribution would not be acceptable, because it makes inefficient use of the footing concrete, tends to overload the soil, and may cause the structure to tilt.

Elastic and Plastic Soil-Pressure Distributions

The soil-pressure diagrams in Fig. 15-5 are based on the assumption that the soil pressure is linearly distributed under a footing. This is a satisfactory assumption at service-load levels and for footings on rock or dense glacial till. For yielding soils, the soil-pressure distribution will approach a uniform (plastic) distribution over part of the base of the footing in such a way that the resultant load on the footing and the resultant of the soil pressure coincide as is required for equilibrium.

For the design of concentrically loaded footings, the distribution of soil pressures is taken to be uniform over the entire contact area, as shown in Fig. 15-5a. For the structural design of eccentrically loaded footings, such as those for retaining walls or bridge abutments, the pressure distribution is a linearly varying distribution like those in Figs. 15-5b, 15-5c, and 15-5d with the resultant of the soil pressure coincident with the resultant of the applied loads.

The examples in this chapter are limited to the predominant case of concentric loads and a uniform or linear soil-pressure distribution over the entire contact area.

Load and Resistance Factors for Footing Design

ACI Code load and resistance factors are given in ACI Code Tables 5.3.1 and 21.2.1. The examples in this chapter are based on those load and resistance factors.

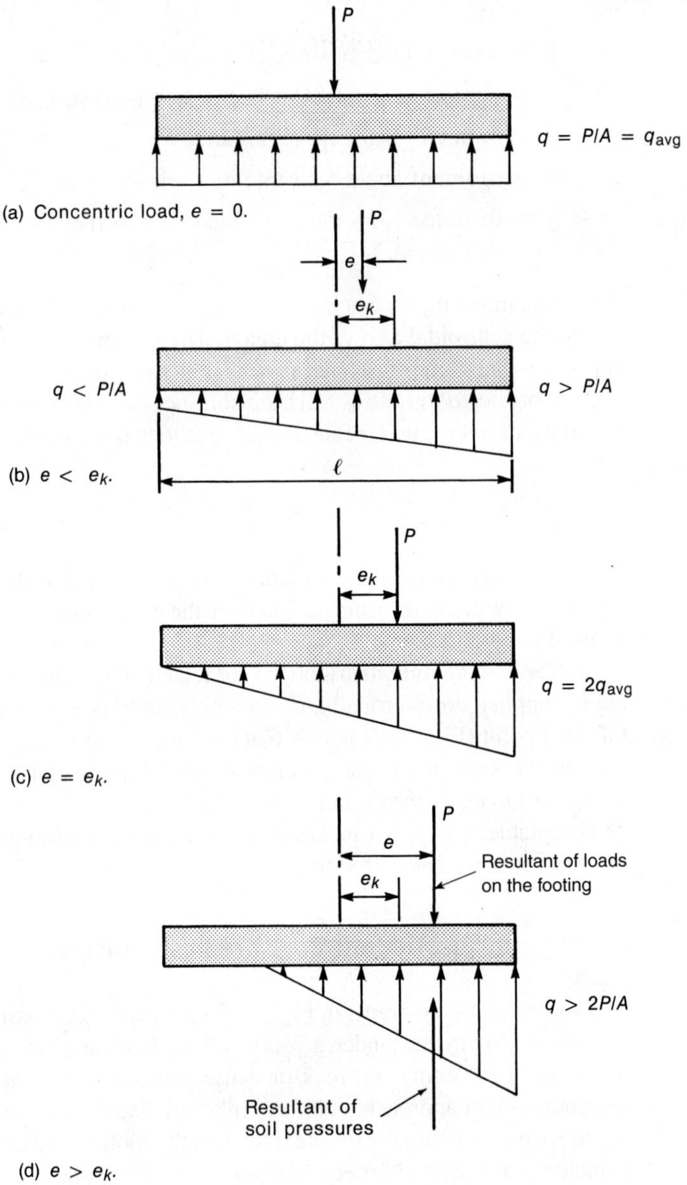

Fig. 15-5
Pressures under an eccentrically loaded footing.

Gross and Net Soil Pressures

Figure 15-6a shows a 2-ft-thick spread footing with a column at its center and with its top surface located 2 ft below the ground surface. There is no column load at this stage. The total downward load from the weights of the soil and the footing is 540 psf. This is balanced by an equal, but opposite, upward pressure. As a result, the net effect on the concrete footing is zero. There are no moments or shears in the footing due to this loading.

When the column load P_c is added, the pressure under the footing increases by $q_n = P_c/A$, as shown in Fig. 15-6b. The total soil pressure is $q = 540 + q_n$. This is referred to as the *gross soil pressure* and must not exceed the allowable soil pressure, q_a. When moments and shears in the concrete footing are calculated, the upward and

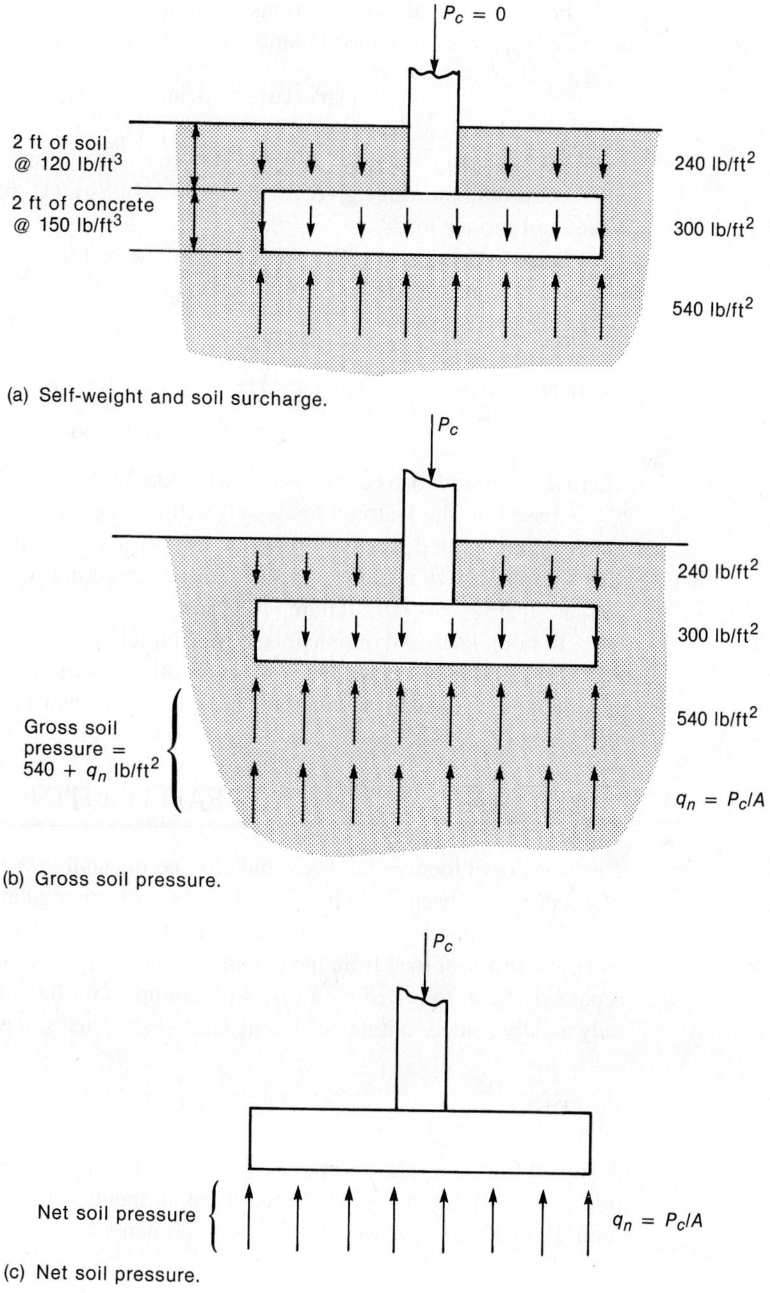

Fig. 15-6
Gross and net soil pressures.

downward pressures of 540 psf cancel out, leaving only the *net soil pressure*, q_n, to cause internal forces in the footing, as shown in Fig. 15-6c.

In design, the area of the footing is selected so that the *gross soil pressure* does not exceed the allowable soil pressure. The flexural reinforcement and the shear strength of the footing are then calculated by using the *net soil pressure*. Thus, the area of the footing is selected to be

$$A = \frac{D(\text{structure, footing, surcharge}) + L}{q_a} \tag{15-6}$$

where D and L refer to the unfactored service dead and live loads.

For service-load combinations including wind, W, most codes allow a 33 percent increase in q_a. For such a load combination, the required area would be

$$A = \frac{D(\text{structure, footing, surcharge}) + L + W}{1.33q_a} \tag{15-7}$$

but not less than the value given by Eq. (15-6). In Eqs. (15-6) and (15-7), the loads are the unfactored service loads.

Once the area of the footing is known, the rest of the design of the footing is based on soil stresses due to the factored loads.

Factored Net Soil Pressure, q_{nu}.

The factored net soil pressures used to design the footing are:

$$q_{nu} = (\text{factored load})/A \tag{15-8}$$

where the factored loads come from ACI Code Table 5.3.1. The factored net soil pressure, q_{nu}, is based on the factored loads and will exceed q_a in most cases. This is acceptable, because the factored loads are roughly 1.5 times the service loads, whereas the factor of safety implicit in q_a is 2.5 to 3. Hence, the factored net soil pressure will be less than the pressure that would cause failure of the soil.

If both load and moment are transmitted to the footing, it is necessary to use Eq. (15-4) (if the load is within the kern) or other relationships (as illustrated in Fig. 15-5d) to compute q_{nu}. In such calculations, the factored loads would be used.

15-3 STRUCTURAL ACTION OF STRIP AND SPREAD FOOTINGS

The behavior of footings has been studied experimentally at various times. Our current design procedures have been strongly affected by the tests reported in [15-4], [15-5], and [15-6].

The design of a footing must consider bending, development of reinforcement, shear, and the transfer of load from the column or wall to the footing. Each of these is considered separately here, followed by a series of examples in the ensuing sections. In this section, only axially loaded footings with uniformly distributed soil pressures, q_{nu}, are considered.

Flexure

A spread footing is shown in Fig. 15-7. Soil pressures acting under the crosshatched portion of the footing in Fig. 15-7b cause the moments about axis A–A at the face of the column. From Fig. 15-7c, we see that these moments are

$$M_u = (q_{nu}bf)\frac{f}{2} \tag{15-9}$$

where $q_{nu}bf$ is the resultant of the soil pressure on the crosshatched area and $f/2$ is the distance from the resultant to section A–A. This moment must be resisted by reinforcement placed as shown in Fig. 15-7c. The maximum moment will occur adjacent to the face of the column on section A–A or on a similar section on the other side of the column.

In a similar manner, the soil pressures under the portion outside of section B–B in Fig. 15-7a will cause a moment about section B–B. Again, this must be resisted by flexural reinforcement perpendicular to B–B at the bottom of the footing; the result is two layers of steel, one each way, shown in section A–A in Fig. 15-7c.

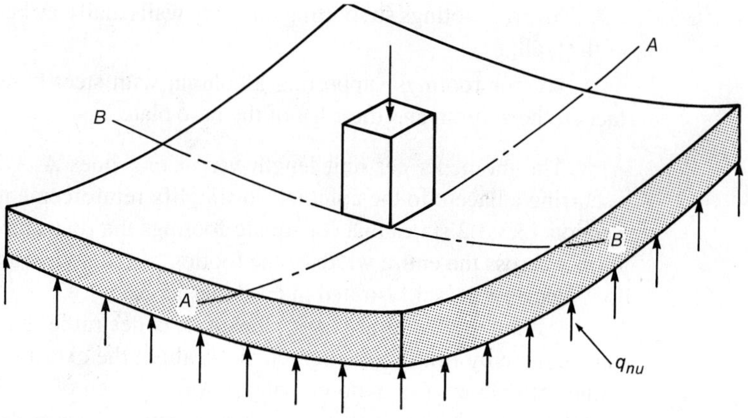

(a) Footing under load.

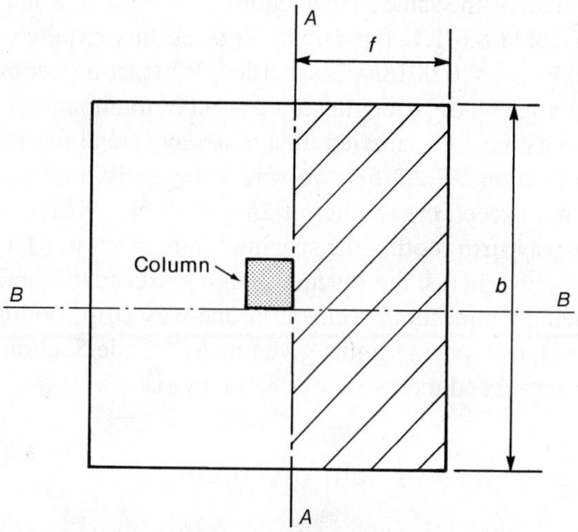

(b) Tributary area for moment at section A-A.

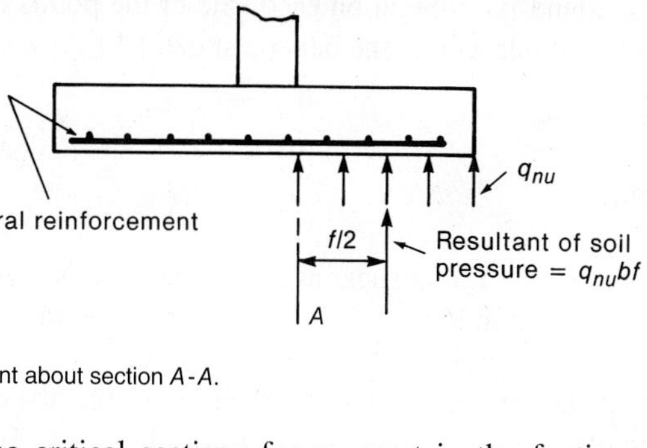

(c) Moment about section A-A.

Fig. 15-7
Flexural action of a spread footing.

The critical sections for moment in the footing are taken as follows (ACI Code Table 13.2.7.1):

1. for footings supporting square or rectangular concrete columns or walls, at the face of the column or wall;

2. for footings supporting circular or regular polygonal columns, at the face of an imaginary square column with the same area (Code Section 13.2.7.3);

3. for footings supporting masonry walls, halfway between the middle and the edge of the wall;

4. for footings supporting a column with steel base plates, halfway between the face of the column and the edge of the base plate.

The moments per unit length vary along lines *A–A* and *B–B*, with the maximum occurring adjacent to the column. To simplify reinforcement placing, however, ACI Code Section 13.3.3.2 states that for square footings the reinforcement shall be distributed uniformly across the entire width of the footing. A banded arrangement is used in rectangular footings, as will be illustrated in Example 15-3.

Although a footing is not a beam, it is desirable that it be ductile in flexure. This can be done by limiting ε_t, net tensile strain in the extreme tension reinforcement, to the value ≥ 0.005 as was done in Chapter 4 for design of beams. When calculating ϕ, ACI Code Table 21.2.2 states that $\phi = 0.9$ for *tension-controlled sections*, i.e., sections where $\varepsilon_t \geq 0.005$ at nominal flexural-strength conditions.

For footings of uniform thickness, the minimum area of flexural tensile reinforcement shall be the same as that required for shrinkage and temperature reinforcement in ACI Code Table 8.6.1.1. For Grade-40 steel, this requires $A_{s,\min} = 0.0020bh$; for Grade-60 steel, $A_{s,\min} = 0.0018bh$ is specified. Whether the required area of flexural reinforcement is governed by the strength requirement or minimum area requirement, the following spacing limits must be satisfied. In a two-way spread footing, the spacing limits given in ACI Code Section 8.7.2.2 for two-way slabs apply, that is, at the critical section the spacing shall not exceed the smaller of $2h$ and 18 in. (where h is the thickness of the footing). In a one-way strip footing the spacing limits given in ACI Code Section 7.7.2.3 for one-way slabs apply, that is, the spacing shall not exceed the smaller of $3h$ and 18 in. For shrinkage and temperature reinforcement in one-way strip footings (placed along the length of the footing), the spacing limits given in ACI Code Section 7.7.2.4 apply, that is, the spacing shall not exceed the smaller of $5h$ or 18 in.

Development of Reinforcement

The footing reinforcement is chosen by assuming that the reinforcement stress reaches f_y along the maximum-moment section at the face of the column. The reinforcement must extend far enough on each side of the points of maximum bar stress to develop this stress. In other words, the bars must extend ℓ_d from the critical section or be hooked at the outer ends.

Shear

A footing may fail in shear as a wide beam, as shown in Fig. 15-8a or as a result of punching, as shown in Fig. 15-8b. These are referred to as one-way shear and two-way shear and are discussed more fully in Section 13-10.

Concern has been expressed about the shear strength of deep, lightly reinforced concrete members [15-7], [15-8], and [15-9]. Tests of members similar to one-way footings [15-9] suggest that, if the ratio of the length, a, of the portion of the footing that projects outward from the column or wall, to the depth of the footing, d, does not exceed 3, the crack-restraining effect of the soil pressure under the footing tends to offset the strength reduction due to size. Shear reinforcement is required in all footings for which V_u exceeds ϕV_c.

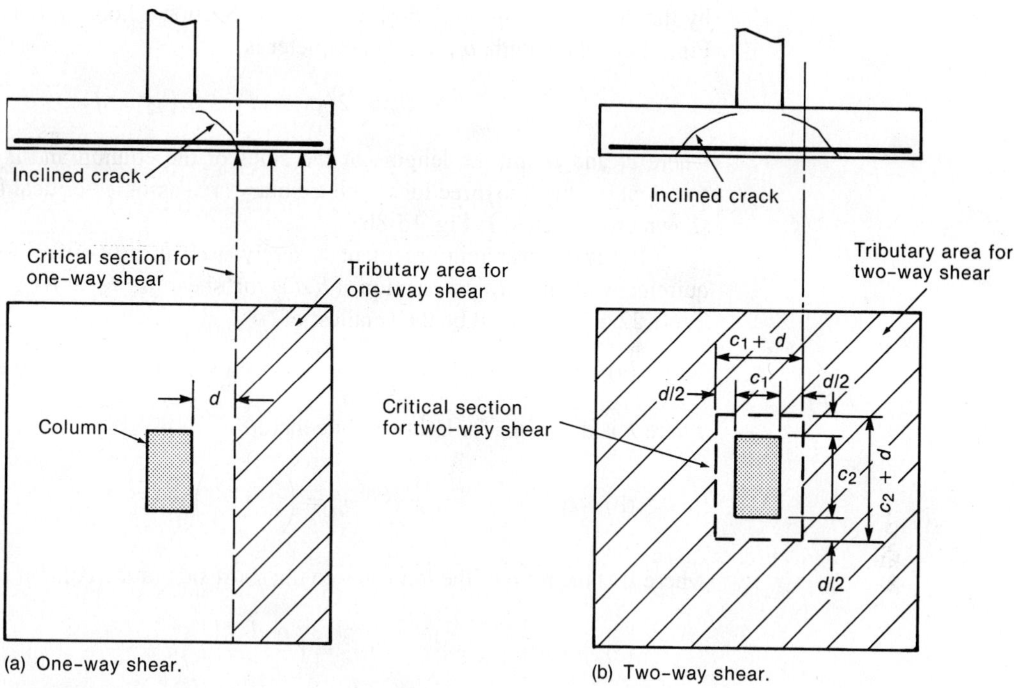

Fig. 15-8
Critical sections and tributary areas for shear in a spread footing.

One-Way Shear

A footing failing through one-way shear is designed as a beam with (ACI Code Section 22.5)

$$V_u \leq \phi(V_c + V_s) \tag{15-10}$$

where

$$V_c = 2\lambda\sqrt{f'_c}b_w d \tag{15-11}$$

Recall that λ is a factor used for lightweight concrete and was defined in Chapter 6. For normal-weight concrete, which is commonly used in footings, $\lambda = 1.0$. Shear reinforcement is very seldom used in strip footings or spread footings, due to the difficulty in placing it, and due to the fact that it is usually cheaper and easier to deepen the footing than it is to provide shear reinforcement. Hence, $V_s = 0$ in most cases. The inclined crack shown in Fig. 15-8a intercepts the bottom of the member about d from the face of the column. As a result, the critical section for one-way shear is located at d away from the face of the column or wall, as shown in plan view in Fig. 15-8a. For footings supporting columns with steel base plates, the critical section is d away from a line halfway between the face of the column and the edge of the base plate. The shear V_u is q_{nu} times the tributary area shown shaded in Fig. 15-8a.

Two-Way Shear

Research [15-6], [15-7] has shown that the critical section for punching shear is at the face of the column, while the critical loaded area is that lying outside the area of the portion punched through the slab. To simplify the design equations, the critical-shear perimeter for design purposes has been defined as lying $d/2$ from the face of the column, as shown by the dashed line in Fig. 15-8b (ACI Code Section 22.6.4.1). For the column shown in Fig. 15-8b, the length, b_o, of this perimeter is

$$b_o = 2(c_1 + d) + 2(c_2 + d) \tag{15-12}$$

where c_1 and c_2 are the lengths of the sides of the column and d is the average effective depth in the two directions. The tributary area assumed critical for design purposes is shown crosshatched in Fig. 15-8b.

Because shear reinforcement is rarely used in a two-way footing, the design requirement at the critical section $(b_o d)$ for shear is, $v_u \leq \phi v_c$, and from ACI Code Table 22.6.5.2, v_c shall be the smallest of

(a)
$$v_c = 4\lambda\sqrt{f'_c} \tag{15-13}$$

where λ is the factor for lightweight concrete,

(b)
$$v_c = \left(2 + \frac{4}{\beta}\right)\lambda\sqrt{f'_c} \tag{15-14}$$

where β is the ratio of the long side to the short side of the column (c_1/c_2 in Fig. 15-8b),

(c)
$$v_c = \left(\frac{\alpha_s d}{b_o} + 2\right)\lambda\sqrt{f'_c} \tag{15-15}$$

where b_o is the critical perimeter from Eq. (15-12) and α_s is 40 for columns in the center of the footing, 30 for columns at an edge of a footing, and 20 for columns at a corner of the footing.

Transfer of Load from Column to Footing

Columns apply a concentrated load on footings. This load is transmitted by bearing stresses in the concrete and by stresses in the dowels or column bars that cross the joint. The design of such a joint is considered in ACI Code Section 16.3. The area of the dowels can be less than that of the bars in the column above, provided that the area of the dowels is at least 0.005 times the column area (ACI Code Section 16.3.4.1) and is adequate to transmit the necessary forces. Such a joint is shown in Fig. 15-9. Generally, the column bars stop at the bottom of the column, and dowels are used to transfer forces across the column–footing joint. Dowels are used because it is awkward to embed the column steel in the footing, due to its unsupported height above the footing and the difficulty in locating it accurately. Figure 15-9a shows an 18 in. × 18 in. column with $f'_c = 5000$ psi and eight No. 8 bars. The column is supported on a footing made of 3000-psi concrete. There are four No. 7 Grade-60 dowels in the connection. The dowels extend into the footing a distance equal to the compression development length of a No. 7 bottom bar in 3000-psi concrete (19 in.) and into the column a distance equal to the greater of

1. the compression lap-splice length for a No. 7 bar in 5000-psi concrete (26 in.), and
2. the compression development length of a No. 8 bar in 5000-psi concrete (17 in.).

This joint could fail by reaching various limit states, including

1. crushing of the concrete at the bottom of the column, where the column bars are no longer effective,
2. crushing of the concrete in the footing below the column,

3. bond failure of the dowels in the footing, and
4. failure in the column of the lap splice between the dowels and column bars.

Bearing Strength

The total capacity of the column for pure axial load is 900 kips, of which 197 kips is carried by the steel and the rest by the concrete, as shown in Fig. 15-9b. At the joint, the area of the dowels is less than that of the column bars, and the force transmitted by the dowels is $\phi A_{sd} f_y$, where A_{sd} is the area of the dowels and ϕ is that for tied columns. As a result, the load carried by the concrete has increased. In Fig. 15-9, the dowels are hooked so that they can be supported on and tied to the mat of footing reinforcement when the footing concrete is placed. The hooks cannot be used to develop compressive force in the bars (ACI Code Section 25.4.1.2).

The maximum bearing load on the concrete is defined in ACI Code Table 22.8.3.2 as $\phi(0.85 f'_c A_1)$. If the load combinations from ACI Code Table 5.3.1 are used, ACI Code Table 21.2.1 gives $\phi = 0.65$ for bearing and A_1 is the area of the contact surface.

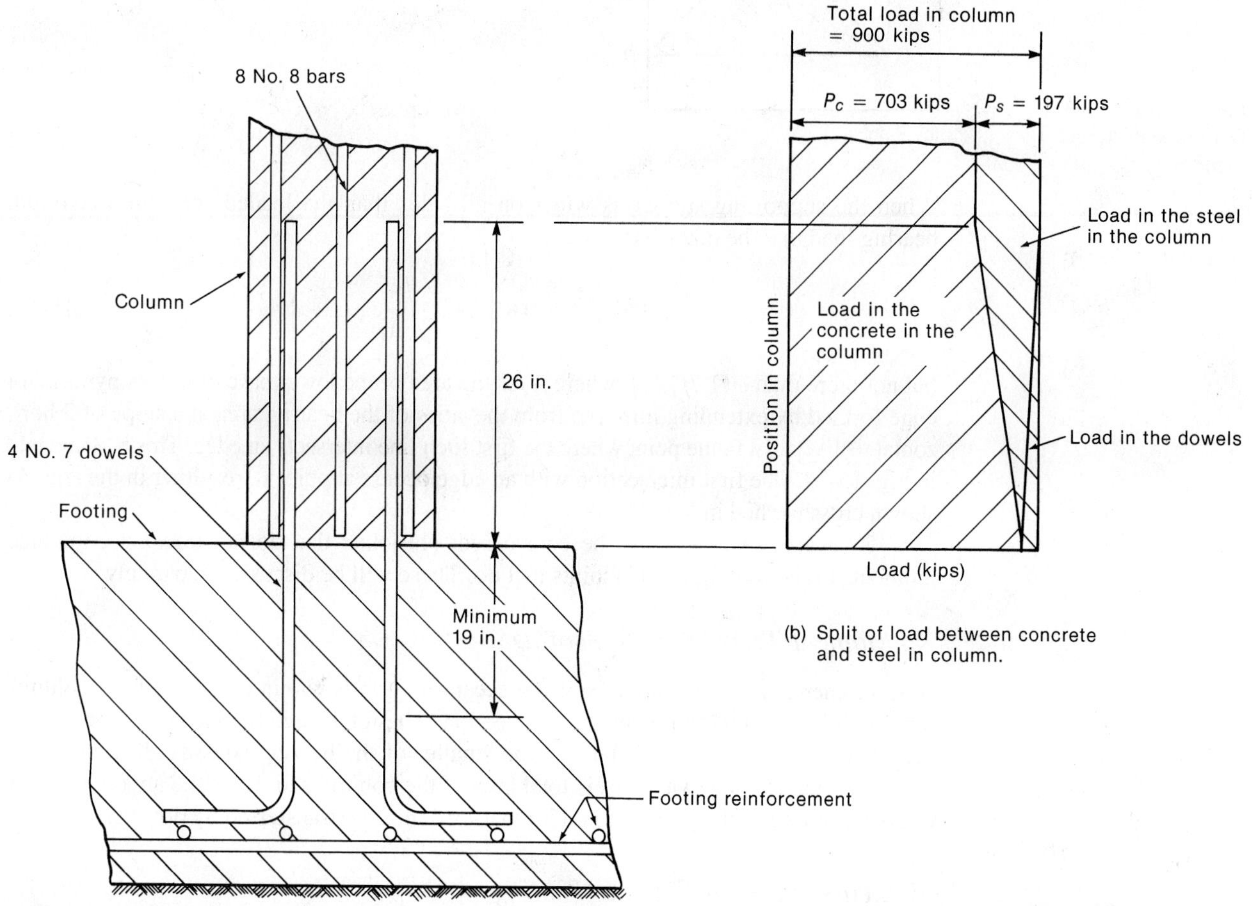

(a) Column–footing joint.

(b) Split of load between concrete and steel in column.

Fig. 15-9
Column–footing joint.

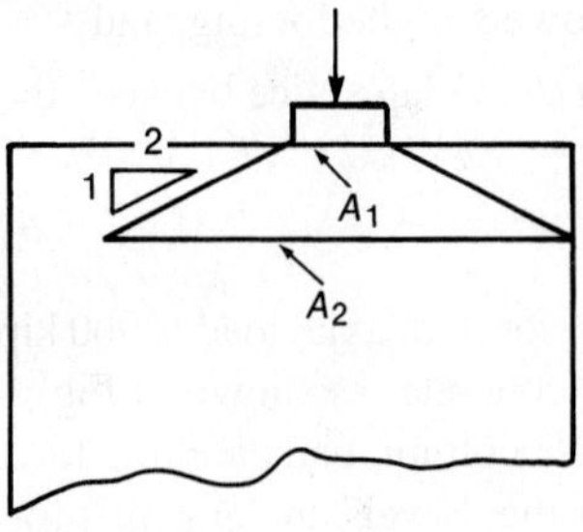

(a) Side view.

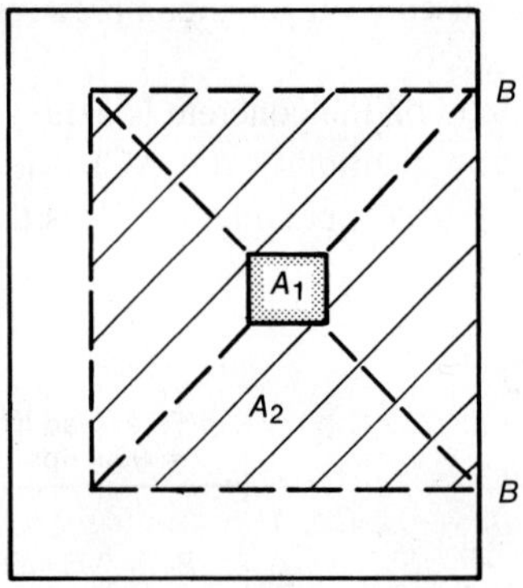

(b) Plan.

Fig. 15-10
Definition of A_1 and A_2.

When the supporting surface is wider on all sides than the loaded area, the maximum bearing load may be taken as

$$\phi(0.85f'_c A_1)\sqrt{\frac{A_2}{A_1}} \qquad (15\text{-}16)$$

but not more than $\phi(1.7f'_c A_1)$, where A_2 is the area of the lower base of a right pyramid or cone formed by extending lines out from the sides of the bearing area at a slope of 2 horizontal to 1 vertical to the point where the first such line intersects an edge. This is illustrated in Fig. 15-10. The first intersection with an edge occurs at point B, resulting in the area A_2 shown crosshatched in Fig. 15-10b.

Two distinct cases must be considered: (1) joints that do not transmit computed moments to the footing and (2) joints that do. These will be discussed separately.

No Moment Transferred to Footing

If no moments are transmitted, or if the eccentricity falls within the kern of the column, there will be compression over the full section. The total force transferred by bearing is then calculated as $(A_g - A_{sd})$ times the smaller of the bearing stresses allowed on the column or the footing, where A_g is total area of the column and A_{sd} is the area of the bars or dowels crossing the joint. Any additional load must be transferred by dowels.

Moments Are Transferred to Footing

If moments are transmitted to the footing, bearing stresses will exist over part, but not all, of the column cross section. The number of dowels required can be obtained by considering the area of the joint as an eccentrically loaded column with a maximum concrete stress equal to the smaller of the bearing stresses allowed on the column or the footing. Sufficient reinforcement must cross the interface to provide the necessary axial load and moment capacity. Generally, this requires that all the column reinforcement or similar-sized dowel bars must cross the interface. This steel must be spliced in accordance with the requirements for column splices.

Practical Aspects

Three other aspects warrant discussion prior to the examples. The minimum cover to the reinforcement in footings cast against the soil is 3 in. (ACI Code Table 20.6.1.3.1). This allows for small irregularities in the surface of the excavation and for potential contamination of the bottom layer of concrete with soil. Sometimes, the bottom of the excavation for the footing is covered with a lean concrete seal coat, to prevent the bottom from becoming uneven after rainstorms and to give a level surface for placing reinforcement.

The minimum depth of the footing above the bottom reinforcement is 6 in. for footings on soil and 12 in. for footings on piles (ACI Code Sections 13.3.1.2 and 13.4.2.1). ACI Code Section 24.3.2, covering the distribution of flexural reinforcement in beams and one-way slabs, does not apply to footings.

15-4 STRIP OR WALL FOOTINGS

A wall footing cantilevers out on both sides of the wall as shown in Figs. 15-1a and 15-11. The soil pressure causes the cantilevers to bend upward, and as a result, reinforcement is required at the bottom of the footing, as shown in Fig. 15-11. The critical sections for design for flexure and anchorage are at the face of the wall (section *A–A* in Fig. 15-11). One-way shear is critical at a section a distance *d* from the face of the wall (section *B–B* in Fig. 15-11). The presence of the wall prevents two-way shear. Thicknesses of wall footings are chosen in 1-in. increments, widths in 2- or 3-in. increments.

EXAMPLE 15-1 Design of a Wall Footing

A 12-in.-thick concrete wall carries a service (unfactored) dead load of 10 kips per foot and a service live load of 12.5 kips per foot. From the geotechnical report, the allowable soil pressure, q_a, is 5000 psf for shallow foundations. Design a wall footing to be based 5-ft below the final ground surface, using $f'_c = 3000$ psi normal-weight concrete and $f_y = 60{,}000$ psi. The density of the soil is 120 lb/ft^3. Most strip footings on soil have one mat of reinforcement.

1. Estimate the size of the footing and the factored net pressure. Consider a 1-ft strip of footing and wall. Allowable soil pressure is 5 ksf; allowable net soil pressure is 5 ksf − weight/ft^2 of the footing and of the soil over the footing. Because the thickness of the footing is not known at this stage, it is necessary to guess a thickness for a first trial.

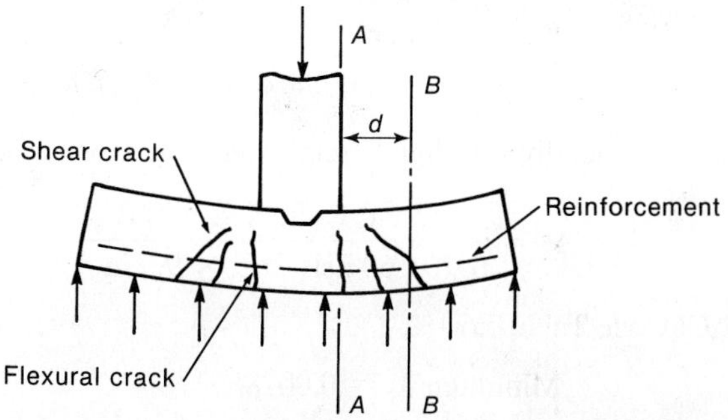

Fig. 15-11
Structural action of a strip footing.

Generally, the thickness will be 1 to 1.5 times the wall thickness. We shall try a 12-in.-thick footing. Therefore, $q_n = 5 - (1 \times 0.15 + 4 \times 0.12) = 4.37$ ksf, and we have:

$$\text{Area required} = \frac{10 \text{ kips} + 12.5 \text{ kips}}{4.37 \text{ ksf}}$$
$$= 5.15 \text{ ft}^2 \text{ per foot length of wall}$$

Try a footing 5 ft 2 in. = 62 in. (5.17 ft) wide.
Using the load factors in ACI Code Table 5.3.1:

$$\text{Factored net pressure, } q_{nu} = \frac{1.2 \times 10 + 1.6 \times 12.5}{5.17} = 6.19 \text{ ksf}$$

In the design of the concrete and reinforcement, we shall use $q_{nu} = 6.19$ ksf

2. Check the shear. Shear usually governs the thickness of footings. Only one-way shear is significant in a wall footing. Check it at d away from the face of the wall (section B–B in Fig. 15-11)

$$d = 12 \text{ in.} - 3 \text{ in. cover} - \tfrac{1}{2} \text{ bar diameter} \simeq 8.5 \text{ in.}$$

The tributary area for shear is shown crosshatched in Fig. 15-12a.

$$V_u = 6.19\left(\frac{16.5}{12} \times 1\right) \text{ft}^2 = 8.51 \text{ kips/ft}$$

$$\phi V_c = \phi \times 2\lambda\sqrt{f'_c}b_w d = 0.75 \times 2 \times 1\sqrt{3000} \times 12 \times 8.5$$
$$= 8380 \text{ lbs/ft} = 8.38 \text{ kips/ft}$$

where, from ACI Code Table 21.2.1, $\phi = 0.75$ for shear design.

Because $V_u > \phi V_c$, the footing depth is too small. If V_u is larger or considerably smaller than ϕV_c, choose a new thickness and repeat steps 1 and 2. **Try a 13-in.-thick footing 5 ft 2 in. wide.** A 13-in.-thick footing has $d \cong 9.50$ in. and $\phi V_c = 9.37$ kips/ft. Because ϕV_c exceeds $V_u = 8.51$ kips/ft, **a 13-in.-thick footing 5 ft 2 in. wide is adequate for shear.**

3. Design the flexural reinforcement. The critical section for moment is at the face of the wall (section A–A in Fig. 15-11). The tributary area for moment is shown crosshatched in Fig. 15-12b.

The required moment is

$$M_u = 6.19 \times \frac{(25/12)^2}{2} \times 1 \text{ kip-ft/ft} = 13.4 \text{ kip-ft/ft of length}$$

We want,

$$M_u \le \phi M_n = \phi A_s f_y jd$$

Footings are generally very lightly reinforced. Therefore, assume that $j = 0.95$. Therefore,

$$A_s = \frac{13.4 \text{ k-ft/ft} \times 12 \text{ in./ft}}{0.9 \times 60 \text{ ksi}(0.95 \times 9.5 \text{ in.})} = 0.330 \text{ in.}^2/\text{ft}$$

From ACI Code Table 7.6.1.1,

$$\text{Minimum } A_s = 0.0018bh$$
$$= 0.0018 \times 12 \times 13 = 0.281 \text{ in.}^2/\text{ft}$$

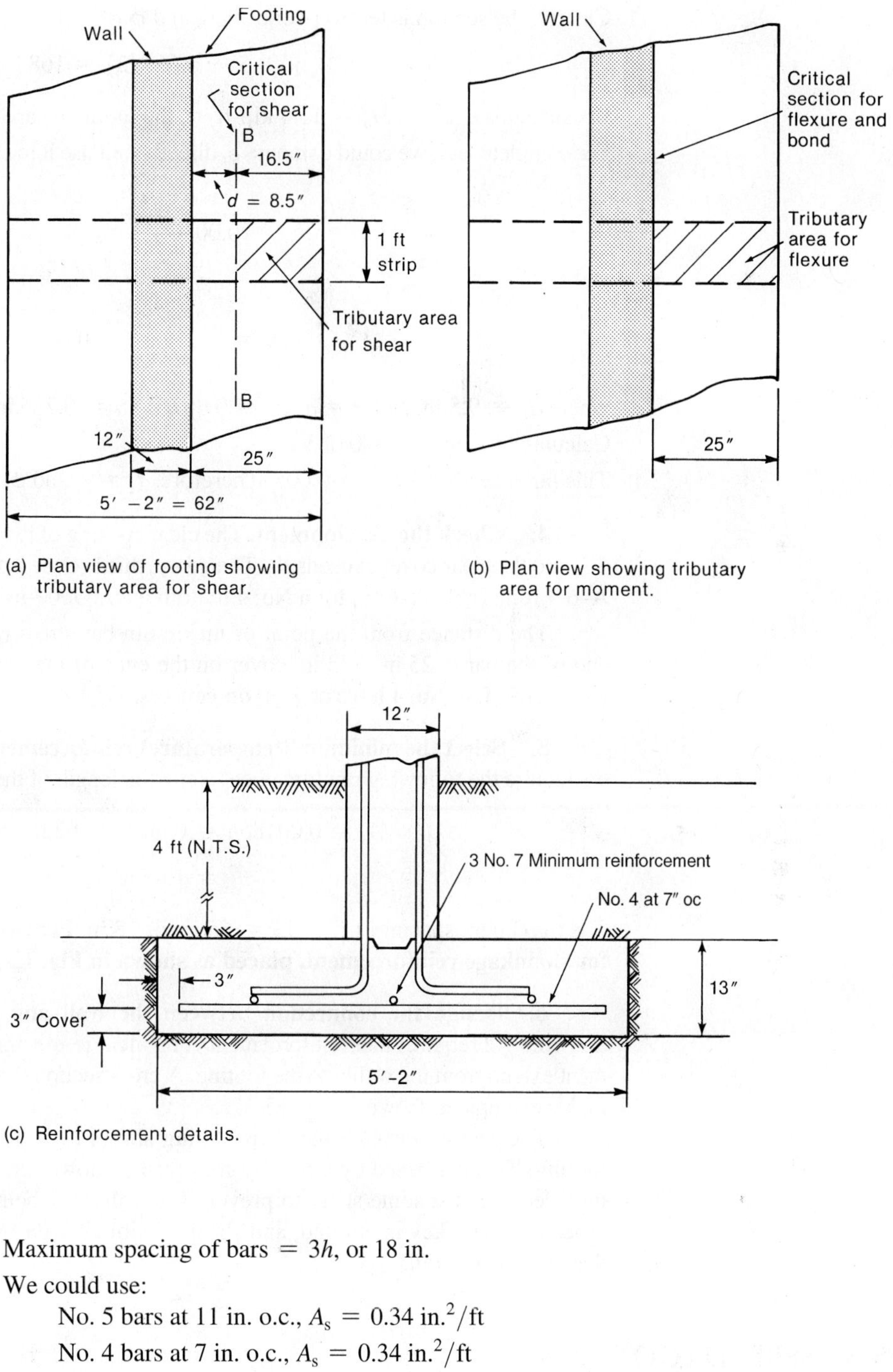

Fig. 15-12
Strip footing—Example 15-1.

Maximum spacing of bars $= 3h$, or 18 in.

We could use:

No. 5 bars at 11 in. o.c., $A_s = 0.34 \text{ in.}^2/\text{ft}$

No. 4 bars at 7 in. o.c., $A_s = 0.34 \text{ in.}^2/\text{ft}$

Try No. 4 bars at 7 in. o.c., $A_s = 0.34 \text{ in.}^2/\text{ft}$

Because the calculation of A_s was based on an assumed j-value, recompute the moment capacity:

$$a = \frac{0.34 \times 60}{0.85 \times 3 \times 12} = 0.667 \text{ in.}$$

Clearly, the section is tension-controlled, and $\phi = 0.9$.

$$\phi M_n = 0.9 \times 0.34 \times 60(9.5 - 0.667/2) = 168 \text{ k-in./ft} = 14.0 \text{ kip-ft/ft}$$

Because this exceeds $M_u = 13.4$ kip-ft/ft, the moment capacity is OK.

For completeness, we could compute ε_t directly and use it to check ϕ. From similar triangles

$$\frac{c}{0.003} = \frac{d_t - c}{\varepsilon_t}$$

and

$$\varepsilon_t = \frac{(d_t - c)}{c} \times 0.003$$

where $d_t \cong 9.5$ in. and $c = a/\beta_1 = 0.667/0.85 = 0.785$ in.

Calculations show $\varepsilon_t = 0.033$.

This far exceeds ε_t limit $= 0.005$. Therefore, $f_s = f_y$ and $\phi = 0.90$.

4. Check the development. The clear spacing of the bars being developed exceeds $2d_b$ and the clear cover exceeds d_b. Therefore, this is Case 2 development in Tables 8-1 and A–6. From Table A–6, ℓ_d for a No. 4 bottom bar in 3000-psi concrete is 21.9 in.

The distance from the point of maximum bar stress (at the face of the wall) to the end of the bar is 25 in. − 3 in. cover on the ends of the bars = 22 in. This is more than $\ell_d = 21.9$. **Use No. 4 bars at 7 in. on centers.**

5. Select the minimum (temperature) reinforcement. By ACI Code Table 7.6.1.1 we require the following reinforcement along the length of the footing.

$$A_s = 0.0018bh = 0.0018 \times 62 \text{ in.} \times 13 \text{ in.}$$
$$= 1.45 \text{ in.}^2$$

The maximum spacing is 5 × 13 = 65 in. or 18 in. **Provide three No. 7 bars (1.80 in.^{2}) for shrinkage reinforcement, placed as shown in Fig. 15-12c.**

6. Design the connection between the wall and the footing. ACI Code Section 16.3.4.2 requires that reinforcement equivalent to the minimum vertical wall reinforcement extend from the wall into the footing. A cross section through the wall footing designed in this example is shown in Fig. 15-12c.

The section through the strip footing shows a shear key in the top surface of the footing. This is formed by a two-by-four pushed down into the surface. The shear key is intended to resist some shear to prevent the wall from being dislocated laterally. Sometimes the shear key is omitted, and shear friction is used to resist dislocation of the wall during construction. ■

15-5 SPREAD FOOTINGS

Spread footings are square or rectangular pads that spread a column load over an area of soil large enough to support the column load. The soil pressure causes the footing to deflect upward, as shown in Fig. 15-7a, causing tension in two directions at the bottom. As a result, reinforcement is placed in two directions at the bottom, as shown in Fig. 15-7c. Two examples will be presented: a square axially loaded footing and a rectangular axially loaded footing.

EXAMPLE 15-2 Design of a Square Spread Footing

A square spread footing supports an 18-in.-square column supporting a service dead load of 400 kips and a service live load of 270 kips. The column is built with 5000-psi concrete and has eight No. 9 longitudinal bars with $f_y = 60{,}000$ psi. Design a spread footing to be constructed by using 3000-psi normal-weight concrete and Grade-60 bars. It is quite common for the strength of the concrete in the footing to be lower than that in the column. Dowels may be required to carry some of the column load across the column–footing interface. The top of the footing will be covered with 6 in. of fill with a density of 120 lb/ft^3 and a 6-in. basement floor (Fig. 15-13). The basement floor loading is 100 psf. For shallow foundations, the allowable bearing pressure on the soil is 6000 psf. Use load and resistance factors from ACI Code Tables 5.3.1 and 21.2.1.

1. Compute the factored loads and the resistance factors, ϕ. Because the statement of the problem mentioned only dead and live loads, we will assume these as the only applicable loads. In effect, we are assuming that the wind loads and roof loads are small compared to the dead loads. This reduces the set of load combinations to the following:

$$U = 1.4(D)$$

$$U = 1.2(D) + 1.6(L)$$

From these equations, the design loads are

$$U = 1.4 \times 400 = 560 \text{ kips}$$

and

$$U = 1.2 \times 400 + 1.6 \times 270 = 912 \text{ kips}$$

Strength-reduction factors ϕ are given in ACI Code Tables 21.2.1 and 21.2.2. For shear, $\phi = 0.75$. For flexure, ϕ will be a function of the strain in the extreme-tension layer of bars, but ϕ will probably be 0.9 for footings.

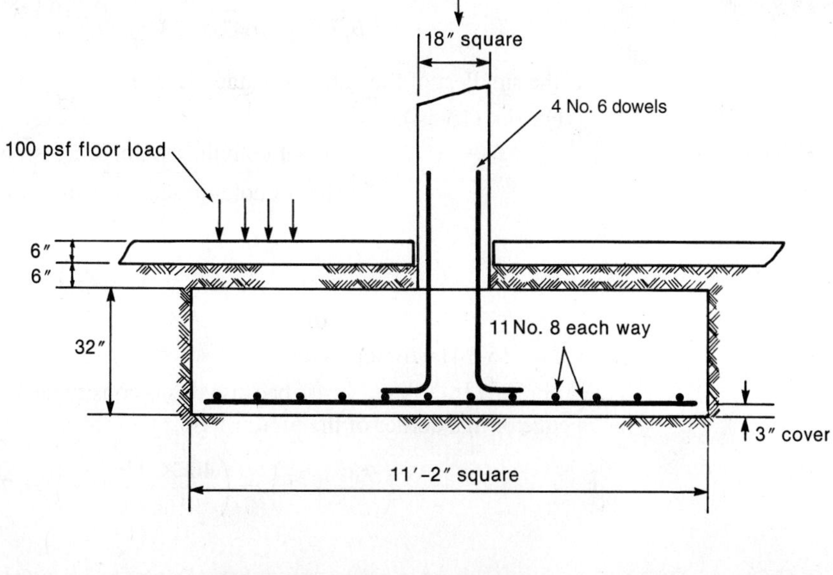

Fig. 15-13
Spread footing—
Example 15-2.

2. Estimate the footing size and the factored net soil pressure. Allowable net soil pressure q_n = 6 ksf − (weight/ft^2 of the footing and the soil and floor over the footing and the floor loading). Estimate the overall thickness of the footing at between one and two times the width of the column, say, 27 in.:

$$q_n = 6.0 - \left(\frac{27}{12} \times 0.15 + 0.5 \times 0.12 + 0.5 \times 0.15 + 0.100\right)$$
$$= 5.43 \text{ ksf}$$

$$\text{Area required} = \frac{400 \text{ kips} + 270 \text{ kips}}{5.43 \text{ ksf}} = 123 \text{ ft}^2$$
$$\approx 11.1 \text{ ft square.}$$

Try a footing 11 ft 2 in. square by 27 in. thick:

$$\text{Factored net soil pressure} = \frac{1.2 \times 400 + 1.6 \times 270}{11.17^2}$$
$$= 7.31 \text{ ksf}$$

3. Check the thickness for two-way shear. Generally, the thickness of a spread footing is governed by two-way shear. The shear will be checked on the critical perimeter at $d/2$ from the face of the column and, if necessary, the thickness will be increased or decreased. Because there is reinforcement in both directions, the average d will be used:

$$\text{Average } d = 27 \text{ in.} - (3 \text{ in. cover}) - (1 \text{ bar diameter})$$
$$\cong 23 \text{ in.}$$

The critical shear perimeter (ACI Code Section 11.11.1.2) is shown dashed in Fig. 15-14a. The tributary area for two-way shear is shown crosshatched. We have

$$V_u = 7.31 \text{ ksf}\left[11.17^2 - \left(\frac{41}{12}\right)^2\right] \text{ft}^2 = 827 \text{ kips}$$

Length of critical shear perimeter:

$$b_o = 4(18 + 23) \text{ in.} = 164 \text{ in.}$$

Thus, the factored shear stress on the critical section is:

$$v_u = \frac{V_u}{b_o \cdot d} = \frac{827 \text{ k}}{164'' \times 23''} = 0.219 \text{ ksi} = 219 \text{ psi}$$

ϕv_c is the smallest of the values obtained from Eqs. (15-13), (15-14), and (15-15).

For Eq. (15-14),

$$\beta = \frac{\text{larger column section dimension}}{\text{shorter column section dimension}} = 1.0$$

and

$$\left(2 + \frac{4}{\beta}\right) = 6 > 4$$

Thus, Eq. (15-14) will not govern.

For Eq. (15-15), $\alpha_s = 40$, because this is considered to be an interior connection (not at an edge or in a corner of the footing). So,

$$\left(\frac{\alpha_s d}{b_o} + 2\right) = \left(\frac{40 \times 23}{164} + 2\right) = 7.6 > 4$$

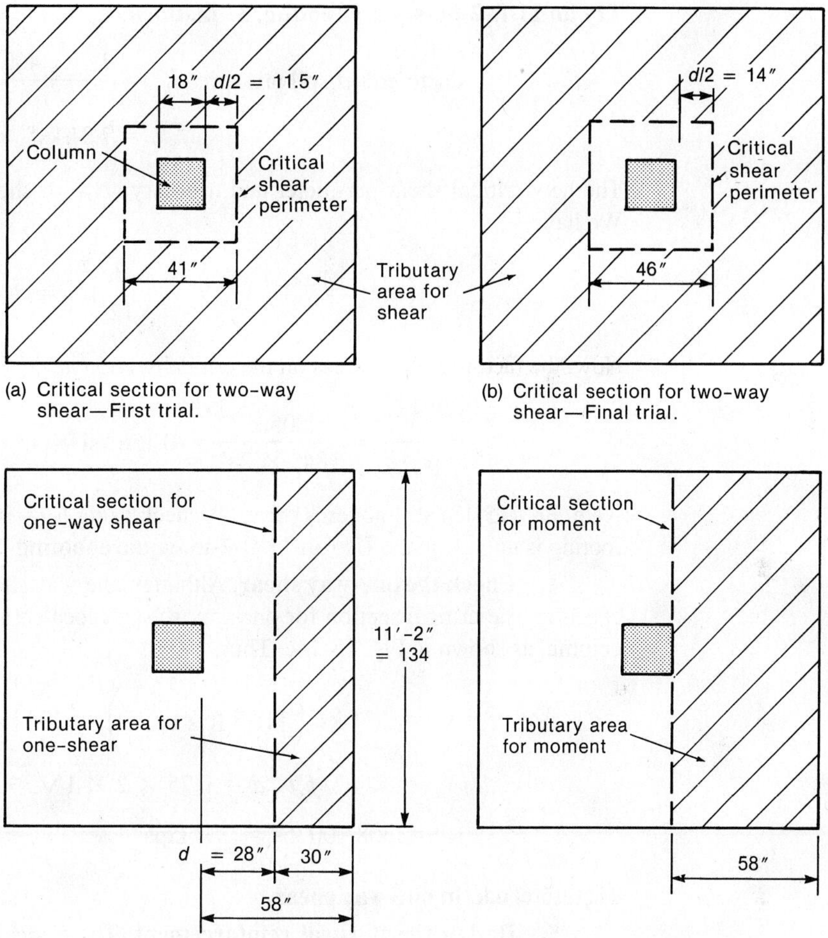

(a) Critical section for two-way shear—First trial.

(b) Critical section for two-way shear—Final trial.

(c) Critical section for one-way shear.

(d) Critical section for moment.

Fig. 15-14
Critical sections—
Example 15-2.

Thus, Eq. (15-15) will not govern.

So, using Eq. (15-13) to find v_c:

$$v_c = 4\lambda\sqrt{f'_c} = 4 \times 1\sqrt{3000 \text{ psi}} = 219 \text{ psi}$$

and $\phi v_c = 0.75 \times 219 = 164$ psi.

Because ϕv_c is less than v_u, the footing is not thick enough. Try $h = 32$ in., $d = 28$ in., and $b_o = 184$ in. The footing is thicker, and it weighs more. Hence, a larger area may be required:

$$q_n = 6.0 - \left(\frac{32}{12} \times 0.15 + 0.5 \times 0.12 + 0.5 \times 0.15 + 0.100\right)$$

$$= 5.37 \text{ ksf}$$

$$\text{Area required} = \frac{400 + 270}{5.37}$$

$$= 125 \text{ ft}^2$$

$$= 11.2 \text{ ft square}$$

Try an 11-ft-2-in.-square footing, 32 in. thick:

$$\text{Factored net soil pressure, } q_{nu} = \frac{1.2 \times 400 + 1.6 \times 270}{11.17^2}$$
$$= 7.31 \text{ ksf}$$

The new critical shear perimeter and tributary area for shear are shown in Fig. 15-14b. We have

$$V_u = 7.31\left[11.17^2 - \left(\frac{46}{12}\right)^2\right] = 805 \text{ kips}$$

Now, the factored shear stress on the critical section is:

$$v_u = \frac{V_u}{b_o \cdot d} = \frac{805 \text{ k}}{184'' \times 28''} = 0.156 \text{ ksi} = 156 \text{ psi} < \phi v_c \text{ (o.k.)}$$

Note, Eq. (15-13) still governs for v_c. A check using $h = 30$ in. shows that a 30-in.-thick footing is not adequate. **Use an 11-ft-2-in.-square footing, 32 in. thick.**

4. Check the one-way shear. Although one-way shear is seldom critical, we shall check it. The critical section for one-way shear is located at d away from the face of the column, as shown in Fig. 15-14c. Thus,

$$V_u = 7.31 \text{ ksf}\left(11.17 \text{ ft} \times \frac{30}{12}\text{ft}\right) = 204 \text{ kips}$$
$$\phi V_c = \phi 2\lambda\sqrt{f'_c}b_w d = 0.75 \times 2 \times 1\sqrt{3000} \times 134 \times 28$$
$$= 308{,}000 \text{ lbs} = 308 \text{ kips}$$

Therefore, **o.k. in one-way shear.**

5. Design the flexural reinforcement. The critical section for moment and anchorage of the reinforcement is shown in Fig. 15-14d. The ultimate moment is

$$M_u = 7.31\left[11.17 \times \frac{(58/12)^2}{2}\right] = 954 \text{ kip-ft}$$

Assuming that $j = 0.95$ and $\phi = 0.90$, the area of steel required is

$$A_s = \frac{954 \times 12}{0.9 \times 60(0.95 \times 28)} = 7.97 \text{ in.}^2$$

The average value of d was used in this calculation for simplicity. The same reinforcement will be used in both directions:

$$\text{Minimum } A_s \text{ (ACI Table 8.6.1.1)} = 0.0018bh$$
$$= 0.0018 \times 134 \times 32 = 7.72 \text{ in.}^2 \text{ (does not govern)}$$
$$\text{Maximum spacing (ACI Code Section 8.7.2.2)} = 18 \text{ in.}$$

Try eleven No. 8 bars each way, A_s = 8.69 in.^{2}. Recompute ϕM_n as a check.

$$a = \frac{8.69 \times 60}{0.85 \times 3 \times 134} = 1.53 \text{ in.}$$

Clearly, the beam is tension-controlled, and $\phi = 0.90$.

$$\phi M_n = 1070 \text{ kip-ft}$$

This exceeds $M_u = 954$ kip-fts

6. Check the development. The clear spacing of the bars being developed exceeds $2d_b$ and the clear cover exceeds d_b. Therefore, this is Case 2 development in Tables 8-1 and A-6. From Table A-6, ℓ_d for a No. 8 bottom bar in 3000-psi concrete is 54.8 d_b. The development length is

$$\ell_d = 54.8\, d_b \psi_e / \lambda$$

where $\psi_e = 1.0$ for uncoated reinforcement and $\lambda = 1.0$ for normal-weight concrete. Accordingly, we have

$$\ell_d = 54.8 \times 1.00 \text{ in.}$$
$$= 54.8 \text{ in.}$$

The bar extension past the point of maximum moment is $(58 \text{ in.} - 3 \text{ in.}) = 55$ in. This is o.k. **Use eleven No. 8 bars each way; $A_s = 8.69$ in.^{2}.**

7. Design the column–footing joint. The column–footing joint is shown in Fig. 15-13. The factored load at the base of the column is

$$1.2 \times 400 + 1.6 \times 270 = 912 \text{ kips}$$

The maximum bearing load on the bottom of the column (ACI Code Table 22.8.3.2) is $\phi 0.85 f'_c A_1$, where A_1 is the area of the contact surface between the column and the footing and f'_c is for the column. When the contact-supporting surface on the footing is wider on all sides than the loaded area, the maximum bearing load on the top of the footing may be taken as

$$0.85\phi f'_c A_1 \sqrt{\frac{A_2}{A_1}}, \text{ but not more than } 1.7\phi f'_c A_1 \tag{15-16}$$

where A_2 is the area of the lower base of a right pyramid or cone as defined in Fig. 15-10. ACI Code Table 21.2.1 defines ϕ equal to 0.65 for bearing. By inspection, $\sqrt{A_2/A_1}$ for the footing exceeds 2—hence, the maximum bearing load *on the footing* is $0.85 \times 0.65 \times 3 \times 18^2 \times 2 = 1070$ kips. The allowable bearing on the base of the column is

$$\phi(0.85 f'_c A_1) = 0.65 \times 0.85 \times 5 \times 18^2$$
$$= 895 \text{ kips}$$

Thus, the maximum load that can be transferred by bearing is 895 kips, and dowels are needed to transfer the excess load. Accordingly, we have

$$\text{Area of dowels required} = \frac{912 - 895}{\phi f_y} = 0.44 \text{ in.}^2$$

where $\phi = 0.65$ has been used. This is the ϕ value from ACI Code Table 21.2.2 for compression-controlled tied columns and Table 21.2.1 for bearing. The area of dowels must also satisfy ACI Code Section 16.3.4.1:

$$\text{Area of dowels} \geq 0.005 A_g = 0.005 \times 18^2 = 1.62 \text{ in.}^2$$

Try four No. 6 dowels ($A_s = 1.76$ in.^{2}); dowel each corner bar. The dowels must extend into the footing a distance equal to the compression-development length (Table A-7) for

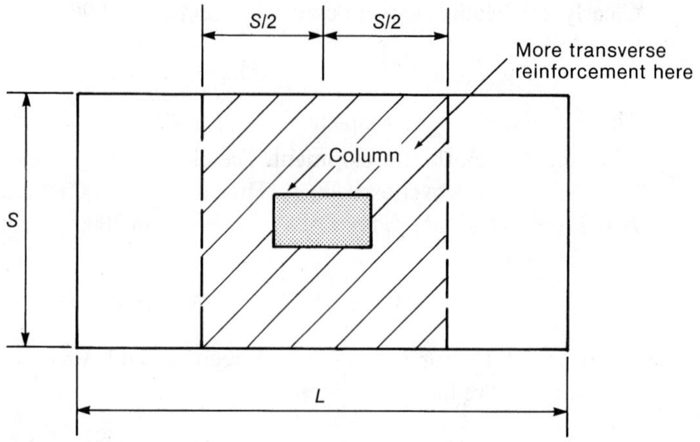

Fig. 15-15
Rectangular footing.

a No. 6 bar in 3000-psi concrete, or 16 in. The bars will be extended down to the level of the main footing steel and hooked 90°. The hooks will be tied to the main steel to hold the dowels in place. From ACI Code Section 25.5.5.4, the dowels must extend into the column a distance equal to the greater of a compression splice (Table A-13) for the dowels (23 in.) or the compression-development length (Table A-7) of the No. 9 column bars for $f'_c = 5000$ psi (20 in.). **Use four No. 6 dowels; dowel each corner bar. Extend dowels 23 in. into column.** (See Fig. 15-13.) ■

Rectangular Footings

Rectangular footings may be used when there is inadequate clearance for a square footing. In such a footing, the reinforcement in the short direction is placed in the three bands shown in Fig. 15-15, with a closer bar spacing in the band under the column than in the two end bands. The band under the column has a width equal to the length of the short side of the footing, but not less than the width of the column (if that is greater) and is centered on the column. The reinforcement in the central band shall be $2/(\beta + 1)$ times the total reinforcement in the short direction, where β is the ratio of the long side of the footing to the short side (ACI Code Section 13.3.3.3). The reinforcement within each band is distributed evenly, as is the reinforcement in the long direction.

EXAMPLE 15-3 Design of a Rectangular Spread Footing

Redesign the footing from Example 15-2, assuming that the maximum width of the footing is limited to 9 ft. Steps 1 through 3 of this example would proceed in the same sequence as in Example 15-2, leading to a footing 9 ft wide by 14 ft long by 33 in. thick. Using the load factors from ACI Code Table 5.3.1, the factored net soil pressure is 7.24 ksf.

1. Check the one-way shear. One-way shear may be critical in a rectangular footing and must be checked. In this case, it required a 1 in. increase in the thickness of the footing. The critical section and tributary area for one-way shear are shown in Fig. 15-16a.

We have

$$V_u = 7.24\left(\frac{46}{12} \times 9\right) = 250 \text{ kips}$$

$$\phi V_c = 0.75 \times 2 \times 1\sqrt{3000} \times 108 \times 29 = 257{,}000 \text{ lbs} = 257 \text{ kips}$$

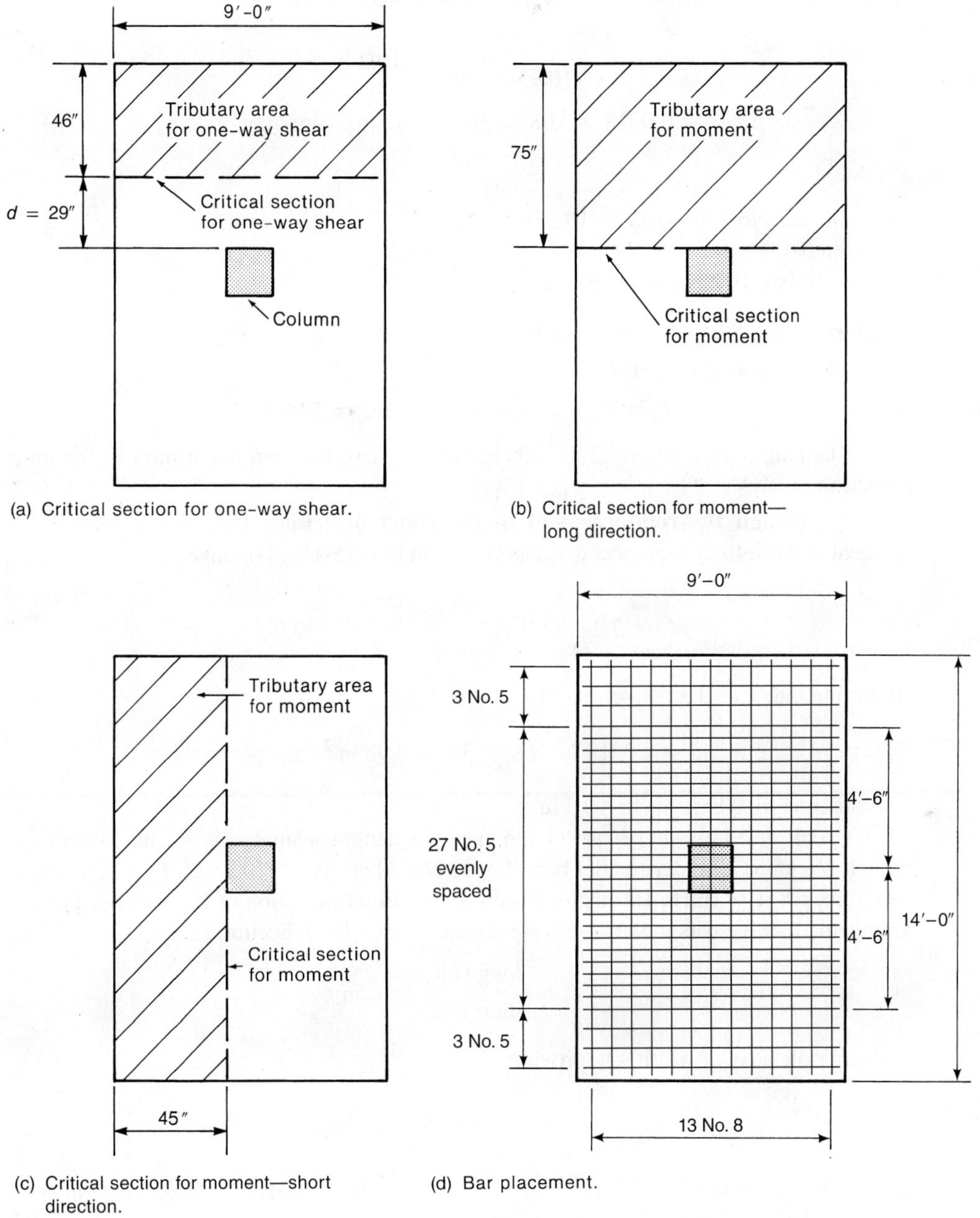

Fig. 15-16
Rectangular footing—Example 15-3.

This is just o.k. in one-way shear.

2. Design the reinforcement in the long direction. The critical section for moment and reinforcement anchorage is shown in Fig. 15-16b. The ultimate moment is

$$M_u = 7.24\left(9 \times \frac{(75/12)^2}{2}\right) = 1270 \text{ kip-ft}$$

Assuming that $\phi = 0.9$ and $j = 0.95$, the area of steel required is

$$A_s = \frac{1270 \times 12}{0.9 \times 60\,(0.95 \times 29)} = 10.2 \text{ in.}^2$$

$$A_{s,\min} = 0.0018 \times 108 \times 33 = 6.42 \text{ in.}^2 \text{ (does not govern)}$$

We could use

thirteen No. 8 bars, $A_s = 10.3$ in.^{2}
ten No. 9 bars, $A_s = 10.0$ in.^{2}
eight No. 10 bars, $A_s = 10.2$ in.^{2}

Try thirteen No. 8 bars—$\phi M_n = 1290$ kip-ft (with $\phi = 0.9$).
Check development; from Table A-6:

$$\ell_d = 54.8 \times 1.0 \times 1.0 \times 1.0 = 54.8 \text{ in.}$$

The length available is 72 in.—therefore, o.k. **Use thirteen No. 8 bars in the long direction.**

3. Design the reinforcement in the short direction. The critical section for moment and reinforcement anchorage is shown in Fig. 15-16c. We have

$$M_u = 7.24\left(14.0 \times \frac{(45/12)^2}{2}\right) = 713 \text{ kip-ft}$$

Assuming that $j = 0.95$, $A_s \geqq 5.75$ in.^{2},

$$A_{s,\min} = 0.0018 \times 168 \times 33 = 9.98 \text{ in.}^2 \text{ (this governs)}$$

Try thirteen No. 8 bars: $A_s = 10.3$ in.^{2}.

Check development. $\ell_d = 54.8$ in. and the length available is 42 in.—therefore, not o.k. We must consider smaller bars. Try 33 No. 5 bars, $A_s = 10.2$ in.^{2}, $\ell_d = 27.4$ in.—therefore, o.k. **Use thirty-three No. 5 bars in the short direction of the footing.** For the arrangement of the bars in the transverse direction (ACI Code Section 15.4.4.2),

$$\beta = \frac{\text{long side}}{\text{short side}} = 1.56$$

In the middle strip of width 9 ft, provide

$$\frac{2}{1.56 + 1} \times 33 \text{ bars} = 25.8 \text{ bars}$$

Provide twenty-seven No. 5 bars in the middle strip, and provide three No. 5 bars in each end strip. The final design is shown in Fig. 15-16d. ■

Footings Transferring Vertical Load and Moment

On rare occasions, footings must transmit both axial load and moment to the soil. The design of such a footing proceeds in the same manner as that for a square or rectangular footing, except for three things. First, a deeper footing will be necessary, because there will be shear stresses developed both by direct shear and by moment. The calculations concerning this are discussed in Chapter 13. Second, the soil pressures will not be uniform, as discussed in Section 15-2 and shown in Fig. 15-5b. Third, the design for two-way shear must consider moment and shear. (See Section 13-11.)

The uneven soil pressures will lead to a tilting settlement of the footing, which will relieve some of the moment if the moment results from compatibility at the fixed end. The tilting can be reduced by offsetting the footing so that the column load acts through the center of the footing base area. If the moment is necessary for equilibrium, it will not be reduced by rotation of the foundation.

EXAMPLE 15-4 Design a Rectangular Footing for a Column Subjected to Axial Load and Bending

Assume the footing has a width of 10 ft and a shape similar to that in Fig. 15-15. Also, assume it will be supporting a 16 in. by 16 in. column that is carrying the following loads.

$$P_D = 180 \text{ kips and } P_L = 120 \text{ kips}$$

$$M_D = 80 \text{ k-ft and } M_L = 60 \text{ k-ft}$$

The length and depth of the footing are to be determined. The geotechnical report indicates that for shallow foundations, the allowable soil bearing pressure is 4000 psf. Design the footing assuming $f'_c = 3500$ psi (normal-weight concrete) and $f_y = 60$ ksi.

1. Factored loads and capacity reduction factors. Assume load combination (b) in Table 5.3.1 governs.

$$P_u = 1.2(180 \text{ k}) + 1.6(120 \text{ k}) = 408 \text{ kips}$$

$$M_u = 1.2(80 \text{ k-ft}) + 1.6(60 \text{ k-ft}) = 192 \text{ k-ft} = 2300 \text{ k-in.}$$

For shear strength, use $\phi = 0.75$ and for flexural strength, use $\phi = 0.9$.

2. Estimate footing length and depth. Estimate the overall thickness, t, of the footing to be between 1.5 and 2.0 times the size of the column, so **select $t = 26$ in.** Assuming a final design section similar to that shown in Fig. 15-13, the net permissible bearing pressure is:

$$q_n = 4.0 - \left[\frac{26}{12} \times 0.15 + 0.5 \times 0.12 + 0.5 \times 0.15 + 0.100\right]$$

$$= 3.44 \text{ ksf}$$

The (nonfactored) loads acting on the footing are:

$$P_a = 180 \text{ k} + 120 \text{ k} = 300 \text{ kips}$$

$$M_a = 80 \text{ k-ft} + 60 \text{ k-ft} = 140 \text{ k-ft}$$

Assume the soil pressure distribution at the base of the footing is similar to that should in Fig. 15-5b. Assuming a footing width, S, and a footing length, L, similar to the footing in Fig. 15-15, an expression for the maximum bearing pressure under the footing is:

$$q(\max) = \frac{P_a}{SL} + \frac{M_a\left(\frac{L}{2}\right)}{\frac{1}{12}SL^3} = \frac{P_a}{SL} + \frac{6M_a}{SL^2}$$

Setting $q(\max)$ equal to q_n (3.44 ksf) and using $S = 10$ ft, we can use the following expression to solve for the required footing length, L.

$$3.44 \text{ ksf} = \frac{300 \text{ k}}{(10 \text{ ft}) \cdot L} + \frac{6(140 \text{ k-ft})}{(10 \text{ ft}) \cdot L^2}$$

Solving for L yields, $L = 11.0$ ft, or -2.23 ft. Using the positive solution, **select a footing length of 12 ft**.

3. Calculate factored soil pressures. (Note: In the prior step the *acting* loads were used with the allowable soil bearing pressure to select the size of the footing. Now, the *factored* loads will be used to determine the soil pressures that will subsequently be used to determine the factored moment and shear used for the design of the footing.) Using $P_u = 408$ kips and $M_u = 192$ k-ft, the minimum and maximum factored soil pressures, as shown in Fig. 15-17(a), are:

$$q_u(\text{min, max}) = \frac{P_u}{SL} \pm \frac{6M_u}{SL^2}$$

$$q_u(\text{max}) = \frac{408\text{ k}}{(10\text{ ft})(12\text{ ft})} + \frac{6 \times 192\text{ k-ft}}{(10\text{ ft})(12\text{ ft})^2}$$

$$= 3.40\text{ ksf} + 0.80\text{ ksf} = 4.20\text{ ksf, and}$$

$$q_u(\text{min}) = 3.40\text{ ksf} - 0.80\text{ ksf} = 2.60\text{ ksf}$$

4. Check footing thickness for two-way shear. The critical shear perimeter is located $d/2$ away from each column face, as shown in Fig. 15-17b. Assume the average effective depth for the footing is:

$$d \cong t - 4\text{ in.} = 26\text{ in.} - 4\text{ in.} = 22\text{ in.}$$

Thus, each side of the critical shear perimeter has a length of 38 in. Using the average factored shear stress inside the critical perimeter ($q_u(\text{avg}) = 3.40$ ksf), the net factored shear to be transferred across the critical perimeter is:

$$V_u(\text{net}) = 408\text{ k} - 3.40\text{ ksf}\left[\left(\frac{38}{12}\text{ft}\right)\left(\frac{38}{12}\text{ft}\right)\right] = 374\text{ kips}$$

The critical perimeter, $b_o = 4 \times 38 = 152$ in., so the factored shear stress on the critical section is:

$$v_u = \frac{V_u}{b_o \cdot d} = \frac{374\text{ k}}{152'' \times 22''} = 0.112\text{ ksi} = 112\text{ psi}$$

Because we have a square column, the coefficient β in Eq. (15-14) is equal to 1.0, and thus, that equation does not govern for the value of V_c. Also, because this is not a large critical perimeter ($b_o = 4 \times 38 = 152$ in.) compared to d ($b_o < 20d$), Eq. (15-15) will not govern. Thus, using Eq. (15-13) to find v_c:

$$v_c = 4\lambda\sqrt{f'_c} = 4 \times 1 \times \sqrt{3500} = 237\text{ psi,}$$

and, $\phi v_c = 0.75 \times 237 = 177\text{ psi} > v_u$ (o.k.).

Thus, the footing depth satisfies the strength requirement for the net shear force.

5. Check for combined transfer of shear and moment. We must now check the maximum shear stress for the combined transfer of shear and moment using Eq. (13-30) of this text.

$$v_u = \frac{V_u}{b_o d} \pm \frac{\gamma_v M_u c}{J_c} \qquad (13\text{-}30)$$

From the previous step, it can be seen that v_u(net) exceeds 40% of ϕv_c. Therefore, the coefficient γ_f, and thus γ_v, cannot be adjusted as permitted in ACI Code Table 8.4.2.3.4. Using Eq. (13-32), with $b_1 = b_2 = 38$ in., we get:

$$\gamma_f = \frac{1}{1 + \frac{2}{3}\sqrt{b_1/b_2}} = \frac{1}{5/3} = 0.60$$

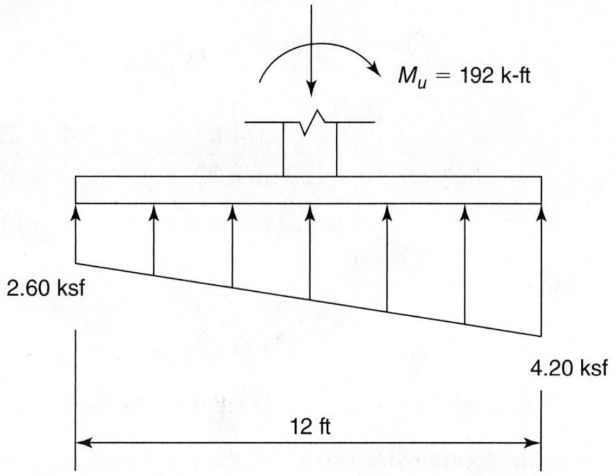

(a) Soil pressure distribution due to factored column loads.

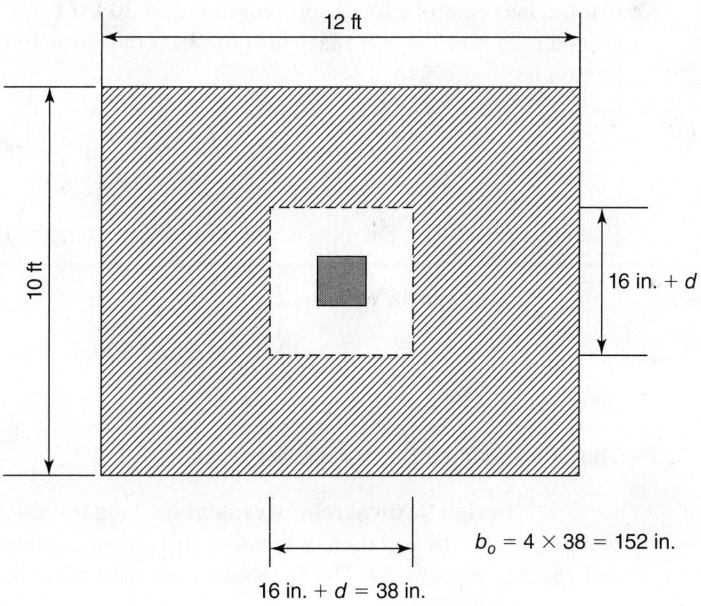

(b) Critical perimeter for two-way shear.

Fig. 15-17
Soil pressure distribution due to factored loads and critical shear perimeter.

And, using Eq. (13-31):

$$\gamma_v = 1 - \gamma_f = 0.40$$

The value of J_c to be used in Eq. (13-30) is given by Eq. (13-34) for an interior connection as ($b_1 = b_2 = 38$ in.):

$$J_c = 2\left(\frac{b_1 d^3}{12} + \frac{d b_1^3}{12}\right) + 2(b_2 d)\left(\frac{b_1}{2}\right)^2$$

$$= 2(33{,}700 + 101{,}000) + 604{,}000 = 872{,}000 \text{ in.}^4$$

Using $c = b_1/2 = 19$ in., Eq. (13-30) results in a maximum combined shear stress of:

$$\nu_u = \frac{V_u(\text{net})}{b_o d} + \frac{\gamma_v M_u c}{J_c}$$
$$= \frac{374 \text{ k}}{(152 \text{ in.})(22 \text{ in.})} + \frac{0.4 \times 2300 \text{ k-in.} \times 19 \text{ in.}}{872{,}000 \text{ in.}^4}$$
$$= 0.112 \text{ ksi} + 0.020 \text{ ksi} = 0.132 \text{ ksi} = 132 \text{ psi}$$

Using Eq. (15-15):

$$\phi v_c = \phi 4\lambda\sqrt{f'_c} = 0.75 \times 4 \times 1\sqrt{3500 \text{ psi}}$$
$$= 177 \text{ psi} > v_u \text{ (o.k.)}$$

Thus, the footing size is o.k. for the combined transfer of shear and moment.

6. Check for one-way shear. The critical section for checking one-way shear strength is shown in Fig. 15-18a. To simplify this check, it is conservative to assume that the maximum factored soil pressure of 4.20 ksf from Fig. 15-17a acts on the entire shaded region in Fig. 15-18a. Thus, the factored shear force to be resisted at the critical section is:

$$V_u = (4.20 \text{ ksf})\left(\frac{42}{12}\text{ft} \times 10\text{ft}\right) = 147 \text{ kips}$$

From text Eq. (6-8b), the reduced one-way shear strength at the crucial section is:

$$\phi V_c = \phi\, 2\lambda\sqrt{f'_c}\, b_w d$$
$$= 0.75 \times 2 \times 1\sqrt{3500 \text{ psi}} \times 120 \text{ in.} \times 22 \text{ in.} = 234{,}000 \text{ lbs}$$
$$= 234 \text{ kips} > V_u \text{ (o.k.)}$$

Thus, the footing is o.k. for one-way shear.

7. Design flexural reinforcement for long direction. The critical section for flexural design and the variation of factored soil pressure to be used are given in Figs. 15-18b and 15-18c, respectively. The factored design moment at the critical section is:

$$M_u = (3.49 \text{ ksf} \times 10 \text{ ft}) \times \frac{(5.33 \text{ ft})^2}{2} + (4.20 \text{ ksf} - 3.49 \text{ ksf}) \times (10 \text{ ft}) \times \frac{(5.33 \text{ ft})^2}{3}$$
$$= 496 \text{ k-ft} + 67.3 \text{ k-ft} = 563 \text{ k-ft} = 6760 \text{ k-in.}$$

Assuming a moment arm, jd, equal to $0.95d$, and using $\phi = 0.9$, the required area of bottom flexural reinforcement in the long direction is:

$$A_s \cong \frac{M_u}{\phi f_y(jd)} = \frac{6760 \text{ k-in.}}{0.9 \times 60 \text{ ksi} \times (0.95 \times 22 \text{ in.})} = 5.99 \text{ in.}^2$$

From ACI Code Table 8.6.1.1 the required minimum area of steel is:

$$A_{s,\min} = 0.0018 \times b \times h = 0.0018 \times 120 \text{ in.} \times 26 \text{ in.} = 5.62 \text{ in.}^2 \text{ (does not govern)}$$

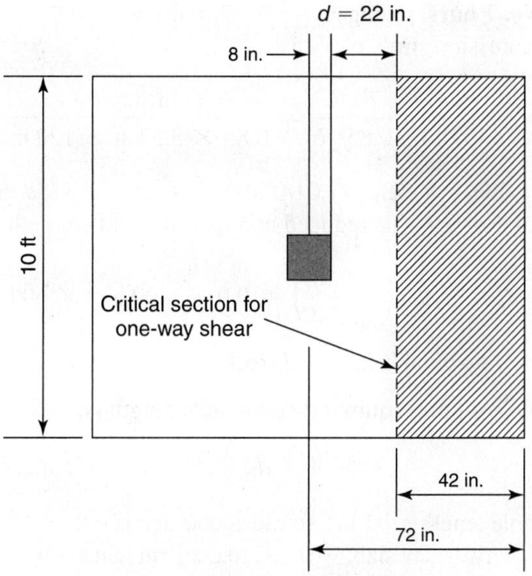

(a) Critical section for one-way shear.

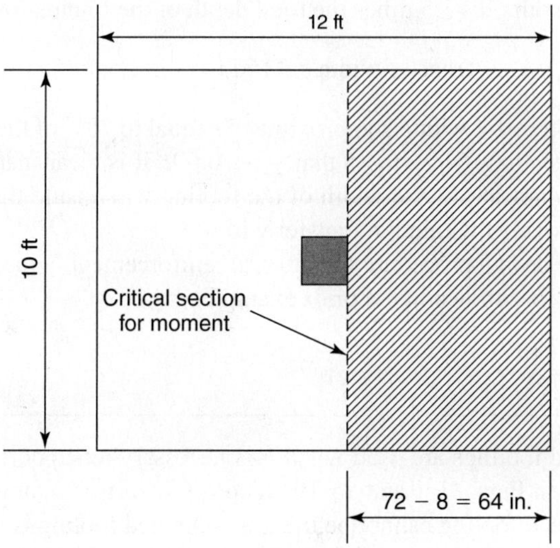

(b) Critical section for design moment.

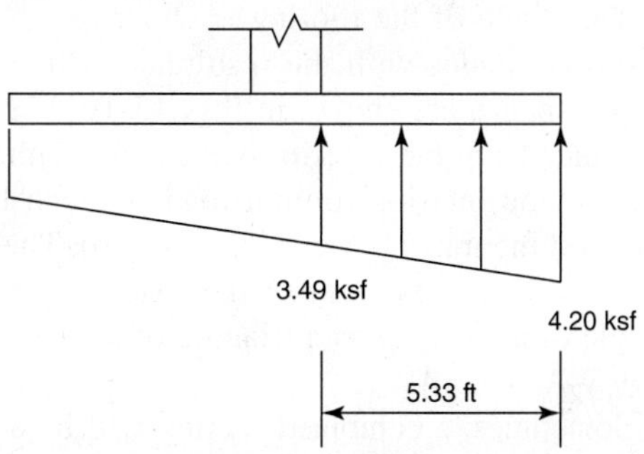

(c) Soil pressure causing moment in footing.

Fig. 15-18
Critical sections for one-way shear and bending.

Try ten No. 7 bars, giving $A_s = 10 \times 0.60\text{ in.}^2 = 6.00\text{ in.}^2$ For this steel area, the depth of the compression stress block is:

$$a = \frac{A_s f_y}{0.85 f'_c b} = \frac{6.00\text{ in.}^2 \times 60\text{ ksi}}{0.85 \times 3.5\text{ ksi} \times 120\text{ in.}} = 1.01\text{ in.}$$

With this value, $c = a/\beta_1 = 1.01/0.85 = 1.19\text{ in.} < 0.375d$. So, this is a tension-controlled section and $\phi = 0.9$. The reduced nominal moment strength is:

$$\phi M_n = \phi A_s f_y\left(d - \frac{a}{2}\right) = 0.9 \times 6.00\text{ in.}^2 \times 60\text{ ksi}(22\text{ in.} - 0.50\text{ in.})$$

$$= 6960\text{ k-in.} > M_u\text{ (o.k.)}$$

From Table A-6, the required development length is:

$$\ell_d = 54.8 \times d_b = 54.8 \times 0.875\text{ in.} = 48.0\text{ in.}$$

The available length is 61 in., so the anchorage is o.k.

As in a two-way slab, we need to confirm that enough reinforcement is placed within the transfer width near the column to resist the percentage of moment transferred to the column by flexural reinforcement, $\gamma_f \times M_u$. The size of the transfer strip is equal to the column width plus 1.5 times the total depth of the foundation on each side of the column.

$$\text{Transfer width} = 16\text{ in.} + 2(1.5 \times 26\text{ in.}) = 94.0\text{ in.}$$

Thus, the transfer width is approximately equal to 78% of the foundation width (94/120 $\approx$ 0.78). Recalling from step 5 that $\gamma_f = 0.60$, it is clear that a uniform distribution of the ten No. 7 bars across the width of the footing will ensure that the required flexural resistance is provided within the transfer width.

The design procedure for flexural reinforcement in the transverse direction would be similar to that shown in the prior example. ■

15-6 COMBINED FOOTINGS

Combined footings are used when it is necessary to support two columns on one footing, as shown in Figs. 15-1 and 15-19. When an exterior column is so close to a property line that a spread footing cannot be used, a combined footing is often used to support the edge column and an interior column. References [15-10] and [15-11] discuss the design of combined footings.

The shape of the footing is chosen such that the centroid of the area in contact with soil coincides with the resultant of the column loads supported by the footing. Common shapes are shown in Fig. 15-19. For the rectangular footing in Fig. 15-19a, the distance from the exterior end to the resultant of the loads is half the length of the footing. If the interior column load is much larger than the exterior column load, a tapered footing may be used (Fig. 15-19b). The location of the centroid can be adjusted to agree with the resultant of the loads by dividing the area into a parallelogram or rectangle of area A_1, plus a triangle of area A_2, such that $A_1 + A_2$ = required area and $\bar{y}_1 A_1 + \bar{y}_2 A_2 = y_R A$.

Sometimes, a combined footing will be designed as two isolated pads joined by a *strap* or stiff beam, as shown in Fig. 15-19c. Here, the exterior footing acts as a wall footing, cantilevering out on the two sides of the strap. The interior footing can be designed as a two-way footing. The strap is designed as a beam and may require shear reinforcement in it.

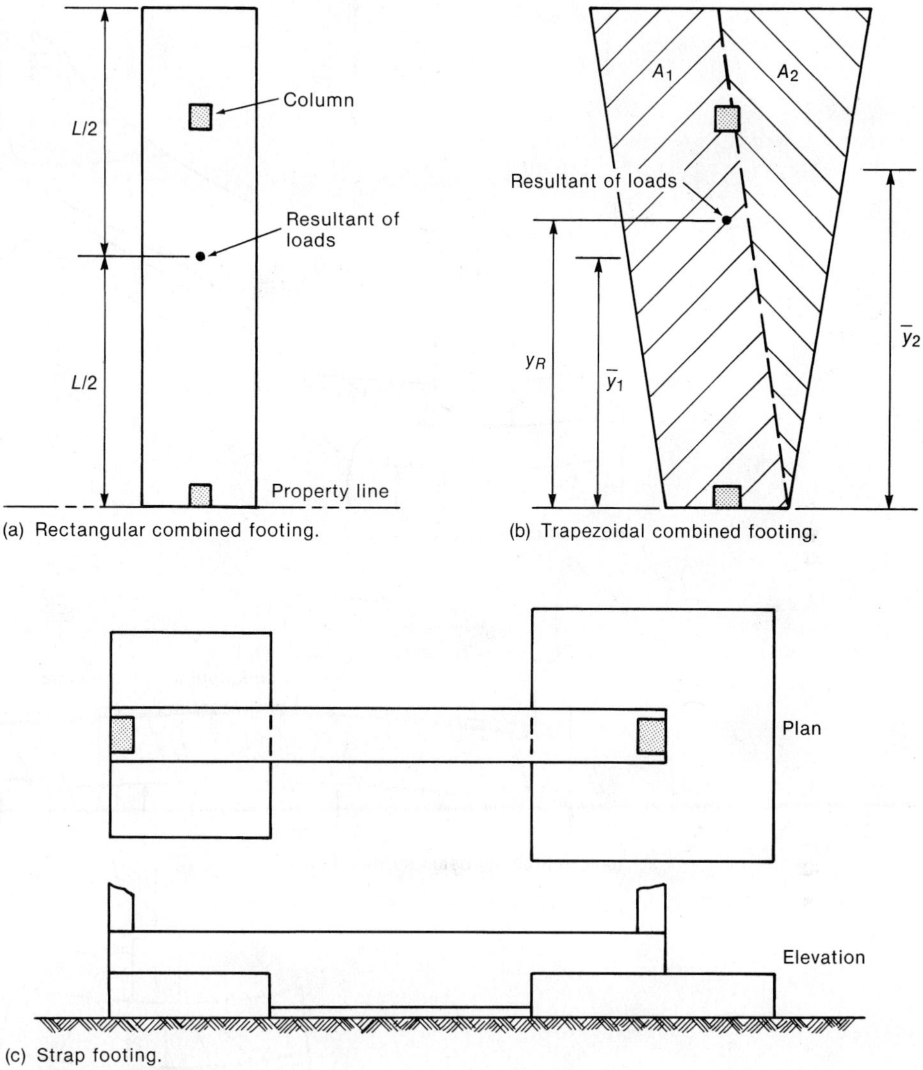

Fig. 15-19
Types of combined footings.

For design, the structural action of a combined footing is idealized as shown in Fig. 15-20a. The soil pressure is assumed to act on longitudinal beam strips, *A–B–C* in Fig. 15-20a. These transmit the load to hypothetical crossbeams, *A–D* and *B–E*, which transmit the upward soil reactions to the columns. For the column placement shown, the longitudinal beam strips would deflect as shown in Fig. 15-20b, requiring the reinforcement shown. The deflected shape and reinforcement of the crossbeams are shown in Fig. 15-20c. The crossbeams are generally assumed to extend a distance $d/2$ on each side of the columns.

EXAMPLE 15-5 Design of a Combined Footing

A combined footing supports a 24 in. × 16 in. exterior column carrying a service dead load of 200 kips and a service live load of 150 kips, plus a 24-in.-square interior column carrying service loads of 300 kips dead load and 225 kips live load. The distance between the columns is 20 ft, center to center. For shallow foundations, the allowable soil bearing

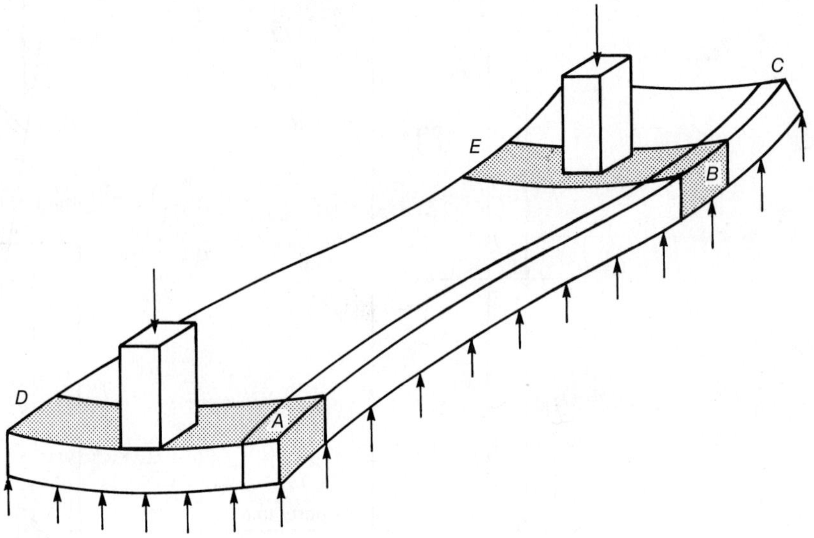

(a) Load distribution.

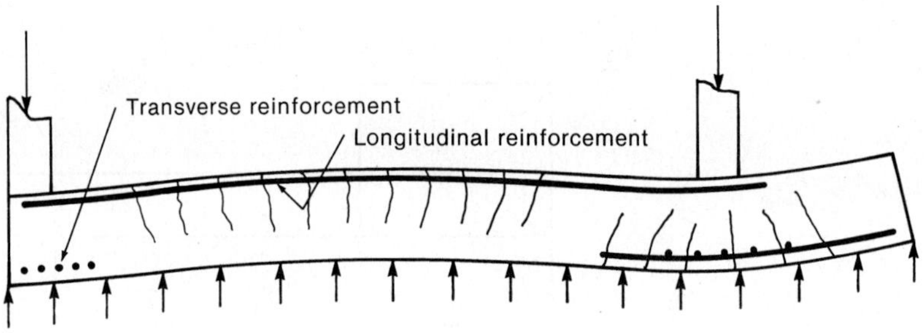

(b) Longitudinal beam strips.

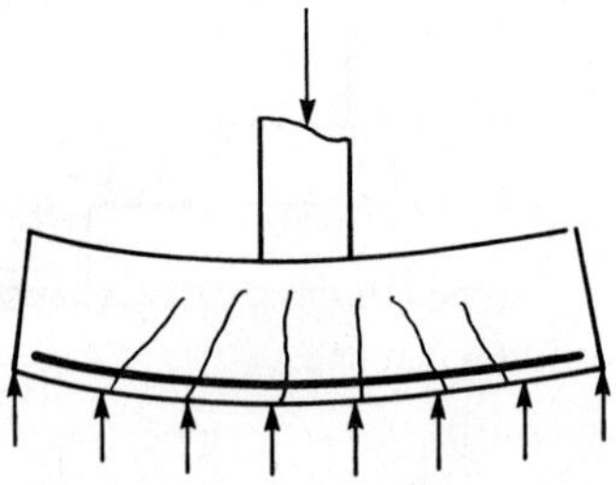

(c) Transverse beam strips.

Fig. 15-20
Structural action of combined footing.

pressure is 5000 psf. The basement floor is 5 in. thick and supports a live load of 100 psf. The density of the fill above the footing is 120 lb/ft^3. Design the footing, assuming that $f'_c = 3000$ psi (normal-weight concrete) and $f_y = 60{,}000$ psi.

Use the load and resistance factors from ACI Code Tables 5.3.1 and 21.2.1.

1. Estimate the size and the factored net pressure. Allowable net soil pressure $q_n = 5$ ksf $-$ (weight/ft^2 of the footing and the soil, the floor over the footing, and the basement floor loading). This can be calculated, as in the previous examples, by using the actual densities and thicknesses of the soil and concrete, or it can be approximated by using an average density for the soil and concrete. This will be taken as 140 lb/ft^3. Therefore,

$$q_n = 5.0 - (4 \times 0.14 + 0.10) = 4.34 \text{ ksf}$$

$$\text{Area required} = \frac{200 + 150 + 300 + 225}{4.34} = 202 \text{ ft}^2$$

The resultant of the column loads is located at

$$\frac{8\text{ in.} \times 350\text{ kips} + 248\text{ in.} \times 525\text{ kips}}{350\text{ kips} + 525\text{ kips}} = 152\text{ in.}$$

from the exterior face of the exterior column (Fig. 15-21a). To achieve uniform soil pressures, the centroid of the footing area will be located at 152 in. from exterior edge. Thus, the footing will be 304 in. = 25 ft 4 in. long. The width of the footing will be 7.96 ft—say, 8 ft.

The factored net pressure is

$$q_{nu} = \frac{1.2(200 + 300) + 1.6(150 + 225)}{25.33 \times 8} = 5.92\text{ ksf}$$

Subsequent design will be based on q_{nu}.

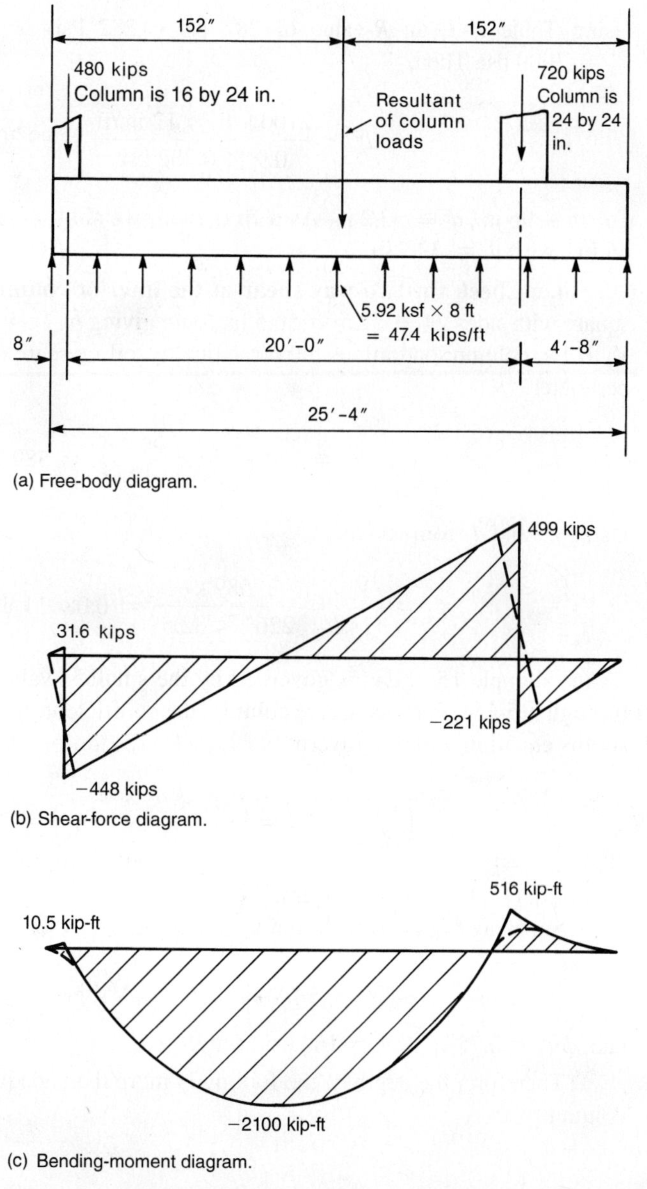

(a) Free-body diagram.

(b) Shear-force diagram.

(c) Bending-moment diagram.

Fig. 15-21 Bending-moment and shearing-force diagrams—Example 15-5.

2. Calculate the bending-moment and shear-force diagrams for the longitudinal action. The factored loads on the footing and the corresponding bending-moment and shearing-force diagrams for the footing are shown in Fig. 15-21. These are plotted for the full 8-ft width of the footing.

3. Calculate the thickness required for maximum positive moment. Because the cross section is so massive (8 ft wide by 2 to 3 ft deep), the use of more than about 0.5 percent reinforcement will lead to very large numbers of large bars. As a first trial, we shall select the depth, assuming 0.5 percent reinforcement. Because this member acts as a beam, the minimum flexural reinforcement ratio, from ACI Code Section 9.6.1.2, is $3\sqrt{f'_c}/f_y \geq 200/f_y = 0.0033$. From Eq. (5-23a), trying a width of 96 in. and $\rho = 0.005$:

$$bd^2 \geq \frac{M_u}{\phi R}$$

Using Table A-3, an R-value of 282 psi (0.282 ksi) is selected for $\rho = 0.005$ and $f'_c = 3000$ psi. Then,

$$bd^2 \geq \frac{2100 \text{ k-ft} \times 12 \text{ in./ft}}{0.9 \times 0.282 \text{ ksi}} = 99{,}300 \text{ in.}^3$$

For $b = 96$ in., $d \cong 32.2$ in. **As a first trial, we shall assume an overall thickness of 36 in., with d = 32.5 in.**

4. Check the two-way shear at the interior column. The critical perimeter is a square with sides $24 + 32.5 = 56.5$ in. long, giving $b_o = 4 \times 56.5 = 226$ in. The shear, V_u, is the column load minus the force due to soil pressure on the area within the critical perimeter:

$$V_u = 720 - 5.92\left(\frac{56.5}{12}\right)^2 = 589 \text{ kips}$$

Using b_o and d from above,

$$v_u = \frac{V_u}{b_o \cdot d} = \frac{589 \text{ k}}{226'' \times 32.5''} = 0.0802 \text{ ksi} = 80.2 \text{ psi}$$

As in Example 15-2, ϕv_c is governed by the smallest value obtained from Eqs. (15-13) through (15-15). For this square column, the coefficient, β, in Eq. (15-14) is equal to 1.0, so this equation will not govern. For Eq. (15-15), $\alpha_s = 40$, so

$$\left(\frac{\alpha_s d}{b_o} + 2\right) = \left(\frac{40 \times 32.5}{226} + 2\right) = 7.75 > 4$$

Thus, Eq. (15-15) will not govern.

So, using Eq. (15-13) to find v_c:

$$v_c = 4\lambda\sqrt{f'_c} = 4 \times 1\sqrt{3000} = 219 \text{ psi}$$

and, $\phi v_c = 0.75 \times 219 = 164$ psi $> v_u$ (o.k.).

Therefore, the depth of the footing is more than adequate for two-way shear at this column.

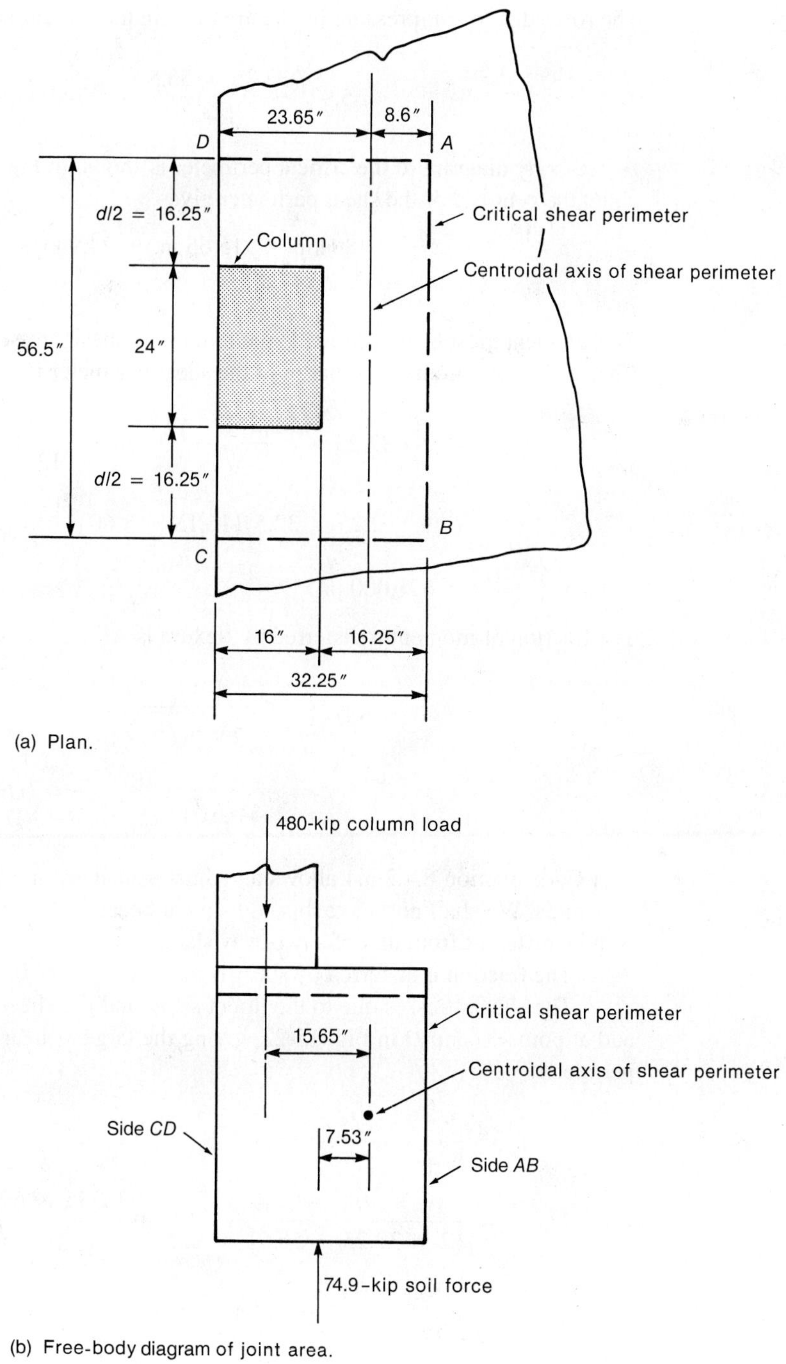

Fig. 15-22
Shear calculations at exterior column—Example 15-5.

5. Check the two-way shear at the exterior column. The shear perimeter around the exterior column is three-sided, as shown in Fig. 15-22a. The distance from line A–B to the centroid of the shear perimeter is

$$\frac{2(32.5 \times 32.25) \times 32.25/2}{32.5(2 \times 32.25 + 56.5)} = 8.60 \text{ in.}$$

The force due to soil pressure on the area within the critical perimeter is

$$5.92\left(\frac{32.25 \times 56.5}{144}\right) = 74.9 \text{ kips}$$

A free-body diagram of the critical perimeter is shown in Fig. 15-22b. Summing moments about the centroid of the shear perimeter gives

$$\begin{aligned} M_u &= 480 \text{ kips} \times 15.65 \text{ in.} - 74.9 \text{ kips} \times 7.53 \text{ in.} \\ &= 6950 \text{ kip-in.} \end{aligned}$$

This moment must be transferred to the footing by shear stresses and flexure, as explained in Chapter 13. The moment of inertia of the shear perimeter is

$$\begin{aligned} J_c = {} & 2\left[\left(32.25 \times \frac{32.5^3}{12}\right) + \left(32.5 \times \frac{32.25^3}{12}\right) \right. \\ & \left. + (32.25 \times 32.5)(16.13 - 8.60)^2\right] + (56.5 \times 32.5)8.60^2 \quad \text{(13-36)} \\ = {} & 621{,}000 \text{ in.}^4 \end{aligned}$$

The fraction of moment transferred by flexure is

$$\gamma_f = \frac{1}{1 + \frac{2}{3}\sqrt{b_1/b_2}} \quad \text{(13-27)}$$

$$= \frac{1}{1 + \frac{2}{3}\sqrt{32.25/56.5}} = 0.665$$

ACI Code Section 8.4.2.3.4 allows an adjustment in γ_f at edge columns in two-way slab structures. We shall not make this adjustment because the structural action of this footing is quite different from that of a two-way slab.

The fraction transferred by shear is $\gamma_v = 1 - \gamma_f = 0.335$.

The shear stresses due to the direct shear and the shear due to moment transfer will add at points C and D in Fig. 15-22, giving the largest shear stresses on the critical shear perimeter:

$$\begin{aligned} v_u &= \frac{V_u}{b_o d} + \frac{\gamma_v M_u c}{J_c} \\ &= \frac{480 - 74.9}{(2 \times 32.25 + 56.5) \times 32.5} + \frac{0.335 \times 6950(32.25 - 8.60)}{621{,}000} \\ &= 0.103 + 0.089 \\ &= 0.192 \text{ ksi} \end{aligned}$$

ϕv_c is computed as the smallest value from Eqs. (15-13) to (15-15): For Eq. (15-14),

$$\beta = \frac{\text{larger column section dimension}}{\text{shorter column section dimension}} = \frac{24}{16} = 1.5$$

and

$$\left(2 + \frac{4}{\beta}\right) = 4.67 > 4$$

Thus, Eq. (15-14) will not govern.

For Eq. (15-15), $\alpha_s = 30$, because this is considered to be an exterior connection. So,

$$\left(\frac{\alpha_s d}{b_o} + 2\right) = \left(\frac{30 \times 32.5}{121} + 2\right) = 10.1 > 4$$

Thus, Eq. (15-15) will not govern.

So, using Eq. (15-13) to find ϕv_c is

$$\phi v_c = \phi 4\lambda\sqrt{f'_c} = 0.75 \times 4 \times 1\sqrt{3000 \text{ psi}}$$
$$= 164 \text{ psi} = 0.164 \text{ ksi}$$

Thus, ϕv_c is less than v_u, and the selected thickness is not adequate for shear at the exterior column. It should be noted that the stress-calculation procedure used here results in finding a maximum stress at a point, as opposed to a maximum stress acting on a significant portion of the critical shear section. In such cases, the ACI Committee 421 [15-12] has suggested that some engineering judgment could be used to permit the calculated stress, v_u, to exceed the reduced nominal strength represented by ϕv_c by 10 to 15 percent. In this case, v_u does exceed ϕv_c by approximately 15 percent, but the author has decided to be conservative and increase the total thickness of the footing. Additional calculations show that an overall depth of 40 in., with $d = 36.5$ in., is required for punching shear at the exterior column. For the final choice, the total moment transferred to the footing is 7550 kip-in. of which 2520 kip-in. is transferred by shear. **Use a combined footing 25 ft 4 in. by 8 ft in plan, 3 ft 4 in. thick, with effective depth 36.5 in.**

6. One-way shear in the longitudinal direction. One-way shear is critical at d from the face of the interior column, ($d + 12$ in. from the center of the column).

$$V_u = 499 \text{ kips} - \left(\frac{12 + 36.5}{12}\right)\text{ft} \times 47.4 \text{ k/ft} = 307 \text{ kips}$$

And $V_u/\phi = 307/0.75 = 410$ kips.

The combined footing is 96 in. wide and $d = 36.5$ in., so

$$V_c = 2\lambda\sqrt{f'_c}b_w d = 2 \times 1\sqrt{3000} \times 96 \times 36.5 = 384 \text{ kips}$$

Thus, shear reinforcement is required, but probably the minimum area requirement will govern. The minimum A_v/s is:

$$\frac{A_v}{s} \geq \frac{0.75\sqrt{f'_c}}{f_y}b_w, \text{ but not less than } \frac{50}{f_y}b_w \text{ (second term governs)}$$

$$\frac{A_v}{s} \geq \frac{50 \times 96}{60{,}000} = 0.080 \text{ in.}^2\text{/in.}$$

Try No. 4, eight leg stirrups at 18 in. o.c., $A_v/s = 1.60/18 = 0.089$ in.^{2}/in.

Note: we should make a check of the A_v/s provided in the transverse direction. The spacing between the eight legs in the transverse direction is approximately 13 in. and taking an 18 in. strip, we get:

$$\frac{A_v}{s} = \frac{0.20}{13} = 0.0154 \text{ in.}^2\text{/in.} \geq \frac{50 \times 18}{60{,}000} = 0.015 \text{ in.}^2\text{/in. (o.k.)}$$

Now, checking in the longitudinal direction for $\phi V_n = \phi(V_c + V_s)$, where:

$$V_s = \frac{A_v f_{yt} d}{s} = \frac{1.60 \times 60 \times 36.5}{18} = 195 \text{ kips}$$

Then,

$$\phi(V_c + V_s) = 0.75(384 + 195) = 434 \text{ kips} \geq V_u = 307 \text{ kips (o.k.)}$$

Use No. 4, eight leg stirrups at a spacing of 18 in. o.c.

7. Design the flexural reinforcement in the longitudinal direction.

(a) Midspan (negative moment):

$$A_s = \frac{M_u}{\phi f_y jd}$$

Estimate $j = 0.95$, because ρ will be very small.

$$A_s = \frac{2100 \text{ k-ft} \times 12 \text{ in./ft/}}{0.9 \times 60 \text{ ksi} \times 0.95 \times 36.5 \text{ in.}} = 13.5 \text{ in.}^2$$

$$\text{Min. } A_s = \frac{3\sqrt{f'_c}}{f_y} bd, \text{ but not less than } \frac{200}{f_y} bd \text{ (second term governs)}$$

$$= \frac{200 \times 96 \times 36.5}{60{,}000}$$

$$= 11.7 \text{ in.}^2 \text{ (does not govern)}$$

Try eighteen No. 8 bars, $A_s = 14.2 \text{ in.}^2$.

$$\phi M_n = 0.9 \times 14.2 \times 60(36.5 - 3.49/2) = 26{,}700 \text{ kip-in.} = 2220 \text{ kip-ft}$$

This exceeds M_u. **Use eighteen No. 8 top bars at midspan.**

(b) Interior column (positive moment). The positive moment at the interior support requires that $A_s = 3.3 \text{ in.}^2$, which is significantly less than $A_{s,\min}$. **Use fifteen No. 8 bottom bars; $A_s = 11.9 \text{ in.}^2$** at the interior column.

8. Check the development of the top bars. ℓ_d for a No. 8 top bar in 3000-psi concrete (Table A-6) is

$$\ell_d = 71.2 \times d_b = 71.2 \text{ in.} = 5.93 \text{ ft}$$

In a normally loaded beam (loaded on the top surface and supported on the bottom surface), it is necessary to check ACI Code Section 9.7.3.8.4 at the positive-moment points of inflection to determine whether the rate of change of moment, and thus of bar force, exceeds the bond strength of the bottom bars. In this footing, the loading, the supports, and the shape of the moment diagram are all inverted from those found in a beam carrying gravity loads. As a result, this check is made for the top steel.

We shall extend all the bars into the column regions at both ends. At points of inflection (1.10 ft from the center of the interior column and under the exterior column), $V_u = 450$ kips, and by ACI Code Section 9.7.3.8.4,

$$\frac{M_n}{V_u} + \ell_a \geq \ell_d \tag{8-21}$$

$$\frac{M_n}{V_u} = \frac{2220/0.9}{450} = 5.48 \text{ ft}$$

where ℓ_a must be at least $5.93 - 5.48 = 0.45$ ft (5.4 in.) to satisfy Eq. (8-21).

At the exterior support, ℓ_a is the extension beyond the center of the support. This cannot exceed 5 in., which is not enough to satisfy Eq. (8-21). Therefore, the top bars will all have to be hooked at the exterior end. This is also necessary to anchor the bars transferring the unbalanced moment from the column to the footing.

At the interior column, extend the bars to the interior face (away from exterior column) of the column. This will give ℓ_a in excess of 5.4 in., therefore, o.k.

9. Check the development of the bottom bars. ℓ_d for a No. 8 bottom bar in 3000-psi concrete is

$$\ell_d = 54.8 \times d_b = 54.8 \text{ in.}$$

Extend the bottom bars to within 3 in. from the interior end of the footing. In the other direction the bottom bars must extend d past the point of inflection (approximately 4.2 ft from the column centerline). Cut off the bottom bars at 4 ft 5 in. from the centerline of the interior column toward the exterior column.

10. Design the transverse "beams." Transverse strips under each column will be assumed to transmit the load from the longitudinal beam strips into the column, as shown in Fig. 15-20a. The width of the beam strips will be assumed to extend $d/2$ on each side of the column. The actual width is unimportant, because the moments to be transferred are independent of the width of the transverse beams. Figure 15-23 shows a section through the transverse beam under the interior column. The factored load on this column is 720 kips. This is balanced by an upward net force of 720 kips/8 ft = 90 kips/ft. The maximum moment in this transverse beam is

$$M_u = \frac{90 \times 3^2}{2} = 405 \text{ kip-ft}$$

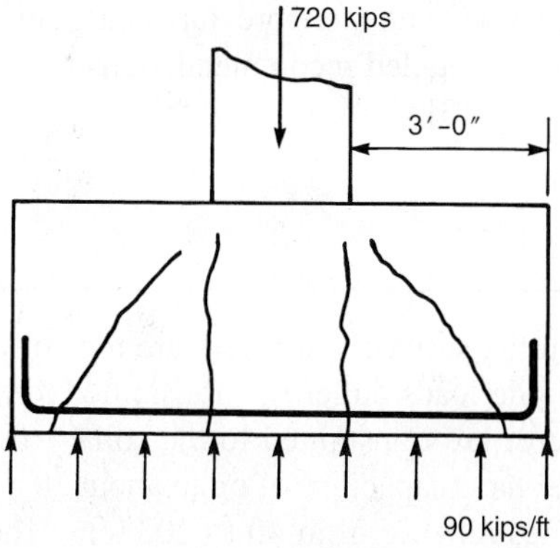

Fig. 15-23
Transverse beam at interior column—Example 15-5.

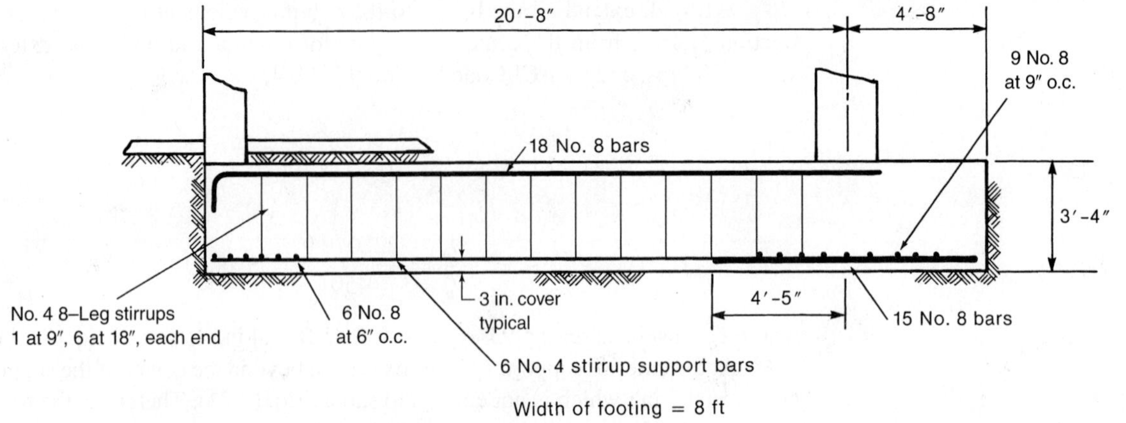

Fig. 15-24
Combined footing—Example 15-5.

Assuming that $j = 0.95$ and $d = 35.5$ in., the required A_s is

$$A_s = \frac{405 \times 12}{0.9 \times 60 \times 0.95 \times 35.5} = 2.67 \text{ in.}^2$$

$$A_{s,\min} = \frac{200}{60{,}000} \times (24 + 2 \times 35.5/2) \times 35.5 = 7.04 \text{ in.}^2 \text{ (this governs)}$$

Use nine No. 8 transverse bottom bars at the interior column; $A_s = 7.11$ **in.**^{2}. $A_{s,\min} = 3.99$ in.^{2} also controls at the exterior column. **Use six No. 8 transverse bottom bars at the exterior column.** Because two-way shear cracks would extend roughly the entire width of the footing, we shall hook the transverse bars at both ends for adequate anchorage outside the inclined cracks, as shown in Fig. 15-23.

11. Design the column-to-footing dowels. This is similar to step 7 in Example 15-2 and so will not be repeated here. The complete design is detailed in Fig. 15-24. ■

15-7 MAT FOUNDATIONS

A mat foundation supports all the columns in a building, as shown in Fig. 15-1g. A mat foundation would be used when buildings are founded on soft or irregular soils in locations where pile foundations cannot be used. Design is carried out by assuming that the foundation acts as an inverted slab. The distribution of soil pressure is affected by the relative stiffness of the soil and foundation, with more pressure being developed under the columns than at points between columns. Detailed recommendations for the design of such foundations are given in [15-10] and [15-13].

15-8 PILE CAPS

Piles may be used when the surface soil layers are too soft to properly support the loads from the structure. The pile loads either are transmitted to a stiff bearing layer some distance below the surface or are transmitted to the soil by friction along the length of the pile. Treated timber piles have capacities of up to about 30 tons. Precast concrete and steel piles often have capacities ranging from 40 to 200 tons. The most common pile caps for high-strength piles comprise groups of 2 to 6 piles, although groups of 25 to 30 piles have been used. The center-to-center pile spacing is a function of the type and capacity of the piles, but frequently is on the order of 3 ft. It is not uncommon for a pile to be several inches to a foot away from its intended location, due to problems during pile driving. The design of pile caps is discussed in References [15-14] and [15-15].

The structural action of a four-pile group is shown schematically in Fig. 15-25. The pile cap is a special case of a "deep beam" and can be idealized as a three-dimensional truss or strut-and-tie model [15-15] and [15-16], with four compression struts transferring

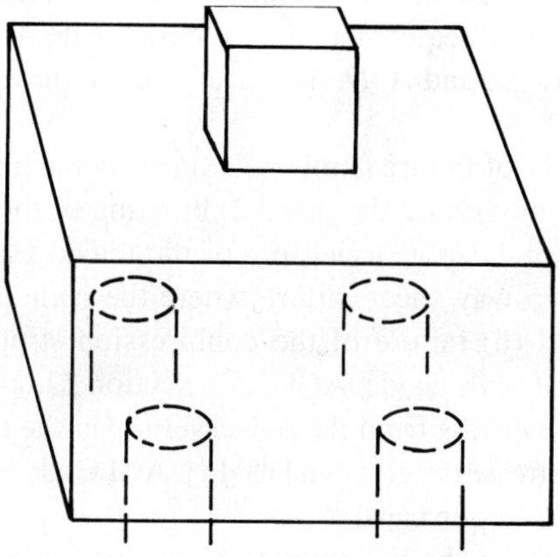

(a) Pile cap.

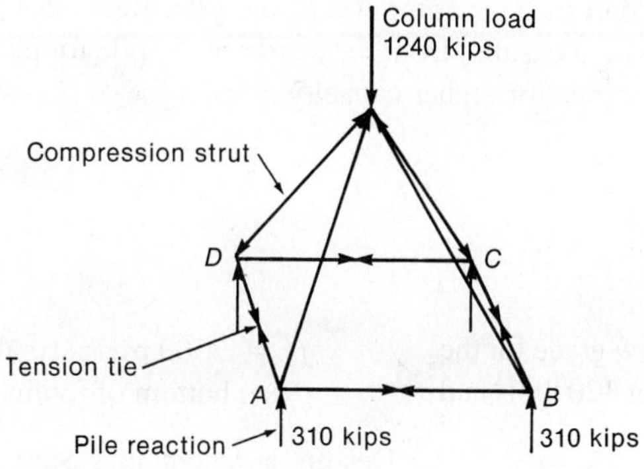

(b) Internal forces in pile cap.

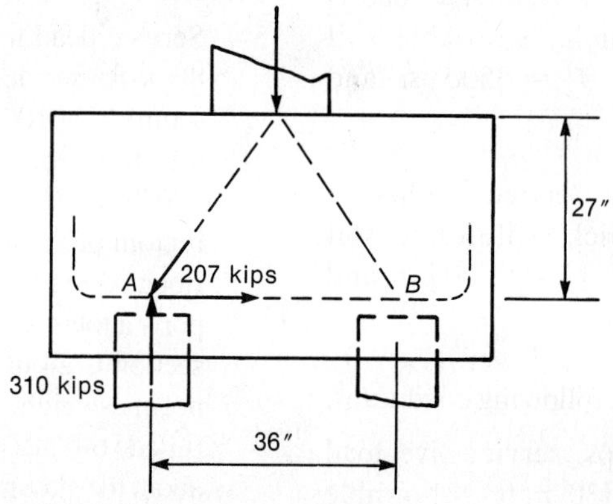

(c) Force in tie A–B

Fig. 15-25
Forces in a pile cap.

load from the column to the tops of the piles and four tension ties equilibrating the outward components of the compression thrusts (Fig. 15-25b). The tension ties have constant force in them and must be anchored for the full horizontal tie force outside the intersection of the pile and the compression strut (outside points *A* and *B* in Fig. 15-25c). Hence, the bars either must extend ℓ_d past the centerlines of the piles or must be hooked outside this point. Tests of model pile caps are reported in [15-16].

For the pile cap shown in Fig. 15-25, the total horizontal tie force in one direction can be calculated from the force triangle shown in Fig. 15-25b. The factored downward force is 1240 kips, equilibrated by a vertical force of 310 kips in each pile. Considering the joint at *A* shown in Fig. 15-25c, we find that the horizontal steel force in tie *A–B* required to equilibrate the force system is 207 kips, corresponding to $\phi A_s f_y$, which, taking $\phi = 0.75$ for shear, requires six No. 8 bars for tie *A–B* and the same number for each of the ties *B–C*, *C–D*, and *A–D*. The use of strut-and-tie models in design is discussed in Chapter 17.

The modes of failure (limit states) for such a pile cap include (1) concrete crushing under the column or over the pile, (2) bursting of the side cover where the pile transfers its load to the pile cap, (3) yielding of the tension tie, (4) anchorage failure of the tension tie, (5) two-way shear failure where the cone of material inside the piles punches downward, and (6) failure of the compression struts. Least understood of these is the two-way shear-failure mode. ACI Code Section 22.6.4.1 is not strictly applicable, because the failure cone differs from the one assumed in the code. Some guidance in selecting allowable shear stresses is given in [15-14]. ACI Code Section 13.4.2 also gives guidance on the shear design of pile caps.

In some cases, the minimum thickness of a pile cap is governed by the development length of the dowels from the pile cap into the column. Also, to prevent bursting of the side cover due to load transfer at the top of the pile, Reference [15-14] recommends an edge distance of 15 in. measured from the center of the pile for plies up to 100-ton capacity, and 21-in. edge distance for higher capacity piles.

PROBLEMS

Design wall footings to be supported 3 ft below grade for the following conditions. Assume a soil density of 120 lb/ft^3 and normal-weight concrete for both problems.

15-1 Service dead load is 6 kips/ft, service live load is 8 kips/ft. Wall is 12 in. thick. Allowable soil pressure, q_a, is 4000 psf. $f'_c = 3500$ psi and $f_y = 60{,}000$ psi.

15-2 Service dead load is 12 kips/ft, service live load is 8 kips/ft. Wall is 16 in. thick. Allowable soil pressure, q_a, is 6000 psf. $f'_c = 3500$ psi and $f_y = 60{,}000$ psi.

Design square spread footings for the following conditions.

15-3 Service dead load is 350 kips, service live load is 275 kips. Soil density is 130 lb/ft^3. Allowable soil pressure is 4500 psf. Column is 18 in. square. $f'_c = 3500$ psi (normal weight), and $f_y = 60{,}000$ psi. Place bottom of footing at 5 ft below floor level.

Design a rectangular spread footing for the following conditions:

15-4 Service dead loads are: axial = 350 kips, moment = 80 k-ft; service live loads are: axial = 250 kips, moment = 100 k-ft. The moments are acting about the strong axis of the column section. Soil density is 120 lb/ft^3. Allowable soil pressure is 5500 psf. Place the bottom of the foundation at 4.5 ft below the basement floor. Assume the basement floor is 6 in. thick and supports a total service load of 80 psf. Assume the column section is 24 in. × 16 in., is constructed with 5000-psi normal-weight concrete and contains six No. 8 bars (Grade 60) placed to give maximum bending resistance about the strong axis of the column section. Assume the footing will use $f'_c = 3500$ psi and $f_y = 60$ ksi.

REFERENCES

15-1 International Code Council, *2012 International Building Code*, Washington, D.C., 2012.

15-2 *AASHTO LRFD Bridge Design Specifications*, Sixth Edition with 2013 Interim Revisions, American Association of State Highway and Transportation Officials, Washington, D.C., 2013.

15-3 *Minimum Design Loads for Buildings and Other Structures* (ASCE/SEI 7-10), American Society of Civil Engineers, Reston, VA, 2010.

15-4 Arthur N. Talbot, *Reinforced Concrete Wall Footings and Column Footings, Bulletin 67*, University of Illinois Engineering Experiment Station, Urbana, IL., 1913, 96 pp.

15-5 Frank E. Richart, "Reinforced Concrete Wall and Column Footings," *ACI Journal, Proceedings*, Vol. 45, October 1948, pp. 97–128; November 1948, pp. 327–360.

15-6 ACI-ASCE Committee 326, "Shear and Diagonal Tension," Part 3, "Slabs and Footings," *ACI Journal, Proceedings*, Vol. 59, No. 3, March 1962, pp. 353–396.

15-7 Z. P. Bazant and J.-K. Kim, "Size Effect in Shear Failure of Longitudinally Reinforced Beams," *ACI Journal, Proceedings*, Vol. 81, No. 5, May 1984, pp. 128–141.

15-8 Michael P. Collins and Dan Kuchma, "How Safe are our Large, Lightly Reinforced Concrete Beams, Slabs, and Footings?" *ACI Structural Journal, Proceedings*, Vol. 96, No. 4, July–August 1999, pp. 482–490.

15-9 A. Uzel, "Shear Design of Large Footings," *University of Toronto Thesis Collection*, Faculty Advisor–M.P. Collins, Department of Civil Engineering, CIENG 2003, 180 pp.

15-10 ACI Committee 336, "Suggested Analysis and Design Procedures for Combined Footings and Mats, ACI 336.2R-88," Reapproved 2002, *ACI Manual of Concrete Practice*, American Concrete Institute, Farmington Hills, MI.

15-11 F. Kramrisch and P. Rogers, "Simplified Design of Combined Footings," *Proceedings ASCE, Journal of the Soil Mechanics Division*, Vol. 87, No. SM5, October 1961, pp. 19–44.

15-12 ACI Committee 421, "Shear Reinforcement for Slabs," (ACI 421.1R), *ACI Structural Journal*, Vol. 89, No 5, September–October 1992, pp. 587–589.

15-13 *Design and Performance of Mat Foundations—State of the Art Review*, SP-152, American Concrete Institute, Farmington Hills, MI, 1995.

15-14 *CRSI Design Handbook*, 11th Edition, Concrete Reinforcing Steel Institute, Schaumberg, IL, 2012.

15-15 Gary J. Klein, "Example 9—Pile Cap," *Examples for the Design of Structural Concrete with Strut-and-Tie Models*, SP-208, American Concrete Institute, Farmington Hills, MI, 2002, pp. 213–223.

15-16 Perry Adebar, Daniel Kuchma, and Michael P. Collins, "Strut-and-Tie Models for the Design of Pile Caps: An Experimental Study," *ACI Structural Journal*, Vol. 87, No. 1, January–February 1990, pp. 81–92.

16

Shear Friction, Horizontal Shear Transfer, and Composite Concrete Beams

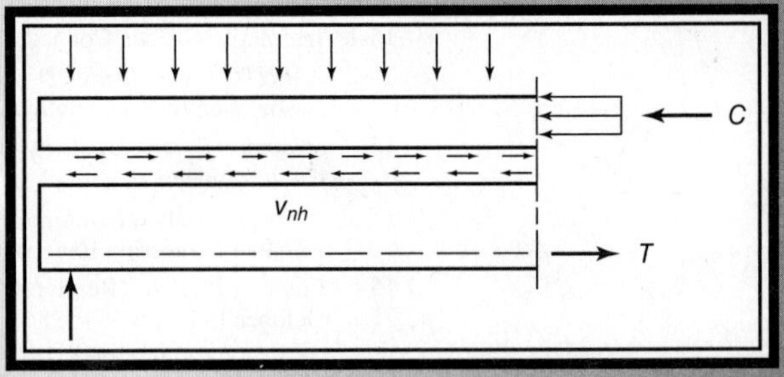

16-1 INTRODUCTION

This chapter considers the shear strength of interfaces between members or parts of members that can slip relative to one another. Among other cases, this includes the interface between a beam and a slab cast later than the beam, but expected to act in a composite manner.

16-2 SHEAR FRICTION

From time to time, shear must be transferred across an interface between two members that can slip relative to one another. The shear-carrying mechanism is known variously as *aggregate interlock*, *interface shear transfer*, or *shear friction*. The last of these terms is used here. The interface on which the shears act is referred to as the *shear plane* or *slip plane*.

Three methods of computing shear transfer strengths have been proposed in the literature. These include: (a) *Shear-friction models*, (b) *Cohesion plus friction models*, and (c) *Horizontal shear models* as occur in composite beams.

Behavior in Shear-Friction Tests

Several types of test specimens have been used to study shear transfer. The most common are the so-called *push-off specimens* similar to that shown in Fig. 16-1, which were tested in one of three ways:

1. Mattock [16-1], [16-2] and his associates reported tests of push-off specimens in which lateral expansion or contraction and crack widths were not held constant or otherwise controlled during the tests except by the reinforcement perpendicular to the slip plane. These specimens were tested either *as constructed* (uncracked) or *precracked* along the shear plane shown by the vertical dashed line in Fig. 16-1. Only the failure load and the amount of transverse reinforcement were reported.

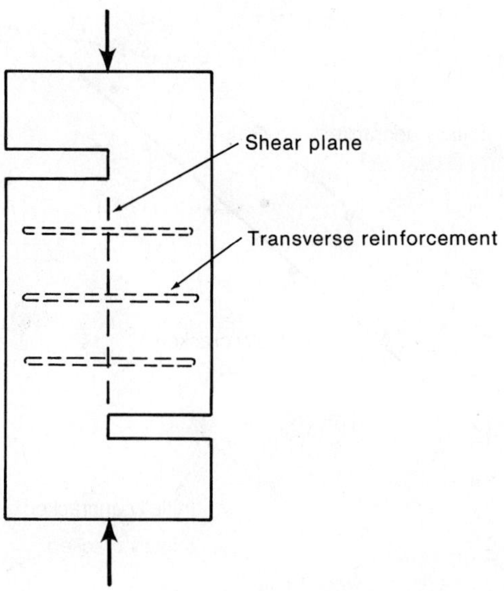

Fig. 16-1
Shear-transfer specimen.

2. Walraven [16-3], [16-4] tested push-off specimens but held the crack width constant during the tests. Loads and slip were reported.

3. A third type of test, reported by Loov and Patnaik [16-5] and Hanson [16-6], measured the shear transferred between the web and the slab in a composite beam. This is referred to as *horizontal shear*.

An excellent review of tests and equations for shear transfer is presented by Ali and White [16-7].

Cohesion and Friction

The results of 13 tests of uncracked specimens, and 21 tests of specimens with previously cracked shear planes are plotted in Fig. 16-2, with open circles and solid circles, respectively. The test results from [16-1] and [16-2] plotted in Fig. 16-2 suggest that:

(a) The strengths of cracked and uncracked specimens can be represented by equations of the form

$$v_n = c + \sigma \tan \theta \qquad (16\text{-}1a)$$

$$v_n = c + \mu\sigma \qquad (16\text{-}1b)$$

where c is a *cohesion-like* term equal to the intercepts of the sloping lines on the vertical axis in Fig. 16-2, plus a *friction-like* term $\sigma \tan \theta$ that is equal to the product of σ, the compressive stress on the shear plane, and the coefficient of friction, μ. The plots in Fig. 16-2, can be idealized as straight sloping lines that intercept the vertical axis at "cohesions" of $c \approx 505$ and $c \approx 255$ psi, for the uncracked and precracked specimens, respectively, and a coefficient of friction of about 0.95. The curves terminate at a ceiling of about 1300 psi.

(b) The initial strengths of the uncracked specimens were higher than those of the precracked specimens. The uncracked and precracked specimens reached similar upper limits on shear transfer as shown in Fig. 16-2.

In Fig. 16-2, the shear stress at failure, v_u, is plotted against $\rho_v f_y$, where ρ_v is the ratio of the area of the transverse reinforcement across the shear plane to the area of the

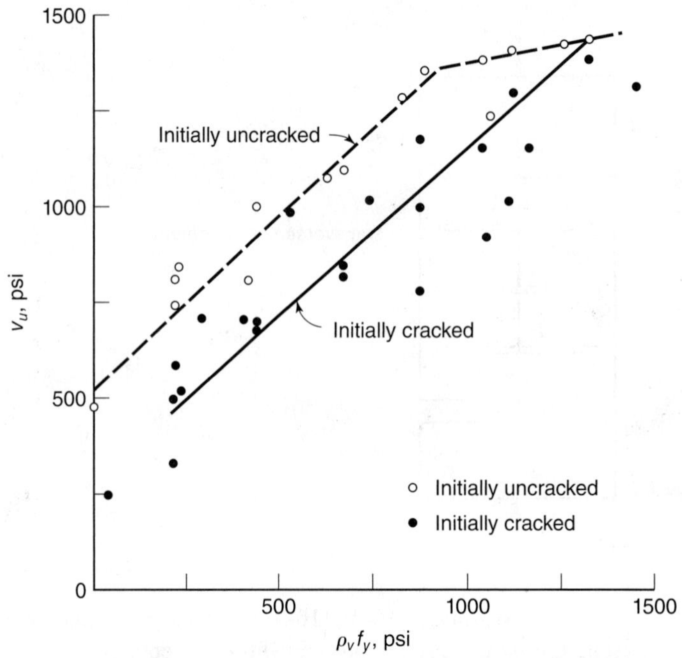

Fig. 16-2
Variation shear strength v_u with web reinforcement ratio $\rho_v f_y$. (From [16-1] and [16-2].)

shear plane. The crack widths, w, were not held constant or otherwise controlled during these test series. The shear stress at failure increases with the value of $\rho_v f_y$, which is taken as a measure of the clamping force due to the tension in the reinforcement crossing the shear plane. In Mattock's tests of initially uncracked push-off tests, the first cracks were a series of diagonal tension cracks, as shown in Fig. 16-3. With further deformation, the compression struts between these cracks rotated at their ends (points A and B) such that point B moved downward relative to A. At the same time, the distance AC, measured across the crack, increased, stretching the transverse reinforcement. The tension in the reinforcement was equilibrated by an increase in the compression in the struts. Failure occurred when the bars yielded or the struts were crushed.

When a shear is applied to an initially cracked specimen or to a specimen formed by placing a layer of concrete on top of or against an existing layer of hardened concrete,

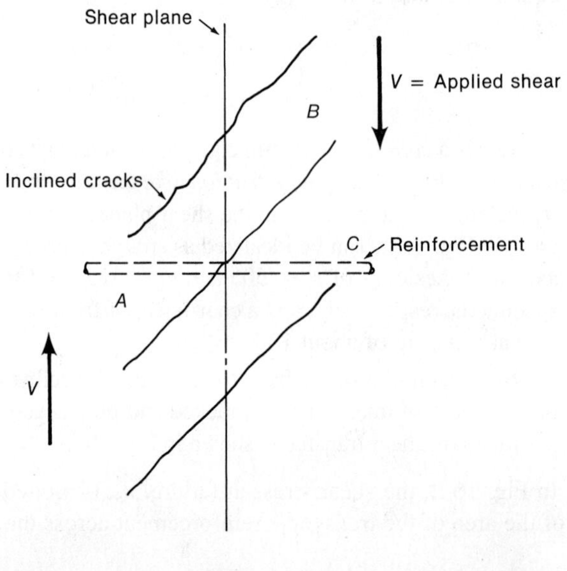

Fig. 16-3
Diagonal tension cracking along a previously uncracked shear plane. (From [16-8].)

relative slip of the layers causes a separation of the surfaces, as shown in Fig. 16-4a. If the reinforcement across the crack is developed on both sides of the crack, it is elongated by the separation of the surfaces and hence is stressed in tension. For equilibrium, a compressive stress is needed as shown in Fig. 16-4b. Shear is transmitted across the crack by (a) friction resulting from the compressive stresses [16-4], [16-5], and (b) interlock of aggregate roughness on the cracked surfaces, combined with dowel action of the reinforcement crossing the surface.

Shear-Friction Model

The original and simplest design model is the *shear-friction model* [16-1], [16-8], shown in Fig. 16-4, which ignores the cohesion-like component, c, and assumes that the shear transfer is due entirely to friction. This model is the basis of the shear-friction design procedure in ACI Code Section 22.9. Because the cohesion component is ignored, unusually high values of the coefficient of friction must be used to fit the test data. Design is based on the shear-friction equation, given in terms of forces as

$$V_n = A_{vf} f_y \mu \qquad (16\text{-}2)$$

(ACI Eq. 22.9.4.2)

or in terms of stresses,

$$v_n = \rho_v f_y \mu \qquad (16\text{-}3)$$

where ρ_v is defined above and μ is the appropriate value of the *coefficient of friction* given in ACI Code Table 22.9.4.2.

Permanent Compression Force

A permanent compressive force, N_u, perpendicular to the slip plane, causes a normal stress on the slip plane in the concrete. If N_u is compressive, the normal stress N_u/A_{cv} adds to the compressive stress on the concrete due to the reinforcement. If tensile forces,

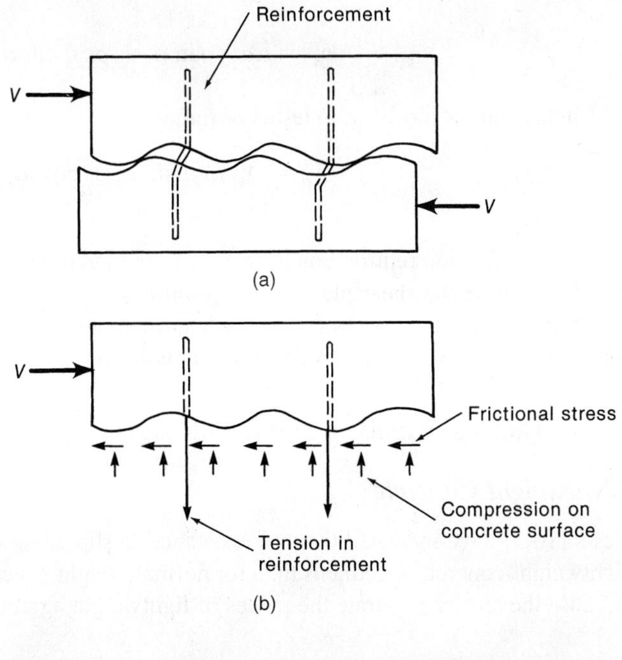

Fig. 16-4
Shear-friction model.
(From [16-6].)

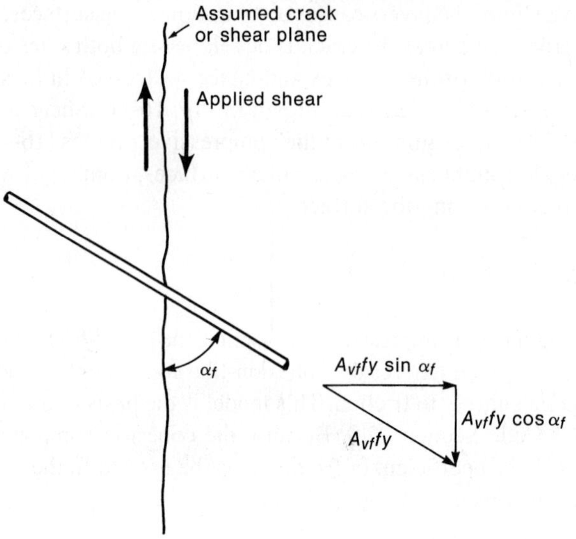

Fig. 16-5
Force components in a bar inclined to the shear plane. (From [16-1].)

N_u, act on the shear plane, they must be equilibrated by tensile reinforcement provided in addition to the shear-friction reinforcement

$$A_n = \frac{N_u}{\phi f_y} \tag{16-4}$$

Inclined Shear-Friction Reinforcement

When the shear-friction reinforcement is inclined to the slip plane such that slip along this plane produces tensile stresses in the reinforcement, as shown in Fig. 16-5, the component of bar force perpendicular to the slip plane causes a compressive force on the plane equal to $A_{vf} f_y \sin \alpha_f$. In addition, there is a force component parallel to the crack equal to $A_{vf} f_y \cos \alpha_f$.

In stress units, the nominal shear-friction resistance at failure from Eq. (16-3) is

$$v_n = \mu \rho_v f_y \sin \alpha_f + \rho_v f_y \cos \alpha_f \tag{16-5}$$

which has units of stress or, in terms of forces,

$$V_n = A_{vf} f_y (\mu \sin \alpha_f + \cos \alpha_f) \tag{16-6}$$

(ACI Eq. 22.9.4.3)

Shear-friction reinforcement that is inclined to the shear plane, such that the anticipated slip along the shear plane cause compression in the inclined bars, is not desirable. This compression tends to force the crack surfaces apart, leading to a decrease in the shear that can be transferred. Such steel is not included as shear-friction reinforcement.

Other Factors Affecting Shear Transfer

Lightweight Concrete

Tests [16-2] have indicated that the resistance to slip along the shear plane is smaller for lightweight-concrete specimens than for normal-weight concrete specimens. This happens because the cracks penetrate the pieces of lightweight aggregate rather than following the perimeters of the aggregate. The resulting crack faces are smoother than for normal-weight concrete [16-2]. This effect has been accounted for by multiplying the coefficient of friction, μ, by the correction factor for lightweight concrete, λ, given in ACI Table 22.9.4.2. Originally, λ was introduced in the ACI Code to account for the smaller tensile strength of lightweight aggregate concrete. Here, it is used for convenience to account for the smoother crack surfaces that occur in lightweight concrete.

Coefficients of Friction

ACI Code Table 22.9.4.2 gives coefficients of friction, μ, as follows:

1. for concrete placed monolithically: 1.4λ

2. for concrete placed against hardened concrete with surface intentionally roughened, as specified in ACI Code Table 22.9.4.2(b): 1.0λ

3. for concrete placed against hardened concrete not intentionally roughened: 0.6λ

4. for concrete anchored to as-rolled structural steel by studs or by reinforcing bars [see ACI Code Table 22.9.4.2(d)]: 0.7λ

where $\lambda = 1.0$ for normal-weight concrete and 0.75 for all-lightweight concrete. The code allows λ to be determined based on volumetric proportions of lightweight and normal-weight aggregates, as specified in ACI Code Section 19.2.4.2, but it shall not exceed 0.85.

The first term in Eq. (16-1) represents the portion of the shear transferred by surface protrusions and by dowel action. Equation (16-5) with $\alpha_f = 90°$, as plotted in Fig. 16-6, applies only for $\rho_v f_y$ greater than 200 psi, so the radial lines in Fig. 16-6 are not plotted below this value. For Grade-60 reinforcement, this limit requires a minimum reinforcement ratio, $\rho = 0.0033$.

Design Rules in the ACI Code

ACI Code Section 22.9 presents design rules for cases "where it is appropriate to consider shear transfer across a given plane, such as an existing or a potential crack, an interface between dissimilar materials, or an interface between two concretes cast at different times." Typical examples are shown in Fig. 16-7.

In design, a crack is assumed to exist along the shear plane, and reinforcement is provided across that crack. The amount of reinforcement is computed (ACI Code Section 22.9.4) from

$$\phi V_n \geq V_u \tag{16-7}$$

$$V_n = A_{vf} f_y \mu \tag{16-2}$$

(ACI Eq. 22.9.4.2)

where $\phi = 0.75$ for designs carried out by using the load factors in ACI Code Table 5.3.1 and where μ is the coefficient of friction, depending on the surfaces in contact.

In cases 2 and 3 defined previously, the surface must be clean and free of *laitance* (a weak layer on the top surface of a concrete placement due to bleed water collecting at the surface). ACI Code Table 22.9.4.2(d) requires that, in case 4, the steel must be clean and free of paint. Case 2 applies to concrete placed against hardened concrete that has been roughened to a "full amplitude" (wave height) of approximately $\frac{1}{4}$ in. but does not specify a "wavelength" for the roughened surface. This was done to allow some freedom in satisfying

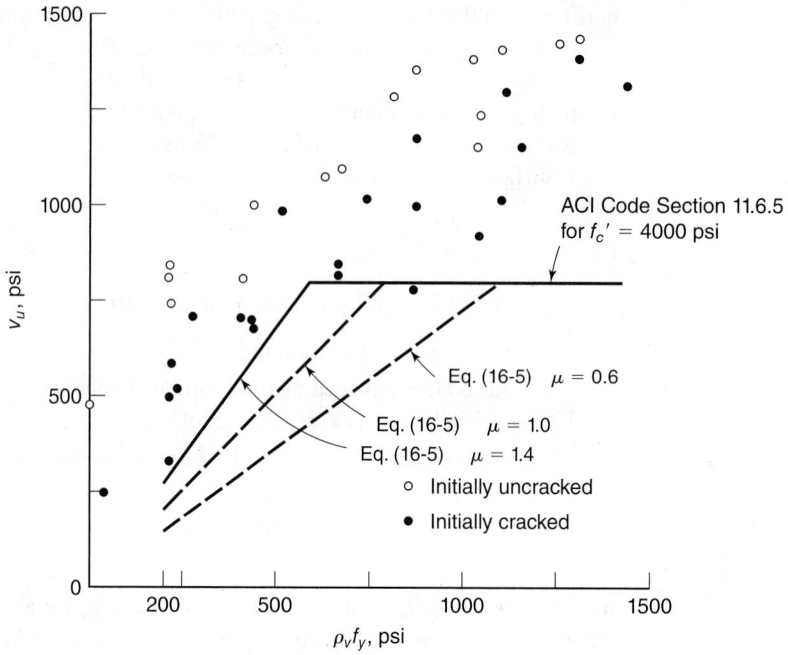

Fig. 16-6
Comparison of Eq. (16-2) with test data from [16-1], [16-2], and [16-6].

this requirement. It was intended, however, that the wavelength be on the same order of magnitude as the full amplitude, say $\frac{1}{4}$ to $\frac{3}{4}$ in.

Upper Limit on Shear Friction

For normal-weight concrete either cast monolithically or placed against hardened concrete with intentionally roughened surfaces, as noted previously, ACI Code Table 22.9.4.4 sets the upper limit on V_n from Eqs. (16-2) and (16-6) to the smallest of $0.2f_c'A_c$, $(480\text{ psi} + 0.08f_c')A_c$ and $(1600\text{ psi})A_c$, where A_c is the area of the concrete section resisting shear transfer. The limits represented by the second two expressions given here are larger than previously permitted limits in the ACI Code and represent a reexamination of test data as presented in references [16-10] and [16-11]. For all other cases, including lightweight concrete and concrete placed against not intentionally roughened surfaces, V_n shall not exceed the smaller of $0.2f_c'A_c$ and $(800\text{ psi})A_c$. In cases where a lower-strength concrete is cast against a higher-strength concrete, or vice versa, the value of f_c' for the lower-strength concrete shall be used to evaluate the limits given here. These upper limits on shear-friction strength are necessary, because Eqs. (16-2) and (16-6) may become unconservative in some cases.

Cohesion-plus-Friction Model

For the usual case of transverse reinforcement perpendicular to the shear plane, Mattock [16-2] rewrote Eq. (16-2) as

$$V_n = K_1 A_{cv} + 0.8 A_{vf} f_y \tag{16-8}$$

where $K_1 = 400$ psi for normal-weight concrete, 200 psi for all-lightweight concrete, and 250 psi for sand-lightweight concrete. The first term on the right-hand side of Eq. (16-8) represents the shear transferred by "cohesion," which is caused by pieces of aggregate bearing on the surfaces of the slip plane, by the shearing off of surface protrusions, and

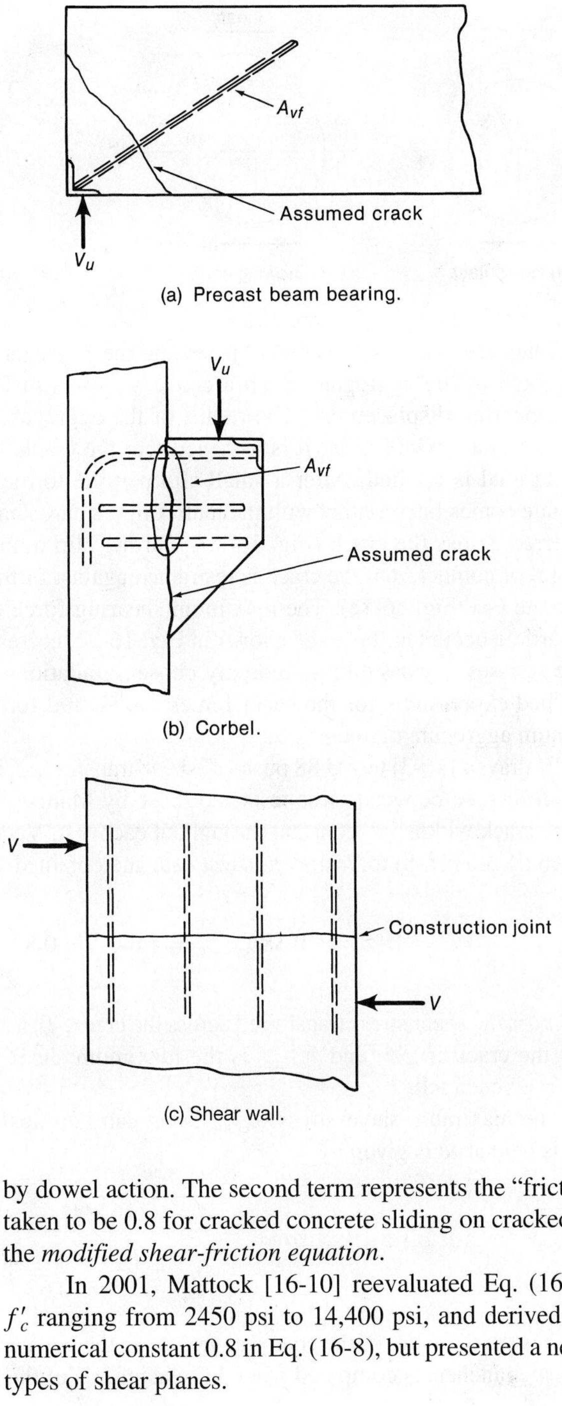

Fig. 16-7
Examples of shear friction.
(From [16-9].)

by dowel action. The second term represents the "friction," with the coefficient of friction taken to be 0.8 for cracked concrete sliding on cracked concrete. Equation (16-8) is called the *modified shear-friction equation.*

In 2001, Mattock [16-10] reevaluated Eq. (16-8) on the basis of 199 tests, with f'_c ranging from 2450 psi to 14,400 psi, and derived a set of equations that retained the numerical constant 0.8 in Eq. (16-8), but presented a new family of values of K_1 for various types of shear planes.

Walraven Model

Walraven [16-3] idealized concrete as a series of size-graded spherical pieces of coarse aggregate embedded in a matrix of hardened concrete paste. A crack was assumed to cross the matrix between pieces of aggregate until it reached a piece of aggregate, at which time it followed the aggregate–matrix interface around the aggregate. The strength

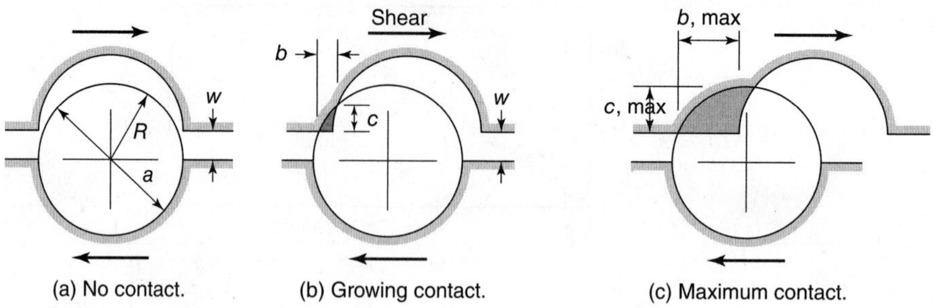

Fig. 16-8
Walraven crack model.
(From [16-3].)

of the interface between the cement paste and the aggregate was assumed to be less than the strength of the aggregate. Such a crack is shown in Fig. 16-8a, after cracking, but before shearing displacement. The radius of the aggregate particle is R, the diameter is a, and the crack width is w. It is assumed that the crack width is held constant while a shearing load is applied. After a small slip parallel to the crack, the spherical piece of aggregate comes into contact with the matrix in the dark shaded area, allowing shear to be transferred across the crack (Fig. 16-8b). Further slip mobilizes rigid–plastic stresses at the point of contact, and the crack is restrained against further slip until the crack surfaces deteriorate (see Fig. 16-8c). The maximum shearing force transferred by a single aggregate particle occurs at the stage shown in Fig. 16-8c, corresponding to the largest area of plastic stresses. By assuming randomly chosen gradations of aggregate sizes, Walraven developed expressions for the shear forces transferred for various crack widths, w, and maximum aggregate diameters, a.

Walraven [16-3] tested 88 push-off shear-transfer specimens similar to Fig. 16-1. The major difference between these tests and those by Mattock and others was that Walraven kept the crack widths, w, constant throughout each test. Vecchio and Collins [16-12] fitted Eqs. (16-5) and (16-8) to Walraven's test data and obtained

$$v_{ci} = 0.18 v_{ci\,\max} + 1.64 f_{ci} - 0.82 \frac{f_{ci}^2}{v_{ci\,\max}} \tag{16-9}$$

where v_{ci} is the shear stress transferred across the crack, f_{ci} is the compression stress required across the crack in psi, and $v_{ci\,\max}$ is the maximum shear stress that can be transmitted across a given crack.

The maximum shear stress, $v_{ci\,\max}$, that can be transferred across a crack when its width is held at w is given by

$$v_{ci\,\max} = \frac{2.16\sqrt{f'_c}}{0.3 + \left(\dfrac{24w}{a + 0.63}\right)} \tag{16-10}$$

where a is the diameter of the coarse aggregate in the cracked concrete in inches. The crack width, w in inches, is computed from the spacing, s_θ, of the inclined crack as

$$w = \varepsilon_1 s_\theta \tag{16-11}$$

where ε_1 is the principal tensile strain, which is assumed to act perpendicular to the crack, and s_θ is the spacing of the cracks, measured perpendicular to the cracks. Equation (16-11) assumes that all the strain is concentrated in the crack.

Walraven's tests were limited to concrete strengths, f'_c, from 2900 to 8200 psi. For high-strength concretes with cylinder strengths in excess of 10,000 psi, the cracks tend

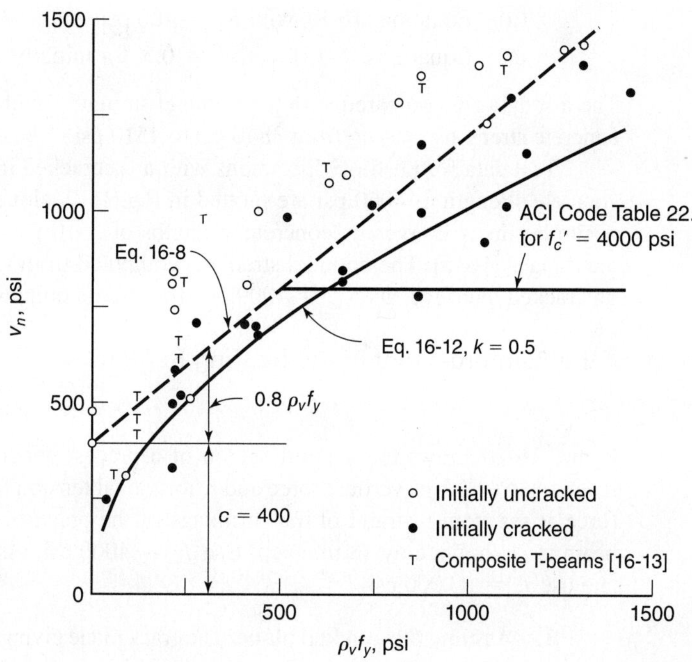

Fig. 16-9
Comparison of test results from [16-1] and [16-5] and Eqs. (16-8) and (16-12).

to cross the individual pieces of aggregate rather than going around them. As a result, the crack surface was smoother than for weaker concretes. For this case, the effective size of the aggregate, *a*, decreases, approaching zero. Angelakos, Bentz, and Collins [16-13] handled this by arbitrarily reducing the effective aggregate size, *a*, in Eq. (16-10) from the nominal diameter, *a*, to zero as the concrete strength increases from 8500 psi to 10,000 psi. For concrete strengths of 10,000 psi or higher, they took *a* equal to zero.

Ali and White [16-7] carried out similar analyses, assuming an undulating crack path, to illustrate the force transfer that develops when the two sides of the crack come into bearing.

Loov and Patnaik—Composite Beams

From tests of composite beams, Loov and Patnaik [16-5] derived the following equation for shear transfer across cracks and for horizontal shear in composite beams (see Fig. 16-11, discussed later):

$$v_{ci\,\max} = \lambda k\sqrt{\sigma f'_c} + \rho_v f_y \cos \alpha_f \tag{16-12}$$

In this equation, λ accounts for lightweight concrete and k is a constant equal to 0.5 for concrete placed against hardened concrete and 0.6 for concrete placed monolithically. The square-root term allows this equation to fit the test data more closely than do other models. Equation (16-12) is plotted in Fig. 16-9 for comparison with test results and with Eq. (16-8).

Because the reinforcement is assumed to yield in order to develop the necessary forces, the yield strength of the steel is limited to 60,000 psi. Each bar must be anchored on both sides of the crack to develop the bar. The steel must be placed approximately uniformly across the shear plane, so that all parts of the crack are clamped together.

Comparison of Design Rules with Test Results

In Fig. 16-9, test data for push-off tests of initially uncracked specimens with reinforcement perpendicular to the shear plane [16-1] are compared with

(a) Equation (16-8), with $K_1 = 400$ psi, and
(b) Equation (16-12), with $k = 0.5$, for initially cracked concrete

The test data are compared with the nominal strengths, with $\phi = 1.0$. The specimens had concrete strengths ranging from 3840 psi to 4510 psi.

Test data for push-off specimens with a precracked interface [16-1] and an average concrete strength of 4060 psi are plotted in Fig. 16-9, along with data from tests of composite beams with average concrete strengths of 5710 psi for the webs and 5160 psi for the flanges [16–5]. The nominal strengths computed from Eq. (16-12), with $k = 0.5$ for a precracked interface, and $f'_c = 4000$ psi, fit the data quite well.

EXAMPLE 16-1 Design of the Reinforcement in the Bearing Region of a Precast Beam

Figure 16-10 shows the support region of a precast concrete beam. The factored beam reactions are 62 kips vertical force and a horizontal tension force of 12 kips. The horizontal force arises from restraint of the shrinkage of the precast beam. The cross section of the beam is 12 in. wide by 18 in. deep. Use $f'_c = 4000$ psi, assume normal-density concrete, and use $f_y = 60{,}000$ psi.

1. Assume the cracked plane. The crack plane giving the maximum area, A_c, is a vertical crack. This will tend to overestimate the cohesion component. Assume that the support region of the beam is enclosed by a 6-in. × 6-in. structural steel angle, as shown in Fig. 16-10, and assume that the crack is at 60° to the horizontal. It intercepts the end of the beam at 10.2 in. above the bottom and has a length of 12 in. We shall take $A_c = 12 \text{ in.} \times 12 \text{ in.} = 144 \text{ in.}^2$.

2. Compute the area of steel required. Resolving the forces onto the inclined plane gives a normal force of 20.6 kips compression and a shear force of 59.7 kips for a 60° plane. Each assumed crack angle will result in a different combination of normal and shear forces. However, it is quick and conservative to assume that the shear force is equal to the vertical reaction of 62 kips and that the normal force is equal to the horizontal reaction $N_u = -12$ kips (negative in tension). In addition, we shall assume that $\alpha_f = 90°$.

From Eq. (16-2),

$$V_n = A_{vf} f_y \, \mu$$

We want $V_n \geq \phi V_u$, so

$$A_{vf} \geq \frac{V_u}{\phi f_y \, \mu}$$

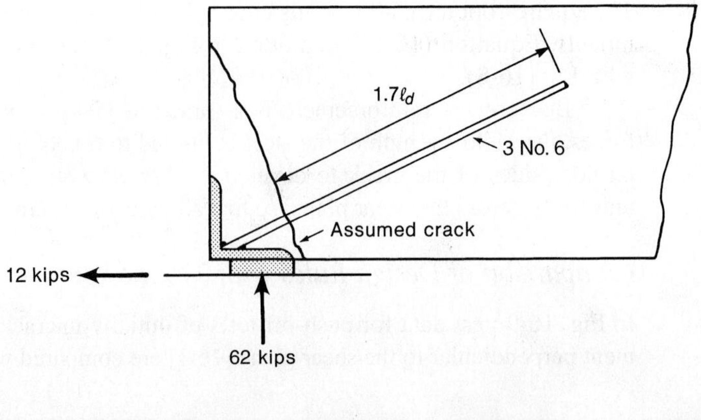

Fig. 16-10
Example 16-1.

where $V_u = 62$ kips, $\mu = 1.4\lambda$ (the crack plane is in monolithically placed concrete), and $\lambda = 1.0$ (normal-weight concrete is used). We thus have

$$A_{vf} = \frac{62 \text{ kips}}{0.75 \times 60 \text{ ksi} \times 1.4} = 0.984 \text{ in.}^2$$

We must confirm that the required value of V_n, V_u/ϕ, does not exceed the upper limit given in ACI Code Table 22.9.4.4. The required value of V_n is

$$V_n = V_u/\phi = 62 \text{ kips}/0.75 = 82.7 \text{ kips}$$

The value for the stress acting on the effective concrete area resisting shear transfer, A_c, is limited in ACI Code Table 22.9.4.4 to the smallest of

$$0.2f'_c = 0.2 \times 4000 \text{ psi} = 800 \text{ psi}$$

$$480 \text{ psi} + 0.08f'_c = 480 + 0.08 \times 4000 = 800 \text{ psi}$$

or

$$1600 \text{ psi}$$

Thus, the limiting stress is 800 psi, and the upper limit on V_n is

$$V_n \leq (800 \text{ psi})A_c = 800 \text{ psi} \times 144 \text{ in.}^2 = 115 \text{ kips}$$

The required value for V_n does not exceed this value, so we do not need to increase the effective section area resisting the factored shear force.

The tensile force must also be transferred across the crack by reinforcement, as calculated using Eq. (16-4),

$$A_n = \frac{12 \text{ kips}}{0.75 \times 60 \text{ ksi}} = 0.267 \text{ in.}^2$$

Therefore, the total steel across the crack must be $(0.984 + 0.267) \text{ in.}^2 = 1.25 \text{ in.}^2$

Provide three No. 6 bars across the assumed crack, $A_s = 1.32$ in.^{2}. These bars must be anchored on both sides of the crack. This is done by welding them to the bearing angle and by extending them $1.7\ell_d$, as recommended in [16-14]. ■

16-3 COMPOSITE CONCRETE BEAMS

Frequently, precast beams or steel beams have a slab cast on top of them and are designed assuming that the slab and beam act as a monolithic unit to support loads. Such a beam-and-slab combination is referred to as a *composite beam*. This discussion will deal only with composite beams where the beam is precast concrete or other concrete cast at an earlier time than the slab.

Shored or Unshored Construction

When the slab concrete is placed, the precast beam can either be shored or unshored. In typical shored construction, the precast beam is placed and must support its own weight. Shores are then added to support the beam and initially resist the weight of the cast-in-place slab. When the strength of the slab concrete is high enough to resist expected stresses, the shores are removed and the slab dead load is resisted by the composite beam and slab. If the precast beam is not shored when the slab is placed, the beam supports its own weight plus the weight of the slab and the slab forms. ACI Code Section 4.12.3 allows either construction process and requires that each element be strong enough to support all loads it supports by itself. If the beam is shored, ACI Code Section 26.11.2.1(c) requires that the shores be left in place until the composite section has a strength adequate to support all loads and to limit deflections and cracking.

Tests have shown that the ultimate strength of a composite beam is the same whether the member was shored or unshored during construction. For this reason, ACI Code Sections 22.3.3.2 and 22.5.4.2 allow strength computations to be made that consider only the final composite member.

Horizontal Shear

In the beam shown in Fig. 16-11a, there are no horizontal shear stresses transferred from the slab to the beam. They act as two independent members. In Fig. 16-11b, horizontal shear stresses act on the interface, and as a result, the slab and beam act in a composite manner. The ACI Code provisions for horizontal shear are given in ACI Code Section 16.4. Although the mechanism of horizontal shear transfer and that of shear friction are similar, if not identical, there is a considerable difference between the two sets of provisions. The difference results from the fact that Eq. (16-8) and ACI Code Sections 16.4.3 and 22.9 are all empirical attempts to fit test data. Equation (16-8) is valid for relatively short shear planes with lengths up to several feet, but is believed to give shear strengths that are too high for long shear-transfer regions if the maximum stress is localized. It is also unconservative for low values of $\rho_v f_y$, as used in Eq. (16-3).

The term *horizontal shear stress* is used to describe the shear stresses acting on the interface in Fig. 16-11b. In a normal composite beam, these stresses are horizontal. If the beam were vertical, however, the term *horizontal shear* would still be used to distinguish between this shear and the orientation of shear stresses in a beam. Tests of horizontal shear in composite beams are reported in [16-3], [16-5], [16-6], [16-15], and [16-16]. The tests reported in [16-6] included members with and without shear keys along the interface. The presence of shear keys stiffened the connection at low slips but had no significant effect on its strength.

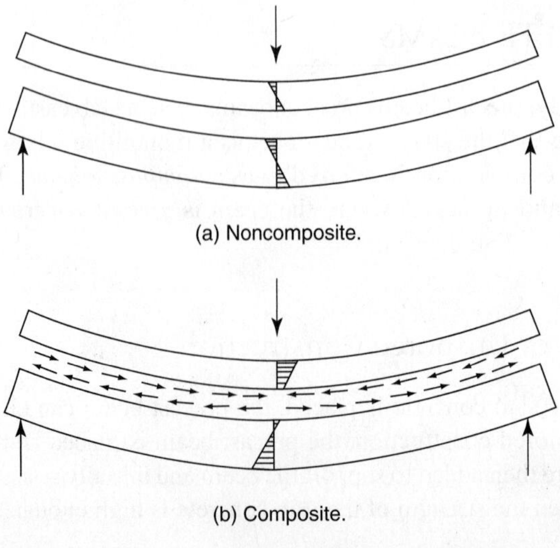

Fig. 16-11
Horizontal shear transfer in a composite beam.

Computation of Horizontal Shear Stress

From strength of materials, the horizontal shear stresses, v_h, on the contact surface between an uncracked elastic precast beam and a slab can be computed from

$$v_{uh} = \frac{VQ}{I_c b_v} \tag{16-13}$$

where

V = shear force acting on the section in question

Q = first moment of the area of the slab or flange about the centroidal axis of the composite section

I_c = moment of inertia of the composite section

b_v = width of the interface between the precast beam and the cast-in-place slab

Equation (16-13) applies to uncracked elastic beams and is only an approximation for cracked concrete beams.

The ACI Code gives two ways of calculating the horizontal shear stress.

ACI Code Section 16.4.3.1 defines the nominal horizontal shear force, V_{nh}, to be transferred as

$$\phi V_{nh} \geq V_u \tag{16-14}$$

Setting $\phi V_{nh} = V_u$ and using expression for shear stress from Chapter 6 gives

$$v_{nh} = \frac{V_u/\phi}{b_v d} \tag{16-15a}$$

This is based on the observation that the shear stresses on opposite pairs of sides of an element located at the top of the web are equal in magnitude, as shown in Fig. 6-3a, but are arranged to give couples in opposite directions. For an element taken from directly over the beam web at the interface between the web and flange, the shear stresses on the top and bottom sides of the element are v_{nh}, and the shear stresses on the vertical sides of the element are

$$v_{nv} = \frac{V_u/\phi}{b_v d} \tag{16-15b}$$

where V_u is the factored shear force acting on the cross section of the beam as obtained from a shear-force diagram for the beam. If the shear stresses on the left and right faces of the element form a counter-clockwise couple, those on the bottom and top of the element must form a clockwise couple. For equilibrium, the shear stresses on four sides of an element must be equal in magnitude. Thus,

$$v_{nh} = v_{nv} \tag{16-16}$$

Calculation of Horizontal Shear Stresses from C and T Forces

Alternatively, ACI Code Section 16.4.5 allows horizontal shear to be computed from the change in compressive or tensile force in the slab in a segment of any length. Figure 16-12 illustrates this clause. At midspan, the force in the compression zone is C, as shown in

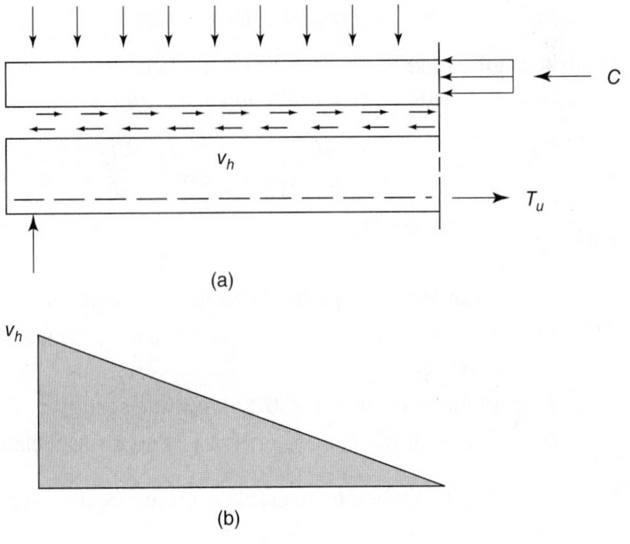

Fig. 16-12
Horizontal shear stresses in a composite beam.

Fig. 16-12a. All of this force acts above the interface. At the end of the beam, the force in the flange is zero. Thus, the horizontal shear force to be transferred across the interface between midspan and the support is

$$V_h = C \tag{16-17}$$

A similar derivation could be carried out by considering the force in the reinforcement,

$$V_{uh} = T_u \tag{16-18}$$

where T_u is the tensile force in the reinforcement related to M_u.

ACI Code Section 16.4.5.2 says that when ties are provided to resist the horizontal shear calculated using Eqs. (16-17) or (16-18), the spacing of the ties should approximately reflect the distribution of shear forces along the member. This is specified because the slip of the slab relative to the web at the onset of a horizontal shear failure is too small to allow much redistribution of horizontal shear stresses. In [16-5], the measured slip at maximum horizontal shear stress was on the order of 0.02 in., and the slip at failure ranged from 0.08 to 0.3 in. We shall calculate the horizontal shear stresses by dividing the beam into a series of segments and computing the value of C_u or T_u at both ends of each segment. Then,

$$v_{uh} = \frac{T_{u1} - T_{u2}}{b_v \ell_s} \tag{16-19}$$

where T_{u1} and T_{u2} are the larger and smaller tension forces on the two ends of a segment, and ℓ_s is the length of the segment. The tension force can be taken as $T_u = M_u/jd$. If jd is assumed to be constant, T_u varies as the M_u diagram. For a uniformly loaded simple beam, the M_u diagram is a parabola, and the moments at the $\frac{1}{4}$ and $\frac{1}{8}$ points of the span are 0.75 and 0.438 times the maximum moment, respectively. These fractions of the moment diagram can be calculated from the shear-force diagram because the change in moment from one section to another is equal to the area of the shear-force diagram between the two sections.

The two procedures give similar results, as will be seen in Example 16-2. The limits on V_{nh} (V_{uh}/ϕ) from ACI Code Table 16.4.4.2 are given in Table 16-1.

TABLE 16-1 Calculation of V_{nh}

Contact Surfaces	Ties	V_{nh}
Intentionally roughened	None	$80b_v d$
Not roughened	Minimum ties	$80b_v d$
Intentionally roughened	$A_v f_{yt}$	$\left(260 + \dfrac{0.6A_v f_{yt}}{b_v s}\right)\lambda b_v d$ but not more than $500b_v d$

In composite beams, the contact surfaces must be clean and free from laitance. The words *intentionally roughened* imply that the surface has been roughened with a "full amplitude" of $\frac{1}{4}$ in., as discussed in connection with shear friction. When the factored shear force, $V_u = \phi V_{nh}$, at the section exceeds $\phi(500b_v d)$ psi, ACI Code Section 16.4.4.1 requires design be based on shear friction, in accordance with ACI Code Section 22.9. This limit reflects the range of test data used to derive ACI Code Table 16.4.4.2.

ACI Code Section 16.4.6 requires that the ties provided for horizontal shear be not less than the minimum stirrups required for shear, given by

$$A_v = \frac{0.75\sqrt{f'_c}\,b_w s}{f_{yt}} \text{ and } \geq \frac{50\,b_w s}{f_{yt}} \tag{16-20}$$

The tie spacing shall not exceed four times the least dimension of the supported element, which is usually the thickness of the slab, but not more than 24 in. The ties must be fully anchored both in the beam stem and in the slab.

Deflections

The beam cross section considered when calculating deflections depends on whether the beam was shored or unshored when the composite slab is placed. If it is shored so that the full dead load of both the precast beam and the slab is carried by the composite section, ACI Code Section 24.2.5.1 allows the designer to consider the loads to be carried by the full composite section when computing deflections. The modulus of elasticity should be based on the strength of the concrete in the compression zone, while the modulus of rupture should be based on the strength of the concrete in the tension zone. For nonprestressed beams constructed with shores, it is not necessary to check deflections if the overall height of the composite section satisfies ACI Code Table 9.3.1.1.

ACI Code Section 24.2.5.2 covers the calculation of deflections for unshored construction of nonprestressed beams. If the thickness of the precast member satisfies ACI Table 9.3.1.1, it is not necessary to consider deflections. If the thickness of the composite section satisfies the table, but the thickness of the precast member does not, it is not necessary to compute deflections occurring after the section becomes composite, but it is necessary to compute the instantaneous deflections and that part of the sustained load deflections occurring prior to the beginning of effective composite action. The latter can be assumed to occur when the modulus of elasticity of the slab reaches 70 to 80 percent of its 28-day value, usually about 4 to 7 days after the slab is placed.

ACI Code Section 24.2.5.3 states that if deflections are computed, they should account for the curvatures induced by the differential shrinkage between the slab and the precast beam. Shrinkage of the slab relative to the beam causes the slab to shorten relative to the beam. Because the slab and beam are joined together, this relative shortening causes the beam to deflect downward, adding to the deflections due to loads. Some of the shrinkage of the concrete in the beam will have occurred before the beam is erected in the structure. All of the slab shrinkage occurs after the slab is cast. As the slab shrinks relative to the beam, tensile stresses are induced in the slab and compressive stresses in the beam. These are redistributed to some degree by creep of the concrete in the slab and beam. This effect can be modeled by using an *age-adjusted effective modulus*, E_{caa}, and an *age-adjusted transformed section* in the calculations, as discussed in Section 3-6 and in [16-17] and [16-18].

EXAMPLE 16-2 Design of a Composite Beam

Precast, simply supported beams that span 24 ft and are spaced 10 ft on centers are composite with a slab that supports an unfactored live load of 100 psf, a partition load of 20 psf, and a superimposed dead load of 10 psf. Design the beams and the composite beam and slab. Use $f'_c = 3000$ psi for the slab, 5000 psi for the precast beams, and $f_y = 60{,}000$ psi. Use load factors from ACI Code Table 5.3.1.

1. Select the trial dimensions. For the end span of the slab, ACI Code Table 9.3.1.1 gives the minimum thickness of a one-way slab as $h = \ell/24 = 120\text{ in.}/24 = 5$ in. For a simply supported beam, the table gives $h = \ell/16 = (24 \times 12)/16 = 18$ in. Deflections of the composite beam may be a problem if the overall depth is less than 18 in. However, for unshored construction, ACI Code Section 24.2.5.2 requires that deflections of the precast member be considered if its overall depth is less than that given by Table 9.3.1.1. To avoid this, we shall try an 18-in.-deep precast beam 12 in. wide to allow the steel to be in one layer, plus a 5-in. slab. For the precast beam, $d = 18 - 2.5 = 15.5$ in.

2. Compute the factored loads on the precast beam. Because the floor will be constructed in an unshored fashion, the precast beam must support its own dead load, the dead load of the slab, the weight of the forms for the slab, assumed to be 10 psf, and some construction live load, assumed to be 50 psf. Thus, we have the following data:

Dead loads:

$$\text{Beam stem } w = \frac{12 \times 18}{144} \times 0.150\text{ kcf} = 0.225\text{ kip/ft}$$

$$\text{Slab } w = 5\text{ in.}/12 \times 10\text{ ft} \times 0.150\text{ kcf} = 0.625\text{ kip/ft}$$

$$\text{Forms } w = 10\text{ ft} \times 0.010\text{ ksf} = 0.100\text{ kip/ft}$$

$$\text{Total} = 0.950\text{ kip/ft}$$

Live load:

$$w = 10\text{ ft} \times 0.050\text{ ksf} = 0.500\text{ kip/ft}$$

The ACI Code does not specifically address the load factors for this construction-load case. We shall take $U = 1.2D + 1.6L$, giving

$$w_u = 1.2 \times 0.950 + 1.6 \times 0.500 = 1.94\text{ kips/ft}$$

3. Compute the size of the precast member required for flexure.

$$M_u = \frac{1.94 \times 24^2}{8} = 140\text{ kip-ft}$$

We shall select a steel percentage close to, but less than, the tension-controlled limit in the precast beam, so that $\phi = 0.9$, and later check whether that provides enough steel for flexure in the composite section. From Table A-3 for 5000 psi, the highest value of ρ corresponding to tension-controlled behavior is 0.021. We shall try $\rho = 0.02$. For this value of ρ, $R = 1030$ psi $= 1.03$ ksi, and from Eq. (5-23a),

$$bd^2 \geq \frac{M_u}{\phi R} = \frac{140 \text{ k-ft} \times 12 \text{ in./ft}}{0.9 \times 1.03 \text{ ksi}} = 1810 \text{ in.}^3$$

For $b = 12$ in., this requires that $d = 12.3$ in. and $h = 12.3 + 2.5 = 14.8$ in. Therefore, the size chosen for deflection control is also adequate for strength.

For the 12-by-18-in. beam, $d = 18 - 2.5 = 15.5$, and $jd \cong 0.9d$, so

$$A_s = \frac{M_u}{\phi f_y jd} = \frac{140 \times 12}{0.9 \times 60 \times 0.9 \times 15.5}$$
$$= 2.23 \text{ in.}^2$$

Try a 12-by-18-in. precast beam with three No. 9 bars; $A_s = 3.00$ in.^{2}. Then

$$a = \frac{3.00 \times 60}{0.85 \times 5 \times 12} = 3.53 \text{ in.}$$
$$c = a/\beta_1 = 3.53/0.8 = 4.41 \text{ in.}$$

This is less than 3/8 d, which is the tension-controlled limit. Hence, $\phi = 0.9$, and we have

$$\phi M_n = \phi A_s f_y \left(d - \frac{a}{2} \right) = 185 \text{ k-ft}$$

Therefore, o.k.

4. **Check the capacity of the composite member in flexure.** The factored loads on the composite member are as follows:

Dead load:

$$\text{Precast stem } w = 0.225 \text{ kip/ft}$$
$$\text{Slab } w = 0.625 \text{ kip/ft}$$
$$\text{Superimposed dead load } w = 0.100 \text{ kip/ft}$$
$$\text{Total } w = 0.950 \text{ kip/ft} \qquad \text{Factored} = 1.2 \times 0.950 = 1.14 \text{ kips/ft}$$

Live load:

$$\text{Floor load } w = 10 \text{ ft} \times 0.100 \text{ ksf} = 1.0 \text{ kip/ft}$$
$$\text{Partitions } w = 10 \text{ ft} \times 0.020 \text{ ksf} = 0.2 \text{ kip/ft}$$
$$\text{Total } w = 1.20 \text{ kips/ft} \qquad \text{Factored} = 1.6 \times 1.20 = 1.92 \text{ kips/ft}$$
$$\text{Total factored load} = 3.06 \text{ kips/ft}$$

$$M_u = \frac{3.06 \times 24^2}{8} = 220 \text{ kip-ft}$$

Compute ϕM_n. From ACI Code Table 6.3.2.1 (assuming $\ell = \ell_n$ for simply supported beam), the effective flange width is 84 in. The overall height is 18 + 5 = 23 in.; $d = 23 - 2.5 = 20.5$ in. Assuming rectangular beam action, where f'_c in the slab is 3000 psi, we have

$$a = \frac{3.00 \times 60}{0.85 \times 3 \times 84} = 0.84 \text{ in.}$$

Because a is less than the flange thickness, rectangular beam action exists. Also, c/d is much less than 3/8 for the tension-controlled limit; thus, $\phi = 0.9$ for flexure, and we have

$$\phi M_n = \phi A_s f_y\left(d - \frac{a}{2}\right)$$
$$= 271 \text{ kip-ft}$$

Therefore, the steel chosen is adequate to resist the moments acting on the composite section. **Use a 12-by-18-in. precast section with three No. 9 longitudinal bars and a 5-in. cast-in-place slab.**

5. Check vertical shear. V_u at d from the support = 3.06 kips/ft × (12 − 20.5/12) ft = 31.5 kips.

Using the smaller of the two concrete strengths in Eq. (6-8b),

$$V_c = 2\lambda\sqrt{f'_c} = 2 \times 1 \times \frac{\sqrt{3000}}{1000} \times 12 \times 20.5 = 26.9 \text{ kips}$$

and using $\phi = 0.75$ for shear,

$$\phi V_c = 0.75 \times 26.9 = 20.2 \text{ kips}$$

Because $V_u > \phi V_c$, we need stirrups. Thus,

$$V_s = \frac{V_u}{\phi} - V_c = \frac{31.5}{0.75} - 26.9$$
$$= 15.1 \text{ kips at } d \text{ from the support}$$

Try Grade-60, No. 3 U stirrups, and let $A_v = 0.22$ in.^{2}. Then

$$s = \frac{A_v f_{yt} d}{V_s} = \frac{0.22 \times 60 \times 20.5}{15.1} \quad \text{(from Eq. 6-14)}$$
$$= 17.9 \text{ in. on centers}$$

$$\text{Maximum spacing} = d/2 = 10.25 \text{ in., say, 10 in.}$$

$$\text{Minimum stirrups (50 psi governs): } s = \frac{A_v f_{yt}}{50 b_w} = \frac{0.22 \times 60{,}000}{50 \times 12} \quad \text{(from Eq. 6-21)}$$
$$= 22.0 \text{ in. on centers}$$

We will select the stirrups after considering horizontal shear.

6. Compute the horizontal shear. Horizontal shear may be computed according to ACI Code Section 16.4.3 or Section 16.4.5. We shall do the calculations both ways and compare the results. We shall assume that the interface is clean, free of laitance, and intentionally roughened.

ACI Code Section 16.4.3: From Eq. (16-14), $\phi V_{nh} \geq V_u = 31.5$ kips at d from the support.

From ACI Code Table 16.4.4.2, an intentionally roughened surface without ties is adequate for

$$\phi V_{nh} = \phi 80 b_v d = 0.75 \times 80 \times 12 \times 20.5 = 14.8 \text{ kips}$$

Therefore, ties are required.

From ACI Code Table 16.4.4.2, if minimum ties are provided according to ACI Code Section 16.4.6 and the interface is intentionally roughened,

$$\phi V_{nh} = \phi(260 + 0.6\rho_v f_{yt})\lambda b_v d$$

The maximum tie spacing allowed by ACI Code Section 16.4.7.2 is 4 × 5 in. = 20 in. but not more than 24 in. The maximum spacing for vertical shear governs. Assume that the ties are No. 3 two-leg stirrups at the maximum spacing allowed for shear, 10 in. Then

$$\rho_v = \frac{A_v}{b_v s} = \frac{0.22}{12 \times 10}$$
$$= 0.00183$$

and

$$\phi V_{nh} = 0.75(260 + 0.6 \times 0.00183 \times 60{,}000) \times 1.0 \times 12 \times 20.5$$
$$= 60.1 \text{ kips}$$

Thus, minimum stirrups provide more than enough ties for horizontal shear. We shall use closed stirrups to better anchor them into the top slab. **Use No. 3, Grade-60 closed stirrups at 10 in. on centers throughout the length of the beam.**

ACI Code Section 16.4.5: From Eq. (16-19),

$$v_{uh} = \frac{T_{u1} - T_{u2}}{b_v \ell_s}$$

Consider the section of the flange between the midspan and the quarter point of the span. At midspan, $T_{u1} = M_u/jd = M_u/0.9d = 143$ kips. The distance from the support to the quarter point is $24/4 = 6$ ft. The moment at the quarter point is

$$M_u = Rx - wx^2/2 = (3.06 \text{ kip/ft} \times 12 \text{ ft}) \times 6 \text{ ft} - 3.06 \times 6^2/2$$
$$= 220 - 55.1 = 165 \text{ kip-ft}$$
$$T_{u2} = M_u/0.9d = 107 \text{ kips}$$

The average horizontal shear stress between the midspan and the quarter point is

$$v_{nh} = \frac{(143 - 107) \text{ kips} \times 1000}{12 \text{ in.} \times (6 \times 12) \text{ in.}} = 41.7 \text{ psi}$$

Because this is less than 80 psi, minimum ties are sufficient between the midspan and the quarter points.

Now consider the portion of the beam between the quarter point and the eighth point:

M_u at the eighth point = 96.4 kip-ft

T_{u1} (at the quarter point) = 107 kips, and T_{u2} (at the eighth point) = 62.7 kips

The average horizontal shear stress between the quarter and the eighth point of the span is

$$v_{nh} = \frac{(107 - 62.7) \times 1000}{12 \times (3 \times 12)} = 103 \text{ psi}$$

Because this value exceeds 80 psi, stirrups are required to satisfy ACI Code Section 16.4.3.1. Minimum ties, according to ACI Code Section 16.4.6, are equivalent to No. 3 two-legged stirrups at 10 in. on centers, with $\rho_v = 0.00183$. For minimum ties, ACI Code Table 16.4.4.2 gives

$$v_{nh} = (260 + 0.6\rho_v f_{yt}) \text{ psi} = (260 + 0.6 \times 0.00183 \times 60{,}000)$$
$$= 326 \text{ psi}$$

Minimum stirrups are more than enough. Consider the portion of beam between the support and the eighth point. At the eighth point, $T_{u1} = 62.7$ kips. At the support, $T_{u2} =$ zero. The average horizontal shear stress on this segment is

$$v_{nh} = \frac{(62.7 - 0) \times 1000}{12 \times (3 \times 12)} = 145 \text{ psi}$$

Thus, minimum ties are satisfactory. We shall use closed stirrups to better anchor them into the top slab. **Use No. 3, Grade-60 closed stirrups at 10 in. on centers throughout the length of the beam.** ■

REFERENCES

16-1 J. A. Hofbeck, I. A. Ibrahim, and Alan H. Mattock, "Shear Transfer in Reinforced Concrete," *ACI Journal, Proceedings*, Vol. 66, No. 2, February 1969, pp. 119–128.

16-2 Alan H. Mattock and Neil M. Hawkins, "Shear Transfer in Reinforced Concrete—Recent Research," *Journal of the Prestressed Concrete Institute*, Vol. 17, No. 2, March–April 1972, pp. 55–75.

16-3 J. C. Walraven, "Fundamental Analysis of Aggregate Interlock," *Journal of the Structural Division, Proceedings of the American Society of Civil Engineers*, Vol. 107, No. ST11, November 1981, pp. 2245–2271.

16-4 H. W. Reinhardt and J. C. Walraven, "Cracks in Concrete Subject to Shear," *Journal of the Structural Division, Proceedings of the American Society of Civil Engineers*, Vol. 108, No. ST1, January 1982, pp. 225–244.

16-5 Robert E. Loov and A. K. Patnaik, "Horizontal Shear Strength of Composite Concrete Beams with a Rough Interface," *PCI Journal*, Vol. 39, No. 1, January–February 1994, pp. 48–69.

16-6 Norman W. Hanson, "Precast-Prestressed Concrete Bridges," Part 2, "Horizontal Shear Connections," *Journal of the Research and Development Laboratories, Portland Cement Association*, Vol. 2, No. 2, May 1960, pp. 38–58.

16-7 M. A. Ali and R. N. White, "Enhanced Contact Model for Shear Friction of Normal and High-Strength Concrete," *ACI Structural Journal*, May–June 1999, Vol. 96, No. 3, pp. 348–360.

16-8 ACI-ASCE Committee 426, "The Shear Strength of Reinforced Concrete Members," *Proceedings ASCE, Journal of the Structural Division*, Vol. 99, No. ST6, June 1973, pp. 1091–1187.

16-9 Mast, R. F., "Auxiliary Reinforcement in Precast Connections," *Proceedings, Journal of the Structural Division* ASCE, Vol. 94, ST6, June 1968, pp. 1485–1504.

16-10 A. H. Mattock, "Shear Friction and High-Strength Concrete," *ACI Structural Journal*, Vol. 98, No. 1, January–February 2001, pp. 50–59.

16-11 L. F. Kahan and A. D. Mitchell, "Shear Friction Tests with High-Strength Concrete," *ACI Structural Journal*, Vol. 99, No. 1 January–February 2002, pp. 98–103.

16-12 F. J. Vecchio and M. P. Collins, "The Modified Compression Field Theory," *ACI Journal, Proceedings*, Vol. 83, No. 2, March–April 1986, pp. 219–231.

16-13 Dino Angelakos, Evan C. Bentz, and Michael P. Collins, "The Effect of Concrete Strength and Minimum Stirrups on the Shear Strength of Large Members," *ACI Structural Journal*, Vol. 98, No. 3, May–June 2001, pp. 290–296.

16-14 *PCI Design Handbook—Precast and Prestressed Concrete*, Seventh Edition (MNL-120-10), Prestressed Concrete Institute, Chicago, IL, 2010.

16-15 Paul H. Kaar, L. B. Kriz, and Eivind Hognestad, "Precast-Prestressed Bridges: (1) Pilot Tests of Continuous Girders," *Journal of the Research and Development Laboratories, Portland Cement Association*, Vol. 2, No. 2, May 1960, pp. 21–37.

16-16 J. C. Saemann and George W. Washa, "Horizontal Shear Connections between Precast Beams and Cast-in-Place Slabs," *ACI Journal, Proceedings*, Vol. 61, No. 11, November 1964, pp. 1383–1409.

16-17 Amin Ghali and Rene Favre, *Concrete Structures: Stresses and Deformations*, Chapman & Hall, New York, 1986, 348 pp.

16-18 Walter H. Dilger, "Creep Analysis of Prestressed Concrete Structures Using Creep-Transformed Section Properties," *PCI Journal*, Vol. 27, No. 1, January–February 1982, pp. 99–118.

17 Discontinuity Regions and Strut-and-Tie Models

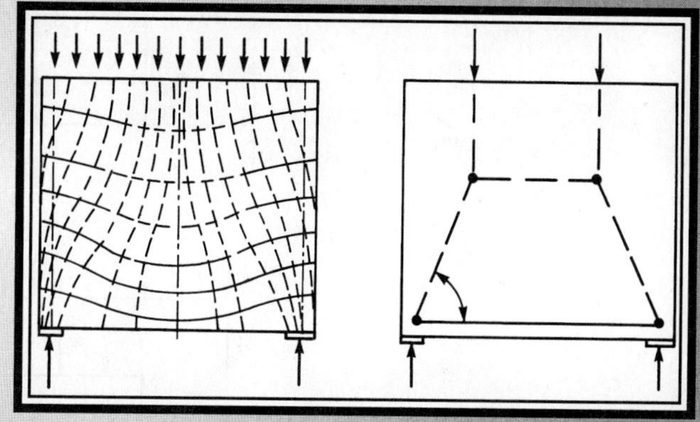

17-1 INTRODUCTION

Definition of Discontinuity Regions

Structural members may be divided into portions called *B-regions*, in which beam theory applies, including linear strains and so on, and other portions called *discontinuity regions*, or *D-regions*, adjacent to discontinuities or disturbances, where beam theory does not apply. D-regions can be *geometric discontinuities*, adjacent to holes or abrupt changes in cross section, or *statical discontinuities*, which are regions near concentrated loads and reactions. Corbels, dapped ends, and joints are affected by both statical and geometric discontinuities. Up to this point, most of this book has dealt with B-regions.

For many years, D-region design has been by "good practice," by rule of thumb or empirical. Three landmark papers by Professor Schlaich of the University of Stuttgart and his coworkers [17-1], [17-2], [17-3] changed this. This chapter will present rules and guidance for the design of D-regions, based largely on these and other recent papers.

Saint Venant's Principle and Extent of D-Regions

St. Venant's principle suggests that the localized effect of a disturbance dies out by about one member-depth from the point of the disturbance. On this basis, D-regions are assumed to extend one member-depth each way from the discontinuity. This principle is conceptual and not precise. However, it serves as a quantitative guide in selecting the dimensions of D-regions.

Figure 17-1 shows D-regions in a number of structures, some of which have B-regions (bending regions) between two D-regions. Figure 17-2 shows examples of D-regions. The D-regions in Figs. 17-2b and 17-2c extend one member-width from the discontinuity as suggested by St. Venant's principle. Occasionally, D-regions are assumed to fill the overlapping region common to two members meeting at a joint. This definition is used in the traditional definition of a joint region.

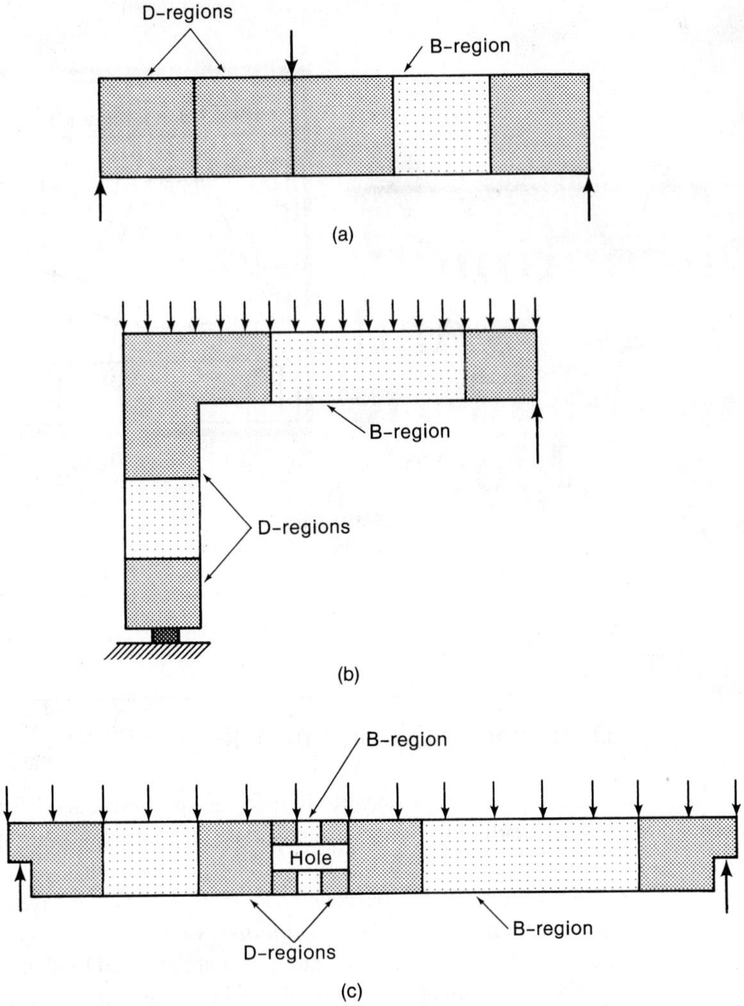

Fig. 17-1
B-regions and D-regions.

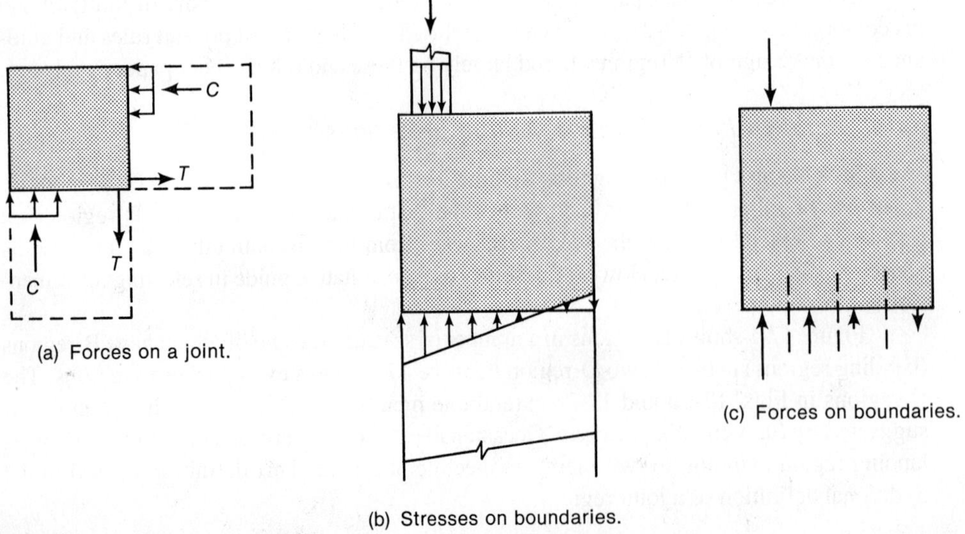

Fig. 17-2
Forces on boundaries of D-regions.

Behavior of D-Regions

Prior to any cracking, an elastic stress field exists, which can be quantified with an elastic analysis, such as a finite-element analysis. Cracking disrupts this stress field, causing a major reorientation of the internal forces. After cracking, the internal forces can be modeled via a *strut-and-tie model* consisting of concrete compression struts, steel tension ties, and joints referred to as nodal zones. If, due to transverse tension, the compression struts are narrower at their ends than they are at midsection, the struts may, in turn, crack longitudinally. For struts without reinforcement crossing their longitudinal axis, this may lead to failure. On the other hand, struts with transverse reinforcement to restrain the cracking can carry additional load and may fail by crushing. Failure may also occur by yielding of the tension ties, failure of the bar anchorage, or failure of the nodal zones. As always, failure initiated by yield of the steel tension ties tends to be more ductile and is desirable.

Strut-and-Tie Models

A strut-and-tie model for a deep beam is shown in Fig. 17-3. It consists of two concrete compressive *struts*, longitudinal reinforcement serving as a tension *tie*, and joints referred to as *nodes*. The concrete around a node is called a *nodal zone*. The nodal zones transfer the forces from the inclined struts to other struts, to ties and to the reactions.

ACI Code Section 6.2.4.3 allows members and D-regions to be designed using strut-and-tie models according to the requirements in ACI Code Chapter 23, *Strut-and-Tie Models*. Background for this code chapter is summarized in [17-4]. Examples in [17-5] and [17-6] were solved using the requirements in Code Chapter 23 by members of ACI Committee 318 Subcommittee E and ACI Committee 445. Additional strut-and-tie examples are available in a recent ACI publication [17-7].

A strut-and-tie model is a model of a portion of the structure that satisfies the following:

(a) it embodies a system of forces that is in equilibrium with a given set of loads, and

(b) the factored-member forces at every section in the struts, ties, and nodal zones do not exceed the corresponding design member strengths for those sections.

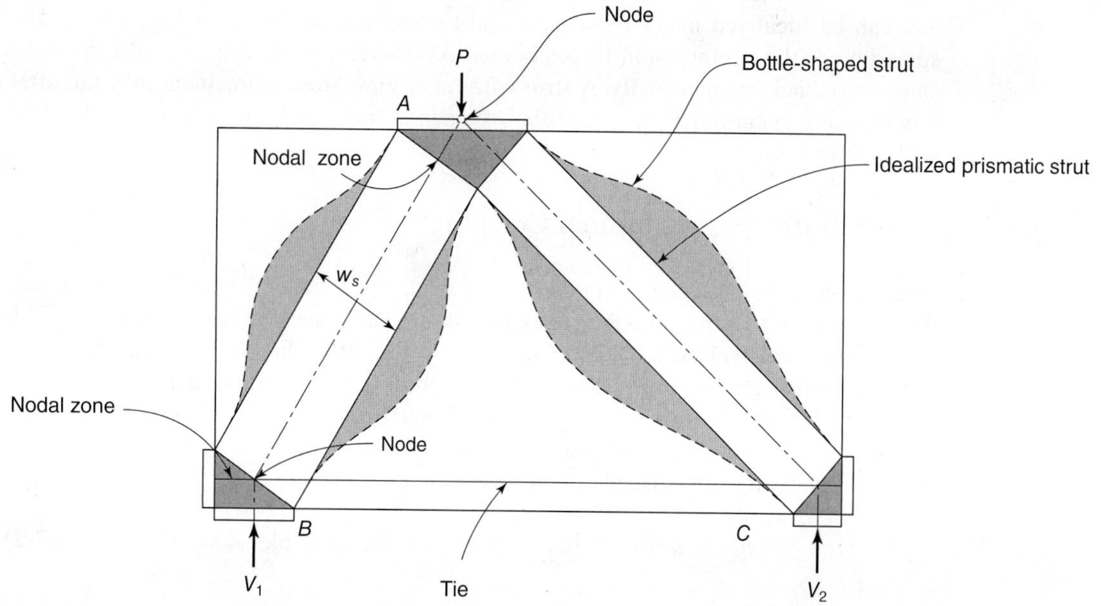

Fig. 17-3
Strut-and-tie model of a deep beam.

The *lower-bound theorem of plasticity* states that the capacity of a system of members, supports, and applied forces that satisfies both (a) and (b) is a lower bound on the strength of the actual structure. For the lower-bound theorem to apply,

(c) The structure must have sufficient ductility to make the transition from elastic behavior to plastic behavior that redistributes the factored internal forces into a set of forces that satisfy items (a) and (b).

The combination of factored loads acting on the structure and the distribution of factored internal forces is a *lower bound on the strength of the structure*, provided that no element is loaded or deformed beyond its capacity. Strut-and-tie models should be chosen so that the internal forces in the struts, ties, and nodal zones are somewhere between the elastic distribution and a fully plastic set of internal forces.

Before we embark on a design example, we will review the strengths of struts, ties, and nodal zones, and factors affecting the layout of strut-and-tie models. In most applications of strut-and-tie models the internal forces, F_u, due to the factored loads, and the struts, ties, and nodal zones are proportioned using

$$\phi F_n \geq F_u \tag{17-1}$$

When design is based on a strut-and-tie model, the load and resistance factors in ACI Code Tables 5.3.1 and 21.2.1 will be used.

17-2 STRUTS

In a strut-and-tie model, the struts represent concrete compressive stress fields with the compression stresses acting parallel to the strut. Although they are frequently idealized as prismatic or uniformly tapering members, as shown in Fig.17-4a, inclined struts generally vary in cross section along their length, as shown in Figs. 17-4b and 17-4c. Transverse tension causes the compression stress fields to be wider at midlength of the strut than at the ends. Struts that change in width along the length of the member are sometimes called *bottle-shaped* due to their shape, as shown in Fig. 17-4b, and can be idealized using *local* strut-and-tie models, as shown in Fig. 17-4c. The spreading of the compression forces is due to transverse tension in the strut that may cause it to crack longitudinally. A strut without transverse reinforcement may fail after this cracking occurs. If transverse reinforcement is provided, the strength of the strut will be increased.

Strut Failure by Longitudinal Cracking

Figure 17-5a shows one end of a bottle-shaped strut. The width of the bearing area is a, and the thickness of the strut (into the page) is t. At midlength the strut has an effective width b_{ef}. Schlaich and Weischede [17-1] assumed that the bottle-shaped region at one end of a strut extends approximately $1.5b_{ef}$ from the end of the strut and in their examples used $b_{ef} = \ell/3$ but not less than a, where ℓ is the length of the strut from face to face of the nodes. For short struts, the limit that b_{ef} not be less than a often governs. We shall assume that in a strut with bottle-shaped regions at each end

$$b_{ef} = a + \ell/6 \text{ but not more than the available width.} \tag{17-2}$$

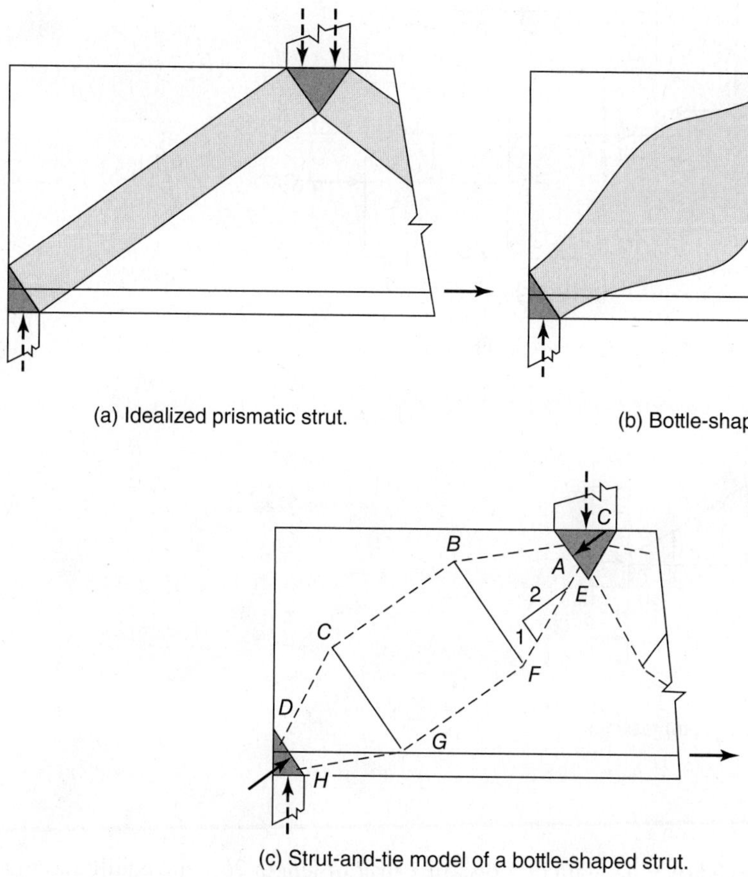

Fig. 17-4
Bottle-shaped strut.

Figures 17-4c and 17-5b show local strut-and-tie models for the bottle-shaped region. It is based on the assumption made in [17-8], that the longitudinal projection of the inclined struts is equal to $b_{ef}/2$. The transverse tension force T at one end of the strut is

$$T = \frac{C}{2}\left(\frac{b_{ef}/4 - a/4}{b_{ef}/2}\right)$$

or

$$T = \frac{C}{4}\left(1 - \frac{a}{b_{ef}}\right) \tag{17-3}$$

The force T causes transverse stresses in the concrete, which may cause cracking. The transverse tensile stresses are distributed as shown by the curved line in Fig. 17-5c. Analyses by Adebar and Zhou [17-9] suggest that the tensile stress distributions at the two ends of a strut are completely separate when ℓ/a exceeds about 3.5 and overlap completely when ℓ/a is between 1.5 and 2. Assuming a parabolic distribution of transverse

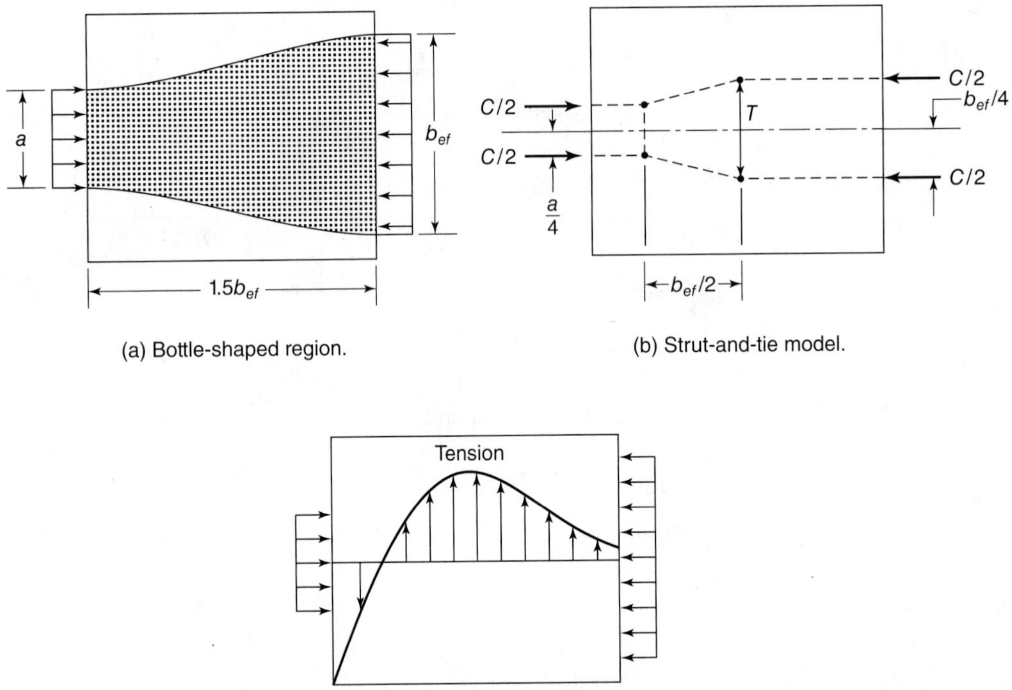

(a) Bottle-shaped region.

(b) Strut-and-tie model.

(c) Transverse tensions and compressions.

Fig. 17-5
Spread of stresses and transverse tensions in a strut.

tensile stresses spread over a length of $1.6b_{ef}$ in a strut of length $2b_{ef}$ and equilibrating a tensile force of $2T$ indicates that the minimum load, C_n, at cracking is 0.51 to 0.57 of f'_c for a strut with $a/b_{ef} = 1/2$. This analysis and [17-2] and [17-8] suggest that longitudinal cracking of the strut may be a problem if the bearing forces on the ends of the strut exceed:

$$C_n = 0.55at\,f'_c \tag{17-4}$$

where at is the loaded area at the end of the strut.

In tests of cylindrical specimens loaded axially through circular bearing plates with diameters less than that of the cylinders, failure occurred at 1.2 to 2 times the cracking loads [17-9].

The maximum load on an unreinforced strut in a wall-like member such as the deep beam in Fig. 17-3, if governed by cracking of the concrete in the strut, is given by Eq. (17-4). This assumes that the compression force spreads in only one direction. If the bearing area does not extend over the full thickness of the member, there will also be transverse tensile stresses through the thickness of the member, as shown in Fig. 17-6. This would require a reanalysis of the support region to design the transverse tie shown in Fig. 17-6a.

Compression Failure of Strut

The crushing strength of the concrete in a strut is referred to as the *effective strength*,

$$f_{cu} = \nu f'_c \tag{17-5}$$

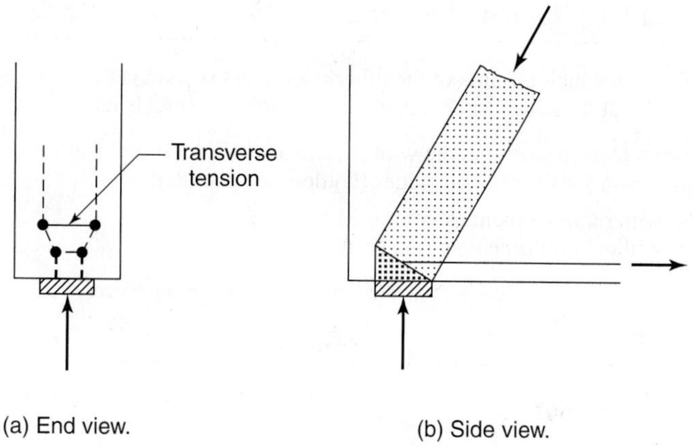

Fig. 17-6
Transverse spread of forces through the thickness of a strut.

where ν is an *efficiency factor* having a value between 0 and 1.0. ACI Code Section 23.4.3 replaces f_{cu} with the *effective compressive strength*, f_{ce}. Various sources give differing values of the efficiency factor [17-3] and [17-10] through [17-15]. The major factors affecting the effective compression strength are:

1. **The concrete strength.** Concrete becomes more brittle and ν tends to decrease as the concrete strength increases.

2. **Load duration effects.** The strength of concrete in beams and columns tends to be less than the cylinder strength, f'_c. Various reasons are given for this lower strength, including the observed reduction in compressive strength under sustained load, the weaker concrete near the tops of members due to vertical migration of bleed water during the placing of the concrete, and the different shapes of compression zones and test cylinders. For flexural members, ACI Code Section 22.2.2.4.1 accounts for this, in part, by taking the maximum stress in the equivalent rectangular stress block as $0.85f'_c$. For struts, load duration effects are accounted for in the ACI Code by rewriting Eq. (17-5) as $f_{ce} = 0.85\beta_s f'_c$. Nodal zones are treated similarly except that β_s is replaced by β_n.

3. **Tensile strains transverse to the strut,** which result from tensile forces in the reinforcement crossing the cracks [17-13] to [17-17]. In tests of uniformly strained concrete panels, such strains were found to reduce the compressive strength of the panels, as discussed in Section 3-2. The AASHTO Specification bases f_{ce} on this concept [17-17].

4. **Cracked struts.** Struts crossed by cracks inclined to the axis of the strut are weakened by the cracks. ACI Code Section 23.4.1 presents the nominal compressive strength of a strut as:

$$F_{ns} = f_{ce}A_c \qquad (17\text{-}6a)$$
(ACI Eq. 23.4.1a)

where subscript n = nominal, s = strut, A_c is the cross-sectional area at the end of the strut, and f_{ce} is:

$$f_{ce} = 0.85\beta_s f'_c \qquad (17\text{-}7a)$$
(ACI Eq. 23.4.3)

Values of β_s are given in Table 17-1.

TABLE 17-1 Values of β_s for Struts (ACI Code Table 23.4.3)

For struts in which the area of the midsection cross section is the same as the area at the nodes, such as the compression zone of a beam.	$\beta_s = 1.0$
For struts located such that the width of the midsection of the strut is larger than the width at the nodes (bottle-shaped struts):	
(a) with reinforcement satisfying 23.5.	$\beta_s = 0.75$
(b) without reinforcement satisfying 23.5.	$\beta_s = 0.60\lambda$
For struts in tension members or the tension flanges of members	$\beta_s = 0.40$
For all other cases	$\beta_s = 0.60\lambda$

Explanation of Types of Struts Described in Table 17-1

$\boldsymbol{\beta_s = 1.0}$ applies to a strut equivalent to a rectangular stress block of depth, a, and thickness, b, as occurs in the compression zones of beams or eccentrically loaded columns. In such cases β_s is equal to 1.0, and the corresponding neutral axis depth is $c = a/\beta_1$. The strut force is equal to the resultant compressive force in the rectangular stress block, $C = f_{ce}ab$, and acts at $a/2$ from the most compressed face of the beam or column as shown in Fig. 17-7.

$\boldsymbol{\beta_s = 0.75}$ applies to bottle-shaped struts similar to those in Fig. 17-4b that contain reinforcement crossing potential longitudinal splitting cracks. The opening of a splitting crack is restrained by the reinforcement, allowing the strut to carry additional load after the splitting cracks develop. For this case $\beta_s = 0.75$. If there is no reinforcement to restrain the opening of a splitting crack, the strut is assumed to fail upon cracking, or shortly after, and a lower value of β_s is used.

The strength of the reinforcement required to restrain the crack is taken equal to the tension force that is lost when the concrete cracks. This is computed using a localized strut-and-tie model of the cracked strut as shown in Fig. 17-5b. The slope of the load-spreading struts is taken as approximately 2 units parallel to axis of strut, to 1 unit perpendicular to the strut axis, resulting in a required tension force as given by Eq. (17-3).

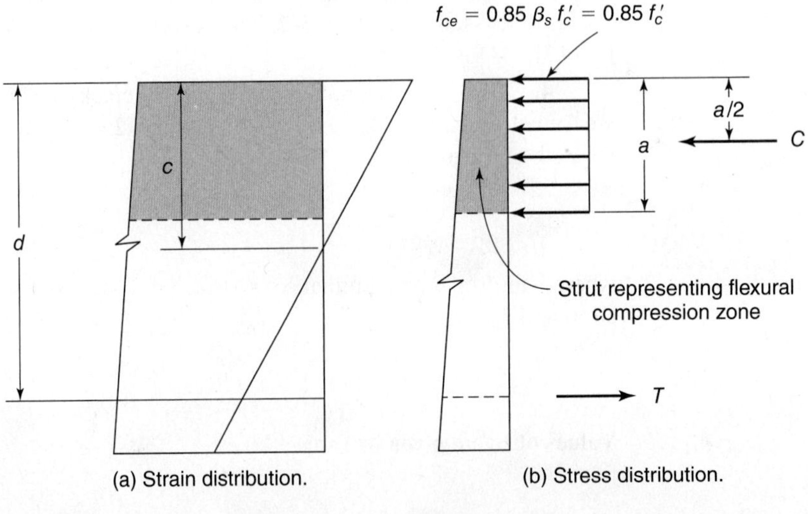

Fig. 17-7
Strut representing the compression stress block in a beam.

Changing to a nominal strength format and representing the nominal strength of the tension tie, T_n, as $A_s f_y$, the required strength of transverse reinforcement at each end of the strut is:

$$A_s f_y \geq \frac{C_n}{4}\left(1 - \frac{a}{b_{ef}}\right) \tag{17-8}$$

where C_n is the nominal compressive strength of the original strut and a is the width of the bearing area at the end of the strut, as shown in Fig. 17-5a. The width of the bottle-shaped strut, b_{ef}, is computed using the distance from one of the parallel longitudinal struts, which are used to model the original strut, to the central axis of the original strut at midlength, shown as $b_{ef}/4$ in Fig. 17-5b. From Eq. (17-8), the transverse reinforcement required to resist spreading of the compression force at both ends of the strut, that is, the total reinforcement required over the length of the strut is:

$$A_s \geq \frac{C_n}{2f_y}\left(1 - \frac{a}{b_{ef}}\right) \tag{17-9}$$

If the reinforcement crosses the strut at an angle α, the reinforcement area on the left side of Eq. (17-9) should be multiplied by sin α. This reinforcement will be referred to as *crack-control reinforcement.*

For concrete compressive strengths of 6000 psi and smaller, ACI Code Section 23.5.3 allows the amount of crack-control reinforcement to be determined using:

$$\Sigma \frac{A_{si}}{bs_i} \sin \alpha_i \geq 0.003 \tag{17-10}$$

where A_{si} refers to the bar area of crack-control reinforcement placed adjacent to the two faces of a member at an angle α_i to the strut axis, as shown in Fig. 17-8. This equation is to be satisfied for either reinforcement placed in only one direction (i.e., vertical or horizontal) or for orthotropic reinforcement, as shown in Fig. 17-8. The arrangement of the crack-control reinforcement is specified in ACI Code Section 23.5.3.1.

Equation (17-10) was written in terms of a reinforcement ratio rather than the tie force to simplify the presentation. For concrete strengths exceeding 6000 psi, the ACI Code requires that the area of crack-control reinforcement be calculated using a procedure similar to that used to derive Eq. (17-9). Orthotropic reinforcement, similar to that shown in Fig. 17-8, can be used to satisfy this requirement.

$\boldsymbol{\beta_s = 0.60\lambda}$ In mass concrete members such as pile caps for more than two piles, it may be difficult to place the crack-control reinforcement. ACI Code Table 23.4.3 specifies a lower value of f_{ce} in such cases. Because the struts are assumed to fail shortly after longitudinal cracking occurs, β_s is multiplied by the correction factor, λ, for lightweight concrete when such concrete is used. Values of λ are defined in ACI Code Table 19.2.4.2. It is 1.0 for normal-weight concrete.

$\boldsymbol{\beta_s = 0.40}$ is used in proportioning struts in strut-and-tie models used to design the reinforcement for the tension flanges of ledger beams (Fig. 4-5), box girders and the like. It accounts for the fact that flexural tension cracks will tend to be wider than cracks in beam webs.

$\boldsymbol{\beta_s = 0.60\lambda}$ applies to all other types of struts not previously covered in ACI Code Table 23.4.3. This includes struts in the web of a beam where more or less parallel cracks divide the web into parallel struts. It also includes struts likely to be crossed by cracks at an angle to the struts.

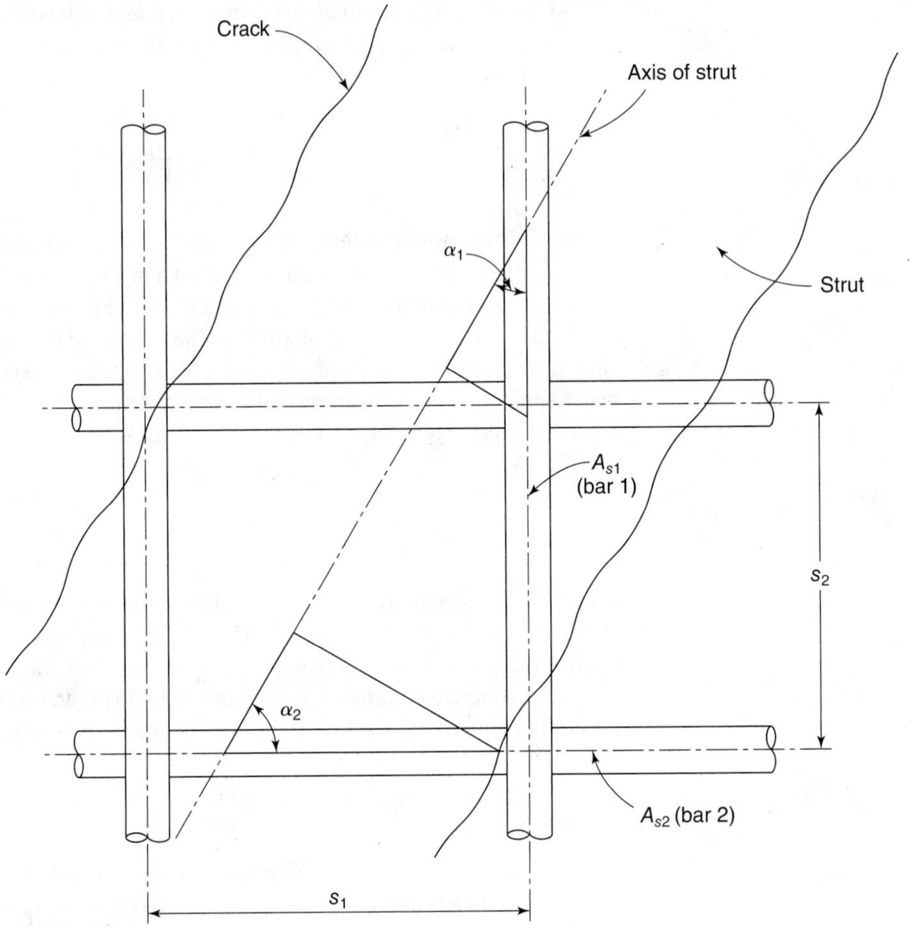

Fig. 17-8
Crack-control reinforcement crossing a strut in a cracked web.

17-3 TIES

The second major component of a strut-and-tie model is the tie. A tie represents one or several layers of reinforcement in the same direction. Design is based on

$$\phi F_{nt} \geq F_{ut} \tag{17-11}$$

where the subscript t refers to "tie," and F_{nt} is the nominal strength of the tie, taken as

$$F_{nt} = A_{ts} f_y + A_{tp}(f_{se} + \Delta f_p) \tag{17-12}$$

(ACI Eq. 23.7.2)

The second term in the parentheses on the right-hand side of Eq. (17-12) is for prestressed ties. It drops out if the member or element does not contain prestressed reinforcement.

ACI Code Section 23.8.1 requires that the axis of the reinforcement in a tie coincide with the axis of the tie in a strut-and-tie model. In the layout of a strut-and-tie model, ties consist of the reinforcement plus a prism of concrete concentric with the longitudinal reinforcement making up the tie. The width of the concrete prism surrounding the tie is referred to as the *effective width* of the tie, w_t. ACI Commentary Section R23.8.1 gives limits for w_t. The lower limit is a width equal to twice the distance from the surface of the concrete to the centroid of the tie reinforcement. In a hydrostatic *C–C–T* nodal zone (defined in Section 17-4), the stresses on all faces of the nodal zone should be equal. As a result, the upper limit on the width of a tie is taken equal to

$$w_{t,\max} = F_{nt}/(f_{ce}b) \tag{17-13}$$

The concrete is included in the tie to establish the widths of the faces of the nodal zones acted on by ties. The concrete in a tie does not resist any load. It aids in the transfer of loads from struts to ties or to bearing areas through bond with the reinforcement. The concrete surrounding the tie steel increases the axial stiffness of the tie by tension stiffening. Tension stiffening may be used in modeling the axial stiffness of the ties in a serviceability analysis.

Ties may fail due to lack of end anchorage. The anchorage of the ties in the nodal zones is a critical part of the design of a D-region using a strut-and-tie model. Ties are normally shown as solid lines in strut-and-tie models.

17-4 NODES AND NODAL ZONES

The points at which the forces in struts-and-ties meet in a strut-and-tie model are referred to as *nodes*. Conceptually, they are idealized as pinned joints. The concrete in and surrounding a node is referred to as a *nodal zone*. In a planar structure, three or more forces must meet at a node for the node to be in equilibrium, as shown in Fig. 17-9. This requires that

$$\Sigma F_x = 0 \qquad \Sigma F_y = 0 \qquad \text{and} \qquad \Sigma M = 0 \tag{17-14}$$

The $\Sigma M = 0$ condition implies that the lines of action of the forces must pass through a common point, or must be able to be resolved into forces that act through a common point. The two compressive forces shown in Fig. 17-9a meet at an angle and are not in equilibrium unless a third force is added, as shown in Fig. 17-9b or 17-9c. Nodal zones are classified as *C–C–C* if three compressive forces meet, as in Fig. 17-9b, and as *C–C–T* if one of the forces is tensile as shown in Fig. 17-9c. *C–T–T* joints may also occur.

Hydrostatic Nodal Zones

Two common ways of laying out nodal zones are illustrated in Figs. 17-10 and 17-11. The prismatic compression struts in Figs. 17-3 and 17-4a are assumed to be stressed in uniaxial compression. A section perpendicular to the axis of a strut is acted on only by compression stresses, while sections at any other angle have combined compression and shear stresses.

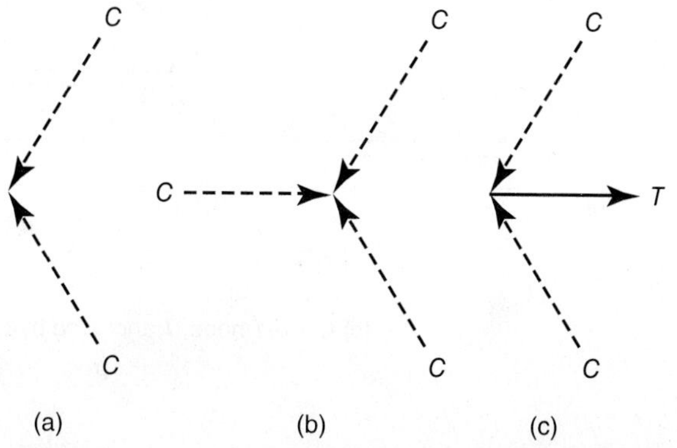

Fig. 17-9
Forces acting on nodes.

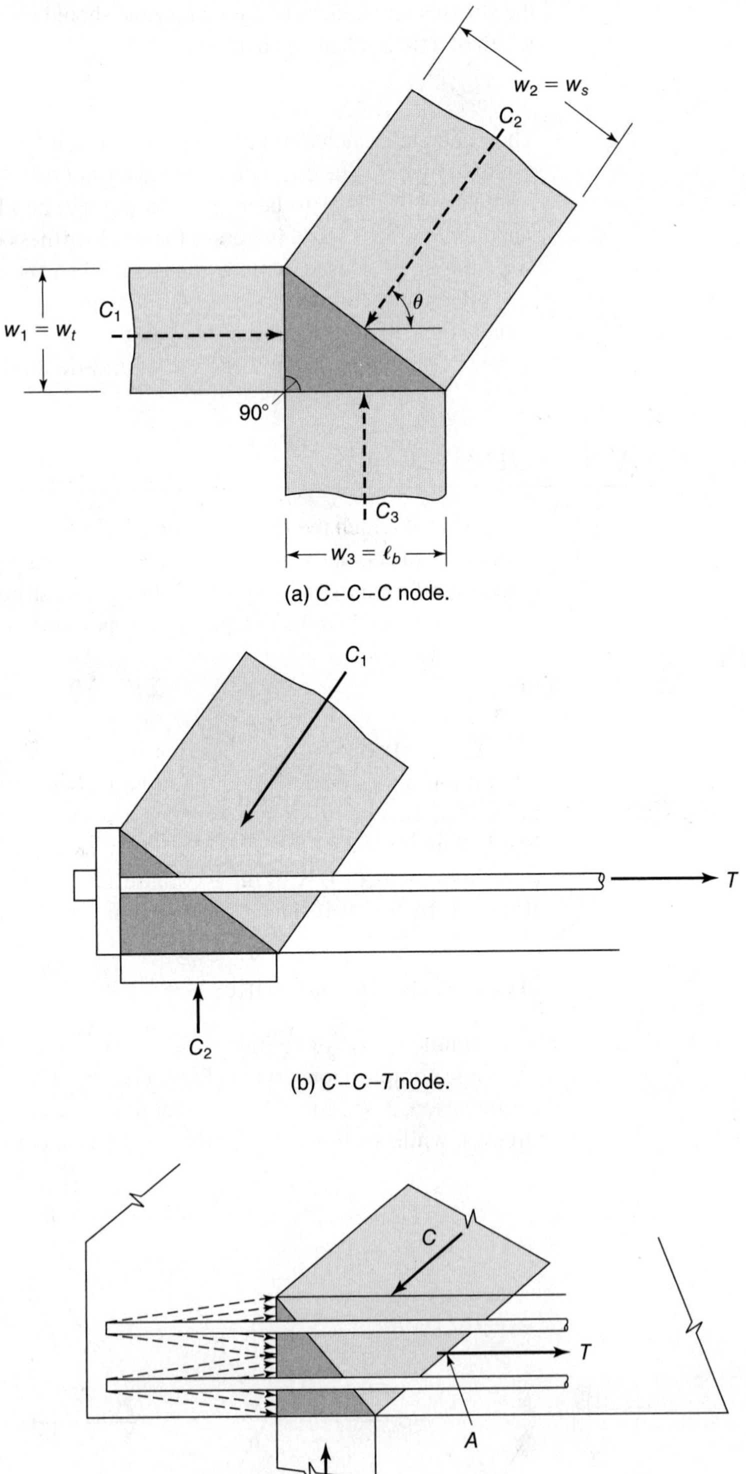

(a) C–C–C node.

(b) C–C–T node.

(c) C–C–T node, T anchored by bond.

Fig. 17-10
Hydrostatic nodal zones in planar structures.

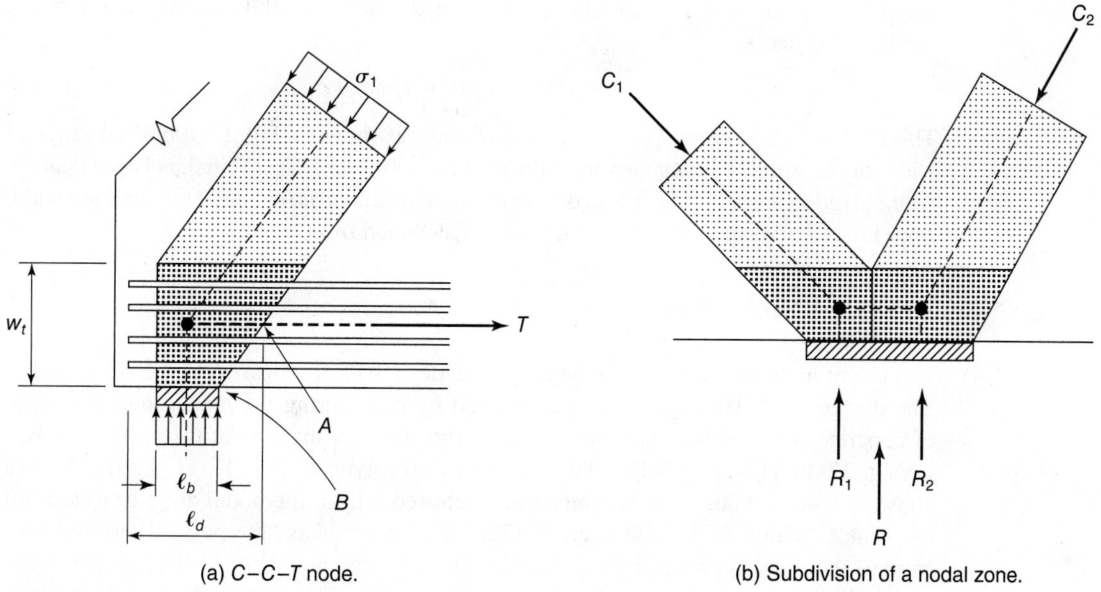

Fig. 17-11
Extended nodal zones.

One way of laying out nodal zones is to orient the sides of the nodes at right angles to the axes of the struts or ties meeting at that node, as shown in Fig. 17-10, and to have the same bearing pressure on each face of the node. When this is done for a *C–C–C* node, the ratio of the lengths of the sides of the node, $w_1 : w_2 : w_3$, is the same as the ratio of the forces in the three members meeting at the node, $C_1 : C_2 : C_3$, as shown in Fig. 17-10a. Nodal zones laid out in this fashion are sometimes referred to as *hydrostatic nodal zones* because the in-plane stresses in the node are the same in all directions. In such a case, the Mohr's circle for the in-plane stresses reduces to a point. If one of the forces is tensile, the width of that side of the node is calculated from a *hypothetical bearing plate* on the end of the tie, which is assumed to exert a bearing pressure on the node equal to the compressive stress in the struts at that node, as shown in Fig. 17-10b. Alternatively, the reinforcement may extend through the nodal zone to be anchored by bond, hooks, or mechanical anchorage before the reinforcement reaches point *A* on the right-hand side of the *extended nodal zone*, as shown by Fig. 17-10c. Such a nodal zone approaches being a hydrostatic *C–C–C* nodal zone. However, the strain incompatibility resulting from the tensile steel strain adjacent to the compressive concrete strain reduces the strength of the nodal zone. Thus, this type of joint should be designed as a *C–C–T* joint.

Geometry of Hydrostatic Nodal Zones

Because the stresses are assumed to be equal on all faces of a hydrostatic nodal zone, equations can be derived relating the lengths of the sides of the nodal zone to the forces acting on each face of the nodal zone. Figure 17-10a shows a hydrostatic *C–C–C* node. For a nodal zone with a 90° corner, as shown, the horizontal width of the bearing area is $w_3 = \ell_b$. The height of the vertical side of the nodal zone is $w_1 = w_t$. The angle between the axis of the inclined strut and the horizontal is θ. The width of the third side, $w_2 = w_s$, can be computed as

$$w_s = w_t \cos\theta + \ell_b \sin\theta \tag{17-15}$$

This equation also can be applied to a *C–C–T* node, as shown in Fig. 17-10b. If the required width of the strut, w_s, computed from the strut force by using Eq. (17-6a), is larger than the width given by Eq. (17-15), it is necessary to increase either w_t or ℓ_b or both until the width from Eq. (17-15) equals or exceeds the width calculated from the strut forces.

Extended Nodal Zones

The use of hydrostatic nodes can be tedious in design, except possibly for *C–C–C* nodes. The design of nodal zones can be simplified by considering the nodal zone to consist of concrete lying within extensions of the struts and ties meeting at the joint as shown in Fig. 17-11 [17-3], [17-18]. Two examples are given in Fig. 17-11. Figure 17-11a shows a *C–C–T* node. The bars must be anchored within the nodal zone or to the left of point *A*, which ACI Code Section 23.8.3(b) describes as "the point where the centroid of the reinforcement in the tie leaves the extended nodal zone." The length, ℓ_d, in which the bars of the tie must be developed is shown. The vertical face of the node is acted on by a stress equal to the tie force *T* divided by the area of the vertical face. The stresses on the three faces of the node can all be different, provided that

1. The resultants of the three forces coincide.
2. The stresses are within the limits given in Table 17-2.
3. The stress is constant on any one face.

Equation 17-15 can be used to compute the widths perpendicular to the axis of the struts, as shown in Fig. 17-12, even though this equation was derived for hydrostatic nodal zones. This equation is useful in adjusting the width of an inclined strut if the original width is found to be inadequate.

An extended nodal zone consists of the node itself, plus the concrete in extensions of the struts, bearing areas, and ties that meet at a joint. Thus in Fig. 17-11a the darker

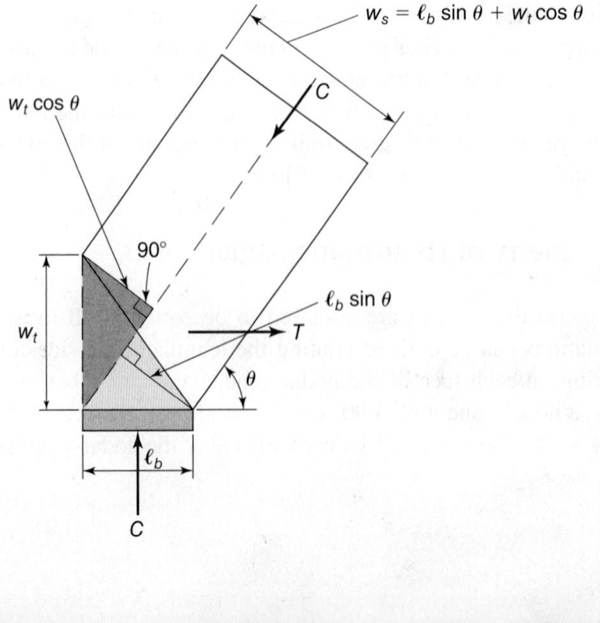

Fig. 17-12
Width of inclined strut at a *C–C–T* nodal zone.

shaded region indicates the nodal zone extends into the area occupied by the struts and ties at this node. The advantage of the nodal zone in Fig. 17-11a comes from the fact that ACI Section Code 23.8.3(b) allows the length available to develop the tie bars to extend out to point A, rather than point B at the edge of the bearing plate. This extended anchorage length recognizes the beneficial effect of the compression from the reaction and the struts for improving bond between the concrete and the tie reinforcement.

Strength of Nodal Zones

Nodal zones are assumed to fail by crushing of the concrete. Anchorage of the tension ties is a required design consideration. If a tension tie is at least partially anchored in a nodal zone there is a strain incompatibility between the tensile strains in the bars and the compressive strain in the concrete of the node. This tends to weaken the nodal zone. ACI Code Section 23.9.1 defines the strength for nodal zones as:

$$F_{nn} = f_{ce} A_{nz} \qquad \text{(17-6b)}$$

(ACI Eq. 23.9.1)

where A_{nz} is the area of the face of the node that the strut or tie acts on, taken perpendicular to the axis of the strut or tie, or the area of a section transferring forces through the nodal zone, and f_{ce} is the effective compression strength of the concrete

$$f_{ce} = 0.85\,\beta_n f'_c \qquad \text{(17-7b)}$$

(ACI Eq. 23.9.2)

Values for β_n are given in Table 17-2.

TABLE 17-2 Values of β_n for Nodal Zones (ACI Code Table 23.9.2)

For C-C-C nodes	$\beta_n = 1.00$
For C-C-T nodes	$\beta_n = 0.80$
For C-T-T nodes	$\beta_n = 0.60$

Subdivision of Nodal Zones

Frequently, it is easier to lay out the size and location of nodal zones if they are subdivided into parts, each of which is assumed to transfer a particular component of the load through the nodal zone. In Fig. 17-11b, the reaction R has been divided into two components R_1, which equilibrates the vertical component of C_1, and R_2, which equilibrates C_2. This subdivision simplifies the layout of the struts and nodes, but the stress on the vertical face between the two subnodes must be less than the limit defined in Eq. (17-7b) and Table 17-2. Subdivision is useful in dealing with the dead load of a beam, which can be assumed to be applied as a series of equivalent concentrated loads, each of which is transferred to a reaction by an individual strut as shown in Fig. 17-13a. In this figure, the dead load is transferred to the support by four inclined struts, which are supported by the portion of the support nodal zone labeled as V_s. The horizontal width of the part of the nodal zone labeled V_s is the sum of the horizontal widths of the four struts that support the dead loads on this half of the beam. The vertical truss members represent the stirrup forces.

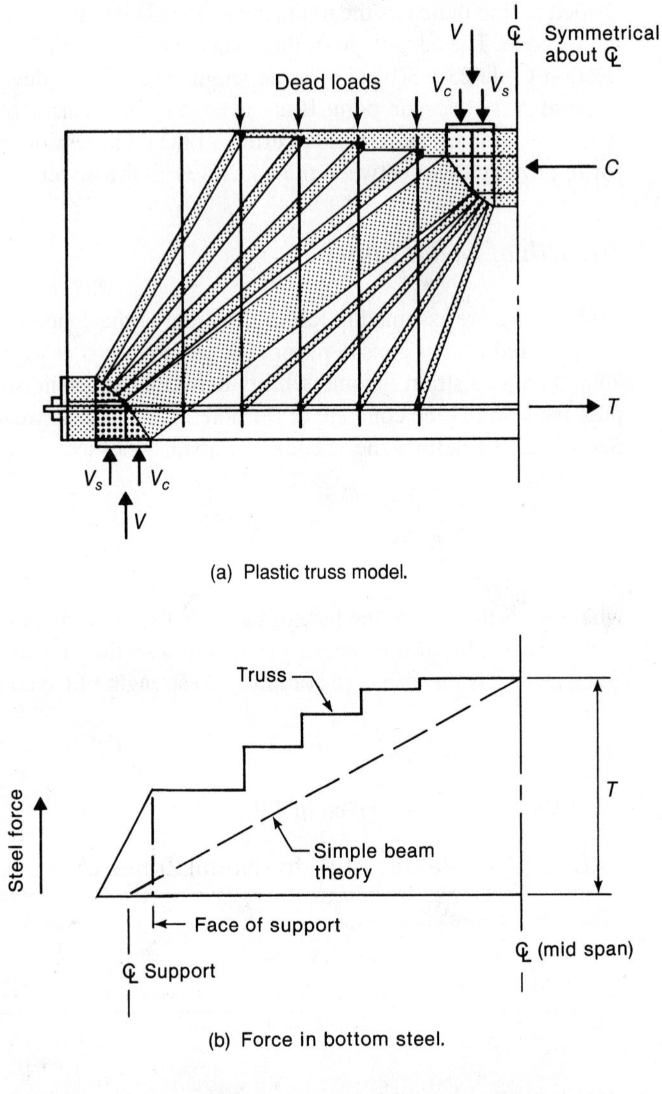

Fig. 17-13
Strut-and-tie model of a deep beam with dead load and stirrups.

Resolution of Forces Acting on a Nodal Zone

If more than three forces act on a nodal zone in a planar strut-and-tie model, it is usually advantageous to subdivide the nodal zone so that only three forces remain on any part of the node. Figure 17-14a shows a hydrostatic nodal zone that is in equilibrium with four strut forces meeting at point D. The nodal zone for point D can be subdivided as shown in Fig. 17-14b. For subnode *E–F–G*, the two forces acting on faces *E–F* and *E–G* can be resolved into a single inclined force (50.6 kips) acting between the two sub-nodes. That inter-nodal force must also be in equilibrium with the forces acting on faces *A–B* and *B–C* of sub-node *A–B–C*. The overall force equilibrium for node *D* is demonstrated in Fig. 17-14c.

Anchorage of Ties in Nodal Zones

A challenge in design using strut-and-tie models is the anchorage of the tie forces at the nodal zones at the edges or ends of a strut-and-tie model. This problem is independent of the type of analysis used in design. It occurs equally in structures designed by elastic analyses or strut-and-tie models. In fact, one of the advantages of strut-and-tie models

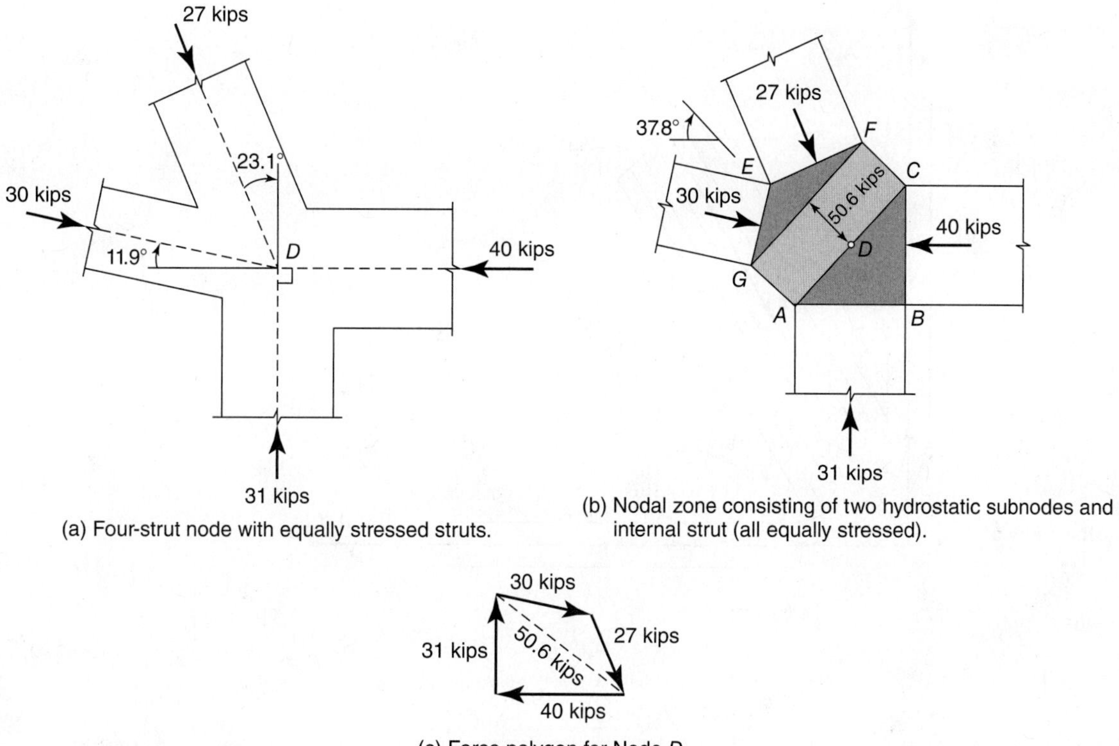

(a) Four-strut node with equally stressed struts.

(b) Nodal zone consisting of two hydrostatic subnodes and internal strut (all equally stressed).

(c) Force polygon for Node *D*.

Fig. 17-14
Resolution of forces acting on a nodal zone.

comes from the attention that the strut-and-tie model places on the anchorage of ties as described in ACI Code Sections 23.8.2 and 23.8.3. For nodal zones anchoring one tie, the tie must be developed by bond, by hooks, or by mechanical anchorage between the free end of the bar and the point at which the centroid of the tie reinforcement leaves the compressed extended portion of the nodal zone. This corresponds to point *A* in Fig. 17-11a. If the bars are anchored by hooks, the hooks should be confined within reinforcement extending into the member from the supporting column, if applicable.

Nodal Zone Anchored by a Bent Bar

Sometimes the two tension ties in a *C–T–T* node are both provided by a bar bent through 90° as shown in Fig. 17-15. The compressive force in the strut can be anchored by bearing and shear stresses transferred from the strut to the bent bar. Such a detail must satisfy the laws of statics and limits on the bearing stresses on the concrete inside the bent bar. A design procedure is given in a recent article by Klein [17-19].

Strut Anchored by Reinforcement

Sometimes, diagonal struts in the web of a truss model of a flexural member are anchored by longitudinal reinforcement that, in turn, is supported by a stirrup, as shown in Fig. 17-16. Reference [17-13] recommends that the length of longitudinal bar able to support the strut be limited to six bar diameters each way from the center of the strut.

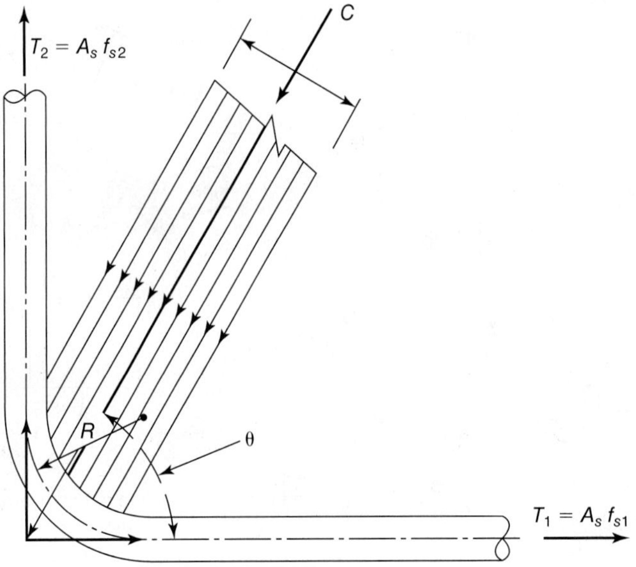

Fig. 17-15
C–T–T node anchored by a bent bar.

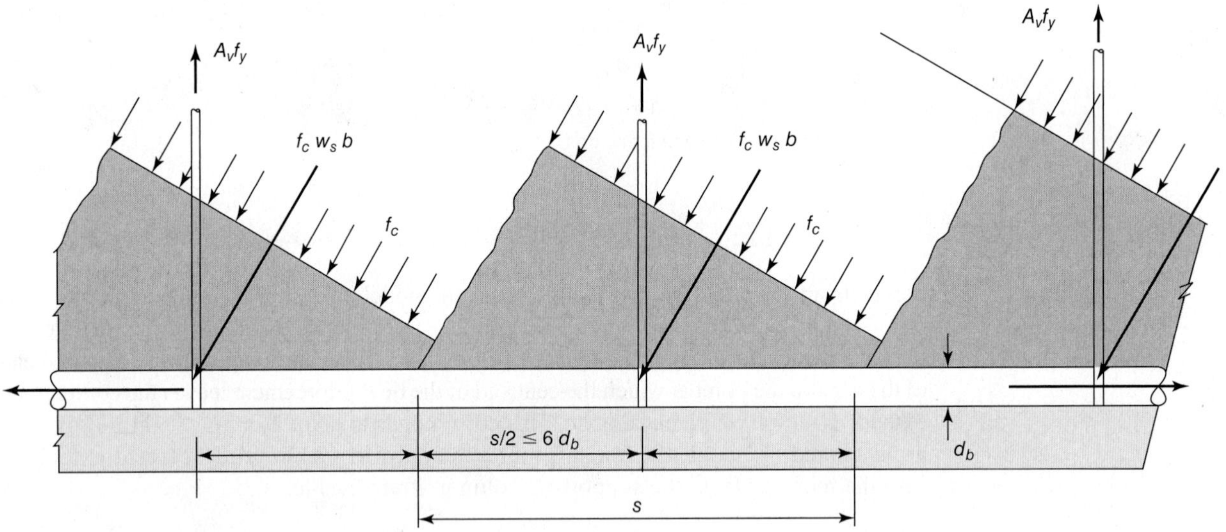

Fig. 17-16
Struts anchored by stirrups and longitudinal bars.

17-5 OTHER STRUT-AND-TIE ELEMENTS

Compression Fans

A compression fan is a series of compression struts that radiate out from a concentrated applied force to distribute that force to a series of localized tension ties, such as the stirrups. Fans are shown over the reaction and under the load in Fig. 17-17.

Compression Fields

A compression field is a series of parallel compression struts combined with appropriate tension ties and compression chords. Compression fields are shown between the compression fans in Fig. 17-17.

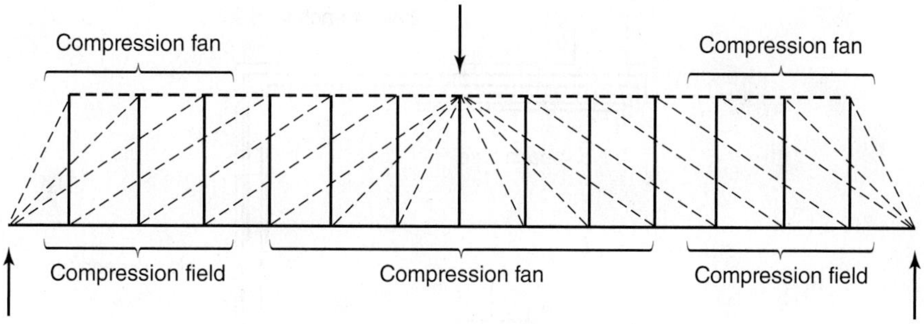

Fig. 17-17
Compression fans and compression fields.

Force Whirls, U-Turns

In Fig. 17-18 the column load causes the stresses shown by the shaded areas at the bottom of the D-region. The stresses on the bottom edge, *A–I*, have been computed with the use of the formula $\sigma = (P/A) + (My/I)$. The neutral axis (axis of zero strains) is at *G*, 26.7 in. from the right edge of the wall. The widths of *E–G* and *G–I* have both been chosen as 26.7 in. so that the upward force at *F* equals the downward force at *H*. The widths of the other two parts (*A–C* and *C–E*) were chosen so that the forces in them were equal. The forces at *F* and

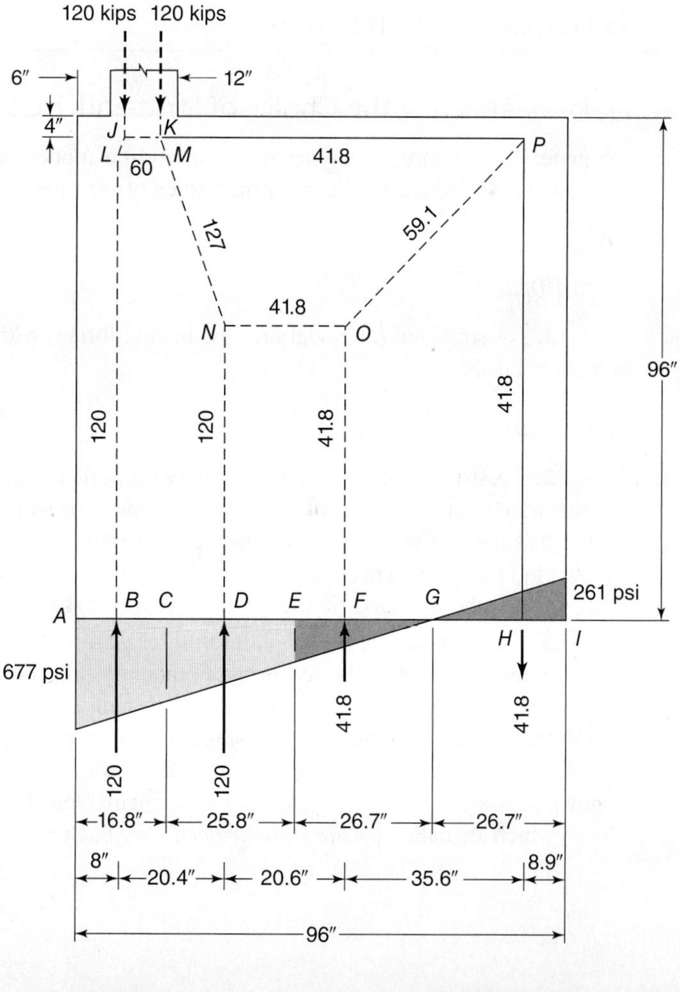

Fig. 17-18
Column supported near one end of a wall, showing a strut-and-tie model with a force whirl or U-turn.

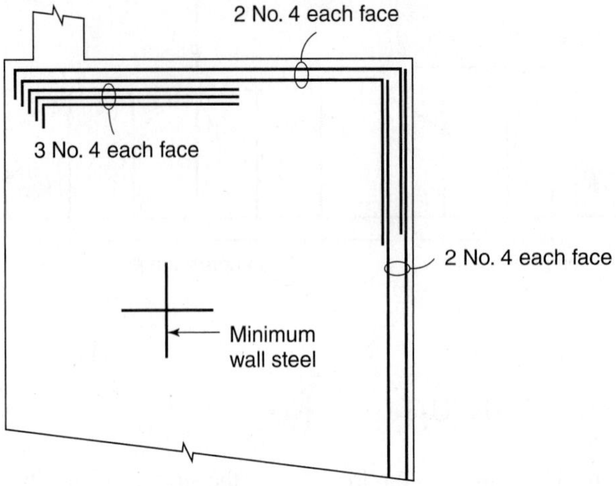

Fig. 17-19
Reinforcement in the wall of Fig. 17-18.

H are equal and cause the force whirl made up of compression members *F–O* and *O–P*, and tension member *H–P*. The strut-and-tie model plotted in Fig. 17-18 is solved in reference [17-18], and the reinforcement computed for that model is shown in Fig. 17-19.

17-6 LAYOUT OF STRUT-AND-TIE MODELS

Factors Affecting the Choice of Strut-and-Tie Models

A general procedure for laying out strut-and-tie models was presented in the previous sections. Additional guidelines for the choice of strut-and-tie models include the following areas.

Equilibrium

1. A strut-and-tie model must be in equilibrium with the loads and there must be a clear load path.

Direction of Struts and Ties

2. A strut-and-tie model for a simply supported beam with unsymmetrically applied concentrated loads consists of either an arch made of straight line segments or a hanging cable that has the same shape as the bending-moment diagram for the loaded beam, as shown in Fig. 17-20. This is also true for uniformly loaded beams, except that the moment diagram and the strut-and-tie model have parabolic shapes.

3. A strut-and-tie model should represent a realistic flow of forces from the loads through the D-region to the reactions. Frequently this can be determined by observation. From an elastic stress analysis, such as a finite element analysis, it is possible to derive the stress trajectories in an uncracked D-region, as shown in Figs. 17-25a, 17-26a, and 17-27a for deep beams. Principal compression stresses act parallel to the dashed lines, which are known as *compressive stress trajectories*. Principal tensile stresses act parallel to the solid lines, which are called *tensile stress trajectories*. Such a diagram shows the flow of internal

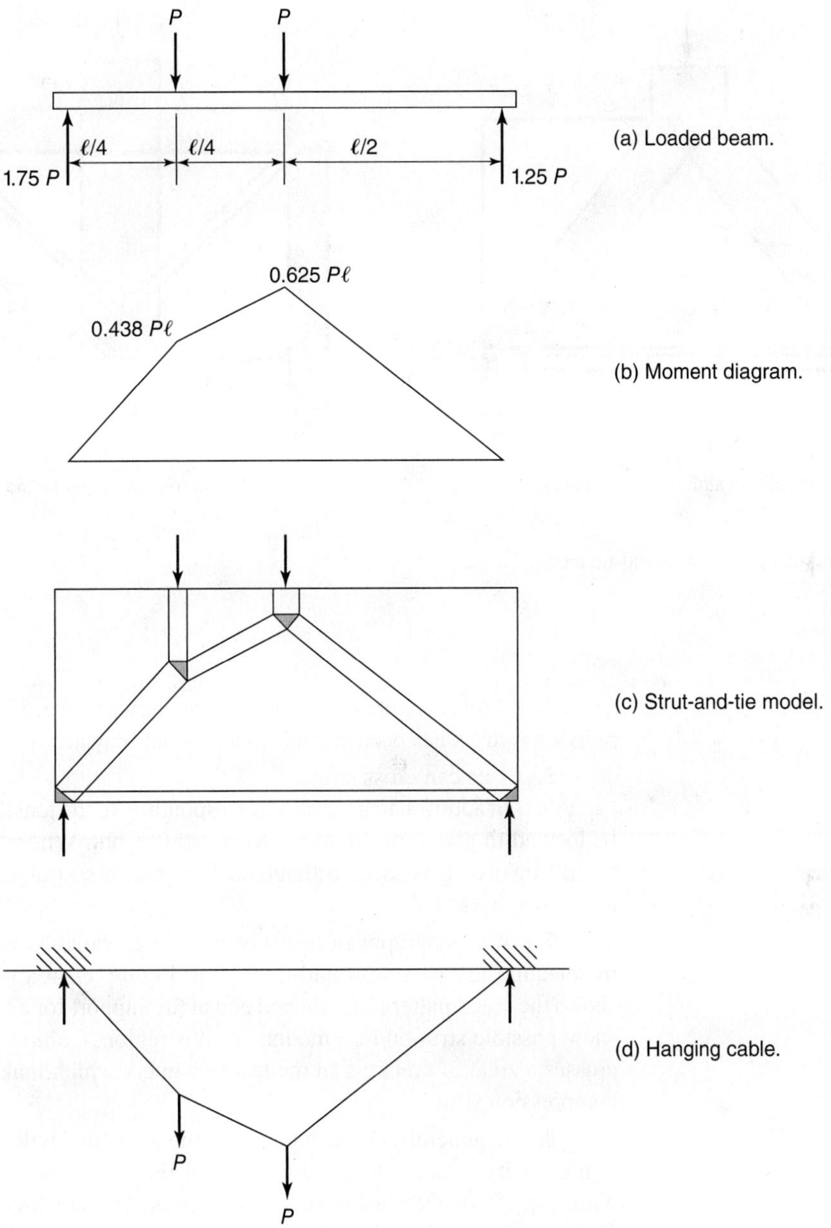

Fig. 17-20
Statical equivalence of various types of structures.

forces and is a useful, but by no means an essential step in laying out a strut-and-tie model. The compressive struts should roughly follow the direction of the compressive stress trajectories, as shown by the refined and simple strut-and-tie models in Figs. 17-25c, 17-26c, and 17-27c. Generally, the strut direction should be within $\pm 15^\circ$ of the direction of the compressive stress trajectories [17-1].

Because a tie consists of a finite arrangement of reinforcing bars that usually are placed orthogonally in the member, there is less restriction on the conformance of ties with the tensile stress trajectories. However, they should be in the general direction of the tension stress trajectories.

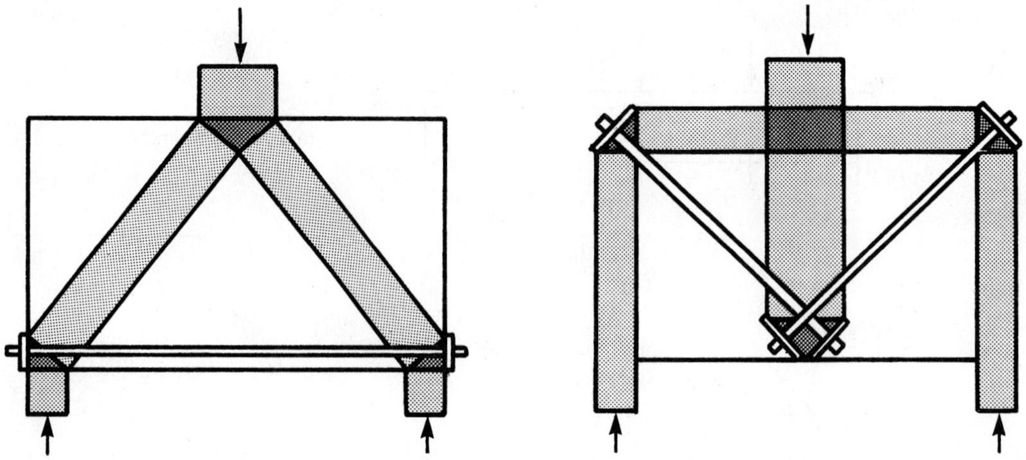

(a) Correct strut-and-tie model.

(b) Incorrect strut-and-tie model.

Fig. 17-21
Suitable and unsuitable strut-and-tie models.

4. Struts cannot cross or overlap, as shown in Fig. 17-21(b), because the width of the individual struts has been calculated assuming they are stressed to the maximum.

5. Ties can cross struts.

6. In addition to generally corresponding to the tensile stress trajectories, ties should be located to give a practical reinforcement layout. Wherever possible, the reinforcement should involve groups of orthogonal bars that are straight, except for hooks needed to anchor the bars.

7. If photographs of test specimens are available, the crack pattern may assist one in selecting the best strut-and-tie model. Figure 17-40a (which will be discussed later) shows the crack pattern in a dapped end at the support for a precast beam. Figures 17-40b–e show possible strut-and-tie models for this region. Compression strut *B–D* in Fig. 17-40d crosses a zone of cracking in the test specimen, which makes it an unlikely location for a compression strut.

8. It generally is assumed that the structure will have enough plastic deformation capacity to adapt to the directions of the struts and ties chosen in design if they are within $\pm 15°$ of the elastic stress trajectories. The crack-control reinforcement from ACI Code Section 23.5 is partially intended to allow the load redistribution needed to accommodate this change in angles.

9. The loads will try to follow the path involving the least forces and deformations. Because the tensile ties are more deformable than the compression struts, the model with the least and shortest ties is the best. Thus the strut-and-tie model in Fig. 17-21a is a better model than the one in Fig. 17-21b because Fig. 17-21a more closely approaches the elastic stress trajectories in Fig. 17-25a. Schlaich et al. [17-2] and [17-3] propose the following criterion for guidance in selecting a good model:

$$\Sigma F_i \ell_i \epsilon_{mi} = \text{minimum}$$

where F_i, ℓ_i, and ϵ_{mi} are the force, length, and mean strain in strut or tie i, respectively. Because the strains in the concrete are small, the struts can be ignored in the summation.

10. Angles, θ, between the struts and attached ties at a node, as shown in Fig. 17-12, should be large (on the order of 45°) and never less than the 25° specified in ACI Code Section 23.2.7. The size of compression struts is sensitive to the angle θ between the strut and the reinforcement in a tie.

Suitable Strut-and-Tie Layouts

A possible strut-and-tie model for the beam shown in Fig. 17-22 consists of two trusses, one utilizing the lower steel as its tension tie, the other using the upper steel. For an ideally plastic material, the capacity would be the sum of the shears transmitted by the two trusses, $V_1 + V_2$. Tests [17-10] have shown, however, that the upper layer of steel has little, if any, effect on the strength. In part this is due to the very low angle θ_2 in the upper truss. When this beam is loaded, the bottom tie yields first.

11. As the angle, θ, between the struts and ties decreases, it is often desirable to include web reinforcement in addition to the confinement reinforcement required by ACI Code Section 23.5. The following equation has been used in European design standards [17-18] to require stirrup reinforcement in beams that approach the ACI Code lower limit on the angle between struts and ties:

$$F_{nw} \approx \frac{2a/jd - 1}{3 - N_n/F_n} F_n \qquad (17\text{-}16)$$

Here F_{nw} is the yield force $\Sigma A_v f_y$ in the web reinforcement in the shear span; F_n is the concentrated load and also the reaction; N_n is the axial force acting on the beam, if any; a is the distance between the axes of the load and reaction; and jd is the lever arm between the resultant compression force and the resultant force in the longitudinal tie. For a beam with $N_n = 0$, and having a/jd equal to 2, Eq. (17-16) requires that all the shear be carried by shear reinforcement. At $a/jd = 0.5$, all the shear is resisted by the compression strut.

12. Sometimes a better representation of the real stress flow is obtained by adding two possible simple models, each of which is in equilibrium with a part of the applied load, provided the struts do not overlap or cross.

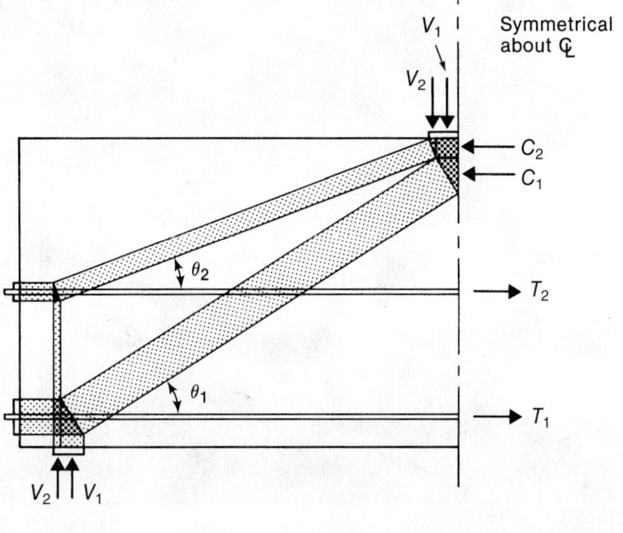

Fig. 17-22
Strut-and-tie model for beam with horizontal web reinforcement at midheight. (From [17-10].)

17-7 DEEP BEAMS

Definition of Deep Beam

The term *deep beam* is defined in ACI Code Section 9.9.1.1 as a member:

(a) loaded on one face and supported on the opposite face so that compression struts can develop between the loads and the supports and

(b) having either

(i) clear spans, ℓ_n, equal to or less than four times the overall member height, h, or

(ii) regions loaded with concentrated loads within $2h$ from the face of the support. The ACI Code does not quantify the magnitude of the concentrated load needed for the beam to act as a deep beam. ACI Committee 445 has suggested that a concentrated load that causes 30 percent or more of the reaction at the support of the beam in question would qualify.

The ACI Code requires that deep beams be designed via nonlinear analyses or by strut-and-tie models. Typically, deep beams occur as *transfer girders*, which may be single span (Fig. 17-23) or continuous (Fig. 17-24). A transfer girder supports the load from one or more columns, transferring it laterally to other columns. Deep-beam action also occurs in some walls and in pile caps. Although such members are not uncommon, satisfactory design methods only recently have been developed.

Analyses and Behavior of Deep Beams

Elastic analyses of deep beams in the uncracked state are meaningful only prior to cracking. In a deep beam, cracking will occur at one-third to one-half of the ultimate load. After cracks develop, a major redistribution of stresses is necessary. The results of elastic analyses are

Fig. 17-23
Single-span deep beam. (Photograph courtesy of J. G. MacGregor.)

Fig. 17-24
Three-span deep beam, Brunswick building, Chicago. (Photograph courtesy of J. G. MacGregor.)

of interest primarily because they show the distribution of stresses, which cause cracking and hence give guidance as to the direction of cracking and the flow of forces after cracking. In Figs. 17-25a, 17-26a, and 17-27a, the dashed lines are *compressive stress trajectories* drawn parallel to the directions of the principal compressive stresses, and the solid lines are *tensile-stress trajectories* parallel to the principal tensile stresses. Cracks would be expected to occur perpendicular to the solid lines (i.e., parallel to the dashed lines).

In the case of a single-span beam supporting a concentrated load at midspan (Fig. 17-25a), the principal compressive stresses act roughly parallel to the dashed lines joining the load and the supports and the largest principal tensile stresses act parallel to the bottom of the beam. The horizontal tensile and compressive stresses on a vertical plane at midspan are shown in Fig. 17-25b. For a beam with aspect ratio about 1.0 the resultant compressive force is closer to the tension tie than a straight-line distribution of stresses would predict. Although it cannot be seen from the figures, it is important to note that the flexural stress at the bottom is constant over much of the span. The stress trajectories in Fig. 17-25a can be represented by the strut-and-tie model given in Fig. 17-25c. Again, dashed lines represent compression struts and solid lines denote tension ties. The angle θ varies approximately linearly from 68° (2.5:1 slope) for $\ell/d = 0.80$ or smaller, to 40° (0.85:1 slope) at $\ell/d = 1.8$. If such a beam were tested, the crack pattern would be similar to the pattern shown in Fig. 17-25d. Note that each of the three tension ties in Fig. 17-25c (*AB*, *CD*, and *EF*) has cracked. At failure, the shaded region in Fig. 17-25d would crush, or the anchorage zones at *E* and *F* would fail. The strut-and-tie model in Fig. 17-25c could be simplified further to the model shown in Fig. 17-25e. However, this simple model does not fully explain the formation of the inclined cracks.

An uncracked, elastic, single-span beam supporting a uniform load has the stress trajectories shown in Fig. 17-26a. The distribution of horizontal stresses on vertical sections at midspan and the quarter point are plotted in Fig. 17-26b. The stress trajectories can be represented by the simple strut-and-tie model in Fig. 17-26c or the slightly more complex model in Fig. 17-26d. In the first case, the uniform load is divided into two parts, each represented by its resultant. In the second case, four parts were used. The angle θ varies from about 68° for $\ell/d = 1.0$ or smaller to about 55° for $\ell/d = 2.0$ [17-1]. The crack pattern in such a beam is shown in Fig. 17-26e.

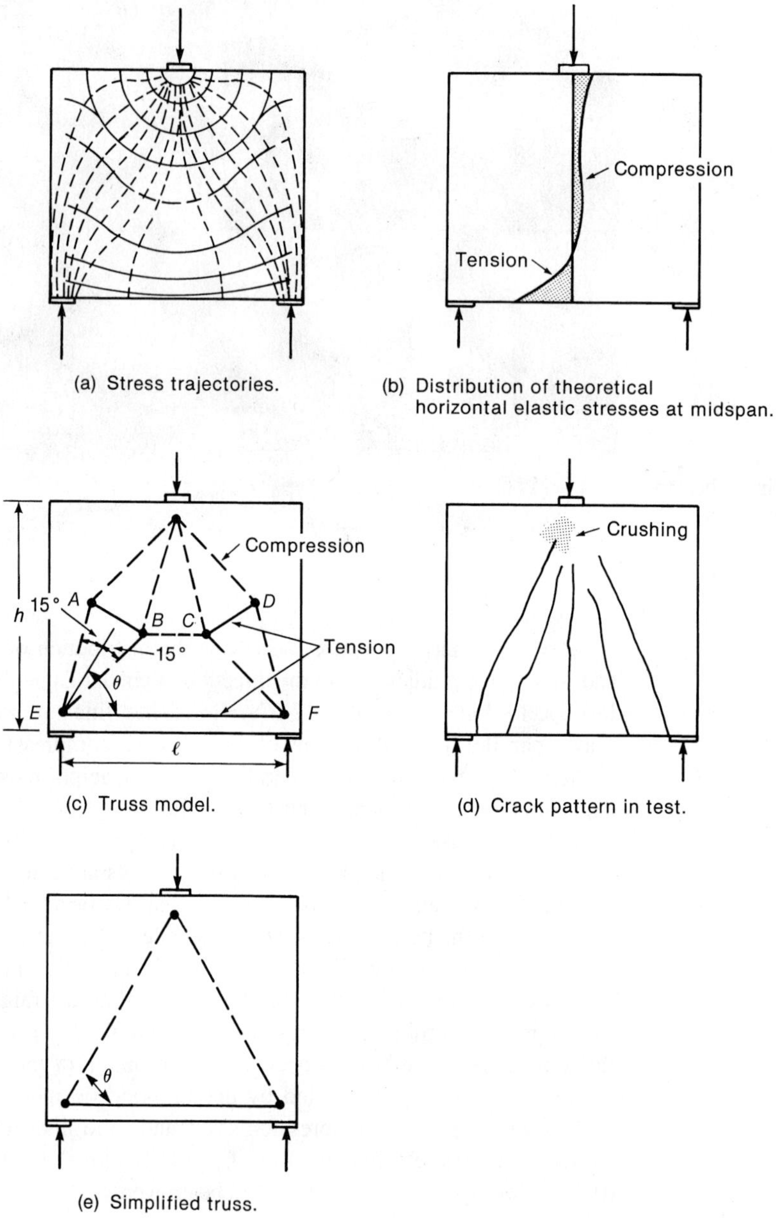

Fig. 17-25
Single-span deep beam supporting a concentrated load. (Adapted from [17-1].)

Figure 17-27a shows the stress trajectories for a deep beam supporting a uniform load acting on ledges at the lower face of the beam. The compression trajectories form an arch, with the loads hanging from it, as shown in Figs. 17-27b and 17-27c. The crack pattern in Fig. 17-27d clearly shows that the load is transferred upward by reinforcement until it acts on the compression arch, which then transfers the load down to the supports.

The strut-and-tie models in Figs. 17-25, 17-26, and 17-27 all suggest that the force in the bottom longitudinal tension ties will be constant along the length of the deep beam. This implies that this force must be anchored at the joints over the reactions. Failure to do so is a major cause of distress in deep beams.

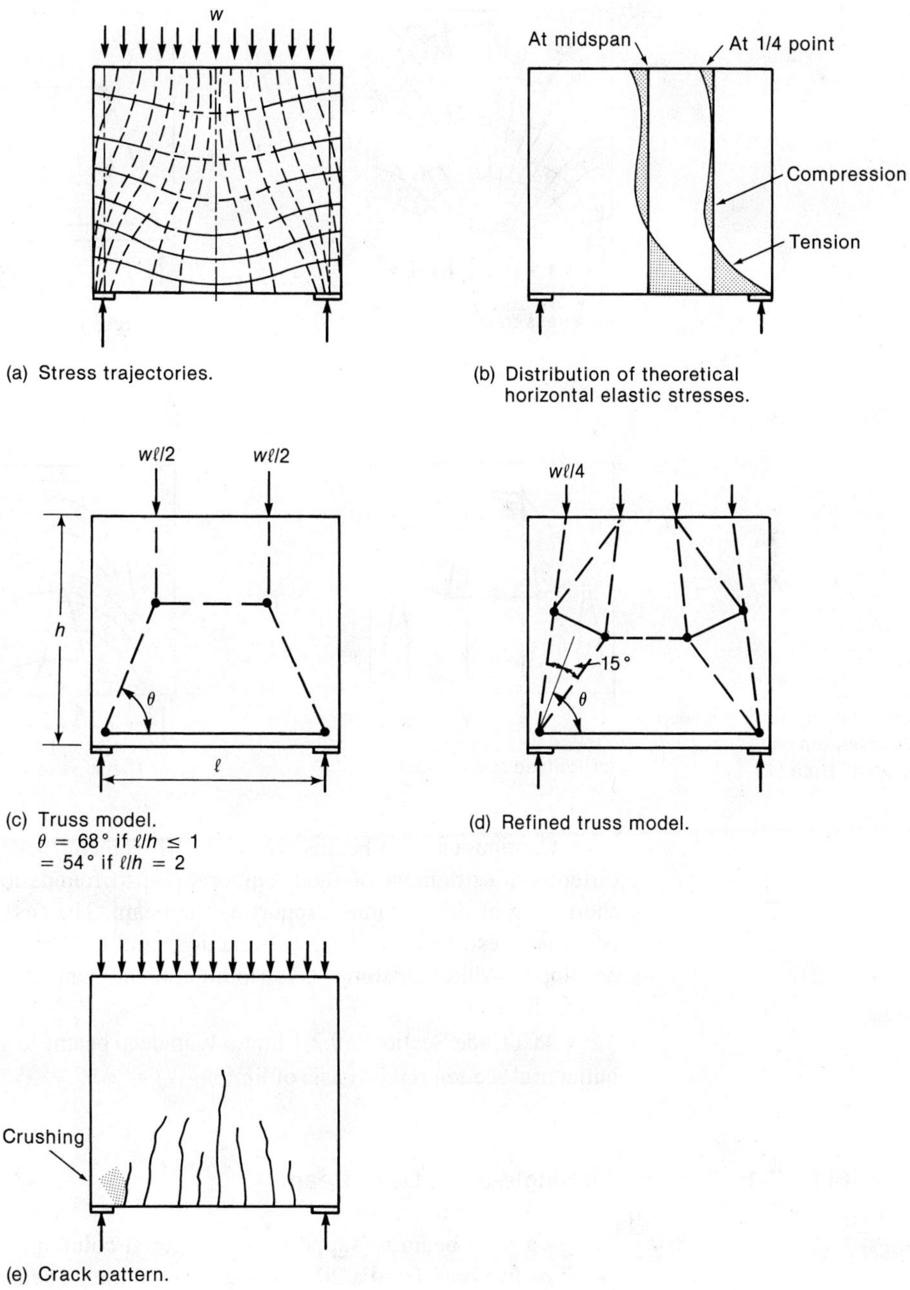

Fig. 17-26
Uniformly loaded deep beam.
(Adapted from [17-1].)

Design Using Strut-and-Tie Models

The design of a deep beam using a strut-and-tie model involves laying out a truss that will transmit the necessary loads. Once a satisfactory truss has been found, the joints and members of the truss are detailed to transmit the necessary forces. The overall dimensions of the beam must be such that the entire truss fits within the beam and has adequate cover. A detailed example that demonstrates the design of a (simply supported) deep beam using the requirements in ACI Code Chapter 23 is given in reference [17-6].

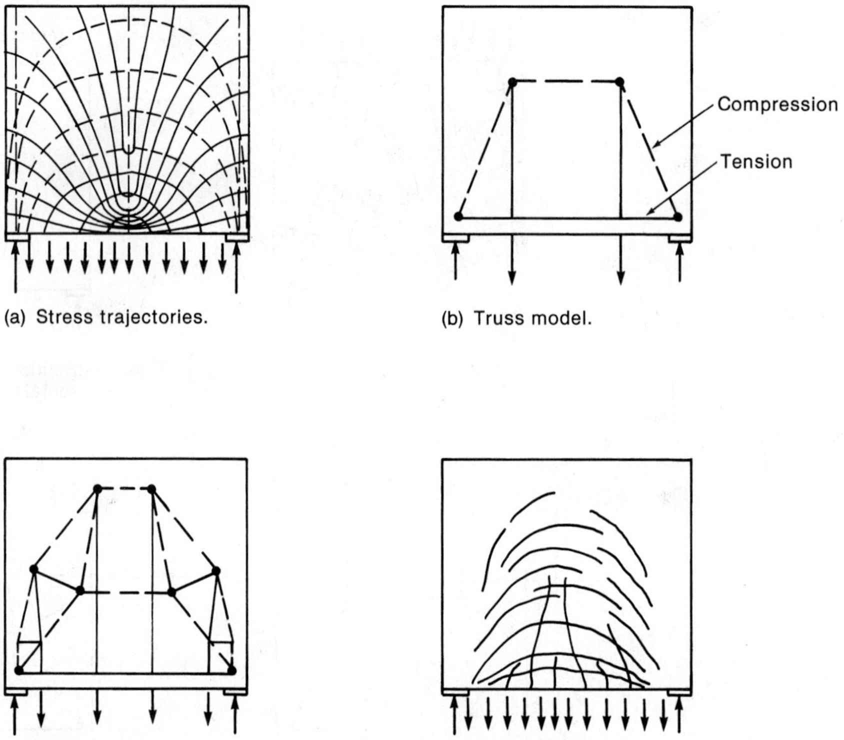

(a) Stress trajectories. (b) Truss model.

(c) Refined truss model. (d) Crack pattern.

Fig. 17-27
Deep beam uniformly loaded on the bottom edge.
(Adapted from [17-1].)

Continuous deep beams are very stiff elements and, as such, are very sensitive to differential settlement of their supports due to foundation movements, or differential shortening of the columns supporting the beam. The first stage in the design of such a beam is to estimate the range of reactions and use this to compute shear and moment envelopes. Although some redistribution of moment and shear may occur, the amount will be limited.

ACI Code Section 9.9.2.1 limits V_n in deep beams to $10\sqrt{f'_c}\,b_w d$. One can obtain an initial trial section on the basis of limiting $V_u = \phi V_n \cong \phi\,(7 \text{ to } 10)\,\sqrt{f'_c}\,b_w d$.

EXAMPLE 17-1 Design of a Single-Span Deep Beam

Design a deep beam to support an unfactored column load of 300 kips dead load and 340 kips live load from a 20-in.-by-20-in. column. The axes of the supporting columns are 6.5 ft and 10 ft from the axis of the loading column as shown in Fig. 17-28. The supporting columns are also 20 in. square and are assumed to act as simple supports. Use 4000-psi normal-weight concrete and Grade-60 steel. Use the load and resistance factors from ACI Code Tables 5.3.1 and 21.2.1.

1. Select a strut-and-tie model.

(a) Select shape and flow of forces: The assumed strut-and-tie model is shown in Fig. 17-28. It consists of two inclined struts, three nodal zones, and one tie. We shall assume that a portion of the column load equal to the left reaction flows down through strut A–B to the reaction at A. The rest of the column load is assumed to flow down strut B–C to the reaction at the right end of the beam.

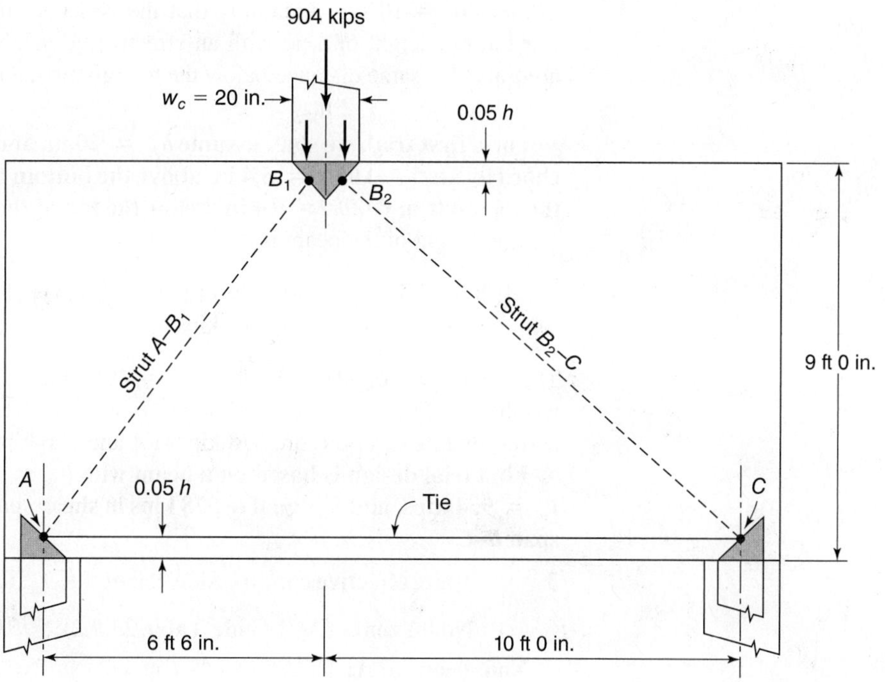

Fig. 17-28
First trial strut-and-tie model of a single-span deep beam—Example 17-1.

(b) Compute the factored load and the reactions: The factored load from the column is

$$P_u = 1.2D + 1.6L = 1.2 \times 300 + 1.6 \times 340 = 904 \text{ kips}$$

The reactions are 548 kips at A and 356 kips at C. The dead load of the beam will be added in after the size of the beam is known.

(c) Design format: In design we will use $\phi V_n = V_u$ where, from ACI Code Table 21.2.1, $\phi = 0.75$, for strut-and-tie models.

2. Estimate the size of the beam. This will be done in two different ways.

(a) Limit ϕV_n to $\phi \times (0.7 \text{ to } 1.0) \times 10\sqrt{f'_c}\, b_w d$: ACI Code Section 9.9.2.1 limits the shear in a deep beam to $10\sqrt{f'_c}\, b_w d$. To allow a little leeway and to leave some capacity for the beam dead load, we will select the beam using maximum shear forces of $\phi 7\sqrt{f'_c}\, b_w d$ to $\phi 10\sqrt{f'_c}\, b_w d$. The maximum factored shear ignoring the dead load of the beam is 548 kips, so

$$b_w d = \frac{V_n}{\phi \times (7 \text{ to } 10)\sqrt{f'_c}} = \frac{548{,}000 \text{ lb}}{0.75(7 \text{ to } 10)\sqrt{4000}} = 1650 \text{ to } 1160 \text{ in.}^2$$

Try b_w equal to the width of the columns, $b_w = 20$ in. This gives d from 82.5 to 58.0 in. Assuming that $h \approx d/0.9$ gives h from 64.4 to 91.7 in. Try a 20-by-96-in. beam.

(b) Select height to keep the flattest strut at an angle of ≈40° from the tie: Note that a limit of 40° is more restrictive than the limit of 25° in ACI Code Section 23.2.7. The strut-and-tie forces begin to grow rapidly for angles less than 40°. For the 120-in. shear span, the minimum height center to center of nodes is

120 tan 40° = 101 in. Assuming that the nodes at the bottom of this strut are located at midheight of a tie with an effective tie height of $0.1h$. We shall locate the node at B the same distance below the top, giving the required height of the beam as $h \approx 101/0.9 = 112$ in.

For a first trial, we shall assume $b_w = 20$ in. and $h = 108$ in. with the lower chord located at $0.05h = 5.4$ in. above the bottom of the beam and the center of the top node at $0.05h = 5.4$ in. below the top of the beam.

The weight of the beam is

$$[(20 \text{ in.}/12 \text{ in./ft.}) \times (108/12) \text{ ft} \times (16.5 + 1.67) \text{ ft} \times 0.150 \text{ kips/ft}^3)$$
$$= 40.9 \text{ kips}$$

The factored weight is $1.2 \times 40.9 = 49.1$ kips. To simplify the calculations we shall add this to the column load giving a total factored load of 954 kips. The corresponding reactions are 578 kips at A and 376 kips at C.

First trial design is based on a beam with $b_w = 20$ in. and $h = 108$ in., with $P_u = 954$ kips, and V_u equal to 578 kips in shear span A–B, and 376 kips in shear span B–C.

3. **Compute effective compression strengths, f_{ce}, for the nodal zones and struts.**

(a) **Nodal zones (ACI Code Table 23.9.2):**

Nodal zone at A: This is a C–C–T node.

$$f_{ce} = 0.85\beta_n f'_c = 0.85 \times 0.80 \times 4000 = 2720 \text{ psi}$$

Nodal zone at B: This is a C–C–C node.

$$f_{ce} = 0.85\beta_n f'_c = 0.85 \times 1.0 \times 4000 = 3400 \text{ psi}$$

Nodal zone at C. This is a C–C–T node.

$$f_{ce} = 2720 \text{ psi.}$$

(b) **Struts (ACI Code Table 23.4.3):**

Strut A–B: This strut has room for the width of the strut to be bottle-shaped. We will provide reinforcement satisfying ACI Code Section 23.5.3.

$$f_{ce} = 0.85\beta_s f'_c = 0.85 \times 0.75 \times 4000 = 2550 \text{ psi}$$

Strut B–C: This also is a bottle-shaped strut, and $f_{ce} = 2550$ psi

The effective compression strengths of the struts govern. **Use $f_{ce} = 2550$ psi** throughout.

4. **Locate nodes—first trial**

Nodal zones at A and C: In step 2(a), we assumed that the nodes at A and C are located at $0.05h = 5.4$ in. above the bottom of the beam. Use this location in the first trial strut-and-tie model.

Nodal zone at B: Assume node B is located at $0.05h = 5.4$ in. below the top of the beam. The node at B is subdivided into two subnodes, as shown in Fig. 17-29, one assumed to transmit a factored load of 578 kips to the left reaction and the other transmitting 376 kips to the right reaction. Each of the subnodes is assumed to receive load from a vertical strut in the column over B. Both of the subnodes are C–C–C nodes. The effective strength of the nodal zone at B is $f_{ce} = 3400$ psi. For

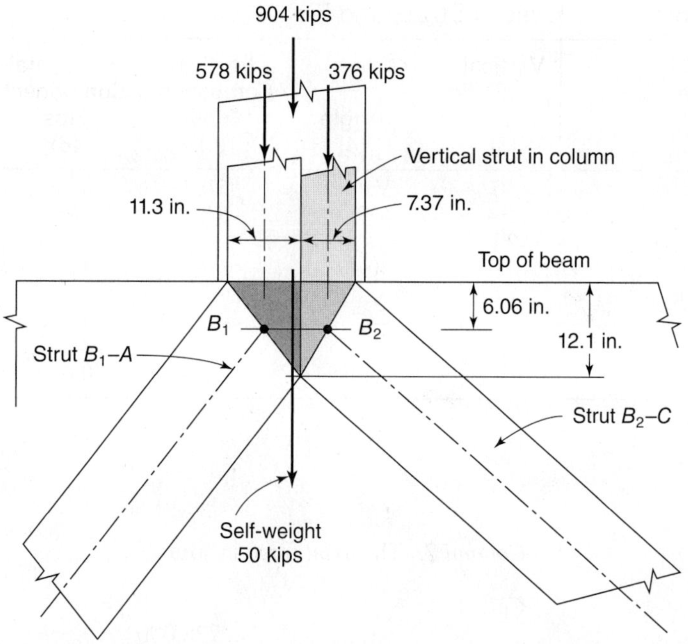

Fig. 17-29
Struts and nodal zones adjacent to the column at B—Example 17-1.

struts A–B_1 and B_2–C, $f_{ce} = 2550$ psi. We shall use $f_{ce} = 2550$ psi for the faces of the nodal zone loaded by bottle-shaped struts, and $f_{ce} = 3400$ psi for the face of the nodal zone that is loaded by compression stresses in the column and for the vertical struts in the column over point B.

$$w_{B1} = \frac{578{,}000}{0.75 \times 3400 \times 20} = 11.3 \text{ in.}$$

$$w_{B2} = 7.37 \text{ in.}$$

The total width of the vertical struts in the column at B is $\Sigma w_{BS} = 18.7$ in.

This will fit in the 20-in.-square column. If it did not, ACI Code Section 23.4.1(b) allows us to include the capacity of the column bars, $\phi A_s f_y$, when computing the capacity of the strut.

Because the vertical load components acting at B_1 and B_2 are different, the vertical division of node B is not located at the center of the column. Assuming equal widths for the unloaded portions of the column outside the two vertical struts in Fig. 17-29, the resultant force in the left-hand strut acts at point B_1, which is 3.7 in. to the left of the center of the column, and the right-hand strut acts at point B_2, located 5.7 in. to the right of the center.

The horizontal distance from A to B is 78 in.; that from B to C is 120 in.

5. Compute the lengths and widths of the struts and ties and draw the first trial strut-and-tie model

(a) Geometry and forces in struts and tie:

Strut A–B_1: The calculations for strut A–B_1 in Table 17-3 are

Column 4: The angle between the axis of the strut and horizontal $= \arctan \dfrac{97.2}{74.3}$ $= 52.6°$

Column 5: The vertical component is equal to the shear $= 578$ kips

Column 6: The horizontal component $= \dfrac{\text{Vertical component}}{\tan 52.6°} = \dfrac{578}{1.307} = 442$ kips

TABLE 17-3 Geometry and Forces in Struts and Tie

Strut or Tie (1)	Horizontal Projection in. (2)	Vertical Projection in. (3)	Angle (4)	Vertical Component kips (5)	Horizontal Component kips (6)	Axial Force kips (7)	Width of Strut in. (8)
Vertical* at *A	0.0	—	Vertical	578	0	578	15.1
***A–B*$_1$**	78.0 − 3.7 = 74.3 in.	108 − 5.4 − 5.4 = 97.2 in.	52.6	578	442	728	19.0
B*$_2$*–C	120 − 5.7 = 114.3 in.	97.2	40.4	376	442	580	15.2
A–C	—	—	—	—	442 at *A* 442 at *C*	442 at *A* 442 at *C*	$w_t = 10.8$
Vertical* at *C	0.0	—	Vertical	376	0	376	9.83

Column 7: The axial force in strut A–$B_1 = \dfrac{578 \text{ kips}}{\sin 52.6°} = 728$ kips

Column 8:

$$w_s = \frac{728{,}000}{0.75 \times 2550 \times 20} = 19.0 \text{ in.}$$

Vertical strut at *A*: The width of the strut under the node at *A* is

$$w_{sA} = \frac{578{,}000}{0.75 \times 2550 \times 20} = 15.1 \text{ in.}$$

Check the width of strut A–B_1 at A, using Eq. (17-15), where angle θ is 52.6°, $\ell_b = 15.1$ in., and $w_t = 10.8$ (calculated below).

$$w_s = w_t \cos\theta + l_b \sin\theta = 10.8 \cos 52.6° + 15.1 \sin 52.6° = 18.6 \text{ in.}$$

Because the 18.6 in. is less than the 19.0 in. required from column 8 of Table 17-3, we should increase w_t or ℓ_b to increase w_s to at least 19.0 in. By increasing the width, ℓ_b, of the vertical strut under node A from 15.1 in. to 15.7 in., the 19.0 width of strut A–B_1 can be accommodated.

Vertical strut at *C*: The width of the strut under the node at *C* is

$$w_{sC} = \frac{376{,}000}{0.75 \times 2550 \times 20} = 9.83 \text{ in.}$$

Strut B_2–*C*: The calculations for strut B_2–C are given in Table 17-3. The bearing length at Node C must be increased to 10.8 in. to reach a strut width of 15.2 in. at Node C.

Tie *A–C*: The force in tie A–C is calculated in Table 17-3. The axial (horizontal) force in the tie is equal to the horizontal forces in the struts at the two ends of the tie.

The axial force in tie based on the geometry at $A = 442$ kips

The axial force based on the geometry at $C = 442$ kips

These should be the same. Slight differences may occur due to approximations in the location of the nodes. The agreement is a partial check on the solution. The effective width of the tie A–C is

$$w_t = \frac{442{,}000}{0.75 \times 2720 \times 20} = 10.8 \text{ in.}$$

The upper limit on the height of the centroid of the tie reinforcement is $10.8/2 = 5.4$ in.

(b) Check whether the beam is a deep beam: ACI Code Section 9.9.1.1 defines a deep beam as one with shear span-to-height ratio less than or equal to 2. This code section also requires that there be a compression strut between the load and the reaction. From step 5(a) in Table 17-3, the horizontal projection of the left shear span is 74.3 in. The height is 108 in. The shear span-to-depth ratio is $74.3/108 = 0.69$. This is less than 2, so the shear span is deep. The right shear span is also deep.

6. Compute tie forces, reinforcement, and the effective width of the tie

(a) Trial choice of reinforcement in tie *A–C*: From Table 17-3, the force in tie *A–C* is 442 kips.

$$A_s = \frac{442 \text{ kips}}{0.75 \times 60 \text{ ksi}} = 9.82 \text{ in.}^2$$

We could use ten No. 9 bars, $A_s = 10.0$ in.^{2}, development length, $\ell_d = 53.3$ in. (Table A-6), minimum web width if bars are in two layers is $b_{w,\min} = 14.5$ in. (Table A-5).

Thirteen No. 8 bars, $A_s = 10.3$ in.^{2}, development length, $\ell_d = 47.4$ in.; $b_{w,\min}$ for three layers is 13.5 in.

Eighteen No. 7 bars, $A_s = 10.8$ in.^{2}, development length, $\ell_d = 41.5$ in.; $b_{w,\min}$ is 15.0 in. for three layers.

Try thirteen No. 8 bars. No. 8 bars will be easier to anchor than No. 9 bars. The bars could fit into two layers, although there would be relatively little free space for inserting vibrators when consolidating the concrete. We will use one layer of five bars and two layers of four bars and we will use No. 10 bars as spacers between the layers.

(b) Minimum flexural reinforcement: ACI Code Section 9.9.3.2 requires that deep beams have at least the minimum flexural reinforcement required by ACI Code Section 9.6.1. This is

$$A_{s,\min} = \frac{3\sqrt{f'_c}}{f_y} b_w d \qquad \text{but not less than } 200 b_w d / f_y$$

$$= \frac{3\sqrt{4000}}{60{,}000} \times 20 \times 102.6 = 6.49 \text{ in.}^2$$

but not less than $\dfrac{200 b_w d}{f_y} = 6.84$ in.^{2}.

The reinforcement provided in the tie is o.k.

(c) Effective width of tie *A–C*, w_t: ACI Commentary Section R23.8.1(b) suggests that the tie be spread over a maximum width equal to the tie force divided by the effective compression strength for the nodes at *A* and *C*. The maximum width calculated here is $w_t = 10.8$ in. The maximum height of the centroid of the reinforcements is $10.8/2 = 5.4$ in. above the bottom of the beam.

If the bars are placed in three layers with No. 4 stirrups and No. 10 bars as spacers between the layers, the centroid of the tie reinforcement is

$$y = \frac{5(2.5) + 4(4.77) + 4(7.04)}{13}$$

$$= 4.60 \text{ in. above the bottom}$$

This corresponds to a width of tie, $w_t = 9.20$ in.

For tie *A–C*, try thirteen No. 8 bars with one layer of five bars and two layers of four bars with 1.27 in. (No. 10 bar) vertical spaces between layers.

(d) Anchor the tie reinforcement at *A*: The tie reinforcement must be anchored for the full yield strength at the point where the axis of the tie enters the struts at *A* and *C*. Assuming an extended nodal zone with node *A* directly over the center of the column, the length available to anchor the tie is (see Fig. 17-30):

Distance from the center of the 20 in. column to the outside face of the cut-off or hooked bars $= 10 - (1.5 + 0.5) = 8.0$ in. The anchorage length available is

$$= 8.0\text{ in.} + \left(\frac{15.1}{2}\text{ in.}\right) + (4.6/\tan 52.6°)\text{ in.} = 19.1\text{ in.}$$

Because $\ell_d = 47.4$ in, the length available to develop anchorage of straight bars is insufficient. We must use hooks or some similar method to anchor the bars.

Try a standard 90° hook with $\ell_{dh} = 19$ in. (Table A-8) and multipliers from:

ACI Table 25.4.3.2 for tail cover $= 2.0$ in., and side cover $= 1.5 + 0.5 + 1.0 = 3.0$ in. (assuming the bars in the column are No. 8 bars and the tie bars are inside the column bars) $\times\ 0.7$

ACI Table 25.4.3.2 We shall ignore the effect of ties in the joint, thus, $\times\ 1.0$.

$\ell_{dh} = 19.0 \times 0.7 \times 1.0 = 13.3$ in. Check whether this will fit into a 20-in.-square column.

We could use thirteen No. 8 bars with 90° hooks in three layers, with 2-in. tail cover on the lower layer and (1.5 in. cover) + (0.5 in. tie) + 1.0 in. bar + (1.0 in. space between hooks) = 4.0 in. cover on the tails of the second layer, leaving 15.1 in. for

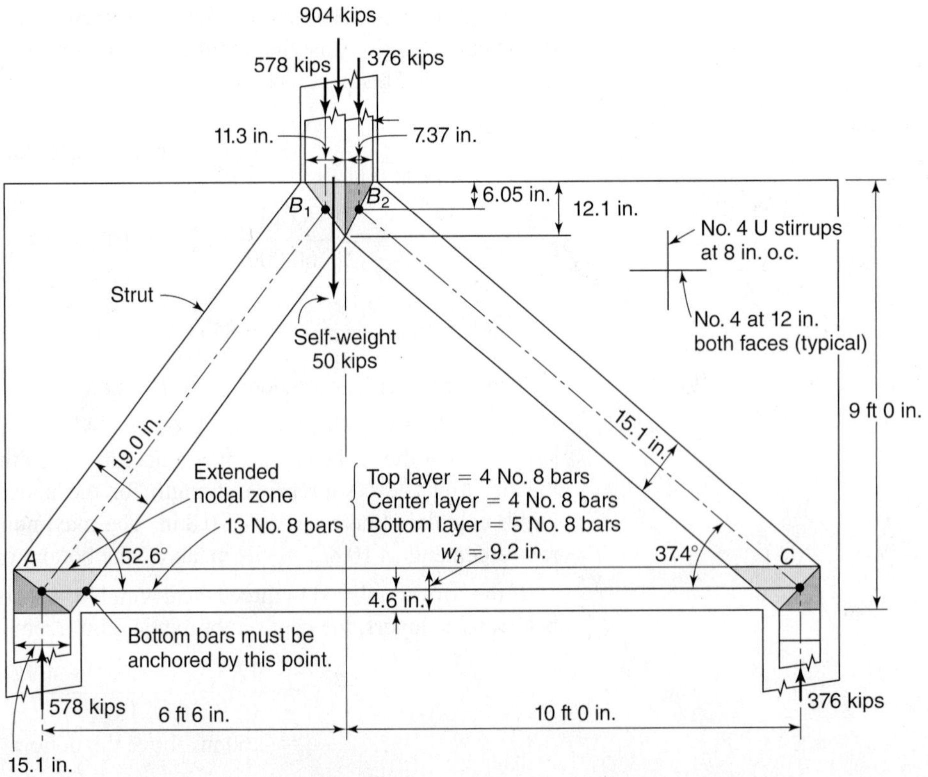

Fig. 17-30
Final design single-span deep beam—Example 17-1.

the hooks on the second layer, and 4.0 in. + 1.0 + 1.0 in. = 6.0 in. cover on the tails of the hooks in the third layer, leaving 13.1 in. for the third layer of hooks (should be o.k.).

Use thirteen No. 8 bars, five in the bottom layer, and two layers of four bars, with 90° hooks inside the column reinforcement.

Alternatively, we could lap splice the tie bars with the legs of horizontal hairpin bars. Assuming the five bars in the lower layer have 90° hooks the remaining bars could be anchored using two No. 8 horizontal hairpin bars lying flat per layer, with the legs of each hairpin lap spliced $1.3\ell_d = 61.6$ in., say 62 in. From ACI Code Table 25.3.1 the total out-to-out width of a No. 8 hairpin bar is $8d_b = 8.0$ in. Two of these would fit between the stirrups in the 20-in. width of the beam.

(e) Check height of node *B*: Summing horizontal forces on a vertical section through the entire beam and the nodal zone at B shows that the horizontal force in the struts meeting at B equals the horizontal force in tie A–C which will be taken equal to $\phi A_s f_y$, thus, $\phi F_{nn} = 0.75 \times 10.3 \text{ in.}^2 \times 60 \text{ ksi} = 464$ kips. Total depth of the nodal zone at B is $464{,}000/(0.75 \times 2550 \times 20) = 12.1$ in. The top nodes, B_1 and B_2, are located at $12.1/2 = 6.05$ in. below the top of the beam. (See Fig. 17-30.)

7. Second trial strut-and-tie model. This step is required if there is a significant change in the geometry of the assumed truss. In this case there was not much change and this step will be skipped.

8. Compute the required crack-control reinforcement.

(a) Minimum reinforcement from ACI Code Section 9.9.3: ACI Code Section 9.9.3.1(a) requires vertical reinforcement not less than $0.0025b_w s$ at a maximum spacing of $d/5 = 103/5 = 20.6$ in. but not more than 12 in.

Try two No. 4 legs at 8 in. on centers each face. $A_v/b_w s = \dfrac{2 \times 0.20 \text{ in.}^2}{20 \text{ in.} \times 8 \text{ in.}} = 0.0025$, exactly right.

ACI Section 9.9.3.1(b) requires horizontal reinforcement not less than $0.0025b_w s$ at a maximum spacing of $d/5 = 20.6$ but not more than 12 in.

Try No. 5 bars at 12 in. on centers each face. $A_{vh}/b_w s = \dfrac{2 \times 0.31}{20 \times 12} = 0.0026$, o.k.

(b) Minimum reinforcement from ACI Code Section 23.5: ACI Code Section 23.5.3 requires an orthogonal grid of bars at each face if $\beta_s = 0.75$ is used. The amount of steel is either computed using a local strut-and-tie model for the end of a bottle-shaped strut, or if f'_c is not greater than 6000 psi, the steel is adequate if it satisfies:

$$\frac{\Sigma A_{si}}{bs_i} \sin \alpha_i \geq 0.003 \tag{17-10}$$

(ACI Eq. 23.5.3)

Left End: The angle between the axis of the strut and the horizontal steel is,

$$\alpha_i = \arctan \frac{97.3}{74.3} = 52.6°$$

Vertical Steel: We selected No. 4 bars at 8 in. on centers vertically on two faces.

The angle between the vertical steel and the axis of the struts is $90° - 52.6° = 37.4°$. Therefore,

$$A_s/(b_w s_i) \sin \alpha_i = 2\left(\frac{0.20}{20 \times 8}\right) \sin 37.4° = 0.00152$$

Horizontal Steel: The angle between the strut and the horizontal is 52.6°. Thus, $\alpha_i = 52.6°$. We selected No. 5 bars at 12 in. on centers, horizontally on two faces, to get

$$A_s/(b_w s_i)\sin\alpha_i = 2\left(\frac{0.31}{20 \times 12}\right)\sin 52.6° = 0.00205$$

Total Steel:

$$\Sigma A_s/(b_w s_i)\sin\alpha_i = 0.00152 + 0.00205 = 0.00357 \text{ (o.k.)}$$

Using the same mesh of reinforcement for the right-hand side of the beam gives

$$\Sigma(A_s/b_w s_i)\sin\alpha_i = 0.00358$$

(c) Alternate solution for crack-control reinforcement using a local strut-and-tie model: The region in which the strut stresses spread out across the width of the strut will be modeled using a local strut-and-tie model. Figure 17-28 shows strut A–B_1. The axial force in the strut is 728 kips. Half of this (364 kips) will be assumed to act at the center points of each half of the width of the end of the strut (at the quarter points of the width). The resulting strut-and-tie model is similar to Fig. 17-4c. ACI Code Section 23.5.1 assumes the resultant forces spread at a slope of 2 longitudinally and 1 transverse. Thus, the transverse force at one end of the strut is $364/2 = 182$ kips. The sum of the transverse forces at the two ends of the strut is $182 + 182 = 364$ kips. Using No. 4 vertical bars at 8 in. on centers at an angle of 52.6° to the crack and No. 5 horizontal bars at 12 in. on centers at an angle of 37.4° to the crack, there will be $(108 \text{ in.})/(12 \text{ in./bar}) = 9$ horizontal bars per face, which can resist a force of

$$2 \times 9 \times 0.31 \times 60 \times \sin 52.6° = 266 \text{ kips}$$

perpendicular to the crack.

In the left shear span, there will be $(78/8) = 9.75$ (use 9) vertical bars per face, which can resist a force of

$$2 \times 9 \times 0.20 \times 60 \times \sin 37.4° = 131 \text{ kips}$$

perpendicular to the crack, for a total of $266 + 131 = 397$ kips. This exceeds 364 kips.

9. Summary of design.

(a) Use a 20-in.-by-108-in. beam with $f'_c = 4000$ psi and $f_y = 60{,}000$ psi.

(b) Provide thirteen No. 8 bars in three layers, five bars in the bottom layer, plus two layers of four bars each. Use No. 10 bars as spacers between the layers. Use 90° hooks on all the No. 8 bars, and place the bars inside the vertical reinforcement from the supporting columns.

(c) Use No. 4 at 8 in. on centers vertically on each face and No. 5 bars at 12 in. on centers horizontally on each face. ■

17-8 CONTINUOUS DEEP BEAMS

Reactions of Continuous Deep Beams

A deep beam is a very stiff element, and as a result, the reactions are strongly affected by the flexural and shear stiffnesses of the beam and by differential settlements of the supports. At one extreme, a very flexible member on rigid supports will have reactions similar to those computed from an elastic-beam analysis. At the other extreme, the three reactions for a flexurally stiff, two-span beam supported on axially soft columns will approach three equal reactions. In the case of short deep beams, a proper analysis will include both shearing deflections and flexural deflections.

EXAMPLE 17-2 Design a Two-Span Continuous Deep Beam

A beam spans between three supporting columns with centerlines 20 ft apart and supports a column at the middle of each span, as shown in Fig. 17-31. Both upper columns support a factored load of $P_u = 1500$ kips. The supporting columns are 24 in. × 20 in., 24 in. × 45 *in.*, and 24 in. × 20 in. The loading columns are 24 in. × 33 in. Assume that the beam is also 24 in. wide. Use a strut-and-tie model to design the beam. Use $f'_c = 4000$-psi normal-weight concrete and $f_y = 60$ ksi, and use the load and strength-reduction factors from ACI Code Tables 5.3.1 and 21.2.1.

1. General—select a strut-and-tie model

(a) Select the first trial strut-and-tie model: In general, the best strut-and-tie models minimize the amount of reinforcement. This will occur when compression forces are transmitted by struts directly to the nearest supports. The left end of an appropriate strut-and-tie model is shown in Fig. 17-31. It consists of three triangular trusses, each consisting of two compression struts and one tie. The load supported by each triangular truss depends on the division of the applied loads between the three reactions.

The load from the column at *B* is divided into three parts, each roughly a third of the column load. One part is transferred to the reaction at *A* by strut $A–B_1$. The change in direction of this load at *A* results in tension in the tie $A–C_1$ at *A*. The second part of the column load is transferred to the reaction at C_1 by strut $B_2–C_1$. This also gives a tension force in tie $A–C_1$. The tension forces at the two ends of tie $A–C_1$ should be equal. The final portion of the column load at *B* is transferred to the reaction at C_2 by strut $B_3–C_2$. For equilibrium at B_3, a tie $B_3–B'_3$ is required.

(b) Compute the reactions: The reactions were estimated by using an elastic analysis of the reactions from an elastic propped-cantilever beam, loaded with a concentrated load, P_u, at midspan, supported on a hinged support at the exterior end and a fixed interior support. The second span will be taken as a mirror image of the first. Each span will have a reaction of 0.312 P_u at the exterior supports and a reaction

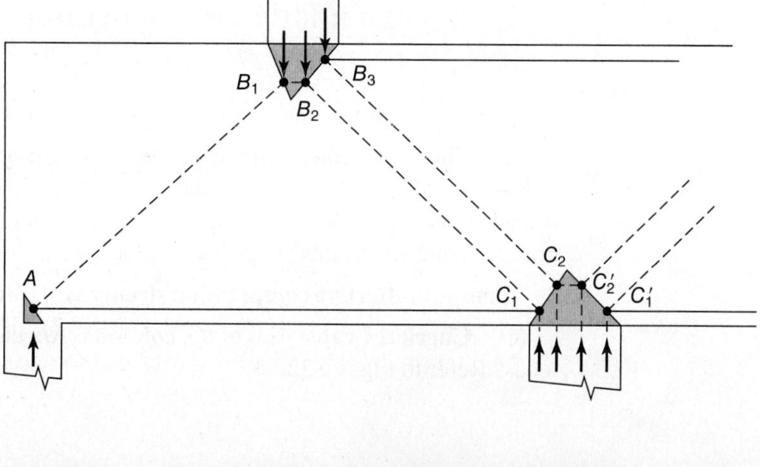

Fig. 17-31
Initial strut-and-tie model, two-span deep beam—Example 17-2.

of 0.688 P_u from each of the spans at the interior support, as shown in Fig. 17-32. In design, it usually is necessary to consider several distributions of loads and reactions with or without support settlements and to provide concrete reinforcement for the highest forces in each strut or tie.

(c) Design equation: Design will be based on

$$\phi F_n \geq F_u \tag{17-1a}$$

2. Estimate the size of the beam.

(a) Based on maximum allowable shear stress: ACI Code Section 9.9.2.1 limits the nominal shear in a deep beam to $\phi V_n = \phi 10\sqrt{f'_c} b_w d$. As a first trial, we shall limit V_n to the range from 0.6 to 0.9 times the maximum:

$$\text{Trial } \phi V_n = (0.6 \text{ to } 0.9) \times \phi 10\sqrt{f'_c}\, b_w d$$

Here, from ACI Code Table 21.2.1, $\phi = 0.75$ for strut-and-tie models. Ignoring the weight of the beam, the maximum shear, V_u is 1030 kips (0.688 × 1500 kips). We will add the weight of the beam later.

$$d = \frac{1{,}030{,}000}{0.75 \times (0.6 \text{ to } 0.9) \times 10\sqrt{4000} \times 24} = 151 \text{ to } 101 \text{ in.}$$

and $h \approx d/0.9 = 168$ to 112 in.

(b) Based on limiting the strut angle to not less than 40°: Assuming that the nodes at A, C_1, and B_3 are $0.05h$ above the bottom or below the top of the beam and that nodes B_1, B_2, C_2, and C'_2 are $3 \times 0.05h = 0.15h$ from the top or bottom of the beam, the first trial vertical projection of strut A–B_1 is

$$h - 0.05h - 0.15h = 0.8h$$

and the approximate horizontal projection is $a = 109$ in. (center-to-center of columns minus 33/3 in.), where the loading columns have a width of 33 in.

Thus,

$$\frac{0.8h}{109 \text{ in.}} \geq \tan 40^\circ$$

or,

$$h \geq 114 \text{ in.}$$

Try b_w = 24 in. and h = 120 in. The dead load of one 20-ft span is

$$W = (2 \text{ ft} \times 10.0 \text{ ft} \times 21 \text{ ft}) \times 0.150 \text{ kips/ft}^3 = 63.0 \text{ kips}$$
$$W_u = 1.2 \times 63.0 = 75.6 \text{ kips, say 76 kips}$$

Add the factored beam weight to the given factored loads, as shown in Fig. 17-32.

(c) Check whether this is a deep beam: $a = 109$ in., $h = 120$ in., and $a/h = 109/120 = 0.91$. ACI Code Section 9.9.1.1 defines a deep beam as one that is loaded on the top surface, is supported on the bottom of the beam, and has a/h not greater than 2. Thus, this beam is a deep beam.

3. Compute effective compression strengths, f_{ce}, for the nodal zones and struts.

(a) Check the capacities of the columns: Are the columns large enough for the loads? Refer to Fig. 17-32.

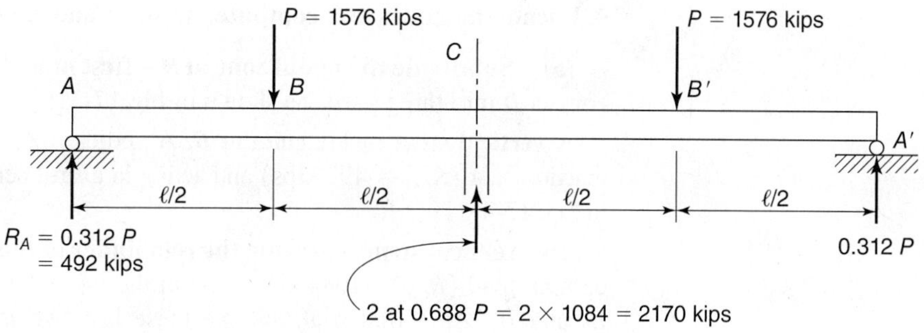

Fig. 17-32
Loads and assumed reactions, two-span deep beam—Example 17-2.

Columns at *A* and *A′*. The columns at *A* and *A′* are 24 in. by 20 in. in section. The factored loads in the columns at *A* and *A′* are $P_u = 492$ kips. From ACI Code Sections 22.4.2.1 and 22.4.2.2, the maximum axial load capacity of a 24 × 20-in. tied column with 1 percent longitudinal reinforcement is

$$\phi P_{n(\max)} = 0.80\,\phi[0.85 f'_c(A_g - A_{st}) + f_y A_{st}]$$

where ϕ is the strength-reduction factor for compression-controlled tied columns ($\phi = 0.65$). So, for the columns at *A* and *A′*, we obtain

$$\phi P_{n(\max)} = 0.8 \times 0.65[0.85 \times 4000(480 - 4.8) + 60{,}000 \times 4.8]$$
$$\phi P_{n(\max)} = 990{,}000 \text{ lbs} = 990 \text{ kips}$$

Because $\phi P_n > P_u$, the columns at *A* and *A′* are adequate with 1 percent steel.

Columns at *B* and *B′*. Similar calculations give the factored capacity $\phi P_{n(\max)} =$ 1630 kips, which exceeds both the 1500-kip load at the bottoms of the loading columns and the 1576-kip load in the part of the beam loaded by the column, which includes the factored dead load of the deep beam.

Again, $\phi P_n > P_u$. The columns over *B* and *B′* are adequate with 1 percent steel.

Column at *C*. The factored load on each half of column *C* is $P_u = 1084$ kips, for a total load of 2170 kips. Assuming a 45-in.-by-24-in. tied column with 1 percent steel, the column capacity is 2230 kips, which is more than the factored load of 2170 kips. If this were not true, it would be necessary to enlarge the column or to increase its capacity by increasing the column reinforcement.

(b) Compute the effective compression strengths of the nodal zones—first model:

Nodal zones *A* and *B* anchor one tension tie. They are *C–C–T* nodes. From ACI Code Table 23.9.2, $\beta_n = 0.80$ and $f_{ce} = 0.85 \times 0.80 \times 4000$ psi $= 2720$ psi.

Nodal zone *C* anchors two ties, one from each side of the line of symmetry. Because it can be subdivided into two *C–C–T* nodes, we shall base f_{ce} on the *C–C–T* nodal zones. From ACI Code Section A.5.2.2, $f_{ce} = 2720$ psi.

(c) Compute the effective compression strengths of the struts—first model: For struts, Eq. (17-7a) gives $f_{ce} = 0.85\beta_s f'_c$.

Struts A–B_1, B_2–C_1, and B_3–C_2 are all "bottle-shaped." The struts can spread laterally into the unstressed concrete adjacent to the struts on at least one side. Crack-control reinforcement will be provided to allow from ACI Code Table 23.4.3, $\beta_s = 0.75$ and $f_{ce} = 0.85 \times 0.75 \times 4000$ psi $= 2550$ psi.

4. Locate the nodes and compute the strut-and-tie forces—first model

(a) Subdivide the nodal zone at *B*—first model: We will subdivide the nodal zone at *B* into three parts, as shown in Fig.17-31:

A vertical strut inside column *B*. A vertical force equal to the nominal vertical reaction at *A* (R_{An} = 492 kips) and acting in a strut centered over node B_1, as shown in Fig. 17-31.

Two vertical struts carrying the remainder of the column load. The remaining column load (R_{Cn} = 1084 kips) acts in the two vertical struts centered over nodes B_2 and B_3. As a first trial, assume these last two forces are equal and are each 0.5×1084 kips = 542 kips.

(b) Compute the widths and locations of the vertical struts in column *B*—first model:

$$\phi F_n \geq F_u \qquad w_s = \frac{F_u}{\phi f_{ce} b}$$

where $f_{ce} = 0.85\beta_s f'_c$ for struts and $f_{ce} = 0.85\beta_n f'_c$ for nodes.

Struts in column over node *B*. For the vertical struts in the column over *B*, ACI Code Table 23.4.3 specifies $\beta_s = 1.0$ because these struts have parallel vertical stress trajectories:

$$f_{ce} = 0.85 \times 1.0 \times 4000 = 3400 \text{ psi}$$

Nodal zone at *B*. The node at *B* anchors one tension tie. From ACI Code Table 23.9.2, $\beta_n = 0.80$ and

$$f_{ce} = 0.85 \times 0.80 \times 4000 = 2720 \text{ psi}$$

We shall assume that this value of f_{ce} controls the strength of the entire nodal zone at *B*.

The width of the vertical strut above B_1 is

$$w_s = \frac{492{,}000}{0.75 \times 2720 \times 24} = 10.0 \text{ in.}$$

At each of B_2 *and* B_3, F_n = 542 kips and w_s = 11.1 in.

The total width of the three struts is (10.0 + 11.1 + 11.1) = 32.2 in. These fit into the 33-in.-wide column, leaving a space of 0.8 in. We shall assume that half of this space occurs on each edge of the column.

Assuming that the sum of the widths of the three struts is centered on the axis of the column at *B*, the locations of the axes of the three vertical struts are as follows:

Node B_1 is 11.1 in. to the left of the centerline of column *B*,
Node B_2 is 0.55 in. to the left, and
Node B_3 is 10.6 in. to the right.

(c) Width of the strut in column *A*—first model: By observation, this is the same width as the column strut at B_1, 10.0 in. This fits inside the 22-in. column width at *A*.

(d) Widths and locations of the vertical struts in the column under *C*—first model: The total factored vertical reaction at *C* from one span is 1084 kips. Assume that each vertical strut in the column at *C* resists half of this, 542 kips. f_{ce} for the struts under *C* is 3400 psi, and for the nodal zone is 2720 psi. The nodal zone governs, so use f_{ce} = 2720 psi both for the nodal zone and for the struts in the column under node C_1. The width of each of the four vertical struts at *C* is 11.1 in., for a total width of 44.4 in. This will fit into the 45-in. column, with a gap of 0.3 in. on each side. Node C_1 is 16.7 in. left of the column centerline. Node C_2 is 5.5 in. to the left. (See Fig. 17-31.)

(e) Assume the vertical positions of the nodes—first model: We will assume that the nodal zones at B and C each consist of two layers, each with a height of $0.10h$, as shown in Fig. 17-31 and Table 17-4. This is arbitrarily based on an assumed depth of flexural compression, a, equal to $0.2h$. Assume that the tie reinforcement is spread over the height of the layer representing the tie; then the centroid of tie A–C_1 can be taken to be $0.05h = 6.0$ in. above the bottom of the beam, and that of tie B_3–B_3' to be 6.0 in. below the top of the beam. The centroid of the second layer is at $0.15h = 18.0$ in. above or below the top of the beam.

(f) Summary of locations of the nodes—first model: The node locations are given in Table 17-4.

Based on the results obtained with the first model, a second strut-and-tie model with a revised geometry will be analyzed.

5. Compute forces in the struts and ties due to factored loads—second model.

(a) Forces in struts and ties—second model: These are computed in Table 17-5. We have

$$\textit{Column 4:}\ \text{Angle} = \arctan \frac{96.0}{108.9} = 41.4°$$

Column 5: Vertical component is the factored shear in shear span A–B, or 492 kips.

TABLE 17-4 Summary of Locations of the Nodes—First Model

Point	Left of Center of Column A, in.	Left of Center of Column B, in.	Left of Center of Column C, in.	Below Top of Beam, in.	Above Bottom of Beam, in.
A	0	—	—	—	6.0
B_1	—	11.1	—	18.0	—
B_2	—	0.55	—	18.0	—
B_3	—	−10.6	—	6.0	—
C_1	—	—	16.9	—	6.0
C_2	—	—	5.8	—	18.0

TABLE 17-5 Geometry and Forces in Struts and Ties—Second Model

Strut or Tie (1)	Horizontal Projection in. (2)	Vertical Projection in. (3)	Angle θ (4)	Vertical Component kips (5)	Horizontal Component kips (6)	Axial Force kips (7)
Strut A–B_1	$120 - 11.1 = 108.9$	$120 - 6.0 - 18.0 = 96.0$	41.4°	492	558	744
Strut B_2–C_1	$120 + 0.55 - 16.7 = 103.9$	$120 - 6.0 - 18.0 = 96.0$	42.7°	542	587	799
Strut B_3–C_2	$120 - 10.6 - 5.5 = 103.9$	$120 - 6.0 - 18.0 = 96.0$	42.7°	542	587	799
Tie A–C_1	—	—	—	—	558 at A 587 at C_1	
Tie B_3–B_3'	—	—	—	—	—	587

$$\textit{Column 6:}\ \text{Horizontal component} = \frac{\text{Vertical component}}{\tan 41.4} = 558\ \text{kips}$$

$$\textit{Column 7:}\ \text{Axial force} = \frac{\text{Vertical component}}{\sin 41.4} = 744\ \text{kips}$$

(b) Balance the tie forces if necessary—second model: From Table 17-5, the tie force from strut A–B_1 is 558 kips, and that from strut B_2–C_1 is 587 kips. These should be the same. If not, several strategies are possible. We will try to maintain the reactions chosen earlier. Because of this, we will not change the load in strut A–B_1.

Reduce the horizontal component of the force in strut B_2–C_1 to 558 kips, so that the tie forces are the same at the two ends of tie A–C_1. This will be done by reducing the vertical force in B_2–C_1 and increasing the vertical force in B_3–C_2 to keep the reaction at C equal to 1084 kips. We thus have

$$\text{Vertical force at } B_2 = \frac{558}{587} \times 542 = 515\ \text{kips}$$

and

$$\text{Vertical force at } B_3 = 1084 - 515 = 569\ \text{kips}$$

The geometry at node B changes: For the struts in the column over B, $f_{ce} =$ 2720 psi; for the vertical strut at B_1, $w_s = 10.0$ in.,

w_s at $B_2 = 10.5$ in.
w_s at $B_3 = 11.6$ in.

Total $w_s = 10.0 + 10.5 + 11.6 = 32.1$ in. This fits into a 24-by-33-in. column with 0.45 in. space on each side. From the geometry of the vertical struts at B (see Fig. 17-33a), B_1 is 11.1 in. left of the centerline of the column, B_2 is 0.8 in. left, and B_3 is 10.3 in. right.

After the tie forces have been balanced, the vertical forces at C_1 and C_2 are 515 and 569 kips, giving struts of widths 10.5 in. below C_1 and 11.6 in. below C_2, which locate C_1 at 16.9 in. left of the center of the column at C and C_2 at 5.8 in. left (Fig. 17-33b). After the tie forces are balanced, the axial forces in B_2–C_1 and B_3–C_2 are 759 and 839 kips, respectively. These values will be used in Table 17-6.

The balanced model with updated node locations will be called the *second model.*

(c) Compute the widths of struts and ties due to factored loads—second model: The calculations in Table 17-6 proceed as follows:

Column (2) is based on Table 17-5, the axial force in B_2–C_1 is 759 kips and that in B_3–C_2 is 839 kips, both after the balancing of the tie force in A–C_1.

Column (3) is the value of f_{ce} for strut A–B_1 at end A.

Column (4) is the value of f_{ce} for the nodal zone at A. The values not crossed out govern.

Column (5) is the computed width of the strut, for the governing value of f_{ce} from columns (3) and (4).

There are two entries for each strut in Table 17-6, to accommodate potentially different values of f_{ce} at the two ends of the struts.

(d) Effective width of ties—second model: ACI Commentary Section R23.8.1, limits the effective width of the tie, w_t, to:

- Twice the minimum height from the bottom or top of the beam to the centroid of the tie reinforcement,

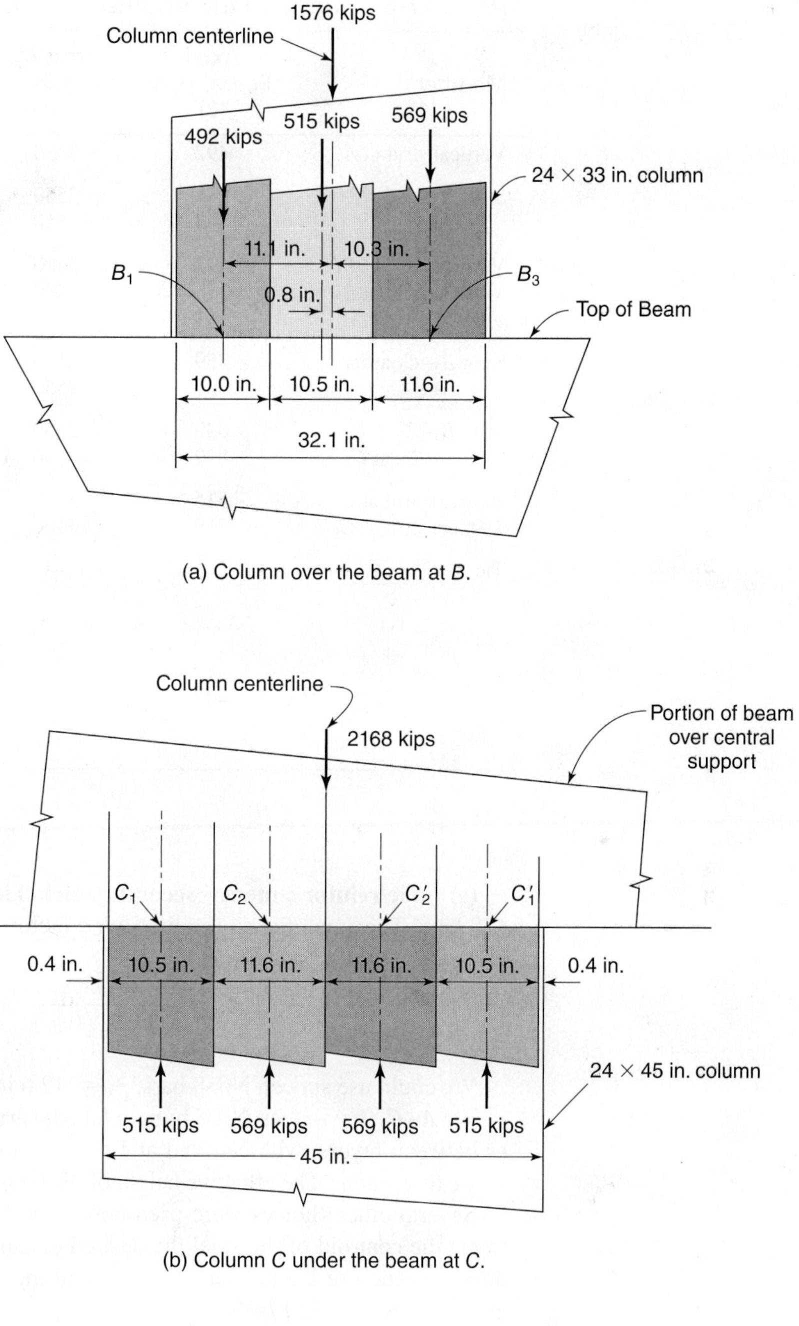

Fig. 17-33
The widths of vertical struts in the columns at *B* and *C*—Two-span deep beam—Example 17-2.

- The width, w_t, computed from the tie force by assuming that the tie is stressed to the value f_{ce} for the nodal zones at the ends of the tie.

From part 5 (b), the effective width of the concrete concentric with tie A–C_1 is

$$w_{t,\max} = \frac{558{,}000}{0.75 \times 2720 \times 24} = 11.4 \text{ in.}$$

where 2720 psi is f_{ce} for a C–C–T node. The height of the centroid of such a tie would be at $11.4/2$ in. $= 5.70$ in.

TABLE 17-6 Compute the Widths of Struts and Ties—Second Model

Member (1)	Axial Force, kips (2)	Strut f_{ce} psi (3)	Node f_{ce} psi (4)	Width in. (5)
Vertical strut at A	492	~~3400~~	2720	10.0
Strut A–B_1 at A	744	2550	~~2720~~	16.2
at B_1	744	2550	~~2720~~	16.2
Vertical strut at B_1	492	~~3400~~	2720	10.0
Vertical strut at B_2	515	~~3400~~	2720	10.5
Vertical strut at B_3	569	~~3400~~	2720	11.6
Strut B_2–C_1 at B_2	759	2550	~~2720~~	16.5
at C_1	759	2550	~~2720~~	16.5
Strut B_3–C_2 at B_3	839	2550	~~2720~~	18.3
Strut B_3–C_2 at C_2	839	2550	~~2720~~	18.3
Vertical strut at C_1	515	~~3400~~	2720	10.5
Vertical strut at C_2	569	~~3400~~	2720	11.6
Tie A–C_1	558	—	2720	$w_t = 11.4$ in. $w_t/2 = 5.70$ in. $A_s = 12.4$ in.^{2}
Tie B_3–B_3'	617		2720	$w_t = 12.6$ in. $w_t/2 = 6.3$ in. $A_s = 13.7$ in.^{2}

(e) Tie reinforcement—second model: Tie A–C_1 forces at A and C_1 are each 558 kips. The area of steel required for a factored tie force of 558 kips is

$$A_s = \frac{558 \text{ kips}}{0.75 \times 60 \text{ ksi}} = 12.4 \text{ in.}^2$$

We could use sixteen No. 8 bars, $A_s = 12.6$ in.^{2}.

Tie A–C_1, try sixteen No. 8 bars in four layers of four bars each, with 1-in. spaces between layers, with centroid at $1.5 + 0.5 + 1.0 + 1.0 + 1.0 + 0.5 = 5.5$ in. above the bottom. The effective height of the tie is 11.0 in.

Several other choices were tried before the final selection of the reinforcement to get the centroid of the steel the desired distance from the bottom of the beam to near the center of the tie. The anchorage of the tie reinforcement in the nodal zone at A is shown in Fig 17-35.

6. Recompute w_t and revise the location of the nodes—second model:

Nodes A and C_1 will be located 5.5 in. above the bottom of the beam, and the effective height of tie A–C_1 is 11.0 in. Node C_2 will be taken at 1.5 times the effective height of tie A–C_1 above the bottom of the beam—that is, at $1.5 \times 11.0 = 16.5$ in. above the bottom.

Tie $B3$–$B3'$: The revised axial force in tie B_3–B_3' is $\dfrac{569}{\tan 42.7°} = 617$ kips,

$$A_s = \frac{617 \text{ kips}}{0.75 \times 60 \text{ ksi}} = 13.7 \text{ in.}^2$$

$$w_t = \frac{617{,}000}{0.75 \times 2720 \times 24} = 12.6 \text{ in.}$$

and

$$w_t/2 = 6.30 \text{ in.}$$

(a) Tie $B3 - B3'$:—Use eighteen No. 8 bars ($A_s = 14.2$ in.^{2}) in four layers; a top layer of six bars plus three layers of four bars each, with transverse No. 11 bars as spacers between layers. The centroid of the steel at B_3 is 6.11 in. below the top of the beam, and nodes B_1 and B_2 are $1.5(2 \times 6.11) = 18.3$ in. below the top. Placing only four or six bars in a layer also allows vertical spaces to ease the vibrating of the concrete.

Based on the results to this point, a slightly revised (third) strut-and-tie model will be used to complete the analysis and design of this continuous beam.

(b) Check agreement between the widths of the struts and the nodal zones—third model: Equation (17-15) relates the widths of the struts, the bearing lengths, and the widths of ties at nodal zones.

Node A–Strut $A-B_1$:

Minimum $w_s = \ell_b \sin \theta_A + w_t \cos \theta_A$

Bearing length: $\ell_b = 10.0$ in.

Height of tie [from steel arrangement chosen in step 5(e)]: $w_t = 11.0$ in.

The modified angle between strut and horizontal: $\theta_A = 41.6°$

In order to maintain the reactions chosen earlier, we will not change the load in strut A–B_1.

$$\begin{aligned} w_s &= 10.0 \sin 41.6° + 11.0 \cos 41.6° \\ &= 6.64 + 8.23 = 14.9 \text{ in.} \end{aligned}$$

Because $w_s = 16.2$ in. from Table 17-6 exceeds this value, the strut A–B_1 is not adequate. Increase ℓ_b to 13.0 in., giving minimum strut width = 16.9 in. This exceeds the $w_s = 16.2$ from Table 17-6 and therefore is adequate.

Node C, Struts $B2$–$C1$ and $B3$–$C2$:

Bearing length: $\ell_b = 10.5 + 11.6 = 22.1$ in.

Total height of nodal zone from steel chosen is equal to $2\ (2 \times 6.11) = 24.4$ in.

The revised angle is 42.9°

$$\begin{aligned} w_s &= 22.1 \sin 42.9° + 24.4 \cos 42.9° = 15.0 + 17.9 \\ &= 32.9 \text{ in.} \end{aligned}$$

Because the widths of the two struts combined from Table 17-6, which is 16.5 + 18.3 = 34.8 in., exceeds the calculated width available at node C, w_s is too small, and the height of the nodal zone at C must be increased. This means the width of the tie $B_3 - B_3'$ must be increased by changing the arrangement of the No. 8 bars. We will change to five layers of bars with four bars per layer, giving us a total of twenty No. 8 bars. Using No. 11 bars as spacers between the layers, this moves the centroid of the reinforcement in tie $B_3 - B_3'$ to 7.32 in. below the top of the beam. Thus, the total width of the strut, w_t, is 14.6 in. and the total height of Node C is twice that value, 29.3 in. Recalculating the width available at node C for the inclined strut gives $w_s = 36.5$ in, which exceeds the required value of 34.8 in.

(c) Draw the strut-and-tie model—third model: Figure 17-36 shows the left span of the third model. The widths of the struts and nodal zones are compatible.

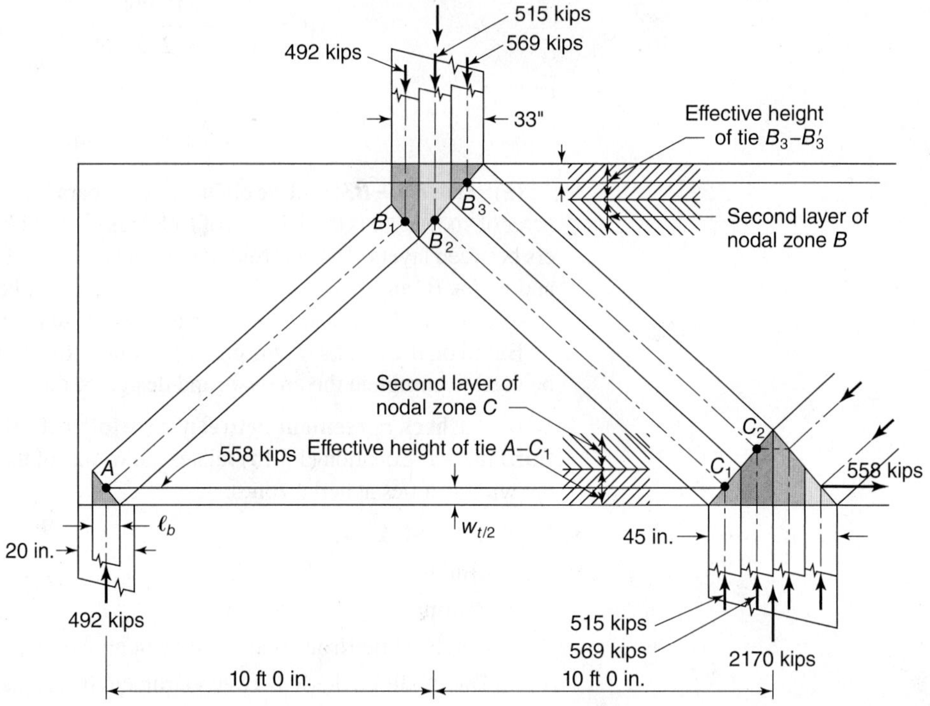

Fig. 17-34
Selection of height of ties—two-span deep beam—Example 17-2.

7. **Select the reinforcement and details.**

(a) Minimum flexural reinforcement: ACI Code Section 9.9.3.2 requires that the flexural reinforcement satisfy the minimum from ACI Code Section 9.6.1, which in turn requires that

$$A_{s,\min} = \frac{3\sqrt{f'_c}}{f_y} b_w d, \text{ but not less than } \frac{200\ psi}{f_y} b_w d$$

$$A_{s,\min} = 8.69 \text{ in.}^2, \text{ but not less than } 9.16 \text{ in.}^2$$

where $d = 120$ in. -5.5 in. $= 114.5$ in. The tie reinforcement chosen previously exceeds these values, so the minimum reinforcement requirement does not govern.

(b) Anchor the ties—third model:

Tie A–C_1 at A: ACI Code Section 23.8.3 requires the tie force at the face of the support to be developed by the point where the centroid of the tie reinforcement leaves the extended nodal zone. In our case, this is where the centroid of the 16 bars composing the tie leaves the bottom side of the inclined strut in the extended nodal zone. (See Fig. 17-35). This occurs at

$$\left(\frac{13 \text{ in.}}{2}\right) + \frac{5.50}{\tan 41.6^\circ} = 12.7 \text{ in.}$$

to the right of the center of the column at A. The total width available to anchor the tie reinforcement measured from the exterior face of column A is $(20/2$ in. $+$ 12.7 in.$) = 22.7$ in. The development length for a straight Grade-60 No. 8 bottom bar, in 4000-psi concrete, is 47.4 in. The length of a 90° hook on the same bars is 19 in. The distance from the exterior face of the column to the tail of the bend

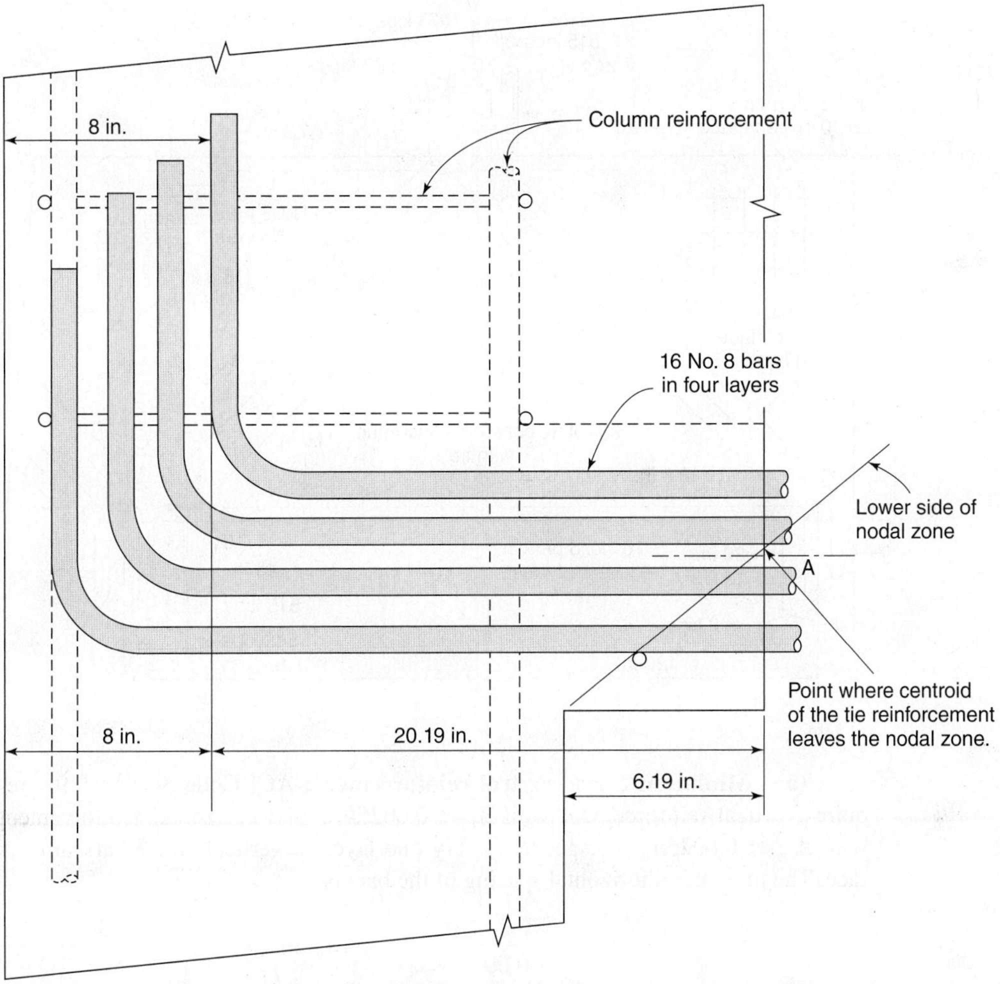

Fig. 17-35
Anchorage of reinforcement from tie *A–C* at node *A*—Two-span deep beam—Example 17-2.

on the top row of bars is (1.5 in. cover + 1/2 in. stirrup + 3 bars and 3 spaces of 1 in. each) = 8 in. (see Fig. 17-35), leaving 14.7 in. to accommodate development length or hook length. The hook development length can be reduced by the factor in ACI Code Table 25.4.3.2 to 0.7 × 19 in. = 13.3 in., if the side cover to the hook is set equal to 2.5 in. and the cover to the tail of the hook is at least 2 in. We will place the reinforcement for the tie inside the column reinforcement. Thus, there is just enough space to develop the bars with hooks.

Tie A–C_1 at C_1: The tie A–C_1 will be anchored at C_1 by extending it continuously through the support at *C*.

Tie B_3–B'_3 at B_3: The development length of No. 8 top bar in 4000-psi concrete is 61.7 in. (Table A-6). Anchor tie B_3–B'_3 by extending the bars through the node at B_3. Node B_3 is 10.3 in. right of the center of the column at *B*. Extend the bars 61.7 in. − 10.3 *in.* =51.4 in.—say, 52 in.—past the centers of the columns at *B* and *B*′.

8. Compute the required minimum crack-control reinforcement and load-spreading reinforcement.

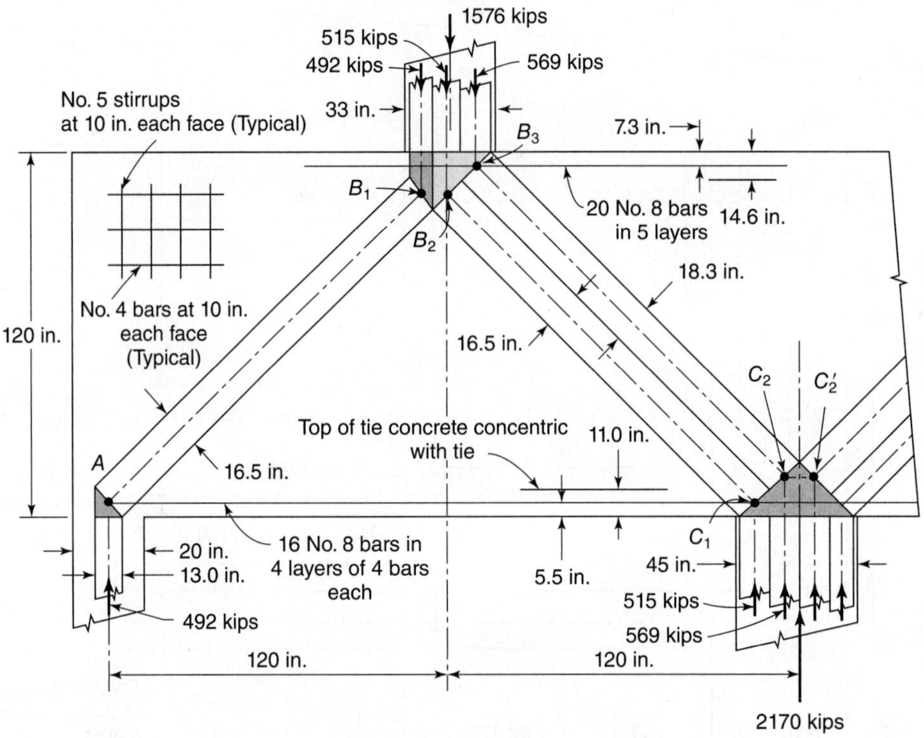

Fig. 17-36
Final design, two-span deep beam—Example 17-2.

(a) Minimum crack-control reinforcement: ACI Code Section 9.9.3 requires vertical reinforcement with $A_v \geq 0.0025 b_w s$ and horizontal reinforcement with $A_{vh} \geq 0.0025 b_w s$, respectively. Try one layer of vertical No. 5 bars in each face. The maximum horizontal spacing of the bars is

$$s_h = \frac{2 \times 0.31}{0.0025 \times 24} = 10.3 \text{ in.}$$

and for one layer of horizontal No. 5 bars on each face, the maximum vertical spacing is

$$s_v = \frac{2 \times 0.31}{0.0025 \times 24} = 10.3 \text{ in.}$$

Thus, to provide the reinforcement required by ACI Code Section 9.9.3 use No. 5 U stirrups at 10 in. o.c. plus horizontal No. 5 bars at 10 in. on each face.

(b) Reinforcement required for bottle-shaped struts: For bottle-shaped struts, satisfy ACI Code Section 23.5.3.

$$\Sigma \frac{A_{si}}{b_w s_i} \sin \alpha_i \geq 0.003 \tag{17-10}$$

We will use layers of reinforcement near both faces of the web of the deep beam.

For strut A–B_1, the angle between the horizontal reinforcement and the strut is 41.6°, and $\sin \alpha_i = 0.664$. For the vertical reinforcement, $\alpha_i = 48.4°$, and $\sin \alpha_i = 0.748$. Thus,

$$\Sigma \frac{A_{si}}{bs} \sin \alpha_i = \frac{2 \times 0.31}{24 \times 10} \times 0.748 + \frac{2 \times 0.31}{24 \times 10} \times 0.664 = 0.00365\text{—o.k.}$$

Provide No. 5 vertical bars on each face at 10 in. on centers and No. 5 horizontal bars on each face at 10 in. on centers.

9. Check other details.

(a) Check the stress on a vertical section through the nodal zone at *C*: Node *C* has a height of 29.3 in. A vertical section through this node is loaded *directly* in compression by the horizontal component of the compressive force in strut B_2–C_1 and is *effectively loaded* in compression by anchoring the tension force in tie A–C_1, similarly to as shown in Fig. 17-10c. Thus, the effective compressive stress on the dividing line between the two halves of this node is as follows:

Horizontal anchor force in tie A–C_1: 558 kips
Horizontal component of the force in strut B_3–C_2: 617 kips

$$\text{Stress on vertical plane through nodal zone: } \frac{558 + 617 \text{ kips}}{29.3 \text{ in.} \times 24 \text{ in.}} \times 1000 = 1670 \text{ psi}$$

f_{ce} for nodal zone *C* is 2720 psi—therefore, o.k.

(b) Check lateral buckling: ACI Code Section 9.2.3 requires that lateral stability of a deep beam be considered. The dimensions of this beam (2 ft wide by 10 ft high) will prevent lateral buckling of the beam itself. We shall assume the bracing of the building prevents relative lateral displacement of the top and bottom of the deep beam.

10. Summary of Design.

- **Use a 24 in.-by-10-ft beam with $f'_c = 4000$ psi and $f_y = 60{,}000$ psi.**
- **Bottom flexural reinforcement: Provide sixteen No. 8 bottom longitudinal bars in four layers of four bars each for full length of beam, with bar layers to be spaced by transverse No. 8 bars. Anchor the No. 8 longitudinal bars into the exterior columns with 90° hooks.**
- **Top flexural reinforcement: Provide twenty No. 8 top bars in five layers spaced by transverse No. 11 bars with four bars per layer. Anchor these by extending them 52 in. past the centers of the columns at *B* and *B′*.**
- **Crack-control reinforcement: Provide No. 5 vertical bars on each face at 10 in. on centers and No. 5 horizontal bars on each face at 10 in. on centers.** ■

17-9 BRACKETS AND CORBELS

A *bracket* or *corbel* is a short member that cantilevers out from a column or wall to support a load. The corbel is generally built monolithically with the column or wall, as shown in Fig. 17-37. The term *corbel* is generally restricted to cantilevers having shear span-to-depth ratios, *a/d*, less than or equal to 1.

Structural Action

A strut-and-tie model for a corbel supported by a column is shown in Fig. 17-37. Within the corbel itself, the structural action consists of an inclined compression strut, *A–C*, and a tension tie, *A–B*. Shears induced in the columns above and below the corbel are resisted by tension in the column bars and ties and by compression forces in struts between the ties.

In tests, [17-20], [17-21] corbels display several typical modes of failure, the most common of which are yielding of the tension tie; failure of the end anchorages of the tension tie, either under the load point or in the column; failure of the compression strut by

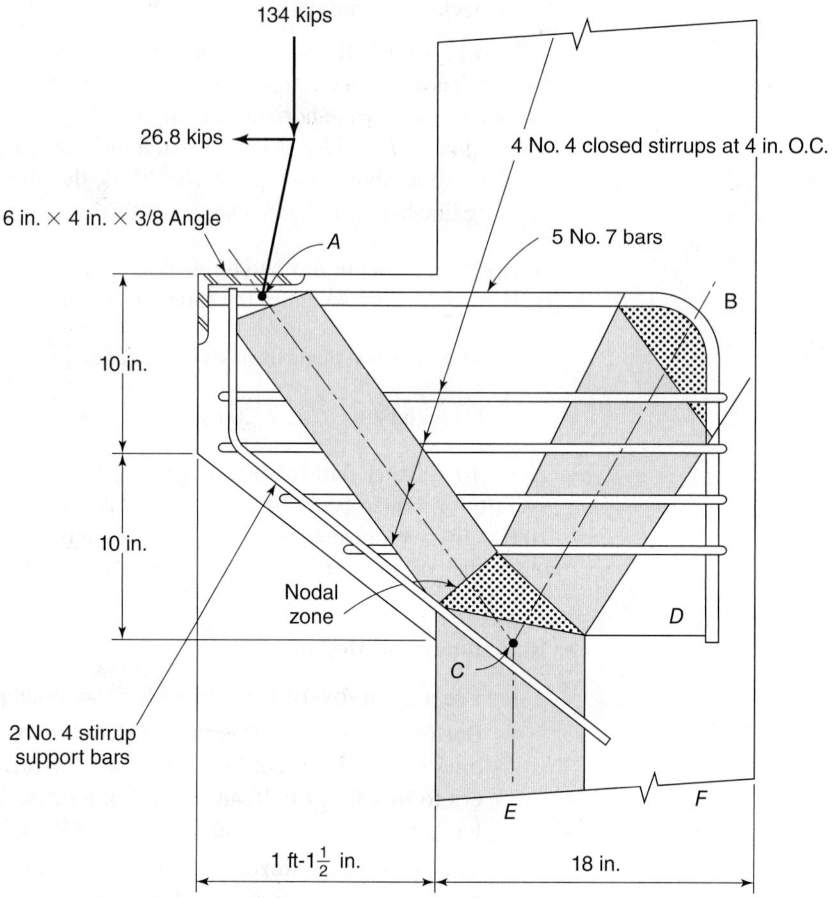

Fig. 17-37
Final strut-and-tie model of a corbel—Example 17-3.

crushing or shearing, and local failures under the bearing plate. If the tie reinforcement is hooked downward, as shown in Fig.17-38a, the concrete outside the hook may split off, causing failure. The tie should be anchored by welding it to a crossbar or plate. Bending the tie bars in a horizontal loop at the outer face of the corbel is also possible, but may be difficult to do because of bends in two directions and may require extra cover. If the corbel is too shallow at the outside end, there is a danger that cracking will extend through the corbel, as shown in Fig. 17-38b. For this reason, ACI Code Section 16.5.2.2 requires the depth measured at the outside edge of the bearing area must be at least one-half the depth at the face of the column.

Design of Corbels

ACI Code Section 23.2.9 allows corbels having *a/d* between 1 and 2 to be designed using strut-and-tie models, where *a* is the distance from the load to the face of the column, and *d* is the depth of the corbel below the tie, measured at the face of the column. Corbels having *a/d* between 0 and 1 may be designed either using strut-and-tie models, or by the closely related traditional ACI design method, which is based in part on the strut-and-tie model and part on shear friction. This procedure was limited to *a/d* ratios less than or equal to 1.0 because little test data was available for longer corbels. Regardless of the design method used, the general requirements in ACI Code Sections 16.5.2, 16.5.6, and 23.2.9 must be satisfied.

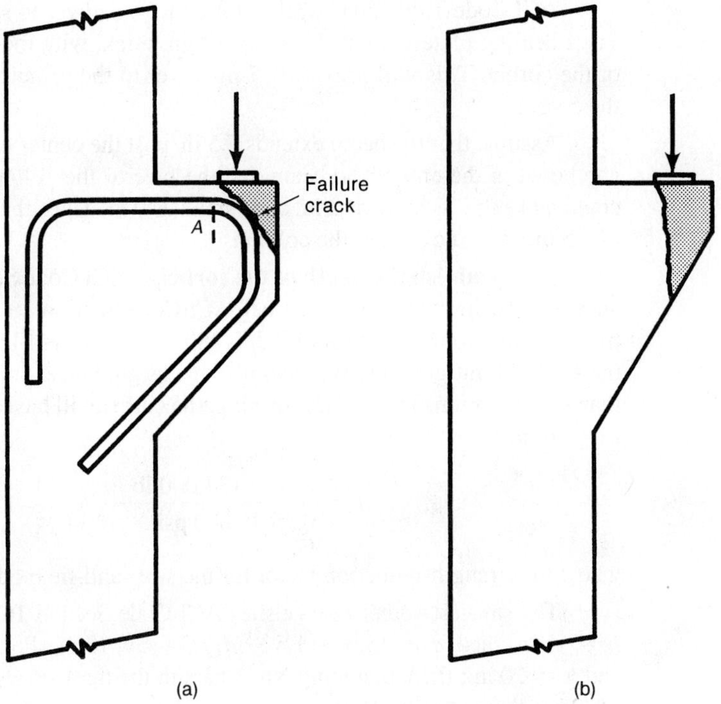

Fig. 17-38
Failure of corbels due to poor detailing.

Two closely related design procedures for corbels will be presented: design using strut-and-tie models, and design according to ACI Code Section 16.5. The strut-and-tie method is a little more versatile than the ACI method, but both give essentially the same results within the range of application of the ACI Code.

EXAMPLE 17-3 Design of a Corbel via a Strut-and-Tie Model

Design a corbel to support the reaction from a precast beam (Fig. 17-37). The end of the beam is 12 in. wide. The column is 16 in. square. The unfactored beam reaction is 60 kips dead load and 39 kips live load. The beam is partially or fully restrained against longitudinal shrinkage. Use f'_c = 5000-psi normal-weight concrete and f_y = 60,000 psi. Use the load factors and strength-reduction factors from ACI Code Tables 5.3.1 and 21.2.1.

1. Compute the factored load—corbel.

$$U = 1.4D = 1.4 \times 60 = 84 \text{ kips}$$

$$U = 1.2D + 1.6L = 1.2 \times 60 + 1.6 \times 39 = 134 \text{ kips}$$

Thus, the factored vertical load on the corbel is 134 kips.

2. Compute the distance from the face of the column to the beam reaction.

Assume a 12-in.-wide bearing plate. From ACI Code Table 22.8.3.2, the maximum bearing stress is $0.85f'_c = 0.85 \times 5000 = 4250$ psi. Using $\phi = 0.65$ as the strength-reduction factor for bearing, the required width of the bearing plate is

$$\frac{134{,}000 \text{ lb}}{0.65 \times 4250 \text{ psi} \times 12} = 4.04 \text{ in.}$$

ACI Code Table 20.6.1.3.1 requires 1.5-in. cover to stirrups or main reinforcement. Try a bearing plate 6 in. by 12 in. by 1.5 in. thick, with the top surface flush with the top of the corbel. This will provide 1.5-in. cover to the principal reinforcement at the top of the corbel.

Assume that the beam extends 7.5 in. past the center of the bearing plate. The design gap between the end of the beam and the face of the column is 1 in. Erection tolerances could make this as large as 2 in. or as small as 0 in. Thus, the beam reaction could act as far as 9.5 in. from the face of the column.

3. Establish the depth of the corbel. ACI Code Section 16.5 does not give guidance in choosing the size of a corbel. ACI Code Section 16.5.2.4 limits the interface shear transfer stress to the smallest of $0.2f'_c$, $480 + 0.08f'_c$, and 1600 psi. For 5000-psi concrete, the second limit governs. We shall use this as guidance, but use a corbel somewhat larger than the minimum. To simplify forming, the corbel will have the same width as the column, $b = 16$ in. Then

$$d = \frac{134{,}000 \text{ lb}}{0.75 \times 880 \text{ psi} \times 16 \text{ in.}} = 12.7 \text{ in.}$$

where the strength-reduction factor for the strut-and-tie model, $\phi = 0.75$, is used.

The smallest corbel that satisfies ACI Code Section 16.5.2.4 would have $b = 16$ in., $h = 15$ in., and $d = 15.0 - 1.5 - d_b/2$—say, 13 in. For conservatism, try $b = 16$ in. and $h = 20$ in.; then, assuming No. 8 bars in the tie *A–B*, $d = 20$ in. $-$ (1.5-in. cover) $-$ (1/2 bar diameter) $= 18$ in.

Try $b = 16$ in., $h = 20$ in., $d = 18$ in.

4. Select design method. ACI Commentary Section R16.5.1.1 allows corbels with any a/d value to be designed using a strut-and-tie model (Code Chapter 23). Corbels with a/d less than 1.0 can also be designed by using the traditional ACI corbel design method in ACI Code Section 16.5. Here, $a/d = 9.5/18 < 1.0$, so either design method may be used. We shall use the strut-and-tie method.

5. Select a first trial strut-and-tie model. ACI Code Section 16.5.3.5 requires corbels to be designed for a shear, V_u, equal to the factored beam reaction and for a factored tensile force, N_{uc}, equal to the actual tension force acting on the corbel, but not less than $0.2\,V_u$. This force represents the forces induced by restrained shrinkage of the structure supported by the corbel [17-22].

Design the corbel for $V_u = 134$ kips and $N_{uc} = 26.8$ kips: (see Fig. 17-37). The forces V_u and N_{uc} can be resolved into an inclined force that intercepts the centroid of the tie reinforcement at node *A*, which is located at $9.5 + 0.2 \times (1.5 + 0.5)$ in. $= 9.9$ in., from the face of the column.

Figure 17-37 shows the final strut-and-tie model. Assuming (a) that the column reinforcement is No. 9 bars with No. 3 ties and (b) that the nodes *B* and *D* are in the plane of the column reinforcement located at 1.5-in. cover + 0.375 in. ties + 1.270/2 = 2.5 in. from the right-hand face of the column, gives the distance from *A* to *B* as 9.9 in. + 16 in. $-$ 2.5 in. = 23.4 in.

Locate node *C*: Node *C* is located at a distance a / 2 from the left side of the column, where a is the depth of the stress block resisting the force in the strut *C–E*. ACI Code Section 23.9.2 gives the effective compression strength of the nodal zone as

$$f_{ce} = 0.85\beta_n f'_c \tag{17-7b}$$

The node at C anchors three compression struts and a tension tie, to be discussed later. From ACI Code Table 23.9.2, a nodal zone anchoring one tie has $\beta_n = 0.80$, so that

$$f_{ce} = 0.85 \times 0.80 \times 5000 \text{ psi} = 3400 \text{ psi}$$

and

$$a = \frac{P_{u,C-E} \text{ kips} \times 1000 \text{ lb/kip}}{0.75 \times 3400 \text{ psi} \times 16 \text{ in.}} = 0.0245 P_{u,C-E} \text{ in.}$$

where $P_{u,C-E}$ is in kips and the 0.75 is the ϕ factor for strut-and-tie models from ACI Code Table 21.2.1. Summing moments about D gives

$$\Sigma M_D = 0 = 134 \text{ kips} \times 23.4 \text{ in.} + 26.8 \text{ kips} \times 18 \text{ in.} - P_{u,C-E} \times (13.5 \text{ in.} - a/2)$$

For $a = 0.0245 P_{u,C-E}$, this becomes

$$0 = 295{,}000 - 1100 P_{u,C-E} + (P_{u,C-E})^2$$

$$P_{u,C-E} = \frac{1100 \pm \sqrt{1100^2 - 4 \times 1 \times 295{,}000}}{2} = 637 \text{ or } 463 \text{ kips}$$

Thus, $P_{u,C-E} = 463$ kips and $a = 0.0245 \times 463 = 11.3$ in. Node C is $11.3/2 = 5.65$ in. from the left edge of the column adjacent to C, and C–D is 13.5 in. − 5.65 in. = 7.85 in.

The nodal zone at C takes up 11.35 in. out of the 13.5 in. width between the compression face of the column and node D. For comparison, the rectangular stress block in a beam would have a depth of about $0.3d$. This indicates that the column is too small. We shall increase the width of the column to 18 in. The distance from A to B increases to 25.4 in.

6. Recompute *a* for the new depth of column—second model. Summing moments about D gives

$$\Sigma M_D = 0 = 134 \times 25.4 + 26.8 \times 18 - P_{u,C-E} \times (15.5 - a/2)$$
$$0 = 317{,}000 - 1265 P_{u,C-E} + (P_{u,C-E})^2$$

which gives $P_{u,C-E} = 345$ kips. Thus, $P_{u,C-E}$ drops from 464 kips to 345 kips, corresponding to $a = 8.45$ in., and $a/2 = 4.23$ in. The distance C–D is 15.5 in. − 4.23 in. = 11.3 in.

7. Solve for forces in the struts and ties—second model.

Node *A*: Figure 17-39a shows the forces acting at node A.

Strut A—C has a horizontal projection of $25.4 - 11.3 = 14.1$ in. and a vertical projection of 18 in. It supports a factored vertical force component of 134 kips, a horizontal force component of

$$\frac{14.1}{18} \times 134 \text{ kips} = 105 \text{ kips}$$

and an axial force of $\sqrt{134^2 + 105^2} = 170$ kips compression. The angle between A–C and the tie A–B is $\arctan(18/14.1) = 51.9°$.

***Tie A–B*:** Summing horizontal forces at node A gives $P_{u,A-B} = 105 + 26.8 = 132$ kips tension.

Node *B*: Figure 17-39b shows the forces acting at node B.

Strut B–C: Member B–C has a vertical projection of 18 in. and a horizontal projection of 15.5 in. − a/2, where $a/2 = 4.23$ in., making the horizontal projection of

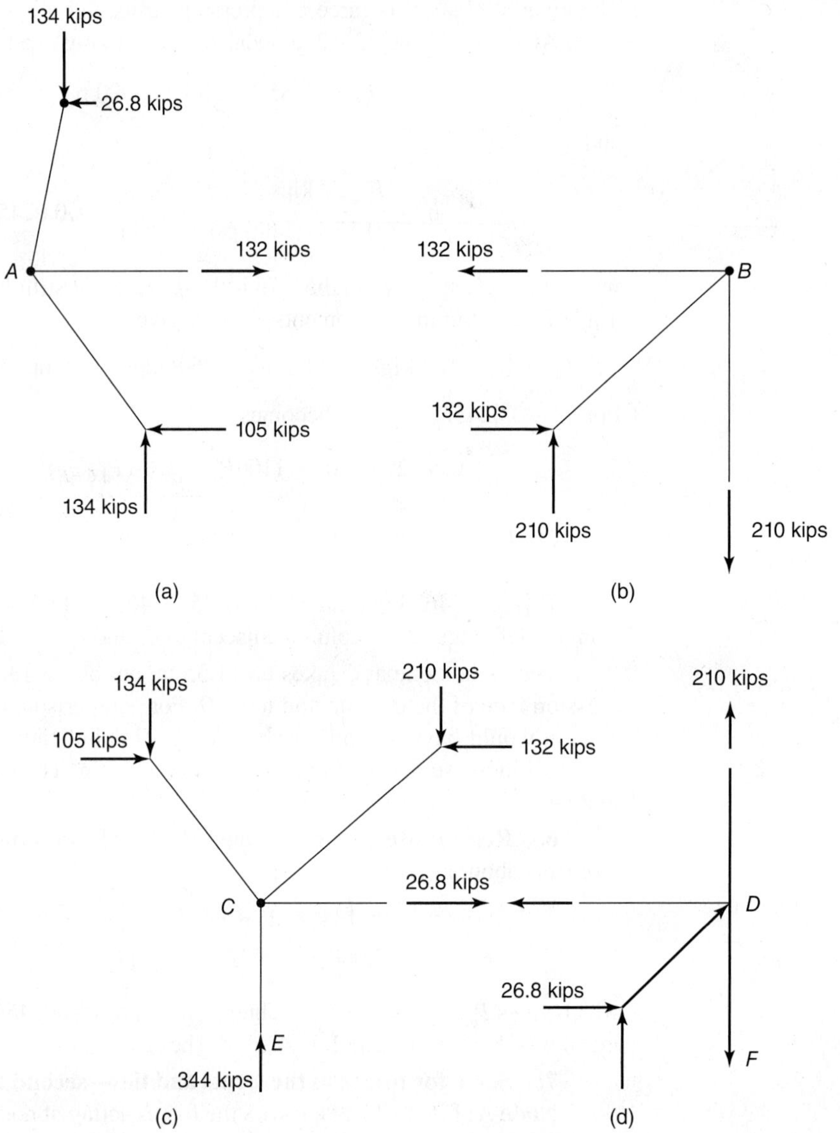

Fig. 17-39
Calculation of forces in struts and ties—Corbel—Example 17-3.

B–C 11.3 in. Summing horizontal forces at joint B gives the horizontal component of the force in B–C as 132 kips. The vertical force component in B–C is

$$\frac{18}{11.3} \times 132 = 210 \text{ kips}$$

and the axial force is 248 kips.

The angle between B–C and tie B–D is $\arctan(11.27/18) = 32.1°$.

***Tie B–D*:** Summing vertical forces at node B gives $P_{u,B-D} = 210$ kips tension.

Node *C*: Figure 17-39c shows the forces acting at node C. The sum of vertical forces is 134 kips down + 210 kips down + 344 kips up = 0. This should total zero—o.k. Summing horizontal forces gives a tension of 26.8 kips in tie C–D. The angle between strut A–C and tie C–D is 51.9° and the angle between strut B–C and tie C–D is 57.9°.

Node *D*: Figure 17-39d shows the forces at node D. An inclined strut D–E is required in the column for equilibrium. A tension tie D–F is also required.

8. Compute reinforcement required for ties.

Tie *A–B*:

$$A_s = \frac{132 \text{ kip}}{0.75 \times 60 \text{ kip}} = 2.93 \text{ in.}^2$$

Possible choices are (a) four No. 8 bars, $A_s = 3.16$ in.^{2}, which will fit into a width of 11.5 in. and have a basic hook-development length of 17 in., and (b) five No. 7 bars, $A_s = 3.00$ in.^{2}, which will fit into a width of 13 in. and have a basic hook-development length of 14.8 in. Use five No. 7 bars for tie *A–B*. Hook these with the vertical tails of the hooks in the plane of the right-hand layer of column steel with a cover of 1.5 + 0.5 in. from the right-hand side of the column.

Tie *B–D*:

$$A_s = \frac{210 \text{ kips}}{0.75 \times 60 \text{ ksi}} = 4.67 \text{ in.}^2$$

The reinforcement for tie *B–D* will be selected by considering the statics and resistance of the column as a whole. The longitudinal column reinforcement will be sized to provide the required reinforcement for this tie. It may be necessary to enlarge the column further.

Tie *C–D*:

$$A_s = \frac{26.8 \text{ kips}}{0.75 \times 60 \text{ ksi}} = 0.60 \text{ in.}^2$$

Use two No. 4 closed column ties, $A_s = 0.80$ in.^{2}.

9. Compute f_{ce} and the widths of the struts—third model. These calculations are summarized in Table 17-7.

10. Draw the strut-and-tie model to scale—third model. This is done to see whether the struts fit within the space available. Figure 17-37 shows that they do with very little extra space.

11. Provide reinforcement to confine struts. When f_{ce} was computed for the struts, they were all assumed to be bottle-shaped. Such struts must have transverse reinforcement satisfying ACI Code Sections 23.5.1 or 23.5.3.

TABLE 17-7 Widths of Struts and Ties—Example 17-3

Member	Axial Force kips	f_{ce} for Strut psi	f_{ce} for Node psi	w_s in.
A–C at *A*	170	0.85 × 0.75 × 5000 = 3190	0.85 × 0.8 × 5000 = 3400	4.44
A–C at *C*	170	3190	= 3400	3.33
B–C at *B*	248	3190	0.85 × 0.60 × 5000 = 2550	6.08
B–C at *C*	248	3190	3400	4.86
C–E at *C*	344	4250	3190	6.74

NOTE: Two rows are provided for each strut, to allow different values of f_{ce} at the two ends of a strut. Also, use an effective corbel width of 12 in. under the bearing plate at node *A*. At all other nodes use a width of 16 in.

Strut *A–B*: Using, ACI Code Section 23.5.3 requires that the strut be crossed by steel satisfying

$$\Sigma \frac{A_{si}}{bs_i} \sin \alpha_i \geq 0.003$$

where A_{si} is the area of transverse steel at an angle α_i to the axis of the strut, which cannot be taken as less than 40°. The computed angle is 51.9°. Use No. 4 closed stirrups at $s_i = 4$ in., spread over the length and width of strut *A–C* (and *B–C*)

$$\Sigma \frac{A_{si}}{bs_i} \sin \alpha_i = \frac{2 \times 0.2}{16 \times 4} \sin 51.9° = 0.00492$$

This exceeds the required 0.003. **Use four No. 4 two-legged closed stirrups at 4 in. o.c.** Place the first one at 4 in. below the centroid of tie *A–B*.

12. Satisfy the detailing requirements—third model. ACI Code Section 16.5 presents a number of detailing requirements for corbels.

16.5.2.2—The depth under the outer edge of the bearing plate shall not be less than half the height at the face of the column—o.k. This requirement was introduced to prevent failures like the one shown in Fig. 17-38b.

16.5.6.3—Probably the most important detailing requirement is that the tie *A–B* be anchored for the tension tie force at the front face of the corbel (at *A*). Because the tie in the truss in Fig. 17-37 is assumed to be stressed to f_y in tension over the whole distance from the loading plate to the column, it must be anchored outside the loading plate for that tension. This is done in one of several ways. It can be anchored

(a) Welding the bars making up the tie to a transverse angle or bar, which may also serve as a bearing plate.

(b) Welding the bars making up the tie to a transverse bar of the same diameter as the bars making up the tie.

(c) Bending the bar in a horizontal loop.

Although the tension tie could also be anchored by bending the bars in a vertical bend, as illustrated in Fig. 17-38a, this is discouraged, because failures have occurred when this detailing was used, as shown in that figure. We shall assume that the $(6 \times 12 \times 1.5)$-in. bearing plate welded across the ends of the five No. 7 bars, as chosen in step 8, will anchor the tie. Because welding is required, the five No. 7 bars must be specified as Grade-60W steel. Alternatively, a steel angle could be used to anchor tie *A–B* at *A*, as shown in Fig. 17-37. The plate is easier to weld, and easier to concrete under, but the angle provides resistance to damage to the outer edge of the bearing area.

16.5.2.3—requires that the bearing area of the load either

(a) not project beyond the start of the bend of the top tie bar, if it is anchored at node *A* by a hook, or

(b) not project past the interior face of the transverse anchor bar, if that detail is used.

Welding the $6 \times 4 \times \frac{3}{8} \times 12$-in. bearing angle across the five No. 7 bars with the center of the top of the angle under the center of the reaction bearing plate will anchor the tie.

13. Check the moments in the column—third model. The loads on the corbel cause a moment of

$$134 \times (9.9 + 18/2) = 2530 \text{ kip-in.}$$

about the center of the column on a line joining nodes A and B. This moment should be divided between the columns above and below the joint in proportion to the stiffnesses of each of these columns. ■

Alternate Design Method for Corbels

ACI Code Section 16.5 presents an alternate design procedure for brackets and corbels. It is based in part on the strut-and-tie truss model and in part on shear friction. The design procedure is limited to a/d ratios of 1.0 or less. At the time it was included in the code there was little test data for longer brackets.

For this design method, the section at the face of the support is designed to resist the shear V_u, the horizontal tensile force N_{uc}, and a moment of $[V_u a + N_{uc}(h - d)]$, where the moment has been calculated relative to the tension steel at the face of the column in Fig. 17-37. The maximum shear strength, V_n, shall not be taken greater than the smallest of $0.2f'_c b_w d$, $(480 + 0.08f'_c)\, b_w d$, and $1600 b_w d$ lb, for normal-weight concrete.

In design, the size of the corbel is selected so that $V_u \leq \phi\, V_n$, based on the maximum shear strength. If a high value of V_n is used, cracking at service loads may lead to serviceability problems. The designer then calculates the following:

1. The area, A_{vf}, of shear-friction steel required is,

$$V_n = A_{vf} f_y \mu \tag{17-17}$$

(ACI Eq. 22.9.4.2)

2. The area, A_f, of flexural reinforcement required to support a moment of $[V_u a + N_{uc}(h - d)]$, based on Chapter 5 of this book.

3. The area, A_n, of direct-tension reinforcement required to resist the tension force N_{uc}, where

$$\phi A_n f_y \geq N_{uc} \tag{17-18}$$

In all these calculations, ϕ is taken equal to the value for shear, 0.75, which is also the value for strut-and-tie models.

The resulting area of tensile steel, A_{sc}, and the area and placement of the horizontal reinforcement within the corbel are specified in ACI Code Sections 16.5.5.1, 16.5.5.2, and 16.5.6.6. In the corbel tests reported in [17-20], [17-21], the best behavior was obtained in corbels that had some horizontal stirrups in addition to the tension tie shown in Fig. 17-37. Accordingly, ACI Code Section 16.5.5.1 requires that two reinforcement patterns be considered and the one giving the greater area, A_{sc}, be used

1. A tension tie having area $A_{sc} = A_f + A_n$, plus horizontal stirrups having area $A_f/2$
2. A tension tie having area $A_{sc} = (2A_{vf}/3) + A_n$, plus horizontal stirrups having area $A_{vf}/3$.

The horizontal stirrups are to be placed within $\frac{2}{3}d$ below the tension tie.

EXAMPLE 17-4 Design of a Corbel—Traditional ACI Code Method

Design a corbel to transfer a precast-beam reaction to a supporting column. The factored shear to be transferred is 134 kips. The column is 16 in. square. The beam being supported is restrained against longitudinal shrinkage. Use $f'_c = 5000$-psi normal-weight concrete and $f_y = 60{,}000$ psi. Use ACI Code Tables 5.3.1 and 21.2.1.

1. Compute the distance, *a*, from the column to V_u. Assume a 12-in.-wide bearing plate. From ACI Code Table 22.8.3.2, the allowable bearing stress is

$$\phi 0.85f'_c = 0.65 \times 0.85 \times 5 \text{ ksi} = 2.76 \text{ ksi}$$

The required width of the bearing plate is

$$\frac{134 \text{ kips}}{2.76 \times 12 \text{ in.}} = 4.04 \text{ in.}$$

Use a 12-in. × 6-in. bearing plate. Assume that the beam overhangs the center of the bearing plate by 7.5 in., that a 1-in. gap is left between the end of the beam and the face of the column, and that the distance a is assumed to be 9.5 in.

2. Compute the minimum depth, *d*. Base this calculation on ACI Code Section 16.5.2.4:

$$\phi V_n \geq V_u$$

V_n is limited to the smallest of $0.2f'_c b_w d$, $(480 + 0.08f'_c)b_w d$, and $1600\, b_w d$. For 5000-psi concrete, the second equation governs. Thus,

$$\text{minimum } d = \frac{V_u}{\phi \times 880 b_w}$$

$$= \frac{134{,}000 \text{ lb}}{0.75 \times 880 \times 16} = 12.7 \text{ in.}$$

Hence, the smallest corbel we could use is a corbel with $b = 16$ in., $h = 15$ in., and $d = 13$ in. For conservatism, we shall use $h = 20$ in. and $d = 20 \text{ in.} - \left(1\frac{1}{2} \text{ in. cover} + \frac{1}{2} \text{ bar diameter}\right)$, which equals 18 in. The corbel will be the same width as the column (16 in. wide), to simplify forming.

3. Compute the forces on the corbel. The factored shear is 134 kips. Because the beam is restrained against shrinkage, we shall assume the normal force to be (ACI Code Section 16.5.3.5)

$$N_{uc} = 0.2V_u = 26.8 \text{ kips}$$

The factored moment is

$$M_u = V_u a + N_{uc}(h - d)$$

$$= 134 \text{ kips} \times 9.5 \text{ in.} + 26.8 \text{ kips}(20 \text{ in.}-18 \text{ in.})$$

$$= 1330 \text{ kip-in.}$$

4. Compute the shear friction steel, A_{vf}. From Eq. (17-17),

$$\phi V_n \geq V_u$$

$$A_{vf} = \frac{V_n}{\mu f_y} = \frac{V_u}{\phi \mu f_y}$$

where $\mu = 1.4\lambda$ for a shear plane through monolithic concrete and $\lambda = 1.0$ for normal-weight concrete. Therefore,

$$A_{vf} = \frac{134 \text{ kips}}{0.75(1.4 \times 1.0)60 \text{ ksi}} = 2.13 \text{ in.}^2$$

5. Compute the flexural reinforcement, A_f. A_f is computed from Eq. (5-16) (with A_s replaced by A_f):

$$M_u \leq \phi A_f f_y\left(d - \frac{a}{2}\right)$$

Here, $\phi = 0.75$ (ACI Code Table 21.2.1) and

$$a = \frac{A_f f_y}{0.85 f'_c b}$$

As a first trial, we shall assume that $(d - a/2) = 0.9d$. Thus,

$$A_f \geq \frac{M_u}{\phi f_y(0.9d)}$$

$$\geq \frac{1330 \text{ k-in.}}{0.75 \times 60 \times 0.9 \times 18} = 1.82 \text{ in.}^2$$

Because this is based on a guess for $(d - a/2)$, we shall compute a and recompute A_f:

$$a = \frac{1.82 \times 60}{0.85 \times 5 \times 16} = 1.61 \text{ in.}$$

$$A_f \geq \frac{1330 \text{ kip-in.}}{0.75 \times 60 \text{ ksi } (18 - 1.61/2) \text{ in.}}$$

$$\geq 1.72 \text{ in.}^2$$

Therefore, use $A_f = 1.72$ in.^{2}.

6. Compute the reinforcement, A_n, for direct tension. From ACI Code Section 16.5.4.3,

$$A_n = \frac{N_{uc}}{\phi f_y} = \frac{26.8 \text{ kips}}{0.75 \times 60 \text{ ksi}}$$

$$= 0.60 \text{ in.}^2$$

7. Compute the area of the tension-tie reinforcement, A_{sc}. From ACI Code Section 16.5.5.1, A_{sc} shall be the larger of

$$(A_f + A_n) = 1.72 + 0.60 = 2.32 \text{ in.}^2\text{, or}$$

$$\left(\frac{2A_{vf}}{3} + A_n\right) = 1.42 + 0.60 = 2.02 \text{ in.}^2$$

Minimum A_{sc} (ACI Code Section 16.5.5.1):

$$A_{sc,(\min)} = \frac{0.04 f'_c}{f_y} b_w d = 0.96 \text{ in.}^2$$

Therefore, $A_{sc} = 2.32$ in.^{2}. Try three No. 8 bars, giving $A_{sc} = 2.37$ in.^{2}.

8. Compute the area of horizontal stirrups.

$$0.5(A_{sc} - A_n) = 2.32 - 0.60 = 0.86 \text{ in.}^2$$

Select three No. 4 double-leg stirrups, area = 1.20 in.^{2}; ACI Code Section 16.5.6.6 requires that these be placed within (2/3)d, measured from the tension tie.

9. Establish the anchorage of the tension tie into the column. The column is 16-in. square. Try a 90° standard hook. From ACI Code Section 25.4.3.1,

$$\ell_{dh} = \left(\frac{f_y \psi_e \psi_c \psi_r}{50\lambda \sqrt{f'_c}}\right) d_b$$

Referring to Table 25.4.3.2, for normal weight concrete, $\lambda = 1.0$. We are not using epoxy coating ($\psi_e = 1.0$) and we will not assume closely space stirrups are used to confine the hook inside the column ($\psi_r = 1.0$). The hook will be placed inside the column bars, so we should have enough side cover and back cover to use $\psi_c = 0.7$. So,

$$\ell_{dh} = \left(\frac{60{,}000 \times 1 \times 0.7 \times 1}{50 \times 1 \times \sqrt{5000}}\right) \times 1.0 \text{ in.} = 11.9 \text{ in. (o.k.)}$$

Therefore, **use three No. 8 bars hooked inside the column** (inside the column bars).

10. Establish the anchorage of the outer end of the bars. The outer end of the bars must be anchored to develop $A_{sc} f_y$. This can be done by welding the bars to a transverse plate, angle, or bar. If the horizontal force, N_{uc}, is required for equilibrium of the structure, some direct connection would be required from the beam base plate to the tension tie. This is not the case here, and a welded cross-angle will be provided, as shown in Fig. 17-37.

11. Consider all other details. To prevent cracks similar to those shown in Fig. 17-38b, ACI Code Section 16.5.2.2 requires that the depth at the outside edge of the bearing area be at least 0.5*d*. ACI Code Section 16.5.6.3 requires anchorage of the tension tie outside the bearing area. Finally, two No. 4 bars are provided to anchor the front ends of the stirrups. All of these aspects are satisfied in the final corbel layout that is similar to Fig. 17-37. ■

Comparison of the Strut-and-Tie Method and the Traditional ACI Method for Corbel Design

The strut-and-tie method required more steel in the tension tie and less confining reinforcement than the traditional ACI method. The strut-and-tie method explicitly considered the effect of the corbel on the forces in the column. The strut-and-tie method could also be used for corbels that have *a*/*d* greater than the limit of 1.0. For $a/d > 1$, the confining stirrups would be more efficient in restraining the splitting of the strut if they were vertical.

17-10 DAPPED ENDS

The ends of precast beams are sometimes supported on an end projection that is reduced in height, as shown in Fig. 17-40. Such a detail is referred to as a *dapped end*. Although several design procedures exist, the best method of design is by means of strut-and-tie models. Tests and analysis of such regions are reported in [17-23], [17-24], and [17-25].

Four common strut-and-tie models for dapped-end regions are compared to the crack pattern observed in tests in Fig. 17-40. Cracking originates at the reentrant corner of the notch, point *A* in Fig. 17-40a. The strut-and-tie models in Figs. 17-40b to 17-40d all involve a vertical tie *B–C* at the end of the deeper portion of the beam and an inclined strut *A–B* over the reaction. In tests, specimens with tie *B–C* composed of closed vertical stirrups with 135° bends around longitudinal bars in the top of the beam performed better than specimens with open-topped stirrups [17-24]. The horizontal component of the

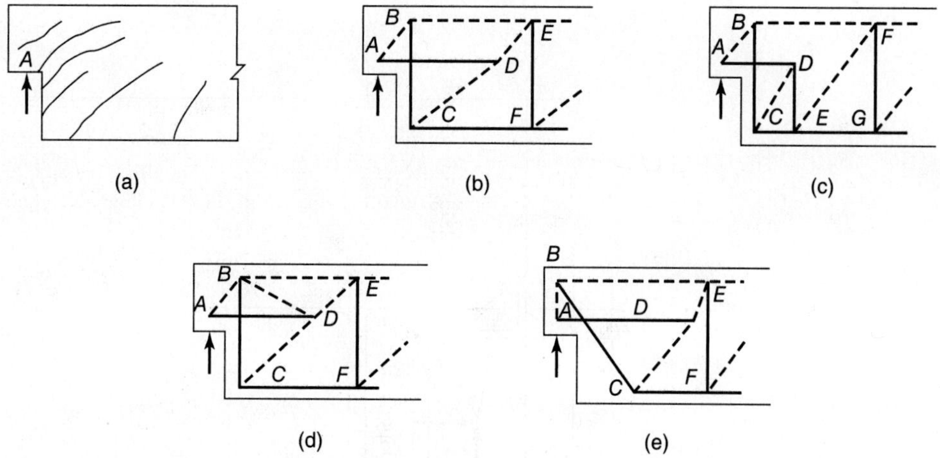

Fig. 17-40
Comparison of strut-and-tie models for dapped ends.

compressive force in *A–B* is equilibrated by the tension tie *A–D*. The three strut-and-tie models differ in the manner in which the horizontal tie is anchored at *D*. The model in Fig. 17-40c has the advantage that the force in tie *C–E* is lower, and hence easier to anchor, than the corresponding force in tie *C–F* in Fig. 17-40b. In Fig. 17-40d, tie *A–D* is anchored by strut *B–D*, which will be crossed by cracks, as shown in Fig. 17-40a. This suggests that Fig. 17-40d is not a feasible model.

The strut-and-tie model in Fig. 17-40e has an inclined hanger tie *B–C* and a vertical strut over the reaction. Care must be taken to anchor the tie *B–C* at its upper end. It is customary to provide a horizontal tie at *A* to resist any tensile forces due to restrained shrinkage of the precast beam. In tests [17-24], dapped ends designed by using the models in Fig. 17-40b or 17-40c performed just as well as ends designed via the model in Fig. 17-40e. A compound model, designed by assuming that half the reaction was resisted by each of these two types of strut-and-tie models, also performed well in tests.

In laying out a dapped-end support, it is good practice to have the depth of the extended part of the beam be at least half of the overall height of the beam. The extended part of the beam should be deep enough that the inclined compression strut *A–B* at the support is no flatter than 45°. Otherwise, the forces in this strut and in the tie that meets it at the support become too large to deal with in a simple manner. Great care must be taken to anchor all the bars in the vicinity of the dap.

EXAMPLE 17-5 Design of a Dapped-End Support

A precast beam supports an unfactored dead load of 2 kips/ft and an unfactored live load of 2.5 kips/ft on a 20-ft span. The beam is 30 in. deep by 15 in. wide and is made from 3000-psi normal-weight concrete and Grade-60 reinforcement. The longitudinal steel is four No. 8 bars. Design the reinforcement in the support region.

1. Isolate the D-Region; compute the reactions and the forces on the boundaries of the D-region

(a) Design equation: Design will be based on

$$\phi F_n \geq F_u \tag{17-1a}$$

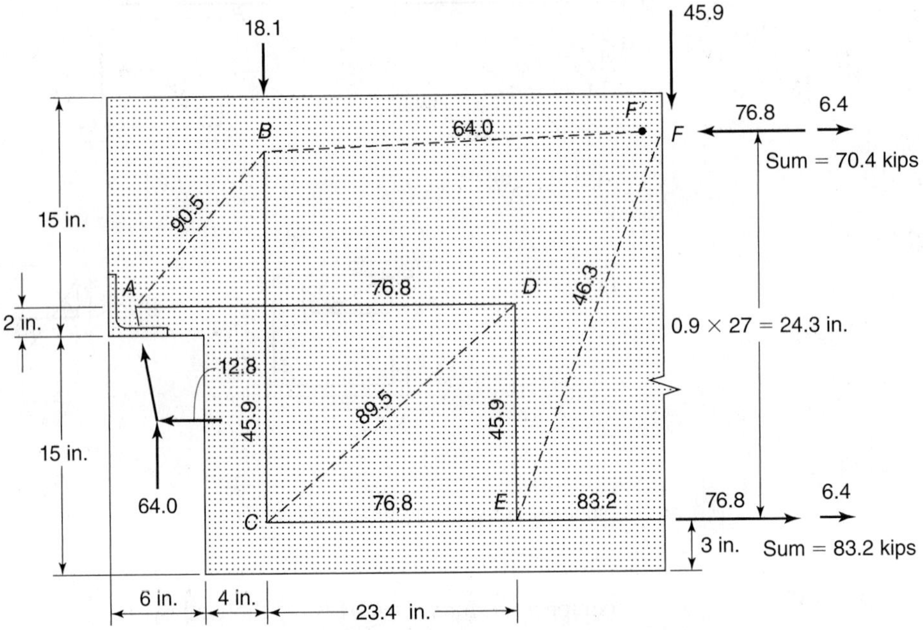

Fig. 17-41
Strut-and-tie model for a dapped end—Example 17-5.

where, from ACI Code Table 21.2.1, $\phi = 0.75$, F_n is the nominal resistance, and F_u is the factored-load effect.

(b) Compute factored loads: $U = 1.2 \times 2 + 1.6 \times 2.5 = 6.4$ kips per ft. This gives a vertical reaction of 64.0 kips and a horizontal reaction of $0.2 \times 64.0 = 12.8$ kips.

(c) Isolate the D-region: The D-region will be assumed to extend 30 in. from the lower corner of the full depth portion of the beam. Assuming the center of the support is 2 in. from the end of the beam (Fig. 17-41), and assuming the loads do not extend beyond the center of the support, the dead and live loads in the D-region are replaced by a single force equal to the sum of the factored dead and live loads acting on the D-region equal to $(34/12)$ ft $\times$ 6.4 kips/ft $=$ 18.1 kips is applied to the beam at convenient place in the D-region. We shall apply it to the beam at node *B*.

Moment equilibrium at the right end of the D-region gives a bending moment of 155.5 kip-ft = 1870 kip-in. acting on the vertical section to the right of node *F*. This can be subdivided into a flexural compression and a flexural tension of $1870/(0.9 \times 27) = 76.8$ kips at nodes *F* and *E*, respectively. The horizontal reaction force of 12.8 kips causes tensions of 6.4 kips at nodes *E* and *F* giving forces $76.8 - 6.4 = 70.4$ kips at *F* and $76.8 + 6.4$ kips $= 83.2$ kips at *E*.

2. Select a strut-and-tie model. A strut-and-tie model similar to that in Fig. 17-40c will be used.

3. Compute the effective compressive strength of the nodal zones and struts. The effective compressive strength of the nodal zones, f_{ce}, is given in ACI Code Section 23.9.2 as

$$f_{ce} = 0.85\beta_n f'_c \tag{17-7b}$$

Use ACI Code Table 23.9.2 to determine β_n values for nodes.

Nodal zone *A*: Nodal zone *A* anchors compressive forces from the reaction and from strut *A–B* and tension from the tie *A–D*. So, $\beta_n = 0.80$, and

$$f_{ce} = 0.85 \times 0.80 \times 3000 = 2040 \text{ psi}$$

Nodal zone *B*: Nodal zone *B* is a *C–C–T* node. So, $\beta_n = 0.80$, and $f_{ce} = 2040$ psi.

Nodal zone *F*: Nodal zone *F* is a *C–C–C* node. So, $\beta_n = 1.0$, and $f_{ce} = 2550$ psi.

Nodal zones *C*, *D*, and *E*: These nodes all anchor more than one tie. So, $\beta_n = 0.60$, and $f_{ce} = 1530$ psi at all of them.

Use ACI Code Table 23.4.3 to determine β_s values for struts.

Struts *A–B*, *C–D*, and *E–F*: We shall assume that all three are bottle-shaped struts with $\beta_s = 0.75$. Thus,

$$f_{ce} = 0.85 \times 0.75 \times 3000 = 1910 \text{ psi}$$

Strut *B–F*: This strut is the compression zone of the beam, so $\beta_s = 1.0$. Thus, $f_{ce} = 0.85 \times 1.0 \times 3000 = 2550$ psi.

4. Estimate the locations of the nodes.

Node *A*: Node *A* is located above the reaction. The size of the bearing plate at the support will be based on the bearing strength from ACI Code Table 22.8.3.2, but must not be less than the nodal zone strength, F_{nn}, from ACI Code Section 23.9.2. From ACI Code Table 22.8.3.2, the bearing strength is $0.85f'_c$. The required bearing area is

$$\frac{64{,}000 \text{ lb}}{0.65 \times 0.85 \times 3000 \text{ psi}} = 38.6 \text{ in.}^2$$

where $\phi = 0.65$ for bearing, but not less than the area based on the strength of nodal zone A given by

$$\frac{64{,}000}{0.75 \times 2040} = 41.8 \text{ in.}^2$$

in which $\phi = 0.75$ for strut-and-tie models.

Use a 4 by 4 by 1/2 in. angle, 15 in. long at the bearings. Bearing area = 60 in.^{2}

We shall assume that the reaction acts 2 in. from the end of the member. **Node *A*** is located at the intersection of the inclined reaction and the bar *A–D*, which we shall assume is located 2 in. above the bottom of the extended part of the beam.

Node *B*: This node is located at the upper ends of strut *A–B* and tie *B–C*.

Node *C*: This node is at the lower right corner of the 30 in. deep portion of the beam.

Node *D*: This node is to the right horizontally from node *A*.

Node *E*: This node is directly below node *D*.

5. Compute the strut-and-tie forces. We shall ignore the slight slope in strut *B–F*. This will lead to a slightly higher force in tie *B–C* and a slight lack of closure in the force diagram. Use can be made of a scale drawing of the strut-and-tie model in Fig. 17-41 in computing the internal forces. The drawing should be large enough to scale lengths from.

Node *B*: This node is located at the upper end of strut *A–B*, which is assumed to be acting at a 45° angle. The vertical reaction from node *A* is 64 kips, resulting in the total force in strut *A–B* of 90.5 kips. Thus, strut *A–B* produces a vertical force of 64 kips and a horizontal force of 64 kips at node *B*. The factored dead and live load force acting on node *B*, from step 1(c), is a vertical downward load of 18.1 kips. For equilibrium at the joint, the force in the strut B–F must be equal to 64 kips and the force in tie B–C must be equal to $64 - 18.1 = 45.9$ kips.

Node A: Strut A–B exerts a horizontal force component of 64 kips, acting to the left at node A. A horizontal reaction of 12.8 kips also acts to the left at node A. Summing horizontal forces at node A results in a force in tie A–D equal to $64 + 12.8 = 76.8$ kips.

Node C and D: The tie B–C applies a vertical upward force of 45.9 kips at node C. Thus, the vertical force component of the inclined strut C–D must be 45.9 kips. Similarly, at node D tie A–D applies a horizontal force of 76.8 kips, acting to the left. Thus, the horizontal force component in strut C–D must be 76.8 kips. From these values, the total force in strut C–D is

$$\sqrt{(45.9)^2 + (76.8)^2} = 89.5 \text{ kips}$$

From Fig. 17-41, the vertical distance from node C to node D is 14.0 in. Thus, the horizontal distance between these nodes must be $14.0 \times 76.8/45.9 = 23.4$ in. To complete the equilibrium at node C, the force in tie C–E must be 76.8 kips. To complete the equilibrium at node D, the force in tie D–E must be 45.9 kips.

Node E: To satisfy horizontal equilibrium at this node, strut E–F must have a horizontal force component of $83.2 - 76.8 = 6.40$ kips. For vertical equilibrium, strut E–F must have a vertical force component of 45.9 kips. Thus, the total force in strut E–F is

$$\sqrt{(6.40)^2 + (45.9)^2} = 46.3 \text{ kips}$$

Node F: Summing horizontal and vertical forces shows that node F is in equilibrium.

6. Compute the strut widths and check whether they will fit. The first estimate of strut-and-tie forces is shown in Fig. 17-41. It is now necessary to compute the widths of the struts to see whether they will fit into the available space without overlapping. From step 3, the effective concrete strength for struts A–B, C–D, and E–F will be taken as $f_{ce} = 1910$ psi. The struts will be assumed to have a thickness equal to the thickness of the beam, 15 in. (except for A–B and B–F). In tests, the cover over the sides of the stirrups spalled off at node B. As a result, we shall assume that struts A–B and B–F are 12 in. thick.

Strut A–B: From step 3 for nodes A and B, $f_{ce} = 2040$ psi. The f_{ce} for strut A–B governs, and we have

$$\text{width: } \frac{90{,}500 \text{ lb}}{0.75 \times 1910 \text{ psi} \times 12 \text{ in.}} = 5.26 \text{ in.}$$

Strut B–F: From step 3 for nodal zone F, $f_{ce} = 2550$ psi and for strut B–F, $f_{ce} = 2550$ psi. The f_{ce} for strut B–F will be used, and we have

$$\text{width: } \frac{64{,}000}{0.75 \times 2550 \times 12} = 2.79 \text{ in.}$$

Strut C–D: From step 3 for nodes C and D, $f_{ce} = 1530$ psi and for strut C–D, $f_{ce} = 1910$ psi. The $f_{ce} = 1530$ psi governs, and

$$\text{width: } \frac{89{,}500}{0.75 \times 1530 \times 15} = 5.20 \text{ in.}$$

7. Provide reinforcement to control cracking. When f_{ce} was selected for the struts, they were assumed to be bottle-shaped struts. To use $\beta_s = 0.75$ such struts must have transverse reinforcement satisfying ACI Code Section 23.5.3. The amount of this steel is given by the following equation:

$$\sum \frac{A_{si}}{bs_i} \sin \alpha_i \geq 0.003 \tag{17-10}$$

Try two horizontal No. 5 U-shaped bars enclosing the strut *A–B*. In Eq. (17-10), we will use the total area provided by both bars in the numerator and in the denominator we will use the projected vertical length of the strut, 15 in., in place of s_i. The angle between the axis of the strut and a horizontal bar is 45°, and sin 45° is 0.707. Substituting into Eq. (17-10) gives:

$$\frac{4 \times 0.31}{15 \times 15} \times 0.707 = 0.00390$$

This is more than the required value of 0.003. **Use two No. 5 U-shaped horizontal bars in the upper portion of the dap extension.**

For strut *C–D*, try three No. 4 vertical U-shaped stirrups and two No. 4 horizontal stirrups. As above, use the total vertical and horizontal steel areas in the numerator of Eq. (17-10) and replace s_i in the denominator with the projected horizontal and vertical lengths of strut *C–D*. The angle between the axis of the strut and a horizontal bar is:

$$\text{Arctan}\left(\frac{14.0}{23.4}\right) = 30.9°$$

and sin 30.9° = 0.513. The angle between the axis of the strut and the vertical bars is 59.1° and sin 59.1° = 0.858. So,

$$\frac{4 \times 0.20}{15 \times 14} \times 0.513 + \frac{6 \times 0.20}{15 \times 23.4} \times 0.858 = 0.00488 \text{ (o.k.)}$$

Strut *E–F* is crossed by both the U-shaped No. 5 bars in the upper half of the beam and the U-shaped No. 4 bars in the lower half of the beam. It is clear that this is sufficient without doing additional calculations.

Use two No. 4 U-shaped horizontal bars in the lower portion of the beam near the dap and U-shaped vertical stirrups at a spacing of 8 in. within the D-region.

8. Compute the steel required in the ties.

Tie *A–D*:

$$A_s = \frac{\text{tie force}}{\phi f_y} = \frac{76.8 \text{ kips}}{0.75 \times 60 \text{ ksi}}$$
$$= 1.71 \text{ in.}^2$$

Use four No. 6 bars welded to the angle. Development length $\ell_d = 42.7$ in. (Table A-6 for top bars). Theoretically, the bars should be anchored toward midspan from node *D*. Extend the bars 43 in. past *D*.

Tie *B–C*:

$$A_s \geq \frac{45.9 \text{ kips}}{0.75 \times 60 \text{ ksi}} = 1.02 \text{ in.}^2$$

Use four No. 4 closed stirrups. Use longitudinal No.4 bars inside the top corners of the stirrups.

Tie *D–E*: The force requirement is the same as for tie *B–C*, so four No. 4 double legged stirrups can be used. We shall arbitrarily spread them out over a longer length of the beam than was done for tie *B–C*. Also, their upper anchorage is less critical, so normal U-shaped stirrups with 135° hooks can be used. **Use No. 4 U-shaped stirrups at a spacing of 6 in. on centers throughout the D-region**. Note that this essentially provides more vertical steel across strut *C–D*.

Tie *C–E*: Assume that at midspan, the bottom steel is four No. 8 bars for flexure. This is enough for the 76.8 kip force in *C–E*. However, it is necessary to develop this force within the node at *C*. The length available for development of the bars within the node is 9.7 in., as

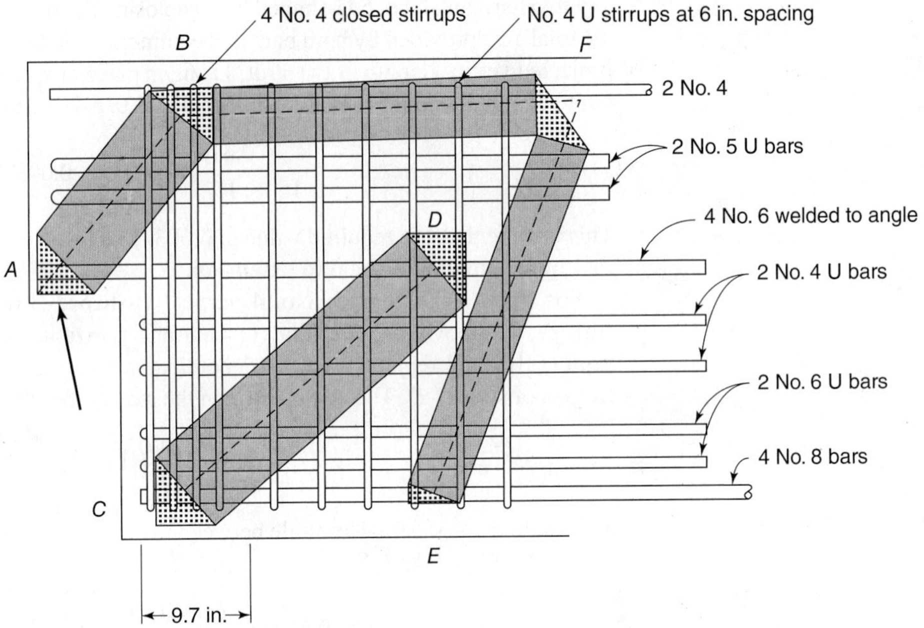

Fig. 17-42
Final strut-and-tie model for a dapped end showing the congestion due to the widths of the struts and reinforcement—Example 17-5.

shown in Fig. 17-42. The development length of a No. 8 bar is 54.8 in. (Table A-6). The force that can be developed in the straight No. 8 bars is

$$\frac{9.7\text{ in.}}{54.8\text{ in.}} \times 4 \times 0.79\text{ in.}^2 \times 60\text{ ksi} = 33.6\text{ kips}$$

This is not enough. Either provide some sort of mechanical anchor for these bars, or provide horizontal U bars to anchor the force. We shall provide horizontal U bars. The area required is

$$A_s = \frac{76.8}{0.75 \times 60} = 1.71\text{ in.}^2$$

Use two No. 6 U bars with bends adjacent to the end of the beam; total area is 1.76 in.^{2}. Place these above the No. 8 bars with clear spaces of 1 in. between the bars. Lap splice these $1.3 \times 32.9 = 42.8$ in.—say, 3 ft 8 in.—into the beam.

The final reinforcement is shown in Fig. 17-44.

9. Check the stresses on the sides of the nodes. It is generally not necessary to check bearing stresses on nodes, but we shall check the height of node *C*.

Height of node *C*: The steel should be at a height approximately equal to the height of the tension tie, to anchor the tie force based on concrete stressed at $f_{ce} = (0.85 \times 0.60 f'_c) = 1530$ psi.

$$\text{Nodal area required} = \frac{76{,}800\text{ psi}}{0.75 \times 1530} = 66.9\text{ in.}^2$$

The width is 15 in., so a height of 4.46 in. is required. The No. 8 bars and No. 6 U bars take more than this—therefore, o.k. ■

17-11 BEAM–COLUMN JOINTS

In Chapters 4, 5, 10, and 11, beams and columns were discussed as isolated members on the assumption that they can somehow be joined together to develop continuity. The design of the joints requires a knowledge of the forces to be transferred through the joint and the likely ways in which this transfer can occur. Most of the design requirements for beam–column joints are given in ACI Code Chapter 15. A few key requirements from Chapter 15 and elsewhere in the Code are:

1. ACI Code Section 15.4.3 requires enclosure of splices of continuing bars and of the end anchorages of bars terminating in connections of primary framing members, such as beams and columns.

2. ACI Code Section 15.4.2 requires a minimum amount of lateral reinforcement (ties or stirrups) in beam–column joints if the joints are not restrained on all four sides by beams or slabs of approximately equal depth. The amount required is the same as the minimum stirrup requirement for beams.

3. ACI Code Section 16.5.6.4 requires primary tension reinforcement in beams to be anchored in, or through, the supporting member.

4. ACI Code Section 9.7.3.8.2 requires that, in frames forming the primary lateral load-resisting system, a portion of the positive-moment steel should be anchored in the joint to develop the yield strength, f_y, in tension at the face of the support.

None of these sections gives specific guidelines for design. Design guidance can be obtained from [17-1], [17-24], and [17-26]. Extensive tests of beam–column joints are referred to in [17-26].

Corner Joints: Opening

In considering joints at the intersection of a beam and column at a corner of a rigid frame, it is necessary to distinguish between joints that tend to be *opened* by the applied moments (Fig. 17-43) and those that tend to be *closed* by the applied moments (Fig. 17-45). Opening joints occur at the corners of frames and in L-shaped retaining walls. In bridge abutments, the joint between the wing-walls and the abutment is normally an opening joint.

The elastic distribution of stresses before cracking is illustrated in Fig. 17-43b. Large tensile stresses occur at the reentrant corner and in the middle of the joint. As a result, cracking develops as shown in Fig. 17-43c. A free-body diagram of the portion outside the diagonal crack is shown in Fig. 17-43d. The force T is necessary for equilibrium. If reinforcement is not provided to develop this force, the joint will fail almost immediately after the development of the diagonal crack. A strut-and-tie model of the joint is shown in Fig. 17-43e.

Figure 17-44a compares the measured efficiency of a series of corner joints reported in [17-27] and [17-28]. The *efficiency* is defined as the ratio of the failure moment of the joint to the moment capacity of the members entering the joint. The reinforcement was detailed as shown in Figs. 17-44b to 17-44e. The solid curved line corresponds to the computed moment at which diagonal cracking is expected to occur in such a joint. Typical beams have reinforcement ratios of about 1 percent. At this reinforcement ratio, the joint details shown in Figs. 17-44d and 17-44e can transmit at most 25 to 35 percent of the moment capacity of the beams.

Nilsson and Losberg [17-27] have shown experimentally that a joint reinforced as shown in Fig. 17-44b will develop the needed moment capacity without excessive deformations. The joint consists of two hooked bars enclosing the corner and diagonal

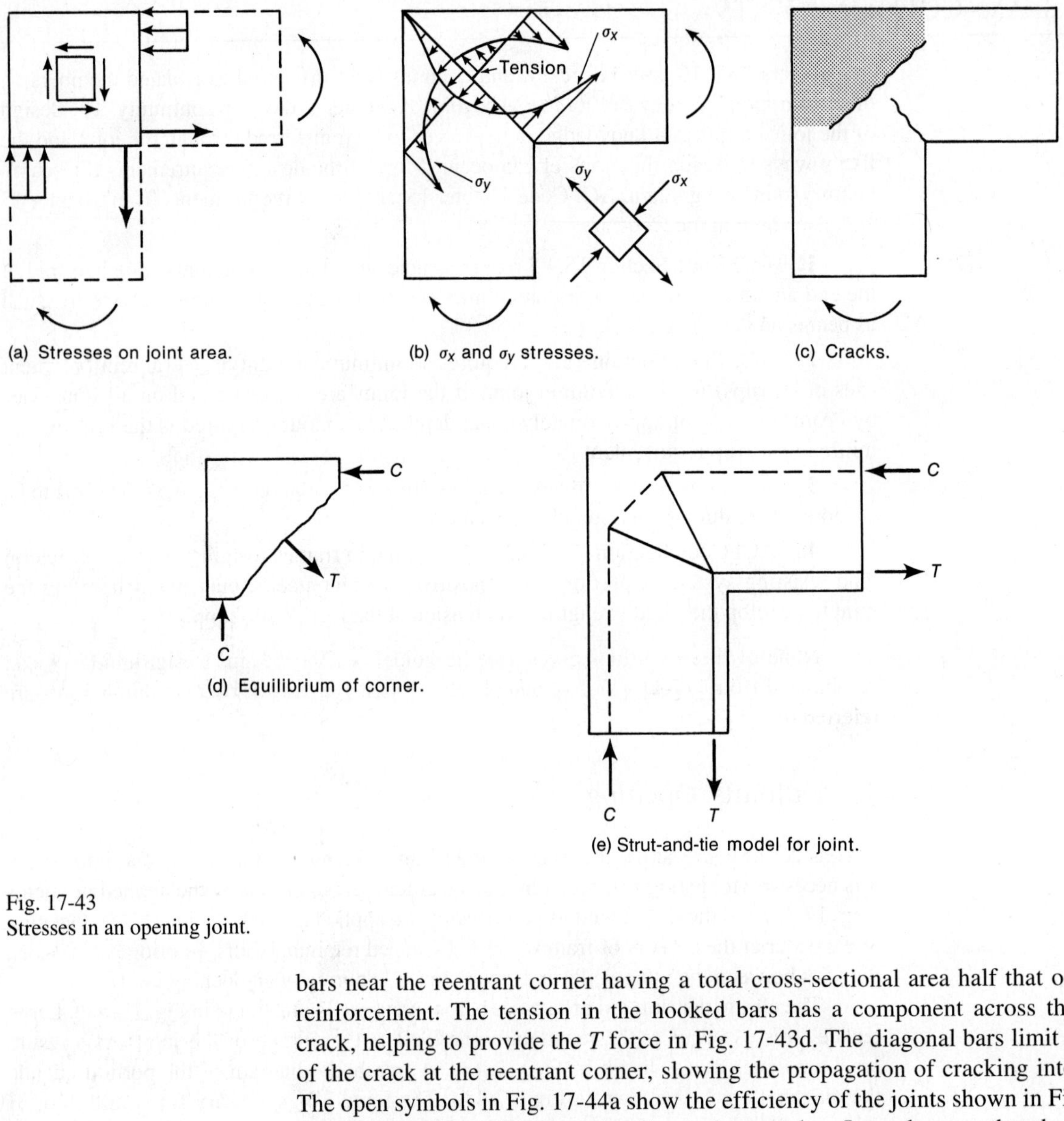

Fig. 17-43
Stresses in an opening joint.

bars near the reentrant corner having a total cross-sectional area half that of the beam reinforcement. The tension in the hooked bars has a component across the diagonal crack, helping to provide the T force in Fig. 17-43d. The diagonal bars limit the growth of the crack at the reentrant corner, slowing the propagation of cracking into the joint. The open symbols in Fig. 17-44a show the efficiency of the joints shown in Figs. 17-44b and 17-44c, with and without the diagonal corner bar. It can be seen that the corner bar is needed to develop the full efficiency in the joint.

Corner Joints: Closing

The elastic stresses in a closing corner joint are exactly opposite to those in an opening corner joint. The forces at the ends of the beams load the joint in shear as shown in Fig. 17-45a. As a result, cracking of such a joint occurs as shown in Fig. 17-45b, with a major crack on the diagonal. Such joints generally have efficiencies between 80 and 100 percent. Problems arise from the bearing inside the bent bars in the corner, because these bars must transmit a force of $\sqrt{2}A_s f_y$ to the concrete on the diagonal of the joint. For this reason, it may be desirable to increase the radius of this bend above minimum values given in ACI Code Section 25.3. Designing these bars as a curved-bar node is discussed by Klein [17–19].

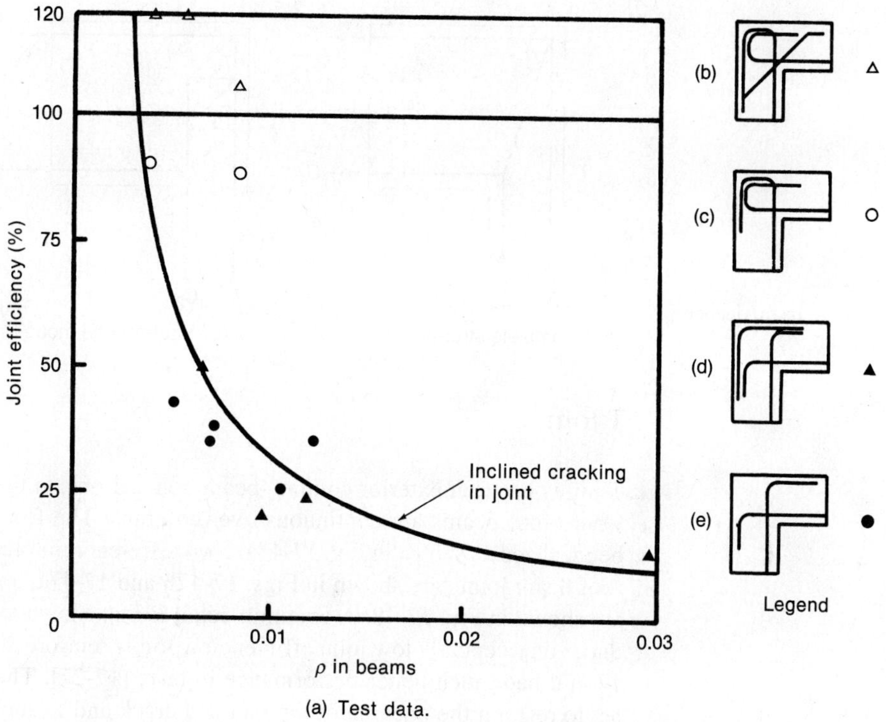

(a) Test data.

Fig. 17-44
Measured efficiency of opening joints.

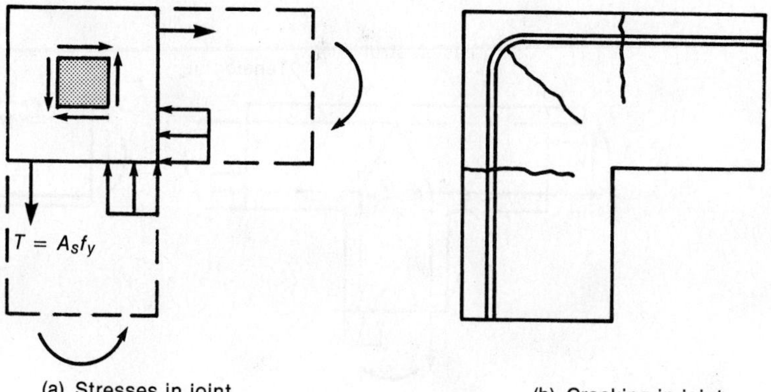

(a) Stresses in joint.

(b) Cracking in joint.

Fig. 17-45
Closing joints.

Frequently, the depth of the beam will be greater than that of the column, as shown in Fig. 17-46a. In such a case, the internal lever arm in the beam is larger than that in the column, and as a result, the tension force in the column steel will be larger than that in the beam. In the case shown, T_2 is three times T_1. For simplicity, the effects of the shears in the beam and column have been omitted in drawing Fig. 17-46. Although the strut-and-tie model in Fig. 17-46a is in overall equilibrium, the bar force jumps suddenly by a factor of 3 at A. A strut-and-tie model that accounts for the change in bar force in such a region is shown in Fig. 17-46b. Stirrups are required in the joint to achieve the increase in tension force in the column reinforcement. It can be seen from this strut-and-tie model that the stirrups in the joint region must provide a tie force of

$$\Sigma T_3 = T_2 - T_1$$

The reinforcement is detailed as shown in Fig. 17-46c.

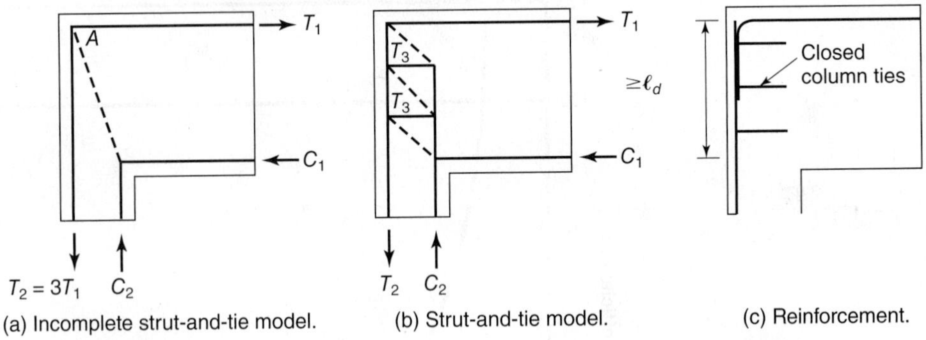

Fig. 17-46
Closing joint, beam deeper than column.

T Joints

T joints occur at exterior column–beam connections, at the base of retaining walls, and where roof beams are continuous over columns. The forces acting on such a joint can be idealized as shown in Fig. 17-47a. Two different reinforcement patterns for column-to-roof beam joints are shown in Figs. 17-47b and 17-47c, and their measured efficiencies are shown in Fig. 17-48. A common detail is that shown in Fig. 17-47b. This detail produces unacceptably low joint efficiencies. Joints reinforced as shown in Figs. 17-47c and 17-47d had much better performance in tests [17-27]. The hooks in these two patterns act to restrain the opening of the inclined crack and to anchor the diagonal compressive strut in the joint.

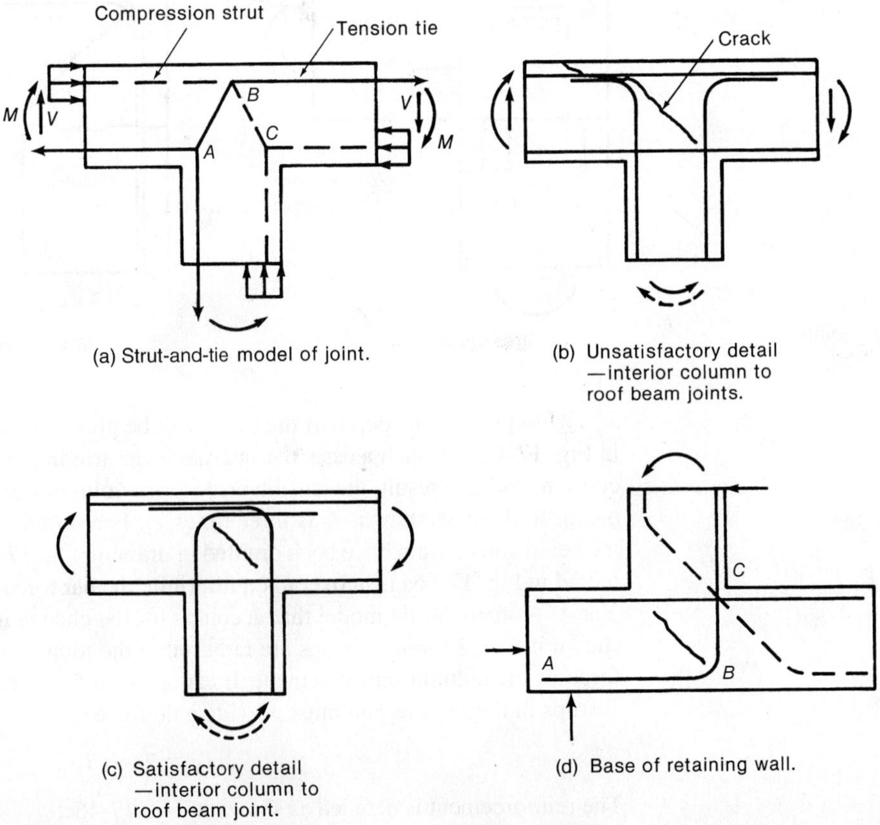

Fig. 17-47
T joints.

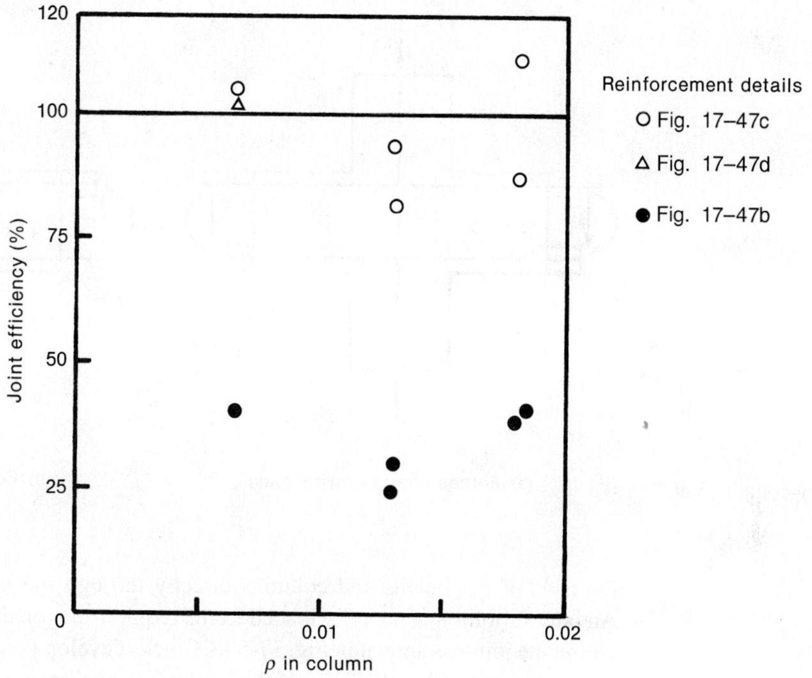

Fig. 17-48
Measured efficiency of T joints.

In the case of a retaining wall, the detail shown in Fig. 17-47d is satisfactory to develop the strength of the wall, provided that the toe is long enough to develop bar *A–B*. The diagonal bar, shown in dashed lines, can be added if desired, to control cracking at the base of the wall at *C*.

Beam–Column Joints in Frames

The function of a beam–column joint in a frame is to transfer the loads and moments at the ends of the beams into the columns. Again, force diagrams can be drawn for such joints. The exterior joint in Fig. 17-49 has the same flow of forces as the T joint in Fig. 17-47a and cracks in the same way. For such joints the reinforcement pattern shown in Fig. 17-47b should not be used. An interior joint under gravity loads transmits the tensions and compressions at

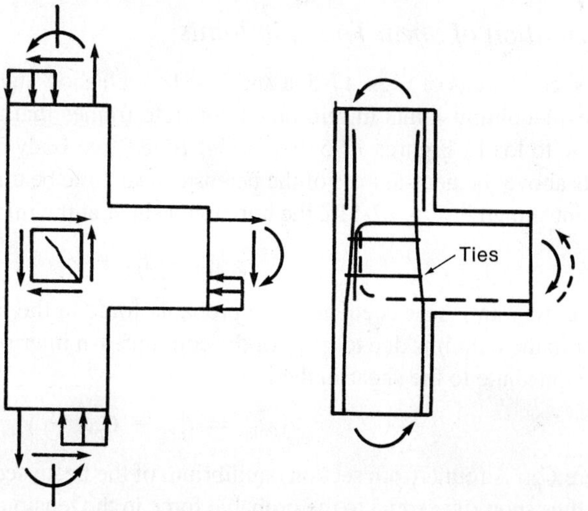

Fig. 17-49
Exterior beam–column joint.

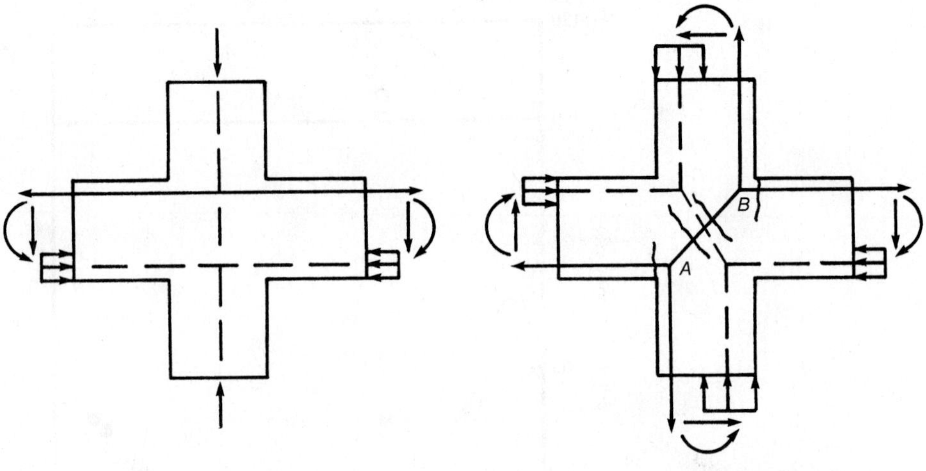

Fig. 17-50
Interior beam–column joint.

(a) Forces due to gravity loads.

(b) Forces due to lateral loads.

the ends of the beams and columns directly through the joint, as shown in Fig. 17-50a. An interior joint in a laterally loaded frame requires diagonal tensile and compressive forces within the joint, as shown in Fig. 17-50b. Cracks develop perpendicular to the tension diagonal in the joint and at the faces of the joint where the beams frame into the joint.

Classification of Joints

The ACI Committee 352 report [17-26] on the design of reinforced concrete beam–column joints divides joints into two groups depending on the deformations the joints are subjected to:

(a) Structures that are not apt to be subjected to large inelastic deformations and do not need to be designed according to ACI Code Chapter 18 are referred to as *nonseismic* structures. Such structures have *Type-1* beam–columns joints, and

(b) Structures that must be able to accommodate large inelastic deformations and as a result must satisfy ACI Code Chapter 18 are referred to as *seismic* structures. Such structures have *Type-2* beam–column joints.

Calculation of Shear Forces in Joints

The shaded areas of Figs. 17-51a and 17-51c each show the upper half of the joint regions in beam–column joints in reinforced concrete frames that are deflecting to the left in response to loads. Figures 17-51b and 17-51d are free-body diagrams of the portions of the joints above the neutral axes of the beams entering the beam–column joints. For the exterior joint shown in Fig. 17-51a, the horizontal shear at the midheight of the joint is given by:

$$V_{u,\,\text{joint}} = T_{pr} - V_{\text{col}} \tag{17-19a}$$

where the joint shear is equal to the probable force in the top steel in the joint, minus the shear in the columns due to sway of the columns. An interior joint has beams on both sides that contribute to the shear in the joint.

$$V_{u,\,\text{joint}} = T_{pr1} + C_{pr2} - V_{\text{col}} \tag{17-19b}$$

where C_{pr2} is found from section equilibrium of the beam section at the left side of the joint, and thus should be equal to the probable force in the tension reinforcement at the bottom of that beam. The column shears, V_{col}, can be obtained from a frame analysis; for most practical

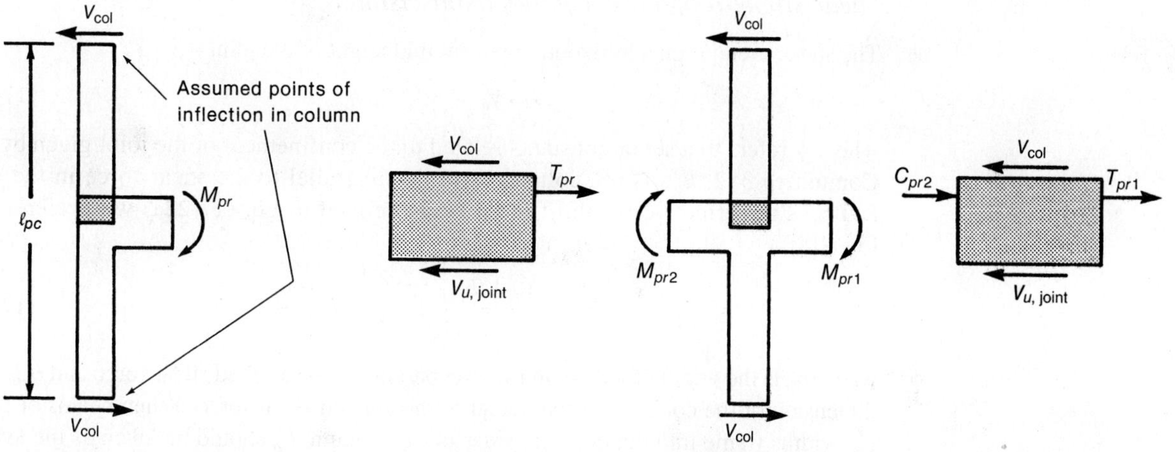

(a) Exterior column–beam joint. (b) Top half of exterior joint. (c) Interior column–beam joint. (d) Top half of interior column–beam joint.

Fig. 17-51
Calculation of shear in joints.

cases, they are estimated from the free-body diagrams in Figs. 17-51a and 17-51c, where points of contraflexure are assumed at the midheight of each story. The force T_{pr} is the tension in the reinforcement in the beam at its probable capacity. Thus,

$$T_{pr} = \alpha A_s f_y \tag{17-20}$$

The factor α is intended to account for the fact that the actual yield strength of a bar is larger than the specified strength and that at large deformations the bar may be strained into the strain-hardening range of behavior (Fig. 3-31). It is taken to be at least 1.0 for Type-1 frames, where only limited ductility is required, and at least 1.25 for Type-2 frames, which require considerable ductility.

The ACI Committee 352 design procedure for Type-1 (nonseismic) joints consists of three main stages:

1. Provide confinement to the joint region by means of beams framing into the sides of the joint or by a combination of the confinement from the column longitudinal bars and from ties in the joint region. Confinement allows the compression diagonal to form within the joint and reinforcement intercepts the inclined cracks. For the joint to be properly confined, the beam steel must be inside the column steel.
2. Limit the shear in the joint.
3. Limit the bar size in the beams to a size that can be developed in the joint.

For best joint behavior, the longitudinal column reinforcement should be uniformly distributed around the perimeter of the column core. For Type-1 joints, ACI Committee 352 recommends that at least two layers of transverse reinforcement (ties) be provided between the top and the bottom levels of the longitudinal reinforcement in the deepest beam framing into the joint. The vertical center-to-center spacing of the transverse reinforcement should not exceed 12 in. in frames resisting gravity loads and should not exceed 6 in. in frames resisting nonseismic lateral loads. In nonseismic regions, the transverse reinforcement can be closed ties, formed either by U-shaped ties with cap ties or by U-shaped ties that are lap spliced within the joint.

The hoop transverse reinforcement can be omitted within the depth of the shallowest beam entering an interior joint, provided that at least three-fourths of the column width is covered by the beams on each side of the column.

Shear Strength of Type-1 Joints (Nonseismic)

The shear strength on a horizontal plane at midheight of the joint is:

$$V_n = \gamma\sqrt{f'_c}b_j h_{col} \tag{17-21a}$$

where γ refers to a set of constants related to the confinement of the joint given by ACI Committee 352, h_{col} is the column dimension parallel to the shear force in the joint, and b_j is the effective width of the joint as defined in Eq. (17-22a) with reference to Fig. 17-53.

$$b_j = \frac{b_b + b_{col}}{2} \leq b_b + h_{col} \leq b_{col} \tag{17-22a}$$

where b_b is the width of the beam running parallel to the applied shear force and b_{col} is the dimension of the column perpendicular to the applied shear force. When beams of different widths frame into the opposite sides of the column, b_b should be taken as the average width of the two beams.

Values for the quantity γ are given in Fig. 17-52 for various classifications of Type-1 joints. These values have been empirically derived from test results. If lightweight concrete is used in the joint, the shear capacity should be multiplied by 0.85 for *sand-lightweight* concrete and by 0.75 for *all-lightweight* concrete.

The nominal shear strength of the joint defined in Eq. (17-21a) must satisfy the normal strength requirement that $\phi V_n \geq V_u$, where $\phi = 0.75$ and V_u is computed from Eq. (17-19a or b). If this is not satisfied, either the size of the column will need to be increased or the amount of shear being transferred to the joint will need to be decreased.

Beam reinforcement terminating in a Type-1 joint should have standard 90° hooks with a development length, ℓ_{dh} given by ACI Code Section 25.4.3.1 (Table A-8). The critical section for developing tension in the beam reinforcement is taken at the face of the joint in the case of Type-1 joints. If ℓ_{dh} is too large to fit into the joint (column), it is necessary to either decrease the size of the bar or increase the size of the column.

Shear Strength of Type-2 Joints (Seismic)

The shear strength on a horizontal plane at midheight of the joint is:

$$V_n = \gamma \cdot \lambda\sqrt{f'_c} \times A_j \tag{17-21b}$$

where γ refers to a set of constants related to the configuration and confinement of the joint given in ACI Code Table 18.8.4.1 and A_j is the effective area of the joint, similar to the product of b_j and h_{col} used in Eq. (17-21a), but it cannot exceed the area of the

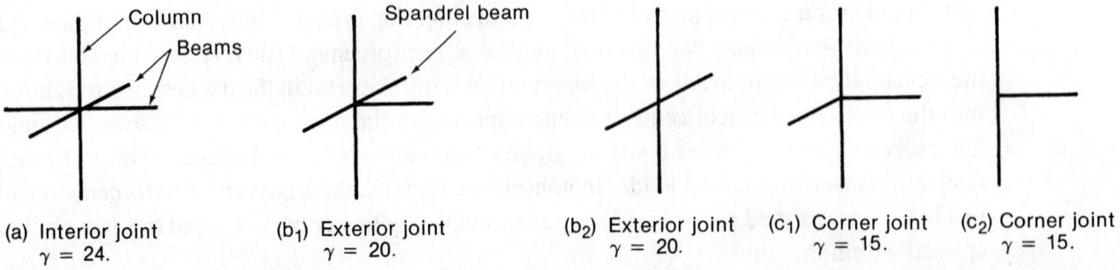

Fig. 17-52
Classification of joints—ACI 352. (γ values are for Type-1 joints.)

column. Assuming the column width exceeds the beam width, the definition of b_j for seismic design, as given in ACI Code Section 18.8.4.3 is:

$$b_j \leq b_b + h_{col}, \text{ and } b_j \leq b_b + 2x \tag{17-22b}$$

where x is the smaller of the distances measured from the beam longitudinal axis to a side face of the column.

Values for the quantity γ, which have been empirically derived from test results, are given below for various classifications of Type-2 joints.

$\gamma = 20$ for confined interior joints

$\gamma = 15$ for exterior joints confined on three faces or on two opposite faces

$\gamma = 12$ for corner or other Type-2 joints.

For a joint to qualify as an interior joint, the beams on the four faces of the joint must cover at least three-quarters of the width and depth of the joint face, where the depth of the joint is taken as the depth of the deepest beam framing into the joint. Joints that have an interior configuration (beams framing into all four faces), but the beams do not satisfy these size requirements, should be evaluated using the γ value for exterior joints.

For joints with an exterior configuration (beams framing into three faces or two opposite faces), the width of the beams on the two opposite joint faces must be at least three-quarters of the width of the joint face and the depth of the shallower of these two beams must be at least three-quarters of the depth of the deeper beam. Joints that have an exterior configuration, but the beams do not satisfy these size requirements, should be evaluated using the γ value for corner joints.

If lightweight concrete is used in the joint, a footnote in ACI Code Table 18.8.4.1 states that λ shall be set equal to 0.75.

The nominal shear strength of the joint defined in Eq. (17-21b) must satisfy the normal strength requirement that $\phi V_n \geq V_u$, where V_u is computed from Eq. (17-19a or b). However, Eq. (17-20) must be used to calculate the T_{pr} and corresponding C_{pr} values with α set equal to at least 1.25 to account for overstrength of the reinforcement and the high probability that those bars will go into strain hardening if plastic hinges form in the beams adjacent to the column faces. Because the joint shear loads are based on increased beam capacities, the appropriate ϕ factor for this *capacity-design* approach is 0.85. If this shear strength requirement is not satisfied, either the size of the column will need to be increased or the amount of shear being transferred to the joint will need to be decreased.

Rules for developing beam reinforcement terminating in a Type-2 joint are given in ACI Code Section 18.8.5.1. For bars terminating in a standard 90° hook, the development length, ℓ_{dh} is given by

$$\ell_{dh} = \frac{f_y \cdot d_b}{65 \cdot \lambda\sqrt{f'_c}} \tag{17-23}$$

but not less than 8d, and 6 in.

This equation considers the beneficial effect of anchoring the bar in the well-confined joint core, and also the detrimental effect of subjecting the bar to load reversals during earthquake loading. For simplicity, the critical section for developing the hooked bars is at the face of the joint, although several researchers state that effective anchorage starts at the face of the joint core.

For beam reinforcement extending through a beam–column joint, ACI Code Section 21.7.2.3 requires that the dimension of the column parallel to the beam bars must be greater than or equal to 20 times the diameter of the largest beam bar for normal weight concrete (26 for lightweight concrete). This length is not sufficient to fully anchor the bars in tension, but rather is intended to delay a potential breakdown in bond between the bars and the concrete in the joint when the beam reinforcement is subjected to load reversals during earthquake loading.

EXAMPLE 17-6 Design of Type-1 Joint Reinforcement

An exterior joint in a braced frame is shown diagrammatically in Fig. 17-54a. The normal-weight concrete and steel strengths are 5000 psi and 60,000 psi, respectively. The story-to-story height is 12 ft 6 in.

1. Check the distribution of the column bars and lay out the joint ties. For Type-1 joints, no specific column–bar spacing limits are given by ACI Committee 352. The column bars should be well distributed around the perimeter of the joint. Figure 17-54b shows a preferred arrangement of column bars and ties.

Because the frame is braced, the frame is not the primary lateral load-resisting mechanism. Hence, the spacing of the joint ties can be $s \leq 12$ in., with at least two sets of ties between the top and bottom steel in the deepest beam. The required area of these ties will be computed in step 3.

2. Calculate the shear force on the joint. We will check this in the direction perpendicular to the edge. A free-body cut through the joint is similar to Fig. 17-51b. The column axial loads have been omitted to simplify Fig. 17-51. The initial beam design used four No.11 top bars, so the shear in the joint is

$$V_{u,\text{joint}} = T_{pr} - V_{\text{col}} \tag{17-19a}$$

where

$$\begin{aligned} T_{pr} &= A_s \alpha f_y \text{ and } \alpha = 1.0 \text{ for a Type-1 joint} \\ &= 4 \times 1.0 \times 1.56 \text{ in.}^2 \times 60 \text{ ksi} = 374 \text{ kips} \end{aligned}$$

To compute V_{col}, consider the free-body diagram in Fig. 17-51a where $\ell_{pc} = 12.5$ ft. The nominal moment capacity of the beam is

$$\begin{aligned} M_n &= A_s f_y\left(d - \frac{a}{2}\right) = 4 \times 1.56 \times 60\left(25.5 - \frac{4.40}{2}\right) \\ &= 8720 \text{ kip-in.} = 727 \text{ kip-ft} \end{aligned}$$

$$V_{\text{col}} = \frac{M_n}{12.5} = 58.2 \text{ kips}$$

Therefore,

$$V_{u,\text{joint}} = 374 - 58.2 = 316 \text{ kips}$$

3. Check the shear strength of the joint. From Fig. 17-53 and Eq. (17-22a), the width of the joint is

$$\begin{aligned} b_j &\leq \tfrac{1}{2}(20 + 24) = 22 \text{ in.} \\ &\leq 20 + 22 = 42 \text{ in.} \end{aligned}$$

Use $b_j = 22$ in. The thickness of the joint, h_{col}, is equal to the column dimension parallel to the shear force in the joint. Use $h_{\text{col}} = 22$ in. Using Eq. (17-21a):

$$V_n = \gamma\sqrt{f'_c}\, b_j h_{\text{col}}$$

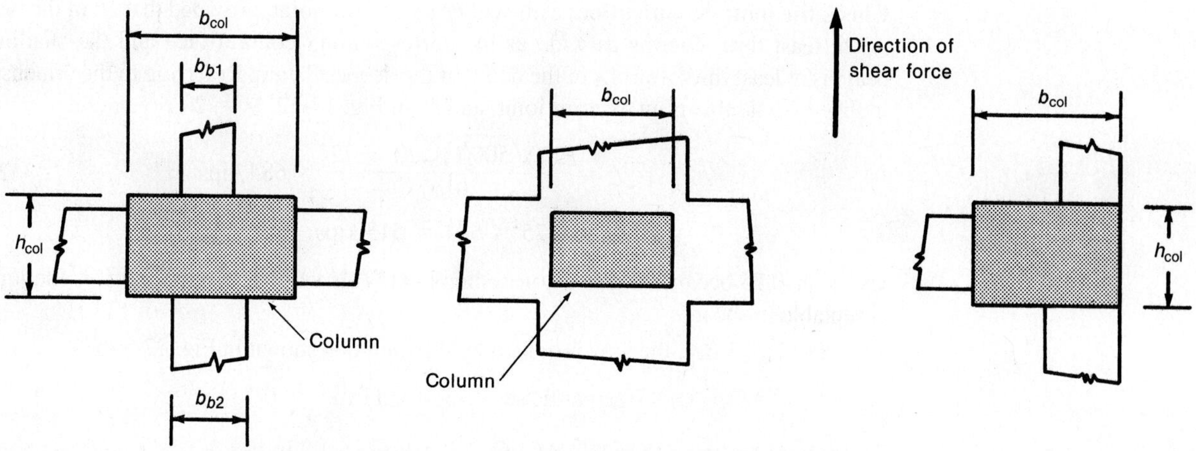

Fig. 17-53
Beam and column dimensions used to calculate width of joint, b_j.

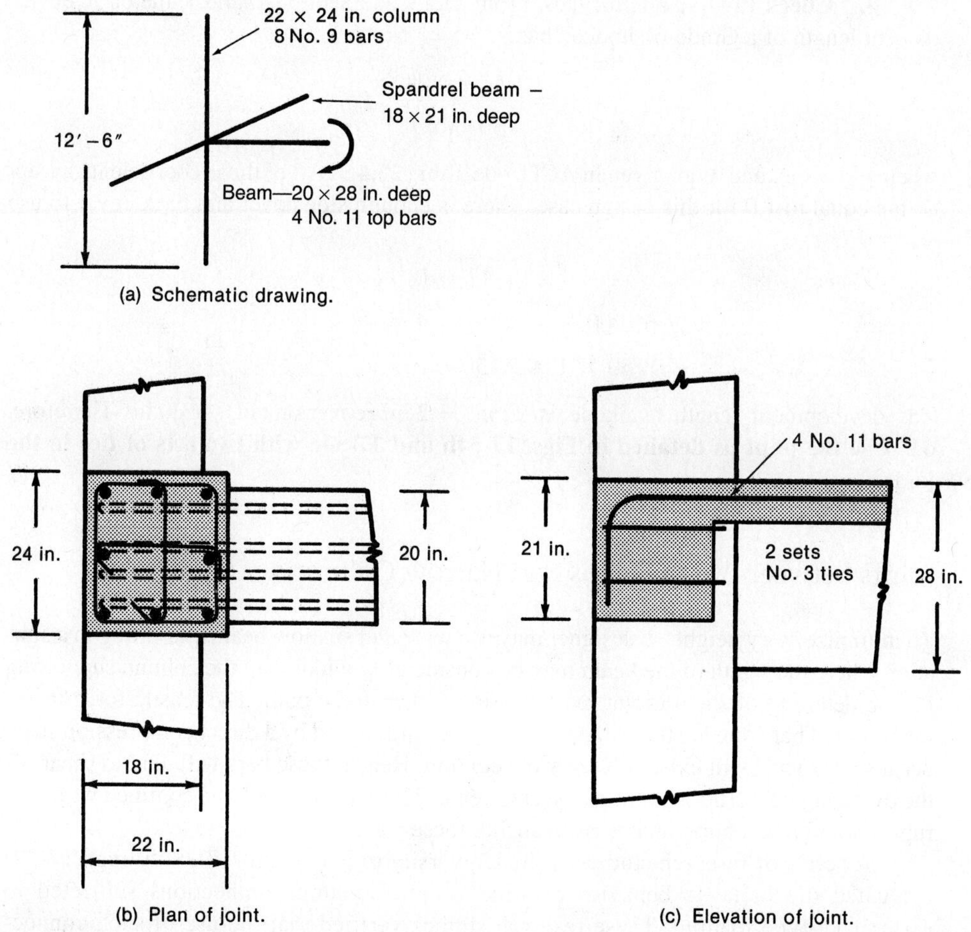

Fig. 17-54
Joint design—Example 17-6.

Check the joint classification: This will be an exterior joint, provided that all of the beams are at least three-fourths as wide as the corresponding column face and the shallowest beam is at least three-fourths of the depth of the deepest beam. Referring to the dimensions in Fig. 17-54a, this is an exterior joint, and from Fig. 17-52, $\gamma = 20$, so that

$$V_n = \frac{20\sqrt{5000} \times 22 \times 22}{1000} = 684 \text{ kips}$$

$$\phi V_n = 0.75 \times 684 = 513 \text{ kips}$$

Use $\phi = 0.75$ because this is a nonseismic joint. This value exceeds $V_{u,\,\text{joint}}$, so the joint is acceptable in shear.

For No. 3 ties, the area provided by the joint ties shown in Fig. 17-54b is

$$\text{Area per tie set} = 3 \times 0.11 \text{ in.}^2 = 0.33 \text{ in.}^2$$

The required spacing to satisfy ACI Code Section 15.4.2 is

$$s = \frac{A_v f_{yt}}{53 \text{ psi} \times b_j} = \frac{0.33 \times 60{,}000}{53 \times 22} = 17.0 \text{ in.}$$

but not more than 12 in. (to satisfy ACI 352). **Provide two sets of ties in the joint.**

4. **Check the bar anchorages.** From ACI Code Section 25.4.3.1, the basic development length of a Grade-60 hooked bar is:

$$\ell_{dh} = \left(\frac{f_y \cdot \psi_e \psi_c \psi_r}{50 \cdot \lambda \sqrt{f'_c}}\right) d_b$$

where ψ_e, ψ_c, ψ_r and λ are given in ACI Code Table 25.4.3. All of those coefficients except ψ_c are equal to 1.0 for this design case. There is enough side cover and back cover to use $\psi_c = 0.7$.

Thus,

$$\ell_{dh} = \left(\frac{60{,}000 \times 1 \times 0.7 \times 1}{50 \times 1 \times \sqrt{5000}}\right) 1.41 \text{ in.} = 16.7 \text{ in.}$$

The development length available is 22 in. − 2 in. (cover on tail) = 20 in. Therefore, o.k. **Use the joint as detailed in Figs. 17-54b and 17-54c with two sets of ties in the joint.** ■

Joints between Wide Beams and Narrow Columns

To minimize story heights, a designer may use wide and shallow beams, resulting in a situation where the width of the beam may be considerably wider than the column supporting it. The detailing of such a joint must provide a clear force path. The tensile force in the longitudinal bars outside the column will not be equilibrated by a direct compression strut, because this strut will exist only over the column. Hence, these bars will tend to shear off the overhanging portions of the transverse beam. This region should be confined with stirrups that provide a horizontal truss to anchor these bars.

A series of research studies at the University of Michigan [17-29 through 17-32] evaluated the inelastic behavior of wide beam-to-column connections subjected to earthquake-type loading. These research studies verified that the use of such connections is feasible in seismic zones and discussed required reinforcing details to transfer moments and shears from the wide beams into the columns for both exterior and interior connections.

Use of Strut-and-Tie Models in Joint Design

Although strut-and-tie models can be used to design joint regions, the models sometimes become complex when the effects of shears in the beams and columns are included. For some purposes, it is adequate to consider only the flexural forces in the steel and concrete within the joint when drawing the strut-and-tie models, as was done in Figs. 17-45e, 17-48, and 17-50. If it is necessary to include the effects of beam and column shears, it is generally necessary to consider a strut-and-tie model that includes portions of the beams between the zero-shear locations and portions of the columns.

17-12 BEARING STRENGTH

Frequently, the load from a column or beam reaction acts on a small area on the top of a wall or a pedestal. As the load spreads out in the wall or pedestal, transverse tensile stresses develop, which may cause splitting and, possibly, failure of the wall or pedestal.

This section starts with a review of the internal stresses and forces developed in such a region, followed by a review of the ACI design requirements for bearing strength. The discussion of internal forces is based in large part on the excellent design manual by Schlaich [17-1].

Internal Forces near Bearing Areas

Figure 17-55a shows the stress trajectories due to a concentrated load, F, acting at the center of the top of an isolated wall of length ℓ, height h, and thickness t, where $\ell = h$ and t is small compared to ℓ and h. As before, compressive-stress trajectories are shown by dashed lines, tensile by solid lines. The horizontal stresses across a vertical line under the load are shown in Fig. 17-55b. These tend to cause a splitting crack directly under the load. For the case shown, where the width of the load is 0.1ℓ, the maximum tensile stress is about 0.4ρ, where $\rho = F/(\ell \times t)$.

For calculation purposes, this state of stress can be idealized as shown in Fig. 17-55c. Here, the uniform compressive stress on the bottom surface is replaced by two concentrated forces, $F/2$ at the quarter-points of the base points D and E. The inclined struts act at an angle θ, which, for load size $w/h = 0.1$ is 65° for $\ell/h = 1.0$ or smaller. Flatter angles will occur if ℓ/h increases, reaching about 55° for a load with $w/h = 0.10$ on a wall having $\ell/h = 2.0$. For design, a slope of 2 vertical to 1 horizontal may be assumed, as done in Fig. 17-4c.

For equilibrium, the truss needs a horizontal tension tie, shown by the solid line B–C in Fig. 17-55c. This is made up of well-anchored horizontal steel placed over the zone of horizontal tensile stress (Fig. 17-55b) and having $A_s f_y$ sufficient to serve as the horizontal tie in the truss.

A similar situation occurs when a series of concentrated loads acts on a continuous wall, as shown in Fig. 17-55d. The horizontal force in the diagonal struts must be equilibrated by the tension tie B–C and a compression strut, C–B'. For force equilibrium on a vertical plane between C and B', the force in tie A–A' must be equal and opposite to C–B'. Elastic analyses suggest that the tension in B–C is roughly twice the compression in C–B'. Reinforcement should be placed in the direction of the tension ties B–C and A–A'.

The two cases shown in Fig. 17-55 involved concentrated loads acting on thin walls. If the concentrated loads were to act on a square pedestal, similar stresses would develop in a three-dimensional fashion. Although the stresses would not be as large, it may be necessary to reinforce for them.

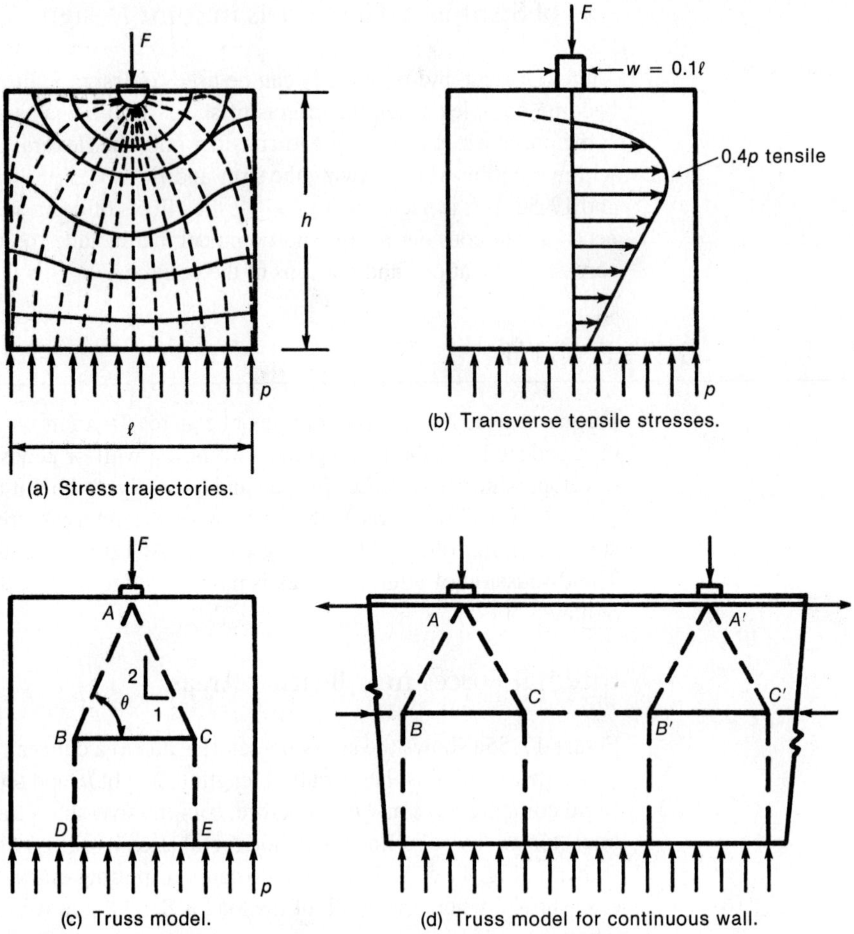

Fig. 17-55
Internal forces near bearing areas. (Adapted from [17-1].)

ACI Code Requirements for Bearing Areas

The ACI Code treats bearing on concrete in ACI Code Section 22.8, for dealing with normal situations, and ACI Code Section 25.9, for dealing with post-tensioning anchorage zones. Sections 25.9.3 and 25.9.4 require consideration of the spread of forces in the anchorage zone and requires reinforcement where this leads to large internal stresses.

ACI Code Section 22.8 is based on tests by Hawkins [17-33] on unreinforced concrete blocks supported on a stiff support and loaded through a stiff plate. A section through such a test is shown in Fig. 17-56a. As load is applied, the crack labeled 1 occurs in the center of the block at a point under the load. This then progresses to the surface, as crack 2. The resulting conical wedge is forced into the body, causing circumferential tension in the surrounding concrete. When this occurs, the radial crack 3 forms (Fig. 17-56b), the block breaks, and a "bearing failure" occurs. Two solutions are available: (1) Provide reinforcement to replace the tension lost when crack 1 formed, or (2) limit the bearing stresses so that internal cracking does not occur. ACI Code Section 22.8 follows the latter course.

The permissible bearing stress is set at $0.85f'_c$ if the bearing area is equal to the area of the supporting member; it can be increased to

$$f_b = 0.85f'_c\sqrt{\frac{A_2}{A_1}}, \text{ but not more than } 1.7f'_c \qquad (17\text{-}24)$$

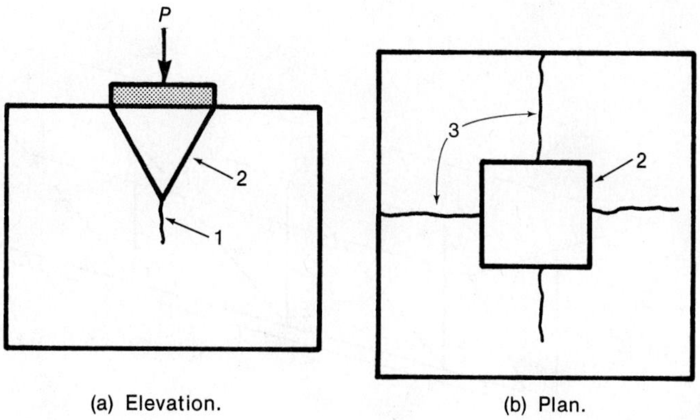

Fig. 17-56
Failure in a bearing test.

if the support has a larger area than the actual bearing area. In Eq. (17-24), A_1 refers to the actual bearing area, and A_2 is the area of the base of a frustrum of a pyramid or cone with its upper area equal to the actual bearing area and having sides extending at 2 horizontal to 1 vertical until they first reach the edge of the block, as illustrated in Fig. 15-10.

The maximum bearing load, B_{max}, is computed from

$$B_{max} = \phi f_b A_1 \tag{17-25}$$

where from ACI Code Table 21.2.1 ϕ is 0.65, f_b is from Eq. (17-24), and A_1 is the bearing area.

The 2:1 rule used to define A_2 does not imply that the load spreads at this rate; it is merely an empirical relationship derived by Hawkins [17-33].

17-13 T-BEAM FLANGES

Figure 4-39 illustrates the spread of forces in the compression flange of a T-beam. The forces in T-beam flanges can be examined more closely in strut-and-tie models of the beam web and flange. Figure 17-57a shows a strut-and-tie model of half of a simply supported beam loaded with a concentrated load at midspan (J). The height of the truss is assumed to be equal to the flexural moment arm, jd. The horizontal components of the compression forces in the diagonal struts in the web apply loads to the top flange at B, D, and so on, as shown in Fig. 17-57b. Compression struts and transverse tension ties in the flange act to spread this compression across the width of the flange. If the web struts are inclined at 45°, except at the reaction and the concentrated load, the horizontal components of the forces acting on the flange at D, F, and H are all equal to the shear V_u. If the flange struts are at a 2:1 slope, the transverse tensions arising from the forces acting on the flange at D, F, and H are $T = V_u/4$ and require transverse reinforcement with a capacity of

$$\phi A_s f_y = V_u/4$$

For the force at F, for example, this steel would be distributed over the length of flange extending about $jd/2$ each way from the location of the transverse tie resulting from the force acting on the flange at F. If transverse steel were needed for cross-bending of the flanges by loads on the overhanging flanges, it would be added to the steel needed to spread the compressive forces.

The strut-and-tie model in Fig. 17-57b indicates that there will not be any compression in the flange to the left of B and that there will be a concentration of horizontal compressive

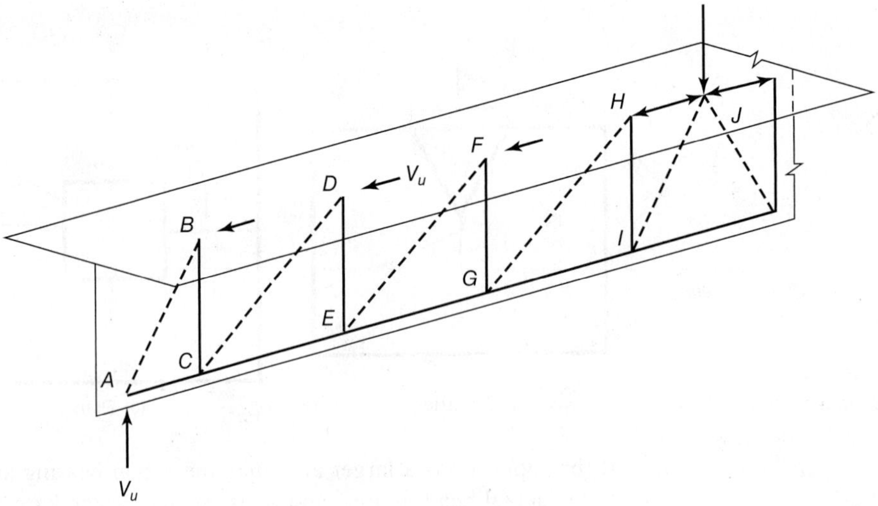

(a) Strut-and-tie model of beam web.

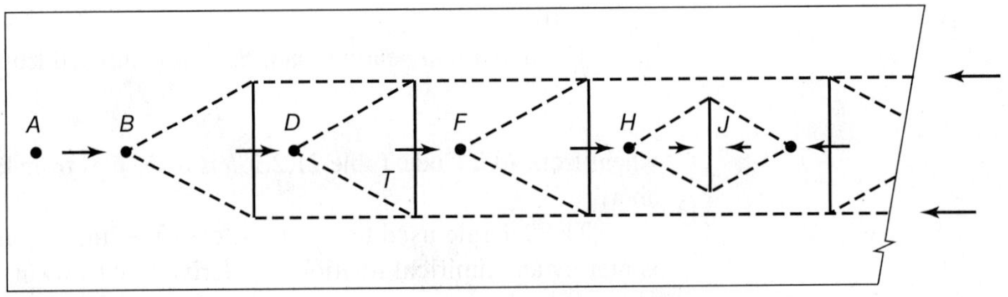

(b) Strut-and-tie model of compression flange.

Fig. 17-57
Strut-and-tie models of a T-beam with the flange in compression.

force in the vicinity of the load at J, because the horizontal component in strut I–J cannot spread across the flange.

The situation for a tension flange is more extreme, as is shown in Fig. 17-58. Here, a cantilever T-beam is loaded with a single concentrated load equal to V_u. Again, when the web struts are at 45°, except at the concentrated load and the support, the horizontal components of the strut forces acting on the flange at C, E, G, and I are equal to V_u. Figure 17-58b is drawn by assuming that the flexural tensile reinforcement in the flange is spread evenly over the width of the flange and hence can be represented by two ties at the quarter points of the width of the flange. The force acting on the flange at G is spread by compression struts to engage longitudinal steel at G' and G''. If the struts are at 2:1, a transverse tension tie G'–G'' is required to resist a transverse tension of $V_u/4$. The longitudinal forces acting on the flange at A and C are too close to the free end to be spread by 2:1 struts. As a result, the transverse-tie forces, and hence the amounts of transverse reinforcement, are larger in this region than in the rest of the beam.

Figure 17-58b was drawn by assuming that the longitudinal tension steel in the flange was evenly spread over the width of the flange. Using more steel close to the web to resist the longitudinal forces introduced into the flange at A and C allows the amount of transverse steel needed at the end of the beam to be reduced.

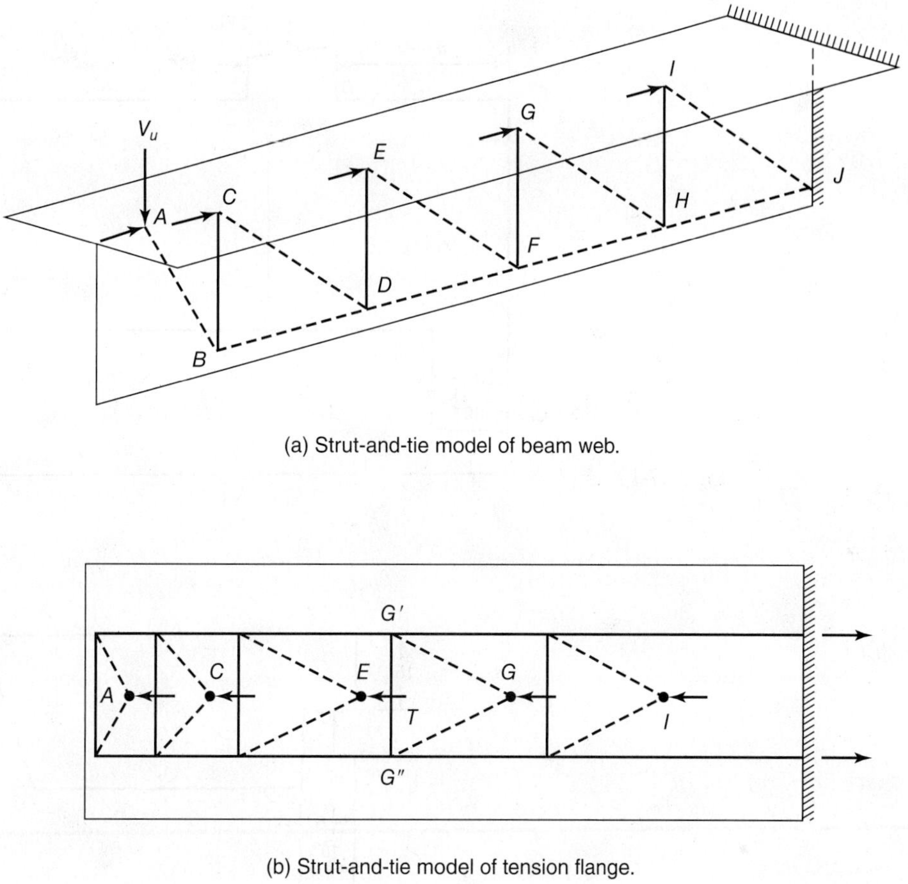

Fig. 17-58
Strut-and-tie models of a T-beam with the flange in tension.

PROBLEMS

17-1 The deep beam shown in Fig. 17-1 supports a factored column load of 1450 kips. The beam and all columns are 24 in. wide. Draw a strut-and-tie model assuming horizontal and vertical confinement reinforcement will be provided to justify using $\beta_s = 0.75$ for inclined struts. Including the weight of the beam, check the strength of the nodes and struts, and design reinforcement for the bottom tension tie. Use normal-weight concrete with $f'_c = 4000$ psi and use $f_y = 60$ ksi.

17-2 Continue Problem 17-1 and design the required horizontal and vertical reinforcement from ACI Code Section 23.5.3 to justify the use of $\beta_s = 0.75$ for the inclined struts. Also check that the reinforcement you provide satisfies the minimum reinforcement percentages specified in ACI Code Sections 9.9.3.1(a) and (b). Use $f_{yt} = 60$ ksi.

17-3 Design a corbel to support a factored vertical load of 120 kips acting at 5 in. from the face of a column. You should include a horizontal load equal to 20 percent of the factored vertical load. The column and corbel are 14 in. wide. The concrete in the column and corbel was cast monolithically. Use 5000-psi normal-weight concrete and $f_y = f_{yt} = 60$ ksi.

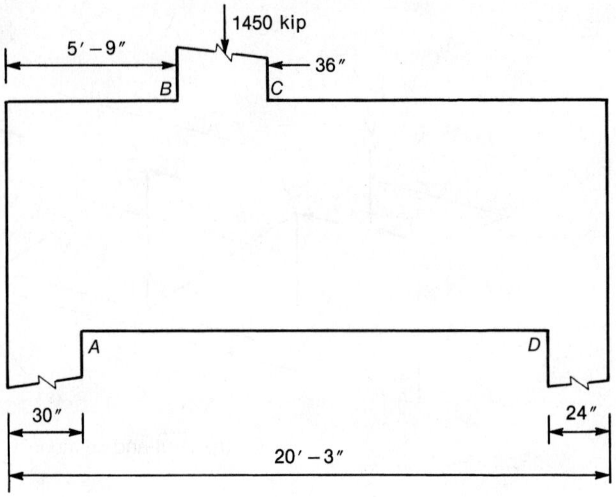

Fig. P17-1

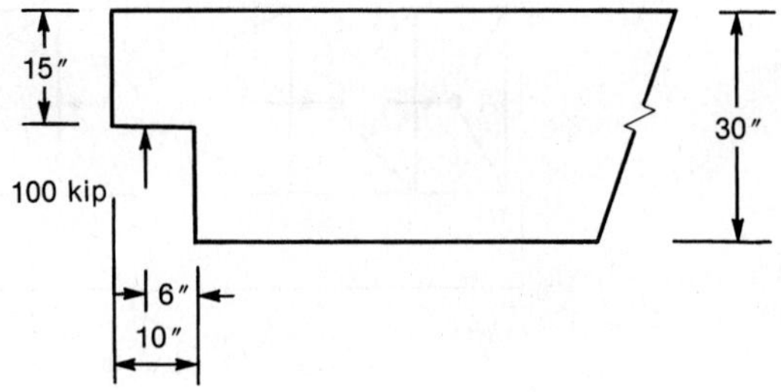

Fig. P17-5

17-4 Repeat Problem 17-3, but design the corbel using the traditional ACI design procedure given in ACI Code Section 16.5.

Video Solution

17-5 Figure P17-5 shows the dapped support region of a simple beam. The factored vertical reaction is 100 kips, and you should include a factored horizontal reaction of 20 kips at same location. Use normal-weight concrete with $f'_c = 5000$ psi, and reinforcement with $f_y = 60{,}000$ psi (weldable). The beam is 16 in. wide.

(a) Define the D-region in the beam.

(b) Draw a strut-and-tie model for the D-region defined in part(a).

(c) For your model in part (b), check the strength of nodes and struts, and design reinforcement of each tie and inclined strut.

REFERENCES

17-1 Jörg Schlaich and Dieter Weischede, *Detailing of Concrete Structures* (in German), Bulletin d'Information 150, Comité Euro-International du Béton, Paris, March 1982, 163 pp.

17-2 Jörg Schlaich, Kurt Schäfer, and Mattias Jennewein, "Toward a Consistent Design of Structural Concrete," *Journal of the Prestressed Concrete Institute*, Vol. 32, No. 3, May–June 1987, pp. 74–150.

17-3 Jörg Schlaich and Kurt Schäfer, "Design and Detailing of Structural Concrete Using Strut-and-Tie Models," *The Structural Engineer*, Vol. 69, No. 6, March 1991, 13 pp.

17-4 James G. MacGregor, "Derivation of Strut-and-Tie Models for the 2002 ACI Code," ACI Publication, SP–208, *Examples for the Design of Structural Concrete with Strut-and-Tie Models*, American Concrete Institute, Farmington Hills, MI, 2002, pp. 7–40.

17-5 Karl-Heinz Reineck, Editor, *Examples for the Design of Structural Concrete with Strut-and-Tie Models*, ACI Publication, SP–208, American Concrete Institute, Farmington Hills, MI, 2002.

17-6 J. K. Wight and G. J. Parra-Montesinos, "Strut and Tie Model for Deep Beam Design Using ACI Appendix A of the 2002 ACI Building Code," *Concrete International*, American Concrete Institute, May 2003, pp. 63–70.

17-7 Karl-Heinz Reineck and Lawrence C. Novak, Editors, *Further Examples for the Design of Structural Concrete with Strut-and-Tie Models*, ACI Publication, SP-273, American Concrete Institute, Farmington Hills, MI, 2010.

17-8 David M. Rogowsky and Peter Marti, "Detailing for Post-Tensioning," *VSL Report Series*, No. 3, VSL International Ltd., Bern, 1991, 49 pp.

17-9 Perry Adebar and Zongyu Zhou, "Bearing Strength of Compressive Struts Confined by Plain Concrete," *ACI Structural Journal*, Vol. 90, No. 5, September–October 1993, pp. 534–541.

17-10 David M. Rogowsky and James G. MacGregor, "Design of Deep Reinforced Concrete Continuous Beams," *Concrete International: Design and Construction*, Vol. 8, No. 8, August 1986, pp. 49–58.

17-11 *fib Model Code for Concrete Structures 2010*, International Federation for Structural Concrete (fib), Lausanne, Switzerland, 2013.

17-12 M. P. Nielsen, M. N. Braestrup, B. C. Jensen, and F. Bach, *Concrete Plasticity, Beam Shear—Shear in Joints—Punching Shear*, Special Publication of the Danish Society for Structural Science and Engineering, Technical University of Denmark, Lyngby/Copenhagen, 1978, 129 pp.

17-13 CSA Technical Committee on Reinforced Concrete Design, *A23.3-04 Design of Concrete Structures*, Canadian Standards Association, Mississauga, Ontario, January 2006, 214 pp.

17-14 Michael P. Collins and Denis Mitchell, "Design Proposals for Shear and Torsion," *Journal of the Prestressed Concrete Institute*, Vol. 25, No. 5, September–October 1980, 70 pp.

17-15 Julio Ramirez and John E. Breen, "Evaluation of a Modified Truss-Model Approach for Beams in Shear," *ACI Structural Journal*, Vol. 88, No. 5, September–October, 1991, pp. 562–571.

17-16 Frank Vecchio, and Michael P. Collins, *The Response of Reinforced Concrete to In-Plane Shear and Normal Stresses*, Publication 82–03, Department of Civil Engineering, University of Toronto, March 1982, 332 pp.

17-17 AASHTO, *LRFD Bridge Specifications*, Sixth Edition with 2013 Interim Revisions, American Association of State Highway and Transportation Officials, Washington, 2013.

17-18 FIP Recommendations, *Practical Design of Structural Concrete*, FIP Commission 3, Practical Design, September 1996, Publ. SETO, London, September 1999. (distributed by *fib* Lausanne.)

17-19 Gary J. Klein, "Curved-Bar Nodes, A Detailing Tool for Strut-and-Tie Models," *Concrete International*, American Concrete Institute, September 2008, pp. 42–27.

17-20 Ladislav Kriz and Charles H. Raths, "Connections in Precast Concrete Structures—Strength of Corbels," *Journal of the Prestressed Concrete Institute*, Vol. 10, No. 1, February 1965, pp. 16–47.

17-21 Alan H. Mattock, K. C. Chen, and K. Soongswany, "The Behavior of Reinforced Concrete Corbels," *Journal of the Prestressed Concrete Institute*, Vol. 21, No. 2, March–April 1976, pp. 52–77.

17-22 *PCI Design Handbook—Precast and Prestressed Concrete*, Seventh Edition (MNL-120-10), Precast/Prestressed Concrete Institute, Chicago, IL, 2010.

17-23 Alan H. Mattock and T. Theryo, *Strength of Members with Dapped Ends*, Research Project 6, Prestressed Concrete Institute, 1980, 25 pp.

17-24 William D. Cook and Denis Mitchell, "Studies of Disturbed Regions near Discontinuities in Reinforced Concrete Members," *ACI Structural Journal*, Vol. 85, No. 2, March–April 1988, pp. 206–216.

17-25 Alan H. Mattock, "Strut-and-Tie Models for Dapped-End Beams," Concrete International, Vol. 34, No. 2, Feb. 2012, pp. 35–40.

17-26 ACI-ASCE Committee 352, "Recommendations for Design of Beam-Column Joints in Monolithic Reinforced Concrete Structures," *ACI* 352R-02, Reapproved 2010, *ACI Manual of Concrete Practice*, American Concrete Institue, Farmington Hills, MI.

17-27 Ingvar H. E. Nilsson and Anders Losberg, "Reinforced Concrete Corners and Joints Subjected to Bending Moment," *Proceedings ASCE, Journal of the Structural Division*, Vol. 102, No. ST6, June 1976, pp. 1229–1254.

17-28 P. S. Balint and Harold P. J. Taylor, *Reinforcement Detailing of Frame Corner Joints with Particular Reference to Opening Corners*, Technical Report 42.462, Cement and Concrete Association, London, February, 1972, 16 pp.

17-29 T. R. Gentry and J. K. Wight, "Wide Beam-Column Connections under Earthquake-Type Loading," *Earthquake Spectra*, EERI, Vol. 10, No. 4, November 1994, pp. 675–703.

17-30 J. M. LaFave and J. K. Wight, "Reinforced Concrete Exterior Wide Beam-Column-Slab Connections Subjected to Lateral Earthquake Loading," *ACI Structural Journal*, Vol. 96, No. 4, July–August 1999, pp. 577–585.

17-31 C. G. Quintero-Febres and J. K. Wight, "Experimental Study of Reinforced Concrete Interior Wide Beam-Column Connections Subjected to Lateral Loading," *ACI Structural Journal*, Vol. 98, No. 4, July–August 2001, pp. 572–582.

17-32 J. M. LaFave and J. K. Wight, "Reinforced Concrete Wide-Beam Construction vs. Conventional Construction: Resistance to Lateral Earthquake Loads," *Earthquake Spectra*, EERI, Vol. 17, No. 3, August 2001, pp. 479–505.

17-33 Neil M. Hawkins, "The Bearing Strength of Concrete Loaded through Rigid Plates," *Magazine of Concrete Research*, Vol. 20, No. 62, March 1968, pp. 31–40.

18

Walls and Shear Walls

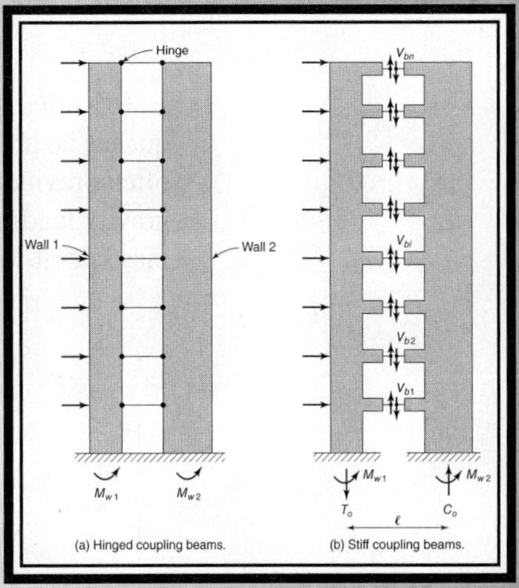

18-1 INTRODUCTION

Definitions—Walls and Wall Loadings

ACI Code Section 2.3 defines a wall as follows:

"*Wall*—a vertical element designed to carry axial load, lateral load, or both, with a horizontal length-to-thickness ratio greater than three, used to enclose or separate spaces."

ACI Section 2.3 also defines the term *structural walls*:

"*Structural wall*—Wall proportioned to resist combinations of shears, moments, and axial forces in the plane of the wall. A shear wall is a structural wall."

Major factors that affect the design of structural walls include the following:

(a) The structural function of the wall relative to the rest of the structure.
- The way the wall is supported and braced by the rest of the structure.
- The way the wall supports and braces the rest of the structure.

(b) The types of loads the wall resists.
(c) The location and amount of reinforcement.

Two frequent characteristics of walls are their slenderness, height-to-thickness ratio, which is generally higher than for columns, and the reinforcement ratios, generally about a fifth to a tenth of those in columns.

Types of Walls

Structural walls can be classified as:

(a) *Bearing walls*—walls that are laterally supported and braced by the rest of the structure that resist primarily in-plane vertical loads acting downward on the top of the wall (see Fig. 18-1a). The vertical load may act eccentrically with respect to the wall thickness, causing *weak-axis bending*.

(b) *Shear walls*—walls that primarily resist lateral loads due to wind or earthquakes acting on the building are called *shear walls* or *structural walls*. These walls often provide lateral bracing for the rest of the structure. (See Fig. 18-1b.) They resist gravity loads transferred to the wall by the parts of the structure tributary to the wall, plus lateral-loads (lateral shears) and moments about the *strong axis* of the wall.

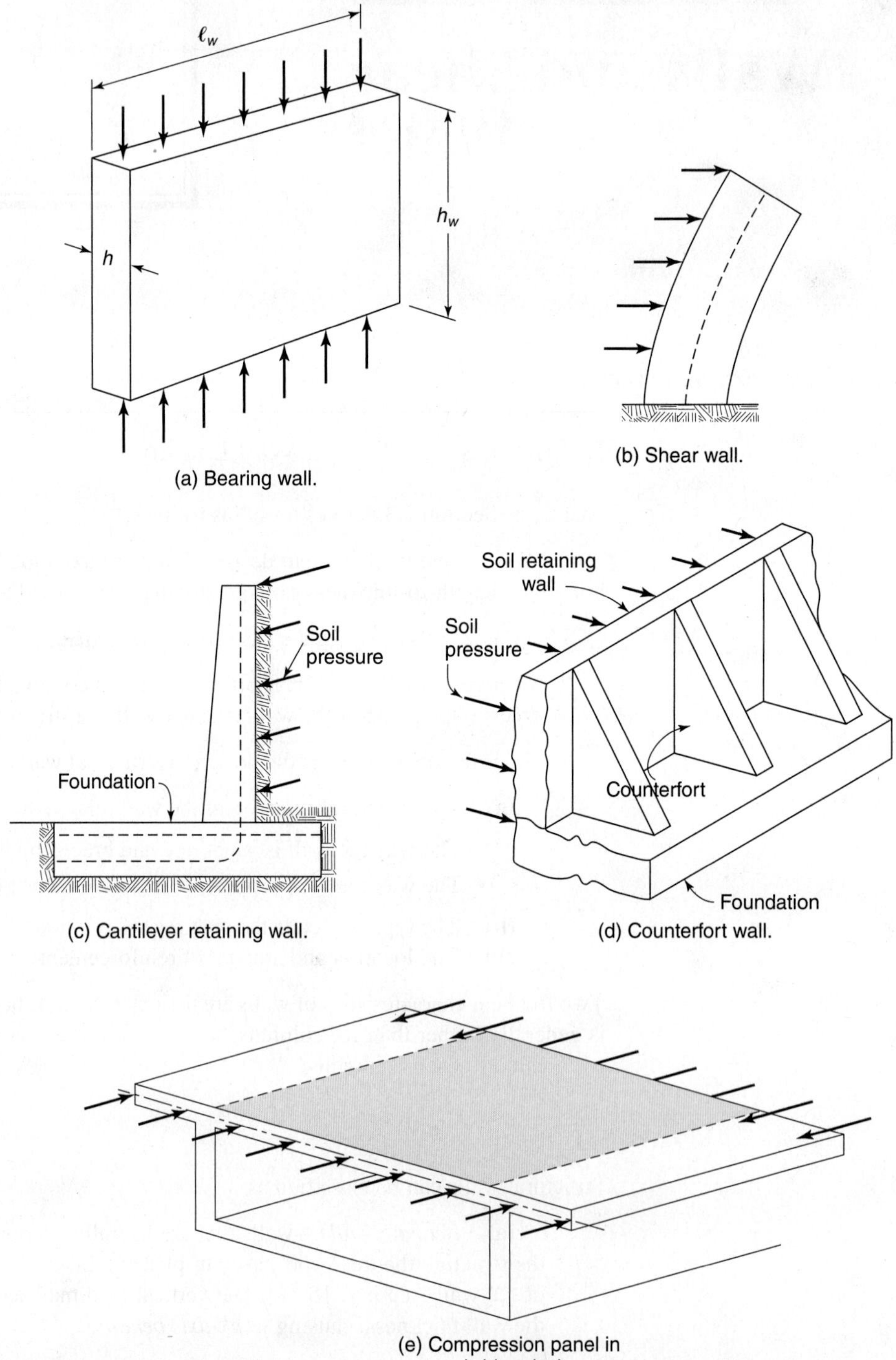

Fig. 18-1
Types of walls.

(c) *Nonbearing walls*—walls that do not support gravity in-plane loads other than their own weight. These walls may resist shears and moments due to pressures or loads acting on one or both sides of the wall. Examples are basement walls and retaining walls used to resist lateral soil pressures. (See Fig. 18-1c and 18-1d.)

(d) *Tilt-up walls*—are very slender walls that are cast in a horizontal position adjacent to the structure. They are then tilted into their intended vertical position and fastened to the foundation, to the roof or floor diaphragm, and to the adjacent panels. They are designed to resist vertical and lateral loads.

(e) Although they are not walls as such, plates that resist in-plane compression, such as the compression flanges or the decks of box girder bridges, display some of the characteristics of walls. (See Fig. 18-1e.)

One-Way and Two-Way Walls

Walls may be supported and restrained against lateral deflections along one to four sides. *Cantilever retaining walls* are generally supported solely along the lower edge of the wall. Such walls act as vertical flexural cantilevers that resist lateral loads from the adjoining soil. *Bearing walls* are generally laterally supported and restrained against deflection along two opposite sides, usually the top and bottom supports. Cantilever retaining walls and bearing walls transfer load in one direction: to supports at the top and bottom of the wall, for example, or to supports at the east and west edges. In the terminology used for one-way and two-way slabs, these are referred to as *one-way walls*. One-way walls may be designed as wide columns spanning between the top and bottom supports, using ACI Code Chapter 10, or they may be designed using ACI Code Chapter 11. Walls that transfer load in more than one direction are called *two-way walls*. Walls supported on three sides may occur in open-topped, rectangular tanks; in storage bins for bulk materials; or in counterfort retaining walls (see Fig. 18-1d). Walls supported on four sides are used to resist forces or pressures applied perpendicular to the walls.

Wall Assemblies

Shear walls may be planar walls, standing in one vertical plane, or three-dimensional assemblies of planar walls or wall segments. The latter occur as elevator shafts in buildings where four or more vertical walls enclose a stairwell or a group of elevators. Figure 18-2 is a photograph of a wall assembly that will enclose an elevator shaft and will brace a steel-frame building that will be built around it. The frame will be attached to the wall assembly by welding or otherwise fastening the frame to steel plates embedded in the wall concrete during construction of the walls.

In the design of a wall assembly, it is necessary to consider the transfer of shear forces from the wall segments serving as webs of the assemblies, to wall segments that act as flanges.

Notation

The orientation of the walls in a vertical direction leads to ambiguity in the notation for height and width. While most of the notation will be defined where it is first used, a few key symbols are defined here. Some of these are illustrated in Fig. 18-1a.

h_b is the height of a beam

h_z is the height of a location in a building

Fig. 18-2
Wall assembly in a building under construction. (Photograph courtesy of J. K. Wight.)

h_w is the overall height of a wall

ℓ_c length of compression member (column or wall) in a frame, measured from center to center of the joints in the frame

ℓ_w is the horizontal length of a wall

h is the thickness of a wall

18-2 BEARING WALLS

Walls used primarily to support gravity loads in buildings are referred to as *bearing walls*. Design of bearing walls normally follows ACI Code Section 11.5.3, which was derived specifically to apply to walls subjected to axial loads and moments due to the axial loads acting at an eccentricity of one-sixth of the thickness of the wall from the midplane of the wall (i.e., at the kern of the wall). The resulting moments are referred to as *weak-axis bending* moments. ACI Code Section 11.5.2.1 allows the design of bearing walls to be carried out either by:

1. Using the column design and slenderness requirements in ACI Code Sections 22.2, 22.4 and 11.8, or

2. The so-called *simplified design method* in ACI Code Section 11.5.3.

Axial-Load Capacity—The axial-load strength of slender walls given in ACI Code Section 11.5.3.1 is based on the results of 54 wall tests reported by Oberlender and Everard [18-1]. Their test results showed no effect of the reinforcement ratios. Concentrically loaded test specimens with $h_w/h = 28$, and eccentrically loaded specimens with $h_w/h = 16$ or more, exhibited buckling failures.

About the same time, Kripanarayanan [18-2] discussed the strength of slender walls based on analytical studies. He observed that reinforcement amounts of $\rho = 0.75$ to 1.0 percent were needed for the reinforcement to affect the failure loads of slender walls.

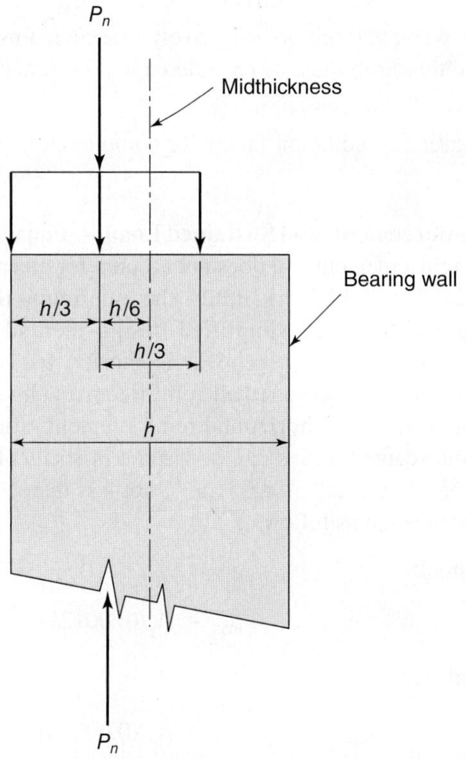

Fig. 18-3
Cross section through the top of a bearing wall showing the flexural compression zone.

Equation (18-1) (ACI Code Eq. 11.5.3.1) was derived in a two-step procedure. First, the capacity of a short wall was derived. Then this was multiplied by a factor reflecting the effects of slenderness on the axial-load capacity.

The largest eccentricity at which a load can be applied to a plain concrete wall without developing tensile stresses is at one-sixth of the wall thickness from the midthickness of the wall (at the kern point of the section). This load case can be approximated by a rectangular stress block extending from the compressed face of the wall for a distance of two-thirds of the thickness of the wall as shown in Fig. 18-3. The force per horizontal length of wall, ℓ_w, is

$$P_{n,\text{short}} = 0.85 f'_c \times \frac{2}{3} h \times \ell_w = 0.567 f'_c \times h \times \ell_w$$

This was rounded off to $0.55 f'_c h \ell_w$ and then multiplied by the term in the square brackets in Eq. (18-1) to account for the slenderness of the wall. The slenderness term was derived to give reasonable agreement with the slenderness effects in ACI Code Section 6.6.4. The equation for the axial-load capacity of a bearing wall is

$$\phi P_n = 0.55 \phi f'_c A_g \left[1 - \left(\frac{k \ell_c}{32h} \right)^2 \right] \qquad \text{(18-1) (ACI Eq. 11.5.3.1)}$$

where

ℓ_c is the clear, vertical distance between lateral supports

k is the effective length factor for a wall, taken as

0.8 if the wall is braced against translation at both ends and the top or bottom (or both) is restrained against rotation

1.0 if both ends are effectively hinged

2.0 for walls that are not effectively braced against lateral translation at the top, and therefore must be considered to be freestanding

h is the overall thickness of the wall

ϕ is the strength-reduction factor for compression-controlled sections, taken equal to 0.65.

Thickness, Reinforcement, and Sustained Loads—Equation (18-1) is not affected by the amount of wall reinforcement and does not account for creep under sustained axial loads.

ACI Code Table 11.3.1.1 limits the minimum thicknesses of bearing walls designed using the so-called simplified design method to 1/25 of the unsupported height or length of the wall, whichever is shorter, but not less than 4 in. ACI Code Eq. (11.5.3.1) does not consider wall reinforcement. Instead, ACI Code Table 11.6.1 gives minimum vertical and horizontal reinforcement ratios.

These reinforcement ratios can be written in terms of the maximum spacing of the bars. Thus, for No. 5 or smaller bars with f_y not less than 60 ksi, the maximum horizontal and vertical spacings are as follows:

Vertical steel:

$$s_{h,\max} = A_v/(0.0012h) \tag{18-2a}$$

Horizontal steel:

$$s_{v,\max} = A_h/(0.0020h) \tag{18-2b}$$

If the reinforcement is in two layers, A_v is the total area of vertical bars within the spacing s_h, and similarly for the horizontal bars.

ACI Code Table 11.6.1 requires more reinforcement horizontally than vertically. This reflects the greater chance that vertical cracks in walls might form as a result of restrained horizontal shrinkage or temperature stresses, compared with a lower chance that horizontal cracks will form as a result of restrained vertical stresses. Generally, if shrinkage occurs in the vertical direction, the shrinkage stresses are dissipated by vertical compression stresses in the wall.

EXAMPLE 18-1 Compute the Capacity of a Bearing Wall

A wall with a vertical height between lateral supports of 16 ft and a horizontal length of 25 ft between intersecting walls supports a uniformly distributed factored gravity load of 41 kips/ft, including the self-weight of the wall. The wall is supported on a strip footing that prevents lateral movement of the bottom of the wall. The wall supports a wooden-frame roof deck, which acts as a diaphragm to restrain lateral displacement of the top of the wall. Is an 8-in.-thick wall adequate if $f'_c = 4000$ psi? If so, select reinforcement for the wall. Use the resistance factors from ACI Code Table 21.2.1.

1. Check whether the wall thickness is sufficient. ACI Code Table 11.3.1.1 limits the thickness of walls designed by the empirical design method to the larger of 4 in. and 1/25 of the shorter of the unsupported height or the length. Thus, the minimum thickness is $(16 \times 12)/25$ in. $= 7.68$ in. An 8-in.-thick wall satisfies the minimum thickness requirement. **Use an 8-in. wall.**

2. Compute the capacity of a 1-ft-wide strip of wall.

$$\phi P_n = 0.55\phi f'_c A_g\left[1 - \left(\frac{k\ell_c}{32h}\right)^2\right] \quad (18\text{-}1)$$

ACI Code Table 11.5.3.2 gives $k = 1.0$ for the end restraints described in the statement of the problem. Walls will seldom have spiral reinforcement. As a result, the wall is an "other" type of member, and ACI Code Table 21.2.1 specifies $\phi = 0.65$. We thus have

$$\phi P_n = 0.55 \times 0.65 \times 4000 \text{ psi} \times 8 \text{ in.} \times 12 \text{ in.}\left[1 - \left(\frac{1.0 \times 16 \times 12}{32 \times 8}\right)^2\right]$$

$$= 137{,}000 \text{ lb} \times [0.438] \text{ per foot of wall length}$$

$$= 60.1 \text{ kips/ft}$$

This value exceeds the applied factored gravity load of 41.0 kips per ft; thus, the wall has adequate capacity.

3. Select reinforcement. ACI Code Table 11.6.1 requires minimum areas of $0.0012A_g$ and $0.0020A_g$ for vertical and horizontal reinforcement, respectively. ACI Code Section 11.7.2.3 allows the reinforcement to be placed in one layer, or "curtain," because the wall thickness is less than 10 in. ACI Code Section 11.7.2.1 gives the maximum bar spacing for cast-in-place walls as the smaller of 3 times the wall thickness—3 times 8 in. = 24 in.—and an upper limit of 18 in. The maximum spacing of the reinforcement is as follows:

Horizontal spacing of vertical reinforcement—from ACI Code Table 11.6.1:

$$s_{h,\max} = A_v/(0.0012h) \quad (18\text{-}2a)$$

If the required vertical steel is placed in a single layer of vertical No. 4 bars, $A_v = 0.20$ in.^{2}, and the spacing is

$$s_{h,\max} = 0.20 \text{ in.}^2/(0.0012 \times 8 \text{ in.}) = 20.8 \text{ in. on centers. Use 18 in. on centers.}$$

Vertical spacing of horizontal reinforcement—from ACI Code Table 11.6.1:

$$s_{v,\max} = A_h/(0.0020h) \quad (18\text{-}2b)$$

Using a single layer of No. 4 bars, the spacing of the minimum horizontal reinforcement is $s_{v,\max} = 0.20/(0.0020 \times 8) = 12.5$ in. on centers. Use 12 in. on center.

Because the vertical reinforcement is not specifically used in the strength design, ACI Code Section 11.7.4.1 does not require the vertical bars to be tied in nonseismic regions provided that

(a) the area of vertical steel is less than 0.01 times the gross area of the wall, or

(b) the steel is not used as compression steel.

The vertical steel provided has $A_s = 0.2$ in.$^2/(8 \times 18)$ in.$^2 = 0.0014$ times the gross area of the wall.

It is good practice to provide a No. 5 bar vertically at each end of each curtain of wall steel.

Use an 8-in.-thick wall with f'_c = 4000 psi, with one curtain of bars at the center of the wall with No. 4 vertical Grade-60 bars at 18 in. o.c. and No. 4 horizontal Grade-60 bars at 12-in. o.c. Add one No. 5 vertical bar at each end of the wall. ■

18-3 RETAINING WALLS

Reinforced concrete walls are used to resist the horizontal soil pressures when the surface of the ground is higher on one side of the wall than on the other. These are called *retaining walls*. The most common type is the *cantilever retaining wall* shown in Fig. 18-1c, which consists of a vertical cantilever wall that resists horizontal earth pressure and a footing that resists the moments at the base of the cantilever and transfers the forces to the ground. ACI Code Section 11.1.4 specifies that flexural design of retaining walls should be in accordance with the flexural design provisions of ACI Code Sections 22.2 through 22.4, with minimum horizontal wall reinforcement in accordance with ACI Code Section 11.6.

The lateral soil pressure acting on the wall and the bearing capacity of the soil under the wall footing are obtained by consultation with a geotechnical engineer, especially if the wall resists surcharge loads acting on the upper ground surface. It is important to provide drainage through or around the wall to minimize hydrostatic pressure behind the wall.

18-4 TILT-UP WALLS

Industrial buildings and warehouses are frequently constructed from tall thin concrete walls that are cast in a horizontal position on a slab on grade that will become the floor slab for the completed building. After the walls have gained adequate strength, a crane is used to tilt them up to their final vertical position at which time they are connected together. Such walls and buildings are referred to as *tilt-up walls* and *tilt-up buildings*. These constitute a form of construction that is becoming common throughout North America. Most aspects of the design of tilt-up buildings are covered by the report of the ACI Committee 551, *Tilt-up Concrete Construction* [18-3]. ACI Code Section 11.8, *Alternative method for out-of-plane slender wall analysis*, presents design requirements for the very slender walls used in tilt-up construction. Guidance is also found in the *Concrete Design Manual* [18-4]. Special consideration should be given to construction loads acting on these walls.

Nathan [18-5] analyzed thin one-way walls. His design charts are used widely in the design of tilt-up buildings.

18-5 SHEAR WALLS

The term *shear wall* is used to describe a wall that resists lateral wind or earthquake loads acting parallel to the plane of the wall in addition to the gravity loads from the floors and roof adjacent to the wall. Such walls are referred to as *structural walls* in ACI Code Chapter 18.

The strength and behavior of short, one- or two-story shear walls (as shown in Fig. 18-4a) generally are dominated by shear. These walls typically have a height-to-length aspect ratio (h_w/ℓ_w) of less than or equal to 2 and are called *short* or *squat walls*. Such walls can be designed by either the sectional strength requirements given in ACI Code Chapter 22 or the strut-and-tie method given in ACI Code Chapter 23. If the wall is more than three or four stories in height, lateral loads are resisted mainly by flexural action of the vertical cantilever wall (Fig. 18-4b) rather than shear action. Shear walls with h_w/ℓ_w greater than or equal to 3 are referred to as slender or flexural walls. These walls typically are designed using the provisions of ACI Code Chapters 11 and 22. Shear walls with h_w/ℓ_w ratios between 2 and 3 exhibit a combination of shear and flexural behavior and are also designed following the provisions of ACI Code Chapters 11 and 22.

Although shear walls may be simple planar walls, several wall segments commonly are connected together to act as a three-dimensional unit. Such *wall assemblies* have

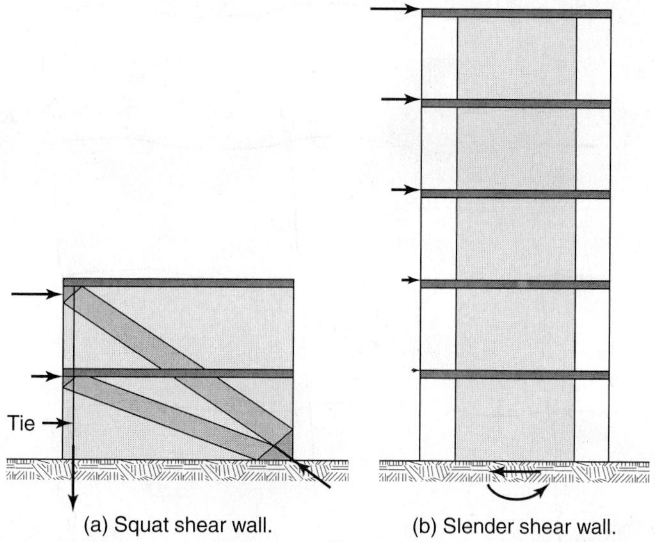

Fig. 18-4
Shear walls.

regular or irregular C, T, L, or H-shaped cross sections with webs and flanges that may enclose spaces in buildings, such as stairwells or elevator shafts.

18-6 LATERAL LOAD-RESISTING SYSTEMS FOR BUILDINGS

Three common systems for resisting wind or earthquake lateral loads are given here.

1. ***Moment-resisting frames*** are made up of interconnected beams and columns. Lateral loads are resisted by bending of the beams and columns (see Fig. 18-5). Such frames undergo relatively large lateral deflections. Typically the deflected shape of a moment-resisting frame is as shown in Fig. 18-5. If all stories have beams and columns with sizes proportional to the shear in the story, the lateral deflection of each story would be similar. To simplify construction, however, the sizes of the beams and columns selected for the lower stories are commonly used throughout, or are changed only every third or fourth story. Hence, the beams and columns in a building tend to be oversized in the upper stories. Moment-resisting frames are used for buildings up to 8 to 10 stories.

In a moment-resisting frame, deflections of the columns and beams both contribute to the sway deflections of the frame. If we isolate the shaded portion of the frame in Fig. 18-5, the deflection of A relative to B is given by

$$\Delta_{AB} = \frac{V\ell^3}{24E_cI_c} + \frac{V\ell^2L}{24E_bI_b} \tag{18-3}$$

where L is the span of the beam, center to center of supports, and ℓ is the height of the column from midheight of the story above the beam to midheight of the story below the beam. The first term on the right-hand side of Eq. (18-3) is the deflection due to the bending of the column. The second term is due to the bending of the beam. Shear deformations are not included in Eq. (18-3) because they are very small compared to bending deformations in a typical frame member.

Assuming $L = 2\ell$ and $E_cI_c = E_bI_b$, the portion of the relative horizontal deflection due to bending of the beam is *two-thirds* of the total deflection. Frequently in tall frames, it

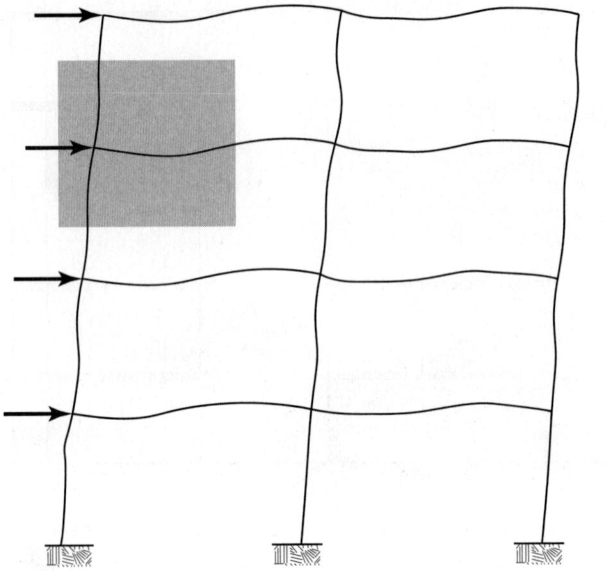

(a) Deflected frame.

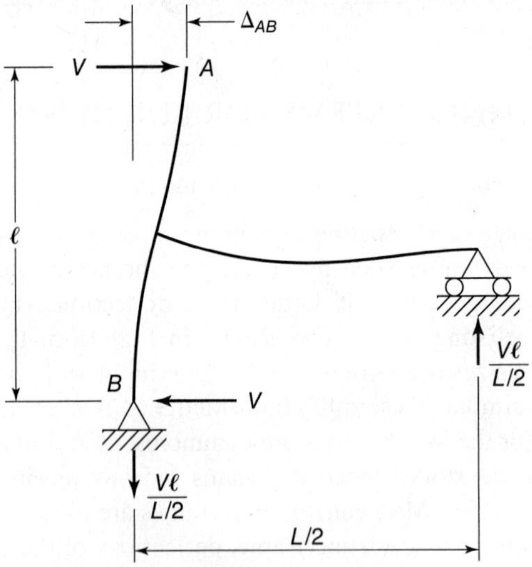

(b) Beam–column subassembly from part (a).

Fig. 18-5
Moment-resisting frame.

may be impossible to make the beams stiff enough to prevent large deflections. In such cases, walls or other stiffening elements are used to control lateral deflections.

2. ***Bearing-wall systems*** are used for apartment buildings or hotels. A bearing-wall building has a series of parallel transverse shear walls between rooms or apartments. The walls resist lateral loads by flexural action and deflect as vertical cantilevers.

3. ***Shear wall–frame buildings,*** shown schematically in Fig. 18-6, are used in buildings ranging from about 8 to about 30 stories. The lateral load is resisted in part by the wall and in part by the frame. Some of the most slender shear-wall–frame structures ever built are described by Grossman [18-6]. He presents some of the wind modeling rationale and two case histories of buildings with heights up to 10 times the least width at ground level.

4. ***Very tall concrete buildings*** Structural systems for very tall concrete buildings are discussed in [18-7]. The bottom stories of such a building are shown in Fig. 17-24. The closely spaced columns in the upper stories transfer vertical loads much like a continuous closed *tube* would. At the top of the second floor the vertical loads are transferred to a continuous deep beam called a *transfer beam*. It, in turn, transfers the vertical loads to the 10 large columns on the perimeter of the ground floor. In this region the more-or-less uniform compression in the tube is disrupted.

18-7 SHEAR-WALL–FRAME INTERACTION

The division of lateral load between the wall and frame in a shear-wall–frame building can be analyzed by using a frame analysis and a model similar to that in Fig. 18-6. The frame members in the model represent the sums of the stiffnesses of the columns and beams in the building in the bays parallel to the plane of the wall. Similarly, the wall in the model represents the sum of the walls in the structure. The wall and frame are connected by axially stiff link beams at every floor. The link beams in the model shown in Fig. 18-6 may or may not be hinged. In computing internal forces and moments due to the factored loads, the flexural stiffnesses, *EI*, from ACI Code Section 6.6.3.1.1 may be used. The model in Fig. 18-6 may be acceptable for buildings that are symmetrical in plan and have rigid floor diaphragms. A three-dimensional model is required for an unsymmetrical building, and where a designer wants to account for diaphragm flexibility.

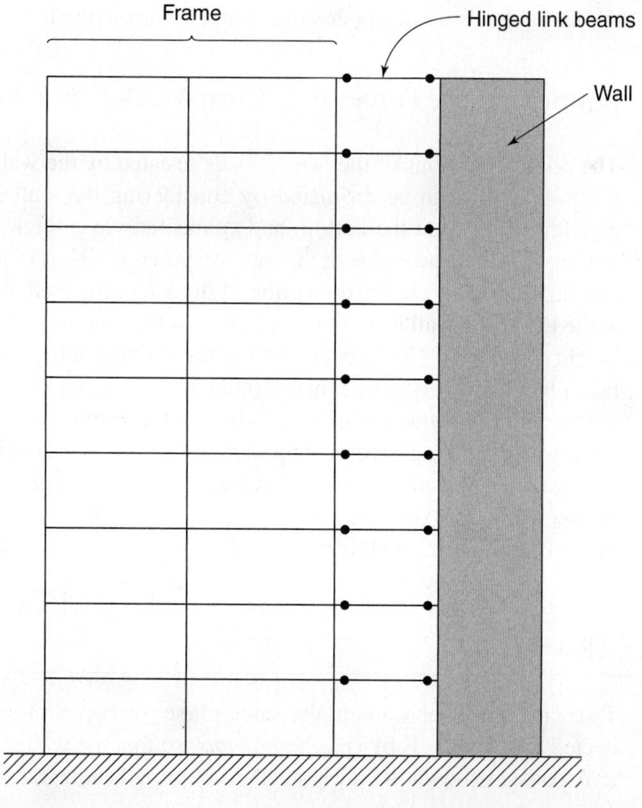

Fig. 18-6
Analytical model of a shear-wall–frame building.

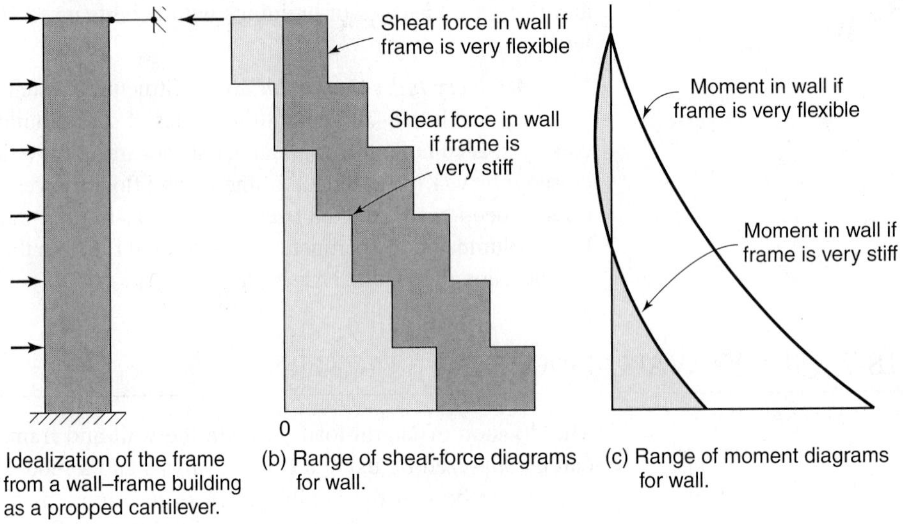

Fig. 18-7
Effect of frame stiffness on shear and moment in the shear wall.

The lateral-force analysis of shear wall–frame buildings must account for the different deformed shapes of the frame and the wall. Due to the incompatibility of the deflected shapes of the wall and the frame, the fractions of the total lateral load resisted by the wall and frame differ from story to story. Near the top of the building, the lateral deflection of the wall in a given story tends to be larger than that of the frame in the same story and the frame pushes back on the wall. This alters the forces acting on the frame in these stories. At some floors the forces change direction, as shown schematically in Fig. 18-7 by the range of possible moment diagrams in the wall. As a result, the frame resists a larger fraction of the lateral loads in the upper stories than it does in the lower stories.

Bounds on the Forces in a Shear-Wall–Frame Building

The relative portions of the lateral loads resisted by the walls and frames in a shear-wall–frame building can be estimated by considering the wall and the frame as two vertical cantilevers, fixed at the bottom and connected via a single extensionally rigid link beam joining the wall and frame at the top, as shown in Fig. 18-7a [18-8]. If the frame is so stiff that it prevents horizontal deflection of the top of the wall, the reaction of the loaded frame at the top of the wall is $3/8wh_w$, where w is the lateral load per foot of height and h_w is the height of the wall. This is equivalent to the reaction at the pinned end of a uniformly loaded beam having a constant EI that is fixed at one end and pinned at the other. As the lateral stiffness of the frame decreases relative to the lateral stiffness of the wall, the reaction at the top of the wall decreases, approaching zero for a very flexible frame combined with a stiff wall. As a result, the shear-force and bending-moment diagrams for the wall can vary between the approximate limits shown in Fig. 18-7b and 18-7c. The sum of the shear forces in the frame and the wall in a given story must equal the shear due to the applied loads.

18-8 COUPLED SHEAR WALLS

Two or more shear walls in the same plane (or two wall assemblies) are sometimes connected at floor levels by *coupling beams*, so that the walls act as a unit when resisting lateral loads, as shown in Fig. 18-8. The coupling beams frame into the edges of the walls, as

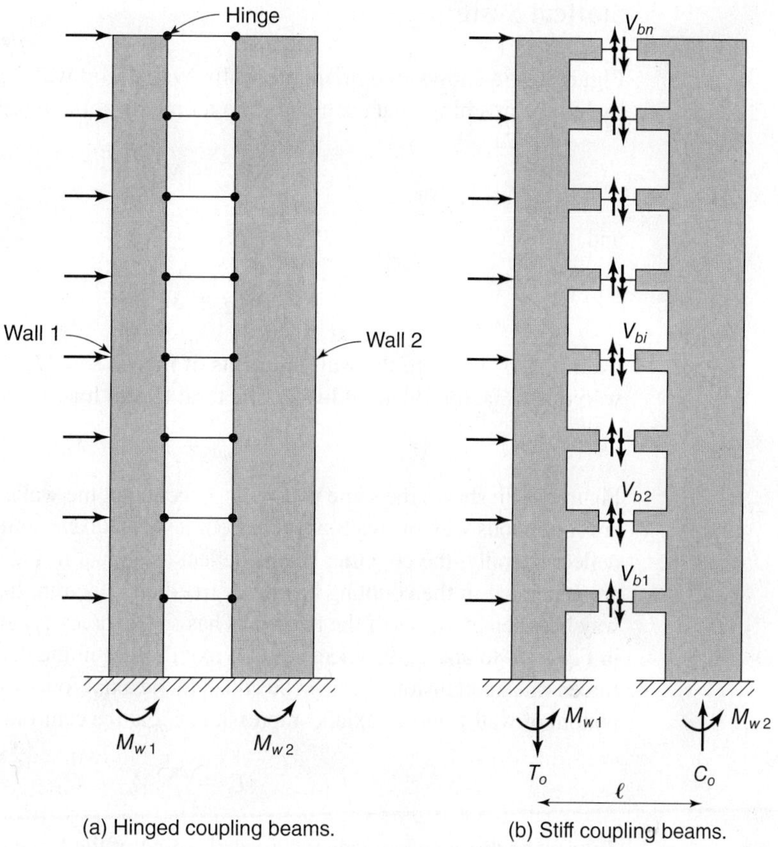

Fig. 18-8
Coupled walls.

shown in this figure. This discussion will be limited to the case of two walls separated by a single vertical line of openings, which are spanned by reinforced concrete *coupling beams*. Walls with more than two lines of openings, as shown in Fig. 18-2, are handled similarly to what is discussed here for two coupled walls. Other coupled wall systems may need special attention, especially if the widths and heights of the line of openings are irregular.

When the coupled wall deflects, the axes of both parts of the wall at A and A' deflect laterally and rotate through an angle θ, as shown in Fig. 18-9. This results in shearing deflections of the coupling beams that join the two walls. Localized cracking of the beam-to-wall joint reduces the angle the coupling beam must go through where it is attached to the wall. It is customary to assume the effect of these localized deflections and reduced stiffness of the coupling beam can be represented by moving the assumed connection point in from the face of the wall by approximately $h_b/2$, where h_b is the height of the coupling beam [18-8]. Thus, we shall assume the walls are joined by coupling beams spanning from B to B'. The downward deflection of point B is

$$\Delta_B = \left(\frac{b_w}{2} - \frac{h_b}{2}\right)\tan\theta \qquad (18\text{-}4)$$

where b_w is the width of the wall.

The moments and axial forces in the two wall segments in Fig. 18-9 must be in equilibrium with the axial forces, shears, and moments in the entire coupled wall system. The signs of the moments and shears may change over the height of the building, similar to those shown by Fig. 18-7b and 18-7c.

Statical System

Figure 18-8a shows two prismatic walls, wall 1 and wall 2, connected at each floor level by link beams hinged at each end. The moments at the bases of the two walls are equal to

$$M_{w1} = M_o \frac{I_{w1}}{I_{w1} + I_{w2}} \tag{18-5}$$

and

$$M_{w2} = M_o \frac{I_{w2}}{I_{w1} + I_{w2}} \tag{18-6}$$

where I_{w1} and I_{w2} are the wall moments of inertia and M_o is the moment at the base of the wall due to factored lateral loads. The total lateral load moments in the walls equal

$$M_{w1} + M_{w2} = M_o \tag{18-7}$$

Figure 18-8b shows the same two walls, except that the walls are now coupled by beams that are continuous with the walls at every floor level and have some flexural stiffness. As the walls deflect laterally, the coupling beams deflect as shown in Fig. 18-9, and shears and moments are generated in the coupling beams. A free-body diagram through the coupling beams halfway between the faces of the two walls has shear forces V_{bi} in each coupling beam, as shown in Figs. 18-8b and 18-9. There are also axial forces in the beams. For equilibrium of vertical forces, an axial tension, T_o, must be added to the axial forces in the walls at the centroid of the bottom of wall 1 and an axial compression, C_o, at the centroid of the bottom of wall 2, where

$$T_o = \sum_{i=1}^{n} V_{bi} = C_o \tag{18-8}$$

Taking the distance between the lines of action of the forces T_o and C_o as ℓ, we find that the total moment at the base of the coupled wall system is

$$M_o = M_{w1} + M_{w2} + T_o \times \ell \tag{18-9}$$

Equation 18-9 is plotted in Fig. 18-10 for a coupled wall consisting of two walls with $EI_{w1} = 2EI_{w2}$ [18-9]. The vertical axis is the slenderness of the beam, h_b/ℓ_b, where h_b and

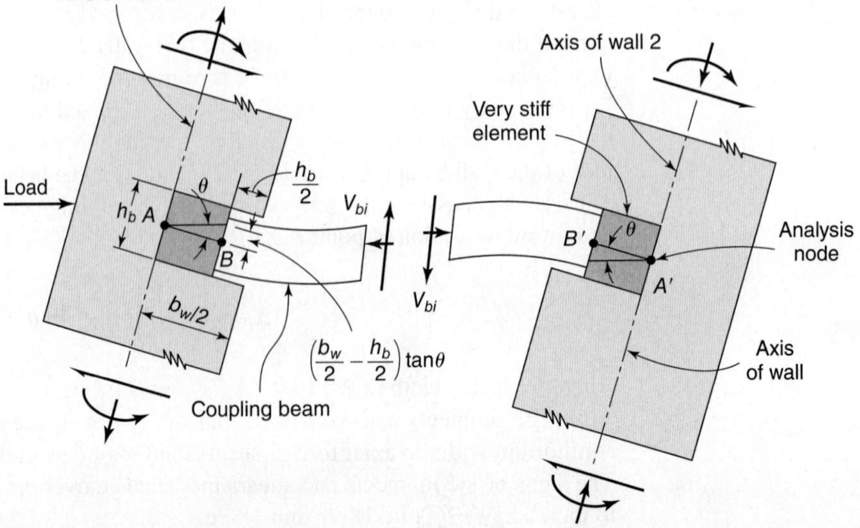

Fig. 18-9
Effect of shear wall deflections on the forces in a coupling beam (coupling beam shown thinner for clarity).

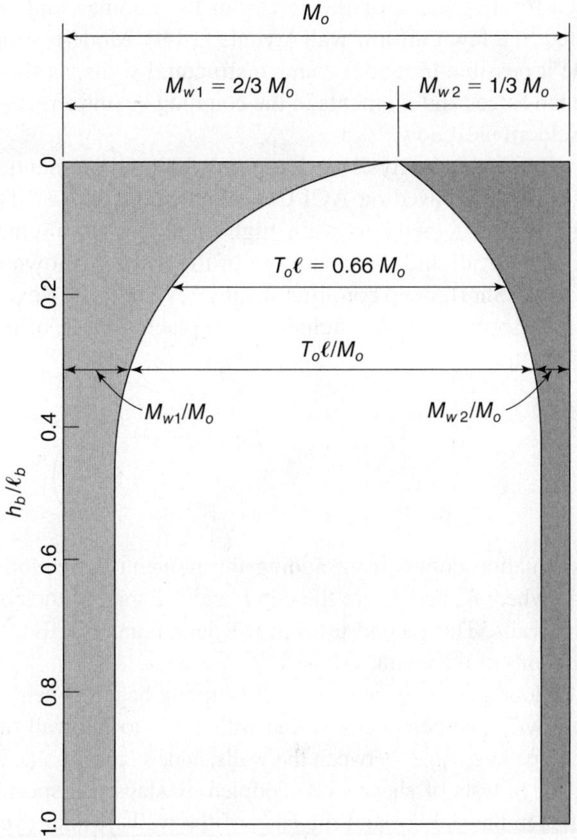

Fig. 18-10
Effect of the stiffness of the coupling beams on the moments in walls 1 and 2, $I_{w1} = 2I_{w2}$. (From [18-9].)

ℓ_b are the height and the adjusted span of the coupling beams, respectively. The beam slenderness is used here as a measure of the stiffness of the coupling beams. A coupling beam with h/ℓ_b equal to zero has no flexural stiffness, and as a result, the wall moments are divided in proportion to the ratio of the wall stiffnesses, as given by Eqs. (18-5) and (18-6). As the flexural stiffness of the coupling beams increases, the shears in them increase. As a result, the fraction of the overturning moment resisted by the axial force couple $T_o \times \ell$ increases asymptotically. A major effect of the coupling beams is to reduce the moments M_{w1} and M_{w2} at the bases of the two walls. This makes it easier to transmit the wall reactions to the foundation. The coupling beams also act to reduce the lateral deflections. If the beams are perfectly rigid, the two walls act as one wall. Figure 18-10 illustrates trends only. The actual division of M_o into M_{w1}, M_{w2}, and $T_o \times \ell$ depends on more variables than just h_b/ℓ_b.

Coupling beams may be rectangular beams, T-beams, or portions of the floor slab [18-10], [18-11]. In seismic regions, short-coupling beams may have diagonally placed reinforcement. This is discussed in Chapter 19.

For seismic design, the Canadian concrete code [18-10] defines a *coupled wall* as one in which $T_o \times \ell$ is at least 66 percent of M_o. In Fig. 18-10, this occurs when h_b/ℓ_b is about 0.2. If $T_o \times \ell$ is less than 66 percent of M_o, the wall is called a *partially coupled wall.*

Analysis of Coupled Walls

Before modern structural analysis programs, the individual coupling beams shown in Figs. 18-8b and 18-9 were replaced for analysis by a series of closely spaced *laminae*, each having a unit height and stiffness I_b/h_s, where h_s is the story height. This allowed a closed-form solution of the forces in the laminae and the walls. Such analyses were limited to a few uniform wall layouts [18-9]. Modern structural-analysis programs have made it possible to model coupled structural walls, as shown in Fig. 18-8b, and to compute the forces and moments in the coupling beams directly. As a result, laminar analyses are seldom used now.

In a frame analysis to determine factored moments for design, the member stiffnesses may be based on ACI Code Table 6.6.3.1.1(a). The coupling beams are joined to hypothetical members with high values of the moment of inertia, I, between the face of the wall and the centerline of the wall, as shown by the dark shaded regions in Fig. 18-9. Short, deep coupling beams develop both flexural and shear deflections. The shear deflections can be included by replacing the I_b of the coupling beam between the walls with

$$I_{\text{eff}} = \frac{I_b}{1 + 2.8\left(\frac{h_b}{\ell_b}\right)^2} \tag{18-10}$$

This equation comes from adding the moment deflections and shear deflections of the beam, where h_b and ℓ_b are the depth and the span of the coupling beam from face to face of the walls. The second term in the denominator of Eq. (18-10) accounts for the shear deflections of the beam.

Floor slabs may serve as soft coupling beams. Their stiffness can be based on a slab with a width perpendicular to the wall equal to the wall thickness plus half of the width of the opening, $\ell_b/2$, between the walls, added on each side of the opening [18-11], [18-12], [18-13]. In tests of shear walls coupled by slabs, the specimens failed by punching-shear failures in the slab around the ends of the walls. Under cyclic loads, the stiffness of slabs serving as coupling beams decreased rapidly.

Figures 18-11a, 18-11b, and 18-11c show the distributions of moments in the walls, the axial-force couple, and the shears in the coupling beams for a typical coupled wall [18-4] where $I_{w1} = I_{w2}$. Typically, the maximum shear in the coupling beams occurs at about one-third of the height above the base. The sawtooth shape of Fig. 18-11a results from the end moments in the coupling beams.

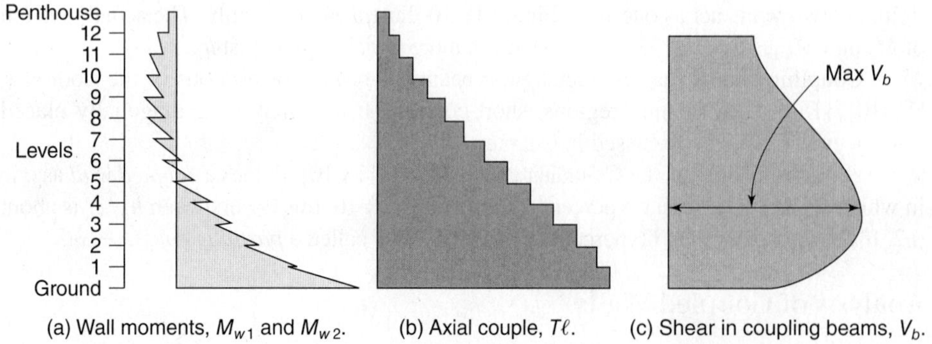

Fig. 18-11
Typical distribution of wall moments, axial-force couple, and shear in the coupling beams, $I_{w1} = I_{w2}$. (From [18-4].)

18-9 DESIGN OF STRUCTURAL WALLS—GENERAL

Layout of Building

Major considerations in selecting a structural system for a multistory building with structural walls are the following:

(a) The building must have enough rigidity to withstand the service loads without excessive deflections or vibrations.

(b) It is desirable that the wall be loaded with enough vertical load to resist any uplift of parts of the wall foundations due to lateral loads.

(c) The locations of frames and walls should minimize torsional deformations of the building about the vertical axis of the building. (See also Section 19-3.)

(d) The walls must have adequate strength in shear, and in combined flexure and axial loads.

(e) The wall thickness or cover on the reinforcement may be governed by the fire code.

Diaphragms

Lateral loads are transferred to the lateral-force resisting system by the floor and roof slabs serving as horizontal *diaphragms* that act as wide, flat beams in the plane of the floor or roof system. The diaphragms must have a tension flange, a compression flange, and a web capable of transmitting the lateral loads. Because the direction of the lateral load changes back and forth during wind or earthquake cycles, the compression and tension flanges of the diaphragm must be able to accommodate changes in the sign of the loading. Notches or other discontinuities in the tension and compression flanges of the diaphragm should be avoided or reinforced to transmit the forces around the discontinuities. Because of the importance of diaphragms, the 2014 edition of the ACI Code has added a chapter (Chapter 15) to address the design and detailing of concrete floor diaphragms.

Distribution of Walls in a Building Floor Plan

A common design recommendation is to minimize the separation, commonly referred to as the eccentricity, between the center of mass (geometric centroid of the floor plate) and the center of lateral resistance (CR) provided by the shear walls and moment resisting frames in the lateral-load system. Because lateral loads are assumed to act through the center of mass (CM), any eccentricity between the CM and CR will result in the generation of torsional moments. A central-core wall system, similar to that shown in Fig. 18-12, commonly is used to minimize eccentricity between the CM and CR.

When a building structure is subjected to large lateral displacements due to earthquake ground motions, the stiffnesses of the lateral-load resisting members are likely to change in a nonuniform fashion. As a result, the CR is likely to be relocated and the eccentricity between the CM and CR may increase. To account for this, the International Building Code [18-14] specifies a minimum eccentricity in the two principal directions that must be added to any calculated eccentricity. For structures where substantial torsional moments may be generated, a wide distribution of shear walls around the perimeter of the floor plan would be the most efficient method for resisting that torsion.

Distribution of Story Forces (Story Shears) to Walls

In the following analysis of the distribution of lateral loads to structural members, only isolated shear walls will be considered as lateral-load resisting elements. It should be clear

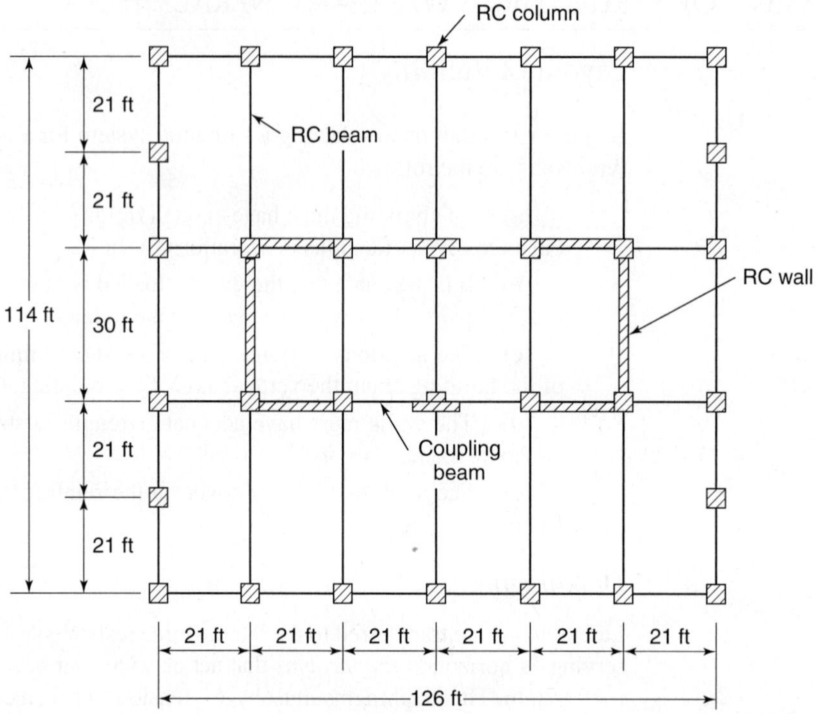

Fig. 18-12
General building floor plan.

that this analysis could be expanded to include moment-resisting frames and wall assemblies, such as T-shapes, L-shapes, U-shapes, and so on.

Consider the floor plan and isolated shear walls shown in Fig. 18-13. We will assume that the slab (diaphragm) connecting these walls is stiff in-plane but has a low flexural stiffness. Thus, the walls are not coupled, but they all should have the same lateral displacement under the acting lateral loads. We will assume the walls have a very low stiffness when bent about their weak axis, and thus, we will only consider the stiffness of the walls when they are bent about their strong axes. We initially will assume that the walls are *slender*. Thus, only flexural stiffnesses will be considered. However, modifications to account for shear deformations in *short* walls will be discussed at the end of this subsection. Finally, for this analysis of the distribution of lateral forces, we initially will assume that the walls are uncracked and thus will use the gross moment of inertia for the walls. The effect of flexural cracking can be accounted for in a second iteration of this analysis by assuming that the moment of inertia for a flexurally cracked wall is equal to one-half of the gross moment of inertia. If lateral displacements are to be calculated, the moment of inertia values should be reduced by 30 percent to correspond to those recommended in ACI Code Table 6.6.3.1.1(a):

Walls uncracked: $0.7\,I_g$
Walls cracked: $0.35\,I_g$

Returning to Fig. 18-13 and using the assumptions discussed here, we now want to calculate the shear forces induced in each wall due to the lateral forces, V_x and V_y. If there was no eccentricity between the CM and CR, the total lateral force V_x would be distributed to the four walls along the north and south edges of the floor plate in proportion to their

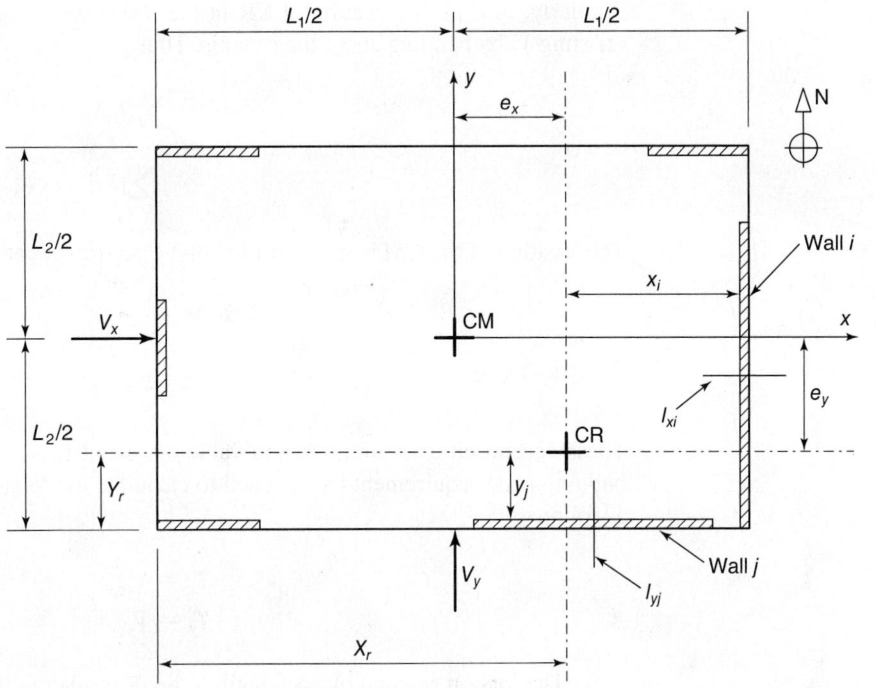

Fig. 18-13
Eccentricity of center of resistance (CR) with respect to center of mass (CM).

moments of inertia about their strong axis (i.e., their y-axis), as shown in Fig. 18-13. Thus, the lateral force resisted by wall j would be

$$V'_{xj} = \left[\frac{I_{yj}}{\sum_n I_{yn}}\right] V_x \tag{18-11a}$$

where n is the number of walls resisting V_x in bending about their strong axis. Similarly, V_y would be resisted by the two walls along the east and west edges of the floor plate, and the lateral force resisted by wall i would be

$$V'_{yi} = \left[\frac{I_{xi}}{\sum_m I_{xm}}\right] V_y \tag{18-11b}$$

where m is the number of walls resisting V_y in bending about their strong axis (x-axis).

If there is an eccentricity between the CM and CR or a minimum eccentricity is specified by a design code, then the effects of torsion must be considered. To find the CR for the floor plate in Fig. 18-13, we initially will assume an origin at the southwest corner of the plate and measure distances from that origin in terms of X and Y. To find the location of the CR in the Y-direction, we will consider only the walls resisting V_x through bending about their y-axis. Following this procedure, the value for Y_r is

$$Y_r = \frac{\sum_j I_{yj} Y_j}{\sum_j I_{yj}} \tag{18-12a}$$

Similarly, to find the location of CR in the X-direction, we will consider only the walls resisting V_y by bending about their x-axis. Thus,

$$X_r = \frac{\sum_i I_{xi} X_i}{\sum_i I_{xi}} \tag{18-12b}$$

The location of the CM is given in Fig. 18-13, so the eccentricities from CM to CR are

$$e_y = \frac{L_2}{2} - Y_r \tag{18-13a}$$

$$e_x = \frac{L_1}{2} - X_r \tag{18-13b}$$

These eccentricities or an increased value of eccentricity required to satisfy a governing building code requirement can be used to calculate the torsion caused by the lateral loads as follows:

$$T_x = V_x e_y \tag{18-14a}$$

$$T_y = V_y e_x \tag{18-14b}$$

The torsion resisted by each wall in the floor plan will be related to the lateral stiffness of the wall for bending about its strong axis multiplied by the distance in the x- or y-direction from the wall to the CR, as measured perpendicular to the weak axis of the wall. Thus, as stated previously, if a significant torsion is to be resisted, the use of widely distributed walls is more effective in resisting torsion. An equivalent torsional stiffness for all of the walls in the floor system (acting as a unit) can be calculated as the sum of the torsional resistance from each wall multiplied by their respective perpendicular distance to the CR. This torsional stiffness can be expressed as

$$K_t = \sum_i I_{xi} x_i^2 + \sum_j I_{yj} \cdot y_j^2 \tag{18-15}$$

With this equivalent torsional stiffness for all of the walls acting as a combined system, we can determine how much shear is induced in each wall when resisting the torsional moments. However, there may be some question regarding what torsional moments should be used when determining the total shear force in each wall. Typically, a structure is analyzed for lateral loads (wind or especially seismic) acting in only one principal direction and then reanalyzed for the lateral loads acting in the other principal direction. These two results normally are considered separately in the design. In this case, the torsion that is generated by either of the lateral loads acting in one principal direction is resisted by *all* of the walls—not just those with their principal axes perpendicular to the lateral load. Thus, it is not clear how much of the torsion generated due to lateral loading in the second principal direction should be included when considering the effect of lateral loading in the first principal direction. Assuming that some percentage of the effect of lateral loading in the second principal direction should be considered, the author presents the following expressions. The first expression is for the shear induced in walls that have their strong axis perpendicular to the lateral force V_x.

$$V''_{xj} = \left[\frac{I_{yj} y_j}{K_t}\right](T_x + \alpha T_y) \tag{18-16a}$$

Similarly, for the walls that have their strong axis perpendicular to the lateral force V_y,

$$V''_{yi} = \left[\frac{I_{xi}x_i}{K_t}\right](T_y + \alpha T_x) \tag{18-16b}$$

For both expressions, the author recommends using $\alpha = 0.25$ to reflect the low probability of having the maximum lateral forces acting simultaneously in both principal directions.

We now can combine the results from Eqs. (18-11) and (18-16) to obtain the total shear resisted by walls with their strong axis perpendicular to the lateral force V_x as

$$V_{xj} = V'_{xj} + V''_{xj} \tag{18-17a}$$

For walls with their strong axis perpendicular to the lateral force V_y, we have

$$V_{yi} = V'_{yi} + V''_{yi} \tag{18-17b}$$

For all of the analysis results given up to this point, only flexural stiffnesses of the walls have been considered. For walls with aspect ratios (h_w/ℓ_w) less than 3, the effect of shear deformations starts to become more significant. For such walls, it is recommended to use a modified moment of inertia to reflect the increased importance of shear deformations. A value recommended in the *PCA Design Handbook* [18-15] is

$$I_{\text{mod}} = \frac{I_e}{1 + \dfrac{24I_e}{A_w h_w^2}} \tag{18-18}$$

where I_e is the effective flexural moment of inertia that has been selected based on the prior discussions, and A_w is the wall area, $h \times \ell_w$.

Wall Foundations

The foundations at the base of the wall must be able to transfer the shear, moment, and axial force as required from the base of the wall to the supporting soil or rock. For the combination of a vertical load, P_u, and moment, M_u, acting on a footing, the moment can be expressed as the vertical load multiplied by an eccentricity, $P_u \cdot e$. It is desirable to have the vertical load acting within the kern region for the footing, $(e \leq L/6)$, to avoid the development of tension (uplift stresses), as are shown in Fig. 18-14a. If the footing size becomes excessive, possible solutions are:

(a) Replace the rectangular footing with an H-shaped footing, to increase the radius of gyration $\left(r = \sqrt{\dfrac{Inertia}{Area}}\right)$ of the footing.

(b) Use pile or caisson foundations.

(c) Use coupled shear walls to widen the footprint of the wall and divide the moment to be transferred into the three components shown in Fig. 18-8b.

(d) Attach the base of the wall to horizontal outrigger beams in the basement that extend the foundation to receive the vertical loads from adjacent frame columns so that there is no uplift; or the uplift is counteracted (see Fig. 18-14b).

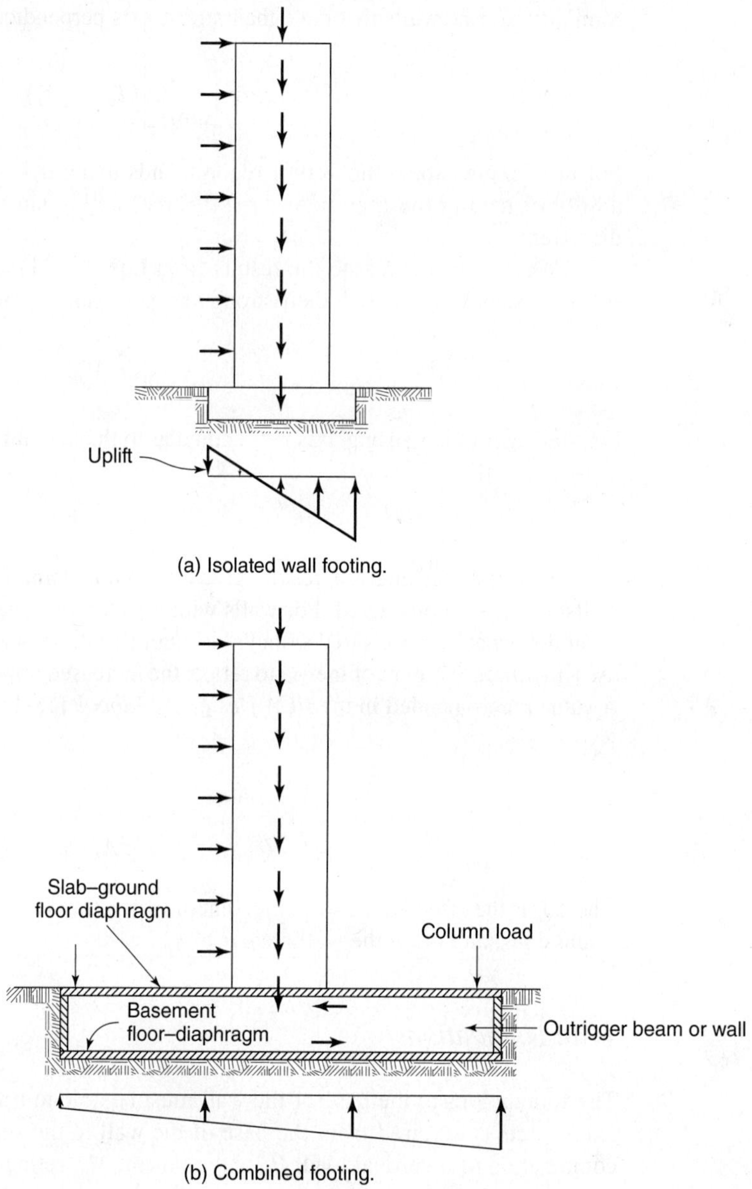

Fig. 18-14
Wall foundations.

(e) Attach the base of the wall to the ground-floor diaphragm and basement floor to provide a horizontal force couple to react to the moment at the base of the wall.

(f) Use a mat foundation.

Wyllie [18-16] and Paulay and Priestly [18-17] discuss the design of shear-wall foundations.

Required Size of Wall

In choosing a structural wall section for a given building, the wall must

(a) have enough strength to resist the factored moments, shears, and axial loads acting on it; and

(b) have enough stiffness to limit the lateral deflections.

There is no widely accepted way of doing this. A *rough estimate* of the minimum wall stiffness, EI, required to limit the lateral deflections to an acceptable value can be made by considering the walls as a vertical cantilever with a constant EI over the height, loaded with a constant wind load over the height, as shown in Fig. 18-14a. Wall thicknesses, and thus the structure's EI would generally decrease as the moments and shears decrease near the top of the structure, we shall assume it is constant over the building height and equal to the sum of the EI values of the walls in the bottom story. This cantilever will be loaded with a constant, uniform service wind load of w_s lb per foot of height. The wind load is the product of the length and height of the building and the wind pressure per foot of height, taken equal to the algebraic sum of the windward wind pressures and leeward wind suctions evaluated at the top of the building. It is customary to limit the relative horizontal deflection of the top and bottom of a story to a fraction of the height of the story, expressed as $\Delta = h_s/(200 \text{ to } 500)$, where Δ/h_s is called the *story drift*. This limit is expressed in terms of the slope of the wall relative to the vertical in any story throughout the height of the building. The largest slope in a shear wall or shear-wall–frame building is generally at the top of the structure and is given by [18-6]:

$$(\text{slope at top}) = (\text{rotation of base of wall}) + (\text{area of } M/EI \text{ diagram from base to top}) \tag{18-19}$$

If the base is fixed against rotation and assuming that $EI = EI_{\text{base}}$ is constant over the height of the building, h_w, the following gives the slope at the top relative to the undeflected position:

$$\text{Slope at top} = \frac{w_s h_w^3}{6E_c I_g} \tag{18-20}$$

Here,

h_w is the height of the wall

E_c is the modulus of elasticity of concrete, psi (because all other units are in pounds and feet, the modulus of elasticity in psi is multiplied by $(144 \text{ in.}^2/\text{ft}^2)$ to change the psi units to psf)

I_g is the uncracked moment of inertia of the wall or walls at the bottom of the structure (an uncracked wall is taken to be representative of service-load levels, because codes limit service-load deflections and the walls are generally not severely cracked under service loads); ACI Code Table 6.6.3.1.1(a) gives $I = 0.70I_g$ ft^4 for an uncracked wall

w_s is the unfactored wind load per vertical foot of wall, evaluated at the top of the wall, in lb/ft of height

Limits on Story Drift

ASCE/SEI 7 [18-18] does not limit the lateral deflection or the story drift. The National Building Code of Canada [18-19] limits the maximum *story drift* under unfactored loads to less than 1/500 of the story height. The maximum acceptable drift depends on the ability for occupants of the building to perceive the motion of the building during a major windstorm. This storm may be chosen as a rare event, such as the maximum annual windstorm or the maximum windstorm in one tenancy of the building—say, every 3 to

8 years. Grossman also discussed design for wind-induced movement of slender concrete buildings [18-6].

If a wall having height h_w and constant wall stiffness EI is fixed at the base, the maximum drift will be in the top story. Setting the service-load story drift from Eq. (18-20) equal to 1/500, we can compute the minimum total I_g for the walls parallel to the direction of the wind load as

$$\Sigma(0.70I_g) = \frac{500w_s h_w^3}{6E_c} \tag{18-21}$$

Once the minimum ΣI_g has been estimated, walls can be selected to give ΣI_g equal to or greater than what is required by Eq. (18-21). Although this analysis is intentionally simplified, it gives an order-of-magnitude estimate of the wall stiffness required to limit story drift to an acceptable value. It does not consider torsional loadings on the structure, and it assumes that the wind pressure is constant over the height of the building. At the stage in design where the sizes of shear walls are initially chosen, these details are not yet known. A better estimate of Δ/h can be obtained by calculating the second term on the right-hand side of Eq. (18-19) more accurately based on a better estimate of the EI values over the height of the structure.

Minimum Wall Thickness

The minimum thicknesses given in ACI Code Section 11.3.1 are intended to apply only to bearing walls designed via the simplified design method from ACI Code Section 11.5.3. The author would limit the thickness of walls with rectangular cross sections to the absolute minimum of 1/20 of the unsupported height of the wall. Preferably, the wall thickness should not be less than 1/15 of the unsupported height. The wall thickness must be large enough to allow the concrete to be placed without honeycombing.

Reinforcement in Shear Walls

Distributed and Concentrated Reinforcement

The reinforcement in a shear wall is generally made up of:

(a) *Distributed* horizontal and vertical reinforcement spread uniformly over the length between the boundary elements and over the height of the wall.

(b) *Concentrated* vertical reinforcement is located in boundary elements at or near the edges of the wall and is tied in much the same way that column cages are.

Minimum Wall Reinforcement

In addition to the distributed reinforcement required by ACI Code Table 11.6.1 for walls subjected to low in-plane shear ($V_u \leq 0.5 \cdot \phi V_c$), several other ACI sections require minimum amounts of distributed horizontal and vertical steel that may apply to walls. These are given in Table 18-1.

TABLE 18-1 Minimum Reinforcement in Walls Compared to Other Thin Members[a]

Reason	ACI Code Section	Requirement	Maximum Spacing
Shrinkage and temperature in slabs	7.6.1.1	Slabs where Grade-60 deformed bars or welded-wire fabric (plain or deformed) are used: 0.0018	Five times the slab thickness, no farther apart than 18 in.
Deep beams	9.9.3.1	The area of web reinforcement perpendicular to the span shall not be less than $0.0025 b_w s$	s shall not exceed $d/5$ or 12 in.
	9.9.3.1	The area of web reinforcement parallel to the span not be less than $0.0025 b_w s_2$	s_2 shall not exceed $d/5$ or 12 in.
Walls, low shear $(V_u \leq 0.5 \cdot \phi V_c)$	11.6.1	Minimum ratio of vertical reinforcement area to gross concrete area shall be: (a) 0.0012 for deformed bars not larger than No. 5 with a specified yield strength not less than 60,000 psi	Vertical and horizontal reinforcement shall not be spaced farther apart than three times the wall thickness, nor farther apart than 18 in.
		Minimum ratio of horizontal reinforcement area to gross concrete area shall be: (a) 0.0020 for deformed bars not larger than No. 5 with a specified yield strength not less than 60,000 psi	
Walls, high shear $(V_u > 0.5 \cdot \phi V_c)$	11.6.2	Ratio ρ_t of horizontal shear reinforcement area to gross concrete area shall not be less than 0.0025	Spacing of horizontal shear reinforcement shall not exceed $\ell_w/5$, $3h$, or 18 in.
		Ratio ρ_ℓ of vertical shear reinforcement shall not be less than $\rho_\ell = 0.0025 + 0.5(2.5 - h_w/\ell_w) \times (\rho_t - 0.0025)$ nor 0.0025	Spacing of vertical shear reinforcement shall not exceed $\ell_w/3$, $3h$, or 18 in.

[a]If more than one of these sections apply, the sections requiring the largest minimum area and the smallest spacing shall govern.

For deformed bars not larger than No. 5, ACI Code Table 11.6.1 requires minimum areas for distributed vertical reinforcement in walls to be equal to $0.0012A_g$ and the minimum area of distributed horizontal reinforcement in walls to be equal to $0.0020A_g$. This steel can be placed in two layers or *curtains* of distributed vertical and horizontal reinforcement with the bars in the two curtains tied together at intervals, but not enclosed in ties. A single reinforcement curtain is allowed at the midplane of walls having thicknesses of 10 in. or less (ACI Code Section 11.7.2.3). This steel cannot be tied because there is only one curtain. Thicker walls require two curtains of reinforcement, each consisting of not less than half of the total minimum reinforcement required in each direction. Each of these two layers is placed not more than one-third of the wall thickness from the surface. It is desirable to have the steel in two layers close to the sides of the wall because the internal lever arm, jd, for weak-axis bending is much smaller if the reinforcement is at the center of the wall. ACI Code Section 11.7.2.2 requires that distributed vertical and horizontal bars be spaced not further apart than three times the wall thickness or 18 in., whichever is less. For high shear, ACI Code Section 11.7.3.1 limits the maximum spacing of horizontal shear reinforcement to be the smallest of $\ell_w/5$, $3h$, and 18 in. and ACI Section 11.7.2.2 limits the maximum spacing of vertical shear reinforcement to be the smallest of $\ell_w/3$, $3h$, and 18 in.

Shear walls subjected to large moment reversals, as likely would occur during strong earthquake ground motions, may require larger minimum vertical-reinforcement areas than given by the previous requirements in order to prevent possible fracture of the vertical reinforcement [18-20].

Ties for Vertical Reinforcement

ACI Code Section 11.7.4.1 specifies that the distributed vertical reinforcement *need not* be enclosed by lateral ties if

(a) the vertical reinforcement area is not greater than 0.01 times the gross concrete area or

(b) the vertical reinforcement is not required as compression reinforcement.

Many designers interpret part (b) to mean that wall steel satisfying (a) that is not enclosed in transverse ties should be ignored in strength calculations if

- it is stressed in compression under static loads, or
- it is stressed by cyclic loads that cause compression in the bars.

Distributed wall steel placed in two separate curtains can be tied through the thickness of the wall using stirrups or through-the-wall cross-ties engaging reinforcement on both faces of the wall. Although the ACI Code does not specify what fraction of the bars should be tied in this manner, it is customary for such ties to engage every second or third bar each way on both faces [18-21].

In many cases the distributed vertical reinforcement has enough moment capacity to resist the wind load moments. If not, concentrated reinforcement is provided in boundary elements at the ends of the walls or at the intersections of walls. These elements should contain vertical steel satisfying the minimum reinforcement requirements from ACI Code Section 10.6.1.1 based on the area of the boundary elements, rather than on the gross area of the wall. This steel is to be enclosed by ties satisfying ACI Code Section 25.7.2. As a result, the steel is assumed to be restrained against buckling, and thus able to resist compressive bar forces. The boundary elements may be the same thickness as the rest of the wall, or they may be thicker.

Although ACI Code Chapter 11 does not require concentrated reinforcement at the ends of the walls, it is good practice to at least use larger bars at the extreme ends of the wall.

Transfer of Wall Load through Floor Systems

In tall buildings, the strength of the concrete in the walls may be higher than the strength of the concrete in the floor system. One reason for the use of high-strength concrete is to reduce the lateral deflections and the wind-induced sway vibrations of the building. On the basis of tests of column–floor joints, ACI Code Section 15.3.1 allows an increase in the effective strength of the floor concrete in locations where the floor concrete is confined on all sides. This effect is smaller in edge and corner column joints than in interior joints, because there is less confinement of the joint concrete at edge or corner columns. The lack of confinement is even more pronounced in wall–floor joints, because a greater length of joint concrete is unconfined. In the absence of tests, it is recommended that the strength of wall-to-floor joints be based on the lower of the wall and floor concrete strengths.

18-10 FLEXURAL STRENGTH OF SHEAR WALLS

Flexure—Nominal Strength, Factored Loads, and Resistance Factors

Cross sections in a wall are designed to satisfy

$$\phi M_n \geq M_u \tag{18-22}$$

$$\phi N_n \geq N_u \tag{18-23}$$

$$\phi V_n \geq V_u \tag{18-24}$$

where M_n is the nominal resistance based on the specified material strengths, M_u is the required resistance computed from the factored loads, and so on. The strength-reduction factor, ϕ, comes from ACI Code Table 21.2.3 for flexure and axial loads and from ACI Code Table 21.2.1 for shear. The factored loads are from ACI Code Table 5.3.1.

Strength-Reduction Factors for Flexure and Axial Load

The strength-reduction (ϕ) factors for combined flexure and axial loads for a shear wall vary, depending on the maximum steel strains anticipated at ultimate load. As explained in Chapters 5 and 11, the calculation of strength-reduction factors, ϕ, is based on the strain, ε_t, in the layer of steel at the depth, d_t, which is located farthest from the extreme-compression fiber.

Tension-controlled limit for a rectangular wall. The strength-reduction factor, ϕ, can be computed directly from the computed strain in the extreme-tension layer of reinforcement, ε_t. ACI Code Table 21.2.2 defines the tension-controlled limit load as the load causing a strain distribution having a maximum strain of 0.003 *in compression* on the most compressed face of the member, when the steel strain in the extreme layer of tension reinforcement, ε_t, reaches 0.005 in tension. From similar triangles, the neutral axis will be located at $c/d_t = 0.375d_t$ from the compressed face, where d_t is the distance from the extreme-compression fiber to the centroid of the layer of bars farthest from the compression face of the member. Thus, when c is less than or equal to $0.375d_t$, the wall is tension-controlled, and $\phi = 0.9$.

Compression-controlled limit for a rectangular wall. The compression-controlled limit corresponds to a strain distribution with the neutral axis at $0.6d_t$. So, when c is greater than or equal to $0.6d_t$, the wall is compression-controlled, and $\phi = 0.65$.

Flexural Strength of Rectangular Walls with Uniform Curtains of Vertical Distributed Reinforcement

Code guidance on the use of vertical distributed reinforcement in walls loaded cyclically is ambiguous. In seismic regions, the loads resisted by vertical bars that are not tied, are ignored. ACI Code Section 11.7.4.1 requires that vertical bars be tied (a) if the total distributed vertical reinforcement, A_s, exceeds $0.01A_g$, or (b) if the vertical reinforcement is included as compression reinforcement in the calculations. The following strength analysis applies to walls with two curtains of reinforcement with ties through the wall to the other curtain of bars. The equations in this subsection also apply if the area of steel is less than 0.01.

ACI Code Section 18.7.5.2 suggests that, in seismic regions, column reinforcement would be adequately tied if the center-to-center spacing between longitudinal bars being held by crossties or the legs of hoops does not exceed 14 in. Given this guidance, the author

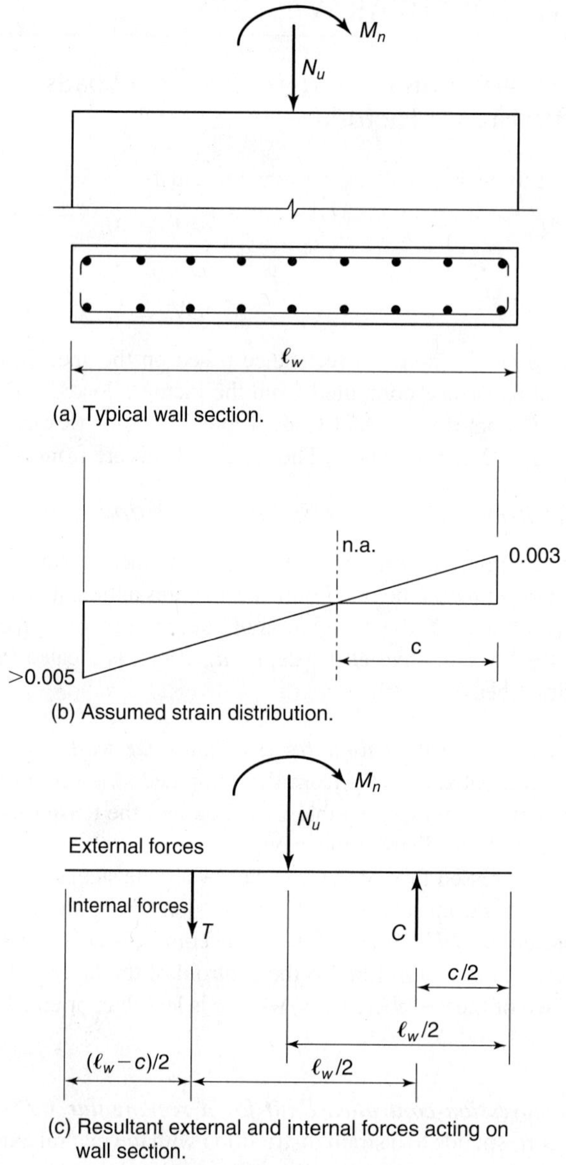

Fig. 18-15
Wall with uniform distribution of vertical reinforcement subjected to axial load and bending.

believes that the vertical reinforcement in a *nonseismic* wall can be taken as "tied" if at least every second bar in a curtain of reinforcement is tied through the wall to a bar in the other curtain of steel, near the opposite face.

Figure 18-15a shows a rectangular wall section with a uniform distribution of vertical steel. We will assume that the wall is subjected to a factored axial load, N_u, and we want to find the nominal flexural strength, M_n, for this wall using the assumed strain distribution in Fig. 18-15b. We will use a procedure developed by A. E. Cardenas and his colleagues [18-22] and [18-23]. They made the following assumptions at nominal strength conditions for shear wall sections similar to that in Fig. 18-15a.

1. All steel in the tension zone yields in tension.
2. All steel in the compression zone yields in compression.
3. The tension force acts at middepth of the tension zone.

4. The total compression force (sum of steel and concrete contributions) acts at middepth of the compression zone.

From those assumptions and using A_{st} to represent the total area of longitudinal (vertical) reinforcement, we can obtain the following expressions for the vector forces in Fig. 18-15c.

$$T = A_{st} f_y \left(\frac{\ell_w - c}{\ell_w} \right) \tag{18-25a}$$

$$C_s = A_{st} f_y \left(\frac{c}{\ell_w} \right) \tag{18-25b}$$

$$C_c = 0.85 f'_c h \beta_1 c \tag{18-25c}$$

and

$$C = C_s + C_c \tag{18-25d}$$

The percentage of total longitudinal reinforcement is

$$\rho_\ell = \frac{A_{st}}{h \ell_w} \tag{18-26a}$$

and the longitudinal reinforcement index is

$$\omega = \rho_\ell \frac{f_y}{f'_c} \tag{18-26b}$$

Finally, Cardenas et al. defined an axial stress parameter as,

$$\alpha = \frac{N_u}{h \ell_w f'_c} \tag{18-27}$$

where N_u represents the factored axial load, positive in compression.

Enforcing section equilibrium leads to

$$C_c + C_s - T = N_u$$

$$0.85 f'_c h \beta_1 c + A_{st} f_y \left(\frac{c}{\ell_w} \right) - A_{st} f_y \left(\frac{\ell_w - c}{\ell_w} \right) = N_u$$

Combining some terms and dividing both sides of this force equilibrium expression by $h \ell_w f'_c$ results in

$$0.85 \beta_1 \frac{c}{\ell_w} - \left(1 - \frac{2c}{\ell_w} \right) \frac{A_{st}}{h \ell_w} \frac{f_y}{f'_c} = \frac{N_u}{h \ell_w f'_c}$$

Substituting the definitions from Eqs. (18-26) and (18-27), we can solve for the distance to the neutral axis from the compression edge of the wall,

$$c = \left(\frac{\alpha + \omega}{0.85 \beta_1 + 2\omega} \right) \ell_w \tag{18-28}$$

With this value for c, we can use Eq. (18-25) to find all of the section forces. Then, summing moment about the compression force, C, in Fig. 18-15c, we get the following expression for the nominal moment strength of the wall section.

$$M_n = T\left(\frac{\ell_w}{2}\right) + N_u\left(\frac{\ell_w - c}{2}\right) \tag{18-29}$$

Cardenas and his colleagues were able to show good agreement between measured moment strengths from tests of shear walls and strengths calculated using Eq. (18-29).

The critical load case for evaluating the nominal moment strength of a structural wall normally corresponds to load combinations (f) and (g) in ACI Code Table 5.3.1.

$$U = 0.9D + 1.0W \qquad \text{(ACI Eq. 5.3.1f)}$$

$$U = 0.9D + 1.0E \qquad \text{(ACI Eq. 5.3.1g)}$$

If service-level wind forces are specified by the governing building code, use a load factor of 1.6 for W in ACI Eq. (5.3.1f).

Either of these load cases will minimize the factored wall axial load, N_u and thus minimize the wall nominal moment strength. Also, it can be shown for essentially all structural walls that $c < 0.375d$. Thus, the wall section is tension-controlled, and $\phi = 0.9$.

Moment Resistance of Wall Assemblies, Walls with Flanges, and Walls with Boundary Elements

Frequently, shear walls have webs and flanges that act together to form H-, C-, T-, and L-shaped wall cross sections referred to as wall *assemblies*. The effective flange widths can be taken from ACI Code Table 6.3.2.1. In regions subject to earthquakes, ACI Code Section 18.10.5.2 limits the flange widths to the smaller of

(a) half the distance to an adjacent web or

(b) 25 percent of the total height of the wall.

We shall use the limiting flange thicknesses from ACI Code Section 18.10.5.2 for both seismic and nonseismic walls.

Frequently, the thickness is increased at the ends of a wall to give room for concentrated vertical reinforcement that is tied like a tied column (see ACI Code Section 25.7.2). The increased thickness also helps to prevent buckling of the flanges. Regions containing concentrated and tied reinforcement are known as *boundary elements*, regardless of whether or not they are thicker than the rest of the wall.

Nominal Moment Strength of Walls with Boundary Elements or Flanges

In this subsection, we will discuss structural walls that have longitudinal reinforcement concentrated at the edges to increase their nominal moment strength. A few typical examples of such walls are shown in Fig. 18-16. The longitudinal reinforcement in boundary elements similar to those in Figs. 18-16a and 18-16b will need to be tied with transverse reinforcement that satisfies ACI Code Section 25.7.2 if the walls are in regions of low or no seismic risk. The confinement requirements are more stringent for boundary elements of structural walls in regions of high seismic risk, as will be discussed in Chapter 19.

When calculating the nominal moment strength, M_n, for walls similar to those in Fig. 18-16, the contribution from the vertical reinforcement in the web is usually ignored because its contribution to M_n is quite small compared to the contribution from the vertical reinforcement concentrated at the edges of the wall. For some flanged sections or wall assemblies, as shown in Fig. 18-16c, ignoring the vertical reinforcement in the web may be

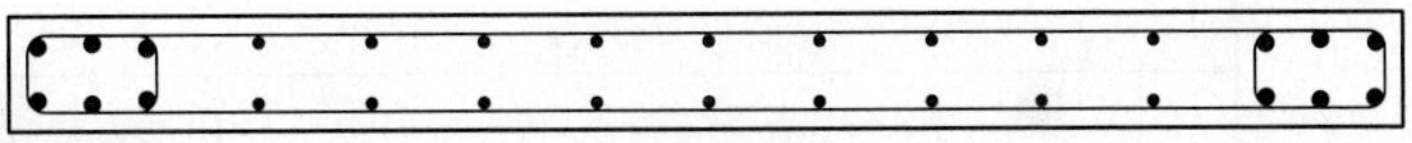

(a) Boundary element within dimensions of wall.

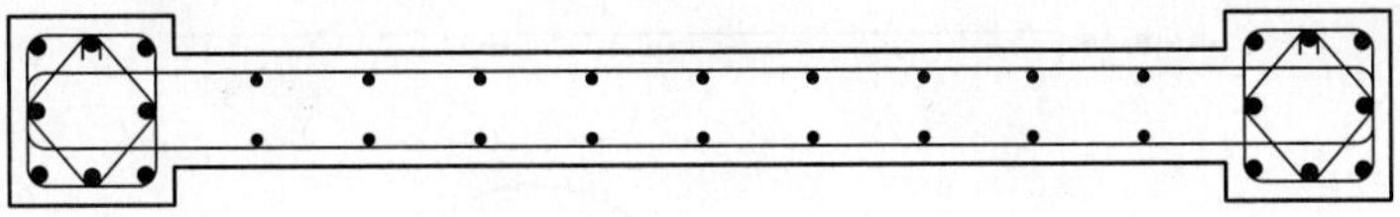

(b) Wall with enlarged boundary element.

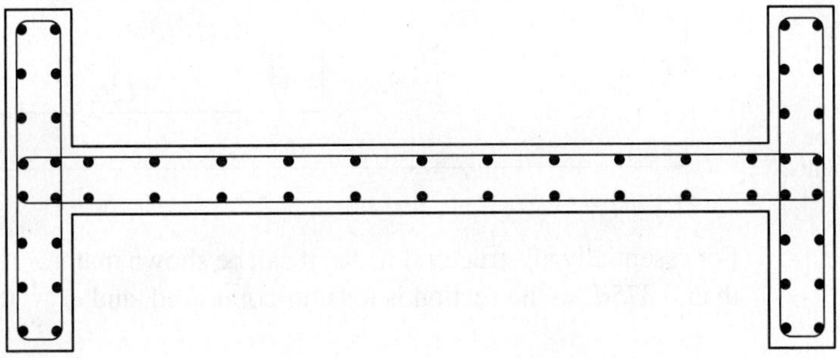

(c) Wall with reinforcement concentrated in flanges.

Fig. 18-16
Structural walls with concentrated reinforcement at their edges.

too conservative. An alternative procedure for analyzing such wall assemblies in flexure is given in the next subsection.

The model used to analyze the nominal moment strength of a structural wall with boundary elements is shown in Fig. 18-17. For the boundary element in tension, the tension force is

$$T = A_s f_y \tag{18-30}$$

where A_s is the total area of longitudinal steel in the boundary element. The longitudinal steel in the compression boundary element is ignored. Using the compression stress-block model from Chapter 4, the compression force is

$$C = 0.85 f'_c ba \tag{18-31}$$

where b is the width of the boundary element. Enforcing section equilibrium for the vertical forces in Fig. 18-17 results in

$$a = \frac{T + N_u}{0.85 f'_c b} \tag{18-32}$$

Normally, the compression stress block is contained within the boundary element, as shown in Fig. 18-17. If the compression force required for section equilibrium is large enough to cause the value of a to exceed the length of the boundary element shown as b' in Fig. 18-17, then a section analysis similar to that discussed in Section 4-8 for flanged sections will be required.

As discussed in the prior subsection, the critical load case for evaluating the nominal moment strength of a structural wall normally corresponds to load combinations (f) or (g) in ACI Code Table 5.3.1. Either of these load cases will minimize the factored axial load, N_u, and thus minimize the wall nominal moment strength. Summing the moment about the compression force in Fig. 18-17 leads to the following expression for M_n.

$$M_n = T\left(d - \frac{a}{2}\right) + N_u\left(\frac{\ell_w - a}{2}\right) \tag{18-33}$$

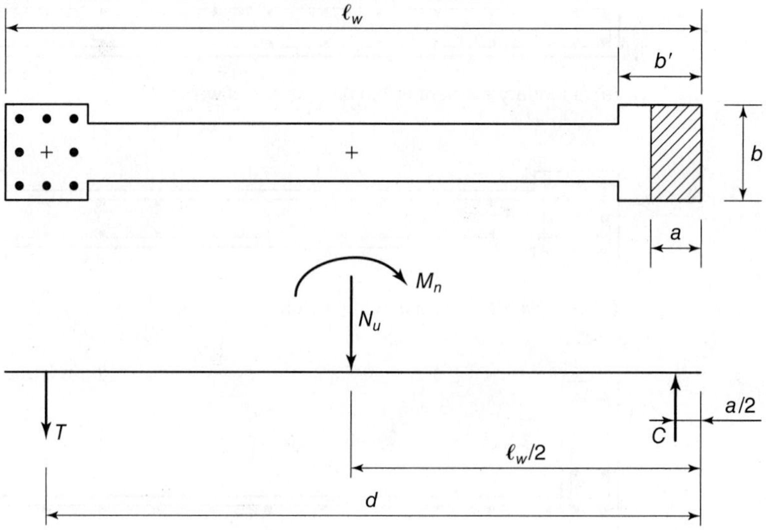

Fig. 18-17
Flexural strength model for wall with boundary elements.

For essentially all structural walls, it can be shown that the neutral-axis depth, c, is well less than $0.375d$, so the section is tension-controlled, and $\phi = 0.9$.

Nominal Moment Strength of Wall Assemblies

Paulay and Priestley [18-17] present the following method of computing the required reinforcement in a wall assembly. Figure 18-18 shows a plan of a wall consisting of a web and two flanges. The wall is loaded with a factored axial load, N_u, and a moment, M_{ua}, that causes compression in flange 1, and a shear, V_u, parallel to the web. The axial load and the moment act through the centroid of the area of the wall. The moment can be replaced by an eccentric axial load located at

$$e_a = \frac{M_{ua}}{N_u} \tag{18-34}$$

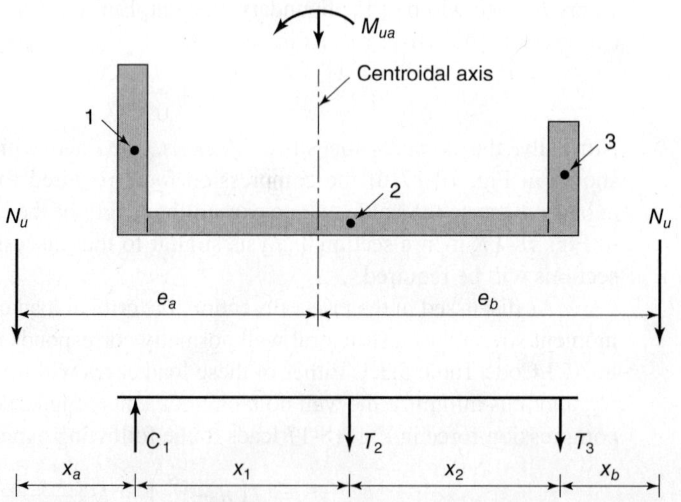

Fig. 18-18
Reinforcement forces acting on a wall segment. (From *Seismic Design of Reinforced Concrete and Masonry Buildings*, Paulay and Priestley, Copyright John Wiley & Sons, 1992. Reprinted by permission of John Wiley & Sons, Inc.)

from the centroid, which is equivalent to it acting at $x_a = e_a - x_1$ from the centroid of the flange that is in compression.

Because of the shear, V_u, the vertical reinforcement in the web will be assumed to be equal to the minimum steel required by ACI Code Section 11.6.2. The calculations will be simplified by assuming that all of the web steel yields in tension. We can assume this because under cyclic loads, the neutral axis will alternately be close to the left and the right ends of the web. The steel in the web resists an axial force of

$$T_2 = A_{s2} f_y \tag{18-35}$$

where A_{s2} is the area of distributed steel in wall element 2, the web. Summing moments about the centroid of the compression flange gives the following equation for the required tension, T_3, in the right-hand flange:

$$T_3 \approx \frac{N_u x_a - T_2 x_1}{x_1 + x_2} \tag{18-36}$$

In a similar manner, the force in the tension reinforcement, required in the left-hand flange for the axial load and the moment M_{ub}, causing compression in the right-hand flange can be computed as

$$T_1 \approx \frac{N_u x_b - T_2 x_2}{x_1 + x_2} \tag{18-37}$$

The required area of concentrated vertical reinforcement in the flanges can be computed by dividing the tension forces from Eqs. (18-36) and (18-37) by the product, $\phi \times f_y$.

Biaxially loaded walls. A wall is said to be biaxially loaded if it resists axial load plus moments about two axes. One method of computing the strength of such walls is the equivalent eccentricity method presented in Section 11-7. In this method, a fraction between 0.4 and 0.8 times the weak-axis moment is added to the strong-axis moment. The wall is then designed for the axial load and the combined biaxial moment treated as a case of uniaxial bending and compression.

Strictly speaking, the elastic moment resistance of an unsymmetrical wall should be computed allowing for moments about both principal axes of the cross section. This is not widely done in practice. It is generally assumed that cracking of the walls and the proportioning of the vertical wall reinforcement can be done considering moments about one orthogonal axis at a time.

Shear Transfer between Wall Segments in Wall Assemblies

For the flanges to work with the rest of the cross section of a wall assembly, so-called *vertical shear stresses* must exist on the interface between the flange and web, even when the wall and the wall segments are constructed monolithically. The stresses to be transferred are calculated in the same manner as for a composite beam, by using Eqs. (16-13) and (16-15). The reinforcement should satisfy ACI Code Section 22.9, *Shear Friction.*

18-11 SHEAR STRENGTH OF SHEAR WALLS

The design of structural walls for shear in nonseismic regions is covered in ACI Code Section 11.5.4, *In-plane shear.* The basis for this code section was a series of tests of one-third-size, planar, shear walls, reported in [18-22], [18-23], and [18-24]. The test specimens were divided between flexural shear walls with ratios of $M_u/(V_u\ell_w)$ of 1.0, 2.0, and higher, and short shear walls with $M_u/(V_u\ell_w) = 0.50$. The basic shear-design equations are similar to those for the shear design of prestressed concrete beams:

$$\phi V_n \geq V_u \tag{18-24}$$

$$V_n = V_c + V_s \tag{18-38}$$

(ACI Eq. 11.5.4.4)

$$V_s \geq \left(\frac{V_u}{\phi} - V_c\right) \tag{18-39}$$

ACI Code Section 11.5.4.3 limits V_n to a maximum value of $10\sqrt{f'_c}\,hd$, where d shall be taken as $0.8\ell_w$ unless a strain–compatibility analysis is used to define the centroid of the tension force in bending. For walls with concentrated vertical reinforcement in boundary elements at the edges of the walls, d may be measured from the extreme compression edge to the centroid of the concentrated vertical reinforcement near the tension edge.

V_c for Shear Walls

As with beam design, the concrete contribution to shear strength is set approximately equal to the value of shear that causes shear (inclined) cracking in a structural wall. For walls subjected to axial compression, a designer is permitted to use Eq. (18-40), unless a more detailed analysis is made, as will be discussed in the next paragraph,

$$V_c = 2\lambda\sqrt{f'_c}\,hd \tag{18-40}$$

where λ is the factor for lightweight aggregate concrete. It is taken as 1.0 for normal-weight concrete and shall be used as defined in ACI Code Table 19.2.4.2 for lightweight concrete. For walls subjected to axial tension, a designer must use the greater of 0.0 or Eq. (18-41),

$$V_c = 2\left(1 + \frac{N_u}{500A_g}\right)\lambda\sqrt{f'_c}\,hd \tag{18-41}$$

where N_u is negative for tension and N_u/A_g shall be expressed in psi units.

For structural walls subjected to axial compression, ACI Code Table 11.5.4.6 permits V_c to be taken as the smaller of

$$V_c = 3.3\lambda\sqrt{f'_c}\,hd + \frac{N_u d}{4\ell_w} \tag{18-42}$$

or,

$$V_c = \left[0.6\lambda\sqrt{f'_c} + \frac{\ell_w\left(1.25\lambda\sqrt{f'_c} + 0.2\dfrac{N_u}{\ell_w h}\right)}{\dfrac{M_u}{V_u} - \dfrac{\ell_w}{2}}\right]hd \tag{18-43}$$

but Eq. (18-43) shall not apply if $(M_u/V_u - \ell_w/2)$ is negative.

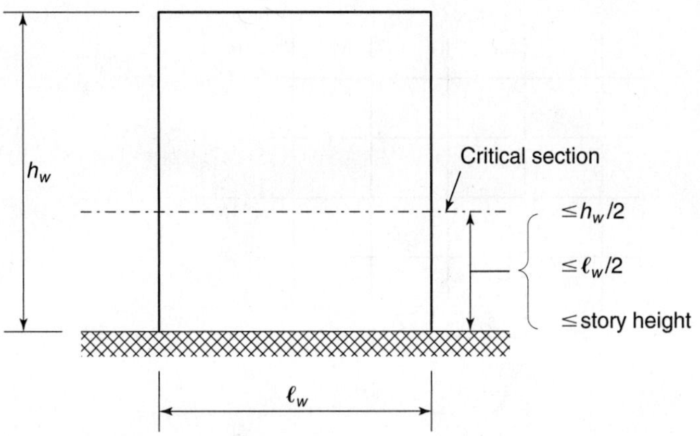

Fig. 18-19
Location of critical section for checking flexural-shear strength.

Equation (18-42) corresponds to the shear force at the initiation of *web-shear* cracking and normally will govern for short walls. Equation (18-43) normally will govern for slender walls and corresponds to the shear force at the initiation of *flexural-shear* cracking at a section approximately $\ell_w/2$ above the base of the wall. The value for M_u/V_u is to be evaluated at a section above the base of the wall, and the distance to that section is to be taken as the smallest of $\ell_w/2$, $h_w/2$, and one story height (Fig. 18-19). The value of V_c computed at that section may be used throughout the height of the shear wall.

Shear Reinforcement for Structural Walls

Shear reinforcement for structural walls always consists of evenly distributed vertical and horizontal reinforcement. In many cases, shear cracks in walls are relatively shallow (i.e., their inclination with respect to a horizontal line is less than 45°), so vertical reinforcement will be just as effective—if not more effective—as horizontal reinforcement in controlling the width and growth of such cracks. However, the shear-strength contribution from wall reinforcement is based on the size and spacing of the horizontal reinforcement.

In many cases, only minimum amounts of shear reinforcement are required in structural walls, and those minimum amounts are a function of the amount of shear being resisted by the structural wall. If the factored shear force, V_u, is less than $\phi V_c/2$, then the distributed vertical and horizontal wall reinforcement must satisfy the requirements of ACI Code Section 11.6.1, as summarized in Table 18-1. If a designer uses Grade-60 reinforcement and bar sizes not larger than No. 5, then the minimum percentage of vertical steel is 0.0012, and the minimum percentage of horizontal steel is 0.0020. Referring to Fig. 18-20 for notation definitions, the percentage of vertical (longitudinal) steel is

$$\rho_\ell = \frac{A_{v,\text{vert}}}{hs_1} \tag{18-44}$$

and the percentage of horizontal (transverse) steel is

$$\rho_t = \frac{A_{v,\text{horiz}}}{hs_2} \tag{18-45}$$

For both the horizontal and vertical steel, ACI Code Sections 11.7.2.1 and 11.7.3.1 limit the bar spacing to the smaller of $3h$ and 18 in.

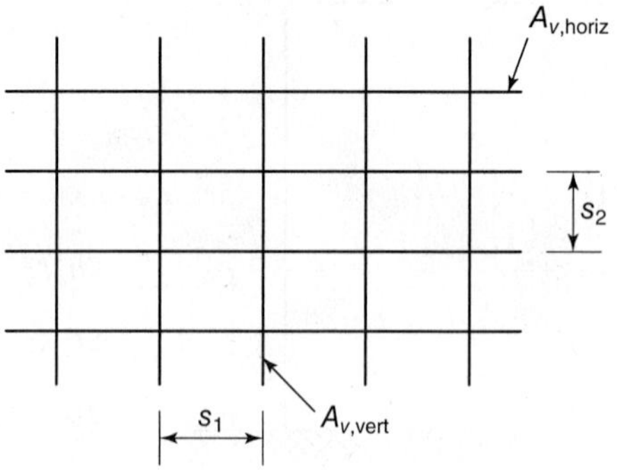

Fig. 18-20
Distribution of vertical and horizontal shear reinforcement in the web of a shear wall.

For walls resisting a higher factored shear force (i.e., $\phi V_c/2 < V_u \leq \phi V_c$), the vertical and horizontal steel in the wall must satisfy the minimum percentages and the maximum spacing requirements of ACI Code Section 11.6.2, as summarized in Table 18-1. The minimum percentage of horizontal (transverse) reinforcement, ρ_t, is 0.0025 and is to be placed at a spacing that does not exceed the smallest of $\ell_w/5$, $3h$, and 18 in. Vertical reinforcement is to be placed at a spacing that does not exceed the smallest of $\ell_w/3$, $3h$, and 18 in. The percentage of vertical (longitudinal) steel, ρ_ℓ, shall not be less than the *larger* of 0.0025 and that calculated using ACI Code Eq. (11.6.2):

$$\rho_\ell = 0.0025 + 0.5\left(2.5 - \frac{h_w}{\ell_w}\right)(\rho_t - 0.0025) \quad (18\text{-}46)$$

For walls with $h_w/\ell_w \geq 2.5$, this equation will not govern. In shorter walls where the horizontal reinforcement percentage, ρ_t, exceeds 0.0025, the value of ρ_ℓ calculated in Eq. (18-46) does not need to exceed the amount (percentage) of horizontal reinforcement required for shear strength, as given next.

If walls are required to resist a factored shear force that exceeds ϕV_c, then horizontal reinforcement must be provided to satisfy the strength requirement expressed in Eq. (18-39). The shear strength, V_s, provided by the horizontal reinforcement is given by ACI Code Eq. (11.5.4.8):

$$V_s = \frac{A_v f_{yt} d}{s} \quad (18\text{-}47)$$

In this equation, A_v is the same as $A_{v,\text{horiz}}$ shown in Fig. 18-20, and s is the same as s_2 in that figure. As was done for shear-strength requirements in Chapter 11, the designer has the option to select a bar size and spacing that satisfies the strength requirement in Eq. (18-39). In addition to satisfying the shear-strength requirement, a designer must check that both the horizontal and vertical wall reinforcement satisfy the minimum reinforcement percentages and maximum spacing limits discussed in the prior paragraphs.

Shear Strength of Structural Walls Resisting Seismic Loads

Chapter 19 discusses *Design for Earthquake Effects*, but the author thought it was essential to include here the modified shear-strength requirements given in ACI Code Chapter 18 for structural walls. Although the equation that defines the nominal shear strength in ACI

Code Chapter 18 appears to be significantly different from equations given in this section, the final values for V_n are not substantially different.

The nominal shear strength, V_n, for a wall designed to resist shear forces due to earthquake ground motions is given by ACI Code Eq. (18.10.4.1):

$$V_n = A_{cv}\,(\alpha_c \lambda \sqrt{f'_c} + \rho_t f_y) \tag{18-48}$$

In this equation, A_{cv} is taken as the width of the web of the wall, h, multiplied by the total length of the wall, ℓ_w. This area is larger than the effective shear area, hd, used for the V_c equations given in ACI Code Table 11.5.4.6.

The first term inside the parenthesis of Eq. (18-48) represents the concrete contribution to shear strength, V_c. The coefficient, α_c, represents the difference between the expected occurrence of *flexure-shear* cracking in slender walls and *web-shear* cracking in short walls. The value of α_c is taken as 2.0 for walls with $h_w/\ell_w \geq 2.0$ and as 3.0 for walls with $h_w/\ell_w \leq 1.5$. A linear variation for the value of α_c is to be used for walls with h_w/ℓ_w ratios between 1.5 and 2.0.

The second term inside the parenthesis of Eq. (18-48) represents the shear-strength contribution from the horizontal wall reinforcement, V_s. Multiplying the definition of ρ_t given in Eq. (18-45) by the definition of $A_{cv}(h\ell_w)$ and the yield strength, f_{yt}, results in the following equivalent value for V_s.

$$V_{s,\text{equiv}} = \frac{A_{v,\text{horiz}}}{hs_2} f_{yt} h \ell_w$$

$$V_{s,\text{equiv}} = \frac{A_{v,\text{horiz}} f_{yt} \ell_w}{s_2} \tag{18-49}$$

This equation is similar to Eq. (18-47), which gives the value for V_s used in ACI Code Chapter 11. However, because $\ell_w > d$, the value of V_s used in ACI Code Chapter 18 exceeds the value of V_s used in ACI Code Chapter 11.

The maximum allowable value for the nominal shear strength, V_n, from Eq. (18-48) is limited to $8A_{cv}\sqrt{f'_c}$ in ACI Code Section 18.10.4.4. Again, this value is similar, but not the same as the upper limit of $10\sqrt{f'_c}\,hd$ given in ACI Code Section 11.5.4.3.

The requirements for minimum vertical and horizontal reinforcement percentages and for maximum permissible spacing are the same as those given in ACI Code Sections 11.7.2.1 and 11.7.3.1, which were discussed previously. The only modification is given in ACI Code Section 18.10.4.3, which states that for walls with $h_w/\ell_w \leq 2.0$, the value of ρ_ℓ (vertical steel) shall not be less than the value of ρ_t. This requirement reflects the fact that in short walls, the vertical reinforcement is equal to or more efficient than the horizontal reinforcement in controlling the width and growth of inclined shear cracks.

The discrepancies noted here between the shear-strength equations given in ACI Code Chapters 11 and 18 can create a dilemma for structural designers. The shear strength of walls designed to resist lateral wind forces should be determined from the equations in ACI Code Chapter 11. However, if the same wall is designed to also resist equivalent lateral forces due to earthquake ground motions, the shear strength must be checked using the equations in ACI Code Chapter 18. A unification of these different shear-strength equations is a major goal of the Code Committee as it works toward future editions of the ACI Code.

Shear Transfer across Construction Joints

The shear force transferred across construction joints can be designed using shear friction. The clamping force is the sum of the tensile reinforcement force components of all the tensile bar forces and permanent compressive forces acting on the joint. This includes all the reinforcement perpendicular to the joint, regardless of the primary use of this steel.

EXAMPLE 18-2 Structural Wall Subjected to Lateral Wind Loads

The moment and shear strengths of the structural wall shown in Fig. 18-21 are to be evaluated for the combined gravity and lateral loads applied to the wall. The wall is 18 ft long and is 10 in. thick. A uniform distribution of vertical and horizontal reinforcement is used in two layers, one near the front face of the wall and the other near the back face. The vertical reinforcement consists of No. 5 bars at 18 in. on centers in each face, and the horizontal reinforcement consists of No. 4 bars at 16 in. on centers in each face. Normal-weight concrete with a compressive strength of 4000 psi is used in the wall. All of the wall reinforcement is Grade-60 steel ($f_y = f_{yt} = 60$ ksi).

The gravity loads applied at each floor level, as shown in Fig. 18-21, are due to dead load. The live loads are not shown but are assumed to be equal to approximately one-half of the dead loads. The lateral wind loads are based on strength-level wind forces and did include the directionality factor.

1. Make an initial check of wall reinforcement. The percentage and spacing of the vertical and horizontal reinforcement will be checked before we do the strength calculations. The percentage of horizontal reinforcement is found using Eq. (18-45):

$$\rho_t = \frac{A_{v,\text{horiz}}}{hs_2} = \frac{2 \times 0.20\ \text{in.}^2}{10\ \text{in.} \times 16\ \text{in.}} = 0.0025$$

This satisfies the minimum requirement in ACI Code Section 11.6.2, so it should be acceptable unless a larger amount is required to satisfy shear-strength requirements. The maximum center-to-center spacing for the horizontal reinforcement is the smallest of $\ell_w/5$

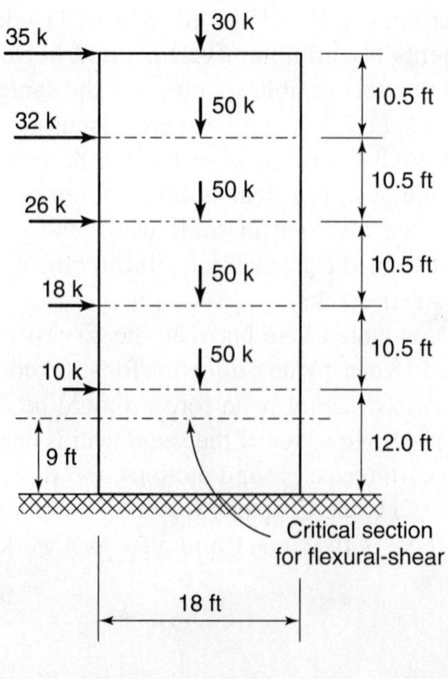

Fig. 18-21
Structural wall for
Example 18-2.

(3.6 ft = 43.2 in.), $3h$ (30 in.), and 18 in. Thus, the provided spacing for the horizontal reinforcement is o.k.

Although it is good practice to use larger vertical bars at the edges of the wall, say No. 6 or No. 7 bars, we will calculate the percentage of vertical reinforcement assuming only No. 5 bars are used. From Eq. (18-44),

$$\rho_\ell = \frac{A_{v,\text{vert}}}{hs_1} = \frac{2 \times 0.31 \text{ in.}^2}{10 \text{ in.} \times 18 \text{ in.}} = 0.00344$$

Because the wall has an aspect ratio, $h_w/\ell_w = 3.0 > 2.5$, Eq. (18-46) will not govern for the minimum percentage of vertical reinforcement. Thus, the minimum percentage of vertical reinforcement is 0.0025 (ACI Code Section 11.6.2), which is less than what is provided. The spacing limit for the vertical reinforcement is the smallest of $\ell_w/3$ (6 ft = 72 in.), $3h$ (30 in.), and 18 in. Thus, the provided spacing of vertical reinforcement is o.k.

2. Check moment strength. The moment at the base of the wall is equal to the sum of the products of the lateral forces times their respective distances to the base of the wall:

$$\begin{aligned} M(\text{base}) &= 35 \times 54 + 32 \times 43.5 + 26 \times 33 + 18 \times 22.5 + 10 \times 12 \\ &= 4670 \text{ kip-ft} \end{aligned}$$

The appropriate load factor for strength-level wind forces is 1.0. Thus, the factored moment at the base of the wall is

$$M_u(\text{base}) = 1.0 \times 4670 = 4670 \text{ kip-ft}$$

The analysis given in [18-23] will be used to evaluate the moment strength of the wall. Using ACI Code Eq. (5.3.1f), the factored axial load is

$$N_u = 0.9N_D = 0.9 \times 230 \text{ kips} = 207 \text{ kips}$$

For 4000-psi concrete, $\beta_1 = 0.85$. Other required parameters are

$$\omega = \rho_\ell \frac{f_y}{f'_c} = 0.00344 \frac{60 \text{ ksi}}{4 \text{ ksi}} = 0.0516 \qquad (18\text{-}26b)$$

and

$$\alpha = \frac{N_u}{h\ell_w f'_c} = \frac{207 \text{ kip}}{10 \text{ in.} \times 216 \text{ in.} \times 4 \text{ ksi}} = 0.0240 \qquad (18\text{-}27)$$

With these parameters, we can use Eq. (18-28) to find the depth to the neutral axis:

$$c = \left(\frac{\alpha + \omega}{0.85\beta_1 + 2\omega}\right)\ell_w$$

$$c = \left(\frac{0.0240 + 0.0516}{0.85 \times 0.85 + 2 \times 0.0516}\right)216 \text{ in.} = 19.8 \text{ in.}$$

If we assume that the effective flexural depth, d, is approximately equal to $0.8\ell_w = 173$ in., it is clear that c is significantly less than $0.375d$. Thus, this is a tension-controlled section, and we will use a strength-reduction factor, ϕ, equal to 0.9.

To calculate the nominal moment strength, we first must determine the tension force at nominal strength conditions. The value of A_{st} for the vertical steel can be calculated as

$$A_{st} = 2A_b \frac{\ell_w}{s_1} = 2 \times 0.31 \text{ in.}^2 \times \frac{216 \text{ in.}}{18 \text{ in.}} = 7.44 \text{ in.}^2$$

Then, from Eq. (18-25a),

$$T = A_{st}f_y\left(\frac{\ell_w - c}{\ell_w}\right)$$

$$T = 7.44 \text{ in.}^2 \times 60 \text{ ksi}\left(\frac{216 \text{ in.} - 19.8 \text{ in.}}{216 \text{ in.}}\right) = 405 \text{ kips}$$

Now, referring to the vector forces in Fig. 18-15c, we can use Eq. (18-29) to calculate the nominal moment strength of the structural wall.

$$\begin{aligned} M_n &= T\left(\frac{\ell_w}{2}\right) + N_u\left(\frac{\ell_w - c}{2}\right) \\ &= 405 \text{ k}\left(\frac{216 \text{ in.}}{2}\right) + 207 \text{ k}\left(\frac{216 \text{ in.} - 19.8 \text{ in.}}{2}\right) \\ &= 64{,}000 \text{ kip-in.} = 5340 \text{ kip-ft} \end{aligned}$$

Using the strength-reduction factor, ϕ, we get

$$\phi M_n = 0.9 \times 5340 = 4800 \text{ kip-ft}$$

Because ϕM_n is larger than M_u (4670 kip-ft), the wall has adequate flexural strength.

3. **Check shear strength.** The factored shear at the base of the wall is

$$V_u = (10 + 18 + 26 + 32 + 35) = 121 \text{ kips}$$

Because this wall is slender, $h_w/\ell_w = 3.0$, we can assume that Eq. (18-43) will govern for the shear-strength contribution from the concrete, V_c. However, we will check both Eqs. (18-42) and (18-43) to demonstrate how each is used. For normal-weight concrete, $\lambda = 1.0$. Using $d = 0.8\ell_w$ (173 in.) and putting all of the quantities into units of pounds and inches, the value of V_c from Eq. (18-42) is

$$\begin{aligned} V_c &= 3.3\lambda\sqrt{f'_c}\,hd + \frac{N_u d}{4\ell_w} \\ &= 3.3 \times 1\sqrt{4000} \times 10 \times 173 + \frac{207{,}000 \times 173}{4 \times 216} \\ &= (361{,}000 + 41{,}400) \text{ lbs} = 402 \text{ kips} \end{aligned}$$

For Eq. (18-43), we need to evaluate the ratio of M_u/V_u at the critical section above the base of the wall. The distance to that section is the smallest of $\ell_w/2$ (9 ft), $h_w/2$ (27 ft), and one story height (12 ft). In this case, $\ell_w/2$ governs, as shown in Fig. 18-21. The factored moment at that section can be found from the following as

$$\begin{aligned} M_u(\text{crit. sect.}) &= M_u\,(\text{base}) - V_u\,(\text{base})\frac{\ell_w}{2} \\ &= 4670 \text{ kip-ft} - 121 \text{ kip} \times 9 \text{ ft} = 3580 \text{ kip-ft} \end{aligned}$$

Thus, the ratio of $M_u/V_u = 3580/121 = 29.6$ ft. Using this value in Eq. (18-43) and expressing all of the quantities in pounds and inches, we have

$$V_c = \left[0.6\lambda\sqrt{f'_c} + \frac{\ell_w\left(1.25\lambda\sqrt{f'_c} + 0.2\dfrac{N_u}{\ell_w h}\right)}{\dfrac{M_u}{V_u} - \dfrac{\ell_w}{2}}\right]hd$$

$$= \left[37.9\text{ psi} + \frac{216\text{ in.}(79.1\text{ psi} + 19.2\text{ psi})}{356\text{ in.} - 108\text{ in.}}\right] \times 10\text{ in.} \times 173\text{ in.}$$

$$= 214{,}000\text{ lbs} = 214\text{ kips}$$

This value governs for V_c, as expected. Using $\phi = 0.75$, as defined in ACI Code Table 21.2.1 for shear and torsion, $\phi V_c = 0.75 \times 214 = 161$ kips. This exceeds V_u, so no calculation of the value of V_s for the provided horizontal (transverse) reinforcement is required. It is clear that the value of $0.5\phi V_c$ is less than V_u, so the requirements for horizontal and vertical reinforcement in ACI Code Section 11.6.2 will govern for this wall. Those requirements were checked in step 1 and were found to be at or above the ACI Code required values.

Thus, the 10-in. thick structural wall with the indicated vertical and horizontal reinforcement has adequate moment and shear strength and the reinforcement satisfies all of the ACI Code requirements. ■

EXAMPLE 18-3 Structural Wall Subjected to Equivalent Lateral Earthquake Loads

The structural wall shown in Fig. 18-22 is to be analyzed for a combined gravity load and equivalent lateral load due to earthquake ground motions. The dimensions of the wall and the section reinforcement (Fig. 18-22b) are taken from a paper by Wallace and Thomsen [18-25], in which the authors analyzed the need for special confinement reinforcement for the boundary elements at the edges of the wall. The details for special confinement reinforcement will be discussed in Chapter 19.

This wall is one of several walls resisting lateral loads for the structure that was analyzed by Wallace and Thomsen [18-25]. The lateral earthquake force, E, assigned to this wall is 205 kips. The dead and reduced live load at the base of the wall are $N_D = 1000$ kips and $N_L = 450$ kips. Use $f'_c = 4000$ psi (normal-weight concrete) and $f_y = f_{yt} = 60$ ksi for all reinforcement.

1. Check moment strength at base of wall. ACI Code Eq. (5.3.1g) will be used to find the factored loads for this analysis.

$$U = 0.9D + 1.0E$$

Seismic lateral loads, as defined in [18-18], typically have an inverted triangular distribution over the height of the building. Therefore, we will assume that the variable x shown in Fig. 18-22a, which is taken as the assumed distance from the base of the wall to the centroid of the lateral force, has a value of $\frac{2}{3}h_w$. Thus, the factored moment at the base of the wall is

$$M_u = 1.0 \times 205\text{ k} \times \tfrac{2}{3} \times 120\text{ ft}$$

$$= 16{,}400\text{ kip-ft}$$

The nominal moment strength of the wall will be analyzed by the procedure shown in Fig. 18-17 for a wall with a vertical reinforcement concentrated in a boundary element. The tension strength for the concentrated reinforcement is

$$T = A_s f_y = 10 \times 1.27\text{ in.}^2 \times 60\text{ ksi} = 762\text{ kips}$$

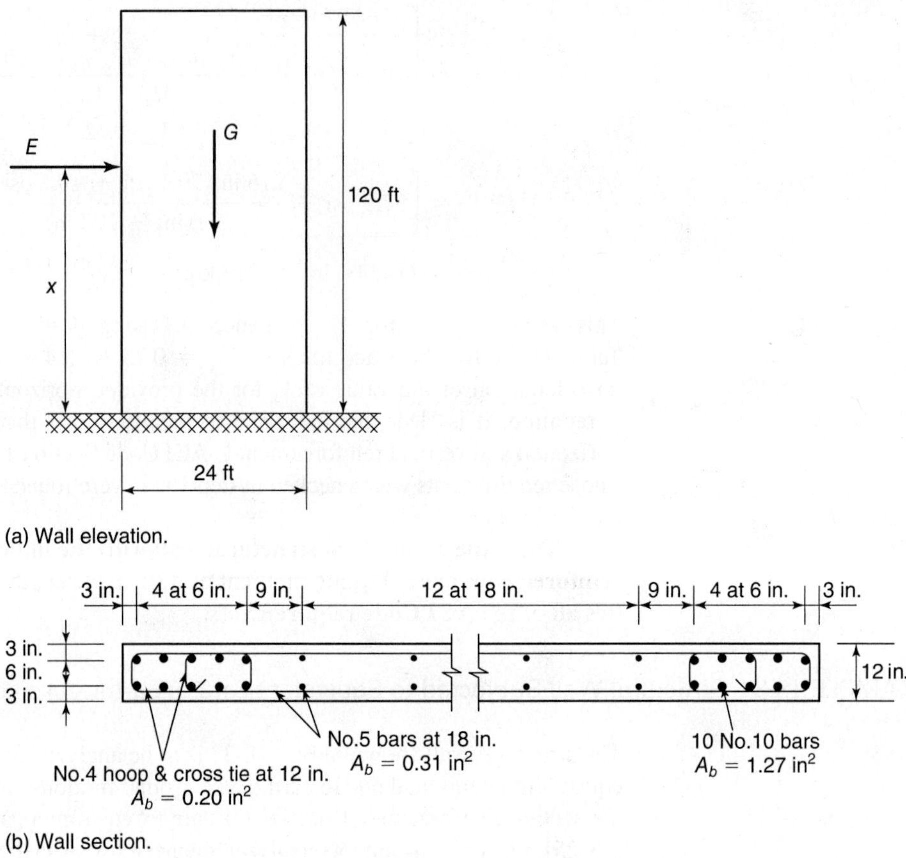

Fig. 18-22 Structural wall for Example 18-3.

The distance from the compression edge to the centroid of the tension reinforcement is

$$d = \ell_w - (3 \text{ in.} + 2 \times 6 \text{ in.}) = 288 - 15 = 273 \text{ in.}$$

The factored axial load for this load case is

$$N_u = 0.9N_D = 0.9 \times 1000 \text{ k} = 900 \text{ kips}$$

Now, we will use Eq. (18-32) to determine the depth of the compression stress block. In this case, $b = h = 12$ in., giving

$$a = \frac{T + N_u}{0.85f'_c b} = \frac{762 \text{ k} + 900 \text{ k}}{0.85 \times 4 \text{ ksi} \times 12 \text{ in.}} = 40.7 \text{ in.}$$

The distance to the neutral axis is, $c = a/\beta_1 = 40.7/0.85 = 47.9$ in., which is well less than $0.375d = 0.375 \times 273 = 102$ in. Therefore, this is a tension-controlled section, and $\phi = 0.9$. Eq. (18-33) will be multiplied by ϕ to find ϕM_n:

$$\phi M_n = \phi\left[T\left(d - \frac{a}{2}\right) + N_u\left(\frac{\ell_w - a}{2}\right)\right]$$

$$= 0.9\left[762 \text{ k}\left(273 - \frac{40.7}{2}\right)\text{in.} + 900 \text{ k}\left(\frac{288 - 40.7}{2}\right)\text{ in.}\right]$$

$$= 0.9[193{,}000 \text{ k-in.} + 111{,}000 \text{ k-in.}]$$
$$= 273{,}000 \text{ kip-in.} = 22{,}800 \text{ kip-ft}$$

Because ϕM_n exceeds M_u, the wall has adequate moment strength.

2. Capacity-based design shear. In seismic design, the design shear force, V_u, is not often based on the factored loads but rather on the *probable flexural strength*, M_{pr}, of the member in question. This is referred to as the *capacity-based design* procedure and is intended to ensure a ductile flexural response in the member—as opposed to a brittle shear failure—if the member is loaded beyond its elastic range of behavior during an earthquake.

For the wall in this example, the probable flexural strength should be based on a *probable* axial load that the wall will be carrying at the time of an earthquake. A value for such an axial load is not defined in either the ACI Code or ASCE/SEI 7-10 [18-18]. The author will assume that a reasonable value for the probable axial load is the sum of the unfactored dead load plus the unfactored reduced live load.

$$N_{pr} = N_D + N_L = 1000 \text{ k} + 450 \text{ k} = 1450 \text{ kips}$$

With this axial load, the wall moment strength will be reevaluated and referred to as the probable moment strength, M_{pr}. First, the depth of the compression stress block is

$$a = \frac{T + N_u}{0.85 f'_c\, b} = 54.2 \text{ in.}$$

In addition to using this probable axial load in the calculation of M_{pr}, we should include all of the wall reinforcement, including the reinforcement in the web of the section and the reinforcement in the compression boundary, both of which were ignored previously in the calculation of M_n. Further, in the definition of M_{pr} given in ACI Code Section 2.2, there is a requirement to use 1.25 f_y for all longitudinal reinforcement when calculating M_{pr}. To properly make this calculation we will need to follow the section equilibrium and strain compatibility procedure used in Chapter 11 for the analysis of column sections that contain multiple layers of reinforcement.

Referring to the wall section in Fig. 18-22, there are 13 pairs of No. 5 bars used as vertical wall reinforcement in the web. For this calculation, we will divide these bars into three lumped areas of steel, one containing five pairs of bars at the middle of the web and two containing four pairs of bars at each of the outer portions of the web. Including the ten No. 10 bars for both the compression and tension edges of the wall, Table 18-2 lists the area and distance from the compression edge of the wall for the five groupings of longitudinal reinforcement used in the analysis of M_{pr}.

Using the steel areas and locations given in Table 18-2, along with a modified yield strength of 75 ksi, the author obtained the following results,

$$c = 59.2 \text{ in.}$$
$$\varepsilon_t = 0.0108 > \varepsilon_y \; (0.00259 \text{ for } f_y = 75 \text{ ksi})$$
$$M_{pr} = 40{,}600 \text{ k-ft}$$

TABLE 18-2 Steel Areas and Locations in Shear Wall for Calculation of M_{pr}

Group No.	1	2	3	4	5
A_{si} (in.^{2})	12.7	2.48	3.10	2.48	12.7
d_i (in.)	273	225	144	63	15

As indicated by these calculations, the strain in the extreme layer of tension reinforcement exceeds the yield strain. We do not need to show that this is a tension-controlled section because we will not be using a capacity reduction factor as part of this calculation. However, we would like to show this is an *under-reinforced* section that will exhibit plastic behavior if overloaded.

The final step is to find the lateral shear force required to develop this moment at the base of the wall. As originally discussed by Bertero et al. [18-26], the lateral load distribution over the height of a shear wall that is part of a complete structural system will tend to be close to a uniform distribution at peak response of the structure when it is subjected to large earthquake motions. Therefore, to more conservatively predict the capacity-based shear force acting on the wall when the base moment reaches M_{pr}, the author recommends using a value of $x = 0.5h_w$ (Fig. 18-22a) for this calculation. Using that value, the capacity-based design shear is

$$V_u(\text{cap-based}) = \frac{M_{pr}}{0.5h_w} = \frac{40{,}600 \text{ k-ft}}{0.5 \times 120 \text{ ft}} = 677 \text{ kips}$$

For this particular structural wall, the capacity-based design shear is approximately more than three times the design shear used to check the flexural strength.

3. Check shear strength. Equation (18-48), which is the same as ACI Code Eq. (18.10.4.1), will be used to check the shear strength of the wall. For this wall, the value of A_{cv} is equal to the gross wall area, so

$$A_{cv} = h\ell_w = 12 \text{ in.} \times 288 \text{ in.} = 3460 \text{ in.}^2$$

Because this is a slender wall, $\alpha_c = 2.0$. Also, because the wall is constructed with normal-weight concrete, $\lambda = 1.0$. Equation (18-45) will be used to determine ρ_t for the distributed horizontal reinforcement in Fig. 18-22b as

$$\rho_t = \frac{A_{v,\text{horiz}}}{hs_2} = \frac{2 \times 0.31 \text{ in.}^2}{12 \text{ in.} \times 18 \text{ in.}} = 0.00287$$

This is greater than the minimum required percentage of 0.0025 from ACI Code Section 11.6.2, and the 18-in. spacing satisfies the limits in ACI Code Section 11.7.3.1. Using the values calculated here and units of pounds and inches, the nominal shear strength of the wall from Eq. (18-48) is

$$\begin{aligned} V_n &= A_{cv}\left(\alpha_c\lambda\sqrt{f'_c} + \rho_t f_{yt}\right) \\ &= 3460 \text{ in.}^2 \,(2 \times 1\sqrt{4000} + 0.00287 \times 60{,}000) \\ &= 3460 \text{ in.}^2 \,(126 \text{ psi} + 172 \text{ psi}) \\ &= 1.03 \times 10^6 \text{ lbs} = 1030 \text{ kips} \end{aligned}$$

ACI Code Section 11.5.4.3 limits the value of V_n to $8A_{cv}\sqrt{f'_c} = 1750$ kips. Thus, we can use the value calculated here. ACI Code Table 21.2.1 states that for shear and torsion, $\phi = 0.75$. So, the reduced, nominal shear strength of the wall is

$$\phi V_n = 0.75 \times 1030 \text{ k} = 773 \text{ kips}$$

Because $\phi V_n > V_u$(cap-based), the wall shear strength is o.k.

Thus, the 12-in. thick structural wall with the indicated vertical and horizontal reinforcement has adequate moment and shear strength, and the reinforcement satisfies all of the ACI Code requirements. ■

18-12 CRITICAL LOADS FOR AXIALLY LOADED WALLS

Buckling of Compressed Walls

The critical stress for buckling of a hinged column or a one-way wall, with a rectangular cross sectional area $(b \times h)$ is

$$\sigma_{cr} = \frac{P_{cr}}{bh} = \frac{\pi^2 EI}{(k\ell)^2}\left(\frac{1}{bh}\right) \tag{18-50}$$

where b is the width of the wall, and $k\ell$ is the *effective length* of the wall. The flexural stiffness of a wall section of width b and thickness h is.

$$EI = \frac{Ebh^3}{12}$$

Buckling of Compressed Walls

The corresponding stiffness of a slab or a two-way wall, per unit of width of wall, is

$$D = \frac{Eh^3}{12(1 - \mu^2)} \tag{18-51}$$

For an axially loaded and uncracked concrete column with little or no reinforcement the critical stress, σ_{cr}, is a function of the tangent modulus of elasticity, E_T, evaluated at the critical stress, σ_{cr}. Substituting E_T into Eq. (18-50), replacing ℓ^2 with b^2, replacing EI with the plate stiffness, D, and including an appropriate edge restraint factor, K, gives the critical stress for a two-way wall:

$$\sigma_{cr} = \frac{P_{cr}}{bh} = K\frac{\pi^2 E_T}{12(1 - \mu^2)}\left(\frac{h}{b}\right)^2 \tag{18-52}$$

In this equation, b is the effective width of the panel, taken equal to the smaller of the width and height of the wall between lateral supports, h is the thickness of the wall, μ is Poisson's ratio, and K is an *edge restraint factor* that varies depending on the degree of fixity of the edges of the plate. It is similar to the effective length term k^2 in Eq. (18-50). The term K is plotted in Fig. 18-23. Values of K at buckling for other types of compressed plates are given in books on structural stability [18-27]. Equations (18-50) and (18-52) have been included to show the similarity between the critical loads for columns and walls.

Properties of Concrete Affecting the Stability of Concrete Walls

Compressive Stress–Strain Curve for the Concrete

The tangent modulus of the concrete in the wall will be used to estimate the tangent modulus buckling loads. A good approximation of the stress–strain curve for the wall is the second-degree parabola given by Eq. (18-53) and shown in Fig. 18-24a, with a peak stress of $f_{2,\max} = 0.90f'_c$. The apex of the parabola will be taken at a strain of $\varepsilon_{co} = 0.002$ and

$$f_2 = f_{2,\max}\left[2\left(\frac{\varepsilon_2}{\varepsilon_{co}}\right) - \left(\frac{\varepsilon_2}{\varepsilon_{co}}\right)^2\right] \tag{18-53}$$

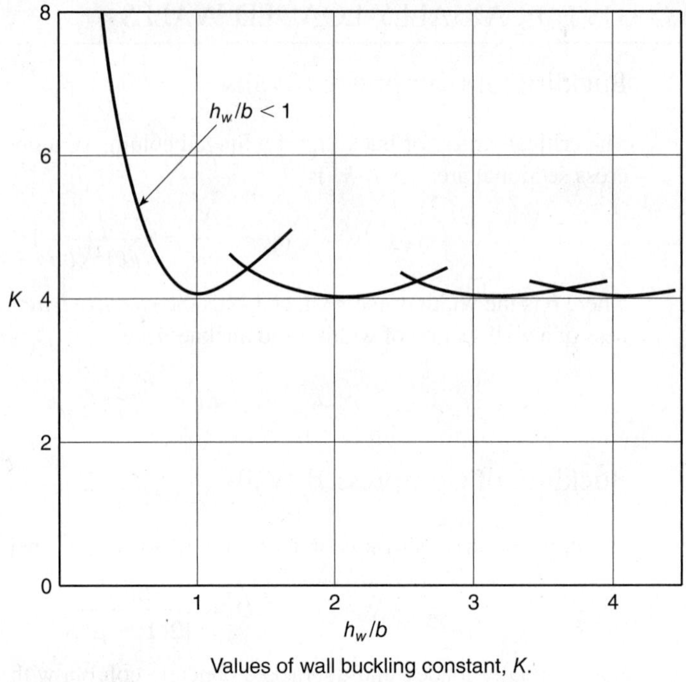

Fig. 18-23
Wall buckling factor, K, for buckling of a compressed plate analogous to a compressed wall in a hollow rectangular bridge pier. (From [18-27].)

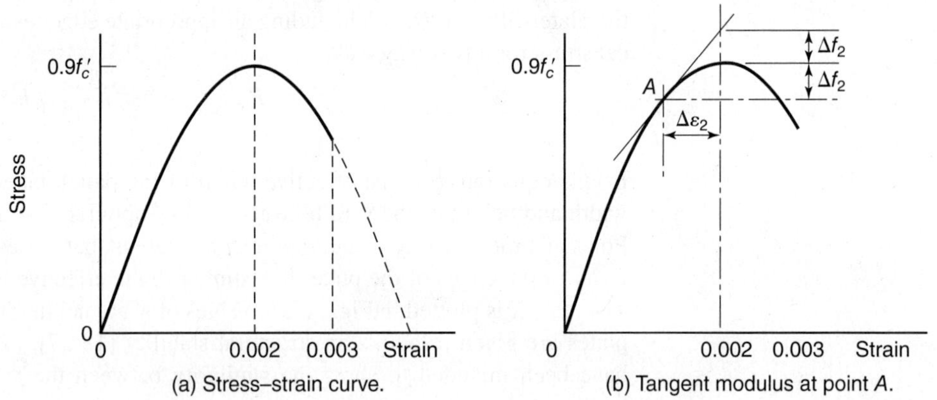

Fig. 18-24
Second-order parabolic stress–strain curve for compressed concrete.

where f_2 is the compressive stress in the most compressed wall of the box pier; ε_2 is the corresponding strain, assumed constant over the thickness of the wall. The strain at any level of stress in Eq. (18-53) is given by

$$\varepsilon_c = \varepsilon_{co}\left(1 - \sqrt{1 - \frac{f_2}{f_{2,\max}}}\right) \qquad (18\text{-}54)$$

The tangent modulus of elasticity, E_T, is the slope of the tangent to the stress–strain curve at some particular stress, in this case of point A in Fig. 18-24. It is computed using

$$E_T = d\sigma/d\varepsilon$$

evaluated at the critical stress, σ_{cr}. For example, the tangent modulus at point A in Fig. 18-24b is calculated as

$$E_T = \frac{2\Delta f_2}{\Delta \varepsilon_2} \tag{18-55}$$

The 2 in Eq. (18-55) comes from the fact that the tangent to a second-order parabola intersects the vertical axis through ε_{co} at 2 times the stress increment between f_2 and $f_{2,\max}$, as shown in Fig. 18-24b. For example, the initial tangent modulus at $f_2 = 0$ for the stress–strain curve given by Eq. (18-55) is for $f'_c = 4000$ psi:

$$E_T = \frac{2 \times 3600 \text{ psi}}{0.0020} = 3{,}600{,}000 \text{ psi}$$

This is very close to the value obtained from ACI Code Section 19.2.2.1:

$$E = 57{,}000\sqrt{f'_c} = 3{,}600{,}000 \text{ psi}$$

The amount of reinforcement in a wall is generally so small that it can be ignored when determining E_T.

Post Buckling Behavior of Concrete plates

After a uniaxially loaded wall panel hinged on four edges buckles, the strips of wall along the edges parallel to the load continue to resist a portion of the compressive in-plane load as shown in Fig. 18-25. At the same time, the load resisted by the center portion of the buckled plate decreases. Failure generally comes from excess bending of the buckled parts. The net effect is that the plate does not suddenly fail. Instead, strips along the edges of the wall continue to support a stress that may approach the buckling stress. This is offset by a drop in the stress resisted by the center of the plate.

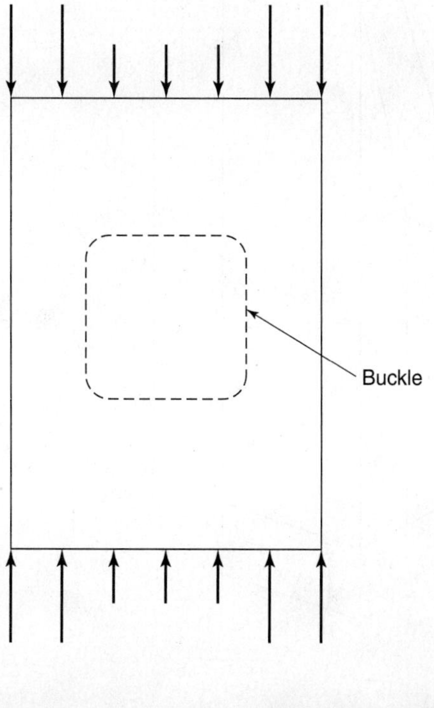

Fig. 18-25
Stresses in a plate after buckling.

EXAMPLE 18-4 Compute the Buckling Load of a Wall

A bridge pier consists of a hollow, rectangular concrete tube similar to the one shown in Fig. 18-26. Side *A–B–C–D–A* has the largest slenderness ratios of any of the sides of the box pier. It has $h_w/h = 25$. The concrete strength is 4000 psi. The applied loads cause axial load and bending about the weak axis of the box as shown in Fig. 18-26. The stress–strain curve of the concrete is assumed to be given by Eqs. (18-53) and (18-54), with a peak compressive stress, $f_{2,\max}$, equal to $0.9f'_c$.

Compute the buckling load for a side wall having a slenderness ratio h_w/h equal to 25.

1. Assume a stress–strain curve for the concrete. We will use the second-degree parabolic stress–strain curve given by Eq. (18-53) with a peak stress and strain of $f_{2,\max} = 0.90f'_c$ and $\varepsilon_{co} = 0.002$:

$$f_2 = f_{2,\max}\left[2\left(\frac{\varepsilon_2}{\varepsilon_{co}}\right) - \left(\frac{\varepsilon_2}{\varepsilon_{co}}\right)^2\right]$$

The strain at any level of stress in Eq. (18-54) is given by

$$\varepsilon_c = \varepsilon_{co}\left(1 - \sqrt{1 - \frac{f_2}{f_{2,\max}}}\right)$$

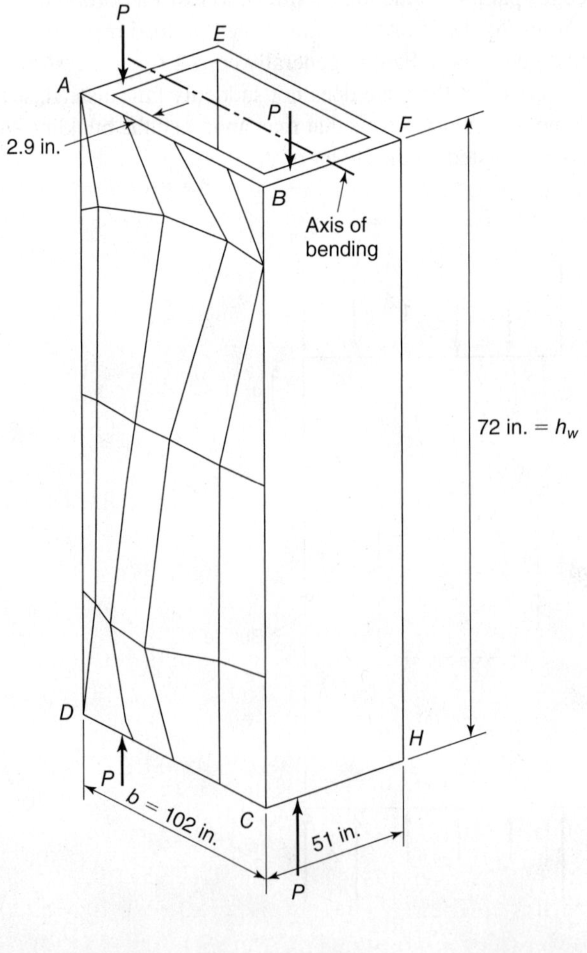

Fig. 18-26
Geometry, dimensions, and buckled shape of the compression flange of a hollow pier test specimen. (Note (a) that the concrete end blocks are not shown and (b) the deflections are plotted to an enlarged scaling.) (From [18-21]).

TABLE 18-3 Critical Slenderness Ratios for a Wall Subjected to Various of $f_2/f_{2,max}$

$f_2/f_{2,max}$ =	0.85	0.90	0.925	0.942	0.95	0.975
ε_c =	0.00095	0.00113	0.00124	0.00133	0.00138	0.00156
E_T psi =	1,025,100	825,900	710,600	622,100	576,400	405,000
h_w/h	34	30	27	**25**	24	20

Origin: Equation 18-52, Example 18-4.

Assume the wall thickness is small compared to the overall distance from the extreme-compression fiber to the neutral axis in the pier. This allows us to assume a uniform stress distribution across the section at the middle of the compressed wall. We shall ignore the effect of reinforcement on the stiffness of the concrete.

2. Compute the tangent modulus of elasticity for various values of $f_2/f_{2,max}$. We shall take the maximum stress at a crushing failure of the wall equal to 0.90×4000 psi = 3600 psi. Values of ε_c will be computed for a series of values of f_2 and, hence, of $f_2/f_{2,max}$

$$E_T = 2(\Delta f_2)/(\Delta \varepsilon_2) \tag{18-55}$$

where Δf_2 is the increase in stress from f_2 to $f_{2,max}$ and $\Delta\varepsilon_2$ is the increase in strain as the stress increases from f_2 to $f_{2,max}$, and the strain increases from ε_2 to ε_{co}.

3. Compute the critical wall slenderness ratios for buckling at various stress ratios, $f_2/f_{2,max}$. Use f_2 to $f_{2,max}$ ratios of 0.85, 0.90, 0.925, 0.95, and 0.975.

The equation for the critical stress of a wall hinged on all four sides and subjected to in-plane load in one direction is:

$$\sigma_{cr} = \frac{P_{cr}}{bh} = K\frac{\pi^2 E_T}{12(1-\mu^2)}\left(\frac{h}{b}\right)^2 \tag{18-52}$$

A plate with height h_w (parallel to the load) and width $b = \ell$ will buckle in a series of square bulges, alternately on one side of the plate, and then the other. One bulge will form if $h_w/\ell = 1$, two bulges if $h_w/\ell = 2$ and so on. At integer values of h_w/ℓ, the buckling constant, $K = 4$.

Substituting $\mu = 0.20$; $0.90f'_c = 3600$ psi; $\sigma_{cr} = 0.85f'_c$ and assuming that $K = 4$, the critical slenderness ratios, h_w/h, are computed in Table 18-3 for the most highly stressed side of the box (side A–B–C–D–A) in Fig. 18-26. This is done for a series of values of $f_2/f_{2,max}$ stress ratios.

The calculated values of h_w/h in Table 18-3 indicate that a side wall with $h_w/h = 25$ would buckle at a stress of 94 percent of the computed flexural capacity.

In the bridge pier tests described in [18-21] and [18-28], the failure strengths of piers with wall thickness ratios between 15 and 35 were reduced by plate buckling. For $h_w/h = 25$, Table 18-3 predicts, a buckling stress of 94 percent of the flexural capacity. This suggests that plates, this thin or thinner, should not be used in situations where such walls are highly stressed in compression. It is difficult to apply this directly to the stability of an assemblage of plates such as those that might enclose an elevator shaft, because the floor slabs and elevator door sills brace the wall at various points along the height of the wall. ■

EXAMPLE 18-5 Compute the Buckling Load of a Compressed Wall Flange

Stability of Compressed Flanges.

Figure 18-27a shows a portion of the projecting flange A–C–D–F–A of an H-shaped wall. When such a flange is loaded in compression, it may buckle in a series of up-and-down

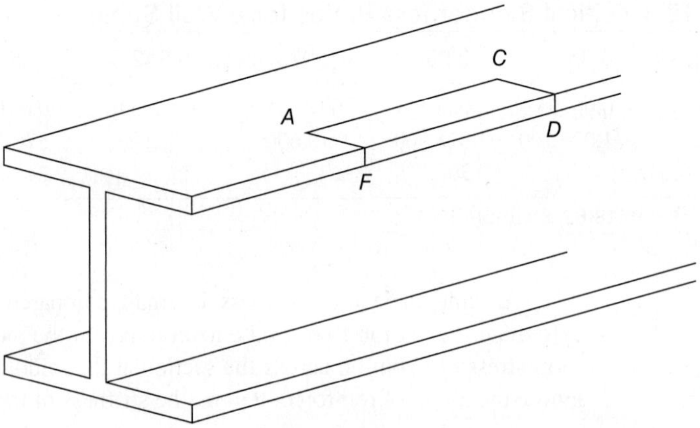

(a) Location of portion of the flange considered.

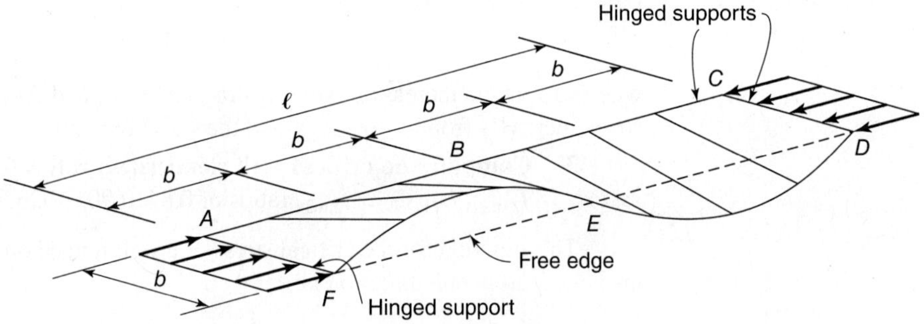

(b) Deformed shape of buckled flange.

Fig. 18-27
Buckling of an outstanding flange of a wall assemblage.

bulges at the free edge of the flange, as shown in Fig. 18-27b. For analysis, the flange can be idealized as a rectangular plate, connected to the web by a hinge *A–B–C* along one long edge of length ℓ, free along the other long edge *D–E–F*, and hinged along the two short edges, *A–F* and *C–D*. The flange projects a distance b from the web and has a thickness t. The buckling load of such a flange can be computed from Eq. (18-52). In this case, books on elastic stability show that K is nearly constant and is equal to 0.5 when $b/t \geq 4$ [18-27].

Substituting $K = 0.5$ into Eq. (18-52), the critical slenderness ratio for an outstanding flange is found to range from 7 to 12 as the average compressive stress in the flange, f_2, increases from 0.85 to 0.975 times $f_{2,\text{max}}$. The calculations are summarized in Table 18-4. The values of the critical slenderness ratios, b/t for $f_2/f_{2,\text{max}} = 0.90$ and 0.925 have been rounded off to the nearest integer, in this case to 10. The analysis given in Table 18-4 predicts that an outstanding flange with a width-to-thickness ratio of $b/t = 10$ is in danger of buckling at flange stresses of 92.5 percent of the maximum stress $f_{2,\text{max}}$ in the stress–strain curve. This may limit the flexural capacity of flanged shear walls. These are similar to the limits on the flange widths allowed in codes for the design of steel structures. ■

TABLE 18-4 Critical Slenderness Ratios, b/t, for a Wall Flange Subjected to Various Compression Stress Ratios, $f_2/f_{2,\text{max}}$

$f_2/f_{2,\text{max}}$	0.85	0.90	0.925	0.95	0.975
$\varepsilon_c =$	0.000946	0.001130	0.001240	0.001380	0.001556
E_T, psi =	1,025,000	825,900	710,600	578,000	405,000
$b/t =$	12	10	10	8	7

Origin: Equation 18-52, Example 18-5.

PROBLEMS

18-1. Check the moment and shear strength at the base of the structural wall shown in Fig. P18-1. Also, show that the given horizontal and vertical reinforcement satisfies all of the ACI Code requirements regarding minimum reinforcement percentage and maximum spacing. The given lateral loads are strength-level wind forces, considering both direct lateral forces and the effects of any torsion. Use a load factor of 1.0 for the wind load effects. The given vertical loads represent dead loads, and you can assume that the vertical live loads are equal to 60 percent of the dead loads. Assume the wall is constructed with normal-weight concrete that has a compressive strength of 4500 psi. Assume all of the steel is Grade 60.

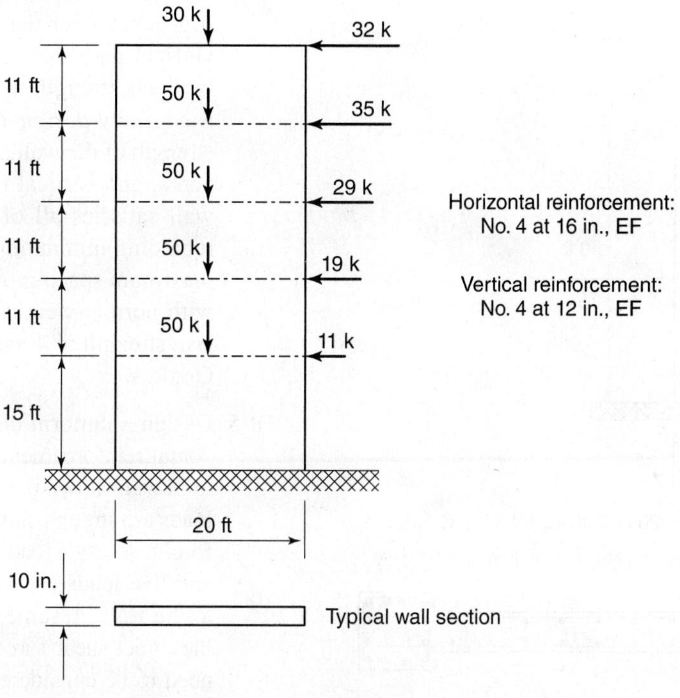

Fig. P18-1

18-2. Design a uniform distribution of vertical and horizontal reinforcement for the structural wall shown in Fig. P18-2. Your design must satisfy all of the ACI Code strength requirements, as well as the requirements for minimum reinforcement percentage and maximum spacing. The given lateral loads are strength-level wind forces, considering both direct lateral forces and the effects of any torsion. Use a load factor of 1.0 for the wind load effects. Assume the wall is constructed with normal-weight concrete that has a compressive strength of 4000 psi. Assume all of the steel is Grade 60.

Video Solution

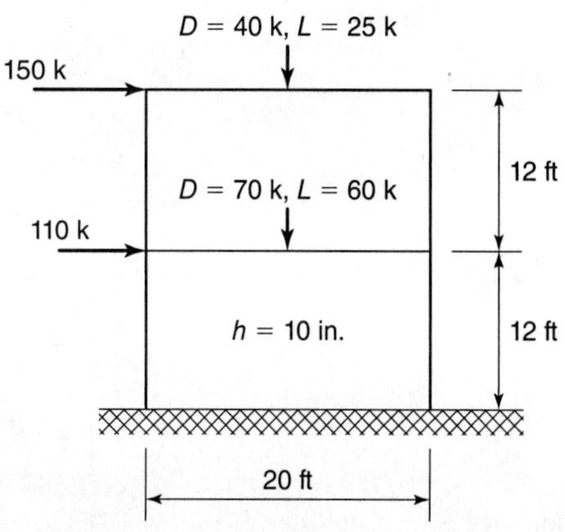

Fig. P18-2

18-3. The structural wall shown in Fig. P18-3 is subjected to gravity loads ($D = 150$ kips and $L = 100$ kips) and an equivalent, static, lateral earthquake load, $E = 150$ kips (torsional effects included). Check the moment strength at the base of the structural wall assuming that the distance, x, from the base of the structure to the lateral force, E, is equal to two-thirds of the wall height. Use a *capacity-design* approach to check the shear strength of the wall and assume the distance, x, is equal to one-half of the wall height for this check. Also, show that the given horizontal and vertical reinforcement in the web of the structural wall satisfies all of the ACI Code requirements regarding minimum reinforcement percentage and maximum spacing. Assume the wall is constructed with normal-weight concrete that has a compressive strength of 5000 psi. Assume all of the steel is Grade 60.

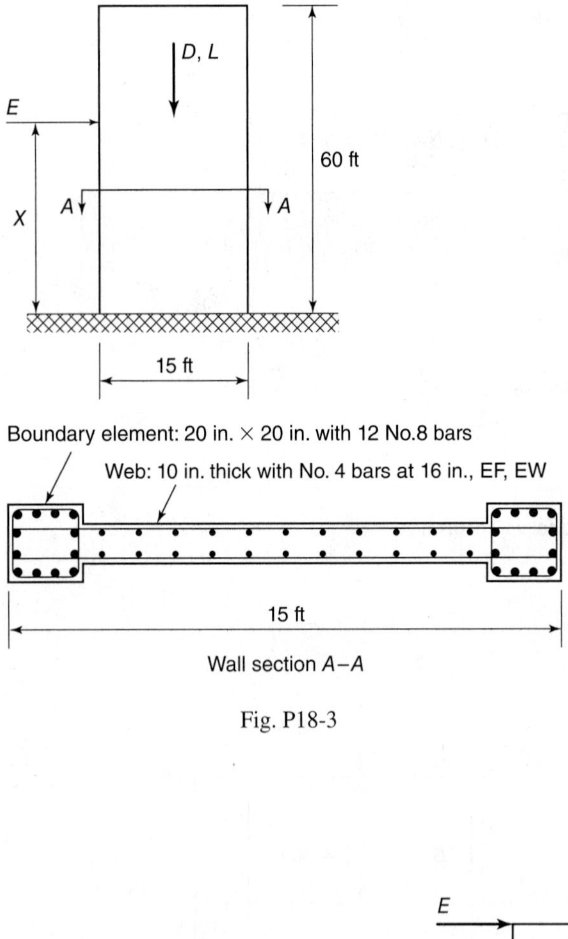

Fig. P18-3

18-4. The structural wall shown in Fig. P18-4 is subjected to gravity loads ($D = 80$ kips and $L = 40$ kips) and a single, equivalent, static, lateral earthquake load ($E = 220$ kips) at the top of the wall (at the roof level). For the given uniform distribution of vertical and horizontal reinforcement, check the moment strength at the base of the wall and use a *capacity-design* approach to check the shear strength of the wall. Also, show that the given horizontal and vertical reinforcement in the structural wall satisfies all of the ACI Code requirements regarding minimum reinforcement percentage and maximum spacing. Assume the wall is constructed with normal-weight concrete that has a compressive strength of 4000 psi. Assume all of the steel is Grade 60.

18-5. Design a uniform distribution of vertical and horizontal reinforcement for the first story of the structural wall shown in Fig. P18-5. The given horizontal loads are strength-level static-equivalent earthquake forces, so use a load factor of 1.0. The vertical dead and live loads are from the tributary area adjacent to the wall. Assume the given lateral loads include the direct shear force and any torsional effects that need to be considered. Use a compressive strength of 4000 psi and Grade-60 reinforcement. Your final design should satisfy all ACI Code requirements for minimum reinforcement percentages and maximum spacing.

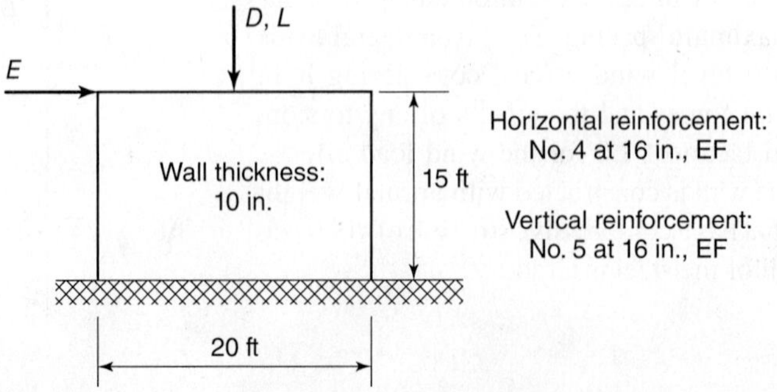

Fig. P18-4

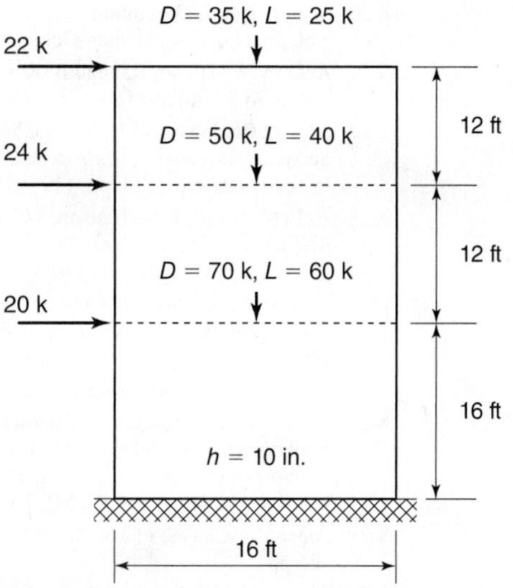

Fig. P18-5

REFERENCES

18-1 G. D. Oberlender and N. J. Everard, "Investigation of Reinforced Concrete Walls," *ACI Journal, Proceedings*, Vol. 74, No. 6, June 1977, pp. 256–263.

18-2 K. M. Kripanarayanan, "Interesting Aspects of the Empirical Wall Design Equation," *ACI Journal, Proceedings*, Vol. 74, No. 5, May 1997, pp. 204–207.

18-3 ACI Committee 551, "Tilt-Up Concrete Construction Guide," (ACI 551.1R-05), *ACI Manual of Concrete Practice*, American Concrete Institute, Farmington Hills, MI.

18-4 *Concrete Design Manual*, Cement Association of Canada, Ottawa, Ontario, January 2006.

18-5 Noel D. Nathan, "Slenderness of Prestressed Concrete Columns," *PCI Journal*, Vol. 28, No. 2, March–April 1983, pp. 50–77.

18-6 Jacob S. Grossman, "Slender Concrete Structures—The New Edge," *ACI Structural Journal*, Vol. 87, No. 1, January–February 1990, pp. 39–52.

18-7 (Former) ACI Committee 442, "Response of Concrete Buildings to Lateral Forces," (ACI 442R-88), *ACI Manual of Concrete Practice*, American Concrete Institute, Farmington Hills, MI, 1989.

18-8 Iain A. MacLeod, *Shear Wall-Frame Interaction*, Special Publication SP011.01D, Portland Cement Association, Skokie, IL, 1971, 62 pp.

18-9 A. R. Santhakumar and Thomas Paulay (Supervisor), *Ductility of Coupled Shearwalls*, Ph.D. Thesis, University of Canterbury, Christchurch, New Zealand, October 1974.

18-10 CSA Technical Committee on Reinforced Concrete Design, A23.3-04, *Design of Concrete Structures*, Canadian Standards Association, Mississauga, Ontario, January 2006.

18-11 Alexander Coull and J. R. Choudhury, "Analysis of Coupled Shear Walls," *ACI Journal, Proceedings*, Vol. 64, No. 9, September 1967, American Concrete Institute, Detroit, pp. 587–593.

18-12 Joseph Schwaighofer and Michael P. Collins, "Experimental Study of the Behavior of Reinforced Concrete Coupling Slabs," *ACI Journal, Proceedings*, Vol. 74, No. 3, March 1977, pp. 123–127.

18-13 Thomas Paulay and R. G. Taylor, "Slab Coupling of Earthquake-Resisting Shearwalls," *ACI Journal, Proceedings*, Vol. 78, No. 2, March–April 1981, pp. 130–140.

18-14 *International Building Code*, International Code Council, Washington, D.C. 2012.

18-15 *Notes on ACI 318-05 Building Code Requirements for Structural Concrete*, Portland Cement Association, Skokie, IL, 2005.

18-16 Loring A. Wyllie, Jr., "Chapter 7, Structural Walls and Diaphragms—How They Function," *Building Structural Design Handbook*, Wiley–Interscience, New York, 1987, pp. 188–215.

18-17 Thomas Paulay and M. J. Nigel Priestley, *Seismic Design of Reinforced Concrete and Masonry Buildings*, Wiley Interscience, New York, 1992, 744 pp.

18-18 *Minimum Design Loads for Buildings and Other Structures (ASCE/SEI 7–10)*, American Society of Civil Engineers, Reston, VA, 2010, 608 pp.

18-19 *National Building Code of Canada*, 2005, Canadian Commission on Building and Fire Codes, National Research Council of Canada, Ottawa, Ontario, Canada.

18-20 Sharon Wood, "Minimum Tensile Reinforcement Requirements in Walls," *ACI Structural Journal*, Vol. 86, No. 4, September–October 1989, pp. 582–591.

18-21 Andrew W. Taylor, Randall B. Rowell, and John E. Breen, "Behavior of Thin-Walled Concrete Box Piers," *ACI Structural Journal*, Vol. 92, No. 3, May–June 1995, pp. 319–333.

18-22 Alex E. Cardenas and Donald D. Magura, "Strength of High-Rise Shear Walls—Rectangular Cross Section," *Response of Multistory Concrete Structures to Lateral Forces*, SP-36, American Concrete Institute, Detroit, 1973, pp. 119–150.

18-23 A. E. Cardenas, J. M. Hanson, W. G. Corley, and E. Hognestad, "Design Provisions for Shear Walls," *ACI Journal, Proceedings*, Vol. 70, No. 3, March 1973, pp. 221–230

18-24 B. Felix, J. M. Hanson, and W. G. Corley, "Shear Strength of Low-Rise Walls with Boundary Elements," *Reinforced Concrete Structures in Seismic Zones*, SP-53, American Concrete Institute, Farmington Hills, MI, 1977, pp. 149–202.

18-25 J. W. Wallace and J. H. Thomsen IV, "Seismic Design of RC Structural Walls, Part II: Applications," *Journal of Structural Engineering*, ASCE, Vol. 121, No. 1, January 1995, pp. 88–101.

18-26 V. Bertero, A. E. Aktan, F. Charney and R. Sause, "Earthquake Simulator Tests and Associated Experimental, Analytical, and Correlation Studies of One-Fifth Scale Model," *Earthquake Effects on Reinforced Concrete Structures: U.S.-Japan Research*, SP-84, J.K. Wight, Editor, American Concrete Institute, Farmington Hills, MI, 1985, pp. 375–424.

18-27 Alexander Chajes, *Principles of Structural Stability Theory*, Prentice-Hall, Englewood Cliffs, NJ, 1974, 336 pp.

18-28 Andrew W. Taylor and John E. Breen, "Design Recommendations for Thin-Walled Box Piers and Pylons," *Concrete International*, American Concrete Institute, Vol. 16, No. 12, December 1994, pp. 36–41.

19

Design for Earthquake Resistance

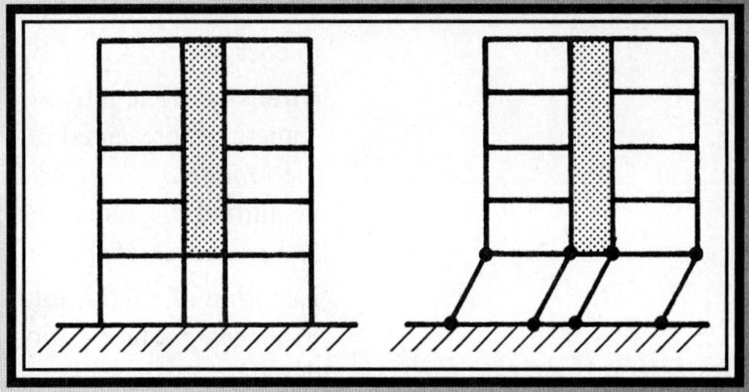

19-1 INTRODUCTION

Plate tectonics theory visualizes the earth as consisting of a viscous, molten magma core with a number of lower-density rock plates floating on it. The exposed surfaces of the plates form the continents and the bottoms of the oceans. As time goes by, the plates move relative to each other, breaking apart in some areas and jamming together in others. Where the plates are moving apart, this movement causes cracks (or rifts) to form, generally in the ocean beds. In some cases, molten magma flows out of these rifts. The regions where the plates are either moving into each other or are sliding adjacent to each other are referred to as fault zones. Compression and shear stresses are generated in the plates and strain energy builds up in at the edges of the plates. At some point in time, the stresses and strain energy at a locked fault surpass the limiting resistance to rupture or slip along the fault. Once started, energy is released rapidly, causing intense vibrations to propagate out from the fault. Three main types of stress waves travel through the rock layers: primary (compression) waves, secondary (shear) waves, and surface waves—each at different speeds. As a result, the effects of these seismic waves and local soil conditions will lead to different ground motions at various sites. Earthquakes may involve regions of slip and/or offsets along surface faults.

Earthquake ground motions impart vertical and horizontal accelerations, a, to the base of a structure. If the structure was completely rigid, forces of magnitude $F = ma$ would be generated in it, where m is the mass of the structure. Because real structures are not rigid, the actual forces generated will differ from this value depending on the period of the building and the dominant periods of the earthquake ground motions. The determination of the seismic force, E, is made more complicated because recorded earthquake ground motions contain a wide range of frequencies and maximum values of base acceleration.

Definitions

Size of earthquake—The size of an earthquake typically is quantified in terms of total energy released.

Magnitude—The *magnitude* of an earthquake is an estimate of the total energy released during the earthquake event, often given by the *Richter magnitude, M* [19-1]. An increase in magnitude by one digit, from 6 to 7, for example, involves an increase in energy released of $10^{1.5}$ times.

Intensity—The intensity of an earthquake is a measure of the shaking at a given site, sometimes presented in terms of observed damage. A commonly used scale is called the *Modified Mercalli, MM Scale* [19-2]. Because these verbal descriptions are not quantifiable, a more precise engineering measure of intensity similar to the *Housner Spectral Intensity* [19-3] may be used.

Location of earthquake—The location where an earthquake is initiated is called the *hypocenter.* The location of the hypocenter is defined by the latitude, longitude, and depth below the surface. The *epicenter* is the point on the surface of the earth directly over the hypocenter.

Seismic Design Requirements

Procedures for the analysis and design of structures to resist the effects of earthquake ground motions are in a continuous state of development. In addition to the work of the ACI Code Committee, several other regulatory bodies [19-4], [19-5] and research development groups [19-6], [19-7] constantly are evaluating and updating code-type analysis and design requirements. Therefore, significant changes to code requirements continue to appear at a rapid rate. For this reason, some of the design requirements noted in this chapter will be modified within a few years. However, the design philosophy and general design procedures for reinforced concrete members are well established and will not change significantly over time. This chapter concentrates on those general principles and gives the latest ACI Code requirements for the earthquake-resistant design of reinforced concrete members. A more in-depth discussion of the inelastic behavior of reinforced concrete members and earthquake-resistant design procedures for reinforced concrete buildings is given by Sozen and Garcia [19-8], [19-9], and by Moehle [19-10].

19-2 SEISMIC RESPONSE SPECTRA

The effect of the size and type of vibration waves released during a given earthquake can be organized so as to be more useful in design in terms of a *response spectrum* for a given earthquake or family of earthquakes. Figure 19-1a shows a family of inverted, damped pendulums, each of which has a different period of vibration, T. To derive a point on a response spectrum, one of these hypothetical pendulum structures is analytically subjected to the vibrations recorded during a particular earthquake. The largest acceleration of this pendulum structure during the entire record of a particular earthquake can be plotted as shown in Fig. 19-1b. Repeating this for each of the other pendulum structures shown in Fig. 19-1a and plotting the peak values for each of the pendulum structures produces an *acceleration response spectrum.*

Generally, the vertical axis of the spectrum is normalized by expressing the computed accelerations in terms of the acceleration due to gravity. If, for example, the ordinate of a point on the response spectrum is 2 for a given period T, it means that the peak acceleration of the pendulum structure for that value of T and for that earthquake was twice that due to gravity. The random wave content of an earthquake causes the derived acceleration response spectrum to plot as a jagged line, as shown in Fig. 19-2c. The spectra in Fig. 19-1b has been smoothed.

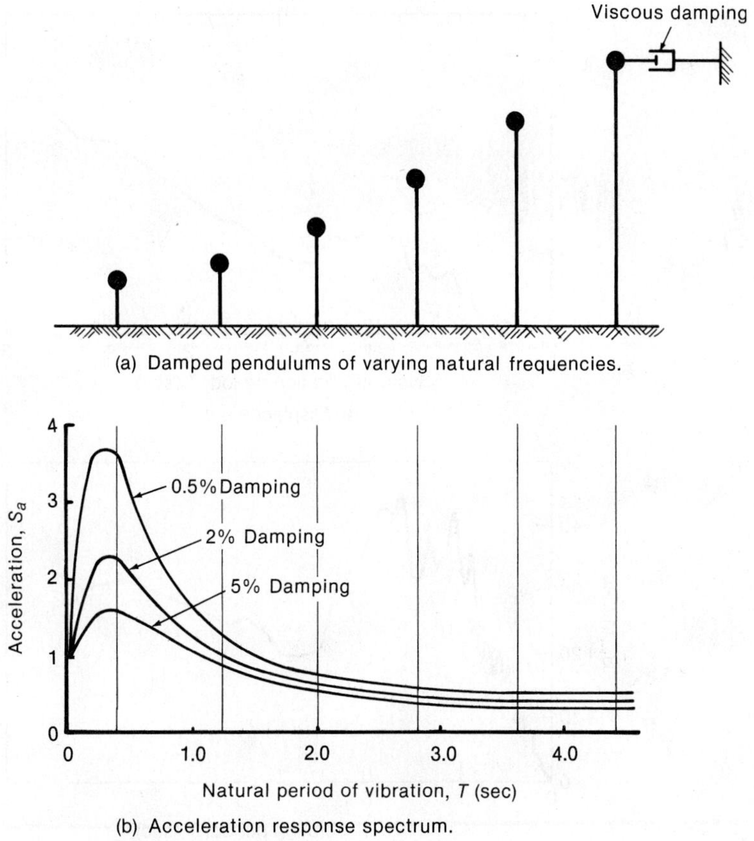

Fig. 19-1
Earthquake response spectrum.

Velocity and Displacement Spectra

Following the procedure used to obtain an acceleration spectrum, but plotting the peak *velocity* relative to the ground during the entire earthquake against the periods of the family of pendulum structures, gives a *velocity response spectrum.* A plot of the maximum *displacements* of the structure relative to the ground during the entire earthquake is called a *displacement response spectrum.* These three spectra for a particular earthquake measured on rock or firm soil sites are shown in Fig. 19-2.

Factors Affecting Peak Response Spectra

Period of Building

Lateral seismic forces are closely related to the fundamental period of vibration of the building. At periods less than about 0.5 sec, the maximum effect for a structure on a firm soil site results from the magnification of the acceleration, as shown by the highest spikes in Fig. 19-2c. For structures with medium periods (from about 0.5 sec to about 2.0 sec), the largest structural response appears in Figure 19-2b, the velocity response spectrum. Finally, at long periods (above about 2.0 sec), the dominant structural response appears in the displacement spectrum.

The period of the first mode of vibration, referred to as the *fundamental or natural period*, can be estimated from empirical equations given in ASCE/SEI 7 [19-4], or from Rayleigh's method [19-11]. For concrete structures, the period, T, in seconds can be estimated from the formula

$$T = C_t \cdot h_n^x \tag{19-1a}$$

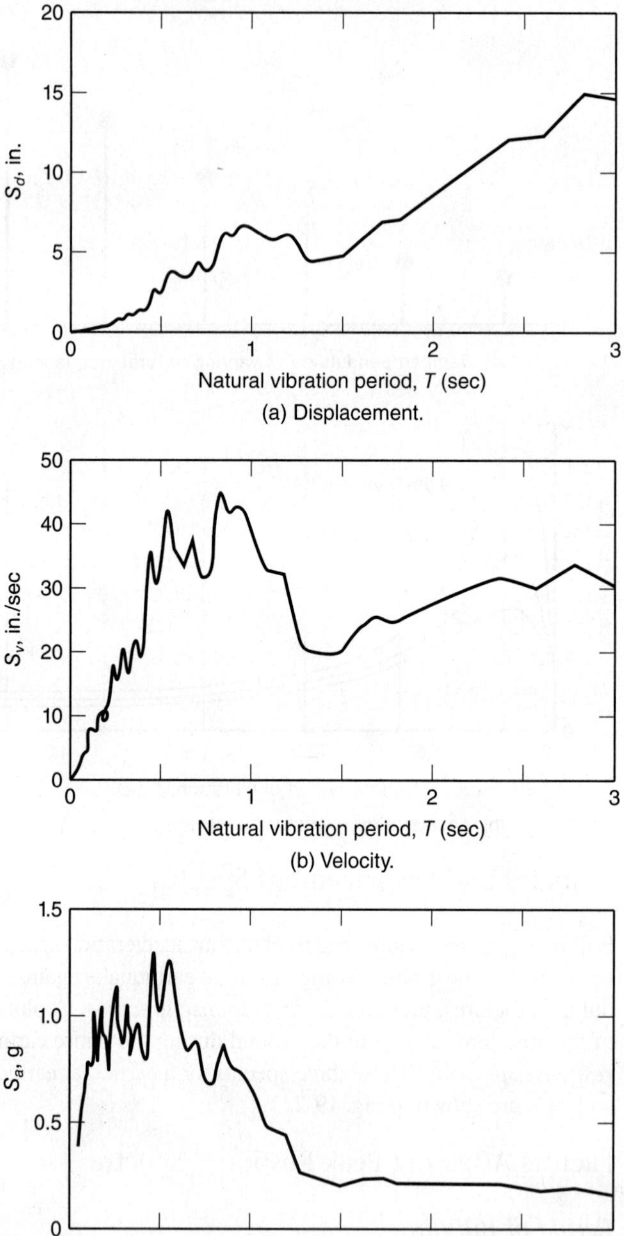

Fig. 19-2
Displacement, velocity, and acceleration spectra for a given earthquake.

where

T = the period in seconds

h_n = the height of the building above exterior grade in feet

C_t and x = 0.016 and 0.9, respectively, for concrete buildings with moment-resisting frames providing 100 percent of the required lateral force resistance

C_t and x = 0.020 and 0.75, respectively, for all other concrete structures

Alternatively, the fundamental period of buildings not exceeding 12 stories and consisting entirely of concrete moment-resisting frames with a story height of at least 10 ft can be estimated from

$$T = 0.1N \tag{19-1b}$$

where N is the number of stories above the exterior grade.

From a series of studies following the 1985 Chilean Earthquake, Wight et al. [19-12] reported that for buildings where the lateral-force resisting system consisted of a high percentage of structural walls, the fundamental period of such buildings could be estimated as

$$T \cong 0.05N \tag{19-1c}$$

Effect of Damping on Response Spectrum

Each of the curves in Fig. 19-1b corresponds to a particular degree of damping. Damping is a measure of the dissipation of energy in the structure and is due to cracking, sliding friction on the cracks, and slip in connections to nonstructural elements. As the damping increases, the ordinates of the response spectrum decrease. Typically, a reinforced concrete building will have 1 to 2 percent of critical damping prior to the building being exposed to an earthquake. As cracking and structural and nonstructural damage develop during the earthquake, the damping increases to about 5 percent. By definition, *critical damping* acts to quickly damp out structural vibrations.

Effect of Ductility on Seismic Forces

As an undamped elastic pendulum is deflected to the right, energy is stored in it in the form of strain energy. The stored energy is equal to the shaded area under the load-deflection diagram shown in Fig. 19-3a. When the pendulum is suddenly released, this energy reenters the system as velocity energy and helps drive the pendulum to the left. This pendulum will oscillate back and forth along the load-deflection diagram shown.

If the pendulum were to develop a plastic hinge at its base, the load-deflection diagram for the same lateral deflection would be as shown in Fig. 19-3b. When this pendulum is suddenly released, only the energy indicated by the triangle *a–b–c* reenters the system as velocity energy, the rest being dissipated primarily by crack development and reinforcement yielding.

Studies of hypothetical elastic and elastic–plastic buildings subjected to a number of different earthquake records suggest that the maximum lateral deflections of elastic and elastic–plastic structures are roughly the same for moderate to long period structures. Figure 19-4 compares the load-deflection diagrams for an elastic structure and an elastic–plastic structure subjected to the same lateral deflection, Δ_u. The ratio of the maximum deflection, Δ_u, to the deflection at yielding for the inelastic structure, Δ_y, is called the *displacement ductility ratio*, μ:

$$\mu = \frac{\Delta_u}{\Delta_y} \tag{19-2}$$

From Fig. 19-4, it can be seen that, for a ductility ratio of 4, the lateral load acting on the elastic–plastic structure would be $\frac{1}{\mu} = \frac{1}{4}$ of that on the elastic structure at the same maximum deflection. Thus, if a structure is ductile, it can be designed for lower seismic forces.

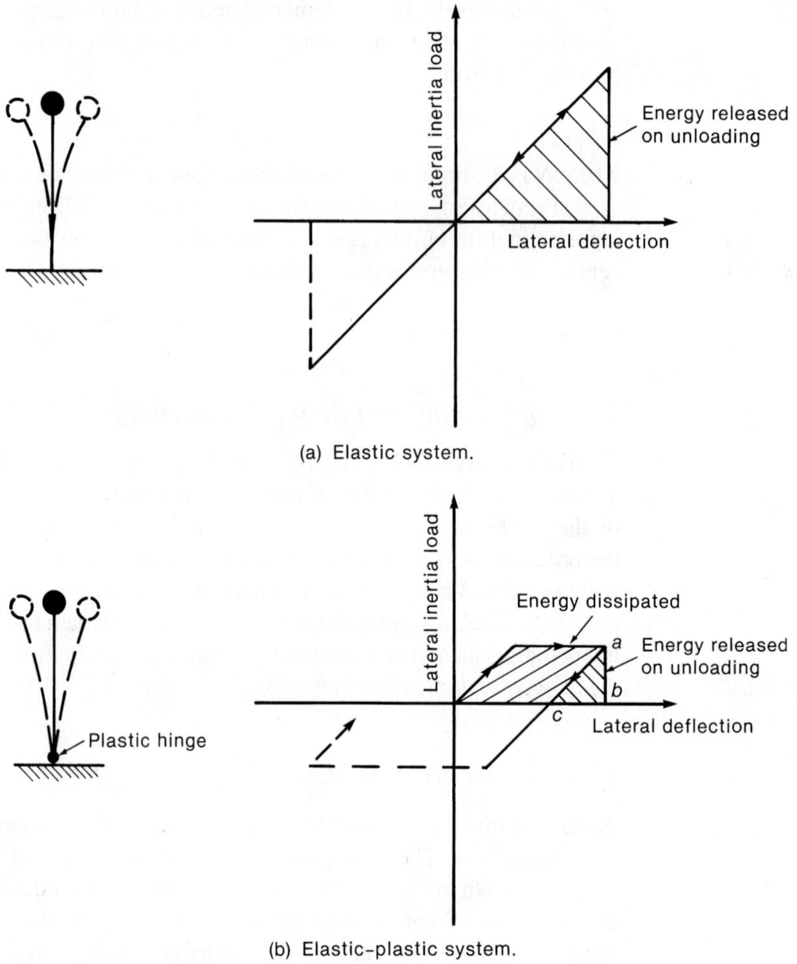

Fig. 19-3
Energy in vibrating pendulums. (From [19-13], copyright John Wiley & Sons, Inc.)

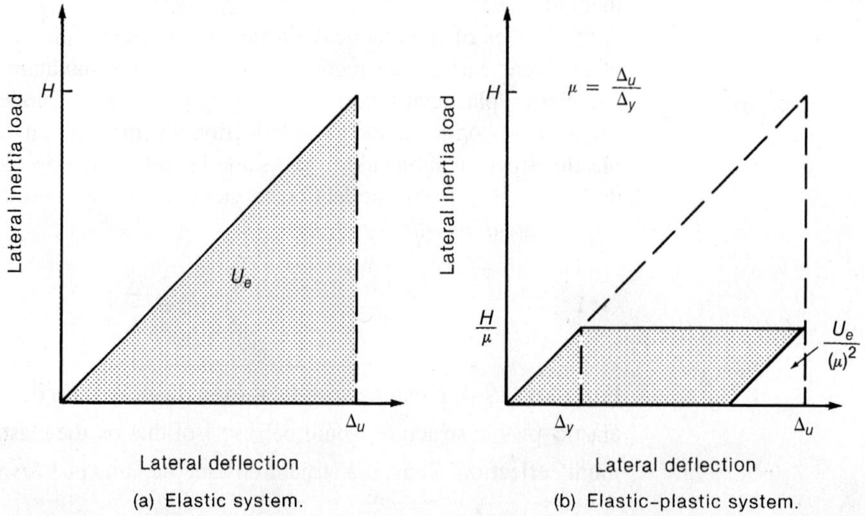

Fig. 19-4
Effect of ductility ratio, μ, on lateral force and strain energy in structures deflected to the same Δ_u. (From [19-13], copyright John Wiley & Sons, Inc.)

Effect of Foundation Soil Stiffness on Response Spectrum

The response spectrum at bedrock normally plots below the response spectrum of a structure that is founded on layers of subsoil between the bedrock and the building footings. For structures on alluvial layers of soil, the increase in the ordinates of the response spectrum is a function of the amplication caused by the various soil layers. Since 1995, seismic design codes have required that designers recognize the effects of foundation soils on seismic response. Geotechnical tests of the subsoil at the site help the designer estimate the degree to which soil–structure interaction will modify the seismic effects on the structure.

19-3 SEISMIC DESIGN REQUIREMENTS

Seismic Design Categories

Seismic design codes [19-4], [19-5] require that all building structures be assigned to a particular *seismic design category*. This assignment is made on the basis of three key parameters, i.e., the expected intensity of the *seismic ground motions*, the *site classification*, and the *building importance factor*. The ASCE/SEI 7 Standard [19-4] gives several maps of spectral response accelerations at periods of 0.2 sec. and 1.0 sec. for all locations throughout the USA. The seismic response accelerations, along with the site classification, are used to establish the design response spectrum for a structure. Site classification is a function of the local soil properties where the structure is to be located. Classifications vary from Class A—hard rock, to Class E—soft clay and Class F—soil requiring a special site response analysis. The building importance factors are related to *building occupancy categories*, which are defined in [19-4]. These categories range from Category I—buildings and other structures that represent a low hazard to human life in the event of failure, to Category IV—buildings and other structures designated as essential facilities. Buildings in Occupancy Category IV have the highest importance factor, and vice versa.

Based on the design response spectrum, which is a function of the expected seismic ground motions and the local site classification, and the building importance factor, ASCE/SEI 7 [19-4] assigns buildings to a specific seismic design category ranging from A to F. Seismic design category A is for structures sited on firm soils where expected maximum ground motions are quite low. Seismic design categories B, C, D, E, and F represent structures where either larger ground motions are expected, or the site consists of softer soil conditions, or the building has a higher importance factor. ACI Code Chapter 18 now gives specific design requirements for concrete buildings based on their seismic design category, as will be discussed in Section 19-6.

Lateral Force-Resisting Structural Systems

The magnitude of the lateral design force for concrete structures is a function of the design response spectrum for the building site and the type of structural system used to resist those forces. As discussed previously, more ductile structural systems can be safely designed for lower seismic forces than systems with limited ductility. This is handled in ASCE/SEI 7 [19-4] by defining a *response modification coefficient*, *R*, which is larger for more ductile structural systems. The general structural systems defined in [19-4] include *bearing wall systems*, *moment-resisting frame systems*, and *dual systems* consisting of a combination of shear walls and moment-resisting frames that work together to resist lateral loads. In general, bearing wall systems are less ductile than either moment-resisting frames or dual systems. Also, depending on the level of structural detailing, which ranges from *ordinary* to *special*, a variety of different *R*-factors are assigned to moment-resisting frames and dual systems. Buildings assigned to high seismic design categories will require special structural detailing, as is discussed in Section 19-6.

Effect of Building Configuration

One of the most important steps in the design of a building for seismic effects is the choice of the building configuration—that is, the distribution of masses and stiffnesses in the building and the choice of load paths by which lateral loads will eventually reach the ground. In recent years, seismic design codes [19-4], [19-5] have classified buildings as *regular* or *irregular*. Irregularities include many aspects of structural design that are conducive to seismic damage. Irregular buildings require a more detailed structural analysis, design provisions to reduce the impact of each irregularity, and more detailing requirements than do regular buildings. Irregularities are classified as *plan irregularities* or *vertical irregularities*, as summarized here.

Plan Irregularities

1. Torsional Irregularities. Ideally, a building subjected to earthquakes should be symmetrical—or, at least, the distance between the center of mass (the point through which the seismic forces act on a given floor) and the center of resistance should be minimized. If there is an eccentricity, as illustrated in Fig. 19-5a, the building will undergo torsional deflections, as shown. The column at *A* in Fig. 19-5a will then experience larger lateral displacements than the column at *B*. The location of the center of resistance is affected by the presence of both structural and "nonstructural" elements.

The computed relative deflection of the top and bottom of a story is referred to as the *story drift*, δ_{max}. A category 1a torsional irregularity exists when the maximum story drift

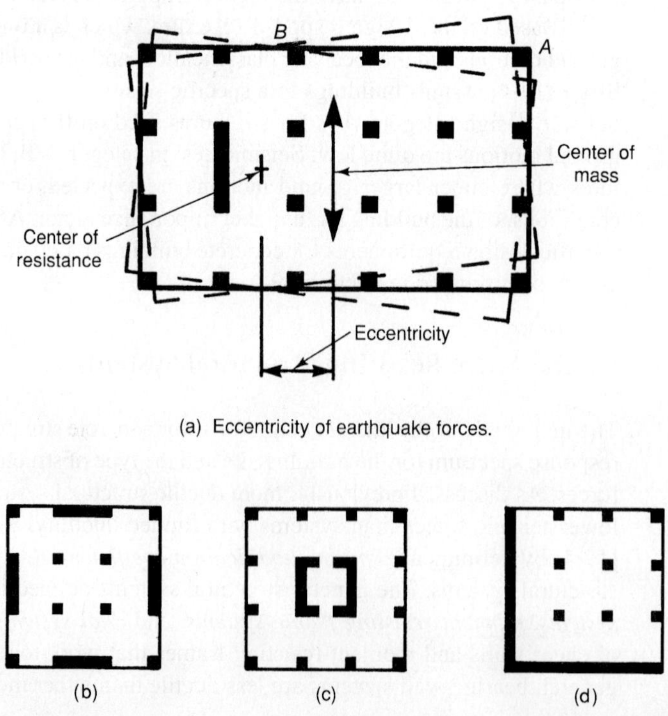

Fig. 19-5
Eccentricities and torsional deformations.

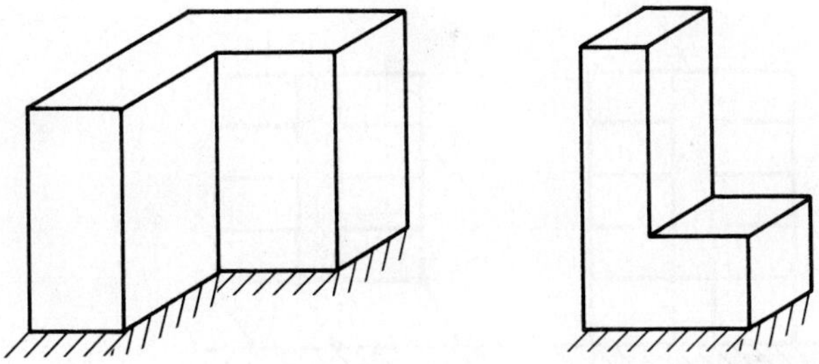

Fig. 19-6
Geometric irregularities.

at one point in a floor level is more than 1.2 times the average story drift for the same floor level [19-4]. This definition applies only to buildings with rigid or semirigid diaphragms.

A category 1b torsional irregularity exists when the ratio of maximum to average elastic computed drifts exceeds 1.4 [19-4].

Irregular buildings should have significant torsional resistances and stiffnesses. As discussed in Chapter 18, because the individual walls in Fig. 19-5b are farther from the center of resistance than those in Fig. 19-5c, they provide more torsional resistance. The core of the building in Fig. 19-5c is almost a closed tube, which tends to be stiffer in torsion than disconnected walls. The plan in Fig. 19-5d, which has been used for corner buildings, is particularly unsuitable. It has a large eccentricity and very little torsional resistance.

2. Reentrant Corner Irregularity. If the plan has reentrant corners and the floor system projects beyond the reentrant corner by more than 15 percent of the plan dimension of the building in the same direction, the building is said to have a *reentrant corner irregularity*. For the building on the left in Fig. 19-6, one solution is to separate the two wings by a joint that is wide enough so that the wings can vibrate separately without banging together. If this is not practical, the region joining the two parts must be strengthened to resist the tendency to pull apart.

3. Diaphragm Discontinuity Irregularity. Figure 19-7 shows a plan of a floor diaphragm transmitting seismic forces to shear walls at each end of a building. The diaphragm acts as a deep thin beam that develops tension and compression on its edges. Abrupt discontinuities or changes in the diaphragms, such as a notch in a flange, may lead to significant damage.

If there are abrupt changes in the stiffness of the diaphragms, including a cutout or open areas comprising more than 50 percent of the diaphragm or cross-sectional area, or 50 percent from one story to the next, the building is said to have a *diaphragm discontinuity irregularity*. The loss of cross-sectional area for the discontinuity shown in Fig. 19-7 is smaller than 50 percent and would not qualify as a diaphragm irregularity.

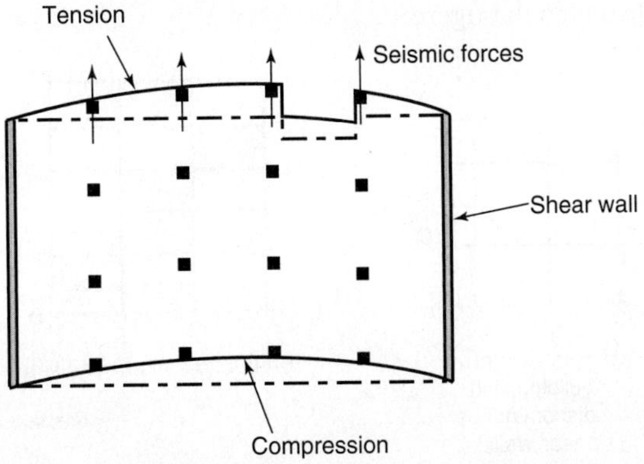

Fig. 19-7
Diaphragm discontinuities.

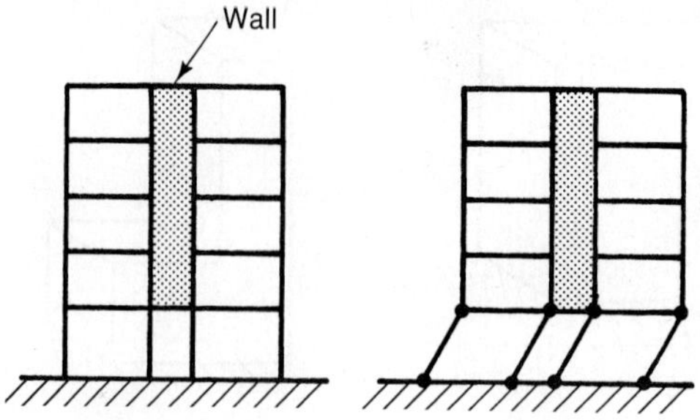

Fig. 19-8
Soft story due to discontinued shear walls.

Vertical Irregularities

Vertical irregularities are abrupt changes in the geometry, strength, or stiffness of a structure from floor to floor.

1. Stiffness irregularity—soft story. A category 1a soft story is one in which the lateral stiffness of the story under consideration is between 70 and 80 percent of that of the story above or below it. This becomes an *extreme soft-story irregularity* (1b) if the lateral stiffness of the story is from 60 to 70 percent of those of the adjacent stories [19-4]. The soft story created by terminating or greatly reducing the stiffness of the shear walls of the ground story, shown in Fig. 19-8, concentrates lateral deformations at that level.

2. Weight (Mass) irregularity. A mass irregularity exists where the effective mass of any story exceeds 150 percent of the effective mass of an adjacent story.

3. Vertical geometric discontinuity. This type of irregularity occurs when the horizontal dimension of the lateral-force-resisting system in any story is more than 130 percent of that in an adjacent story.

4. In-Plane discontinuity in vertical, lateral-force-resisting elements. An in-plane discontinuity is considered to exist where an in-plane offset of the primary lateral-force-resisting elements exists, as shown in Figs. 19-9a and 19-9b.

5. Discontinuity in lateral strength: weak story. A weak story exists if the lateral resistance of a story is less than 80 percent of that of the story above. The lateral resistance of a story is the total strength of all lateral-force-resisting elements in the story.

Variations in the column stiffnesses attract forces to the stiffer columns. Because of their different free lengths, the lateral stiffness of column *D* in Fig. 19-10 would be approximately four times that of column *B* for the same cross section. Initially, column *D* would be called upon to resist four times the shear force in column *B*. Frequently, such a column will fail in shear above the wall. Sometimes the change in column stiffness is caused by the restriction of free movement caused by nonstructural elements, such as the masonry walls shown shaded in the figure.

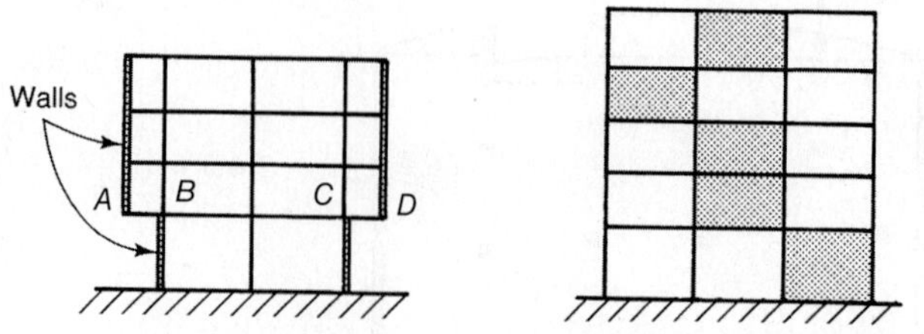

(a) Side view of building with discontinuous shear walls.

(b) In-plane discontinuities.

Fig. 19-9
Discontinuous shear walls.

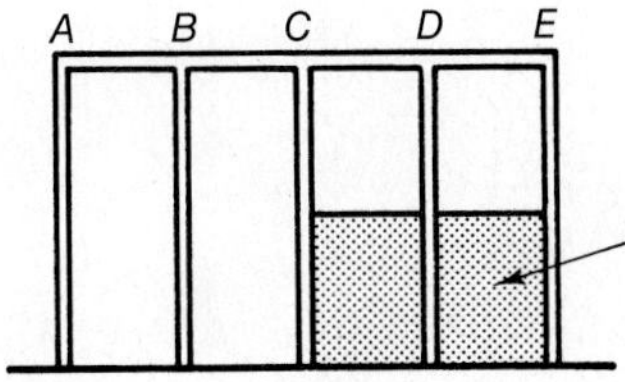

Fig. 19-10
Differences in column stiffnesses.

19-4 SEISMIC FORCES ON STRUCTURES

The ACI Code does not specify earthquake ground motions for a given site or give details about how structures should be analyzed for seismic actions. These details are provided in the general building code for the area [19-5]. Currently, general building codes allow different levels of seismic analysis. The three analysis procedures permitted by ASCE/SEI 7 [19-4] are the *equivalent lateral force* procedure, the *modal spectrum analysis* procedure, and an *inelastic response history analysis* procedure.

Vertical and Horizontal Components of E

American earthquake design codes [19-4], [19-5] published since 1997 have divided the seismic load effect, E, into horizontal and vertical force components, E_h and E_v, as follows:

$$E = E_h + E_v \tag{19-3}$$

$$E = \rho Q_E + 0.2 S_{DS} D \tag{19-4}$$

where ρ is a redundancy factor, commonly equal to 1.0 for most structures, and Q_E is the effect of horizontal seismic forces. E_v is the vertical component of seismic forces, taken equal to $0.2\,S_{DS}D$, where S_{DS} is the design spectral acceleration at a short period, such as 0.2 sec, and D is the dead load.

Equivalent Lateral Force Method for Computing Earthquake Forces

Typically, the equivalent lateral force method is permitted for regular buildings up to about 20 stories. Sometimes it can be used for irregular buildings if special attention is given to the types of irregularities.

Seismic Base Shear, V

The seismic base shear is calculated as

$$V = C_s W \tag{19-5}$$

where C_s is the seismic response coefficient for the building, and W is the effective seismic weight of the building.

Seismic Response Coefficient, C_s

The upper plateau for the seismic response coefficient, C_s, as shown in Fig. 19-11, is given by the following expression:

$$C_s = \frac{S_{DS}}{R/I_e} \tag{19-6}$$

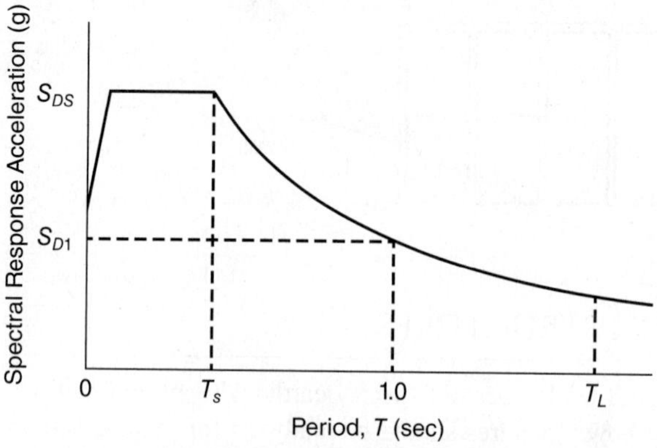

Fig. 19-11
Variation of seismic response coefficient, C_s, with period, T.

where S_{DS} is the design spectral response acceleration parameter in the short period range, R is the response modification factor that accounts for a reduction of the design seismic lateral load for ductile structures capable of dissipating enemrgy through inelastic deformations, and I_e is the earthquake importance factor based on the use and occupancy of the building.

For longer period structures, the C_s coefficient is given by the following two expressions, which are also represented in Fig. 19-11,

$$C_s = \frac{S_{D1}}{T(R/I_e)} \text{ for } T \leq T_L \tag{19-7}$$

and

$$C_s = \frac{S_{D1}T_L}{T^2(R/I_e)} \text{ for } T \leq T_L \tag{19-8}$$

where S_{D1} = the design spectral response acceleration parameter for a structural period of 1.0 sec., T is the fundamental period of the structure, and T_L is the transition period for long-period structures. For most locations, a lower limit for C_s is given as,

$$C_s = 0.044 S_{DS} I_e \geq 0.01 \tag{19-9}$$

The values of S_{D1}, S_{DS}, and T_L can all be determined using maps provided by the U.S. Geological Survey (USGS), which are duplicated in reference [19-4]. Also, if the exact longitude and latitude of the building is known, the USGS website [19-14] will provide the parameters required to evaluate these variables.

Effective Seismic Weight of the Building, W

The effective seismic weight, W, of the building used to compute V is intended to represent the gravity loads likely to be present when earthquake occurs. In ASCE/SEI 7 [19-4], W is calculated as follows:

W = 100 percent of the unfactored dead load,
+ the partition load based on the weight of the partitions (or a minimum weight of 10 psf),
+ in areas used for storage, at least 25 percent of the unfactored live load,
+ where the design show load exceeds 30 psf, not less than 20 percent of the unfactored snow load on a roof,
+ the unfactored load from the full contents of any tanks, and
+ the weight of permanent equipment

TABLE 19-1 Response-Modification Coefficient, *R*, for Seismic Resistance

Basic Structural System (RC)	Seismic Force Resisting System	Response Modification Coefficient *R*
Bearing wall system	Special reinforced concrete shear walls	5
Building frame system	Special reinforced concrete shear walls	6
Moment-resisting frame system	Special moment frames (SMF)	8
	Intermediate moment frames (IMF)	5
	Ordinary moment frames (OMF)	3
Dual system with a SMF capable of resisting at least 25 percent of prescribed seismic forces	Special reinforced concrete shear walls	7
	Ordinary reinforced concrete shear walls	6
Dual system with a IMF capable of resisting at least 25 percent of prescribed seismic forces	Special reinforced concrete shear walls	6.5
	Ordinary reinforced concrete shear walls	5.5

Source: Abridged from [19-4].

Response-Modification Factor, R

The response-modification factor, *R*, reflects

(a) the ability of the structure to dissipate energy through inelastic action, as shown in Fig. 19-4b, and

(b) the redundancy of the structure.

It is assumed that the level of ductility governs the reduction in seismic forces for the various families of lateral-force-resisting systems. Typical values for reinforced concrete structures are given in Table 19-1.

Distribution of Lateral Forces over the Height of a Building

The base shear, V, from Eq. (19-5) is distributed as a series of lateral forces at each floor level and at the roof. In general, the distribution of the lateral forces is assumed to be similar to the deflected shape for the first mode of vibration, which corresponds to an inverted v distribution of lateral forces for short period structures. The coefficient, k, in Eq. (19-8) is used to account for the higher modes of vibration.

Thus, the lateral force at a given floor level, x, is

$$F_x = C_{vx} \cdot V \tag{19-10}$$

where

$$C_{vx} = \frac{w_x h_x^k}{\sum_{i=1}^{n} w_i h_i^k} \tag{19-11}$$

in which F_x, w_x, and h_x are the lateral force, weight, and height, respectively, at level x above grade; $i = 1$ refers to the first level of the building above grade; and $i = n$ refers to

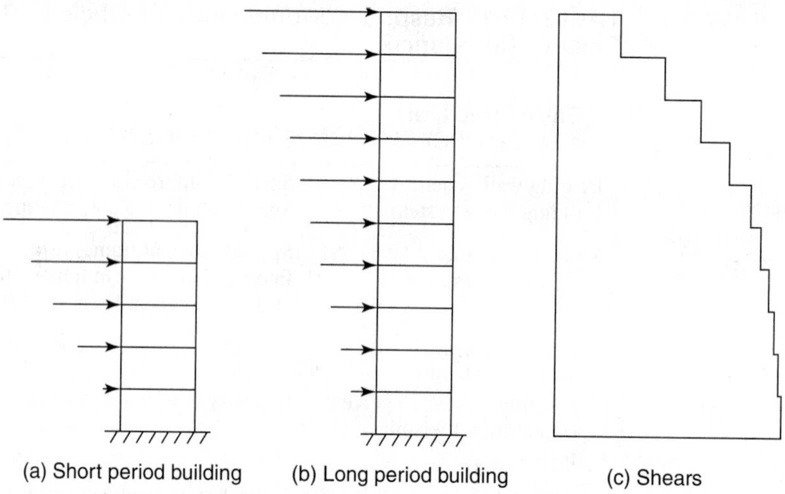

Fig. 19-12
Distribution of equivalent lateral forces and shears.

the top level (roof). The coefficient, k, is an exponent related to the structural period. It is taken equal to 1.0 for structural periods at or below 0.5 seconds, and it is equal to 2.0 for periods at or above 2.5 seconds. For structural periods between 0.5 and 2.5 seconds, k is determined by a linear interpolation between 1.0 and 2.0. Typical lateral load distributions are shown in Fig. 19-12.

Story Shear

The shear, V_x, in any story x is the sum of the lateral forces, F_i, acting *above* that story:

$$V_x = \sum_{i=x}^{n} F_i \tag{19-12}$$

Story Torsion

As discussed in Chapter 18, the story shear, V_x, in a story is assumed to act horizontally through the center of mass of the story, resulting in a torsional moment, T, equal to the product of the seismic forces times the horizontal eccentricity, e_x or e_y, between the center of mass and the center of resistance, measured perpendicular to the line of action of the seismic force. (See Fig. 18-13.) To account for accidental torsion, a torsional moment, T_{ax}, equal to the story shear times a distance e_{ay} is added. ASCE/SEI 7 [19-4] takes e_{ay} equal to $\pm 0.05L_b$, where L_b is the horizontal length of the building perpendicular to the assumed direction of the applied forces. Thus,

$$T_x = V_x \times e_y + V_x \times e_{ay} \tag{19-13}$$

where T_x is the torsional moment due to earthquake forces in the x direction, e_y is the distance in the y direction between the center of mass and the center of resistance, and e_{ay} is 5 percent of the building length in the y direction. An accidental increase in the torsional effects may occur if, for example, a corner column, such as A in Fig. 19-5a, cracks and loses some of its stiffness before the other columns crack. When this occurs, the center of resistance moves toward the stiffer columns (to the left in Fig. 19-5a), thereby increasing the torsional effects. Each element in the building is to be designed for the most severe effects of the accidental torsions due to forces in the x direction and the y direction.

Analysis

The total lateral forces are used in a linearly elastic structural analysis of the frame. For regular structures, independent two-dimensional models may be used. For concrete buildings, cracked-section properties are assumed in the analysis. For irregular structures, three-dimensional analyses must be used. Where the diaphragms are flexible relative to the lateral-force-resisting members, that flexibility must be represented in the analysis.

In summary, two concepts are important here. First, the force developed in the structure does not have a fixed value, but instead results from the stiffness of the structure and its response to ground motions. Second, if a structure is detailed so that it can respond in a ductile fashion, the earthquake forces are reduced from the elastic values.

19-5 DUCTILITY OF REINFORCED CONCRETE MEMBERS

Factors affecting the ductility of reinforced concrete beams under monotonically applied loadings have been discussed in Section 4-2. (See Figs. 4-11 and 4-12.) The flexural ductility of a beam increases as the tension–reinforcement ratio ρ goes down and as the compression–reinforcement ratio, ρ' goes up.

When a reinforced concrete member is subjected to load, flexural and shear cracks develop, as shown in Fig. 19-13a. When the load is reversed, these cracks close and new cracks form [19-15]. After several cycles of loading, the member will resemble Fig. 19-13b. The left end of the beam is divided into a series of blocks of concrete held together by the reinforcing cage. If the beam cracks through its depth, as shown in Fig. 19-13b, shear is transferred across the crack at low rotations by dowel action of the longitudinal reinforcement and grinding friction along the crack. If the concrete cover crushes, the longitudinal bars will buckle unless restrained by closely spaced stirrups or hoops. The hoops also provide confinement of the core concrete, increasing the beam's ductility.

It should be noted that the beam's displacement ductility ratio discussed earlier, Δ_u/Δ_y, is defined in Eq. (19-2) in terms of the deflection Δ at the end of the beam, similar to

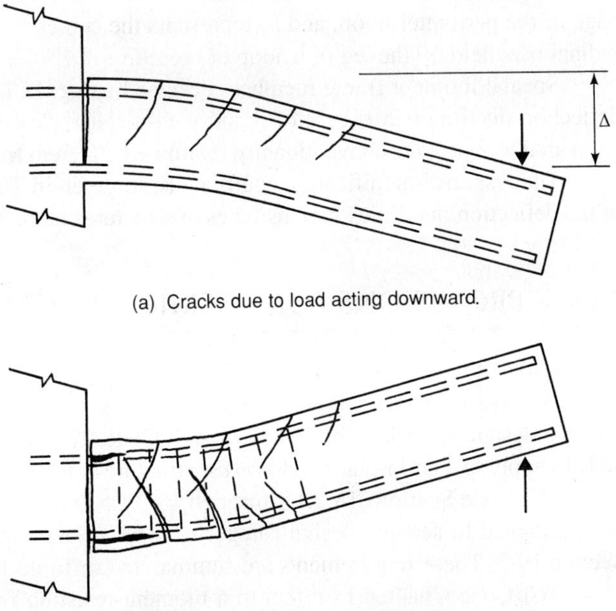

(a) Cracks due to load acting downward.

(b) Cracks due to load acting upwards.

Fig. 19-13
Beam subjected to cyclic loads.

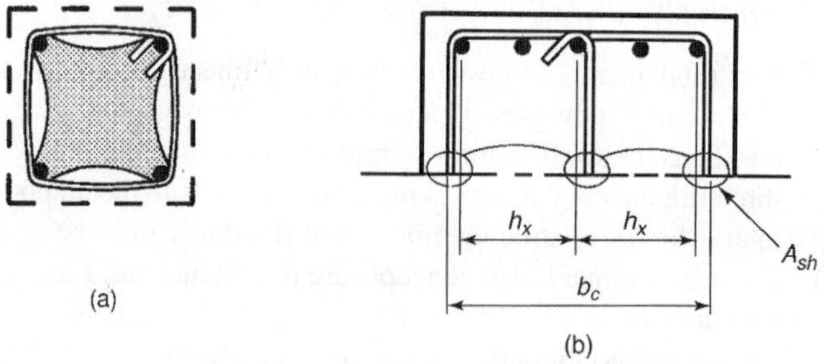

Fig. 19-14
Confinement by hoops.

that shown in Fig. 19-13. Because most of the deformation is concentrated in the cracked plastic hinging region, the rotational ductility, θ_u/θ_y, measured over the length of the plastic hinging zone and the curvature ductility, ϕ_u/ϕ_y, measured at the section of maximum moment for the beam are larger than the required deflection ductility, Δ_u/Δ_y.

In Section 3-4, it was shown that concrete subjected to triaxial compressive stresses increases in both strength and ductility (Fig. 3-15). In a spiral column, the lateral expansion of the concrete inside the spiral stresses the spiral in tension, and this in turn causes a confining pressure on the core concrete leading to an increase in the strength and ductility of the core (Fig. 11-4). ACI Code Chapter 18 requires that beams, columns, and the ends of shear walls have *hoops* in regions where the flexural reinforcement is expected to yield. Hoops are closely spaced closed ties or continuously wound ties or spirals, the ends of which have 135° hooks with six-bar-diameter (but not less than 3 in.) extensions into the confined core. The hoops must enclose the longitudinal reinforcement and give lateral support to those bars in the manner required for column ties in ACI Code Section 25.7.2. Although hoops can be circular, they most often are rectangular, as shown in Fig. 19-14, because most beams and columns have rectangular cross sections. The core concrete shown shaded in Fig. 19-14a is confined by the hoop. As a result, it tends to be more ductile and a little stronger than the unconfined concrete. In addition to confining the core concrete, the hoops restrain the buckling of the longitudinal bars and act as shear reinforcement. In Fig. 19-14b, A_{sh} represents the sum of the hoop and crosstie leg areas, b_c represents the dimension of the confined core, measured from outside edge to outside edge of the perimeter hoop, and h_x represents the center-to-center spacing between longitudinal bars held by the leg of a hoop or crosstie.

Special moment frame members designed using ACI Code Chapter 18 can achieve deflection ductilities in excess of 5 and well-detailed flexural walls can achieve about 4, compared to 1 or 2 for conventionally reinforced concrete frame members.

The response-modification coefficients, R, given in Table 19-1 are in part a measure of the deflection ductilities various types of structures can attain.

19-6 GENERAL ACI CODE PROVISIONS FOR SEISMIC DESIGN

Applicability

Seismic design provisions presented in ACI Code Chapter 18 were extended in the 2002 code to apply to cast-in-place and *precast* structures.

ACI Code Sections 18.2.1.1 through 18.2.1.5 give the design requirements for structures assigned to seismic design categories (SDC) B through F, which were discussed in Section 19-3. These requirements are summarized in Table 19-2.

ACI Code Chapter 18 refers to a moment-resisting frame designed by using Code Chapters 1 to 17 and 19 to 26 as an ordinary moment frame (OMF), and to a moment-resisting

TABLE 19-2 ACI Code Sections Applicable to Various Seismic Design Categories*

Seismic Design Category (SDC)	Moment-Resisting Frames	Beams, Columns, and Joints	Structural Walls and Coupling Beams	Diaphragms	Foundations
A	None	None	None	None	None
B	18.3	None	None	None	None
C	18.4	None	None	None	None
D, E, F	18.6, 18.7, and 18.14	18.8	18.10	18.12	18.13

*In addition to ACI Code Chapters 1 through 17 and 19 through 26.

frame designed using those Code Chapters plus ACI Code Section 18.4, which requires special detailing, as an intermediate moment frame (IMF). A moment-resisting frame designed using the same Code Chapters plus ACI Code Sections 18.2.4 through 18.2.8, and 18.6 through 18.9 is called a *special moment frame (SMF)*.

Materials

The compressive strength of the concrete shall not be less than 3000 psi (ACI Code Table 19.2.1.1). Because some high-strength lightweight concretes display brittle crushing failures, the strength of lightweight concrete shall not exceed 5000 psi unless good behavior of the particular high-strength lightweight aggregate concrete is documented.

Reinforcement resisting earthquake-induced stresses in frame members and the boundary elements of walls shall comply with ASTM A 706, *Standard Specification for Low-Alloy Steel Deformed and Plain Bars for Concrete Reinforcement*. ASTM A 615 Grade-60 steel also may be used if ACI Code Section 20.2.2.5 parts (i) and (iii) are satisfied.

Load Factors, Load Combinations, and Strength-Reduction Factors

Design will be based on

$$\phi S_n \geq S_u \tag{19-14}$$

where S_u is the sum of the factored load effects for an applicable load combination. S_n is the nominal resistance of the member and ϕ is the applicable strength-reduction factor from ACI Code Tables 21.2.1 and 21.2.2.

Load and Resistance Factors—ACI Code

ACI Code Table 5.3.1 presents two load combinations that include earthquake loads, E:

$$U = 1.2D + 1.0E + 1.0L + 0.2S \qquad \text{(ACI Eq. 5.3.1e)}$$

and

$$U = 0.9D + 1.0E \qquad \text{(ACI Eq. 5.3.1g)}$$

where D is an unfactored dead load effect, L is an unfactored live load effect, S is an unfactored snow load effect, E is an unfactored earthquake load effect. (An *effect* is the result of a force acting on a structure.)

ACI Eq. (5.3.1g) is used when the dead load stabilizes a structure subjected to overturning loads or stress reversals.

ACI Code Section 21.2.4 defines special reduction factors, ϕ, for three types of *shear-sensitive* members encountered in seismic design. In 21.2.4.1 for structural members with a nominal shear strength less than the shear corresponding to the development of the nominal flexural strength of the member, the factor ϕ shall be taken as 0.60. In 21.2.4.2 for diaphragms, the ϕ factor for shear strength shall not exceed the lowest ϕ factor in shear used for the vertical components of the primary lateral-force-resisting system. In 21.2.4.3, the ϕ factor for shear in joints and diagonally reinforced coupling beams is set at 0.85. The shear forces in these members are determined by a *capacity design* procedure, as discussed in Example 18-3 and the next section, and the ϕ factor was selected to be consistent with prior editions of the ACI Code.

Capacity Design

The reinforcing details required to ensure adequate ductility of hinging regions in a laterally loaded structure tend to be tedious to design and expensive to place. This complexity can be reduced if the structure is designed so that only a few cross sections form hinges under the seismic loads, while the rest of the structure has enough reinforcement to remain elastic. Consider, for example, a special structural wall with the shear and moment diagrams shown in Fig. 19-15. This wall is loaded by vertical loads totaling N_u and horizontal loads totaling E. In a nonseismic design, the size of the wall and the wall reinforcement would be chosen to have the desired stiffness plus a strength equal or greater than the effect of the factored loads.

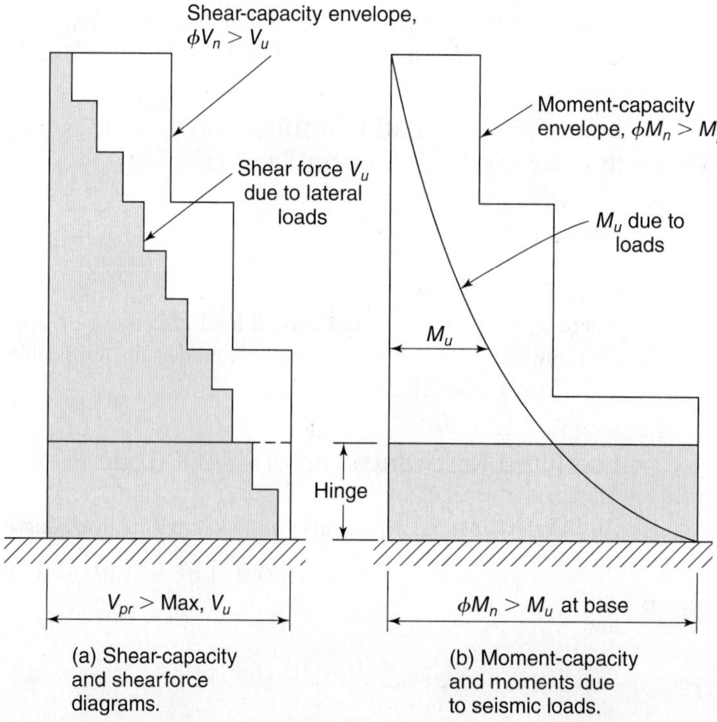

Fig. 19-15
Capacity design of a shear wall.

Because this wall acts as a vertical cantilever, the first plastic hinge is expected to occur at the base of the wall. In *capacity design* this section is designed to hinge at a lateral load level that will allow the hinge to be detailed for ductile behavior under flexure and compressive axial loads. At the same time, however, the sections away from the hinging region are designed to remain elastic throughout the loading history, thereby avoiding the need for seismic detailing at those sections.

Because a shear failure of the hinge region would not be ductile, the shear strength at the base of the wall is chosen to exceed the shear expected at flexural hinging of the wall, as shown by the moment and shear envelopes in Fig. 19-15 and as was demonstrated in Example 18-3. Also, the moment strengths of all sections not chosen to be hinges exceed the moments from the assumed hinging mechanism in the structure. This process is called *capacity design*.

The structure is not merely designed to resist the applied load effects; instead it is proportioned so that the moment and shear strengths of all nonyielding elements in the structure exceed the loads corresponding to yielding of the critical elements that the designer has selected.

Strong-Column–Weak-Beam Design

If plastic hinges form in columns, the stability of the structure may be compromised, as shown in Fig. 19-8. As a result, the design philosophy for ductile moment-resisting frames attempts to force the structure to respond in what is referred to as strong-column–weak-beam action in which the plastic hinges induced by the seismic forces form at the ends of the beams, as shown in Fig. 19-16. To achieve this goal, ACI Code Section 18.7.3.2 requires the following strength check at all beam-to-column connections,

$$\sum M_{nc} \geq (6/5)\sum M_{nb} \tag{19-15}$$

where $\sum M_{nc}$ is the sum of nominal flexural strength of columns framing into the joint, evaluated at the faces of the joint, and $\sum M_{nb}$ is the sum of nominal flexural strengths of

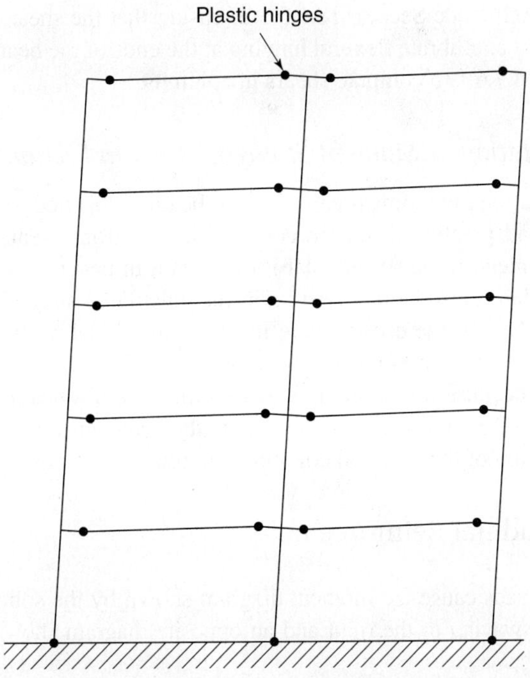

Fig. 19-16
Strong-column–weak-beam behavior.

the beams framing into the joint, evaluated at the faces of the joint. Because beam yielding may occur in both directions as the structure responds to earthquake ground motions, this equation is to be evaluated for beam moments ac ting in both directions in the vertical plane of the frame under consideration. As noted earlier, the beam hinging regions are to be detailed to maintain their shear capacity as the beam flexural hinges undergo yielding in both positive and negative bending.

19-7 BEAMS IN SPECIAL MOMENT FRAMES

Geometric Limits on Beam Cross Sections

ACI Commentary Section R18.6.1 defines a flexural member as a member proportioned to resist primarily flexure and having either no axial load or a factored axial compressive force less than $(A_g f'_c/10)$. ACI Code Section 18.6.2 defines geometric limitations on the span-to-depth ratio $(\ell_n \geq 4d)$ to avoid deep beam action, except that this limit does not apply to coupling beams in shear walls. The widths of flexural members in special moment frames shall not be less than (a) 0.3 times the depth of the beam, or (b) 10 in., or (c) not more than the width of the supporting member, c_2, plus a distance on each side of the supporting member equal to the smaller of the dimension, c_2, or 0.75 times the perpendicular dimension of the supporting member, c_1.

Classification of Moment Strengths

Two levels of moment strengths are used in seismic design:

M_n = *nominal moment strength*, calculated using the specified yield strength, f_y, and the specified concrete strength, f'_c.

M_{pr} = *probable moment strength*, calculated by using $1.25f_y$ because the average yield strength tends to be greater than f_y and because beam longitudinal reinforcement will likely go into strain hardening in plastic hinging zones. The probable moment strength is used in ACI Code Section 18.6.5.1 to ensure that the shear strengths of beams exceed the shears that equilibrate flexural hinging at the ends of the beams. It is also used in ACI Code Section 18.7.6.1 to compute shears in columns.

Computation of Moment Strength of Beam Sections

In calculating the moment capacities of beams subjected to earthquake forces, it has been widespread practice to ignore compression reinforcement of the beam and any tension reinforcement in the flange (slab) for a beam in negative bending (tension at top). Ignoring the slab reinforcement results in the calculated strength of the beam being less than it would be if these effects were included in the calculations [19-16]. This overstrength in flexure, uses up some of the cushion of shear strength provided from a capacity-design procedure. To avoid an under-estimation of a beam's flexural strength, ACI Code Section 18.7.3.2 requires that the steel in the beam tension flanges be considered during the computation of the required column strengths.

Longitudinal Reinforcement

Seismic loads cause the moment diagram shown by the solid line in Fig. 19-17b when the frame is swaying to the right and an opposite diagram shown by the dashed line when the

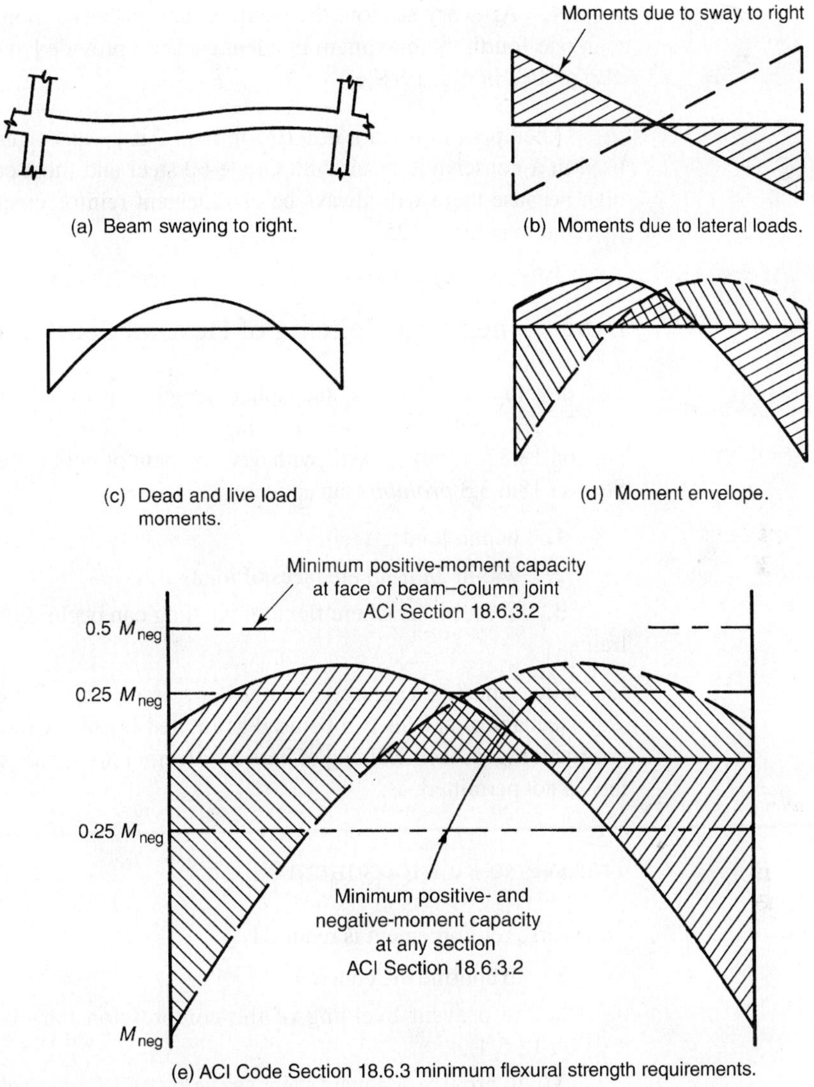

Fig. 19-17
Moment diagram due to gravity loads and seismic loads.

frame is swaying to the left. To this must be added the dead- and live-load moments shown in Fig. 19-17c, giving the moment envelope in Fig. 19-17d. The maximum moments in the span normally occur at the face of a column. In addition to providing adequate moment strength, flexural reinforcement must satisfy the following detailing requirements from ACI Code Section 18.6.3 to provide adequate ductility:

1. At least two bars must be provided continuously top and bottom.

2. The areas of each of the top and bottom reinforcement at every section shall not be less than given by (ACI Code Section 9.6.1.2), nor less than $200b_w d/f_y$. The reinforcement ratio, $\rho = A_s/bd$, shall not exceed 0.025 for either the top or bottom reinforcement.

3. The positive-moment strength of the beam section at the face of the beam–column joint shall not be less than half the negative-moment strength. (See Fig. 19-17e.) This provides $\rho' \geq 0.5\rho$, which allows the beam to develop large curvatures at hinging regions and greatly improves the ductility of the ends of the beams.

4. At every section, the positive- and negative-moment capacity shall not be less than one-fourth the maximum moment capacity provided at the face of either joint. This is also plotted in Fig. 19-17e.

The upper limit on ρ of 0.025 in item 2 is greater than the ρ that would normally be used in a nonseismic beam with Grade-60 steel and most concrete strengths. It is set this high because there will always be confinement reinforcement and compression steel with ρ' equal to at least 0.25ρ.

Development and Splicing of Flexural Reinforcement

The development lengths and splice lengths specified in ACI Code Sections 25.4.2 and 25.5.2 apply to frames resisting seismic forces except as altered in ACI Code Section 18.8.5.1, which deals with development of bars in beam–column joints. ACI Code Section 18.6.3.3 *prohibits* lap splices

1. within joints;

2. within $2h$ from the faces of joints; and

3. in locations where flexural yielding can occur due to lateral deformations of the frame.

Lap splices must be enclosed by hoops or spirals at spacing equal to the smaller of 4 in. or $d/4$. Mechanical splices can be used as limited by ACI Code Section 18.6.3.4. Welding is not encouraged; tack welding of bars for assembly purposes embrittles the bars locally, and thus is not permitted.

Transverse Reinforcement

Transverse reinforcement is required

1. to confine the concrete,

2. to prevent buckling of the compression bars in plastic hinging areas (ACI Section 18.6.4),

3. to provide adequate shear strength (ACI Code Section 18.6.5), and as mentioned in the preceding section, and

4. to confine lap splices.

Confinement Reinforcement

Hoops for confinement and to control buckling of the longitudinal reinforcement are required

1. over a length equal to $2h$ from the face of supports and

2. within $2h$ on each side of other locations where plastic hinging can result due to lateral deformations of the frame.

The spacing of the hoops is specified in ACI Code Section 18.6.4.4. In the rest of the beam, either stirrups or hoops are required at a maximum spacing of $d/2$.

ACI Code Section 2.3 defines a *seismic hook* as a hook on a stirrup, hoop, or crosstie having a bend not less than 135° with a six-diameter (but not less than 3 in.) extension that engages the longitudinal reinforcement and projects into the confined concrete in the interior of the stirrup or hoop.

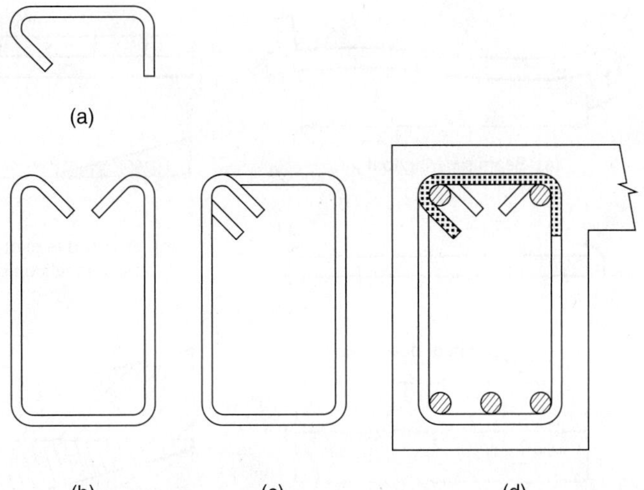

Fig. 19-18
Hoops and crossties.

A *crosstie* is defined as a continuous reinforcing bar having a seismic hook at one end and a hook not less than 90° with at least a six-diameter extension at the other end, as shown in Fig. 19-18a. Both hooks shall engage peripheral longitudinal bars. The 90° hooks of two successive crossties engaging the same longitudinal bars are alternated end for end, except as allowed in ACI Code Section 18.6.4.3.

A *hoop* is a closed tie, as shown in Fig. 19-18c, or a continuously wound tie. A closed tie can be made up of several reinforcing bars, each having seismic hooks at one or both ends. This allows the use of a number of bars or interlocking sheets of welded wire fabric to make up a cage of hoops and longitudinal bars for a beam or column. In flexural members, ACI Code Section 18.6.4.3 allows hoops to be made up of a crosstie as shown in Fig. 19-18a plus a U-shaped stirrup with seismic hooks at each end, as shown in Fig. 19-18b. If the longitudinal bars secured by the crossties are confined by a slab on only one side of the beam, as shown in Fig. 19-18d, the 90° hooks on the crossties must be placed on that side.

Shear Reinforcement

When the frame is displaced laterally through the inelastic deformations required to develop the ductility of the structure, the reinforcement at the ends of the beam will yield unless the moment strength is several times the moment due to seismic loads. The yielding of the reinforcement sets an upper limit on the moments that can be developed at the ends of the beam. The design shear forces, V_e, are based on the shears due to factored dead and live loads (Fig. 19-19c) plus the shears due to hinging at the two ends of the beam for the frame swaying to the right or to the left, as shown in Fig. 19-19a. M_{pr} is the probable moment strength of the members, based on the dimensions and reinforcement at the joint and assuming a tensile strength of $1.25f_y$. For a rectangular beam without axial loads, ACI Code Section 18.6.5.1 requires that beams be designed for the sum of

$$V_{\text{sway}} = \frac{M_{pr1} + M_{pr2}}{\ell_n} \tag{19-16}$$

and

$$V_g = \frac{w_u \ell_n}{2} \tag{19-17}$$

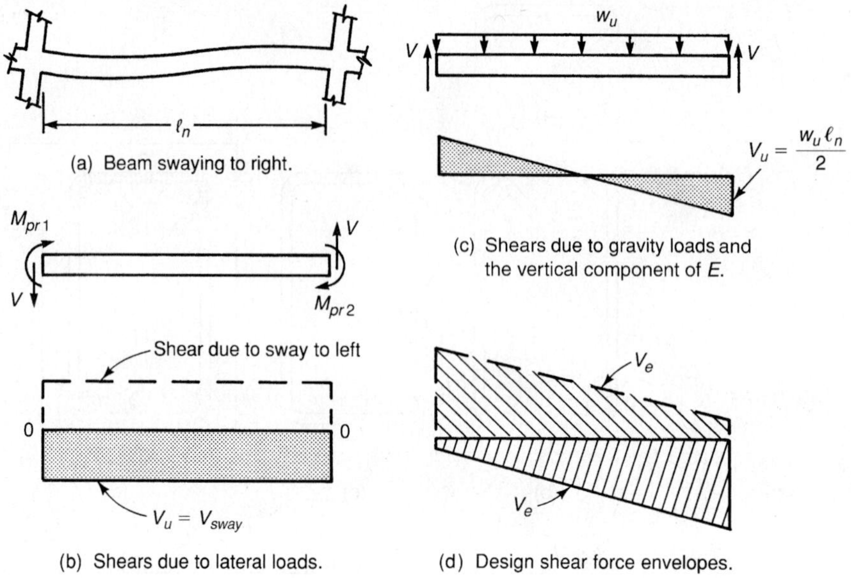

Fig. 19-19
Shear force diagrams due to gravity loads and seismic loads.

Thus, the total design shear is

$$V_e = V_g \pm V_{\text{sway}} \tag{19-18}$$

where

$$M_{pr} = 1.25 f_y A_s \left(d - \frac{a}{2} \right) \tag{19-19}$$

where

$$a = \frac{1.25 f_y A_s}{0.85 f'_c b}$$

The beam is then designed for the resulting shear force envelope with $V_u = V_e$ in the normal way, except that if

(a) the shear, V_{sway}, due to the moments M_{pr1} and M_{pr2} is half or more of the total shear, V_e, within the span *and*

(b) the factored axial compressive force (if any) including earthquake effects is less than $(A_g f'_c/20)$,

then V_c is taken equal to zero. The damage to the hinging area due to repeated load reversals greatly reduces the ability of the cross section to resist shear, requiring more transverse reinforcement [19-15]. Hoops and stirrups provided to satisfy ACI Code Section 18.6.4 also can serve as shear reinforcement.

EXAMPLE 19-1 Design of Flexural Member in a Special Moment-Resisting Frame

The beam shown in Fig. 19-20a and b is a typical floor beam in the special moment-resisting frame of an office building. The beam supports a uniform unfactored dead load of

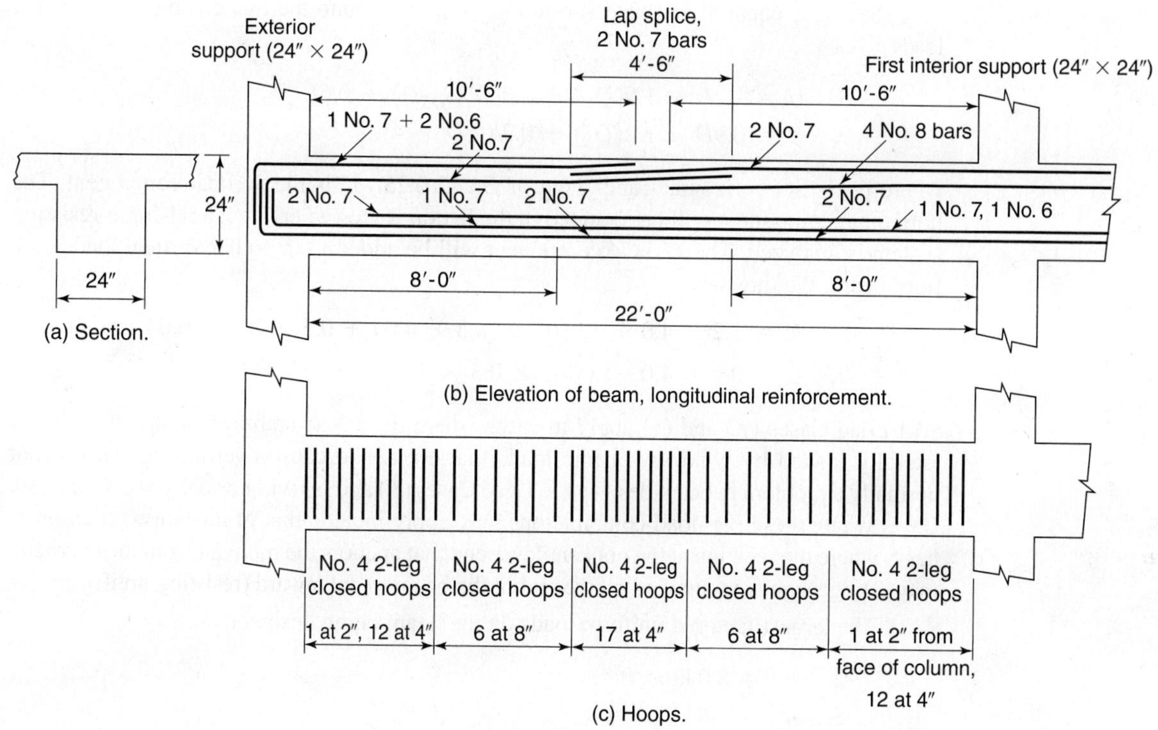

Fig. 19-20
Beam—Example 19-1.

4.0 kips/ft and a uniform unfactored live load of 2 kips/ft. The normal-weight concrete and reinforcement strengths are 4000 psi and 60,000 psi. Design the reinforcement.

1. Select the level of seismic design. The seismic-force calculations will not be completed in detail. We shall arbitrarily choose $S_{DS} = 0.30$ g.

2. Compute factored load combinations. ACI Table 5.3.1 presents seven load combinations for the structural design of buildings. Two of these apply specifically to earthquake loads. Often, the first selection of the beam reinforcement is made by using the nonseismic load combinations, because the beam must be able to resist everyday loads while it waits for an earthquake. Following this, the seismic detailing requirements from ACI Code Chapter 18 will be used to choose steel at other locations where hinges will form under seismic loads. By substituting the live-load factor 0.5 from ACI Section 5.3.3 in ACI Load Cases (c) and (e), the factored loads can be computed as follows:

1. Load Case (a) $U = 1.4 \times 4.0 = 5.60$ kips/ft
2. Load Case (b) $U = 1.2 \times 4.0 + 1.6 \times 2.0 = 8.0$ kips/ft
3. Load Case (c) $U = 1.2 \times 4.0 + 0.5 \times 2.0 = 5.8$ kips/ft

Seismic Load Combinations

5. Load Case (e) $U = 1.2D + 1.0E + 0.5L + 0.2S$
7. Load Case (g) $U = 0.9D + 1.0E$

Setting S equal to zero and substituting Eq. (19-4) into the two combinations with E loads gives

5. $U = 1.2D + 1.0(Q_E) + 1.0(0.2S_{DS}D) + 0.5L$

7. $U = 0.9D + 1.0(Q_E) + 0.2S_{DS}D$

where Q_E is the horizontal component of E and $0.2S_{DS}D$ is the vertical component. The horizontal component will be included in the lateral forces when the lateral-force-resisting system is analyzed. The vertical component will be added to the other vertical loads on a floor beam. We then have

5. $U = 1.2 \times 4.0 + 1.0(0.2 \times 0.3 \times 4.0) + 0.5 \times 2.0 = 6.04 \text{ kips/ft}$

7. $U = 0.9 \times 4.0 - 1.0(0.2 \times 0.3 \times 4.0) = 3.36 \text{ kips/ft}$

ACI Load Cases (f) and (g) apply in cases where dead load stabilizes a structure that is laterally loaded by wind or seismic loads that are apt to cause overturning. This is not normally a problem in beam design, and Load Cases (f) and (g) will normally not be applied. Cases where these two load combinations may apply include that of a beam with an overhang, where the reaction at the noncantilever end that anchors the moment from the overhang may change sign from upward on the end of the beam to downward (resisting uplift).

The largest factored uniform loads on the beam being designed are

2. $w_u = 8.0 \text{ kips/ft}$

5. $w_u = 6.04 \text{ kips/ft}$

3. Does the beam satisfy the definition of a special moment frame beam? ACI Code Sections 18.6.1 and 18.6.2 require beams to have

(i) very little, if any, axial load—o.k.

(ii) a clear span not less than four times the effective depth, $\ell_n/d = 22 \times 12/21.5 = 12.3$—o.k.

(iii) a width

(a) not less than 10 in.—o.k;

(b) not more than width of column—o.k.

Thus, the beam satisfies the requirements of a beam in a special moment frame. If it did not, it would be necessary to change the dimensions of the beam.

4. Calculate the design moments. At all sections, the beam must have $\phi M_n \geq M_u$, where M_u is the moment due to the factored gravity loads plus the vertical component of the seismic loads. We shall use ACI Code Table 6.5.2 to compute the gravity-load moments at critical sections of an exterior span for maximum negative and positive moments. Normally, this would be part of the frame analysis.

Exterior negative gravity-load moment for Load Case (b): $\dfrac{w\ell_n^2}{16} = \dfrac{8.0 \times 22^2}{16}$

$M_u = -242 \text{ kip-ft}$

Midspan positive gravity-load moment Load Case (b): $M_u = 277 \text{ kip-ft}$

Interior-support negative gravity-load moment Load Case (b): $M_u = -387 \text{ kip-ft}$

The reinforcement chosen for each cross section, the ultimate moment, and the nominal and probable moment strengths are listed in Table 19-3. The various sections are referred to in the calculations by the case number from the first column of that table.

TABLE 19-3 Reinforcement and Moment Strengths, End Span

Case	Location	Sway Direction	M_u kip-ft	Reinforcement	A_s in.^{2}	ϕM_n kip-ft	M_{pr} kip-ft
1.	Interior End Negative M	Right	−387	4 No. 8 plus 2 No. 7	4.36	390	531 cw
2.	Exterior End Negative M	Left	−242	3 No. 7 plus 2 No. 6	2.68	247	339 ccw
3.	Exterior End Positive	Right	+124	3 No. 7	1.80	169	233 cw
4.	Interior End Positive	Left	+195	3 No. 7 plus 1 No. 6	2.24	208	286 ccw
5.	Midspan, Positive M	Either	+ 277	5 No. 7 plus 1 No. 6	3.44	313	n.a.

Note: "cw" and "ccw" stand for "clockwise" and "counterclockwise," respectively.

5. Calculate the steel required for flexure.

Case 1, Table 19–3—Interior support, negative moment, sway to right. Assuming rectangular beam action, the required reinforcement for $M_u = -387$ kip-ft is computed. Assume one layer of steel, $d = 24 - 2.5 = 21.5$ in. We shall ignore compression steel (if any).

$$A_s = \frac{M_u}{\phi f_y jd}$$

Assume that $j = 0.90$ and $\phi = 0.9$. Then

$$A_s = \frac{387 \times 12}{0.9 \times 60 \times 0.9 \times 21.5}$$

$$= 4.44 \text{ in.}^2$$

For a first estimate, try six No. 8 bars, $A_s = 4.74$ in^2. These will fit into one layer. ACI Code Section 18.8.2.3 requires the bar sizes be limited so that the column width or depth parallel to the bars is at least $20d_b$. The columns are 24 in. square. This sets the maximum bar diameter as $24/20 = 1.2$ in. Thus, No. 9 bars are the largest that can be used, and the No. 8 bars are o.k.

However, the moment strength of a beam with six No. 8 bars is 12 percent higher than is needed. We thus have

$$a = \frac{A_s f_y}{0.85 f'_c b} = \frac{4.74 \times 60}{0.85 \times 4 \times 24}$$

$$= 3.49 \text{ in.}$$

$$\phi M_n = \phi A_s f_y (d - a/2)$$

$$= 0.9 \times 4.74 \times 60(21.5 - 3.49/2)$$

$$= 5060 \text{ kip-in.} = 421 \text{ kip-ft}$$

Try four No. 8 bars plus two No. 7 bars, $A_s = 4.36$ in^2. These bars give $\phi M_n =$ 390 kip-ft, which just satisfies $M_u = -387$ kip-ft.

ACI Code Section 18.6.3.1

Check whether $A_s \geq A_{s,\min}$.

$$A_{s,\min} = \frac{3\sqrt{f'_c}}{f_y} b_w d, \text{ but not less than } \frac{200 b_w d}{f_y}$$

$$= 1.63 \text{ in.}^2, \text{ but } \geq 1.72 \text{ in.}^2\text{—}A_{s,\min} = 1.72 \text{ in.}^2$$

$$A_s = 4.36 \text{ in.}^2 > A_{s,\min} = 1.72 \text{ in.}^2\text{—o.k.}$$

$$\text{Check whether } \rho = \frac{4.36}{24 \times 21.5} = 0.0085 \leq 0.025\text{—o.k.}$$

Check whether section is tension controlled.

$$a = 3.21 \text{ in.}$$

$$c = a/\beta_1 = 3.77 \text{ in.}$$

Because $c < 0.375\,d$, the section is tension controlled and $\phi = 0.9$.

Case 1—Interior support, negative moment. Use four No. 8 plus two No. 7 top bars at the interior support. $A_s = 4.36$ in.^{2}, $\phi M_n = -390$ kip-ft.

Case 2—Exterior support, negative moment. From step 4, $M_u = -242$ kip-ft. This moment is required to support the factored gravity loads at times other than during an earthquake. **Try three No. 7 bars plus two No. 6 bars,** $A_s = 2.68$ in.^{2}, $\phi M_n = 247$ kip-ft—o.k. At these two sections, Cases 1 and 2, the gravity loads control the steel selection. For the rest of the span, longitudinal reinforcement is provided to satisfy detailing rules in ACI Code Section 18.6.3.

Case 3—Exterior support, positive moment. ACI Section 18.6.3.2 requires the positive-moment strength at the face of the support to be at least 0.5 times the ϕM_n for the negative-moment steel at the face of the support: $0.5 \times 247 = 124$ kip-ft. Design for 124 kip-ft. **Try three No. 7 bars,** $A_s = 1.80$ in.^{2}, $a = 1.32$ in., and $\phi M_n = 169$ kip-ft. From step 4, the minimum A_s was 1.72 in.^{2}. As a result, it is not possible to use fewer than three No. 7 bars.

Case 4—Interior support, positive moment. ACI Section 18.6.3.2 requires that the positive-moment strength at the face of the joint not be less than 0.5 times the negative-moment strength ϕM_n provided by the negative-moment reinforcement at the face of the same joint. Thus, the minimum positive-moment strength is $0.5 \times 390 = 195$ kip-ft. Design the interior support for a positive moment of $\phi M_n = 195$ ft-kips. **Use three No. 7 bars plus one No. 6, with the No. 6 bar cut off after passing midspan, as shown in Fig. 19-20b.** Before the cutoff, $A_s = 2.24$ in.^{2}, and $\phi M_n = 208$ kip-ft.

Case 5—Midspan, positive moment: From step 4, the maximum positive moment at midspan is 277 kip-ft. **Try five No. 7 bars plus one No. 6 bars,** $A_s = 3.44$ in.^{2}, $\phi M_n = 313$ kip-ft; A_s satisfies the minimums. The extreme-tensile strain, ε_t, exceeds 0.005, so the beam is tension controlled and ϕ can be taken as 0.9.

Minimum positive- and negative-moment strengths: ACI Section 18.6.3.2 requires that the minimum positive- and negative-moment strengths at any section along the beam not be less than 0.25 times the maximum negative-moment strength provided at either joint: 0.25×390 kip-ft $= 97.5$ kips-ft. Two No. 7 bars are adequate as minimum steel.

The steel chosen for flexure and to suit the detailing requirements is shown in Fig. 19-20b.

6. Compute the probable moment strengths, M_{pr}. The seismic shears in the beam are computed by assuming that plastic hinges form at each end of the beam with the reinforcement stressed to $1.25f_y$.

Moments for frame swaying to the right (see Fig. 19-21a).

Case 1—Probable interior-end negative moment. Use four No. 8 plus two No. 7 bars, and $A_s = 4.36$ in.^{2} top steel. The depth of the stress block, a, and the moment strength, M_{pr}, for the top steel, with $1.25\,f_y$, are

$$a = \frac{(1.25 \times 60 \times 4.36)}{0.85 \times 4 \times 24} = 4.01 \text{ in.}$$

$$M_{pr} = \frac{(1.25 \times 60) \times 4.36(21.5 - a/2)}{12} = 531 \text{ kip-ft}$$

(clockwise on the interior end of the beam).

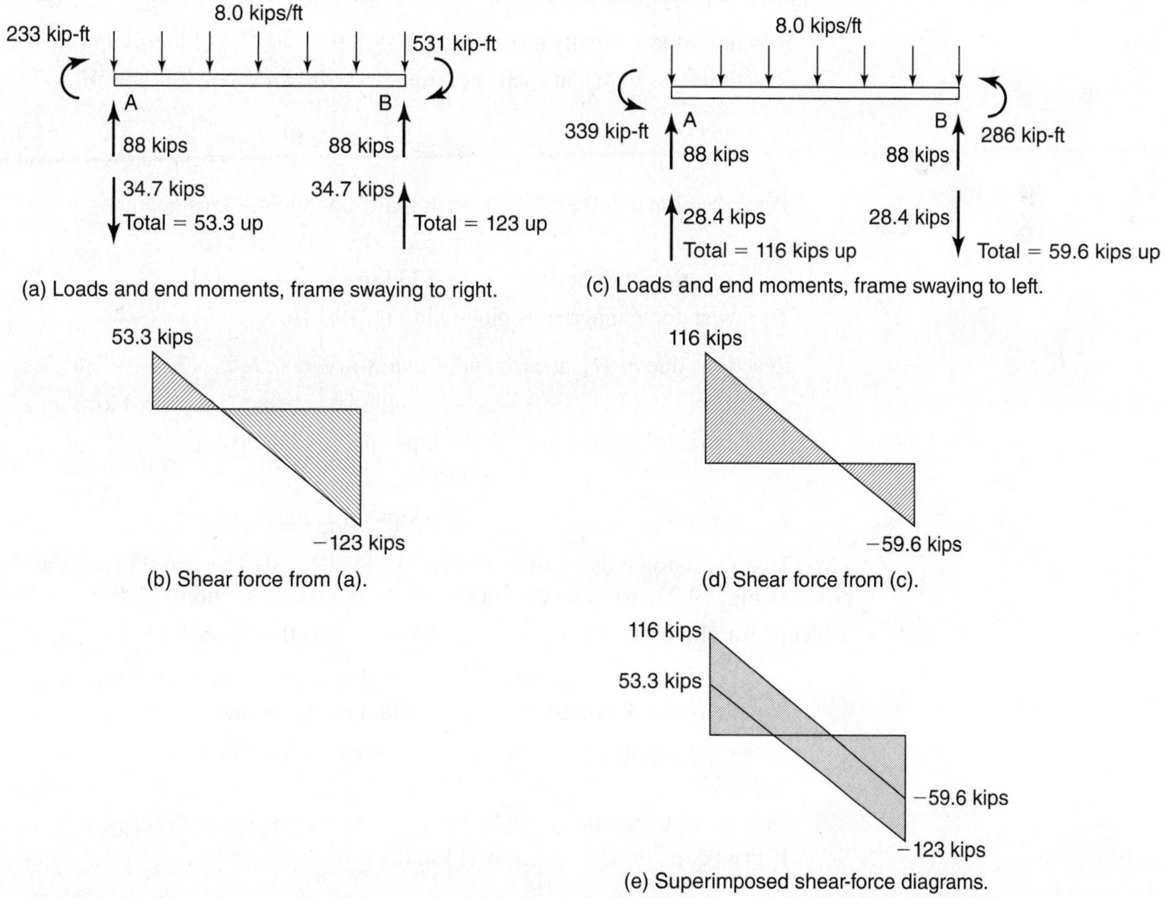

Fig. 19-21
Computation of shear forces—Example 19-1.

Case 3—Probable exterior-end positive-moment strength for bottom steel. Use three No. 7 bars, $A_s = 1.80$ in.^{2}. Then

$$M_{pr} = \frac{(1.25 \times 60) \times 1.80\left(21.5 - \dfrac{1.25 \times 60 \times 1.80}{1.7 \times 4 \times 24}\right)}{12} = 233 \text{ kip-ft}$$

(clockwise on the exterior end of the beam).

Moments for frame swaying to the left (see Fig. 19-21c).

Case 2—Probable exterior-end negative-moment strength. Use three No. 7 and two No. 6 bars, and $A_s = 2.68$ in.^{2}. Probable moment strength is $M_{pr} = 339$ kip-ft, counter clockwise on the exterior end of the beam.

Case 4—Probable interior-end positive-moment strength. Use three No. 7 bars plus one No. 6 bar and, $A_s = 2.24$ in.^{2}. Probable moment capacity is $M_{pr} =$ 286 kip-ft counter clockwise on interior end of beam.

7. Compute the shear-force envelope and design the stirrups. Figure 19-21a shows the moments and uniform load for Load Case (e) acting on the beam with the frame swaying to the right. The reactions consist of two parts:

Hinges develop at both ends of the beam, with hinging moments of $M_{pr} = 233$ kip-ft at the exterior end and 531 kip-ft at the interior-end negative-moment region.

Reactions due to gravity loads: $w_u\ell_n/2 = 8.0 \times 22/2 = 88$ kips upward at each end.

Reactions due to M_{pr} at each end, *frame swaying to right*: $(M_{pr1} + M_{pr2})/\ell_n$
$= (233 + 531)/22$
$= 34.7$ kips down at left end

Total reaction at left end $= 88$ kips up $\pm$ 34.7 kips down
$= 53.3$ kips up at left end

Total reaction at right end $= 123$ kips up at right end.

The shear-force diagram is plotted in Fig. 19-21b.

Reactions due to M_{pr} at each end, *frame swaying to left*: $(339 + 286)/22$
$= 28.4$ kips up at left end.

Total reaction at left end $= 88$ kips up $\pm$ 28.4 kips up
$= 116$ kips up at left end
$= 59.6$ kips up at right end

This shear-force diagram is plotted in Fig. 19-21d. The two diagrams are superimposed in Fig. 19-21e to show the maximum shear at every section.

Stirrups for shear: ACI Code Section 18.6.5.2 states that V_c shall be taken equal to zero if

(a) the shear V_{sway} due to plastic hinging at the two ends of the beam exceeds half or more of the maximum shear, V_u, within the span; and

(b) the factored axial compression force, including earthquake effects, is less than $A_g f'_c/20$.

Otherwise, V_c has the regular value $V_c = 2\lambda\sqrt{f'_c}\, b_w d$—or the effects of axial load on V_c can be included by using ACI Eq. (22.5.6.1).

The end reactions for gravity loads are 88 kips upward at both ends of the beam, independently of the sway direction. For sway to the right, shown in Figs. 19-21a and 19-21b, the shear due to sway moments plus the gravity reactions, total 53.3 kips at A and 123 kips at B. $V_{sway} = 34.7$ kips does not exceed half of the required maximum shear strength within the span. For sway to the left (see Figs. 19-21c and 19-21d), $V_{sway} = 28.4$ kips, and the total end reactions are 59.6 kips and 116 kips. Again, V_{sway} does not exceed half of the required maximum shear strength within the span. Thus, at all sections, V_c takes its usual value for beams.

Exterior end: Maximum shear $V_u = 116$ kips

$$V_c = 2\lambda\sqrt{f'_c}\, b_w d = 65.3 \text{ kips}$$

$$V_s = V_u/\phi - V_c = 116/0.75 - 65.3 = 89.4 \text{ kips}$$

ACI Code Section 22.5.1.2 sets the maximum $V_s = 8\sqrt{f'_c}\, b_w d = 261$ kips—o.k. Then

$$\frac{A_v}{s} = \frac{89.4 \text{ kips}}{f_{yt}\, d} = 0.069 \text{ in.}^2/\text{in.}$$

Try No. 4, two-leg stirrups, $A_v = 0.40 \text{ in.}^2$, and $s = 0.40/0.069 = 5.80$ in. The hoops will be selected after the confinement stirrups are calculated.

Hoops for confinement: ACI Code Section 18.6.4.1 requires hoops over a distance of $2h = 48$ in. measured from the face of the column. Every corner and alternate longitudinal bars in the beam must be at the corner of a stirrup in accordance with ACI Code Section 25.7.2.3. The stirrups and hoops have the shapes shown in Fig. 19-22. ACI Code Section 18.6.4.4 requires the first hoop to be 2 in. from the face of the column and the subsequent spacing of hoops to be the smallest of:

(a) $d/4 = 5.38$ in.

(b) 6 times the minimum primary longitudinal bar diameter $= 6 \times 0.75$ in. $=$ 4.5 in.

(c) 6 in.

Try No. 4 two-leg hoops, one at 2 in. from the face of the column at each end, plus twelve at 4 in. o.c., total 50 in. each end. The plastic hinge zone ends at 48 in. from the face of the support.

Interior end: $V_u = 123$ kips

$$V_s = V_u/\phi - V_c = 123/0.75 - 65.3 = 99 \text{ kips}$$

Using No. 4 two-leg stirrups at a spacing of 4-in. gives,

$$V_s = \frac{A_v f_{yt} d}{s}$$

$$= \frac{0.40 \text{ in.}^2 \times 60 \text{ ksi} \times 21.5 \text{ in.}}{4 \text{ in.}} = 129 \text{ kips (o.k.)}$$

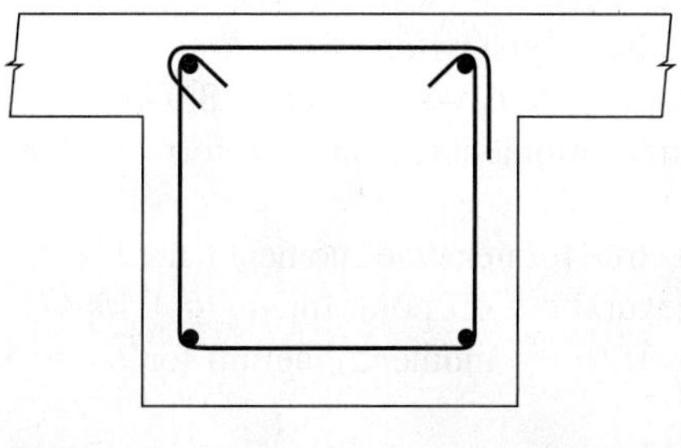

Fig. 19-22
No. 4 two-leg closed hoop—Example 19-1.

The maximum shear, V_u, at the end of the hinging region, 48 in. from the interior end, is 123 kips $-$ 4 ft $\times$ 8 kips/ft $=$ 123 kips $-$ 32 kips $=$ 91 kips. Also,

$$V_s = V_u/\phi - V_c = (91/0.75) - 65.3 = 56.0 \text{ kips}$$

From Eq. (6-14),

$$\frac{A_v}{s} = \frac{V_s}{f_{yt}d} = \frac{56}{60 \times 21.5} = 0.0434 \text{ in.}^2/\text{in.}$$

For No. 4 two-leg stirrups, $0.40/0.0434 = 9.22$ in. Maximum spacing from ACI Code Section 18.6.4.6 is $d/2 = 10.75$ in.

Use No. 4 two-leg hoops, one at 2 in. from the face of the column at each end, plus twelve at 4 in. o.c., plus ten No. 4 two-leg closed stirrups at 8 in.

ACI Code Section 18.6.3.1 requires that at least two bars be made continuous, top and bottom. At the top we will accomplish this by lap splicing two No. 7 bars. The moment in the midspan region will either be positive or a very small negative value. Thus, the stress in the top bars will either be compression or a very small tension value. For these conditions, ACI Code Table 25.5.2.1 allows a Class A lap splice of length 1.0 ℓ_d, where $\ell_d = 61.7\, d_b = 54.0$ in (Table A-6). ACI Section 18.6.3.3 requires that the lap splice be enclosed in hoops spaced at a maximum spacing of the smaller of $d/4 = 5.38$ in. or 4 in. Thus, we will provide No. 4 closed hoops at a 4 in. spacing along the 54.0 in. length of the lap splice for the top No. 7 bars. Sufficient continuous bottom steel is already in place to satisfy ACI Code Section 18.6.3.1. Any possible splicing of bottom steel should not be done at midspan because this is the location of maximum positive bending, i.e., maximum tension stress in the bottom bars.

Summary — From each column face use No. 4 three-leg hoops, one at 2 in. from the face of the column, plus twelve at 4 in. (covers plastic hinging zones), plus six No. 4 two-leg closed stirrups at 8 in. Then, in the midspan region use seventeen No. 4 two-leg stirrups at a 4 in. spacing. (Fig. 19-20)

Summary—Lap splice two No. 7 top bars at midspan for 54 in.

8. Compute the cutoff points for the flexural reinforcement. The bar cutoffs are calculated by assuming that the ends of the beam are hinging at a moment of $\pm M_{pr}$, due to the frame swaying to the right, and that the moment at the cut-off point is ϕM_n for the bars remaining after the cut-off point.

Figure 19-23 shows one-half of an interior span of a bending-moment diagram between midspan at A and the right-hand support at H. If the beam is loaded only with concentrated loads at midspan and the supports, the moment diagram is the sloping straight line A–H. If the beam supports a factored uniform load of $w_u = 8.0$ kips/ft from Load Case (b), the moment diagram is curved, as shown by the dashed line in Fig. 19-23. Line B–C–D represents a group of bars with a moment strength corresponding to the height 0–B.

If $w_u = 0$, the bars must extend from B to C, with the flexural cut-off point at C. If there is a uniform load of w_u—in this case, 8.0 kips/ft—the cut-off point is at D. This shows that, in a positive-moment region, a uniform load increases the length to the cutoff by the distance C–D.

The opposite is true for negative-moment bars: line G–E represents the distance from the support to the flexural cut-off point for $w_u = 0$ kips/ft. Line G–F represents the distance from the support to the moment diagram for $w_u = 8.0$ kips/ft. The distance from

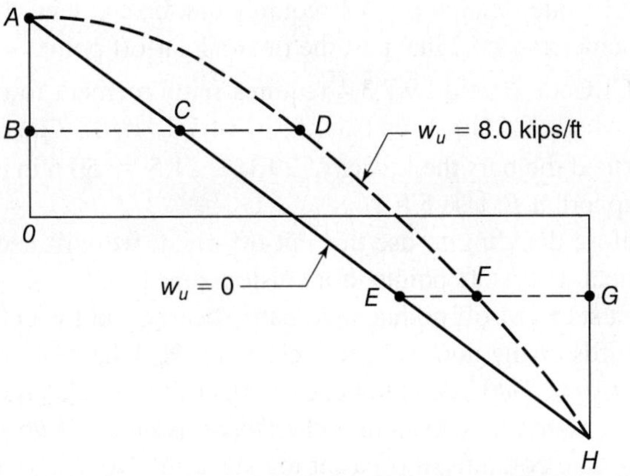

Fig. 19-23
Effect of uniform load on cut-off points.

G to F is shorter than the distance from G to E. This shows that the negative-moment bars must extend farther from the support if there is no uniform load w_u than they would have to if the full w_u were present.

Frame swaying to the right—top (negative-moment) bars at interior support. There are four No. 8 bars plus two No. 7 top bars at the interior (right) support. (See Fig. 19-20.) These are in tension when the frame sways to the right and the ACI Code Section 18.6.3.1 requires two continuous bars top and bottom. We will use two of the No. 7 top bars as continuous bars; two of No. 8 bars will be cut off when they are no longer needed for negative moment at the right end. The probable moment capacity right end support, M_{pr} is -531 kip-ft. The negative-moment capacity after the bars are cut off is $\phi M_n = -256$ kip-ft (two No. 8 and two No. 7). Figure 19-24 is a free-body diagram of the interior end of the span. The uniform load has been taken as 8.0 kips/ft., and x is the distance in feet from the right end of the beam to the flexural cut-off point.

Summing moments about O gives

$$4x^2 - 123x + 275 = 0$$

$$x = \frac{-b \pm \sqrt{b^2 - 4ac}}{2a} = 2.43 \text{ ft}$$

The theoretical flexural cut-off point for the two No. 8 bars is located 2.43 ft = 29.1 in. from the face of the support.

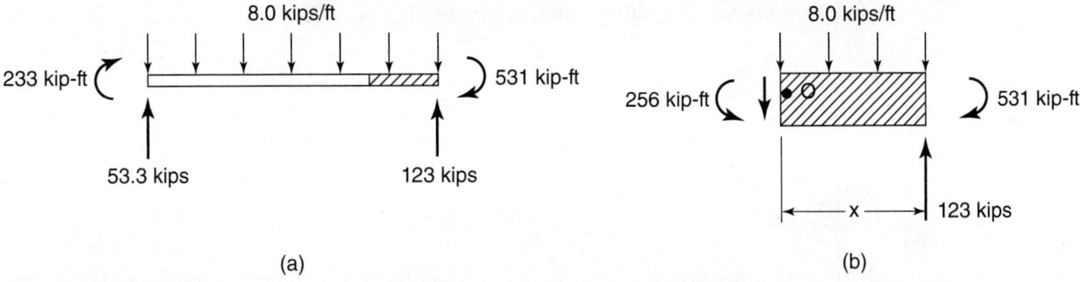

Fig. 19-24
Calculation of cut-off points—frame swaying to the right—Example 19-1.

ACI Code Section 9.7.3.3 requires that bars extend the longer of $d = 21.5$ in., and 12 bar diameters ($=12$ in.) past the flexural cut-off point.

ACI Code Section 9.7.3.4 requires reinforcement to extend ℓ_d from the face of the column, where, for No. 8 top bars, $\ell_d = 61.7d_b = 61.7$ in. (Table A-6).

Extend the bars the longer of $29.1 + 21.5 = 50.6$ in., or $\ell_d = 61.7$ in. past the face of the support at B. Use 5 ft 2 in.

Before deciding to use this cut-off point, we will need to check the shear strength requirement at cut-off points in a tension zone. ACI Code Section 9.7.3.5 requires extra stirrups at such cut-off points, unless the shear, V_u, at the cut-off point is less than or equal to two-thirds of the nominal shear capacity, V_n, where $V_n = V_c + V_s$. A quick check will show that $(0.667)\phi V_n$ does not exceed V_u at the calculated cut-off point for the top No. 8 bars. Thus, rather than extending the closer spacing of two-leg No. 4 stirrups further from the face of the columns, it is easier to extend the No. 8 bars past the point of inflection before terminating them, following the requirements of ACI Code Section 9.7.3.8.4. A similar decision was made when determining the cut-off points for the negative- and positive-moment reinforcement at other points in the span.

19-8 COLUMNS IN SPECIAL MOMENT FRAMES

ACI Code Section 18.7 applies to columns in special moment frames resisting earthquake-induced axial forces, moments, and shear. Columns in frames in regions of high seismic risk must satisfy two geometric requirements: the smallest dimension through the centroid of the column must be at least 12 in., and the ratio of the shortest to the longest cross-sectional dimension shall not be less than 0.4. These limits ensure a minimum robustness and produce a cross section that can be confined using practical hoop layouts, which would be difficult with highly rectangular columns.

Required Capacity and Longitudinal Reinforcement

Because gravity loads must be transferred through the columns before and after a major seismic event, damage to the columns should be minimized. As discussed previously in Section 19-6, the use of a strong-column–weak-beam design philosophy, expressed in Eq. (19-15) from ACI Code Section 18.7.3.2, is normally used to achieve this goal.

$$\Sigma M_{nc} \geq (6/5)\Sigma M_{nb} \tag{19-15}$$

M_{nc} and M_{nb} are the nominal moment strengths of the column and beam sections, evaluated at the faces of a beam-to-column joint, as shown in Fig. 19-25.

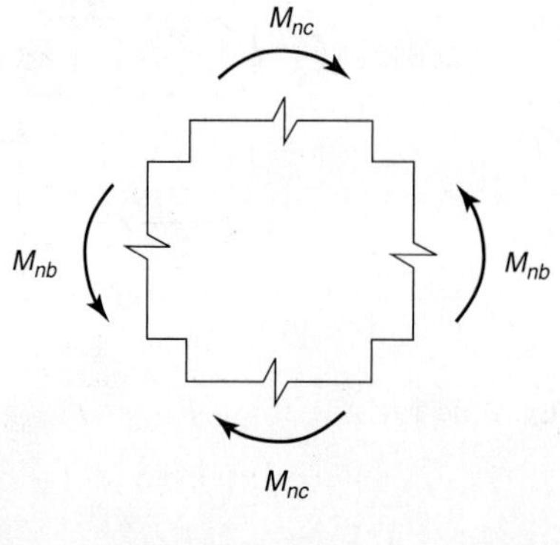

Fig. 19-25
Nominal moment strengths at beam-column joint.

If the column flexural strength requirement in Eq. (19-15) is not achieved at a joint, the strength and stiffness of the columns in question are to be disregarded when calculating the lateral strength and stiffness of the structure. Further, those columns must have transverse reinforcement satisfying ACI Code Sections 18.7.5.2 through 18.7.5.4 over their entire length.

Longitudinal reinforcement is designed for the axial loads and moments in the same way as in a nonseismic column. It may range from $\rho = 0.01$ to 0.06. Generally, it is difficult to place and splice much more than 2 to 3 percent reinforcement in a column.

Because the cover concrete will probably spall in plastic hinging regions, which may form near the ends of the column during a major seismic event, longitudinal bars that are to be lap spliced must be spliced in the center half of the column height. Such splices must be designed as tension splices, because the alternating moments due to the cyclic loads alternately stress the bars on each side of the column in tension and compression. Furthermore, there is frequently a possibility of uplift forces. The considerable length required for tension lap splices may require small-diameter bars or mechanical splices.

Transverse Reinforcement

Confinement Reinforcement

Transverse reinforcement in the form of spirals or hoops must be provided over a height of ℓ_o from each end of the column to confine the concrete and restrain the longitudinal bars from buckling. The height ℓ_o is the largest of (ACI Code Section 18.7.5.1)

(a) the depth of the column, h at the face of the joint,
(b) one-sixth of the height of the column, and
(c) 18 in.

Within the length ℓ_o, ACI Code Section 18.7.5.3 requires that the spacing of the transverse reinforcement shall not exceed

(a) one-quarter of the minimum dimension, b or h, of the column cross section;
(b) six times the diameter of the longitudinal bar diameter; and
(c) the distance

$$s_o = 4 + \left(\frac{14 - h_x}{3}\right) \qquad (19\text{-}20)$$

(ACI Eq. 18.7.5.3)

where h_x is the spacing between longitudinal bars that are held by hoop or crosstie legs (Fig. 19-14b). ACI Code Section 18.7.5.2 states that the value of h_x shall not exceed 14 in. and Code Section 18.7.5.3 states that s_o shall not exceed 6 in. and need not be taken less than 4 in. The hoops and crossties must also satisfy ACI Code Section 25.7.2.3, which requires every corner bar and alternate longitudinal bar to be held by the corner of a hoop or crosstie.

Requirements for the minimum amount of transverse reinforcement in columns of special moment frames are given in ACI Code Table 18.7.5.4. Minimum areas are given for the use of either a combination of rectilinear hoops and crossties, or the use of spirals (or circular hoops). The minimum required areas for both types of transverse reinforcement are a function of the level of the axial load and the concrete strength. When using concrete strengths greater than 10,000 psi, ACI Code Section 18.7.5.4 defines a concrete strength factor, k_f, as:

$$k_f = \frac{f'_c}{25{,}000} + 0.6 \geq 1.0 \qquad (19\text{-}21)$$

The same Code section also defines a confinement effectiveness factor, k_n, as:

$$k_n = \frac{n_\ell}{n_\ell - 2} \tag{19-22}$$

where n_ℓ is defined as the number of bars around the perimeter of a column core that are supported by legs of hoops or crossties. Referring to Fig. 19-14b, if all of the column faces were similar to the one shown in the figure, n_ℓ would be equal to $4 \times 3 = 12$, and $k_n = 12/10$.

For a combination of rectilinear hoops and crossties, the area of legs confining the concrete core in one direction is given as A_{sh}, as defined in Fig. 19-14b. The core dimension perpendicular to the legs defining A_{sh} is shown as b_c in Fig. 19-14b, measured outside edge to outside edge of the perimeter hoop. Assuming spacing, s, along the column vertical axis between sets of hoops and crossties, ACI Code Table 18.7.5.4 defines the required minimum area of transverse reinforcement as a ratio, $A_{sh}/(s \cdot b_c)$, which must be checked in both principal directions of the column section. When the design axial load is approximately at or below the balanced point ($P_u \le 0.3A_g f'_c$), and when concrete strengths are less than or equal to 10,000 psi, only the following two requirements are to be satisfied:

$$\frac{A_{sh}}{s \cdot b_c} \ge 0.3\left(\frac{A_g}{A_{ch}} - 1\right)\frac{f'_c}{f_{yt}} \tag{19-23}$$

and

$$\frac{A_{sh}}{s \cdot b_c} \ge 0.09\frac{f'_c}{f_{yt}} \tag{19-24}$$

where A_g is the gross area of the column section and A_{ch} is the area of the concrete core, taken as the product of the two perpendicular core dimensions, b_c. When either the design axial load is likely to be above the balanced point ($P_u > 0.3A_g f'_c$), i.e., in the compression-controls region of behavior, or the column concrete strength exceeds 10,000 psi, a third minimum area requirement must also be satisfied.

$$\frac{A_{sh}}{s \cdot b_c} \ge 0.2k_f k_n \frac{P_u}{A_{ch} f_{yt}} \tag{19-25}$$

For columns with spiral transverse reinforcement, the required amount of transverse reinforcement is expressed as a reinforcement ratio, ρ_s, which was defined in Chapter 11 as the volume of spiral reinforcement divided by the volume of core concrete, both measured for one complete turn of the spiral around the core as,

$$\rho_s = \frac{4A_{sp}}{s \cdot D_c} \tag{11-21}$$

where A_{sp} is the area of the spiral bar, D_c is the diameter of the core measured to the outside edges of the spiral, and s is the spacing or pitch of the spiral for one turn around the core. As with the rectilinear transverse reinforcement, for low axial load ($P_u \le 0.3A_g f'_c$) and for concrete strengths less than 10,000 psi, only two requirements are to be satisfied,

$$\rho_s \ge 0.45\left(\frac{A_g}{A_{ch}} - 1\right)\frac{f'_c}{f_{yt}} \tag{19-26}$$

and

$$\rho_s \ge 0.12\frac{f'_c}{f_{yt}} \tag{19-27}$$

where A_g is the gross area of the column section and A_{ch} is the area of the concrete core. When either the design axial load exceeds $0.3A_g f'_c$ or the column concrete strength exceeds 10,000 psi, a third requirement must also be satisfied

$$\rho_s \geq 0.35k_f \frac{P_u}{A_{ch} f_{yt}} \tag{19-28}$$

Because the outward pressure against the side of rectilinear transverse reinforcement tends to cause the perimeter legs to deflect (Fig. 19-14a), hoops are less efficient that spirals at confining the concrete core. Thus, although the expressions given above are similar for the two types of transverse reinforcement, the requirements for rectilinear transverse reinforcement result in use of approximately one-third more reinforcement than used for spirals.

Columns supporting discontinued shear walls are extremely susceptible to seismic damage. ACI Code Section 18.7.5.6 requires hoops or spirals over the full height of such members. These hoops must extend into the wall from the face of the column and the footing or other member under the column.

Shear Reinforcement

The transverse reinforcement also must be designed for shear. The design shear force V_{sway} is computed by assuming inelastic action in either the columns or the beams.

(a) The shear corresponding to plastic hinges at each end of the column is given by

$$V_{sway} = \frac{M_{prc,\,top} + M_{prc,\,btm}}{\ell_u} \tag{19-29}$$

where $M_{prc,\,top}$ and $M_{prc,\,btm}$ are the probable moment capacities at the top and bottom of the column and ℓ_u is the clear height of the column. These are obtained from an interaction diagram for the probable strength, $P_n - M_{pr}$ of the column, using the factored loads on the member for the load combination under consideration.

(b) It need not be more than

$$V_{sway} = \frac{\Sigma M_{prb,\,top} DF_{top} + \Sigma M_{prb,\,btm} DF_{btm}}{\ell_u} \tag{19-30}$$

where $\Sigma M_{prb,\,top}$ and $\Sigma M_{prb,\,btm}$ are the sum of the probable moment capacities of the beams framing into the joints at the top and bottom of the column for the frame swaying to the left or right, and DF_{top} and DF_{btm} are the moment-distribution factors at the top and bottom of the column being designed. This reflects the strong-column–weak-beam philosophy and Eq. (19-15), which requires that the beams are weaker than the columns.

(c) but not less than the factored shear from a frame analysis.

Transverse reinforcement is designed for shear according to ACI Code Section 22.5.1, and V_c may be increased to allow for the effect of axial loads, except that, within the length ℓ_o, defined in the discussion of confinement reinforcement, V_c shall be taken equal to zero when the earthquake-induced shear force makes up half or more of the maximum shear force in the lengths ℓ_o and *if* the factored compression force is less than $A_g f'_c/20$ (ACI Code Section 18.7.6.2.1). Columns with such low axial loads essentially behave like a beam. Thus, the concrete contribution to shear, V_c, is set equal to zero in potential plastic hinging zones at the ends of a column, just as was done for plastic hinging zones at the end of beams.

It should be noted that, although the axial load increases V_c, it also increases the rate of shear degradation [19-17]. For this reason, it may be prudent to ignore V_c when a major portion of the shear results from earthquake loads.

EXAMPLE 19-2 Design of a Column

The column supporting the interior end of the beam designed in Example 19-1 is 24 in. square and is constructed of 4000-psi normal-weight concrete and 60,000-psi for longitudinal and transverse steel. The floor-to-floor height is 12 ft, with 24-in.-deep beams in each floor, giving a clear column height of 10 ft. The column size and the floor-to-floor heights are the same in the stories above and below the column being designed. The unfactored moments, shears, and axial loads from an elastic analysis for earthquake loads are given in Table 19-4. Design the reinforcement in the column. Use load factors from ACI Code Table 5.3.1.

1. Select the design procedure, method of analysis, and the load combinations to be used. In some building codes the design method depends on the size of the design earthquake and it is necessary to choose the method to be used to analyze the seismic effects.

2. Compute the factored loads and moments. An examination of the relative size of the loads suggested that the governing load combinations (a) and (b), and the two seismic load combinations, (e) and (g) in ACI Code Table 5.3.1.

Load combination (a) $U = 1.4D$

Load combination (b) $U = 1.2D + 1.6L$

Load combination (e) $U = 1.2D + 1.0E + 0.5L$

Load combination (g) $U = 0.9D + 1.0E$

The calculations are summarized in Table 19-5.

3. Does the column satisfy the definition of a column in a special moment frame? ACI Code Sections 18.7.1 and 18.7.2 lists three requirements for a member to be designed as a column in a special moment frame:

(a) Column resists earthquake-induced forces—o.k.

(b) Shortest cross-sectional dimension is not less than 12 in.—o.k.

(c) Ratio of cross-sectional dimensions is not less than 0.4—o.k.

Therefore, design the column according to ACI Code Section 18.7. If these were not satisfied, it would be necessary to modify the column dimensions.

4. Initial selection of column steel. As a first trial, we shall select a 24 × 24 in. column with twelve No. 8 bars, $A_{st} = 9.48$ in.^{2}:

$$\rho_g = \frac{9.48}{24 \times 24} = 0.0165$$

TABLE 19-4 Unfactored Axial Forces, Moments, and Shears in Column

	Dead Load	Live Load	Earthquake
Axial load (kips)			
Column in story above	510	140	±5
Column being designed	560	154	±5
Column in story below	610	168	±6
Moments (kip-ft[a])			
Top of column	−4	−1	±195
Bottom of column	−4	−1	±210
Shears (kips)	0	0	40

[a]Counterclockwise moment on the end of a member is positive.

TABLE 19-5 Factored Forces and Moments on Columns

	Axial Load (kips)	Top Moment[a] (kip-ft)	Bottom Moment (kip-ft)	Shear (kips)
Column in story above				
Load combination (a)	714			
Load combination (b)	836			
Load combination (e)	687			
Load combination (g)	454			
Column being designed				
Load combination (a)	784	−6	−6	
Load combination (b)	918	−6	−6	
Load combination (e)				
Sway to right	749 + 5	195 − 5	210 − 5	40
Sway to left	749 − 5	−195 − 5	−210 − 5	40
Load combination (g)				
Sway to right	504 + 5	195 − 5	210 − 5	40
Sway to left	504 − 5	−195 − 5	−210 − 5	40
Column in story below				
Load combination (a)	854			
Load combination (b)	1000			
Load combination (e)	822			
Load combination (g)	543			

[a]Counterclockwise moment on the end of the column is positive.

ACI Code Section 18.7.4.1 limits ρ_g to not less than 0.01 or more than 0.06—o.k. No. 8 bars were chosen to avoid excessive splice lengths. Interaction diagrams for $\phi P_n - \phi M_n$ and for $P_n - M_{pr}$ are given in Fig. 19-27.

5. Check whether the column strengths satisfy $\Sigma M_{nc} \geq 1.2\,\Sigma M_{nb}$. ACI Code Section 18.7.3.2 requires that the nominal flexural strengths of the columns satisfy

$$\Sigma M_{nc} \geq 1.2\,\Sigma M_{nb} \tag{19-15}$$

where ΣM_{nc} is the sum of the nominal flexural strengths for the two columns meeting at a floor joint corresponding to the factored axial loads in the columns and ΣM_{nb} is the sum of the nominal flexural strengths of the beams meeting at the joint.

For the frame swaying to the right, the nominal flexural strength values at the ends of the beam meeting at the top of the column can be obtained from the ϕM_n values calculated in step 4 of Example 19-1, by dividing those values by $\phi = 0.9$. The resulting M_{nb} values are shown in Fig. 19-26a, and are used here to obtain,

$$\begin{aligned} 1.2\Sigma(M_{nb}) &= 1.2 \times (231 \text{ kip-ft} + 433 \text{ kip-ft}) \\ &= 797 \text{ kip-ft} \end{aligned}$$

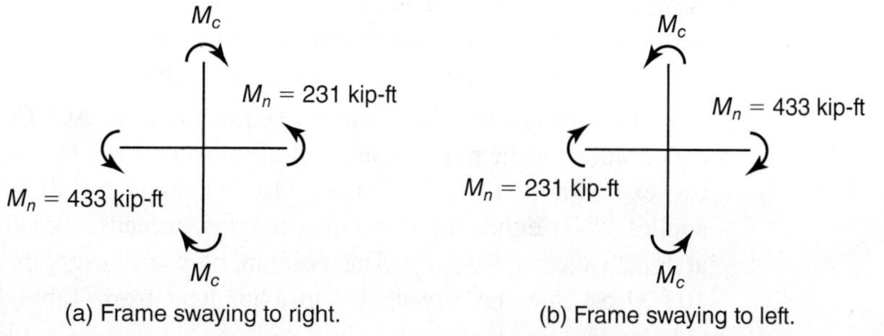

Fig. 19-26
Moments at a beam–column joint—Example 19-2.

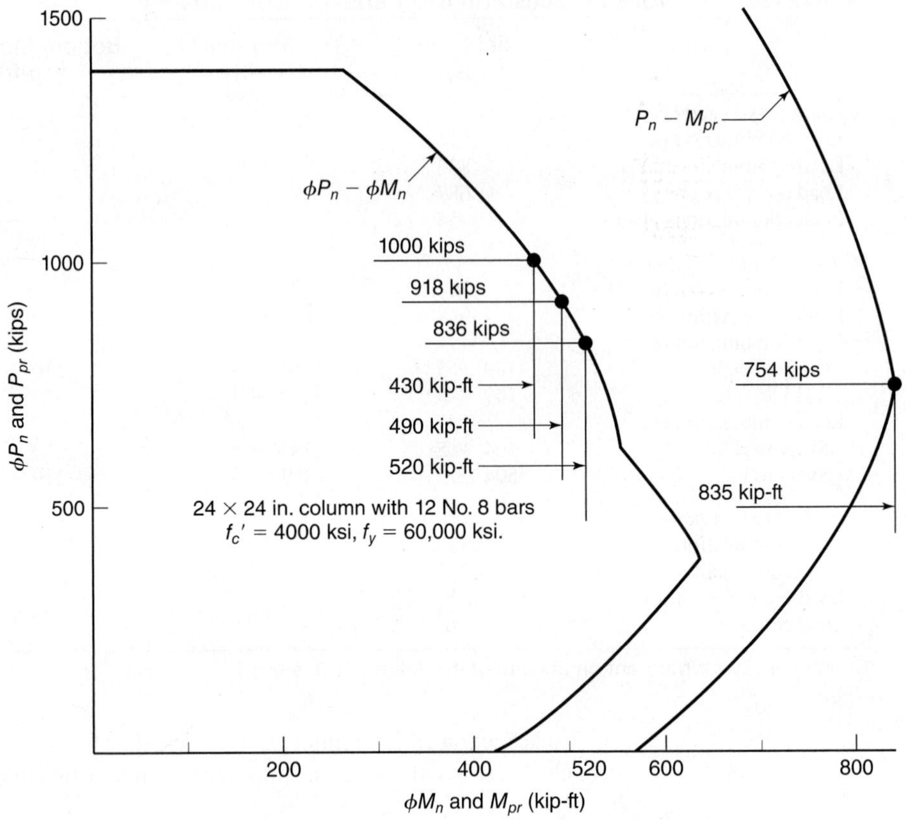

Fig. 19-27
Interaction diagrams for ϕP_n–ϕM_n and for P_n–M_{pr}.

For load combination (b), the axial load in the column in the story above the column being designed is 836 kips. From the interaction diagram for ϕP_n–ϕM_n in Fig. 19-27, the moment strength corresponding to an axial load of 836 kips is $\phi M_n = 520$ kip-ft. The axial load in the column being designed for the same load combination is 918 kips, corresponding to a moment strength $\phi M_n = 490$ kip-ft. For both of these points $\phi = 0.65$. Thus, at the top of the column,

$$\Sigma M_{nc} = (520 + 490)/0.65 = 1550 \text{ kip-ft}$$

This column cross section satisfies the requirement that $\Sigma M_{nc} \geq 1.2\Sigma M_{nb}$ at the top of the column.

Assuming that the beams at the bottom of the column are the same as those at the top, $1.2\Sigma M_{nb} = 797$ kip-ft. The reduced moment strength of the column being designed is 490 kip-ft. The axial load in the column in the story below the column being designed is 1000 kips, corresponding to a moment strength $\phi M_n = 430$ kip-ft. Here, $\Sigma M_{nc} = (430 + 490)/0.65 = 1490$ kip-ft, which is still greater than $1.2\Sigma M_{nb}$—o.k.

6. Design the confinement reinforcement. ACI Code Table 18.7.5.4 gives three expressions for the required amount of transverse reinforcement in tied columns. The first two expressions, Eqs. (19-23) and (19-24), apply for all tied columns. Equation (19-25) applies when either the concrete strengths exceeds 10,000 psi or when the design axial load exceeds $0.3A_g f_c'$. The column concrete strength (4000 psi) does not exceed 10,000 psi, but the largest design axial load from Table 19-3 (918 kips) does exceed

$0.3A_g f'_c$ ($0.3 \times 24^2 \times 4 = 691$ kips). For Eq. (19-25), we need to calculate values for k_f and k_n using Eqs. (19-21) and (19-22).

$$k_f = \frac{f'_c}{25{,}000} + 0.6 = \frac{4000}{25{,}000} + 0.6 = 0.76 \text{ (use 1.0)} \tag{19-21}$$

For Eq. (19-22), $n_\ell = 12$, because all of the bars are supported by a corner of the transverse reinforcement (Fig. 19-28).

$$k_n = \frac{n_\ell}{n_\ell - 2} = \frac{12}{12 - 2} = 1.2 \tag{19-22}$$

The dimension of the core in both directions is, $b_c = 24 - 2 \times 1.5 = 21$ in. Using that value, we will solve Eqs. (19-23) through (19-25) for the ratio of the required area divided by the spacing,

$$\frac{A_{sh}}{s} \geq 0.3 \times b_c\left(\frac{A_g}{A_{ch}} - 1\right)\frac{f'_c}{f_{yt}} = 0.3 \times 21 \text{ in.}\left(\frac{24 \times 24}{21 \times 21} - 1\right)\frac{4 \text{ ksi}}{60 \text{ ksi}}$$

$$\frac{A_{sh}}{s} \geq 0.129 \text{ in}^2\text{/in.}$$

and

$$\frac{A_{sh}}{s} \geq 0.09 \times b_c \times \frac{f'_c}{f_{yt}} = 0.09 \times 21 \text{ in.} \times \frac{4 \text{ ksi}}{60 \text{ ksi}}$$

$$\frac{A_{sh}}{s} \geq 0.126 \text{ in.}^2\text{/in.}$$

and

$$\frac{A_{sh}}{s} \geq 0.2k_f k_n \frac{b_c \cdot P_u}{A_{ch} f_{yt}} = 0.2 \times 1.0 \times 1.2 \frac{21 \text{ in.} \times 918 \text{ k}}{441 \text{ in.}^2 \times 60 \text{ ksi}}$$

$$\frac{A_{sh}}{s} \geq 0.175 \text{ in.}^2\text{/in. (this governs)}$$

ACI Code Section 18.7.5.3 sets the maximum spacing as the smaller of

(a) 0.25 times the minimum cross-sectional dimension, 0.25×24 in. $= 6$ in.

(b) Six times the longitudinal bar diameter, 6×1 in. $= 6$ in.

(c) the distance:

$$s_o \leq 4 + \frac{14 - h_x}{3} \tag{19-20}$$

From Fig. 19-28, h_x is approximately one-third of $h - 2(2.5'')$, i.e., 0.333×19 in. $= 6.33$ in. (which is less than the maximum permissible spacing of 14 in.). With this value for h_x, $s_o = 6.6$ in., which is above the limit of 6 in. Thus, use $s = 6$ in.

For $s = 6$ in., the required $A_{sh} = 0.175 \times 6 = 1.05$ in.^{2} Using No. 5 hoops arranged as shown in Fig. 19-28 gives

$$A_{sh} = A_b\left(2 + 2\frac{\sqrt{2}}{2}\right) = 0.31 \text{ in.}^2 (3.414) = 1.06 \text{ in.}^2 \text{ (o.k.)}$$

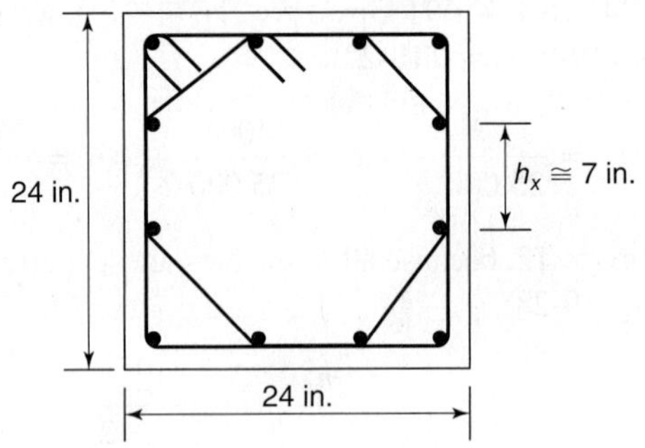

Fig. 19-28
Hoops—Example 19-2.

Thus, use No. 5 hoops at $s = 6$ in. arranged as shown in Fig. 19-28.

ACI Code Section 18.7.5.1 requires hoop reinforcement over a length ℓ_o adjacent to each end of the column, where ℓ_o is the largest of

(a) the depth of the member at the joint face: 24 in.,
(b) one-sixth of the clear height of the column, or 120 in./6 = 20 in., and
(c) 18 in.

Thus, $\ell_o = 24$ in. Use $s = 6$ in. within ℓ_o and throughout the rest of the height of the column, ACI Code Section 18.7.5.5 permits hoops at 6 in.

7. **Design the shear reinforcement.** The design shear force V_e shall be:

(a) The shear corresponding to plastic hinges at each end of the column, given by

$$V_{\text{sway}} = \frac{M_{prc,\,\text{top}} + M_{prc,\,\text{btm}}}{\ell_u} \tag{19-29}$$

(b) Need not be more than

$$V_{\text{sway}} = \frac{\Sigma M_{prb,\,\text{top}} DF_{\text{top}} + \Sigma M_{prb,\,\text{btm}} DF_{\text{btm}}}{\ell_u} \tag{19-30}$$

(c) Not less than the factored shear from a frame analysis.

For load combination (e), the corresponding factored axial loads in the column being designed are 744 and 754 kips for sway to the left and right, respectively. From the interaction diagram for $P_n - M_{pr}$ in Fig. 19-27, the maximum value of M_{pr} for the column is 835 kip-ft. Substituting into Eq. (19-29) gives

$$V_{\text{sway}} = \frac{835 \text{ kip-ft} + 835 \text{ kip-ft}}{10 \text{ ft}} = 167 \text{ kips}$$

From step 5 of Example 19-1, the probable moment capacities of the beams framing into the joints at the top and bottom of the interior column are 531 kip-ft and 286 kip-ft. Because the columns in the stories above and below and the column being designed all have the same stiffness, DF_{top} and DF_{btm} are both 0.5. Substituting into Eq. (19-30) gives

$$V_{\text{sway}} = \frac{(531 + 286) \text{ kip-ft} \times 0.5 + (531 + 286) \text{ kip-ft} \times 0.5}{10 \text{ ft}} = 81.7 \text{ kips}$$

The shear V_e shall not be less than the factored shear from the analysis, 40 kips. Therefore, the design shear $V_e = 81.7$ kips.

As stated earlier, it may be prudent to set V_c equal to zero if the earthquake-induced shear represents half or more of the total design shear, regardless of axial loads. For the column being designed, this is true; hence, V_c is set equal to zero. We thus have

$$V_s = \frac{V_u}{\phi} - V_c, V_c = 0$$

$$V_u = V_e = 81.7 \text{ kips}$$

Therefore,

$$V_s = \frac{81.7}{0.75} = 109 \text{ kips}$$

$$\frac{A_v}{s} = \frac{V_s}{f_{yt}d} = \frac{109}{60 \times 21.5}$$

$$= 0.0845 \text{ in.}^2/\text{in.}$$

For $s = 6$ in., $A_v = 0.51$ in.^{2}. The hoops for confinement have $A_v = 1.06$ in.^{2}—o.k. Outside of the lengths ℓ_o, V_c is given by ACI Eq. (22.5.6.1):

$$V_c = 2\left(1 + \frac{N_u}{2000A_g}\right)\lambda\sqrt{f'_c}\, b_w d \qquad \text{(6-13a)}$$

where N_u is conservatively taken as the lowest value from the load combinations. Thus, $N_u = 499$ kips.

$$V_c = 2\left(1 + \frac{499 \text{ kips} \times 1000}{2000 \times 576 \text{ in.}^2}\right)1.0\sqrt{4000} \times 24 \text{ in.} \times 21.5 \text{ in.}$$

$$= 93{,}500 \text{ lbs.} = 93.5 \text{ kips}$$

Because V_c exceeds V_u outside the length ℓ_o, stirrups are only required at a spacing of $d/2$. Thus, spacing will be controlled by confinement requirements.

Provide No. 5 hoops as shown in Fig. 19-28 at 2 in. from end of column and five at 6 in. on centers at each end; provide similar No. 5 hoops at 6 in. on centers over the rest of the height.

8. Design lap splices for the column bars. ACI Code Section 18.7.4.3 requires that splices be in the midheight of the column, be designed as tension splices, and be enclosed within transverse reinforcement conforming to ACI Code Sections 18.7.5.2 and 18.7.5.3. Thus, over the length of the lap splice the spacing between the layers of transverse reinforcement must be limited to 6 in. o.c.

ACI Code Table 10.7.5.2.2 requires a Class B tension lap splice if all the bars are spliced at the same location. From Table A-6 for a vertical No. 8 bar,

$$\ell_d = 47.4 \text{ in.}$$

A Class B splice has a length of $1.3\ell_d = 1.3 \times 47.4$ in. $= 61.7$ in. ACI Code Section 10.7.5.2.1 allows this length to be multiplied by 0.83 if the ties throughout the splice length have an effective area of not less than $0.0015hs$, which, for the $s = 6$ in. is 0.22 in.^{2}. The hoops have an area of 1.06 in.^{2}; therefore, they are adequate to allow this reduction. The lap length becomes 0.83×61.7 in. $= 51.2$ in.—say, 4 ft 4 in.

Lap splice all vertical bars with a 4 ft 4 in. lap splice at midheight of the column. ■

19-9 JOINTS OF SPECIAL MOMENT FRAMES

The flow of forces within beam–column joints and the design of such joints has been discussed in Section 17-11, and an example of the design of an exterior nonseismic joint is given in Example 17-6. Code provisions for joints in special moment-resisting frames (SMFs) are given in ACI Code Section 18.8. A more extensive discussion of design recommendations for beam–column connections is given by ACI Committee 352 [19-18].

ACI Code Section 18.8.2.1 requires that joint forces be calculated by taking the stress in the flexural reinforcement in the beams as $1.25f_y$. This is analogous to using the probable strength in the calculations of shear in columns and beams in special-moment frames.

ACI Code Section 18.8.2.3 limits the diameter of the longitudinal beam reinforcement that passes through a joint to $\frac{1}{20}$ of the width of the joint parallel to the beam bars. When hinges form in the beams, the beam reinforcement is stressed to the actual yield strength of the bar on one side of the joint and is stressed in compression on the other side. This results in very large bond stresses in the joint, possibly leading to slipping of the bar in the joint. The minimum bonded length of such a bar in a joint is thus $20d_b$, which is considerably less than is required by the development-length equations in ACI Code Section 25.4. The minimum bonded length was selected from test results of joints tested under cyclic loads to limit, but not entirely eliminate slip of the beam bar in the joint. Based on research by Leon [19-19] and others, better stiffness retention and energy dissipation is obtained for interior beam–column connections when the column dimension is increased to at least 24 times the diameter of beam bars passing through the joint.

ACI Committee 352 [19-18] uses the same limit for the diameter of column bars passing through a beam–column joint (i.e., the column–bar diameter should not exceed 1/20 of the overall depth of the shallowest beam framing into the joint). Although this design recommendation has not been adopted by the ACI Code Committee, based on research results [19-20], the author recommends that designers should attempt to satisfy this limit when selecting the size of column bars.

ACI Code Section 18.8.3.1 requires hoop reinforcement around the column reinforcement in all joints in special moment-resisting frames. In joints confined on all four sides by beams satisfying ACI Code Section 18.8.3.2, the amount of hoop reinforcement is reduced, and its spacing is less restrictive within the depth of the shallowest beam entering the joint.

ACI Code Table 18.8.4.1 gives upper limits on the shear strength of joints. As indicated in Section 17-11, these are lower than the joint shear strengths recommended in nonseismic joints. This reflects the possible damage to joints resulting from cyclic loads.

ACI Code Section 18.8.5 gives special development lengths for hooks and straight bars in joints. These are shorter than the development lengths given in ACI Code Section 25.4.3 because the effects of the joint confinement by hoops have already been included.

EXAMPLE 19-3 Design an Interior Beam–Column Joint

Design the interior beam–column joint connecting the beams and columns from Examples 19-1 and 19-2. Beams, which are 24 in. by 24 in. in section, frame into the 24-in.-by-24-in. column on all four sides.

1. Define the size of the joint. The joint has width, depth, and vertical height of 24 in. The area of a horizontal section through the joint, A_j (ACI Code Section 18.8.4.3), is $A_j = 24 \times 24 = 576 \text{ in.}^2$.

ACI Code Section 18.8.2.3 requires the length of the joint measured parallel to the flexural steel causing the joint shear to be at least 20 times the diameter of those bars, (20 × 1 in.)—o.k.

From ACI Committee 352 [19-18], the column–bar diameter should be less than 1/20 of the total depth of the shallowest beam framing into the joint, so 1/20 × 24 = 1.2 in. Thus, the selected No. 8 bars are o.k.

2. Determine the transverse reinforcement for confinement. ACI Code Section 18.8.3.1 requires confinement steel within the joint. Because the joint has beams on all four sides, ACI Code Section 18.8.3.2 sets the amount of confinement steel as half of the confinement steel required in the ends of the columns, given by Eqs. (19-23) through (19-25). In the column example, Eq. (19-25) governed (see step 6 of Example 19-2) and required that $A_{sh}/s = 0.175$ in.^{2}/in. Within the height of the joint, we require that

$$\frac{A_{sh}}{s} = 0.5 \times 0.175 \text{ in.}^2/\text{in.} = 0.0875 \text{ in.}^2/\text{in.}$$

The vertical spacing of the hoops from ACI Code Section 18.8.3.2 is permitted to be 6 in.

The clear distance between the top and bottom beam steel is 18 in. Provide three sets of hoops, the first at 3 in. below the top steel. The required A_{sh} is 6 × 0.0875 = 0.525 in.^{2}. Use No. 4 hoops in the arrangement shown in Fig. 19-28, so A_{sh} = 0.68 in.^{2}.

3. Compute the shear in the joint and check the shear strength. Figure 19-29 is a series of free-body diagrams of the joint for the frame swaying to the right. The beams entering the joint have probable moment capacities of −531 kip-ft and +286 kip-ft. At

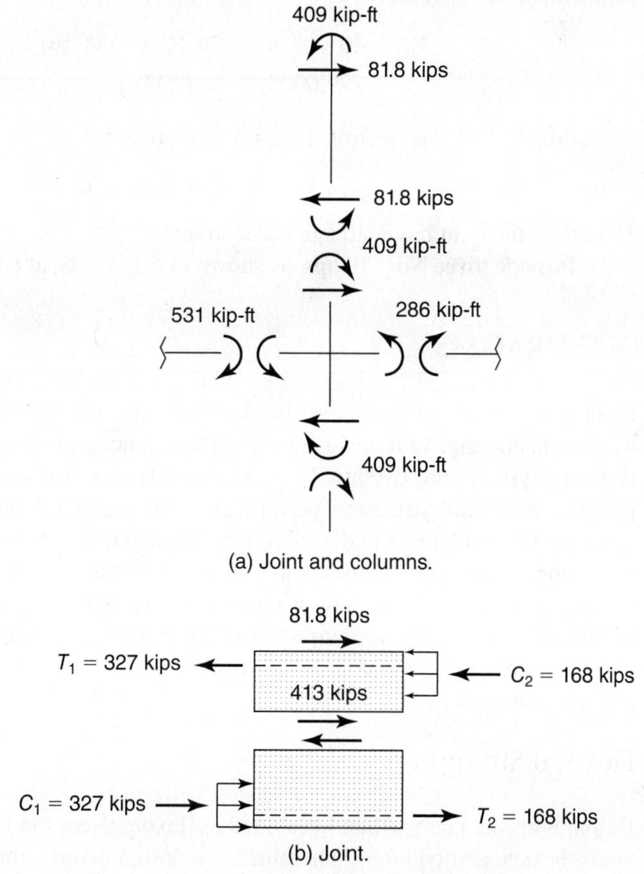

Fig. 19-29
Free-body diagrams of joint—Example 19-3.

the joint, the stiffnesses of the columns above and below the joint are the same, giving distribution factors of $DF = 0.5$ for each column. Thus, the moment in the column above is

$$M_c = 0.5(531 + 286) = 409 \text{ kip-ft}$$

The shear in the column above is

$$V_{\text{sway}} = \frac{409 + 409}{10 \text{ ft}} = 81.8 \text{ kips} \quad \text{(See step 7 of Example 19-2.)}$$

The area of the top steel at the interior support is four No. 8 bars plus two No. 7 bars, so $A_s = 4.36 \text{ in.}^2$ The force in the steel in the beam on the left of the joint is

$$T_1 = 1.25A_s f_y = 1.25 \times 4.36 \text{ in.}^2 \times 60 \text{ ksi}$$
$$= 327 \text{ kips}$$

The compression force in the beam to the left is $C_1 = T_1 = 327$ kips. Similarly, T_2 and C_2 in the beam to the right of the joint are $1.25 \times 2.24 \times 60 = 168$ kips.

Summing horizontal forces gives the shear in the joint as

$$V_j = V_{\text{sway}} - T_1 - C_2$$
$$= 81.8 \text{ kips to the right} - 327 \text{ kips to the left} - 168 \text{ kips to the left}$$
$$= 413 \text{ kips to the left}$$

From ACI Code Table 18.8.4.1, the nominal shear strength of an interior joint confined on all four sides is

$$V_n = 20\sqrt{f'_c}\,A_j = 20 \times \sqrt{4000} \text{ psi} \times 576 \text{ in.}^2$$
$$= 729{,}000 \text{ lbs.} = 729 \text{ kips}$$

From ACI Code Section 21.2.4.3, $\phi = 0.85$ for seismic joints. So,

$$\phi V_n = 0.85 \times 729 = 619 \text{ kips}$$

Therefore, the joint has adequate shear strength.

Provide three No. 4 hoops, as shown in Fig. 19-28, at 6 in. on centers in the joint. ■

19-10 STRUCTURAL DIAPHRAGMS

Floor slabs, roof slabs, and cast-in-place toppings on precast concrete floors may all be used as diaphragms to transfer horizontal forces acting in the building to the lateral-force-resisting system and, eventually, to the foundations. Because of the importance of diaphragms to overall structural performance, the ACI Code has developed a new chapter (Chapter 12) to address the design of diaphragms. In effect, they act as deep flexural members lying in horizontal planes. Cast-in-place toppings serving as diaphragms may or may not be composite with the floor or roof members. ACI Code Section 18.12.6 allows the use of 2-in.-thick composite topping slabs or $2\frac{1}{2}$-in. noncomposite toppings. ACI Code Section 18.12.5 requires that noncomposite toppings be designed for the forces transferred as if they are acting alone.

Flexural Strength

Diaphragms are analyzed and designed in flexure using the same assumptions as used for walls, beams, and columns. Therefore, the design assumptions and general procedures in ACI Code Sections 22.2 and 22.3 shall apply for diaphragms, except that the nonlinear strain distributions discussed in ACI Code Section 9.9.1.2 for deep beams do not need to be applied to diaphragms. Designing diaphragms as a normal flexural member with distributed reinforcement represents a significant change in philosophy from earlier design practice that assumed moments were resisted entirely by tension and compression chord forces acting at opposite edges of the diaphragm. Now, all of the longitudinal reinforcements satisfying the requirements of ACI Code Section 18.12.7 is assumed to contribute to the flexural strength of the diaphragm.

Although the revised analysis and design procedure removes the need to use concentrated reinforcement in chords at the edges of a diaphragm, there may be some structural layouts where it is necessary to concentrate longitudinal reinforcement at the diaphragm edges to obtain adequate flexural strength. For such designs, ACI Code Section 18.12.7.5 requires that if the calculated concrete stress in the compression chord exceeds $0.2f'_c$, confining transverse reinforcement that satisfies ACI Code Sections 18.7.5.2(a) through (e) and Code Section 18.7.5.3, except the spacing limit of 18.7.5.3(a) shall be one-third of the least dimension of the collector. The amount of transverse reinforcement shall be in accordance with ACI Code Table 18.12.7.5. Furthermore, this transverse reinforcement must be used over the portion of the compression chord where the calculated concrete stress exceeds $0.15f'_c$.

Shear Strength

ACI Code Section 18.12.9 gives expressions for the nominal shear resistance of diaphragms. The nominal shear strength of monolithic structural diaphragms shall not exceed

$$V_n = A_{cv}(2\lambda\sqrt{f'_c} + \rho_t f_y) \tag{19-31}$$

(ACI Eq. 18.12.9.1)

This is equivalent to $V_n = (V_c + V_s)$, where V_c and V_s have been divided by the shear area, A_{cv}, defined as the area bounded by the thickness of the diaphragm and the length of the section in the direction of the shear force being considered.

For cast-in-place composite-topping-slab diaphragms and cast-in-place noncomposite topping slabs on precast floors or roofs, V_c is taken equal to zero above the joints between the precast elements, to reflect the likelihood that the topping will be cracked by shrinkage or other effects during construction and service. The required slab steel in this case is computed using

$$V_n = A_{vf} f_y \mu \tag{19-32}$$

(ACI Eq. 18.12.9.3)

where A_{vf} is the total area of *shear friction* reinforcement in the topping slab, including both the distributed reinforcement and concentrated reinforcement (if any) at the edges of the diaphragm that is oriented perpendicular to the joints in the precast elements. At least one-half of A_{vf} shall be distributed along the potential shear plane. The coefficient of friction, μ, is taken as 1.0λ, where λ is the factor for lightweight concrete given in ACI Code Table 19.2.4.2. The value of V_n shall not exceed the limits given in ACI Code Section 18.12.9.2.

Effect of Diaphragm Stiffness on Lateral-Load Distribution

Figure 19-30 illustrates the effect of diaphragm stiffness on the distribution of lateral loads to the lateral-load-resisting elements. The building shown in the figure has three walls of equal lateral stiffness. If the diaphragm is essentially rigid in plane and there is no torsion, all three walls will displace by the same amount, and each wall will resist one-third of the total lateral load, as shown in Fig. 19-30a. On the other hand, if the diaphragm is very flexible relative to the walls, the two end walls will each resist approximately a quarter of the lateral shear and the center wall will resist half of it, as shown in Fig. 19-30b.

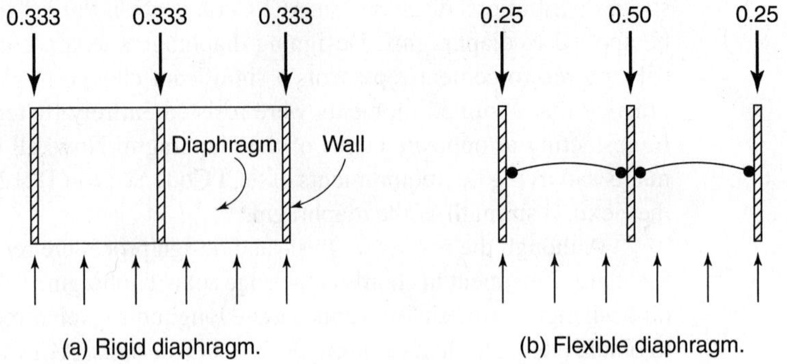

Fig. 19-30
Plan view of a building showing the effect of diaphragm stiffness on distribution of lateral loads to walls in a building.

The following derivations are presented to give an idea of the factors affecting the relative stiffnesses of the walls and diaphragms. The lateral stiffness, K_ℓ, of a cantilever of height ℓ that is fixed at the base is

$$K_\ell = \frac{V}{\Delta} \tag{19-33}$$

where Δ is the lateral deflection on the top of the cantilever due to the load V at the top, equal to

$$\Delta = \frac{V\ell^3}{3EI} + \frac{1.2V\ell}{AG} \tag{19-34}$$

The first term represents the flexural deflections, the second the shear deflections. Substituting Eq. (19-34) into (19-33) and taking $G = E/2$ gives

$$K_\ell = \frac{1}{\dfrac{\ell^3}{3EI} + \dfrac{2.4\ell}{AE}} \tag{19-35}$$

A similar expression can be derived for the lateral stiffness of a piece of diaphragm between two walls.

Benjamin [19-21] has shown that if the stiffness of the diaphragm exceeds about two times that of the walls, the diaphragm will act as a rigid diaphragm in transmitting loads to the walls.

19-11 STRUCTURAL WALLS

Structural walls or *shear walls* are frequently used to resist a major fraction of the design seismic shears. These are designed according to ACI Code Section 18.10. The factored shears, moments, and axial forces to be considered in the design of the wall are obtained from a frame analysis. Chapter 18 discussed the layout and design of structural walls.

Design of Shear Walls

Chapter 18 identified two types of shear walls: slender shear walls, and short (or squat) shear walls. Slender shear walls are designed using the theory of flexure for members subjected to axial loads and bending. The design of short walls may be based, in part, on strut-and-tie models.

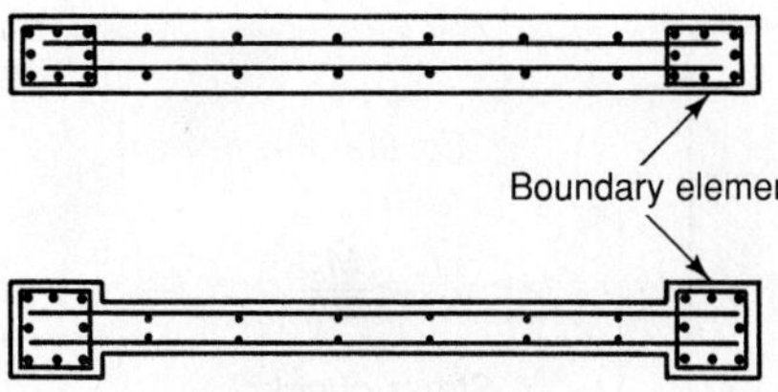

Fig. 19-31
Plan views of structural walls with boundary elements.

1. The first step in the design of a shear wall is to select the size and shape of the wall, using stiffness, building geometry, and the required moment and shear strengths. At this stage, the geometry of the wall probably is chosen.

2. The foundations for the walls may be required to transfer very large overturning moments to the soil or rock under the building. ACI Code Section 18.13 presents requirements for the design of foundations resisting earthquake forces. ACI Code Section 18.13.2 requires that special attention be given to the anchorage of the wall reinforcement into the foundations.

3. Once the concrete section is established, it is necessary to investigate the need for boundary elements. Boundary elements are regions at the ends of the cross section of the wall that are reinforced as columns, with the reinforcement enclosed by hoop reinforcement, as shown in Fig. 19-31. Boundary elements strengthen and confine the edges of the walls to resist stress reversals and prevent reinforcement buckling near the edges. They generally are thicker than the walls although they may have the same thickness as the wall. ACI Code Sections 18.10.6.2 and 18.10.6.3 give two methods of determining the need for special confinement reinforcement in boundary elements.

(a) ACI Code Section 18.10.6.2 applies to walls that are effectively continuous from the base of the structure to the top of the wall and are designed to have a single critical section for axial loads and bending at the base of the wall. Boundary elements are required if

$$c \geq \frac{\ell_w}{600(1.5\delta_u/h_w)} \qquad \text{(19-36)} \quad \text{(ACI Eq. 18.10.6.2)}$$

where

c = depth from the neutral axis to the extreme compression fiber
ℓ_w = horizontal length of entire wall or of a segment of wall considered in the direction of the shear force
h_w = height of the entire wall, or the segment of wall considered
δ_u = design displacement, defined as the total lateral displacement deflection of the top of the building for the design-basis earthquake

Equation (19-36) essentially represents a check on the maximum strain at the compression edge of the wall at the base of the structure [19-22]. The quantity δ_u/h_w, referred to as the building drift, shall not be taken less than 0.005. Larger building drifts would lead to larger curvatures in the wall sections at the base of the structure and thus, to larger strains at the edges of the wall. Furthermore, for a given value of curvature, a larger value for the depth to the neutral axis, c, would result in a larger strain at the extreme compression fiber. Thus, by defining an upper limit for the neutral-axis depth as a function of the calculated building drift, Eq. (19-36) essentially is requiring that special confinement reinforcement must be used when the maximum strain experienced at the compression edge of the wall exceeds a certain value, which was suggested to be 0.004 in reference [19-22].

Where boundary elements are required by Eq. (19-36), the special boundary reinforcement shall extend upward from the base of the structure for a distance not less than

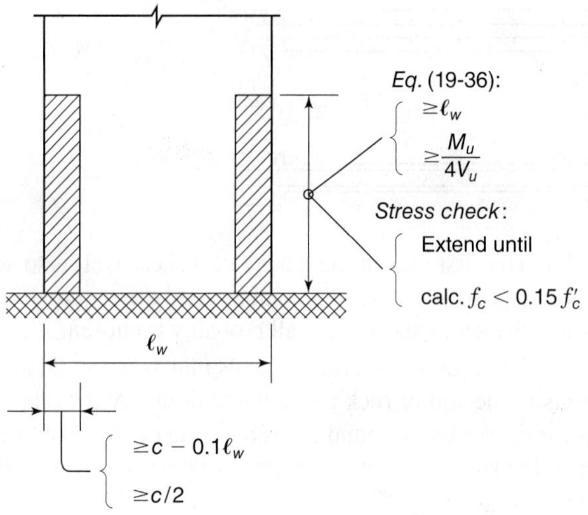

Fig. 19-32
Size of boundary region where special confinement steel is required.

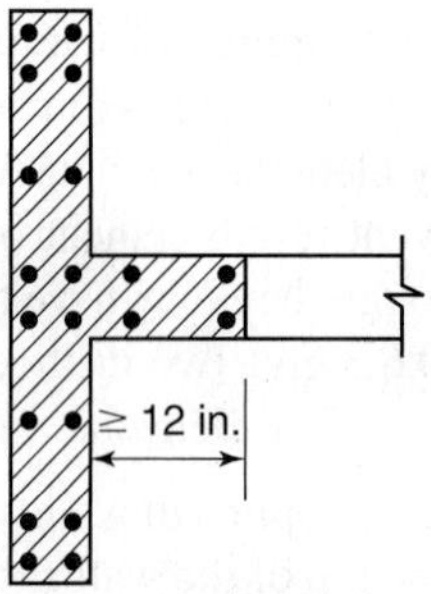

Fig. 19-33
Required extension of special confinement region into web of flanged wall.

the larger of ℓ_w or $M_u/4V_u$. That vertical dimension and the minimum required horizontal extension [ACI Code Section 18.10.6.4(a)] of the confined boundary zone from the compression edge of the wall are shown in Fig. 19-32. ACI Code Section 18.10.6.4(d) also requires that if a specially confined boundary zone is required for a wall with a compression flange, the boundary zone must extend at least 12 in. into the adjacent web, as shown in Fig. 19-33.

The transverse reinforcement in the specially confined boundary elements must satisfy the requirements for column confinement reinforcement given in ACI Table 18.10.6.4(f) and the transverse reinforcement spacing limit of ACI Code Section 18.7.5.3(a) shall be one-third of the least dimension of the boundary element.

(b) In ACI Code Section 18.10.6.3, stresses in the uncracked wall are analyzed by using

$$f_c = \frac{N_u}{A} \pm \frac{M_u y}{I} \tag{19-37}$$

where A and I are based on the gross concrete section. If the computed maximum compressive stress in the extreme fiber exceeds $f_c = 0.2f'_c$ at any point, ACI Code Section 18.10.6.3 requires confined boundary elements over that portion of the height where the extreme-fiber stress exceeds $f_c = 0.15f'_c$. Figures 19-32 and 19-33 show the required vertical and horizontal dimensions of the specially confined boundary elements.

If specially confined boundary elements are not required when checking by either Eq. (19-36) or (19-37), then the reinforcement details at the edges of the wall must satisfy the requirements given in ACI Code Section 18.10.6.5. Example 19-4 will demonstrate the use of Eqs. (19-36) and (19-37) for a wall that was analyzed for required flexural and shear strength in Example 18-3.

Design for flexure should follow ACI Code Sections 22.2 and 22.3 except that the nonlinear strains mentioned in ACI Code Section 9.9.1.2 shall not apply. In addition, the axial-load capacity computed by ACI Code Section 22.4.2 does not apply.

Design for Shear

As was discussed in Section 18-11, the design of structural walls for shear is given in ACI Code Section 18.10.4. The basic design equation is essentially the same as the $\phi(V_c + V_s)$ procedure used in beam design. The basic shear (v_c) stress carried by the concrete is $2\sqrt{f'_c}$ psi, except for short stubby walls, for which a higher stress is allowed. The horizontal reinforcement in the wall must be anchored in the boundary elements, as specified in ACI Code Section 18.10.6.4(h).

Strength-Reduction Factor

Because walls generally have axial load ratios N_u/A_gF_c between 0.1 and 0.3, they are normally tension controlled and the ϕ-factor for flexure is 0.9. The ϕ-factor for shear is 0.75 unless the nominal shear strength is less than the shear corresponding to development of the nominal flexural strength of the wall. In such a case, ϕ is taken as 0.6 (ACI Code Section 21.2.4.1).

EXAMPLE 19-4 Check Need for Boundary Elements in Structural Wall Analyzed in Example 18-3.

All of the loads, wall dimensions, and material properties are given in Example 18-3. For that example, the structural wall was found to have adequate flexural and shear strength. We now want to use Eqs. (19-36) and (19-37) to determine if specially confined boundary elements are required for this structural wall.

1. Use Eq. (19-36) to determine need for specially confined boundary elements. As stated earlier, this equation represents a check on the maximum strain at the compression edge of the structural wall at the maximum lateral drift of the structure. The building drift in Eq. (19-36) is represented by the ratio, δ_u/h_w. ACI Code Section 18.10.6.2 requires that the minimum value for the drift ratio is 0.005. The structural wall selected for Example 18-3 came from a paper by Wallace and Thomsen [19-23]. For the structure that they analyzed, the minimum drift ratio specified by the ACI Code would represent a fairly strong ground motion. Therefore, we will use a drift ratio of 0.005 for this check of the need for specially confined boundary elements.

ACI Load Combination, 5.3.1(e) will be used to determine the factored axial load. We will assume this is a standard occupancy building, so the load factor for live load can be reduced to 0.5. Thus,

$$N_u = 1.2N_D + 1.0N_E + 0.5N_L$$
$$= 1.2 \times 1000\text{ k} + 1.0 \times 0.0 + 0.5 \times 450\text{ k} = 1430\text{ kips}$$

From Example 18-3, the flexural tension force is

$$T = A_s f_y = 12.7\text{ in.}^2 \times 60\text{ ksi} = 762\text{ kips}$$

With these numbers, the depth of the compression stress block can be calculated.

$$a = \frac{T + N_u}{0.85 f'_c b} = \frac{762\text{ k} + 1430\text{ k}}{0.85 \times 4\text{ ksi} \times 12\text{ in.}} = 53.7\text{ in.}$$

From this value of a, the depth to the neutral axis is

$$c = a/\beta_1 = 53.7/0.85 = 63.2 \text{ in.}$$

This value for c is to be compared to the limiting value from Eq. (19-36):

$$c(\text{limit}) = \frac{\ell_w}{600(1.5\delta_u/h_w)}$$

$$= \frac{288}{600 \times 1.5 \times 0.005} = 64.0 \text{ in.}$$

Because the calculated value for c (63.2 in.) is less than the limit from Eq. (19-36), specially confined boundary elements are not required for this wall. However, we used an assumed value for δ_u/h_w and found a limiting value for the neutral-axis depth from Eq. (19-36) that was very close to the value calculated for flexural analysis. Thus, if more information was available regarding the design earthquake and the structural properties of the building, a more exact determination of the design displacement, δ_u, would be recommended.

2. Use Eq. (19-37) to determine the need for specially confined boundary elements. It is clear that this equation represents a check on the maximum compressive stress, as opposed to a check on the maximum compression strain from Eq. (19-36). From step 1, the factored axial load is $N_u = 1430$ kips. Also from step 1 of Example 18-3, the factored design moment at the base of this wall is $M_u = 16{,}400$ kip-ft $= 197{,}000$ kip-in. The wall section properties are

$$A = \ell_w h = 288 \times 12 = 3460 \text{ in.}^2$$

$$I = \frac{1}{12} h \ell_w^3 = \frac{12 \times (288)^3}{12} = 23.9 \times 10^6 \text{ in.}^4$$

and

$$y = \ell_w/2 = 288/2 = 144 \text{ in.}$$

With these values, the maximum compression stress at the base of the wall is

$$f_c = \frac{N_u}{A} + \frac{M_u y}{I}$$

$$= \frac{1430 \text{ k}}{3460 \text{ in.}^2} + \frac{197{,}000 \text{ k-in.} \times 144 \text{ in.}}{23.9 \times 10^6 \text{ in.}^4}$$

$$= 0.41 \text{ ksi} + 1.19 \text{ ksi} = 1.60 \text{ ksi}$$

This value is greater than the limiting value given in ACI Code Section 18.10.6.3, defined as $0.20 f'_c = 0.20 \times 4 \text{ ksi} = 0.80$ ksi. For this case, confined boundary elements are required according to Eq. (19-37). The required dimensions for the boundary elements are shown in Fig. 19-32. The horizontal dimension from the edge of the wall is the larger of

$$c - 0.1\ell_w = 63.2 - 0.1 \times 288 = 34.4 \text{ in. (governs)}$$

or

$$c/2 = 63.2/2 = 31.6 \text{ in.}$$

The vertical dimension is a function of the maximum calculated compression stress at the edge of the wall. Equation (19-37) can be used to calculate the maximum compression stress at the base of each successive story, proceeding up from the base. When a story level is reached where the calculated compressive stress, f_c, is less than $0.15 f'_c$, the special confinement reinforcement can be stopped. ■

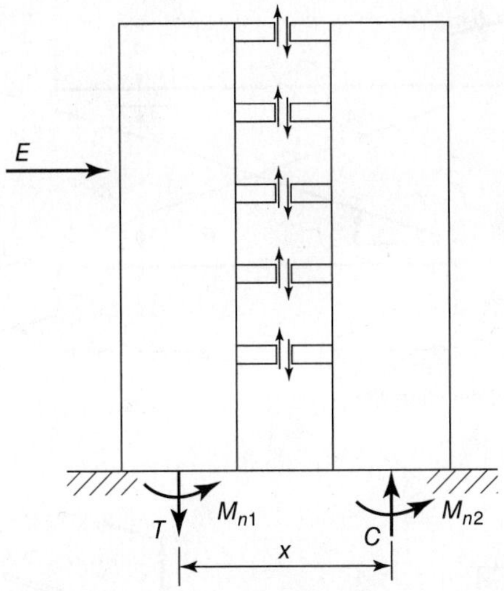

Fig. 19-34
Overturning moment resistance of coupled shear wall.

Coupled Walls and Coupling Beams

Frequently, two shear walls are *coupled* by beams spanning across a doorway or similar opening. Depending on the stiffness of the coupling beams, the walls act as either two independent cantilevers when the coupling-beam stiffness approaches zero or as one solid cantilever if the coupling-beam stiffness is high. The coupling beams transmit shear forces from one wall to the other as shown in Fig. 19-34. Thus, the overturning moment resistance at the base of the structure is composed of moments at the base of each wall (M_{n1} and M_{n2}) plus the net tension-compression couple ($T \cdot x$) due to shear forces in the coupling beams. The *degree of coupling* can be defined as:

$$\text{Degree of coupling} = \frac{T \cdot x}{M_{n1} + M_{n2} + T \cdot x} \tag{19-38}$$

where $T =$ sum of shear forces in coupling beams and x is the distance between the centroids of the coupled walls. Typical coupled wall systems have a degree of coupling between 0.3 and 0.5.

During a major earthquake coupling beams may be subjected to large cyclic reversals of shear deformations that can lead to a rapid deterioration of their shear strength and stiffness. Park and Paulay [19-13] discuss tests of short coupling beams (length/depth ≤ 1.5) that demonstrated that the diagonal reinforcement pattern shown in Fig. 19-35 transmits cyclic shear load and maintains member stiffness much better than conventional reinforcement consisting of top and bottom longitudinal steel, and vertical stirrups. This diagonal steel acts as a truss to develop forces T_u and C_u, which transmit a moment and shear, where:

$$T_u = C_u = \phi A_s f_y \tag{19-39}$$

$$V_u = 2T_u \sin\alpha = 2\phi A_s f_y \sin\alpha \tag{19-40}$$

$$M_u = (\phi A_s f_y \cos\alpha)(h - 2d') \tag{19-41}$$

ACI Code Section 18.10.7.2 requires that diagonal reinforcement similar to that in Fig. 19-35 must be used for coupling beams with $\ell_n/h < 2$ and with a design shear of

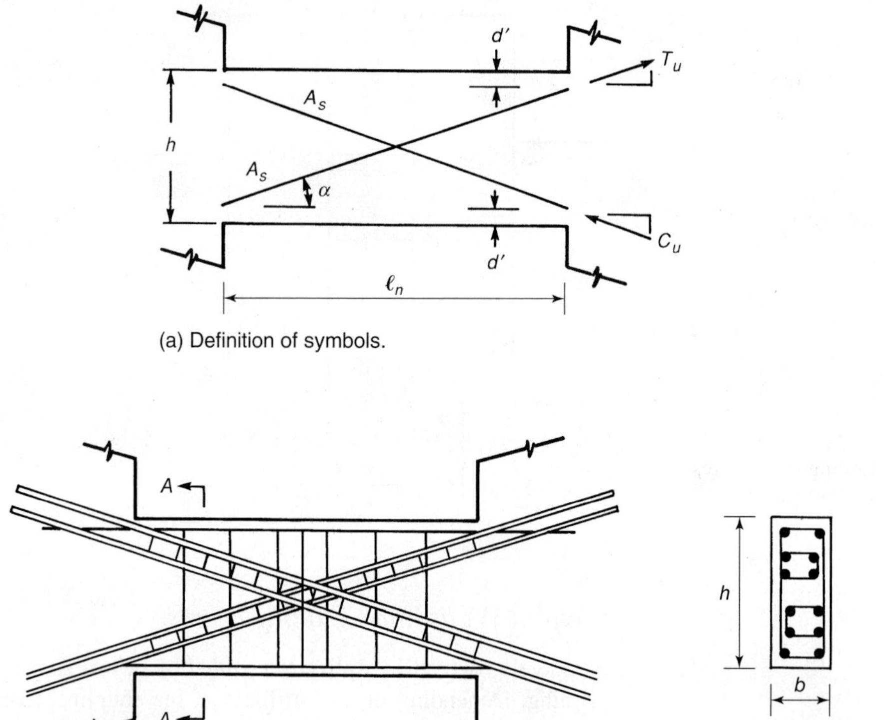

(a) Definition of symbols.

(b) Reinforcement in a coupling beam.

(c) Section *A–A*.

Fig. 19-35
Diagonally reinforced coupling beams.

$V_u > 4\lambda\sqrt{f'_c}A_{cw}$, where A_{cw} is the coupling beam area $b \times h$. ACI Code Section 18.10.7.1 states that coupling beams with $\ell_n/h > 4$ shall be designed as a flexural member of a special moment frame following ACI Code Section 18.6. For coupling beams not governed by ACI Code Sections 18.10.7.1 or 18.10.7.2, Code Section 18.10.7.3 gives a designer the option of either using diagonal reinforcement, as shown in Fig. 19-35, or using conventional reinforcement, as governed by ACI Code Sections 18.6.3 through 18.6.5. Although the ACI Code does not discuss a combination of these design procedures for such coupling beams, the author believes that for coupling beams with aspect ratios of $4 > \ell_n/h > 2$ a combination of diagonal and conventional reinforcement would be more efficient than using only one or the other type of reinforcement.

As a coupling beam span to depth ratio (ℓ_n/h) increases from 1.5 to 4.0, the author recommends that the percentage of moment resistance assigned to the diagonal reinforcement should be reduced from 100 percent to zero. Top and bottom longitudinal reinforcement should be added to the beam to maintain the required flexural strength. In a similar manner, the maximum shear stress (psi units) permitted in a coupling beam should be decreased from $10\sqrt{f_c'}$ to $6\sqrt{f_c'}$ as the percentage of shear resisted by the diagonal reinforcement decreases from 100 percent to zero. Closed vertical stirrups should be added to the beam to maintain the required shear strength. Because of the large shear deformation reversals expected in a coupling beam, the concrete contribution to shear resistance is ignored. The lower maximum shear stress limit of $6\sqrt{f_c'}$ for coupling beams with conventional transverse reinforcement is based on test results reported in [19-15].

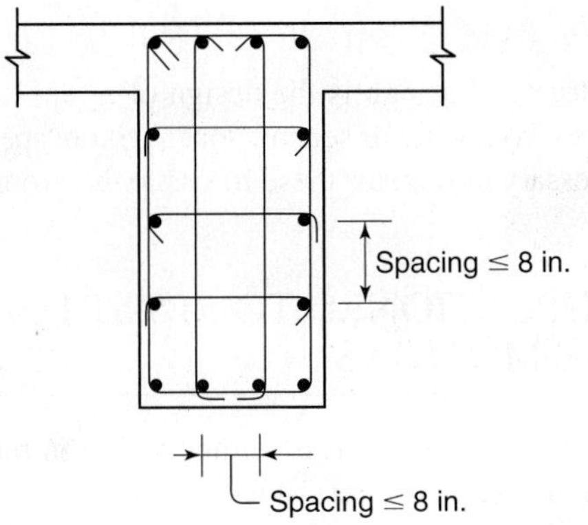

Fig. 19-36
Alternate procedure for confinement reinforcement in coupling beams.

When using diagonal reinforcement, ACI Code Section 18.10.7.4 allows designers to use one of two procedures for confining the diagonals, which act alternatively in compression and tension during the response of a coupled wall system to earthquake ground motion. Part (c) of ACI Code Section 18.10.7.4 gives detailing rules for transverse reinforcement to be placed around each diagonal (Fig. 19-35c). This transverse reinforcement is intended to provide confinement to the set of diagonal bars (minimum of four bars), similar to that required for columns. The area of transverse reinforcement must satisfy the requirements in ACI Code Section 18.10.7.4(c). Limits on the spacing between legs of crossties and for the spacing of this transverse reinforcement along the diagonal bars also are given in the same Code Section. An alternate procedure for confining the diagonals, which is given in ACI Code Section 18.10.7.4(d), is to use transverse reinforcement to confine the *entire* coupling-beam section, as shown in Fig. 19-36. This transverse reinforcement must be provided in both the vertical and horizontal directions and must satisfy the requirements for column confinement reinforcement given in ACI Code Section 18.10.7.4(d). Many structural engineers have indicated a preference for this second option because of the construction difficulties associated with placing the confinement reinforcement directly around the groups of diagonal bars.

Construction of coupling beams with diagonal reinforcement can be difficult and time-consuming, especially if the diagonal bars need to be threaded through heavy confinement steel at the boundaries of adjacent shear walls. Recent research studies [19-24] of coupling beams constructed with high-performance fiber-reinforced concrete (high-performance meaning a strain-hardening behavior in tension) have indicated that the percentage of moment and shear assigned to the diagonal reinforcement can be significantly reduced and the need for confinement of that diagonal reinforcement can be virtually eliminated. The high-performance fiber-reinforced concrete (HPFRC) behaves like a confined concrete and interacts with a nominal amount of transverse reinforcement to effectively confine the reduced amount of diagonal reinforcement. The HPFRC also contributes to the shear strength and stiffness of coupling beams subjected to large cyclic reversals of shear deformations.

Coupled walls may need specially confined boundary elements at the outside ends only or at both ends of both walls, depending on the stiffness of the coupling beams. This would be determined by an analysis according to ACI Code Section 18.10.6.2 and either Eq. (19-36) or (19-37), using the moments and axial forces assigned to each part of the coupled wall.

Wall Foundations

A major stage in structural design is the design of a foundation for the wall. Frequently, overturning moments from wind or seismic loads require special structures in the basement of the building necessary to transfer these forces to the ground.

19-12 FRAME MEMBERS NOT PROPORTIONED TO RESIST FORCES INDUCED BY EARTHQUAKE MOTIONS

ACI Code Section 18.14 provides less stringent design requirements for members that are not part of the designated lateral-load-resisting system in structures subjected to severe earthquakes. Such members must be able to resist the factored axial forces due to gravity loads and the moments and shears induced in them when the frame is deflected laterally through twice the elastically calculated lateral deflections under factored lateral loads. In some older structures building frame members were designed by assuming that the frame members supported only gravity loads, while shear walls resisted all of the lateral loads. In the 1994 Northridge earthquake, columns in a number of buildings of this type failed when subjected to the lateral displacements imposed by that earthquake [19-25]. After that earthquake, ACI Code Section 18.14 was made considerably more stringent.

19-13 SPECIAL PRECAST STRUCTURES

Since 2002, the ACI Code has included earthquake-resistant design requirements for precast frame members and precast structural walls. Requirements for special moment frames constructed with precast concrete elements are given in ACI Code Section 18.9. This section gives a designer the option of using ductile connections (Code Section 18.9.2.1) or strong connections (Code Section 18.9.2.2) between precast elements. The first option assumes that at least part of the inelastic behavior during the overall structural response to strong ground motions will occur in the connections between precast elements. The second option assumes that the connections will remain elastic and that plastic hinging zones will occur at other locations in the precast frame elements. Based on the probable member strengths at the plastic hinge locations, a *capacity-based design* procedure can be used to determine the required strength for the connections between the precast elements.

Special precast structural walls are required to satisfy all of the design provisions for special cast-in-place structural walls given in ACI Code Section 18.10 and discussed previously in Section 19-11. Intermediate precast structural walls also are permitted and must satisfy the design requirements in ACI Code Section 18.5.

19-14 FOUNDATIONS

ACI Code Section 18.13 deals with foundations for seismic structures, including footings, mat foundations, pile caps, piles, piers, and caissons. The major emphasis is on the pull-out strength of reinforcement extending from the structure into the foundations. Minimum confinement reinforcement and shear reinforcement is required in piles, piers, and caissons.

PROBLEMS

19-1. Use Eq. (19-36) to check the need for specially confined boundary elements in the structural wall described in Problem 18-3. Assume the design displacement, δ_u, at the top of the wall is 0.6 ft. Use all of the dimensions, loading information, and material properties given in Problem 18-3. If a specially confined boundary element is required, define the required vertical and horizontal dimensions for the boundary element.

19-2. Use Eq. (19-37) to check the need for specially confined boundary elements in the structural wall described in Problem 18-3. Use all of the dimensions, loading information, and material properties given in Problem 18-3. If a specially confined boundary element is required, define the required vertical and horizontal dimensions for the boundary element.

REFERENCES

19-1 B. Gutenberg and C. F. Richter, "Earthquake Magnitude, Intensity, Energy and Acceleration," *Bulletin of the Seismological Society of America*, Vol. 46, No. 2, 1956.

19-2 J. H. Hodgson, *Earthquakes and Earth Structure*, Englewood Cliffs, Prentice-Hall, NJ, 1964.

19-3 G. W. Housner, "Spectrum Intensities of Strong Motion Earthquakes," *Proceedings of the Symposium on Earthquakes and Blast Effects on Structures,* Earthquake Engineering Research Institute, Oakland, CA, 1952.

19-4 *Minimum Design Loads for Buildings and Other Structures* (ASCE/SEI 7–10), American Society of Civil Engineers, Reston, VA, 2010, 608 pp.

19-5 *International Building Code*, International Code Council, Washington, D.C. 2012.

19-6 *Seismic Design Recommendations*, Structural Engineers Association of California, (www.seaoc.org/bluebook), Sacramento, CA, 2008.

19-7 *NEHRP Recommended Provisions for Seismic Regulations for New Buildings and Other Structures*, FEMA 450-1, Part 1: Provisions and FEMA 450-2 Part 2: Commentary, 2003 Edition, Building Seismic Safety Council, National Institute of Building Sciences, Washington, D.C.

19-8 M. A. Sozen, "Seismic Behavior of Reinforced Concrete Buildings," *Earthquake Engineering*, Y. Bozorgnia and V. V. Bertero, Editors, CRC Press, New York, 2004, pp. 13–1 to 13–41.

19-9 L. E. Garcia and M. A. Sozen, "Earthquake-Resistant Design of Reinforced Concrete Buildings," *Earthquake Engineering*, Y. Bozorgnia and V. V. Bertero, Editors, CRC Press, New York, 2004. pp. 14–1 to 14–85.

19-10 Jack Moehle, *Design of Reinforced Concrete Buildings*, McGraw-Hill, 2014, pp. 778.

19-11 A. K. Chopra, *Dynamics of Structures*, Second Edition, Prentice-Hall, NJ, 2001, pp. 844.

19-12 J. K. Wight, S. L. Wood, J. P. Moehle, and J. W. Wallace, "On Design Requirements for Reinforced Concrete Structural Walls," *Mete A. Sozen Symposium: A Tribute from his Students*, J. K. Wight and M. E. Kreger, Editors, Special Publication SP-162, American Concrete Institute, Farmington Hills, MI, 1996, pp. 431–456.

19-13 Robert Park and Thomas Paulay, *Reinforced Concrete Structures*, Wiley-Interscience, New York, 1975, 768 pp.

19-14 USGS Website: *earthquake.usgs.gov/hazards/designmaps/.*

19-15 C. F. Scribner and J. K. Wight, "Strength Decay in R/C Members Under Load Reversals," *Journal of the Structural Division*, ASCE, Vol. 106, No. ST4, April 1980, pp. 861–876.

19-16 C. W. French and J. P. Moehle, "Effect of Floor Slab on Behavior of Slab-Beam-Column Connections," *Design of Beam-Column Joints for Seismic Resistance*, Special Publication SP-123, American Concrete Institute, Farmington Hills, MI, 1991, pp. 225–258.

19-17 J. K. Wight and M. A. Sozen, "Shear Strength Decay of RC Columns under Shear Reversals," *Journal of the Structural Division*, ASCE, Vol. 101, No. ST5, May 1975, pp. 1053–1065.

19-18 ACI-ASCE Committee 352, "Recommendations for Design of Beam-Column Joints in Monolithic Reinforced Concrete Structures," ACI 352R-02, (Reapproved 2010), *ACI Manual of Concrete Practice*, American Concrete Institute, Farmington Hills, MI.

19-19 R. T. Leon, "Interior Joints with Variable Anchorage Lengths," *Journal of Structural Engineering*, ASCE, Vol. 115, No. 9, September 1989, pp. 2261–2275.

19-20 C. G. Quintero-Febres and J. K. Wight, "Experimental Study of Reinforced Concrete Interior Wide Beam-Column Connections Subjected to Lateral Loading," *ACI Structural Journal*, Vol. 98, No. 4, July–August 2001, pp. 572–582.

19-21 J. R. Benjamin, *Statically Indeterminate Structures*, McGraw-Hill, New York, 1959, 350 pp.

19-22 J. W. Wallace, "Seismic Design of RC Structural Walls, Part I: New Code Format," *Journal of Structural Engineering*, ASCE, Vol. 121, No. 1, January 1995, pp. 75–87.

19-23 J. W. Wallace and J. H. Thomsen IV, "Seismic Design of RC Structural Walls, Part II: Applications," *Journal of Structural Engineering*, ASCE, Vol. 121, No. 1, January 1995, pp. 88–101.

19-24 Lequesne, G. Parra-Montesinos and J. K. Wight, "Test of a Coupled Wall with High-Performance Fiber-Reinforced Concrete Coupling Beams," *Thomas T.C. Hsu Symposium on Shear and Torsion in Concrete Structures*, ACI Special Publication 265, pp. 1–18.

19-25 *Northridge Earthquake January 17, 1994: Preliminary Reconnaissance Report*, J. F. Hall, Technical Editor, Earthquake Engineering Research Institute, Pub. 94–01, March 1994, 96 pp.

APPENDIX A

Design Aids

Tables

Table A-1	Areas, Weights, and Dimensions of Reinforcing Bars
Table A-1M	Areas, Weights, and Dimensions of Reinforcing Bars—SI Units
Table A-2	Welded-Wire Reinforcement
Table A-3	Values of Flexural Resistance Factor, R (psi), for Grade-60 Reinforcement
Table A-3M	Values of Flexural Resistance Factor, R (MPa), for Grade-420 Reinforcement
Table A-4	Total Area (in.^{2}) of Multiple U.S. Reinforcing Bars
Table A-4M	Total Area (mm^2) of Multiple U.S.-Metric Reinforcing Bars
Table A-5	Minimum Beam Width (in.) for Multiple U.S. Bars per Layer; Interior Exposure
Table A-5M	Minimum Beam Width (mm) for Multiple U.S.-Metric Bars per Layer; Interior Exposure
Table A-6	Basic Tension Development-Length Ratio, ℓ_d/d_b (in./in.)
Table A-6M	Basic Tension Development-Length Ratio, ℓ_d/d_b (mm/mm)
Table A-7	Basic Compression Development Length, ℓ_{dc} (in.)
Table A-7M	Basic Compression Development Length, ℓ_{dc} (mm)
Table A-8	Basic Development Lengths for Hooked Bars, ℓ_{dh} (in.)
Table A-8M	Basic Development Lengths for Hooked Bars, ℓ_{dh} (mm)
Table A-9	Minimum Thicknesses of Nonprestressed Beams or One-Way Slabs Unless Deflections Are Computed

Table A-10 Maximum Number of Bars That Can Be Placed in Square Columns with the Same Number of Bars in Each Face, Based on Normal (Radial) Lap Splices; Minimum Bar Spacing

Table A-11 Maximum Number of Bars That Can Be Placed in Circular Columns, Based on Normal (Radial) Lap Splices; Minimum Bar Spacing

Table A-12 Number of Bars Required to Provide a Given Area of Steel

Table A-13 Lap-Splice Lengths for Grade-60 Bars in Columns (in.)

Table A-14 Moment-Distribution Factors for Slabs without Drop Panels

Table A-15 Moment-Distribution Factors for Slabs with Drop Panels; $h_1 = 1.5h$

Table A-16 Moment-Distribution Factors for Slabs with Drop Panels; $h_1 = 1.5h$

Table A-17 Stiffness and Carryover Factors for Columns

Figures

Fig. A-1 Bending-moment envelope for typical interior span (moment coefficients: $-1/11$, $+1/16$, $-1/11$).

Fig. A-2 Bending-moment envelope for exterior span with exterior support built integrally with a column (moment coefficients: $-1/16$, $+1/14$, $-1/10$).

Fig. A-3 Bending-moment envelope for exterior span with exterior support built integrally with a spandrel beam or girder (moment coefficients: $-1/24$, $+1/14$, $-1/10$).

Fig. A-4 Bending-moment envelope for exterior span with discontinuous end unrestrained (moment coefficients: 0, $+1/11$, $-1/10$).

Fig. A-5 Standard bar details.

Fig. A-6a-c Nondimensional interaction diagram for rectangular tied columns with bars in two faces: $f'_c = 4000$ psi

Fig. A-7a-c Nondimensional interaction diagram for rectangular tied columns with bars in two faces: $f'_c = 5000$ psi

Fig. A-8a-c Nondimensional interaction diagram for rectangular tied columns with bars in two faces: $f'_c = 6000$ psi

Fig. A-9a-c Nondimensional interaction diagram for rectangular tied columns with bars in four faces: $f'_c = 4000$ psi

Fig. A-10a-c Nondimensional interaction diagram for rectangular tied columns with bars in four faces: $f'_c = 5000$ psi

Fig. A-11a-c Nondimensional interaction diagram for tied columns with bars in four faces: $f'_c = 6000$ psi

Fig. A-12a-c Nondimensional interaction diagram for circular *spiral* column: $f'_c = 4000$ psi

Fig. A-13a-c Nondimensional interaction diagram for circular *spiral* column: $f'_c = 5000$ psi

Fig. A-14a-c Nondimensional interaction diagram for circular *spiral* column: $f'_c = 6000$ psi

TABLE A-1 Areas, Weights, and Dimensions of Reinforcing Bars

			Nominal Dimensions[a]	
Bar Size Designation No.[b]	Grades[c]	Weight (lb/ft)	Diameter (in.)	Cross-Sectional Area (in.^{2})
3	40, 60	0.376	0.375	0.11
4	40, 60	0.668	0.500	0.20
5	40, 60	1.043	0.625	0.31
6	40, 60, 75	1.502	0.750	0.44
7	60, 75	2.044	0.875	0.60
8	60, 75	2.67	1.000	0.79
9	60, 75	3.40	1.128	1.00
10	60, 75	4.30	1.270	1.27
11	60, 75	5.31	1.410	1.56
14	60, 75	7.65	1.693	2.25
18	60, 75	13.60	2.257	4.00

[a]The nominal dimensions of a deformed bar are equivalent to those of a plain round bar having the same weight per foot as the deformed bar.

[b]Bar numbers are based on the number of eighths of an inch included in the nominal diameter.

[c]Grade is nominal yield strength in ksi.

TABLE A-1M Areas, Weights, and Dimensions of Reinforcing Bars—SI Units

			Nominal Dimensions[a]	
Bar Size Designation No.[b]	Grades[c]	Nominal Mass (kg/m)	Diameter (mm)	Cross-Sectional Area (mm^2)
10	300, 420	0.560	9.5	71
13	300, 420	0.994	12.7	129
16	300, 420	1.552	15.9	199
19	300, 420, 520	2.235	19.1	284
22	420, 520	3.042	22.2	387
25	420, 520	3.973	25.4	510
29	420, 520	5.060	28.7	645
32	420, 520	6.404	32.3	819
36	420, 520	7.907	35.8	1006
43	420, 520	11.38	43	1452
57	420, 520	20.24	57.3	2581

[a]The nominal dimensions of a deformed bar are equivalent to those of a plain round bar having the same mass per meter as the deformed bar.

[b]Bar-designation numbers are the nominal diameter, rounded off to the nearest mm. The sequence of the bar-designation numbers is the sequence of U.S. Customary diameters rounded off to the nearest mm.

[c]Grade is nominal yield strength in MPa.

TABLE A-2 Welded-Wire Reinforcement

(a) Wires

Wire Size Number[a] Smooth	Wire Size Number[a] Deformed	Nominal Diameter (in.)	Area ($in.^2$ per ft of width for center-to-center spacing, in.) 4	6	10	12
W31	D31	0.628	0.93	0.62	0.372	0.31
W11	D11	0.374	0.33	0.22	0.132	0.11
W10	D10	0.356	0.30	0.20	0.12	0.10
W9	D9	0.338	0.27	0.18	0.108	0.09
W8	D8	0.319	0.24	0.16	0.096	0.08
W7	D7	0.298	0.21	0.14	0.084	0.07
W6	D6	0.276	0.18	0.12	0.072	0.06
W5.5	—	0.264	0.165	0.11	0.066	0.055
W5	D5	0.252	0.15	0.10	0.06	0.05
W4	D4	0.225	0.12	0.08	0.048	0.04
W3.5	—	0.211	0.105	0.07	0.042	0.035
W2.9	—	0.192	0.087	0.058	0.035	0.029
W2.5	—	0.178	0.075	0.05	0.03	0.025
W2.1	—	0.162	0.063	0.042	0.025	0.021
W1.4	—	0.135	0.042	0.028	0.017	0.014

[a]Wire size number is 100 times the wire area in $in.^2$.

(b) Common-Stock Welded-Wire Reinforcement

Style Designation[a]	Steel Area $in.^2$/ft Longitudinal	Steel Area $in.^2$/ft Transverse	Approximate Weight (lb/100 ft^2)
6 × 6—W2.9 × W2.9	0.058	0.058	42
4 × 4—W2.1 × W2.1	0.062	0.062	44
6 × 6—W4 × W4	0.080	0.080	58
4 × 4—W2.9 × W2.9	0.087	0.087	62
6 × 6—W5.5 × W5.5	0.110	0.110	80
4 × 4—W4 × W4	0.120	0.120	85
4 × 4—W5.5 × W5.5	0.165	0.165	119

[a]The numbers in the style designation refer to (longitudinal wire spacing × transverse wire spacing)–(longitudinal wire size × transverse wire size).

TABLE A–3 Values of Flexural Resistance Factor, *R* (psi) for Grade-60 Reinforcement

$R = \omega f'_c(1 - 0.59\,\omega)$ $\qquad \omega = \rho f_y / f'_c$

	Concrete Compressive Strength, f'_c (psi)				
ρ	3000	4000	5000	6000	7000
0.0033	190	192			
0.0035	201	203	205		
0.0039	223	226	228	229	
0.0042	240	243	245	246	247
0.005	282	287	289	291	292
0.006	335	341	345	347	349
0.007	385	394	399	403	405
0.008	435	446	453	457	461
0.009	483	497	506	511	515
0.010	529	547	558	565	570
0.011	574	596	609	617	623
0.012	618	644	659	669	676
0.013	660	690	708	720	729
0.014	**701**	736	757	771	781
0.015	**741**	781	804	820	832
0.016	**779**	824	851	869	882
0.017	**815**	867	897	918	932
0.018	**851**	908	942	965	982
0.019	**884**	**948**	987	1012	1030
0.020	**917**	**988**	1030	1058	1079
0.021	**948**	**1026**	1073	1104	1126
0.022	**977**	**1063**	**1114**	1149	1173
0.023		**1099**	**1155**	1193	1219
0.024		**1134**	**1195**	1236	1265
0.025		**1168**	**1235**	**1279**	1310
0.026		**1201**	**1273**	**1321**	1355
0.027		**1233**	**1310**	**1362**	**1399**
0.028		**1264**	**1347**	**1402**	**1442**
0.029		**1293**	**1383**	**1442**	**1485**
0.030			**1418**	**1481**	**1527**
0.032			**1485**	**1558**	**1609**
0.034			**1549**	**1631**	**1689**
0.036				**1701**	**1767**
0.038				**1769**	**1842**
0.040					**1915**
0.042					**1985**

Note: Regular numbers indicate section reinforcement ratios for tension-controlled behavior and $\rho \geq \rho$ (min). Bold numbers indicate transition behavior and thus, $0.65 \leq \phi \leq 0.9$. Numbers stop when section is in compression-controlled region of behavior.

TABLE A–3M Values of Flexural Resistance Factor, R (MPa) for Grade-420 Reinforcement

$R = \omega f'_c(1 - 0.59\,\omega)$ $\qquad \omega = \rho f_y / f'_c$

	Concrete Compressive Strength, f'_c (MPa)					
ρ	20	25	30	35	40	50
0.0033	1.34	1.35	1.36			
0.0035	1.41	1.42	1.43	1.43		
0.0038	1.52	1.54	1.55	1.55	1.56	
0.0042	1.67	1.69	1.70	1.71	1.72	1.73
0.005	1.97	2.00	2.01	2.03	2.03	2.05
0.006	2.33	2.37	2.40	2.41	2.43	2.45
0.007	2.69	2.74	2.77	2.79	2.81	2.84
0.008	3.03	3.09	3.14	3.17	3.19	3.23
0.009	3.36	3.44	3.50	3.54	3.57	3.06
0.010	3.68	3.78	3.85	3.90	3.94	3.99
0.011	3.99	4.12	4.20	4.26	4.31	4.37
0.012	4.29	4.44	4.54	4.61	4.67	4.74
0.013	4.58	4.76	4.87	4.96	5.02	5.11
0.014	**4.86**	5.06	5.20	5.30	5.37	5.47
0.015	**5.13**	5.36	5.52	5.63	5.71	5.83
0.016	**5.39**	5.65	5.83	5.96	6.05	6.19
0.017	**5.64**	**5.94**	6.14	6.28	6.39	6.54
0.018	**5.87**	**6.21**	6.44	6.60	6.72	6.89
0.019	**6.10**	**6.48**	6.73	6.91	7.04	7.23
0.020	**6.32**	**6.73**	**7.01**	7.21	7.36	7.57
0.021		**6.98**	**7.29**	7.51	7.67	7.90
0.022		**7.23**	**7.56**	**7.80**	**7.98**	8.23
0.023		**7.46**	**7.82**	**8.09**	**8.28**	8.56
0.024		**7.68**	**8.08**	**8.37**	**8.58**	8.88
0.025		**7.90**	**8.33**	**8.64**	**8.87**	9.20
0.026		**8.11**	**8.57**	**8.91**	**9.16**	9.51
0.027			**8.81**	**9.17**	**9.44**	**9.82**
0.028			**9.04**	**9.43**	**9.72**	**10.13**
0.029			**9.26**	**9.68**	**9.99**	**10.43**
0.030			**9.48**	**9.92**	**10.26**	**10.73**
0.032				**10.40**	**10.78**	**11.31**
0.034				**10.84**	**11.27**	**11.87**
0.036					**11.75**	**12.42**
0.038						**12.95**
0.040						**13.47**
0.042						**13.97**

Note: Regular numbers indicate section reinforcement ratios for tension-controlled behavior and $\rho \geq \rho$ (min). Bold numbers indicate transition behavior and thus, $0.65 \leq \phi \leq 0.9$. Numbers stop when section is in compression-controlled region of behavior.

TABLE A-4 Total Area (in.^{2}) of Multiple U.S. Reinforcement Bars

	Number of Bars					
Bar No.	1	2	3	4	5	6
3	0.11	0.22	0.33	0.44	0.55	0.66
4	0.20	0.40	0.60	0.80	1.00	1.20
5	0.31	0.62	0.93	1.24	1.55	1.86
6	0.44	0.88	1.32	1.76	2.20	2.64
7	0.60	1.20	1.80	2.40	3.00	3.60
8	0.79	1.58	2.37	3.16	3.95	4.74
9	1.00	2.00	3.00	4.00	5.00	6.00
10	1.27	2.54	3.81	5.08	6.35	7.62
11	1.56	3.12	4.68	6.24	7.80	9.36

TABLE A-4M Total Area (mm^2) of Multiple U.S. Metric Reinforcing Bars

	Number of Bars					
Bar No.	1	2	3	4	5	6
10	71	142	213	284	355	426
13	129	258	387	516	645	774
16	199	398	597	796	995	1190
19	284	568	852	1140	1420	1700
22	387	774	1160	1550	1930	2320
25	510	1020	1530	2040	2550	3060
29	645	1290	1930	2580	3220	3870
32	819	1640	2460	3280	4090	4910
36	1010	2010	3020	4020	5030	6040

TABLE A-5 Minimum Beam Width (in.) for Multiple U.S. Bars per Layer; Interior Exposure[a]

	Diameter	Number of Bars in Single Layer				
Bar No.	(in.)	2	3	4	5	6
4	0.50	7.0	8.5	10.0	11.5	13.0
5	0.625	7.0	8.5	10.5	12.0	13.5
6	0.75	7.0	9.0	11.0	12.5	14.0
7	0.875	7.5	9.0	11.0	13.0	15.0
8	1.00	7.5	9.5	11.5	13.5	15.5
9	1.128	8.0	10.0	12.5	14.5	17.0
10	1.27	8.0	10.5	13.0	15.5	18.0
11	1.41	8.5	11.0	14.0	17.0	19.5

[a]Clear cover of 1.5 in.; No. 3 double-leg stirrup; ¾ in. maximum-size aggregate.

TABLE A–5M Minimum Beam Width (mm) for Multiple U.S. Metric Bars per Layer; Interior Exposure[a]

	Diameter	Number of Bars in Single Layer				
Bar No.	(mm)	2	3	4	5	6
13	12.7	180	220	260	300	330
16	15.9	180	220	270	310	350
19	19.1	180	230	280	320	360
22	22.2	190	230	280	330	380
25	25.4	190	240	290	350	400
29	28.7	210	260	320	370	440
32	32.3	210	270	340	400	460
36	35.8	220	280	360	440	500

[a]Clear cover of 40 mm; No. 10 double-leg stirrup; 19 mm maximum-size aggregate.

TABLE A-6 Basic Tension Development-Length Ratio, ℓ_d/d_b (in./in.)

$$\ell_d = \text{(table value)} \times \frac{\psi_e}{\lambda} \times d_b, \text{ but not less than 12 in.}^a$$

Bar No.	$f'_c = 3000$ psi Bottom Bar	$f'_c = 3000$ psi Top Bar	$f'_c = 3750$ psi Bottom Bar	$f'_c = 3750$ psi Top Bar	$f'_c = 4000$ psi Bottom Bar	$f'_c = 4000$ psi Top Bar	$f'_c = 5000$ psi Bottom Bar	$f'_c = 5000$ psi Top Bar	$f'_c = 6000$ psi Bottom Bar	$f'_c = 6000$ psi Top Bar
Case 1: Clear spacing of bars being developed or spliced not less than d_b, clear cover not less than d_b, and stirrups or ties not less than the ACI Code minimum, throughout ℓ_d										
or										
Case 2: Clear spacing of bars being developed or spliced not less than $2d_b$ and clear cover not less than d_b.										
$f_y = 60{,}000$ psi, uncoated bars, normal-weight concrete										
3 to 6	43.8	57.0	39.2	50.9	37.9	49.3	33.9	44.1	31.0	40.3
7 to 18	54.8	71.2	49.0	63.7	47.4	61.7	42.4	55.2	38.7	50.3
$f_y = 40{,}000$ psi, uncoated bars, normal-weight concrete										
3 to 6	29.2	38.0	26.1	34.0	25.3	32.9	22.6	29.4	20.7	26.9
Other Cases										
$f_y = 60{,}000$ psi, uncoated bars, normal-weight concrete										
3 to 6	65.7	85.4	58.8	76.4	56.9	74.0	50.9	66.2	46.5	60.5
7 to 18	82.2	106.8	73.5	95.6	71.1	92.6	63.6	82.8	58.1	75.5
$f_y = 40{,}000$ psi, uncoated bars, normal-weight concrete										
3 to 6	43.8	57.0	39.2	51.0	38.0	49.4	33.9	44.1	31.1	40.4

[a] ψ_e coating factor; λ, lightweight-concrete factor.

TABLE A-6M Basic Tension Development-Length Ratio, ℓ_d/d_b (mm/mm)

$$\ell_d = (\text{table value}) \times \frac{\psi_e}{\lambda} \times d_b, \text{ but not less than 300 mm}^a$$

Bar No.	f'_c = 20 MPa Bottom Bar	f'_c = 20 MPa Top Bar	f'_c = 25 MPa Bottom Bar	f'_c = 25 MPa Top Bar	f'_c = 30 MPa Bottom Bar	f'_c = 30 MPa Top Bar	f'_c = 35 MPa Bottom Bar	f'_c = 35 MPa Top Bar	f'_c = 40 MPa Bottom Bar	f'_c = 40 MPa Top Bar
Case 1: Clear spacing of bars being developed or spliced not less than d_b, clear cover not less than d_b, and stirrups or ties not less than the ACI Code minimum, throughout ℓ_d										
or										
Case 2: Clear spacing of bars being developed or spliced not less than $2d_b$ and clear cover not less than d_b.										
f_y = 420 MPa, uncoated bars, normal-weight concrete										
10 to 19	44.7	58.1	40.0	52.0	36.5	47.5	33.8	43.9	31.6	41.1
22 to 57	55.2	71.8	49.4	64.2	45.1	58.6	41.8	54.3	39.1	50.8
f_y = 300 MPa, uncoated bars, normal-weight concrete										
10 to 19	31.9	41.5	28.6	37.1	26.1	33.9	24.1	31.4	22.6	29.4
Other Cases										
f_y = 420 MPa, uncoated bars, normal-weight concrete										
10 to 19	67.1	87.2	60.0	78.0	54.8	71.2	50.7	65.9	47.4	61.7
22 to 57	85.4	111	76.4	99.3	69.7	90.6	64.5	83.9	60.4	78.5
f_y = 300 MPa, uncoated bars, normal-weight concrete										
10 to 19	47.9	62.3	42.9	55.7	39.1	50.9	36.2	47.1	33.9	44.0

[a] ψ_e coating factor; λ, lightweight-concrete factor.

TABLE A-7 Basic Compression Development Length, ℓ_{dc} (in.)[a]

$$\ell_{dc} = \ell_{dc} \times \psi_r/\lambda$$

Bar No.	3000	4000	5000 psi and up
f_y = 60,000 psi			
3	8	8	8
4	11	9	9
5	14	12	11
6	16	14	14
7	19	17	16
8	22	19	18
9	25	21	20
10	28	24	23
11	34	27	25
14	37	32	30
18	49	43	41
f_y = 40,000 psi			
3	8	8	8
4	8	8	8
5	9	8	8
6	11	9	9

[a] Lengths may be reduced if excess reinforcement is anchored or if the splice is enclosed in confining reinforcement. See ACI Code Table 25.4.9.3 and Section 25.4.10. Reduced length shall not be less than 8 in.

TABLE A-7M Basic Compression Development Length, ℓ_{dc} (mm)[a]

$$\ell_{dc} = \ell_{dc} \times \psi_r/\lambda$$

Bar No.	f_c' (MPa) 20	25	30	35	40
	$f_y = 420$ (MPa)				
10	235	210	192	177	168
13	305	273	249	231	216
16	376	336	307	284	266
19	446	399	364	337	315
22	517	462	422	390	365
25	587	525	479	444	415
29	681	609	556	515	481
32	751	672	613	568	531
36	845	756	690	639	598
43	1010	903	824	763	714
57	1338	1197	1093	1012	946
	$f_y = 300$ (MPa)				
10	200	200	200	200	200
15	252	225	205	200	200
20	335	300	274	254	240

[a]Lengths may be reduced if excess reinforcement is anchored or if the splice is enclosed in confining reinforcement. See ACI Code Table 25.4.9.3 and Section 25.4.10. Reduced length shall not be less than 200 mm.

TABLE A-8 Basic Development Lengths for Hooked Bars, ℓ_{dh} (in.)

$\ell_{dh} = \ell_{dh} \times$ (Factors in ACI Code Table 25.4.3.2)[a]

Normal-weight concrete, $f_y = 60{,}000$ psi

Standard 90° or 180° Hooks

Bar No.	f_c' (psi) 3000	4000	5000	6000
3	8.2	7.1	6.4	5.8
4	11	9.5	8.5	7.8
5	13.7	11.9	10.6	9.7
6	16.4	14.2	12.7	11.6
7	19.2	16.6	14.8	13.6
8	22	19	17	15.5
9	25	21	19	17.5
10	28	24	22	20
11	31	27	24	22
14	37	32	29	26
18	49	43	38	35

[a]ℓ_{dh} is defined in Fig. 8-12a. The development length of a hook, ℓ_{dh}, is the product of ℓ_{dh} from this table and factors relating to epoxy coating, cover, presence of stirrups, and type of concrete, given in ACI Code Table 25.4.3.2. The resulting length, ℓ_{dh}, shall not be less than the larger of eight bar diameters or 6 in.

TABLE A-8M Basic Development Lengths for Hooked Bars, ℓ_{dh} (mm)

$\ell_{dh} = \ell_{dh} \times$ (Factors in ACI Code Table 25.4.3.2)[a]
Normal-weight concrete, $f_y = 400$ MPa
Standard 90° or 180° hooks

Bar No.	f'_c (MPa) 20	25	30	35	40
10	224	200	183	169	158
13	291	260	237	220	206
16	358	320	292	270	253
19	425	380	347	321	300
22	492	440	402	372	348
25	559	500	456	423	395
29	648	580	529	490	459
32	716	640	584	541	506
36	805	720	657	609	569
43	962	860	785	727	680
57	1275	1140	1041	963	901

[a]ℓ_{dh} is defined in Fig. 8-12a. The development length of a hook, ℓ_{dh}, is the product of ℓ_{dh} from this table and factors relating to epoxy coating, cover, presence of stirrups, and type of concrete, given in ACI Code Table 25.4.3.2. The resulting length, ℓ_{dh}, shall not be less than the larger of eight bar diameters or 150 mm.

TABLE A-9 Minimum Thicknesses of Nonprestressed Beams or One-Way Slabs Unless Deflections Are Computed

Exposure	Member	Minimum Thickness, h: Simply Supported	One End Continuous	Both Ends Continuous	Cantilever	Source
Not supporting or attached to partitions or other construction likely to be damaged by large deflections	Solid one-way slabs	$\ell/20$	$\ell/24$	$\ell/28$	$\ell/10$	ACI Code Table 7.3.1.1
	Beams or ribbed one-way slabs	$\ell/16$	$\ell/18.5$	$\ell/21$	$\ell/8$	ACI Code Table 9.3.1.1
Supporting or attached to partitions or other construction likely to be damaged by large deflections	All members: $\omega \leq 0.12$[a] and $\frac{\text{sustained load}}{\text{total load}} < 0.5$	$\ell/10$	$\ell/13$	$\ell/16$	$\ell/4$	[9-20]
	All members: $\frac{\text{sustained load}}{\text{total load}} > 0.5$	$\ell/6$	$\ell/8$	$\ell/10$	$\ell/3$	[9-20]

[a]$\omega = \rho f_y / f'_c$.

TABLE A-10 Maximum Number of Bars That Can Be Placed in Square Columns with the Same Number of Bars in Each Face, Based on Normal (Radial) Lap Splices; Minimum Bar Spacing[a]

b (in.)	A_g (in.[2])		Bar No. 5	6	7	8	9	10	11
10	100	n_{max}	8	4	4	4	4		
		A_{st}	2.48	1.76	2.40	3.16	4.00		
		ρ_g	0.025	0.018	0.024	0.032	0.040		
12	144	n_{max}	12	8	8	8	4	4	4
		A_{st}	3.72	3.52	4.80	6.32	4.00	5.08	6.24
		ρ_g	0.026	0.024	0.033	0.044	0.028	0.035	0.043
14	196	n_{max}	16	12	12	12	8	8	4
		A_{st}	4.96	5.28	7.20	9.48	8.00	10.16	6.24
		ρ_g	0.025	0.027	0.037	0.048	0.041	0.052	0.032
16	256	n_{max}	—	16	16	12	12	8	8
		A_{st}		7.04	9.60	9.48	12.00	10.16	12.48
		ρ_g		0.028	0.038	0.037	0.047	0.040	0.049
18	324	n_{max}	—	20	20	16	16	12	12
		A_{st}		8.80	12.00	12.64	16.00	15.24	18.72
		ρ_g		0.027	0.037	0.039	0.049	0.047	0.058
20	400	n_{max}	—	—	20	20	16	16	12
		A_{st}			12.0	15.80	16.00	20.32	18.72
		ρ_g			0.030	0.039	0.040	0.051	0.047
22	484	n_{max}	—	—	24	24	20	16	16
		A_{st}			14.40	18.96	20.00	20.32	24.96
		ρ_g			0.030	0.039	0.041	0.042	0.052
24	576	n_{max}	—	—	28	28	24	20	16
		A_{st}			16.80	22.12	24.00	25.40	24.96
		ρ_g			0.029	0.038	0.042	0.044	0.043
26	676	n_{max}	—	—	32	28	24	20	20
		A_{st}			19.20	22.12	24.00	25.40	31.20
		ρ_g			0.028	0.033	0.036	0.038	0.046
28	784	n_{max}	—	—	36	32	28	24	20
		A_{st}			21.60	25.28	28.00	30.48	31.20
		ρ_g			0.028	0.032	0.036	0.039	0.040
30	900	n_{max}	—	—	—	36	32	28	24
		A_{st}				28.44	32.00	35.56	37.44
		ρ_g				0.032	0.036	0.039	0.042
32	1024	n_{max}	—	—	—	40	32	28	28
		A_{st}				31.60	32.00	35.56	43.68
		ρ_g				0.031	0.031	0.035	0.043

[a]Based on 1-in.-maximum-size aggregate.

Source: From [11-10], the *Design Handbook*, reprinted with permission of the American Concrete Institute.

TABLE A-11 Maximum Number of Bars That Can Be Placed in Circular Columns, Based on Normal (Radial) Lap Splices; Minimum Bar Spacing[a]

Diameter (in.)	A_g (in.^{2})		Bar Size 5	6	7	8	9	10	11
12	113	n_{max}	8	7	6	6	—	—	—
		A_{st}	2.48	3.08	3.60	4.74			
		ρ_g	0.022	0.027	0.032	0.042			
14	154	n_{max}	11	10	9	8	7	—	—
		A_{st}	3.41	4.40	5.40	6.32	7.00		
		ρ_g	0.022	0.029	0.035	0.041	0.046		
16	201	n_{max}	14	13	12	11	9	7	6
		A_{st}	4.34	5.72	7.20	8.69	9.00	8.89	9.36
		ρ_g	0.022	0.029	0.036	0.043	0.045	0.044	0.047
18	254	n_{max}	—	16	14	13	11	9	8
		A_{st}		7.04	8.40	10.27	11.00	11.43	12.48
		ρ_g		0.028	0.033	0.040	0.043	0.045	0.049
20	314	n_{max}	—	—	17	16	13	11	10
		A_{st}			10.20	12.64	13.00	13.97	15.60
		ρ_g			0.033	0.040	0.041	0.044	0.050
22	380	n_{max}	—	—	20	18	16	13	12
		A_{st}			12.00	14.22	16.00	16.51	18.72
		ρ_g			0.032	0.037	0.042	0.043	0.049
24	452	n_{max}	—	—	22	21	18	15	13
		A_{st}			13.20	16.59	18.00	19.05	20.28
		ρ_g			0.029	0.037	0.040	0.042	0.045
26	531	n_{max}	—	—	25	23	20	17	15
		A_{st}			15.00	18.17	20.00	21.59	23.40
		ρ_g			0.028	0.034	0.038	0.041	0.044
28	616	n_{max}	—	—	28	26	22	19	17
		A_{st}			16.80	20.54	22.00	24.13	26.52
		ρ_g			0.027	0.033	0.036	0.039	0.043
30	707	n_{max}	—	—	—	28	25	21	19
		A_{st}				22.12	25.00	26.67	29.64
		ρ_g				0.031	0.035	0.038	0.042
32	804	n_{max}	—	—	—	31	27	23	21
		A_{st}				24.29	27.00	29.21	32.76
		ρ_g				0.031	0.034	0.036	0.041

[a]Based on No. 4 spirals or ties, 1-in.-maximum-size aggregate, and $1\frac{1}{2}$—*in* clear cover to spirals.

Source: This table is an abridged version of a table in [11-10] and is printed with the permission of the American Concrete Institute.

TABLE A-12 Number of Bars Required to Provide a Given Area of Steel[a]

	Bar No.				
	5	6	7	8	9
Area (in.^{2})	0.31	0.44	0.60	0.79	1.00
1.24	**4**				
1.76		**4**			
1.86	6				
2.40			**4**		
2.48	**8**				
2.64		6			
3.16				**4**	
3.52		**8**			
3.60			6		
3.72	**12**				
4.00					**4**
4.40		10			
4.74				6	
4.80			**8**		
5.28		**12**			
6.00			10		6
6.32				**8**	
7.04		**16**			
7.20			**12**		
7.90				10	
8.00					**8**
8.40			14		
8.80		**20**			
9.48				**12**	
9.60			**16**		
10.00					10
10.80			18		

	Bar No.				
	8	9	10	11	14
Area (in.^{2})	0.79	1.00	1.27	1.56	2.25
10.16			**8**		
11.06	14				
12.00		**12**			
12.48					**8**
12.64	**16**				
12.70			10		
13.50					6
14.00		14			
14.22	18				
15.24			**12**		
15.60				10	
15.80	**20**				
16.00		**16**			
17.78			14		
18.00		18			**8**
18.72				**12**	
20.00		**20**			
20.32			**16**		
21.84				14	
22.00		22			
22.50					10
22.86			18		
24.00		**24**			
24.96					**16**

[a]Bold figures denote combinations that will give an equal number of bars in each side of a square column.

TABLE A-13 Lap-Splice Lengths for Grade-60 Bars in Columns (in.)

f_c' (psi)	Bar No. 5	6	7	8	9	10	11
Compression lap splices							
Lap splice length = (length from table) × (factors in note[a])							
<3000	26	31	35	40	46	51	56
≥3000	19	23	26	30	34	38	42
Tension lap splices							
Lap splice length = (length from table) × ψ_e/λ (note[b])							
Class A tension lap splice: half or fewer of the bars spliced at any location and $0 \le f_s \le 0.5f_y$ in tension (ACI Code Table 10.7.5.2.2)							
3000	27.4	32.9	48.0	54.8	61.8	69.6	77.3
4000	23.7	28.4	41.5	47.4	53.5	60.2	66.8
5000	21.2	25.4	37.1	42.4	47.8	53.8	59.8
6000	19.4	23.3	33.9	38.7	43.7	49.1	54.6
Class B tension lap splices: more than half of the bars spliced at any section and/or f_s greater than $0.5f_y$ in tension (ACI Code Table 10.7.5.2.2)							
3000	35.6	42.7	62.3	71.2	80.4	90.5	100
4000	30.8	37.0	53.9	61.6	69.5	78.3	86.9
5000	27.5	33.0	48.2	55.1	62.1	70.0	77.7
6000	25.2	30.2	44.0	50.3	56.7	63.9	70.9

[a]Compression lap splices may be multiplied by 0.83 or 0.75 if enclosed by ties or spirals satisfying ACI Code Section 10.7.5.2.1.

[b]ψ_e = coating factor, λ = lightweight − concrete factor.

TABLE A-14 Moment-Distribution Factors for Slabs without Drop Panels[a]

FEM (uniform load w) = $Mw\ell_2\ell_1^2$ K (stiffness) = $kE\ell_2t^3/12\ell_1$

Carryover factor = *COF*

c_1/ℓ_1		c_2/ℓ_2 = 0.00	0.05	0.10	0.15	0.20	0.25
0.00	*M*	0.083	0.083	0.083	0.083	0.083	0.083
	k	4.000	4.000	4.000	4.000	4.000	4.000
	COF	0.500	0.500	0.500	0.500	0.500	0.500
0.05	*M*	0.083	0.084	0.084	0.084	0.085	0.085
	k	4.000	4.047	4.093	4.138	4.181	4.222
	COF	0.500	0.503	0.507	0.510	0.513	0.516
0.10	*M*	0.083	0.084	0.085	0.085	0.086	0.087
	k	4.000	4.091	4.182	4.272	4.362	4.449
	COF	0.500	0.506	0.513	0.519	0.524	0.530
0.15	*M*	0.083	0.084	0.085	0.086	0.087	0.088
	k	4.000	4.132	4.267	4.403	4.541	4.680
	COF	0.500	0.509	0.517	0.526	0.534	0.543
0.20	*M*	0.083	0.085	0.086	0.087	0.088	0.089
	k	4.000	4.170	4.346	4.529	4.717	4.910
	COF	0.500	0.511	0.522	0.532	0.543	0.554
0.25	*M*	0.083	0.085	0.086	0.087	0.089	0.090
	k	4.000	4.204	4.420	4.648	4.887	5.138
	COF	0.500	0.512	0.525	0.538	0.550	0.563
$x = (1 - c_2/\ell_2^3)$		1.000	0.856	0.729	0.613	0.512	0.421

[a]c_1 and c_2 are the widths of the column measured parallel to ℓ_1 and ℓ_2.

Source: [13-16].

TABLE A-15 Moment-Distribution Factors for Slabs with Drop Panels; $h_1 = 1.5h^a$

FEM (uniform load w) = $Mw\ell_2\ell_1^2$ K (stiffness) = $kE\ell_2 t^3/12\ell_1$

Carryover factor = *COF*

c_1/ℓ_1		c_2/ℓ_2 = 0.00	0.05	0.10	0.15	0.20	0.25	0.30
0.00	*M*	0.088	0.088	0.088	0.088	0.088	0.088	0.088
	k	4.795	4.795	4.795	4.795	4.795	4.795	4.795
	COF	0.542	0.542	0.542	0.542	0.542	0.542	0.542
0.05	*M*	0.088	0.088	0.089	0.089	0.089	0.089	0.090
	k	4.795	4.846	4.896	4.944	4.990	5.035	5.077
	COF	0.542	0.545	0.548	0.551	0.553	0.556	0.558
0.10	*M*	0.088	0.088	0.089	0.090	0.090	0.091	0.091
	k	4.795	4.894	4.992	5.039	5.184	5.278	5.368
	COF	0.542	0.548	0.553	0.559	0.564	0.569	0.573
0.15	*M*	0.088	0.089	0.090	0.090	0.091	0.092	0.092
	k	4.795	4.938	5.082	5.228	5.374	5.520	5.665
	COF	0.542	0.550	0.558	0.565	0.573	0.580	0.587
0.20	*M*	0.088	0.089	0.090	0.091	0.092	0.093	0.094
	k	4.795	4.978	5.167	5.361	5.558	5.760	5.962
	COF	0.542	0.552	0.562	0.571	0.581	0.590	0.600
0.25	*M*	0.088	0.089	0.090	0.091	0.092	0.094	0.095
	k	4.795	5.015	5.245	5.485	5.735	5.994	6.261
	COF	0.542	0.553	0.565	0.576	0.587	0.598	0.609
0.30	*M*	0.088	0.089	0.090	0.092	0.093	0.094	0.095
	k	4.795	5.048	5.317	5.601	5.902	6.219	6.550
	COF	0.542	0.554	0.567	0.580	0.593	0.605	0.618

[a] *h*, Slab thickness; h_1, total thickness in drop panel.

Source: [13-16].

TABLE A-16 Moment-Distribution Factors for Slabs with Drop Panels; $h_1 = 1.5h^a$

FEM (uniform load w) = $Mw\ell_2\ell_1^2$ K (stiffness) = $kE\ell_2 t^3/12\ell_1$

Carryover factor = *COF*

c_1/ℓ_1		c_2/ℓ_2 = 0.00	0.05	0.10	0.15	0.20	0.25
0.00	*M*	0.093	0.093	0.093	0.093	0.093	0.093
	k	5.837	5.837	5.837	5.837	5.837	5.837
	COF	0.589	0.589	0.589	0.589	0.589	0.589
0.05	*M*	0.093	0.093	0.093	0.093	0.094	0.094
	k	5.837	5.890	5.942	5.993	6.041	6.087
	COF	0.589	0.591	0.594	0.596	0.598	0.600
0.10	*M*	0.093	0.093	0.094	0.094	0.094	0.095
	k	5.837	5.940	6.024	6.142	6.240	6.335
	COF	0.589	0.593	0.598	0.602	0.607	0.611
0.15	*M*	0.093	0.093	0.094	0.095	0.095	0.096
	k	5.837	5.986	6.135	6.284	6.432	6.579
	COF	0.589	0.595	0.602	0.608	0.614	0.620
0.20	*M*	0.093	0.093	0.094	0.095	0.096	0.096
	k	5.837	6.027	6.221	6.418	6.616	6.816
	COF	0.589	0.597	0.605	0.613	0.621	0.628
0.25	*M*	0.093	0.094	0.094	0.095	0.096	0.097
	k	5.837	6.065	6.300	6.543	6.790	7.043
	COF	0.589	0.598	0.608	0.617	0.626	0.635

[a] *h*, Slab thickness; h_1, total thickness in drop panel.

Source: [13-16].

TABLE A-17 Stiffness and Carryover Factors for Columns

$K_c = k\frac{EI_c}{\ell_c}$

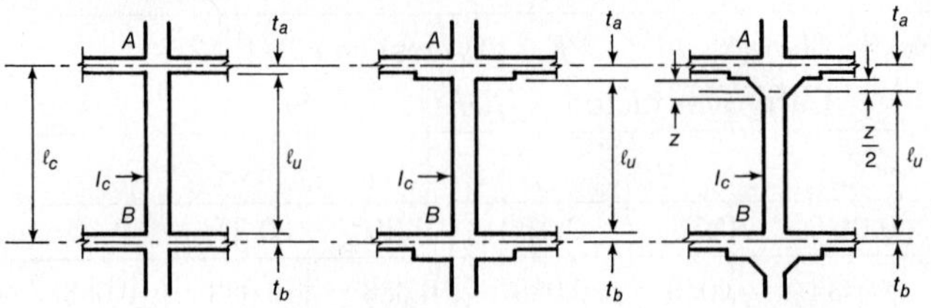

t_a/t_b		ℓ_c/ℓ_u = 1.05	1.10	1.15	1.20	1.25	1.30	1.35	1.40	1.45
0.00	k_{AB}	4.20	4.40	4.60	4.80	5.00	5.20	5.40	5.60	5.80
	C_{AB}	0.57	0.65	0.73	0.80	0.87	0.95	1.03	1.10	1.17
0.2	k_{AB}	4.31	4.62	4.95	5.30	5.65	6.02	6.40	6.79	7.20
	C_{AB}	0.56	0.62	0.68	0.74	0.80	0.85	0.91	0.96	1.01
0.4	k_{AB}	4.38	4.79	5.22	5.67	6.15	6.65	7.18	7.74	8.32
	C_{AB}	0.55	0.60	0.65	0.70	0.74	0.79	0.83	0.87	0.91
0.6	k_{AB}	4.44	4.91	5.42	5.96	6.54	7.15	7.81	8.50	9.23
	C_{AB}	0.55	0.59	0.63	0.67	0.70	0.74	0.77	0.80	0.83
0.8	k_{AB}	4.49	5.01	5.58	6.19	6.85	7.56	8.31	9.12	9.98
	C_{AB}	0.54	0.58	0.61	0.64	0.67	0.70	0.72	0.75	0.77
1.0	k_{AB}	4.52	5.09	5.71	6.38	7.11	7.89	8.73	9.63	10.60
	C_{AB}	0.54	0.57	0.60	0.62	0.65	0.67	0.69	0.71	0.73
1.2	k_{AB}	4.55	5.16	5.82	6.54	7.32	8.17	9.08	10.07	11.12
	C_{AB}	0.53	0.56	0.59	0.61	0.63	0.65	0.66	0.68	0.69
1.4	k_{AB}	4.58	5.21	5.91	6.68	7.51	8.41	9.38	10.43	11.57
	C_{AB}	0.53	0.55	0.58	0.60	0.61	0.63	0.64	0.65	0.66
1.6	k_{AB}	4.60	5.26	5.99	6.79	7.66	8.61	9.64	10.75	11.95
	C_{AB}	0.53	0.55	0.57	0.59	0.60	0.61	0.62	0.63	0.64
1.8	k_{AB}	4.62	5.30	6.06	6.89	7.80	8.79	9.87	11.03	12.29
	C_{AB}	0.52	0.55	0.56	0.58	0.59	0.60	0.61	0.61	0.62
2.0	k_{AB}	4.63	5.34	6.12	6.98	7.92	8.94	10.06	11.27	12.59
	C_{AB}	0.52	0.54	0.56	0.57	0.58	0.59	0.59	0.60	0.60
2.2	k_{AB}	4.65	5.37	6.17	7.05	8.02	9.08	10.24	11.49	12.85
	C_{AB}	0.52	0.54	0.55	0.56	0.57	0.58	0.58	0.59	0.59
2.4	k_{AB}	4.66	5.40	6.22	7.12	8.11	9.20	10.39	11.68	13.08
	C_{AB}	0.52	0.53	0.55	0.56	0.56	0.57	0.57	0.58	0.58
2.6	k_{AB}	4.67	5.42	6.26	7.18	8.20	9.31	10.53	11.86	13.29
	C_{AB}	0.52	0.53	0.54	0.55	0.56	0.56	0.56	0.57	0.57
2.8	k_{AB}	4.68	5.44	6.29	7.23	8.27	9.41	10.66	12.01	13.48
	C_{AB}	0.52	0.53	0.54	0.55	0.55	0.55	0.56	0.56	0.56
3.0	k_{AB}	4.69	5.46	6.33	7.28	8.34	9.50	10.77	12.15	13.65
	C_{AB}	0.52	0.53	0.54	0.54	0.55	0.55	0.55	0.55	0.55
3.5	k_{AB}	4.71	5.50	6.40	7.39	8.48	9.69	11.01	12.46	14.02
	C_{AB}	0.51	0.52	0.53	0.53	0.54	0.54	0.54	0.53	0.53
4.0	k_{AB}	4.72	5.54	6.45	7.47	8.60	9.84	11.21	12.70	14.32
	C_{AB}	0.51	0.52	0.52	0.53	0.53	0.52	0.52	0.52	0.52
4.5	k_{AB}	4.73	5.56	6.50	7.54	8.69	9.97	11.37	12.89	14.57
	C_{AB}	0.51	0.52	0.52	0.52	0.52	0.52	0.51	0.51	0.51
5.0	k_{AB}	4.75	5.59	6.54	7.60	8.78	10.07	11.50	13.07	14.77
	C_{AB}	0.51	0.51	0.52	0.52	0.51	0.51	0.51	0.50	0.49
6.0	k_{AB}	4.76	5.63	6.60	7.69	8.90	10.24	11.72	13.33	15.10
	C_{AB}	0.51	0.51	0.51	0.51	0.50	0.50	0.49	0.49	0.48
7.0	k_{AB}	4.78	5.66	6.65	7.76	9.00	10.37	11.88	13.54	15.34
	C_{AB}	0.51	0.51	0.51	0.50	0.50	0.49	0.48	0.48	0.47
8.0	k_{AB}	4.78	5.68	6.69	7.82	9.07	10.47	12.01	13.70	15.54
	C_{AB}	0.51	0.51	0.50	0.50	0.49	0.49	0.48	0.47	0.46
9.0	k_{AB}	4.80	5.71	6.74	7.89	9.18	10.61	12.19	13.93	15.83
	C_{AB}	0.50	0.50	0.50	0.49	0.48	0.48	0.47	0.46	0.45

Source: [13-15], courtesy of the Portland Cement Association.

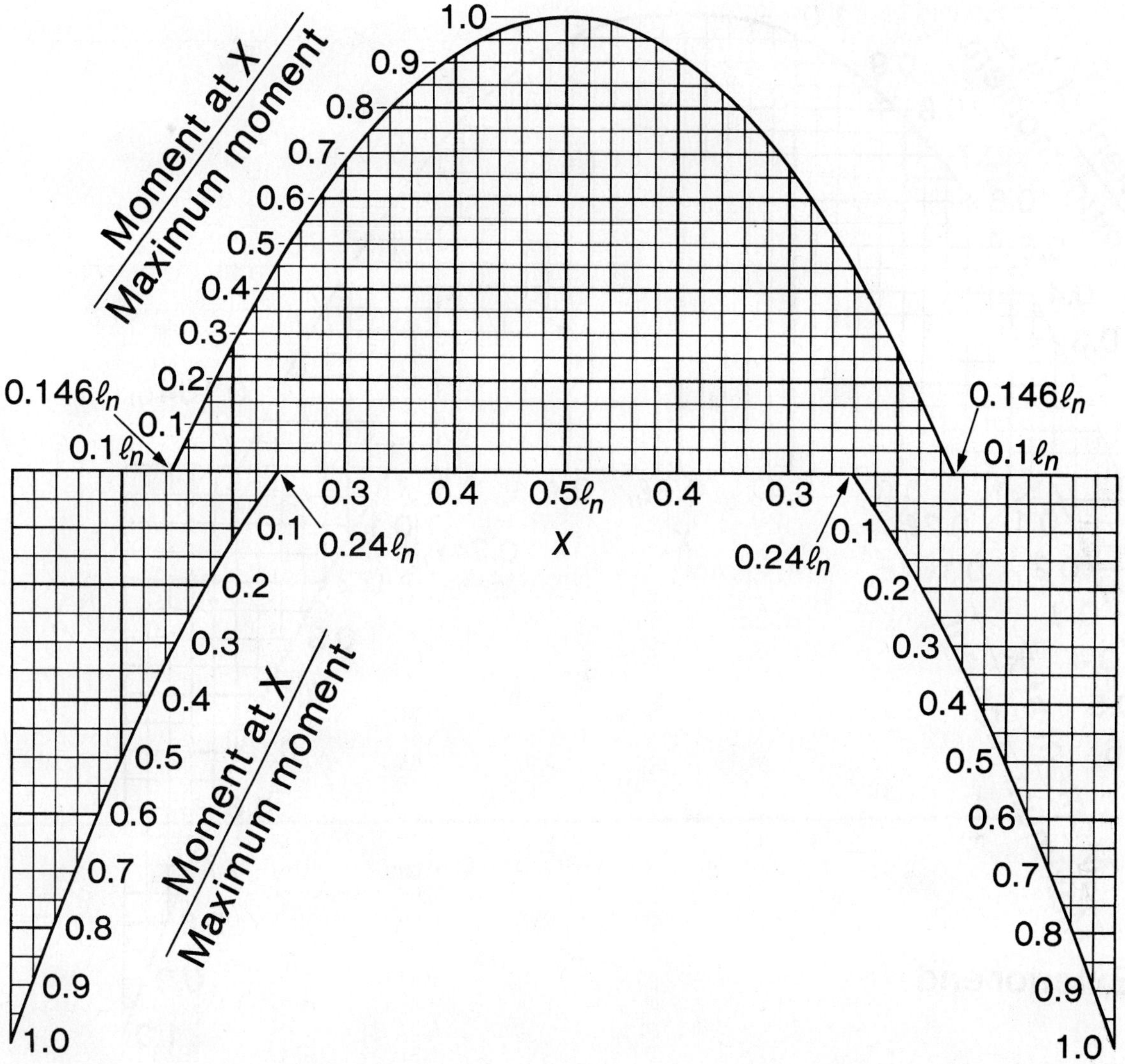

Fig. A-1
Bending-moment envelope for typical interior span (moment coefficients: $-1/11$, $+1/16$, $-1/11$).

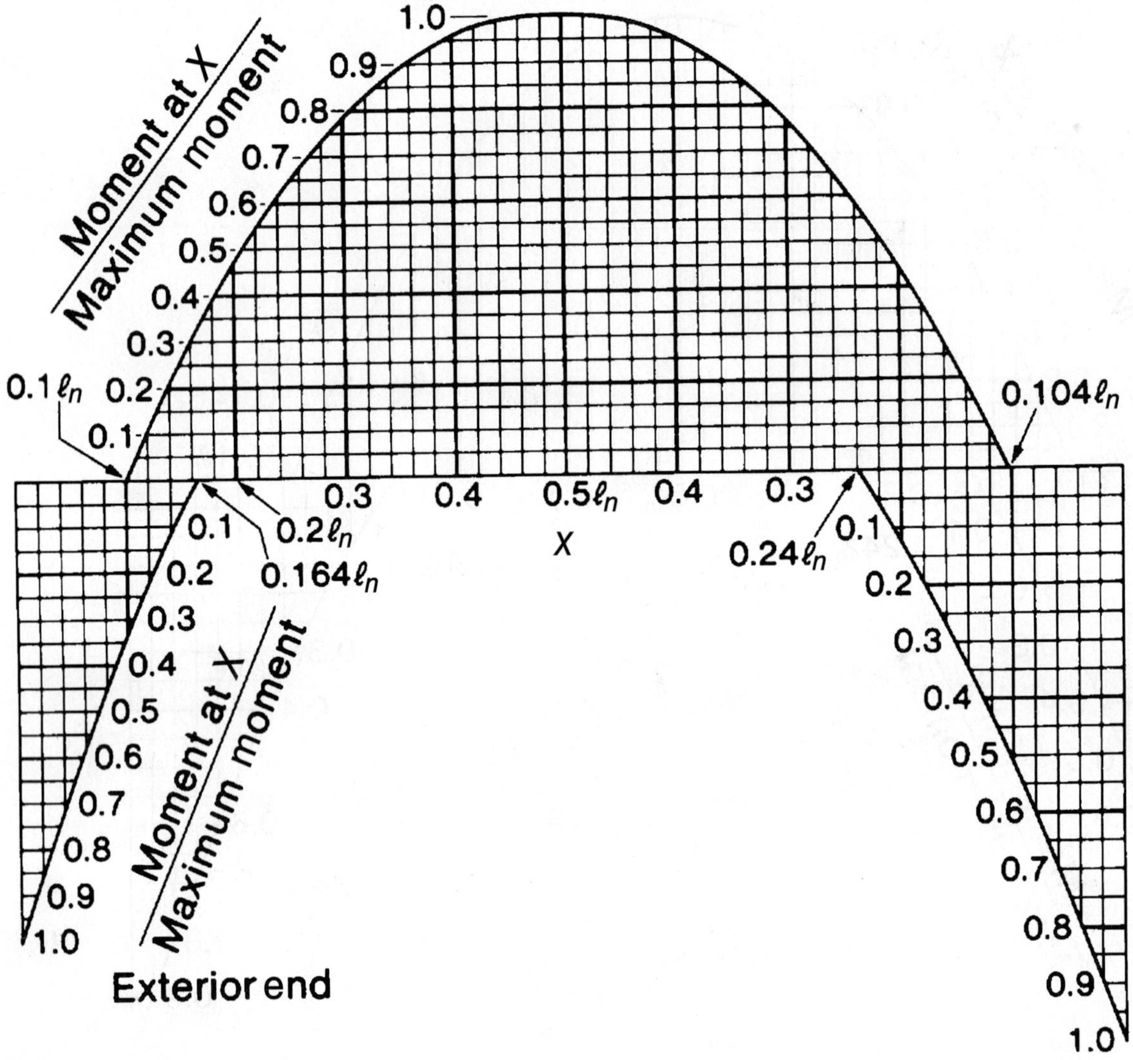

Fig. A-2
Bending-moment envelope for exterior span with exterior support built integrally with a column (moment coefficients: $-1/16$, $+1/14$, $-1/10$).

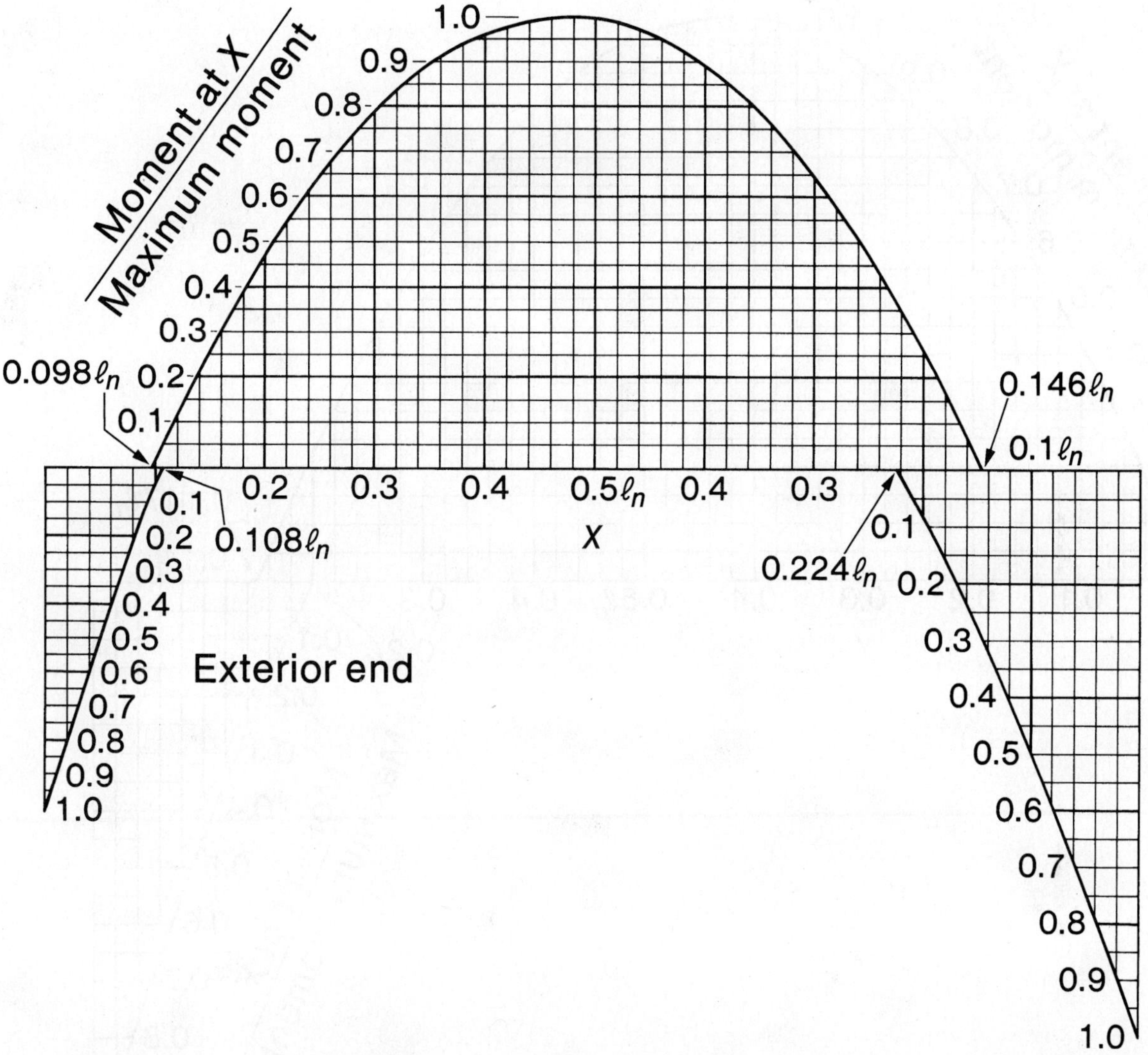

Fig. A-3
Bending-moment envelope for exterior span with exterior support built integrally with a spandrel beam or girder (moment coefficients: $-1/24$, $+1/14$, $-1/10$).

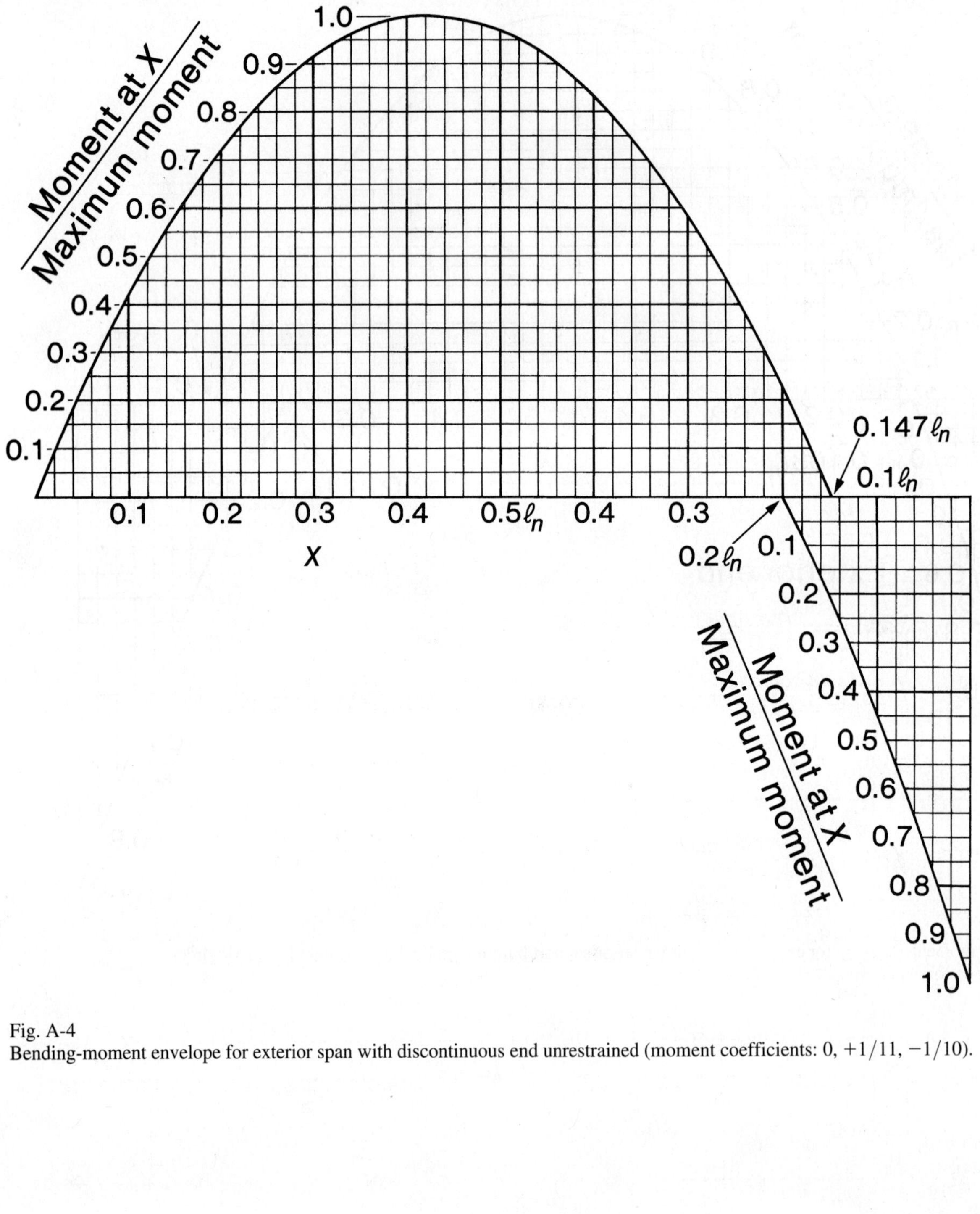

Fig. A-4
Bending-moment envelope for exterior span with discontinuous end unrestrained (moment coefficients: 0, $+1/11$, $-1/10$).

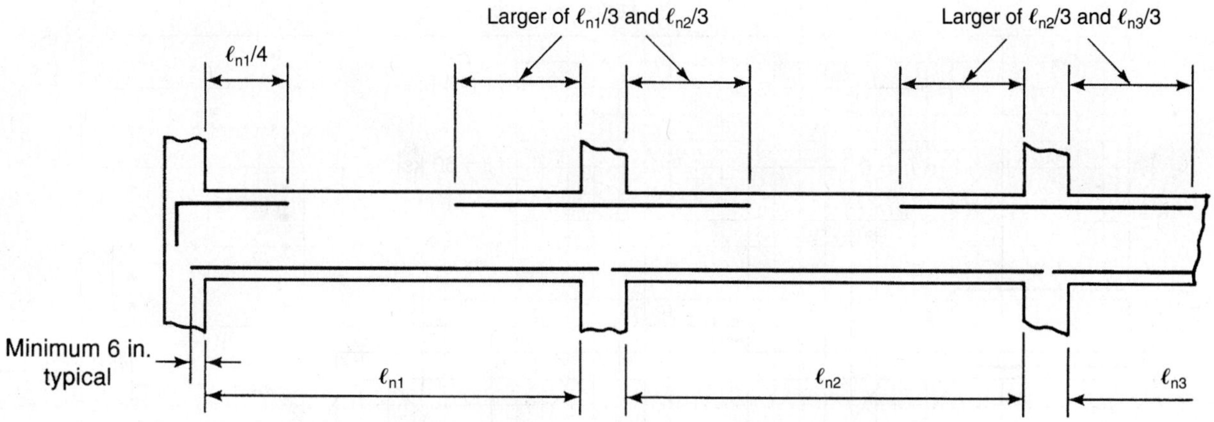

(a) Beam with closed stirrups. If closed stirrups are not provided, see ACI Code Section 9.7.7.2.

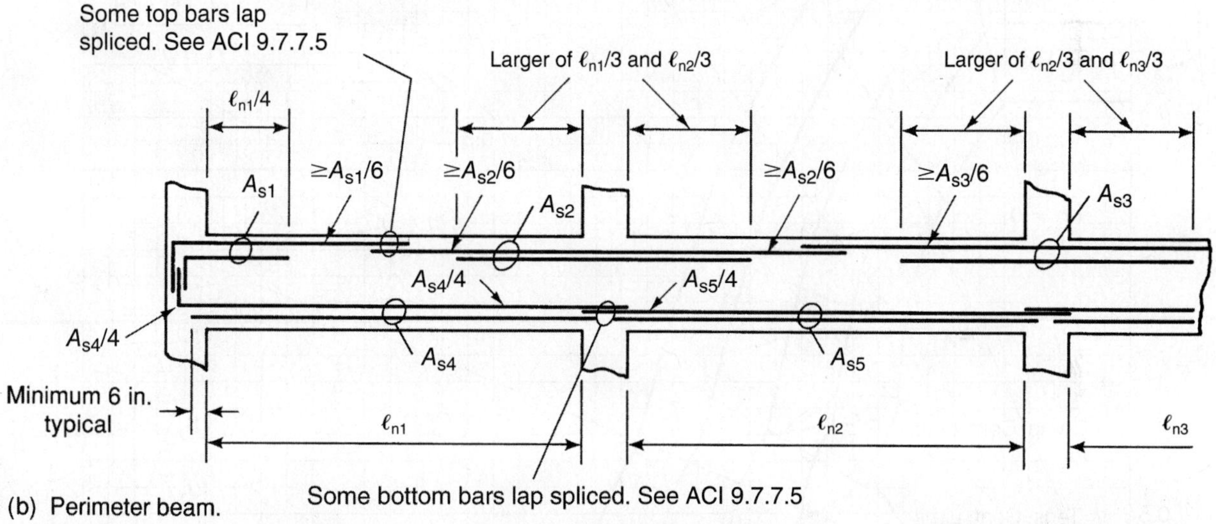

(b) Perimeter beam.

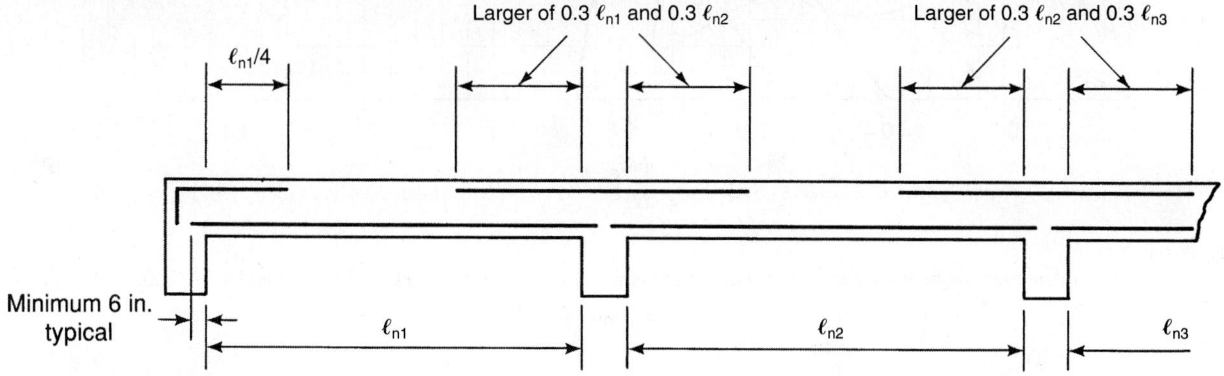

(c) One-way slab.

Fig. A-5
Standard bar details.

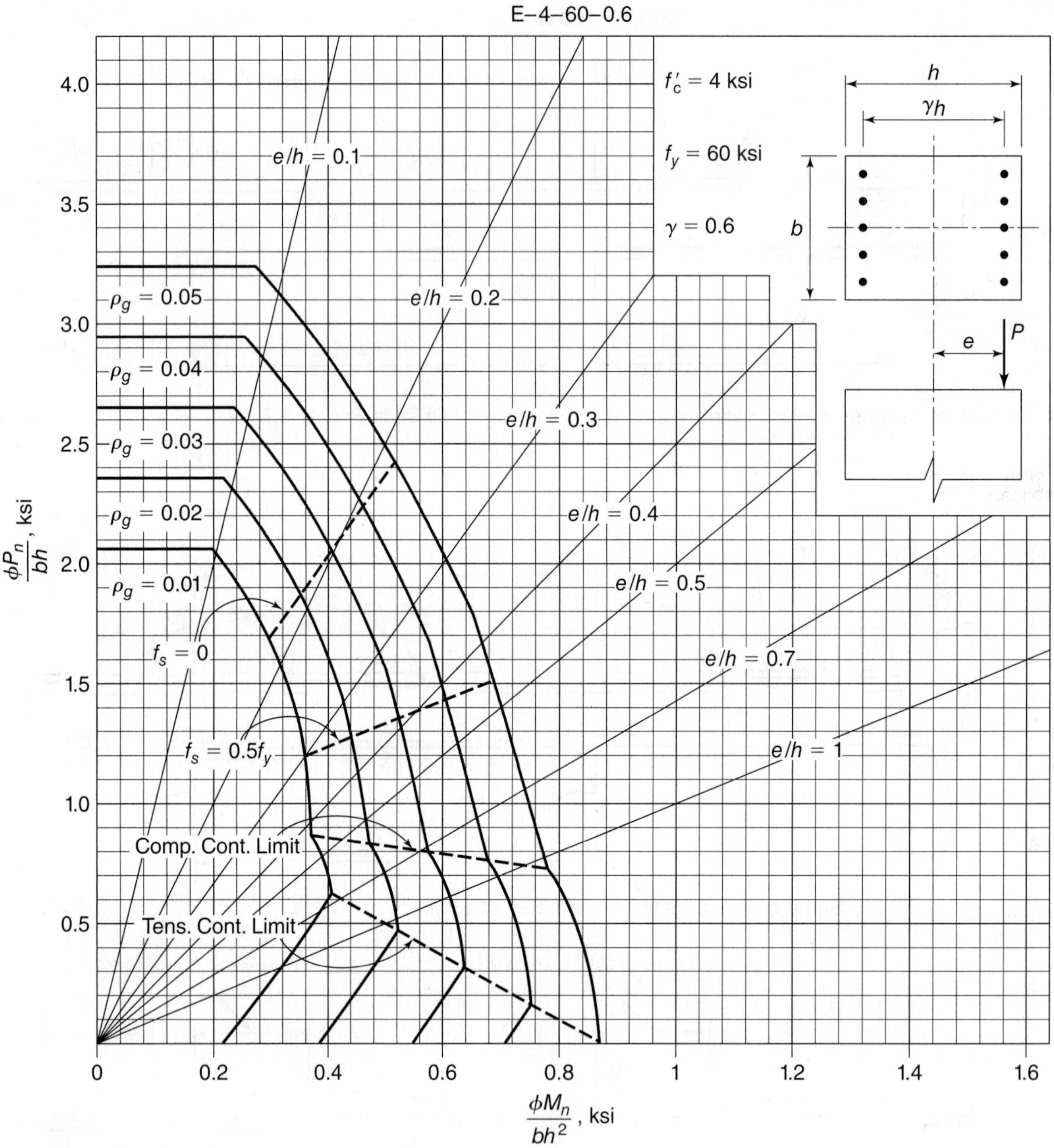

Fig. A-6a
Nondimensional interaction diagram for rectangular tied columns with bars in two faces: $f'_c = 4000$ psi and $\gamma = 0.60$.

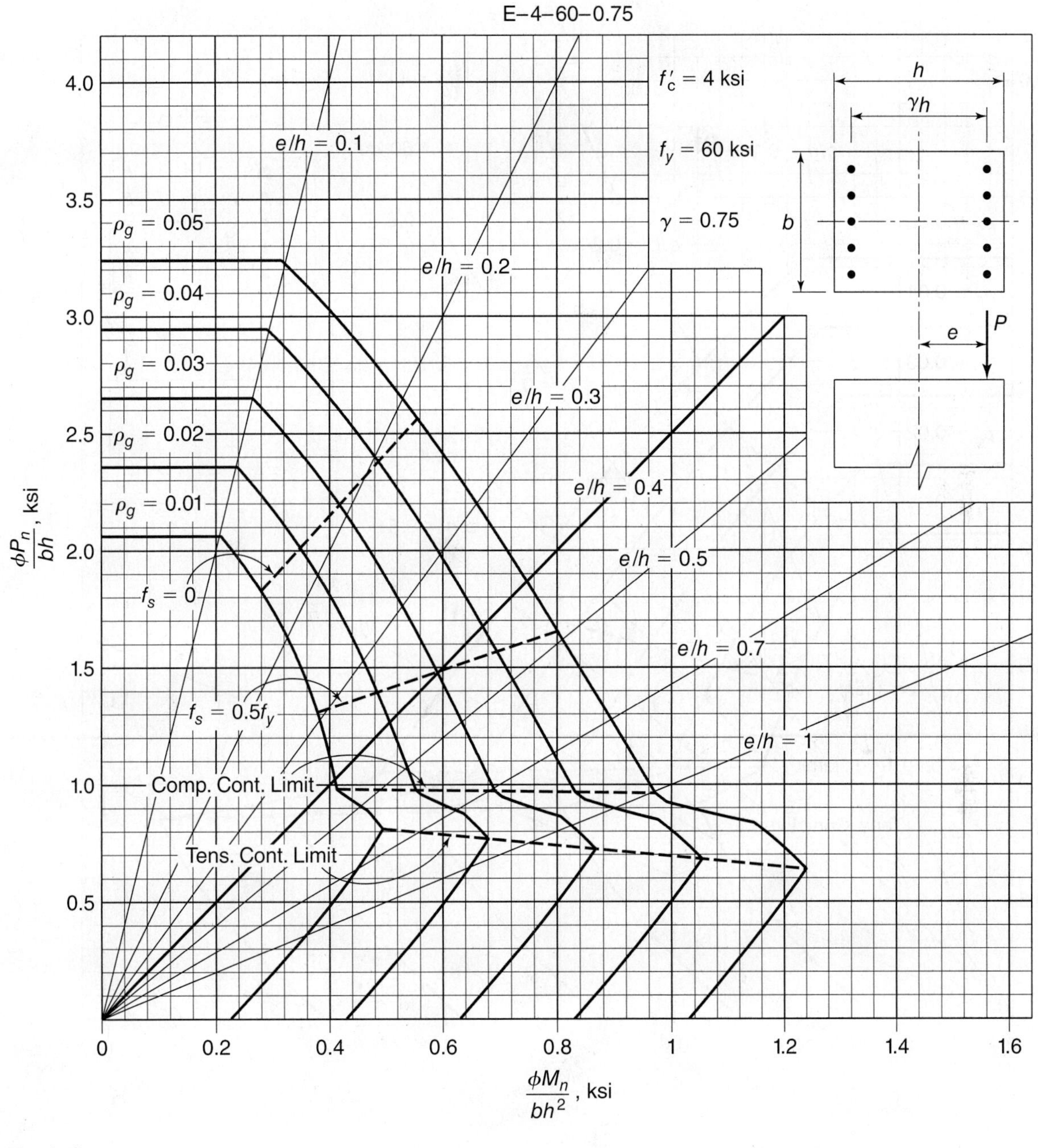

Fig. A-6b
Nondimensional interaction diagram for rectangular tied column with bars in two faces: f'_c = 4000 psi and γ = 0.75.

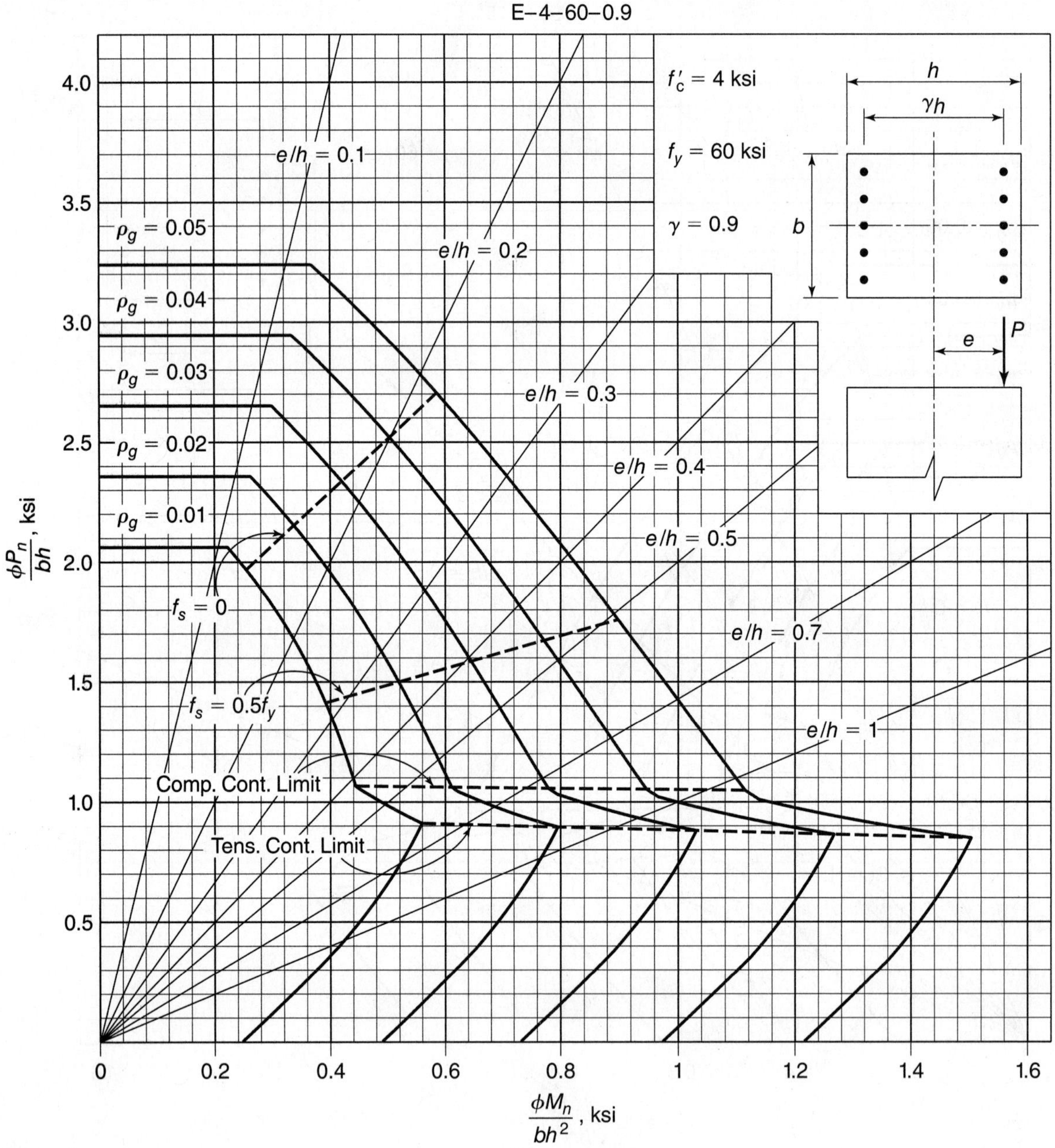

Fig. A-6c
Nondimensional interaction diagram for rectangular tied column with bars in two faces: $f'_c = 4000$ psi and $\gamma = 0.90$.

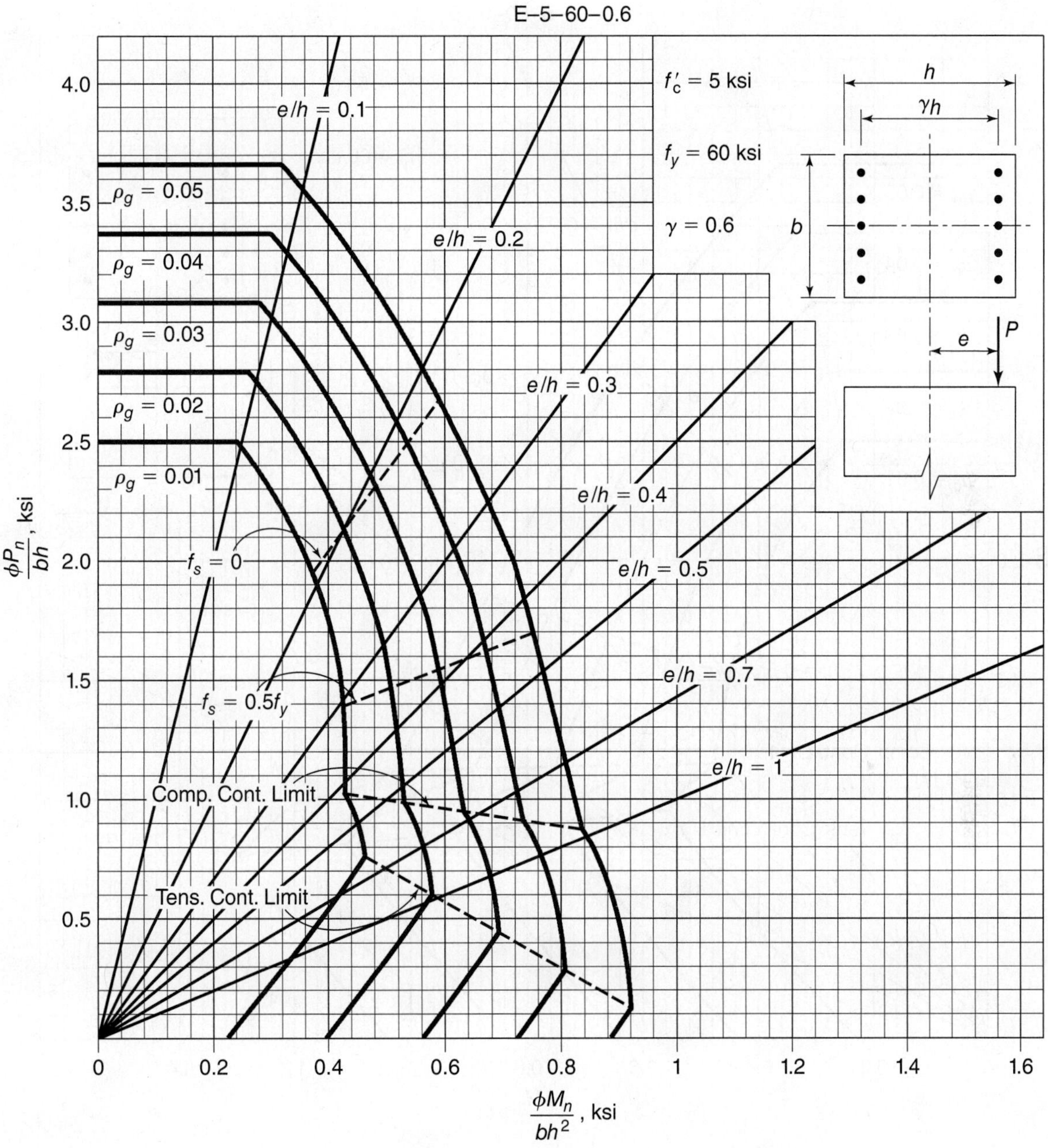

Fig. A-7a
Nondimensional interaction diagram for rectangular tied column with bars in two faces: $f'_c = 5000$ psi and $\gamma = 0.60$.

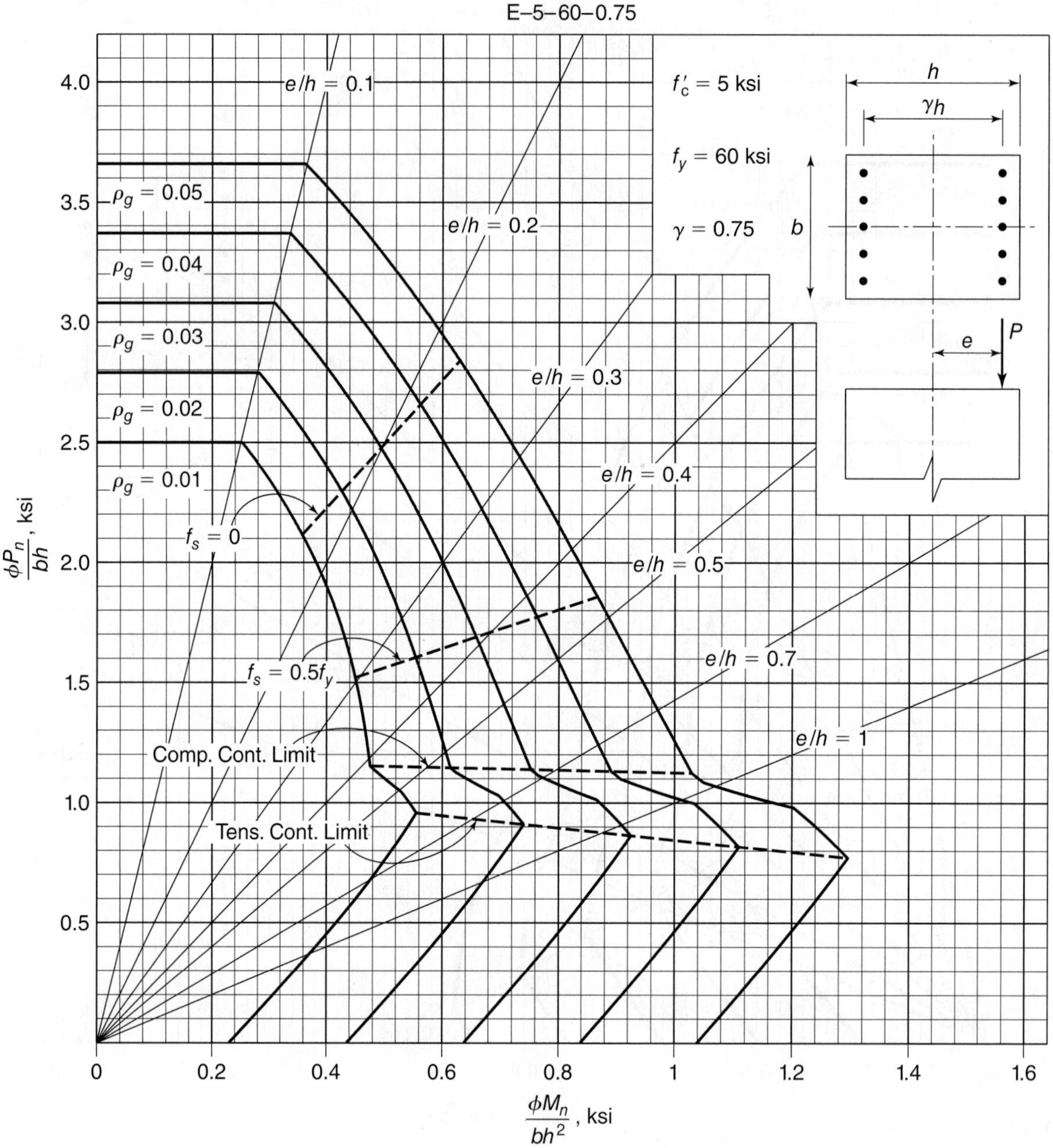

Fig. A-7b
Nondimensional interaction diagram for rectangular tied column with bars in two faces: $f'_c = 5000$ psi and $\gamma = 0.75$.

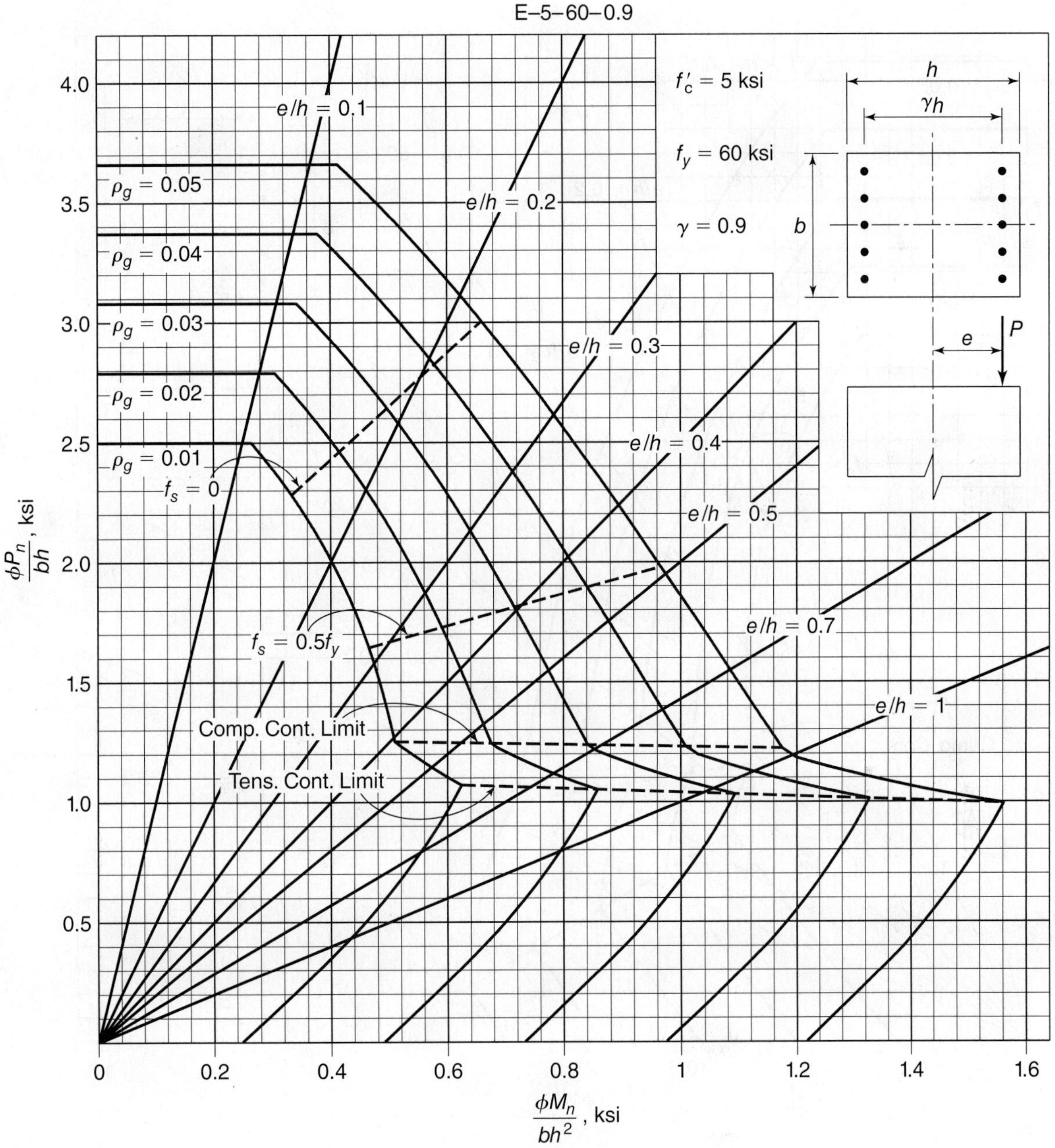

Fig. A-7c
Nondimensional interaction diagram for rectangular tied column with bars in two faces: $f'_c = 5000$ psi and $\gamma = 0.90$.

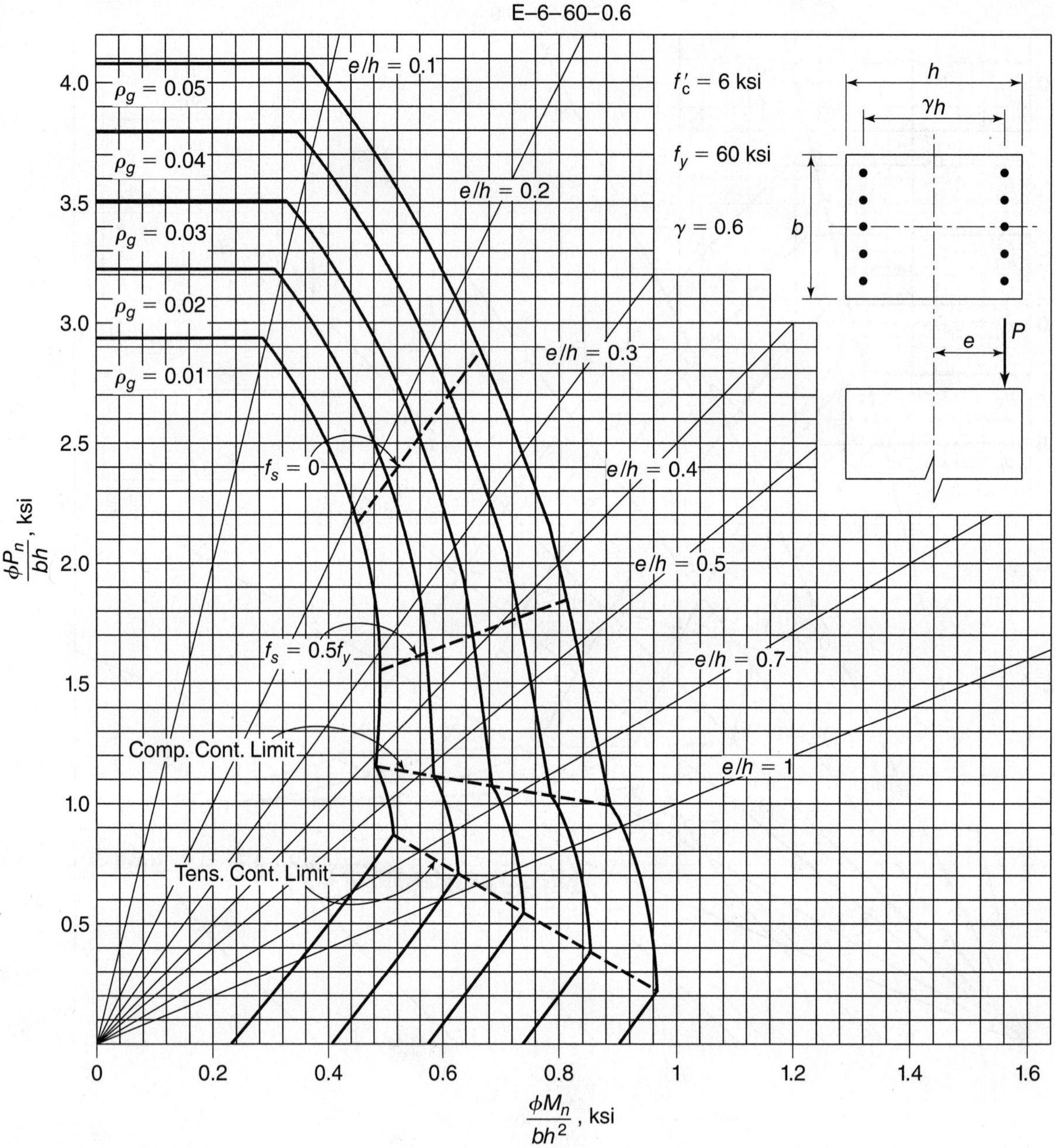

Fig. A-8a
Nondimensional interaction diagram for rectangular tied column with bars in two faces: $f'_c = 6000$ psi and $\gamma = 0.60$.

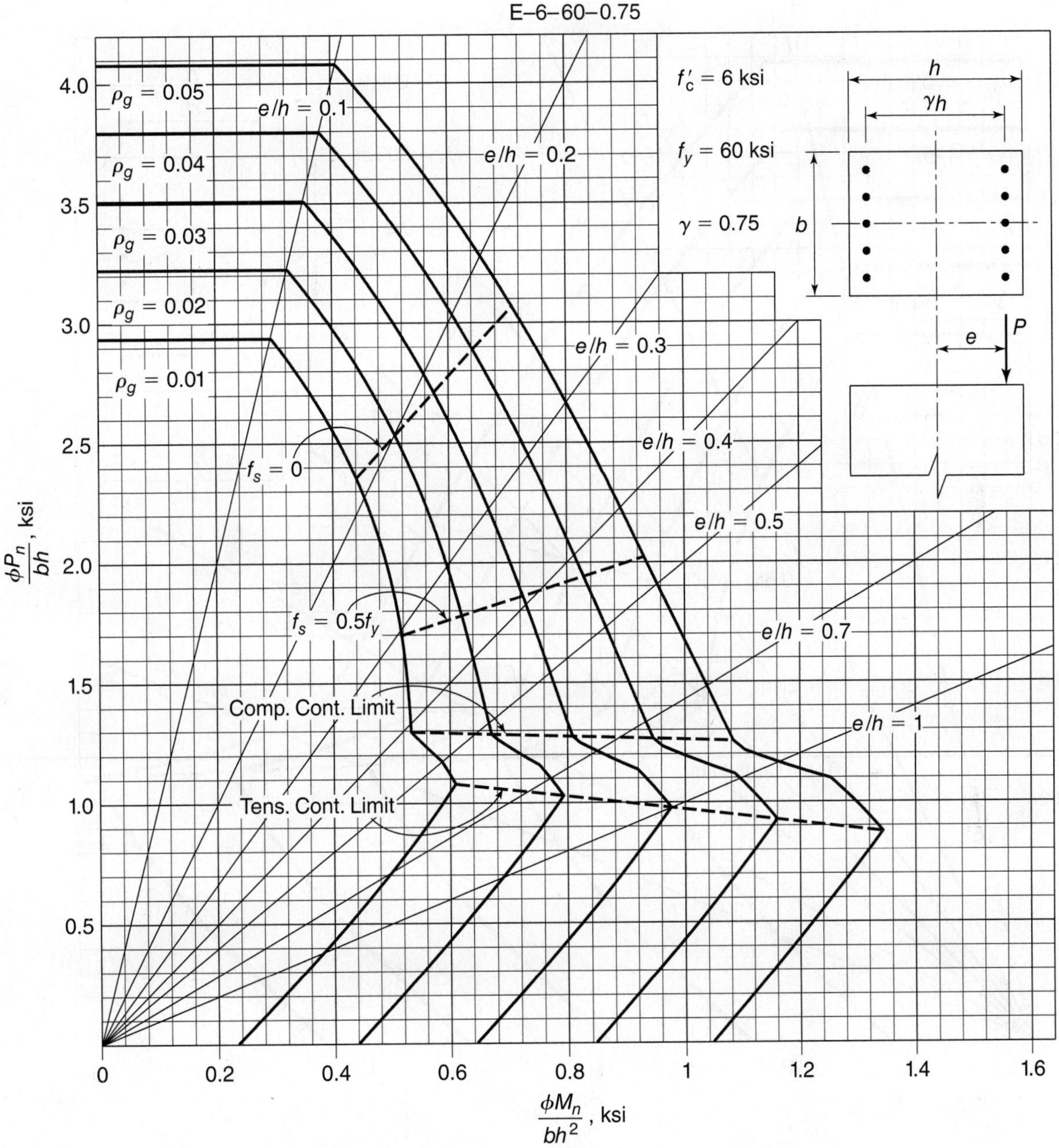

Fig. A-8b
Nondimensional interaction diagram for rectangular tied column with bars in two faces: $f'_c = 6000$ psi and $\gamma = 0.75$.

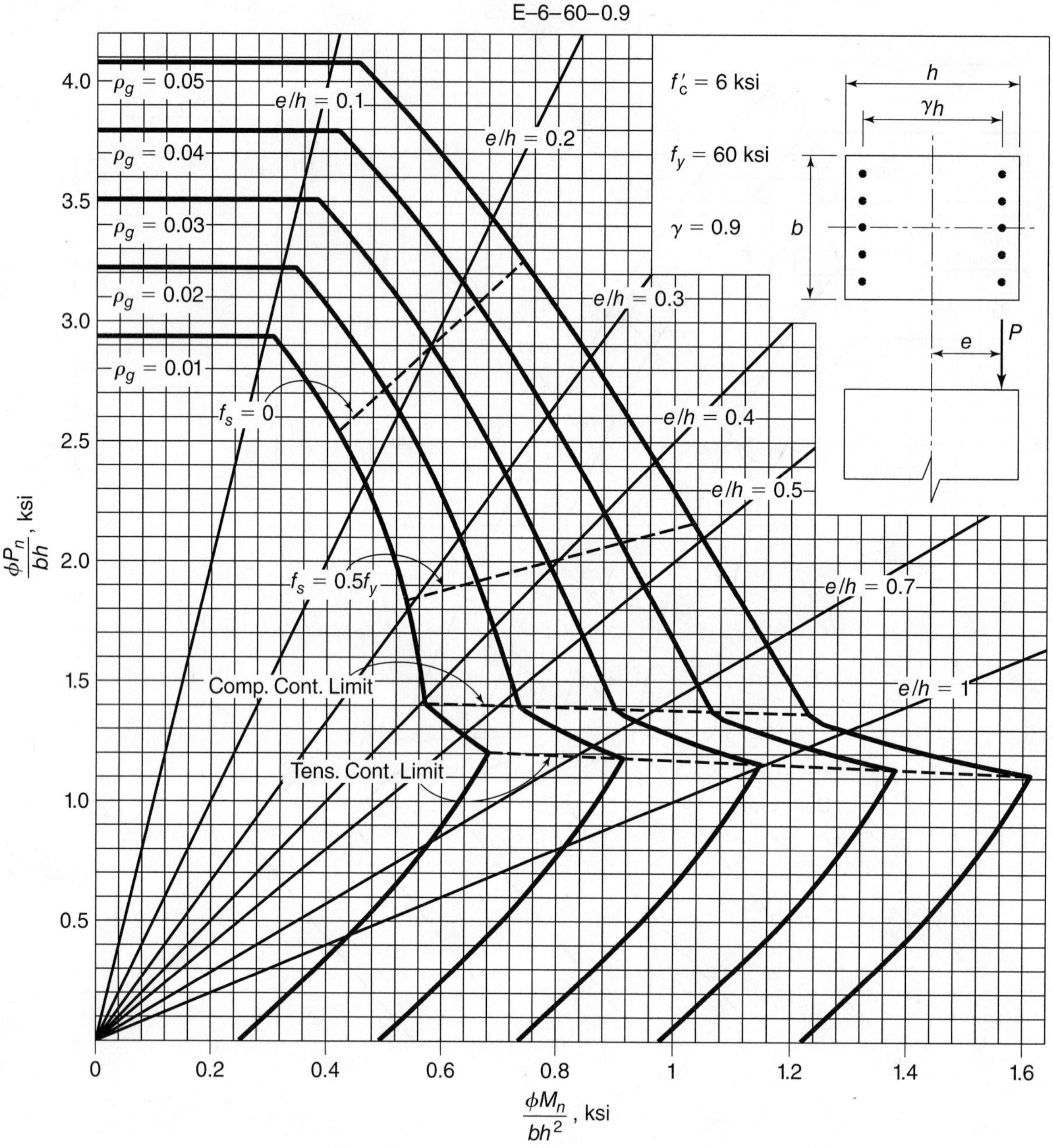

Fig. A-8c
Nondimensional interaction diagram for rectangular tied column with bars in two faces: f'_c = 6000 psi and γ = 0.90.

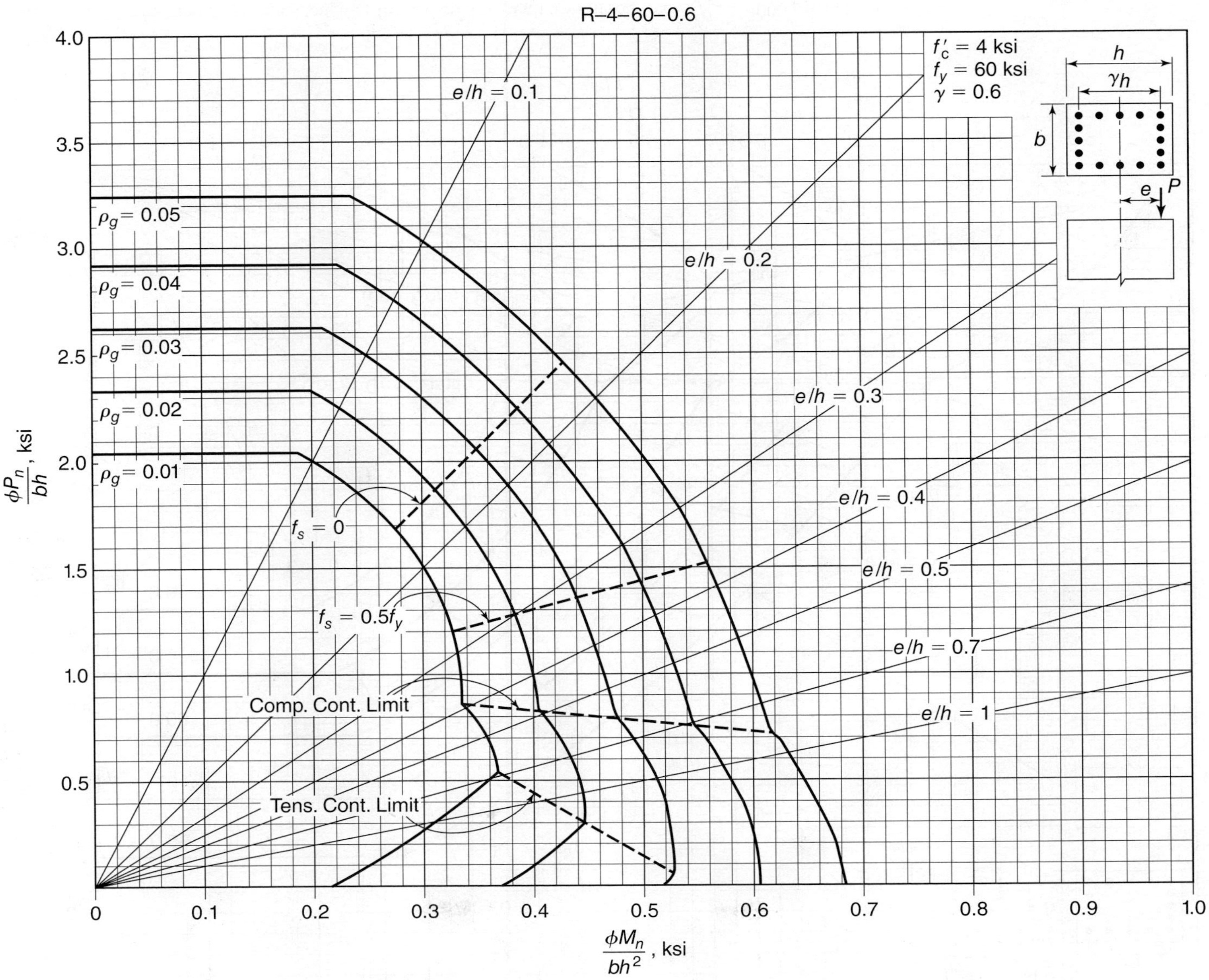

Fig. A-9a
Nondimensional interaction diagram for rectangular tied column with bars in four faces: $f'_c = 4000$ psi and $\gamma = 0.60$.

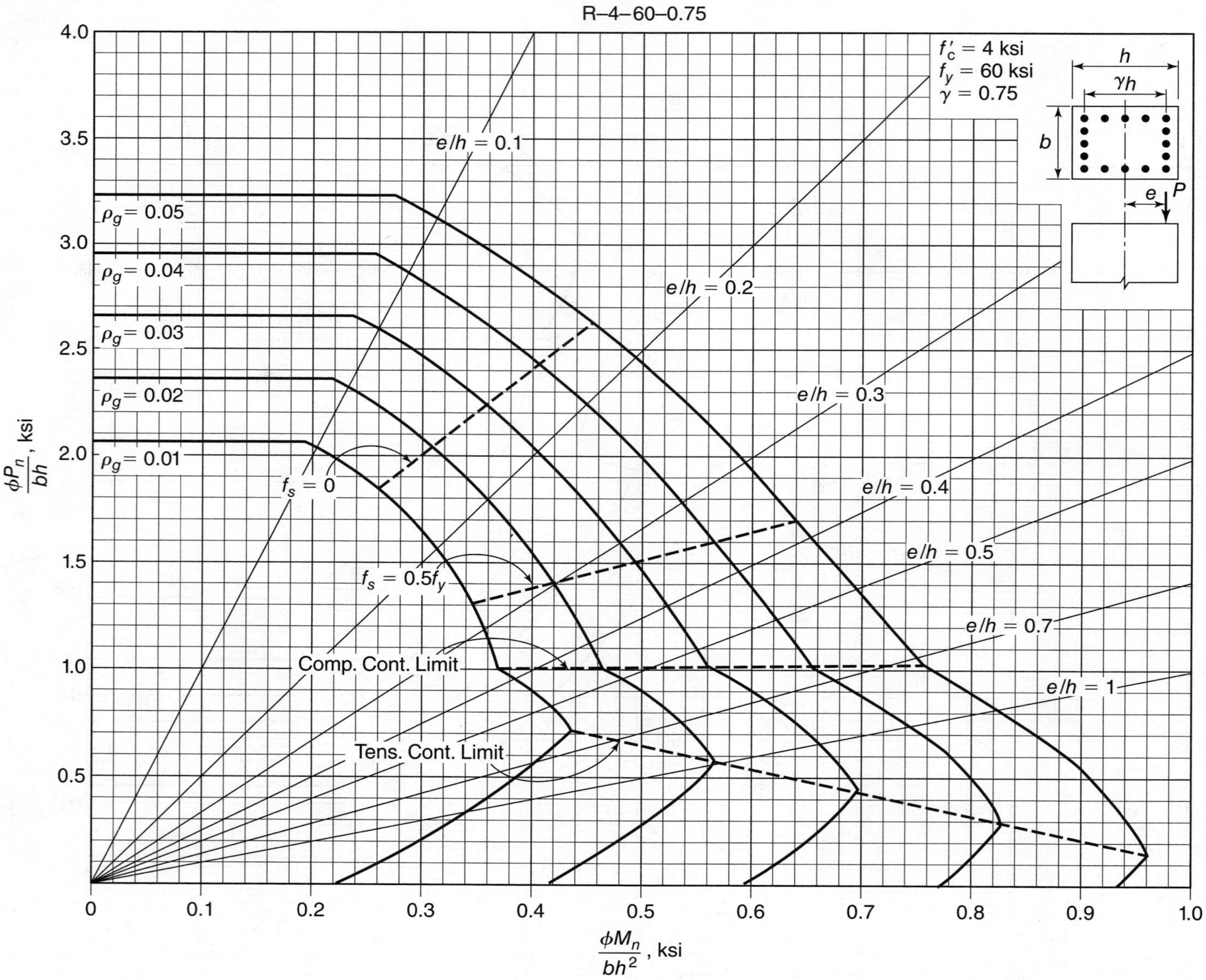

Fig. A-9b
Nondimensional interaction diagram rectangular for tied column with bars in four faces: f'_c = 4000 psi and γ = 0.75.

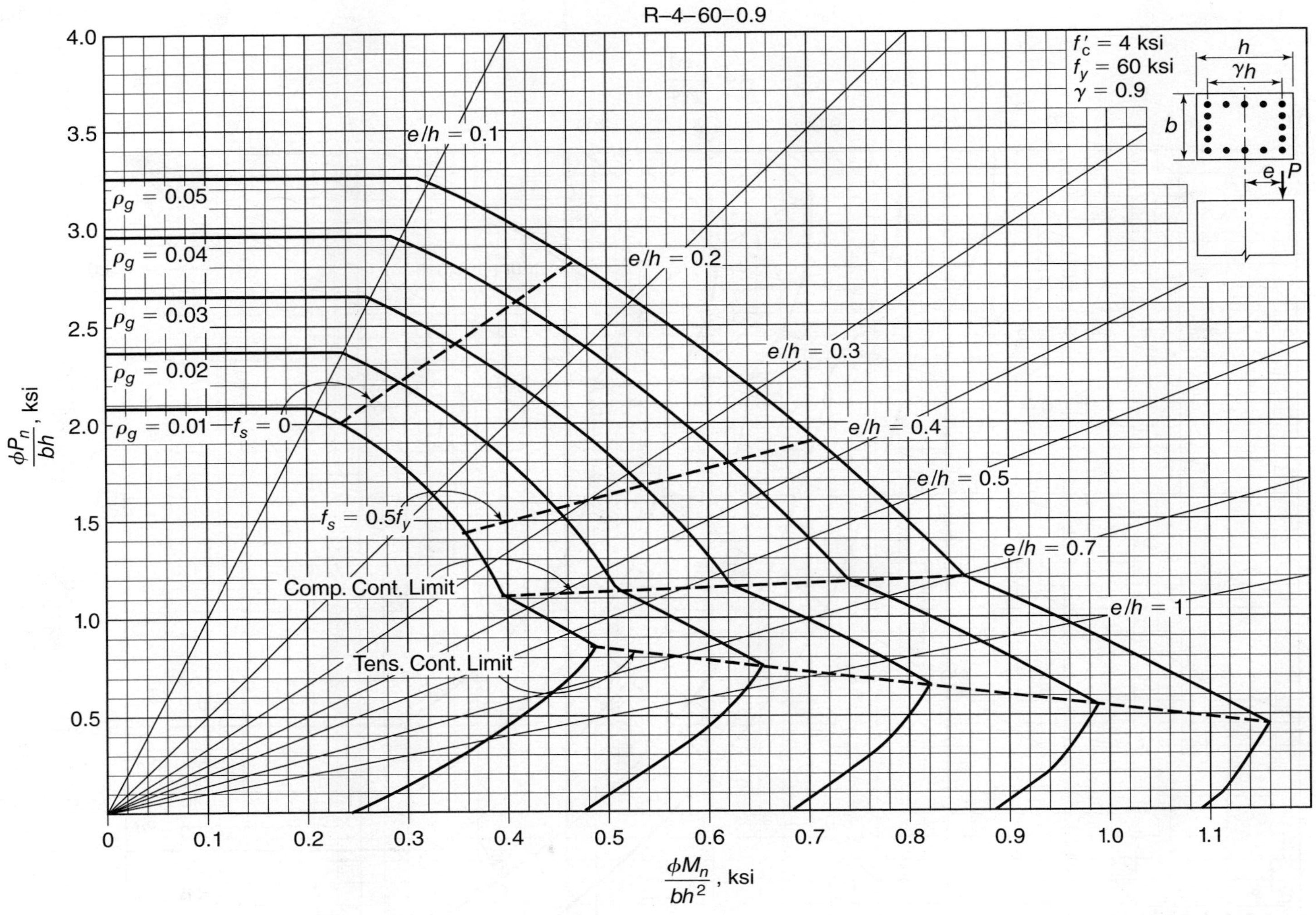

Fig. A-9c
Nondimensional interaction diagram for rectangular tied column with bars in four faces: $f'_c = 4000$ psi and $\gamma = 0.90$.

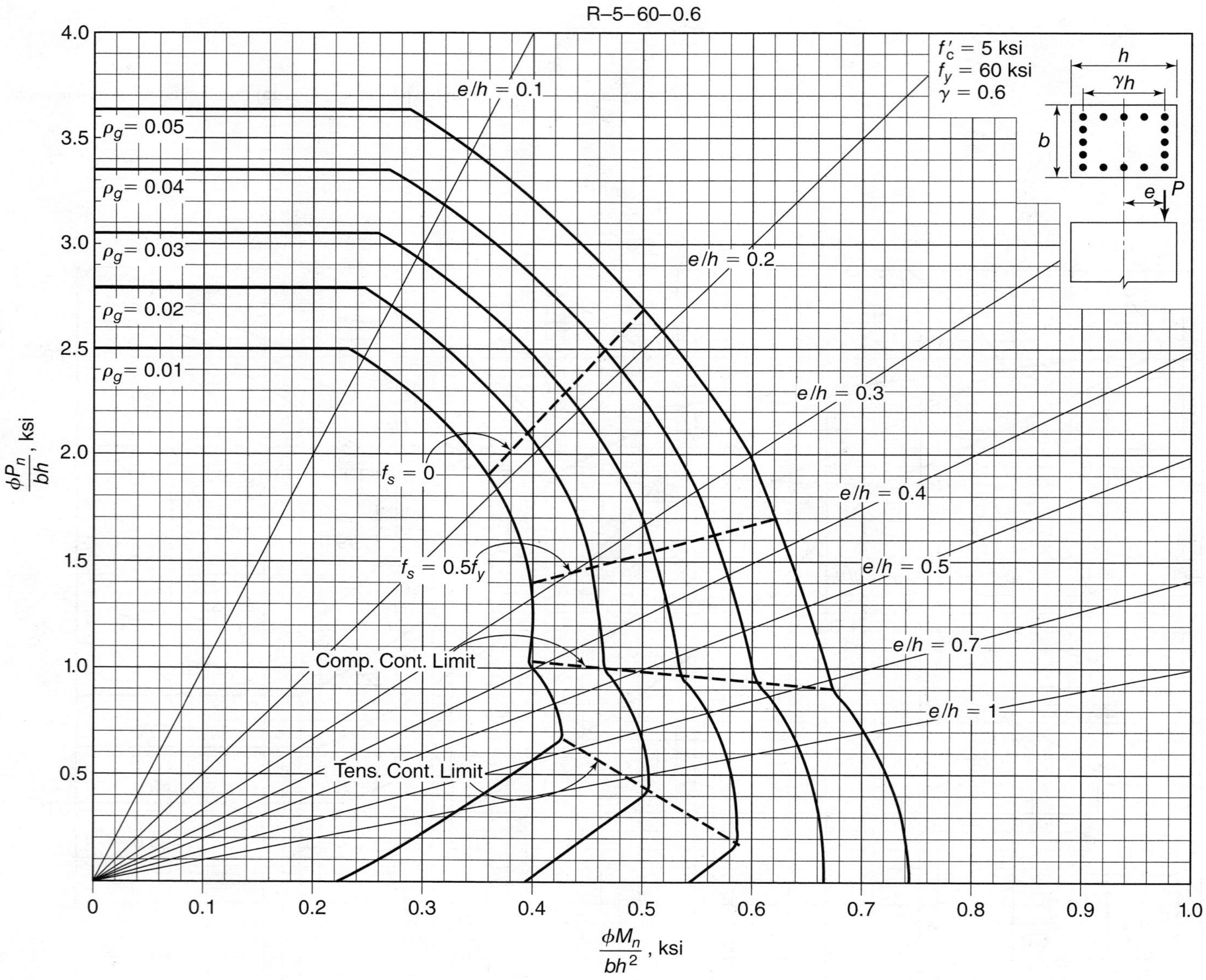

Fig. A-10a
Nondimensional interaction diagram for rectangular tied column with bars in four faces: $f'_c = 5000$ psi and $\gamma = 0.60$.

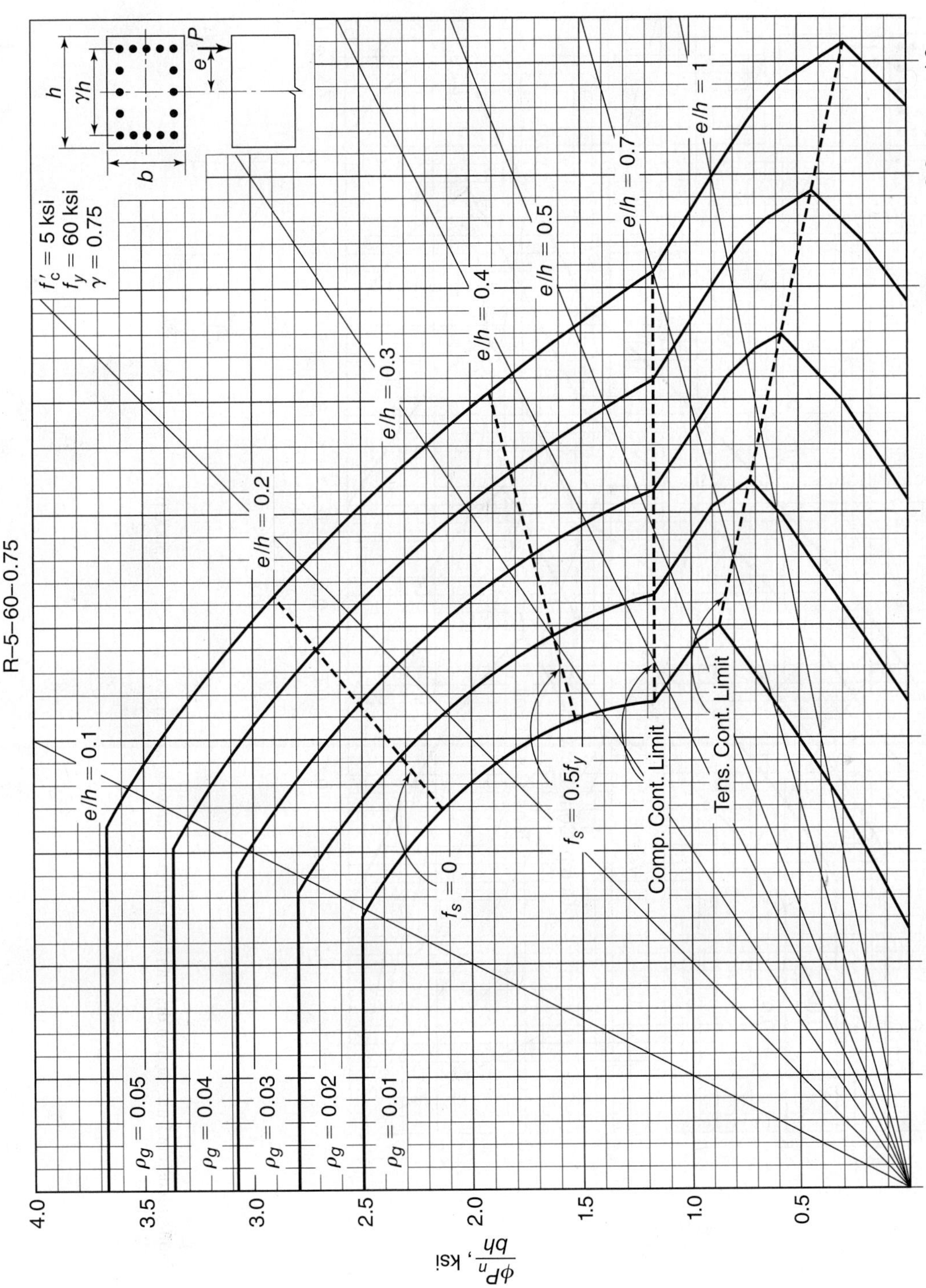

Fig. A-10b
Nondimensional interaction diagram for tied column with bars in four faces: f'_c = 5000 psi and γ = 0.75.

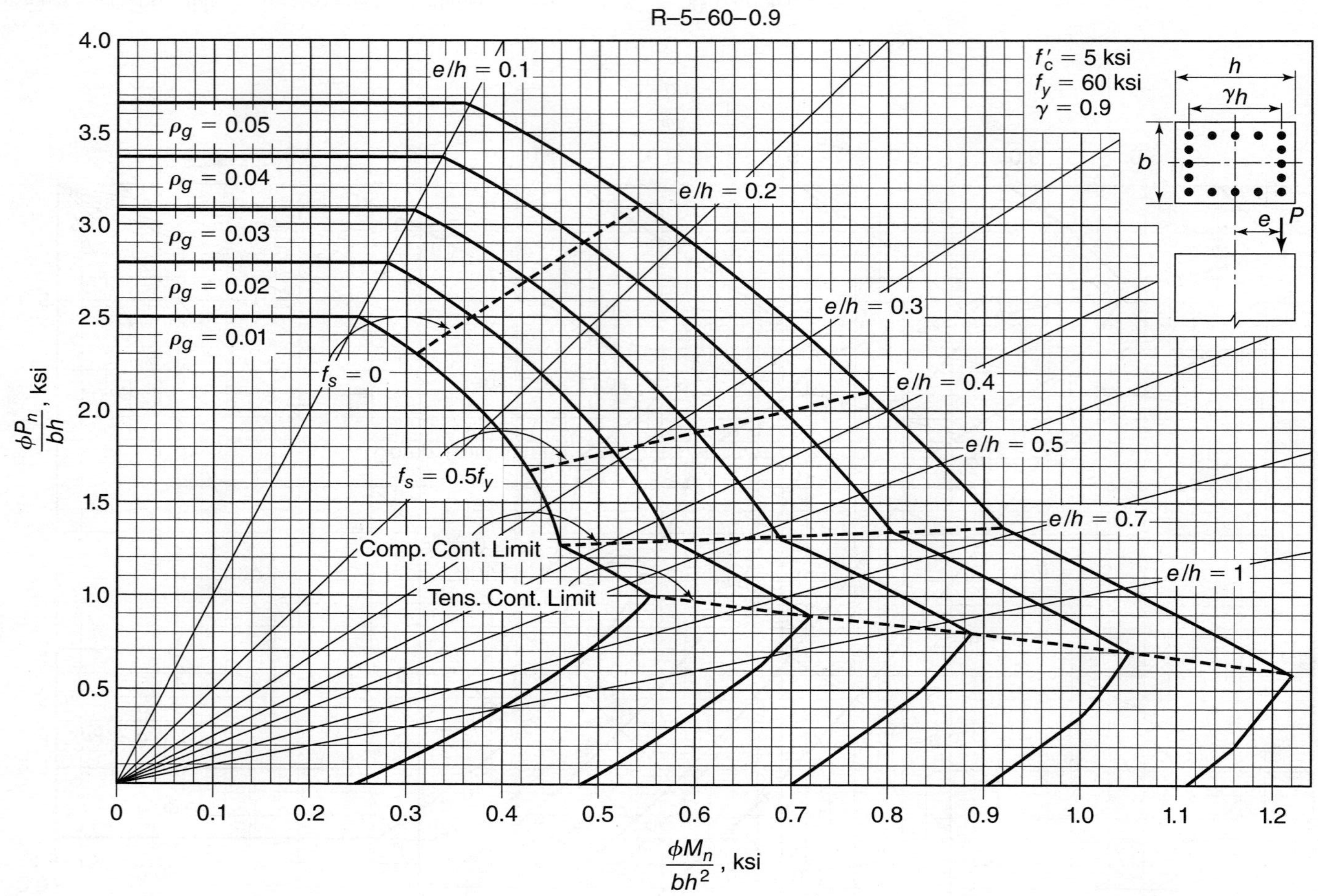

Fig. A-10c
Nondimensional interaction diagram for tied column with bars in four faces: $f'_c = 5000$ psi and $\gamma = 0.90$.

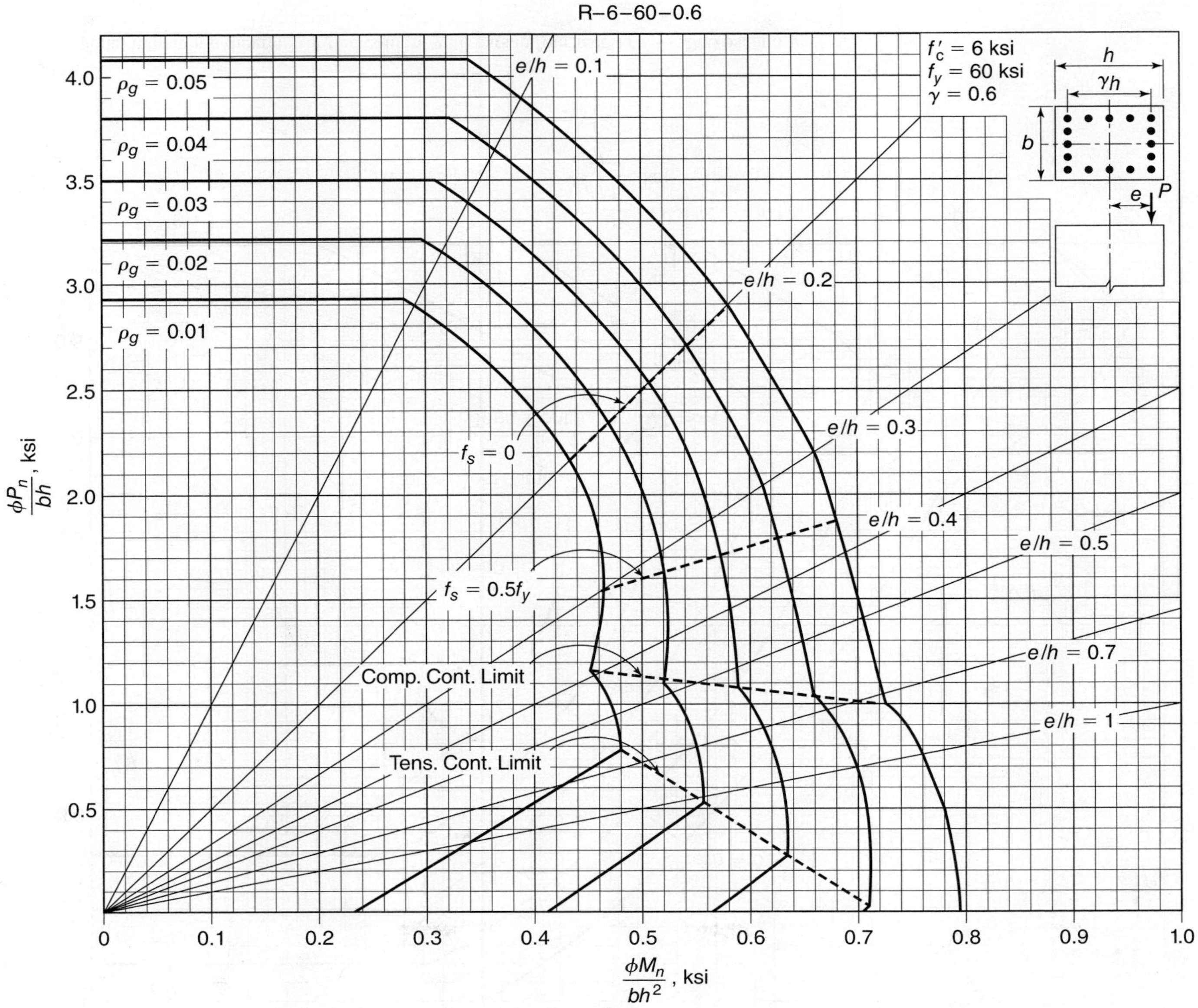

Fig. A-11a
Nondimensional interaction diagram for tied column with bars in four faces: $f'_c = 6000$ psi and $\gamma = 0.60$.

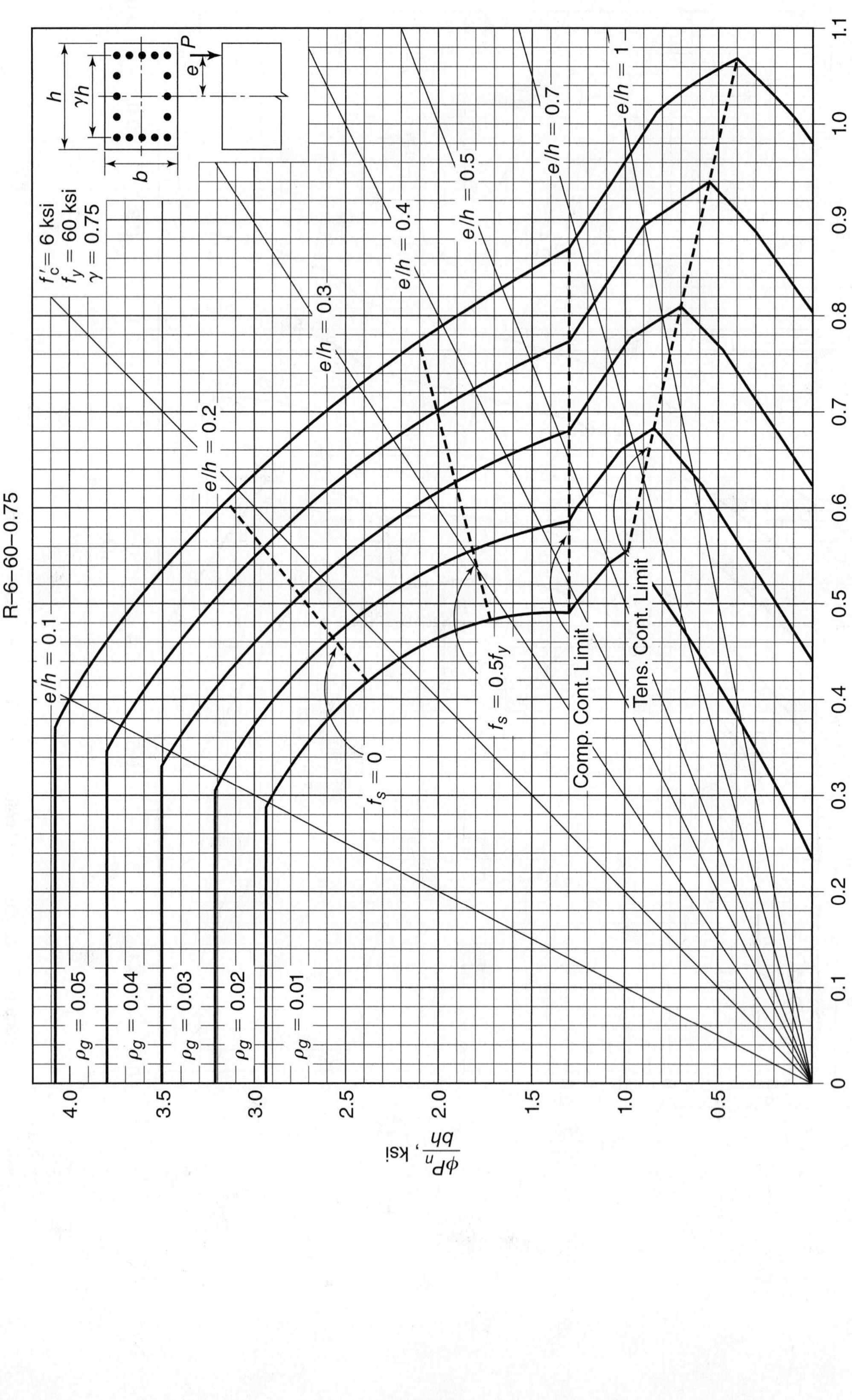

Fig. A-11b
Nondimensional interaction diagram for tied column with bars in four faces: $f'_c = 6000$ psi and $\gamma = 0.75$.

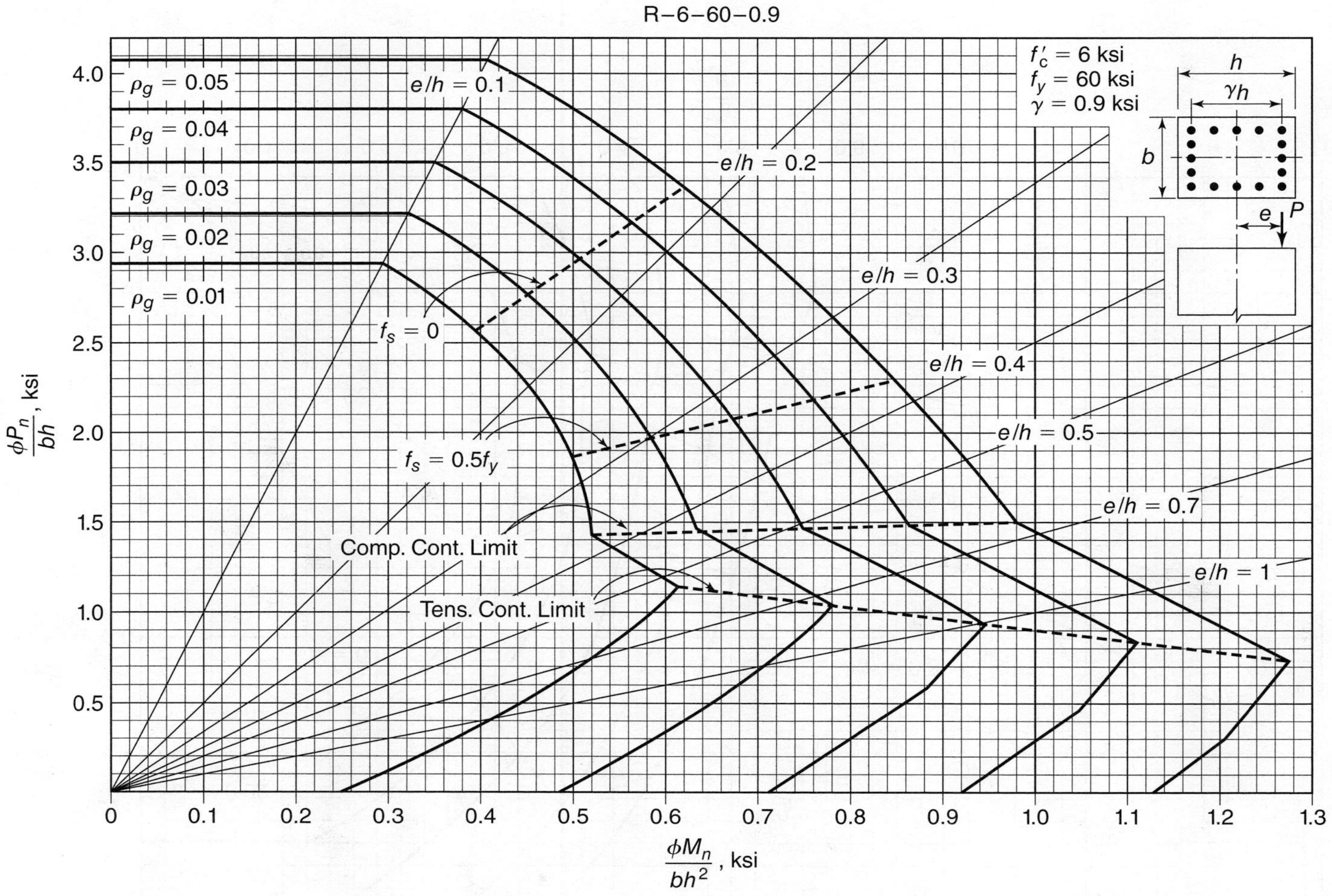

Fig. A-11c
Nondimensional interaction diagram for tied column with bars in four faces: $f'_c = 6000$ psi and $\gamma = 0.90$.

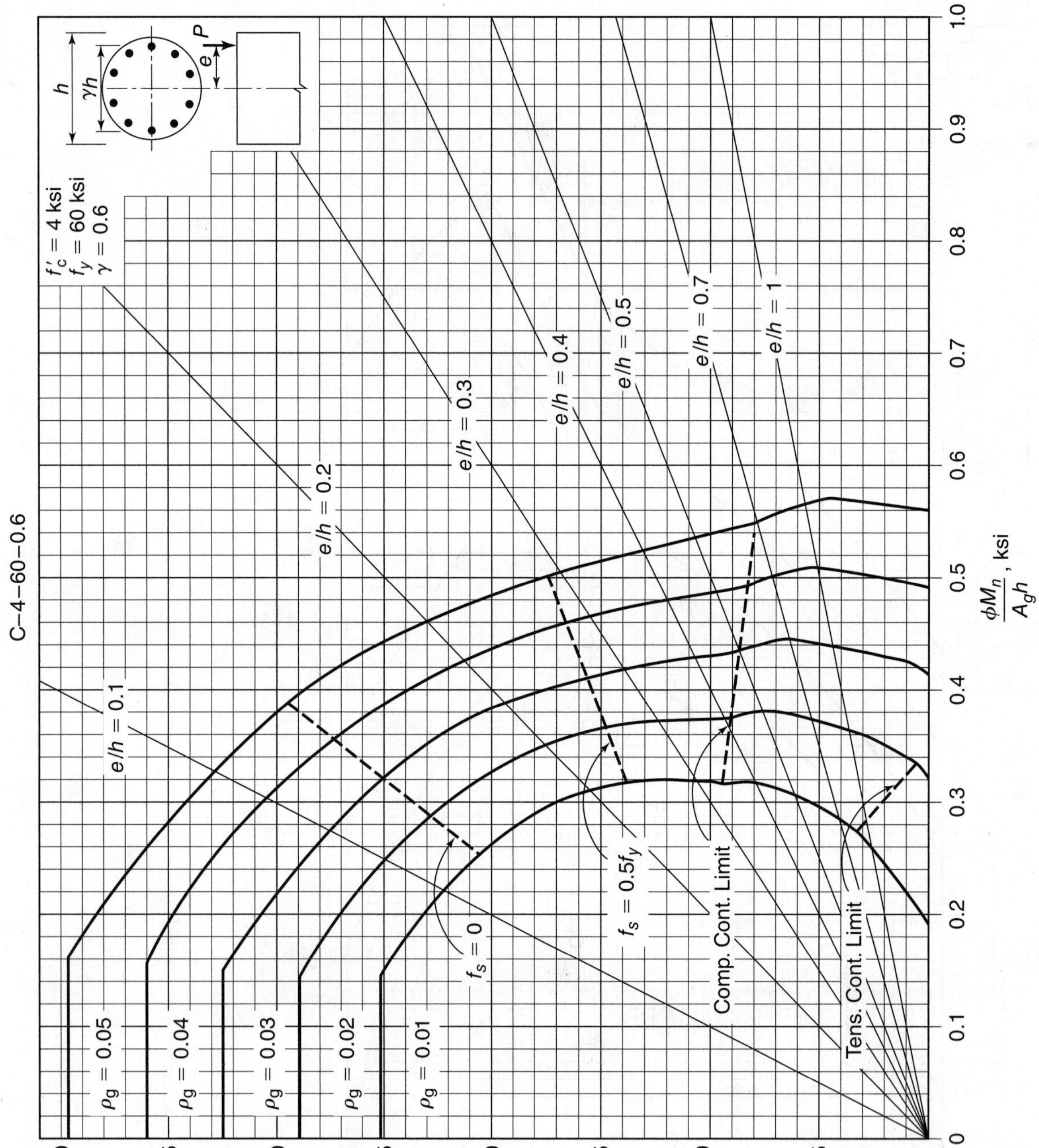

Fig. A-12a
Nondimensional interaction diagram for circular spiral column: f'_c = 4000 psi and γ = 0.60.

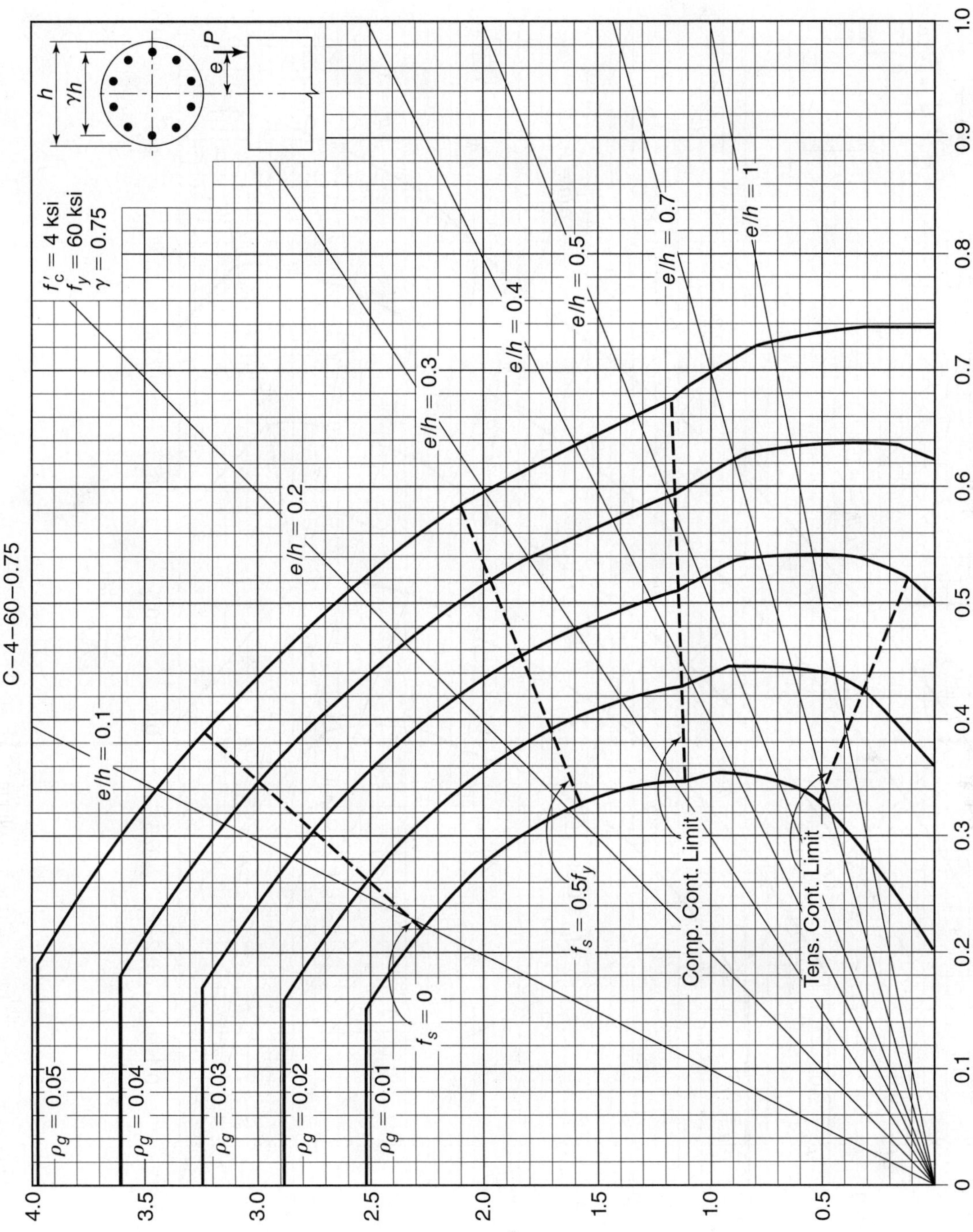

Fig. A-12b
Nondimensional interaction diagram for circular spiral column: $f'_c = 4000$ psi and $\gamma = 0.75$.

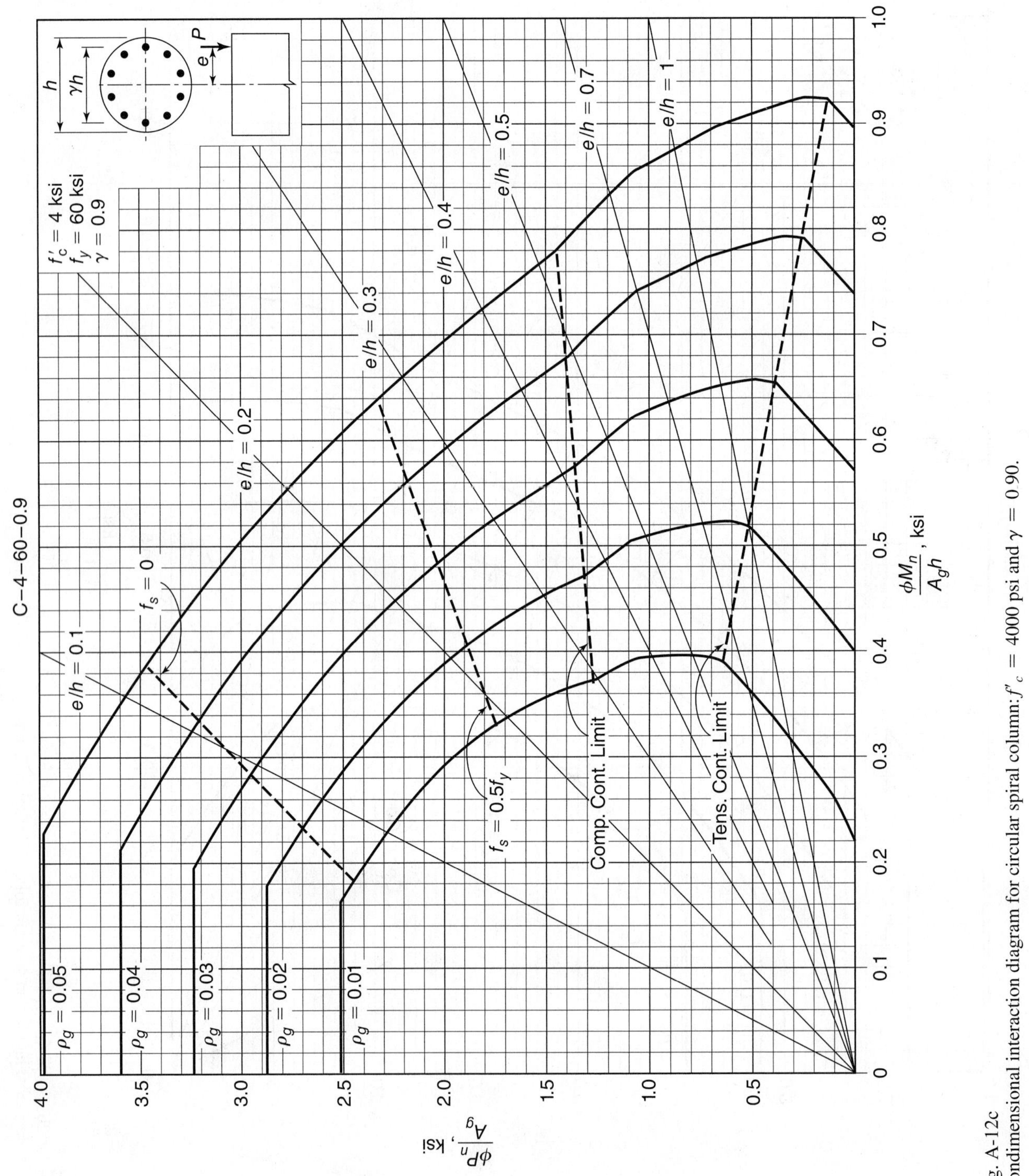

Fig. A-12c
Nondimensional interaction diagram for circular spiral column: $f'_c = 4000$ psi and $\gamma = 0.90$.

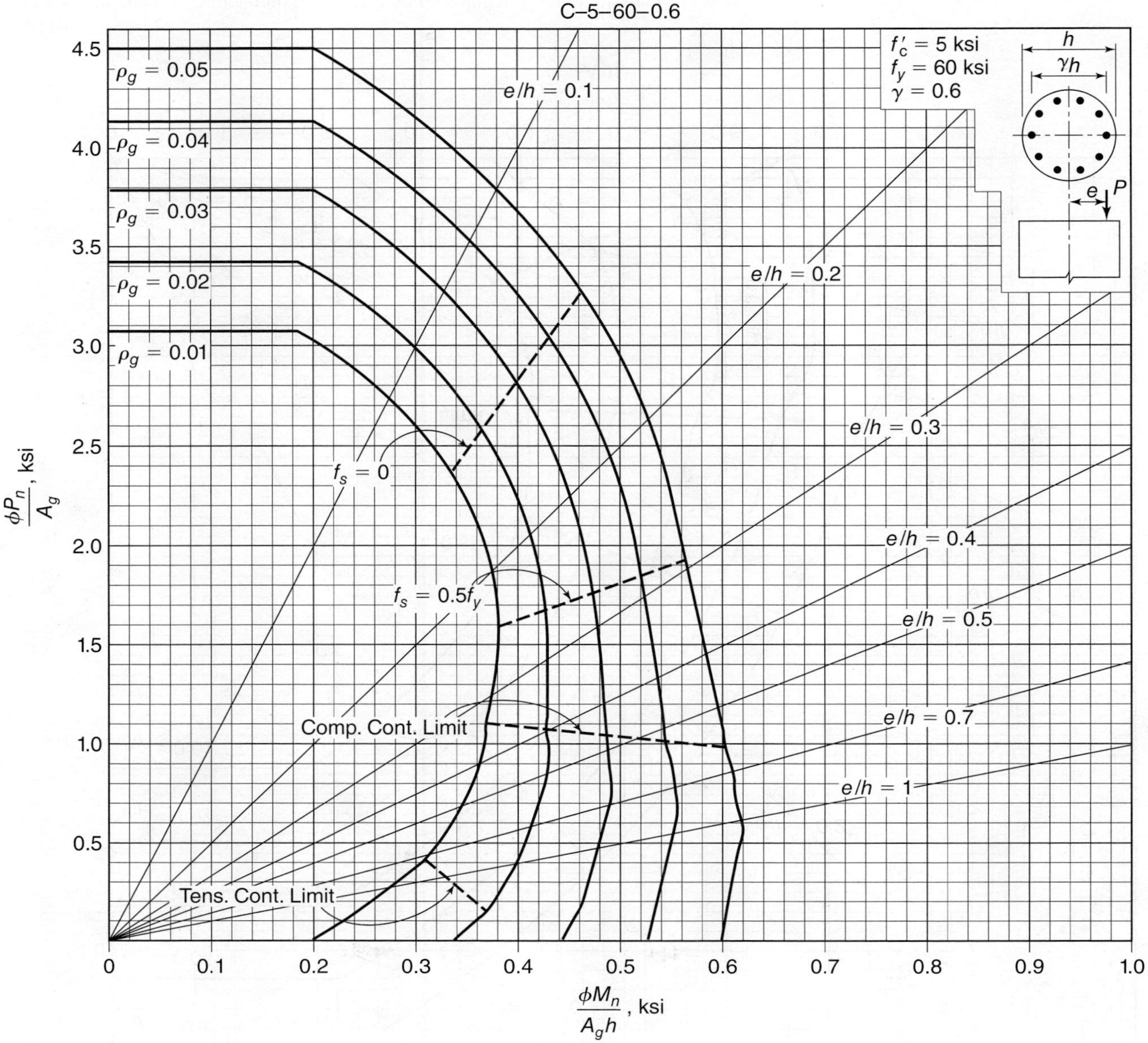

Fig. A-13a
Nondimensional interaction diagram for circular spiral column: f'_c = 5000 psi and γ = 0.60.

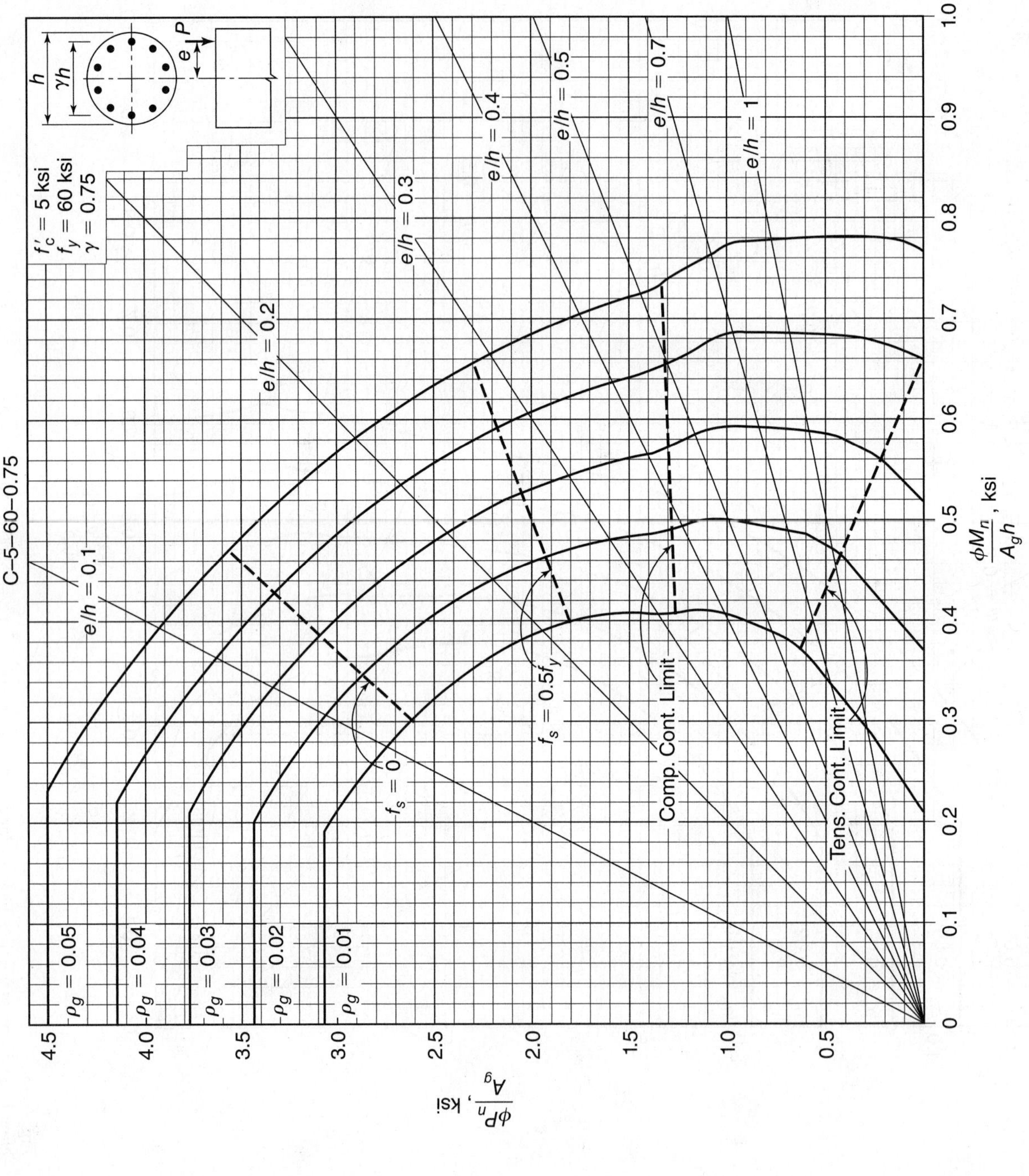

Fig. A-13b
Nondimensional interaction diagram for circular spiral column: f'_c = 5000 psi and γ = 0.75.

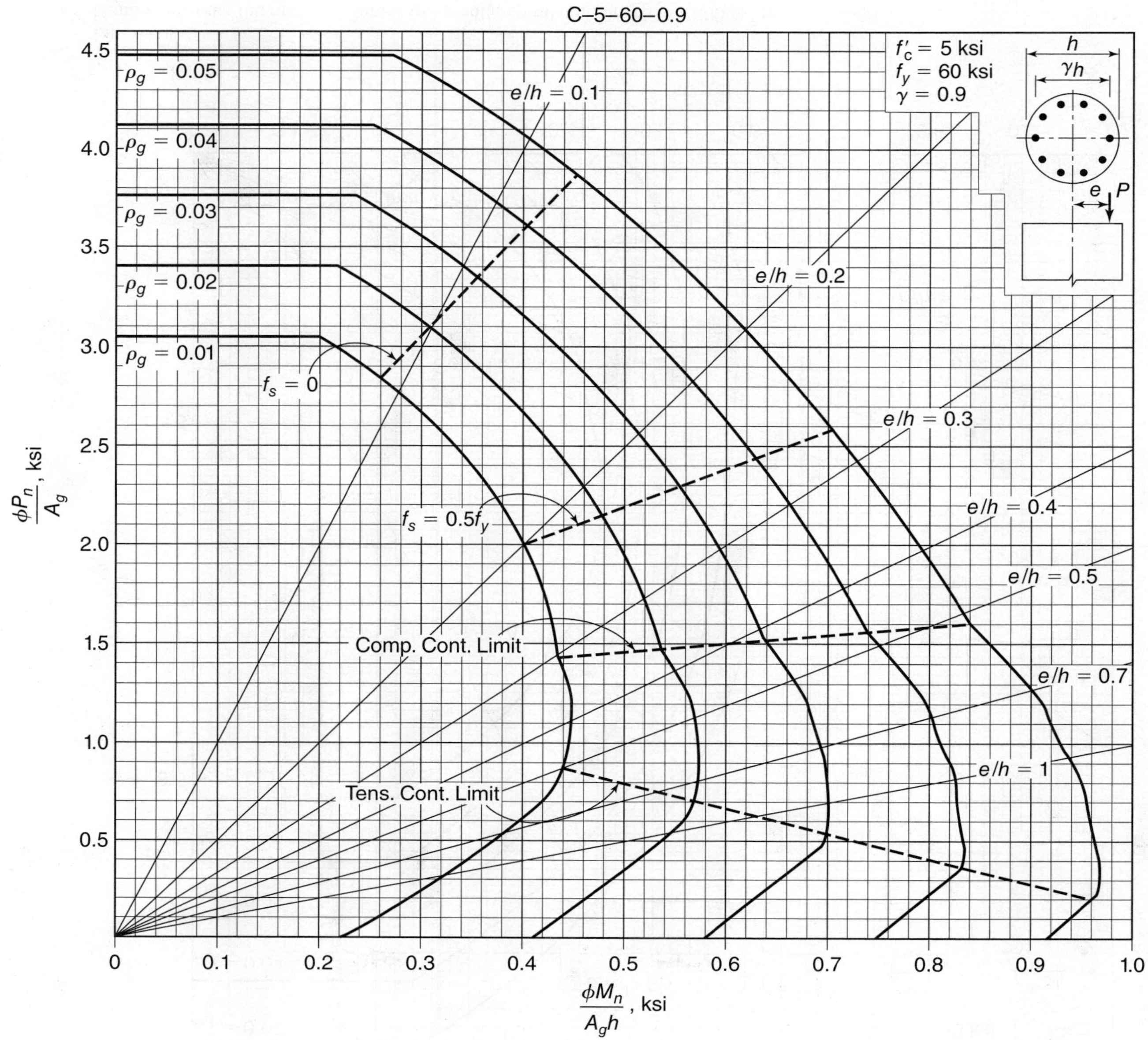

Fig. A-13c
Nondimensional interaction diagram for circular spiral column: f'_c = 5000 psi and γ = 0.90.

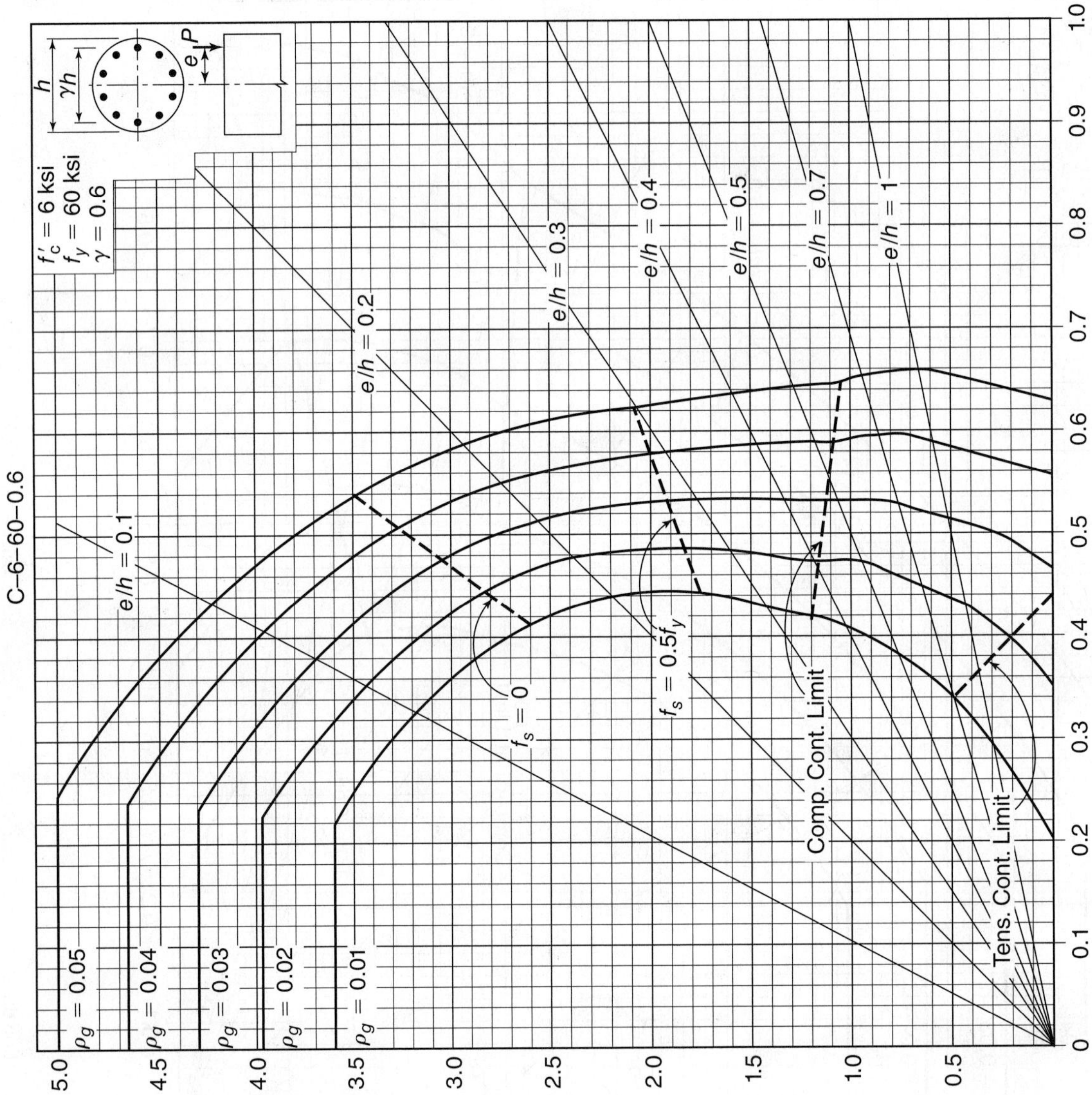

Fig. A-14a
Nondimensional interaction diagram for circular spiral column: f'_c = 6000 psi and γ = 0.60.

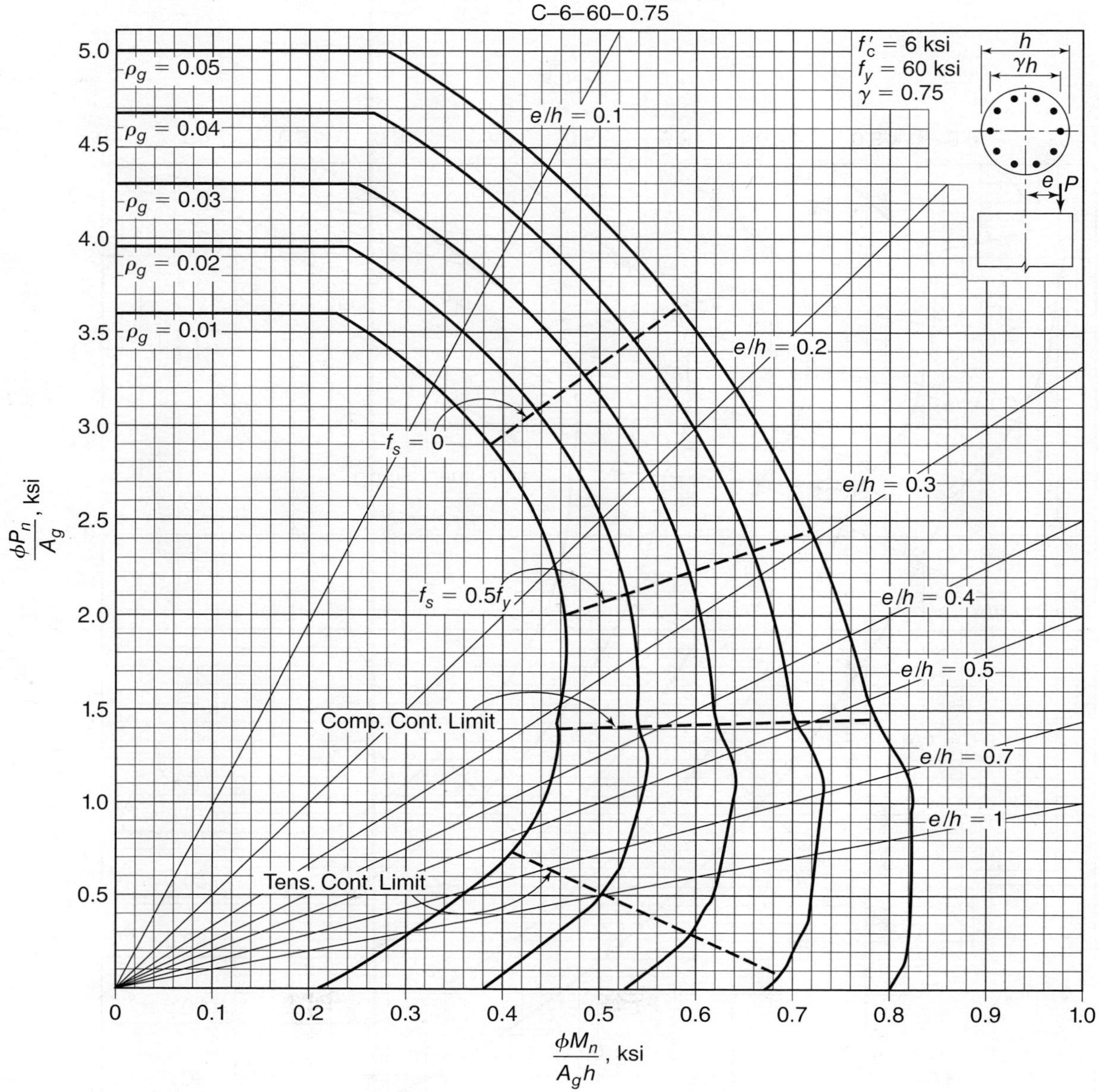

Fig. A-14b
Nondimensional interaction diagram for circular spiral column: f'_c = 6000 psi and γ = 0.75.

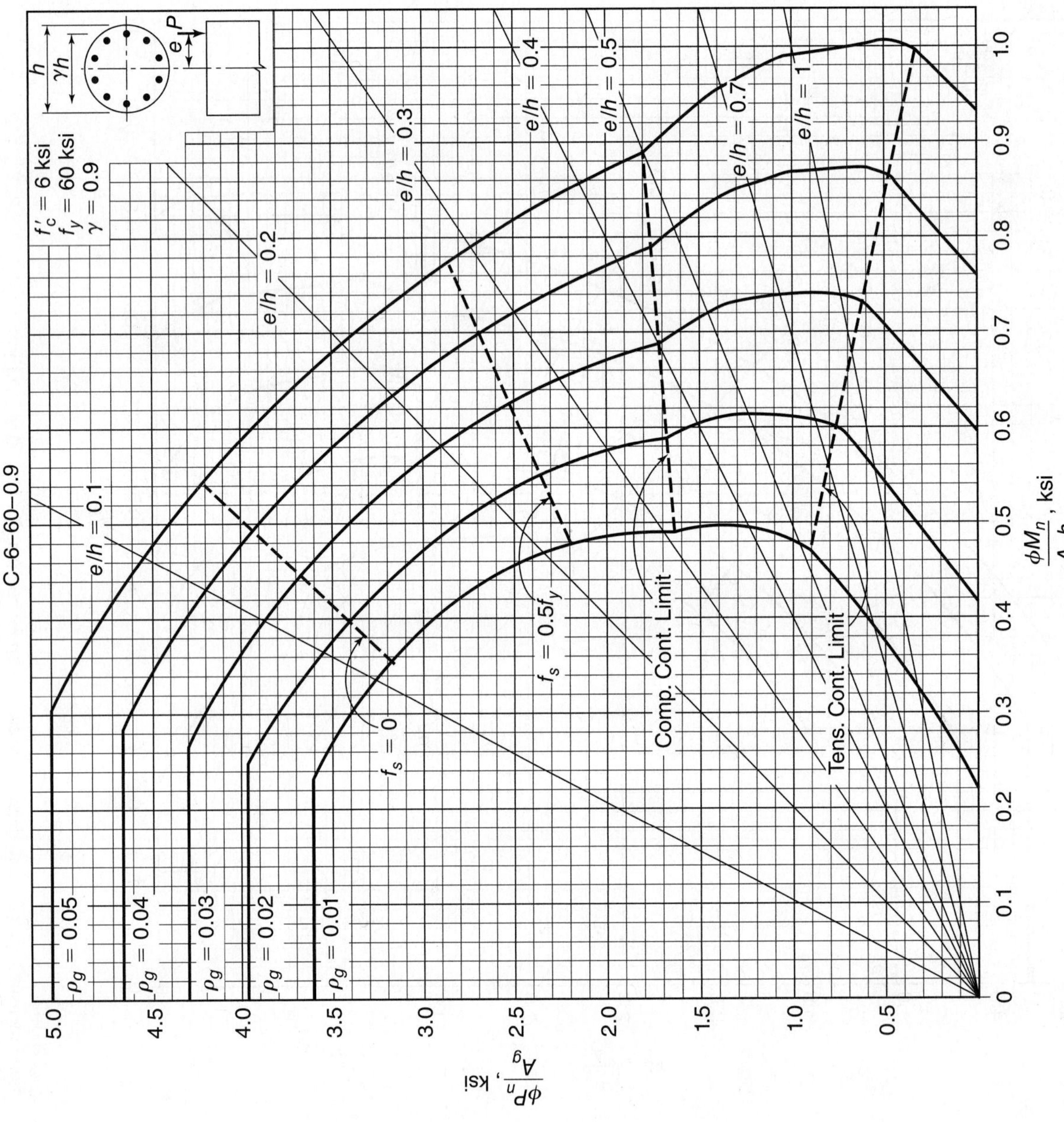

Fig. A-14c
Nondimensional interaction diagram for circular spiral column: $f'_c = 6000$ psi and $\gamma = 0.90$.

APPENDIX B

Notation

Note: U.S. Customary units are used to define common units of terms. Alternate definitions with SI units are also possible. Some seldom-used notation is defined where it occurs in the text.

a = depth of equivalent rectangular stress block, in.

a = shear span, distance between concentrated load and face of support, in.

A_b = area of an individual bar, in.^{2}

A_c = area of concrete section resisting shear transfer, in.^{2}

A_c = the effective area at one end of a strut in a strut-and-tie model, taken perpendicular to the axis of the strut, in.^{2}

A_{ch} = area of core of reinforced compression member measured to outside edges of transverse reinforcement, in.^{2}

A_{cp} = area enclosed by outside perimeter of concrete cross section, in.^{2}

A_{cv} = gross area of concrete section bounded by web thickness and length of section in the direction of shear force considered, in.^{2}

A_g = gross area of section, in.^{2} (For a hollow section, A_g is the area of concrete only and does not include the area of the void(s).)

A_I = influence area (see Section 2-8)

A_ℓ = total area of longitudinal reinforcement to resist torsion, in.^{2}

A_n = area of a face of a nodal zone or a section through a nodal zone, in.^{2}

A_o = gross area enclosed by shear flow path, in.^{2}

A_{oh} = area enclosed by centerline of the outermost closed transverse torsional reinforcement, in.^{2}

A_{ps} = area of prestressed reinforcement in a tie, in.^{2}

A_s = area of nonprestressed tension reinforcement, in.^{2}

A'_s = area of compression reinforcement, in.^{2}

A'_s = area of compression reinforcement in a strut, in.^{2}

A_{sf} = area of tension reinforcement balancing the compression force in the overhanging flanges of a T beam, in.^{2}

A_{si} = area of surface reinforcement in the ith layer crossing a strut, in.^{2}

A_{sk} = area of skin reinforcement per unit height in one side face, in.^{2}/ft

A_{st} = area of nonprestressed reinforcement in a tie, in.^{2}

A_{sw} = area of tension reinforcement balancing the compression force in the web of a T beam, in.^{2}

A_t = area of one leg of a closed stirrup resisting torsion within a distance s, in.^{2}

A_{tr} = total cross-sectional area of all transverse reinforcement that is within the spacing s and crosses the potential plane of splitting through the reinforcement being developed, in.^{2}

A_v = area of shear reinforcement within a distance s, in.^{2}

A_{vf} = area of shear-friction reinforcement, in.^{2}

A_1 = loaded area in bearing, in.^{2}

A_2 = area of the lower base of the largest frustrum of a pyramid, cone, or tapered wedge contained wholly within the support, having for its upper base the loaded area and having side slopes of 1 vertical to 2 horizontal, in.^{2}

b = width of compression face of member, in.

b_e = effective compressive flange width of a T beam, in.

b_o = perimeter of critical section for two-way shear in slabs and footings, in.

b_w = web width, or diameter of circular section, in.

b_1 = length of critical shear perimeter for two-way shear, measured parallel to the span ℓ_1, in.

b_2 = length of critical shear perimeter for two-way shear measured perpendicular to b_1, in.

c = spacing or cover dimension, in.

c = distance from extreme-compression fiber to neutral axis, in.

c_c = clear cover from the nearest surface in tension to the surface of the flexural tension reinforcement, in.

c_1 = size of rectangular or equivalent rectangular column, capital, or bracket, measured parallel to the span ℓ_1, in.

c_2 = size of rectangular or equivalent rectangular column, capital, or bracket, measured parallel to ℓ_2, in.

C = compressive force in cross section; with subscripts, c = concrete, s = steel

C = cross-sectional constant to define torsional properties, in.^{4}

C_m = factor relating the actual moment diagram of a slender column to an equivalent uniform moment diagram (Chapter 12)

C_m = moment coefficient (Chapter 5)

C_v = shear coefficient

d = effective depth = distance from extreme-compression fiber to centroid of tension reinforcement, in.

d' = distance from extreme-compression fiber to centroid of compression reinforcement, in.

d_b = nominal diameter of bar, wire, or prestressing strand, in.

d_t = distance from extreme-compression fiber to extreme-layer of tension steel, in.
D = dead loads, or related internal moments and forces
D = diagonal compression force in the web of a beam, lb
e = eccentricity of axial load on a column = M/P
E_c = modulus of elasticity of concrete, psi
EI = flexural stiffness of compression member, lb−in.^{2}
E_s = modulus of elasticity of reinforcement, psi
f_c = compressive stress in concrete
f'_c = specified compressive strength of concrete, psi
$\sqrt{f'_c}$ = square root of specified compressive strength of concrete, psi
f_{ce} = effective compressive strength of the concrete in a strut or a nodal zone, psi
f'_{cr} = required average compressive strength for f'_c to meet statistical acceptance criteria, psi
f_{ct} = splitting tensile strength of concrete, psi
f_r = modulus of rupture of concrete, psi
f_s = calculated stress in reinforcement at service loads, ksi
f'_s = stress in compression reinforcement, psi
f_{si} = stress in the ith layer of surface reinforcement, psi
f_y = specified yield strength of nonprestressed reinforcement, psi
F = loads due to weight and pressures of fluids with well-defined densities and controllable heights
F_n = nominal strength of a strut, tie, or nodal zone, lb
F_{nn} = nominal strength of a face of a nodal zone, lb
F_{ns} = nominal strength of a strut, lb
F_{nt} = nominal strength of a tie, lb
F_u = factored force acting on a strut, tie, bearing area, or nodal zone in a strut-and-tie model, lb
h = height of member, in.
h = overall thickness of member, in.
h_t = effective height of tie, in.
h_w = total height of wall from base to top, in.
H = loads due to weight and pressure of soil, water in soil, or other materials, or related internal moments and forces
I = moment of inertia of section; in.^{4} subscripts, b = beam, c = column, s = slab
I_{cr} = moment of inertia of cracked section transformed to concrete, in.^{4}
I_e = effective moment of inertia for computation of deflection, in.^{4}
I_g = moment of inertia of gross concrete section, neglecting reinforcement, in.^{4}
I_{se} = moment of inertia of the reinforcement in a column about centroidal axis of gross member cross section, in.^{4}
jd = distance between the resultants of the internal compressive and tensile forces on a cross section, in.
J_c = property of assumed critical section for two-way shear, analogous to polar moment of inertia, in.^{4}
k = effective length factor for compression members

K = flexural stiffness: moment per unit rotation; used with subscripts, b = beam, c = column, ec = equivalent column, s = slab

K_t = torsional stiffness of torsional member; moment per unit rotation

K_{tr} = transverse reinforcement index (Chapter 8)

ℓ = span length of beam or one-way slab, generally center to center of supports; clear projection of cantilever, in.

ℓ_a = additional embedment length at support or at point of inflection, in.

ℓ_a = length in which anchorage of a tie must take place, in.

ℓ_b = length of bearing area parallel to the plane of the strut-and-tie model, in.

ℓ_c = length of compression member in a frame, measured from center to center of the joints in the frame

ℓ_d = development length, in.

ℓ_{dc} = development length in compression, in.

ℓ_{dh} = development length of standard hook in tension, measured from critical section to outside end of hook, in.

ℓ_n = clear span measured face to face of supports:

ℓ_n = clear span for positive moment, negative moment at exterior support, or shear

ℓ_n = average of adjacent clear spans for negative moment

ℓ_o = minimum length, measured from joint face along axis of structural member, over which transverse reinforcement must be provided, in.

ℓ_u = unsupported length of compression member

ℓ_w = total length of structural wall

ℓ_1 = length of span of two-way slab in direction that moments are being determined, measured center to center of supports

ℓ_2 = length of span of two-way slab transverse to ℓ_1, measured center to center of supports

L = live loads, or related internal moments and forces

L_r = specified live load for roof, or related internal forces or moments

m = moment per unit width; subscripts, x, y, and xy for twisting moment; kip-ft/ft

m_r = resisting moment per unit width, kip-ft/ft

M_a = maximum moment in member at the stage for which deflections are being computed (Chapter 9)

M_c = factored moment to be used for design of a slender compression member, lb-in.

M_{cr} = cracking moment

M_n = nominal moment strength, lb-in.

M_o = total factored static moment, lb-in.

M_{pr} = probable flexural moment strength

M_s = moment in column due to loads causing appreciable sidesway

M_u = moment due to factored loads

M_1 = smaller factored end moment on a compression member; negative if the member is bent in single curvature, positive if bent in double curvature

M_{1ns} = factored end moment on a compression member at the end at which M_1 acts, due to loads that cause no appreciable sidesway

M_{1s} = factored end moment on a compression member at the end at which M_1 acts, due to loads that cause appreciable sidesway, calculated by using a first-order elastic frame analysis

M_2 = larger factored end moment on a compression member, always positive, lb-in.

M_{2ns} = factored end moment on a compression member at the end at which M_2 acts, due to loads that cause no appreciable sidesway

M_{2s} = factored end moment on a compression member at the end at which M_2 acts, due to loads that cause appreciable sidesway, calculated using a first-order elastic-frame analysis

n = modular ratio $= E_s/E_c$

n = number of bars in a layer being spliced or developed at a critical section

N_u = factored axial load normal to cross section, positive for compression and negative for tension

p_{cp} = outside perimeter of the concrete cross section, in.

p_h = perimeter of the centerline of the outermost closed transverse torsional reinforcement, in.

P_b = nominal axial load strength at balanced strain conditions

P_c = critical axial load

P_E = buckling load of an elastic, hinged-end column

P_n = nominal axial load strength at given eccentricity

P_o = nominal axial load strength at zero eccentricity

P_u = axial force due to factored loads

q_D = dead load per unit area of slab, lb/ft^2

q_L = live load per unit area of slab, lb/ft^2

q_u = factored load per unit area

Q = stability index for a story (Chapter 12)

r = radius of gyration of cross section of a compression member, in.

R = flexural resistance factor

R = rain load, or related internal moments and forces

s = standard deviation

s = spacing of shear or torsion reinforcement measured along the longitudinal axis of the structural member, in.

s_i = spacing of reinforcement in the ith layer adjacent to the surface of a strut-and-tie model, in.

S = snow load, or related internal forces and moments

s_{sk} = spacing of skin reinforcement, in.

t = thickness of wall in hollow section, in.

T = cumulative effect of temperature, creep, shrinkage, differential settlement, and shrinkage-compensating cement

T_n = nominal torsional moment strength, lb-in.

T_u = factored torsional moment at section, lb-in.

U = strength required to resist factored loads or related internal moments and forces

v = shear stress

v_c = nominal shear stress carried by concrete, psi

v_n = nominal shear strength in stress units, psi
V_c = nominal shear force carried by concrete, lb
V_e = design shear force in beams or columns due to dead, live, and seismic loads, lb
V_n = nominal shear strength, lb
V_s = nominal shear strength provided by shear reinforcement, lb
V_{sway} = the shear in the span of a beam or column, resulting from a plastic hinging mechanism in the columns and beams of the frame due to seismically induced sidesway deflections of the frame, lb
V_u = factored shear force at section, lb
w_c = weight of concrete, lb/ft^3
w_D = dead load per unit length of beam, lb/ft
w_L = live load per unit of length of beam
w_s = effective width of a strut, in.
w_t = effective width of a tie, in.
w_u = total factored load per unit length of beam or one-way slab
x = shorter overall dimension of rectangular part of cross section, in.
y = longer overall dimension of rectangular part of cross section, in.
y_t = distance from centroidal axis of cross section, neglecting reinforcement, to extreme fiber in tension, in.
α = angle defining orientation of reinforcement
α_f = ratio of flexural stiffness of beam section to flexural stiffness of a width of slab bounded laterally by centerlines of adjacent panels (if any) on each side of the beam
α_{fm} = average value of α_f for all beams on edges of a slab panel
α_i = angle between the axis of a strut and the bars in the ith layer of reinforcement crossing that strut
α_s = coefficient used to compute V_c in slabs and footings
α_{f1} = α_f in direction of ℓ_1
α_{f2} = α_f in direction of ℓ_2
β = ratio of clear spans in long to short direction of two-way slabs
β = ratio of long side to short side of footing
β = ratio of long side to short side of column or concentrated load
β_{dns} = ratio used to account for reduction of stiffness of columns due to sustained axial loads
β_{ds} = ratio used to account for reduction of stiffness of columns due to sustained lateral loads
β_n = factor to account for the effect of the anchorage of ties on the effective compressive strength of a nodal zone
β_s = factor to account for the effect of cracking and confining reinforcement on the effective compressive strength of the concrete in a strut
β_t = ratio of torsional stiffness of edge beam section to flexural stiffness of a width of slab equal to span length of beam, center to center of supports
β_1 = ratio of depth of rectangular stress block, a, to depth to neutral axis, c
γ = ratio of the distance between the outer layers of reinforcement in a column to the overall depth of the column

γ_f = fraction of unbalanced moment transferred by flexure at slab–column connections

γ_v = fraction of unbalanced moment transferred by eccentricity of shear at slab–column connections

δ = moment-magnification factor for frames braced against sidesway, to reflect effects of member curvature between ends of compression member

δ_s = moment-magnification factor for frames not braced against sidesway, to reflect lateral drift resulting from lateral and gravity loads

ϵ = strain

ϵ_c = strain in concrete

ϵ_{cu} = maximum useable compressive strain for concrete

ϵ_s = strain in steel

ϵ_t = net tensile strain in extreme layer of tension steel at nominal strength

λ = correction factor related to unit weight of concrete

λ_Δ = multiplier for additional long-term deflection

μ = coefficient of friction

ν = ratio of effective concrete strength in web of beam or compression strut to f'_c

ξ = time-dependent factor for sustained load deflections

ρ = ratio of nonprestressed tension reinforcement

ρ' = ratio of nonprestressed compression reinforcement

ρ_b = reinforcement ratio corresponding to balanced strain conditions

ρ_g = ratio of total reinforcement area to cross-sectional area of column

ρ_ℓ = ratio of area of distributed vertical wall reinforcement to area of concrete perpendicular to that reinforcement

ρ_s = ratio of volume of spiral reinforcement to total volume of concrete core (out-to-out spirals) of a spirally reinforced compression member

ρ_t = ratio of area distributed transverse reinforcement to gross concrete area perpendicular to that reinforcement

ρ_w = ratio of A_s to $b_w d$

ϕ = strength-reduction factor

θ = angle of compression struts in web of beam

ψ_c = factor used to modify hooked bar development length based on side and back cover

ψ_e = factor used to modify development length based on reinforcement coating

ψ_r = factor used to modify hooked bar and compression bar development lengths based on confining reinforcement

ψ_s = factor used to modify development length based on reinforcement size

ψ_t = factor used to modify development length based on reinforcement location

ω = reinforcement index

Index

A

ACI, *see* American Concrete Institute
Actions, 27–37. *See also* Loads
Adjacent span loading, 179
Age-adjusted effective modulus, 83
Age-adjusted transformed section, 83, 868
Aggregates, 52, 256–257
Air entrainment, 50–51, 91
Alkali silica reaction, 91
Allowable stress design for footings, 808
Alternate span loading, 179
American Concrete Institute (ACI), 9–10, 20–27, 40–41, 182–184, 287–288, 331–356, 857–858, 918–921, 1028–1030
  building codes, 10, 20–27, 40
  concrete, specifications for, 37–38
American Society of Testing and Materials (ASTM), 51, 54, 56, 59, 93–95, 98–99
  cement, specifications for, 51–52
  concrete, specifications for, 52–53, 91
  steel reinforcement specifications, 90–92
  welded-wire reinforcement specifications, 97–98
Analysis strip, 177
Anchorage, 265–267, 331–332, 373–385
  bars, 365–366, 861
  design for, 375
  failure of, 265–267
  headed, 378
  hooked, 373–377
  mechanical, 378
  nodal zones, 862–864
  stirrups, 265–267, 331–332
  struts, 864
  ties, 862–864
  torsional reinforcement, 331–332
Arches, actions of, 248–249
Area (A), 31, 32, 176–177, 181–182, 307–308, 326–330
  floor systems, 176–177, 181–182
  influence (A_I), 31, 181–182
  within perimeter of a section (A_{cp}), 326
  shear flow, enclosed (A_o), 307–308, 322
  tributary (A_T), 31, 176–177
Area, steel reinforcement
  balanced reinforcement, 132–133
  compression reinforcement (A'_s), 107, 116–117, 142–143
  effects of on strength and ductility, 117–119, 142–144
  flexure of beam sections, 105, 113, 117–118, 131–132, 142–143, 159, 204
  tension reinforcement (A_s), 107, 113, 117–119, 131–132, 159, 204
  longitudinal reinforcement (A_l), 328–330
Autogenous shrinkage, 74
Average shear stress, 247–248
Axial loads, 27, 258, 295–299, 496–498, 505, 508, 512, 962–964, 1000, 1003–1008
  capacity, 501

Axial loads *(Continued)*
columns, 496–498, 499–502
compression, 298
maximum, 508
pure, 505, 512
shear in, 298
tension, 297
walls, 962–964, 1003–1008
Axial load-moment curves, *see* Interaction diagrams

B

B–regions (Bernoulli), 873
Balanced conditions, 131–132
Balanced failure, 131, 503, 507
Balanced reinforcement ratio (ρ_b), 132
Balanced strain diagrams, 131
Bar-location factor (ψ_t), 370
Bar-size factor (ψ_s), 370
Bar-spacing factor (c_b), 366–370
Bars, 195–196, 198–203, 229–231, 331, 365–373, 520, 724, 726–728, 1072, 1077–1086, 1093
anchorages for, 365–373, 728
bundled, 372
centroids of, 201
coated, 370, 372
covers, 198–199, 228, 726
cut-off, 195, 385–395, 728
development lengths for, 365–373
effective depth (*d*) of, 199–202
locations in beams, 196–197
slabs, 229–231, 724–729, 1093
spacing, 198–199, 333–334, 527, 727
structural integrity requirements, 395–413
tables for, 1072, 1077–1086
torsional reinforcement, 331–334
trusses (bent-up), 197
Basement walls, shrinkage of, 76
Beam-action (one way) shear, 687, 817
Beam–column joints, 939–951, 1046
corners, 939–942
frames, in, 943–944
moment-resisting frames and, 967–968
nonseismic design of, 944–946
reinforcement, 947–950
seismic design of, 946–947
shear strength of, 944–948
T-shaped, 942–943
Beam-to-slab stiffness ratio (α_f), 640–642
Beams, 1–3, 4, 105–299, 408–412, 458–477, 480–481, 616–618, 755–765, 825, 827–835, 896–899, 953–954, 1077–1078, 1082. *See also* Concrete beams; Reinforced concrete
actions, statics of, 108–111
composite, 825, 827–835
compression-controlled sections, 132–141
compression reinforcement of, 119, 141–149
continuous, 182–194, 177, 180, 289–290, 464494
cracked moment of inertia (I_{cr}) for, 189
cracking in, 243–246, 249–250
cross sections, 133–134, 992
coupling, 1065–1067
deep, 875, 878, 892
doubly-reinforced sections, 116, 145–153
edge, 616–618, 635–636
elastic analysis of stresses for, 409–412
failure of in shear, 243–259
fiber-reinforced concrete (FRC), 260
flanged sections, 133–165
flexural design of sections, 173–238
flexure of, 109–172
forming for, 2–4
internal forces in, 254–255, 259
inverted L-, 152, 155
moment-curvature relationships for, 109–113
over-reinforced sections, 136
plain, 5
prestressed, 1
service-load stresses in, 424–426
shear (V) in, 243–299, 833, 835–843
shear strength (V_c) of, 255–259
simply supported, 274–281
singly reinforced sections, 111, 124–130, 138–142
size of, 256–257
slabs designed with, 755–765
spandrel, 152–153, 183
stirrups for, 196–197, 249, 254–255, 274–284
structural analysis for, 188–194
strut-and-tie models for, 896–902, 953–954
T-, 152–153, 155, 163–165, 468–489, 953–954
tables for, 1077–1078, 1082
tapered, 294–295
tension-controlled sections, 132–141
transformed sections, 420–424
uncracked, 246
under-reinforced sections, 121, 125, 133–139
web reinforcement for, 252, 258–260, 269–270
Bearing strength, 819–820, 951–953
Bearing walls, 959–960
Bending-moment diagrams, 408–412, 1089–1090
Bending moments, 16, 108–109
Biaxial loads, 62–66, 545–557
Bolsters (chairs), 197
Bond, 43–44, 359–363
average, 360
cracks, 44, 63
force transfer as, 359–360
pull-out test for, 362–363
splitting loads, 366–367
stresses (μ), 359–363
transfer mechanisms, 364–365
true (in-and-out), 359, 361–362
Boundary elements in shear walls, 988–990
Braced (nonsway) frames, 465–466, 562–563
design of, 582–593
effective length (*kl*) of columns for, 562–563, 577
end restraints, effects of on, 582–584
moment magnifier design for, 566–568
reinforced concrete floor systems and, 465–467

relative stiffness (ψ) of columns for, 583, 614
sustained loads, effects on, 574–575
Brackets, *see* Corbels
Biaxially loaded columns
Bresler reciprocal load method for, 548, 555–557
equivalent-eccentricity method for, 547, 554
strain-compatibility method for, 547
Buckling, 561–563, 575, 962
concrete properties and, 1003
creep, 575
edge restraint factor for, 1003
Euler load, 562
slender columns, 561–563, 575–576
walls, 959–1008
Building codes, 10, 20–27, 50
Button-head anchorages, 100

C

Cantilever retaining walls, 961
Capacity design, 21, 1030–1031
Capitals, 6, 624, 725
Carbonation shrinkage, 74
Carryover factors (COF), 660–661
Cement, 7–8, 43, 50–51, 93
Centroids of reinforcement bars, 201
Checkerboard loading, 179
Chemical attack causes of breakdown in concrete, 91
Circular columns, 519–520, 714
Circulatory torsion, 310–314
Coating factor (ψ_e), 370
Coefficient of variation (V), 47–48
Cohesion-plus-friction model, 858–859
Collapse mechanism, 21
Columns, 495–557, 636–638, 659, 802–804, 811–812, 1073–1120
axially loaded, 499–502, 561–563
bar spacing requirements for, 527
biaxially loaded, 545–557
circular, 519–520, 714
critical shear sections around, 710–712
eccentrically loaded, 498–499
edge (exterior), 659, 710–712, 719–724
fan yield-line patterns at, 802–804
interaction diagrams for, 504–523, 1072–1120
interior, 642, 659, 717
load transfers to footings from, 811–812
material properties for, 525–526
maximum axial load for, 508, 509
moment-resisting frames and, 1036–1046
pin-ended, 564–582
polar moment of inertia (J_c) for critical section around, 710–712
reinforced concrete, 504–523
reinforcement ratio for, 525–526
second maximum load of, 498
shear and moment transfer to, 705–724
shear strength at slab–column connections and, 701
short, 495, 523–543
size estimation of, 525–526
slabs supported by, 635–638, 644–659
slender, 495, 526–527, 559–621
spiral, 496–502, 526–527, 539–540
splices for reinforcement of, 527–531
stiffness and carryover factors for, 1088
strength, steel and concrete contributions to, 544–545
strength-reduction factor (ϕ) for, 507–508
tied, 495, 496–502, 526, 535–538
torsional members and, 666–675
unsymmetrical, 521
Combined footings, 806, 838–848
Companion-action loads, 23, 29
Compatibility torsion, 323–326, 489
Composite beams, 861, 863–872
deflections in, 867–868
design of, 868–872
horizontal shear in, 865–867
Loov and Patnaik shear transfer equation for, 861
shored and unshored construction of, 863–864
Compression, 43–46, 67–72, 142–151, 195–219, 289–290, 855–856
axial, 298
concrete behavior in, 43–46
microcracking from, 43–44
mode failure changes from, 145
permanent force (N_u), 855–856
stress–strain curves, 43–44, 67–73
transverse, 153–155
Compression-controlled beam sections, 132–142
Compression-controlled limit (CCL), 134
Compression fans, 890
Compression fields, 289–290, 890–891
Compression lap splices, 415
Compression reinforcement, 107, 118–119, 142–151
doubly-reinforced beam sections area of (A'_s), 112, 142
ductility increased from, 117, 144
fabrication ease from, 145
flexural behavior, effect of on, 144–145
nominal moment (M_n) strength and, 146–149
strength, effect of on, 140–141
sustained load deflections reduced by, 144
ties for, 149
Compressive strength (f'_c), 46–58, 60–61, 63–65, 92, 467
aggregates and, 52
biaxial loadings and, 62–63
building code requirements for, 50
cement type and, 51
coefficient of variation (V) of, 47–48
concrete, 46–58, 60, 63–65
control data for, 49
core tests for, 54–57
curing conditions, 53
distribution of, 47–49
effective, 457
equivalent specified strength (f'_{ceq}), 56–58
pozzolans for, 52

Compressive strength *(Continued)*
rate of loading and, 54
standard deviation (s) of, 47–48
standard tests for, 46–47
stress–strain curves and, 43–44
supplementary materials for, 51–52
tensile strength and, 59–66
water/cement ratio for, 50–51
water quality and, 52
Compressive stress block, 110
Concentrated loads, 801–805, 818, 896
columns and, 801–805, 818
strut-and-tie models and, 890–892
Concrete, 8, 43–101, 255–256, 544–545. *See also* Reinforced concrete
age of, 53
air entrainment, 50–51, 91
chemical attack causes of breakdown, 91
compression, behavior of in, 43–46, 67–72
compressive strength (f'_c), 46–58, 60–61, 63–65
corrosion of steel in, 90
creep and, 78–85, 87
critical stress of, 44
discontinuity limit of, 44
durability of, 90–91
fiber-reinforced, 90
high-strength, 47, 85–87, 273
lightweight, 49, 87, 256
maturity of, 53–54
microcracking, 43–44
multiaxial loading of, 59–66
normal-weight, 67–73
shear strength (V_c) of beams and, 255–258
shotcrete, 93
shrinkage, 73–78, 87
strength of in structures, 57–58, 563–564
stress–strain curves for, 43–44, 46, 67–73, 1003–1004
temperature effects on, 91–92
tensile strength of, 59–66, 72–73, 255
tests for strength of, 46–47, 54–57
thermal expansion and contraction, 85
volume changes, time-dependent, 73–85
walls and, 1003–1008
Concrete beams, 1–4, 8, 197, 260–287, 439–447, 852, 863–872
composite, 852, 863–872
deflections of, 439–447, 867
flexural stiffness (EI) of, 440–443
load–deflection behavior of, 440
moment of inertia (I) for, 440, 441–443
reinforced, 1–4, 8, 197, 260–287
Concrete Reinforcing Steel Institute (CRSI), 41
Connections, 939–950, 705–729
beam-column, *see* Beam-column joints
slab-column, 705–729
Construction loads, 34, 765–766
Continuity reinforcement, 395–397
Continuous beams, 188–194, 281–284, 464–494
deep, 908–921
design of, 468–494
floor system design using, 184–194, 467
moment coefficient (C_m) for, 185–191
moment redistribution for, 492–494
reinforced concrete for, 464–468
stirrup design for, 281–284
structural analysis of, 188–194
strut-and-tie models for, 909
Control data for concrete compressive strength, 49
Corbels (brackets), 6, 921–932
ACI code method of design for, 929–932
structural action of, 921–922
strut-and-tie model design of, 923–929
Core of spiral columns, 498
Core tests, 54–57
Corner joints, 939–942
Coupled shear walls, 970–974
Coupling beams, 1065–1066
construction of, 1067
high-performance fiber-reinforced concrete (HPFRC), 1067
with diagonal reinforcement, 1065–1067
Cover, 228–229, 239, 726–727
Cracked moment of inertia (I_{cr}), 191
Cracking point, 114–116
Cracks, 14, 43–44, 63–65, 247–248, 253–254, 265, 267–268, 334–335, 430–439, 876–878
angle (θ), 326
beam sections, 248–249, 253–254
bond, 43–44, 432
concrete, 43–44
control of, 434–436, 439
development of, 432–434
flexural, 246–247
flexure-shear, 253
heat-of-hydration, 431
inclined, 246, 249–250, 253–254
load-induced, 430–439
longitudinal, 876–878
map, 431–432
microcracking, 43–44
mortar, 43–44
plastic slumping, 431
reinforced concrete, 63–65
reinforcement for, 249–250, 437
serviceability and, 419–461
shear between, 247–248
skin reinforcement for, 439
temperature reinforcement for, 438
torsion and, 334, 335
web-shear, 253
widths, 14, 334–335, 436–438
Creep, 78–85, 87, 575
age-adjusted effective modulus for, 83
buckling, 575
calculation of, 81–83, 85
coefficient, 79–80
compliance function, 82
high-strength concrete, 87
restrained, 83–85
unrestrained concrete, 78–83
variability of, 80–81
Critical loads for walls, 1003–1008
Critical sections, 337–338, 690–691, 707–708, 815–816
beams, 337–338
footings, 815–816
torsion and, 337–338
two-way slabs, 689–690, 710
Critical compression stress, 43–44
Crushing strength of webs, 335–336

Curing concrete, 53
Curvature relationships to moments in slabs, 633–634
Cut-off bars, 195, 385–389
bending-moment diagrams for, 408–412
development of, 393–395
equations of moments for, 399–407
flexural reinforcement and, 385–395
inflection, point of, 392
location of, 195, 386–389
maximum force, points of, 389–390
positive-moment regions, 390–395
required-moment diagram for, 386
shear, effect of on, 388–389
structural integrity requirements, 395–413

D

D-regions, *see* Discontinuity regions
Dapped ends, 932–938
Dead loads (*D*), 24, 29–30
Deep beams, 250, 896–921
continuous, 908–921
discontinuity regions in, 896–921
elastic analysis of, 896–898
strut-and-tie models design of, 899–900
Deflections, 15, 144–145, 212, 439–458, 573–574, 767–771, 867–868
allowable, 449
buckling, 561–563
compression reinforcement effects on, 142–144
concrete beams, 439–447, 867–868
design considerations for, 447–457
dimensions for control of, 197–198
first-order, 564–565
flexural stiffness (*EI*) and, 440–447, 458, 573
frames, 458, 573–574
instantaneous, 443–447
lateral, 458, 560
nonstructural elements, damage to from, 447–449
second-order, 573–574
serviceability limit states (SLS) from, 448–449, 459
singly-reinforced beam sections, 197–198
slender columns, 573–574
span-to-depth limits for control of, 450–457
structural stability and, 14
sustained-load, 144, 446–447
two-way slabs, 767–771
vertical, 458
visual appearance of, 447
Depth, 113, 133, 202–203, 250
effective (d), 113, 133, 202–203
shear (d_v), 250
Design, *see* Flexural design; Seismic design; Structural design
Design loads for continuous floor systems, 467
Design pressure (p), 35
Development length, 365–373
Diaphragms, 975, 1021, 1058–1060
discontinuity irregularities for seismic design, 1021
earthquake loads and, 1021, 1058–1060
flexural strength of, 1058–1059
shear strength (V_c), 1059
stiffness effects on load distribution, 1059–1060
walls and, 975
Direct-design method for slabs, 644–659
Direct loads, load factors for, 23
Directionality factor (K_d), 35
Discontinuity limit, 44
Discontinuity regions, 873–956
beam–column joints, 939–951
bearing strength, 951–953
behavior of, 875
continuous deep beams, 908–921
corbels (brackets), 921–932
dapped ends, 932–938
deep beams, 896–908
Saint Venant's principle, 873–874
strut-and-tie models for, 873–956
T-beam flanges, 953–955
Discontinuous corners of slabs, 800–801
Distribution factors (DF), 660
Doubly-reinforced beam sections, 113, 142, 220–228
compression reinforcement and, 142–151
flexural analysis of, 149
flexural behavior of, 111, 142
flexural design of, 220–228
nominal moment (M_n) strength of, 145–148
strength-reduction factors (ϕ) for, 149
ties for, 149–150
Drop panels, 6, 661–662, 724–729
Drying shrinkage, *see* Shrinkage
Ductility, 117–119, 144, 1017–1019
beam section variables, effects on, 117–119
compression reinforcement and, 119, 144
earthquake loads and, 1017–1019
Dynamic loads, 28

E

Earthquake loads (*E*), 24–25, 947–948, 999–1002
beam–column joints and, 1056–1058
beam cross sections, geometric limits of, 1032
columns and, 1046–1055
design for resistance of, 1013–1069
distribution of, 1025–1027
ductility and, 1027–1028
equivalent lateral force method for, 1023–1025
flexural members and, 1032–1046
foundations and, 1068
load factors for, 24
load-resisting systems for, 947–948, 999–1002
moment-resisting frames for, 1032–1058
nonductile frame members for, 1068
precast structures for, 1068
reinforced concrete and, 1027–1028
reinforcement for, 1028–1055
resisting moments for, 1028–1029
response-modification factor (*R*) for, 1025

Earthquake loads *(Continued)*
seismic design, 1014–1027, 1028–1032
seismic response spectra, 1014–1019, 1023
shear walls and seismic load resistance, 999–1002
structures, seismic forces on, 1017–1019
walls and, 960, 966, 967
Eccentrically loaded columns, 525
Eccentricity (e) of load for interaction diagrams, 525, 554
Edge beams, 635–636, 651
Edge (exterior) columns, 659, 710
Effective compressive strength, 467
Effective depth (d), 133, 199–203
Effective flange width, 153–154
Effective length (kl), 562, 577
Elastic analysis, 420–429, 779–781, 896–898
beam sections, 420–429
deep beams, 896–908
flexural stiffness (EI) for, 420–429
modular ratio for, 420
modulus of elasticity (E) for, 420
serviceability and, 420–429
two-way slabs, 779–781
Elastic distribution of soil pressure, 810–811
Elasticity (E), moduli of, 67, 420
End restraints, effects on braced frames, 582–593
Equilibrium, 13, 124, 323–326, 561–563, 875
columns, states of, 561–563
internal forces for flexural analysis, 124
structural stability and loss of, 13
strut-and-tie models and, 875–876
torsion, 323–326
Equivalent eccentricity method for biaxial columns, 547–548, 554–555
Equivalent-frame methods for slabs, 659–687
columns, properties of, 666
computers used for analysis, 681–684
flexural stiffness (EI) for, 660–661
live load arrangement for, 675
moment distribution for, 676
slab–beams, properties of, 661–662
Equivalent lateral force method for earthquake loads, 1023–1025
Equivalent specified strength (f'_{ceq}), 56–58
Euler buckling load, 562
Extended nodal zones, 886–887
External pressure coefficient (C_p), 36

F

Factored loads (U), 17, 20–21, 23–27, 106, 180–183, 190
computation of effects, 25–27
design moment (M_u), 106, 187
flexure and, 106
floor system design for, 180–182
load combinations and, 23–25, 184–188
moment coefficient (C_m) for, 182–188
structural design and, 17, 20–22
Failure, 14, 18–19, 33–34, 131, 144, 243, 255, 449, 498, 501–502, 559
balanced, 131–132, 503, 507
compression, 144–145, 243
consequences of, 18–19
mode, 145
probability of (P_f), 19–20
shear, 248, 258, 804
stability, 565
structural safety and, 19–20
Fan-shaped yield patterns, 801–804
Fatigue, 14, 96–97, 460
Fiber-reinforced concrete (FRC), 88–90, 260
Fiber-reinforced polymer (FRP) reinforcement, 99–100
Fillers, 490
Finite-element analysis for slabs, 781–783
Fire resistance of structures, 6
First-order analysis, 564
Fixed-end moments (FEM), 660–661
Flanged beam sections, 152–165
area of tension reinforcement (A_s) for, 161
effective width of, 153–155
flexural analysis of, 159–163
inverted L-beam, 152, 154
nominal moment (M_n) strength of, 155–158
overhanging portions, 155
spandrel beam for, 152–153
strength-reduction factors (ϕ) for, 159
T-beams, 152–153, 155, 163–165
Flat (plates) slabs, 6, 624–625, 652–659, 677–681
direct-design method for, 652–659
equivalent-frame method for, 677–681
system, 624–625
Flexural cut-off points, 407–412
Flexural design, 119–124, 173–239
beam sections, 173–239
continuous beams, 184–194
doubly-reinforced beam sections, 220–228
flexural theory simplifications for, 119–124
floor sections, 173–194
one-way slabs, 188–191, 228–239
singly-reinforced beam sections, 195–220
Flexural reinforcement, 371, 1034
Flexural-resistance factor (R), 211, 1075–1076
Flexural stiffness (EI), 440–441, 445, 573, 596–598, 660–661
concrete beams, 439–441
lateral reduction factor for, 596
serviceability and, 420–421
slabs, 660–661
slender columns, 573, 596–598
sway (unbraced) frames, 596–598
ultimate limit state, 596–597
Flexural strength, 105, 117–119, 985–991, 1058–1059
design for, 106–107
diaphragms, 1058–1059
ductility and, 117–119
required, 106–107
shear walls, 985–991
Flexure, 27, 105–168, 626–628, 814–816
analysis of reinforced concrete, 106
balanced conditions for, 131–132
beam sections, 105–168

code definitions for, 132–142
compression reinforcement and, 142–151
equilibrium of internal forces for, 124
failure in, 626–628
footings, 814–816
loads and, 27, 105–108
nominal moment (M_n) strength, 106, 116, 124–130, 136–138, 146–149, 155–158
slabs, 105–106, 626–628
strength-reduction factors (ϕ), 106–107
stress–strain relationships for, 120–122, 124
structural design and, 27
symbols and notation for, 107–108
theory, 108–124
Flexure-shear cracking in walls, 995
Flexure theory, 108–124
assumptions of, 111
beam action, statics of, 108–111
bending moments, 108
compressive stress block, 109–110
cracking point, 114–116
internal resisting moments for, 108–111
moment–curve relationships for, 114
simplifications in for design, 119–124
stress blocks, 123–124
stress–strain relationships for, 120–122
uncracked-elastic range, 114
Whitney stress block, 122–124
yield point, 112, 116–117
Floor systems, 173–194, 467–468, 489–490, 771–773, 984. *See also* Girders; Slabs
column loads transferred through, 467–468
continuous beams for, 184–194
design loads for, 467
factored load combinations for, 184–188
flexural design of, 173–194
girders for, 175, 489–490
influence lines for, 177–178
joist, 490–492
live load reductions for, 181–182
load paths in, 175–176
moment coefficient (C_m) for, 182–188
Mueller-Breslau principle for, 178–180
one-way, continuous, 173–194
pattern loadings in, 177–180
post-tensioning, 771–773
shear coefficient (C_v) for, 182–184
slabs for, 188–191, 771–773
spandrel beams for, 184
structural analysis for, 188–194
tributary areas in, 176–177
wall load transfer through, 984
Fly ash, 51–52
Footings, 4, 806–850
allowable stress design of, 808
combined, 806, 838–848
flexure of, 814–816
limit-states design of, 808–810
load transfers in, 818–821, 832–833
mat (raft) foundations, 806, 848
pile caps, 806, 848–850
rectangular, 830–832
reinforcement, 814–816
shear in, 816–818
soil pressure under, 806–814
spread, 806, 814–821, 824–838
strip (wall), 806, 814–824
Force whirls (U-turns), 891–892
Forms, construction of, 1–3, 7
Foundations, 979–980, 1068
Frames, 458, 465–467, 562–563, 577–578, 582–618, 943–944, 967–968, 1068
beam–column joints in, 943–944, 1056–1058
braced (nonsway), 465–467, 562–563, 576, 582–593
deflections of, 458, 573–574
design of, 582–618
earthquake loads and, 999–1002, 1068
effective length of columns for, 562–563
first-order analysis for, 601
moment-resisting, 967–968, 1032–1058
moments in, 577–578, 596–597
second-order analysis for, 596–600
slender columns and, 563–564, 577, 582–618
stability index (Q) for, 576
sway (unbraced), 465–467, 562–563, 576, 593–618
Friction, 852–863
ACI code design rules for, 857–858
coefficients of (μ), 857
cohesion and, 853–855
cohesion-plus-friction model for, 858–859
inclined reinforcement, 856
lightweight concrete and, 856–857
permanent compression force (N_u) for, 855–856
push-off specimens for, 852–853
shear-friction model for, 855–856
upper limits on, 858
Walraven model for, 859–861

G

Girders, 175–176, 284–287, 489–490
compatibility torsion and, 489–490
load paths and, 175–176
shear reinforcement of, 284–287
Gravity loads, 603–618
Gross moment of inertia (I_g), 189
Gross soil pressure, 812–814
Gust-effect factor, (G), 36

H

Hanger reinforcement, 292–294
Heat-of-hydration cracks, 431
High-alumina cement, 93
High-performance fiber-reinforced concrete (HPFRC), 1067
High-strength concrete, 47, 85–87, 273
Hollow members, torsion in, 307–310
Hoops, 1027, 1028, 1034, 1043
Horizontal shear, 864–866
Hydrostatic nodal zones, 883–886

I

Impact factor, 28
Imposed deformation, 27–28, 37

Inclined cracks, 246, 249–250, 253–254
Inertia (*I*), moment of, 440, 441–443
Influence area (A_I), 31, 181–182
Influence lines, 179–180
Instantaneous deflections, 443–446
Interaction diagrams, 502–523, 566, 1072
  balanced failures in, 503, 507
  compression-controlled failures, 482
  computational method for, 509–519
  eccentricity (*e*) of load for, 502, 521
  nondimensional, 520, 1072
  significant points of, 505–508
  strain compatibility solution for, 504–505
  strain limits for, 507
  transition failures in, 507
Internal forces, 124, 254–255
  beams, 254–255
  equilibrium of, 124
  shear failure and, 254–255
Internal moments, *see* Moments (*M*)
Inverted L-beam, 152

J

Joints, *see* Beam–column joints
Joists, 4, 490–492

K

Kinematically admissible mechanism, 21, 785

L

Lap splices, 413–415, 1034, 1072
Lateral deflections, 458
Lateral earth pressure, 24–25
Lateral loads, *see* Earthquake loads; Windloads
Lateral stiffness-reduction factor, 596
Leaning columns, 594
Lightweight-aggregate-concrete factor (λ), 371
Lightweight concrete, 49, 87–88, 256, 856–857
Limit design, 21
Limit states, 13–15, 419–420, 448–450, 459
  deflections and, 448–450, 459
  design, 15
  reinforced concrete design and, 13–15
  serviceability (SLS), 14, 419–420, 448–449, 459
  special, 14
  ultimate, 13–14
  vibrations, 459
Limit-states design of footing, 808–810
Limiting slenderness (*kl*/*r*) ratio, 575–576
Live loads (*L*), 24, 28–33, 176–182, 448, 675
  arrangement of for structural analysis, 675
  checkerboard, 179
  deflection, 448
  floor systems, 176–182
  impact factor for, 28
  influence area for (A_I) 31, 181–182
  load factors for, 24
  pattern loadings for, 177–180
  reductions for, 181–182
  specifications for, 29
  sustained (quasi-permanent), 29
  tributary area (A_T) for, 31, 176–177
  use and occupancy, due to, 30–33
Load and resistance factor design (LRFD), 21, 811–812
Load–deflection behavior, 439–440
Load-induced cracks, 430–437
Loadings, 27–37
Loads, 15–17, 18, 20–37, 62–66, 176–180, 408, 606–608, 801–803, 818–821, 800, 894. *See also* Earthquake Loads; Wind loads
  accidental, 28
  ACI building codes for, 20–27
  actions, 27–37
  axial, 27
  biaxial, 62–65
  classification of, 28–29
  combinations, 23–24, 419
  companion-action, 23, 29
  concentrated, 801–803, 818–821, 892, 895
  construction, 34
  dead (*D*), 23, 29–30
  direct, 23
  dynamic, 28
  factored (*U*), 17, 20, 22–23, 25–27
  factors, 16–17, 20, 23–25
  fixed, 28
  footings, 818–821, 832–833
  gravity, 606–618
  imposed deformation, 27–28, 37
  internal resisting moment for, 16
  lateral, 24–25
  live (*L*), 24, 28–33, 176–182
  multiaxial, 62–66
  occupancy categories for, 32
  permanent, 28
  rain (*R*), 33–34
  required strength for, 20–21, 22
  roof (L_r), 23, 33–34
  snow (*S*), 33
  static, 28
  structural design and, 15–37
  strut-and-tie models, spreading regions of, 892, 895
  sustained, 28–29
  triaxial, 65–66
  variable, 28
  vertical, 832–833
  walls for resistance of lateral, 967–968
  working (service), 419
Long columns, *see* Slender columns
Longitudinal cracking, failure of struts by, 876–882
Longitudinal reinforcement, 255, 328–336, 1032–1034, 1046–1047
Lower-bound plastic theorem, 21

M

Map cracking, 431–432
Mat (raft) foundations, 806–807, 848
Materials, 7, 43–101, 565
  choices for structures, 7
  compressive strength (f'_c), 46–58, 60–61, 63–66, 86
  concrete, 43–92
  fiber-reinforced concrete, 88–89
  fiber-reinforced polymer (FRP) reinforcement, 99–100

high-alumina cement, 93
high-strength concrete, 49, 85–87
lightweight concrete, 49, 87
prestressing steel, 100–101
shotcrete, 92
steel reinforcement, 97, 93–99
stress–strain curves, 43–44, 46, 67–73
volume changes, time-dependent, 7, 73–86
Mechanical splices, 415
Microcracking, 43–44
Mill tests, 95
Modular ratio, 420
Modulus of rupture (f_r) test, 59–60
Mohr rupture envelope, 66–67
Mohr's circle of compatibility, 289
Moment coefficient (C_m) for floor systems, 178–183
Moment curve relationships, 111–114
Moment-resisting frames, 967–968, 1031, 1036–1046, 1056
beam–column joints and, 1034
columns in, 1046–1055
earthquake loads and, 1032–1055
flexural members in, 1036–1046
reinforcement for, 1032–1055
walls, 967–968
wind loads and, 967–968
Moments (M), 16–17, 105–111, 116, 124–130, 189, 304, 338–357, 440, 441–443, 566–571, 596–600, 629–638, 644–659, 675–677, 705–724, 781–783, 1032
analysis of in two-way slabs, 629–633
bending, 16, 108–109
curvature and, 633–638
distribution of, 633–638, 644–659, 675–677
equivalent factor (C_m), 570–571
factored design (M_u), 106
finite-element analysis for, 781–783
first-order (M_o), 566–568
flexure and, 105–107, 124–130
inertia (I), 189, 440, 441–443
internal resisting, 16, 108–110
magnifiers for columns, 566–568, 602
maximum, 602–603
minimum, 603
negative, 107
nominal (M_n) strength, 16–17, 105–107, 116, 124–130
nonsway (M_{ns}), 595
positive, 107
second-order (M_c), 566–568, 596–600
shear and torsion combined with, 337–357
shear transfer and, 705–724
slabs and, 629–638, 644–659, 675–677, 705–724
slender columns and, 566–571, 596–600
static (M_o), 632, 644–647
sway (M_s), 596
torsional (T), 304, 323–324, 332, 337
transfer of, 659, 705–724
yield (M_y), 116
Mortar cracks, 43–44
Mueller-Breslau principle, 178–180
Multiaxial loads, 62–66
biaxial, 62–65
compressive strength (f'_c), and, 63–65
concrete strength under, 62–66
Mohr rupture envelope for, 65–66
triaxial, 65–66

N

Net soil pressure, 812–814
Nichols' analysis of moments, 626, 630–633
Nodal zones, 875, 883–890
anchorages in, 888–889
extended, 886–887
forces acting on, resolution of, 888–889
hydrostatic, 883–886
strength of, 887
subdivision of, 887
Nodes, 875, 883
Nominal moment (M_n) strength, 16–17, 105, 116, 124–130, 136–138, 146–149, 151–154, 985, 988–990
beam sections, 116, 124–130, 138, 146–149, 155–158
compression reinforcement and, 146–149
flexure and, 105, 116, 124–130, 136–138
reinforcement ratio (ρ) and, 136–138
shear walls, 985, 988–990
Nomographs, effective length factors, 585
Notation, 107–108, 961, 1121–1127

O

One-way slabs, 5, 105–106, 188–191, 228–239
concrete covers for, 228–229
flexural behavior of, 105–106
flexural design of, 188–191, 228–229
floor systems and, 188–191
reinforcement of, 229–231
thickness of, 228–229
One-way walls, 961
Over-reinforced beam sections, 138

P

Pans, 490
Pattern loadings in floor systems, 176–180
Perimeter of a section, (ρ_{cp}), 326
Permanent loads, 28
Pile caps, 806, 848–850
Pin-ended columns, 564–582
deflections of, 564–568
design of, 578–582
failure of materials and stability of, 565–566
moments for, 566–571
sustained loads, effects of on, 565–566
symmetrically loaded, 566–568
unequal end moments of, 568–571
Plastic design, 21
Plastic distribution of soil pressure, 811
Plastic mechanisms, structural stability and, 14
Plastic slumping cracks, 431

Poisson's ratio, 73
Polar moment of inertia (J_c), 710–712
Ponding failure 14, 34, 449
Post-tensioning, 771–773
Pozzolans, 52
Prestress losses, 100
Prestressed beams, 1
Prestressing steel, 100–101
Principal stresses, 246, 310
Probability of failure (P_f), 19–20
Probable moment strength (M_{pr}), 1032
Progressive collapse, 13–14, 395
Pull-out test, 362–363
Punching (two-way) shear, 687–689, 693–694, 817–818

R

Radius of curvature (r), 633
Rain loads (R), 33–34
Rectangular footings, 830–838
Reentrant corner irregularities for seismic design, 1021
Reinforced concrete, 1–11, 13–15, 106, 111–114, 197, 260–287, 315–316, 464–468, 1027–1028
  ACI building codes for, 10
  analysis versus design, 106
  beams, 1–4, 197, 260–287
  building elements, 6
  cement and, 7–8
  compressive strength (f'_c) of cracked, 63–65
  continuity in, 464–468
  design specifications for, 9, 260–263
  ductility of, 1027–1028
  earthquake loads and, 1027–1028
  factors effecting choice of, 6–7
  flexure theory for, 111–114
  floor systems, 467–468
  footings, 4
  frames, 465–467
  joists, 2
  limit states, 13–15
  mechanics of, 1–2
  shear, design and analysis for, 260–287
  slabs, 4–6, 197
  structures, 1, 464–468
  torsion, subjected to, 315–326
Reinforcement, 93–99, 113, 132–151, 195–209, 229–231, 249–252, 258–260, 270, 274–284, 292–294, 359–417, 437–439, 699–705, 724–729, 814–821, 856, 948–950, 982–985, 993–994, 1032. *See also* Area (A)
  anchorages, 265–267, 373–385
  ASTM specifications for, 93–95, 98–99
  bars, 195–196, 198–203, 365–372, 385–395, 727
  beam–column joints, 939–951
  beam sections, 195–220
  bolsters (chairs), 197
  bond, 359–365, 366–367
  code definitions for, 132–142
  compression, 142–151
  concentrated, 982, 984
  continuity, 395–413
  cracks and, 250–251, 437
  coupling beams, 1065–1066
  design of, 204–209, 701–703
  development of, 359–373
  distributed, 982
  extreme layer of tension, 133
  fabrication ease, 145
  fatigue strength of, 96–97
  fiber-reinforced polymer (FRP), 99–100
  flexural, 371, 385–395, 814–816, 1034
  footings, 814–816
  hanger, 292–294
  hoops, 1027, 1028, 1034
  inclined shear friction, 856
  location of (placing sequence), 195–199, 726
  longitudinal, 328–336, 1032–1034, 1046–1047
  minimum, 115, 203, 982–984
  seismic design, 1031–1046
  shear, 249–250, 388–389, 993–994, 1035–1036, 1049
  shearheads, 700
  shrinkage, 230, 438
  skin (web face), 439
  slabs, 229–231, 699–701, 724–729
  spiral, 135
  splices for, 413–415
  stirrups (transverse bars), 196–197, 249, 258–260, 274–284, 327–328, 699–700
  structural integrity requirements, 395–413, 436–438, 729
  studs, 700–701
  temperature, 230, 438
  torsional, 315–326
  transverse, 1034, 1047–1049
  upper limits on, 136–138
  walls, 982–984, 993–994
  web, 249, 250–252, 255–256, 439
  welded-wire, 97–99, 372–373
Reinforcement ratio (ρ), 132, 136–138, 255, 525–526
  balanced (ρ_b), 132
  columns and, 525–526
  longitudinal (ρ_w), 255
  nominal moment (M_n) strength and, 136–138
  over-reinforced beams sections, 138
Relative humidity effects on shrinkage, 73–75
Relative stiffness (ψ), 583, 586–587
Restrained columns, 593–595
Retaining (nonbearing) walls, 960–961, 966
Rigidity, *see* Stiffness
Roof loads (L_r), 23–24, 33–34

S

Safety index (β), 19–20
Saint Venant's principle, 873
Sand-heap analogy, 307
Secant modulus of elasticity, 67–69
Second peak load in spiral columns, 498
Second-order analysis, 564–565, 596–600
Seismic design, 1019–1022, 1028–1032, 1037
  ACI code for, 1028–1032
  building configurations and, 1019–1023
  capacity design for, 1030–1031
  categories, 1019

columns, 1046–1055
lateral force-resisting systems for, 1019–1020
load and resistance factors for, 1029–1030
plan irregularities for, 1020–1021
reinforcement for, 1032–1046
strong column–weak beam design for, 1031–1032
structures, seismic forces on, 1023–1027
vertical irregularities for, 1022–1023
walls, 1060–1068
weight (W) for, 1024
Seismic response spectra, 1014–1019, 1023–1025
acceleration response, 1014–1015
coefficient (C_s), 1023–1024
damping effects on, 1017
displacement response, 1015–1016
ductility effects on, 1017–1018
period of building for, 1015–1017
soil effects on, 1019
velocity response, 1015
Self-straining effects, 25
Self-straining forces, 37
Serviceability, 14, 419–420, 597
concrete beams, 439–447
cracks, 430–439
deflections, 439–447
elastic analysis of stresses for, 420–429
fatigue, 460
frames, 458
limit states (SLS), 14, 419–420, 447–449, 464, 597
service-load stresses for, 424–425
transformed beam sections, 420–424
vibrations, 458–459
Shake-down limit, 46
Shear (*V*), 182–184, 243–301, 309–314, 337, 338, 687–705, 817–818, 852–867, 991–1006, 1035–1036. *See also* Failure; Friction
area within shear flow (A_o), 307–308
average stress, 247–248
axially loaded members and, 295–299
B-regions (Bernoulli) of, 873
beam-action (one way), 687, 754–755, 817
beam sections, 243–301
coefficient (C_v), 182–184
column–slab connections and, 705–724
compression failure, 250–252
compression field theories for, 289–290
cracks, between, 247–248
critical section for, 337–338, 689–690, 695–696, 817–818
D-regions (discontinuity) of, 873–874
depth for calculations, (d_v), 248
equivalence of on planes, 290–291
floor systems, 184–188
flow (q), 249, 307–308, 323
footings, failure in, 817–818
friction, 852–863
girder design for, 294
hanger reinforcement for, 292–294
horizontal, 864–867
inclined cracking and, 250–251, 253–254
maximum, 271–272
models for, 243–301
Mohr's circle of compatibility for, 288
moment and torsion combined with, 337–357
moment transfer, and, 705–724
punching (two-way), 693–694, 817–818
reinforcement, 249–250, 479–482, 699, 993–994, 1035–1036
seismic base, 1023
shear friction method for, 290
smeared-crack models for, 289
stirrups for, 196–197, 254–255, 258–259, 262–267, 327–328, 699–700
story (V_x), 975–979
strength-reduction factor (ϕ) for, 270
stresses(τ), 245–248, 304–310
strut-and-tie models for, 288
traditional ACI method for, 289
transfer, 852–867
tributary areas for, 691–692
two-way slabs, 687–705
whole-member design methods for, 288
Shear caps, 725–726
Shear-friction model, 855–856
Shear-force envelope, 272, 274
Shear strength (V_n), 255–258, 687–705, 946, 991–1002, 1057
beam–column joints, 939–951
beam sections, 255–260
diaphragms, 1059
seismic load resistance from, 994–995, 1037
shear walls, 991–1002
two-way slabs, 687–705
Shear-wall–frame buildings, 969–970
Shear walls, 961, 966–967, 970–974, 982–1002, 1060–1068
boundary elements in, 988–990
coupled, 970–974, 1065–1067
flexural strength of, 985–991
nominal moment (M_n) strength of, 987, 988–991
reinforcement in, 982–984, 990–991
seismic design of, 1060–1068
shear strength (V_c) of, 991–1002
strength-reduction factors (ϕ) for, 985, 1063
structural behavior of, 980–981
wall assemblies and, 966–967, 990–991
Shearheads, 700
Short columns, 495, 523–543
Shotcrete, 93
Shrinkage, 73–78, 87, 230–231, 438
basement walls, 76
calculation of, 76–78
carbonation, 74
drying, 73–74
floor slabs, 85
high-strength concrete, 87
reinforcement for slabs, 229–231
relative humidity effects on, 75–76
temperature reinforcement and, 438
Sidesway buckling, 603
Silica fume, 52
Simply supported beams, 274–281

Singly-reinforced beam sections, 114, 124–130, 138–142, 195–219
area of tension reinforcement (A_s) for, 210
concrete cover spacing for, 198–199
deflections, dimensions for control of, 198
effective depth (d) of, 199–203
flexural analysis of, 146
flexural behavior of, 111–114
flexural design of, 195–242
flexural-resistance factor (R) for, 211
nominal moment (M_n) strength of, 128–130
rectangular compression zones, with, 195–242
reinforced concrete, construction of, 197
reinforcement of, 195–197, 203–206
strength requirements for, 203–204
unknown dimensions, design of with, 209–220
Skin reinforcement, 439
Slab-beams, 661–666
Slabs, 5–6, 76–77, 105, 197–198, 228–239, 624–776, 779–805, 1086–1087
direct-design method for, 644–659
elastic analysis of, 779–783
equivalent-frame method for, 639, 659–681
finite-element analysis for, 781–783
flat (plates), 6, 624, 652–659, 662–664
flexural design of, 197–198, 228–239
floor system design using, 76, 191–194, 768–771
moment-distribution factors for, 1086–1087
one-way, 5, 106, 188–194, 228–239
plate rigidity (D) of, 780–781
reinforcement for, 229–231, 767
shrinkage of, 76–77
structural analysis of, 188–194
thickness of, 779
two-way, 5, 175, 624–805
waffle, 624
yield-line analysis of, 628, 783–803
Slag, ground granulated blast-furnace, 52
Slender beams, 260–287
Slender columns, 495, 526–527, 559–622. *See also* Frames
braced (nonsway) frames and, 582–593
buckling, 562–563, 575
deflections of, 566–567
effective length (kl) of, 562–563
equilibrium states of, 561–562
flexural rigidity (EI) of, 571–574
interaction diagrams for, 566
limiting slenderness (kl/r) ratio for, 575–576
moments for, 565, 566–571, 595
pin-ended, 564–582
restrained, 582–587
size estimation of, 579
structures, 563–564
sway frames and, 593–595
torsional critical load for, 619–621
Smeared-crack models, 289
Snow loads (S), 33
Soap-film analogy, 304–307
Soil pressure, 807–814
elastic distribution of, 810–812
footing design and, 807–814
gross, 812–814
limit states and, 809
net, 812–814
plastic distribution of, 811
Solid members, torsion in, 304–307, 309
Spandrel beams, 152, 184
Span-to-depth limits, 450–457
Span-to-depth ratio (a/d), 256
Special moment-resisting frames (SMF), *see* Moment-resisting frames
Spiral columns, 496–502, 520, 526–527, 539–540
behavior of, 497–499
design of, 539–540
requirements for, 534–535
size estimation of, 526
Spiral reinforcement, 135
Splices, 413–415, 527–531
column reinforcement using, 527–531
compression lap, 415
mechanical, 415
seismic design and, 1034
tension lap, 413–415
welded, 415
Split-cylinder (f_{ct}) test, 59–60
Spread footings, 4, 807–808, 814–821, 824–838
St. Anthony Falls Bridge, 3
Stability failure, 565–566
Stability index (Q), 576
Standard deviation (s), 47–48
Static loads, 28
Static yield strength, 96
Statically admissible structure, 21
Steel reinforcement, 97, 99, 544–545, 727
ASTM specifications for, 93–95, 98–99
corrosion of, 90–91
fatigue strength of, 96–97
hot-rolled deformed bars, 93–99
mechanical properties of, 95–96
prestressing, 100–101
required area (A_s) for slabs, 723
yield strength of, 95
welded-wire, 97–99
Stiffness, 7, 319–321, 596–598, 640–641, 780–781, 1022, 1059–1060. *See also* Flexural stiffness (*EI*)
beam-to-slab ratio (α_f), 640–641
diaphragm effects on load distribution, 1059–1060
importance of for structural rigidity, 7
irregularities for seismic design, 1022
plate rigidity (D), 780–781
relative (ψ) at joint, 583
serviceability limit state, 597–598
torsional (K_t), 319–320
Stirrups, 197, 249, 254–255, 265–267, 277–284, 331–332, 699–700
anchorage for, 265–267, 331–332
design of, 281–284
failure of reinforced concrete beams due to yielding, 262–269
location of, 195–197
shear reinforcement using, 249–250, 701–705

torsional reinforcement using, 322, 326, 330–334
web reinforcement using, 249, 258–260
yielding of, 262–265
Story drift (δ_{max}), 981, 1020
Story shear (V_x), 1026
Story torsion (T_x), 1026
Straight-line theory, *see* Elastic analysis
Strain-compatibility method for biaxial columns, 547
Strain limits, 507
Strength, 22–24, 57–58, 59–66, 96–97, 149, 203–204, 255–258, 499–502, 544–545, 687–705, 819–821. *See also* Compressive strength (f'_c); Flexural strength; Nominal moment strength (M_n); Shear strength (V_n)
beam sections, 203–204, 256–257
bearing, 819–821
columns, 544–545
compression reinforcement effect on, 149–150
concrete in structures, 58, 544–545
crushing of the web, 268, 288
design, 20–22
fatigue, 96–97
flexural, 105, 119–124, 985–991
loads and, 59–66
nodal zones, 875–876
steel reinforcement contributions to, 544–545
structural design, 20–21
Strength-reduction factors (ϕ), 16, 20–21, 27, 106–107, 133, 149, 159, 270, 507–508, 985, 1063
beam sections, 106–107, 142, 149, 159, 270
code definition for, 139–140
columns, 507–508
flexure, 106, 139, 149, 159
shear and, 270
shear walls, 985, 1063
Stress, 122–124, 245–248, 304–310, 360–363, 420–429
blocks, 122–124
bond (μ), 359–363
elastic analysis of, 420–429
principal, 245
service-load, 424–429
shear (τ), 245–248, 304–310
torsion, due to, 304–310
trajectories, 246
Stress–strain curves, 43–44, 46, 67–73, 1003–1005
compression, 43–44, 46, 67–72
concrete properties and, 43, 67–73, 1003–1005
equations for, 69–72
modified Hognestad, 69
moduli of elasticity (E), 67, 1007
normal-weight concrete, 67–69, 72–73
Poisson's ratio and, 73
tension, 72–73
wall buckling loads determined from, 1006–1008
Stress–strain relationships, 112–113, 120, 124
Strip (wall) footings, 807–808, 814–824
Strong column–weak beam design for earthquake loads, 1031–1032
Structural analysis, 13, 188–194
Structural design, 10, 12–40, 1023–1027, 1071–1120
accuracy of calculations for, 40–41
ACI building code procedures, 20–27
dimensions and tolerances for, 40
economic factors for, 12, 37–38
handbooks and aids for, 41
inspection and, 40
limit states, 15
load effects and factors, 15–18, 22–37
plastic, 21
process, 12–40
reinforced concrete specifications for, 9
safety, 17–20
seismic forces and, 1023–1027
seismic requirements for, 1019–1023
strength, 20–21
Structural continuity, 395–413, 729
ACI requirements for, 397–407
bending-moment diagrams for, 408–412
cracking and, 430–439
equations of moments for, 399–407
flexural cut-off points, 407–412
slabs, 729
standard cut-off points, 412–413
Structures, 1–7, 12–40, 464–468, 563, 1068
beams, 1–3
building elements, 5
continuity in, 464–468
design process for, 12–40
economic factors for, 6
fire resistance of, 6
forms and shoring, 7
load effects and factors, 15–17, 22–37
material choices for, 7
members for, 3–4, 13
precast, 1068
reinforced concrete, 1–3, 8–9, 458–459
rigidity of, 7
safety of, 17–20
slender columns in, 563–564
stability of, 14
time-dependant volume changes in, 7
Strut-and-tie models, 288, 873–970
compression fans in, 890–891
compression fields of, 890–891
corbels (brackets), 921–932
deep beams, 896–921
direction of struts and ties in, 892–895
discontinuity regions and, 873–875
equilibrium of, 892
force whirls (U-turns), 891–892
layout of, 892–895
longitudinal cracking, failure by, 878
nodal zones, 875, 883–890
nodes, 875, 883
struts, 876–882, 890–895, 900–901
T-beam flanges, 953–955
ties, 882–883, 888–889, 892–893
Struts, 876–882, 890–895, 900–901
design of, 876–882
direction of, 892–893
Studs, 700

Sustainable/green construction
  aesthetics and occupant comfort, 39
  and cement industry, 39–40
  durability factor, 39
  economic impact, 39
  in terms of CO_2 emissions, 39
Sustained loads, 28–29, 144, 574
  beams, compression reinforcement of, 144
  characteristics of, 28–29
  pin-ended columns, effects of on, 574
Sway (unbraced) frames, 465–467, 593–618
  design of, 593–595, 601–618
  direct $P-\Delta$ analysis of, 601–602
  flexural stiffness (*EI*) of, 596–598
  gravity loads and, 603–618
  iterative $P-\Delta$ analysis of, 598
  lateral stiffness-reduction factor for, 596
  moments, calculation of in, 596–600
  reinforced concrete floor systems and, 467–468
  second-order analysis of, 596–600
  sidesway buckling, 603
  slender columns and, 593–618
  statics of, 593–595

T

T-beams, 152, 154, 464–494, 953–955
  compression zone for, 152–153
  continuous, design of, 464–494
  effective flange width for, 153–155
  flanges, 953–955
  flexural analysis of, 159–163
  strut-and-tie model for, 953–955
T-joints, 942
Tangent modulus of elasticity, 67
Tapered beams, 294–295
Temperature, 85, 91–92, 236–237, 438–439
  cold, 92
  concrete, effects of on, 85, 91–92
  high (fire), 91
  reinforcement, 236–237, 438
  shrinkage and, 438
  slabs, effects of on, 236–237, 438
  thermal expansion and contraction, 85
Tensile-controlled limits, 507, 816
Tensile strength, 7, 59–65, 72–73, 255
  compressive strength (f'_c) and, 60–61
  concrete, 59–61, 71–72, 255
  low, importance of, 7
  modulus of rupture (f_r), 59–60
  split-cylinder (f_{ct}), 60
  standard tests for, 59–60
  stress–strain curves for, 72–73
Tension, 72–73, 107, 113, 117–119, 133, 195, 246, 297
  area of reinforcement (A_s), 107, 119, 129–131, 137–138, 204
  axial, 297
  extreme layer of reinforcement, 133
  singly-reinforced beam sections, 204
  stress–strain curves, 72–73
Tension chord, failure of, 268–269
Tension-controlled beam sections, 132–142
Tension-controlled limit (TCL), 134–136
Tension lap splices, 413–415
Tests for concrete, 46, 54–58
  compressive strength (f'_c), 46–47, 56–59
  core, 54–58
  modulus of rupture (f_r), 60–61
  split-cylinder (f_{ct}), 60
  standard cylinder, 46, 59–60
  tensile strength, 59–65
Thermal expansion and contraction of concrete, 85
Thin-walled tube/plastic space truss
  shear flow area (A_o) of for, 329
Thin-walled tube theory, 309–310
Threshold temperature, 53
Tied columns, 496, 497–499, 526, 531–533, 535–538
  behavior of, 497–499
  design of, 535–538
  requirements for, 531–533
  size estimation of, 526
Ties, 149, 882–883, 888–889, 892–895, 984
  anchorage of in nodal zone, 888–889
  design of, 882–883
  direction of, 892–893
  strut-and-tie model layout using, 888–889, 892–895
  vertical reinforcement, as, 984
Tilt-up walls, 961, 966
Torsion, 304–358, 489, 1021, 1026
  ACI code design for, 337–357
  anchorages for, 331–332
  circulatory, 310–314
  compatibility, 323–326, 489
  crack width limit for, 334–335
  critical section for, 337–338
  cross sections for, 304–305, 337
  earthquake loads and, 1021, 1026
  equilibrium, 323–326
  girder design and, 489–490
  hollow members, 307–309
  irregularities for seismic design, 1020–1021
  moment combined with, 336
  principal stresses due to, 310
  pure, 315
  reinforced concrete members subjected to, 315–316
  reinforcement, 328–330, 331
  shear and moment combined with, 336
  shearing stresses due to, 304–310
  skew bending theory for, 315
  soap-film analogy for, 304–305
  solid members, 304–307
  stirrups, area of for, 327–328
  story (T_x), 1026
  thin-walled tube analogies, 316–323
  thin-walled tube theory for, 309–310
  threshold, 332
  warping, 310–314
  web crushing limit for, 335–336
Torsional critical load, 619–621
Torsional members and columns, 666–675
Torsional moments (*T*), 304, 326
Torsional stiffness (K_t), 326
Transfer beam, 969
Transfer width, 708
Transformed beam sections, 421–424
Transition failures, 507
Transition-zone section, 134–135
Transverse reinforcement, 490, 1034–1036, 1047–1049. *See also* Hoops; Stirrups

confinement, 1034–1035, 1047–1049
index (K_{tr}), 371
shear, 1035–1036, 1049
Triaxial loads, 65–66
Tributary area (A_T), 31, 176–177, 691–692
Trump International Hotel and Tower, 2
Truss (bent-up) bars, 197
Two-way slabs, 5, 105, 624–776
beams in two directions, 755–765
capitals for, 624, 725
columns, supported by, 636–638, 644–659
construction loads on, 765
deflections in, 767–771
design of, 639–643, 729–755
direct-design method for, 644–659
discontinuous corners of, 800–801
drop panels for, 624, 724–725
edge beams for, 641–642
elastic analysis of, 779–783
equivalent-frame methods for, 644, 659–681
fan-shaped yield patterns of, 801–804
finite-element analysis for, 781–783
flexural failure in, 115, 626–628
minimum thickness of, 642–643
moments in, 627–628, 629–638, 644–659, 705–724
Nichols' analysis of moments for, 626, 630–633
post-tensioning, use of, 771–773
radius of curvature (r) for, 633
reinforcement of, 699–701, 719–724
shear and moment transfer in, 705–724
shear caps for, 725–726
shear strength of, 687–705
walls, supported by, 635–636
without beams, 729–755
yield-line analysis of, 628, 783–804
Two-way walls, 961

U

Ultimate limit states, 13–14, 596–597
Unbraced frames, *see* Sway frames
Uncracked-elastic range, 114
Under-reinforced beam sections, 112, 116, 124–130
Uniqueness plastic theorem, 22
Unsymmetrical beam sections, 167–168
Unsymmetrical columns, 521
Upper-bound plastic theorem, 21–22

V

Variable loads, 28–29
Velocity pressure (q), 35–36
Vertical deflections, 458
Vertical loads, 832–833
Vibrations, 14, 458–459
Virtual work method for yield-line analysis, 786–789
Volume changes of concrete, 7, 73–85

W

Waffle slabs, 624
Walls, 635–636, 959–1008, 1060–1068
assemblies, 961, 966
axially loaded, 991, 1003–1008
bearing, 962–965
buckling of, 1003
building floor plans for, 975
cantilever retaining, 961
concrete properties of, 1003–1008
critical loads for, 1003–1008
design of, 975–984
diaphragms for, 975
floor systems, load transfer through, 984
foundations for, 960–961
lateral load-resisting systems for, 967–969
minimum thickness of, 982
one-way, 961
reinforcement in, 964–965, 982–984
retaining (nonbearing), 961, 966
seismic design of, 1060–1068
seismic load resistance, 994–995
shear, 959, 960, 961, 966, 969–974, 1060–1068
shear-wall–frame buildings, 968–969
size requirements of, 980–982
slabs supported by, 635–636
story drift limits for, 981–982
story forces (shears) and, 975–979
ties for vertical reinforcement of, 982
tilt-up, 961, 966
two-way, 961
Walraven model, 859–861
Warping torsion, 310–314
Water/cement (w/c) ratio, 50–51
Water quality and concrete strength, 52
Weak story in seismic design, 1022
Web-shear cracking in walls, 993
Web width (b_w), 115, 202, 206
Webs, 250, 258–260, 261, 262–270, 335–336, 439
crushing strength of, 335–336
minimum specifications for, 270
reinforcement, 254, 255–260, 262–270, 439
stirrups for, 249, 254–255
Wedge anchors, 100
Weight (mass) for seismic design, 1022
Welded splices, 413
Welded-wire reinforcement, 97–99, 372–373, 1074
ASTM specifications for, 98
development length of, 365–373
Whitney stress block, 122–124
Whole-member design methods, 288
Wind loads, 24, 25, 34–36, 614–616, 981–982, 996–999
design pressure (p) for, 35
directionality factor (K_d) for, 35
external pressure coefficient (C_p) for, 36
load factors for, 24
moment-resisting frames for, 967–968
velocity pressure (q) for, 35–36
walls, load-resisting systems for, 967–969
Wind turbine foundation, 3
Working (service) loads, 419
Working-stress design, 9, 421

Y

Yield-line analysis of slabs, 628, 783–804
  applications of, 790–800
  axes location for, 784–786
  concentrated loads and, 801–804
  criterion for, 783–784
  discontinuous corners and, 800–801
  equilibrium method for, 786
  failure in flexure from, 628
  virtual work method for, 786–789
  yield-line patterns, 784–786, 800–804
Yield moment (M_y), 116
Yield points, 116–117
Yielding, 112, 116, 262–265
  ductility and, 117
  flexural behavior and, 111–114
  stirrups, 262–265
  under-reinforced beam sections, 112, 116